**Handbook of
Heterogeneous Catalysis**

*Edited by
Gerhard Ertl, Helmut Knözinger,
Ferdi Schüth, and Jens Weitkamp*

Handbook of Heterogeneous Catalysis

8 Volumes

Volume 1

Edited by
Gerhard Ertl, Helmut Knözinger, Ferdi Schüth,
and Jens Weitkamp

Second, Completely Revised and Enlarged Edition

WILEY-VCH Verlag GmbH & Co. KGaA

The Editors

Prof. Dr. Gerhard Ertl
Fritz-Haber-Institut der
Max-Planck-Gesellschaft
Faradayweg 4-6
14195 Berlin
Germany

Prof. Dr. Helmut Knözinger
Department of Chemistry and Biochemistry
University of Munich
Butenandtstr. 5-13
81377 München
Germany

Prof. Dr. Ferdi Schüth
Max-Planck-Institut
für Kohlenforschung
Kaiser-Wilhelm-Platz 1
45470 Mülheim
Germany

Prof. Dr. Jens Weitkamp
Institute of Chemical Technology
University of Stuttgart
Pfaffenwaldring 55
70569 Stuttgart
Germany

Cover
Photo of shaped catalysts
courtesy of Süd-Chemie AG, Munich, Germany.

All books published by Wiley-VCH are carefully produced. Nevertheless, authors, editors, and publisher do not warrant the information contained in these books, including this book, to be free of errors. Readers are advised to keep in mind that statements, data, illustrations, procedural details or other items may inadvertently be inaccurate.

Library of Congress Card No.: applied for

British Library Cataloguing-in-Publication Data
A catalogue record for this book is available from the British Library.

Bibliographic information published by the Deutsche Nationalbibliothek
Die Deutsche Nationalbibliothek lists this publication in the Deutsche Nationalbibliografie; detailed bibliographic data are available on the Internet at <http://dnb.d-nb.de>.

© 2008 WILEY-VCH Verlag GmbH & Co. KGaA, Weinheim

All rights reserved (including those of translation into other languages). No part of this book may be reproduced in any form – by photoprinting, microfilm, or any other means – nor transmitted or translated into a machine language without written permission from the publishers. Registered names, trademarks, etc. used in this book, even when not specifically marked as such, are not to be considered unprotected by law.

Composition Laserwords Private Ltd., Chennai, India

Printing Betz-Druck GmbH, Darmstadt

Bookbinding Litges & Dopf GmbH, Heppenheim

Cover Design Schulz Grafik-Design, Fußgönheim

Printed in the Federal Republic of Germany
Printed on acid-free paper

ISBN: 978-3-527-31241-2

Preface

In the past century, catalysis and in particular heterogeneous catalysis have become the basis of huge industries, such as petroleum refining, natural gas processing, the chemical industry including the manufacture of polymers, and environmental protection. In ca. 85 to 90% of all chemical processes, use is made of a catalyst. Catalysis helps to accelerate chemical transformations and to make them significantly more selective for the desired products, more energy-efficient and/or environmentally more benign. In ca. 80 to 85% of all catalytic processes, heterogeneous catalysis is being employed, i.e., the catalyst is a solid. Today, the worldwide trade volume for solid catalysts is estimated to amount to ca. 15 billion US $ per year, and the value created by using these catalysts is about 100 to 1000 times as high. The manufacture of solid catalysts has thus become an important branch of industrial chemistry. It continues to be based on a large amount of empirism and know-how, but a more rational scientific understanding of the function of solid surfaces and the complex steps involved during heterogeneous catalysis is emerging at an unprecedented pace.

Heterogeneous catalysis is an interdisciplinary field that demands the cooperation of experts from a multitude of traditional branches of natural and engineering sciences. It is based on solid-state chemistry and physics, materials science and surface science, but various other disciplines have steadily contributed to an improved understanding of and progress in heterogeneous catalysis. Among these are reaction kinetics and mechanisms, theoretical chemistry, solid-state spectroscopy, analytical chemistry and chemical reaction engineering. It is this pronounced interdisciplinarity that contributes to the fascination of heterogeneous catalysis. At the same time, however, the rapidly expanding literature covering the entire field tends to be scattered over an unusually broad variety of sources which are difficult to overlook even for the expert. Noteworthy early attempts to bundle and summarize the then available knowledge on catalysis were the *Handbuch der Katalyse* edited by G.M. Schwab in the period from 1941 to 1943 and P.H. Emmett's book series *Catalysis* published between 1954 and 1960. A recent treatise on heterogeneous catalysis which attempted to describe and order all facets of the field from the scientific fundamentals to the chemical engineering of catalytic processes was the *Handbook of Heterogeneous Catalysis*, the 1st edition of which was published in 1997 by VCH in five volumes.

Here is the completely revised and significantly enlarged 2nd edition of the *Handbook of Heterogeneous Catalysis*. Not only were the contributions from the 1st edition updated, so as to cover the recent literature up to ca. 2005 or 2006, but numerous chapters from the 1st edition were deeply revised or entirely re-written in an effort to have rapidly developing fields adequately discussed from today's view. Moreover, the reader will find a fairly large number of completely new chapters dealing with topics that were not yet addressed in the 1st edition. Altogether, the eight volumes of the 2nd edition contain nearly 200 chapters authored by ca. 330 experts. To ensure a quick and efficient access to the wealth of information offered by the new Handbook, an in-depth and detailed subject index was added which will guide the reader by means of keywords with up to three hierarchical levels.

A handbook of this scope and size can only be produced in a large and cooperative effort by many contributors. In the first place, we would like to express our gratitude to all authors for having brought in their expertise, for their enthusiastic cooperation and for their patient willingness to adapt their manuscripts to the layout asked for by the publisher and the editors. Likewise, our sincere thanks go to the Publisher's numerous employees, who were involved in the production of the new Handbook, for their competent and professional cooperation. The editors are particularly grateful to Dr. Rainer Münz from Wiley-VCH who managed and steered this project in a most efficient and cooperative manner from the very first planning till the delivery of the eight volumes.

Handbook of Heterogeneous Catalysis, 2nd Ed.
Edited by G. Ertl, H. Knözinger, F. Schüth, and J. Weitkamp
Copyright © 2008 Wiley-VCH Verlag GmbH & Co. KGaA, Weinheim
ISBN: 978-3-527-31241-2

We are confident that the 2nd edition of the *Handbook of Heterogeneous Catalysis* offers useful information to a broad and diverse readership, both in industry and at the academia. It is our hope that the new Handbook may contribute to a further promotion of heterogeneous catalysis and that it stimulates and deepens the cooperation between its various sub-fields. Much remains to be achieved by heterogeneous catalysis in the century that has just begun. May the new Handbook also help to attract young scientists to devote their professional career to a fascinating and interdisciplinary area, for the benefit of mankind.

December 2007

Gerhard Ertl
Helmut Knözinger
Ferdi Schüth
Jens Weitkamp

Outline

Vol. 1	1	Introduction	1
	2	Preparation of Solid Catalysts	57
Vol. 2	3	Characterization of Solid Catalysts	721
Vol. 3	4	Model Systems	1259
	5	Elementary Steps and Mechanisms	1375
	6	Macrokinetics and Transport Processes	1693
Vol. 4	7	Activity Loss	1829
	8	Special Catalytic Systems	1873
	9	Laboratory Testing of Solid Catalysts	2019
	10	Reaction Engineering	2075
Vol. 5	11	Environmental Catalysis	2265
	12	Inorganic Reactions	2501
Vol. 6	13	Energy-Related Catalysis	2677
Vol. 7	14	Organic Reactions	3123
Vol. 8	15	Polymerization Reactions	3733
	16	Reactions on Immobilized Biocatalysts	3831
		Index	3867

Handbook of Heterogeneous Catalysis, 2nd Ed.
Edited by G. Ertl, H. Knözinger, F. Schüth, and J. Weitkamp
Copyright © 2008 Wiley-VCH Verlag GmbH & Co. KGaA, Weinheim
ISBN: 978-3-527-31241-2

Contents

Volume 1

Preface V

List of Contributors XIX

1 Introduction 1

1.1 Principles of Heterogeneous Catalysis 1
James A. Dumesic, George W. Huber, and Michel Boudart
1.1.1 Introduction 1
1.1.2 Definitions of Catalysis and Turnover 1
1.1.3 Steps in a Heterogeneous Catalytic Reaction 2
1.1.4 Desired Characteristics of a Catalyst 3
1.1.5 Reaction Schemes and Adsorbed Species 4
1.1.6 Conditions for Catalyst Optimality 6
1.1.7 Catalyst Design 10
1.1.8 Catalyst Development 11
1.1.9 Bridging Gaps in Heterogeneous Catalysis 12
1.1.10 A Philosophical Note 13
References 14

1.2 Development of the Science of Catalysis 16
Burtron H. Davis
1.2.1 Early Concepts: Berzelius, Liebig, and Faraday 16
1.2.2 Wilhelm Ostwald 19
1.2.3 The Concepts of Kinetics and Intermediate Compounds 20
1.2.4 Negative Catalysis: Autocatalysis 24
1.2.5 Adsorption 24
1.2.6 Active Site: Geometric or Electronic? 26
1.2.7 Selected Systems 29
1.2.7.1 Ammonia Synthesis 29
1.2.7.2 Acid Catalysis 30
1.2.7.3 Zeolites 31
1.2.7.4 Ions in Catalysis 31
1.2.7.5 Hydrogenation 32
1.2.7.6 Oxidation 33
1.2.8 Summary 34
References 35

1.3 The Development of Industrial Heterogeneous Catalysis 37
Uwe Dingerdissen, Andreas Martin, Daniel Herein, and Hans Jürgen Wernicke
1.3.1 Introduction 37
1.3.2 The Raw Materials Situation 38
1.3.2.1 Inorganic Base Chemicals 38
1.3.2.2 Petroleum Refining 39
1.3.2.3 Coal and Natural Gas Processing 41
1.3.3 The Base Chemicals 42
1.3.3.1 Olefins and Aromatics 43
1.3.3.2 Alcohols, Ethers, Aldehydes, and Ketones 43
1.3.3.2.1 Alcohols 43
1.3.3.2.2 Ethers 44
1.3.3.2.3 Aldehydes 44
1.3.3.3 Acids, Esters, Anhydrides, and Epoxides 44
1.3.3.3.1 Acids, Esters, Anhydrides 44
1.3.3.3.2 Epoxides 45
1.3.3.4 Halogenated Chemicals 45
1.3.4 The Fine Chemicals 46
1.3.4.1 Catalytic Hydrogenation 46
1.3.4.1.1 Hydrogenation of Carbon–Carbon Multiple Bonds 46
1.3.4.1.2 Hydrogenation of Carbon–Oxygen and Carbon–Nitrogen Multiple Bonds 47
1.3.4.1.3 Reductive Amination 48
1.3.4.1.4 Nitrile Hydrogenation 48
1.3.4.1.5 Hydrogenation of Nitro Groups 48
1.3.4.2 Catalysis on Solid Acids and Bases 48
1.3.4.2.1 Aromatic Substitution (Friedel–Crafts Alkylation and Acylation) 49
1.3.4.2.2 Amination 49
1.3.4.2.3 Rearrangement, Isomerization 49
1.3.4.3 Oxidation Catalysis 49

Handbook of Heterogeneous Catalysis, 2nd Ed.
Edited by G. Ertl, H. Knözinger, F. Schüth, and J. Weitkamp
Copyright © 2008 Wiley-VCH Verlag GmbH & Co. KGaA, Weinheim
ISBN: 978-3-527-31241-2

1.3.4.3.1	Oxidation of Alcohols and Aldehydes 49		2.3.1.1	Introduction 81
1.3.4.3.2	Epoxidation and Aromatic Ring Hydroxylation 50		2.3.1.2	The Concept of Fused Catalysts 82
1.3.5	Environmental Catalysis 50		2.3.1.3	Thermodynamic and Kinetic Considerations 85
1.3.5.1	Cleaning of Automobile Emissions 50		2.3.1.4	Sulfuric Acid Catalyst 87
1.3.5.2	Removal of Harmful Emissions from Stationary Sources 52		2.3.1.5	Metallic Glasses 88
			2.3.1.6	Tribochemistry 90
1.3.5.3	Wastewater Treatment 53		2.3.1.7	Mesostructure of Fused Catalyst Materials 91
1.3.5.4	Fuel Cell Catalysis 53			References 91
1.3.5.5	Photocatalytic Applications 54			
	References 54		2.3.2	Skeletal Metal Catalysts 92
				Andrew James Smith and Mark Sebastian Wainwright
2	**Preparation of Solid Catalysts 57**		2.3.2.1	Introduction 92
2.1	Development of Industrial Catalysts 57		2.3.2.2	General Aspects 93
	Ewald F. Gallei, Michael Hesse, and Ekkehard Schwab		2.3.2.2.1	Alloy Preparation 93
			2.3.2.2.2	Activation Using Alkali Leaching 93
2.1.1	Introduction 57		2.3.2.2.3	Storage and Handling 94
2.1.2	Types and Properties of Technical Catalysts 57		2.3.2.2.4	Advantages of Skeletal Metal Catalysts 94
2.1.3	Characterization of Technical Catalysts 60		2.3.2.3	Skeletal Nickel Catalysts 94
2.1.4	High-Throughput Experimentation (HTE) 62		2.3.2.3.1	Alloy Preparation of Skeletal Nickel Catalysts 94
2.1.5	Identification of Technical Catalysts 63		2.3.2.3.2	Properties of Skeletal Nickel Catalysts 95
2.1.6	Production of Technical Catalysts 65		2.3.2.3.3	Uses of Skeletal Nickel Catalysts 95
2.1.7	Conclusions 65		2.3.2.3.4	The Promotion of Skeletal Nickel Catalysts 95
	References 66		2.3.2.4	Skeletal Copper Catalysts 96
			2.3.2.4.1	Alloy Preparation of Skeletal Copper Catalysts 96
2.2	Computer-Aided Design of Solid Catalysts 66		2.3.2.4.2	Properties of Skeletal Copper Catalysts 97
	Martin Holena and Manfred Baerns		2.3.2.4.3	Uses of Skeletal Copper Catalysts 97
2.2.1	Introduction 66		2.3.2.4.4	Promotion of Skeletal Copper Catalysts 98
2.2.2	Theoretical Methods 66		2.3.2.5	Skeletal Cobalt Catalysts 99
2.2.3	Rational Methods for the Design of Catalytic Experiments 68		2.3.2.6	Other Skeletal Metal Catalysts 99
2.2.3.1	The Statistical Design of Experiments 68			References 99
2.2.3.2	Optimization Methods for Empirical Objective Functions 69		2.3.3	Precipitation and Coprecipitation 100
				Ferdi Schüth, Michael Hesse, and Klaus K. Unger
2.2.3.2.1	Evolutionary Methods 70			
2.2.3.2.2	Other Stochastic Methods 72		2.3.3.1	Introduction 100
2.2.3.2.3	Deterministic Methods 72		2.3.3.2	General Principles Governing Precipitation from Solutions 101
2.2.3.3	Concluding Remarks 73			
2.2.4	Data Analysis and Data Mining 74		2.3.3.2.1	Physico-Chemical Considerations 102
2.2.4.1	Statistical Methods: An Overview 74		2.3.3.2.2	Chemical Considerations 104
2.2.4.2	Artificial Neural Networks 75		2.3.3.2.3	Process Considerations 105
2.2.4.3	Concluding Remarks 77		2.3.3.3	Influencing the Properties of the Final Product 106
2.2.5	Conclusions 79			
	References 79		2.3.3.3.1	Influence of Raw Materials 106
			2.3.3.3.2	Influence of Concentration and Composition 108
2.3	Bulk Catalysts and Supports 81			
2.3.1	Fused Catalysts 81		2.3.3.3.3	Solvent Effects 108
	Robert Schlögl		2.3.3.3.4	Influence of Temperature 109

2.3.3.3.5	Influence of pH 110		2.3.6	Ordered Mesoporous Materials 178
2.3.3.3.6	Influence of Aging 110			*Freddy Kleitz*
2.3.3.3.7	Influence of Additives 111		2.3.6.1	Introduction 178
2.3.3.4	Prototypical Examples of Precipitated Catalysts and Supports 111		2.3.6.2	Ordered Mesoporous Molecular Sieves: MCM-41 179
2.3.3.4.1	Silica as Support Material 111		2.3.6.3	Synthesis of Ordered Mesoporous Materials 180
2.3.3.4.2	Active Aluminas 113		2.3.6.3.1	Synthesis Strategies for Mesostructure Formation 180
2.3.3.4.3	Ni/Al_2O_3 Catalysts by Coprecipitation 115		2.3.6.3.2	Inorganic Polymerization and Self-Assembly with Surfactants 182
2.3.3.4.4	V/P/O Catalysts for Butane Oxidation to Maleic Anhydride 116		2.3.6.3.3	Synthesis Pathways and Structural Diversity 187
2.3.3.5	Conclusions 116 References 116		2.3.6.3.4	Pore Size Tailoring and Structure Engineering 193
			2.3.6.3.5	Removal of the Template 197
2.3.4	Sol–Gel Process 119 *Miron V. Landau*		2.3.6.4	Functionalization of Ordered Mesoporous Materials 200
2.3.4.1	Introduction 119		2.3.6.4.1	Functionalization Strategies 200
2.3.4.2	Physico-Chemical Basis and Principles of Sol–Gel Processing for the Preparation of Porous Solids 121		2.3.6.4.2	Surface Properties 201
			2.3.6.4.3	Surface Functionalization 201
2.3.4.3	Application of Sol–Gel Processing for the Preparation of Solid Catalysts 128		2.3.6.4.4	Framework Functionalization 203
			2.3.6.5	Non-Siliceous Mesostructured and Mesoporous Materials 205
2.3.4.3.1	Bulk Catalytic-Phase Materials: Xerogels and Aerogels 129		2.3.6.5.1	Transition Metal Oxides 205
2.3.4.3.2	Catalytic Materials and Modifiers Entrapped in Porous Matrices 134		2.3.6.5.2	Alumina 206
			2.3.6.5.3	Other Non-Siliceous Compositions 207
2.3.4.3.3	Catalytic Coatings, Films, and Membranes 147		2.3.6.6	Hard Templating (Nanocasting) 208
			2.3.6.7	Morphology Control 211
2.3.4.4	Summary 155 References 156		2.3.6.8	Concluding Remarks 213 References 214
2.3.5	Hydrothermal Zeolite Synthesis 160 *Christine E. A. Kirschhock, Eddy J. P. Feijen, Pierre A. Jacobs, and Johan A. Martens*		2.3.7	Pillared Clays 219 *Georges Poncelet and José J. Fripiat*
			2.3.7.1	Introduction 219
			2.3.7.1.1	Historical Overview 219
2.3.5.1	Introduction 160		2.3.7.1.2	Which Clays or Smectites can be Pillared? 220
2.3.5.2	Zeolitization in General 161			
2.3.5.2.1	Key Steps in Zeolite Synthesis 161		2.3.7.1.3	Al-Pillared Vermiculites and Phlogopite Micas 222
2.3.5.2.2	Key Parameters Governing Zeolitization 162		2.3.7.1.4	Reviews 222
2.3.5.2.3	Precursor Species 165		2.3.7.2	Pillaring Mechanisms 222
2.3.5.2.4	Templates 167		2.3.7.2.1	Basic Pillaring Mechanism: Exchange Precipitation 222
2.3.5.2.5	Rationalization of Crystallization Conditions for Zeolites A, X, Y, and EMC-2 169		2.3.7.2.2	New Pillaring Mechanism: Template 223
2.3.5.2.6	Zeolite Growth Models 170		2.3.7.3	Pillars 224
2.3.5.3	Synthesis of Industrial Zeolites 174		2.3.7.3.1	Homo-Atomic Pillars 224
2.3.5.3.1	Industrial Zeolites Crystallized from the Na_2O-Al_2O_3-SiO_2-H_2O System: Zeolites A, X, Y, Mordenite, and ZSM-5 174		2.3.7.3.2	Heteroatomic Pillars 227
			2.3.7.3.3	Mixed Al–Fe Pillars 228
			2.3.7.3.4	Miscellaneous 228
2.3.5.3.2	Industrial Zeolite Crystallization in the Presence of Organic Compounds 175		2.3.7.4	Pillared Clays 228
			2.3.7.4.1	Preliminary Remarks 228
2.3.5.4	Conclusions and Outlook 176 References 176		2.3.7.4.2	Main Physical Characteristics 229

2.3.7.4.3	Chemical Properties 232		2.3.9.6.1	Promoters 268
2.3.7.4.4	Chemical Modifications or Functionalization of the Pillaring Species 235		2.3.9.6.2	Structure–Activity Relationships 268
			2.3.9.6.3	Effect of Promoters 268
2.3.7.5	Catalytic Properties 236		2.3.9.6.4	Addition of Promoters 269
2.3.7.5.1	A Brief Summary of Previous Studies 236		2.3.9.7	Calcination 269
2.3.7.5.2	An Overview of Catalytic Reactions Investigated during the Past Decade 237		2.3.9.7.1	Events Occurring During Calcination 269
			2.3.9.7.2	Effect of Bed Volume and Packing 270
2.3.7.6	Conclusions 242		2.3.9.7.3	Effect of Calcination Temperature on Physical Properties 272
	References 242			
			2.3.9.7.4	Effect of Calcination Temperature on Catalytic Performance 272
2.3.8	Chemistry and Applications of Porous Metal–Organic Frameworks 247 *Ulrich Müller, Markus M. Schubert, and Omar M. Yaghi*			
			2.3.9.7.5	Effect of Holding Time 272
			2.3.9.7.6	Calcination: A Summary 273
			2.3.9.8	Sulfation of Thermally Treated Crystalline Oxides 273
2.3.8.1	Introduction 247			
2.3.8.2	Terminology and Structure 248		2.3.9.8.1	Sulfation with Aqueous Solutions (Solid–Liquid Phase) 273
2.3.8.3	Synthesis 250			
2.3.8.4	Characterization 252		2.3.9.8.2	Sulfation via Gas Phase (Solid–Gas Phase) 273
2.3.8.5	Emerging Applications 252			
2.3.8.5.1	Adsorption Properties 252		2.3.9.8.3	Sulfation of Zirconia using Solid Sulfate Precursors 274
2.3.8.5.2	Diffusional Properties 253			
2.3.8.5.3	Gas Purification 254		2.3.9.9	Addition of Noble Metals 274
2.3.8.5.4	Gas Separation 254		2.3.9.10	Activation 274
2.3.8.5.5	Gas Storage 254		2.3.9.11	Summary 274
2.3.8.5.6	Catalysis 257			References 275
2.3.8.6	Conclusions 260			
	References 261		2.3.10	Catalysis by Ion-Exchange Resins 278 *Bruce C. Gates*
2.3.9	Oxo-Anion Modified Oxides 262 *Friederike C. Jentoft*		2.3.10.1	Introduction 278
			2.3.10.2	Classes of Ion-Exchange Resin Catalysts: The Importance of Porosity 278
2.3.9.1	Introduction 262			
2.3.9.1.1	Classification 262		2.3.10.3	Examples of Industrial Catalysis by Ion-Exchange Resins 279
2.3.9.1.2	Variety of Materials and Focus 262			
2.3.9.2	Target Properties 263		2.3.10.4	Synthesis of Ion-Exchange Resins 279
2.3.9.2.1	Structure–Activity Relationships 263		2.3.10.5	Structure, Influence of Water, and Acidity in Ion-Exchange Resins 280
2.3.9.2.2	Effect of Sulfate and Other Oxo-Anions 264			
			2.3.10.6	Accessibility, Swelling, and Mass Transfer Effects 281
2.3.9.3	Preparation Routes 265			
2.3.9.4	Formation of Primary Solid 265		2.3.10.7	Mechanisms of Reactions Catalyzed by Sulfonic Acid Resins 282
2.3.9.4.1	Formation of Pure Hydrous Zirconia 266			
2.3.9.4.2	Formation of Sulfate-Containing Hydrous Zirconia 266		2.3.10.8	Kinetics of Reactions Catalyzed by Ion-Exchange Resins 283
			2.3.10.9	Deactivation and Stability of Ion-Exchange Resin Catalysts 284
2.3.9.4.3	Template-Directed Formation of Primary Solid 267			
			2.3.10.10	Modified and Metal-Containing Ion-Exchange Resin Catalysts 284
2.3.9.4.4	Commercially Available Precursors 267			
2.3.9.5	Anion-Modification of Primary Solid 267		2.3.10.11	Hybrid Organic/Inorganic Catalysts akin to Ion-Exchange Resins 285
2.3.9.5.1	Addition of Oxo-Anions to Primary Solid in Liquid Medium 267			
			2.3.10.12	Summary and Outlook 285
2.3.9.5.2	Sulfation of Primary Solid via Gas Phase 268			References 285
2.3.9.5.3	Introduction of Oxo-Anions Using Solid Precursors 268		2.3.11	Flame Hydrolysis 286 *Dieter Kerner and Matthias Rochnia*
2.3.9.6	Addition of Promoters (Optional) 268			

2.3.11.1	Manufacture of Pyrogenic Oxides 286		2.3.13.5.7	HPA Encaged in Supercages 334
2.3.11.2	Physico-Chemical Properties of Pyrogenic Oxides 288		2.3.13.5.8	Heteropolyanions Intercalated in Layered Double Hydroxides 335
2.3.11.3	Preparation of Formed Supports 291		2.3.13.6	Lacunary and Metal-Substituted Heteropoly Compounds 335
2.3.11.4	Applications 292		2.3.13.6.1	Lacunary Silicotungstate 335
2.3.11.5	Conclusions 293		2.3.13.6.2	Transition Metal-Substituted Polyanions (TMSP) 336
	References 293		2.3.13.6.3	Heteropolyanion-Supported Metals 337
2.3.12	Solid-State Reactions 295 *Bernard Delmon and Michel Devillers*		2.3.13.6.4	Microstructured POM toward a Shape-Selective Catalyst 337 References 338
2.3.12.1	Why Prepare Catalysts from Solid Precursors? 295			
2.3.12.1.1	Some General Concepts and Parameters in Solid-State Reactions 297		2.3.14	Transition Metal Carbides, Nitrides, and Phosphides 342 *S. Ted Oyama*
2.3.12.1.2	The Objective of this Chapter in the Handbook 303		2.3.14.1	Introduction 342
2.3.12.2	Description of Preparative Methods 303		2.3.14.2	General Properties 342
2.3.12.2.1	Dry Methods 304		2.3.14.3	Structure 342
2.3.12.2.2	Wet Methods 304		2.3.14.3.1	Binary Compounds of C and N 343
2.3.12.2.3	Chemical Complexation Methods 307		2.3.14.3.2	Binary Compounds of Phosphorus 344
2.3.12.2.4	Advanced Methods for Obtaining Solid Precursors of Flexible Composition 312		2.3.14.4	Preparation 345
2.3.12.2.5	Specific Methods: Example of Hydrotalcite-Type Catalysts 314		2.3.14.4.1	The Reactions of Metals or Metal Compounds with Gas-Phase Reagents 345
2.3.12.3	Conclusions and Prospects 315 References 316		2.3.14.4.2	Decomposition of Metal Halide Vapors 346
			2.3.14.4.3	Decomposition of Metal Compounds 346
2.3.13	Heteropoly Compounds 318 *Kwan-Young Lee and Makoto Misono*		2.3.14.4.4	Temperature-Programmed Methods 346
2.3.13.1	Structure and Catalytic Properties 318		2.3.14.4.5	Utilization of High-Surface-Area Supports 347
2.3.13.2	Heteropolyacids: Acid Forms in the Solid State and in Solution 320		2.3.14.4.6	Reaction Between Metal Oxide Vapor and Solid Carbon 347
2.3.13.2.1	Keggin-Type HPAs 320		2.3.14.4.7	Liquid-Phase Methods 347
2.3.13.2.2	Wells–Dawson-Type HPAs 323		2.3.14.4.8	Amorphous Materials 347
2.3.13.3	Salts of Heteropolyacids: Cation-Exchanged Forms 324		2.3.14.5	Catalytic Properties 347
2.3.13.3.1	Acid Catalysis of Salts of Heteropolyacids 324		2.3.14.5.1	Acid and Base Catalysis 348
2.3.13.3.2	Oxidation Catalysis of Salts of Heteropolyacids 325		2.3.14.5.2	Ammonia Synthesis and Decomposition 349
2.3.13.3.3	Bifunctional Catalysts 326		2.3.14.5.3	Aromatization 349
2.3.13.3.4	Shape Selectivities of Acidic Salts of Hetero- polyacids 327		2.3.14.5.4	Hydrogenation and Dehydrogenation 349
2.3.13.3.5	Preparation 328		2.3.14.5.5	Hydrotreating 351
2.3.13.4	Mixed-Coordinated Heteropoly Compounds 328		2.3.14.5.6	Isomerization and Hydroisomerization (Reforming) 352
2.3.13.5	Supported Heteropoly Compounds 329		2.3.14.5.7	Methanation and Fischer–Tropsch 353
2.3.13.5.1	Oxide Supports 329		2.3.14.5.8	Oxidation 353
2.3.13.5.2	Mesoporous Supports 330		2.3.14.5.9	Reforming (Steam, Dry, and Autothermal) 354
2.3.13.5.3	Carbon Supports 332		2.3.14.5.10	Water-Gas Shift and Reverse Water-Gas Shift 354
2.3.13.5.4	HPA Salt Supports 333		2.3.14.6	Perspective 354 References 355
2.3.13.5.5	Polymer Supports 333			
2.3.13.5.6	Clay Supports 334			

2.3.15	Carbons 357		2.4.1.5.2	Deposition–Precipitation by Changing the pH 435
	Robert Schlögl		2.4.1.5.3	Changing the Valency of Precursor Ions 454
2.3.15.1	Introduction 357		2.4.1.5.4	Removal of a Complexing Agent 455
2.3.15.2	The Complexity of Carbon 357		2.4.1.5.5	Electrochemical Deposition–Precipitation 456
2.3.15.3	Electronic Structure of Carbon Allotropes 358		2.4.1.5.6	Deposition–Precipitation of Gold 456
2.3.15.4	Basic Structures 361		2.4.1.5.7	Deposition–Precipitation within Pre-Shaped Support Bodies 460
2.3.15.5	Nanostructured Carbons 363		2.4.1.6	Production of Catalyst Supports and Synthetic Clay Minerals 462
2.3.15.6	Carbolites and Fullerenes 364		2.4.1.7	Concluding Remarks 464
2.3.15.7	Disordered Graphites 365			References 465
2.3.15.8	Nanocrystalline Carbon: Carbon Black 369			
2.3.15.9	Formation of Carbon: General Pathways 375		2.4.2	Ion Exchange and Impregnation 467
2.3.15.10	Formation of Solid Carbon: Mechanistic Aspects 376			*Eric Marceau, Xavier Carrier, Michel Che, Oliver Clause, and Christian Marcilly*
2.3.15.11	Catalytic Formation of Carbon from Molecules 378		2.4.2.1	Deposition of Active Component Precursors 467
2.3.15.12	Carbon on Noble Metal Catalysts 380		2.4.2.2	Ion Exchange: Physicochemical Aspects 469
2.3.15.13	Carbon Formation in Zeolites 383		2.4.2.3	Impregnation: Physical Aspects 473
2.3.15.14	Graphitization of Carbons 385		2.4.2.4	Support and Precursor during the Preparation Steps: Chemical Aspects 475
2.3.15.15	Reaction of Oxygen with Carbon 387		2.4.2.5	Applications 477
2.3.15.16	Surface Chemistry of Carbon 393		2.4.2.6	Conclusions 482
2.3.15.17	Chemical Quantification of Oxygen Groups 395			References 482
2.3.15.18	Non-Oxygen Heteroelements on Carbon Surfaces 397		2.4.3	Solid-State Ion Exchange in Zeolites 484
2.3.15.19	Chemical Analysis of Surface Oxygen Groups 400			*Hellmut G. Karge*
2.3.15.20	Reactivity in Fluid Phase 405		2.4.3.1	Introduction 484
2.3.15.21	Spectroscopy of Functional Groups: XPS 410		2.4.3.2	Comparison of Conventional and Solid-State Ion Exchange 485
2.3.15.22	Vibrational Spectroscopy of Functional Groups 414		2.4.3.3	Experimental Procedure for Solid-State Ion Exchange in Zeolites 486
2.3.15.23	Advanced Thermal Desorption 416		2.4.3.4	Methods of Monitoring Solid-State Ion Exchange 486
2.3.15.24	Synopsis: Carbon and Catalysis 418		2.4.3.4.1	Infrared Spectroscopy 486
	References 419		2.4.3.4.2	ESR Spectroscopy 487
2.4	Supported Catalysts 428		2.4.3.4.3	Mössbauer Spectroscopy 487
2.4.1	Preparation of Supported Catalysts by Deposition–Precipitation 428		2.4.3.4.4	Solid-State NMR Spectroscopy 487
	John W. Geus and A. Jos van Dillen		2.4.3.4.5	X-Ray Diffraction 487
2.4.1.1	Introduction 428		2.4.3.4.6	Extended X-Ray Absorption Fine Structure (EXAFS) and X-Ray Absorption Near Edge Structure (XANES) 487
2.4.1.2	Supported Catalysts 428			
2.4.1.3	Production of Supported Catalysts 430		2.4.3.4.7	Temperature-Programmed Evolution (TPE) of Volatile Gases 487
2.4.1.3.1	Selective Removal of One or More Component 430			
2.4.1.3.2	Application on Separately Produced Supports 431		2.4.3.4.8	Thermogravimetric Analysis (TGA) 488
2.4.1.4	The Theory of Nucleation 433		2.4.3.5	Systems Investigated for Solid-State Ion Exchange in Zeolites 488
2.4.1.5	Deposition–Precipitation on Suspended Supports 435		2.4.3.5.1	Solid-State Ion Exchange with Salts of Alkaline Metals 488
2.4.1.5.1	Survey of Deposition–Precipitation Procedures 435			

2.4.3.5.2	Solid-State Ion Exchange with Salts of Alkaline Earth Metals 490	2.4.4.4.5	Electron Deficiency 515	
2.4.3.5.3	Solid-State Ion Exchange with Salts of Lanthanum 491	2.4.4.4.6	Is Bifunctionality due to "Shuttling" between Sites or to "Collapsed Sites"? 516	
2.4.3.5.4	Solid-State Ion Exchange with Salts of Copper 493	2.4.4.4.7	Metal Oxidation by Protons, Selective "Leaching" 517	
2.4.3.5.5	Solid-State Ion Exchange with Salts of Iron 494	2.4.4.5	Effects of Zeolite Geometry on Catalysis 517	
2.4.3.5.6	Solid-State Ion Exchange with Salts of Manganese 495	2.4.4.5.1	Transport and Transition State Restrictions 517	
2.4.3.5.7	Solid-State Ion Exchange with Salts of Noble Metals 498	2.4.4.5.2	Collimation of Molecules in Pores 517	
		2.4.4.5.3	Types of Pore Diffusion in Zeolites; Effects on Apparent Activation Energy 518	
2.4.3.5.8	Miscellaneous 499			
2.4.3.6	Some Considerations on Thermodynamic and Kinetic Aspects of SSIE 499	2.4.4.5.4	Beneficial Damage of Zeolite Matrix by Growing Metal Particles 519	
2.4.3.6.1	Effect of Temperature and Salt Concentration in Salt/Zeolite Mixtures 499	2.4.4.6	Conclusions 519 References 520	
2.4.3.6.2	The Role of Water in Solid-State Ion Exchange 499	2.4.5	Grafting and Anchoring of Transition Metal Complexes to Inorganic Oxides 522 *Frédéric Averseng, Maxence Vennat, and Michel Che*	
2.4.3.6.3	Kinetics of Solid-State Ion Exchange 500			
2.4.3.6.4	A Possible Mechanism of Solid-State Ion Exchange 501			
2.4.3.7	Modified Solid-State Ion Exchange and Related Processes 502	2.4.5.1	Introduction 522	
		2.4.5.2	Grafting versus Anchoring: Definitions, Nomenclature, and Basic Principles 522	
2.4.3.7.1	Contact-Induced Ion Exchange 502	2.4.5.2.1	A Brief Overview of Earlier Definitions 522	
2.4.3.7.2	Incipient Wetness Technique of Impregnation 503	2.4.5.2.2	Proposed New Definitions 523	
2.4.3.7.3	Gas-Phase-Mediated Processes Related to Solid-State Ion Exchange 503	2.4.5.2.3	Nomenclature, Basic Principles, and Oxide Support Characteristics 524	
		2.4.5.3	Preparation: Context and Reactions 527	
2.4.3.7.4	Oxidative and Reductive Solid-State Ion Exchange 503	2.4.5.3.1	Grafting 527	
2.4.3.8	A Tabulated Survey of the Systems Studied in SSIE and Related Processes 504	2.4.5.3.2	Anchoring 528	
		2.4.5.4	Characterization of Anchored/Grafted Complexes 528	
2.4.3.9	Concluding Remarks 504 References 504	2.4.5.5	Grafted and Anchored Materials as Heterogeneous Catalysts: Selected Overview 529	
2.4.4	Metal Clusters in Zeolites 510 *Wolfgang M. H. Sachtler and Z. Conrad Zhang*	2.4.5.5.1	Grafting 529	
		2.4.5.5.2	Anchoring 533	
2.4.4.1	Introduction 510	2.4.5.6	Conclusions 536 References 537	
2.4.4.2	Nano Clusters of Metal versus Macroscopic Metals 511			
2.4.4.3	Preparation of Mono- or Bimetal Clusters in Zeolite Cavities 512	2.4.6	Supported Catalysts from Chemical Vapor Deposition and Related Techniques 539 *Mizuki Tada and Yasuhiro Iwasawa*	
2.4.4.3.1	Ion Exchange 512			
2.4.4.3.2	Calcination 512	2.4.6.1	Chemical Vapor Deposition (CVD) Technique 539	
2.4.4.3.3	Reduction 513			
2.4.4.4	Interaction of Metal Clusters and Zeolite Protons 513	2.4.6.2	CVD Precursors and Equipment 539	
		2.4.6.3	CVD Catalysts 540	
2.4.4.4.1	Brønsted Acids in Zeolites 513	2.4.6.3.1	Gold (Au) 540	
2.4.4.4.2	Metal Clusters as Lewis Bases 513	2.4.6.3.2	Cobalt (Co) 540	
2.4.4.4.3	Chemical Anchoring 514	2.4.6.3.3	Chromium (Cr) 541	
2.4.4.4.4	Heterogeneous Catalysis by Single Metal Atom Sites 515	2.4.6.3.4	Copper (Cu) 542	
		2.4.6.3.5	Iron (Fe) 542	

2.4.6.3.6	Gallium (Ga) 542		2.4.8.3.3	The Practical Use of Mills: Some Comments 575
2.4.6.3.7	Molybdenum (Mo) 542		2.4.8.4	The Use of Mechanochemistry for the Synthesis of Catalysts 577
2.4.6.3.8	Niobium (Nb) 544		2.4.8.5	The Influence of Mechanical Activation on Catalytic Properties 578
2.4.6.3.9	Nickel (Ni) 544			
2.4.6.3.10	Osmium (Os) 545		2.4.8.5.1	The Influence of Mechanical Activation on Catalytic Activity 578
2.4.6.3.11	Palladium (Pd) Films and Particles 545			
2.4.6.3.12	Pd/Zeolites 545		2.4.8.5.2	The Influence of Mechanical Activation on Catalytic Selectivity 579
2.4.6.3.13	Platinum (Pt) 545			
2.4.6.3.14	Rhenium (Re) 546		2.4.8.6	Pseudocatalytic and Catalytic Reactions During Mechanical Activation 580
2.4.6.3.15	Rhodium (Rh) 548			
2.4.6.3.16	Ruthenium (Ru) 548		2.4.8.7	Conclusions 580
2.4.6.3.17	SiO$_2$ 548			References 581

2.4.6.3.18 Tin (Sn) 549
2.4.6.3.19 Titanium (Ti) 549
2.4.6.3.20 Vanadium (V) 550
2.4.6.3.21 Tungsten (W) 551
2.4.6.3.22 Zirconium (Zr) 551
2.4.6.3.23 Bimetals 551
2.4.6.3.24 Mixed Metal Oxides and Sulfides 552
2.4.6.4 CVD-Related Techniques 552
2.4.6.5 Concluding Remarks 553
References 553

2.4.7 Spreading and Wetting 555
Helmut Knözinger and Edmund Taglauer
2.4.7.1 Introduction 555
2.4.7.2 Theoretical Considerations 556
2.4.7.2.1 Thermodynamics of Wetting and Spreading 556
2.4.7.2.2 Qualitative Discussion of the Dynamics of Interfacial Processes 558
2.4.7.3 Supported Metal Catalysts 561
2.4.7.3.1 Sintering and Redispersion 561
2.4.7.3.2 Strong Metal Support Interactions (SMSI) 563
2.4.7.3.3 Bimetallic Catalysts 564
2.4.7.4 Supported Oxide Catalysts 565
2.4.7.4.1 Molybdenum-Based Catalysts 566
2.4.7.4.2 Vanadium-Based Catalysts 568
2.4.7.4.3 Tungsten-Based Catalysts 569
2.4.7.4.4 Zeolites as Supports 569
2.4.7.5 Solid-State Ion Exchange of Zeolites 569
References 569

2.4.8 Mechanochemical Methods 571
Bernd Kubias, Martin J. G. Fait, and Robert Schlögl
2.4.8.1 Introduction 571
2.4.8.2 The Effect of Mechanical Activation on the Reactivity of Solids 572
2.4.8.3 Mills 573
2.4.8.3.1 Stress Types 573
2.4.8.3.2 Types and Mode of Action of Mills 574

2.4.9 Immobilization of Molecular Catalysts 583
Reiner Anwander
2.4.9.1 Introduction and Scope 583
2.4.9.2 Metal Complex–Support Interaction 584
2.4.9.3 Physically Immobilized ("Entrapped") Metal Complexes 586
2.4.9.3.1 Intrazeolite Assembly: Ship-in-a-Bottle Catalysts 586
2.4.9.3.2 Entrapment in Sol–Gel and Polysiloxane Matrices 587
2.4.9.3.3 Microencapsulation and Polymer-Incarcerated Methods 588
2.4.9.3.4 Supramolecular Complexation: Mechanical Immobilization 589
2.4.9.4 Tethered Metal Complexes 589
2.4.9.4.1 Multifunctional Surfaces 589
2.4.9.4.2 Spatial Restrictions Imparted by Periodic Mesoporous Silica 590
2.4.9.4.3 Site Isolation via Spatial Patterning of the Silica Surface 591
2.4.9.5 Grafted Metal Complexes: Surface Organometallic Chemistry (SOMC) 592
2.4.9.5.1 Alkyl-Based SOMC 593
2.4.9.5.2 Alkoxide-Based SOMC 595
2.4.9.5.3 Amide-Based SOMC 596
2.4.9.5.4 Molecular Model Oxo–Surfaces 597
2.4.9.6 Metal–Organic Assemblies: Self-Immobilized Homogeneous Catalysts 598
2.4.9.6.1 Metal–Organic Frameworks 599
2.4.9.6.2 Matrix-Embedded Catalysts 601
2.4.9.6.3 Liquid Crystal Assemblies 605
References 608

2.4.10 Zeolite-Entrapped Metal Complexes 614
Stefan Ernst
2.4.10.1 Introduction 614

2.4.10.2	Synthesis of Zeolite-Entrapped Metal Complexes 614	2.4.12.4	Conclusion 652
			References 652
2.4.10.2.1	Flexible-Ligand Method 615		
2.4.10.2.2	Ship-in-the-Bottle Method 617	2.5	Formation of the Final Catalyst 655
2.4.10.2.3	Zeolite Synthesis Method 618	2.5.1	Reactions During Catalyst Activation 655
2.4.10.3	Characterization 618		*Bernard Delmon*
2.4.10.3.1	Electron Paramagnetic Resonance Spectroscopy 619	2.5.1.1	Introduction 655
		2.5.1.2	Particular Features of Chemical Reactions of Supported Solids 657
2.4.10.3.2	Electron Excitation by Ultraviolet/Visible Light 619	2.5.1.2.1	Characteristics of Reactions of Solids: Possible Consequences in the Kinetics of Activation 657
2.4.10.3.3	X-Ray Photoelectron Spectroscopy 621		
2.4.10.3.4	Infrared Spectroscopy 622	2.5.1.2.2	Interaction Between Supported Phases and Supports 658
2.4.10.3.5	Nuclear Magnetic Resonance Spectroscopy 622	2.5.1.3	Activation of Supported Catalysts by Calcination 660
2.4.10.3.6	Raman Spectroscopy 622		
2.4.10.3.7	Stability Analysis 623	2.5.1.4	Activation of Supported Catalysts by Reduction 661
2.4.10.3.8	Oxygen Adsorption 623		
2.4.10.3.9	Cyclovoltammetry 623	2.5.1.4.1	General: Effect of Precursor Dispersion 662
2.4.10.4	Catalysis by Zeolite-Entrapped Transition Metal Complexes 623		
		2.5.1.4.2	Overall Effects: Influence of Calcination Temperature in Precursor Preparation 662
2.4.10.4.1	Selective Oxidations on Zeolite-Encaged Transition Metal Complexes 624		
		2.5.1.4.3	Role of Precursors 663
2.4.10.4.2	Selective Hydrogenations on Zeolite-Encaged Transition Metal Complexes 626	2.5.1.4.4	Influence of the Amount of Deposited Precursor Loaded on the Support 663
		2.5.1.4.5	Effect of the Formation of Compounds between the Precursor and the Support 664
2.4.10.4.3	Catalysis by Zeolite-Encapsulated Chiral Complexes 627		
2.4.10.4.4	Miscellaneous Reactions 627	2.5.1.4.6	Nature of the Support 664
2.4.10.5	Conclusions 628	2.5.1.4.7	Effect of Modifiers (Promotors) 665
	References 628	2.5.1.4.8	Influence of the Activation Conditions 666
		2.5.1.4.9	Final Remarks 666
2.4.11	Supported Liquid Catalysts 631	2.5.1.5	Reduction–Sulfidation 667
	Anders Riisager,	2.5.1.5.1	Fundamental Data 667
	Rasmus Fehrmann, and Peter Wasserscheid	2.5.1.5.2	Influence of the Sequence of Steps in Reduction-Sulfidation 668
2.4.11.1	Introduction 631		
2.4.11.2	Historical Development 631	2.5.1.5.3	Influence of the Activation Temperature 669
2.4.11.3	The Sulfuric Acid Catalyst 633		
2.4.11.4	Solid Phosphoric Acid (SPA) Catalyst 635	2.5.1.5.4	Influence of the Sulfiding Molecule 670
2.4.11.5	Supported Ionic Liquid-Phase (SILP) Catalysts 637	2.5.1.5.5	Other Results 670
		2.5.1.5.6	Outlook 670
2.4.11.6	Conclusions 642	2.5.1.6	Activation of Other Catalysts 671
	References 642	2.5.1.7	Conventional Activation and the Real State of Catalysts during Catalysis 672
2.4.12	Immobilization of Biological Catalysts 644		
	Marion B. Ansorge-Schumacher	2.5.1.8	Conclusions 672
2.4.12.1	Introduction 644		References 673
2.4.12.2	Immobilization by Attachment 645		
2.4.12.2.1	Carrier-Less Crosslinking 645	2.5.2	Catalyst Forming 676
2.4.12.2.2	Adsorption onto a Carrier 647		*Ferdi Schüth and Michael Hesse*
2.4.12.2.3	Covalent Binding to a Carrier 649	2.5.2.1	Introduction 676
2.4.12.3	Immobilization by Entrapment 650	2.5.2.2	The Physico-Chemical Background of Forming Processes 677
2.4.12.3.1	Entrapment in a Matrix 650		
2.4.12.3.2	Entrapment in Membranes 651	2.5.2.3	Spray Drying 679

2.5.2.3.1	The Spray-Drying Process	680	2.6.1.2.2	EUROPT-2	706

2.5.2.3.1   The Spray-Drying Process   680
2.5.2.3.2   Atomizers   681
2.5.2.3.3   Dryer Configuration   683
2.5.2.4   Extrusion   683
2.5.2.4.1   Basic Considerations   683
2.5.2.4.2   Equipment   685
2.5.2.4.3   Paste Composition and Additives   686
2.5.2.4.4   Drying and Calcination   689
2.5.2.5   Tableting   690
2.5.2.5.1   Basic Considerations   690
2.5.2.5.2   Equipment   691
2.5.2.5.3   Formulation of the Feed   692
2.5.2.6   Granulation   694
2.5.2.6.1   Basic Considerations   695
2.5.2.6.2   Equipment   696
2.5.2.7   Miscellaneous Techniques   697
2.5.2.7.1   Oil Drop Coagulation   697
2.5.2.7.2   Pastillation   698
2.5.2.7.3   Coating of Preshaped Bodies   698
2.5.2.8   Perspectives   698
             References   698

2.6   Standard Catalysts   700
2.6.1   Non-Zeolitic Standard Catalysts   700
             Geoffrey C. Bond
2.6.1.1   Introduction   700
2.6.1.2   EUROCAT Metal Catalysts   701
2.6.1.2.1   EUROPT-1   701
2.6.1.2.2   EUROPT-2   706
2.6.1.2.3   EUROPT-3 and EUROPT-4   706
2.6.1.2.4   EURONI-1   708
2.6.1.3   Other EUROCAT Catalysts   710
2.6.1.3.1   Vanadia-Titania EUROCAT Oxides   710
2.6.1.3.2   EUROTS-1 Zeolite   712
2.6.1.4   Gold Reference Catalysts   712
2.6.1.5   Other Programs   713
2.6.1.6   Summary and Conclusions   714
             References   714

2.6.2   Zeolite Standard Catalysts and Related Activities of the International Zeolite Association   715
             Michael Stöcker and Jens Weitkamp
2.6.2.1   Introduction   715
2.6.2.2   Atlas of Zeolite Framework Types   715
2.6.2.3   Verified Syntheses of Zeolitic Materials   716
2.6.2.4   Preparation of the Standard Large-Pore, Acidic Zeolite Catalyst La,Na-Y   717
2.6.2.5   Apparatus and Procedure for Catalyst Testing   717
2.6.2.6   Comparison of the Results Achieved in Five Different Laboratories   718
2.6.2.7   Conclusions   718
             References   719

List of Contributors

Zarah Ainbinder[†]
formerly Central Research and Development
E. I. du Pont de Nemours and Company
Wilminton, DE
USA

Peter W. Albers
AQura GmbH
AQ-EM
Rodenbacher Chaussee 4, post code 915 D115
63457 Hanau
Germany

Masakazu Anpo
Osaka Prefecture University
Department of Applied Chemistry
1-1 Gakuencho, Naka-ku
Sakai, Osaka 599-8531
Japan

Marion B. Ansorge-Schumacher
TU Berlin
Institute of Chemistry
Department of Enzyme Technology
Straße des 17. Juni 124
10623 Berlin
Germany

Reiner Anwander
University of Bergen
Department of Chemistry
Allégaten 41
5007 Bergen
Norway

Heiko Arnold
Technische Universität Darmstadt
Fachbereich Chemie
Petersenstr. 20
64287 Darmstadt
Germany

Frédéric Averseng
Université Pierre et Marie Curie-Paris 6
Laboratoire de Réactivité de Surface–UMR 7609 CNRS
Case 178, 4 place Jussieu
75252 Paris Cedex 05
France

Sean A. Axon
Johnson Matthey Technology Centre
P.O. Box 1
Belasis Avenue
Billingham Cleveland
TS23 1LB
UK

Manfred Baerns
Ruhr-Universität Bochum
Lehrstuhl für Technische Chemie
Universitätsstr. 150
44801 Bochum
Germany

currently:
Fritz-Haber-Institute of Max-Planck Society
Department of Inorganic Chemistry
Faradayweg 4-6
14195 Berlin
Germany

Alfons Baiker
ETH Zurich
Department of Chemistry and Applied Biosciences
Hönggerberg, HCI
8093 Zurich
Switzerland

Handbook of Heterogeneous Catalysis, 2nd Ed.
Edited by G. Ertl, H. Knözinger, F. Schüth, and J. Weitkamp
Copyright © 2008 Wiley-VCH Verlag GmbH & Co. KGaA, Weinheim
ISBN: 978-3-527-31241-2

Nicolae Barsan
University of Tübingen
Faculty of Chemistry
IPTC
Auf der Morgenstelle 8
72076 Tübingen
Germany

Mark A. Barteau
University of Delaware
Department of Chemical Engineering
Colburn Laboratory
Newark, DE 19716
USA

Yann Batonneau
Université de Poitiers
UMR 6503 CNRS
Catalyse par les métaux
Laboratoire de catalyse en chimie organique
Bâtiment de chimie - 1er étage nord–108
40, av. du recteur Pineau
86022 Poitiers cedex
France

Colin Baudouin
AXENS–Catalysts & Adsorbents
89, Boulevard Franklin Roosvelt
92508 Rueil Malmaison
France

Frank Bauer
Leibniz-Institut für Oberflächenmodifizierung e.V.
Permoserstraße 15
04318 Leipzig
Germany

Peter Baumeister
CHEMGO Organica AG
Leonhardsgraben 36, Postfach
4003 Basel
Switzerland

Jeffrey S. Beck
ExxonMobil Research and Engineering
Corporate Strategic Research
1545 Route 22
Annandale, NJ 08801
USA

Robert G. Bell
University College London
Davy Faraday Research Laboratory
3rd Floor
Kathleen Lonsdale Building
Gower Street
London WC1E 6BT
UK

Giuseppe Bellussi
Eni S.p.A.
Direzione Ricerca e Sviluppo Tecnologico
Divisione Refining & Marketing
Centro Ricerche di San Donato Milanese (MI)
Via Maritano 26
20097 San Donato Milanese
Italy

Rob J. Berger
Delft University of Technology
Faculty of Applied Sciences
DelftChemTech, Catalysis Engineering
Julianalaan 136
2628 BL Delft
The Netherlands

Gérard Bergeret
Université Lyon 1–CNRS
Institut de Recherches sur la Catalyse et l'Environment de Lyon
2 Avenue Albert Einstein
69626 Villeurbanne Cedex
France

Jaap A. Bergwerff
Albemarle Catalysts Company b.v.
Research Center Catalysts
P.O. Box 37650
1030 BE Amsterdam
The Netherlands

Flemming Besenbacher
University of Århus
Interdisciplinary Nanoscience Center (iNANO)
and Department of Physics and Astronomy
Ny Munkegade, building 1521
8000 Århus C
Denmark

Michèle Besson
IRCELYON
Institut de Recherches sur la Catalyse et l'Environment de Lyon
2 Avenue Albert Einstein
69626 Villeurbanne Cedex
France

Martin Bewersdorf
Evonik Degussa GmbH
Process Technology & Engineering
Rodenbacher Chaussee 4
63457 Hanau-Wolfgang
Germany

Hans-Ulrich Blaser
Solvias AG
P.O. Box
4002 Basel
Switzerland

Jochen H. Block†
formerly Max-Planck-Gesellschaft
Fritz-Haber-Institut
Berlin
Germany

Paula L. Bogdan
UOP LLC
50 East Algonquin Road
Des Plaines, IL 60017
USA

Geoffrey C. Bond
59 Nightingale Road
Rickmansworth
Hertfordshire WD3 7 BU
UK

Silvia Bordiga
University of Torino
Dipartimento di Chimica IFM
and NIS Centre of Excellence
Via Pietro Giuria 7
10125 Torino
Italy

Michel Boudart
Stanford University
Department of Chemical Engineering
381 North South Mall
Stanford, CA 94305
USA

François Bozon-Verduraz
Université Paris 7 – Denis Diderot
LCMDC, case 7090
Bât. 44-45, 5eme étage
2, place Jussieu
75251 Paris Cedex 05
France

Sigmar Bräuninger
BASF AG
Chemicals Research and Engineering
67056 Ludwigshafen
Germany

Susanne Brosda
University of Patras
Department of Chemical Engineering
1, Caratheodory St
26504 Patras
Greece

Stephen H. Brown
ExxonMobil Research and Engineering
Corporate Strategic Research
1545 Route 22
Annandale, NJ 08801
USA

Grigorii A. Bunimovich
Matros Technologies, Inc.
14963 Green Circle Drive
Chesterfield, MI 63017
USA

James D. Burrington
The Lubrizol Corporation
29400 Lakeland Blvd.
Wickliffe, OH 44092
USA

Tilman Butz
Universität Leipzig
Fakultät für Physik und Geowissenschaften
Linnestr. 5
04103 Leipzig
Germany

Nelson Cardona Martinez
University of Puerto Rico – Mayagüez
Department of Chemical Engineering
P.O. Box 9046
Mayagüez, PR 00681-9046
USA

Jürgen Caro
Universität Hannover
Institut für Physikalische Chemie und Elektrochemie
Callinstr. 3-3A
30167 Hannover
Germany

Xavier Carrier
Université Pierre et Marie Curie-Paris 6
Laboratoire de Réactivité de Surface–UMR 7609 CNRS
Case 178, 4 place Jussieu
75252 Paris Cedex 05
France

John L. Casci
Johnson Matthey Technology Centre
P.O. Box 1
Belasis Avenue
Billingham TS23 1LB
UK

Karl Josef Caspary
Uhde GmbH
Research and Development Division
Friedrich-Uhde-Str. 15
44141 Dortmund
Germany

Richard Catlow
University College London
Davy Faraday Research Laboratory
3rd Floor
Kathleen Lonsdale Building
Gower Street
London WC1E 6BT
UK

Michel Che
Université Pierre et Marie Curie-Paris 6
Laboratoire de Réactivité de Surface–UMR 7609 CNRS
Case 178, 4 place Jussieu
75252 Paris Cedex 05
France

Wu-Cheng Cheng
W. R. Grace & Co.–Conn.
7500 Grace Drive
Columbia, MD 21044
USA

Alessandro Cimino
Università La Sapienza
Dipartimento di Chimica
Piazzale Aldo Moro, 5
00185 Roma
Italy

Michael C. Clark
ExxonMobil Research and Engineering
Process Engineering
3225 Gallows Road
Fairfax, VA 22037
USA

Peter Claus
Technische Universität Darmstadt
Technische Chemie II
Petersenstr. 20
64287 Darmstadt
Germany

Oliver Clause
Institut Français du Pétrole
Catalysis and Separation Direction
BP 3
69390 Vernaison
France

Cinzia Cristiani
Politecnico di Milano
CMIC–Dipartimento di Chimica, Materiali, Ingegneria
Chimica "Giulio Natta"
Piazza Leonardo da Vinci 32
20133 Milano
Italy

Weilin Dai
Fudan University
Department of Chemistry
Shanghai 200433
P.R. China

Srinivas Darbha
National Chemical Laboratory
Pune 411 008
India

Abhaya K. Datye
University of New Mexico
Department of Chemical & Nuclear Engineering
MSC01 1120, Albuquerque, NM 87131-0001
USA

Burtron H. Davis
University of Kentucky
Center for Applied Energy Research
2540 Research Park Drive
Lexington, KY 40511-8410
USA

Bernard Delmon
Université Catholique de Louvain
Catalyse et Chimie des Matériaux divisés
Place Croix du Sud 2/17
1348 Louvain-la-Neuve
Belgium

Olaf Deutschmann
Universität Karlsruhe (TH)
Institute for Chemical Technology and Polymer Chemistry
Engesserstr. 20
76131 Karlsruhe
Germany

Michel Devillers
Université catholique de Louvain
Chimie des Matériaux inorganiques et organiques
SC/CHIM/CMAT, Place Louis Pasteur 1
1348 Louvain-la-Neuve
Belgium

Dirk De Vos
Katholieke Universiteit Leuven
Centre for Surface Chemistry and Catalysis
Kasteelpark Arenberg 23
3001 Leuven
Belgium

Simon Diezi
ETH Zurich
Department of Chemistry and Applied Biosciences
Hönggerberg, HCI
8093 Zurich
Switzerland

Uwe Dingerdissen
Leibniz-Institut für Katalyse
Richard-Willstätter-Str. 12
12489 Berlin
Germany

Roland Dittmeyer
Karl-Winnacker-Institut der DECHEMA
Technische Chemie
Theodor-Heuss-Allee 25
60486 Frankfurt
Germany

Frank Döbert
Technische Universität Darmstadt
Fachbereich Chemie
Petersenstr. 20
64287 Darmstadt
Germany

Mark E. Dry
University of Cape Town
Department of Chemical Engineering
Private Bag
7701 Cape Town
Republic of South Africa

James A. Dumesic
University of Wisconsin
Department of Chemical and Biological Engineering
1415 Engineering Drive
Madison, WI 53706
USA

Gerhart Eigenberger
Universität Stuttgart
Institut für Chemische Verfahrenstechnik
Böblingerstr. 72
70199 Stuttgart
Germany

Gerhard Emig
Universität Erlangen-Nürnberg
Technische Chemie I
Egerlandstr. 3
91058 Erlangen
Germany

Stefan Ernst
University of Kaiserslautern
Department of Chemistry–Chemical Technology
P.O. Box 3049
67653 Kaiserslautern
Germany

Gerhard Ertl
Fritz-Haber-Institut der
Max-Planck-Gesellschaft
Faradayweg 4-6
14195 Berlin
Germany

Douglas H. Everett
The Universoty of Bristol
School of Chemistry
Department of Physical Chemistry
Bristol BS8 1TS
UK

Martin J. G. Fait
Leibniz Institut für Katalyse an der Universität Rostock,
Außenstelle Berlin
Bereich Hochdurchsatztechnologien
Richard-Willstätter-Str. 12
12489 Berlin
Germany

Rasmus Fehrmann
Technical University of Denmark
Department of Chemistry
Building 207
2800 Kgs. Lyngby
Denmark

Eddy J. P. Feijen
Oleon NV
Assenedestraat 2
9940 Ertvelde
Belgium

Thomas Fetzer
BASF AG
Carl-Bosch-Str. 38
67056 Ludwigshafen
Germany

Gerhard Fink
Max-Planck-Institut für Kohlenforschung
Kaiser-Wilhelm-Platz 1
45470 Mülheim-Ruhr
Germany

Laura Forni
University of Bologna
Department of Industrial Chemistry and Materials
Viale Risorgimento, 4
40136 Bologna
Italy

Pio Forzatti
Politecnico di Milano
CMIC–Dipartimento di Chimica, Materiali, Ingegneria
Chimica "Giulio Natta"
Piazza Leonardo da Vinci 32
20133 Milano
Italy

Hans-Joachim Freund
Fritz-Haber-Institut der Max-Planck-Gesellschaft
Abteilung Chemische Physik
Faradayweg 4-6
14195 Berlin
Germany

Jose Fripiat
retired, formerly University of Louvain
95 Montserrat
04330 Coyoacan, DF
Mexico

Terje Fuglerud
Hydro Polymers AS
Herøya Research Park
P.O. Box 2560
3908 Porsgrunn
Norway

Pär Gabrielsson
Haldor Topsøe A/S
Nymøllevej 55
2800 Lyngby
Denmark

Ewald F. Gallei
BASF AG
67056 Ludwigshafen
Germany

Pierre Gallezot
Université Lyon 1–CNRS
Institut de Recherches sur la Catalyse et l'Environment de
Lyon
2 Avenue Albert Einstein
69626 Villeurbanne Cedex
France

Bruce C. Gates
University of California, Davis
Department of Chemical Engineering and Materials Science
Davis, CA 95616
USA

Jürgen Garche
Center for Solar Energy and Hydrogen Research
Baden-Warttemberg
Helmholtzstr. 8
89075 Ulm
Germany

Hubert A. Gasteiger
General Motors Corporation
Fuel Cell Activities
10 Carriage Street
Honeoye Falls, NY 14472
USA

Johann Gaube
Technische Universität Darmstadt
Fachbereich Chemie
Petersenstr. 20
64287 Darmstadt
Germany

Helmut Gehrke
Uhde GmbH
Research and Development Division
Neubeckumer Str. 127
59320 Ennigerloh
Germany

John W. Geus
Utrecht University
Room West 524, Hugo R. Kruytbuilding
Padualaan 8
3584 CH Utrecht
The Netherlands

Lynn F. Gladden
University of Cambridge
Department of Chemical Engineering
Pembroke Street
Cambridge
CB2 3RA
UK

Roger Gläser
University of Leipzig
Institute of Chemical Technology
Linnéstr. 3
04103 Leipzig
Germany

D. Wayne Goodman
Texas A&M University
Department of Chemistry
College Station, TX 77842
USA

Robert K. Grasselli
University of Delaware
Center for Catalytic Science and Technology
Newark, DE 19716
USA

and

Technische Universität München
Department of Chemistry
Lichtenbergstr. 4
85748 Garching
Germany

Gianpiero Groppi
Politecnico di Milano
CMIC–Dipartimento di Chimica, Materiali, Ingegneria
Chimica "Giulio Natta"
Piazza Leonardo da Vinci 32
20133 Milano
Italy

Henrik Guldberg Pedersen
Haldor Topsøe A/S
Nymøllevej 55
2800 Lyngby
Denmark

Bernhard Gutsche
Cognis GmbH
Process Development
P. O. Box 13 01 64
40551 Düsseldorf
Germany

Jerzy Haber
Polish Academy of Sciences
Institute of Catalysis and Surface Chemistry
Ul. Niezapominajek 8
30-239 Krakow
Poland

E. Thomas Habib, Jr.
W. R. Grace & Co.–Conn.
7500 Grace Drive
Columbia, MD 21044
USA

Matthias W. Haenel
Max-Planck-Institut für Kohlenforschung
Kaiser-Wilhelm-Platz 1
45470 Mülheim an der Ruhr
Germany

Christian Hagelüken
Umicore AG & Co. KG
Precious Metals Refining
Rodenbacher Chaussee 4
63457 Hanau
Germany

Silje Fosse Håkonsen
Norwegian University of Science and Technology
Department of Chemical Engineering
7491 Trondheim
Norway

Ahmad Hammad
University of Patras
Department of Chemical Engineering
1, Caratheodory St
26504 Patras
Greece

John Bøgdil Hansen
Haldor Topsoe, A/S
Nymollevej, 55
2800 Kgs. Lyngby
Denmark

Poul Lenvig Hansen
Haldor Topsoe, A/S
Nymollevej, 55
2800 Kgs. Lyngby
Denmark

Stephan T. Hatscher
BASF AG
Carl-Bosch-Str. 38
67056 Ludwigshafen
Germany

R. A. W. Haul
National Chemical Research Laboratory
Pretoria
South Africa

Max Heinritz-Adrian
Uhde GmbH
Gas Technology Division
Friedrich-Uhde-Str. 15
44141 Dortmund
Germany

Stig Helveg
Haldor Topsøe, A/S
Nymøllevej, 55
2800 Kgs. Lyngby
Denmark

Emiel J. M. Hensen
Eindhoven University of Technology
Department of Chemical Engineering and Chemsitry
P.O. Box 513
5600 MB Eindhoven
The Netherlands

Daniel Herein
Leibniz-Institute for Catalysis
Richard-Willstätter-Str. 12
12489 Berlin
Germany

Michael Hesse
BASF AG
G-CCP/KR–E 100
67056 Ludwigshafen
Germany

Kai-Olaf Hinrichsen
Technische Universität München
Lehrstuhl I für Technische Chemie
Lichtenbergstr. 4
85747 Garching b. München
Germany

Arend Hoek
Shell Global Solutions Int.
XTL department
Badhuisweg 3
1031 CM Amsterdam
The Netherlands

Martin Holena
Institute of Computer Science of the
Academy of Sciences of Czech Republic

currently:
Leibniz-Institute for Catalysis at University of Rostock
Branch Berlin
Richard-Willstätter-Str. 12
12489 Berlin
Germany

Anders Holmen
Norwegian University of Science and Technology
Department of Chemical Engineering
7491 Trondheim
Norway

Annett Horn
University of Rostock
Department of Chemistry
Albert-Einstein-Str. 3 a
18059 Rostock
Germany

Russell Francis Howe
University of Aberdeen
Chemistry Department
Meston Building
Aberdeen AB24 3UE
UK

George W. Huber
University of Massachusetts–Amherst
Chemical Engineering Department
112 Goessmann Laboratory
686 North Pleasant Street
Amherst, MA 01003
USA

Michael Hunger
University of Stuttgart
Institute of Chemical Technology
Pfaffenwaldring 55
70569 Stuttgart
Germany

Laurent G. Huve
Shell RTCA
Dept. GSRH
P.O. Box 38000
1030 BN Amsterdam
The Netherlands

Hicham Idriss
The University of Auckland
Department of Chemistry
Room 527 A
Private Bag 92019
Auckland
New Zealand

Yasuhiro Iwasawa
The University of Tokyo
Department of Chemistry
Graduate School of Science
7-3-1 Hongo, Bunkyo-ku
Tokyo 113-0033
Japan

Pierre A. Jacobs
K.U. Leuven
Centrum Oppervlaktechemie en Katalyse
Kasteelpark Arenberg 23 bus 2461
3001 Leuven
Belgium

Friederike C. Jentoft
Fritz-Haber-Institut der Max-Planck-Gesellschaft
Abteilung Anorganische Chemie
Faradayweg 4-6
14195 Berlin
Germany

Gregor Jenzer
Shell Deutschland Oil GmbH
Rhineland Refinery
Ludwigshafenerstr. 1
50389 Wesseling
Germany

Hervé Jobic
Université de Lyon
IRCELYON, Institut de Recherches sur la Catalyse et l'Environment de Lyon
CNRS
UMR 5256
2 Avenue Albert Einstein
69626 Villeurbanne Cedex
France

Gert Jonkers
Eindhoven University of Technology
Department of Chemical Engineering and Chemsitry
P.O. Box 513
5600 MB Eindhoven
The Netherlands

Arvids Judzis, Jr.
CDTECH
10100 Bay Area Boulevard
Pasadena, TX 77507
USA

Eckhard Jüngst
Lurgi AG
Lurgiallee 5
60295 Frankfurt am Main
Germany

Wilfried Kalchauer
Wacker Chemie AG
Wacker Silicones—Basic Materials
Johannes Hess Str. 24
84489 Burghausen
Germany

Takashi Kamegawa
Osaka Prefecture University
Department of Applied Chemistry
1-1 Gakuencho, Naka-ku
Sakai, Osaka 599-8531
Japan

Charles J. Kappenstein
University of Poitiers
LACCO UMR 6503
Laboratoire de Catalyse par les Métaux
40 Avenue du Recteur Pineau
86022 Poitiers Cedex
France

Freek Kapteijn
Delft University of Technology
Faculty of Applied Sciences
DelftChemTech, Catalysis Engineering
Julianalaan 136
2628 BL Delft
The Netherlands

Hellmut G. Karge
Fritz Haber Institute of the Max Planck Society
Faradayweg 8
14195 Berlin
Germany

Jörg Kärger
Universität Leipzig
Abteilung Grenzflächenphysik
04103 Leipzig
Germany

Alexandros Katsaounis
Technical University of Crete
Department of Environmental Engineering
Laboratory of Air Waste Treatment Technology
73100 Chania
Greece

Wilhelm Keim
RWTH Aachen
Institut für Technische und Makromolekulare Chemie
Worringerweg 1
52074 Aachen
Germany

Dieter Kerner
Degussa GmbH
Abt. AS-FA
Rodenbacher Chaussee 4
63457 Hanau-Wolfgang
Germany

Kari I. Keskinen
Neste Jacobs Oy
Process Engineering
P.O. Box 310
06101 Porvoo
Finland

Alain Kiennemann
Université Louis Pasteur Strasbourg
ECPM-LMSPC UMR CNRS 7515
25 rue Becquerel
67087 Strasbourg Cedex 2
France

Johoo Kim
Lehigh University
Center for Advanced Materials
and Nanotechnology
Department of Chemistry
305C Sinclair Laboratory, 7 Asa Drive
Bethlehem, PA 18015-3172
USA

Christine E. A. Kirschhock
K.U. Leuven
Centrum Oppervlaktechemie en Katalyse
Kasteelpark Arenberg 23
3001 Leuven
Belgium

Masaaki Kitano
Osaka Prefecture University
Department of Applied Chemistry
1-1 Gakuencho, Naka-ku
Sakai, Osaka 599-8531
Japan

Lioubov Kiwi-Minsker
École polytechnique fédérale de Lausanne
EPFL-ISIC-GGRC, Station 6
1015 Lausanne
Switzerland

Wolfgang Kleist
Technische Universität München
Department of Chemistry
Lichtenbergstr. 4
85747 Garching
Germany

Freddy Kleitz
Université Laval
Department of Chemistry
Pavillon Alexandre Vachon
G1K 7P4, Québec, QC
Canada

Elias Klemm
Chemnitz University of Technology
Institute of Chemistry
Straße der Nationen 62
09111 Chemnitz
Germany

Heinz-Josef Kneuper
BASF AG
Carl-Bosch-Str. 38
67056 Ludwigshafen
Germany

Helmut Knözinger
Ludwig-Maximilians-Universität München
Department Chemie und Biochemie
Butenandtstr. 5-13 (Haus E)
81377 München
Germany

Karl Kochloefl
Schwarzenbergstr. 15
83026 Rosenheim
Germany

Bruce E. Koel
Lehigh University
Center for Advanced Materials
and Nanotechnology
Department of Chemistry
305C Sinclair Laboratory
7 Asa Drive
Bethlehem, PA 18015-3172
USA

Klaus Köhler
Technische Universität München
Department of Chemistry
Lichtenbergstr. 4
85747 Garching
Germany

Stein Kolboe
University of Oslo
Institute of Chemistry
P.O. Box 1033 Blindern
0315 Oslo
Norway

Evgenii V. Kondratenko
Institute for Applied Chemistry Berlin-Adlershof (ACA)
Richard-Willstätter-Str. 12
12489 Berlin
Germany

Diek C. Koningsberger
Utrecht University
Inorganic and Catalysis Group, Faculty of Science
P.O. Box 80083
3508 TB Utrecht
The Netherlands

Martin Köstner
Evonik Röhm GmbH
Business Unit Methacrylates
Kirschenallee
64293 Darmstadt
Germany

Stefan Kotrel
BASF AG
Chemicals Research and Engineering
67056 Ludwigshafen
Germany

Herman W. Kouwenhoven
Rozewerf 17
1156 CX Marken
The Netherlands

Udo Kragl
University of Rostock
Department of Chemistry
Albert-Einstein-Str. 3 a
18059 Rostock
Germany

A. Outi I. Krause
Helsinki University of Technology
Chemical Technology
P.O. Box 6100
02015 TKK
Finland

Katharina Krischer
Technische Universität München
Physik Department, E19
James-Franck-Str. 1
85748 Garching
Germany

Norbert Kruse
Université Libre de Bruxelles
Chemical Physics of Materials (Catalysis – Tribology)
Campus Plaine, CP 243
1050 Brussels
Belgium

Bernd Kubias
Fritz-Haber-Institut der Max-Planck-Gesellschaft
Abteilung Anorganische Chemie
Faradayweg 4-6
14195 Berlin
Germany

Shefali Kumar
University of Rostock
Department of Chemistry
Albert-Einstein-Str. 3 a
18059 Rostock
Germany

Steinar Kvisle
Hydro Polymers AS
Herøya Research Park
P.O. Box 2560
3908 Porsgrunn
Norway

Milos Kraus[†]
formerly Institute of Chemical Process Fundamentals
CSAD
Praha
Czech Republik

Miron V. Landau
Ben-Gurion University of the Negev
Chemical Engineering Department
Ben-Gurion Av. 1, P.O. Box 653
84105 Beer-Sheva
Israel

Armin Lange de Oliveira
hte Aktiengesellschaft
Kurpfalzring 104
69123 Heidelberg
Germany

Jeppe Vang Lauritsen
University of Århus
Interdisciplinary Nanoscience Center (iNANO) and
Department of Physics and Astronomy
Ny Munkegade, building 1521
8000 Århus C
Denmark

Kwan-Young Lee
Korea University
Department of Chemical and Biological Engineering
Anam-Dong, Sungbuk-Du
Seoul 136-701
Korea

Andreas Liese
Hamburg University of Technology (TUHH)
Institute of Technical Biocatalysis
Denickestr. 15
21073 Hamburg
Germany

Vladimir A. Likholobov
Institute of Hydrocarbons Processing
Omsk Scientific Center of the Siberian Branch of the Russian Academy of Sciences
Neftezavodskaya street, 54
Omsk 644040
Russia

Karl Petter Lillerud
University of Oslo
Institute of Chemistry
P.O. Box 1033 Blindern
0315 Oslo
Norway

Suljo Linic
University of Michigan
Department of Chemical Engineering
Herbert H. Dow Building
Ann Arbor, MI 48109
USA

Egbert S. J. Lox
Umicore N.V./S.A.
Group Research and Development
Kasteelstraat 7
2250 Olen
Belgium

Theo M. Maesen
Chevron, Energy Technology Company
100 Chevron Way
Richmond, CA 94802-0627
USA

Tamas Mallat
ETH Zurich
Department of Chemistry and Applied Biosciences
Hönggerberg, HCI
8093 Zurich
Switzerland

Michael D. Mantle
University of Cambridge
Department of Chemical Engineering
Pembroke Street
Cambridge
CB2 3RA
UK

Leo E. Manzer
Catalytic Insights, LLC'
714 Burnley Road
Wilminton, DE 19803
USA

Eric Marceau
Université Pierre et Marie Curie-Paris 6
Laboratoire de Réactivité de Surface–UMR 7609 CNRS
Case 178, 4 place Jussieu
75252 Paris Cedex 05
France

Christian Marcilly
Institut Français du Pétrole
Catalysis and Separation Direction
BP 3
69390 Vernaison
France

Anderson L. Marsh
Lebanon Valley College
Department of Chemistry
Neidig-Garber Science Center
Annville, PA 17003
USA

Johan A. Martens
K.U. Leuven
Centrum Oppervlaktechemie en Katalyse
Kasteelpark Arenberg 23
3001 Leuven
Belgium

Andreas Martin
Leibniz-Institut für Katalyse
Richard-Willstätter-Str. 12
12489 Berlin
Germany

Germain Martino
80 av. F. Lefebvre
78300 Poissy
France

Yurii S. Matros
Matros Technologies, Inc.
14963 Green Circle Drive
Chesterfield, MI 63017
USA

Masaya Matsuoka
Osaka Prefecture University
Department of Applied Chemistry
1-1 Gakuencho, Naka-ku
Sakai, Osaka 599-8531
Japan

Max P. McDaniel
Chevron-Phillips Chemical Co.
Bartlesville Research Center
Routes 60 & 123A
Bartlesville, OK 74004
USA

Ian McKenzie
STFC Rutherford Appleton Laboratory
ISIS Pulsed Muon Facility
OX11 0QX Chilton, Didcot, Oxon
U.K.

Gerhard Mestl
Catalytic Technologies
Süd-Chemie AG
Waldheimer Str. 13
83052 Bruckmühl
Germany

Johannes K. Minderhoud
Shell RTCA
Dept. GSRH
P.O. Box 38000
1030 BN Amsterdam
The Netherlands

Ivano Miracca
Snamprogretti S.p.A.
Onshore Technologies
Viale A. De Gasperi, 16
20097 San Donato Milanese
Italy

Makoto Misono
Kogakuin University
Graduate School of Engineering
Department of Applied Chemistry
Tokyo 163-8677
Japan

Johannes C. Mol
Institute of Molecular Chemistry
Universiteit van Amsterdam
Nieuwe Achtergracht 166
1018 WV Amsterdam
The Netherlands

Árpád Molnár
University of Szeged
Department of Organic Chemistry
Dóm tér 8
6720 Szeged
Hungary

Giuliano Moretti
Sapienza University of Rome
Department of Chemistry
Piazzale Aldo Moro 5
00185 Roma
Italy

Boris L. Moroz
G. K. Boreskov Institute of Catalysis
Siberian Branch of the Russian Academy of Sciences
Lavrent'eva avenue, 5
Novosibirsk 630909
Russia

L. Moscou
Formerly Akzo Chemical
Research Centre Amsterdam
P.O. Box 15
1000 AA Amsterdam
The Netherlands

Mark D. Moser
UOP LLC
50 East Algonquin Road
Des Plaines, IL 60017-5017
USA

Jacob A. Moulijn
Delft University of Technology
Faculty of Applied Sciences
DelftChemTech, Catalysis Engineering
Julianalaan 136
2628 BL Delft
The Netherlands

Ulrich Mueller
BASF Aktiengesellschaft
GCC/PZ–M 301
67056 Ludwigshafen
Germany

Martin Muhler
Ruhr-Universität Bochum
Lehrstuhl für Technische Chemie
44780 Bochum

Mario J. Nappa
Central Research and Development
E. I. du Pont de Nemours and Company
Wilminton, DE
USA

Fritz Näumann
BASF Catalysts LLC
25 Middlesex/Essex Turnpike
P.O. Box 770
Iselin, NJ 08830-0770
USA

Stoyan Nedeltchev
Bulgarian Academy of Sciences
Institute of Chemical Engineering
"Acad. G. Bontchev" Str. Bl. 103
1113 Sofia
Bulgaria

Wolfgang Nehb
Lurgi AG
Lurgiallee 5
60295 Frankfurt am Main
Germany

Alexander V. Neimark
The State University of New Jersey
Department of Chemical and Biochemical
Engineering Rutgers
98 Brett Road
Piscataway, NJ 08854-8058
USA

Matthew Neurock
University of Virginia
Department of Chemical Engineering
School of Engineering and Applied Science
P.O. Box 400741
Charlottesville, VA 22904-4741
USA

Poul Erik Højlund Nielsen
Haldor Topsoe, A/S
Nymollevej, 55
2800 Kgs. Lyngby
Denmark

Hans Niemantsverdriet
Schuit Institute of Catalysis
Eindhoven University of Technology
5600 MB Eindhoven
The Netherlands

Jens K. Nørskov
Center for Atomic-scale Materials Design
Department of Physics NanoDTU
Technical University of Denmark Building 311
2800 Lyngby
Denmark

Cyril T. O'Connor
University of Cape Town
Faculty of Engineering and the Built Environment
University Private Bag
Rondebosch 7701
Republic of South Africa

Unni Olsbye
University of Oslo
Institute of Chemistry
P.O. Box 1033 Blindern
0315 Oslo
Norway

Yücel Önal
Technische Universität Darmstadt
Technische Chemie II
Petersenstr. 20
64287 Darmstadt
Germany

S. Ted Oyama
Virginia Polytechnic Institute
Department of Chemical Engineering
Environmental Catalysis and Nanomaterials Laboratory
Blacksburg, VA 24061
USA

Bernd Pachaly
Wacker Chemie AG
Wacker Silicones–Elastomers
Johannes Hess Str. 24
84489 Burghausen
Germany

George K. Papadopoulos
National Technical University of Athens
School of Chemical Engineering
9 Heroon Polytechniou Str., Zografou Campus
157 80 Athens
Greece

Valentin M. Parmon
Boreskov Institute of Catalysis
Prospekt Akademika Lavrentieva, 5
630090 Novosibirsk
Russia

Carlo Perego
Eni S.p.A.
Direzione Ricerca e Sviluppo Tecnologico
Divisione Refining & Marketing
Centro Ricerche di San Donato Milanese (MI)
Via Maritano 26
20097 San Donato Milanese
Italy

Ralf Peters
Forschungszentrum Jülich GmbH
Institut für Energieforschung–Brennstoffzellen (IEF-3)
Helmholtzring 8
52425 Jülich
Germany

Robert A. Pierotti
The Georgia Institute of Technology
School of Chemistry
Atlanta, GA 30332
USA

Catherine Pinel
IRCELYON
Institut de Recherches sur la Catalyse et l'Environment de Lyon
2 Avenue Albert Einstein
69626 Villeurbanne Cedex
France

Georges Poncelet
retired, formerly University of Louvain
145 rue Latinis
1030 Bruxelles
Belgium

Roel Prins
ETH Zurich
Institute for Chemical and Bio-Engineering
Wolfgang-Pauli-Str. 10, HCI E 125
8093 Zurich
Switzerland

Ulf Prüße
Federal Agricultural Research Centre
Institute of Technology and Biosystems Engineering
Bundesallee 50
38116 Braunschweig
Germany

Benoît Pugin
Solvias AG
P.O. Box
4002 Basel
Switzerland

Ljubisa R. Radovic
Penn State University
Department of Energy and Mineral Engineering
205 Hosler Building
University Park, PA 16802
USA

and

Universidad de Concepcion
Department of Chemical Engineering
Chile

Kuppuswamy Rajagopalan
W. R. Grace & Co.–Conn.
7500 Grace Drive
Columbia, MD 21044
USA

Dave E. Ramaker
George Washington University
Department of Chemistry
2121 I street N.W.
Washington, DC 20052
USA

Paul Ratnasamy
National Chemical Laboratory
Pune 411 008
India

Andreas Reitzmann
SÜD-CHEMIE AG
Research & Development, Catalytic Technologies
Waldheimer Str. 15
83052 Bruckmühl
Germany

Tom J. Remans
Shell Global Solutions Int.
XTL department
Badhuisweg 3
1031 CM Amsterdam
The Netherlands

Liping Ren
Shanghai Research Institute of Petrochemical Technology
1658 Pudung Beilu
Shanghai 201208
P.R. China

Albert Renken
École polytechnique fédérale de Lausanne
EPFL-ISIC-GGRC, Station 6
1015 Lausanne
Switzerland

Fabio H. Ribeiro
Purdue University
School of Chemical Engineering
Forney Hall of Chemical Engineering
West Lafayette, IN 47907
USA

Anders Riisager
Technical University of Denmark
Department of Chemistry
Building 207
2800 Kgs. Lyngby
Denmark

Michael Rinner
Evonik Degussa GmbH
Process Technology & Engineering
Rodenbacher Chaussee 4
63457 Hanau-Wolfgang
Germany

Terry G. Roberie
W. R. Grace & Co.–Conn.
7500 Grace Drive
Columbia, MD 21044
USA

Matthias Rochnia
Degussa GmbH
Abt. AS-FA-AE-FT
Rodenbacher Chaussee 4
63457 Hanau-Wolfgang
Germany

José A. Rodriguez
Brookhaven National Laboratory
Chemistry Department
Upton, NY 11973
USA

Emil Roduner
Universität Stuttgart
Institut für Physikalische Chemie
Pfaffenwaldring 55
70569 Stuttgart
Germany

Frank Rosowski
BASF Aktiengesellschaft
GCC/O - M 301
67056 Ludwigshafen
Germany

Felix Rössler
DSM Nutritional Products AG
Wurmisweg 576
4303 Kaiseraugst
Switzerland

Harald Rößler
Cognis GmbH
Process Development
P. O. Box 13 01 64
40551 Düsseldorf
Germany

Frank Rößner
Carl von Ossietzky University of Oldenburg
Faculty of Mathematic and Natural Sciences
Institute of Pure and Applied Chemistry
Industrial Chemistry II
26111 Oldenburg
Germany

Jens Richard Rostrup-Nielsen
Haldor Topsøe A/S
Nymøllevej 55
2800 Lyngby
Denmark

Jean Rouquerol
Centre de Thermodynamique et de Microcalorimétrie du CNRS
26 rue du 141° RIA
13003 Marseille
France

Agnieszka M. Ruppert
Utrecht University
Inorganic Chemistry and Catalysis group
Department of Chemistry
P.O. Box 80083
3508 TB Utrecht
The Netherlands

Wolfgang M. H. Sachtler
Northwestern University
Department of Chemistry
Center for Catalysis and Surface Science
V. N. Ipatieff Laboratory
Evanston, IL 60208
USA

Nishit Sahay
CDTECH
10100 Bay Area Boulevard
Pasadena, TX 77507
USA

Domenico Sanfilippo
Snamprogretti S.p.A.
Onshore Technologies
Viale A. De Gasperi, 16
20097 San Donato Milanese
Italy

Don S. Santilli
University of California, Davis
Department of Chemical Engineering and Materials Science
Davis, CA 95616
USA

Patrick Sarrazin
IFP–Institut Français du Pétrole
Refining-Petrochemicals Technology Business Unit
1 & 4, av. de Bois-Préau
92852 Rueil-Malmaison
France

Jörg Sauer
Evonik Degussa GmbH
Process Technology & Engineering
Paul-Baumann-Str. 1
45764 Marl
Germany

Elena R. Savinova
Université Louis Pasteur
LMSPC-UMR 7515 du CNRS-ULP-ECPM
25, rue Becquerel
67087 Strasbourg Cedex 2
France

Domenica Scarano
University of Torino
Dipartimento di Chimica IFM and NIS Centre of Excellence
Via Pietro Giuria 7
10125 Torino
Italy

Klaus D. Schierbaum
Heinrich-Heine-Universität Düsseldorf
Materialwissenschaft IpkM, Geb. 25.23.01
Universitätsstr. 1
40225 Düsseldorf
Germany

Robert Schlögl
Fritz-Haber-Institut der Max-Planck-Gesellschaft
Abteilung Anorganische Chemie
Faradayweg 4-6
14195 Berlin
Germany

Friedrich Schmidt
Lug ins Land 52
83024 Rosenheim
Germany

Anita Schnyder
Solvias AG
P.O. Box
4002 Basel
Switzerland

Markus M. Schubert
BASF Aktiengesellschaft
GCC/PZ–M 301
67056 Ludwigshafen
Germany

Matthias Schulz
BASF Catalysts LLC
25 Middlesex/ Essex Turnpike
P.O. Box 770
Iselin, NJ 08830-0770
USA

Adrian Schumpe
Braunschweig University of Technology
Institute of Technical Chemistry
Hans-Sommer-Str. 10
38106 Braunschweig
Germany

Stephan Andreas Schunk
hte Aktiengesellschaft
Kurpfalzring 104
69123 Heidelberg
Germany

Ferdi Schüth
Max-Planck-Institut für Kohlenforschung
Kaiser-Wilhelm-Platz 1
45470 Mülheim an der Ruhr
Germany

Ekkehard Schwab
BASF AG
GCC/PH–M 301
67056 Ludwigshafen
Germany

Meinhard Schwefer
Uhde GmbH
Research and Development Division
Friedrich-Uhde-Str. 15
44141 Dortmund
Germany

Andrew J. Sederman
University of Cambridge
Department of Chemical Engineering
Pembroke Street
Cambridge
CB2 3RA
UK

Swan Tiong Sie
8 Laan van vogelenzang
1217 PH Hilversum
The Netherlands

Teresa Siemieniewska
Technical University of Wrocław
Institute of Chemistry and Technology
of Petroleum and Coal
ul. Gdańska 7/9
50-344 Wrocław
Poland

Peter L. Silverston
University of Waterloo
Department of Chemical Reaction
Waterloo, Ontario, N2L 3G1
Canada

Kenneth S. W. Sing
1 Rosemullion Court
Cliff Road
Budleigh Salterton
Devon EX9 6JU
UK

A. P. Singh
National Chemical Laboratory
Catalysis Division
Pune 411 008
India

Sara E. Skrabalak
University of Illinois at Urbana-Champaign
School of Chemical Sciences
Chemical and Life Sciences Building, Room A422
600 S. Mathews Ave.
Urbana, IL 61801
USA

Ben Slater
University College London
Davy Faraday Research Laboratory
3rd Floor
Kathleen Lonsdale Building
Gower Street
London WC1E 6BT
UK

Berend Smit
Centre Européen de Calcul Atomique (CECAM)
Ecole Normale Supérieure
46 Allée d'Italie
69007 Lyon
France

and

Univeristy of California
Department of Chemical Engineering
101B Gilman Hall
Berkeley, CA 94720-1462
USA

Andrew James Smith
University of New South Wales
School of Chemical Sciences & Engineering
UNSW Sydney NSW 2052
Australia

C. Morris Smith
ExxonMobil Chemical Company
Technology Licensing
4500 Bayway Drive
Baytown, TX 77520
USA

Zbigniew Sojka
Jagiellonian University
Faculty of Chemistry
ul. Ingardena 3
30-060 Kraków
Poland

Gabor A. Somorjai
University of California, Berkeley
Department of Chemistry
D58 Hildebrand Hall
Berkeley, CA 94720
USA

James J. Spivey
Louisiana State University
Department of Chemical Engineering
S. Stadium Drive
Baton Rouge, LA 70803
USA

Guiseppe Spoto
University of Torino
Dipartimento di Chimica IFM and NIS Centre of Excellence
Via Pietro Giuria 7
10125 Torino
Italy

Heinz Steiner
Solvias AG
P.O. Box
4002 Basel
Switzerland

David L. Stern
ExxonMobil Research and Engineering
Corporate Strategic Research
1545 Route 22
Annandale, NJ 08801
USA

Michael Stöcker
SINTEF Materials and Chemistry
P.O. Box 124 Blindern
0314 Oslo
Norway

Per Stoltze
Aalborg University
Niels Bohrs Vej 8
6700 Esbjerg
Denmark

Frank S. Stone
University of Bath
Department of Chemistry
Claverton Down
Bath BA2 7AY
UK

Sebastian Storck
BASF Aktiengesellschaft
GCC/O - M 301
67056 Ludwigshafen
Germany

Wim H. J. Stork[†]
formerly Shell Research and Technology Centre
Amsterdam
The Netherlands

Kenneth S. Suslick
University of Illinois at Urbana-Champaign
School of Chemical Sciences
Chemical and Life Sciences Building, Room A422
600 S. Mathews Ave.
Urbana, IL 61801
USA

Mizuki Tada
The University of Tokyo
Department of Chemistry
Graduate School of Science
7-3-1 Hongo, Bunkyo-ku
Tokyo 113-0033
Japan

Edmund Taglauer
Max-Planck-Institut für Plasmaphysik
Max-Planck-Gesellschaft
EURATOM Association
Boltzmannstr. 2
85748 Garching
Germany

Masato Takeuchi
Osaka Prefecture University
Department of Applied Chemistry
1-1 Gakuencho, Naka-ku
Sakai, Osaka 599-8531
Japan

Miachel A. Tenhover
873 Wala Drive
Oceanside, CA 92058
USA

Doros N. Theodorou
National Technical University of Athens
School of Chemical Engineering
9 Heroon Polytechniou Str., Zografou Campus
157 80 Athens
Greece

Nadine Thielecke
Federal Agricultural Research Centre
Institute of Technology and Biosystems Engineering
Bundesallee 50
38116 Braunschweig
Germany

Matthias Thommes
Quantachrome Instruments
1900 Corporate Drive
Boynton Beach, FL 33426
USA

Yvonne Traa
University of Stuttgart
Institute of Chemical Technology
Pfaffenwaldring 55
70569 Stuttgart
Germany

Ferruccio Trifirò
University of Bologna
Department of Industrial Chemistry and Materials
Viale Risorgimento, 4
40136 Bologna
Italy

Klaus K. Unger
Am alten Berg 40
64342 Seeheim
Germany

Herman van Bekkum
Delft University of Technology
DelftChemTech
Self-Assembling Systems
Julianalaan 136
2628 BL Delft
The Netherlands

Annelies E. van Diepen
Delft University of Technology
Faculty of Applied Sciences
DelftChemTech, Catalysis Engineering
Julianalaan 136
2628 BL Delft
The Netherlands

A. Jos van Dillen
Utrecht University
Room West 524, Hugo R. Kruytbuilding
Padualaan 8
3584 CH Utrecht
The Netherlands

Piet W. N. M. van Leeuwen
Universiteit van Amsterdam
Van't Hoff Institute for Molecular Sciences
Nieuwe Achtergracht 166
1018 WV Amsterdam
The Netherlands

Rutger A. van Santen
Eindhoven University of Technology
Department of Chemical Engineering and Chemsitry
P.O. Box 513
5600 MB Eindhoven
The Netherlands

J. A. Rob van Veen
Shell RTCA
Dept. GSCT/5
P.O. Box 38000
1030 BN Amsterdam
The Netherlands

Ronnie T. Vang
University of Århus
Interdisciplinary Nanoscience Center (iNANO)
and Department of Physics and Astronomy
Ny Munkegade, building 1521
8000 Århus C
Denmark

Constantinos G. Vayenas
University of Patras
Department of Chemical Engineering
1, Caratheodory St
26504 Patras
Greece

Maxence Vennat
Université Pierre et Marie Curie-Paris 6
Laboratoire de Réactivité de Surface–UMR 7609 CNRS
Case 178, 4 place Jussieu
75252 Paris Cedex 05
France

Götz Veser
University of Pittsburgh
Department of Chemical Engineering
1249 Benedum Hall
Pittsburgh, PA 15261
USA

Thierry Visart de Bocarmé
Université Libre de Bruxelles
Chemical Physics of Materials (Catalysis–Tribology)
Campus Plaine, CP 243
1050 Brussels
Belgium

Bernd Vogel
Evonik Röhm GmbH
Process Development Department
Kirschenallee
64293 Darmstadt
Germany

Bipin V. Vora
UOP LLC
25 E. Algonquin Road
Des Plaines, IL 60016-6101
USA

Klaus-Dieter Vorlop
Federal Agricultural Research Centre
Institute of Technology and Biosystems Engineering
Bundesallee 50
38116 Braunschweig
Germany

Eckhart Wagner
BASF AG
Carl-Bosch-Str. 38
67056 Ludwigshafen
Germany

Joos Wahlen
Katholieke Universiteit Leuven
Centre for Surface Chemistry and Catalysis
Kasteelpark Arenberg 23
3001 Leuven
Belgium

Mark Sebastian Wainwright
University of New South Wales
School of Chemical Sciences & Engineering
UNSW Sydney NSW 2052
Australia

Wei Wang
Lanzhou University
State Key Laboratory of Applied Organic Chemistry
730000 Gansu
P.R. China

Peter Wasserscheid
Universität Erlangen-Nürnberg
Lehrstuhl für Chemische Reaktionstechnik
Egerlandstr. 3
91058 Erlangen
Germany

Bert M. Weckhuysen
Utrecht University
Inorganic Chemistry and Catalysis group,
Department of Chemistry
P.O. Box 80083
3508 TB Utrecht
The Netherlands

James Wei
Princeton University
School of Engineering and Applied Sciences
Princeton, NJ 08544-5263
USA

Udo Weimar
University of Tübingen
Faculty of Chemistry
IPTC
Auf der Morgenstelle 8
72076 Tübingen
Germany

Jens Weitkamp
University of Stuttgart
Institute of Chemical Technology
Pfaffenwaldring 55
70569 Stuttgart
Germany

Hans Jürgen Wernicke
Süd-Chemie AG
Lenbachplatz 6
80333 München
Germany

Joachim Werther
Hamburg University of Technology
Institute of Solids Process Engineering and
Particle Technology
Denickestr. 15
21073 Hamburg
Germany

Dorit Wolf
Evonik Degussa GmbH
Business Line Catalysts
Rodenbacher Chaussee 4
63457 Hanau-Wolfgang
Germany

Scott M. Woodley
University College London
Davy Faraday Research Laboratory
3rd Floor
Kathleen Lonsdale Building
Gower Street
London WC1E 6BT
UK

Richard F. Wormsbecher
W. R. Grace & Co.–Conn.
7500 Grace Drive
Columbia, MD 21044
USA

Stephan Würkert
3M ESPE AG, MTD
P. O. Box 11 61
82224 Seefeld
Germany

Omar M. Yaghi
UCLA
Department of Chemistry and Biochemistry
607 East Charles E. Young Drive
Los Angeles, CA 90095-1569
USA

Adriano Zecchina
University of Torino
Dipartimento di Chimica IFM
and NIS Centre of Excellence
Via Pietro Giuria 7
10125 Torino
Italy

Z. Conrad Zhang
Northwestern University
Department of Chemistry
Center for Catalysis and Surface Science
V. N. Ipatieff Laboratory
Evanston, IL 60208
USA

Michael S. Ziebarth
W. R. Grace & Co.–Conn.
7500 Grace Drive
Columbia, MD 21044
USA

Jürgen Zühlke
BASF Aktiengesellschaft
GCC/O - M 301
67056 Ludwigshafen
Germany

1
Introduction

1.1
Principles of Heterogeneous Catalysis[1]

James A. Dumesic, George W. Huber, and Michel Boudart*

1.1.1
Introduction

Heterogeneous catalysis is of vital importance to the world's economy, allowing us to convert raw materials into valuable chemicals and fuels in an economical, efficient, and environmentally benign manner. For example, heterogeneous catalysts have numerous industrial applications in the chemical, food, pharmaceutical, automobile and petrochemical industries [1–5], and it has been estimated that 90% of all chemical processes use heterogeneous catalysts [6]. Heterogeneous catalysis is also finding new applications in emerging areas such as fuel cells [7–9], green chemistry [10–12], nanotechnology [13], and biorefining/biotechnology [14–18]. Indeed, continued research into heterogeneous catalysis is required to allow us to address increasingly complex environmental and energy issues facing our industrialized society.

Discussing the principles of heterogeneous catalysis is difficult, because catalysts are used for a wide range of applications, involving a rich range of surface chemistries. Moreover, the field of heterogeneous catalysis is highly interdisciplinary in nature, requiring the cooperation between chemists and physicists, between surface scientists and reaction engineers, between theorists and experimentalists, between spectroscopists and kineticists, and between materials scientists involved with catalyst synthesis and characterization. Furthermore, industrial catalysts are complex materials, with highly optimized chemical compositions, structures, morphologies, and pellet shapes; moreover, the physical and chemical characteristics of these materials may depend on hidden or unknown variables. Accordingly, principles of heterogeneous catalysis are typically formulated from studies of model catalysts in ideal reactors with simplified reactants under mild pressure conditions (e.g., 1 bar), rather than from catalytic performance data obtained with commercial catalysts in complex reactors using mixed feed streams under industrial reaction conditions. The principles derived from these more simplified studies advance the science of heterogeneous catalysis, and they guide the researcher, inventor, and innovator of new catalysts and catalytic processes.

1.1.2
Definitions of Catalysis and Turnover

The definition of a catalyst has been discussed many times [19]. For example, a catalyst is a material that converts reactants into products, through a series of elementary steps, in which the catalyst participates while being regenerated to its original form at the end of each cycle during its lifetime. A catalyst changes the kinetics of the reaction, but does not change the thermodynamics. Another definition is that a catalyst is a substance that transforms reactants into products, through an uninterrupted and repeated cycle of elementary steps in which the catalyst participates while being regenerated to its original form at the end of each cycle during its lifetime [20].

The main advantage of using a heterogeneous catalyst is that, being a solid material, it is easy to separate from the gas and/or liquid reactants and products of the overall catalytic reaction. The heart of a heterogeneous catalyst involves the active sites (or active centers) at the surface of the solid. The catalyst is typically a high-surface area material (e.g., 10–1000 $m^2\ g^{-1}$), and it is usually desirable to maximize the number of active sites per reactor volume. Identifying the reaction intermediates – and hence the

---

1 A list of abbreviations/acronyms used in the text is provided at the end of the chapter.
* Corresponding author.

Handbook of Heterogeneous Catalysis, 2nd Ed.
Edited by G. Ertl, H. Knözinger, F. Schüth, and J. Weitkamp
Copyright © 2008 Wiley-VCH Verlag GmbH & Co. KGaA, Weinheim
ISBN: 978-3-527-31241-2

References see page 14

mechanism – for a heterogeneous catalytic reaction is often difficult, because many of these intermediates are difficult to detect using conventional methods (e.g., gas chromatography or mass spectrometry) because they do not desorb at significant rates from the surface of the catalyst (especially for gas-phase reactions).

Heterogeneous catalysts typically contain different types of surface sites, because crystalline solids exhibit crystalline anisotropy. Equilibrated single crystals expose different faces with different atomic structures so as to minimize total surface energy. It would be surprising, in fact, if different crystallographic planes exposing sites with different coordination environments possessed identical properties for chemisorption and catalytic reactions. Moreover, most catalytic solids are polycrystalline. Furthermore, in order to achieve high surface areas, most catalysts contain particles with sizes in the nanometer length scale. The surfaces of these nanoscopic particles contain sites associated with terraces, edges, kinks, and vacancies [21]. If the catalyst contains more than one component (as is generally the case), the surface composition may be different from that of the bulk and differently so for each exposed crystallographic plane. Solids normally contain defects of electronic or atomic nature; in addition, they contain impurities which are either known or unknown in the bulk, but are mostly unknown at the surface. Finally, the surface atomic structure and composition may change with time-on-stream as the catalytic reaction proceeds. In short, it is normal to expect that a catalytic surface exposes a variety of surface sites, in contrast to displaying a single type of active site. Indeed, it is so normal today to expect such complexity that it seems surprising that, in 1925, when Taylor formulated his principle of active sites or active centers, the report created so much attention and remains one of the most often cited in heterogeneous catalysis [22]. The relative importance of surface structure – as influenced by crystalline anisotropy, surface defects, and surface composition – underlines the difficulty of identifying the active sites, either simple or complex, that are responsible for turning over the catalytic cycle. The identification and counting of active sites in heterogeneous catalysis became the "Holy Grail" of heterogeneous catalysis in 1925, and the situation remains the same today.

The activity of a catalyst is defined by the number of revolutions of the catalytic cycle per unit time, given in units of *turnover rate* (TOR) or *turnover frequency* (TOF). In cases where the rate is not uniform within the catalytic reactor or within the catalyst pellets, it is useful to report the rate as a *site time yield* (STY), defined as the overall rate of the catalytic reaction within the reactor normalized by the total number of active sites within the reactor, again in units of reciprocal time. Catalysis by solid materials has been observed quantitatively at temperatures as low as 78 K and as high as 1500 K; at pressures between 10^{-9} and 10^3 bar; with reactants in the gas phase or in polar or non-polar solvents; with or without assistance of photons, radiation or electron transfer at electrodes; with pure metals as unreactive as gold and as reactive as sodium; with multicomponent and multiphase inorganic compounds and acidic organic polymers; and at STYs as low as 10^{-5} s^{-1} (one turnover per day) and as high as 10^9 s^{-1} (gas kinetic collision rate at 10 bar). TOFs of commonly used heterogeneous catalysts are commonly on the order of one per second. The life of the catalyst can be defined as the number of turnovers observed before the catalyst ceases to operate at an acceptable rate. Clearly, this number must be larger than unity, otherwise the substance used is not a catalyst but a reagent. Catalyst life can either be short, as in catalytic cracking of oil, or very long, corresponding to as many as 10^9 turnovers in ammonia synthesis.

1.1.3
Steps in a Heterogeneous Catalytic Reaction

During an overall catalytic reaction, the reactants and products undergo a series of steps over the catalyst, including:

1. Diffusion of the reactants through a boundary layer surrounding the catalyst particle.
2. Intraparticle diffusion of the reactants into the catalyst pores to the active sites.
3. Adsorption of the reactants onto active sites.
4. Surface reactions involving formation or conversion of various adsorbed intermediates, possibly including surface diffusion steps.
5. Desorption of products from catalyst sites.
6. Intraparticle diffusion of the products through the catalyst pores.
7. Diffusion of the products across the boundary layer surrounding the catalyst particle.

Accordingly, different regimes of catalytic rate control can exist, including: (i) film diffusion control (Steps 1 and 7); (ii) pore diffusion control (Steps 2 and 6); and (iii) intrinsic reaction kinetics control (Steps 3 to 5) of catalyst performance. In addition to mass transfer effects, heat transfer effects can also occur in heterogeneous catalysis for highly exothermic or endothermic reactions (especially in combustion or steam reforming).

Figure 1 shows a general effect of temperature on the reaction rate for a heterogeneous catalyst. At low temperatures, diffusion through the film and pores is fast compared to rates of surface reactions, and the overall reaction rate is controlled by the intrinsic reaction

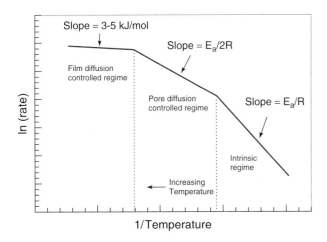

Fig. 1 General effects of temperature on catalytic activity. The intrinsic activation energy is equal to E_a, and R is the gas constant.

kinetics. As the temperature is increased, the rates of surface reactions typically increase more rapidly than the rates of diffusion, and the overall rate of the catalytic process becomes controlled by intraparticle diffusion. The apparent activation energy in this regime is equal to the intrinsic activation energy divided by two. As the temperature is increased further, mass transfer through the external boundary layer becomes the controlling step. The onset of diffusion limited regimes can be altered by changing the reactor design, the catalyst pore structure, the catalyst particle size, and the distribution of the active sites in the catalyst particles. Values of various dimensionless groups can be calculated to estimate the extents to which transport phenomena may control catalytic performance for specific operating conditions [23–31]; however, these calculations are most reliable for cases where the intrinsic reaction kinetics are known. In these cases, it is possible to make catalysts with structures designed to provide adequate rates of diffusion and yet offering high surfaces areas, leading to high rates of reaction per reactor volume, such as the design of specific pore size distributions (e.g., bimodal distributions containing large pores leading to high accessibility of the active sites within the interior of the catalytic pellet, and small pores that branch from the larger pores leading to high surface areas), the formulation of unique pellet shapes (that lead to high accessibility of the active sites but do not cause large pressure drops through the catalytic reactor), and the synthesis of catalyst pellets containing a spatial distribution of the active material within the catalyst pellet [32]. In some cases, transport effects can be used to improve the selectivity of a catalyst, such as in the case of shape-selective catalysis in zeolites [33–36]. In the following sections, we focus on various factors controlling the intrinsic reaction kinetics of catalysts, and we refer the reader to other articles for further discussion on transport effects in heterogeneous catalysis [23–31].

1.1.4
Desired Characteristics of a Catalyst

The following list provides several of the key attributes of a good catalyst:

- The catalyst should exhibit good selectivity for production of the desired products and minimal production of undesirable byproducts.
- The catalyst should achieve adequate rates of reaction at the desired reaction conditions of the process (remembering that achieving good selectivity is usually more important than achieving high catalytic activity).
- The catalyst should show stable performance at reaction conditions for long periods of time, or it should be possible to regenerate good catalyst performance by appropriate treatment of the deactivated catalyst after short periods.
- The catalyst should have good accessibility of reactants and products to the active sites such that high rates can be achieved per reactor volume.

The first three key attributes of a good catalyst are influenced primarily by the interactions of the catalyst surface with the reactants, products, and intermediates of the catalytic process. In addition, other species may form on the catalyst surface (e.g., hydrogen-deficient carbonaceous deposits denoted as coke) that are not directly part of the reaction scheme (or mechanism) for the overall catalytic process.

The principle of Sabatier states that a good heterogeneous catalyst is a material that exhibits an intermediate strength of interaction with the reactants, products, and intermediates of the catalytic process [37, 38]. Interactions of the catalyst surface with the various adsorbed species of the reaction mechanism that are too weak lead to high activation energies for surface reactions and thus low catalytic activity, whereas interactions of the catalyst with adsorbed species that are too strong lead to excessive blocking of surface sites by these adsorbed species, again leading to low catalytic activity.

The principle of Sabatier is elegant in its simplicity and generality, but it is deceptively difficult to use in practice. In particular, this principle applies to a catalyst in its working state, and the nature of the catalyst surface can be expected to be dependent on the nature of the catalytic reaction conditions. For example, one may begin the catalytic reaction with the heterogeneous catalyst

---

References see page 14

in a given oxidation state (e.g., containing zero-valent metal particles following treatment of the catalyst in H_2 at elevated temperature); however, the nature of the surface can be changed dramatically upon interaction with strongly adsorbed species, such as the formation of carbonaceous deposits (coke), and formation of oxides, carbides, nitrides, or sulfides upon interaction with O, C, N, or S species, respectively [39–44]. In this case, the interactions of these oxide, carbide, nitride, or sulfide surfaces with the adsorbed species enter into the reaction mechanism. Of even greater complexity is the fact that a variety of different types of sites are typically present on a catalyst surface (e.g., sites having different coordination and/or chemical composition), and a majority of the observed catalytic activity may be caused by the contributions from a small fraction of the sites present on the catalyst surface. In this case, the adsorbed species interact with these special surface sites (e.g., steps and defect sites on a metal nanoparticle, or sites present at the metal–support interface of a supported metal catalyst). Another factor that complicates catalyst design is that the strengths of interaction of the surface with adsorbed species typically depend on the surface coverages by adsorbed species. For example, the interaction of a transition metal surface with adsorbed CO may be very strong at low surface coverages (e.g., binding energy of nearly 200 kJ mol^{-1}), suggesting that these surfaces would be completely covered and thus poisoned by adsorbed CO at moderate pressures and temperatures; however, these surfaces may carry out catalytic reactions in the presence of gaseous CO at these pressures and temperatures because the differential heat of CO adsorption decreases significantly (e.g., to binding energies near 100 kJ mol^{-1}) as the surface coverage by adsorbed CO increases [45, 46]. Accordingly, there is a relationship between activity and the interaction of the surface with adsorbed species at the surface coverage regime appropriate for the catalytic reaction conditions.

The aforementioned complications caused by the presence of different types of sites on the surface, and the effects on the surface binding energies caused by changes in surface coverages, clearly make it difficult to interpret the performance of a heterogeneous catalyst in quantitative detail. Tools are certainly available to address these complications, such as kinetic Monte Carlo simulations combined with results from density functional theory (DFT) calculations [47–50]. Yet, from a different point of view, the presence of different types of sites and the effects of surface coverage may well contribute to the robustness of the heterogeneous catalyst for operation over a wide range of reaction conditions. In general, the presence of different types of sites and the effects of surface coverage both contribute to surface non-uniformity (different types of sites producing *a prior* non-uniformity, and effects of surface coverage causing induced non-uniformity). At a selected set of reaction conditions, an optimal set of surface binding energies exists that satisfy the principle of Sabatier (as discussed below). Accordingly, the performance of a heterogeneous catalyst with a non-uniform surface will be dominated by the subset of the sites having surface binding energies closest to the optimal values. At higher temperatures, other sites having stronger binding energies with adsorbed species will become the dominant contributors to the observed catalytic activity, whereas sites having weaker binding energies with adsorbed species will control catalytic activity at lower temperatures. Thus, while the effects of surface non-uniformity make it more difficult to predict the performance of a heterogeneous catalyst from a molecular-level understanding, these effects may serve to broaden the range of reaction conditions over which the catalyst can operate effectively. In this respect, our desire to design catalysts having very high selectivity is guided by the synthesis of uniform catalysts, where each site has the optimal properties for production of the desired reaction product. This strategy leads to the idea of highly selective, single-site catalysis as discussed by Thomas et al. [51]. In contrast, the design of catalysts that operate over a wide range of reaction conditions is guided by the synthesis of non-uniform catalysts, such that different subsets of sites control catalyst performance at different reaction conditions. The disadvantage of using non-uniform catalysts, however, is that different sites may display different selectivities for the production of various products, and control over catalytic selectivity may thus be limited [22].

1.1.5
Reaction Schemes and Adsorbed Species

We now explore further the principle of Sabatier using a specific example: water-gas shift over a metal catalyst (e.g., Cu). This reaction ($CO + H_2O \rightleftarrows CO_2 + H_2$) is of importance for the production of H_2 from steam-reforming of fossil fuels, and for controlling the CO : H_2 ratio in synthesis gas mixtures used in methanol and Fischer–Tropsch synthesis processes. For this example, we consider the reaction scheme shown in Fig. 2, where * represents a surface site. The stoichiometric numbers, $\sigma_{i,1}$ and $\sigma_{i,2}$, indicate the number of times that step i occurs to give the overall reaction for reaction schemes 1 and 2, respectively.

In this sequence of steps, the water-gas shift reaction can take place via the formation of carboxyl species (COOH) or through the formation of formate species (HCOO) [52]. In the absence of a catalyst, the rate of water-gas shift via this mechanism is negligible, because the

		$\sigma_{i,1}$	$\sigma_{i,2}$
1.	$CO + {}^* \rightleftarrows CO^*$	1	1
2.	$H_2O + {}^* \rightleftarrows H_2O^*$	1	1
3.	$H_2O^* + {}^* \rightleftarrows OH^* + H^*$	2	1
4.	$CO^* + OH^* \rightleftarrows COOH^* + {}^*$	1	0
5.	$COOH^* + OH^* \rightleftarrows CO_2^* + H_2O^*$	1	0
6.	$CO_2^* \rightleftarrows CO_2 + {}^*$	1	1
7.	$2H^* \rightleftarrows H_2 + 2^*$	1	1
8.	$CO^* + OH^* \rightleftarrows HCOO^{**}$	0	1
9.	$HCOO^{**} \rightleftarrows CO_2^* + H^*$	0	1
overall	$CO + H_2O \rightleftarrows CO_2 + H_2$		

Fig. 2 Assumed reaction mechanism for water-gas shift reaction over Cu. Adapted from Ref. [52].

reaction intermediates (e.g., OH*, H*, COOH*, HCOO*) are at very low concentrations in the gas phase. For example, the enthalpy change for step 3 in the gas phase is approximately 500 kJ mol^{-1}. However, adsorption of the reaction intermediates onto the catalyst surface allows these steps to take place with small enthalpy changes. In the case of copper, the binding energies of H and OH are approximately 250 and 280 kJ mol^{-1} on Cu(111), such that the enthalpy change for step 3 on the catalyst surface is now slightly exothermic. According to the principle of Sabatier, a good catalyst is a material that adsorbs reaction intermediates with intermediate strength. However, we now must distinguish between reactive intermediates and spectator species on the catalyst surface.

In the above reaction scheme, we see that the water-gas shift reaction can take place through adsorbed carboxyl species or formate species. Results from DFT calculations indicate that adsorbed formate species have lower energy compared to adsorbed carboxyl species on copper surfaces, suggesting that path 2 for water-gas shift ($\sigma_{i,2}$) would be favored versus path 1 ($\sigma_{i,1}$) based on thermodynamic arguments. However, the activation energy for step 4 is considerably lower than that for step 8, and the primary path for water-gas shift over copper involves the formation and subsequent reaction of adsorbed carboxyl species. Accordingly, the most stable species on the catalyst surface are not necessarily the most reactive species. This idea leads us to distinguish between a *most abundant surface intermediate* (MASI) and a *most abundant reactive intermediate* (MARI). In certain cases, the MASI and the MARI may be the same species, but in other cases (such as in this case of water-gas shift on copper), the MASI is a spectator species that does not participate in the overall reaction. In this latter case, the spectator species inhibits the overall reaction by blocking surface sites, and it serves no useful role in the overall reaction scheme. For purposes of elucidating catalytic reaction schemes it is essential to distinguish between reactive intermediates and spectator species. This distinction is of paramount importance in spectroscopic studies of adsorbed species on catalyst surfaces, where the detection of a specific adsorbed species using a spectroscopic method (e.g., the detection by infra-red (IR) studies of adsorbed ethylidyne species on platinum surfaces during ethylene hydrogenation [53]) does not guarantee that this species is a reactive intermediate. Instead, these spectroscopic studies must be conducted under dynamic conditions (e.g., so-called operando measurements, where spectroscopic and reaction kinetics data are collected simultaneously) to determine that the time constant for the formation or disappearance of the surface species is the same as the time constant for the overall catalytic reaction [54, 55].

The overall catalytic reaction is given by a linear combination of elementary steps, and the enthalpy change for the overall reaction, ΔH, is given by:

$$\Delta H = \sum_i \sigma_i \Delta H_i \qquad (1)$$

where ΔH_i are the enthalpy changes for elementary steps i. From the principle of Sabatier, it is now clear that the overall value of ΔH should be composed of approximately equal contributions from each of the values of ΔH_i, giving rise to a relatively flat potential energy diagram of energy versus reaction coordinate in moving from reactants, through adsorbed intermediates, to products. Specifically, any value of ΔH_j that is very negative must be balanced by a value of ΔH_k that is very positive, such that the surface will become highly covered (and poisoned) by the adsorbed species produced in step j, and the activation energy for step k will be high. Both of these effects lead to low catalytic activity. We note that the reaction mechanism can certainly contain steps with positive values of ΔH_i, because the intermediates produced in such a step can be consumed by following steps having negative values of ΔH_i. This situation is termed "kinetic coupling", where the conversion of an unfavorable step is increased by its combination with a favorable step that consumes the unstable products of the first step. The highest value of ΔH_i for a surface reaction that can be tolerated can be estimated from transition state theory. The value of $\Delta H_{i,\max}$ depends on the overall rate of the reaction (TOF), the surface coverage by the surface species that reacts in this step (θ_A), a frequency

---

References see page 14

factor ν (of the order of 10^{13} s^{-1}), and the temperature T, as given by:

$$\text{TOF} = \nu \exp\left(-\frac{\Delta H_{i,\max}}{RT}\right) \theta_A \quad (2)$$

For a reaction operating at 500 K with a TOF of 1 s^{-1}, the maximum value of ΔH_i that can be tolerated for a species with high surface coverage (θ_A approaching unity) corresponds to 125 kJ mol^{-1}, which is still a rather high value. In practice, the highest value of ΔH_i that could be tolerated would be lower than this value of 125 kJ mol^{-1}, because the surface coverage by the reactive intermediate would typically be lower than unity and the above analysis assumes that the activation energy for the reverse of step i (i.e., the exothermic direction for this step) is equal to zero. This situation where the overall enthalpy change is shared fairly equally between the various steps of the reaction scheme is a necessary condition for high catalytic activity, but it is not a sufficient condition, because we have not yet considered the transition states for the various elementary steps.

The aforementioned reaction scheme for water-gas shift involving the formation of carboxyl species contains seven steps, thereby requiring the determination (or estimation) of 13 rate constants to describe the reaction kinetics completely; that is, a forward and reverse rate constant for each step ($k_{\text{for},i}$ and $k_{\text{rev},i}$) constrained by the relationship that these rate constants must give the proper equilibrium constant for the overall reaction, K_{eq}, as given below:

$$\prod_i \left(\frac{k_{\text{for},i}}{k_{\text{rev},i}}\right)^{\sigma_i} = K_{\text{eq}} \quad (3)$$

However, it is a rare case that all of these rate constants are kinetically significant. Thus, while we generally have the desire to know the values for as many rate constants as possible, we typically need to know only the values of a limited number of these rate constants to describe the performance of the catalyst for the reaction conditions of interest. Unfortunately, at the outset of research on a given catalyst process, we usually do not know which rate constants will be kinetically significant. Accordingly, an important objective of research into a given catalytic process is to identify which steps are kinetically significant, such that further research can focus on altering the nature of the catalyst and the reaction conditions to enhance the rates of these kinetically controlling steps. This situation is illustrated in Fig. 3 for the above case of water-gas shift involving carboxyl species, according to which the rate is controlled by steps 3 and 4, whereas steps 1, 2, 5, 6, and 7 are quasi-equilibrated.

The net rate of step 3 in Fig. 3, is twofold faster than the net rates of all other steps, because the stoichiometric number of step 3 is equal to 2 whereas all other stoichiometric numbers are equal to 1. Importantly, the net rate of each step divided by its stoichiometric number is equal to the net rate of the overall catalytic reaction. This equality is due to the principle of kinetic steady state, as stated by Bodenstein (see Ref. [38]), according to which determining the rate of one single reaction (typically the overall reaction) allows one to calculate the net rates of all the other individual reactions. The Bodenstein principle is an important foundation of our thinking about how catalytic cycles turn over. This principle also shows that the notion of a "slow" step in a catalytic cycle at the steady state is a misnomer, because all steps proceed at the same net rate.

1.1.6
Conditions for Catalyst Optimality

It can be shown that the net rate of the overall catalytic reaction is controlled by kinetic parameters which depend only on the properties of the transition states for the kinetically significant steps relative to the reactants (and possibly the products) of the overall reaction [56]. The overall rate is also controlled by an additional kinetic parameter for each surface species that is abundant on

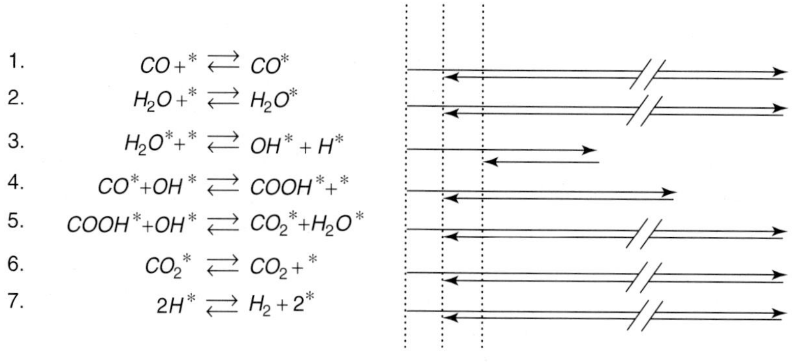

1. $CO + {}^* \rightleftarrows CO^*$
2. $H_2O + {}^* \rightleftarrows H_2O^*$
3. $H_2O^* + {}^* \rightleftarrows OH^* + H^*$
4. $CO^* + OH^* \rightleftarrows COOH^* + {}^*$
5. $COOH^* + OH^* \rightleftarrows CO_2^* + H_2O^*$
6. $CO_2^* \rightleftarrows CO_2 + {}^*$
7. $2H^* \rightleftarrows H_2 + 2{}^*$

Fig. 3 Rates of forward and reverse steps in the water-gas shift reaction on Cu.

the catalyst surface. Specifically, the net rate of the overall reaction is determined by the kinetic parameters as well as by the fraction of the surface sites, θ^*, that is available for the formation of the transition states; the value of θ^* is determined by the extent of site blocking by abundant surface species.

To illustrate how to determine the optimal activity of a catalyst, we consider an example in which the reaction scheme contains a single rate-controlling step and a single abundant surface species. According to results obtained using DeDonder relations (discussed in Section 5.2.1.10) [56], we may write this reaction scheme in terms of a quasi-equilibrated step involving the transition state for the rate-controlling step, TS_i, and a second equilibrated step involving the formation of the most abundant surface species, A^*, as given below and:

1. Reactants $+ 2^* \rightleftharpoons TS_i$ (4)
2. $A + {}^* \rightleftharpoons A^*$ (5)

The overall rate of the reaction, r_{net}, as will be discussed in Section 5.2.1.12, is now given by:

$$r_{net} = \frac{v^{\ddagger}}{\sigma_1} \exp\left(\frac{\Delta S_1^{o\ddagger}}{R} - \frac{\Delta H_1^{o\ddagger}}{RT}\right) F(a_i) \theta_*^2 \left(1 - z_{tot}^{1/\sigma_1}\right) \quad (6)$$

$$\theta_* = \frac{1}{1 + \exp\left(\frac{\Delta S_2^o}{R} - \frac{\Delta H_2^o}{RT}\right) a_A} \quad (7)$$

where $F(a_i)$ is a function of the activities (a_i) of the reactants and/or products of the overall reaction. Neglecting entropy effects, as we change the nature of the catalyst for constant reaction conditions, the primary items in the above equations that change are $\Delta H_1^{o\ddagger}$ and ΔH_2^o. (Note, we implicitly assume that the reaction mechanism does not change.) Accordingly, the overall rate of the reaction for different catalysts is given by:

$$r_{net} = \frac{C_1 \exp\left(-\frac{\Delta H_1^{o\ddagger}}{RT}\right)}{\left(1 + C_2 \exp\left(-\frac{\Delta H_2^o}{RT}\right)\right)^2} \quad (8)$$

$$C_1 = \frac{v^{\ddagger}}{\sigma_1} \exp\left(\frac{\Delta S_1^{o\ddagger}}{R}\right) F(a_i)\left(1 - z_{tot}^{1/\sigma_1}\right) \quad (9)$$

$$C_2 = \exp\left(\frac{\Delta S_2^o}{R}\right) a_A \quad (10)$$

We next consider that the surface properties of the catalyst are described in terms of some fundamental catalyst property, x. This property x could be a heat of adsorption of one of the reactants [57], the heat of formation of a bulk compound that can be correlated with a heat of adsorption [58], the position of the catalytic element along a horizontal series in the Periodic Table, an electronic property of the catalyst such as Pauling's d-band character of the metal [59], or the d-band center of the metal [60]. The optimal catalyst can thus characterized by the following relationship:

$$\frac{dr_{net}}{dx} = 0 = \frac{-C_1 \exp\left(-\frac{\Delta H_1^{o\ddagger}}{RT}\right) \frac{d\Delta H_1^{o\ddagger}}{RT dx}}{\left(1 + C_2 \exp\left(-\frac{\Delta H_2^o}{RT}\right)\right)^2}$$

$$+ \frac{2C_1 \exp\left(-\frac{\Delta H_1^{o\ddagger}}{RT}\right) C_2 \exp\left(-\frac{\Delta H_2^o}{RT}\right) \frac{d\Delta H_2^o}{RT dx}}{\left(1 + C_2 \exp\left(-\frac{\Delta H_2^o}{RT}\right)\right)^3} \quad (11)$$

This relationship may be simplified to give:

$$\frac{d\Delta H_1^{o\ddagger}}{dx} = \frac{2C_2 \exp\left(-\frac{\Delta H_2^o}{RT}\right) \frac{d\Delta H_2^o}{dx}}{\left(1 + C_2 \exp\left(-\frac{\Delta H_2^o}{RT}\right)\right)} = 2\theta_A \frac{d\Delta H_2^o}{dx} \quad (12)$$

Thus, for the optimal catalyst, the surface coverage by the most abundant surface species is equal to:

$$\theta_A = \frac{\frac{d\Delta H_1^{o\ddagger}}{dx}}{2\frac{d\Delta H_2^o}{dx}} = \frac{\omega_1^{\ddagger}}{2\omega_2} \quad (13)$$

where the values of ω_i are defined as:

$$\omega_1^{\ddagger} = \frac{d\Delta H_1^{o\ddagger}}{dx} \quad (14)$$

$$\omega_2 = \frac{d\Delta H_2^o}{dx} \quad (15)$$

In the above derivation, we assume that $\omega_1^{\ddagger}$ and ω_2 have the same sign, such that variations in x change the enthalpy of the transition state and the MASI in the same direction. We also assume that $(d^2\Delta H_1^{o\ddagger})/dx^2$ and $(d^2\Delta H_2^{o\ddagger})/dx^2$ are small or zero. This assumption is valid if we are searching for improved catalysts over a small range of x, which typically occurs when testing catalysts. In fact, when we vary x over a large range, then the mechanism of the catalytic reaction would probably change.

References see page 14

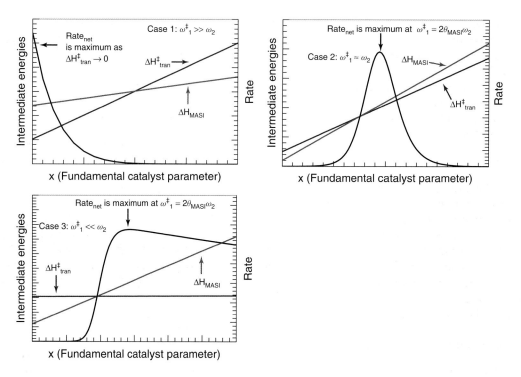

Fig. 4 Reaction rates and energies of transition state and most abundant surface intermediate (MASI) as functions of fundamental catalyst parameter "x".

First, we will consider Case 1 shown in Fig. 4, where $\omega_1^\ddagger \gg \omega_2$. In this case, a maximum rate does not exist and the parameter x should be adjusted to its lowest possible value (i.e., the strongest bonding to the surface) which would decrease the enthalpy of the transition state as much as possible. In this case the optimal catalyst operates at high surface coverage. We now consider Case 2, where $\omega_1^\ddagger \approx \omega_2$; that is, x changes the enthalpies of the transition state and the MASI by similar amounts. This situation is probably more physically realistic, and if the MASI is in fact a reactive intermediate (i.e., if it is a MARI), then the family of catalysts described by the variation of x follows the Brønsted–Evans–Polanyi–Semenov relationship, which relates the thermodynamics and kinetics of the system. Here, there is a clear maximum in the rate versus x, as shown in Fig. 4. The plot of rates versus x appears as a volcano-type curve which decreases symmetrically on both sides. In the case where $\omega_1^\ddagger$ is equal to ω_2, the surface coverage by the MASI on the optimal catalyst is equal to 0.5. For Case 3, $\omega_1^\ddagger \ll \omega_2$, corresponding to the situation where x changes the enthalpy of the MASI more significantly than the transition state. The optimal catalyst in this case has a low surface coverage by the MASI. A maximum rate occurs in this case, as shown in Fig. 4, provided that $|\omega_1^\ddagger| > 0$; however, the plot of rate versus x is not symmetric with respect to x. We note that as x increases for these three cases, the number of vacant sites on the catalyst increases.

Cases 2 and 3 clearly illustrate the principle of Sabatier, in which a maximum rate occurs at some moderate level of interaction of the catalyst surface with the intermediates and adsorbed species. While Case 1 appears to contradict Sabatier's principle, the situation where $\omega_1^\ddagger \gg \omega_2$ is highly unlikely. In particular, this situation corresponds to the case where the catalyst interacts more strongly with the transition state than with any of the reactive intermediates. However, if the activation energies for the elementary steps of the mechanism are positive, then the reactants and/or products of the elementary step involving the rate-controlling transition state are more strongly adsorbed on the surface than is the transition state, leading to the situation described by Case 2 or 3.

We may generalize the above expression for catalyst optimality to the case where the surface contains several abundant surfaces species, A*, B*, C*, and D*, leading to the following expression:

$$\omega_1^\ddagger = 2(\omega_A \theta_A + \omega_B \theta_B + \omega_C \theta_C + \omega_D \theta_D) \quad (16)$$

In this case, the nature of the optimal catalyst is controlled by the change in the binding energy of the transition state with respect to x ($\omega_1^\ddagger$) compared to the changes in the bindings energies of species A, B, C, and

D ($\omega_A, \omega_B, \omega_C, \omega_D$) weighed by their respective surface coverages at the steady state.

A bridge between the thermodynamics and kinetics of a reaction is provided by the Brønsted–Evans–Polanyi–Semenov relationship, which states that there is a linear relationship between the activation energy E_{act} of an elementary step and the heat of reaction if entropy effects are neglected [38]:

$$E_{act} = E_0 + \alpha \Delta H \qquad (17)$$

where ΔH is the enthalpy of reaction, α is the transfer coefficient that varies between zero and one, and E_0 is a constant. In other words, if we neglect entropic effects, the activation energy of an elementary step in the exothermic direction is lower when the heat of reaction becomes more favorable (i.e., ΔH becomes more negative). DFT calculations have recently shown that Brønsted–Evans–Polanyi–Semenov relationships are generally upheld in chemical reactions on catalyst surfaces [61–64].

We now consider the following catalytic reaction:

$$A^* \longrightarrow B^* + C^* \qquad (18)$$

If we use the gas phase A species as the zero energy level (as shown in Fig. 5) we can define the activation energy as:

$$\begin{aligned}E_{act} &= E_0 + \alpha(H_{Bgas} + B.E._B + H_{Cgas} + B.E._C \\ &\quad - H_{Agas} - B.E._A) = E_0 + \alpha(B.E._B \\ &\quad + B.E._C - B.E._A) + \alpha \Delta H_{gas}\end{aligned} \qquad (19)$$

where the $B.E.$ terms are the binding energies of the various species on the surface.

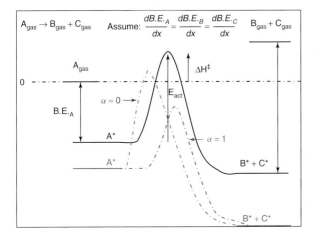

Fig. 5 Schematic potential energy diagram of energy versus reaction coordinate, showing the relationship between the energy of transition state, $\Delta H^{\neq}$, and changes in the energies of adsorbed species.

We now define the standard enthalpy for the formation of the transition state from gaseous species A as:

$$\begin{aligned}\Delta H_1^{o\ddagger} &= E_{act} + B.E._A = E_0 + \alpha(\Delta H_{gas}) \\ &\quad + \alpha(B.E._B + B.E._C) + (1-\alpha)B.E._A\end{aligned} \qquad (20)$$

and we differentiate the enthalpy of the transition state with respect to x, leading to the following result:

$$\frac{d \Delta H_1^{o\ddagger}}{dx} = \alpha \left(\frac{dB.E._B}{dx} + \frac{dB.E._C}{dx} \right) + (1-\alpha) \frac{dB.E._A}{dx} \qquad (21)$$

This relationship shows how the change of the transition state enthalpy with respect to $x(\omega_1^{\ddagger})$ is related to the changes in the binding energies $B.E.$ of the adsorbed species (ω_2). If α is equal to zero, then the change of the transition state enthalpy depends only on the change of the binding energy of A, and $\omega_1^{\ddagger} = \omega_2$. In this case, the transition state is an early transition state that chemically looks similar to A [65]. If α is equal to 1, then the transition state is a late transition state that resembles the products of the reaction, and the change of the transition state enthalpy is related to changes of the binding energies of B and C. If $(dB.E._A)/dx = (dB.E._B)/dx = (dB.E._C)/dx$, then the late transition state gives rise to the situation where $\omega_1^{\ddagger} = 2\omega_2$.

The above examples show how it is possible to maximize the activity of the catalyst. However, it is often more important to optimize the selectivity of the catalyst. Similar types of analyses can be carried out for these cases. In general, these situations are classified as being series selectivity challenges, such as A → B → C, where B is the desired product, and/or parallel selectivity challenges, such as A → B coupled with A → C. In these cases, if we want to optimize the selectivity of the catalyst, we search for some catalyst property, x, that decreases the enthalpy of the transition state for the desired reaction more than it decreases the enthalpies of the transition states for the undesired reactions.

An undesired reaction that leads to progressive blocking of surface sites leads to deactivation of the catalyst with respect to time-on-stream. More generally, various mechanisms exist by which a catalyst can undergo deactivation, such as: (i) poisoning; (ii) thermal degradation (sintering); (iii) leaching of the active site; and (iv) attrition [66]. The first three of these mechanisms are chemical in nature, whereas the last mechanism is physical (e.g., the catalyst pellet breaks apart). Some catalysts do not show any measurable deactivation over periods of years, such as in ammonia synthesis. However, other catalysts lose an important fraction of their activity

References see page 14

after less than a minute of contact with feed, as in catalytic cracking. In the latter case, if deactivation is caused by coking, the catalyst must be regenerated by continuous regeneration in an oxidizing atmosphere.

1.1.7
Catalyst Design

Given that the performance of a catalyst is controlled by a limited number of kinetic parameters, it is unclear why it is so difficult to design a catalyst from molecular-level concepts. As noted above, during the early stages of research into a catalytic process, first we do not know which steps in the reaction mechanism are kinetically significant, and which species are most abundant on the catalyst surface under reaction conditions. Second, we do not often know the structure of the active site and its dependence on the nature of the reaction conditions. Third, we do not usually know how the activity and selectivity for the catalytic reaction depend on the structure of the active sites. Fourth, we do not typically know during these early stages the rates of various modes of catalyst deactivation (e.g., sintering, phase changes, deposition of carbonaceous deposits on the surface, etc.), and we do not know whether the catalyst can be regenerated following deactivation. Finally, we must ensure that the texture of the catalyst and the geometry of the reactor are designed in such a way that mass transport of reactants and products to and from the active sites is sufficiently rapid that high rates of reaction per unit volume of reactor can be achieved.

Because of these difficulties, the field of heterogeneous catalysis is highly interdisciplinary in nature, and involves close collaboration between experts in such areas as catalyst synthesis, catalyst characterization, surface spectroscopy, chemical kinetics, chemical reaction engineering and, most recently, in theoretical calculations of catalyst structure and performance using density function theory. These broad studies can be grouped into three levels, as shown in Fig. 6.

All studies of heterogeneous catalysis begin at the Materials Level. High-surface area catalytic materials must be synthesized with specific structures and textures, the latter referring to such features as the sizes of the various phase domains and the details of the pore structure. Clearly, the synthesis of catalytic materials must be guided by detailed characterization studies to determine the structures, compositions, and textures of the materials that have been prepared. These characterization studies should be conducted after the catalyst has been subjected to various treatment steps (such as those treatments employed during activation of the catalytic material), and it is most desirable to carry out characterization studies of the catalyst under the actual reaction conditions of the catalytic process. Indeed, the properties of a heterogeneous catalyst are inherently dynamic in nature, and these properties often change dramatically with changes in the reaction conditions (e.g., phase changes, surface reconstructions, changes in surface versus bulk composition, etc.) [67].

The central level of research and development of heterogeneous catalysts involves the quantification of catalyst performance (this is known as the Catalyst Performance Level). These studies can be carried out in a preliminary fashion over a wide range of catalytic materials (e.g., high-throughput studies) to identify promising catalysts

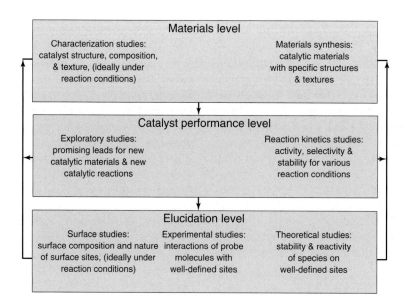

Fig. 6 Levels of study in heterogeneous catalysis research.

and reaction conditions for further studies. The performance of the catalyst is then documented in greater detail by determining catalytic activity, selectivity, and stability with respect to time-on-stream for various reaction conditions. These measurements must be made at various conversions when multiple reaction pathways exist, because catalytic selectivities in these cases are different, depending on whether the desired products are formed in primary versus secondary reactions, or in series versus parallel pathways. We note here that various definitions of catalytic activity are used, depending on the nature of the study. For practical studies, catalytic activities can be reported as rates per gram of catalyst or per unit surface area. However, for more detailed studies or for research purposes, it is often desirable to report catalytic activities as rates per surface site (i.e., TOFs), with the number of surface sites measured most often by selective adsorption measurements (e.g., adsorption of H_2 or CO to titrate metal sites, adsorption of ammonia or pyridine to titrate acid sites). In some cases it is possible to report catalytic activity as rate per active site (also called TOF), when it is possible to distinguish active sites from the larger number of surface sites using special probe molecules (e.g., dissociative adsorption of N_2 to titrate sites for ammonia synthesis [68]; selective poisoning by adsorbates that compete with the reactants of the catalytic reaction [69]); or by transient isotopic tracing [70].

For the purposes of catalyst development, it is probably sufficient to work at the Materials Level and the Catalyst Performance Level. However, research into heterogeneous catalysis is dominated by studies conducted at a third level – the Elucidation Level – where the aim is to identify the fundamental building blocks of knowledge which can be assembled to build a molecular-level understanding of catalyst performance in order to guide further investigations to improve catalyst performance. At the Elucidation Level the studies are designed to determine the surface composition and nature of the surface sites on the catalyst [71–74]. Clearly, these investigations must be conducted with the catalyst under controlled conditions (e.g., under ultra-high vacuum, after treatment with H_2, after calcination, etc.) and, where possible, such measurements should be made with the catalyst under reaction conditions. Moreover, the studies may be carried out on real catalytic materials and on more well-defined surfaces (e.g., single crystals, or model samples formed by depositing known amounts of materials onto well-defined supports) [73, 75, 76]. Most measurements at the Elucidation Level involve studies of the interactions of specific probe molecules with the catalyst surface. These probe molecules may be the reactants, intermediates, or products of the catalytic reaction, or they may be more simple species chosen to monitor a specific functionality of the surface. Alternatively, a molecule may be used as a probe because it has an advantageous feature for spectroscopic identification (e.g., CO for infrared studies, a ^{13}C-containing molecule for NMR studies). These studies of the interaction of probe molecules with surfaces are designed to determine the surface concentrations of different types of surface site, to determine the nature of the adsorbed species formed on the surface sites, and to determine the reactivities of the surface sites by monitoring the adsorbed species on the surface versus time, versus temperature or, most commonly, during a temperature ramp (e.g., temperature-programmed desorption).

The third pillar of studies at the Elucidation Level involves the use of DFT calculations to assess the structures, stabilities, and reactivity of species adsorbed onto the surface sites (with the sites being composed of clusters of atoms or as periodic slabs of atoms) [77–81]. These studies are used to help interpret the results obtained from spectroscopic studies of catalyst surfaces (e.g., to predict the vibrational spectra of species adsorbed in different orientations on different sites), to calculate heats of adsorption for various intermediates in a reaction mechanism (e.g., to predict which species are expected to be abundant on the catalyst under reaction conditions), to estimate the energy changes for possible steps in a reaction mechanism (thereby eliminating from further consideration steps with very positive energy changes), and to determine activation energy barriers for steps that are suspected as being kinetically significant in the reaction scheme. Indeed, a key feature of these theoretical studies is the ability to predict how the surface properties are expected to change as the nature of the surface is altered (e.g., by changing the surface structure, or by adding possible promoters). This in turn will provide feedback to the Materials Level with regards to new materials that should be synthesized and which are likely to lead to an improved catalyst performance. In addition, these theoretically based studies provide information about highly reactive intermediates which might be difficult to obtain by direct experimental measurements. Most importantly, studies conducted at the Elucidation Level provide a scientific basis about the working catalysis that may, in future, be used to design different reaction pathways.

1.1.8
Catalyst Development

Catalyst development typically involves testing a large number of catalysts with a feedback loop, as it is currently difficult to design catalysts *a priori*. In this respect, catalyst development studies involve examining a large number of catalysts, for which recent advances in high-throughput

References see page 14

testing have attracted considerable attention [82–87]. Catalyst development through the testing of a wide range of materials was first practiced in 1909 by Mittasch at BASF who, according to Timm [88], issued the following directive to his team who at the time were developing the synthesis of ammonia:

- The search for a suitable catalyst necessitates carrying out experiments with a number of elements, together with numerous additives.
- The catalytic substances must be tested at high pressures and temperatures, just as in the case of Haber's experiments.
- A very large number of tests will be required.

Ten years later, the number of tests conducted had exceeded 10 000, and more than 4000 catalysts had been studied. This extraordinary effort was also extraordinarily successful. What has changed since then, however, is the way in which the systematic search is assisted. Today, armed with an arsenal of principles, concepts, instrumentation and computers, it is possible to identify and to improve new catalytic materials in a much shorter time and with a smaller number of trial samples, especially with the possibility of advanced characterization methods (especially *in-situ* techniques) and insights from theoretical calculations (e.g., DFT calculations). The practical merit of this "assisted catalyst design" is clear, while its scientific dividend is the possibility of learning as the design proceeds, with the building of a data bank of rate constants and the formulation of more precise models of active sites. With new theoretical insights or principles, quantitative bases of catalyst preparation and reproducibility of catalyst behavior, the future of heterogeneous catalysis still looks very bright.

The path to the design of an optimal catalytic process would be clear if the activity, selectivity and stability of the catalyst were to move in the same direction upon an increase in a single process variable, such as temperature. However, this simple behavior is not typically observed, and choices must be made in every instance. For example, while the activity of a catalyst may increase with temperature, its stability usually decreases with temperature. In addition, the relationship between catalytic activity and selectivity is typically very complex, and is not understood in detail until the surface chemistry of the catalytic process has been elucidated. Accordingly, selectivity, stability and activity must be considered together, and trade-offs may have to be negotiated, perhaps by using multi-functional reactors with catalytic distillation or catalytic membranes. Success in heterogeneous catalysis begins with chemistry, but ends with catalytic reaction engineering.

1.1.9
Bridging Gaps in Heterogeneous Catalysis

The above description of research and development into heterogeneous catalysis as being interdisciplinary in nature, involving studies at the levels of materials, catalyst performance and elucidation, can also be cast in the form of building bridges between various types of studies and different types of material. As depicted in Fig. 7, we often talk about bringing together the field of surface science (which traditionally is focused on studies of single crystal surfaces at low pressures) with the field of heterogeneous catalysis (which traditionally is focused on studies of high-surface area catalytic materials surfaces under high-pressure reaction conditions). More recently, we have talked about "bridging the materials gap", as we have attempted to use experimental results from studies of well-defined model materials to interpret the performance of more complex, high-surface area catalytic materials. Traditionally, these model materials have been single crystals, cut at various angles to expose surfaces containing different types of sites, such as surfaces with different symmetries and atoms present at terraces, steps, and kinks [89]. More recently, however, these model materials have become highly sophisticated, such as the deposition of nanoparticles with specific sizes and geometries on well-defined support surfaces (e.g., metal nanoparticles supported on thin films of oxides deposited on single crystal metal surfaces, or non-metallic nanoparticles supported directly on single crystal metal surfaces) [73, 75, 76]. We also talk about "bridging the pressure gap", as we attempt to use experimental results from studies conducted at low pressures (less than 10^{-6} Torr) to interpret the performance of catalysts under high-pressure reaction

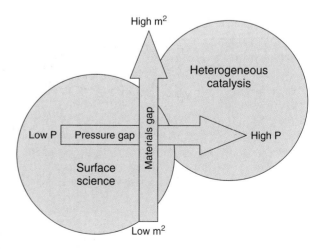

Fig. 7 Bridging the gap between surface science and heterogeneous catalysis.

conditions. The origin for this pressure gap comes from the fact that, whereas some spectroscopic techniques can be employed to study the surface and bulk properties of catalysts under high-pressure reaction conditions (e.g., FTIR, Raman, XRD, EXAFS, Mössbauer spectroscopy), other spectroscopic and characterization techniques (e.g., XPS, TEM) are most easily conducted with the sample at low pressures (e.g., $<10^{-6}$ Torr) [90]. These latter techniques are typically associated with use of electrons to probe the sample, with the electrons interacting strongly with molecules in the gas phase. This pressure gap can be bridged directly by designing advanced instrumentation, such that the distance traversed by the electrons in the gas phase is minimized [91, 92]. In addition, the pressure gap can be bridged indirectly by using molecular-based models (e.g., kinetic Monte Carlo calculations, microkinetic models), first to describe the experimental results obtained at low pressures, and then to extrapolate this information to high-pressure reaction conditions.

The past few years have witnessed an explosion in the area of nanotechnology, in which researchers have learned – and are continuing to learn – how to engineer materials at the nanometer length scale. The field of heterogeneous catalysis has been involved in the synthesis of nanomaterials for many years (e.g., the synthesis of zeolites). Indeed, the scheme depicted in Fig. 6 shows that essentially all studies of heterogeneous catalysis begin at the Materials Level. Recent advances in nanotechnology offer new routes for catalyst synthesis (e.g., atomic layer deposition, self-assembly methods) [93–96] and, importantly, also for catalyst characterization (e.g., techniques such as scanning tunneling microscopy that allow atomic-scale imaging of materials at elevated temperatures and pressures) [73, 97]. However, as an as increasing number of research groups become involved in nanotechnology, it is possible that an "applications gap" will be created in heterogeneous catalysis, where new materials are formed without clear applications for catalytic processes. Clearly, this gap can be bridged by realizing that research and development into heterogeneous catalysis involves the combination of studies at the levels of materials, catalyst performance, and elucidation. As advances in nanotechnology allow us to create new materials (the Materials Level) and to characterize these materials in greater detail (Materials and Elucidation Levels), we are positioned to take full advantage of these advances by conducting studies at the Catalyst Performance Level.

1.1.10
A Philosophical Note

Today, we live in an era in which it is possible to employ a vast range of advanced experimental techniques and theoretical methods to elucidate in detail the surface chemistry of catalytic processes. This situation, with respect to the hierarchy of theoretical methods that can be employed to describe the reaction kinetics for a catalytic process, is illustrated schematically in Fig. 8. At the lowest level, we use empirical rate expressions to fit reaction kinetics data over a range of process conditions;

References see page 14

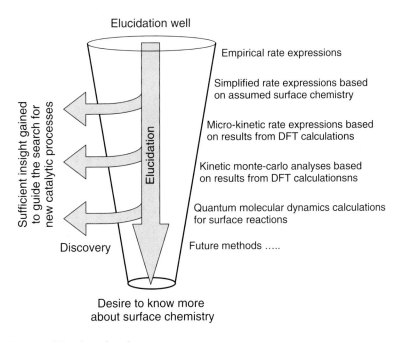

Fig. 8 The catalyst "elucidation well" and catalyst discovery.

however, these models typically have questionable success in predicting catalyst performance outside the range of experimental conditions used to fit the model.

We then move to rate expressions based on assumed surface chemistry. These models should have a better predictive value, although it is often difficult to determine which types of assumption should be made. Accordingly, we turn to results from DFT calculations to build micro-kinetic models that describe catalyst performance without the need to make prior assumptions about which steps are kinetically significant and which species are abundant on the catalyst surface. These micro-kinetic models, however, are typically based on the mean-field approximation, and they thus make simplified assumptions about (or neglect) the effects of surface coverage and lateral interactions between adsorbed species. Accordingly, these restrictive assumptions are relaxed when using kinetic Monte Carlo methods to describe reaction kinetics based on binding energies and lateral interaction terms determined from DFT calculations. Indeed, today's research groups are beginning to combine quantum mechanics and molecular dynamics calculations to describe a variety of surface processes. Who knows what new computational methods are on the horizon?

The sequential use of the aforementioned methods to describe reaction kinetics in greater detail is depicted in Fig. 8, as the digging of an "elucidation well". As we dig deeper by using more sophisticated methods, we learn more about the details of the surface chemistry. Indeed, we are driven to dig deeper by our desire to learn as much as possible about the fundamental principles that control catalyst performance. However, this desire to know as much as possible must be balanced by our need to discover new catalysts and new catalytic processes. Clearly, as we learn more about the fundamentals of the catalytic process (i.e., as we dig deeper), we should have better insight to guide our search for better catalysts. Luckily, we need not dig to the deepest levels to begin the search for better catalysts. As noted at the beginning of this chapter, our industrialized society is facing increasingly complex environmental and energy issues for sustained growth. Thus, while our scientific curiosity for fundamental knowledge drives us to dig deeper toward detailed elucidation of catalytic phenomena, we must also continue to look horizontally as we use our newfound insight to develop new catalysts and catalytic processes for the benefit of society.

List of Abbreviations

DFT	density functional theory
EXAFS	extended X-ray absorption fine structure
FTIR	Fourier transform infrared
STY	site time yield
TEM	transmission electron microscopy
TOF	turnover frequency
TOR	turnover rate
XPS	X-ray photoelectron spectroscopy
XRD	X-ray diffraction

References

1. J. M. Thomas, W. J. Thomas, *Principles and Practice of Heterogeneous Catalysis*, VCH, Weinheim, 1997, 669 pp.
2. J. N. Armor, *Appl. Catal., A* **2001**, *222*, 407.
3. I. Chorkendorff, J. W. Niemantsverdriet, *Concepts of Modern Catalysis and Kinetics*, Wiley-VCH, Weinheim, 2003, 452 pp.
4. R. J. Farrauto, C. H. Bartholomew, *Fundamentals of Industrial Catalytic Processes*, Chapman & Hall, London, 1997, 754 pp.
5. R. A. van Santen, P. W. N. M. v. Leeuwen, J. A. Moouljn, B. A. Averill, *Catalysis: An Integrated Approach*, Elsevier Science B.V., Amsterdam, 1999, 574 pp.
6. National Research Council Panel on New Directions in Catalytic Sciences and Technology, *Catalysis Looks to the Future*, National Academy Press, Washington D.C., 1992, p. 1.
7. W. Vielstich, A. Lamm, H. Gasteiger, *Handbook of Fuel Cells: Fundamentals, Technology, Applications*, Wiley, Chichester, 2003, 2690 pp.
8. S. Park, J. M. Vohs, R. J. Gorte, *Nature* **2000**, *404*, 265.
9. S. Ha, R. Larsen, R. I. Masel, *J. Power Sources* **2005**, *144*, 28.
10. A. Corma, H. Garcia, *Chem. Rev.* **2003**, *103*, 4307.
11. G. Centi, *Catal. Today* **2003**, *77*, 287.
12. R. A. Sheldon, *Green Chem.* **2005**, *7*, 267.
13. A. Borgna, L. Balzano, J. E. Herrera, W. E. Alvarez, D. E. Resasco, *J. Catal.* **2001**, *204*, 131.
14. H. van Bekkum, P. Gallezot, *Top. Catal.* **2004**, *27*, 1.
15. G. W. Huber, J. W. Shabaker, J. A. Dumesic, *Science* **2003**, *300*, 2075.
16. G. W. Huber, J. N. Chheda, C. J. Barrett, J. A. Dumesic, *Science* **2005**, *308*, 1446.
17. I. K. Mbaraka, B. H. Shanks, *J. Catal.* **2005**, *229*, 365.
18. S. Varadarajan, D. J. Miller, *Biotechnol. Progr.* **1999**, *15*, 845.
19. E. K. Rideal, H. S. Taylor, *Catalysis in Theory and Practice*, Macmillan, London, 1919, Chapter 2.
20. M. Boudart, in *Perspectives in Catalysis*, J. M. Thomas, K. I. Zamaraev (Eds.), Blackwell, Oxford, 1992, p. 183.
21. G. A. Somorjai, *Introduction to Surface Chemistry and Catalysis*, John Wiley, New York, 1994, 667 pp.
22. H. S. Taylor, *Proc. Roy. Soc. (London)* **1925**, *A108*, 105.
23. J. B. Anderson, *Kagaku Kogaku (Chem. Eng. Jpn.)* **1962**, *147*, 191.
24. J. J. Carberry, *AICHE* **1961**, *7*, 350.
25. G. F. Froment, K. B. Bischoff, *Chemical Reactor Analysis and Design*, Wiley, New York, 1990, 664 pp.
26. C. N. Satterfield, *Mass Transfer in Heterogeneous Catalysis*, MIT Press, Cambridge, MA, 1970, 267 pp.
27. D. E. Mears, *Ind. Eng. Chem. Proc. Des. Dev.* **1971**, *10*, 541.
28. D. E. Mears, *J. Catal.* **1971**, *20*, 127.
29. J. M. Smith, *J. Chem. Eng. Japan* **1973**, *6*, 191.
30. P. B. Weisz, *Z. Phys. Chem.* **1954**, *11*, 1.
31. P. B. Weisz, C. D. Prater, *Adv. Catal.* **1957**, *6*, 143.

32. R. Aris, in *Catalyst Design: Progress and Perspectives*, L. L. Hegedus (Ed.), Wiley Interscience, New York, 1987, Chapter 7.
33. P. A. Jacobs, J. A. Martens, J. Weitkamp, H. K. Beyer, *Faraday Discuss. Chem. Soc.* **1981**, *72*, 353.
34. W. O. Haag, R. M. Lego, P. B. Weisz, *Faraday Discuss. Chem. Soc.* **1981**, *72*, 317.
35. E. G. Derouane, P. Dejaifve, Z. Gabelica, *Faraday Discuss. Chem. Soc.* **1981**, *72*, 331.
36. J. M. Thomas, G. R. Millward, S. Ramdas, L. A. Busil, M. Audier, *Faraday Discuss. Chem. Soc.* **1981**, *72*, 345.
37. P. Sabatier, *La catalyse en chimie organique*, Bérange, Paris, 1920, 388 pp.
38. M. Boudart, *Kinetics of Chemical Processes*, Blackwell, Oxford, Stoneham, MA, 1991, 246 pp.
39. J. V. Lauritsen, M. Nyberg, J. K. Norskøv, *J. Catal.* **2004**, *224*, 94.
40. B. Hinnemann, J. K. Norskøv, H. Topsøe, *J. Phys. Chem. B* **2005**, *109*, 2245.
41. H. H. Hwu, J. G. Chen, *Chem. Rev.* **2005**, *105*, 185.
42. R. B. Levy, M. Boudart, *Science* **1973**, *181*, 547.
43. M. K. Neylon, S. Choi, H. Kwon, K. E. Curry, L. T. Thompson, *Appl. Catal., A* **1999**, *183*, 253.
44. S. T. Oyama, *Catal. Today* **1992**, *15*, 179.
45. S. G. Podkolzin, J. Shen, J. J. D. Pablo, J. A. Dumesic, *J. Phys. Chem. B* **2000**, *104*, 4169.
46. R. M. Watwe, B. E. Spiewak, R. D. Cortright, J. A. Dumesic, *Catal. Lett.* **1998**, *51*, 139.
47. C. G. M. Hermse, A. P. van Bave, A. P. J. Jansen, L. A. M. M. Barbosa, P. Sautet, R. A. van Santen, *J. Phys. Chem. B* **2004**, *108*, 11035.
48. M. Neurock, S. A. Wasileski, D. Mei, *Chem. Eng. Sci.* **2004**, *59*, 4703.
49. S. Raimondeau, P. Aghalayam, A. B. Mhadeshwar, D. G. Vlachos, *Ind. Eng. Chem. Res.* **2003**, *42*, 1174.
50. F. J. Garcia, E. E. Wolf, *Chem. Eng. Sci.* **2004**, *59*, 4723.
51. J. M. Thomas, C. R. A. Catlow, G. Sankar, *Chem. Commun.* **2002**, 2921.
52. A. A. Gokhale, *Water–Gas Shift Reaction and Fischer–Tropsch Synthesis on Transition Metal Surfaces*, PhD Thesis, University of Wisconsin, 2005.
53. P. S. Cremer, X. Su, Y. R. Shen, G. A. Somorjai, *J. Am. Chem. Soc.* **1996**, *118*, 2942.
54. H. Topsøe, *J. Catal.* **2003**, *216*, 155.
55. I. E. Wachs, *Catal. Today* **1996**, *27*, 437.
56. J. A. Dumesic, *J. Catal.* **1999**, *185*, 496.
57. O. Beeck, *Disc. Faraday Soc.* **1950**, *8*, 118.
58. W. J. M. Rootsaert, W. M. H. Sachtler, *Z. Physik. Chem.* **1960**, *26*, 16.
59. J. H. Sinfelt, *Bimetallic Catalysts*, Wiley, New York, 1983 p. 14.
60. M. Boudart, *J. Am. Chem. Soc.* **1950**, *72*, 1050.
61. T. Bligaard, J. K. Norskøv, S. Dahl, J. Matthiesen, C. H. Christensen, J. Sehested, *J. Catal.* **2004**, *224*, 206.
62. J. K. Norskøv, T. Bligaard, A. Logadottir, S. Bahn, L. B. Hansen, M. Bollinger, H. Bengaard, B. Hammer, Z. Sljivancanin, M. Mavrikakis, Y. Xu, S. Dahl, C. J. H. Jacobsen, *J. Catal.* **2002**, *209*, 275.
63. V. Pallassana, M. Neurock, *J. Catal.* **2000**, *191*, 301.
64. Z. P. Liu, P. J. Hu, *J. Chem. Phys.* **2001**, *114*, 8244.
65. R. A. van Santen, J. W. Niemantsverdriet, *Chemical Kinetics and Catalysis*, Plenum Press, New York, 1995, p. 233.
66. C. H. Bartholomew, *Appl. Catal., A* **2001**, *212*, 17.
67. G. A. Somorjai, *Annu. Rev. Phys. Chem.* **1994**, *45*, 721.
68. H. Topsøe, J. A. Dumesic, N. Topsøe, H. Bohlbro, in *Proceedings of the Seventh International Congress in Catalysis*, T. Seiyama, K. Tanabe (Eds.), Elsevier, Amsterdam, 1981, p. 247.
69. H. Knözinger, *Adv. Catal.* **1976**, *25*, 183.
70. J. G. Goodwin Jr., S. Kim, W. D. Rhodes, *Catal.* **2004**, *17*, 320.
71. D. W. Goodman, *J. Catal.* **2003**, *216*, 213.
72. G. A. Somorjai, K. R. McCrea, J. Zhu, *Top. Catal.* **2002**, *18*, 157.
73. H.-J. Freund, M. Baumer, J. Libuda, T. Risse, G. Rupprechter, S. Shaikhutdinov, *J. Catal.* **2003**, *216*, 223.
74. G. Ertl, *J. Mol. Catal. A: Chem.* **2002**, *182–183*, 5.
75. J. V. Lauritsen, M. Nyberg, J. K. Norskov, B. S. Clausen, H. Topsoe, E. Laegsgaard, F. Besenbacher, *J. Catal.* **2004**, *224*, 94.
76. M. S. Chen, D. W. Goodman, *Science* **2004**, *306*, 252.
77. J. Greeley, J. K. Norskøv, M. Mavrikakis, *Annu. Rev. Phys. Chem.* **2002**, *53*, 319.
78. M. Neurock, *J. Catal.* **2003**, *216*, 73.
79. B. Hammer, J. K. Norskøv, *Adv. Catal.* **2000**, *45*, 71.
80. S. Linic, H. Piao, K. Adib, M. A. Barteau, *Angew. Chem. Int. Ed.* **2004**, *43*, 2918.
81. M. T. M. Koper, R. A. van Santen, M. Neurock, in *Catalysis and Electrocatalysis at Nanoparticle Surfaces*, A. Wieckowski, E. R. Savinova, C. G. Vayenas (Eds.), Marcel Dekker, Inc., New York, 2003, p. 1.
82. A. Hagemeyer, P. Strasser, J. Anthony, F. Volpe, *High-Throughput Screening in Chemical Catalysis: Technologies, Strategies and Applications*, Wiley-VCH Verlag, Weinheim, Germany, 2004, 339 pp.
83. A. Hagemeyer, R. Borade, P. Desrosiers, S. Guan, D. M. Lowe, D. M. Poojary, H. Turner, H. Weinberg, X. Zhou, R. Armbrust, G. Fengler, U. Notheis, *Appl. Catal., A* **2002**, *227*, 43.
84. R. J. Hendershot, C. M. Snively, J. Lauterbach, *Chem. Eur. J.* **2005**, *11*, 806.
85. J. M. Serra, E. Guillon, A. Corma, *J. Catal.* **2005**, *232*, 342.
86. S. Senkan, *Angew. Chem. Int. Ed.* **2001**, *40*, 312.
87. J. W. Saalfrank, W. F. Maier, *Angew. Chem. Int. Ed.* **2004**, *43*, 2028.
88. B. Timm, *Proceedings of the 8th International Congress on Catalysis Berlin 1984*, Verlag Chemie, Frankfurt, 1984, Vol. I, p. 7.
89. P. L. J. Gunter, J. W. Niemantsverdriet, F. H. Ribeiro, G. A. Somorjai, *Cat. Rev. - Sci. Eng.* **1997**, *39*, 77.
90. J. W. Niemantsverdriet, *Spectroscopy in Catalysis*, Wiley, Weinheim, Germany, 1993, 288 pp.
91. P. L. Hansen, J. B. Wagner, S. Helveg, J. R. Rostrup-Nielsen, B. S. Clausen, H. Topsøe, *Science* **2002**, *295*, 2053.
92. D. Teschner, A. Pestryakov, E. Kleimenov, M. Haevecker, H. Bluhm, H. Sauer, A. Knop-Gericke, R. Schloegl, *J. Catal.* **2005**, *230*, 186.
93. A. Corma, F. Rey, J. Rius, M. J. Sabater, S. Valencia, *Nature* **2004**, *431*, 287.
94. Y. Yin, R. M. Rioux, C. K. Erdonmez, S. Hughes, G. A. Somorjai, A. P. Alivisatos, *Science* **2004**, *304*, 711.
95. M. J. Pellin, P. C. Stair, G. Xiong, J. W. Elam, J. Birrell, L. Curtiss, S. M. George, C. Y. Han, L. Iton, H. Kung, M. Kung, H. H. Wang, *Catal. Lett.* **2005**, *102*, 127.
96. A. Gervasini, P. Carniti, J. Keranen, L. Niinisto, A. Auroux, *Catal. Today* **2004**, *96*, 187.
97. F. Besenbacher, E. Laegsgaard, I. Stensgaard, *Mater. Today* **2005**, *8*, 26.

1.2
Development of the Science of Catalysis

*Burtron H. Davis**

1.2.1
Early Concepts: Berzelius, Liebig, and Faraday

Prior to the introduction of the concept by Berzelius during the period of 1835–1836, catalysis was an experimental fact and a subject of much debate. The first application of catalysis, to produce ethanol by fermentation, is lost to antiquity. However, by the middle ages the use of sulfuric acid to catalyze the production of diethyl ether was widespread and indeed, written records of this synthesis date back to 1552 [1].

The reception by the scientific community of the concept of catalysis is well illustrated by the studies of Mitscherlich and Berzelius. In 1834, Mitscherlich reported that when alcohol was run into dilute sulfuric acid at 140 °C, ether and water could be distilled from the mixture [2]. He extended his observations by stating that decompositions and combinations of this kind were very frequent. Mitscherlich introduced the term "contact" to describe these actions, and summarized a number of reactions that were caused by contact – the formation of ether, the oxidation of ethanol to acetic acid, the fermentation of sugar, the production of sugar from starch by boiling sulfuric acid, the hydrolysis of ethyl acetate by alkali, and the formation of ethene from ethanol by heating with acid.

Berzelius, from 1821 on, summarized and reviewed critically the scientific investigations conducted worldwide in his "Annual Report" (Jahresberichte) [3]. The generalizations in Berzelius' reviews added as much or more to his reputation as his own discoveries, and these were many and important. In his annual review of 1835, Berzelius covered a number of reactions which take place in the presence of a substance which remains unaffected. Some roots of catalysis are provided in Fig. 1, which emphasizes those considered by Berzelius; a number of additional examples of reactions that predate Berzelius' definition can be found elsewhere [4]. Trofast [3] presents an English language version of Berzelius' conclusions:

> "This is a new power to produce chemical activity belonging to both inorganic and organic nature, which is surely more widespread than we have hitherto believed and the nature of which is still concealed from us. When I call it a new power, I do not mean to imply that it is a capacity independent of the electrochemical properties of the substance. On the contrary, I am unable to suppose that this is anything other than a kind of special

---

* Corresponding author.

> manifestation of these, but as long as we are unable to discover their mutual relationship, it will simplify our researches to regard it as a separate power for the time being. It will also make it easier for us to refer to it if it possesses a name of its own. I shall therefore, using a derivation well-known in chemistry, call it the catalytic power of the substances, and the decomposition by means of this power catalysis, just as we use the word analysis to denote the separation of the component parts of bodies by means of ordinary chemical forces. Catalytic power actually means that substances are able to awaken affinities which are asleep at this temperature by their mere presence and not by their own affinity."

Thus, although the same concept was put forth from two perspectives within a two-year period, one perspective had a much more significant and lasting impact. The concepts of Berzelius attracted much criticism that was directed, for the most part, not at catalysis, but rather at the concept of *catalytic force*. Berzelius has therefore been credited with introducing the concept of catalysis, even though one could easily conclude that Mitscherlich predated him by two years. Mitscherlich was not completely ignored, however, since "contact catalysis" was utilized into the middle of the 20th century.

The theories which were advanced during the 19th century to explain the mechanism of catalysis may be grouped into three classes: the chemical, vibrational, and physical. Proponents of the chemical theory view the catalyst to operate through the continuous formation and decomposition of unstable intermediate products. The vibration theory was applied especially to fermentation, and was at its zenith under the leadership of Liebig. The ferment (enzyme) was then viewed to possess a particular internal motion that could be transmitted to neighboring molecules, and thereby induce reactions of them. The physical theory explains the phenomena as being due to the condensation, and the increase in concentration, of the reacting substances at the surface of the catalyst, such increase in concentration being brought about by capillary forces.

It has long been recognized that a catalyst has the same chemical composition at the end of the reaction that it effected; however, it has been equally recognized that its physical state may have undergone a profound alteration [5]. Thus, when ammonia is decomposed by contact with heated metals there is nearly always an alteration of the physical state of the metal, and this has been attributed to the formation and decomposition of intermediate compounds, metallic nitrides in this particular case. While these views are obvious today, it would be a demanding task to trace their history "to the original statement of the concept".

Whereas Berzelius defined "catalysis", "catalytic force", and "catalytic action", it remained for Armstrong [6] to define catalyst:

> "... so little has been done to ascertain the nature of the influence of the contact-substance, or catalyst, as I would term it, the main

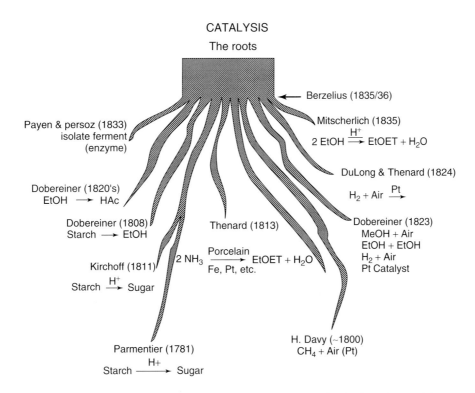

Fig. 1 Schematic representation of some of the studies that serve as a background for the formulation of the concept of catalysis by Berzelius.

object in view being the study of the product of the reaction, that the importance of the catalyst is not duly appreciated."

This is an outstanding example of an accurate historical fact, and a presentation, on the one hand, of an obvious definition, and on the other hand, an insightful statement of needs for future research.

Adsorption is at the heart of any heterogeneously catalyzed reaction. In 1800, adsorption was absorption, and scientists were aware of it. The concepts of adsorption are illustrated in an admirably lucid account of its status at that time by de Saussure in a presentation to the Geneva Society on April 16, 1812 [7]. The questions that de Saussure asked in the introduction of his paper are equally valid for a manuscript that is being written today. He begins with a statement that no accurate experimental data were available on the question of whether a gas, when it penetrates into the pores of a solid body, undergoes any diminution of bulk (volume) when no chemical bonding takes place between the gas and the solid. de Saussure, assuming that such contraction of volume does take place, is led to a number of insightful questions: What influence has the size of the pores on this condensation? Are all gases equally condensed by the same bodies? What influence has the density of the gas on this condensation? When equal quantities of two gases are in contact with a solid, do they adsorb in equal quantities? Do the mixed gases, when condensed in the solid, enter into combinations which they would not form in the free state? He understood the need for careful experimental results. Also, he clearly outlined and defined many scientific aspects of adsorption.

de Saussure confirmed that adsorption in all cases, except for oxygen, was complete at the end of 24 to 36 hours. He also demonstrated that the adsorption of a gas is exothermic, liberating a quantity of heat during the condensation that is often sensible to the feel, and sufficient to raise by several degrees the temperature indicated by a thermometer in contact with the solid (charcoal). Aware that condensation of a gas was exothermic and evaporation was endothermic, de Saussure speculated that the solid should become cooler during evaporation, and demonstrated experimentally that this was indeed the case.

de Saussure speculated that charcoal should be activated by evacuation as well as by heating, and showed this to be the case. He compared the amount adsorbed on the whole and ground charcoal. The pulverized sample adsorbed less gas, and de Saussure attributed this to the amount of preadsorbed water. The rate of adsorption of different gases was reported to be the same in all solids of similar chemical properties, even though the amounts of gases condensed in the various solids may vary greatly.

References see page 35

de Saussure observed that exposure of a solid saturated with one gas to a second gas would cause adsorption of the second gas, resulting in the desorption of some fraction of the first gas.

He stated that the affinity of the gas and the solid, as well as the condensibility of the gas, must be considered in addition to the porosity of the solid. Furthermore, the forces of affinity and of condensibility act to oppose each other. de Saussure considered the condensation of gases in solid bodies to be analogous with the rise of liquids in capillary tubes. Both are in the inverse ratio of the size of the interior diameters of the tubes or pores.

Mellor [5] devotes considerable space to Faraday's "condensation theory" of catalysis. Taylor wrote [8]: "Faraday was the first to indicate the zone of adsorbed material as the reaction space of a heterogeneous catalytic action."

Robertson [9] and Schwab [10], among others who survey the concepts of catalysis, attribute the adsorption theory to Faraday. However, Faraday's view of catalysis was a limited extension of de Saussure's views that are described above.

Faraday, during about a three-month period, investigated the recombination of oxygen and hydrogen as they remained in contact with platinum following their liberation during electrolytic studies [11]. He concluded that this property belonged to platinum "... *at all times*, and was *always effective* when the surface was *perfectly clean*." The platinum could be activated by heating in acid or by employing mechanical cleaning; other approaches, such as treating or heating with alkali, were effective for only part of the time. Faraday was convinced that the phenomenon which he observed had a satisfactory explanation based on known principles, and did not require the assumption of any new state or new property. Faraday clearly showed that the presence of hydrogen sulfide, phosphine or ethene, among others, would prevent the platinum from exhibiting its properties of causing the combination of hydrogen and oxygen.

Faraday wrote that Döbereiner [12] considered the effect to be an electric action in which hydrogen, being very highly positive, represents the zinc of the usual electric cell arrangement, and like it, attracts oxygen and combines with it. Faraday stressed the need for a perfectly clean and metallic surface. He considered the effect to be produced by most, if not all, solid bodies, weakly perhaps in many of them, but rising to a high degree in platinum. He wrote that he was prepared to admit that the sphere of action of particles extends beyond those other particles with which they are immediately and evidently in union, and in many cases produces effects rising into considerable importance, and he thought that this kind of attraction was a determining cause of Döbereiner's effect. Faraday compared the spongy platinum to a hygroscopic body which becomes moist due to the condensation of the water vapor with which it is in contact. He then considers that a gas, even when compressed at high pressure, still has a low density and possesses sufficient vacant volume that another gas can be added in the void space. He wrote that the two gases, hydrogen and oxygen, will not react even when compressed unless the platinum surface is present to suppress, or remove, their elasticity and/or the action of the metal in condensing them against its surface by an attractive force. He compares the hydrogen/oxygen case to that where Hall found CO_2 and lime to remain combined under pressure at temperatures at which they would not have remained combined if the pressure had been removed. The course of events may be stated, according to Faraday by these principles:

- the deficiency of elastic power and the attraction of the metal for the gases are such that the gases are so far condensed as to be brought within the action of their mutual affinities at the existing temperature
- the deficiency of elastic power, not merely subjecting them more closely to the attractive influence of the metal, but also bringing them into a more favorable state for union, by abstracting a part of that power (upon which depends their elasticity) which elsewhere in the mass of gases is opposing their combination

The consequence of their combination is the production of the vapor of water and an elevation of temperature. The attraction of Pt for water is not as great as for the gases, and so the water quickly diffuses through the remaining gases. Fresh portions of the gases come into juxtaposition with the metal, combine, and the vapor formed also diffuses, allowing new portions of gas to be acted upon. In this way the process advances and is accelerated by the evolution of heat. The platinum is not considered to cause the combination of any particles with itself, but only associating them closely around it; the compressed particles are free to move from the platinum, being replaced by other particles, as a portion of dense air upon the surface of the globe, or at the bottom of a deep mine, is free to move by the slightest impulse into the upper and rarer parts of the atmosphere. Thus, apart from introducing the need for a clean surface, and what may be, in the most optimistic view, a small step in the direction of stating the concept of chemisorption, Faraday held views that were essentially the same as those of de Saussure. This is not meant to degrade Faraday's work; in fact, he applied extraordinary concepts and visionary thought in an experimental program that extended over only three months.

During the period 1840–1880, Liebig's view of catalysis prevailed. Liebig matched Berzelius in scientific stature, published his own journal, and had the advantage that most chemists had been trained by him, either directly or indirectly by his students. During the period between 1836 and 1846, Liebig was engaged in research into putrefaction, and in developing agriculture science where organic ferments were a subject for active investigation. The isolation of an enzyme from barley malt in 1832 [13] ignited these intense research efforts.

Liebig was a proponent of the vibration theory of fermentation and catalysis. Pasteur [14] proved that alcoholic fermentation and putrefaction are caused by the presence of certain low forms of life, namely bacteria. Liebig attacked this view, and with vigor:

"To suppose that putrefaction or fermentation is caused by the physiological action of such creatures can only be compared with the idea entertained by a child who would explain the rapid current of a river through a mill-wheel by supposing that the mill-wheel, by its force, drives the water down the stream."

However, Liebig's criticism did push Pasteur to conduct more experiments, and to firmly establish his views. When considering the decomposition of hydrogen peroxide, Liebig wrote [15]:

"Yet it is singular that the cause of the sudden separation of the component parts of peroxide of hydrogen has been viewed as different from those of common decomposition, and has been ascribed to a new power termed the catalytic force. *Now, it has not been considered, that the presence of the platinum and silver serves here only to accelerate the decomposition; for without the contact of these metals, the peroxide of hydrogen decomposes spontaneously, although very slowly."*

Thus, Liebig presented a major concept that was to be a part of the theory of catalysis advanced later by Ostwald, although it did not have the framework of kinetics that developed during the 40-year period separating the two.

After considering several examples of such decompositions, Liebig [16] concludes that:

"No other explanation of these phenomena can be given, than that a body in the act of combination or decomposition enables another body, with which it is in contact, to enter into the same state. It is evident that the active state of the atoms of one body has an influence upon the atoms of a body in contact with it; and if these atoms are capable of the same change as the former, they likewise undergo that change; and combinations and decompositions are the consequence. But when the atoms of the second body are not capable of such an action, any further disposition to change ceases from the moment at which the atoms of the first body assume the state of rest, that is when the changes of transformations of this body are quite completed."

"In that state in which they exist within the pores or upon the surface of solid bodies, their repulsion ceases, and their whole chemical action is exerted. Thus, combinations which oxygen cannot enter into, decompositions which it cannot effect while in the state of gas, take place with the greatest facility in the pores of platinum containing condensed oxygen. When a jet of hydrogen gas, for instance, is thrown upon spongy platinum, it combines with the oxygen condensed in the interior of the mass; at their point of contact water is formed, and as the immediate consequence heat is evolved; the platinum becomes red hot and the gas is inflamed... In finely pulverized platinum, and even in spongy platinum, we therefore possess a perpetuum mobile – a mechanism like a watch which runs out and winds itself up – a force which is never exhausted – competent to produce effects of the most powerful kind, and self-renewed ad infinitum. Most phenomena, formerly inexplicable, are satisfactorily explained by these recently discovered properties of porous bodies. The metamorphosis of alcohol into acetic acid, by the process known as the quick vinegar manufacture, depends upon principles, at a knowledge of which we have arrived by a careful study of these properties."

1.2.2
Wilhelm Ostwald

During the 50 years following the definition of catalysis by Berzelius, very little progress was made in its understanding. Berzelius had organized enzyme catalysis, as well as organic and inorganic catalysis, into a single discipline, and it remained that way for the next half-century. During this period, as in the period preceding its definition, catalysis was a science based on trial and error – Edisonian research – and reports of the observations. However, during this period great advances in the understanding of chemistry were taking place, and these provided the framework for an explosion of soundly based concepts of catalysis. Organic chemistry was gaining a structure through many developments that included van't Hoff–Le Bel's three-dimensional molecules, the recognition of four-valent carbon, the development of reliable chemical formulas based upon Liebig's method of elemental analysis of organic compounds, the recognition of stereoisomers, and the isolation of enantiomers. Physical chemistry was introduced and developed, primarily by Arrhenius, Ostwald and van't Hoff; its acceptance was documented with the appearance of Ostwald's two-volume book and the *Zeitschrift für Physikalische Chemie*, edited by van't Hoff and Ostwald. Kinetics became a subject for study following the measurements by Wilhelmy in 1850 of the acid-catalyzed inversion of cane sugar [17], and attained its first mathematical maturity in the studies of Harcourt and Esson [18]. Mayer introduced the concept of the conservation of energy in 1842, and others placed the concept on a firm scientific basis [19]. Thermodynamics was established, and Gibbs provided advanced concepts of statistical thermodynamics and the phase rules. Boltzmann provided his theories of the distribution of energy, the theory of approach to equilibrium, and entropy and probability, among others.

Ostwald first came to catalysis through his studies on the acceleration of homogeneous reactions by acids. These findings were widely accepted at the time although,

References see page 35

ultimately, they were shown to be incorrect because Ostwald believed that the acid, in acting as a catalyst, did not enter into the chemical change which it influences but rather acted by its mere presence (contact catalysis).

Ostwald, when reviewing a paper in which Stohmann [20] utilized the vibrational theory of catalysis, wrote:

> "The abstractor has several objections to make to this definition. First, the assumption of a 'condition of movement of the atoms in a molecule' is hypothetical and therefore not suitable for purposes of definition.... If the abstractor were to formulate for himself the problem of characterizing the phenomena of catalysis in a general way, he would consider the following expression as probably most suitable: Catalysis is the acceleration of a chemical reaction, which proceeds slowly, by the presence of a foreign substance. It would then be necessary to give the following explanations... This acceleration occurs without alteration of the general energy relations, since after the end of the reaction the foreign body can again be separated from the field of the reaction, so that the energy used by the addition can once more be obtained by the separation, or the reverse. However, these processes, like all natural ones, must always occur in such a direction that the free energy of the entire system is decreased.... The existence of catalytic processes is to me therefore a positive proof that chemical processes cannot have a kinetic nature."

With this review, the dominant concept of catalysis passed from Liebig to Ostwald. In these comments, Ostwald showed that a catalyst could not change the equilibrium, and it was he, more than anyone, who brought down the vibrational theory as not being amenable to experimental verification. At the same time, his definition emphasized an equally unverifiable concept: a catalyst cannot start a reaction which is not taking place without it. In 1901, Ostwald [21] included as catalytic phenomena:

- the release of supersaturation
- catalysis in homogeneous mixtures
- catalysis in heterogeneous systems
- enzyme action.

Ostwald had not accepted the kinetic theory at the time of his definitions of catalysis, and incorrectly saw the phenomena disproving the atomic hypothesis championed by, among others, Boltzmann. Like Liebig before him, Ostwald saw catalysis theory by analogy: a catalyst acts like oil on a machine, or as with a whip on a tired horse [22]. His investigations into catalysis were recognized with the award of Nobel prize for chemistry in 1909.

In Ostwald we have an example which demonstrates that concepts that are shown, ultimately, to have little validity – or even to be incorrect – may have great impact in developing science. Indeed, in many instances it is the drive and forceful dominance by the individual that establishes the concept, and not the novelty of the concept. Thus, most authors associate the concept of a catalyst speeding up an existing reaction with Ostwald, not realizing that Liebig advanced the same suggestion some 50 years earlier. With the acceptance of physical chemistry, and the training of a sufficient number of scientists to make use of the advances in the field, catalysis was set to make many advances beginning at about 1900, and the Ostwald school was in the driver's seat.

1.2.3
The Concepts of Kinetics and Intermediate Compounds

Döbereiner believed that platinum black carried oxygen over to hydrogen, while de la Rive suspected that a layer of platinum oxide formed on the surface of platinum. Fusinieri [23] went so far as to propose that waves of oxidation and reduction were responsible for the reaction of hydrogen and oxygen; he even reported the visual observation of such wave action. Although Faraday was unable to form a distinct idea of the power of Fusinieri's theory, the latter defended his proposals, citing evidence to show that those with good eyesight could see these waves. Berthollet suggested that platinum hydrides were formed in the presence of hydrogen, and that these reacted with oxygen to produce hydrogen.

Brodie [24] explained the decomposition of peroxides on the basis of chemical affinity and of catalysis in terms of coupled chemical reactions. For the reaction (his uncorrected equations are used in the following)

$$I_2 + Ba_2O_2 = 2BaI + O_2$$

Brodie considered the reaction to take place by the decomposition and reformation of water, according to the two equations:

$$I_2 + BaO_2 + H_2O = 2HI + Ba_2O + O_2$$
$$2HI + Ba_2O = 2BaI + H_2O$$

Viewing water as a catalyst provides the cyclic concept of catalytic action.

Brodie [24] also noted that:

> "Views as to the polarization of oxygen, and the cause of the decompositions effected by the alkaline peroxides, which to a great extent are identical with the preceding, and in which the same language and the same notation are employed, have recently been put forward, with considerable pretension, as new and originating with himself, by Schönbein, Professor of Chemistry at Basle. This chemist can scarcely be aware of the memoir referred to, as in his numerous publications he makes no allusion to it. A reclamation of priority of ten years ought not to be required, but I am compelled to call the attention of chemists to these circumstances in order that I myself may not be considered to appropriate without acknowledgment the ideas of discoveries of another."

The intermediate compound theory has guided the thoughts of many individuals whose contributions moved catalysis to the forefront. For example, Sabatier [25] wrote:

> "Having arrived at the end of my career as a chemist, I have thought that it might be of interest, and at the same time of some utility, to relate how I was led to develop the direct hydrogenation method by finely divided metals, which won for me the Nobel Prize in 1912... It is the faith in the theory of temporary compounds furnished by the catalyst which has constantly guided me in these various labors; it is these inductions to which I owe all my new results."

The concept of the intermediate compound remains dominant today in the many "volcano plots", and is frequently associated with the names of Sabatier and/or Balandin. Thus, the rate increases with the heat of adsorption (or formation) until a sufficiently high value is obtained so that the species is adsorbed so strongly that the rate decreases; this produces the well-known volcano plot [26] (Fig. 2). As with any principle or correlation, it is frequently utilized in situations where its use is questionable. Thus, without the data point for Au, the volcano curve in Fig. 3 would be a straight line.

Polanyi [27] utilized this concept in the development of his "transition state" theory of catalysis, assuming that free homopolar valencies protrude from the surface of a solid, and that a reaction $AB + CD = AC + BD$ would be catalyzed on the surface of such a solid (see Fig. 4). For the reaction of $H + H_2$, Polanyi provided the curve shown in Fig. 5, and concluded that the catalytic effect would be

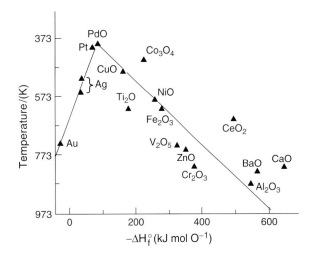

Fig. 3 Example of a volcano plot with little data to support one side of the mountain. (Reproduced from an anonymous reference.)

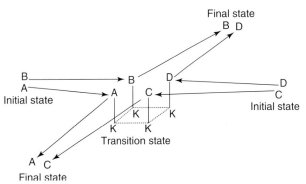

Fig. 4 Polanyi's representation of his "transition state" theory. (Redrawn from Ref. [27].)

strongest when there is adsorption of moderate overall energy. Surface valencies forming links of 50 000–51 000 gcal, corresponding to QY = 0–2000, will be the most efficient.

The highlights in chemical kinetics have been summarized by Laidler in his outstanding history of physical chemistry (Table 1) [28]. The kinetic ideas introduced by Wilhelmy, and also by Harcourt and Esson, were put forth by van't Hoff in a book, *Études de dynamique chimique*, so that a compilation of the concepts of both chemical thermodynamics and kinetics was available. van't Hoff introduced what is now considered to be the order of the reaction. Arrhenius also introduced the concept of an intermediate, Z, in the reaction scheme:

$$A + C \underset{k_3}{\overset{k_1}{\rightleftharpoons}} Z$$

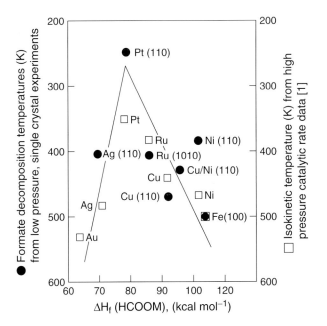

Fig. 2 Example of a typical volcano plot correlating the formate decomposition rate for both pressure and single crystal studies. (Reproduced from Ref. [26].)

References see page 35

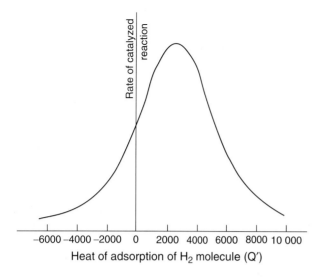

Fig. 5 Relationship between the rate of a catalyzed reaction and the heat of adsorption of hydrogen: an early "volcano-type" plot. (Redrawn from Ref. [27].)

$$Z \xrightarrow{k_3} B + C$$

Here, the reaction that determines the velocity of the process is the decomposition of Z, and the assumptions of the low and steady-state concentrations of Z permits the rate of production of product B, catalyzed by C, to be written as:

$$\frac{d[B]}{dt} = k_3[Z] = k_3 K_1[A][C]; \quad K_1 = \frac{k_1}{k_2}$$

Langmuir [29] derived his isotherm from a consideration of adsorption/desorption kinetics, and developed kinetic equations for a number of cases of catalysis. The adsorption equation developed by Langmuir was advanced independently by Michaelis and Menten [30] to describe the kinetics of enzyme catalysis, and illustrates that, by 1916, biocatalysis had again developed into a study separated from much of catalysis science. The systematic studies performed by Hinshelwood extended the use of Langmuir isotherms to what is now known as "Langmuir–Hinshelwood kinetics". Hougen and coworkers subsequently provided the basis for kinetic applications in chemical engineering [31].

One development that provided a conceptual framework for the kinetics of catalytic reactions was the introduction of the activated reaction rate theory by Eyring and coworkers [32, 33]. Eyring's theory provided a basis for the reaction coordinate shown for the homogeneous and heterogeneous reaction in Fig. 6 [34]. In this figure – and even more so in many other instances – the Ostwald view that a catalyst simply speeds up a reaction that is already occurring is evident, as the activated complexes for the thermal and catalytic reaction are depicted at a common point on the reaction coordinate. Only recently has it become widely accepted that the role of the catalyst is, in almost all cases, to alter the reaction mechanism so that the thermal pathway which requires a high activation energy is transformed to two or more steps, each with a lower activation energy.

The concept of utilizing a stoichiometric number to define (or at least limit the number of steps that

Tab. 1 Highlights in chemical kinetics (From Ref. [28].)

Date	Author	Contribution
1850	Wilhelmy	Rate-concentration dependence
1865	Harcourt and Esson	Time course of reactions
1884	van't Hoff	Differential method; temperature dependence
1889	Arrhenius	"Arrhenius equation"
1891	Ostwald	Theory of catalysis
1899	Chapman	Theory of detonation
1913	Chapman	Steady-state treatment
1914	Marcelin	Potential-energy surfaces
1917	Trautz; W. C. McC. Lewis	Collision theory
1918	Nernst	Atomic chain mechanism
1921	Langmuir	Surface reactions
1921–1922	Lindemann; Christiansen	Unimolecular reactions
1927–1928	Semenov; Hinshelwood	Branching chains
1931	Eyring and M. Polanyi	Potential-energy surface for H + H_2
1934	Rice and Herzfeld	Organic chain mechanisms
1935	Eyring; Evans and M. Polanyi	Transition-state theory
1949	Porter and Norrish	Flash photolysis
1954	Eigen	Relaxation methods
1980	J. C. Polanyi	Spectroscopy of transition species

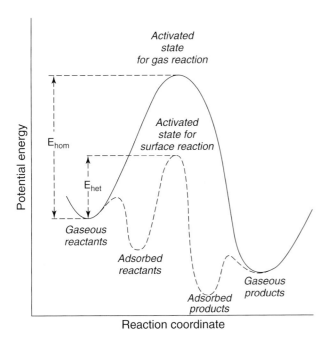

Fig. 6 Potential-energy curves for a reaction proceeding homogeneously (solid curve) and on a surface (dotted curve). These curves illustrate retention of the Ostwald concept, that a catalyst only speeds up a homogeneous reaction, but does not change the mechanism. (Reproduced from Ref. [34].)

define) the rate was introduced by Horiuti [35], and many of the mathematical derivations relating to this method were developed by his group. The stoichiometric number is defined as the number of times that a step of the reaction mechanism occurs. The first efforts to utilize the technique to define the mechanism for ammonia synthesis led to contradictory results; Enomoto and Horiuti [36] reported a stoichiometric number of 2, whereas the results of a combined kinetic and isotopic study indicated that it was 1 [37]. Similar problems have been encountered in applying the stoichiometric number to other catalytic systems; thus, while it is an attractive concept and provides assistance in eliminating some potential mechanisms from consideration, the experimental data available today usually limit its usefulness.

The concept of turnover – the number of catalytic reactions per catalytic site per unit time – came into common usage in enzyme catalysis. Only recently has this concept been adapted to heterogeneous catalysis, particularly by Boudart and his students. Where the number of active sites can be defined precisely (e.g., H^+ in a sulfuric acid solution), the concept has great utility in a definition of catalysis; however, it has been applied too frequently in instances where the measure of the number of catalytic sites is uncertain, or worse.

Jost [38] attributes:

"... the discovery of chain reactions to Bodenstein in 1913, and the coining of the word to Christiansen.... E. Cremer, in her thesis with Bodenstein in 1927 on the H_2-Cl_2 reaction, first noted the occurrence of chain branching and instability caused by branching. The implications of this concept, however, were not fully recognized either by her or by Bodenstein until Semenoff's later publications."

Kemball, however, wrote that the idea of an atomic chain based on the repeated reactions was proposed by Nernst in 1916 [39]. In any event, catalysis and chain reactions have features of a common concept.

One area of kinetics that has attracted much attention is the compensation effect; here, a high activation energy is compensated for by an increase in the pre-exponential term, the effect being derived from a one-page note by Cremer and Schwab [40]. The initial experimental data demonstrating the compensation effect was provided by Constable [41]. The debate continues as to the cause of the linear relationship between ln A and E; these include views that it is based upon one of several theoretical explanations, is merely an empirical observation, or even to question its existence [42, 43].

Today, the progress of a heterogeneous catalytic reactions can be resolved into at least five distinct steps:

(i) diffusion of the reactants to the catalyst
(ii) formation of the adsorption complex (reactant-surface)
(iii) the chemical change on the surface
(iv) decomposition of the adsorption complex (product-surface)
(v) diffusion of the reaction products from the catalyst

Unless all five steps have the same activation energy, the one step with the highest activation energy determines the rate.

Steps (i) and (v) received prominent attention during the early application of kinetics to catalysis. Nernst, among others, considered the rate-controlling step of the dissolution of a solid to be due to the diffusion of the molecule or ions through the nearly saturated boundary layer that surrounds each particle. By analogy, during the late 1800s and early 1900s, Nernst, Bodenstein, and others proposed that the rate of a catalytic reaction was determined by the diffusion of reactants and/or products through the layer(s) of adsorbed molecules. While this particular view of catalysis gradually fell from favor, diffusion remains an important concept in catalysis.

Thiele, while investigating the theory of catalysis at home in his spare time, developed a connection

References see page 35

between the rate of reaction and the diffusional limitations imposed by the porosity of a catalyst [44]. This theory was amplified by many, particularly Wheeler [45], whose studies proved to be highly original and the results monumental. Wheeler's most significant results included:

- emphasizing the apparent decrease in the activation energy, pressure and temperature gradients expected in catalyst pellets
- the influence of catalyst poisons on activity and selectivity
- the role of diffusion in altering the observed kinetics from, for example, series to parallel reactions
- an estimate of the percentage of internal catalyst surface utilized for many reactions (Fig. 7) [46]

1.2.4
Negative Catalysis: Autocatalysis

Negative catalysis was one of the concepts that added to the confusion about catalysis. The findings of Turner [47] and Faraday [11] clearly showed that some substances retarded the effectiveness of a catalyst. Mellor [48] noted that:

> "The interesting feature is that these gases may be regarded as catalytic agents, which inhibit the action of another catalytic agent. This phenomenon will be called negative catalysis."

Subsequently, Mellor explained that negative catalysis in ester hydrolysis has been explained, with more or less success, by assuming that:

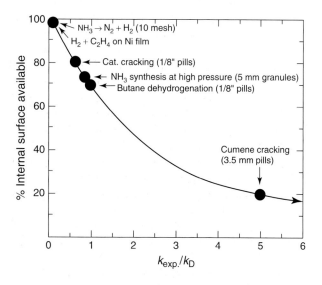

Fig. 7 Wheeler's estimate for the percent of internal catalyst surface available for effecting a number of important industrial reactions. (Reproduced from Ref. [46].)

(i) The degree of ionization of the ester or catalyzer is diminished, or else the catalyzer combines with the "foreign substance" so that the quantity of the available catalytic agent is diminished.
(ii) The combination of the ester with the retarding salt by which the active mass of the ester is diminished.

However, it was gradually recognized that negative catalysis was merely the result of catalyst poisoning. Likewise, the chemical concepts slowly developed to show that a strong adsorptive bond between the poison and the catalyst caused the effect to be highly specific. Due in a large part to the extensive studies of Maxted [49], it was understood that the effect depended on definite types of electronic configuration of both the catalyst and the poison. A material would only be regarded as a "poison" if it exerted an appreciable effect on catalysis even when present in very small concentrations. Maxted was careful to distinguish between catalyst activity loss due to poisoning and that of "... mechanically covering up of a catalyst surface by less specifically held coatings, such as the cloaking of a catalyst by a layer of gums or waxes or by a deposit of carbon in organic reactions at high temperatures." Today, the study of these topics has become so extensive that they are important components of International Symposia on Catalyst Deactivation; the books based on these symposia contain many reports that provide finer details and more complex examples of these two basic concepts.

The term "autocatalysis" was coined by Ostwald to identify a class of catalytic reactions which are slow during an initial induction period, after which the rate accelerates rapidly. This effect will occur when one of the products can act as a catalyst for the reaction. Thus, the acid formed as a product of the hydrolysis of an ester can act as a hydrolysis catalyst. Another instance where autocatalysis will be observed is during the reduction of a metal oxide with hydrogen, provided that the metal formed catalyzes the reduction. Another example occurs in the methanol-to-gasoline conversion, where a product (in this case the C_{3+} alkenes) combines with methanol more rapidly than the hydrocarbons can be formed from only methanol.

1.2.5
Adsorption

During the early 1900s, it was gradually recognized that there are two types of adsorption. Langmuir's studies [29] emphasized the type which has become known as *chemisorption*, whereas the value of the other type – commonly known as *physisorption* – was emphasized by the introduction of the Brunauer, Emmett and Teller (BET) method for measuring surface area. Taylor was primarily responsible for the concept of activated adsorption, and the need for an activation energy for the transformation from the physisorbed state to the chemisorbed state [50]. This proposal was made by Taylor in an effort to more closely correlate the specific activity of the catalysts with their adsorption characteristics.

Wheeler [45] advanced the use of a physisorbed state by introducing the concept that, by taking into account the formation of multilayers of physically adsorbed gas and capillary condensation in capillaries, one could calculate the size distribution of pores present in a catalyst. Barrett et al. [51] provided a detailed procedure for quantitative calculations. Wheeler subsequently pioneered the development of the theory of interpreting the influence of pore size on many characteristics of catalytic activity and selectivity. For example, the activation energy for a reaction completely controlled by pore diffusion will have only one-half the true activation energy; likewise, a series reaction A → B → C may be made to appear as parallel reactions by pore diffusion control. Moreover, Wheeler pointed out that the catalyst preparation conditions can be utilized to obtain a material with the optimum surface area and porosity for the production of a desired product. The isomerization of xylenes to produce greater than 90% of the para-isomer, in contrast to the 20–25% present at equilibrium, simply by controlling the pore opening of the HZSM-5 catalyst, is one example of the application of the concept selectivity control by diffusion. The concept of diffusion in catalysis has now advanced to become a scientific discipline that represents an independent field of study.

Washburn [52] developed a theory based upon mercury penetration to provide a method for measuring porosity (especially for larger pores), and this concept was converted into an experimental reality by the investigations of Ritter and Drake [53].

The classification into five types of physical adsorption isotherms was accomplished by Brunauer and colleagues (see [54]). The desorption hysteresis associated with each adsorption isotherm came to be related to the type of pores that the material possessed, and while a single source cannot be cited for this concept, de Boer and coworkers had much to do with publicizing the importance of the shapes of the hysteresis loops.

During 1914–15, Langmuir put forth three concepts important to chemisorption and its role in catalysis: kinetics, his checkerboard model of the surface, and chemical forces and bonding [55]. In developing his concepts, Langmuir described heterogeneous catalysis as a chemical drama that occurred in a single layer and by the same chemical forces that held molecules and solids together. Langmuir's concepts dominated the field for decades, and remain important today. The kinetic nature of the adsorption and desorption steps at equilibrium for a given pressure led to the Langmuir Isotherm:

$$\theta = \frac{kP}{1 + kP}$$

where θ is the surface coverage, k is the ratio of the adsorption and desorption rate constants, and P is the pressure of the adsorbent. Langmuir's notebook contain his "Fundamental Laws of Heterogeneous Reactions":

- The surface of a metal contains atoms spaced according to a surface lattice (number per cm^2).
- Adsorption films consist of atoms or molecules held to the atoms forming the surface lattice by chemical forces [55].

Langmuir's model implies that the amount of gas (e.g., CO) adsorbed onto a metal (e.g., Pt) permits a calculation of the surface of the metal exposed to the gas; however, Langmuir did not emphasize this concept. He did, however, introduce the "checkerboard" model for the surface of the catalyst.

During the 1930s, Brunauer and Emmett utilized the combination of physical and chemical adsorption to obtain a measure of the fraction of the surface of an iron synthetic ammonia catalyst that consists of metallic iron, alkali oxide promoter, and a structural promoter such as silica or alumina. Thus, low-temperature nitrogen adsorption provided a measure of the total surface, chemisorbed CO a measure of free metallic iron, and CO_2 adsorption the surface concentration of alkali oxide promoter, and – by difference – the fraction covered by the structural promoter. Significant improvements in the technique have occurred since these initial investigations, but little has been altered concerning the concepts involved.

In 1935, Lennard–Jones [56] provided a pictorial scheme for the adsorption of, in this example, hydrogen (Fig. 8). This scheme illustrates that the transition from van der Waals adsorption to that of the chemisorbed state requires an activation energy. This concept allows for the activated adsorption championed by Taylor; at the same time, in some special cases the activation energy may be so low that adsorption may occur with little or no activation. deBoer [57] considered adsorption from the viewpoint of an adsorption time, whereby the average stay of a molecule on the surface should depend upon the interaction energy. Thus, as the interaction energy increases, the average residence time increases; hence, as the energy increases the residence time increases from essentially no adsorption to physical adsorption to chemisorption.

Surface mobility became an important concern following Langmuir's studies on chemisorption. Mobility, as it applies to physical adsorption, means that the adsorbate

References see page 35

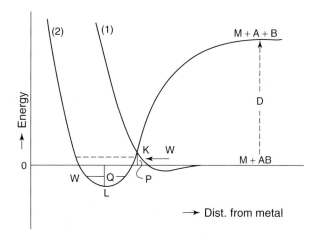

Fig. 8 Schematic representation of the Leonard-Jones potential, illustrating the activation energy needed to transfer from the physically adsorbed to the chemisorbed state. (Redrawn from Ref. [56].)

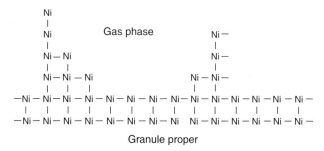

Fig. 9 Taylor's representation of heterogeneity and the presence of various types of active site on a nickel catalyst. (Reproduced from Ref. [62].)

can move so freely on the surface that its state is that of a two-dimensional gas. For chemisorption, at least three definitions have been advanced [58]. Beeck [59] suggested that adsorption be classified as mobile or immobile according to whether the heat of adsorption fell with increasing coverage or remained constant. Thus, if the adsorbate is mobile it should adsorb on high-energy sites first, resulting with a decrease in the heat of adsorption with increasing surface coverage. The field emission microscope [60] provided the first direct method of studying surface mobility; indeed, it is surprising that the developer of this technique, E. Müller, did not receive greater recognition [61].

1.2.6
Active Site: Geometric or Electronic?

Langmuir's checkerboard model provided an initial impetus to relating physical/chemical properties of the solid and the catalytic characteristics. Taylor's name is usually the one associated with the concept of an active site [62], although he was not the first to propose a special site of unusually high activity. However, his simple model and his stature focused much attention to the active site. Thus, Taylor attributed special activity to those atoms which, because of the uneven geometry of the surface (Fig. 9), have many atoms whose coordination to other catalyst atoms is very low (unsaturation), and these are the atoms that were attributed to provide the seat of most catalytic conversions. This view focused attention on the heterogeneity of the surface of almost all catalysts, and the fact that the total surface would not be equally active in effecting chemical reactions. Constable assumed that these active sites followed an exponential distribution [41]. The application of modern surface techniques have provided the data to place the concept of active sites on a firm foundation. At about the same time that he advanced the active site, Taylor [63] made an observation that is equally insightful, but has received little attention:

"... an oxide catalyst surface is to be regarded as composed, not of a single catalyst, but of two catalysts, metal ions and oxide ions and the nature of the changes induced in the adsorbed reactant is determined by the charge of the ion on which the reactant molecule is adsorbed."

This statement contains the rudiments of an acid–base theory of metal oxide catalysis.

A liberal historian could trace the concept of active sites back at least to Loew [64], who suggested that when a molecule of the reactant contacts the catalyst, the "sharp corners" of the catalyst break up the molecule into atoms, and these are more reactive. Hence, the finer the particle, the greater the number of corners, and the greater the activity of the catalyst.

Shortly following Taylor's active-site hypothesis, Balandin proposed what has become known as the multiplet hypothesis. Balandin proposed that binding between two atoms of an adsorbed molecule can be broken if they are attracted by two different catalyst atoms; a bond may be formed between atoms of an adsorbed molecule bound to the same catalyst atom [65]. This hypothesis was applied by Balandin to many reactions, including the dehydrogenation of cyclohexane; here, the spatial arrangement of the catalyst surface atoms must be such that metal–hydrogen bonding can result (Fig. 10) [66]. Those metals with little mismatch between the spacing of the reactant hydrogen atoms and the catalyst surface atoms were considered to be active catalysts, whereas those with a large mismatch would be inactive. Balandin expanded his scope to include the activation enthalpy of the formation/decomposition of the multiplet complex, and conducted extensive studies to obtain a measure of the sum of energies of disrupted bonds in reactants and

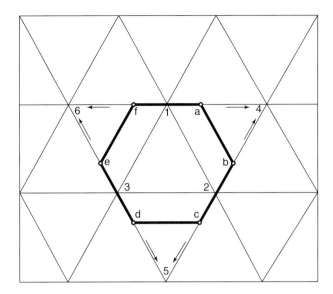

Fig. 10 Schematic showing the location of surface atoms of the catalyst (intersections of light lines) and the adsorbed cyclohexane (heavy lines); the multiplet theory indicated that the spacings had to match in order to form the appropriate bonding. (Reproduced from Ref. [66].)

newly formed ones in products. This view attracted much attention and, because of Balandin's position within the USSR scientific community, probably survived in its original form much longer than could have been anticipated.

In 1934, Gwathmey began a series of studies to show, at least qualitatively, that different metal crystal faces possess different catalytic activities. Gwathmey's experiments, which were based on his captivation by the diffraction experiments of Davisson and Germer in 1927 [67], allowed him to demonstrate the anisotropy of different crystal faces in chemical reactions, as well as the view that anisotropy of surface behavior was widespread, if not universal.

Subsequently, in 1938, Kobozev [68] extended Balandin's concept and introduced the hypothesis of active ensembles. As originally proposed, the theory lacked detail and was not readily susceptible to experimental verification. Later, during the 1960s, Van Hardeveld and Hartog [69] placed the Kobozev concept upon a more quantitative basis when they made calculations for the dependence of the number of metal atoms of a specified coordination number upon the crystal size. The studies conducted by Van Hardeveld and Hartog resulted from their initial failure to reproduce the report by Eischens and Jacknow [70] of infrared (IR) bands due to the adsorption of nitrogen on nickel. Only when they reduced a fresh catalyst *in situ* could they obtain the sought-after IR band, and they then proceeded to combine quantitative IR data and models of the surface of various sized metal catalyst particles to develop their quantitative model. The situation

became even more complex with the introduction of the concept of facile and demanding reactions [71]; furthermore, it was considered that a single atom should be adequate for a facile reaction, whereas two or many more atoms would be required for a catalytic site adequate for demanding reactions. Thus, while some reactions should be dependent on surface geometry, others may not.

The pioneering studies of Beeck, using evaporated metal films, led to the concept of catalytic activity being related to the number of holes in the d-band and/or the geometry (Figs. 11 and 12) (e.g., [72, 73]). The concept of relating catalytic activity to the electronic theories of solid-state properties was given impetus by the theoretical report by Dowden and the accompanying experimental data which showed that the catalytic activity of nickel decreased linearly as the nickel was alloyed with copper to cause a parallel decrease in the number of d-band holes. Furthermore, the activity approached zero for the alloy composition where the number of d-band holes approached zero [74]. Although these findings provided impetus for many studies during the following years, it was subsequently shown that surface enrichment caused the composition of the outermost layer of the NiCu alloy used by Dowden and Reynolds to differ dramatically from that of the bulk.

Schwab [75], using Hume–Rothery alloys, showed that the addition of a second element to increase the occupation ratios of the first Brillouin zone (conduction band) provided a corresponding increase in the activation

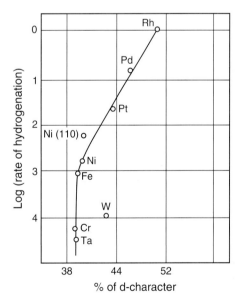

Fig. 11 Correlation of the catalytic activity for ethene hydrogenation with the percentage d-character of the catalyst. (Reproduced from Ref. [72].)

References see page 35

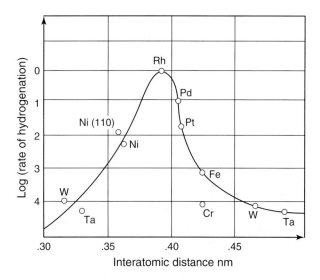

Fig. 12 Correlation of the catalytic activity for ethene hydrogenation with the surface atom interatomic distance. (Reproduced from Ref. [73].)

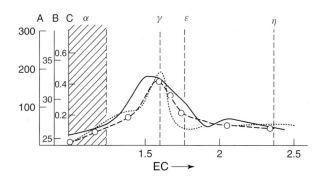

Fig. 13 Alteration energy for the conversion of formic acid as electron levels are filled by varying the composition of binary metal catalysts [A –, hardness (kg mm^{-2}); B –o–, activation energy (kcal mol^{-1}); C ···, resistance (Ωcm·10^4)]. (Redrawn from Ref. [75].)

energies of donor reactions (Fig. 13). Schwab therefore concluded that electron holes must be considered to be involved in the catalysis.

The pioneering studies of Farnsworth [76] (who struggled from his first publication in 1929 until the 1950s to develop adequate instrumentation to obtain data showing the likelihood of differences in atomic spacing within the surface layer and the bulk, as well as the relaxation of the surface layer inwards) and Germer [77] showed that the surface layer in many instances undergoes reorganization to a more stable configuration that differs dramatically from the structure of the bulk. Furthermore, the gas molecules may adsorb and arrange themselves into a regular two-dimensional lattice of their own, and this may be dependent upon surface coverage.

In addition, chemisorption may cause reconstruction of the surface arrangement [78]. The pioneering IR studies of Eischens and Pliskin [79] provided definitive evidence that CO adsorbed onto a Pt surface in more than one form, contrary to the checkerboard model of Langmuir, and that, by analogy with carbonyl cluster molecules, these corresponded to the bridged and linear forms of adsorbed CO.

The staggering number of developments in the theoretical physics of both metals and semiconductors during the mid-1900s led their application in catalysis. For example, most of Volume 7 of *Advances in Catalysis* was devoted to some aspect of this issue. Wagner [80] showed that zinc oxide could accommodate excess zinc atoms that were associated with interstitial positions of the lattice, and that the conductivity of the solid depended upon the pressure of the oxygen in contact with the solid, as predicted by the theory. Doping ZnO with a metal of valence greater than Zn^{2+} requires an increase in free electrons, whereas doping with a univalent cation will decrease the electron concentration. Parravano and Boudart [81] pointed out that, while the literature data and the activity for hydrogen/deuterium exchange could be shown to be compatible with the predictions of the semiconductor theory, the limitations of the verification of the concept were numerous. These included deviations from the Arrhenius plots for calculating the activation energy, and evidence for a compensation effect in the low-temperature range where activation energies could be calculated. Likewise, the data for the changes in activity for CO oxidation with doping of NiO led to contradictory conclusions [81]. Even so, it was concluded that "... catalytic behavior of semiconducting oxides has been modified in opposite directions by the addition of impurities which modify their electrical characteristics in opposite directions" [81].

During the late 1940s, Garner et al. [82] conducted extensive studies of the catalytic oxidation of CO on cuprous oxide and, at the same time, followed the conductivity of the solid, which resembled that of the solid following saturation with CO. These authors suggested that oxygen and CO react on the surface to produce a carbonate complex which then reacts with adsorbed CO to produce CO_2:

$$2O^-_{ads} + CO_{ads} \longrightarrow CO_3^{2-}{}_{ads}$$

and $\quad CO_3^{2-}{}_{ads} + CO_{ads} \longrightarrow 2CO_2$

Vol'kenshtein [83] attempted to formulate adsorption on semiconductors in a manner that took into account not only the Fermi level but also the surface states and the gas phase and adsorbed electronic structure of the adsorbate. This concept led Vol'kenshtein to the conclusion that the act of adsorption could create new sites for adsorption. Thus, there may be different types of level even for the same molecules adsorbed on the same (from its chemical nature) surface. A "homogeneous" surface could therefore become "heterogeneous" by the act of adsorption.

Studies to elucidate the electronic aspects of catalysis dominated the 1950s, and continue even today. Subsequently, two models have emerged: (i) the *atomistic model* (a surface molecule of one or more atoms); and (ii) the *band model*. The atomistic model essentially ignores the solid and concentrates attention on the ensemble of one or a few surface atoms. In contrast, the band model describes the surface in terms of surface states and localized energy levels available at the surface. Vol'kenshtein serves as an example of an early proponent of the valence band model, while Knor [84] summarizes the other approach. Morrison [85] surveyed the application of solid state theories to catalysis in 1977 and in 1989.

Today, the *geometric effect* is one of two dominant themes of catalysis by metals, the other being the *electronic effect*. The geometric and electronic effects are usually discussed as separate topics, although it appears that – just as with acid–base catalysis – the two effects are interrelated so that, at best, one can only document that one of the effects plays a dominant role in the catalysis. In fact, it appears that it is necessary to accept the concept that the initial catalytic material, and especially with metallic catalysts, may interact with the reactant(s) to redesign itself. This concept can be illustrated by the observation of Beeck initially [86], by the retention of ^{14}C-labeled acetylene during hydrogenation of ethylene [87], as well as by the surface science of, for example, Somorjai and Ertl, that different metal catalysts will strongly – and essentially irreversibly – bond different amounts of a reactant. Thus, one can view the metallic catalyst as redesigning itself by doping its surface, and in some cases the bulk, with sufficient reactant so as to attain an electronic state where the bonding is sufficiently weakened that chemical reactions may occur.

1.2.7
Selected Systems

1.2.7.1 Ammonia Synthesis
Le Chatelier's principle indicates that ammonia formation is favored at high pressure. Convinced of his principle, Le Chatelier had his assistant compress a mixture of hydrogen and nitrogen to a very high pressure, and then apply an electrical spark. An explosive reaction occurred which killed the assistant, and was the result of a faulty experimental procedure whereby air was permitted to mix with the nitrogen–hydrogen mixture. Hence, it was the oxygen–hydrogen reaction, rather than ammonia formation, that caused the explosion. This was just one of many hundreds of unsuccessful early attempts to synthesize ammonia.

Larson [88] credits Perman with the initial synthesis of ammonia using an iron catalyst. However, it was the interactions of Nernst and Haber that led to the successful demonstration of the catalytic synthesis of ammonia and the results that were to generate the excitement needed for rapid progress. Haber determined experimentally the amount of ammonia present at equilibrium. Nernst computed, by using the heats of reaction and other chemical constants, the equilibrium amounts for many reactions. All values calculated by Nernst were in agreement with the experimental values, except for the experimental value for ammonia reported by Haber. In diplomatic terms, a very spirited debate arose over the correct value for ammonia [89]. It is sufficient here to state that Haber's additional measurements at high pressure, spurred by Nernst's comments, resulted in showing that the equilibrium concentration of ammonia was sufficiently high to indicate industrial significance. Haber's most active catalyst, osmium, was not of industrial significance. One of Haber's novel concepts was to use a process in which the unconverted hydrogen and nitrogen, after removal by condensation of the ammonia produced, were recycled to the reactor. In this connection, Haber introduced the concept of "space-time-yield" [90]. Even so, Haber's studies would only have resulted in a victory for ego and academics if he had not been able to interest BASF in the process.

Mittasch, a student of Ostwald, guided the research at BASF that led to the iron synthetic ammonia catalyst that is still used today, albeit in an improved form. Mittasch was led, just as Sabatier, to a successful catalyst following the intermediate compound concept; in this case, the catalyst was viewed briefly to form a nitride intermediate with subsequent reduction by hydrogen. While concepts were utilized by Mittasch, the investigations were systematic and extensive. During about a two-year period, 6500 experiments with about 2500 different catalysts were conducted [90], truly an example of Edisonian research. A magnetite sample from Sweden proved to be a surprisingly active catalyst and, because of its relative high density, the concept of "compact while porous" developed, but was subsequently proved to be wrong. Haber considered, temporarily, that the high density of osmium and similar metals was the reason for their effectiveness as an ammonia synthesis catalyst. During extensive studies by Mittasch and coworkers, the concepts of promoter action of catalysis were developed and, building on the meager earlier results, defined the options for the role of promoters during the 1920s (Fig. 14) [91].

References see page 35

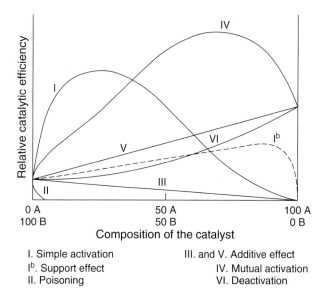

Fig. 14 The five ways in which a promoter may impact catalytic activity as the concentration of the promoter is varied. (Reproduced from Ref. [91].)

I. Simple activation
I^b. Support effect
II. Poisoning
III. and V. Additive effect
IV. Mutual activation
VI. Deactivation

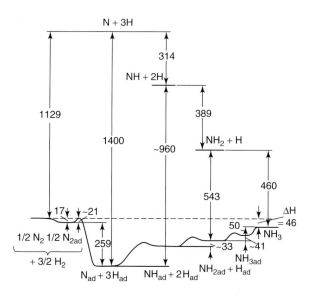

Fig. 15 Energy profile of the progress of the ammonia synthesis on Fe (units of energy are kJ mol^{-1}). (Reproduced from Ref. [94].)

The development of the United States synthetic ammonia industry provides an outstanding example of exceptional accomplishments in the development of the technology first, and then in scientific understanding, whilst simultaneously showing exceptionally disastrous political actions. Within six years of commercial operations in Germany, Larson and coworkers at the Fixed Nitrogen Laboratory (FNL) had developed a suitable catalyst. Sir Hugh Taylor [92], in considering the scientific studies conducted by Emmett, Brunauer and coworkers at the FNL following the development scheme, wrote:

"These authors have given us the most detailed kinetic study ever made by a single reaction, with all aspects of the reaction studied, adsorptions, kinetics, influence of reactant concentrations as a function of composition and mode of preparations. The treatment is so comprehensive that it is possible to present an almost complete account of the phenomenon of surface catalysis by reference to this one example alone."

The concept of the dissociation of nitrogen as the slow step was demonstrated using both kinetic and isotopic techniques. The catalysts were carefully characterized. Kinetic concepts were developed and these, together with the concept of "virtual pressure", which was introduced by Temkin and Pyzhev [93], provides a kinetic model that is still valid today. However, the more detailed model that has resulted from many surface science studies provides a much sounder basis, with greater detail for the concepts of how the catalyst operates (Fig. 15) [94].

1.2.7.2 Acid Catalysis

Acids have been a dominant theme in the study of reactions, and acid catalysis predates its definition. The dissociation theories advanced by Arrhenius, and applied by Ostwald, did much to aid in developing the understanding of acid catalysis. These authors dominated investigations to show that the catalytic power of a solution of an acid is directly proportional to its electrical conductivity, and independent of the nature of the anion. This indicated that the catalyst is the hydrogen-ion, and that its effect is directly proportional to its concentration in the solution. However, by 1900 acid catalysis was in a state of confusion, primarily because of the failure to recognize the role of the undissociated acid and the salt effects. The concepts of acid–base catalysis have been presented during the past 50 years in a series of books by Bell [95]. Unfortunately, all too frequently scientists forget that acid–base catalysis is the title of Bell's first book, and treat acid catalysis in isolation from the conjugate base.

Hydrocarbon conversions have been dominant in the study of acid catalysis. Early studies on this topic involved acids such as the one formed by an aluminum chloride/hydrogen chloride mixture; however, the combination of the corrosion and the difficulty of catalyst recovery has limited its use to special situations, and these two factors led to the failure of the cracking process introduced by Gulf Oil that was based on this catalyst. The applications of acid catalysis were first realized in petroleum processing by Eugene Houdry, with his introduction of catalytic cracking. Actually, the Houdry process provided much greater advances in catalyst regeneration and process control than it did in introducing new catalytic concepts, and the same was also true of the introduction of fluidized catalytic cracking. Acid-treated clays were the first successful acidic heterogeneous cracking catalyst to be used, but they were subsequently replaced by a synthetic amorphous silica–alumina catalyst. The silica–alumina catalyst is acidic because charge-balance requires that a proton, or other cation, be present for each aluminum ion that is substituted into the SiO_2 framework. The credit for developing this concept for the source of acidity in the amorphous silica–alumina catalyst is usually attributed to Thomas [96]. In time, the amorphous catalyst was replaced by crystalline zeolite catalysts.

In June 1932, Ipatieff and Pines carried out experiments, using $AlCl_3$ as the catalyst and HCl as a promoter, to show that alkylation of an alkane with an alkene is possible. These studies, augmented by those of Komarewsky, Grosse and other Universal Oil Products (UOP) scientists, led to the commercial processes. Louis Schmerling explained these reactions in terms of the carbonium mechanism, and the catalyst was shown to be effective for hydrocarbon isomerizations, with side reactions being inhibited by the presence of hydrogen [97].

Acid catalysis is also involved in a more complicated process, namely naphtha reforming. Here, two functions are involved – a metallic function for hydrogenation, and an acidic function to effect hydrocarbon isomerizations. Thus, the concept of bifunctional catalysis was introduced [98]. Here, an alkene is dehydrogenated to an alkene at the metallic function, after which the alkene migrates, either on the surface or through the gas phase, to an acid site where it undergoes isomerization. It then migrates again to the metallic function, where it is hydrogenated to an isomerized alkane. The introduction of bi- or poly-functional catalysis led to the concept of the need for the two types of site to be located sufficiently close to each other so that transport between them would not be the rate-limiting step of the overall process.

1.2.7.3 Zeolites

Flanigen [99] provides an historical perspective of the introduction of zeolites into catalysis. The incubation period of the concepts needed for the application of zeolites in catalysis began in 1756, with the discovery of the first natural zeolite mineral. Following determination of the structure in 1930 [100], the colloidal scientist McBain coined the term "molecular sieve", to indicate that the materials act on the molecular scale [101]. The pioneering studies by Barrer on the synthesis and adsorption properties of zeolites attracted attention to the subject, and Barrer subsequently offered the first classification of zeolites based on molecular size and accomplished the first definitive synthesis of zeolites [102]. Inspired by Barrer's findings, by 1954 Milton, Breck and coworkers at Union Carbide synthesized several commercially significant zeolites (including A, X, and Y). Initially, the zeolites found application as superior drying materials and for separations. However, the introduction of the "ISOSIV" process was an application of the concept of molecular sieving, and offered the possibility of separating normal and isoalkanes.

By 1959, the concept of a zeolite as an acid catalyst – and specifically as an isomerization catalyst – had been taken to the commercial level [99]. However, the acidic nature and the possible potential of their use as cracking catalysts predated this; for example, Houdry's team had presented data for catalytic cracking for a zeolite catalyst at the 1950 Faraday Society meeting. While zeolites were recognized to have a high catalytic activity, they could not be kept in this state for a sufficient amount of time, and extensive studies conducted at Mobil Oil eventually led to the utilization of materials exchanged by rare earths. It was the practical concept of Plank and Rosinski to test both the fresh zeolite and the material produced by high-temperature steaming to simulate the regeneration process that led to the explosive introduction of zeolite-based cracking catalysts [103]. The steamed material exhibited a dramatic increase in the production of gasoline, together with a dramatic decrease in coke make, over that of the best amorphous silica–alumina catalyst then in use. Since the development of the pioneering concepts and the development of many commercial processes based on zeolites, the field has grown into a specialization in its own right, with the formation of the International Zeolite Association and the introduction of scientific journals to record the dramatic growth in both concepts and commercial applications of zeolites.

This provided the intellectual and economic driving force that has since led to the production of at least a hundred crystalline materials, many of which have unique catalytic properties. As one example, the ZSM-5 zeolite was found to convert methanol to gasoline range hydrocarbons, high in aromatics, with a very high selectivity [104]. This property was considered to be related to the unique structure of the catalyst, and this has resulted in the advancement of numerous additional concepts.

1.2.7.4 Ions in Catalysis

The introduction of the concept of ions, and the success of Ostwald in relating homogeneous catalytic action to the hydrogen ion, made it likely that the concept would be extended to heterogeneous catalysis at a early date. However, it appears that this did not occur prior to 1905, when Kirkby proposed that the hydrogen–oxygen reaction was connected with the corpuscular discharge emitted

References see page 35

by hot platinum, which he supposed would produce ions [105]. In 1940, Emmett and Teller [106] considered four earlier attempts [107] to explain the catalytic action of surfaces by the formation of ions which then react with each other. By focusing their attention on metallic catalysts, Emmett and Teller concluded that it was not likely that surface catalytic reactions had an ionic nature. Indeed, they pointed out that in many cases it reduces to a question of definition as to whether or not one is dealing with ions. For surface phenomena it is much more difficult to prove the existance of an ion than it is for the gaseous state or for solutions. Thus, use of the concept of ionic intermediates in catalysis expanded for metal oxide catalysts, and probably as reaction intermediates where they are not appropriate, although the concept did not prosper for metal-catalyzed reactions. Couper and Eley [108] later calculated, on a theoretical basis, that all possible ionization processes for the adsorption of hydrogen onto the surfaces of transition metals are prohibitively endothermic. This concept was conceived on a train during travel to a scientific meeting!

The concept of alkyl cations provided a basis for the understanding of hydrocarbon cracking. The landmark publication on rearrangements in acyclic hydrocarbon and other systems by Whitmore established the usefulness of carbocations in explaining such conversions. Three groups independently adapted the concepts advanced by Whitmore, and utilized carbocations to explain the differences between the products obtained from thermal and acid-catalyzed cracking [109]. With time, pentavalent carbon cations (carbonium ions) were also identified, and these are utilized together with trivalent carbocations (carbenium ions) in explaining catalytic cracking. Olah was a leader in advancing this area (particularly the use of the superacid $HSbF_6$ and in applying NMR techniques for the precise identification of cation structures), and in 1994 received the Nobel Prize in recognition of his efforts.

1.2.7.5 Hydrogenation

Studies of the reactions of hydrogen have contributed dramatically to the development of concepts of catalysis. The early fascination by the conversion of hydrogen and oxygen over platinum and other solids was instrumental in laying the groundwork for the definition of catalysis. In fact, the reaction between hydrogen and oxygen was sufficiently advanced that a book was written on this subject in 1934 [110].

The role of hydrogenation in advancing the concepts of catalysis received major boosts from the investigations of both Sabatier and Ipatieff. Sabatier and his coworkers were instrumental in developing the field of hydrogenation of organic materials. French scientists had reported previously that the porous metals adsorb acetylene, with spontaneous destruction. On the basis of scientific courtesy, therefore, Sabatier did not feel that he could work with acetylene, and consequently his initial studies were with ethylene. Sabatier was a co-recipient of the Nobel Prize in 1912 for his studies on hydrogenating organic compounds in the presence of finely divided metals.

Ipatieff was trained as an artillery officer in the Russian army, and began research after becoming a Professor of Chemistry at the Artillery Academy. In 1905, he introduced his high-pressure autoclave, taking advantage of his artillery training to develop a tight seal for the reaction vessel [111]. With his new high-pressure reactor, Ipatieff made significant advances in the hydrogenation of aromatics, the high-pressure polymerization of alkenes, and the hydrogenation of sugars. In spite of his involvement in administrative duties following the Russian Revolution, he continued his research studies, including the destructive hydrogenation of hydrocarbons. In 1929, after many of his associates and coworkers had been arrested, some being shot without trial, Ipatieff left Russia for Germany, never to return, and was prohibited from contact with his children. In 1930, efforts were commenced that were to lead him to UOP and Northwestern University in the United States, where he arrived at the age of 62 years and subsequently accomplished more research than most persons do during an entire career. Among his most prominent achievements was the development of alkylation (see Section 1.2.7.2).

In 1929, serendipity led to the discovery of the catalytic conversions of ortho-para hydrogen (o-p-H_2) [112]. Convinced of the concept that metallic platinum would catalyze, at liquid air temperature, the transformation of a room-temperature mixture of o-p-H_2 to one that was enriched in the para compound, the experiment was run repeatedly without success. Since Pt *had* to be a good catalyst, the investigators reasoned that it must have been poisoned by an impurity in the hydrogen. Thus, Harteck applied the classic gas clean-up technique, by passing the gas over charcoal at low temperature. Harteck and Bonhoffer were surprised to find that charcoal was an effective catalyst and, having a source of near-pure para-H_2, they immediately began their pioneering studies on the catalysis of the o-p-H_2 conversion. The intense academic interest caused the numbers of published reports to increase very rapidly, but to decline just as rapidly as *all* of the curiosity-driven questions were answered. In 1935, no research director would have advocated funding this research area as one with a commercial potential, and would have been correct. However, in 1960, with space travel seemingly iminent and the use of liquid hydrogen both as a fuel and potential coolant, the o-p-H_2 conversion assumed immense commercial importance, and the number of publications on the topic showed a second

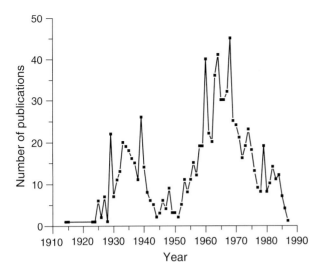

Fig. 16 Graphical plot of the numbers of reports on ortho-para-hydrogen catalytic conversion from 1910 to 1988.

rapid increase (Fig. 16). The academic studies performed during the 1930s were directed toward deciding which concept – dissociative adsorption or magnetic effects acting on the undissociated molecule – were responsible for the transformation. In 1933, Bonhoeffer et al. [113] reviewed their findings using metallic, charcoal and salt catalysts, and concluded that the two mechanisms applied, but at different temperatures. At low temperatures the conversion is monomolecular, it occurs in the van der Waals adsorption layer, and is most likely catalyzed by magnetic effects. However, at high temperatures the reaction involves atomic adsorption, and this has become known as the Bonhoeffer–Farkas mechanism.

The discovery in 1932 of the deuterium isotope, along with methods to enrich its concentration, ignited a period of very fruitful research in catalysis. The Farkas brothers, with their previous experience of o-p-H_2 studies, were at the forefront of this effort. However, because of the Nazi takeover of the Government, they were forced to leave Germany in 1933, and moved to Sir Eric Rideal's laboratory in Cambridge. During a one-year period following their arrival in England, the brothers published 17 reports on the catalysis of hydrogen. These publications illustrated a personality trait of Rideal, in that he provided the brothers with laboratory space and facilities for their pioneering research, but was a co-author on only one of the reports – which was an unusual gesture, then or now. The first report by the Farkas brothers demonstrated the occurrence of the thermal and catalytic reaction

$$H_2 + D_2 = 2HD$$

They showed that the addition of D_2 to ethene in the presence of a nickel catalyst was accompanied above 60 °C by the exchange reaction

$$C_2H_4 + D_2 \longrightarrow C_2H_3D + HD$$

This mechanism differed from the one advanced earlier that involved a half-hydrogenated intermediate during D_2-benzene exchange [114]. In 1939, Adalbert Farkas summarized the published results and concluded that the dissociative adsorption of hydrogen and the hydrocarbon could explain the observed findings. This report heightened interest in these reactions and initiated the debate concerning the concepts of the dissociative or the associative mechanism. During the subsequent investigations, the Rideal–Eley mechanism was advanced, where interchange between a chemisorbed atom and hydrogen adsorbed in the van der Waals layer. The experimental data for this mechanism was shown later by Rideal to be erroneous, although the Rideal–Eley concept survived and is applicable in some cases even today. The measurements made by Rideal and his student Trapnell, along with the classic studies by Beeck and many others showed that as chemisorption continued, the heat of adsorption decreased, and that this altered the state from that of the irreversible adsorption at low coverages to one of reversible adsorption at higher coverages. The status today is that the concept of dissociative adsorption describes the mechanism; however, a few examples of associative adsorption persist even today.

Today, the hydrogenation of fats has became an important industry, and studies to understand the process have led to data that define the series nature of the reaction mechanism, namely the sequential hydrogenation of triene to diene to monoene, and finally to saturated fat. Burwell [115] presented a classic report that showed the hydrogenation of butenes to be as (or even more) rapid than that of butyne, although inhibition of butene hydrogenation by the adsorbed butyne allowed near-complete hydrogenation of the butyne prior to hydrogenation of the butenes to form butane (Fig. 17). The classic book by Bond on catalysis by metals [116] includes many other examples where hydrogenations have been utilized to develop concepts of catalysis.

1.2.7.6 Oxidation

Catalytic oxidation dates at least from 1794, when Mrs. Fulhame advanced the concept that the presence of water was necessary to accelerate the oxidation of hydrocarbons [117]. Likewise, written documentation of the oxidation of sulfur to sulfur trioxide in the presence of saltpeter and then to sulfuric acid dates to the latter half of the 15th century [118]. The patent for the manufacture

References see page 35

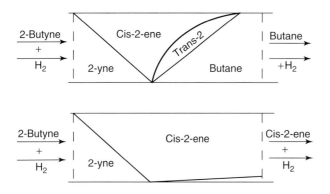

Fig. 17 Upper: hydrogenation of 2-butyne to form 2-butene and then the further hydrogenation to produce only butane at the exit of the reactor. Lower: selective poisoning of the catalyst to produce near-pure 2-butene. (Redrawn from Ref. [115].)

of sulfuric acid based upon the oxidation of SO_2 using air and a finely divided platinum metal predates the definition of catalysis by Berzelius. A publication by Knietsch [119] revealed concepts that had been learned during intensive studies at BASF: (i) that the Winkler concept that stoichiometric mixture of SO_2 and oxygen, undiluted by other gases, had to be fed to the catalyst was incorrect; (ii) the need to remove some of the heat of reaction from the catalyst was recognized; and (iii) it was claimed that the principles governing the manufacture of the acid were now well understood.

The Hopcalite catalyst was unique as the discovery and development studies were conducted in universities, and under a tight time-frame required to develop an effective gas mask for protection against a poison gas, CO, during World War I [120]. The catalyst that had sufficient activity to work at room temperature was a combined effort by workers at The Johns Hopkins University and the University of California, Berkeley. After much study following World War I, Frazer was able to explain the mechanism of the catalysis by a two-point attachment of the O to the metal atom and the C to an O of manganese dioxide. The two-point attachment of the CO to the catalyst surface would require stretching of the CO bond from 1.15 to 1.86 Å.

The Arabian alchemists are credited with a process for the production of nitric acid as early as the 8th century [121], while in 1788, Rev. Milner had reported that nitric oxide resulted from passing ammonia over heated MnO_2 [122]. The use of a platinum catalyst was accomplished by Henry in 1824 [123]. Kuhlmann reported a number of catalysts, including Pt, that catalyzed the oxidation of ammonia to nitric acid [124]. Although a heated platinized asbestos catalyst was used in 1891 by Warren [125], it remained for Ostwald [126] to make a detailed study of ammonia oxidation, to settle on a platinum catalyst, and for it also to be a catalyst for building a commercial plant near Bochum, Germany. The concept of using a platinum gauze catalyst was offered by Kaiser [127], and has been considered to be the most important innovation in ammonia oxidation to date [128]. During 1920–30, extensive studies were conducted by Andrussow [129], Bodenstein [130], and Rasching [131] to understand the reaction mechanism, and consequently three mechanisms were proposed: (i) the imide mechanism; (ii) the nitroxyl mechanism; and (iii) the hydroxylamine mechanism (NH_2OH intermediate). The hydroxylamine mechanism, as proposed by Bodenstein, became widely accepted.

An excellent example of the admonition of Armstrong to study the catalyst is present in the understanding of oxidation with bismuth–molybdenum oxide catalysts. Studies using carbon-13 and deuterium labeling had established conclusively that the formation of an allylic intermediate by the abstraction of an α-hydrogen was the rate-limiting step [132]. Making use of ^{18}O isotopic tracer studies, Keulks and Matsuzaki [133] showed that, for a pre-reduced catalyst, the low-temperature reoxidation that incorporates oxygen is accompanied by the Bi^0 to Bi^{3+} conversion, whereas the corresponding high-temperature reoxidation is accompanied by the Mo^{4+} to Mo^{6+} conversion. The higher-temperature reoxidation process is related to the rate-limiting step for propene oxidation to acrolein at temperatures below 400 °C. This step is identified as reoxidation of the Mo centers by lattice oxygen diffusion. The oxygen inlet site is identified as reduced Bi, and the oxygen outlet site is associated with Mo. Grasselli [134] combined the results from his group and published the results to arrive at a detailed mechanistic cycle for the selective ammoxidation or oxidation of propene over bismuth molybdate (Fig. 18).

1.2.8
Summary

In catalysis, as in many areas of science, a concept will be associated with one or a few individuals. Frequently, the concept will have been advanced in a similar form and at an earlier time, and in many instances the concept will be associated with the name of the forceful individual who provides a driving force to cause widespread acceptance of the concept. One of the major reasons for the more-or-less haphazard assignment of credit for concepts is the scant attention paid to the history of the subject. Thus, in catalysis, credit for the development of a concept is likely to be assigned on the basis of common usage.

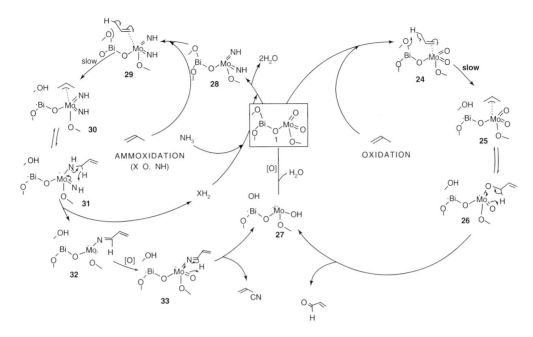

Fig. 18 Mechanism of selective propene ammoxidation and oxidation over bismuth molybdate catalyst. (Redrawn from Ref. [134].)

References

1. S. J. Green, *Industrial Catalysis*, Ernest Benn, Ltd., 1928.
2. E. Mitscherlich, *Pagg. Ann.* **1834**, *31*, 273.
3. A. Trofast, in *Proceedings of Swedish Symposium of Catalysis, 12th*, R. Larsson (Ed.), Liber Laeromedel, Lund, Sweden, 1981, pp. 9–17.
4. J. R. Partington, *A History of Chemistry*, MacMillan & Co., Inc., London, 1964, pp. 261–264.
5. J. W. Mellor, *Chemical Statics and Dynamics*, Longmans, Green, and Co., London, 1904, p. 246.
6. H. E. Armstrong, *Chemical News*, September 11, **1885**, pp. 130–156.
7. T. de Saussure, *Ann. Philos.* **1815**, *6*, 214, 331.
8. H. S. Taylor, in *Colloid Chemistry. Theoretical and Applied*, J. Alexander (Ed.), The Chemical Catalog Co., Inc., New York, Vol III, 1931.
9. (a) A. J. B. Robertson, *Platinum Metals Rev.* **1975**, *19*, 64; (b) M. Farinelli, A. L. B. Gale, A. J. B. Robertson, *Ann. Sci.* **1974**, *31*, 19.
10. G.-M. Schwab, in *Catalysis. Science and Technology*, J. R. Anderson, M. Boudart (Eds.), Springer-Verlag, Berlin, 1981, Vol. 2, pp. 1–11.
11. M. Faraday, *Phil. Trans. Roy. Soc.* **1834**, pt. 1, 55–71.
12. J. W. Döbereiner, as described in P. L. Dulong, L. J. Thénard, *Ann de Chim. et de phys.* **1813**, *23*, 440; *Ann de Chim. et de phys.* **1813**, *24*, 380.
13. J. Persoz, A. Payen, *Ann. Chim. Phys.* **1833**, *53*, 73.
14. (a) L. Pasteur, *Compt. Rend.* **1858**, *46*, 179; (b) L. Pasteur, *Compt. Rend.* **1859**, *48*, 640, 735; (c) L. Pasteur, *Compt. Rend.* **1860**, *50*, 1083.
15. From *Liebig's Complete Works on Chemistry*, Justus Liebig, M.D., Ph.D., F.R.S., Professor of Chemistry in the University of Giessen (Edited by Lyon Playfair, from the last London edition, much improved). T. B. Peterson and Brothers, Philadelphia, no publication date, Chapter II, p. 88.
16. J. Liebig, *Familiar Letters on Chemistry*, Taylor, Walton & Maberly, London, 3rd edn, Revised and Much Enlarged, 1851, Letter X, pp. 125–138.
17. F. Wilhelmy, *Ann. Phys.* **1850**, *81*, 413, 499.
18. (a) A. F. Harcourt, W. Esson, *Proc. Roy. Soc.* **1865**, *14*, 470; (b) A. F. Harcourt, W. Esson, *Phil. Trans.* **1866**, *156*, 193; (c) A. F. Harcourt, W. Esson, *Phil. Trans.* **1867**, *157*, 117.
19. B. H. Davis, *Appl. Catal.* **1992**, *82*, N–15.
20. (a) W. Ostwald, *Z. Phys. Chem.* **1894**, *15*, 705–706; (b) W. Ostwald, translation in D. D. Runes (Ed.), *A Treasury of World Science*, Philosophical Library, New York, 1962, pp. 782–784.
21. W. Ostwald, *Nature* **1902**, *65*, 522–526.
22. J. R. Partington, *A History of Chemistry*, MacMillan & Co., Ltd., London, 1964, Vol. 4, pp. 595–600.
23. G. Fusinieri, *Giornale di Fisica* **1825**, *8*, 259.
24. B. C. Brodie, *Philos. Trans. Royal Soc., London* **1862**, *151*, 837.
25. P. Sabatier, *Ind. Eng. Chem.* **1926**, *18*, 1005.
26. M. A. Barteau, *Catal. Lett.* **1991**, *8*, 175.
27. M. Polanyi, *J. Soc. Chem. Ind.* **1935**, *54*, 123.
28. K. J. Laidler, *The World of Physical Chemistry*, Oxford University Press, Oxford, 1993, p. 233.
29. I. Langmuir, *J. Am. Chem. Soc.* **1916**, *38*, 2221; *J. Am. Chem. Soc.* **1918**, *40*, 1361.
30. L. Michaelis, M. L. Menten, *Biochem. Z.* **1913**, *49*, 333.
31. S. W. Weller, *Catal. Rev.-Sci. Eng.* **1992**, *34*, 227–280.
32. S. Glastone, K. J. Laidler, H. Eyring, *The Theory of Rate Processes*, McGraw-Hill Book Co., Inc., New York, 1941.
33. H. Eyring, M. Polanyi, *Z. Physik. Chem.* **1931**, *B12*, 279.
34. K. Laidler, in *Catalysis*, P. H. Emmett (Ed.), Reinhold Publishing Corp., New York, Vol. 1, 1954, p. 236.
35. J. Horiuti, *Adv. Catal.* **1957**, *9*, 339.
36. (a) S. Enomoto, J. Horiuti, *J. Res. Inst. Catal., Hokkaido Univ.* **1953**, *2*, 87; (b) S. Enomoto, J. Horiuti, *J. Res. Inst. Catal., Hokkaido Univ.* **1955**, *3*, 185.
37. C. Bokhoven, M. J. Gorgels, P. Mars, *Trans. Faraday Soc.* **1959**, *55*, 339.

38. W. Jost, in *Annual Reviews of Physical Chemistry*, H. Eyring (Ed.), Annual Reviews, Inc., Palo Alto, California, Vol. 17, 1966, pp. 1–14.
39. C. Kemball, in *Biographical Memoirs of Fellows of The Royal Society*, The Royal Society, London, Vol. 21, 1975, pp. 517–547.
40. E. Cremer, G. M. Schwab, *A. Physik. Chem.* **1929**, *A144*, 234.
41. F. H. Constable, *Proc. Roy. Soc. London* **1925**, *A108*, 355.
42. E. Cremer, *Adv. Catal.* **1955**, *7*, 75.
43. M. P. Suárez, A. Palermo, C. M. Aldao, *J. Thermal Anal.* **1994**, *41*, 807–816.
44. R. Randhava, in *Heterogeneous Catalysis, Selected American Histories*, B. H. Davis, W. P. Hettinger (Eds.), ACS Symposium Series, Vol. 222, 1983, pp. 173–177.
45. A. Wheeler, *Adv. Catal.* **1951**, *3*, 249; A. Wheeler, *Catalysis*, P. H. Emmett (Ed.), Van Nostrand, New York, 1955, Vol. 2.
46. G. A. Mills, in *Heterogeneous Catalysis, Selected American Histories*, B. H. Davis, W. P. Hettinger (Eds.) ACS Symposium Series, Vol. 222, 1983, pp. 179–182.
47. E. Turner, *Edin. Phil. J.* **1824**, *11*, 99, 311; *Pogg. Ann.* **1824**, *2*, 210.
48. J. W. Mellor, *Chemical Statics and Dynamics*, Longmans, Green, and Co., London, 1904, pp. 258, 285.
49. E. B. Maxted, *Adv. Catal.* **1951**, *3*, 129–177.
50. H. S. Taylor, *J. Am. Chem. Soc.* **1931**, *53*, 578.
51. E. P. Barrett, L. G. Joyner, P. P. Halenda, *J. Am. Chem. Soc.* **1951**, *73*, 373.
52. W. E. Washburn, *Proc. Natl. Acad. Sci. USA* **1921**, *7*, 115.
53. H. L. Ritter, L. C. Drake, *Ind. Eng. Chem., Anal. Ed.* **1945**, *17*, 787.
54. S. Brunauer, *The Adsorption of Gases and Vapors*, Princeton University Press, Princeton, 1942.
55. G. L. Gaines, G. Wise, in *Heterogeneous Catalysis. Selected American Histories*, B. H. Davis, W. P. Hettinger (Eds.), ASC Symposium Series, Vol. 222, 1983, pp. 13–22.
56. J. E. Lennard–Jones, *Trans. Faraday Soc.* **1932**, *28*, 333.
57. J. H. deBoer, *The Dynamic Character of Adsorption*, The Clarendon Press, Oxford, 1953.
58. A. W. Adamson, *Physical Chemistry of Surfaces*, John Wiley & Sons, New York, 3rd edn, 1976, pp. 652–654.
59. O. Beeck, *Adv. Catal.* **1950**, *2*, 151.
60. R. Germer, *Adv. Catal.* **1955**, *7*, 93.
61. E. W. Müller, T. T. Tsong, *Field Ion Microscopy*, Elsevier Publishing Co., Amsterdam, 1969.
62. (a) H. S. Taylor, *Proc. Royal Soc.* **1925**, *A108*, 105; (b) H. S. Taylor, *J. Phys. Chem.* **1926**, *30*, 145.
63. H. S. Taylor, *Colloid Symposium Monograph*, H. B. Weiser (Ed.), The Chemical Catalog Co., Inc., New York, 1926, pp. 19–28.
64. O. Loew, *J. Prakt. Chem.* **1875**, *11*, 372.
65. (a) A. A. Balandin, *Adv. Catal.* **1969**, *19*, 1; (b) D. V. Sokolsky, B. V. Erofeev, *Chemtech* **1973**, *3*, 728.
66. B. M. W. Trapnell, *Adv. Catal.* **1951**, *3*, 4.
67. H. Leidheiser, Jr., in *Heterogeneous Catalysis. Selected American Histories*, B. H. Davis, W. P. Hettinger, Jr. (Eds.), ACS Symposium Series, Vol. 222, 1983, pp. 121–130.
68. N. I. Kobozev, *Acta Physicochim. USSR* **1938**, *9*, 805.
69. R. Van Hardeveld, F. Hartog, *Surf. Sci.* **1969**, *15*, 189.
70. R. P. Eischens, J. Jacknow, *Proceedings 3rd International Congress of Catalysis*, W. M. H. Sachtler, G. C. A. Schuit, P. Zwietering (Eds.), North-Holland Publishing Co., Amsterdam, 1964, p. 627.
71. M. Boudart, *Proceedings 6th International Congress of Catalysis*, G. C. Bond, P. B. Wells, F. C. Tompkins (Eds.), The Chemical Society, London, 1977, pp. 1–9.
72. O. Beeck, A. W. Ritchie, *Disc. Faraday Soc.* **1950**, *8*, 159.
73. O. Beeck, *Modern Phys.* **1945**, *17*, 61.
74. (a) D. A. Dowden, *J. Chem. Soc.* **1950**, 242; (b) D. A. Dowden, P. W. Reynolds, *Disc. Faraday Soc.* **1950**, *8*, 184–190.
75. (a) G. M. Schwab, *Disc. Faraday Soc.* **1950**, *8*, 166; (b) G. M. Schwab, in *Catalysis*, J. R. Anderson, M. Boudart (Eds.), Springer-Verlag, Berlin, **1981**, Vol. 2, pp. 1–11.
76. H. E. Farnsworth, *Adv. Catal.* **1957**, *9*, 493.
77. L. H. Germer, *Surf. Sci.* **1966**, *5*, 147–151.
78. J. W. May, *Ind. Eng. Chem.* **1965**, *57*, 18.
79. R. P. Eischens, W. A. Pliskin, *Adv. Catal.* **1957**, *9*, 662.
80. C. Wagner, *Z. Physik. Chem.* **1933**, *B22*, 181.
81. G. Parravano, M. Boudart, *Adv. Catal.* **1955**, *7*, 47.
82. W. E. Garner, T. J. Gray, F. S. Stone, *Disc. Faraday Soc.* **1950**, *8*, 246.
83. F. F. Vol'kenshtein, in *Scientific Selection of Catalysts*, A. A. Balandin, et al. (Eds.), English Translation, Israel Program for Scientific Translations, Jerusalem, 1968, pp. 70–81.
84. Z. Knor, *Adv. Catal.* **1972**, *22*, 51.
85. (a) S. R. Morrison, *The Chemical Physics of Surfaces*, Plenum Press, New York, 1977; (b) M. J. Madou, S. R. Morrison, *Chemical Sensing with Solid State Devices*, Academic Press, New York, 1989.
86. O. Beeck, *Disc. Faraday Soc.* **1950**, *8*, 118.
87. G. F. Taylor, S. J. Thomson, G. Webb, *J. Catal.* **1968**, *12*, 191.
88. A. T. Larson, *J. Chem. Educ.* **1926**, *3*, 284–290.
89. K. Tamaru, in *Catalytic Ammonia Synthesis*, J. R. Jennings (Ed.), Plenum Press, New York, 1991, pp. 118.
90. E. Farber, *Chymia* **1966**, *11*, 157.
91. A. Mittasch, *Adv. Catal.* **1950**, *2*, 81.
92. H. S. Taylor, *Am. Sci.* **1946**, October, 553.
93. M. Temkin, V. Pyzhev, *Acta Physicochem. USSR* **1940**, *12*, 327.
94. G. Ertl, in *Catalytic Ammonia Synthesis*, J. R. Jennings (Ed.), Plenum Press, New York, 1991, p. 128.
95. R. P. Bell, *The Proton in Chemistry*, 3rd edn, Cornell University Press, 1973.
96. C. L. Thomas, in *Heterogeneous Catalysis*, B. H. Davis, W. P. Hettinger, Jr. (Eds.), ACS Symposium Series, Vol. 222, 1983, pp. 241–245.
97. V. N. Haensel, L. Schmerling, *Adv. Catal.* **1948**, *1*, 27.
98. G. A. Mills, H. Heinmann, T. H. Milliken, A. G. Oblad, *Ind. Eng. Chem.* **1953**, *45*, 134.
99. E. M. Flanigen, in *Introduction to Zeolite Science and Practice*, H. van Bekkum, E. M. Flanigen, J. C. Jander (Eds.), Elsevier, Amsterdam, 1991, pp. 13–34.
100. (a) W. H. Taylor, *Z. Kristallogr.* **1930**, *74*, 1; (b) L. Pauling, *Proc. Natl. Acad. Sci. USA* **1930**, *16*, 453; (c) L. Pauling, *Z. Kristallogr.* **1930**, *74*, 213.
101. J. W. McBain, *The Sorption of Gases and Vapors by Solids*, Routledge and Sons, London, 1932, Chapter 5.
102. (a) R. M. Barrer, *J. Soc. Chem. Ind.* **1945**, *64*, 130; (b) R. M. Barrer, *J. Chem. Soc.* **1948**, 2158.
103. C. J. Plank, in *Heterogeneous Catalysis. Selected American Histories*, B. H. Davis, W. P. Hettinger, Jr. (Eds.), ACS Symposium Series, Vol. 222, 1983, pp. 253–271.
104. G. T. Kerr, *Sci. Am.* **1989**, *261*, 100.
105. (a) P. J. Kirkby, *Phil. Mag.* **1905**, *10*, 467; (b) A. J. B. Robertson, *Platinum Metals Rev.* **1983**, *27*, 31–39.

106. P. H. Emmett, E. Teller, in *12th Report on the Committee on Catalysis, National Research Council, Committee on Contact Catalysis*, John Wiley & Sons, Inc., 1940, pp. 68–81.
107. (a) A. K. Brewer, *J. Phys. Chem.* **1928**, *32*, 1006; (b) J. E. Nyrop, *The Catalytic Action of Surfaces*, Levin and Munkgaard, Copenhagen, Denmark, **1937**; (c) L. V. Pisarzhevskii, *Ukrain. Khem. Zhur.* **1925**, *1*, Sci. part. 1–18; (d) L. V. Pisarzhevskii, *Acta Physiochim. URSS* **1937**, *6*, 555–574; (e) O. Schmidt, *Chem. Rev.* **1933**, *12*, 363.
108. A. Couper, D. D. Eley, *Disc. Faraday Soc.* **1950**, *8*, 172.
109. (a) C. L. Thomas, in *Heterogeneous Catalysis. Selected American Histories*, B. H. Davis, W. P. Hettinger, Jr. (Eds.), ACS Symposium Series, Vol. 222, 1983, pp. 241–245; (b) H. H. Voge, in *Heterogeneous Catalysis. Selected American Histories*, ACS Symposium Series, Vol. 222, 1983, pp. 235–240; (c) R. C. Hansford, in *Heterogeneous Catalysis. Selected American Histories*, ACS Symposium Series, Vol. 222, 1983, pp. 247–252.
110. C. N. Hinshelwood, A. T. Williamson, *The Reaction Between Hydrogen and Oxygen*, The Clarendon Press, Oxford, 1934.
111. H. Pines, in *Heterogeneous Catalysis. Selected American Histories*, B. H. Davis, W. P. Hettinger, Jr. (Eds.), ACS Symposium Series, Vol. 222, 1983, pp. 23–32.
112. A. Farkas, in *Heterogeneous Catalysis. Selected American Histories*, B. H. Davis, W. P. Hettinger, Jr. (Eds.), ACS Symposium Series, Vol. 222, 1983, pp. 89–117.
113. K. F. Bonhoeffer, A. Farkas, K. W. Rummel, *Z. Physik. Chem.* **1933**, *B21*, 225.
114. J. Horiuti, M. Polanyi, *Nature* **1933**, *132*, 931.
115. R. L. Burwell, Jr., *C & E News* **1966**, August 22, 56–67.
116. G. C. Bond, *Catalysis by Metals*, Academic Press, 1962.
117. (a) Mrs. Fulhame, *An Essay on Combustion*, London, **1794**; (b) J. W. Mellor, *J. Phys. Chem.* **1903**, *7*, 557.
118. A. M. Fairlie, *Sulfuric Acid Manufacture*, Reinhold Publishing Corp., New York, 1936, pp. 23–26.
119. R. Knietsch, *Ber.* **1901**, *34*, 4069.
120. A. B. Lamb, W. C. Bray, J. C. W. Frazer, *Ind. Eng. Chem.* **1920**, *12*, 213.
121. *Kirk-Olhmer Encyclopedia of Chemical Technology*, 3rd edn, John Wiley & Sons, New York, 1979, p. 853.
122. J. R. Partington, L. H. Parker, *The Nitrogen Industry*, Van Nostrand Co., 1923, p. 270.
123. W. Henry, *Philos. Trans. Roy. Soc.* **1824**, *114*, 266.
124. F. Kuhlmann, *Compt. Rend.* **1838**, *7*, 1107.
125. H. N. Warren, *Chem. News* **1891**, *63*, 290.
126. W. Ostwald, *Chem. Zeitung* **1903**, *27*, 457.
127. K. Kiaser, U.S. Patent 987,375, 1911.
128. B. S. Beshty, Presentation at ACS division of history of chemistry, Philadelphia, Fall Meeting, 1983.
129. (a) L. Andrusow, *Z. Angew. Chem.* **1926**, *39*, 321; (b) L. Andrusow, *Z. Angew. Chem.* **1927**, *40*, 166; (c) L. Andrusow, *Z. Angew. Chem.* **1928**, *41*, 206.
130. M. Bodenstein, *Angew. Chem.* **1926**, *40*, 174.
131. F. Rasching, *Angew. Chem.* **1927**, *40*, 118.
132. (a) W. H. M. Sachtler, N. H. deBoer, *Proc. Int. Congr. Catal.* **1965**, *3*, 252; (b) C. R. Adams, T. Jennings, *J. Catal.* **1964**, *3*, 549; (c) C. R. Adams, T. Jennings, *J. Catal.* **1962**, *2*, 63.
133. G. W. Keulks, T. Matsuzaki, in *Adsorption and Catalysis on Oxide Surfaces*, M. Che, G. C. Bond (Eds.), Elsevier Science Publishers, Amsterdam, 1985, pp. 297–307.
134. R. K. Grasselli, in *Adsorption and Catalysis on Oxide Surfaces*, M. Che, G. C. Bond (Eds.), Elsevier Science Publishers, Amsterdam, 1985, pp. 275–283.

1.3
The Development of Industrial Heterogeneous Catalysis

Uwe Dingerdissen*, Andreas Martin, Daniel Herein, and Hans Jürgen Wernicke

1.3.1
Introduction

The catalyst industry is almost as old as the chemical industry, with the history of its underlying technologies dating back more than 150 years. Yet, today the discipline of catalysis is specific to the facets of the industry that it serves.

One of the earliest pioneers in the development of industrial catalysts was Mittasch. While Haber discovered the ammonia synthesis catalyst based on osmium, Mittasch developed the industrial ammonia catalyst based on iron oxide by preparing and testing several thousand catalyst samples at BASF.

Since that time, it has been found that catalysts must satisfy a variety of requirements, including selectivity, activity, thermal and mechanical stability, poison resistance, and last – but not least – commercial feasibility. Today, the reactivation and/or regeneration of catalysts is an additional issue of commercial interest, as catalyst production must be not only reproducible but also economically and ecologically acceptable.

Until the 1950s, catalysts were predominantly developed and produced for captive use, but since then their production as specialty materials, independent of the catalytic process used, has been established step by step in the United States, Western Europe, and Japan. Today, more than 150 international companies produce some 1000 catalysts, based on about 100 fundamental types [1]. In addition to the classic fields of catalysis – namely, the petrochemical and chemical industries – catalysis has steadily penetrated new industrial areas such as environmental catalysis, polymer production, power generation, fuel cell technology, specialties, and fine chemicals for pharmaceuticals.

Currently, the annual worldwide value of chemicals (including petrochemicals) produced by catalytic reactions is about US$ 2–4 trillion. Industrial catalysis contributes both directly and indirectly through processes and products to about 25% of the Gross National Product (GNP) in developed countries [2]. Approximately 80% of all chemical conversions require catalysts, and about 20% of the value of all commercial products manufactured in the US is derived from processes involving catalysis [3].

References see page 54
* Corresponding author.

In 2004, the total global market for catalysts was valuated at approximately US$ 13 billion.

The driving forces, and the motivation to create new or replace old catalytic or even non-catalytic by catalytic processes, is based on obtaining [4]:

- more environmentally benign processes in terms of reducing undesired byproducts
- the use of more benign feedstocks
- lower-cost feedstocks
- less capital-intensive processes
- less energy-intensive processes
- reaction capabilities not commercially feasible heretofore
- higher reactor efficiencies by higher space-time-yields.

1.3.2
The Raw Materials Situation

Nowadays, the raw materials situation in the chemical industry depends to a high degree on the side products of the energy industry – that is, the petroleum, gas and coal industries. Therefore, changes in the raw materials situation and resultant first chemical derivatives such as olefins, aromatics and synthesis gas are only expected with the changing situation in the feedstock supply. There is little doubt that shifts to new energy systems will occur in time, although the point in time at which this happens to a substantial degree remains an open issue. Besides the interest in solar energy, wind energy and biomass, a renewed interest in the large resources of as yet non-economically utilizable hydrocarbons (i.e., heavy oil, tar sand and gas hydrates) and for coal is observed. Currently, gas and oil provide about 70% of the world's energy needs.

Although crude oil and natural gas are non-renewable, finite resources, analysts estimate that the current world oil production is more than adequate to meet a 60% to 200% increase in demand until the year 2030. Subsequently, oil production will decline slowly until the resources are depleted [5], though the needs of rapidly developing nations such as India and China might accelerate this process. Currently, new oil fields are being discovered at a rate which is twofold lower than the needs of current consumption (Fig. 1) [6].

A high volatility and steadily increasing oil price continue to push new developments in the petrochemical and chemical industries towards new base feedstocks and related new – mostly catalytic – processes. This will support the increasing use of natural gas, gas condensates and – in the long term – of regenerative sources such as biomass. Moreover, these changes will be accompanied by the need for environmentally more friendly technologies.

1.3.2.1 Inorganic Base Chemicals

The existence of sulfuric acid has been known since the late Middle Ages, at which time it was obtained in small quantities in glass vessels by burning sulfur with saltpeter in the presence of moisture. The production of larger amounts only became possible following the introduction of the lead chamber process by Roebuck, in 1746. The lead chamber process used nitric acid as a catalyst to oxidize SO_2 to SO_3 in the presence of water. In 1831, Phillips patented the oxidation of SO_2 to SO_3

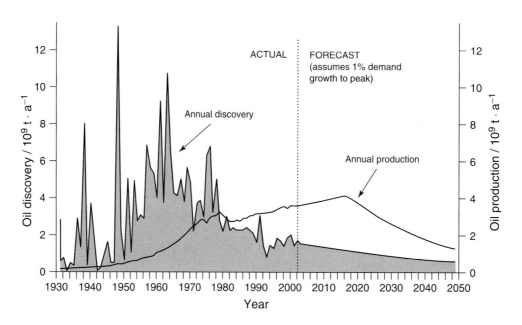

Fig. 1 Estimated global annual discovery rates of oil. (Adopted from Ref. [6].)

using platinum on acid-resistant materials as a catalyst, and this process continued to be used on an industrial basis until 1872, when intensive development of the so-called "contact process" began. In the following years, catalysts consisting mainly of silica-supported V_2O_5 and K_2SO_4 were substituted for the more easily poisoned Pt. Today, the oxidation of SO_2 into SO_3 over vanadium-based supported liquid phase (SLP) catalysts is used almost exclusively in the industrial production of sulfuric acid [7].

The first major breakthrough of modern industrial catalysis was the fixing of nitrogen by the Haber–Bosch process [8]. This process, in which ammonia is synthesized from its basic elements, was commercialized in 1913, and replaced the previous dependence on guano as a source of nitrogenous material. The Haber–Bosch process also removed the dependence on the high-temperature (and thermally inefficient) arc process used to produce cyanamid. While Bosch and his coworkers successfully constructed and operated the high-pressure equipment, Mittasch investigated numerous ammonia synthesis catalysts and scaled-up his synthesis to commercial dimensions. Currently, the industrial iron-based ammonia synthesis process uses a promoted catalyst (K-, Ca-, and Al-oxide as structural and electronic promoters) operating at 400 to 700 °C, with pressures of approximately 3×10^7 Pa (300 bar). Today, ammonia is the base from which virtually all nitrogen-containing products are derived, such that in 1990 the worldwide demand for ammonia reached 100×10^6 t a^{-1}, with ammonia – in terms of quantity produced – being the sixth largest chemical produced worldwide. The dominance of iron-based catalysts has recently been challenged by promoted ruthenium/graphite catalysts applied by the BP group [9]. As ruthenium shows a remarkable improvement in activity, ammonia production is technically feasible at pressures as low as 4×10^6 Pa.

The oxidation of ammonia to nitric acid was first proposed in 1906 by Ostwald, and the process was commercialized in 1915, using platinum as catalyst [10]. This commercialization was accelerated by military demand for nitrates used in the production of explosives. In the first industrial process designed by Ostwald, pure platinum gauzes served as the catalysts. As ammonia accounts for approximately 90% of the production costs of nitric acid [11], the efficiency of ammonia oxidation represents a key factor in nitric acid production. Consequently, the pure platinum gauze catalyst was replaced with platinum–rhodium alloys (Rh content from 5–10%), which offer a higher selectivity. Despite many attempts to develop other types of catalyst to reduce nitric acid production costs, platinum–rhodium and platinum–palladium–rhodium alloys still serve as the main catalysts [12]. The search for alternative, oxide-based catalysts began in 1902, when Ostwald patented catalysts based on transition metal oxides (mainly Co- and Mn-based oxides). During the 1960s, intensive studies resulted in a re-evaluation of the cobalt oxide systems, which subsequently were partly operated on an industrial basis for several years [12]. Between 1985 and 1990, these developments resulted in the industrial operation of some two-bed catalytic systems, combining oxide and platinum-based alloy catalysts. However, currently the platinum alloy-based catalyst process is used exclusively for nitric acid production.

1.3.2.2 Petroleum Refining

Petroleum refining includes not only distillation and extraction but also the synthesis of usable and saleable products from crude oil, for a variety of needs such as transportation fuels or chemical feedstocks.

Most modern, heterogeneously catalyzed processes in petroleum refining were developed between 60 and 70 years ago. The catalysts are mainly based on acid aluminas and zeolites (see, e.g. Refs. [13, 14]), and are continuously improved to meet stricter fuel specifications. In 2004, the total value of catalysts used in crude oil processing was approximately US$ 2.8 billion. A typical refinery is a complex system of units for the thermal, catalytic, or extractive treatments of the crude oil fractions (Fig. 2) [15].

Refinery operations aim at a maximum output of either high-octane, low-sulfur gasoline or high-cetane diesel and jet fuel. Most of the post-distillation processes are heterogeneously catalyzed, including catalytic cracking, hydrocracking and different hydroprocessing steps to remove sulfur, basic nitrogen compounds and metals, and also to reduce aromatics.

Fluid catalytic cracking (FCC) of heavy petroleum fractions is among the most important of the catalytic processes, with annual worldwide capacity amounting to approximately 750×10^6 t. Catalytic crackers contribute between 20% and 50% of the blending components in the gasoline pool of a refinery.

As early as the 1930s, it had been found that if heavy oil fractions were heated over clay-type materials, the cracking reactions would lead to significant yields of lighter hydrocarbons (the Houdry fixed-bed process). Subsequently, the improvements in catalyst activity progressed steadily, with acid-treated clays being replaced by synthetic amorphous silica–alumina catalysts, which were commonly used until 1960. However, a major breakthrough occurred during the early 1960s at Mobil Oil Co., when it was discovered that zeolites (particularly X and Y zeolites) provided tremendous improvements in

References see page 54

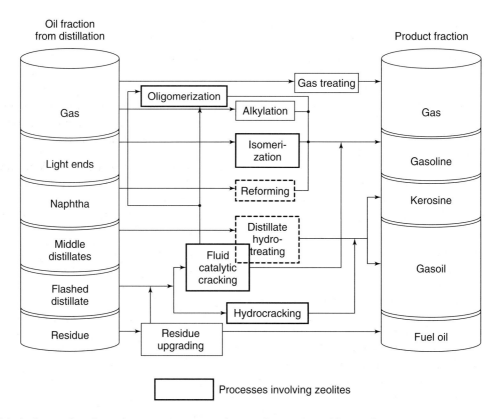

Fig. 2 Simplified scheme of a refinery showing processes involving zeolites. (Adopted from Ref. [14].)

both yield and octane number. In order to achieve the best results, the original faujasite zeolites (Y-type) were modified and stabilized by dealumination and/or rare-earth ion exchange [so-called US-Y (ultrastable) or DA-Y (dealuminated)].

During the cracking reactions, coke is deposited on the catalyst surface. Hence, for their reactivation, the catalysts must periodically be treated with air to burn off the carbonaceous deposits. Nowadays, a fully continuous process is used for catalytic cracking which comprises a riser reactor for cracking and a fluidized-bed reactor for regeneration (FCC). In the cracking reactions, which are carried out at approximately 500–550 °C, the FCC catalysts typically consist of acid-treated clays (filler), synthetic silica–aluminas or silica–magnesia as binders, and zeolites as active components. The catalysts must meet certain requirements such as good cracking activity for large molecules, a low rate of coke and gas formation, hydrothermal stability, abrasive strength, resistance to poisoning by metals (mainly V and Ni), and high regeneration temperatures. Promoters improve selectivities and ease carbon burn-off. The first FCC unit went on stream at Standard Oil of New Jersey's refinery in Baton Rouge, Louisiana in May 1942. As of 2001, the global value of FCC catalysts was approximately 10% of the global catalyst market (i.e., US$ 1.1 billion).

Hydrodesulfurization and Hydrodenitrogenation Hydrodesulfurization (HDS) and hydrodenitrogenation (HDN) are both important refinery steps, not only to process heavy crudes but also to protect downstream processing units against sulfur and basic nitrogen poisoning. Typically, Co–Mo and Ni–Mo (W) supported (γ-alumina) catalysts are used for HDS. The oxidic catalyst precursors or regenerated batches must be presulfided before use, whereupon the process converts organic sulfur compounds (sulfides, disulfides, thiols, thiophenes) to hydrogen sulfide and hydrocarbons. The hydrogen sulfide is recovered and oxidized to elemental sulfur via the Claus process.

Hydrodenitrogenation is the removal of various nitrogen-containing compounds (e.g., indole, quinoline) which otherwise would inhibit acidic catalysts (in either FCC units or hydrocrackers). In HDN, similar catalysts are used as in HDS, but under somewhat increased pressure.

Hydrocracking Hydrocracking is a combination of hydrogenation/hydrotreating and cracking to produce light fractions from vacuum gas oil and residues in a one- or two-stage catalytic process.

The history of this process dates back to the late 1920s, when a plant for the commercial hydrogenation of lignite was commissioned at Leuna in Germany. Tungsten sulfide was used as a catalyst in this process, in which high reaction pressures ($2-3 \times 10^7$ Pa) and a reaction temperature of 400 °C were applied. By the end of the 1950s, the development of improved catalysts enabled the operation of this process at considerably lower pressure (ca. $70-150 \times 10^5$ Pa).

Nowadays, bifunctional catalysts are predominantly used. Cracking and isomerization require acidic catalyst components, and hydrogenation a noble metal component (e.g., Pt on aluminosilicates or zeolites). The reaction is carried out at temperatures between 300 and 600 °C and pressures of up to 15×10^6 Pa.

Reforming, Isomerization, and Alkylation Reforming, isomerization, and alkylation are the main processes to provide high-octane gasoline.

The endothermic *catalytic reforming process* – yet another basic step in petroleum refining – has been used in refining for almost 50 years, the historic driving force being the synthesis of toluene for TNT production during the Second World War. Initially, a catalytic process for the dehydrogenation of methylcyclohexane was developed, but later it was realized that this reaction route could also help to transform cycloalkanes to aromatics used as blending components for gasoline to achieve high anti-knock qualities. Processes such as "Platforming", "Houdryforming" and "Catforming" were introduced simultaneously during the early 1950s; these were developed to obtain aromatics as selectively as possible. The catalysts used contained Pt on acidic aluminas, and the reaction was performed at temperatures of 700–800 K and pressures up to 30×10^5 Pa. The main reaction routes were as follows: (i) dehydrogenation of cyclohexanes to aromatics; (ii) dehydroisomerization of alkylcyclopentanes to aromatics; (iii) dehydrocyclization of alkanes to aromatics; (iv) isomerization of *n*-alkanes to branched compounds; and fragmentation. The introduction of bimetallic catalysts (addition of Re) in 1967 by Chevron allowed the reaction pressure to be decreased, which favored the formation of reformate, for thermodynamic reasons.

Due to their toxicity, the aromatics content of gasoline was restricted during the early 1990s. This legislative-driven decision, and the introduction of new anti-knock additives [e.g., methyl-*tert*-butyl ether (MTBE); ethyl-*tert*-butyl ether (ETBE)] or (bio)ethanol blending (oxyfuels) were the reasons that catalytic reforming partly lost its impact in gasoline manufacture.

During the early 1980s, MTBE usage grew in response to octane demand which resulted initially from the phase-out of lead from gasoline, and later from the rising demand for premium gasoline. Maximum growth rates of 44% per year for MTBE were achieved between 1992 and 1997. As a result of its widespread use, and some problems with underground tanks, MTBE was detected in water supplies in the US as a contaminant. This led to a ban on MTBE in several US states (California, Connecticut, Kentucky, Missouri, and New York, which accounted for 44% of all MTBE consumed in the US) and its eventual phasing out. Today, the only oxygenate replacement considered in the US is ethanol. As yet, similar legislative regulations are not foreseen for the rest of the world, although a switch from MTBE to ETBE is observable for Europe.

Catalytic isomerization is used to increase the octane number of the C_5/C_6 fraction by skeletal branching. Process development began during the Second World War, when there was a need to isomerize *n*-butane to isobutane for alkylation with C_3 and C_4 olefins. Suitable catalysts must be active at temperatures as low as possible, favoring branched compounds on a thermodynamic basis. The early catalysts were based on aluminum chloride, but today the catalysts consist of Pt supported on chlorinated alumina or acidic zeolites such as H-Mordenite. Recent developments have used sulfated metal oxides (e.g., sulfated zirconia) as acidic carriers.

Alkylation in petroleum refining refers to the acid-catalyzed conversion of isobutane with light alkenes (mostly butenes, sometimes propene or pentenes) to a mixture of isoparaffins with high octane numbers. Either anhydrous hydrogen fluoride or concentrated sulfuric acid are used as catalysts.

Oligomerization represents the synthesis of linear, branched or cyclic compounds from monomers or their mixtures. For alkene oligomerization, catalysts consisting of silicophosphoric acid are used to produce so-called "poly-gasoline". Shell's SHOP process uses chelated Ni-hydride complexes for the selective ethylene oligomerization to α-olefins mainly for chemical use.

Recent developments and future trends in catalysis for oil refining and petrochemistry are summarized in Refs. [16, 17].

1.3.2.3 Coal and Natural Gas Processing

The manufacture of synthesis gas was originally based on the gasification of coke from hard coal and low-temperature coke from brown coal by means of air and steam. In 1931, Bergius demonstrated the liquefaction of coal by liquid-phase hydrogenation, with the coal and solid catalyst suspended in tar oil. The hydrogen required for this process was produced by the water gas shift reaction (chromia-containing Fe_3O_4, discovered in 1920) from synthesis gas. After the Second World War, liquid

References see page 54

and gaseous fossil fuels (oil and natural gas) were also employed as feedstocks.

Nowadays, advanced coal gasification processes are undergoing a renaissance in connection with a partial replacement of oil as a petrochemical feedstock. The amount of synthesis gas produced from coal (only 3% in 1976) had increased to about 16% in 2003. More than half of the capacity can be attributed to the Fischer–Tropsch plants in South Africa (Sasol), while many new projects have been announced in coal-rich areas such as in China and Russia.

The gasification process of coal with steam and oxygen consists of several reactions, for example the exothermic partial combustion and the endothermic water gas formation. The various gasification processes can be characterized by the type of coal used (hard or brown coal), the technology involved (allothermal or autothermal), the reactor type (fixed-bed, fluidized-bed), and the gasification agent (water, oxygen or air). The Winkler, Koppers–Totzek and Lurgi pressure gasification are established industrial processes. The Winkler process (finely ground coal, fluidized-bed, 950 °C, oxygen and steam) was developed in Germany by Leunawerke in 1931, and today is operated in numerous plants worldwide. The first commercial plant using the Koppers–Totzek process (powdered coal at 1400–2000 °C, oxygen and steam) was constructed in Finland in 1952, and the process has subsequently been operated in several countries. The Lurgi pressure gasification (coal granules, fixed-bed, 750 °C at $20-30 \times 10^5$ Pa, oxygen and steam) has been in use since 1930.

The main route to synthetic fuels from coal via synthesis gas was developed by Fischer and Tropsch, and first commercialized in 1938. The Fischer–Tropsch process is based on the hydrogenation of CO at 220–330 °C and $7-30 \times 10^5$ Pa pressure over alkalized iron turnings as catalyst. The earlier catalysts consisted of cobalt and of nickel or manganese oxides. In Germany during the Second World War, large-scale plants using fixed-bed reactors produced liquid fuels via this route; these products were alkanes containing different amounts of oxygenated compounds. After the war, the large-scale use of Fischer–Tropsch synthesis was centered in South Africa, where Sasol operated first a fixed-bed reactor plant (since 1955, capacity 2.5×10^5 tons per year) and later two entrained-bed reactor plants (since 1983, with a total capacity of 4.5×10^6 tons per year).

During the 19th century, which were the early days of the natural gas industry, gas was mainly used for street lamps and, sporadically, for the heating and lighting of buildings. Only after the Second World War natural gas and town gas from coking plants were used in the production of synthesis gas as a feedstock for liquid fuels.

Today, the conversion of natural gas into liquid fuels (gas to liquid fuels process, GTL) remains a challenge to industrial catalysis [18], and this point has been the subject of discussion for decades since the first energy crisis [19]. Although, in the past, gas conversion processes have suffered from a small price differential between the product and natural gas, today a number of factors have changed. For example, it has become unacceptable to flare associated gas from oil fields. Moreover, synfuels obtained by GTL will become increasingly important because they are sulfur- and mostly aromatics-free, and because of their high cetane numbers [20].

Nowadays, the available GTL technology is based exclusively on the indirect conversion of methane (70–90% content in natural gas) into either methanol (see Section 1.3.3.2) or liquid fuels via the synthesis gas route. As the direct conversion of CH_4 into methanol or other liquid fuels is still far from being a feasible industrial process [21, 22], the indirect route via synthesis gas is presently the only viable option. The advantage of using natural gas as feedstock for synthesis gas production instead of coal is the higher H/C ratio in the resulting synthesis gas (up to 4:1 from natural gas compared to 1:1 from coal) [20]. The conversion of natural gas into synthesis gas is based on the autothermal reforming and non-autothermal steam reforming processes. While autothermal reforming (ATR) [23] has been used for industrial synthesis gas production since the late 1950s, new developments (burner design and catalysts) were made during the 1990s, including operation at low steam to carbon ratios, which ensure safe operation [24, 25]. The first large-scale steam reforming process was operated by ICI in 1962, whereby hydrocarbon feeds with boiling points up to 200 °C (naphtha) were used as feedstocks for the process. Recent years have witnessed a progress in steam reforming technology that has resulted in cheaper plants based on better reactor tube materials, and better catalysts and process concepts (e.g., the Lurgi process, Ni catalyst at 750–800 °C). Today, ATR appears to be the cheapest solution for fulfilling the optimum requirements of synthesis gas production for methanol, and Fischer–Tropsch synthesis from natural gas. The manufacture of synfuels from natural gas is technically feasible, as demonstrated in a plant at Bintulu, Malaysia; this was the first natural gas-based Fischer–Tropsch plant outside Africa, and was opened in 1993 by Shell.

1.3.3
The Base Chemicals

The total number of base chemicals is estimated at approximately 300; hence, the development of industrial catalysis can be shown, on an exemplary basis, for only some cases, and more complete descriptions are available elsewhere (e.g., Refs. [26, 27]).

The major trends and drivers for base chemicals are the optimization of existing processes and the partial replacement of crude oil as base feedstock. Therefore, the development of catalysts with improved selectivity and activity to reduce or eliminate the formation of undesired byproducts, to reduce the number of process-related steps, and to increase capacities, remains the primary challenge. The recycling of waste and the avoidance of emissions have each approached high levels, and are close to 100% in some cases. Additionally, hazardous products as reactants or products have been phased out and substituted with less hazardous compounds in order to increase the operational safety of the processes.

1.3.3.1 Olefins and Aromatics

Butadiene In 2003, the global demand for butadiene was 8.7×10^6 tons, with modern industrial processes its production being based exclusively on petrochemicals. Cracking processes for the manufacture of olefins have rendered the coproduct butadiene less expensive, and thus have made the C_4 dehydrogenation process unprofitable; for example, in 1999 only 1.9% of the total butadiene was manufactured by dehydrogenation. Older processes (e.g., in the former German Democratic Republic) used acetylene or ethanol as feedstock. In China, India, Poland and Brazil, the ethanol-based Lebedew process is still used; indeed, this is one process that may become more attractive in the future as part of a global renewable resources strategy.

Isoprene Isoprene, with an annual world production capacity of 8.5×10^5 tons in 1997, is manufactured largely in processes analogous to those employed for butadiene, either by direct isolation from C_5 cracking fractions (i.e., isoprene extraction by sulfolane) or by the catalytic dehydrogenation of C_5 isoalkanes/isoalkenes. Additional synthetic routes to isoprene used include the acetone–acetylene process (used until 1982, KOH as catalyst) and the route from isobutene and formaldehyde with 4,4-dimethyl-(1,3)-dioxane as intermediate.

Styrene The history of styrene can be tracked back to Charles Goodyear in 1839, who discovered the vulcanization of natural rubber with sulfur, whereby natural rubber obtained global economic interest. The next chapter opened in 1876, when *Hevea brasiliensis*, a latex-producing tree indigenous to the Amazon basin, was smuggled by the English biologist Henry Wickham in two stuffed crocodiles out of Brazil to establish the European-colonized rubber plantations of what is now Sri Lanka and Malaysia.

During the First World War, the British blockaded rubber transport to Germany, thereby prompting the German chemical industries to develop styrene and butadiene polymers. Similarly, during the Second World War, the Japanese occupation of the rubber-producing areas in south-east Asia led US chemists to develop and produce massive amounts of synthetic rubbers. The wars affected the demand for synthetic styrene butadiene rubber (SBR) as a substitute for isoprene-based natural rubber, and prompted accelerated technology improvements and capacity expansion. This wartime effort led to the construction of several large-scale factories, which rapidly transformed styrene production into a giant industry.

In 2001, the global demand for styrene was 20.5×10^6 tons. The main production process for styrene, which today accounts for 85% of its commercial production, is the direct catalytic dehydrogenation of ethylbenzene. The process was developed in 1931, using mixed $ZnO/Al_2O_3/CaO$ catalysts. The iron oxide catalysts which are today generally preferred were introduced in 1957, and consist of iron oxide, usually containing Cr_2O_3, and KOH or K_2CO_3. The oxydehydrogenation of ethylbenzene (e.g., the Styro-Plus or SMART SM process) has been piloted, and is soon expected to be commercialized. In this process, the catalysts consist of oxides of V, Mg and Al or phosphates of Ce/Zr, Zr or alkaline earth/Ni.

Styrene production, using ethylbenzene as a starting material, consumes approximately 50% of the world's benzene production. Currently, almost all ethylbenzene is produced commercially by the ethylation of benzene. Historically, the alkylation is carried out in the liquid phase with aluminum chloride as catalyst (first described by Balsohn in 1879 [28], today labeled as Friedel–Crafts alkylation). Today, either alkylation in the vapor phase or in the liquid phase use fixed beds of shape-selective acidic zeolites (see Chapter 14.3 of this Handbook).

1.3.3.2 Alcohols, Ethers, Aldehydes, and Ketones
1.3.3.2.1 Alcohols

Methanol The first synthetic methanol was produced in 1923 by BASF from synthesis gas. The "high-pressure" process, using a zinc oxide/chromium-based catalyst (ZnO/Cr_2O_3), which was operated around $250-350 \times 10^5$ Pa and $320-450\,°C$, remained the dominating technology for over 45 years. During the 1960s, ICI made improvements on the use of a copper/zinc oxide-based catalyst concept, which allows milder operating conditions ("low-pressure" process: $35-55 \times 10^5$ Pa, $200-300\,°C$) and today, is the only process used commercially. The current catalysts used in the "low-pressure" methanol

References see page 54

synthesis are composed of copper oxide and zinc oxide on a carrier of aluminum oxide. As a rule, the proportion of CuO ranges between 40 and 80%, that of ZnO between 10 and 30%, and Al_2O_3 from 5 to 10%. Additives such as MgO may also be present.

Plant capacities of 5000 t day^{-1} and above (such as the new Lurgi Mega-Methanol-Design) require new generations of methanol synthesis catalysts with higher low-temperature activities and increased stability. In 2002, the global demand for methanol was 31.0×10^6 tons.

Ethanol Ethanol is one of the oldest manufactured chemicals, with alcoholic (ethanolic) beverages such as wine and beer being produced primarily through the fermentation of fruit or grain. The first civilizations to be formed around a fixed agricultural life style were the Sumerians, in Mesopotamia, around 4000 BC. The evidence that alcohol was produced by these people has been confirmed by archaeological findings and images on many of their cuneiform tablets, which show images of alcohol being drunk. Although distilled spirits have their origin in China and India in about 800 BC, the distillation process did not make its way to Europe until the 11th century. Today, some 90% of ethyl alcohol is produced by fermentation.

The production of synthetic ethanol on an industrial scale was introduced in 1930 by Union Carbide. In a liquid-phase process, ethylene in sulfuric acid was converted to ethyl sulfates which finally yield ethanol by subsequent hydrolytic cleavage. The direct catalytic hydration of ethylene was first commercialized by Shell in 1947. The addition of water is carried out in the gas phase, generally over acidic catalysts, such as H_3PO_4/SiO_2 ($300\,°C$, 70×10^5 Pa). Indeed, chemically synthesized ethanol is still today manufactured from ethylene by catalytic hydration.

The annual world production of ethanol amounts to approximately 25×10^6 t (in 2003), of which 93% is generated through fermentation, the remainder through ethylene hydration.

2-Propanol In 2000, the annual world production capacity for 2-propanol (isopropyl alcohol; IPA) was about 2.1×10^6 t. Usually, 2-propanol is produced by the hydration of propene over various acidic catalysts such as supported heteropoly acids or mineral acids in the gas phase, and acidic ion-exchangers in the trickle phase. From 1951 until the end of the 1970s, IPA was manufactured by ICI, using WO_3/SiO_2 (heteropoly acids) with ZnO as promoter in a "high-pressure" process at $270\,°C$ and 250×10^5 Pa. In Germany, Condea uses H_3PO_4/SiO_2 in a "low-pressure" process at $170-190\,°C$ and $25-45 \times 10^5$ Pa. A trickle-bed, three-phase variant of the direct hydration, which was first developed by Condea in 1972, uses a solid sulfonic acid ion-exchanger as catalyst at $130-160\,°C$ and $80-100 \times 10^5$ Pa pressure.

1.3.3.2.2 Ethers MTBE, as a major consumer of methanol, was first manufactured commercially in Western Europe (Italy) in 1973, by the reaction of isobutene and methanol on an acid ion-exchange resin catalyst ($30-100\,°C$). The feedstock for MTBE production is raffinate I, a C_4 fraction containing isobutene, isobutane, 1-butene, 1,3-butadiene, *n*-butane and *trans*- and *cis*-2-butene. Only isobutene is converted, and all other components remain unchanged. In 1996, the annual global capacity peaked at 22×10^6 t. Due to its low biodegradability and high water solubility (42 g L^{-1} at $20\,°C$), MTBE is being phased out from gasoline, especially in the US (see also Section 1.3.2.2).

1.3.3.2.3 Aldehydes Formaldehyde was first commercially produced from methanol in 1888. The catalytic oxidation is carried out in the presence of Ag or Cu catalysts or Fe-containing MoO_3 catalysts. In 2001, the global demand for formaldehyde was 4.45×10^6 tons.

1.3.3.3 Acids, Esters, Anhydrides, and Epoxides
1.3.3.3.1 Acids, Esters, Anhydrides

Acetic Acid For many years, acetic acid was mainly manufactured from acetaldehyde and thus, was closely coupled to the acetaldehyde manufacturing process (at the start of the First World War), and underwent the same change in feedstock from acetylene to ethylene. An alternative route to acetic acid was based on the non-catalyzed or Co-acetate-catalyzed oxidation of *n*-butane. The initial discovery of the homogeneously catalyzed methanol carbonylation at high temperature ($250\,°C$) and high pressure (7×10^7 Pa) with a cobalt iodide catalyst was described by Reppe at BASF [29] as early as 1913, but this was not commercialized before 1960 due to the extremely corrosive process conditions. The stimulus to these investigations was to develop an acetic acid process that did not depend on petroleum-based feedstocks. The initial capacity of 3600 t a^{-1} was expanded to $4.5 \times 10^5 \text{ t a}^{-1}$ in 1981 [30].

Today, around 70% of the capacity worldwide is based on the low-pressure Monsanto methanol carbonylation process ($170\,°C$, 30×10^5 Pa, iodine-promoted rhodium or iridium catalyst) introduced in 1970, and improved processes such as the Celanese low water process (1980) and the BP Cativa process (1996). In 2002, the global demand for acetic acid was 6.94×10^6 tons. Recent developments have been based on the direct selective oxidation of ethane to acetic acid [31].

The first vinyl esters were prepared by Klatte in 1912, by the treatment of acetylene with the corresponding carboxylic acid and mercury salts as catalysts in a liquid-phase process. Initially, the reaction was used for the industrial production of vinyl chloroacetate; the polymer obtained from this vinyl ester was used during the First World War as a varnish for airplanes [26].

Until 1965, almost all vinyl acetate was produced by the acetylene gas-phase process [32]. The conversion of acetic acid with acetylene is still a commercially used process, albeit of minor economic importance (market share <10%). The reaction is carried out in the gas phase in the presence of solid catalysts containing zinc salts ($Zn(OAc)_2$/charcoal) at 170–250 °C.

The preferred process today is the gas-phase reaction of ethylene with acetic acid and oxygen, and this has been used industrially since 1968. It was developed independently by National Distillers Products (US) [33], and by Bayer in cooperation with Knapsack and Hoechst (Germany) [34–36]. The reaction is carried out at pressures below 10×10^5 Pa and temperatures between 170 and 200 °C. For the two commercial catalyst systems (Cd, Pd on silica and Au, Pd on silica), the cadmium-containing catalysts are replaced by the gold-containing system, for reasons of reduced toxicity. Two more processes – conversion of acetic acid with acetylene, and oxidative coupling with ethylene – have also been developed as liquid-phase processes, but none of these has reached substantial industrial importance, and they are no longer in use. Currently, approximately 90% of the worldwide vinyl acetate capacity uses the ethylene gas-phase process. In 2004, the global demand for vinyl acetate in 2004 was about 4.4×10^6 tons.

Acrylic acid was first manufactured industrially in 1901 (Rohm & Haas) by the ethylene cyanohydrin process, and this was utilized until 1971. However, in recent years, the oxidation of propene to acrylic acid has become dominant and has mostly replaced the traditional processes (cyanohydrin and Reppe processes, alcoholysis of β-propiolactone, hydrolysis of acrylonitrile). Propene is either converted in a one-step process or, more commonly, in a two-step process. In the single-step direct oxidation, propene is converted into a mixture of acrylic acid and acrolein (200–500 °C, multifunctional Mo-based catalyst with Te compounds as promoters). In the two-step process, propene is oxidized to acrolein, with a minor fraction of acrylic acid in a first reactor system in the presence of steam (330–370 °C, 1–2 $\times 10^5$ Pa, Bi, P, Mo-based catalyst with Fe, Co and other promoters). The outlet of the first reactor is further oxidized to acrylic acid (260–330 °C, Mo, V, W, Fe, Ni, Mn, Cu mixed catalyst) in a second reactor system.

Until the early 1960s, benzene was the only raw material used for the manufacture of maleic anhydride but, with increasing demand, more economical routes based on C_4 compounds were developed. In 1991, 36% of the world capacity was still based on benzene. The first commercial plant (Monsanto) based on n-butane began operation in the US in 1974. Initially, only fixed-bed reactors were used, but these were later converted into fluidized-bed reactors and processes. Both processes use the same vanadium pyrophosphate-based catalysts (Fe, Cr, Ti, Co, Ni, Mo as promoters). In 2004, the global demand for maleic anhydride was about 1.4×10^6 tons.

Until 1960, phthalic anhydride was manufactured almost exclusively from naphthalene from coal tar (started in 1916 by BASF with V_2O_5 catalysts). A reduction in coal coking, and therefore a depletion in naphthalene, led to the use of o-xylene as a more economic feedstock. During the 1990s, about 85% of phthalic anhydride was produced by this process, and this has remained fairly constant up to the present. In the commercial process, o-xylene is oxidized in fixed-bed or fluidized-bed reactors at 375–410 °C over V_2O_5-based catalysts (with TiO_2 and Al and Zr phosphate promoters). In 1998, the global demand for phthalic anhydride was about 3.5×10^6 tons.

1.3.3.3.2 **Epoxides** Ethylene oxide has experienced a dramatic expansion in production since its discovery in 1859, and the first industrial process in 1925 (chlorohydrin process). The direct epoxidation of ethylene with oxygen or air to ethylene oxide was industrialized in 1937. This silver-catalyzed process (Ag/Al_2O_3 with Cs or Ba/Cs or K/Rb promoters at 270–290 °C and 30×10^5 Pa) has displaced the chlorohydrin process, and has been the exclusive manufacturing process since 1975. In 2004, the annual world production capacity of ethylene oxide was about 17×10^6 t.

All attempts to manufacture propylene oxide commercially by the direct oxidation of propene (analogous to ethylene oxide) have been unsuccessful. Thus, the chlorohydrin process, which is no longer economic for ethylene oxide, remains important in propylene oxide manufacture and accounted for 51% of the worldwide production capacity in 1998. The remaining production was based on indirect oxidation processes with hydroperoxides. In 2004, the annual global demand for propylene oxide was 2.8×10^6 t.

1.3.3.4 **Halogenated Chemicals**

The industrial production of vinyl chloride monomer (VCM) began in the 1930s, and reached an annual worldwide capacity of about 31×10^6 t in 2000. The classical production process for VCM is based on the addition of hydrogen chloride to acetylene ($HgCl_2$/charcoal catalyst at

References see page 54

140–200 °C). Today, however, most of these plants have been closed down and VCM is now generally manufactured by the oxychlorination of ethylene with hydrogen chloride and oxygen (supported $CuCl_2$ with rare-earth and alkali metal chlorides at 220–240 °C and $2-4 \times 10^5$ Pa) to 1,2-dichlorethane (EDC) in a fluidized-bed and subsequent thermal dechlorination of EDC.

The chloromethanes (from CH_3Cl to CCl_4), with an annual worldwide production capacity of about 2.3×10^6 tons in 1998, were manufactured since 1923 by the direct non-catalyzed chlorination of methane. Since 1975, however, a process which can be regarded as an indirect oxychlorination of methane (melt of a mixture of $CuCl_2$ and KCl as catalyst and chlorine source) has been developed and used.

1.3.4
The Fine Chemicals

Although there is no universally accepted definition of "fine chemicals", a useful working terminology might be to classify them as chemicals with a price tag of more than 10 US$ per kg, or an annual production capacity of less than 10 000 t. Hence, most catalysts themselves belong to the category of fine chemicals [37].

The fine chemical industry is a growth industry, with the current growth rate of 3 to 5% per annum exceeding gross domestic product growth in both Europe and the US. The catalyst value used for these applications is estimated to be in the magnitude of 10^9 US$ per annum. From the chemical viewpoint, fine chemicals are complex, multifunctional molecules, which are often sensitive to thermal treatment. Mostly, they are processed in liquid phase and often in multi-step syntheses in multi-purpose equipment.

The pressure of time – that is, the rapid development of new fine chemicals or new ecological processes, due to a shorter lifecycle of the products – is of utmost importance. In contrast, the development of a new or alternative catalytic process generally is often time-consuming. Therefore, strategies to generate suitable catalysts in the fine chemicals area often consist of three major steps: (i) the use of so-called "catalyst kits" – that is, available "on-shelf" solutions; (ii) the rapid adaptation of basic, more broadly useable catalyst systems by making minor but effective changes; and (iii) the use of high-throughput synthesis and testing (see Chapter 9.3 of this Handbook). Only in very rare cases are specific catalytic processes developed exclusively for a certain reaction, and this predominantly for rather large-scale processes.

Instead of describing specific reactions, it is possible to classify certain principal reaction categories as selected and characterized here. While some of the cited examples might not be mentioned typically as fine chemicals due to their size, they are well-documented examples which help to express the described technology and are conceptually assignable to representatives of smaller volume.

1.3.4.1 Catalytic Hydrogenation

The selective hydrogenation of carbon–carbon double or triple bonds, the hydrogenation of carbonyl compounds to the corresponding alcohols, the reductive amination, the nitrile hydrogenation, and the reduction of nitro groups to amines have been fundamental and often-practiced reactions for the synthesis and manufacture of fine chemicals, agrochemicals and pharmaceuticals for several decades. Catalytic hydrogenation over supported noble metal catalysts has a longstanding tradition that dates back to its discovery by Sabatier [38, 39] in 1897, and still is a vivid domain for new developments, such as the combination with chiral additives to yield enantioselectivity [40]. Several excellent monographs, papers and books have been produced which review catalytic hydrogenation with respect to the fine chemical synthesis [41–51]. However, only selected aspects of developments achieved over the past 40 years are summarized exemplarily and reported.

1.3.4.1.1 Hydrogenation of Carbon–Carbon Multiple Bonds
The catalysts mainly used in technical applications for fine chemical production usually involve precious metals (i.e., Pt, Rh, Ru, Pd) supported on carbon, alumina, and silica and other special supports. Besides noble metal catalysts, activated base metal catalysts (ABM, such as Raney-Ni or Raney-Cu), or Cu, Co, Ni on various support materials are preferred catalytic materials for the hydrogenation of carbon–carbon multiple bonds.

Palladium, which is widely used and of outstanding performance, is an active and selective component in comparison to other elements of Group VIII. One notable advantage of Pd-based catalysts is their simple adjustability to selectivity and activity by the addition of modifiers. Besides modifiers, metal crystal size and support surface area influence reactivity. Reaction conditions (ambient temperature and low hydrogen pressure) are usually mild for steric unpretending substrates. On occasion, only complex steric factors and substitution patterns demand an increase in the reaction temperature or hydrogen partial pressure. One notable disadvantage in many cases is the ability of Pd to catalyze C–C double bond isomerizations, although in such cases Pt, Ru, and Ni catalysts may be used as alternatives due to their lower isomerization activity.

Presumably, the best-known selective hydrogenation is the reaction of alkynes to alkenes which is connected with Lindlar [52]. The Lindlar catalyst, which was developed in 1952, is based on a Pd system supported on calcium carbonate and modified by lead acetate to improve selectivity. Lindlar-type catalysts are still in use today.

Aromatic ring hydrogenation typically requires more severe reaction conditions than the hydrogenation of other functional groups, whereas the reduction of heteroatomic systems succeeds often under milder conditions. This permits selective hydrogenation of the heteroaromatic ring besides an aromatic ring, which remains unaltered. In addition to total hydrogenation of the aromatic ring system (ease of hydrogenation: benzene to cyclohexane, phenol to cyclohexanol, pyridine to piperidine, pyrrole to pyrrolidine) for which supported Rh, Ru, Pt, Ni, Pd, and Co are active catalysts, partial hydrogenation of aromatic ring systems is of special industrial interest. For example, Asahi Chemicals has been conducting the partial hydrogenation of benzene to cyclohexene since 1990, where a supported Ru-catalyst is used in a biphasic system [53]. Of equally worthy mention is the "Allied Process"; this is the hydrogenation of phenol to cyclohexanone by means of Pd supported on carbon and further modifiers such as alkali metals [54]. Although Pd is the noble metal of choice for the hydrogenation of C=C bonds, it lacks the ability to hydrogenate C–O double bonds, which is consequently used in this process by taking advantage of the faster keto–enol tautomerization.

1.3.4.1.2 Hydrogenation of Carbon–Oxygen and Carbon–Nitrogen Multiple Bonds Consistently, aliphatic carbonyl compounds are hydrogenated to alcohols on Pt, Rh, and Ru catalysts, of which Pt is most often preferred. Usually, Pd catalysts are not used for these reductions. Since 1940 primary alcohols have often been produced via olefin hydroformylation followed by reduction on an industrial scale.

The hydrogenation of α, β-unsaturated aldehydes to the corresponding allylic alcohols, the saturated aldehydes or the saturated alcohols, is an interesting example of the development of selective catalysts. A prominent representative for these types of reaction is the hydrogenation of E- and Z-citral (see Fig. 3). Depending on the catalyst, either the saturated alcohol (i.e., citronellol) or the allylic isomers geraniol and nerol are obtained with good yield. Citronellol is the dominant product when Ni on alumina is used as catalyst, whereas the isomers geraniol and nerol are preferred when using Ru on carbon [55, 56].

In contrast to the aliphatic carbonyl compounds, aromatic aldehydes and ketones are activated for hydrogenation of the carbonyl group, and therefore milder hydrogenation conditions can be applied. Selective reductions by means of Pd catalysts could be achieved [46]. The hydrogenation of furfural to furfuryl alcohol is typically applied in the fragrance industry. Here, commonly used copper chromite catalysts exhibit a selectivity of 80%, whereas newer results showed enhanced selectivity of 98% and conversion of 98% for Raney-type Ni or Cu catalyst modified with Mo-heteropolyacids [57, 58].

Another process worthy of mention is the hydrogenation of aromatic carboxylic acids over chromium-modified zirconia catalysts. The production of benzaldehyde and derivatives via the corresponding acids was realized by Mitsubishi in 1988 with an annual production capacity of 2000 t. Selectivities of up to 96% and conversions of up to 98% were obtained [59].

Other typical representatives of carbonyl hydrogenation include the conversion of sugars on Raney-Ni [46, 60], of fatty esters on Cu–Al–Fe catalysts [61], or maleic anhydride on copper chromite or Pd–Re on carbon [62].

Although enantioselective hydrogenations are usually the domain of homogeneous catalysis, there are some excellent examples of heterogeneous chiral catalysis implemented in industrial processes. The most widely published are the enantioselective hydrogenation of β-ketoesters and α-ketoesters. β-Ketoesters may be hydrogenated to their respective β-hydroxyesters by Raney-type Ni catalysts modified with tartaric acid/NaBr. The enantiomeric excess (*e.e.*) can surpass 90%, as shown

References see page 54

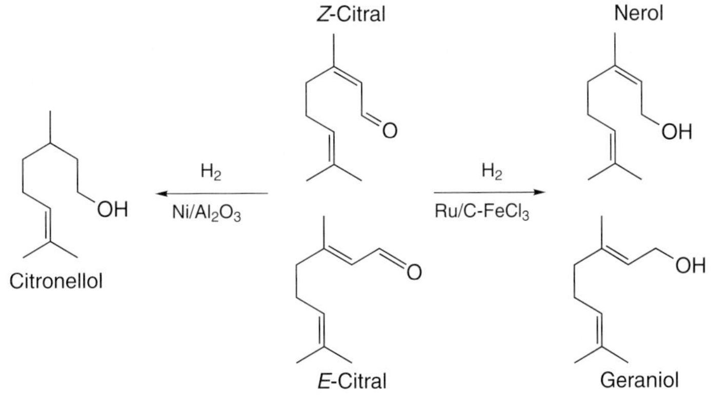

Fig. 3 The hydrogenation of citral.

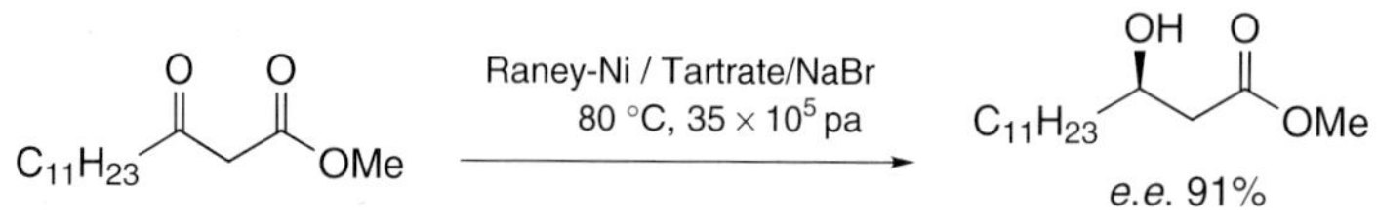

Fig. 4 The enantioselective hydrogenation of an intermediate used in the synthesis of orlistat [40].

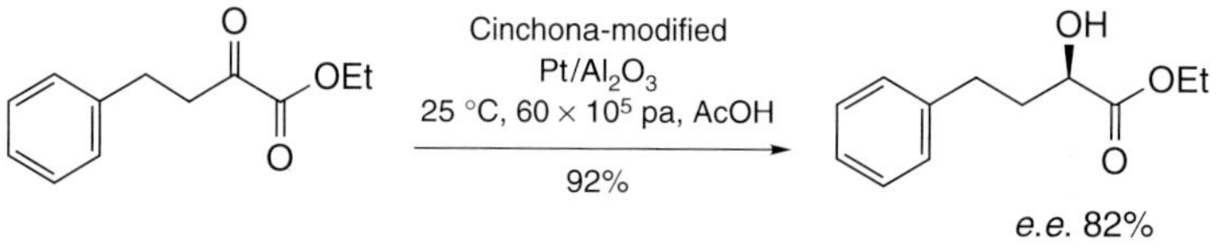

Fig. 5 The enantioselective hydrogenation of an α-ketoester to the chiral α-hydroxyester used in the synthesis of benazepril [40].

by Roche [44, 63] where the β-hydroxyester (Fig. 4) is used as intermediate for the production of the anti-obesity drug, orlistat. α-Ketoesters are reduced by means of a chinchona-modified Pt–Al$_2$O$_3$ catalyst. The respective α-hydroxyester (see Fig. 5) is an interesting intermediate for the synthesis of benazepril, which serves as an angiotensin-converting enzyme (ACE) inhibitor; the reported *e.e.*-value of the synthesis [40] reaches 82%. Recent developments of Degussa have led to a novel commercialized catalyst system (catASium®F214) which exhibits an even higher *e.e.*-value of up to 90%, at very high activity.

1.3.4.1.3 Reductive Amination Reductive amination represents the condensation of an amine with a carbonyl group, followed by reduction of the intermediate imine or enamine to the desired primary, secondary, and potentially tertiary, amine. Alcohols, as far as they do not react via nucleophilic substitution, follow the same scheme by dehydrogenation to the carbonyl compounds, and then undergo reductive amination to the corresponding amine. A typical example of this reaction class might be the reductive amination of the cyclohexanone/cyclohexanol mixture to cyclohexylamine, or the condensation of fatty amines with formaldehyde, which is carried out on Cu-containing catalysts [64–66].

1.3.4.1.4 Nitrile Hydrogenation The reduction of nitriles yields primary, secondary, or tertiary amines. Nitriles are reduced stepwise via interim imines, which react with the amines over a complex reaction cascade to yield additionally secondary and tertiary amines, besides the expected primary amines. The catalyst, reaction conditions and substrate concentration control the product distribution. Pt/C forms predominantly tertiary amines, Rh/C leads mainly to secondary amines, and Ru/C gives the primary amine in case of butyronitrile hydrogenation [67–70]. Isophorone diamine production is an interesting example of the combination of two different transformations, viz. reductive amination and nitrile hydrogenation, which are carried out with Ru–Al$_2$O$_3$, CoO–Mn$_2$O$_3$ [71], or Raney-type Co or Ni catalysts [72, 73].

1.3.4.1.5 Hydrogenation of Nitro Groups The access to aromatic amino derivatives usually occurs by the reduction of aromatic nitro compounds, with known intermediates being the nitroso component and hydroxylamine. An accumulation of the latter is often undesired, and therefore the catalyst development is forced to suppress these unwanted components. Depending on the availability of other reducible groups in the molecule of interest, such as C—C, C—N double or triple bonds or halo groups, many catalytic systems based on Pt/carbon have been adopted to solve this problem. Promoters are typically V, Pd, Pb, Cu, Fe, H$_3$PO$_4$, or formamide [74–79].

1.3.4.2 Catalysis on Solid Acids and Bases
In the synthesis of fine chemicals, in second position of importance after hydrogenation is that of acid–base catalysis. Traditionally, stoichiometric chemistry or homogeneous acid and base catalysis have been applied to a major extent in liquid-phase syntheses. The commonly encountered drawbacks of such syntheses include salt formation and problems of waste disposal. In a recent, excellent review, 127 different catalytic processes (e.g., alkylation, acylation, rearrangement and isomerization, amination, cracking, etherification, esterification, condensation, aromatic substitution and others) were identified which have achieved commercial implementation within the past 25 years, having overcome the problem of salt formation [80]. Interestingly, between 30 and 40 of these processes are related to the production of intermediates and fine chemicals. Meanwhile, the application of solid base catalysts, which began much later than that of solid acid catalysts, is becoming of more interesting to the chemical industry. Worthy of mention is the fact that

bifunctional catalysts with separated acid and base sites that are close to each other are of increasing interest for commercial applications. The solids used for acid catalysis are typically zeolites, clays, acidic mixed-metal oxides, phosphates and ion-exchange resins (polystyrene-based sulfonated resins, nafion). The typical base catalysts are include MgO, Na/NaOH/Al$_2$O$_3$, K/KOH/Al$_2$O$_3$, K/CaO, and Na/K$_2$CO$_3$.

1.3.4.2.1 Aromatic Substitution (Friedel–Crafts Alkylation and Acylation)
Compared to the currently used (and still largely predominant) Lewis acid catalysts (e.g., AlCl$_3$, BF$_3$ and FeCl$_3$) used in fine chemical synthesis, solid catalysts have significant environmental advantages for both alkylations and acylations. For example, the use of shape-selective zeolites in such reactions leads to enhanced yields of the *para*-isomer.

Alkylations are widely used in industrial processes; typically, benzene or higher aromatic compounds are alkylated with olefins such as ethene, propene and butadiene, or alcohols such as methanol or ethanol. Acidic zeolites are the catalysts of choice, but basic catalysts such as MgO, Na/K$_2$CO$_3$ or K/KOH/Al$_2$O$_3$ are also being applied. Besides widespread use in the production of base chemicals and intermediates, a variety of interesting new processes have been developed, including the O-alkylation of phenol with methanol to *o*-cresol and 2,6-xylenol by Asahi Kasai Chemicals in 1984 (catalyst: Fe−V−O/SiO$_2$) [80]. In 1992, Sumitomo commercialized the alkylation of cumene with ethene to *tert*-amylbenzene over a superbase catalyst (K/KOH/Al$_2$O$_3$) with selectivities and conversions in excess of 99% [80].

The Friedel–Crafts acylation represents the most relevant means of generating aromatic ketones. The acylating agents are usually organic acids or anhydrides. During the past 20 years, major research activity has been applied towards solid catalysts (mainly zeolites) as a replacement for the classic Lewis acids, although in many cases rapid deactivation and poor selectivity remain issues. An initial successful industrial application was the Rhodia/Rhône Poulenc acylation process [80, 81] (in 1996) of anisole or veratrole (1,2-dimethoxybenzene), using zeolite H-beta.

1.3.4.2.2 Amination
The synthesis of *tert*-butylamine is an example of the replacement of stoichiometric reactions by heterogeneous catalysis (BASF, 1986) [82–84]. Another example, in 1991, was the synthesis of ethyleneimine by Nippon Shokubai [85, 86].

The direct amination of isobutene to *tert*-butylamine is catalyzed by Brønsted acid materials, especially zeolites, under supercritical conditions. For thermodynamic reasons, the pressure should be as high as useful and the temperature as low as possible. At sufficiently high concentration and strength of the acid sites in the zeolite, the reaction proceeds rapidly enough to reach the thermodynamic equilibrium. Isobutene is reacted with ammonia in a slight excess over a pentasil [83] or beta-zeolite [84] in a fixed-bed reactor at around 270 to 300 °C and 3×10^7 Pa pressure. The selectivity reaches 99%, while the conversion varies between 12% and 17%, depending on the temperature. This process provides a clear advantage over the HCN-based Ritter route (via saponification of the intermediate formamide) by avoiding the production of 3 t Na$_2$SO$_4$ as byproduct per ton *tert*-butylamine.

The conventional process for ethyleneimine, which is applied for the production of amine-based functional polymers for coatings of papers and textiles, is the intramolecular dehydration of monoethanolamine in liquid phase using sulfuric acid and sodium hydroxide (the Wenker process). The drawback of this process is the formation of large amounts of Na$_2$SO$_4$ (4 t t^{-1} ethyleneimine). The new Nippon Shokubai vapor-phase process, by using solid acid–base catalysts (Ba−Cs−Si−P-oxide), avoids such byproducts and achieves a monoethanolamine conversion of approximately 86% and 81% selectivity at ambient pressure and 410 °C.

1.3.4.2.3 Rearrangement, Isomerization
An example is the salt-free Beckmann rearrangement of cyclohexanone oxime, as recently commercialized by Sumitomo Chemicals. The catalyst presumably consists of a zeolite with weak acidic sites [87], and the reaction is carried out at about 300 °C and under reduced pressure ($<10^5$ Pa).

Other reactions such as the Fries, the benzamine, or the pinacol rearrangements, are referred to in the literature, but may not yet be widely performed on a commercial scale.

An example of a terpene-type rearrangement/isomerization is the conversion of 5-vinyl-bicyclo[2.2.1] hepta-2-ene to the 5-ethylidene-bicyclo[2.2.1]hepta-2-ene, a compound which is used for vulcanization purposes and recently commercialized by Sumitomo Chemicals. The reaction is carried out at −30 °C over Na/NaOH/Al$_2$O$_3$ as catalyst, and both the conversion and selectivity are in excess of 99%. This was the first example of using the so-called "superbasic catalysts", which have an extremely high base strength, sometimes higher than $H_0 > -37$.

1.3.4.3 Oxidation Catalysis

1.3.4.3.1 Oxidation of Alcohols and Aldehydes
The first indications that cinnamyl alcohol can be oxidized by air to cinnamaldehyde in the presence of a Pt catalyst

---

References see page 54

date back to 1855 [88]. Pioneering studies conducted by Heyns et al. [89, 90] around 1960 showed that the liquid-phase oxidation of alcohols in the presence of supported noble metal and other transition metal (V, Mn, Co, Ni, Cu, Zn) catalysts was potentially very attractive for the synthesis of fine chemicals. Mild temperatures, oxygen as reagent and water as solvent are other attractive reaction conditions. However, one drawback of this reaction type is catalyst deactivation, which tends to be the major problem. Besides the agglomeration of noble metals, the leaching of metals – particularly of metal promoters – is a known cause of deactivation [91–94]. Deactivation by blocking the surface with adsorbed reaction products has been also reported [95–98]. Another serious obstacle is the demand for a base medium in many cases (pH = 8–11), which leads to neutralization costs and additional work-up steps. Typical catalysts are Pd, Pt, Au, Ru, Rh on carbon, $CaCO_3$ or Al_2O_3 modified with Bi or Pb. The discovery that p-electron metals, such as Pb and Bi as modifiers, improve the rate of conversion and selectivity proved to be a major breakthrough. For example, the rate of oxidation of glucose to gluconate was 20-fold higher on Pd–Bi/C than on a Pd/C catalyst [98].

Although numerous promising attempts have been reported, implementation in industrial processes remains modest. A novel continuous process for carbohydrate oxidation by air on noble metal catalysts was designed by Südzucker [92, 99, 100]. This process was originally developed for the oxidation of isomaltulose, but has been adapted for the oxidation of sucrose, maltose, glucose, and methyl α-D-glucopyranoside, with both selectivity and yield as high as 99%.

One very exciting example of a gas-phase oxidation is part of the BASF citral process which was industrialized during the late 1980s [87, 101]. The synthesis of citral (which is a valuable intermediate in the synthesis of fragrances or vitamins) begins with cheap isobutene and formaldehyde. By condensation, 3-methyl-butane-3-ol (isoprenol) is formed. A part of this alcohol is isomerized. The other part of the isoprenol is regioselectively oxidized to the corresponding isoprenal (yield 95%) by means of a Ag catalyst at 500 °C in an excess of oxygen. This is followed by a Cope rearrangement and isomerization of the alcohol, whereupon the isoprenal yields citral.

1.3.4.3.2 Epoxidation and Aromatic Ring Hydroxylation

Many solid catalysts, which are mainly based on Ti, have been described for epoxidations in which the epoxidizing reagent is an alkyl hydroperoxide. However, from an industrial point of view, aqueous hydrogen peroxide is preferred as a cheaper oxidant. Besides epoxidation, another oxidation reaction worth mentioning for this type of Ti-containing class of catalysts in combination with hydrogen peroxide is that of aromatic ring hydroxylation. The development of TS-1 zeolite by the Enichem research group proved to be a major breakthrough during the mid-1980s [102–106].

1.3.5
Environmental Catalysis

Environmental pollution and destruction on a global scale, as well as the lack of sufficient clean and natural energy sources, have attracted much attention and concern to the vital need for ecologically clean chemical technologies, materials, and processes.

During the past 30 years, legislative demands have been of major influence in the development of catalysts, most notably in legislation concerning automotive exhaust emission, stack gas emission from power stations, and gaseous effluents from processes associated with the refining and chemical industries. During the past few years, the automobile emission catalyst market in particular has undergone extraordinarily expansion.

The annual growth rate of the catalyst market over a 10-year period between 1995 and 2005 has been about 10%. By comparison, the annual growth rate of the emission control catalyst market has been approximately 17%, while other segments have shown growth rates of between 3% and 5% [107] (only European catalyst market). More than 30% of the worldwide market for solid catalysts belongs to the environmental sector, and more than 90% of this can be attributed to the control of emissions from mobile sources.

1.3.5.1 Cleaning of Automobile Emissions

During the past 40 years, the development of a successful automotive emission control technology has been a remarkable achievement, given the demanding nature of the application. Since the 1970s, the introduction of catalysts has resulted in a massive (>90%) reduction of emissions [108, 109], with automotive catalysts for emission control perfectly meeting both European Union (EU) and US standards for CO, hydrocarbons (HC) and NO_x emissions. A comparison of present and future EU emission control standards for spark ignition and diesel engines demonstrates an impressive confidence in further innovation (Table 1) [110].

The early catalytic systems for after-treatment in automobile emission control, as introduced in the US in 1975, comprised simple alumina-supported noble (Pt–Pd) metal systems. Whilst the initial goal was to lower the proportion of unburned hydrocarbons and CO, the subsequent step was to limit the emission levels of NO_x. In turn, changes in fuel formulations, as well as developments in engine operation and control

Tab. 1 European Union standards for emission limits of automotive exhaust gases (all values in g km^{-1})

Emission	EURO I (2003)		EURO IV (2005)		EURO V (2008/10)[a]	
	Gasoline	Diesel	Gasoline	Diesel	Gasoline	Diesel
CO	2.72	2.72	1.0	0.5	0.5	0.5–0.25
HC	0.97	0.97	0.1	0.3	0.05	0.15
NO$_x$	–	–	0.08	0.25	0.04	0.13
PM[c]	0.14	–	–	0.025	–	0.004–0.015[b]

[a] Currently under discussion.
[b] Light/heavy duty diesel.
[c] PM = particulate matter.

systems, led to the introduction of the so-called "three-way catalyst" (TWC) during the early 1980s, these being designed specifically for spark ignition engines. Figure 6 depicts, graphically, the CO, HC and NO$_x$ conversion depending on the equivalence ratio λ (the lambda value); λ is the ratio of the actual engine air-to-fuel ratio (A/F) to the stoichiometric engine A/F, which is 14.7 (the value needed to burn completely all HCs present).

A λ value <1 (rich conditions) leads to incomplete combustion, whereas λ > 1 (lean conditions) gives rise to an exhaust gas that contains more oxidizing reactants than reducing ones.

TWC catalysts can remove the three main pollutants in automobile waste gases, namely CO, HC, and NO$_x$; elimination of the first two components is by oxidation, and the latter by reduction, as shown in the stoichiometric equations:

$$2CO + 2NO \longrightarrow 2CO_2 + N_2$$

$$HC + 2NO \xrightarrow{O_2} CO_2 + \frac{1}{2}H_2O + N_2$$

As can be seen clearly in Fig. 6, it is of utmost importance that the redox balance of the waste gas stream is maintained in order to ensure a high conversion of both reductants and oxidants.

A typical TWC consists of a ceramic honeycomb substrate (e.g., α-cordierite) that is coated with a thin, stabilized γ-Al$_2$O$_3$ washcoat as carrier, and a noble metal component, mainly Pt, Pd, and Rh. The ceramic substrate always contains ceria as the oxygen storage compound; this helps to suppress the unavoidable oxygen fluctuations in the exhaust gas by changing the oxidation state (Ce^{4+} ⇆ Ce^{3+}), and therefore keeps the λ-value in balance.

Due to further economic demands (e.g., The Kyoto Protocol), leaner conditions are more desirable for: (i) complete fuel burning; and (ii) a more economical use of fuel. However, lean conditions tend to push the λ-value into a region where NO$_x$ is not yet removed satisfactorily.

Compression ignition engines (diesel engines) differ from spark ignition engines in terms of the A/F ratio, and also the ignition of the A/F mixture. The exhaust gas of diesel engines has a more complex composition, with particulate matter (PM) components being present as well as gaseous, liquid and solid matter. Since the early 1990s, diesel oxidation catalysts (supported Pt, Pd systems) have been applied to passenger cars in the EU and to medium- and heavy-duty trucks in the US. Following oxidation, the amount of CO, HC and organic components (e.g., aldehydes) is diminished, and most of the PM is also removed. It is crucial, however, that the catalyst selectively catalyzes the above-mentioned reactions but not the

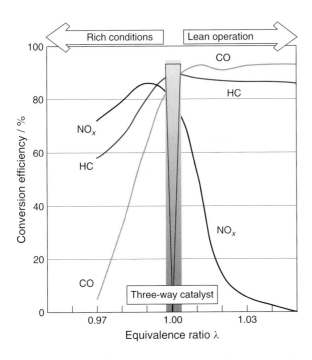

Fig. 6 CO, HC, and NO$_x$ conversion on a three-way catalyst, depending on the equivalence ratio (λ-value).

References see page 54

oxidation of NO to NO_2, nor the oxidation of SO_2 to SO_3. As with TWCs, diesel oxidation catalysts consist of a ceramic monolithic support on which a washcoat is deposited as support for the noble metal component.

Whereas the diesel oxidation catalyst became an accepted technology for limiting the emission of CO, gaseous HC and PM, the major remaining challenge is to remove the NO_x. Intensive research into lean NO_x after-treatment was initiated during the early 1980s, with the twofold purpose of delivering low levels of exhaust emissions combined with exceptionally good fuel economy. The pioneering experiments on cation-exchanged (e.g., Cu, Fe, Mn, V) mordenite, X and Y zeolites were carried out by Volkswagen during the early 1980s [111]. Initially, a Cu-exchanged mordenite was able significantly to reduce the NO_x concentration over a temperature range of 200 to 400 °C in the presence of a hydrocarbon (e.g., C_1-C_4 alkanes and alkenes) and oxygen. Due to the use of a hydrocarbon as reductant, this process became known as Hydrocarbon Selective Catalytic Reduction (HC-SCR). A second generation of these catalysts, which were based on ZSM-5 zeolites, proved to be more steam-resistant compared to mordenites.

Another way of reducing NO_x emission in heavy-duty diesel trucks by catalytic after-treatment is to use ammonia as reductant (NH_3-SCR), a technology which is well known from NO_x removal in stationary units such as power plants. Here, mainly TiO_2-supported V_2O_5- and WO_3-containing catalysts are used. There are some advantages compared to noble metal catalysts, including: (i) resistance to sulfur; and (ii) a lower cost of the non-noble metal systems. As it is unlikely that ammonia would be used in road vehicles, for NO_x removal in heavy-duty diesel trucks an aqueous urea solution would be carried in additional tanks. An acidic precatalyst decomposes the urea to produce ammonia, which is then used for the NH_3-SCR. At present, the so-called "AdBlue™" aqueous urea solution has been launched at certain EU diesel filling stations.

Another concept for the removal of NO_x from gasoline engines was developed by Toyota Laboratories during the 1990s. Toyota introduced NO_x storage catalysts (NSR) which function under lean conditions [112], such that barium carbonate traps NO_x and barium nitrate is formed on the catalyst surface. By switching the regime to rich conditions for a short period of time, the barium carbonate is recovered by its releasing NO_x, which then is converted to nitrogen through a conventional TWC reaction.

NO_x storage under lean conditions and release under rich conditions are illustrated in Fig. 7a and b, respectively [113]. Unfortunately, the severe inhibitory effect of sulfur in the fuel has so far prevented a broader commercialization, although a number of passenger vehicles equipped with such catalysts are currently running, mainly in Japan.

1.3.5.2 Removal of Harmful Emissions from Stationary Sources

Emissions removal from stationary sources mainly concerns the SCR of NO_x [109], the removal of volatile organic compounds (VOC) [114], and the decomposition of NO from nitric acid plants [109].

Titania and titania-silica-doped vanadia, molybdena, tungsta and chromia materials have been applied as SCR catalysts, with the vanadia on titania system receiving most attention. Most of these catalyst formulations are bulk materials shaped in different monolith geometries. A variety of reductants such as CH_4, CO, H_2 and NH_3 have been used, of which ammonia is the most favored. The application of such catalysts is found in power stations, waste incinerators and gas turbines, with most of the industrial experience in SCR applications having been acquired in Japan and Germany.

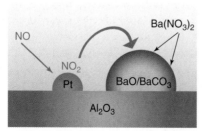

Fig. 7 NO_x storage catalyst concept. (a) Storage under lean conditions; (b) release under rich conditions. (Adopted from Ref. [113].)

The flue gas produced by coal-fired burners in power plants is passed to the SCR reactor, together with NH_3. The overall reactions over highly selective vanadia catalysts may be described by the following stoichiometry:

$$4NO + 4NH_3 + O_2 \longrightarrow 4N_2 + 6H_2O$$

$$6NO_2 + 8NH_3 \longrightarrow 7N_2 + 12H_2O$$

A diversity of VOCs are emitted from various stationary sources, mainly from chemical industries such as solvent preparation, the synthesis of organic chemicals, and miscellaneous processes.

Today, different classes of catalysts are used for gas clean-up, including supported noble metals, transition metal oxides, and zeolites [109]. Most of the VOC removal catalysts are Pt- and Pd-containing systems, while Cu-, V- or Cr-oxides are mainly used for the oxidation of halogenated compounds. VOCs are oxidized in a catalytic incinerator (often also known as the "afterburner", into CO_2 and H_2O, the efficiency of which conversion is about 70–90%.

1.3.5.3 Wastewater Treatment

Besides the removal of VOCs in the gas phase, several processes have also been developed which utilize catalysts for the clean-up of wastewater in liquid phases running under elevated pressure but at moderate temperatures [115]. Most such catalysts are noble metal-containing systems (Pt, Pd) on supports such as oxides of Al, Hf, Zr, and Ti. Various disadvantages such as leaching and sintering can affect the catalyst's lifetime, but in general the noble metal catalysts are the most active in such reactions. For acetic acid or ammonia destruction, the degree of leaching is rather low.

The catalytic wet oxidation wastewater treatment (NSLC Wastewater Treatment System) process developed by Nippon Shokubai uses Pt, Pd on TiO_2–ZrO_2 supports for the removal of acetic acid, phenol, formaldehyde, or glucose. The reaction is carried out at approximately $220\,°C$ and 40×10^5 Pa, and the reactor concept realizes a two-phase flow (segmented gas–liquid flow) through vertical monoliths. In this way, conversion rates in excess of 99% have been reported.

One technique which has been developed for wastewater clean-up from coal gasification, coking plants and the steel industry is known as the Osaka Gas Process. The system utilizes similar noble metal catalysts on TiO_2 or TiO_2–ZrO_2 supports as described above, with the reaction running at $250\,°C$ and ca. 70×10^5 Pa pressure. In this way, highly polluted wastewaters are cleaned with residence times of approximately 25 min, up to 10 mg L^{-1} total organic carbon (TOC) in the feed, and no detectable NO_x and SO_3 emissions are observed. Moreover, a catalyst lifetime of up to 8 years has been reported. A similar process was developed by the Kurita Company, using Pt on titania catalysts. At present, further applications are under development, mainly using Cu, Zn oxides on alumina or Ru on activated carbon.

1.3.5.4 Fuel Cell Catalysis

A fuel cell is an electrochemical equipment similar to a battery, but produces electricity from an external fuel supply such as hydrogen on the anode side and oxygen on the cathode side (as opposed to the limited internal energy storage capacity of a battery). Although the fuel cell effect was discovered 170 years ago, the first applications (as the alkaline fuel cell; AFC) did not become well known until the early 1960s, when NASA began to equip space probes with fuel cells for power generation. Further technological advances during the 1980s and 1990s, such as the use of polymers (Nafion) as the electrolyte, and limitations in the amounts of expensive platinum catalyst required, have made the prospect of fuel cells in consumer applications such as automobiles more or less realistic. One Canadian company, Ballard Power Systems, is a major manufacturer of fuel cells, and conducts research and development for car manufacturers. Both, The Ford Motor Company and DaimlerChrysler are major investors in Ballard, as well as many smaller automobile companies, are customers of Ballard. Among the major automobile manufacturers, only General Motors and Toyota are pursuing an internal development of fuel cells, although both Nissan and Honda initiated similar research programs in 2004.

In addition to applications in the automobile sector, many recent developments have demonstrated the potential of fuel cells in power generation. Examples range from solid oxide fuel cells (SOFC) for stationary units to proton-exchange fuel cells (PEFC) (also known as polymer electrolyte membrane fuel cell; PEMFC) or direct methanol fuel cells (DMFC) for independent power supplies in laptop computers and cameras, as well as for auxiliary units in trucks, sport boats, and others. At present, the development of hydrogen-driven automobiles as an environmentally benign transport systems suffers from a missing distribution system, high investment, the lack of rather low efficiency and high CO_2 emission in conventional H_2 manufacture.

Catalytically active materials are Ni, Ni–Ti, Pt–Pd for the anode and Pt/C, Ag or perovskites and spinels for the cathode (AFC). As the anode, PEFC uses Pt (which might be poisoned by CO), Pt/C, Pt–Ru/C and Pt alloys or Pt/C as cathode material. For molten carbonate fuel cells (MCFC), pure Ni or Ni–Cr is used as anode and Li–NiO$_x$ as cathode.

References see page 54

1.3.5.5 Photocatalytic Applications

TiO$_2$ nanoparticles are ideal and powerful photocatalysts for the elimination of pollutants in air and wastewater, due to their physical properties, chemical stability, non-toxicity, and high reactivity. In particular, the use of clean, safe and abundant solar energy for photocatalytic processes addresses ecological interests [116].

Most applications of titania can be seen in the purification and detoxification of polluted water and air, using a relatively weak ultra-violet (UV) light by oxidation. One example is the photocatalytic decomposition of pentachlorophenol into water, CO$_2$ and HCl, while another application is the titania-catalyzed removal of organic deposits on titania-coated glass and mirror surfaces. In the latter application, two effects are combined: the coating leads to a spreading of water droplets (i.e., the so-called self-cleaning Lotus effect), which is additionally supported by the titania-induced catalytic oxidative removal of organic deposits.

A further field of application is the removal of NO$_x$ from ambient air, particularly in areas of high NO$_x$ concentration (e.g., highways and motorways), by oxidation to nitrate using TiO$_2$ nanoparticles and sunlight. The TiO$_2$ nanoparticles are coated onto concrete building elements, walkway plates, etc. (the so-called NOXER application of Mitsubishi [117]), and the nitrate so formed is easily flushed away by the rain.

One future challenge may be the photocatalytic splitting of water under visible light irradiation. Since the discovery of this effect by Honda and Fujishima [118] during the early 1970s, the photocatalysis of water by TiO$_2$-derived materials has been widely studied, and a breakthrough in this field would clearly open a direct route to the use of solar energy for hydrogen production [118].

References

1. *1998 World Wide Catalyst Product, Process Licensing & Service Directory*, Catalyst Consultants Publishing, Spring House, PA, 1998.
2. I. E. Maxwell, *CatTech* **1997**, *1*, 5; 11th International Congress on Catalysis – 40th Anniversary, J. W. Hightower, W. N. Delgass, E. Iglesia, A. T. Bell (Eds.), *Studies in Surface Science and Catalysis*, Vol. 101, Part A, Elsevier, Amsterdam, **1996**, p. 1.
3. H. Heinemann, in *Handbook of Heterogeneous Catalysis*, G. Ertl, H. Knözinger, J. Weitkamp (Eds.), Vol. 1, VCH, Weinheim 1997, p. 35.
4. J. F. Roth, *Appl. Catal. A: General* **1994**, *113*, 131.
5. Extended Abstracts of the Symposium *Fossil Fuels: Reserves and Alternatives – A Scientific Approach*, Amsterdam, The Netherlands, December 9, 2004. Royal Netherlands Academy of Arts and Sciences (KNAW), 'Het Trippenhuis', Kloveniersburgwal 29, Amsterdam, Organised by the Royal Netherlands Academy of Arts and Sciences, Earth and Climate Council (RAK) and the Clingendael International Energy Programme (CIEP).
6. R. W. Bentley, M. R. Smith, Paper presented at the IAEE (International Association for Energy Economics) International Conference Prague, Czech Republic June 5–7, 2003.
7. S. G. Masters, K. M. Eriksen, R. Fehrmann, *J. Mol. Catal. A: Chemical* **1997**, *120*, 227.
8. S. A. Topham in *Catalysis – Science and Technology*, Vol. 7, J. R. Anderson, M. Boudart (Eds.), Springer, Berlin, Heidelberg, New York, 1985, p. 1.
9. A. I. Foster, P. G. James, J. J. McCarroll, S. R. Tennison, US Patent 4,163,775, assigned to BP, 1979; A. I. Foster, P. G. James, J. J. McCaroll, S. R. Tennison, US Patent 4,250,057, assigned to BP, 1981; J. J. McCarroll, S. R. Tennison, N. P. Wilkinson, US Patent 4,600,571, assigned to BP, 1986.
10. W. Ostwald, *Berg- und Hüttenmännische Rundschau* **1906**, *20*, 12.
11. P. Pickwell, *Chemistry & Industry* **1981**, *4*, 114.
12. V. A. Sadykov, L. A. Isupova, I. A. Zolotarskii, L. N. Bobrova, A. S. Noskov, V. N. Parmon, E. A. Brushtein, T. V. Telyatnikova, V. I. Chernyshev, V. V. Lunin, *Appl. Catal. A: General* **2000**, *204*, 59.
13. I. E. Maxwell, W. H. J. Stork, in *Introduction to Zeolite Science and Practice*, H. van Bekkum, E. M. Flanigen, J. C. Jansen (Eds.), *Studies in Surface Science and Catalysis*, Vol. 58, Elsevier, Amsterdam, 1991, p. 571.
14. S. T. Sie, in *Advanced Zeolite Science and Applications*, J. C. Jansen, M. Stöcker, H. G. Karge, J. Weitkamp (Eds.), *Studies in Surface Science and Catalysis*, Vol. 85, Elsevier, Amsterdam, 1994, p. 587.
15. W. W. Irion, O. S. Neuwirth, in *Ullmann's Encyclopedia of Industrial Chemistry*, 5th Ed., Vol. A18, VCH, Weinheim, 1991, p. 51.
16. G. Martino, in *12th International Congress on Catalysis*, A. Corma, F. V. Melo, S. Mendioroz, J. L. G. Fierro (Eds.), *Studies in Surface Science and Catalysis*, Vol. 130, Elsevier, Amsterdam, 2000, p. 83.
17. J. R. Rostrup-Nielsen, *Catal. Rev.-Sci. Eng.* **2004**, *46*, 247.
18. J. R. Rostrup-Nielsen, *Catal. Today* **1994**, *21*, 257.
19. J. M. Fox, *Catal. Rev.-Sci. Eng.* **1993**, *35*, 169.
20. K. Aasberg-Petersen, J.-H. Bak Hansen, T. S. Christensen, I. Dybkjaer, P. Seier Christensen, C. Stub Nielsen, S. E. L. Winter Madsen, J. R. Rostrup-Nielsen, *Appl. Catal. A: General* **2001**, *221*, 379.
21. J. R. Rostrup-Nielsen, in *Proceedings of the 15th World Petroleum Congress*, Wiley, New York, 1998, p. 169.
22. J. P. Lange, *Ind. Eng. Chem.* **1997**, *36*, 4282.
23. T. S. Christensen, I. I. Primdahl, *Hydrocarbon Process.* **1994**, *73*(3), 39.
24. T. S. Christensen, P. S. Christensen, I. Dybkjaer, J.-H. Bak Hansen, I. I. Primdahl, in *Natural Gas Conversion V*, A. Parmaliana, D. Sanfilippo, F. Frusteri, A. Vaccari, F. Arena (Eds.), *Studies in Surface Science and Catalysis*, Vol. 119, Elsevier, Amsterdam, 1998, p. 883.
25. T. S. Christensen, I. Dybkjaer, I. I. Primdahl, *Ammonia Plant Safety* **1994**, *35*, 205.
26. *Ullmann's Encyclopedia of Industrial Chemistry*, 6th Ed., Wiley-VCH, Weinheim, 2002, 40 Volumes, 30.080 pp.; *Ullmann's Encyclopedia of Industrial Chemistry*, 1st Electronic Ed., 2007.
27. K. Weissermel, H.-J. Arpe, *Industrial Organic Chemistry*, 4th Ed., Wiley-VCH, Weinheim, 2003, 491 pp.
28. R. H. Boundy, R. F. Boyer, *Styrene. Its Polymers, Copolymers, and Derivatives*, American Chemical Society Monograph No. 15, Reinhold Publishing Co., New York, 1952, p. 16.

29. R. P. Lowry, A. Aguiló, *Hydrocarbon Process.* **1974**, *53*(11), 103.
30. A. Aguiló, Ch. C. Hobbs, E. G. Zey, in *Ullmann's Encyclopedia of Industrial Chemistry*, 5th Ed., Vol. A1, VCH, Weinheim, 1985, p. 45.
31. H. Borchert, U. Dingerdissen, WO 98/05619 A1, assigned to Celanese Chemicals Europe GmbH, 1996; H. Borchert, U. Dingerdissen, R. Roesky, DE 197 17 075 A1, assigned to Hoechst AG, 1998; H. Borchert, U. Dingerdissen, R. Roesky, DE 197 17 076 A1, assigned to Hoechst AG, 1998; H. Borchert, U. Dingerdissen, DE 196 30 832 A1, assigned to Hoechst AG, 1998; H. Borchert, U. Dingerdissen, DE 197 45 902 A1, assigned to Hoechst AG, 1999; H. Borchert, U. Dingerdissen, R. Roesky, WO 98/47850 A1, assigned to Hoechst Research & Technology Deutschland GmbH & Co. KG, 1998; H. Borchert, U. Dingerdissen, R. Roesky, WO 98/47851 A1, assigned to Hoechst Research & Technology Deutschland GmbH & Co. KG, 1998; S. Jobson, D. J. Watson, WO 98/05620, assigned to BP Chemicals International Limited, 1998; S. Jobson, D. J. Watson, US Patent 6,040,474, assigned to BP Chemicals Limited 2000; S. Jobson, D. J. Watson, US Patent 6,180,821, assigned to BP Chemicals Limited, 2001; S. Zeyß, U. Dingerdissen, J. Fritch, WO 01/90043, assigned to Aventis Research & Technologies GmbH & Co KG, 2001.
32. R. P. Arganbright, R. H. Evans, *Hydrocarbon Processing and Petroleum Refiner* **1964**, *43*(11), 159.
33. R. E. Robinson, US Patent 3,190,912, assigned to National Distillers and Chemical Corporation, 1965.
34. H. Holzrichter, W. Krönig, B. Frenz, BE Patent 62 7888, assigned to Bayer, 1962.
35. K. Sennewald, W. Vogt, H. Glaser, DE Patent 1 244 766, assigned to Knapsack, 1967.
36. H. Fernholz, H.-J. Schmidt, F. Wunder, DE Patent 1 296 138, assigned to Hoechst, 1969.
37. R. A. Sheldon, H. van Bekkum, *Fine Chemicals through Heterogeneous Catalysis*, Wiley-VCH, Weinheim, 2001, 611 pp.
38. R. L. Augustine, *Catalytic Hydrogenation: Techniques and Applications in Organic Synthesis*, Marcel Dekker, New York, 1965, 200 pp.
39. P. N. Rylander, *Catalytic Hydrogenation in Organic Synthesis*, Academic Press, New York, 1979, 325 pp.
40. H.-U. Blaser, M. Studer, *Chirality* **1999**, *11*, 459.
41. B. Chen, U. Dingerdissen, J. G. E. Krauter, H. G. J. Lansink-Rotgerink, K. Möbus, D. J. Ostgard, P. Panster, T. H. Riermeier, S. Seebald, T. Tacke, H. Trauthwein, *Appl. Catal. A: General* **2005**, *280*, 17.
42. H.-U. Blaser, *Catal. Today* **2000**, *60*, 161.
43. S. Bailey, F. King, in *Fine Chemicals through Heterogeneous Catalysis*, R. A. Sheldon, H. van Bekkum (Eds.), Wiley-VCH, Weinheim, 2001, p. 351.
44. H.-U. Blaser, F. Spindler, M. Studer, *Appl. Catal. A: General* **2001**, *221*, 119.
45. H. J. Rimek, in *Methoden der Organischen Chemie*, J. Houben, E. Müller (Eds.), 4th Ed., Vol. 4,1c (Reduktion 1), Thieme-Verlag, Stuttgart, 1980, p. 105.
46. S. Nishimura, *Handbook of Heterogeneous Catalytic Hydrogenation for Organic Synthesis*, Wiley, New York, 2001, 720 pp.
47. A. Molnar, *J. Mol. Catal. A: Chemical* **2001**, *173*, 185.
48. A. O. King, R. D. Larsen, E. Negishi, in *Handbook of Organopalladium Chemistry for Organic Synthesis*, E. Negishi (Ed.), Wiley, New York, 2002, p. 2719.
49. R. L. Augustine, *Heterogeneous Catalysis for the Synthetic Chemist*, Marcel Dekker, New York, 1995, 640 pp.
50. H.-U. Blaser, C. Malan, B. Pugin, F. Spindler, H. Steiner, M. Studer, *Adv. Synth. Catal.* **2003**, *345*, 103.
51. H.-U. Blaser, A. Indolese, A. Schnyder, H. Steiner, M. Studer, *J. Mol. Catal. A: Chemical* **2001**, *173*, 3.
52. H. Lindlar, *Helv. Chim. Acta* **1952**, *35*, 446.
53. O. Mitsui, Y. Fukuoka, US Patent 4,678,861, assigned to Asahi Chemicals, 1987.
54. G. G. Joris, J. Vitrone, Jr., US Patent 2,829,166, assigned to Allied Signal, 1958.
55. B. Bachiller-Baeza, A. Guerrero-Ruiz, P. Wang, I. Rodríguez-Ramos, *J. Catal.* **2001**, *204*, 450.
56. D. V. Sokol'skii, A. M. Pak, S. R. Konuspaev, M. A. Ginzburg, L. D. Rozmanova, *Trudy Instituta Organicheskogo Kataliza i Elektrokhimii, Akademiya Nauk Kazakhskoi SSR* **1977**, *14*, 3.
57. D. J. Ostgard, M. Berweiler, S. Roder, K. Möbus, A. Freund, P. Panster, in *Catalysis of Organic Reactions*, M. E. Ford (Ed.), Chemical Industries, Vol. 82, Marcel Dekker Inc., New York, 2001, p. 75.
58. L. Baijun, L. Lianhai, W. Bingchun, C. Tianxi, K. Iwatani, *Appl. Catal. A: General* **1998**, *171*, 117.
59. T. Maki, T. Yokoyama, US Patent 4,613,700, assigned to Mitsubishi Kasei Corp., 1986.
60. R. Albert, A. Stratz, G. Vollheim, in *Catalysis of Organic Reactions*, W. R. Moser (Ed.), Chemical Industries, Vol. 5, Marcel Dekker Inc., New York, 1981, p. 421.
61. H. Hirayama, JP Patent 07 206 737, assigned to Showa Denko, 1995; T. Fleckenstein, J. Pohl, F. J. Carduck, EP Patent 300 347, assigned to Henkel KGaA, 1989; A. Y. Kadono, Y. Hattori, M. Horio, F. Nakamura, US Patent 5,763,353, assigned to Kao Corp., 1998; G. Stochniol, K. A. Gaudschun, I. Beul, D. Ostgard, P. Panster, DE Patent 101 08 842, assigned to Degussa AG, 2001; D. Ostgard, B. Bender, M. Berweiler, K. Möbus, G. Stein, US Patent 6,284,703, assigned to Degussa AG, 2001; D. Ostgard, B. Bender, M. Berweiler, K. Möbus, G. Stein, US Patent 6,489,521, assigned to Degussa AG, 2002.
62. V. Pallassana, M. Neurock, G. Coulston, *Catal. Today* **1999**, *50*, 589; G. L. Castiglioni, M. Gazzano, G. Stefani, A. Vaccari, in *Heterogeneous Catalysis and Fine Chemicals III*, M. Guisnet, J. Barbier, J. Barrault, C. Bouchoule, D. Duprez, G. Pérot, C. Montassier (Eds.), *Studies in Surface Science and Catalysis*, Vol. 78, Elsevier, Amsterdam, 1993, p. 275.
63. R. Schmid, M. Scalone, in *Comprehensive Asymmetric Catalysis I-III*, E. N. Jacobsen, H. Yamamoto, A. Pfaltz (Eds.), Springer, Berlin, 1999, p. 1439.
64. A. Baiker, W. Caprez, W. L. Holstein, *Ind. Eng. Chem. Prod. Res. Dev.* **1983**, *22*, 217.
65. J. Becker, J. P. M. Nieder, M. Keller, W. F. Hölderich, *Appl. Catal. A: General* **2000**, *197*, 229.
66. J. Barrault, S. Brunet, N. Essayem, A. Piccirilli, C. Guimon, J. P. Gamet, in *Heterogeneous Catalysis and Fine Chemicals III*, M. Guisnet, J. Barbier, J. Barrault, C. Bouchoule, D. Duprez, G. Pérot, C. Montassier (Eds.), Studies in Surface Science and Catalysis, Vol. 78, Elsevier, Amsterdam, 1993, p. 305.
67. C. D. Bellefon, P. Fouilloux, *Catal. Rev.-Sci. Eng.* **1994**, *36*, 459.
68. H. Greenfield, *Ind. Eng. Chem. Prod. Res. Dev.* **1967**, *6*, 142.
69. M. Besson, J. M. Bonnier, M. Joucla, *Bull. Soc. Chim. Fr.* **1990**, *127*, 5.
70. S. N. Thomas-Pryor, T. A. Manz, Z. Liu, T. A. Koch, S. K. Sengupta, W. N. Delgass, in *Catalysis of Organic Reactions*, F. E. Herkes (Ed.), Chemical Industries, Vol. 75, Marcel Dekker Inc., New York, 1998, p. 195.

71. F. Merger, C. U. Priester, T. Witzel, G. Koppenhoeffer, W. Harder, EP Patent 449 089, assigned to BASF AG, 1991.
72. D. J. Ostgard, S. Roder, B. Jaeger, M. Berweiler, N. Finke, C. Lettmann, J. Sauer, US Patent 6,437,186, assigned to Degussa AG, 2002.
73. J. Ph. Gillet, J. Kervennal, M. Pralus, in *Heterogeneous Catalysis and Fine Chemicals III*, M. Guisnet, J. Barbier, J. Barrault, C. Bouchoule, D. Duprez, G. Pérot, C. Montassier (Eds.), Studies in Surface Science and Catalysis, Vol. 78, Elsevier, Amsterdam, 1993, p. 321.
74. P. Baumeister, H.-U. Blaser, M. Studer, *Catal. Lett.* **1997**, *49*, 219.
75. P. Baumeister, U. Siegrist, M. Studer, EP Patent 0 842 920, assigned to Novartis AG, 1997.
76. M. Studer, P. Baumeister, WO 96/36597, assigned to Ciba Geigy AG, 1996.
77. H.-U. Blaser, U. Siegrist, H. Steiner, M. Studer, in *Fine Chemicals through Heterogeneous Catalysis*, R. A. Sheldon, H. van Bekkum (Eds.), Wiley-VCH, Weinheim, 2001, p. 389.
78. P. Baumeister, H.-U. Blaser, U. Siegrist, M. Studer, in *Catalysis of Organic Reactions*, F. E. Herkes (Ed.), Chemical Industries, Vol. 75, Marcel Dekker Inc., New York, 1998, p. 207.
79. U. Siegrist, P. Baumeister, WO 95/32941, assigned to Ciba-Geigy AG, 1995.
80. K. Tanabe, W. F. Hölderich, *Appl. Catal. A: General* **1999**, *181*, 399.
81. P. Métivier, in *Fine Chemicals through Heterogeneous Catalysis*, R. A. Sheldon, H. van Bekkum (Eds.), Wiley-VCH, Weinheim, 2001, p. 161.
82. W. F. Hölderich, G. Heitmann, *Catal. Today* **1997**, *38*, 227.
83. W. F. Hölderich, V. Taglieb, H. Pohl, R. Kummer, K. G. Baur, DE Patent 363 42 747, assigned to BASF AG, 1987.
84. U. Dingerdissen, K. Eller, R. Kummer, H. J. Lützel, P. Stops, J. Herrmann, WO 97/07088, assigned to BASF AG, 1995.
85. M. Ueshima, H. Tsuneki, *Catal. Sci. Technol.* **1991**, *1*, 357.
86. M. Misono, N. Notori, *Appl. Catal.* **1990**, *64*, 1.
87. W. F. Hölderich, *Catal. Today* **2000**, *62*, 115.
88. A. Strecker, *Liebigs Ann. Chem.* **1855**, *93*, 370.
89. K. Heyns, H. Paulsen, *Angew. Chem.* **1957**, *69*, 600.
90. K. Heyns, H. Paulsen, G. Ruediger, J. Weyer, *Fortschritte der Chemischen Forschung* **1969**, *11*, 285.
91. P. Fordham, M. Besson, P. Gallezot, *Catal. Lett.* **1997**, *46*, 195.
92. C. Bronnimann, Z. Bodnar, P. Hug, T. Mallat, A. Baiker, *J. Catal.* **1994**, *150*, 199.
93. T. Mallat, C. Bronnimann, A. Baiker, *Appl. Catal. A: General* **1997**, *149*, 103.
94. I. Bakos, T. Mallat, A. Baiker, *Catal. Lett.* **1997**, *43*, 201.
95. T. Mallat, A. Baiker, L. Botz, *Appl. Catal. A: General* **1992**, *86*, 147.
96. T. Mallat, Z. Bodnar, A. Baiker, in *Heterogeneous Catalysis and Fine Chemicals III*, M. Guisnet, J. Barbier, J. Barrault, C. Bouchoule, D. Duprez, G. Pérot, C. Montassier (Eds.), Studies in Surface Science and Catalysis, Vol. 78, Elsevier, Amsterdam, 1993, p. 377.
97. T. Mallat, Z. Bodnar, M. Maciejewski, A. Baiker, in *New Developments in Selective Oxidation II*, V. Cortés Corberán, S. Vic Bellón (Eds.), Studies in Surface Science and Catalysis, Vol. 82, Elsevier, Amsterdam, 1994, p. 561.
98. T. Mallat, Z. Bodnar, P. Hug, A. Baiker, *J. Catal.* **1995**, *153*, 131.
99. T. Mallat, A. Baiker, *Catal. Today* **1994**, *19*, 247.
100. C. Bronnimann, Z. Bodnar, R. Aeschimann, T. Mallat, A. Baiker, *J. Catal.* **1996**, *161*, 720.
101. W. F. Hölderich, in *New Frontiers in Catalysis, Proceedings of the 10th International Congress on Catalysis*, L. Guczi, F. Solymosi, P. Tétényi (Eds.), Studies in Surface Science and Catalysis, Vol. 75, Part A, Elsevier, Amsterdam, 1993, p. 127.
102. M. Taramasso, G. Perego, B. Notari, US Patent 4,410,501, assigned to Enichem, 1983.
103. B. Notari, in *Advances in Catalysis*, D. D. Eley, W. O. Haag, B. Gates (Eds.), Vol. 41, Academic Press, San Diego, 1996, p. 253; B. Notari, *Catal. Today* **1993**, *18*, 163.
104. B. Notari, in *Innovation in Zeolite Materials Science*, P. J. Grobet, W. J. Mortier, E. F. Vansant, G. Schulz-Ekloff (Eds.), Studies in Surface Science and Catalysis, Vol. 37, Elsevier, Amsterdam, 1988, p. 413.
105. U. Romano, A. Esposito, F. Maspero, C. Neri, M. G. Clerici, *Chimica e l'Industria* **1990**, *72*, 610.
106. P. Ingallina, M. G. Clerici, L. Rossi, G. Bellussi, in *Science and Technology in Catalysis 1994*, Y. Izumi, H. Arai, M. Iwamoto (Eds.), Studies in Surface Science and Catalysis, Vol. 92, Elsevier, Amsterdam, 1995, p. 31.
107. *European Catalyst Markets*, #3034-82, Frost & Sullivan Ltd., London, 1996, p. 3.
108. R. Burch, *Catal. Rev.* **2004**, *46*, 271.
109. E. S. J. Lox, B. H. Engler, in *Environmental Catalysis*, G. Ertl, H. Knözinger, J. Weitkamp (Eds.), Wiley-VCH, Weinheim, 1999, p. 1.
110. R. Fricke, M. Richter, *Nachrichten aus der Chemie* **2006**, *54*, 520.
111. A. König, W. Held, T. Richter, *Top. Catal.* **2004**, *28*, 99.
112. M. Takeuchi, S. Matsumoto, *Top. Catal.* **2004**, *28*, 151.
113. S. Matsumoto, *CatTech* **2000**, *4*, 102.
114. F. I. Khan, A. Ghoskal, *J. Loss Prevent. Proc. Ind.* **2000**, *13*, 527.
115. F. Lucjk, *Catal. Today* **1999**, *53*, 81.
116. M. Anpo, *Bull. Chem. Soc. Jpn.* **2004**, *77*, 1427.
117. Y. Murata, K. Kamitani, H. Tawara, H. Obata, Y. Yamada, US Patent 6,454,489, assigned to Mitsubishi, 2002.
118. K. Honda, A. Fujishima, *Nature* **1972**, *238*, 37.

2
Preparation of Solid Catalysts

2.1
Development of Industrial Catalysts[1]

Ewald F. Gallei*, Michael Hesse, and Ekkehard Schwab

2.1.1
Introduction

Catalytic chemistry is the heart of industrial chemistry in which chemical reactions are used to transform inexpensive raw materials into high-value products. In order to carry out a chemical transformation or a synthesis, chemical processes are used, and in more than 85% of these cases a catalyst is also employed (Fig. 1). Moreover, in excess of 80% of these catalytic processes utilize solid catalysts.

The amount of byproducts associated with catalyzed chemical transformations is much lower than with their non-catalyzed counterparts (Table 1). Therefore, catalytic processes are both more economic and more environmentally friendly. Furthermore, in every technical process the costs of the catalyst itself must be taken into account.

2.1.2
Types and Properties of Technical Catalysts

The term "catalysis" is used to describe three distinct areas: biocatalysis, homogeneous catalysis, and heterogeneous catalysis.

Biocatalysis, by enzymes or cells, is performed in the liquid phase, usually in a water-based environment. Although this is a rapidly growing area, it is strongly associated with biochemistry and is beyond the scope of this chapter. In homogeneous catalysis [1], the catalyst is dispersed on a molecular level in a liquid phase. The structure of the active center is often known, and the catalytic reaction proceeds very selectively with well-defined substances via organometallic chemistry. The chances for successful rational catalyst design using molecular modeling are much higher than for solid catalysts. The homogeneous catalyst complexes, however, are very sensitive to temperature, air, and moisture, and are normally very difficult to separate from the reaction products without destroying the expensive ligands. Especially with the last point in mind, there are strong tendencies to heterogenize homogeneous catalysts. However, this is difficult as in most cases either the catalyst activity is lower or there is a leaching of the active components.

With respect to separation, *solid catalysts* offer one major advantage. Typically, solid catalysts are much more rugged, and the catalyst, the reactants and the products are present in different phases. The catalytic reaction occurs directly at the interphase between the two phases, and consequently the catalyst and the reactants and/or products can be separated very easily. Thus, with the exception of slurry phase catalysis, a process stage for catalyst separation is not required (Fig. 2).

The disadvantages of heterogeneous catalysts are:

- The active center is normally not known.
- Molecular modeling for rational design is difficult because computer capacity allows calculations involving only up to some 500 atoms.
- The reproducible synthesis of empirically derived catalyst formulations can be difficult.

The development of solid catalysts therefore requires different concepts (see Chapter 3) in comparison with homogeneous catalysts. The fundamental strategies which are applied for the development of industrial catalysts have not changed since the time of Alwin Mittasch, and are described in a number of monographs [2–7].

---

1 A list of abbreviations/acronyms used in the text is provided at the end of the chapter.
* Corresponding author.

Handbook of Heterogeneous Catalysis, 2nd Ed.
Edited by G. Ertl, H. Knözinger, F. Schüth, and J. Weitkamp
Copyright © 2008 Wiley-VCH Verlag GmbH & Co. KGaA, Weinheim
ISBN: 978-3-527-31241-2

References see page 66

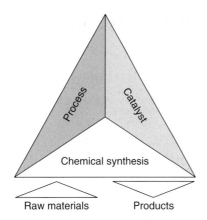

Fig. 1 The combined system process and catalyst.

In order to meet the economic requirements of an industrial catalyst (space-time yield, activity, selectivity, costs), three different structural types of solid catalyst are employed, depending on the chemical reactions to be catalyzed. In this way, it is possible to distinguish between carrier-free, supported, and coated catalysts (Fig. 3):

- *Carrier-free catalysts*, as the name indicates, consist entirely of catalytic material.
- In *supported catalysts*, the catalytically active material – which often has only a limited surface area – is anchored to and dispersed on a carrier with a high surface area. The carrier itself may also be catalytically active. One can consider a carrier as being like a sponge with high porosity and large internal surface. These catalyst types make better use of expensive active materials, such as noble metals. By using special preparation techniques, concentration profiles of the active component across the cross-section of the support can be produced similar to that for coated catalysts.
- *Coated catalysts* comprise a thin layer or shell of active material distributed across the surface of an inert carrier with low or no porosity. Typically, this shell of active material is only a few hundred micrometers thick. The reactants diffuse rapidly through this layer, so that the residence time at the active surface is very

Tab. 1 Comparison of catalyzed and non-catalyzed chemical transformations

Transformation	Reagent used in laboratory method	Byproduct	Reagent used in catalytic process	Byproduct
Hydrogenation	e.g. LiAlH$_4$ reduction	LiAlO$_2$	H$_2$	None
Oxidation	e.g. CrO$_3$ oxidation	Cr^{3+} salts	Air	None
Alkylation	e.g. AlCl$_3$	Al(OH)$_3$, HCl	None (zeolite surface)	None

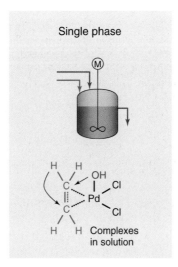
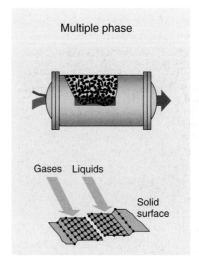

Fig. 2 Homogeneous and heterogeneous catalysis.

  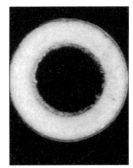

Carrier-free catalyst Supported catalyst Coated catalyst

Fig. 3 Types of heterogeneous catalyst.

short. Such catalysts find widespread application for kinetically rapid reactions such as partial oxidation and selective hydrogenation. In partial oxidations, the short residence time of the products is needed to avoid further oxidation to unwanted compounds such as CO_2; in hydrogenations, the thin layer also allows for the rapid access of hydrogen to the hydrogenation substrates.

Moreover, these three types of heterogeneous catalysts can be manufactured in many different shapes and sizes (Fig. 4). In slurry-phase catalysis the catalysts can be employed as powders suspended in liquids, whereas for use in fixed-bed reactors the usual physical forms include spheres, pellets, rings, extrudates, and honeycombs. The latter are especially useful in operations where a small decrease in pressure is required.

Before attempting to develop or improve a solid catalyst, the properties to be improved or optimized must be defined. In this respect, the most important properties of a catalyst are listed in Fig. 5.

Among the catalysts' properties, activity and selectivity are of course the most important. Other, perhaps less-important factors include suitable shape, chemical and

Fig. 4 Catalysts are available in a variety of different shapes.

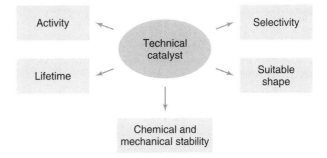

Fig. 5 Catalyst properties.

mechanical stability, and lifetime, which in turn is influenced by the stability. All of these parameters can have an enormous impact on the economics of the operation of a catalytic process. For example, the selection of an unfavorable shape, or using a catalyst with poor mechanical stability, can lead to an early plant shut down and catalyst replacement. If the catalyst is too densely packed, or if the catalyst shape breaks down – either during operations or when initially charging the reactor, which may be up to 10 meters high – the gas or liquid flow through the reactor is impeded and no longer uniform. This leads to increased requirements for compression energy, such that the resultant additional costs may endanger the economics of the plant.

The development of a solid catalyst requires knowledge not only of the performance characteristics but also of the parameters which have the greatest influence on catalyst performance. Thus, the seven most important parameters include (Fig. 6):

- Mass and heat transfer, both in the catalyst particle itself and in the reactor
- Raw materials for catalyst manufacture (impurities)
- Structure of the solid catalyst (porosity, phases)

References see page 66

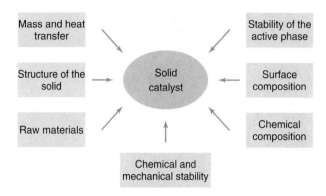

Fig. 6 The properties that control a catalyst's performance.

- Chemical and mechanical stability
- Chemical composition of the bulk
- Chemical composition of the active surface
- Stability of the active phases.

A recent report from BP's Fischer–Tropsch technology team describes an illustrative case study for the development of an industrial catalyst along the guidelines listed above. Their story describes the route from microreactor testing with a few grams of catalyst to a 300-barrel per day demonstration plant which holds 45 tons of catalyst [8].

2.1.3
Characterization of Technical Catalysts

The most important parameters to be taken into account when developing a solid catalyst are those listed above. Some of the instrumental techniques available for investigating these different parameters, together with the underlying physical and chemical structures defining them, are illustrated in Fig. 7.

The catalyst performance test in the laboratory or semi-technical reactor remains the most reliable method for evaluating a catalyst to this day. These tests must frequently be supplemented by kinetic measurements in order to determine the optimum technical reaction conditions. The testing of catalysts and the evaluation of kinetic data is carried out either on the laboratory scale or in pilot plant reactors, as illustrated in Fig. 8.

The catalyst volume employed ranges between 5 mL and several liters. For reactions performed in multi-tube reactors (e.g., partial oxidations) it is highly recommended that the tests be carried out in pilot plants with representative tube dimensions of a technical plant to ensure realistic hydrodynamic flow patterns with the corresponding mass and heat transfer.

In the case of a technical, high-performance catalyst, all of the quantifiable properties such as activity, selectivity and effective lifetime should, of course, achieve the highest possible values. In general, however, these properties are inter-dependent, and so a compromise must often be accepted in order to create an effective, but realistic, catalyst (see Figs. 6 and 15).

Today, many situations persist where there is no representative short-term test available, from which one can make a reliable assessment of the long-term behaviour of a catalyst, and it is for this reason that catalyst development may often take a long time. In order to minimize this period of time, a *correlation analysis* should be employed:

- The physical structure – that is, the pore structure of the catalyst is characterized using techniques such as BET-sorption, mercury porosimetry and scanning electron microscopy.

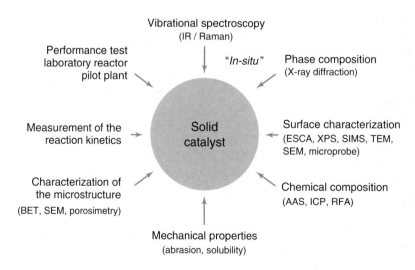

Fig. 7 Instruments for catalyst research.

Laboratory   Pilot plant

Fig. 8 Testing of catalysts.

- The mechanical and chemical stability of a catalyst are defined in terms of parameters such as abrasion resistance, crushing strength, and chemical inertness towards the reaction media.
- The chemical composition is normally determined by means of atomic or inductively coupled plasma (ICP) spectroscopy and X-ray fluorescence.

The characterization of a catalyst's surface and bulk involves techniques such as electron spectroscopy (ESCA or XPS), scanning tunnel electron microscopy (STM), transmission electron microscopy (TEM), scanning electron microscopy (SEM), and electron probe microanalysis (EPMA). Surface and solid state characterization can also be carried out by infrared (IR), Raman, and NMR spectroscopy. The latter methods, in addition, have the advantage that the analysis can be performed under "*in-situ*" conditions. The same is valid for X-ray diffraction to investigate the bulk phase composition.

In addition, a number of recently developed instrumental techniques have led to further breakthroughs in heterogeneous catalysis.

2.1.3.1 *In-Situ* Experiments

In-situ measurements provide a deeper insight into the catalyst solid-state system under operating conditions. For such an investigation, a set-up is used in which X-ray diffraction (XRD) measurements are taken from catalyst samples operated at close to real conditions in a miniature reactor, with the sample being placed in the X-ray beam (see Fig. 9). The presence of active sites on catalytic surfaces is strongly determined by the underlying bulk phase. *In-situ* XRD provides information on the crystal structure of the bulk phase and its stability under reaction conditions.

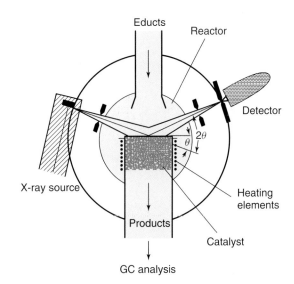

Fig. 9 *In-situ* X-ray diffraction equipment.

One example of using an *in-situ* method is determining the phase changes during the lifetime of a heteropolyacid catalyst (HPC) used for the oxidation of methacrolein to methacrylic acid. Previously, this catalyst demonstrated only a short lifetime, and it was found that the segregation of stable MoO_3 from the catalytic active phase of a metastable heteropoly compound is responsible for the decline in catalyst activity (Figs. 10 and 11). The catalyst (Fig. 11a), within short time, showed a significant increase in MoO_3 content with operation time. By using these findings, it was possible – by means of doping elements – to develop a catalyst (Fig. 11b) that showed a much higher stability and hence a longer lifetime.

References see page 66

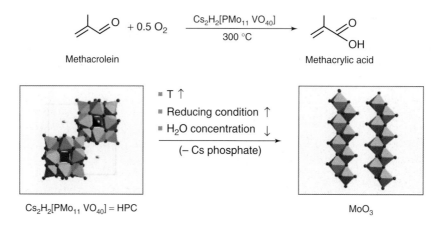

Fig. 10 Structural changes of a heteropolyacid catalyst (HPC).

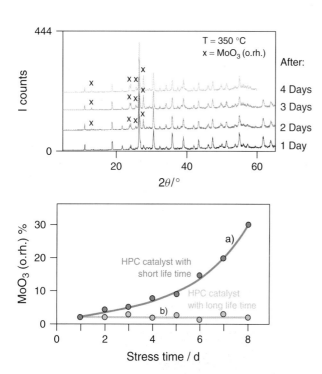

Fig. 11 *In-situ* experiments with two heteropolyacid catalysts (HPCs). (a) With a short lifetime; (b) with a longer lifetime.

2.1.4
High-Throughput Experimentation (HTE)

In order to facilitate and accelerate the discovery of potential catalysts, high-throughput experimentation (HTE) can be used both in the preparation and testing of catalysts. The iterations of model, design, preparation and testing are similar to those of conventional catalyst development (Fig. 12), although the number of catalysts evaluated at each iteration are, by a factor of 10 to 1000, higher in HTE.

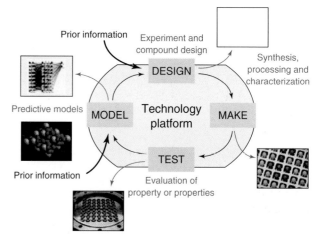

Fig. 12 Iteration cycle in catalyst development. (Illustration courtesy of hte, Heidelberg, Germany.)

To achieve these numbers, the parallel synthesis of catalyst samples must be applied, with pipetting robots being used to mix large numbers of metal salt compositions or to impregnate hundreds of carrier substrates with these solutions (Fig. 13). The thus-prepared manifold samples may then be thermally treated to obtain the catalysts for testing.

For testing, these samples are then transferred to a high-throughput reactor. Such reactors usually consist of many tubes in parallel, which can be filled and analyzed using gas chromatography (GC) or GC-mass spectrometry (MS) separately, but operated in a single reactor system (Fig. 14).

Other approaches include reactor systems with different temperature zones or reactor discs, onto which catalyst powder is fixed as a thin film in small fields, and which are then analyzed by monitoring the heat of reaction on each catalyst field of that disc. In any case, using

Fig. 13 Parallel synthesis of catalysts. (Illustration courtesy of hte, Heidelberg, Germany.)

these systems means generating a very large amount of data. Thus, the information technology (IT) systems managing these data and reducing them in such a way as to identify the best catalysts in each run are of utmost importance.

2.1.5
Identification of Technical Catalysts

Neither the catalyst nor process can be seen as being independent of each other. The chemical synthesis can only be optimized by optimizing the combined system process/catalyst, and for this reason it is essential that, when developing an industrial synthesis, the catalyst development is started at an early stage. This requires close collaboration between the synthetic chemist, catalyst researcher, and process engineer. It has been said that, "...designing catalysts to satisfy process needs is like writing prescriptions to cure illness" [7].

Therefore, before a catalyst development is started, the goals and objectives must be defined, there being four different scenarios:

- The development of a new catalyst for a new process.
- The development of improved catalysts for existing processes.
- The scale-up of laboratory recipes into technical catalyst production.
- Elucidating the reasons in case of catalyst failure.

As mentioned previously (see Fig. 6), the development of a catalyst must consider not only the chemical reaction, but also the mass and heat transfer effects within the catalyst particle and in the reactor. It is therefore evident that a successful optimization of a catalyst system is only possible in combination with a search for the most favorable operating conditions, or by tailoring the catalyst to the existing conditions. However, the latter approach is the less-preferred alternative. The structure of the solid – that is, the pore structure and the type, structure, and distribution of the active phases of the catalyst – have a major influence on its performance,

References see page 66

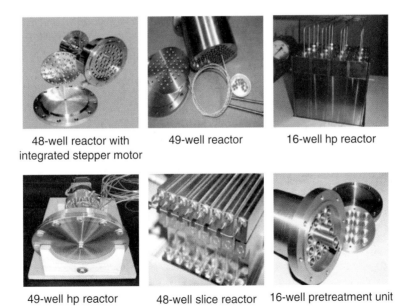

48-well reactor with integrated stepper motor | 49-well reactor | 16-well hp reactor

49-well hp reactor | 48-well slice reactor | 16-well pretreatment unit

Fig. 14 Parallel reactors. (Illustration courtesy of hte Heidelberg and MPI für Kohlenforschung, Germany.)

through the interaction of heat and mass transfer and the chemical reaction.

The choice of raw materials for a catalyst production process can lead to a variety of negative effects; such an example is when poisons are inadvertently introduced during the catalyst's manufacture. In addition, adequate chemical and mechanical stabilities are preconditions for a reasonable lifetime for the catalyst within the technical plant.

The chemical composition of the bulk solid phase determines which active elements are present in the catalyst. Modern high-performance catalysts are mostly multicomponent systems comprising several active phases doped with numerous promotors. However, the integral chemical composition of the solid is not in itself sufficient to ensure high activity and selectivity. Rather, it is necessary to determine a definite chemical composition of the active surface (both, internal and external) in terms of functional active sites. Similarly, a high morphological stability of the active phases is a precondition for an acceptable effective lifetime of a catalyst.

Accordingly, the main objective of catalyst development is optimization of the various different catalyst properties, as illustrated in Fig. 15.

An essential precondition for achieving such an objective is a close cooperation between experts working in very different fields of scientific research – as can be seen in Fig. 16.

Experts in the field of reactor technology must select the appropriate types of reactor, so that the technical process is correctly tailored to the catalytic reactions. An equivalent input must also be obtained from experts within each of the other fields listed above.

As highlighted above, the performance test is the most important method of catalyst evaluation. Normally, all other physical techniques help to determine which structural parameters of the catalyst have changed during operation, or even during the preparation process. The

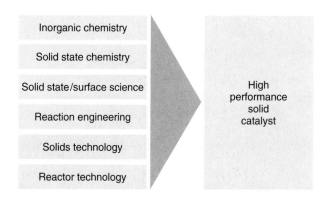

Fig. 16 The development of a solid catalyst.

physical techniques provide an indication of the most favorable structural properties of a high-performance catalyst, so that improvements can be made in terms of the reproducibility of technical production. On the other hand, they are only of limited value in developing a new catalyst starting from scratch. These techniques provide no indication as to how certain catalyst properties can be achieved. Accordingly, when developing a new catalyst one must proceed as shown in Fig. 17.

Starting with a detailed analysis of the problem, a working hypothesis on the basis of the available know-how and relevant information in the literature is set up. A decision must then be made as to which are the most appropriate methods for characterizing and testing the catalysts. In this connection, a considerable range of techniques for microstructure analysis is of particular significance, with the most important point being correlation analysis – that is, the establishment of any relationships between preparation parameters, structural properties and performance data of the catalyst. When a catalyst has been prepared successfully on both the laboratory and semi-technical scales, the recipe must be scaled up to the technical production level – a step which is often very problematic. Naturally, if a high-performance catalyst cannot be prepared reproducibly on a laboratory

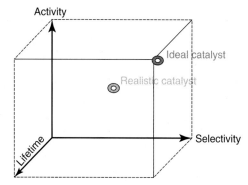

Fig. 15 A compromise must be found between the target values.

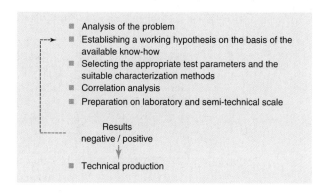

Fig. 17 The stages of catalyst development.

or pilot plant scale, then a new working hypothesis must be established.

2.1.6
Production of Technical Catalysts

Currently, two principal routes exist for the production of technical catalysts (see Fig. 18), and in excess of 80% of all technical catalysts are manufactured according to these procedures.

The first route, yielding carrier-free catalysts, starts with the precipitation of a catalyst precursor, which is filtered, dried, and shaped. Calcination steps may be included after drying or after shaping. The second route starts with a carrier material, which can be impregnated with salt solutions or coated with powders, these powders often stemming from a route one-type preparation. Drying and calcination give the final catalyst. The carriers used can range from powders, granules, or extrudates to tablets of various forms and honeycombs or foam-like structures.

Techniques of inorganic, organometallic and solid-state chemistry are used in the initial preparation of the catalyst in the laboratory. The physical methods of solid-state and surface science are applied to characterize the bulk catalyst and the surfaces where the reactions take place. The successful production of a catalyst on a technical scale requires a comprehensive knowledge in the area of solids handling technology, and this is particularly true during the catalyst scale-up from the laboratory to the technical plant (see Fig. 16). Most reports relating to the synthesis of industrial catalysts have been disclosed in a huge number of patents, a compilation of which (for the most important industrial catalytic processes) has been provided [9]. Unfortunately, however, these studies have not been extended to more recent years.

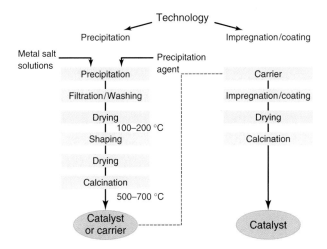

Fig. 18 Preparation routes for technical catalysts.

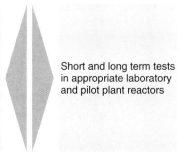

Fig. 19 Correlation analysis of catalysts.

Nonetheless, a number of alternative techniques are employed in catalyst preparation, including fusion and chemical vapor deposition (CVD) or physical vapor deposition (PVD). Recently, flame pyrolysis has also been used for carbon-supported precious metal catalysts in fuel cell applications.

The correlation analysis is described schematically in Fig. 19, where an attempt is made to establish the relationships between the catalyst production process and the resulting chemical and physical properties on the one hand, and the catalyst performance on the other hand.

The wide variety of parameters capable of influencing catalyst performance has already been stressed. Important factors include the production process, the raw materials and, consequently, the chemical composition, the distribution of the active components, the characteristics of the various phases and of the surface, as well as the geometric shape of the catalyst.

2.1.7
Conclusions

From the above discussions, the following conclusions can be drawn for the development of industrial catalysts:

- Catalysts play a key role in the value chains of the chemical industry.
- The development of catalysts is an interdisciplinary task.
- Surface science and modeling are necessary tools, but they alone do not allow "*ab-initio*" development.
- Performance tests in pilot plants are the most reliable method for performance evaluation.
- Technical catalysts are defined not only by the active sites but also by their physical environment and accessibility, as well as their interaction with the operation parameters and reactor systems used.

All in all the, saying of Alwin Mittasch, the grandseigneur of the art of catalysis still holds true today:

References see page 66

*"Chemistry without catalysis would be
a sword without handle,
a light without brilliance,
a bell without sound."*

List of Abbreviations

AAS	atomic absorption spectrometry
CVD	chemical vapor deposition
EPMA	electron probe microanalysis
ESCA	electron spectroscopy for chemical analysis
NMR	nuclear magnetic resonance
PVD	physical vapor deposition
RFA	X-ray fluorescence analysis
SEM	scanning electron microscopy
SIMS	secondary ion mass spectrometry
STM	scanning tunnel electron microscopy
TEM	transmission electron microscopy
XPS	X-ray photoelectron spectroscopy
XRD	X-ray diffraction

References

1. M. Röper, *Chemie in unserer Zeit* **2006**, *40*, 126–135 (in German).
2. A. Mittasch, *Geschichte der Ammoniaksynthese*, Verlag Chemie, Weinheim, 1951.
3. C. H. Collier (Ed.), *Catalysis in Practice*, Reinhold Publishing Corporation, New York, 1957.
4. D. L. Trimm, *Design of Industrial Catalysts*, Chem. Eng. Monographs, Vol. 11, Elsevier, 1980.
5. N. Pernicone, F. Traina, *Applied Industrial Catalysis*, Vol. 3, Academic Press, 1984, pp. 1–24.
6. J. F. Le Page, *Applied Heterogeneous Catalysis*, Chapters 4 & 5, Edition Technip, 1987.
7. J. T. Richardson, *Principles of Catalyst Development*, Plenum Press, New York, 1989.
8. J. J. H. M. Font Freide, T. D. Gamlin, C. Graham, J. R. Hensman, B. Nay, C. Sharp, *Top. Catal.* **2003**, *26*(1–4), 3–12.
9. M. Sittig, *Handbook of Catalyst Manufacture*, Chemical Technology review # 98, Noyes Data Corporation, 1978.

2.2
Computer-Aided Design of Solid Catalysts

*Martin Holena and Manfred Baerns**

2.2.1
Introduction

Our understanding of heterogeneous catalysis has increased significantly during the past 10 years, due mainly to the application of *in-situ* spectroscopies to surface and bulk processes, and to advanced computational methods based on ab-initio quantum chemistry techniques. The results of such investigations with regards to catalysis allow the identification not only of interactions of gaseous or liquid reactants with the solid catalytic surface, but also of changes in the electronic and structural properties of the solid catalytic material. Moreover, elementary catalytic reaction steps can be frequently discovered. The information derived from these processes provides essential indications for designing a catalyst based on certain materials of catalytic impact. Such a procedure often does not lead to an optimum catalyst, however, and this is applicable to both qualitative and quantitative compositions, as well as to the method of preparation. Against this background, empirical methods – which make use of fundamental knowledge – are still required in the quest for an optimal catalyst.

In this chapter we present a brief introductory overview of the results of the theoretical methods mentioned above, but no further mention is made of the various spectroscopic techniques described elsewhere in this handbook [1]. The main emphasis of this chapter is placed on a combination of fundamental and empirical knowledge in preparing new catalysts, as well as on the design of catalytic experiments and the assessment of experimental data. A careful evaluation of the catalytic results by suitable data mining procedures may add to our fundamental understanding of catalysis [2, 3]. Combinatorial methodologies were introduced into the development of solid catalysts during the late 1990, at a time when they were already well known from the viewpoints of organic synthesis and enzymatic catalysis. One major advantage of the combinatorial techniques is their ability to cover of a wide range of variables during the development process; these variables include mostly the composition of the catalytic material, as well as other properties of the solid material, and finally also the reaction conditions to which the catalytic material is exposed (see Section 2.2.3). An extensive overview of related subjects, notably with respect to organic chemistry, has been provided elsewhere [4]. So-called expert systems, which have been described earlier [4–7], have not proven to be especially successful in heterogeneous catalysis; hence, they are not considered within this chapter.

2.2.2
Theoretical Methods

In an earlier review, the use of ab-initio molecular-orbital calculation had already been discussed for the elucidation of catalytic processes; in this way, the method offers a potential for predicting the catalytic

---

* Corresponding author.

properties of solid materials and hence, in the widest sense, for designing catalysts [8]. Likewise, it was pointed out later [9] that an understanding of the catalytic and adsorptive properties of solid materials at atomic and electronic levels is essential in the design of novel catalysts. In this way, computer simulations can significantly contribute to a rational interpretation of experimental results, and they can also suggest modifications of new catalytic materials. In 2004, Motoki and Shiga [10] developed a reaction simulator, which is an intermolecular interaction analyzer based on the theories of paired interacting orbitals and localized frontier orbitals. For specific situations, activities and selectivities, as well as molecular weight in olefin polymerization, can be predicted. By using density functional theory (DFT) calculations, various catalytic performances became predictable; for example, the difference in dehydrogenation activity for cyclohexane by Pt and Ni [11], as well as the activation of C–H, C–C and C–I bonds by Pd and cis-Pd(CO)$_2$I$_2$ [12].

A first-principles method was illustrated by Linic et al. [13] for the identification of bimetallic alloy catalysts. The elementary reaction steps for ethylene epoxidation were derived from surface science experiments, and DFT calculations used for the catalytic chemistry. This molecular-level mechanistic information was used as an input in computational screening of potential bimetallic alloy catalysts that might offer greater selectivity to ethylene oxide than the traditional monometallic silver catalyst. In this way, a formulation of a novel Cu/Ag alloy leading to improved selectivity was found which was later verified experimentally. The main point of discussion with regards to this example was the confirmation of the great value of first-principles studies in catalyst design.

The above-described methods are often referred to as *combinatorial computational chemistry approach*, a term introduced by Kubo et al. [14]. Originally, the term had been applied mainly to the synthesis of organic compounds, but was then extended to include drugs and enzyme catalysis; today, it is successfully applied also to inorganic chemistry and hence, to the preparation of solid catalysts. Examples of this approach include the design of metal catalysts for methanol synthesis [15] and deNO$_x$ chemistry [16].

The selected studies mentioned above indicate that significant progress has been made in computational chemistry, and these advances are progressively coming to bear on catalyst design. In the past, computational chemistry usually served the purpose of elucidating the physico-chemical properties of materials, which then further contributed to their improved design by computation. Unfortunately, first-principle quantum-chemical calculations, which represent the most powerful approach, suffered from certain disadvantages, the main problem being that as the quality of the computational results depends largely on the assumed cluster size of the catalytic material, significant amounts of computer time are needed for large clusters. A variety of simplifications have been introduced, such as the neglect of long-range electrostatic interactions. More recently, Selvam et al. [17] dealt with the implications related to materials development and introduced a new approach, termed the *tight-binding quantum chemical molecular dynamics method*, whereby the computing time was reduced by a factor of 5000. This method can obviously be applied not only to single-site but also to nano-scale catalysis. In their report, the authors described comprehensively the design of materials and an understanding of its performance for a variety of applications, of which only those related to catalysis are mentioned. These include chemical reaction dynamics over organometallic catalysts, electronic states of supported metal catalysts, interface characteristics of precious metals-on-zirconia catalysts, chemical reaction dynamics on a Ziegler–Natta catalyst, adsorption and electron transfer dynamics on a metal surface, and chemical reaction dynamics of materials-synthesis processes. The effective simulation of a number of physical and chemical systems at reaction temperatures is of special importance, as such simulations cannot yet be studied by static first-principles calculations and classical molecular dynamics simulations.

In similar manner to the studies described above, a concept of micro-kinetic analysis of heterogeneous catalysis was pioneered by Stoltze and Norskov [18, 19], as well as by Dumesic et al. [20]. For the kinetics of the elementary reaction steps the kinetic parameters were derived by computational studies. Hereby, the catalytic performance, especially of catalytic metal sites, could be predicted, and this then served as a basis for selecting catalytic elements for the reaction under consideration.

In spite of all the recent progress that has been made in computational chemistry coupled with combinatorial approaches, there is an ongoing demand for experimental studies related to catalyst design. By introducing the concept of combinatorics to computational chemistry, a theoretical high-throughput screening of catalysts has become feasible, as long as sufficient knowledge can be made available for carrying out the computational chemistry calculations. Unfortunately, at present this is still not the case in many situations, and consequently a variety of empirical combinatorial approaches have been established. However, these require the rational design of experiments using all available empirical

References see page 79

and fundamental knowledge, a suitable analysis of the catalytic data obtained from these experiments, and their subsequent use in the design of new experiments to develop improvements in catalytic performance. Moreover, the data analysis comprises extensive data mining in order to extract all available empirical and fundamental knowledge. No mention is made here of the experimental details of any high-throughput experimentation based on the design of experiments, but the interested reader is referred to extensive reference material available on the subject [21, 22] (see also Chapter 9.3).

2.2.3
Rational Methods for the Design of Catalytic Experiments

Many properties and conditions influence the ability of a material to serve as the catalyst for a particular reaction. First, the composition of the material is important, it being defined by the present chemical elements or compounds, in addition to their respective masses or molar fractions within the material. In addition, the material's structure and texture, both of which are mainly determined by the preparation method used, play important roles. The composition and structure of the bulk and the surface are interconnected by further properties of the material, such as acidity, basicity, redox potential, electronic conductivity; all such properties are normally referred to as *descriptors* (cf. [23, 24]). The catalytic performance of a material depends not only on the material itself, but also on the conditions to which it is exposed in the reaction, at a particular temperature, at partial pressures of the reactants, total pressure, and the space velocity of the feed. Taken together, the descriptors of the material and the reaction conditions are termed the *input variables* in the sequence.

In a catalytic experiment, a certain number of catalytic materials are tested, and these are chosen from a very comprehensive set of potential materials. The number of materials to be examined is affected by the following aspects:

- The *ranges* of the individual input variables.
- Whether a particular input variable can have *only particular prespecified values*, or whether it can assume *any value* within the corresponding range, restricted only through the finite discernibility due to experimental error.
- The *experimental error* of a particular input variable. As there always exists some experimental error for a continuous variable, the number of discernible values within its range is always finite, but may be quite high. For example, a range of fractions of 0 to 20% and an experimental error of 0.1% will entail 201 discernible values.
- *Constraints* on input variables, such as the constraint that the fractions of all components should sum up to 100%, or constraints on the number of non-zero fractions. The latter actually express constraints on the number of components in a catalytic material, for example, two to three active components plus one dopant.

2.2.3.1 The Statistical Design of Experiments

The task of finding, for a given set of potential catalytic materials, a subset of representatives which convey the required information about the whole set has, for almost a century, been addressed by methods of *statistical design of experiments* (DOE). A very simple situation called *factorial design* occurs if the required information represents the impact of any combination of possible values of some n independent factors. If for $i = 1, \ldots, n$ the i-th factor can assume f_i different values, the factorial design needs $f_1 \cdot \ldots \cdot f_n$ representatives. As an example, let a catalytic material have a fixed fraction of support chosen from two possibilities, active components chosen from three possibilities, among which one is always present, whereas each of the remaining two either is absent or its ratio to the first one can assume one of three given values, and can have one dopant which either is absent or assumes a given value. Then $n = 4$ independent factors exist, $f_\text{support} = 2$, $f_\text{2nd active component} = f_\text{3rd active component} = 4$, $f_\text{dopant} = 2$, and the factorial design needs:

$$f_\text{support} \cdot f_\text{2nd active component} \cdot f_\text{3rd active component}$$
$$\times f_\text{dopant} = 64 \text{ representatives.}$$

On the other hand, assuming that there is no interaction between different factors and the only information required is about the impact of any single factor in isolation, one needs only $n + 1$ representatives. In the above example, n + 1 = 5.

In both of the above methods, the number of representatives can vary from experiment to experiment, depending on the number of independent factors, and in factorial design also on the number of their possible values. *Computer-aided statistical DOE methods* have been developed only since the 1970s, and these are used to solve more ambitious and computationally much more demanding tasks: the main target here is to choose, for any given set of possible materials, a *representative subset of particular size such that the amount of required information about the whole set is maximal among all subsets of that size*.

Among the computer-aided DOE methods available, the *D-optimal design* is the most frequently used. This relies on a model assuming a dependent variable, such as yield, conversion or selectivity, which depends linearly on input variables, and interactions between them. The information maximized in the D-optimal design is an information measure called the *Fisher information* for that model; more precisely, it is the determinant of the matrix of that information measure (actually, the term "D-optimal" relates to the fact that the determinant is maximized). Moreover, from statistics it is known that maximizing this information is equivalent to minimizing the volume of the confidence ellipsoid for parameter estimates in the underlying linear model. To illustrate these principles again by an example, a catalytic material is considered that consists of a support material and four active components. For this material, it is assumed that the yield y_i of the reaction catalyzed with the i-th catalytic material from the representative subset depends on the fraction $x_{i,0}$ of the support and the fractions $x_{i,0}$, $x_{i,1}$, $x_{i,2}$, $x_{i,3}$ of the active components 1, 2, and 3 in this material. The fraction $x_{i,4}$ of the active component 4 is obtained from the condition $x_{i,0} + x_{i,1} + x_{i,2} + x_{i,3} + x_{i,4} = 1$. Moreover, the yield y_i is also assumed to depend on interactions between any two of the fractions $x_{i,1}$, $x_{i,2}$, $x_{i,3}$, and on interactions between $x_{i,0}$ and any pair of those three active-component fractions. Rewritten in a formula, these assumptions read:

$$y = \beta_0 x_{i,0} + \beta_1 x_{i,1} + \beta_2 x_{i,2} + \beta_2 x_{i,3} + \beta_{01} x_{i,0} x_{i,1}$$
$$+ \beta_{02} x_{i,0} x_{i,2} + \beta_{03} x_{i,0} x_{i,3} + \beta_{12} x_{i,1} x_{i,2}$$
$$+ \beta_{13} x_{i,1} x_{i,3} + \beta_{23} x_{i,2} x_{i,3} + \beta_{012} x_{i,0} x_{i,1} x_{i,2}$$
$$+ \beta_{013} x_{i,0} x_{i,1} x_{i,3} + \beta_{023} x_{i,0} x_{i,2} x_{i,3}$$

where $\beta_0, \beta_1, \ldots, \beta_{023}$ are the unknown model parameters. The information, which is maximized in the D-optimal design, is then the determinant $\det(R'R)$, with R denoting the 13-column matrix with rows

$(x_{i,0}, x_{i,1}, x_{i,2}, x_{i,3}, x_{i,0}x_{i,1} \ldots x_{i,2}x_{i,3}, x_{i,0}x_{i,1}x_{i,2},$

$x_{i,0}x_{i,1}x_{i,3}, x_{i,0}x_{i,2}x_{i,3})$

and R' representing the transpose of R. As was recalled above, maximizing that information is equivalent to minimizing the volume of the confidence ellipsoid for the estimates of $\beta_0, \beta_1, \ldots, \beta_{023}$.

For a comprehensive explanation of various statistical DOE methods, competent monographs are available, both of a general form [25–27] and those addressing specifically chemical applications [28, 29]. Examples of such applications in catalytic research can be found elsewhere [30–34].

2.2.3.2 Optimization Methods for Empirical Objective Functions

Different statistical methods for the design of experiments use different criteria to choose experimentally tested materials from among those considered possible for testing. Nonetheless, the respective criteria are always applied uniformly to the entire set of possible candidate materials. This is to be quite impractical if the most interesting catalytic materials are not distributed uniformly in the parameter space of possible materials, but rather form only one or several small clusters. This often occurs when interest is expressed only in catalytic materials of sufficiently high performance (in terms of yield, conversion, selectivity, etc.). Indeed, high values of such performance measures are typically achieved only in small areas of contiguous compositions (Fig. 1).

In such situations, it is more relevant to use a method designed specifically to seek, for any given objective function, locations at which the function values are maximized. Since each such method equivalently seeks locations in which the negative of the objective function takes its minimal value, the more general terms *optima/extremes of the objective function* are used rather than *maxima* and *minima*. The process of searching such optima by means of a corresponding algorithm is termed *function optimization*, and the involved computer-aided methods *optimization methods*. Before presenting an overview of these types of methods, two introductory points should be clarified:

- Whether a location is an optimum of the objective function in the space of input variables depends on

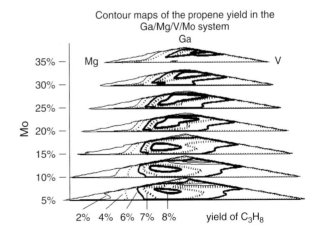

Fig. 1 Example of dependency of yield on catalyst composition. This example shows the dependency of propene yield on the fractions of Mg, V, Ga, and Mo in the oxidative dehydrogenation of propane. The approximation was obtained by means of artificial neural networks on data analyzed in Ref. [35].

References see page 79

other locations with which it is compared. On occasion, the function value in a location is not surpassed by values in locations within a certain neighborhood, yet outside that neighborhood there are locations with a higher function value. For example, varying the fraction of individual elements in the catalytic material within a certain small range does not lead to a material with a better performance, whereas varying some of those fractions outside that range can lead to such a performance. The material is then said to have a *locally optimal performance* or, in mathematical terms, to be a *local optimum* of the function describing the dependency of its performance on the composition. On the other hand, if even varying any of the fractions within the whole range of their admissible values does not lead to a better performance, the considered catalytic material is said to correspond to a *global optimum* of that function.

- In most other applications (including, e.g., reaction kinetics), values of the objective function may be obtained analytically – that is, either as the result of setting the function input into a mathematical expression, or as the solution of an equation described with a mathematical expression (e.g., of a differential equation). In contrast, values of functions describing the dependence of catalyst performance on its composition are obtained *empirically*, through experimental measurements.

2.2.3.2.1 **Evolutionary Methods** Evolutionary optimization methods are stochastic methods, which means that the available information about the objective function is complemented with random influences. The term "evolutionary" refers to the fact that the way of incorporating random influences into the optimization process has in those methods been inspired by *biological evolution*. The most frequently used (and most deeply elaborated) representatives of evolutionary methods are *genetic algorithms* (GAs; see below), in which the incorporation of random influences attempts to mimic the *evolution of a genotype*. Basically, that method comprises:

- randomly exchanging coordinates between two particular locations in the input space of the objective function (*recombination, crossover*);
- randomly modifying coordinates of a particular location in the input space of the objective function (*mutation*); and
- *selecting* the locations for crossover and mutation (parent locations) according to a probability distribution, either uniform or skewed towards locations at which the objective function takes high values (the latter being a probabilistic expression of the survival-of-the-fittest principle).

Detailed treatment of various types of genetic algorithms, as well as of other evolutionary optimization methods, can be found in specialized monographs [36–45]. In this chapter, only those four features of GAs will be highlighted which are particularly important for the development of catalytic materials.

Feature 1 The meaning of the individual coordinates of locations in the input space of the objective function is strongly problem-dependent. In catalyst design, the coordinates typically convey some of the following meanings:

(i) The *qualitative composition* of the catalytic material; that is, of which active components does it consist, whether it contains dopants (and which ones), whether it is supported, and what is its support.
(ii) The *quantitative composition* of the catalytic material; that is, the fractions of the various components mentioned in (i).
(iii) The *preparation of the catalytic material*, its individual steps and their quantitative characterizations, such as temperatures or times they need.
(iv) The *reaction conditions* of the catalyzed reaction.

There is an intimate connection between qualitative and quantitative composition of catalytic materials. The presence of a particular component in the catalytic material is equivalent to the fraction of that component being non-zero. There are consequences for the algorithms accomplishing the operations of recombination and mutation. Namely, the algorithms must guarantee that this equivalence cannot be invalidated through the respective operation. For example, if the presence of a particular component in one of the parent materials, and the absence of that component in the other parent material, are exchanged during recombination, then the fraction of that component has also to be changed at the same time. Similarly, if a mutation eliminates a certain component from the catalytic material, the fraction of that component must be set simultaneously to zero. In this context, it is useful to differentiate between *quantitative mutation* (which modifies only the quantitative composition of the catalytic material, without affecting its qualitative composition) and *qualitative mutation* (which modifies its qualitative composition) (Fig. 2).

Feature 2 Crossover and mutation operations can be applied to many individuals simultaneously; thus, the GA can *follow many optimization paths in parallel*. Moreover, optimization proceeds between subsequent iterations for different paths independently. Because of the biological inspiration of GAs, individual iterations of a GA are

Crossover

Co = 0	Cr = 0	Cu = 40	Mn = 0	Mo = 34	P = 0	Sn = 11	W = 15
Co = 0	Cr = 0	Cu = 32	Mn = 0	Mo = 17	P = 18	Sn = 10	W = 23

↓

Co = 0	Cr = 0	Cu = 40	Mn = 0	Mo = 17	P = 18	Sn = 10	W = 15
Co = 0	Cr = 0	Cu = 32	Mn = 0	Mo = 34	P = 0	Sn = 11	W = 23

Qualitative mutation

Co = 0	Cr = 0	Cu = 32	Mn = 0	Mo = 17	P = 18	Sn = 10	W = 23

↓

Co = 0	Cr = 0	Cu = 49	Mn = 0	Mo = 21	P = 0	Sn = 12	W = 18

Quantitative mutation

Co = 0	Cr = 0	Cu = 32	Mn = 0	Mo = 17	P = 18	Sn = 10	W = 23

↓

Co = 0	Cr = 0	Cu = 48	Mn = 0	Mo = 15	P = 1	Sn = 16	W = 20

Fig. 2 Illustration of operations used in genetic algorithms; the values in the examples are mass fractions of active elements in the catalytic material expressed in mol%. (Reproduced from Ref. [46].)

referred to as *generations*, and all locations in which the value of the objective function is considered in a particular generation (e.g., all catalytic materials of which the performance has been measured in that generation) are denoted as *population*. The fact that GAs follow many optimization paths in parallel is actually the main reason for their attractiveness in high-throughput catalyst development, because a straightforward correspondence can be established between those optimization paths and channels of the high-throughput reactor in which the materials are experimentally tested.

Feature 3 As GAs do not use derivatives, they are not attracted to local optima. On the contrary, the random variables incorporated into recombination, mutation, and selection enable the optimization paths to *leave the attraction area of the nearest local optimum*, and to continue *searching for a global one*.

Feature 4 Due to the biological inspiration of the GAs, a particular distribution of the incorporated random variables cannot be justified mathematically, but its choice is an heuristic task. The most important heuristic parameters describing such a distribution are the *overall probability of any modification* of an individual, the *ratio between the conditional probabilities* of crossover and qualitative or quantitative mutation, conditioned on any modification, and the *distribution of the intensity of quantitative mutation*. The *population size* is also sometimes a matter of heuristic choice, although in the practice of catalyst development it is usually determined by the number of channels in the testing reactor.

As a consequence of the problem-dependency of GA inputs, for solving optimization problems in specific application areas such as in the development of catalytic materials, it is quite difficult to use general GA software, such as Matlab's Genetic Algorithm and Direct Search Toolbox [47]. Indeed, such general GA software optimizes only functions within input spaces of low-level data types, examples of which are vectors of real numbers and bit-strings. However, encoding the qualitative and quantitative composition of catalytic materials, their preparation and reaction conditions with low-level data types is tedious and error-prone. Moreover, it requires a great deal of mathematical erudition.

For the reasons highlighted, it is not surprising that – apart from early attempts to use general GA software to optimize the distribution of active sites of two catalytic components [48, 49] – the application of GAs in this area has followed the way of developing specific algorithms for the optimization of catalytic materials. The first such algorithm for solid catalyst development was implemented by Wolf et al. during the late 1990s [50–52], followed by a modified version in 2001 [53, 54]; similar algorithms were later also developed at other institutions [2, 32, 55–59]. Experience gained with those algorithms shows that any new method of catalyst design can substantially decrease the usefulness of an earlier implemented specific GA after several years. Maintaining a high usefulness of the implementation for a long time requires (during the GA's development) the anticipation of all catalytic materials and their preparation methods for which the implemented algorithm might need to be used in the future. Nevertheless, there is no guarantee that all of these factors can be anticipated during development. In addition, the more possibilities the implementation of a GA attempts to cover, the broader the

References see page 79

class of possible distributions of the incorporated random variables. Thus, according to Feature 4 (see above), the greater becomes the number of possible combinations of parameters of those distributions that need to be heuristically chosen. In order to avoid that problem, a recent proposal was made to generate implementations of specific GAs by a software system, according to a description of the problem in a machine-readable description language [46].

2.2.3.2.2 Other Stochastic Methods Several other stochastic methods have been developed for the optimization of an objective function; these include simulated annealing, tabu search, multilevel single linkage, topographical optimization, stochastic hill climbing, stochastic tunneling, stochastic branch and bound [60, 61]. Like GAs, these methods typically only compare function values at different locations; hence, they tend to find global optima rather than local ones. On the other hand, as function values alone contain only little information, the search for the optimum progresses only very slowly.

Simulated annealing has sometimes been used in heterogeneous catalysis, although less frequently than GAs [49, 62–65]. Basically, the way in which the simulated annealing method searches for the global minimum of an objective function mimics the way in which a crystalline substance reaches the ground state with minimal energy in a process of heating and then slowly cooling. The method works by iteratively proposing changes and either accepting or rejecting each of them. There are various criteria for the acceptance or rejection of proposed changes, the most often encountered being a criterion called *Metropolis*. This accepts the change unconditionally if the objective function decreases; otherwise it accepts it only with an *exponentially distributed probability* that depends on a driving-force parameter being called *temperature*, which is equal to the mean of that exponential distribution. Thus, if the temperature is very large, nearly all changes are accepted and the method simply moves the system through various states, irrespective of the values of the objective function. Consequently, if the method starts at a higher temperature, which is then gradually decreased, it brings the system to a state with a minimum of the objective function, but still allows it to escape that state if the minimum is not global (hence, the temperature is not zero). In particular, when searching for catalyst materials with high performance, the simulated annealing method can leave a locally optimal catalyst, and continues with lower-performance materials, in order to ultimately identify a globally optimal catalyst.

2.2.3.2.3 Deterministic Methods From the overall perspective of function optimization, the plethora of deterministic methods is much more frequently used than stochastic methods [66, 67]. In those methods, the optimization of a function relies solely on the available information about its response surface, with no randomness involved. Hence, the same starting location of the optimization procedure always leads to iteration through the same locations in the input space of the objective functions. Before discussing the role of deterministic methods in the optimization of catalytic materials it should be recalled that, according to the information they use, all such methods could be divided into three large groups.

Group 1 Methods These include methods that use only information about function values. As mentioned above in the context of stochastic optimization, such methods tend to find a global optimum rather than a local one (in contrast to Groups 2 and 3, below), and they are very slow. Their most frequently encountered example is the *simplex method*. This produces a sequence of simplexes; that is, of polyhedra with $(n + 1)$ vertices, where n refers to the dimension of the input space of the objective function. The vertices of the first simplex of the sequence are typically chosen at random, and any further simplex is obtained from the previous one through replacing the vertex in which the value of the objective function was the worst (i.e., the lowest value in the case of maximization, and the highest value in the case of minimization). Other methods of this group, used in the design of catalytic materials, include the *holographic research strategy* and the *sequential weight increasing factor technique*.

Group 2 Methods These include methods which use, in addition to function values, information about first partial derivatives; that is, about the gradient. The gradient of a function has the property that its direction coincides with the direction of the fastest increase of function values, and is opposite to the direction of their fastest decrease. Thus, if the optimization path follows from some location in the input space the direction of the objective function gradient, then the function value increases with the highest possible speed along that path (at least in the immediate neighborhood of that location). Similarly, if the path follows the direction opposite to its gradient, then function values decreases with the highest possible speed. However, following the direction of the gradient (or opposite to the gradient) of the objective function does not generally allow the optimization path to reach its global maximum (or minimum), but only a local one. More precisely, for every maximum (global as well as local) there exists an area that an optimization path will never leave if it starts within that area and follows in each location the direction of the gradient of the objective function. The area is called the *attraction area* of the considered maximum, and the maximum is said to be an *attractor* of that area. Global and local minima are also attractors, although the optimization paths in their attraction areas follow in each location the direction opposite to the gradient. Simple examples of methods of this group are various variants of *steepest descent*, which directly employs the above recalled property of gradients and searches for a minimum of the function in such a way, that the next location in a sequence of iterations is chosen in the direction opposite to the gradient in the current location. Individual variants of this method differ in their particular way of choosing the next location in that direction. More sophisticated representatives of this group include several *methods of conjugate gradients*.

Group 3 Methods These include methods which use, in addition to function values, information about partial derivatives up to the second order. Like gradient-based methods, they search for a maximum or minimum of the objective function only within its attraction area. However, second-order derivatives allow the construction of a quadratic approximation of the function, whereas gradient approximates the function only linearly. Close to a maximum or minimum, a quadratic approximation of a function is more accurate than a linear one, and therefore methods of this group can localize searched optima faster, once they reach their proximity. On the other hand, linear approximations are frequently more accurate far from any optima. Therefore, methods of this group are most often used as combined methods, which switch between the behavior of second-order methods close to an optimum and the behavior of a gradient-based method far from the optimum. The most frequently encountered method of that type is the *Levenberg–Marquardt method*, whereas using only second-order derivatives leads in the most simple case to the *Gauss–Newton method*.

For the optimization of empirical functions in the development of solid catalytic materials, it is, unfortunately, impossible to employ methods using partial derivatives – that is, methods from Groups 2 and 3. The reasons for this impossibility are connected with the *character of catalytic experiments*, and can be summarized as follows:

- As a *mathematical expression* for the dependency of performance of the catalytic material on the various input variables is not known, mathematical expressions for its partial derivatives cannot be obtained either.

- To obtain sufficiently accurate *numerical estimates* of the partial derivatives, small differences between values of the dependent variable must be recorded; typically, differences should not be larger than 0.1% of the function value in the respective location. However, for the dependent variables occurring in catalyst optimization (e.g., yield, conversion, selectivity), such differences commonly lie within the *experimental error*.

- Even if the experimental setting would lead to a lower experimental error (or if sufficient accuracy of the estimates were not required), obtaining numerical estimates of the gradient must be given up for practical reasons. Indeed, to obtain the numerical estimate of the gradient in any location of an n-dimensional input space of the objective function would require empirical evaluation of the function in $(n+1)$ locations located *very close to each other*. For example, imagine that the objective function describes the dependence of yield on the composition of catalytic material on choosing its components from a pool of 15 elements. Then, to estimate its gradient numerically in one single location would require testing 16 materials with nearly the same composition. That is clearly not affordable for reasons of cost and time.

Therefore, in heterogeneous catalysis only deterministic optimization methods of Group 1 have been occasionally used [64, 68–70]. On the other hand, in the context of high-throughput development of solid catalysts, it was pointed out that their best known representative – the simplex method – allows an easy adaptation to produce any prescribed number of sequences of simplexes in parallel [64]. Consequently, the method can propose in each step a *prescribed number of new catalytic materials in parallel*, which provides a straightforward correspondence with their subsequent testing in a multichannel reactor, as in the case of evolutionary algorithms.

2.2.3.3 Concluding Remarks

The choice of computer-aided methods for the design of catalytic experiments depends on whether the space of potential catalytic materials should be investigated in some systematic way, or whether the search for catalysts with the best performance is its only objective. The former case is a task for statistical DOE methods, whereas optimization methods are appropriate for the latter. Within the computer-aided branch of statistical DOE methods, the D-optimal design is most universally employed. An optimization method has to require solely function values to be applicable to the optimization of catalytic materials. Among such methods, GAs are most

---
References see page 79

frequently used, especially those developed specifically for catalytic optimization.

One problem with GAs is the crucial role of their heuristic parameters, a correct tuning of which requires a great deal of preliminary knowledge. In part, that knowledge can be obtained by means of data analysis and data mining methods, which will be surveyed in the following section. Therefore, GAs are frequently *combined and interleaved* with such methods, especially with neural networks [54, 56–58, 68, 71–73].

2.2.4
Data Analysis and Data Mining

The term *data analysis* is traditionally used for statistically assessing the extent to which data support particular relationships between the various objects, properties and phenomena that they describe. Until the early 1970s, researchers hypothesized such relationships, and computers were employed only for their assessment. However, the steadily increasing amount of experimental data, which became available following the advent of database technology, required an automated, data-driven search for such relationships within a new branch of statistics, referred to as *exploratory statistics*. On the other hand, the increasing significance of computers also led to a rapid development of *artificial intelligence* and *machine learning*, which attempt to mimic the way in which human beings themselves hypothesize relationships from observations. The most important types of machine learning methods include rule-based methods, decision trees, inductive logic programming, support vector machines, and artificial neural networks (ANNs). Methods based on ANNs have been frequently used also in heterogeneous catalysis (see Section 2.2.4.2 below). Although methods of exploratory statistics and machine-learning rely on different paradigms, they deal with the same data and employ the same information technologies, especially databases, the world-wide web, and various object-oriented technologies (Fig. 3). Together, they form a new interdisciplinary area known as *data mining*. In the context of heterogeneous catalysis, the distinction between traditional data analysis and data mining should be taken into account when choosing between the two conceptually different approaches to the design of libraries of solid catalysts – the optimization approach and the discovery approach [74].

2.2.4.1 Statistical Methods: An Overview
The simplest use of statistical methods is to provide summary parameters characterizing important statistical properties of input variables and of various measures

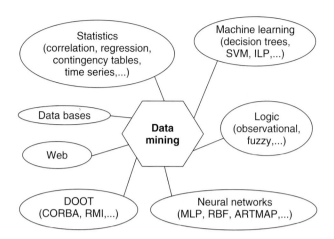

Fig. 3 Main data mining approaches and supporting technologies. ILP, inductive logic programming; MLP, multilayer perceptron; RBF, radial base functions network; ARTMAP, adaptive resonance theory mapping network; DOOT, distributed object-oriented technologies; CORBA, common object request broker; RMI, remote method invocation.

of catalyst performance (yield, conversion, etc.), or relationships between them. Such summary parameters are usually called *descriptive statistics*; their common representatives are *mean*, *median* and other *quantiles, variance, standard deviation, skewness, kurtosis, covariance* and *correlation*.

The two last-mentioned descriptive statistics – covariance and correlation – allow one to summarize the relationship between a performance measure and a particular input variable. The situation becomes substantially more complicated if one is interested in the relationship between a performance measure and a whole set of input variables. Indeed, then not only parameters corresponding to individual variables, but also parameters corresponding to various levels of interactions between them are needed. In such situations, the parameters are usually combined with an assumption about the form of the dependency of the performance measure on the input variables (such as linear dependency, polynomial dependency, or dependency derived from some theoretical model), for once an assumption about its form is made, the parameters already fully determine that dependency. In statistics, this approach is known as *regression* or *response surface modeling*, and the parameters determining the dependency are called *regressors*. For example, the regression of yield y on three component fractions x_1, x_2, x_3 uses in the case of a *linear regression* four parameters $\alpha_0, \alpha_1, \alpha_2, \alpha_3$:

$$y = \alpha_0 + \alpha_1 x_1 + \alpha_2 x_2 + \alpha_3 x_3$$

whereas in the case of a quadratic regression, it uses 10 parameters $\alpha_0, \alpha_1, \alpha_2, \alpha_3, \alpha_{1,1}, \alpha_{1,2}, \ldots, \alpha_{3,3}$:

$$y = \alpha_0 + \alpha_1 x_1 + \alpha_2 x_2 + \alpha_3 x_3 + \alpha_{1,1} x_1^2 + \alpha_{2,2} x_2^2$$
$$+ \alpha_{3,3} x_3^2 + \alpha_{1,2} x_1 x_2 + \alpha_{1,3} x_1 x_3 + \alpha_{2,3} x_2 x_3$$

Descriptive statistics can also not be used in the case of other properties of catalytic materials that depend on a whole set of input variables. Examples of such properties are similarity between different materials, dependence on unobservable factors, or classification of materials according to their catalytic behavior in particular reactions. In order to characterize such properties, methods of *multivariate analysis* are needed. The multivariate approaches which are most relevant to the analysis of catalytic data are:

- *Principal component analysis*, which reduces data dimensionality through concentrating on those linear combinations of input variables that are most responsible for the variability of the data set (unfortunately, such combinations usually do not convey any real meaning).
- *Factor analysis*, which explains input variables as combinations of a smaller number of unobservable factors.
- Analysis of the relationship between factors influencing input variables and those influencing performance measures by means of an approach called *partial least squares*.
- *Cluster analysis*, which entails grouping catalytic materials into clusters or a hierarchy of clusters according to similarity among the values of the input variables or according to a similar catalytic performance, which can be measured with various *similarity measures*.
- *Classification* of new materials, according to values of their input variables, with respect to their usability as catalytic materials in particular reactions. An important feature of classification is that the discrimination between classes relies solely on data with known correct classification, for example, on already tested materials.

Frequently, the primary purpose of data analysis is to check the compatibility of the available data with certain assumptions about the probability distribution governing those data. To this end, methods of *statistical hypotheses testing* are needed. The most commonly tested hypotheses include:

- that the probability distribution of data belongs to a certain family of distributions, for example normal distributions or exponential distributions
- that certain parameters characterizing the distribution, such as mean or variance, have a particular value, or that their value lies within a particular range
- that probability distributions governing two data sets are identical, or that some of their parameters are identical.

One important example of testing hypotheses of the last-mentioned type is an analysis of the influence of varying the values of individual input variables on the performance of a catalytic material by means of an approach called *analysis of variance*.

For detailed information about statistical methods, the reader is referred to comprehensive monographs about statistics [75–78], or chemometrics [79, 80]. Recent applications of statistical methods to heterogeneous catalysis have been reported in [23, 24, 33, 34, 81–88].

2.2.4.2 Artificial Neural Networks

ANNs are distributed computing systems that attempt to implement a greater or smaller part of the functionality characterizing biological neural networks. Their most basic concepts are a *neuron*, the biologically inspired meaning of which is an elementary signal processing unit, and a *connection* between neurons enabling the transmission of signals between them. In addition to signal transmission between different neurons, signal transmission between neurons and the environment can also take place. Neurons, connections between them, and connections between the environment and neurons form together the *architecture* of the ANN. Those neurons that receive signals from the environment but not from other neurons are called *input neurons*, whereas those sending signals to the environment but not to other neurons are called *output neurons*. Finally, neurons that receive and send signals only from and to other neurons are called *hidden neurons* (Fig. 4).

Partitioning of the set of neurons into input, hidden, and output neurons allows a large variety of architectures; however, for almost all types of ANNs frequently encountered in practical applications, the architecture is basically the same. It is a *layered architecture*, in which all signals propagate from layer to layer in the same direction, denoted as forward direction. During that propagation, a *neuron-specific linear transformation* of signals from the preceding layer is performed in each hidden or output neuron, whereas a *network-specific nonlinear function*, called *activation function*, is subsequently applied to the results of the linear transformations in hidden and

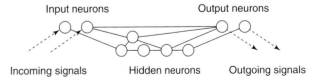

Fig. 4 Simple generic artificial neural network architecture. (Reproduced from Ref. [102].)

References see page 79

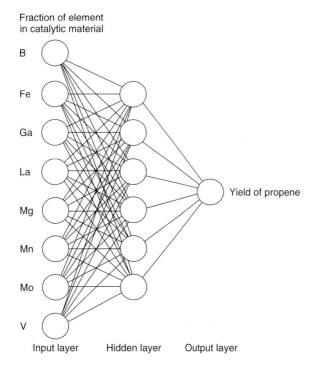

Fig. 5 Example of multilayer perceptron with one layer of hidden neurons, employed to approximate an unknown dependency of propene yield on catalyst composition. (Reproduced from Ref. [102].)

sometimes also in output neurons. Such ANNs are called feedforward neural networks, and their most prominent example is the *multilayer perceptron* (MLP; see Fig. 5), in which the activation functions are S-formed functions known as *sigmoidal functions*.

A MLP with n_I input neurons and n_O output neurons computes a function from the space of n_I-dimensional vectors to the space of n_O-dimensional vectors. The precise form of that function depends on the specific linear transformations connected with individual hidden and output neurons. It is the set of computable functions that accounts for the most useful feature of multilayer perceptrons, which is their *universal approximation property*. This property means that even to a *very general unknown dependency*, and to any prescribed arbitrarily small distance in an appropriate function space, a MLP can always be found, with as few as one layer of hidden neurons, such that the function computed by this MLP lies within the prescribed distance from the unknown dependency. In catalysis, one is most often interested in dependencies of product yields, reactant conversions, and product selectivities on input variables.

In order to obtain a neural network that computes such a function, two crucial steps are needed:

1. To choose an appropriate architecture which, in the case of a MLP, means choosing an appropriate number of hidden layers plus an appropriate number of neurons in each of them.
2. To connect, in a process called *network training*, to each hidden and output neuron the linear transformation that fits the available data best. The quality of the overall fit of all those linear transformations may be assessed using various measures [89]. The most commonly used measure is the mean sum of squared errors; that is, the mean sum of squared distances between the output values that the network computes for a given sequence of inputs, and the output values that for those inputs have been obtained experimentally.

In reality, the choice of the number of hidden neurons cannot be separated from network training, since to assess the appropriateness of a particular architecture for the available data, a number of networks with that architecture must first be trained.

A better fit to the data used for training does not necessarily entail a better approximation of the ultimate unknown dependency. To assess the quality of the desired approximation, an independent set of data called *test data* is needed, obeying the unknown dependency but unseen by the network during training. The phenomenon of *overtraining* – that is, of a good fit to the training data accompanied by a bad fit to the test data – is the main problem faced during neural network training. Up to now, various methods have been developed that reduce that phenomenon, such as *early stopping* or *Bayesian regularization* (Fig. 6). Nevertheless, to fully exclude an influence of overtraining to the choice of the number of neurons, that choice must be based on the fit to the test data. (For details, see Ref. [89].)

The architecture of a trained neural network and the mapping computed by the network inherently represent the knowledge contained in the data used to train the network, i.e., deriving and knowledge about relationships between the input and output variables, for example, about relationships between component fractions and yield. Such a representation is not very human-comprehensible, however, as it is far from the symbolic and modular manner in which people represent knowledge by themselves. Therefore, various *knowledge extraction* methods have been developed with the aim of transforming the network architecture and the computed mapping into a representation that is more acceptable to humans, namely into representation by means of *logical rules*. For dependencies of output variables on input variables, logical rules are of the general form:

- IF the input variables fulfill an input condition C_{input}

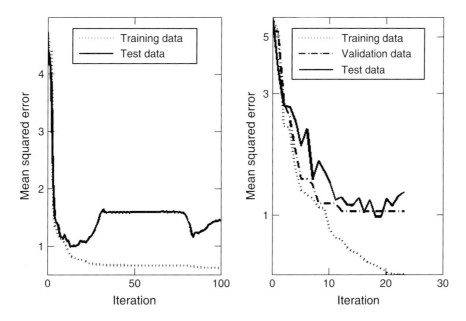

Fig. 6 An example of overtraining encountered when training a MLP with the data analyzed in Ref. [35], using the Levenberg–Marquardt training method (left), and an example of overtraining reduction using the early-stopping method (right). (Reproduced from Ref. [102].)

- THEN the output variables are likely to fulfill an output condition C_{output}.

For examples of such rules, see Fig. 7.

Detailed information about ANNs is available in specialized monographs [90–96]. Applications of multilayer perceptrons to catalysis on solid materials have been reported in [23, 24, 57, 87, 97–107], along with applications of other types of neural network [23, 56, 58, 71, 86, 87, 108].

2.2.4.3 Concluding Remarks

In Section 2.2.4 we dealt with the relationships between various objects, properties and phenomena described by catalytic data, which primarily are relationships between input variables and catalyst performance in particular reactions. The relationships to be investigated may either be hypothesized by the researcher, or searched for based on the available data. In the former case, statistical data analysis methods are used, whereas in the latter, data mining methods are appropriate, notably methods of exploratory statistics or machine learning. Among the latter, ANNs have been frequently used in catalytic research. As the area of machine learning is still undergoing rapid development, other relevant methods may become comparably important in the next few years, including *decision trees* [23, 24, 86, 87] and *support vector machines* [109].

Both, data analysis and data mining are performed in a stepwise manner, and the result of any one particular

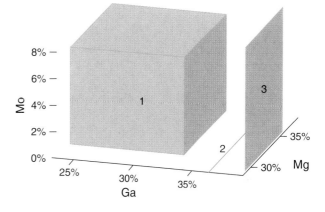

Fig. 7 Visualization of the three-dimensional projection to the dimensions corresponding to Ga, Mg, and Mo of the antecedents (left-hand sides) of three logical rules extracted from an artificial neural network trained with data on oxidative dehydrogenation of propane. (Reproduced from Ref. [102].)
Rule 1:
IF the concentrations fulfill 24% $\leq$ Ga $\leq$ 33% AND 31% $\leq$ Mg $\leq$ 39% AND 0% $\leq$ Mo $\leq$ 7% AND the remaining concentrations are 0%, THEN yield of propene $\geq$ 8%
Rule 2:
IF the concentrations fulfill Ga $\approx$ 36% AND 28% $\leq$ Mg $\leq$ 38% AND the remaining concentrations are 0%, THEN yield of propene $\geq$ 8%
Rule 3:
IF the concentrations fulfill 0% $\leq$ Fe $\leq$ 12% AND Ga $\approx$ 38% AND 29% $\leq$ Mg $\leq$ 36% AND 0% $<$ Mo $<$ 9% AND the remaining concentrations are 0%, THEN yield of propene $\geq$ 8%.

References see page 79

Tab. 1 Application of computational tools in catalyst design, data assessment and optimization

Computational tools for experimental design and data assessment	Catalytic reactions	Potential catalytic elements, compounds and supports	References	Year
First principles, periodic DFT calculations	Ethylene to ethylene oxide	Cu/Ag alloys, α-Al_2O_3	[13]	2004
Combinatorial computational DFT chemistry	Methanol synthesis	Co, Cu, Ru, Pd, Ag, Re, Os, Ir, Pt, Au as metals and cations	[15]	2002
Combinatorial computational chemistry based on DFT	Fischer–Tropsch synthesis	Fe_6 clusters Additives for Fe_5M M = Fe, Si, Mn, Ge, Zr, Mo	[14]	2004
Combinatorial computational chemistry based on DFT	$DeNO_x$ reaction	Rh, Pd, Ag, Ir, Au as clusters	[16]	2004
Combinatorial computational chemistry based on DFT	$DeNO_x$ reaction	Cu^+, Ag^+, Au^+, Fe^{2+}, Co^{2+}, Ni^{2+}, Pd^{2+}, Pt^{2+}, Cr^{3+}, Fe^{3+}, Ir^{3+}, Tl^{3+} ZSM-5	[111]	2000
GA, factorial design	Paraffin (C_5 to C_6) isomerization	Me oxide supports: γ-Al_2O_3, ZrO_2, TiO_2 (+ SO_4^{2-}, BO_3^{3-}, PO_4^{3-}, WO_x. Metals: Pt, Ce, Pd, Sn, Ni, Mn, Nb	[32, 55]	2003
Factorial design, multivariate analysis	Zeolite synthesis	(NaOH, KOH, 1,4-dibromopentane, 1-methylpiperidine, Sylobloc 47 (Grace), aluminium-tri-iso-propoxide	[33]	2003
GA	Low-temperature oxidation of low-concentration propane in air	Pt, Pd, Rh, Ru, Au, Cu, Ag, Mn on TiO_2	[52]	2001
GA, ANN	Oxidative dehydrogenation of propane	Redox MeO of V, Mo, Mn, Fe, Ga; non-reducible MeO of La; acidic & basis MeO of B, Mg.	[54]	2004
GA, ANN	Methanol synthesis	Oxides of Cu, Zn, Al, Sc, B, Zr (at different preparation condi-tions)	[56, 73]	2004
GA, ANN, clustering and factorial analysis	Epoxidation of cyclohexene by *tert*-butylhydroperoxide	Synthesis conditions of Ti-silicate-based catalysts	[58, 86]	2005
GA	Selective oxidation of CO	Au, Cu, Pt and oxides of Mo, Nb, V supported on TiO_2, ZrO_2, CeO_2	[59]	2005
GA, ANN	Oxidative coupling of methane	Oxides of Na, W, P, S, Zr, Mn	[68, 100]	2003, 2001
ANN	Oxidation of butane	MgO, Al_2O_3, SiO_2, TiO_2, ZnO, ZrO_2, SnO_2, Bi_2O_3	[97]	1995
ANN	Decomposition of NO into N_2 and O_2	Cu_x/ZSM−5x: variable & different conditions	[98]	1995
ANN	Propane ammoxidation	Oxides of V, Sb, W, Sn (P, K, Cr, Mo) on Al_2O_3/SiO_2	[99]	1997
ANN	CO_2 hydrogenation to hydrocarbons	Ag, Al, Au, Ce, Co, Cr, Cu, Fe, Ga, La, Mn, Mo, Ni, Pd, Pt, Ru, Si, Ti, W, Zn, Zr; supports ZrO_2, Al_2O_3, ZnO, TiO_2	[101]	2001
GA, ANN	Oxidative dehydrogenation of ethane	Oxides of Ga, Cu, Mn, Mo, W, Sn, Cr, Co, Zr, Ca, La, Au	[53, 72]	2003, 2002
Holographic research strategy and ANN	Total oxidation of methane	Oxides of Co, Zr, Cr, La, Cu, Pt, Pd, Au; Support CeO_2, La_2O_3	[110]	2005
Factorial design, ANN	Selective CO oxidation in excess H_2	Co_x/$SrCO_3x$: variable	[107]	2005

GA, genetic algorithms; ANN, artificial neural networks.

step might then be used in later steps. This is especially apparent in the following two situations:

- The objective of an initial stage of data analysis or data mining is to determine which input variables play the most important role in the investigated relationships, or which of them play in that relationship similar roles. Such an initial stage is called *feature selection*, and this leads ultimately to the *dimensionality reduction* of the space of input variables, allowing subsequently a

more thorough analysis in that less-dimensional space [23, 24, 86].
- The results of data analysis are used in the next step to improve catalyst optimization [2]. In particular, ANNs are frequently combined with GAs [54, 56–58, 68, 71–73] or other optimization methods [49, 68, 110].

2.2.5
Conclusions

Today, computer-aided catalyst design includes a comprehensive arsenal of tools which are used to predict and optimize the catalytic performance of materials, either on a theoretical basis (i.e., by combinatorial computational catalytic chemistry) or on an experimental basis. The latter approach includes the design of experiments and the assessment of experimental data by computer-aided methods originating from statistics, optimization, and machine learning. The fundamentals of the various procedures have been applied in many cases; selected examples, which have been reported in the literature are detailed in Table 1.

In the present authors' opinion, knowledge within the broad field of computer-aided catalyst design has progressed significantly during the past five to 10 years, and this trend will surely continue as it offers a high potential for further development. As a joint effort, combinatorial computational catalytic chemistry and high-throughput experimentation will undoubtedly foster catalyst development in the years to come.

It is also believed that the results of high-throughput experimentation based on combinatorial approaches, together with sophisticated data analysis, will in the long run contribute to a better understanding of catalysis and supplement our fundamental knowledge gained by other means.

References

1. G. Ertl, H. Knözinger, F. Schüth, J. Weitkamp (Eds.), *Handbook of Heterogeneous Catalysis*, 2nd edn, Wiley-VCH, Weinheim, 2007.
2. J. M. Caruthers, J. A. Lauterbach, K. T. Thomson, V. Venkatasubramanian, C. M. Snively, A. Bhan, S. Katare, G. Oskarsdottir, *J. Catal.* **2003**, *216*, 98.
3. O. V. Buyevskaya, A. Brückner, E. V. Kondratenko, D. Wolf, M. Baerns, *Catal. Today* **2001**, *67*, 369.
4. J. Gasteiger (Ed.), *Handbook of Chemoinformatics*, Vol. 3, Wiley-VCH, Weinheim, 2004, pp. 490.
5. T. Hattori, Y. Murakami, *Appl. Catal.* **1989**, *48*, 107.
6. E. Körting, M. Baerns, in *Proceedings of the Symposium of the Materials Research Society*, Vol. 454, Boston, 1997, p. 187.
7. E. Körting, M. Baerns, in *Handbook of Heterogeneous Catalysis*, G. Ertl, H. Knözinger, J Weitkamp (Eds.), Vol. 1, Wiley-VCH, Weinheim, 1997, p. 419.
8. H. Kobayashi, *Catal. Today* **1991**, *10*, 167.
9. S. C. Ammal, S. Takami, M. Kubo, A. Miyamoto, *Bull. Mater. Sci.* **1999**, *22*, 851.
10. T. Motoki, A. Shiga, *J. Comput. Chem.* **2004**, *25*, 106.
11. M. Tsuda, W. A. Dino, S. Watanabe, H. Nakanishi, H. Kasai, *J. Physics. Cond. Mat.* **2004**, *16*, S5721.
12. A. Diefenbach, F. M. Bickelhaupt, *J. Organomet. Chem.* **2005**, *690*, 2191.
13. S. Linic, J. Jankowiak, M. A. Barteau, *J. Catal.* **2004**, *224*, 489.
14. M. Kuobo, T. Kubota, C. Jung, M. Ando, S. Sakahura, K. Yajima, K. Seki, R. Belosludov, A. Endou, S. Takami, A. Myiamoto, *Catal. Today* **2004**, *89*, 479.
15. S. Sakahura, K. Yajima, R. Belosludov, S. Takami, M. Kumo, A. Myiamoto, *Appl. Surface Sci.* **2002**, *189*, 253.
16. A. Endou, C. Jung, T. Kusagaya, M. Kuobo, P. Selvam, A. Myiamoto, *Appl. Surface Sci.* **2004**, *223*, 159.
17. P. Selvam, H. Tsuboi, M. Koyama, M. Kuobo, A. Myiamoto, *Catal. Today* **2005**, *100*, 11.
18. P. Stolze, J. K. Norskov, *Phys. Rev. Lett.* **1985**, *55*, 2502.
19. P. Stolze, J. K. Norskov, *J. Catal.* **1988**, *110*, 1.
20. J. A. Dumesic, D. A. Rudd, L. A. Aparicio, J. E. Bekoske, A. A. Trevino, *The Microkinetics of Heterogeneous Catalysis*, American Chemical Society, Washington, 1993, pp. 316.
21. U. Rodemerck, M. Baerns, in *Basic Principles in Applied Catalysis*, M. Baerns (Ed.), Springer, Berlin, 2003, p. 259.
22. A. Hagemeyer, P. Strasser, A. F. Volpe (Eds.), *High-Throughput Screening in Chemical Catalysis*, Wiley-WCH, Weinheim, 2004, pp. 319.
23. C. Klanner, *Evaluation of Descriptors for Solids*, PhD thesis, Ruhr-University Bochum, Bochum, 2004, pp. 198.
24. C. Klanner, D. Farrusseng, L. Baumes, M. Lengliz, C. Mirodatos, F. Schüth, *Angew. Chem. Int. Ed.* **2004**, *43*, 5347.
25. G. E. P. Box, W. G. Hunter, J. S. Hunter, *Statistics for Experimenters. An Introduction to Design, Data Analysis, and Model Building*, Wiley, New York, 1978, pp. 653.
26. A. C. Atkinson, A. N. Donev, *Optimum Experimental Designs*, Oxford University Press, 1992, pp. 328.
27. P. D. Fukenbusch, *Practical Guide to Designed Experiments. A Unified Modular Approach*, Dekker, New York, 2005, pp. 197.
28. S. N. Deming, S. L. Morgan, *Experimental Designs: A Chemometric Approach*, 2nd edn, Elsevier, Amsterdam, 2005, pp. 454.
29. R. Carlson, *Design and Optimisation in Organic Synthesis*, 2nd edn, Elsevier, Amsterdam, 2005, pp. 596.
30. M. Nele, A. Vidal, D. L. Bhering, J. C. Pinto, V. M. M. Salim, *Appl. Catal. A: General* **1999**, *178*, 177.
31. R. Ramos, M. Menendez, J. Santamaria, *Catal. Today* **2000**, *56*, 239.
32. A. Corma, J. M. Serra, A. Chica, *Catal. Today* **2003**, *81*, 495.
33. M. Tagliabue, L. C. Carluccio, D. Ghisletti, C. Perego, *Catal. Today* **2003**, *81*, 405.
34. R. J. Hendershot, W. B. Rogers, C. M. Snively, B. A. Ogannaike, J. Lauterbach, *Catal. Today* **2004**, *98*, 375.
35. J. N. Cawse, M. Baerns, M. Holena, *J. Chem. Inf. Computer Sci.* **2004**, *44*, 143.
36. D. Goldberg, *Genetic Algorithms in Search, Optimization, and Machine Learning*, Addison-Wesley, Reading, 1989, pp. 412.
37. J. R. Koza, *Genetic Programming: On the Programming of Computers by Means of Natural Selection*, MIT Press, Cambridge, 1992, pp. 840.
38. J. R. Koza, *Genetic Programming II: Automatic Discovery of Reusable Programs*, MIT Press, Cambridge, 1994, pp. 350.

39. J. R. Koza, F. H. Bennett, D. Andre, M. A. Keane, *Genetic Programming III: Darwinian Invention and Problem Solving*, Academic Press, Orlando, 1999, pp. 1154.
40. J. R. Koza, M. A. Keane, M. J. Streeter, W. Mydlowec, J. Yu, G. Lanza, *Genetic Programming IV: Routine Human-Competitive Machine Intelligence*, Kluwer, Dordrecht, 2003, pp. 590.
41. T. Bäck, *Evolutionary Algorithms in Theory and Practice: Evolution Strategies, Evolutionary Programming, Genetic Algorithms*, Oxford University Press, New York, 1996, pp. 314.
42. M. Mitchell, *An Introduction to Genetic Algorithms*, MIT Press, Cambridge, 1996, pp. 224.
43. D. B. Fogel, *Evolutionary Computation: Toward a New Philosophy of Machine Intelligence*, 2nd edn, IEEE Press, New York, 1999, pp. 270.
44. M. L. Wong, K. S. Leung, *Data Mining Using Grammar Based Genetic Programming and Applications*, Kluwer, Dordrecht, 2000, pp. 213.
45. A. A. Freitas, *Data Mining and Knowledge Discovery with Evolutionary Algorithms*, Springer, Berlin, 2002, pp. 264.
46. M. Holena, in *High-Throughput Screening in Chemical Catalysis*, A. Hagemeyer, P. Strasser, A. F. Volpe (Eds.), Wiley-VCH, Weinheim, 2004, p. 153.
47. *Genetic Algorithm and Direct Search Toolbox*. The MathWorks, Inc., Natick, 2004, pp. 268.
48. A. S. McLeod, M. E. Johnston, L. F. Gladden, *J. Catal.* **1997**, *167*, 279.
49. A. S. McLeod, L. F. Gladden, *J. Chem. Inf. Computer Sci.* **2000**, *40*, 981.
50. D. Wolf, O. V. Buyevskaya, M. Baerns, *Appl. Catal. A: General* **2000**, *200*, 63.
51. O. V. Buyevskaya, D. Wolf, M. Baerns, *Catal. Today* **2000**, *62*, 91.
52. U. Rodemerck, D. Wolf, O. V. Buyevskaya, P. Claus, S. Senkan, M. Baerns, *Chem. Eng. J.* **2001**, *82*, 3.
53. G. Grubert, E. Kondratenko, S. Kolf, M. Baerns, P. van Geem, R. Parton, *Catal. Today* **2003**, *81*, 337.
54. U. Rodemerck, M. Baerns, M. Holena, D. Wolf, *Appl. Surface Sci.* **2004**, *223*, 168.
55. J. M. Serra, A. Chica, A. Corma, *Appl. Catal. A: General* **2003**, *239*, 35.
56. Y. Watanabe, T. Umegaki, M. Hashimoto, K. Omata, M. Yamada, *Catal. Today* **2004**, *89*, 455.
57. A. Corma, J. M. Serra, *Catal. Today* **2005**, *107–108*, 3.
58. A. Corma, J. M. Serra, P. Serna, S. Valero, E. Argente, V. Botti, *J. Catal.* **2005**, *229*, 513.
59. R. M. Pereira, F. Clerc, D. Farrusseng, J. C. Waal, T. Maschmeyer, *QSAR Combinatorial Sci.* **2005**, *24*, 45.
60. R. H. J. M. Otten, L. P. P. P. Gineken, *The Annealing Algorithm*, Springer, New York, 1989, pp. 224.
61. Z. B. Zabinsky, *Stochastic Adaptive Search for Global Optimization*, Kluwer, Boston, 2003, pp. 224.
62. B. Li, P. Sun, Q. Jin, J. Wang, D. Ding. *J. Mol. Catal. A: Chemical* **1999**, *148*, 189.
63. A. Eftaxias, J. Font, A. Fortuny, J. Giralt, A. Fabregat, F. Stüber. *Appl. Catal. B: Environmental* **2001**, *33*, 175.
64. A. Holzwarth, P. Denton, H. Zanthoff, C. Mirodatos, *Catal. Today* **2001**, *67*, 309.
65. S. Senkan, *Angew. Chem. Int. Ed.* **2001**, *40*, 312.
66. C. A. Floudas, *Deterministic Global Optimization. Theory, Methods and Applications*, Kluwer, Dordrecht, 2000, pp. 739.
67. J. A. Snyman, *Practical Mathematical Optimization. An Introduction to Basic Optimization Theory and Classical and New Gradient-Based Algorithms*, Springer, New York, 2005, pp. 257.
68. K. Huang, X. L. Zhan, F. Q. Chen, D. W. Lü, *Chem. Eng. Sci.* **2003**, *58*, 81.
69. L. Végvári, A. Tompos, S. Göbölös, J. F. Margitfalvi, *Catal. Today* **2003**, *81*, 517.
70. B. Lin, S. Chavali, K. Camarda, D. C. Miller, *Comput. Chem. Eng.* **2005**, *29*, 337.
71. T. Cundari, J. Deng, Y. Zhao, *Ind. Eng. Chem. Res.* **2001**, *40*, 5475.
72. A. Corma, J. M. Serra, E. Argente, V. Botti, S. Valero, *Chemphyschem* **2002**, *3*, 939.
73. K. Omata, M. Hashimoto, Y. Watanabe, T. Umegaki, S. Wagatsuma, G. Ishiguro, M. Yamada, *Appl. Catal. A: General* **2004**, *262*, 207.
74. C. Klanner, D. Farrusseng, L. Baumes, C. Mirodatos, F. Schüth, *QSAR Combinatorial Sci.* **2003**, *22*, 729.
75. A. F. Siegel, C. J. Morgan, *Statistics and Data Analysis: An Introduction*, 2nd edn, Wiley, New York, 1996, pp. 635.
76. M. Berthold, D. J. Hand, *Intelligent Data Analysis. An Introduction*, 2nd edn, Springer, Berlin, 2002, pp. 460.
77. M. Kantardzic, *Data Mining. Concepts, Models, Methods, and Algorithms*, Wiley, Chichester, 2003, pp. 345.
78. J. K. Taylor, C. Cihon, *Statistical Techniques for Data Analysis*, 2nd edn, Chapman & Hall, Boca Raton, 2004, pp. 273.
79. P. C. Meier, R. E. Zund, *Statistical Methods in Analytical Chemistry*, Wiley, New York, 1993, pp. 345.
80. M. Caria, *Measurement Analysis. An Introduction to the Statistical Analysis of Laboratory Data in Physics, Chemistry and the Life Sciences*, Imperial College Press, London, 2000, pp. 229.
81. D. Wallenstein, B. Kanz, R. H. Harding, *Appl. Catal. A: General* **1999**, *178*, 117.
82. J. T. Richardson, D. Remue, J. K. Hung, *Appl. Catal. A: General* **2003**, *250*, 319.
83. O. M. Wilkin, P. M. Maitlis, A. Haynes, M. L. Turner, *Catal. Today* **2003**, *30*, 309.
84. A. M. Amat, A. Arques, S. H. Bossmann, A. M. Braun, S. Göb, M. A. Miranda, E. Oliveros, *Chemosphere* **2004**, *57*, 1123.
85. D. G. Cantrell, L. J. Gillie, A. F. Lee, K. Wilson, *Appl. Catal. A: General* **2005**, *287*, 183.
86. A. Corma, J. M. Serra, P. Serna, M. Moliner, *J. Catal.* **2005**, *232*, 335.
87. D. Farrusseng, C. Klanner, L. Baumes, M. Lengliz, C. Mirodatos, F. Schüth, *QSAR Combinatorial Sci.* **2005**, *24*, 78.
88. R. J. Hendershot, R. Vijay, B. J. Feist, C. M. Snively, J. Lauterbach, *Measurement Sci. Technol.* **2005**, *16*, 302.
89. M. Holena, M. Baerns, in *Experimental Design for Combinatorial and High Throughput Materials Development*, J. N. Cawse (Ed.), Wiley, New York, 2003, p. 163.
90. J. Zupan, J. Gasteiger, *Neural Networks for Chemists*, Wiley-VCH, Weinheim, 1993, pp. 305.
91. C. M. Bishop, *Neural Networks for Pattern Recognition*, Clarendon Press, Oxford, 1996, pp. 482.
92. M. T. Hagan, H. Demuth, M. Beale, *Neural Network Design*, PWS Publishing, Boston, 1996, pp. 736.
93. B. D. Ripley, *Pattern Recognition and Neural Networks*, Cambridge University Press, Cambridge, 1996, pp. 403.
94. K. Mehrota, C. K. Mohan, *Elements of Artificial Neural Networks*, MIT Press, Cambridge, 1997, pp. 344.
95. S. Haykin, *Neural Networks. A Comprehensive Foundation*, 2nd edn, IEEE, New York, 1999, pp. 600.

96. J. Zupan, J. Gasteiger, *Neural Networks in Chemistry and Drug Design: An Introduction*, Wiley-VCH, Weinheim, 1999, pp. 402.
97. T. Hattori, S. Kito, *Catal. Today* **1995**, *23*, 347.
98. M. Sasaki, H. Hamada, Y. Kintaichi, T. Ito, *Appl. Catal. A: General* **1995**, *132*, 261.
99. Z. Y. Hou, Q. Dai, X. Q. Wu, G. T. Chen, *Appl. Catal. A: General* **1997**, *161*, 183.
100. K. Huang, C. Feng-Qiu, D. W. Lü, *Appl. Catal. A: General* **2001**, *219*, 61.
101. Y. Liu, D. Liu, T. Cao, S. Han, G. Xu, *Computers Chem. Eng.* **2001**, *25*, 1711.
102. M. Holena, M. Baerns, *Catal. Today* **2003**, *81*, 485.
103. J. M. Serra, A. Corma, E. Argente, S. Valero, V. Botti, *Appl. Catal. A: General* **2003**, *254*, 133.
104. J. M. Serra, A. Corma, A. Chica, E. Argente, V. Botti, *Catal. Today* **2003**, *81*, 393.
105. S. Kito, A. Satsuma, T. Ishikura, M. Niwa, Y. Murakami, T. Hattori, *Catal. Today* **2004**, *97*, 41.
106. M. Moliner, J. M. Serra, A. Corma, E. Argente, S. Valero, V. Boti, *Microporous Mesoporous Mater.* **2005**, *78*, 73.
107. K. Omata, Y. Kobayashi, M. Yamada, *Catal. Commun.* **2005**, *6*, 563.
108. G. Zahedi, A. Jahanmiri, M. H. Rahimpor, *Int. J. Chem. Reactor Eng.* **2005**, *3*, A8.
109. M. W. B. Trotter, B. F. Buxton, S. B. Holden, *Measurement Control* **2001**, *34*, 235.
110. A. Tompos, J. F. Margitfalvi, E. Tfirst, L. Végvári, M. A. Jaloull, H. A. Khalfalla, M. M. Elgarni, *Appl. Catal. A: General* **2005**, *285*, 65.
111. K. Yajima, Y. Ueda, H. Tsuruya, T. Kanougi, Y. Oumi, S. S. C. Ammal, S. Takami, M. Kubo, A. Miyamoto, *Appl. Catal. A: General* **2000**, *194*, 183.

2.3
Bulk Catalysts and Supports

2.3.1
Fused Catalysts

*Robert Schlögl**

2.3.1.1 Introduction

A small number of heterogeneous catalysts is prepared by the fusion of mixtures of oxidic or metallic precursors. The obvious group of compounds to be prepared in this way includes the metal alloy catalysts that are applied in unsupported form, such as the noble metal gauze used for the oxidation of ammonia to nitric oxide. Currently, melting of the elements in the appropriate composition is the only way in which bulk amounts of a chemical mixture of the constituent atoms may be produced. The process is well described by thermodynamics, and today a large database is available which includes phase diagrams and detailed structural studies. Metallurgy not only provides the technologies for the preparation and characterization of these products [1] but also enables the synthesis of a large number of bulk alloys with well-defined properties.

One material-saving development in the use of such bulk-phase metallic alloy catalysts is the application of bulk metallic glasses in the form of ribbons with macroscopic dimensions [2–5]. In this class of material, the atomic dispersion in the liquid alloy is preserved in the solid state as a single phase, although the material may be metastable from its composition. This allows the preparation of unique alloy compositions which are inaccessible by equilibrium synthesis. The solidification process, which utilizes rapid cooling (with cooling rates $>10^4$ K s^{-1}) creates "glassy" materials with well-defined short-range order but without long-range order. The difference in free energy between compositional equilibration and crystallization, when stored in the metallic glass, can be used to transform the material in an initial activation step from a glassy state into a nanocrystalline agglomerate with a large internal surface interface between crystallites. This still-metastable state is the active phase in catalysis, and the final transformation into the stable solid-phase mix with an equilibrium composition that terminates the life of such a catalyst.

In oxide materials [6] that are fused for catalytic applications, two additional factors contribute to the unique features of this preparation route. When in their liquid states, many oxides are thermodynamically unstable with respect to the oxygen partial pressure present in ambient air; that is, they decompose into lower-valent oxides and release molecular oxygen into the gas phase. This process can occur rapidly on the timescale of the fusion process – for example, with vanadium pentoxide or manganese oxides – or it may be slow, as with iron oxides. The existence of such decomposition reactions and the control of their kinetics [6] can create a unique quenched solid which is thermodynamically metastable at ambient conditions with respect to its oxygen content. In addition, by controlling the phase nucleation, a local anisotropy of phases – that is, a mixture of particles of different oxide forms interdispersed with each other – can be obtained. These oxides exhibit a complex and reactive internal interface structure which may be useful, either for direct catalytic application in oxidation reactions or in predetermining the micromorphology of resulting catalytic material when the fused oxide is used as precursor.

In both oxide and metal systems, application of the fusion process enables control to be exerted over nanostructuring and elemental composition outside the ranges of thermodynamically stable phases. The prototype example for such a catalyst is the multiply promoted iron oxide

* Corresponding author.

precursor used for the synthesis of ammonia (a detailed description is provided in Chapter 12.1, which discusses the consequences of the metastable oxide precursors for the catalytic action of the final metal catalyst).

Another feature of fused catalytic compounds is the generation of a liquid film during catalytic action. Such supported liquid-phase (SLP) catalysts consist of an inert solid support on which a mixture of oxides is precipitated and transforms into a homogeneous melt under reaction conditions. One such variety is self-supporting, and results from the segregation of components from the fused catalyst precursor to the surface, where they melt at operation temperatures due to the compositional gradient between the bulk and the surface. SLP systems provide a chemically and structurally homogeneous reaction environment which contains molecular clusters as active components that would be non-reactive in crystalline form. The standard example of this type of catalyst is the vanadium oxide contact used for the oxidation of SO_2 to SO_3.

2.3.1.2 The Concept of Fused Catalysts

The main preparation strategies for unsupported catalysts are summarized in Fig. 1. The group of thermochemical methods represents a small minority family in comparison to the wet chemical procedures. For a description of these methods, the reader is referred to individual chapters in this Handbook, wherein these methods are reviewed. Among the thermochemical methods, the fusion processes are rarely applied as they require extensive special equipment which is generally unavailable to research groups, and difficult to use in industry. The preparation of non-supported catalysts by fusion is an expensive and very energy-consuming, high-temperature process. It must compete with the concepts of wet chemical preparation (Fig. 1) by the mixing–precipitating calcining process which can be used in oxidative and reductive modes to obtain oxides and metals. Sol–gel preparations, either without or with templates, are derivatives of this general approach.

Unconventional alternatives among the thermochemical methods are offered by flame hydrolysis and by tribochemical procedures, although these are still at an early stage of development. Fusion products require extensive engineering of their mesostructure, since the geometric surface area of such catalysts is very low due to the high processing temperature and the occurrence of a fluid state of the material.

The term "fused catalyst" is not synonymous with "unsupported catalyst", but rather designates a small subgroup of unsupported catalytic materials. Fused catalysts have passed through the molten state at least once during their preparation, and in this respect they differ fundamentally from catalysts prepared at high temperatures (e.g., carbons, which are produced by gas–solid reaction processes) and have substantial kinetic differences compared to melt-solidification reactions.

The motivation to apply fusion reactions in catalyst synthesis stems from the unique possibility of obtaining mesoscopically structured materials with a substantial degree of compositional and structural disorder. The mesoscopic grain size allows for high thermal and chemical stability in high-temperature or chemically corroding processes. An ordering scheme of catalyst synthesis strategies, in terms of the resulting ordering in microscopic and mesoscopic dimensions, is shown

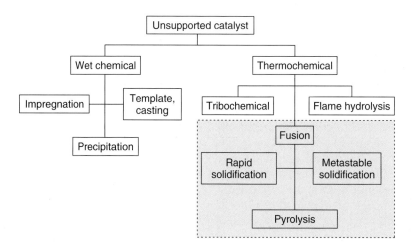

Fig. 1 Principal pathways to generate unsupported catalytic materials. The synthetic pathway of fusion (gray area) can be applied via differing techniques of application of heating or cooling procedures. Rapid and metastable solidification processes begin from a stationary melt, whereas pyrolysis (including metallothermal [7, 8] fusion) transforms the material for only a very short time into the molten state and exhibits fast heating and cooling kinetics; consequently, there is little time for the material to adapt to stable conditions at each temperature level.

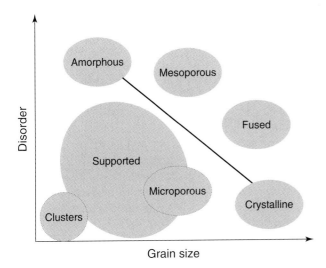

Fig. 2 Discrimination of catalysts with respect to their extent of microscopic and mesoscopic ordering. "Disorder" represents local microscopic ordering, and "grain size" mesoscopic structuring. For supported systems the ordering refers to the active component(s), not to the support.

schematically in Fig. 2. On the trajectory between the amorphous and crystalline systems, the metallic alloys after partial crystallization (see below) form a family of metastable systems. The grain size of a bulk material is zero for an amorphous solid and increases with its microscopic ordering. Systems with unusually large grains and a high extent of microscopic disorder are fused catalysts and mesoporous systems, with their whole wall system being amorphous but forming a microscopically well-ordered structure (e.g., a hexagonal array). Systems with well-defined microscopic ordering and unusually small crystallite or grain sizes form a large family of supported catalysts (the active component on the support) and (supported) cluster catalysts. Microporous materials, with their small crystallites and well-ordered wall systems (zeolites), form an unusual group in this diagram and are frequently used to define and maintain [9, 10] cluster systems. Figure 2 reveals clearly the special combination of ordering attainable with the fusion synthesis strategy, thereby justifying the substantial effort required to produce such systems. An important class of fused catalyst represents the metal aluminum alloys with subsequent leaching of the Al component; these "Raney catalysts" [11, 12] are discussed extensively in Chapter 2.3.2 of this Handbook.

A collection of the major individual reaction steps for the synthesis of unsupported catalysts is listed in Table 1. One fundamental insight from this rather schematic comparison is that differences in the reaction kinetics of the synthesis of a given material will lead to different mesoscopic and macroscopic structures, which in turn considerably affect the catalytic performance. It is necessary to control these analytically difficult-to-describe parameters with much the same precision as the atomic arrangement or the local electronic structure. Whereas these latter parameters influence the nature of the active site, it is the meso/macrostructure which controls the distribution and abundance of active sites on any given material. In fact, in certain cases it is necessary to apply the costly method of fusion, as there is no other way to obtain the desired (and in most cases unknown) optimal meso/macrostructure of the final catalyst.

Whilst details of the chemistry in the precipitation process can be found elsewhere in this Handbook (see Chapters 2.3.3), this chapter focuses on fused catalysts. (In this respect, it may benefit the reader to compare and contrast the following discussion with the contents of Table 1.)

Fused catalysts allow the combination of compounds and elements in atomic dispersions that do not mix either in solution (e.g., oxides) or in the solid state. Melting provides the necessary means to generate an intimate, eventually atomically disperse distribution; a carefully

References see page 91

Tab. 1 Main reaction steps in the precipitation and fusion of catalytic solids

Step	Wet chemical	Fusion
Mixing of atoms	In solution	In melt
Preformation of compounds	Frequently, with solvent ligands	Possible for alloys (E-L-TM), always for compounds (oxides)
Compositional modification	Frequently, by ligand exchange and incorporation of solvent	Possible with volatiles, frequently with compounds by thermochemical reduction
Solidification	Precipitation, difficult to control, very fine particles with molecular homogeneity	Cooling, very important to control, affects chemical structure (exsolution) and long-range ordering
Calcination	Required for ligand removal, complex reaction, difficult to control	Not required
Formulation	Pressing, extrusion precipitation onto supports	Crushing, sieving, production of wires and gauze

controlled solidification can preserve the metastable situation in the melt down to operation temperature. In the melt, the pre-formation of structural units such as oxo complexes or alloy clusters can occur, and the final short-range order of the catalyst is predetermined. Examples are alloys of noble metals with elements located in the main group sections or in the early transition metal groups of the Periodic Table [13, 14] (the early-late transition metal (E-LTM) alloys). In the case of oxides, the partial pressure of oxygen has the chance to equilibrate between the gas-phase environment of the melting furnace and the liquid oxide. With compounds in high formal oxidation states this can lead to thermochemical reduction, as for iron oxide (the reduction of hematite to magnetite and wustite) or for silver oxide (which reduces to the metal). Compounds in low oxidation states, such as MnO, Sb_2O_3 or VO_2, will oxidize to higher oxidation states and thus also change the chemical structure.

The kinetics to reach the equilibrium situation can be quite slow, so that the holding time and mechanical mixing of the melt will crucially affect the extent of the chemical conversion. Early termination of the holding time will lead to metastable situations for the melt, with local heterogeneity in the chemical composition of the final product. On occasion this may be desirable (as in the case of the iron oxide precursor for ammonia synthesis), although it can also be unwanted, as in most intermetallic [15, 16] compounds. The dissolution of, for example, one oxide into another, may also be a prolonged process, and early cooling will lead to a complex situation of disperse binary compounds coexisting with ternary phases. Examples are alumina and calcium oxide promoters in iron oxide melts, where ternary spinel compounds can be formed, provided that sufficient trivalent iron is present. This requires the addition of activated forms of the binary oxides in order to dissolve some of the ions before the thermochemical reduction has removed the trivalent iron in excess of that required for the formation of the matrix spinel of magnetite. These examples illustrate that the starting compounds, their purity and physical form, and the heating program will severely affect the composition and heterogeneity of the resulting material. The scaling-up of such fusion processes is a major problem, as the heat and ion transport determines to a significant extent the properties of the material. The gas phase over the melt and its control are also of major importance, as its chemical potential will determine the phase inventory of the resulting compound.

In addition to the complex cases of mixed oxides, other challenges must also be overcome in alloy production, such as oxide and scale formation [17, 18]. The detrimental effect of the oxide shells around metal particles preventing intermixing is well known, while the compositional changes resulting from the preferential oxidation of one component must also be taken into account. Instability of the product and/or drastic changes in the thermochemical properties of the material after shell formation (such as massive increases in the required fusion temperature in noble metal eutectic mixtures) are common, especially in the case of small-scale preparations. These effects still set limits as to the availability of catalytically desired alloys for practical purposes (e.g., for compounds with Zr, Si, alkali, and Mg).

As well as these more practical problems of catalyst preparation, a number of severe theoretical problems are also associated with the prediction of the chemistry in the fluid state of a compound. The motion of all structural elements (atoms, ions, molecules) is controlled by a statistical contribution from Brownian motion, by gradients of the respective chemical potentials (those of the structural elements and those of all species such as oxygen or water in the gas phase which can react with the structural elements and thus modify the local concentration), and by external mechanical forces such as stirring and gas evolution. In electric fields (as occurs in an arc melting furnace), the field effects will further contribute to non-isotropic motion and thus to the creation of concentration gradients. For the interested reader, an exhaustive discussion of these problems has been published (see Ref. [6] and references therein).

The second step in the process is the cooling of the melt. *Slow cooling* will result in an equilibration of the mix according to the thermodynamic situation, and only in simple cases will this yield the desired compound. In most cases the mixture of structural elements that are stable in the melt will be metastable under ambient conditions, and techniques of supercooling must be applied in order to maintain the desired composition [19].

Rapid cooling with temperature gradients up to about $100\ K\ s^{-1}$ is required to generate metastable crystalline solids. Local heterogeneity (such as concentration gradients or non-dissolved particles) will disturb the equilibrium formation [20] of crystals and lead to unusual geometries of the grain structure. The crystallite size is also affected by the cooling rate, especially at temperatures near the solidus point, where the abundance of (homogeneous) nuclei is determined.

Rapid cooling limits the growth of large crystals as the activation energies for the diffusion and dissolution of smaller crystallites are only available for a short time. Annealing of the solid after initial solidification can be used to modify the crystallite size, provided that no unwanted phase transition occurs in the phase diagram at or below the annealing temperature. A knowledge of the complete phase diagram for the possible multicomponent reaction mixture is mandatory in order to design a temperature–time profile for a catalyst fusion experiment. Unfortunately, however, in many cases these phase diagrams are either not available or are not known with sufficient accuracy, and consequently a series of experiments must be conducted to adjust this most critical step in the whole process. Frequently, empirical relationships between characteristic temperatures in the phase diagram and the critical temperatures for stable-to-metastable phase transformations (e.g., the ratio between an eutectic temperature and the crystallization temperature of a binary system) are used to predict compositions of stable amorphous compounds of metals and metalloids [20, 21].

Cooling rates of between 100 and 10 000 K s^{-1} can lead to a modification of the long-range order of the material. Under such rapid solidification conditions, the time at which the activation energy for motion of the structural elements is overcome is so short that the mean free path length reaches the dimension of the structural unit. The random orientation of the units in the melt is then preserved and the glassy state obtained. Such solids are X-ray amorphous, contain no grain boundary network, and exhibit no exsolution phenomena. They are also chemically and structurally isotropic [6, 22]. However, these solids preserve the energy of crystallization as potential energy in the solid state.

It is possible to transform these glasses into cascades of crystalline states, some of which may also be metastable under the crystallization conditions since the activation energy, for falling into the state of equilibrium, is not high enough. Glassy materials are therefore interesting precursors for the formation of metastable compositions and/or metastable grain boundary structures that are inaccessible by precipitation and calcination. The critical glass-forming temperatures vary widely for different materials; typically, alkali silicates require low cooling rates of several 100 K s^{-1}, while transition metal oxides and E-L TM alloys require rates above 1000 K s^{-1}. Pure elements cannot be transformed into the glassy state, but by utilizing these differences it is possible to obtain composite materials with a glassy phase coexisting with a crystalline phase. Examples of this [23, 24] are restructured metallic glasses and amorphous oxide promoter species dispersed between the iron oxides of the ammonia synthesis catalyst precursor.

The third step in the catalyst preparation process is the thermal treatment known as *calcination*, which is essential in all wet chemical processes. This leads to solvent-free materials and causes chemical reactions to occur between components, with the oxidation states of all elements reaching their desired values. All of this is accomplished during preparation of the fluid phase and during precipitation of the fused catalyst, and hence such a step is rarely required for these catalysts. This feature significantly reduces the difference in energy input to the final catalyst, between fusion and precipitation. There are two important consequences of conditioning of the catalytic material occurring for a fused catalyst in the fluid state, but for the precipitated catalyst in the solid state:

- the temperature levels of conditioning (and hence the range of accessible phases and composites) are different, and so is the chemical composition of the resulting material in particular with respect to volatile components
- in wet chemically prepared systems the calcination reaction occurs as a solid–solid state reaction with diffusion limitations and eventual topochemical reaction control, both of which give rise to spatial heterogeneity in large dimensions relative to the particle size.

In fused systems, the fluid state allows very intimate mixing and hence isotropic chemical reactivity, provided that the composition is either stable during cooling or is quenched so rapidly that no de-mixing occurs. Chemical heterogeneity at any dimensional level can be created or will occur unintentionally, with no gradients between particle boundaries if the cooling process is suitably adjusted to allow partial equilibration of the system.

The final step of catalyst preparation is that of *activation*, and this is required for both types of material. In this step – which often occurs during the initial stages of catalytic operation (*in-situ* conditioning) – the catalyst is transformed into the working state, being frequently chemically and/or structurally different from the as-synthesized state. It is desirable to store free energy in the catalyst precursor which can be used to overcome the activation barriers into the active state in order to initiate the solid-state transformations required for a rapid and facile activation. These barriers can be quite high for solid–solid reactions, and may therefore inhibit the activation of a catalyst.

A special case are catalysts which are metastable in their active state with respect to the catalytic reaction conditions. In this case, a suitable lifetime is only reached if the active phase is regenerated by solid-state reaction [25–29] occurring in parallel to the substrate-to-product conversion. In this case it is of special relevance to store free energy in the catalyst precursor as insufficient solid-state reaction rates will interfere with the substrate-to-product reaction cycle. One class of catalyst in which this effect is of relevance is the oxide materials used for selective oxidation reactions.

2.3.1.3 Thermodynamic and Kinetic Considerations

The following general considerations are intended to illustrate the potential and complications when a fused catalyst material is prepared. The necessary precondition

References see page 91

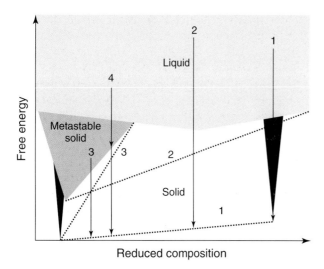

Fig. 3 The thermodynamic situation upon solidification of a multernary system. Vertical lines designate the principal reaction pathways; dashed tangent lines illustrate the compositional changes arising from an equilibrium solidification at the respective pathways. The narrow areas of existence (dark) designate stable phases with a finite phase width; the area designated "metastable" indicates the existence of a single-phase solid which is unstable at ambient conditions.

is that the starting state is a homogeneous phase (the fluid).

A general free energy versus composition diagram [22] for a fused catalyst system containing several components is illustrated in Fig. 3. The composition coordinate may be a projection [30] through a multernary phase diagram. If the melt is cooled slowly it will solidify in the phase (1), with little compositional variation; this path leads to a stable solid with few problems in its preparation and identification. If the melt is cooled suitably to follow the solidus curve further down in free energy (2), it can then be rapidly quenched without any compositional variation; this creates a metastable solid with a large amount of free energy stored in the solid state. The resulting material is a characteristic fused catalyst (or precursor). If the cooling is slowed down, the composition will split in a primary crystallization [30] of the supersaturated solution. The melt is then enriched in one component according to the tangent line (2), and the solid is depleted until it reaches the composition of the metastable solid (3). The enriched melt can either crystallize in (1) or react along pathway (3), provided that enough energy of crystallization is released and the cooling conditions [31] are still adequate. The metastable solid (3) may either be quenched and form a further metastable component of the phase mix, or it can undergo equilibration in the same way as system (2), along the tangent line (3). The cooling conditions and eventual annealing intervals will decide over the branching ratio. A further possibility is the formation (4) of a supersaturated solid solution (the metastable solid in Fig. 3) directly from the melt followed by either quench cooling to ambient temperature and leading to another metastable phase in the mix or by an equilibration-affording pathway (3).

The solidification kinetics and compositional fluctuations in the melt will decide upon the crystallization pathway that can be followed by all of the melt. If local gradients of temperature or composition exist within the system, the crystallization pathway can be locally inhomogeneous and create different metastable solids at different locations in the macroscopic solidified blocks. Thus, it is possible that all pathways of Fig. 3 will be followed simultaneously, the result being an enormous heterogeneity in the primary solid and substantial local variation in the post-solidification reactions. Especially in oxide mixtures which become highly viscous near the solidus line, such origins of heterogeneity are quite possible.

This simple consideration shows that a wide variety of stable and metastable solids can be produced from a homogeneous melt if the solidification conditions are suitably chosen. In this way, a complex solid-phase mix can be obtained which is inaccessible by the wet chemical preparation route. The metastable phase mix may either contain an active phase, or may be used to generate it by using a suitable activation procedure at relatively low temperatures. Stable phases that are catalytically valuable should be accessible by other less complex and costly routes (see Figs. 1 and 2) and so are not considered here.

The kinetic situation is generalized in Fig. 4. For a fused catalyst system, a liquid phase is assumed to coexist with a metastable solid solution. Additional solid phases

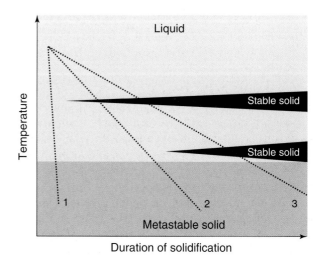

Fig. 4 Kinetics of solidification, illustrating how various cooling programs (pathways 1 to 3) can affect the final inventory of phases which differ in their respective crystallization kinetics. The characteristic times are in the microsecond regime for metallic alloys, but extend into time-scales of days for oxides.

crystallize with retarded kinetics and form lenses in the time–temperature diagram of Fig. 4. Three characteristic cooling profiles are sketched. Rapid quenching (1) leads to only the solid solution without compositional changes and without mesoscopic heterogeneity. Intermediate quenching (2) passes through the solid 1 area and leads to a branching of the solid products between solid 1 and the solid solution with a modified composition (primary crystallization, path 2 in Fig. 3). Slow cooling, eventually with a holding sequence (3), allows the pre-formation of structural units in the melt and leads to the formation of three solids with different compositions. Moving the holding temperature further down into the ranges occupied by the solid phases provides control over the branching of the solid products. It can be seen that rapid cooling of the fused melt leads to a clear situation with respect to the solid, as all free energy is transferred into the solid phase and liberated only in solid–solid reactions. If the cooling rate is intermediate, or if it is not isokinetic in the whole melt, then complex situations are obtained with wide variations in chemical and local compositions of the final solids.

The reduced fused iron oxide for ammonia synthesis is a perfect example illustrating, in its textural and structural complexity, the merit of this preparation strategy by allowing the creation of a metastable porous form of the element iron. The necessary kinetic stabilization of the metastable solid is achieved by the ex-solution of irreducible oxide phases of structural promoters. Some of these precipitate during solidification, whereas others are liberated from the matrix during activation. A prerequisite for the very important secondary ex-solution species is the intimate phase mixture of ternary iron earth alkali oxides, which cannot be achieved by wet chemical precipitation techniques due to the extremely different coordination chemistry of the various cations in solvent media.

2.3.1.4 Sulfuric Acid Catalyst

The reaction of gaseous SO_2 with molecular oxygen in the contact process seems to proceed over two independent mechanisms [32], one of which is the direct oxidation of a vanadium pentoxide–sulfur dioxide adduct by oxygen, and the other proceeding via a redox cycle involving V^{4+} and V^{3+} intermediate species [32–34].

The technical catalyst is a supported liquid-phase system of vanadium pentoxide in potassium pyrosulfate [35, 36]. Other alkali ions influence the activity [37] at the low-temperature end of the operation range, with Cs exhibiting a particular beneficial effect [32]. It is noted that, during the earlier stages of catalyst development phases, the fused vanadium oxides were operated in a regime of solid systems. In this regime, the redox chemistry of the vanadium oxide becomes rate-determining, and this led Mars and van Krevelen [38, 39] to the formulation of their famous rate law of catalyst reoxidation being rate-determining, and not the redox chemistry of SO_2.

It is necessary to work at the lowest possible temperature in order to achieve complete conversion. Only at temperatures below 573 K is the equilibrium conversion of SO_2 practically complete (ca. about 99.5%). The binary phase diagram vanadium–oxygen shows the lowest eutectic for a mixture of pentoxide and the phase V_3O_7 at 910 K. All binary oxides are stable phases, from their crystallization down to ambient temperature. The pyrosulfate promoter is thus an essential ingredient rather than a beneficial additive to the system. Compositions of 33% alkali (equivalent to 16.5% pyrosulfate) solidify at around 590 K, but this temperature is still too high as at about 595 K the activation energy increases sharply, even although the system is still liquid. The liquid state is thought to be essential for the facile diffusion [32, 39] of oxygen to the active sites [32].

The small mismatch between the required and achieved minimum operation temperatures has the severe consequence that a special preabsorption stage must be included in the reactor set-up in order to achieve an essentially complete conversion. In this manner, the partial pressure of the SO_3 product is lowered before the last stage of conversion, rendering acceptable incomplete conversion of the overheated catalyst. If the reason why the catalyst does not operate efficiently down to its solidification point could be eliminated, it might be possible to circumvent the intermediate absorption stage and thus facilitate reactor design to a considerable degree.

Catalyst fusion is essential to bring (and maintain) the pyrosulfate–vanadium oxide system into a homogeneous state, which is the basis for operating the system at the eutectic in the ternary phase diagram. The reaction mechanism, and the fact that the operation point of the catalyst is at the absolute minimum in the V^{5+}–O_2 section of the phase diagram, point to the existence of a supersaturated solution of partly reduced vanadium oxides in the melt. The point at which the activation energy for SO_2 oxidation changes over to a lower (transport-controlled, Mars–van Krevelen regime) value marks the stage at which crystallization of the supersaturated solution begins under catalytic conditions. This hypothesis was verified in pioneering studies conducted by Fehrmann and coworkers using electric conductivity measurements and preparative isolation techniques [35, 36]. These authors isolated crystals of a variety of V^{4+} and V^{3+} ternary alkali sulfates. These precipitates could be redissolved in a regeneration procedure of the catalyst involving heat treatment to 800 K under oxidizing conditions [36]. In a rather elegant

---

References see page 91

in-situ electron paramagnetic resonance (EPR) study, the deactivation mechanism was experimentally confirmed on an industrial supported catalyst in which the phase $K_4(VO)_3(SO_4)_5$ was identified as V^{4+} deactivating species that could also be redissolved by a high-temperature treatment [40].

The accurate analysis of the problem is complicated as, under reaction conditions (presence of oxygen), all redox equilibria between V^{5+} and the lower oxidation states are shifted towards the pentavalent state. The generation of realistic model systems in which, for example, conductivity experiments can be performed, thus requires the exact control of the gas phase in contact with the melt.

In a recent summarizing review using EPR and NMR data of technical catalysts [41], it was established that an oligomeric oxocomplex of pentavalent vanadium with sulfate represents the active phase. Species of reduced vanadium oxocomplexes are present but represent side chains of catalyst deactivation mechanisms interfering with the SLP nature of the active phase. With respect to the need to use a fused catalyst, it is relevant to compare the kinetic data with those from solid supported catalysts [42] containing also pentavalent oxocomplexes supported on titania. This system, which allows *in-situ* vibrational spectra to elucidate the reaction mechanism, is significantly inferior in performance compared to the characteristic fused systems. However, neither the detailed origin of this difference nor the exact nature of the structure of active sites is yet clear.

The real pseudo-binary phase diagram [35] of $V_2O_5/S_2O_7M_2$ with $M = K$ or Na is rather complex in the interesting range around the eutectic, which is shown in Fig. 5. The formation of a complex salt with the composition $3M_2S_2O_7 \cdot xV_2O_5$ interferes with the eutectic and gives rise to two eutectic points with fusion temperatures of 587 K and 599 K. It is interesting to note that the chemistry of vanadium pentoxide in molten alkali sulfates is different from the present case with pyrosulfates, where no vanadium oxo oligomers are formed. This is an indication of a complex formation between pyrosulfate and vanadium oxide in the sense of preformed molecules in the fused melt. The dashed lines at high molar fractions of the pentoxide in Fig. 5 indicate the estimated continuation of the phase boundaries which are inaccessible experimentally, as in this regime glassy oxides with unknown compositions are formed. The dashed lines which point towards the global eutecticum indicate the expected position of the solidus line, without the occurrence of compound formation in the melt. The complexation of molecular motifs within the oxide melt exerts a significant influence on the crystallization behavior. The significant gain in Gibbs free energy of the oxo complex with respect

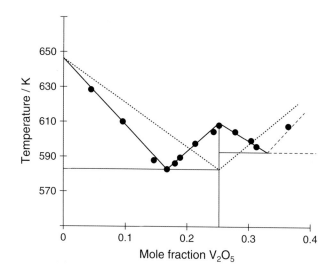

Fig. 5 Section of the pseudobinary phase diagram of the sulfuric acid SLP catalytic material (data from Ref. [16]). The data points were derived from anomalies of the conductivity versus temperature curves of the respective mixtures. At the high compositional resolution and in the range of the global eutectic, the formation of a vanadate-sulfato complex causes the local maximum in the solidus curve. Note that extreme precision in experimental procedures was necessary to derive this result, which illustrates the characteristics of fused systems, that compound formation can also occur in the molten state.

to a random fluid of vanadium oxide "ligands" can be estimated by the gain in melting temperature at 0.25 mol.% pentoxide.

These observations on the sulfuric acid catalyst are fully in line with the general thermodynamic behavior of fused catalyst systems. The metastable solid in Fig. 3 must be replaced in this case by a cascade of the partly reduced vanadium ternary sulfates. The processes sketched above occur under thermodynamic control in a quaternary phase diagram, vanadium–oxygen–sulfur–alkali, as illustrated by the reversibility of the ex-solution of the partly reduced vanadium compounds under suitable partial pressures of oxygen within the melt. This partial pressure is adjusted by the operating temperature. The desired low operation temperature increases the viscosity of the melt and hence increases the diffusion barrier of the gas into the liquid. This, in turn, facilitates the exsolution of reduced vanadium sulfates which further inhibit the oxygen diffusion.

2.3.1.5 Metallic Glasses

Amorphous metals can be prepared in a wide variety of stable and metastable compositions, with all catalytically relevant elements. This synthetic flexibility and the isotropic nature of the amorphous state with no defined surface orientations and no defect structure (as no long-range ordering exists) provoked the search for their application in catalysis [43]. The drastic effect of an average statistical mixture of a second metal component to a catalytically active base metal was illustrated in a model experiment of CO chemisorption on polycrystalline Ni which was alloyed by Zr as a crystalline phase and in the amorphous state. Since CO chemisorbs as a molecule on Ni and dissociates on Zr, it was observed that, on the crystalline alloy, a combination of molecular and dissociative chemisorption in the ratio of the surface abundance occurred. This additive behavior was replaced by a synergistic effect of the Zr in the amorphous state, where molecular adsorption with a modified electronic structure of the adsorbate was observed [44]. This experiment led to the conclusion [23] that, with amorphous metals, a novel class of catalytic materials with tunable electronic properties might be at our disposal.

The first attempts to check this hypothesis [45] revealed a superior catalytic activity of iron in amorphous iron–zirconium alloys in ammonia synthesis compared to the same iron surface exposed in crystalline conventional catalysts. A detailed analysis of the effect subsequently revealed that the alloy, under catalytic conditions, was not amorphous but rather was crystallized into platelets of metastable epsilon-iron supported on Zr-oxide [46, 47].

This was the first proven example of the operation of the principle that free energy stored in the metastable amorphous alloy can be used to create a catalytically active species which is still metastable against phase separation and recrystallization, but which is low enough in residual free energy to maintain the catalytically active state for useful lifetimes.

In Pd–Zr alloys, a different principle of usage for the excess free energy can be found. Amorphous alloys of the composition $PdZr_2$ were activated in several procedures and compared to a Pd on ZrO_2 catalyst for the activity in CO oxidation applications [48, 49]. *In-situ* activation of the amorphous alloy caused crystallization into small nanocrystalline Pd + O solid solution particles and larger pure Pd particles, both of which are embedded into a high interface area of zirconia, being present as a poorly crystalline phase mix of monoclinic and tetragonal polymorphs. This phase mix is still metastable against the formation of large particles of pure Pd and well crystallized large particles of zirconia with minimal common interface area, as it is obtained from conventional impregnation techniques. A detailed analysis of the surface chemistry of the *in-situ*-activated amorphous alloy, which is metastable against the segregation of a thick layer of zirconia in air, revealed that only under crystallization in the reaction mixture is the intimate phase mix between zirconia and Pd present at the outer surface of the material. It was concluded from kinetic data [48] that the intimate contact between zirconia and Pd should facilitate the spillover of oxygen from the oxide to the metal.

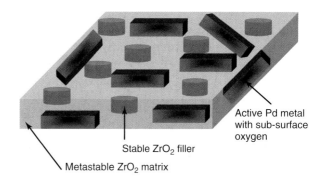

Fig. 6 Schematic arrangement of the surface of a partly crystallized E-L TM amorphous alloy such as Pd–Zr. A matrix of zirconia consisting of the metastable polymorph with crystallites of the stable polymorph holds particles of the L transition metal (Pd) which are structured in a skin of solid solution with oxygen (light) and a nucleus of pure metal (dark). (Figure compiled from experimental data in Refs. [48, 49].)

Figure 6 illustrates, schematically, the advantages of the metastable precursor structure for generating via a large internal oxide–metal interface a continuous supply of activated oxygen to replenish the Pd–O compound that is believed to be the active phase, and to constitute the active surface. It remains speculative as to whether the beneficial effect is really a spillover of oxygen from the oxide through the surface and/or bulk diffusion [48], or whether the structural stabilization of the known [49] oxygen storage phase in the Pd (the solid solution) by the defective zirconia matrix is the reason for the superior catalytic performance.

Most relevant for the oxygen transport should be the defective crystal structure of both catalyst components. The defective structure and intimate contact of crystallites of the various phases are direct consequences of the fusion of the catalyst precursor, and are features which are inaccessible by conventional wet chemical methods of preparation. Possible alternative strategies for the controlled synthesis of such designed interfaces may be provided by modern chemical vapor deposition (CVD) methods, although considerably more chemical control is required than for the fusion of an amorphous alloy.

In a study of the application of Pd–Si amorphous alloys as selective hydrogenation catalysts [3], it was found that *in-situ* activation provides a route to active and selective catalysts, whereas *ex-situ* activation caused crystallization of the system into the thermodynamically stable Pd + SiO_2 system, which is in distinguishable in terms of its activity and poor selectivity from conventional catalysts of the same composition. In this study, it was possible to show conclusively that bulk amorphous alloys are not amorphous on their surfaces as they

References see page 91

undergo, in reactive gas atmospheres, chemically induced phase segregation which starts the crystallization process according to Fig. 3 (pathways 2 and 3).

The function of the fused amorphous alloys is thus to serve as a precursor material for the formation of a metastable active phase characterized by an intimate mixture of phases with different functions. This mixture is preformed during preparation of the metallic melt, and preserved by rapid solidification. The micromorphology consists of quenched droplets allowing subsequent segregation into platelets. *In-situ* activation is the method used to prevent crystallization in the structure with the global free energy minimum. This activation allows transformation of the supersaturated solution from the fusion process only to crystallize until the metastable state of the tangent line (2) in Fig. 3 is reached. At this stage of transformation, the catalytically active state is present. This principle of application of amorphous alloys is also highlighted in reviews [3–5] describing other catalytic applications of this class of fused material.

2.3.1.6 Tribochemistry

The tribochemical synthesis of catalysts by high-energy ball milling is still used in academic environments to obtain bulk catalysts with a designed mesoscopic structuring. The raw materials may be bulk materials synthesized by wet chemical methods or of any other origin. The aim of the treatment is to use strong local chemical overheating induced by mechanical forces for rapid local solidification under mechanical stress [50]. The action of the milling tools on the milled material may be first to break the initial crystals and thus enhance the surface area. This initial process reaches equilibrium (the "milling equilibrium") when the local melting and fusion process, acting in parallel to the disintegration, leads to sintering of the broken grains. Continuation of the milling process beyond the milling equilibrium (this is detected by plotting the evolution of surface area versus milling time) will lead to extensive fusion and thus to a loss in long-range order. In extreme cases phase changes or even complete bulk amorphization [17] can be reached. This structuring process can be combined with a mixing of phases [50, 51], provided that the mechanical properties (hardness and anisotropy of mechanical stability) are comparable for the constituents. Although milling activation is a technical, large-scale process used in the ceramics industry, and suitable equipment exists, few applications remain of this effective process (there is no waste) in catalyst synthesis.

A typical academic study is the modification of a wet chemically synthesized bulk vanadium pyrophosphate (VPP) catalyst [52, 53] used for the selective oxidation of butane [54, 55] to maleic anhydride. The platelet-shaped material is believed to be active only at the basal planes,

and thus thinning of the flakes by thermal, vacuum or milling treatment was attempted. Some results are summarized in Fig. 7. A common precursor material with good performance and after equilibration [53] for 100 h on stream was used for the structural modifications.

It was evident that milling beyond the milling equilibrium (which was reached after 2 min) had an extensive and positive effect, whereas thermal or vacuum treatment had only negative effects on performance. The mesostructure of the strongly activated material showed many defects arising from the breaking of the initially present platelets. The gain in performance per unit mass shown in Fig. 7 is not paralleled by a simple increase in active surface area, as the performance per unit surface area was much reduced by this treatment. This indicated that a more profound structural change had occurred rather than a simple exfoliation of the initially present platelets. The data also illustrated [52] that the activity does not correlate simply with the abundance of basal plane surfaces [55], but rather is related to a small fraction of the accessible surface that is not obviously correlated to a preferred orientation.

Another successful example of a modified mesostructure of a bulk catalyst was described [56] which aimed at activating iron oxide for the reaction of NO and CO in the context of automobile exhaust catalysis. Here, high-energy ball milling effected beyond the milling equilibrium the incorporation of a large number of lattice defects detectable by transmission electron microscopy (TEM) and calorimetry into hematite. These lattice defects prevented

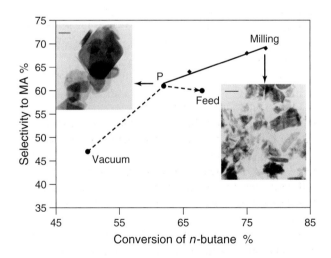

Fig. 7 Tribochemical activation of vanadyl pyrophosphate (VPP) for the selective oxidation of butane to maleic anhydride (MA). Only high-energy ball milling improved the performance of a common precursor material (P) consisting of thin platelets of the VPP phase. After extensive milling (30 min in ethanol), the platelets were disintegrated and broken individually. The effect of milling is compared to other equilibration procedures in vaccum or in feed.

the reduction of hematite to magnetite by CO under operation conditions, as occur in the blast furnace process. The origin of this stabilization should be the additional activation of the diffusion of iron species through the bulk to the surface where they may be reduced by CO. The remaining hematite, being metastable and self-regenerating under excess CO, acted as a highly effective catalyst for CO oxidation and was found to be even more active than supported platinum, with a 10-fold higher specific surface area than the bulk iron oxide.

2.3.1.7 Mesostructure of Fused Catalyst Materials

The aim of fusion and controlled solidification of a catalytic material is the generation of a metastable catalytic material. The thermodynamic instability (at operation conditions) can be caused by a non-equilibrium composition, by a non-equilibrium morphology, or by a combination of both. In the case of the SLP catalysts, the desired effect is to avoid the formation of solidification in order to maintain a disordered state of the active material.

The detection of metastable phases by spectroscopic and local structure-sensitive methods has been described in case studies [3, 48, 49, 57, 58]. The detection of non-equilibrium mesostructures is rather difficult, and less frequently carried out due to the fact that the relevant size range is between local atomic microstructural motives and macroscopic crystal morphologies. For this reason, conventional scanning electron microscopy (SEM) as well as TEM (which reveals only two-dimensional projections) are not ideally suited to the study of such mesostructures. High-resolution SEM with high-voltage probes and field emission instruments or scanning probe microscopies [47] are suitable techniques for retrieving information about the metastable mesostructure. This information is of significant catalytic relevance, as many reactions are structure-sensitive and thus exhibit different kinetics on different surface orientations. The generation of anisotropic particles with the consequence of preferred abundances of selected orientations (i.e., basal planes of platelets) or with large interfaces between different phases in the catalysts are key issues in the process of improving or even tailoring catalytic performance.

Fused materials provide a viable route to bulk amounts of anisotropic particles prepared in a controlled, yet complex, procedure. This is demonstrated with the fused iron catalyst for ammonia synthesis. An alternative to mesoscopic metastable structuring is provided with the surfaces of covalently bonded intermetallic [11, 15] phases. These thermodynamically stable bulk phases retain their rigid bulk structure [59, 60] even under catalytic reaction conditions. Structural motifs such as the site isolation of, for example, Pd atoms in a non-reactive matrix of Pd–Ga [61], are retained as no segregation and formation of subsurface compounds (e.g., Pd hydride) can occur. The metastability of these fused catalysts must be confined to the surface, where unusual terminations (no CO adsorption on Pd surface atoms) give rise to novel selective catalytic functions [13], as exemplified in the stable high performance of PdGa in the selective hydrogenation of acetylene in a large excess of ethylene. It may be expected that this class of material, though not yet frequently used in catalysis, may become a relevant alternative in selective transformations when their synthesis [12, 16, 50] has been simplified.

References

1. A. Cottrell, *An Introduction to Metallurgy*, Jesus College, Cambridge, 1975.
2. F. E. Luborsky (Ed.), *Amorphous Metallic Alloys*, Butterworths, London, 1983.
3. K. Noack, C. Rehren, H. Zbinden, R. Schlögl, *Langmuir* **1995**, *11*, 2018–2030.
4. A. Baiker, *Faraday Discuss. Chem. Soc.* **1989**, *87*, 239–251.
5. A. Molnar, G. V. Smith, M. Bartok, *Adv. Catal.* **1989**, *36*, 329–383.
6. H. Schmalzried (Ed.), *Chemical Kinetics of Solids*, VCH, Weinheim, 1995.
7. R. Abramovici, *Mater. Sci. Eng.* **1985**, *71*, 313–320.
8. Gn. Kozhevni, V. V. Efremkin, *Russian Metallurgy* **1972**, 35–37.
9. D. Bruhwiler, G. Calzaferri, *Microporous Mesoporous Mater.* **2004**, *72*, 1–23.
10. J. C. Fierro-Gonzalez, S. Kuba, Y. L. Hao, B. C. Gates, *J. Phys. Chem. B* **2006**, *110*, 13326–13351.
11. Y. Xu, S. Kameoka, K. Kishida, M. Demura, A. P. Tsai, T. Hirano, *Mater. Transact.* **2004**, *45*, 3177–3179.
12. Y. Xu, S. Kameoka, K. Kishida, M. Demura, A. P. Tsai, T. Hirano, *Intermetallics* **2005**, *13*, 151–155.
13. A. Bahia, I. R. Harris, J. M. Winterbottom, *J. Chem. Technol. Biotechnol.* **1994**, *60*, 347–351.
14. M. M. Jaksic, J. M. Jaksic, *Electrochim. Acta* **1994**, *39*, 1695–1714.
15. A. Bahia, J. M. Winterbottom, *J. Chem. Technol. Biotechnol.* **1994**, *60*, 305–315.
16. O. V. Chetina, T. V. Vasina, V. V. Lunin, *Russian Chem. Bull.* **1994**, *43*, 1630–1633.
17. E. Gaffet, M. Abdellaoui, N. Malhourouxgaffet, *Mater. Transact.* **1995**, *36*, 198–209.
18. H. J. Grabke, V. Leroy, H. Viefhaus, *Isij Int.* **1995**, *35*, 95–113.
19. D. R. Uhlmann, *J. Non-Cryst. Solids* **1972**, *7*, 337.
20. H. A. Davies, B. G. Lewis, *Scripta Met.* **1975**, *9*, 1107–1112.
21. D. Turnbull, *Contemp. Phys.* **1969**, *10*, 473–488.
22. U. Koster, P. Weiss, *J. Non-Cryst. Solids* **1975**, *17*, 359–368.
23. P. Barnickel, A. Wokaun, A. Baiker, *J. Chem. Soc., Faraday Trans.* **1991**, *87*, 333–336.
24. T. Takahashi, M. Kawabata, T. Kai, H. Kimura, A. Inoue, *Mater. Transact.* **2006**, *47*, 2081–2085.
25. V. F. Balakirev, Y. V. Golikov, *Inorganic Mater.* **2006**, *42*, 49–69.

26. L. Giebeler, P. Kampe, A. Wirth, A. H. Adams, J. Kunert, H. Fuess, H. Vogel, *J. Molec. Catal. A. Chemical* **2006**, *259*, 309–318.
27. L. O'Mahony, T. Curtin, D. Zemlyanov, M. Mihov, B. K. Hodnett, *J. Catal.* **2004**, *227*, 270–281.
28. F. Jentoft, S. Klokishner, J. Kröhnert, J. Melsheimer, T. Ressler, O. Timpe, J. Wienold, R. Schlögl, *Appl. Catal. A* **2003**, *256*, 291–317.
29. R. Schlögl, A. Knop-Gericke, M. Hävecker, U. Wild, D. Frickel, T. Ressler, R. E. Jentoft, J. Wienold, A. Blume, G. Mestl, O. Timpe, Y. Uchida, *Top. Catal.* **2001**, *15*, 219–228.
30. M. von Heimendahl, H. Oppolzer, *Scripta Met.* **1978**, *12*, 1087–1094.
31. P. G. Boswell, *Scripta Met.* **1977**, *11*, 701–707.
32. F. J. Dacring, D. A. Berkel, *J. Catal.* **1987**, *103*, 126–139.
33. G. K. Boreskov, G. M. Polyakova, A. A. Ivanov, V. M. Mastikhin, *Dokl. Akad. Nauk.* **1973**, *210*, 626.
34. G. K. Boreskov, A. A. Ivanov, B. S. Balzhinimaev, L. M. Karnatovskaya, *React. Kinet. Catal. Lett.* **1980**, *14*, 25–29.
35. D. A. Karydis, S. Boghosian, R. Fehrmann, *J. Catal.* **1994**, *145*, 312–317.
36. S. Boghosian, R. Fehrmann, N. J. Bjerrum, G. N. Papatheodorou, *J. Catal.* **1989**, *119*, 121–134.
37. L. G. Simonova, B. S. Balzhinimaev, O. B. Lapina, Y. O. Bulgakova, T. F. Soshkina, *Kin. Katal.* **1991**, *32*, 678–682.
38. P. Mars, D. W. van Krevelen, *Chem. Eng. Sci.* **1954**, *8*, 41–59.
39. F. P. Holroyd, C. N. Kenney, *Chem. Eng. Sci.* **1971**, *26*, 1971–1975.
40. K. M. Eriksen, R. Fehrmann, N. J. Bjerrum, *J. Catal.* **1991**, *132*, 263–265.
41. O. B. Lapina, B. S. Balzhinimaev, S. Boghosian, K. M. Eriksen, R. Fehrmann, *Catal. Today* **1999**, *51*, 469–479.
42. J. P. Dunn, H. G. Stenger, Jr., I. E. Wachs, *Catal. Today* **1999**, *51*, 301–318.
43. R. Schlögl, *Rapidly Quenched Metals*, S. Steeb, H. Warlimont (Eds.), Elsevier, 1986, 1723–1727.
44. R. Hauert, P. Oelhafen, R. Schlögl, H.-J. Güntherodt, *Rapidly Quenched Metals* S. Steeb, H. Warlimout (Eds.), Elsevier, 1986, 1493–1496.
45. E. Armbruster, A. Baiker, H. Baris, H.-J. Güntherodt, R. Schlögl, B. Walz, *J. Chem. Soc., Chem. Commun.* **1986**, 299–301.
46. A. Baiker, R. Schlögl, E. Armbruster, H.-J. Güntherodt, *J. Catal.* **1987**, *107*, 221–231.
47. R. Schlögl, R. Wiesendanger, A. Baiker, *J. Catal.* **1987**, *108*, 452–466.
48. A. Baiker, D. Gasser, J. Lenzner, A. Reller, R. Schlögl, *J. Catal.* **1990**, *126*, 555–571.
49. R. Schlögl, G. Loose, M. Wesemann, A. Baiker, *J. Catal.* **1992**, *137*, 139–157.
50. C. C. Koch, J. D. Whittenberger, *Intermetallics* **1996**, *4*, 339–355.
51. C. Suryanarayana, *Prog. Mater. Sci.* **2001**, *46*, 1–184.
52. I. Ayub, D. Su, M. Willinger, A. Kharlamov, L. Ushkalov, V. A. Zazhigalov, N. Kirilova, R. Schlögl, *Phys. Chem. Chem. Phys.* **2003**, *5*, 970–978.
53. A. Kharlamov, L. N. Ushkalov, D. S. Su, I. Ayub, R. Schlögl, *Rep. Natl. Acad. Sci. Ukraine* **2003**, *2*, 94–102.
54. M. Hävecker, N. Pinna, K. Weiss, H. Sack-Kongehl, R. E. Jentoft, D. Wang, M. Swoboda, U. Wild, M. Niederberger, J. Urban, D. S. Su, R. Schlögl, *J. Catal.* **2005**, *236*, 221–232.
55. Y. H. Taufiq-Yap, C. K. Goh, G. J. Hutchings, N. Dummer, J. K. Bartley, *J. Molec. Catal. A. Chemical* **2006**, *260*, 24–31.
56. T. Rühle, O. Timpe, N. Pfänder, R. Schlögl, *Angew. Chem. -Int. Ed.* **2000**, *39*, 4379.
57. H. Yamashita, M. Yoshikawa, T. Funabiki, S. Yoshida, *J. Catal.* **1986**, *99*, 375–382.
58. H. Yamashita, M. Yoshikawa, T. Funabiki, S. Yoshida, *J. Chem. Soc., Faraday Trans. I* **1987**, *83*, 2883–2893.
59. T. F. Fässler, *Chem. Soc. Rev.* **2003**, *32*, 80–86.
60. R. Nesper, *Angew. Chem. -Int. Ed.* **1991**, *30*, 789–817.
61. C. Wannek, B. Harbrecht, *J. Alloys Compounds* **2001**, *316*, 99–106.

2.3.2
Skeletal Metal Catalysts

Andrew James Smith* and Mark Sebastian Wainwright

"It is in the preparation of catalysts that the Chemist is most likely to revert to type and to employ alchemical methods. From all evidence, it seems the work should be approached with humility and supplication, and the production of a good catalyst received with rejoicing and thanksgiving" [1].

2.3.2.1 Introduction

Murray Raney graduated as a Mechanical Engineer from the University of Kentucky in 1909, and in 1915 joined the Lookout Oil and Refining Company in Tennessee with responsibility for the installation of electrolytic cells to produce hydrogen for use in the hydrogenation of vegetable oils. At the time, the industry used a nickel catalyst that was prepared by hydrogen reduction of supported nickel oxide. Raney believed that better catalysts could be produced, however, and in 1921 he formed his own research company. In 1924, Raney produced a 50% nickel–silicon alloy which he treated with aqueous sodium hydroxide to produce a grayish metallic solid that was tested by the hydrogenation of cottonseed oil. Raney found the activity of his catalyst to be fivefold greater than the best catalyst then in use, and consequently applied for a patent which was issued on December 1, 1925 [2].

Subsequently, Raney produced a nickel catalyst by leaching a 50 wt.% Ni–Al alloy in aqueous sodium hydroxide; this catalyst was even more active, and a patent application was filed in 1926 [3]. This class of materials is generically referred to as "skeletal" or "sponge" metal catalysts. The choice of an alloy containing 50 wt.% Ni and 50 wt.% Al was fortuitous and without scientific basis, and is part of the alchemy referred to above. However, it is of interest to note that it is the preferred alloy composition for the production of skeletal nickel catalysts currently in use. In 1963, Murray Raney sold his business to the W. R. Grace and Company, the Davison Division of which produces and markets a wide range of these

---
* Corresponding author.

catalysts. Because Raney catalysts are protected by a registered trademark, only those products produced by Grace Davison are correctly termed "Raney Ni", "Raney Cu", "Raney Cu", etc. Alternatively, the more generic term "skeletal" is used to refer to catalysts in the following text. In addition, "Ni−Al" or "Cu−Al", etc. rather than "Raney alloy" is used to refer generically to the precursor to the catalyst.

Following Raney's development of sponge-metal nickel catalysts by alkali leaching of Ni−Al alloys, other alloy systems were considered. These included iron [4], cobalt [5], copper [6], platinum [7], ruthenium [8], and palladium [9]. Small amounts of a third metal such as chromium [10], molybdenum [11], or zinc [12] have also been added to the binary alloy to promote catalyst activity. The two most common skeletal metal catalysts currently in use are nickel and copper, in either unpromoted or promoted forms. Skeletal copper is less active and more selective than skeletal nickel in hydrogenation reactions, and it also finds use in the selective hydrolysis of nitriles [13]. Therefore, this chapter is mainly concerned with the preparation, properties and applications of promoted and unpromoted skeletal nickel and skeletal copper catalysts which are produced by the selective leaching of aluminum from binary or ternary alloys.

2.3.2.2 General Aspects

2.3.2.2.1 Alloy Preparation

Alloys are prepared both commercially and in the laboratory by melting the active metal and aluminum in a crucible and quenching the resultant melt; the latter is then crushed and screened to the particle size range required for a particular application. The alloy composition is very important, as different phases leach quite differently and this leads to markedly different porosities and crystallite sizes of the active metal. Mondolfo [14] provides an excellent compilation of the binary and ternary phase diagrams for aluminum alloys, including those used for the preparation of skeletal metal catalysts. Alloys of a number of compositions are available commercially for activation either in the laboratory or plant. These include alloys of aluminum with nickel, copper, cobalt, chromium−nickel, molybdenum−nickel, cobalt−nickel, and iron−nickel.

2.3.2.2.2 Activation Using Alkali Leaching

Skeletal catalysts are generally prepared by the selective removal of aluminum from alloy particles, using aqueous sodium hydroxide. The leaching reaction is given by:

$$2M\text{-}Al_{(s)} + 2OH^- + 6H_2O_{(l)} \longrightarrow 2M_{(s)} + 2Al(OH)_4^- + 3H_{2(g)} \quad (1)$$

The dissolution of aluminum in aqueous sodium hydroxide may be represented more simply by:

$$2Al + 2NaOH + 2H_2O \longrightarrow 2NaAlO_2 + 3H_2 \quad (2)$$

The formation of sodium aluminate ($NaAlO_2$) requires that high concentrations (20–40 wt.%) of excess sodium hydroxide are used in order to avoid the formation of aluminum hydroxide, which precipitates as bayerite:

$$2NaAlO_2 + 4H_2O \longrightarrow Al_2O_3 \cdot 3H_2O + 2NaOH \quad (3)$$

and causes the blocking of pores and surface site coverage in the sponge metals formed during leaching. The bayerite deposition leads to a loss of surface area and hence of catalyst activity. Care must be taken during activation of the alloys to vent the large quantities of hydrogen that are produced by reaction, in order to prevent explosions and fires.

The temperature used to leach the alloy has a marked effect on the pore structure and surface area of the catalyst. The surface areas of skeletal catalysts decrease with increasing temperature of leaching due to structural rearrangements that lead to increases in crystallite size analogous to sintering [15]. Leaching of aluminum from certain alloys can be extremely slow at low temperatures, and hence a compromise in the temperature of leaching must be reached in order to produce a catalyst with an appropriate surface area within a reasonable period of time.

One convenient method of producing powdered skeletal catalysts is to use the procedure described by Freel et al. [16] for Raney nickel. The same relative proportions of alloy and leachant can be used to maintain the leach reaction relatively isothermal when larger quantities of catalyst are desired. In this method, an aqueous solution containing 40 wt.% NaOH is added stepwise to 30 g of alloy powder and 150 mL of distilled water in a vessel at 313 K. Alkali additions are made at approximately 2-min intervals. For the first 20-min period the volume of alkali added is 2 mL, and 5-mL additions are made thereafter. A reaction time of around 3 h is sufficient to fully leach 500 μm particles of aluminum−nickel or aluminum−copper alloys at 313 K. After extraction, the catalysts are washed with water at ambient temperature, first by decantation and then by water flow in a vessel until the pH is lowered to around 9. The samples can then be stored in a closed vessel under deaerated distilled water in order to prevent oxidation prior to use.

Activated skeletal catalysts including nickel, copper, cobalt, and molybdenum or chromium-promoted nickel are available commercially.

References see page 99

2.3.2.2.3 Storage and Handling
Skeletal metal catalysts are extremely pyrophoric due to the small sizes of the metal crystallites that form during the leaching process. If the catalysts are allowed to dry in air, the metal particles rapidly oxidize; this generates large amounts of heat such that the particles glow red. The heat may also cause the ignition of combustible materials in the vicinity. It is therefore important that, after preparation, the catalysts be properly stored in a liquid. Although water is generally used as the storage medium, there is the possibility that the hydrolysis of any residual sodium aluminate will occur according to Eq. (3), and that this will lead to the formation of hydrated alumina resulting in catalyst deactivation by surface and pore blocking. For this reason, storage under slightly alkaline conditions (pH 9–11) is preferred. Studies of the storage of catalysts using aliphatic alcohols suggest that isopropanol is a better storage medium than water, although this does not have any significant practical importance.

2.3.2.2.4 Advantages of Skeletal Metal Catalysts
The principal advantage of skeletal catalysts is that they can be stored in the form of the active metal and therefore require no pre-reduction prior to use, as do conventional catalysts which are in the form of the oxide of the active metal supported on a carrier. These catalysts can also be prepared on demand by a simple caustic leaching procedure. They have very high activity as the Brunauer, Emmett and Teller (BET) surface area (typically up to 100 m^2 g^{-1} for skeletal nickel and 30 m^2 g^{-1} for skeletal copper) is essentially the metal surface area. Skeletal catalysts are low in initial cost per unit mass of metal, and therefore provide the lowest ultimate cost per unit mass of active catalyst. The high metal content provides good resistance to catalytic poisoning.

Because alloy composition and leaching conditions can be carefully controlled, skeletal catalysts exhibit excellent batch-to-batch uniformity. The particle size of the catalyst can be easily controlled through crushing and screening; thus, ultrafine powders can be produced for use in slurry-phase reactors whilst large granules can be produced for use in fixed-bed applications. The relatively high densities of skeletal catalysts (particularly nickel) provide excellent settling characteristics compared with supported catalysts when used in slurry-phase reactors. The high thermal conductivity of the all-metal skeletal catalyst is a further advantage.

2.3.2.3 Skeletal Nickel Catalysts

2.3.2.3.1 Alloy Preparation of Skeletal Nickel Catalysts
Skeletal nickel catalysts used industrially are produced from alloys that typically contain 40–50 wt.% nickel, with the 50% composition being most commonly used. The alloys are produced by adding molten aluminum to nickel, which dissolves by a highly exothermic reaction. For laboratory preparations the use of an induction furnace and graphite crucibles provides a very convenient method of preparation. The melt is rapidly quenched in water, leading to alloys with quenched structures consisting of the intermetallics Ni$_2$Al$_3$, NiAl$_3$, and some frozen eutectic. It has been found [17] that the eutectic material leaches more rapidly than NiAl$_3$, which leaches much more rapidly than Ni$_2$Al$_3$.

The phase diagram for the Al–Ni system is shown in Fig. 1. An alloy of composition 42 wt.% nickel corresponds to NiAl$_3$, whilst Ni$_2$Al$_3$ contains approximately 60 wt.% nickel. A study of the selective leaching of essentially pure NiAl$_3$ and Ni$_2$Al$_3$ intermetallics [17] has shown that NiAl$_3$ readily leaches in 20 wt.% aqueous NaOH at temperatures from 274 to 323 K, producing porous nickel which was friable and readily disintegrated. On the other hand, at those temperatures Ni$_2$Al$_3$ was unreactive, requiring temperatures from 343 to 380 K to produce significant extents of leaching. The reaction between Ni$_2$Al$_3$ and the NaOH solution proceeded in two steps. First, a two-phase mixture of Ni$_2$Al$_3$ plus Ni was produced, but after longer times nickel alone was produced. It is apparent that the 50 wt.% Ni alloy used frequently in an industrial scenario represents a composition that is a compromise between the readily leached NiAl$_3$, which produces mechanically weak catalysts, and Ni$_2$Al$_3$, which is more difficult to leach but forms a strong residual material.

Rapid quenching of the alloy melt is desirable as it produces very small crystals within the alloy [18],

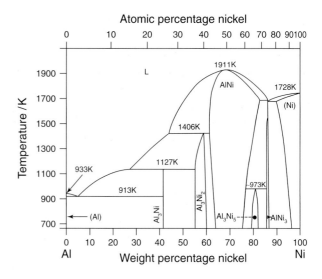

Fig. 1 Al–Ni phase diagram.

providing a large number of grain boundaries through which the leach liquor can penetrate. It is also possible to obtain supersaturation of additive species by using a particularly fast quenching method [19–21].

Some interest has been shown in rapid quenching techniques that produce amorphous metal glasses. Deng et al. [22] reviewed amorphous alloys involving Ni–P, Ni–Co, Ni–B, Ni–Co–B, and Ni. Increased catalytic activity was observed for various hydrogenation reactions, partly due to increased surface area (Ni–P) and partly to chemical promotion (Ni–Co–B). Aluminum species remaining on or near the surface are thought to act as a matrix to stabilize the nickel [23].

Nanocrystalline $NiAl_x$ ($1 < x < 3$) alloys have been prepared via a wet chemistry technique involving nickel(0) and aluminum coordination compounds [24]. The catalysts produced after leaching have a very fine structure and can lead to catalytic activities exceeding those of commercially available skeletal nickel [24, 25].

2.3.2.3.2 Properties of Skeletal Nickel Catalysts Skeletal nickel catalysts have BET surface areas typically in the range 50 to 100 $m^2\ g^{-1}$. The theoretical pore volumes for fully leached $NiAl_3$ and Ni_2Al_3 are 0.48 $cm^3\ g^{-1}$ and 0.17 $cm^3\ g^{-1}$, respectively. The large pore volume for the material prepared from $NiAl_3$ accounts for its low mechanical strength. Typical characteristics of Raney nickel catalysts produced by leaching a 50 wt.% Ni alloy using a NaOH : Al molar ratio of approximately 1.8 : 1 are presented in Table 1. These data show that an increased temperature of leaching leads to an increased pore diameter, increased crystallite size, and a lower total surface area.

One remarkable property of skeletal nickel is its ability to store hydrogen that is produced during the leaching process. It has been shown previously that the amount of hydrogen present in a freshly prepared sample of skeletal nickel can exceed by an order of magnitude the amount that could be chemisorbed on the surface nickel atoms. Many explanations have been proposed to explain this phenomenon, but suffice to say the ability to store hydrogen accounts for the high activity of Raney nickel over a wide range of hydrogenation reactions.

2.3.2.3.3 Uses of Skeletal Nickel Catalysts Raney nickel catalysts are used over a wide range of organic synthesis reactions including:

- hydrogenation of nitro compounds
- hydrogenation of alkenes
- hydrogenation of carbonyl compounds
- hydrogenation of nitriles
- ammonolysis of alcohols
- hydrogenation of alkynes
- hydrogenation of aromatic compounds
- reductive alkylation
- methanation.

Some typical industrial applications of Raney nickel catalysts are listed in Table 2.

Skeletal nickel also finds application as an electrocatalyst, particularly for the hydrogen fuel cell [28–32]. The catalyst is usually embedded in a polytetrafluoroethylene (PTFE) matrix to create the gas diffusion electrode for the fuel cell. Skeletal nickel does not only offer the advantage of low-temperature operation but also removes the need for precious metals.

The 2004 Murray Raney Award was presented to Lessard for his group's studies on electrocatalytic hydrogenation of organic compounds over skeletal metal catalysts [33]. In particular, highly efficient hydrogenation is possible at milder conditions than would otherwise be achievable with conventional catalytic hydrogenation over the same catalysts. Indeed, in the case of skeletal Cu and Co, high selectivity to specific functional groups is possible.

2.3.2.3.4 The Promotion of Skeletal Nickel Catalysts The addition of a second component in metal catalysts is widely used in order to enhance activity and/or selectivity. In the case of skeletal nickel catalysts, it is a simple procedure to add small amounts of a second metal during the alloy preparation stage. Although other metals have been used in laboratory studies, the most common metals used to promote skeletal nickel catalysts employed industrially are Co, Cr, Cu, Fe, and Mo.

References see page 99

Tab. 1 Surface properties of skeletal nickel catalysts produced by leaching a 50 wt.% Ni alloy in aqueous sodium hydroxide solution (Data compiled from Refs. [16, 26, 27])

Activation temperature/K	BET surface area/$m^2\ g^{-1}$	Pore volume/ $cm^3\ g^{-1}$	Average pore diameter/nm	Crystallite size/nm	Surface as Ni/%
323	100	0.064	2.6	3.6	59
380	86	0.125	5.8	5.7	75

Tab. 2 Industrial applications of skeletal nickel catalysts

Reaction	Reactant	Product
Hydrogenation of nitro compounds	2,4-Dinitrotoluene	2,4-Toluenediamine
	2-Nitropropane	Isopropylamine
Hydrogenation of alkenes	Sulfolene	Sulfolane
Hydrogenation of carbonyl compounds	Dextrose	Sorbitol
	2-Ethylhexanal	2-Ethylhexanol
Hydrogenation of nitriles	Stearonitrile	Stearylamine
	Adiponitrile	Hexamethylenediamine
Ammonolysis of alcohols	1,6-Hexanediol	Hexamethylenediamine
Hydrogenation of alkynes	1,4-Butynediol	1,4-Butanediol
Hydrogenation of aromatics	Benzene	Cyclohexane
	Phenol	Cyclohexanol
Reductive alkylation	Dodecylamine + formaldehyde	N,N-dimethyldodecylamine
Methanation	Synthesis gas ($CO/CO_2/H_2$)	Methane

Montgomery [11] has made a detailed study of the functional group activity of promoted Raney nickel catalysts. He prepared catalysts by leaching alloy of the type Al (58 wt.%)/Ni (37–42 wt.%)/M (0–5 wt.%), where M = Co, Cr, Cu, Fe, and Mo, in aqueous sodium hydroxide (NaOH/Al (molar) = 1.80) at 323 K. The activities of the catalysts were measured by the rates of hydrogenation of various organic compounds, including an alkene, a carbonyl compound, a nitro compound, and a nitrile compound. Among the metals tested, molybdenum was found to be the most effective promoter. All of the metals tested were found to increase the activity of Raney nickel in the hydrogenation of a nitrile compound. In fact, the optimum level of promoter present in the precursor alloy was found to be Cr = 1.5 wt.%, Mo = 2.2 wt.%, Co = 2.5–6.0 wt.%, Cu = 4.0 wt.%, and Fe = 6.5 wt.%. The effect of promoters was most apparent for the hydrogenation of a nitrile compound. The results of Montgomery's group [11], showing catalysts with optimum activity, are summarized in Table 3.

More recent investigations with promotion have found that skeletal nickel promoted with Sn is particularly effective for hydrogen production from biomass-derived oxygenated hydrocarbons [34]. The Sn addition decreases byproduct methane formation from C–O bond cleavage, thereby enhancing the selectivity of the nickel catalyst toward hydrogen and providing a performance that compares favorably with that of precious metal catalysts.

The surface of skeletal nickel can also be modified by the addition of organic species; the most important of these is tartaric acid, which is used to create a modified skeletal nickel catalyst capable of enantioselective hydrogenations. An example of this is the synthesis of chiral alcohols from ketones for pharmaceutical manufacture [35].

Tab. 3 Effect of metallic promoters in the hydrogenation of organic compounds using Raney® nickel (Data compiled from Ref. [11])

Promoter (M)	Alloy composition $\frac{M \times 100}{Ni + M + Al}$	Organic compound	Relative activity[a] $\frac{r(M + Ni)}{r(Ni)}$
Mo	2.2	Butyronitrile	6.5
Cr	1.5	"	3.8
Fe	6.5	"	3.3
Cu	4.0	"	2.9
Co	6.0	"	2.0
Mo	2.2	Acetone	2.9
Cu	4.0	"	1.7
Co	2.5	"	1.6
Cr	1.5	"	1.5
Fe	6.5	"	1.3
Fe	6.5	Sodium p-nitrophenolate	2.1
Mo	1.5	"	1.7
Cr	1.5	"	1.6
Cu	4.0	"	1.3
Mo	2.2	Sodium itaconate	1.2

[a] Ratio of reaction rate for promoted catalyst to reaction rate over unpromoted Raney® nickel.

2.3.2.4 Skeletal Copper Catalysts

2.3.2.4.1 Alloy Preparation of Skeletal Copper Catalysts

The earliest study of copper catalysts prepared by the skeletal method was that of Fauconnau [6], who used Devarda's alloy (45 wt.% Cu, 50 wt.% Al, 5 wt.% Zn) and aluminum bronze (90 wt.% Cu, 10 wt.% Al). In a later study, Stanfield and Robins [36] investigated the influence of the composition of the precursor Cu–Al alloy along with leaching conditions. These authors found that a catalyst prepared from a 40 wt.% Cu, 60 wt.% Al alloy was the most active in hydrogenation reactions. The most commonly used alloy has a nominal composition of 50 wt.% Cu and 50 wt.% Al, which corresponds to an almost pure $CuAl_2$ phase with a small amount of Al–$CuAl_2$ eutectic.

2.3.2.4.2 Properties of Skeletal Copper Catalysts

Leaching of aluminum takes place at a sharp reaction front (see Fig. 2a), with the final skeletal copper product consisting of a three-dimensional network of short interconnected copper rods (see Fig. 2b). The temperature and time of leaching have a marked effect on the surface area of skeletal copper catalysts. The surface area decreases with increasing temperature of leaching, whilst prolonged contact with caustic solutions leads to structural rearrangements that cause significant reductions in the surface area [15]. The structural rearrangements involve the dissolution and reprecipitation of copper atoms that work to coarsen the skeletal structure and occur continuously during leaching, as it is the mechanism forming a coherent structure [37]. At longer leach times the result is a reduction in surface area, while aluminum continues to be leached.

The surface properties of skeletal copper catalysts produced by leaching a 50 wt.% Cu alloy in aqueous sodium hydroxide solution at 293 K are listed in Table 4. These data show that the surface area decreases with increasing particle size of the alloy. The effect of temperature of extraction on the surface area and pore structures of completely leached 1000–1180 μm particles of the 50 wt.% Cu alloy are listed in Table 5. These results show that increasing the temperature of leaching from 275 to 363 K leads to a steady decrease in surface area, from 25.4 m² g⁻¹ to 12.7 m² g⁻¹.

2.3.2.4.3 Uses of Skeletal Copper Catalysts

Skeletal copper catalysts are used in a range of selective hydrogenation and dehydrogenation reactions. For example, they are

References see page 99

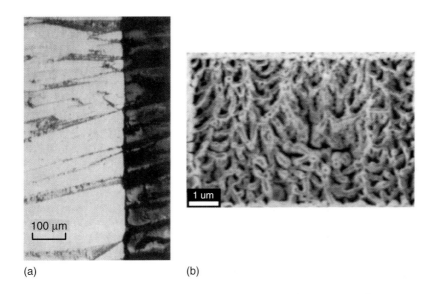

Fig. 2 (a) Skeletal copper leaches with a sharp reaction front, producing (b) a uniform network of short interconnected copper rods. (Compiled from Refs. [37, 38].)

Tab. 4 Surface and pore structure data for particles of $CuAl_2$ alloy that are fully leached at 293 K (From Ref. [39])

Particle size/μm	S_{BET}/m² g⁻¹	Pore diameter $2r_p$/nm	Pore volume/ cm³ g⁻¹	Crystallite size/nm	Copper rod diameter/nm
105–180	31.4	23.6	0.214	8.7	20.6
353–420	25.3	31.6	0.203	8.2	28.5
710–850	23.5	34.2	0.197	8.7	31.5
1000–1180	24.0	33.8	0.197	8.5	31.1
1400–1676	23.9	34.2	0.197	8.5	31.5
2000–2360	21.4	38.2	0.195	8.2	35.4

Tab. 5 The effect of temperature of extraction on the surface area and pore structure of completely leached 1000–1180 μm particles of $CuAl_2$ alloy (From Ref. [39])

Extraction temperature/ K	S_{BET}/ $m^2\ g^{-1}$	Pore diameter $2r_p$/nm	Crystallite size/nm	Copper rod diameter/ nm
275	25.4	30.2	7.5	27.5
293	24.0	33.8	8.5	30.8
308	22.6	55.0	11.2	50.1
323	18.5	61.0	12.0	55.6
343	16.9	75.0	13.7	68.3
363	12.7	107.6	14.6	98.0

Tab. 6 Methanol yields at 493 K and 4.5 MPa for skeletal catalysts (produced using different leaching conditions) compared to coprecipitated catalysts tested under the same conditions (From Ref. [45])

Catalyst[a]	Space velocity/h^{-1}	Methanol yield/kg $L^{-1} h^{-1}$
I	36 000	1.12
I	15 000	0.80
IIa	36 000	0.64
IIb	36 000	0.61
III	12 000	0.60
IV	36 000	0.60
IV	15 000	0.44
V	15 000	0.45

[a] I: Cu–Al–Zn leached in 6.1 M NaOH/0.62 M Na_2ZnO_2, at 303 K.
II: $CuAl_2$ leached in 6.1 M NaOH/0.62 M Na_2ZnO_2; catalyst IIa 274 K; catalyst IIb 303 K.
III: Cu–Al–Zn leached 6.1 M NaOH, at 274 K.
IV, V: Commercial coprecipitated catalysts.

highly specific for the hydrogenation of the 4-nitro group in 2,4-dinitro-1-alkyl-benzene to the corresponding 4-amino derivative. They are also used for the hydrogenation of aldehydes to the corresponding alcohols, the dehydrogenation of alcohols to aldehydes or ketones, the hydrogenation of esters to alcohols, the dehydrogenation of methanol to produce methyl formate, and in the steam reforming of methanol. Thus, skeletal copper catalysts can be used in a wide range of gas-phase and liquid-phase hydrogenation and dehydrogenation processes.

One major industrial process that uses skeletal copper catalysts is the manufacture of iminodiacetic acid as an intermediate for the herbicide, glyphosate. Skeletal copper catalyzes the liquid-phase oxidative dehydrogenation of diethanolamine to iminodiacetate according to:

$$NH(CH_2CH_2OH)_2 + 2OH^-$$
$$\longrightarrow NH(CH_2COO^-)_2 + 4H_{2(g)} \quad (4)$$

The reaction is carried out in slurry-phase batch reactors at elevated temperature (around 435 K) and pressure (900 kPa). Under these conditions, the copper catalysts can suffer deactivation due to structural rearrangement in the high pH environment, as well as some attrition losses. Promoted skeletal copper catalysts (as described below) can provide a greater activity and an increased stability against this deactivation [40].

2.3.2.4.4 Promotion of Skeletal Copper Catalysts The addition of other metals to promote skeletal copper catalysts has been the subject of a number of investigations including the use of V, Cr, Mn, and Cd for the hydrogenation of nitro compounds [41], Cd in the hydrogenation of unsaturated esters to unsaturated alcohols [42], and Ni and Zn for the dehydrogenation of cyclohexanol to cyclohexanone. The use of Cr as a promoter is particularly attractive, as copper chromite catalysts are used in a wide range of industrial applications.

Zinc-promoted skeletal copper has been investigated as a potential replacement to conventional coprecipitated $CuO-ZnO-Al_2O_3$ catalysts for low-temperature methanol synthesis from syngas and the water gas shift (WGS) reaction [43–45]. The comparative performances of skeletal Cu–Zn and commercial catalysts for the methanol synthesis reaction are listed in Table 6. High activity and selectivity has also been identified with zinc-promoted skeletal copper for other reactions, such as the steam reformation of methanol, methyl formate production by dehydrogenation of methanol, and the hydrogenolysis of alkyl formates to produce alcohols.

The promotion of skeletal catalysts can be achieved by adding the promoting metal either to the precursor alloy or to the leach solution. The latter method can achieve a more uniform distribution of promoter across the catalyst particle after leaching [45]. More importantly, addition via the leach solution can overcome solubility issues in dissolving sufficient of the promoting metal into the precursor alloy.

Structural rearrangements referred to above for pure skeletal copper also occur with promoted skeletal copper, resulting in a loss of surface area and activity, albeit to a lesser extent depending on the promoting metal chosen [46]. Chromium promotion is known to produce a catalyst structure which is more resistant to coarsening than zinc promotion [46], and this has been related to the mechanism of incorporation of these two promoters [47].

The effects of chromium promotion on the characteristics of a skeletal copper catalyst are listed in Table 7. Clearly, chromium is acting to provide a much finer structure in the final catalyst, with smaller crystallite size

Tab. 7 Comparison of unpromoted and chromium-promoted skeletal copper catalysts produced from a 50 wt.% Cu alloy leached in NaOH solution at 273 K, with or without 0.02 M Na$_2$CrO$_4$ added to the leach liquor (Data compiled from Ref. [46])

Skeletal catalyst	BET surface area/m^2 g^{-1}	Pore volume cm^3 g^{-1}	Average pore diameter/nm	Crystallite size/nm
Cu	19.5	0.15	30.8	14.7
Cu–Cr$_2$O$_3$	40.7	0.18	17.7	9.2

and mean pore diameter and a correspondingly higher surface area.

2.3.2.5 Skeletal Cobalt Catalysts

Cobalt catalysts have activities in hydrogenation reactions between those of nickel and copper. For example, nickel catalyzes methanation, cobalt catalyzes low-molecular-weight hydrocarbon and higher alcohol synthesis, whilst copper catalyzes methanol synthesis. Skeletal Co has less activity but greater selectivity than skeletal Ni, and is effective in converting nitriles to primary amines in the absence of ammonia. Skeletal cobalt can be readily prepared from a nominal 50 wt.% Co alloy. For example, when particles of a 48.8 wt.% Co, 51.3 wt.% Al alloy were leached in a 40% aqueous sodium hydroxide solution, 97.5% of the Al was leached, and this resulted in porous cobalt with a BET surface area of 26.7 m^2 g^{-1} and a bimodal pore size distribution with pore diameter maxima of 4.8 nm and 20 nm [48]. The crystallite size of the extracted catalyst was 4.7 nm.

2.3.2.6 Other Skeletal Metal Catalysts

Skeletal catalysts from many other metals have successfully been prepared, although these have not found widespread use industrially.

It is possible to use skeletal metal structures as supports for other catalytic metals; typical examples include copper metal coated onto skeletal nickel supports [49], or cobalt coated onto skeletal iron [50]. Such a catalyst offers the strength and high surface area of the skeletal metal structure, but with the catalytic selectivity/activity of the coated metal. This is relevant when the desired catalytic metal cannot be made into a skeletal catalyst directly, or when the desired metal provides insufficient strength for the intended application. Methods such as metal displacement or electroless plating have been used to achieve the coating.

References

1. M. Raney, *Ind. Eng. Chem.* **1940**, *32*, 1199.
2. M. Raney, US Patent 1,563,587, assigned to Murray Raney (US), 1925.
3. M. Raney, US Patent 1,628,190, assigned to Murray Raney (US), 1927.
4. M. Raney, US Patent 1,915,473, assigned to Murray Raney (US), 1933.
5. B. V. Aller, *J. Appl. Chem.* **1957**, *7*, 130.
6. L. Fauconnau, *Bull. Soc. Chim.* **1937**, *4*(5), 58.
7. A. A. Vendenyapin, N. D. Zubareva, V. M. Akimov, E. I. Klabunovskii, N. G. Giorgadze, N. F. Barannikova, *Izv. Akad. Nauk. SSSR. Ser. Khim.* **1976**, *10*, 2340.
8. K. Urabe, T. Yoshioka, A. Ozaki, *J. Catal.* **1978**, *54*, 52.
9. T. M. Grishina, L. I. Lazareva, *Zh. Fiz. Khim.* **1982**, *56*, 2614.
10. R. Paul, *Bull. Soc. Chim. Fr.* **1946**, *13*, 208.
11. S. R. Montgomery, in *Functional Group Activity of Promoted Raney Nickel Catalysts*, W. R. Moser (Ed.), *Catalysis of Organic Reactions*, Dekker, New York, USA, 1981, p. 383.
12. W. F. Marsden, M. S. Wainwright, J. B. Friedrich, *Ind. Eng. Chem. Prod. Res. Dev.* **1980**, *19*, 551.
13. N. I. Onuoha, M. S. Wainwright, *Chem. Eng. Commun.* **1984**, *29*, 1.
14. L. F. Mondolfo, *Aluminium Alloys: Structure and Properties*, Butterworths, London, UK, 1976, 971 pp.
15. A. D. Tomsett, D. J. Young, M. R. Stammbach, M. S. Wainwright, *J. Mater. Sci.* **1990**, *25*, 4106.
16. J. Freel, W. J. M. Pieters, R. B. Anderson, *J. Catal.* **1969**, *14*, 247.
17. M. L. Bakker, D. J. Young, M. S. Wainwright, *J. Mater. Sci.* **1988**, *23*, 3921.
18. M. Hu, M. Qiao, Y. Pei, K. Fan, H. Li, B. Zong, X. Zhang, *Appl. Catal. A* **2003**, *252*, 173.
19. I. Ohnaka, I. Yamauchi, M. Itaya, *J. Jpn. Inst. Metals* **1992**, *56*, 973.
20. I. Yamauchi, I. Ohnaka, Y. Ohashi, *J. Jpn. Inst. Metals* **1993**, *57*, 1064.
21. I. Ohnaka, I. Yamauchi, *Mater. Sci. Eng.* **1994**, *A181–182*, 1190.
22. J.-F. Deng, H. Li, W. Wang, *Catal. Today* **1999**, *51*, 113.
23. H. Lei, Z. Song, X. Bao, X. Mu, B. Zong, E. Min, *Surf. Interface Anal.* **2001**, *32*, 210.
24. R. Richards, G. Geibel, W. Hofstadt, H. Bönnemann, *Appl. Organometal. Chem.* **2002**, *16*, 377.
25. H. Modrow, M. O. Rahman, R. Richards, J. Hormes, H. Bönnemann, *J. Phys. Chem.* **2003**, *107*, 12221.
26. J. Freel, S. D. Robertson, R. B. Anderson, *J. Catal.* **1970**, *18*, 243.
27. S. D. Robertson, R. B. Anderson, *J. Catal.* **1971**, *23*, 286.
28. E. W. Justi, W. Scheibe, A. W. Winsel, German Patent 1,019,361, assigned to Ruhrchemie AG and Steinkohlen-Elektrizitaet AG, 1954.
29. G. Sandstede, E. J. Cairns, V. S. Bagotsky, K. Wiesener, in *History of low temperature fuel cells*, W. Vielstich, H. A. Gasteiger, A. Lamm (Eds.), *Handbook of Fuel Cells - Fundamentals, Technology and Applications*, John Wiley & Sons Ltd, 2003, p. 173.
30. H. Binder, A. Koehling, G. Sandstede, in *Raney Catalysts*, G. Sandstede (Ed.), *From Electrocatalysis to Fuel Cells*, University of Washington Press, Seattle, 1972, p. 15.
31. G. F. McLean, T. Niet, S. Prince-Richard, N. Djilali, *Int. J. Hydrogen Energy* **2002**, *27*, 507.
32. *Fuel Cell Handbook*, National Energy Technology Laboratory, US Department of Energy, Morgantown, West Virginia, 2002, Ch 4.
33. J. Lessard, in *2004 Murray Raney Award: Electrocatalytic Hydrogenation of Organic Compounds at Raney Metal Electrodes*,

J. R. Sowa, Jr. (Ed.), *Catalysis of Organic Reactions*, CRC Press, Taylor & Francis Group, Boca Raton, US, 2005, p. 3.
34. G. W. Huber, J. W. Shabaker, J. A. Dumesic, *Science* **2003**, *300*(5628), 2075.
35. J. Court, M. Lopez, US Patent 6,825,370, assigned to Centre National de La Recherche Scientifique, 2004.
36. J. A. Stanfield, P. E. Robbins, *Actes. Congr. Intern. Catal (2nd), Paris* **1960**, *2*, 579.
37. A. J. Smith, T. Tran, M. S. Wainwright, *J. Appl. Electrochem.* **1999**, *29*, 1085.
38. J. Szot, D. J. Young, A. Bourdillon, K. E. Easterling, *Philos. Mag. Lett.* **1987**, *55*, 109.
39. A. D. Tomsett, *Pore Development in Skeletal Copper Catalysts*, PhD Thesis, University of New South Wales, Sydney, Australia, 1987, 197 pp.
40. T. S. Franczyk, US Patent 5,292,936, assigned to Monsanto Company (US), 1994.
41. K. Wimmer, O. Stichnoth, German Patent 875,519, assigned to BASF AG, 1953.
42. UK Patent 1,029,502, assigned to Kyowa Hakko Kogyo Co. Ltd., 1966.
43. A. Andreev, V. Kafedjiiski, T. Halachev, B. Kuner, M. Kaltchev, *Appl. Catal.* **1991**, *78*, 199.
44. J. R. Mellor, *The Water Gas Shift Reaction: Deactivation Studies*, PhD Thesis, University of Witwatersrand, Johannesburg, South Africa, 1993, 295 pp.
45. H. E. Curry-Hyde, M. S. Wainwright, D. J. Young, in *Improvements to Raney Copper Methanol Synthesis Catalysts through Zinc Impregnation: III Activity Testing*, D. Bibby, C. C. Chang, R. F. Howe, S. Yurchak (Eds.), Methane Conversion - Studies in Surface Science and Catalysis, Vol. 36, Elsevier, Amsterdam, The Netherlands, 1988, p. 239.
46. L. Ma, D. L. Trimm, M. S. Wainwright, in *Promoted Skeletal Copper Catalysts for Methanol Synthesis, Advances of Alcohol Fuels in the World*, 1998, Beijing, China, p. 1.
47. A. J. Smith, T. Tran, M. S. Wainwright, *J. Appl. Electrochem.* **2000**, *30*, 1103.
48. J. P. Orchard, A. D. Tomsett, M. S. Wainwright, D. J. Young, *J. Catal.* **1983**, *84*, 189.
49. D. A. Morgenstern, J. P. Arhancet, H. C. Berk, W. L. Moench, J. C. Peterson, US Patent, 2002019564, assigned to Monsanto Company (US), 2002.
50. S. R. Schmidt, Int. Patent Appl. WO2004091777, assigned to W. R. Grace & Co (US) and Stephen Raymond Schmidt (US), 2004.

2.3.3
Precipitation and Coprecipitation

Ferdi Schüth*, Michael Hesse, and Klaus K. Unger

2.3.3.1 Introduction

Precipitation is the process in which a phase-separated solid is formed from homogeneous solution, after supersaturation with respect to the precipitating solid has been achieved. A number of related phenomena are known,

---
* Corresponding author.

which are often not clearly discriminated. Crystallization from solution is a process, in which the solid is directly obtained in crystalline form. Crystallization typically proceeds at relatively low supersaturation, which is induced mostly by reduction of the temperature or evaporation of the solvent. Precipitation is often used for the description of processes, in which the solid formation is induced by addition of an agent which initiates a chemical reaction or which reduces the solubility (antisolvent). Precipitation normally involves high supersaturation, and thus frequently amorphous intermediates are obtained as the first solids formed. However, no consistent distinction between crystallization and precipitation is made in the literature, although with respect to catalyst synthesis, mostly indeed an agent is added which effects the solid formation – that is, precipitation in the narrow sense of the word is used.

Sol–gel synthesis is also related to precipitation, the key difference being that, in a sol–gel reaction, the formed solid is a container-spanning hydro- or alcogel, while in precipitation a clear phase separation occurs. However, there may also be confusion here in the literature, because often a sol–gel synthesis is referred to in cases where alkoxide precursors are used, regardless of whether a true gel is formed as an intermediate, or a precipitate with clear phase separation occurs. In this chapter, the term "precipitation" will be used in a rather broad sense, but in cases where a true sol–gel synthesis is referred to, this will be clearly indicated. More information on sol–gel processes can be found in Chapter 2.3.4 of this Handbook.

The preparation of catalysts and supports by precipitation or coprecipitation is technically very important [1]. However, precipitation is usually more demanding than several other preparation techniques, due to the necessity of product separation after precipitation and the large volumes of salt-containing solutions generated in precipitation processes. Thus, techniques for catalyst manufacture must produce catalysts with better performance in order to compensate for the higher costs of production in comparison, for instance, to solid-state reactions for catalyst preparation.

Nevertheless, for several catalytically relevant materials – and especially for support materials – precipitation is the most frequently applied method of preparation. These materials include mainly aluminum and silicon oxides. In other systems precipitation techniques are also used, for example in the production of iron oxides, titanium oxides, or zirconias. The main advantages of precipitation when preparing such materials are the possibility of creating very pure materials and the flexibility of the process with respect to final product quality.

Other catalysts, based on more than one component, can be prepared by coprecipitation. According to the

IUPAC nomenclature [2], coprecipitation is the simultaneous precipitation of a normally soluble component with a macrocomponent from the same solution by the formation of mixed crystals, by adsorption, occlusion, or mechanical entrapment. However, in catalyst preparation technology, the term is usually used in a more general sense in that the requirement of one species being soluble is dropped. In many cases, both components to be precipitated are essentially insoluble under precipitation conditions, although their solubility products might differ substantially. Thus, in this chapter the term coprecipitation will be used for the simultaneous precipitation of more than one component. Such systems prepared by coprecipitation include, for example, Ni/Al_2O_3, Cu/Al_2O_3, Cu/ZnO, $Cu/ZnO/Al_2O_3$, and Sn–Sb oxides.

Coprecipitation is very suitable for the generation of a homogeneous distribution of catalyst components, or for the creation of precursors with a definite stoichiometry, which can be easily converted to the active catalyst. If the precursor for the final catalyst is a stoichiometrically defined compound of the later constituents of the catalyst, a calcination and/or reduction step to generate the final catalyst usually creates very small and intimately mixed crystallites of the components. This has been shown for several catalytic systems, and is discussed in more detail later in the chapter. Such a good dispersion of catalyst components is difficult to achieve by other means of preparation, and thus coprecipitation will remain an important technique in the manufacture of solid catalysts, in spite of the disadvantages associated with such processes. These disadvantages are the higher technological demands, the difficulties in following the quality of the precipitated product during the precipitation, and the problems in maintaining a constant product quality throughout the whole precipitation process, if the precipitation is carried out discontinuously.

In order to stress the technical relevance of precipitated catalysts, an overview of industrially used precipitated catalysts and supports is provided in Table 1. Since the catalyst compositions – and even less so the catalyst preparation procedures – for many industrial processes are not disclosed by the respective companies, this list is by no means comprehensive.

2.3.3.2 General Principles Governing Precipitation from Solutions

Precipitation processes are relevant not only for catalysis, but also for other industries, as for example the production of pigments. Despite the tremendous importance of precipitation from solution, many basic questions in this field remain unsolved, and the production of a precipitate with properties that can be adjusted at will is still rather more of an art than a science. This is primarily due to the fact that the key step – the nucleation of the solid from a homogeneous solution – is not only very elusive but also difficult to study using the analytical tools currently available. On the one hand, spectroscopies using local probes are not sufficiently sensitive to study larger arrangements of atoms, whereas on the other hand neither are diffraction methods suitable for analysis, as a nucleus is not large enough to produce a distinctive

References see page 116

Tab. 1 Some industrially relevant catalysts and supports obtained by precipitation and coprecipitation techniques

Material	Use	Important examples
Al_2O_3, mostly γ, in special cases α or η	Support, catalyst	Claus process, dehydration of alcohols to alkenes and ethers, support of hydrotreating catalysts, support for three-way catalyst
SiO_2	Support	Noble metal/SiO_2 for hydrogenation reactions, Ni/SiO_2 for hydrogenation reactions, V_2O_5/SiO_2 for sulfuric acid production, support for Ziegler catalysts
Al_2O_3/SiO_2	Catalyst	Acid-catalyzed reactions, such as isomerization, component of FCC catalysts
Fe_2O_3	Catalyst, catalyst component	Fischer–Tropsch reactions, major component of catalyst for ethylbenzene dehydrogenation to styrene
TiO_2	Support, catalyst, catalyst component	Major component of DeNO$_x$ catalyst, photocatalyst
ZrO_2	Catalyst, support	Acid catalyst after sulfate modification, component of three-way catalyst support
Cu/ZnO	Catalyst	Methanol synthesis, methanol steam reforming
$(VO)_2P_2O_7$	Catalyst	Selective oxidation (e.g., butane to maleic anhydride)
Cu–Cr-oxides	Catalyst	Combustion reactions, hydrogenations
$AlPO_4$	Support, catalyst	Polymerization, acid-catalyzed reactions
Sn–Sb-oxides	Catalyst	Selective oxidation (e.g., isobutene to methacrolein)
Bi-molybdates	Catalyst	Selective oxidation (e.g., propene to acrolein), mostly supported in thin-layer on non-porous carriers

FCC = fluid catalytic cracking.

diffraction pattern. Thus, investigations of crystallization and precipitation processes from solution must often rely on indirect and theoretical methods. A general flow scheme for the preparation of a precipitated catalyst is depicted in Fig. 1.

2.3.3.2.1 Physico-Chemical Considerations The best-defined situation for the formation of a solid from solution is that of homogeneous nucleation; that is, the solid is formed in the bulk of the solution which initially is free of any solid particles. However, in practice in most cases of catalyst precipitation heterogeneous nucleation prevails; that is, the formation of solid in the presence of either seeds of the same solid or other solid material, which can be added at will or may be present as impurities. Nevertheless, with regards to the underlying physical principles, homogeneous precipitation will first be discussed.

In order for a solid to precipitate from homogenous solution, a nucleus has first to form. The formation of a particle is governed by the free energy of agglomerates of the constituents of the solution. The total free energy change due to agglomeration, ΔG, is determined by

$$\Delta G = \Delta G_{bulk} + \Delta G_{interface} + \Delta G_{others} \quad (1)$$

where ΔG_{bulk} is the difference of the free energy between solution species and solid species, $\Delta G_{inerface}$ is the free energy change related to the formation of the interface, and ΔG_{others} summarizes all other contributions, as for instance strain or impurities, which can be neglected here. The agglomeration will be energetically favored if ΔG is negative. At supersaturation conditions, ΔG_{bulk} is always negative but, to create an interface, energy is needed; $\Delta G_{interface}$ is thus positive. For very small particles the total free energy change is positive. If spherical particles are formed, ΔG_{bulk} increases with $4\pi r^3/3$ while the interfacial energy only increases with $4\pi r^2$. There is therefore a critical size r of the agglomerate, from which on ΔG_{bulk} predominates the total free energy change and the total free energy decreases with the particle size. This critical size is the size of the nucleus.

The general process of the formation of a solid from a solution can be described in a simplified form, as indicated in Fig. 2. The most important curve is the nucleation curve, which describes the development of the precursor concentration with time. Such a precursor could, for instance, be the hydrolysis product of the metal ions in solution. Only if the precursor concentration exceeds a critical threshold concentration will a nucleus form, and the precipitation begin. The nucleus is defined as the "smallest solid-phase aggregate of atoms, molecules or ions which is formed during a precipitation and which is capable of spontaneous growth" [2]. As long as the concentration of precursor species stays above the nucleation threshold, new particles are formed, but as soon as the concentration falls below the critical concentration, due to the consumption of precursors by nucleation or by the growth process, then only particle growth of existing particles prevails. Thus, in the framework of this simple concept, which was introduced mainly by the studies of LaMer during the early 1950s [3] (and later used extensively by Matijevic [4],

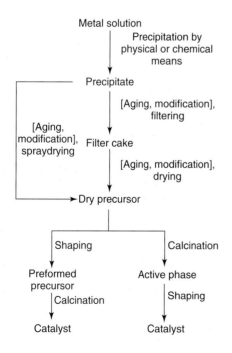

Fig. 1 Preparation scheme for precipitated catalysts. Optional steps are indicated by square brackets.

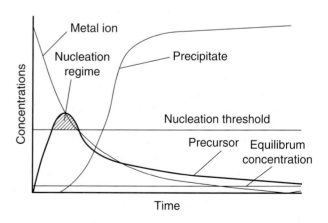

Fig. 2 Simplified scheme for the formation of a solid product from solution. From the metal ions a precursor species is formed, for example by hydrolysis or raising the pH. When the concentration of the precursor species exceeds the nucleation threshold, precipitation of the product begins, consuming the precursor by nucleation and growth. New nuclei are formed only in the shaded area.

who produced a large number of different monodispersed oxides or hydroxides), particles with a narrow particle size distribution will result from a short nucleation burst, and a wide particle size distribution will result from nucleation over a longer period of time.

The size of the particles finally resulting from a precipitation process will be dependent on the area of the shaded section between the nucleation curve and the nucleation threshold. In general, the larger the area, the more particles nucleate and the smaller the resulting particles will be. It should be borne in mind, however, that this is not a direct proportionality, since at higher supersaturation the nucleation rates are higher; that is, more particles per time unit are formed under high supersaturation conditions than in the case where the supersaturation just exceeds the nucleation threshold. The final particle size is determined by the interplay between nucleation and growth (and possibly also later steps, such as Ostwald ripening). The nucleation process is strongly temperature-dependent, since the rate constant for homogeneous nucleation usually does not follow an Arrhenius-type law, but is, for instance, better described by [5]:

$$\frac{dN}{dt} = \beta \exp\left(\frac{-A}{\ln^2 s}\right) \qquad (2)$$

where β is a pre-exponential term, A is the interfacial energy parameter $16\pi\sigma^3\Gamma^2/3(kT)^3$, σ the solid–fluid interfacial energy, Γ the solid molecular volume, and s the supersaturation. However, this is only one possible expression for the nucleation rate, and several others have been proposed.

Over the past few years, evidence has been accumulated that the formation of complex solids does not simply follow the pathways described above, and many of the problems associated with this description are discussed in an excellent review by Horn and Rieger [6]. According to broad evidence from different fields and systems, the nucleus is a much less clearly defined entity than is suggested from classical nucleation theory [7, 8]. A precipitation reaction rather must be considered as a sequence of chemical reactions, each with its own activation threshold, and there may be a number of individual energy barriers which are about equal in height. Inorganic solids – especially if they are rather complex and are formed at high supersaturation (which holds for many solid catalysts) – may often first occur as amorphous precipitates which later recrystallize gradually. For instance, in the precipitation of ZnS, theoretical modeling suggests that the initial stages involve flat clusters, then "bubble" clusters, and even "double bubble" clusters, in which a smaller more or less spherical shell is embedded in a bigger one. These species are generated by a sequence of chemical reactions, in which aquo-ligands of the Zn^{2+} centers are replaced by S^{2-} groups [9]. Even for crystals formed from colloidal particles (which are rather simple model systems), the expectations from classical nucleation theory were not confirmed in experimental observations of the ordering of the colloidal spheres: although nuclei were identified, they had a gradient of order from high in the inside to low close to the surface; moreover, the nuclei were found to be ellipsoidal on average instead of spherical [10]. Theoretical simulation using Monte Carlo methods also revealed a more complex picture of the nucleation process, as comprehensively reviewed by Auer and Frenkel [11]. Thus, although classical nucleation theory provides a conceptually simple picture, which in turn allows an explanation for some of the peculiarities of nucleating systems, more detailed study of each specific system is necessary for a true understanding of the underlying reaction sequences that lead to the formation of solids from solution.

For phenomenological modeling purposes, nucleation and growth processes can be described mathematically by sets of differential equations which balance the concentrations of the various species in the system. The most well-known approach to this problem is the so-called "population balance formalism" introduced during the early 1960s [12]. Although such models can provide valuable insight into the basic ideas of particle formation from solution, it is an extremely simplified concept. The models are usually only formulated for the formation of a single phase. If more than one phase is possible, the model does not provide information on the nature of the phase eventually formed. According to the Ostwald rule of successive phase transformation, initially the thermodynamically most unstable phases are formed which then transform to more stable phases. Another factor, which is relatively difficult to implement, is the lack of information on the decisive solution species for many systems. Usually, the species responsible for nucleation are also considered to be the species that contribute to the particle growth. However, the nucleation might involve relatively complex species, while growth – at least of the primary particles – in many cases is assumed to proceed via monomer addition. The model also completely neglects the role of aggregation and agglomeration processes, which further contribute to growth and can result in the formation of fewer (but larger) particles than predicted for the simple nucleation burst model. Such processes can be of great importance in the formation of technically relevant hydroxides and oxides [13]. However, even if aggregation and agglomeration do occur, narrow particle size distributions (which are often desired) can

References see page 116

be obtained. The uniformity in the final particle size distribution can be reached by size-dependent aggregation rates [14]. In addition, mechanical agitation or other processes can lead to the fragmentation of growing crystals, thus forming secondary nuclei which can alter the particle size distribution.

Another way to induce precipitation without needing a homogeneous nucleation step is to seed the solution. The best results are usually obtained if seeding is performed with the desired phase. If the solution is seeded, usually no nucleation takes place, as the precursor concentration never exceeds a critical threshold. The precipitation rates in seeded systems normally follow Arrhenius-type rate laws. The precipitation of $Al(OH)_3$ in the Bayer process is described by [13]:

$$\frac{-dc}{dt} = k \exp\left(\frac{-E}{RT}\right) A(c - c_{eq})^2 \quad (3)$$

where c is the Al_2O_3 concentration, k the rate constant, E the activation energy (ca. 59 kJ mol^{-1}), R the gas constant, T the temperature, A the seed surface area, and c_{eq} the equilibrium concentration. However, here as well, temperature dependencies can be complicated, as the equilibrium concentration might vary strongly with temperature.

From the facts considered above, it is clear that supersaturation of the solution from which precipitation occurs is a key factor of the precipitation process. Supersaturation can be achieved either by physical means, which usually means cooling down the reaction mixture or evaporating the solvent, or by chemical means – that is, by the addition of a precipitating agent or an antisolvent. The precipitating agent either changes the pH, thus leading to the condensation of precursors to form the hydroxides or the oxides, or it introduces additional ions into the system by which the solubility product for a certain precipitate is exceeded. The influence of such precipitating agents is discussed in Section 2.3.3.2.2.

If catalysts are prepared by coprecipitation, knowledge of the relative solubilities of the precipitates, and the possibility for the formation of defined mixed phases, are essential. If one of the components is much more soluble than the other, there is a possibility that sequential precipitation will occur. This will lead to concentration gradients in the product and to less intimate mixing of the components. If this effect is not compensated by adsorption or occlusion of the more soluble component, the precipitation should be carried out at high supersaturation in order to exceed the solubility product for both components simultaneously. Precipitation of the less-soluble product will proceed slightly faster, and the initially formed particles can act as nucleation sites for the more-soluble precipitate which forms by heterogeneous precipitation. The problem is less crucial if both components form a defined, insoluble species; this is the case for the coprecipitation of nickel and aluminum, which can form defined compounds of the hydrotalcite type (for an extensive review, see Cavani et al. [15] and the summary by Andrew [16]).

2.3.3.2.2 **Chemical Considerations** It is generally desirable to precipitate the target material in such a form, that the counterions of the precursor salts and the precipitation agent, which can be occluded in the precipitate during the precipitation, can easily be removed by a calcination step. If precipitation is induced by physical means, namely cooling or evaporation of the solvent to reach supersaturation of the solution, only the counterion of the metal salt is relevant. If precipitation is induced by the addition of a precipitating agent, then the ions introduced into the system via this route must also be considered. Favorable ions are nitrates, carbonates, or ammonium, which decompose to volatile products during the calcination. For catalytic applications, usually hydroxides, oxohydrates, oxides (in the following, the term ''hydroxides'' is used in a rather general sense, comprising hydroxides and oxides with different degrees of hydration) are precipitated; in some cases carbonates, which are subsequently converted to the oxides or other species in a calcination step, are formed. The precipitation of oxalates as precursors for spinel-type catalysts has also occasionally been reported to give good results [17]. If the ions do not decompose to volatile products, then careful washing of the precipitate is advisable.

In many cases it has been found advantageous to work at low and relatively constant supersaturation, which is achieved homogeneously in the whole solution (precipitation from homogeneous solution; PFHS). This approach can also be employed for deposition–precipitation processes (see Chapter 2.4.1). PFHS can be achieved by using a precipitating agent which slowly decomposes to form the species active in the precipitation. The most commonly employed precursor for the liberation of ammonia is urea, which has been used in many precipitation processes [18]. The ammonia is liberated homogeneously over the whole precipitation vessel, thus avoiding higher concentrations at the inlet point, as may occur if aqueous ammonia is used. In addition, the carbon dioxide released during the urea hydrolysis can keep the solution essentially oxygen-free. These differences can lead to markedly different products [19]. For the preparation of sulfides, thioacetamide might be used.

The precipitation of hydroxides can be performed either starting from an alkaline solution which is acidified, or from acidic solutions by raising the pH value. In the first case, the formation of the solid product proceeds via polyanionic species. These polyanionic species undergo condensation reactions, either via olation reactions

$$E - OH + H_2O - E \longrightarrow E - (OH) - E + H_2O \quad (4)$$

or via oxolation reactions

$$E - OH + HO - E \longrightarrow E - O - E + H_2O \quad (5)$$

A prototypic example for such precipitation reactions from alkaline solutions is that of SiO_2, which is prepared from silicates, as for instance sodium water glass by acidification. In this case, depending on the exact conditions, either precipitation or gelation can occur, but both processes are caused by condensation reactions between the various silicate ions in solution. However, most hydroxides for technical applications are precipitated from acidic solutions by the addition of a precipitating agent (usually ammonia or sodium carbonate). However, if other ions do not adversely influence the catalytic performance, then $Ca(OH)_2$ or NaOH can also be used. Depending on the metal ion and the precipitating agent, either the hydroxides, carbonates or hydroxycarbonates precipitate. Precipitation from acidic solutions mostly proceeds via polycations and – as in the basic case – by either olation or oxolation reactions. Intermediate states are less well known than for the polyanionic species, and only a few defined polycations have been reported in selected cases.

One special case to induce precipitation is the addition of an antisolvent. Here, two miscible liquids are used, in one of which the precursor of the solid has a high solubility, in the other a low solubility. The precursor is dissolved in the liquid in which it has a high solubility, after which the other liquid (the "antisolvent") is added, and this induces precipitation of the solid, typically in the form of relatively small particles. For the synthesis of catalytic materials, especially supercritical CO_2 has been used as the antisolvent [20, 21].

2.3.3.2.3 Process Considerations There are several alternative ways in which the precipitation process can be carried out (Fig. 3) [22]. The simplest implementation is the *batch operation*, where the solution from which the salt is to be precipitated is usually present in the precipitation vessel and the precipitating agent is added. The advantage of this mode of operation is the simple way in which the product can be obtained; the most severe disadvantage is the variation of batch composition during the precipitation process. This can lead to differences between the product formed during the initial stages of the precipitation and the precipitate formed at the end of the process. If a coprecipitation is carried out in this way, it is important to decide which compounds are present in the vessel and which compounds are to be added. If the

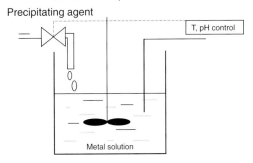

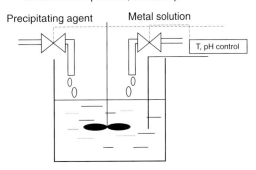

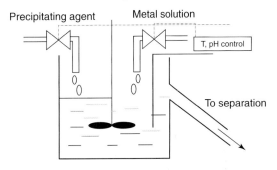

Fig. 3 Possible implementations of precipitation processes (after Ref. [22]). In the batchwise process (a), the pH and all other parameters except for the temperature change continuously during the precipitation due to consumption of the metal species. Coprecipitation should be carried out in the reversed arrangement by sequential precipitation. In process (b), the pH is kept constant but the batch composition and residence time of the precipitate change continuously. In process (c) all parameters are kept constant.

precipitating agent is present in the precipitator and the mixed-metal solutions are added, the product tends to be homogeneous, as the precipitation agent is always present in large excess. If, on the other hand, the precipitating agent is added to a mixed-metal solution, the precipitate with the lower solubility tends to precipitate first, thus resulting in the formation of an inhomogeneous product.

References see page 116

A slightly more complex process is the *simultaneous addition of both reagents* under strict control of the pH and the reagent ratios. If the precipitation is carried out following this procedure, the ratio of the metal salt and precipitating agent remains constant; all other concentrations, however, change during the process. The homogeneity of the product is usually better than in the first process described, but might still vary between the first precipitate and the precipitate formed last. This is due to the different concentrations of the other ions which are not precipitated and might be occluded in the precipitate to a greater extent during the final stages of the procedure. Moreover, the precipitates first formed are aged for a longer time in the solution. Thus, phase transitions might have already occurred, while fresh precipitates are still formed.

These problems are avoided if a *continuous process*, mostly in a continuously stirred-tank reactor (CSTR), is employed for the precipitation; however, this makes higher demands on the process control. In a continuous process all parameters such as temperature, concentration, pH, and residence time of the precipitate can be kept constant or altered at will. Continuous operation is, for instance, used for the precipitation of aluminum hydroxide in the Bayer process. Bayer aluminum hydroxide is the main source for the production of catalytically active aluminas. The precipitation step of the Bayer process is carried out continuously. An aluminum solution supersaturated with respect to $Al(OH)_3$, but not supersaturated enough for homogeneous nucleation, enters the precipitation vessel which already contains precipitate so that heterogeneous nucleation is possible. The nucleation rate must be controlled very carefully to maintain constant conditions. This is usually done by controlling the temperature of the system to within 2–3 K [13].

The continuous process usually allows precipitation at low supersaturation conditions, as the seeds are already present in the precipitation vessel. Thus, no homogeneous precipitation (which needs high levels of supersaturation) is necessary, and nucleation occurs heterogeneously with the associated lower supersaturation levels.

Alternatively, *tubular reactors* may be used for precipitation reactions instead of a CSTR. However, these are rather difficult to operate, as very long reactors may be needed in order to obtain the required product quality. Moreover, clogging of these reactors may lead to pressure build-up, causing severe problems in the practical operation. In laboratory processes, tubular reactors have been used for fundamental studies of precipitation reactions [6, 8, 23, 24]. In combination with a micromixer, a continuous reactor has also proved to be an interesting synthesis tool for production of the Cu–ZnO catalyst for methanol production [25]. The precursor solutions were mixed in a micromixer with precise temperature control and then directed to a settling vessel, before further processing steps were performed. The methanol productivity of catalysts produced with this procedure was quite high, but this was predominantly attributed to the relatively low solution concentrations needed to avoid clogging of the microstructure, and less to use of the continuous process. However, such synthesis schemes may be of interest if the whole process of catalyst production by precipitation could be carried out in a fully continuous fashion using efficient mixing devices, instead of in CSTRs with their relatively broad residence time distribution.

2.3.3.3 Influencing the Properties of the Final Product

Basically, all process parameters – some of which are fixed and some of which are variable – influence the quality of the final product of the precipitation. Usually, precipitates with specific properties are desired; these properties could be the nature of the phase formed, the chemical composition, purity, particle size, surface area, pore sizes, pore volumes, separability from the mother liquor, and many more, including demands imposed by the requirements of downstream processes, such as drying, pelletizing, or calcination. It is therefore necessary to optimize the parameters in order to produce the desired material. The parameters which may be adjusted in precipitation processes, and the properties which are mainly influenced by these parameters, are summarized in Fig. 4. The following discussion provides some general guidelines concerning the influence of various process parameters on the properties of the resulting precipitate. It should be stressed that the stated tendencies are only trends which might vary in special cases, as the exact choice of precipitation parameters usually results from a long, empirically driven optimization procedure and is a well-guarded secret of catalyst manufacturers or the producers of precursors for catalysts. Since for laboratory studies the exact synthetic pathway needs to be known and controlled, requiring information on addition rates, pH, concentrations, feed rates of reagents, etc., it is advisable to employ computer-controlled set-ups for the synthesis of catalysts, as shown in Fig. 5.

2.3.3.3.1 Influence of Raw Materials
As stated above, precursors are usually chosen with counterions that can easily be decomposed to volatile products. These are preferably the nitrates of metal precursors and ammonia or sodium carbonate as the precipitating agent; oxalates have also occasionally been employed. If the precipitation is carried out in the presence of ions which can be occluded, repeated washing steps are necessary if the ions

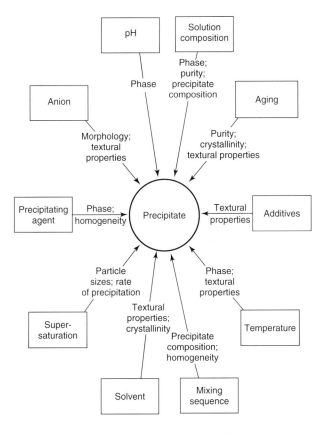

Fig. 4 Parameters affecting the properties of the precipitate, and the main properties influenced.

adversely affect the catalytic performance of the later catalyst. Ions such as chlorides or sulfates act as poisons in many catalytic reactions, and should therefore be avoided in the precipitation. The problem is reduced if supersaturation is reached by physical means; however, higher degrees of supersaturation – and thus more rapid precipitation and smaller particle sizes – are better achieved by changing the pH.

The nature of the ions present in the precipitation solution can strongly influence the properties of the final product. This was demonstrated effectively by Matijevic, who investigated the precipitation of many different metals, primarily as hydroxides [26]. The anions present do not only influence particle morphologies and sizes, but can even result in the formation of different phases. One striking example is given by Matijevic [26]: a solution of 0.0315 M $FeCl_3$ and 0.005 M HCl results in the formation of hematite, $\alpha\text{-}Fe_2O_3$; at higher concentrations (0.27 M $FeCl_3$ and 0.01 M HCl), $\beta\text{-}FeOOH$ is formed; in the presence of nitrate (0.18 M $Fe(NO_3)_3$) and sulfate (0.32 M Na_2SO_4), ferric basic sulfate ($Fe_3(OH)_5(SO_4)_2 \cdot 2H_2O$) precipitates; finally, with phosphate (0.0038 M $FeCl_3$ and 0.24 M H_3PO_4), $FePO_4$ is formed.

The precipitates differ both in the phase formed and also in their morphologies. Depending on the conditions, rather spherical particles are formed on the one hand, or needle-like crystals on the other hand.

References see page 116

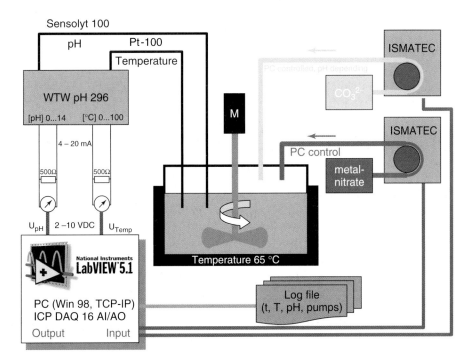

Fig. 5 Schematic drawing for a laboratory set-up for controlled precipitation reactions. (Illustration courtesy of C. Kiener.)

Other examples of the influence of starting materials are the precipitation of MoO_3 [27] or the preparation of $AlPO_4$ [28]. For MoO_3, small particles with relatively high surface area are formed with Na_2MoO_4 as the precursor salt, whereas larger particles with lower surface area precipitate from solutions containing $(NH_4)_6Mo_7O_{24}$. In the $AlPO_4$ system, the type of anion has a strong influence on the recrystallization behavior. Recrystallization to the α-crystobalite form occurs at 1073 K for $AlPO_4$ precipitated from aluminum nitrate. If the sulfate is used, even at calcination temperatures of 773 K, recrystallization to the phase with the tridymite structure takes place. The precipitate from chloride solutions only begins to recrystallize at 773 K, but at these temperatures large fractions of amorphous material are present. The phases formed are the tridymite structure and, at temperatures above 1200 K, the phase analogous to α-crystobalite. As in the case of the iron oxides, textural properties can vary drastically.

2.3.3.3.2 Influence of Concentration and Composition

In most cases it is desirable to precipitate at high concentrations of the metal ions. This increases the space-time yields by decreasing the vessel volume for the same mass of precipitate. Moreover, the higher degrees of supersaturation lead to faster precipitation, such that plant investment is reduced. With respect to the quality of the product obtained, smaller particle sizes and higher surface areas are usually achieved at higher concentrations due to increased nucleation rates at higher supersaturation if homogeneous nucleation takes place, although in some cases also lower particle sizes and thus higher surface areas were obtained in highly diluted systems, for instance with zirconia [29]. If for some reason the precipitation is carried out at low concentrations, for example to produce larger primary particles, it must be performed either in continuous systems where seeds are present in the stationary state, or seeds must be added to the solution.

If catalysts are prepared by coprecipitation, the composition of the solutions determines the composition of the final product. Often, the composition of the precipitate will reflect the solution concentrations, as was shown for CuO–ZnO catalysts for methanol synthesis [30], but this is not necessarily the case. For aluminum phosphates it was found that, at low P:Al ratios, the precipitate composition is identical with the solution composition; however, if the P:Al ratio in the solution comes close to and exceeds unity, the precipitate composition asymptotically approaches a P:Al ratio of 1 [31]. Deviations from solution composition in coprecipitation processes will generally occur if the solubilities of the different compounds differ strongly and precipitation is not complete or, if in addition to stoichiometric compounds, only one component forms an insoluble precipitate. This occurs in the case of aluminum phosphate.

2.3.3.3.3 Solvent Effects

For economic reasons, water is used almost exclusively as the solvent for precipitation processes, at least for bulk catalysts and supports; organic solvents are much more expensive than water. This economic disadvantage is even more severe than would be expected from the price difference, because solubilities for most metal salts are much lower in organic solvents. Thus, to achieve the same space-time yield, larger systems usually have to be employed. Moreover, increased environmental problems are associated with the use of organic solvents. However, some groups have reported that organic solvents can be advantageous for the precipitation of certain materials. The low solubilities of the precursor materials may, for example, result in very low supersaturations, which means a slow crystallization. Hence, the particle size distributions can be altered, or phases closer to the equilibrium phases can be formed. The disadvantages of using organic solvents in precipitation processes must be compensated by superior product qualities which cannot be achieved by other means. Although the possible benefits of precipitation from organic solvents were highlighted during the early 1990s [32], this approach does not seem to have found widespread acceptance, most likely due to the inherent disadvantages described above.

One of the most important systems which can be prepared advantageously from organic solvents is the $(VO)HPO_4 \cdot 0.5H_2O$ precursor for vanadium–phosphorus mixed oxides [33]. This is the best known catalyst for the selective conversion of n-butane to maleic anhydride, and the system is discussed in more detail below.

The possibility of obtaining higher-surface-area precipitates from organic solvents is described by Desmond et al. [34]; polar compounds such as alcohols, aldehydes, esters and glycols are used. Due to low solubilities of the precipitating agent (preferably Na_2CO_3 or K_2CO_3) in the organic solvent (10^{-1} to 10^{-6} mol L^{-1}), Group VIIIA, IB and IIB metal oxides can be prepared with higher surface areas than from aqueous solutions. This is believed to be due to a constant, low concentration of the precipitating agent which is dissolved as the metal oxide precipitates. The patent claims that different structures can be formed as compared to precipitation from aqueous solutions. Iron oxide, for example, precipitates as magnetite (Fe_3O_4), whereas in water it is usually hematite (α-Fe_2O_3) or maghemite (γ-Fe_2O_3) that is formed. One problem associated with this technique is the long precipitation time, perhaps of several days. The materials are reported to exhibit surface areas approximately twofold that of comparative catalysts prepared from water; the yield in Fischer–Tropsch reactions was about 50% higher than for conventionally prepared catalysts. Copper-based catalysts for the production of higher alcohols from syngas, which were precipitated from alcoholic solution, have also been described [35]. Small-particle-size (25 nm) ZrO_2 has been precipitated by the hydrolysis of $ZrOCl_2 \cdot 8H_2O$ in organic solvents, which are thought to prevent the formation of hard agglomerates [36].

Whilst for most catalyst systems, precipitation from aqueous solution is used almost exclusively, in two cases precipitation proceeds from organic solvents due to the very nature of the system. These are the hydrolysis of organic precursors and the formation of polymerization catalysts in non-polar media. In the hydrolysis of organic precursors, alkyls or alkoxides (especially of silicon or aluminum) are hydrolyzed by the action of aqueous bases or water; this process can be carried out in water as the solvent. The reversed system, with the organometallic compound itself as the solvent or an additional solvent, has also been described. This process has the advantage that catalyst particles with predetermined shapes, usually spheres, can be formed directly [37]. Moreover, materials resulting from hydrolysis processes are very pure due to the high purity of the starting materials. The aluminum oxides obtained from the hydrolysis of aluminum alcoholates were initially only a byproduct of detergent production but, due to the high quality of the resulting aluminas (e.g., PURAL®-type aluminas, produced by Sasol), they became the other main product of the process. Catalyst production processes based on the hydrolysis of organic precursors are discussed in greater detail in Chapter 2.3.4. It should be borne in mind that the hydrolysis of organic precursors does not necessarily lead to formation of a gel, but instead precipitation can occur.

If polymerization catalysts (e.g., for alkene polymerization; Ziegler-type catalysts) are prepared by precipitation methods, they may be precipitated from organic solvents, as claimed in several patents [38]. In these patents, the precipitation of titanium–magnesium compounds in tetrahydrofuran (THF) with hexane as precipitating agent is used for catalyst formation. The $MgCl_2$ support can be formed *in situ* by dissolving magnesium alcoholate in a solvent and then adding $TiCl_4$, which directly yields the supported titanium catalyst [39]. One method used to produce a spherical $MgCl_2$ support for Ziegler catalysts consists of dissolving $MgCl_2$ in ethanol, creating an emulsion with silicon oil or paraffin, and the subsequent precipitation of a magnesium chloride/ethanol adduct by the addition of hexane as antisolvent [40]. Many important Ziegler–Natta polymerization initiators are solids, and heterogeneous initiator systems seem to be necessary for the production of isotactic polyalkenes [41]. However, very few data are available in the open literature on such catalyst preparation.

Finally, a special case is the synthesis in reverse micelle systems (microemulsions). In such systems, a water-in-oil microemulsion is created, typically involving surfactants for stabilization of the microemulsion. The water droplets in the oil phase act as reaction environments, in which the precipitation can proceed. Precipitation can be triggered by the addition of a second microemulsion which contains the precipitation agent dissolved in the aqueous phase, or by passing a reactive gas through the microemulsion [42]. Since the aqueous phase droplets are small (some tens of nanometers), the particles of the resultant precipitate also tend to be small. These systems have been used to produce TiO_2 [43], ZrO_2 [44], Fe_2O_3 [45], aluminophosphates [46], Fe/SiO_2 [47] or CeO_2/ZnO [48]. In microemulsions, the oil does not act as a solvent in the strict sense, but rather as a medium for dispersing the water droplets; however, due to the small sizes of the droplets the microemulsions typically appear transparent and isotropic. Whilst this process seems attractive for research purposes, no commercial catalysts have been reported as being formed from microemulsion-based syntheses.

To summarize, the use of organic solvents to precipitate catalysts appears to be of minor importance. Only in special cases are the higher costs and problems involved in such processes justified by the superior properties of the final catalyst product.

2.3.3.3.4 **Influence of Temperature** As nucleation rates are extremely sensitive to temperature changes, the precipitation temperature is a decisive factor for controlling precipitate properties such as primary crystallite sizes, surface areas, and even the phases formed. It is very difficult, however, to state how the precipitation temperature must be adjusted in order to achieve a product with specific properties. Usually, the optimum precipitation temperature must be determined experimentally.

In general, most precipitation processes are carried out above room temperature, often close to 373 K. One obvious reason for this is that precipitation occurs more rapidly, provided that high levels of supersaturation are maintained.

Depending on the kinetics of the different elementary processes involved in the formation of the precipitate, a temperature increase might lead to an increase in crystallite size, as has been observed for the crystallization of pseudoboehmite [49] or iron molybdates [50]. However, in other cases no influence of the precipitation temperature on the crystallite size of the final catalyst was

References see page 116

reported [51], or indeed a decrease was reported, as for the ZnO system [52].

For some systems, the influence of temperature on phase composition can be predicted based on chemical considerations. For instance, the composition of bismuth molybdate catalysts is believed to be determined by the nature of the molybdate anion present in solution [53], which is in turn dependent on the solution temperature. For Ni/SiO$_2$ catalysts the differences between catalysts prepared at high or low temperatures are explained by the formation of nickel hydrosilicate at high temperatures, while at low temperature the main precipitate is nickel hydroxide [54].

A strong influence of the precipitation temperature on catalytic activity has been observed for Cu/ZnO–Al$_2$O$_3$ catalysts for methanol synthesis. Activity passes through a maximum at a precipitation temperature of about 343 K, and this maximum scales well both with total surface area and copper surface area [55]. In this study, the influence of precipitation pH on catalyst activity was also studied (Fig. 6).

2.3.3.3.5 Influence of pH

Since the pH directly controls the degree of supersaturation (at least if hydroxides are precipitated), this should be one of the crucial factors in precipitation processes. As for many other parameters, the influence of pH is not simple, and must be investigated experimentally for any specific system. Even in a relatively well-known system such as iron oxide [56], the effect which the precipitation pH has on the properties of the final product is not yet clear. There is no obvious relationship between the precipitation pH and the textural and catalytic properties of the precipitate. It is relatively well known which phases are formed under specific conditions, due to the importance of iron oxides as pigments, but the influence of pH with respect to catalytic properties is not well studied.

In the aluminum oxide system, the precipitation pH is one of the variables which control the nature of the phase eventually obtained. However, the aging conditions of the initially formed amorphous precipitate are at least equally important. In general, it can be stated that precipitation above pH 8 leads to the formation of bayerite, while precipitation under more acidic conditions favors the subsequent formation of boehmite. Hydrargillite is formed as the product of the Bayer process by seeding a supersaturated alkali-containing aluminum solution. The formation of bayerite is strongly facilitated by the presence of alkali cations which stabilize the structure.

The influence of precipitation pH on the performance of methanol synthesis catalysts was mentioned above, and several investigations have been devoted to this problem [55, 57–60]. In this case, the influence of pH can be clearly traced to the different hydroxycarbonate phases or oxide/hydroxide phases formed at different precipitation pHs. Whilst at low pH, separate Cu- and Zn-containing phases are preferred (e.g., copper hydroxynitrate), at higher pH the favorable mixed hydroxycarbonates (e.g., malachite or aurichalcite) are precipitated, and this results in a high copper dispersion in the final catalyst.

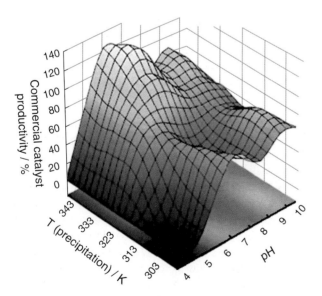

Fig. 6 Plot of methanol synthesis activity (in percent of the activity of a commercial benchmark) for Cu/ZnO-Al$_2$O$_3$ catalysts obtained from precursors precipitated at different pH and precipitation temperature. pH was constant during aging. (Illustration courtesy of C. Baltes.)

2.3.3.3.6 Influence of Aging

The time for which the precipitate is left in the mother liquor under precipitation conditions, or for which it is post-treated under somewhat different conditions, may also have a powerful influence on the performance of a precipitated catalyst. This factor is often neglected, and frequently the time over which precipitation takes place is not even reported in synthetic procedures.

During aging, several processes may take place which affect the precipitate properties. Particle sizes can increase due to Ostwald ripening, which is caused by the higher solubility of small particles with high surface curvature compared to larger particles. Small particles thus dissolve and the material is reprecipitated onto the larger ones, which leads to an overall shift of the sizes to larger particles. Furthermore, while the solid is in solution, any impurities which may be occluded (especially during a fast precipitation reaction) may be redissolved. Finally – and perhaps most importantly in terms of the performance of the final catalysts – recrystallization, either from an amorphous precipitate to a crystalline material or between different crystalline phases, is possible during aging.

Crystallization of initially amorphous precipitates is a quite frequently observed process in precipitation reactions. If the precipitate is formed at high supersaturation, the first precipitate is often amorphous, and conversion to a crystalline structure proceeds while the solid is in the mother liquor [61]. Controlled aging, either in the mother liquor or under different conditions (such as altered pH) can be used in a systematic manner to adjust the phase of the final catalyst. Especially for zirconia, the aging (or "digestion" as it is often called) allows the synthesis of either the monoclinic or the tetragonal phase in a controlled fashion, and fine adjustment of the textural properties of the materials [62].

Aging of the precursors is a crucial step for the later performance of methanol synthesis catalysts [58, 59]. The initially precipitated, amorphous hydroxycarbonates change strongly after the precipitation, and the pH of the solution also varies substantially, if it is not kept constant by the controlled addition of acid or base. In a typical precipitation, amorphous products are first obtained which crystallize to the hydroxcarbonate phases upon aging in the mother liquor. The times for this depend on the conditions, but range typically between 30 min and 2 h. The crystallization is accompanied by a change in pH (unless it is controlled), foaming due to evolved CO_2, and a change in the color of the precipitate. The initial, amorphous precursors lead to low-activity catalysts, whereas the crystallized hydroxycarbonates yield much more active catalysts after calcination and reduction. It is advisable, therefore, that this type of catalyst is left in the mother liquor for some time after precipitation.

2.3.3.3.7 Influence of Additives The properties of precipitates can be strongly influenced by additives, which are not necessarily ingredients of a precipitation reaction. Thus, although anions of precipitation agents might strongly influence the product's properties, they are not considered as additives, because their presence is unavoidable in the precipitation reaction.

The most frequently used additives are organic molecules, the role of which is to control the catalyst's pore structure [63]. These organic molecules can be removed from the precipitate at a later stage, in a calcination step. One very promising route for preparing high-surface-area oxides (M41S) is to use surfactants as additives; this scheme was introduced in 1992 by the Mobil Oil Corporation (now Exxon-Mobil) [64]. The surfactants are able to form liquid crystal-like structures in cooperation with the silicates present in solution [65]. Removal of the surfactant by a calcination step leaves a silica negative of the organic liquid crystal with a relatively perfect hexagonal arrangement of pores, the diameters of which can be adjusted between about 2 and 10 nm. The pore size distribution is very sharp, as verified by sorption analysis [66]. It was shown subsequently that this concept can be generalized to the preparation of other oxides [67]; in the case of the silicate, typical BET surface areas may be as high as 1200 $m^2\,g^{-1}$. Developments in this field have expanded vastly since publication of the First Edition of this Handbook, and the synthesis of ordered mesoporous materials under the influence of liquid crystal forming surfactants is discussed in detail in Chapter 2.3.6.

Surfactant molecules are not only useful for increasing surface areas by forming liquid crystal-like structures; they may also stabilize small particle sizes by being adsorbed onto the newly forming particles during the precipitation process [68]. Today, the use of additives to control catalyst precipitation is mainly restricted to trade secrets, and so cannot be discussed in greater detail at this point.

2.3.3.4 Prototypical Examples of Precipitated Catalysts and Supports

In this section it is shown, for four examples of increasing complexity, how precipitates are formed and how their properties are controlled to produce materials suitable for catalytic applications. The first two examples are: (i) silica, which is used primarily as a support material and usually formed as an amorphous solid; and (ii) alumina, which is also used as a catalytically active material and can be formed in various modifications with widely varying properties as pure precipitated compounds.

The other examples result from coprecipitation processes: (i) Ni/Al_2O_3, which can be prepared by several pathways and for which the precipitation of a certain phase determines the reduction behavior and the later catalytic properties; and (ii) (VO)$HPO_4 \cdot 0.5H_2O$, which is the precursor of the V/P/O catalyst for butane oxidation to maleic anhydride, where even the formation of a specific crystallographic face with high catalytic activity must be controlled.

2.3.3.4.1 Silica as Support Material Silica applied as a support for catalysts is an X-ray amorphous form of silicon dioxide [69]. These silicas are produced by sol–gel processes (silica gel), by precipitation (precipitated silica, although this is sometimes also called silica gel), or flame hydrolysis (fumed silica). Precipitated silica and silica gel have many similarities, the major difference lying in the fact that during the synthesis of silica gel a container-spanning gel is formed, whereas in the synthesis of precipitated silica the solid appears in a

References see page 116

phase separation step between the liquid phase and the precipitate. However, the synthetic processes leading to both forms and the resulting materials have many similarities, and hence are discussed together. For further information on the sol–gel process, the reader is referred to Chapter 2.3.4. Comprehensive information on porous silica can also be found in Ref. [70].

The initial step for both silica gel formation and precipitation is the formation of a sol; that is, the polymerization of silicates to form small primary particles in aqueous solution. These primary particles aggregate and form a porous network. In acid solution, or in the presence of electrolytes, gel formation is then favored – that is, the network grows to extend through the whole container and forms a container-spanning solid. At neutral or alkaline conditions, and in solutions containing substantial amounts of salts, a precipitate is formed. Precipitation can also be induced by the presence of coagulants or flocculants, such as water-miscible organic liquids, polymers or surfactants [70, 71]. The precipitation of silica in commercial production is almost exclusively induced by the addition of sulfuric acid to a sodium silicate solution. Both solutions are fed simultaneously to a stirred vessel containing water, where precipitation occurs. The formation of a gel is avoided by stirring at elevated temperatures.

In the second step, the silica hydrogel or the precipitate is subjected to aftertreatment, such as washing, followed by dehydration to remove water. The product is either a silica xerogel composed of hard porous granules, or the dried silica precipitate. The granules can be milled to a powder and sized to a desired particle size; in most cases some additional shaping operations are necessary, such as granulation. Silica beads can be obtained directly during the course of the sol–gel process by dispersing sol droplets into a water-immiscible liquid, whereby gelling of the droplets into hydrogel beads occurs [72] (see also Chapter 2.5.2 of this Handbook).

In the sol–gel process, sodium silicate (water glass) solution is acidified with sulfuric acid. By lowering the pH of the water glass solution, silicic acids are formed which immediately undergo polycondensation and further growth to colloidal silica particles. Depending on the final pH, the silica concentration, the type and concentration of the electrolyte, the temperature, and other parameters, the dispersion of the colloid particles can be stabilized as a silica sol, agglomeration of the colloidal particles occurs to silica hydrogel, or silica precipitates. The silica hydrogel represents a coherent system composed of a three-dimensional network of agglomerated spheres with sodium sulfate solution as dispersing liquid. The pores in both silica gels and precipitated silicas are preformed by the interstices between the agglomerated nonporous silica globules. The pore structure is thus determined by the size of the globules and their coordination in the agglomerate.

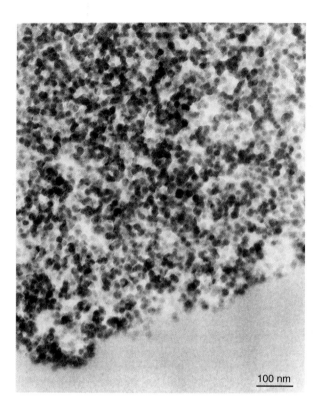

Fig. 7 Transmission electron micrograph of the globular structure of a silica xerogel.

The globular structure of a silica xerogel is shown in Fig. 7. In order to convert the initially formed hydrous silica to the silica xerogel, it is first washed to remove the salt solution and then dried at about 423–473 K. The washing step may influence the pore structure of the resulting material, with silica gels being more sensitive to the conditions of washing than precipitated silicas. The drying conditions also affect the specific surface area and the pore structure parameters of the xerogel. When the dried xerogel is exposed to air, it adsorbs water to about 5–10 wt.%, depending on its specific surface area. During the course of aging, washing and drying, a number of procedures can be applied systematically to vary the pore structure; this procedure is known as "modification", and may include washing with acids or bases or hydrothermal treatment. As a result, the pores are enlarged and the specific surface area is decreased. This step is important when preparing catalyst supports in which transport limitations are minimized.

Large-pore granules of silica xerogel are not suitable as catalyst supports due to the lack of large macropores which are necessary for good mass transfer, and to low stability when packed in a fixed-bed reactor. It is advisable to subject the powdered xerogel or the precipitated silica to pelletization on a rotary plate, or to extrusion. In both cases a binder must be applied to control the porosity and size of macropores, and to obtain sufficient mechanical stability (crushing strength and attrition resistance).

The xerogel or precipitated silica contain inorganic impurities which depend on the purity of the water glass, and which might affect catalytic performance. Most common are sodium, aluminum, and titanium at concentrations exceeding 1000 ppm. These impurities affect the thermal stability of the xerogel and the catalytic properties of the final catalyst. The silica surface is covered by hydroxyl groups and adsorbed water. At full hydroxylation, the concentration of hydroxyl groups amounts to about 8 µmol m^{-2} (four to six OH groups per nm^2). There are different types of Brønsted sites, namely isolated (or free), geminal, and vicinal; these can be monitored using ^{29}Si cross-polarization magic angle spinning nuclear magnetic resonance (CP MAS NMR) spectroscopy [73], infra-red (IR) spectroscopy [74], and other techniques. The Brønsted sites on the silica surface are weakly acidic (p$K_a \approx 7$). The point of zero charge (PZC) of silica is around 2; that is, at a pH >2 the surface is negatively charged. When dispersed in electrolyte solution, silica thus acts as a weak cation exchanger, which makes it possible to impregnate silicas by equilibration with the ion to be loaded. It is worthy of note that silica is soluble in solutions of pH >9.

One major advantage of silica over other support materials is the ease of adjustment and control of the mean pore diameter, the specific surface area and the specific pore volume. Supports for catalytic applications require a sufficiently high specific surface area which is generated by mesopores with mean pore diameters between 5 and 50 nm, and a sufficient amount of macropores of mean pore diameter >50 nm to ascertain a rapid mass transfer. Also important is a high degree of pore connectivity, which requires a high porosity, although enhancing the porosity leads to a reduction in the mechanical strength of the support particles. Thus, a compromise must be found. Silica xerogels usually have a specific pore volume between 0.5 and 1.0 cm^3 g^{-1}, which corresponds to a porosity of about 50–70%. Silica xerogels are typical mesoporous adsorbents with a mean pore diameter between 5 and 50 nm and corresponding specific surface areas of between 50 and 600 m^2 g^{-1}. Macropores of a mean pore diameter of >50 nm are generated during the course of the formulation and the shaping process.

Silicas are used as supports in a variety of catalytic processes. Typical applications are Ni-, Pd-, or Pt-supported catalysts for hydrogenation reactions, and V$_2$O$_5$ supported catalysts in the oxidation of SO$_2$ in gases from roasting sulfides or sulfur to yield SO$_3$. One disadvantage of silica is its limited hydrothermal stability, such that its application is limited in water-vapor containing environments at high temperature.

2.3.3.4.2 **Active Aluminas** Active aluminas are much more widely used in catalytic applications than silicas, as they are not only excellent support materials but also very active as catalysts in their own right for several applications. The chemistry of aluminas is much more complicated than that of silica, as many crystallographic modifications exist, only a few of which are useful as catalysts. A comprehensive survey of aluminas can be found in Ref. [75].

Porous aluminas are manufactured by a controlled dehydration from aluminum hydroxide (Al(OH)$_3$ = Al$_2$O$_3 \cdot$ 3H$_2$O) or aluminum oxide hydrates (AlOOH = AlO$_3 \cdot$ H$_2$O). Aluminum hydroxide exists in various crystalline forms as gibbsite (hydrargillite), bayerite, and nordstrandite [13, 76], all of which are non-porous. Thermal treatment of gibbsite formed in the Bayer process at about 423 K yields microcrystalline boehmite with a specific surface area of about 200 to 400 m^2 g^{-1}. The best Bayer precursor for the preparation of catalysts is the fine crystalline material deposited as a hard crust on the precipitator wall (it must be removed periodically). Further treatment between 673 K and 1273 K yields a series of porous crystalline transition aluminas assigned by Greek letters as chi (χ), kappa (κ), gamma (γ), eta (η), theta (θ), and delta (δ). The most important transition oxides for catalytic applications are γ-alumina and the η-alumina. Calcination above 1273 K yields α-alumina as a crystalline, non-porous product which is occasionally used as a low-surface-area support material.

The pathways for the formation of different aluminas between Al(OH)$_3$ and Al$_3$O$_3$ are shown in Fig. 8. γ-Alumina and η-alumina are also called active aluminas; these materials are seldom phase pure, and contain other transition aluminas as impurities. Their properties depend strongly on the type of starting material, the procedure chosen for thermal treatment, and the operation parameters such as temperature and pressure. Highly active γ-alumina has been prepared by shock calcination, followed by rehydration [77, 78].

In addition to gibbsite, several other routes exist for the manufacture of Al(OH)$_3$ and the consecutive transition oxides. One method is to precipitate Al(OH)$_3$ from aluminum salts by adjusting the pH between 7 and 12 by adding bases. Precipitation at elevated temperatures and high pH leads to the formation of bayerite, whereas at lower pH pseudoboehmite and subsequently boehmite are formed. By heating, these materials can be converted to the active transition aluminas.

A third, very important, source of aluminum oxide hydroxides is the hydrolysis of aluminum alcoholates. Due to the high purity of the alcoholate, almost no impurities, except for carbon residues at low concentrations, are

References see page 116

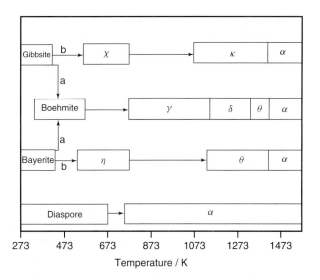

Fig. 8 Sequence of phases formed after thermal treatment of aluminum hydroxides and aluminum oxide hydroxide. Pathway (a) is favored by pressures >0.1 MPa, moist air, heating rates >1 K min^{-1}, and particle sizes >100 μm. Pathway (b) is favored by a pressure of 0.1 MPa, dry air, heating rates <1 K min^{-1}, and particle sizes <10 μm.

present in the product:

$$Al(OR)_3 + (2 + \chi)H_2O \longrightarrow 3ROH + AlOOH \cdot \chi H_2O \tag{6}$$

where R can vary.

By using such procedures, high-purity pseudoboehmite is produced (PURAL®, Sasol). With this type of microcrystalline pseudoboehmite, spheres of γ-alumina can be manufactured. PURAL can also be extruded to pellets with binders and then subjected to calcination to prepare γ-alumina. The extrudation of various aluminum hydroxides and oxide hydroxides with binders to pellets is described in Ref. [79]. The properties of the different aluminum hydroxide and oxide hydroxide precursors for the preparation of active aluminas are summarized in Table 2.

The structure of γ-alumina resembles that of the spinel-type $MgAl_2O_4$; η-alumina has a very similar structure. The difference lies in the concentration of stacking faults which is caused by the difference in the structures of the precursors boehmite and bayerite [80]; the concentration of stacking faults is higher in the η-alumina.

γ-Alumina exhibits Brønsted acidity, Lewis acidity, and Lewis basicity [81]. Towards electrolytes, it acts as a cation and anion exchanger, depending on the pH [82]. γ-Alumina develops mesoporosity with specific surface areas between 50 and 300 m^2 g^{-1} and specific pore volumes up to 0.6 cm^3 g^{-1}. Following special procedures, highly porous γ-alumina can be prepared with a specific pore volume up to 0.8 cm^3 g^{-1} [83]. Ordered mesoporous γ-alumina has recently been produced via a surfactant-assisted process [84]. α-Alumina (corundum) is obtained after high-temperature calcination; this material is essentially non-porous and has a low specific surface area. It is also marketed as a ceramic support with a specific surface area of around 3–5 m^2 g^{-1}, a porosity of 50%, and a pellet size of 2–4 mm.

Aluminas are used in a variety of catalytic applications. In fact, α-, γ- and η-aluminas are all used as support materials, the first in applications where low surface areas are desired, as in partial oxidation reactions. The latter two, especially γ-alumina, are used in applications requiring high surface areas and high thermal and mechanical stabilities. One of the most prominent applications of γ-alumina as a support is in the catalytic converter for pollution control, where an alumina washcoat covers a monolithic support. The washcoat is impregnated with the catalytically active noble metals (see Chapter 11.2 of this Handbook). Another major application area of high-surface aluminas as a support is in the petrochemical industry in hydrotreating plants, with alumina-supported

Tab. 2 Properties of different precursors used for active alumina preparation

Property	Bayer hydroxides	Precipitated hydroxide	Ziegler hydroxides
Raw material	Bauxite	Aluminum salt solution	Aluminum alcoholate
Process	Digestion in NaOH and crystallization	Neutralization with bases at pH 7–12	Hydrolysis
Phase	Gibbsite	Pseudoboehmite, bayerite, nordstrandite, gibbsite	Pseudoboehmite
Size of primary particles/nm	500–15 000	10–1000	4–10
Main impurity	Na$_2$O	Chloride, sulfate, nitrate	Carbon (very low concentration)
Weight loss on calcination/wt.%	35	18–40	22–28
Maximum of pore radius distribution/nm	1.2–2.2	1.2–2.2	3.0–7.5
BET surface area/m^2 g^{-1}	0.5–10	200–400	160–600

catalysts with Co, Ni, and/or Mo being used for this purpose. All noble metals are also available as supported catalysts based on aluminas; such catalysts are used for hydrogenation reactions or sometimes in oxidation reactions. If a high mechanical stability and a low surface area of the catalyst support are desired, then α-alumina rather than γ-alumina might be used. This is especially the case for ammonia catalysts, which often contain alumina as support material.

γ-Alumina is used not only as a support material but also as a catalyst in its native form, for example in the alcohol dehydration reaction, due to the acidic nature of its surface; it is used most prominently as the catalyst in the Claus process (see Chapter 12.4 of this Handbook), the severe conditions of which require the use of high-stability aluminas.

2.3.3.4.3 Ni/Al$_2$O$_3$ Catalysts by Coprecipitation

Alumina-supported nickel catalysts serve as excellent examples of the advantages of, and the problems associated with, coprecipitation processes in the manufacture of catalysts. These catalysts are accessible via several pathways, as impregnation, deposition/precipitation, coprecipitation from alumina gels, and more conventional coprecipitation routes. In the case of coprecipitation, several different routes are possible, the oldest of which date back to the 1920s [85]. Starting from the nitrate solutions of nickel and aluminum, there are at least three different routes:

- The precipitation of a hydrotalcite (HTlc)-like compound with the general formula $[M^{II}_{1-x}M^{III}_x(OH)_2]^{x+}(A^{n-})_{x/n} \cdot nH_2O$, where M^{II} = Mg, Ni, Zn, ..., M^{III} = Al, Fe, Cr, ..., and A^{n-} = CO_3^{2-}, SO_4^{2-}, NO_3^- ..., using Na$_2$CO$_3$ as the precipitating agent. The composition of the nickel–alumina hydroxycarbonate (NiAlCO$_3$-HTlc) may thus vary, but an idealized composition is that of the mineral takovite Ni$_6$Al$_2$(OH)$_{16}$CO$_3 \cdot$ 4H$_2$O. Hydrotalcites as catalytic precursors, amongst them NiAlCO$_3$-HTlc, are discussed in an excellent recent review [86].
- Nickel–aluminum coprecipitates can be prepared in the complete absence of carbonate ions, for example by precipitation with NaOH in carbonate-free solutions [87]. In this case, no mixture of hydroxides is obtained, but nitrates are incorporated and a mixed hydroxynitrate precipitates.
- The same holds for ammonia solutions, from which the precipitation can be carried out by evaporating the ammonia. If the ammonia is removed without excluding the carbonates, the normal hydroxycarbonate forms. If carbonate is excluded, the hydroxynitrate precipitates [87].

An alternative approach is precipitation by combining an alkaline NaAlO$_2$ solution with an acidic nickel nitrate solution [88]. In this case as well, hydrotalcite-like species are probably formed first although, due to the very high nickel contents reported, some nickel hydroxide also forms. Pure hydrotalcite-like structures are reported to form for Ni : Al atomic ratios between 2 and 3 [89] or 4 [90]; outside this range the hydroxides of the excess species form as additional phases. Rapid precipitation tends to form pure Al(OH)$_3$ due to the lower solubility. If nitrates are used as the starting materials, a low precipitation pH leads to the incorporation of nitrate ions [91]. Aging can also lead to recrystallization of some of the pure hydroxides to the hydrotalcite structure.

The formation of hydrotalcite highlights one of the key advantages of coprecipitation techniques, in that it can provide intimate mixing of the catalyst constituents on an atomic scale at very high loadings. The intimate mixing leads to very different behavior with respect to calcination and reduction as compared to impregnated catalysts [92, 93]. Whereas impregnated or mechanically mixed catalysts form surface-mixed oxides of the spinel structure at best, which segregate again on reduction to result in the formation of a separate nickel phase on a γ-alumina support, the coprecipitation technique leads to formation of a high-surface-area mixed-oxide with varying composition between NiO and the spinel-type NiAl$_2$O$_4$, depending on the composition of the precursor. On reduction of this oxide, nickel crystallites are formed which exhibit a high degree of "paracrystallinity"; that is, highly defective nickel crystallites with aluminate groups as the defects [94]. The impregnated catalysts are easier to reduce than the coprecipitated materials, as might be expected from the structure of the precursors [95].

The thermal treatment is one of the factors which control the properties of the final catalyst [93, 95, 96]. The total surface area (in the range of 100 to 300 m^2 g^{-1}) decreases with increasing reduction temperature; however, the nickel surface area (typically 20–50 m^2 g^{-1}) increases, most likely due to a higher degree of reduction. The best precursor with respect to a high surface area is the hydroxycarbonate; the surface areas of catalysts prepared from hydroxychlorides and nitrates are smaller by about a factor of two. Nickel particle sizes are in the order of 5 nm for such catalysts.

Ni/Al$_2$O$_3$ catalysts prepared by coprecipitation exhibit excellent sintering stability, and can be used in many catalytic reactions, the most important being steam reforming (see Chapter 13.11) and methanation.

The preparation of mixed-oxide catalysts from hydroxycarbonates is not restricted to Ni/Al$_2$O$_3$, but is also possible for several other systems. Ni/MgO-Al$_2$O$_3$ catalysts can also be prepared via a hydrotalcite primary coprecipitate, where the mode of nickel introduction

References see page 116

(coprecipitation, coprecipitation involving anionic nickel complexes, or nickel incorporation into a previously precipitated hydrotalcite) has a strong influence on catalytic properties [97]. One other prominent example is the technically very important methanol catalyst based on $CuO/ZnO\text{-}Al_2O_3$. This can also be prepared from a hydrotalcite-like precursor [98], but mostly a hydrozincite or aurichalcite precursor is used [99]. The resulting catalysts after calcination and reduction contain strained copper particles, an effect which seems to be related to the catalytic activity of these materials [100]. On reduction of aurichalcite precursors, epitaxial relationships between the copper and the ZnO were observed. Such preferential formation of certain orientations is also the decisive factor in the final example discussed here.

2.3.3.4.4 V/P/O Catalysts for Butane Oxidation to Maleic Anhydride
The V/P/O $((VO)_2P_2O_7)$ catalyst is a very specialized material which has a unique activity for the oxidation of n-butane to maleic anhydride. Its activity has often been related to the presence of (100) planes in the catalyst, in which the arrangement of surface atoms ideally matches the geometry of the reacting molecules [101, 102], although this view is increasingly questioned as there appears to be an amorphous overlayer on the surface of the equilibrated catalyst [103]. Nevertheless, catalyst particle morphologies in which high fractions of (100) surfaces are present seem to be favorable for the reaction. Since excellent recent reviews on this subject already exist [104, 105], only those important findings which concern the preparation of the precursor are reported here, as this example highlights the problems involved in creating high fractions of certain crystallographic faces. In the case of the V/P/O catalyst there exists a pseudomorphic relationship between the precursor $(VO)HPO_4 \cdot 0.5H_2O$ and the final catalyst. The equilibrium shape of $(VO)_2P_2O_7$ exhibits only about 32% of the optimum (100) faces. The hemihydrate precursor now is special in that it can be precipitated in a morphology that leads to formation of a much higher (100) fraction, as the (001) of the precursor is structurally related to the (100) of the pyrophosphate.

Several pathways have been described for the preparation of the precursor [104]. Usually, V_2O_5 is reductively dissolved in water or an organic solvent such as an alcohol, mostly iso-butanol with benzyl alcohol as additive, to yield V^{IV} species. Synthesis in non-aqueous solvents is advantageous here, as the organic solvent can play a dual role of solvent and reducing agent. To this solution phosphoric acid is added and the solid precursor precipitates. Precipitation is often carried out under reflux. There is general agreement that preferentially (001) planes should be exposed in the precursor, and that a precursor with a high concentration of stacking faults is superior to more ordered materials. This can best be achieved by precipitation in organic solvents, as can be seen by comparison of the X-ray patterns of samples produced either from organic or aqueous media [106].

The influence of several parameters on the morphology of the precipitate formed was examined in a study on the precipitation conditions for the hemihydrate [107]. The surface area of the precursors formed was mainly related to the thickness of the crystals; the thicker the crystals (meaning exposure of undesired faces), the lower the surface area. High surface areas could be achieved with a high-boiling alcohol. The conditions favoring the formation of thin, high-surface-area $(VO)HPO_4 \cdot 0.5H_2O$ platelets ($H_3PO_4 \cdot 85\%$, P:V = 1.1–1.2, iso-butanol/benzyl alcohol, addition of tetraethylorthosilicate) are identical to conditions defined earlier for obtaining the best-performing catalyst [108]. The addition of polyethylene glycol was also reported to result in high-surface-area catalysts with very good catalytic performance [109]. The example of the V/P/O catalyst shows, convincingly, how control of the precipitation conditions, based on a knowledge of the catalytic process, can eventually result in the formation of a highly active catalyst.

2.3.3.5 Conclusions
Precipitation is – and probably will remain – one of the most important methods for the synthesis of solid catalysts. Change in this field is typically not revolutionary, but new methods are being added continuously to the tool box. For the industrial preparation of catalysts by precipitation, the ease and cost of the operation must always be balanced against the catalyst performance. Advanced precipitation methods, such as microemulsion synthesis, in most cases do not yield catalysts of sufficiently superior quality to compensate for the higher efforts required in the synthesis. Moreover, the more complex a synthetic process is, the less robust it is usually, and this represents a major obstacle for industrial implementation. Hence, conventional precipitation and coprecipitation from aqueous solution will most likely continue to be the "workhorse" of catalyst synthesis for many years.

References

1. C. L. Thomas, *Catalytic Processes and Proven Catalysts*, Academic Press, New York, 1970, Chapters 11 and 14; A. B. Stiles, T. A. Koch, *Catalyst Manufacture*, 2nd Ed., Marcel Dekker, New York, 1995, p. 291.
2. Recommendations on Nomenclature for Contamination Phenomena in Precipitation from Aqueous Solution, *Pure Appl. Chem.* **1975**, *37*, 463.
3. V. K. LaMer, R. H. Dinegar, *J. Am. Chem. Soc.* **1950**, *72*, 4847.
4. E. Matijevic, *Acc. Chem. Res.* **1981**, *14*, 22.

5. C. J. J. den Ouden, R. W. Thompson, *J. Colloid Interface Sci.* **1991**, *143*, 77.
6. D. Horn, J. Rieger, *Angew. Chem. Int. Ed.* **2001**, *40*, 4330.
7. H. Haberkorn, D. Franke, T. Fechen, W. Goesele, J. Rieger, *J. Colloid Interface Sci.* **2003**, *259*, 112.
8. F. Schüth, P. Bussian, P. Ågren, S. Schunk, M. Lindén, *Solid State Sci.* **2001**, *3*, 801.
9. S. Hamad, S. Crystol, C. R. A. Catlow, *J. Am. Chem. Soc.* **2005**, *127*, 2580.
10. U. Gasser, E. R. Weeks, A. Schofield, P. N. Pusey, D. A. Weitz, *Science* **2001**, *292*, 258.
11. S. Auer, D. Frenkel, *Adv. Polym. Sci.* **2005**, *173*, 149.
12. A. D. Randolph, M. A. Larson, *AIChE J.* **1962**, *8*, 639; The concepts are described in more detail in: A. D. Randolph, M. A. Larson, *Theory of Particulate Processes*, 2nd Ed., Academic Press, San Diego, **1988**, pp. 251.
13. L. K. Hudson, C. Misra, A. J. Perrotta, K. Wefers, F. S. Williams, Aluminum Oxide, in *Ullmann's Encyclopedia of Industrial Chemistry*, Wiley-VCH, Weinheim, 2005, available on CD or as Web-Edition: http://jws-edck.interscience.wiley.com:8087/index.html.
14. G. H. Bogusch, C. F. Zukoski, IV, *J. Colloid Interface Sci.* **1991**, *142*, 19.
15. F. Cavani, F. Trifiro, A. Vaccari, *Catal. Today* **1991**, *11*, 173.
16. S. P. S. Andrew, in *Preparation of Catalysts I*, B. Delmon, P. A. Jacobs, G. Poncelet (Eds.), Studies in Surface Science and Catalysis, Vol. 1, Elsevier, Amsterdam, 1976, p. 429.
17. P. Peshev, A. Toshev, G. Gyurov, *Mater. Res. Bull.* **1989**, *24*, 33.
18. G. Pass, A. B. Littlewood, R. L. Burwell, Jr., *J. Am. Chem. Soc.* **1960**, *82*, 6281; M. P. McDaniel, R. L. Burwell, Jr., *J. Catal.* **1975**, *36*, 384; C. Sivaraj, P. Kantaro, *Appl. Catal.* **1988**, *45*, 103; C. Sivaraj, B. R. Reddy, B. R. Rao, P. K. Rao, *App. Catal.* **1986**, *24*, 25.
19. K. Tanabe, M. Itoh, K. Morishige, H. Hattori, in *Preparation of Catalysts I*, B. Delmon, P. A. Jacobs, G. Poncelet (Eds.), Studies in Surface Science and Catalysis, Vol. 1, Elsevier, Amsterdam, 1976, p. 65.
20. E. Reverchon, *J. Supercrit. Fluids* **1999**, *15*, 1.
21. G. J. Hutchings, J. K. Bartley, J. M. Webster, J. A. Lopez-Sanchez, D. J. Gilbert, C. J. Kiely, A. F. Carley, S. M. Howdle, S. Sajip, S. Caldarelli, C. Rhodes, J. C. Volta, M. Poliakoff, *J. Catal.* **2001**, *197*, 232.
22. P. Courty, C. Marcilly, *Preparation of Catalysts III*, G. Poncelet, P. Grange, P. A. Jacobs, (Eds.), Studies in Surface Science and Catalysis, Vol. 16, Elsevier, Amsterdam, 1976, p. 485.
23. M. Lindén, S. Schunk, F. Schüth, *Angew. Chem. Int. Ed.* **1998**, *37*, 821.
24. M. Linden, J. Blanchard, S. Schacht, S. A. Schunk, F. Schüth, *Chem. Mater.* **1999**, *11*, 3002.
25. M. Schur, B. Bems, A. Dassenoy, I. Kassatkine, J. Urban, H. Wilmes, O. Hinrichsen, M. Muhler, R. Schlögl, *Angew. Chem. Int. Ed.* **2003**, *42*, 3815.
26. E. Matijevic, *Pure Appl. Chem.* **1978**, *50*, 1193; E. Matijevic, in *Preparation of Catalysts II*, B. Delmon, P. Grange, P. Jacobs, G. Poncelet (Eds.), Studies in Surface Science and Catalysis, Vol. 3, Elsevier, Amsterdam, 1979, p. 555.
27. A. Dalas, C. Kordulis, P. G. Koutsoukos, A. Lycourghiotis, *J. Chem. Soc., Faraday Trans.* **1993**, *89*, 3645.
28. J. M. Campelo, A. Garcia, D. Luna, J. M. Marina, *J. Catal.* **1988**, *111*, 106.
29. W. Stichert, F. Schüth, *Chem. Mater.* **1998**, *10*, 2020.
30. P. Porta, S. de Rossi, G. Ferraris, M. Lo Jacono, G. Minelli, G. Moretti, *J. Catal.* **1988**, *109*, 367; G. C. Shen, S. I. Fujita, N. Takezawa, *J. Catal.* **1992**, *138*, 754.
31. T. T. P. Cheung, K. W. Wilcox, M. P. McDaniel, M. M. Johnson, *J. Catal.* **1986**, *102*, 10.
32. J. T. Wrobletski, M. Boudart, *Catal. Today* **1992**, *15*, 349.
33. J. C. Burnett, R. A. Kleppel, W. D. Robinson, *Catal. Today* **1987**, *1*, 537.
34. M. J. Desmond, M. A. Pepera, US Patent 4686203, 1987, assigned to Standard Oil Company.
35. T. J. Mazanec, G. John, Jr., US Patent 4677091, 1987, assigned to Standard Oil Company.
36. C. Yu, D. Liang, P. Lin, L. Lin, *Ranlia Huaxue Xuebao* **1998**, *26*, 230; C. Yu, D. Liang, P. Lin, L. Lin, *CAN 129*, 114063.
37. T. Kawaguchi, K. Kouhei, *J. Non-Cryst. Solids* **1990**, *121*, 383.
38. M. Matsura, T. Fujita, JP 62257906 1987, assigned to Showa; M. Kioka, N. Norio, EP 279586, 1988, assigned to authors; K. Yamaguchi, N. Kanoh, T. Tanaka, E. Tom, N. Enokido, A. Murakami, S. Yoshida, Ger. Offenl. 2515211, 1975, assigned to authors.
39. R. C. Brady, III, F. C. Stakem, H. T. Liu, A. Noshay, EP 291958, 1988, assigned to Union Carbide Corp.
40. R. Jamjah, G. H. Zohuri, J. Vahezi, S. Ahmadjo, M. Nekomanesh, M. Pourjari, *J Appl. Polym. Sci.* **2006**, *101*, 3829.
41. A. E. Hamielec, H. Tobita, in *Ullmann's Encyclopedia of Industrial Chemistry*, 5th Ed., 1992, Vol. A21, p. 305.
42. V. Pillai, P. Kumar, M. J. Hou, P. Ayyub, D. O. Shah, *Adv. Colloid Interf. Sci.* **1995**, *55*, 241.
43. V. Chabra, V. Pillai, B. K. Mishra, A. Morrone, D. O. Shah, *Langmuir* **1995**, *11*, 3307.
44. R. Palkovits, S. Kaskel, *J. Mater. Chem.* **2006**, *16*, 391.
45. D. Barkhuizen, I. Mabaso, E. Voljoen, C. Welker, M. Claeys, E. Van Steen, J. C. Q. Fletcher, *Pure Appl. Chem.* **2006**, *78*, 1759.
46. M. A. Aramendia, V. Borau, C. Jiménez, J. M. Marinas, F. J. Romero, F. J. Urbano, *J. Mol. Catal. A: Chemical* **2002**, *182*, 35.
47. H. Hayashi, L. Z. Chen, T. Tago, M. Kishida, K. Wakabayashi, *Appl. Catal. A: General* **2002**, *231*, 81.
48. Y. He, B. Yang, G. Cheng, *Catal. Today* **2004**, *98*, 595.
49. D. Damyanov, I. Ivanov, L. Vlaev, *Zh. Prikl. Khim.* **1989**, *62*, 486.
50. F. Traina, N. Pemicone, *Chim. Ind. (Milan)* **1970**, *52*, 1.
51. A. S. Ivanova, E. M. Moroz, G. S. Litvak, *Kinet. Katal.* **1992**, *33*, 1208.
52. A. Vian, J. Tijero, E. Guardiola, *Actas Simp. Iberoam. Catal.*, 9th Volume 2, Soc. Iberoam. Catal., Lisbon, 1984, p. 1601.
53. F. Trifiro, H. Hoser, R. D. Scarle, *J. Catal.* **1972**, *25*, 12; F. Trifiro, P. Forzatti, P. L. Villa, in *Preparation of Catalysts I*, B. Delmon, P. A. Jacobs, G. Poncelet (Eds.), Studies in Surface Science and Catalysis, Vol. 1, Elsevier, Amsterdam, 1976, p. 147.
54. Y. Nitta, T. Imanaka, S. Teranishi, *J. Catal.* **1985**, *96*, 429 and references therein.
55. C. Baltes, S. Vukojevic, F. Schüth, *DGMK Tagungsberichte* **2006**, *2006-4*, 19.
56. R. A. Diffenbach, D. J. Fauth, *J. Catal.* **1986**, *100*, 466.
57. R. A. Hadden, P. J. Lambert, C. Ranson, *Appl. Catal.* **1995**, *122*, L1.
58. B. Bems, M. Schur, A. Dassenoy, H. Junkes, D. Herein, R. Schlögl, *Chem. Eur. J.* **2003**, *9*, 2039.

59. C. Kiener, M. Kurtz, H. Wilmer, C. Hoffmann, H.-W. Schmidt, J.-D. Grunwaldt, M. Muhler, F. Schüth, *J. Catal.* **2001**, *216*, 110.
60. J. L. Li, T. Inui, *Appl. Catal. A: General* **1996**, *137*, 105.
61. W. Gösele, E. Egel-Hess, K. Wintermantel, F. R. Faulhabe, A. Mersmann, *Chem. Ing. Tech.* **1990**, *62*, 544.
62. K. T. Jung, A. T. Bell, *J. Mol. Catal. A: Chemical* **2000**, *163*, 27; G. K. Chua, S. H. Liu, S. Jaenicke, J. Li, *Micropor. Mesopor. Mater.* **2000**, *39*, 381; G. K. Chua, *Catal. Today* **1999**, *49*, 131; W. Stichert, F. Schüth, *J. Catal.* **1998**, *174*, 242; A. Clearfield, *Inorg. Chem.* **1964**, *3*, 146.
63. D. Basmadijan, G. N. Fulford, B. I. Parsons, D. S. Montgomery, *J. Catal.* **1962**, *1*, 547.
64. C. T. Kresge, M. E. Leonowicz, W. J. Roth, J. C. Vartuli, J. S. Beck, *Nature* **1992**, *359*, 710.
65. A. Monnier, F. Schüth, Q. Huo, D. Kumar, D. Margolese, R. S. Maxwell, G. D. Stucky, M. Krishnamurty, P. Petroff, A. Firouzi, M. Janicke, B. Chmelka, *Science* **1993**, *261*, 1299.
66. O. Franke, G. Schulz-Ekloff, J. Rathousky, J. Starek, A. Zukai, *J. Chem. Soc., Chem. Commun.* **1993**, 724; P. J. Branton, P. G. Hall, K. S. W. Sing, *J. Chem. Soc., Chem. Commun.* **1993**, 1257; P. L. Llewellyn, H. Reichert, Y. Grillet, F. Schüth, K. K. Unger, *Micropor. Mater.* **1994**, *3*, 345; P. J. Branton, P. G. Hall, K. S. W. Sing, H. Reichert, F. Schüth, K. Unger, *J. Chem. Soc., Faraday Trans.* **1994**, *90*, 2965; A. V. Neimark, P. I. Ravikovitch, M. Grün, F. Schüth, K. K. Unger, *J. Colloid Interface Sci.* **1998**, *207*, 159; M. Kruk, M. Jaroniec, A. Sayari, *Langmuir* **1997**, *13*, 6267.
67. F. Schüth, *Chem. Mater.* **2001**, *13*, 3184.
68. Y. Gao, G. Chen, Y. Oli, Z. Zhang, Q. Xue, *Wear* **2002**, *252*, 454; F. Jones, H. Cölfen, M. Antonietti, *Biomacromolecules* **2000**, *1*, 556.
69. R. K. Iler, *The Chemistry of Silica and Silicates*, Wiley, New York, 1979, pp. 866; D. Barby, in *Characterization of Powder Surfaces*, G. D. Parfitt, K. S. W. Sing (Eds.), Academic Press, London, 1976, p. 353; K. K. Unger, *Porous Silica*, J. Chromatogr. Libr. Vol. 16, Elsevier, Amsterdam, 1979, p. 294; H. Ferch, A. Kreher, in *Chemische Technologie*, K. Winnacker, L. Küchler (Eds.), Vol. 3, 4th Ed., VCH, Weinheim, 1983, p. 75.
70. C. Setzer, G. von Essche, N. Pryor, in *Handbook of Porous Solids*, F. Schüth, K. S. W. Sing, J. Weitkamp (Eds.), Vol. 3, Wiley-VCH, Weinheim, 2002, p. 1543.
71. J. F. Brinker, G. W. Scherer, *Sol-Gel Science: The Physics and Chemistry of Sol-Gel Processing*, Academic Press, Boston, 1990, pp. 908.
72. C. W. Higginson, *Chem. Eng.* **1974**, 98.
73. G. Engelhardt, D. Michel, *High Resolution Solid State NMR of Silicates and Zeolites*, Wiley, Chichester, 1988, pp. 500.
74. B. A. Morrow, in *Spectroscopic Analysis of Heterogeneous Catalysts. Part A: Methods of Surface Analysis*, J. L. G. Fierro (Ed.), Studies in Surface Science and Catalysis, Vol. 57A, Elsevier, Amsterdam 1990, p. 161.
75. P. Euzen, P. Raybaud, X. Krokidis, H. Toulhoat, J. L. de Loarer, J. P. Jolivet, C. Froidefond, in *Handbook of Porous Solids*, F. Schüth, K. S. W. Sing, J. Weitkamp (Eds.), Vol. 3, Wiley-VCH, Weinheim, 2002, p. 1591.
76. K. Wefers, G. M. Bell, *Oxides and Hydroxides of Aluminum*, Alcoa Research Laboratories, Technical Paper No. 19, 1972; K. Bielefeldt, G. Winkhaus, in *Chemische Technologie*, K. Winnacker, L. Küchler (Eds.), Vol. 3, 4th Ed., VCH, Weinheim, 1983, p. 2.
77. U. Bollmann, *Habilitationsschrift*, Bergakademie Freiberg, FRG, 1992.
78. R. Thome, H. Schmidt, R. Feige, U. Bollmann, R. Lange, S. Engels, EP 518106, 1992, assigned to VAW Aluminum AG.
79. A. Danner, K. K. Unger, in *Preparation of Catalysts IV*, B. Delmon, P. Grange, P. A. Jacobs, G. Poncelet (Eds.), Studies in Surface Science and Catalysis, Vol. 31, Elsevier, Amsterdam, 1987, p. 343; W. Stoepler, K. K. Unger, in *Preparation of Catalysts III*, G. Poncelet, P. Grange, P. A. Jacobs, (Eds.), Studies in Surface Science and Catalysis, Vol. 16, Elsevier, Amsterdam, 1982, p. 643.
80. B. C. Lippens, J. H. de Boer, *Acta Crystallogr.* **1964**, *17*, 1312; H. Sallfeld, B. B. Mehrotra, *Naturwissenschaften* **1966**, *53*, 128.
81. H. Knözinger, P. Ratnasamy, *Catal. Rev.-Sci. Eng.* **1978**, *17*, 31.
82. C. Laurent, H. A. H. Billet, L. de Galaan, *Chromatographia* **1983**, *17*, 253.
83. http://www.engelhard.com/documents/HiQ%20Gamma%20phase%20alumina%20powders%20rev%204.06.pdf, accessed on February 21st, 2007.
84. C. Boissière, L. Nicole, C. Gervais, F. Babonneau, M. Antonietti, H. Amenitsch, C. Sanchez, D. Grosso, *Chem. Mater.* **2006**, *18*, 5238.
85. N. Zelinsky, W. Kommarewsky, *Chem. Ber.* **1924**, *57*, 667.
86. F. Cavani, F. Trifiro, A. Vaccari, *Catal. Today* **1991**, *11*, 173.
87. E. C. Kruissink, L. L. van Reijen, J. R. H. Ross, *J. Chem. Soc., Faraday Trans. I* **1981**, *77*, 649.
88. J. Zielinski, *Appl. Catal.* **1993**, *A94*, 107; J. Zielinski, *J. Mol. Catal.* **1993**, *83*, 197; R. Lamber, G. Schulz-Ekloff, *J. Catal.* **1994**, *146*, 601.
89. E. C. Kruissink, L. E. Alzamora, S. Orr, E. B. M. Doesburg, L. L. van Reijen, J. R. H. Ross, G. van Veen, *Preparation of Catalysts II*, B. Delmon, P. Grange, P. A. Jacobs, G. Poncelet (Eds.), Studies in Surface Science and Catalysis, Vol. 3, Elsevier, Amsterdam, 1979, p. 143.
90. J. Rathousky, G. Schulz-Ekloff, J. Starek, A. Zukal, *Chem. Eng. Technol.* **1994**, *17*, 41.
91. E. C. Kruissink, L. L. van Reijen, J. R. H. Ross, *J. Chem. Soc., Faraday Trans. I* **1981**, *77*, 649.
92. D. C. Puxley, I. J. Kitchener, C. Komodromos, N. D. Parkyns, in *Preparation of Catalysts III*, G. Poncelet, P. Grange, P. A. Jacobs (Eds.), Studies in Surface Science and Catalysis, Vol. 16, Elsevier, Amsterdam, 1983, p. 237.
93. B. Vos, B. Poels, A. Bliek, *J. Catal.* **2001**, *198*, 77.
94. C. J. Wright, C. G. Windsor, D. C. Puxley, *J. Catal.* **1982**, *78*, 257.
95. L. E. Alzamora, J. R. H. Ross, E. C. Kruissink, L. L. van Reijen, *J. Chem. Soc., Faraday Trans. I* **1981**, *77*, 665.
96. E. D. Rodeghiero, J. Chisaki, E. P. Giannelis, *Chem. Mater.* **1997**, *9*, 478.
97. A. Y. Tsygonok, T. Tsunoda, S. Hamakawa, K. Suzuki, K. Takehira, T. Hayakawa, *J. Catal.* **2003**, *213*, 191.
98. J. G. Nunan, P. B. Himmelfarb, R. G. Herman, K. Klier, C. E. Bogdan, G. W. Simmons, *Inorg. Chem.* **1989**, *28*, 3868.
99. P. Porta, S. De Rossi, G. Ferraris, M. Lo Jacono, G. Minelli, G. Moretti, *J. Catal.* **1988**, *109*, 367.
100. M. M. Günther, T. Ressler, B. Bems, C. Büscher, T. Genger, O. Hinrichsen, M. Muhler, R. Schlögl, *Catal. Lett.* **2001**, *71*, 37.

101. J. C. Burnett, R. A. Keppei, W. D. Robinson, *Catal. Today* **1987**, *1*, 537; R. M. Contractor, A. W. Sleight, *Catal. Today* **1987**, *1*, 587; G. Centi, *Catal. Today* **1993**, *16*, 5; G. Centi, *NATO ASI Ser. Ser. C* **1993**, *398*, 93.
102. V. V. Guliants, S. A. Holmes, J. P. Benziger, P. Heaney, D. Yates, I. E. Wachs, *J. Mol. Catal. A: Chemical* **2001**, *172*, 265.
103. H. Bluhm, M. Hävecker, E. Kleimenov, A. Knop-Gericke, A. Liskowski, R. Schlögl, D. S. Su, *Top. Catal.* **2003**, *23*, 99.
104. F. Cavani, F. Trifiro, in *Preparation of Catalysts VI*, G. Poncelet, J. Martens, B. Delmon, P. A. Jacobs, P. Grange (Eds.), Studies in Surface Science and Catalysis, Vol. 91, Elsevier, Amsterdam, 1995, p. 1.
105. N. Ballarini, F. Cavani, C. Cortelli, S. Ligi, F. Pierelli, F. Trifiro, C. Fumagalli, G. Mazzoni, T. Monti, *Top. Catal.* **2006**, *38*, 147.
106. L. M. Comaglia, C. A. Sanchez, E. A. Lombardo, *Appl. Catal.* **1993**, *95*, 117.
107. E. Kesteman, M. Merzouki, B. Taouk, E. Bordes, R. Contractor, in *Preparation of Catalysts VI*, G. Poncelet, J. Martens, B. Delmon, P. A. Jacobs, P. Grange (Eds.), Studies in Surface Science and Catalysis, Vol. 91, Elsevier, Amsterdam, 1995, p. 707.
108. H. S. Horowitz, C. M. Blackstone, A. W. Sleight, G. Teufer, *Appl. Catal.* **1988**, *38*, 193.
109. X. Wang, W. Nie, W. Ji, X. Guo, Q. Yan, Y. Chen, *Chem. Lett.* **2001**, 696.

2.3.4
Sol–Gel Process

*Miron V. Landau**

2.3.4.1 Introduction

Sol–gel processing is one of the routes for the preparation of porous materials by their solidification from a true solution phase [1], and its physico-chemical principles and applications are well described in the literature [2–4]. The method is characterized by the formation of stable colloidal solutions ("sol") in the first step, followed by anisotropic condensation of colloidal particles (micelles) producing polymeric chains with entrapped solution of condensation byproducts, resulting in the formation of a "liogel" or "hydrogel" or "monolith" when external solvent is not used. After washing out the byproducts, the solvent removal produces "xerogels" or "aerogels", depending on the drying mode, with distinct structures of the primary particles and their packing manner (texture) [5, 6]. For clarity of terminology, the sol–gel method should be distinguished from other routes of materials solidification from solutions, such as precipitation and deposition–precipitation [7, 8], crystallization from melts [9], expansion of supercritical solvent [10], supercritical anti-solvent method [11], supramolecular assembling [12, 13], and others. The main peculiarity – which makes the sol–gel route unique and clearly discernible – is the formation of a clear colloidal solution due to primary condensation of dissolved molecular precursors. The second peculiarity is the merging of these colloidal particles during the subsequent gelation stage into polymeric chains by chemical bonding between local reactive groups at their surface. This prevents flocculation, that is a result of isotropic micelle aggregation (Fig. 1). The porous solids (xero- or aerogels) are produced in the next step – desolvation – depending on the drying mode (see Fig. 1).

Both stages are controlled by condensation chemistry that can include, as a first step, the hydrolysis of hydrated metal ions [14, 15] or metal alkoxide molecules [14–16] (hydrolytic sol–gel processing). The condensation chemistry in this case is based on olation/oxolation reactions between hydroxylated species [14, 15]. The hydroxylated species for further condensation can also be formed by a non-hydrolytic route; that is, by reactions between metal chlorides and alcohols with electron-donor substituents [17]. The non-hydrolytic sol–gel processing may also proceed without intermediate formation of hydroxylated species when it is based on esterification of metal chelate complexes with free carboxylic groups and polyalcohols [20, 21]. Another non-hydrolytic/non-hydroxylating sol–gel route relies on direct condensation reactions between metal alkoxides and metal chlorides or acetates, with the elimination of alkylchlorides or esters [17–19].

The characteristics of sol–gel processing allow the application of different strategies for the preparation of solid catalytic materials. A schematic representation is shown in Fig. 2. The gelation of colloidal solution, followed by desolvation of the obtained gel, can be applied for the preparation of three types of materials:

- Bulk uniphasic materials, i.e., mono- or multimetallic xero- or aerogels (I → II).
- Bulk multiphasic materials where molecular moieties or condensed phases are entrapped between polymeric chains of the gel matrices and/or co-gelated from a mixed colloidal solution [(I + III) → IV].
- Porous uni- and multiphasic coatings and nanometric films prepared by conducting the gelation inside a thin film of colloidal solution at the surface of a supporting material (substrate) [I → II → V; (I + III) → V].

In this chapter, the physico-chemical basis of sol–gel processing will be discussed, as well as details of the synthesis parameters that control the properties of the

---
* Corresponding author.

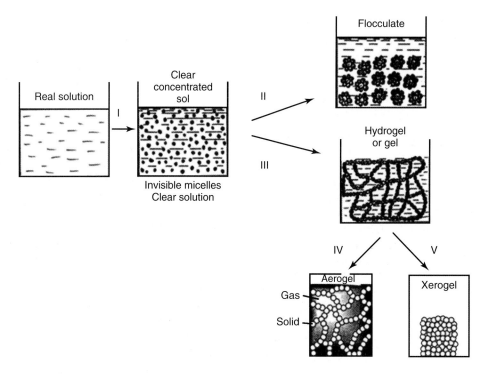

Fig. 1 General scheme of sol–gel processing in the preparation of solid materials. I, colloidization; II, flocculation; III gelation; IV, supercritical fluid processing; V, drying by evaporation.

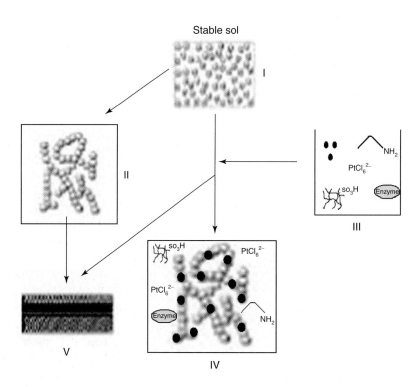

Fig. 2 Schematic diagram showing the different applications of sol–gel processing for the preparation of catalytic materials. I, stable sol; II, gel; III, sources of additional catalytic chemical functionality; IV, gel with included sources of additional catalytic chemical functionality; V, catalytic coating.

obtained materials, namely structure/texture and catalytic performance. Attention is focused on the three types of sol–gel-derived catalysts that can be prepared using the three strategies presented above:

- Bulk inorganic mono- and multimetallic phases: catalysts, catalyst supports [5, 6, 22–28].
- Bulk multiphasic materials: nanostructured inorganic catalytic phases and functionalized molecular moieties including transition metal complexes, polymers and enzymes entrapped in the phase of a solid porous matrix [29–37].
- Uniphasic and multiphasic catalytic coatings, films, and membranes [26, 27, 31, 32].

Only those catalytic materials will be considered where the sol–gel principle is an essential part of the preparation strategy. Deposition, grafting, or anchoring of the catalytic phases or functionalized molecular moieties on sol–gel-derived supports or coatings, as well as those methods that do not use the full sequence of sol–gel processing (e.g., spray-drying of colloidal solutions) are beyond the scope of this chapter.

2.3.4.2 Physico-Chemical Basis and Principles of Sol–Gel Processing for the Preparation of Porous Solids

Sol–gel processing involves a sequence of operations that includes chemical reactions and physical processes (phase separation, dissolution, evaporation, phase transition, etc.) leading to the formation of porous solids from liquid solutions of molecular precursors. The general principles and the physico-chemical bases of the separate steps have been extensively described and analyzed in many comprehensive books and reviews [2–4, 6, 14, 15, 20, 38–48]. The sol–gel processing sequence includes the following seven main stages (see Fig. 3):

- Conversion (activation) of dissolved molecular precursors to the reactive state
- Polycondensation of activated molecular precursors into nanoclusters (micelles) forming a colloidal solution, the sol
- Gelation
- Aging
- Washing
- Drying
- Stabilization.

The first two steps are hardly distinguished in real sol–gel practice because they occur in parallel after mixing the starting reagents. The preparation of a solid catalyst aims at producing a porous solid with a controlled texture, bulk structure, and chemical functionality of its surface. Here, we will briefly discuss all of the sol–gel steps with regards to their impact on these properties. The chemical nature of the selected precursors determines the reactions involved in sol–gel processing, the required additives (solvents, reagents, catalysts), and the conditions needed to control the properties of a porous solid (pH, reaction time, concentrations, temperature, drying mode). The choice of the chemical strategy depends on the possibility to control the rates of activation/condensation reactions which determine the materials texture, cost/availability of corresponding precursors, and their ease of handling. The choice is also determined by the compatibility of functionalizing substances that should be entrapped in the final gel, with solvents and products liberated during the solids preparation. Choice depends also on the physical properties of the corresponding sols, such as stability, viscosity, and the wetting of substrates, all of which are important for the fabrication of catalytic coatings.

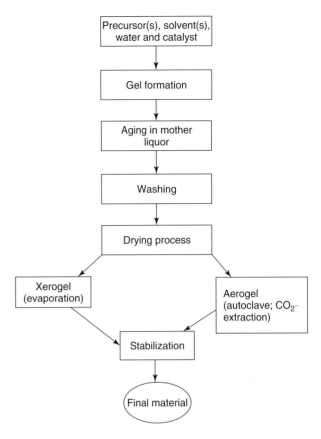

Fig. 3 The sequence of steps involved in sol–gel processing for the synthesis of porous solid catalytic phases. (Adapted from Ref. [6].)

Activation of Sol–Gel Precursors Activation of sol–gel precursors is needed when they are chemically inert

References see page 156

and cannot participate in condensation reactions directly. However, this is not always the case – for example, metal chlorides can react with alkoxides of the same or another metal directly, yielding condensed matter via the non-hydrolytic sol–gel route [20]. In many cases, however, activation is necessary and this most often involves the hydrolysis of solvated metal ions in aqueous solutions or metal alkoxides in alcoholic solutions containing controlled amounts of water. In solvated complexes (hydrates) of metal salts in solution: $[M(OH_2)_n]^{z+}$, where $z+$ is the cation charge and n the coordination number of the metal atom, cations interact with nucleophilic water molecules by transferring electrons from their filled bonding $3a_1$ orbital to empty d-orbitals of metal atoms. This interaction creates a positive partial charge on water protons, increasing their acid strength and ability to participate in hydrolysis (deprotonation) reactions with nucleophilic basic species in the solution. The reaction can proceed in two steps for every water ligand:

$$[M(OH_2)_n]^{z+} + mOH^-$$
$$\rightleftharpoons [M(OH_2)_{(n-m)}(OH)_m]^{(z-m)+} + mH_2O$$

$$[M(OH_2)_{(n-m)}(OH)_m]^{(z-m)+} + OH^-$$
$$\rightleftharpoons [M(OH_2)_{(n-m)}(OH)_{(m-1)}O]^{(z-m-1)+} + H_2O$$

These reactions equilibrate the amounts of complexes with different content and combinations of aqua- (H_2O), hydroxo- (−OH) or oxo- (=O) ligands of general formula $[MO_nH_{(2n-h)}]^{(z-h)+}$, where h is the hydrolysis extent (molar ratio). The reactivity of these species in further condensation reactions, and the structure of polycondensed clusters, are determined by their composition. The principles governing the reactivity of the hydrolyzable metals are reviewed in detail by Livage et al. [14, 15, 45], and the available experimental data for most of transition metals are listed in Ref. [1]. The equilibrium is shifted to the right with increasing solution pH, cation charge density and temperature. The electronegativity of the metal and its coordination number also have an influence in this respect.

Besides the addition of bases, which increases the hydroxide concentration in solution, the hydrolysis of hydrated metal cations may be initiated by addition of another proton scavenger such as propylene oxide [47, 49, 50]. Protonation of the epoxide oxygen by an acidic proton of a water ligand is followed by the epoxide ring opening through nucleophilic attack of the conjugated base (Scheme 1). The use of chloride salts ($A^- = Cl^-$) here is essential, as they are better nucleophiles than water molecules, in contrast to NO_3^- ions, while the ring opening to yield chloropropanol strongly increases the solution pH [50].

Inorganic compounds with a high positive charge on the metal atoms (+IV–VI), present as anions of corresponding salts in aqueous solutions, such as silicon, vanadium, molybdenum or tungsten, should be activated by acidification [2, 14, 15, 45]. In the case of silicates, this causes conversion of the silicate anions to tetrahedrally coordinated precursors $[H_nSiO_4]^{(4-n)-}$ that do not contain coordinative water and can be further condensed by oxolation reactions. Here, the acidification extent n is controlled by the pH, being 2 under strongly alkaline conditions and reaching 4 at pH < 9 [51]. The high-valent metals, such as V(V) or Mo(VI), form polyoxometallate anions with highly symmetrical structures that cannot act as precursors or building blocks for the oxide network. In this case, the neutral species arising from the dissociation of polyanions such as decavanadic acid under acidic conditions should be considered as molecular precursors for further condensation reactions [45]:

$$[V_{10}O_{26}(OH)_2]^{4-} + 4H_3O^+ + 28H_2O$$
$$\rightleftharpoons 10[VO(OH)_3(OH_2)_2]^0$$

Another type of sol–gel precursor – the metal alkoxides (the chemistry and peculiarities of which in sol–gel processing are well described elsewhere [2–4, 52]) – react with water at rates depending on the metals' electronegativity, which in turn determines the partial charge on the metal atoms (δ) [14]. Silica exhibits optimal electronegativity, so that for $Si(OC_2H_5)_4 \delta = +0.32$. This makes the hydrolysis rate slow, but tunable by catalyst addition. Higher partial charges, as for Ti ($\delta = +0.63$) and Zr ($\delta = +0.74$) atoms in the corresponding alkoxides speed up the hydrolysis, which requires careful management of the amount of water present, the temperature, and the concentrations of solvents and/or organic ligands for its control. Lower partial charges do not allow hydrolysis at all, whereas highly positive δ values (as in alkali-metal alkoxides) accelerate the rate too much for a reasonable control. Therefore, $Si(OR)_4$ compounds are widely used as sol–gel precursors for the preparation of silica supports with tunable texture, entrapped catalytic materials and coatings.

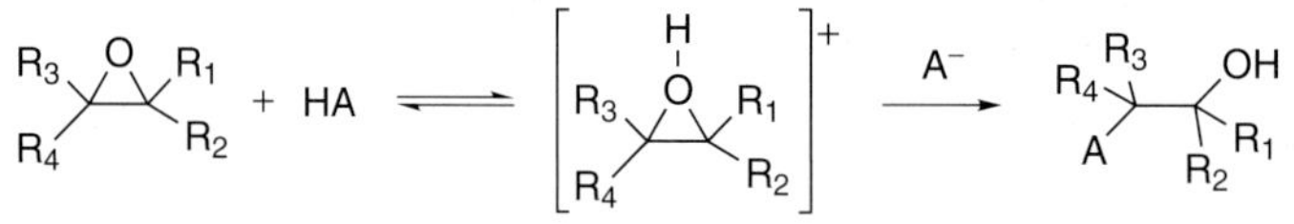

Scheme 1

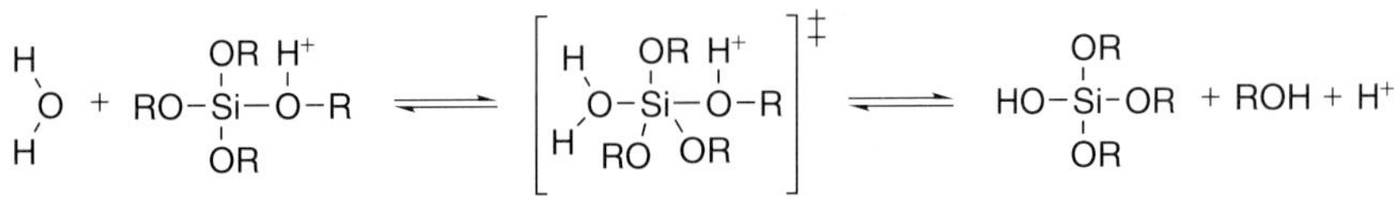

Scheme 2

Their hydrolysis proceeds in the presence of either acid (Scheme 2) or base catalysts including strong Lewis bases, such as F⁻ ions (Scheme 3) via S_N2 nucleophilic substitution reactions with a transition state involving five-coordinated silica atoms. It passes a rate minimum at pH 7 (Fig. 4a). The main variables controlling the hydrolysis rate of the alkoxides, besides pH and temperature, are the hydrolysis number–water/alkoxide ratio, and the structure of R derivatives in the molecules.

References see page 156

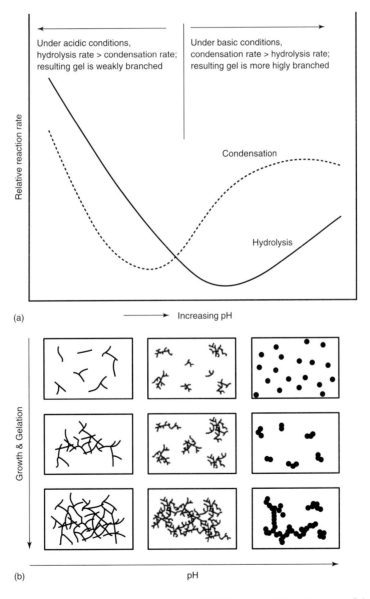

Fig. 4 Dependence on pH of (a) hydrolysis and condensation rates of $Si(OR)_4$ and (b) the structure of the obtained gel. (a) Redrawn from Ref. [46]; (b) Redrawn from Ref. [28].

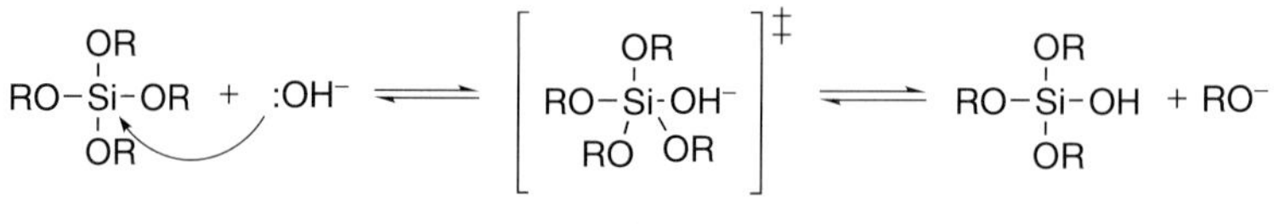

$$\text{RO-Si(OR)}_3 + :\text{OH}^- \rightleftharpoons \left[\text{RO-Si(OR)}_3\text{-OH}^- \right]^{\ddagger} \rightleftharpoons \text{RO-Si(OR)}_2\text{-OH} + \text{RO}^-$$

Scheme 3

Alkoxide sol–gel precursors can be converted to the same hydroxylated activated substances by non-hydrolytic routes. The thermal intermolecular hydroxylation (Scheme 4) was used as a first step for gelation of ZrO_2 at 200–300 °C in a conventional reactor [53] or inside the channels of mesostructured SBA-15 silica [54].

Partially hydroxylated alkoxides can be synthesized by reactions between metal chlorides and alcohols with electron-donor substituents (Scheme 5) or reactions between basic alkoxides and carbonyl compounds such as ketones (Scheme 6) [20].

Metal cations that display unfavorable hydrolysis equilibria in aqueous solutions, such as alkali or alkaline earth metals as well as transition metals, can be activated by complexation with bi- and tridentate organic chelation agents, such as citric acid, oxalic acid, and ethylenediaminetetraacetic acid. This allows them to participate in condensation reactions with polyalcohols, resulting in gelation according to the Pechini method [18, 28].

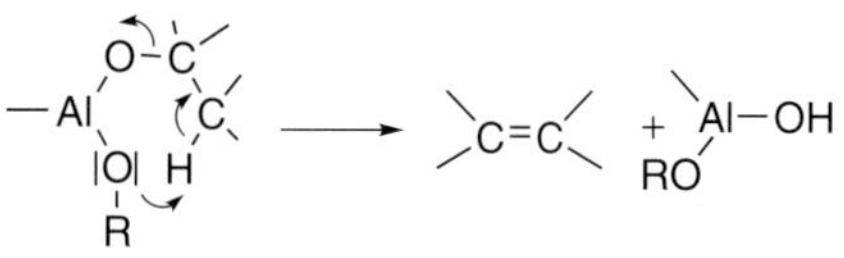

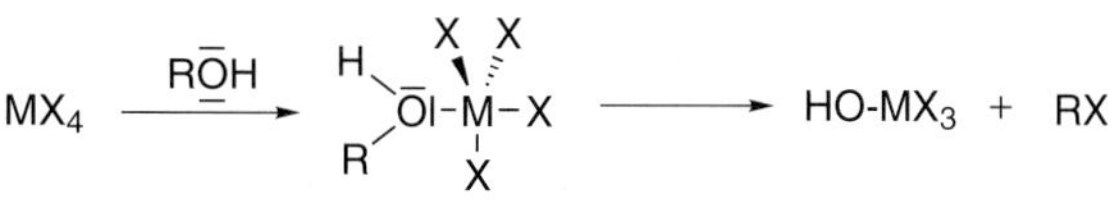

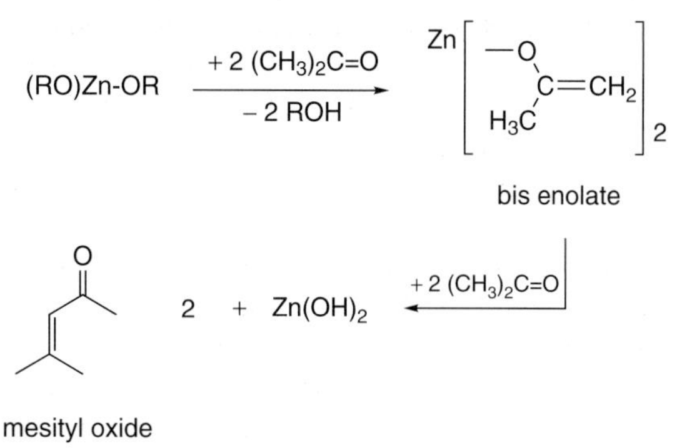

Polycondensation The activated species of the sol–gel precursors continue to react with each other to form colloidal solutions of oligomeric species. The precursors containing hydroxyl groups bonded to metal atoms condense according to olation, oxolation and/or alcoxolation routes catalyzed under acidic and basic conditions [14, 15, 45]. The coordinatively saturated, partially hydrolyzed metal cations are converted by olation through nucleophilic substitution, where the OH group is a nucleophile and H_2O a leaving group (Scheme 7a). It forms one or two M–OH–M bridges between metal cations.

Another route, depending on the extent of hydrolysis, is a two-step reaction involving the nucleophilic addition of hydroxyl species to an oxy-hydroxy-precursor followed by water elimination that gives M–O–M bridges (Scheme 7b). Coordinatively unsaturated ions with deficiency of water ligands react rapidly by oxolation via nucleophilic addition, also forming M–O–M bridges (Scheme 7c). The rates of these reactions depend on solution pH, as the nucleophilic attack is favored by base catalysis and acids favor elimination of the leaving group (H_2O). This is a result of the higher nucleophilicity of anionic precursors $[MO(OR)(OH)_n]^{m-}$ formed in basic solutions compared with neutral species, or protonation of the leaving hydroxyl groups in acidic media. The partially hydrolyzed alkoxides of low reactivity, such as $Si(OR)_4$, react by oxolation under acidic conditions according to oxolation or alcoxolation reactions involving pentacoordinated silicon intermediates (Scheme 8a and b, in that order).

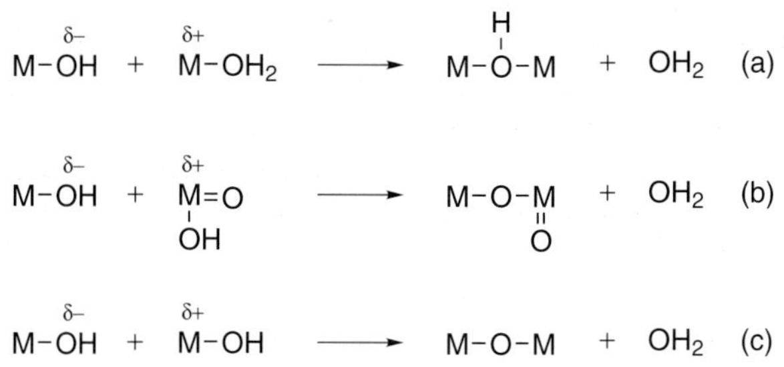

Scheme 4

Scheme 5

Scheme 6

Scheme 7

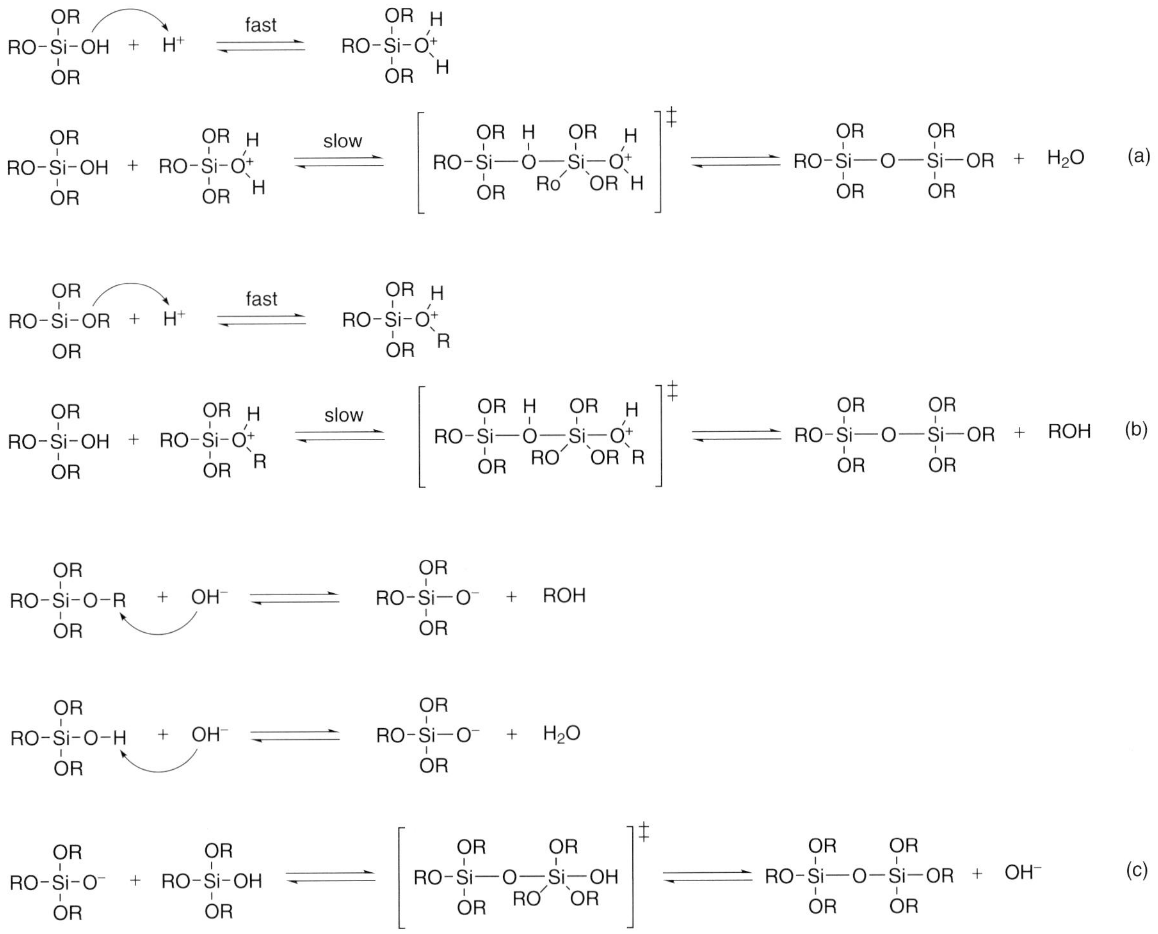

Scheme 8

Under basic conditions, both $Si(OR)_4$ and $Si(OR)_3OH$ molecules are converted to negatively charged oxo-intermediates evolving ROH or H_2O, respectively. These species act as nucleophiles in subsequent nucleophilic addition reactions with activated precursors forming Si—O—Si bridges (Scheme 8c). The salts formed during acidification of aqueous solutions containing high-valence metal anionic precursors (e.g., silicates) decrease the electrostatic repulsion between ions that favors the corner sharing of $[SiO_4]^{4-}$ tetrahedra and the formation of cage polysilicates [45] and increasing the condensation rates. These salt effects can be minimized by acidifying the solutions with proton-exchange resins that remove the counter ions, or by separating metal ions by dialysis.

Another class of reactions leading to polycondensation of oxygen-containing metal precursors is the so-called "aprotic condensation" which proceeds during non-hydrolytic sol–gel processing where water is not required for precursor activation and is not produced by condensation [20, 21, 55]. This includes reactions between alkoxides of dissimilar metals differing in the polarity of M—O bonds, or of alkoxides with metal esters and metal chlorides. The most convenient is the last one that normally is carried out without solvents at temperatures between 313 and 373 K (Scheme 9).

This reaction results in the nucleophilic cleavage of an O—R bond, and for silica the reaction proceeds only with tertiary or benzilic OR radicals that are able to stabilize a carbocation. The alkoxides of Al, Ti, Zr, W, Nb, and Fe with primary carbon groups are reactive enough to participate in condensation reactions. Also, the reaction between $SiCl_4$ and $Si(OR)_4$ (where $R = CH_3$ or C_2H_5) can be successfully conducted in the presence of Lewis acid catalysts such as $FeCl_3$ or $AlCl_3$. The redistribution of OR and Cl ligands that takes place even at room temperatures (M—Cl + M′—OR → M—OR + M′—Cl) results in the situation that the real condensing precursors are mixed chloroalkoxide molecules $M(OR)_xCl_y$. The condensation rate is strongly dependent on the OR/M ratio in mixed precursors [56],

References see page 156

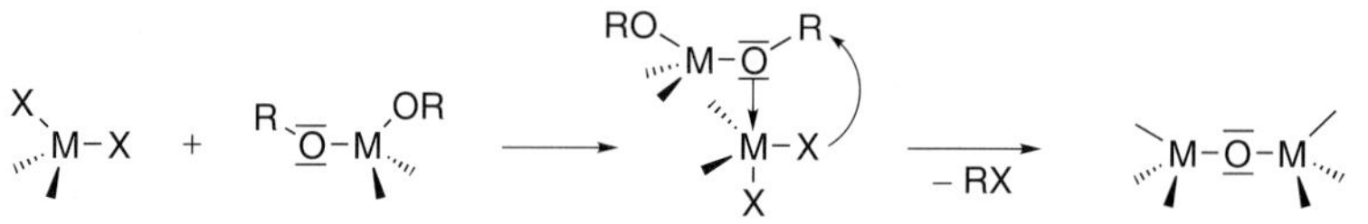

Scheme 9

as well as on temperature, the presence of catalyst, and the structure of the R derivative.

The complexes of metal cations with polydentate chelating agents are condensed with polyalcohols such as ethylene glycol according to acid-catalyzed polyesterification reactions [18, 28]. A balance should be found between the low pH needed for condensation and the ability of the chelating ligand to bind the cation, which increases with increasing pH.

The activation of sol–gel precursors and condensation of activated molecules proceed in parallel, yielding oligomers or polymers with a relatively small number of metal atoms, in the form of packed clusters consisting of polyhedra where metal atoms exist in favorable coordination. With an increased degree of hydrolysis/polycondensation of metal cations, they lose electron-donating water molecules and hydroxyl ligands and form more bonds, leading to a decrease in the nucleophilic power of the OH groups. This, together with increasing positive charge on oligomeric clusters, stops polycondensation following the formation of dimers-tetramers or oligomers containing more than 10 metal atoms, such as $[Al_{13}O_4(OH)_{24}(OH_2)_{12}]^{7+}$ [14, 45]. A similar phenomenon stops the condensation of low-reactivity alkoxides, such as silica. Further controlled increase of temperature and/or pH leads to the condensation of these primary units, that ensures the existence of a colloidal solution (sol) with cluster (micelle) diameters of 5 to 50 nm over periods ranging from several minutes to several days. The same result can be achieved with highly reactive alkoxides or precursors of aprotic condensations by tuning the activation/polycondensation kinetics.

Gelation/Aging/Washing After a period of time called the "gelation time", the sol experiences a gelation transition, converting a liquid solution to a state where it can support a stress elastically [39]. This is the main characteristic of sol–gel processing, which differentiates it from precipitation or flocculation that yields a liquid–solid phase separation. It is a result of the formation of a network, which consists of condensed colloidal clusters entrapping the solution (see Fig. 1). In many cases it is difficult to direct the process to gelation. Increasing the pH (adding OH^-, CO_3^{2-}, NR_3) and the temperature of aqueous salts solutions, or the temperature, amount of water and catalyst in organic alkoxide solutions, often leads to the instantaneous precipitation of an oxidic solid. In such cases, gelation can be achieved by very slow, controlled base addition with a pump [57] or by slow release of a base (OH^-) by hydrolysis of urea added to aqueous salt solution [58]. The same effect is attained by the addition of an epoxide as a proton scavenger (e.g., propylene oxide, 1,2-epoxybutane, glycidol, epichlorohydrin) to the organic solutions of hydrated metal salts, binding the acidic protons of hydrated ions under epoxide ring opening [49]. This results in a uniform and slow increase in pH that controls the olation/oxolation reactions between partially hydrolyzed hydrates of Cr, Fe, Al, Zr, and other metals, forming a stable metal oxide sol and then a gel network. A similar effect is achieved by the addition of chelating complexing agents to metal alkoxide solutions (carboxylic acids, β-diketonates), thereby reducing the hydrolysis rate of metal precursors [14].

The gel structure is determined by the ionic character of the M−O bond, and the relationship between the activation/condensation rates. The Si−O bond, being about 50% covalent, allows a wide distribution of Si−O−Si angles [59], whereas other metals such as Al, Ti and Zr show less flexibility. Therefore, the condensation of silica under certain conditions yields much more linear and open networks of interconnected, "lacy" particles. Other metals tend to form a random bonding between dense micelles forming particulate or colloidal gels [6]. The relationship between the hydrolysis and condensation rates of silicon alkoxides changes with increasing pH (see Fig. 4a). This has a major impact on the architecture of the forming network (Fig. 4b). At low pH, where condensation is a limiting step, the Si−O−Si chains are formed preferentially, followed by their branching and crosslinking. This results, after drying, in the formation of a polymeric gel with a relatively low pore volume and microporous structure. Under basic conditions where the hydrolysis is the rate-determining step, the hydrolyzed species are immediately consumed. This leads to the production of large agglomerates that eventually crosslink, forming a particulate gel with a high pore volume and mesoporous structure after drying.

In aqueous solutions the gelation rate of colloids resulting from silicate acidification to $[H_nSiO_4]^{(4-n)-}$ precursors is lowest at pH $\sim$ 3, and increases by changing the pH in both directions due to acid/base catalysis of oxolation reactions [51]. At higher pH values the oxolation proceeds according to the basic catalytic cycle (Scheme 8c), and since the Q^2 silicon atoms with two μ_2 bridges are more positively charged than Q^1 silicons in terminal $-Si(OH)_3$ groups [45], and thus are predominantly attacked by OH^-, preferentially branched oligomers are formed. At pH < 3 the oxolation proceeds through protonation of terminal Si—OH groups according to the acidic catalytic cycle, which leads to the formation of chain polymers [51]. As a result, the architecture of the forming gel network can be controlled as for the alkoxide precursors, taking into account the cation effects as shown in Fig. 5.

After visible gelation the sol–gel processing proceeds to the aging step, where the structure and properties of the formed network continue to change up to the point that yields the target gel density. This includes four processes: polycondensation; syneresis; coarsening; and phase transformation [39, 46]. The polycondensation reactions continue to occur due to a large concentration of hydroxyl groups, creating more crosslinked structures. The syneresis–shrinkage of the gel with expulsion of liquid from pores, is a result of the creation of new chemical bonds by polycondensation and the tendency to reduce the huge solid–liquid interface. This can be controlled by reducing the temperature and by selecting the correct solvents, which form hydrogen bonds with the hydroxyls. Coarsening or Ostwald ripening is caused by the formation of a solid network with non-uniform surface curvature. The dissolution and reprecipitation of dissolved material into regions of negative curvature causes a growth of the necks between particles, filing the small pores. This reduces the microporosity and surface area due to redistribution of the pore volume between the different pore size ranges. Temperature and solution pH that control the material's solubility as well as replacing the gel's liquid for another one (such as soaking the silica gel in diluted ammonia solution) can effectively influence the coarsening and other aging processes. Washing the pore liquor before drying is used

References see page 156

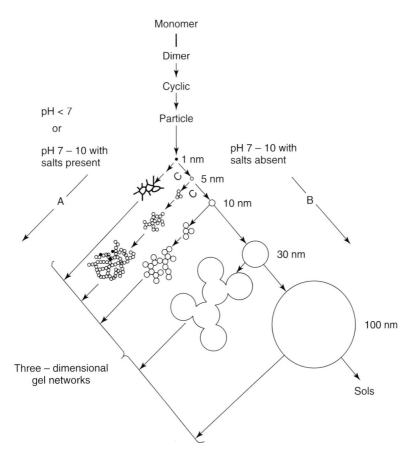

Fig. 5 Structure of silica-gels obtained by polycondensation of acidified silicate precursors in basic and acidic conditions. (Redrawn from Ref. [51].)

not only to remove undesired components but also to change the gel's structure by altering the rates of different routes in the aging progression. The aging conditions, such as temperature, pH, solvent, and the presence of salts, strongly affect the gelation process.

Gel Drying/Desolvation Desolvation of the gel converts the hydro- or alcogel to a porous solid, and is generally carried out by solvent evaporation [39] or supercritical drying [5, 47, 60]. Drying by evaporation, to yield xerogels, consists of three steps: (i) desolvation without pore opening accompanied with the network shrinkage caused by capillary forces; (ii) opening of the pores by liquid flow to the surface after the gel reaches a "leatherhard point" and resists the capillary forces due to greater packing density; and (iii) evaporation of the remaining liquid film in the last step. The shrinkage in the first step is caused by differential pressure across the pores that, for a water-filled pore with radius of 1 nm, reaches 1.5×10^8 Pa [46]. This strongly reduces the pore volume and size in the xerogel compared to the texture of the corresponding wet gel (see Fig. 1). The surface area is also reduced two- to threefold relative to this potential value. One measure to reduce this effect (at least partially) is solvent exchange in the wet gel for organic liquid substances with a lower surface tension before evaporative drying [61].

A more efficient approach that, in principle, allows full conservation of the wet gels texture is supercritical drying, producing aerogels (see Fig. 1). The supercritical liquid has no meniscus in the wet gel pores and does not generate the structure-damaging capillary forces [62]. Two methods for producing the aerogels are widely used in practice, namely supercritical solvent release and supercritical extraction. The first method implies setting the wet gel to the temperature and pressure corresponding to the supercritical state of the liquid phase, followed by its slow release. The second method consists of extracting the liquid phase from the wet gel by passing a flow of supercritical fluid, CO_2 or N_2O that dissolves and extracts the liquid in the gels. The experimental set-up of these techniques has been well described [5, 6, 60].

Under conditions approaching the critical point (647 K, 22.1 MPa), the water becomes a powerful solvent and peptizer that causes "chemical damage" of the gel structure. It is therefore generally replaced by alcohols or ethers, which have a much lower critical temperature (<550 K). Slow release of the solvent from the high-pressure vessel maintained at supercritical temperature yields aerogels. The parameters which affect the aerogel's properties are mainly the high temperature, which favors accelerated aging, phase transitions of the gel and increased hydrophobicity due to the formation of alkoxy groups. The rate of depressurization should also be adjusted in order to avoid the structure stress caused by faster solvent expansion compared with the rate of its escape from the network.

Supercritical extraction of the solvent is conducted with a CO_2 flow in a semicontinuous operation at low temperatures of 310–318 K and pressures of 8–30 MPa. In many cases the original reaction solvent (especially polar water molecules) are replaced by alcohols or liquid CO_2 before extraction, in order to increase the solubility. The main parameters which affect the structure of low-temperature aerogels are the pressure (an increase in which allows CO_2 to approach the density of the solvent) and temperature. With increasing temperature the pore size distribution is shifted to higher pore diameters.

Stabilization of Xerogels and Aerogels In order to ensure a steady operation of catalytic materials prepared by sol–gel processing, the dried xero- or aerogels should be thermally transformed to a more active and stable solid phase. Heat treatment of a dried gel at increasing temperatures causes chemical modification of its surface, crystallographic transformations of the solid matter, and reorganization of the pore geometry [39, 40]. With increasing temperature, first topotactic transformations of the solid matter take place, which are a result of dehydroxylation with water evolution, but without changing the crystallographic packing of oxygen anions. This results in additional shrinkage of the material. Nucleation and the growth of stable crystalline phases occur at higher temperatures. This is always accompanied with sintering and a reduction in surface area and porosity. In addition to the temperature and heating time, the final properties of the stabilized material depend on the heating atmosphere, which either favors the removal of residual organics by burning (O_2, air) or reductive transformations of catalytic phases and entrapped additives (H_2).

2.3.4.3 Application of Sol–Gel Processing for the Preparation of Solid Catalysts

Sol–gel processing is widely used for the preparation of solid catalytic materials, which covers all known applications of solid catalysts. Materials produced include bulk mono- or multimetallic monophasic oxide catalysts, mixed multiphasic catalysts prepared by the entrapment of organic or inorganic catalytic materials in porous matrices, catalytic coatings, and membranes. The sol–gel technique is also employed for the production of preformed porous solids with well-controlled texture that are widely used as catalysts supports, such as silica and alumina. Such widespread use is due to the great flexibility of the sol–gel approach, which permits the fine-tuning of pore and phase structure by adjusting a wide variety of precursors, chemical properties, and preparation parameters. Unfortunately, such an approach also introduces complexity into selection of the optimal chemical route and processing sequence/conditions, depending on the type of target catalytic material. In this section we will highlight those peculiarities of sol–gel processing that are important for the preparation of bulk monophasic catalysts, entrapped multiphasic composites, and catalytic coatings.

2.3.4.3.1 Bulk Catalytic-Phase Materials: Xerogels and Aerogels

A Monometallic Catalytic Materials The sol–gel processing chemistry discussed above allows the preparation of metal-oxide phases with controlled size, shape, and packing mode of primary particles, the structure of which (phase composition) is formed at the final thermolysis/crystallization steps. The requirements to reach maximal surface area and mesoporous structure with high pore volume are generally not compatible with the necessity to synthesize a catalytic material with a proper crystal structure formed at high temperatures, because that goes along with significant degradation of the gel's texture. While this is less important in the production of thermally stable but chemically inert porous silica for use as catalyst supports [2, 51, 63], it becomes critical in the preparation of thermally less-stable transition metal oxides [1]. The careful choice of molecular precursor and sol–gel chemical route, combined with an optimization of activation/condensation kinetics (hydrolysis ratio, catalysts, complexation agents), drying and crystallization conditions, has permitted the preparation of alumina [64] and many crystalline transition metal oxides [1] such as TiO_2, ZrO_2, HfO_2, Nb_2O_5, CrO_x or Fe_2O_3 with a surface area exceeding 200 $m^2 g^{-1}$.

The gelation conditions, structure, texture and performance of many sol–gel-derived monometallic catalytic phases have been analyzed in series of comprehensive reviews [1, 5, 22, 23, 27, 44, 64]. The process is illustrated briefly in the following section with examples of three monometallic oxide phases – alumina, chromia, and titania – which are widely used as solid catalysts and catalyst supports, and where the sol–gel processing provides materials with higher surface areas relative to those of other oxides.

a Alumina Three main sol–gel strategies used for the preparation of porous alumina differ in the nature of the molecular Al-precursors and chemical routes for their activation/condensation: (i) hydrolysis-condensation of Al-alkoxides or salts; and (ii) non-hydrolytic condensation of Al-salt with alkoxide. Alumina xerogels were originally prepared from solutions of aluminum iso-propoxide, dissolving it in excess water ($H_2O/Al = 100$) with acid (HCl, HNO_3) addition to clear the sol, followed by heating the solution to the gelling point, and drying the gel in air at 353 K [65, 66]. The rapid condensation under these conditions yields hydrated boehmite with a surface area of 300 $m^2 g^{-1}$ [66]. This surface area is maintained up to a temperature of ∼723 K, and gradually decreases to 45 $m^2 g^{-1}$ at 1333 K, with progressive dehydration, to form transition aluminas [66]. Increasing the acid/Al molar ratio in the range of 0.05 to 0.25 causes a steady decline in surface area and pore volume. In fact, the surface area and pore volume are decreased for xerogel prepared in this way (acid/Al = 0.05–0.25) and fired at 873 K from 220 to 160 $m^2 g^{-1}$ and from 0.43 to 0.17 $cm^3 g^{-1}$, respectively [65]. The very slow hydrolysis/condensation at room temperature, controlled by the humidity of the air above the Al-*tert*-butoxide–acetone solution in the absence of any catalyst, yielded a similar material, namely crystalline γ-alumina with a surface area of 285 $m^2 g^{-1}$ after calcination at 873 K [67]. The use of a mediate hydrolysis ratio (H_2O/Al-alkoxide) of 10, combined with the addition of acid catalyst and acetylacetone as complexation agent (which controls the rate of Al-*tert*-butoxide hydrolysis) allowed the surface area of γ-alumina xerogel fired at 773 K to be increased to 524 $m^2 g^{-1}$, with a pore volume of 0.4 $cm^3 g^{-1}$ and a pore diameter of 2.8 nm [68]. An alumina xerogel with similar texture characteristics (506 $m^2 g^{-1}$, 0.26 $cm^3 g^{-1}$ after firing at 773 K) was prepared by gelation with acid catalyst in the absence of a chelating agent, but at a lower hydrolysis ratio of 1.1 [69]. Addition of the non-ionic copolymer surfactant Pluronic 123 to the sol formed in Al-*tert*-butoxide–EtOH–H_2O–HCl solution at a hydrolysis ratio of 6 yielded a well-ordered hexagonal mesostructured alumina with a surface area of 410 $m^2 g^{-1}$ and pore volume of 0.8 $cm^3 g^{-1}$ after gelation/crystallization at 313 K and burning out the organics at 673 K [70]. The supercritical drying did not change the texture of alumina gels obtained at fast alkoxide hydrolysis conditions ($H_2O/Al = 100$, HCl), yielding a dry material with a surface area of 350 $m^2 g^{-1}$ [71]. Gelation under mild conditions in the absence of catalysts and H_2O/Al-alkoxide ratios of 3–9 [72–74], or at $H_2O/Al = 1.1$ in the presence of HNO_3 [69], allowed the preparation of aluminas with substantially higher pore volumes of 1.35 to 9 $cm^3 g^{-1}$ and surface areas of 456 to 680 $m^2 g^{-1}$ after calcination at 773 to 1073 K. Separation of the hydrolysis and condensation steps of Al-*tert*-butoxide, conducting the procedure slowly at $H_2O/Al = 0.6$ with subsequent addition of HAc catalyst

References see page 156

to a stable alumina sol, produced crystalline aerogels with a very open texture and >98% porosity [75].

Slow basification of aqueous aluminum nitrate solution by hydrolysis of urea produced a material that was transformed at 573 K to γ-alumina xerogel with a surface area of 425 m^2 g^{-1} [76]. Gels obtained from aqueous solutions of inorganic aluminum salts, as a result of capturing the acidic protons of Al-water ligands by reaction with propylene oxide, displayed surface areas of 660 to 709 m^2 g^{-1} after supercritical drying [77]. These retained a high surface area of 431 m^2 g^{-1} and a pore volume of 2.3 cm^3 g^{-1} after conversion to γ-alumina at 1073 K [77], and even as xerogels had surface areas of 325 m^2 g^{-1} and pore volumes of 1.2 cm^3 g^{-1} after calcination at 873 K [78].

An amorphous alumina xerogel is prepared by the non-hydrolytic method according to:

$$AlCl_3 + 1.5i\text{-}Pr_2O \longrightarrow AlO_yCl_{3-x-y}(i\text{-}PrO)_{x-y}$$
$$\times (i\text{-}Pr_2O)_{1.5-x} + (x+y)i\text{-}PrCl$$

where x denotes the extent of the condensation reaction of AlCl$_3$ with Pr$_2$O yielding the i-propoxy ligand, and y indicates the condensation reaction between chloride and the i-PrO ligand forming Al-O-Al bridges. The amorphous hydrated aluminum oxide has a surface area of 640 m^2 g^{-1} [79], remains amorphous up to 973 K (320 m^2 g^{-1}), and crystallizes to η-alumina at 1123 K, resulting in a loss of surface area to 230 m^2 g^{-1} [79, 80]. Wetting of this xerogel with water decreases the crystallization temperature to 873 K [81].

b Chromia Two sol–gel strategies were used for the preparation of high-surface-area chromia: (i) slow hydrolysis/condensation of inorganic Cr-salts in aqueous solution [57, 58, 82–85]; and (ii) reductive polycondensation of chromate esters formed by slow addition of alcohol to aqueous chromium (VI) oxide [84, 87, 88]. Originally, chromia gel was prepared by the slow basification of aqueous Cr(NO$_3$)$_3$ with ammonia [57, 83]. The implementation of urea hydrolysis for this purpose further slows down the basification rate that permitted an increase in the surface area of chromia xerogel (593 K) obtained at the same gelation time from 193–310 (NH$_4$OH) to 280–340 m^2 g^{-1} (urea), with a pore volume of ~0.2 cm^3 g^{-1} [57, 84]. Supercritical drying of urea-assisted chromia gels with CO$_2$ at 318 K, followed by calcination at 593 K [58], or at high temperature (548–578 K) [83, 84] after replacement of the water solvent with alcohols, yielded chromia aerogels with surface areas of 516 to 785 m^2 g^{-1} and pore volumes of 1.3 to 3.9 cm^3 g^{-1}. Freeze-drying caused partial degradation of this gel structure: when followed by calcination at similar temperatures (500–570 K), this yielded materials with surface areas of 150–163 m^2 g^{-1} and pore volumes of 0.5 cm^3 g^{-1} [82]. A careful identification of the structure of a urea-assisted chromia aerogel showed that it represents a fragile packing of 3- to 5-nm particles consisting of two-dimensional fragments (clusters) of α-CrOOH crystals [58, 85]. This structure is stable up to 650 K in air and up to 773 K in inert atmosphere, being converted to low-surface-area (<200 m^2 g^{-1})α-Cr$_2$O$_3$ as a result of a glow transition [58]. Aerogels obtained from aqueous solutions of inorganic chromium salts (nitrate, chloride) as a result of capturing the acidic protons of the water ligands by reaction with propylene oxide, followed by CO$_2$-supercritical drying at 590 K, exhibited surface areas of 520 m^2 g^{-1} and pore volumes of 1.9 cm^3 g^{-1} [86].

Reaction of chromic acid (a solution of CrO$_3$ in water) with alcohols at 288–298 K causes gelation as a result of the redox reaction proposed in the original publication [84]:

$$2CrO_3 + ROH \longrightarrow Cr_2O_3 + xCO_2 + yH_2O$$

Later, it was supposed that this process is more complicated and includes the polycondensation of chromate ester formed by the reaction of aqueous CrO$_3$ with alcohol [87]. This was partially confirmed by detecting the formate and C$_1$–C$_4$ carboxylate species stable up to 673–723 K in chromia aerogel prepared according to this method [87]. The aerogels obtained after supercritical drying of CrO$_3$/alcohol-derived gels have surface areas of 519 to 594 m^2 g^{-1} and pore volumes of 3.6 cm^3 g^{-1} [84, 87, 88]. Their structure was not identified; only broad diffuse peaks in X-ray diffractograms were mentioned, reflecting very small crystal size and large lattice distortion [84]. After heating to 773 [34] or 973 K [88], the gels crystallize into α-Cr$_2$O$_3$ with a low surface area of ≤100 m^2 g^{-1}. The texture of aerogels derived from ethanol or 1-propanol is thermally more stable [88].

c Titania Two sol–gel protocols were used for the preparation of titania xerogels and aerogels: controlled hydrolysis/condensation of inorganic Ti-salts in aqueous solutions and of Ti-alkoxides in organic solvents. At low temperatures (ice addition), it is possible to stabilize the violently hydrolyzed TiCl$_4$ in aqueous solution as TiOCl$_2$, forming a stable acidic sol that can be further converted to gel at elevated temperatures or by base addition [1]. Xerogels obtained in this way by basification of the sol with NH$_4$OH [89], NaOH [89] or hexamethylenetetramine [90] exhibited surface areas of 390–420, 320 and 470 m^2 g^{-1}, respectively. The use of amine as a base yielded a microporous material, while inorganic bases increased the pore volume by 50–70%

and the pore diameter from 0.5 to 1.7 nm. Heating the xerogels results in a strong decrease of surface area to 20–80 $m^2 g^{-1}$ at temperatures of 673–873 K due to the glow transition caused by crystallization of amorphous material into the anatase structure [90, 91]. The replacement of water with n-butanol, followed by its supercritical release at 598 K, gave a crystalline (rutile + anatase) material with a surface area of 167 $m^2 g^{-1}$ and a pore volume of 2.28 $cm^3 g^{-1}$ [92]. Another method of increasing the surface area of a crystalline titania aerogel is hydrothermal treatment of the hydrogel (reflux, autoclave), which favors titania crystallization at relatively low temperatures. The surface area of such-prepared titania reaches 262 $m^2 g^{-1}$ after firing at 873 K [93].

In the preparation of titania from alkoxide precursors, the main parameters affecting xerogel texture are the hydrolysis ratio (H_2O/Ti) and the concentration of the acid catalyst. Optimization of these parameters, starting from Ti(n-BuO)$_4$–methanol solution, produced an anatase aerogel with a surface area of 220 $m^2 g^{-1}$ and a pore volume of 0.6 $cm^3 g^{-1}$ after calcination at 773 K [94]. The catalyst type used for hydrolysis/condensation of Ti(n-BuO)$_4$ controls the structural evolution of xerogels at elevated temperatures: the HCl-catalyzed gel crystallizes into anatase while acetylacetone allowed the rutile phase to be obtained more easily [95]. Acetylacetone was also shown to inhibit the condensation process due to its chelating action that decreases the Ti-alkoxide hydrolysis rate. This in turn prolongs the gelation time, favoring the formation of less-dense structures [96], and increases the crystallization temperature by 200 K [95]. Hydrogen peroxide acts as a complexing agent in the opposite direction, strongly increasing the alkoxides hydrolysis/condensation rates by the removal and conversion of terminal alkyl groups to hydroxyl groups, possibly with coordination change [97]. This causes gel shrinkage [97] and produces a crystalline gel (anatase/rutile phases) after drying of the gel at 423 K [98]. The addition of polyethylene glycol (PEG) to the gel obtained by hydrolysis/condensation of Ti(i-PrO)$_4$ in ethanol in the presence of H_2O, HCl and acetylacetone increased the surface area of the anatase phase fired at 793 K by a factor of 1.7 [99].

The xerogel prepared by hydrolysis/condensation of Ti(n-BuO)$_4$ in i-PrOH consists of a pure anatase phase with a surface area of 122 $m^2 g^{-1}$ after calcination at 673 K [100]. At 973 K, the rutile phase became predominant with surface areas of 11 $m^2 g^{-1}$ and a crystal size of 35 nm [100]. Crystalline xerogel was obtained using the acidic hydrolysis of tetrabutyltitanate modified with acrylic acid in EtOH, where gelation was induced by subsequent basification with ammonia solution [101]. Drying and removal of bonded organics by photocatalytic oxidation yielded a titania xerogel being pure anatase and exhibiting a surface area of 359 $m^2 g^{-1}$ with a pore volume of 0.36 $cm^3 g^{-1}$. The supercritical drying of wet gels allows the surface area of the aerogels to be increased to 769–840 $m^2 g^{-1}$ and a pore volume of up to 4 $cm^3 g^{-1}$ to be attained [102, 103]. A titania cryogel with a surface area of 511 $m^2 g^{-1}$ was obtained by freeze-drying [104]. Increasing the calcination temperature leads to a proportional decrease in the surface area of these materials due to sintering/crystallization, with the lowest decline for materials dried by low-temperature CO_2-solvent extraction.

B Multimetallic Composite Catalytic Phases Use of the sol–gel strategy to prepare multimetallic crystalline catalytic phases with well-organized solid surfaces is based not only on its potential for mixing the cations at the precursors' molecular scale, but also on the potential for direct crystallization of composite phases. The principles of the sol–gel protocol for preparing multimetallic phases have been discussed comprehensively by Narendar and Messing [105], and are briefly summarized below.

The main advantage of the sol–gel strategy is the possibility of achieving molecular homogenization and spatial fixation of metal cations of different chemical origin in aged gels that crystallize in the stabilization step via high-temperature, solid-state reactions. This reduces the cation diffusion distances, lowers the crystallization temperature of the mixed oxide phases, which in turn allows retention of the texture of multimetallic xerogels and aerogels. The phase separation is determined by the thermodynamic driving force and the kinetics of cation diffusion. At the relatively low temperatures needed for crystallization of multimetallic oxide phases from homogeneous gels, the mixing enthalpy is low and phase separation is essentially determined by the kinetics of long-range cation diffusion. Since this is hindered at low temperatures, the preparation of multimetallic oxide phases requires cation homogenization during the colloidization–gelation steps of sol–gel processing. It also requires a high cation homogeneity to be maintained during the subsequent stabilization steps, including thermolysis/densification that yields amorphous solids, and their further crystallization.

In cases where the precursor activation is carried out by hydrolysis, the main reason for segregating the metal cations at the colloidization–gelation steps is a wide difference of hydrolysis rates for metal atoms with different electrophilicity and ability to undergo coordinative expansion [14, 105]. This favors homocondensation (M_1-O-M_1, M_2-O-M_2) instead of heterocondensation (M_1-O-M_2), yielding separation of $M_1O_x(OH)_y$ and

References see page 156

$M_2O_z(OH)_\zeta$ phases in a gel. In such cases, the initial cation homogenization can be achieved by separating the steps of precursor hydrolysis and condensation by:

- lowering the concentration of water for hydrolysis and controlling carefully the amount of water and the rate of its addition;
- providing a partial prehydrolysis of precursors with lower reactivity and reducing the reactivity of highly reactive precursors by complexation with chelating agents;
- using multimetallic molecular precursors; and
- selecting the optimal acid- or base-catalyzed hydrolysis/condensation routes.

When the metal cations are activated by their chelation with polydentate carboxylic acids, the reasons for cation segregation in multimetallic systems are incomplete chelation, so that part of the metals precipitates as nitrates or hydroxides during the dehydration (condensation) step, and selective precipitation of some metal carboxylates occurs at corresponding pH values. This can be avoided by controlling the carboxylate/metal ratio and pH during the activation step, and also by favoring the crosslinking reactions between uncomplexed carboxylate groups using citric acid and the addition of ethylene glycol that leads to the formation of ester bridges [18, 21, 105]. The non-hydrolytic sol–gel route that does not require precursor activation and includes direct condensation reactions between alkoxides, esters or chlorides of dissimilar metals [20, 55] yields highly homogeneous gels.

After desolvation, homogeneous multimetallic gels are amorphous solids that require further thermolysis-crystallization in order to produce catalytic materials with desired structure(s). To achieve this, the homogeneous distribution of cations must be maintained, and the crystallization process directed towards formation of the required phases at the gel-stabilization steps. Here, the main problems are reactions between the gel network and materials evolved at the thermolysis step: examples include water created as a result of the dehydroxylation of gels obtained by hydrolytic sol–gel routes, or CO and CO_2 formed during carboxylate decomposition, different rates of decomposition of metal–carboxylate complexes of dissimilar metals, and high nucleation rates of thermodynamically less-stable phases [105]. In this situation, water acts as mineralizer, breaking the M_1-O-M_2 bonds, and leading to cation segregation. Cations can react with gases evolved during decarboxylation of the gel, producing carbonate phases of high thermal stability, while less-stable metal carboxylates are transformed to separate oxide phases. This requires a final high-temperature treatment for the crystallization of required composite structures as a result of solid-state reactions, and in general this yields materials with a low dispersion and surface area. Measures that can be taken to improve the structure and texture of composite catalytic phases include reduction of the water or CO_2 pressure at the thermolysis step by optimizing the gel's transport porosity [106] or evacuation [107], rapid heating to favor fast crystallization before cation segregation that leads to formation of thermodynamically more-stable phases, and optimizing the annealing temperature.

a Spinel Compounds These have the general formula $M(II)M(III)_2O_4$, and a crystalline structure where trivalent metal ions occupy octahedral and divalent metal ions tetrahedral positions in a network formed by oxygen ions [108]. The traditional "ceramic route" for their preparation involves solid-state reactions between parent metal oxides that require high temperatures of >1400 K and yield materials with low surface areas of 1 to 5 $m^2 g^{-1}$ [109, 110]. Homogenization of parent metal cations in solids prepared by the sol–gel synthetic protocol allows the crystallization temperature to be decreased to <800 K, maintaining the xero- (aero-) gel texture that, in turn, increases the spinel's surface area to >200 $m^2 g^{-1}$. The mixing of aqueous $Mg(OAc)_2-(C_3H_7)_4NOH$ solution with $Al(OC_4H_9)_3-i\text{-}PrOH$ produced a gel which, after calcination at 823 K, yielded only the crystalline MgO phase in the wide Mg/Al range of 0.15 to 6 [111]. The synthesis of a double Mg–Al precursor as a complex-containing triethanolamine derivative

improves the metal homogenization in the gel formed by its hydrolysis under basic conditions, yielding a mixture of MgO and $MgAl_2O_4$ spinel phases after calcination at >873 K [112]. The pure spinel phase in this case was obtained only at 1473 K, with a surface area of ~5 $m^2 g^{-1}$. Further improvement of Mg–Al homogenization in the gel was achieved due to hydroxylation of Mg- and Al-ions in deionized water by the addition of ammonia, followed by base-catalyzed condensation conducted in a closed vessel which prevented the formation of carbonates [113]. This allowed the required decrease in temperature that was needed for the crystallization of a pure $MgAl_2O_4$ spinel phase to 473 K [113]. The nanocrystalline material prepared by this method retained the small crystal size of <10 nm (X-ray diffraction) up to 1073 K.

Similar results were obtained for the synthesis of $CoAl_2O_4$ [114] and $NiAl_2O_4$ [115] spinels using the commercial double alkoxide precursors $Co(Al(C_3H_7)_4)_2$ and $Ni(Al(C_3H_7)_4)_2$ mixed with $Al(OC_3H_7)_3$ to adjust the Co(Ni)/Al ratio. After hydrolysis/gelation, the air-dried gels were crystallized at 523 (Co) and 773 K (Ni) to the corresponding pure spinel phases with surface areas of 235 and 234 $m^2 g^{-1}$, respectively. Reaction of $CoCl_2$ with preformed AlOOH sol gave a xerogel with lower metal homogenization, leading to an increase in the temperature of $CoAl_2O_3$ spinel crystallization to 673 K [116]. Correspondingly, the material's surface area was decreased to 156 $m^2 g^{-1}$. Gelation achieved by refluxing the mixed solution of $Co(OAc)_2$, Bi- and Fe-nitrates in 2-methoxyethanol in presence of water yielded a xerogel that was transformed at 523 K to the nanocrystalline $CoFe_{1.9}Bi_{0.1}O_4$ spinel phase and sintered at >923 K [117].

b Hydrolcites These are layered double hydroxides of the general formula $[M(II)_{1-x}M(III)_x(OH)_2]^{x+}[A^{n-}]_{x/n} \cdot zH_2O$, where A represents anions (mainly carbonate) and $x = 0.15-0.5$ (M(III)/M(II) = 1–6) [118]. They have relatively low thermal stability, being converted to spinels at >600–773 K. When using the sol–gel synthesis protocol, hydrotalcites should be crystallized at the gel ageing step. Reaction of aqueous AlOOH sol with $MgCO_3$, $Mg(OH)_2$ or MgO under basic conditions at 358 K [119], as well as reaction of prehydrolyzed $Mg(OEt)_2$ with $Al(sec-OBu)_3$ in $BuOH-H_2O$ solution at 343 K [120], yielded gels where only part of the material is present as the hydrotalcite phase. Combining the prehydrolysis of $Mg(OEt)_2$, its subsequent mixing with $Al(OAcOAc)_3$, and base catalysis at the condensation step, prevented metal segregation during the following gelation step under reflux conditions [121]. This yielded a pure hydrotalcite xerogel with a surface area of 211 $m^2 g^{-1}$. Further improvements in the hydrotalcite texture characteristics were achieved by co-gelation of $Mg(OMe)_2$ and $Al(i-OPr)_3$ in MeOH, adjusting the hydrolysis/condensation rate by a slow, controlled addition of water [122]. This produced amorphous binary oxy/hydroxides that were converted to aerogel by supercritical methanol release, yielding a pure hydrotalcite phase with crystal size of 10–20 nm and surface area >500 $m^2 g^{-1}$ after ion exchange with carbonate anions. A pure Ni–Al phase with a hydrotalcite-like crystalline structure and surface area of 127 $m^2 g^{-1}$ was prepared by gelation from $Ni(OAcOAc)_2-Al(i-PrO)_3$ solution under reflux conditions after the addition of base catalyst [123]. Base catalysis (NaOH) for the condensation of mixtures containing prehydrolyzed $Mg(OEt)_2$–Cr-acetylacetonate or Ni- and Cr-acetylacetonates in ethanol produced, under reflux at 353 K, pure Mg–Cr- and Ni–Cr- layered anionic clays with hydrotalcite structure and <20 nm crystal size [124].

c Perovskites These are mixed oxide compounds of the general formula ABO_3 containing a transition metal, such as Ti, Nb, Ta, in the "B" position, with Ca, Sr, Co, Pb, Ce or other RE-metals in the "A" position, and having the same basic structure with isometric symmetry where A ions occupy voids between BO_6 octahedra [125]. The yttrium cobaltate phase $YCoO_3$ with perovskite structure was prepared either by co-gelation of Y and Co ions at 363 K in aqueous solution under slow basification conditions due to urea hydrolysis, or by the formation of a bimetallic carboxylate complex after addition of citric acid [126]. The xerogels obtained by these two methods demonstrated similar behavior: segregated Y_2O_3 and Co_3O_4 phases at temperatures up to 1073 K, and crystallization of a perovskite phase with a crystal size of 18 nm at 1173 K. Using partially hydroxylated basic Pb-acetate salt and $Ti(n-OBu)_4$ in the presence of acetic acid that controls the rates of hydrolysis/polycondensation reactions, and conducting these reactions under slow i-$PrOH-H_2O$ addition, yielded xerogel that formed ~50-nm nanocrystals with perovskite structure at 873 K [127]. The insertion of $PbTiO_3$ seeds into this gel reduced the crystallization temperature to 673 K, yielding nanocrystals of ~20 nm [128].

The pure $PbMg_{0.33}Nb_{0.67}O_3$ perovskite phase was crystallized at 1253 K from xerogel obtained by co-gelation of Pb-acetate, Mg- and Nb-ethoxides in methoxyethanol [129]. Its purity was controlled by the addition of excess Pb and Mg components that prevented formation of the lead niobate pyrochlore phase. Gelation of Pb–Mg–Nb–EDTA solution at basic conditions at Pb : EDTA = 1 : 1 caused precipitation of Pb_2EDTA salt and the formation of 35% pyrochlore phase together with perovskite at 1073 K [105]. Reducing this ratio to 1 : 4 stabilized the homogeneous gel structure, yielding a phase-pure perovskite Pb–Mg-niobate at the same temperature. A further decrease in crystallization temperature of the $PbMg_{0.33}Nb_{0.67}O_3$ phase to 923 K was achieved by gelation from a methanol solution of Nb-ethoxide and Mg-, Pb-acetates in presence of acetylacetone as complexing agent [130]. A pure perovskite phase with the composition $SrBi_2Ta_2O_9$ and a crystal size <50 nm was formed at 1073 K from a xerogel precursor, obtained by gelation from $Ta(EtO)_5$ and Sr-acetate and BiO-nitrate solution in methoxyethanol using an acetic acid catalyst [131]. Acetate ions act as bidentate ligands in this solution, equalizing the hydrolysis/condensation rates of Ta and other ions.

References see page 156

2.3.4.3.2 Catalytic Materials and Modifiers Entrapped in Porous Matrices

The sol–gel entrapment of guest entities in xerogels entities in xerogels or aerogels represents a viable tool for the preparation of advanced catalytic materials. From the viewpoint of phase composition, the target here is opposite to that of the preparation of multimetal, monophasic gels. Entrapment should retain the structure of a guest phase as nanoparticles or isolated molecular substances in the well-defined matrix. Its aim is to add or change the chemical functionality, surface polarity or pore structure of the matrix gel, creating or altering its catalytic performance. These targets could be achieved by three main strategies:

- The insertion of foreign atoms or molecular substances inside the primary micelles at the colloidization stage of sol–gel processing by co-condensation with matrix precursors.
- The adsorption or steric (physical) entrapment of molecular substances at the surface of primary particles, or between them during the gelation/desolvation stages.
- The entrapment of preformed nanoparticles (>1 nm) of catalytic phases.

Depending on the nature and final state of the entrapped guests, this can produce composite catalytic phases which appear either as xerogels or aerogels (i.e., by isomorphous substitution of silicon for aluminum or titanium atoms in the network of SiO_2 matrix particles), it may immobilize (heterogenize) the soluble molecular catalysts (functionalized chemicals, coordinative metal complexes, enzymes) in a porous matrix, or stabilize the nanoparticles of catalytic phases (metals, metal oxides, mixed composites, polymers) in a porous matrix at a high dispersion state. In addition to these conventional applications of catalyst distribution in porous matrices, sol–gel entrapment allows one to control the surface polarity, to create organic–inorganic hybrid materials, and to tailor the pore structure by eliminating entrapped organic templates. This greatly affects the adsorption and diffusion patterns of the reagents and/or products. Co-condensation also permits the spatial separation of different catalytic functionalities in porous matrices, opening a wide range of possibilities for combining chemically non-compatible catalysts (i.e., acids and bases) to conduct one-pot processes that include several reaction steps [132, 133]. Indeed, today many hundreds of reports describe the flexibility of sol–gel entrapment, involving a vast number of reagents and chemical reactions for combining components into composite catalytic materials. As each of the three above-mentioned strategies requires a special approach in the handling of sol–gel processing, their peculiarities are briefly considered below.

A Atoms or Molecular Substances Entrapped by Co-Condensation at the Colloidization Step

a Mixed Metal Oxides Xerogels and Aerogels The insertion of foreign metal atoms into the inorganic network of metal oxides is widely used in the practice of catalyst preparation to create active sites with different functionality, such as acidic (Al in SiO_2) or redox functionality (Ti in SiO_2, V in TiO_2). The foreign atoms can be entrapped in the main oxide (matrix) using sol–gel processing by either of two strategies:

- by mixing the separately prepared colloidal solutions of matrix and dopants, followed by co-gelation of a composite sol
- by co-condensation of the matrix and dopant molecular precursors at the colloidization step.

The second of these approaches is much more flexible, and ensures (in principle) molecular-scale homogeneity. However, special measures are required in order to avoid component segregation, as in case of multimetallic phases preparation, especially when the reactivities of precursors in hydrolysis/condensation reactions are substantially different.

This problem can be solved by addition of the less-reactive precursor to the solution where the second, more-reactive precursor is partially prehydrolyzed, or equalization of the reactivity by selection of precursors with different molecular structure, or by using a complexing agent to control the hydrolysis rate. For the preparation of aluminum-silicates with acidic functionality created by entrapment of tetracoordinated aluminum ions in the silica matrix, prehydrolysis of the less-reactive silicon alkoxide before the addition of aluminum alkoxide suppresses the early formation of Al−OH−Al bridges and subsequent precipitation of pseudoboehmite or bayerite phases [134, 135]. Careful control of the hydrolysis ratio facilitates the formation of an aluminosilicate precursor with tetrahedrally coordinated aluminum already at the colloidization step [135]. The advantage of this synthesis protocol is the absence of cations, needed for the incorporation of aluminum into the silica matrix by conventional co-gelation of silicate and aluminate sols [136]. This allows direct formation of acid sites without the ion exchange or high-temperature calcination needed to remove organic cations. The length of the alkyl chain in the aluminum alkoxide controlling its reactivity determines the relative amounts of tetrahedral, octahedral, or pentacoordinated aluminum in the silica matrix [137]. The use of complexing agents that regulate the alkoxides hydrolysis rates increases the acidity of aluminosilicate gels to a much greater extent than could be achieved by conventional methods due to effective Al—O—Si bond formation [138]. Co-gelation after mixing the separately prepared sols causes segregation of silica and alumina phases after thermal treatment that yields materials with packing of interpenetrated alumina particles and polymeric clusters of silica [139, 140].

A modified procedure was successfully used for entrapment of Ti [141] and Zr [142] molecular species into a silica matrix. This relies on three-step colloidization, first producing partially hydrolyzed silicon alkoxide species:

$$Si(OR)_4 + H_2O \longrightarrow HO-Si(OR)_3 + ROH$$

followed by formation of mixed alkoxides after addition of the second reagent:

$$HO-Si(OR)_3 + Ti(OR')_4$$
$$\longrightarrow (R'O)_3Ti-O-Si(OR)_3 + ROH$$

and the final hydrolysis/condensation of these substances. The fine-tuning of the hydrolysis extent, utilizing at the second hydrolysis step the water released from the initial condensation, favors the entrapment of titanium atoms into the silica network, and high activity/selectivity in partial organics oxidation with a peroxide [143, 144].

If efficient active sites are to be created by stabilizing the molecular species at the surface of a well-defined phase (as in the case of VO_x/TiO_2-anatase catalysts [145]), the better activity in NO selective catalytic reduction (SCR) was achieved by co-condensation of the dopant alkoxide with the surface groups of preformed colloidal particles of the matrix precursor, followed by gelation [146]. A similar approach of modifying the matrix precursor sol with dopant molecules yielded well-dispersed tantalum species at the silica surface, efficient in selective oxidation of pyrimidine thioether [147].

The addition of $CoCl_2$ to the reaction mixture in the non-hydrolytic sol–gel processing of alumina, according to the reaction

$$AlCl_3 + 1.5Al(i\text{-}OPr)_3 \longrightarrow AlO_{1.5} + 3i\text{-}PrCl$$

caused efficient entrapment of Co(II) centers in alumina gels [148]. The material was leach-proof and displayed high activity in olefin epoxidation. Similar results were achieved by the entrapment of Ti-ions in a silica matrix by reaction of $SiCl_4$ with $Ti(i\text{-}PrO)_4$, forming a colorless composite sol [149].

b Hybrid Organic–Inorganic Interphase Catalysts The entrapment of organic molecules entrapment of organic molecules by sol–gel processing in inorganic xerogel or aerogel matrices is made possible by the homo- and co-condensation of metal alkoxides modified with different functional groups. This approach is used to prepare solid catalysts for the insertion of specific chemical functionality sufficient to create various catalytic species and the adjustment of matrix surface polarity (hydrophobicity) and pore structure (matrix network modifiers). Such an approach can also alter the product distribution obtained with entrapped catalysts by tuning the surface chemistry of the surrounding matrix. This technique has the potential to achieve a uniform distribution of organic moieties over a wide loading range, and to prevent their leaching in liquid-phase catalytic runs due to strong covalent bonding to the matrix. Depending on the required function of entrapped organics in the catalytic material, they can be either "non-reactive" (i.e., they do not contain reactive functional groups) or "reactive"; the latter category can be further subdivided as molecules containing reactive organic functional groups or as coordinated metal complexes with reactive centers. Many aspects of the preparation and handling of sol–gel co-condensed hybrid organic–inorganic catalytic materials have been discussed in detail in comprehensive reviews [29, 150–153].

The general synthesis strategy for the entrapment of organics by co-condensation at the colloidization step is based on using organically modified monomeric precursors of the inorganic matrix. These contain at least two alkoxide (OR) groups that enable polymerization or copolymerization with corresponding metal alkoxides: $(OR)_{n\geq 2}A-X$. Here, A is the matrix metal constituent and X the organic derivative, non-reactive (N) or containing reactive functional groups (FG). Depending on their chemical nature, these materials can be synthesized via a variety of routes used to prepare metal-binding ligands with silylated side chains [153, 154]. This chemistry, which is well developed for the synthesis of $(OR)_nSi-X$ compounds, includes hydrosilylation, nucleophilic displacement of alkylhalides, Grignard and coupling reactions, and organically modified sol–gel precursors, many of which are commercially available. The coordinated metal complexes (M), being efficient molecular catalysts for many organic reactions, can be attached or anchored to these monomeric precursors by complexation with silicon-tethered functional groups, typically Lewis bases (FG = $-NR_2, -Ph, -PPh_2, -P(CH_3)_2$, etc. [152, 154]), or by the reaction of a terminal functional group of the metal ligand with a corresponding alkoxide precursor:

$$M\text{-}FG + (OR)_4Ti \longrightarrow M\text{-}FG\text{-}Ti(OR)_3$$

where M = Co-phthalocyanine, FG = SO_2OH, SO_2Cl [155], both giving a compound $(OR)_nA-FG-M$. Other metals (e.g., Al, Ti, Zr, Sn, V, Mo) that also can form sol–gel-derived porous oxide matrices are rarely used to entrap functionalized organics in catalyst preparations due to their limited potential to modify the chemical properties of the corresponding precursors, their brittleness, and their limited porosity.

Although the dialkoxy-derivatives $(OR)_2AFG_2$, $(OR)_2AN_2$ or $(OR)_2A(FG-M)_2$ cannot form a solid network, the chains that result from their condensation can tune the flexibility of the oxide arrangement formed by co-condensation with metal alkoxides [156]. Polycondensation of trialkoxy-derivatives (i.e., $(OR)_3$-Si-FG-M) yields a two-dimensional "pleated sheet" structure [2]. The creation of hard three-dimensional (3D) structures requires co-condensation of organically modified, and not modified metal alkoxides where the latter play the role of crosslinking agents. This is common practice in the preparation of sol–gel-entrapped catalysts with organically modified matrix precursors [29, 151, 154]. The co-condensation of two modified precursors – one with catalytic function and the other containing non-reactive organic molecules – extends the potential for the regulation of a catalyst's performance [167]. Another possibility of obtaining the 3D network is provided by the implementation of polysilylated organic molecules $[(OR)_nSi]_zN$, $[(OR)_nSi]_zFG$ or $[(OR)_nSi]_zFG-M$, the condensation of which results in the formation of polysilsesquioxane structures $[(O_{1.5}Si)_z(N, FG, FG-M)]_m$ with entrapped organic moieties [157]. The co-condensation of polysilylated organics with non-modified metal alkoxides allows the density of organic molecules to be controlled, while the network flexibility is seen to depend also on the structure of organic molecules [151].

The introduction of organic modifiers to molecular alkoxide precursors greatly affects their reactivity in hydrolysis/condensation reactions, as the alkyl substituents are electron donors and decrease the electrophilicity of the metal center. This is one of the reasons for a strong dependence of the density and texture of dried gels on the $Si(OCH_3)_4/FG-(CH_2)_nSi(OR)_3$ ratio used in co-condensation, and its independence of the type of functional group [158]. Under acidic conditions, the inductive effect of the organic modifier promotes the sol–gel reactions relative to non-modified alkoxides, with a reverse situation occurring in a basic environment [153]. Optimization of the sol–gel entrapment conditions, concentration of modified precursor, and prehydrolysis of the less-reactive component allows control of the gel properties.

Methods used for the entrapment of organic catalysts and modifiers in porous solids by co-condensation are listed in Table 1, with the silica matrix being the best-investigated case. The general schemes are illustrated by real examples of the preparation of immobilized catalysts on silica supports. The non-reactive and functionalized organic molecules are inserted by co-condensation

of preformed $(OR)_3Si[FG(N)]$ or $[(OR)_3Si]_zFG(N)$ with $Si(OR)_4$. The coordinated metal complexes can be inserted using one of two strategies, conducting their reactions with functional groups either before (methods 2a, 4a) or after the co-condensation of functionalized metal alkoxides $(OR)_3SiFG$ or $[(OR)_3Si]_zFG$ with $Si(OR)_4$ (methods 2b, 4b). The first approach (methods 2a, 4a) produces more active and leachproof catalysts, if the accessibility of active species is not strongly reduced due to unfavorable gel structure [153].

Enhancement of the catalyst's surface hydrophobicity, to favor the adsorption of organic reagents, can be achieved by the incorporation of non-reactive organics by means of co-condensation. This is not accompanied by any damage of texture parameters that takes place during high-temperature dehydroxylation. The high efficacy of this approach was demonstrated especially in the preparation of Ti- and Zr-doped silica catalysts for selective oxidation of organics with peroxides in liquid phase [160, 168–171]. The structure of the organic molecule entrapped by co-condensation in the Ti–Si aerogel greatly affected epoxide yield in the oxidation of cyclohexene or cyclohexenol (Fig. 6).

The choice of modifier depends on the nature of the substrate: the hydroxypropyl-modified catalyst showed the best performance in alcohol oxidation, whilst for olefin oxidation the highest yield was obtained with phenyl-modified material. Increasing the extent of surface

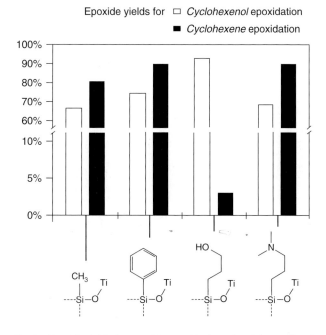

Fig. 6 Epoxide yields in peroxide epoxidation of cyclohexenol and cyclohexene catalyzed by Ti-doped hydrophobic silica aerogels modified with different organic substances. (Reproduced from Ref. [168].)

Tab. 1 Synthesis methods for entrapment of organic catalysts and modifiers in sol–gel-derived catalytic materials with silica matrix by co-condensation approach. N, non-functionalized derivatives; FG, organic derivatives with reactive functional groups; M, coordinative transition metal complexes; ⌒, organosilyl tether (spacer)

Entry	Catalyst or modifier	Synthesis method	Examples of catalysts or catalysts supports	References
1	SiO₂–Si(O)(O)(O)–CH₂–FG	(RO)₃Si–FG (FG,N) + Si(OR)₄ →[H₂O, Catalyst] SiO₂–Si(O)(O)(O)–CH₂–(FG,N)	2 Si(OMe)₄ + (EtO)₂SiMe(CH₂)₃NH₂ →[H₂O/MeOH, sol-gel] Si–(CH₂)₃NH₂	[29]
	SiO₂–Si(O)(O)(O)–CH₂–N		x(EtO)₃Si-CH₃ + y(OMe)₄Si + Zr(OPr)₄ →[H₂O, NH₄OH] Si–CH₃ / Zr	[30]

(continued overleaf)

Tab. 1 (Continued)

Entry	Catalyst or modifier	Synthesis method	Examples of catalysts or catalysts supports	References
2		(a) $(RO)_3Si{-}FG + M \rightarrow (RO)_3Si{-}FG{-}M \xrightarrow{Si(OR)_4, H_2O, Catalyst} SiO_2\text{-supported }Si{-}FG{-}M$ (b) $(RO)_3Si{-}FG \xrightarrow{Si(OR)_4, H_2O, Catalyst} SiO_2\text{-supported }Si{-}FG \xrightarrow{M} SiO_2\text{-supported }Si{-}FG{-}M$	$(EtO_3)SiC_3H_5 + Cr(CO)_6 \longrightarrow (EtO)_3Si{-}C_6H_4{-}Cr(CO)_3 \xrightarrow{+TMOS, \text{ sol-gel}}$ supported arene-Cr(CO)$_3$ $(EtO)_3Si{-}(CH_2)_3{-}NH_2 \xrightarrow{TEOS, \text{ sol-gel}}$ supported amine; then + PhC(O)NCS → supported thiourea; then sol-gel [Rh(COD)Cl]$_2$ → supported Rh(COD)Cl complex	[31] [32]

2.3.4 Sol–Gel Process **139**

(continued overleaf)

Tab. 1 (Continued)

Entry	Catalyst or modifier	Synthesis method	Examples of catalysts or catalysts supports	References
4		(a) $[(RO)_3Si\text{-}]_z FG + M \rightarrow [(RO)_3Si\text{-}]_z FG$		[35]
		(b)		[36]

methylation in Zr-doped silica to 60% led to an increased (up to fourfold, in linear fashion) cyclohexene conversion and hydrogen peroxide efficiency, this being due to improving the substrate's adsorption capacity and decreasing H_2O_2 decomposition [160]. Methylation of the Ti—SiO_2 surface also protects the entrapped Ti-species from leaching during oxidation with aqueous H_2O_2, by reducing water adsorption and the probability of Ti—O—Si bond hydrolysis. This improves the catalyst's recovery and re-use performance, as has been demonstrated for the oxidation of cyclooctene [169]. The entrapment of organics with amine groups enhanced the selectivity for 3-methylcyclohexen-2-ol-1 by up to 98%; this was due to neutralization of the surface acidity of the Ti—SiO_2 catalyst, which reduced the rate of side reactions [171].

The insertion of non-reactive organics by co-condensation creates additional micro- or mesoporosity in xerogels after their removal, increasing the surface area and pore volume [150]. The extractive removal of acetylenic organic fragments from hybrid silsesquioxane gel at low temperature in the presence of catalytic amounts of NH_4F allowed careful control of the silica pore structure [164]. Co-condensation of silica alkoxides substituted by selected molecular templates produces template-shaped cavities, which act as selective recognition spaces or molecular imprints inside the oxide matrix after their removal. This allows a combination of the accelerating action of catalysts with molecular recognition effects, enantiomeric resolution and careful control of the chemical environment of catalytic sites [172, 173]. Imprinting the acidic aluminum-doped silica-gels with templates having a structure which is close to the required transition state of acid-catalyzed reactions gave rise to substrate selectivity, or caused the enantiomers to react at different rates [150, 172]. Entrapment of the bulky 3-(triethoxysilyl)propyl-benzylcarbamate by sol–gel co-condensation with $Si(OEt)_4$, followed by selective removal of the aromatic core of the template molecule, left tethered amine basic sites within an imprinted environment [174]. This led to a high shape-selectivity in the condensation of isophthalaldehyde with malononitrile, by strongly inhibiting the addition of a second malononitrile molecule.

The entrapment of functionalized organics and co-ordinated metal complexes by co-condensation with precursors containing tethered reactive centers allows the preparation of so-called "interphase catalysts" (Fig. 7), where the reactive centers (organic functional groups, FG, or coordinated metal complexes, FG—M, bound to the matrix) are located in the mobile liquid or gas phase. This overcomes the restricted accessibility of active sites to reacting molecules, and provides the mobility that is required for formation of the transition state. Numerous examples of the successful preparation of such catalysts with high activity/selectivity patterns have been described [151, 152]. In many cases, however,

References see page 156

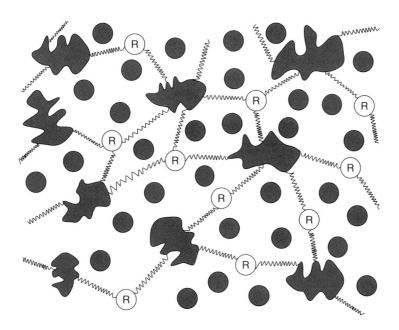

Fig. 7 Representation of an interphase catalyst containing reactive groups immobilized in an inorganic porous matrix. R, functional reactive group (FG) or immobilized coordinative metal complex (FG-M), ▓-support/matrix, ∿- spacer, ●-substrate/solvent. (Reproduced from Ref. [151].)

synergistic effects were observed (i.e., [175–178]), resulting in substantially higher turnover frequencies and product selectivities than those observed with homogeneous analogues of the entrapped catalysts.

B Molecular Substances Adsorbed or Entrapped at the Gelation Step In many instances it is possible to prepare efficient catalytic materials through encapsulation of the chemically functionalized molecules in a porous oxide matrix by physical or steric entrapment at the gelation step of the sol–gel sequence (Fig. 8). This implies the absence of strong covalent bonding between entrapped species and the matrix precursor that excludes co-condensation and the formation of mixed compounds at the colloidization step. The entrapped molecular catalysts, if dissolved in the intermicellar liquid, become adsorbed at the surface of matrix pore walls after the drying step. This bonding is a result of "interfacial coordination chemistry" [179], including the matrix surface groups in the inner sphere of coordinated metal complexes, electrostatic interactions of metal ions with surface SiO^- species that behave similarly to water at higher ligand field strength [51], hydrophobic interactions, H-bonding, acid–base or other interactions, depending on the entrapped matter and matrix precursor functionality [180]. This makes the immobilization of molecular catalysts more efficient, in addition to steric hindrance, with respect to leaching, but it can also cause damage to the catalyst's performance. When the catalyst's activity/selectivity is maintained, the advantage of physical entrapment is the simplicity of the preparation, avoiding the difficulties associated with synthesis of tethered -FG or -FG–M substances.

a Functionalized Organic Catalysts The preparation and performance of molecular catalysts that have been successfully heterogenized by physical entrapment for a wide range of reactions have been reviewed in details [33, 181]. Two synthetic protocols were used, depending on the chemical properties of the entrapped molecules: (i) the addition of dopant solution to prehydrolyzed $Si(OR)_4$ at pH close to 7; or (ii) together with ammonium hydroxide to tetraalkoxysilane precondensed in acidic solution. Both methods, when followed by gelation, aging and drying, yielded a variety of leach-proof recyclable organoceramic catalysts with good activity/selectivity [33].

The main disadvantage of this approach is a need to direct the sol–gel procedure such that microporous silica matrices with pore radius of 1–2 nm are formed for the efficient steric entrapment of dopants. Unfortunately, this restricts the use of such catalysts to substrates that can penetrate the small pore openings, and reduces the overall reaction rates by diffusion limitations, since part of the pore volume is also occupied with dopants. As a result, the loading of entrapped catalysts is limited to dopant/Si ratios of <0.1. As demonstrated previously [182], the catalyst surface area falls from 200–700 $m^2\,g^{-1}$ at $Rh(CO)Cl(L)_2/Si(OEt)_4$ ratios of <0.1 to below 20 $m^2\,g^{-1}$ when this ratio exceeds 0.2. However, in some cases the accessibility of entrapped catalysts can be substantially improved at high loadings of reactive groups, using organically modified silica precursors and selected solvents, which favor swelling of the organic/inorganic hybrid matrix [151, 167].

Another approach is the physical sol–gel entrapment of metal complexes dissolved in ionic liquid which, after extraction, leaves a more open xerogel structure with only strongly adsorbed, accessible catalyst molecules [183]. Careful comparison of co-condensation and physical entrapment strategies applied to immobilization of the same reactive molecule in a silica matrix shows that co-condensation provides catalysts that are not only more active but also leach-proof [175]. In some cases, a better immobilization of the molecular catalyst is achieved by physical entrapment in an alumina matrix obtained by the non-hydrolytic sol–gel route [184].

b Inorganic Molecular Catalysts The bulky molecules of the Keggin-type dodecaheteropolyacids ($H_n[XM_{12}O_{40}]$, n = 3, 4; X = Si, P; M = Mo, W) of about 1 nm in size could be efficiently immobilized in a silica matrix by physical entrapment using the sol–gel method [185, 186].

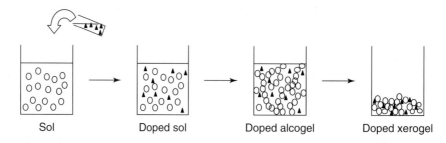

Fig. 8 (Left to right) Steps in forming physically entrapped (adsorbed) molecular catalysts by sol–gel processing. (Adapted from Ref. [153].)

These catalysts demonstrated higher activity and recyclability in many acid-catalyzed reactions compared with their silica-supported analogues. In contrast, small inorganic molecules (e.g., the $PtCl_4$–CO complex) were only weakly adsorbed at the surface of the matrix pores, and easily leached out in catalytic runs, even if they had been entrapped in the silica matrix by sol–gel processing [187]. As shown recently by several groups, strong adsorption at the surface of the porous matrix is critical for the successful immobilization of inorganic molecular catalysts by sol–gel entrapment. Steric factors and matrix microporosity are of secondary importance. The addition of $CrCl_3$ salt to the prehydrolyzed sol of Si- or Zr-alkoxides caused entrapment of Cr-ions in the resulting gels, so that they became immobilized at the silica surface as molecular Cr(VI)-oxide species after oxidation with ozone due to reactions with surface hydroxyls [188]. Analogously, a Cu/SiO_2 catalyst containing adsorbed Cu-ions was prepared by the addition of $CuSO_4$ salt to an acidified $Si(OEt)_4$ solution, followed by decomposition of the salt in the obtained gel at 773 K [189]. Both catalysts displayed high activity in the oxidation of alcohols and CO, and good recyclability. Interaction of $Ru(CO)_{12}$ and $Os(CO)_{12}$-carbonyls with an alumina surface, after sol–gel entrapment using $Al(i\text{-}OPr)_3$ precursor, immobilized them in the form of coordinatively unsaturated complexes $(Al-O-C-Ru(Os)(CO)_{10})$ which proved to be stable against leaching and metal cluster formation [190]. These surface species, when held in a mesoporous matrix, displayed substantially higher activity in hydrogen transfer reactions compared with homogeneous carbonyls, in addition to excellent leach-proof properties and recycling performance.

c Enzymes Bioceramics – which are porous inorganic matrices with entrapped enzymes, catalytic antibodies, or whole cells – represent a viable catalytic tool for the improvement of biotechnological and chemical processes, and have been studied extensively during the past decade. The details of their preparation by the sol–gel entrapment strategy, properties and performance have been reviewed elsewhere [191–195].

All biological catalysts are bulky polymeric molecules with molecular weights of 5000 to 400 000 Da and sizes in the range of 1 to 6 nm for globular conformations. Therefore, the main sol–gel encapsulation strategy in this case is either physical or steric entrapment. The sol–gel entrapment protocol for bioencapsulation should be modified to ensure that condensation of the silica matrix precursors and subsequent processing can be performed in the biocompatible pH range of 5 to 9, compatible redox conditions, and temperatures of <333 K, while minimizing the use of solvents and toxic organic species.

The main problems here are the decomposition of biocatalysts by reactions with alcohols evolved from the hydrolysis of Si–OR groups, deactivation as a result of reactions of silica precursors with reactive centers of the enzymes, and gel shrinkage, which pressurizes the entrapped molecules. The optimized entrapment protocol [194] involves between 5 and 100% prehydrolysis of alkoxysilane precursors, removal of the alcohols by evaporation, mixing with a buffered biological catalyst containing fluoride or amine condensation catalysts, drying the control additives such as formamide and polyols and structure modifiers such as poly(vinyl alcohol) (PVA) or PEG, and directing the silicate/siloxane oligomer growth via protein topology. After aging for 12 to 72 h and washing, controlled desiccation allows extreme shrinkage of the catalyst gel to be avoided.

This procedure, using $Si(OR)_4$ precursors, has allowed preparation of a series of entrapped biological catalysts that included phosphatase, aspartase, glucose oxidase, trypsin, and other enzymes [191, 196]. Being optimized, they expressed 100% activities relative to the natural forms. However, for other enzymes such as lipases – which have versatile biocatalytic activity in synthetic organic chemistry, food and oil processing – it was necessary to use silica gels modified with alkyl groups in order to obtain similar results [192]. The highest activity was observed for catalysts which had been prepared starting from $CH_3\text{-}Si(OCH_3)_3$ and $n\text{-}C_3H_7\text{-}Si(OCH_3)_3$ precursors mixed with $Si(OR)_4$ at a stoichiometric water-to-silane ratio. This provides a lipophilic environment for the enzyme, which is inactive in the absence of an aqueous–lipid interface, and favors the transport of organic substrate inside the matrix pores. Further improvement of the entrapment efficiency, which avoided reactions of the biomolecules with evolved alcohols and allowed the enzyme loadings to be increased to 20–30 wt.%, was achieved by substituting polyalkoxysilanes with water-soluble polyglyceroxysilanes as gel precursors. This was carried out by the transesterification of partially hydrolyzed alkoxysilanes with glycerol before enzyme addition [193, 197]. The glycerol evolved during synthesis of bioceramics is biocompatible and acts as additional drying control additive.

C Nanoparticles of Catalytic Phases Entrapped in Inorganic Porous Matrices The catalysis of chemical reactions by catalytic phases – solids with well-defined atomic structure(s) that contain at their surface(s) groups of atoms in a specific arrangement acting as active sites – requires correct control of their particle size in the range of 1 to 10 nm. One common practice in the

References see page 156

preparation of catalytic materials is stabilization of the corresponding phases, metals, metal oxides, sulfides, mixed oxides, and salts inside the porous inorganic matrices at high dispersion levels. The entrapment of nanoparticles of catalytic phases in such matrices by sol–gel processing is a very useful method to achieve this goal, and involves three main synthesis strategies [28, 198]:

- *Precipitation* of a target catalytic phase or its precursor from the intermicellar liquid of the matrix sol at the gelation/drying step.
- *Co-condensation* of the molecular precursor of a catalytic phase with the matrix precursor at the colloidization step, followed by formation of the desired nanoparticles by chemical/thermal transformations of entrapped substances.
- *Co-gelation* of the mixed sol containing colloidal matrix precursor and preformed colloidal nanoparticles of a target catalytic phase.

The choice of strategy depends on the chemistry of the selected phase and the matrix precursors, where silica and alumina are used in most cases. Another application of gel-included catalytic particles is the formation of pellets that contain catalytic phases as powders with micrometer particle size, such as zeolites. In this case, the gel of silica, alumina or alumina-silicate is implemented as a binder, filling the intraparticle space of the catalytic phase, which is not really entrapped in the gel structure. The principles of controlling the binders structure and adhesion are described in Sections 2.3.4.3.1 and 2.3.4.3.3.

a Metals The application of different sol–gel methods for the entrapment of metallic nanoparticles has been reviewed elsewhere [28, 199–201]. The first strategy, in which the metal precursor is dissolved in the intermicellar liquid and precipitated during gel drying, followed by precipitate reduction, is the simplest to use, but provides poor control of the metal state. The insertion of metal salts at relatively low concentrations, as used in supported metal catalysts, only slightly affects the catalyst's texture, which is determined predominantly by the matrix precursor and the selected sol–gel processing conditions [200, 202–204]. However, the resultant dispersion of the metallic phase depends heavily on the strength of the interaction between metal precursors and matrix, so that selection of the metal compound and gelation conditions becomes critical. The dispersion of noble metals (Pt, Pd) entrapped in silica and alumina matrix varies over a wide range, from 5 to 75%, depending on the metal precursor (metal chloride, chloride of metal-ammonia complex, chloroplatinic acid, acetylacetonate) and the pH of the matrix at the gelation step [202–206]. This was rationalized on the basis of the Zeta-potential values determining preferential adsorption of cationic or anionic species [200].

The addition of reactive organic stabilizers to $Si(OR)_4$-metal salt solutions at the initial mixing step, such as ethylene glycol in case of Ni-salt [207], leads to formation of metal glycoxide complexes, that react with silicon alkoxide, thus strengthening the adsorption of the metal precursor. The organic additives can also reduce the metal directly at room temperature ($Pd(OAc)_2/SiO_2$ [208]; $AuCl_3/ZrO_2$ [209]), or by reaction with evolved CO at elevated temperature ($Ni(NO_3)_2/SiO_2$ [210]), favoring a uniform distribution and high dispersion of the metallic phase. Much better control of size and particle distribution in the porous matrix is achieved by a method whereby a reducible metal precursor is bonded covalently to the matrix precursor, which is then processed by the full sequence of sol–gel steps or co-condensed with a non-modified precursor. This has been demonstrated [210, 211] for Au, Pt, Pd, and Ag compounds attached to N-[3-(trimethoxysilyl)propyl]-ethylene diamine to yield, for example, $(CH_3)_3Si\text{-}C_3H_7\text{-}NH_3^+ \; AuCl_4^-$, followed by hydrolysis condensation/gelation (H_2O, HCl, optionally $Si(OC_2H_5)_4$) and reduction of the metal derivatives with $NaBH_4$. An additional advantage of this method is that, after metal reduction, the tethered amine groups play the role of nanoparticle stabilizers, as in colloidal solutions. This results in a narrow size distribution of the corresponding metal nanoparticles distributed uniformly through the matrix volume.

Entrapment into a silica matrix of the metal carbonyl $Ru_2Co(CO)_{12}$ (which is converted to the metallic state at 100 °C) rather than salts, avoids high-temperature reduction and yields well-dispersed metal particles of 2–3 nm diameter [213]. In some instances, the reagents used and released during the matrix sol–gel processing have a negative effect on the formation of entrapped metal nanoparticles, notably when the metal and matrix precursors are mixed before gelation. However, this problem can be solved by completing the matrix gelation first and then exchanging the swelling solution for a solution of the metal precursor with controlled composition. This approach also helps to avoid the spontaneous silver precipitation caused by reduction with released methanol radicals, or the formation of unfavorable photosensitive AgCl by reaction of $AgNO_3$ with HCl used as condensation catalyst [214]. In addition, the strongly acidic conditions used for condensation of the silica matrix [215] are not compatible with the existence of Ni-citrate in solution. Ni-citrate decomposes under such conditions, with the liberated Ni ions interacting strongly with silica. This problem is solved by first completing the silica matrix gelation, and then inserting nickel citrate into a non-acidic gel.

The addition of specialized precipitation agents such as urea to the $Si(OR)_4$–Ni-salt solution favors homogeneous metal distribution in the matrix at high metal loadings [216].

The reduction of transition metal salts in the presence of organic stabilizers (special ligands, polymers, surfactants) [217] serves as another source of metal nanoparticles of 2 to 10 nm size for entrapment into oxide matrices by sol–gel processing. This particle size, which is predetermined before entrapment, is comparable to that of the matrix precursor colloids. Therefore, the nanoparticles can only be physically entrapped in the matrix gels by co-gelation, followed by removal of the stabilizer and desolvation (Fig. 9). The main problem to be solved here is the uniform mixing (distribution) of metal nanoparticles in the catalyst pores, and their accessibility to reacting molecules in the final composite. In order to obtain a uniform distribution of metal nanoparticles and to depress their aggregation (sintering) inside the matrix, four main measures should be taken:

- the processing temperatures should be kept below the level causing desorption of organic stabilizers
- the matrix hydrophobicity should be increased for better compatibility with organic stabilizers adsorbed at the nanoparticle surface
- there should be supercritical drying of co-gels
- the metal nanoparticles should be coated with protective layers.

Increasing the hydrophobicity of the silica matrix using $CH_3Si(OCH_3)_3$ and $Mg(OC_2H_5)_2$ precursor allows one to maintain the size of the protected $Pd(0)/N(C_8H_{17})Br$ nanoparticles at 2 to 3.5 nm, while in a $Si(OCH_3)_4$-derived matrix Pd is formed as 100- to 200-nm aggregates [218]. Another possible solution is to use an ionic liquid as solvent for sol–gel processing of the silica matrix, and which is compatible with silica and can stabilize metal colloids [219]. Shrinkage of the matrix at the gel desolation step, forming a microporous xerogel structure, can cause phase segregation and the growth of metal particles. The low-temperature CO_2-supercritical drying leaves sufficient mesopore space for efficient entrapment of metal colloids in SiO_2 and TiO_2 matrices [220–222].

The metal colloids can be protected against aggregation in the co-gelled matrix by insertion into nanocapsules with nanometer-thick walls composed of silica [223–225], self-organized protein superstructures [226], or polymers such as PVA [227]. Avoiding aggregation leaves the nanoparticles in colloidal solution (Fig. 10a) that limits their use as solid catalysts. However, co-gelation of the protected metal nanoparticles with colloidal silica yields a continuous porous material, where the distance between nanoparticles can be controlled by the loading of the protected metal colloid (Fig. 10b,c) [228]. The accessibility of sol–gel-entrapped nanoparticles for molecules involved in catalytic reactions was investigated in only a few cases. However, they have been proven to be efficient catalysts in olefin hydrogenation (Pd) [218] and NO reduction by CO (Rh) [101]. Addition of the metal colloid to silica sol

References see page 156

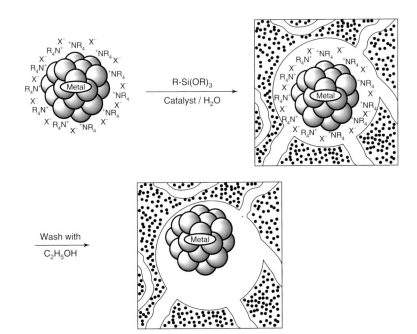

Fig. 9 Scheme for the entrapment of preformed metal colloids in a sol–gel matrix. (Adapted from Ref. [218].)

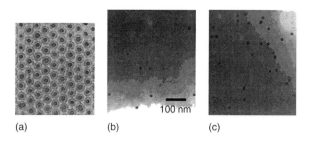

Fig. 10 Silica-encapsulated gold nanoparticles before (a) and after entrapment in a sol–gel silica matrix (b,c). (Adapted from Refs. [223, 228].)

only a few minutes before gelation reduced the interaction between the two colloids, making the nanoparticles more accessible [221].

b Other Inorganics The entrapment of transition metal oxide nanoparticles, or nanoparticles of other inorganic catalytic phases, can be achieved by the same co-condensation or co-gelation strategies as discussed above for metal nanoparticles. Besides uniformity of the catalytic phase distribution in the oxide matrix, which determines its dispersion and can be controlled by selection of the correct, strongly interacting precursors and conditions of sol–gel processing, the main problem is the creation of the desired crystalline modification of the entrapped phase that is optimal for catalysis. The principles of oxide phase formation in multicomponent molecular gels containing matrices and precursors of catalytic phases have been reviewed [105], and discussed in more detail in Section 2.3.4.3.1B. The direct crystallization of the equilibrium metal oxide nanoparticles requires homogeneous distribution of the metal precursors in the mixed gel. It avoids phase separation at the gel thermolysis step, and creates an energetic barrier for rapid nucleation. A highly homogeneous distribution of iron and silicon oxides in corresponding sol–gel-derived composites was achieved by co-gelation of hydrated $FeCl_3 \cdot 6H_2O$ and polysilylated organic precursors in the presence of propylene oxide, forming a hybrid iron–polysilsesquioxane network [230]. The crystallization behavior of the oxides depends also on the sol–gel processing conditions. The addition of cobalt nitrate to the acidic silicon alkoxide gel with microporous polymeric structure resulted in phase segregation and the appearance of large Co_3O_4 crystals after thermolysis at 973 K [231]. Entrapment of iron nitrate to the same gel under basic conditions delayed the crystallization of small clusters of the Fe_2O_3 phase up to 1173 K [232]. Undesired phases of the catalyst can be formed at elevated temperatures needed for crystallization of the entrapped oxide; this is due to reactions of these oxides with the matrix. Insertion of the correct additives can prevent the formation of phases such as potassium in the case of the MoO_3/ZrO_2 catalyst, where the formation of a $Zr(MoO_4)_2$ phase was fully suppressed [233].

The entrapment of other catalytic phases, such as metal phosphides or molybdates, requires careful selection of the corresponding precursors. Successful synthesis of metal phosphine complexes containing alkoxysilyl groups co-condensed with $Si(OCH_3)_4$ allowed the preparation, after hydrogen reduction, of a series of silica-entrapped phosphides of Fe, Ru, Co, Ni, Rh, Pd, and Pt with crystal sizes of <10 nm [234]. The embedding of Co(Ni) molybdates in a silica matrix was achieved by the simultaneous insertion of two metal salts to the silicon alkoxide precursor at the colloidization step [235].

c Functionalized and Templating Polymers The catalytic activity of functionalized polymers as acidic ion-exchange resins in non-polar solvents or gas-phase reactions is limited due to the extremely small surface area of polymer beads of $<0.02\ m^2\ g^{-1}$. This problem was solved for the perfluorosulfonic Nafion resin containing terminal $-CF_2CF_2SO_3H$ groups attached to $[(CF_2CF_2)_nC(OCF_2CF_3F)_mFCF_2O-]_x$ polymeric chains, where $x \sim 1000$ by its entrapment in a porous silica matrix, using the sol–gel co-gelation strategy [236, 237]. Two-step condensation of a hydrolyzed $Si(OR)_4$ precursor was applied: first under acid conditions (HCl) to form a sol, followed by gelation under basic conditions (NaOH) after the addition of a polymer sol. As the polymer/$Si(OR)_4$ ratio is increased, the polymer is fully entrapped in the silica matrix at low loadings up to 20–40 wt.%, whilst at higher loadings the polymer microstructure becomes dominant. After reacidification and drying, the Nafion is distributed in a porous silica matrix as discrete 20- to 60-nm domains, with a specific surface area in the range of 50 to 150 $m^2\ g^{-1}$ polymer. This surface is highly accessible to reacting molecules, and the high efficiency of this composite catalyst was proven in a variety of acid-catalyzed reactions [236–239]. The factors affecting the performance include the extent of polymer aggregation [237] and the matrix pore structure, which could be controlled by tuning the polymer loading and pH during preparation. It seems that the sol–gel entrapment technique is unique in achieving polymer dispersion in porous matrices: attempts to infiltrate the polymer into the pores of regular, preformed silicas with a pore diameter of about 15 nm were unsuccessful, and yielded only polymer films that coated only the very low external surface of the silica pellets [237].

Sol–gel-entrapped polymers can also be used in the preparation of solid catalysts as templates directing the structure of catalytically active inorganic phases or catalyst supports [240, 241]. The co-condensation/gelation of metal alkoxides and other precursors with amphiphilic, heteroatomic copolymers with a tendency for self-assembly in aqueous solutions yielded a variety of mesostructured materials with different symmetries and uniform pore size distribution; however, a detailed discussion is beyond the scope of this chapter. The physical entrapment of polymeric spheres with controlled diameter and spatial arrangement in a silica matrix by the sol–gel route, followed by calcination to remove the polymer, provides ordered macroporous solids for applications as catalyst supports [242].

2.3.4.3.3 Catalytic Coatings, Films, and Membranes

Thick coatings and thin films represent a class of catalytic materials used in structured reactors with catalytic walls: monoliths with parallel channels, ceramic foams [243, 244], and microreactors of different configurations [245, 246], as well as catalytically active membranes, combining chemical reactions with partitioning of the reagents and product separation [247, 248]. Coating can also be used as a step in the fabrication of bulk nanostructured catalytic materials by the production of nanometric layers at the surface of rigid substrates of different shape and composition, such as inorganic or polymeric particles, fibers, carbon nanotubes, mesostructured silicas, and others. If needed, this can be followed by the template removal [249–251].

The goal of the coating process, as well as the type of the coated substrate, determines the requirements with respect to film thickness and morphology. In structured reactors, which are designed mainly for conducting very rapid exothermic reactions, the surface area of ceramic monoliths or microreactor bodies fabricated by lithography, electroforming or etching processes, is low (<1 m^2 g^{-1}). In order for the catalytic wall to reach high efficiency, it should be covered with a highly porous layer of catalytic material of 20–50 μm thickness and containing mesopores for good mass-transfer [244]. This results in a sufficiently high surface area for an efficient catalyst or catalyst support.

For high-surface-area substrates such as ceramic membranes, metals and metal-oxide nanoparticles, mesostructured silicas or carbon nanotubes coated with catalytic phases, the porosity of the layer is not sufficient for catalytic performance. In this case, increasing the thickness of the catalytic film to 5–10 nm can decrease the composite's surface area by filling the pores of the substrate [249]. Two additional targets for the deposition of materials onto porous substrates by sol–gel processing are: (i) to fix large amounts of precursors for catalytic phases inside the substrate's pore system for further compartment crystallization of desired catalytic phases [252, 253]; and (ii) protection of the catalytic phase from sintering at high temperature [254].

Catalytic membranes should be essentially microporous, thus creating selectivity according to the size and shape of reagent and product molecules. This is achieved by the deposition of non-aggregated ceramic particles down to 4 nm in size on the surface of mesoporous supports [247]. The mesoporous ceramic membranes provide a broad range of opportunities with regards to catalytic applications, such as filling the membrane mesopores with catalytic phases, combining the successive inert and active layers of different porosity, and others [255]. The major requirement here is the formation of a continuous catalyst structure with high permeability for selected molecules.

Coating strategies depend on goals of the catalyst preparation, and include: (i) direct coating of the substrate surface with a stable catalyst precursor sol; or (ii) the full sol–gel sequence including gelation inside the liquid layer deposited on the substrate surface, as shown in Fig. 11 for a perovskite coating [256]. Here, the "amorphous" mixed oxide film is obtained by co-gelation of a titanium alkoxide-derived sol and metal carboxylates. The gelation is a critical step which determines the film's densification properties during further processing stages. The process is driven by solvent evaporation, forcing interactions between precursors and crosslinking of micelles or polymeric clusters. Gelation can proceed according to three different routes, depending on the type of gelation behavior, to produce three types of coating [256, 262, 263]:

- *Chemical or "polymeric" gel films*, in which condensation reactions start during or immediately after coating, forming chemical bonds between oligomeric species (micelles) that cannot be redissolved in their parent solvent.
- *Physical gel films* that display gelation through physical aggregation of oligomers due to van der Waals' forces or steric interactions which can be redissolved in their parent solvents.
- *Non-gelling films* displaying a "wet" character after deposition, and which is associated with difficulties of solvent evaporation that occur at the temperature of oligomer decomposition.

The film gelation behavior is determined by the reactivity of the precursor species, sol concentration and composition, solvent choice (dissolution, stability, solution aging behavior, evaporation rate, surface wetting), and the presence of an acid or base catalysts for condensation of alkoxide precursors and sols. The thickness of the film

References see page 156

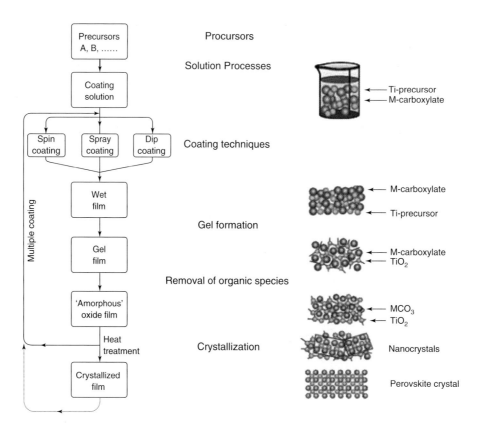

Fig. 11 Flow chart of a typical sol–gel coating process, showing the steps of the formation of $MTiO_3$ perovskite films at the surface of a substrate. (Adapted from Ref. [256].)

deposited in one step strongly increases with increasing viscosity of the sol precursor. The coating quality (uniformity and adhesion) is also improved by the introduction of organic binders, by maintaining the film's integrity after drying before the post-gelation densification, and by plasticizers controlling its rheology [264]. The relationships between coating–drying conditions and sol–gel film-coating properties have been analyzed in terms of fluid mechanics, mass transfer, and solid mechanics [269].

The two techniques of spin- and spray-coating are widely used for the coating of flat substrates (plates) in membrane preparation (Fig. 12a,b). However, they are not efficient for the formation of uniform films in channeled monoliths, micro- and nanotubes, or for microreactor channels with specific configurations, except when the coating is performed on microstructured wafers which are subsequently packed to form the microchannels [271]. Films of colloidal solution (sol) on structured substrates are deposited exclusively by dip-coating (Fig. 12c). The soaking time and withdrawal speed are additional parameters which affect the coating quality. A special rotation technique with proper control of the coating liquid's viscosity should be applied in order to obtain a uniform film thickness on shaped substrates and inside cavities [257]. For uniform coating on the interior surfaces of tubes, the conventional dip-coating process should be modified to achieve an enforced laminar flow. This was achieved, for example, by organizing an air flow in the opposite direction to the direction of tube withdrawal from the colloidal solution [258] (Fig. 12d).

Monolith Coating The coating of low-surface-area ceramic monoliths with an alumina porous layer serving as noble metal support in the preparation of automotive catalysts is generally carried out by washcoating with preformed alumina particles of ~5 μm size that cannot penetrate the ceramics pores [244, 259]. The direct coating of ceramic or metallic substrates with stable alumina- or silica-sols yields physical gel films, which are converted to insufficiently thin (<100 nm) non-porous layers after drying–calcination [244, 260]. These films are produced by means of dip-coating, followed by the removal of excess fluid from the channels and drying under rotation of the monolith around the channel axis. Coating of monoliths with chemically gelled sols, using the same technique, allows the production of highly porous layers, the thickness of which could be adjusted to tens of micrometers by repetitive coating–drying–calcination steps. This approach is not widely used in industry for automotive catalyst preparation, since it is more expensive

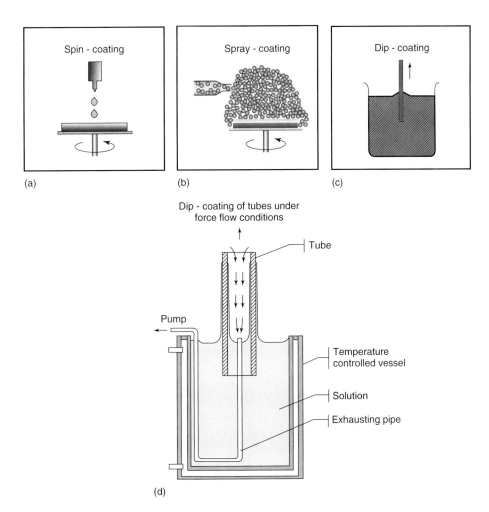

Fig. 12 (a–d) Commonly used techniques for the sol–gel coating of substrates. (a–c, reproduced from Ref. [256]; d, reproduced from Ref. [258].)

than washcoating, especially when the alkoxides are used as precursors. The main advantage of sol–gel coating is the ability to form a uniform layer, without filling up the corners of the channels, and this results in some cost reduction [259].

A procedure has been developed to coat monoliths with a sol of optimal viscosity prepared from a cheap pseudoboehmite (AlOOH) stabilized by urea and nitric acid, that yields a polymeric film [244, 261]. The process produced a coating with a surface area of 250 m^2 g^{-1}. The preparation of thick silica washcoat supports of 20 to 50 μm with surface areas of 60 to 140 m^2 g^{-1} on metallic monoliths by sol–gel coating with silica sols containing potassium silicate was also reported [260]. In order to overcome the inherent low loading when coating a ceramic monolith with Ce-doped alumina-sol (due to the large amounts of solvent required), the proposal was made to coat the substrate with hybrid systems, such as sol–gel powders dispersed in sols [265]. This method was proven to combine an effective loading procedure with fine-tuned coating characteristics: 50–150 μm thickness, high integrity, porosity, pore size distribution in the range of 0.001 to 10 μm, and stability.

Later, it was found that catalytically active components could be added to the coating sol, thereby allowing preparation of the catalytic film in one step. A successful example was the coating of an α-Al$_2$O$_3$ foam monolith with a LiLaNiO$_3$/γ-Al$_2$O$_3$ layer of 10 μm thickness, yielding an efficient and stable catalyst for the oxidative conversion of methane to syngas at high space velocities, and with a surface area of 104 m^2 g^{-1} [266]. The addition of metal salts to a boehmite sol did not cause micelle aggregation, but the addition of PVA as a binder and PEG as plasticizer to the sol improved the uniformity of the catalytic film. The Ni/La$_2$O$_3$ catalyst, when deposited on a cordierite monolith as a layer by coating together with alumina, and using the sol–gel method, displayed substantially higher selectivity to hydrogen in

References see page 156

the partial oxidation of ethanol at high gas velocities compared with the washcoated catalyst of the same composition [267]. The selection of a suitable condensation chemistry for direct coating of the substrates with multicomponent catalytic phases is essential for the coating quality and performance: the acid-catalyzed condensation of Ti–Ru precursors (Ti(OPr)$_4$; RuCl$_3$) in a mixed solution produced thin, continuous, crack-free coatings of RuO$_2$–TiO$_2$ solid solutions, but these could not be used on electrocatalysis electrodes [268]. In contrast, coating with a base-catalyzed solution yielded thick, cracked coatings containing the anatase phase which performed well in the electrocatalytic reaction.

Microreactors Microreactors as miniaturized devices can have a shorter response time, efficient heat management, and an inherent safety of operation. They often contain structured micromachined channels of 10 to 500 μm size, mainly in metallic bodies, and can be coated with highly porous ceramic films as supports for catalytically active components or directly with catalytic phases [246, 270]. Reduction of the channel dimensions by one to two orders of magnitude relative to monolithic or ceramic foam substrates creates some specific requirements to catalytic coating of their walls by sol–gel methods. Dip-coating becomes inadequate for the formation of uniform liquid films, and consequently it became necessary to develop special methods for *sol infiltration* into microchannels. The high *selectivity of film deposition* entirely on the channel walls should exclude contamination of the microreactor body, thus enabling an efficient packing of the reactor elements. Substantially more stringent requirements with regards to the *adhesion of the catalyst layer* are dictated by the high possibility for blocking the microchannels due to delamination of the films. Nonetheless, adapting the sol–gel coating methods to these requirements is possible by adjusting the parameters of the coating procedure.

Closed microchannels can be filled by gravitation by using sols of low viscosity, keeping the reactor in the vertical position [272–274] or under limited pressure, and injecting the sols via a syringe [275, 276]. In both cases, the excess of coating sol should be slowly withdrawn after a short time by blowing with air or another gas flow, using forced circulation, vacuum aspiration, or another fluid not miscible with the sol [277]. This is followed by film drying and calcination. This technique does not guarantee high uniformity of the coating, due to movement of the insufficiently withdrawn precursor to the channel inlet and outlet as a result of thermal expansion during the drying step. It can also result in contamination of the reactor body.

An original approach was proposed to solve these problems, in which the microchannels (opened from one side) were coated using a micropipetting method that spread the precursor along the channel as a result of capillary action [276]. Selective coating of the hydrophilic hydroxylated surface of silicon channel walls with aqueous alumina sol containing platinum salt was achieved as a result of hydrophobization of the top reactor surface by coating with a paraffin film. A schematic representation of this surface-selective infiltration procedure is shown in Fig. 13. After coating, the clean reactor top is sealed by anodic, diffusion, or compressive bonding.

The film's adhesion can be substantially improved by controlling the layer thickness to less than 3 μm; this avoids the relief of the tensile stress developed in the film due to volume shrinkage at the drying step [276]. The film's shrinkage can be reduced by decreasing the condensation rate in the sol through addition of a stabilizer (i.e., acetylacetone to aluminasol), increasing the sol aging time before coating, or selecting a catalyst (acid or base) that provides a high degree of gel condensation. All such changes yield more polymer-like structures of the Al$_2$O$_3$, SiO$_2$ and TiO$_2$ films with better adhesion and an absence of cracks [273, 281]. The problem of film delamination was reduced by the design and fabrication of microchannels with smooth corners that reduced the tensile stress at the film–substrate interface [276]. The film adhesion could be strongly improved with an adherent layer formed at the metallic substrate surface before film deposition. This was made possible by pre-coating with a nanometric metal oxide layer [275] or thermal generation of a metal oxide layer at an alloy surface (i.e., Fe-based alloy with 5% Al [273]). The surface area and pore size distribution of catalytic coatings in microchannels is mainly a function of the post-coating thermal treatment, as in the case of bulk gels. The film, after drying, should be calcined at an optimal temperature that is high enough for gel dehydration, decomposition and elimination of organics, but not too high to prevent the sintering of gel particles and film recrystallization. For an alumina film on stainless steel, coated from an aluminum propoxide-derived sol, this temperature was 773 K, and the procedure yielded 152 m^2 m^{-2} enhancement of the surface area [273].

Successful examples of optimizing the sol–gel coating parameters for the fabrication of efficient catalytic microreactors can be found in the literature: examples include Pd/Al$_2$O$_3$/silicon for partial methane oxidation [274], serine protease cucumicin enzyme immobilized in an aminated silica layer deposited at the inner surface of silica capillaries for polypeptide hydrolysis [278], Fe–Co/Al$_2$O$_3$/silicon for syngas conversion to alkanes [37], and Pt/Al$_2$O$_3$/silicon for CO oxidation [275].

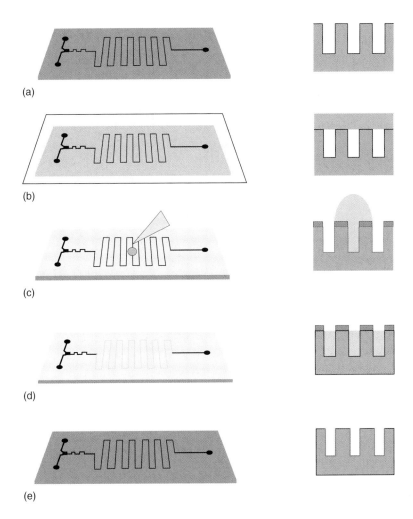

Fig. 13 Schematic illustration of the surface-selective infiltration procedure for the coating of channels. (a) initial microreactor; (b) top surface of the reactor coated with the hydrophobic material; (c) a measured drop of precursor sol placed at the reactor surface; (d) the sol is infiltrated into the channel; (e) the catalyst layer is solidified and the hydrophobic layer removed by calcination. (Reproduced from Ref. [276].)

Thin, Nanometric Catalytic Films Thin catalytic films ranging from several nanometers to several hundred nanometers thickness are prepared by sol–gel processing on flat surfaces of non-porous substrates (Si-plates, glasses) or inside the macro- (10^{-6} to 10^{-4} m) or mesopores (10^{-9} to 10^{-8} m) of powdered or preformed materials (metallic or polymeric fibers, organic membranes, polymeric colloidal crystals or gels, regular or ordered inorganic mesostructured materials). Such films can also be fabricated at the surface of discrete entities, such as inorganic nanoparticles of $<10^{-7}$ m, polymeric latex spheres, or organic crystals of 10^{-6} to 10^{-4} m.

Nanometric films on flat surfaces contain much less catalytic material compared with the thick porous films discussed above. They are used in cases when the film's efficiency in catalytic reactions is limited by its thickness, for example, by diffusion limitations, saturation of the light-harvesting efficiency in photocatalytic reactions, or resistance in electrocatalytic devices. For instance, when the TiO_2 film thickness exceeds a value of about 250 nm, its photocatalytic activity remains unchanged with increasing thickness [281]. This makes fine-tuning of the thickness of catalytic coatings on the nanometer scale important.

This is possible by tuning the viscosity of the coating solution by changing the precursor's concentration, by the addition of H_2O_2 to form a metal alkoxide-peroxo complex, by variation of the sol aging time before coating [282], or by the addition of a plasticizer (PEG) [283]. The thickness of a TiO_2 film deposited by one-step dip-coating on flat substrates was increased linearly, from 30 to 100 nm, when increasing the sol viscosity from

References see page 156

4 to 12 N·s m^{-2} [282]. In order to produce a nanometric film of uniform thickness, it is important to keep the substrate withdrawal rate from the precursor sol during dip-coating in the range of 1 to 2 mm s^{-1} [284]. The porosity of thin catalytic films on non-porous substrates is critical for their catalytic performance. Therefore, the final calcination temperature of the coated layers which is needed for formation of the required phase composition of the primary particles should be decreased to a level that does not cause intraparticle or interparticle sintering. For TiO$_2$ films which require calcination at 673–873 K for crystallization to the active anatase phase, this can be achieved by the addition of metal additives such as iron that, in addition, are efficient promoters of photocatalytic activity [283], or by hydrothermal treatment of the precursor sol. The latter method produces a 20-nm anatase nanocrystal sol that yields a uniform and stable 200-nm-thick coating at the surface of soda lime glass after heating at only 115 °C [285]. Such reduction of the final calcination temperature is important also for the coating of substrates with low thermal stability, such as the above-mentioned glass, or organic polymeric materials.

Coating of catalytically active discrete entities with nanometric films is used for promoting the catalytic phases, and for their protection against aggregation/sintering. Non-active particulates as shape-directing templates for catalytic phases can be used for the fabrication of porous hollow spheres, open-ended tubes, or hollow structures with a grained shape for their potential use as catalysts or catalyst supports. For promotion, the formation of a continuous film is not needed, and can in fact be detrimental. The coating of TiO$_2$ powder consisting of 87-nm size anatase nanocrystals with Fe$_2$O$_3$ or Al$_2$O$_3$ sols at only 0.1 wt.% loadings substantially improved its photooxidizing power in the decomposition of phenols under ultraviolet (UV) irradiation [286]. The coating of 10- to 13-nm TiO$_2$ nanoparticles with a Ce-salt urea aqueous sol yielded agglomerated particles with an average size of 70 nm, where the TiO$_2$ was located in the center, surrounded by discrete 10-nm CeO$_2$ nanocrystals (Fig. 14) [287]. Doping with CeO$_2$ in this way strongly increased the oxygen sensitivity of the TiO$_2$ catalyst.

A continuous film (SiO$_2$) with a thickness which is five- to tenfold higher than the nanoparticle diameter is needed for the stabilization of nanoparticles of catalytic phases (metals, metal oxides or other compounds) against sintering at high temperatures [288–290, and references therein]. The main scientific challenge here is to direct the polymeric chains formed during gelation to the nanoparticle's surface, thus yielding a film and retaining the structure of discrete particles instead of their entrapment in a bulk gel body. This can be achieved by treatment of the nanoparticle surface with specific

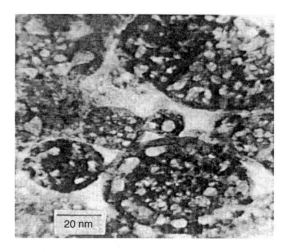

Fig. 14 Morphology of nano-CeO$_2$ coated titanium oxide. (Reproduced from Ref. [287].)

reagents to promote selective gelation of silica at their surface; an example is 3-aminopropyltrimethoxysilane, which is used in the coating of gold nanoparticles [288]. Another original approach utilizes the gelation of a tetraethyl orthosilicate aqueous sol inside the droplets of a water-in-oil microemulsion where nanoparticles of a rhodium–hydrazine complex were created by previous treatment of aqueous RhCl$_3$ solution droplets with hydrazine [290]. The following H$_2$-reduction of the obtained precipitate at 723 K yielded porous silica spheres of ∼30 nm with a surface area of 246 m^2 g^{-1}, with every single sphere containing at its center a metallic rhodium particle of ∼4 nm size (Fig. 15b). This material had a high thermal stability up to 1173 K. In contrast, gelation of a tetraethoxysilane (TEOS)-derived sol with an entrapped Rh–hydrazine complex at similar Rh-loading, but without using a microemulsion, yielded a material with rather similar surface area (231 m^2 g^{-1}), dispersion of metallic particles (Fig. 15a) and catalytic activity in NO reduction by CO, which demonstrated a strong loss of metal dispersion and activity after calcination at 1173 K [290]. This difference is a result of the uniform distance between the metal nanoparticles in the selectively coated material leaving sufficient thickness of the protection layer for every particle.

Coating catalytically inactive discrete entities with sol–gel-derived thin films of catalytic phases, followed by template removal, represents a viable route for the preparation of advanced particulate catalytic materials. Various of these materials were successfully fabricated and their physico-chemical properties well characterized: examples include silica on polymer spheres [291] and ammonium D,L-tartrate crystallites [292], titania on polymer spheres [293] and carbon nanotubes [294], vanadium-, tungsten-, ruthenium-, iridium and molybdenum oxides

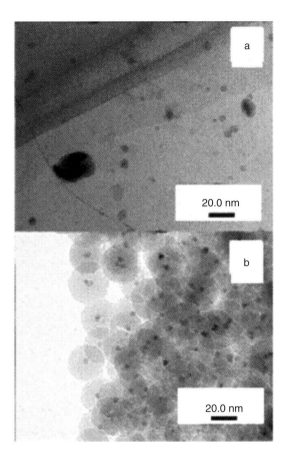

Fig. 15 Transmission electron microscopy images of Rh/SiO$_2$ catalysts. (a) Rh entrapped in silica-gel as RhCl$_3$; (b) Rh-particles coated with silica-gel, using the microemulsion method. (Adapted from Ref. [289].)

on carbon nanotubes [294], and other catalytic phases with versatile particulate templates, including those of biological origin [250, 251]. Unfortunately, the potential of such materials – which still require substantial fundamental research effort – will only become clear in the future.

Coating the macropores of monolithic substrates with nanometric layers is not suitable for the fabrication of thick porous films to be used as catalyst supports. For instance, the penetration of alumina colloidal particles from a sol into the macropores of ceramic monoliths yielded a material with reduced porosity [244]. However, this approach can be used successfully to functionalize the surfaces of structured macroporous materials, such as fibrous metallic filters or organic membranes, with catalytic phases or supports. In such cases the creation of thick layers would reduce the permeability due to pore blocking. The method would also be useful for preparing replicated structured porous catalysts using monolithic templates and the nanocoating technique [250, 251]. In both cases, the high integrity of the film is essential to avoid macropore blocking and to retain the template-directed shape of the material. Coating of metal alloy fibers, which form sintered filters with pores of several to tens of micrometers size, by nanometric layers of ZrO$_2$ [295], SiO$_2$ [296] or Al$_2$O$_3$ [297] increases their surface area from <1 to 12–38 m^2 g^{-1}, thus enabling the preparation of efficient catalysts after sulfation, or the further deposition of Co, Pt, or Pd. Such coating is carried out by impregnation of the substrate with an excess of starting colloidal solution, followed by direct drying after withdrawal. In this case the formation of a stable uniform coating of the metallic fibers can be achieved by optimizing the nature of the sol precursor, its concentration, polymerization rate, viscosity, the contacting time before substrate withdrawal, and the surface chemistry of the substrate. This should provide good wetting and coating of the fibers with thin liquid films, preventing blockage of the macropores with the bulk gel.

The development of nanometric catalytic layers by sol–gel processing inside the macropores of monolithic materials with subsequent matrix removal producing a positive replica (the "nano-coating approach"; see Fig. 16) can provide advantages, compared with the complete filling of macropores before template elimination, giving a negative replica ("nano-casting"; Fig. 16). The second approach used to fabricate thick macro-mesoporous membranes and films [250, 251] may have disadvantages due to the higher layer thickness, especially when the filling is essentially microporous. Conducting the sol–gel process with metal-alkoxide precursors within the pores of polymeric gel, which have diameters ranging from 100 nm to several micrometers, yielded coatings of Ti- and Zr-oxides with a thickness of 100 to 150 nm [298, 299]. The template removal provided materials with a "coral-like" network and a surface area of up to 100 m^2 g^{-1}. Tubular structures consisting of TiO$_2$ or ZnO with a predetermined diameter of less than 200 nm were synthesized by coating an alumina membrane with corresponding sols, followed by thermal treatment and membrane removal in an aqueous base [300]. The nanocoating of macroporous monoliths is an area of research that is presently undergoing very close examination.

Mesopores of high-surface-area supports can be loaded with a precursor for the catalytic phase in the colloidal state or even before colloidization; formation of the oligomeric species is then conducted inside the pores, the goal being their subsequent deposition on the pore walls. This strategy, which is referred to as "internal gelation" [252], resembles the nanocasting approach and differs from the coating techniques considered above

References see page 156

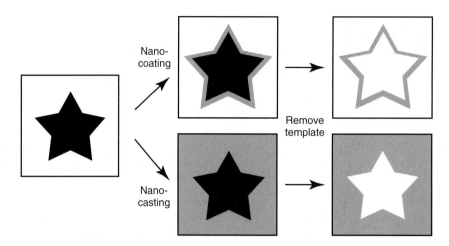

Fig. 16 Templating steps involved in nano-coating and nano-casting strategies for the preparation of sol–gel-derived porous solids. (Reproduced from Ref. [251].)

by conducting the gelation step in the whole volume of the substrate pores, which are filled with a colloidal solution. This method for depositing catalytic materials with high dispersion on the surface of supports can be used to prepare catalysts at loading levels well beyond the support's capacity for the chemical bonding of precursor molecules. As their interaction with the support surface may be poor, conducting the gelation process inside the mesopores is essential for fixation of the precursor inside the pores at the drying step, and for the uniform distribution of nanoparticles. The coating integrity is not important in this case. However, filling the pores with discrete nanoparticles of catalytic phases is preferable, due to the higher surface area of nanoparticle ensembles compared to coated layers of identically sized phases [249].

The addition of a stable Pt–Ru-sol solution (prepared by the sonication of metal acetylacetonates dissolved in an ethanol-acetic acid mixture) to carbon black by incipient wetness impregnation allowed the catalytic material to be fixed in the matrix mesopores at the subsequent drying–polymerization step; this was followed by Pt–RuO$_2$ phase formation at 673 K [301]. The catalytic anode prepared by coating the glassy disk electrode with such a highly loaded Pt–RuO$_2$/C composite performed much better in methanol electrooxidation than did the state-of-the-art commercial catalysts. Filling the mesopores of mesostructured SBA-15 and SBA-16 silica with a methanolic solution of NiCl$_2$ containing propylene oxide and water, at a gelation time of 4 h, allowed fixation of about 25% NiO in the mesopores in the form of nanocrystals of 3–7 nm size. These were distributed uniformly along the channels after calcination at 623 K [252]. The NiO loading can be increased to about 70 wt.% by repeating this procedure.

A similar strategy, using a mixed solution of Mn- and Ce-chlorides in propylene oxide-methanol with a sol stabilizer, was successfully used to coat the walls of cylindrical pores in mesostructured SBA-15 silica with an amorphous Mn–Ce-oxide layer of 2–3 nm thickness. After thermal treatment at 700 °C, this layer was converted to nanocrystals of Mn$_2$O$_3$ and CeO$_2$ phases of the same size [253]. The material after matrix removal was active for full mineralization of organics in the catalytic wet oxidation. The internal gelation of a TiO$_2$ precursor inside the channels of mesostructured silica, initiated by immersion of the SBA-15 matrix filled with titanium n-butoxide solution in water, yielded composites with up to 60 wt.% loading of TiO$_2$ anatase nanocrystals of ∼6 nm size [302, 303]. These composite catalysts displayed high activity in the photocatalytic purification of waste water [302] and combustion of ethylacetate [303].

Catalytic Membranes The implementation of catalytic membranes permits separate reagent feeding and product separation at the micro/nanoscale. This can increase product yields due to a better contact between reagents and the solid catalyst surface, by limiting side reactions as a result of controlled reagent addition, and by shifting the reaction equilibrium by selectively extracting one of the products. According to the working mechanism, catalytic membranes are generally defined as *contactors*, *distributors*, and *extractors*, and are fabricated as three different types: catalytic layers; inherently catalytic membranes; and catalysts embedded in the inert membranes (Fig. 17) [304]. Several aspects of the catalytic membrane reactors have been described in a number of comprehensive reviews [304–307].

The preparation of *catalytic layers* at the inert membrane surface does not require additional catalyst properties

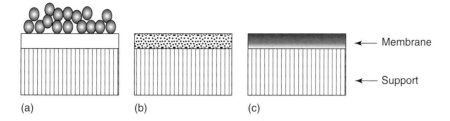

Fig. 17 The three possible membrane/catalyst combinations used in membrane catalytic reactors. (a) A catalyst bed at the surface of an inert membrane; (b) catalytic material dispersed in the pores of an inert membrane; (c) an inherently catalytic membrane. (Reproduced from Ref. [304].)

beyond those discussed above for coating structured substrates and microreactor walls using the sol–gel approach. Examples of optimized procedures for the preparation and processing of sols for coating membranes with VMgO, ZrO$_2$ or CeO$_2$-Al$_2$O$_3$ catalytic materials have been reported [308]. In this case, even the strict requirement for the existence of a continuous porous structure of the coating is not critical, as all the specific distribution/extraction functions are localized in the supporting membrane.

This is not the case in the preparation of *catalytically active membranes*. Such membranes, when fixed on macroporous substrates or formed by extrusion in tubular or plate shape, should be inherently continuous, with a narrow pore size distribution at <10-nm pore diameters, where the Knudsen diffusion of reagents/products dominates, and with a phase composition required for best catalytic performance. In addition, the absence of large meso-macropores is important for high mechanical strength. These requirements restrict the conditions of sol–gel processing for the strategies of membrane preparation:

- the casting of thick layers on macroporous substrates, including consecutive coating with layers of different pore size and catalytic functionality
- the infiltration of catalytic material in macroporous inert membranes
- casting the gels in special forms for the production of membrane films
- the extrusion of plasticized mass with sol–gel-derived catalytic material
- the coating of polymeric membranes, followed by template removal.

The full arsenal of sol–gel-processing strategies and conditions should be adjusted to control the size of the colloids and the shape of the aggregates (as described in Section 2.3.4.3), to meet the requirements listed above. Some measures are a reduction of the amount of molecular units condensing to stable particles of nanometric size, the exclusion of particle aggregation at the sol stage, the reduction of precursor reactivity, and the use of binders at the sol stage and others. Examples of such an optimization, adapted to zirconia, titania, alumina, LaOCl, Pd/SiO$_2$ and other catalytic phases, are presented for the preparation of microporous [309] and mesoporous catalytic membranes [310]. When preparing catalytic membranes by the template coating strategy, gelation is conducted inside the macropores of a preformed film (i.e., cellulose), which is then removed by calcination [311]. This procedure resembles the internal gelation for catalyst deposition inside the mesopores of supports. In this case, additional porosity – which cannot be controlled by the parameters of sol–gel processing – arises as voids after template removal. The fabrication of catalytic membranes by sol–gel techniques can include other steps, such as chemical vapor deposition and the impregnation of sol–gel-derived materials with solutions of active metal compounds. The Rh/TiO$_2$/SiO$_2$ membrane with high permselectivity in the catalytic isomerization of 1-butene was prepared in such a way [312].

2.3.4.4 Summary

Sol–gel processing, which originally was developed as a tool for controlling the texture of pure metal oxide phases, has become a universally acclaimed technique for preparing the full spectrum of modern catalytic materials, ranging from multimetallic composite oxide phases to organic–inorganic hybrid interphase catalysts, heterogenized enzymes, and catalytic coatings and films. This universality results from the combination of soft- and hard-matter chemistry and technology employed during sol–gel processing. Today, sol–gel processing presents the synthetic chemist with a wide range of opportunities not only for defining the chemical functionalization of porous materials, but also allowing the careful control of their texture. The possibility of handling a material as a viscous liquid before its gelation/solidification brings an added value to this emerging technology – namely, the opportunity to control the shape of a catalytic material at the

---

References see page 156

nanoscale by its insertion into mesoporous matrices, or by the formation of thin films. At present, sol–gel processing is developing in line with progress in soft- and hard-matter chemistry and catalytic functionalization of materials, and today is regarded as one of the chemist's most important tools in the preparation of advanced catalysts.

References

1. M. V. Landau, in *Handbook of Porous Solids*, F. Schüth, K. S. W. Sing, J. Weitkamp (Eds.), Wiley-VCH, Weinheim, 2002, Vol. 3, p. 1677.
2. C. J. Brinker, G. W. Scherer, *Sol-Gel Science. The Physics and Chemistry of Sol-Gel Processing*, Academic Press, Inc., Boston, San Diego, NY, 1990, 908 pp.
3. A. Pierre, *Introduction in Sol-Gel Processing*, Kluwer, Boston, 1998, 394 pp.
4. J. D. Wright, N. A. J. M. Sommerdijk, *Sol-Gel materials: Chemistry and Applications*, Taylor and Francis, London, 2001, 125 pp.
5. M. Schneider, A. Baiker, *Catal. Rev. -Sci. Eng.* **1995**, *37*, 15.
6. A. C. Piere, G. M. Pajonk, *Chem. Rev.* **2002**, *102*, 4243.
7. F. Schüth, K. Unger, in *Preparation of Solid Catalysts*, G. Ertl, H. Knözinger, J. Weitkamp (Eds.), Wiley-VCH, Weinheim, 1999, p. 60.
8. J. W. Geus, A. J. van Dillen, in *Preparation of Solid Catalysts*, G. Ertl, H. Knözinger, J. Weitkamp (Eds.), Wiley-VCH, Weinheim, 1999, p. 460.
9. P. Afanasiev, A. Thiollier, M. Breysse, J. L. Dubois, *Topics Catal.* **1999**, *8*, 147.
10. B. Helfgen, M. Turk, K. Schaber, *J. Supercrit. Fluid* **2003**, *26*, 225.
11. G. Muhrer, C. Lin, M. Mazzotti, *Ind. Eng. Chem. Res.* **2002**, *41*, 3566.
12. U. Ciesla, F. Schüth, *Micropor. Mesopor. Mater.* **1999**, *27*, 131.
13. A. E. C. Palmquist, *Curr. Opin. Colloid Interface Sci.* **2003**, *8*, 145.
14. J. Livage, M. Henry, C. Sanchez, *Prog. Solid State Chem.* **1988**, *18*, 259.
15. J. P. Jolivet, M. Henry, J. Livage, *Metal Oxide Chemistry and Synthesis – From Solution to Solid State*, J. Wiley & Sons, Ltd, Chichester, New York, Weinheim, 2000, 321 pp.
16. Yu. M. Rodionov, E. M. Slyusarenko, V. V. Lunin, *Russian Chem. Rev.* **1996**, *65*, 797.
17. P. A. Lessing, *Ceram. Bull.* **1989**, *68*, 1002.
18. M. Kakihana, M. Yoshimura, *Bull. Chem. Soc. Jpn.* **1999**, *72*, 1427.
19. A. Vioux, D. Leclercq, *Heterogeneous Chem. Rev.* **1996**, *3*, 65.
20. A. Vioux, *Chem. Mater.* **1997**, *9*, 2292.
21. G. Predieri, D. Cauzzi, *Chimica & Industria* **2000**, 1.
22. G. M. Pajonk, *Appl. Catal. A* **1991**, *72*, 217.
23. G. M. Pajonk, *Catal. Today* **1997**, *35*, 319.
24. G. M. Pajonk, *Colloid. Polym. Sci.* **2003**, *281*, 637.
25. S. J. Monaco, E. I. Ko, *CHEMTECH*, June **1998**, 23.
26. D. A. Ward, E. I. Ko, *Ind. Eng. Chem. Res.* **1995**, *34*, 421.
27. H. H. Kung, E. I. Ko, *Chem. Eng. J.* **1996**, *64*, 203.
28. B. L. Cushing, V. L. Kolesnichenko, C. J. O'Connor, *Chem. Rev.* **2004**, *104*, 3893.
29. U. Schubert, *New J. Chem.* **1994**, *18*, 1049.
30. Z-lin Lu, E. Lindner, H. A. Mayer, *Chem. Rev.* **2002**, *102*, 3543.
31. R. A. Caruso, *Top. Curr. Chem.* **2003**, *226*, 91.
32. Y. Kobayashi, T. Ishizaka, Y. Kurokawa, *J. Mater. Sci.* **2005**, *40*, 263.
33. J. Blum, D. Avnir, H. Schumann, *CHEMTECH*, February **1999**, 32.
34. Y. Izumi, Silica-occluded Heteropolyacids, in *Fine Chemicals through Heterogeneous Catalysis*, R. A. Sheldon, H. van Bekkum (Eds.), Wiley-VCH, Weinheim, New York, 2001, p. 100.
35. W. Jin, J. D. Brennan, *Anal. Chim. Acta* **2002**, *461*, 1.
36. S. P. Watton, C. M. Taylor, G. M. Kloster, S. C. Bowman, *Prog. Inorg. Chem.* **2003**, *51*, 333.
37. R. Ciriminna, M. Pagliaro, *Curr. Org. Chem.* **2004**, *8*, 1851.
38. H. D. Gesser, P. C. Goswami, *Chem. Rev.* **1989**, *89*, 765.
39. L. L. Hench, J. K. West, *Chem. Rev.* **1990**, *90*, 33.
40. A. C. Pierre, *Ceramics Int.* **1997**, *23*, 229.
41. J. Fricke, T. Tillotson, *Thin Solid Films* **1997**, *297*, 212.
42. L. L. Murrel, *Catal. Today* **1997**, *35*, 225.
43. J. B. Miller, E. I. Ko, *Catal. Today* **1997**, *35*, 269.
44. N. Hüsing, U. Schubert, *Angew. Chem. Int. Ed.* **1998**, *37*, 22.
45. J. Livage, *Catal. Today*, **1998**, *41*, 3.
46. E. I. Ko, in *Preparation of Solid Catalysts*, G. Ertl, H. Knözinger, J. Weitkamp (Eds.), Wiley-VCH, Weinheim, 1999, p. 85.
47. T. F. Baumann, A. E. Gash, G. A. Fox, J. H. Satcher, L. W. Hrubesh, in *Handbook of Porous Solids*, F. Schüth, K. S. W. Sing, J. Weitkamp (Eds.), Wiley-VCH, Weinheim, 2002, Vol. 3, p. 2014.
48. C. Moreno-Castilla, F. J. Maldonado-Hodar, *Carbon* **2005**, *43*, 455.
49. C. N. Chervin, B. J. Clapsaddle, H. W. Chiu, A. E. Gash, J. H. Satcher, S. M. Kauzlarich, *Chem. Mater.* **2005**, *17*, 3345.
50. A. E. Gash, T. M. Tillotson, J. H. Satcher, J. F. Poco, L. W. Urubesh, R. L. Simpson, *Chem. Mater.* **2001**, *13*, 999.
51. R. K. Iller, *The Chemistry of Silica*, John Wiley & Sons, New York, 1979, 866 pp.
52. A. Singh, D. C. Bradley, R. C. Mehrotra, I. Rothwell, *Alcoxo and Aryloxo Derivatives of Metals*, Academic Press, New York, 2001, 704 pp.
53. M. Inoue, H. K. Ominami, T. Inui, *Appl. Catal. A* **1993**, *97*, L25.
54. M. V. Landau, L. Titelman, V. Vradman, P. Wilson, *Chem. Commun.* **2003**, 594.
55. V. Lafond, P. H. Mutin, A. Vioux, *J. Mol. Catal. A* **2002**, *182*, 81.
56. P. Arnal, R. J. P. Corriu, D. Leclercq, P. H. Mutin, A. Vioux, *Chem. Mater.* **1997**, *9*, 694.
57. J. Sonnemans, H. de Keijer, P. Mars, *J. Colloids Interface Sci.* **1975**, *51*, 335.
58. M. Abecassis-Wolfowich, H. Rotter, M. V. Landau, E. Korin, A. I. Erenburg, D. Mogilyansky, E. Gartstein, *J. Non-Cryst. Solids* **2003**, *318*, 95.
59. R. L. Mozzi, B. E. Warren, *J. Appl. Crystallogr.* **1969**, *2*, 164.
60. R. J. Ayen, P. A. Jacobucci, *Rev. Chem. Eng.* **1988**, *5*, 157.
61. D. M. Smith, R. Desphande C. J. Brinker, in *Better Ceramics through Chemistry V*, M. J. Hampden-Smith, W. G. Klemperer, C. J. Brinker (Eds.), Materials Research Society, Pittsburgh, 1992, Vol. 271, pp. 553, 567.
62. C. E. Bunker, H. W. Rollins, Y.-P. Sun, in *Supercritical fluid technology in material science and engineering: synthesis, properties and applications*, Y.-P. Sun (Ed.), Marcel Dekker, New York, 2002, p. 1.
63. C. Setzer, G. van Essche, N. Pryor, in *Handbook of Porous Solids*, F. Schüth, K. S. W. Sing, J. Weitkamp (Eds.), Wiley-VCH, Weinheim, 2002, Vol. 3, p. 1543.

64. P. Euzen, P. Raybaud, X. Krokidis, H. Toulhoat, J.-L. Le Loarer, J.-P. Jolivet, C. Froidefond, in *Handbook of Porous Solids*, F. Schüth, K. S. W. Sing, J. Weitkamp (Eds.), Wiley-VCH, Weinheim, 2002, Vol. 3, p. 1591.
65. B. E. Yoldas, *Am. Ceram. Soc. Bull.* **1975**, *54*, 286; *J. Mater. Sci.* **1975**, *10*, 1856.
66. M. Nguefack, A. F. Popa, S. Rossignol, C. Kappenstein, *Phys. Chem. Chem. Phys.* **2003**, *5*, 4279.
67. S. Kureti, W. Weisweiler, *J. Non-Cryst. Solids* **2002**, *303*, 253.
68. L. Le Bihan, F. Dumeignil, E. Payen, J. Grimblot, *J. Sol-Gel Sci. Technol.* **2002**, *24*, 113.
69. D. J. Suh, T.-J. Park, *Chem. Mater.* **1997**, *9*, 1903.
70. K. Niesz, P. Yang, G. A. Somorjai, *Chem. Commun.* **2005**, 1986.
71. S. Keysar, Y. Cohen, S. Shagal, S. Slobodiansky, G. S. Grader, *J. Sol-Gel Sci. Technol.* **1999**, *14*, 131.
72. J. Walendziewski, M. Stolarski, *React. Kinet. Catal. Lett.* **2000**, *71*, 201.
73. A. A. Khaleel, K. J. Klabunde, *Chem. Eur. J.* **2002**, *8*, 3991.
74. J. N. Armor, E. J. Carlson, *J. Mater. Sci.* **1987**, *22*, 2549.
75. J. F. Poco, J. H. Satcher, L. W. Hrubesh, *J. Non-Cryst. Solids* **2001**, *285*, 57.
76. M. I. F. Macedo, C. C. Osawa, C. A. Bertran, *J. Sol-Gel Sci. Technol.* **2004**, *30*, 135.
77. T. F. Baumann, A. E. Gash, S. C. Chinn, A. M. Sawvel, R. S. Maxwell, J. H. Satcher, *Chem. Mater.* **2005**, *17*, 395.
78. L. Gan, Z. Xu, Y. Feng, L. Chen, *J. Porous. Mater.* **2005**, *12*, 317.
79. G. S. Grader, Y. De Hazan, D. Bravo-Zhivotovskii, G. E. Shter, *J. Sol-Gel Sci. Technol.* **1997**, *10*, 127.
80. Y. De Hazan, G. E. Shter, Y. Cohen, C. Rottman, D. Avnir, G. S. Grader, *J. Sol-Gel Sci. Technol.* **1999**, *14*, 233.
81. G. S. Grader, G. E. Shter, D. Avnir, H. Frenkel, D. Sclar, A. Dolev, *J. Sol-Gel Sci. Technol.* **2001**, *21*, 157.
82. J. Kirchnerova, D. Klvana, J. Chaouki, *Appl. Catal. A* **2000**, *196*, 191.
83. A. Erenburg, E. Gartstein, M. V. Landau, *J. Phys. Chem. Solids* **2005**, *66*, 81.
84. J. N. Armor, E. J. Carlson, W. C. Conner, *Reactivity of Solids* **1987**, *3*, 155.
85. H. Rotter, M. V. Landau, M. Carrera, D. Goldfarb, M. Herskowitz, *Appl. Catal. B* **2004**, *47*, 104.
86. A. E. Gash, T. M. Tillotson, J. H. Satcher, L. W. Hrubesh, R. L. Simpson, *J. Non-Cryst. Solids* **2001**, *285*, 22.
87. D. J. Suh, *J. Non-Cryst. Solids* **2004**, *350*, 314.
88. T. Skapin, *J. Non-Cryst. Solids* **2001**, *285*, 128.
89. V. Yu. Gavrilov, G. A. Zenkovetz, *Kinet. Katal.* **1988**, *31*, 168.
90. U. Trüdinger, G. Müller, K. K. Unger, *J. Chromatogr.* **1990**, *535*, 111.
91. V. Yu. Gavrilov, G. A. Zenkovetz, *Kinet. Katal.* **1993**, *34*, 357.
92. Z. S. Hu, J. X. Dong, G. X. Chen, *Powder Technol.* **1999**, *101*, 205.
93. M. Iwasaki, M. Hara, S. Ito, *J. Mater. Sci. Lett.* **1998**, *17*, 1769.
94. L. K. Campbell, B. K. Na, E. I. Ko, *Chem. Mater.* **1992**, *4*, 1329.
95. X. Ding, X. Liu, *Materials Sci. Eng.* **1997**, *A224*, 210.
96. A. Ponton, S. Barboux-Doeuff, C. Sanchez, *J. Non-Cryst. Solids* **2005**, *351*, 45.
97. B. E. Yoldas, *J. Mater. Sci.* **1986**, *21*, 1087.
98. Z. Wang, J. Chen, X. Hu, *Mater. Lett.* **2000**, *43*, 87.
99. A. A. C. Magalhaes, D. L. Nunes, P. A. Robles-Dutenhefner, E. M. B. de Sousa, *J. Non-Cryst. Solids* **2004**, *348*, 185.
100. C. Su, B.-Y. Hong, C.-M. Tseng, *Catal. Today* **2004**, *96*, 119.
101. L. Mao, Q. Li, H. Dang, Z. Zhang, *Mater. Res. Bull.* **2005**, *40*, 201.
102. D. J. Suh, T.-J. Park, *Chem. Mater.* **1996**, *8*, 509.
103. G. M. Anikumar, A. D. Damodaran, K. G. Warrier, *Ceram. Trans.* **1997**, *73*, 93.
104. T. Boiadjieva, G. Cappelletti, S. Ardizzone, S. Rondinini, A. Vertova, *Phys. Chem. Chem. Phys.* **2003**, *5*, 1689.
105. Y. Narendar, G. L. Messing, *Catal. Today* **1997**, *35*, 247.
106. J. C. Huling, G. L. Messing, *J. Non-Cryst. Solids* **1992**, *147/148*, 213.
107. S. C. Zhang, G. L. Messing, *J. Mater. Res.* **1990**, *5*, 1806.
108. K. E. Sickafus, S. M. Wills, N. W. Grimes, *J. Am. Ceram. Soc.* **1999**, *82*, 3279.
109. D. Ganguli, M. Chatterjee, *Ceramic Powder Preparation: A Handbook*, Kluwer Academic Publishers, Boston, 1977, 221 pp.
110. A. Cimino, F. Pepe, *J. Catal.* **1972**, *25*, 362.
111. M. Bolognini, F. Cavani, D. Scagliarini, C. Flego, C. Perego, M. Saba, *Micropor. Mesopor. Mater.* **2003**, *66*, 77.
112. N. Thanabodeekij, M. Sathupunya, A. M. Jamieson, S. Wongkasemjit, *Mater. Characterization* **2003**, *50*, 325.
113. M. K. Naskar, M. Chatterjee, *J. Am. Ceram. Soc.* **2005**, *88*, 38.
114. C. O. Arean, M. P. Mentruit, E. E. Platero, F. X. Llabres, I. Xamena, J. B. Parra, *Mater. Lett.* **1999**, *59*, 22.
115. C. O. Arean, M. P. Mentruit, A. J. Lopez Lopez, J. B. Parra, *Colloids Surfaces A* **2001**, *180*, 253.
116. S. Chemlal, A. Larbot, M. Persin, J. Sarrazin, M. Sghyar, M. Rafiq, *Mater. Res. Bull.* **2000**, *35*, 2515.
117. W. C. Kim. S. W. Lee, S. J. Kim, S. H. Yoon, C. S. Kim, *J. Magnet. Magn. Mater.* **2000**, *215–216*, 217.
118. R. Allmann, *Acta Crystallogr. Sect. B* **1986**, *24*, 972.
119. D. Stamires, P. O'Connor, W. Jones, M. Brady, US Patent 6,555,496, assigned to Akzo Nobel, 2003.
120. J. A. Wang, A. Morales, X. Bokhimi, O. Novaro, T. Lopez, R. Gomez, *Chem. Mater.* **1999**, *11*, 308.
121. T. Lopez, P. Bosch, E. Ramos, R. Gomez, O. Novaro, D. Acosta, F. Figueras, *Langmuir* **1996**, *12*, 189.
122. B. M. Choudary, V. S. Jaya, B. R. Reddy, M. L. Kantam, M. M. Rao, S. S. Madhavendra, *Chem. Mater.* **2005**, *17*, 2740.
123. M. Jitanu, M. Balasoiu, M. Zaharescu, A. Jitanu, A. Ivanov, *J. Sol-Gel Sci. Technol.* **2000**, *19*, 453.
124. M. Jitanu, M. Zaharescu, M. Balasoiu, A. Jitanu, *J. Sol-Gel Sci. Technol.* **2003**, *26*, 217.
125. C. N. R. Rao, B. Raveau, *Transition Metal Oxides*, VCH, New York, 1995, 373 pp.
126. O. S. Buassi-Monroy, C. C. Luhrs, A. Shavez-Chavez, C. R. Michel, *Mater. Lett.* **2004**, *58*, 716.
127. V. Umar, R. Marimuthu, S. S. Patil, Y. Ohya, Y. Takahashi, *J. Am. Ceram. Soc.* **1996**, *79*, 2775.
128. J. Tartaj, J. F. Fernandez, M. E. Villafuerte-Castrejon, *Mater. Res. Bull.* **2001**, *36*, 479.
129. W.-F. A. Su, *Mater. Chem. Phys.* **2000**, *62*, 18.
130. H. Beltran, E. Cordoncillo, P. Escribano, J. B. Carda, A. Coats, A. R. West, *Chem. Mater.* **2000**, *12*, 400.
131. W. Wang, D. Jia, Y. Zhou, F. Ye, *Mater. Res. Bull.* **2002**, *37*, 2517.
132. F. Gelman, J. Blum, H. Schumann, D. Avnir, *J. Sol-Gel Sci. Technol.* **2003**, *26*, 43.
133. F. Gelman, J. Blum, D. Avnir, *J. Am. Chem. Soc.* **2002**, *124*, 14460.
134. B. E. Yoldas, *J. Non-Cryst. Solids* **1984**, *63*, 150.
135. G. A. Pozarnsky, A. V. McCormick, *J. Non-Cryst. Solids* **1995**, *190*, 212.
136. A. Corma, J. Perez-Pariente, *Appl. Catal.* **1990**, *63*, 145.
137. A. May, M. Asomna, T. Lopez, R. Gomez, *Chem. Mater.* **1997**, *9*, 2395.

138. M. Toba, F. Mizukami, S. Niwa, T. Sano, K. Maeda, H. Shoji, *J. Mater. Chem.* **1994**, *4*, 1131.
139. C. K. Karula, M. Rokosz, L. F. Allard, R. J. Kudla, M. S. Chattha, *Langmuir* **2000**, *16*, 3818.
140. Q. Wei, D. Wang, *Mater. Let.* **2003**, *57*, 2015.
141. H. Talmon, T. Sone, M. Mikami, M. Okazaki, *J. Colloid Interface Sci.* **1997**, *188*, 493.
142. M. Nogami, K. Nagasaka, *J. Non-Cryst. Solids* **1989**, *109*, 79.
143. M. A. Holland, D. M. Pickup, G. Mountjoy, E. S. C. Tsang, G. W. Wallidge, R. J. Newport, M. E. Smith, *J. Mater. Chem.* **2000**, *10*, 2495.
144. D. M. Pickup, G. Mountjoy, M. A. Holland, G. W. Wallidge, R. J. Newport, M. E. Smith, *J. Phys.: Condens. Matter* **2000**, *12*, 9751.
145. G. Busca, L. Lietti, G. Ramis, F. Berti, *Appl. Catal. B* **1998**, *18*, 1
146. M. Schneider, M. Maciejewski, S. Tschudin, A. Wokaun, A. Baiker, *J. Catal.* **1994**, *149*, 326.
147. V. Cimpeanu, V. Parvulescu, V. I. Parvulescu, M. Capron, P. Grange, J. M. Thompson, C. Hardacre, *J. Catal.* **2005**, *235*, 184.
148. O. J. de Lima, A. T. Papacidero, L. A. Rocha, H. C. Sacco, E. J. Nassar, K. J. Giuffi, L. A. Bueno, Y. Messaddeq, S. J. L. Ribeiro, *Mater. Characterization* **2003**, *50*, 101.
149. V. Lafond, P. H. Mutin, A. Vioux, *J. Mol. Catal. A* **2002**, *182–183*, 81.
150. J. J. E. Moreau, M. W. C. Man, *Coord. Chem. Rev.* **1998**, *178–180*, 1073.
151. E. Linder, T. Schneller, F. Auer, H. A. Mayer, *Angew. Chem. Int. Ed.* **1999**, *38*, 2154.
152. Z. Lu, E. Linder, H. A. Mayer, *Chem. Rev.* **2002**, *102*, 3543.
153. S. P. Watton, C. M. Taylor, G. M. Kloster, S. C. Bowman, *Prog. Inorg. Chem.* **2003**, *51*, 333.
154. U. Schubert, N. Huesing, A. Lorenz, *Chem. Mater.* **1995**, *7*, 2010.
155. T. Stuchinskaya, N. Kundo, L. Gogina, U. Schubert, A. Lorenz, V. Maizlish, *J. Mol. Catal. A* **1999**, *140*, 235.
156. J. Livage, C. Sanchez, F. Babonneau, in *Chemistry of Advanced Materials: An Overview*, L. V. Interrante, M. J. Hampden-Smith (Eds.), Wiley-VCH, New York, 1998, p. 389.
157. R. Corriu, D. Leclercq, *Angew. Chem. Int. Ed. Engl.* **1996**, *35*, 1420.
158. N. Husing, U. Schubert, *Chem. Mater.* **1998**, *10*, 3024.
159. L. Sarussi, J. Blum, D. Avnir, *J. Sol-Gel Sci. Technol.* **2000**, *19*, 17.
160. M. Morandin, R. Gavagnin, F. Pinna, G. Strukul, *J. Catal.* **2002**, *212*, 193.
161. R. Ophir, Y. Shvo, *J. Mol. Catal. A* **1999**, *140*, 259.
162. D. Cauzzi, M. Lafranchi, G. Marzolini, G. Predieri, A. Tripicchio, M. Costa, R. Zanoni, *J. Organomet. Chem.* **1995**, *488*, 115.
163. M. Jurado-Gonzalez, D. L. Ou, B. Ormsby, A. C. Sullivan, J. R. H. Wilson, *Chem. Commun.* **2001**, 67.
164. R. J. P. Corriu, J. J. E. Moreau, P. Thhepot, M. W. C. Man, *Chem. Mater.* **1996**, *8*, 100.
165. O. Kroecher, R. A. Koeppel, A. Baiker, *Chem. Commun.* **1999**, 2303.
166. D. Cauzzi, M. Costa, L. Consalvi, M. A. Pellinghelli, G. Predieri, A. Tripicchio, M. Costa, R. Zanoni, *J. Organomet. Chem.* **1997**, *541*, 377.
167. E. Linder, T. Schneller, F. Auer, P. Wegner, H. A. Mayer, *Chem. Eur. J.* **1997**, *3*, 1833.
168. A. Baiker, J.-D. Grunwaldt, C. A. Müller, L. Schmidt, *Chimia* **1998**, *52*, 517.
169. M. Buechler-Skoda, R. Gill, D. Wu, C. Nguen, G. Larsen, *Appl. Catal. A* **1999**, *185*, 301.
170. A. Karli, G. Larsen, *Catal. Lett.* **2001**, *77*, 107.
171. M. Dusi, C. A. Mueller, T. Mallat, A. Baiker, *Chem. Commun.* **1999**, 197.
172. M. Tada, Y. Iwasawa, *J. Mol. Catal. A* **2003**, *199*, 115.
173. W. F. Mayer, J. Heilmann, *Angew. Chem. Int. Ed. Engl.* **1996**, *35*, 554.
174. A. Katz, M. E. Davis, *Nature* **2000**, *403*, 286.
175. J. Zhao, J. Han, Y. Zhang, *J. Mol. Catal. A* **2005**, *231*, 129.
176. E. Linder, A. Jager, F. Auer, W. Wielandt, P. Wegner, *J. Mol. Catal. A* **1998**, *129*, 91.
177. L. Schmid, M. Rohr, A. Baker, *Chem. Commun.* **1999**, 2303.
178. P. Gancitano, R. Ciriminna, M. L. Testa, A. Fidalgo, L. M. Ilharco, M. Pagliaro, *Org. Biomol. Chem.* **2005**, *3*, 2389.
179. C. Lepetit, M. Che, *J. Mol. Catal. A* **1995**, *100*, 147.
180. B. Dunn, J. I. Zink, *Chem. Mater.* **1997**, *9*, 2280.
181. D. Avnir, L. C. Klein, D. Levy, U. Schubert, A. B. Wojcik, in *The Chemistry of Organosilicon Compounds*, Y. Apeloig, Z. Rapoport (Eds.), Wiley-VCH, Chichester, UK, 1998, Vol. 2, p. 2317.
182. U. Schubert, K. Rose, H. Schmidt, *J. Non-Cryst. Solids* **1988**, *105*, 165.
183. S. J. Craythorne, A. R. Crozier, F. Lorenzini, A. C. Marr, P. C. Marr, *J. Organomet. Chem.* **2005**, *690*, 3518.
184. M. A. Caiut, S. Nakagaki, O. J. De Lima, C. Mello, C. A. P. Leite, *J. Sol-Gel Sci. Technol.* **2003**, *28*, 57.
185. A. Molnar, C. Keresszegi, B. Török, *Appl. Catal. A* **1999**, *189*, 217.
186. Y. Izumi, in *Fine Chemicals through Heterogeneous Catalysis*, R. A. Sheldon, H. van Bekkum (Eds.), Wiley-VCH, Weinheim, 2001, p. 100.
187. O. Israelsohn, K. P. Volhardt, J. Blum, *J. Mol. Catal. A* **2002**, *184*, 1.
188. M. Gruttadauria, L. F. Liotta, G. Deganello, R. Noto, *Tetrahedron* **2003**, *59*, 4997.
189. E. M. B. de Sousa, A. P. Guimaraes, N. D. S. Mohallem, R. M. Lago, *Appl. Surf. Sci.* **2001**, *183*, 216.
190. N. Eliau, D. Avnir, M. S. Eisen, J. Blum, *J. Sol-Gel. Sci. Technol.* **2005**, *35*, 159.
191. D. Avnir, S. Braun, O. Lev, M. Ottolenghi, *Chem. Mater.* **1994**, *6*, 1605.
192. M. T. Reetz, *Adv. Mater.* **1997**, *9*, 943.
193. I. Gill, A. Ballesteros, *Trends Biotechnol.* **2000**, *18*, 282.
194. I. Gill, *Chem. Mater.* **2001**, *13*, 3404.
195. W. Jin, J. D. Brennan, *Anal. Chim. Acta* **2002**, *461*, 1.
196. S. Braun, S. Shtelzer, S. Rappoport, D. Avnir, M. Ottolenghi, *J. Non-Cryst. Solids* **1992**, *147/148*, 739.
197. I. Gil, A. Ballesteros, *J. Am. Chem. Soc.* **1998**, *120*, 8587.
198. K. M. K. Yu, C. M. Y. Yeung, D. Thompsett, S. C. Tsang, *J. Phys. Chem. B* **2003**, *107*, 4515.
199. M. A. Cauqui, J. M. Rodrigues-Izquierdo, *J. Non-Cryst. Solids* **1992**, *147/148*, 724.
200. R. D. Gonzalez, T. Lopez, R. Gomez, *Catal. Today* **1997**, *35*, 293.
201. M. L. Toebes, J. A. van Dillen, K. P. de Jong, *J. Mol. Catal. A* **2001**, *173*, 75.
202. W. Zou, R. D. Gonzalez, *J. Catal.* **1995**, *152*, 291.
203. M. Schneider, D. G. Duff, T. Mallat, M. Wildberger, A. Baiker, *J. Catal.* **1994**, *147*, 500.
204. M. Schneider, M. Wildberger, M. Maciewsci, D. G. Duff, T. Mallat, A. Baiker, *J. Catal.* **1994**, *148*, 625.
205. T. Lopez, M. Asomoza, P. Bosch, E. Garcia-Figueroa, R. Gomez, *J. Catal.* **1992**, *138*, 463.

206. D. H. Kim, S. I. Woo, O.-B. Yang, *Appl. Catal. B* **2000**, *26*, 285.
207. A. Ueno, H. Suzuki, Y. Kotera, *J. Chem. Soc., Faraday Trans. I* **1983**, *79*, 127.
208. J. M. Tour, J. P. Cooper, S. L. Pendalwar, *Chem. Mater.* **1990**, *2*, 647.
209. R. Vacassy, L. Lemaire, J.-C. Valmalette, J. Dutta, H. Hofmann, *J. Mater. Sci. Lett.* **1998**, *17*, 1665.
210. S. Bharathi, O. Lev, *Chem. Commun.* **1997**, 2303.
211. S. Bharathi, N. Fishelson, O. Lev, *Langmuir* **1999**, *15*, 1929.
212. F. C. Fonseca, G. F. Goya, R. F. Jardim, N. L. V. Carreno, E. Longo, E. R. Leite, R. Muccillo, *Appl. Phys. A* **2003**, *76*, 621.
213. F. Gelman, D. Avnir, H. Schumann, J. Blum, *J. Mol. Catal. A* **2001**, *171*, 191.
214. P.-W. Wu, B. Dunn, V. Doan, B. J. Schwartz, E. Yablonovich, M. Yamane, *J. Sol-Gel. Sci. Technol.* **2000**, *19*, 249.
215. R. Takahashi, S. Sato, T. Sodesawa, M. Kato, S. Takenaka, S. Yoshida, *J. Catal.* **2001**, *204*, 259.
216. S. Tomiyama, R. Takahashi, S. Sato, T. Sodesawa, S. Yoshida, *Appl. Catal. A* **2003**, *241*, 349.
217. J. H. Fendler (Ed.), *Nanoparticles and Nanostructured Films*, Wiley-VCH, Weinheim, 1998, 468 pp.
218. M. T. Reetz, M. Dugal, *Catal. Lett.* **1999**, *58*, 207.
219. K. Anderson, S. C. Fernandez, C. Hardacre, P. C. Marr, *Inorg. Chem. Commun.* **2004**, *7*, 73.
220. M. L. Anderson, C. A. Morris, R. M. Stroud, C. I. Merzbacher, D. R. Rolison, *Langmuir* **1999**, *15*, 674.
221. C. A. Morris, M. L. Anderson, R. M. Stroud, C. I. Merzbacher, D. R. Rolison, *Science* **1999**, *284*, 622.
222. Y. Tai, M. Watanabe, K. Kaneko, *Adv. Mater.* **2001**, *13*, 1611.
223. L. M. Liz-Marzan, M. Giersing, P. Mulvaney, *Langmuir* **1996**, *12*, 4329.
224. V. V. Hardikar, E. Matijevic, *J. Colloid Interface Sci.* **2000**, *221*, 133.
225. Y. Lu, Y. Yin, Z.-Y. Li, Y. Xia, *Nano Lett.* **2002**, *2*, 785.
226. J. M. Wallace, J. K. Rice, J. J. Pietron, R. M. Stroud, J. W. Long, D. R. Rolison, *Nano Lett.* **2003**, *3*, 1463.
227. M. Richard-Plouet, J.-L. Guilloe, Y. Frere, L. Danicher, *J. Sol-Gel. Sci. Technol.* **2002**, *25*, 207.
228. Y. Kobayashi, M. A. Correa-Duarte, L. M. Liz-Marzan, *Langmuir* **2001**, *17*, 6375.
229. M. Kishida, T. Tago, T. Hatsuta, K. Wakabayashi, *Chem. Lett.* **2000**, 1108.
230. L. Zhao, B. J. Clapsaddle, J. H. Satcher, D. W. Schaefer, K. J. Shea, *Chem. Mater.* **2005**, *17*, 1358.
231. L. S. Sales, P. A. Robles-Dutenhefner, D. L. Nunes, N. D. S. Mohallem, E. V. Gusevskaya, E. M. B. Sousa, *Mater. Characterization* **2003**, *50*, 95.
232. P. Fabrizioli, T. Burgi, M. Burgener, S. van Doorslaer, A. Baiker, *J. Mater. Chem.* **2002**, *12*, 619.
233. S. N. Koc, G. Gurdag, S. Geisser, M. Guraya, M. Orbay, M. Muhler, *J. Mol. Catal. A* **2005**, *225*, 197.
234. C. M. Lukehart, S. B. Milne, S. R. Stock, *Chem. Mater.* **1998**, *10*, 903.
235. D. Cauzzi, M. Deltratti, G. Opredieri, A. Tiripicchio, A. Kaddouri, C. Mazzoccia, E. Tempesti, A. Armigliato, C. Vignali, *Appl. Catal. A* **1999**, *182*, 125.
236. M. A. Harmer, W. E. Farneth, Q. Sun, *J. Am. Chem. Soc.* **1996**, *118*, 7708.
237. M. A. Harmer, Q. Sun, A. J. Vega, W. E. Farneth, A. Heidekum, W. F. Hoelderich, *Green Chem.* **2000**, *2*, 7.
238. I. Ledneczki, A. Molnar, *Synth. Commun.* **2004**, *34*, 3683.
239. I. Ledneczki, M. Daranyi, F. Fulop, A. Molnar, *Catal. Today* **2005**, *100*, 437.
240. S. Förster, T. Plantenberg, *Angew. Chem. Int. Ed.* **2002**, *41*, 688.
241. G. J. de A. A. Soler-Illia, E. L. Crepaldi, D. Grosso, C. Sanchez, *Curr. Opinion Colloid Interface Sci.* **2003**, *8*, 109.
242. A. Stein, R. C. Schroden, *Curr. Opinion Solid State Mater. Sci.* **2001**, *2*, 553.
243. J. W. Geus, J. C. Giezen, *Catal. Today* **1999**, *47*, 169.
244. T. A. Nijhuis, A. E. W. Beers, T. Vergunst, I. Hoek, F. Kapteijn, J. A. Moulijn, *Catal. Rev.* **2001**, *43*, 345.
245. K. Hass-Santo, O. Gorke, P. Pfeifer, K. Schubert, *Chimia* **2002**, *56*, 605.
246. H. Pennemann, V. Hessel, H. Löwe, *Chem. Eng. Sci.* **2004**, *59*, 4789.
247. A. Julbe, C. Guizard, A. Larbot, L. Cot, A. Giroir-Fender, *J. Membrane Sci.* **1993**, *77*, 137.
248. A. Julbe, D. Farrusseng, C. Guizard, *J. Membrane Sci.* **2001**, *181*, 3.
249. M. V. Landau, L. Vradman, P. M. Rao, A. Wolfson, M. Herskowitz, *C. R. Chimie* **2005**, *8*, 679.
250. R. A. Caruso, M. Antonietti, *Chem. Mater.* **2001**, *13*, 3272.
251. R. A. Caruso, *Topics Curr. Chem.* **2003**, *226*, 91.
252. D. Kantorovich, L. Haviv, L. Vradman, M. V. Landau, in *Nanoporous Materials IV*, M. Jaroniec, A. Sayari (Eds.), Studies in Surface Science and Catalysis, Vol. 156, Elsevier, Amsterdam, 2005, p. 147.
253. M. Abecassis-Wolfovich, M. V. Landau, A. Brenner, M. Herskowitz, *J. Catal.* **2007**, *247*, 201.
254. I. Yuranov, L. Kiwi-Minsker, A. Renken, *Appl. Catal. B* **2003**, *43*, 217.
255. J. Coronas, J. Santamaria, *Catal. Today* **1999**, *51*, 377.
256. R. W. Schwartz, T. Schneller, R. Waser, *C. R. Chimie* **2004**, *7*, 433.
257. S. F. Kistler, L. E. Scriven, *Int. J. Numerical Methods in Fluids* **1984**, *4*, 207.
258. M. A. Aegerter, M. Puetz, G. Gasparro, N. Al-Dahoudi, *Opt. Mater.* **2004**, *26*, 155.
259. C. K. Narula, J. E. Allison, D. R. Bauer, H. S. Gandhi, *Chem. Mater.* **1996**, *8*, 984.
260. M. F. M. Zwinkels, S. G. Järas, P. G. Menon, in *Preparation of Catalysts VI*, G. Poncelet, J. Martens, B. Delmon, P. A. Jacobs, P. Grange (Eds.), Studies in Surface Science and Catalysis, Vol. 91, Elsevier, Amsterdam, 1995, p. 85.
261. A. E. Duisterwinkel, *Clean coal combustion with in-situ impregnated sol-gel sorbent*, PhD Thesis, Delft University of Technology, Delft, 1991.
262. R. W. Schwartz, J. A. Voigt, C. D. Buchheit, T. J. Boyle, *Ceram. Trans., Ferroic Mater.: Design, Prep., Charact.* **1994**, *43*, 145.
263. R. W. Schwartz, T. L. Reichert, P. G. Clem, D. Dimos, D. Liu, *Integr. Ferroelectr.* **1997**, *18*, 275.
264. R. J. R. Uhlhorn, M. H. B. J. Huisint Veld, K. Keizer, A. J. Burggraaf, *J. Mater. Sci.* **1992**, *27*, 527.
265. C. Agrafiotis, A. Tsetsekou, C. J. Stournaras, A. Julbe, L. Dalmazio, C. Guizard, *J. Eur. Ceram. Soc.* **2002**, *22*, 15.
266. R. Ran, G. Xiong, W. Yang, *J. Mater. Chem.* **2002**, *12*, 1854.
267. D. K. Liguras, K. Goundani, X. E. Verykios, *J. Power Sources* **2004**, *130*, 30.
268. M. Aparicio, L. C. Klein, *J. Sol-Gel Sci. Technol.* **2004**, *29*, 81.
269. R. A. Carncross, K. S. Chen, P. R. Schunk, C. J. Brinker, A. J. Hurd, *Ceram. Trans.* **1997**, *69*, 153.
270. K. Jensen, *Chem. Eng. Sci.* **2001**, *56*, 293.
271. S. Schimpf, M. Lucas, C. Mohr, U. Rodemerck, J. Bruckner, J. Radnik, H. Hofmeister, P. Claus, *Catal. Today* **2002**, *72*, 63.

272. R. Wunsch, M. Fichtner, O. Görke, K. Haas-Santo, K. Schubert, *Chem. Eng. Technol.* **2002**, *25*, 7.
273. K. Haas-Santo, M. Fichtner, K. Schubert, *Appl. Catal. A* **2001**, *220*, 79.
274. O. Younes-Metzler, J. Svagin, S. Jensen, C. H. Christensen, O. Hansen, U. Quaade, *Appl. Catal A* **2005**, *284*, 5.
275. S. Srinivas, A. Dhingra, H. Im, E. Gulari, *Appl. Catal. A* **2004**, *274*, 285.
276. H. Chen, L. Bednarova, R. S. Besser, W. Y. Lee, *Appl. Catal. A* **2005**, *286*, 186.
277. P. Case, C. Remy, P. Woehl, Eur. Pat. Appl., European Patent 1,415,706, assigned to Corning, 2004.
278. M. Miyazaki, J. Kaneno, M. Uehara, M. Fujii, H. Shimizu, H. Maeda, *Chem. Commun.* **2003**, 648.
279. V. S. Nagineni, S. Zhao, A. Potluri, Y. Liang, U. Siriwardane, N. V. Seetala, J. Fang, J. Palmer, D. Kuila, *Ind. Eng. Chem. Res.* **2005**, *44*, 5602.
280. Q. Fu, C.-B. Cao, H.-S. Zhu, *Thin Solid Films* **1999**, *348*, 99.
281. J. U. Shang, W. Li, Y. Zhu, *J. Mol. Catal. A* **2003**, *202*, 187.
282. R. S. Sonawane, S. G. Hegde, M. K. Dongare, *Mater. Chem. Phys.* **2002**, *77*, 744.
283. R. S. Sonawane, B. B. Kale, M. K. Dongare, *Mater. Chem. Phys.* **2004**, *85*, 52.
284. T. Watanabe, A. Nakajima, R. Wang, M. Minabe, S. Koizumi, A. Fujishima, K. Hashimoto, *Thin Solid Films* **1999**, *351*, 260.
285. Y. J. Yun, J. S. Chung, S. Kim, S. H. Hahn, E. J. Kim, *Mater. Lett.* **2004**, *58*, 3703.
286. T. K. Kim, M. N. Lee, S. H. Lee, Y. C. Park, C. K. Jung, J.-H. Boo, *Thin Solid Films* **2005**, *475*, 171.
287. M. Zhang, X. Wang, F. Wang, W. Li, *Mater. Sci. Technol.* **2002**, *18*, 345.
288. L. M. Liz-Marzan, M. Giersig, P. Mulvaney, *Langmuir* **1996**, *12*, 4329.
289. M. Ikeda, T. Tago, M. Kishida, K. Wakabayashi, *Chem. Commun.* **2001**, 2512.
290. M. Kishida, T. Tago, T. Hatsuta, K. Wakabayashi, *Chem. Lett.* **2000**, 1108.
291. H. Bamnolker, B. Nitzan, S. Gura, S. Margel, *Mater. Sci. Lett.* **1997**, *16*, 1412.
292. F. Miyaji, S. A. Davis, J. P. H. Charmant, S. Mann, *Chem. Mater.* **1999**, *11*, 719.
293. H. Shiho, N. Kawahashi, *Colloid Polym. Sci.* **2000**, *278*, 270.
294. B. C. Satishkumar, A. Govindaraj, M. Naqth, C. N. R. Rao, *J. Mater. Chem.* **2000**, *10*, 2115.
295. P. Stefanov, G. Atanasova, R. Robinson, A. Julbe, C. Guizard, P. O'Hara, Ts. Marinova, *Surf. Interface Anal.* **2002**, *34*, 84.
296. I. Yuranov, L. Kiwi-Minsker, A. Renken, *Appl. Catal. B* **2003**, *43*, 217.
297. W. Fei, S. C. Kuiry, Y. Sohn, S. Seal, *Surf. Eng.* **2003**, *19*, 189.
298. R. A. Caruso, M. Giersing, F. Willing, M. Antonietti, *Langmuir* **1998**, *14*, 6333.
299. R. A. Caruso, M. Antonietti, M. Giersig, H.-P. Hentze, J. Jia, *Chem. Mater.* **2000**, *13*, 1114.
300. B. B. Lakshmi, P. K. Dorhout, C. R. Martin, *Chem. Mater.* **1997**, *9*, 2544.
301. H. B. Suffredini, V. Tricoli, L. A. Avaca, N. Vatistas, *Electrochem. Commun.* **2004**, *6*, 1025.
302. R. van Grieken, J. Aguado, M. J. Lopez-Munoz, J. Marugan, *J. Photochem. Photobiol.* **2002**, *A148*, 315.
303. Xueguang Wang, M. V. Landau, H. Rotter, L. Vradman, A. Wolfson, A. Erenburg, *J. Catal.* **2004**, *222*, 565.
304. A. Julbe, D. Farrusseng, C. Guizard, *J. Membrane Sci.* **2001**, *181*, 3.
305. J. Zaman, A. Chakma, *J. Membrane Sci.* **1994**, *92*, 1.
306. J. N. Armor, *Catal. Today* **1995**, *25*, 199.
307. J. Coronas, J. Santamaria, *Catal. Today* **1999**, *51*, 377.
308. C. G. Guizard, A. C. Julbe, A. Ayral, *J. Mater. Chem.* **1999**, *9*, 55.
309. A. Julbe, C. Guizard, A. Larbot, L. Got, A. Giroir-Fendler, *J. Membrane Sci.* **1993**, *77*, 137.
310. A. Larbot, J. P. Fabre, C. Guizard, L. Got, *J. Membrane Sci.* **1988**, *39*, 203.
311. R. A. Caruzo, J. H. Schattka, *Adv. Mater.* **2000**, *12*, 1921.
312. P. A. Sermon, F. P. Cetton, *J. Sol-Gel Sci. Technol.* **2005**, *33*, 113.

2.3.5
Hydrothermal Zeolite Synthesis

Christine E. A. Kirschhock, Eddy J. P. Feijen, Pierre A. Jacobs*, and Johan A. Martens

2.3.5.1 Introduction

The crystallization of natural zeolites involves geological timescales. Large crystals of zeolite minerals grow in cavities of basic volcanic or metamorphic rocks in the presence of mineralizing solutions. Vast sediments of natural zeolite microcrystallites originate from the alteration of volcanic glass and from sedimentary rocks of marine origin. Extensive deposits of natural zeolites are present in all oceans and find many applications, but they are less suitable as catalysts owing to the presence of impurities that are difficult to eliminate. In general, synthetic zeolites are preferred for catalytic applications.

During the period between 1940 and 1950, Barrer and his coworkers duplicated the natural zeolite crystallization process under more severe conditions of alkalinity and temperature, and were able to crystallize the first synthetic zeolites in reaction times which ranged from hours to days [1]. Following the pioneering studies of Barrer [2] and Milton [1], many new zeolite types and framework compositions have been synthesized as a result of significant research activity, especially by oil companies. Since the mid-1970s, much research effort has been devoted to unravel the mechanisms of nucleation and crystal growth of zeolites. Concepts for tailor-made zeolite crystallizations have been developed, and the "art" of zeolite synthesis became a real scientific issue [3, 4].

In this chapter, zeolite synthesis is rationalized in a broad chemical sense, and illustrated with industrial zeolite crystallization processes. Aluminosilicate zeolites and hydrothermal synthesis form the central topic of the chapter, as these materials and their conditions of synthesis cover the majority of industrial zeolite products

---

* Corresponding author.

and manufacturing processes. However, with appropriate changes the theory presented here is also applicable to zeolites with other framework elements.

Aluminosilicate zeolite synthesis has been described in detail in several books [2, 5–8a], in proceedings of conferences organized by the International Zeolite Association [8b] and reviews [3, 4, 9, 10]. A practical guide containing a collection of verified zeolite synthesis recipes is also available [11].

Although the particle size of current zeolite powders is around 1 μm, the proliferation of applications of zeolites has fostered research and development in the area of nanosized zeolites [12], as well as on the synthesis of larger zeolite crystals [13]. Today, while the literature describing the hydrothermal synthesis of zeolites is vast, the references cited herein are those that are most appropriate to illustrate this chapter.

2.3.5.2 Zeolitization in General

2.3.5.2.1 Key Steps in Zeolite Synthesis

The hydrothermal synthesis of aluminosilicate zeolites corresponds to the conversion of a mixture of silicon and aluminum compounds, alkali metal cations, organic molecules and water via a supersaturated solution into a microporous crystalline aluminosilicate. This complex chemical process is called *zeolitization*. Colloidal silica, waterglass, pyrogenic silica or silicon alkoxides such as tetramethyl and tetraethyl orthosilicate are common sources of silicon. These raw materials differ with respect to the degree of polymerization of the silicate. Aluminum can be introduced via compounds such as gibbsite, pseudo-boehmite, aluminate salts, aluminum alkoxides or as metal powder. Cationic and neutral organic molecules are added as solvents or as structure-directing agents (SDAs). Besides silica, alumina and water, a zeolite synthesis mixture typically comprises an alkali or alkaline earth metal, an organic compound, or a combination of an alkali metal and an organic compound [14].

When the reactants are mixed, an aluminosilicate hydrogel or precipitate is usually formed. For example, the addition of a solution of sodium aluminate to a silica sol increases the ionic strength of the solution, resulting in immediate gelling; that is, in destabilization of the sol and formation of a network of colloidal aluminosilicate particles. Although in particular systems a direct gel to crystal transformation occurs [15, 16], in the majority of syntheses the crystallization process is mediated by the liquid phase (Fig. 1).

Dissolution of the amorphous solid phase catalyzed by the mineralizing agent provides the solution with silicate and aluminate monomers and oligomers, which condense into crystalline zeolite phases. Elucidation of

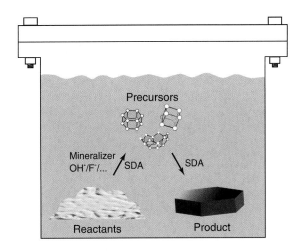

Fig. 1 Schematic representation of the zeolite formation process, from reactants to the final crystalline product. The process is depicted inside a laboratory-scale autoclave, as used for most experimental zeolite syntheses. SDA, structure-directing agent.

the nature of the crucial species involved, and of the growth mechanism, has been a scientific challenge for decades [17]. The principal growth models are: (i) nucleation followed by crystal growth by addition of a preassembled elementary unit; and (ii) growth by aggregation (Fig. 2).

Combinations of the two growth modes occur frequently. Similar kinetic growth models referred to as monomer–cluster and cluster–cluster growth, respectively, are handled in sol–gel science [18]. The nucleation and crystal growth model holds for zeolitization processes from inorganic gels. The aggregation model pertains to systems with high concentrations of organic molecules.

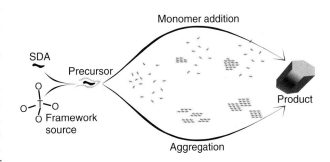

Fig. 2 Principal growth models of zeolites. Zeolites grow from preassembled aluminosilicate precursors. Depending on the conditions and composition, the final crystalline product is formed either by monomer addition to a growing nucleus, or by successive aggregation following a self-organization mechanism.

References see page 176

2.3.5.2.2 Key Parameters Governing Zeolitization

A Molar Composition The composition of the synthesis mixture is not an independent parameter influencing zeolitization for it reflects other parameters indirectly, such as the alkalinity and the nature and amount of organic templates, the impact of which will be dealt with in more detail later on. The molar composition should be regarded as one of the most important factors determining the outcome of the crystallization, because every zeolite has a specific compositional range, which can be either very broad or extremely narrow. Such compositional ranges – commonly called the *crystallization field* – are usually represented graphically in a ternary diagram, such as that illustrated in Fig. 3 [19].

The chemical composition of a specific synthesis is usually expressed as molar ratios of oxides:

$$a\, SiO_2 : Al_2O_3 : bM_xO : c\, N_yO : dR : eH_2O$$

in which M and N represent (alkali) metal ions and R the organic template, while $a-e$ are molar ratios. To indicate the crystallization field, the preferred ranges for the relative molar amounts of SiO_2, Al_2O_3, M_2O, N_2O, R and H_2O can be added to the oxide formula. Next to the nature of the templates used, the ratios $SiO_2 : Al_2O_3$, $M_xO(N_yO) : SiO_2$, $R : SiO_2$ and $H_2O : SiO_2$ are often handled to characterize a zeolitization process. The compositional ratios influence nucleation and crystallization kinetics, the nature of the crystalline phases obtained, the framework composition, the Si and Al distribution in the lattice, and the crystal size and morphology [20–26]. For example, the organic-free synthesis of ZSM-5 and mordenite was shown to be dependent on the $Na_2O:SiO_2$ as well as on the $SiO_2 : Al_2O_3$ ratio, ZSM-5 being preferred at high $SiO_2 : Al_2O_3$ and low $Na_2O : SiO_2$ ratios in the gel [20].

To date, the use of aqueous media has been very successful for growing zeolites. Nevertheless, it is possible to grow zeolites in other, non-aqueous media such as alcohols (e.g., hydroxysodalite in glycol or ethanol) [27].

The potential use of organic solvents in zeolitization is influenced by their viscosity and their relative electric permittivity, with the latter property greatly influencing the solvating capacity [27]. Synthesis in non-aqueous media is one of the techniques used to grow large zeolite crystals [13]. A synthesis method to prepare millimeter-scale and morphologically well-defined crystals in non-aqueous media has been reported by Kuperman et al. [28]. Solvents used for the synthesis of large crystals include pyridine, triethylamine, and polyethylene glycol. Next to a mineralizer and an optional templating species, water was added in stoichiometric amounts necessary to dissolve the silicon and aluminum sources. HF-pyridine and HF-alkylamines are claimed to be novel mineralizers by acting as reservoirs for anhydrous HF in the organic solvent, and thereby allowing the amount of reactant water to be controlled.

B Mineralizing Agent Most zeolite crystallizations are performed under alkaline conditions, with the *pH* of the synthesis solution, which is generally between 9 and 13, being of key importance. The OH^- anions fulfill a crucial catalytic function in the field of inorganic crystallizations, defined as the *mineralizing* or *mobilizing function*. The rates of nucleation as well as of crystallization are influenced by the alkalinity of the medium. Furthermore, the alkalinity has an impact on the Si : Al ratio of the zeolite product, and even on the aspect ratio of the final zeolite crystals. An overview of the impact of alkalinity on the zeolitization process is shown in Scheme 1. The *pH* is known to affect different stages in the process of zeolite formation, as well as different product features. The alkalinity itself, however, changes during the zeolitization process.

The role of the mineralizing agent is to depolymerize and/or hydrolyze the amorphous aluminosilicate particles at an appropriate rate. Soluble and adequate precursor complexes have Si^{IV} and Al^{III} atoms in tetrahedral coordination and contain condensable ligands [29]. The mineralizing agent is indispensable in order to reach a supersaturated state which makes nucleation and crystal

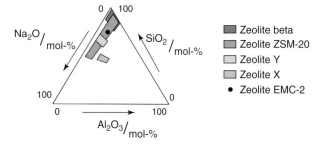

Fig. 3 Ternary diagram indicating the crystallization fields of different zeolites. (After Ref. [19].)

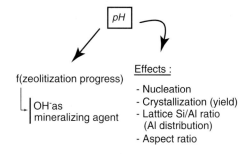

Scheme 1

growth possible [29, 30]. In general, an increased pH will accelerate crystal growth and shorten the induction period before the formation of viable nuclei, by enhancing the precursor concentration [30]. This alkalinity effect on the crystallization kinetics is explained schematically in Fig. 4 [31].

The dissolution of the gel, catalyzed by OH^-, proceeds via a nucleophilic S_N2 mechanism (as depicted in Scheme 2a). The fivefold coordination state of Si^{IV} in the transition state weakens the siloxane bonds [27], and condensation occurs via an attack of a nucleophilic deprotonated silanol group on a neutral species (as depicted in Scheme 2b) [27].

During zeolite crystallization from aluminosilicate gels under basic conditions, changes in pH occur [32, 33]; this pH variation for the synthesis of EU-1 [32] is illustrated graphically in Fig. 5. First, the pH falls as a result of the hydrolysis of the amorphous aluminosilicate, consuming hydroxyl ions, but upon the formation of crystals through condensation reactions the pH increases. In the example shown in Fig. 5 this occurs after 6 h. During the condensation reaction, hydroxy ligands are

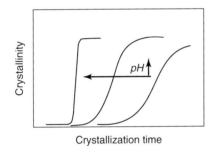

Fig. 4 Increasing alkalinity of the synthesis mixture reduces the induction period and accelerates crystal growth, increasing the tangent of the steepest part of the crystallization curve. (After Ref. [31].)

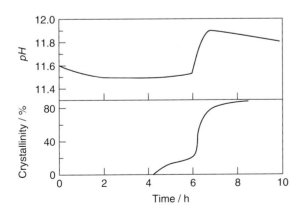

Fig. 5 pH-variation and crystallization curve for the zeolite EU-1. (After Ref. [32].)

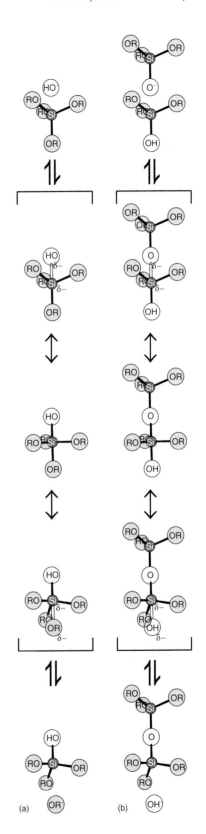

Scheme 2

expelled upon formation of the oxo (–O–) bridges of the zeolite framework (see Scheme 2b). The *pH* jump coincides with the period of autocatalytic crystal growth (Fig. 5, lower part). Indeed, it is even possible to use this *pH* change as a simple and rapid method to follow the course of crystallization [32].

Furthermore, the alkalinity of aluminosilicate gels influences the Si:Al ratio of the crystalline product. The average Si:Al ratio in the zeolite shows a tendency to decrease with increasing *pH* of the synthesis hydrogel [29, 30]; this is depicted in Fig. 6 for the crystallization of chabazite and edingtonite-type zeolites from hydrogels with different $SiO_2:Al_2O_3$ ratios. This is readily explained by a nucleophilic condensation mechanism which involves $Si-O^-$ and/or $Si-OH$ functions. At increasing *pH*, the silicate species are more deprotonated. The reaction rate of the condensation of silicate species is decreased because the dissociated hydroxy ligands are poor leaving groups. Whereas in the range of effective *pH* conditions, Al^{III} is present in solution as the aluminate anion ($Al(OH)_4^-$), the condensation rate of silicates with the aluminate anion is less influenced by *pH* [29, 30].

For faujasite-type zeolites, the very slow linear crystal growth rates for siliceous polytypes crystallized from inorganic hydrogels, within a reasonable time frame, appear to place an upper limit to the Si:Al lattice ratios for the crystals, with a maximum Si:Al ratio of ca. 3.0 [2]. The lower limit for the lattice Si:Al ratio of aluminosilicate zeolites crystallized in alkaline media is 1.0, as the formation Al–O–Al bonds is avoided according to the Löwenstein rule [34]. Aluminosilicate zeolites with an Si:Al ratio of 1.0 are crystallized in very alkaline media where, for kinetic reasons, the silicate monomer condenses only with the aluminate.

The choice of a mineralizing agent is not limited to OH^- anions. Indeed, F^- anions can fulfill this catalytic function as well, allowing zeolitization in acid media [10]. The use of F^- anions increases the solubility of certain tri- and tetravalent elements (e.g., Ga^{III}, Ti^{IV}) through complexation [27, 29]. However, when present at higher concentrations, F^- anions will inhibit the condensation reactions as they are retained in the coordination sphere of the T-atoms [27]. According to Guth et al. [29], an increasing F^- concentration for Si and Al complexes tends to change the tetrahedral into octahedral coordination. The amount of F^- used in zeolite synthesis is limited. Generally, no highly supersaturated solutions are involved, avoiding sudden massive nucleation and in this way generating a large number of small crystals.

Therefore, the crystallization process is more easily controlled with respect to crystal size and morphology, and consequently the crystals formed in these solutions contain fewer structural defects. Unfortunately, this procedure requires longer crystallization periods, however [29]. When using F^- anions, the incorporation of elements such as Co^{2+} with a poor solubility in highly alkaline solutions is possible [29].

C Temperature and Time In general, as crystallization is an activated process, within certain limits temperature has a positive influence on zeolitization. Crystal growth rates are currently expressed as linear growth rates calculated as $k = 0.5\delta l/\delta t$, with *l* being the crystal length. Increasing temperatures yield steeper crystallization curves, shifted towards shorter crystallization times, as illustrated in Fig. 7 for the synthesis of mordenite.

The temperature also influences the nature of the crystallized zeolite phase. Increasing temperatures give rise to denser phases, as under hydrothermal synthesis conditions the fraction of water in the liquid phase

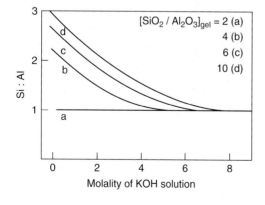

Fig. 6 Influence of alkalinity of the hydrogel on Si:Al ratio of the products, chabazite and edingtonite-type zeolites, from hydrogels containing various relative amounts of SiO_2 and Al_2O_3. (After Ref. [2].)

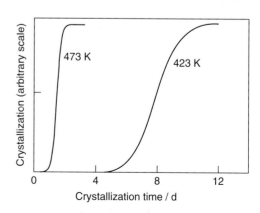

Fig. 7 Impact of temperature on crystallization kinetics of zeolite mordenite. (After Ref. [2], p. 145.)

(which stabilizes the porous products by filling the pores) decreases. Therefore, there exists an upper temperature limit for the formation of each specific zeolite, and of zeolites in general [2, 35]. The use of non-volatile microvoid-filling molecules substituting the water would, in principle, allow the high-temperature synthesis of open, porous structures [2].

Since during crystallization, the solid product usually consists of a mixture of zeolite and unreacted amorphous solids, partially transformed into precursor species, the crystallinity of the product will increase in time. However, since zeolites are metastable phases, zeolite crystallization is governed by the occurrence of successive phase transformations; this is known as the *Ostwald rule of successive phase transformation*. The thermodynamically least favorable phase crystallizes first, and is successively replaced in time by more stable and often denser phases [2]. Typically, the following crystallization sequence exists:

- Na–Y (FAU framework type) → Na–P (GIS framework type) [2, 37]
- Na–Y → ZSM-4 (MAZ framework type) (Fig. 8) [36].

2.3.5.2.3 Precursor Species After reaching the gel point upon mixing of the reactants, the hydrogel is usually aged at room temperature or at a slightly increased temperature but below that applied during the crystallization. This treatment of the hydrogel is known as "ageing" or "ripening", and is often crucial for obtaining the desired crystalline phase and accelerating the crystallization [37]. After ageing, the synthesis mixture is heated to the appropriate crystallization temperature, which most often ranges from 333 to 473 K [2, 30]. Crystallization above the boiling point of water requires the use of closed systems, allowing a high pressure of water vapor to build up.

Crystallization temperatures up to 470 K or even higher allow a high yield of crystals to be achieved in an acceptable period of time.

When solid or gelatinous Si-sources are used, one of the important reactions occurring during the ripening period is the (partial) depolymerization of the solid silicate phase, which is catalyzed by hydroxyl ions [2]. This dissolution leads to an increase in the concentration of silica in the liquid phase of the gel. By using ^{29}Si NMR, Ginter et al. [38] studied the evolution of the gel upon ageing during the crystallization of zeolite Na–Y. A slow dissolution of gel solids was observed at room temperature, but this was accelerated by increasing the temperature. Upon dissolution, the initial products formed are monomeric silicate anions (Q^0). In the Q^n notation n represents the number of siloxane bonds of the Si atom. After ageing for 24 h, the relative NMR signal intensity of the low connectivity bands, Q^0, Q^1 and Q^2, was characteristic for sodium silicate solutions with a low silicate ratio (SiO_2: Na_2O), which is tantamount to high alkalinity. Further ageing (for 36–48 h) resulted in the formation of silicate species with a higher degree of condensation (Q^2 and Q^3), suggesting that the silicate ratio in solution has increased [38]. This evolution is depicted schematically in Fig. 9.

Following depolymerization of the silica sol, the monomeric silicate anions released into solution further copolymerize into oligomeric species. These oligomers are negatively charged and undergo rapid intramolecular as well as intermolecular exchange of silicate monomers [39, 40]. The condensation and hydrolysis of silicate oligomers occurs through a nucleophilic mechanism catalyzed by hydroxyl ions (see Scheme 2b) [41].

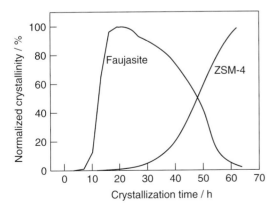

Fig. 8 Example of Ostwald's rule of successive phase transformations: in time, the less-stable faujasite phase is replaced by the more-stable ZSM-4 (hydrogels containing TMA^+ cations). (After Ref. [36].)

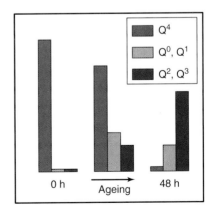

Fig. 9 Schematic evolution of different Si populations during gel ageing for the synthesis of zeolite Na–Y. Non-aged gels show the exclusive presence of polymeric Si (Q^4). Upon ageing, monomeric Si (Q^0) is formed by hydrolysis, and later transformed into species with higher connectivity (Q^1–Q^3). (After Ref. [38].)

References see page 176

Both the acidity and nucleophilicity increase with the degree of silicate connectivity in the oligomer ($Q^3 > Q^2 > Q^1 > Q^0$).

In alkaline solutions, aluminum is present as the tetrahedral aluminate anion, $Al(OH)_4^-$. The silicate molecules condense with this monomeric $Al(OH)_4^-$ to produce aluminosilicate structures [42, 43]. The condensation rate of silicate oligomers and the aluminate anion increases with increasing silicate ratio; that is, with increasing degree of oligomerization of the silicate [42]. The aluminate anion condenses preferentially, with the larger silicate species in solution containing the strongest nucleophiles. The preference of aluminum for condensation with large silicate anions is also explained by the low electrostatic repulsion between large silicate species and the $Al(OH)_4^-$ anion, since the average negative charge per silicon decreases with the size of the oligomer [44]. Once these aluminosilicate complexes are formed, further polymerization progresses only slowly [44].

The silicate and aluminosilicate polyanion solution chemistry is extremely complex [39, 42, 45], and ^{29}Si and ^{27}Al NMR spectroscopic techniques have each contributed to the identification of these silicate and aluminosilicate polyanion structures [2, 39, 42, 44, 45]. In silicate oligomers, ring- and cage-like structures are more stable compared to chain-like structures [45]. The ^{29}Si NMR technique is suitable for the detection and analysis of silicate species containing up to 12 Si atoms [39]. Furthermore, the concise study of mixed silicon–aluminum anions is hampered by the strong line-broadening caused by the quadrupolar ^{27}Al nucleus. The identification and characterization of aluminosilicate polyanions of extended character would contribute significantly to the understanding of the processes initiating nucleation and crystal growth, as well as of the impact of the various crystallization parameters. Among the siliceous species which could be identified are a number of the so-called "secondary building units" (SBUs) [2, 45] (see Fig. 10), which Meier [46] proposed as basic building units of zeolite framework structures. The presence of such units in the liquid phase of hydrogels is rather interesting, for Barrer [2] proposed that mechanisms of nucleation and crystal growth involve the condensation of specific polyanions rather than Q^0-monomers [45]. The SBU is a structural rather than a chemical concept, as the nature of preassembled silicate units is dictated by chemistry rather than crystallography; hence, the correspondence of some precursor species and SBUs is fortuitous [47].

Katovic et al. [37] studied the competitive crystallization of zeolite Na–X and cubic Na–P, denoted with Pc. The crystallization yield and nature of the zeolite phase obtained depend on the ageing period of the hydrogel.

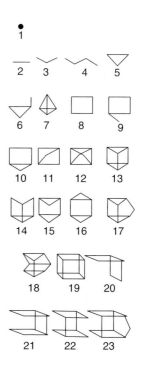

Fig. 10 Silicate polyanions identified by ^{29}Si liquid-state NMR in alkaline solutions. (After Ref. [47].)

Ageing leads to a shortening of the induction period of the crystallization of both zeolite Na–X and Pc. It is explained by an increase in the number of nuclei formed in solution upon ageing [37]. During polycondensation of the silicate and aluminate precursors, a predominantly amorphous aluminosilicate hydrogel is formed, containing very small semi-crystalline particles with a structure resembling that of Pc. The growth of these gel-occluded semi-crystalline particles, however, is impeded because the material flux inside the hydrogel matrix is slow [37]. During ageing, structural changes take place in the solid part of the gel, as detected with X-ray diffraction (XRD). The Raman/infra-red (IR) spectra of the gel suggest a slow formation of six-membered aluminosilicate rings, their ordering into sodalite cages, and a possible formation of quasi-crystalline particles resembling the faujasite structure [48]. Dissolution of the gel during ageing results in the release of these semi-crystalline nanoparticles into the solution. The enhanced crystallization of zeolite Na–from an aged gel can, therefore, be attributed to the increased number of nuclei formed inside the gel matrix, and explains why zeolite Na–X-crystallizes only from aged gels [37]. The competitive formation of double six-rings (D6R), as building units for zeolite Na–X, versus double four-rings (D4R), as building units for zeolite Pc, will be discussed later in this chapter.

The impact of ageing on the crystallization of zeolite Na–X and Na–A is illustrated graphically in Fig. 11. The

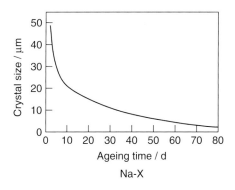
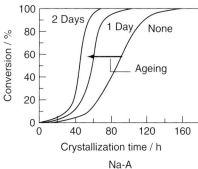

Fig. 11 Left: Dependence of crystal size and crystallization kinetics on the ageing period, as demonstrated for the crystallization of Na−X. Right: The crystallization curves of zeolite Na−A for to non-aged, 1-day and 2-day aged gels. (After Ref. [49].)

Na−X zeolite crystal size decreases substantially when the gel is aged for longer; in addition, ageing of the gel enhances the concentration of precursor species and the concentration of nuclei, and the average final crystal size is reduced when a larger number of nuclei develop into crystals. The presence of an increased concentration of nuclei is also reflected in enhanced crystallization kinetics.

2.3.5.2.4 Templates Templates (which sometimes are also known as structure directing agents, SDAs) are species in the reaction mixture which, in general kinetic and thermodynamic terms, contribute to the formation of the zeolite lattice during the zeolitization process in several ways. Mainly, they influence gelation, precursor formation, nucleation, and crystal growth processes. The template organizes aluminosilicate oligomers into a particular geometry and, as a result, provides precursor species for nucleation and growth of the "templated" zeolite structure [39, 50]. The incorporation of templates in the zeolite micropores lowers the chemical potential of the zeolite by lowering the interfacial energy. Template inclusion in the zeolite micropores contributes to the stability by new interactions (hydrogen bonding, electrostatic and van der Waals interactions), and also controls the formation of a particular topology through geometrical factors (form and size) [29]. It is evident that changes in charge density provoked by geometrical or physical properties of the template will be reflected in the chemical composition (Si: Al ratio) of a given topology.

The capability to predict which template is required for a given structure and composition has long been sought [29, 51]. Hence, in the synthesis of silicate-based zeolites, template design has been used successfully to force the formation of zeolites with multidimensional channel systems [52, 53]. In selecting possible templates, some general criteria must be borne in mind regarding the templating potential of a candidate molecule; these include solubility in the solution, stability under synthesis conditions, steric compatibility, and framework stabilization [29]. The possibility of removing the template without destroying the framework can also be an important practical issue [29].

Neutral molecules as well as cations or ion pairs are able to fulfill these structure- and composition-directing functions of a template. Each of these three types can be either organic or inorganic in nature.

A Charged Molecules The majority of structure-directing agents operated in zeolite synthesis are positively charged. These cations not only function as structure- and composition-directing agents, but also positively influence the rate of crystallization [2, 42]. Inorganic as well as organic cations are frequently used and comprise species such as Na^+, Li^+, Cs^+, K^+, Rb^+, Ca^{2+}, Sr^{2+}, tetraalkylammonium cations (TMA^+, TEA^+, TPA^+, dihydroxyethyl-dimethylammonium, etc.), dialkyl and trialkyl amines, and phosphonium compounds [27, 30, 54]. The neutral crown-ethers form charged complexes with an alkali metal cation in the central cavity [29, 33].

In aqueous solution, cations are known to influence the ordering of water molecules, and on the basis of this influence the concept of *structure-making* and *structure-breaking* agents [39, 42, 55] has been introduced.

- *Structure-making* cations are small cations (e.g., Na^+ and Li^+) which interact strongly with water molecules because of their high charge density. As a result of this interaction, the original hydrogen bonds in the water network are broken and the water molecules are strongly organized around the cations [39]. The strong interaction of the cation M with the oxygen atom of water molecules (M^+-O) leads to the formation of a network of hydrogen-bonded water molecules [42], dominated by the 1st shell coordination around the cation (see Fig. 12).

References see page 176

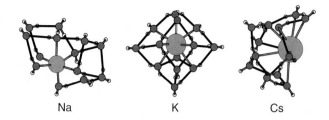

Fig. 12 M(H$_2$O)$_{13}^+$ alkali-ion–water microcluster as optimized for geometry. The small Na$^+$ clearly has a strong re-organizing effect on the water structure, dominated by the first shell octahedral arrangement of water around the ion. Cs$^+$ does not markedly disturb the inherent water–water interaction found in pure water. The strength of the effect of K$^+$ on the local water structure is situated between Na$^+$ and Cs$^+$. (Based on Ref. [55].)

- *Structure-breaking* properties usually are assigned to large cations such as tetramethylammonium [56], K$^+$ and Cs$^+$ which interact also with the water molecules, but rather form transient cage- or ring-like structures around the cation [57]. The interaction of the larger cations with the water molecules is rather weak, and the structures that are formed are dominated by inter-water hydrogen-bonding [39, 56].

It has been proposed that the organized network of water molecules surrounding the cations can be (partially) replaced by silicate and aluminate tetrahedra. Oxygen atoms of silica replace the oxygen atoms of water molecules and form cage-like structures [27, 42]. When using ^{29}Si NMR, McCormick et al. [58] indeed observed an increase in the amount of cage-like structures for alkali metal-containing silicate solutions, with increasing cation size. As for a fixed silicate ratio, changes in the alkalinity of the mixture are negligible by using alkali cations with different sizes, this effect was attributed to cation–anion interactions. It could also be related to the observed clathrate-like water structures surrounding these ions.

An example of the similarity of the oxygen positions in the water-cation cluster around potassium compared to the situation of potassium within the silicate, D6R is illustrated in Fig. 13.

It remains unclear whether water molecules are successively replaced by T-atoms, or if the preference of K$^+$ for D6R is only due to the energetically favorable charge distribution.

The selective formation and stabilization of particular silicate structures in quaternary ammonium ion-containing silicate solutions has been observed many times. For example, the presence of the trimethylamine group results in the formation of the cubic octamer [59–61], also referred to as D4R. Other alkylamines and alkylammonium species show different structure-directing effects, resulting in different silicate polyanion species. For example, tetrapropylammonium has been observed to favor the formation of oligomeric silicate species based on annealed five-rings [62]. The formation of phase-specific structures mainly depends on hydrophobic–hydrophilic and steric interactions of the organic cations with their surrounding. The structure-directing effect of organic amines and ammonium compounds can be inhibited by the presence of alkali metal ions [63]. The disturbed interaction between the organic cations and the water molecules by the hydration of the alkali ions might be responsible for this inhibition [63]. Another explanation could be the competition among templating species consisting of alkali and organic cations, resulting in a mixture of molecular silicate species.

A large number of zeolite types was successfully synthesized in fluoride medium [10]. In several zeolite syntheses in acidic media, the F$^-$ anion has a structure-directing role and is incorporated into the D4R unit of the framework. The addition of germanium to the synthesis mixture favors the formation of D4R units even further.

B Neutral Molecules In the present context, water is a very important *neutral* molecule acting as a templating

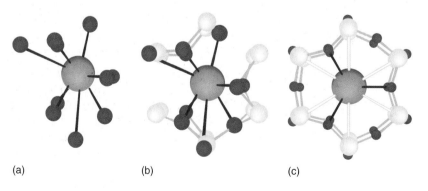

Fig. 13 (a) K(H$_2$O)$_{13}^+$ microcluster as optimized for geometry; (b) Si positions as found in the D6R overlaid onto K(H$_2$O)$_{13}^+$; (c) Si- and O-positions in D6R.

species. Next to its templating effect by interaction with cations (as discussed above) and its solvating and hydrolyzing ability (which are of key importance in zeolite synthesis), it enhances the formation of a zeolitic structure during crystal growth by filling the micropore system as a guest molecule, thereby stabilizing the porous oxide frameworks. Here, the template acts on a supramolecular level, assuming the role of void-filling, rather than intervening at the molecular scale in the precursor formation. By the selection of suitably shaped precursor species, this supramolecular template effect can direct the synthesis toward the formation of a desired structure. A similar behavior has been demonstrated in the cases of synthesis of LTA- or EMT-type zeolites using organic charge transfer complexes [64] or crown-ethers [65], respectively.

The concept of void-filling has successfully been taken one step further: Catlow and coworkers have developed a computer program (Zebedde) which suggests the optimum organic template molecule for a given zeolite cage [66].

Other neutral molecules acting as templates include ethers, alcohols, and di- and tri-ols [27, 29]. Amines usually are present in the protonated form and should be classified as charged template species.

C Ion Pairs Alkali and alkaline earth metal salts (viz., NaCl, KCl, KBr, CaF_2, $BaCl_2$, $BaBr_2$) can be occluded as ion pairs, possibly next to water molecules into the zeolite micropore system, thereby stabilizing the zeolite [2, 29, 30]. As mentioned above, these non-volatile space-filling species can be used in high-temperature zeolite synthesis applications. When salts are present, and are catalyzing the zeolitization, they may cause the formation of specific zeolites, thereby improving crystallization yield and even crystallinity [30].

2.3.5.2.5 Rationalization of Crystallization Conditions for Zeolites A, X, Y, and EMC-2 Based on the wealth of accumulated data, the formation of some well-known zeolite frameworks in relation to the synthesis conditions can be rationalized (Fig. 14). In the synthesis of zeolites A (LTA-type), X (FAU-type), Y (FAU-type) and EMC-2 (EMT-type), the sodium cation assumes a particular structure-directing role in combination with co-templates and depending on the aluminum content. Sodium has been observed to show a marked preference for the formation of four-rings (4R) [67]. In a solution with a high content of alumina, this results in the specific formation

References see page 176

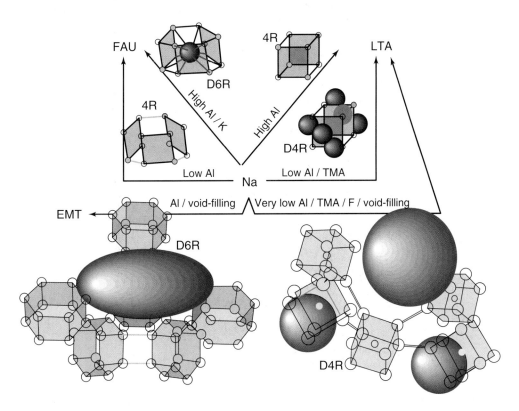

Fig. 14 Different zeolite topologies formed from synthesis gels containing sodium ions, depending on the presence of co-templates and the Al-content.

of the D4R unit, containing four Al and four Si atoms alternating on the nodes. The D4R has been identified as the growth unit for zeolite A (LTA-type) [68], and is present in these solutions. Lowering the content of Al, however, results in the formation of Y-type zeolites (FAU-type), which have been proposed to be assembled from the D6R unit [65] which can be formed from three four-rings, 4Rs. Four-ring units containing less than two Al atoms clearly are not condensing exclusively into D4R but rather prefer the formation of D6R, which results in the formation of a FAU structure. In order to obtain the FAU topology with very high Al content, the immediate condensation of two 4Rs into D4R has to be prevented. This can occur via the addition of K^+, which preferentially is sited inside the D6R in the final crystal [69] and which has been observed to be involved in D6R formation in numerous zeolite syntheses [70, 71]. The common use of Sr^{2+} for templating structures containing D6R is not a surprise considering the diagonal relationship of K^+ and Sr^{2+} in the Periodic Table. On the other hand, if an LTA topology with lowered Al content is desired, a co-template forcing the formation of D4R, such as the tetramethylammonium ion, should preferably be added [10, 72]. Further agents for preferred D4R formation are the F^- ion (which almost exclusively is found inside D4R units) and Ge-ions (which are preferentially incorporated into D4R moieties of the framework [10, 64]). The LTA structure can be formed with almost no Al in the lattice when a void-filling template is used which is chosen to fit nicely into the α-cages of the structure [64]. The void-filling concept can also be used to redirect the assembly of D6R into EMC-2 zeolite (EMT-type) rather than into the formation of FAU-type topology. Crown-ethers arrange D6R units in ellipsoidal cages being part of the EMT structure.

2.3.5.2.6 Zeolite Growth Models

A The Nucleation and Growth Process

a Achievement of Supersaturation A solution with specific concentration and temperature can exist in stable, metastable, or labile conditions (as depicted in Fig. 15). The stable and metastable areas are separated by the solubility curve, which determines the equilibrium saturation concentration of a component c^* [73]. The boundary between the metastable and labile region is not always well defined. The degree of supersaturation, S, is defined as the ratio of the actual and the equilibrium concentration ($S = c/c^*$) [73]. In the stable region, neither nucleation nor crystal growth can occur, while in the labile region nucleation as well as crystal growth are possible. In the metastable area, only crystal growth can occur.

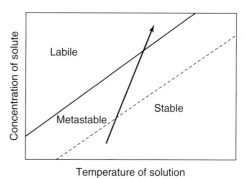

Fig. 15 Solubility–supersolubility diagram. The arrow indicates how supersaturation is reached in the hydrothermal gel method.

During ageing – and especially upon increasing the temperature – the concentration of adequate aluminosilicate precursor species (as discussed in Section 2.3.5.2.5) increases in time by dissolution of the amorphous solid phase and interconversions of dissolved aluminosilicate oligomers in solution. A stable solution is transformed into a metastable solution, and finally into a labile one. This transformation is indicated with an arrow in the solubility–supersolubility diagram (Fig. 15).

b Nucleation Primary nucleation from a supersaturated solution can be either homogeneous or heterogeneous [73]. The latter condition is induced by impurities or foreign particles present in the solution, while the former occurs spontaneously. It is evident that heterogeneous nucleation can be suppressed by filtration of the different solutes. Secondary nucleation is induced by crystals and is relevant to the technique of seeding (as discussed below).

During the period preceding the formation of viable nuclei, different types of germ-nuclei (embryos) form by chemical aggregations of the aluminosilicate precursor species, and disappear again through depolymerization [2]. As a result of such fluctuations, the germ-nuclei will grow in time, and form eventually different types of nuclei, with dimensions having the critical size to become viable; that is, nuclei on which crystal growth occurs spontaneously [2]. The concentration of these species is highest in the boundary layer surrounding the amorphous solid particles, as a result of the enhanced local concentration by dissolution. Presumably, nucleation occurs preferentially in these boundary regions [42].

The net free energy of formation of a nucleus, consisting of j structural units can be expressed as [30]:

$$\Delta G = A j^{2/3} - B j \qquad (1)$$

in which A and B are constants. The first term $A j^{2/3}$ reflects the interfacial free energy between the nucleus and the solution. It is proportional to the area of the interface between the nucleus and the solution, and has a positive value, corresponding to the destabilization of the nucleus owing to surface tension. The second term represents the free energy of formation of a nucleus containing j structural units. According to Eq. (1), the nuclei will become viable if $\delta \Delta G/\delta\ j = 0$. Further addition of structural units will decrease ΔG, and favor crystal growth [30]. For a spherical nucleus, the free energy of formation depends on the degree of supersaturation, S, the density, ρ and the surface energy, σ of the nucleus [73]:

$$\Delta G = \frac{16\pi\sigma^3(MW)^2}{3(RT\rho \ln S)^2} \quad (2)$$

From Eq. (2), it is evident that from a saturated solution ($S = 1$, $\ln S = 0$) no spontaneous nucleation can occur. For a supersaturated solution with $S > 1$, ΔG has a finite negative value and spontaneous nucleation is possible [73].

Nucleation is an activated process, and the nucleation rate J (the number of viable nuclei formed per unit time) is commonly described by an Arrhenius rate equation [73]:

$$J = A \exp\left(\frac{-\Delta G}{RT}\right) \quad (3)$$

Equation (3) predicts an exponential increase in the nucleation rate with temperature and, consequently, in the degree of supersaturation, as the concentration in solution increases exponentially with temperature. The onset of nucleation occurs at a critical degree of supersaturation, $S_{crit.}$ (Fig. 16, curve a) [74]. The experimentally observed nucleation rates typically exhibit a maximum with increasing degree of supersaturation (Fig. 16, curve b) [74]. The decrease at high concentrations is caused by the enhanced viscosity of the medium, inhibiting molecular migrations.

The rate of nucleation can be determined from crystal size distribution measurements in the final crystallization

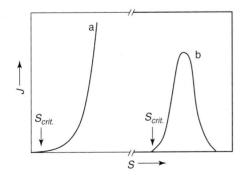

Fig. 16 Nucleation rate, J, plotted versus degree of supersaturation, S. Curve 'a' is plotted according to the Arrhenius rate equation; curve 'b' is based on experimental observations. (Reproduced from Ref. [72], with permission of Elsevier Science.)

product and size increase measurements of the largest crystals during the course of crystallization [49]. During crystallization of Na−X, the majority of nuclei are formed at the early stages of the crystallization; that is, during the autocatalytic stage of the crystallization process, corresponding to the exponential part of the crystallization yield curves (for an explanation, see Section 2.3.5.2.6Ac) [74]. As nucleation and crystal growth consume the same precursor species, the nucleation rate is expected to go through a maximum and to decline again after a certain period of time when the consumption of precursor species by crystal growth limits their availability [2, 73].

The degree of supersaturation influences the crystal size and morphology. In highly alkaline synthesis gels, nucleation and crystallization kinetics are enhanced, and rapid and massive nucleation is to be expected as well as an increased crystal growth rate. As a result, the crystals are not well defined morphologically. At high alkalinity, the crystals sometimes partially redissolve, and in this way the crystal morphology can be totally changed.

c *Crystal Growth* Zeolite crystal growth occurs at the crystal–solution interface by condensation of dissolved preassembled building units onto the crystal surface [2, 29, 42]. Likewise, nuclei grow to crystals by multiple addition of precursor species. Experimental "crystallization curves" express the yield of crystals against time, and these curves usually exhibit an S-shaped profile [2, 73–75]. The inflection point of the sigmoid curves separates an autocatalytic growth period from a stage of delayed crystal growth owing to a depletion of the solution with precursor species. The autocatalytic nature of the first stage of the crystallization reflects the intrinsic property of self-acceleration of a crystallization process, provided that the source of nutrients is not limiting. The experimental crystallization curves can be described by Kholmogorov's equation [74]:

$$Z = 1 - \exp(kt^n) \quad (4)$$

in which n and k are constants. This equation describes the evolution of the ratio, Z, of the mass of zeolite formed at time t to the final zeolite mass. The constant n reflects the nucleation kinetics; the condition $n = 4$ is characteristic for a constant nucleation rate, but values higher or lower than 4 represent increasing or decreasing nucleation rates in time, respectively.

Many zeolite structures are strongly anisotropic. Since micropores extend in specific crystallographic directions, for catalytic applications it is important to control the crystal morphology. The growth rate of zeolite crystals in different crystallographic directions can differ strongly

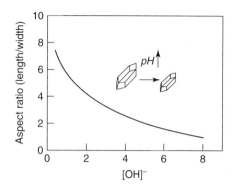

Fig. 17 Influence of alkalinity of the hydrogel on the crystal aspect ratio for the synthesis of silicalite-1 crystals. (After Ref. [31].)

and depend on the synthesis conditions. The alkalinity might also alter the shape of the crystals, as is illustrated by its effect on the aspect ratio (length:width) of silicalite-1 crystals in Fig. 17. The aspect ratio falls for an increased alkalinity of the solution as the length of the crystals decreases relative to their width. For the growth of silicalite-1 zeolite, a separate analysis of the growth rates along the crystal length and width reveals that the rate of crystal growth along the c-axis is relatively independent of the hydroxide concentration. The rate of crystal growth, measured along the width (a and b crystallographic directions) was found to depend heavily on the reaction mixture alkalinity; this rate was also found to decrease in a non-linear fashion with the pH. The role of alkalinity in crystal growth along the a-axis and b-axis is fundamentally different from that for the c-axis. Attempts were made to relate this pH effect with condensation patterns of presumed precursor species [34, 76], but a definitive molecular explanation for the observed crystal growth kinetics is still lacking.

d Seeding Seeding is a technique in which the supersaturated system is inoculated with small particles of material to be crystallized [73]. These seed particles (crystals) increase in size as crystalline material is deposited on them [30]. By a seeding operation, the nucleation stage is bypassed and the induction period eliminated [42].

As the surface area provided by the seed crystals is larger than that provided by fresh nuclei, the seeding technique provides favorable conditions for measuring linear growth rates [2]. Examples of seeds are Na−X crystals for the crystallization of Na−Y-type zeolites, aged hydrogels, or mother liquor from a previous crystallization [2].

The effect of seeding on the crystallization kinetics of Na−Y is illustrated graphically in Fig. 18 [75]. The impact of the amount of seed crystals added is shown in Fig. 18a; increasing the amount of seeds clearly enhances crystallization, as shown by the shift of the crystallization curve towards shorter crystallization times. In another series of experiments, the same weight of seed crystals was added, but with a different crystal size (Fig. 18b). Here, the zeolitization proceeds faster with the smallest crystals, thus exposing the largest surface area.

B The Aggregation Model Studies of synthesis mixtures at successive states of zeolitization revealed, in some cases, a behavior similar to the *cluster–cluster* growth mechanism observed in general sol–gel chemistry [77–79]. The detection of particles with discrete sizes in the nanometer range by dynamic light-scattering (DLS), small angle X-ray scattering (SAXS), and small angle neutron scattering (SANS) [80–85] supports the view of nanoscopic clusters aggregating into larger units, and eventually condensing into zeolite product. For the case of silicalite-1 zeolite formation with tetrapropylammonium (TPA) as template, a detailed study of the molecular and supramolecular steps of the formation of the final zeolite, starting from silicate monomer, has been performed [86, 87]. The aggregation sequence is shown schematically in Fig. 19.

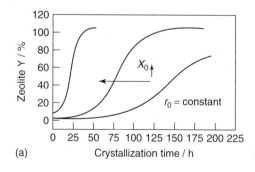
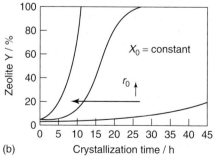

Fig. 18 The addition of seed crystals (Na−X nuclei) affects the crystallization kinetics of Na−Y. (a) Increasing the total amount of seed crystals (X_0) with constant radius (r_0) favors crystallization. (b) A constant amount of seed crystals with decreasing radius also enhances the formation of crystalline material (next to Na−Y, Na−P was sometimes co-crystallized). (After Ref. [74].)

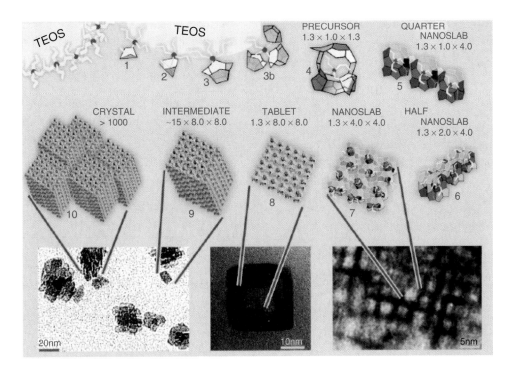

Fig. 19 Formation of silicalite-1 type zeolite from clear solutions [86].

The formation of polysilicate species in close contact with the TPA template, and already containing characteristics of the silicalite-1 framework topology, was proposed long ago [88–91]. Key to the confirmation of the aggregation model for silicalite-1 formation was the use of a so-called *clear solution* [92], which is a solution of TPA silicate that is spectroscopically much better accessible than hydrogel systems. A clear solution is obtained by the reaction of tetraethylorthosilicate (TEOS) in concentrated, aqueous TPA hydroxide solution. Silicalite-1 crystallizes from clear solution upon heating. Depending on the duration of synthesis, subcolloidal and colloidal zeolite particles can be obtained from this synthesis.

Due to the structure-directing effect of the TPA template, the dominant silica species present in the clear solution already are related to the framework of the silicalite-1 zeolite. Ordered condensation of these solution-borne precursors leads to the formation of slab-shaped nanoparticles which, upon heating, finally condense and give rise to the formation of bulk zeolite material [86, 87].

Other studies have shown the involvement of the nanosized zeolite particles during growth [93]. A computational evaluation of rival crystallization models showed the proposed aggregation mechanism to be consistent [70].

A number of other zeolites also can be prepared as nanoparticles, including the zeosils with MTW and BEA topology [94], MEL [95, 96], LTA [97] and FAU [98, 99]. In the synthesis of nanosized FAU-type zeolite it was observed that irregularly shaped 10- to 20-nm-sized FAU particles are formed in the gel phase, and these subsequently aggregate into larger crystals at later stages during synthesis [98]. This is a noteworthy observation, as FAU synthesis occurred in the absence of any organic structure-directing agent, whereas the silicalite-1 synthesis usually is studied with excess organic template present.

The concept of the formation of extended structures during zeolite formation was first discussed in the literature well before nanoscopic species were shown experimentally to play an important role in some zeolite syntheses. Vaughan [54] proposed the formation of "extended sheet and columnar substructures" in aluminosilicate gels, while Feijen et al. suggested the involvement of faujasite sheets in the crown-ether-mediated synthesis of EMT-1 and EMT-2 zeolites [24].

In some zeolite syntheses, layered precursor phases are obtained which only condense into bulk zeolite upon calcination [100]. These examples indicate that the precursors and building units present in the synthesis mixtures may have the tendency to order during synthesis, without immediate condensation into bulk zeolite. During this ordering process, structure-directing agent molecules not occluded into the already formed structures take over the important role of charge-compensation between the negatively charged layers. The existence of ordered liquid

References see page 176

phases was detected in recent *in-situ* studies of silicalite-1 formation from clear solution [87], and this transient laminar phase was proposed to be a key intermediate during formation of this zeolite following the aggregation route [87].

2.3.5.3 Synthesis of Industrial Zeolites

The industrial importance of zeolites can be attributed to their high surface areas, in combination with their chemical nature. Indeed, both material features make them suited to roles as cation exchangers (zeolite A in laundry powder), adsorbents (zeolite A- and X-types mainly) or as catalysts (ultrastable zeolite Y, mordenite, MCM-22, zeolite beta, ferrierite, ZSM-5, ZSM-22, ZSM-23, and TS-1) [101]. The aluminosilicate zeolites mentioned here are currently prepared on an industrial scale.

Figure 20 shows, in schematic form, the set-up for the industrial preparation of a sodium aluminosilicate zeolite [102]. In a first reactor, a sodium aluminate solution is prepared by mixing an aluminum source (e.g., gibbsite) with a NaOH solution. This alkaline solution is mixed with a silicon-containing solution (e.g., waterglass or colloidal silica), and the mixed solution is then heated to the crystallization temperature, using steam. After crystallization, the zeolite suspension is filtered to remove the product, which is then washed and finally spray-dried. The alkaline mother liquor is concentrated through heating, and recirculated to the crystallization reactor; in this way no waste products are formed. Zeolites are usually crystallized in batch reactors under one of the following conditions [103]:

- 90–100 °C; atmospheric pressure; $pH > 10$
- 140–180 °C; autogenous water pressure (0.5–1 MPa, closed vessel); $pH > 10$
- 100–180 °C; water + "amine"; autogenous pressure; $pH > 10$.

Open-tank crystallization vessels and autoclaves are used, with capacities of up to 38 m³ and with a volumetric product yield ranging from 10 to 120 kg m^{-3} [103, 104].

2.3.5.3.1 Industrial Zeolites Crystallized from the Na$_2$O-Al$_2$O$_3$-SiO$_2$-H$_2$O System: Zeolites A, X, Y, Mordenite, and ZSM-5 The industrially relevant A, X, and Y zeolites are crystallized from sodium aluminosilicate hydrogels. The molar compositional ranges for the crystallization of these three zeolite types are best illustrated by the compositional diagrams of Fig. 21, where sodium silicate is used as the silicon source [5]. The water content of the gel is between 90 and 98 mol%. The diagrams show that a decreasing aluminum fraction in the gel promotes the formation of zeolite X and eventually zeolite Y instead of zeolite A. Zeolite X, and especially zeolite Y, are known to contain more silicon and less aluminum compound to the zeolite product A. The water content of these synthesis hydrogels is critical, as a reduction to 60–85% water leads to the formation of the dense hydroxysodalite phase instead of a zeolite (Fig. 21b).

The crystallization time is heavily influenced by temperature; for example, in the case of zeolite X it is reduced from 800 h at 25 °C to 6 h at 100 °C [5]. At temperatures exceeding 150 °C, the formation of zeolite X and Y is inhibited, and zeolite P becomes

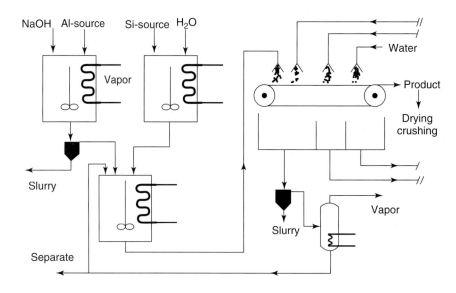

Fig. 20 Schematic representation of an industrial sodium aluminosilicate zeolite manufacturing process. (After Ref. [102].)

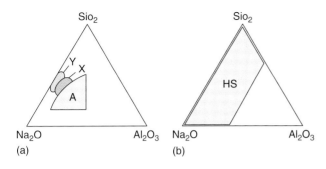

Fig. 21 Ternary diagrams for the crystallization of different zeolites from an inorganic sodium aluminosilicate gel with a water content of (a) 90–98 mol% and (b) 60–85 mol%. The Si source is sodium silicate; HS, hydroxysodalite. (After Ref. [5].)

the dominant product [5]. Crystallization is also greatly influenced by the nature of the reactants used to create the gel. Typically, the use of a colloidal silica sol might provoke the crystallization of zeolite Y from gel compositions which yield zeolite P in the soluble silicate system [5]. In the industrial manufacture of zeolite Y, the seeding technique is currently applied to facilitate crystallization.

Figure 22 shows, schematically, a typical manufacturing process for zeolite A, which is used in gas drying, in the separation of iso- from normal-alkanes, and in detergent water softening [103]. Zeolites ZSM-5 and mordenite are also crystallized from diluted sodium aluminosilicate hydrogels [6, 105]. Increasing the $Na_2O:SiO_2$ ratio or decreasing the $SiO_2:Al_2O_3$ ratio favors the phase transformations from ZSM-5 to mordenite [105]. The production

of MFI-type zeolites in the absence of an organic SDA requires high temperatures (>200 °C), a typical framework Si:Al ratio being 13 [106, 107]. Zeolites of the MFI-type are currently synthesized in the presence of quaternary ammonium compounds or other organic SDA molecules [6], as discussed below.

2.3.5.3.2 Industrial Zeolite Crystallization in the Presence of Organic Compounds In order to prepare zeolite ZSM-5, the TPA cation is the most commonly used organic template. However, the ZSM-5 zeolite is also known to be formed in the presence of triethylpropylammonium (TEPA) and tripropylmethylammonium (TPMA) compounds, as well as amines, alcohols, and many other organic molecules [6].

The ZSM-5 zeolite crystals synthesized in the presence of TPA^+ exhibit an aluminum zoning that is opposite to that present in crystals synthesized from inorganic hydrogels [107] (Fig. 23) and reflected in the catalytic performances [107]. The typical hydrogel composition, according to examples from the original patent by Argauer and Landolt [108] is given by: 29 $SiO_2:Al_2O_3:Na_2O$:9–17 TPA_2O:450–480 H_2O.

Many hydrogel compositions, however, have been found to yield ZSM-5-type zeolites [6]. In particular, the $H_2O:SiO_2$ ratio does not seem to be critical for the crystallization of ZSM-5, and indeed this ratio has been varied between 7 to 122 [6].

Another important member of the MFI family of materials is a titanosilicate, designated as TS-1. This material is applied as a partial oxidation/oxygenation

References see page 176

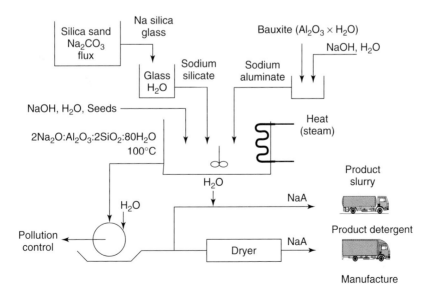

Fig. 22 The manufacturing process for type A zeolite. (After Ref. [103].)

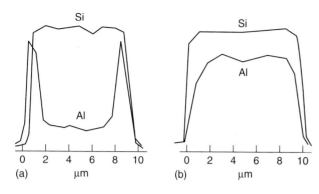

Fig. 23 Aluminum zoning in MFI-type crystals synthesized in the presence of (a) TPA$^+$ and (b) in the absence of any organic molecule. (After Ref. [107].)

catalyst, and is also synthesized in the presence of TPA$^+$ cations. The molar gel composition, according to the invention by Taramasso et al. [109], is shown in Table 1. The success of the TS-1 synthesis depends very much on the purity of the reactants, as traces of alkali metals impede the insertion of Ti into the silicalite-1 framework [110, 111].

2.3.5.4 Conclusions and Outlook

Synthetic aluminosilicate zeolites are technical materials of central importance to a growing area of industrial production and services. The hydrothermal gel method remains the most important synthesis procedure for traditional as well as new types of aluminosilicate zeolites. The hydrothermal zeolitization process essentially is a solution-mediated equilibration between the solid phase of the hydrogel and the crystalline zeolite phase. Based on a wealth of accumulated data, the formation of well-known zeolite frameworks in relation to synthesis conditions can be rationalized using the concept of preassembled units. The precise location of the template action, for example, in the amorphous phase, in solution, or on the growing crystal surface is unknown, simply because of an absence of adequate *in-situ* experimental techniques and the heterogeneous and dynamic nature of the system. It is likely that the mechanism depends on the zeolite, and for the same zeolite type even on the synthesis conditions employed. The chemical details of the formation process of zeolite crystals has intrigued several generations of scientists, and will likely continue to do so for many years.

Acknowledgments

The authors acknowledge the Flemish government for supporting a Concerted Research Action (GOA), the Federal government for an IAP network, and FWO for sponsoring of specific actions in this area.

References

1. R. M. Milton, in *Zeolite Synthesis*, M. L. Occelli, H. E. Robson (Eds.), ACS Symposium Series, Vol. 398, American Chemical Society, Washington, 1989, p. 1.
2. R. M. Barrer, *Hydrothermal Chemistry of Zeolites*, Academic Press, London, 1982, 360 pp.
3. C. S. Cundy, P. A. Cox, *Chem. Rev.* **2003**, *103*, 663.
4. C. S. Cundy, P. A. Cox, *Microporous Mesoporous Mater.* **2005**, *82*, 1.
5. D. W. Breck, *Zeolite Molecular Sieves: Structure, Chemistry and Use*, Wiley, New York, 1973, 771 pp.
6. J. A. Martens, P. A. Jacobs, Synthesis of high-silica aluminosilicate zeolites, in *Studies in Surface Science and Catalysis*, Vol. 33, Elsevier, Amsterdam, 1987, 390 pp.
7. K. Byrappa, M. Yoshimura, *Handbook of Hydrothermal Technology*, Noyes, New York, 2001, 875 pp.
8. (a) K. J. Balkus, in *Progress in Inorganic Chemistry*, K. D. Karlin (Ed.), Vol. 50, John Wiley & Sons, New York, 2001, p. 217; (b) www.iza-online.org.
9. G. J. A. A. Soler-Illia, C. Sanchez, B. Lebeau, J. Patarin, *Chem. Rev.* **2002**, *102*, 4093.
10. P. Caullet, J. L. Paillaud, A. Simon-Masseron, M. Soulard, J. Patarin, *C. R. Chimie* **2005**, *8*, 245.
11. H. Robson (Ed.), *Verified Syntheses of Zeolitic Materials*, Elsevier, Amsterdam, 2001, 266 pp.
12. L. Tosheva, V. P. Valtchev, *Chem. Mater.* **2005**, *17*, 2494.
13. Z. A. D. Lethbridge, J. J. Williams, R. I. Walton, K. E. Evans, C. W. Smith, *Microporous Mesoporous Mater.* **2005**, *79*, 339.
14. G. Kühl, in *Verified Syntheses of Zeolitic Materials*, H. Robson (Ed.), Elsevier, Amsterdam, 2001, p. 19.
15. S. Mintova, N. H. Olson, V. Valtchev, T. Bein, *Science* **1999**, *283*, 958.
16. D. P. Serrano, R. van Grieken, *J. Mater. Chem.* **2001**, *11*, 2391.
17. F. Schüth, *Curr. Opin. Solid State Mater. Sci.* **2001**, *5*, 389.
18. C. J. Brinker, G. W. Scherer, *Sol-Gel Science*, Academic Press, London, 1990, 908 pp.
19. Z. Gabelica, N. Dewaele, L. Maistriau, B. J. Nagy, E. G. Derouane, in *Zeolite Synthesis*, M. L. Occelli, H. E. Robson (Eds.), ACS Symposium Series, Vol. 398, American Chemical Society, Washington, 1989, p. 518.
20. F.-Y. Dai, M. Suzuki, H. Takahashi, Y. Saito, in *Proceedings of the 7th International Zeolite Conference*, Y. Murakami, A. Iijima, J. W. Ward (Eds.), Elsevier, Amsterdam, 1986, p. 223.
21. R. M. Barrer, W. Sieber, *J. Chem. Soc., Dalton Trans.* **1977**, *10*, 1020.

Tab. 1 Molar gel composition and crystallization conditions for the synthesis of zeolite TS-1 [107]

Parameter	Value
$SiO_2 : TiO_2$	5–200
$OH^- : SiO_2$	0.1–1.0
$H_2O : SiO_2$	20–200
$TPA^+ : SiO_2$	0.4–1.0
Crystallization time	6–30 days
Crystallization temperature	403–473 K

22. M. A. Camblor, A. Mifsud, J. Perez-Pariente, *Zeolites* **1991**, *11*, 792.
23. G. Zi, T. Dake, Z. Ruiming, *Zeolites* **1988**, *8*, 453.
24. E. J. P. Feijen, K. De Vadder, M. H. Bosschaerts, J. L. Lievens, J. A. Martens, P. J. Grobet, P. A. Jacobs, *J. Am. Chem. Soc.* **1994**, *116*, 2950.
25. U. Mueller, K. Unger, *Zeolites* **1988**, *8*, 154.
26. J. Dwyer, K. Karim, W. J. Smith, N. E. Thompson, R. K. Harris, D. C. Apperley, *J. Phys. Chem.* **1991**, *95*, 8826.
27. J. C. Jansen, Introduction to zeolite science and practice, in *Studies in Surface Science and Catalysis*, Vol. 58, H. van Bekkum, E. M. Flanigen, J. C. Jansen (Eds.), Elsevier Science, Amsterdam, 1991, p. 77.
28. A. Kuperman, S. Nadimi, S. Oliver, G. A. Ozin, J. M. Garcés, M. M. Olken, *Nature* **1993**, *365*, 239.
29. J. L. Guth, P. Caullet, A. Seive, J. Patarin, F. Delprato, in *Guidelines for Mastering the Properties of Molecular Sieves*, D. Barthomeuf, E. G. Derouane, W. Hölderich (Eds.), NATO ASI Series, Vol. 221, Plenum Press, New York, 1990, p. 69.
30. R. M. Barrer, *Zeolites* **1981**, *1*, 130.
31. D. T. Hayhurst, A. Nastro, R. Aiello, F. Crea, G. Giordano, *Zeolites* **1988**, *8*, 416.
32. J. L. Casci, B. M. Lowe, *Zeolites* **1983**, *3*, 186.
33. F. Delprato, L. Delmotte, J. L. Guth, L. Huve, *Zeolites* **1990**, *10*, 546.
34. W. Löwenstein, *Am. Mineral.* **1954**, *39*, 92.
35. B. Stringham, *Econ. Geol* **1952**, *47*, 661.
36. F. G. Dwyer, P. Chu, *J. Catal.* **1979**, *59*, 263.
37. A. Katovic, B. Subotic, I. Smit, Lj. A. Despotovic, M. Curic, in *Zeolite Synthesis*, M. L. Occelli, H. E. Robson (Eds.), ACS Symposium Series, Vol. 398, American Chemical Society, Washington, 1989, p. 124.
38. D. M. Ginter, C. J. Radke, A. T. Bell, Zeolites: Facts, figures, future, in *Studies in Surface Science and Catalysis*, P. A. Jacobs, R. A. van Santen (Eds.), Elsevier Science, Amsterdam, 1989, p. 161.
39. J. P. Gilson, in *Zeolite Microporous Solids: Synthesis, Structure, and Reactivity*, E. G. Derouane, F. Lemos, C. Naccache, F. R. Ribeiro (Eds.), NATO ASI Series, Vol. 352, Plenum Press, New York, 1992, p. 19.
40. C. T. G. Knight, R. J. Kirkpatrick, E. Oldfield, *J. Magn. Reson.* **1988**, *79*, 31.
41. J. Livage, Advanced zeolite science and applications, in *Studies in Surface Science and Catalysis*, Vol. 85, J. C. Jansen, M. Stöcker, H. G. Karge, J. Weitkamp (Eds.), Elsevier, Amsterdam, 1994, p. 1.
42. A. V. McCormick, A. T. Bell, *Catal. Rev.-Sci. Eng.* **1989**, *31*, 97.
43. J. L. Guth, P. Caullet, P. Jacques, R. Wey, *Bull. Soc. Chim. Fr.* **1980**, *3-4*, 121.
44. G. Harvey, L. S. D. Glasser, in *Zeolite Synthesis*, M. L. Occelli, H. E. Robson (Eds.), ACS Symposium Series, Vol. 398, American Chemical Society, Washington, 1989, p. 49.
45. A. T. Bell, in *Zeolite Synthesis*, M. L. Occelli, H. E. Robson (Eds.), ACS Symposium Series, Vol. 398, American Chemical Society, Washington, 1989, p. 66.
46. W. M. Meier, in *Molecular Sieves*, Society of Chemical Industry, London, 1968, p. 10.
47. C. T. G. Knight, S. D. Kinrade, *J. Phys. Chem. B* **2002**, *106*, 3329.
48. P. K. Dutta, D. C. Shieh, M. Puri, *J. Phys. Chem.* **1978**, *91*, 2332.
49. S. P. Zhadanov, N. N. Samulevich, in *Proceedings of the 5th International Zeolite Conference*, L. V. C. Rees (Ed.), Heyden, London, 1980, p. 75.
50. B. M. Lok, T. R. Cannan, C. A. Messina, *Zeolites* **1983**, *3*, 282.
51. H. Gies, B. Marler, *Zeolites* **1992**, *12*, 42.
52. R. F. Lobo, M. Pan, I. Chan, R. C. Medrud, S. I. Zones, P. A. Crozier, M. E. Davis, *J. Phys. Chem.* **1994**, *98*, 12040.
53. S. I. Zones, M. M. Olmstead, D. S. Santilli, *J. Am. Chem. Soc.* **1992**, *114*, 4195.
54. D. E. W. Vaughan, Catalysis and adsorption by zeolites, in *Studies in Surface Science and Catalysis*, Vol. 65, G. Öhlmann, H. Pfeifer, R. Fricke (Eds.), Elsevier Science, Amsterdam, 1991, p. 275.
55. R. K. Iler, *The Chemistry of Silica*, Wiley, New York, 1979, 896 pp.
56. J. Turner, A. K. Soper, J. L. Finney, *Mol. Phys.* **1990**, *70*, 679.
57. F. Schulz, *J. Phys. Chem* **2002**, *3*, 98.
58. A. V. McCormick, A. T. Bell, C. J. Radke, in *Perspectives in Molecular Sieve Science*, W. H. Flank, T. E. Whyte (Eds.), ACS Symposium Series, Vol. 368, American Chemical Society, Washington, 1988, p. 222.
59. S. D. Kinrade, C. T. G. Knight, D. L. Pole, R. Syvitski, *Inorg. Chem.* **1998**, *37*, 4272.
60. S. D. Kinrade, C. T. G. Knight, D. L. Pole, R. Syvitski, *Inorg. Chem.* **1998**, *37*, 4278.
61. S. Caratzoulas, D. Vlachos, M. Tsapatsis, *J. Phys. Chem. B* **2005**, *109*, 10429.
62. C. E. A. Kirschhock, R. Ravishankar, F. Verspeurt, P. J. Grobet, P. A. Jacobs, J. A. Martens, *J. Phys. Chem. B* **1999**, *103*, 4960.
63. I. Hasegawa, S. Sakka, in *Zeolite synthesis*, M. L. Occelli, H. E. Robson (Eds.), ACS Symposium Series, Vol. 398, American Chemical Society, Washington, 1989, p. 140.
64. A. Corma, F. Rey, J. Rius, M. J. Sabater, S. Valencia, *Nature* **2004**, *431*, 287.
65. F. Delprato, L. Delmotte, J. L. Guth, L. Huve, *Zeolites* **1990**, *10*, 546.
66. (a) D. W. Lewis, D. J. Willock, C. R. A. Catlow, J. M. Thomas, G. J. Hutchings, *Nature* **1996**, *382*, 604; (b) D. W. Lewis, G. Sankar, J. K. Wyles, J. M. Thomas, C. R. A. Catlow, D. Willock, *Angew. Chem., Int. Ed. Engl.* **1997**, *36*, 2675; (c) D. J. Willock, D. W. Lewis, C. R. A. Catlow, G. J. Hutchings, J. M. Thomas, *J. Mol. Catal. A: Chemical* **1997**, *A119*, 415.
67. M. T. Melchior, D. E. W. Vaughan, C. F. Pictroski, *J. Phys. Chem.* **1995**, *99*, 6128.
68. S. Sugiyama, S. Yamamoto, O. Matsuoka, H. Nozoye, J. Yu, G. Zhu, S. Qiu, O. Terasaki, *Microporous Mesoporous Mater.* **1999**, *28*, 1.
69. T. Gibbs, D. W. Lewis, *Chem. Commun.* **2002**, *22*, 2660.
70. R. W. Thompson, in *Verified Syntheses of Zeolitic Materials*, H. Robson (Ed.), Elsevier, Amsterdam, 2001, p. 21.
71. T. Ohsuna, B. Slater, F. Gao, J. Yu, Y. Sakamoto, G. Zhu, O. Terasaki, D. E. W. Vaughan, S. Qiu, C. R. A. Catlow, *Chem. Eur. J.* **2004**, *10*, 5031.
72. A. Firouzi, F. Atef, A. G. Oertli, G. D. Stucky, B. F. Chmelka, *J. Am. Chem. Soc.* **1997**, *119*, 3596.
73. P. A. Jacobs, in *Zeolite Microporous Solids: Synthesis, Structure, and Reactivity*, E. G. Derouane, F. Lemos, C. Naccache, F. R. Ribeiro (Eds.), NATO ASI Series, Vol. 352, Plenum Press, New York, 1992, p. 3.
74. E. J. P. Feijen, J. A. Martens, P. A. Jacobs, Zeolites and Related Microporous Materials: State of the Art 1994, in *Studies in Surface Science and Catalysis*, Vol. 84, J. Weitkamp,

H. G. Karge, H. Pfeifer, W. Hölderich (Eds.), Elsevier, Amsterdam, 1994, p. 3.
75. H. Kacirek, H. Lechert, *J. Phys. Chem.* **1975**, *79*, 1589.
76. R. A. Van Santen, J. Keijsper, G. Ooms, A. G. T. G. Kortbeek, in *Proceedings of the 7th International Zeolite Conference*, Y. Murakami, A. Iijima, J. W. Ward (Eds.), Kodansha, Tokyo, and Elsevier, Amsterdam, 1986, p. 169.
77. P.-P. E. A. de Moor, T. P. M. Beelen, R. A. van Santen, *Microporous Mater.* **1997**, *9*, 117.
78. W. H. Dokter, H. F. van Garderen, T. P. M. Beelen, R. A. van Santen, W. Bras, *Angew. Chem.* **1995**, *34*, 73.
79. T. P. M. Beelen, W. H. Dokter, H. F. van Garderen, R. A. van Santen, in *Synthesis of Porous Materials, Zeolites, Clays, and Nanostructures*, M. I. Occelli, H. Kessler (Eds.), Dekker Inc., New York, 1997, p. 59.
80. B. J. Schoeman, J. Sterte, J.-E. Otterstedt, *Zeolites* **1994**, *14*, 568.
81. J. M. Fedeyko, J. D. Rimer, R. F. Lobo, D. G. Vlachos, *J. Phys. Chem. B* **2004**, *108*, 12271.
82. S. Yang, A. Navrotsky, D. J. Wesolowski, J. A. Pople, *Chem. Mater.* **2004**, *16*, 210.
83. J. N. Watson, A. S. Brown, L. E. Iton, J. W. White, *J. Chem. Soc. Faraday Trans.* **1998**, *94*, 2181.
84. B. J. Schoeman, *Zeolites* **1997**, *18*, 97.
85. O. Regev, Y. Cohen, E. Kehat, Y. Talmon, *Zeolites* **1994**, *14*, 314.
86. C. E. A. Kirschhock, V. Buschmann, S. P. B. Kremer, R. Ravishankar, J. V. Houssin, B. L. Mojet, R. A. van Santen, P. J. Grobet, P. A. Jacobs, J. A. Martens, *Angew. Chem. Int. Ed.* **2001**, *40*, 2637.
87. C. E. A. Kirschhock, S. P. B. Kremer, J. Vermant, G. Van Tendeloo, P. A. Jacobs, J. A. Martens, *Chem. Eur. J.* **2005**, *11*, 4306.
88. E. J. J. Groenen, A. G. T. G. Kortbeek, M. Mackay, O. Sudmeijer, *Zeolites* **1986**, *6*, 403.
89. P. A. Jacobs, E. G. Derouane, J. Weitkamp, *Chem. Soc., Chem. Commun.* **1981**, *12*, 591.
90. C. D. Chang, A. T. Bell, *Catal. Lett.* **1991**, *8*, 305.
91. S. L. Burkett, M. E. Davis, *J. Phys. Chem.* **1994**, *98*, 4647.
92. A. E. Persson, B. J. Schoeman, J. Sterte, J.-E. Otterstedt, *Zeolites* **1994**, *14*, 557.
93. V. Nikolakis, E. Kokkoli, M. Tirrell, M. Tsapatsis, D. G. Vlachos, *Chem. Mater.* **2000**, *12*, 845.
94. P.-P. E. A. de Moor, T. P. M. Beelen, R. A. van Santen, K. Tsuji, M. E. Davis, *Chem. Mater.* **1999**, *11*, 36.
95. F. Testa, R. Szostak, R. Chiappetta, R. Aiello, A. Fonseca, J. B. Nagy, *Zeolites* **1997**, *18*, 106.
96. S. Mintova, N. Petkov, K. Karaghiosoff, T. Bein, *Mater. Sci. Eng.* **2002**, *C19*, 111.
97. S. Mintova, N. H. Olson, V. Valtchev, T. Bein, *Science* **1999**, *283*, 958.
98. V. Valtchev, *J. Phys. Chem. B* **2004**, *108*, 15587.
99. Q. Li, D. Creaser, J. Sterte, *Chem. Mater.* **2002**, *14*, 1319.
100. (a) Y. X. Wang, H. Gies, B. Marler, U. Müller, *Chem. Mater* **2005**, *17*, 43; (b) J. F. M. Denayer, R. A. Ocakoglu, I. C. Arik, C. E. A. Kirschhock, J. A. Martens, G. V. Baron, *Angew. Chem. Int. Ed.* **2005**, *117*, 404; (c) A. Corma, V. Fornés, M. S. Galletero, H. Gárcia, C. J. Gómes-García, *Phys. Chem. Chem. Phys.* **2001**, *3*, 1218; (d) M. A. Camblor, A. Corma, M.-J. Diaz-Cabanas, *J. Phys. Chem. B* **1998**, *102*, 44; (e) L. Schreyeck, P. Caullet, J. C. Mougenel, J.-L. Guth, B. Marler, *Microporous Mater.* **1996**, *6*, 259.
101. J.-P. Gilson, M. Guisnet, *Zeolites for Cleaner Technologies*, Imperial College, London, 2002, 388 pp.
102. P. Kripylo, K.-P. Wendlandt, F. Vogt, *Heterogene Katalyse in der Chemischen Technik*, Deutscher Verlag für Grundstoffindustrie, Leipzig, Stuttgart, 1993, 342 pp.
103. D. E. W. Vaughan, *Chem. Eng. Prog.* **1988**, *84*(2), 25.
104. N. Y. Chen, Th. F. Degnan Jr., C. Morris-Smith, *Molecular Transport and Reaction in Zeolites: Design and Application of Shape Selective Catalysts*, VCH, Weinheim, 1994, p. 20.
105. F.-Y. Dai, M. Suzuki, H. Takahashi, Y. Saito, in *Zeolite Synthesis*, M. L. Occelli, H. E. Robson (Eds.), ACS Symposium Series, Vol. 398, American Chemical Society, Washington, 1989, p. 245.
106. U. Müller, A. Brenner, A. Reich, K. K. Unger, in *Zeolite Synthesis*, M. L. Occelli, H. E. Robson (Eds.), ACS Symposium Series, Vol. 398, American Chemical Society, Washington, 1989, p. 346.
107. A. Tissler, P. Polanek, U. Girrbach, U. Müller, K. K. Unger, Zeolites as Catalysts, Sorbents and Detergent Builders, in *Studies in Surface Science and Catalysis*, Vol. 46, H. G. Karge, J. Weitkamp (Eds.), Elsevier Science, Amsterdam, 1989, p. 399.
108. R. J. Argauer, G. R. Landolt, US Patent 3, 702, 886, assigned to Mobil Oil, 1972.
109. M. Taramasso, G. Perego, B. Notari, US Patent 4, 410, 501, assigned to Enichem, 1983.
110. D. R. C. Huybrechts, R. F. Parton, P. A. Jacobs, Chemistry of Microporous Crystals, in *Studies in Surface Science and Catalysis*, Vol. 60, T. Inui, S. Namba, T. Tatsumi (Eds.), Kodansha, Tokyo, and Elsevier, Amsterdam, 1991, p. 225.
111. G. Bellussi, V. Fattore, Zeolite Chemistry and Catalysis, in *Studies in Surface Science and Catalysis*, Vol. 69, P. A. Jacobs, N. I. Jaeger, L. Kubelková, B. Wichterlová (Eds.), Elsevier, Amsterdam, 1991, p. 79.

2.3.6
Ordered Mesoporous Materials

*Freddy Kleitz**

2.3.6.1 Introduction

Porous materials were originally defined in terms of their adsorption properties. From this, porous materials are distinguished by their pore size. According to the IUPAC definition [1], porous solids are divided into three classes: microporous (<2 nm), mesoporous (2–50 nm), and macroporous (>50 nm) materials. In addition, the term "nanoporous", referring to pores in the nanometer size range, is nowadays increasingly being used. Ideally, a porous material should have a narrow pore size distribution which is critical for size-specific applications, and a readily tunable pore size allowing flexibility for host–guest interactions. In addition, it should have high chemical, thermal and mechanical stabilities, with high surface area and large pore volumes [2, 3]. Furthermore, the material should have appropriate particle size and

---
* Corresponding author.

morphology. Well-known examples of microporous solids are zeolites, which are crystalline aluminosilicates [4–7]. Some related compounds usually possess similar structures with different framework compositions, such as aluminophosphates. Due to their crystalline network, zeolites and related materials exhibit an extremely narrow and uniform pore size distribution which allows their successful exploitation for size-specific applications in adsorption, molecular sieving, host–guest chemistry and shape-selective catalysis [8–13]. In particular, zeolites have attracted much attention as solid acids and redox catalysts. However, a major drawback for application is their limited pore size (<1.5 nm), which excludes size-specific processes involving large molecules. Larger pore sizes are required for heavy oil cracking and catalytic conversion of large molecules, and other applications as separation media or host for large molecules and macromolecules. Also, severe mass transfer limitations are observed for microporous catalysts when exploited in liquid-phase systems. In contrast, sol–gel-derived porous oxides are non-crystalline solids which, however, offer distinct advantages in terms of processing – that is, tailoring on a macroscopic size scale to create membranes, monoliths, or fibers. In addition, they can be tailored on the molecular size scale [4, 14]. Such amorphous porous gels or aerogels and porous glasses can be made as mesoporous solids. In contrast to zeolites, they have disordered pore systems and exhibit therefore a rather broad pore size distribution. These materials are commonly used in separation processes and as catalyst supports.

Numerous attempts have been made to extend the hydrothermal synthesis procedures used to prepare zeolite-type materials to the mesopore range, but the success was limited [13]. Further attempts to achieve mesoporous materials with narrow pore size distribution were based on controlling the pore size of amorphous silica-aluminas by preparing them in the presence of alkyltrimethylammonium cations [15]. Disordered silica-aluminas with pores in the mesopore region and a quite regular distribution were obtained with high surface areas. Other ordered non-crystalline mesoporous oxide molecular sieves were synthesized by the intercalation of layered materials (pillaring) such as double hydroxides, titanium and zirconium phosphates or clays. However, these exhibit broad mesopore-size distributions [2, 13]. The introduction of supramolecular assemblies (i.e., micellar aggregates, rather than molecular species) as structure-directing agents (SDAs) allowed the synthesis of a new family of mesoporous silica and aluminosilicate compounds which was designated as M41S. These solids are characterized by ordered arrangements of mesopores exhibiting narrow pore size distribution. This discovery proved to be a major breakthrough which subsequently opened up a whole field of research and new possibilities in many areas of chemistry and materials science.

2.3.6.2 Ordered Mesoporous Molecular Sieves: MCM-41

Originally, the synthesis of an ordered mesoporous material was described in a patent filed in 1969 [16], but due to a lack of analytical techniques these early scientists were unable to recognize the remarkable features of their product. In 1992, a similar material was obtained by Mobil Corporation scientists [17, 18]. MCM-41 (Mobil Composition of Matter No. 41) exhibits a highly ordered hexagonal array of unidimensional cylindrical pores with a narrow pore size distribution (Fig. 1), although the walls are amorphous. Other related phases, such as cubic (MCM-48) and lamellar (MCM-50) were also reported in early publications. The characteristic approach for the synthesis of ordered mesoporous materials is the use of liquid crystal-forming templates that enable the specific formation of pores with predetermined size. In contrast to zeolite synthesis, where single molecules act as templates, aggregates of surfactant molecules are used to obtain the mesoporous materials. At approximately the same time as the discovery of MCM-41, an alternative approach to mesoporous materials was described by Yanagisawa et al. [19], who used kanemite, a layered silicate, as the silica source. The material is designated as FSM-n (Folded Sheet Mesoporous Materials-n), where n is the number of carbon atoms in the chain of the surfactant used. MCM-41 and FSM-16 are very similar, though with some slightly different adsorption and surface properties [20].

The standard use of transmission electron microscopy (TEM), powder X-ray diffraction (XRD) and gas physisorption analysis allows independent and reliable characterization of mesoporous materials. The hexagonal arrangement of uniform pores can be visualized by TEM, as shown in Fig. 1. The pore sizes usually vary between 2 and 10 nm depending on the alkylammonium surfactant chain length, the presence of organic additives, or post-treatments. The pore wall thickness is usually evaluated to be between 0.7 and 1.1 nm. MCM-41-type mesoporous silicas are highly porous and commonly show BET (Brunauer-Emmett-Teller [21]) surface areas exceeding 1000 m^2 g^{-1} and pore volumes up to 1 cm^3 g^{-1}. The N$_2$ sorption isotherm measured for MCM-41 is distinctive, being a type IV isotherm which is generally observed for mesoporous materials [1], but with an unusually sharp capillary condensation step (see Fig. 1). MCM-41 exhibits long-range periodicity in two directions. The regular arrangement of the pores can be considered as a type of "super-structure" with long-range ordering. Along the

References see page 214

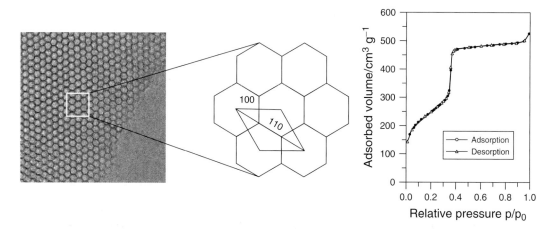

Fig. 1 Left: Transmission electron microscopy image of MCM-41. (Image courtesy Dr. Y. Sakamoto, Stockholm University, Sweden.) Center: Representation of the hexagonal lattice. Right: Nitrogen physisorption isotherm measured at 77 K of calcined MCM-41. (Adapted from Ref. [24].)

c-axis, which is the axial direction of the pores, no periodicity is observed. The long-range ordering of MCM-41 can be evidenced using XRD, where Bragg reflections can be detected at low 2θ angles. The only reflections observed are due to the ordered hexagonal array of parallel pore walls and can be indexed as a hexagonal $p6mm$ structure. No reflections at higher 2θ are observed.

Subsequent developments proved that ordered mesoporous materials could also be formed in an acidic medium, or by using neutral amines, non-ionic surfactants, or block copolymers as the SDAs. In addition, heteroatoms, such as Al or Ti, could be incorporated into the mesostructured silicate framework, as is done with zeolitic frameworks, and surface functionalization was achieved by incorporating organic and organometallic groups. The concepts of mesoporous materials were also extended to hybrid organic–inorganic frameworks and fully non-siliceous frameworks (e.g., titania, alumina, carbon). However, the removal of the template and the stability of the inorganic networks represent, in this latter case, major issues [22]. Numerous excellent reviews have been published which cover all aspects of the ordered mesoporous materials, such as their synthesis, characterization, modification, and their applications in catalysis or in other fields [13, 22–37].

2.3.6.3 Synthesis of Ordered Mesoporous Materials

2.3.6.3.1 Synthesis Strategies for Mesostructure Formation

Ordered mesoporous materials with various compositions are accessible by following different routes. The three major synthesis strategies used to construct ordered mesoporous materials are illustrated in Fig. 2, according to the description given by Schüth [22]. Since the discovery of MCM-41, there has been impressive progress in the development of many new mesoporous solids based on templating mechanisms related to the one used for the original MCM-41 synthesis. The majority of these routes consist in the use of organic precursor species which allow the formation of liquid crystals, such as molecular surfactants or block copolymers. Two general pathways exist:

- The surfactant-inorganic composite mesophase composite forms cooperatively from the species present in solution (i.e., cooperative self-assembly), which are not in a liquid crystalline state prior to mixing of the precursors.
- A liquid crystalline precursor phase is used which is infiltrated with the inorganic species (so-called true liquid crystal templating).

The self-assembly process of the template is followed by (or synchronized with) the formation of the inorganic network deposited around the "self-assembled substrate". Inorganic replication occurs at the accessible interfaces built by the preorganized or self-assembled templates. In many cases, a cooperative self-assembly takes place *in situ* between the templating species and the mineral network precursors, with synchronized self-assembly and inorganic network formation, yielding the highly organized mesoscopic architectures.

The first ordered mesoporous materials (e.g., MCM-41 and FSM-16) were prepared from ionic surfactants, such as quaternary alkylammonium ions. There, the formation of the inorganic–organic hybrid mesophase is based on electrostatic interactions between the positively charged surfactant molecules and the negatively charged silicate species in solutions. Depending on the synthesis

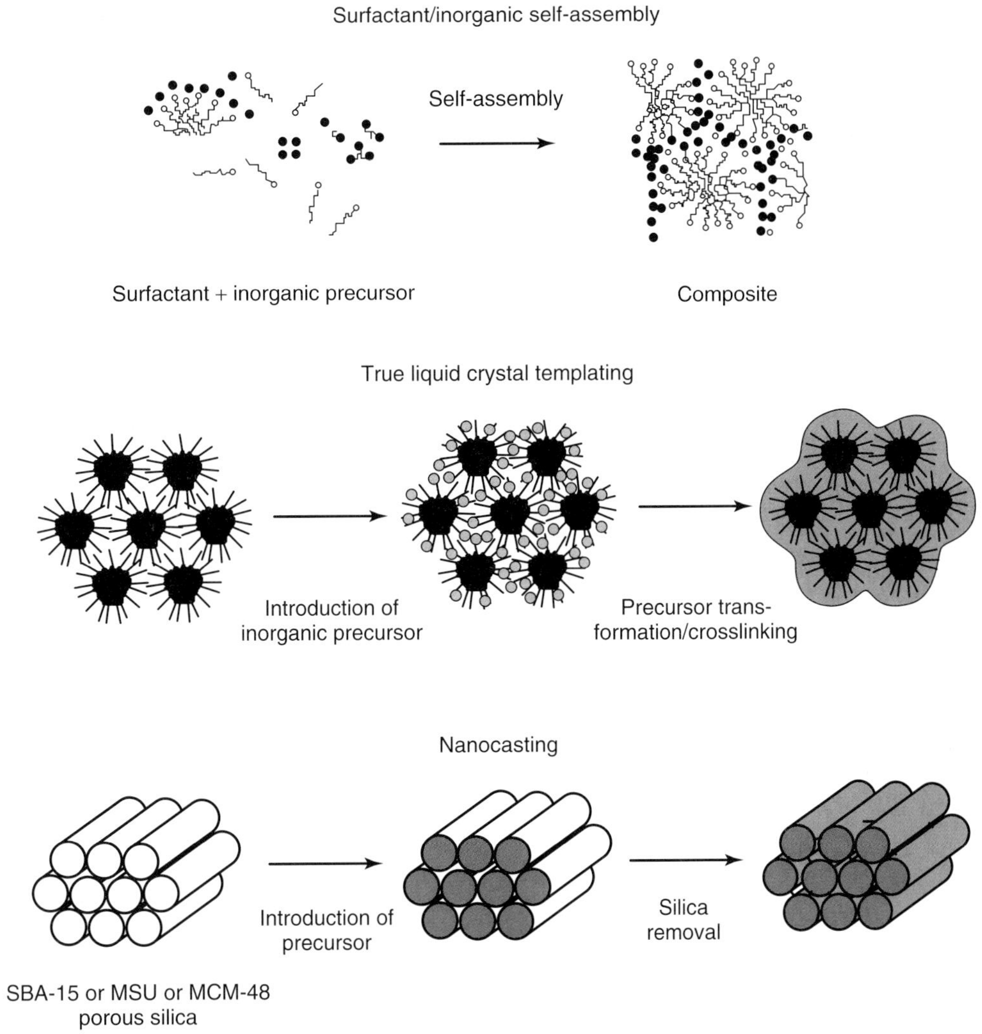

Fig. 2 Schematic representation of the pathways leading to ordered mesoporous materials, as described by Schüth. (Reprinted with permission from Ref. [22]; © 2001, American Chemical Society.)

conditions, the silicon source or the type of surfactant used, many other mesoporous materials can be synthesized following the cooperative assembly pathways. The cooperative formation mechanism was extended by Stucky et al. to a whole series of other electrostatic assembly mechanisms and counterion-mediated pathways. Many silica-based mesoporous materials, as well as non-silica frameworks, have been reported to be formed via these electrostatic assembly pathways. Pinnavaia et al. developed two additional approaches for the synthesis of mesoporous materials on the basis on non-ionic organic–inorganic interactions (H-bondings or dipolar). These authors used neutral surfactants such as primary amines and poly(ethylene oxides) to prepare HMS (Hexagonal Mesoporous Silica) and MSU (Michigan State University) materials. Later, a two-dimensional (2-D) hexagonal mesoporous material (denoted as SBA-15 for Santa Barbara No. 15) was also formed via this pathway. SBA-15 silica exhibits a thick walls of 3 to 7 nm and large pore sizes that are adjustable between 6 and about 15 nm. The amphiphilic triblock copolymer poly(ethylene oxide)-*block*-poly(propylene oxide)-*block*-poly(ethylene oxide)$_x$, $(EO)_{20}$-$(PO)_{70}$-$(EO)_{20}$ (EO = ethylene oxide; PO = propylene oxide) was employed as the SDA in highly acidic media. In addition to these cooperative pathways, the true liquid crystal templating (TLCT) pathway and the nanocasting route that uses already formed ordered mesoporous materials as hard templates, have also been developed during the past 10 years. In the case of the TLCT, an already preformed surfactant liquid crystalline (meso)phase is used, which is subsequently loaded with the precursor for the

References see page 214

inorganic materials [38]. This pathway had already been suggested as one possible mechanism for the formation of MCM-41 in the original publication by the Mobil group. The nanocasting route, on the other hand, which was first reported by Ryoo et al. in 1999 [39], is a clearly distinct method. Here, no "soft" surfactant template is used, but instead the pore system of an ordered mesoporous silica is used as a hard template serving as a true mold for preparing compositions such as metals, carbons, or transition metal oxides. This nanocasting route is very suitable for creating frameworks that are difficult to access using liquid crystal-based templates (i.e., reactive oxides, polymers, carbons, non-oxides, metals, etc.). This method – which is highly versatile for the preparation of non-siliceous mesostructured and mesoporous materials – will be discussed in detail in Section 2.3.6.6.

2.3.6.3.2 Inorganic Polymerization and Self-Assembly with Surfactants
In the case of ordered mesoporous oxides, the templating relies on supramolecular arrays: micellar systems formed by surfactants or block copolymers. The formation of an inorganic–surfactant composite mesophase is controlled by the chemistry of both the inorganic and organic parts of the system, and the way in which the surfactant molecules interact with the inorganic species. MCM-41-type materials generally result from a process of silica polymerization during which (liquid crystal) supramolecular templating is simultaneously proceeding. The overall mesophase formation is driven by a cooperative assembly of organic surfactant molecules and inorganic solution species.

A Inorganic Polymerization (the Case of Silica) The methods employed to prepare mesoporous oxide molecular sieves are similar to those commonly used also for sol–gel-derived oxides [14]. Both types of material consist of non-crystalline oxidic frameworks, the main difference being the degree of order since porous sol–gel oxides are completely disordered. As mesoporous silicas were the first to be developed and the most commonly employed, the sol–gel process will be described here only for silicates. The sol–gel process related to other compositions is detailed in Ref. [40]. The silicon source usually plays an important role by determining the reaction conditions. For non-molecular silicon sources, silica is obtained as a gel formed from a non-homogeneous solution, and the gel is subsequently treated hydrothermally. In the case of molecular silicon sources, water, surfactant, and catalyst are first combined to form a homogenous micellar solution. To this solution, the silicon alkoxide is added and mesophase formation follows. In both cases, the first step for inorganic polymerization is the formation of silanol groups. This occurs either by a neutralization reaction or by hydrolysis of the alkoxysilane. In this latter case, hydrolysis of monomeric tetrafunctional silicon alkoxide precursors proceeds as in Scheme 1. Hydrolysis is generally catalyzed by a mineral acid (HCl) or a base (NaOH, NH_3).

The hydrolysis rate is furthermore affected by the nature of the silicon precursor. For example, the more hydrophobic or sterically hindered the precursor is, the slower the hydrolysis rate. Depending on the synthesis conditions, hydrolysis may be only partial or go to completion. In both cases, the silicic acid formed will undergo self-condensation or condensation with an alkoxysilane molecule producing siloxane bonds (Si–O–Si) and alcohol or water (Scheme 2). This type of condensation reactions leads to the formation of oligomeric species, which form chains, rings or branched structures and continues to build larger polymeric silicates [14].

The overall reaction proceeds as a polycondensation to form soluble higher molecular-weight polysilicates; this resulting colloidal dispersion is the *sol*. The polysilicates eventually link together to form a three-dimensional (3-D) network which spans the container and is usually filled with solvent molecules, called the *gel*, or precipitates as precipitated silica. Sol–gel polymerization proceeds in several, possibly overlapping, steps: polymerization to form primary particles, growth of the particles, and finally aggregation [14]. Each step strongly depends on pH, temperature, concentration, and co-solvent effects. Hydrolysis and condensation depend most strongly on the nature and the concentration of the catalysts, resulting in a pH dependence. The pH dependence of the polymerization reactions, which has been recognized for colloidal silica–water systems can be employed in the synthesis of silicate–surfactant mesophases. In this regard, the polymerization process can be divided into three pH domains, as described by Iler [4]: pH < 2; pH = 2–7; and pH > 7. pH 2 corresponds approximately to the isoelectric point (IEP) of silica (the electrical mobility of the silica particles equals zero), and the point of zero

$$\text{hydrolysis}$$
$$\equiv Si-OR + H_2O \longrightarrow \equiv Si-OH + ROH$$

Scheme 1

$$\text{alcohol condensation (alcoxolation)}$$
$$\equiv Si-OR + HO-Si\equiv \longrightarrow \equiv Si-O-Si\equiv + ROH$$

$$\text{water condensation (oxolation)}$$
$$\equiv Si-OH + HO-Si\equiv \longrightarrow \equiv Si-O-Si\equiv + H_2O$$

Scheme 2

charge (PZC, when the surface charge is zero). Below pH 2, the silicate species are positively charged, while above pH 2 they are negatively charged. pH 7 serves as a boundary, as above this value solubility of the silicates increases strongly with the pH and the condensed species are likely to be ionized.

B Template-Assisted Synthesis A successful approach to generate tailored pores is based on the use of templates or "imprints" following biological models. The first success in templating was achieved in 1949 by the use of bioorganic molecules to create artificial antibodies [41]. The principles of templating have since then been adapted widely for the synthesis of many organic or inorganic materials. In most cases, templating comprises the use of a synthesis solution and a template molecule or an assembly of molecules. The solutions from which the templated solid is produced contain precursors which allow some form of solidification that can be precipitation from solution, sol–gel synthesis of inorganic materials, redox processes leading to metal deposition, or polymerization of organic monomers. During this time, morphological construction occurs by direct imprinting of the shape and texture of the template. A template is thus described as a central structure around which a network forms. The cavity created after the removal of the template retains morphological and stereochemical features of the central structure [42, 43]. Zeolites and zeotype materials are synthesized in the presence of a single small organic template or SDA, typically quaternary alkyl ammonium ions [44]. The porosity is subsequently created by the removal of the template. However, examples of templating with exact geometrical correspondence are rare for zeolites, as the internal cavities usually do not conform rigorously to the molecule shape. Individual organic molecules do not act as true templates, but more typically direct structures by participating in the ordering of the reagents, and fill space in the porous product. Consequently, it results in a rather indirect correlation of the shape and size of the organic molecule to the structure and volume of the cavity created in the inorganic framework. Initially, supramolecular assemblies [45] were considered to be promising templates. In general, these assemblies are the result of an association of a large number of small molecular building blocks into a specific phase with specific macroscopic characteristics and a well-defined microscopic organization, for example liquid crystalline (LC) phases. Supramolecular engineering rapidly provided access to the controlled generation of well-defined polymolecular architectures in layers, films, membranes, micelles, or mesophases [46, 47]. The generation of ordered mesoporous materials is possible via templating by self-assembled liquid crystalline phases.

In this case, there is a geometric correlation between the surfactants array size and shape and the final pore size and geometry in the mesophase. Two main types of organic template (SDAs) are used in the elaboration of ordered mesoporous solids: molecular-based surfactant systems and polymeric templates.

Surfactants consist of a hydrophilic region, for example ionic, non-ionic, zwitterionic or polymeric groups, often called the "head", and a hydrophobic region, the "tail", for example alkyl or polymeric chains. This amphiphilic character enables surfactant molecules to associate in supramolecular micellar arrays [24]. Single amphiphilic molecules tend to associate into aggregates in aqueous solution due to hydrophobic effects. Above a certain critical concentration of amphiphiles, the formation of an assembly, such as a spherical micelle, is favored. Within these nanometric aggregates, the surfactant molecules are arranged such that the heads form the outer surface facing the water and the tails are clustered together, pointing toward the center. (Note: the term *aggregate* is used for the supramolecular array formed upon self-assembly of single surfactant molecules.) Micellar aggregates organize according to different shapes (spherical or cylindrical micelles, lamellae, etc.), which permits the coexistence of two incompatible phases (polar and apolar). Some typical micellar structures are represented in Fig. 3 [48].

The formation of micelles, their shape, and their aggregation into liquid crystals all depend on the surfactant concentration. At very low concentration, the surfactant is present as free molecules dissolved in solution and adsorbed at interfaces. At slightly higher concentration, called the critical micelle concentration (CMC), the individual molecules form small spherical aggregates. At higher concentrations, spherical micelles eventually coalesce to form elongated cylindrical, rod-like micelles. These transition concentrations depend strongly on temperature, surfactant chain length and surfactant counterion binding strength. With further increasing concentrations, liquid crystalline phases form. Initially, the rod-like micelles aggregate in hexagonal, close-packed arrays, but as the concentration increases further, the cubic phases form followed by lamellar phases [24, 43]. Details of this sequence might vary, depending on the surfactant, but in general it is valid for most of the systems. The architecture of the final materials will rely directly on the nature of the surfactant molecules; that is, the morphology of the micellar aggregates and the interactions at the inorganic–organic interface. Furthermore, efficient template removal and faithful imprinting have been shown to depend largely on the nature of the interactions between the template and the embedding matrix, and the ability of the matrix

References see page 214

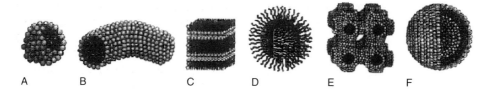

Fig. 3 Diversity of the micellar structures. A, spherical micelle; B, cylindrical micelle; C, bilayer lamellae; D, reverse micelle; E, bicontinuous cubic phase; F, vesicular-liposomes. (Reprinted with permission from Ref. [48]; © 1984, John Wiley & Sons.)

to adapt to the template. The intimate template–matrix association required for supramolecular templating of inorganic mesophases is generally facilitated by the flexibility of amorphous inorganic networks with low structural constrains (e.g., small inorganic oligomers), and by the large radius of curvature of the organic template. Ideally, following removal of the core molecules from the surrounding matrix, the shape of the voids that remain reflects the shape of the template.

The first syntheses of mesoporous materials were carried out using ionic surfactants such as cationic alkytrimethylammonium ($C_n TA^+$, $n = 8-20$), anionic alkylsulfonates ($C_n SO_3^{2-}$, $n = 12-18$) or long-chain alkylphosphates as SDAs. Two main limitations, however, are evident for the resulting mesostructured materials: a typical wall thickness of 0.8–1.3 nm, which is a limitation regarding stability for catalysis (thermal and hydrothermal stability), and a quite limited pore size. The only way to exceed 5 nm in pore size consisted in employing swelling agents (such as trimethylbenzene), although poorly reproducible syntheses led to low-quality materials. Thus, more versatile templates were intensively sought. Amphiphilic block copolymers belong to an important family of surfactants that are widely used in detergents and emulsifiers. The amphiphilic block copolymers are able to self-assemble, much like their molecular surfactant counterparts, in a variety of morphologies (spherical or cylindrical micelles, lamellar structures, hexagonal structures, gyroids, body-centered micellar cubic, etc.) [49]. The organized systems formed by these amphiphilic block copolymers are excellent templates for the structuring of inorganic networks. Diblock or triblock copolymers (Fig. 4) are the most often used, in which hydrophilic blocks (polyethylene oxide, PEO, or polyacrylic acid, PAA) are associated with hydrophobic blocks (polypropylene oxide, PPO or polystyrene, PS, for example).

The self-assembly characteristics of these block copolymers can be tuned by adjusting solvent compositions, synthesis temperature, molecular weight or polymer architecture. One of the most commonly employed triblock copolymers is poly(ethylene oxide)$_x$-poly(propylene oxide)$_y$-poly(ethylene oxide)$_x$ [EO)$_x$-(PO)$_y$-(EO)$_x$; trade name: Pluronics] which shows an ability to form LC structures. These can be used under acidic conditions to synthesize, via the H-bonding-based pathways, a variety of different mesoporous solids with large pores, large wall thicknesses, and various framework compositions.

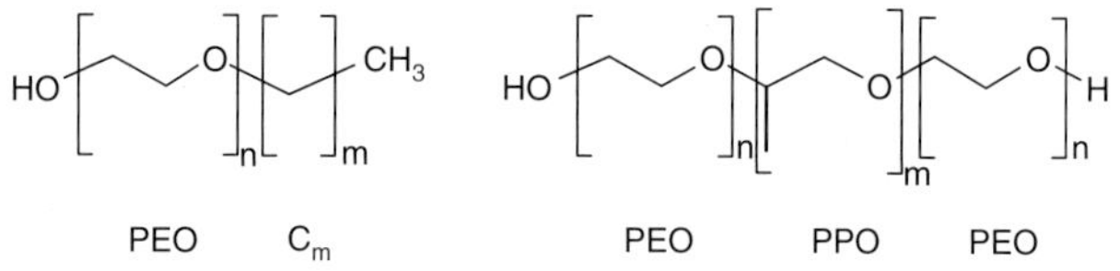

Fig. 4 Examples of poly(alkylene oxide) ethers used as structure-directing agents. PEO, polyethylene oxide; PPO, polypropylene oxide.

C Surfactant Packing Structural properties of the surfactant molecules and micellar solutions are crucial when preparing mesoporous structures. Predictions about the inorganic–surfactant phase behavior can be made based on models developed for dilute surfactant systems. Surfactants in solution assemble into structures, the geometry of which can be described by the surfactant packing parameter. The packing parameter concept is based on a model that relates the geometry of the individual surfactant molecule to the shape of the supramolecular aggregate structures most likely to form. According to Israelachvili et al. [50], the preferred shape of the surfactant molecule aggregates above the CMC depends on the effective mean molecular parameters which establish the value of the packing parameter, g. The dimensionless surfactant packing parameter g is defined by:

$$g = \frac{V}{a_0 l_c} \quad (1)$$

where V is the total volume of the hydrophobic chains plus any cosurfactant organic molecules between the chains, a_0 is the effective head group area, and l_c is the critical length of the hydrophobic tail. (Note: l_c is also called kinetic length of the surfactant tail and represents normally 80–90% of the fully extended hydrocarbon chain. The exact value of l_c depends on the extension of the chain; to ensure chain fluidity, l_c must verify that $l_c < l$, where l is the length of the fully extended chain.)

The parameter g depends on the molecular geometry of the surfactant. The number of carbons in the hydrophobic chain, the degree of chain saturation and the size and charge of the polar head group influence the value of g. Additionally, pH, ionic strength of the solution, cosurfactant effects and temperature can be included in V, a_0 and l_c [51]. Spherical aggregates are preferentially formed by surfactants bearing large polar head groups, but if the head groups can pack tightly then the aggregation number will increase, and rod-like or lamellar packing will be favored (Fig. 5).

Furthermore, for ionic surfactants the value of a_0 is strongly dependent on the degree of dissociation of the head groups and the ionic strength of the solution. The packing parameter g increases with increasing hydrophobic chain volume, smaller head group area and decreasing kinetic tail length [52]. In principle, the larger the value of g, the lower the curvature in the aggregate. Therefore, changes in micellar curvature may be achieved by altering the surfactant chain length, introducing electrolytes or by adding polar and non-polar organic additives.

The model described above was expanded to silica–surfactant mesophases by Stucky et al., who included the inorganic components to predict the final mesostructures, these including a 3-D hexagonal $P6_3/mmc$ phase [53–55]. With increasing g value, a phase transition reflects a decrease in surface curvature from the cubic ($Pm\bar{3}n$) phase over the hexagonal to a cubic phase with ($Ia\bar{3}d$) space group to finally a lamellar phase (Table 1).

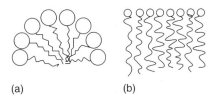

Fig. 5 (a) Surfactants with large head groups provoke a high surface curvature. (b) Surfactants with a large hydrophobic chain volume and relatively small head groups generate a small surface curvature.

Tab. 1 Surfactant packing parameter g and predicted mesophases. (Adapted from Ref. [54].)

g	Expected structure	Space group	Name
<1/3	Hexagonal (hcp)	$P6_3/mmc$	SBA-2
1/3	Cubic	$Pm\bar{3}n$	SBA-1
1/2	Hexagonal	$p6mm$	MCM-41, FSM-16, SBA-3
1/2–2/3	Cubic	$Ia\bar{3}d$	MCM-48
1	Lamellar	$P2$	MCM-50

In contrast to ionic surfactants, block copolymer-based surfactants can exhibit a wider range of architectures (linear, branched, star, etc.). The size, shape and curvature of the aggregates formed by these types of amphiphile in water is usually determined by the degree of polymerization of each block, the volume fraction of each block, and the incompatibility between the blocks (i.e., the Flory–Huggins parameter) [56]. In general, micellization behavior is driven by hydrophilic/hydrophobic character, block size and chain conformation. The aggregation behavior of triblock copolymers, which is well-described under thermodynamic equilibrium conditions, is known to depend not only on the block copolymer concentration, but more importantly also strongly on the temperature [57–59].

It is noteworthy that the thermodynamics of surfactant and block copolymer mesophases can be useful to explain solvent–surfactant and possibly inorganic–surfactant interactions. However, the behavior of the inorganic components is more complex, and will also depend markedly on the kinetics of inorganic polymerization, which can "freeze" a given mesostructure.

D Formation of the Mesostructure The different formation mechanisms suggested for the generation of various ordered mesoporous materials have been summarized by Patarin et al. [60]. The original synthesis of mesoporous molecular sieves of the M41S family was performed in aqueous alkaline solution (pH > 8). Originally, Beck and coworkers [17, 18] proposed a "liquid-crystal templating" (LCT) mechanism (Fig. 6) on the basis of similarities between liquid crystalline surfactant phases and the mesostructured materials. In this hypothesis, the structure is defined by the organization of surfactant molecules into LC phases which serve as templates for the building of M41S structures. The inorganic silicate species would occupy the space between a preformed hexagonal lyotropic crystal phase, and are deposited on the micellar rods of the LC phase (pathway 1 in Fig. 6). However, since the LC structures, which are formed in surfactant solutions, are very sensitive to the characteristics of the solution, the same authors proposed that the addition of inorganics could mediate the ordering of the surfactants into specific mesophases (pathway 2). In both pathways, however, the inorganic species interact electrostatically with the charged surfactant head groups and condense into a continuous framework which could be regarded as a hexagonal array of surfactant micellar rods embedded in an inorganic matrix.

References see page 214

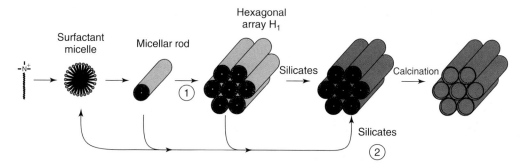

Fig. 6 Liquid crystal templating mechanism (LCT) according to Beck et al. Pathway 1 is liquid crystal-initiated; pathway 2 is silicate-initiated. (Adapted and reprinted with permission from Ref. [18]; © 1992, American Chemical Society.)

Further investigations brought additional support in favor of pathway 2 [61]. Monnier et al. [62] investigated how MCM-41 forms at concentrations where only spherical micelles are present (1 wt.%), and also proposed a silicate-initiated mechanism. These authors introduced a "charge density matching" model which was based on the cooperative organization of inorganic and organic molecular species into 3-D structures. In this model, three steps are involved in the formation of the silica-surfactant hybrid mesophase:

- Multidentate binding of the silicate poly-ions to the cationic head groups through electrostatic interactions leading to a surfactant–silica interface
- Preferential silicate polymerization in the interface region
- Subsequent charge density matching between the surfactant and the silicate.

Before silicate addition, the surfactant molecules are in dynamic equilibrium between spherical or cylindrical micelles and single molecules. However, upon addition of a silicon source, the multicharged silicate species displace the surfactant counteranions to form organic–inorganic ion pairs, which reorganize first into the silicatropic mesophase, followed by silica crosslinking. Growing silica species cooperatively attach to an increasing number of surfactant molecules and, when the amount of surfactants is large enough, the organized structures precipitate. The nature of the resulting mesophase is controlled by the multidentate interaction via interface packing density. In conclusion, both surfactant and inorganic soluble species direct the synthesis of mesostructured MCM-41-type materials. The key feature in the synthesis of mesostructured materials is to achieve a well-defined segregation of the organic and inorganic domains at the nanometric scale. Here, the nature of the hybrid interface plays a fundamental role. The most relevant thermodynamic factors affecting the formation of a hybrid interface have been proposed first by Huo et al. [63, 64]

in their description of the charge density matching model. The free energy of the mesostructure formation (ΔG_{ms}) is composed of four main terms, which represent respectively, the contribution of the inorganic–organic interface (ΔG_{inter}), the inorganic framework (ΔG_{inorg}), the self-assembly of the organic molecules (ΔG_{org}), and the contribution of the solution (ΔG_{sol}):

$$\Delta G_{ms} = \Delta G_{inter} + \Delta G_{inorg} + \Delta G_{org} + \Delta G_{sol} \quad (2)$$

In the cooperative assembly route, the template concentrations may be well below those necessary to obtain liquid crystalline assemblies or even micelles. Thus, the creation of a compatible hybrid interface between the inorganic walls and the organic templates (ΔG_{inter}) is essential for the generation of a well-ordered hybrid structure with appropriate curvature. In contrast, in the case of LC templating, the contribution due to the organization of the amphiphilic molecules (ΔG_{org}) would prevail. From the kinetic point of view, the formation of an organized hybrid mesostructure is considered to be the result of a balance of organic–inorganic phase separation, organization of the SDA, and inorganic polymerization. Hence, two aspects are essential to fine-tune the mesophase formation, namely the reactivity of the inorganic precursors (polymerization rate, isoelectric point, etc.) and the interactions involved to generate the hybrid interface.

A generalized cooperative mechanism of formation was proposed by Huo et al. [63, 64] based on the specific electrostatic interactions between an inorganic precursor I and a surfactant head group S. A number of different strategies for obtaining a variety of ordered mesoporous materials have been identified. The hybrid mesophases obtained are strongly dependent on the interactions between the surfactants and the inorganic precursors. In the case of ionic surfactants, the formation of the mesostructured material is mainly governed by electrostatic interactions (Fig. 7). In the simplest case, the charges of the surfactant (S) and the mineral species

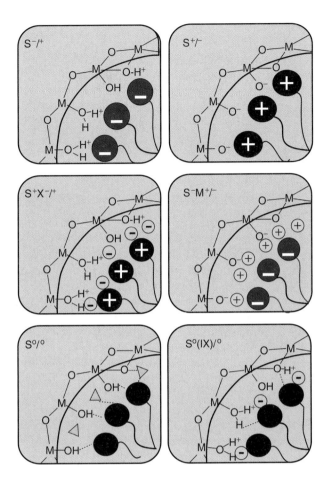

Fig. 7 The different inorganic–organic hybrid interfaces. (Reprinted with permission from Ref. [30]; © 2002, American Chemical Society.)

(I) are opposite under the synthesis conditions. Along with S^+I^-, the cooperative interaction between inorganic and organic species based on charge interactions can be achieved by using the reverse charge matching S^-I^+. With these two main direct synthesis routes identified, S^+I^- and S^-I^+, two other synthesis paths, both of which are considered to be indirect, also yield hybrid mesophases from the self-assembly of inorganic and surfactant species.

Synthesis routes involving interactions between surfactants and inorganic ions with similar charges are possible through the mediation of ions with the opposite charge ($S^+X^-I^+$ or $S^-M^+I^-$). The $S^+X^-I^+$ pathway takes place under acidic conditions, in the presence of halide anions ($X^- = Cl^-$, Br^-) and the $S^-M^+I^-$ route is characteristic of basic media, in the presence of alkali metal ions ($M^+ = Na^+$, K^+). Besides the syntheses based on ionic interactions, the assembly approach has been extended to pathways using neutral (S^0) [65] or non-ionic surfactants (N^0) [66]. In the approaches denoted (S^0I^0) and (N^0I^0), hydrogen-bonding is considered to be the main driving force for the formation of the mesophase.

E True Liquid Crystal Templating Attard et al. performed silica sol–gel polymerization specifically in lyotropic liquid crystals, creating an alternative to conventional precipitation methods [38]. The TLCT pathway leads to mesostructured hexagonal, cubic or lamellar silicas, shaped as gels, monoliths or membranes. Metallic mesoporous materials are also obtained by chemical or electrochemical reduction of metal salts, dissolved in the hydrophilic region of the LC phase [67]. The TLCT method allows a cast replica of the LC phase to be obtained; hence it should, in principle, be possible to control the final phase by knowing the phase diagram of the surfactant. However, inhomogeneity may sometimes result from inadequate diffusion of the inorganic precursors within the preformed LC phase, limiting somewhat the use of this pathway.

F Evaporation-Induced Self-Assembly Evaporation-induced self-assembly (EISA) was introduced by Brinker et al. [68] to describe synthesis methods used to produce mesostructured materials starting from dilute solution, followed by solvent evaporation. EISA can be considered as an LC templating-based method (after solvent evaporation, a hybrid LC mesophase is formed) [69]. Starting from solutions below the CMC allows thin films or gels to be obtained with excellent homogeneity, as the high dilution prevents uncontrolled inorganic polymerization. Surfactant and block copolymer-based mesoporous silica films and monoliths have been obtained using EISA. This method is also particularly versatile for preparing non-silica systems, where condensation must be thoroughly controlled (see Section 2.3.5), and processed as thin films in most cases.

2.3.6.3.3 Synthesis Pathways and Structural Diversity
Silica-based materials are the most widely studied systems for several reasons, namely that a wide variety of structures is possible, there is precise control of the hydrolysis–condensation reactions due to lower reactivity, and enhanced thermal stability and a wide variety of functionalization methods available. The most popular ordered mesoporous silicas are listed in Table 2, together with details of the type of hybrid interface interaction involved, pore size ranges and mesostructure symmetries. Examples of powder XRD spectra obtained for silica mesophases, with their respective pore topology, are depicted in Fig. 8. Due to the appealing physicochemical properties of the triblock copolymer-based

References see page 214

Tab. 2 Structural diversity of some of the most common mesoporous silica materials

Name	Possible structure-directing agent	Type of interaction	Structure	Pore size/nm	References
MCM-41	$C_n(CH_3)_3 N^+ Br^-$ or Cl^- $8 \leq n \leq 20$	S^+I^-	2-D hexagonal $p6mm$	1.5–10	[17, 18]
MCM-48	$C_n(CH_3)_3 N^+ Br^-$ or Cl^- $12 \leq n \leq 20$	S^+I^-	3-D cubic $Ia\bar{3}d$	1.5–4.6	[18, 62, 80]
KIT-1	$C_{16}(CH_3)_3 N^+ Cl^-$	S^+I^-	Disordered	3.4	[70]
SBA-1	$C_n(C_2H_5)_3 N^+ Br^-$ or Cl^- $12 \leq n \leq 18$	$S^+X^-I^+$	3-D cubic $Pm\bar{3}n$	1.5–3.0	[63, 64]
SBA-3	$C_n(CH_3)_3 N^+ Br^-$ or Cl^- $12 \leq n \leq 18$	$S^+X^-I^+$	2-D hexagonal $p6mm$	1.5–3.5	[54, 55]
SBA-12	Brij 76 $C_{18}EO_{10}$	$(S^0H^+)(X^-I^+)$	3-D hexagonal (intergrowth)	3.0–5.0	[71, 92]
SBA-15	P123 $EO_{20}PO_{70}EO_{20}$ Blend of organics and P123	$(N^0H^+)(X^-I^+)$	2-D hexagonal $p6mm$	4.0–15	[92, 93, 119, 157]
SBA-16	F127 $EO_{106}PO_{70}EO_{106}$ Blends of F127 and P123 Blend of F127 and butanol	$(N^0H^+)(X^-I^+)$	3-D cage-like cubic $Im\bar{3}m$	4.7–12	[93, 111, 130, 142]
KIT-6	Blend of P123 and butanol	$(N^0H^+)(X^-I^+)$	3-D cubic $Ia\bar{3}d$	4.0–11.5	[120, 138]
FDU-1	B50-6600 $EO_{39}BO_{47}EO_{39}$	$(N^0H^+)(X^-I^+)$	3-D cage-like cubic $Fm\bar{3}m$	8.0–14	[99, 114]
FDU-12 or KIT-5	F127 $EO_{106}PO_{70}EO_{106}$ Blend of F127 and trimethyl-benzene (TMB)	$(N^0H^+)(X^-I^+)$	3-D cage-like cubic $Fm\bar{3}m$	6.0–12.5	[117, 140]
MCF	Blend of P123 and TMB	$(N^0H^+)(X^-I^+)$	Disordered	20–42	[118, 119]
MSU-X	Tergitol 15-S-n $C_{11-15}EO_n$ $7 \leq n \leq 30$ Triton X-100, X-114 $C_8Ph(EO)_n$ $8 \leq n \leq 10$ P64 $EO_{13}PO_{30}EO_{13}$	N^0I^0	Disordered	2.0–5.8	[66, 91]
HMS	Alkylamines (C_8–C_{18})	S^0I^0	Disordered	2.3–4.0	[65]
MSU-H	P123 $EO_{20}PO_{70}EO_{20}$	N^0I^0	2-D hexagonal $p6mm$	7.5–12	[100]

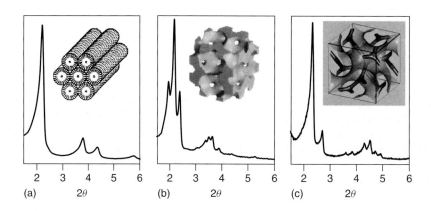

Fig. 8 Examples of powder X-ray diffraction patterns obtained for ordered mesoporous silica mesophases, shown with their pore topology. (a) MCM-41 with $p6mm$ symmetry; (b) SBA-1 with $Pm\bar{3}n$ symmetry; (c) MCM-48 with $Ia\bar{3}d$ symmetry.

silicas for applications that require improved stability or involve large guest molecules, current research is increasingly oriented towards the development of ordered mesoporous materials prepared with triblock copolymers as SDAs.

A Silica Polymorphs from the Alkaline Route (S^+I^-)
Mesoscopic polymorphs of silica synthesized under alkaline conditions have attracted the most scientific interest so far. Under these conditions, anionic silicates I^-, and cationic surfactant molecules S^+, associate

cooperatively and organize to form hexagonal, lamellar, or cubic structures. This synthesis in alkaline medium can lead to three well-defined structures: MCM-41, MCM-48, and MCM-50 [18, 42]. Vartuli et al. [42] showed that the surfactant-to-silica molar ratio is a critical variable in the formation of M41S materials. By using tetraethoxysilane (TEOS) with cetyltrimethylammonium chloride [CTAB; $(C_{16}H_{33}(CH_3)_3NCl)$], these authors found that increasing the surfactant-to-silica molar ratio from 0.5 to 2.0 resulted in hexagonal (<1), cubic $Ia\bar{3}d$ (1–1.5), lamellar (1.2–2) and uncondensed cubic octamer (>2) composite structures. Moreover, structural phase transformations to lower energy configurations may be induced during hydrothermal treatment as changes in the charge density of the silicate occur upon condensation. In addition to the lamellar to hexagonal phase transition described by Monnier et al. [62], transition from hexagonal to lamellar [54] and hexagonal to cubic [74] geometries upon condensation (i.e., transition to lower curvature) can also be observed. The phase transitions during hydrothermal treatment may generally be explained by further condensation of the silicate framework and subsequent restructuring which induces changes in the packing of the surfactant. In addition, structural transformations may also be coupled with temperature-dependent changes in hydration, silicate solubility, counterion binding and the migration of organic cosolvent in the system.

The original synthesis of MCM-41 was carried out in aqueous alkaline solutions at elevated reaction temperatures (150 °C) with non-molecular silicon sources such as fumed silicas or water glasses. In 1997, Grün et al. [72] proposed a very efficient novel route in the synthesis of MCM-41 by rapid hydrolysis of a molecular silicon source (TEOS) with ammonia as the catalyst. This method provides a convenient route to high-quality product in a short period of time (1 h) at room temperature, and yields MCM-41 without sodium traces. The synthesis results in a highly reproducible product with respect to the specific surface area and pore structural parameters. Later, this procedure was modified to yield a MCM-41 with increased structural order by aging the material at 90 °C in the mother liquor for several days [73]. MCM-48 can be prepared, via hydrothermal synthesis, either by adjusting the silica-to-CTAB ratio and varying the synthesis conditions [42, 74, 75], or at room temperature by using gemini-type surfactants [54]. The preparation of MCM-48 ($Ia\bar{3}d$) was also demonstrated by adjustment of the packing parameter g with the addition of organic additives [76–78] or by employing mixtures of different surfactants as SDAs [79–81]. The structure of MCM-48 belongs to the $Ia\bar{3}d$ space group, which has also been found in the binary water/CTAB system [82]. The structure is considered to be bicontinuous with a simplified representation of two 3-D mutually intertwined, unconnected networks of rods (Fig. 8c). This structure contains a 3-D channel network with channels running along the [111] and [100] axis directions [83]. The unit cell parameter measured for the cubic MCM-48 usually ranges between ca. 8 and 10 nm.

In the as-synthesized form, MCM-50 is a lamellar structure; this phase can be represented by sheets or bilayers of surfactant molecules with the hydrophilic head group pointing towards the silicate at the interface. Removal of the template results in collapse of the lamellar structure, unless the material is stabilized.

B Silica Polymorphs from the Acidic Route ($S^+X^-I^+$)

Huo et al. [63, 64] showed that mesoporous silicas are accessible at low pH. The formation of silica is possible by the cooperative assembly of cationic inorganic species with cationic surfactants. These syntheses are carried out under strongly acidic conditions (usually 1–3 M HCl or HBr), where silicate species are positively charged. 2-D hexagonal $p6mm$, 3-D hexagonal $P6_3/mmc$, cubic $Pm\bar{3}n$ and lamellar structures could all be obtained in the presence of HCl [54, 55, 64]. Over the pH range employed, silica polymerization proceeds through the condensation of cationic intermediates and resulting in gel and glass morphologies very different from those obtained in basic medium [4, 14]. It is suggested that the main driving force for self-assembly at low pH is electrostatic interface energy, as well as hydrogen bonding. At high concentrations of HCl, the cationic hydrophilic region of the surfactant is surrounded with a peripheral negative charge (S^+X^-). The positively charged species (I^+) resulting from hydrolysis of the silicon sources at low pH are attracted electrostatically to the anionic portion of the surfactant ion pair, thus forming a triple layer where the halide ions coordinate through Coulombic interactions to the protonated silanol groups. The formation of this triple layer ($S^+X^-I^+$), where the halide anions in the electrical double layer around the micellar aggregate serve as charge mediators, is considered to account for formation of the mesophase [64]. During the cooperative assembly of the soluble molecular species in the silica synthesis under concentrated acidic conditions, precipitation and polymerization occur. In this process, the protons associated with the silicates, along with associated excess halide ions, are excluded until an almost neutral inorganic framework remains. This quasi-neutral polysilicic acid framework is hydrogen-bonded to the S^-X^+ complex. The mechanism for the $S^+X^-I^+$ synthesis is clearly not the same as that for the S^+I^- route. Generally, similar interplanar d-spacings are observed

References see page 214

for mesostructured materials prepared at low and high pH. However, the physisorption results indicate that the materials prepared in acidic medium have smaller pores, thicker pore walls, and a very high adsorption capacity [55]. Moreover, the overall framework charge is different from the base-derived mesophase, due to the different precipitation conditions and charge balance requirements. The 2-D hexagonal materials ($p6mm$) prepared according to the acidic route are sometimes labeled SBA-3 (Santa Barbara No. 3) [54, 64, 84]. More recently, Tatsumi showed that four types of well-ordered mesophases, 3-D hexagonal $P6_3/mmc$, a cubic $Pm\bar{3}n$ (SBA-1), 2-D hexagonal $p6mm$ (SBA-3), and the cubic $Ia\bar{3}d$ phase can actually be synthesized in the same surfactant/water systems, but in the presence of different acids [85]. These mesoporous materials were synthesized by using cetyltriethylammonium bromide (CTEABr) as the surfactant and TEOS as the silicon source with different acids: H_2SO_4, HCl, HBr, and HNO_3. These studies were also the first to demonstrate the possible synthesis of a cubic $Ia\bar{3}d$ MCM-48-like mesophase under acidic conditions. Although easily accessible, these materials have been investigated to a much less degree than their base-derived counterparts.

C Anionic Surfactant-Templated Mesoporous Silicas: AMS-*n* Materials A rather new development is the preparation of the mesoporous silica materials described by Che et al. [86, 87] using anionic surfactants in combination with organosilane groups as costructure-directing agents (CSDAs). In this system, the mesoporous silica materials are thought to be formed through electrostatic interactions between negatively charged carboxylic head groups of anionic surfactants, such as N-acyl-alanine, glycine, or N-acyl-glutamic acid, and positively charged amine or ammonium groups of 3-aminopropyltrimethoxysilane (APS) or N-trimethoxysilylpropyl-N,N,N-trimethylammonium chloride (TMAPS). The organosilane additives APS and TMAPS co-condense with TEOS to form ultimately the silica framework. The species involved are represented schematically in Fig. 9. According to these authors, the interaction mechanism may be generalized as $S^-N^+ \sim I^-$, where S^- is the anionic surfactant, N^+ represents the positively charged amine moiety in the aminosilane group, and I^- is the condensing silica framework.

This new family of mesoporous materials has been named AMS-*n* (anionic surfactant-templated mesoporous silicas). High-resolution TEM and XRD studies of the silica materials generated in this system have revealed

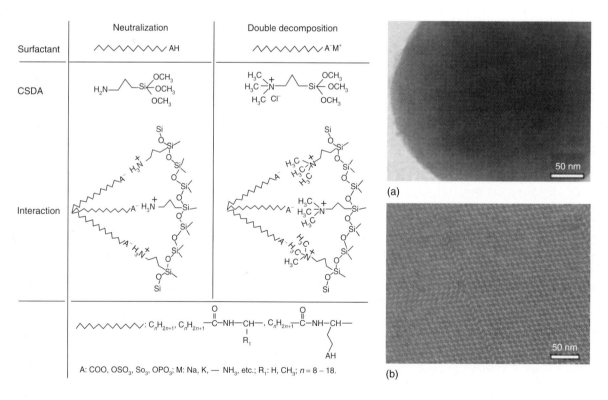

Fig. 9 Left: Schematic representation of the two possible types of interaction involved in the formation of AMS-*n* silica materials. (Reprinted with permission from Ref. [86]; © 2002, Nature Publishing Group.) Right: Representative high-resolution TEM images of (a) AMS−6 $Ia\bar{3}d$ material with an image taken along the [100] direction, $a = 10.73$ nm; and (b) AMS−8 $Fd\bar{3}m$ with an image taken along the [110] direction, $a = 17.79$ nm. (Original images courtesy Dr. A. Garcia-Bennett, Uppsala University, Sweden.)

an impressive structural diversity as a function of synthesis conditions (e.g., pH, reagent ratios, temperature, and time) and the nature of the surfactant [87–89]. Highly ordered mesoporous silica with either 3-D hexagonal (AMS-1), 3-D cubic $Pm\bar{3}n$ (AMS-2), 2-D hexagonal (AMS-3), 3-D cubic $Ia\bar{3}d$ (AMS-6), 3-D cubic $Fd\bar{3}m$ (AMS-8) or tetragonal $P4_2/mnm$ structures may be obtained, to list some of the mesophases achieved. The materials have high surface areas (300–900 m^2 g^{-1}) and pore volumes (0.22–0.8 cm^3 g^{-1}), with pore sizes ranging from 2.3 to ca 7.5 nm. The different structures are obtained in syntheses performed at pH ranging from 5.0 to 9.5 and subsequent aging treatment at 60–100 °C for 1 to 4 days. For example, the cubic $Ia\bar{3}d$ AMS-6 silica is obtained with N-acyl-alanine surfactant and APS as the CSDA after curing for 2 days. It should be noted that the resulting highly ordered sample is composed of thin, spherical particles of sizes ranging from 10 to 80 nm (Fig. 9a). Also of particular interest is the 3-D cubic AMS-8 ($Fd\bar{3}m$) material that can be prepared at 60 °C by using sodium N-lauroyl-L-glycine surfactant, together with TMAPS as the additive. In this synthesis, an additional aging step at 100 °C for 3 days is necessary. This very promising AMS-8 material was identified as a bimodal pore system consisting of a 3-D network of large cage-like pores of 7.6 nm and smaller cages of 5.6 nm, all of which were interconnected (these values were extracted from an HRTEM analysis) [88].

D Non-Ionic Routes (Hydrogen-Bonding Interactions S^0I^0, N^0I^0 or $(N^0H^+)(X^-I^+)$) In general, syntheses employing non-ionic SDAs are carried out under acidic or neutral conditions, and the hybrid inorganic–organic mesophase assembly is governed by weak van der Waals interactions or hydrogen bondings. Tanev and Pinnavaia developed the neutral templating route, which is based on hydrogen bonding and self-assembly involving neutral amine surfactants and neutral inorganic precursors [65]. According to the S^0I^0 approach, primary amines with alkyl chains from C$_8$ to C$_{18}$ are used to prepare a mesoporous product, named HMS, from precursor silica species at pH 7. Silicate species are suggested to interact with the micellar aggregates through hydrogen bonding between the hydroxyl groups of hydrolyzed silicate species and the polar amine head groups. Particularly versatile for the design of large-pore mesoporous materials is the use of non-ionic block copolymer-based amphiphiles as the SDAs [90]. In 1995, Pinnavaia's group was the first to use non-ionic poly(ethylene oxide) monoethers (polyoxyethylene-based oligomeric amphiphiles), as well as Pluronic-type surfactants in neutral aqueous media to direct the formation of MSU-type silicas [66]. A neutral templating pathway based on hydrogen-bonding interactions was proposed. Worm-like mesopore structures denoted MSU-X, with uniform diameters ranging from 2.0 to 5.8 nm, were obtained by varying the size and structure of the surfactant molecules and using TEOS as a silicon source. Some advantages of using such neutral non-ionic rather than ionic surfactants were immediately recognized: (i) thicker inorganic walls (1.5–4 nm) are obtained, resulting in improved hydrothermal stability; (ii) tailoring of the pore diameter is easier; and (iii) template removal by solvent extraction is facilitated. However, the first materials originating from this route presented disordered structures. Optimization of the precipitation conditions of MSU-X is possible by using a two-step method:

- First, the hybrid mesophase assembly is favored, with the inorganic network condensation being temporarily "quenched".
- Second, an extended inorganic network condensation is subsequently triggered separately by a pH increase or by the addition of F$^-$ as a condensation catalyst.

By using this strategy, the synthesis yields excellent organization of the mesophase [91]. Major progress in the preparation of mesoporous silicas was made by Zhao et al. in 1998 [92, 93] who used poly(ethylene oxide)-poly(propylene oxide)-poly(ethylene oxide) triblock copolymers (Pluronic-type) for the synthesis under acidic conditions of a large-pore material called SBA-15 and other related materials. These syntheses were based on the use of organic silica sources such as TEOS or tetramethoxysilane (TMOS) in combination with diluted acidic aqueous solution of Pluronic-type triblock copolymer (<7 wt.% in water, ideally 2.5 wt.%) such as P123 (EO$_{20}$-PO$_{70}$-EO$_{20}$, EO = ethylene oxide, PO = propylene oxide) and F127 (EO$_{106}$-PO$_{70}$-EO$_{106}$). However, here the mechanism of formation is proposed to rather be $(N^0H^+)(X^-I^+)$, as the block copolymer could be positively charged under the reaction conditions. The use of triblock copolymers led to a considerable expansion in accessible range of pore sizes. In most cases, mesoporous silicas obtained with triblock copolymers possess uniform large pores with dimensions well above 5 nm in diameter and thick walls. The latter provided high thermal stability and improved the hydrothermal stability when compared with mesoporous silicas synthesized with low-molecular-weight ionic surfactants [94]. Hexagonally ordered SBA-15 can be prepared with pore sizes between 4.5 nm and 15 nm with wall thicknesses of 3.0 to 6.5 nm. SBA-15 usually exhibits a large surface area (ca. 800–1000 m^2 g^{-1}) and pore volumes up to 1.5 cm^3 g^{-1}.

References see page 214

Other important related phases such as SBA-12 (3-D hexagonal $P6_3/mmc$) and SBA-16 (cubic $Im\bar{3}m$) were obtained with different non-ionic block copolymers or non-ionic oligomeric surfactants (Brij-type) [93]. TEM images and the pore topology of SBA-15 and SBA-16 are presented in Fig. 10. Several of the triblock copolymer-templated silicas are also accessible through syntheses using less-expensive inorganic silica sources, such as sodium silicates [95–99]. It is worthy of mention that as the condensation of silicate species under neutral conditions (pH 7) occurs too rapidly to be finely controlled, most syntheses are performed under acidic conditions. The hexagonally ordered mesoporous MSU-H silica prepared with sodium silicate and P123 as SDA is one of the few examples of successful synthesis carried out under almost neutral conditions [100].

SBA-15 silica, being more or less the large pore equivalent to MCM-41, is of growing interest for a wide range of applications (e.g., selective sorbent, support for catalysts and biomolecules, nanoreactor, etc.), and is therefore the subject of increasing numbers of studies. Originally, SBA-15 was thought to be a large-pore equivalent of MCM-41 mesoporous silica with unconnected mesoporous cylindrical channels. However, several studies have shown that the pore size distribution of the hexagonal SBA-15 is rather bimodal, in which the larger, hexagonally ordered structural pores are connected by smaller pores (micropores) located inside the silica walls [101–104]. TEM imaging of a platinum replica of the pore structure of SBA-15 allowed visualization of the presence of these complementary pores [102, 105]. These pores do not seem to be well ordered, and may likely result from the fact that the PEO blocks of the copolymer penetrate inside the silica framework. The presence of complementary porosity in the walls of SBA-15-type silicas is one of the fundamental differences from the smaller-pore MCM-41-type systems, and it is of major importance for further development of block copolymer-directed silicas. In being more complex than that of MCM-41, the formation mechanism of SBA-15 and other related phases is currently undergoing intense study, using techniques such as in-situ small-angle X-ray scattering (SAXS/XRD) methods [106–108], NMR [109], electron paramagnetic resonance (EPR) [110], and TEM [109].

Among triblock copolymer-templated materials, mesoporous silicas consisting of interconnected large, cage-like pores (pore diameter >5 nm) with a 3-D organization are also of growing importance. For example, the large-pore SBA-16 silica, which is synthesized in acidic media, consists of spherical cavities of 8 to 10 nm diameter arranged in a body-centered cubic (bcc) array (with $Im\bar{3}m$ symmetry), and connected through mesoporous openings of 2.0 to 3.5 nm [111]. The aqueous synthesis route to achieve the highly ordered cubic $Im\bar{3}m$ mesoporous silica is based on the use of F127 (EO_{106}-PO_{70}-EO_{106}), which exhibits a high hydrophilic to hydrophobic volume ratio (high EO/PO ratio), implying high curvature of the micelles. These types of cage-like mesophase with cubic symmetry are usually more difficult to prepare than their 2-D hexagonal counterparts (e.g., SBA-15), and are often only synthesized in a narrow range of synthesis mixture composition. However, such materials with interconnected 3-D pore structures are expected to be superior to materials with one-dimensional channels for applications dealing with selectively tuned adsorption, enhanced diffusion, or other host–guest interactions [36, 112, 113]. Thus, an increasing number of research investigations are being devoted to the development and application of large-pore 3-D cubic phases [93, 111, 114–117]. In addition, Stucky et al. have described the preparation of a family of materials designated as mesostructured cellular foams (MCF) by employing triblock copolymer species stabilized by oil-in-water microemulsions [118, 199]. These materials exhibit a 3-D mesostructure with pores consisting of large, interconnected spherical cells with uniform windows.

Another type of ordered mesoporous material with 3-D interconnected pore structure was synthesized by Kleitz et al. [120], this time by using a blend of P123

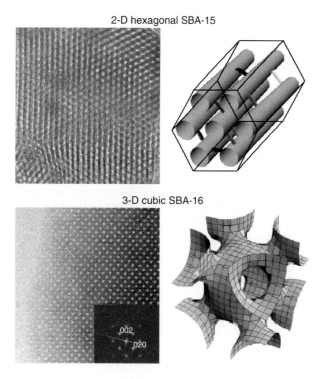

Fig. 10 Transmission electron microscopy images of SBA-15 and SBA-16, as indicated, with the respective representation of the pore topology. (Adapted from Refs. [92, 110].)

and n-butanol as SDAs in combination with a precise adjustment of the acidic medium concentration. This material, designated as KIT-6, exhibits a structure with cubic $Ia\bar{3}d$ symmetry, being thus the large-pore equivalent of the MCM-48 silica. The pore network topology of this material can be represented as an interpenetrating bicontinuous network of interconnected channels (the gyroid structure) [121]. The porosity of KIT-6 is quite similar in nature to that of SBA-15. KIT-6 silica has high pore volume and large accessible pores tailored between 4 and 12 nm, and complementary intrawall porosity is also observed. Several other routes to produce large-pore cubic $Ia\bar{3}d$ analogues are also available [122–124].

Finally, it should be mentioned at this point that large-pore ordered mesoporous silicas were also obtained from solutions containing preformed liquid crystalline mesophases, at high block copolymer concentrations (>20 wt.%) [125], following the TLCT mechanism.

2.3.6.3.4 Pore Size Tailoring and Structure Engineering

Structural and textural control is crucial for the design of functional porous solids. In particular, a fine control of the mesopore size is important because of potential applications as catalysts, molecular sieves, and hosts for quantum size effect materials. Furthermore, processing of the hybrid mesophases prior to calcination is determining for pore size, structure, and porosity. Processing affects wall thickness and framework density, which in turn determines hydrothermal and thermal stability, being of major importance for applications under demanding conditions such as catalysis or membranes.

A Surfactant Chain Length   One simple means of tuning the pore size of the surfactant-directed inorganic material is to change the length of the surfactant carbon chain [18, 126]. Usually, a linear relationship is observed between the pore sizes and the length of the carbon chain of a molecular template. With C_nTAB ($n = 8$, 10, 12, 14, 16, and 18), the pore size of the as-synthesized MCM-41 material is shown to increase by about 0.45 nm when n is increased by two carbon atoms. The shortest chain surfactant from which mesophases could be made is usually $n = 8$. Long-chain surfactants ($n > 20$) are not commercially available, and are virtually insoluble in water; moreover, the pore expansion is generally accompanied by a loss of order [18]. Changes in the molecular geometry and chain length of non-ionic oligomeric surfactant and block copolymers allow similar fine tuning of the pore size. In these cases, the adjustment of pore size can be achieved continuously by varying the concentration of the templating agent (or eventually the polymer-to-silica molar ratio), and by changing the composition of the copolymers or the block size [66, 92, 93, 98, 127]. Specifically, the ratio of hydrophilic to hydrophobic blocks (EO/PO, x/y) can determine the structure of the mesophase, with block copolymers that possess longer hydrophilic chains, such as F127, leading to materials having highly curved, cage-like pores [93]. The EO/PO ratio of the copolymer has also a major influence on the pore size and wall thickness of the resulting materials. In general, for a triblock copolymer such as EO_x-PO_y-EO_x, an hexagonal phase is generated if the ratio x/y is between 0.1 and 1.0. Lower ratios usually lead to a lamellar phase, while values higher than 1.0 (e.g., EO_{70-106} vs. PO_{70}) afford cubic mesostructures with spherical pores. Alfredsson et al. used the effects of different block lengths and their relative volume fraction in the synthesis of mesoporous silicas in order to tune pore dimensions and structure [128, 129]. For different structure-directing copolymers that lead normally to an hexagonal phase under given synthesis conditions, a marked increase in the PO chain length, with constant EO chain length, results in larger mesopores. In contrast, longer EO blocks, with the PO chains remaining constant, lead to mesoporous silica with much thicker walls. Another example of the role of the volume ratio (geometry) of the block copolymer in the synthesis of ordered mesoporous materials is the preparation of the cage-like silica material called FDU-1 [99]. This interesting material, which has very large pore dimensions and a large pore volume, is synthesized using a different triblock copolymer containing polybutylene oxide blocks, denoted B50-6600 (EO_{39}-BO_{47}-EO_{39}). FDU-1 exhibits a 3-D interconnected face-centered cubic (fcc) pore structure with intergrowths of hexagonal close-packed (hcp) phase [114].

In addition, by employing copolymer blends (P123 mixed with F127) jointly with a control of the synthesis temperature and time, a tailoring of the cage dimensions and pore openings of SBA-16 silica is also possible [130].

B Time and Temperature   Another method used to control the pore size of mesoporous solids is based on restructuring upon prolonged hydrothermal treatment. The application of aging treatments at different temperatures and for extended periods (e.g., from 24 h to several days) has proven to be efficient for tailoring the mesophases. This can be done either directly in the mother liquor [131, 132] or at a different pH in water [54] or alcohol. For example, Khushalani et al. [131] showed that siliceous MCM-41-type mesophases could be restructured at elevated temperatures in the alkaline mother liquor, resulting in the expansion in pore size from 3.7 to 5.9 nm. Generally, these post-synthesis treatments

References see page 214

improve the thermal stability of samples obtained at room temperature, and seem to afford materials with higher structural quality, upon the promotion of wall polymerization. However, the increased value of the unit cell that is sometimes observed is not always connected to an increase in pore size, but could also be due to thicker walls. The silicate framework consists of a large amount of silanols, which can further condense and rearrange. Improved stability may arise from increased condensation within the framework, leading to fewer silanol groups and thicker walls. Furthermore, the pronounced contraction of the unit cell that occurs upon calcination of as-synthesized MCM-41 materials at elevated temperatures is usually prevented for hydrothermally aged samples [133, 134].

In the case of ordered mesoporous materials prepared with triblock copolymer, the synthesis temperature is known to have a marked influence on the mesostructure, and this by changing the micelle hydrophobicity. The solution reactions leading to mesophase formation are usually performed within an ideal range of 35 to 45 °C, depending on the block copolymer employed. Increasing the synthesis temperature within this range, and beyond, renders the EO groups more hydrophobic, leading to micelles with a larger hydrophobic core volume and smaller hydrophilic regions. As a consequence, higher synthesis temperatures result in materials with increased pore size and slightly reduced wall thickness [128, 129].

For all block copolymer-templated materials (e.g., SBA-15, KIT-6, FDU-1, SBA-16), the time and the temperature of the hydrothermal aging which is then applied after the first synthesis step are also particularly important. The aging temperatures are usually lower than those employed for zeolite synthesis, ranging from as low as 40 °C up to 150 °C, in which the temperature range of 90 to 100 °C is the most commonly employed. Aging temperatures in excess of 150 °C usually lead to a pronounced decrease in the structural quality of the materials. An extended aging time of 12 h up to several days is usually required to produce silica materials of satisfactory quality. It should be noted that the hydrothermal treatment time can be shortened to 2 hours or even less if microwave conditions are employed [135]. Increasing the aging temperature in the preparation of SBA-15 always leads to a larger pore diameter. In addition, larger pore volumes are obtained and the nature of the wall porosity is modified [92, 101, 136] (Fig. 11). A similar effect of pore size increase occurs upon longer hydrothermal treatment times. For example, the mesopore size of SBA-15 can easily be tailored in diameters ranging from 4.5 nm to 11.5 nm by increasing the aging temperature from 60 °C to 130 °C and applying a hydrothermal aging time ranging from a few hours to 4 days [136, 137]. Similarly, the mesopore size of cubic $Ia\bar{3}d$ KIT-6 silica can be tailored within the same range of diameters, as exemplified by representative sorption results shown in Fig. 11 [138].

Ryoo and Fajula confirmed that the pore structure of SBA-15 depends on the synthesis and processing (thermal treatment) of the initial as-synthesized hybrid mesophase [136]. For example, at a temperature of >60 °C the PEO blocks become even less hydrophilic and dehydrate. Therefore, the wall microporosity can be tuned by modifying the silica–template interactions upon thermal treatment (i.e., by modifying the degree of polymerization of the inorganic species while enhancing segregation between the template and the inorganic phase). Aging at 80–100 °C seems to help segregation of the PEO blocks and the inorganic framework, by promoting condensation of the latter, leading to materials with less microporosity detectable. In fact, it is proposed that SBA-15 silica prepared between 35 °C and 60 °C exhibits micropores on the pore surface, forming a so-called "corona" [102], and with apparently no connection between the mesopores. At 100 °C, SBA-15 possesses some micropores and connections between the hexagonally ordered mesopores, whereas at 130 °C the material no longer has any micropores but has larger connections. This evolution of porosity is shown schematically in Fig. 11.

The case of cage-like materials such as SBA-16 or FDU-1 is also very interesting. Similar to materials with cylindrical pores, the size of the spherical mesopores of these materials can be changed by applying different aging temperatures and times. Pore enlargement occurs together with a pronounced increase in pore volume. Here, the mesopore size of SBA-16-type silica must be assessed using XRD modeling methods or physisorption data evaluated with models based on spherical pore geometry [139–143]. The mesopore size of SBA-16, for example, can be increased from about 6.5 nm to more than 11 nm while the aging temperature ranges from 45 °C to 130 °C over a 24-h period. However, it is not only the main mesocage diameter that increases, but also the pore windows or pore openings of the cages that are substantially enlarged upon prolonged hydrothermal treatment at elevated temperatures; this is represented in Fig. 12 for the case of SBA-16. Enlargement of the cage apertures, the precise size of which is not easily determined, seems to be a general feature observed for most cage-like silicas [96, 114, 130, 140, 142].

C Effects of Electrolytes and pH Adjustment Further improvements of the long-range order of mesoporous silica molecular sieves or fine tuning of the pore size and mesostructure can also be made by adjusting the pH; this is achieved by adding the required amount of acid or base during the synthesis. In particular, this has

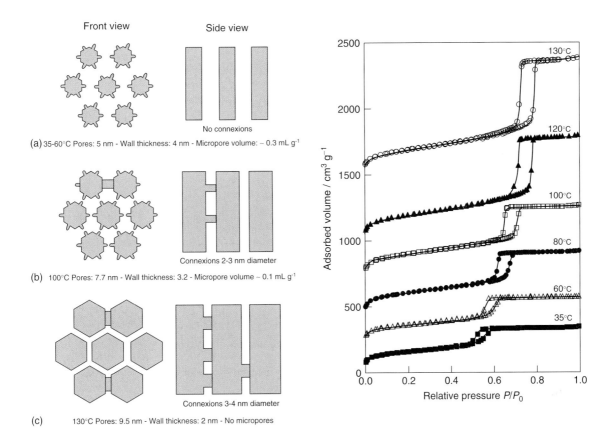

Fig. 11 Left: Schematic representation of the pore topology of SBA-15 silica synthesized: (a) between 35 °C and 60 °C; (b) at ca. 100 °C; and (c) at 130 °C. (Adapted and reprinted with permission from Ref. [136]; © 2003, Royal Society of Chemistry. Original figure courtesy Dr. Minkee Choi, KAIST, Deajeon, South Korea.) Right: N_2 adsorption–desorption isotherms for cubic $Ia\bar{3}d$ KIT-6 silica samples synthesized with different hydrothermal treatment temperatures as indicated (the isotherms are plotted with offsets). (Reprinted with permission from Ref. [138]; © 2005, American Chemical Society.)

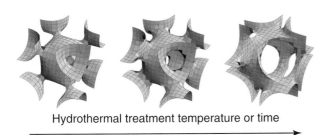

Hydrothermal treatment temperature or time

Fig. 12 Schematic representation of the structure tailoring of SBA-16 by increasing the hydrothermal treatment time and/or temperature. (Adapted and reprinted with permission from Ref. [130]; © 2004, American Chemical Society. Original figures courtesy Dr. Y. Sakamoto, Stockholm University, Sweden.)

been demonstrated for MCM-41 and MCM-48 syntheses [80, 144, 145], and is based on the powerful influence that the solution pH has on the degree of condensation and polymerization of the inorganic oligomeric species, the charge density of the polyelectrolyte inorganic species involved, and the surfactant packing parameter g.

Salt addition also has a significant effect on the synthesis mixture and, consequently, also on the properties of the resulting mesoporous solids. Salts affect the packing parameter g and the interaction conditions between the surfactants and silica. The surface charge density of the surfactant phase is affected by adsorbed counterions. When the charge density of the silica phase is reduced by the presence of screening electrolytes in the synthesis mixture, the surfactant molecules self-assemble into a mesophase with a lower surface charge density, and phase transitions can be observed caused by a change in the mesophase surface curvature. Furthermore, the addition of salts (e.g., NaCl) is suggested to increase the hydrothermal stability of the material [146]. The role of the salt, in moderating wall–surfactant interaction, is believed to be an important factor that drives a restructuring process of the silica framework. For example, the improved hydrothermal stability observed upon the addition of salt to the preparation of MCM-48

References see page 214

is claimed to originate from a local restructuring of the silica walls.

Electrolyte addition is also a possible means of affecting the phase behavior of block copolymer-based mesoporous silicas. In general, inorganic salts have a strong influence on the block copolymer micelles. The CMC and critical micellar temperature (CMT) of the block copolymer micelles are both decreased upon salt addition. Salting-out electrolytes (lyotropic ions) such as KCl, NaCl or K_2SO_4, which are not adsorbed in the copolymer micelles, dehydrate the hydrophilic portion of the copolymer and induce a pronounced decrease in the preferential interfacial curvature of the micelles. In contrast, salting-in electrolytes (hydrotropic ions) adsorbed in the micelles tend to inhibit micelle growth, and may increase the preferential interfacial curvature [123, 147]. Inorganic salting-out electrolytes, in addition to dehydrating the PEO blocks, also contribute to a pronounced increase in the organic–inorganic interfacial energy. This effect seems to reduce the interpenetration degree of the PEO chains inside the inorganic structure, thus leading ultimately to materials with lower micropore volumes [148]. Yu et al. were the first to use the salting-out effects caused by the addition of electrolytes in combination with triblock copolymers to generate various well-defined mesostructures [149, 150]. In contrast, the salting-in effect observed upon the addition of NaI was employed successfully to prepare large-pore cubic $Ia\bar{3}d$ silica [124]. In the case of salt additions, it is important to note that it is not only the aggregation behavior of copolymer micelles that is affected, but also that the presence of electrolytes will influence hydrolysis and condensation, as well as the kinetics of aggregation of the silica precursors.

D Organic Additives The solubilization of hydrophobic additives inside the core of the micellar assembly is another method employed to increase the pore size of mesoporous silicas. Trimethylbenzene (TMB) has been the most widely used additive [18, 54, 118, 119], although aliphatic hydrocarbons such as hexane have also been used [151]. In the early studies conducted by the Mobil research group, it was shown that the pore size of MCM-41 could be varied in controlled manner between 1.5 nm and 10 nm by the addition of TMB [18]. A near-linear increase in pore size was observed with increasing oil concentration (TMB). In general, hydrocarbons or hydrophobic aromatics (TMB, toluene [151]) are considered to be swelling agents that are solubilized in the core of the surfactant aggregate. On the other hand, cosurfactant molecules, such as alcohol or amines, are found to accumulate in the palisade layer of the aggregates, and may therefore have more complex effects. Both, the phase behavior and the d-spacing of mesoscopically ordered silicates, can be affected by adding either short-chain n-alcohols [152], n-amines [153] or polar benzene derivatives [154] as cosurfactants. In particular, the addition of short-chain amines, such as butylamine, can decrease the pore size of the inorganic structures prepared under basic conditions [155]. In contrast, variation of the cohesive properties of the solvent by performing the synthesis in mixed solvents can also be used to decrease the d-spacing of the resulting material, and allows fine-tuning of the pore size [156]. Cosolvents change the solution thermodynamics, which either alters the packing or the number of surfactant molecules in the micelles. Many of these modifications used in the synthesis of MCM-41 can also be applied to HMS materials and others.

By using TMB as a swelling agent, the pore size of block copolymer-based mesoporous silica can be enlarged significantly. Silica materials with pores reaching 30 nm or more were obtained, however, in the form of disordered mesocellar foams (MCF). A pore size expansion of SBA-15 with retention of the 2-D hexagonal ordered mesostructure, seems to be restricted to sizes reaching 15 nm, or slightly above, at the maximum [119, 157]. When using TMB again, Fan et al. developed an interesting approach to modify the bimodal porosity of SBA-15. Under specific synthesis conditions, the introduction of TMB into the P123 micelles was found to provoke a pronounced enlargement of the intrawall complementary porosity. With high-temperature hydrothermal treatment (130–140 °C) and the addition of TMB to the initial synthesis mixture, the wall microporosity almost vanished, leaving only very large connecting mesopores, also called "mesotunnels", with random distribution. The size of these tunnels was determined at about 8 nm using HRTEM [158]. Under these conditions, SBA-15 apparently becomes a fully interconnected 3-D material.

The complexity of PEO-PPO-PEO/water/cosolute systems has been revealed by Alexandridis et al., who established correlations between the cosolvent/cosolute polarity and its location within each block, or at the different interfaces [58, 159]. With this in mind, it is in principle possible to design different mesostructures, and this has been achieved by using n-butanol as a cosolute in the preparation of various highly ordered mesoporous silicas [120, 125, 160]. However, in the case of syntheses based on triblock copolymers, adjustment of the HCl concentration used for the syntheses is a prerequisite. At HCl concentrations of 0.3–0.5 M, it became possible to enrich the phase behavior of the triblock copolymers in water in the presence of polymerizing silica species, with the help of slower silica condensation kinetics. It is likely that the lower concentrations of the acid catalyst provide more flexibility and facilitate the synthesis of mesoporous silica in more designed ways. The addition of butanol as a cosurfactant, coupled with the low HCl concentrations, allowed the cubic $Ia\bar{3}d$ KIT-6 mesophase to be obtained in high phase purity [138]. When using this method, it is also possible to synthesize large-pore cubic phases ($Ia\bar{3}d$, $Im\bar{3}m$ and $Fm\bar{3}m$) over a wide range of starting mixture compositions, and diagrams of phase domains in the different $SiO_2-EO_xPO_yEO_x-BuOH-H_2O-HCl$ systems are available [138, 161]. As shown in the diagrams plotted in Fig. 13, different mesophases are generated as a function of the molar amounts of TEOS and butanol introduced in the synthesis mixture.

Further modification of the mesophase behavior with the simultaneous introduction of organic functional groups is achieved by performing co-condensation reactions (see Section 2.3.6.4) involving TEOS and organo-alkoxysilanes. Particularly remarkable here are the syntheses of the large-pore cubic $Ia\bar{3}d$ phase, either by the cocondensation reaction of TEOS and mercaptoalkoxysilane derivatives [162], or by using a vinyl-containing silane in the presence of NaCl [123].

E Stability and Zeolitization The stability of the materials is influenced by wall thickness and also the degree of silica polymerization/condensation. Several approaches exist to improve this property, including salt addition during the hydrothermal synthesis [163], hydrophobic modification of the surface by silylation [164, 165], post-synthesis grafting of inorganic compounds to increase wall thickness or chemically stabilize the surface [166], and zeolitization of the walls of mesoporous silica. Zeolitization, in particular, is an interesting development regarding the thermal or hydrothermal stability and surface acidity of the materials. Two possible methods have been proposed: (i) the walls of an ordered mesoporous silica are subsequently converted into zeolitic units [167]; or (ii) mesoporous silica is assembled from solution containing zeolitic structural elements [168–170]. Of these two methods, the second seems to be the most successful. Alternatively, the walls of mesoporous silicas can be coated with zeolitic fragments [171], leading to materials that should combine the acidity of zeolites with the improved mass transfer of mesoporous materials.

2.3.6.3.5 Removal of the Template When the material has reached a sufficient degree of condensation, the templating molecules are no longer needed and can be removed to open the porous structure. As some hybrid mesophases may contain 45 to 55% of organic substance by weight, the removal procedure of the organics is of utmost importance in the preparation process. Moreover, as this step can cause considerable variation in the final properties of the desired materials, several studies have focused specifically on template removal [133, 134, 172–176].

References see page 214

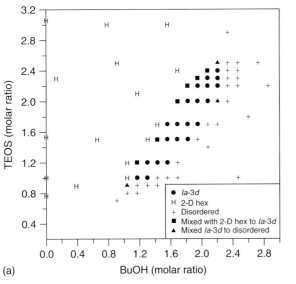
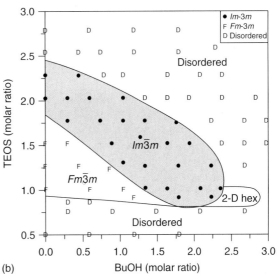

Fig. 13 Diagrams of mesophase structures synthesized using blends of triblock copolymer and n-butanol. The diagrams are established according to XRD measurements. Left: Each sample is prepared with a molar ratio 0.017 P123/x TEOS/y BuOH/1.83 HCl/195 H_2O. Right: Each sample is prepared with a molar ratio of 0.0035 F127/x TEOS/y BuOH/0.91 HCl/117 H_2O. (Reprinted with permission from Refs. [138, 161]; © 2005, 2006, American Chemical Society.)

A Calcination The most common method used in laboratories to remove the template is *calcination*, whereby the as-synthesized materials are alternatively heated in flowing nitrogen, oxygen or air, to burn away the organics. Any necessary charge-compensating counterions are supplied from the decomposition of organics. Usually, the heating rates required are slow, with typical heating ramps of $1\,°C\,min^{-1}$ up to $550\,°C$. This is followed by an extended period of heating at a temperature plateau (4–8 h). In addition, the calcination of as-synthesized mesophase containing large amounts of carbonaceous species can leave carbon deposits or coke as a contaminant in the porous materials, and pore blocking may occur as a consequence. In general, when the template is removed by calcination, an increase in XRD reflection intensities is observed, the mesostructure often shrinks, and the mesoscopic order may be strongly affected. The reported conditions for calcination are found to vary widely; for example, MCM-41 was originally calcined at $540\,°C$ in N_2 for 1 h and then in O_2 for 6 h [18]. However, a standard general procedure that can be used to calcine mesostructured materials is performed under air at $550\,°C$ during 5 h, this temperature being reached at a heating rate of $1\,°C\,min^{-1}$. The first investigations into calcination were described in the early reports on MCM-41 from the Mobil research group [18], which studied the removal of the alkylammonium template by thermogravimetric analysis combined with temperature-programmed amine desorption analysis (TPAD). Evolving species were observed which corresponded to the decomposition of $C_{16}H_{33}(CH_3)_3N^+$ into hexadecene and trimethylamine. It was proposed that, as the siloxy groups were stronger bases (silanols are weaker acids), they could promote the Hofmann elimination, which is also suggested in the case of MFI-type zeolites [177]. Template decomposition based on Hofmann degradation was confirmed by Corma et al. [172], who carried out an *in-situ* IR study of the thermal desorption of the template.

Although calcination has the advantage of being the most efficient method for the complete removal of organics in one step, the process has two major consequences, in that there is: (i) a decrease in the unit cell of the material; and (ii) an increase of the surface hydrophobicity, with both effects being caused by silanol condensation. The ratio Q^3/Q^4 (measured with ^{29}Si solid-state MAS NMR) is commonly used to estimate the degree of condensation. The reported framework condensation increases from Q^3/Q^4 of about 0.67 in the as-synthesized MCM-41 precursor to about 0.25 after calcination. Removal of the template by calcination from mesostructured silica materials was further investigated by Kleitz et al. [133, 134], who combined high-temperature XRD, thermogravimetry-differential thermal analysis, and mass spectrometry. The *in-situ* XRD studies showed a strong increase in X-ray scattering contrast for the low-angle reflections when the template was removed from the inside of the pores. The analyses proved that removal of the surfactant in the case of MCM-41 or MCM-48, both of which were synthesized under alkaline conditions, is a stepwise mechanism which takes place first via the Hofmann degradation at lower temperatures ($150–250\,°C$), followed by oxidation and combustion reactions above $250\,°C$. On the other hand, materials synthesized following the acidic route showed different behaviors, depending on the type of SDA employed. The acidic conditions used to synthesize SBA-3-type materials seemed to favor reactions of oxidation after an initial evaporation of water and hydrochloric acid. Furthermore, despite having thicker walls, the materials obtained via the acidic synthesis route showed the highest lattice shrinkage. The case of SBA-15 was also interesting with regards to template removal. Here, two main processes occur: at low temperature ($80\,°C$), desorption of physisorbed water takes place, followed at higher temperature ($145\,°C$) by exothermic decomposition of the template [92, 93, 134]. Residual carbonaceous species and water are removed from the structure upon heating from $300\,°C$ up to $550\,°C$. During this latter secondary process, a large contraction of the hexagonal unit cell is observed, as a result of further framework condensation. SBA-15 materials prepared with aging temperatures below $90\,°C$ show a contraction of their network of about 10–15%, whereas higher aging temperatures seem to lead to better framework stability. This framework contraction has a direct influence not only on the mesopore size but also on the intrawall porosity. High-temperature calcination treatments (up to $900\,°C$) were claimed to diminish dramatically, or even suppress totally, the complementary wall porosity [178]. In order to improve the efficiency of the calcination treatment and limit the detrimental exothermic effects, a brief pre-extraction step using an ethanol/HCl mixture is often performed in the case of SBA-15-type materials.

B Solvent Extraction and Acid Treatments An alternative method for surfactant removal is to extract the organic template. This can be achieved either by liquid extraction [65, 179], acid treatment [180], oxygen plasma treatment [180], or supercritical fluid extraction [181]. Dried as-synthesized MCM-41 samples are usually extracted in acidic solutions, alcohols or salt solutions. For example, Hitz et al. [179] showed that MCM-41 samples could be extracted in acidic media for 1 h at $78\,°C$, whereupon up to 73% of the template could be removed by extraction with solutions of an acid or salt in ethanol. These authors showed that, when extracting with acidic ethanol, ion exchange of the countercations for protons could be achieved simultaneously. The use of strong acid or small cations proved to be more efficient for extraction of the template in ethanol, which suggested that the size – and thus the mobility – of the cations in the close-packed micellar aggregates is one of the factors determining the extent of extraction. Acids with a low acid dissociation constant (e.g., CH_3COOH) are less efficient. Moreover, it seems that more polar solvents are superior for dissolving the template ions. Accordingly, it is widely suggested that an ion-exchange mechanism occurs during solvent extraction of M41S-type materials, and the presence of cationic species in the extraction liquid for charge balance is therefore essential for the ion exchange. Various acidic media are used for surfactant extraction, with ethanolic HCl solutions being the most frequently employed. The HMS- or SBA-type frameworks are considered to be relatively neutral, and the resulting framework–surfactant interactions are weak. Such weak electrostatic interactions or hydrogen bonding are more favorable for surfactant extraction, even in the absence of cationic species, as counterions are not needed. It is, for example, possible to remove large amounts of cationic surfactant from an SBA-3 mesophase by extraction in boiling ethanol for a short time [182]. The templates from mesophases obtained with long-chain amines [65], as well as transition metal-based mesophases obtained from the ligand-assisted method [25] are readily extracted. Block copolymers could also be extracted from SBA-15 using acidic ethanol solutions for short periods and at low temperatures [101]. In addition, an increase in the aging temperature of SBA-15 seems to lead to a more efficient removal of the template by solvent extraction. Indeed, the high temperatures dehydrate the PEO chains, which will interact less strongly with the inorganic framework. Clearly, the possibility of extracting the template molecules depends heavily on the nature of the interactions between the template and inorganics, and the efficiency of this method relies on a balance between extraction time and temperature as well as on the composition and concentration of the extraction solution. The advantages of this method are that the physico-chemical properties are not drastically modified, and framework contraction is very limited. Furthermore, by using solvent extraction, it is possible to extract the structure-directing species without affecting organic groups that are eventually present in the silica framework. However, in the case of SBA-15 materials, solvent extraction cannot provide a complete elimination of the templating species, and intrawall microporosity is often not fully liberated.

Another method of removing the template from triblock copolymer-templated silicas is to perform an oxidation of the organics with hydrogen peroxide in the presence of nitric acid under controlled microwave conditions [183]. When compared with conventionally calcined samples, materials which have been microwave-digested with H_2O_2/HNO_3 show a much higher pore volume and, most importantly, a much higher density of silanol groups on the pore surface. This method is fast (<15 min), enabling complete removal of the organics and the generation of a material with high pore volume and a more hydrophilic surface, but it cannot be used with functionalized organic–inorganic hybrids.

Finally, Yang et al. [175, 184] have found that the template can be removed from triblock copolymer-directed silicas in a sequential manner. These authors developed a method that permits first, the liberation of the main mesopores, and then the intrawall porosity. The copolymer template is fragmentized by hydrolysis and ether cleavage with aqueous sulfuric acid. As discussed earlier, in the as-synthesized form of SBA-15, the PPO groups are placed predominantly in the mesopore region, whereas the PEO groups are embedded into the silica walls. Upon conventional calcination, removal of the PEO moieties from the walls usually creates micropores that eventually connect the mesopore channels. The ether cleavage with sulfuric acid (48 wt.% in water) solution removes the accessible polymer located in the large mesopore system, whereas the PEO groups are essentially unaffected. Mild calcination at 200 °C then allows removal of the remaining PEO groups in the walls and makes microporosity accessible. This acid treatment prevents any pronounced shrinkage of the mesostructure, which in turn affords materials with much larger pore dimensions than the regularly calcined SBA-15. More importantly, the strongly acidic medium induces a more pronounced condensation of silica, but also permits the generation of a high density of silanols on the surface, as opposed to the calcination method. This may be important for certain catalytic applications, as demonstrated by Palkovits et al. Acid-treated SBA-15 materials, after mild calcination, were used as catalysts in the vapor-phase Beckmann rearrangement of cyclohexanone oxime to ε-caprolactam. These acid-treated samples exhibited the highest activities, corresponding to a very high total silanol concentration, whereas SBA-15 samples that were conventionally calcined at 550 °C showed lower activity, most likely because of fewer silanol groups [185]. In terms of porosity, the acid treatment of SBA-15 prepared with aging at 95 °C is somewhat comparable to preparing SBA-15 at higher temperatures (>130°C), leading to large mesopore sizes, but with clear differences in the wall microporosity. Sulfuric acid treatment is a good method if materials are required which exhibit large pore volumes and high stability, while conserving a relatively high wall microporosity and hydrophilic surfaces. In

---

References see page 214

particular, the method seems to be suitable for the removal of polyether templates, and may be used for some organically functionalized materials depending on the nature of the attached organic groups [123]. Furthermore, simultaneous template removal and pore topology control is possible when performing sulfuric acid treatment of SBA-16-type silicas that are prepared under relatively low HCl concentrations (0.4 M). Under these conditions, the pore apertures of the cage-like silicas are widely enlarged, resulting in 3-D cubic materials with large (*pseudo*)-cylindrical pores [186].

2.3.6.4 Functionalization of Ordered Mesoporous Materials

Ordered mesoporous silica is not often used itself as a catalyst. More frequently, functionalities are introduced by the incorporation of active sites inside the silica walls, or by the deposition of catalytic species on the inner surface of the mesopores. The number of methods available for the functionalization of ordered mesoporous materials is vast, and many studies have been dedicated to potential applications of functionalized mesoporous materials, especially in heterogeneous catalysis. It is not intended that this section serves as a comprehensive survey of the topic; rather, it will provide only a brief overview of the various strategies developed to modify mesoporous materials, accompanied by a few representative examples. For further information, the reader should consult comprehensive reviews on the topic [13, 25–27, 31–34, 37, 187, 188]. The preparation of catalytic materials through the deposition of highly dispersed metal and metal oxide particles will not be discussed at this point.

2.3.6.4.1 Functionalization Strategies
In contrast to microporous zeolites, mesoporous materials are not strongly acidic by themselves. Instead, these materials must often be functionalized with active sites for applications. A wide choice of different textural properties, pore sizes and pore shape, with high surface areas and large pore volumes, makes ordered mesoporous silicas highly attractive as supports. There are many methods available to modify ordered mesoporous materials but, depending on the location of the incorporated functionalities, two aspects of modifications can be distinguished: (i) surface-restricted functionalization; and (ii) framework functionalization with active species or functional groups placed inside the framework walls. In both cases, it is possible to introduce all manner of functional groups, including inorganic active species (heteroelements, clusters, nanoparticles, oxide layers), organic functional groups (acid, base, ligands, chiral entities), molecular organometallic catalysts, and polymers (fillings, coatings, dendrimers). Furthermore, the introduction of functional groups or modifications may be performed either directly during the synthesis of the materials, or subsequently by post-synthesis procedures.

Among strategies for functionalization, the association of organic and inorganic units within a single material is particularly attractive because it allows the possibility of combining the enormous functional diversity of organic chemistry with the advantages of a thermally stable and mechanically robust inorganic scaffold. The inorganic framework ensures a stable mesoscopically ordered structure. On the other hand, the organic species integrated to the material permit fine-tuning of both the interfacial and bulk properties of the material (e.g., hydrophobicity, porosity, site accessibility, physical properties). Two main strategies exist for the preparation of mesoporous organic–inorganic hybrid materials:

- Modification of the pore surface of a purely inorganic material by post-synthesis methods. This is usually carried out by covalent grafting methods, but impregnation or adsorption are also sometimes employed.
- Direct incorporation of the organic functionalities by condensation reaction of functional organosilica precursors in a so-called "one-pot" synthesis.

For the second approach, two methods are again possible. The first method is to prepare ordered mesoporous organosilicas by co-condensation of TEOS (or TMOS) and terminal trialkoxyorganosilanes of the type $(R'O)_3Si-R$ (where R' is either methyl or ethyl and R is a non-hydrolyzable organic group). With this method, the functional groups are most often introduced not only onto the surface but also partly inside the framework walls. The second method is based on using bis-silylated organosilicas as single precursors, which allows the direct incorporation of organic groups as bridging components specifically into the pore walls. Organic functions are thus incorporated as bridges between silicon atoms within the wall. The advantages and drawbacks of these different strategies, and the methods used for characterization, have been discussed extensively elsewhere [27, 34, 37, 189–191].

Functionalization of ordered mesoporous silicas is also performed by introducing various inorganic heteroelements or oxidic compounds either selectively onto the surface or into the framework of the mesoporous material. The incorporation of metal centers into a silica framework is usually achieved by post-synthesis treatment, or by direct mixing of adequate precursors in the initial synthesis mixture. Again, methods can be distinguished which result in the positioning of inorganic components preferentially on the mesopore surface, mostly as metal oxides, and substitution of silicon atoms by metal ions within the framework, similarly to zeolites. In general, pore surface functionalization is achieved by post-synthesis grafting of metal centers such as Al, Ti, V, Cr, and Mo onto the surface silanols [33, 35, 192–194]. On the other hand, modification of the mesoporous framework by substitution of silicon atoms is possible via one-pot direct synthesis [13, 25, 33, 35]. Both methods are employed to generate acid or redox active sites, and the thus-obtained materials may be used for different catalytic reactions. However, the outcome of the two different methods is often not identical. Further, fine control and proper characterization of the localization and distribution of the heteroelements throughout the silica framework are not always straightforward. While the direct method typically results in a relatively homogeneous incorporation of the heteroelements in the material, the post-synthesis treatments primarily modify the wall surface and lead to higher concentrations of the heteroelement on the mesopore surface. Less-ordered mesostructures also sometimes result from the direct synthesis route.

The elaboration of methods to modify ordered mesoporous materials in a sequential and spatially controlled manner is an emerging concept. In particular, several strategies permit spatial control of the material functionalization and a selective positioning of active sites [35, 187, 190, 195]. Selective passivation of the silanol groups on the external surface of ordered mesoporous silica by reaction with small amounts of diphenyldichlorosilane was the first example of spatially controlled modification [196]. After such a passivation step, the interior of the pore system remained free to be selectively functionalized. This strategy has subsequently been implemented for the creation of enantioselective catalysts based on silica [197]. Currently, other new approaches towards the spatially resolved positioning of functional groups inside mesoporous materials, and the fabrication of novel systems based on multiple functions operating cooperatively, are being developed worldwide [198]. Although detailed characterization of these materials remains rather difficult. Both, XRD and electron microscopy have been shown suitable for the analysis of spatial modifications. For example, XRD patterns are dependent on the location of guest species in the pore system [199–201], while HRTEM allows direct visualization of guest species introduced into the materials [202, 203].

2.3.6.4.2 **Surface Properties** Many of these modification and functionalization methods use the silanol groups (SiOH) that are present at the silica mesopore surface as anchoring sites for metal species or silane coupling agents. Therefore, information concerning surface – OH sites is important for surface modifications such as silylation or organosilane coupling. At present, there are indications that silanol density in ordered mesoporous silica (e.g., MCM-41 or SBA-15) is much lower than that of conventional hydroxylated silica. Moreover, unmodified ordered mesoporous silicas have some distribution of surface sites. In general, the silanol density of most mesoporous silicas ranges between one and three SiOH per nm^2 [34, 204, 205]. By comparison, the silanol density of regular hydroxylated silica is about four to six SiOH per nm^2 [206]. For SBA-15, a silanol density closer to one SiOH per nm^2 is observed if the material is pretreated (activated) at 200 °C under vacuum [207]. It is clear, however, that the population of silanols and their distribution on the mesopore surface will depend greatly on thermal treatment conditions, activation under vacuum or dehydroxylation–rehydroxylation processes. By carrying out the adsorption of polar and non-polar probe molecules monitored by Fourier transform infrared (FTIR) spectroscopy, a relatively hydrophobic character of MCM-41 was suggested [18, 208]. Following removal of the template, MCM-41 adsorbs a much larger amount of organic substances than water, which indicates that the internal surfaces may be quite hydrophobic, even though silanol groups are present. However, there is evidence of surface heterogeneity with hydrophobic and hydrophilic patches, at least for the MCM-41-type materials [209]. The nature of the silanol groups is also relevant: at least three different types of silanol group can be distinguished in MCM-41 silica by using pyridine adsorption and spectroscopic examinations [204]: single silanol (SiO)$_3$SiOH, hydrogen-bonded (SiO)$_3$SiOH−OHSi(SiO)$_3$, and geminal silanol groups (SiO)$_2$Si(OH)$_2$. However, only single and geminal silanols seem to be accessible to a silylating agent such as trimethylchlorosilane. Surface silylation allows the passivation of the silanol groups to obtain a more homogeneous surface. Such grafting of alkylorganosilanes on the mesopore surface to passivate silanol groups is also a known strategy that is employed to improve the structural stability of mesoporous silicas, because it provides a more hydrophobic nature to the walls [164]. It should be noted that, although grafting refers frequently to the subsequent modification of the pore surface via the covalent bonding of organic groups [27, 30, 37], the term is also used to describe the chemical surface fixation of inorganic active species (e.g., oxides, metal cations, etc.) [35, 210].

2.3.6.4.3 **Surface Functionalization** A significant number of techniques has been developed to attach organic functionalities to the walls of mesoporous silica. Organic

---

References see page 214

functions can be anchored covalently (or tethered) on the oxide walls, leading to hybrid mesostructured materials that combine the properties of the inorganic mesoporous framework and tunable organic surface properties. The numerous methods available for surface modification of ordered mesoporous materials are often based on standard techniques that were developed for the surface functionalization of conventional silicas and used, for example, in chromatography [27, 37, 211, 212]. Surface silylation is relatively straightforward when using trialkylchloro- or trialkylalkoxysilanes, and was described in the early reports of the Mobil group [18]. Alternatively, hexamethyldisilazane is also an efficient silylating agent. In this first approach, grafting takes place via a condensation reaction between organoalkoxysilanes of the type $(R'O)_3SiR$, and the free surface silanol groups. Chlorosilanes $(ClSiR_3)$ or silazanes $(HN(SiR_3)_2)$ are also suitable organosilane precursors. In principle, functionalization with a variety of organic groups can be realized in this way by varying the nature of the organic residue R. However, the mesoporous hosts must be thoroughly dried before addition of the organosilane precursors in order to avoid self-condensation in the presence of water. Post-synthesis surface modification of mesoporous silica with basic functional groups is achieved by using condensation of surface silanols with 3-aminopropyltriethoxysilane for example, resulting in $-NH_2$ groups placed on the pore surface. Trimethoxysilylpropylethylenediamine is also frequently employed. Acid functionalities may also be incorporated by post-synthesis grafting. As examples, MCM-41 [213] and SBA-15 [214] can be first functionalized by post-synthesis modification with organosilanes bearing propylthiol groups, which are subsequently converted into propylsulfonic acid groups under mild oxidative conditions with H_2O_2. In all cases, however, the concentration and repartition of the organic functionalities is dictated by the surface silanols and their accessibility. The grafting ratio is known to depend on the precursor reactivity, being also limited by diffusion and steric effects. The most frequently used functional groups covalently linked to a silica surface are summarized in Fig. 14. These functions may serve not only as acid or base catalysts and interaction sites, but also are useful for further immobilizing various larger regio- and enantioselective molecular catalysts (e.g., organometallic complexes, chiral catalysts), as detailed in Refs. [27, 31, 32, 35, 37, 187]. In the case of anchored molecular catalysts, optimum binding and a designed active site environment are required. Moreover, depending on the length of the spacer connecting the catalyst to the surface or any specific functional group included in the tether, the activity and selectivity of a catalyst may be tuned [35, 215]. By using post-synthesis functionalization, complex architectures may be constructed within the mesopore system.

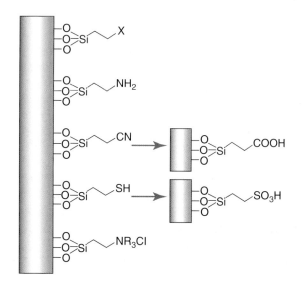

Fig. 14 Representation of the most frequently used functional groups covalently linked to a silica surface and employed for anchoring molecular catalysts. (Adapted from Ref. [35].)

As a particularly interesting example, Acosta et al. [216] were able to build quite large, melamine-like structures which resembled dendrimers, confined inside the pores of amino-functionalized SBA-15.

In 1996, Burkett et al. [217] and Macquarrie [218] introduced the direct formation of organic–inorganic hybrid solids by the co-condensation of tetraalkoxysilanes and organically functionalized silanes in the presence of SDAs. This direct incorporation of organic groups, which was also called the "one-pot method", produces mesoporous materials with the organic moieties pending on the walls [27, 37]. In particular, Macquarrie used 3-aminopropyltriethoxysilane as the organosilica precursor to prepare, in one step, a modified HMS silica with pending amine groups. In this pathway, the organoalkoxysilane actually plays two roles as it contributes as a building block in the inorganic structure, co-condensing with the tetraalkoxysilane precursor, and it also provides the organic functionality. Here again, various functional groups can be introduced by using this pathway (vinyl, phenyl, aminopropyl, imidazole, cyanopropyl, mercaptopropyl, more complex ligands, etc.) [27, 37]. By using this approach, however, the choice of the organosilane precursor is restricted by the synthesis conditions. For example, syntheses carried out under alkaline conditions and hydrothemal conditions will limit the choice of precursors in terms of their Si–C bond stability. Furthermore, template removal can only be carried out by extraction using organic solvents or acidic media. The mercaptopropyl groups are especially interesting as these may subsequently be oxidized with nitric acid or

H_2O_2 to yield sulfonic acid groups pending in the mesopores. Stucky showed that oxidation of the mercaptopropyl groups could even be achieved *in situ* during the material formation with the addition of H_2O_2 directly in the acidic reaction medium [219]. In this way, the direct synthesis method for the preparation of sulfonic acid-functionalized ordered mesoporous SBA-15 materials could be improved significantly in terms of synthesis time and processing conditions. SBA-15 mesoporous silica can also be functionalized with arenesulfonic acid groups by means of such a one-step simple synthesis approach [191, 220] involving, for example, the co-condensation of TEOS and 2-(4-chlorosulfonylphenyl)-ethyltrimethoxysilane. The different principles of mesopore surface functionalization achieved either via the post-synthetic grafting method or by co-condensation are illustrated schematically in Fig. 15.

2.3.6.4.4 Framework Functionalization The two principal methods for efficient framework modification are: (i) the inclusion of organic groups within the silica walls according to the route developed for the preparation of periodic mesoporous organosilica; and (ii) the isomorphous substitution of silicon atoms with heteroelements. A third approach, which is exploited much less often, is emerging in connection with the concept of spatial design of mesostructured materials, and involves the possibility of placing entities selectively inside the complementary intrawall pores of triblock copolymer-templated silicas (e.g., SBA-15, SBA-16, KIT-6). To date, only a few attempts have been reported, with the introduction of Pd particles [198b, 221] or polymeric units [222]. The following section will focus on the two first methods, both of which are widely implemented.

The use of silsequioxanes such as [(RO)$_3$Si—R'—Si(OR)$_3$] is widespread in sol–gel chemistry for the preparation of disordered porous aerogels and xerogels hybrids with high surface areas [223]. When combined with a SDA (CTAB or P123), similar bridged silsequioxanes can also be used in direct one-pot syntheses to generate organically modified mesoporous silica. In contrast to porous gels, the resulting materials are characterized by an ordered mesopore system (2-D hexagonal, 3-D hexagonal, 3-D cubic phases) exhibiting a narrow pore size distribution. This allowed the construction of a new class of mesostructured materials which are often designated as periodic mesoporous organosilicas (PMOs). In these mesoporous materials, the organic groups R' are homogeneously incorporated as bridges between silicon atoms inside the mesoporous silica walls, which permits the addition of new functions without pore blocking. Since the alkyl-C bond is not affected during synthesis, the organic groups are usually retained in the final material. The first PMOs were synthesized in 1999 by three independent research groups using either 1,2-bis(trimethoxysilyl)ethane (BTME) [224] or 1,2-bis(triethoxysilyl)ethene [225, 226] as single organosilica precursors in the presence of alkylammonium surfactants as templates for the mesostructure formation. Later, the preparation of periodic organic–inorganic mesoporous hybrids was also extended to triblock copolymer systems, thereby affording PMOs with much larger pores [227, 228]. A great variety of bridging groups have been introduced (R' = methylene, ethylene, phenylene, vinylene, benzene, etc.), and even rather bulky functional groups can be incorporated into the framework in this way. As a striking example, the 1,4,8,11-tetraazacyclotetradecane chelate, a functional group with excellent binding ability toward transition metal centers, could be placed inside the walls of large-pore silica prepared with P123 [229]. A more detailed overview of the wide variety of different bis- or multi-silylated organosilica precursors employed for the synthesis of PMOs is provided by Hoffmann et al. in a comprehensive review of the topic [37].

References see page 214

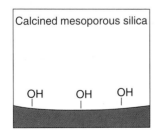
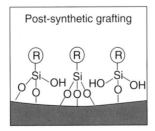
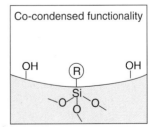

Fig. 15 Schematic illustration of the principles employed for functionalization of the mesopore surface. The regular silica mesopore surface exhibits a certain concentration of available Si—OH sites, depending on treatment conditions. Functionalization may be achieved subsequently on such a mesoporous silica by the post-grafting of functional organosilanes (through siloxane coupling on the surface). The other method consists of co-condensation of the tetraalkoxysilane silicon source and a functional organosilane in a one-pot synthesis. Note the difference between the two methods with regards to the nature of the anchoring site for the functional group. (Adapted from Ref. [37].)

Interestingly, combined meso- and molecular-scale periodicity can be observed when using 1,4-bis(triethoxysilyl) benzene (i.e., R′ = phenyl) as the organosilica precursor. By using alkyltrimethylammonium surfactant under alkaline conditions, Inagaki et al. [230] demonstrated the formation of a particularly well-ordered 2-D hexagonal mesostructure which showed additional structural periodicity with a spacing of 0.76 nm along the channel direction. Evidence for the crystalline nature of the wall structure was obtained from powder XRD data, which showed diffraction peaks at medium scattering angles in addition to the low angles typical of the 2-D hexagonal structure (Fig. 16a). The crystal-like organization of the walls was confirmed by HRTEM observations which showed lattice fringes along the pore axis. In the material, the benzene rings are integral parts of the wall structure. A model picture of the pore structure of PMO with crystal-like walls is shown in Fig. 16b. Other organosilica precursors, such as 4′,4-bis(triethoxysilyl)biphenyl or 1,4-bis[(E)-2-(triethoxysilyl)vinyl]benzene, are also suitable to provide PMO products with crystal-like organization of the organic bridges. Moreover, it is worth noting that the organic group incorporated in the pore walls can itself be functionalized by a post-treatment. Inagaki et al. showed, for example, the possibility of post-synthesis sulfonation of the above benzene-bridged PMO material, leading to a sulfonic acid-functionalized mesoporous solid. The PMO materials are considered as particularly promising candidates for technical applications in catalysis, adsorption, and chromatography.

The doping of mesoporous silicas with heteroelements has attracted enormous interest for the production of new catalytic materials. In general, the surface of purely siliceous mesoporous materials is only weakly acidic, and therefore, much effort has focused on the introduction of heteroatoms, particularly aluminum, in the siliceous framework to modify the composition of the inorganic walls. The isomorphous substitution of heteroatoms [13, 25, 33, 35, 231, 232] in mesoporous silica frameworks is especially important with respect to catalytic applications, as the substitution of silicon allows fine-tuning of the acidity or the creation of redox properties, as observed in amorphous alumosilicates or zeolites. These mesoporous materials are usually synthesized by incorporating a tetrahedrally coordinated tri- or tetravalent element, such as Al [18, 233, 234], B [235], Fe [236, 237], Ga [238, 239], or Ti [240, 241] in the silica framework. In addition to the hydrophobic/hydrophilic properties of the surface, the resulting catalytic behavior is strongly influenced by the environment and the stabilization of the metal introduced. MCM-41 materials containing trivalent elements (prepared by one-pot synthesis, post-impregnation, ion exchange, etc.) have been reported as having interesting catalytic activities [13, 25]. Aluminum incorporation seems to be of particular interest if, as in zeolites, this results in the formation of ion-exchange sites and Brønsted acidity. Tetrahedrally coordinated aluminum atoms can be substituted for silicon atoms in the MCM-41 framework, but the degree of substitution depends on the aluminum precursor and method of preparation [231]. Generally, such an incorporation of aluminum in the framework of MCM-41 is reported to increase the acid site concentration, ion-exchange capacity, and hydrothermal stability of the material [242]. However, in general, the strength of acid sites was found to be similar to that of amorphous aluminosilicates. As the local environment around the acid sites corresponds to amorphous materials, they do exhibit a weaker acidity than zeolites, and correspond to silica-alumina in terms

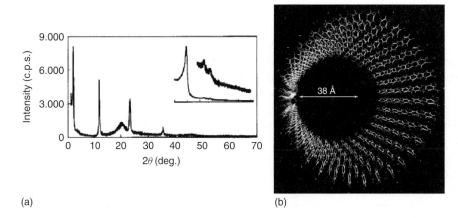

Fig. 16 Powder X-ray diffraction pattern (a) and model pore system (b) of a mesoporous benzene-bridged organosilica with crystalline walls. The benzene rings are aligned in a circular manner along the pore, connected at both sides by silica layers. The benzene and silica layers are arranged alternatively along the pore axis with intervals of 0.76 nm. (Reproduced with permission from Ref. [230]; © 2002, Nature Publishing Group.)

of the number of acid sites and acid strength distribution. Nevertheless, they contain larger pores than zeolites and thus are more suitable for catalytic reactions involving substrates which are too large to enter the pores of zeolites, for example bulky hydrocarbons [243–245]. In order to generate catalytic activity, and especially to create redox properties, many other metal ions have been incorporated into the silica framework, including Ti, Zr, V, Cr, Mn, Co, Cu, La, and Ce [25, 34, 231]. In particular, Ti-MCM-41 was shown to be a promising catalyst for the selective oxidation and epoxidation of organic molecules using peroxides. The incorporation of Ti into MCM-41 or SBA-15 is generally achieved via the direct synthesis procedure which involves addition of a titanium source, such as titanium isopropoxide in ethanol, to the gel for hydrothermal synthesis [246, 247]. It should be noted that the post-synthesis addition of Ti is also possible by grafting techniques [248–250], for example using $TiCl_4$, titanium alkoxide, or metallocene complexes. However, metal incorporation does not lead to the formation of defined sites as in zeolites, but rather to a wide variety of different sites with different local environments. Considering their catalytic properties, the transition metal-containing materials are also much closer in nature to substituted amorphous silica than framework-substituted crystalline zeolites.

2.3.6.5 Non-Siliceous Mesostructured and Mesoporous Materials

Shortly after the discovery of mesoporous M41S-type materials, the use of surfactant species to organize various non-siliceous mesostructured oxides was explored over a wide range of conditions [63]. Since then, a large number of mesoporous oxides, sulfides, phosphates and metals have been reported [22, 23, 251]. However, compared to silica-based networks, non-siliceous ordered mesoporous materials have attracted less attention, due to the relative difficulty of applying the principles employed to create mesoporous silicon oxides to non-silicate species. Moreover, other framework compositions are more susceptible to redox reactions, hydrolysis, or phase transformations. In particular, template removal, which is needed to achieve porous materials, has been shown to be a critical point [252]. In contrast to silicates, other compositions are usually more sensitive to thermal treatments, and calcination can result in a breakdown of the structural integrity. Hydrolysis, redox reactions or phase transformations to the thermodynamically preferred denser crystalline phases account for this lower thermal stability. Therefore, removal of the surfactant by thermal treatment is more difficult in the case of non-siliceous mesostructured materials, and many of the transition metal-based mesostructured materials synthesized in the presence of cationic surfactants were shown to collapse during thermal treatments.

In 1993, it was suggested that it might be possible to synthesize non-siliceous materials by substituting the silicate by metal oxides that are able to form polyoxoanions [62]. Mesostructured surfactant composites of tungsten oxide, molybdenum oxide and antimony oxide were thus obtained. This approach was extended to the charge reverse system (S^-I^+) by using polyoxocations and to the mediated combinations $S^+X^-I^+$ and $S^-M^+I^-$ [63, 252]. Several authors subsequently reported the syntheses of mesostructured vanadia with lamellar and hexagonal phase and vanadium phosphorus oxide with hexagonal, cubic ($Ia\bar{3}d$), and lamellar phases [253–255]. However, due to a poor thermal stability, none of these structures could be obtained as template-free mesoporous solids. The poor thermal stability observed is most probably due to the different oxo chemistry of the metals compared to silicon. Several oxidation states of the metal centers are assumed to be responsible for oxidation and/or reduction during calcination [252]. In addition, incomplete condensation of the framework is also possible [63].

2.3.6.5.1 Transition Metal Oxides The first porous transition metal oxide was reported by Antonelli et al. in 1995 [256], based on titania. The material was prepared by using an anionic surfactant with phosphate head groups and titanium alkoxy-precursors stabilized by bidentate ligands. After calcination at 350 °C, hexagonally ordered porous TiO_2 with surface area of about 200 $m^2\,g^{-1}$ and narrow pore size distribution around 3.2 nm was obtained, including phosphate groups in the framework. However, the thermal stability of this material was not very high. Later, other materials based on titanium [257–259], zirconium [260, 261] niobium oxide [262], and tantalum oxide [263] were synthesized. For the niobium and tantalum oxides, the redox stability problem was solved by using a ligand-assisted templating pathway, with subsequent extraction of the template, thereby avoiding calcination. The ligand-assisted pathway is based on covalent bonding between the inorganic species and the surfactant head groups (S-I) [262]. The synthesis of Nb- and Ta-TMS1 (Transition metal oxide Mesoporous molecular Sieve No. 1) involves the hydrolysis of long-chain amine complexes of niobium or tantalum alkoxides. The respective phase is formed by self-assembly of the metal alkoxide–amine complexes. The template can be removed by extraction, leading to open porous structures with surface areas of up to 500 $m^2\,g^{-1}$. One important area of application of these

References see page 214

mesoporous transition metal oxides is in photocatalysis. For example, Takahara and Domen [264] reported the photocatalytic activity of mesoporous tantalum oxide. By loading NiO into the Ta-TMS material, the photocatalytic activity for water decomposition could be improved, despite the walls of Ta-TMS being amorphous. Stucky and coworkers [265] developed an interesting synthesis procedure for a very wide range of non-siliceous oxides (e.g., TiO_2, ZrO_2, Nb_2O_5, Ta_2O_5, etc.) by using amphiphilic polyalkylene oxide triblock copolymers in non-aqueous solutions. The mesoporous solids obtained were thermally stable, ordered and showed a large pore size up to 14 nm. Whereas the pore walls of the materials described before were amorphous, these mesoporous materials contained nanocrystalline domains within relatively thick framework walls. The mesoporous oxides are believed to form through a mechanism that combines block copolymer self-assembly with alkylene oxide complexation of the inorganic metal species.

Non-silica systems present very fast inorganic condensation kinetics. There are several strategies aimed at retarding condensation, such as complexation, acidification or controlled hydrolysis [30]. The most successful is to selectively favor hydrolysis and to limit condensation by using a highly acidic medium (pH < 1) and working under EISA conditions [265, 266]. This strategy has been particularly applied to prepare films and aerosols, and ordered mesoporous titania [267] and zirconia [268] thin films have been obtained via this pathway. Whilst silica walls maintain their amorphous nature upon thermal treatment, crystalline phases can arise in transition metal-based frameworks. However, the growth of nanocrystals beyond the wall thickness leads to a deterioration in the mesostructure. With a block copolymer template, the walls are thicker, and the first studies by Yang et al. [265] revealed the presence of partially crystallized inorganic walls. Materials with high crystallite fractions were obtained later by Sanchez et al., who combined a structuring approach using block copolymer and the EISA method, associated with layer deposition, to produce nanocrystalline titania and zirconia thin films [269, 270] and nanocrystalline perovskite layers [271]. These materials were thought to show particular promise for applications in photocatalysis, electronics, or energy conversion.

2.3.6.5.2 **Alumina** Alumina, in particular, is of major importance in heterogeneous catalysis, for several reasons. Its use as a catalyst support is widespread as it is considered superior to silica because of a higher hydrolytic stability and its different PZC that facilitates loading with different metal species. Furthermore, it is inexpensive and available in a wide range of surface area and porosities. The surface area of traditional alumina materials manufactured by precipitation varies usually from a few $m^2\ g^{-1}$ up to 300 $m^2\ g^{-1}$, depending on its phase and texture [272]. Unfortunately, the pore size distribution is often broad, and this can be a major drawback in catalysis. Consequently, substantial effort has been focused on the possibility of fine-controlling the porosity of mesoporous alumina, introducing long-range order into the structure, and improving thermal stability [22, 273]. Mesoporous alumina has been synthesized by a variety of methods using anionic, cationic and neutral surfactants. Bagshaw and Pinnavaia reported the first synthesis of MCM-41-related mesoporous alumina using non-ionic poly(ethylene oxide) as the SDA and aluminum tri-sec-butoxide as the aluminum source [274]. The low-angle XRD pattern obtained exhibited a single broad peak which revealed a disordered pore structure resembling that of MSU silica. A surface area around 400 to 500 $m^2\ g^{-1}$ was reported. Similar worm-like pore structures were also obtained by Cabrera et al. [275], but these authors followed a pathway in conditions under which the aluminum species in solution were anionic, and they used a cationic surfactant in combination with triethanolamine in water. By using this pathway (which was quite similar to that used to synthesize MCM-41), the authors were able to adjust the pore sizes to between 3.3 and 6 nm by carefully varying the ratio among surfactant, water, and triethanolamine. The approach developed by Stucky et al. [265], as mentioned previously (see Section 2.3.6.5.1), also permitted access to alumina materials with quite large pore sizes. However, these materials exhibited a disordered pore structure, and their pore size distribution appeared to be bimodal, with the strongest contribution from pores centered around 14 nm.

Later, neutral diblock and triblock copolymers were also used in a multistep synthesis of mesoporous alumina with aluminum salts as metal source. Interestingly, the size of the pores was found to be independent of the size of the SDAs [276]. In turn, it became evident that in the alumina system the use of surfactant-type molecules for the synthesis does not necessarily follow a real templating route. The possibility should also be considered that the surfactant does not act as a porogen, but rather stabilizes small alumina particles. Furthermore, it seems that the presence of water is an issue in the treatment of mesoporous alumina materials. Indeed, it was observed that washing the sample after filtration with distilled water led to a collapse of the mesostructure, and hence the use of alcohols is recommended for the washing step. Structural collapse caused by the presence of water was also observed after calcination or template extraction [273]. Template removal is a particularly critical step in the case of mesoporous alumina.

Nevertheless, a major success in the preparation of ordered mesoporous alumina was finally achieved by the group of Sanchez. By combining triblock copolymer structure-direction and the EISA method, these authors showed that it was possible to prepare thermally stable nanocrystalline γ-alumina films which exhibited ordered mesoporosity with a 3-D face-centered cubic structure [277]. The ordered mesoporous alumina films were found to be stable up to temperatures of 900 °C. The pores were ellipsoidal with the shorter pore axis ranging from 3 to 6 nm in size, and the larger pore axis, as deduced from the pore aspect ratio, ranged from 13 to 24 nm (Fig. 17).

2.3.6.5.3 Other Non-Siliceous Compositions

Among non-siliceous mesoporous materials, phosphates [259–261, 278], sulfides [279–281] and selenides [282], metals, and many others may also be of interest as catalysts or catalyst supports. As an example, hexagonal and cubic mesoporous tin(IV) phosphates, as described by Serre et al. [283], were used for NO removal in the presence of C_2H_4 and O_2. Titanium phosphate materials were used for liquid-phase oxidation of cyclohexane in the presence of H_2O_2 [259]. Also of interest, mesostructured zirconium-based composites with hexagonal phase could be readily produced with zirconium sulfate as the inorganic precursor and alkylammonium surfactants [260]. Zirconium sulfate proved to be an ideal precursor because the negatively charged solution species interact favorably with the cationic surfactant molecules. However, the presence of sulfate groups in the inorganic framework prevents full condensation, and leads to major framework destruction because of the removal of the sulfate upon calcination. By using a post-synthesis treatment with phosphoric acid, materials with improved thermal stability were obtained [260, 284] which retained their mesoporosity upon thermal removal of the template. A thermal stability up to 500 °C is due to crystallization delay caused by the phosphate groups in the structure, so that the disordered wall structure which is favorable for mesoporous materials is retained. These porous solids, which are referred to as "zirconium oxo-phosphates", show structures that are analogous to MCM-41 (see Fig. 18a), or analogous to MCM-48 with cubic $Ia\bar{3}d$ symmetry [285]. Hexagonally ordered mesoporous titanium oxo-phosphates were also described [286]. Another interesting development was proposed by Kanatzidis, who described the coassembly of Ge_4S_{10} units together with Fe_4S_4 clusters, resulting in a well-ordered hexagonal structure. These thiogermanate materials incorporating Fe_4S_4 units may find applications as photocatalysts or for artificial photosynthesis [287].

Several authors have also reported mesoporous aluminophosphates. Crystalline, microporous aluminophosphate, silicoaluminophosphates and their transition metal-containing counterparts have found use in heterogeneous catalysis in hydrocarbon oxidation reactions and in methanol conversion. Mesostructured aluminophosphates were prepared under a wide range of synthetic conditions, with the pH varying from highly acidic to alkaline. A variety of different surfactants were employed such as anionic, neutral, and cationic compounds [288–290]. In many cases, the mesostructure was either disturbed or lost during calcination, but Cabrera et al. succeeded in obtaining a more thermally stable mesoporous aluminophosphate with pore sizes ranging from 1.3 to 3.7 nm as a function of the reaction conditions [291].

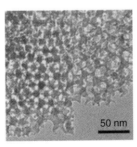

Fig. 17 TEM images of nanocrystalline γ-alumina films exhibiting ordered mesoporosity with a 3-D face-centered cubic structure. The pores are ellipsoidal, with the shorter pore axis ranging from 3 to 6 nm, and the larger pore axis (as deduced from the pore aspect ratio) ranging from 13 to 24 nm. (Adapted and reprinted with permission from Ref. [277]; © 2005, Wiley-VCH Verlag. Original images courtesy Drs. M. Kümmel and D. Grosso, Université Pierre et Marie Curie, Paris, France.)

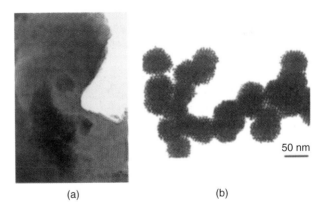

Fig. 18 (a) TEM image of calcined zirconium oxo-phosphate with hexagonal pore arrangement. (Reprinted with permission from Ref. [284]; © 1999, American Chemical Society.) (b) TEM image of a mesoporous Pt sample obtained via the true liquid crystal templating pathway. (Reprinted with permission from Ref. [292]; © 1997, Wiley-VCH Verlag.)

With regards to the direct preparation of metals by supramolecular templating strategies, Attard et al. succeeded relatively early in the synthesis of hexagonally ordered noble metals obtained in a mesoporous form [67, 292]. The syntheses are based on the TLCT mechanism, where a lyotropic LC phase is infiltrated with the metal precursor ions, which are then reduced within the liquid crystal. As a reducing agent, either less-noble metals are used or other agents, such as hydrazine. The surfactant, for example octaethylene glycol monohexadecyl ether, is removed by washing. The materials are obtained either as powders composed of particles with sizes of some hundred nanometers, or as mesostructured metallic thin films. Mesoporous metals, such as platinum or nickel, are hexagonally ordered (Fig. 18b) with pore sizes of approximately 3 nm and surface areas of 20 to 85 $m^2\,g^{-1}$, depending on the mass density (this is usually very high for metals) and the thickness of the walls. Subsequently, the strategy was adapted to other metals, metal alloys, and non-metals [293, 294].

2.3.6.6 Hard Templating (Nanocasting)

Despite remarkable progress in the synthesis of non-siliceous ordered mesoporous materials, liquid crystal-like templating methods were considered unsuitable for the synthesis of materials with composition and structure requiring high-temperature treatments. Especially, non-siliceous compositions such as carbon, polymers or some transition metal-based materials are very sensitive to thermal treatment conditions and redox reactions. A possible solution to this problem was identified by a South Korean research group, who used a new approach of hard templating, which they referred to as "nanocasting". Hence, Ryoo and coworkers succeeded in producing highly ordered mesoporous carbons with high surface areas and unprecedented narrow pore size distribution [39, 295, 296]. In contrast to the soft (LC) templating methods described above, the hard templating principle is based on the use of ordered mesoporous silica itself as a form of rigid nanoscopic mold. Hence, several of the ordered mesoporous silicas described in Section 2.3.6.3.3 were used as rigid templates for the preparation of a series of ordered mesoporous carbon materials and mesostructured noble metals. The ordered mesoporous carbons obtained via the hard templating pathway are designated as CMK-x materials (Carbon Molecular Sieves Korean Advanced Institute of Science and Technology No. x) [39, 295, 297]. This pathway, which essentially was pioneered by the group of Ryoo, may be considered as a major breakthrough in materials science. Indeed, it has emerged as one of the most successful alternatives for synthesizing highly ordered, non-siliceous mesoporous materials. Following the nanocasting preparation of mesoporous carbons and metal nanostructures, other mesostructured materials, such as transition metal oxides and sulfides, have been successfully obtained by using similar casting approaches [297–299].

The principle of the hard templating (nanocasting) strategy is shown schematically in Fig. 19, and the procedure may be described as follows. First, suitable precursors are incorporated (by sorption, ion-exchange, covalent grafting, etc.) inside the pores of a highly ordered mesoporous silica solid template such as SBA-15 or MCM-48. The precursor infiltration may be repeated to achieve high loadings, as this will facilitate the rigidification of the templated framework. When sufficient solidification has been achieved within the host pore system, and eventual heat treatment has been carried out to form a desired phase, the silica matrix may be selectively removed and the shape-reversed molded structures are obtained. Most importantly for successful replication, the resulting material composition must be stable either in diluted HF or NaOH solution, and the precursor must not react with silica at elevated temperatures. Because a mold with a 3-D bridged structure is necessary to maintain a stable replica, only experiments with silicas having interconnected mesopores are successful. In the case of SBA-15, for example, the complementary pores connect the

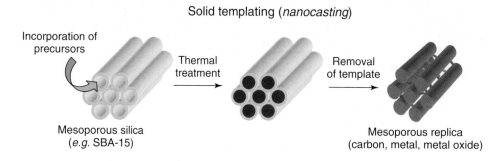

Fig. 19 Schematic representation of the process of nanocasting using ordered mesoporous silica material as a hard template for the preparation of non-siliceous ordered mesoporous materials.

hexagonally packed linear mesopores, thus providing the crosslinking required to obtain a stable replica. This crucial point is illustrated in Fig. 20. At present, ordered mesoporous 3-D cubic MCM-48, 2-D hexagonal SBA-15, and the 3-D cubic KIT-6 silicas are the most frequently utilized hard templates. Conventional MCM-41 silica does not exhibit interconnected channel topology, and is therefore not suitable as a template for the casting pathway [300].

Originally, ordered mesoporous carbon CMK-1 was prepared using MCM-48 as a template, with sucrose (as the carbon source) being impregnated in the presence of sulfuric acid. After a drying step, the obtained sample was pyrolyzed at 900 °C under vacuum or nitrogen to produce pure carbon. The byproducts formed during carbonization generally leave the channels in the gas phase. Finally, the silica framework was dissolved with either a NaOH solution, or with HF. Upon template removal, however, a change in the mesostructure occurred such that the symmetry of the resulting CMK-1 was different from that of the MCM-48 silica [39]. Interestingly, the specific particle morphology of MCM-48 is replicated during nanocasting (Fig. 21). The first ordered mesoporous carbon to be prepared as a faithful inverse replica of the template was synthesized employing SBA-15 as a template. This material, named CMK-3, consisted of uniformly sized amorphous carbon rods arranged in a 2-D hexagonal pattern, as shown in Fig. 21b.

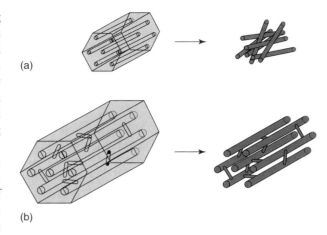

Fig. 20 Schematic illustration of the formation of (a) disordered carbon using a hard template exhibiting disconnected pores (e.g., MCM-41), and (b) ordered mesoporous CMK-3 carbon obtained using a hard template exhibiting an interconnected pore system (e.g., SBA-15). (Reprinted with permission from Ref. [296]; © 2001, Wiley-VCH Verlag. Original figure courtesy Dr. M. Choi, KAIST, Deajeon, South Korea.)

References see page 214

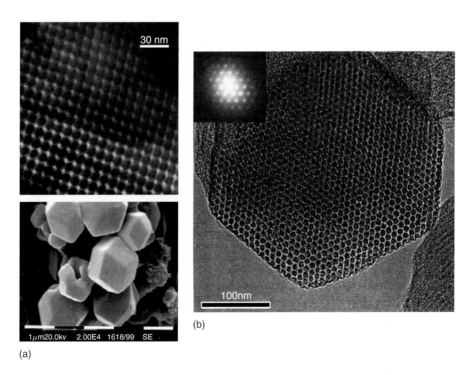

Fig. 21 (a) TEM (upper) and SEM (lower) images of ordered mesoporous CMK-1 carbon. (Reprinted with permission from Ref. [39]; © 1999, American Chemical Society.) (b) Typical TEM image of ordered mesoporous carbon molecular sieve, CMK-3, consisting of a 2-D hexagonal arrangement of carbon rods. The structure of the CMK-3 carbon is exactly an inverse replica of SBA-15. (Adapted from Ref. [295].)

As the connecting pores of SBA-15 are also filled with carbon precursor, they form structure-supporting links between the carbon rods [295]. Interestingly, this preparation actually proved that the mesopores of SBA-15 are interconnected through their walls via smaller pores.

Mesoporous carbons can be prepared with a variety of pore shapes, connectivity and pore wall thickness, depending on the pore structure and pore diameter of the parent silica templates. Because the pores are formed by dissolution of the silica framework, the size of the pores are dictated by the silica wall thickness. Pore size control of mesoporous CMK-3 is therefore achieved by using ordered mesoporous silica templates synthesized with tailored wall thicknesses [301]. Furthermore, the composition and microstructure of the carbon framework is influenced by the nature of the carbon precursor. Various suitable carbogenic precursors (monomeric or polymeric) are available, including sucrose, furfuryl alcohol, phenol–formaldehyde resin, acenaphthene, polydivinylbenzene, acetylene, acrylonitrile, and pyrrole [296–298, 302–304]. Normally, an appropriate polymerization catalyst (aluminum sites, oxalic acid, p-toluenesulfonic acid, $FeCl_3$, etc.) is added to the system.

Most interestingly, by varying the filling degree of the carbon precursor in the pores of mesoporous silica and the processing conditions, the structure of the resulting carbon can be varied. The carbon structures formed within the mesopores of the silica templates can be tuned to be either rod-type or tube-type structures, depending on the amount of the precursors and the pyrolysis conditions. If the pore system of large-pore silicas such as SBA-15 (or KIT-6) is only coated by the carbon precursor instead of being completely filled, a surface-templated mesoporous carbon named CMK-5 [305] is generated instead of the volume-templated CMK-3-type material. The structure of CMK-5 mesoporous carbon is composed of hexagonal arrays of carbon nanopipes that were originally formed inside the cylindrical nanotubes of the SBA-15 template. Removal of the silica results in a bimodal pore system observed for the CMK-5 material, where one type of pore is generated in the inner part of the channels that is not completely filled by carbon, and the other porosity arises from the spaces previously occupied by the silica walls of the SBA-15. A TEM image of CMK-5 is shown in Fig. 22. CMK-5 has an extremely high surface area (which may reach 2500 $m^2\ g^{-1}$) and a large pore volume that make it suitable for applications in adsorption and catalysis. Such a CMK-5 mesoporous carbon is most easily synthesized by using aluminum-modified SBA-15 silica and furfuryl alcohol as a carbon precursor. Carbonization under vacuum pyrolysis conditions usually favors the formation of carbon as a film coated on the mesopore walls, rather than complete volume filling. The cubic $Ia\bar{3}d$ surface-templated equivalent (designated as CMK-9) is obtained using KIT-6 silica as the solid template [120].

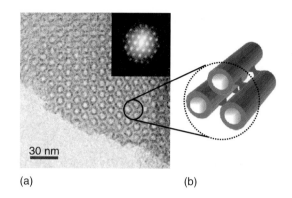

Fig. 22 (a) TEM image of CMK-5 viewed along the direction of the ordered nanoporous carbon tubes and the corresponding Fourier diffractogram. (Reprinted with permission from Ref. [305]; © 2001, Nature Publishing Group.) (b) Schematic representation of the carbon structure. The structural model indicates that the carbon nanopores are rigidly interconnected into a highly ordered hexagonal array by carbon spacers.

The nanocasting approach was extended to the preparation of mesostructured metals and metal oxides of various compositions, which are usually difficult to prepare by direct templating methods due to poor redox stability upon template removal. Mesostructured metals (e.g., Pt or Ag) were prepared by infiltrating the silica hard template with metal precursor complexes and subsequent reduction. The resulting nanostructures grew in the confined space of the pores, further interconnecting with each other [306, 307]. Provided that the loadings are sufficiently high, the nanocrystals crosslink properly to form a continuous framework. Similarly, ordered mesoporous metal oxides (e.g., In_2O_3, Co_3O_4, Fe_2O_3, CeO_2, Cr_2O_3) are also accessible by direct replication of SBA-15, SBA-16 and KIT-6 [308–312]. In addition, in most of the cases reported the framework walls of these mesoporous materials are composed of crystalline nanoparticles.

In principle, the CMK-type carbon materials might also serve as templates to construct ordered pore structures with other compositions such as metal oxides. This is indeed possible [297, 299], and has been nicely illustrated by Roggenbuck and Tiemann, who exploited repeated nanocasting, by using CMK-type carbon as a template for the preparation of magnesium oxide [313]. Following removal of the CMK-3 carbon template by a standard calcination, a high surface area mesoporous MgO material with ordered $p6mm$ pore symmetry could be produced.

2.3.6.7 Morphology Control

Morphology control is thought to be helpful for obtaining efficient solid-phase catalysts, because diffusion limitations are dependent upon the size and morphology of the primary particles of the catalyst. Mesoporous silicas obtained from the early syntheses were typically finely divided powders consisting of small particles (<10 μm) with no well-defined morphology. However, the true liquid crystal templating mechanism introduced by Attard et al. [38] allowed the preparation of monoliths [125]. Since 1996, a wide variety of other shapes, including thin films, spheres, fibers, tubes, monoliths and many other more complex morphologies have been reported [23, 24, 314–318]. Some examples of different morphologies are depicted in Fig. 23. The motivations for the development of materials with defined morphology are very diverse, and include catalysis, optics, electronics, photonics, separation, or biotechnology [28, 319, 320].

Some solids with defined macroscale morphology can be designed by the process conditions, such as dip-coating, spin-coating or emulsion templating. Others are formed spontaneously by a self-organization processes based on kinetic regimes in which equilibrium phases are replaced by high-order organizational states determined by local minimum. However, the origin of the different morphologies is often not established, and the mechanisms of formation proposed are sometimes hypothetical. Macrostructures can be considered as matter organized on different length scales, while control may be achieved on the molecular, supramolecular, and colloidal scales. Such materials are often described as being "hierarchical", as they actually consist of primary building blocks associated with more complex secondary structures integrated in the next size level [321]. By combining the sol–gel technique and self-assembly, Yang et al. [322] showed that the synthesis of mesoporous silica following the acid route $S^+X^-I^+$ is particularly well suited to the preparation of a large variety of different morphologies. This is most likely based on the weaker interactions between surfactant and silicate surface and less-charged framework walls, compared to the base-derived composites, thus allowing more flexibility for the mesophase. Ordered thin films of thicknesses between 0.2 and 1.0 μm have been prepared under acidic conditions by heterogeneous nucleation at the air–water [323] and oil–water [84] interfaces as free-standing films, and at

References see page 214

Fig. 23 (a) SEM image of mesoporous silica spheres obtained by a modified Stöber synthesis. (Reprinted with permission of Ref. [330]; © 1999, Elsevier.) (b) SEM image of a mesoporous silica fiber synthesized with P123. (Reprinted with permission of Ref. [338]; © 1998, American Chemical Society.) (c) Representative TEM image of an ordered mesoporous silica thin film. (Reprinted with permission of Ref. [317]; © 2000, American Chemical Society.) (d) SEM image of mesoporous silica fibers synthesized with tetrabutoxysilane and CTAB in a static acidic system. (Original SEM image obtained by H. Bongard and F. Kleitz, Max-Planck-Institut für Kohlenfoschung, Mülheim, Germany.)

the solid–liquid interface [324]. Furthermore, Ogawa prepared a self-supporting transparent hexagonal-phase thin film, by slow evaporation of a sol mixture containing TMOS, water, CTAC and a mineral acid [325]. Brinker and coworkers [326] synthesized mesostructured thin films with hexagonal, lamellar or cubic phases by using dip-coating or spin-coating techniques. Slow evaporation of the solvent induced the surfactant concentration to cross the CMC and then provoke coassembly of the silicate and the surfactant (EISA mechanism). Efficient solvent evaporation triggers the mesophase formation. Usually, the films remain as flexible hybrid mesophases, presenting low inorganic condensation. Therefore, post-processing can help to improve ordering or even change the nature of the phase. The dip- and spin-coating techniques were also applied using non-ionic surfactant and triblock copolymers under acidic conditions, to prepare oriented mesoporous thin films with 2-D hexagonal, 3-D hexagonal, and cubic structures through the EISA mechanism [106, 266–268, 327].

Through biphasic emulsion chemistry, the preparation of hollow spheres of mesoporous silica has been achieved under acidic conditions, with a decisive control of the shearing rate [84]. In contrast, Bruinsma et al. [328] reported the synthesis of hollow spheres by spray-drying techniques, based on very rapid solvent evaporation and the retention of preformed shapes. Particles with spherical morphology were also very successfully prepared under alkaline conditions. For example, Huo et al. [329] reported the preparation of hard transparent spheres from an emulsion at room temperature, while Grün et al. [72, 330] modified the well-known Stöber synthesis of monodisperse spheres [331] and succeeded in preparing monodispersed mesoporous MCM-41 and MCM-48 spheres in the presence of ethanol and ammonium hydroxide (Fig. 23a). A concept of pseudomorphic transformation has been applied to amorphous preshaped spherical silica particles (5 to 800 μm), such as LiChrospher 60 or 100 and Nucleosil 100-5, to produce ordered mesoporous MCM-41- and MCM-48-type silica and aluminosilica particles with the same spherical morphology. This pseudomorphic transformation method is based on using a mild alkaline solution to dissolve–reprecipitate silica in a controlled manner. Silica is dissolved progressively and locally, and reprecipitated at the same rate in the presence of a quaternary ammonium surfactant (CTAB), without modifying the global morphology. Micrometer-sized spheres are obtained via this pathway [332, 333].

In general, microporous solids such as zeolites and zeotypes, show well-defined crystal morphology that corresponds to their order at the atomic level. On the other hand, it is rather unusual to obtain ordered mesoporous materials with well-defined crystal shapes. The preparation of mesoporous materials exhibiting single crystal particle morphology has attracted some interest, but success has been limited. One such example was reported by Kim and Ryoo [78], who synthesized MCM-48 crystals ($Ia\bar{3}d$) with a truncated rhombic dodecahedral shape, generated with the aid of an inorganic salt. Subsequently, other materials with single crystal particle morphology, such as silica and organosilica hybrid cubic mesoporous crystals ($Pm\bar{3}n$) with well-defined decaoctahedron shapes, were also described [111, 334, 335]. It appears that the addition of inorganic salts would play a crucial role in the morphology of the particles of ordered mesoporous silica. For example, Yu et al. showed that it was possible to control the particle shape of triblock copolymer-based silicas in the presence of inorganic salts. By using the salt effect of K_2SO_4, these authors demonstrated an elegant synthesis of large-pore cubic mesoporous silica single crystals, equivalent to SBA-16, with particles of uniform, rhombodecahedral shape [150]. Some examples of mesoporous silica single crystals are depicted in Fig. 24. Zhao and Stucky [336] were also able to control the morphology of SBA-15 silica, preparing SBA-15 rods of uniform length (1–2 μm) in relatively high yields by adding KCl to the synthesis mixture.

In 1996, Lin et al. [337] reported a remarkable synthesis of hollow tubes of 0.3 to 3 μm diameter and with mesopores channels along the tubular axis. This organization could be achieved by careful control of the surfactant-to-water ratio and the rate of silica condensation at high pH. The methods of preparation of porous silica gel fibers by spinning a high-viscosity silica sol or by freezing a solution of silicic acid in water were applied under acidic conditions to produce mesoporous silica fibers (MSF). Later, Bruinsma et al. [328] introduced the spinning process to synthesize mesoporous fibers with a cationic alkyltrimethylammonium surfactant. This dry-spinning process is based on slow solvent evaporation, which drives mesophase formation, and an increased viscosity facilitated by the addition of a polymer (PEO). The fibers, which are drawn from the mixture into continuous filaments, present pore channels that are oriented parallel to the fiber axis. MSF with larger pores can be prepared in a similar manner by drawing a gel strand from a highly viscous amphiphilic block copolymer–silicate solution [338] (see Fig. 23b). In contrast, the spontaneous formation of mesoporous silica fibers at the oil–water interface is also possible. The formation of well-defined mesoporous silica fibers using specific halide-mediated, acidic conditions, occurs preferentially when the oil–water interface is held static (Fig. 23d). If stirring is applied the fiber morphology is no longer produced and spherical particles are formed instead [339]. These fibers are optically transparent in the visible region of the spectrum, and can be doped with a variety of dyes. Moreover, optical and TEM investigations

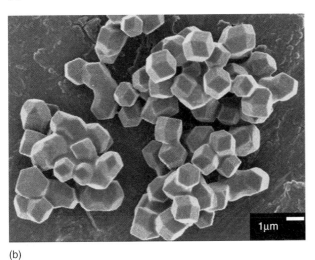

Fig. 24 SEM images of single crystals. (a) Mesoporous SBA-1 silica prepared with alkylammonium surfactant under acidic conditions. (Reprinted with permission from Ref. [111]; © 2000, Nature Publishing Group.) (b) Large pore mesoporous silica with cubic $Im\bar{3}m$ structure prepared with triblock copolymer template in the presence of K_2SO_4 salt. (Reprinted with permission from Ref. [150]; © 2002, American Chemical Society.)

suggested that these synthesized fibers had a circular inner architecture, consisting of hexagonally organized channels running circularly around the fiber's axis [340].

2.3.6.8 Concluding Remarks

In this chapter, an attempt has been made to illustrate the broad horizons encompassed in the world of ordered mesoporous materials. It is impossible, however, to produce a comprehensive survey of the many different aspects of these exciting materials, and the topics discussed provide only a brief overview of this immense diversity in terms of synthesis strategies, functionalization methods, and other aspects of processing. Today, ordered mesoporous solids clearly form their own specific class of very exciting molecular sieves. These periodic materials consist of extended inorganic or inorganic–organic hybrid arrays with exceptional long-range ordering, tunable textural and surface properties, and controlled pore size and geometry. The different aspects of experimental preparation, processing and application of such ordered mesoporous materials are vast in number, and the large body of related publications, patents and reports continues to grow. Since the first reports during the early 1990s, much intensive scientific effort has resulted in almost 7000 publications, of which over 4500 date within the period 2002 to 2006. In particular, the demand for porous solids with specific framework and surface compositions, which may exhibit structures controlled on several length scales or integrated multiple functionalities, has led to considerable development, and has greatly expanded the scope of research, by bridging chemistry with physics and/or biology. In turn, the field of mesoporous materials is now truly interdisciplinary, merging fields such as materials chemistry and engineering, heterogeneous catalysis, sol–gel science, coordination chemistry, surfactant technology and solid-state physics. Today, the properties of ordered mesoporous materials are investigated for industrial and commercial applications, not only in catalysis and separation, but also in drug delivery techniques, in biomedicals, sensing, electronic technologies and electro-optical devices, among other prospects.

An increasing knowledge of these mesoporous and mesostructured materials has led to the almost continuous development of new processes and techniques for their synthesis and modification, overcoming previous limitations. The impressive diversity of structure and composition is continually demonstrated, with new mesophases being discovered on a regular basis in original systems, while the variation of composition is almost infinite. As a striking example, highly ordered polymers and carbon frameworks with 2-D hexagonal mesostructures or other complex, interconnected 3-D structures may now be produced through direct soft templating methods, using triblock copolymer as SDAs [341, 342]. Nanocasting approaches remain extremely versatile pathways for preparing carbons with different properties (e.g., graphitic-like, nitrogen-containing, surface-functionalized, tubular-type [297, 299, 303]) and other exciting compositions such as ternary oxides or non-oxides (e.g., SiC [343]), all of which have major potential as either catalysts or catalyst supports.

References see page 214

Nanocasting offers, in addition to compositional variety, the possibility of selective and spatial functionalization. One such example was demonstrated by Lu et al. [344], who produced magnetically separable mesoporous carbon-based hydrogenation catalysts via sequential and selective positioning of different entities in CMK-type materials. Also, the surface functionalization of ordered mesoporous carbons by the selective deposition of polymers (e.g., polystyrene) was achieved to create materials with the surface properties of polymers, but with the physical properties of the carbon framework, for advanced applications [345]. But, many other developments are creating similar levels of excitement, including the creation of combined polyfunctional mesoporous materials [346–348], the manufacture of stimuli-responsive functional systems, with eventual dynamic control of porosity [349], and the repercussion of chirality from a molecular level to micrometer length scale [350].

References

1. K. S. W. Sing, D. H. Everett, R. H. W. Haul, L. Moscou, R. A. Pierotti, J. Rouquerol, T. Siemieniewska, *Pure Appl. Chem.* **1985**, *57*, 603.
2. P. Behrens, *Adv. Mater.* **1993**, *5*, 127.
3. T. J. Barton, L. M. Bull, W. G. Klemperer, D. A. Loy, B. McEnaney, M. Misono, P. A. Monson, G. Pez, G. W. Scherer, J. C. Vartuli, O. M. Yaghi, *Chem. Mater.* **1999**, *11*, 2633.
4. K. K. Iler, *The Chemistry of Silica*, Wiley, New York, 1979, 896 pp.
5. H. van Bekkum, E. M. Flanigen, J. C. Jansen (Eds.), *Introduction to Zeolite Science and Practice, Studies in Surface Science and Catalysis*, Vol. 58, Elsevier, Amsterdam, 1991, 754 pp.
6. R. M. Barrer, *Hydrothermal Chemistry of Zeolites*, Academic Press Inc., London, 1982, 348 pp.
7. P. A. Jacobs, J. A. Martens (Eds.), *Synthesis of Aluminosilicate Zeolites, Studies in Surface Science and Catalysis*, Vol. 33, Elsevier, Amsterdam, 1987, 406 pp.
8. D. W. Breck, *Zeolite Molecular Sieves*, John Wiley & Sons, New York, London, 1974, 784 pp.
9. A. Dyer, *An Introduction to Zeolite Molecular Sieves*, John Wiley & Sons, New York, London, 1988, 164 pp.
10. W. F. Hölderich, M. Hesse, F. Näumann, *Angew. Chem. Int. Ed.* **1988**, *27*, 226.
11. J. C. Jansen, M. Stöcker, H. G. Karge, J. Weitkamp (Eds.), *Advanced Zeolite Science and Applications, Studies in Surface Science and Catalysis*, Vol. 85, Elsevier, Amsterdam, 1994, 708 pp.
12. A. Corma, *Chem. Rev.* **1995**, *95*, 559.
13. A. Corma, *Chem. Rev.* **1997**, *97*, 2373.
14. C. J. Brinker, G. W. Scherer, *Sol-Gel Science*, Academic Press, New York, 1990, 912 pp.
15. M. R. S. Manton, J. C. Davidtz, *J. Catal.* **1979**, *60*, 156.
16. V. Chiola, J. E. Ritsko, C. D. Vanderpool, US Patent 3 556 725, 1971, assigned to Sylvania Electric Products Inc.
17. C. T. Kresge, M. E. Leonowicz, W. J. Roth, J. C. Vartuli, J. S. Beck, *Nature* **1992**, *359*, 710.
18. J. S. Beck, J. C. Vartuli, W. J. Roth, M. E. Leonowicz, C. T. Kresge, K. D. Schmitt, C. T.-W. Chu, D. H. Olson, E. W. Sheppard, S. B. McCullen, J. B. Higgins, J. L. Schlenker, *J. Am. Chem. Soc.* **1992**, *114*, 10834.
19. T. Yanagisawa, T. Shimizu, K. Kuroda, C. Kato, *Bull. Chem. Soc. Jpn.* **1990**, *63*, 988.
20. J. C. Vartuli, C. T. Kresge, M. E. Leonowicz, A. S. Chu, S. B. McCullen, I. D. Johnson, E. W. Sheppard, *Chem. Mater.* **1994**, *6*, 2070.
21. S. Brunauer, P. H. Emmett, E. Teller, *J. Am. Chem. Soc.* **1938**, *60*, 309.
22. F. Schüth, *Chem. Mater.* **2001**, *13*, 3184.
23. U. Ciesla, F. Schüth, *Micropor. Mesopor. Mater.* **1999**, *27*, 131.
24. M. Lindén, S. Schacht, F. Schüth, A. Steel, K. K. Unger, *J. Porous Mater.* **1998**, *5*, 177.
25. J. Y. Ying, C. P. Mehnert, M. S. Wong, *Angew. Chem. Int. Ed.* **1999**, *38*, 56.
26. A. Stein, B. J. Melde, R. C. Schroden, *Adv. Mater.* **2000**, *12*, 1403.
27. A. Sayari, S. Hamoudi, *Chem. Mater.* **2001**, *13*, 3151.
28. B. J. Scott, G. Wirnsberger, G. D. Stucky, *Chem. Mater.* **2001**, *13*, 3140.
29. F. Schüth, W. Schmidt, *Adv. Mater.* **2002**, *14*, 629.
30. G. J. A. A. Soler-Illia, J. Patarin, B. Lebeau, C. Sanchez, *Chem. Rev.* **2002**, *102*, 4093.
31. D. E. de Vos, M. Dams, B. F. Sels, P. A. Jacobs, *Chem. Rev.* **2002**, *102*, 3615.
32. A. P. Wight, M. E. Davis, *Chem. Rev.* **2002**, *102*, 3589.
33. D. Trong On, D. Desplantier-Giscard, C. Danumah, S. Kaliaguine, *Appl. Catal. A: General* **2003**, *253*, 545.
34. A. Stein, *Adv. Mater.* **2003**, *15*, 763.
35. A. Taguchi, F. Schuth, *Micropor. Mesopor. Mater.* **2005**, *77*, 1.
36. M. Hartmann, *Chem. Mater.* **2005**, *17*, 4577.
37. F. Hoffmann, M. Cornelius, J. Morell, M. Fröba, *Angew. Chem. Int. Ed.* **2006**, *45*, 3216.
38. G. S. Attard, J. C. Glyde, C. G. Göltner, *Nature* **1995**, *378*, 366.
39. R. Ryoo, S. H. Joo, S. Jun, *J. Phys. Chem. B* **1999**, *103*, 7743.
40. J. Livage, M. Henry, C. Sanchez, *Prog. Solid State Chem.* **1988**, *18*, 259.
41. F. H. Dickey, *Proc. Natl. Acad. Sci. USA* **1949**, *35*, 227.
42. J. C. Vartuli, K. D. Schmitt, C. T. Kresge, W. J. Roth, M. E. Leonowicz, S. B. McCullen, S D Hellring, J. S. Beck, J. L. Schlenker, D. H. Olson, E. W. Sheppard, *Chem. Mater.* **1994**, *6*, 2317.
43. N. K. Raman, M. T. Anderson, C. J. Brinker, *Chem. Mater.* **1996**, *8*, 1682.
44. M. E. Davis, R. F. Lobo, *Chem. Mater.* **1992**, *4*, 756.
45. J.-M. Lehn, *Science* **1985**, *227*, 849.
46. S. Mann, G. A. Ozin, *Nature* **1996**, *382*, 313.
47. G. A. Ozin, *Acc. Chem. Res.* **1997**, *30*, 17.
48. D. F. Evans, H. Wennerström (Eds.), *The Colloidal Domain: Where Physics, Chemistry, Biology and Technology Meet*, Wiley-VCH Publishers, Weinheim, Germany, 1984, 672 pp.
49. S. Förster, M. Antonietti, *Adv. Mater.* **1998**, *10*, 195.
50. (a) J. N. Israelachvili, D. J. Mitchell, B. W. Ninham, *J. Chem. Soc., Faraday Trans.* **1976**, *72*, 1525; (b) J. N. Israelachvili, D. J. Mitchell, B. W. Ninham, *Biochim. Biophys. Acta* **1977**, *470*, 185.
51. J. N. Israelachvili, *Intermolecular and Surfaces Forces*, Academic Press, London, 1992, 450 pp.
52. S. T. Hyde, *Pure Appl. Chem.* **1992**, *64*, 1617.
53. A. Firouzi, D. Kumar, L. M. Bull, T. Besier, P. Sieger, Q. Huo, S. A. Walker, J. A. Zasadzinski, C. Glinka, J. Nicol, D. Margolese, G. D. Stucky, B. F. Chmelka, *Science* **1995**, *267*, 1138.

54. Q. Huo, D. I. Margolese, G. D. Stucky, *Chem. Mater.* **1996**, *8*, 1147.
55. Q. Huo, R. Leon, P. M. Petroff, G. D. Stucky, *Science* **1995**, *268*, 1324.
56. F. S. Bates, G. H. Frederickson, *Phys. Today*, **1999**, *52*, 32.
57. G. Wanka, H. Hoffmann, W. Ulbricht, *Macromolecules* **1994**, *27*, 4145.
58. P. Alexandridis, T. A. Hatton, *Colloids Surfaces A: Physicochem. Eng. Aspects* **1995**, *96*, 1.
59. C. Booth, D. Attwood, *Macromol. Rapid Commun.* **2000**, *21*, 501.
60. J. L. Patarin, B. Lebeau, R. Zana, *Curr. Opin. Colloid Interface Sci.* **2002**, *7*, 107.
61. C.-Y. Chen, S. L. Burkett, H.-X. Li, M. E. Davis, *Micropor. Mater.* **1993**, *2*, 27.
62. A. Monnier, F. Schüth, Q. Huo, D. Kumar, D. Margolese, R. S. Maxwell, G. D. Stucky, M. Krishnamurty, P. Petroff, A. Firouzi, M. Janicke, B. F. Chmelka, *Science* **1993**, *261*, 1299.
63. Q. Huo, D. I. Margolese, U. Ciesla, P. Feng, T. E. Gier, P. Sieger, R. Leon, P. M. Petroff, F. Schüth, G. D. Stucky, *Nature* **1994**, *368*, 317.
64. Q. Huo, D. I. Margolese, U. Ciesla, D. G. Demuth, P. Feng, T. E. Gier, P. Sieger, A. Firouzi, B. F. Chmelka, F. Schüth, G. D. Stucky, *Chem. Mater.* **1994**, *6*, 1176.
65. P. T. Tanev, T. J. Pinnavaia, *Science* **1995**, *267*, 865.
66. S. A. Bagshaw, E. Prouzet, T. J. Pinnavaia, *Science* **1995**, *269*, 1242.
67. G. S. Attard, P. N. Bartlett, N. R. B Coleman, J. M. Elliot, J. R. Owen, J. H. Wang, *Science* **1997**, *278*, 838.
68. C. J. Brinker, Y. Lu, A. Sellinger, H. Fan, *Adv. Mater.* **1999**, *11*, 579.
69. D. Grosso, F. Cagnol, G. J. A. A. Soler-Illia, E. L. Crepaldi, H. Amenitsch, A. Brunet-Bruneau, A. Bourgeois, C. Sanchez, *Adv. Funct. Mater.* **2004**, *14*, 309.
70. R. Ryoo, J. M. Kim, C. H. Ko, C. H. Shin, *J. Phys. Chem.* **1996**, *100*, 17718.
71. Y. Sakamoto, I. Diaz, O. Terasaki, D. Zhao, J. Pérez-Pariente, J. M. Kim, G. D. Stucky, *J. Phys. Chem. B* **2002**, *106*, 3118.
72. M. Grün, I. Lauer, K. K. Unger, *Adv. Mater.* **1997**, *9*, 254.
73. M. Grün, K. K. Unger, A. Matsumoto, K. Tsutsumi, *Micropor. Mesopor. Mater.* **1999**, *27*, 207.
74. A. A. Romero, A. D. Alba, W. Zhou, J. Klinowski, *J. Phys. Chem. B* **1997**, *101*, 5294.
75. J. Xu, Z. Luan, H. He, W. Zhou, L. Kevan, *Chem. Mater.* **1998**, *10*, 3690.
76. Q. Huo, D. I. Margolese, G. D. Stucky, *Chem. Mater.* **1996**, *8*, 1147.
77. K. W. Gallis, C. C. Landry, *Chem. Mater.* **1997**, *9*, 2035.
78. J. M. Kim, S. K. Kim, R. Ryoo, *Chem. Commun.* **1998**, 259.
79. F. Chen, L. Huang, Q. Li, *Chem. Mater.* **1997**, *9*, 2685.
80. (a) R. Ryoo, S. H. Joo, J. M. Kim, *J. Phys. Chem. B* **1999**, *103*, 7435; (b) M. Kruk, M. Jaroniec, R. Ryoo, S. H. Joo, *Chem. Mater.* **2000**, *12*, 1414.
81. W. Zhao, Q. Z. Li, *Chem. Mater.* **2003**, *15*, 4160.
82. X. Auvray, C. Petipas, R. Anthore, I. Rico, A. J. Lattes, *J. Phys. Chem.* **1989**, *93*, 7458.
83. (a) R. Schmidt, M. Stöcker, D. Akporiaye, E. H. Tørstad, A. Olsen, *Micropor. Mater.* **1995**, *5*, 1; (b) V. Alfredsson, M. W. Anderson, *Chem. Mater.* **1996**, *8*, 1141.
84. S. Schacht, Q. Huo, I. G. Voigt-Martin, G. D. Stucky, F. Schüth, *Science* **1996**, *273*, 768.
85. S. Che, S. Lim, M. Kaneda, H. Yoshitake, O. Terasaki, T. Tatsumi, *J. Am. Chem. Soc.* **2002**, *124*, 13962.
86. S. Che, A. E. Garcia-Bennett, T. Yokoi, K. Sakamoto, H. Kunieda, O. Terasaki, T. Tatsumi, *Nature Mater.* **2003**, *2*, 801.
87. A. E. Garcia-Bennett, O. Terasaki, S. Che, T. Tatsumi, *Chem. Mater.* **2004**, *16*, 813.
88. A. E. Garcia-Bennett, K. Miyasaka, O. Terasaki, S. Che, *Chem. Mater.* **2004**, *16*, 3587.
89. A. E. Garcia-Bennett, N. Kupferschmidt, Y. Sakamoto, S. Che, O. Terasaki, *Angew. Chem. Int. Ed.* **2005**, *44*, 5317.
90. G. J. A. A. Soler-Illia, E. L. Crepaldi, D. Grosso, C. Sanchez, *Curr. Opin. Colloid Interface Sci.* **2003**, *8*, 109.
91. C. Boissière, A. Larbot, A. van der Lee, P. Kooyman, E. Prouzet, *Chem. Mater.* **2000**, *12*, 2902.
92. D. Zhao, J. Feng, Q. Huo, N. Melosh, G. H. Frederickson, B. F. Chmelka, G. D. Stucky, *Science* **1998**, *279*, 548.
93. D. Zhao, Q. Huo, J. Feng, B. F. Chmelka, G. D. Stucky, *J. Am. Chem. Soc.* **1998**, *120*, 6024.
94. K. Cassiers, T. Linssen, M. Mathieu, M. Benjelloun, K. Schrijnemakers, P. van der Voort, P. Cool, E. F. Vansant, *Chem. Mater.* **2002**, *14*, 2317.
95. J. M. Kim, G. D. Stucky, *Chem. Commun.* **2000**, 1159.
96. J. R. Matos, L. P. Mercuri, M. Kruk, M. Jaroniec, *Langmuir* **2002**, *18*, 884.
97. S. H. Joo, R. Ryoo, M. Kruk, M. Jaroniec, *J. Phys. Chem. B* **2002**, *106*, 4640.
98. M. Choi, W. Heo, F. Kleitz, R. Ryoo, *Chem. Commun.* **2003**, 1340.
99. C. Z. Yu, Y. H. Yu, D. Y. Zhao, *Chem. Commun.* **2000**, 575.
100. (a) S. S. Kim, T. R. Pauly, T. J. Pinnavaia, *Chem. Commun.* **2000**, 1661; (b) S. S. Kim, A. Karkambar, T. J. Pinnavaia, M. Kruk, M. Jaroniec, *J. Phys. Chem. B* **2001**, *105*, 7663.
101. M. Kruk, M. Jaroniec, C. H. Ko, R. Ryoo, *Chem. Mater.* **2000**, *12*, 1961.
102. M. Impéror-Clerc, P. Davidson, A. Davidson, *J. Am. Chem. Soc.* **2000**, *122*, 11925.
103. R. Ryoo, C. H. Ko, M. Kruk, V. Antoschuk, M. Jaroniec, *J. Phys. Chem. B* **2000**, *104*, 11465.
104. P. L. Ravikovitch, A. V. Neimark, *J. Phys. Chem. B* **2001**, *105*, 6817.
105. Z. Liu, O. Terasaki, T. Ohsuna, K. Hiraga, H. J. Shin, R. Ryoo, *ChemPhysChem.* **2001**, *4*, 229.
106. D. Grosso, A. R. Balkende, P. A. Albouy, A. Ayral, H. Amenitsch, F. Babonneau, *Chem. Mater.* **2001**, *13*, 1848.
107. K. Flodström, C. V. Teixeira, H. Amenitsch, V. Alfredsson, M. Lindén, *Langmuir* **2004**, *20*, 4885.
108. A. Y. Khodakov, V. L. Zholobenko, M. Impéror-Clerc, D. Durand, *J. Phys. Chem. B* **2005**, *109*, 22780.
109. K. Flodström, H. Wennerström, V. Alfredsson, *Langmuir* **2004**, *20*, 680.
110. (a) S. Ruthstein, V. Frydman, S. Kababya, M. Landau, D. Goldfarb *J. Phys. Chem. B* **2003**, *107*, 1739; (b) S. Ruthstein, V. Frydman, D. Goldfarb, *J. Phys. Chem. B* **2004**, *108*, 9016.
111. Y. Sakamoto, M. Kaneda, O. Terasaki, D. Y. Zhao, J. M. Kim, G. D. Stucky, H. J. Shin, R. Ryoo, *Nature* **2000**, *408*, 449.
112. Y. Park, T. Kang, J. Lee, P. Kim, H. Kim, J. Yi., *Catal. Today* **2004**, *97*, 195.
113. O. C. Gobin, Q. Huang, H. Vinh-Thang, F. Kleitz, M. Eić, S. Kaliaguine, *J. Phys. Chem. C* **2007**, *111*, 3059.
114. J. R. Matos, M. Kruk, L. P. Mercuri, M. Jaroniec, L. Zhao, T. Kamiyama, O. Terasaki, T. J. Pinnavaia, Y. Liu, *J. Am. Chem. Soc.* **2003**, *125*, 821.
115. B. L. Newalkar, S. Komarneni, *Chem. Commun.* **2002**, 1774.
116. S. S. Kim, A. Karkambar, T. J. Pinnavaia, M. Kruk, M. Jaroniec, *J. Phys. Chem. B* **2001**, *105*, 7663.

117. J. Fan, C. Yu, F. Gao, J. Lei, B. Tian, L. Wang, Q. Luo, B. Tu, W. Zhou, D. Zhao, *Angew. Chem. Int. Ed.* **2003**, *42*, 3146.
118. P. Schmidt-Winkel, W. W. Lukens, Jr., D. Zhao, P. Yang, B. F. Chmelka, G. D. Stucky, *J. Am. Chem. Soc.* **1999**, *121*, 254.
119. J. S. Lettow, Y. J. Han, P. Schmidt-Winkel, P. Yang, D. Zhao, G. D. Stucky, J. Y. Ying, *Langmuir* **2002**, *16*, 8291.
120. F. Kleitz, S. H. Choi, R. Ryoo, *Chem. Commun.* **2003**, 2136.
121. Y. Sakamoto, T. W. Kim, R. Ryoo, O. Terasaki, *Angew. Chem. Int. Ed.* **2004**, *42*, 5231.
122. X. Liu, B. Tian, C. Yu, F. Gao, S. Xie, B. Tu, R. Che, L.-M. Peng, D. Zhao, *Angew. Chem. Int. Ed.* **2002**, *41*, 3876.
123. Y. Q. Wang, C. M. Yang, B. Zibrowius, B. Spliethoff, M. Lindén, F. Schüth, *Chem. Mater.* **2003**, *15*, 5029.
124. K. Flodström, V. Alfredsson, N. Källrot, *J. Am. Chem. Soc.* **2003**, *125*, 4402.
125. P. Feng, X. Bu, D. J. Pine, *Langmuir* **2000**, *16*, 5304.
126. M. Kruk, M. Jaroniec, A. Sayari, *J. Phys. Chem. B* **1997**, *101*, 583.
127. J. M. Kim, Y. Sakamoto, Y. K. Hwang, Y.-U. Kwon, O. Terasaki, S.-E. Park, G. D. Stucky *J. Phys. Chem. B* **2002**, *106*, 2552.
128. P. Kipkemboi, A. Fogden, V. Alfredsson, K. Flodström, *Langmuir* **2001**, *17*, 5398.
129. K. Flodström, V. Alfredsson, *Micropor. Mesopor. Mater.* **2003**, *59*, 167.
130. T.-W. Kim, R. Ryoo, M. Kruk, K. P. Gierszal, M. Jaroniec, S. Kamiya, O. Terasaki, *J. Phys. Chem. B* **2004**, *108*, 11480.
131. K. Kushalani, A. Kuperman, G. A. Ozin, K. Tanaka, J. Garces, M. M Olken, N. Coombs, *Adv. Mater.* **1995**, *7*, 842.
132. A. Corma, Q. Kan, M. T. Navarro, J. Pérez-Pariente, F. Rey, *Chem. Mater.* **1997**, *9*, 2123.
133. F. Kleitz, W. Schmidt, F. Schüth, *Micropor. Mesopor. Mater.* **2001**, *44–45*, 92.
134. F. Kleitz, W. Schmidt, F. Schüth, *Micropor. Mesopor. Mater.* **2003**, *65*, 1.
135. B. L. Newalkar, S. Komarneni, U. T. Turaga, H. Katsuki, *J. Mater. Chem.* **2003**, *13*, 1710.
136. A. Galarneau, H. Cambon, F. DiRenzo, R. Ryoo, M. Choi, F. Fajula, *New J. Chem.* **2003**, *27*, 73.
137. P. F. Fulvio, S. Pikus, M. Jaroniec, *J. Mater. Chem.* **2005**, *15*, 5049.
138. T.-W. Kim, F. Kleitz, B. Paul, R. Ryoo, *J. Am. Chem. Soc.* **2005**, *127*, 7601.
139. P. I. Ravikovitch, A. V. Neimark, *Langmuir* **2002**, *18*, 1550.
140. F. Kleitz, D. Liu, G. M. Anilkumar, I. S. Park, L. A. Solovyov, A. N. Shmakov, R. Ryoo, *J. Phys. Chem. B* **2003**, *107*, 14296.
141. M. Thommes, B. Smarsly, M. Groenewolt, P. I. Ravikovitch, A. V. Neimark, *Langmuir* **2006**, *22*, 756.
142. F. Kleitz, T. Czuryszkiewicz, L. A. Solovyov, M. Lindén, *Chem. Mater.* **2006**, *18*, 5070.
143. M. Jaroniec, L. A. Solovyov, *Chem. Commun.* **2006**, 2242.
144. R. Ryoo, J. M. Kim, *Chem. Commun.* **1995**, 711.
145. K. J. Edler, J. W. White, *Chem. Mater.* **1997**, *9*, 1226.
146. S. Jun, J. M. Kim, R. Ryoo, Y.-S. Ahn, M. H. Han, *Micropor. Mesopor. Mater.* **2000**, *41*, 119.
147. A. Kabalnov, U. Olsson, H. Wennerström, *J. Phys. Chem.* **1995**, *98*, 6220.
148. B. L. Newalkar, S. Komarneni, *Chem. Mater.* **2001**, *13*, 4573.
149. C. Z. Yu, B. Tian, J. Fan, G. D. Stucky, D. Y. Zhao, *Chem. Commun.* **2001**, 2726.
150. C. Z. Yu, B. Tian, J. Fan, G. D. Stucky, D. Y. Zhao, *J. Am. Chem. Soc.* **2002**, *124*, 4556.
151. M. Lindén, P. Ågren, S. Karslsson, P. Bussian, H. Amenitsch, *Langmuir* **2000**, *16*, 5831.
152. P. Ågren, M. Linden, J. Rosenholm, R. Schwarzenbacher, M. Kriechbaum, H. Amenitsch, P. Laggner, J. Blanchard, F. Schüth, *J. Phys. Chem. B* **1999**, *103*, 5943.
153. P. Ågren, M. Lindén, S. Karlsson, J. B. Rosenholm, J. Blanchard, F. Schüth, H. Amenitsch, *Langmuir* **2000**, *16*, 8809.
154. A. Lind, J. Andersson, S. Karlsson, M. Lindén, *Colloids Interfaces A.* **2001**, *183*, 415.
155. F. Kleitz, J. Blanchard, B. Zibrowius, F. Schüth, P. Ågren, M. Lindén, *Langmuir* **2002**, *18*, 4963.
156. M. T. Anderson, J. E. Martin, J. Odinek, P. P. Newcomer, *Chem. Mater.* **1998**, *10*, 311.
157. J. Sun, H. Zhang, D. Ma, Y. Chen, X. Bao, A. Klein-Hoffmann, N. Pfänder, D. S. Su, *Chem. Commun.* **2005**, 5343.
158. J. Fan, C. Yu, L. Wang, B. Tu, D. Zhao, Y. Sakamoto, O. Terasaki, *J. Am. Chem. Soc.* **2001**, *123*, 12113.
159. (a) P. Holmqvist, P. Alexandridis, B. Lindman, *Macromolecules* **1997**, *30*, 6788; (b) P. Holmqvist, P. Alexandridis, B. Lindman, *J. Phys. Chem. B* **1998**, *102*, 1149.
160. F. Kleitz, L. A. Solovyov, G. M. Anilkumar, S. H. Choi, R. Ryoo, *Chem. Commun.* **2004**, 1536.
161. F. Kleitz, T. W. Kim, R. Ryoo, *Langmuir* **2006**, *22*, 440.
162. B. Tian, X. Liu, L. A. Solovyov, Z. Liu, H. Yang, Z. Zhang, S. Xie, F. Zhang, B. Tu, C. Yu, O. Terasaki, D. Zhao, *J. Am. Chem. Soc.* **2003**, *126*, 865.
163. R. Ryoo, S. Jun, *J. Phys. Chem. B* **1997**, *101*, 317.
164. K. A. Koyano, T. Tatsumi, Y. Tanaka, S. Nakata, *J. Phys. Chem. B* **1997**, *101*, 9436.
165. J. M. Kisler, M. L. Gee, G. W. Stevens, A. J. O'Connor, *Chem. Mater.* **2003**, *15*, 619.
166. A. S. O'Neil, R. Mokaya, M. Poliakoff, *J. Am. Chem. Soc.* **2002**, *124*, 10636.
167. K. R. Kloetstra, H. van Bekkum, J. C. Jansen, *Chem. Commun.* **1997**, 2281.
168. Y. Liu, W. Zhang, T. J. Pinnavaia, *Angew. Chem. Int. Ed.* **2001**, *40*, 1255.
169. Z. Zhang, Y. Han, L. Zhu, R. Wang, Y. Yu, S. Qiu, D. Zhao, F.-S. Xiao, *Angew. Chem. Int. Ed.* **2001**, *40*, 1258.
170. S. P. B. Kremer, C. E. A. Kirschhock, A. Aerts, K. Villani, J. A. Martens, O. I. Lebedev, G. Van Tendeloo, *Adv. Mater.* **2003**, *15*, 1705.
171. D. Trong On, S. Kaliaguine, *Angew. Chem. Int. Ed.* **2002**, *41*, 1036.
172. A. Corma, V. Fornés, M. T. Navarro, J. Pérez-Pariente, *J. Catal.* **1994**, *148*, 569.
173. M. T. J. Keene, R. D. M. Gougeon, R. Denoyel, R. H. Harris, J. Rouquerol, P. L. Llewellyn, *J. Mater. Chem.* **1999**, *9*, 2843.
174. B. Tian, X. Liu, C. Yu, F. Gao, Q. Luo, S. Xie, B. Tu, D. Zhao, *Chem. Commun.* **2002**, 1186.
175. C. M. Yang, B. Zibrowius, W. Schmidt, F. Schüth, *Chem. Mater.* **2004**, *16*, 2918.
176. (a) R. Zaleski, J. Wawryszczuk, A. Borowka, *Micropor. Mesopor. Mater.* **2003**, *62*, 47; (b) J. Goworek, A. Borówka, R. Zaleski, R. Kusak, *J. Thermal. Anal. Cal.* **2005**, *79*, 555.
177. M. Soulard, S. Bilger, H. Kessler, J. L. Guth, *Zeolites* **1991**, *11*, 107.
178. H. J. Shin, R. Ryoo, M. Kruk, M. Jaroniec, *Chem. Commun.* **2001**, 349.
179. S. Hitz, R. Prins, *J. Catal.* **1997**, *168*, 194.
180. F. Schüth, *Ber. Bunsen.-Ges. Phys. Chem.* **1995**, *99*, 1306.
181. S. Kawi, M. W. Lai, *Chem. Commun.* **1998**, 1407.
182. P. T. Tanev, T. J. Pinnavaia, *Chem. Mater.* **1996**, *8*, 2068.

183. B. Tian, X. Liu, C. Yu, F. Gao, Q. Luo, S. Xie, B. Tu, D. Zhao, *Chem. Commun.* **2002**, 1186.
184. C. M. Yang, B. Zibrowius, W. Schmidt, F. Schüth, *Chem. Mater.* **2003**, *15*, 3739.
185. R. Palkovits, C. M. Yang, S. Olejnik, F. Schüth, *J. Catal.* **2006**, *243*, 93.
186. C. M. Yang, W. Schmidt, F. Kleitz, *J. Mater. Chem.* **2005**, *15*, 5112.
187. D. M. Ford, E. E. Simanek, D. F. Shantz, *Nanotechnology* **2005**, *16*, S458.
188. H. Yoshitake, *New. J. Chem.* **2005**, *29*, 1107.
189. M. H. Lim, A. Stein, *Chem. Mater.* **1999**, *11*, 3285.
190. F. Schüth, *Annu. Rev. Mater. Res.* **2005**, *35*, 209.
191. J. A. Melero, R. van Grieken, G. Morales, *Chem. Rev.* **2006**, *106*, 3790.
192. R. Mokaya, *Angew. Chem. Int. Ed.* **1999**, *38*, 2930.
193. T. Klimova, E. Rodríguez, E. Martínez, J. Ramírez, *Micropor. Mesopor. Mater.* **2001**, *44–45*, 357.
194. M. Baltes, K. Cassiers, P. Van Der Voort, B. M. Weckhuysen, R. A. Schoonheydt, E. F. Vansant, *J. Catal.* **2001**, *197*, 160.
195. V. Dufaud, M. E. Davis, *J. Am. Chem. Soc.* **2003**, *125*, 9403.
196. D. S. Shephard, W. Z. Zhou, T. Maschmeyer, J. M. Matters, C. L. Roper, J. M. Thomas, *Angew. Chem. Int. Ed.* **1998**, *37*, 2719.
197. B. F. G. Johnson, S. A. Raynor, D. S. Shephard, T. Maschmeyer, J. M. Thomas, *Chem. Commun.* **1999**, 1167.
198. (a) R. K. Zeidan, S.-J. Hwang, M. E. Davis, *Angew. Chem. Int. Ed.* **2006**, *45*, 6332; (b) C. M. Yang, H. A. Lin, B. Zibrowius, B. Spliethoff, F. Schüth, S. C. Liou, M. W. Chu, C. H. Chen, *Chem. Mater.* **2007**, *19*, 3205.
199. J. Sauer, F. Marlow, F. Schüth, *Phys. Chem. Chem. Phys.* **2001**, *3*, 5579.
200. L. A. Solovyov, S. D. Kirik, A. N. Shmakov, V. N. Romannikov, *Micropor. Mesopor. Mater.* **2001**, *44–45*, 17.
201. L. A. Solovyov, O. V. Belousov, R. E. Dinnebier, A. N. Shmakov, S. D. Kirik, *J. Phys. Chem. B* **2005**, *109*, 3233.
202. J. Sauer, F. Marlow, B. Spliethoff, F. Schüth, *Chem. Mater.* **2002**, *14*, 217.
203. J. M. Thomas, P. A. Midgley, *Chem. Commun.* **2004**, 1253.
204. X. S. Zhao, C. Q. Lu, A. K. Whittaker, G. J. Millar, H. Y. Zhu, *J. Phys. Chem. B* **1997**, *101*, 6525.
205. M. Widenmeyer, R. Anwander, *Chem. Mater.* **2002**, *14*, 1827.
206. L. T. Zhuralev, *Langmuir* **1987**, *3*, 316.
207. C. Nozaki, C. G. Lugmaier, A. T. Bell, T. D. Tilley, *J. Am. Chem. Soc.* **2002**, *124*, 13194.
208. P. L. Llewellyn, F. Schüth, Y. Grillet, F. Rouquerol, J. Rouquerol, K. K. Unger, *Langmuir* **1995**, *11*, 574.
209. M. F. Ottaviani, A. Galarneau, F. Desplantiers-Giscard, F. DiRenzo, F. Fajula, *Micropor. Mesopor. Mater.* **2001**, *44–45*, 1.
210. M. S. Morey, J. D. Bryan, S. Schwarz, G. D. Stucky, *Chem. Mater.* **2000**, *12*, 3435.
211. D. Brunel, *Micropor. Mesopor. Mater.* **1999**, *27*, 329.
212. R. Anwander, *Chem. Mater.* **2001**, *13*, 4419.
213. D. Das, S. Cheng, *Chem. Commun.* **2001**, 2178.
214. B. Sow, S. Hamoudi, M. H. Zahedi-Niaki, S. Kaliaguine, *Micropor. Mesopor. Mater.* **2005**, *79*, 129.
215. A. Bleloch, B. F. G. Johnson, S. V. Ley, A. J. Price, D. S. Shephard, A. W. Thomas, *Chem. Commun.* **1999**, 1907.
216. E. J. Acosta, C. S. Carr, E. E. Simanek, D. F. Shantz, *Adv. Mater.* **2004**, *16*, 985.
217. E. L. Burkett, S. D. Sims, S. Mann, *Chem. Commun.* **1996**, 1367.
218. D. J. Macquarrie, *Chem. Commun.* **1996**, 1961.
219. D. Margolese, J. A. Melero, S. C. Christiansen, B. F. Chmelka, G. D. Stucky, *Chem. Mater.* **2000**, *12*, 2448.
220. J. A. Melero, G. D. Stucky, R. van Grieken, G. Morales, *J. Mater. Chem.* **2002**, *12*, 1664.
221. I. Yuranov, L. Kiwi-Minsker, P. Buffat, A. Renken, *Chem. Mater.* **2004**, *16*, 760.
222. M. Choi, F. Kleitz, D. Liu, H. Y. Lee, W.-S. Ahn, R. Ryoo, *J. Am. Chem. Soc.* **2005**, *127*, 1924.
223. P. Gómez-Romero, C. Sanchez (Eds.), *Functional Hybrid Materials*, Wiley-VCH, Weinheim, 2004.
224. S. Inagaki, S. Guan, Y. Fukushima, T. Ohsuna, O. Terasaki, *J. Am. Chem. Soc.* **1999**, *121*, 9611.
225. B. J. Melde, B. T. Holland, C. F. Blanford, A. Stein, *Chem. Mater.* **1999**, *11*, 3302.
226. T. Asefa, M. J. MacLachlan, N. Coombs, G. A. Ozin, *Nature* **1999**, *402*, 867.
227. H. Zhu, D. J. Jones, J. Zajac, J. Rozière, R. Dutartre, *Chem. Commun.* **2001**, 2568.
228. O. Muth, C. Schellbach, M. Fröba, *Chem. Commun.* **2001**, 2032.
229. R. J. P. Corriu, A. Mehdi, C. Reyé, C. Thieuleux, *Chem. Commun.* **2002**, 1382.
230. S. Inagaki, S. Guan, T. Ohsuna, O. Terasaki, *Nature* **2002**, *416*, 304.
231. R. Ryoo, C. H. Ko, R. F. Howe, *Chem. Mater.* **1997**, *9*, 1607.
232. A. Sayari, *Chem. Mater.* **1996**, *8*, 1840.
233. Z. Luan, C. F. Cheng, W. Zhou, J. Klinowski, *J. Phys. Chem.* **1995**, *99*, 1018.
234. A. Tuel, S. Gontier, *Chem. Mater.* **1996**, *8*, 114.
235. A. Sayari, C. Danumah, I. L. Moudrakovski, *Chem. Mater.* **1995**, *7*, 813.
236. Z. Y. Yuan, S. Q. Liu, T. H. Chen, J. Z. Wang, H. X. Li, *Chem. Commun.* **1995**, 973.
237. B. Echchahed, A. Moen, D. Nicholson, L. Bonneviot, *Chem. Mater.* **1997**, *9*, 1716.
238. C. F. Cheng, H. He, W. Zhou, J. Klinowski, J. A. S. Gonçalves, L. F. Gladden, *J. Phys. Chem.* **1996**, *100*, 390.
239. T. Takeguchi, J. B. Kim, M. Kang, T. Inui, W. T. Cheuh, G. L. Haller, *J. Catal.* **1998**, *175*, 1.
240. P. T. Tanev, M. Chibwe, T. J. Pinnavaia, *Nature* **1994**, *368*, 321.
241. K. A. Koyano, T. Tatsumi, *Chem. Commun.* **1996**, 145.
242. H. Kosslick, G. Lischke, B. Parlitz, W. Storek, R. Fricke, *Appl. Catal. A: General* **1999**, *184*, 49.
243. A. Corma, M. S. Grande, V. Gonzalez-Alfaro, A. V. Orchilles, *J. Catal.* **1996**, *159*, 375.
244. A. Corma, V. Fornes, M. T. Navarro, J. Perez-Pariente, *J. Catal.* **1994**, *148*, 569.
245. A. Corma, A. Martínez, V. Martínez-Soria, J. B. Montón, *J. Catal.* **1995**, *153*, 25.
246. M. D. Alba, Z. Luan, J. Klinowski, *J. Phys. Chem. B* **1996**, *102*, 857.
247. M. Morey, A. Davidson, G. D. Stucky, *Micropor. Mater.* **1996**, *6*, 99.
248. B. J. Aronson, C. F. Blanford, A. Stein, *Chem. Mater.* **1997**, *9*, 2842.
249. Z. Luan, E. M. Maes, P. A. W. van der Heide, D. Zhao, R. S. Czernszewicz, L. Kevan, **1999**, *11*, 3680.
250. T. Machmeyer, F. Rey, G. Sankar, J. M. Thomas, *Nature* **1995**, *378*, 159.
251. A. Sayari, P. Liu, *Micropor. Mater.* **1997**, *12*, 149.
252. U. Ciesla, D. Demuth, R. Leon, P. Petroff, G. D. Stucky, K. K. Unger, F. Schüth, *Chem. Commun.* **1994**, 1387.
253. T. Abe, A. Taguchi, M. Iwamoto, *Chem. Mater.* **1995**, *7*, 1429.

254. P. Liu, L. Moudrakovski, J. Liu, A. Sayari, *Chem. Mater.* **1997**, *9*, 2513.
255. H. Hatayama, M. Risono, A. Taguchi, N. Mizuno, *Chem. Lett.* **2000**, 884.
256. D. M. Antonelli, J. Y. Ying, *Angew. Chem. Int. Ed.* **1995**, *34*, 2014.
257. V. S. Stone, Jr., R. J. Davis, *Chem. Mater.* **1998**, *10*, 1468.
258. D. M. Antonelli, *Micropor. Mesopor. Mater.* **1999**, *30*, 315.
259. A. Bhaumik, S. Inagaki, *J. Am. Chem. Soc.* **2001**, *123*, 691.
260. U. Ciesla, S. Schacht, G. D. Stucky, K. K. Unger, F. Schüth, *Angew. Chem. Int. Ed.* **1996**, *35*, 541.
261. M. S. Wong, J. Y. Ying, *Chem. Mater.* **1998**, *10*, 2067.
262. D. M. Antonelli, J. Y. Ying, *Angew. Chem. Int. Ed.* **1996**, *35*, 426.
263. D. M. Antonelli, J. Y. Ying, *Chem. Mater.* **1996**, *8*, 874.
264. Y. Takahara, J. N. Kondo, T. Takata, D. Lu, K. Domen, *Chem. Mater.* **2001**, *13*, 498.
265. (a) P. Yang, D. Zhao, D. I. Margolese, B. F. Chmelka, G. D. Stucky, *Nature* **1998**, *396*, 152; (b) P. Yang, D. Zhao, D. I. Margolese, B. F. Chmelka, G. D. Stucky, *Chem. Mater.* **1999**, *11*, 2813.
266. G. J. A. A. Soler-Illia, E. Scolan, A. Louis, P. A. Albouy, C. Sanchez, *New. J. Chem.* **2001**, *25*, 156.
267. D. Grosso, G. J. A. A. Soler-Illia, F. Babonneau, C. Sanchez, P. A. Albouy, A. Brunet-Bruneau, A. R. Balkenende, *Adv. Mater.* **2001**, *13*, 1085.
268. E. L. Crepaldi, G. J. A. A. Soler-Illia, D. Grosso, P. A. Albouy, C. Sanchez, *Chem. Commun.* **2001**, 1582.
269. E. L. Crepaldi, G. J. A. A. Soler-Illia, D. Grosso, C. Sanchez, *Angew. Chem. Int. Ed.* **2002**, *42*, 347.
270. E. L. Crepaldi, G. J. A. A. Soler-Illia, D. Grosso, F. Cagnol, F. Ribot, C. Sanchez, *J. Am. Chem. Soc.* **2003**, *125*, 9770.
271. D. Grosso. C. Boissière, B. Smarsly, T. Brezesinski, N. Pinna, P. A. Albouy, H. Amenitsch, M. Antonietti, C. Sanchez, *Nature Mater.* **2004**, *3*, 787.
272. F. Schüth, K. Unger, in G. Ertl, H. Knözinger, J. Weitkamp (Eds.), *Preparation of Solid Catalysts*, Wiley-VCH, Weinheim, 1999, pp. 77–80.
273. J. Cejka, *Appl. Catal. A: General* **2003**, *254*, 327.
274. S. Bagshaw, T. J. Pinnavaia, *Angew. Chem. Int. Ed.* **1996**, *35*, 1102.
275. S. Cabrera, J. El Haskouri, J. Alamo, A. Beltrán, D. Beltrán, S. Mendioroz, M. Dolores Marcos, P. Amorós, *Adv. Mater.* **1999**, *11*, 379.
276. Z. Zhang, T. J. Pinnavaia, *J. Am. Chem. Soc.* **2002**, *124*, 12294.
277. M. Kuemmel, D. Grosso, C. Boissière, B. Smarsly, T. Brezesinski, P. A. Albouy, H. Amenitsch, C. Sanchez, *Angew. Chem.* **2005**, *117*, 4665.
278. B. Z. Tian, X. Y. Liu, B. Tu, C. Z. Yu, J. Fan, L. M. Wang, S. H. Xie, G. D. Stucky, D. Y. Zhao, *Nature Mater.* **2003**, *2*, 159.
279. K. K. Rangan, S. J. L. Billinge, V. Petkov, J. Heising, M. G. Kanatzidis, *Chem. Mater.* **1999**, *11*, 2629.
280. M. J. MacLachlan, N. Coombs, G. A Ozin, *Nature* **1999**, *397*, 681.
281. K. K. Rangan, P. N. Trikalitis, M. G. Kanatzidis, *J. Am. Chem. Soc.* **2000**, *122*, 10230.
282. P. N. Trikalitis, K. K. Rangan, T. Bakas, M. G. Kanatzidis, *Nature* **2001**, *410*, 6712.
283. C. Serre, A. Auroux, A. Gervasini, M. Hervieu, G. Férey, *Angew. Chem. Int. Ed.* **2002**, *41*, 1594.
284. U. Ciesla, M. Fröba, G. D. Stucky, F. Schüth, *Chem. Mater.* **1999**, *11*, 227.
285. F. Kleitz, S. J. Thomson, Z. Liu, O. Terasaki, F. Schüth, *Chem. Mater.* **2002**, *14*, 4134.
286. J. Blanchard, P. Trens, M. Hudson, F. Schüth, *Micropor. Mesopor. Mater.* **2000**, *39*, 163.
287. P. N. Trikalitis, T. Bakas, V. Papaefthymiou, M. G. Kanatzidis, *Angew. Chem. Int. Ed.* **2000**, *39*, 4558.
288. M. Fröba, M. Tiemann, *Chem. Mater.* **1998**, *10*, 3475.
289. D. Zhao, Z. Luan, L. Kevan, *Chem. Commun.* **1997**, 1009.
290. T. Kimura, Y. Sagahara, K. Kuroda, *Chem. Mater.* **1999**, *11*, 508.
291. S. Cabrera, J. El Haskouri, C. Guillem, A. Beltrán-Porter, D. Beltrán-Porter, S. Mendioroz, M. Dolores Marcos, P. Amorós, *Chem. Commun.* **1999**, 333.
292. G. S. Attard, C. G. Göltner, J. M. Corker, S. Henke, R. H. Templer, *Angew. Chem. Int. Ed.* **1997**, *36*, 1315.
293. P. A. Nelson, J. M. Elliott, G. S. Attard, J. R. Owen, *Chem. Mater.* **2002**, *14*, 524.
294. I. Nandhakumar, J. M. Elliott, G. S. Attard, *Chem. Mater.* **2001**, *13*, 3840.
295. S. Jun, S. H. Joo, R. Ryoo, M. Kruk, M. Jaroniec, Z. Liu, T. Ohsuna, O. Terasaki, *J. Am. Chem. Soc.* **2000**, *122*, 10712.
296. R. Ryoo, S. H. Joo, M. Kruk, M. Jaroniec, *Adv. Mater.* **2001**, *13*, 677.
297. A. H. Lu, F. Schüth, *Adv. Mater.* **2006**, *18*, 1793.
298. B. Tian, X. Y. Liu, H. F. Yang, S. H. Xie, C. Z. Yu, B. Tu, D. Zhao, *Adv. Mater.* **2003**, *15*, 1370.
299. H. F. Feng, D. Y. Zhao, *J. Mater. Chem.* **2005**, *15*, 1217.
300. Y. Sakamoto, T. Ohsuna, K. Hiraga, O. Terasaki, C. H. Ko, H. J. Shin, R. Ryoo, *Angew. Chem. Int. Ed.* **2000**, *39*, 3107.
301. J. S. Lee, S. H. Joo, R. Ryoo, *J. Am. Chem. Soc.* **2002**, *124*, 1156.
302. J. Lee, S. Yoon, T. Hyeon, S. M. Oh, K. B. Kim, *Chem. Commun.* **1999**, 2177.
303. J. Lee, S. Han, T. Hyeon, *J. Mater. Chem.* **2004**, *14*, 478.
304. A. H. Lu, A. Kiefer, W. Schmidt, F. Schüth, *Chem. Mater.* **2004**, *16*, 100.
305. S. H. Joo, S. J. Choi, I. Oh, J. Kwak, Z. Liu, O. Terasaki, R. Ryoo, *Nature* **2001**, *412*, 169.
306. H. J. Shin, R. Ryoo, Z. Liu, O. Terasaki, *J. Am. Chem. Soc.* **2001**, *123*, 1246.
307. Y. J. Han, J. M. Kim, G. D. Stucky, *Chem. Mater.* **2000**, *12*, 2068.
308. Y. Wang, C. M. Yang, W. Schmidt, B. Spliethoff, E. Bill, F. Schüth, *Adv. Mater.* **2005**, *17*, 53.
309. A. Rumplecker, F. Kleitz, E. L. Salabas, F. Schüth, *Chem. Mater.* **2007**, *19*, 485.
310. H. Yang, Q. Shi, B. Tian, Q. Lu, F. Gao, S. Xie, J. Fan, C. Yu, B. Tu, D. Zhao, *J. Am. Chem. Soc.* **2003**, *125*, 4724.
311. B. Tian, X. Liu, L. A. Solovyov, Z. Liu, H. Yang, Z. Zhang, S. Xie, F. Zhang, B. Tu, C. Yu, O. Terasaki, D. Zhao, *J. Am. Chem. Soc.* **2004**, *126*, 865.
312. K. Jiao, B. Zhang, B. Yue, Y. Ren, S. Liu, S. Yan, C. Dickinson, W. Zhou, H. He, *Chem. Commun.* **2005**, 5618.
313. J. Roggenbuck, M. Tiemann, *J. Am. Chem. Soc.* **2005**, *127*, 1096.
314. G. A. Ozin, *Chem. Commun.* **2000**, 419.
315. S. H. Tolbert, A. Firouzi, G. D. Stucky, B. F. Chmelka, *Science* **1997**, *278*, 264.
316. P. Yang, T. Deng, D. Zhao, P. Feng, D. Pine, B. F. Chmelka, G. M. Whitesides, G. D. Stucky, *Science* **1998**, *282*, 2244.
317. H. Miyata, K. Kuroda, *Chem. Mater.* **2000**, *12*, 49.
318. H. Fan, S. Reed, T. Bear, R. Schunk, G. P. López, C. J. Brinker, *Micropor. Mesopor. Mater.* **2001**, *44–45*, 625.

319. R. C. Hayward, P. Alberius-Henning, B. F. Chemlka, G. D. Stucky, *Micropor. Mesopor. Mater.* **2001**, *44–45*, 619.
320. M. Vallet-Regi, A. Rámila, R. P. del Real, J. Pérez-Pariente, *Chem. Mater.* **2001**, *13*, 308.
321. S. Mann, S. L. Burkett, S. A. Davis, C. E. Fowler, N. H. Mendelson, S. D. Slims, D. Walsh, N. T. Whilton, *Chem. Mater.* **1997**, *9*, 2300.
322. H. Yang, N. Coombs, G. A. Ozin, *Nature* **1997**, *386*, 692.
323. H. Yang, N. Coombs, I. Sokolov, G. A. Ozin, *Nature* **1996**, *381*, 589.
324. H. Yang, A. Kuperman, N. Coombs, S. Mamiche-Afara, G. A. Ozin, *Nature* **1996**, *379*, 703.
325. M. Ogawa, *Chem. Comm.* **1996**, 1149.
326. Y. Lu, R. Ganguli, C. A. Drewien, M. T. Anderson, C. J. Brinker, W. Gong, Y. Guo, H. Soyez, B. Dunn, M. H. Huang, J. I. Zink, *Nature* **1997**, *389*, 364.
327. D. Zhao, P. Yang, N. Melosh, J. Feng, B. F. Chmelka, G. D. Stucky, *Adv. Mater.* **1998**, *10*, 1380.
328. P. J. Bruinsma, A. Y. Kim. J. Liu, S. Baskaran, *Chem. Mater.* **1997**, *9*, 2507.
329. Q. Huo, J. Feng, F. Schüth, G. D. Stucky, *Chem. Mater.* **1997**, *9*, 14.
330. K. Schumacher, M. Grün, K. K. Unger, *Micropor. Mesopor. Mater.* **1999**, *27*, 201.
331. W. Stöber, A. Funk, E. Bohn, *J. Colloid Interface Sci.* **1968**, *26*, 62.
332. T. Martin, A. Galarneau, F. DiRenzo, F. Fajula, D. Plee, *Angew. Chem. Int. Ed.* **2002**, *41*, 2590.
333. A. Galarneau, J. Iapichella, K. Bonhomme, F. DiRenzo, P. Kooyman, O. Terasaki, F. Fajula, *Adv. Funct. Mater.* **2006**, *16*, 1657.
334. S. Guan, S. Inagaki, T. Ohsuna, O. Terasaki, *J. Am. Chem. Soc.* **2000**, *122*, 5660.
335. A. Sayari, S. Hamoudi, Y. Yang, I. L. Moudrakovski, J. R. Ripmeester, *Chem. Mater.* **2000**, *12*, 3857.
336. C. Z. Yu, J. Fan, B. Z. Tian, G. D. Stucky, D. Y. Zhao, *Adv. Mater.* **2002**, *14*, 1742.
337. H.-P. Lin, C.-Y. Mou, *Science* **1996**, *273*, 765.
338. P. Yang, D. Zhao, B. F. Chmelka, G. D. Stucky, *Chem. Mater.* **1998**, *10*, 2033.
339. (a) F. Marlow, M. D. McGehee, D. Zhao, B. F. Chmelka, G. D Stucky, *Adv. Mater.* **1999**, *11*, 632; (b) F. Kleitz, F. Marlow, G. D. Stucky, F. Schüth, *Chem. Mater.* **2001**, *17*, 2136.
340. F. Marlow, B. Spliethoff, B. Tesche, D. Zhao, *Adv. Mater.* **2000**, *12*, 961.
341. S. Tanaka, N. Nishiyama, Y. Egashira, K. Ueyama, *Chem. Commun.* **2005**, 2125.
342. F. Zhang, Y. Meng, D. Gu, Y. Yan, C. Yu, B. Tu, D. Zhao, *J. Am. Chem. Soc.* **2005**, *127*, 13508.
343. P. Krawiec, D. Geiger, S. Kaskel, *Chem. Commun.* **2006**, 2469.
344. A. H. Lu, W. Schmidt, N. Matoussevitch, H. Bönnemann, B. Spliethoff, B. Tesche, E. Bill, W. Kiefer, F. Schüth, *Angew. Chem. Int. Ed.* **2004**, *43*, 4303.
345. M. Choi, R. Ryoo, *Nature Mater.* **2003**, *2*, 473.
346. F. Goettmann, D. Grosso, F. Mercier, F. Matthey, C. Sanchez, *Chem. Commun.* **2004**, 1240.
347. S. Huh, H. T Chen. J. W. Wiench, M. Pruski, V. S. Y. Lin, *Angew. Chem.* **2005**, *117*, 1860.
348. (a) J. Alauzun, A. Mehdi, C. Reyé, R. J. Corriu, *J. Am. Chem. Soc.* **2006**, *128*, 8718; (b) M. Jaroniec, *Nature* **2006**, *442*, 638.
349. N. K. Mal, M. Fujiwara, Y. Tanaka, *Nature* **2003**, *421*, 350.
350. S. Che, Z. Liu, T. Ohsuna, K. Sakamoto, O. Terasaki, T. Tatsumi, *Nature* **2004**, *429*, 281.

2.3.7
Pillared Clays

*Georges Poncelet and José J. Fripiat**

2.3.7.1 Introduction

2.3.7.1.1 Historical Overview Clays have played an important role in the history of acid catalysis. For example, the acid leaching of natural smectites led to amorphous alumino-silicates with a large acid surface able to initiate the carbocation chemistry which has so many diverse applications in catalysis. The original Houdry catalyst was prepared in such a manner. Due to the variability in composition of naturally occurring smectites, synthetic aluminosilicates obtained through the coprecipitation of aluminum and of silicon salts were substituted for the acid-leached clays as soon as the economic constraints of World War II had been released. During the 1960s, the amorphous synthetic aluminosilicates were replaced by zeolites when it was realized that the latter were more active and more stable acid catalysts, with high stability being of particular importance in the regeneration process. Nonetheless, the unaltered smectites were used as acid catalysts in soft chemistry processes, as the interlayer space (see below) can be kept swollen up to 250 °C by using an adequate solvent. The studies of Purnell et al. [1] have provided many examples of reactions catalyzed by un-pillared clays.

Perhaps the most well-recognized problem with smectites is that of reversibility of the swelling. For example, in the presence of water vapor the interlayer distance may reach 0.6 nm, and correspond to the intercalation of two layers of water molecules on a surface area of about 800 m^2 g^{-1}. Such a hydrated surface is not thermally stable, however, and the interlayer space begins to collapse above ∼100 °C.

The intercalation of alkylammonium cations (and especially of tetraalkylammonium), as pioneered by Barrer [2], is a straightforward process, because the clay lattice is negatively charged and a large surface area becomes available in the swollen smectite derivatives. The ammonium cations may withstand a moderate heat treatment.

References see page 242
* Corresponding author.

Thus, use of the beautiful and highly flexible architecture of the natural smectites seemed to be excluded under the harsh conditions of most catalytic reactions involved in petroleum chemistry. Amazingly, the idea of keeping the clay layers "permanently expanded" by intercalating robust inorganic pillars arose relatively late, with "paternity" of the idea being accredited to Vaughan et al. [3] and to Brindley and Sempels [4], who had in fact combined their efforts. Lahav et al. [5] subsequently published details of their pioneering studies after the patents had been posted, but not issued.

These pioneering investigations were at the origin of a worldwide interest in the so-called pillared clays; these resemble zeolites in the sense that both types of solid are microporous and can easily be transformed into acid catalysts. To some extent, those working with pillared clays were – at least during the early, enthusiastic stages, challenging those working with zeolites. From a theoretical standpoint, if large distances could be obtained between the clay sheets, then this would allow for the chemistry of bulky molecules within a large, interlayer gallery. For pillared clays, this would be a major asset in the treatment of heavy oils.

Although, to date, such a goal has not been achieved, zeolite chemists have since synthesized architectures with larger pores, yet the future of pillared clays remains restricted to certain specialized areas.

Clay is a phyllosilicate composed of tetrahedral and octahedral building units that are organized around at least one (theoretically) infinite tetrahedral layer. Thus, lamellar structures such as double hydroxides and silicic acids are not clays. Only the pillared structures with at least one non ambiguous $d001$ (X-ray) reflections and a surface area of at least 100 m^2 g^{-1} after calcination at 200 °C will be considered here.

2.3.7.1.2 Which Clays or Smectites can be Pillared?

A number of excellent books are available on the subject of clay mineralogy [6], and consequently the subject will not be described in great detail in this chapter. However, the further use of mineralogical names and an understanding of the mechanism of pillaring requires that a brief summary of smectite structure be included at this point.

All smectites derive from two non-swelling lamellar minerals, namely, pyrophyllite and talc (Fig. 1). The only structural difference between pyrophyllite and talc is the complete occupation of the octahedral holes in the latter, whereas in the former two out of three octahedral positions are occupied. For this reason, pyrophyllite and talc are referred to as "dioctahedral" and "trioctahedral" minerals, respectively, and the clays derived from them retain this distinction. Whilst this point is of some academic importance, it should also be noted that

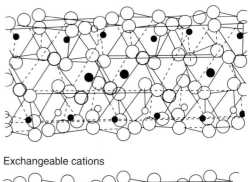

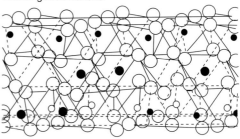

Fig. 1 Smectite basic structure. Only two superimposed sheets are shown. Each sheet comprises two layers of tetrahedra "sandwiching" one octahedral layer. Here, two octahedral cavities out of three are occupied (dioctahedral structure). The gallery, or interlayer space, is the space between sheets where the exchangeable cations are located. The basal 001 plane is made from the oxygen atoms of the tetrahedral layer. The c axis is perpendicular to 001.

trioctahedral minerals are generally more thermally stable than their dioctahedral counterparts.

As noted above, the layers of talc and pyrophyllite cannot be separated. Clay minerals differ from these precursors by the presence of isomorphic substitutions in either the tetrahedral or octahedral layer, or in both. If the substitution introduces a cation of lower valency, then a negative charge appears in the lattice which, according to Pauling's principle, must be balanced within the shortest possible distance by a cation. As clay synthesis occurs in aqueous media in Nature, these cations become hydrated and serve as nuclei for the stepwise hydration of the surface, as a dehydrated clay is exposed to increasing relative humidity. The formation of a single monolayer, or of a two- and sometimes a three-layer hydrate, can be followed by the change in the $d001$ X-ray spacing (see Fig. 2). The thickness of a sheet is about 0.96 nm; the interlayer distance is $\Delta = (d001-0.96)$ nm.

The presence of electric charges in clay lattices, generated by isomorphic substitutions, is at the origin of cation exchange capacity, just as in zeolites where the charges are created by Al for Si substitutions. In principle, any cation in the interlayer can be replaced by any other, and since the swelling is not limited by a tridimensional architecture (as in zeolites), even very bulky complex

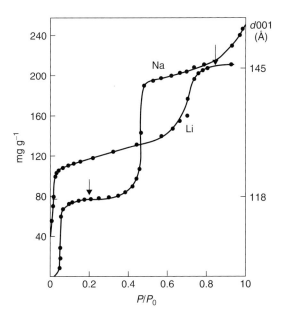

Fig. 2 Example of the stepwise swelling of a sodium or lithium (Llano) vermiculite upon water adsorption at 25 °C; left ordinate, weight gain; right ordinate, $d001$ spacing. (Reprinted from Ref. [7], with permission.)

cations can expand the interlayer gallery over distances which can reach about 2 nm (see below).

Depending on the nature of the isomorphic substitutions, clays with different electric charge are obtained. A brief list of the most common clays, together with what is considered the typical corresponding chemical formula, is provided in Table 1. On the basis of their chemical constitution, mother Nature has provided the mineralogist with an endless number of clay family members. If control of the chemical composition is needed, then the synthesis of clays from pure reagents is possible, but at a cost larger than that for most zeolites because the synthesis conditions (temperature and pressure) are more demanding.

In summary, the properties of clays stem from a textural characteristic, the lamellar (or bidimensional) nature of their crystal lattice, and from a crystallo-chemical characteristic, the negative charge of the lattice due to isomorphic substitutions.

The hydration or complexation of the charge-balancing cation provokes swelling of the lamellar structure. It should be noted that there is an optimum density of charge for optimal swelling: if the charge is too small (<0.5 meq g^{-1}), the cation hydration or complexation energy is lower than the adhesion energy between the platelets. (Throughout this section, meq is used to denote milli-equivalent, i.e., 10^{-3} equivalent charge). If the charge is too large (>3 meq g^{-1}), the electrostatic attraction between the negative planes and the middle plane containing the exchangeable cations is larger than the swelling force. As far as the two families of clays shown in Table 1 are concerned, the highly charged non-swelling end members are the micas muscovite (dioctahedral) or phlogopite (trioctahedral). As with pyrophyllite and talc – but for exactly the opposite reasons – micas do not swell unless special treatments are applied. Note that

References see page 242

Tab. 1 Some properties of commonly occurring clays[a]

Family	Name	Typical composition
Dioctahedral parent: pyrophyllite $x=0$ mica: muscovite, $K[Si_3,Al]^{IV}[Al_2]^{VI}(OH)_2O_{10}$	Montmorillonite Beidellite	$M_x^+[Si_4]^{IV}[Al_{2-x}R_x^{2+}]^{VI}(OH)_2O_{10}$ $x \sim 1 R^{2+} = Mg^{2+}$ $M_x^+[Si_{4-x},Al]^{IV}[Al_2]^{VI}(OH)_2O_{10}$ $x \sim -0.5$
Trioctahedral parent: talc $x=y=0$ mica: phlogopite, $K_x[Si_{4-x}Al_x]^{IV}[(Mg, Fe^{2+})_3]^{VI}$ $O_{10}(OH, F)_2$	Hectorite Saponite Vermiculite	$M_x^+[Si_4]^{IV}[Mg_{3-x},R_x^*]^{VI}(OH)_{2-y} y\, FO_{10}$ $x = 0.7, R^+ = Li^+, y = -$variable $M_x^+[Si_{4-x},Al_x]^{IV}[R_3^{2+}]^{VI}(OH)_2O_{10}$ x ca. $0.6, R^{2+} = Mg^{2+}, Fe^{2+}$ $M_{x-y}^+[Si_{4-x}, Al_x]^{IV}[R_{3-y}^{2+}, R_y^{3+}]^{VI}(OH)_2O_{10}$ $x \sim 1, y =$ variable

[a] M^+ represents a monovalent exchangeable cation.
Superscripts IV and VI indicate the tetrahedral and octahedral layers, respectively.
Hydration of the clay is not accounted for. In one example, pillaring of rectorite is reported.
Rectorite is an interstratified mica-smectite with tetrahedral Al for Si substitution.
Laponite is a synthetic, poorly crystallized hectorite.

vermiculite (which is listed as clay in Table 1) has a lattice charge such that swelling is limited.

2.3.7.1.3 Al-Pillared Vermiculites and Phlogopite Micas

Phlogopites are trioctahedral sheet silicates where the octahedral positions are occupied mostly by divalent cations (Mg^{2+}) and with Al for Si substitutions in the tetrahedral layers. The high degree of substitutions within the octahedral and tetrahedral layers is at the origin of the high net negative layer charge, as in micas. The model-phlogopite has the following general composition: $K_2Mg_6(Si_6Al_2)O_{20}(OH,F)_4$. Natural phlogopites generally contain Mg^{2+}, Fe^{2+}, Al^{3+}, and Fe^{3+} in the octahedral layers, and many also have substantial amounts of structural fluorine anions replacing hydroxyls, which confers a high resistance to weathering, hardness, and thermal stability. The dominant charge-balancing cations are potassium ions; these are located between the layers, with adjacent layers being stacked in such a way that the potassium ion is equidistant to 12 oxygen atoms, six for each tetrahedral layer [8]. In their original state, natural micas do not swell in the presence of water because the hydration energy of the interlayer potassium ions does not overcome the adhesion energy between layers [9]. Hence, ion exchange does not occur.

Thus, to allow pillaring, the main obstacle to be overcome is the high selectivity for potassium that makes the exchange by other cations very difficult. Even a Na-exchanged mica, when brought into contact with an Al-pillaring solution, retains selectively Al^{3+} instead of the bulkier Keggin species (Al_{13}^{7+}) present in the pillaring solution [10].

Vermiculites constitute intermediate minerals in the natural sequence of weathering of trioctahedral micas to smectites. They also have, as phlogopites, Al for Si substitutions in the tetrahedral layers and Fe and Al for Mg substitutions in the octahedral layers. Their lattice charge is intermediate between those of micas and of smectites. Hydrated cations (most often Mg^{2+}) have replaced to a large extent the original potassium ions. Consequently, vermiculites are swelling minerals.

Straight pillaring of vermiculite suspensions with Al_{13}-solutions results in materials with 1.4-nm interlayer spacing at room temperature [11], instead of 1.8 nm for Al-pillared smectites (montmorillonites, saponites, etc.), thus with a gallery height (0.4 nm) which is half that of smectites (0.8 nm) [10, 11]. This difference has been attributed to the location of the negative charge on the basal oxygens of the tetrahedral layers, which prevents selective uptake of the Keggin-type Al_{13}^{7+} cations. When pillaring was preceded by a treatment with ornithine, only a small fraction of the vermiculite was pillared [12].

Vermiculites and phlogopites have been successfully pillared with Al-solutions when the layer charge density of the minerals was reduced before pillaring. The charge-reduced minerals brought into contact with Al-pillaring solutions provide materials exhibiting 1.8-nm spacings that are thermally stable up to temperatures in excess of 800 °C, namely, an improvement of about 200–300 °C relative to the Al-pillared smectites. From a catalytic point of view, the higher number of Al for Si tetrahedral substitutions compared, for instance, with beidellites and saponites, both smectites with such substitutions, makes those materials very attractive acid catalysts because (as will be discussed later) more Brønsted acid sites can be generated upon pillaring. The procedure developed to prepare well-organized pillared vermiculites and phlogopites and their characterization have been reported elsewhere [13, 14]. (Note: the textural stability of these materials is illustrated in Fig. 11.)

Briefly, the conditioning treatment consisted of four operations carried out in the following stepwise sequence: (1) controlled acid leaching with diluted nitric acid; (2) calcination in air at 600 °C for 4 h; (3) acid leaching with a diluted solution of a complexing acid; and (4) saturation of the residual exchange positions of the charge-reduced minerals with sodium (or calcium) ions. At the end of this "accelerated weathering" process, the overall negative charge of the minerals is reduced by about one third. The charge reduction mainly occurs at steps (1) and (2), whereas step (3) is aimed at cleaning the surface. Step (4) is essential to achieve well-organized pillared materials. The charge-reduced vermiculites and phlogopites are then dispersed in the Al-pillaring solutions.

2.3.7.1.4 Reviews

Since 1996, several reviews – some covering the traditional aspects, including the catalytic applications – have been published [15, 17, 19, 21, 23]. Others have insisted on their nanoporosity [20], whereas Schoonheidt and Jacobs [18] have approached the comparison between pillared clays and zeolites from the dimensionality point of view. Fundamental aspects of acidity were treated by Lambert and Poncelet [16], while Auroux [22] has shown the benefit of microcalorimetry in the distinction of the acid sites.

2.3.7.2 Pillaring Mechanisms

2.3.7.2.1 Basic Pillaring Mechanism: Exchange Precipitation

At first sight, the pillaring mechanism is simple, as it would consist essentially of the exchange of the charge-balancing cation of the clay (preferably Na^+) by a cationic oligomer P^+ made from n cations bound by oxo or hydroxo bridges and with a total charge v^+ [24]. If M is the symbol representing the clay, the following exchange occurs in the gallery:

$$\text{MNa}_v + P_n^{v+} \longrightarrow \text{M}P_n + v\text{Na}^+ \quad (1)$$

The gallery height, or basal spacing $d001$, increases by a length somewhat less than the radius of the oligomer. If the exchange was the only reaction involved in the pillaring mechanism, the experimental procedure would indeed be simple, as is the case when exchanging sodium by a quaternary alkylammonium, for instance. However, the oligomer itself results from a condensation reaction such as

$$nP^{z+} \longleftrightarrow P_n^{v+} + (nZ - v)\text{H}^+ + x\text{H}_2\text{O} \quad (2)$$

involving hydronium ions and water. Therefore, the overall pillaring reaction necessarily involves Eqs. (1) and (2), and the chemical properties of the clay surface, such as its acidity, play a role.

Finally, if a successful pillaring mechanism has been achieved – as shown by the increase in the $d001$ reflection in the X-ray diffraction pattern of the air-dried pillared material – it remains necessary to create the microporosity. This most important step is performed by removing the hydration water filling the gallery by thermal activation. This step will be successful insofar as P^{v+} does not depolymerize, and remains anchored in one way or another to the surface. The nature and the extent of the crosslinking action of P_n^{v+} on the adjacent microcrystals forming the galleries depend on the nature of the surface of both the clay and the pillar. The solid, stabilized by a thermal treatment in the temperature range 400–500 °C, is the actual pillared clay. Its structure is essentially turbostratic as the long-range crystalline order is only along the c axis perpendicular to the layers. There is no long-range order along the a and b unit cell axes. Since the pillaring oligomers are cationic, it might be assumed that they will be distributed on the surface as far as possible from one another in order to reduce the mutual electrostatic repulsion. Again, the way in which the pillars are distributed is also a function of an additional factor which becomes apparent when considering the typical composition of the clays shown in Table 1.

In a dioctahedral clay such as montmorillonite, the lattice electric charge results from the substitution of octahedral Al^{3+} by Mg^{2+}. Thus, the lattice charge is buried in the octahedral layer and smeared out on the oxygen surface of the 001 plane (see Fig. 1). In such a case, the lattice charge is not localized. In contrast, in beidellite, for example, the origin of the lattice charge is replacement of tetrahedral Si^{4+} by Al^{3+}, and the charge is localized in the corresponding tetrahedron, as is the case in zeolites. Based on studies with ^{29}Si high-resolution solid-state NMR (^{29}Si MAS NMR), it is known that the distribution of the Al substitutions in layer lattice minerals such as micas and vermiculite is homogeneous [25], bringing them to the largest possible distance from one another. In beidellite and saponite, according to ^{29}Si MAS NMR, it appears that at least the Loewenstein rule applies (this is the avoidance rule, which states that two tetrahedral Al cannot be next-neighbors). The homogeneous dispersion observed for a lamellar structure richer in tetrahedral substitutions such as vermiculite is more demanding since, in addition, it requires the largest possible distance between substitutions.

In lattices with localized electric charges, the distribution of small cations must coincide with that of the tetrahedral isomorphic substitutions. The distribution of the oligomeric cations should be the same, and as the homogeneous distribution advocated by Herrero [25] is also that which would minimize the repulsion between pillars, there should be no essential difference between clays with localized and de-localized lattice charges insofar as the pillar distribution is concerned. However, as shown below, the charge localization increases the surface reactivity, and will influence the crosslinking mechanism between the clay platelets.

2.3.7.2.2 New Pillaring Mechanism: Template In the basic pillaring mechanism, the main interaction between the cationic hydroxy-polymer and the clay surface is electrostatic, as long as the material has not been calcined. Hydrogen bonding and van der Waals interactions play a secondary role and the pillars are discrete molecular units.

When it was shown that crystalline mesoporous silica can be obtained by ordered condensation of silica oligomers around a template, the intercalation of mesoporous fragments within the clay interlamellar volume was logically attempted. It had already been shown by Moini and Pinnavaia [26] that crystallized synthetic imogolite can be intercalated.

Zhou et al. [27] prepared mesoporous silica pillared clays with porous montmorillonite. The heterostructures were synthesized by using pillaring and template-assembly techniques. The purified montmorillonite is modified by cetyltrimethylammonium bromide (CTAB) and mixtures of dodecylamine (DDA), and tetraethylorthosilicate (TEOS) at montmorillonite/DDA/TEOS molar ratios of 1 : 20 : 150 were allowed to react for 8 h. The resulting intercalates were then centrifuged and air-dried. Finally, after calcination at 540 °C for 4 h to remove the organic templates, novel silica porous montmorillonite heterostructures (PMHs) were obtained. These possess mesopores (average pore size $\sim$2.17 nm), a super-gallery

---

References see page 242

height of 2.75 nm, a large BET specific surface area of 822 m² g⁻¹, and high thermal stability up to 800 °C. The mechanism of pore formation through gallery template-assembly was postulated, and the higher thermal stability of PMHs was ascribed to the uniformity of the pillars.

Zhou et al. [28] prepared first a bentonite pillared with a mixed sol of Si and Ti hydroxides, which was subsequently treated with quaternary ammonium surfactants. The surfactant micelle acts as a template, similar to their role in MCM-41 synthesis. Because both the surfactant micelles and the sol particles are positively charged, the formation of a meso-phase in the galleries between the clay layers that bear negative charges is greatly favored. Besides, the sol particles do not bind the clay layers strongly as do other types of pillar precursor, so that the treatment with surfactants can result in a radical structural modification of the sol-pillared clays. This allows the pore structure of these porous clays to be tailored by the choice of surfactant. The surfactant treatment also results in increases of the porous volume and an improvement of the thermal stability.

Pires et al. [29] prepared several pillared clays by using a polyalcohol [ethylene glycol or poly(vinyl alcohol)] or a poly(ethylene oxide) surfactant as an interlayer gallery template and an aluminum oligomer species as the pillaring agent. The use of polyalcohols or non-ionic surfactants (e.g., Tergitol) produced materials which, in general, presented larger basal spacing than those found for the solids prepared without additives. The initial positive effect on the expansion of the clay interlayers was not totally retained after calcination of the materials; most probably, at the end, the basal spacing is still ruled by the intercalated aluminum species.

From this short review on a subject that deserves more attention, it is clear that further research is needed before claiming that the hetero structures obtained by the use of surfactants present substantial advantages over the classical pillared clays, especially regarding the thermal stability.

2.3.7.3 Pillars

Among the elements shown in Table 2 that have been used as pillars, the most common are aluminum and silicon; all of the others are transition metals, some of which are used alone but, on most occasions, also in combination with aluminum or zirconium.

2.3.7.3.1 Homo-Atomic Pillars

The chemical elements used in the preparation of mononuclear pillared clays are shown in Table 3, together with information on the nature of the clay. The preparation process will be detailed hereafter.

Tab. 2 Chemical elements used for the preparation of pillared clays

									13 Al	14 Si
22 Ti	23 V	24 Cr	25 Mn	26 Fe	27 Co	28 Ni	29 Cu	30 Zn	31 Ga	
40 Zr	41 Nb	42 Mo								

Tab. 3 Homonuclear pillars

Element	Clay	References
Al	Montmorillonite	[30–33]
Al	Beidellite	[30, 33, 34]
Al	Saponite	[33, 35, 36]
Al	Hectorite	[29, 33, 35, 37]
Al	Laponite	[33, 38]
Zr	Montmorillonite, Beidellite, Saponite	[28, 31, 39, 40]
Ti	Montmorillonite	[31, 41, 42]
Ta	Montmorillonite	[43]
Cr	Montmorillonite	[44]
Fe	Montmorillonite	[43, 45]

A Hydroxyaluminum Polymer The complexity of the aqueous solution chemistry of Al^{3+} and the precipitation of aluminum hydroxide colloids in the smectites interlayer was a popular subject among clay scientists during the 1960s, mainly due to the resemblance of these mixed compounds with natural interstratified minerals. In 1978, Lahav et al. [5] were the first to publish X-ray recordings showing the swelling of montmorillonite up to ∼1.8 nm obtained by mixing an aged suspension of a hydrolyzed aluminum cation with $R = OH:Al$ varying from 1 to 2.3. The total Al content, the R ratio, and the aging time were each shown to be important parameters. Prior to the publication by Lahav et al., and as mentioned in Section 2.3.7.1.1, in 1974 a team of scientists at W. R. Grace began some extensive studies of pillared clays, but it was only in 1980 that Vaughan and Lussier [46a] disclosed their patents dealing with Al, Zr, and Ti pillared clays (for a review, see [46b]). In the meantime, progress in analytical techniques had afforded a better understanding of the nature of the pillars.

Using the polyhydroxyaluminum polymer as an illustration, it was presumed that the pillar was the so-called Al_{13} cation, $[Al_{13}O_4(OH)_{24}(H_2O)_{12}]^{7+}$, the structure of which in a hydrated sulfate salt had been determined by Johannson et al. [47] (Fig. 3). As early as 1972, in a partially hydrolyzed aluminum solution, Akitt and coworkers [48] had recognized the signature of Al_{13} in a NMR peak attributable to fourfold coordinated aluminum (Al^{IV}), typical of the Keggin cation. Since the Keggin cation contains the four planes of oxygen atoms

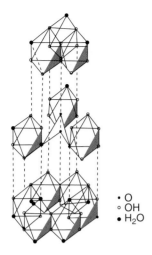

Fig. 3 Magnified structure of the Al_{13} cation, showing the central Al tetrahedron surrounded by 12 Al octahedra.

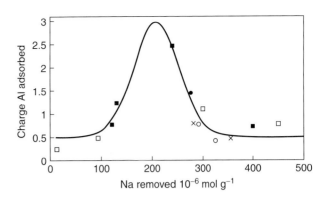

Fig. 4 Exchange Al_{13}-Na^+ in (□) saponite and (■), (◇) and (x) hectorite, using a pure Al_{13} solution. The charge per Al is the ratio Na^+ removed : Al uptaken. (Adapted from Ref. [35].)

required to prop apart the clay interlayer by ~0.8 nm, it was tempting to suggest that pillaring with Al_{13} was just a cation exchange process. It was thus assumed that the Al_{13} polymer retains its integrity throughout the exchange process, in spite of the fact that Al_{13} results from a complex set of reversible hydrolysis reactions [49] which can eventually occur on the clay surface.

The importance of aging the polymer solution, as outlined previously by Lahav et al. [5] and repeatedly reported by others, had shed some doubt on oversimplifying the pillaring mechanism by a simple cation exchange. An original investigation by Schoonheydt et al. [35] has shed interesting light on the complexity of the phenomenon.

A pure solution of Al_{13} chloride was prepared according to the recipe initially proposed by Schönherr et al. [50]. From the analysis of the solution by ^{27}Al NMR (e.g., in measuring quantitatively the intensity of the resonance due to the tetrahedral Al in Al_{13}, and the release of sodium), real exchange isotherms were obtained for the first time [35].

Figure 4 (the details of which are calculated from the results displayed in Fig. 8 of Ref. [35]) shows the variation of the average charge per Al adsorbed with respect to the Na released. It is obtained simply as the ratio of equiv. Na^+ released : mol Al fixed. Up to a loading of about 20% of the cation exchange capacities (CEC) – which are 0.55 meq g^{-1} and 0.66 meq g^{-1} for hectorite and saponite, respectively – the charge is slightly lower than +0.55. This is the theoretical charge per Al in Al_{13}^{7+}. It may be that the first part of the curve shown in Fig. 4 corresponds to the exchange on the external surface.

In the gallery, the exchange is accompanied by a partial or even total depolymerization of Al_{13}. The charge distribution represented by a Gaussian function would be random because of the randomness of the depolymerization process. As soon as about two-thirds of the Na content corresponding to the CEC is desorbed, reaggregation of oligomers into Al_{13} occurs. The interlayer spacing of about 0.8 nm is obtained when the Al_{13} content exceeds the CEC – which means, of course, that the charge per Al must decrease. It is possible that washing might help further hydrolysis and/or polymerization.

Figure 5 shows the composition of solutions with different OH : Al ratios prepared at room temperature or refluxed for 24 h. In spite of the high relative content in Al_{13}^{7+} of the solution with $R = 2$ prepared at room temperature, the achievement of the 1.8-nm spacing requires aging, washing (eventually by dialysis) and also a pretreatment of the solution at 50 °C, in agreement with Plee et al. [30]. With a multicomponent solution, there is obviously a competition of more than one cationic species to replace Na^+ (the initial cation) in the interlayer space. The complexity of the pillaring solution is even greater in chlorhydrol solution, which has been used successfully (as claimed in Refs. [46, 51, 52]). Chlorhydrol is the trade-name of the reaction product of Al metal with an aluminum chloride solution. In a chlorhydrol solution with $R = 1.8$, the amounts of monomeric Al, of Al_{13}, and of polymeric Al are about one-third [53]. The polymer in question is made only from octahedrally coordinated aluminum.

However, when either a solution of chlorhydrol is used or the refluxed solution with compositions shown in Fig. 5b, swelling of the clay to 1.8 nm is achieved easily. The main difference between the chlorhydrol or refluxed solution and the solutions obtained at room temperature by partial hydrolysis ($R \sim 2$) is in the small relative abundance of monomeric Al. Meanwhile, it seems that we have also to accept that, on the acid clay surface, the "polymeric" fractions present in the chlorhydrol

---

References see page 242

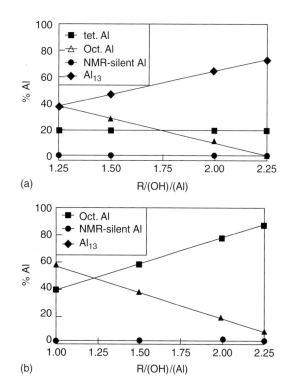

Fig. 5 Relative quantities of Al species in solution as a function of $R =$ OH:Al in a 0.1 M Al^{3+} solution prepared at (a) room temperature and (b) under reflux. (Reproduced from Ref. [37], with permission.)

or in the "refluxed" solution can undergo structural rearrangements with some reconstruction of Al_{13} or of Al_{13} polymer, as suggested in Fig. 4. Indeed, in the pillared dry material the observed $Al^{IV}:Al^{VI}$ ratio is roughly that expected in Al_{13} (see Section 2.3.7.3).

It is interesting to note that Figueras et al. [54], in what they termed "competitive adsorption", had prepared a pillaring solution containing NH_4^+ with NH_4^+/Al ratios as high as 10. Relatively large concentrations of foreign cations decreased the amount of alumina species which was fixed, as expected from simple equilibrium considerations. The addition of ammonium increased the pH (to ~6) and almost completely depleted the concentration in monomeric aluminum. The result was an increased crystallinity of the pillared materials [54]. Thereafter, Michot and Pinnavaia [32] have improved the crystallinity of alumina-pillared montmorillonite by protecting the Al_{13} polymer by a non-ionic surfactant. Aging is no longer necessary.

B Gallium and Mixed Gallium Aluminum Hydroxypolymers The chemistry of gallium resembles that of aluminum. It has been shown by Bradley et al. [55, 56] that, in a pure Ga^{3+} solution, the equivalent of the Al_{13} polymer is synthesized upon limited hydrolysis. The existence of a mixed (Ga,12Al) polymer (where Ga is in tetrahedral coordination) has also been noted [56]. The ^{69}Ga ($I = 3/2$) NMR resonance line is, unfortunately, broader than the ^{27}Al line, which makes detailed study of the pillaring mechanism more difficult.

Both, Vieira-Coelho and Poncelet [57] and Bradley et al. [55, 56], have shown that the Ga or mixed Ga/Al polymeric solutions, when prepared as described above for Al polymeric solutions, must also be aged and the intercalated smectite washed in order to obtain a successful pillaring, very much as reported for Al_{13}.

C Zirconia Hydroxypolymers Historically, the zirconia hydroxypolymers obtained through the hydrolysis of $ZrOCl_2 \cdot xH_2O$ were among the first to be used for pillaring clays because they required less optimization than the alumina polymers [46a,b]. The generally accepted structure of the pillaring agent is a $[Zr(OH)_2 \cdot (H_2O)_4]^{8+}$ tetramer (Fig. 6) [58], and this structure is present in solid zirconium chloride [59]. The starting material for producing the polymer is the zirconyl chloride, which is hydrolyzed by NaOH and/or Na_2CO_3. An increase in pH increases the degree of polymerization.

Farfan-Torres et al. [39] have used a 0.1 M NaOH solution to produce hydrolysis of either the $ZrOCl_2$ solution or the $ZrOCl_2$-clay suspension. As long as the final pH of the pillaring solution was below 2, a stable intercalation with 0.74 nm basal spacing was obtained. It is important to note that the clay suspension to which the pillaring solution was added, was made from a 1:1 mixture of acetone and water. It is evident that, at the low pH used for pillaring zirconia, hydronium was a serious competitor for the clay exchange sites and that some acid dissolution of the lattice was likely. Farfan-Torres et al. [39] have measured a residual CEC close to 88% of the initial CEC of an unpillared (Westone-L) montmorillonite, after pillaring at pH 1.9 and fixation of ~24 wt.% of ZrO_2 (or 0.2 mol ZrO_2). The corresponding charge brought by each zirconium in the polymer must be as low as 0.2^+.

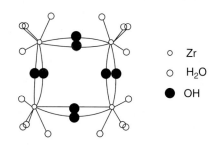

Fig. 6 Structure of $[Zr(OH)_2 \cdot 4H_2O]^+$ tetramer. (Reproduced from Ref. [58], with permission.)

The extent of polymerization that the tetramer should undergo to explain such a lowering of the charge is large, and it forces the clay to work as a powerful proton sink. The strong acidity observed by temperature-programmed desorption (TPD) of ammonia from zirconia-pillared montmorillonite may be partially attributable to the surface protonation [39].

D Titania Pillars Pillaring smectites with titanium polyhydroxide was among the claims of the original W. R. Grace patents held by Vaughan et al. [3]. The partial hydrolysis of $TiCl_4$ [59] or alkoxide [41, 60] is performed in HCl solution, and potentiometric titrations suggest the existence of a polynuclear species $[(TiO)_8(OH)_{12}]^{4+}$, the structure of which is unknown [61]. The procedure used to obtain this polymer is a slow hydrolysis of $TiCl_4$ in 6.0 M HCl until solutions with final concentrations in the range 0.11 to 1.0 M Ti are obtained [59]. Again, these solutions must be aged.

The polymer obtained from the hydrolysis of Ti alkoxide is also probably made from octahedra sharing edges [41]. At a constant hydrolysis ratio (HCl : Ti alkoxide), the loading in TiO_2 decreases with increasing molecular weight of the alkoxide (from ethoxide to n-butoxide), suggesting a simultaneous reduction in the extent of polymerization.

E Silica and Silicate Pillars As early as 1980, Endo et al. [62] attempted to pillar smectite with silica obtained through the decomposition of silicon acetyl acetonate. A spacing corresponding to a single SiO_2 layer was obtained. Fetter et al. [63] have reviewed the various attempts to use organosilicon compounds for obtaining thermally stable pillared smectites. The same authors provided details of the preparation of the pillaring solutions using either 2(trichlorosilylethyl)pyridine (TCSEP) or 3-aminopropyltrimethoxysilane (APTMS). A limited hydrolysis was obtained at 0 °C for TCSEP in water, or at room temperature for APTMS in methanol and a 16% HCl solution. The hydrolysis product of TCSEP was aged for 3 days, while the aged (27 h) hydrolysis product of APTMS was added to acetone and refluxed for 1 h. It was necessary to store these solutions in the dark before use.

The results of a ^{29}Si NMR study suggest that the hydrolysis of TCSEP is a complicated process comprising a mixture of compounds such as hydrated monomer, hydrated and trimeric species, and higher polymers with a resonance line near -67 ppm, also observable in the pillared solid.

Moini and Pinnavaia [26] have succeeded in intercalating tubular imogolite in montmorillonite; the synthetic imogolite was brought to a pH higher than the zero point charge, such that it was positively charged and played the role of a supercation.

F Transition Metal Oxide Pillars The possibility of pillaring clays with transition metal oxides which should exhibit electron-transfer properties has been investigated by Pinnavaia and coworkers [44, 64], who studied chromia-pillared clay as early as 1985. Chromium nitrate $(Cr(NO_3)_3 \cdot 9H_2O)$ is hydrolyzed by Na_2CO_3 and aged for more than one day at 95 °C, the final pH being < 2, and the ratio meq CO_3^{2-} per mole Cr^{III} being about 2. Accordingly, the $4A_{2g} \rightarrow 4T_{1g}$ and $4A_{2u} \rightarrow 4T_{2g}$ transitions are red shifted, indicating the formation of higher chromia polymer. Aging is essential for obtaining large gallery heights. After calcination at 500 °C, the $d001$ reflection can still reach 2.2 nm.

Attempts have been made to pillar smectites by refluxing Cr (III) acetate with a Na-montmorillonite suspension [65]. Linear oligomeric chromium polyhydroxoacetate is believed to be the pillaring agent. However, in order to obtain pillared material, calcination under gaseous NH_3 is necessary, and the crystallinity is worse than that obtained in the procedure proposed by Pinnavaia et al.

The oligomer resulting from the hydrolysis of chromium nitrate has been identified by X-ray absorption fine-structure spectroscopy (EXAFS) as a tetranuclear species with edge-bridged octahedral chromia [38].

Tantalum-pillared clay has been obtained using a pillaring solution obtained through the hydrolysis of $Ta(OC_2H_5)_5$ in an acidified ethanolic solution [43]. The clay is slowly added to this solution.

2.3.7.3.2 Heteroatomic Pillars Pillared clays offer a unique opportunity of tuning the catalytic properties through the intercalation of pillars with different functionalities. Previously, aluminum-pillared beidellite [30] has been shown to combine the Brønsted acidity of the clay surface with the Lewis acidity of the pillars, though more sophisticated developments have been reported recently (see Table 4). Guo et al. [66] have pillared rectorite with mixed zirconium, silicon oxide modified by SO_4^{2-}. After being calcined at 500 °C, (SO_4^{2-}/SiZrR) exhibited superacidity within a mesoporous structure: the Hammett acidity function H_0 was smaller than -11.93 and the most probable pore size was at ca. 2.3 nm. The superacid sites were attributed to Lewis acid sites, and proved to be the dominant active centers.

In the domain of desulfurization catalysis, Colin et al. [67] have studied molybdenum supported on Al-pillared clay and on Zr-pillared clay; these materials were tested in the hydrogenation of naphthalene. The molybdenum sulfide (MoS_2) catalysts supported on Zr-pillared clays were found to be more active than MoS_2 supported on Al-pillared clays and on alumina.

References see page 242

Tab. 4 Mixed pillars or coated pillars

Elements	Clay	References
Al, Ga	Montmorillonite, Beidellite, Rectorite	[69–73]
Si, Al	Montmorillonite, Fluorohectorite	[74–76]
Si, Zr		[66]
Si, Ti	Montmorillonite, Saponite	[77–79]
Al, Zr	Saponite	[30]
Al, Cr		[80]
Al, Re		[81, 82]
Al, Cu		[83]
Al, Ce	Montmorillonite	[71, 84]
Al, La	Montmorillonite, Rectorite	[85–87]
Al, V	Montmorillonite, Saponite	[88, 89]
Al, Fe		[90]
Ti, V	Montmorillonite	[27, 91]
Zr, Mn		[92]
Ni, Mo, Al		[93]
Al, Mo		[67]
Mn, Co, Al		[68]
Cr, S, Cr		[94]
Zr, S		[95]
Mo compl.	Saponite, Montmorillonite	[96]
Co, Ni, Zn, Zr	Montmorillonite	[97]

Vicente et al. [68] have characterized Mn- and Co-supported catalysts prepared by wet impregnation of Al-pillared clays with Mn- and Co-complexes, using complex salts of the metals, (Co^{2+} acetylacetonate, tris(ethylenediamine) Co^{3+} chloride, Mn^{2+} acetylacetonate and Mn^{3+} acetate dihydrate) as precursors. The incorporation of these metals into the clays by cointercalation of Al–Mn and Al–Co solutions was also considered, but this method was not useful for the preparation of catalysts as only very small amounts of Mn or Co were fixed. Dispersed Mn_2O_3, Mn_3O_4 and Co_3O_4 particles were present in the calcined catalysts.

2.3.7.3.3 Mixed Al–Fe Pillars Pillaring smectites with iron polyhydroxy polymers which would be similar to those of aluminum is difficult as far as hydrolysis conditions are concerned. The base-to-metal ratio must be about 2, the pH about 1.7, and the X-ray reflections are broad and weak (see Table III in Ref. [45]). In contrast, mixed Al/Fe^{3+} pillars have been prepared and the intercalated smectite is thermally stable up to 500 °C, if the Fe : Al ratio is equal to or less than 0.2 [90]. The ^{27}Al NMR spectra in the pillaring solutions with Fe : Al < 0.5 show the Al^{IV} signature of the Keggin cation at 63 ppm [98]. Thus, the pillaring solution would be made from these cations in which Al^{3+} would be partially replaced by Fe^{3+}. The nature of the substitutions is not known, but in view of the ubiquitous isomorphic replacement of Al by Fe in the octahedral layers of micas, it is likely that some of the octahedral Al is replaced by Fe in the Keggin structure. The pillaring solutions are prepared by either: (i) contacting an aged Al_{13} solution prepared by the method of Lahav et al. [5] with a 0.1 M $FeCl_3$ solution at pH between 4 and 4.5 (0.1 M Na_2CO_3); or (ii) co-hydrolysis of 0.1 M $AlCl_3$ to which 0.1 M $FeCl_3$ is added. The hydrolysis is carried out after the addition of 0.1 M Na_2CO_3 solution. In both cases, the hydrolysis products are aged for one week at room temperature and at 120 °C for 4 h before use.

2.3.7.3.4 Miscellaneous To date, reviews of the pillaring mechanism have been restricted to materials in which the nature of the polymer has been investigated. Pillars other than those reviewed above have been used successfully; these include binary oxides such as $LaNiO_x$ [99], lanthanum–aluminum [85], and silicon–titanium [79] pillars.

2.3.7.4 Pillared Clays

2.3.7.4.1 Preliminary Remarks Based on a brief description of the structure of the clay (see Fig. 1), an oversimplified representation of a pillared clay would be that shown in Fig. 7. However, such a representation would be grossly inaccurate. When a smectite is air-dried, the vector normal to the clay platelets is scattered by ±18° with respect to the averaged direction of the c axis [100]. Therefore, there is a significant orientational disorder which can be considered as a distribution of the orientation of tactoids made from a few parallel plates [101, 102]. Moreover, to this orientational disorder must be added a translational disorder, as the turbostratic character is evident from the shape of the (hkl) X-ray reflection. Thus, the aggregate of platelets to be pillared is more likely to resemble that shown in Fig. 8 (see also micrograph in Ref. [105]).

In brief, the real situation is far from the "rosy" picture shown in Fig. 7. As a consequence, when questions about the specific surface area or of the porosity of pillared clays are addressed, the precipitation of the pillaring agent outside of the gallery, the contribution to the surface area of the "dangling" sheets, etc. also has to be accounted for.

The coherent scattering domain is always small as evidenced by the observation of three broad 001 reflections, at the most. From the viewpoint of crystallinity, zeolites and pillared clays belong to two different worlds. Yet, in spite of these remarks (which are evident to those who have worked in the field), one sometimes still finds researchers who are amazed by the fact that calculations of microporous volume or of surface area, using a model such as that in Fig. 7, do not agree with the experimental results. A more realistic representation is that shown

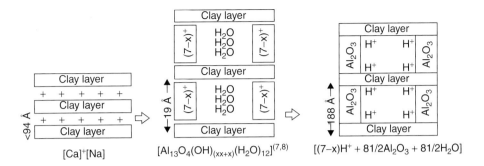

Fig. 7 Early representation of clay pillaring with chlorhydrol. (Reproduced from Ref. [46b], with permission.)

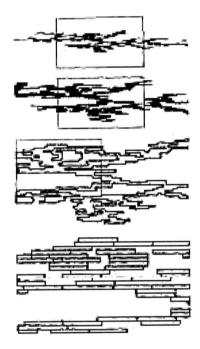

Fig. 8 Computer simulation of the formation of clay tactoids by a process of aggregation of 10^3 particles. The length/thickness ratio of each particle is 9. In the three topmost diagrams, each rectangle encloses a portion of tactoid shown enlarged in the diagram immediately below. (Reproduced from Refs. [103, 104], with permission.)

in Fig. 8. It is important to stress these points before reviewing the properties of the pillared clay.

2.3.7.4.2 Main Physical Characteristics The $d001$ spacing, the pore size distribution, the specific surface area and its distribution versus the pore radius are essential physical characteristics of the solid. As thermal activation is always necessary at the final step in the preparation of pillared clay, the variation of these physical parameters with temperature contains important information. A temperature-programmed calcination is too rarely investigated.

Whatever the nature of the pillar, the achievement of a pillared clay with a gallery of at least ∼0.8 nm and specific surface area (calculated from the Langmuir adsorption isotherm of N_2 at liquid nitrogen temperature) of at least 200 m^2 g^{-1} after calcination at 500 °C, requires a critical concentration of pillaring agents. This critical concentration depends upon the nature of the pillaring agent and of the clay. An interesting aspect is shown in Fig. 9 where, clearly, pillaring the *montmorillonite* (fraction < 2 μ of a Wyoming Na-bentonite) requires more Al^{13} polymers, and higher $R = $ OH : Al, than pillaring a synthetic beidellite.

Amazingly, the total Al adsorbed is already near saturation (2 mmol Al fixed per g clay) for an initial content in the solution, (the ordinate in Fig. 9), of about 10 meq g^{-1} clay. Clearly, at this content the full expansion is not reached in the beidellite while it is in the montmorillonite. Figure 10 represents a good illustration of the large enhancement of the N_2 adsorption capacity resulting from pillaring beidellite.

A strong link between the $d001$ basal spacing and the surface area has been generally observed for all categories of pillared clays. However, the link between the loading in pillaring agents, the $d001$ spacing, the surface area, and the porosity is not obvious. As mentioned in Section 2.3.7.3.1, what is pillared is an assembly of tactoids and, consequently, one must account for the fact that adjacent platelets separated by an interlayer distance Δ are overlapping one another by a variable fraction [104].

The adequate technique of measuring the porosity, the pore-size distribution and the surface area for pillared material is not evident, either. The team led by Vansant has devoted a great deal of effort to these aspects that are so important for the separation techniques. Only two recent references [31, 106] are given here; the results in Table 5 are reproduced from Ref. [31].

A report by Yamanaka and Hattori [107] showed that in a montmorillonite pillared with 48 wt.% Fe_2O_3, an unknown (but not negligible) fraction of the iron oxide

References see page 242

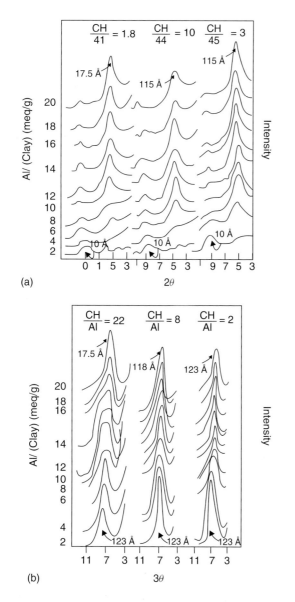

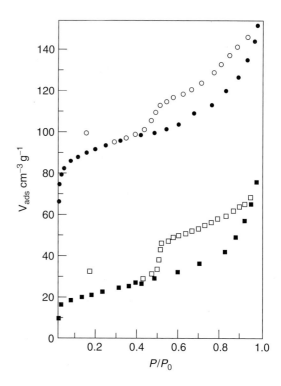

Fig. 9 (a) 001 reflection of calcined (500 °C) films of pillared montmorillonite as a function of the initial Al : clay ratio and for R = OH/Al ratios between 1.2 and 2. The composition of such solutions can be found in Fig. 5a. (Reproduced from Ref. [110], with permission.) (b) Same situation as in Fig. 9a, but with the smectite replaced by beidellite.

Fig. 10 N_2 adsorption–desorption isotherm at -183 °C on (a) beidellite and (b) pillared beidellite calcined at 300 °C. (Reproduced from Ref. [110], with permission.)

Tab. 5 Adsorption capacities (mmol g^{-1} at 0.45 bar) of various pillared montmorillonites [31]

	Al	Zr	Ti	Fe
At -79 °C				
N_2	0.05	0.09	0.16	0.03
O_2	0.02	0.03	0.07	0.01
CO_2	0.95	2.4	1.46	0.29
At -196 °C				
N_2	2.5	3.5	2.8	
O_2	3.38	5	5.9	
A	3.04	4.4	5.2	
Physical characteristics				
$\Delta d001$ (nm)	0.78	0.40	1.6	
Specific area (m^2 g^{-1})^a	341	280	410	198

^{a}From Langmuir isotherm, except BET for the last column.

locates on the external surface. Indeed, N_2 adsorption isotherms of Brunauer type II are observed instead of the Langmuir isotherm, which would be expected from the observation of an interlayer spacing of 0.64 nm. It is for the same reason that for Fe-pillared montmorillonite in Table 5, it is the BET surface area which is given.

In Table 6 are listed the interlayer spacings when the loadings were given. The data where the $d001$ reflection appears as a shoulder on the rapidly decreasing scattering intensity away from the direct X-ray beam, were skipped.

As can be observed, the data obtained with the Al-pillared montmorillonite, beidellite, laponite, and saponite are in good agreement. The mixed aluminum–iron pillar provides spacing close to that of aluminum, whereas the data for iron are more debatable. It seems that an appreciable fraction of Fe_2O_3 is precipitated on the outside surfaces, but in contrast pillaring with chromia

Tab. 6 Observed interlayer spacing A = $(d001 - 9.6)$ Å, nature of the pillar, and of the clay. Loading is oxide: clay wt% or clay unit cell (uc)

Clay[a]	Pillar	Loading	Temp./°C	Δ/Å	References
Mon	Al	18.7	500	7.8	[54]
Mon	Al	10	300	8	[17]
Mon	Al	6	500[f]	5.7	[22]
Bei	Al	10	300	8	[17]
Bei	Al	8.5–34	400	8	[52]
Sap	Al	10–13.2	400	7.9–8.6[b]	[14]
Hoc	Al	≈11	250	8.6	[14, 16]
Sap	Al	9.2	500	8.2	[45]
Lap	Al	13	400	7	[32]
Mon	Zr	16.2–25.1	400	7.4	[28]
Mon	Zr	19.1	500	4	[54]
Mon	Ti	20	500	16–19[c]	[54]
Mon	Ti	29–49	400	10–12[b]	[38]
Mon	Cr	3.5 Cr_2O_3/uc	350	13.4	[30]
Mon	Fe	6–9 Fe_2O_3/uc	350	13–15[d]	[56]
Mon	Fe	33.2–39.6	350	≈18	[53]
Mon	Ta	17–24	400	13.9–15.9[b]	[33]
Mon	Al,Fe	Al_2O_3 11.6, Fe_2O_3 8.6	400	≈7	[34]
Mon	Al,La	Al_2O_3 5, La_2O_3 1.7–1.9	autoclave; 120 °C	≈16[c]	[44]
Sap	Ti,Si	TiO_1 10.6 SiO_2 9.9	500	22.4	[45]
Mon	Al,Si	Si : Al ratio, 0.36–0.76, total loading Al + Si ≈ 1.5/uc	500	7.6[e]	[57]
Fhec	Al,Si	see above	500	10.2[f]	[57]

[a] Mon = montmorillonite; Lap = laponite or synthetic hectorite; Bei = beidellite; Sap = saponite; Fhec = fluorchectorite.
[b] The lowest spacing corresponds to the lowest loading.
[c] Probable interstratification.
[d] Unstable in moist atmosphere.
[e] Sharpness of basal reflection not indicated.
[f] Calcined under vacuum or inert atmosphere.

Tab. 7 Molecular sieving by Al- and a Cr-pillared clay freeze-dried and calcined at 350 °C [19, 30]; amount adsorbed (mmol) g^{-1} clay at saturation

Kinetic diameter (Å)	4	5.8	6.2	9.2	10.2
Molecule	N_2	C_6H_6	$(CH_3)_4C$	$1,3,5$-$(C_2H_5)_6C_6H_3$	$(F_6C_4)_3N$
Al-pillared[a] 2.87 Al/uc $\Delta d001 = 9.5$ Å	2.91	1.61	1.02	0.71	0.43
Cr-pillared[b] 3.53 Cr/uc $\Delta d001 = 11.4$ Å	3.63	2.58	1.56	0.91	0.76

[a] BET surface area: 280 m^2 g^{-1}.
[b] BET surface area: 350 m^2 g^{-1}.

and Ta_2O_5 leads to less ambiguous results. It should be noted that, upon calcination in air, the interlayer spacing of the chromia-pillared montmorillonite collapses [44]. Saponite with mixed Si–Ti pillars is an example of very good thermal stability.

It should be emphasized that interlayer spacings in the order of 1.2 to 1.6 nm require pillars with about six or more oxygen layers. These pillaring species are likely to be nanosized colloidal particles on the positive side of their zero point charge. Some examples of molecular sieving are shown in Table 7. It can be observed that the kinetic molecular diameter, the corresponding amounts of adsorbed molecules, and the interlayer spacing (A d001) follow the same trend. A fractal analysis of the adsorption process of pillared clays by Van Damme and Fripiat [108] has shown that the values in Table 6 are compatible with an Euclidean fractal dimension of 2 for Al-pillared montmorillonite. It must be emphasized that the preparation conditions of the pillared clay are highly critical for the ultimate size of the molecules which have access to the gallery [109].

The development of the surface area and of the microporosity upon Al-pillaring a vermiculite or a phlogopite is illustrated in Fig. 11.

2.3.7.4.3 Chemical Properties

Smectite surfaces are acidic, as a result of the high degree of dissociation of the cation hydration water [111–113] and of the bidimensional structure. The exchangeable cation–clay association is similar to a cation–anion association, but the "anion" (namely, the clay lattice) has an infinite radius of curvature. Thus, the electric field created by the cation strongly polarizes a hydration water molecule in the first coordinated shell. One of the protons is probably delocalized [114] on the network of water filling in the space between the cations, while the OH^- remains coordinated to the cation. In the pillared clay, there is always a residual CEC, the value of which depends on the nature of the pillar and the loading. In a dioctahedral clay such as beidellite, Brønsted acid centers can be created (as in Y and other zeolites) by calcining the ammonium-exchanged form up to about 400 °C. Upon de-ammoniating the $NH_4^+-Si-O-Al-$ cluster in the tetrahedral layer, the OH bridge $Si-OH-Al$ is obtained, as shown by an OH stretching band at 3440 cm^{-1} in the infrared (IR) spectra recorded at increasing temperature (Fig. 12).

A similar behavior is also observed upon moderate acid leaching. These observations are in line with the well-established fact that an Al–Si isomorphic substitution is a weak point in a tetrahedral lattice. Therefore, in addition to the Brønsted acidity attributable to the cation hydration water, stronger Brønsted sites can be induced in the *001* plane of dioctahedral smectites, as shown in Fig. 13. These two types of acid site exist independently of the pillars.

It might have been anticipated that the behavior of a trioctahedral smectite rich in isomorphic substitutions (e.g., saponite) would be similar to that observed for beidellite. It is, indeed, observed that acid leaching [115] produces OH groups not present in the Na-exchanged form, although its stretching frequency, instead of being at 3440 cm^{-1}, is observed at 3594 cm^{-1}, whilst a band which is observed in the sodium form at 3716 cm^{-1} shifts to 3738 cm^{-1}.

Chevalier et al. [116] have observed that the band at 3594 cm^{-1} disappears upon pyridine adsorption, but these authors did not provide information on the thermal stability of the pyridinium.

Besides the Brønsted acidity, the exchangeable cation – especially if it is multivalent and partially dehydrated – serves as a source of Lewis acidity.

Pillaring the clay will modify the surface acidity in at least two ways: (i) the polycationic pillars displace the cations originally present on the surface of the clay; and (ii) the pillars' own acidity may create $Si-OH-Al$ bridges in smectites with Al, Si substitutions in the tetrahedral layer. These OH may provide an anchoring point to the pillar. These two points are fundamental to the nature of the cross-inking mechanisms.

In clay suspension in contact with the pillaring solution, what may be considered to happen is simply a flocculation provoked by contraction of the double layer [117]. Such a flocculation process likely affects the texture of the tactoids shown in Fig. 8 to an unknown extent. Once the polycations are within the interlayer space, electrostatic forces maintain the clay sheets at the distance corresponding

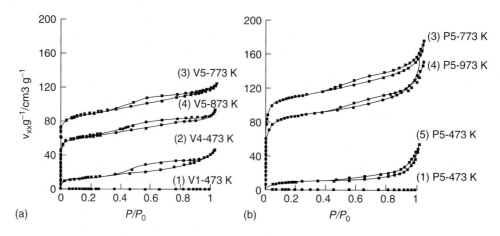

Fig. 11 N_2 adsorption–desorption isotherms at −196 °C of (1) initial vermiculite (a) and phlogopite (b); after the charge reduction step and Na-exchange (2), and after calcination at 500 °C (3) and at 720 °C (4) of the Al-pillared material [13, 14].

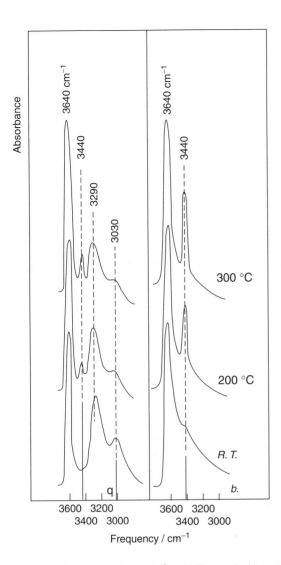

Fig. 12 Infrared spectra of (a) NH_4^+-beidellite and (b) H_3O^+-beidellite, after outgassing at room temperature, 200 °C, and 300 °C. (Reproduced from Ref. [110], with permission.)

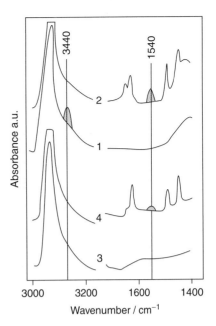

Fig. 13 Infrared spectra of (1) Al-pillared beidellite calcined at 400 °C and (3) Al-pillared calcined montmorillonite before the adsorption of pyridine. Spectra (2) and (4) were obtained after adsorption. (Reproduced from Ref. [115], with permission.)

to the size of the pillars. This, of course, affects the fraction of the platelets facing one another, and at that stage crosslinking *per se* is not achieved in the sense that the poly-cation pillars can be back-exchanged. Actual cross-linking occurs upon thermal activation.

As thermal activation is known to: (i) dehydroxylate the pillars; (ii) activate the eventual attack of the surface by protons; and (iii) remove the clay and pillars' hydration water, crosslinking is clearly a complex process. The process depends – as shown above – on the nature of the smectites as well as on the nature of the pillars, and consequently several mechanisms of crosslinking must exist. The most trivial mechanism would result from the action of the short-range van der Waals forces between the pillar and the clay surface, but it must not be forgotten that, in one way or another, the clay lattice charge must be balanced, irrespective of the residual charge (if any) of the dehydroxylated pillar. If the pillar dehydroxylation is represented by

$$P_n^{v+}(OH, H_2O) \longrightarrow P_n^{v'+} + (v - v')H^+ \quad (3)$$

and if v' is zero, then cation back-exchange is impossible.

The protons must: (i) remain on the surface as H_3O^+; (ii) react with the lattice by opening Si–O–M bridges, as observed for M = Al; or (iii) disappear into the lattice. Initially, it was observed [46] that the surface charge of an Al_{13} crosslinked montmorillonite can be partially restored by calcining in the presence of ammonia. That the initial charge is only partially retrieved is most probably due to a partial dehydroxylation by reaction of the "free" protons with lattice OH^-. This explanation is supported by the observation that an ammonium montmorillonite dehydroxylates at least 100 °C lower than a potassium montmorillonite [118]. If the pillar has no charge, the van der Waals forces alone suffice to explain the structural stability of the pillared material. Such a material appears as a nanocomposite resulting from the dispersion of particles with diameters in the order of 1 nm within the gallery, but there is no crosslinking if we reserve that denomination to covalent (or partially covalent) bonds between elements in the layer and elements in the pillar.

References see page 242

Covalent bonding has been noted in Al_{13}-pillared beidellite [119] and fluorophlogopite [120] by observing the ^{27}Al and ^{29}Si MAS NMR spectra in the calcined materials. In both cases, an inversion of lattice tetrahedron is suggested, with the apex oxygen pointing towards the gallery where it binds with the pillar. In the case of the fluorophlogopite studied by Pinnavaia et al. [120] it is, of course, a silicon tetrahedron which is inverted. The ^{29}Si $Q^3(Si-O-Al)$ line (at -92.3 ppm) splits into two components at -93.5 and 95.5 ppm. In the case of beidellite, where the crosslinking is most likely initiated by the reaction of the $Si-OH-Al$ bridging hydroxyl on the surface of the clay with an $HO-Al$ group on the pillar, an up-field shift of the ^{29}Si $Q^3(Si-1Al)$ line from -90.8 to -92.6 ppm has been reported [119], but the most noticeable change is the weakening (by ~50%) and up-field shift of the line attributable to the fourfold (Al^{IV})-coordinated aluminum atom in the lattice. This observation prompted Plee et al. [119] to suggest the inversion of a lattice Al tetrahedron and the formation of $Al^{IV}-O-Al^{VI}$ (pillar) linkages, as shown in Fig. 14.

A cross-linking mechanism of another type was suggested to occur by Tennakoon et al. [121], between alumina pillars and montmorillonite. Based on the absence of perturbation of the ^{29}Si NMR spectrum of the tetrahedral layer (also observed in pillared hectorite in Ref. [119]), these authors suggested that the OH of the octahedral layer would be implied in the formation of a covalent bond with the pillar. Before commenting on this hypothesis, the thermal behavior of Al-pillared montmorillonite and beidellite as revealed by differential thermal analysis (DTA) should be summarized.

The maximum endothermic peak resulting from the loss of physically adsorbed water which is observed at about 100 °C, extends up to about 250 °C in unpillared Na-beidellite and montmorillonite [115]. It is followed, in pillared beidellite, by two well-defined endothermic peaks at 330 °C and 540 °C assigned to pillar dehydration and partial dehydroxylation, and to the loss of lattice OH, respectively. In pillared montmorillonite, a continuous exothermic effect extends from about 250 °C to 700 °C, with maxima at 330 °C and 630 °C. The fact that the pillar and the lattice dehydroxylation temperatures differ by at least 200 °C sheds a serious doubt, from our point of view, on the hypothesis of Tennakoon et al. [121]. In addition, noticeable modification of the ^{27}Al NMR line attributable to octahedral Al has not been observed.

A report by Lambert et al. [122] on Al-pillared saponite is particularly illuminating as to the question of crosslinking, with a (Otay)-montmorillonite serving as a reference. These authors observed: (i) a significant decrease in the intensity of the ^{29}Si $Q^3(Si-1Al)$ line, but no shift of this line as the pillar density in the gallery increased; and (ii) a line assigned to the inverted

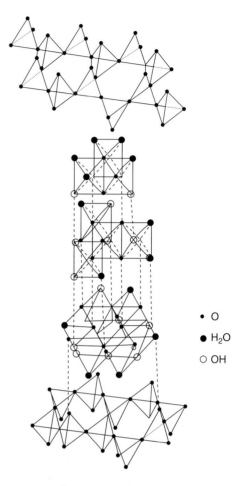

Fig. 14 Hypothetical structure of thermally modified alumina pillars sharing three oxygen atoms with three inverted Al^{IV} tetrahedra belonging to two tetrahedral layers of a beidellite. (Reproduced from Ref. [119]; copyright 1985, American Chemical Society.)

Al tetrahedron at ~56 ppm in double rotation (DOR) ^{27}Al spectra. Some of these findings, which are summarized in Table 8, support the conclusion in Ref. [119] concerning the inversion of Al tetrahedra in beidellite.

Similar observations were also reported by Zheng et al. [123], who pillared rectorite with alumina or silica pillars, and also by Malla and Komarneni [79]. Neither splitting of the ^{29}Si resonance nor an appreciable shift is observed in pillared and calcined montmorillonite.

We can summarize the crosslinking question by stating that there is no experimental evidence for the existence of covalent links between the pillars and the *001* surface of the clay lattice, except in the case of clays with tetrahedral substitutions. For experimental reasons, the crosslinking mechanism has not been studied for pillars other than Al pillars. It might be predicted, however, that tetrahedral inversion operates in tetrahedrally Al-substituted clay with any pillar since, without exception, they all contain OH

Tab. 8 NMR data supporting the hypothesis of a lattice Al tetrahedron inversion in Al-pillared saponite. R is the octahedral Al/tetrahedral Al ratio; in saponite Al^{VI} is only in the pillar. (Source: Ref. [7])

Sample	Al pillar/ mmol g^{-1}	Integr. Intens. Q^3 (Si–1Al)^a	$R = Al^{VI}/Al^{IVb}$
Na-saponite	0.0	29	0.0
D	2.1	21	1.13
A	2.7	14	1.37
C	4.5	7	3.03

^{a}Calcined at 500 °C.
^{b}Uncalcined.

groups which should react with the clay surface OH groups.

The final question to be developed in this section, which is devoted to the chemical properties, is of course the modification of acidity obtained upon pillaring and calcining. As outlined above, the pillared clay treatment with ammonia drives the protons that result from the pillar dehydroxylation from the lattice into the gallery. Thus, Brønsted acidity is potentially present and it should be available to proton acceptor molecules. The strength of this acidity is unknown, however. In contrast, the Si–OH–Al linkages on the surface of pillared beidellite or saponite, and of montmorillonite containing a few Al–Si substitutions in the tetrahedral layer [117], protonate pyridine.

As anticipated, pillars such as the zirconia, titania, and alumina pillars are a source of Lewis acidity due to the presence of coordinatively unsaturated sites.

For example, in alumina-pillared saponite, the band near 1450 cm^{-1} that is attributable to the pyridine-Lewis site adduct is still observable after desorption at 520 °C, whereas pyridinium disappears after desorption at 350 °C [79]. According to Sun et al. [124] and Del Castillo et al. [41], Ti-pillared montmorillonite has some Brønsted acidity, as shown by observation of the IR band at 1540 cm^{-1} assigned to pyridinium, up to 250 °C. Amazingly, Fetter et al. [63] report a rather strong acidity, indicated by NH$_3$ retention up to 500 °C, on silica-pillared montmorillonite. These authors assign the origin of this strong acidity to the residual protons, but do not explain why the acidity of the SiO$_2$-pillared clay is larger than that of comparable alumina- or zirconia-pillared samples. The surface protons removed from the bulk of the lattice may also be at the origin of the Brønsted acidity in titania- and zirconia- pillared montmorillonites studied by Farfan-Torres et al. [125]. However, in a recent report, Bodoardo et al. [126] challenged this possibility in suggesting that some of the OH-stretching bands observed in pillared montmorillonite can be indicative of acidic AlOH or SiOH.

In particular, a band at 3660 cm^{-1} which shifts by 224 cm^{-1} upon adsorbing benzene, is assigned to an acidic AlOH in the layer of the pillar.

Unfortunately, what is too-often overlooked in these discussions is the relative importance of the edge surfaces in the tactoids (see Fig. 8). On this point, the existence of Si–OH–Al bridges resulting either from the condensation of dangling bonds and/or of the reaction of the dangling bonds with components of the pillaring solution, could create Brønsted acid sites. In fact, it has been estimated that the edge surface area represents about one-fifth of the gallery area.

2.3.7.4.4 Chemical Modifications or Functionalization of the Pillaring Species The chemical properties of a pillar can be modified without affecting markedly the physical properties of the pillared material. For instance, the preparation of a titania-pillared montmorillonite which is stable up to 600 °C, with a specific surface area of $\sim$330 m^2 g^{-1} and a basal spacing of 2.2 nm, has been described by Sterte [59]. On the other hand, it is well documented [127] that the sulfate doping of the TiO$_2$, ZrO$_2$, HfO$_2$, SnO$_2$, and Al$_2$O$_3$ surface yields "superacids" containing, most probably, Brønsted and Lewis acidity. There is also a consensus that the Lewis acidity created by the electron-withdrawing power of SO$_4$ on the neighbor metallic cation is predominant. The origin of the Brønsted acidity is more controversial.

Among the five oxides cited above, some may be used as pillars and, therefore, it is amazing that very few reports have been devoted to the modification of pillar acidity by sulfation. Admaiai et al. [128] added ammonium sulfate to the aged pillaring TiCl$_4$/HCl solution, and up to 5% SO$_4$ could be fixed in that way. The specific surface area of the 2.6% SO$_4$–TiO$_2$-pillared material is reduced to a constant value of $\sim$200 m^2 g^{-1} upon calcination between 200 and 600 °C, while the basal spacing decreased by $\sim$0.2 nm. The ammonia retention by the sulfated TiO$_2$-pillared material calcined at 600 °C is almost twice that of the unsulfated pillared clay. From the IR spectra of chemisorbed pyridine, these authors conclude that the acidity enhancement is mainly due to Lewis acidity. Similar results are reported by Boudali et al. [129].

Farfan-Torres and coworkers [125, 130] have prepared sulfated ZrO$_2$-pillared clays in a similar manner. As for SO$_4$–TiO$_2$, an excess of sulfate (SO$_4$: Zr $\sim$ 0.7) collapsed the pillared structure at $\sim$500 °C, but the acidity of the sulfated material (SO$_4$/Zr = 0.35) is also practically doubled with respect to the unsulfated material. The

Brønsted acidity is greater in the SO_4–ZrO_2-pillared clay than in the 50% TiO_2-pillared material, but the unsulfated zirconia-pillared clay is also richer in Brønsted sites than the titania analogue.

In another vein, one may also consider the introduction of a small amount of alumina in a silica pillar, as a dopant, as it will most probably implement Brønsted sites on the surface of a pillar that otherwise would be neutral. This aspect has been studied by Sterte [131] in pillared montmorillonites and fluorohectorites. In both cases, and using a pillar with Si:Al ratio 2.2, the total acidity judged from NH_3 retention at 250 °C was more than doubled with respect to that measured with alumina pillars.

2.3.7.5 Catalytic Properties

2.3.7.5.1 A Brief Summary of Previous Studies
This section briefly summarizes some of the main trends of earlier research performed with pillared clays as catalysts. As outlined previously, with the exception of the reaction of ethene oxide and ethanol producing glycolether on Al-pillared montmorillonite [132], no large-scale application of pillared clays exists. This application is far from the initial goals referred to in Section 2.3.7.1.1 (e.g., the cracking of heavy hydrocarbons), mainly due to a lack of thermal and hydrothermal stability of the pillared material, which cannot withstand the severity of the regeneration treatments applied to cracking catalysts.

Molina et al. [133] have compared, in decane hydro-isomerization-hydrocracking, alumina-pillared synthetic beidellite (with Si:Al = 10.9) and montmorillonites relatively rich in Al–Si substitutions, with Si:Al of 13.3 for N7 and 19 for N4 (both clays from Niquelandia, Brasil), and 132 (Westone L montmorillonite). Pt (1 wt.%) was introduced into all samples, while two ultrastable Y zeolites with Si:Al ratios of 13.5 (CBV720) and 36 (CBV760), respectively, served as references. A Ga-pillared WL montmorillonite was also used for comparison.

As may be observed in Fig. 15, which shows the conversion-temperature curves, the activities of the Al-pillared beidellite (Si:Al = 10.9) and N7 (Si:Al = 13.3) lie between the activity of the ultrastable Y CBV720 (Si:Al = 13.5) and CBV760 (Si:Al = 36). AlPN4 (Si:Al = 19), and Ga- and Al-pillared montmorillonite are less efficient; the order of activity AlPB > AlN7 > AlN4 > GaPW > AlPW follows the Brønsted sites' content of the clay lattice – that is, the number of Al for Si substitutions in the tetrahedral layer. The isomerization selectivity at 60% conversion follows almost the same order, namely, 89.7%, 83.6%, 55.2%, 35.2% (AlPW), and 20.7% (GaPW). This study concluded that USY- and Al-pillared beidellite are similar isomerization catalysts, although on USY (containing 1 wt.% Pt) the hydrogenolysis is absent (no CH_4 formed),

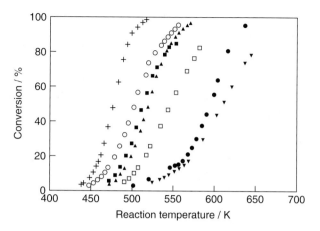

Fig. 15 Decane conversion versus reaction temperature on Al-pillared clays, Ga-pillared montmorillonite GaPW and USY CBV720 (Si:Al = 13.5) (+), and CBV760 (Si:Al = 36) (▲); (○) Al-pillared beidellite; (■) Al-N7; (●) Al-N4; (♦) Ga-montmorillonite; (▼) Al-montmorillonite. (Reproduced from Ref. [133], with permission.)

primary hydrocracking is obtained, and the cracked products are essentially branched. On the pillared clay, with 1 wt.% Pt also, hydrogenolysis is superimposed on hydrocracking.

The isomerization of m-xylene, which is a less complex and acidity-demanding reaction, has been studied by Molina et al. [134] on similar samples, but without Pt. After 10 h on stream, the conversion on AlPB was about 25%, namely, half of that observed on USY (with Si:Al = 5), but the isomerization/disproportionation ratio was 8 on AlPB and less than 1 on USY, making AlPB a better catalyst than USY in m-xylene isomerization. In an earlier study of hydrocracking, Occelli and Renard [135] concluded that pillared montmorillonites display a good activity when they are used as supports for metals (Mo and Ni) in composite catalysts containing zeolites. Cumene cracking was carried out at 250 °C as a test reaction for Brønsted acidity on alumina-pillared montmorillonite precalcined between 250 and 500 °C [136]. Chemisorbed ammonia is removed from Al-pillared and other pillared montmorillonites at between 350 and 400 °C. Interestingly, cumene conversion fell sharply when the pretreatment temperature of the pillared clay was above 400 °C.

The conversion of cyclohexane to benzene over chromia-pillared montmorillonite [44] prereduced at 550 °C in H_2, is about 100% at 550 °C during 2 h when the content in chromia is equivalent to 1.24 Cr per unit cell (uc). This falls to half this value on clay containing 3.53 Cr per uc. The weight-hourly space velocity was in the range of 1 to 3 h^{-1}. Compared to chromia dispersed on alumina, the use of the Cr-pillared clay does not represent any real advantage.

From this brief review of the use of pillared clays as catalysts in hydrocarbon chemistry, it clearly appeared that, at best, the pillared clays could reach some of the performances of zeolites, but not exceed them.

Although, surprisingly, the domain of fine chemistry catalysis has not been extensively explored, it has received much attention with regards to reactions occurring in the interlayer space of unpillared clays [1]. Gutierrez et al. [137, 138] studied the conversion of pinacol **1** and of 2,3-diphenyl-2,3-butanediol **2** into pinacolone **3**, 1,2-diphenyl-2-methyl-1-butanone **4**, and 1,2-diphenyl-2-methyl-1-propanone **5**. These authors used a slurry of calcined Al-, Zr-, or Cr-pillared montmorillonite in either **1** or **2**, and the catalyst : reagent ratio was about 6 : 1. For reaction **1** the temperature was 100 °C, but for reaction **2** was 160 °C, and both reactions were conducted in concentrated sulfuric acid. A selectivity higher than 90% was obtained in the transformation of **1** → **2** on Zr- and Cr-pillared montmorillonite precalcined at 400 °C, the secondary product being 2,3-dimethyl-1,3-butadiene. With regards to the transformation of **2**, in concentrated sulfuric acid the yield of **4** was about 50 %. On calcined Al- and Zr-pillared clays, the selectivity toward ketone **4** was more than 80% after 2 h, while after 17 h both **4** and **5** were obtained in about equal proportions. Gutierrez et al. pointed out that on HY the ketone is a minor component, whilst on acid Amberlyst-15, **5** is the major component.

Nonetheless, in the above example of "soft" chemistry the acidity plays the major role. Studies conducted by Rightor et al. [45] and Barrault et al. [90, 139] tested Fe- and mixed Al,Fe-pillared montmorillonite in the transformation of syngas into light alkenes. The reaction conditions employed by Rightor et al. were as follows: $H_2 : CO = 2$, pressure 8.6 bar (8.6×10^5 Pa), 275 °C, gas hourly space velocities (GHSV) = 2100 h^{-1}. After 21 h on stream, the conversion was 1.3%. The Schulz–Flory distribution was obeyed for hydrocarbons from C1 to C6. The results of Barrault et al. [90] were quite different on mixed Al–Fe pillars, although the conditions were not comparable. At 450 °C, with a $H_2 : CO$ ratio close to 1 and a pressure of 0.1 MPa, the selectivity towards alkenes decreased, but the total conversion increased as the iron content of the pillars increased. The Schulz–Flory law was not obeyed because of an unusual selectivity for the C2–C6 fraction. This deviation was attributed to a shape-selectivity effect.

Choudary et al. [140] reported the synthesis of chiral sulfoxides on Ti-pillared montmorillonite using diethyl-tartrate (DET) and *t*-butylhydroperoxide (TBHP) as chiral auxiliary. For example, Ti-pillared clay was added to dry CH_2Cl_2 + DET, and the prochiral sulfide added. The mixture was cooled at −20 °C, dried azeotropically, and TBHP in toluene then introduced. The reaction was run for 3 h at −20 °C, whereupon chiral sulfoxide was produced with a yield of 85%. According to Choudary et al., the yield was higher in the heterogeneous system (at −20 °C) than with the homogeneous catalyst. Without apparent reason, only an acid-leached montmorillonite was suitable, which makes the need for a better characterization of the catalyst especially evident.

Finally, it should be noted that pillared clays, as zeolites, are suitable supports for metal catalysts. An example of the selective hydrogenation of 3-butenonitrile and 2-butenonitrile on palladium supported on Ti-pillared montmorillonite was reported [141], although apparently there was no gain in using this catalyst when, for example, its performance was compared to that of Raney nickel.

2.3.7.5.2 An Overview of Catalytic Reactions Investigated during the Past Decade The diversification of the nature of chemical species added to clay with the aim of introducing pillars into the clay's interlamellar space, as well as the type of clay used to host these species, is reflected in numerous reactions. Some have been investigated either as simple test reactions to correlate the activity of the materials with particular characteristics or properties of the solids (most often acidity, but also porous architecture), and some to evaluate the potential applications of more complex catalytic systems.

The main different classes of catalytic reactions are briefly presented below on a non-exhaustive basis. In addition, some preliminary remarks should be made at this point:

- Several review articles which partly deal with the use of pillared clays in heterogeneous catalysis have been published in recent years [16, 21, 142–146].
- In most reports concerned with the study of the catalytic properties of the pillared clays, an absence of comparative data obtained on existing catalysts makes it difficult to identify the actual efficiency of a particular pillared material relative either to known industrial counterparts, or, if not available, to catalysts which have shown interesting activities.
- In several publications, the term "pillared clay" appears to be used incorrectly. The fact that a swelling clay is brought into contact with a solution of the chemical element(s) which is (are) intended to be introduced into the interlamellar space does not necessarily provide a material which meets the criteria of a pillared clay. This is particularly the case if no interlayer spacing, no data on the thermal stability, nor on the textural properties (specific surface area, microporous volume) – that is, the main features which characterize pillared clays – are given.

References see page 242

It is worthwhile recalling that the first pillared clays were obtained using either partially neutralized solutions of an aluminum salt (also called hydroxy-aluminum solutions), or chlorohydrol [3]. These pillaring aluminum solutions contain the Al_{13}^{7+} species with the Keggin structure, the dimension of which corresponds to the increment of the interlamellar space of the host clay as observed by X-ray diffraction, from about 1.0–1.2 nm (depending on the charge-balancing counterion and its hydration level) to about 1.7–1.8 nm after calcination.

Ga is the nearest element for which the chemistry in solution is similar to that of aluminum, therefore providing pillared materials with very close characteristics to those of Al-pillared clays. The difficulties in obtaining well-organized pillared materials begins when using solutions of Cr, Fe, Zr, Ti, etc., because these elements exhibit different hydrolysis behaviors and species in solution. Such difficulties can partly be overcome by using mixed pillaring solutions in which Al is one of these species (preferably the dominant), or by starting with particular compounds of those elements.

Considering as "pillared clay" a material obtained by dispersing a clay suspension in a solution of any element(s), with surface areas barely higher than the external surface area of the parent clay, is highly questionable.

It has been shown, some 20 years ago, that clays with isomorphic Al for Si substitutions within the silica tetrahedral layers (beidellite) generated, upon pillaring with hydroxy-aluminum solutions, materials with higher contents of stronger Brønsted acid sites and higher catalytic activities compared with Al-pillared materials obtained with clays with substitutions in the octahedral layers (montmorillonite, hectorite) [30, 34]. The formation, in the pillared clay, of Si−OH−Al strong acid sites in the tetrahedral layer due to the opening of Si−O−Al linkages by protons was invoked to account for the superior catalytic performances of Al-pillared beidellite relative to Al-pillared montmorillonite [34]. Indeed, such strong acid sites cannot form in octahedrally substituted clays. The influence of the location of the isomorphic substitutions on proton acidity and catalytic properties was verified in the case of Al-pillared clays prepared with clays with various degrees of tetrahedral substitutions, using as a test reaction the hydroisomerization of alkanes (decane, octane, heptane) on Pt-impregnated Al-pillared montmorillonites from various origins, and Al-pillared saponites, magnesian clays with Al for Si tetrahedral substitutions from Spanish deposits, and from Ballarat [40, 133, 147, 148]. The beneficial influence of tetrahedral substitutions on the development of strong Brønsted acidity and/or higher catalytic efficiency has been confirmed [149–152]. From these results, it could be anticipated that clay minerals with more tetrahedral substitutions than in synthetic or naturally occurring beidellite (a clay which until now has not been available in sizeable amounts) and saponites (large deposits are known, e.g., in Spain), such as vermiculites and phlogopites (magnesian micas), with high tetrahedral layer charge, would provide even more active catalysts, provided that a suitable pillaring procedure could be developed. As mentioned above, a successful Al-pillaring method for these minerals has been recently reported [13, 14]. Preliminary catalytic results of octane hydroconversion on Pt-impregnated samples of Al-pillared vermiculite and phlogopite showed clearly that these materials were much more active than the Al-pillared montmorillonites and saponites, and substantially more efficient than most zeolites [153]. That only a limited number of catalytic reactions requiring proton acidity has been performed on tetrahedrally substituted (saponites, beidellites) pillared materials is therefore discouraging.

Also, rather few studies have been devoted to the preparation of pillared clays from concentrated clay suspensions (e.g., [154–157]) and pillaring solutions, or to scale up laboratory preparations of pillared clays to large batches (e.g., [158, 159]).

During the past decade, over 600 articles relating to pillared clays have been published, and more than 200 of these have been partly or totally focused on the catalytic properties of the materials. The main reactions or groups of reactions will be briefly presented as follows:

- Selective reduction of nitrogen oxides (NO and NO_x). This reaction, which was scarcely investigated before 1994, has received much attention, for environmental reasons. About 30% of the reports have been devoted to the catalytic reduction of NO and NO_x, either by light hydrocarbons such as methane, ethene, propene, [92, 160–176], or by ammonia [177–192]. The large majority of these studies used Ti- and mixed Fe-Ti-pillared clays as base catalysts; these catalysts were eventually modified by partial exchange/impregnation with solutions of the following elements: Cu, Co, Ni, Ag, Zr, Ce, Fe, VO^{2+}. In some studies, reduction was operated with carbon monoxide and hydrogen, which required the deposition of a noble metal to activate hydrogen as, for instance, Rh/Al-pillared clay [193], and Pd/TiO_2-pillared clay [194]. Independent of the catalyst composition, a high reduction yield was generally achieved. Ti-pillared clays are good examples of catalytic systems in which the distribution of Ti as $TiOH_x$ pillar species within the interlamellar space and/or TiO_x clusters deposited at the external surface is not known accurately.
- Wet oxidation with hydrogen peroxide of pollutants in waste waters was investigated on different pillared

clays according to the nature of the pollutant. Mixed Fe,Al-pillared montmorillonites were found to be highly efficient and relatively stable catalysts for phenol oxidation, with 80% conversion at 70 °C, owing to the presence of isolated Fe species in the pillars [195–200]. Nitrophenol [201], and dyes [202–205] were successfully decomposed on Cu–Al-pillared clays and on Fe-pillared clays [206].

- Photocatalytic decomposition of organic molecules in waste water [78, 207–209], among which phenol [210], dyes [211], carboxylic acids [212], and the decomposition of carbon monoxide [213] have been performed almost exclusively on Ti-pillared clays; this was, of course, due to the photocatalytic activity of TiO_2. Different clays were used in the preparation of the catalyst, including montmorillonite, saponite, and fluor-mica, with Ti being introduced via usual procedures as well as by sol–gel methods.

- Oxidation of organic compounds has been carried out on different types of pillared clay. The oxidation of acetone and methylethylketone used Pt/Al-pillared clays [214]. Acetone oxidation was also investigated on MnOx/Al- and Zr-pillared montmorillonite [215], and mixed Cr, Al-pillared saponite [216]. Oxidation of cyclohexanol was tested on a V- and Mo-doped mixed Fe,Al-pillared clays [217]. Benzene and cyclohexane were oxidized on Ti-pillared mica, montmorillonite and saponite [218], and on Pd-impregated Al-, Ce-, and La-pillared laponite [219]. Chlorinated hydrocarbons (methylene chloride) were efficiently decomposed on mixed Al,Cr-pillared bentonites [220, 221], whereas Pd/Ti-pillared montmorillonite was used in the oxidation of trichloroethene [222].

- Dehydration of alcohols: Al-pillared montmorillonites have been evaluated in the dehydration of ethanol, isopropanol, and pentanol [223–228]. Dehydration of 2-propanol, butanol, and pentanol has also been carried out on mixed Ce, Al-pillared montmorillonite [84], Ta-pillared montmorillonite [229], and Al- and Zr-pillared saponites [230], respectively. Dehydration of glucose to organic acids was performed on Fe-, Cr-, and Al-pillared clays [231]. The dehydration of methanol to dimethylether was shown to be efficient on Zr-pillared montmorillonite, and the addition of Cu resulted in a promising catalyst for the dehydrogenation of methanol to methyl formate [232]. Selective dehydration of 1-phenylethanol to 3-oxo-2,4-diphenylpentane was investigated on Ti-pillared clays [233].

- Dehydrogenation: Fe-, Al-, and mixed Al, Cr-pillared montmorillonite and saponite were reported to be suitable catalysts for the dehydrogenation of ethylbenzene to styrene [234–236]. It was also observed that, in the case of mixed Al, Cr-pillared saponite, the selectivity to styrene increased when the Cr content decreased [237]. Coimpregnated Al-pillared clays also catalyzed the reaction [238]. H-, Fe-, and Mn-exchanged Al-pillared clays were active and selective catalysts for the dehydrogenation of styrene [239]. Oxidative dehydrogenation of ethane and of propane was successfully performed on Cr-pillared montmorillonite [240] and V/Al-pillared saponite (saponite from Ballarat) [241], respectively. Cu-doped Zr-pillared clays were efficient catalysts for the dehydrogenation of methanol to methyl formate, although the acid–base properties of these catalysts contributed to the formation of dimethylether and to the degradation of methyl formate to carbon dioxide and monoxide, and to formaldehyde. Undoped Zr-pillared clays were also active in the dehydration of methanol to dimethylether and hydrocarbons [242].

- Epoxidation: Ti-pillared montmorillonites were tested in the liquid phase epoxidation of allylic alcohol with t-Bu hydroperoxide in the presence of (+)-DET, with a yield of 50% at 25 °C, comparable to that observed on conventional polymer-supported Ti(IV) catalysts [243, 244].

- Miscellaneous: Mixed Fe,Cr-pillared clays were also evaluated in the acylation of alcohols with acetic acid [245]; aldol condensation on Al-pillared clays [246]; cyclization of citronellal to menthone and isomenthone in 1,2-dichloroethane on mixed Al,Fe-pillared clays [247]; hydration of propene to acetone via intermediate isopropanol oxidative dehydrogenation on Al-, and mixed Fe,Al- and Ru,Al-pillared bentonite, with the activity being related to the nature and number of redox and acid sites [248]. Hydroformylation of 1,1-diarylethenes and vinylnaphthalenes was investigated on rhodium supported on pillared clays and other inorganic supports, with yields comparable or even higher than those observed with the classical homogeneous rhodium-based catalysts [249]. Hydroxylation of phenol by hydrogen peroxide was achieved on Cu-doped Al-pillared montmorillonite [250], on Ti-pillared montmorillonite [251], and by microwave irradiation on mixed Fe,Al-pillared montmorillonite, with conversions close to 70% [252]. Selective nitration of chlorobenzene in the para position was tested on transition metal (Fe, Cr, Mn) oxide-pillared montmorillonites, with Fe-pillared clay being more active and Cr-pillared clay more selective [253]. Para selectivities of about 85% were also obtained on Zr-, Ti-, and mixed Zr,Ti-pillared montmorillonites, in the activity sequence Zr,Ti-pillared clay > Zr-pillared clay > Ti-pillared clay [254].

- The hydrogenation of CO (Fischer–Tropsch; FT) has been carried out on CoO/Al-pillared montmorillonites. The best catalysts exhibited a good shape selectivity and higher intrinsic activity compared with a standard

References see page 242

FT catalyst [255]. In the case of Rh/Al-pillared clays prepared with an organic or inorganic metal precursor, higher activities and selectivities for oxygenates were obtained with the organic precursor [256]. Hydrogenation of benzene was performed on Ni/Al-pillared montmorillonite [257] and saponite [258], whereas hydrogenation of phenyl alkyl acetylenes was studied on Pd/Al-pillared clay [259].

- Hydrotreatment of naphthalene was investigated on MoS_2 supported on Al- and Zr-pillared clays [260]. The higher hydrogenation activity achieved on the MoS_2/Zr-pillared clay was attributed to interactions of MoS_2 with ZrO_2. Hydrotreatment of petroleum heavy fractions has been carried out on Ni- and Mo-impregnated Al-, Ti-, and mixed Al, La-pillared bentonites. The Al, La-pillared clay reduced the asphaltenes by 64% versus 11% for a reference Ni-Mo/alumina [261]. Montmorillonite and laponite pillared with Sn, Cr, and Al were tested in the hydrocracking of coal extracts and petroleum distillation residue and compared with Ni-Mo/alumina [262–264]. Better performances were achieved with Cr-pillared montmorillonite and Sn-pillared laponite [262].
- Hydrodesulfurization (HDS) of thiophene has been investigated on chromia-pillared montmorillonite [94, 265, 266], synthetic saponites with framework Co, Ni, Zn, and Mg [267], and on W, Re, NiW, and NiRe supported on pillared clay, USY zeolite, active carbon, and alumina [268]. For the latter catalysts, the activity increases when Ni is incorporated into W for the four supports. The incorporation of Ni to Re has no effect when the support is carbon or alumina, whereas the activity decreases with pillared clay and USY as support. The same reaction was studied on mixed Ru, Al-pillared montmorillonite [269] and Ni–Mo supported on Al-, and Ti-pillared montmorillonite [93]. Bifunctional Pt/H-ZSM-5, Rh/Al-pillared hectorite (sulfided and non-sulfided), Rh/Al-pillared montmorillonite, and Pt/MCM-41 showed higher activity for the HDS of benzothiophene than a commercial Co–Mo/alumina catalyst. In these catalysts, the noble metal activates hydrogen to form hydrogen spillover, and the acid sites activate thiophene [270]. Al- and Ti-pillared montmorillonites prepared by microwave irradiation and impregnated with Co–Mo were compared with an industrial Co–Mo/alumina catalyst in the HDS of heavy vacuum gas oil (with 2.21% sulfur). The Co–Mo/Al-pillared clay, with the highest surface area, was the most active catalyst [271].
- Cracking of cumene, which probably is the most popular reaction used to correlate acidity and activity, has been carried out on Cr-pillared montmorillonite [272], Al-pillared saponites [273], Al- and Ti-pillared montmorillonite and beidellite [274], Ti-pillared montmorillonite, saponite, and rectorite [275], and Al(Ce)-pillared montmorillonite; the results obtained were compared with H-ZSM-5 [276]. The results pointed to a higher efficiency of pillared materials prepared with clays with tetrahedral substitutions. High cracking of 1-heptene occurred on Al-pillared beidellite and saponite, whereas no reaction was observed on Al-pillared montmorillonite and hectorite, again pointing to the stronger acidity of the former catalysts [277]. Surprisingly, no such superior performance of Al-pillared saponite was mentioned in comparative trials of polyethene cracking carried out on Al-pillared montmorillonite and saponite [278]. Pillared saponite performed better than a pillared montmorillonite in the cracking of polyethene, although both pillared clays could fully degrade the polymer. They were less active, however, than a USY-based commercial cracking catalyst. Owing to their weaker acidity, the pillared clays produced more liquid and less coke was formed [279].
- Alkylations, another class of reactions which has been investigated classically on acid catalysts, was studied on different forms of pillared materials. Cation-exchanged Al-pillared clays [280] and Fe-pillared [281] clays were tested in the alkylation of benzene with propene. Other studies used octane and dodecene as alkylating agent on Al-montmorillonites with different extents of tetrahedral substitution [282], acid-activated Al-pillared montmorillonite [283], and Al-pillared saponite [284]. More recently, the alkylation of benzene with dodecene was compared on a series of ten different catalysts, including an Al-pillared clay. A molar ratio of 10:1 benzene:dodecene favored the formation of linear dodecylbenzenes. With decreasing ratios, however, the formation of didodecylbenzenes was increased. The rate of benzene alkylation decreased with the chain length of the alkene, in the order: 1-octene > 1-decene > 1-tetradecene [285]. Alkylation of benzene with primary alcohols was also investigated on pillared clays [286]. Toluene by methanol alkylation was studied on Al- and mixed Al, La-pillared smectites [287], and Zr-pillared montmorillonite and vermiculite [288], but the activity was low [288]. Phenol has been alkylated with methanol on Ni-exchanged Zr-pillared montmorillonite, and a positive correlation was observed between alkylation activity and the acidity of the pillared materials. The samples with a lesser pillar density were more selective toward anisole [289]. Liquid-phase benzylation of o-xylene with benzyl chloride has been carried out on Fe-, Al-, and mixed Fe, Al-pillared montmorillonites doped with 10% of either Mo, V, or Cr. The V-containing catalyst had the maximum of acidity as found by ammonia TPD.

High selectivities toward the monobenzylated product were obtained in all cases, with the Fe-catalysts being the most active. Catalytic activity and strong acidity appeared to be correlated [290].

- Disproportionation, another acid-catalyzed reaction, has been investigated on Ga-substituted Al-pillared saponite using 1,2,4-trimethylbenzene. The activity was proportional to the amount of Ga in the tetrahedral layer [291]. On Al-pillared montmorillonite and Y zeolite, the activity was stabilized by spill-over hydrogen [292].
- Transalkylation of diisopropylbenzene with excess benzene occurred with high conversion and selectivity to isopropylbenzene on Cu-impregnated Al-pillared clay [293].
- Isomerization of o- and m-xylene was studied on Al- and Cr-pillared montmorillonites [294], and Al-pillared beidellite and montmorillonite [295]. Isomerization reached higher conversion over the Al-pillared clay relative to the Cr-pillared clay. A higher conversion of m-xylene was achieved on Al-pillared beidellite than on Al-pillared montmorillonite. Katoh et al. [296] observed a remarkable enhancement of the acidic and catalytic properties of sulfate-promoted Zr-pillared montmorillonite prepared with a refluxed solution of zirconium oxychloride, in the isomerization of 1-butene and cyclopropane. The transformation of 1-butene was also investigated on Al- and Zr-pillared montmorillonite and laponite, and the results were compared with data obtained on H-Y zeolite and discussed in terms of acidity measurements [297]. Cyclopropane isomerization was tested on different pillared montmorillonites (with Al, Zr, Ti, Fe, Cr, Sn pillars). On Al-pillared clay, oligomerization followed by isomerization was attributed to Lewis and Brønsted acidity. Only negligible isomerization of hexane was noted using Al-, Fe-, Cu-, Zn-pillared clays (a montmorillonite-type, and a mixture of non-swelling kaolinite and illite) [298].
- Hydroisomerization-hydrocracking of methylcyclohexane has been studied on NiSMM (synthetic nickel mica-montmorillonite), Zr-pillared NiSMM, and Al-pillared NiSMM. High conversions mainly to branched aliphatic hydrocarbons (isoalkanes) were obtained [299]. Hydroisomerization-hydrocracking of normal alkanes (heptane, octane, decane) has also been performed on Pt-impregnated Al-pillared natural clays (bifunctional catalysts) including montmorillonites and saponites from various deposits, hectorite (Hector California), and Al-pillared vermiculites and phlogopites, in order to provide evidence of the influence of the type of clay (with octahedral and tetrahedral substitutions) on the acid and catalytic properties. The isomerization activity of Pt/Al-pillared samples followed the sequence: Al-pillared montmorillonites = Al-pillared hectorite < Al-pillared beidellite = Al-pillared saponites < Al-pillared vermiculites = Al-pillared phlogopites [40, 133, 147, 148, 153, 158]. It was also clear that, in octane hydroisomerization, the latter two catalysts were more efficient than USY and ZSM-5 zeolites and equivalent to a Beta zeolite tested under similar conditions. The most active zeolite for this reaction was Pt/H-ZSM-22 which, at maximum isomerization (229 °C), gave a

References see page 242

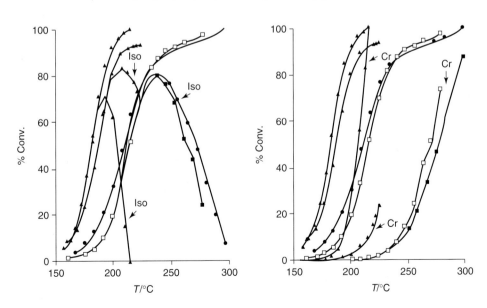

Fig. 16 Conversion of octane versus temperature over H-beta (▲), pillared vermiculite (○), pillared phlogopite (□), and acid-treated pillared vermiculite (△). Iso: yields of C8 isomers; Cr: yields of cracking products. (Reproduced from Ref. [153], with permission.)

yield in C8-isomers of 79% (selectivity of 92%). Under similar reaction conditions, Pt/acid-treated Al-pillared vermiculite reached an isomer conversion of 82%, with a selectivity of 91%. A batch of this catalyst was submitted to a long-term run at 207 °C, and no deactivation at all was noted after more than 200 h on stream. Throughout the run, the conversion was 88% with a yield of isomers of 83% (selectivity of 94%) [153]. Representative results showing the variation of the total conversion and of the yields of C8 isomers and of the cracking products as a function of reaction temperature are illustrated in Fig. 16 for Pt/H-Beta zeolite, Pt/Al-pillared vermiculite, Pt/Al-pillared phlogopite, and for acid-treated Al-pillared vermiculite.

- Oligomerization (polymerization) of ethene was discussed in terms of protonic acidity of aluminum and oligosilsesquioxane complex cations-pillared montmorillonite [300] and of Al-pillared clay and zeolites [301].
- Metathesis of propene was investigated on Mo/Al-pillared montmorillonite [302].
- Reforming of methane with carbon dioxide to syngas was carried out on Ni deposited on Zr-pillared laponite. Interesting results (high initial conversion, long-term stability) were achieved on samples with adequate porosity [303].

2.3.7.6 Conclusions

Pillared clays have an interesting architecture as the distance between the layers can be tuned to between ∼0.6 and ∼1.2 nm. Chemical functions, such as Brønsted or Lewis acid sites, can be modified by using a variety of techniques. However, three main warnings should be addressed to those who are not familiar with the subject:

- Pillared clays are poorly crystallized systems.
- The thermal stability becomes questionable above 600 °C, when maintained for at least 24 h at this temperature and in the presence of steam.
- Al-pillared vermiculite and phlogopite are an exception, as they are much better crystallized and
- significantly more stable above 800 °C.

Although the field of pillared clays has been investigated to a much lesser degree than that of zeolites, it is hoped that interesting applications will continue to be discovered. From what has been described in the preceding sections, it appears that – potentially – the most interesting domains are in separation technology and fine chemical catalysis. In the field of hydrocarbon chemistry, pillared clays should not compete against, but rather complement, zeolites.

Acknowledgments

The authors greatly appreciate the help provided by F. Somers in editing the second edition of this chapter.

References

1. (a) J. A. Ballantine, M. Davies, I. Patel, J. H. Purnell, M. Rayanakom, K. J. Williams, J. M. Thomas, *J. Mol. Catal.* **1984**, *26*, 37; (b) J. A. Ballantine, J. H. Purnell, M. Rayanakom, K. J. Williams, J. M. Thomas, *J. Mol. Catal.* **1985**, *30*, 373, and references therein.
2. R. M. Barrer, *Phil. Trans. R. Soc. London* **1984**, *A311*, 333.
3. (a) D. E. W. Vaughan, R. J. Lussier, J. S. Magee, US Patent 3775395, 1973; (b) D. E. W. Vaughan, R. J. Lussier, J. S. Magee, US Patent 3838037, 1974; (c) D. E. W. Vaughan, R. J. Lussier, J. S. Magee, UK Patent 1483466, 1979.
4. G. W. Brindley, R. E. Sempels, *Clay Miner.* **1977**, *12*, 229.
5. N. Lahav, U. Shani, J. Shabtai, *Clays Clay Miner.* **1978**, *26*, 107.
6. G. Brown, *Phil. Trans. R. Soc. London* **1984**, *A311*, 221.
7. J. Hougardy, W. E. E. Stone, J. J. Fripiat, *J. Magn. Res.* **1977**, *25*, 563 and references therein.
8. R. E. Grim, *Clay Mineralogy*, McGraw-Hill, New York, 1953, p. 65.
9. A. C. D. Newman, G. Brown, in *Chemistry of Clays and Clay Minerals*, A. C. D. Newman (Ed.), Mineralogical Society 6, Longman, New York, 1987, p. 75.
10. J. B. d'Espinose de la Caillerie, J. J. Fripiat, *Clays Clay Miner.* **1991**, *39*, 270.
11. P. H. Hsu, *Clays Clay Miner.* **1992**, *40*, 300.
12. L. J. Michot, D. Tracas, B. S. Lartiges, F. Lhote, C. H. Pons, *Clay Miner.* **1994**, *29*, 133.
13. F. del Rey-Perez-Caballero, G. Poncelet, *Microporous Mesoporous Mater.* **2000**, *41*, 169.
14. F. del Rey-Perez-Caballero, G. Poncelet, *Microporous Mesoporous Mater.* **2000**, *37*, 313.
15. A. Clearfield, *Advanced Catalysts and Nanostructured Materials. Modern Synthetic Methods.* W. R. Moser (Ed.), Academic Press Inc., 1996, p. 345.
16. J.-F. Lambert, G. Poncelet, *Topics Catal.* **1997**, *4*, 43.
17. J. T. Kloprogge, *J. Porous Mater.* **1998**, *5*, 5.
18. R. A. Schoonheydt, K. Y. Jacobs, *Stud. Surf. Sci. Catal.*, Elsevier, Amsterdam **2001**, *137*, 299.
19. J. G. Cao, S. M. Yu, *Kuangwu Yu Huagong*, **2001**, *30*, 1.
20. S. Yamanaka, K. Inumaru, *Mater. Integration* **2000**, *13*, 1.
21. A. Gil, L. M. Gandia, M. A. Vicente, *Catal. Rev. -Sci. Eng.* **2000**, *42*, 145.
22. A. Auroux, *Topics Catal.* **2002**, *19*, 205.
23. P. Cool, E. F. Vansant, in *Handbook of Layered Materials*, S. M. Auerbach, K. A. Carrado, P. K. Dutta, K. Prabir (Eds.), Marcel Dekker Inc., New York, 2004, p. 261.
24. C. F. Baes, Jr., R. E. Mesmer, *The Hydrolysis of Cations*, Wiley, New York, 1976.
25. C. P. Herrero, in *Proceedings International Clay Conference 1985*, L. G. Schultz, H. Van Olphen, F. A. Mumpton (Eds.), The Clay Mineral Society, 1987, p. 24.
26. A. Moini, T. J. Pinnavaia, *Solid State Ionics* **1988**, *26*, 119.
27. C. H. Zhou, Q. W. Li, Z. H. Ge, X. N. Li, Z. M. Ni, *GaodengXuexiao Huaxue Xuebao* **2003**, *24*, 1351.
28. C. H. Zhou, X. N. Li, D. S. Tong, Y. F. Zhu, Z. H. Ge, *Chin. Chem. Lett.* **2005**, *16*, 261.

29. J. Pires, J. Francisco, A. Carvalho, M. B. de Carvalho, A. R. Silva, C. Freire, B. de Castro, *Langmuir* **2004**, *20*, 2861.
30. D. Plee, L. Gatineau, J. J. Fripiat, *Clays Clay Miner.* **1987**, *35*, 81.
31. A. Molinard, E. F. Vansant, *Adsorption*, Kluwer, Dordrecht, 1994.
32. L. J. Michot, T. J. Pinnavaia, *Chem. Mater.* **1992**, *4*, 1433.
33. X. T. Yuan, J. Yu, H. Z. Liu, W. J. Li, *Huaxue Xuebao* **2004**, *62*, 1049.
34. G. Poncelet, A. Schutz, in *Chemical Reactions in Organic and Inorganic Constrained Systems* R. Setton (Ed.), D. Reidel Publishers, 1986, p. 165.
35. R. A. Schoonheydt, H. Leeman, A. Scorpion, J. Lenotte, P. J. Grobet, *Clays Clay Miner.* **1994**, *42*, 518.
36. H. Y. Zhu, Z. Ding, G. Q. Lu, *Stud. Surf. Sci. Catal., Elsevier, Amsterdam* **2000**, *129*, 425.
37. R. A. Schoonheydt, J. Van den Eynde, H. Tubbax, H. Leeman, M. Stuyckens, J. Lenotte, W. E. E. Stone, *Clays Clay Miner.* **1993**, *41*, 598.
38. K. Bornholdt, J. M. Corker, J. Evans, J. M. Rommey, *Inorg. Chem.* **1991**, *30*, 2.
39. E. M. Farfan-Torres, O. Dedeycker, P. Grange, *Stud. Surf. Sci. Catal., Elsevier, Amsterdam* **1990**, *63*, 337.
40. S. Moreno, R. Sun Kou, R. Molina, G. Poncelet, *J. Catal.* **1999**, *182*, 174.
41. H. L. Del Castillo, P. Grange, *Appl. Catal. A: General* **1993**, *103*, 23.
42. M. Nakatsuji, R. Ishii, Z. M. Wang, K. Ooi, *J. Colloid Interface Sci.* **2004**, *272*, 158.
43. G. Guiu, P. Grange, *Mol. Cryst. Liq. Cryst.* **1994**, *245*, 221.
44. M. S. Tzou, T. J. Pinnavaia, *Catal. Today* **1988**, *2*, 243.
45. E. G. Rightor, M. S. Tzou, T. J. Pinnavaia, *J. Catal.* **1991**, *130*, 29.
46. (a) D. E. W. Vaughan, R. J. Lussier, *Preprints, 5th International Conference on Zeolites*, Naples, Italy, L. V. Rees (Ed.), Heyden & Sons, 1980, p. 94; (b) D. E. W. Vaughan, *Catal. Today* **1988**, *2*, 187.
47. G. Johannson, G. Lundgren, L. G. Sillen, R. Soderquist, *Acta Chem. Scand.* **1960**, *14*, 403.
48. (a) J. W. Akitt, W. Greenwood, B. L. Khandelwal, G. R. Lester, *J. Chem. Soc., Dalton Trans.* **1972**, 604; (b) J. W. Akitt, A. Farting, *J. Magn. Reson.* **1988**, *32*, 345.
49. J. Y. Bottero, D. Tchoubar, J. M. Cases, R. Flessinger, *J. Phys. Chem.* **1982**, *85*, 3667.
50. S. Schönherr, H. Gorz, D. Muller, W. Gessner, *Z. Anorg. Allg. Chem.* **1981**, *476*, 188.
51. M. L. Occelli, R. M. Tindwa, *Clays Clay Miner.* **1983**, *31*, 22.
52. T. J. Pinnavaia, M. S. Tzou, S. D. Landau, H. R. Raythata, *J. Mol. Catal.* **1984**, *27*, 195.
53. J. Sterte, *Catal. Today* **1988**, *2*, 219.
54. F. Figueras, Z. Klapyta, P. Massiani, Z. Montassir, D. Tichit, F. Fajula, G. Guegen, J. Bousquet, A. Auroux, *Clays Clay Miner.* **1990**, *38*, 257.
55. S. M. Bradley, R. A. Kydd, R. Yamdagni, *J. Chem. Soc., Dalton Trans.* **1990**, *413*, 2653.
56. S. M. Bradley, R. A. Kydd, R. Yamdagni, *Magn. Res. Chem.* **1990**, *28*, 741.
57. A. Vieira-Coelho, G. Poncelet, *Appl. Catal. A: General* **1991**, *77*, 303.
58. G. J. J. Bartley, *Catal. Today* **1988**, *2*, 233.
59. J. Sterte, *Clays Clay Miner.* **1986**, *34*, 658.
60. S. Yoneyama, S. Hafa, S. Yamanaka, *J. Phys. Chem.* **1989**, *93*, 4833.
61. H. Einaga, *J. Chem. Soc., Dalton Trans.* **1976**, 1971.
62. T. Endo, M. Mortland, T. J. Pinnavaia, *Clays Clay Miner.* **1980**, *28*, 105.
63. G. Fetter, D. Tichit, P. Massiani, R. Dutarte, F. Figueras, *Clays Clay Miner.* **1994**, *42*, 61.
64. T. J. Pinnavaia, M. S. Tzou, S. D. Landau, *J. Am. Chem. Soc.* **1985**, *107*, 2783.
65. A. Jimenez-Lopez, J. Maza-Rodriguez, P. Oliveira-Pastor, P. Maireles-Torres, E. Rodriguez-Castellon, *Clays Clay Miner.* **1993**, *41*, 328.
66. X. K. Guo, G. P. Qin, N. Shen, *Huaxue Xuebao* **2004**, *62*, 208.
67. L. J. A. Colin, J. A. De los Reyes, A. Vazquez, A. Montoya, *Appl. Surf. Sci.* **2005**, *240*, 8.
68. M. A. Vicente, C. Belver, R. Trujillano, V. Rives, A. C. Alvarez, J.-F. Lambert, S. A. Korili, L. M. Gandia, A. Gil, *Appl. Catal. A: General* **2004**, *267*, 47.
69. S. A. Bagshaw, R. P. Cooney, *Chem. Mater.* **1995**, *7*, 1384.
70. X. Z. Tang, W. Q. Xu, Y. F. Shen, S. L. Suib, *Chem. Mater.* **1995**, *7*, 102.
71. M. J. Hernando, C. Pesquera, C. Blanco, I. Benito, F. Gonzalez, *Appl. Catal. A: General* **1996**, *141*, 175.
72. M. J. Hernando, C. Pesquera, C. Blanco, I. Benito, F. Gonzalez, *Chem. Mater.* **1996**, *8*, 7.
73. K. B. Brandt, R. A. Kydd, *Chem. Mater.* **1997**, *9*, 567.
74. G. Fetter, D. Tichit, L. C. de Ménorval, F. Figueras, *Appl. Catal.* **1995**, *126*, 165.
75. J. Sterte, J. Shabtai, *Clays Clay Miner.* **1987**, *35*, 429.
76. A. Gil, G. Guiu, P. Grange, M. Montes, *J. Phys. Chem.* **1995**, *99*, 301.
77. J. H. Choy, J. H. Park, J. B. Yoon, *Bull. Korean Chem. Soc.* **1998**, *19*, 1185.
78. Z. Ding, H. Y. Zhu, P. F. Greenfield, G. Q. Lu, *J. Colloid Interface Sci.* **2001**, *238*, 267.
79. P. V. Malla, S. Komarneni, *Clays Clay Miner.* **1993**, *41*, 472.
80. D. Zhao, Y. Yang, X. Guo, *Zeolite* **1995**, *15*, 58.
81. E Booij, J. T. Klopprogge, J. A. R. van Veen, *Appl. Clay Sci.* **1996**, *11*, 155.
82. E. Booij, J. T. Klopprogge, J. A. R. van Veen, *Clays Clay Miner.* **1996**, *44*, 774.
83. N. Frini, M. Crespin, M. Trabelsi, D. Messad, H. Van Damme, F. Bergaya, *Appl. Clay Sci.* **1997**, *12*, 281.
84. G. Fetter, P. Salas, L. A. Velazquez, P. Bosch, *Ind. Eng. Chem. Res.* **2000**, *39*, 1944.
85. J. Sterte, *Clays Clay Miner.* **1991**, *39*, 167.
86. H. Y. Zhu, E. F. Vansant, J. A. Xia, G. Q. Lu, *J. Porous Mater.* **1997**, *4*, 17.
87. L. Caballero, J. M. Dominguez, J. L. De Los Santos, A. Montoya, J. Navarrete, in *Synthesis of Porous Materials*, M. L. Occelli, H. Kessler (Eds.), Marcel Dekker Inc., 1997, *69*, 491.
88. K. Bahranowski, R. Dula, J. Komorek, T. Romotowski, E. M. Serwicka, *Stud. Surf. Sci. Catal., Elsevier, Amsterdam* **1995**, *91*, 747.
89. M. A. Vicente, C. Belver, R. Trujillano, M. A. Banares-Munoz, V. Rives, S. A. Korili, A. Gil, L. M. Gandia, J.-F. Lambert, *Catal. Today* **2003**, *78*, 181.
90. F. Bergaya, J. Barrault, in *Pillared Layered Structures*, I. V. Mitchell (Ed.), Elsevier, 1990, p. 167.
91. T. Nakao, M. Nogami, *J. Ceram. Soc. Jap.* **2005**, *113*, 435.
92. T. Grzybek, J. Klinik, D. Olszewska, H. Papp, J. Smarzowski, *Polish J. Chem.* **2001**, *75*, 857.
93. C. E. Ramos-Galvan, G. Sandoval-Robles, A. Castillo-Mares, J. M. Dominguez, *Appl. Catal. A: General* **1997**, *150*, 37.
94. M. Sychev, V. H. J. de Beer, J. A. R. van Santen, *Microporous Mater.* **1997**, *8*, 255.

95. S. Ben Chaabene, L. Bergaoui, A. Ghorbel, *Colloids and Surfaces, A: Physicochem. Eng. Aspects* **2004**, *251*, 109.
96. F. Costa, C. J. R. Silva, A. M. Fonseca, I. C. Neves, A. P. Carvalho, J. Pires, M. B. Carvalho, *Mater. Sci. Forum*, **2004**, *455–456*, 569.
97. C. Flego, L. Galasso, R. Millini, I. Kiricsi, *Appl. Catal. A: General* **1998**, *168*, 323.
98. D. G. Zhao, G. J. Wang, Y. Y. Xie, X. Guo, Q. B. Wang J. Y. Ren, *Clay Clay Miner.* **1993**, *41*, 317.
99. S. P. Skaribas, P. J. Pomonis, P. Grange, B. Delmon, in *Multifunctional Mesoporous Inorganic Solids*, C. A. C. Sequeira, M. J. Hudson (Eds.), Kluwer, 1993, p. 127.
100. M. Gutierrez-Lebrun, J. M. Gaite, *J. Magn. Res.* **1980**, *40*, 105.
101. A. Banin, N. Lahav, *Israel J. Chem.* **1968**, *6*, 235.
102. J. J. Fripiat, J. M. Cases, M. Francois, M. Letellier, *J. Colloid Interf. Sci.* **1982**, *89*, 378.
103. H. Van Damme, P. Levitz, J. J. Fripiat, J. F. Alcover, L. Gatineau, F. Bergaya, in *Physics of Finely Divided Matter*, N. Boccara, H. Daoud (Eds.), Springer Verlag, Berlin, 1985, p. 24.
104. J. J. Fripiat, R. Setton, *J. Appl. Phys.* **1987**, *61*, 1811.
105. M. L. Occeli, *Catal. Today* **1988**, *2*, 334.
106. J. Heylen, N. Maes, A. Molinard, E. F. Vansant, in *Separation Technology*, E. F. Vansant (Ed.), 1994, p. 355.
107. S. Yamanaka, M. Hattori, *Catal. Today* **1988**, *2*, 261.
108. H. Van Damme, J. J. Fripiat, *J. Chem. Phys.* **1985**, *82*, 2785.
109. E. Kikuchi, T. Masuda, *Catal. Today* **1988**, *2*, 297.
110. A. Schutz, W. E. E. Stone, G. Poncelet, J. J. Fripiat, *Clays Clay Miner.* **1987**, *35*, 251.
111. M. M. Mortland, J. J. Fripiat, J. Chaussidon, J. Uytterhoeven, *J. Phys. Chem.* **1963**, *67*, 248.
112. M. M. Mortland, K. V. Raman, *Clays Clay Miner.* **1968**, *16*, 398.
113. J. J. Fripiat, A. Jelli, G. Poncelet, J. André, *J. Phys. Chem.* **1965**, *69*, 2185.
114. R. Touillaux, P. Salvador, C. Vandermeersche, J. J. Fripiat, *Israel J. Chem.* **1968**, *6*, 337.
115. D. Plee, A. Schutz, G. Poncelet, J. J. Fripiat, *Stud. Surf. Sci. Catal., Elsevier, Amsterdam* **1985**, *20*, 343.
116. S. Chevalier, R. Franck, H. Suquet, J. F. Lambert, D. Barthomeuf, *J. Chem. Soc., Faraday Trans.* **1994**, *90*, 667.
117. E. J. W. Verwey, J. Th. G. Overbeek, *Theory of the Stability of Lyophobic Colloids*, Elsevier, New York, 1948.
118. B. Chourabi, J. J. Fripiat, *Clays Clay Miner.* **1981**, *29*, 260.
119. D. Plee, F. Borg, L. Gatineau, J. J. Fripiat, *J. Am. Chem. Soc.* **1985**, *107*, 2362.
120. T. J. Pinnavaia, S. D. Landau, M. Tzou, I. D. Johnson, M. Lipsicas, *J. Am. Chem. Soc.* **1985**, *107*, 7222.
121. D. T. B. Tennakoon, W. Jones, J. M. Thomas, *J. Chem. Soc., Faraday Trans.* **1986**, *82*, 3081.
122. J.-F. Lambert, S. Chevalier, R. Franck, H. Suquet, D. Barthomeuf, *J. Chem. Soc., Faraday Trans.* **1994**, *90*, 675.
123. J. Zheng, Y. Hao, L. Hao, Y. Zhang, Z. Hue, *Zeolites* **1992**, *12*, 374.
124. G. Sun, F. S. Yan, H. H. Zhu, Z. H. Liu, *Stud. Surf. Sci. Catal., Elsevier, Amsterdam* **1987**, *31*, 649.
125. E. M. Farfan-Torres, E. Sham, P. Grange, *Catal. Today* **1992**, *15*, 515.
126. S. Bodoardo, F. Figueras, E. Garrone, *J. Catal.* **1994**, *147*, 223.
127. K. Arata, *Adv. Catal.* **1990**, *37*, 165, and references therein.
128. F. Admaiai, A. Bernier, P. Grange, *Stud. Surf. Sci. Catal., Elsevier, Amsterdam* **1993**, *75*, 1629.
129. J. K. Boudali, A. Ghorbel, D. Tichit, B. Chiche, R. Dutarte, F. Figueras, *Microporous Mater.* **1994**, *2*, 525.
130. E. M. Farfan-Torres, P. Grange, *Catal. Sci. Techn.* **1991**, *1*, 103.
131. J. Sterte, *New Types of Cracking Catalysts Based on Cross-linked Smectites*, Dept. Eng. Chem., Chalmers Biblioteks Tryckeri, Göteborg, 1987, Vol. 1, p. 27.
132. M. P. Atkins, in *Pillared Layered Structures*, I. V. Mitchell (Ed.), Elsevier Applied Science, 1990, p. 159.
133. R. Molina, S. Moreno, A. Vieira-Coelho, J. A. Martens, P. A. Jacobs, G. Poncelet, *J. Catal.* **1994**, *148*, 304.
134. R. Molina, A. Schutz, G. Poncelet, *J. Catal.* **1994**, *145*, 79.
135. M. L. Occelli, R. J. Renard, *Catal. Today* **1988**, *2*, 309.
136. M.-Y. He, Z. Liu, E. Min, *Catal. Today* **1988**, *2*, 321.
137. E. Gutierrez, E. Ruiz-Hitzky, *Mol. Cryst. Liq. Cryst.* **1988**, *161*, 453.
138. E. Gutierrez, A. J. Aznar, E. Ruiz-Hitsky, *Stud. Surf. Sci. Catal., Elsevier, Amsterdam* **1988**, *41*, 211.
139. J. Barrault, C. Zivkov, F. Bergaya, L. Gatineau, N. Hassoun, H. Van Damme, D. Mari, *J. Chem. Soc., Chem. Commun.* **1988**, 1403.
140. B. M. Choudary, S. Shobha-Rani, N. Narender, *Catal. Lett.* **1993**, *119*, 299.
141. A. Lamesh, H. Del Castillo, P. Vanderwegen, L. Daza, G. Jannes, P. Grange, *Stud. Surf. Sci. Catal., Elsevier, Amsterdam* **1993**, *41*, 299.
142. Z. Ding, J. T. Kloprogge, R. L. Frost, G. Q. Lu, H. Y. Zhu, *J. Porous Mater.* **2001**, *8*, 273.
143. N. K. Mitra, S. Maitra, *Trans. Indian Ceram. Soc.* **2001**, *60*, 121.
144. P. X. Wu, D. Q. Ye, C. B. Ming, *Kuangwu Yanshi Huaue Tongbao* **2002**, *21*, 228.
145. P. Cool, E. F. Vansant, G. Poncelet, R. A. Schoonheydt, in *Handbook of Porous Solids*, Wiley-VCH Verlag GmbH & Co., Vol. 2, Weinheim, 2002, p. 1250.
146. T. Kloprogge, L. V. Duong, R. L. Frost, *Environ. Geol.* **2005**, *47*, 967.
147. S. Moreno, R. Sun Kou, G. Poncelet, *J. Catal.* **1996**, *162*, 198.
148. S. Moreno, R. Sun Kou, G. Poncelet, *J. Phys. Chem.* **1997**, *101*, 1569.
149. R. Swarnakar, K. B. Brandt, R. A. Kydd, *Appl. Catal. A: General* **1996**, *142*, 61.
150. K. B. Brandt, R. A. Kydd, *Appl. Catal. A: General* **1997**, *165*, 103.
151. F. Kooli, J. Bovey, W. Jones, *J. Mater. Chem.* **1997**, *7*, 153.
152. M. Yao, Z. Y. Liu, K. X. Wang, C. J. Li, Y. C. Zou, H. J. Sun, *J. Porous Mater.* **2004**, *11*, 229.
153. F. del Rey-Perez-Caballero, M. L. Sanchez-Henao, G. Poncelet, *Stud. Surf. Sci. Catal. Elsevier, Amsterdam* **2000**, *130*, 2417.
154. W. Diano, R. Rubino, M. Sergio, *Microporous Mater.* **1994**, *2*, 179.
155. L. Storaro, M. Lenarda, R. Ganzerla, A. Rinaldi, *Microporous Mater.* **1996**, *6*, 55.
156. G. Fetter, G. Heredia, L. A. Velazquez, A. M. Maubert, P. Bosch, *Appl. Catal. A: General* **1997**, *162*, 41.
157. R. Molina, S. Moreno, G. Poncelet, *Stud. Surf. Sci. Catal., Elsevier, Amsterdam* **2000**, *130*, 983.
158. S. Moreno, E. Gutierrez, A. Alvarez, N. G. Papayannakos, G. Poncelet, *Appl. Catal. A: General* **1997**, *165*, 103.
159. J. L. Valverde, A. De Lucas, F. Dorado, R. Sun-Kou, P. Sanchez, I. Asencio, A. Garrido, A. Romero, *Ind. Eng. Chem. Res.* **2003**, *42*, 2783.
160. Y. Ralph, W. Li, *J. Catal.* **1996**, *155*, 414.
161. W. Li, M. Sirilumpen, R. T. Yang, *Appl. Catal. B: Environ.* **1997**, *11*, 347.

162. M. Sirilumpen, R. T. Yang, S. Tharapiwattananon, *J. Mol. Catal.* **1999**, *137*, 273.
163. B.-S. Kim, S.-H. Lee, Y.-T. Park, S.-W. Ham, H.-J. Chae, I.-S. Nam, *Korean J. Chem. Eng.* **2001**, *18*, 704.
164. G. A. Konin, A. N. Il'ichev, V. A. Matyshak, T. I. Khomenko, V. N. Korchak, V. A. Sadykov, V. P. Doronin, R. V. Bunina, G. M. Alikina, T. G. Kuznetsova, E. A. Paukshtis, V. B. Fenelonov, V. I. Zaikovskii, A. S. Ivanova, S. A. Beloshapkin, A. Ya. Rozovskii, V. F. Tretyakov, J. R. H. Ross, J. P. Breen, *Topics Catal.* **2001**, *16*, 193.
165. A. Bahamonde, F. Mohino, M. Rebollar, M. Yates, P. Avila, S. Mendioroz, *Catal. Today* **2001**, *69*, 233.
166. A. De Stefanis, M. Dondi, G. Perez, A. A. G. Tomlinson, *Chemosphere* **2000**, *41*, 1161.
167. V. A. Sadykov, R. V. Bunina, G. M. Alikina, V. P. Doronin, T. P. Sorokina, D. L. Kochubei, B. N. Novgorodov, E. A. Paukshtis, V. B. Fenelonov, A. Yu. Derevyankin, A. S. Ivanova, V. I. Zaikovskii, T. G. Kuznetsova, B. A. Beloshapkin, V. N. Kolomiichuk, L. M. Plasova, V. A. Matyshak, G. A. Konin, A. Ya Rozovskii, V. F. Tretyakov, T. N. Burdeynaya, M. N. Davydova, J. R. H. Ross, J. P. Breen, F. C. Meunier, *Mater. Res. Soc. Symp. Proc. 581* (Nanophase and Nanocomposite Materials III) **2000**, 435.
168. L. A. Galeano, S. Moreno, *Rev. Colomb. Quim.* **2000**, *31*, 57.
169. J. L. Valverde, F. Dorado, P. Sanchez, I. Asencio, A. Romero, *Ind. Eng. Chem. Res.* **2003**, *42*, 3871.
170. J. L. Valverde, A. De Lucas, P. Sanchez, F. Dorado, A. Romero, *Appl. Catal., B: Environ.* **2003**, *43*, 43.
171. J. L. Valverde, A. De Lucas, F. Dorado, R. Sun Kou, P. Sanchez, I. Asencio, A. Garrido, A. Romero, *Ind. Eng. Chem. Res.* **2003**, *42*, 2783.
172. J. L. Valverde, A. De Lucas, P. Sanchez, F. Dorado, A. Romero, *Stud. Surf. Sci. Catal., Elsevier, Amsterdam* **2002**, *142*, 723.
173. J. H. Choy, H. Jung, Y. S. Han, J. B. Yoon, Y. G. Shul, H. J. Kim, *Chem. Mater.* **2002**, *14*, 3823.
174. J. L. Valverde, A. De Lucas, F. Dorado, A. Romero, P. B. Garcia, *Ind. Eng. Chem. Res.* **2005**, *44*, 2955.
175. V. A. Sadykov, T. G. Kuznetsova, V. P. Doronin, E. M. Moroz, D. A. Ziuzin, D. Kochubei, B. N. Novgorodov, V. N. Kolomiichuk, G. M. Alikina, R. V. Bunina, E. A. Paukshtis, V. B. Fenelonov, O. B. Lapina, I. V. Yudaev, N. V. Mezentseva, A. M. Volodin, V. A. Matyshak, V. V. Lunin, A. Ya. Rozovskii, V. F. Tretyakov, T. N. Burdeynaya, J. R. H. Ross, *Topics Catal.* **2005**, *32*, 29.
176. C. Belver, M. A. Vicente, A. Martinez-Arias, M. Fernandez-Garcia, *Appl. Catal. B: Environ.* **2004**, *50*, 227.
177. D. P. Chen, M. C. Hausladen, R. T. Yang, *J. Catal.* **1995**, *151*, 135.
178. R. T. Yang, W. Li, *J. Catal.* **1995**, *155*, 414.
179. J. P. L. Chen, R. T. Yang, *Chem. Eng. Commun.* **1996**, *153*, 161.
180. L. S. Cheng, R. T. Yang, N. Chen, *J. Catal.* **1996**, *164*, 70.
181. H. L. Del Castillo, A. Gil, P. Grange, *Catal. Lett.* **1996**, *36*, 237.
182. K. Bahranowski, J. Janas, T. Machej, E. M. Serwicka, L. A. Vartikian, *Clay Miner.* **1997**, *32*, 665.
183. S. Perathoner, A. Vaccari, *Clay Miner.* **1997**, *32*, 123.
184. K. Sato, T. Fujimoto, S. Kanai, Y. Kintaichi, M. Inaba, M. Haneda, H. Hamada, *Appl. Catal. B: Environ.* **1997**, *13*, 27.
185. R. Q. Long, R. T. Yang, *J. Catal.* **2000**, *196*, 73.
186. L. Chmielarz, P. Kustrowski, M. Zbroja, A. Rafalska-Lasocha, B. Dudek, R. Dziembaj, *Appl. Catal. B: Environ.* **2003**, *45*, 103.
187. S. C. Kim, J. K. Kang, D. S. Kim, D. K. Lee, *Stud. Surf. Sci. Catal., Elsevier, Amsterdam* **2003**, *146*, 713.
188. L. K. Boudali, A. Ghorbel, P. Grange, *Catal. Lett.* **2003**, *86*, 251.
189. D. K. Lee, S. C. Kim, S. J. Kim, J. K. Kang, D. S. Kim, S. S. Oh, *Stud. Surf. Sci. Catal., Elsevier, Amsterdam* **2002**, *142*, 895.
190. L. K. Boudali, A. Ghorbel, P. Grange, S. M. Jung, *Stud. Surf. Sci. Catal., Elsevier, Amsterdam* **2002**, *143*, 873.
191. D. K. Lee, S. C. Kim, *Stud. Surf. Sci. Catal., Elsevier, Amsterdam* **2004**, *154*, 2973.
192. L. Chmielarz, P. Kustrowski, M. Zbroja, W. Lasocha, R. Dziembaj, *Catal. Today* **2004**, *90*, 43.
193. C. Philippopoulos, N. Gangas, N. Papayannakos, *J. Mater. Sci. Lett.* **1996**, *15*, 1940.
194. G. Qi, R. T. Yang, L. T. Thompson, *Appl. Catal. A: General* **2004**, *259*, 261.
195. J. Barrault, C. Bouchoule, K. Echachoui, N. Frini Srasra, M. Trabelsi, F. Bergaya, *Appl. Catal. A: General* **1998**, *158*, 269.
196. J. Barrault, C. Bouchoule, J. M. Tatibouet, M. Abdellaoui, A. Majeste, A. Louloudi, N. Papayannakos, N. H. Gangas, *Stud. Surf. Sci. Catal., Elsevier, Amsterdam* **2000**, *130*, 749.
197. J. Barrault, M. Abdellaoui, C. Bouchoule, A. Majeste, J. M. Tatibouet, A. Louloudi, N. Papayannakos, N. H. Gangas, *Appl. Catal. B: Environ.* **2000**, *27*, L225.
198. C. Catrinescu, C. Teodosiu, M. Macoveanu, J. Miehe-Brendle, R. Le Dred, *Environ. Eng. Manag. J.* **2000**, *1*, 567.
199. E. Guelou, J. Barrault, J. Fournier, J. M. Tatibouet, *Appl. Catal. B: Environ.* **2003**, *44*, 1.
200. M. N. Timofeeva, S. T. Khankhasaeva, S. V. Badmaeva, A. L. Chuvilin, E. B. Burgina, A. B. Ayupov, V. N. Panchenko, A. V. Kulikova, *Appl. Catal. B: Environ.* **2005**, *59*, 243.
201. L. Chirchi, I. Mrad, A. Ghorbel, *J. Chim. Phys. Physico-Chim. Biol.* **1997**, *94*, 1869.
202. S. C. Kim, S. S. Oh, G. S. Lee, J. K. Kang, D. S. Kim, D. K. Lee, *Stud. Surf. Sci. Catal. Elsevier, Amsterdam* **2003**, *146*, 633.
203. S. C. Kim, D. S. Kim, G. S. Lee, J. K. Kang, D. K. Lee, Y. K. Yang, *Stud. Surf. Sci. Catal., Elsevier, Amsterdam* **2002**, *142*, 683.
204. S. C. Kim, D. K. Lee, *Stud. Surf. Sci. Catal., Elsevier, Amsterdam* **2004**, *154*, 2958.
205. S. C. Kim, D. K. Lee, *Catal. Today* **2004**, *97*, 153.
206. S. Perathoner, G. Centi, *Topics Catal.* **2005**, *33*, 207.
207. Z. Ding, H. Y. Zhu, P. F. Greenfield, G. Q. Lu, *J. Colloid Interface Sci.* **2001**, *238*, 267.
208. C. Ooka, H. Yoshida, K. Suzuki, T. Hattori, *Appl. Catal. A: General* **2004**, *260*, 47.
209. C. Ooka, H. Yoshida, K. Suzuki, T. Hattori, *Microporous Mesoporous Mater.* **2004**, *67*, 143.
210. S. F. Cheng, S. J. Tsai, Y. F. Lee, *Catal. Today* **1995**, *26*, 87.
211. S. V. Awate, K. Suzuki, *Adsorption* **2001**, *7*, 319.
212. T. Kaneko, H. Shimotsuma, M. Kajikawa, T. Hatamachi, T. Kodama, Y. Kitayama, *J. Porous Mater.* **2001**, *8*, 295.
213. T. Takahama, T. Sako, M. Yokoyama, S. Hirao, *Nippon Kagaku Kaishi* **1994**, *7*, 618.
214. A. Gil, M. A. Vicente, J.-F. Lambert, L. M. Gandia, *Catal. Today* **2001**, *68*, 41.
215. L. M. Gandia, M. A. Vicente, A. Gil, *Appl. Catal. B: Environ.* **2002**, *38*, 295.
216. A. Gil, M. A. Vicente, R. Toranzo, M. A. Banares, L. M. Gandia, *J. Chem. Tech. Biotech.* **1998**, *72*, 131.
217. K. Manju, S. Sugunan, *React. Kinet. Catal. Lett.* **2005**, *85*, 37.

218. K. Shimizu, T. Kaneko, T. Fujishima, T. Kodama, H. Yoshida, Y. Kitayama, *Appl. Catal. A: General* **2002**, *225*, 185.
219. J. J. Li, J. Zheng, Z. P. Hao, X. Y. Xu, Y. H. Zhuang, *J. Mol. Catal. A: Chem.* **2005**, *225*, 173.
220. L. Storaro, R. Ganzerla, M. Lenarda, R. Zanoni, *J. Mol. Catal.* **1995**, *97*, 139.
221. L. Storaro, R. Ganzerla, M. Lenarda, R. Zanoni, A. Jimenez Lopez, P. Olivera-Pastor, E. R. Rodriguez Castellon, *J. Mol. Catal.* **1997**, *115*, 329.
222. K. Bahranowski, A. Gawel, R. Janik, J. Komorek, T. Machej, A. Michalik, E. M. Serwicka, W. Wlodarczyk, *Polish J. Chem.* **2003**, *77*, 5.
223. J. R. Jones, J. H. Purnell, *Catal. Lett.* **1994**, *28*, 283.
224. M. Sychev, V. H. J. de Beer, R. A. van Santen, R. Pridhod'ko, V. Goncharuk, *Stud. Surf. Sci. Catal.*, Elsevier, Amsterdam **1994**, *84A*, 267.
225. A. K. Lavados, P. N. Trikalitis, P. J. Pomonis, *J. Mol. Catal.* **1996**, *106*, 241.
226. M. Perissinotto, M. Lenarda, L. Storaro, R. Ganzerla, *J. Mol. Catal.* **1995**, *121*, 103.
227. M. Raimondo, A. De Stefanis, G. Perez, A. A. G. Tomlinson, *Appl. Catal. A: General* **1998**, *171*, 85.
228. H. E. Lin, A. N. Ko, *J. Chinese Chem. Soc.* **2000**, *47*, 509.
229. G. Guiu, P. Grange, *J. Catal.* **1997**, *168*, 463.
230. F. Kooli, W. Jones, *J. Mater. Chem.* **1998**, *8*, 2119.
231. K. Luvrani, G. L. Rorrer, *Appl. Catal.* **1994**, *109*, 147.
232. S. C. Kim, J. K. Kang, D. S. Kim, D. K. Lee, *Stud. Surf. Sci. Catal.*, Elsevier, Amsterdam **2003**, *146*, 713.
233. A. Gil, H. L. Del Castillo, J. Masson, J. Court, P. Grange, *J. Mol. Catal.* **1996**, *107*, 185.
234. G. Perez, A. De Stefanis, A. A. G. Tomlinson, *J. Mater. Chem.* **1997**, *7*, 351.
235. L. Huerta, A. Meyer, E. Choren, *Microporous Mesoporous Mater.* **2003**, *57*, 219.
236. T. Oberto, A. Meyer, E. Gonzalez, *Ciencia (Maracaibo)* **2002**, *10*, 185.
237. M. A. Vicente, A. Meyer, E. Gonzalez, M. A. Banares-Munoz, L. M. Gandia, A. Gil, *Catal. Lett.* **2002**, *78*, 99.
238. E. Gonzalez, A. Moronta, *Appl. Catal. A: General* **2004**, *258*, 99.
239. A. B. Boricha, H. M. Mody, A. Das, H. C. Bajaj, *Appl. Catal. A: General* **1999**, *340–341*, 349.
240. P. Oliveira-Pastor, J. Maza-Rodriguez, A. Jimenez Lopez, A. Guerrero-Ruiz, J. L. G. Fierro, *Stud. Surf. Sci. Catal.*, Elsevier, Amsterdam **1994**, *82*, 11.
241. A. I. Panizo, C. Belver, M. A. Banares-Munoz, O. Guerrero, M. A. Banares, M. A. Vicente, *Afinidad* **2004**, *61*, 161.
242. M. R. Sun Kou, S. Mendioroz, P. Salerno, V. Munoz, *Appl. Catal. A: General* **2003**, *240*, 273.
243. L. K. Boudali, A. Ghorbel, H. Amri, F. Figueras, *C. R. Acad. Sci. Paris, Ser. IIc: Chim.* **2001**, *4*, 67.
244. L. K. Boudali, A. Ghorbel, F. Figueras, C. Pinel, *Stud. Surf. Sci. Catal.*, Elsevier, Amsterdam **2000**, *130*, 1643.
245. M. Akcay, *Appl. Catal. A: General* **2004**, *269*, 157.
246. A. Azzouz, E. Dimitriu, V. Hulea, C. Catrinescu, G. Carga, *Prog. Catal.* **1996**, *5*, 9.
247. M. R. Cramarossa, L. Forti, U. M. Pagnoni, M. Vidali, *Synthesis* **2001**, *1*, 52.
248. M. Lenarda, R. Ganzerla, L. Storaro, S. Enzo, R. Zanoni, *J. Mol. Catal.* **1994**, *92*, 201.
249. M. Lenarda, R. Ganzerla, L. Riatto, L. Storaro, *J. Mol. Catal. A: Chem.* **2002**, *187*, 129.
250. K. Bahranowski, M. Gasior, E. Jagielska, J. Podobinski, E. M. Serwicka, L. A. Vartikian, *Proceedings, Rosenquist Symposium*, Oslo, 1996, p. 7.
251. H. L. Del Castillo, A. Gil, P. Grange, *Clays Clay Miner.* **1996**, *44*, 706.
252. S. Letaief, B. Casal, P. Aranda, M. A. Martin-Luengo, E. Ruiz-Hitzky, *Appl. Clay Sci.* **2003**, *22*, 263.
253. T. Mishra, K. M. Parida, *J. Mol. Catal.* **1997**, *121*, 91.
254. D. Das, H. K. Mishra, K. Parida, A. K. Dalai, *Indian J. Chem., Sect. A: Inorg., Bio-inorg., Phys., Theor. & Anal. Chem.* **2002**, *41*, 2238.
255. K. Sapag, S. Rojas, M. L. Granados, J. L. G. Fierro, S. Mendioroz, *J. Mol. Catal. A: Chem.* **2001**, *167*, 81.
256. S. Mendioroz, B. Asenjo, P. Terreros, P. Salerno, V. Munoz, *Stud. Surf. Sci. Catal.*, Elsevier, Amsterdam **2000**, *130*, 3741.
257. A. Louloudi, N. Papayannakos, *Appl. Catal. A: General* **2000**, *204*, 167.
258. A. Louloudi, J. Michalopoulos, N. H. Gangas, N. Papayannakos, *Appl. Catal. A: General* **2003**, *242*, 41.
259. N. Marin-Astorga, G. Alvez-Manoli, P. Reyes, *J. Mol. Catal. A: Chem.* **2005**, *226*, 81.
260. L. J. A. Colin, J. A. De los Reyes, A. Vazquez, A. Montoya, *Appl. Surf. Sci.* **2005**, *240*, 48.
261. G. Sandoval-Robles, R. Ramos-Gomez, S. Robles, R. Garcia-Alamilla, *Inform. Technol.* **2001**, *12*, 71.
262. S. D. Bodman, W. R. McWhinnie, V. Begon, M. Millan, I. Suelves, M. J. Lazaro, A. A. Herod, R. Kandiyoti, *Fuel* **2003**, *82*, 2309.
263. S. D. Bodman, W. R. McWhinnie, V. Begon, I. Suelves, M. J. Lazaro, T. J. Morgan, A. A. Herod, R. Kandiyoti, *Fuel* **2002**, *81*, 449.
264. M. E. Gyftopoulou, M. Millan, A. V. Bridgwater, D. Dugwell, R. Kandiyoti, J. A. Hriljac, *Appl. Catal. A: General* **2005**, *282*, 205.
265. M. Sychev, N. Kostoglod, E. M. van Oers, V. H. J. de Beer, R. A. van Santen, J. Kornatowski, M. Rozwadowski, *Stud. Surf. Sci. Catal.*, Elsevier, Amsterdam **1995**, *94*, 39.
266. M. Sychev, V. H. J. de Beer, A. Kodentsov, E. M. van Oers, R. A. van Santen, *J. Catal.* **1997**, *168*, 245.
267. R. G. Leliveld, W. C. A. Huyben, A. J. van Dillen, J. W. Geus, D. C. Koninsberger, *Stud. Surf. Sci. Catal.*, Elsevier, Amsterdam **1997**, *106*, 137.
268. G. Alvarez, R. Garcia, R. Cid, N. Escalona, F. J. Gil-Llambias, *Bol. Soc. Chil. Quim.* **2001**, *46*, 363.
269. M. J. Perez-Zurita, G. P. Quintana, J. G. Biomorgi, C. E. Scott, *Prepr. Symp. -Am. Chem. Soc., Div. Fuel Chem.* **2003**, *48*, 90.
270. M. Sugioka, T. Kurosaka, *J. Japan Petrol. Inst.* **2002**, *45*, 342.
271. M. J. Martinez-Ortiz, G. Fetter, J. M. Dominguez, J. A. Melo-Banda, R. Ramos-Gomez, *Microporous Mesoporous Mater.* **2003**, *58*, 73.
272. S. Vijayakumar, C. Vijaya, K. Rengaraj, B. Sivasanker, *Bull. Chem. Soc. Jpn.* **1994**, *7*, 3107.
273. S. Chevalier, R. Frank, J.-F. Lambert, D. Barthomeuf, H. Suquet, *Appl. Catal.* **1995**, *110*, 153.
274. R. Swarnakar, K. B. Brandt, R. A. Kydd, *Appl. Catal. A: General* **1996**, *142*, 61.
275. F. Kooli, J. Bovey, W. Jones, *J. Mater. Chem.* **1997**, *7*, 153.
276. U. Kürschner, V. Seefeld, B. Parlitz, W. Gessner, H. Lieske, *React. Kinet. Catal. Lett.* **1998**, *65*, 17.
277. J. Alcaraz, J. Triphahn, S. Wojnicki, M. Cohn, J. Holmgren, *International Symposium on Acid-Base Catalysis III*, Rolduc, April 20–24, 1997 (paper 27).

278. G. Manos, I. Y. Yusof, N. H. Gangas, N. Papayannakos, *Energy & Fuels* **2002**, *16*, 485.
279. K. Gobin, G. Manos, *Polym. Degrad. Stabil.* **2004**, *82*, 267.
280. A. Geatti, M. Lenarda, L. Storaro, R. Ganzerla, M. Perissinotto, *J. Mol. Catal.* **1997**, *121*, 111.
281. B. M. Choudary, M. L. Kantam, M. Sateesh, K. K. Rao, P. L. Santhi, *Appl. Catal. A: General* **1997**, *149*, 257.
282. E. Min, *Stud. Surf. Sci. Catal.*, Elsevier, Amsterdam **1994**, *83*, 443.
283. R. Mokaya, W. Jones, *J. Chem. Soc. Chem. Commun.* **1994**, 929.
284. B. Casal, J. Merino, E. Ruiz-Hitzky, E. Gutierrez, A. Alvarez, *Clay Miner.* **1997**, *32*, 41.
285. G. D. Yadav, N. S. Doshi, *Org. Proc. Res. Dev.* **2002**, *6*, 263.
286. M. Raimondo, G. Perez, A. De Stefanis, A. A. G. Tomlinson, O. Ursini, *Appl. Catal. A: General* **1997**, *164*, 119.
287. L. Caballero, J. M. Dominguez, J. L. De Los Santos, A. Montoya, J. Navarrete, in *Synthesis of Porous Materials*, M. L. Occelli and H. Kessler (Eds.), Marcel Dekker Inc., 1997, p. 491.
288. C. Vijaya, K. Rengaraj, B. Sivasankar, *Indian J. Chem. Technol.* **1998**, *5*, 281.
289. B. G. Mishra, G. R. Rao, *Microporous Mesoporous Mater.* **2004**, *70*, 43.
290. M. Kurian, S. Sugunan, *Indian J. Chem., Section A: Inorg., Bio-inorg., Phys., Theor. & Anal. Chem.* **2003**, *42*, 2480.
291. T. Jiang, T. Sun, J. Shen, D. Jiang, E. Min, *Gaodeng Xuexiao Huaxue Xuebao* **1994**, *15*, 93.
292. T. Matsuda, H. Seki, E. Kikuchi, *J. Catal.* **1995**, *154*, 41.
293. R. Russu, A. Russu, O. Cira, O. V. Albu, *Prog. Catal.* **1995**, *4*, 31.
294. J. Krajovic, P. Hudec, F. Grejtak, *React. Kinet. Catal. Lett.* **1995**, *54*, 87.
295. R. Molina, A. Schutz, G. Poncelet, *J. Catal.* **1994**, *145*, 79.
296. M. Katoh, H. Fujisawa, T. Yamaguchi, *Stud. Surf. Sci. Catal.*, Elsevier, Amsterdam **1994**, *90*, 263.
297. A. P. Carvalho, A. Martins, J. M. Silva, J. Pires, H. Vasques, M. B. de Carvalho, *Clays Clay Miner.* **2003**, *51*, 340.
298. M. F. Molina, S. Moreno, *Rev. Colomb. Quim.* **2001**, *30*, 133.
299. E. S. Olson, R. K. Sharma, *Energy & Fuels* **1996**, *10*, 587.
300. J. Bovey, W. Jones, *J. Mater. Chem.* **1995**, *5*, 2027.
301. S. Bodoardo, R. Chiapetta, F. Fajula, E. Garrone, *Microporous Mater.* **1995**, *3*, 613.
302. A. Gil, M. Montes, *Ind. Eng. Chem. Res.* **1997**, *36*, 1431.
303. Z. P. Hao, H. Y. Zhu, G. Q. Lu, *Appl. Catal. A: General* **2003**, *242*, 275.

2.3.8
Chemistry and Applications of Porous Metal–Organic Frameworks

Ulrich Müller*, Markus M. Schubert, and Omar M. Yaghi

2.3.8.1 Introduction

As early as 1965, a preliminary report by Tomic [1] on novel solids was introduced which, today, would be categorized and addressed as metal–organic frameworks (MOFs), coordination polymers, or supramolecular structures. In the aforementioned contribution, details were described of the simple synthesis of coordination polymers based on metals such as zinc, nickel, iron, aluminum, as well as thorium and uranium, employing bivalent to tetravalent aromatic carboxylic acids. Moreover, at such an early date, interesting features of these compounds such as their high thermal stability and high metal content were already under investigation.

Some decades later, a major stimulation to the field was provided by the publication of the synthesis, structure and porosity of MOF-2 and, more significantly, of MOF-5 in 1998 and 1999 [2], in addition to the concept of reticular design with totally different carboxylate linkers in 2002 [3–5]. In the meantime, numerous reviews have addressed this rapidly growing area of research, with the most comprehensive being provided by Kitagawa [6] and Yaghi [7]. Of particular interest is the progress achieved in studying the structures, properties, and possible application of MOFs as storage media [7, 8]. Comparisons with oxides, molecular sieves, porous carbons and heteropolyanion salts have been reported by Barton and coauthors [9], and today several hundred different types of MOF have been identified. The self-assembling of metal ions, which act as coordination centers, linked together with a variety of polyatomic organic bridging ligands, has resulted in tailored nanoporous host materials which are robust solids with high thermal and mechanical stabilities (Fig. 1).

Interestingly, unlike other solid matter – for example, zeolites, carbons, and oxides – a number of coordination compounds are known with a high flexibility of the framework and shrinkage or expansion due to interactions with guest molecules. The most striking difference to state-of-art materials is probably the total lack of non-accessible bulk volume in MOF structures. Although high surface areas are already recognized with activated carbons and zeolites, it is the absence of any dead volume in MOFs which principally provides them, on a weight-specific basis, with the highest porosities and record surface areas (Fig. 2). For example, in the case of MOF-177, values of 4500 m^2 g^{-1} are reported [5], and for MIL-101 up to 5900 m^2 g^{-1} [10]. Of course, properties such as a drastically increased velocity of molecular traffic of molecules in these open structures are also closely related to the regularity of pores of nanometer size.

Thus, the combination of as-yet unrealized level of porosity, surface area, pore size and wide chemical inorganic–organic composition has recently brought these materials to the attention of many research groups both in both academia and industry, with over 1000 publications on "metal–organic frameworks" and "coordination polymers" produced on an annual basis [6].

---
* Corresponding author.

References see page 261

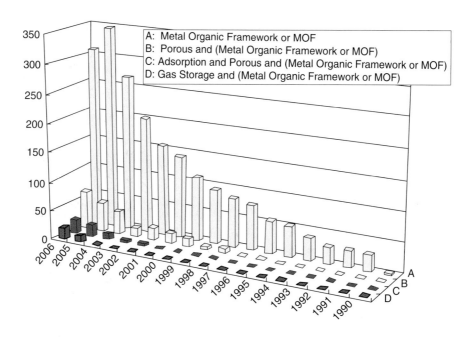

Fig. 1 Overview of publications on metal–organic frameworks (MOFs).

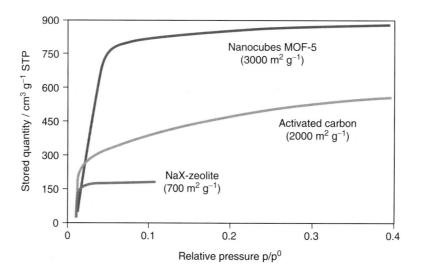

Fig. 2 Outstanding surface areas and porosities on metal–organic frameworks.

This chapter aims to describe how MOF materials can be synthesized using verified synthesis methods, as well as by a totally novel electrochemical approach [11]. Today, with large-scale samples being available, the testing of MOFs in the fields of catalysis and gas processing is possible, and is expanding. Details will be provided on the catalytic activation of alkynes (the formation of methoxypropene from propyne, vinylester synthesis from acetylene) [12], while further examples such as olefin polymerization, the Diels–Alder reaction, transesterification [6] or cyanosilylation [13] are detailed elsewhere in the literature.

The removal of impurities in natural gas (traces of tetrahydrothiophene in methane), the pressure swing separation of rare gases (krypton and xenon) and the storage of hydrogen (3.3 wt.% at 2.5 MPa/77 K on Cu–BTC–MOF) will underline the perspectives for future industrial uses of MOFs. Whenever possible, comparisons will be made to state-of-art applications in order to outline the possibilities of processes which might benefit from the use of MOFs.

2.3.8.2 Terminology and Structure

The design and synthesis of MOFs has yielded a large number of structures which can be classified as being

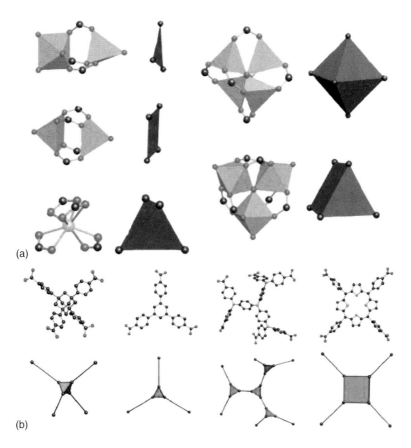

Fig. 3 Examples of: (a) inorganic metal carboxylate clusters and their corresponding SBU geometry; (b) organic units and their corresponding SBU geometry.

composed of two components; namely, transition metal oxide units and organic units that are linked together to form extended porous frameworks. Most often, the metal oxide units are either zero-dimensional (0-D) or one-dimensional (1-D), corresponding to discrete or rod-like geometric units named secondary building units (SBUs). The organic units, when more than two-connected, can also be viewed as SBUs. In essence, a MOF structure is composed of inorganic and organic SBUs (Fig. 3a,b), and the concept of SBUs has allowed the classification of more than 2000 MOF structures. Techniques have now been developed for reducing MOF crystal structures down to their underlying topologies (nets). Indeed, it is found that most MOF structures fall in one of a handful of nets which are referred to as "default" nets. MOFs based on these nets are the most likely outcome of solution synthesis using metal ions and organic units.

In the interpretation and prediction of MOF structures, the SBUs are considered as the "joints" and the organic links as the "struts" of the underlying net. MOFs based on discrete shapes (triangles, squares, tetrahedra, etc.) have been synthesized and studied. An illustrative example of this is MOF-5, where the metal–carboxylate structure (Fig. 4) is an octahedral SBU linked by benzene units to produce a primitive cubic net. The simplification of MOF structures in this way has led to a full enumeration and description of the principal topological possibilities available for the assembly of various discrete SBU geometries.

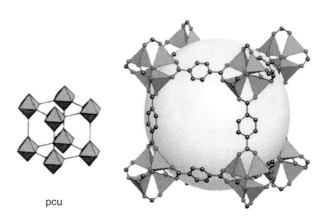

Fig. 4 MOF-5 structure and its corresponding underlying primitive cubic (pcu) net topology.

References see page 261

Once the reaction conditions for a given inorganic SBU have been determined, it is then possible to add to the synthesis preparation the organic link desired. This procedure was followed to yield the isoreticular (IR) analogues of MOF-5, as shown for IRMOF-1 to -16 in Fig. 5.

The large majority of structural studies have been performed on MOFs of *discrete* SBUs. Recently, MOFs were introduced that are constructed from *infinite* rod-shaped SBUs (see Fig. 6). These rod-shaped metal–carboxylate SBUs were found to provide means of accessing MOFs that do not interpenetrate due to the intrinsic packing arrangement of such rods in the crystal structure. Specifically: (i) an account was provided of the geometric principles of basic rod packing and nets based on them; (ii) the principal topological possibilities, and the most likely structures that could form from rods, were identified; (iii) a number of new MOFs were adduced with structures constructed according to these geometric principles; and (iv) it was shown that, similar to MOFs of discrete SBUs, MOFs based on rod SBUs may also have stable architectures and permanent porosity.

Another outcome of the SBU concept is the ability to produce MOFs with permanent porosity. As the SBUs are based on carboxylates, the metal centers are chelated to provide for robust structures wherein each metal ion is locked into position by the carboxylates. The practical outcome of this concept is embodied in the applications presented in this contribution.

2.3.8.3 Synthesis

Usually, the synthesis of MOFs is straightforward, using dissolvable salt precursors of the inorganic metal compound, viz. metal nitrates, sulfates, acetates, etc. The organic ingredients, which mostly are the carboxylic mono-, di-, tri- and tetracarboxylic acids, are supplied in an organic polar miscible solvent, typically an amine (triethylamine) or amide (diethylformamide, dimethylformamide). After combination of these inorganic and organic components under stirring, the metal–organic structures are forming by self-assembly at temperatures between almost room temperature and up to 200 °C under solvothermal conditions, and within a few hours. A typical scheme for this semi-technical process is provided in Fig. 7, indicating not only the different steps of preparation but also recycling of the solvent and further processing of the dried powders into shaped particles.

It should be pointed out, however, that the filtering and drying of metal–organic compounds in wet processing must be carried out with great care, as – due to their high porosity and surface area – they may easily carry 50–150 wt.% of occluded solvent; this is an order of magnitude higher than in a zeolite or base metal oxide preparation. Hence, it is advisable first to remove the

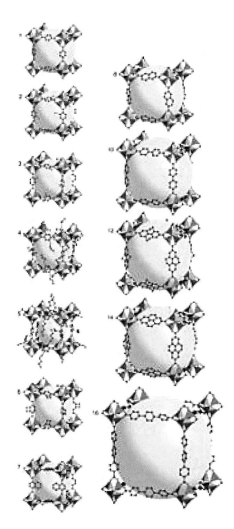

Fig. 5 Examples of isoreticular (having the same underlying net topology) metal–organic frameworks (IRMOFs).

large proportion of adsorbed solvent and water under gentle conditions of pressure and temperature, prior to high thermal activation.

In addition, the crystallization of MOFs can be considered as an esterification reaction between an inorganic base metal salt and an organic acid. Thus, the preparation of, for example MOF-5, can be expressed stoichiometrically as:

$$4[Zn(NO_3)_2 \cdot 4H_2O] + 3[BDC] + 8[OH^-]$$
$$\longrightarrow [Zn_4O(BDC)_3] + 8[NO_3^-] + 23[H_2O]$$

where BDC is terephthalic acid and $Zn_4O(BDC)_3$ is the MOF-5 unit composition. It is clearly seen, that the equilibrium can be shifted to the MOF-product side by working on the concentration profiles of, for example, the solvent, liberated water and nitrate, respectively.

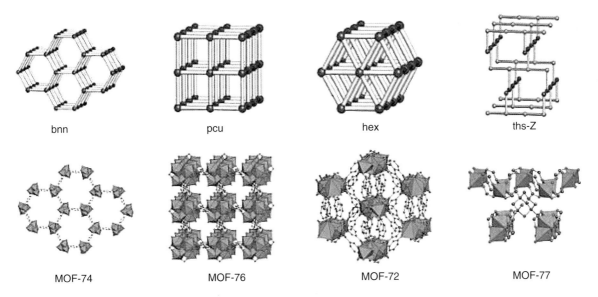

Fig. 6 Examples of MOFs with infinite rod SBUs, and their underlying net topology.

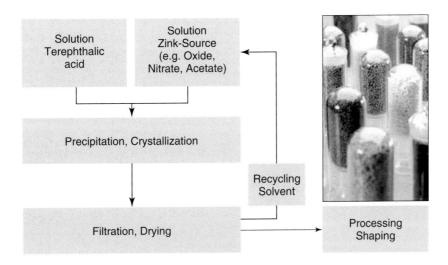

Fig. 7 Simplified flowsheet of metal–organic framework (MOF) synthesis procedure.

By taking these equilibrium conditions into account, it became possible to improve the MOF quality; that is, to increase the surface area from 2600 m² g⁻¹ up to 3400 m² g⁻¹.

As esterification reactions can be easily driven in both directions, it becomes clear that the stability and integrity of MOFs in their use can depend on polar protic environments and the pH values applied, and these must be considered on a case-by-case basis.

On a larger production scale, attention must be paid to safety issues whenever high nitrate concentrations are involved, especially in addition to the build-up of large surface area volumes in adiabatic reactor vessels. Thus, an alternative salt-free procedure has been developed at BASF, starting via an electrochemical route and employing bulk metal sacrificial anodes with dissolved carboxylates in an electrochemical cell [11, 14]. Simple recovery of the solid formed by filtration directly yields the final MOF powder after drying. This procedure is especially beneficial with MOF-structures having open metal ligand sites, as no anions from added salts can block the access to the sites.

Recently, the details have been summarized of some verified synthesis procedures on a larger, laboratory scale [14].

References see page 261

2.3.8.4 Characterization

As MOFs are both crystalline and highly porous materials, most frequently the technique of X-ray diffraction (XRD) is used to characterize the crystallinity and phase purity, adsorption measurements are performed to check for the porosity, and more elaborate studies make use of neutron scattering to detect sorption sites [15].

Commercially available equipment using nitrogen sorption at 77 K or argon uptake at 87 K are applied, and equivalent surface areas calculated according to, for example, the Langmuir equation. However, it should be considered that the underlying model of independent, equivalent and non-infringing sorption sites might be different on a molecular level. Previously, many reports have described localized rather than bulk volume adsorption phenomena on metal–organic materials, with some even clearly differentiating between various crystallographic site and adsorption strengths [15–18].

For applications, it is simply a convenient and brief means of comparing the higher sorption capacities of MOFs over state-of-the-art sorbents, although care must be taken whenever applications might rely upon gravimetric or volumetric scale at a later date.

As mass transfer can heavily influence results in catalysis, it is clearly necessary to have information at hand on crystal sizes and size distribution of the MOF samples that are routinely collected by scanning electron microscopy (Fig. 8).

As the metal component content arrives at values in between 20–40 wt.%, it is also desirable to check the local metal cluster arrangements and environments. using more refined methodology such as X-ray photon spectroscopy (XPS), extended X-ray absorption fine structure (EXAFS), and X-ray absorption near edge structure (XANES).

Checking adsorbates in the inner voids of MOFs can be performed by UV-Visible and Raman spectroscopy [5].

2.3.8.5 Emerging Applications

To date, only a limited number of possible uses have been discussed for MOF compositions, and it appears that none has yet been realized on an industrial basis. However, when examining the specific properties of MOFs more closely, it becomes clear that – in principle – the high porosity and absence of hidden volumes in these new frameworks render them valuable for volume-specific applications such as adsorption, separations, purification purposes, and catalysis. For example, typical drop-in technologies might use MOFs initially, perhaps as substitutes for zeolitic molecular sieves, activated carbons and base metal oxides in existing plants. This promises a better performance and variable costs, without any fixed capital expenditure.

Some of the preliminary findings with MOFs, which are already being exploited at the pilot plant stage, are described in the following sections.

2.3.8.5.1 Adsorption Properties
In accordance with the values of surface area (with some of up to almost 5900 $m^2\,g^{-1}$ [10]), it can be expected that molecules other than nitrogen or argon will also be adsorbed to a considerable extent. Figure 9 shows a comparison for the adsorption of selected alkanes on MOF-5 and NaX, respectively; clearly, the MOF materials exhibit a much higher uptake capacity. As noted previously, this high performance is based on the intrinsic material property that metal–organic complexes do not have any hidden dead-space within their structure.

Interestingly, even small-pore MOFs have been prepared which are capable of separating molecules either by size or kinetic diameter [19–21]; for example, while water uptake continues to occur, slightly larger molecules such as N_2, O_2, CO_2, and methane are occluded from adsorption [20].

MOF-5 from Zn-Nitrate
50–100 μm

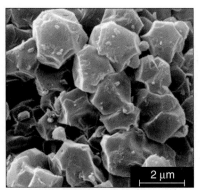

MOF-5 from ZnO
2 μm

Fig. 8 Morphology of MOF-crystals derived from different zinc sources.

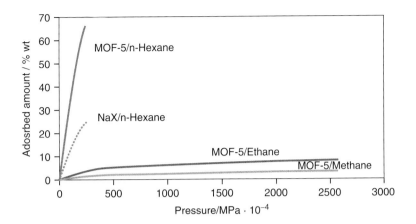

Fig. 9 Hydrocarbon uptake of MOF-5 compared to NaX zeolite.

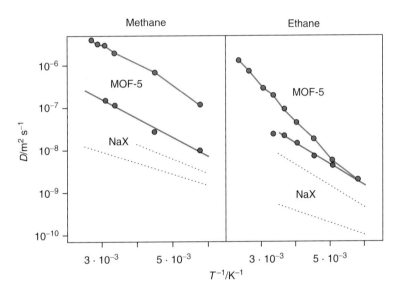

Fig. 10 Comparison of diffusion properties of MOF-5 over NaX zeolite. Arrhenius presentation of the intracrystalline (lower) and effective (upper) self-diffusion coefficients of methane and ethane in MOF-5. The region between the dotted lines shows the range of intracrystalline self-diffusion coefficients in NaX zeolite for loadings from about 30 mg g^{-1} up to 90 mg g^{-1} methane and 110 mg g^{-1} ethane, respectively. The solid lines correspond to activation energies of 8.5 kJ mol^{-1} (methane) and 9.5 kJ mol^{-1} (ethane) for adsorbate self-diffusion in MOF-5.

In addition, fully reversible so-called "gate-pressure" effects become observable when adsorption and desorption curves are monitored over a larger span of pressures [22–24]. Huge hysteresis effects are registered which no longer can be simply attributed to subtle phase-change effects of the adsorbate state, as occurs with some zeolites [25]. Moreover, a progressive opening up of pores and the entire lattice in the flexible MOFs is observed. In future, it will be interesting to see whether the other material properties, such as electronic conductivity, optical appearance, magnetism, and also thermal properties, also undergo changes, depending on the various stages in the gated pressure regime.

As a consequence, applications might become possible whenever these lattice-dependent memory effects can be included by design, and controlled.

2.3.8.5.2 Diffusional Properties As expected, for ethane and benzene (see Fig. 10), diffusion in MOF-5 is clearly faster – by orders of magnitude – than in zeolite NaX [26]. This is mainly considered to be a consequence of the difference between the diameters of the two large nanoporous cavities in MOF-5 (1.1–1.3 nm) over the smaller NaX supercage (1.2 nm), with an even lower

References see page 261

entrance window size (0.7 nm). The effective self-diffusion coefficient of benzene in the MOF-5 structure is only slightly less than that in neat liquids at the same temperature (C_6H_6: 2.5×10^{-9} m^2 s^{-1}).

For both, catalysis and gas processing, this is an important observation and promises rapid molecular transport and low diffusional resistance as additional benefits whenever MOF materials rather than zeolites, are used. In industrial applications of MOFs, this might contribute to providing permanent benefits in terms of variable energy costs, for example when comparing pressure-swing unit operations with state-of-the-art solids.

2.3.8.5.3 Gas Purification

The removal of ppm traces of sulfur components from gases might represent one possible field where MOFs could be used beneficially. In particular, MOF structures with accessible open metal ligand sites would be well-suited to strongly (>30 kJ mol^{-1}) chemisorb electron-rich, odor-generating molecules such as amines, phosphines, oxygenates, alcohols, water, or sulfur-containing molecules.

One such example which was evaluated experimentally in continuous breakthrough trials was the removal of an odorant, tetrahydrothiophene (THT), from natural gas. At room temperature, traces (10–15 ppm) of sulfur were fully captured down to less than 1 ppm using an electrochemically prepared Cu–MOF (Fig. 11) [14]. The overall capacity of the MOF material (70 g THT L_{MOF}^{-1}) outperformed, by about an order of magnitude, commercially available activated carbon materials as adsorbents, viz. Norit (type RB4) and CarboTech (type C38/4).

2.3.8.5.4 Gas Separation

Unlike gas purification, in gas separation processes the mixtures usually consist of components with concentrations in the same order of magnitude.

Very often, either distillation or pressure- and/or thermal-swing adsorption/desorption can be used to separate the different mixtures. Examples of existing technologies include nitrogen/oxygen air separation, nitrogen/methane, noble gases (e.g., Kr/Xe, etc.), some of which are well known for their use with zeolitic adsorbents.

Recently, the separation of Kr/Xe by pressure-swing, as well as the purification of methane in natural gas, was piloted on MOF-adsorbents, and details of this have been reported elsewhere [27].

Furthermore, the separation of propane and propene (50/50 vol.%) at 298 K and up to 0.5 MPa pressure was benchmarked against zeolitic molecular sieves (see Fig. 12). The separation was found to function best with MOFs; indeed, among the samples tested a Cu–MOF was shown to have a much higher dynamic saturation (14.9 wt.%) compared to a 13X molecular sieve (11.9 wt.%).

2.3.8.5.5 Gas Storage

Due to the unique structure of the MOFs, and especially to the absence of dead space, it has become possible for the first time to increase volume-specific gas storage above previously known levels. This storage effect can be very pronounced, and depends on the type, temperature and pressure of the gas, as well as on the specific MOF material being used.

The mechanism of increased storage in MOF-filled gas canisters over empty gas bottles is easy to understand, when the underlying principles are considered. The filling

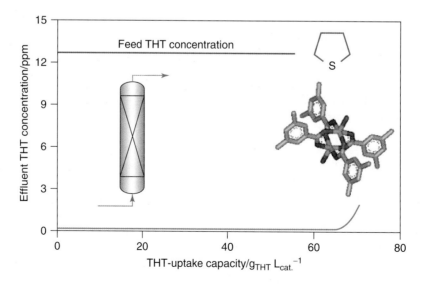

Fig. 11 Feedstream purification and removal of tetrahydrothiophene from natural gas.

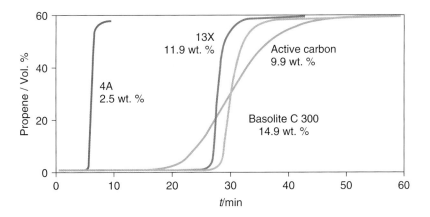

Fig. 12 Separation of propane/propene on different adsorbents.

of a conventional gas container is simply applying physical effects that are dependent upon the pVT-characteristic of the gas under investigation. However, in the case of MOF-filled gas canisters, the above-mentioned principle is greatly enhanced by an additional adsorption effect inside the MOF. As these frameworks are free from dead-volume, there is almost no notable loss of storage capacity due to space-blocking by any non-accessible solids volume. In summary, pVT-filling plus adsorption contribute to an enhanced storage capacity per volume.

This situation is exemplified in Fig. 13 by the storage of propane at room temperature, with pressures up to 10^6 Pa for a container filled with tabletted MOF material. A nonlinear uptake curve can be monitored as the pressure is increased; release will then occur on the same curve when the valve is opened and the pressure is reduced. At a given pressure of about 1 MPa it can be clearly seen that about a three-fold capacity is reached over the state-of-the-art filling curve. It is important to note at this point that the benefit of enhanced capacity with MOF-filled vessels will only be gained if the gas is in a true gaseous state, and not in liquid-phase.

Recently, exceptionally high uptakes of acetylene were reported by Kitagawa, in this case clearly exceeding the packing density usually obtained in today's storage devices [28].

Of course, the most challenging storage is with hydrogen, as the underlying possibility of using hydrogen as a fuel for mobile or portable fuel-cell applications attracts great interest in hydrogen storage possibilities. Unlike metal hydrides, MOF-storage for hydrogen is fully reversible, avoids complicated heat treatments, and the recharging proceeds within a timeframe of seconds to minutes. In MOF-5, the distinct location of hydrogen has been elucidated using inelastic neutron scattering [15], while density functional theory (DFT) studies have further calculated that it might be possible to store about 16–20 molecules of H_2 per Zn_4O-cluster [18].

The need for alternative fuel sources and energy carriers, in addition to the target values of the US Department of Energy, was recently reviewed [8]. Storage data have been reported of up to approximately 7.5 wt.% of hydrogen on MOFs (e.g., MOF-177) at 9 MPa and 77 K [27–29]. Unsurprisingly, many other research groups have also investigated this challenge, and Férey reported a 3.8% weight-specific calculation on MIL-53 [22] derived from aluminum salts and BDC. Pillaring with secondary amine (triethylenediamine) linkers as a strategy was employed by Kim [30] and Seki [31], and this resulted in hydrogen uptakes of up to 2 wt.% at 0.1 MPa and 87 K. Similar values were achieved with MOF-505 by Yaghi's group [32], where Cu-paddle-wheels are connected by 3,3′,5,5′-biphenyltetracarboxylic acid. Doubly interpenetrated nets of zinc frameworks built by NTB-linkers (4,4′,4″-nitrilotrisbenzoic acid) by Suh [33] were reported to reach 1.9 wt.% of hydrogen uptake at 77 K. However, it has not yet been foreseen as to whether high-surface-area materials such as MOF-177 [5], MIL-100 [34] or MOF-5 and isoreticular members [2–4], materials with an average surface of between 1000–1500 $m^2\,g^{-1}$ [22, 33], or even small-pore MOFs [16, 35], will prove to be the most promising storage medium. Neither can it be concluded whether divalent or trivalent metal clusters [36] are the most favorable. Whatever the outcome, such comparison with NaX-zeolite indicates very clearly the superior behavior of MOFs over microporous inorganic media [37].

Based on the results of prototype equipment (77 K and up to 4 MPa), it can be seen, how different MOF materials contribute differently to volume-specific hydrogen storage (Fig. 14; Table 1). These data from a large-scale prototype

References see page 261

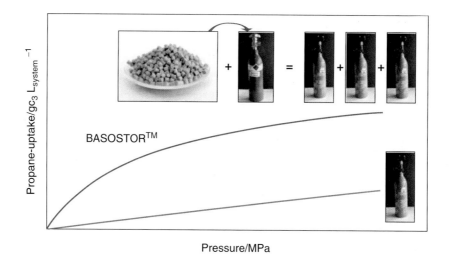

Fig. 13 Gas storage of propane in shaped MOF (Basostor™).

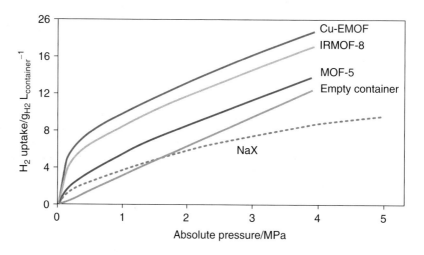

Fig. 14 Hydrogen storage capacities measured in prototype trials.

compare reasonably well with those reported in the literature [8].

For many volume-limited mobile and portable fuel-cell applications, it would be industrially much more relevant to compare storage data on a volume-specific rather than on a weight-specific storage capacity basis. Typically, the packing densities of MOF powders are approximately 0.2 to 0.4 g cm^{-3}, increasing to 0.5–0.8 g cm^{-3} when shaped into tablets or extrudates. The density of this material is so low that weight limitation in an application is not relevant; this is in direct contrast, of course, to the alternative use of metal hydrides as storage media.

The most important issue here, of course, is the *quantity* of hydrogen which, within a reasonable timescale, can be discharged from the storage media, and within this context MOFs have a fully reversible uptake- and release-behavior. As the storage mechanism is based predominantly on physisorption, there are no huge activation energy barriers to be overcome when liberating the stored hydrogen. A simple pressure reduction by controlling the valve opening is sufficient to draw off hydrogen from MOFs within a few seconds.

Energy density values of 1.1 kWh L^{-1}, as requested in the European hydrogen and fuel cell Strategic Research Agenda and Deployment Strategy [38], are equivalent to a volumetric hydrogen stored capacity of about 33 g H$_2$ L^{-1}. As has been demonstrated, almost two-thirds of this value can be achieved by storing hydrogen in MOFs at 77 K and at a moderate pressure of 4–5 MPa. However, due to the low degree of interaction possibilities of hydrogen, and its low heat of adsorption, significant amounts of storage at room temperature have not yet been achieved.

Tab. 1 Summary of hydrogen adsorption in MOFs [8]

Material[a]	Free/fixed diameters[a,c] / Å	Accessible volume fraction[c]	Apparent surface area[d] /m^{2}g^{-1}	Pore volume[c] / cm^{3}g^{-1}	H$_2$ uptake/ wt.%	Conditions	References
Zn$_4$O(bdc)$_2$, IRMOF-1	7.8/15.2	0.59	3362	1.19	1.32	77 K, 1 atm	[22]
					1.0	RT, 20 bar	[23]
					1.65	RT, 48 atm	[24]
Zn$_4$O(R^6-bdc)$_2$, IRMOF-6	5.9/15.2	0.50	2630	0.93	1.0	RT, 10 bar	[23, 30]
Zn$_4$O(ndc)$_3$, IRMOF-8	8.4/18.0	0.66	1466	0.52	1.50	77 K, 1 atm	[22]
					2.0	RT, 10 atm	[23]
Zn$_4$O(hpdc)$_2$, IRMOF-11	6/12.4	0.40	1911	0.63	1.62	77 K, 1 atm	[22]
Zn$_4$O(tmbdc)$_3$, IRMOF-18	5.4/13.8	0.42	1501	0.53	0.89	77 K, 1 atm	[22]
Zn$_4$O(ntb)$_2$, MOF-177	9.6/11.8	0.63	4526	1.61	1.25	77 K, 1 atm	[22]
Al(OH)(bdc), MIL-53(Al)	6.4/6.4	0.29	1590, 1020[i]	–	3.8	77 K, 16 bar	[36, 45]
Cr(OH)(bdc), MIL-53(Cr)	6.6/6.6	0.29	1500, 1026[g]	–	3.1	77 K, 16 bar	[36, 46]
Mn(HCO$_2$)$_2$	3/4.7	0.10	297[h]	–	0.9	77 K, 1 atm	[26]
Cu$_2$(hfipbb)$_2$(H$_2$hfipbb)	3/4.7	0.03	–	–	1.0	RT, 48 atm	[24]
Ni(cyclam)(bpydc)	6.1/7.6	0.18	817	0.37	1.1	77 K, 1 atm	[47]
Zn$_2$(bdc)$_2$(dabco)	7.8/9.5	0.45	1450[i]	–	2.0	77 K, 1 atm	[48]
Ni$_2$(bpy)$_2$(NO$_3$)$_4$ (M)	2.4/4.0	0.05	–	0.181[i]	0.8	77 K, 1 atm	[49]
Ni$_2$(bpy)$_3$(NO$_3$)$_4$ (E)	2.1/4.2	0.05	–	0.149[j]	0.7	77 K, 1 atm	[49]
Ni$_2$(btc)$_3$(3-plc)$_6$(pd)$_2$	8.5/10.7	0.30	–	0.63	2.1	77 K, 14 bar	[49]
Zn$_4$O(L^1)$_2$	3.8/7.8	0.21	502[e]	0.20	1.12	RT, 48 atm	[28]
Zn$_4$O(L^2)$_3$	3.8/5.4	0.17	396[f]	0.13	0.98	RT, 48 bar	[28]
Cu$_2$(pzdc)$_2$(pyz), CPL-1	3.4/5.0	0.04	–	–	0.2	89 K, 1 atm	[40]
Cu$_2$(bptc), MOF-505	6.7/10.1	0.37	1646	0.63	2.48	77 K, 1 atm	[50]

[a]Acronyms: bdc-benzene-1,4-dicarboxylate, R^6-bdc = 1,2-dihydrocyclobutylbenzene-3,6-dicarboxylate, ndc = naphthalene-2,6-dicarboxylate, hpdc = 4,5,9,10-tetrahydropyrene-2,7-dicarboxylate, tmbdc = 2,3,5,6-tetramethylbenzene-1,4-dicarboxylate, btb = benzene-1,3,5-tribenzoate, hfipbb = 4,4'-(hexafluoroisopropylidene)bisbenzoate, cyclam = 1,4,8,11-tetraazacyclotetradecane, bpydc = 2,2'-bipyridyl-5,5'-dicarboxylate, dabco-1,4-diazablcyclo[2.2.2]octane, bpy = 4,4'-bipyridline, btc = benzene-1,3,5-tricarboxylate, 3-plc = 3-picoline, pd-1,2-propanediol, L^1 = 6,6'-dichloro-2,2'-diethoxy-1,1'-binaphthyl-4,4'-dibenzoate, L^2 = 6,6'-dichloro-2,2'-dibenzyloxy-1,1'-binaphthyl-4,4'-dibenzoate, bptc = biphenyl-3,3',5,5'-tetracarboxylate.
[b]Calculations were performed using the Census^2 software package. Crystallographic data for the evacuated frameworks was used where available.
[c]Free and fixed diameters correspond to the largest spheres that can pass through the apertures and fit in the largest pores of the frameworks, respectively.
[d]Calculated using a probe radius of 1.45 Å, which corresponds to the kinetic diameter of H$_2$.
[e]Calculated from N$_2$ adsorption data collected at 77 K using the Langmuir model except where indicated.
[f]BET surface area from N$_2$ at 77 K.
[g]BET surface area from CO$_2$ at 195 K.
[h]BET surface area from CO$_2$ at 273 K.
[i]Calculated from N$_2$ adsorption data collected at 77 K using the Dubirin-Radushkevich method except where indicated.
[j]Methanol used as adsorbate.

When the situation is clarified as to where exactly the hydrogen molecules are favorably bound and attached in the different MOFs, it should be possible to identify the most promising structures. In this respect, molecular modeling tools may become equally important as elaborate experimental syntheses [37]. Depending on the temperature of a possible application, highly porous MOFs might be favorable for low temperatures, whereas rather small-pore materials [16, 35], or highly attractive and flexible materials [22, 39], may be preferred for room temperature hydrogen storage.

More recent reports have addressed biomimetic approaches; an example is that of Co$_4$(μ_4-O)(carboxylate)$_4$ units [40] being related to the structure of active centers, such as Fe in hemoglobin. These materials have relatively high sorption enthalpies of 10 kJ mol^{-1} for hydrogen, and similarly very high enthalpies for O$_2$, CO, or CH$_4$.

2.3.8.5.6 **Catalysis** It is surprising, that although research efforts into the synthesis of materials has

References see page 261

Tab. 2 Summary of catalytic reactions over metal-organic frameworks

MOF type	Reaction type	Reactants	References
Ag@[$Zn_4O(BDC)_3$] (MOF-5)	Epoxidation	Propylene + O_2	[41]
Pt@[$Zn_4O(BDC)_3$] (MOF-5)	H_2O_2-Synthesis	$H_2 + O_2$	[42]
Cu@[$Zn_4O(BDC)_3$] (MOF-5)	Methanol synthesis	Syngas	[43]
Pd@[$Zn_4O(BDC)_3$] (MOF-5)	Hydrogenation	Cyclooctene + H_2	[43]
[$Mn(BTC)_2(H_2O)_6$]·[$Mn(phen)_2(H_2O)_2$]$_2$·($H_2O)_{24}$, [$Cu(BTC)_2(H_2O)_6$]·[$Mn(phen)_2(H_2O)_2$]$_2$·($H_2O)_{22}$	Oxidation	Phenol + H_2O_2	[64]
$Ti(O^iPr)_4$/[$Cd_3Cl_6(L1)_3$]·$(DMF)_4(MeOH)_6(H_2O)_3$	Addition to carbonyls	$ZnEt_2$ + Aromatic aldehydes	[65]
[$Ti_2(^iPrO)_2Cl_2(L2)$]	Diels–Alder reaction	Acrolein + 1,3-cyclohexadiene	[66]
$Ti(O^iPr)_4$/[$Zr(L3)$]	Addition to carbonyls	$ZnEt_2$ + Aromatic aldehydes	[67]
[$Zr(L4)$], [$Zr(L5)$]	Hydrogenation	Aromatic ketones + H_2	[68]
[$Zr(L6)$], [$Zr(L7)$]	Hydrogenation	β-Ketoesters + H_2	[69]
[$Rh_2(M^{2+}TCPP)_2$] (M^{2+} = Cu, Ni, Pd)	Hydrogenation	Propene + H_2	[46]
		1-Butene + H_2	[47]
[$Zn_2(BPDC)_2(L8)$]*10DMF*8H_2O	Epoxidation	2,2-Dimethyl-2H-chromene + 2-(tert-butylsulfonyl)iodosylbenzene	[48]
[$Zn_3O(L9$-H$)$]·($H_3O)_2(H_2O)_{12}$ (D-POST-1)	Transesterification	Acetic acid 2,4-dinitrophenyl ester + EtOH	[70]
[$Sm(L10$-H$_2)(L10$-H$_3)(H_2O)_4$]*($H_2O)_x$	Esterification	Meso-2,3-dimethylsuccinic anhydride + MeOH	[55]
[$Zn_4O(BDC)_3$] (MOF-5)	Alkoxylation	Dipropylene glycol + PO, Methyl dipropylene Glycol + EO, Acrylic acid + EO	[54]
Zn-DHBDC-MOF, Zn-BTC-MOF	Polyalkylene carbonate formation	PO + CO_2	[71]
[$Ln(OH)(H_2O)$(naphthalenedisulfonate)], Ln = Nd, Pr, La	Oxidation	Linanool + H_2O_2	[72]
[$In_4(OH)_6(BDC)_3$]	Oxidation	Methylphenylsulfide, (2-ethylbutyl)phenylsulfide + H_2O_2	[52]
Pd-($L11$)-MOF	Oxidation	Benzyl alcohol + air	[73]
[$Cu_3(BTC)_2$]	Oxidation	Polyphenols + H_2O_2	[74]
[$Sc_2[BDC]_3$]	Oxidation	Methylphenylsulfide + H_2O_2	[53]
[$In_2(BDC)_3(bpy)_2$], [$In_2(BDC)_2(OH)_2(phen)_2$], [$In(BTC)(H_2O)(bpy)$], [$In(BTC)(H_2O)(phen)$]	Acetalization	Benzaldehyde + Trimethylorthoformate	[75]
[$Cu_3(BTC)_2$]	Cyanosilylation	Benzaldehyde + Cyanotrimethylsilane	[13]
[$Cd(4,4'$-bpy$)_2(H_2O)_2$]($NO_3)_2$·($H_2O)_4$	Cyanosilylation	Benzaldehyde, Aromatic imines + Cyanotrimethylsilane	[76, 77]
[$Sm(L10$-H$_2)(L10$-H$_3)(H_2O)_4$]*($H_2O)_x$	Cyanosilylation	Benzaldehyde + Cyanotrimethylsilane	[55]
[$Zn_4O(BDC)_3$] (MOF-5)	Addition to Alkynes	4-tert-butylbenzoic acid + C_2H_2, MeOH + propyne	[12]
[$Ti_2O_2(L12)$]	Carbonyl-ene reaction	Acetoxy-acetaldehyde + isopropenylenzene	[78]
Li[Al($L12$)]	Michael reaction	Cyclohexenone + dibenzyl malonate	[78]
Ti-(2,7-dihydroxynaphthalene)-MOFs	Ziegler–Natta polymerization	Ethylene, Propylene	[79]
[$Cu_3(BTC)_2$]	Isomerization	α-Pinene oxide	[80]
[$Cu_3(BTC)_2$]	Cyclization	Citronellal	[80]
[$Cu_3(BTC)_2$]	Rearrangement	2-Bromopropiophenone	[80]
[$Rh_2(L13)$]	Hydrogenation	Ethylene, Propylene + H_2	[81]
[Rh(BDC)], [Rh(fumarate)]	H-D exchange, Hydrogenation	Propylene + H_2	[50]

Tab. 2 (Continued)

MOF type	Reaction type	Reactants	References
[Ru(1,4-diisocyanobenzene)$_2$]Cl$_2$	Hydrogenation	1-Hexene + H$_2$	[82, 83]
[Rh(diisocyanobiphenyl)$_2$(H$_2$O)$_{2.53}$]Cl	Hydrogenation	1-Hexene + H$_2$	[84]
[In$_4$(OH)$_6$(BDC)$_3$]	Hydrogenation	1-Nitro-2-Methylnaphthalene, Nitrobenzene + H$_2$	[52]
[Ru$_2$(BDC)$_2$], [Ru$_2$(BPDC)$_2$], [Ru$_2$(BDC)$_2$(dabco)], [Ru$_2$(BPDC)$_2$(dabco)]	Hydrogenation	Ethylene + H$_2$	[51]
[Rh$_2$(fumarate)$_2$], [Rh$_2$(BDC)$_2$], [Rh$_2$(H$_2$TCPP)$_2$]	Hydrogenation H-D Exchange	Ethylene + H$_2$ Propylene + H$_2$	[46]

Abbreviations: BDC = 1,4-Benzenedicarboxylate; BTC = 1,3,5-Benzenetricarboxylate; bpy = 4,4'-bipyridine; phen = phenanthroline; BPDC = Biphenyldicarboxylate; DHBDC = 2,5-Dihydroxy-1,4-benzenedicarboxylate; TCPP = 4,4′,4″,4‴-(21H, 23H-porphine-5,10,15,20-tetrakis benzoicdicarboxylate).

afforded hundreds of different new MOFs during recent years, very few examples are available for catalytic applications. Selected samples from the literature are summarized in Table 2, and these may be subdivided into at least three different types.

First, MOFs have simply been employed as a support material for catalytically active species, such as precious metal particles. Advantages may be expected from the high intrinsic surface area and the large diffusion coefficients of reactants within the MOF. For example, BASF recently filed patent applications, where silver or platinum were supported on MOF-5 [41, 42]. The resulting catalysts were active for the epoxidation of propylene and the direct synthesis of H_2O_2 from the elements, respectively. Likewise, others deposited Pd and Cu particles by a chemical vapor deposition (CVD) method on MOF-5 [43]. The latter showed a remarkable activity for methanol synthesis from syngas, as it showed high conversion in the absence of ZnO. From the recent appearance of thermally and chemically more stable framework materials, such as Al–BDC–MOF [44] or Mg–NDC–MOF [45], a new stimulus for such strategies might be expected.

Second, the framework structure is used as a medium in order to heterogenize molecular transition metal catalysts. The latter may be either part of a rather complex linker unit, or they can be introduced subsequently by anchoring on functional groups in the basic framework. For instance, Mori and coworkers reported details on Rh–MOFs with porphin-based linker units which showed a high activity for the hydrogenation of olefins [46, 47]. Recently, the enantioselective epoxidation of olefins was carried out on a framework with chiral manganese–salen struts [48]. Further examples of adequate synthesis strategies, with the focus being set on the development of enantioselective structures for heterogeneous catalysis, have been developed in the research group of Lin during recent years [49].

Third, in some studies more simply structured MOFs are also directly employed as catalysts for a variety of reactions. Some reactions, such as hydrogenations, are expected to proceed at free metal coordination sites, available for example in paddle-wheel structures, which are typical for Ru–or Rh–MOFs [46, 50, 51]. However, other reactions most likely utilize unique active sites, which are formed by the interaction of the individual node and linker elements. For example, the first promising results were published for oxidation reactions [52, 53], alkoxylation [54] or acid-catalyzed reactions, such as esterification [55] or addition reactions [12]. In the near future, ring-opening reactions starting from epoxides [56], as well as multisite catalysis, might become possible on Co–MOFs, while some of them can also be simply tuned by varying linkers or cobalt configuration. With a growing understanding of the coupling of CO_2 and epoxides on dinuclear zinc complexes [57], it can be foreseen that even reaction mechanisms on zinc paddlewheels containing MOFs may be elucidated in the near future.

Clearly, the industrial use of MOFs in catalysis must also be applied to investigating time-on-stream behavior, recycle, and reuse. Thus, additional studies will need to be performed in order to identify viable and profitable applications for the long term.

2.3.8.6 Conclusions

In this chapter, we have discussed the point that MOFs or coordination polymers are not merely a new class of porous materials in terms of having combined inorganic and organic chemistry classifications. Rather, from an industrial viewpoint they offer – both in principle and for the first time – many interesting and promising features, including:

- ultimate nanoporosity with a total absence of blocked volume in solid matter
- fully exposed metal sites in solids
- world record values of weight-specific surface areas
- combined flexibility plus robust frameworks
- a high mobility of guest species in regular framework nanopores
- fast-growing numbers (hundreds) of novel inorganic–organic compositions.

Clearly, many applications might – and surely will – be tested and reported in the literature, as soon as verified recipes for syntheses using MOFs are available. Although the procedures provided in Ref. [14] will allow the reader to prepare these new compounds in laboratory-scale quantities, their industrial synthesis at BASF has now advanced to the pilot scale and commercial efforts.

Unlike many other novel materials, such as carbon polymorphs, fullerenes, bucky-balls, and carbon nanotubes (CNT), the preparation and fabrication of MOF materials does not necessarily require additional capital investment into new synthesis technology. Rather, a simple adaptation of conventionally available precipitation and crystallization manufacturing methods is sufficient. Likewise, the shaping of MOF powders into industrially widespread geometries of tablets, extrudates, honeycombs, etc., can also be carried out without major obstacle.

The examples provided here for catalysis and for gas processing and storage show that there is still plenty of scope for future research efforts. It should be noted, however, that – unlike state-of-the-art solid catalysts – the metal sites in MOFs are fully exposed, and so provide an ultimately high degree of metal dispersion. Compared to zeolites, the metal content of MOFs is almost 10-fold higher, with the majority of the metal species being transition metals. Preliminary results in catalysis suggest that the use of solid MOFs will compare favorably with organometallic homogeneous compounds or single-site catalysis.

In summary, the aforementioned issues should lead to a rapidly growing, prosperous and widespread innovation in materials science, in areas of both academia and industry. Attention will be focused on identifying superior performance by applying MOFs over state-of-the-art technologies, and seeking improvement. Such an approach will surely lead to sustainable innovation and value-added growth for Society in general.

References

1. E. A. Tomic, *J. Appl. Polym. Sci.* **1965**, *9*, 3745.
2. H. Li, M. Eddaoudi, M. O'Keeffe, O. M. Yaghi, *Nature* **1999**, *402*, 276.
3. M. Eddaoudi, J. Kim, N. Rosi, D. Vodak, J. Wachter, M. O'Keefe, O. M. Yaghi, *Science* **2002**, *295*, 469.
4. O. M. Yaghi, M. Eddaoudi, H. Li, J. Kim, N. Rosi, WO 2002/088148, assigned to University of Michigan, 2002.
5. H. K. Chae, D. Y. Siberio-Pérez, J. Kim, Y. B. Go, M. Eddaoudi, A. J. Matzger, M. O'Keeffe, O. M. Yaghi, *Nature* **2004**, *427*, 523.
6. S. Kitagawa, R. Kitaura, S. Noro, *Angew. Chem. Int. Ed.* **2004**, *43*, 2334.
7. J. L. C. Rowsell, O. M. Yaghi, *Microporous Mesoporous Mater.* **2004**, *73*, 3.
8. J. L. C. Rowsell, O. M. Yaghi, *Angew. Chem.* **2005**, *117*, 4748 and *Angew. Chem. Int. Ed.* **2005**, *44*, 4670.
9. T. J. Barton, L. M. Bull, W. G. Klemperer, D. A. Loy, B. McEnaney, M. Misono, P. A. Monson, G. Pez, G. W. Scherer, J. C. Vartuli, O. M. Yaghi, *Chem. Mater.* **1999**, *11*, 2633.
10. G. Férey, C. Mellot-Draznieks, C. Serre, F. Millange, J. Dutour, S. Surblé, I. Margiolaki, *Science* **2005**, *309*, 2040.
11. U. Mueller, H. Puetter, M. Hesse, H. Wessel, WO 2005/049892, assigned to BASF Aktiengesellschaft, 2005.
12. U. Mueller, M. Hesse, L. Lobree, M. Hoelzle, J. D. Arndt, P. Rudolf, WO 2002/070526, assigned to BASF Aktiengesellschaft, 2002.
13. K. Schlichte, T. Kratzke, S. Kaskel, *Microporous Mesoporous Mater.* **2004**, *73*, 81.
14. U. Mueller, M. Schubert, F. Teich, H. Puetter, K. Schierle-Arndt, J. Pastré, *J. Mater. Chem.* **2006**, *16*, 626.
15. N. L. Rosi, J. Eckert, M. Eddaoudi, D. T. Vodak, J. Kim, M. O'Keefe, O. M. Yaghi, *Science* **2003**, *300*, 1127.
16. Y. Kubota, M. Takata, R. Matsuda, R. Kitaura, S. Kitagawa, K. Kato, M. Sakata, T. C. Kobayashi, *Angew. Chem.* **2005**, *117*, 942 and *Angew. Chem. Int. Ed.* **2005**, *44*, 920.
17. T. Yildirim, M. R. Hartman, *Phys. Rev. Lett.* **2005**, *95*, 215504.
18. T. Mueller, G. Ceder, *J. Phys. Chem. B* **2005**, *109*, 17974.
19. L. Pan, B. Parker, X. Huang, D. H. Olson, J. Y. Lee, J. Li, *J. Am. Chem. Soc.* **2006**, *128*, 4180.
20. T. K. Maji, G. Mostafa, H. Ch. Chang, S. Kitagawa, *Chem. Commun.* **2005**, 2436.
21. L. Pan, D. H. Olson, L. R. Ciemnolonski, R. Heddy, J. Li, *Angew. Chem. Int. Ed.* **2006**, *45*, 616.
22. G. Férey, M. Latroche, C. Serre, F. Millange, T. Loiseau, A. Percheron-Guégan, *Chem. Commun.* **2003**, 2976.
23. R. Kitaura, K. Fujimoto, S. Noro, M. Kondo, S. Kitagawa, *Angew. Chem.* **2002**, *114*, 141 and *Angew. Chem. Int. Ed.* **2002**, *41*, 133.
24. X. Zhao, B. Xiao, A. J. Fletcher, K. M. Thomas, D. Bradshaw, M. J. Rosseinsky, *Science* **2004**, *306*, 1012.
25. U. Muller, K. K. Unger, in *Characterization of Porous Solids*, K. K. Unger, J. Rouquerol, K. S. W. Sing, H. Kral (Eds.), Studies in Surface Science and Catalysis, Vol. 39, Elsevier, Amsterdam, 1988, p. 101.
26. F. Stallmach, S. Groeger, V. Kuenzel, J. Kaerger, O. M. Yaghi, M. Hesse, U. Mueller, *Angew. Chem.* **2006**, *118*, 2177 and *Angew. Chem. Int. Ed.* **2006**, *45*, 2123.
27. U. Mueller, M. Hesse, H. Puetter, M. Schubert, D. Mirsch, EP 1,674,555, assigned to BASF Aktiengesellschaft, 2005.
28. R. Matsuda, R. Kitaura, S. Kitagawa, Y. Kubota, R. V. Belosludov, T. C. Kobayashi, H. Sakamoto, T. Chiba, M. Takata, Y. Kawazoe, Y. Mita, *Nature* **2005**, *436*, 238.
29. A. G. Wong-Foy, A. J. Matzger, O. M. Yaghi, *J. Am. Chem. Soc.* **2006**, *128*, 3494.
30. D. N. Dybtsev, H. Chun, K. Kim, *Angew. Chem.* **2004**, *116*, 5143 and *Angew. Chem. Int. Ed.* **2004**, *43*, 5033.
31. K. Seki, W. Mori, *J. Phys. Chem. B* **2002**, *106*, 1380.
32. B. Chen, N. W. Ockwig, A. R. Millward, D. S. Contreras, O. M. Yaghi, *Angew. Chem.* **2005**, *117*, 4823 and *Angew. Chem. Int. Ed.* **2005**, *44*, 4745.
33. E. Y. Lee, S. Y. Jang, M. P. Suh, *J. Am. Chem. Soc.* **2005**, *127*, 6374.
34. G. Férey, C. Serre, C. Mellot-Draznieks, F. Millange, S. Surblé, J. Dutour, I. Margiolaki, *Angew. Chem.* **2004**, *116*, 6456 and *Angew. Chem. Int. Ed.* **2004**, *43*, 6296.
35. D. N. Dybtsev, H. Chun, S. H. Yoon, D. Kim, K. Kim, *J. Am. Chem. Soc.* **2004**, *126*, 32.
36. C. Serre, F. Millange, S. Surblé, G. Férey, *Angew. Chem.* **2004**, *116*, 6446 and *Angew. Chem. Int. Ed.* **2004**, *43*, 6286.
37. Q. Yang, Ch. Zhong, *J. Phys. Chem. B* **2005**, *109*, 11862.
38. S. Barrett, *Fuel Cells Bulletin*, 2005, May, 12.
39. K. Uemura, R. Matsuda, S. Kitagawa, *J. Solid State Chem.* **2005**, *178*, 2420.
40. S. Ma, H.-C. Zhou, *J. Am. Chem. Soc.* **2006**, *128*, 11734.
41. U. Mueller, L. Lobree, M. Hesse, O. M. Yaghi, M. Eddaoudi, WO 03/101975, assigned to BASF Aktiengesellschaft/University of Michigan, 2003.
42. U. Mueller, O. Metelkina, H. Junicke, T. Butz, O. M. Yaghi, US 2004/081611, assigned to BASF Aktiengesellschaft/University of Michigan, 2004.
43. S. Hermes, M.-K. Schröter, R. Schmid, L. Khodeir, M. Muhler, A. Tissler, R. W. Fischer, R. A. Fischer, *Angew. Chem.* **2005**, *117*, 6394 and *Angew. Chem. Int. Ed.* **2005**, *44*, 6237.
44. T. Loiseau, C. Serre, C. Huguenard, G. Fink, F. Taullele, M. Henry, T. Bataille, G. Férey, *Chem. Eur. J.* **2004**, *10*, 1373.
45. M. Dinca, J. R. Long, *J. Am. Chem. Soc.* **2005**, *127*, 9376.
46. W. Mori, T. Sato, T. Ohmura, C. N. Kato, T. Takei, *J. Solid State Chem.* **2005**, *178*, 2555.
47. T. Sato, W. Mori, C. N. Kato, E. Yanaoka, T. Kuribayashi, R. Ohtera, Y. Shiraishi, *J. Catal.* **2005**, *232*, 186.
48. S.-H. Cho, B. Ma, S. T. Nguyen, J. T. Hupp, T. E. Albrecht-Schmitt, *Chem. Commun.* **2006**, 2563.
49. H. L. Ngo, W. Lin, *Top. Catal.* **2005**, *34*, 85.
50. S. Naito, T. Tanibe, E. Saito, T. Miyao, W. Mori, *Chem. Lett.* **2001**, 1178.
51. T. Ohmura, W. Mori, H. Hiraga, M. Ono, Y. Nishimota, *Chem. Lett.* **2003**, *32*, 468.

52. B. Gomez-Lor, E. Guiterrez-Puebla, M. Iglesias, M. A. Monge, C. Ruiz-Valero, N. Snejko, *Inorg. Chem.* **2002**, *41*, 2429.
53. J. Perles, M. Iglesias, M.-A. Martín-Luengo, M. Á. Monge, C. Ruiz-Valero, N. Snjeko, *Chem. Mater.* **2005**, *17*, 5837.
54. U. Mueller, M. Stoesser, R. Ruppel, E. Baum, E. Bohres, M. Sigl, L. Lobree, O. M. Yaghi, M. Eddaoudi, WO 03/035717, assigned to BASF Aktiengesellschaft, 2003.
55. O. R. Evans, H. L. Ngo, W. Lin, *J. Am. Chem. Soc.* **2001**, *123*, 10395.
56. F. Molnar, G. A. Luinstra, M. Allmendinger, B. Rieger, *Chem. Eur. J.* **2003**, *9*, 1273.
57. H. S. Kim, J. J. Kim, S. D. Lee, M. S. Lah, D. Moon, H. G. Jang, *Chem. Eur. J.* **2003**, *9*, 678.
58. J. L. C. Rowsell, A. R. Millward, K. S. Park, O. M. Yaghi, *J. Am. Chem. Soc.* **2004**, *126*, 5666.
59. L. Pan, M. B. Sander, X. Huang, J. Li, M. Smith, E. Bittner, B. Bockrath, J. K. Johnson, *J. Am. Chem. Soc.* **2004**, *124*, 1308.
60. F. Millange, C. Serre, G. Férey, *Chem. Commun.* **2002**, 822.
61. E. Y. Lee, M. P. Suh, *Angew. Chem. Int. Ed.* **2004**, *43*, 2798.
62. X. Zhao, B. Xiao, A. J. Fletcher, K. M. Thomas, D. Bradshaw, M. J. Rosseinsky, *Science* **2004**, *306*, 1012.
63. B. Kesanli, Y. Cui, M. R. Smith, E. W. Bittner, B. C. Bockrath, W. Lin, *Angew. Chem.* **2005**, *117*, 74 and *Angew. Chem. Int. Ed.* **2005**, *44*, 72.
64. L.-G. Qiu, A.-J. Xie, L.-D. Zhang, *Adv. Mater.* **2005**, *17*, 689.
65. C.-D. Wu, A. Hu, L. Zhang, W. Lin, *J. Am. Chem. Soc.* **2005**, *127*, 8940.
66. T. Sawaki, T. Dewa, Y. Aoyama, *J. Am. Chem. Soc.* **1998**, *120*, 8539.
67. H. L. Ngo, A. Hu, W. Lin, *J. Mol. Catal. A: Chemical* **2004**, *215*, 177.
68. A. Hu, H. L. Ngo, W. Lin, *J. Am. Chem. Soc.* **2003**, *125*, 11490.
69. A. Hu, H. L. Ngo, W. Lin, *Angew. Chem. Int. Ed.* **2003**, *42*, 6000.
70. J. S. Seo, D. Whang, H. Lee, S. I. Jun, J. Oh, Y. J. Jeon, K. Kim, *Nature* **2000**, *404*, 982.
71. U. Mueller, G. Luinstra, O. M. Yaghi, O. Metelkina, M. Stoesser, WO 2004/037895, assigned to BASF Aktiengesellschaft/The Regents of the University of Michigan, 2004.
72. E. Guitérrez-Puebla, C. Cascales-Sedano, B. Gómez-Lor, M. M. Iglesias-Hernandez, M. A. Monge-Bravo, C. Ruiz-Valero, N. Snejko, ES 2,200,681, assigned to Consejo Superior de Investigaciones Cientificas Serrano, 2004.
73. B. Xing, M.-F. Choi, B. Xu, *Chem. Eur. J.* **2002**, *8*, 5028.
74. S. De Rosa, G. Giordano, T. Granato, A. Katovic, A. Siciliano, F. Tripicchio, *J. Agric. Food Chem.* **2005**, *53*, 8306.
75. B. Gómez-Lor, E. Guitérrez-Puebla, M. Iglesias, M. A. Monge, C. Ruiz-Valero, N. Snjeko, *Chem. Mater* **2005**, *17*, 2568.
76. M. Fujita, Y. J. Kwon, S. Washizu, K. Ogura, *J. Am. Chem. Soc.* **1994**, *116*, 1151.
77. O. Ohmori, M. Fujita, *Chem. Commun.* **2004**, 1586.
78. S. Takizawa, H. Somei, D. Jayaprakash, H. Sasai, *Angew. Chem. Int. Ed.* **2003**, *42*, 5711.
79. J. M. Tanski, P. T. Wolczanski, *Inorg. Chem.* **2001**, *40*, 2026.
80. L. Alaerts, E. Séguin, H. Poelman, F. Thibault-Starzyk, P. A. Jacobs, D. E. De Vos, *Chem. Eur. J.* **2006**, *12*, 7353.
81. T. Sato, W. Mori, C. N. Kato, T. Ohmura, T. Sato, K. Yokoyama, S. Takamizawa, S. Naito, *Chem. Lett.* **2003**, *32*, 854.
82. R. Tannenbaum, *Chem. Mater.* **1994**, *6*, 550.
83. R. Tannenbaum, *J. Molec. Catal. A: Chemical* **1996**, *107*, 207.
84. I. Feinstein-Jaffe, A. Efraty, *J. Molec. Catal.* **1987**, *40*, 1.

2.3.9
Oxo-Anion Modified Oxides

*Friederike C. Jentoft**

2.3.9.1 Introduction

2.3.9.1.1 Classification In the interaction of two or more catalyst components, two extremes can be found: the formation of a compound or a solid solution constitutes the most intimate and homogeneous type of interaction, while the formation of a non-wetting surface species on an inert support can be seen as the minimal type of interaction. For supported catalysts, the goal is usually the dispersion of the active species; the structural integrity of the support remains unchanged, although strong interactions may occur at the interface of the support and the dispersed phase (see Chapter 3.2.5). In this sense, anion-modified oxides such as sulfated or tungstated zirconia or sulfated titania could simply be understood as supported systems, specifically as mounted acids. In a typical preparation route, however, the second component (sulfate, tungstate) is added early in the preparation and is already present when the support or, better, "matrix oxide" crystallizes during the thermal treatment. As a consequence, the textural and structural properties (zirconia features a particularly vivid phase chemistry) of the matrix oxide are severely influenced by the additive. If further components such as promoters are added, the situation will become more complex. Although the product might be considered a surface-functionalized oxide, the systems are characterized by a strong interaction of the functionalizing agent and the matrix, leading to a mutual directing of their structures. The final product ideally contains only the matrix oxide as crystalline phase. This characteristic distinguishes these systems from coprecipitated catalysts such as Ni/Al_2O_3 and Cu/ZnO, which also feature precursors with intimate mixing of the components, but the final product obtained after a reduction step consists of separate phases, which are individually detectable by diffraction. Oxo-anion-modified oxides can thus be viewed as a separate class of catalysts.

2.3.9.1.2 Variety of Materials and Focus The lead system discussed in this chapter is sulfated zirconia but other varieties of this type of catalyst, that is, other anions and other oxides are also discussed. Sulfated zirconia was first described in a patent by Holm and Bailey in 1962 [1]. Profound interest in the system was raised years later through two articles by Hino and Arata, who reported

---

* Corresponding author.

sulfated zirconia to be active for *n*-butane isomerization at room temperature; that is, under conditions that thermodynamically favor the desired branched isomer [2, 3]. Commercialization of sulfated zirconia by UOP has been reported [4, 5], emphasizing the significance of this catalyst. Sulfated zirconia can be employed for a number of different reactions, most of which are acid-catalyzed processes; e.g. alkylation, condensation, etherification, acylation, esterification, nitration, and oligomerization [6].

Both catalyst components can be varied, i.e. (i) zirconia can be mixed with other oxides [7], or other matrix oxides or mixtures thereof can be employed and (ii) other oxo-anions can be used. For example, sulfated TiO_2, HfO_2, SnO_2 and Fe_2O_3 are each active for the skeletal isomerization of *n*-butane to isobutane at low temperatures [8]. Sulfated SiO_2 is active for the dehydration of ethanol [8]. ZrO_2 combined with tungstate is a catalyst for alkane isomerization [9, 10] and cracking [11], ZrO_2 with molybdate for conversion of hexane or benzoylation of toluene with benzoic anhydride [10] and ZrO_2 with borate for benzoylation of anisole with benzoyl chloride [12]. Materials obtained by treating ZrO_2 with SeO_4^{2-}, TeO_4^{2-} and CrO_4^{2-} are active for alkane dehydrogenation [8]. As a selection of combinations containing neither sulfate nor zirconia, tungstated SnO_2, TiO_2 and Fe_2O_3 may be named [10]; numerous other combinations can be imagined.

The activity of sulfated zirconia for butane isomerization can be further improved by orders of magnitude through the addition of cations of Fe and Mn [13, 14], or, e.g., Co or Ni [15]. Equally, tungstated zirconia can be promoted with, for example, Fe [16]. In general, sulfated or tungstated zirconia catalysts deactivate rapidly [17, 18], but by adding a noble metal to the catalyst, and hydrogen to the feed, the isomerization activity can be stabilized [16, 19, 20]. A complete catalyst thus consists of up to four types of components: the matrix oxide(s), the oxo-anion(s), promoters (optionally), and noble metal(s).

2.3.9.2 Target Properties

Three types of correlations have been made, namely between (i) performance and a measured or known physical quantity, (ii) performance and a preparation parameter or an observable that is characteristic of the preparation procedure, and (iii) physical quantities and a preparation parameter. Correlations of the first type allow for a targeted synthesis but still one must identify relationships between physical quantities and preparation parameters.

2.3.9.2.1 Structure–Activity Relationships
Unless they are noble-metal doped and operated in the presence of H_2, sulfated zirconia catalysts usually do not exhibit a stable performance. Thus, for any correlation it should be specified whether the maximum or steady-state activity has been evaluated.

The phase composition of zirconia has been identified as important. Zirconia occurs as three polymorphs, with the monoclinic phase being the room-temperature-stable modification. Above 1443 K, the tetragonal modification is preferred, whereas beyond 2643 K the cubic phase becomes stable [21]. The monoclinic phase of zirconia has been reported as inactive [22], or at least a factor of four to five less active than tetragonal zirconia [23]. Consistent with insignificant activity of the monoclinic phase, maximum *n*-butane conversion (or isomerization rate) was found to be proportional to the fraction of tetragonal phase [24, 25]. However, the presence of the tetragonal phase is only a necessary but not a sufficient requirement for good catalytic activity [26]; that is, not all tetragonal materials are automatically good catalysts. Sulfated zirconia that does not exhibit a distinct diffraction pattern has been claimed to be more [27] or less [28] active than conventional sulfated zirconia.

Catalytic activity is related to the surface area, for example in the case of butane isomerization [29, 30] or toluene cracking [31]. For the liquid-phase acylation of anisole with benzoyl chloride, the conversion is linearly related to the surface area, and increases monotonously with pore diameter in the range of 4.5 to 10.5 nm [32].

The sulfur content is also an essential parameter. For toluene benzylation, conversion increases linearly in the range of 0.01 to 0.023% S per $m^2 \, g^{-1}$, which should correspond to densities of two to four S atoms per nm^2 [33]. However, the conversion of methylcyclopentane passes through a maximum at an intermediate sulfate content [34]. The *n*-butane isomerization activity of Pt-doped sulfated zirconia increases with increasing sulfur content up to a level of ca. one S atom per nm^2, and then remains constant [35]. Other sources report distinct maxima in the isomerization rate of sulfated zirconia for sulfur contents of 170 μg m^{-2} (3 S atoms per nm^{-2}) [37], or 1–2 wt.% S [38], or 2.6 wt.% S [30]. There is evidence that a disulfate surface structure ($S_2O_7^{2-}$) is relevant for good performance in *n*-butane isomerization, and this condensation product is only formed at high sulfate surface density upon dehydration [38–42].

The highest butane isomerization rates per m^2 are observed at a Brønsted-to-Lewis site ratio of 1 after activation (as determined by pyridine adsorption) [36]. The Brønsted-to-Lewis sites ratio increases with increasing sulfate content. Unfortunately, the correctness of the number of Lewis sites as obtained by pyridine adsorption has been questioned, as this strong nucleophile may substitute for sulfate as a ligand of Zr ions and the Lewis acid sites will be overdetermined [43]. Correlations of the

References see page 275

reaction rate to the total number of acid sites as measured by ammonia adsorption are not always convincing; for example, a poor correlation was obtained for an acylation reaction [44], but good correlations were observed for 2-propanol dehydration and cumene dealkylation [45]. On the other hand, the n-butane isomerization rate was found to be proportional to the number of Brønsted acid sites as determined by pyridine adsorption [46].

Therefore, a preparative goal should be to maximize the fraction of tetragonal zirconia which, in the overwhelming proportion of reports, has been found more active in comparison to monoclinic or amorphous zirconia. A second goal is to maximize the surface area. Attempts have been made simply to support sulfated zirconia (or other oxo-anion modified oxides) on oxides with a higher surface area (such systems are not described here, but further information is available elsewhere [47–51]). The target with respect to the sulfate content is less clear. There appears to be an optimum, because Brønsted and Lewis sites are both necessary and the fraction of Brønsted sites increases with increasing sulfate content. On the other hand, a disulfate structure seems to be desirable, and such a condensed species forms preferably with increasing sulfate surface density. Furthermore, the described properties are not independent of each other due to the strong interaction between oxo-anion and matrix oxide.

2.3.9.2.2 Effect of Sulfate and Other Oxo-Anions

The sulfate content influences the structural and textural properties of the final product. While pure zirconia, depending upon temperature, crystallizes into a mixture of monoclinic and tetragonal zirconia, the tetragonal phase becomes increasingly predominant as the sulfate content is raised [52]. No general statement can be made as to the amount of sulfate necessary to obtain solely tetragonal zirconia, because the phase composition depends also strongly on the calcination conditions (*vide infra*).

The surface area was found to be a linear function of the sulfate content up to about 5 wt.% SO_4^{2-}, with a slope of 30 m^2 g^{-1} per wt.% SO_4^{2-} for ZrO_2, TiO_2 and SnO_2, and a slope of 8 m^2 g^{-1} per wt.% SO_4^{2-} for Al_2O_3 and Fe_2O_3 [31]. The linear correlation can be interpreted as the stabilization of a certain area per sulfate group, namely 0.5 nm^2 for oxides with the stoichiometry MO_2, or 0.14 nm^2 for the M_2O_3 type. Other studies confirm the surface area increase up to a content of 4–6 wt.% SO_4^{2-}; at higher loadings, the data are not consistent [37, 53]. Anions such as sulfate, tungstate, and molybdate have the same effect, as can be seen in Fig. 1 [52, 54]. The relative weights of WO_3, MoO_3, and SO_4^{2-} explain the offset of the curves with respect to content. The attainable surface area, however, is considerably higher for WO_3/ZrO_2 and MoO_3/ZrO_2. For sulfated zirconia, the surface area decreases at higher calcination temperatures, and the maximum shifts to lower sulfate contents, as Fig. 2 demonstrates.

At very high sulfate contents, the surface area decreases (see Fig. 1), and the size of the tetragonal crystallites increases (based on analysis of the line width with the Scherrer formula) [55, 56]. No systematic trend in the porosity with the sulfur content could be found [52]. A maximum in the number of Brønsted sites (as measured by pyridine adsorption) is found at around four S atoms per nm^2 [57].

A high sulfate content is desirable because the active condensed sulfate structure seems to form only when

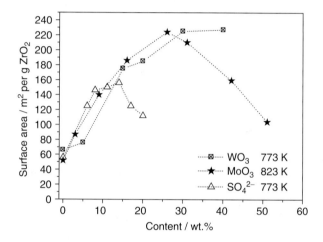

Fig. 1 Surface area as a function of modifier content, with modifier type as a parameter. (After Ref. [54].)

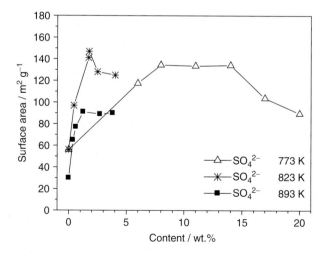

Fig. 2 Surface area as a function of sulfate content, with calcination temperature as a parameter. (After Refs. [35, 52, 54].)

a sufficient sulfate density is reached on the surface; furthermore, the surface area increases with sulfate content. However, very high sulfate contents have negative effects. The sulfate content may be controlled through the sulfation procedure only if it is ensured that there will be no sulfur loss during calcination.

2.3.9.3 Preparation Routes

The typical preparation route of sulfated zirconia and related materials follows the classical sequence according to IUPAC [58]: formation of a primary solid with all components, thermal treatment, and activation. The principles of preparing a precursor for the matrix oxide by precipitation or sol–gel synthesis are presented in Chapters 2.3.3 and 2.3.4. of this handbook. The procedures for dispersing noble metals are discussed in Chapter 2.4.2. However, oxo-anions, promoters, and noble metals can be associated at different stages of the preparation (see Fig. 3). The main theme of this chapter is to develop an understanding for the interaction of oxide precursor, oxo-anion, and promoters during formation of the primary solid or during thermal treatment and for the ensuing solid state chemistry, and from this knowledge to recommend suitable compositions and preparation procedures.

2.3.9.4 Formation of Primary Solid

To produce a suitable precursor for a high-surface-area zirconia, the starting point is usually a zirconium species dissolved in liquid phase. The zirconium reagent may be either an inorganic salt or an organyl.

Zirconium salt solutions have a number of industrial uses [59, 60] and their chemistry has been studied extensively [61–73]. In general, zirconium(IV) cations have a strong tendency to hydrolyze in aqueous medium; that is, the solutions are not stable and typically turn cloudy, which makes it difficult to exert control. Attempts have been made to identify the complexes that are formed in the hydrolysis reaction, as well as the species formed in subsequent reactions. Research has focused on solutions containing anions such as chloride, nitrate, and perchlorate, which are neither strongly complexing nor bridging. A discrete tetrameric complex, $[Zr_4(OH)_8(H_2O)_{16}]^{8+}$ is present; it can be neutralized through attachment of eight chloride ions [62, 63]. The tetrameric complex has been identified in zirconium oxychloride solutions [64, 68, 71], in zirconium perchlorate solutions [66], and in zirconium nitrate solutions [65]. Upon increase of temperature or pH, the tetramer will oligomerize presumably in a first step to an octamer [64, 71] of the constitution $[Zr_8(OH)_{20}(H_2O)_{24}Cl_{12}]$ [68], and then to higher oligomers, which will finally aggregate and precipitate. At 373 K, 3 to 7 nm primary particles were observed [67]; secondary particles (aggregates of primary particles) reach 175 to 250 nm [70, 72]. The particle growth rate can be decreased through addition of HCl [71]. Hence, the precipitation chemistry can be determined by concentration, pH, and temperature.

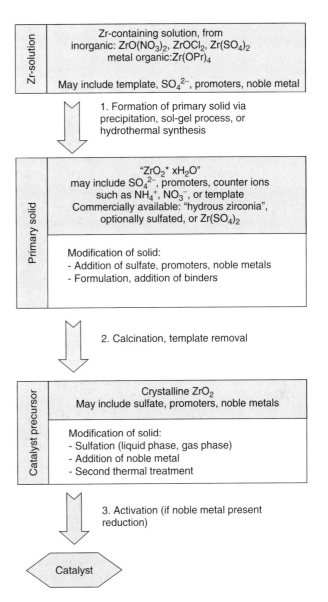

Fig. 3 General preparation scheme for sulfated zirconia as an example for an oxo-anion modified oxide.

The chemistry of zirconium sulfate solutions is different because sulfate not only strongly complexes with zirconium [74, 75] but also potentially acts as bridging ligand and promotes polymerization [75, 76]. While cationic or neutral complexes prevail in chloride, nitrate, and perchlorate solutions, anionic [77, 78] mixed

References see page 275

hydroxosulfato complexes [76], also of polynuclear type, are formed in sulfate solutions. Polynuclear complexes can form through different bridging mechanisms; that is, hydroxo bridges (formed in an "olation" reaction), oxo bridges ("oxolation"), and sulfato bridges. Sulfate-bridged complexes of the constitution $[Zr_n(OH)_{n+1}(SO_4)_{2n}]^{(n+1)-}$ have been proposed [76], or complexes of the type $[Zr(OH)_2(SO_4)_x(OH_2)_y]_n^{-n(2x-2)}$, which include water as a ligand [80]. Hence, it is not hydrous zirconia that is precipitated from these solutions, but sulfates; and the large number of possible sulfates [61, 81, 82], especially basic sulfates [61, 83–85], suggests that numerous complexes of different constitution may exist in solution. Consistently, no particular species have been reported as being stable over a range of conditions in zirconium sulfate solutions. The time frame for changes in these solutions [83] indicates that equilibration is slow: precipitation in 0.5 M $Zr(SO_4)_2$ solutions was observed only after 2 weeks [86], and precipitation in 0.2 mM $Zr(SO_4)_2$ could be delayed by 10 h, 2 days, or 4 days by dissolving the salt in 1, 2, or 4 mM HNO_3, respectively [77]. Heating promotes hydrolysis [87]. The kinetics of formation of larger particles in zirconium sulfate solutions have been investigated, and growth was accelerated at higher temperature, higher zirconium concentration and lower acid concentration [88]. From the available data it becomes clear that not only concentration, temperature and pH, but also aging, will play a role in the formation of the primary solid.

The chemistry in zirconium aqueous solutions depends strongly on the complexation with the available anions. Zr complexes may be either cationic or anionic, which also determines the choice of surfactant in a templated synthesis. In general, for each metal cation and each oxo-anion, their joint solution chemistry must be considered, and adding the oxo-anion during the formation of the primary solid may affect the properties of the matrix oxide significantly.

2.3.9.4.1 **Formation of Pure Hydrous Zirconia** Starting compounds used for the formation of hydrous zirconia are $ZrOCl_2$ [89], $ZrO(NO_3)_2$ [36], and zirconium alkoxides [28, 90]. Precipitating agents are aqueous ammonia [89] and solutions of either KOH or NaOH [91, 92]. The precipitates are of a gelatinous nature [89], and a crystalline zirconium hydroxide with a defined stoichiometry does not exist [93]. Hydrolysis of Zr alkoxides, which are soluble in for example ethanol, 1-propanol, 2-propanol or cyclohexane [44], can be initiated by water [94, 95], acids [90], or bases [33]. For other oxides, e.g. $TiCl_4$ or $SnCl_4$ are used as starting compounds.

The precipitation can be conducted either by adding the zirconium solution to the base, or by adding the base to the zirconium solution. A more dramatic effect than the method of addition has the subsequent ageing or digesting of the precipitate [89]. Figure 4 shows how the surface area and the fraction of tetragonal phase increase with increasing digestion temperature. Moreover, the surface area correlates with the fraction of tetragonal phase. The surface area increases with digestion time, for example from 48 $m^2 g^{-1}$ without digestion to 248 $m^2 g^{-1}$ after a 96-h digestion at 373 K and calcination at 773 K [89]. In addition, the pore size distribution becomes narrower [28]. Digestion at pH 14 (KOH or NaOH) produces materials with surface areas stable up to 773 K, while undigested materials or those digested at pH 11–9.4 (NH_4OH) lose surface area upon thermal treatment [91]. The acid site distribution is also altered; ammonia desorbs at higher temperatures from previously digested calcined zirconia, but the adsorbed amount per m^2 is unaffected [92].

Hydrous zirconia can also be prepared via hydrolysis of zirconium alkoxides. Variations of the water/alkoxide ratio between 2 and 32 were found to have only a minor influence on the surface area of calcined zirconia (773 K) in comparison to the digestion time [95]. The zirconia surface area was reported to increase until about 200 h of digestion time at pH 9, reaching values larger than 300 $m^2 g^{-1}$. Digestion at pH 1 or 3 produces surface areas smaller than 100 $m^2 g^{-1}$; in general the fraction of the monoclinic phase in the calcined material (773 K) increases with decreasing pH of the suspension [28, 95]. Zirconia gels can also be produced in supercritical CO_2 by hydrolyzing zirconium alkoxides with acetic acid [90].

2.3.9.4.2 **Formation of Sulfate-Containing Hydrous Zirconia** Sulfate can already be added during the formation

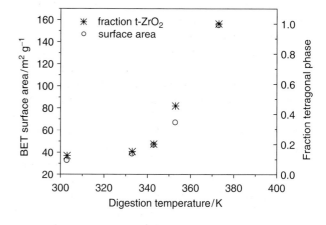

Fig. 4 Surface area and fraction of tetragonal phase as a function of digestion temperature. Basic precipitation, calcination at 873 K. (After Ref. [89].)

of the primary solid. In a synthesis via gel formation from alkoxides, sulfate can be introduced during the gelling step [53, 96, 97] or during peptization [98]. Tungstate can equally be co-gelled [99]. Reverse microemulsions can also be used as a synthesis medium for sulfated zirconia nanoparticles [100].

A large amount of sulfate can be incorporated into the primary solid due to the ability of sulfate to act as a bridging ligand between zirconium cations. The amount of sulfate can exceed 50 wt.% after application of sulfuric acid for peptization following a sol–gel synthesis; this will lead to sulfate contents of approximately 20 wt.% after calcination at 823 K [98]. For calcination temperatures below 873 K, samples co-gelled with sulfate exhibit higher surface areas than those sulfated with ammonium sulfate at a later stage [96]. The surface and bulk sulfate concentrations increase with increasing H_2O/Zr ratio [101]. The drying method has also been investigated, with a xerogel (regular drying) being more active in n-butane isomerization than an aerogel (supercritical drying) [102].

2.3.9.4.3 Template-Directed Formation of Primary Solid
Supports such as silica can be synthesized in the form of mesoporous molecular sieves [103] (see also Chapter 2.3.6). These materials are characterized by a very high surface area, which would be desirable also for zirconia, and can be achieved via a template-assisted hydrothermal synthesis. Surfactant molecules assemble to micelles, which direct the growth of the oxide matrix from an organometallic precursor. The surfactant is later removed in a calcination step. The difficulty with zirconium is that, in the strongly acidic solutions required to restrict hydrolysis, it is present in the form of cationic complexes (see Section 2.3.9.4); the typically used surfactants are protonated under these conditions and hence assembly cannot take place [104]. One solution to this problem is the addition of sulfate to generate anionic zirconium species, but full condensation is apparently prevented and the framework may be partially destroyed when the sulfate is removed by calcination. Using hexadecyltrimethyl ammonium bromide as surfactant, it was possible to obtain an MCM-41-like structure, which collapsed during calcination [105]. Treatment of the as-synthesized composite with phosphoric acid stabilizes the structure throughout calcination, with S and P contents of 1 and 10 wt.%, respectively, in the final product. By using zirconium propoxide and ammonium sulfate instead of zirconium sulfate, an ordered pore system with S contents of up to 8.5 wt.% could be obtained [80]. Further procedures either with [106–108] or without [109] sulfate in the surfactant-containing synthesis mixtures have been published. The structural homogeneity and the catalytic properties of these materials vary. Inactivity in n-hexane isomerization was found combined with high activity for benzene alkylation with propene [106]. A lower activity in n-butane isomerization than for conventional tetragonal zirconia was observed for mesostructured zirconia [108], but a higher activity has also been reported [109].

2.3.9.4.4 Commercially Available Precursors
Hydrous and sulfated hydrous zirconia materials are available, e.g., from MEL Chemicals or Sigma-Aldrich [110], and are frequently employed in both, academia and industry to produce sulfated zirconia catalysts [15, 111–113]. Zirconium compounds contain Hf as an impurity, e.g. hydrous zirconia contains typically about 1 wt.% Hf [112]. Zirconium sulfate, titanium sulfate and tin sulfate can also be used as precursors for the corresponding sulfated metal oxides [31, 114, 115].

2.3.9.5 Anion-Modification of Primary Solid

2.3.9.5.1 Addition of Oxo-Anions to Primary Solid in Liquid Medium
Different sulfating agents are used for sulfation in aqueous solution: H_2SO_4, $(NH_4)_2SO_4$, HSO_3Cl [116], $(NH_4)_2S_2O_8$ [117, 118], and $Zr(SO_4)_2$ [119]. The procedures used are incipient wetness impregnation [35], soaking in excess impregnating solution and filtering (sometimes referred to as percolation) [52, 120], and soaking in excess impregnation solution followed by solvent evaporation [121]. Typical precursors for other oxo-anions are ammonium metatungstate $(NH_4)_6H_2W_{12}O_{40} \cdot xH_2O$ [8] or ammonium metamolybdate [122].

The surface sulfur content (calculated from total content and surface area with the assumption that all sulfate is on the surface) in the calcined and reduced catalysts increases linearly with the concentration of the H_2SO_4 used in an incipient wetness procedure [35]. When an excess of acid is applied and the solid is filtered, the total sulfate content in the calcined product increases first steeply (<0.5 M), and then linearly with H_2SO_4 concentration [52, 123]. A similar course is found for the sulfate surface density [118]. The surface sulfate content of the calcined materials (as measured by photoelectron spectroscopy) increases initially more steeply with increasing H_2SO_4 molarity than the total sulfate content [52]. The sulfate content increases with the amount of sulfuric acid used for the impregnation [34]. More sulfate can be deposited (solvent evaporation) than adsorbed (filtration) [124].

As a natural consequence of the described dependence of surface area on sulfate content, and the possible ways to modify the sulfate content by way of the impregnation

---

References see page 275

step, the surface area correlates with the amount of sulfate applied. These correlations all show maxima in the surface area, either at an intermediate H_2SO_4 amount [34], or at intermediate concentrations e.g., at 0.5 M H_2SO_4 [120] or 0.2–0.25 M H_2SO_4 [123]. A maximum surface area at 0.25 M H_2SO_4 coincides with the maximum in butane conversion [94]. The pore diameter decreases with the addition of sulfate, and remains constant at about 8 nm for H_2SO_4 concentrations larger than 0.05 M [120]. Not only the sulfate concentration but also the pH plays a role: the maximum surface area is attained when hydrous zirconia is soaked in a sulfate-containing solution of pH 6–9, and these samples also produce the maximum isobutane yield under steady-state conditions [125]. The precursors, which result in different pH, unless it is adjusted, are also important; for example, catalysts obtained via H_2SO_4 impregnation give a higher butane conversion than those from $(NH_4)_2S_2O_8$ impregnation [118]. Some authors prefer H_2SO_4 as sulfating agent [10], and others $(NH_4)_2SO_4$ [6].

2.3.9.5.2 Sulfation of Primary Solid via Gas Phase
Treatment of hydrous zirconia with SO_2 enhances the activity for n-butane isomerization as much as impregnation with sulfuric acid, whereas treatment with H_2S or SO_3 does not produce active materials [2]. Silica gel can be treated with SO_2Cl_2, which reacts with the surface OH groups, in order to generate sulfate attached to silica [8].

2.3.9.5.3 Introduction of Oxo-Anions Using Solid Precursors
Calcination of a kneaded mixture of hydrous zirconia and ammonium sulfate at 773 K produces a material that is active for butene double bond isomerization and benzoylation of anisole [126]. Kneading hydrous zirconia with H_2WO_4, followed by calcination, yields a catalyst for pentane conversion [8].

2.3.9.6 Addition of Promoters (Optional)

2.3.9.6.1 Promoters
Of the main group elements, Al [117, 127–133] and Ga [24, 134] act as promoters of sulfated zirconia isomerization catalysts. First row transition metals such as Ti [127], V [127], Cr [127, 135], Mn [10, 15, 136–138], Fe [10, 15, 127, 135–137, 139–142], Co [15, 127, 135], and Ni [15, 127, 132, 135, 139, 143, 144] have attracted considerable attention. Nb is a promoter for alkane isomerization [145]. Silver and copper in the form Ag^0 and Cu^0 promote the activity of sulfated zirconia for n-pentane isomerization [146]. Ag and Cu are also suitable as additives to tungstated zirconia [147]. Ce and other lanthanides are also promoters of sulfated zirconia [112, 145].

Combinations of promoters have also been tested. In particular, the combination of Fe and Mn was studied in detail for isomerization [14, 127, 136, 137, 139, 148–171], and also for cracking [172–174] and alkylation reactions [175], since a synergistic effect between the two promoters was reported. Other successful combinations include Cr–Fe and V–Fe [127].

2.3.9.6.2 Structure–Activity Relationships
A number of relationships concerning performance and promoters have been reported. The efficacy of the first row transition metal promoters follows a trend in the Periodic Table, that is, under the selected conditions, the maximum activity decreases in the order Mn > Fe ≫ Co > Ni ≫ Zn [15]. The conversion of n-pentane increases with decreasing ionic radius of the promoter cation in eightfold coordination; this relationship applies to yttrium and lanthanides [112].

The promoter content can be optimized [117, 144], and, e.g., when combining Group V, VI, and VII elements, the sum of all promoters is ideally between 0.1 and 4.5 wt.% metal [13]. The relationship between the conversion to isobutane and promoter content depends on the promoter type [139]. The steady-state n-butane isomerization rate was found to increase with an increasing ratio of the lattice constant c/a of the tetragonal phase for a family of Fe- or Mn-promoted catalysts [176].

2.3.9.6.3 Effect of Promoters
Promoters such as Mn or Fe stabilize the desired tetragonal phase of zirconia and, at high content, the cubic phase [113]. Stabilization of the high-temperature phases is a result of the incorporation of the promoter cations into the zirconia lattice [113]. This phenomenon is well known for ions of e.g. Y, Ca, or Mg; because of its wide application as a ceramic material yttria-stabilized zirconia has been studied extensively [177]. Many cations can be dissolved in the zirconia lattice [178–181]. The lattice parameters of the stabilized tetragonal or cubic phase vary, depending on promoter type and content. The lattice may either contract or expand, with a linear relationship between certain lattice parameters and the molar fraction of promoter, corresponding to Vegard's law [181]. For example, Mn or Fe cause a contraction [113]. In light of the formation of solid solutions it can be understood why pentane conversion and ionic radius of the promoter are correlated [112]. The valence of the promoter ions differs from that of zirconium, and charges must be compensated in the lattice, for example by creating oxygen vacancies to balance the incorporation of ions with a valence less than +IV. Hence, the promoters alter the geometric and electronic structure of the zirconia matrix.

Promotion sometimes results in a slight increase in surface area [24, 134], but not on all occasions [144]. The differences in surface area seem to be more pronounced when the calcination produces an unpromoted sulfated zirconia with significant amounts of monoclinic phase, while the promoters stabilize the tetragonal phase with its typically higher surface area (see Fig. 4). A 0–67% higher sulfate content is observed in promoted compared to unpromoted sulfated zirconia samples [24, 127, 134]. The promoters affect the reducibility of the oxo-anions on zirconia. In the presence of Mn, Fe, or Ga, the reduction temperature of sulfate is decreased [134, 182, 183] and, in the presence of Ga, tungstate is more easily reducible [184].

While the choice of promoter type and content is straightforward, it is unclear whether the synthesis should target the dissolution of the promoter ions in the zirconia lattice or the formation of surface species. Evidence exists for both scenarios, that is either incorporated or surface promoter species may be catalytically relevant, as discussed in Ref. [113].

2.3.9.6.4 Addition of Promoters
Promoters can be introduced at the solution stage before formation of the primary solid, they can be impregnated onto the primary solid, or they can be added to the crystalline oxide, always followed by calcination. It would seem that the first method should produce preferably incorporated species whereas the last method should produce preferably surface species. The literature though is not in agreement as to which method produces predominantly which species [113]. Figure 5 shows a more pronounced effect of the promoters on the lattice constant for preparation via coprecipitation in comparison to incipient wetness impregnation, suggesting that, indeed, incorporation is furthered when zirconium and promoter ion species are well interspersed in the primary solid.

Coprecipitation from a mixed solution of zirconium and promoter ions has been applied to introduce, for example, Al, Ga, Ti, Cr, Mn, Fe, Co, or Ni [24, 117, 127, 134]; the sulfate can be added later [127]. If the promoters are introduced by impregnating a solid and the oxo-anion is already present, then the solubility of the oxo-anion in the impregnation solution must be considered. Even after calcination, the sulfate in sulfated zirconia is water-soluble and will be partially removed in aqueous medium [113, 185]. One frequently applied method is that of incipient wetness [14, 15], in which the solubility problem is suppressed. The order of adding sulfate and promoters to the primary solid via incipient wetness impregnation does not result in significant differences in n-butane isomerization activity [186]. If Al- or Ga-promoted sulfated zirconia are prepared by

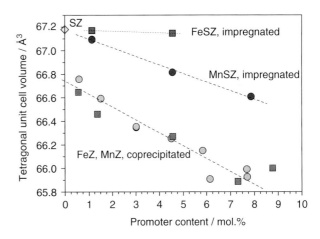

Fig. 5 Tetragonal unit cell volume as a function of promoter content with promoter type and promoter addition method as parameters. Circles, Mn promotion; squares, Fe promotion.

first introducing the promoters and then the sulfate, coprecipitation yields superior catalysts in comparison to wet impregnation [24].

Calcination of physical mixtures of sulfated zirconia with $FeSO_4$, sulfated Fe_2O_3 or Fe_2O_3 at 723–873 K also results in catalysts with promoted activity in n-butane isomerization [187].

2.3.9.7 Calcination

2.3.9.7.1 Events Occurring During Calcination
During calcination, a largely disordered hydroxide-type precursor is converted into a crystalline metal oxide. The chemical reactions that occur are dehydration and condensation reactions such as dehydroxylation, which leads to the formation of oxo-bridges (oxolation) between the metal cations [188]. Further components that are able to cleave out volatile compounds may be decomposed; for example, NH_3 may be released from ammonium, NO_x from nitrates, and SO_2 and O_2 from sulfate. Also from other oxo-anions volatile species may be produced, e.g. MoO_3. Redox processes can occur through reaction with the oxygen of the calcination atmosphere, or through re-reaction with volatilized species, affecting the valences of the catalyst components. Surfactant molecules would be combusted. Structure and morphology change considerably during calcination. The oxide crystallizes and particles coalesce, leading to a loss in surface area and a change in porosity.

Thermogravimetric (TG) and differential thermal analysis (DTA) data for an ammonium sulfate-treated hydrous zirconia are presented in Fig. 6 (for further data relating to zirconia-based materials, see Refs. [91, 136,

References see page 275

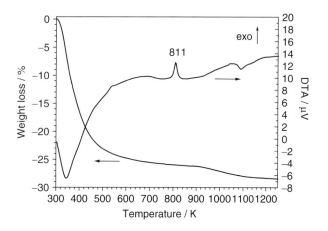

Fig. 6 Thermogravimetry and differential thermal analysis data for sulfated hydrous zirconia (MEL XZO 682). Heating rate 7 K min^{-1}.

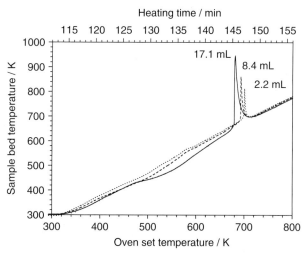

Fig. 7 Actual sample temperature of hydrous zirconia (center of bed) as a function of oven set temperature with precursor volume as a parameter. Heating rate 3 K min^{-1}; synthetic air flow 200 mL min^{-1}. Samples were placed in quartz boats of different volume [204].

189–191]). The TG curve is characterized by a large weight loss of more than 25% that extends to 873 K, and a second, smaller weight loss that sets in at about 900 K. The first loss is connected with the evaporation of water and the decomposition of ammonium, and the second loss to the decomposition of sulfate. The DTA curve shows the water loss and sulfate decomposition to be endothermic, and additionally shows an exothermic peak at about 800 K. The origin of this "glow exotherm" has been debated, whereby crystallization [192, 193], the loss of internal surface hydroxyl groups [194], and the coalescence of small particles to larger ones [195, 196] have been proposed as heat-generating processes.

The enthalpies derived from thermal analysis for, presumably, the crystallization of tetragonal or monoclinic zirconia show a wide spread, ranging from −4.3 to −58.6 kJ mol^{-1} [91, 191–193, 197–201]. The energetic difference between the two phases is small; the monoclinic is about 5 kJ mol^{-1} more stable [199, 202]. These discrepancies in enthalpy can be ascribed to variations in the initial and final states. The properties of "hydrous zirconia" depend strongly on the procedure of preparation, and this state is usually not closely investigated. The final state is also insufficiently characterized, in that only the detected phases are considered but not amorphous fractions or properties such as crystallite sizes and surface areas.

From the thermal analysis data it is obvious that, during calcination, considerable mass and heat transfer occurs which, particularly when working on a preparative scale, may be limited. The chemistry will be influenced by the mass transfer conditions, e.g. the condensation chemistry will be affected by the water vapor partial pressure in the pores and the surrounding gas phase. Surface diffusion and hence zirconia crystallite growth is reported to be accelerated by water vapor [203]. Any volatile species can potentially be re-deposited on the material's surface.

2.3.9.7.2 Effect of Bed Volume and Packing The actual sample temperature of hydrous zirconia during the heat-up phase of the calcination program is presented in Fig. 7. Significant deviations from the planned ramp are apparent, which differ depending on the sample volume employed [204, 205]. The sample temperature lags behind the oven set temperature, and more so with increasing sample volume, because of the endothermic evaporation of water. Retention of water vapor in a tightly packed bed will delay condensation reactions, which should lead to higher surface areas. At oven temperatures of 675 to 700 K, the bed temperature exhibits a brief, but significant, overshoot. For a 17.1-mL volume, the overshoot amounts to 260 K, with an increase of 45 K s^{-1}. The curves indicate explosion kinetics; heat cannot be dissipated as rapidly as it is produced. The onset shifts to lower temperature with increasing volume, probably due to a restricted heat transfer with the small surface-to-volume ratio and more complete enclosure of the bed in the larger quartz boats.

This violent reaction is designated in the literature as the "glow phenomenon", in reference to the emission of visible light, and was first reported by Berzelius in 1812 [206, 207]. For zirconia, it was mentioned already in 1818 [208] but was more thoroughly investigated only decades later [209, 210]. The phenomenon is also observed during the formation of chromium oxide, iron oxide and titania, and is thus of general relevance; however, it does not occur during alumina formation [195, 210]. Additives of any type shift the glow event to a higher temperature [189, 190, 196, 197, 204, 205] and subdue or even suppress it [193, 195]; for sulfate, the exotherm temperature shifts linearly with content [190]. Estimation of the expected temperature overshoot from enthalpy changes due to crystallization or surface area loss, under the assumption of quasi-adiabatic conditions, shows that either process can account for the phenomenon [205].

The actual temperature curves during calcination for a sulfated hydrous zirconia material promoted with 0.5 wt.% Mn are shown in Fig. 8. The addition of sulfate and Mn shifts the onset of the overheating to about 100 K higher temperature. The nitrogen adsorption and desorption isotherms for the obtained materials are shown in Fig. 9. The hysteresis loop becomes more pronounced with increasing bed volume. The surface area increases slightly, from 92 to 96 to 106 m^2 g^{-1}, consistent with the idea of inhibition of oxolation and coalescence by water retained in a larger bed. In Fig. 10, the catalytic activity data are presented for the same set of samples. The rates increase significantly with the volume used for the calcination, excluding surface area as the sole explanation for the differences.

These data show that variations in bed size and packing during calcination can produce different materials from a single precursor batch [204, 205]. Although the nominal conditions (heating ramp, holding temperature and time, atmosphere) are identical in these calcinations, the samples experience quite different actual conditions. During the initial phase, the water vapor partial pressure will affect the condensation chemistry. Crystallization starts before overheating is observed. Surface area and

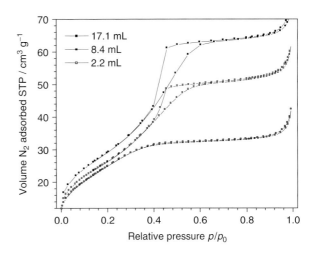

Fig. 9 Adsorption and desorption branches of nitrogen adsorption isotherms at 77 K. Sulfated zirconia promoted with 0.5 wt.% Mn.

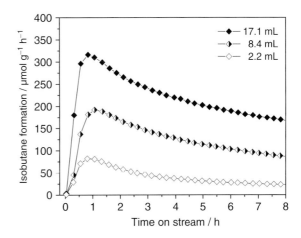

Fig. 10 Isobutane formation rate versus time on stream for sulfated zirconia promoted with 0.5 wt.% Mn. Activation: 0.5 h at 723 K in N$_2$. Reaction conditions: 500 mg catalyst, 1 kPa n-butane in N$_2$ at atmospheric pressure, 338 K, 80 mL min^{-1} total flow.

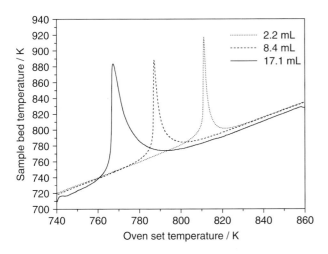

Fig. 8 Actual sample temperature of hydrous zirconia promoted with 0.5 wt.% Mn (center of bed) as a function of oven set temperature with precursor volume as a parameter. Heating rate 3 K min^{-1}; synthetic air flow 200 mL min^{-1}; holding time 3 h at 923 K. Samples were placed in quartz boats of different volume [204].

diffractogram of a sample that was cooled directly after the glow phenomenon largely correspond to those of samples that underwent the complete heating program, which indicates that structure and morphology are developed during overheating [205]. During the temperature overshoot, the desired maximum temperature may be exceeded, which allows for a chemistry that is not possible within the planned program. Enhanced ion mobility within the forming solid, as well as volatilization and re-anchoring with a restructuring of surface species, can also be envisioned. Because of the fairly rapid cooling down following the overshoot, non-equilibrium states might be

References see page 275

quenched. The subsequent holding time at high temperature does not equalize differences evoked during the heat-up phase. Consistently, attempts have been made to relate high n-pentane isomerization activity to a sharp and high exothermal peak [191]. In part, the reported lack of reproducibility [211] of these catalysts may originate from underestimating the importance of the details of the calcination procedure. Samples calcined in different set-ups are not comparable on an absolute scale, and the data discussed below illustrate such trends.

2.3.9.7.3 Effect of Calcination Temperature on Physical Properties

The calcination temperature has a major influence on phase composition, surface area, and sulfate content. Upon heating, the metastable tetragonal (cubic) phase is formed first. The higher the calcination temperature, the more likely is the formation of monoclinic zirconia. Crystalline zirconia appears at about 663 K in non-sulfated samples [212]. The crystallite size of the tetragonal phase and the fraction of the monoclinic phase increase with increasing calcination temperature [89, 212]. In the presence of sulfate, the tetragonal phase is stabilized, and even at temperatures of 933 K, purely tetragonal material can be obtained [212]. However, typically, formation of the monoclinic phase sets in at calcination temperatures above ca. 873 K [37]. The crystallite size of the metal oxide in sulfated zirconia, titania and tin oxide prepared via sulfate decomposition, was found to be constant up to the calcination temperature that provides the largest surface area [31].

The surface area of pure zirconia decreases with increasing calcination temperature above 773 K, with a smaller effect for materials with a small initial surface area [95, 101]. The surface area drops significantly once the temperature exceeds that of the glow exotherm [213]. Such a decrease is also observed for oxo-anion modified zirconia, for example for sulfated zirconia (from 135 m^2 g^{-1} at 773 K to 40 m^2 g^{-1} at 1173 K [53]), and for tungstated zirconia (from 100 m^2 g^{-1} at 973 K to 35 m^2 g^{-1} at 1223 K [99]). A linear correlation of surface area and temperature is seen between 773 and 1223 K [188]. The trends for different metal oxides are shown in Fig. 11. The sulfate content is also reduced by calcination, and this occurs in parallel with a loss of surface area, as Fig. 11 demonstrates. The published data vary as to the temperature where these losses start; values also depend on the initial sulfur content, and range from 473 K to 873 K [36, 37, 101, 212, 214]. It has been claimed that sulfate in certain crystallographic locations desorbs preferentially, and

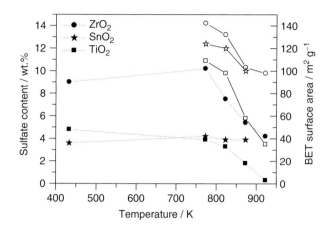

Fig. 11 Sulfate content (solid symbols) and BET surface area (open symbols) as a function of calcination temperature after drying at 433 K with oxide type as a parameter. (After Ref. [124].)

highly uncoordinated Zr^{4+} sites, which are strong Lewis acid sites, are generated at calcination above 773 K [215]. With increasing calcination temperature, the number of OH groups on the surface decreases [183].

2.3.9.7.4 Effect of Calcination Temperature on Catalytic Performance

The best performance in n-butane isomerization is exhibited by sulfated zirconia that has been calcined at 823–873 K [187], 873 K [53], 893 K [216], or 875–900 K [214]. A minimum temperature of 773 K is required to obtain an active material [215]. Much higher temperatures are necessary for tungstated zirconia, about 1100–1200 K [99].

The initial activity of the obtained sulfated metal oxide catalyst in cumene cracking depends on the calcination temperature employed in decomposing the commercial Zr(SO$_4$)$_2$ or Ti(SO$_4$)$_2$ precursor. The activity increases with increasing calcination temperature, reaching a maximum at 900 K for the titania and 1000 K for the zirconia catalyst [114]. This increase in activity was found to be parallel to an increase in surface area [31].

The optimal temperature in the presence of promoters has been evaluated for sulfated zirconia promoted with Fe and Mn; the results are shown in Fig. 12 [217]. Because the activity of these catalysts changes rapidly, several data points representing different times on stream have been selected. The best performance is obtained after calcination at 923 K [166, 217].

2.3.9.7.5 Effect of Holding Time

Little is known about the effect of holding time. Neither zirconia crystallization nor crystallite growth progress further after 2 h at ≥773 K [218]. If the temperature is high enough to decompose sulfate, e.g. 873 K, then the sulfate content will

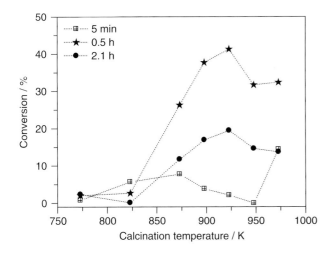

Fig. 12 Conversion at different times on stream for sulfated zirconia promoted with 1.5 wt.% Fe and 0.5 wt.% Mn. Calcination: 5 g raw material supported on frit in vertical open tube, 3 h heating phase to desired temperature, 4 h holding time, free air convection. Activation: 1.5 h at 723 K in N_2. Reaction conditions: 1.5 g catalyst, 0.5 kPa n-butane in N_2 at atmospheric pressure, 373 K, 80 mL min^{-1} total flow.

decrease with time [219]. Holding time and temperature have a combined effect, thus one can be selected if the other is adapted. For example, calcination for 12 h at 823 K delivers a promoted sulfated zirconia catalyst that yields equal conversion in alkane isomerization as one calcined for 4 h at 848 K [13].

2.3.9.7.6 Calcination: A Summary
The calcination is a critical step, but it is difficult to name optimal conditions for two reasons.

First, phase composition, surface area, and oxo-anion content are interdependent properties. During calcination, excess sulfate or tungstate – that is, more than can be accommodated in a chemisorbed monolayer – will be converted to oxides, and more so as the surface area shrinks with increasing temperature. In the case of sulfate, oxide formation means volatilization and, in the case of tungstate, the excess remains in the sample in the form of WO_3 [99]. The sulfate content thus will be diminished; typical sulfur losses during calcination (2 h 873 K) depend on the sulfur loading and amount to 20–50% of the initial content [56]. Concurrent with sulfate loss and surface area shrinkage, monoclinization of zirconia may occur.

Second, the number of parameters influencing the outcome of calcination is high. Because of the condensation chemistry when starting from a hydrous oxide precursor, and the strongly exothermic crystallographic and morphological changes, both heat and mass transfer conditions are extremely important. The restriction of heat transfer and gas exchange seems favorable, contrary to IUPAC recommendations to use moving beds [58]. The most important point is to realize that simply stating the maximum temperature and holding time is insufficient; the bed size and packing, gas exchange conditions and heating rates must be specified and controlled to render calcinations reproducible.

2.3.9.8 Sulfation of Thermally Treated Crystalline Oxides
It has often been claimed that adding sulfate to crystalline ZrO_2 does not yield an active catalyst [37], or yields only about one order of magnitude less active catalysts [214]. Specifically, the activity of the resultant materials drops when the hydrous oxide is pretreated at a temperature that will induce crystallization [2]. However, the situation is different for other oxides; for example, materials that are active in acid catalysis can be generated from Fe_2O_3 [220, 221].

2.3.9.8.1 Sulfation with Aqueous Solutions (Solid–Liquid Phase)
The impregnation of zirconia, followed by calcination at 873 K, produces a material that is active in n-butane isomerization, with the activity related to the surface area [29]. Incipient wetness impregnation with ammonium sulfate and calcination at a temperature of at least 773 K also results in a sample active in n-butane isomerization [96].

2.3.9.8.2 Sulfation via Gas Phase (Solid–Gas Phase)
SO_3 treatment at 573 K has been used to sulfate Fe_2O_3, with positive results in cyclopropane isomerization; also, H_2S and SO_2 treated Fe_2O_3 samples become active after oxidation at 723–773 K [221]. Equally, exposure of ZrO_2 to H_2S or SO_2, followed by treatment in an excess of O_2 at 723 K, yields virtually identical IR spectra [119]. Moreover, the spectra correspond to those obtained after aqueous solution impregnations using H_2SO_4, $(NH_4)_2SO_4$, or $Zr(SO_4)_2$ and suitable activation. At low coverage, monosulfate species are observed, while at more than 250 µmol S g^{-1} an S_2O_7-type species appears, indicating that a high sulfate density can be achieved via gas-phase reaction. It is possible to generate an active material via gas-phase sulfation of predominantly monoclinic zirconia using H_2SO_4 vapor [214].

SO_3 can also be employed as a sulfating agent. Exposure of zirconia supported on silica at 673 K generates a catalyst that is active in the transalkylation of benzene and diethylbenzene, though only after hydrating the material [222]. The sulfur content of sulfated zirconia can be increased by treatment with SO_3 at 673 K, with an increase from 0.44 to 0.64 mmol g^{-1} or 6 wt.% having been reported [24]. In a series of samples, the total amount of sulfate after SO_3 treatment was found to be proportional to the fraction of monoclinic material in the sample, with loadings reaching 1.15 mmol g^{-1} or 12 wt.%. SO_3 treatment enhances the butane isomerization activity of sulfated zirconia, and zirconia becomes active. A sample prepared via SO_3 sulfation, and containing only 30% tetragonal zirconia, was even found to be catalytically more active than a sample created by the calcination of amorphous hydrous zirconia. Previous investigations have confirmed that monoclinic zirconia may be sulfated to give an active catalyst, but only at high sulfate loading and at the expense of low sulfate stability and rapid deactivation [223].

Calcination of pure hydrous zirconia leads to a lower surface area, and a monoclinic fraction is more easily formed. The successful sulfation of crystalline materials requires a high temperature; hence, it appears that a reaction of the sulfur species with the support must be initiated.

2.3.9.8.3 Sulfation of Zirconia using Solid Sulfate Precursors
Kneading calcined zirconia (773 K) with ammonium sulfate, followed by a second calcination at 773 K, produces a material that is active for butene double-bond isomerization and the benzoylation of anisole [126].

2.3.9.9 Addition of Noble Metals
Noble metals can be added at different stages of the preparation, for example, during alkoxide hydrolysis [224], before calcination [225, 226], or after a calcination step [227]. Normal impregnation methods are employed; non-water-soluble noble metal salts can be kneaded together with the primary solid [225]. Noble metals are also effective when not located directly on the oxo-anion modified oxide [187].

2.3.9.10 Activation
Reports on the optimal activation temperature vary significantly. For sulfated zirconia, temperatures of 523–573 K [216], 573 K [228], 590 K [229], 673 K [94, 228], 723 K [55], or 923 K [214] have been reported as being optimal for high n-butane isomerization activity. The individual materials may favor different activation temperatures, but atmosphere, heating rates and holding times may also play a role.

The performance of promoted sulfated zirconia catalysts, for example with Fe and Mn as promoters, appears to be less sensitive to the temperature, particularly if the sample is activated in air and not in inert gas [152]. Temperatures of 623–723 K have been reported to generate the highest maximum activities [152, 166].

Noble metal-containing sulfated zirconia catalysts require a reduction step, with the inherent danger of reducing and volatilizing the sulfate. In the presence of Pt, the sulfate reduction temperature is shifted by about -200 K into the range 733 to 773 K [230]. The butane isomerization activity of Pt-doped sulfated zirconia decreases to zero with increasing reduction temperature in the range 583 to 693 K [46]. The situation will be more complicated if promoters are present; for example, Mn-promoted sulfated zirconia becomes inactive after treatment in 50% H_2 at 703 K [231].

Similar to sulfate, tungstate is much more easily reduced in the presence of Pt; W^{5+} species are formed at room temperature, but the systems are active [232]. Pt in Pt-doped Fe-promoted tungstated zirconia is reduced to metal at room temperature and hence, under hydroisomerization conditions, is also expected to be metallic, rendering an extra reduction step unnecessary [16].

2.3.9.11 Summary
The catalysts discussed in this chapter contain four types of components, namely matrix oxide, oxo-anion, promoter, and noble metal. If the oxide and oxo-anion precursors are associated in solution, then the metal cation's tendency for hydrolysis and for complexation with the oxo-anion will determine the nature of the primary solid. Alternatively, oxo-anions can be introduced to a solid hydrous or crystalline oxide.

If promoters are added, then the solid-state chemistry of the matrix oxide and its ability to accommodate foreign cations in its lattice come into play; on the other hand, oxo-ions will compete to form stable compounds with the promoter ions, e.g., $Al_2(SO_4)_3$. In the first case, the geometric and electronic structures of the matrix oxide can be tuned, although not independently. Thus, zirconia is an interesting matrix oxide for oxo-anion modification because, by varying its properties via promoter incorporation, the reactivity, e.g. the reducibility, of the oxo-anion can be determined.

All preparations require a high-temperature thermal treatment after associating the components. The calcination conditions are difficult to control because the reactions occurring in the sample consume and produce significant amounts of heat and gases; thus, heat- and mass-transfer conditions must be carefully considered. The thermal treatment represents a key step in the formation of oxo-anion modified oxide catalysts. As a consequence of the close interaction of the individual catalyst components, properties such as phase composition, surface area and oxo-anion content are interdependent, and determine further properties, such as acid site distribution. Therefore, the targeted synthesis of oxo-anion modified oxides with specific properties is challenging.

References

1. V. C. F. Holm, G. C. Bailey, US Patent 3, 032, 599, assigned to Phillips Petroleum Co., 1962.
2. M. Hino, S. Kobayashi, K. Arata, *J. Am. Chem. Soc.* **1979**, *101*, 6439.
3. M. Hino, K. Arata, *J. Chem. Soc., Chem. Commun.* **1980**, 851.
4. F. Schmidt, *Appl. Catal. A: General* **2001**, *221*, 15, and references therein.
5. N. Lohitharn, J. G. Goodwin, Jr., *J. Catal.* **2007**, *245*, 198.
6. G. D. Yadav, J. J. Nair, *Micropor. Mesopor. Mater.* **1999**, *33*, 1, and references therein.
7. X. Song, Y. Sayari, *Catal. Rev. -Sci. Eng.* **1996**, *38*, 329, and references therein.
8. K. Arata, M. Hino, *Mater. Chem. Phys.* **1990**, *26*, 213, and references therein.
9. C. D. Chang, F. T. DiGuiseppi, J. G. Santiesteban, US Patent 5, 780, 382, assigned to Mobil Oil Corporation, 1998.
10. K. Arata, *Appl. Catal. A: General* **1996**, *146*, 3, and references therein.
11. C. D. Chang, S. Han, R. A. Morrison, J. G. Santiesteban, US Patent 5, 999, 643, assigned to Mobil Oil Corporation, 1999.
12. P. T. Patil, K. M. Malshe, P. Kumar, M. K. Dongare, E. Kemnitz, *Catal. Commun.* **2002**, *3*, 411.
13. E. J. Hollstein, J. T. Wei, C. Y. Hsu, US Patent 4, 918, 041, assigned to Sun Refining and Marketing Co., 1990.
14. C.-Y. Hsu, C. R. Heimbuch, C. T. Armes, B. C. Gates, *J. Chem. Soc., Chem. Commun.* **1992**, 1645.
15. F. C. Lange, T.-K. Cheung, B. C. Gates, *Catal. Lett.* **1996**, *41*, 95.
16. S. Kuba, B. C. Gates, R. K. Grasselli, H. Knözinger, *Chem. Commun.* **2001**, 321.
17. R. Ahmad, J. Melsheimer, F. C. Jentoft, R. Schlögl, *J. Catal.* **2003**, *218*, 365.
18. S. Kuba, P. Lukinskas, R. Ahmad, F. C. Jentoft, R. K. Grasselli, B. C. Gates, H. Knözinger, *J. Catal.* **2003**, *219*, 376.
19. H. Liu, G. D. Lei, V. Adeeva, W. M. H. Sachtler, *J. Mol. Catal.* **1995**, *100*, 35.
20. S. Kuba, P. Lukinskas, R. K. Grasselli, B. C. Gates, H. Knözinger, *J. Catal.* **2003**, *216*, 353.
21. J. Luo, R. Stevens, *J. Am. Ceram. Soc.* **1999**, *82*, 1922.
22. C. Morterra, G. Cerrato, F. Pinna, M. Signoretto, *J. Catal.* **1995**, *157*, 109.
23. W. Stichert, F. Schüth, S. Kuba, H. Knözinger, *J. Catal.* **2001**, *198*, 277.
24. J. A. Moreno, G. Poncelet, *J. Catal.* **2001**, *203*, 453.
25. X. Li, K. Nagaoka, R. Olindo, J. A. Lercher, *J. Catal.* **2006**, *238*, 39.
26. X. Li, K. Nagaoka, L. J. Simon, J. A. Lercher, S. Wrabetz, F. C. Jentoft, C. Breitkopf, S. Matysik, H. Papp, *J. Catal.* **2005**, *230*, 214.
27. M. A. Risch, E. E. Wolf, *Appl. Catal. A: General* **1998**, *172*, L1.
28. C. Breitkopf, A. Garsuch, H. Papp, *Appl. Catal. A: General* **2005**, *296*, 148.
29. T. Riemer, D. Spielbauer, M. Hunger, G. A. H. Mekhemer, H. Knözinger, *J. Chem. Soc., Chem. Commun.* **1994**, 1181.
30. M. G. Cutrufello, U. Diebold, R. D. Gonzalez, *Catal. Lett.* **2005**, *101*, 5.
31. D. Fraenkel, *Ind. Eng. Chem. Res.* **1997**, *36*, 52.
32. S. Melada, M. Signoretto, F. Somma, F. Pinna, G. Cerrato, G. Meligrana, C. Morterra, *Catal. Lett.* **2004**, *94*, 193.
33. M. K. Mishra, B. Tyagi, R. V. Jasra, *Ind. Eng. Chem. Res.* **2003**, *42*, 5727.
34. D. Fărcașiu, J. Q. Li, S. Cameron, *Appl. Catal. A: General* **1997**, *154*, 173.
35. J. C. Yori, J. M. Parera, *Appl. Catal. A: General* **1995**, *129*, L151.
36. P. Nascimento, C. Akratapoulou, M. Oszagyan, G. Coudurier, C. Travers, J. F. Joly, J. C. Vedrine, *Proceedings 10th International Congress on Catalysis*, July 19–24, 1992, Budapest, Hungary, *New Frontiers in Catalysis*, L. Guczi, F. Solymosi, P. Tetényi (Eds.), Elsevier, Amsterdam, 1993, p. 1185.
37. F. R. Chen, G. Coudurier, J.-F. Joly, J. C. Védrine, *J. Catal.* **1993**, *143*, 616.
38. A. Hofmann, J. Sauer, *J. Phys. Chem. B*, **2004**, *108*, 14652.
39. B. S. Klose, F. C. Jentoft, R. Schlögl. *J. Catal.* **2005**, *233*, 68.
40. B. S. Klose, Dissertation, Technische Universität Berlin, 2005, http://opus.kobv.de/tuberlin/volltexte/2005/1177/.
41. X. Li, K. Nagaoka, L. J. Simon, R. Olindo, J. A. Lercher, A. Hofmann, J. Sauer, *J. Am. Chem. Soc.* **2005**, *127*, 16159.
42. C. Breitkopf, H. Papp, X. Li, R. Olindo, J. A. Lercher, R. Lloyd, S. Wrabetz, F. C. Jentoft, K. Meinel, S. Förster, K.-M. Schindler, H. Neddermeyer, W. Widdra, A. Hofmann, J. Sauer, *Phys. Chem. Chem. Phys.* **2007**, *9*, 3600.
43. F. Babou, G. Coudurier, J. C. Védrine, *J. Catal.* **1995**, *152*, 341.
44. K. Biró, F. Figueras, S. Békássy, *Appl. Catal. A: General* **2002**, *229*, 235.
45. J. R. Sohn, S. H. Lee, *Appl. Catal. A: General* **2004**, *266*, 89.
46. K. Ebitani, J. Konishi, H. Hattori, *J. Catal.* **1991**, *130*, 257.
47. F.-S. Xiao, *Top. Catal.* **2005**, *35*, 9, and references therein.
48. V. M. Benítez, C. R. Vera, C. L. Pieck, F. G. Lacamoire, J. C. Yori, J. M. Grau, J. M. Parera, *Catal. Today* **2005**, *107–108*, 651.
49. R. Akkari, A. Ghorbel, N. Essayem, F. Figueras, *J. Sol-Gel Sci. Tech.* **2005**, *33*, 121.
50. R. Akkari, A. Ghorbel, N. Essayem, F. Figueras, *J. Sol-Gel Sci. Tech.* **2006**, *38*, 185.
51. X. Yang, R. E. Jentoft, F. C. Jentoft, *Catal. Lett.* **2006**, *106*, 195.
52. M. A. Ecormier, K. Wilson, A. F. Lee, *J. Catal.* **2003**, *215*, 57.
53. D. A. Ward, E. I. Ko, *J. Catal.* **1994**, *150*, 18.
54. B.-Y. Zhao, X.-P. Xu, H.-R. Ma, D.-H. Sun, J.-M. Gao, *Catal. Lett.* **1997**, *45*, 237.
55. H. K. Mishra, K. M. Parida, *Appl. Catal. A: General* **2002**, *224*, 179.
56. A. F. Bedilo, K. J. Klabunde, *J. Catal.* **1998**, *176*, 448.
57. D. J. Zalewski, S. Alerasool, P. K. Doolin, *Catal. Today* **1999**, *53*, 419.
58. J. Haber, J. H. Block, B. Delmon, *Pure Appl. Chem.* **1995**, *67*, 1257.
59. A. L. Hock, *Chemistry & Industry* **1974**, *2*, 864.
60. C. E. Morris, T. L. Vigo, C. M. Welch, *Textile Res. J.* **1981**, 90.
61. W. B. Blumenthal, *Ind. Eng. Chem.* **1954**, *46*, 528.
62. A. Clearfield, P. A. Vaughan, *Acta Crystallogr.* **1956**, *9*, 555.
63. G. M. Muha, P. A. Vaughan, *J. Chem. Phys.* **1960**, *33*, 194.
64. S. Hannane, F. Bertin, J. Bouix, *Bull. Soc. Chim. Fr.* **1990**, *127*, 43.
65. L. M. Toth, J. S. Lin, L. K. Felker, *J. Phys. Chem.* **1991**, *95*, 3106.
66. M. Åberg, J. Glaser, *Inorg. Chim. Acta* **1993**, *206*, 53.
67. K. Matsui, H. Suzuki, M. Ohgai, *J. Am. Ceram. Soc.* **1995**, *78*, 146.
68. A. Singhal, L. M. Toth, J. S. Lin, K. Affholter, *J. Am. Chem. Soc.* **1996**, *118*, 11529.

69. A. Singhal, L. M. Toth, G. Beaucage, J.-S. Lin, J. Peterson, *J. Colloid Interface Sci.* **1997**, *194*, 470.
70. K. Matsui, M. Ohgai, *J. Am. Ceram. Soc.* **1997**, *80*, 1949.
71. M. Z.-C. Hu, J. T. Zielke, J.-S. Lin, C. H. Byers, *J. Mater. Res.* **1999**, *14*, 103.
72. K. Lee, A. Sathyagal, P. W. Carr, A. V. McCormick, *J. Am. Ceram. Soc.* **1999**, *82*, 338.
73. P. Riello, A. Minesso, A. Craievich, A. Benedetti, *J. Phys. Chem. B.* **2003**, *107*, 3390.
74. R. E. Connick, W. H. McVey, *J. Am. Chem. Soc.* **1949**, *71*, 3182.
75. S. Ahrland, D. Karipides, B. Norén, *Acta Chem. Scand.* **1963**, *17*, 411.
76. A. Clearfield, *Rev. Pure Appl. Chem.* **1964**, *14*, 91.
77. E. Matijevic, *Acc. Chem. Res.* **1981**, *14*, 22.
78. R. Ruer, *Z. Anorg. Chem.* **1904**, *42*, 87.
79. F. G. Baglin, D. Breger, *Inorg. Nucl. Chem. Lett.* **1976**, *12*, 173.
80. U. Ciesla, M. Fröba, G. Stucky, F. Schüth, *Chem. Mater.* **1999**, *11*, 227.
81. M. Falinski, *Ann. Chim.* **1941**, *16*, 237.
82. M. Chatterjee, J. Ray, A. Chatterjee, D. Ganguli, *J. Mater. Sci. Lett.* **1989**, *8*, 548.
83. J. D'Ans, H. Eick, *Z. Elektrochem.* **1951**, *55*, 19.
84. I. J. Bear, W. G. Mumme, *Rev. Pure Appl. Chem.* **1971**, *21*, 189.
85. P. J. Squattrito, P. R. Rudolf, A. Clearfield, *Inorg. Chem.* **1987**, *26*, 4240.
86. O. Hauser, *Z. Anorg. Allg. Chem.* **1905**, *45*, 185.
87. I. G. Atanov, L. M. Zaitsev, *Russ. J. Inorg. Chem.* **1967**, *12*, 188.
88. H. Cölfen, H. Schnablegger, A. Fischer, F. C. Jentoft, G. Weinberg, R. Schlögl, *Langmuir* **2002**, *18*, 3500.
89. G. K. Chuah, S. Jaenicke, S. A. Cheong, K. S. Chan, *Appl. Catal. A: General* **1996**, *145*, 267.
90. R. Sui, A. S. Rizkalla, P. A. Charpentier, *Langmuir* **2006**, *22*, 4390.
91. G. K. Chuah, S. Jaenicke, B. K. Pong, *J. Catal.* **1998**, *175*, 80.
92. G. K. Chuah, S. Jaenicke, T. H. Xu, *Surf. Interface Anal.* **1999**, *28*, 131.
93. E. Riedel, *Anorganische Chemie*, 5th Ed., Walter deGruyter, Berlin, 2002, p. 776.
94. B. Li, R. D. Gonzalez, *Ind. Eng. Chem. Res.* **1996**, *35*, 3141.
95. G. K. Chuah, S. H. Liu, S. Jaenicke, J. Li, *Micropor. Mesopor. Mater.* **2000**, *39*, 381.
96. D. A. Ward, E. I. Ko, *J. Catal.* **1995**, *157*, 321.
97. D. Tichit, B. Coq, H. Armendariz, F. Figuéras, *Catal. Lett.* **1996**, *38*, 109.
98. V. Pârvulescu, S. Coman, P. Grange, V. I. Pârvulescu, *Appl. Catal. A: General* **1999**, *176*, 27.
99. R. A. Boyse, E. I. Ko, *J. Catal.* **1997**, *171*, 191.
100. H. Althues, S. Kaskel, *Langmuir* **2002**, *18*, 7428.
101. S. Melada, S. A. Ardizzone, C. L. Bianchi, *Micropor. Mesopor. Mater.* **2004**, *73*, 203.
102. S. Melada, M. Signoretto, S. A. Ardizzone, C. L. Bianchi, *Catal. Lett.* **2001**, *75*, 199.
103. C. T. Kresge, M. E. Leonowicz, W. J. Roth, J. C. Vartuli, *Nature* **1992**, *359*, 710.
104. F. Schüth, *Chem. Mater.* **2001**, *13*, 3184.
105. U. Ciesla, S. Schacht, G. D. Stucky, K. K. Unger, F. Schüth, *Angew. Chem.* **1996**, *108*, 597; U. Ciesla, S. Schacht, G. D. Stucky, K. K. Unger, F. Schüth, *Angew. Chem. Int. Ed. Eng.* **1996**, *35*, 541.
106. V. N. Romannikov, V. B. Fenelonov, E. A. Paukshtis, A. Yu. Derevyankin, V. I. Zaikovskii, *Micropor. Mesopor. Mater.* **1998**, *21*, 411.
107. D. M. Antonelli, *Adv. Mater.* **1999**, *11*, 487.
108. X. Yang, F. C. Jentoft, R. E. Jentoft, F. Girgsdies, T. Ressler, *Catal. Lett.* **2002**, *81*, 25.
109. Y. Sun, L. Yuan, W. Wang, C.-L. Chen, F.-S. Xiao, *Catal. Lett.* **2003**, *87*, 57.
110. www.zrchem.com, www.sigma-aldrich.com.
111. G. Hausinger, A. Reimer, J. Schönlinner, F. Schmidt, German patent application 100 33 477, Süd-Chemie AG, 2002.
112. R. D. Gillespie, M. Cohn, US Patent 6,706,659, assigned to UOP LLC, 2004.
113. F. C. Jentoft, A. Hahn, J. Kröhnert, G. Lorenz, R. E. Jentoft, T. Ressler, U. Wild, R. Schlögl, C. Häßner, K. Köhler, *J. Catal.* **2004**, *224*, 124, and references therein.
114. K. Arata, M. Hino, N. Yamagata, *Bull. Chem. Soc. Jpn.* **1990**, *63*, 244.
115. E. Escalona Platero, M. Peñarroya Mentruit, C. Otero Areán, A. Zecchina, *J. Catal.* **1996**, *162*, 268.
116. G. D. Yadav, A. D. Murkute, *J. Catal.* **2004**, *224*, 218.
117. Y. Xia, W. Hua, Z. Gao, *Appl. Catal. A: General* **1999**, *185*, 293.
118. R. L. Marcus, R. D. Gonzalez, E. L. Kugler, A. Auroux, *Chem. Eng. Commun.* **2003**, *190*, 1601.
119. M. Bensitel, O. Saur, J.-C. Lavalley, B. A. Morrow, *Mater. Chem. Phys.* **1988**, *19*, 147.
120. J. M. Parera, *Catal. Today* **1992**, *15*, 481.
121. D. Fărcaşiu, J. Q. Li, *Appl. Catal.* **1995**, *128*, 97.
122. C. D. Chang, F. T. DiGuiseppi, J. G. Santiesteban, US Patent 5,780,382, assigned to Mobil Oil Corporation, 1998.
123. J. B. Laizet, A. K. Søiland, J. Leglise, J. C. Duchet, *Top. Catal.* **2000**, *10*, 89.
124. A. Corma, A. Martínez, C. Martínez, *Appl. Catal. A: General* **1996**, *144*, 249.
125. D. F. Stec, R. S. Maxwell, H. Cho, *J. Catal.* **1998**, *176*, 14.
126. V. Quaschning, J. Deutsch, P. Druska, H.-J. Niclas, E. Kemnitz, *J. Catal.* **1998**, *177*, 164.
127. C. Miao, W. Hua, J. Chen, Z. Gao, *Catal. Lett.* **1996**, *37*, 187.
128. Z. Gao, Y. Xia, W. Hua, C. Miao, *Top. Catal.* **1998**, *6*, 101.
129. Y. D. Xia, W. M. Hua, Y. Tang, Z. Gao, *Chem. Commun.* **1999**, 1899.
130. W. Hua, Y. Xia, Y. Yue, Z. Gao, *J. Catal.* **2000**, *196*, 104.
131. P. Canton, R. Olindo, F. Pinna, G. Strukul, P. Riello, M. Meneghetti, G. Cerrato, C. Morterra, A. Benedetti, *Chem. Mater.* **2001**, *13*, 1634.
132. M. Perez-Luna, J. A. Toledo-Antonio, F. Hernandez-Beltrán, H. Armendariz, A. Garcia Borquez, *Catal. Lett.* **2002**, *83*, 201.
133. M. Hino, K. Arata, *React. Kin. Catal. Lett.* **2004**, *81*, 321.
134. M. Signoretto, S. Melada, F. Pinna, S. Polizzi, G. Cerrato, C. Morterra, *Micropor. Mesopor. Mater.* **2005**, *81*, 19.
135. J. C. Yori, J. M. Parera, *Appl. Catal. A: General* **1996**, *147*, 145.
136. R. Srinivasan, R. A. Keogh, B. H. Davis, *Appl. Catal. A: General* **1995**, *130*, 135.
137. T. Yamamoto, T. Tanaka, S. Takenaka, S. Yoshida, T. Onari, Y. Takahashi, T. Kosaka, S. Hasegawa, M. Kudo, *J. Phys. Chem. B* **1999**, *103*, 2385.
138. R. E. Jentoft, A. Hahn, F. C. Jentoft, T. Ressler, *J. Synchrotron Rad.* **2001**, *8*, 563.
139. M. A. Coelho, D. E. Resasco, E. C. Sikabwe, R. L. White, *Catal. Lett.* **1995**, *32*, 253.
140. E. A. García, E. H. Rueda, A. J. Rouco, *Appl. Catal. A: General* **2001**, *210*, 363.
141. M. Hino, K. Arata, *Catal. Lett.* **1996**, *34*, 125.

142. J. M. M. Millet, M. Signoretto, P. Bonville, *Catal. Lett.* **2000**, *64*, 135.
143. J. C. Yori, J. M. Parera, *Appl. Catal. A: General* **1995**, *129*, 83.
144. M. Perez-Luna, J. A. Toledo Antonio, A. Montoya, R. Rosas-Salas, *Catal. Lett.* **2004**, *97*, 59.
145. A. Corma, J. M. Serra, A. Chica, *Catal. Today* **2003**, *81*, 495.
146. M. L. Occelli, D. A. Schiraldi, A. Auroux, R. A. Keogh, B. H. Davis, *Appl. Catal. A: General* **2001**, *209*, 165.
147. C. T. Kresge, C. D. Chang, J. G. Santiesteban, D. S. Shihabi, S. A. Stevenson, J. C. Vartuli, US Patent 5, 902, 767, assigned to Mobil Oil Corporation, 1999.
148. A. Jatia, C. Chang, J. D. MacLeod, T. Okubo, M. E. Davis, *Catal. Lett.* **1994**, *25*, 21.
149. V. Adeeva, G. D. Lei, W. M. H. Sachtler, *Appl. Catal. A: General* **1994**, *118*, L11.
150. A. S. Zarkalis, C.-Y. Hsu, B. C. Gates, *Catal. Lett.* **1994**, *29*, 235.
151. V. Adeeva, J. W. de Haan, J. Jänchen, G. D. Lei, V. Schünemann, L. J. M. van de Ven, W. M. H. Sachtler, R. A. van Santen, *J. Catal.* **1995**, *151*, 364.
152. E. C. Sikabwe, M. A. Coelho, D. E. Resasco, R. L. White, *Catal. Lett.* **1995**, *34*, 23.
153. J. E. Tábora, R. J. Davis, *J. Chem. Soc., Faraday Trans.* **1995**, *91*, 1825.
154. K. T. Wan, C. B. Khouw, M. E. Davis, *J. Catal.* **1996**, *158*, 311.
155. R. Srinivasan, R. A. Keogh, A. Ghenciu, D. Fărcaşiu, B. H. Davis, *J. Catal.* **1996**, *158*, 502.
156. M. A. Coelho, W. E. Alvarez, E. C. Sikabwe, R. L. White, D. E. Resasco, *Catal. Today* **1996**, *28*, 415.
157. A. S. Zarkalis, C.-Y. Hsu, B. C. Gates, *Catal. Lett.* **1996**, *37*, 1.
158. M. Benaïssa, J. G. Santiesteban, G. Diaz, M. José-Yacamán, *Surf. Sci.* **1996**, *364*, L591.
159. X. Song, K. R. Reddy, A. Sayari, *J. Catal.* **1996**, *161*, 206.
160. C. Morterra, G. Cerrato, S. Di Ciero, M. Signorotto, A. Minesso, F. Pinna, G. Strukul, *Catal. Lett.* **1997**, *49*, 25.
161. W. E. Alvarez, H. Liu, D. E. Resasco, *Appl. Catal. A: General* **1997**, *162*, 103.
162. A. Sayari, Y. Yang, X. Song, *J. Catal.* **1997**, *167*, 346.
163. V. Adeeva, H.-Y. Liu, B.-Q. Xu, W. M. H. Sachtler, *Top. Catal.* **1998**, *6*, 61.
164. S. Rezgui, A. Liang, T.-K. Cheung, B. C. Gates, *Catal. Lett.* **1998**, *53*, 1.
165. F. C. Jentoft, *Erdöl, Erdgas, Kohle* **1998**, *114*, 441.
166. S. X. Song, R. A. Kydd, *Catal. Lett.* **1998**, *51*, 95.
167. D. R. Milburn, R. A. Keogh, D. E. Sparks, B. H. Davis, *Appl. Surf. Sci.* **1998**, *126*, 11.
168. M. Scheithauer, E. Bosch, U. A. Schubert, H. Knözinger, T.-K. Cheung, F. C. Jentoft, B. C. Gates, B. Tesche, *J. Catal.* **1998**, *177*, 137.
169. T. Tanaka, T. Yamamoto, Y. Kohno, T. Yoshida, S. Yoshida, *Jpn. J. Appl. Phys.* **1999**, *38*, 30.
170. A. Sayari, Y. Yang, *J. Catal.* **1999**, *187*, 186.
171. R. E. Jentoft, B. C. Gates, *Catal. Lett.* **2001**, *72*, 129.
172. T.-K. Cheung, J. L. d'Itri, F. C. Lange, B. C. Gates, *Catal. Lett.* **1995**, *31*, 153.
173. T.-K. Cheung, F. C. Jentoft, J. L. d'Itri, B. C. Gates, *Chem. Eng. Sci.* **1997**, *52*, 4607.
174. T.-K. Cheung, B. C. Gates, *Top. Catal.* **1998**, *6*, 41, and references therein.
175. A. S. Chelappa, R. C. Miller, W. J. Thomson, *Appl. Catal. A: General* **2001**, *209*, 359.
176. A. H. P. Hahn, T. Ressler, U. Wild, R. E. Jentoft, F. C. Jentoft, in preparation.
177. P. Li, I. W. Chen, J. E. Penner-Hahn, *Phys. Rev. B* **1993**, *48*, 10063, 10074, 10082.
178. J. Stöcker, R. Collongues, *Compt. Rend.* **1957**, *245*, 695.
179. H. J. Stöcker, *Ann. Chim.* **1960**, *5*, 1459.
180. J. Stöcker, *Bull. Soc. Chim.* **1961**, 78.
181. M. Yashima, N. Ishizawa, M. Yoshimura, *J. Am. Ceram. Soc.* **1992**, *75*, 1541, 1550.
182. D. Das, H. K. Mishra, A. K. Dalai, K. M. Parida, *Catal. Lett.* **2004**, *93*, 185.
183. B. S. Klose, F. C. Jentoft, R. Schlögl, I. R. Subbotina, V. B. Kazansky, *Langmuir* **2005**, *21*, 10564.
184. X.-R. Chen, C.-L. Chen, N.-P. Xua, S. Han, C.-Y. Mou, *Catal. Lett.* **2003**, *85*, 177.
185. X. Li, K. Nagaoka, J. A. Lercher, *J. Catal.* **2004**, *227*, 130.
186. E. A. Garcia, M. A. Volpe, M. L. Ferreira, E. H. Rueda, *Lat. Am. Appl. Res.* **2005**, *35*, 281.
187. K. Arata, H. Matsuhashi, M. Hino, H. Nakamura, *Catal. Today* **2003**, *81*, 17.
188. C. J. Norman, P. A. Goulding, I. McAlpine, *Catal. Today* **1994**, *20*, 313.
189. R. Srinivasan, D. Taulbee, B. H. Davis, *Catal. Lett.* **1991**, *9*, 1.
190. S. Chokkaram, R. Srinivasan, D. R. Milburn, B. H. Davis, *J. Colloid Interface Sci.* **1994**, *165*, 160.
191. T. Tatsumi, H. Matsuhashi, K. Arata, *Bull. Chem. Soc. Jpn.* **1996**, *69*, 1191.
192. J. Livage, K. Doi, C. Mazières, *J. Am. Ceram. Soc.* **1968**, *51*, 349.
193. A. Keshavaraja, N. E. Jacob, A. V. Ramaswamy, *Thermochim. Acta* **1995**, *254*, 267.
194. B. Djuricic, S. Pickering, D. McGarry, P. Glaude, P. Tambuyser, K. Schuster, *Ceram. Int.* **1995**, *21*, 195.
195. M. Sorrentino, L. Steinbrecher, F. Hazel, *J. Colloid Interface Sci.* **1969**, *31*, 307.
196. R. Srinivasan, B. H. Davis, *J. Colloid Interface Sci.* **1993**, *156*, 400.
197. K. Haberko, A. Ciesla, A. Pron, *Ceram. Int.* **1975**, *1*, 111.
198. R. Srinivasan, M. B. Harris, S. F. Simpson, R. J. De Angelis, B. H. Davis, *J. Mater. Res.* **1988**, *3*, 787.
199. P. D. L. Mercera, J. G. van Ommen, E. B. M. Doesburg, A. J. Burggraaf, J. R. H. Ross, *Appl. Catal.* **1990**, *57*, 127.
200. I. Molodetsky, A. Navrotsky, M. J. Paskowitz, V. J. Leppert, S. H. Risbud, *J. Non-Cryst. Solids* **2000**, *262*, 106.
201. S. Xie, E. Iglesia, A. T. Bell, *Chem. Mater.* **2000**, *12*, 2442.
202. I. Molodetsky, A. Navrotsky, M. Lajavardi, A. Brune, *Z. Phys. Chem.* **1998**, *207*, 59.
203. Y. Murase, E. Kato, *J. Am. Ceram. Soc.* **1983**, *66*, 196.
204. A. Hahn, T. Ressler, R. E. Jentoft, F. C. Jentoft, *Chem. Commun.* **2001**, 537 and electronic support information.
205. A. H. P. Hahn, R. E. Jentoft, T. Ressler, G. Weinberg, R. Schlögl, F. C. Jentoft, *J. Catal.* **2005**, *236*, 324.
206. J. Berzelius, *J. Chem. Phys.* (Schweigger, Nürnberg) **1812**, *6*, 119.
207. J. Berzelius, *Ann. Phys.* (Gilbert, Leipzig) **1812**, *42*, 276.
208. J. Berzelius, *J. Chem. Phys.* (Schweigger, Nürnberg) **1818**, *12*, 51.
209. R. Ruer, *Z. Anorg. Allg. Chem.* **1905**, *43*, 282.
210. L. Wöhler, *Koll.-Zeitschr.* **1926**, *38*, 97.
211. R. A. Keogh, R. Srinivasan, B. H. Davis, *J. Catal.* **1995**, *151*, 292.
212. R. A. Comelli, C. R. Vera, J. M. Parera, *J. Catal.* **1995**, *151*, 96.
213. B. H. Davis, *Catal. Today* **1994**, *20*, 219.
214. C. R. Vera, J. M. Parera, *J. Catal.* **1997**, *165*, 254.

215. C. Morterra, G. Cerrato, M. Signoretto, *Catal. Lett.* **1996**, *41*, 101.
216. S. X. Song, R. A. Kydd, *J. Chem. Soc. Faraday Trans.* **1998**, *94*, 1333.
217. F. C. Jentoft, B. C. Gates, unpublished results.
218. A. V. Chadwick, G. Mountjoy, V. M. Nield, I. J. F. Poplett, M. E. Smith, J. H. Strange, M. G. Tucker, *Chem. Mater.* **2001**, *13*, 1219.
219. X. Li, K. Nagaoka, L. J. Simon, R. Olindo, J. A. Lercher, *Catal. Lett.* **2007**, *113*, 34.
220. M. Hino, K. Arata, *Chem. Lett.* **1979**, 477.
221. T. Yamaguchi, T. Jin, K. Tanabe, *J. Phys. Chem.* **1986**, *90*, 3148.
222. I. J. Dijs, J. W. Geus, L. W. Jenneskens, *J. Phys. Chem. B* **2003**, *107*, 13403.
223. C. Morterra, G. Cerrato, F. Pinna, M. Signoretto, *J. Catal.* **1995**, *157*, 109.
224. M. Signoretto. F. Pinna, G. Strukul, G. Cerrato, C. Morterra, *Catal. Lett.* **1996**, *36*, 129.
225. M. Hino, K. Arata, *Catal. Lett.* **1995**, *30*, 25.
226. G. Larsen, L. M. Petkovic, *Appl. Catal. A: General* **1996**, *148*, 155.
227. S. Vijay, E. E. Wolf, *Appl. Catal. A: General* **2004**, *264*, 117.
228. M. Risch, E. E. Wolf, *Appl. Catal. A: General* **2001**, *206*, 283.
229. S. Y. Kim, J. G. Goodwin, Jr., D. Galloway, *Catal. Today* **2000**, *63*, 21.
230. C. R. Vera, J. C. Yori, C. L. Pieck, S. Irusta, J. M. Parera, *Appl. Catal. A: General* **2003**, *240*, 161.
231. R. E. Jentoft, A. H. P. Hahn, F. C. Jentoft, T. Ressler, *Phys. Chem. Chem. Phys.* **2005**, *7*, 2830.
232. S. Kuba, P. Concepción Heydorn, R. K. Grasselli, B. C. Gates, M. Che, H. Knözinger, *Phys. Chem. Chem. Phys.* **2001**, *3*, 146.

2.3.10
Catalysis by Ion-Exchange Resins

*Bruce C. Gates**

2.3.10.1 Introduction

Ion-exchange resins, typified by sulfonated crosslinked polystyrene, have been used widely in industry for many decades to catalyze reactions that are also catalyzed by mineral acids. These solid polymeric catalysts may be thought of as anchored sulfonic acid ($-SO_3H$) groups and solid analogues of sulfuric acid (or, more precisely, of toluenesulfonic acid). They were the first synthetic analogues of soluble molecular species to be used on a large scale as catalysts, having been applied in industry more than 60 years ago for esterification. Ion-exchange resins were the forerunners of what is now a large class of "anchored" catalysts that are solid analogues of soluble molecular or ionic species.

---

* Corresponding author.

Sulfonic acid resins are finding increasing application as replacements for soluble mineral acid catalysts, primarily because they minimize the detrimental environmental effects of acidic waste streams. Ion-exchange resin catalysts also offer the processing advantages of ease of separation from products and minimization of corrosion. These advantages are offset by their cost (they are much more expensive, per acid group, than mineral acids such as H_2SO_4 and HCl) and their relative lack of stability (see below).

The term "ion-exchange resin" refers to an organic polymeric resin that exchanges ions with a liquid that contains ions. Anion-exchange resins exchange anions, and cation-exchange resins exchange cations. For example, a resin with anchored $-SO_3H$ groups, in water, undergoes dissociation of these groups, giving (hydrated) protons and anchored sulfonate anions; cations from a solution in contact with the resin replace some of the protons. Thus, an aqueous solution containing Ca^{2+} ions will become enriched in hydrogen ions and depleted in Ca^{2+} ions as a result of contacting with the sulfonic acid resin. Sulfonic acid resins are used on a large scale in the purification of hard water. When anions such as Cl^- are replaced by OH^- from an anion-exchange resin and cations such as Ca^{2+} are replaced by H^+, the net effect is replacement of calcium chloride with water (an example of water softening).

The ion-exchange resins that are most important as industrial catalysts are sulfonic acid resins, and these are the principal topic of this chapter, the goals of which are to provide a summary of the synthesis and structural properties of these catalysts, with a summary of the most important catalytic applications and descriptions of the performance. The coverage of this subject here is not extensive, and earlier reviews [1–4] have provided more depth. The syntheses of these catalysts are described here only in broad terms, as key details remain proprietary. The inferences about catalyst structures are based on physical characterization methods that are also glossed over here.

Information closely complementing the content of the present chapter is available elsewhere in this Handbook (Chapter 13.10), which describes the synthesis of methyl *tert*-butyl ether (MTBE) and related ethers, which are manufactured on a large scale with sulfonic acid resin catalysts, and also in Chapter 10.6, in which details are described of the catalytic distillation that is applied with such catalysts in the manufacture of these ethers.

2.3.10.2 Classes of Ion-Exchange Resin Catalysts: The Importance of Porosity

When they were first prepared, ion-exchange resins were gels such as crosslinked polystyrene, and materials without true pores; transport within such materials results from the movement of species between the chains of polymer, and is affected by swelling of the gel, which separates the chains. Both hydrocarbons (which have an affinity for the crosslinked polystyrene) and polar solvents (which have an affinity for functional groups such as $-SO_3H$ groups) cause the gels to swell. The lower the degree of crosslinking of the gel, the more it can swell. Lightly crosslinked resins [e.g., poly(styrene-divinylbenzene) with 2% divinylbenzene] can swell to many times their unswelled volumes and become almost indistinct from liquids.

Because highly crosslinked resins are more stable than their lightly crosslinked counterparts, the former are preferred in catalytic applications. However, because gel-form resins with high crosslink densities may swell so little as to be almost impermeable to many reactants, they are generally ineffective catalysts, with reactions occurring only near their peripheral surfaces. In order to render a large fraction of the anchored catalytic groups accessible in highly crosslinked resins, research groups some decades ago developed methods for the synthesis of ion-exchange resins which incorporated true pores. These macroporous resins are referred to as "macroreticular", and many of the important catalytic applications involve such resins; such resins are usually highly crosslinked, and may contain up to 20% divinylbenzene, and sometimes even more.

2.3.10.3 Examples of Industrial Catalysis by Ion-Exchange Resins

Examples of important industrial reactions catalyzed by ion-exchange resins are listed in Table 1; additional reactions inferred by Chakrabarti and Sharma [1] to be practiced industrially are listed in the review by these authors (who typically cited patent literature but not references to specific processes and manufacturers). The list of Table 1 is representative of lists provided by industrial manufacturers of ion-exchange resin catalysts. Details of applications of ion-exchange resin catalysts are provided in the review by Harmer and Sun [2].

Macroporous sulfonic acid resins predominate in these applications. The reactants are principally organic compounds that are relatively strong bases, chiefly those with oxygen-containing functional groups, such as alcohols. In a few processes, the reactants are unfunctionalized hydrocarbons, including alkenes but not alkanes.

2.3.10.4 Synthesis of Ion-Exchange Resins

Ion-exchange resins are typically synthesized from the hydrocarbon monomers (usually styrene and divinylbenzene) by copolymerization, and the catalytic functional groups are added in a later step, such as sulfonation of the copolymer with concentrated sulfuric acid. Formation of the hydrocarbon copolymers by free-radical emulsion polymerization results in near-spherical resin beads, and this is the common form of the gel-form catalysts, which are available commercially in a variety of particle size ranges.

Macroporous catalysts, comprising aggregates of approximately spherical gel-form microparticles, are also supplied as spherical beads. Synthesis of the macroporous resins involves copolymerization of the monomers in the presence of a poorly swelling solvent such as heptane. After the polymer has been formed, this solvent is removed, leaving void spaces that are the pores.

The conditions of the sulfonation have usually been chosen to provide approximately one $-SO_3H$ group per benzene ring in the resin (which corresponds to approximately 5 meq of $-SO_3H$ groups per gram dry

---

References see page 285

Tab. 1 Industrial reactions catalyzed by ion-exchange resins

Reactant(s)	Product(s)	Catalyst
Methanol + isobutene	Methyl *tert*-butyl ether (MTBE)	Sulfonic acid resin
Ethanol + isobutene	Ethyl *tert*-butyl ether (ETBE)	Sulfonic acid resin
Methanol + isoamylene	*tert*-Amyl methyl ether (TAME)	Sulfonic acid resin
Isobutene + water	*tert*-Butyl alcohol	Sulfonic acid resin
tert-Butyl alcohol	Isobutene + water	Sulfonic acid resin
Phenols + alkenes	Alkylphenols	Sulfonic acid resin
Phenol + acetone	Bisphenol A + water	Sulfonic acid resin
Alkene (e.g., *n*-butenes) + carboxylic acid (e.g., acetic acid)	Ester (e.g., *s*-butyl acetate) + water	Sulfonic acid resin
Methanol + CO	Acetic acid	Anion-exchange resin with pyridinium groups anchoring rhodium carbonyl complex

Tab. 2 Physical properties of sulfonic acid resin catalysts. In the calculation of pore diameters, the macropores were modeled as cylinders and the microparticles as spheres [5]

Styrene content/ wt.%	Divinyl-benzene content/wt.%	Ethylvinyl-benzene content/wt.%	$-SO_3H$ group content/ meq g^{-1}	Skeletal density/ g cm^{-3}	Apparent density/ g cm^{-3}	BET surface area/m^2 g^{-1}	Average micro-particle diameter/nm	Average macropore diameter/nm
85.9	8.0	6.1	5.46 ± 0.12	1.500 ± 0.005	0.482 ± 0.005	24 ± 3	167 ± 21	235 ± 30
71.8	16.0	12.2	5.25 ± 0.06	1.426 ± 0.005	0.460 ± 0.005	23 ± 3	183 ± 24	256 ± 34
85.9	8.0	6.1	5.47 ± 0.14	1.410 ± 0.005	0.495 ± 0.005	11 ± 3	387 ± 105	477 ± 130
85.9	8.0	6.1	5.54 ± 0.18	1.473 ± 0.005	0.709 ± 0.005	12.5 ± 3	326 ± 78	234 ± 56

resin). Because $-SO_3H$ groups deactivate the ring for further substitution, forcing conditions are required to produce higher degrees of sulfonation. It is inferred that such conditions are applied in the manufacture of some of today's sulfonic acid resin catalysts, which have acid group densities in excess of 5 meq g^{-1} (such details remain proprietary). Chakrabarti and Sharma [1] suggested that commercial catalysts with degrees of sulfonation up to 1.2–1.8 $-SO_3H$ groups per aromatic ring may be available.

The gel-form polymers are translucent and amber in color; the porous polymers are generally gray or brown and opaque. Variations in the conditions of synthesis and monomer compositions lead to a wide range of physical properties of macroporous resins, as illustrated by the data in Table 2 for catalysts composed of the monomers styrene, divinylbenzene, and ethylvinylbenzene [5].

Although the predominant cation-exchange resin catalysts are functionalized crosslinked polystyrene, many reports exist of catalysis by more strongly acidic polymers that consist of $-SO_3H$ groups anchored to perfluorinated backbones. However, these resins are relatively expensive, and there is a lack of information regarding their large-scale application as catalysts. Cation-exchange resins with relatively weak acidic groups – including carboxylic acid, phosphonic acid, phosphinic acid, and arsonic acid groups – have been prepared, but their applications in catalysis are insignificant.

Anion-exchange resin catalysts with groups including quaternary ammonium, quaternary phosphonium, and pyridinium groups have been reported, but there are few large-scale catalytic applications.

2.3.10.5 Structure, Influence of Water, and Acidity in Ion-Exchange Resins

The role of water in catalysis by ion-exchange resins is often crucial. When excess water is present in a sulfonic acid resin, the sulfonic acid groups are fully dissociated, and for all practical purposes the catalyst can be regarded as an aqueous solution of a strong (dissociated) acid in the presence of an inert solid polymer. The catalytic properties are essentially those of strong aqueous mineral acids. Such catalysis is referred to as specific acid catalysis (catalysis by hydrated protons), and the subject ties directly to an extensive literature that is decades old [6, 7].

On the other hand, when only a small amount of water (or another strongly polar solvent) is present, the $-SO_3H$ groups remain undissociated and are themselves the catalytic sites. This type of catalysis is referred to as general acid catalysis (catalysis by undissociated acid groups), and one seeks parallels to catalysis in solution by compounds such as p-toluenesulfonic acid. The catalytic properties of undissociated $-SO_3H$ groups are much different from those of hydrated protons, in part because their acid–base properties are much different. Thus, sulfonic acid resins in the absence of water may be highly active for some reactions, but almost inactive in the presence of significant amounts of water.

Infrared spectroscopy has been used to characterize the interactions of the resin $-SO_3H$ groups with each other [8]. In a dry resin with enough flexibility of the polymer network (not too high a crosslink density) and a sufficiently high concentration of $-SO_3H$ groups, these groups bond strongly with each other. The structures include those that are dimer-like, trimer-like, etc., and they form part of a hydrogen-bonded network of which the following structure is representative [8]:

When water is added to the dry resin, it intrudes into the network, breaking the bonds between the acid groups and becoming hydrogen-bonded within the network. Even trace amounts of water can inhibit some catalytic reactions strongly; this is because the water competes for the catalytic sites with reactants that are weaker bases, and because water interrupts the network and limits the opportunities for reactants to be (hydrogen-)bonded within it. Infrared spectroscopy has been used to determine the structure of water bonded to $-SO_3H$ groups in the resins [8]; the following structure, which shows one water molecule hydrogen-bonded to three $-SO_3H$ groups, is representative (the water is bound extremely tightly within this structure, and cannot be removed, even by heating the resin under vacuum):

When enough water is present (approximately 2 molecules per $-SO_3H$ group [8]), the acid groups become dissociated. Rapid removal of water by evacuation with heating (e.g., at 393 K) from a highly hydrated resin is effective in removing all of the water [8].

Polar molecules other than water (e.g., alcohols) can bond similarly to the $-SO_3H$ groups in the network, and such bonding may facilitate catalytic reactions. It is evident that the $-SO_3H$ groups in the network are not generally single, site-isolated catalytic sites; rather, it is the network that functions as a catalyst. The $-SO_3H$ groups are both proton donors and proton acceptors; the concerted action of proton-donor groups with proton-acceptor groups is also important in catalysis by some enzymes.

When the network of $-SO_3H$ groups is broken up by the presence of too much reactant, such as an alcohol, the activity of the catalyst declines; thus, alcohols – like water – act as reaction inhibitors. If the alcohols are reactants, then they can cause self-inhibition of catalysis by breaking up the network of $-SO_3H$ groups; at sufficiently high concentrations, alcohols even cause dissociation of the $-SO_3H$ groups. Such results have been illustrated for the dehydration reactions of alcohols [9].

Sulfonic acid groups are strongly acidic, and addition of the base water (or cations such as sodium, for example) lowers the acid strength and, for the typical catalytic reaction, also lowers the catalytic activity. The activities of sulfonic acid resins for reactions taking place in a network of $-SO_3H$ groups are typically higher than those of resins with isolated $-SO_3H$ groups, and the difference reflects the interactions of the acid groups and their roles as both proton acceptors and proton donors.

The result of these interactions has occasionally been shown to confer reactivity associated with very strong acids. For example, in the reaction of isobutyl alcohol to give butenes and water, some isobutane and tars were observed in addition to butenes and water when the partial pressure of the alcohol in the gas phase was <0.2 bar (2×10^4 Pa) (this is consistent with hydride abstraction by a cationic intermediate) [9]. When the alcohol partial pressure was increased, however, the formation of isobutane and tars ceased as the acidity was leveled, and the reaction was inhibited by the alcohol itself.

2.3.10.6 Accessibility, Swelling, and Mass Transfer Effects

Catalytic groups in the interior regions of polymer particles are effective only when the reactants reach them. Accessibility to these groups is facilitated when the resin is swelled. Swelling is favored by strongly polar compounds such as water or methanol, both of which interact strongly with $-SO_3H$ groups, and by hydrocarbons which interact strongly with the hydrocarbon polymer backbone. If the resin crosslinking is too extensive, however, then the swelling is limited and the reactants may not find efficient access to the interior catalytic groups. Highly crosslinked resins are usually preferred only when they offer stability advantages (see below).

When gel-form catalysts with moderate degrees of crosslinking (e.g., 8% divinylbenzene) operate in the presence of excess liquid water, they become highly swelled and the acid groups are dissociated. The catalysis is then modeled as occurring in a homogeneous aqueous solution of (hydrated) protons within the polymer beads. It is possible to measure the distribution coefficients of the reactants (and products) in the resin phase relative to the surrounding liquid phase, and to represent the reaction as a liquid-phase reaction taking place in the presence of the reactants and products at the concentrations prevailing in the swelled resin. These concentrations may differ substantially from those in the surrounding liquid, especially when some components cause selective swelling the resin. Thus, the reaction is considered to be pseudo-homogeneous in the resin interior; such a reaction model is best for aqueous reactant mixtures, and has found only minimal application in non-aqueous media.

Because gel-form resins (unless they are only lightly crosslinked) usually do not swell substantially when the reactants are non-aqueous, the interior volume is often not fully accessible to the reactants (i.e., the effectiveness factors are low). For example, the hydrocarbons propene and benzene (which react to give alkylation products) barely cause swelling of a gel-form resin that is 8% crosslinked, and the resin has such a low effectiveness factor as to be almost entirely ineffective [10].

References see page 285

In fact, macroporous resins are preferred with such reactants, because most of the catalytic groups are accessible as they are either near the interior surfaces or inside the polymer network and not far from such surfaces. In order to account rigorously for the performance of such catalysts, one should monitor both a *macro*effectiveness factor (representing the accessibility of the true interior surface, via macropores) and a *micro*effectiveness factor (representing the accessibility of the catalytic groups in the interior of the gel-form microparticles comprising the resin beads). Although microeffectiveness factors would have to account for the swelling of the microparticles, such factors are rarely known because they are difficult to measure. The uncertainty arises in part because it is difficult to prepare resins with independently varied crosslink densities and microparticle sizes [5], and in part because there are gradients in crosslink densities and acid group concentrations within the microparticles.

Although the microparticles often may not swell substantially (because they are highly crosslinked), the overall effectiveness factors of macroporous resin particles may often be close to unity, because a large fraction of the catalytic groups are close to the true internal surfaces. Typically, these macroporous catalysts (which usually have high degrees of crosslinking) working with reactants that do not cause so much swelling are modeled as true surface catalysts, although this is an oversimplification.

2.3.10.7 Mechanisms of Reactions Catalyzed by Sulfonic Acid Resins

When the resin particles are caused to swell with water and reactants, the $-SO_3H$ groups are dissociated and the catalysis essentially matches that in aqueous solutions of mineral acids; this is a well-established topic in homogeneous catalysis [6, 7], and so will not be reviewed at this point.

When only a little water is present and the $-SO_3H$ groups are undissociated, these groups act as catalytic sites on solids. However, these solids are not those of typical catalysts with two-dimensional surfaces, but instead are flexible, three-dimensional matrices. The catalytic reaction mechanisms often involve multiple $-SO_3H$ groups, with reactants bonded into a network of these groups. A proton is donated from one $-SO_3H$ group to a reactant, and a proton is then donated back to the network from an intermediate species as the reaction occurs. This process is often concerted and referred to as "push–pull" catalysis. Such mechanisms are common on solid acids, ranging from Al_2O_3 to zeolites. The $-SO_3H$ groups act simultaneously as proton donors and proton acceptors. The rates of such concerted reactions may often be much greater than the rates of reactions catalyzed by individual, independent groups such as $-SO_3H$, and the activation energies are often markedly less than those of reactions in which a proton must be fully donated to a reactant, after which a protonated intermediate must donate a proton back to the catalyst.

Evidence of the involvement of networks of $-SO_3H$ groups has been obtained not only from infrared spectra showing reactants bonded in the network of these groups [8], but also from measurements of rates of catalytic reactions as a function of the fraction of the groups that are $-SO_3H$ groups (with the remainder being their salts, such as $-SO_3Na$). For example, the dehydration reaction of *tert*-butyl alcohol to produce isobutene is characterized by an order of reaction in $-SO_3H$ groups of approximately 4 when the density of these groups is relatively high, and this is consistent with a mechanism such as that shown schematically in Fig. 1 [11]. This strong dependence on $-SO_3H$ group content, which pertains with some generality to reactions catalyzed by sulfonic acid resins in the (near) absence of water and other strongly polar compounds, indicates the benefit of using forcing conditions of sulfonation to provide high densities of $-SO_3H$ groups in the catalyst.

When the $-SO_3H$ groups in the resin catalyzing the dehydration of *tert*-butyl alcohol were highly diluted with catalytically inactive groups (e.g., $-SO_3Na$), the increase in the rate of the catalytic reaction was approximately linear with respect to the concentration of $-SO_3H$ groups, which indicated that the groups acted alone (as site-isolated groups). Consequently, the catalytic activity was much less than when the reaction occurred in the presence of a network of $-SO_3H$ groups acting in concert [11].

It is clear that for catalysis to occur efficiently in a network of $-SO_3H$ groups, the latter must be anchored to polymers that are sufficiently flexible to allow them to interact; thus, the crosslink density of the resin has an important influence on the catalysis. (It should be noted that crosslink densities are usually not uniform within resins.)

The simplified mechanisms associated with Eley–Rideal and Langmuir–Hinshelwood models are often good approximations of reactions catalyzed by resin-bound $-SO_3H$ groups. For example, the dimerization of isobutene catalyzed by a macroporous resin was inferred to proceed according to a mechanism whereby an isobutene molecule from the fluid phase reacted in a rate-determining step with a chemisorbed isobutene [12]. Evidence for such a mechanism was found from the kinetics of the reaction (for a summary, see Section 2.3.10.8). The nature of the chemisorbed isobutene is not fully understood, although it might form (in a limiting case) *tert*-butyl cations that are more-or-less solvated by

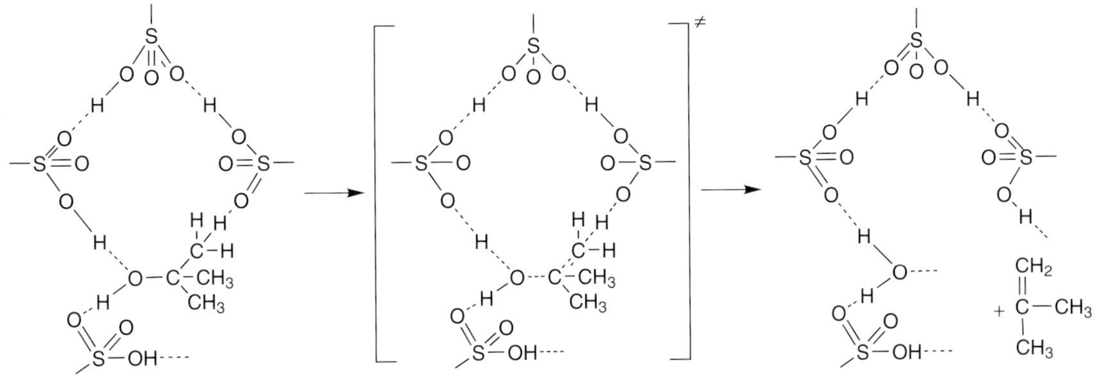

Fig. 1 Schematic representation of the mechanism of catalytic dehydration of *tert*-butyl alcohol within the network of $-SO_3H$ groups of a sulfonic acid resin [20].

the network of sulfonic acid groups lacking a positive charge [9].

Eley–Rideal mechanisms are common in catalysis by solid acids; one reactant (predominantly the more polar if there is more than one) is bonded to the catalytic groups, and the other reactant then combines with it. In isobutene dimerization, there is only the one reactant; in benzene propylation, it has been inferred – on the basis of the reaction kinetics – that propene is chemisorbed and then combines with benzene from the adjacent fluid phase [10].

When the reactants are highly polar and strongly bonded to $-SO_3H$ groups, then Langmuir–Hinshelwood mechanisms are common. For example, in the dehydration of methanol to produce dimethyl ether, two methanol molecules bonded in the network of $-SO_3H$ groups are believed to react to yield the products; the evidence for this inference is in the kinetics of the catalytic reaction and infrared spectra of the alcohol bonded to the catalyst [13, 14]. The mechanism seems similar to that represented for the dehydration of *tert*-butyl alcohol to produce isobutene (see Fig. 1), and also to the mechanisms of dehydration of other alcohols [9].

2.3.10.8 Kinetics of Reactions Catalyzed by Ion-Exchange Resins

Kinetics reported for a number of reactions catalyzed by sulfonic acid ion-exchange resins have provided the bases to infer mechanisms such as those stated above. For example, the dehydration reactions of methanol [15] and of ethanol [16], catalyzed by gel-form sulfonic acid resins (in the absence of mass transfer limitations, as indicated by a lack of effect of catalyst particle size on rate) to form the respective ethers, are represented by rate equations of the following form:

$$r = \frac{k(K_A P_A)^2}{(1 + K_A P_A + K_W P_W)^2} \quad (1)$$

where r is reaction rate, P is partial pressure of the gas-phase reactants and product water, and the subscripts A and W refer to alcohol and water, respectively; k is a rate constant; and the Ks are adsorption equilibrium constants according to the Langmuir formalism. This equation, which represents only the forward reaction, is consistent with a rate-determining combination of two alcohol molecules bonded adjacent to each other within the hydrogen-bonded network of $-SO_3H$ groups. The denominator term indicates competition between alcohol and water for bonding to sites in the network and the inhibition of the reaction by water and self-inhibition by alcohol. The lack of a denominator term for ether implies that it competes negligibly for these bonding sites.

The K-values determined from the kinetics data nicely match values determined from adsorption equilibrium measurements [16]. This is one of the few examples of such good agreement. The agreement points to the catalyst as ideal in the sense of constituting a near-uniform set of catalytic sites that are less than saturated with reactants and products, according to the Langmuir model.

The kinetics of dimerization of isobutene catalyzed by a macroporous sulfonic acid resin was found to be of the form

$$r = \frac{k K_{IB} C_{IB}^2}{(1 + K_{IB} C_{IB})} \quad (2)$$

where C is concentration of the liquid-phase reactant, the subscript IB refers to isobutene, and the other terms are as defined for Eq. (1). The form of this equation corresponds to Eley–Rideal kinetics, consistent with the mechanism stated above.

Extensive kinetics measurements of the reaction of methanol and isobutene to give MTBE led to a rate equation expressed in terms of activities of reactants in

---
References see page 285

the liquid phase [17]; the data imply self inhibition of reaction by methanol [18]. These data are likely to be useful in reactor design for MTBE production. (This is evidently one of the few industrial catalytic reactions for which quantitative kinetics are available, and they form part of the basis for detailed process optimization calculations [19].)

Kinetics data have been reported for the dehydration of *tert*-butyl alcohol catalyzed by a macroporous sulfonic acid resin in the presence of liquid reactants [20]. When water concentrations were relatively low, the rate equation was of a form suggested by Langmuir–Hinshelwood models, demonstrating competition between the alcohol and product water for the catalytic sites. However, when sufficient water was present to cause dissociation of the $-SO_3H$ groups, then the reaction became simply first-order in the alcohol – and much slower. This limiting case corresponds to essentially homogeneous catalysis by dissociated acid. The data show that, at about 353 K, the pseudo-first-order rate constant for the reaction per acid group catalyzed by the network of $-SO_3H$ groups was about 40-fold greater than the pseudo-first-order rate constant per acid group for the reaction catalyzed by hydrated protons formed by dissociation of the $-SO_3H$ groups. By inference, this comparison illustrates the advantage of a network of undissociated acid groups over a conventional aqueous acid catalyst.

2.3.10.9 Deactivation and Stability of Ion-Exchange Resin Catalysts

Stability limitations of ion-exchange resin catalysts prevent their practical application at high temperatures. Sulfonic acid resins in the presence of water undergo desulfonation and thus loss of catalytic sites. The typical sulfonic acid resin is desulfonated at a significant rate at temperatures exceeding approximately 393 K, although certain trade literature indicates that some resins can be operated at temperatures up to about 443 or even 463 K (these resins are probably highly crosslinked).

When metal ions are present in feed streams (e.g., as impurities such as tramp iron) to reactors containing sulfonic acid resin catalysts, ion exchange occurs, removing catalytic sites and also poisoning the catalysts. The detrimental effect on catalytic activity of a small amount of metal ions interrupting the network of $-SO_3H$ groups may be severe, as stated above for the example of *tert*-butyl alcohol dehydration. Catalysts poisoned by exchange cations can be regenerated by ion exchange with aqueous acid.

Bases such as ammonia or amines can similarly poison sulfonic acid resin catalysts and, again, detrimental effects of breakup of the network of $-SO_3H$ groups can be expected to be severe.

The formation of high-molecular-weight side products of catalytic reactions, such as oligomers formed from isobutene in the manufacture of MTBE, can foul ion-exchange resin catalysts and lead to their deactivation.

Ion-exchange resins offer little resistance to abrasion, and do not withstand use in stirred reactors; hence, they are more generally used in fixed-bed reactors. If a resin that can swell substantially is used in a fixed-bed reactor, precautions must be taken to ensure that it is not allowed to come in contact with fluids that would cause it to swell markedly, as this might cause the reactor to burst. Likewise, an increased particle volume would hinder flow through the bed and cause an increased pressure drop across the reactor.

Commercial ion-exchange resins may contain low-molecular-weight components that bleed from the particles during operation in the presence of liquid reactants. These components may include monomers or oligomers, or even unreacted sulfonating agents. Such materials may slough into the products and, although present in only trace amounts, they may be significantly detrimental to product specifications, such as color.

Further details of deactivation and regeneration are available in the review by Chakrabarti and Sharma [1].

2.3.10.10 Modified and Metal-Containing Ion-Exchange Resin Catalysts

A variety of modified forms of sulfonic acid resins have been used as catalysts. For example, resins containing palladium particles in addition to the acidic groups are available commercially and may be of importance on an industrial basis. These resins can be formed by ion-exchange with a palladium salt followed by reduction of the palladium. The catalysts are bifunctional, and simultaneously catalyze hydrogenation (by palladium) and condensation and dehydration (by the acidic groups).

Sulfonic acid resins incorporating $-SO_3Ag$ groups are bifunctional catalysts in which the latter groups increase the activity for reactions of alkenes by virtue of the selective bonding of alkenes to silver cations [21]. The local concentration of the alkenes near the catalytically active acid groups is increased because of this affinity.

Resins in which some of the $-SO_3H$ groups are replaced by groups containing an SH functionality are more active than the sulfonic acid resin alone in the synthesis of bisphenol A from phenol and acetone [22, 23]. The SH groups are inferred to act analogously to mercaptan promoters in solution catalysis.

In attempts to increase the acid strengths of sulfonic acid resins and to mimic the properties of superacids in solution, researchers have added Lewis acids such as $AlCl_3$ and SbF_5 to macroporous resins [24]. Although these resins are indeed superacidic, and catalyze reactions such as alkane isomerization at room temperature, they are too unstable to be of practical use as catalysts, evidently losing volatile acids such as HCl.

Among the few large-scale applications of anion-exchange resins as catalysts is a recently announced process for the carbonylation of methanol to produce acetic acid [25]. The catalyst is a rhodium carbonyl, which is expected to exist predominantly in the form of an anionic species (such as $[Rh(CO)_2I_2]^-$), and presumably is ion-paired to pyridinium groups in a crosslinked polymer. The catalyst functions in similar manner to the soluble rhodium carbonyl complex, which is also an industrial catalyst for methanol carbonylation.

2.3.10.11 Hybrid Organic/Inorganic Catalysts akin to Ion-Exchange Resins

Sulfonated fluorocarbon resin catalysts have not found important applications, being limited by their cost and non-porous structures that do not swell in weakly polar media. However, the appeal of the strong acidity of these resins and the prospects that they offer for the catalysis of reactions requiring stronger acidity than the conventional sulfonic acid resins (such as alkene/alkane alkylation) have led to development of new porous materials that provide the advantages of strong acidity combined with the accessibility of macroporous catalysts. The new catalysts are resin/silica composites [26] consisting of nanometer-sized sulfonated fluorocarbon resin particles entrapped within a highly porous silica matrix. These materials provide much improved access to the strongly acidic groups (many of which are near surfaces), the catalysts being much more effective than the gel-form sulfonated fluorocarbon resins for a number of reactions. Data presented for the propylation of benzene at 353 K, for example, show that these hybrid catalysts are highly active (even more active than a conventional macroporous sulfonated crosslinked polystyrene catalyst, which is less strongly acidic) [26]. According to Harmer and Sun [2], these hybrid catalysts are available commercially and have been used successfully in two (unspecified) industrial processes.

2.3.10.12 Summary and Outlook

Sulfonic acid resins have been utilized for more than 50 years as industrial catalysts; they continue to be important and to find increasing applications, due mainly to their low cost and environmental benefits. The broad class of acidic resin catalysts continues to be developed and improved, with innovations producing more stable and more highly acidic catalysts, as well as catalysts with an enhanced accessibility of the acidic groups.

Furthermore, the results of recent investigations into precisely structured polymeric supports for sulfonic acid groups [27], and of the precise placement of $-SO_3H$ and $-SH$ groups on supports, suggest that advanced versions of these catalysts will continue to be developed [28].

Acknowledgments

This work was supported by the U.S. Department of Energy, Office of Energy Research, Office of Basic Energy Sciences, contract DE-FG02-04ER15600.

References

1. A. Chakrabarti, M. M. Sharma, *Reactive Polymers* **1993**, *20*, 1.
2. M. A. Harmer, Q. Sun, *Appl. Catal. A Gen.* **2001**, *221*, 45.
3. H. Widdecke in D. C. Sherrington, P. Hodge (Eds.), *Syntheses and Separations Using Functional Polymers*, Wiley, Chichester, 1988, p. 149.
4. B. Corain, M. Zecca, K. Jerabek, *J. Mol. Catal. A Gen.* **2001**, *177*, 3.
5. K. M. Dooley, J. A. Williams, B. C. Gates, R. L. Albright, *J. Catal.* **1982**, *74*, 361.
6. M. L. Bender, *Mechanisms of Homogeneous Catalysis from Protons to Proteins*, Wiley, New York, 1971.
7. B. C. Gates, *Catalytic Chemistry*, Wiley, New York, 1992.
8. G. Zundel, *Hydration and Intermolecular Interaction*, Academic Press, New York, 1969.
9. R. Thornton, B. C. Gates, *J. Catal.* **1974**, *34*, 275.
10. R. B. Wesley, B. C. Gates, *J. Catal.* **1974**, *34*, 288.
11. B. C. Gates, J. S. Wisnouskas, H. W. Heath, Jr., *J. Catal.* **1972**, *24*, 320.
12. W. O. Haag, *Chem. Eng. Prog. Symp. Ser.* **1967**, *63*, 140.
13. R. Thornton, B. C. Gates, *Proceedings, 5th International Congress on Catalysis* **1973**, *1*, 357.
14. E. Knözinger, H. Noller, *Z. Phys. Chem. (Frankfurt)* **1967**, *55*, 59.
15. B. C. Gates, L. N. Johanson, *AIChE J.* **1971**, *17*, 981.
16. R. L. Kabel, L. N. Johanson, *AIChE J.* **1962**, *8*, 621.
17. (a) A. Rehfinger, U. Hoffman, *Chem. Eng. Sci.* **1990**, *45*, 1605; (b) A. Rehfinger, U. Hoffman, *Chem. Eng. Sci.* **1990**, *45*, 1619.
18. (a) F. Ancillotti, M. Massi Mauri, E. Pescarollo, *J. Catal.* **1977**, *46*, 49; (b) F. Ancillotti, M. Massi Mauri, E. Pescarollo, *J. Mol. Catal.*, **1978**, *4*, 37.
19. J. F. Burri, V. I. Manousiouthakis, *Computers Chem. Eng.* **2004**, *28*, 2509.
20. B. C. Gates, W. Rodriguez, *J. Catal.* **1973**, *31*, 27.
21. S. Affrossman, J. P. Murray, *J. Chem. Soc. B* **1996**, 1015.
22. R. A. Reinicker, B. C. Gates, *AIChE J.* **1974**, *20*, 933.
23. B. B. Gammill, G. R. Ladewig, G. E. Ham, U.S. Patent 3,634,341, 1972.
24. K. M. Dooley, B. C. Gates, *J. Catal.* **1985**, *96*, 347.
25. N. Yoneda, S. Kusano, M. Yasui, P. Pujado, S. Wilcher, *Appl. Catal. A Gen.* **2001**, *221*, 253.
26. M. A. Harmer, W. E. Farneth, Q. Sun, *J. Am. Chem. Soc.* **1996**, *118*, 7708.
27. Y. Xu, W. Gu, D. L. Gin, *J. Am. Chem. Soc.* **2004**, *126*, 1616.
28. R. K. Zeidan, V. Dufaud, M. E. Davis, *J. Catal.* **2006**, *239*, 299.

2.3.11
Flame Hydrolysis

Dieter Kerner* and Matthias Rochnia

2.3.11.1 Manufacture of Pyrogenic Oxides

Synthetic nanoscale metal oxides can be manufactured via gas to particle conversion processes such as the decomposition of suitable precursors in hot-wall and plasma reactors or through laser ablation [1]. Due to their low production rates, relatively high energy consumption and limited scalability, these processes have only minor technical relevance in manufacturing commercial quantities of nanoscale particles.

The most common – and by far most important – process is the flame hydrolysis or so-called AEROSIL® process [2], invented in 1942 by H. Kloepfer [3]. The invention was driven by the requirement to develop a white filler (silicon dioxide) for reinforcement purposes, in analogy to carbon black, which is widely used in rubber applications. The term "flame hydrolysis" or high-temperature hydrolysis in general describes a process in which a gaseous mixture of a metal chloride precursor, hydrogen and air is reacting in a continuously operated flame reactor [4, 5]. This reactor design allows the high throughput of metal chlorides – essential for a commercial production of nanoscale metal oxides – combined with particle properties such as defined particle size distributions, nonporous and spherical primary particles and a high purity. The manufacturing process is briefly described in the following section, but further details are provided in Ref. [6] and the patents and papers cited therein.

Silicon tetrachloride ($SiCl_4$) – the most common precursor for the production of pyrogenic silica – is vaporized, mixed with dry air and hydrogen, and then fed into the flame reactor (burner). During the combustion, hydrogen and oxygen form water which quantitatively hydrolyzes the $SiCl_4$, thus forming nanoscale primary particles of SiO_2. The particle size can be adjusted by the flame parameters in order to generate tailored particles with specific properties. The generated heat of reaction is carried off through heat exchangers downstream of the flame reactor before entering the gas–solid separation system. Filters or cyclones separate the silica from the offgas containing hydrochloric acid. Residual hydrochloric acid adsorbed on the large surface of the silica particles is removed by an air and steam treatment (deacidification) in a fluidized bed or rotary kiln reactor. A simplified flow sheet of the flame hydrolysis process and the general chemical reaction taking place in this reactor are shown in Fig. 1.

The general mechanism of formation and growth of the silica particles occurring in the flame reactor [7–9] is dominated by three processes:

- The chemical reaction (hydrolysis) of $SiCl_4$ with water forms spherical primary particles through nucleation.
- Further growth of these particles is accomplished by the reaction of additional $SiCl_4$ on the surface of the already-generated particles.
- Through the means of coagulation (collision of primary particles) and coalescence (sintering), primary particles form aggregated structures [8, 10–13].

Accumulations of primary particles can be described by the expressions "aggregate" and "agglomerate". An aggregate is a cluster of particles held together by strong chemical bonds, whereas agglomerates are defined as loose accumulations of particles sticking together by, for example, hydrogen bonds and van der Waals forces. The main processes leading to particle formation and growth are illustrated schematically in Fig. 2.

The growth of particles to aggregates and agglomerates in relation to the residence time in the flame reactor is shown graphically in Fig. 3; these data are supported by transmission electron microscopy (TEM) images taken at 20, 50, and 90 ms residence time. The formation of primary particles is already completed after about 5–10 ms, whereas the extent of aggregation and agglomeration increases with the residence time in the reactor [14]. Theoretical investigations support these experimental results [15, 16]: in Fig. 3 the aggregate values printed in bold type represent calculated values. Further agglomeration in the downstream process leads to products in the range of several micrometers, although the primary particles are still in the nanometer size range.

The size of the primary particles and the degree of aggregation and agglomeration can be influenced by the flame temperature, the hydrogen to oxygen ratio, the silicon tetrachloride concentration, and the residence time in the nucleation and aggregation zone of the flame reactor [17].

In principle, any vaporizable silicon-containing precursor, such as methyltrichlorosilane (MTCS) or trichlorosilane (TCS), can be decomposed using the flame hydrolysis process. However, the above-described process is not limited to silicon-containing raw materials. In general, any volatile precursor – either in the liquid or solid phase – can be decomposed at high temperatures to form highly dispersed nanoscale metal oxides. Based on the respective metal chlorides, aluminum oxide, titanium dioxide, iron oxide, and zirconium dioxide (or mixtures thereof) are available on a commercial scale [18]. Of these materials, pyrogenic silica is by far the most widely produced pyrogenic oxide [19]. The annual worldwide production capacity for pyrogenic silica was estimated to

---
* Corresponding author.

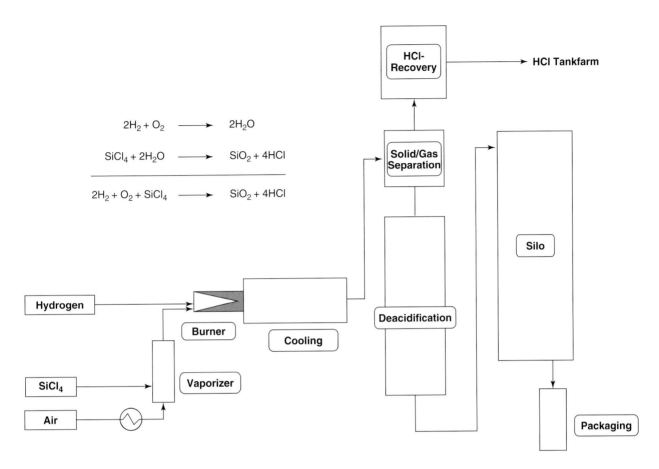

Fig. 1 Schematic diagram of the pyrogenic silica process.

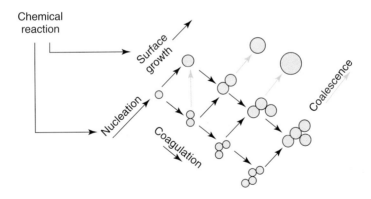

Fig. 2 Particle formation and growth.

be in excess of 120 000 tonnes in 2002 [20], the main producers being Degussa (AEROSIL®), Cabot (CAB-O-Sil®), Wacker (HDK®), and Tokuyama (Reolosil®).

Spray pyrolysis is another process by which pyrogenic oxides may be generated. Whereas the term flame hydrolysis describes a gas-to-particle process, spray pyrolysis is an example of a droplet-to-particle process, whereby droplets containing precursors are formed mechanically by liquid atomization and then pyrolyzed in flames [21]. Combining both processes significantly broadens the variety of pyrogenic oxides available through flame hydrolysis.

References see page 293

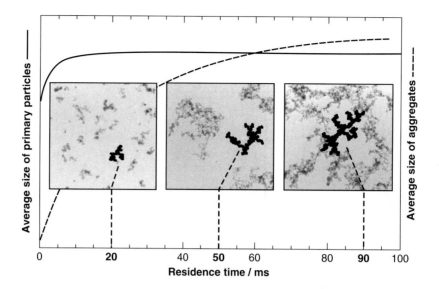

Fig. 3 Average primary particle and aggregate size in relation to the residence time in the flame reactor. The scales are indexed and different for primary particles (nm) and aggregates (×100 nm).

2.3.11.2 Physico-Chemical Properties of Pyrogenic Oxides

The large specific surface area of pyrogenic oxides is one of their key properties. The surface area is determined mainly by the size of the primary particles formed during the nucleation process in the flame reactor. By varying certain process parameters (see Section 2.3.11.1), the size of the primary particles can be adjusted in the range of approximately 7 to 40 nm, corresponding to pyrogenic oxides with specific surface areas between 300 m^2 g^{-1} and 50 m^2 g^{-1}. These oxides consist of amorphous, spherical particles without an inner surface area. Materials with a specific surface area above 300 m^2 g^{-1} show a certain amount of microporosity rather than a further reduced primary particle size.

The primary particle size distribution of pyrogenic silica grades with different surface areas is shown graphically in Fig. 4.

TEM images of two different pyrogenic silica grades are shown in Figs. 5 and 6. Pyrogenic silica with a specific surface area of 300 m^2 g^{-1} (Fig. 6) shows a highly aggregated and agglomerated structure, consisting of primary particles in the 7-nm range; the TEM image of the 50 m^2 g^{-1} grade (Fig. 5) shows a much lesser degree of aggregation.

The bulk density of freshly produced pyrogenic silica is in the range of 10 to 20 g L^{-1}. Since such a voluminous, fluffy powder is difficult to handle, dedicated compactors and packaging machines are used to increase the bulk density up to 150 g L^{-1}.

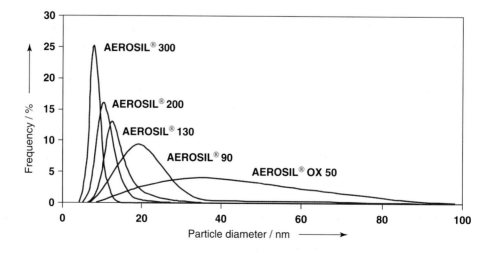

Fig. 4 Primary particle size distribution of pyrogenic silica grades with different specific surface areas from 50 to 300 m^2 g^{-1}.

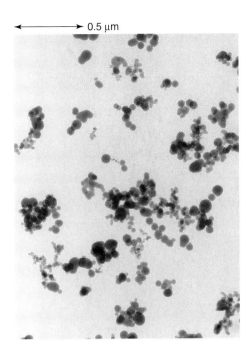

Fig. 5 Transmission electron microscopy image of pyrogenic silica with a specific surface area of 50 m² g⁻¹.

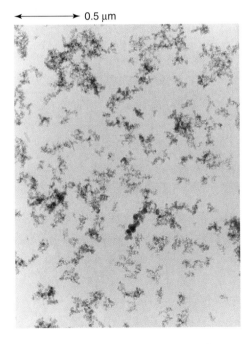

Fig. 6 Transmission electron microscopy image of pyrogenic silica with a specific surface area of 300 m² g⁻¹.

The fundamental building blocks of pyrogenic silica – the SiO_4 tetrahedra – are arranged in random fashion, and therefore a distinct and well-defined X-ray pattern (as obtained with any crystalline silica modification occurring in nature, such as quartz or christobalite), cannot be observed [6]. The X-ray photograph of pyrogenic silica is characterized by an absence of any diffraction pattern, indicating an entirely amorphous material. In contrast to the crystalline forms, this material does not cause silicosis.

Unlike pyrogenic silica, pyrogenic alumina is characterized by a crystalline structure consisting of the thermodynamically metastable γ- and δ-forms instead of the stable α-form [22]. The primary particle size of pyrogenic alumina is in the range of 13 nm, corresponding to a specific surface area of about 100 m² g⁻¹ (AEROXIDE® Alu C; Degussa AG). Pyrogenic alumina can be transformed to α-Al_2O_3 at temperatures above 1200 °C, which is associated with a loss in specific surface area. This type of heat treatment results in improved mechanical properties, such as higher hardness and abrasiveness.

Pyrogenic titania (AEROXIDE® TiO_2 P25; Degussa AG) has an average primary particle size of about 21 nm and a specific surface area of about 50 m² g⁻¹ [22]. This product differs significantly from pigmentary TiO_2, which is produced in substantial quantities through precipitation processes. The average particle diameter of precipitated titania is about 0.3 µm, with surface areas in the region of 10 m² g⁻¹. The precipitation route leads to the thermodynamically stable rutile modification. In pyrogenic TiO_2, in contrast, the thermodynamically metastable anatase phase is the dominant modification (ca. 80%). A lattice transformation towards higher proportions of rutile is observed at temperatures above 700 °C, and is associated with a reduction in the specific surface area.

Pyrogenic zirconium dioxide (ZrO_2) consists predominantly of the thermodynamically stable monoclinic and, to a lesser degree, of the metastable tetragonal phase. ZrO_2 obtained by the flame hydrolysis process has an average primary particle size of 30 nm and a specific surface area of about 40 m² g⁻¹ [22]. At a temperature of approximately 1170 °C the conversion from a monoclinic to a tetragonal structure is observed, followed by another transformation at about 2370 °C from tetragonal to cubic.

In particular, the transformation from monoclinic to tetragonal is associated with an increase in density (monoclinic 5.68 g cm⁻³; tetragonal 6.10 g cm⁻³), and has a negative effect on the thermal resistance of, for example, molded ceramic parts, up to the point where the formation of microcracks is observed [23]. By adding dopants (besides CaO, MgO and CeO_2, Y_2O_3 is the most common stabilization aid), high-temperature modifications can be stabilized far below their original transformation point [23–25].

A general overview of relevant physico-chemical data relating to the above-mentioned pyrogenic oxides is provided in Table 1.

References see page 293

Tab. 1 Physico-chemical properties of pyrogenic oxides

Property	AEROSIL® OX 50	AEROSIL® 200	AEROXIDE® AluC	AEROXIDE® TiO₂ P25	VP ZrO₂
Specific surface area by BET/m² g⁻¹ [a]	50 ± 15	200 ± 25	100 ± 15	50 ± 15	40 ± 10
Average size of primary particles/nm	40	12	13	21	30
Tapped density/g L⁻¹ [b]	ca. 130	ca. 50	ca. 50	ca. 130	ca. 80
Density/g mL⁻¹ [c]	ca. 2.2	ca. 2.2	ca. 3.2	ca. 3.7	ca. 5.4
Purity/wt.% [d]	$SiO_2 \geq 99.8$	$SiO_2 \geq 99.8$	$Al_2O_3 \geq 99.6$	$TiO_2 \geq 99.5$	$ZrO_2 \geq 96$
Loss on drying/wt.% [e]	≤ 1.5	≤ 1.5	≤ 5	≤ 1.5	≤ 2
Loss on ignition/wt.% [f,g]	≤ 1	≤ 1	≤ 3	≤ 2	≤ 3
Isoelectric point at pH	2	2	9	6.5	8.2
Phase composition	Amorphous	Amorphous	γ- and δ- modification	80% anatase, 20% rutile	Monoclinic and tetragonal

[a] According to DIN 68131.
[b] Density of the powder according to DIN ISO 787/XI, JIS K 510108 (not screened).
[c] Density of the solid SiO₂ particles, determined with a pycnometer.
[d] Based on material ignited for 2 h at 1000 °C.
[e] According to DIN ISO 787/II, ASTM D 280, JIS K 5101/21.
[f] According to DIN 55921, ASTM D 1208, JIS K 5101/23.
[g] Based on material dried for 2 h at 105 °C.

In addition to their comparatively large specific surface area, their well-defined, spherical primary particles, and their high chemical purity, the surface chemistry of pyrogenic oxides represents another key aspect for their use in catalytic applications. Siloxane (Si—O—Si) and silanol groups (Si—OH) can be described as the main functional groups (Fig. 7) on the surface of pyrogenic silica.

Although a hydrophobic effect can be attributed to the siloxane group, the hydrophilic character of the silanol groups dominates the surface chemistry and makes pyrogenic silica wettable. The density of the silanol groups can be determined, for example, by the reaction with lithium aluminum hydride, and ranges between 1.8 and 2.7 Si—OH nm⁻² [6]. When exposed to humid conditions, water is adsorbed onto the surface of pyrogenic silica, allowing additional siloxane groups to react and the subsequent generation of additional silanol groups. This reaction can be reversed (dehydroxylation) by heating pyrogenic silica up to temperatures above 150 to 400 °C. Changes in the moisture balance at different temperatures of pyrogenic silica can easily be detected and followed using infrared spectroscopy [26–28]. The hydroxyl groups are acidic, resulting in an isoelectric point at $pH = 2$.

Pyrogenic Al₂O₃ has basic hydroxyl groups at its surface, corresponding to an isoelectric point at $pH = 9$. The total dehydroxylation of alumina results in the presence of aluminum ions located at the surface, coordinated by only five rather than six oxygen atoms, thus representing Lewis acid centers. These centers can either add pyridine or can be rehydroxylated by the adsorption of water [29].

Depending on the coordination of the hydroxyl groups at the surface of titania, an acidic as well as a basic character of these groups can be observed. While the acidic sites accumulate ammonia and can be esterified with diazomethane, the basic sites can be detected by exchange reactions with anions. The existence of equimolar amounts of acidic and basic hydroxyl groups is also reflected by an isoelectric point at $pH = 6.5$ [30].

Zirconia shows a similar surface chemistry but, relative to titania, it has more basic rather than acidic sites, and this results in an isoelectric point at $pH = 8.2$. The zeta potentials of the above-mentioned pyrogenic oxides are illustrated graphically in Fig. 8.

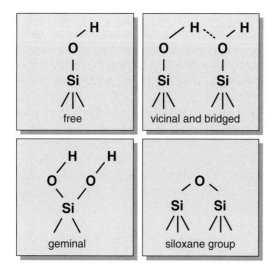

Fig. 7 Functional groups on the surface of pyrogenic silica.

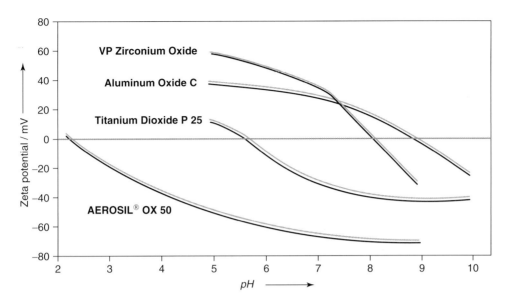

Fig. 8 Zeta potential of pyrogenic oxides as a function of pH.

2.3.11.3 Preparation of Formed Supports

In general, pyrogenic oxides are fine and fluffy powders, but they tend to be rather inconvenient for use in catalysis unless the powder does not form part of the final product. Shaping processes can help to create more convenient supports, especially for use in fixed-bed reactors. The requirements for an excellent support are [31]:

- a well-defined chemical composition
- a high purity
- a well-defined surface chemistry
- no sintering at high temperatures
- good abrasion resistance and crushing strength
- a well-defined porosity, pore size distribution and pore volume
- easy separation from the reactants.

Pyrogenic oxides in powder form fulfill only part of these requirements. For a long time catalysts prepared by the impregnation of oxide powders were only of academic interest because, on a technical scale, problems arose when separating the powder from the products. However, these disadvantages can be overcome by size enlargement – that is, by compaction of the powder.

The basic options for preparing formed supports include [32]:

- agglomeration (drum agglomerator, inclined-disk agglomerator)
- spray-drying or spray-granulation
- pressure compaction or extrusion.

Agglomeration processes are not applicable to pyrogenic oxides because of their fluffiness.

In the *spray-drying process*, a suspension of pyrogenic oxides in water is fed via a spraying device into the chamber of a spray-dryer, and microgranulates of pyrogenic silica [33, 34], alumina [35] and titania [36] have been reported when using this technique. The properties of the resulting spheres can be controlled by the solid content of the pyrogenic oxide in the suspension, the type of spraying devices (nozzles, discs, etc.), and the residence time and temperature in the spraying chamber. Typically, spheres in the range of 10 to 150 μm can be produced. In addition, the pore size and pore size distribution can be adjusted by selecting oxides with different particle size distributions and surface areas, respectively.

On the laboratory scale, *pressure compacting* (to make tablets) is more prominent for basic investigations into the processability and performance of newly developed recipes.

Extrusion is preferred on the pilot and production scales, but basically both processes follow the same schematic procedure.

In the compacting process, binders and other auxiliaries such as plasticizers or pore-building substances are needed to achieve supports with the desired properties. Ettlinger et al. described a simple way of making tablets from pyrogenic silica, alumina and titania by using silica sol as a binder and polyfunctional alcohols as plasticizers [37]. The corresponding pyrogenic oxide was mixed with water, silica sol and glycerol, pelletized, dried at ambient temperature, and calcined at 550 °C. The tablets consisted of 50–60% void volume, and the initial

References see page 293

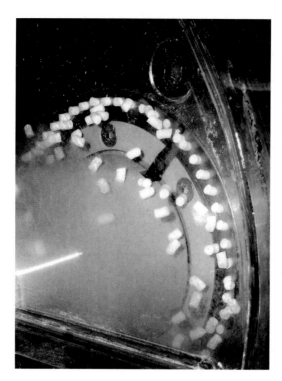

Fig. 9 Extruder (left) and the extrudates (right).

surface area of the powder was reduced by less than 20%. The extruder and the corresponding extrudates prepared are illustrated in Fig. 9.

The use of a silica sol as a binder can be of disadvantage in systems other than silica, as it may change the surface properties of the other oxides. This can be overcome by using small amounts of decomposable substances such as organic acids [38] and graphite [31]. Such tablets of pyrogenic titania have a crushing strength of 150–200 N and a pore volume of 45–53%, which corresponds to 0.3–0.4 cm^3 g^{-1}. The initial specific surface area is reduced by approximately 10%. Additionally, by using corresponding salt solutions such as zirconium oxychloride or nickel chloride, an inorganic surface treatment of the pyrogenic titania with other oxides can be made. The use of urea as a pore-building agent in order to achieve higher porosities of 0.6–1.0 cm^3 g^{-1} has also been proposed [31, 39].

Deller et al. described the use of various auxiliaries such as kaolin, graphite, sugar, starch, urea, and wax as binders and pore-building agents for preparing pellets of pyrogenic silica [40, 41], alumina [42], zirconia [43], and silica-alumina mixed oxides [44]. A detailed description of the extrusion process of pyrogenic titania is provided in Refs. [45, 46]. The procedure comprises four crucial steps, namely kneading of the raw materials, extrusion, drying of the green bodies, and calcination.

2.3.11.4 Applications

Very often, pyrogenic oxides are used as model substances because of their unique properties such as purity and a well-defined and easily accessible surface. A review up to 1980 is provided in Ref. [47] concerning the investigation of silica (26 references), alumina (11 references) and titania (four references). As can be seen from these data, these basic investigations centered mainly on pyrogenic silica rather than on the other oxides.

Among commercial applications, silica supports play an important role in the synthesis of vinyl acetate monomers. Previously, Wunder et al. [48] claimed the use of Pd- and/or Au-impregnated cylindrical supports made from either pyrogenic silica or a mixture of silica and alumina. Formed silica and/or alumina are also used as supports in catalysts where the support is impregnated with Pd/K/Cd, Pd/K/Ba or Pd/K/Au giving a selectivity in excess of 90% [49, 50]. Bankmann et al. [51] have described improved vinyl acetate catalysts based on formed silica and silica/alumina with various shapes. Silica supports in the form of pellets, beads or of a globular shape impregnated with phosphoric acid are used in the hydration of olefins in a fixed-bed reactor [52]. In a rather basic study of the gas-phase polymerization of ethene, it was shown that the activity of catalysts using pyrogenic silica as support for metallocenes was ten-fold higher than with other silica supports [53].

To date, pyrogenic alumina has played only a minor role in catalysis, although more recent studies have centered on the use of pyrogenic alumina in automotive catalysts. Modern, three-way catalysts consist of an alumina washcoat containing one or more of the elements Pt, Rh, Pd and so-called storage components for NOx and oxygen, respectively. Liu and Anderson investigated the stored NOx stability [54, 55]. NOx is stored under lean-burn conditions on an alumina-supported alkaline earth oxide component (10% BaO on Al_2O_3), is released during intermittent rich/stoichiometric periods, and reduced by hydrogen, CO, or hydrocarbons over the noble metal component. In a similar manner, oxygen can be stored under oxygen-rich conditions and released under oxygen-lean conditions using CeO/ZrO_2 on alumina. Oxygen storage in fresh, thermally aged catalysts as well as catalysts treated by oxychlorination has been studied [56].

During recent years, titania in general has gained much interest in the field of photocatalysis, where it plays the role not only of the support but also the catalyst. A wide variety of reviews, containing many hundreds of references, has been published during the past 10 years [57–60]. Subsequently, Mills and Lee [61] reported a web-based overview of current commercial applications.

One major field of application is the treatment of (waste) water and air by the photodegradation of inorganic compounds (e.g., ammonia, nitrates) and organic substances (e.g., chlorinated aliphatic and aromatic compounds), as well as volatile organic compounds (VOCs) in the air. Even 2,4,6-trinitrotoluene (TNT) can be completely destroyed under aerobic conditions by the use of pyrogenic titania (AEROXIDE® TiO_2 P 25; Degussa AG) [62]. Another field is the use of titania as a sensitizer in the photodissociation of water. Here, the preliminary investigations were conducted during the early 1970s by Fujishima and Honda [63], followed by the studies of Graetzel et al. during the early 1980s [64–68]. Graetzel further improved the titania catalyst by depositing RuO_2 or Pt on the surface, or doping the titania with Nb_2O_5. These authors also used sensitizers such as $Ru(bpy)_3^{2+}$, $Ru(bpy)_2(4,4'$-tridecyl-2,2'-bpy$)^{2+}$ and 8-hydroxy-orthoquinoline.

Surprisingly, the photocatalytic activity of pyrogenic titania is higher than expected, most likely due to the specific mixed crystal structure. The commercial product AEROXIDE® TiO_2 P 25 (Degussa AG) consists of approximately 80% by weight of anatase and 20% of rutile [69], and is often regarded as the "reference" in photocatalytic investigations [59, 70]. Both crystal forms are tetragonal, but with different dimensions of the elementary cell. According to Hurum et al. [71], the rutile acts as an antenna to extend the photoactivity into visible wavelengths. In addition to the increase in the spectral range, an increased charge separation (relative to the pure-phase material) reduces recombination and increases the efficiency of the catalyst. The extraordinary photocatalytic performance of AEROXIDE® TiO_2 P 25 in comparison to other nanoscale titania particles has been published in several reports: it is, for example, useful in the degradation of humic acid [72], of phenol and salicylic acid [73], of 1,4-dichlorobenzene [74], and in the photocatalytic reduction of Hg(II) [75]. It is also used in the oxidation of primary alcohols to aldehydes [76] or in the photopolymerization of methyl methacrylate [77]. Its use in cement can also help in reducing environmental pollution [78, 79].

Pyrogenic titania is useful not only in photocatalysis but also in other catalytic applications. It serves as the base material for DeNOx catalysts [80–82] and in catalysis for selective hydrogenations [51]. A broad field is the use in Fischer–Tropsch catalysts [83–87].

2.3.11.5 Conclusions

Because of their unique properties, pyrogenic oxides provide many advantages when used as catalyst supports, such as purity, a well-defined surface chemistry, and particle size and aggregate size. Forming of catalysts can be carried out using conventional compacting processes such as spray-drying or spray-granulation, pressure compaction or extrusion, which leads to a well-defined porosity, pore-size distribution, and pore volume. Formed supports show good crushing strengths and a favorable sintering behavior, even at high temperatures.

Although pyrogenic oxides are very often used as model substances in fundamental research, an increasing number of commercial applications also exist. Pyrogenic silica, alumina and titania are the main supports in commercial applications, and can be used in the mass production of chemicals as well as in more-sophisticated special applications. A typical example of mass production is the application of pyrogenic silica in the production of vinyl acetate monomers, or its use in Fischer–Tropsch catalysis. Additional, sophisticated applications include the use of alumina in automotive exhaust catalysts or the use of pyrogenic titania in various photocatalytic processes.

References

1. T. Kodas, M. J. Hampden-Smith, *Aerosol Processing of Materials*, Wiley-VCH, New York, 1999, 712 pp.
2. AEROSIL® is a registered trademark of Degussa AG.
3. H. Kloepfer, DE Patent 762 723, assigned to Degussa, 1942.
4. S. Pratsinis, *Prog. Energy Combust. Sci.* **1997**, *24*, 197–219.
5. H. Kammler, L. Mädler, S. Pratsinis, *Chem. Eng. Technol.* **2001**, *24*, 583–596.
6. Degussa A. G, Düsseldorf, *Technical Bulletin Fine Particles Nr. 11*, 2003.
7. D. Schaefer, A. Hurd, *Aerosol Sci. Technol.* **1990**, *12*, 876–890.
8. S. Pratsinis, *J. Colloid Interface Sci.* **1988**, *124*, 416–427.
9. G. Ulrich, *Combust. Sci. Technol.* **1971**, *4*, 47–57.

10. S. K. Friedlander, *Smoke, Dust and Haze – Fundamentals of Aerosol Behavior*, John Wiley & Sons, New York, 1977, 338 pp.
11. T. Johannessen, S. Pratsinis, H. Livbjerg, *Chem. Eng. Sci.* **2000**, *55*, 177–191.
12. W. Koch, S. Friedlander, *J. Colloid Interface Sci.* **1990**, *140*, 419–427.
13. J. Landgrebe, S. Pratsinis, *J. Colloid Interface Sci.* **1990**, *139*, 63–86.
14. O. Arabi-Katbi, S. Pratsinis, P. Morrison, C. Megaridis, *Combustion and Flame* **2001**, *124*, 560–572.
15. F. Kruis, K. Kusters, S. Pratsinis, B. Scarlett, *Aerosol Sci. Technol.* **1993**, *19*, 514–526.
16. A. Schild, A. Gutsch, H. Mühlenweg, S. E. Pratsinis, *J. Nanoparticle Res.* **1999**, *1*, 305–315.
17. L. White, G. Duffy, *Ind. Eng. Chem.* **1959**, *51*, 232–238.
18. A. Liu, P. Kleinschmit, DE Patent 36 11 449 A1, assigned to Degussa, 1986.
19. M. Ettlinger, in *Ullmann's Encyclopedia of Industrial Chemistry*, 5th Ed., Vol. A 23, VCH Verlagsgesellschaft, Weinheim, 1993, pp. 635–642.
20. D. Kerner, N. Schall, W. Schmidt, R. Schmoll, J. Schütz, in *Winnacker-Küchler, Chemische Technik-Prozesse und Produkte*, R. Dittmeyer, W. Keim, G. Kreysa, A. Oberholz (Eds.), Vol. 3, Wiley-VCH, Weinheim, 2005, pp. 853–855.
21. L. Mädler, H. Kammler, R. Mueller, S. Pratsinis, *Aerosol Sci.* **2002**, *33*, 369–389.
22. Degussa A. G, Düsseldorf, *Technical Bulletin Pigments Nr. 56*, 2001.
23. *Grundzüge der Keramik*, Skript zur Vorlesung Ingenieurkeramik I, Professur für nichtmetallische Werkstoffe 2001, ETH Zürich, 37–43.
24. G. Dressler, P. Minuth, in *Ullmann's Encyklopädie der technischen Chemie*, 4th Ed., Vol. 24, Verlag Chemie, Weinheim, 1983, pp. 694–696.
25. H.-J. Bargel, G. Schulze, in *Werkstoffkunde*, 8th Ed., Springer-Verlag, Berlin, 2004, p. 312–314.
26. L. Zhuravlev, *Colloids Surfaces* **2000**, *A173*, 1–38.
27. R. K. Iler, *The Chemistry of Silica*, Wiley-Interscience, New York, 1979, 896 pp.
28. A. P. Legrand, *The Surface Properties of Silicas*, Wiley & Sons, New York, 1998, 494 pp.
29. J. Peri, *J. Phys. Chem.* **1965**, *69*, 220–230.
30. H.-P. Boehm, *Angew. Chemie* **1966**, *78*, 617–652.
31. B. Despeyroux, K. Deller, H. Krause, *Chemische Industrie* **1993**, *10*, 48.
32. C. E. Capes, in *Kirk-Othmer Encyclopedia of Chemical Technology*, 3rd edn, Vol. 21, John Wiley & Sons, New York, 1978, pp. 77–105.
33. H. Biegler, G. Kallrath, DE Patent 1 209 108, assigned to Degussa, 1966.
34. K. Deller, H. Krause, J. Meyer, D. Kerner, W. Hartmann, H. Lansink-Rotgerink, EP Patent 0 725 037, assigned to Degussa, 1996.
35. J. Meyer, P. Neugebauer, M. Steigerwald, US Patent 6,743,269, assigned to Degussa, 2004.
36. H. Gilges, D. Kerner, J. Meyer, EP Patent 1 078 883 A1, assigned to Degussa, 2001.
37. M. Ettlinger, H. Ferch, D. Koth, E. Simon, DE Patent 31 32 674 A1, assigned to Degussa, 1983.
38. M. Ettlinger, H. Ferch, D. Koth, E. Simon, DE Patent 32 17 751 A1, assigned to Degussa, 1983.
39. K. Deller, R. Klingel, H. Krause, DE Patent 38 03 894 A1, assigned to Degussa, 1989.
40. K. Deller, R. Klingel, H. Krause, DE Patent 38 03 895 C1, assigned to Degussa, 1989.
41. K. Deller, M. Förster, H. Krause, DE Patent 39 12 504 A1, assigned to Degussa, 1990.
42. K. Deller, R. Klingel, H. Krause, DE Patent 38 03 897 A1, assigned to Degussa, 1989.
43. K. Deller, M. Ettlinger, R. Klingel, H. Krause, DE Patent 38 03 898 A1, assigned to Degussa, 1989.
44. K. Deller, R. Klingel, H. Krause, K.-P. Bauer, DE Patent 38 03 899 C1, assigned to Degussa, 1989.
45. M. Bankmann, R. Brand, B. Engler, J. Ohmer, *Catal. Today*, **1992**, *14*, 225–242.
46. R. Brand, B. Engler, M. Foerster, W. Hartmann, P. Kleinschmit, E. Koberstein, J. Ohmer, R. Schwarz, DE Patent 40 12 479 A1, assigned to Degussa, 1991.
47. D. Koth, H. Ferch, *Chem.-Ing.-Tech.* **1980**, *52*, 628–634.
48. F. Wunder, G. Roscher, K. Eichler, DE Patent 38 03 900 A1, assigned to Hoechst, 1989.
49. R. Abel, K.-F. Wörner, EP Patent 0 634 214 A1, assigned to Hoechst, 1995.
50. H. Krause, H. Lansink-Rotgerink, O. Feuer, T. Tacke, P. Panster, EP Patent 0 916 402 A1, assigned to Degussa, 1998.
51. M. Bankmann, B. Despeyroux, H. Krause, J. Ohmer, R. Brand, in *Studies in Surface Science and Catalysis*, Vol. 75, Elsevier, Amsterdam, 1993, pp. 1781–1784.
52. R. W. Cockman, G. Haining, P. Lusmann, A. D. Melville, EP Patent 0 578 441 A2, assigned to BP Chemicals, 1993.
53. M. Walden, Dissertation, Universität Essen, 2002.
54. Z. Liu, J. Anderson, *J. Catal.* **2004**, *224*, 18–27.
55. Z. Liu, J. Anderson, *J. Catal.* **2004**, *228*, 243–253.
56. R. Daley, S. Christou, A. Efstathiou, J. Anderson, *Appl. Catal. B: Environmental* **2005**, *60*, 119–129.
57. M. Hoffmann, S. Martin, W. Choi, D. Bahnemann, *Chem. Rev.* **1995**, *95*, 69–96.
58. A. Linsebigler, G. Lu, J. Yates, *Chem. Rev.* **1995**, *95*, 735–758.
59. A. Mills, S. Le Hunte, *J. Photochem. Photobiol. A: Chemistry* **1997**, *108*, 1–35.
60. A. Fujishima, T. Rao, D. Tryk, *J. Photochem. Photobiol. C: Photochem. Rev.* **2000**, *1*, 1–21.
61. A. Mills, S.-K. Lee, *J. Photochem. Photobiol. A: Chemistry* **2002**, *152*, 233–247.
62. Zh. Wang, Ch. Kutal, *Chemosphere* **1995**, *30*, 1125–1136.
63. A. Fujishima, K. Honda, *Nature* **1972**, *238*, 37–38.
64. E. Borgarello, J. Kiwi, E. Pelizzetti, M. Visca, M. Graetzel, *Nature* **1981**, *289*, 158–160.
65. E. Borgarello, J. Kiwi, E. Pelizzetti, M. Visca, M. Graetzel, *J. Am. Chem. Soc.* **1981**, *103*, 6324–6329.
66. M. Graetzel, *Acc. Chem. Res.* **1981**, *14*, 376–384.
67. V. Houlding, M. Graetzel, *J. Am. Chem. Soc.* **1983**, *105*, 5695–5696.
68. M. Graetzel, *Dechema Monographien* **1987**, *106*, 189–204.
69. Degussa A. G, Düsseldorf, Schriftenreihe Fine Particles Nr. 80, 2003.
70. A. Mills, A. Lepre, N. Elliott, Sh. Bhopal, I. Parkin, S. O'Neill, *J. Photochem. Photobiol. A: Chemistry* **2003**, *160*, 213–224.
71. D. Hurum, A. Agrios, K. Gray, T. Rajh, M. Thurnauer, *J. Phys. Chem. B* **2003**, *107*, 4545–4549.
72. C. Uyguner, M. Bekbolet, *Int. J. Photoenergy* **2004**, *6*, 73–80.
73. K. Chhor, J. Bocquet, C. Colbeau-Justin, *Mater. Chem. Phys.* **2004**, *86*, 123–131.
74. J. Papp, S. Soled, K. Dwight, A. Wold, *Chem. Mater.* **1994**, *6*, 496–500.

75. X. Wang, S. Pehkonen, A. Ray, *Electrochimica Acta* **2004**, *49*, 1435–1444.
76. M. Malati, N. Seger, *J. Oil Col. Chemists Assoc.* **1981**, *64*, 231–233.
77. C. Dong, X. Ni, *J. Macromolec. Sci. A* **2004**, *A41*(5), 547–563.
78. L. Cassar, C. Pepe, WO Patent 98/05601, assigned to Italcementi, 1998.
79. M. Lackhoff, X. Prieteo, N. Nestle, F. Dehn, R. Niessner, *Appl. Catal. B: Environmental* **2003**, *43*, 205–216.
80. H. Hellebrand, H. Schmelz, N. Landgraf, DE Patent 39 38 155 A1, assigned to Siemens, 1990.
81. R. Brand, E. Engler, W. Honnen, E. Koberstein, J. Ohmer, EP Patent 0 385 164, assigned to Degussa, 1990.
82. R. Brand, B. Engler, W. Honnen, P. Kleine-Möllhoff, E. Koberstein, DE Patent 37 40 289 A1, assigned to Degussa, 1989.
83. M. A. Vannice, R. L. Garten, US Patent 4,042,614, assigned to Exxon, 1976.
84. I. E. Wachs, R. A. Fiato, C. C. Chersich, US Patent 4,559,365, assigned to Exxon, 1984.
85. K. Thampi, J. Kiwi, M. Graetzel, *Nature* **1987**, *327*, 506–508.
86. R. A. Fiato, S. Miseo, US Patent 4,749,677, assigned to Exxon, 1988.
87. S. Plecha, C. H. Mauldin, L. E. Pedrick, US Patent 6,117,814, assigned to Exxon, 2000.

2.3.12
Solid-State Reactions

Bernard Delmon* and Michel Devillers

2.3.12.1 Why Prepare Catalysts from Solid Precursors?

The methods most commonly used to prepare catalysts are precipitation and impregnation, in both of which the catalytically active material is transferred from a liquid phase, usually an aqueous solution, to a solid. By contrast, other catalysts are obtained from solid precursors. So-called solid-state reactions – namely, solid-to-solid reactions in which both the starting material (the catalyst precursor) and the catalyst are solids – offer convenient methods to prepare several industrial catalysts, and especially those containing two or more metallic elements or their oxides. The reason for the conspicuous efficiency of these methods in preparing phases containing two or several metallic elements is the existence of certain special features of solid-state reactions, compared to liquid-to-solid or gas-to-solid reactions. In this chapter we briefly outline these peculiarities, and present the most frequent types of solid-state processes used in the preparation of catalysts. More recently developed approaches are also outlined.

The requirement for more than one component in a catalyst arises from many needs: those needs linked to the polyfunctionality often required for the different steps in a reaction; the need to enhance the rate of some reaction steps and to inhibit unwanted side reactions; the wish to increase thermal stability; and the necessity to take advantage of observed synergetic effects. From a fundamental point of view, the presence of several metal elements in a common structure permits adjustment of the local electronic properties, imposes well-defined coordinations, limits the extent of oxidation–reduction phenomena, and may also stabilize the entire catalyst by retarding sintering. Mixed-oxide catalysts are used in this way, or as precursors of active catalysts, for a wide range of important industrial processes, a representative selection of which is provided in Table 1.

The preparation of atomically homogeneous multicomponent catalysts by conventional methods, however, is often difficult due to three main reasons:

- The deposition of active precursors, either as a precipitate or onto a support, often produces separate moieties, instead of a single phase of uniform chemical composition.
- The segregation of the active components can occur during thermal treatment of the catalyst precursor.
- The support can selectively interact with one of the active components at one or the other stage of the formation of the mixed oxide.

A good example of the third case is the reaction of iron with silica in silica-supported iron molybdate, a fact which made impossible the use of silica as a support in the case of catalysts used for the oxidation of methanol to formaldehyde. Some of the methods used to overcome these difficulties in catalyst preparation form the subject basis of this chapter, which deals with a small group of solid-to-solid reactions that are relevant to the preparation of certain mixed-oxide catalysts.

In principle, liquids are ideal precursors when atomically homogeneous mixtures are desired. The reason for this is that the ordinary liquid state has no long-range structure, and hence permits a statistical distribution over the entire volume of its contents, at the molecular scale. However, segregation almost invariably occurs when liquids are reacted to give solids (e.g., by precipitation, when one element usually precipitates before the other component or components). In the use of solid-state reactions, the guiding principle is that – unlike in a liquid – the diffusion of atoms or ions in a solid is very difficult, and is practically frozen at relatively low temperatures. Whenever a solid of complex composition

---

* Corresponding author.

Tab. 1 Some important mixed-oxide catalysts used in industry

Catalyst	Active phases	Industrial processes
Copper chromite	$CuCr_2O_4$	CO conversion (low temp.), oxidations, hydrogenation, hydrogenation-hydrogenolysis
Zinc chromite	$ZnCr_2O_4$, ZnO	Methanol synthesis (high pressure)
Copper/zinc chromite	$Cu_xZn_{1-x}Cr_2O_4$, CuO_y	Methanol synthesis (low pressure)
Iron molybdate	$Fe_2(MoO_4)_3$, MoO_3	Methanol to formaldehyde
Iron oxide/potassium oxide	$KFeO_2$, $K_2Fe_{22}O_{34}$	Ethylbenzene to styrene
Bismuth molybdate, other oxides + promoter oxides	Mixed oxides: Mo, Bi, Sb, Sn, Co, etc.	Propene to acrolein Propene to acrylonitrile
Zinc ferrite	$ZnFe_2O_4$	Oxidative dehydrogenation
Vanadia-molybdena	V_2O_5–MoO_3 solid solution	Benzene/butene to maleic anhydride
Vanadium phosphate	Mixed $(VO)_x(PO_4)_y$	Butane to maleic anhydride
Chromia-alumina	$Cr_xAl_{2-x}O_3$	Dehydrogenation of light alkanes
Mixed oxides	Perovskite-type	Catalytic combustion
Barium hexaaluminate	$BaAl_{12}O_{19}$	Catalytic combustion (support)

cannot be obtained by a direct reaction, an indirect method can be attempted. This is possible if a solid can be obtained of a composition which differs from the desired solid (e.g., a hydroxide, salts of organic acids, etc.), but which contains the metallic elements in the proportions required by the catalyst. The key reaction is then the solid-state transformation of this hydroxide or salt to the desired oxide. Keeping the material in the solid state throughout the transformation permits a powerful control (Fig. 1) which prevents all unwanted atomic restructuring in the bulk during catalyst preparation. In practice, two goals can be achieved here:

As mentioned above, when specific compositions can be prepared with homogeneity at the molecular level, but in a thermally unstable state (as salts, hydroxides, complexes, or vitreous compounds), these can be decomposed into the ultimately desired mixed-oxide catalyst.

Solid-state chemistry offers an additional advantage in that specific structures and even textures (as for the layers of phyllosilicate in pillared clays; see Chapter 2.3.7) can be preserved intact throughout these solid-state transformations.

Today, however, the potential of processes starting from solid precursors has broadened. On the one hand, advances in the elucidation of physico-chemical processes occurring during the reactions of a solid reactant to a solid product have opened a new perspective. Sometimes, a single starting phase yields solid products containing two or several phases, each of them very highly dispersed and in intimate contact with each other. This was observed in catalyst preparation. On the other hand, the identification of synergetic actions between catalyst components progressively convinced catalysis scientists that the synergy actually corresponded to a cooperation between distinct phases (initially called "remote control" [1–3]; see also Chapter 5.3.2). This is typically the case of oxidation, ammoxidation, and related processes. The use of solid precursors is therefore not limited to the preparation of homogeneous phases containing several metallic elements. Such a technique can provide intimate mixtures of phases in close contact with each other, and enhances the efficiency of phase cooperation [4]. This also allows the preparation of two or several multi-element oxides in a highly dispersed form that can be intimately mixed with each other. This maximizes the cooperative effect, by optimizing (by adequate doping) the efficiency of each partner in the synergetic action. This occurs in the preparation of the so-called "iron molybdate" (for the oxidation of methanol to formaldehyde), which is an intimate mixture of MoO_3 and iron molybdate.

Antimony oxide is essential in many selective oxidation reactions, and the use of solid precursors very likely produces highly dispersed Sb_2O_4 in intimate contact with the other component. Recent investigations on the very complex catalysts presently in development for the ammoxidation of propane (see Chapter 14.11.8) suggest a synergy between two distinct phases, each of them of a complex composition. If both of these could be optimized in terms of elemental composition and textural properties, then intimately mixing them could yield catalysts with very attractive properties.

Now, considering both highly dispersed solids of uniform composition at the atomic scale and mixtures of intimately interdispersed phases, the use of solid precursors offers an important additional advantage. The transformation of solids usually allows various types of control of the texture of the solid products that other processes do not

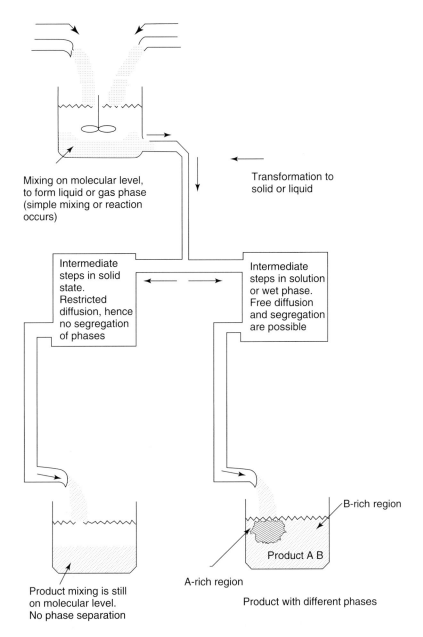

Fig. 1 Schematic representation of the vital role of solid-state reactions in preparing single-phase catalysts.

allow. An adequate selection of reaction conditions leads to highly dispersed solids possessing a well-defined porosity. The preparation methods described in this section are cases that can be considered as typical, and have been selected to highlight not only the possible goals that can be reached, but also the procedures used to reach them.

2.3.12.1.1 Some General Concepts and Parameters in Solid-State Reactions
When multicomponent solid systems are used to prepare a catalyst, homogenization of the precursors (mixing at the molecular level) is extremely important. The activity of the finished catalyst should not differ in the different parts of a catalyst charge, nor from batch to batch of the preparation. Two fundamental aspects of solid-state reactions involved in the preparation of catalysts are nucleation and the growth from solution of the nuclei or elementary particles into distinct solid phases in the surrounding liquid medium. Both of these are involved in the preparation of many catalysts and supports, as well as in the deposition or dispersion of an active catalyst component on the support surface. The problem here is that nucleation is, by essence, a process which favors

References see page 316

segregation; thus, it goes against the goal of preserving homogeneity at a molecular or atomic scale. Supersaturation of the medium essentially controls both nucleation and growth processes, thus determining ultimately the size, structure, and phases of elementary catalyst particles. The most important parameter is the relative contribution of surface energy in the whole energy context. This surface energy is considerable in small particles of any phase in contact with vacuum or any other phase (gas, liquid, solid) in comparison to that of large particles or flat surfaces (this is expressed by the Gibbs–Thomson law). Nevertheless, nucleation can take place because of statistical fluctuations, the efficiency of which in triggering nucleation increases dramatically in a narrow range of supersaturation. The kinetics of nucleation are extremely important when the synthesis of highly dispersed materials is considered. A detailed discussion of these aspects and the relevant literature are provided in reviews by Marcilly [5] and Marcilly and Franck [6]. The importance of nucleation and growth phenomena in solid-state reactions in catalyst preparation will be discussed briefly in the next section. Their vital role, and the different ways in which they can be controlled during the preparation of supported catalysts, is discussed elsewhere in this Handbook. General surveys of solid-state reactions involved in catalyst preparation or production are provided by Courty et al. [7], Courty and Marcilly [8, 9], and Delmon and de Keyser [10]. Several aspects of this subject are also described and updated in the Proceedings of the International Symposia on Scientific Bases for the Preparation of Heterogeneous Catalysts, held in 1975, 1978, and subsequently at four-year intervals, at Louvain-la-Neuve, Belgium [11–18].

The thermodynamics of the formation and transformation of a solid phase into another are characterized by two aspects, both of which explain the difficulty of producing solids of homogeneous composition. The more important of these aspects is nucleation; the other aspect is the tendency of certain components of the solid to diffuse to, or away from, surfaces. These aspects, however, cannot be considered in isolation. Chemical reactions involve the breaking of bonds and the formation of new ones, which involves kinetically limited processes. In many cases, diffusion brings about additional kinetic limitations. The final result is the combination of the effects of all these processes.

Nucleation in solid-state reactions essentially obeys laws similar to those that rule the formation of solids from liquid or gases. The only difference is the enhanced role of diffusion limitations. The simpler case of nucleation from liquids will be sufficient for explaining the problems encountered in the formation of homogeneous solids in all cases.

A form of phase diagram with two intensive parameters is shown in Fig. 2; here, they are the temperature (T) and

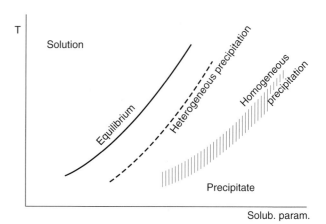

Fig. 2 Schematic representation of the precipitation conditions for a given compound.

any parameter (indicated as *solub. param.*) on which the solubility or dissociation constant depends. For example, this could be the pH or the concentration of a reactant or a precipitating agent. It is also possible to use as coordinates two of these *solub. param.* parameters, holding the temperature constant. This diagram indicates the limits of concentration of a given compound or of a third component (or temperature) for which either a liquid solution or a solution containing a precipitated solid is observed. The line labeled "equilibrium" corresponds to the exact equilibrium curve (as determined by thermodynamics), separating the monophasic domain (upper left) from the biphasic domain (lower right). The special kinetics of nucleation ensure that, as explained above, the new phase is not formed for a small supersaturation unless a patch of surface of that phase is already formed; in that case, deposition can take place. In the absence of any solid surface, precipitation will only occur in supersaturation conditions represented by the curve "homogeneous precipitation". In truth this is not a curve, but rather a band, as actual deposition will depend on the rate at which the *solub. param.* changes. The upper left-hand border of this band corresponds to so-called critical nucleation conditions, namely those below which no nucleation occurs. The dotted curve represents the conditions where precipitation can occur when a foreign surface is present, namely heterogeneous precipitation. The fact that this curve lies at an intermediary position is explained by the fact that the surface energy necessary for forming the precipitate is diminished in comparison to homogeneous precipitation, but is not equal to zero as in the presence of a flat surface of the precipitating compound, that is at equilibrium.

Let us now consider a solution containing more than one compound. There will be as many curves of

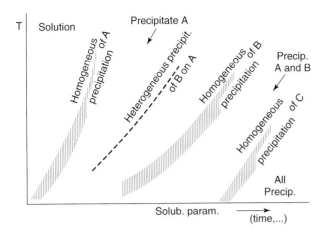

Fig. 3 Schematic representation of precipitation from a solution containing the compounds A, B, and C.

equilibrium, heterogeneous precipitation, and homogeneous precipitation as compounds are present. Figure 3 represents homogeneous precipitation for components A, B, and C – a similar figure would represent the situation for the other curves of Fig. 2. The consequence is that the various compounds in a solution precipitate sequentially, namely they bring about a segregation and create a solid of heterogeneous composition when the *solub. param.* changes (e.g., with time, in the direction of the arrow in Fig. 3). This can be very easily seen in coprecipitated silica-alumina, where pure silica is very frequently observed and considerable fluctuations of the Si/Al ratio by factors of 2 to ∞ are measured over a distance of several tens of nanometers for the best, and several micrometers for most samples [19, 20].

This is illustrated in Figs. 4 to 7, taken from data obtained with 17 different samples. These samples were prepared by coprecipitation (see Chapter 2.3.3) or by the sol–gel method (see Chapter 2.3.4), where the heterogeneity comes from the different reactivity of the starting oxides [21]. The laboratory samples were obtained from the same laboratory, using always the same procedure [20]. With respect to the industrial samples, most were probably obtained by coprecipitation, but perhaps some of them using the sol–gel or modified sol–gel techniques. In these investigations, the samples were dispersed in water with an ultrasonic vibrator, and one drop of the resultant suspension was deposited on a thin carbon film supported on a standard copper grid. After drying, the samples were observed and analyzed by transmission electron microscopy (TEM) on a JEOL-JEM 100 °C TEMSCAN equipped with a KEVEX energy dispersive spectrometer for electron-probe microanalysis (EPMA). The accelerating potential used was 100 kV. The first observations concerned the general aspect of the sample, examining at the whole grid. Then, different

References see page 316

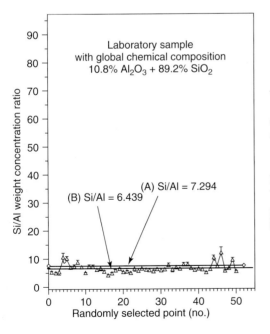
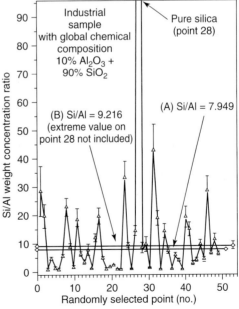

Fig. 4 Nanoscale fluctuations of the Si/Al weight concentration ratio for laboratory- and industry-prepared samples of alumina content around 10% (similar data were obtained for Si/Al around 5% or 50–60%). (A) is the value of Si/Al calculated on the basis of the global silica and alumina contents. (B) is the average value calculated from the Si/Al ratios measured experimentally on all the analyzed points (except for the extreme values, that could not be taken into account for calculation). See text for the physical meaning of the A-value [21].

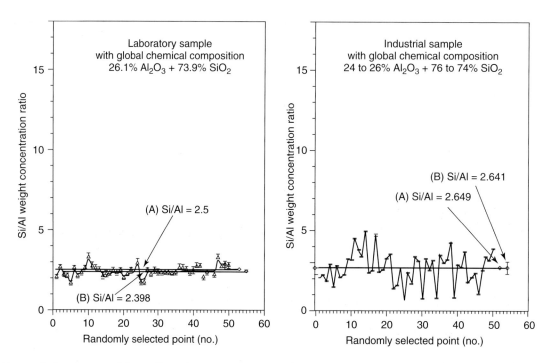

Fig. 5 Nanoscale fluctuations of the Si/Al weight concentration ratio for samples of alumina content around 25–26%. See legend of Fig. 4 for other details [21].

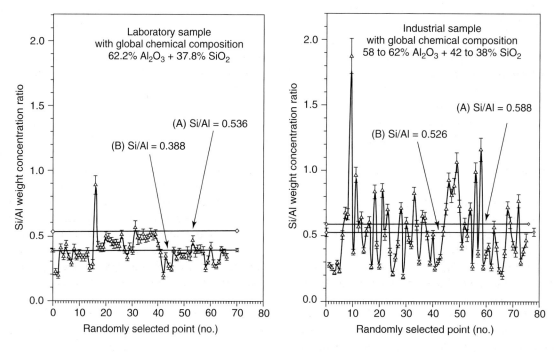

Fig. 6 Nanoscale fluctuations of the Si/Al weight concentration ratio for samples of comparable contents of silica and alumina. See legend of Fig. 4 for other details [21].

points of each particle were analyzed (two or three); the same procedure was used for the analysis of a total of 10 to 25 particles for each sample. The diameter of the static spot was 0.2 µm, while the duration for each analysis (accumulation) varied between 60 and 240 s. The details of the procedure are reported in Ref. [21].

In general, those samples prepared according to the original sol–gel method (from alkoxides) in the laboratory

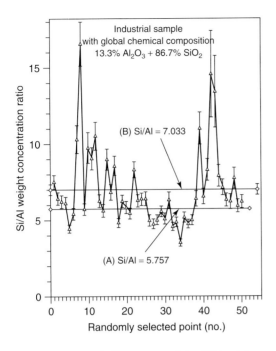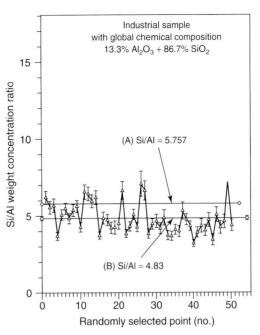

Fig. 7 Differences in nanoscale fluctuations of the Si/Al weight concentration ratio between batches for industrial samples of identical composition. See legend of Fig. 4 for other details [21].

are more homogeneous, but one sample (14.8% Al_2O_3; Fig. 4a of Ref. [21], not represented here) contained particles of pure silica. Except perhaps for compositions around 25%, the degree of heterogeneity was similar for all sol–gel samples. It is understandable that industrial silica-aluminas examined in identical ways can exhibit substantially different degrees of homogeneity (Fig. 7).

Several programs have been conducted for improving homogeneity in coprecipitation (e.g., Refs. [22] and references therein, and [23]). The approach was typically a chemical engineering one, but data showing important progress in improving homogeneity have not been published. It seems that the attention was finally focused on a better control of pore size uniformity.

Comment should perhaps be made at this point with regards to the other technique used for obtaining the results shown in Figs. 5–7. The co-hydrolysis of alkoxides was apparently first studied during the 1960s to create high-temperature-resistant ceramics for the space programs. This became known as the "sol–gel method", in spite of the fact that coprecipitation also produces a gel. At present, the name tends to be used in industry also for classical coprecipitation and, in industry or university laboratories, for modified reactions involving alkoxides together with other reactants (most often nitrates) and even for other methods. There is a common obstacle to reaching homogeneity in all these methods, in that the reaction leading to precipitation does not occur at the same rate for the two (or several) components supposed to coprecipitate. In the case of the original alkoxide method, mildly chelating agents, or, alternatively the choice of different alkoxide groups for each cation, can make the reactivity of one component approach that of the other. Success is limited, however, the crucial step being the introduction of the solution containing the reactant that triggers the precipitation (e.g., sodium hydroxide in classical coprecipitation, or acidified water in the sol–gel technique). In spite of a very energetic mixing action, the compositions at the scale of several nanometers (and often micrometers) differ from place to place, and reactions occur at different rates, producing gel particles not only of different composition but also often of different sizes.

There are two possible approaches to improve the situation. The first approach relies on the basic features of the reactions involving solids. It is possible to take advantage of the difference in the form of the kinetic laws corresponding to nucleation, on the one hand, and the growth of precipitated crystallites, on the other hand. The basis for these attempts is represented schematically in Fig. 8 [24], corresponding to one given compound. On the right-hand side of the figure, the rate of nucleation will be very high in comparison to growth, and very tiny particles will be obtained. The situation represented on the left-hand side corresponds to large particles and even, below the critical limit for nucleation, to Ostwald ripening. At very high supersaturations of all compounds

---

References see page 316

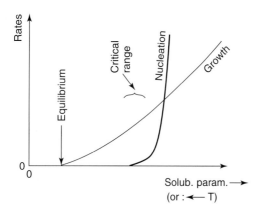

Fig. 8 Qualitative comparison of the dependency of the rates of nucleation and growth of precipitate particle as a function of the solubility parameter (increasing from left to right) or T (opposite direction).

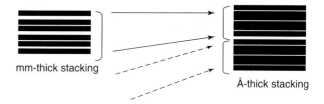

Fig. 9 The "mille-feuille" concept of perfect mixing of plastic material.

present in a solution, finely interdispersed particles will be formed. However, this will not lead to the production of atomically homogeneous solids. Although extremely rapid mixing can produce such finely interdispersed phases, rapid cooling is generally unsuccessful, because diffusion in the liquid phase will considerably perturb the processes leading to nucleation. In a recent approach, a semi-batch reactor method based on high-speed mixing with an impeller was proposed to control the nucleation and growth mechanisms of high surface area pure and doped CeO_2 nanopowders [25].

The second approach is the one resting on the "unit operations" of chemical engineering. It is well known that the process of mixing continues to resist spectacular improvements, despite the invention of many devices including high-energy jets from a nozzle, cyclones, and turbines. Yet, a molecular-scale homogeneity is perhaps possible, using a concept borrowed from ideas developed for microdevices. The speculative idea is to: (i) make thin films of two (or several) components (the black and white bars in the figure) to mix, or of the imperfectly mixed solid or highly viscous materials; (ii) intercalate sheets of type 1 between sheets of type 2, as a sort of "mille-feuille"; and (iii) repeat the operations several times (Fig. 9). Perfectly conducted, the process based on this principle would need 21 to 23 successive operations!

Other possible approaches include the relatively frequent use of ultrasound (e.g., [23]) or that of microwaves (e.g., [26], for the preparation of VPO catalysts). These techniques can be used alone or in addition to those already mentioned, but comparative studies should first be made. In the case of microwaves, it should be made clear that, in addition to supplying heat and favoring the intense movement of molecules, microwaves are able selectively to activate chemical bonds [26]. Adequately "tuning" the wavelength might allow the selective activation of reactions between two different alkoxides in the sol–gel technique. A partial conclusion – or, more precisely, a remark – is that the above section outlines the perspective that the use of many different contributions of chemical engineering can open. A reasonable conclusion is that the combination of purely chemical and chemical engineering approaches offers the best prospects.

When such strategies are used to produce approximately homogeneous solids, the other phenomenon mentioned above can occur, namely *heterogeneous precipitation* of one compound on the surface of the already formed precipitate of another phase. This was shown in Fig. 3, for the precipitation of compound B taking place when A had already precipitated. This has given rise to severe misinterpretations of the physico-chemical characterization of precipitated solids, especially when surface-sensitive techniques such as X-ray photoelectron spectroscopy (XPS) were used.

As indicated above, elementary thermodynamic considerations show that the creation of a surface is accompanied by a positive free-energy change. This leads to another phenomenon, one which takes place inside an already-formed mixed oxide. In order to minimize the positive free-energy change, the solid component with the lowest free energy tends to migrate from the bulk to the surface [27, 28]. Since the migration of a constituent of a solid from its interior to its surface usually involves overcoming an activation energy barrier, such a process necessarily requires a higher temperature. As a rule of thumb, the Tammann temperature (taken approximately as half the melting temperature, in Kelvin) is generally believed to be that over which the atoms or ions in the bulk are sufficiently mobile for bulk-to-surface migrations, while the Hüttig temperature (about one-third the melting temperature) is enough to make the species, which are already on the surface, sufficiently mobile to undergo agglomeration or sintering. As working temperatures in catalytic processes often exceed the Hüttig temperatures of metals or oxides used as catalysts, one objective of catalyst preparation is to anchor the active species onto the catalyst or support and thus retard or prevent their free motion or migration under the process conditions. This is why homogeneous compound oxides may be useful for stabilizing the coordination of an oxide element and maintaining dispersion.

2.3.12.1.2 The Objective of this Chapter in the Handbook

In conforming with the title of this chapter, the objective here is to describe approaches to catalyst preparations using solid precursors. This excludes in particular the original sol–gel method (see Chapter 2.3.4) which, in common with precipitation processes, is faced by the fundamental problems described in Section 2.3.12.1.1. Unless a method is found for the direct preparation of a catalytic phase (usually an oxide) that contains all of the required metallic elements in the necessary proportions, a more indirect method should be selected. In general, however, a direct preparation from the gas or liquid phase will be impossible, for the reasons given above. Thus, it becomes necessary to resort to a two-step or multiple-step approach, using at least one solid precursor.

The presentation will essentially concern oxides, which is by far the most frequent case. For other catalysts (e.g., carbides [29], nitrates [30, 31], oxynitrides [32], phosphides [33]), mixed oxides usually serve as precursors, and homogeneity is yet more crucial. Even bimetallic catalysts have been prepared by the reduction of precursors, using techniques presented in this section of the Handbook [34].

Several cases must be distinguished when an atomically homogeneous multi-element catalyst is required:

(i) Direct preparation by various types of reaction (e.g., precipitation), though very few examples exist. In practice, this approach is only possible if an oxoanion (e.g., a molybdate or a vanadate) reacts with a cation to produce the precipitate. The stoichiometry is dictated by the chemistry of the reaction, and has little flexibility.
(ii) A two-step process, through preparation of another type of homogeneous solid of exactly the same composition (e.g., carbonate, hydroxide, oxalate or other precursors, for making finally, in a second step, an oxide) and thermally activated solid-state reaction of this precursor.
(iii) The preparation of a solid, approximately homogeneous, followed by a solid-state transformation, the hope being that a higher homogeneity will be achieved during this second step.

As category (i) forms the subject of other parts of this Handbook, attention is given here to categories (ii) and (iii). Category (ii) corresponds to two different cases – those where the precursor is crystalline, and those where it is amorphous. However, the following discussion is organized according to the technique used, considering whether the precursor is crystalline or not, and without distinguishing between cases (ii) and (iii).

Thus, comments will be made on the potential of the technique to produce atomically homogeneous solid precursors.

The details concerning the transformation of precursors to final catalysts will be mentioned in each case. A general comment must first be made at this point: that the strategy which consists of starting from a homogeneous solid precursor and transforming it to the desired catalyst is only successful if diffusions in the solid state that potentially lead to segregation are made difficult. This requires that only those precursors which decompose to the final solid at low temperature, below the Tammann temperatures, be used. On the other hand – and for reasons which cannot be presented here in full detail – it is advantageous that the solid precursor decomposes with the production of large quantities of gases. The main reason for this is that a coupling between the solid-state transformation and the formation of pores is necessary, and the formation of pores is obviously promoted by the evolution of gases. Coupling effects in catalyst preparation have been discussed elsewhere [35, 36].

In practice, hydroxides, carbonates, hydroxocarbonates, nitrates, and salts of carboxylic acids (formates, oxalates, acetates, citrates, etc.) satisfy both requirements. However, there are – potentially – many other possible compounds, some examples of which will be outlined.

2.3.12.2 Description of Preparative Methods

In very rare cases, catalysts can be prepared by using dry methods, namely reactions in the solid state. However, the most commonly used methods to prepare mixed-oxide catalysts start from a liquid (Fig. 10). As complete mixing at the molecular level is possible in solution, a solution of all the concerned salts in the appropriate amounts is often taken as the common starting material. One can essentially distinguish two steps in the procedure: (i) the preparation of a homogeneous solid precursor; and (ii) the careful decomposition of this precursor to the oxide. For the first step, it is crucial to avoid the segregation of the oxide species into separate solid phases. This is usually achieved by:

- Physical methods such as evaporation to dryness, drying by vaporization, freeze-drying, and crystallization; for the reasons explained above, the operation should be extremely rapid.
- A physico-chemical method, such as some gelation methods.
- Chemical methods such as precipitation, coprecipitation, oil-drop, oil-up, and complexation. Oil-drop and oil-up involve processes pertaining to categories (ii) and (iii) above.

References see page 316

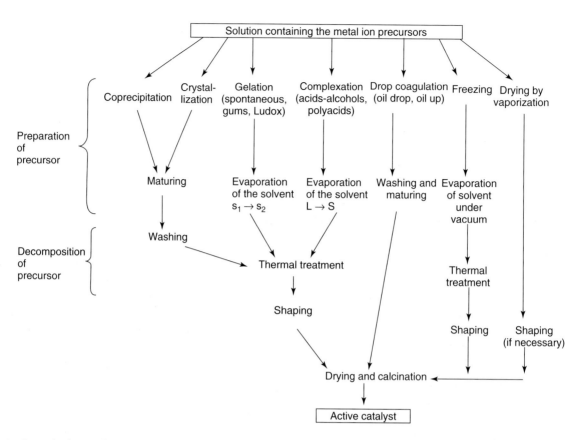

Fig. 10 General scheme of various methods for the preparation of unsupported catalysts. (From Ref. [10].)

The considerable development of heterogeneous catalysts has led to innumerable variations around the most frequently used techniques, in addition to combinations of techniques. Examples include micro-emulsions + oil drop, supercritical precipitation [17, 18], the impregnation of high-surface-area solids that later can be eliminated by combustion, and complexation + spray-drying (see Refs. [11–16, 37–40]).

2.3.12.2.1 Dry Methods One common method of preparing mixed-oxide catalysts is to grind or ball-mill the components together, either dry or in a wet slurry (see also Chapter 2.4.8); this is followed by calcination to temperatures of 600–1000 °C. The entire operation may need to be repeated several times. In the past, catalysts for the styrene process were generally prepared in this way from KOH and oxides of Fe and Cr and other (proprietary) promoter components. The solid–solid reaction during calcination can be accelerated by "priming" or catalyzing by using a small quantity of the product premixed with the reactant oxides. Catalysts used in the synthesis of ammonia are prepared by mixing the components in the molten state. They correspond to a very special case, where very high dispersion is not necessary, and very high temperatures can therefore be used. A "fluid energy mill", operated with steam injection at 400–700 °C and 5–15 bar ($5-15 \times 10^5$ Pa) steam pressure, is applied in some cases. Some other rarely used techniques include the dry chemical oxidation of alloys, and electrolytic deposition.

The calcination step may sometimes have a more subtle importance than is generally realized. For example, at a calcination temperature of 600 °C, a promoter such as Li can be mixed to modify a catalyst such as TiO_2, whereas at 930–1030 °C it can be incorporated into the lattice structure of TiO_2. For the former case, optimum performance in the oxidative coupling of methane (OCM) was shown at 17 wt.% Li_2O loading, and catalytic activity for OCM fell by a factor of five after 2 days on stream [41, 42]. For the catalyst calcined at 930–1030 °C, the optimal Li_2O loading was only 1–2 wt.%, it was active for OCM at higher temperatures, and its stability over several days was much better [43].

In general, the methods described here produce catalytic solids with a low surface area.

2.3.12.2.2 Wet Methods
A Coprecipitation The most common method for mixed-oxide catalyst preparation is crystallization or

precipitation or coprecipitation in solution of a precursor form (hydroxide, oxide, insoluble salt) of the catalyst (see also Chapter 2.3.3). Other specific steps, for example either addition of an extra component or its removal by partial extraction, may sometimes be necessary to adjust the final catalyst composition and to ensure homogeneity.

One of the most studied cases – and also the most successful – was that of hydroxo-carbonate for preparing catalysts containing copper and zinc or chromium with additional elements (e.g., Co or Al). The reason for these studies was the practical importance of catalysts used for producing synthesis gas and methanol. The data in Table 2 [9], which provide the composition of various precursors and their structures, suggest details of those mixed oxides which could be formed by decomposition. Additional data can be found elsewhere [9, 44–47].

As discussed previously, precipitation (or simple crystallization) involves two main steps – nucleation and growth. At the nucleation stage, very small crystals of the solid phase are formed; these have a high specific surface area and high surface energy, and hence are unstable. Below a certain critical nuclei size, the free energy of formation of the solid phase is less than its surface energy, and the new phase is thermodynamically unstable. Above this critical size, the solid particles can grow. According to the conditions, this growth step is mainly controlled by interface or diffusion phenomena. The first case generally corresponds to the formation of complicated solid structures, whereas the second case is more common in the precipitation processes used in catalyst preparation. An overview is provided elsewhere [5, 6, 24], and an excellent detailed treatment of nucleation and growth phenomena is also available [48]. At this point, however, only more general statements or rules of thumb can be suggested.

For *nucleation*:

- the higher the supersaturation, the higher will be the nucleation rate
- any interface may play the role of a heterogeneous nucleus by lowering the surface free-energy of the new phase, and thus increase the rate of nucleation

- in general, higher temperatures lower the nucleation rate by increasing the critical size of the nuclei – the upper left boundary of the homogeneous nucleation band reflects this trend.

These conditions are represented schematically in Fig. 2.

For *growth*:

- the higher the supersaturation, the higher will be the growth rate, but it is rapidly limited by diffusion processes
- as an increase of temperature enhances the diffusion rates, the higher the temperature, the higher will be the growth rate of the new phase.

By combining the above guidelines, and taking into consideration the data shown in Fig. 8, some general rules for obtaining a fine precipitate can be formulated:

(i) A vigorous stirring of the solution, while adding the chemical compound triggering precipitation is beneficial in two ways: first, each elementary volume of solution reaches rapidly the highest degree of supersaturation; second, it comes into contact several times with the agitator and the walls of the vessel, thereby also promoting heterogeneous nucleation.
(ii) A rapid addition of the precipitating agent ensures a rapid reaching of the highest degree of supersaturation in the whole volume of the solution; hence, a maximum nucleation rate is obtained. For the same reason, the best precipitating agent is the one giving the precipitate with the lowest solubility product.
(iii) Although the situation may differ widely from case to case, precipitation is often advantageously made at the lowest practical temperature, as this often favors nucleation over growth.
(iv) As demonstrated recently in the case of $Ce_{1-x}Zr_xO_2$ solid solutions, crystalline mixed-oxide nanoparticles can also be obtained by hydroxide coprecipitation, followed by redispersion in an aqueous medium by sonication, using nitric acid as peptizing agent [49].

In the *coprecipitation process*, the multi-element product can be of three types:

- Metals combined in an insoluble single compound (e.g., $BaTiO_3$) can be obtained by precipitation of a hydrated oxalate ($BaTiO(C_2O_4)_2 \cdot 4H_2O$) and its thermal decomposition. To illustrate this approach with an example more directly related to catalysis, a cerium zirconyl oxalate corresponding to the stoichiometry $Ce_2(ZrO)_2(C_2O_4)_5$ has been described as precursor

Tab. 2 Hydroxocarbonate precursors for Cu/Zn/Cr catalysts

Phase	Formula
Hydrotalcite-type phase	$M^{2+}_6 M^{3+}_2 (OH)_{16} CO_3 \cdot 4H_2O$
	$M^{2+} = Cu^{2+}, Co^{2+}, Zn^{2+}, Mg^{2+}$
	$M^{3+} = Al^{3+}, Cr^{3+}, Fe^{3+}$
Malachite-type phase (rosasite)	$Cu^{2+}_{2-x} Zn^{2+}_x (OH)_2 CO_3$
Copper-zinc hydroxocarbonate (aurichalcite)	$Cu^{2+}_{5-x} Zn^{2+}_x (OH)_6 (CO_3)_2$

References see page 316

for CeO_2-ZrO_2 solid solutions [50]. Other instances are precipitations as chromates, molybdates, and tungstates.

- Some metals precipitate into a single-phase solid solution or a single-phase mixed structure; for example, precipitation by sodium carbonate of aluminates or hydroxoprecursors of aluminates of Cu, Fe, Ni, Zn, and Mg. Divalent metal oxalates can also be precipitated in this way [51].
- Some metals precipitate more or less simultaneously, but in separate phases – for example, metal hydroxides precipitated by NaOH, KOH, and (NH_3)aq. Normally, aqueous ammonia is preferable for precipitation, especially when the washing off all alkali from the catalyst is a prerequisite for preparing the final catalyst. An exception to this is when (NH_3)aq can form ammine complexes, as with many transition metals (Ni, Co, Cu, Zn).

Non-exhaustive lists of mixed salts which had been claimed to be obtained by coprecipitation were presented in the first edition of this Handbook (see Chapter 2.1.6.2C; Tables 3–7) [52].

B Spray-Drying and Spray Calcination Spray-drying is a versatile method which can produce mixtures of phases with a high degree of interdispersion and a relatively high degree of homogeneity, although there is, in principle, no atomic-level homogeneity. This is essentially an improved evaporation method.

Very often, slow evaporation of a homogeneous solution of two precursor salts, for example $MgSO_4$ and $Al_2(SO_4)_3$, leads to segregations of the two salts in the solid phase. In order to prevent this, it is advantageous rapidly to eliminate the solvent, either by spray-drying or hot-petroleum drying. Spray-drying is also frequently used for removing water when the active phase has been produced. In certain cases, the temperature in the equipment can be raised sufficiently to permit the reaction between the salts to produce an oxide combining two metallic elements. This has been carried out from solutions of $MgSO_4$ and $Al_2(SO_4)_3$ to produce highly dispersed $MgAl_2O_4$. The success of the method, in this case, lies in the very small size of the crystallites obtained by rapid drying and their excellent interdispersion, which promotes solid-state reactions at higher temperatures.

The spray-drying technique is used industrially to produce powders for the fabrication of ceramics [53]. It is mainly single-metal oxides that are made in this way, but has been claimed that $MnFe_2O_4$, $NiFe_2O_4$, $NiAl_2O_4$, dopant-stabilized zirconium oxide and titanates can also be obtained. Powders of surface area up to $20\,m^2\,g^{-1}$ can be produced, the powder generally being in the form of aggregates of 0.2–0.4 μm in the best cases, but often of over 100 μm in diameter [53].

By using this technique it is possible to prepare fine spherical catalyst particles in the range of 10 to 100 μm diameter, as are required for typical fluidized-bed catalytic processes. In this technique, which is used for large-scale catalyst manufacture, the feed is generally a dilute hydrogel or sol that is sprayed from the top of a tower while hot air is blown in a cocurrent or countercurrent direction in order to dry the droplets before they reach the bottom of the tower. The fine droplets are produced or atomized by pumping the hydrogel or sol under pressure either through nozzles, or onto wheels discs that are rotated at high speed. The method is also convenient for embedding crystalline particles of micron or submicron size in an amorphous matrix of 10 to 100 μm diameter. The best-known examples are the embedding of multicomponent bismuth molybdates in a 50% silica matrix for the ammoxidation of propene to acrylonitrile, and the incorporation of <1–3 μm zeolite crystallites in an amorphous silica-alumina/clay matrix for fluid catalytic cracking (FCC) catalysts.

The use of spray-drying or spray calcination to prepare atomically homogeneous mixed oxides is slowly beginning to emerge from its development stage. Flame spray pyrolysis of precursors mixtures was recently shown to provide Pt/ceria-zirconia materials which exhibited higher specific surface areas than did conventional catalysts obtained by precipitation or incipient wetness impregnation [54].

C Hot-Petroleum Drying Hot-petroleum drying, which constitutes an alternative to spray-drying, was developed originally by Reynen and Bastius [55], initially for small-scale preparations. However, the development of what industry incorrectly calls the sol–gel method – namely integrated conventional precipitation and gelling – has led to hot-petroleum drying becoming a large-scale preparation method. The principle of the technique involves dropping a fluid material containing the active elements into kerosene. This fluid material may be a gel, a solution which is in the course of becoming a gel, a stabilized emulsion of a salt solution and kerosene, or an aqueous solution of salts. The temperature of the kerosene bath is maintained at about 170 °C, and it may be vigorously agitated if necessary. The water evaporates and the powder or (according to processes) the small solid spheres obtained are easily filtered and dried at around 250 °C. The product is a free-flowing powder which consists of a homogeneous mixture of salts.

The hot-petroleum drying method, in contrast to the spray-drying variant (called spray-calcination) does not permit the solid-state reactions needed to form the oxides, but the highly dispersed and intermixed salts can be thermally decomposed without phase segregation. As with spray-drying, this method also depends on the evaporation of water being faster than the segregation phenomena. Thus, the controlling factors are the temperature of the hot kerosene and the rate of water evaporation from the droplets. The advantages of this method are the simple set-up, the fact that kerosene can be recycled, and the lower energy consumption than for spray-drying.

D Freeze-Drying Although freeze-drying has the same purpose as both above-mentioned methods, it has a different strategy. In order to limit segregation, the action now concerns the diffusion process, which is slowed down either by rapid cooling (freeze-drying) or by increasing the viscosity of the solution (e.g., by using some of the methods described below, or by the addition of hydroxycellulose or gums). This increase in viscosity also plays a role in the more elaborate amorphous citrate process (see below).

Undercooling is the driving force in freeze-drying. An aqueous salt solution is introduced dropwise into an immiscible liquid (hexane or a petroleum fraction such as kerosene) that is cooled below $-30\,^\circ$C. The individual droplets are frozen instantaneously, and the solid particles decanted or filtered off. The frozen liquid is then sublimed in a vacuum to obtain a homogeneous powder of fairly uniform particle size. Important parameters in freeze-drying are the cooling rate and the final temperature of the salt solution. These can be controlled to some extent, but only on a small scale, and hence the method is not well suited to the large-scale manufacture of catalysts.

2.3.12.2.3 Chemical Complexation Methods

A General Comments Quite generally, the methodology for preparing finely divided heterogeneous catalysts also benefits from the development of the so-called "soft chemistry" strategies implemented for the synthesis of multicomponent inorganic materials. In that context, "precursor" routes based on the thermal decomposition of suitable precursors containing organic groups, under rather low temperatures (below 500 $^\circ$C) are widely used, essentially for the synthesis of oxides. Sulfides, nitrides, or carbides can be obtained in a similar way, either starting with appropriate precursors or by adapting the decomposition medium. Because these routes allow crystalline materials to form under conditions that are significantly milder than those employed in conventional solid-state synthesis ("ceramic method"), the final materials often present relatively high surface areas. To select the appropriate temperature for the decomposition process, in agreement with the thermal degradation scheme of the starting materials, a compromise between the high surface areas and the extent of residual carbonaceous surface contamination must be found. However, in most cases, when precursors containing only carbon, oxygen, and nitrogen are used the choice of appropriate conditions allows finely divided materials with very high surface purity to be obtained.

Two major classes of precursor are encountered. On the one hand, well-established methods are based on the use of polymeric precursors of ill-defined stoichiometry at the molecular scale, containing the desired metallic elements in appropriate amounts, together with a sacrificial matrix that ensures the efficient mixing at the molecular level. The widely used "citrate methods" belong to this category, although in some cases (Al [56], Ti [57], V [58], Mn [59], Co [60], Mo [61], and La [62]) well-defined citrate complexes have recently been identified and crystallized. On the other hand, in some other cases, coordination compounds with defined stoichiometries can be obtained, and even characterized from a structural point of view when single crystals are available. The presence of bridging and/or chelating organic ligands in these precursors was shown to avoid metal segregation during oxide formation. Carboxylates and polyaminocarboxylates (PAC) such as EDTA (ethylenediaminetetraacetate) or its heavier analogues such as DTPA (diethylenetriaminepentaacetate) or TTHA (triethylenetetraminehexaacetate), give rise to a wide variety of such complexes [63, 64]. More particularly, when different metals are involved in the formulation of the final oxide, the main advantage of these routes is the potential use of heterometallic single-source precursors, because they provide a much greater control of the metal stoichiometry in the final oxide. More particularly, several examples of alkoxysiloxy metal complexes or other siloxane compounds have been described as precursors to silica-based multicomponent oxides containing for instance magnesium [65], chromium [66], molybdenum [67, 68], tungsten [68], vanadium [69], or zirconium [70]. Ideally, when bulk mixed oxides are concerned, precursor complexes containing the number of metal atoms corresponding to the stoichiometry of the final oxide are required to optimize this approach, but this often represents an ambitious challenge. Otherwise, intimate mixtures of identical (or as similar as possible, with respect to the thermal behavior) homometallic precursors of different metals to be combined can be used ("multiple precursors method").

Depending on the functionality of the PAC acid and the total number of possible N and O coordination sites, various geometric arrangements and coordination numbers can be achieved to fit the size and geometry requirements of the cations involved. In some specific cases, slightly modified ligands constitute promising alternatives. This is the case, for example, with the elements of Groups 4 (Ti) and 5 (V, Nb, Ta), for which the

---

References see page 316

availability of well-defined soluble precursors is restricted. Peroxo-carboxylates [71–74] and peroxo-PACs [75–77] of these metals are available as water-soluble molecular precursors for the manufacture of catalysts. Because of the presence of the peroxo groups, however, care must be exercised when handling these compounds on a large scale.

B Crystalline Complex Salts Oxalates [78], tartrates [79] and citrates [80] of various metals can form well-crystallized complex salts (e.g., Mg(NH$_4$)$_4$(Al$_x$Cr$_{1-x}$(C$_2$O$_4$)$_3$)$_2$ · nH$_2$O) which, on thermal decomposition at moderate temperatures, can yield mixed oxides. Although these complex salts permit substitution over large concentration ranges (or even in all proportions in some cases of elements of the same ionic volumes), their stoichiometry is still rather rigid and this imposes a serious limitation on any wide applicability of this method. This limitation is conspicuous when comparing the chemical formulas of crystallized citrates that were mentioned in the First Edition of this Handbook (see Chapter 2.1.6, Table 7 [52], and Ref. [53]) with the very flexible composition of compounds that will be described in the next subsections. This difference concerns the nature of the associated elements and the stoichiometry. Very few crystallized citrates contain elements of primary catalytic importance, and almost none of them contains two elements of interest. On the other hand, stoichiometry is determined in a rigid way by the crystallographic structure. Oxalato-complexes, which have been the object of detailed studies [78–80], correspond to the formula:

$$M^I_2(NH_4)_4[M^{III}(C_2O_4)_3]_2 \cdot nH_2O$$

where M^I and M^{III} represent a single metal or sometimes a mixture of simple ions of identical valency, namely I or III, respectively. This flexibility in stoichiometry, however, is restricted to within, respectively, the group of valency I or that of valency III. The method was also adapted for complex ions such as the vanadyl cation VO^{2+}. In general, the problem is further complicated due to the possibility of non-homogenization or phase separation during the chain of downstream unit operation involved in catalyst production. The amorphous complex method (which is described below) provides an escape from most of these difficulties.

C Amorphous Precursor "Citrate" Method: General The principle of this method consists of preparing a stable amorphous or glassy precursor which, on the basis of its amorphous structure, tolerates an enormously wide range of composition. The starting solution is a mixture of metallic salts and a hydroxyacid such as citric, malic, tartaric, lactic, or glycolic acid [7, 81–83].

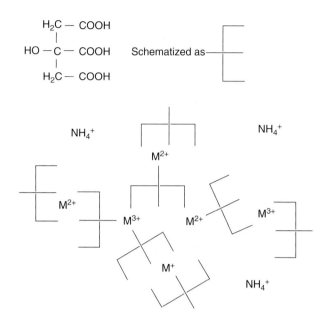

Fig. 11 Schematic representation of the 3-D structure of the amorphous citrate precursor. Only the hydrocarbon skeleton of citric acid is represented schematically with its three acidic functions and one hydroxyl function. The coordination sphere of each cation is completed by water molecules. Extra ammonium and negative counter-ions are involved for charge balance.

The metals are preferably in the form of nitrates or ammonium salts, which will not leave any residues on thermal decomposition at a later stage. In addition, the decomposition of the nitrate and ammonium ions promotes surface area and porosity, due to the large amount of gas that is evolved during the decomposition. Typical concentrations are 1 g equivalent organic acid per equivalent of total metal content. The solution is concentrated by rapid evaporation under vacuum, for instance in a Rotavapor, until its viscosity exceeds 500 cP at ambient temperature. A very stable syrup is obtained in this way which can be dehydrated under vacuum at about 80 °C to a rigid-foam (meringue) precursor, which is amorphous to X-rays, hygroscopic, and exhibits the color of the metallic ions present. The amorphous precursor appears to be some sort of a three-dimensional polymer in which the multifunctional organic acids are linked to two or more cations (Fig. 11). The thermal decomposition of the amorphous mass may proceed more or less continuously or in two distinct steps at 80–100 °C and 250–400 °C. Because of the danger of violent reactions, this decomposition step must be carried out very carefully, taking all necessary precautions. The preparation of oxides containing metallic ions highly active in oxidation requires very stringent precautions, and is generally not advisable. In particular, the possible detonation of copper and silver salts should be borne in mind. In these cases, the freshly formed metal particles could be powerful catalysts for some unexpected reactions, for example with ammonia. In addition, secondary reactions can occur even at room temperature, resulting in the formation of a small amount of lethal cyanogen gas. Clearly, this is not a method to be tried by students or inexperienced researchers!

The mechanisms of reactions occurring during the preparation of mixed oxides by the citrate methods have been outlined in some detail [84]. In the dehydration under vacuum, extensive loss of nitrate ions (in the form of various nitrogen oxides and ammonium nitrate) occur, together with the loss of water. Although easier in the presence of citric acid, the decomposition of nitrate does not involve extensive reaction with the latter below 100 °C. The semi-decomposed precursor obtained at about 140 °C is essentially a dehydrated, amorphous, highly porous mixed citrate. The higher temperature decomposition stage, mainly between 225 and 400 °C, consists of the burning of citric acid by oxygen. This is facilitated because of the presence of nitrate ions in the initial precursor. As indicated above, this stage is catalyzed by the presence of metals of which the oxides are oxidation catalysts. Additional details for this procedure may be found in Refs. [9] and [84]. Some carbon remains in the oxides, as in most other preparation methods, although the use of ozone has been shown to diminish the C content, a fact illustrated only recently [85].

Mixed oxides obtained using the above method are very lightweight powders of apparent density <0.05 g cm^{-3}, and are amorphous to X-rays, free flowing, and consist of aggregates of particles of 20 to 100 nm diameter. Upon further heating, these oxides crystallize into various well-defined structures, depending on the nature and composition of the starting materials: solid solutions, spinels, perovskites, garnets [7]. Variants of the citrate method consist of simply adding citric acid or other hydroxymultifunction acids to solutions used for preparing catalysts by other methods (precipitation, impregnation, sol–gel, etc.). Although not leading to such excellent results as the original citrate method, the outcome of such a use of hydroxyacid corresponds to a substantial improvement. A modified citrate method (commonly called the "Pechini method") has been proposed in which a polyalcohol is added to the initial solution [86]. The polyalcohol is thought to promote reticulation in the transformation of the starting solution to the rigid foam, with the reticulation or polymer formation being due to the esterification reaction between citric acid and the polyalcohol. This method has been used for the preparation of barium titanates doped by a large variety of ions, and can be easily applied for preparing catalysts.

The chemical complexation methods – and particularly those using amorphous precursors – have found important and innovative applications in the preparation of a wide variety of catalysts and supports, and also of various perovskite-type catalysts and barium hexaaluminates, as required for high-temperature (>1200 °C) applications such as catalytic combustion. Thus, some applications of these materials are described in the following two subsections.

D   Citrate Method for the Preparation of CeO$_2$–ZrO$_2$ Solid Solutions as Supports   Ceria–zirconia solid solutions have attracted attention because of their thermal stability and ability to store oxygen or, at least, to increase oxygen mobility in complete oxidation catalysts [87, 88]. These materials were investigated over a whole range of compositions; in addition, their high heat capacity constitutes a major advantage for use in reverse-flow reactors [89]. Extensive studies have been conducted on the Ce–Zr–O system since 1995 [89–97], and their range of stabilities for application were found to correspond to phase metastability in the phase diagram [98]. Hence, it would be advantageous to start from very homogeneous precursors, in order to avoid the presence of large regions where the composition would correspond to major thermodynamic stability and correlatively particles of low surface area due to rapid growth of the nuclei. Data obtained with samples prepared by the citrate and sol–gel methods after calcination at 700 °C [89] are compared in Fig. 12. Measurements with X-ray diffraction (XRD) indicate the presence of two phases, possessing the tetragonal and the monoclinic structures, respectively. Observation of the XRD lines of both phases does not show any broadening that might suggest heterogeneity of composition. A variety of techniques, using notably the MSI-CERIUS software, indicated that both polymorphic phases were solid solutions, and not monometallic oxides [89]. In contrast, the samples prepared using the sol–gel method exhibited asymmetric XRD lines, in addition to the presence of stoichiometric Ce$_{0.5}$Zr$_{0.5}$O$_2$, as found previously in other studies (see references cited in Ref. [89]). The results of a detailed study of these samples [by XPS, energy dispersive X-ray (EDX), electron microscopy, and the observation of modifications of the XRD diagrams as a function of temperature during calcination of the precursor], in addition to the technique mentioned previously, are reported in Ref. [89].

E   Citrate Method for the Preparation of Unsupported and Supported Perovskite Catalysts   A general introduction to perovskites and their catalytic properties is provided

References see page 316

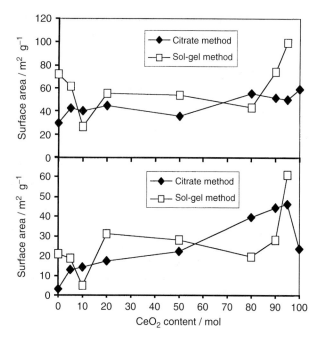

Fig. 12 Specific surface area of CeO$_2$-ZrO$_2$ oxides (Ce = 0–100%), prepared by the citrate and sol–gel method and calcined at (upper) 450 °C and (lower) 700 °C.

Tab. 3 Methods of preparation of high-surface-area perovskites (Adapted from Ref. [99].)

Description	Preparation method
Control of evaporation (decomposition) process	Freeze-drying, spray-drying, mist decomposition, combustion synthesis
Use of precursor material of a composition adapted to the desired catalyst	Coprecipitated oxalate decomposition, amorphous citrate decomposition, decomposition of solid solutions of cyanides and hydroxides

Tab. 4 Specific surface area of LaCoO$_3$, prepared by various methods (Adapted from Ref. [99].)

Preparation method	Final heat treatment[a]	Surface area/m^{2}g^{-1}
Ceramic methods	1250 °C, 48 h	1.7
Oxalate decomposition	1000 °C, 12 h	4.5
	900 °C, 15 h	
Acetate decomposition	850 °C, 5 h	2.5
	850 °C, 5 h	2.2
Mist decomposition	700 °C[b]	11.8
Citrate decomposition	700 °C, 4 h	1.3
	600 °C, 2 h	11.3
	700 °C, 4 h	8.5
Cyanide decomposition	650 °C, 4 h	37.5
Freeze-drying	500 °C, 10 h	36.2
Combustion synthesis	500 °C, 10 h	34.6

[a] Temperature and period of calcination in air sufficient for obtaining a single perovskite phase.
[b] A mist of mixed aqueous solution of cobalt and lanthanum nitrates was treated successively in a flow-type reactor equipped with three furnaces (170 °C, 300 °C, 700 °C); the residence time in a final furnace (700 °C) was about 11 s.

in the reviews of Tejuca et al. [94] and Yamazoe and Teraoka [99]. In general, perovskites have been prepared by calcining a solid mixture of constituent metal oxides or carbonates (the method used in the ceramic industry), or by evaporating an aqueous solution of constituent metal nitrates or acetates and by subsequent calcination. The resulting oxides have relatively small surface areas. Some methods which can produce higher specific surface areas are listed in Table 3. The guiding principle here is to attain as thorough a dispersion as possible, almost to a molecular level, of constituent components in the precursors prior to calcination. Once this is achieved, lower calcination temperatures can convert the precursors to the required perovskite structure of larger surface areas (see Table 4). Among recent examples are the preparations of La$_{1-x}$Ce$_x$MnO$_3$ ($x \leq 0.4$) [100] and La$_{1-x}$Ce$_x$Mn$_{1-y}$Co$_y$O$_z$ ($x \leq 0.3$, $y \leq 0.7$) [101] homogeneous solid solutions. Nitrate coprecipitation apparently failed to provide homogeneous solid solutions [102].

Teraoka et al. [103, 104] have applied the amorphous citrate process to prepare unsupported (or neat) and supported perovskites of the type LaMn, LaCo, LaMnCu, LaCoFe, LaCaCo, LaCaMn, LaSrMn, LaSrCo, LaSrCoCu, and LaSrCoFe. The use of the citrate method for preparing atomically homogeneous oxides on supports has been a very important extension of the citrate method over the years. Nitrates of constituent metals of the required perovskite were dissolved in water and mixed with an aqueous solution of citric acid (molar ratio of citric acid to total metals, 1:1). Water was evaporated from the mixed solution using a rotary evaporator at 60–70 °C until a sol was obtained. The sol was further dehydrated at the same temperature under vacuum for 5 h. A stabilized alumina of composition La$_2$O$_3 \cdot 19$Al$_2$O$_3$ was used as the support for the perovskites.

One of the supports mentioned in the previous subsection (Ce$_{0.8}$Zr$_{0.2}$O$_2$) has been used for supporting up to 20 wt.% LaCoO$_3$ or La$_{0.8}$Ce$_{0.2}$CoO$_3$ [105]. The sensitivity of the XRD technique does not allow such solid conclusions as those obtained for unsupported phases, because the supported phase represents only a relatively small fraction of the solid investigated. At a loading up to 15 wt.%, no difference appears between impregnation by citrates and by nitrates, but for a loading of 20 wt.% a segregation of cobalt oxide is more important in the case of impregnation by nitrates in comparison with the use of citrates. In neither case is there any evidence of a reaction between the supported perovskite and the support or of other types of decomposition of the supported phase. The activity of the catalysts obtained from citrates showed a slight advantage over the other series, but the conclusion of a critical discussion was that this could not be related directly to the quality of the supported solid solution.

It is of interest to mention here that Teraoka et al. also prepared the perovskites mentioned in the results described above by using an acetate process rather than the citrate process. The specific surface areas of the perovskites prepared by the citrate and acetate processes are compared in Table 5. The amorphous citrate precursor decomposed in three steps, with the third decomposition occurring at 550–650 °C and already inducing the crystallization of perovskite-type oxides. The acetate process required a higher calcination temperature of 850 °C, and consequently produced samples the surface areas of which were only one-third to one-seventh of those of samples created by the citrate process. The higher-area perovskites from the citrate process had a higher catalytic activity for methane oxidation. The citrate process was also more effective than the acetate process in the preparation of supported perovskite catalysts, a fact confirmed in a more recent study [105].

Many attempts have been made to modify the citrate method, with most combining the original technique with other approaches (see Section 2.3.12.2.3G). As an example, the solid precursor was decomposed under nitrogen and subsequently "annealed" in air [106] (i.e., submitted to oxidation), in addition to that normally occurring due to the use of nitrates in the citrate process. This might perhaps lead to further investigation, because it could be supposed that: (i) the presence of residual carbon after the first step may hinder sintering; and (ii) the succession of steps might modify a segregation of phases if this had been the beneficial effect.

F Complex Cyanide Method   Geus and coworkers [107, 108] have applied another method based on chemical complexes; this is the complex cyanide method used to prepare both monocomponent (Fe or Co) and multicomponent Fischer–Tropsch catalysts. A large range of insoluble complex cyanides are known in which many metals can be combined; for example, hexacyanoferrate(II) (also called ferrocyanide) and hexacyanoferrate(III) (also called ferricyanide) can be combined with nickel, cobalt, copper, and zinc ions. Bimetallic precursors associating Cu cyano complexes with ammine or ethylenediamine complexes of Zn(II) were also described as model precursors to Cu–Zn oxide catalysts [109]. Soluble complex ions of molybdenum (IV) ($[Mo(CN)_8]^{4-}$) which can produce insoluble complexes with metal cations have also been identified.

Deposition–precipitation can be performed by the injection of a solution of a soluble cyanide complex of one desired metal into a suspension of a suitable support in a solution of a simple salt of the other desired metal. By adjusting the cation composition of the simple salt solution, with a same cyanide, it is possible to adjust the composition of the precursor from a monometallic oxide (the case when the metallic cation is identical to that contained in the complex) to oxides containing one or several foreign elements.

Complex metal cyanides decompose at fairly high temperatures. In fact, decomposition in an oxidizing atmosphere (up to 330 °C) results in the corresponding oxides which, in an inert or a reducing atmosphere (up to 630 °C), results in the metals or alloys. In either case, reaction of the oxides or of the metals/alloy with the

References see page 316

Tab. 5 Specific surface area of perovskite-type catalysts, prepared by the citrate process and the acetate process (Adapted from Ref. [103].)

Sample	Citrate process		Acetate process	
	Calcination temp./°C	Surface area/$m^2 g^{-1}$	Calcination temp./°C	Surface area/$m^2 g^{-1}$
$LaMnO_3$	650	44.8	850	7.3
$LaCoO_3$	600	11.3	850	2.2
$LaCo_{0.4}Fe_{0.6}O_3$	550	22.7	850	3.3
$LaMn_{0.6}Cu_{0.4}O_3$	600	33.0	850	7.5
$La_{0.8}Sr_{0.2}MnO_3$	600	36.4	850	8.2
$LaCo_{0.8}Cu_{0.2}O_3$	550	24.7	–	–
$La_{0.6}Ca_{0.4}MnO_3$	600	33.0	–	–
$La_{0.4}Ca_{0.6}CoO_3$	600	14.3	–	–

support is minimized. The complex cyanide method can thus establish homogeneity of the constituent ions on a molecular scale, as well as prevent undesired reactions of the active precursors with the support.

G Hybrid Complexation Methods The shortcomings of several preparation techniques has led to the use of hybrid approaches. The idea here is to use complexing agents to improve catalyst homogeneity [110], and this has been applied to different preparation techniques mentioned above. Whilst the success of the citrate technique meant that citric acid, hydroxy-acids and complexing molecules were used very frequently, many different mixtures have been used as precursors, with amounts of complexing agent varying from small to stoichiometric excesses. It should be noted, conversely, that so-called "modified citrate methods", rather than starting exclusively from citrates, also use organometallic salts, sulfates, chlorides, etc. (see, for example the preparation of TiO_2 and TiO_2-ZrO_2 oxides, in Refs. [111, 112]).

In some cases, the advantage of these hybrid approaches is to modify the amount of carbon residue that can most likely contribute to the inhibition of sintering [113]. Both, sulfates and chlorides leave residues, the consequence of which might be modifications to the catalyst's activity and selectivity.

2.3.12.2.4 Advanced Methods for Obtaining Solid Precursors of Flexible Composition

A Specific Approaches A number of alternative methods exist by which the segregation of phases can be avoided during the decomposition of a solid precursor to form an oxide.

One such method uses soluble gums or polymers, wherein the principle is the same as in previously described methods. A solution containing all of the desired elements is made very viscous by dissolving gums or functional polymers; hydroxymethylcellulose and similar structures have been used for this purpose. Usually, the viscous solution can be concentrated (e.g., by evaporation) with little or no crystallization or segregation, and then calcined to a mixed oxide containing all of the elements, finely interdispersed.

Another very interesting method, albeit of limited application, is a type of self-gelation. This occurs when amorphous hydrated precipitates have a strong hydrophilic character that favors their interaction with the solution, and subsequent transformation through tridimensional reticulation to a homogeneous hydrogel that retains the solution in its net. The only well-characterized example here is that of iron molybdate gels (pure or with additives), which were developed to produce catalysts for the oxidation of methanol to formaldehyde [114–119]. The data in

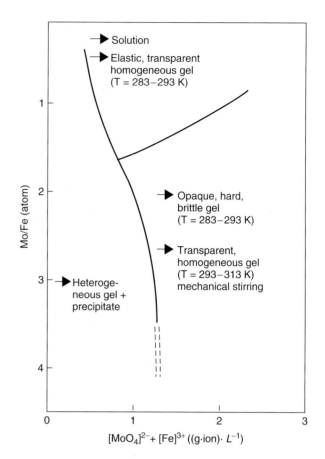

Fig. 13 Formation of gels from iron molybdate precipitates [9]. A very fine colloidal precipitate forms in all cases, but undergoes different changes as suggested by the arrows, according to concentration and Mo:Fe proportions, as well as other conditions mentioned in the figure. In the upper part domain, the colloidal precipitate seems to dissolve to form a solution, the viscosity of which increases until the formation of a gel. In the right-hand domain, the gel forms directly from the colloidal system, but can become transparent if the reaction takes place in the temperature range 20–40 °C.

Fig. 13 [9] show that gel formation occurs only within a narrow range of composition and operating conditions. The main parameters involved are the Mo/Fe ratio, the nature of the starting salts, the nature and amount of additives, the temperature, the concentrations, and the intensity of stirring. Aging increases reticulation and homogeneity; consequently the hydrogel becomes hard and brittle (the fractured surfaces exhibit the aspect of glass fractures also commonly found in hydrogels). Depending on the concentrations and Mo/Fe proportions, a precipitate first forms, and then partly or totally dissolves to produce a gel. A true metastable solution is formed transiently for atomic ratios Mo:Fe ≤ 1.5 [9]. The aged gel is then dried to yield a brown, transparent, homogeneous xerogel containing less than 1 wt.% water, and produces the activated catalyst after calcination. This method of preparation permits the introduction of many elements as doping agents. Likewise, the gel may incorporate a large excess of molybdenum compared to the stoichiometry of $Fe_2(MoO_4)_3$.

At present it is unclear whether this method can be extended to other systems. However, it may be speculated that all elements producing high-molecular-weight polyanions could form similar gels when combined with cations carrying three or more positive charges.

B Possible Approaches Based on Specific Interactions

Speculative approaches would be to use specific colloidal or molecular interactions for preparing precursors of flexible composition. The simplest idea would be to flocculate colloids constituted of species M, N, P, etc., carrying opposite charges in order to obtain three dimensional -M^+-N^--M^+-P^--M^+ clusters, though very few publications exist which deal with this possible method. It seems that current ideas correspond to very complicated approaches, such as the synthesis of organometallic molecules that contain all of the desired metals in the desired proportions, that would be calcined or transformed in other ways in very mild conditions to a solid solution or a compound containing these two or several elements.

A relatively practical approach could be derived from the sol–gel process. If the alkoxides contained several elements in the correct proportions, then uniformity of composition could be achieved for these elements. The difference from the conventional sol–gel method would be to start from "building blocks", namely complex alkoxides containing two or several elements, instead of individual "bricks". Conceptually, this is the way in which zeolites are constructed. Zeolites of different compositions and structure can be made because the building blocks are different. Figure 14 illustrates, schematically, a series of building blocks that contain several elements, and the bidimensional or tridimensional linking of these building blocks to make a multi-element material [120]. In zeolite synthesis, colloidal and van der Waals forces attach the building blocks to each other. Formally, chemical links might do the same as they do in the citrate process.

Major advances have been made since the mid-1990s in concepts related to colloidal processes, molecular recognition, new polymers and, generally speaking, processes which allow specific arrangements in space of chemical entities and chemical functions. This, in turn, has led to new ideas for obtaining precursors of very flexible composition.

Since the main problem in sol–gel process is the lack of selectivity when forming bonds between different alkoxides, one idea advanced by specialists in this field was to attach different functions to the parts that contain the different elements to be associated. These functions could be selected in such a way that they only react with a single matching type of function. This is shown schematically in Fig. 15, where the arrows and respective selectively matching clefts symbolize molecular recognition or similar selective interactions. In this way, a sort of elaborate "Lego" structure can potentially be constructed according to this principle [120]. An embodiment of such ideas for making a variety of materials – including possibly surface-active structures – is provided in Ref. [121].

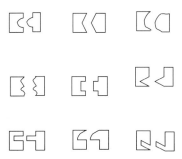

Fig. 14 Schematic representation of precursors made of building blocks possessing *terminal functions*, each attaching specifically to other terminal functions of other building blocks.

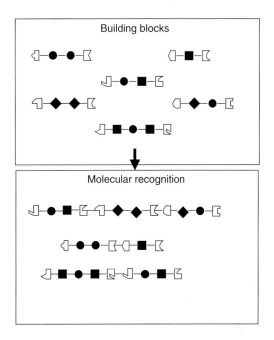

Fig. 15 Schematic representation of the use of molecular recognition for selectively attaching elements or building blocks to other elements or building blocks.

References see page 316

C The Use of Multi-Functional Polymers Approaches involving the use of polymer–metal complexes based on structurally simple organic polymers such as poly(ethylene glycol), poly(vinyl alcohol), poly(acrylic acid), polyamide or polyimide derivatives, have been described to synthesize many oxide-type inorganic materials [122, 123]. When nitrate ions are simultaneously present, auto-ignited combustion may occur and lead to multicomponent oxides [124]. Recently, investigations were undertaken for using more elaborate polymer structures: these consisted of polyampholytes resulting from the association of monomers bearing carboxylic acid and amine moieties. The objective was to synthesize polymer–metal complexes, which were converted into solid materials upon appropriate thermal treatment. Depending on the substituents present on the main chain, these polymers offer the advantage of tunable properties such as the solubility in various polar media, the number and nature of functional groups of each type, the presence of selected substituents able to induce controlled steric hindrance, cation complexation, or superstructures in the solid state. In the case of polybetaine structures, the simultaneous presence of opposite electric charges along the chains, within a given pH window corresponding to the zwitterionic state, facilitates the interaction with both the anions and the cations of a soluble inorganic salt, in the sense that electrostatic interactions operate together with metal complexation, if any (Fig. 16). This approach has been validated to obtain several transition metal molybdates, starting from inorganic precursors and a copolymer matrix made from diallylammonium and (functionalized) maleic or maleamic acid units [125, 126]. After removal of the solvent, homogeneous hybrid materials containing up to equimolar amounts of metals with respect to the "repeat unit" of the copolymer can be obtained. An appropriate thermal treatment under oxidizing conditions generates the final oxide powder with very low residual amounts of carbonaceous species.

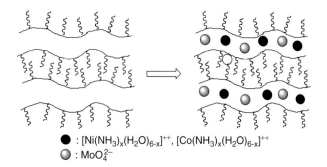

● : $[Ni(NH_3)_x(H_2O)_{6-x}]^{++}$, $[Co(NH_3)_x(H_2O)_{6-x}]^{++}$
○ : MoO_4^{2-}

Fig. 16 Idealized scheme for the incorporation of the inorganic salts in the ordered polymer matrix.

2.3.12.2.5 Specific Methods: Example of Hydrotalcite-Type Catalysts The example of hydrotalcite catalyst can serve to illustrate the fact that solids of given structures containing several elements in rather flexible proportions can be obtained by different techniques. This example will also show that special techniques can be designed for the preparation of specific precursors.

Hydrotalcite (HT) is a clay mineral which, on crushing, becomes a fine powder, similar to talc. It is a hydroxocarbonate of Mg and Al, of the general formula $Mg_6Al_2(OH)_{16}CO_3 \cdot 4H_2O$, and occurs in nature in foliated and rolled plates and/or fibrous masses. Anionic clays of the HT type have been of importance in catalysis since 1970. Their interesting properties for catalytic applications are:

- a large surface area
- basic properties
- formation of homogeneous mixtures of oxides with very small crystal size which, on reduction, form small and thermally stable metal crystallites
- a "memory effect", which allows reconstruction under mild conditions of the original HT structure when the product of the thermal treatment is contacted with aqueous solutions containing various anions.

Cavani et al. [127], who reviewed this area extensively in 1991, emphasize that all the stages of preparation of a catalyst with a HT-type precursor need precise chemical control in order to avoid inhomogeneities and chemical segregation.

To date, three different methods of preparation have been attempted for HT-type catalysts:

- Precipitation (increasing pH method or coprecipitation either at low or at high supersaturation)
- Hydrothermal synthesis and treatments, also ageing
- Ion-exchange methods.

As several cases and details for these preparation methods have been provided by Cavani et al. [127], only the preparation by Trifiro et al. [128] of a Ni catalyst for steam reforming will be cited here as an illustrative example. The HT-type precursors were prepared by coprecipitation at pH 8 by adding a solution of the nitrates of Mg and Al (or Cr), in the correct proportions, to a stirred solution of $NaHCO_3$. The precipitate was washed with distilled water until its Na_2O content was <0.02%, and then dried at 90 °C. By heating the precipitate in air or vacuum in the range 350 to 900 °C, various catalysts could be prepared with average Ni crystallite sizes ranging from 2 to 100 nm. These catalysts contained a NiO phase, probably containing traces of Al^{3+} ions, a spinel-type phase, and an alumina-type phase (doped with small amounts of Ni^{2+} ions), probably grafted onto the spinel-type phase. This last phase contributes significantly to the surface area of the catalyst samples, but it has little or no influence on their thermal stability or catalytic activity.

Applications of HT-type catalysts, prepared by the above methods, have been reported in recent years for basic catalysis (polymerization of alkene oxides, aldol condensation), steam reforming of methane or naphtha, CO hydrogenation as in methanol and higher-alcohol synthesis, conversion of syngas to alkanes and alkenes, hydrogenation of nitrobenzene, oxidation reactions, and as a support for Ziegler–Natta catalysts.

2.3.12.3 Conclusions and Prospects

In conclusion, the preparation of catalysts which contain several elements distributed homogeneously in the solid at the subnanometer scale remains a major challenge, and perhaps is deserving of more intense research than is presently being undertaken. Considered independently, neither chemical engineering nor synthetic chemistry approaches seem able to provide simple solutions, although a breakthrough may reasonably be expected from cooperative studies between both fields of scientific activity. Even imperfect homogeneity might be a valuable objective for a lack of real molecular dispersion, because phase cooperation is now coming to the forefront of concepts in catalysis, notably for oxidation reactions. Most methods used with the objective of ultimate atomic dispersion may actually achieve a very fine inter-dispersion of different phases and therefore provide – together with other methods – more active and/or selective catalysts because contacts between phases would be more numerous.

In this chapter, the main topic has been the use of solid-state transformations to prevent segregation of the composition of solid precursors in a final oxide catalyst. In general, although crystalline precursors do not permit the desired composition to be achieved, they can be excellent in some specific cases. Amorphous precursors (especially those using polyfunctional hydroxy acids) of almost any possible composition can be prepared, and provide the corresponding oxides. The initial aim of the method was to prepare solid solutions or oxides of two or several elements in well-defined structures, sometimes containing elements substituted within the normal lattice structure. Only one other method has been claimed to offer similar flexibility, namely the sol–gel method. Whilst the two methods have still to be compared in detail, it seems that the kinetics of hydrolysis are altered very much from one alkoxide to another, with the consequence that the different elements become incorporated into the gel sequentially during gel formation. This phenomenon has the same consequences as the successive precipitation of different elements mentioned above (see Fig. 2), namely that the homogeneity of the gel at a molecular scale is not perfect.

When the objective is to prepare an extremely intimate mixture of two or more different oxides, each of them constituted by very small grains or particles, techniques such as spray calcination, oil drop, or freeze-drying, complexation methods (and generally all methods described in Sections 2.3.12.2.2B and C above) and the sol–gel method seem, in principle, to have the same potential. The use of a particular technique is then dictated by other considerations, such as availability, stability and the costs of starting materials (especially for alkoxides) or ease of decomposition.

Finally, a word of caution should be added. The successful transformation of precursors to a homogeneous oxide or a finely interdispersed mixture of oxides demands that diffusion leading to segregation be made as difficult as possible. In particular, the formation of liquid phases during decomposition and calcination of the precursor should be completely avoided. However, what cannot be avoided is surface contamination.

Another important consideration in preparing mixed-oxide catalysts is the spontaneous monolayer dispersion of oxides and salts onto the surfaces of support substrates on calcination. Both, temperature and duration of calcination are important here, as discussed in reviews by Xie and Tang [129] and by Knözinger and Taglauer [130]. If this dispersion step is inadequate or incomplete, then the resulting oxide layer – and any reduced metal surface from it – will not be reproducible: from the same catalyst system therefore, it is possible to prepare different catalysts at different times and, of course, in different laboratories. Spreading and wetting phenomena in the preparation of supported catalysts is discussed in Chapter 2.4.7 of this Handbook.

Heterogeneous catalysts are not chemicals in the ordinary sense of the word; rather, they are performance chemicals or surface-active materials. Naturally, the performance of the catalyst will depend not so much on the initial composition or surface of the starting material, as on the real surface, which is formed and stabilized and then changed dynamically under the prevailing process conditions. Here, one must take into account known and controlled process parameters such as temperature, pressure, reactant concentration and space velocity, as well as variable factors such as feed composition, and unpredictable or unsuspected factors such as impurities and poisons in the feed [131, 132].

References see page 316

References

1. L. T. Weng, B. Delmon, *Appl. Catal. A* **1992**, *81*, 141.
2. B. Delmon, *Heterog. Chem. Rev.* **1994**, *1*, 219.
3. B. Delmon, *Surf. Rev. Lett.* **1995**, *2*, 25.
4. L. T. Weng, P. Patrono, E. Sham, P. Ruiz, B. Delmon, *J. Catal.* **1991**, *132*, 360.
5. C. Marcilly, *Rev. Inst. Fr. Petr.* **1984**, *39*, 189.
6. C. Marcilly, J.-P. Franck, *Rev. Inst. Fr. Petr.* **1984**, *39*, 337.
7. P. Courty, H. Ajot, C. Marcilly, B. Delmon, *Powder Tech.* **1973**, *7*, 21.
8. P. Courty, C. Marcilly, in *Preparation of Catalysts I*, B. Delmon, P. A. Jacobs, G. Poncelet (Eds.), Elsevier, Amsterdam, 1976, pp. 119–145.
9. P. Courty, C. Marcilly, in *Preparation of Catalysts III*, G. Poncelet, P. Grange, P. A. Jacobs (Eds.), Elsevier, Amsterdam, 1983, pp. 485–519.
10. B. Delmon, N. de Keyser, in *Chemical and Physical Aspects of Catalytic Oxidation*, J. L. Portefaix, F. Figueras (Eds.), CNRS, Lyon, 1978, pp. 491–534.
11. B. Delmon, P. A. Jacobs, G. Poncelet (Eds.), *Preparation of Catalysts I, Studies in Surface Science and Catalysis*, Vol. 1, Elsevier, Amsterdam, 1976.
12. B. Delmon, P. Grange, P. A. Jacobs, G. Poncelet (Eds.), *Preparation of Catalysts II, Studies in Surface Science and Catalysis*, Vol. 3, Elsevier, Amsterdam, 1979.
13. G. Poncelet, P. Grange, P. A. Jacobs (Eds.), *Preparation of Catalysts III, Studies in Surface Science and Catalysis*, Vol. 16, Elsevier, Amsterdam, 1983.
14. B. Delmon, P. Grange, P. A. Jacobs, G. Poncelet (Eds.), *Preparation of Catalysts IV, Studies in Surface Science and Catalysis*, Vol. 31, Elsevier, Amsterdam, 1987.
15. G. Poncelet, P. A. Jacobs, P. Grange, B. Delmon (Eds.), *Preparation of Catalysts V, Studies in Surface Science and Catalysis*, Vol. 63, Elsevier, Amsterdam, 1991.
16. G. Poncelet, J. Martens, P. A. Jacobs, P. Grange, B. Delmon (Eds.), *Preparation of Catalysts VI, Studies in Surface Science and Catalysis*, Vol. 91, Elsevier, Amsterdam, 1995.
17. B. Delmon, P. A. Jacobs, R. Maggi, J. A. Martens, P. Grange, G. Poncelet (Eds.), *Preparation of Catalysts VII, Studies in Surface Science and Catalysis*, Vol. 118, Elsevier, Amsterdam, 1988.
18. E. Gaigneaux, D. E. De Vos, P. Grange, P. A. Jacobs, J. A. Martens, P. Ruiz, G. Poncelet (Eds.), *Scientific Bases for the Preparation of Heterogeneous Catalysts, Studies in Surface Science and Catalysis*, Vol. 143, Elsevier, Amsterdam, 2002.
19. M. Ruwet, B. Delmon, unpublished results.
20. P. G. Rouxhet, R. E. Sempels, *J. Chem. Soc., Faraday Trans. I* **1974**, *70*, 2021.
21. C. Sarbu, B. Delmon, *Appl. Catal. A* **1999**, *185*, 85.
22. B. Brizzo, G. Bellussi, in *Oxide-based Systems at the Crossroads of Chemistry*, A. Gamba, C. Colella, S. Coluccia (Eds.), *Studies in Surface Science and Catalysis* Vol. 140, Elsevier, Amsterdam, 2002, 401.
23. S. C. Emerson, C. F. Coote, H. Boote III, J. C. Tufts, R. LaRoque, W. R. Moser, in *Scientific Bases for the Preparation of Heterogeneous Catalysts*, B. Delmon, P. A. Jacobs, R. Maggi, J. A. Martens, P. Grange, G. Poncelet, (Eds.), *Studies in Surface Science and Catalysis*, Vol. 118, Elsevier, Amsterdam 1998, p. 773.
24. M. Haruta, B. Delmon, *J. Chim. Phys.* **1986**, *83*, 859.
25. X.-D. Zhou, W. Huebner, H. U. Anderson, *Chem. Mater.* **2003**, *15*, 378.
26. U. R. Pillai, E. Sahle-Demessie, R. S. Varma, *Appl. Catal. A* **2003**, *252*, 1.
27. S. H. Overbury, P. A. Bertrand, G. A. Somorjai, *Chem. Rev.* **1975**, *75*, 547.
28. G. A. Somorjai, *Chemistry in Two Dimensions: Surfaces*, Cornell University Press, Ithaca, NY, 1981.
29. T. Xiao, H. Wang, A. P. E. York, V. C. Williams, M. L. H. Green, *J. Catal.* **2002**, *209*, 318.
30. K. Hada, J. Tanabe, S. Omi, M. Nagai, *J. Catal.* **2002**, *207*, 10.
31. A. Boisen, S. Dahl, C. J. H. Jacobsen, *J. Catal.* **2002**, *208*, 180.
32. P. Perez-Romo, C. Potvin, M.-J. Manoli, M. M. Chehimi, G. Djéga-Mariadassou, *J. Catal.* **2002**, *208*, 187.
33. X. Wang, P. Clark, S. T. Oyama, *J. Catal.* **2002**, *208*, 321.
34. C.-M. Chen, J.-M. Jehng, *Appl. Catal. A* **2004**, *267*, 103.
35. B. Delmon, in *Reactivity of Solids*, K. Dyrek, J. Haber, J. Nowotny (Eds.), Elsevier, Amsterdam, and PWN, Warszawa, 1982, pp. 327–369.
36. B. Delmon, *J. Chim. Phys.* **1986**, *83*, 875.
37. S. Eriksson, U. Nylén, S. Rojas, M. Boutonnet, *Appl. Catal. A* **2004**, *265*, 207.
38. A. Muto, T. Bhaskar, Y. Kaneshiro, Y. Sakata, Y. Kusano, K. Murakami, *Appl. Catal. A* **2004**, *275*, 173.
39. J. Knert, A. Drochner, J. Ott, H. Vogel, H. Fuess, *Appl. Catal. A* **2004**, *269*, 53.
40. M. T. Le, W. J. M. Van Well, I. Van Driessche, S. Hoste, *Appl. Catal. A* **2004**, *267*, 227.
41. A. M. Efstathiu, D. Boudouvas, N. Vamvouka, X. E. Verikios, *J. Catal.* **1993**, *140*, 1.
42. D. Papagiorgio, A. M. Efstathiou, X. E. Verikios, *Appl. Catal. A* **1994**, *111*, 41.
43. G. S. Lane, E. Miro, E. E. Wolf, *J. Catal.* **1989**, *119*, 161.
44. P. Gherardi, O. Ruggeri, F. Trifiro, A. Vaccari, G. Del Piero, G. Manara, B. Notari, in *Preparation of Catalysts III*, G. Poncelet, P. Grange, P. A. Jacobs (Eds.), Elsevier, Amsterdam, 1983, pp. 723–733.
45. G. Petrini, F. Montino, A. Bossi, F. Garbassi, in *Preparation of Catalysts III*, G. Poncelet, P. Grange, P. A. Jacobs (Eds.), Elsevier, Amsterdam, 1983, pp. 735–744.
46. E. B. M. Doesburg, R. H. Höppener, B. de Koning, Xu Xiaoding, J. J. F. Scholten, in *Preparation of Catalysts IV*; B. Delmon, P. Grange, P. A. Jacobs, G. Poncelet (Eds.), Elsevier, Amsterdam, 1987, pp. 767–776.
47. B. S. Rasmussen, P. E. Højlund Nielsen, J. Villadsen, J. B. Hansen, in *Preparation of Catalysts IV*, B. Delmon, P. Grange, P. A. Jacobs, G. Poncelet (Eds.), Elsevier, Amsterdam, 1987, pp. 785–794.
48. W. D. Kingery, *Introduction to Ceramics*, Wiley, New York, 1960, pp. 291–305.
49. A. S. Deshpande, N. Pinna, P. Beato, M. Antonietti, M. Niederberger, *Chem. Mater.* **2004**, *16*, 2599.
50. T. Masui, Y. Peng, K.-I. Machida, G.-Y. Adachi, *Chem. Mater.* **1998**, *10*, 4005.
51. P. K. Gallager, H. M. O'Bryan, F. Schrey, F. R. Monforte, *Am. Ceram. Soc. Bull.* **1969**, *48*, 1053.
52. P. G. Menon, B. Delmon, in *Handbook of Heterogeneous Catalysis*, G. Ertl, H. Knözinger, J. Weitkamp (Eds.), Vol. I, Wiley-VCH, Weinheim, 1997, Section 2.1.6, pp. 100–110.
53. W. F. Kladnig, W. Karner, *Ceram. Bull.* **1990**, *69*, 814.
54. W. J. Stark, J.-D. Grunwaldt, M. Maciejewski, S. E. Pratsinis, A. Baiker, *Chem. Mater.* **2005**, *17*, 3352.
55. P. Reynen, H. Bastius, *Powder Metall. Int.* **1976**, *8*, 91.
56. M. Matzapetakis, C. P. Raptopoulou, A. Terzis, A. Lakatos, T. Kiss, A. Salifoglou, *Inorg. Chem.* **1999**, *38*, 618.

57. Y.-F. Deng, Z.-H. Zhou, H.-L. Wan, *Inorg. Chem.* **2004**, *43*, 6266.
58. M. Kaliva, T. Giannadaki, A. Salifoglou, *Inorg. Chem.* **2002**, *41*, 3850.
59. M. Matzapetakis, N. Karligiano, A. Bino, M. Dakanali, C. P. Raptopoulou, V. Tangoulis, A. Terzis, J. Giapintzakis, A. Salifoglou, *Inorg. Chem.* **2000**, *39*, 4044.
60. N. Kotsakis, C. P. Raptopoulou, V. Tangoulis, A. Terzis, J. Giapintzakis, T. Jakusch, T. Kiss, A. Salifoglou, *Inorg. Chem.* **2003**, *42*, 22.
61. Z.-H. Zhou, H.-L. Wan, K.-R. Tsai, *Inorg. Chem.* **2000**, *39*, 59.
62. R. Baggio, M. Perec, *Inorg. Chem.* **2004**, *43*, 6965.
63. H. Wullens, D. Leroy, M. Devillers, *Int. J. Inorg. Mater.* **2001**, *3*, 309.
64. H. Wullens, N. Bodart, M. Devillers, *J. Solid State Chem.* **2002**, *167*, 494.
65. J. W. Kriesel, T. D. Tilley, *J. Mater. Chem.* **2001**, *11*, 1081.
66. K. L. Fujdala, T. D. Tilley, *Chem. Mater.* **2001**, *13*, 1817.
67. K. L. Fujdala, T. D. Tilley, *Chem. Mater.* **2004**, *16*, 1035.
68. J. Jarupatrakorn, M. P. Coles, T. D. Tilley, *Chem. Mater.* **2005**, *17*, 1818.
69. A. Haoudi, P. Dhamelincourt, A. Mazzah, M. Drache, P. Conflant, *Int. J. Inorg. Mater.* **2001**, *3*, 357.
70. J. W. Kriesel, M. S. Sander, T. D. Tilley, *Adv. Mater.* **2001**, *13*, 331.
71. D. Bayot, B. Tinant, M. Devillers, *Catal. Today* **2003**, *78*, 439.
72. D. Bayot, B. Tinant, M. Devillers, *Inorg. Chem.* **2005**, *44*, 1554.
73. M. Dakanali, E. T. Kefalas, C. P. Raptopoulou, A. Terzis, G. Voyiatzis, I. Kyrikou, T. Mavromoustakos, A. Salifoglou, *Inorg. Chem.* **2003**, *42*, 4632.
74. M. Kaliva, C. P. Raptopoulou, A. Terzis, A. Salifoglou, *Inorg. Chem.* **2004**, *43*, 2985.
75. D. Bayot, B. Tinant, B. Mathieu, J.-P. Declercq, M. Devillers, *Eur. J. Inorg. Chem.* **2003**, *4*, 737.
76. D. Bayot, B. Tinant, M. Devillers, *Inorg. Chem.* **2004**, *43*, 5999.
77. D. Bayot, M. Devillers, *Chem. Mater.* **2004**, *16*, 5401.
78. J. Paris, R. Paris, *Bull. Soc. Chim. Fr.* **1965**, 138.
79. Y. Saikali, J. M. Paris, *C. R. Acad. Sci., série C* **1967**, *265*, 1041.
80. G. Paris, G. Szabo, R. A. Paris, *C. R. Acad. Sci., série C* **1968**, *266*, 554.
81. C. Marcilly, B. Delmon, *C. R. Acad. Sci., série C* **1969**, *268*, 1795.
82. P. Courty, B. Delmon, *C. R. Acad. Sci., série C* **1969**, *268*, 1874.
83. C. Marcilly, P. Courty, B. Delmon, *J. Am. Ceram. Soc.* **1970**, *53*, 56.
84. B. Delmon, J. Droguest, in *Fine Particles, 2nd International Conference*, W. E. Kuhn, J. Ehretsmann (Eds.) The Electrochemical Soc., Princeton, 1974, pp. 242–255.
85. C. L. Pieck, C. R. Vera, C. A. Querini, J. M. Parera, *Appl. Catal. A* **2005**, *278*, 173.
86. M. P. Pechini, N. Adams, US Patent 3330697, 1967.
87. R. Prasad, L. A. Kennedy, E. Ruckenstein, *Catal. Rev. Sci. Eng.* **1984**, *26*, 1.
88. L. D. Pfefferle, W. C. Pfefferle, *Catal. Rev. Sci. Eng.* **1987**, *29*, 219.
89. M. Alifanti, B. Baps, N. Blangenois, J. Naud, P. Grange, B. Delmon, *Chem. Mater.* **2003**, *15*, 395.
90. K. Sekizawa, K. Eguchi, H. Arai, *J. Catal.* **1993**, *142*, 655.
91. H. Arai, T. Yamada, K. Eguchi, T. Seiyama, *Appl. Catal. A* **1986**, *26*, 265.
92. J. G. McCarty, H. Wise, *Catal. Today* **1990**, *8*, 231.
93. T. Seiyama, *Catal. Rev. Sci. Eng.* **1992**, *34*, 281.
94. L. G. Tejuca, J. L. G. Fierro, J. M. D. Tascon, *Adv. Catal.* **1989**, *36*, 237.
95. J. Kirchnerova, M. Alifanti, B. Delmon, *Appl. Catal. A* **2002**, *231*, 65.
96. D. Terribile, A. Tovarelli, C. de Leitenburg, A. Primavera, G. Dolcetti, *Catal. Today* **1999**, *47*, 133.
97. P. E. Marti, A. Baiker, *Catal. Lett.* **1994**, *26*, 71.
98. Y. Takeda, S. Nakai, T. Kojima, N. Imanishi, G. Q. Shen, O. Yamamoto, M. Mori, C. Asakawa, T. Abe, *Mater. Res. Bull.* **1991**, *26*, 153.
99. N. Yamazoe, Y. Teraoka, *Catal. Today* **1990**, *8*, 175.
100. M. Alifanti, J. Kirchnerova, B. Delmon, *Appl. Catal. A* **2003**, *245*, 231.
101. M. Alifanti, R. Auer, J. Kirchnerova, F. Thyrion, P. Grange, B. Delmon, *Appl. Catal. B: Environmental* **2003**, *41*, 71.
102. M. F. Wilkes, P. Hayden, A. K. Bhattacharya, *J. Catal.* **2003**, *219*, 286.
103. Y. Teraoka, H.-M. Zhang, N. Yamazoe, *Proceedings, 9th International Congress on Catalysis*, Calgary, 1988, pp. 1984–1991.
104. H.-M. Zhang, Y. Teraoka, N. Yamazoe, *Appl. Catal.* **1988**, *41*, 137.
105. M. Alifanti, N. Blangenois, M. Florea, B. Delmon, *Appl. Catal. A* **2005**, *280*, 255.
106. L. Chen, X. Sun, Y. Liu, Y. Li, *Appl. Catal. A* **2004**, *265*, 123.
107. E. Boellaard, A. M. van der Kraan, J. W. Geus, in *Preparation of Catalysts VI*, G. Poncelet, J. Martens, P. A. Jacobs, P. Grange, B. Delmon (Eds.) Elsevier, Amsterdam, 1995, pp. 931–940.
108. E. Boellaard, *Supported Iron (Alloy) Catalysts from Complex Cyanides*, PhD Thesis, University of Utrecht, 1994.
109. M. Jia, A. Seifert, W. R. Thiel, *J. Catal.* **2004**, *221*, 319.
110. S. Lambert, C. Cellier, P. Grange, J.-P. Pirard, B. Heinrichs, *J. Catal.* **2004**, *221*, 335.
111. H. Zou, Y. S. Lin, *Appl. Catal. A* **2004**, *265*, 35.
112. Y. Wan, J. Ma, W. Zhou, Y. Zhu, X. Song, H. Li, *Appl. Catal. A* **2004**, *277*, 55.
113. Y. Zhang, G. Xiong, W. Yang, S. Sheng, in *Natural Gas Conversion VI*, J. J. Spivey, E. Iglesia, T. H. Fleisch (Eds.), Studies in Surface Science and Catalysis, Vol. 133, Elsevier, Amsterdam 2001, p. 21.
114. Ph. Courty, H. Ajot, B. Delmon, *C. R. Acad. Sci., série C* **1973**, *276C*, 1147.
115. Ph. Courty, H. Ajot, B. Delmon (Institut Français du Pétrole), French Patent 1600128, 1970; Belgian Patent 743578, 1970.
116. Ph. Courty, H. Ajot, B. Delmon (Institut Français du Pétrole), French Patent 2031818, 1970; Belgian Patent 745024, 1970.
117. Ph. Courty, H. Ajot, B. Delmon (Institut Français du Pétrole), German Patent 1965176, 1970.
118. (a) Institut Français du Pétrole: French Patent 2060 171, 1969; (b) Institut Français du Pétrole: French Patent 2082444, 1970; (c) US Patent 3716497, 1973; (d) US Patent 3846341, 1974; (e) US Patent 3975302, 1976; (f) US Patent 4000085, 1976; (g) US Patent 4141 861, 1979.
119. Ph. Courty, H. Ajot, B. Delmon, unpublished results (1967).
120. B. Delmon, in *Handbook of Catalyst Preparation*, J. Regalbuto (Ed.), Taylor & Francis, Boca Raton, Fl., Chapter 18, 2006, pp. 449–463.
121. H. Fan, J. Brinker, in *Mesoporous Crystals and Related Nano-Structured Materials*, O. Terasaki (Ed), Studies in Surface Science and Catalysis, Vol. 148, Elsevier, Amsterdam 2004, p. 213.

122. J. C. W. Chien, B. M. Gong, Y. Yang, I. Cabrera, J. Effing, H. Ringsdorf, *Adv. Mater.* **1990**, *2*, 305.
123. K. Naka, Y. Tanaka, K. Yamasaki, A. Ohki, Y. Chujo, S. Maeda, *Bull. Chem. Soc. Jpn.* **2001**, *74*, 571.
124. K.-B. Park, H.-J. Kweon, Y.-S. Hong, S.-J. Kim, K. Kim, *J. Mater. Sci.* **1997**, *32*, 57.
125. F. Rullens, M. Devillers, A. Laschewsky, *J. Mater. Chem.* **2004**, *14*, 3421.
126. F. Rullens, N. Deligne, A. Laschewsky, M. Devillers, *J. Mater. Chem.* **2005**, *15*, 1668.
127. P. Cavani, F. Trifiro, A. Vaccari, *Catal. Today* **1991**, *11*, 173.
128. F. Trifiro, A. Vaccari, O. Clause, *Catal. Today* **1994**, *21*, 185.
129. Y. C. Xie, Y. Q. Tang, *Adv. Catal.* **1991**, *37*, 1.
130. H. Knözinger, E. Taglauer, *Catalysis*, Vol. 10, The Royal Society of Chemistry, Cambridge, 1993, pp. 1–40.
131. P. G. Menon, *Catal. Today* **1991**, *11*, 161.
132. P. G. Menon, *Chem. Rev.* **1994**, *94*, 1021.

2.3.13
Heteropoly Compounds

*Kwan-Young Lee and Makoto Misono**

2.3.13.1 Structure and Catalytic Properties

Heteropolyanions are polymeric oxoanions (polyoxometalates) formed by the condensation of more than two types of oxoanion [1]. The term "heteropoly compound" is used for the acid forms and the salts; other heteropolyacid (HPA)-related compounds are organic and organometallic complexes of polyanions.

Heteropolyanions are composed of oxides of addenda atoms (V, Nb, Mo, W, etc.) and heteroatoms (P, Si, etc.). Various elements which are listed in Table 1 may serve as the addenda atoms and heteroatoms [2]. The structures are classified into several groups based on the similarity of the composition and structure, for example: Keggin-type, $XM_{12}O_{40}^{n-}$; Silverton-type, $XM_{12}O_{42}^{n-}$; Dawson-type, $X_2M_{18}O_{62}^{n-}$; Strandberg-type, $X_2M_5O_{23}^{n-}$; Anderson-type, $XM_6O_{24}^{n-}$; and Lindqvist-type, $XM_6O_{24}^{n-}$, where X is a heteroatom and M is an addenda atom [3]. Structural isomers and lacunary polyanions are also known. Keggin-type anions, of which the catalytic activity has been extensively examined, have been the main subject of HPAs. Recently, the numbers of studies conducted on catalysis with heteropolyanions having other structures such as Dawson-type have gradually increased. For HPAs in the solid state, the primary structure (polyanions), the secondary (three-dimensional arrangement of polyanions, counter cations, water of crystallization, etc.) and the tertiary structures (particle size, pore structure, etc.) are differentiated (Fig. 1) [4, 5]. Not only the primary and secondary structures, but also tertiary and higher-order structures are influential in the catalytic function. In this respect, the preparation process and resulting structure are very important.

There are three prototypes of heterogeneous catalysis with heteropoly compounds, as shown in Fig. 2 [4, 5]. Actual cases may be intermediate and vary by the type of heteropoly compound(s), reacting molecule(s), and reaction conditions. Ordinary heterogeneous catalysis is of the surface type, where the catalytic reaction takes place on a two-dimensional surface. Bulk type I is the reaction in the pseudoliquid phase. The secondary structure (Fig. 1b) of certain HPAs is flexible and polar molecules are readily absorbed in interstitial positions of the solid bulk to form the pseudoliquid phase. Bulk type II has been demonstrated for several catalytic oxidations at relatively high temperatures. The reaction fields for the bulk types are three-dimensional.

For acid catalysis, the rates of bulk-type reactions show close correlations with the bulk acidity, while the catalytic activities for surface-type reactions are related to the surface acidity, which is sensitive to the surface composition and often changes randomly. Similarly, in the case of oxidation catalysis, good correlations exist between the oxidizing ability of the catalyst and the catalytic activity for oxidation in both bulk-type and surface-type reactions. Acid and redox bifunctionality is another characteristic of HPAs. For example, the acidity and oxidizing ability function cooperatively for the oxidation of methacrolein, but competitively for the oxidative dehydrogenation of isobutyric acid [5]. Interestingly, the former is of the surface-type and the latter of bulk-type.

Heteropoly compounds are useful as catalysts due to the following unique characteristics [5–8]:

- acidic properties and oxidizing ability are systematically controllable
- polyanions are well-defined oxide clusters; hence, catalyst design at the molecular level is possible
- a unique reaction environment, such as the pseudoliquid phase, is available, and catalytically useful coordination is possible.

Some HPAs in the solid state are thermally quite stable and suitable for vapor-phase reactions conducted at high temperatures. The thermal stability is in the order $H_3PW_{12}O_{40} > H_4SiW_{12}O_{40} > H_3PMo_{12}O_{40} > H_4SiMo_{12}O_{40}$, and can be enhanced by the formation of appropriate salts [5–11].

Attempts to utilize HPAs as catalysts have a long history [5], with one of the initial investigations – a bench-scale test for the alkylation of aromatics – being reported in 1971. The first industrial process using a HPA catalyst was launched in 1972, for the hydration of propene in the

---
* Corresponding author.

Tab. 1 Known addenda atoms (□) and heteroatoms (o) incorporated into heteropolyanions

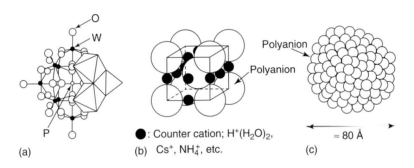

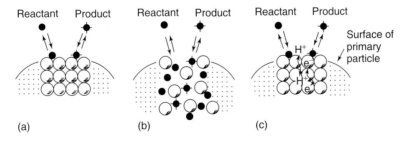

Fig. 1 (a) Primary structure, Keggin structure, $PW_{12}O_{40}^{3-}$; (b) secondary structure, $H_3PW_{12}O_{40} \cdot 6H_2O$; (c) example of a tertiary structure, a primary particle of $Cs_3PW_{12}O_{40}$.

Fig. 2 Three types of heterogeneous catalysis for heteropoly compounds. (a) Surface type; (b) bulk type I (pseudoliquid); (c) bulk type II.

liquid phase. The essential role of the Keggin structure for the oxidation of methacrolein was indicated in a patent in 1975. Systematic basic research into heterogeneous catalysis began in the mid-1970s, the aim being to elucidate the quantitative correlations between the acid-redox properties and catalytic performance of HPAs. Pseudoliquid phase (bulk-type I catalysis) was reported in 1979, and bulk-type II behavior in 1983 [5]. Similar efforts for homogeneous systems were also carried out during the same period. During the 1980s, several new large-scale

References see page 338

industrial processes (e.g., the oxidation of methacrolein, the hydration of isobutene and n-butene, and the polymerization of tetrahydrofuran) were initiated by using HPA catalysts [4, 12]. In the 1990s, the catalytic activity for acid-catalyzed reactions was greatly enhanced by supporting it on mesoporous materials and the formation of an acidic Cs salt, $Cs_{2.5}H_{0.5}PW_{12}O_{40}$ [13]. The direct synthesis of acetic acid from ethene has been commercialized using a bifunctional $Pd-H_4SiW_{12}O_{40}/SiO_2$ [14]. Precise control of pore size has been possible through the understanding of the microstructure, which results in unique shape selectivity observed for various reactions [15]. For example, the $Pt-Cs_{2.1}H_{0.9}PW_{12}O_{40}$ possessed ultramicropores with widths varying between 0.43 and 0.50 nm, and exhibited shape selectivities for oxidation [16, 17], hydrogenation, and the skeletal isomerization of n-butane [18]. Additional attractive and important aspects include the oxidative stability and the possibility of introducing various elements into polyanions and the counter cations. These properties are suitable for the activation and functionalization of lower alkanes, including methane, alkenes, alcohols. Typical reactions catalyzed by HPAs are listed in Tables 2–6.

2.3.13.2 Heteropolyacids: Acid Forms in the Solid State and in Solution

2.3.13.2.1 Keggin-Type HPAs
Heteropolyacids are also Brønsted acids in the solid state, their acidities depending on the constituent elements and their structure. $H_3PW_{12}O_{40}$ (Keggin-type) and its acidic salts are especially strong acids. The color changes of indicators

Tab. 2 Homogeneous acid-catalyzed reactions

Reaction	Catalyst	Remarks	References
Hydration of olefins	$H_3PW_{12}O_{40}$	$T = 313–423$ K	[184a]
n ⟨furan⟩ ⟶ HO[(CH$_2$)$_4$-O]$_n$H (PTMG)	$H_3PW_{12}O_{40} \cdot nH_2O$ ($n = 0–6$)	$T = 333$ K, MW ≈ 3000	[184c]
AcO-sugar-OAc + ROH ⟶ AcO-sugar-OR	$H_3PW_{12}O_{40}$	$T = 298$ K, yield = 60–98% (2 h)	[184d]
⟨alkene⟩ + CH_3OH ⟶ ether	$H_6P_2W_{18}O_{62}$	$T = 315$ K, selectivity ≈ 100%	[184e]
3 C_2H_5CHO ⟶ trimer	$H_3PW_{12}O_{40}$, $H_3PMo_{12}O_{40}$	$T = 298$ K	[184f]

Tab. 3 Heterogeneous acid-catalyzed reactions

Reaction	Catalyst	Remarks	References
$CH_3COOH + C_2H_5OH \longrightarrow CH_3COOC_2H_5 + H_2O$	$H_3PW_{12}O_{40}$	$T = 423$ K, selectivity = 91% (90% conv.)	[185a]
Alkylation of aromatics	$H_3PW_{12}O_{40}$	$T = 303–373$ K,	[185b]
Alkylation of alkanes	$Cs_{2.5}H_{0.5}PW_{12}O_{40}$	$T = 293$ K, $P = 5$ kg cm^{-2}	[185d]
Isomerization of alkanes	$Pd_{1.5}PW_{12}O_{40}$	$T = 483$ K	[185e]
$CH_3OH(CH_3OCH_3) \longrightarrow C_1–C_6$ Hydrocarbons	$H_3PW_{12}O_{40}$	$T = 348$ K	[185f,g]
	$H_3PW_{12}O_{40}$	$T = 573$ K	
⟨isobutene⟩ + CH_3OH ⟶ MTBE	$Cs_{2.5}H_{0.5}PW_{12}O_{40}$	$T = 563$ K, selectivity ($C_2–C_4$ alkenes) = 74%	[185h]
⟨benzene⟩ + HNO_3 ⟶ ⟨nitrobenzene⟩	$H_6P_2W_{18}O_{62}$	$T = 323$ K	[185i]
⟨arene⟩ + $(CH_3CO)_2O$ ⟶ ⟨aryl ketone⟩	$Cs_{1.5}H_{1.5}PMo_{12}O_{40}$	$T = 413$ K, selectivity = 97% (94% conv.)	[185j]
	$Cs_{2.5}H_{0.5}PMo_{12}O_{40}$	$T = 573$ K	[185k]
$C_2H_4 + CH_3COOH \longrightarrow CH_3COOC_2H_5$	$H_4SiW_{12}O_{40}/SiO_2$	$T = 430$ K	[191]

Tab. 4 Homogeneous oxidation reactions

Reaction	Catalyst	Remarks	References
PhCH$_2$CH$_3$ + H$_2$O$_2$ → PhCOCH$_3$	H$_3$PMo$_{12}$O$_{40}$	reflux, selectivity = 100%	[186a]
2,6-dimethylphenol + O$_2$ → 2,6-dimethylbenzoquinone	H$_5$PMo$_{10}$V$_2$O$_{40}$	$T = 333$ K, O$_2$ = 1 atm, yield = 80%	[186b]
cyclopentene, cyclohexene + H$_2$O$_2$ → dialdehydes	H$_3$PMo$_6$W$_6$O$_{40}$	$T = 303$ K	[35]
1-alkene + H$_2$O$_2$ → epoxide	H$^+$/WO$_4^{2-}$/PO$_4^{3-}$/QX; (CP)$_3$PW$_{12}$O$_{40}$, CP = cetylpyridinium ion	$T = 343$ K, yield = 82%; $T = 333$ K, yield = 76%	[186c], [186d]
propene + O$_2$ → acetone + isopropanol	H$_7$PW$_9$Fe$_2$NiO$_{37}$	$T = 423$ K, turnover number = 9730 (3 h)	[186e]
cyclohexane + t-BuOOH → cyclohexanol	PW$_{11}$CoO$_{39}^{5-}$	$T = 298$ K	[165c]

Tab. 5 Heterogeneous oxidation reactions

Reaction	Catalyst	Remarks	References
CH$_2$=C(CH$_3$)CHO + O$_2$ → CH$_2$=C(CH$_3$)COOH	CsH$_3$PVMo$_{11}$O$_{40}$	$T = 553$ K, selectivity = 80–85%	[90, 187a]
isobutane + O$_2$ → methacrylic acid	H$_3$PMo$_{12}$O$_{40}$	$T = 623$ K, selectivity = 45%	[187b]
CH$_3$CH(CH$_3$)COOH → CH$_2$=C(CH$_3$)COOH	H$_5$PV$_2$Mo$_{10}$O$_{40}$	$T = 573$ K, selectivity = 72% (52% conv.)	[94, 187c]
n-butane + O$_2$ → maleic anhydride	H$_5$PV$_2$Mo$_{10}$O$_{40}$	$T = 583$ K, selectivity = 55%	[187d]
C$_2$H$_4$ + O$_2$ → CH$_3$COOH	Pd–Te/H$_4$SiW$_{12}$O$_{40}$/SiO$_2$	$T = 423$ K	[14, 191]

Tab. 6 Heteropoly-metal multifunctional catalysts

Reaction	Catalyst	References
RCH=CH$_2$ + 1/2 O$_2$ → RCHOCH$_3$	PMo$_{12-n}$V$_n$O$_{40}^{3+n}$ + PdSO$_4$	[22, 188a]
C$_6$H$_6$ + 1/2 O$_2$ → (C$_6$H$_5$)$_2$	PMo$_{12-n}$V$_n$O$_{40}^{3+n}$ + Pd(OCOCH$_3$)$_2$	[22, 188a]
PhNO$_2$ + 3 CO + CH$_3$OH → PhNHCO$_2$CH$_3$ + 2 CO$_2$	H$_5$PMo$_{10}$V$_2$O$_{40}$ + PdCl$_2$	[188b, c]
cyclohexene + O$_2$ → cyclohexenone	[(n-C$_4$H$_9$)$_4$N]$_5$Na$_3$[(1,5-cod)·IrP$_2$W$_{15}$Nb$_3$O$_{62}$]	[174]
HC≡C–OH + H$_2$ → H$_2$C=CH–OH	Li$_4$SiMo$_{12}$O$_{40}$ + RhCl(PPh$_3$)$_3$	[188d]
CH$_3$OH + CO → CH$_3$COOH	RhPW$_{12}$O$_{40}$/SiO$_2$	[188e]

show that solid $H_3PW_{12}O_{40}$ (anhydrous) is a very strong acid with Hammett acidity function $H_0 < -8$ [19a,b], and a superacid, $H_0 < -13.16$ [19c]. Thermal desorption of pyridine shows that a significant amount of pyridine (equal to the number of protons of the whole bulk) remains bonded for $H_3PW_{12}O_{40}$ at 573 K, while pyridinium ions adsorbed on $SiO_2-Al_2O_3$ are completely removed at the same temperature [20, 21]. This strong acidity of the heteropolyacid can be attributed to the large size of the polyanion, having a low and delocalizable surface charge density. The acid strength in solution has been measured by various methods, with the general trends being $W^{6+} > Mo^{6+} > V^{5+}$ (for the addenda atom) [22] and $P^{5+} > Si^{4+}, Ge^{4+} > B^{3+} > Co^{2+}$ (for the heteroatom or the charge of the polyanion) [23–25]. The latter order is also observed for solid HPAs [26]. These results indicate that the acid strength can be controlled by the addenda atoms and the heteroatoms. On the other hand, the amount of acidic sites (not the acid strength) on the solid surface is sensitive to the kind of counter cations, as well as to the resulting tertiary structures (see below).

The heteropolyanion stabilizes protonated intermediates by coordination in solution and in the pseudoliquid phase as well as on the surface, thus lowering the activation energy and accelerating reactions. Several protonated intermediates including the protonated ethanol dimer and monomer [27], the protonated pyridine dimer [21, 28], and protonated methanol [29] have been detected in the pseudoliquid phase directly by use of X-ray diffraction (XRD), infrared (IR) or solid-state NMR. In solid-state ^{1}H NMR, the chemical shift for the protonated ethanol dimer $(C_2H_5OH)_2H^+$ is 9.5 ppm downfield from tetramethylsilane, which lies in the range of superacids reported by Olah et al. [27]. This fact also supports the strong acidity of heteropolyacids.

Heteropolyacids are much more active than mineral acids for several types of homogeneous reactions in both organic solvents and aqueous solution [4, 11]. The enhancement is generally greater in organic solvents. For the hydration of isobutene in a concentrated aqueous HPA solution (above 1.5 mol dm^{-3}), the reaction rate is about 10-fold greater than for mineral acids [30]. This rate enhancement is attributed to the combination of stronger acidity, the stabilization of protonated intermediates, and an increased solubility of alkenes [30]. In this case, the selectivity is also much improved with HPA catalysts.

The catalytic behavior of HPA depends on the basicity of reactants [31]. For the decomposition of weakly basic isobutyl propionate, the catalytic activities of the heteropolyacids are 60- to 100-fold higher than those of H_2SO_4 and p-toluenesulfonic acid, and the catalytic activity decreases in the order of $H_3PW_{12}O_{40} > H_4SiW_{12}O_{40} \approx H_4GeW_{12}O_{40} > H_5BW_{12}O_{40} > H_6CoW_{12}O_{40}$. This is also the order of the acid strength in solution, which increases with the decrease in the negative charge of the polyanion [24, 25, 32, 33]. Izumi et al. [32] indicated that the catalytic activities in nonaqueous solutions were related to the acidity of the heteropolyacid solution and the softness of the polyanion. In dilute aqueous solutions, however, the softness becomes more important, as most heteropolyacids are almost completely dissociated [32]. For some reactions involving basic molecules such as alcohols, no large differences in activity among these catalysts, including H_2SO_4 and p-toluenesulfonic acid, are observed [31]; this is most likely due to the leveling effect.

Heteropolyacids are good multielectron oxidants due to the presence of Mo^{6+}, W^{6+}, and V^{5+} and the charge delocalizability in the polyanion structure. Generally, the oxidizing ability decreases in the order V- > Mo- > W-containing heteropolyanions [22]. The reduction potential decreases linearly with a decrease in the valence of the central atom or an increase in the negative charge of the heteropolyanions: $PW_{12}O_{40}^{3-} > GeW_{12}O_{40}^{4-} \approx SiW_{12}O_{40}^{4-} > FeW_{12}O_{40}^{5-} \approx BW_{12}O_{40}^{5-} > CoW_{12}O_{40}^{6-} > CuW_{12}O_{40}^{7-}$ [22, 23]; this property is well reflected in their catalytic activities. For cyclohexene oxidation with hydrogen peroxide, the catalytic activity increased in the order $B^{3+} < Si^{4+} < C^{4+} \leq P^{5+}$ [34, 35]. Cetylpiridinium salts of HPA can be used as phase-transfer catalysts, and catalyze the epoxidation of alkenes very efficiently in a two-phase system comprising aqueous H_2O_2 and $CHCl_3$ [36].

The typical method for the preparation of heteropolyacids is the acidification of aqueous solutions of oxoanions of addenda atoms and heteroatoms [1b]:

$$12WO_4^{2-} + HPO_4^{2-} + 23H^+ \longrightarrow [PW_{12}O_{40}]^{3-} + 12H_2O$$

Acidification is generally achieved by the addition of mineral acids. The rates of formation are high, so that the polyanions can be crystallized from stoichiometrically acidified mixtures of the components at room temperature. Although the stoichiometry is often a good guide to the preparation, in some cases an excess heteroatom, the addition of small amounts of oxidants, or careful control of temperature and pH is necessary.

The hydrolysis and condensation are usually rapid, and depend on the pH as well as on the solvent. For example, $PW_{12}O_{40}^{3-}$ begins to decompose above pH 2 and coexists with $PW_{11}O_{39}^{7-}$ and $PW_9O_{34}^{9-}$ [37]. In this respect, pure $PW_{12}O_{40}^{3-}$ should be isolated as the free acid by extraction by ether, the so-called "etherate" method. If a highly acidic solution of the heteropolyanion is shaken with excess diethyl ether, three phases separate: a top ether layer; a middle aqueous layer; and a bottom heteropoly-etherate.

The bottom layer is drawn off, shaken with excess ether to remove entrained aqueous solution, and separated again. The etherate is decomposed by the addition of water, the ether is removed, and the aqueous solution of the heteropolyacid is allowed to evaporate until crystallization occurs [1].

Using this general method, heteropolyacids containing various types of heteroatom can be prepared: $H_3PW_{12}O_{40} \cdot 30H_2O$ is prepared from $Na_2WO_4 \cdot 2H_2O$ and Na_2HPO_4 [38]; $H_4SiW_{12}O_{40} \cdot 22H_2O$ from Na_2WO_4 and $Na_2SiO_3 \cdot 9H_2O$ [39]; $H_4GeW_{12}O_{40} \cdot 7H_2O$ from $Na_2WO_4 \cdot 2H_2O$ and GeO_2 [31]; $H_5BW_{12}O_{40} \cdot 15H_2O$ from $Na_2WO_4 \cdot 2H_2O$ and H_3BO_3 [40]; and $H_6CoW_{12}O_{40} \cdot 19H_2O$ from $Na_2WO_4 \cdot 2H_2O$ and $Co(OCOCH_3)_2 \cdot 4H_2O$ [31, 41]. The number of water molecules included in the crystal is variable in all these cases. The structure can be identified using IR, NMR, and XRD [1b, 5].

2.3.13.2.2 Wells–Dawson-Type HPAs
Nowadays, even though more than 100 heteropolyacids of different compositions and structures are known [1b, 42], only the Keggin-structure HPAs are well described in respect of their physico-chemical and catalytic properties as described above; this is due to the simple synthesis procedure and the thermal stability of Keggin-structure HPAs. Nevertheless, in recent years an increasing amount of attention has been focused on obtaining new quantitative data for the acid-catalytic properties of HPAs having other structures and compositions (Fig. 3).

The achievements in the field of acid catalysis by heteropolyacids having versatile structures have been recently reviewed by Timofeeva [43], where $H_xPW_{11}LO_{40}$ (L = Zr^{IV}, Ti^{IV}, Th^{IV}), $H_xZW_{12}O_{40}$ (Z = Si^{IV}, P^V, B^{III}, Co^{III}, Ge^{IV}), $H_{21}B_3W_{39}O_{132}$, $H_6P_2W_{18}O_{62}$, $H_6P_2W_{21}O_{71}$, $H_6As_2W_{21}O_{69}$ were reviewed and the data on the acidity of the HPAs generalized. Timofeeva concluded that the catalytic effect of HPAs in homogeneous acidic-type reactions depends mainly on three factors: the acid strength; heteropolyanion structure; and type of reaction (nature of reagents). The catalytic activity is more dependent on the HPA structure rather than its composition, and a prediction of catalytic activity is possible only for those HPAs having similar structures.

The acidic strength of heteropolyacids has been determined more quantitatively by calorimetry of NH_3 adsorption [26, 44]. When comparing the initial heats of NH_3 adsorption, acidity was seen to decrease in the series:

$$H_3PW_{12}O_{40} > H_4SiW_{12}O_{40} > H_6P_2W_{21}O_{71}(H_2O)_3$$
$$> H_6P_2W_{18}O_{62}$$

These data indicate that the Keggin-type heteropolyacids are much stronger acids than the Dawson-type HPAs. Furthermore, the heats of NH_3 adsorption confirm that HPAs are stronger acids than zeolites or simple metal oxides.

Recently, the scientific literature concerning the structure, hydrolytic stability in solution, thermal stability in the solid state, redox-acid properties and applications of HPAs with Wells–Dawson structure has been reported [45]. Wells–Dawson HPAs possess the formula $[(X^{n+})_2M_{18}O_{62}]^{(16-2n)-}$, where X^{n+} represents a central atom (phosphorus (V), arsenic (V), sulfur (VI), fluorine) surrounded by a cage of M addenda

References see page 338

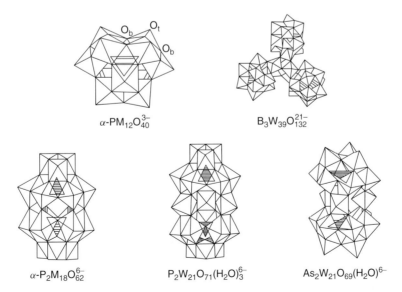

Fig. 3 The structures of heteropoly anions (M = Mo^{VI}, W^{VI}). Top left: Keggin-type, Bottom left: Dawson-type [43].

atoms, such as tungsten (VI), molybdenum (VI) or a mixture of elements, each of them comprising MO_6 octahedral units. The addenda atoms are partially substituted by other elements, such as vanadium, transition metals, lanthanides, halogens, and inorganic radicals. Wells–Dawson acids (phosphotungstic $H_6P_2W_{18}O_{62} \cdot 24H_2O$, phosphomolybdic $H_6P_2Mo_{18}O_{62} \cdot nH_2O$ and arsenicmolybdic $H_6As_2Mo_{18}O_{62} \cdot nH_2O$) possess superacidity and a remarkable stability, both in solution and in the solid state. Although the application of the acids in heterogeneous gas-phase reactions is less developed, a patented method has been devised to oxidize alkanes to carboxylic acids on a supported Wells–Dawson catalyst that combines acid and redox properties. Wells–Dawson anions possess the ability to accept or release electrons through an external potential, or upon exposure to visible and ultraviolet (UV) radiation (electro- and photochemical reactions). Additionally, Wells–Dawson HPAs catalyze the oxidation of organic molecules with molecular oxygen, hydrogen peroxide and iodosylarenes; epoxidation and hydrogenation in homogeneous and heterogeneous liquid-phase conditions. The ability of transition metal-substituted Wells–Dawson HPAs to be reduced and reoxidized without degradation of the structure is promising in the application of those HPAs replacing metalloporphyrin catalysts in redox and electrochemical reactions.

In order to determine the nature, amount and acid strength of phosphotungstic Wells–Dawson-type HPAs, isopropanol chemisorption and temperature-programmed surface reaction analyses have been studied. The properties were compared to those of Keggin-type HPAs, bulk WO_3, and monolayer-supported tungsten oxide over titania [46]. The number of available sites for isopropanol adsorption of Keggin and Wells–Dawson heteropolyacids is significantly higher than that of bulk WO_3, and even of supported tungsten oxide monolayers. This observation indicates that the alcohol is adsorbed both on the surface and the bulk, in agreement with the pseudoliquid phase behavior. Moreover, the temperature of surface decomposition of adsorbed isopropoxy species towards propene is significantly lower on the HPAs than that of the other tungsten oxide-based catalysts.

2.3.13.3 Salts of Heteropolyacids: Cation-Exchanged Forms

Salts of heteropolyacids can be divided into two groups according to Niiyama et al. [5, 10]: (i) the salts of small cations such as Na^+ and Cu^+ (group A salts); and (ii) salts of large cations such as Cs^+, K^+, and NH_4^+ (group B salts). The behavior of group A salts is similar to that of the acid forms, but is very different in several respects from that of group B salts. Group A salts can exhibit pseudoliquid behavior and are very soluble in water and other polar solvents, whereas group B salts are generally insoluble and have high surface areas. Organic salts also exist, but very few studies on their thermal stability and catalytic function have been reported. The thermal stability of group B salts is high (no decomposition occurs below their melting points, for example; the melting points of K and Cs salts of $H_3PMo_{12}O_{40}$ are reported to be 913 and 963 K, respectively [9]). This is most likely due to an absence of protons and water in the lattice of B salts, as the decomposition begins with the formation of water from oxygen of the polyanion and protons (or water) in the lattice.

2.3.13.3.1 Acid Catalysis of Salts of Heteropolyacids

The group B salts of $H_3PW_{12}O_{40}$ are also strong acids unless they are stoichiometrically neutralized. These salts catalyze various types of reaction in the liquid phase as solid catalysts. For the alkylation of 1,3,5-trimethylbenzene with cyclohexene, the activity of $Cs_{2.5}H_{0.5}PW_{12}O_{40}$ (abbreviated as $Cs_{2.5}W$) is much higher than those of SO_4^{2-}/ZrO_2 and Nafion. Here, $Cs_{2.5}W$ is not soluble and the reactions take place on the surface of the solid.

For $Cs_xH_{3-x}PW_{12}O_{40}$ (a typical group B salt), the catalytic activity changes dramatically with the extent of neutralization by Cs (Fig. 4). $Cs_{2.5}W$ shows the highest catalytic activity, which is even higher than that of the acid form. This is in contrast to Na salts (group A salts), where the activity as well as the acid strength decreases monotonically with the Na content [47]. The acid strength of $Cs_{2.5}W$ is almost the same as that of $H_3PW_{12}O_{40}$ according to temperature-programmed desorption (TPD)

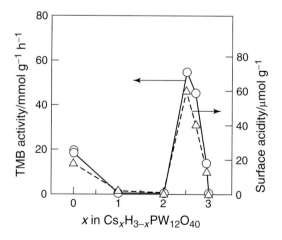

Fig. 4 Changes in catalytic activity for the alkylation of 1,3,5-trimethylbenzene (TMB) with cyclohexene and surface acidity (number of protons on the surface) as a function of Cs content in $Cs_xH_{3-x}PW_{12}O_{40}$.

of NH$_3$ [48]. Cs$_{2.5}$W consists of very fine particles (particle size 8–10 nm, surface area 100–200 m^2 g^{-1}). Solid-state ^{31}P NMR studies revealed that all protons in the salts are distributed randomly, indicating that near-uniform acidic salts are formed [4]. The same ^{31}P NMR spectrum is observed after heat treatment of a sample prepared by the impregnation of H$_3$PW$_{12}$O$_{40}$ on the Cs$_3$W salt, indicating formation of the same solid solution by the diffusion of Cs ion and proton. By assuming a uniform distribution of protons, the surface acidity (the number of protons on the surface) of the Cs salts can be estimated from the proton content in the bulk and the specific surface area. When the catalytic activities are plotted against the surface acidity estimated in this way, a good linear correlation is found [19b], indicating that the high activity of Cs$_{2.5}$W is due to its high surface acidity. It is also noteworthy that the activities per unit acidity are much greater for Cs salts than for SiO$_2$–Al$_2$O$_3$, SO$_4^{2-}$/ZrO$_2$, and zeolites. This indicates the presence of additional effects such as acid–base bifunctional acceleration by the cooperation of acidic protons and the basic polyanions.

The acid strength of the acid forms in the solid state usually reflects that in solution, but the acidic properties of the salts in the solid state are more complicated in general. Five mechanisms are possible for the generation of acidity in metallic salts [5]:

- protons in acidic salts, as for Cs salts as described above
- partial hydrolysis of polyanions to form weakly acidic protons
- acidic dissociation of water coordinated with multivalent metal ions
- Lewis acidity of multivalent metal ions
- protons formed by the reduction of metal ions with hydrogen.

2.3.13.3.2 Oxidation Catalysis of Salts of Heteropolyacids

The redox properties can also be controlled by the formation of salts. However, the relationship between these properties and the catalytic activity for oxidation is not sufficiently clarified. For example, it was reported that the activity order for alkali salts was reversed by the reaction temperature [49]. Nevertheless, good correlations are obtained if the concept of surface- and bulk-type catalysis is appropriately considered [4, 50–52]. The reduction by H$_2$ of free acid and group A salts proceeds both on the surface and in the bulk due to the diffusion of protons and electrons, whereas reduction by CO occurs only near the surface. Hence, the rate of reduction by H$_2$ represents the oxidizing ability of the catalyst bulk (r[H$_2$]), while the rate of reduction by CO expresses the oxidizing ability of the surface (r[CO]). Figure 5a shows the correlation of the rates of oxidation of acetaldehyde (surface-type) with the surface oxidizing ability (r[CO]). In Fig. 5b, a similar relationship between the rate of oxidative dehydrogenation of cyclohexene (bulk-type) and the bulk oxidizing ability (r[H$_2$]) is presented. Only a poor correlation exists if the former reaction (surface)

References see page 338

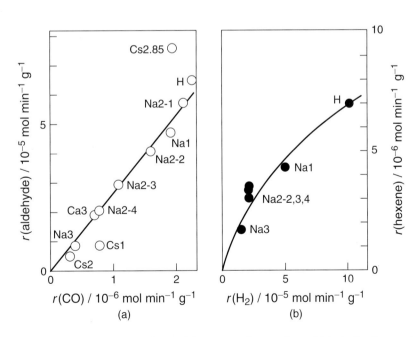

Fig. 5 Correlations between catalytic activity and oxidizing ability. (a) Oxidation of acetaldehyde (surface-type) and surface oxidizing ability; (b) oxidative dehydrogenation of cyclohexene (bulk-type) and bulk oxidizing ability [4, 38]. r(aldehyde) and r(hexene) are the rates of catalytic oxidation of acetaldehyde and cyclohexene, respectively.

is plotted against the bulk oxidizing ability. In this way, the catalytic activity and the oxidizing ability of Na salts decrease monotonically with the Na content for both reactions. However, for the oxidations over $Cs_xH_{3-x}PMo_{12}O_{40}$, activity patterns are similar to that in Fig. 4 due to differences in the surface area and secondary structures [4].

Although the redox properties of $H_{3+x}PMo_{12-x}V_xO_{40}$ have been examined by many research groups in relation to selective partial oxidations, their oxidizing abilities remain controversial [5]. One reason for this is that these catalysts are unstable at high temperatures. Several contributions by Mizuno have shown the possibility of the acidic salts of heteropolyacids as oxidation catalysts [53–55]. Lee and Misono [56] have also reported that the acidic Cs salts of HPA could reveal not only high conversion but also highly selective catalytic activity for oxidative dehydrogenation of isobutyric acid. $Cs_{2.75}Mo_{11}V$ was very efficient: 97% conversion and 78% selectivity to methacrylic acid (MAA) at 623 K, as compared to 40% conversion and 52% selectivity for $Mo_{11}V$ (acid form) [56]. $Mo_{11}V$ is significantly decomposed during the pretreatment in O_2 at 623 K, while $Cs_{2.75}Mo_{11}V$ decomposes only slightly. The improved thermal stability probably enabled the use of the controlled redox ability. For the reaction, unique dependency on the substituted amount of Cs has also been shown. $Cs_{2.75}PMo_{11}V$ was found to be the best catalyst for the production of methacrylic acid, while acetone was the exclusive product over $Cs_nPMo_{11}V$ ($n > 3$), from which it could be considered that HPA might be modified to have basic properties.

It has been intended to investigate the oxidation catalysis related to the basic properties more precisely. The reactivity of the acidic Cs salts ($Cs_nPMo_{11}V$, $n = 0-4$) for ethanol oxidation was investigated by Lee et al. [57]. For this reaction, $Cs_nPMo_{11}V$ showed a unique activity pattern with the substituted Cs amount. Acid catalysis was dominant as compared to oxidation catalysis with $n < 3$, which can be explained by the high surface acidity similar to acidic Cs salts of tungstophosphoric acids [58]. Oxidation catalysis was recessive in $n < 3$, but for $n \geq 3$ it was remarkably improved compared to the case of $n < 3$, which resulted in 100% of acetaldehyde selectivity and a threefold increased yield (38%).

Several types of characterization were carried out to account for the high oxidation activity of $Cs_nPMo_{11}V$ at $n > 3$. XRD patterns of $Cs_nPMo_{11}V$ ($n > 3$) were identical to that of $Cs_3PMo_{11}V$, the structure of which was interpreted as a bcc type. Considering the lattice parameter of $Cs_nPMo_{11}V$ and the sizes of Cs^+ and Keggin ions, up to three Cs^+ ions per Keggin anion can be incorporated into the stable bcc structure. Thus, it was inferred that the excess Cs ions above $n = 3$ cannot exist in the bulk but on the surface, thus playing an important role in oxidation catalysis. Subsequently, X-ray photoelectron spectroscopy (XPS) was performed to confirm the surface Cs species, with results indicating that Cs species different from bulk Cs were formed on the surface above $n = 3$. Oxidation activity is generally influenced by the basicity of the catalysts. CO_2 adsorption was carried out to examine the surface basicity of the Cs salts, whereupon the amount of adsorbed CO_2 increased above $n = 3$ and the increased uptake per gram coincided with that of Cs_2O; from this it could be speculated that the surface Cs species were Cs_2O. Based on the results of this study, it can be concluded that acidic Cs salts of molybdovanadophosphoric acids with $n > 3$ served as very active and selective catalysts for ethanol oxidation, the high oxidation activity of which was attributable to the surface basicity caused by surface Cs species being not incorporated in the bulk structure [57].

2.3.13.3.3 Bifunctional Catalysts

Noble metal salts of heteropolyacids function in novel ways as catalysts. $Pd_{1.5}PW_{12}O_{40}$, when supported on SiO_2, catalyzes the skeletal isomerization of hexane in the presence of hydrogen more efficiently than the parent compound $H_3PW_{12}O_{40}$ [59, 60]. The isomerization rate depends reversibly on the hydrogen partial pressure. The role of hydrogen is thought to be twofold: it dissociates on the surface of metallic palladium and spilled-over hydrogen atoms not only become acidic protons but also suppress coke formation, thus ensuring a higher steady activity and longer catalytic life.

Acid sites of $Ag_3PW_{12}O_{40}$ are generated upon the reduction with hydrogen at 300 °C [59, 61]; the catalytic activity (e.g., for o-xylene isomerization) develops following the reduction. The generation of acid sites has been confirmed by pyridine adsorption and solid-state ^1H NMR [61].

Recently, bifunctional catalysts consisting of a metal and a heteropolyacid have attracted much attention. Sano et al. [14] have developed a bifunctional catalyst consisting of Pd, $H_4SiW_{12}O_{40}$ and SiO_2 for the selective oxidation of ethene to acetic acid in the presence of water, proposing a mechanism that involves the hydration of ethene on the acid sites and the subsequent oxidation of ethanol on Pd, which is different from the Wacker-type reactions.

Okuhara et al. [62–64] have reported the hydroisomerization of benzene and skeletal isomerization of n-pentane and n-heptane using bifunctional catalysts such as Pt–$Cs_{2.5}H_{0.5}PW_{12}O_{40}$ (Pt–$Cs_{2.5}$W) and Pd–$Cs_{2.5}H_{0.5}PW_{12}O_{40}$ (Pd–$Cs_{2.5}$W), with the catalysts being supported on SiO_2. The catalytic activity for the formation of methylcyclopentane in the hydroisomerization of benzene was in the order of Pt–$Cs_{2.5}$W/SiO_2 > Pt–$Cs_{2.5}$W > Pt–WO_3/ZrO_2 > Pt–SO_4^{2-}/ZrO_2 [62]. The highest activity of Pt−$Cs_{2.5}$W/SiO_2 among these bifunctional catalysts would be due to its high ability for hydrogenation of benzene and high activity for skeletal isomerization of cyclohexane. In the skeletal isomerization of *n*-heptane, Pt−$Cs_{2.5}$W/SiO_2 was found to be more selective than Pt−H−β zeolite, which is known to be the most efficient catalyst for this reaction, though it was less active than Pt−H−β zeolite. Pt−$Cs_{2.5}$W was prepared by a titration method [65]. An aqueous solution of $H_2PtCl_6 \cdot 6H_2O$ (0.04 mol dm^{-3}) was added to an aqueous solution of $H_3PW_{12}O_{40}$ (0.08 mol dm^{-3}) at room temperature, after which an aqueous solution of Cs_2CO_3 (0.10 mol dm^{-3}) was added dropwise to the mixture at a rate of 0.1 cm^3 min^{-1}, with vigorous stirring, at room temperature. The obtained suspension was allowed to stand overnight at room temperature, and evaporated at 318 K to obtain a solid. Silica-supported Pt−$Cs_{2.5}$W was prepared from the aqueous solutions of $Cs_{2.5}H_{0.5}PW_{12}O_{40}$ and $H_2PtCl_6 \cdot 6H_2O$ using SiO_2 [66].

The skeletal isomerization of *n*-heptane in the presence of hydrogen has been studied over silica-supported bifunctional catalysts, Pd−$H_4SiW_{12}O_{40}$/SiO_2 with different loadings of $H_4SiW_{12}O_{40}$ [64]. The Pd catalysts with low loadings of $H_4SiW_{12}O_{40}$, such as 2 wt.% Pd-10, 15 and 20 wt.% $H_4SiW_{12}O_{40}$/SiO_2, exhibited very high selectivity towards branched heptanes comparable to that of Pd−H−β zeolite, and demonstrated higher activity than did Pd−H−β zeolite in the skeletal isomerization of *n*-heptane. Pd−H−β zeolite is the best catalyst known for the skeletal isomerization of *n*-alkane in the presence of hydrogen at atmospheric pressure at 180 °C. These bifunctional heteropoly catalysts were far superior to Pt−SO_4^{2-}/ZrO_2 and Pd−WO_3/ZrO_2 catalysts in selectivity and activity. Selectivity towards branched heptanes was sensitive to the loading amount of $H_4SiW_{12}O_{40}$; the selectivity increased with a decrease in $H_4SiW_{12}O_{40}$ loading amount. The acid strength/acid amount decreased, while the dispersion of Pd on $H_4SiW_{12}O_{40}$/SiO_2 increased with a decrease in the $H_4SiW_{12}O_{40}$ loading amount. These results, and the changes in product distribution with contact time, indicated that *n*-heptane was isomerized via a consecutive reaction pathway over Pd−$H_4SiW_{12}O_{40}$/SiO_2. Here, *n*-heptane was dehydrogenated on Pd sites, the resulting *n*-heptene was isomerized to *mono-*, *di-*, and *tri-*branched heptenes via carbenium ions at acidic sites on $H_4SiW_{12}O_{40}$/SiO_2, and the branched heptenes were hydrogenated on Pd sites. Cracking products are considered to be produced from dibranched heptenes.

2.3.13.3.4 Shape Selectivities of Acidic Salts of Heteropolyacids Okuhara [67] reported shape selectivities of heteropoly compounds with different cations and compositions. Indeed, in his recent review, a variety of studies is described on the pore-structure, the nano-structure of the heteropoly compounds, and the mechanism for pore formation.

As in the case of porous heteropoly compounds, Gregg and Tayyab [68] reported first that stoichiometric ammonium salts of HPAs were porous. Moffat and coworkers [69, 70] subsequently reported the porosity of alkaline salts and ammonium salts, while Okuhara and coworkers [71–73] noted that an acidic cesium salt, $Cs_{2.5}H_{0.5}PW_{12}O_{40}$ ($Cs_{2.5}$W), possesses pores with a bimodal distribution. Many reports have been made regarding catalysis of the catalyst, $Cs_{2.5}$W and its Pt-promoted form because of their high catalytic performances [74–76]. Subsequently, microporous heteropoly compounds with uniform pores were synthesized by the partial substitution of Cs^+ for H^+ of $H_3PW_{12}O_{40}$. It was shown that $Cs_{2.1}H_{0.9}PW_{12}O_{40}$ ($Cs_{2.1}$W) exhibited shape selectivity for the decomposition of esters and for alkylation reactions in solid–liquid reaction systems [71, 77]. In addition, the Pt-promoted $Cs_{2.1}$W exhibited shape selectivities for hydrogenation and oxidation reactions in gas–solid systems [71, 73, 78–81]. Recently, Okuhara and coworkers [81, 82] also found that the micropore-size of heteropoly compounds can be controlled by choosing the counter cations. Nanomaterials consisting of Cs or ammonium salts of heteropoly compounds have been reported by several research groups [83–86].

Micropores and mesopores are defined as pores having a width of less than 2 nm, and 2 to 50 nm, respectively [87]. $M_{2.1}H_{0.9}PW_{12}O_{40}$ ($M_{2.1}$W; M = Cs, Rb, and K) produced curves of micropore size-distribution similar to that of H-ZSM-5, without any peak in the mesopore region. The pore sizes were 0.51, 0.53 and 0.55 nm for $Cs_{2.1}$W, $Rb_{2.1}$W and $K_{2.1}$W, respectively – that is, the pore-width increased as the cation size decreased (Cs > Rb > K). The primary particles of $Cs_{2.1}$W are microcrystallites consisting of heteropolyanion ($PW_{12}O_{40}^{3-}$) and Cs^+ with bcc structure [72]. These microcrystallites themselves are non-porous, and the micropores are presumed to be spaces formed between the crystallites. Pt−$Cs_{2.1}$W has the micropores, but no mesopores. The width (ca. 0.5 nm) of micropores of Pt−$Cs_{2.1}$W is slightly smaller than that of H-ZSM-5. Proposed models for $Cs_{2.5}$W and $Cs_{2.1}$W are illustrated in Fig. 6 [88].

It has been demonstrated that Pt−$Cs_{2.1}$W was active only for the oxidation of smaller molecules such as CH_4, CO and *n*-butane, but was essentially inactive for those of isobutane and benzene [73, 81, 89]. Furthermore, Pt−$Rb_{2.1}$W exhibited shape selectivity for the hydrogenation of aromatic compounds [76, 78, 79].

References see page 338

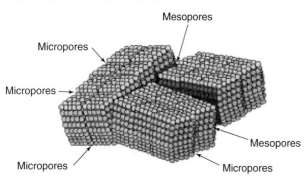

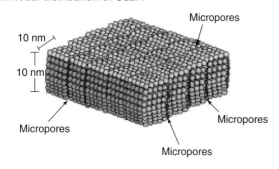

Fig. 6 Model for pores of $Cs_{2.5}H_{0.5}PW_{12}O_{40}$ (upper) and $Cs_{2.1}H_{0.9}PW_{12}O_{40}$ (lower) [88].

2.3.13.3.5 Preparation
Isolation of the polyanions from solution is generally achieved by the addition of an appropriate counter cation, usually alkali metal, ammonium or tetraalkylammonium ion [1].

Acidic Cs salts of $H_3PW_{12}O_{40}$ are prepared from aqueous solutions of heteropolyacids and Cs_2CO_3. The compound $H_3PW_{12}O_{40} \cdot 20H_2O$, which is prepared as described above, is evacuated at 323 K to form $H_3PW_{12}O_{40} \cdot 6H_2O$. Commercial Cs_2CO_3 is dehydrated by evacuation at 693 K for 2 h, prior to use for the preparation of an aqueous solution of Cs_2CO_3. The acidic cesium salts ($Cs_xH_{3-x}PW_{12}O_{40}$, Cs_xW) are prepared by the titration of aqueous solution of $H_3PW_{12}O_{40}$ (0.08 mol dm^{-3}) with an aqueous solution of Cs_2CO_3 ([Cs^+] = 0.25 mol dm^{-3}), where the aqueous solution of Cs_2CO_3 is added dropwise to the $H_3PW_{12}O_{40}$ solution at room temperature at a controlled rate (1 mL min^{-1}) [48, 58]. The white colloidal solution obtained is evaporated to a solid at 323 K. The surface area changes considerably with the Cs content and the method of preparation. For example, the surface area of $H_3PW_{12}O_{40}$ is 5 m^2 g^{-1}, whereas for $Cs_xH_{3-x}PW_{12}O_{40}$ prepared as above it is 1, 1, 135, and 156 m^2 g^{-1} for x = 1, 2, 2.5, and 3, respectively, after the treatment at 573 K.

Pt-promoted acidic Cs salts of $H_3PW_{12}O_{40}$, Pt–Cs_x $H_{3-x}PW_{12}O_{40}$ are prepared from the aqueous solution of $H_3PW_{12}O_{40} \cdot nH_2O$, an aqueous solution of $Pt(NH_3)_4(OH)_2$, and the aqueous solution of Cs_2CO_3 with the titration method [78, 79]. The content of Pt (in the case of $Pt(NH_3)_4(OH)_2$) influences the pore-structure. Rb_xW and Pt–Rb_xW are prepared similarly to those of Cs_xW using an aqueous solution of Rb_2CO_3 instead of Cs_2CO_3 [68, 81]. These are calcined in air at 573 K for 3 h prior to use as catalyst.

2.3.13.4 Mixed-Coordinated Heteropoly Compounds
Molybdenum-vanadium mixed-coordinated HPAs are active for several selective oxidations in both homogeneous and heterogeneous systems. The main component of commercial catalysts for the oxidation of methacrolein to methacrylic acid is an acidic Cs salt of $H_{3+x}PMo_{12-x}V_xO_{40}$ ($x = 1–2$) [12, 90, 91]. For the oxidative dehydrogenation of isobutyric acid over $Cs_{2.75}H_{0.25+x}PMo_{12-x}V_xO_{40}$ ($Cs_{2.75}Mo_{12-x}V_x$), the activity increased in the order of $Cs_{2.75}Mo_{11}V > Cs_{2.75}Mo_{10}V_2 > Cs_{2.75}Mo_{12}$ and the methacrylic acid (MAA) selectivity was $Cs_{2.75}Mo_{11}V \geq Cs_{2.75}Mo_{10}V_2 > Cs_{2.75}Mo_{12}$ [56]. Although various studies of mixed-coordinated HPA catalysts have been reported, their stability and redox properties remain controversial [92–94]. The states and roles of V of the $H_4PMo_{11}VO_{40}$ ($PMo_{11}V$) catalyst in the partial oxidation of isobutane have been analyzed by electron paramagnetic resonance (EPR), ^{51}V and ^{31}P NMR, IR spectroscopy, and redox titration, which was also compared with the $H_3PMo_{12}O_{40}$ (PMo_{12}) catalyst [189]. Thermal treatment of $PMo_{11}V$ at 623 K in O_2 caused the elimination of V from the Keggin anion and formed undefined polymeric and a monomeric V species. It was shown, based on the redox titration and EPR spectra, that the average valency of V in used catalysts significantly changed with the oxygen partial pressure, while that of Mo remained almost unchanged for both $PMo_{11}V$ and PMo_{12} [189]. The compound $Na_5PMo_{10}V_2O_{40}$, supported on active carbon, is active for the oxidative dehydrogenation of benzylic alcohols and amines without overoxidation of benzaldehyde and benzylamine [95]. The mechanism of the aerobic oxidative dehydrogenation of α-terpinene to p-cymene in solution with mixed-coordinated $PMo_{10}V_2O_{40}^{5-}$ has been reported [96].

Molybdenum–tungsten mixed-coordinated HPA shows a remarkable synergistic effect for, for example, the formation of dialdehydes from alkenes [35, 97]. Compounds of the type $H_3PMo_{12-x}W_xO_{40}$ ($Mo_{12-x}W_x$; $x = 0–12$), prepared by the conventional method, are statistical mixtures of 13 mixed-coordinated heteropolyacids [35, 98]. For the oxidation of cyclopentene

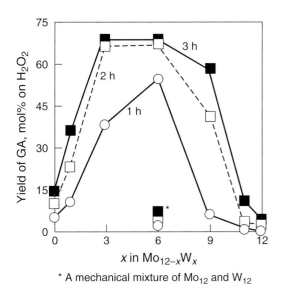

Fig. 7 Synergistic effect of Mo–W mixed-coordinated heteropolyacids for the oxidation of cyclopentene with hydrogen peroxide [26]; the reaction times shown are 1 h, 2 h, and 3 h. GA, glutaraldehyde.

by H_2O_2 in *tri-n*-butylphosphate solution, Mo_6W_6 shows the highest yield of aldehyde, much higher than Mo_{12} and W_{12} alone or a mechanical mixture of them (Fig. 7). The selectivity of concurrent formation of diol increases with the ratio of W to Mo, reflecting the acidity. The synergistic effect is explained quantum chemically by the increase in the stability of reduced states of mixed-coordinated polyanions [99]. However, there is a possibility that active peroxo species are formed by partial degradation of the starting Keggin anion and the formation is easier with mixed coordinated HPA.

Reproducible preparation of $H_{3+x}PMo_{12-x}V_xO_{40}$ ($x = 1$–3, $Mo_{12-x}V_x$) was established by Tsigdinos and Hallada [100]. An aqueous solution of Na_2HPO_4 and sodium metavanadate is acidified with concentrated sulfuric acid, and to this mixture is added an aqueous solution of $Na_2MoO_4 \cdot 2H_2O$ (the detailed procedure is available in the literature; it should be noted that an excess of sodium metavanadate is used).

Molybdenum–tungsten mixed-coordinated HPAs, $H_3PMo_{12-x}W_xO_{40}$, are prepared from $Na_2HPO_4 \cdot 12H_2O$, Na_2WO_4, and $Na_2MoO_4 \cdot 2H_2O$ solutions. After the solutions have been mixed at 353 K for 1 h, hydrochloric acid is added at room temperature. The HPAs are then extracted with diethyl ether and crystallized at room temperature; $H_3PMo_{12}O_{40}$ and $H_3PW_{12}O_{40}$ can also be used as the starting materials. The mixed aqueous solution is held at 353 K for 6 h, followed by extraction by diethyl ether and recrystallization. The HPAs prepared by these methods are mixtures of mixed-coordinated heteropolyanions having a statistical distribution of x.

The synthesis of specifically Mo- or V-substituted positional isomers of tungstophosphates is achieved by the method of Massart et al. [101] and Domaille and Watunya [102], though their catalytic activities are not as high. A requirement for these syntheses is isolation, or the in-place generation, of an appropriate defect polytungstate anion that is subsequently resubstituted with V or Mo: the ion $[PVW_{11}O_{40}]^{4-}$ is derived from in-place generation of $[PW_{11}O_{39}]^{7-}$; $[\alpha\text{-}1,2,3\text{-}PV_3W_9O_{40}]^{6-}$ is prepared from preformed $[\alpha\text{-}PW_9O_{34}]^{9-}$; $[\beta\text{-}$ and $\gamma\text{-}PV_2W_{10}O_{40}]^{5-}$ are isolated from $[PW_{10}O_{36}]^{7-}$ [102]; $[\alpha\text{-}PW_9Mo_2O_{39}]^{3-}$ is prepared from $[\beta\text{-}PW_9O_{34}]^{9-}$; $[\alpha\text{-}PW_9Mo_3O_{40}]^{3-}$ is derived from $[\alpha\text{-}PW_9Mo_2O_{39}]^{3-}$ [101]. Two important preparative details are emphasized [103]:

- reproducible syntheses require accurate pH control by a calibrated pH meter
- vanadium (V) species are strong oxidants, particularly at low pH, and metallic spatulas are attacked, leaving an intense heteropoly "blue" vanadium (IV) product (ceramic spatulas are recommended).

Product identification relies on ^{31}P NMR of their solutions. Details of ^{31}P NMR measurements and preparation methods are available in the literature [104].

2.3.13.5 Supported Heteropoly Compounds

Supported heteropoly compounds, where heteropoly compounds are dispersed on the surface of porous solids, are important for applications, as the surface area of HPAs is usually low. The structure, stability, and catalytic properties are very much dependent on the support materials, the extent of loading, and the method of preparation [5]. As supports for heteropoly compounds, SiO_2, Al_2O_3, and TiO_2, active carbons, ion-exchange resins, mesoporous materials, and high-surface-area salts of HPA have been used. A silane coupling reagent has been tested for binding HPA on oxide supports [105]. Insolubility is also an important property when HPA is applied to liquid-phase reactions.

2.3.13.5.1 Oxide Supports
Basic solids such as Al_2O_3 and MgO tend to decompose HPAs [5, 106, 107], although significant activities have been reported in some cases. On the other hand, SiO_2 and some carbons are relatively inert. In the cases of $H_3PW_{12}O_{40}$ and $H_3PMo_{12}O_{40}$ on SiO_2, HPAs are dispersed as thin layers on the support up to a certain quantity of HPA, but above that quantity they form thick layers or separate particles [52]. Detailed studies of $H_4SiMo_{12}O_{40}$ supported on SiO_2 have been reported [108]. Both, spectroscopic and catalytic tests

References see page 338

indicate that the Keggin structure is maintained and catalytic properties corresponding to the parent HPA are revealed at high loading levels [108a,c]. However, at a very low level of loading, strong interactions between the HPA and surface silanol groups suppress the acidity, and the redox catalysis predominates. It was also shown that the state in solution influences the dispersion after being supported on SiO_2 [108b]. The thermal stability of HPA on SiO_2 is usually comparable or slightly lower than that of the parent HPA [108c, 109], whereas it was reported that $H_3PMo_{12}O_{40}$ was stable up to 853–873 K when supported on SiO_2 [110]. Besides decomposition, reformation of the Keggin structure takes place under certain (wet) conditions [5, 111–113]. $H_4SiW_{12}O_{40}$ supported on SiO_2 has been reported recently by Okuhara et al. to be superior in activity and selectivity for Friedel-Crafts-Type reactions of p-xylene with γ-butyrolactone or vinylacetic acid [190]. The catalytic activity showed good correlation with the acid. Silica-bound $Cs_{2.5}H_{0.5}PW_{12}O_{40}$, which is prepared by the hydrolysis of ethyl orthosilicate in the presence of colloidal $Cs_{2.5}H_{0.5}PW_{12}O_{40}$ in ethanol, is catalytically more active than Amberlist-15 and HZSM-5 with respect to the turnover frequency, as based on the unit acid site [114].

Molybdo(vanado)phosphoric heteropolyacids [HPMoV_x ($x = 0, 1, 2, 3$)] of Keggin structure supported on oxide supports (SiO_2, TiO_2, Al_2O_3) have been used as catalysts for ethane to acetic acid oxidation in the range of reaction temperature from 523 to 673 K [115]. HPMoV_x ($x = 0, 1, 2, 3$) were supported on SiO_2, Al_2O_3 and TiO_2 by the incipient wetness method, with subsequent calcination at 623 K for 2 h. Vanadium atoms located in the Keggin structure enhanced the catalyst's oxidative activity, while the introduction of a vanadyl group exchanged into the cationic position (H(VO)PMoV_x) resulted in a decrease in oxidative activity. The nature of the support (acidic or base centers on the surface) influenced both ethane conversion and distribution of side products (CO_2 and ethene). Titania-supported HPMoV_x resulted in higher yields of acetic acid, when compared to silica-supported catalysts, although among the side products CO_2 predominated over titania-supported samples. Silica- and titania-supported HPA preserved the Keggin structure, which was confirmed by Fourier transform (FT)–IR, while alumina-supported HPA has not shown any IR bands characteristic of the Keggin structure.

Depending on the surface basicity, the Keggin structure may be transformed to lacunary HPA (at pH 5–8) or to mixed oxide system (pH > 8). The isoelectric points were estimated to be about 2.2, 5.6 and 8.1 for SiO_2, TiO_2 (from isopropoxide) and Al_2O_3 (from isopropoxide), respectively [116]. From the value of the isoelectric point, it can be concluded that OH groups on SiO_2 surface have acidic character, OH groups on TiO_2 show only slight acidity, while the surface of Al_2O_3 shows a basic character. By considering the acid–base character of the support, it can be inferred that the oxidative activity of HPMoV_x/SiO_2 and HPMoV_x/TiO_2 was due to the presence of heteropoly anions of regular or defect structure, while the low activity of oxidation on HPMoV_x/Al_2O_3 resulted from the formation of mixed Mo–V–P oxides on the alumina surface. The presence of water vapor in the reaction mixture was required for both acetic acid desorption and modification of the catalyst surface.

The activity of the Pt-promoted systems prepared on the basis of Keggin $H_3PW_{12}O_{40}$ (HPW) and $H_3PW_{11}ZrO_{40}$ (HPW$_{11}$Zr), and Dawson $H_6P_2W_{18}O_{62}$ (HP$_2$W$_{18}$) and $H_6P_2W_{21}O_{71}$ (HP$_2$W$_{21}$) heteropolyacids supported on zirconia in the isomerization of n-hexane was studied [117]. The Pt/HPW/ZrO$_2$ catalytic system shows high activity in n-hexane isomerization, with the yield of iso-hexanes close to 80% and selectivity 96–98% at 363 K. The differential thermal analysis (DTA) and FTIR spectroscopic studies showed that partially distorted grafted Keggin ions are responsible for the catalytic activity. However, these species cannot be formed by dispersion of HP$_2$W$_n$ and HPW$_{11}$Zr heteropolyacids because of their insufficient stability. The support of these systems on ZrO$_2$ is accompanied by significant structural transformations. The Pt/HP$_2$W$_n$/ZrO$_2$ or HPW$_{11}$Zr/ZrO$_2$ catalytic systems show the best performance (yield close to 80%) at higher temperatures (303–313 K) than Pt/HPW/ZrO$_2$. However, these systems show higher selectivity due to a lower density of active sites and inhibition of bimolecular alkylation-cracking side reactions.

Butylation of p-cresol by tert-butanol was catalyzed by 12-tungstophosphoric acid supported on zirconia (HPA/ZrO$_2$) under flow conditions [118]. Catalysts prepared with different HPA loading (5–30 wt.%) were calcined at 1023 K and acidity was estimated by TPD of NH_3. Some 15% of HPA/ZrO$_2$ showed the highest acidity, and was found to be the most active catalyst in the butylation of p-cresol. Under the optimized conditions, conversion of p-cresol was found to be 61 mol.%, with product selectivity for 2-tert-butyl-p-cresol (TBC) 81.4%, 2,6-di-tert-butyl-p-cresol (DTBC) 18.1% and tert-butyl-p-tolyl ether (ether) 0.5%. Studies of time on stream performed as a function of time for 100 h showed that the loss in activity in terms of conversion of p-cresol was 6%.

Conventional methods such as incipient wetness impregnation are applicable to the preparation of supported HPAs [11].

2.3.13.5.2 Mesoporous Supports Many research activities have been conducted aiming to couple the advanced characteristics of mesoporous materials with superacidic properties of the heteropolyacid catalysts, the goal being to develop highly active and shape-selective solid acid catalysts.

Soled et al. [119] have prepared supported forms of insoluble cesium- and ammonium-acid salts. The silica-supported Cs-salt was prepared by sequential impregnation and *in-situ* reaction on the support. First, cesium carbonate was impregnated by aqueous incipient wetness onto silica powder or extrudates, dried at 383 K, and calcined at 573 K. Subsequently, the 12-tungstophosphoric acid was impregnated by a similar aqueous impregnation route, dried at 383 K, and calcined at 573 K. Cs-salts supported on silica extrudates prepared by the *in-situ* reaction/precipitation route showed the egg-white morphology [119].

Using a grafting technique, Wang et al. [120] have prepared improved supported Cs-HPA salts that are uniformly dispersed on a mesoporous silica carrier. This was confirmed by transmission electron microscopy (TEM) and supported by the catalytic performance for a model alkylation reaction (Fig. 8). Experiments assessing the stability of supported HPA catalysts indicate that not only their resistance to leaching by solvents but also the thermal stability of these catalysts are also significantly improved. The catalysts thus prepared possess the potential advantages offered by the MCM-41 materials, such as ordered structure, high surface area, and controllable and uniform pore size distributions for improved mass transfer or added shape selectivity.

The grafting method to prepare Cs-HPA salts supported on MCM-41 is as follows [121]: Aqueous Cs_2CO_3 solutions (0.03–1.0 M) and HPA solutions (0.03–1.0 M) using the Keggin-type $H_3PW_{12}O_{40}$ dissolved in an organic solvent (e.g., butanol) are initially prepared. The organic solvent for the HPA solutions is chosen to be one in which the heteropolyacid is soluble, while Cs_2CO_3 is less soluble or even insoluble. Two steps are then utilized to mitigate Cs mobility during catalyst preparation in order to yield the desired high dispersion of the Cs-HPA salt on mesoporous silica. First, Cs_2CO_3 is impregnated by aqueous incipient wetness onto mesoporous silica, dried at 383 K overnight, and then calcined at temperature ranges between 573 and 973 K for 2 h. It is believed that this calcination step facilitates the anchoring of Cs to the silica surface, most likely via a reaction between surface silanols and Cs_2CO_3. Following this, HPA is impregnated using a similar incipient wetness technique with various organic solvents including ethanol, 1-propanol, and 1-butanol. The use of an organic solvent rather than water for this impregnation mitigates Cs migration because it likely avoids the hydrolysis of Si-O-Cs bonds and subsequent dissolution of CsOH. The catalysts are then dried at 383 K overnight and calcined at 573 K for 2 h.

The tungstovanadogermanic ($H_5GeW_{11}VO_{40}$) heteropolyacid with Keggin structure has been chemically anchored to a modified SBA-15 surface [122]. A two-step synthesis has been employed to synthesize a new type of heteropolyacid on amino-functionalized SBA-15. The material was characterized using solid-state ^{29}Si MAS NMR, FT-IR spectroscopy, XRD, TG-DTA and N_2 adsorption experiments. Preservation of the structure of the HPA and dispersion of the cluster on the support were verified; spectroscopic data suggested that the HPA clusters were attached firmly to the surface. The protonic conductivity

References see page 338

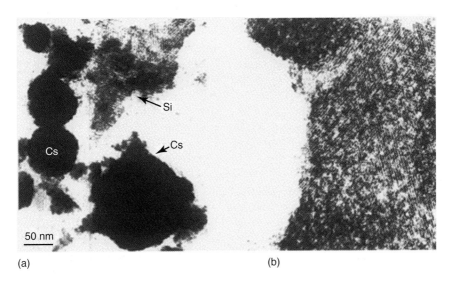

Fig. 8 Transmission electron micrographs of (a) Cs2.5-TPA/MS prepared by impregnation method [119] and (b) Cs2.5-HPA/MS synthesized by grafting method [120a].

of this composite was 3.09×10^{-3} S cm^{-1} with the 85% HPA at 296 K [123].

The crystallization of mesoporous molecular sieves of SBA-3 structure carried out from the gel containing HPAs (pH < 2) resulted in the formation of new mixed mesostructures, which included the Keggin units in mesoporous material walls [124]. Taking into consideration that the Keggin units are stable in acidic pH, the mesoporous material of SBA-3 structure synthesized at a pH below 2 [125–127] was applied as the heteropolyacid matrix. Correct mesostructure was formed up to 20 wt.% of HPW in the system. Template removal by means of calcination preserved the mesoporous structure, while washing of calcined samples with polar solvents removed the Keggin unit from the walls, and consequently the mesostructure collapsed.

The effect of the cation on the immobilization of 12-molybdophosphoric and 12-tungstophosphoric acids on hexagonal mesoporous materials was studied [128]. Hexagonal mesoporous silicate (HMS) materials containing Ti, Zr, and Al ions were prepared using the neutral template route [129]. IR and ^{1}H and ^{31}P NMR spectroscopic data revealed that the species present on the surface of HMS and TiHMS mesoporous materials after impregnation with solutions of 12-molybdophosphoric and 12-tungstophosphoric acid are intact heteropoly anions with a preserved Keggin structure. Partial degradation of the anion was observed for zirconium-containing silicate (ZrHMS) supports, the extent of which was stronger for supported molybdophosphoric acid in comparison with that of tungstophosphoric acid.

Phosphotungstic acid, $H_3PW_{12}O_{40}$ (PW) was supported on the synthesized mesoporous aluminophosphate (AlPO) at various loadings, namely 10, 20, and 40 wt.% to yield PW/AlPOs [130]. The catalysts were characterized by XRD, thermogravimetric analysis (TGA) and FTIR. Mesoporous aluminophosphate has been synthesized hydrothermally using cetyltrimethylammonium bromide (CTAB) as structure-directing agent in the presence of tetramethylammonium hydroxide (TMAOH) at room temperature. The supported catalyst was prepared by the impregnation of phosphotungstic acid ($H_3PW_{12}O_{40} \cdot 12H_2O$) on mesoporous AlPO [131]. XRD patterns confirmed that HPA retains the Keggin structure on the AlPO surface and forms finely dispersed HPA species. HPA crystal phases were not developed, even at a HPA loading as high as 40 wt.%. Catalytic activities of both parent AlPO and HPA-loaded AlPOs were studied by carrying out t-butylation of phenol with tert-butanol in the vapor phase in the temperature range 363 to 523 K. The products obtained were o-tert-butyl phenol (o-t-BP), p-tert-butyl phenol (p-t-BP) and tert-butylphenyl ether (t-BPE), with high selectivity towards p-t-BP. The effects of feed ratio and of the weight hourly space velocity (WHSV) on the conversion of phenol were studied, and finally time-on-stream studies were also carried out to compare the activities of the catalysts. The PW/AlPO catalysts were found to yield higher conversions than the parent AlPO due to an increase in the Brønsted acid site density. Hence, PW/AlPO compositions are promising catalysts exclusively for acid-type conversion of large organic molecules.

2.3.13.5.3 Carbon Supports
Active carbons are excellent supports for the insolubilization of HPAs [132]. It has been reported that proton transfer from HPAs to the carbon support tightens their binding [133].

For heterogenization, different types of catalytic filamentous carbon (CFC) have been used as supports for HPAs, such as $H_6P_2W_{21}O_{71}$ $(H_2O)_3$ (P$_2$W$_{21}$), $H_6As_2W_{21}O_{69}$ (H_2O) (As$_2$W$_{21}$), Dawson-type α-$H_6P_2W_{18}O_{62}$ (P$_2$W$_{18}$) and Keggin-type $H_3PW_{12}O_{40}$ (PW) [134]. The support properties, which affect the interaction between HPAs and supports were studied. The concentration of Brønsted acid sites in PW and PW/support was determined by electron spin resonance (ESR) studies of the stable nitroxyl radical (4-hydroxy-2,2′,6,6′-tetramethylpyperidin-N-oxyl (TEMPOL)) adsorbed from hexane. The reaction of 2,6-di-tert-butyl-4-methylphenol (DMBP) with toluene as tert-butyl acceptors was studied using bulk and supported PW and P$_2$W$_{18}$. It was found that P$_2$W$_{18}$ was more active than PW. It was also shown that the catalytic activities of HPAs supported on SiO$_2$ and CFC were approximately equal. The activity of the surface protons in HPA/support increased monotonically as the total HPA loading increased, reaching a maximum for the bulk HPA. HPA/SiO$_2$ and HPA/CFC catalysts were prepared by impregnation of SiO$_2$ and CFC with an aqueous (or MeOH) solution and dried overnight in a desiccator.

Scanning electron microscopy (SEM) data (Fig. 9) showed that PW molecules are adsorbed onto the outer surface of filamentous carbon. Individual HPA molecules penetrated into holes, localized at defects of packages of graphite sheets in filaments, and became new centers for further adsorption (Fig. 9a). Various HPA forms are observed on the CFC-2 (obtained with Cu–Ni catalysts [135]) surface: individual molecules less than 20 Å in size (Fig. 9b), clusters of approximately 50 Å in size (Fig. 9c), and large crystallites larger than 150 Å (Fig. 9a); the distribution of these forms was seen to depend on the HPA loading.

The adsorption of $H_3PW_{12}O_{40}$ from water and organic solvents (acetone and acetonitrile) by carbon mesoporous materials [Sibunit and fluorinated Sibunit (F-Sibunit, superstoichiometrical fluorocarbon material)] was also

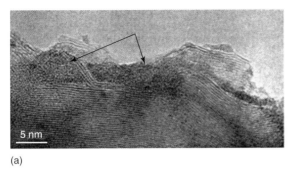
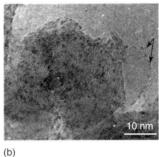

Fig. 9 Scanning electron micrographs of $H_3PW_{12}O_{40}$ supported on CFC-2 [134].

studied [136]. TEM data (Fig. 10) showed that the HPA is adsorbed on the surface of F-Sibunit in two main forms: isolated molecules less than 20 Å in size and clusters of ca. 35–50 Å in size. HPA aggregates larger than 100–150 Å were not observed, which might be explained by the pronounced hydrophobic properties of the F-Sibunit surface, preventing the formation of the large hydrophilic aggregates.

Keggin-type HPAs were immobilized in the unique network structure of resorcinol-formaldehyde (RF)-derived carbon gels [137]. Two methods were employed for immobilization: one by impregnating 12-tungstophosphoric acid in a wet RF hydrogel followed by drying and carbonization of the support; and the other by synthesizing 12-molybdophosphoric acid in a slurry mixture of carbon gel particles and deionized water. It was found that significant amounts of the acids could be immobilized on carbon gel supports by both methods, and large amounts of the acids remained on the carbon supports, even after long-term hot-water extraction. The obtained complexes were found to show catalytic activity in the synthesis of methyl-*tert*-butyl ether from methyl alcohol and *tert*-butyl alcohol, indicating that they could be practically used as solid acid catalysts in various reactions which involve water.

Fig. 10 Transmission electron micrograph of 8% heteropolyacid (HPA) supported on F-Sibunit [136].

2.3.13.5.4 HPA Salt Supports Potassium and cesium salts of HPAs which have high surface areas are useful supports for HPAs. Increased stability, as well as improved yield in the selective oxidation of acrolein, are observed when $H_{3+x}PV_xMo_{12-x}O_{40}$ compounds are supported on $K_3PMo_{12}O_{40}$ [92, 93]. It was inferred that free acids covered epitaxially the surface of the support, but the possible formation of a solid solution of $H(K,Cs)_3PW_{12}O_{40}$ [138] and $H(K,Cs)_3PMo_{12}O_{40}$ [139] due to the diffusion of K^+, Cs^+ and H^+ need also to be considered.

2.3.13.5.5 Polymer Supports High activity was reported for HPAs supported on ion-exchanged resins [140] and HPAs doped in polymers [141, 142].

Polyacetylene can be doped with large anions of $H_3PMo_{12}O_{40}$; such doping increases not only the conductivity of the polymer but also the catalytic activity. The HPA is distributed almost uniformly over the cross-section of the polymer film. For the conversion of ethanol, the catalyst exhibits acid–base activity as well as redox activity. In pulse reaction experiments, it has been shown that the rate of formation of ethene and diethyl ether increased 10-fold and the rate of formation of acetaldehyde increased 40-fold with the supported material [141]. Lee and coworkers [142] showed that the oxidation rate of ethanol was remarkably increased by using a $H_3PMo_{12}O_{40}$-blended porous polysulfone (PSF) film. The selectivity for acetaldehyde was high (>90% selectivity at ≈10% conversion, compared with 40% selectivity at 2% conversion without polymer at 433–463 K).

The doping of polyacetylene with HPAs is performed via chemical oxidation in acetonitrile containing $H_3PMo_{12}O_{40}$ and a minute amount of suitable complexing reagent. Anodic oxidation in an $H_3PM_{12}O_{40}$-acetonitrile electrolyte is also possible using a polyacetylene film as the anode. After doping, the films are washed

References see page 338

with dry acetonitrile and again vacuum-dried; the samples are stored in vacuum-sealed glass tubes [141].

For blending porous polysulfone (PSF) and HPAs, the choice of solvent is important; dimethylformamide (DMF) has been recommended [142]. A HPA–PSF film is prepared by casting the HPA–PSF–DMF solution on a glass plate, followed by drying in air for 4–5 h and subsequent evacuation for 2 h. The thickness of the HPA–PSF film is usually 0.1 mm [142].

2.3.13.5.6 **Clay Supports** Yadav et al. [143] have synthesized $Cs_{2.5}H_{0.5}PW_{12}O_{40}$ supported on montmorillonite-K-10 clay with complete intact Keggin anion structure. This catalyst was found to exhibit excellent activity for the decomposition of cumene hydroperoxide exclusively into phenol and acetone. The catalyst is stable and reusable, providing 100% conversion with exclusive selectivity to 100% phenol and acetone during all runs. The K-10 clay was shown to be a suitable support for HPAs in a variety of industrially important reactions [144–149].

The K-10 clay is first impregnated with an aqueous solution containing a Cs^+ precursor (cesium chloride), dried at 383 K for 12 h, and then calcined at 573 K for 3 h. HPA is then impregnated onto this support from a methanolic solution. Thus, 20% (w/w) $Cs_{2.5}H_{0.5}PW_{12}O_{40}$/K-10 is dried at 383 K for 12 h and then calcined at 573 K for 3 h.

2.3.13.5.7 **HPA Encaged in Supercages** Sulikowski et al. [150, 151] and Mukai et al. [152–155] showed, independently, that molecules of Keggin-type HPAs can be encaged in the supercages of Y-type zeolite using the "ship-in-a-bottle" approach. The size of the supercage of zeolite Y is 1.3 nm in diameter, and the supercages are interconnected by windows that are 0.74 nm in diameter. As the molecular sizes of HPAs are about 1.0 nm, then assuming a spherical shape, this might be a good support for HPAs (see Fig. 11). Thus, if the molecules could be encaged successfully in the supercages of zeolite Y, highly dispersed HPA catalysts can be prepared. Mukai et al. also inferred that the encaged HPA did not decompose or leak from supercages even in the liquid-phase reactions, indicating that highly dispersed HPA catalysts can be prepared by encaging HPAs into supercages of zeolite Y. According to Mukai et al., even though using MoO_3 as the Mo source might not be very efficient (as the solubility of MoO_3 in water is extremely low), only the preparation method, which uses MoO_3 and H_3PO_4 for PMo_{12} synthesis, led to a significant amount of encaged PMo_{12}. Applying water-soluble Mo sources (e.g., Na_2MoO_4) and using electromembrane technology proved not to be successful in producing encaged HPAs [152]. However, by adjusting the reaction temperature, the stability of the resulting encaged catalyst

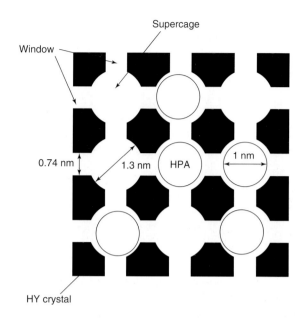

Fig. 11 Schematic illustration of the encagement of heteropolyacid (HPA) molecules in HY crystal [154].

could be enhanced to higher levels. By adding t–butyl alcohol (TBA) to the solution used for catalyst synthesis, an active and stable encaged catalyst can be prepared, even at low synthesis temperatures.

The framework of the Y-type zeolite used contains aluminum atoms. As aluminum atoms show basicity, they may accelerate the decomposition of HPAs, or disturb the formation of HPA anions. Therefore, Y-type zeolite was dealuminated in a steam atmosphere for 10 h. All of the zeolites were ion-exchanged with a 10 wt.% NH_4Cl solution before use. The conventional preparation method by Mukai et al. is as follows [153, 154]: 2.0 g of HY and 7.2 g of molybdenum (VI) oxide are mixed in 70 g of deionized water. This mixture is stirred for 24 h at room temperature. 0.48 g phosphoric acid is added, and the resultant mixture stirred for 3 h at a temperature of 368 K. The synthesized sample ($PMo12$ encaged HY) is filtered and dried at 383 K.

Recently, Kakuta et al. [156] suggested a hydrothermal synthesis method to prepare heteropolyacids encaged in US-Y. A hydrothermal technique using an autoclave was adopted to achieve the formation of HPAs in the supercages of US-Y, where not only was the encagement guaranteed but the concentrations of encaged HPAs were also controllable. Silicomolybdic acid (SMA) and phosphomolybdic acid (PMA) were synthesized by refluxing treatments with phosphoric or nitric acids. FT-IR and extended X-ray absorption fine structure (EXAFS) analyses revealed the presence of HPAs even after washing several times in hot water, indicating that the HPA molecules were anchored in the supercages. The preparation method was as follows. A certain amount of molybdenum trioxide (ca. twofold the final concentration desired) and 3 g of US-Y were mixed with 50 mL distilled water, and the mixture was retained in an autoclave at 383 K for 24 h. This hydrothermal condition facilitates the synthesis of molybdenum species in the supercage of US-Y. Following the addition of nitric acid for SMA formation (or of phosphoric acid for PMA formation) to the molybdenum species encaged in US-Y, the slurry was stirred and refluxed at 353 K for 3 h to build up HPA molecules in the supercages. The as-prepared sample was then washed with hot water (353 K) several times to remove the HPAs formed on the outer surface of US-Y, as well as unreacted material. Finally, the sample was dried at 383 K overnight.

2.3.13.5.8 Heteropolyanions Intercalated in Layered Double Hydroxides

Layered silicate clays intercalated by pillaring polyoxocations are precursors to an important class of microporous catalysts. Smectite clay was the only host structure known to be pillarable by purely inorganic oxo ions. Layered double hydroxides (LDH) pillaring oxo ions were reported by Pinnavaia and coworkers [157, 158] (see also Chapter 2.3.7).

The pillaring of LDH by Keggin-type heteropolyanions and their lacunary species has been attempted. The gallery heights are reported to be about 10 Å. The reactivity of $[XM_{12}O_{40}]^{n-}$ species for intercalative ion exchange depends strongly on both the net charge and polyhedral framework of the ion [158]. Since the cross-section of a Keggin anion (10 Å diameter) is ≈ 80 Å^2 and the area per unit positive charge is 16.6 Å^2 for the layer surface of an LDH, $[Zn_2Al(OH)_6]NO_3 \cdot 2H_2O$, Keggin ions with a negative charge less than -5 cannot be electrically balanced by the host LDH [158]. However, further investigations are needed to establish the catalytic properties, structures, and thermal stabilities of the LDH-pillared HPAs.

The ion-exchange reactions are carried out by dropwise addition of a boiling suspension of $[Zn_2Al(OH)_6]NO_3 \cdot 2H_2O$ to a 40% excess of the polyoxometalates in aqueous solution at room temperature. A nitrogen atmosphere is used to avoid possible reaction of the LDH with atmospheric CO_2. The final products are stored under nitrogen [158]. The host LDH is prepared by the reaction of freshly precipitated aluminum hydroxide with an aqueous zinc chloride solution at pH 6.2, according to the general method of Taylor [159]. The reaction time at 373 K is 7 days.

2.3.13.6 Lacunary and Metal-Substituted Heteropoly Compounds

Metals can be coordinated with a heteropolyanion to form metal-coordinated polyanions, which show unique catalytic activities for various reactions. Lacunary Keggin anions, α-$XM_{11}O_{39}^{n-}$, have a "defect" Keggin structure, in which one addenda atom and one terminal oxygen are missing. These lacunary heteropolyanions, which are formed at relatively high pH, are used as ligands of 3d metal ions. The lacunary HPA can serve as a catalyst by itself, and the vacancy of the lacunary HPA can be the binding site of the metal. The lacunary anion can be used as a ligand for transition metal ions; that is, transition metal-substituted polyanions (TMSP; Fig. 12a). A polyanion can be used to support transition metals, that is, heteropolyanion-supported transition metal (Fig. 12b).

2.3.13.6.1 Lacunary Silicotungstate

Mizuno et al. [161, 162] found a widely usable green route to the production of epoxides using a silicotungstate compound, $[\gamma\text{-}SiW_{10}O_{34}(H_2O)_2]^{4-}$. Epoxides are an important class of industrial chemicals that have been used as chemical intermediates. This catalyst is synthesized by protonation of a divacant, lacunary, Keggin-type polyoxometalate of $[\gamma\text{-}SiW_{10}O_{36}]^{8-}$ (Fig. 13), and exhibits high catalytic performance for the epoxidation of various alkenes, including propene, with a hydrogen peroxide (H_2O_2) oxidant at 305 K. The effectiveness of this catalyst is evidenced by $\geq 99\%$ selectivity to epoxide, $\geq 99\%$ efficiency of H_2O_2 utilization, high stereospecificity, and easy recovery of the catalyst from the homogeneous reaction mixture. The high regioselectivity of the catalyst for the epoxidation would be caused by the steric hindrance of the active site [162].

Di-vanadium-substituted silicotungstate, $[\gamma\text{-}1,2\text{-}H_2SiV_2W_{10}O_{40}]^{4-}$ with the $\{VO\text{-}(\mu\text{-}OH)_2\text{-}VO\}$ core could catalyze epoxidation of alkenes using only one equivalent

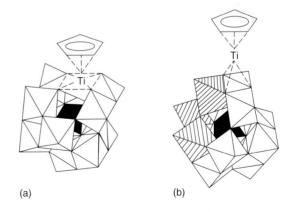

Fig. 12 (a) Transition metal-substituted heteropolyanion $[PW_{12}O_{39}(TiCp)]^{4-}$; (b) heteropolyanion-supported transition metal $[TiCp\cdot SiW_9V_3O_{40}]^{4-}$ [65]. The hatched octahedra represent VO_6, and the central phosphate is shown by black tetrahedra. $Cp = C_5H_5$.

References see page 338

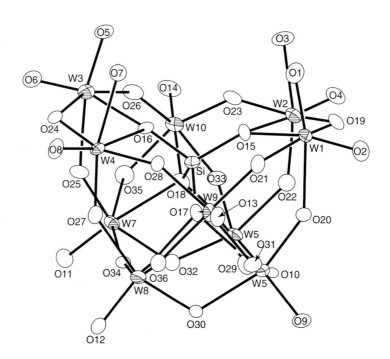

Fig. 13 Structure of a lacunary Keggin-type silicodecatungstate [162].

H_2O_2 with the high epoxide yield and high efficiency of H_2O_2 utilization [163]. This system shows a quite different performance from that of the epoxidation systems including $[\gamma\text{-SiW}_{10}O_{34}(H_2O)_2]^{4-}$.

2.3.13.6.2 Transition Metal-Substituted Polyanions (TMSP)

The vacancy of the lacunary heteropolyanions is the binding site; for example, $PW_{11}O_{39}^{7-}$ acts as a pentadentate and tetradentate ligand of Co and Cu, respectively [1b, 160, 164]. Complexes of anions such as $XM_{11}O_{39}^{n-}$ (X = P, Si; M = Mo, W, V, Nb) with metal ions are used as inorganic analogues of metalloporphyrins (inorganic synzyme) and have important advantages as catalysts – that is, a high resistance to oxidative degradation and thermal stability.

Hill and coworkers [165] have reported that TMSP is effective for the epoxidation of alkenes and the oxygenation of alkanes by use of t-BuOOH and PhIO as oxidants. The characteristics of TMSP in alkene epoxidation, compared with those of metalloporphyrins, Schiff base complexes, and triflate salts, are as follows [165]:

Reactivity:

$$PW_{11}Co^{II}O_{39}^{5-} > PW_{11}Mn^{II}O_{39}^{5-} \geq Fe^{III}(TDCPP)Cl$$
$$> Fe^{III}(TPP)Cl > M(OTf)_2$$

Selectivity to epoxide:

$$PW_{11}Co^{II}O_{39}^{5-} \approx PW_{11}Mn^{II}O_{39}^{5-} > M(OTf)_2$$
$$> Fe^{III}(TDCPP)Cl > Fe^{III}(TPP)Cl$$

Stability:

$$TMSP > Fe^{III}(TDCPP)Cl > Fe^{III}(TPP)Cl > M(OTf)_2$$

(TDCPP = tetrakis-2,6-dichlorophenylporphyrin; TPP = tetraphenylporphyrin; OTf = triflate ion).

Thus, the substituted heteropolyanion is stable and active even in the presence of oxidants such as t-BuOOH or PhIO. Note that the heteropolyanion is unstable with respect to hydrogen peroxide. Based on the high stability, TMSP can be used for alkane hydroxylation [165b]. Mansuy et al. [166] have reported that $P_2W_{17}O_{61}(Mn^{3+}\cdot Br)^{8-}$ is oxidation-resistant and more active for the epoxidation of cyclooctene with PhIO than those containing Fe^{3+}, Co^{2+}, Ni^{2+}, or Cu^{2+}. The oxygenations of cyclohexane, adamantane, and heptane and the hydroxylation of naphthalene, are also catalyzed by TMSP. Ruthenium has been used as the catalytic center in TMSP for the oxidations of cyclohexane, adamantane, and styrene with PhIO or t-BuOOH [167].

The preparation of TMSPs (Keggin-type and Dawson-type) is described in detail by Walker and Hill [168] and by Lyon et al. [169]. The tetrakis(tetra-n-butylammonium) salt of $PMo_{11}O_{39}^{7-}$, $(n\text{-Bu}_4N)_4H_3PMo_{11}O_{39}$ (**1**) is prepared and metalated to form the corresponding Keggin-type TMSP. The compound **1** is prepared as follows: $\alpha-H_3PMo_{12}O_{40}$ is dissolved in water, the pH of the solution is adjusted to 4.3 with Li_2CO_3, and a precipitate is then formed by adding solid $n\text{-Bu}_4NBr$ (14 equiv.) to the solution, with vigorous stirring. The crude product is collected, washed with H_2O, evaporated to dryness, and allowed to air-dry overnight at room temperature. In contrast to recrystallization from protic media [170], the light yellow-green crystalline product **1** can be obtained by repeated recrystallization from organic solvents, without decomposition. Recrystallization under non-protic conditions involves dissolving the crude product in CH_3CN by stirring without heating, followed by slow evaporation of the solvent at room temperature. After three recrystallizations, the complex has >99% purity, as confirmed by ^{31}P NMR [168].

From the lacunary polyanion, the Mn-substituted polyanion, $(n\text{-}Bu_4N)_4HPMo_{11}Mn^{II}(CH_3CN)O_{39}$, can be prepared [168]: **1** is dissolved in acetonitrile with stirring, after which toluene is added to produce a homogeneous light yellow-green solution. This solution is poured into a separation funnel to which is added an aqueous solution of $MnSO_4 \cdot H_2O$ (1.5 equiv.). Vigorous shaking of the heterogeneous mixture for ca. 1 min results in an immediate color change of the organic layer to brown, indicating that **1** has rapidly extracted the manganese ion into the organic layer. The water layer is drawn off, and the remaining organic layer placed in a crystallizing dish and allowed to evaporate slowly at room temperature. The crude brown crystalline product is separated by filtration and recrystallized from acetonitrile. Unlike Co^{II}, Mn^{II}, or Cu^{II}, metalation with Zn^{II} induces the decomposition of **1** [168].

The bis-(μ-oxo)-bridged di-iron site on the lacunary polyoxometallates (POM) has been used as a catalyst for the oxygenation of alkenes in homogeneous reaction media using molecular oxygen as an oxygen donor [171]. It is remarkable that selectivity to cyclooctene oxide and turnover number (TON) reached 98% and 10 000, respectively, for the oxygenation of cyclooctene. Not only cyclooctene, but also cyclodecene, 1-octene, 2-octene, 2-heptene, and 2-hexene, were catalytically oxygenated with high TONs and high selectivity to the corresponding epoxides.

The bis-(μ-oxo)-bridged di-iron on the lacunary POM was synthesized as its tetrabutylammonium salt by the reported procedure [172]. A solution of $K_8[\gamma\text{-}SiW_{10}O_{36}] \cdot 12H_2O$ in deionized water is quickly adjusted to pH 3.9 with concentrated nitric acid; an aqueous solution of $Fe(NO_3)_3 \cdot 9H_2O$ in water is then added, and the color of the solution turns pale yellow. The solution is then stirred for 5 min, after which the addition of an excess of tetra-n-butylammonium nitrate results in a yellow-white precipitate. The precipitate is collected by filtration and purified by twice dissolving it in acetonitrile and adding water to reprecipitate the product. The polyoxometalate $\gamma\text{-}SiW_{10}\{Fe^{3+}(OH_2)\}_2O_{38}^{6-}$ has a γ-Keggin structure with C_{2v} symmetry, and the two iron centers occupy adjacent, edge-sharing octahedra. UV/Visible, Mössbauer, EPR, and magnetic susceptibility data showed that the two high-spin Fe^{3+} centers were equivalent and antiferromagnetically coupled [172].

2.3.13.6.3 Heteropolyanion-Supported Metals

The organometallic cations can be supported on reduced heteropolyanions, or more negatively charged polyanions such as $SiW_9M_3O_{40}^{7-}$ and $P_2W_{15}M_3O_{62}^{9-}$ (M = V^{5+}, Nb^{5+}). The latter has three full units of negative surface charges that enable the tight and covalent bonding of transition metals [164].

Finke and coworkers have reported that a $P_2W_{15}Nb_3O_{62}^{9-}$-supported Ir catalyst, $[(n\text{-}C_4H_9)_4N]_5Na_3[(1,5\text{-}cod)Ir \cdot P_2W_{15}Nb_3O_{62}]$ (cod = cyclooctadiene), is active for both hydrogenation [173] and oxygenation [174] of cyclohexene. The turnover frequency (TOF) is 100-fold higher than that of its parent Ir compound, $[(1,5\text{-}cod)IrCl]_2$. The supported structure probably remains intact during the reaction, but is unstable under H_2 atmosphere [175].

A heteropolyanion-supported metal, $[(n\text{-}C_4H_9)_4N]_5Na_3[(1,5\text{-}cod)Ir \cdot P_2W_{15}Nb_3O_{62}]$ (**2**) is prepared from $(Bu_4N)_9P_2W_{15}Nb_3O_{62}$ and $[(1,5\text{-}cod)Ir(CH_3CN)_2]BF_4$ [176]. The choice of the metal–ligand combination $Ir(1,5\text{-}cod)^+$ and synthesis of $(Bu_4N)_9P_2W_{15}Nb_3O_{62}$ are described in the literature [177, 178]. The synthesis and subsequent storage of **2** require strict oxygen-free conditions. The key to obtaining **2** as a pure solid is the use of mixed Bu_4N^+/Na^+ salts and at least two reprecipitations from paper-filtered homogeneous CH_3CN solution using EtOAc. Bu_4NBF_4 is very soluble in EtOAc, and thus is removed by this process. Analytically pure **2** is obtained as a bright yellow, very air-sensitive powder.

2.3.13.6.4 Microstructured POM toward a Shape-Selective Catalyst

The Keggin-type polyoxometalate, $SiW_{12}O_{40}^{4-}$, and Cr(III) trinuclear cation, $Cr_3O(OOCH)_6(H_2O)_3^{3+}$, assembled in the presence of K^+ to create a channeled complex $K_3[Cr_3O(OOCH)_6(H_2O)_3]SiW_{12}O_{40} \cdot 16H_2O$ with an opening of 0.5×0.8 nm (Fig. 14) [179–181] was highly selective, and even a difference of one methylene group in the organic guest molecule was discriminated. Polar organic molecules with a longer methylene chain and non-polar molecules were completely excluded. The distinctive guest inclusion was successfully applied to the oxidation of methanol from a mixture of alcohols.

Mizuno et al. also reported a zeotype polyoxometalate-macrocation ionic crystal of $Cs_5[Cr_3O(OOCH)_6(H_2O)_3][\alpha\text{-}CoW_{12}O_{40}] \cdot 7.5H_2O$ with the shape-selective adsorption of water and separation of only water from an ethanol/water azeotropic mixture [182].

References see page 338

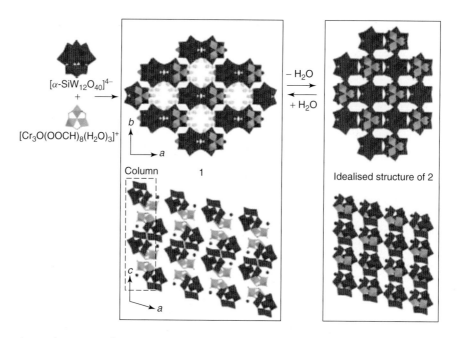

Fig. 14 Synthesis and crystal structure of $K_3[Cr_3O(OOCH)_6(H_2O)_3][\alpha\text{-}SiW_{12}O_{40}]\cdot 16H_2O$ (1) and the ideal closest-packing model of guest-free phase (2). The water of crystallization is omitted for clarity; the small dark (blue) spheres symbolize K^+ ions [180].

More recently, these authors also reported the complexation of Dawson-type polyoxometalates of $[\alpha\text{-}P_2W_{18}O_{62}]^{6-}$, $[\alpha_2\text{-}P_2W_{17}V_1O_{62}]^{7-}$, and $[\alpha\text{-}P_2W_{15}V_3O_{62}]^{9-}$ with the macrocation $([Cr_3O(OOCH)_6(H_2O)_3]^+)$ forms ionic crystals of $(NH_4)_4[Cr_3O(OOCH)_6(H_2O)_3]_2[\alpha\text{-}P_2W_{18}O_{62}]\cdot 15H_2O$ (**1a**), $(NH_4)_5[Cr_3O(OOCH)_6(H_2O)_3]_2$ $[\alpha_2\text{-}P_2W_{17}V_1O_{62}]\cdot 15H_2O$ (**2a**), and $(NH_4)_7[Cr_3O(OOCH)_6(H_2O)_3]_2$ $[\alpha\text{-}P_2W_{15}V_3O_{62}]\cdot 15H_2O$ (**3a**), respectively [183]. The compounds **1a** to **3a** show the honeycomb packing, and the symmetry of the constituent ions reflects the crystal structures. The lengths of the a-axes of **1a** to **3a** are almost the same, while those of the c-axes decrease in the order of **1a** ≥ **2a** > **3a**, with the increase in the anion charges of the polyoxometalates. The water of crystallization in **1a–3a** is desorbed by the evacuation at 373 K to form the respective guest-free phases **1b–3b**. The compounds **1b–3b** are crystalline, and the lengths of the a-axes of **1b–3b** are almost the same, while those of the c-axes decrease in the order of **1b** ≥ **2b** > **3b**. The compounds **1b–3b** possess voids running perpendicular to the c-axis. The sizes of the voids decrease in the order of **1b** > **2b** > **3b**, and are comparable to those of water and methanol. The crystal structures, water sorption kinetics, and alcohol sorption properties of **1b–3b** change systematically with the anion charges.

References

1. (a) G. A. Tsigdinos, *Topics Curr. Chem.* **1978**, *76*, 1; (b) M. T. Pope, *Heteropoly and Isopoly Oxometallates*, Springer-Verlag, Berlin, 1983; (c) M. T. Pope, A. Müller, *Angew. Chem. Int. Ed. Engl.* **1991**, *30*, 34; (d) V. W. Day, W. G. Klemperer, C. Schwartz, R.-C. Wang, in *Surface Organometallic Chemistry: The Molecular Approach to Surface Science and Catalysis*, J. M. Basset, B. C. Gates (Eds.), Reidel, Dordrecht, 1988, p. 173.
2. M. Misono, M. Hashimoto, *Shokubai (Catalyst)* **1992**, *34*, 152.
3. Y. Jeannin, M. Fournier, *Pure Appl. Chem.* **1987**, *59*, 1529.
4. M. Misono, *Proceedings 10th International Congress on Catalysis*, 1992, Akademiai Kiado: Budapest, 1993, p. 69.
5. (a) M. Misono, *Catal. Rev. Sci. Eng.* **1987**, *29*, 269; (b) M. Misono, *Catal. Rev. Sci. Eng.* **1988**, *30*, 339.
6. (a) T. Okuhara, N. Mizuno, M. Misono, *Adv. Catal.* **1996**, *41*, 113; (b) X. Song, A. Sayari, *Catal. Rev. Sci. Eng.* **1996**, *38*, 329; (c) A. Corma, H. Garcia, *Catal. Today* **1997**, *38*, 257.
7. I. V. Kozhevnikov, *Chem. Rev.* **1998**, *98*, 171.
8. T. Okuhara, *Chem. Rev.* **2002**, *102*, 3641.
9. K. Eguchi, N. Yamazoe, T. Seiyama, *Nippon Kagaku Kaishi* **1981**, 336.
10. H. Niiyama, Y. Saito, S. Yoshida, E. Echigoya, *Nippon Kagaku Kaishi* **1982**, 569.
11. Y. Izumi, K. Urabe, M. Onaka, *Zeolite, Clay, and Heteropolyacid in Organic Reactions*, Kodansha, Tokyo, and VCH, Weinheim, 1992.
12. M. Misono, N. Nojiri, *Appl. Catal.* **1990**, *64*, 1.
13. T. Okuhara, N. Mizuno, M. Misono, *Appl. Catal. A: General* **2001**, *222*, 63.
14. K. Sano, H. Uchida, S. Wakabayashi, *Catal. Survey Jpn.* **1999**, *3*, 55.
15. M. Misono, *Chem. Commun.* **2001**, 1141.
16. Y. Yoshinaga, K. Seki, T. Nakato, T. Okuhara, *Angew. Chem. Int. Ed.* **1997**, *36*, 2833.
17. Y. Yoshinaga, T. Okuhara, *J. Chem. Soc., Faraday Trans. I* **1998**, *94*, 2235.
18. T. Okuhara, R. Watanabe, Y. Yoshinaga, *ACS Symp. Ser.* **2000**, *738*, 369.
19. (a) M. Otake, T. Onoda, *Shokubai* **1975**, *17*, 13; (b) H. Hayashi, J. B. Moffat, *J. Catal.* **1982**, *77*, 473;

(c) M. Misono, T. Okuhara, *CHEMTECH*, **1993**, November 23.
20. M. Misono, K. Sakata, Y. Yoneda, W. Y. Lee, *Proceedings, 7th International Congress on Catalysis, 1980*, Kodansha, Tokyo, and Elsevier, Amsterdam, 1981, p. 1047.
21. M. Misono, N. Mizuno, K. Katamura, A. Kasai, Y. Konishi, K. Sakata, T. Okuhara, Y. Yoneda, *Bull. Chem. Soc. Jpn.* **1982**, *55*, 400.
22. I. V. Kozhevnikov, K. I. Matveev, *Appl. Catal.* **1983**, *5*, 135.
23. J. J. Altenau, M. T. Pope, R. A. Prados, H. So, *Inorg. Chem.* **195**, *14*, 417.
24. D. E. Katsoulis, M. T. Pope, *J. Am. Chem. Soc.* **1984**, *106*, 2737.
25. T. Okuhara, C. Hu, M. Misono, *Bull. Chem. Soc. Jpn.* **1994**, *67*, 1156.
26. F. Lefebvre, F. X. Liu-Cai, A. Auroux, *J. Mater. Chem.* **1994**, *4*, 125.
27. K. Y. Lee, T. Arai, S.-I. Nakata, S. Asaoka, T. Okuhara, M. Misono, *J. Am. Chem. Soc.* **1992**, *114*, 2836.
28. M. Hashimoto, M. Misono, *Acta Crystallogr.* **1994**, *C50*, 231.
29. (a) J. G. Highfield, J. B. Moffat, *J. Catal.* **1985**, *95*, 108; (b) J. G. Highfield, J. B. Moffat, *J. Catal.* **1986**, *98*, 245.
30. A. Aoshima, S. Yamamatsu, T. Yamaguchi, *Nippon Kagaku Kaishi* **1987**, 976.
31. C. Hu, M. Hashimoto, T. Okuhara, M. Misono, *J. Catal.* **1993**, *143*, 437.
32. Y. Izumi, K. Matsuo, K. Urabe, *J. Mol. Catal.* **1983**, *18*, 299.
33. L. Barcza, M. T. Pope. *J. Phys. Chem.* **1975**, *79*, 92.
34. N. Mizuno, S. Yokota. I. Miyazaki, M. Misono, *Nippon Kagaku Kaishi* **1991**, 1066.
35. K. Y. Lee, K. Itoh, M. Hashimoto, N. Mizuno, M. Misono, in *New Developments in Selective Oxidation II*, V. C. Corberán, S. V. Bellón (Eds.), Elsevier, 1994, p. 583.
36. (a) Y. Matoba, H. Inoue, J. Akagi, T. Okabayashi, Y. Ishii, M. Ogawa, *Syn. Commun.* **1984**, *14*, 865; (b) Y. Ishii, Y. Kazumasa, T. Ura, H. Yamada, T. Yoshida, M. Ogawa, *J. Org. Chem.* **1988**, *53*, 3587, 5549; (c) Y. Ishii, K. Yamawaki, T. Yoshida, T. Ura, M. Ogawa, *J. Org. Chem.* **1987**, *52*, 1868; (d) T. Oguchi, Y. Sakata, N. Takeuchi, K. Kaneda, Y. Ishii, M. Ogawa, *Chem Lett.* **1989**, 2053.
37. P. Souchay, *Ions Minéraux Condensés*, Masson & Cie, Paris, 1969.
38. J. C. Bailar, *Inorg. Synth.* **1939**, *1*, 132.
39. C. Rocchiccioli-Deltcheff, M. Fournier, R. Franck, R. Thouvenot, *Inorg. Chem.* **1983**, *22*, 207.
40. P. Souchay, *Bull. Soc. Chim. Fr.* **1951**, *18*, 365.
41. L. C. W. Baker, T. P. McCutcheon, *J. Am. Chem. Soc.* **1956**, *78*, 4503.
42. A. Muller, F. Peters, M. T. Pope, D. Gatteschi, *Chem. Rev.* **1998**, *98*, 239.
43. M. N. Timofeeva, *Appl. Catal. A: General* **2003**, *256*, 19.
44. A. Auroux, J. C. Vedrine, *Stud. Surf. Sci. Catal.* **1985**, *20*, 311.
45. L. E. Briand, G. T. Baronetti, H. J. Thomas, *Appl. Catal. A: General* **2003**, *256*, 37.
46. L. A. Gambaro, L. E. Briand, *Appl. Catal. A: General* **2004**, *264*, 151.
47. T. Okuhara, A. Kasai, N. Hayakawa, Y. Yoneda, M. Misono, *J. Catal.* **1983**, *83*, 121.
48. T. Okuhara, T. Nishimura, H. Watanabe, M. Misono, *J. Mol. Catal.* **1992**, *74*, 247.
49. M. Akimoto, K. Shima, H. Ikeda, E. Echigoya, *J. Catal.* **1981**, *72*, 83.
50. M. Misono, N. Mizuno, T. Komaya, *Proceedings, 8th International Congress on Catalysis, Berlin, 1984*, Verlag Chemie-Dechema, 1984, Vol. 5, p. 487.
51. N. Mizuno, T. Watanabe, M. Misono, *J. Phys. Chem.* **1985**, *89*, 80.
52. N. Mizuno, T. Watanabe, H. Mori, M. Misono, *J. Catal.* **1990**, *123*, 157.
53. N. Mizuno, M. Tateishi, M. Iwamoto, *J. Chem. Soc., Chem. Commun.* **1994**, 1411.
54. N. Mizuno, M. Tateishi, M. Iwamoto, *Appl. Catal. A: General* **1994**, *118*, L1.
55. N. Mizuno, M. Tateishi, M. Iwamoto, *Appl. Catal. A: General* **1995**, *128*, L165.
56. K.-Y. Lee, S. Oishi, H. Igarashi, M. Misono, *Catal. Today* **1997**, *33*, 183.
57. J.-I. Yang, D.-W. Lee, J.-H. Lee, J. C. Hyun, K.-Y. Lee, *Appl. Catal. A: General* **2000**, *194–195*, 123.
58. T. Nishimura, T. Okuhara, M. Misono, *Appl. Catal.* **1991**, *73*, L7.
59. Y. Ono, *Perspectives in Catalysis*, J. M. Thomas, K. I. Zamaraev (Eds.), Blackwell: London, 1992.
60. S. Suzuki, K. Kogai, Y. Ono, *Chem. Lett.* **1984**, 699.
61. T. Baba, M. Nomura, Y. Ono, in *Acid-Base Catalysis II*, H. Hattori, M. Misono, Y. Ono (Eds.), Elsevier, Amsterdam, and Kodansha, Tokyo, 1994, p. 419.
62. A. Miyaji, T. Okuhara, *Catal. Today* **2003**, *81*, 43.
63. A. Miyaji, T. Echizen, K. Nagata, Y. Yoshinaga, T. Okuhara, *J. Mol. Catal. A: Chem.* **2003**, *201*, 145.
64. A. Miyaji, R. Ohnishi, T. Okuhara, *Appl. Catal. A: General* **2004**, *262*, 143.
65. K. Na, T. Okuhara, M. Misono, *J. Catal.* **1997**, *170*, 96.
66. T. Okuhara, T. Yamada, K. Seki, K. Johkan, T. Nakato, *Micropor. Mesopor. Mater.* **1998**, *21*, 637.
67. T. Okuhara, *Appl. Catal. A: General*, **2003**, *256*, 213.
68. S. T. Gregg, M. M. Tayyab, *J. Chem. Soc., Faraday Trans. I* **1978**, *74*, 348.
69. J. B. McMonagle, J. B. Moffat, *J. Coll. Interface Sci.* **1984**, *101*, 479.
70. D. Lapham, J. B. Moffat, *Langmuir* **1991**, *7*, 2273.
71. T. Okuhara, T. Nishimura, M. Misono, *Chem. Lett.* **1995**, 155.
72. T. Okuhara, H. Watanabe, T. Nishimura, K. Inumaru, M. Misono, *Chem. Mater.* **2000**, *12*, 2230.
73. Y. Yoshinaga, T. Suzuki, M. Yoshimune, T. Okuhara, *Topics Catal.* **2002**, *19*, 179.
74. P. B. Venute, *Micropor. Mater.* **1994**, *2*, 297.
75. M. Misono, I. Ono, G. Koyano, A. Aoshima, *Pure Appl. Chem.* **2000**, *72*, 1305.
76. T. Okuhara, T. Nakato, *Catal. Surv. Jpn.* **1999**, *2*, 31.
77. T. Okuhara, T. Nishimura, M. Misono, *Stud. Surf. Sci. Catal.* **1996**, *101*, 559.
78. Y. Yoshinaga, K. Seki, T. Nakato, T. Okuhara, *Angew. Chem. Int. Ed. Engl.* **1997**, *36*, 2833.
79. Y. Yoshinaga, T. Okuhara, *J. Chem. Soc., Faraday Trans.* **1998**, *94*, 2235.
80. T. Okuhara, T. Yamada, K. Seki, K. Johkan, T. Nakato, *Micropor. Mesopor. Mater.* **1998**, *21*, 637.
81. M. Yoshimune, Y. Yoshinaga, T. Okuhara, *Chem. Lett.* **2002**, 330.
82. M. Yoshimune, Y. Yoshinaga, T. Okuhara, *Micropor. Mesopor. Mater.* **2002**, *51*, 165.
83. T. Ito, K. Inumaru, M. Misono, *J. Phys. Chem. B* **1997**, *191*, 9958.
84. K. Inumaru, T. Ito, M. Misono, *Micropor. Mesopor. Mater.* **1998**, *21*, 629.

85. A. Kaliadima, L. A. Perez-Maqueda, E. Matijevic, *Langmuir* **1997**, *13*, 3733.
86. L. A. Perez-Maqueda, E. Matijevic, *Chem. Mater.* **1998**, *10*, 1430.
87. J. Rouquerol, D. Avnir, C. W. Fairbridge, D. H. Everett, J. H. Haynes, N. Pernicone, J. D. F. Ramsay, K. S. Sing, K. K. Unger, *Pure Appl. Chem.* **1994**, *66*, 1739.
88. T. Okuhara, *Catal. Today* **2002**, *73*, 167.
89. M. Yoshimune, R. Ohnishi, T. Okuhara, *Micropor. Mesopor. Mater.* **2004**, *75*, 33.
90. S. Nakamura, H. Ichihashi, in *Proceedings, 7th International Congress on Catalysis, Tokyo, 1980*, Kodansha, Tokyo, and Elsevier, Amsterdam, 1981, p. 755.
91. N. Shimizu, M. Ueshima, M. Wada, *Shokubai* **1988**, *30*, 555.
92. J. B. Black, N. J. Claydon, P. L. Gai, J. D. Scott, E. M. Serwicka, J. B. Goodenough, *J. Catal.* **1987**, *106*, 1.
93. (a) K. Bruckman, J. Haber, E. M. Serwicka, *Faraday Disc. Chem. Soc.* **1989**, *87*, 173; (b) K. Bruckman, J. Haber, E. Lalik, E. M. Serwicka, *Catal. Lett.* **1988**, *1*, 35; (c) K. Bruckman, J. Haber. E. M. Serwicka, N. Yurchenko, T. P. Lazarenko, *Catal. Lett.* **1990**, *4*, 181.
94. M. Akimoto, H. Ikeda, A. Okabe, E. Echigoya, *J. Catal.* **1984**, *89*, 196.
95. R. Neumann, M. Levin, *J. Org. Chem.* **1991**, *56*, 5707.
96. R. Neumann, M. Levin, *J. Am. Chem. Soc.* **1992**, *114*, 7278.
97. H. Furukawa. T. Nakamura, H. Inagaki, E. Nishikawa, C. Imai, M. Misono, *Chem. Lett.* **1988**, 877.
98. M. A. Schwegler, M. Floor, H. van Bekkum, *Tetrahedron Lett.* **1988**, *29*, 823.
99. J. K. Burdett, C. K. Nguyen, *J. Am. Chem. Soc.* **1990**, *112*, 5366.
100. G. A. Tsigdinos, C. J. Hallada, *Inorg. Chem.* **1968**, *7*, 437.
101. R. Massart, R. Contant, J.-M. Fruchart, J.-P. Ciabrini, M. Fournier, *Inorg. Chem.* **1977**, *16*, 2916.
102. P. J. Domaille, G. Watunya, *Inorg. Chem.* **1986**, *25*, 1239.
103. P. J. Domaille, *Inorg. Synth.* **1990**, *27*, 96.
104. A. Tézé, G. Hervé, *Inorg. Synth.* **1990**, *27*, 85.
105. M. Kamada, Y. Kera, *Chem. Lett.* **1991**, 1831.
106. J. A. R. van Veen, P. A. J. M. Hendricks, R. R. Andrea, E. J. M. Romers, A. E. Wilson, *J. Phys. Chem.* **1990**, *94*, 5282.
107. K. Nowinska, R. Fiedorow, J. Adamiec, *J. Chem. Soc. Faraday Trans.* **1991**, *87*, 749.
108. (a) J.-M. Tatibouet, M. Che, M. Amirouche, M. Fournier, C. Rocchiccioli-Deltcheff, *J. Chem. Soc., Chem. Commun.* **1988**, 1260; (b) M. Fournier, R. Thouvenot, C. Rocchiccioli-Deltcheff, *J. Chem. Soc., Faraday Trans.* **1991**, *87*, 349; (c) C. Rocchiccioli-Deltcheff, M. Amirouche, G. Herve, M. Fournier, M. Che, J.-M. Tatibouet, *J. Catal.* **1990**, *126*, 591.
109. M. J. Bartoli, L. Monceaux, E. Bordes, G. Hecquet, P. Courtine, *Stud. Surf. Sci. Catal.* **1992**, *72*, 81.
110. S. Kasztelan, E. Payen, J. B. Moffat, *J. Catal.* **1990**, *125*, 45.
111. C. R. Rocchiccioli-Deltcheff, M. Amirouche, M. Che, J. M. Tatibouet, M. Fournier, *J. Catal.* **1990**, *125*, 292.
112. A. Ogata, A. Kazusaka, A. Yamazaki, M. Enyo, in *Acid-Base Catalysis*, K. Tanabe, H. Hattori, T. Yamaguchi, T. Tanaka (Eds.), VCH, Weinheim, 1989, p. 249.
113. Y. Konishi, K. Sakata, M. Misono, Y. Yoneda. *J. Catal.* **1982**, *77*, 169.
114. Y. Izumi, M. Ono, M. Ogawa, K. Urabe, *Chem. Lett.* **1993**, 825.
115. M. Sopa, A. Waclaw-Held, M. Grossy, J. Pijanka, K. Nowinska, *Appl. Catal. A: General* **2005**, *285*, 119.
116. J. Winkler, S. Marme, *J. Chromatogr. A* **2000**, *888*, 51.
117. A. V. Ivanov, T. V. Vasina, V. D. Nissenbaum, L. M. Kustov, M. N. Timofeeva, J. I. Houzvicka, *Appl. Catal. A: General* **2004**, *259*, 65.
118. B. M. Devassy, G. V. Shanbhag, F. Lefebvre, S. B. Halligudi, *J. Mol. Catal. A: Chem.* **2004**, *210*, 125.
119. S. Soled, S. Miseo, G. McVicker, W. E. Gates, A. Gutierrez, J. Paes, *Catal. Today* **1997**, *36*, 441.
120. (a) S. Choi, Y. Wang, Z. Nie, J. Liu, C. H. F. Peden, *Catal. Today* **2000**, *55*, 117; (b) Y. Wang, K.-Y. Lee, S. Choi, C. H. F. Peden, *International Symposium on Acid-Base Catalysis IV*, Matsuyama, Japan 2001, p. 69; (c) C. H. F. Peden, Y. Wang, K.-Y. Lee, S. Choi, *Abstract of 221st ACS National Meeting*, San Diego, USA. 2001, PETR 43.
121. (a) Y. Wang, C. H. F. Peden, S. Choi, US Patent 6,472,344, 2002; (b) Y. Wang, C. H. F. Peden, S. Choi, US Patent 6,815,392, 2004.
122. H. Jin, Q. Wu, P. Zhang, W. Pang, *Solid State Sciences*, **2005**, *7*, 333.
123. H. Jin, Q. Wu, W. Pang, *Mater. Lett.* **2004**, *58*, 3657.
124. K. Nowinska, R. Fórmaniak, W. Kaleta, A. Waclaw, *Appl. Catal. A: General* **2003**, *256*, 115.
125. Q. Huo, D. I. Margolese, U. Ciesla, D. G. Demuth, P. Feng, T. E. Gier, P. Sieger, A. Firouzi, B. F. Chmelka, F. Schüth, G. D. Stucky, *Chem. Mater.* **1994**, *6*, 1176.
126. Q. Huo, D. I. Margolese, G. D. Stucky, *Chem. Mater.* **1996**, *8*, 1147.
127. U. Ciesla, F. Schüth, *Micropor. Mesopor. Mater.* **1999**, *27*, 131.
128. S. Damyanova, L. Dimitrov, R. Mariscal, J. L. G. Fierro, L. Petrov, I. Sobrados, *Appl. Catal. A: General* **2003**, *256*, 183.
129. S. Gotier, A. Tuel, *Zeolites* **1995**, *15*, 601.
130. K. U. Nandhini, B. Arabindoo, M. Palanichamy, V. Murugesan, *J. Mol. Catal. A: Chem.* **2004**, *223*, 201.
131. I. V. Kozhevnikov, K. K. Kloetstra, A. Sinnema, H. W. Zandbergen, H. van Bekkum, *J. Mol. Catal. A: Chem.* **1996**, *114*, 287.
132. Y. Izumi, K. Urabe, *Chem. Lett.* **1981**, 663.
133. M. A. Schwegler, P. Vinle, M. van der Eijk, H. van Bekkum, *Appl. Catal.* **1992**, *80*, 41.
134. M. N. Timofeeva, M. M. Matrosova, T. V. Reshetenko, L. B. Avdeeva, A. A. Budneva, A. B. Ayupov, E. A. Paukshtis, A. L. Chuvilin, A. V. Volodin, V. A. Likholobov, *J. Mol. Catal. A: Chemical* **2004**, *211*, 131.
135. L. B. Avdeeva, O. V. Goncharova, D. I. Kochubey, V. I. Zaikovskii, L. M. Plyasova, B. N. Novgorodov, S. K. Shaikhutdinov, *Appl. Catal. A: General* **1996**, *141*, 117.
136. M. N. Timofeeva, A. B. Ayupov, V. N. Mitkin, A. V. Volodin, E. B. Burgina, A. L. Chuvilin, G. V. Echevsky, *J. Mol. Catal. A: Chemical* **2004**, *217*, 155.
137. S. R. Mukai, T. Sugiyama, H. Tamon, *Appl. Catal. A: General* **2003**, *256*, 99.
138. (a) T. Nishimura, H. Watanabe, T. Okuhara, M. Misono, *Shokubai* **1991**, *33*, 420; (b) H. Watanabe, T. Nishimura, S. Nakata, T. Okuhara, M. Misono, *63rd National Meetings of Chem. Soc. Jpn.*, March 1992; (c) N. Mizuno, M. Misono, *Chem. Lett.* **1987**, 967.
139. M. Misono. N. Mizuno, H. Mori, K. Y. Lee, J. Jiao, T. Okuhara, in *Structure-Activity and Selectivity Relationships in Heterogeneous Catalysis*, R. K. Grasselli, A. W. Sleight (Eds.), Elsevier, Amsterdam, 1991, p. 87.
140. T. Baba, Y. Ono, *Appl. Catal.* **1986**, *22*, 321.
141. J. Pozniczek, I. Kulszewicz-Bajer, M. Zagorska, K. Kruczala, K. Dyrek, A. Bielanski, A. Pron, *J. Catal.* **1991**, *132*, 311.
142. I. K. Song, S. K. Shin, W. Y. Lee, *J. Catal.* **1993**, *144*, 348.

143. G. D. Yadav, N. S. Asthana, *Appl. Catal. A: General* **2003**, *244*, 341.
144. G. D. Yadav, N. Kirthivasan, *J. Chem. Soc., Chem. Commun.* **1995**, 203.
145. G. D. Yadav, N. Kirthivasan, *Appl. Catal. A: General* **1997**, *154*, 29.
146. G. D. Yadav, V. V. Bokade, *Appl. Catal. A: General* **1996**, *147*, 299.
147. G. D. Yadav, N. S. Doshi, *Catal. Today* **2000**, *60*, 263.
148. G. D. Yadav, N. S. Doshi, *Org. Proc. Res. Dev.* **2002**, *6*, 263.
149. G. D. Yadav, N. S. Doshi, *Appl. Catal. A: General* **2002**, *236*, 129.
150. B. Sulikowski, J. Haber, A. Kubacka, K. Pamin, Z. Olejniczak, J. Ptaszyñski, *Catal. Lett.* **1996**, *39*, 27.
151. K. Pamin, A. Kubacka, Z. Olejniczak, J. Haber, B. Sulikowski, *Appl. Catal. A: General* **2000**, *194–195*, 137.
152. S. R. Mukai, M. Shimoda, L. Lin, H. Tamon, T. Masuda, *Appl. Catal. A: General* **2003**, *256*, 107.
153. S. R. Mukai, L. Lin, T. Masuda, K. Hashimoto, *Chem. Eng. Sci.* **2001**, *56*, 799.
154. S. R. Mukai, T. Masuda, I. Ogino, K. Hashimoto, *Appl. Catal. A: General* **1997**, *165*, 219.
155. S. R. Mukai, I. Ogino, L. Lin, Y. Kondo, T. Masuda, K. Hashimoto, *React. Kinet. Catal. Lett.* **2000**, *69*, 253.
156. M. H. Tran, H. Ohkita, T. Mizushima, N. Kakuta, *Appl. Catal. A: General* **2005**, *287*, 129.
157. T. Kwon, G. A. Tsigdinos, T. J. Pinnavaia, *J. Am. Chem. Soc.* **1988**, *110*, 3653.
158. T. Kwon, T. J. Pinnavaia, *Chem Mater.* **1989**, *1*, 381.
159. R. M. Taylor, *Clay Miner.* **1984**, *19*, 591.
160. N. Mizuno, M. Misono, *J. Mol. Catal.* **1994**, *86*, 319.
161. K. Kamata, K. Yonehara, Y. Sumida, K. Yamaguchi, S. Hikichi, N. Mizuno, *Science* **2003**, *300*, 964.
162. K. Kamata, Y. Nakagawa, K. Yamaguchi, N. Mizuno, *J. Catal.* **2004**, *224*, 224.
163. Y. Nakagawa, K. Uehara, N. Mizuno, *Inorg. Chem.* **2005**, *44*, 14.
164. R. K. Ho, W. G. Klemperer, *J. Am. Chem. Soc.* **1978**, *100*, 6772.
165. (a) M. Faraj, C. L. Hill, *J. Chem. Soc., Chem. Commun.* **1987**, 1487; (b) C. L. Hill, *Activation and Functionalization of Alkanes*, Wiley, New York, 1989; (c) C. L. Hill, R. B. Brown, Jr., *J. Am. Chem. Soc.* **1986**, *108*, 536.
166. D. Mansuy, J.-F. Bartoli, P. Battioni, D. K. Lyon, R. G. Finke, *J. Am. Chem. Soc.* **1991**, *113*, 7222.
167. R. Neumann, C. A. Gnim, *J. Chem. Soc., Chem Commun.* **1989**, 1324.
168. L. A. C.- Walker, C. L. Hill, *Inorg. Chem.* **1991**, *30*, 4016.
169. D. K. Lyon, W. K. Miller, T. Novet, P. J. Domaille, E. Evitte, D. C. Johnson, R. G. Finke, *J. Am. Chem. Soc.* **1991**, *113*, 7209.
170. M. Fournier, R. C. R. Massart, *Acad. Sci. Paris, Ser. C* **1974**, *279*, 875.
171. Y. Nishiyama, T. Hayashi, Y. Nakagawa, N. Mizuno, *Angew. Chem. Int. Ed.* **2001**, *40*, 3639.
172. C. Nozaki, I. Kiyoto, Y. Minai, M. Misono, N. Mizuno, *Inorg. Chem.* **1999**, *38*, 5724.
173. D. K. Lyon, R. G. Finke, *Inorg. Chem.* **1990**, *29*, 1789.
174. N. Mizuno, D. K. Lyon, R. G. Finke, *J. Catal.* **1991**, *128*, 84.
175. M. Pohl, R. G. Finke, *Organometallics* **1993**, *12*, 1453.
176. R. G. Finke, D. K. Lyon, K. Nomiya, S. Sur, N. Mizuno, *Inorg. Chem.* **1990**, *29*, 1787.
177. (a) M. Green, T. A. Kuc, S. H. Taylor, *J. Chem. Soc. A* **1971**, 2334; (b) R. R. Schrock, J. A. Osborne, *J. Am. Chem. Soc.* **1971**, *93*, 3089.
178. (a) D. J. Edlund, R. J. Saxton, D. K. Lyon, R. G. Finke, *Organometallics* **1988**, *7*, 1692; (b) R. G. Finke, D. K. Lyon, K. Nomiya, T. R. J. Weakley, *Acta Crystallogr. C* **1990**, *46*, 1592.
179. S. Uchida, M. Hashimoto, N. Mizuno, *Angew. Chem. Int. Ed.* **2002**, *41*, 2814.
180. S. Uchida, N. Mizuno, *Chem. Eur. J.* **2003**, *9*, 5850.
181. R. Kawamoto, S. Uchida, N. Mizuno, *J. Am. Chem. Soc.* **2005**, *127*, 10560.
182. S. Uchida, N. Mizuno, *J. Am. Chem. Soc.* **2004**, *126*, 1602.
183. S. Uchida, R. Kawamoto, T. Akatsuka, S. Hikichi, N. Mizuno, *Chem. Mater.* **2005**, *17*, 1367.
184. (a) Y. Onoue, Y. Mizutani, S. Akiyama, Y. Izumi, *Chemtech.* **1978**, *8*, 432; (b) A. Aoshima, S. Yamamatsu, T. Yamaguchi, *Nippon Kagaku Kaishi* **1990**, 233; (c) A. Aoshima, S. Tonomura, S. Yamamatsu, *Polymers Adv. Tech.* **1990**, *2*, 127; (d) EU Patent, EP 263027 (Nippon Fine Chemical Co.); (e) G. Maksimov, I. V. Kozhevnikov, *React. Kinet. Catal. Lett.* **1989**, *39*, 317; (f) S. Sato, C. Sakurai, H. Furuta, T. Sodesawa, F. Nozaki, *J. Chem. Soc., Chem. Commun.* **1991**, 1327.
185. (a) Y. Izumi, R. Hasebe, K. Urabe, *J. Catal.* **1983**, *84*, 402; (b) T. Nishimura, T. Okuhara, M. Misono, *Chem. Lett.* **1991**, 1695; (c) Y. Izumi, N. Natsume, H. Takamine, I. Tamaoki, K. Urabe, *Bull. Chem. Soc. Jpn.* **1989**, *62*, 2159; (d) T. Okuhara, M. Yamashita, K. Na, M. Misono, *Chem. Lett.* **1994**, 1450; (e) Y. Ono, M. Taguchi, Gerile, S. Suzuki, T. Baba, *Stud. Surf. Sci. Catal.* **1985**, *20*, 167; (f) Y. Ono, T. Baba, J. Sakai, T. Keii, *J. Chem. Soc., Chem. Commun.* **1982**, 400; (g) K. Na, T. Okuhara, M. Misono, *Chem. Lett.* **1993**, 1141; (h) T. Okuhara, T. Hibi, K. Takahashi, S. Tatematsu, M. Misono, *J. Chem. Soc., Chem. Commun.* **1984**, 697; (i) S. Shikata, T. Okuhara, M. Misono, *Sekiyu Gakkaishi* **1994**, *37*, 632; (j) Japanese Patent 1991 300150; (k) Y. Izumi, M. Ogawa. W. Nohara, K. Urabe, *Chem. Lett.* **1992**, 1987.
186. (a) D. Attanasio, D. Orru, L. Suber, *J. Mol. Catal.* **1989**, *57*, L1; (b) M. Lissel, H. Jansen, R. Neumann, *Tetrahedron Lett.* **1992**, *33*, 1795; (c) C. Venturello, R. D'Aloisio, *J. Org. Chem.* **1988**, *53*, 1553; (d) Y. Ishii, K. Yamawaki, T. Ura, H. Yamada, T. Yoshida, M. Ogawa, *J. Org. Chem.* **1988**, *53*, 3587; (e) J. E. Lyons, P. E. Ellis, Jr., V. A. Durante, *Stud. Surf. Sci. Catal.* **1991**, *67*, 99.
187. (a) M. Ueshima, H. Tsuneki, A. Shimizu, *Hyomen* **1985**, *23*, 69; (b) Japanese Patent 145249, 1988 (Asahi Chem. Ind.); (c) Japanese Patent 106839, 1991 (Sumitomo Chem.); (d) G. B. McGarvey, J. B. Moffat, *J. Catal.* **1991**, *132*, 100; (e) G. Centi, J. L. Nieto, C. Iapalucci, K. Brückman, E. M. Serwicka, *Appl. Catal.* **1989**, *46*, 197.
188. (a) K. I. Matveev, I. V. Kozhevnikov, *Kinet. Katal.* **1980**, *21*, 1189; (b) Y. Izumi, Y. Satoh, H. Kondoh, K. Urabe, *J. Mol. Catal.* **1992**, *72*, 37; (c) Y. Izumi, Y. Satoh, K. Urabe, *Chem. Lett.* **1990**, 795; (d) K. Urabe, Y. Tanaka, Y. Izumi, *Chem. Lett.* **1985**, 1595; (e) R. W. Wegman, *J. Chem. Soc. Chem. Commun.* **1994**, 947.
189. K. Inumaru, A. Ono, H. Kubo, M. Misono, *J. Chem. Soc., Faraday Trans.* **1998**, *94*, 1765.
190. Y. Kamiya, Y. Ooka, C. Obara, R. Ohnishi, T. Fujita, Y. Kurata, K. Tsuji, T. Nakajyo, T. Okuhara, *J. Mol. Catal. A: Chem.* **2007**, *262*, 77.
191. T. Sugii, R. Ohnishi, J. Zhang, A. Miyaji, Y. Kamiya, T. Okuhara, *Catal. Today* **2006**, *116*, 179.

2.3.14
Transition Metal Carbides, Nitrides, and Phosphides

S. Ted Oyama*

2.3.14.1 Introduction

The combination of elemental carbon, nitrogen, and phosphorus with transition metals produces compounds known as carbides, nitrides, and phosphides, respectively, which have a rich catalytic chemistry. The carbides and nitrides have similar properties and crystal structures, while the phosphides are somewhat distinct. However, they are still compounds between electropositive metals and electronegative main group elements and so share many common characteristics. In this chapter these similarities and differences will be contrasted, with attention focused on general properties, structure, preparation, and catalytic activity. A lack of space does not permit the coverage of spectroscopic properties, surface science studies, and theoretical investigations.

2.3.14.2 General Properties

Many monographs and reviews have described the nature, structure and synthesis of carbides and nitrides [1–6], and phosphides [7, 8]. The bonding in the carbides, nitrides, and phosphides ranges from ionic for the alkali and alkaline earth metals, metallic or covalent for the transition elements, and covalent for the main group elements. The focus of this chapter concerns the metal-rich compounds, MX or M_2X (X = C, N, P), of the transition metals (M), which have metallic properties. Metal-deficient compositions are rare in the carbides and nitrides, except among the alkali and alkaline earth metals (e.g., Na_2C_2, CaC_2, Mg_3N_2), but they are common among the phosphides (e.g., MoP_2, NiP_3). However, phosphorus-rich compositions are semiconducting and are considerably less stable than the metal-rich compounds, and will not be considered here.

The metallic carbides and nitrides are well known for their strength and hardness, and are routinely used in cutting tools, drill bits, and rocket nozzles. In ferrous alloys the carbides are the components responsible for the toughness of steels. However, they also have interesting optical, electronic, and magnetic properties and have been applied in optical coatings, electrical contacts, and diffusion barriers. Because of its gold color, TiN is widely used for wear-resistant coatings, and decorative coatings on watches and jewelry. Thin-film resistors based on Ta_2N are currently being mass-produced by magnetron sputtering.

The metallic phosphides also have interesting physical properties. Titanium phosphide (TiP) is a hard-wearing metallic conductor that is highly resistant to oxidation, and has been suggested as a diffusion barrier for Al/W metallization. Zinc phosphide (Zn_3P_2) has been studied as a novel optoelectronic material. Phosphides are important in the study of magnetism because their interatomic spacing and anion electronegativity places them in an intermediate position between oxides and metals [9]. A complex composition, $MnFeP_{0.45}As_{0.55}$, has been reported to have a very high Curie temperature of 300 K, making it a candidate for magnetic refrigeration at room temperature [10]. Some metal-rich molybdenum phosphides are superconducting, with Mo_3P having the highest critical temperature (T_c) of 7 K among binary metal phosphides.

Basically, these materials combine the electronic properties of metals and the physical properties of ceramics. Thus, they are good conductors of heat and electricity, are hard and strong, and have high thermal and chemical stabilities (Tables 1 and 2) [4]. In general, the carbides and nitrides are more refractory than the phosphides, possessing higher heats of formation, melting points, and microhardness values. Interestingly, they also have a strong metallic nature, with lower electrical resistivity and higher heat capacity. This is related to their compact crystal structure. In all materials stability decreases in going to the right in the Periodic Table, as antibonding levels are filled.

2.3.14.3 Structure

Although the physical and chemical properties of carbides, nitrides and phosphides are similar, they differ substantially in their crystal structure. In most carbides and nitrides, the carbon and nitrogen atoms reside in the interstitial spaces between metal host atoms to form relatively simple, compact metal lattices: face-centered cubic (fcc), hexagonal closed packed (hcp) or simple hexagonal. For these compounds, geometric considerations predict stable structures when the ratio of non-metal to metal radii (r_X/r_M) is between 0.41 and 0.59 (Hägg's rule) [11]. The exact metallic arrangement (fcc, hcp, or simple hexagonal) is largely determined by the number of outer *sp* electrons [12]. For the phosphides, however, the atomic radius of phosphorus (0.109 nm) is substantially larger than those of carbon (0.071 nm) or nitrogen (0.065 nm), and the radius ratio is too large for octahedral coordination around the non-metal to be favorable. For this reason, in phosphides (also borides and silicides) the non-metal atom is usually found at the center of a triangular prism [7]. Different arrangements of these building blocks give rise to different structures.

---
* Corresponding author.

Tab. 1 The physical properties of carbides and nitrides[a]

Ceramic properties		Metallic properties	
Melting point/K	2000–3600	Electrical resistivity/$\mu\Omega$ cm	10–80
Microhardness/kg mm^{-2}	1100–3200	Hall coefficient/10^{11} m^3 °C^{-1}	−200 to 30
Young's modulus/GPa	200–700	Magnetic susceptibility/10^6 emu mol^{-1}	6–30
−Heat of formation/kJ mol^{-1}	−50 to 170	Heat capacity/J mol^{-1} K^{-1}	30–80

[a] Principally, Groups 4 to 8, first row.

Tab. 2 The physical properties of metal-rich phosphides[a]

Ceramic properties		Metallic properties	
Melting point/K	1100–1800	Electrical resistivity/$\mu\Omega$ cm	900–25 000
Microhardness/kg mm^{-2}	600–1100	Hall coefficient/10^{11} m^3 °C^{-1}	Unavailable
Young's modulus/GPa	Unavailable	Magnetic susceptibility/10^6 emu mol^{-1}	110–620
−Heat of formation/kJ mol^{-1}	30–180	Heat capacity/J mol^{-1} K^{-1}	20–50

[a] Principally, Groups 4 to 8, first row.

Carbides, nitrides, and phosphides can manifest considerable deviations from ideal stoichiometry, with both metal and non-metal vacancies. They also can form mixed-element compositions; the compounds readily incorporate small amounts of boron, silicon, sulfur, or oxygen. Of these elements, because of its size and electronegativity, only oxygen is found in a large proportion [13]. In fact, it is often difficult to exclude oxygen [3], and contamination by this element means that the interpretation of research results must be carried out with caution.

2.3.14.3.1 Binary Compounds of C and N The best-known metal carbides and nitrides are the binary compounds of the early transition metals (Groups 4 to 6), where the structural unit is the M$_6$X octahedron. In moving to the right in the Periodic Table to the Group 6–9 metals, the r_X/r_M ratio increases and the M$_6$X trigonal prism is favored (Fig. 1).

The monocarbides of Group 4 metals (Ti, Zr, Hf) adopt the cubic rocksalt structure with an fcc metal arrangement. The Group 5 metal carbides (V, Nb, Ta) also crystallize in this structure, although metal-rich compositions can form the hexagonal M$_2$X phases where the non-metal atoms (X) randomly occupy octahedral sites in the fcc lattice. The Group 6 metals (Cr, Mo, W) can form several different phases [14]. For example, for Mo several different monocarbides can form with different stacking, δ-MoC with the rocksalt structure and the well-known ABCABC stacking, γ-MoC isostructural with simple hexagonal WC with AAAA stacking, η-MoC with an ABCACB stacking, and γ'-MoC with AABB stacking. In

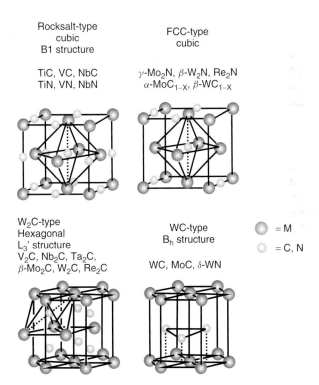

Fig. 1 Common structures in transition metal carbides and nitrides.

addition, several metal-rich Mo$_2$C compounds can form that differ in their carbon atom ordering that include γ-Mo$_2$C (also known as β-Mo$_2$C in the catalytic literature) with perfect hexagonal ABAB stacking.

References see page 355

The mononitrides of the Group 4 and 5 metals also form the ABCABC rocksalt structure, which is also common with the δ-nitrides of Ti, Zr, Hf, V, Nb, Ta and Mo, which are nitrogen-deficient with a stoichiometry close to M_2X. The nitrides of Group 5 differ from the carbides in being able to form M_5X square pyramids and the metal arrangement changes from fcc to tetragonal for VN, and to hexagonal for ε-TaN and η-NbN [6]. In Group 6 the fcc arrangement is increasingly disfavored, with CrN having very low stability and Mo nitride only existing with a stoichiometry of close to Mo_2N. Hexagonal ABAB stacking is found in β-V_2N, β-Nb_2N, β-Ta_2N, ε-Cr_2N, ζ-Mn_2N, and ε-Fe_2N [6].

Binary compounds of the mid and late transition metals (Groups 7–10) are formed almost exclusively only with the first row elements, and adopt complex structures. The compounds in this group include the carbides and nitrides of iron, cobalt, and nickel, with chromium also included, and are well known for imparting hardness and strength to steels.

The late second and third row metals do not form stable carbides and nitrides for reasons of electronic stability. The increased electron population resulting from compound formation can only be accommodated in the antibonding portion of their band structure, resulting in unstable materials.

2.3.14.3.2 **Binary Compounds of Phosphorus** The monophosphides, MP, are structurally closely related to the NaCl-type, NiAs-type, and WC-type structures, which are characterized by both types of atom coordinating six nearest neighbors (Fig. 2). In the NaCl-type structure, the six neighbors of each atom are arranged at the corners of an octahedron. In NiAs, the metal atoms also have an octahedral environment, while the non-metal atoms have triangular prismatic surroundings. In the WC-type structure the nearest neighbors of both atoms are situated at the corners of a triangular prism. These structures are thus conveniently described in terms of PM_6 octahedra and PM_6 triangular prisms. Because of the relatively large size of phosphorus, trigonal prismatic coordination of the P by the metal atoms is mostly present, and the bonding is probably partly metallic and partly covalent. Therefore, most of these compounds are hard, chemically inert, and resistant to oxidation at high temperatures.

The monophosphides TiP, ZrP, and HfP have the same γ-MoC-type structure. In this structure, the metal atoms are octahedrally coordinated by phosphorus; half of the P atoms are octahedrally coordinated, and the other half have trigonal prismatic coordination by metal atoms. TiP contains both PTi_6 octahedral and PTi_6 triangular prisms.

In the other phosphides of Group 5 and 6 metals, the structural environment is triangular prismatic. The isomorphous β-TaP and β-NbP have a NbAs-type structure, which can be described as a variation of a WC-type structure in which half of the non-metals are displaced by half a lattice spacing. The structure of VP is of the NiAs-type, with the metal and phosphorus coordination being octahedral and trigonal prismatic, respectively. This coordination scheme is similar to that adopted in the MnP-type phosphide. The structure of MoP

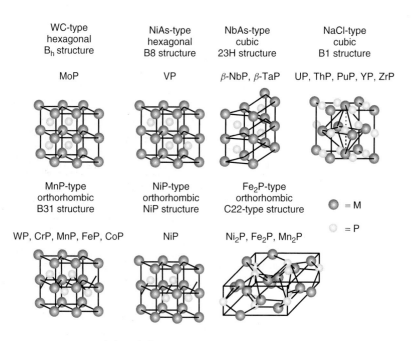

Fig. 2 Common structures in transition metal phosphides.

is of the ordered WC-type. Trigonal prismatic coordination of the P atom is found in VP (NiAs-type), MoP (WC-type), TiP, ZrP, HfP (γ-MoC type) structures, but no short P–P distances are found.

In the other elements of the Groups 6–10, the monophosphides such as Cr, Mn, Fe, Co, Ru, and W are categorized as an isomorphous orthorhombic series. They adopt the MnP and NiP structures, both of which can be regarded as distortions of the NiAs structure. The phosphorus atoms in the MnP structure form chains, while the phosphorus atoms in the NiP structure form pairs.

The hemiphosphides of formula M_2P adopt mostly tetrakaidecahedral (TKD) coordination. TKD coordination is nine-fold, and consists of a trigonal prismatic coordination with three additional metal atoms placed near the centers of the vertical faces of the prism. Examples are Co_2P and Ru_2P with the $PbCl_2$(C23)-type structure, Re_2P with a closely related structure, and Fe_2P, Mn_2P, and Ni_2P with the C22-type structure. Mo_3P, Cr_3P, Mn_3P, Ni_3P, and Fe_3P are an isomorphous series of tetragonal structures of the Fe_3P-type.

2.3.14.4 Preparation

Many methods [1–3, 15–17] have been developed for the preparation of carbides and nitrides, and these will be described in this section. Attention will be focused on methods that produce powders, particles and supported forms, which have high surface area (Table 3).

2.3.14.4.1 The Reactions of Metals or Metal Compounds with Gas-Phase Reagents
A direct method of preparation of carbides is the reaction of a metal or metal compound with a carbon-containing gas, generally at isothermal conditions. The compound can be a metal hydride, oxide, or halide, and is generally a solid. Flowing ammonia has been applied to the synthesis of the nitrides of iron [18], cobalt and nickel [19] – compounds with very low enthalpies of formation. Ammonia is preferred over molecular nitrogen to nitride metals or oxides because it is a stronger nitriding agent. The reduction of phosphates with hydrogen has been used to prepare phosphides [20].

References see page 355

Tab. 3 The preparation of high-surface-area materials

Reaction/method	Examples	References
1. Reactions with gas-phase reagents		
a. Carbides $M + 2CO \rightarrow MC + CO_2$	Cr_3C_2, Mo_2C, TcC, Fe_3C, Co_2C, Ni_3C	[18]
b. Nitrides $MO + NH_3 \rightarrow MN + H_2O$	Mo_2N, TiN, ZrN, HfN, VN, NbN, W_2N	[19]
c. Phosphides $MPO_x + PH_2 \rightarrow MP + H_2O$		[20]
2. Decomposition of metal halide vapors		
a. Carbides $MCl + H_xC_y \rightarrow MC + HCl + \cdots$	TiC, TaC, HfC	[4]
b. Nitrides $MCl + N_2/H_2 \rightarrow MN + HCl + \cdots$	TiN, VN, Re_2N, Fe_2N, Fe_4N, Cu_3N	[16, 21]
c. Phosphides $MCl + PH_3 \rightarrow MP + HCl + H_2$	MoP, WP, Fe_2P, Ni_2P, FeP, RuP	[22]
3. Decomposition of metal compounds		
a. Carbides $W(CO)_n + H_xC_y \rightarrow WC + H_2O + CO$	HfC, VC, WC	[23]
b. Nitrides $Ti(NR_2)_4 + NH_3 \rightarrow TiN + \cdots$	TiN, Zr_3N_4	[16, 21]
c. Phosphides $TiCl_4(PH_2C_6H_{11})_2 \rightarrow TiP + PH_3 + C_6H_{10}$	TiP, CoP	[24, 25]
4. Temperature-programmed methods		
a. Carbides $MoO_3 + CH_4 + H_2 \rightarrow Mo_2C + \cdots$	Mo_2C, MoC_{1-x}, WC, WC_{1-x} NbC	[26–29]
b. Nitrides $WO_3 + NH_3 \rightarrow W_2N + \cdots$	Mo_2N, MoN, VN, TiN_xO_y	[30, 31]
c. Phosphides $MoPO_4 + H_2 \rightarrow MoP$	MoP, WP	[32, 33]
5. Utilization of high-surface-area supports		
a. Carbides $Mo(CO)_6/Al_2O_3 \rightarrow Mo_2C/Al_2O_3$	Mo_2C/Al_2O_3	[34]
b. Nitrides $TiO_2/SiO_2 \rightarrow TiN/SiO_2$	TiN/SiO_2	[35]
c. Phosphides $NiPO_4/SiO_2 + H_2 \rightarrow Ni_2P/SiO_2$	WP/SiO_2, MoP/C, Ni_2P/USY $Ni_2P/MCM-41$	[36–39]
6. Reaction between metal oxide vapor and solid carbon		
a. Carbides V_2O_5 (gas) $+ 7C$ (solid) $\rightarrow 2VC + 5CO$	Mo_2C, VC, WC	[40]
7. Liquid-phase methods		
a. Carbides $MoCl_4(thf)_2 + LiBEt_3H \rightarrow Mo_2C$	Mo_2C, W_2C	[41]
b. Nitrides $[(Me_3Si)_2N]_3La + NH_3 \rightarrow LaN$	CoN, Ni_3N_2, TiN, LaN	[42]
c. Phosphides $Fe(CO)_5 + P(C_8H_{17})_3 \rightarrow FeP$	FeP, Fe_2P, MnP, Ni_2P, CoP, Zn_3P_2	[43, 44]
8. Electroless plating		
a. Phosphides $NiCl_2 + NaH_2PO_4 \rightarrow Ni-P/SiO_2$	$Ni-P/SiO_2$	[45, 46]

2.3.14.4.2 Decomposition of Metal Halide Vapors
Carbides are prepared from volatile metal chlorides or oxychlorides by reacting them with gaseous hydrocarbons in the vicinity of a localized heat source at 1400–2900 K (the Van Arkel process). Because the chloride starting materials can be obtained in pure form, the reaction is one of the most convenient and versatile for producing small amounts of pure carbides. The reaction is thermodynamically favorable [17]. The Van Arkel process and variations are used to make many metal nitrides such as BN, AlN, TiN, ZrN, HfN, CrN, Re_2N, Fe_2N, Fe_4N, and Cu_3N [16]. Although NH_3 can be used as a nitriding gas, it tends to decompose at the temperatures needed for nitride formation (1300–1900 K), and hence a mixture of H_2 and N_2 is often used. Phosphides have been prepared by the reaction of metal chlorides with phosphine [22].

2.3.14.4.3 Decomposition of Metal Compounds
Various compounds aside from the chlorides discussed in the previous section decompose to produce carbides and nitrides. For example, the reaction of molybdenum and tungsten carbonyls with hydrogen can be used to produce Mo_2C and W_2C [47]. A soluble oxide of Mo prepared by dissolution of MoO_3 in H_2O_2 has been complexed with sucrose and heated in a carburizing atmosphere to form β-Mo_2C [48]. The material retains excess carbon. The decomposition of tungsten bipyridine chloride in inert gas followed by activation in hydrogen produces a WC of up to 240 m^2 g^{-1} [23]. However, the sample also retains substantial carbon, which likely contributes to the observed area.

The synthesis reactions can be carried out in the gas phase if the compounds are volatile, or in the solid state, otherwise. Many nitrides can be obtained from the reaction of oxyhalides ($VOCl_3$, CrO_2Cl_2) or ammonium-oxo complexes (NH_4VO_3, NH_4ReO_4) [16]. A high-surface-area TiN of 100–200 m^2 g^{-1} was made by the nitridation of $TiCl_4$-tetrahydrofuran or $TiCl_4$-bipyridine or $TiCl_4$-ammonia complexes in a stream of NH_3 at 973–1273 K (700–1000 °C) [21].

2.3.14.4.4 Temperature-Programmed Methods
The temperature-programmed reaction (TPR) method of preparation consists of treating a precursor compound in a reactive gas stream while raising the temperature in a uniform manner. The precursor can be an oxide, sulfide, nitride, or other compound, while the reactive gas can be a mixture of hydrocarbon (e.g., methane) and hydrogen for carbides or ammonia for nitrides. The precursor is generally a phosphate for phosphides and the reactive gas is hydrogen, although recently the phosphidation of Ni metal with phosphine has been reported [49]. The use of the temperature program allows an optimal balance between synthesis and sintering rates, and results in products with high specific surface areas. The use of TPR has been described for the synthesis of Mo_2C, WC, NbC, VC, TaC, Mo_2N, W_2N, NbN, and Re_3N [26–31]. The nitrogen-deficient phase β-$Mo_2N_{0.78}$ is obtained by heating γ-Mo_2N in He [50].

The precursor can be quite important in the mechanism of synthesis and the properties of the products. In the case of Mo_2N [30], the solid-state transformation of the precursor oxide to the nitride has been shown to be pseudomorphous and topotactic. Pseudomorphism refers to the retention of the external size and shape of crystallites in a transformation, while topotacticity refers to the existence of a crystallographic relationship between the reactant and product phases. In the case of Mo_2N this occurs by a process of atomic substitution of N for O which results in minimal disruption of the atomic arrangement of the crystals. This gives rise to a porous product of high surface area (225 m^2 g^{-1}). The use of other precursors gives rise to high-surface-area products when the transformation is pseudomorphic, as occurs for MoO_3, $(NH_4)_6Mo_7O_{24}·4H_2O$, and H_xMoO_3, but not for MoO_2 and $(NH_4)_2MoO_4$ [51]. The nitridation of V_2O_5 to VN is pseudomorphic and topotactic, producing solids of surface area 90 m^2 g^{-1} of similar morphology as the starting material [31, 52]. The topotactic nature of the transformation may depend on reaction conditions [53]. It is also possible to obtain high-surface-area carbides by first nitriding an oxide and then carburizing the nitride, and this approach has been used to produce α-MoC_{1-x} from γ-Mo_2N by a topotactic transformation which retains the same fcc metal arrangement of the nitride [54].

Generally, the generation of a high surface area requires high space velocities of synthesis gases and low temperature ramp rates. This is to reduce the partial pressure of water vapor (which inhibits the synthesis) and to reduce the final temperature of reaction (which leads to sintering). A factorial analysis of synthesis parameters has been reported for Mo_2N [55]. For the carbides, the use of high temperatures and sometimes carburizing agents such as ethane [56], propane [57], or butane [58, 59] which are more prone to decompose, results in a greater deposition of surface carbon. This can be seen from ^{13}C NMR signals of surface carbon [58]. However, the use of higher hydrocarbons decreases the temperature of reaction. For Mo_2C, TPR synthesis with methane/H_2 requires close to 1000 K (727 °C), with ethane/H_2 around 900 K (627 °C), and with butane/H_2 around 823 K (550 °C) [58]. The surface carbon can be removed by treatment with hydrogen [26, 48], although the precise conditions need to be determined for each compound.

The first report of the use of TPR for phosphides was for MoP using a phosphate precursor [32]. Since then, the method has been applied to many compounds, including WP [33], Fe_2P, CoP, and Ni_2P [60]. Supported materials will be described in the next section.

2.3.14.4.5 Utilization of High-Surface-Area Supports
The use of supports offers the advantage of better usage of the active component and higher control of surface area and pore size distribution. Highly dispersed carbide phases may be prepared by depositing an oxide precursor on a high-surface-area support and carburizing it. This was demonstrated for molybdenum carbide supported on alumina [34]. The nitrogen-deficient phase β-$Mo_2N_{0.78}$ supported on alumina is obtained by the TPR of supported MoO_3 with N_2/H_2 mixtures [61]. Other supports used for Mo_2C include carbon, ZrO_2, SiO_2, and TiO_2 [62]. The ZrO_2 was found to interact strongly with the Mo_2C and to lead to its oxidation. Supported molybdenum nitride prepared from molybdenum sulfide on SiO_2, Al_2O_3, TiO_2 and ZrO_2 has been reported [63]. Carbon deposition in carbide formation remains a problem. Temperature-programmed experiments have shown that sequential nitridation and carburization produced a cleaner Mo_2C/Al_2O_3 than direct carburization [64, 65].

The preparation of highly dispersed phosphide phases on carriers such as SiO_2 [36, 60], γ-Al_2O_3 [66], carbon [37], zeolites [38], and mesoporous supports [39] has been reported. The preparations mostly employ temperature-programmed reduction of phosphate precursors, but simple reduction of supported metal carbonyl precursors supported on η-Al_2O_3 has also been carried out [67]. When the precursor is a phosphate, supports that interact weakly with the phosphate are desirable. For this reason SiO_2, carbon, siliceous zeolites such as dealuminated USY [38], and siliceous mesoporous supports such as MCM-41 [39] have been found to be effective. In contrast, Al_2O_3 forms aluminum phosphate, and the formation of the phosphide requires very high temperatures because of the difficulty of extracting the phosphorus from the support [66]. A typical synthesis of Ni_2P on SiO_2 requires 923 K (650 °C), whereas on Al_2O_3 it requires 1123 K (850 °C) [66]. It has also been found that excess phosphorus is necessary for formation of the stoichiometric phosphide on the support [39, 68]. In the case of Ni_2P/SiO_2, a Ni/P ratio of 1/2 was found to produce the most active catalyst [68], although the optimal value should depend on the preparation method and the surface area of the support.

2.3.14.4.6 Reaction Between Metal Oxide Vapor and Solid Carbon
A novel method for the preparation of ultra-high-surface-area carbides [40] involves the reaction of solid carbon with vaporized metal oxide precursors such as MoO_3 or WO_2. The synthesis uses high-specific-surface-area-activated carbons, and the final product appears to retain a "memory" of the porous structure of the starting material. The carbon acts like a skeleton around which the carbides are formed, and catalytically active samples with surface areas between 100 and 400 $m^2\,g^{-1}$ are generated.

2.3.14.4.7 Liquid-Phase Methods
The synthesis of Mo_2C and W_2C via chemical reduction methods [41] is carried out by mixing tetrahydrofuran (thf) suspensions of the metal chlorides, $MoCl_4(thf)_2$, $MoCl_3(thf)_3$ and WCl_4 at 263 K with lithium triethylborohydride, $LiBEt_3H$, to generate black colloidal powders (1–2 μm diameter) composed of smaller 2 nm-sized particles. These powders are subsequently heat-treated at 723–773 K to produce the hemicarbide phases.

Nitrogen-rich nitrides are produced by the thermal decomposition of amides prepared from precipitation from liquid ammonia [42]. An example is CoN from $Co(NH_2)_3$, although the product is not crystalline.

Nanorods of FeP prepared by the liquid-phase reaction of $Fe(CO)_5$ in COP (octylphosphine, $P(C_8H_{17})_3$) at 573 K were characterized by X-ray diffraction (XRD), X-ray photoelectron spectroscopy (XPS), scanning electron microscopy (SEM) and magnetic measurements [43]. Nanoparticles of Fe_2P were prepared by a solvothermal method in a Teflon autoclave at 423–473 K using $FeCl_2 \cdot 6H_2O$ and yellow phosphorus in an ethylenediamine solvent [44]. These nanorods and nanoparticles were found to have interesting magnetic properties.

2.3.14.4.8 Amorphous Materials
Amorphous metal–P alloys are prepared by the rapid quenching of melts, or by a process known as "electroless chemical deposition" involving the autocatalytic reduction of Ni^{2+} ions by hypophosphite ions $H_2PO_2^-$. Unsupported amorphous Ni–P alloys can be prepared by both methods [69], but supported alloys can only be prepared by the electroless chemical deposition method [45]. Amorphous Raney Ni–P alloys are obtained from amorphous Ni–Al–P alloys by leaching the Al component with alkali solutions [70, 71]. The unsupported amorphous alloys have low surface areas of 0.1 to 1 $m^2\,g^{-1}$, the Raney alloys have intermediate surface areas of 20 to 90 $m^2\,g^{-1}$, and the supported materials have surface areas of 100 to 200 $m^2\,g^{-1}$. A combination of sol–gel processing and electroless deposition produces a Ni–P/SiO_2 aerogel of 254 $m^2\,g^{-1}$ [72]. In general, the materials have higher activity than metallic Ni in hydrogenation reactions [73], but suffer from instability at high temperatures, generally crystallizing a low P-content phosphide.

2.3.14.5 Catalytic Properties
The catalytic properties of carbides and nitrides have been detailed in several older reviews [74, 75], and more recently in a review on hydroprocessing [76]. The catalytic properties of phosphides have also been covered

---
References see page 355

briefly [77], and this section will concentrate on recent developments in this field.

The carbides and nitrides show excellent reactivity for a variety of reactions that include ammonia synthesis and decomposition, amination, hydrogenolysis, hydrogenation, dehydrogenation, hydrazine decomposition, isomerization, methanation, aromatization, hydroprocessing, reforming, and water-gas shift. Much attention has been placed on the carbides and nitrides because of early studies which reported that WC had platinum-like properties [78]. Although the similarity was based on selectivity rather than activity, the concept has persisted and remains an important guiding principle [79–81].

Although the phosphides have been studied less extensively, they also show excellent activity in reactions involving hydrogen transfer. In particular, these compounds show good tolerance to sulfur and nitrogen heteroatoms, and activities that often are similar to that of Pt-group metals; hence, these materials are beginning to attract considerable attention.

2.3.14.5.1 Acid and Base Catalysis

Acid and base properties are fundamental to many solids, and have been studied extensively. The surfaces of carbides and nitrides have been found to have both acid and base characteristics, although much depends on the oxygen content of the surface. Deliberately adding oxygen increases the acid character, as will be discussed below and in the section on hydroisomerization.

The acid–base properties of Mo_2C [82] and Mo_2N [83] prepared by TPR, followed by passivation and re-reduction at 673 K (400 °C) for Mo_2C and 753 K (480 °C) for Mo_2N, have been reported. The surfaces were characterized by TPD of ammonia and CO_2 as respective probes of acid and base sites, and comparisons were made to HZSM-5, USY, MgO, and 1% Pt/SiO_2 (Table 4).

Comparison of the TPD results indicated that the acid sites on Mo_2C were weaker than in HZSM-5, and that the base sites were weaker than in MgO. Mo_2N had more basic sites than Mo_2C, some of them weak, and also had some intermediate-strength acid sites. These properties gave rise to different catalytic behavior. The acid–base properties were attributed to polarization of the surface due to electron donation from metal to non-metal.

The dehydrogenation of isopropyl alcohol to acetone at 433 K and 10^5 Pa (1 bar) was used to probe the acid–base properties of Mo_2C [82]. Mo_2C was found to have a higher activity than the MgO, with conversion of about 40% and selectivity to acetone of about 80% and propene of 20%. Experiments in which NH_3 and CO_2 were co-fed reduced the activity of acetone formation, indicating that both acid and base sites were utilized in the transformation.

As yet, the acid–base properties of transition metal phosphides have not been studied extensively. A number of phosphides supported on η-Al_2O_3 have been tested for the dimerization of isobutene [67], and although rates normalized to surface area or active site number are unavailable, the conversions at 373–423 K (100–150 °C) followed the order Fe > Mo > Pt > Pd > Ru > Rh > W > Co. The products consisted of an 80:20 mixture of 2,4,4-trimethyl-1-pentene and 2,4,4-trimethyl-2-pentene, which were indicated to be very close to those of the acid-catalyzed dimerization of isobutene.

Tab. 4 Acid–base characterization of catalysts

Catalyst	Surface area/$m^2\ g^{-1}$	Desorption of NH_3		
		Below 523 K/ molec cm^{-2}	Between 523–823 K/ molec cm^{-2}	Above 823 K/ molec cm^{-2}
Mo_2N	135	5.6×10^{13}	6.2×10^{13}	–
Mo_2C	50	2.3×10^{14}	–	–
USY	625	2.0×10^{14}	3.2×10^{13}	6.7×10^{13}
HZSM-5	625	4.4×10^{14}	1.9×10^{14}	–

Catalyst	Surface area/$m^2\ g^{-1}$	Desorption of CO_2		
		Below 423 K/ molec cm^{-2}	Between 423–673 K/ molec cm^{-2}	Above 673 K/ molec cm^{-2}
Mo_2N		1.9×10^{13}	5.4×10^{13}	–
Mo_2C	50	–	4.4×10^{13}	–
MgO	212	–	2.1×10^{14}	1.3×10^{14}
1% Pt/SiO_2	244	–	2.4×10^{12}	–

2.3.14.5.2 Ammonia Synthesis and Decomposition
The synthesis of ammonia using a doubly-promoted iron catalyst became a commercial process in 1913, and is known as the Haber–Bosch process. Even today, this technique still ranks as one of the most important industrial processes for the manufacture of a bulk commodity.

The activity of molybdenum carbides and nitrides was found to be higher than Ru, but slightly lower than the doubly-promoted iron catalyst based on active sites titrated by CO chemisorption [75]. A reinvestigation of the system [84] at 673 K and 0.1 MPa has confirmed that the order of activity is β-Mo_2C > α-MoC_{1-x} > γ-Mo_2N, and has further found that the addition of Cs as a promoter reduces the surface area but increases the areal rate (Table 5).

Recently, bimetallic molybdenum-containing nitride catalysts, such as Co_3Mo_3N, have been reported [85, 86] which are more active than the doubly-promoted iron catalyst. A comparison at 673 K (400 °C) and 0.1 MPa is presented in Table 6 [86].

The effect of pressure was studied up to 3.1 MPa, and it was found that the Co_3Mo_3N catalyst promoted with Cs (2 mol.%) had much better performance than the remainder of the catalysts, including the commercial doubly-promoted iron catalyst, because it was not inhibited as much by NH_3.

Ammonia decomposition has been studied over a number of carbides and nitrides. Ammonia decomposition rates at 843 K and 10^5 Pa were 0.5–1 μmol m^{-2} s^{-1} for TaC, compared to 2 μmol m^{-2} s^{-1} for V_8C_7, 3 μmol m^{-2} s^{-1} for Mo_2C, 4 μmol m^{-2} s^{-1} for Mo_2N [87], and 0.2 μmol m^{-2} s^{-1} for VN [88]. Kinetic rate constants were reported to be between those for Pt and Fe, which was taken as an indication that VN had noble metal-like behavior [88].

2.3.14.5.3 Aromatization
The dehydrocondensation reaction of methane to form aromatics such as benzene and naphthalene represents a means of converting natural gas to liquids, and has received considerable attention [89] following its initial report [90]. The most-studied catalyst is ZSM-5 loaded with molybdenum, and considerable efforts have been spent to characterize the active phase. The studies indicate that the Mo enters the zeolite channels in highly dispersed form [91]. Extended X-ray absorption fine structure (EXAFS) studies indicate the formation of $(Mo_2O_5)^{2+}$ ditetrahedral species interacting with two cation-exchange sites [92]. Early studies suggested the formation of active Mo_2C species [93, 94] by reduction of oxidic species during the observed induction period, which was supported by subsequent EXAFS studies [95]. Brønsted acid sites also appear to be important in the reaction [96].

2.3.14.5.4 Hydrogenation and Dehydrogenation
Carbides, nitrides, and phosphides are highly active in hydrogen-transfer reactions, and have been explored extensively for hydrogenation and dehydrogenation reactions. In the hydrogenation of benzene to cyclohexane, molybdenum carbide had higher activity than Mo metal, and its steady-state turnover rate was comparable to that of Pt or Ru [97]. The carbide deactivated over time. Similar results were obtained with various other carbide catalysts [98]. Activities were comparable to 5% Ru/Al_2O_3, but the catalysts deactivated. Mo_2C showed the highest initial activity and deactivation rate, while α-WC and mixed Mo–Ta and Mo–Nb catalysts had lower activity but improved stability.

The dehydrogenation of n-butane was studied over carbide and nitride catalysts at 723 K (450 °C) and 101 kPa for a reactant mixture containing 4% n-C_4H_{10}, 60% H_2, and 36% He [53] (Table 7).

The order of activity was ranked as follows: γ-Mo_2N > W_2C ~ WC > W_2N > WC_{1-x} > Mo_2C > VN ~ VC ≫ NbN ~ NbC. Butane conversion activities for the nitrides

Tab. 5 Comparison of ammonia synthesis rates on molybdenum carbides and nitrides

Sample	Surface area/m^2 g^{-1}	Rate/μmol m^{-2} h^{-1}
γ-Mo_2N	146	0.3
α-MoC_{1-x}	134	4.2
β-Mo_2C	32	19.2
γ-Mo_2N-Cs2	64	2.5
α-MoC_{1-x}-Cs2	87	9.6
β-Mo_2C-Cs2	a	a

Cs2 = 2 mol.%.
a Forms metal.

Tab. 6 Comparison of ammonia synthesis rates on bimetallic nitrides

Sample	Surface area/m^2 g^{-1}	Rate/μmol m^{-2} h^{-1}
Fe–K_2O–Al_2O_3	14	24
Co_3Mo_3N	21	31
Co_3Mo_3N–Cs2	16	62
Co_3Mo_3N–Cs10	10	59
Co_3Mo_3N–K30	8	46
Ni–Mo–N	20	14
Fe–Mo–N	7	20
Mo_2N	121	1

Cs2 = 2 mol.%; Cs10 = 10 mol.%; K30 = 30 mol.%.

Tab. 7 Reaction performance of carbides and nitrides in the dehydrogenation of butane

Catalyst	Surface area/m^2 g^{-1}	Selectivity/%			Rate/ nmol m^{-2} s^{-1}
		Dehydrogenation	Isomerization	Hydrogenolysis	
VN	14	100	<0.1	0	6.4
VC	11	100	<0.1	<0.1	5.8
NbN	33	a	a	a	0.01
NbC	30	a	a	a	0.01
γ-Mo$_2$N	21	54	2	44	171
β-Mo$_2$C	63	65	0	35	32
β-W$_2$N	81	40	41	19	72
WC	12	32	2	66	95
W$_2$C	18	25	2	73	112
WC$_{1-x}$	48	39	4	57	57
Pt–Sn/Al$_2$O$_3$	228	99	0.3	0.7	–

^{a}Reaction rate too low to measure selectivity.

and carbides ranged from 0.4×10^{12} to 10×10^{12} molecules cm^{-2} s^{-1}, with corresponding turnover frequencies (TOFs) based on oxygen chemisorption of up to 10^{-2} s^{-1}. Activities for Pt powder at the same temperature have been reported as 1.0×10^{12} and 3.2×10^{12} molecules cm^{-2} s^{-1} for butane hydrogenolysis and isomerization, respectively [99]. The TOF for the Pt–Sn/γ-Al$_2$O$_3$ was 0.063 s^{-1}. Thus, the best carbide and nitride catalysts have activity comparable to that of Pt.

The hydrogenation of propene on phosphorus-containing tungsten oxynitrides has also been studied [100] (Table 8); notably, the addition of phosphorus enhanced the TOF.

The hydrogenation of ethene, propene, and acetylene on ruthenium phosphide supported on η-Al$_2$O$_3$ has also been examined [67]. The active catalyst surface area is unknown, but the reactions are reported not to proceed at 393–398 K, but then to light-off and produce hydrogenation products. The same study reported that the hydrogenation of nitrobenzene to aniline also proceeds over the supported phosphides of Ru and Rh, but with substantially lower activity compared to a supported Pd reference.

Tab. 8 Surface area and propene hydrogenation rate of P-containing tungsten oxynitrides

Sample	Surface area/m^2 g^{-1}	TOF at 353 K/s^{-1}	TOF at 423 K/s^{-1}
WN$_{1.08}$O$_{0.48}$	74	–	0.018
WN$_{1.28}$O$_{1.67}$P$_{0.09}$	87	0.012	–
WN$_{1.10}$O$_{0.48}$P$_{0.11}$	64	0.25	–
WN$_{1.00}$O$_{0.44}$P$_{0.16}$	33	0.15	–
WN$_{1.09}$O$_{0.8}$P$_{0.28}$	44	0.13	–

TOF, turnover frequency.

The hydrogenation of butadiene has been studied on a catalyst prepared by reducing nickel phosphate on alumina at 873 K and claimed to be Ni$_2$P/Al$_2$O$_3$ [101]. At 333 K and approximately 4×10^4 Pa the phosphide favored 1,4-addition to form 2-butene, while a sample reduced at 673 K (400 °C) and believed to be Ni/Al$_2$O$_3$ favored 1,2-addition to form 1-butene. The possibility of secondary isomerization was not considered. The phosphide had lower specific rate than the metal, but the active surface areas were not reported.

The liquid-phase hydrogenation of benzene to cyclohexane was studied at 373 K (100 °C) and 1 MPa with four types of Ni–P amorphous catalyst [70]. The TOF based on H$_2$ chemisorption indicated that the amorphous Raney catalysts had comparable activity to Raney Ni, but supported samples had higher activities (Table 9).

The gas-phase hydrogenation of cyclopentadiene to cyclopentene and cyclopentane was investigated at 323 K and 10^5 Pa with amorphous Ni–P/SiO$_2$ catalysts [72]. The TOF for Ni–P/SiO$_2$ on an aerogel support was 0.46 s^{-1},

Tab. 9 Catalytic activity of amorphous alloys in the hydrogenation of benzene

Sample	Composition	S_{Ni}/m^2 g^{-1}	TOF/s^{-1}
Raney Ni–P	Ni$_{91}$P$_9$	38	2.42×10^{-2}
Raney Ni–P (cryst)^a	Ni$_{93}$P$_7$	15	1.25×10^{-2}
Raney Ni	Ni	43	1.80×10^{-2}
Ni–P/SiO$_2$^{b}	Ni$_{86}$P$_{14}$	21	3.47×10^{-2}
Ni–W–P/SiO$_2$^{c}	Ni$_{81.5}$W$_{1.2}$P$_{17.3}$	19	4.08×10^{-2}
Ultrafine Ni–P	Ni$_{86}$P$_{14}$	5.2	7.60×10^{-3}

^{a}Obtained by heating Raney Ni–P at 673 K for 2 h in N$_2$ flow.
^{b}Ni loading = 11.5 wt.%.
^{c}Ni loading = 11.2 wt.%; W/Ni molar ratio = 1.5%.
TOF, turnover frequency.

for Ni–P/SiO$_2$ on a standard silica support was 0.42 s^{-1}, and for a Ni/SiO$_2$ standard was 0.40 s^{-1}. Although the activities were comparable, the selectivity to cyclopentene was slightly higher (>90%) on the amorphous Ni–P catalysts than on the metallic Ni catalyst (>80%).

2.3.14.5.5 Hydrotreating A global tightening in the allowed sulfur content in fuels, together with increased restrictions on the release of nitrogen oxides, have resulted in considerable efforts to develop better catalysts for hydrodesulfurization (HDS) and hydrodenitrogenation (HDN). The chief challenge in ultra-deep HDS is the removal of sulfur from unreactive sulfur compounds, of which the family of dibenzothiophene molecules alkylated next to the sulfur atom is characteristic. A representative member of this family is 4,6-dimethyldibenzothiophene (4,6-DMDBT), which has been used extensively as a model compound.

The first application of carbides to HDS showed that α-MoC$_{1-x}$ had activity similar to Mo–S/Al$_2$O$_3$ in the reaction of thiophene [102], and a subsequent study with various carbides and nitrides found that β-Mo$_2$C and γ-Mo$_2$N had good activity in the HDN of quinoline [103]. This was followed by a comprehensive evaluation of the performance of different carbide and nitride catalysts in the simultaneous HDS of dibenzothiophene and HDN of quinoline at 643 K and 3.1 MPa and a liquid-hourly space velocity (LHSV) of 5 h^{-1} [104]. The order of activity in HDS was Ni–Mo–S/Al$_2$O$_3$ > Mo$_2$C > WC > Mo$_2$N > NbC > VC > VN > TiN (Fig. 3), while the order in HDN was Mo$_2$C > Ni–Mo–S/Al$_2$O$_3$ > WC > Mo$_2$N > NbC $\sim$ VN > VC > TiN (Fig. 4). The comparisons were based on equal sites (70 μmol) loaded in the reactor, as counted by CO chemisorption at room temperature for the carbides and nitrides and by low-temperature, pulse O$_2$ chemisorption for the sulfide, so that the comparisons provided information on the intrinsic activity of the catalysts.

The reactivity follows the order Group 6 > Group 5 > Group 4, and it is notable that Mo$_2$C has higher activity than a commercial Ni–Mo–S/Al$_2$O$_3$ catalyst (Shell 324). The same study also found that VN has excellent hydrodeoxygenation activity for benzofuran. Other studies have confirmed the excellent activity in HDS of Mo$_2$C [105]. A study with an actual feed showed that it had superior activity in HDS to a commercial Ni–Mo–S/Al$_2$O$_3$ catalyst, and had good stability for over 300 h [106]. The active surface of the carbide contained sulfur and was suggested to be a carbosulfide phase [105]. The catalytic activity of supported molybdenum nitride and carbide were investigated in the HDS of thiophene at 693 K and 10^5 Pa [105]. The activity (μmol thiophene mol Mo^{-1} s^{-1}) followed the order:

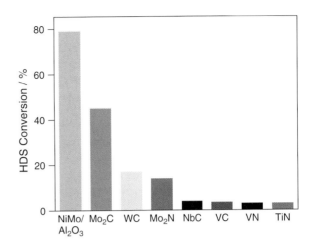

Fig. 3 Catalytic activity in the hydrodesulfurization (HDS) of dibenzothiophene.

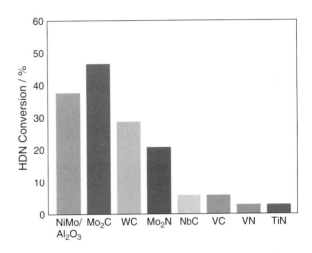

Fig. 4 Catalytic activity in the hydrodenitrogenation (HDN) of quinoline.

Mo$_2$C/Al$_2$O$_3$ > Mo$_2$N/Al$_2$O$_3$ > MoS$_2$/Al$_2$O$_3$. A kinetic study of Mo$_2$C/Al$_2$O$_3$ in the HDS of 4,6-DMDBT at 613 K and 4 MPa reports an unusual result [107]. In the sulfur-removal reactions for 4,6-DMDBT as for DBT [108], two pathways have been identified: a hydrogenation (HYD) pathway, and a direct desulfurization (DDS) pathway. Although usually unhindered molecules such as thiophene or dibenzothiophene undergo HDS by the DDS route, the hindered 4,6-DMDBT molecule proceeds by the HYD route because prior hydrogenation removes the planarity of the molecule and makes the sulfur atom more accessible. In the case of Mo$_2$C/Al$_2$O$_3$, the HDS of 4,6-DMDBT was observed to proceed by the DDS route at conversions lower than 45% with a HYD/DDS selectivity of 0.7 [107].

References see page 355

Although less active than Mo_2C, numerous studies have also examined the activity of Mo_2N, and much attention has been placed on supported Mo_2N/Al_2O_3 prepared by temperature-programmed reduction. The nitrogen-deficient phase β-$Mo_2N_{0.78}/Al_2O_3$ has been studied for the desulfurization of dibenzothiophene between 523 and 623 K and 7 to 14 MPa [61]. The catalyst shows higher activity than molybdenum sulfide and good stability, although some sulfur incorporation was reported (this was believed to occur in the passivation layer).

The first reports of the application of phosphides in HDN [109] and simultaneous HDS and HDN have appeared only recently [32]. In HDN, Ni_2P was found to have higher activity for quinoline denitrogenation than a commercial phosphorus-containing sulfided Ni–Mo/Al_2O_3 catalyst. However, deviations in the kinetics of the reaction network suggested that the two systems worked in fundamentally different manners; that is, the activity of the Ni–Mo sulfide was not due to the formation of Ni_2P on its surface.

Recent studies have shown that MoP [32], WP [33], Ni_2P, and Co_2P [60, 109, 110] are active for HDS and HDN of petroleum feedstocks. A summary of reactivity in the HDS of dibenzothiophene and the HDN of quinoline (Fig. 5) is provided under the same conditions as for the carbides and nitrides (see Figs. 3 and 4) based on 250 µmol of sites counted by CO uptake loaded in the reactor.

Among the phosphides, Ni_2P has been found to be the most active [32, 38] and several reports have been published describing its HDS properties for treating thiophene [66, 111], dibenzothiophene (DBT) [38, 39, 68], and 4,6-DMDBT [38]. For example, a Ni_2P catalyst supported on a low-surface-area SiO_2 provided a DBT conversion of 100%, which was much greater than that of a commercial Ni–Mo–S/γ-Al_2O_3, with an HDS conversion of 76% [68]. Subsequently, it was found that Ni_2P operated principally by the hydrogenation route with higher first-order rate constants per active site than Co–Mo–S/γ-Al_2O_3 and Ni–Mo–S/γ-Al_2O_3 [38] (Table 10).

Catalytic activity depends on the phosphorus content of the Ni_2P catalysts [66, 68]; when the P content is low, and P-deficient phases such as $Ni_{12}P_5$ are formed, then the activity is also low [39, 66]. High activity is found with excess phosphorus, but too much can block sites and reduce activity, so a careful balance is important. Studies of Ni_2P catalysts after reaction [60, 68] indicates that the bulk phase is unaltered, but that EXAFS shows the presence of a Ni–S distance, indicating the formation of a surface phosphosulfide. Treatment of Ni_2P/SiO_2 with H_2S/H_2 increases the activity by 10% for thiophene HDS compared to treatment in H_2, indicating that sulfur activates the surface [111]. The sulfur content of the catalysts is small, and insufficient to cover the entire surface. On Ni_2P/SiO_2 and Ni_2P/Al_2O_3 the compositions after reaction were $Ni_{2.0}P_{1.0}S_{0.017}$ and $Ni_{2.0}P_{1.0}S_{0.050}$ [66].

A study of silica-supported phosphides was conducted for the HDN of o-methylaniline at 643 K and 3 MPa in the presence and absence of H_2S [112]. The most active catalysts were MoP/SiO_2 and Co_2P/SiO_2, while the ternary compounds were less active (Table 11). The activity of all catalysts was reduced by the presence of H_2S, most likely due to competitive adsorption.

2.3.14.5.6 Isomerization and Hydroisomerization (Reforming)

In this section, isomerization will refer

Tab. 10 First-order rate constants for the reactions of 4,6-DMDBT

	Co-Mo-S/ γ-Al_2O_3	Ni-Mo-S/ γ-Al_2O_3	Ni_2P/ USY	Ni_2P/ SiO_2
$k(HYD)/s^{-1}$	2.2	6.7	12.7	21.6
$k(DDS)/s^{-1}$	1.8	2.1	2.5	2.1

Tab. 11 Activity of supported phosphides in the hydrodenitrogenation of *ortho*-methylaniline

Sample	Conversion/%	TOF/s^{-1}
Co_2P/SiO_2	45	11.2×10^{-3}
MoP/SiO_2	26	12.7×10^{-3}
WP/SiO_2	9	Not determined
NiMoP/SiO_2	25	2.6×10^{-3}
CoMoP/SiO_2	25	6.4×10^{-3}
Ni-Mo-S/Al_2O_3	36	2.6×10^{-3}

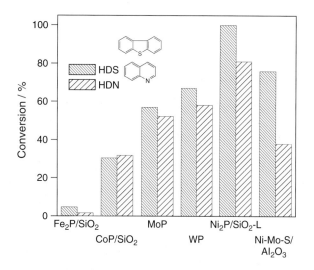

Fig. 5 Catalytic activity in the hydrodesulfurization (HDS) of dibenzothiophene and the hydrodenitrogenation (HDN) of quinoline.

TOF, turnover frequency.

to hydrocarbon skeletal changes that are simply catalyzed by acid or base sites, whereas hydroisomerization or reforming will refer to the same transformations catalyzed by bifunctional catalysts that involve hydrogenation/dehydrogenation sites. Another difference is that the former process is usually carried out simply in the presence of the hydrocarbon, whereas the latter process usually requires hydrogen in the reactant stream. Carbides and nitrides tend to deactivate with hydrocarbon feedstocks, and usually the catalysts are operated in the reforming mode.

The reforming reaction of neopentane was studied over WC and β-W$_2$C prepared by the reduction of WO$_3$ in CH$_4$–H$_2$ mixtures [113]. The catalysts chemisorbed between 0.2 and 0.4 monolayers of CO, and so were likely free of polymeric carbon. The catalysts had very high activity. Over the fresh catalysts, hydrogenolysis predominated, but over the oxygen-covered catalysts hydrogenolysis was inhibited and isopentane was formed – a product observed only on Pt, Ir, and Au catalysts. Thus, chemisorbed oxygen induced noble-metal like behavior on tungsten carbide.

The chemisorbed oxygen served to reduce the binding energy of hydrocarbon intermediates on the surface, and also to decrease hydrogenolysis and deactivation rates [114]. The product distributions of n-heptane and 3,3-dimethylpentane, and the isotope distribution of n-heptane-1-C^{13}, indicated that isomerization involves methyl-shift pathways that require olefin intermediates. The results also suggested that the reactions involved bifunctional (dehydrogenation, carbenium ion) pathways [115].

A study of the isomerization of C6–C8 alkanes at pressures ranging from 10^5 to 6×10^5 Pa has shown that oxygen-modified molybdenum carbide has comparable activity to supported Pt catalysts, but a much higher selectivity [40]. The main products are monomethyl isomers and dibranched molecules, with little production of cyclic or cracked products. No deactivation was observed for tens of hours and, importantly, a high isomerization selectivity could be obtained at high conversions. The product distribution suggested a bond-shift mechanism via a metallacyclobutane intermediate. A recent study has shown the existence of carbenes that give rise to metathesis products by the formation of metallacyclobutanes on molybdenum carbide surfaces [116].

The hydroisomerization of 1-butene at atmospheric pressure has been studied over metal phosphides supported on η-Al$_2$O$_3$ [67]. Active surface areas were not reported, and so rates cannot be calculated, although conversion at 418–448 K followed the order Ru > Mo > Pd > Rh > Ni > Pt, and comparison of product selectivities with the pure metals indicated that phosphidation reduced activity and changed the course of reaction from hydrogenation to isomerization. Only with the phosphides of Pd – and, to a lesser extent Ru and Pt – was some hydrogenation to butane observed.

2.3.14.5.7 Methanation and Fischer–Tropsch The hydrogenation of carbon monoxide has been studied over carbide catalysts [57, 117]. Very similar TOFs were obtained for β-Mo$_2$C (2.7×10^{-2} s^{-1}), α-MoC$_{1-x}$ (4.6×10^{-2} s^{-1}), and γ-Mo$_2$N (5.0×10^{-2} s^{-1}) at 53 kPa and 570 K [117], which were close to those of Ru/Al$_2$O$_3$ (3.0×10^{-2} s^{-1}) [118]. However, the kinetics for β-Mo$_2$C were positive order in both CO and H$_2$, whereas for Ru they were negative for CO and positive for H$_2$ [117]. This was most likely due to the existence of different adsorption sites in the materials. On β-Mo$_2$C, where the intermetallic distance is wider than in Ru, adsorption of CO probably occurs on on-top sites, whilst the adsorption of hydrogen occurs on hollow sites. On Ru, there is competitive adsorption only on on-top sites, and the stronger adsorption of CO leads to inhibition.

In the Fischer–Tropsch synthesis, tungsten and molybdenum carbides produce mainly light alkanes, whereas the formation of alcohols is related to the surface stoichiometry and to the extent of carburization [119, 120]. Promotion of Mo$_2$C with K$_2$CO$_3$ has been found greatly to enhance the selectivity to alcohols composed of linear C1–C7 [120].

The Fischer–Tropsch reaction has been studied over low-surface-area Mo$_2$C, Ru/Mo$_2$C and Co/Mo$_2$C at 473 to 513 K (200–240 °C) and 2×10^6 and 5×10^6 Pa [121]. It was found that Mo$_2$C produces principally light hydrocarbons and a small amount of alcohols, while the addition of Ru and Co increases the rate but suppresses alcohol formation (Table 12). The hydrocarbon distribution was consistent with the Anderson–Schulz–Flory distribution, except for the case of Co/Mo$_2$C.

2.3.14.5.8 Oxidation The catalytic activity of carbides in the oxidation of carbon monoxide, ammonia, and oxidative coupling of methane has been reviewed [122]. In general, the activity of the carbides is less than that of the metals, but greater than that of the oxides. Carbides, nitrides, and phosphides are each susceptible to oxidation, and are not suitable catalysts for catalytic oxidation. However, reports exist of their successful use in oxidation at low temperatures or with liquid hydrocarbons [123].

The partial oxidation of methane at 1123 K (850 °C) was attempted on Mo$_2$C/Al$_2$O$_3$ with Ni, Cu, or K additives, but the catalysts were found gradually to be deactivated due to oxidation of the Mo [124].

References see page 355

Tab. 12 Rate and selectivity in the Fischer–Tropsch reaction

Sample	Rate/mmol h^{-1} g^{-1}	TOF^a/s^{-1}	Selectivity/%						α
			CO$_2$	C1	C2–C4	C5+	C1OH	C2 + OH	
Mo$_2$C	3.17	1.8×10^{-2}	31.4	37.9	50.6	7.1	4.4	10	0.38
Ru/Mo$_2$C	8.63	3.4×10^{-2}	30.4	46.7	45.4	7.9	0	0	0.37
Co/Mo$_2$C	11.9	9.9×10^{-2}	29.4	50.4	44.0	5.6	0	0	0.35

Conditions: T = 513 K, P = 20 bar, volumetric space velocity = 3150 h^{-1}.
^{a}From the surface area and assuming 10^{19} sites m^{-2}.
TOF, turnover frequency.

2.3.14.5.9 Reforming (Steam, Dry, and Autothermal)

The conversion of natural gas, of which the principal component is methane, is an important current area of investigation because of the projected long-term availability of this resource compared to petroleum feedstocks. The industrial use of natural gas proceeds by its initial conversion to carbon monoxide and hydrogen (syngas) by the steam-reforming (SR) process.

A study of unsupported, high-surface-area Mo$_2$C and WC has shown that these are highly efficient catalysts for the SR process, and also for the dry-reforming (DR) and partial oxidation (PO) reactions, with activities comparable to those of elemental iridium and ruthenium [125]. At ambient pressure, the catalysts were deactivated in all three processes by oxidation to form MoO$_2$ and WO$_2$. However, at moderate pressure (8×10^5 Pa) they were stable for all three reactions, despite the use of high temperatures of 1220 K (947 °C) for SR, 1120 K (847 °C) for DR, and 1170 K (897 °C) for PO.

2.3.14.5.10 Water-Gas Shift and Reverse Water-Gas Shift

The water-gas shift reaction is an important component of the industrial production of hydrogen. It is used to raise the H$_2$/CO molar ratio in the product of steam reformers.

$$H_2O + CO \rightleftharpoons H_2 + CO_2 \quad \Delta H^{\circ}_{298} = -41 \text{ kJ mol}^{-1} \quad (1)$$

The water-gas shift reaction was studied over Mo$_2$C at 568 K and 10^5 Pa using a simulated methane steam reformer effluent containing 62.5% H$_2$, 31.8% H$_2$O, and 5.7% CO [126]. The catalyst was highly stable and did not produce CH$_4$. Its areal activity (0.22 μmol m^{-2} s^{-1}) was comparable to that of a commercial Zn–Cu–Al catalyst (0.69 μmol m^{-2} s^{-1}).

The hydrogenation of carbon dioxide has been studied over carbide catalysts [127–129]. The main products at atmospheric pressure were CO and H$_2$, indicating the occurrence of the reverse water-gas shift reaction. On molybdenum carbide supported on alumina, it was found that a slight carbon deficiency (Mo$_2$C$_{0.96}$) gave rise to a higher rate than a stoichiometric carbide (Mo$_2$C) [128]. Turnover rates based on CO uptakes for CO$_2$ methanation at 573 K, 10^5 Pa, with a CO$_2$/H$_2$ feed of 1/3 have been reported for various supported molybdenum carbide and nitride catalysts [65] (Table 13). The catalysts were prepared from 12 wt.% MoO$_3$ by direct carburization or nitridation, or by carburization followed by nitridation. In the latter case it was found that the phase formed depended on the preparation temperature.

It is clear that the carbides are more active than the nitrides, and that the activity increases with the preparation temperature.

2.3.14.6 Perspective

The catalytic properties of carbides and nitrides have been examined over a broad range of reactions, and both materials continue to be of great interest. In general, carbides are slightly more active than nitrides, with activity comparable to that of the platinum-group metals. The greatest limitations to the use of carbides and nitrides in industrial applications are most likely cost and reliability. Typically, their preparation requires relatively high temperatures and gas flow rates, although this can be handeled by passivation and re-reduction. The question mark on reliability is a result of their propensity to deposit coke, both in the preparation and during the reaction, and the lack of ready methods for regeneration.

The catalytic properties of phosphides have been explored in only a handful of reactions, most notably

Tab. 13 Catalytic activity of molybdenum nitride and carbide catalysts in the reverse water-gas shift reaction

Sample	Preparation temp./K	Pre-nitridation temp./K	Turnover rate/s^{-1}
Mo$_2$N/Al$_2$O$_3$	973	None	0.0068
MoC$_x$/Al$_2$O$_3$	973	None	0.0910
α-MoC$_x$/Al$_2$O$_3$	773	973	0.0154
β-Mo$_2$C/Al$_2$O$_3$	973	973	0.0683
η-Mo$_3$C$_2$/Al$_2$O$_3$	1173	973	0.173

that of hydroprocessing, where they show great promise, although the control of the surface composition might be problematic. Not only is the metal to phosphorus ratio important, but surface sulfidation tends also to occur. From a cost standpoint, however, the phosphides can be prepared at lower temperatures than the carbides and nitrides, and this may provide financial benefits.

References

1. P. Schwarzkopf, R. Kieffer, *Refractory Hard Metals*, Macmillan, New York, 1953.
2. E. K. Storms, *The Refractory Carbides*, Academic Press, New York, 1967.
3. L. E. Toth, *Transition Metal Carbides and Nitrides*, Academic Press, New York, 1971.
4. S. T. Oyama. Ed. *The Chemistry of Transition Metal Carbides and Nitrides*, Blackie Academic and Professional, Glasgow, 1996.
5. H. O. Pierson, *Handbook of Refractory Carbides and Nitrides*, Noyes Publications, Westwood, New Jersey, 1996.
6. P. Ettmayer, W. Lengauer, Nitrides: Transition Metal Solid State Chemistry, in *Encyclopedia of Inorganic Chemistry*, R. B. King, Ed., Wiley, New York, 1994, p. 2498.
7. B. Aronsson, T., Lundström, S., Rundqvist, *Borides, Silicides and Phosphides*, Methuen, London and Wiley, New York, 1965.
8. D. E. C. Corbridge, *Studies in Inorganic Chemistry*, Vol. 10, 4th Ed., Elsevier, Amsterdam, 1990.
9. B. F. Stein, R. H. Walmsley, *Phys. Rev. B* **1996**, *148*, 933.
10. O. Tegus, E. Brück, K. H. J. Buschow, F. R. de Boer, *Nature* **2002**, *415*, 150.
11. Hägg, G., *Z. Phys. Chem.* **1931**, *12*, 33.
12. S. T. Oyama, *J. Solid State Chem.* **1992**, *96*, 442.
13. D. P. Thomson, The Role of Oxygen in Non-Oxide Engineering Ceramics, in *The Physics and Chemistry of Carbides, Nitrides, and Borides*, R. Freer, Ed., NATO ASI Series, Series E: Applied Sciences, Vol. 185, Kluwer, Dordrecht, 1990, p. 423.
14. H. W. Hugosson, O. Eriksson, L. Nordström, U. Jansson, L. Fast, A. Delin, J. M. Wills, B. Johansson, *J. Appl. Phys.* **1999**, *86*, 3758.
15. R. Jusa, in *Advances in Inorganic Chemistry and Radiochemistry*, H. J. Eméleus, A. G. Sharpe, Eds., Vol. 9, Academic Press, New York, 1966, p. 81.
16. F. Benesovsky, R. Kieffer, P. Ettmayer, Nitrides, in *Kirk Othmer, Encyclopedia of Chemical Technology*, 3rd Ed., Vol. 15, John Wiley, New York, 1981, p. 871.
17. S. T. Oyama, in *Handbook of Heterogeneous Catalysis*, G. Ertl, H. Knözinger, J. Weitkamp, Eds., VCH, Weinheim, 1997, p. 132.
18. E. Lehrer, *Z. Electrochemistry* **1930**, *36*, 383.
19. R. Juza, W. Sachsze, *Z. Anorg. Chem.* **1947**, *253*, 95.
20. J. Gopalakrishnan, S. Pandey, K. K. Rangan, *Chem. Mater.* **1997**, *9*, 2113.
21. S. Kaskel, K. Schlichte, G. Chaplais, M. Khanna, *J. Mater. Chem.* **2003**, *13*, 1496.
22. G. Cordier, H. Schafer, M. Stelter, *Z. Nat. B.* **1986**, *41*, 1416.
23. S. Wanner, L. Hilaire, P. Wehrer, J. P. Hindermann, G. Maire, *Appl. Catal. A: General* **2000**, *203*, 55.
24. C. H. Winter, T. S. Lewkebandara, J. W. Proscia, *Chem. Mater.* **1995**, *7*, 1053.
25. P. G. Edwards, P. W. Read, M. B. Hursthouse, K. M. Abdul Malik, *J. Chem. Soc. Dalton Trans.* **1994**, 975.
26. J. S. Lee, S. T. Oyama, M. Boudart, *J. Catal.* **1987**, *106*, 125.
27. S. T. Oyama, J. C. Schlatter, J. E. Metcalfe, J. Lambert, *Ind. Eng. Chem. Res.* **1988**, *27*, 1639.
28. V. L. S. Teixeira da Silva, E. I. Ko, M. Schmal, S. T. Oyama, *Chem. Mater.* **1995**, *7*, 179.
29. V. L. S. Teixeira da Silva, M. Schmal, S. T. Oyama, *J. Solid State Chem.* **1996**, *123*, 168.
30. L. Volpe, M. Boudart, *J. Solid State Chem.* **1985**, *59*, 332.
31. R. Kapoor, S. T. Oyama, *J. Solid State Chem.* **1992**, *99*, 303.
32. W. Li, B. Dhandapani, S. T. Oyama, *Chem. Lett.* **1988**, 207.
33. Clark, P., Li, W., Oyama, S. T., *J. Catal.* **2001**, *200*, 140.
34. J. S. Lee, S. T. Oyama, M. Boudart, *J. Catal.* **1987**, *106*, 125.
35. R. E. Partch, Y. Xie, S. T. Oyama, E. Matijević, *J. Mater. Res.* **1993**, *8*, 2014.
36. Stinner, C., Tang, Z., Haouas, M., Weber, Th., Prins, R., *J. Catal.* **2002**, *208*, 456.
37. Y. Shu, S. T. Oyama, *Carbon* **2005**, *43*, 1517.
38. J. H. Kim, X. Ma, C. Song, Y.-K. Lee, S. T. Oyama, *Energy & Fuels* **2005**, *19*, 353.
39. A. Wang, L. Ruan, Y. Teng, X. Li, M. Lu, J. Reng, Y. Wang, Y. Hu, *J. Catal.* **2005**, *229*, 314.
40. M. J. Ledoux, C. Pham-Huu, A. P. E. York, E. A. Blekkan, P. Delporte, P. del Gallo, Study of the isomerization of C6 and C6+ alkanes over molybdenum oxycarbide catalysts, in *The Chemistry of Transition Metal Carbides and Nitrides*, S. T. Oyama (Ed.), Blackie Academic & Professional, Glasgow, 1996, pp. 373–397.
41. D. Zheng, M. J. Hampden-Smith, *Chem. Mater.* **1992**, *4*, 968.
42. O. Schmitz-Dumont, *Z. Elektrochem.* **1956**, *60*, 866.
43. C. Qian, F. Kim, L. Ma, F. Tsui, P. Yang, J. Liu, *J. Am. Chem. Soc.* **1994**, *126*, 1195.
44. F. Luo, H.-L. Su, W. Song, Z.-M. Wang, Z.-G. Yan, C.-H. Yan, *J. Mater. Chem.* **2004**, *14*, 111.
45. Y. Chen, *Catal. Today* **1998**, *44*, 3.
46. H. Li, W. Wang, H. Li, J.-F. Deng, *J. Catal.* **2000**, *194*, 211.
47. J. Hojo, T. Oku, A. Kato, *J. Less-Common Met.* **1978**, *59*, 85.
48. D. C. LaMont, A. J. Gilligan, A. R. S. Darujati, A. S. Chellappa, W. J. Thompson, *Appl. Catal. A: General* **2003**, *255*, 239.
49. S. Yang, R. Prins, *Chem. Commun.* **2005**, 4178.
50. K. Hada, J. Tanabe, S. Omi, M. Nagai, *J. Catal.* **2002**, *207*, 10.
51. C. H. Jaggers, J. N. Michaels, A. M. Stacy, *Chem. Mater.* **1990**, *2*, 150.
52. S. T. Oyama, R. Kapoor, H. T. Oyama, D. J. Hofmann, E. Matijević, *J. Mater. Res.* **1993**, *8*, 1450.
53. H. Kwon, S. Choi, L. T. Thompson, *J. Catal.* **1999**, *184*, 236.
54. L. Volpe, M. Boudart, *J. Sol. St. Chem.* **1985**, *59*, 348.
55. J.-G. Choi, R. L. Curl, L. V. Thompson, *J. Catal.* **1994**, *146*, 218.
56. A. Hanif, T. Xiao, A. P. E. York, J. Sloan, M. L. H. Green, *Chem. Mater.* **2002**, *14*, 1009.
57. M. Saito, R. B. Anderson, *J. Catal.* **1980**, *63*, 483.
58. T.-C. Xiao, A. P. E. York, K. S. Coleman, J. B. Claridge, J. Sloan, J. Charnock, M. L. H. Green, *J. Mater. Chem.* **2001**, *11*, 3094.
59. T. C. Xiao, A. P. E. York, C. V. Williams, H. Al-Megren, A. Hanif, X. Zhou, M. L. H. Green, *Chem. Mater.* **2000**, *12*, 3896.
60. X. Wang, P. Clark, S. T. Oyama, *J. Catal.* **2002**, *208*, 321.
61. S. Gong, H. Chen, W. Li, B. Li, T. Hu, *J. Molec. Catal. A: Chemical* **2005**, *225*, 213.

62. S. Naito, A. Takada, S. Tokizawa, T. Miyao, *Appl. Catal. A: General* **2005**, *289*, 22.
63. J. Y. Lee, T. H. Nguyen, A. Khodakov, A. A. Adesina, *J. Molec. Catal. A: Chemical* **2004**, *211*, 191.
64. T. Miyao, I. Shishikura, M. Matsuoka, M. Nagai, S. T. Oyama, *Appl. Catal. A: General* **1997**, *165*, 419.
65. M. Nagai, K. Oshikawa, T. Murakami, T. Miyao, S. Omi, *J. Catal.* **1998**, *180*, 14.
66. S. J. Sawhill, K. A. Layman, D. R. Van Wyk, M. H. Engelhard, C. Wang, M. E. Bussell, *J. Catal.* **2005**, *231*, 300.
67. E. L. Muetterties, J. C. Sauer, *J. Am. Chem. Soc.* **1974**, *96*, 3410.
68. S. T. Oyama, X. Wang, Y. K. Lee, K. Bando, F. G. Requejo, *J. Catal.* **2002**, *210*, 207.
69. S.-H. Ko, T.-C. Chou, T.-J. Yang, *Ind. Eng. Chem. Res.* **1995**, *34*, 457.
70. H. Li, W. Wang, B. Zong, E. Min, J.-F. Deng, *Chem. Lett.* **1998**, 371.
71. X. Yu, H. Li, J.-F. Deng, *Appl. Catal. A: General* **2000**, *199*, 191.
72. W.-J. Wang, M.-H. Qiao, H. Li, J.-F. Deng, *Appl. Catal. A* **1998**, *166*, L243.
73. H. Yamashita, M. Yoshikawa, T. Funabiki, S. Yoshida, *J. Chem. Soc. Faraday Trans. I* **1986**, *82*, 1771.
74. S. T. Oyama, G. L. Haller, *Catalysis, Specialist Periodical Reports* **1982**, *3*, 333.
75. S. T. Oyama, *Catal. Today* **1992**, *15*, 179.
76. E. Furimsky, *Appl. Catal. A: General* **2003**, *240*, 1.
77. S. T. Oyama, *J. Catal.* **2003**, *216*, 343.
78. R. B. Levy, M. Boudart, *Science* **1973**, *181*, 547.
79. J. G. Chen, *Chem. Rev.* **1996**, *96*, 1477.
80. H. H. Hwu, J. G. Chen, *Chem. Rev.* **2005**, *105*, 185.
81. J. Chen, J. Eng, S. P. Kelty, *Catal. Today* **1998**, *43*, 137.
82. S. K. Bej, C. A. Bennett, L. T. Thompson, *Appl. Catal. A: General* **2003**, *250*, 197.
83. S. K. Bej, L. T. Thompson, *Appl. Catal. A: General* **2004**, *264*, 141.
84. R. Kojima, K. Aika, *Appl. Catal. A: General* **2001**, *219*, 141.
85. C. J. H. Jacobsen, *Chem. Commun.* **2000**, 1057.
86. R. Kojima, K. Aika, *Appl. Catal. A: General* **2001**, *218*, 121.
87. J.-G. Choi, J. Ha, J.-W. Hong, *Appl. Catal. A: General* **1998**, *168*, 47.
88. S. T. Oyama, *J. Catal.* **1992**, *133*, 358.
89. Y. Xu, M. Ichikawa, *Catal. Today* **2001**, *71*, 51.
90. L. Wang, L. Tao, M. Xie, G. Xu, J. Huang, Y. Xu, *Catal. Lett.* **1993**, *21*, 35.
91. L. Chen, L. Lin, Z. Xu, X. Lin, T. Zhang, *J. Catal.* **1995**, *157*, 190.
92. W. Li, G. D. Meitzner, R. W. Borry, III, E. Iglesia, *J. Catal.* **2000**, *191*, 373.
93. F. Solymosi, J. Cserenyi, A. Szoke, T. Bansagi, A. Dszko, *J. Catal.* **1997**, *165*, 150.
94. D. Wang., J. H. Lunsford, M. P. Rosynek, *J. Catal.* **1997**, *169*, 345.
95. S. Liu, L. Wang, R. Ohnishi, M. Ichikawa, *J. Catal.* **1999**, *18*, 175.
96. Y. Shu, D. Ma, X. Liu, X. Han, Y. Xu, X. Bao, *J. Phys. Chem. B* **2000**, *104*, 8245.
97. J. S. Lee, M. H. Yeom, K. Y. Park, I.-S. Nam, J. S. Chung, Y. G. Kim, S. H. Moon, *J. Catal.* **1991**, *128*, 126.
98. C. Marques-Alvarez, J. B. Claridge, P. E. Andrew, J. Sloan, M. H. L. Green, *Stud. Surf. Sci. Catal.* **1997**, *106*, 145.
99. L. Guczi, A. Sárkány, P. Tétényi, *Faraday Trans.* **1974**, *70*, 1971.
100. P. Pérez-Romo, C. Potvin, J.-M. Manoli, G. Djéga-Mariadassou, *J. Catal.* **2002**, *205*, 191.
101. F. Nozaki, R. Adachi, *J. Catal.* **1975**, *40*, 166.
102. J. S. Lee, M. Boudart, *Appl. Catal.* **1985**, *19*, 207.
103. J. C. Schlatter, S. T. Oyama, J. E. Metcalfe, III, J. M. Lambert, Jr, *Ind. Eng. Chem. Res.* **1988**, *27*, 1648.
104. S. Ramanathan, S. T. Oyama, *J. Phys. Chem.* **1995**, *99*, 16365.
105. P. A. Aegerter, W. W. C. Quigley, J. G. Simpson, D. D. Ziegler, J. W. Logan, K. R. McCrea, S. Glazier, M. E. Bussell, *J. Catal.* **1996**, *164*, 109.
106. D. J. Sajkowski, S. T. Oyama, *Appl. Catal. A: General* **1996**, *134*, 339.
107. P. Da Costa, C. Potvin, J.-M. Manoli, J.-L. Lemberton, G. Pérot, G. Djéga-Mariadassou, *J. Molec. Catal. A: Chemical* **2002**, *184*, 323.
108. M. Houalla, N. K. Nag, A. V. Sapre, D. H. Broderick, B. C. Gates, *AIChE J.* **1978**, *24*, 1015.
109. W. R. A. M. Robinson, J. N. M. van Gestel, T. I. Korányi, S. Eijsbouts, A. M. van der Kraan, J. A. R. van Veen, V. H. J. de Beer, *J. Catal.* **1996**, *161*, 539.
110. T. I. Korányi, *Appl. Catal. A: General* **2002**, *237*, 1.
111. S. J. Sawhill, D. C. Phillips, M. E. Bussell, *J. Catal.* **2003**, *215*, 208.
112. V. Zuzaniuk, R. Prins, *J. Catal.* **2003**, *219*, 85.
113. F. A. Ribeiro, R. A. Dalla Betta, M. Boudart, J. E. Baumgartner, E. Iglesia, *J. Catal.* **1991**, *130*, 86.
114. F. H. Ribeiro, M. Boudart, R. A. Dalla Betta, E. Iglesia, *J. Catal.* **2001**, *130*, 498.
115. E. Iglesia, J. E. Baumgartner, F. H. Ribeiro, M. Boudart, *J. Catal.* **2001**, *131*, 523.
116. M. Siaj, P. H. McBreen, *Science* **2005**, *309*, 588.
117. G. S. Ranhotra, G. W. Haddix, A. T. Bell, J. A. Reimer, *J. Catal.* **1987**, *108*, 24.
118. L. Leclercq, K. Imura, S. Yoshida, T. Barbee, M. Boudart, in *Preparation of Catalysts II*, B. Delmon (Ed.), p. 627, Elsevier, New York, 1978.
119. L. Leclercq, A. Almazouari, M. Dufour, G. Leclercq, in *The Chemistry of Transition Metal Carbides and Nitrides*, S. T. Oyama (Ed.), Blackie Academic & Professional, Glasgow, 1996, pp. 345–361.
120. H. C. Woo, K. Y. Park, Y. G. Kim, I. S. Nam, J. S. Chung, J. S. Lee, *Appl. Catal.* **1991**, *75*, 267.
121. A. Griboval-Constant, J.-M. Giraudon, G. Leclercq, L. Leclercq, *Appl. Catal. A: General* **2004**, *260*, 35.
122. N. I. Ilchenko, Yu. I. Pyatnitsky, Carbides of transition metals as catalysts for oxidation reactions, in *The Chemistry of Transition Metal Carbides and Nitrides*, S. T. Oyama (Ed.), Blackie Academic & Professional: Glasgow, 1996; pp. 311–326.
123. G. P. Khirnova, M. G. Bulygin, E. A. Blyumberg, *Petrol. Chem. U.S.S.R.* **1981**, *21*, 49.
124. Q. Zhu, B. Zhang, J. Zhao, S. Ji, J. Yang, J. Wang, H. Wang, *J. Molec. Catal. A: General* **2004**, *213*, 199.
125. J. B. Claridge, A. P. E. York, A. J. Brungs, C. Marques-Alvarez, J. Sloan, S. C. Tsang, M. L. H. Green, *J. Catal.* **1998**, *180*, 85.
126. J. Patt, D. J. Moon, C. Phillips, L. Thompson, *Catal. Lett.* **2000**, *65*, 193.
127. M.-D. Lee, J.-F. Lee, C.-S. Chang, *Appl. Catal.* **1991**, *72*, 267.
128. M. Nagai, T. Kurakami, S. Omi, *Catal. Today* **1998**, *45*, 235.
129. M. Nagai, K. Oshikawa, T. Kurakami, T. Miyao, S. Omi, *J. Catal.* **1998**, *180*, 14.

2.3.15
Carbons

Robert Schlögl*

2.3.15.1 Introduction

The element carbon plays a multiple role in catalysis. In this chapter, the structures of solid carbon in concept are introduced, and the enormous variability in the structural properties of catalytically relevant carbon are investigated. The concepts of surface reactivity are then examined, focusing on the role of defects and their saturation by functional groups. Finally, our present understanding of the function of carbon in catalysis is discussed by studying selected examples of catalytic processes.

Carbon-containing molecules are, in most catalytic applications, the substrates of the process under consideration. In homogeneous catalysis, such molecules occur as the most prominent constituent of ligand systems. However, deposits and polymers of carbon atoms often occur as poisons on catalysts, and carbon deposition is indeed a major problem in solid acid–base-catalyzed reactions. In hydrogenation reactions, carbon deposits are thought to act as modifiers of catalyst activity and to provide selectivity by controlling the hydrogenation–dehydrogenation activity of the metal part of the catalyst. In fact, in some cases the carbon deposits are even thought to constitute part of the active site.

Carbon is a prominent catalyst support material, as it allows the anchoring of metal particles onto a substrate which, after suitable preparation, does not exhibit solid acid–base properties. Carbon is in fact a catalyst in its own right, as it enables the activation of oxygen and chlorine for selective oxidation, chlorination, and dechlorination reactions.

These multiple roles occur in parallel with a complex structural chemistry which gives rise to families in which the element shows vastly different chemical modifications. These differences arise from the mesoscopic or nanostructural features of arrays of carbon atoms with essentially only two local geometric structures. It is thus a special characteristic of carbon chemistry that many of these modifications cannot be obtained as phase-pure materials, as the medium-range ordering of materials is difficult to control. This limits our comprehensive knowledge of the physical and chemical properties to two archetypal allotropes with substantial long-range ordering of the carbon atoms, namely graphite and diamond.

The reason for the poor structural definition of these materials can be found in the processes of their formation, because it is difficult to control the polymerization reactions. Such processes also occur as unwanted side reactions in catalytic transformations with small organic molecules. Therefore, the nature of carbon deposits reflects the complexity of these "organic" carbon materials. A major aim of this chapter is to describe the structural and chemical complexity of "coke" or "soot", in order to provide an understanding of the frequently observed complexity of their chemical reactivity in the reactivation of catalysts, which are invariably aimed at the oxidative removal of deposits.

During the past 20 years, the number of families of carbon materials with complex structures has increased substantially, notably with the advent and production of nanostructured carbons. Fullerenes, nanotubes, and nanocarbons are materials of catalytic relevance, and will also be discussed in terms of their structures and reactivities.

The surface chemistry of carbon, which in turn determines its interfacial properties, is a complex field, since a huge variety of surface functional groups are known to exist. By far the most important of these groups is oxygen, although other prominent surface heteroatoms include halogens, hydrogen, and nitrogen. In general however, this chapter will focus on the description and characterization of oxygen surface functional groups.

2.3.15.2 The Complexity of Carbon

The chemistry of carbon is highly complex, both in the molecular and solid states. The reason for this complexity lies in the pronounced tendency of the element to form homonuclear bonds in three bonding geometries. These can occur either uniformly or in almost unlimited combinations, thereby allowing for a very large number of homonuclear structures of elemental carbon. The moderate electronegativity of carbon results in strong covalent interactions with all main group elements. This adds enormously to the molecular (heteronuclear molecules) and solid-state structural complexity of carbon.

The literally unlimited number of covalent bonding situations for carbon atoms in a solid hardly allows for samples with perfect structural order to complicate the determination of structural, chemical, and physical reference properties. Processes of defect healing, solution–precipitation or crystal growth according to Ostwald ripening are limited for elemental carbon to extreme conditions that are not relevant to catalysis. Segregation and condensation processes lead to the carbon solids of interest discussed here, with the catalytic synthesis of carbon nanostructures by catalytic chemical vapor deposition (CCVD) representing the main process by which structural control can be achieved by designing not only the catalyst but also the reaction conditions.

The poor structural definition of carbon as an element is illustrated by the fact that even the crystal structure of graphite is only an approximate model [1], as no single crystal with few enough defect concentrations has been found to allow a regular three-dimensional (3-D) X-ray structure determination. Figure 1 illustrates the challenge when determining the structure of graphite from diffracted intensities. Both, the image and the diffraction pattern illustrate the ordering (spots in diffraction and graphene layers in the image), but likewise the high abundance of disorder [cloudy contrast in the thin sections of the transmission electron microscopy (TEM) image and frequent rupture in the graphene layers; wide profiles, satellite spots (arrows) plus anisotropic random ordering (ring) in the diffraction pattern]. The electron diffraction pattern represents the fortunate case for structure analysis, as it allows the selection of a small, defect-poor region from a graphite flake. The predominant types of disorder arise from the composition of graphene layers from small basic structural units (BSUs) with in-plane defects around their interconnects, and the arbitrary rotation of the graphene layers with respect to each, thereby other violating the ABAB stacking order.

In the context of catalysis it is important to note that "black carbon", "coke", "soot" or "carbonaceous deposit" is by no means a graphitic or even a homogeneous material with reproducible properties. This leads to the wide variations in application profiles of synthetic carbon blacks (there are several thousand varieties), and also to difficulties when determining the oxidation properties of "coke" on catalyst surfaces. Graphitic carbon (as seen in Fig. 1) always requires, for its comparable high state of ordering, treatment at very high temperatures in excess of 2500 K. Lower grades of graphitic order (planar graphene layers, no heteroatoms, well-developed in-plane ordering, turbostratic stacking of BSUs) can occur at lower temperatures from sources of high pregraphitic structures such as polycyclic aromatic compounds, or from systems which allowing for a liquid "mesophase" during carbonization.

2.3.15.3 Electronic Structure of Carbon Allotropes

The carbon atom, with its $2s^2 2p^2$ electron configuration, can form a maximum of four bonds of bond orders between 1 and 3. Three motifs of connectivity result, as depicted in Fig. 2. The valence orbitals hybridize, forming in the elemental state, sp, sp^2, or sp^3 hybrid orbitals. The data in Fig. 2 illustrate that the local bonding geometry and the carbon–carbon bond distances vary significantly with the connectivity as the bond order (and bond strength) increases from the sp^3 (single bond) over the sp2 (double bonds or aromatic bonds) to the sp state (triple bond). Carbon atoms also bond to heteroatoms among which, in the present context, hydrogen (single bonds), oxygen and nitrogen (single and double bonds) are the most important partners.

The different bonding motifs shown in Fig. 2 correspond to significantly different local electronic structures of the carbon atom. X-ray absorption spectroscopy (XAS) at the carbon K-edge (C1s) is a most sensitive tool to probe the electronic structure due to the involvement of the valence band as final state. Much more popular, but somewhat insensitive, is photoelectron spectroscopy [X-ray photoelectron spectroscopy (XPS), electron spectroscopy for chemical analysis (ESCA)], but this cannot discriminate the connectivity on the basis of chemical shifts. Nonetheless, XPS is very useful for analyzing for carbon–heteroatomic bonds. The XAS data from thin-film

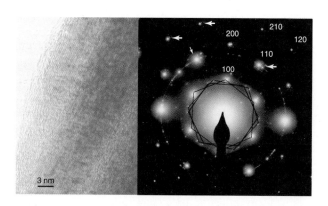

Fig. 1 Transmission electron microscopy image and electron diffraction pattern of a natural graphite single crystal. The image shows the view along the c-axis representing the stacking of sp2 graphene layers with its disorder. The effect of disorder is also shown in the diffraction pattern highlighting the anisotropy in ordering (high in hk0 and low 001).

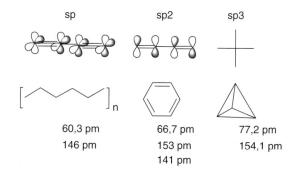

Fig. 2 Hybridizations, connectivity, orbital orientations and structural motifs for the carbon atoms relevant in solid carbon forms. The top numbers give the atom radius in the respective bonding situations and the bottom row indicates the bond distances. For sp^2 the top value applies to chain-like structures and the lower value to aromatic ring structures larger than benzene such as occurring in basic structural units of graphene.

samples of fullerene (C$_{60}$), graphite, and diamond are presented in Fig. 3. The band assignment is shown in regions of resonances, and the multiple spectra reveal the polarization dependence of π and σ contributions to the spectral weight. As expected from Fig. 2, the resonances for the π electron systems show sharp features for the lowest unoccupied orbital (LUMO) and LUMO+1 and broad features for the structures involved in the delocalized components [2–4] of the valence electron density. The weak contribution of C–H in graphite, and its strong counterpart in the hydrogen-assisted CVD-synthesized diamond [5, 6], indicates the abundance of terminal carbon atoms and may be regarded as proxy feature for defects.

The tetrahedral connectivity leads to space-filling carbon polymers which are realized as defect-free ideal structures in diamond. Figure 4 shows a high-resolution transmission electron microscopy (HRTEM) image of the catalytically only potentially relevant form of ultradispersed diamond (UDD) [7]; this is available in small quantities from catalytic shock-wave synthesis. This high-surface-area material exhibits many defects at its surface (roughness) and in the bulk, despite the high binding energy and dense packing of the atoms

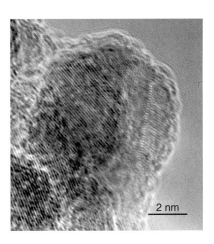

Fig. 4 High-resolution TEM image of nanoparticles of ultra-disperse diamond (UDD). Bulk and surfaces exhibit numerous defects despite the overall crystallinity giving rise to a well-developed powder-X-ray diffraction pattern of diamond. The defects create the spotty contrast and the "stepped" structure, both being imaging artifacts caused by non-resolved deviations from the periodicity of the lattice.

which should prevent the existence of many defects in a thermodynamically metastable material [8, 9]. It is assumed that many of these defects are stabilized by hydrogen, as is the case [10] with single crystal surfaces of this material.

The trigonal connectivity leads to sheets of carbon that are, in the defect-free variety, composed of hexagons polymerized to planar graphene layers. An edge-on view of a stack of such graphene sheets and their perfect parallel orientation is presented in Fig. 1. The idealized structure of graphite (see Fig. 5) will result if these layers are regularly stacked in the third dimension.

In the trigonal connectivity, the sp^2 hybrid orbitals leave one electron per carbon atom in a non-bonding state. These electrons can interact with each other, forming so-called "weak" bonds which exhibit their plane of electron density orthogonal to the plane of the strong bonds (Fig. 2). Hence, they can either interact as isolated or conjugated localized double bonds (as in alkene molecules), as delocalized aromatic bonds (as in aromatic hydrocarbon molecules), or in graphite, where they carry metallic properties. It should be noted here that the simplification made in the context of carbonaceous deposits, that graphite was a large polycyclic hydrocarbon molecule, is incorrect, as its electronic structure cannot be constructed by extrapolation from polycyclic hydrocarbon molecules.

Any combination between these electronic interactions exists in defective carbons. A continuous variation of the geometry between the prototypes sp^2 and sp^3 of carbon connections (see Figs. 2 and 3) occurs by

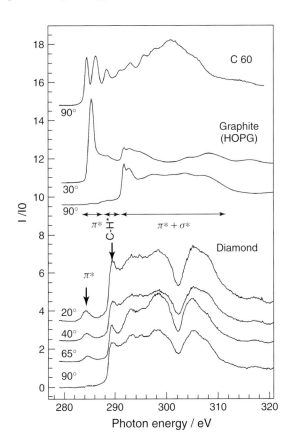

Fig. 3 X-ray absorption spectra at the carbon K edge for three types of solid carbon. For a discussion of the technique and the band assignment, see text.

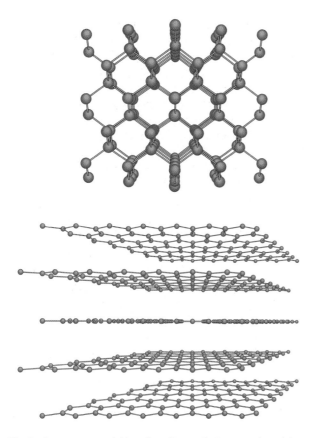

Fig. 5 Atom connectivities for diamond (top) and graphite (bottom). The common structural motif are layers, that are buckled in diamond allowing a three-dimensional network, whereas they are flat in graphite and give rise to the interlayer space referred to in the literature as "van der Waals gap".

bending the planarity of the graphene sheets as the topological consequence of a point defect giving rise to a non-six-membered carbon ring. The resulting fractional hybridization sp^{2-x} results in centers of varying double bond character [11]. In molecular chemistry, this effect of varying the degree of double bond localization is referred to as "mesomerism", and is observed experimentally in certain polycyclic aromatic molecules [12, 13]. Attempts to understand the synthesis of fullerenes have brought about studies [14] on the electronic structure of bowl-shaped "aromatic" molecules that can be seen as models for local distortions of graphenes by point defects.

One consequence of the existence of weak bonds in sp^2 carbons is the action of strong dispersive forces (or van der Waals bonds) in successive layers, resulting in the cohesion of graphene layers. These interactions provide only a moderate mechanical stability of the perfect layer system in graphite (see Fig. 5) against shear forces in the a–b plane of the platelet crystals resulting from the planar motif of the sp^2 bond configuration. Graphite is mechanically weak as a catalyst support, but may be used due to its excellent tribological properties as an inert formulation additive for many catalytically relevant oxide systems.

The van der Waals forces provide energy for the chemisorption of both polar and non-polar molecules on graphite surfaces [15–18]. The in-plane bonding strength of the planar connectivity leads to a very poor chemical reactivity of graphene layers on their basal planes, allowing only for dispersive physisorption. The adsorption of xenon or methane on graphite, and its temperature-driven surface phase transitions, are an example of studies of molecular interactions [19, 20] that are not disturbed by specific (chemisorptive) bonding. Chemically, significant reactivity is confined to edges and holes in planar graphene sheets, thereby exposing dangling bonds that become saturated in functionalized groups which determine the chemical behavior of the graphene (see below).

The weak bonding between graphene layers allows graphite to serve as a host for intercalation of a very wide variety of molecular and sheet-forming compounds, ranging from free chlorine to most elemental halides, oxo acids, metal oxides electron-donating organic molecules, and to electropositive metals such as alkaline, earth alkaline and rare earth metals. Hydrogen and oxygen cannot be intercalated into graphite (for a comprehensive review, see Refs. [21, 22]). In the context of catalytic studies, it is likely that alkali metals are intercalated into graphitic carbon where they catalyze their gasification, or where they may be stored for gradual release in non-oxidizing atmospheres. The formation of intercalates is sometimes detectable by a change in the c-axis lattice parameter, but as catalytically relevant graphite is often nanocrystalline (and hence not detectable by X-ray diffraction) it is difficult to identify [23–25] the occurrence of intercalation in disordered systems. One prominent case [26] of intercalation of alkali into carbon is the modification of the catalytic properties of graphite supports by Cs in the promoted Ru-ammonia synthesis catalyst.

The trigonal connectivity leads not only to hexagons but also enables the incorporation of triangles, pentagons, heptagons and larger cyclic structures as local defects into the basic honeycomb structure [27–29]. A consequence of these defects is the strong localization of double bonds around these defects, and the deviation of the carbon sheet from planarity [30–34]. Three-dimensional bends always require the incorporation of non-six-membered structures if they are continuous and not realized by breaking the graphene; they are realized in fullerenes, in curved ribbons or sheets of nanocarbons and in bends of graphene units constituting defective graphite crystals (see also Fig. 1). These units are locations of chemical reactivity, since the resulting transformation of "inert" aromatic electronic structure into "reactive" double bond character gives rise to an enhanced reactivity pattern of addition and insertion reactions (see below and Section 2.3.15.16, Surface chemistry) of alkene molecules as compared to the typical substitution chemistry of aromatic structures.

The linear connectivity of sp hybridization leads to chain-like polymers with conjugated triple bonds formed by the two sets of orthogonal nonbonding orbitals, as shown in Fig. 2. The resulting structure of cumulated electron density creates "explosive" chemical reactivity [35] towards restructuring into the trigonal or tetragonal connectivity, and towards the addition of heteroatoms. A redistribution of the weakly bonding electrons, as found in molecular ketenes, does not occur in pure carbon materials. Rather, the inherent chemical instability of this arrangement [36] causes the materials to occur only rarely in pure form after rapid quenching from high-temperature carbon sources, or in meteorites. Although these phases (e.g., chaotit) are thermodynamically metastable, they may occur as a part of more complex carbon polymers [37, 38].

Figure 3 illustrates how different bonding situations are reflected in electron spectroscopy. The preceding discussion indicates that the extent and exact nature of the weak bonding in carbon may serve as an excellent and continuous descriptor for the local structure of carbon allowing the prediction of stability and reactivity from this quantity that can be determined both in bulk and surface modes. Figure 3 depicts the surface-sensitive XAS [39] of C_{60}, graphite and diamond (as CVD films [40, 41] with a trace impurity of graphite). The carbon 1s XAS method is chosen here as being complementary to the prominently used bulk-sensitive electron energy loss spectroscopy (EELS) technique [42–46]. Both methods resolve clearly the presence of delocalized electrons from the π bonds shown in the connectivity diagrams. The polarization dependence of the excitation with synchrotron light allows, for single-crystalline surfaces, determination of the orientation of the π bond relative to the carbon–carbon backbone bonding which accounts for the high-energy structures above 290 eV depicted in Fig. 3 [2, 47]. Fullerene and graphite, although often described collectively under the label "perfect sp^2 systems", are dissimilar in their spectra. This is due mainly to the molecular character of fullerene, which gives rise to discrete bands of weakly interacting molecular orbitals, whereas in graphite the complete overlap of molecular states to broad bands of density of states gives rise to only a few structural features in the spectrum [3, 48, 49]. The sequence of electronic states with energy is, however, similar for the two carbons with the same local connectivity [50, 51].

2.3.15.4 Basic Structures

The two allotropes of carbon with particularly well-defined properties are hexagonal graphite, as a thermodynamically stable modification at ambient conditions, and its high-pressure, high-temperature allotrope cubic diamond. Although both well-crystallized forms will only very rarely be encountered in catalytic systems, it is important to recall some details about their properties as these also prevail in the catalytically relevant nanostructured forms of carbon. The idealized crystal structures are displayed in Fig. 5. The structural consequences of the different connectivities can be clearly seen.

Diamond Diamond is an isotropic material consisting of corner-sharing tetrahedra, equivalent to a face-centered, close-packed structure of carbon atoms with additional atoms in the centers of every other octant of the cubic unit cell. The ideal tetrahedral arrangement of carbon atoms leads to four equal bonds of 154 pm in length. The long-range motif can be described as layers of bent carbon hexagons with one additional interlayer bond per carbon atom. The bent hexagons are equivalent to cyclohexane with the in-plane bonds in equatorial and the interlayer bonds in axial positions. The resulting crystals are highly symmetric (space group $Fd3m$) with a preferred (111) cleavage plane. The historical classification of diamond [9] follows optical absorption properties, which are predominantly controlled by the abundance of substitutional nitrogen and boron atoms.

The synthesis conditions of both natural and artificial diamond require extreme conditions and the presence of impurities acting as catalysts. For the synthesis of diamond and its properties, the reader is referred to selected reports [6, 7, 52–55].

The structural properties are reflected in the physical properties selected in Table 1. Diamond is a super hard, electrical insulator with an extremely good thermal conductivity. Phase transitions into the liquid state or

Tab. 1 Selected properties of hexagonal graphite and cubic diamond. Anisotropic data for graphite are given in directions first parallel to, and then perpendicular to, the graphene layers

Property	Graphite (hex)	Diamond
Density/g/cm^{-3}	2.266	3.514
C—C bond length/pm	142	154
Enthalpy of sublimation/kJ mol^{-1}	715	710
Enthalpy of formation/kJ mol^{-1}	0.00	1.90
Enthalpy of oxidation/kJ mol^{-1}	393.51	395.41
Melting point/K at kbar	2600, 9	4100, 125
Thermal expansion/1 K^{-1}	-1×10^{-6}// $+29 \times 10^{-6}$	1×10^{-6}
Electrical resistivity/Ω cm^{-1}	5×10^{-7}//1	1020
Mohs hardness	0.5//9	10

References see page 419

sublimation occur outside the range of conditions for catalysis (see Table 1). The surface of, in particular, artificial diamond is usually terminated by hydrogen, giving rise to alkane-like reconstructions [5, 6, 10, 56–58], and thus high chemical stability. The chemical reactivity of natural diamond is very limited, occurring mainly through a catalyzed transformation into graphite [18, 59, 60]. From this modified surface, the oxidation and halogenation reactions will occur [5, 53, 61]. An extensive report is available [9] concerning structural, morphological and kinetic aspects.

In studies of carbon deposits on catalysts, it must be taken into account that the conversion of reactive forms of carbon into diamond can occur under the beam of an electron microscope as a particular form of beam damage [62, 63]. An example of such artifact formation is presented in Fig. 6. The artificial object in Fig. 6, which is not particularly distinct in its diamond structure in the center, is formed as consequence of the beam irradiation and is typical for "catalytic" carbon. Thus, care must be taken when diamond is found analytically by electron microscopy in carbon on catalytic materials. It is, however, not impossible to find diamond as genuine constituent in the form of nanodiamond in systems which have undergone rapid and drastic temperature changes. Onion-like objects with metals instead of "diamond" in their centers frequently occur after the deactivation of metal nanoparticles by carbon deposition.

Graphite The trigonal connectivity leads to a stable planar mesostructure of carbon with strong bonds between the carbon atoms within the planes, as seen from the reduced bond distance of 142 pm. The interplanar distance of 335 pm is significantly larger than the in-plane bonds, leading to a significant reduction of the bulk density of graphite as compared to that of diamond (see Table 1). All physical properties are strongly anisotropic, with a metallic character within the plane directions and an insulator behavior [64, 65] perpendicular to the planes.

The electrical properties are unusual in that a small band overlap between valence and conduction band of the order of 0.004 eV allows a very small number of carriers (0.02 electrons per C atom) to move freely with extremely small scattering probability. This yields good electrical in-plane conductivity. Perpendicular to the graphene planes, the conductivity is of the zero-band semiconductor type with an inverse temperature coefficient as for the metallic in-plane conductivity [65, 66]. As defects increase in the graphene layers, the conduction mechanism gradually changes from metallic to semiconducting [28], with only moderate changes in the absolute resistivity, which now cannot be measured as an anisotropic quantity. A high density of defect states on both sides of the Fermi level [67] accounts for the loss of anisotropy (this is the reason why carbon brushes and carbon filament lamps work). The determination of temperature dependence of the electrical resistivity is a suitable method by which to determine the degree of graphitic medium-range structure in nanocrystalline carbon materials. This method, as well as an analysis of the electron paramagnetic resonance (EPR) signal from defects, allows one to follow the structural effects of heat treatment during the graphitization of non-graphitic precursors [68, 69].

For analytical purposes, it is important to know that anisotropic large graphitic particles are difficult to find in NMR studies as they react sensitively only to defective and semiconducting carbon. Great care must be executed when determining the structural distribution of carbonaceous deposits from NMR data due to the exceptionally large variation in the detection sensitivities of carbon species [70, 71] in magnetic resonance arising from relaxation problems in graphitic carbon, and from the coexistence of metallic and semiconducting species within the sample.

The interplanar bonding forces are of the order of 5 kJ mol^{-1}, and are so weak that small changes in the electronic structure of adjacent planes allow the incorporation of a very large number of heteroelements and molecular compounds. The electronic changes can be either: (i) a weak chemical reduction of the graphene layers by donor compounds such as alkali atoms; or (ii) a weak oxidation by bromine, halide salts or mineral acids. All of these compounds form intercalation structures with complicated structural and physical properties, and have undergone extensive investigation [21, 22, 72].

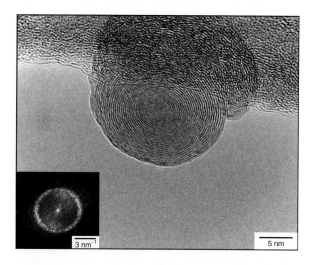

Fig. 6 TEM image of reactive nanostructured carbon after 45 min irradiation at 200 keV. The onion carbon fixed onto a base of nanostructured sp^2 carbon ("coke") is stable and exhibits in its inner 2 nm a core structure of non-graphitic, diamond-like character. Such images are not typical of natural carbon formation but require the assistance of a high-energy beam.

The graphene layers in perfect graphite are very large relative to their thickness, and are free of intralayer defects. Furthermore, they are stacked in perfect periodicity either in ABAB or in ABCABC stacking sequences, giving rise to hexagonal or rhombohedral graphite [73, 74]. Graphite in catalytic situations differs considerably from this ideal structural type, as several types of defect give rise to different families of modifications.

Apart from reactivity towards intercalation, graphite is rather inert to other chemicals, except to strong oxidizing agents which form covalent compounds such as graphite fluoride or graphite oxide [75]. The total oxidation of graphite to CO and CO_2 is a highly complex gas–solid reaction which is strongly influenced by the defect structure of the graphite [76–78]. The reason for this is the chemical inertness of the ideal graphene layer, which cannot be attacked by molecular or atomic oxygen at thermal energies. Thus, reaction occurs only at defect sites, such as missing atoms or non-hexagons, and on the perimeter of the graphene layer which collectively form the prism faces of graphite crystal [37, 79]. The number of reactive centers for oxidation in graphite is small, as these faces occupy less than 0.1% of the graphite crystal faces. Reactivity thus depends heavily on the additional sites provided by in-plane defects.

In summary, graphite and diamond are two modifications of carbon with similar strong bonds between the carbon atoms (difference in bond length 8.5%). The differing hybridization leads to a different orientation of the covalent interaction. The strong structural anisotropy of graphite as compared to the isotropic diamond gives rise to fundamental physical and structural differences. Their signatures in X-ray absorption [39, 80, 81] and X-ray emission [82, 83] spectra can also be used to discriminate the two modifications in cases where absolute minute amounts of material in disordered forms are present, as occurring in carbon deposition processes on catalysts.

Carbynes Carbynes are a metastable form of solid carbon consisting of columns of cumulenes – that is, carbons connected only with double bonds. These columns are interconnected to a quasiperiodic structure by occasional crosslinks leading to the occurrence of alternating single and triple bonds. The highly energetic material of silver-white color can be obtained under extreme non-equilibrium conditions, such as in the product of quenched graphite after shock-wave treatment, or at the cups of graphite electrodes for nanocarbons synthesis. Its formation seems to require temperatures above 2300 K, but the material can survive in ambient conditions for thousands of years as a mineral. A compilation of data [35, 36] allows one to check for the occurrence of this form of carbon under certain catalytic conditions, where hydrogen-free transparent pure carbon of non-diamond structure is identified.

2.3.15.5 Nanostructured Carbons

The complexity of carbon chemistry arises from the infinite number of combinations of mesoscopic structural elements of graphite and diamond, together with polymers of simple organic molecules and rarely of carbyne, in even a small particle of carbon material. The many classes of resulting materials can be discriminated by stating the structure type of their predominant structural unit (graphitic or sp^2, diamond-like or sp^3) and their nanostructure, meaning the connectivity in real space from molecular dimensions up to a few nanometers. As it is inherently difficult to describe this structure (see below), only coarse classifications that are based only on functional properties and not on structural details are found in literature. An overview is provided in Fig. 7.

In the literature, it has been suggested that the term "polytypism" is introduced for the families [84] of complex nanocarbon materials. In agreement with the classification of Fig. 7, one assumes three allotropes of carbon, as characterized by sections in the phase diagram, from which non-equilibrium species called "polytypes" are derived.

It is fair to state that most families of carbon are based upon the sp^2-derived motif, reflecting the spontaneous formation of this building block as most stable form of carbon under ambient conditions. This does not mean, however, that all these forms are similar to graphite or are "graphitic" carbon. The arrows in Fig. 7 indicate that admixtures of structural features from different families are often found in an individual carbon particle, rendering the discrimination of species a complex endeavor that can be mastered when complementary techniques of characterization such as electronic structure analysis, surface elemental analysis and electron microscopy are applied [85]. The benefit of such studies is an insight into interfacial behavior of carbon, such as the redox effects of surface functional groups on catalyst supports with respect to the activation of transition metal species. Also highly relevant are these structural discriminations for the prediction of stability, and hence for the design of scavenging or regeneration procedures [86] (oxidative regeneration).

"Soot" or "coke" are not individual forms of carbon but rather are mixtures of forms from Fig. 7 characterized by large amounts of heteroatomic impurities reaching levels well above 10 wt.%. The family of macroscopic graphites is also discussed here, as it will be shown below (and can be seen in Fig. 1) that even large macroscopic portions

---
References see page 419

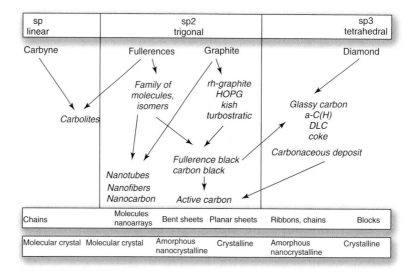

Fig. 7 Families of nanostructured carbon materials. The top line indicates the connectivity followed by the parent macroscopic allotropes of carbon. The ranges of materials within the dashed zones indicates families with similar functional properties. Their exact structures are often not known. A sharp discrimination is also not possible. Often, mixtures of such forms can be found. The arrows mark structural relationships between the families. The lower lines contain information about the shape and size of the basic structural units and the type of long-range order that can be formed in secondary or higher structural dimensions (quasi-periodicity).

of these materials are not "single crystals" (as often and incorrectly stated) with "highly oriented pyrolytic graphite" (HOPG), but rather complex arrangements of mosaic crystals composed of nanostructured building blocks with well-recognizable defect structures [87, 88] separating the individual building blocks [89]. A set of suggestions for the correct terminology of the structural and functional properties of nanostructured carbon can be found elsewhere [90].

Some basic characterization data for archetypical materials are listed in Tables 2 and 3. Cell parameters for crystalline forms of carbon are shown in Table 2, while characteristic Raman resonance data are summarized in Table 3. It should be noted that, in the context of catalytic carbon, it is highly advisable to use local characterization methods as many carbon species do not form particles sufficiently large for integral characterization by X-ray diffraction (XRD) or integral Raman [91] scattering. Even then, many nanostructured carbons still occur in mixtures of polytypes. The Raman signature indicates that, by the occurrence of several signals originating fundamentally from the C—C bond vibration, mixtures of structural motifs contribute to the polytype; however, this must not be mistaken for the coexistence of polytypes. The Raman analysis of carbon nanostructures is involved, and can feature a multiplicity of sharp lines in the frequency range indicated when the usual mixture of tubes, "graphitic carbon", and fibers is present [43, 64, 92, 93].

Tab. 3 Selected Raman resonances useful for material characterization

Material	Frequency/cm^{-1}	Lineshape
Graphite (hex)	1582	Sharp
Diamond	1332	Sharp
C60 (oligomeric)	1471 (1452)	Sharp
Carbon black	1355	Broad
a-CH	1332 + 1500	Sharp + broad
Glassy carbon	1359 + 1600	Broad + broad
Nanotubes (mixture swnt, mwnt)	1566 + 1592	Sharp + sharp

2.3.15.6 Carbolites and Fullerenes

Both polytypes have in common that they arise from high-temperature evaporation–quench processes of graphitic material. In the context of catalysis, it is possible to encounter such carbon when rapid quench situations

Tab. 2 Unit cell parameters for selected carbon materials

Material	a-axis/pm	c-axis/pm	γ/deg
Graphite (hex)	246.2	678.8	120
Carbine (I)	894.0	1536.0	120
Carbine (II)	824.0	768.0	120
Carbolite	11 928	1062	120
Diamond (cubic)	356.7	–	–
C60 (300 K)	1415.2	–	–

occur with volatile organic matter; colored films are often misinterpreted as arising from polycyclic aromatic compounds instead of considering the interference coloring of self-assembled nanocarbons which differ in their hydrogen content from the latter carbon forms.

Carbolites are orange-colored transparent solids with a very low density of 1.46 g cm^{-3} [94]. They are obtained in pure forms, similar to fullerene formation from graphite arc discharges using argon or argon–hydrogen as quench gas, but at higher pressures than for fullerene formation [95–97]. The structural properties are listed in Tables 2 and 3. Models of short chains of carbyne structure with assumed interchain crosslinks are controversial. The material is quite stable under ambient conditions.

The solid *fullerenes* [98–100] are well-defined molecular solids from a family of spherical, all-carbon molecules, the smallest monomers being C_{60}, C_{70}, C_{72} and C_{84}, all of which can be obtained in bulk quantities as purified materials. Besides the traditional graphite arc process [101–103] and laser ablation [100, 104, 105] techniques, solid fullerenes are also readily formed in combustion processes designed for both the intended [106–108] and unintended [86] generation of these molecular carbon species.

In contrast to all other polytypes, and in parallel with many polycyclic aromatic compounds, these molecules are soluble in solvents such as toluene or trichlorobenzene. They form deeply colored solutions that can be purified by liquid-phase chromatography. The respective molecular solid requires careful sublimation to remove solvent and gas adducts [109]. Fullerenes are both air-sensitive [110–114] and light-sensitive [11, 115], and decompose slowly by polymerization (becoming insoluble after storage in air) [116, 117].

Fullerenes are formed from curved and closed graphene sheets modified by the incorporation of pentagons in between hexagons. Stability rules state that two pentagons can never be direct neighbors, which results automatically in the well-known "football" structural motif of pentagons fully surrounded by hexagons. In stable fullerene forms not only one but also several rings of hexagons can surround one pentagon, giving rise to a large family of molecules with an even larger family of structural isomers [27, 49, 104].

The majority of predicted larger molecular structures [98] has not been observed as pure compounds, although evidence for the formation of larger cage molecules has been obtained from TEM investigations of the residues of evaporated graphite after the extraction [95, 118–122] of the small, soluble fullerenes [123]. Such cage molecules may be present in amorphous carbon materials that have been subjected to sufficiently high temperatures for skeletal rearrangement. These temperatures are reached in many combustion processes, and also under non-stoichiometric oxidation reactions of aromatic and aliphatic hydrocarbons [107, 124]. These studies have shown that fullerenes can occur in many conventional reactions [125, 126] involving substoichiometric oxidation reactions of hydrocarbons of almost any structure (see below), and that fullerenes and their polymers are not a curiosity among the large number of nanostructured carbons obtained from high-temperature processes of graphite.

Some morphologies of polymeric derivatives of fullerenes are shown in Fig. 8. The samples [86, 127] were obtained from various oxidation processes of polymers and fullerene black. These materials are all highly beam-sensitive, and are transformed readily into graphitic ribbons (as also shown in Fig. 8). If such morphologies are detected in the context of catalytic processes, then conditions of large gradients in carbon chemical potential and rapid transport kinetics must be present in the system.

2.3.15.7 Disordered Graphites

A large group of carbon materials arise from nanostructuring of bulk graphite, and thus represent "disorder" varieties. The average structure of this catalytically most relevant group of materials is shown schematically in Fig. 9, which summarizes the stacking schemes along the crystallographic c-axis.

Figure 9A shows the stacking of large graphene layers, these being continuous up to lateral sizes of several hundred micrometers. The most frequently observed regular stacking of carbon atoms within the sheets are the hexagonal ABA stacking, as in natural graphite, and the rhombohedral ABCA stacking. The regular stacking of graphene layers which are only very weakly bound to each other (by induced dipole moments of the aromatic π electrons) is given by a closest packing of the carbon atom spheres. This gives rise to the dimensions for ideal hexagonal graphite reported in Table 2, and to a perfect interlayer distance of 353.9 pm. It should be noted that a wide variation in the interlayer distance (up to 400 pm and down to 290 pm) can be observed in highly defective and strained nanoparticles. In many disordered forms of carbon, the interlayer distance gives rise to a broad maximum in the pair correlation function as probed by XRD, and often is incorrectly mistaken as a broad Bragg diffraction peak. Its correct analysis [128, 129] can yield valuable micromorphological information regarding the size and shape of the basic structural units or graphene units, including their average defect disposition.

The rhombohedral graphite polytype [73, 74] is slightly less stable than hexagonal graphite, and converts during oxidation or even during mechanical treatment

References see page 419

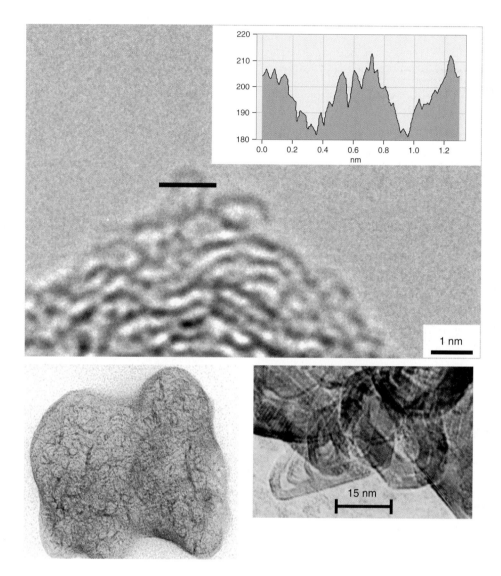

Fig. 8 TEM image of highly beam-sensitive nanocarbons created by oxidation of reactive precursors. The top image shows a single fullerene molecule (see inset with the diameter measurement of 0.6 nm) on a carbon black particle obtained from incomplete combustion and rapid quench. The lower images are typical for reaction products from fullerene-containing precursors after mild oxidation in ambient air. The highly ordered graphitic ribbons in the lower right image may arise from beam artifacts.

(ultrasonication, milling) into the hexagonal form. It can be discriminated by its powder X-ray diffractogram in the region of the cross-lattice reflections (h 0, l0), which are generally sensitive to the stacking order of a graphite material. The presence of rhombohedral reflections can be used as an absolute criterion to identify natural graphite.

The structure of graphitic material is conveniently described by the concept of BSUs. In its original form, the units were identified in electron microscopy images of poorly graphitized carbon materials [130, 131] and characterized by vibrational spectroscopy. The BSUs in crystalline graphite (see Fig. 1) are graphenes denoting large (many micrometers) patches of the hexagonal carbon network. A graphene sheet is planar and, ideally, exhibits no defects in the carbon interconnection topology; rather, it exhibits edge atoms that need to be saturated for dangling bonds by heteroatoms. The topology of terminating graphenes can either be "zigzag" or "armchair" (Fig. 10), with consequences for the electronic structure and geometric hindrance of heteroatoms (oxygen versus hydrogen, see below).

The low interlayer binding allows for a broad distribution of stacking variants coexisting with a broad distribution of orientation angles, which gives rise to a poorly ordered graphite but with a large and stable BSU. A convenient tool for the surface superlattice formation [29,

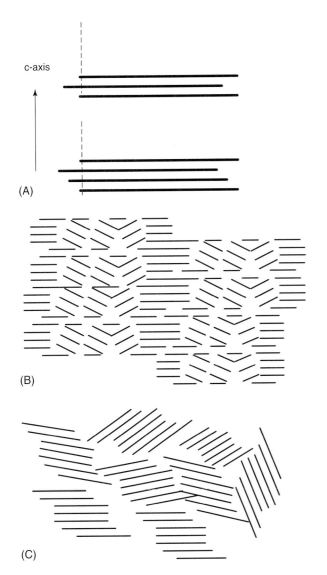

Fig. 9 Stacking variants of graphitic carbon. (A) Ideal stacking of large graphene units in hexagonal (AB) or rhombohedral (ABC) sequence. (B) Partly graphitized hard carbon with stacks of small graphene units in ideal order packed irregularly as inclined mosaic units; heteroatom defects and few sp^3 bridging functions keep the metastable nanostructure in place. (C) Irregular stacks of molecular basic structural units in a pre-graphitic stage typical for carbon black from sub-stoichiometric oxidation processes.

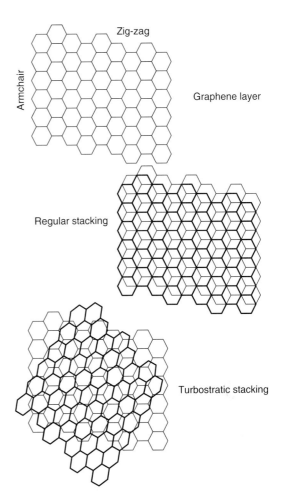

Fig. 10 (Top) A single graphene sheet with its terminations. The lower sketches describe ideal and turbostratic stacking. It should be pointed out that a continuous range of angles between successive graphene layers is possible, and that no periodic stacking sequence at all (see Fig. 9A) can coexist.

132–136], as indicated by the bottom scheme in Fig. 10, is the scanning tunneling microscopy (STM) analysis of flat areas of carbon. This can also be achieved on catalytically relevant nanocarbons [118, 137] to provide valuable information regarding stability and (oxidative) reactivity.

The unidimensional disorder of graphene stacking has a two-dimensional (2-D) counterpart in the relative angular orientation of the hexagons within adjacent graphene layers. Figure 10 shows, schematically, the two limiting possibilities of stacking in 2-D registry (relative displacement of carbon atoms 0.5 and 0.5 unit cell lengths, hexagonal stacking) and without registry referred to as the "turbostratic" disorder. This type of crystalline disorder is characteristic of all layered materials with weak interlayer bonding. Its detection indicates well-developed graphene layers with very little chemical impurity, which would force the stacking of layers into a three-dimensional (3-D) disordered state with no parallel orientation of adjacent BSUs. Non-graphitic carbons frequently exhibit this type of disorder, with the consequence that all XRD peaks other than the (002) line are either absent or very broad.

These disordered carbons constitute a transition form between graphite and carbon blacks, which are turbostratic materials with the additional complication that the lateral size of the BSUs becomes very small, with

References see page 419

typical sizes below 5 nm. As discussed for Fig. 9, in such small and chemically impure BSUs, the tendency to form planar arrangements becomes minimal, and 3-D cluster arrays [118, 138, 139] dominate the nanostructuring of the BSUs.

In graphite, the termination is achieved predominantly via hydrogen, thereby allowing a close contact of planar BSUs. This is seen in Fig. 1, and in more detail in Fig. 11. In Fig. 11, decoration of the surface of graphite crystals with "debris" from stray BSUs can also be seen as weak contrasts. These surface impurities have severe consequences on the reactivity, as they are themselves highly reactive, exposing many non-basal sites and masking the reactivity of the true surface of the graphitic crystal. The facile movements of these stray BSUs, for example under the forces of STM and the Moiré patterns between these loose BSUs and the still in-plane fixed BSUs in the surface, give rise to numerous artifacts of imaging carbon atoms in scanning probe techniques that once were misinterpreted but now are fully understood [29, 140].

The mechanical prehistory of the sample and possible preoxidation will strongly affect the abundance of this surface debris, and thus affect the chemical properties of such samples. In the context of graphitic materials as formulation additives, this surface disposition will minimize contact between matrix and additive, and hence increase the mobility of the additive under hydrostatic pressure or under mechanical shear forces. The TEM images in Fig. 11 show, again, the high abundance of defects in the graphene BSUs; these appear as bends in the center image, and as strain contrast in the bottom (lower resolution) image.

Figure 12 shows the ab-plane view on a basal plane of a graphite crystal. The hexagonal motif of sp^2 can be seen clearly, and the blurred hexagons are locations of point defects. As this TEM image is a projection of the whole transparent crystal, the defective sites occur frequently, without indicating that the surface is so highly defected. The cloudy, long-range contrast arises from random orientation of the large BSUs (larger than the field of view in Fig. 12) with respect to stacking (see schemes in Figs. 9 and 10). The imaging of graphite with TEM rather than with scanning probe techniques, removes the misconception of atomically defect-free surfaces of graphite [88], and highlights the complex surface morphology of graphite relevant to the binding of catalytic materials and for hosting heteroatoms as functional groups.

In contrast to the situation with graphite formed at high temperature, the BSUs of low-temperature carbons are of nanometer size in the non-graphitic nanocarbons, and resemble in chemical terms polycyclic aromatic compounds rather than carbon-only structures. The

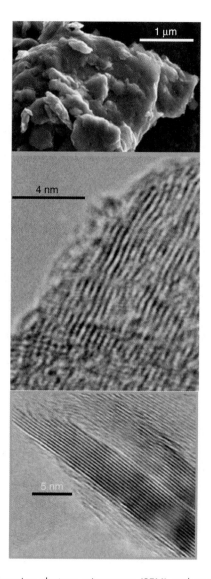

Fig. 11 Scanning electron microscopy (SEM) and transmission electron microscopy (TEM) images of natural graphite surfaces viewed along the a–b plane. The stacking of basic structural units (BSUs) of strongly varying size can be seen. Note the surface being covered with fragments of crystals (SEM) and with stray BSUs (TEM), giving rise to many scanning probe microscopy artifacts when they are slid over the underlying solid basal plane.

influence of the termination on electronic structure is substantial, as the topology of carbon connections (aromaticity) is modified with respect to a large assembly of regular hexagons. The consequences of these modified bonding situations are twofold:

- The BSUs are no longer flat but rather begin to curve due to double-bond localization at the sites of heteroatomic, non-hydrogen bonding.
- The large volume of heteroatoms other than hydrogen, with respect to that of carbon, causes problems with

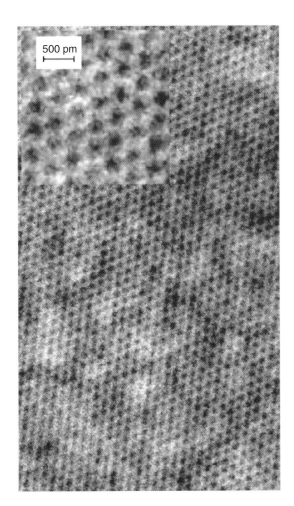

Fig. 12 High-resolution TEM image of a natural graphite crystal viewed along the c-axis. The hexagonal motif of the sp² connectivity can be seen. The large BSUs exhibit point defects (blurred hexagons) and turbostratic disorder (diffuse large contrast).

the close stacking of such units in a parallel fashion: it becomes energetically favorable to maximize the van der Waals interactions between heteroatoms and the sp^2 electronic system.

With the support of adventitious hydrogen bonding between neighboring BSUs, an arrangement of clusters of staggered BSUs in strong inclination to each other becomes energetically more favored than flat stacks, where the dispersive forces are outweighed by the repulsive interactions of the terminating atoms.

In Fig. 9B, regular arrays of parallel and inclined stacks of BSUs can be seen. This alignment arises from mixtures of graphenes and aromatic BSUs, and is characteristic of incompletely graphitized (chemically still impure) synthetic carbon which is used frequently as catalyst support materials. As the chemistry of the edges of the BSU stacks is different from that of the basal planes, a patterned surface chemistry will result. The patterning depends on the average angle of inclination, which can be measured by ω-scans around the (001) XRD peaks. In synthetic graphite, which is post-annealed after its formation under high temperature and at hydrostatic pressure [141] and called "highly oriented pyrolytic graphite" (HOPG), this angle is as low 1° for top-quality samples used as graphite monochromators. The in-plane coherence of the BSUs is weak, giving rise to frequent ruptures and turbostratic rotations of the fragments, as shown in Figs. 1 and 10. At average angles of 5° in XRD, these materials are known as "pyrographite".

The absence of any preferential alignment, as depicted in Fig. 9C, is observed between the BSU stacks in pyrocarbons and some carbon black materials. Such a secondary structure is often observed when the process of carbon generation occurs at low temperatures, or in large temperature gradients. A substantial number of covalent bonds still exists between individual BSUs via local sp^3 connectivities, locking in space random orientations between otherwise flat BSUs. The surface chemistry of these hard carbons is characterized by a hydrophilic nature, caused by the many terminating oxygen functionalities on these materials.

Kish graphite is a well-ordered (by XRD, it is better than HOPG) variety of synthetic graphite obtained from exsolution of carbon out of liquid iron. Under certain conditions, excess carbon – which is insoluble as FeC_x – crystallizes in a well-ordered platelet morphology and can be isolated by dissolution of the iron matrix in mineral acids. The macroscopic size of the BSUs allows the formation of graphene stacks of very high structural integrity. Under certain cooling conditions of the iron melt, these flakes can grow to 1 mm in diameter. They represent, by their extraordinarily high degree of ordering, the closest known approximation to graphite single crystals. In catalytic systems with iron as active species, the microkish graphite particles are also of highly crystalline perfection, rendering them highly resistant to chemical transformation in oxidizing or reducing atmospheres, and hence resistant to regeneration procedures.

2.3.15.8 Nanocrystalline Carbon: Carbon Black

Carbon blacks appear as chainlike aggregates of spherical particles of 10 to 50 nm diameter [142–144]. This morphology represents the second hierarchy level of nanostructuring most evident in electron microscopy studies. A typical image from a sample of carbon black from a diesel engine is shown in Fig. 13. These materials are similar in their structuring and chemical behavior to synthetic carbon blacks, and represent compact carbon

References see page 419

produced by incomplete oxidation in many catalytic processes. Deposits from non-oxidative processes appear morphologically much more graphitic, with stacks of small BSUs, as shown in Figs. 1 and 11.

After contact with fluid phases, the chains of functionalized carbon black become further densified, giving rise to compact flake-like appearance which represent the third hierarchical level of structuring of nanocarbons. The distinction of these platelets from true graphitic platelets by TEM requires careful deaggregation in, for example, dilute ammonia solution, or sonication in non-aqueous media.

The strong tendency of the primary spherical carbon particles to reform the chain-like agglomerates after chemical or mechanical disruption serves as the origin of many applications of carbon blacks, for example in well-dispersing catalysts supports or as fillers in numerous polymer composites. Elemental analysis reveals a significant hydrogen- and non-hydrogen atom content of carbon black. These non-carbon atoms are responsible for the aggregation of the spherical units as they create surface functional groups which allow for hydrogen bonding and solvation interaction through ubiquitous water films.

Highly impure forms of carbon black are often called "soot", which refers to their likely origin of incomplete combustion [43, 145, 146]. Such materials exhibit a typical nanocrystalline ordering as the XRD mechanism changes from Bragg diffraction to elastic scattering with the different physics of the scattering process [81, 128, 129, 147–149]. The broad peak in the range of the graphite (002) is no longer a Bragg peak, but represents the absolute maximum of the pair correlation function which is dominated by the coplanar arrangement of small graphene layers with sizes comparable to those of large aromatic hydrocarbons.

The origin of the spherical morphology of carbon blacks is an energy-optimized arrangement of small BSUs with termination irregularities, as discussed above for the prize of turbostratic disorder. The spherical mesostructure maximizes the contacts between heterostructures within one particle under simultaneous exploitation of the dispersive forces between BSUs. Figure 14 represents models [131, 144, 150, 151] and HRTEM images of the primary structure of carbon blacks and their BSUs. The shell structure of stacks of tiny graphene units (top left in Fig. 14) optimizes the interstack interaction via heteroatoms located at the prism edges of each graphene layer. The shells are filled by molecular aggregates polycyclic aromatic hydrocarbons (PAH), or grow around a nucleus of inorganic ash.

The interactions between stacks of BSUs (Fig. 14, bottom left) are mediated in many carbon blacks through hydrogen bonds of adjacent OH groups. The outer surface of a black particle also contains significant numbers of OH groups [17] which control the aggregation of the primary spherical particles to fractal larger aggregates; this is shown in Fig. 13, and is often called the "structure" of carbon black. Modification of the ion strength (e.g., in aqueous solution) of the environment of secondary carbon aggregates affects the strength of the interaction and influences the aggregation behavior [85, 152]. Thermal activation of the hydroxyl groups creates dangling bonds and a reorientation of the BSU stacks so as to form covalent crosslinking within each spherical particle and within the secondary structure. The process creates sp^3 bonds and leads to a decrease in chemical reactivity and an increase in mechanical strength. This important aspect of the reactivity of nanocarbons with a high fractional prismatic surface area has the consequence that, even under oxidizing conditions, thermal treatment can transform the reactive carbon black material into a highly stable glassy carbon (see below).

The process of structural reorientation during the combustion of carbon is most relevant for regeneration processes of catalysts, leading to "carbon ash" materials that are non-reactive towards the regeneration conditions. In a stability study on carbon black [153], a correlation was found between the surface oxygen content measured by XPS as proxy for the number of defects or the inverse of a "graphitic" structure, and the oxidative stability under catalysis-relevant conditions. It is clear from Fig. 15 that

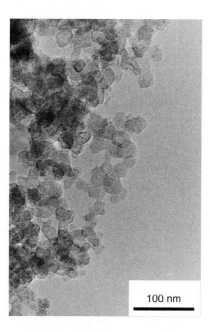

Fig. 13 TEM image of a carbon black sample from combustion in a diesel engine. The typical chain morphology of the secondary structure of carbon black can be seen, as well as the beginning of agglomeration to the (dark) platelet structures representing the third level of hierarchical structuring. The platelets, although being agglomerates of agglomerates, are not transparent to 200-keV electrons.

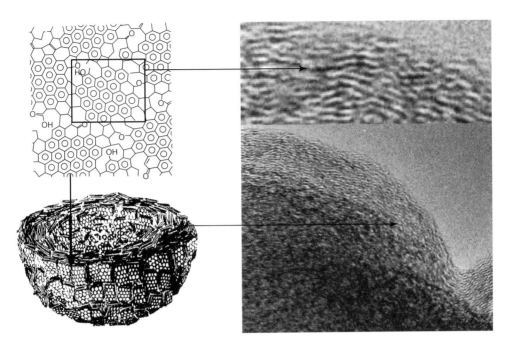

Fig. 14 Basic structural units (BSUs) and their three-dimensional arrangement in carbon black. The individual units (box in top-left scheme) exhibit irregular terminations (see Fig. 10), are covalently crosslinked, and carry functional groups preventing both the planar array of graphitic structures. The images reveal size, stacking and bending of the individual BSUs occurring as black and white stripes in the top-right image.

the defect structure of the carbon is highly relevant to its reactivity. For strategies of the gasification of nanocarbons, it becomes further relevant to avoid thermal cleavage of functional groups, as the reactivity of the solid will then decrease.

The reactivity can be estimated qualitatively [76, 79, 86, 153–155] by electron microscopy classification of the agglomeration of the BSUs in a given nanoparticle of carbon. Figure 16 exemplifies how the relative abundance of clusters of parallel-aligned BSUs indicates the reactivity as a trend. The thicker the compact stacks of BSUs, the lower is the reactivity of the whole particle as the abundance of sites capable of initiating oxidative attack is smaller. For this reason, it is highly desirable to accurately characterize the structure of the BSUs and their supramolecular arrangement. The reaction of carbon with oxygen is either wanted (regeneration) or unwanted (support), and this can be controlled by suitable nanostructuring during formation of the carbon. In conditions of unwanted carbon formation, care should be taken to control the kinetics of supramolecular assembly so as to achieve the morphology depicted on the right-hand side of Fig. 16.

Fullerene black [70, 97, 108, 118, 119, 127] is of a similar constitution and secondary structure as normal carbon black, but with the important difference that its BSUs are composed of non-planar graphene layers. Consequently, the stacks are not straight but are curved, giving rise to a more disordered and less-dense carbon material (see Fig. 8, bottom left). The chemical reactivity is much higher [110, 156] as the stability of non-planar graphene layers is reduced. In addition, the irregular shape of each graphene layer prevents the formation of parallel stacks to a large extent. A fully amorphous array with a large

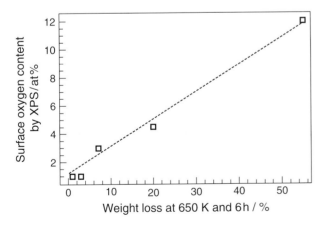

Fig. 15 Correlation of carbon black stability measured as weight loss in combustion against its content of surface oxygen groups. The stability was measured isothermally by thermogravimetry under 5% oxygen in inert gas. The samples were several types of carbon black from combustion processes.

References see page 419

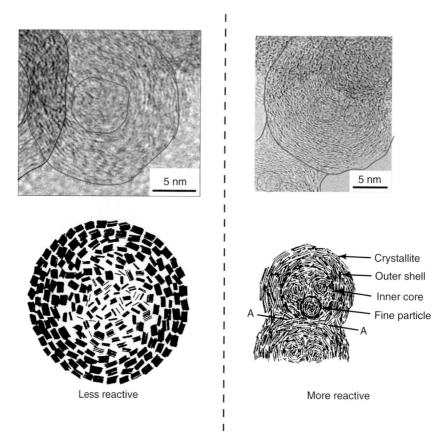

Fig. 16 Typology of reactive nanocarbons from their microstructure. The more packages of parallel inclined BSUs that can be found in a particle, the less will be the accessible fraction of surface functional groups and the less reactivity can be expected.

surface fraction of prism faces will result, and this will additionally increase the chemical reactivity [95, 111]. The low stability in the electron beam of the TEM instrument is an indicator of the low structural stability of BSUs with a significant content of non-six-membered carbon rings.

It should be noted that, in fullerenoid carbons, this origin of enhanced reactivity is not caused by the presence of peripheral defects as heteroatoms, but arises from internal defects of the non-six-membered carbon rings of the BSUs. Figure 17 summarizes these interactions in nanocrystalline carbons constituting a large fraction of carbonaceous deposits on catalytic materials. The central scheme in Fig. 17 indicates how a simple point defect in a growing graphene sheet leads to fullerenoid structures by minimization of the total energy through minimization of the number of dangling bond into a cyclic perimeter. Thus, the fullerenoid structures as depicted in Fig. 17A are by no means material curiosities, but will occur readily in the polymeric carbon that is frequently encountered in deposits from catalytic processes. The upper scheme in Fig. 17 contrasts the fullerenoid curvature to that originating from functionalization of the perimeter of a non-defective graphene with heteroatoms (see above and Fig. 14). Weakly curved graphene BSUs will form small stacks which are limited in their thickness by the internal strain caused by the mismatch of atomic dimensions, and thus counteracting the inert-graphene dispersive forces. Figure 17B shows, schematically, how blocks of BSUs will result, as observed in reality in Fig. 16 (top left image). Non-planar carbons exhibit a high reactivity towards polymerization under weakly oxidizing conditions, and a high reactivity towards total oxidation under strongly oxidizing conditions [76, 79, 156].

The materials described in Fig. 17A and B result from weak interactions between the individual graphene building blocks. A hard and stable carbon with sp^2 and sp^3 local bond geometries is formed after crosslinking planar or non-planar stacks of BSUs with bonds, atoms or even molecules, as shown schematically in Figs. 14 and 17C. The resulting hard carbon black is non-graphitizable by thermal treatment up to 3000 K. The sp^3 bond cannot be rehybridized to the sp^2 configuration. These hard carbons are referred to in catalytic systems as carbonaceous deposit, carbon ash, or "hard coke" [71, 82, 157–160]. The material is chemically heterogeneous, structurally amorphous, mechanically hard, and exhibits

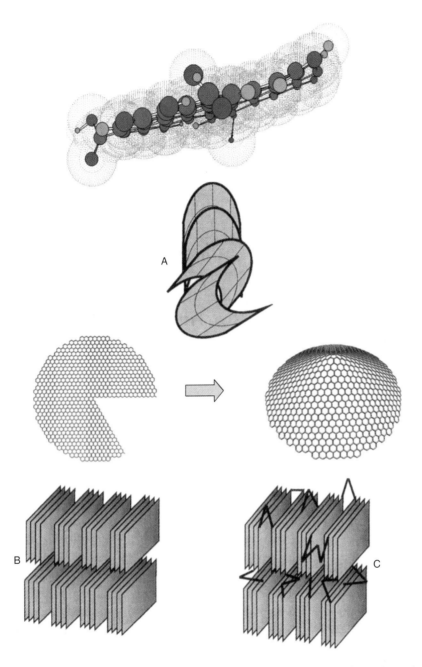

Fig. 17 Origin of non-graphitic BSUs. (A) Curved structures due to peripheral or central defects of a single graphene unit: top scheme: curvature as consequence of defect saturation by oxygen atoms (light gray), central scheme: formation of a fullerenoid structure due to one single missing carbon atom at the apex (five-membered ring). (B) Packages of molecular BSUs units with few peripheral heteroatoms other than hydrogen. (C) Packages of BSUs with covalent interlinks created by sp^3 carbons and/or heteroatomic bridges.

no graphitic properties, although the large majority of all carbon–carbon bonds are sp^2 hybridized.

Higher contents of heteroatoms and the presence of nitrogen, sulfur and oxygen are characteristic properties of coke, the product from carbonization of coal or other hydrocarbon sources with larger molecular constituents. Carbonization is a polycondensation process in which small hydrocarbon molecules such as methane, propene or benzene are liberated. Heteroatoms in the form of ammonia, water and hydrogen disulfide are removed, and irregular stacks of small graphene layers (originating from the large organic molecules) are crosslinked by molecular fragments under preservation of the irregular orientation of the stacks. The resulting structure is similar

References see page 419

to "catalytic hard carbon", with the same features as described in Fig. 17C.

There exist families of carbon with a still higher content of heteroatoms enumerated in Fig. 7. In a-CH [161] or diamond-like carbon (DLC) [63, 162], the hydrogen content is as high as CH0.4 and the carbon atoms are arranged in ribbons or chains, with no discernible stack structure of graphene layers.

Glassy carbon [163], with its pronounced electronic structure much different from that of graphite [164], is a technical product prepared via the carbonization of organic polymers; alternatively, in the electronics industry glassy carbon is produced by plasma-treating organic precursors. It consists of ribbons of sp^2 carbon with a significant contribution of non-hexagonal rings, allowing a 3-D entangled structure. This creates the high mechanical hardness [165] and resistance against graphitization, which renders this material excellent mechanical stability [166] at high thermal and mechanical load (only under non-oxidizing conditions). Turbostratic stacks of 3 nm width and 10 nm length form the continuously bent BSUs of this carbon, which can be formed in catalytic reactions allowing polymerization of molecular precursors [83, 121] followed by carbonization at around 1000 K (in regeneration cycles, for example). Although the content of sp^3 bonds is low in this material, the hydrogen content is significant as it is required to saturate the high prismatic surface area of the ribbons. Vibrational spectroscopy is a versatile tool to identify the presence of the interlinking bonds, both with Raman for carbon–carbon bonds and with infrared (IR) spectroscopy for the heteroatomic bonds [167]. Such carbons of "metallic" hardness can be found on catalysts and reactors likewise, and are sometimes difficult to analyze due to their apparent non-carbon properties and their frequently strong adhesion to the substrate.

Active carbon [59, 138, 168–173] or charcoal [169, 174] is an important modification of carbon in catalysis. It consists of carbonized biopolymer material, being activated in a second step of synthesis. This procedure creates a high specific surface area by oxidative generation of micropores of enormously variable size and shape distribution [175]. A more controlled activation is achieved by the addition of phosphoric acid or zinc chloride [176] to the raw product. The additive is incorporated during carbonization into the hard carbon, and is subsequently removed by leaching; this creates the empty voids in a more narrow pore size distribution as achievable by oxidation. Other activation strategies employ liquid oxidants such as nitric acid to create large pores with a very wide size distribution [177]. Today, a large body of patent literature describes numerous ways to structurally activate carbon [178] from mostly natural sources [179–181].

A large number of oxygen functional groups are created during the activation process by the saturation of dangling bonds with oxygen. This creates a rich surface chemistry [17, 59, 85, 172, 182–184] which is used for selective adsorption [148, 183, 185]. In addition, it determines the ion-exchange properties that are relevant for catalyst loading with active components [152, 185–189]. A more detailed description of the surface chemical aspects most relevant to the catalytic uses of nanocrystalline and activated carbons follows.

The microstructure of activated carbon may be described by the archetypes depicted in Fig. 17, and may even contain significant amounts of graphitic structure [189] depending on the temperature regime of activation. The micromorphology is often characteristic of the biological origin of the raw material, with shapes of cell arrays and even whole plant organs being detectable by scanning electron microscopy (SEM) imaging. Active carbon contains significant amounts of inorganic matter (iron silicates, silicates, calcium oxides, etc.) being analyzed as ash content after complete combustion. Activated carbons are very heterogeneous materials and are difficult to characterize in terms of microstructure and reactivity. Several reviews related to this class of large-scale technical carbon are available [85, 152, 190].

To summarize, a wide variety of forms of carbon can be found under catalytic conditions. In particular, during deposition from organic feedstock and during thermal post-treatments in regeneration cycles a very complex carbon material with unexpected chemical and structural properties can result, causing severe problems in subsequent regeneration cycles. The group of carbons specifically employed in catalytic systems is summarized in Table 4, together with its most important physical parameter, the specific geometric surface area. This ranges over three orders of magnitude, and reaches values between geometric surfaces of micron-sized particles up to the surface area of the number of carbon atoms in the sample. These latter values are non-representative if they are interpreted in terms of a monolayer coverage of the nitrogen probe molecules. Highly porous carbons exhibit

Tab. 4 Types of carbon support and their geometric surface areas

Carbon	Surface area/$m^2\ g^{-1}$
Natural graphite	0.1–20
Synthetic graphite	0.3–300
Graphitized carbon blacks	20–150
Carbon blacks	40–250
Activated carbon from wood	300–1000
Activated carbon from peat	400–1200
Activated carbon from coal (coke)	200–1000
Activated carbon from coconut shells	700–1500

pronounced capillary condensation phenomena requiring a full BET isothermal analysis for the determination of the true geometric surface area. Although this is widely known and described in detail in other chapters of this Handbook, it is still customary to present these very high numbers of BET surface area. Care must be taken when comparing data based on surface area measurements in carbon chemistry, as some sources determine the monolayer sorption capacity, whereas others do not. For this reason, surface area values determined with various adsorbents [169], such as nitrogen, CO_2 or Kr, may differ significantly as these probe molecules exhibit quite different condensation properties under the conditions of adsorption. The most useful contribution of sorption data [191] to evaluate the nature and size distribution of porosity in carbon materials is not discussed at this point.

It should be noted that, in different research areas, the use of terminology to describe the structure and properties of carbons varies widely. However, within the carbon-science community a clear system of terminology and definitions has been elaborated which is recommended whenever applicable [90].

2.3.15.9 Formation of Carbon: General Pathways

The technical synthesis of graphite, diamond and a variety of other forms of sp^2 carbons is described in a comprehensive review [192]. As the unintended formation of carbon in deactivation processes and the modification of primary carbon surfaces during chemical treatment (in catalytic service and during oxidative reactivation) are frequently encountered problems in catalytic carbon chemistry, it seems appropriate to discuss some general mechanistic ideas. These concepts stem from the analysis of combustion processes (flame chemistry) and their carbon formation (frequently called "sooting" [155], without meaning solid carbon rich in heteroatoms) and from controlled-atmosphere electron microscopy. In a review on carbon formation on acidic surfaces [159] it is stressed that the mechanisms of sooting would depend on the reaction conditions, and vary from condensation to chains to the formation of polyaromatic intermediates. It should be noted that exactly the same dual pathways exist also in combustion chemistry, and that notions are justified about process similarity between combustion and catalyst deactivation by carbon deposition from molecular (reactant) precursors without flame.

The source of all carbon relevant to the present context is the feedstock of hydrocarbon molecules (aliphatic, aromatic, with and without heteroatoms); the possibilities for their conversion into black carbon are summarized in Fig. 18. The chemical route comprises polymerization [193–195] via several steps into aromatic hydrocarbons with final thermal dehydrogenation. The reaction steps [126] depend on the fuel and the combustion conditions used. Even when, during the first steps of aggregation, condensation into small C_2–C_4 chain-like fragments is the dominant process, this scheme does not continue as intuitively inferred from Fischer–Tropsch chemistry, but rather is taken over by cyclic condensation reactions according to the Diels–Alder reaction scheme.

A critical intermediate is the formation of first rings such as benzene and alkyl-substituted benzenes and cyclopentene units [196]. The following steps of condensation and growth involve the critical participation of radicals and ions, leading to a large variety of aromatic and polycyclic compounds [124, 197, 198]. At this stage, a liquid-crystalline phase is often formed immediately before final solidification. In this phase, large aromatic molecules can self-organize into parallel stacks and form well-ordered precursors [199] for graphitic structures with large planar graphene layers. This phase is referred to as mesophase [199, 200], and can be observed using polarized light optical microscopy.

The physical pathway in Fig. 18 requires activation of the molecules by a high external energy input, for example through high-temperature flames, through projectiles such as accelerated and free electrons, or the presence of a plasma source [201, 202]. These conditions can also occur in analytical instruments such as electron spectrometers and microscopes, which have the potential to create black carbon during the analytical inspection of hydrocarbon layers [203]. In all of these cases the molecules are fragmented and dehydrogenated in one process, and represent a source of free carbon atoms [101, 102, 201] or small fragments such as C_2 dimers. These fragments condense into chains of carbon atoms that are also observed in plasmas and in the interstellar medium [83, 97], and have a strong tendency to form

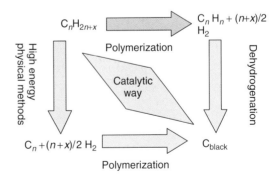

Fig. 18 Reaction pathways from organic molecules to solid carbon. The "chemical" synthesis (top, left to right) encompasses both dehydrogenation and polymerization. The "physical" synthesis (left side) involves fragmentation of feedstock by high-energy projectiles (ions, atoms, electrons) followed by polymerization.

References see page 419

small ring molecules by nucleophilic self-substitution of the end groups. Saturation of the dangling bonds of the primary rings by polymerization leads to graphene layers. Primary condensates giving rise to non-six-membered carbon rings are very rapidly cut to size through the evolution of radical rests; alternatively, they may redissolve under the attack of smaller units. These processes occur very rapidly compared to the slow activation along the chemical pathway. Only a limited time is available for orientation of the rings into stacks, and crosslinking into non-planar units can easily occur (see Fig. 17). Hard, non-graphitizable carbons are the final products of these processes, giving rise to the poor reactivity of the "physical carbon" as opposed to "chemical carbon" generated through slower multiple condensation.

A large number of intermediate pathways are possible when catalytic reactions interfere with the polymerization–dehydrogenation steps. Here, a common scenario is the catalytic dehydrogenation of hydrocarbons on nickel surfaces, followed by dissolution of the activated carbon atoms and exsolution of graphene layers after exceeding the solubility limit of carbon in nickel. Such processes have been observed experimentally [204, 205] and used to explain the shapes of carbon filaments. In the synthetic routes to nanotubes (see below) [206–208], the catalytic action of *in-situ*-prepared metal particles was applied to create a catalyst for the dehydrogenation and for shape-controlled [209] synthesis.

2.3.15.10 Formation of Solid Carbon: Mechanistic Aspects

A schematic correlation between reaction pathways and the principal structural properties of common carbon forms for materials with predominant sp^2 connectivity (as shown in Fig. 7) is presented in Fig. 19. The literature concerning structure formation [97, 146, 159, 193, 210–214] is somewhat controversial [215, 216], and allows only a conceptual survey of chemically plausible pathways relevant to the conditions of catalytic reactions in carbon formation.

Three main synthetic routes to carbon can be identified, none of which produces a single polymorph or subfamily of solid carbon. In all cases mixtures are produced, as indicated by the central component of Fig. 19. This complexity highlights the strong influence that kinetic aspects have on formation reactions. The realization of a perfect sp^2 local configuration is always energetically favorable, and is reached in graphene layers with only hexagonal constituents. The saturation of the peripheral dangling bonds of graphenes can occur either by homonuclear crosslinking (resulting in macroscopic planes or irregular arrays of stacked graphenes) or by heteronuclear bonding (resulting in either polymers or surface functional groups). Metastable configurations contain mixtures of six-membered rings with other ring geometries causing strain in the sp^2 bonds. A wide variety of non-planar structures results, referred to as nanocarbons. These structures are chemically reactive and disappear in oxidizing or hydrogenating atmospheres faster than do planar structures exposed to the same conditions. This explains the additional strong influence of gas phase composition on the evolution of structure and reactivity of carbon materials during their gasification.

The synthesis routes shown at the top of Fig. 19 represent the chemistry of carbon formation via initial fragmentation, as initially derived for combustion processes [196, 215, 217]. In high-temperature combustion

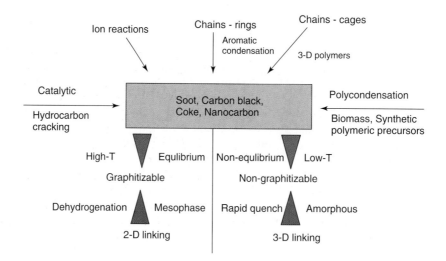

Fig. 19 Reaction pathways towards the formation of mixed carbons with predominant sp^2 connectivity. The arrows denote the pathways of formation, which may act together or occur as individual modes of operation; the triangles indicate the key structural properties of the product family denoted in the central box.

reactions (above 1200 K), the general reaction occurs via the "physical" pathway indicated in Fig. 18, with initial fragmentation into neutral species and ions. C_2 fragments or oligomers react to form carbon chains which react further to predominantly aromatic molecular fragments (with dangling bonds). In highly oxidizing atmospheres, 3-D cage structures are formed which exhibit both sp^2 and sp^3 bonds. The building blocks condense to graphene BSUs with a planar or bent layer structure in cooperation with ionic fragments acting through their high chemical energy as "etchants" for "wrongly condensed" primary condensation intermediates.

When comparing the models and reaction pathways in the literature, it should be considered that even qualitative results depend on the type of combustion process studied. As the whole process is under strong kinetic control, it is not surprising that energy production, the transport properties of the flowing system and homogeneous–heterogeneous exchange processes will affect the findings. Sooting flames at low temperatures with low transport may be a suitable model for carbon deposition in heterogeneous cases, for example at surfaces from reacting hydrocarbon intermediates. In a detailed analysis of the flame chemistry of very fuel-rich systems [107, 124, 197, 198], a three-zone model of a flame was derived. The elementary building blocks for carbon were formed in the hottest oxidation zone, while a surrounding zone with luminous constituents was identified as the location of chain growth into polyalkyne molecules of the form C_nH_2. In the third dark zone, these molecules rearranged into polycyclic hydrocarbons, and finally into graphene layers. Ample evidence for this model comes from mass spectral analysis of the gas phase in various types of flame. This diagnostic method [218] does not provide any indication about the reactive particles since, by the time of their detection, they are energetically cold and represent the ground state of a mere species rather than the reaction intermediate. For this reason, it is difficult to see that long linear chains, as are found under certain conditions of carbon condensation characterized by low density of fragments [83, 97], could be important intermediates under conditions of high abundance of reactive intermediates, such as when adsorbates on catalytic surfaces react, since long-chain structures are rarely found in carbon products.

Scenarios in which initial condensation processes [126, 196] are modified by interaction of thermoions [146, 150] and/or aromatic radicals [219] and several independent repair mechanisms for a maximum yield of six-membered condensed rings, seem more plausible [218]. The kinetic boundary conditions are of major importance in all of these processes, as large local temperature gradients are required to allow a significant accumulation of carbons from reactive intermediates. At constant high temperatures the unstable species decompose effectively back into their constituents. This effect was studied in detail in several types of combustion reaction [220, 221], and is the guiding principle of all technical processes for carbon black synthesis [192].

Another important aspect is the nature of the hydrocarbon source in relation to the efficiency of carbon formation. The idea that the structure of the molecule should be relevant for carbon formation (aromatic molecules are better than aliphatic structures) is incorrect in all situations where a pool of primary building blocks is formed (in flames, under oxidative polymerization conditions and under physical activation). However, it is confirmed that a decrease in the average C–C bond energy allows easier chemical fragmentation, which enhances the rate of carbon formation [222]. Also, under low-temperature formation conditions there seem to be effects of the structure of the precursor on the final carbon particle [223].

The growth of carbon particles cannot occur in one synthetic step by condensation of uncharged small molecular units. However large an aromatic molecule may become [124, 126, 219, 224], it will be too small to create a micron-sized carbon particle. Three major routes can lead to such particles. The most unlikely yet proposed case (by reason of intuition) [49] is a continuous polymerization process at the surface of a fullerene molecule. The starting point would be long-chain carbon molecules, since these exist [225] in low-density, high-energy situations. Both, the unfavorable kinetics [226, 227] and an absence of any structural evidence that fullerene-like structural elements (as identified by characteristic C–C bond lengths) occur in the radial distribution function of black carbon material [147, 228], are strong indicators that carbon particles, despite their spherical morphology, grow in a heterogeneous manner rather than a homogeneous manner. The explanation for this morphology was provided above (see Section 2.3.15).

In a combined kinetic, statistical and microscopic analysis of the growth process [146, 214], a model of surface growth was proposed. According to this, the polyaromatic molecules are dehydrogenated and the dangling bonds saturated by the cocondensation of alkyne units. Such a process, whilst being assumed to work with neutral species, also serves as a likely model for the here most relevant heterogeneous condensation mechanism whereby not only a carbon surface but any catalyst surface may be precovered with reactive carbon fragments to serve as a starting unit. In *in-situ* electron microscopic studies, such processes [62, 63, 229] with defective carbon surfaces have been observed and discussed. A mass growth of carbon particles with 90% of the total carbon

References see page 419

inventory could be explained in this way, which is in excellent agreement with experimental flame studies. The prediction of this model for a structure-less agglomerate of aromatic molecules surrounded by a concentric array of graphene layers [217] is also in excellent agreement with HRTEM observations of homogeneously produced carbon particles. Such cases are also seen quite clearly in Fig. 16. The spherical texture is assumed to be the consequence of droplet formation of aromatic molecules (mesophase) in the initial stages of supersaturation [199, 200] required for homogeneous nucleation. These droplets are not residues from the possible liquid initial fuel, but rather occur as a consequence of the very rapid initial polymerization of the highly active small carbon fragments.

The third – and most likely – mechanism of carbon formation is a combination of the generation of graphene units with and without curvature, according to the availability of heteroatoms and of reactive repairing species with a spatiotemporal decoupled aggregation process forming the secondary structure; this, in turn, may be annealed in temperature gradients of the reacting system. Such a scenario is not amenable to a direct mechanistic evaluation, despite such claims [209, 230], due to the loss of data regarding formation chemistry and aggregation chemistry in post-reaction studies, as these are usually performed by analyzing the reaction products. This problem is identical to any catalytic mechanistic study performed with reactor exit analyses only. The *in-situ* study of carbon formation is, despite progress with mass spectrometry [231–233], photoemission [223, 234, 235] and light-scattering techniques [97, 124, 194, 218, 236], still incomplete and has not provided a full picture of the initiation, formation and parameter dependence of the process. Great care should be taken with *ex-situ* and post-mortem studies of carbon deposits when they are used to infer reaction mechanisms [64, 97, 195, 209, 212, 224] for carbon deactivation. A useful compilation of the partly difficult-to-access literature can be found in [237] dealing with the mechanistic aspects of carbon formation in combustion processes.

In summary, all processes that homogeneously generate carbon from molecular precursors and from homogeneous activation, lead to spherical particles composed of small BSUs of molecular dimensions which may either be flat, as graphene layers, or curved. These BSUs are still very much of a molecular nature, carry a significant number of peripheral non-sp^2 and non-carbon atoms, and give rise to molecular reactivity. The mechanisms of their secondary structure formation were discussed in Section 2.3.15. The resultant materials are also non-graphitic, despite the literature frequently referring to carbon deposits as "graphitic". These may be polycyclic aromatic in nature, they may contain fullerenoid structural elements, and in many cases they are still chemically reactive. Their reactivity shows two avenues of oxidation and supramolecular ordering, with temperature and the oxidizing potential of the surrounding phase being the key control variables. Detailed descriptions of these reaction avenues are provided in the following sections, while portraits of such nanocarbons have been discussed earlier, with Figs. 4, 6, 8, and 13 representing morphologies all of which may found with "catalytic carbon".

It is also possible, however, that much better-ordered carbon is observed in catalysts which have undergone a mechanism of prepolymerization of hydrocarbons prior to carbon formation. Such carbons, with much larger continuous graphene layers as BSUs, can be obtained as materials from the pyrolytic polymerization [169, 238–240] of prepolymerized hydrocarbons. Synthetic materials such as kapton or polyacrylonitrile (PAN), or biopolymers [178] such as coconut shells, wood coal and rice husks, can be used. Carbonization or "coking" involves the chemical removal of all heteroatoms such as nitrogen or sulfur, and crosslinking of the polymeric blocks via the dangling bonds created by removal of the heteroatoms. In this way, the texture of the precursor material is preserved in the final product. As no fragmentation of the precursor is required, this method of carbon generation is much less energy-intensive. Moreover, it leads to a great variety of structural properties that give rise to many materials of the classes of "activated carbon" and "carbon black" (see Section 2.3.15.8). The possibility of retaining a portion of the heteroatoms, such as silicates in biomass carbon, makes it feasible to produce extra strong carbons (silicate-reinforced). These may be useful for anchoring catalytically active metal components such as noble metal particles. However, the exact chemistry of carbonization is complex and depends on the nature of the precursor and on the gas phase chosen (inert, oxidizing, steam, reducing). For this reason, no general description is possible of this difficult-to-control process, which occurs between 500 K and 1000 K. However, details of structural information and relevant literature [59, 137, 144, 171, 172, 178, 180, 200, 241–247] are available in Section 2.3.15.9.

2.3.15.11 Catalytic Formation of Carbon from Molecules

The catalytic generation of carbon, as indicated in Figs. 18 and 19, is of significance as the deactivation reaction in many processes involves the presence of hydrocarbons and of transition metal reaction sites. These either catalyze dehydrogenation reactions or form solid acid–base centers catalyzing polymerization reactions. Much data have been compiled [204, 205, 207, 208] concerning the morphology of carbon deposited as a function of deposition temperature. The data in Fig. 20 show that the predominant carbon form at most catalytically

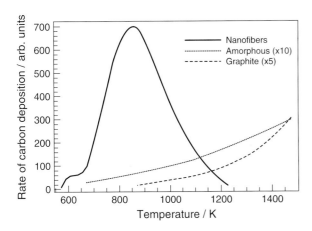

Fig. 20 Relative rates of carbon polytypes developing on metal particles. Under all temperatures of deposition the evolution of nanofibers (in older literature, "filaments" or "whiskers") is the dominating process, followed by the occurrence of graphitic flakes and to a small extent of amorphous soot-like morphologies. This finding is universal for all transition metals. Plotted from data of Refs. [101, 204, 206].

useful temperatures is filamentous carbon. The term "filaments", which was used before 1990, is synonymous with "carbon whiskers", and designates nanofibers and nanotubes of carbon. The carbon source may be CO, methane, acetylene or other hydrocarbons [205, 206, 230, 248]. The metals for which this correlation is valid are either nickel or iron on all types of conventional support, and also as binary or Cu alloys. It is important to note that forms of carbon other than filaments rarely grow under the conditions required for steam reforming or Fischer–Tropsch synthesis [249–251].

The sequence of events for the growth of these nanocarbons, which can destroy the entire catalyst bed, is now well understood [211, 230, 252–255]. Electron micrographs [204, 256] of the catalyst particles show that they are detached from their support and form either the front (or occasionally the center) of a growing carbon filament. In all cases, the shape of the particle changes during its operation as catalyst for the growth of nanocarbons. The growth of an anisotropic carbon particle requires structure sensitivity from the polymerization process of carbon atom to graphene fragments. Otherwise, a shell of graphite would be produced around a core of metal, inhibiting its catalytic function (for the desired reaction as well as for nanocarbon growth). Such morphologies have been observed frequently with noble metals [257], but less frequently with iron systems at moderate temperatures in carbon-rich environments or at high temperatures and gas compositions which are lean in terms of carbon source.

An example of a dense graphite layer covering a nanoparticle of Pt after contact with n-hexane can be seen in Fig. 21. The well-ordered graphene layers reproduce the shape of the Pt particle in a perfect way, and produce a continuously bent structure with substantial strain in the graphene layers to retain a tight fit against the parent metal surface. The fact that very little production of amorphous carbon with uncontrolled polymerization of the BSUs is observed further illustrates the relevance of the metal catalyst particle, which not only generates the source for the solid carbon formation but also controls the geometry of the polymerization process.

The structure sensitivity of the catalyst is generated during an inductive period in which carbon is dissolved into the metal. A flat surface with a subsurface carbide forms, which dissociates the molecular carbon source. The activated atoms diffuse through this layer and migrate to other faces where they exsolute, forming graphene layers. The reconstruction was observed with electron microscopy, which revealed a deformation of spherical metal particles into droplet-like conical particles. The base forms the active surface for dissociative adsorption, and is always directed away from the surface from which the filament grows. The conical mantels are faceted in various directions and allow the growth of graphene layers with different velocities and orientation. It remains a current question of research as to whether solution–dissolution

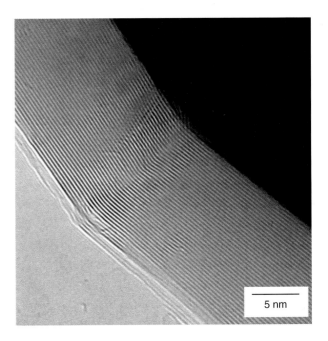

Fig. 21 High-resolution TEM of a section from a graphite capsule around a Pt particle (black) after it was contacted to hydrogen and n-hexane at 353 K. Not the low temperature of the structure-forming process, indicating clearly the catalytic function of metal directing the condensation of carbon fragments.

References see page 419

alone or also surface dissociation contributes to the non-disputable transport of the activated carbon to specific growth facets of the catalyst particle. The important role of carbide formation may well be explained by the kinetic stabilization of a particular active surface structure that does not exist on a pure metal.

A summary of typical situations of nanocarbon growth is shown in Fig. 22. The images reveal the operation of the scenarios described above, and show that carbon is indeed transported through a metal particle; the images also highlight the non-equilibrium shape of the metal particles and clearly rule out, by their multiple topotactic growth features of carbon, that the catalyst might be fluid during carbon growth. It is clear that varying abundances of growth and activation facets will result in varying nanocarbon morphologies, such as filaments, herringbones, and tubes (see Section 2.3.15.5). From the multiple growth directions of a coherent nanocarbon (Fig. 22A) it becomes clear that the growth must have occurred in multiple steps, with interruptions and reorientations of the catalyst particle between producing multiple directional changes in growth and the "crumbly" morphology. It is suspected that the size of the catalyst particles is so large that insufficient supersaturation with activated carbon can be achieved to allow for the instantaneous (whisker) growth of a large portion of the carbon filaments.

The coadsorption of impurities such as oxygen, ammonia, water or sulfur strongly affects the growth kinetics by controlling the reconstruction of the catalyst species. This sensitivity of the carbon generation process on surface orientation can be used to advantage when controlling the shape and rate of carbon filament growth. In this way, inhibitions as well as an enhanced production of carbon filaments, can be achieved. Additives in the gas phase (oxygen, sulfides, phosphorous compounds) or promotion with alkali or boron oxide all affect the metal faceting. Likewise, additives of alloys of molybdenum, silicon or tungsten influence the migration kinetics and the thermodynamic solubility limits of the carbon atoms in the metal particles (10–300 nm diameter), which is often believed to be the rate-limiting step in the overall process. A number of models differ in the character of the carbide phase, which may be a solid solution or a distinct phase such as cementite in the Fe−C system. An additional source of discrepancy [206, 248, 258, 259] is the origin of the carbon concentration gradient, allowing continuous transport through the metal particle. Thermal gradients were excluded by not only several independent experiments but also convincing circumstantial evidence.

2.3.15.12 Carbon on Noble Metal Catalysts

In the catalytic reactions of hydrocarbon conversions on noble metals (hydrogenation, isomerization over platinum, palladium or rhodium are typical examples), it was found that submonolayer quantities of carbonaceous deposits exert a dominating influence [260] on the [261, 262] sequence of events. This is also known in practical applications, where it is noted that a short prereaction time in the hydrocarbon atmosphere is required to obtain a selective [263, 264] performance. The beneficial effect of the carbon overlayer [265] is believed to arise from a moderation of the dehydrogenation activity of noble metal surfaces. Dehydrogenation reactions [266] are the fastest to be catalyzed by platinum, whereas skeletal rearrangements and hydrogenolysis reactions are about an order of magnitude slower on various single-crystal surfaces of platinum at atmospheric pressure and at

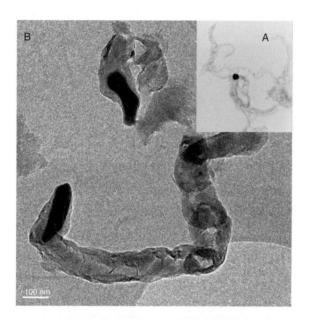
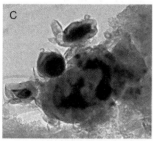
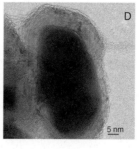

Fig. 22 Typical morphologies of nanocarbon grown over Ni particles. Ethylene in hydrogen was used as feed gas, the growth temperature was 973 K. (A) A TEM image of a 50-nm large, multiply facetted metal particle with a spiral of carbon structures generated from it. (B) A high-resolution SEM image of a facetted particle generating bamboo-type carbons, proving that at least some carbon must be transported through the metal to the inner facet of the catalyst particle. (C) Deactivated metal particles with shell-type carbon overlayers. (D) A high-resolution close-up, revealing the exact topotaxy of the graphene layers around the metal surface. Note the high quality of the ordering, and compare to Fig. 21.

573 K [262, 265]. The beneficial role of carbon would be to allow the hydrocarbon substrate not to be adsorbed directly onto the metal; instead, if it is bound on a carbon material onto which hydrogen atoms can spill over, it would be expected that hydrogenation would occur more selectively relative to all other processes. This idea of an intermediate inert surface with intimate contact to the hydrogen-activating metal, which was developed during the analysis of the coking behavior of supported Ni, Rh, Pd and Pt catalysts [267–269], requires complete chemical inertness of the carbon against hydrogenation, as well as a suitable electronic structure [270] to permit chemisorption of the organic substrate. In addition, the beneficial effect of structure sensitivity, which is prominent in platinum metal catalysis [271], is lost. A compromise would be the synergistic action of patches of carbon (or a holey, polymeric overlayer) over a suitable structured metal surface, providing the necessary hydrogen activation. Such a scenario has been proposed earlier [272], using a silica modifying layer.

Combined carbon radiotracer and temperature-programmed desorption experiments on platinum [266] have revealed that, for temperatures up to 473 K, a carbon species with the composition $C_n H_{1.5n}$ prevails, which can be removed in excess hydrogen. At higher exposition temperatures, the carbon : hydrogen ratio falls sharply to below 1 : 1 and the resulting carbon is no longer reactive in hydrogen. This more graphitic carbon fulfils the conditions required to act as a stable adsorption intermediate. Above 750 K, the layer converts into structurally identified graphite which is, however, inactive for hydrogen transfer, as was shown by hydrogen isotopic exchange studies [273]. All reactions carried out with hydrocarbons below 473 K require a permanent readsorption of the hydrogen-rich carbon intermediate, whereas platinum treated at high temperature is permanently modified and passivated for hydrocarbon reactions above 750 K.

The structure of the hydrogen-poor carbon deposit has not been clarified by surface science, and is referred to as the "carbonaceous overlayer". The structural nature of the hydrogen-rich carbon deposit, however, was amenable to a suite of surface-science techniques, and the unique structure of the alkylidines [274] was derived. These adsorbate species, which have molecular counterparts in organometallic chemistry [35], consist of at least two carbon atoms one of which is bonded to three metal atoms. The other carbon atoms are fully hydrogenated alkyl groups with only C–C single bonds in-between. On reacting platinum surfaces, no structure-sensitive effect exerted via adsorption of the organic substrate is expected because 80–95 atom.% of all platinum atoms are covered by the carbonaceous deposit, which can be of a different structure. Polymerization and dehydrogenation of hydrocarbons produce, according to the general schemes of Figs. 18 and 19, hydrogen-poor molecular species with multiple bonds towards the surface. The alkylidine species is a well-defined limiting structure which occurs instantaneously on clean metal surfaces [261] upon exposure to pure small alkene molecules. With longer residence times and higher temperatures, these still-hydrogenated species carbonize to form graphene layers. If they are inhibiting the metal from contact with the reactants (see Fig. 21), then they are deactivating. If, however, the graphene units allow participation of the metal, in the transformation of reactants for example by providing a suitable rate of activated hydrogen, then these layers may well form part of the active site and are by no means detrimental to the catalyst function. The function of such deposits may be of multiple [264, 275–280] nature: First, they may act as a dynamic cover for unselective active sites, and be compressed laterally upon substrate adsorption and reexpanded after desorption of the product. Second, they may serve as a weakly interacting surface for substrate adsorption [269].

One way to clarify the role of the adsorbate is the *in-situ* investigation of platinum catalysts with electron spectroscopy, which is capable of determining quantitatively the abundance and chemical structure of the carbon species present. This has been carried out for a platinum black [271] material used after various activation procedures in *n*-hexane conversion reactions. The pristine catalysts were found to be significantly oxidized, with the surface oxides being reduced by alkane decomposition into methane, carbon, and ethylene. Selectivities to higher hydrocarbons and skeletal isomers of hexane were observed only after an initial induction period. Surface analysis at this stage revealed the presence of Pt metal, of adsorbed OH groups, and of a variety of carbon species [264, 281] including carbon dissolved subsurface, graphite, amorphous hydrogenated carbon, and several partial oxidation products of the polymer. The total abundance of carbon, as measured by XPS, varied between 15 atom.% for purified Pt and 48 atom.% after catalysis. The complex scenario of a working surface is summarized in the upper scheme of Fig. 23. The abundance of the various carbon species is likely to be much larger than estimated from the atomic percentage in XPS represented in Fig. 23, as the electron mean free path though such species is much larger than is estimated from the general equations [282, 283].

The initial contact of the metal and hydrocarbons results in a combination of surface and subsurface species, and also of stable graphene fragments with metastable alkylidine species. Only a small fraction of the carbon deposit is of a graphitic nature, with most carbon species remaining in a state of carbonization with

---

References see page 419

still significant peripheral heteroatoms. The nature of the catalytically active carbon deposit may be considered as a multilayer of porous hydrogenated carbon (the islands seen in Fig. 23); this remains sufficiently stable due to its network structure, and provides active centers for hydrocarbon adsorption through hydrogen bonding or through dispersive interactions. The pores (or free patches of metal seen in Fig. 23) provide access for the reactants to the metal surface.

The dissolution of carbon atoms below the surface is depicted schematically in Fig. 23B. This phenomenon, which occurs as a dynamic effect only during the reaction of noble metals with hydrocarbons, exerts significant control over the stability of surface terminations, and thus over the selectivity of metal–hydrocarbon interactions. The signature of this carbon form is an XPS binding energy below that of graphite, but above that of carbide. The electronic structure is indeed not that of an ionic or covalent carbide with large charge transfer, but rather that of an interstitial atomic species, as in an alloying species. Fundamental studies on the selectivity control and the effects of subsurface carbon in more

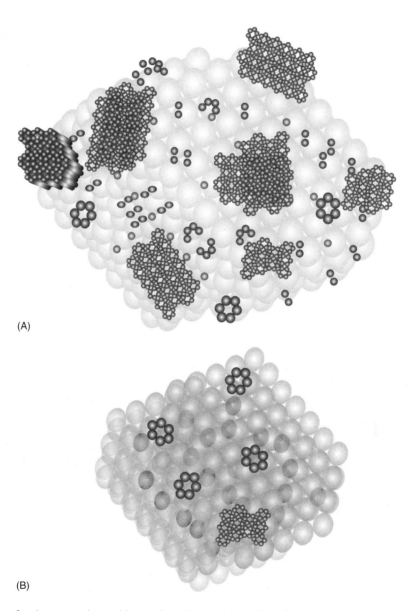

Fig. 23 Representation of carbon on and in noble metal catalyst particles. (A) Surface species: individual "atoms" oxygen and OH groups, "dimers" ethylidine and related species, "C4" alkoxy- and chain fragments, "C6" pregraphitic rings, condensates graphene units and graphitic elements. (B) Atoms subsurface of metal; carbon, oxygen and hydrogen can coexist without forming compounds; at the surface segregation processes begin to form carbonaceous units. The representation is idealistic in the low coverage that was chosen to allow facile distinction of species.

complex catalytic processes [270, 284–286] have been carried out.

A realistic impression of the working surface of a noble metal catalyst is presented in Fig. 24. This image (from Ref. [286]) highlights the complicated interpenetration of carbon into the metal surface. Seen are the carbon (002) lattice fringes of graphene layers formed at temperatures below 373 K at the surface, and the Pd (100) fringes in the deeper part of the image. The circles denote positions at which local precision determinations of the lattice constant were taken to ascertain the assignment of phases. In addition, the data reveal that a gradient of lattice expansion from the interface into the bulk indicates dissolution of the carbon atoms deeper into the lattice such that their location cannot be imaged directly at the resolution of the instrument being used.

2.3.15.13 Carbon Formation in Zeolites

Zeolites and other oxide-based solid acid catalysts are used in hydrocarbon conversion reactions, with enormous economic [287, 288] impact. Their lifetime is, however, very limited due to the deposition of carbon, which is referred to in the literature as "coke formation". The definition of coke in zeolites is very much wider than in the sense of the present chapter, as it is understood that coke is "... a carbonaceous deposit which is deficient in hydrogen compared with coke forming molecules" [160, 289]. Two types of coke [82, 158, 240, 280, 290] are

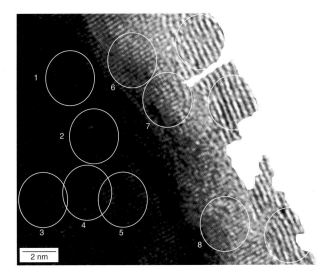

Fig. 24 High-resolution TEM image of a cross-section through a Pd foil that was used as catalyst for pentyne hydrogenation. The blank area was filled with the image of the glue used to prepare the specimen. The circles 1 to 8 indicate the presence of Pd lattice fringes as measured from their local FFT spectrum; whereas the three circles at the surface are typical of graphitic lattice distances. Note the porous structure of the carbon and the interpenetration of the carbon atoms into the Pd lattice at the interface (circles 7, 8).

discriminated by several techniques, amongst which are liquid phase extraction, thermal desorption [157], electron microscopy with energy loss near edge structure (ELNES) [82], and other spectroscopies such as vibrational methods [291–293] and NMR [71, 294–296].

One coke type, known as "white coke", covers polymeric alkanes, alkoxides and oligo-olefins, including naphthalene-type aromatic molecules and derivatives with alkyl chains; the second coke type – "black coke" – covers larger pericondensed aromatic molecules and alkylated derivatives. The latter form would be referred to in carbon chemistry as a precursor to coke in the sense of Figs. 18 and 19.

One significant problem in oxide coke analysis is determination of the C:H ratio, which is the most important integral value for a structural characterization. The main problem is the interference of adsorbed or structural water, and the structural change occurring during temperature-programmed analysis allowing for carbonization and the formation of irreversible coke, which is structurally identical with black carbon from polymerization (see Fig. 19). The solutions to this problem, which contributed much to the controversial descriptions of coke on oxide catalysts, are described in detail in an excellent review [297]. The absence of large amounts of graphitic carbon in zeolite coke was shown using XAS at the carbon K-edge [298]. The spectra from an industrial zeolite which was treated with aliphatic and complex aromatic hydrocarbon feedstock to generate coke showed, in both cases, the predominant presence of catacondensed aromatic molecules (naphthenes). It should be noted that results obtained with this spectroscopic technique, which is applicable "*in situ*", agree well with those obtained after traditional chemical extraction [159, 240] methods, which are always subject to doubts about transformations during elution and work-up. The independence of the coke properties from the nature of the hydrocarbon feedstock being either methanol (in the MTG process), small hydrocarbons (e.g., ethylene or propylene), aromatic molecules (e.g., mesitylene) or heavy oil fractions was demonstrated for a technical zeolite sample [299]. The data showed that, under the influence of the solid acid catalyst [280, 288, 294], the feed is broken into small units from which catacondensed aromatics (with alkyl side chains or as bridging units) are formed. This process, which is similar to carbon formation in homogeneous systems (flames), occurs at comparatively very low temperatures (as low as 475 K). After this polymerization stage, and according to the classification of Fig. 19, a step of dehydrogenation up to 700 K follows in which all molecules which escape the micropore system react into pericondensed

References see page 419

polycyclic aromatic molecules. Those molecules that remain in the micropores retain their shape and are not polymerized, which explains the simultaneous presence of both types of coke in zeolites [160, 289, 292, 297]. If the conditions of operation or regeneration are chosen such that the whole micropore system is liberated from hydrocarbons, or if the catalytic reaction occurs only at the external surface of a zeolite, then only aromatic coke in possibly varying degrees of condensation will be found. A strong dependence on the diffusion kinetics [300, 301] of the type of coke present on zeolites may be one origin of the conflicting reports about speciation found in the literature. In subsequent oxidative regeneration steps, the pericondensed molecules at the outer surface carbonize and are eventually oxidized, rendering free all of the outer surface according to selective adsorption experiments. The catacondensed deposits in the micropores cannot fully carbonize due to spatial constraints, and so remain partly intact as molecular pore blockers and causing a significant reduction in the micropore accessibility [292, 297].

The sequence of events in coke formation was studied in the model reaction [288, 290, 302] of H-Y zeolite with propene at 723 K. Under these drastic conditions, the soluble white coke formed rapidly within 20 min and was converted into insoluble coke within 6 h under inert gas, without losing carbon atoms in the deposit. Due to the larger pores in the Y-zeolites compared to the ZSM-type zeolites used in the other studies mentioned to date, the structure of the aromatic molecules was somewhat different. The soluble coke in this system consisted of alkyl cyclopentapyrenes (C_nH_{2n-26}, Type A) as the hydrogen-rich primary product, and of alkyl benzoperylenes (C_nH_{2n-32}, Type B) and alkyl coronenes (C_nH_{2n-36}, Type C) as matured components. The temporal evolution of the various products is shown graphically in Fig. 25. It is clear that the soluble coke fractions are precursors for the insoluble coke, and that within the soluble coke fraction the final steps of dehydrogenation–polymerization are very slow compared to the initial formation of smaller aromatic molecules from propene. The sequential formation of precursors with decreasing C : H ratio follows from the shift of the maximum in the abundance of each fraction on the time axis.

The formation of insoluble coke (Fig. 25) occurs clearly in several kinetically well-discernible stages which can be attributed to initial growth at the outside of the zeolite crystallites, and in a slow diffusion-limited growth inside the micropore system. The diffusion limitation will affect the migration of soluble coke molecules rather than the motion of the small precursor molecules. It can be further seen that the acidity [158] of the zeolite is an important controlling parameter, which is summarized here rather crudely as the Si : Al ratio. Clearly, the more acid sites available in the system, the faster the kinetics of coking and the greater the absolute amount of carbon deposition.

The role of the acid sites [280, 294, 303] may be either catalytic, or they are consumed irreversibly during coke formation. In the propene coking study [160], the abundance of coke molecules remained always significantly below the number of supercages in the zeolite, which pointed to a catalytic function responsible for coking, together with the fact that insoluble coke formation was not saturated within 30 h on stream. This was studied in great detail in a combined catalysis Fourier transform (FT)-IR study [304] which revealed the constant abundance of Brønsted acid sites in mordenite with a Si : Al ratio of 12 during coke formation in ethyl benzene disproportionation. The fact that an IR-active band at 1600 cm^{-1} can be assigned to coke shows that the structure of zeolite coke is not comparable to coke from coal (in which this band is also present, indicating carbon–oxygen functions at non-aromatic C=C bonds) carbonization. This was supposed in a study of coke formation from hexane and hexene in faujasite zeolites [291]. In this study, it was found further that alkenic hydrocarbons were much more active than aliphatic, though this may be interpreted as an inhibition of the initial cleavage of the aliphatic feed molecules, and which becomes important for long-chain molecules.

The degree of ethylbenzene conversion [304] fell with time on stream, and the coke IR band increased proportionally, without any change in the Brønsted acid band at 3600 cm^{-1}. Moreover, after complete deactivation of ethylbenzene conversion, the acid sites were still active for coke formation with a small precursor molecules as ethylene, which increased the coke band further when fed after ethylbenzene. These results were in line with observations from propene coking, but conflicted with those of other studies from the same group [297]. The latter claimed either the involvement of extremely hygroscopic Lewis acid sites (which are hydrolyzed by minute traces of water from hydrocarbon oxidation with lattice oxygen) or the consumption of acid sites during alkene polymerization (which may be an experimental artifact due to the extreme loss in IR transmittance of the samples in these highly reactive coking gas mixtures). It remains an intrinsic disadvantage of this type of study that IR can detect only various forms of soluble coke (the shifts and varying positions of the coke band are a direct consequence of the successive dehydrogenation–polymerization of polyaromatic molecules), and does not indicate the formation, transformation and absolute abundance of insoluble coke, which is detectable only in Raman or NMR experiments.

The extensive study of coking on solid acid systems has brought about an additional pathway to

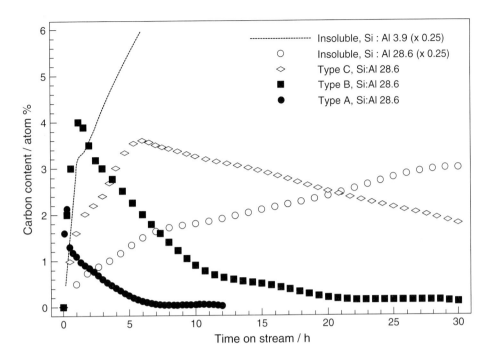

Fig. 25 Abundance of various types of coke on zeolite H-Y during reaction with propene. The soluble fractions (A–C, black symbols) represent different aromatic structures which form during the first polymerization steps. The insoluble coke is the final product of polymerization which is affected by the solid-state acidity of the zeolite, represented here as the Si : Al ratio. Data taken from Ref. [302].

those described previously on carbon formation (see Sections 2.3.15.9–2.3.15.11). This pathway, which leads to a hydrocarbon polymer with predominantly or exclusively chain-like structure that is saturated with many hydrogen atoms is, in the sense of Fig. 19, still a precarbon polymer. However, the literature uses the incorrectly applied term "coke" which, in carbon chemistry is a hard, carbon-rich material of aromatic molecular structure but with a significant disorder of the graphene BSUs with respect to the planar stacking.

2.3.15.14 Graphitization of Carbons

The various types of carbon obtained by one of the three routes shown in Fig. 19 can be discriminated with respect to their reactivity upon thermal treatment. The lower part of the Fig. 19 shows the two categories of carbon that will either graphitize (further) upon thermal treatment (annealing), or remain non-graphitic even at temperatures up to 3000 K. The main reason for this distinction of reactivity is the dimensionality of the crosslinking of constituting BSUs. If these are linked in 3-D networks then, by thermal treatment, little possibility exists for rearrangement of the cross-linking into the 2-D network required to form the large planar graphene layers that constitute graphite. The larger the fraction of 3-D links in a carbon, the quicker is its formation, as there is insufficient time to reach the equilibrium situation of an all-sp^2 carbon connectivity (which excludes 3-D networks). The synthesis conditions of rapid quenching from high temperatures are typical for this carbon. Alternatively, the reaction temperature is very low and thus insufficient to provide the activation energy for a rearrangement of the C–C bonds. Catalytic situations, charcoals and activated carbons are the usual materials of this class.

The synthesis conditions of high temperatures, with dehydrogenation of hydrocarbons being rate-determining, and the intermediate formation of a liquid phase or even a liquid–crystalline phase (mesophase) of 2-D stacking of PAH molecules, are favorable for reaching the equilibrium situation. Pregraphitic structures are then formed which only require annealing for better stacking order and polycondensation of the BSUs by a final dehydrogenation step. The sequence of graphitization events [199, 305] is summarized in Fig. 26. The end of the carbonization process leaves molecular structures consisting of polycyclic aromatic molecules, with no heteroatoms other than hydrogen and some alkyl terminating groups. Their alignment is parallel to each other, as in van der Waals crystals of polycyclic molecules. The intermolecular dispersive forces between parallel ring systems are responsible for the alignment. At higher temperatures, all non-six-membered structures and most of the alkyl groups [141] are removed. At around 1900 K, the large

References see page 419

aromatic molecules begin condensing to graphene layers, although as some of the interconnections are not achieved by *peri*-condensation but rather by the formation of C—C single bonds, considerably strained sheets with significant buckling will result. This state of graphitization can be observed experimentally with HRTEM [305]. The final step – the aromatization of all C—C bonds – occurs only when the temperature is well above 2000 K, since only at such temperatures does the parallel alignment of the graphene layers and the formation of a regular stacking occur. It is then that the abundance of defective structures within the graphene layers becomes small enough to allow the dispersive forces between the graphene layers to direct the orientation of the BSUs.

The sequences of events with the existence of intermediate stages of ordering between graphite and

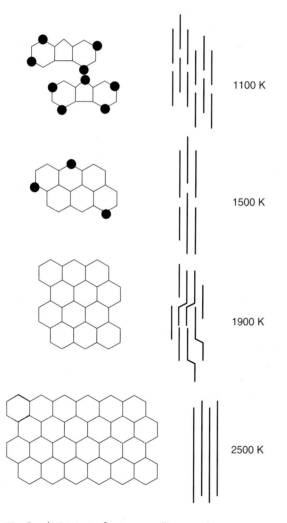

Fig. 26 Graphitization of nanocrystalline graphene precursors (see Figs. 9–11). Chemical structures (dots represent heteroatoms or methyl groups) and stacking order are given as a function of characteristic temperatures.

aromatic molecules has been derived conclusively using XRD [128, 129, 147, 228, 306]. The evolution of interplanar spacing with carbonization temperature, and the kinetics of its equilibration, have served as tools for the analysis. The aromatization was detectable from a shrinking in the a-axis parameter, reflecting the shortening of the average C—C bond distance. Both changes led to a shrinking in the cell volume, which is reflected in the relationship between the lattice parameters shown in Fig. 27. The graphitization reaction proceeds through a series of metastable states [307] with an average change in the carbon–carbon bonding. This is unavoidable, as in the stages of carbonization [308] a significant number of covalent bonds between the BSUs are formed that stabilize a 3-D network of BSUs and not a 2-D stack. The network forms the porous structure of non-graphitic carbon [169, 171, 184] that needs to be broken through the simultaneous opening of C—C bonds, allowing restructuring of the BSUs. Thus, graphitization is not a simple reorientation of turbostratic units into well-stacked units, as is frequently assumed.

Good candidates for such solid-state phase transformations are carbons produced by polymerization of pitch, tar, or other high-boiling fractions of fossil fuels [199, 305]. Pyrocarbons from methane, which are generated under conditions of strongly understoichiometric oxidation, are also suitable candidates for highly ordered synthetic graphitic materials [89, 141]. In catalytic situations, most carbonaceous deposits, carbon blacks and synthetic graphite belong to this class of carbons.

Graphitization occurs between 1100 K and 2500 K, with significant variations depending on the number and quality of structural defects in the material. In fullerenoid material, the sequence of events of graphitization occurs at temperatures of about 1000 K below [83, 95, 120, 127] that for other carbons. The incorporation of large amounts of potential energy in the form of ring tension by the presence of non-six-membered sp^2 carbon rings drastically lowers the activation energy for the solid structural transformations, and allows graphitization at combustion temperatures. A similar low-temperature graphitization is also possible if the carbonaceous deposit on a hydrogenation catalyst [281] has stored chemical energy in the form of highly reactive organic molecular structures. Nanostructured carbons as tubes and fibers that are graphitic in their structure can also be converted into planar graphitic structures [72, 309, 310] at relatively mild temperatures as, again, chemical energy is stored in the bent structure allowing facile breaking into BSUs rearranging into graphene stacks. In the context of catalytic carbon, these studies are of relevance because, if such nanostructured forms of carbon are initially formed (during catalyst deactivation), then their removal by burning may be severely hindered by simultaneous

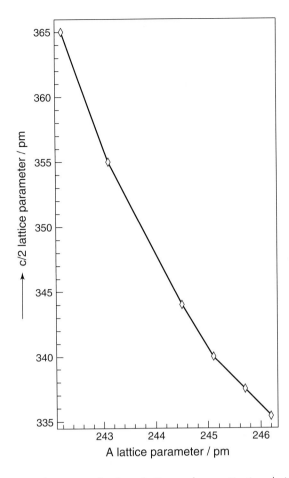

Fig. 27 The extent of polymerization and aromatization during graphitization is reflected in the average lattice parameters which can be extracted from X-ray diffraction data. The correlation between the axis parameters indicates the shrinking of the cell volume. Plotted from data of Ref. [306].

transformation of some of the reactive carbon into graphitic "carbon ash" using the excess temperature of oxidation of some of the deposited carbon as energy source.

2.3.15.15 Reaction of Oxygen with Carbon

This apparently simple reaction is of multiple relevance in catalytic carbon chemistry. The reaction is used to activate carbon by the increase of surface area due to hole burning. It is further used to create surface functional groups. Finally, it is used to remove undesired carbon deposits in regeneration procedures.

The reaction is a gas–solid interface reaction requiring two independent types [311] of electronically different [79, 312] sites. One site should be electron-rich to activate molecular oxygen (site **A**); the other site should be electron-deficient to react with activated oxygen atoms (site **C**). A minimal sequence of elementary steps is composed of the following processes:

$$2A + O_2 \longrightarrow O_2(ads) \qquad (1)$$

$$O_2(ads) + 2e^- \longrightarrow O_2^{2-}(ads) \qquad (2)$$

$$O_2^{2-}(ads) + 2e^- \longrightarrow 2O_2^{2-}(ads) \qquad (3)$$

$$O^{2-}(ads) + C \longrightarrow CO(ads) + 2e^- + A \qquad (4)$$

$$CO(ads) \longrightarrow CO(gas) + C \qquad (5)$$

$$O^{2-} + CO(ads) \longrightarrow CO_2(ads) + 2e^- + A \qquad (6)$$

$$CO_2 + (ads) \longrightarrow CO_2(gas) + C \qquad (7)$$

This sequence of events was suggested by several authors [37, 78, 154, 182, 311] with, however, different assignments to the nature of the two sites ranging from "ionic to covalent" [311], or to a distribution of sites with varying energy of oxygen chemisorption [313]. In these models it was assumed that the steps (5) to (7) are rate-determining.

Steps 1–3 designate the activation of molecular oxygen into an oxidizing atomic species. This process requires the presence of weakly bound electrons in order to split the dioxygen double bond. Such catalytic sites are available on the basal planes of graphitic carbon. The notion that almost free electrons are also available at structural defect sites [314, 315] is only valid for extremely fast reaction rates, as otherwise stable surface complexes with oxygen, water or hydrogen will deactivate [313, 316, 317] such structural defects. The site **A** is thus at graphene planes of sufficient size to exhibit delocalized electrons more weakly bound than in aromatic molecules [182]. An efficient activation on the perfect plane without any defect can only be achieved through the help of a catalyst [318] (alkali, transition metals) enhancing the extremely low sticking coefficient [75, 319–321] of oxygen on basal planes of graphite. The oxygen atoms will diffuse on the surface [322], as shown by isotope tracer studies using their excess dissociation energy, until they either recombine and desorb or find a structural defect **C** where they can form a covalent bond to carbon [step (4)]. This reaction is an oxidation reaction in which the directly bonding carbon atom **C** loses two electrons which flow back into the reservoir of almost free electrons of the graphene layer. It is a matter of ongoing debate to decide on the influence of the diffusion of activated oxygen on the overall kinetics [37, 78, 239, 323]. Due to the extremely different reaction conditions and usually intentionally small conversions, at which experiments are made, it is difficult to decide this question at present. The main problem is the integrating effect of the large number of reactive sites present in an oxidizing carbon sample,

References see page 419

which precludes the identification of the influence of any particular reaction step, due to the convolution of the elementary step kinetic parameters with the energy distribution of the reactive sites. It is useful to consider the branching reaction of activated oxygen reacting with its carbon substrate or with an adsorbed species in a catalytic fashion [323–325]; from the rates of the two processes and the number of edge sites known an estimation of the delivery function of the basal plane for activated oxygen can be made.

The resulting carbon–oxygen complexes [59, 148, 200, 321, 322, 326] can either desorb using its energy of formation for activation, or survive due to insufficient activation energy. The amount of activation energy depends on the carbon skeleton onto which site **C** is bonded. It is a structure-sensitive reaction [327–329]. If the primary carbon–oxygen complex remains long enough at the surface where activated oxygen atoms are available (or if a sufficient flux of activated oxygen is available), then complete oxidation will occur, according to step (6). The desorption of the oxidation product leaves in any case an empty site for a new attack of an oxygen atoms [steps (5) and (7)]. The oxidation reactions [steps (4) and (6)] leave an active site each for oxygen adsorption and dissociation. These sites are, however, not those at which the primary oxygen atoms are formed which changed their positions to diffuse from their location of generation on a graphene layer to the oxidation site. From this scenario, which was developed from extensive experimental observations [79, 182, 312, 314, 317, 330–332], it is evident that efficient oxidation requires the presence of extended graphene layers for oxygen activation [39, 182, 331, 333] and sufficient defect sites [319, 334] for reaction of the activated oxygen. It was found necessary in an extensive formulation of hypothetical elementary steps to invoke the existence of stationary and mobile initial oxidation products [step (4)] in order to explain the varying selectivity between CO and CO_2. The experimental evidence showed different TDS peaks for carbon–oxygen compounds, but not for di-oxygen [312, 314, 322], which led the authors to consider the possibility that the reaction products are mobile, but not the activated oxygen. The process of oxygen activation was assumed to be possible only at the reaction sites themselves (**C** in the reaction scheme). The sensitivity of the overall reaction kinetics to the presence of graphene layers for oxygen activation is lost and a fully structure-insensitive process occurs if the activation of molecular oxygen occurs by non-chemical means (glow discharge [314, 335] or radiolysis [315]). In the other extreme of a dioxygen reaction on a perfect graphene layer, reaction of activated oxygen will always occur at the unavoidable external prism surfaces and very efficiently at the few intra-layer defects.

One consequence of the reaction mechanism is that carbon burns with molecular oxygen in a layer-by-layer fashion at a decreasing rate as the perimeter of the graphene stack becomes progressively smaller with increasing burn-off. The defects present in variable abundance in each layers contribute significantly to the rate, with an increasing tendency towards burn-off as the perimeter of the etch pits enlarges with the ongoing reaction. The resulting net rate is an addition of a constant decreasing term and a fluctuating term with an unpredictable overall behavior with burn-off. The unusual dominating role of in-plane defects was established by microscopic studies on graphite oxidation [318, 336]. This effect precludes a definition of reactivity of carbon materials as their average structure is not sufficiently accurately determined to account for the defect distribution in the carbon which controls the activity. This was established to be valid not only for graphite but also for a wide range of carbons [59, 312, 313, 325, 337] containing most of the structural features shown in Figs. 4, 8, and 11. A non-linear compensation effect of the kinetic parameters was discussed, as several groups of carbons with different average structures (graphite, carbon blacks, coals) are differently affected by the rate-enhancing effect of defects, causing simultaneously a decrease in apparent activation energy and an increase in the pre-exponential factor. The correlation of the kinetic parameters of all carbons results [338] in a non-linear relationship between the rate (thought to be intrinsic, but is in fact extrinsic as it is defect-controlled) and the activation energy due to the disregarded structural influence.

The plausible model with a period of increasing defect creation followed by a period of rapid gasification was challenged for HOPG as substrate by *in-situ* spectroscopic experiments, revealing that the average degree of graphitic chemical bonding does not change with burn-off [39]. The local electronic structure of graphitic sp^2 configurations and the extended structural feature of parallel graphene planes is also not drastically changed with burn-off [39]. This finding is in sharp contrast to the defect-hypothesis discussed above. The spectra showed that the abundance of such defective and non-graphitic sites with several types of oxygen functional group was so low that the differences in spectral weight during gasification showed an increase in graphitic ordering rather than a deterioration in ordering. This unique feature of carbon oxidation, as compared to the oxidation of other elements, arises from the fact that the carbon oxides are all volatile and do not form a product at the reacting surface. Disintegration of the well-ordered surface as preparation for oxidation at every surface site, as found for the reaction of atomic oxygen [314, 335], was not observed with molecular oxygen. Experiments with *in-situ* XRD [339] revealed that defective parts of graphite crystals oxidize preferentially, leaving behind a more perfect structure than at the beginning of oxidation, until the total mass loss erodes the ordered parts of the crystal. These data are in agreement with the general experience of solid-state chemistry, that an increase in defect density creates an increase in gas–solid reactivity.

A series of single-crystal experiments using natural graphite flakes was conducted [79, 87, 330] to elucidate the complex influence of the 2-D geometry on the anisotropic reaction rate. A series of 256 optical micrographs of the gasification of single graphite flake is shown in Fig. 28. The drift-corrected [339] images reveal, by the relative density of contours, how the reaction rate varies with location and time. Over wide parts the reaction proceeds in a topotactic mode, as seen by the self-similar shapes of the contours. However, some parts do react very fast after a slow-reacting hindering part had been gasified, and some small particles separate off during general shrinkage of the particle. Figure 28 shows clearly that graphite oxidation is highly anisotropic and is strongly dominated by the reactivity of the edges, as the crystal's perimeter is reduced faster than its width. This finding is in good agreement with the XAS data discussed above, and indicates that whilst the defects at basal planes do contribute to oxidation [66, 135, 336] they do not have a dominant effect on the reaction rate. The image further shows that graphitization during carbon removal is detrimental for catalyst regeneration. These data also suggest that carbon can be effectively stabilized against oxidation by sealing the edges with heteroelements such as phosphate and boron [37, 340].

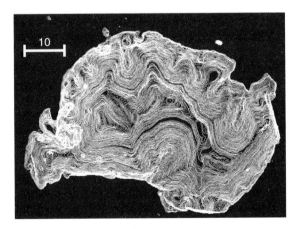

Fig. 28 Drift-corrected superposition of optical micrographs showing the combustion of a single graphite flake in synthetic air at 977. The density of the contours is inversely proportional to the rate of gasification. The anisotropy of reaction follows from the fact that the flake disappeared by recession of the perimeter and not by thinning (this have resulted in a black hole in this contrast).

A rate measurement using a single flake of graphite is shown in Fig. 29. The CO_2 evolution begins rapidly, passes through a minimum when the surface defects are all removed, and then stabilizes at substantial burn-off into the topotactic mode, as discussed for Fig. 28. The morphological studies at higher resolution displayed in the insets of Fig. 29 show [330] that, after the removal of the rough initial surface layer giving rise to the rate maximum, the terraces become atomically flat with very few exceptions which are stabilized by subsurface defects [318] which are more difficult to oxidize than the matrix. The resulting structure has sharp needles and small towers. These observations confirm the complex influence of the defects on the oxidation kinetics that can be either accelerated or decelerated by various types of defect. Elemental impurities such as boron are likely candidates for such inhibitory effects, as boron compounds are well known for their inhibitory effect on carbon oxidation [316, 341, 342]. Other known inhibiting groups include C−H and certain C−O surface complexes [343, 344].

The *in-situ* XRD data further excluded [339] the possibility that oxygen intercalates between the graphene layers and starts the reaction in a 3-D fashion at the internal surface. In-depth oxidation occurs only at structural defects with 3-D character, such as at grain boundaries and interlayer defects [69, 238, 339]. The contribution of this extra-oxidation activity to the overall rate is small (as discussed above), but severe consequences arise for the mechanical stability of all carbon materials under oxidative load. At the very early stages of oxidation, the materials may disintegrate into small particles of individual graphitic crystallites. Such mechanical disintegration has important implications for catalyst support materials, where the active component falls from the support at the very start of any oxidative load on the system, for example after a short pulse of air from a temporary leak into the feed stream. A number of reviews [26, 345, 346] of non-catalyzed and alkali-ion-catalyzed [319, 347–350] gasification at defect sites have summarized the present status of research in this highly relevant area.

In any practical consideration of carbon application or intended removal (regeneration), the oxidation reactivity is of importance. The results of initial oxidation experiments are summarized in Fig. 30. Different carbon materials exhibiting the sp^2 connectivity reveal a wide range of oxidation temperatures, the values of which depend on the definition of reactivity, which can be either the first uptake of oxygen, the 50% line of oxygen consumption, or even the 50% line of weight loss. However, care must be taken when comparing published values [331, 351–356],

References see page 419

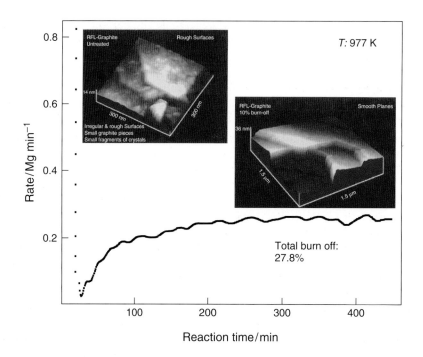

Fig. 29 Rate of CO_2 formation during initial combustion of a single graphite flake. A high-sensitivity ion-molecule reaction mass spectrometer was used as detector of the ppb traces of CO_2 in the N_2 carrier gas. The insets display two scanning tunneling microscopy (STM) images of the initial state and of the interrupted state of this combustion experiment. The different states of the surface definition are clearly visible. Note in the initial state the many nanoflakes that give rise to imaging artifacts of high-resolution STM of graphite.

as all of these definitions are in currently use and the differences in temperatures are quite drastic.

The reactivity of the molecular fullerene solid resembles the expected pattern for a homogeneous material. Only a small prereactivity at 700 K indicates that a fullerene–oxygen complex is formed as an intermediate stoichiometric compound [113, 114, 357]. At 723 K, the formation of this compound and the complete oxidation are in a steady state, with the consequence of a stable rate of oxidation which is almost independent of the burn-off [79, 125] of the fullerene solid. This solid transforms prior to oxidation into a disordered polymeric material. The process is an example of the alternative reaction scenario sketched above for the graphite oxidation reaction. The simultaneous oxidation of many individual fullerene molecules, leaving behind open cages with radical centers, is the main reason for the polymerization of carbons with molecular dimensions in which the activation of oxygen can occur due to the localized electronic structure at the sp^2 centers without immediate reaction to CO_x.

The molecular nature of this carbon allows the spectroscopic identification [114] of this intermediate and several oxo-complexes serving as precursor compounds [4]. On this basis, the reactivity of fullerenes in oxidation is understood in detail [117, 358, 359] and the reaction scenario is secured by experimental verification.

The other two sp^2 carbons oxidize, according to the data in Fig. 30, at quite different temperatures and with complex reactivity patterns. They show extensive prereaction activity. A significant oxygen consumption initiates at lower temperatures than the main reaction, indicating gasification at extra-reactive sites. The spread in temperature of this prereaction activity is very significant.

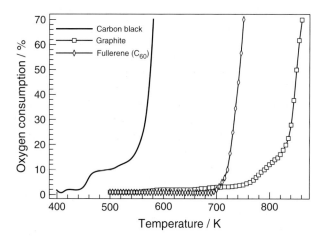

Fig. 30 Reactivity of different forms of carbon against oxidation in 5% oxygen in Ar (100%) as a function of temperature. The carbon black was FW-1 (Degussa), the graphite natural graphite powder AF (Kropfmühl).

Indeed, it has been shown recently that the gasification activity at low temperatures is not a surface effect but can lead to a complete combustion of all the carbon [76, 156]. This means that, for practical applications in catalysis, the minimum corrosion temperature of carbon in an oxidative atmosphere can be very much lower than the bulk gasification temperature measured by thermogravimetry, which may be too insensitive [356] to indicate the prereaction activity. The distribution of the prereactivity over reaction temperature is not a characteristic property of a carbon material, but depends in the same way as the main reactivity on the origin, storage and thermal history of the carbon. The oxidation rate as a function of the geometric carbon surface area [total surface area (TSA)] is a sensitive function [354, 360] of extrinsic sample properties.

The concept of the existence of an active surface area (ASA) representing a small fraction of the TSA was developed [77] in order to overcome this highly complicated [323] situation. In full analogy with the concept of actives sites in catalysis, it was concluded that a surface area measurement with di-oxygen under defined conditions should measure the active sites in combustion (**C** sites in the reaction scheme). This technique allows the detection of the average change of the abundance of active sites with burn-off. The change in ratio between TSA and ASA by over an order of magnitude [77] for a graphitized carbon black highlights the relevance of surface modifications for the oxidation reactivity. A numerical example of the ratio TSA:ASA of $76:0.24$ m^2 g^{-1} (ratio 316) at the start of burn off of the graphitized carbon black, and of $128:2.12$ m^2 g^{-1} (ratio 60) at 35% burn-off, illustrates the accuracy required to obtain meaningful results. This problem of large ratios may be the reason for the deviating numbers for carbon reactivity still quoted by different authors in the literature [79, 86, 226, 351, 354]. The same concept was further very successful in eliminating all of the extrinsic influences on the carbon reactivity over ranges of materials, with the consequence that a universal rate for the oxidation of sp^2 carbon was established, deviating for various classes of materials only by a factor of about 2 [37, 78, 361].

The method of determination of the ASA creates some carbon–oxygen surface complexes which are non-reactive and which cause the still significant dependence of the turnover frequency (TOF) on burn-off or material property. A further improvement of this situation, reducing the dependence of the TOF on burn off to below 10%, was achieved by introducing the concept of reactive surface area (RSA) [327] or active site density [350]. A simultaneous *in-situ* measurement of rate and abundance of active sites allows determination of the RSA without any assumptions on reactivity or energy spread of active sites to be incorporated in the method. There is now good agreement that the oxidation of carbon [153, 352, 356, 360, 362] is strongly dependent on its defect concentration, and thus its morphology providing the structural basis for the reactive sites as exemplified for a simple case in Figs. 28 and 29.

The oxidation of carbon may also be catalyzed, although two fundamentally different cases should be discriminated. Transition metal oxides and carbides were found to be efficient local sources of atomic oxygen, increasing its abundance much above the non-catalyzed case. Streams of diffusing oxygen atoms created decorated pathways of non-selective oxidation of basal plane sites, as detected by TEM [363].

Alkali and earth alkaline metal compounds (hydroxides and bicarbonates) were found to be efficient catalysts [345, 349], reducing the 50% weight loss temperature of all types of carbon material by up to 200 K [346]. Two principally different modes of action have been proposed for this technologically and scientifically interesting effect. First, it was assumed that alkali graphite intercalation compounds [364] could form and modify the surface electronic structure [348] of the top graphene layer to enhance the sticking coefficient of molecular oxygen on carbon. This may, however, more effectively result from the action of the strong positive point electrostatic charge [365] of an alkali cation residing chemisorbed at the carbon surface. Oxygen molecules still in the gas phase would become polarized, which would greatly enhance their sticking probability at the surface.

The second effect is related to the set of elementary steps [25, 26, 345, 366] required to form a surface–oxygen complex. Many of these structures involve the presence of charged oxygen species requiring protons as counterions (carboxylic acids, alcohol, phenols, etc.). Alkali ions provide ideal counterions in the situation of gasification where few protons are available. In addition, while alkali ions are able to accelerate the formation of salts of surface functional groups [17, 346], they cannot act as site blockers – as can hydrogen, which may form very stable surface hydrogen bonds and reduce the number of active sites. Finally, the alkali salts are all strong Brønsted bases which helping in the desorbing of surface functional groups which form during carbon oxidation. One point which remains unclear here, but which may be referred to frequently in the literature, is the proven fact that CO_2 as a weak acid will tend to react with alkali species, forming carbonates. It must be assumed that either traces of water and/or the high local reaction temperature destroy the carbonates, as these also are not detected in post-mortem studies [348] of interrupted gasification systems. By using isotope labeling and other kinetic experiments, it was established [25, 323, 366] that one very important function

---

References see page 419

in alkali-catalyzed gasification is the enhancement of the amount of "exchangeable oxygen" between gas-phase molecules and surface complexes. An alkali–oxygen cluster [367] is envisaged at the active species, which is loaded with oxygen from the gas phase and unloaded by spillover of oxygen to defects at the carbon surface. A fraction of carbon–oxygen–alkali always remains non-exchangeable [347, 348, 365], and this is seen as the "backbone" of the system, keeping the alkali species at the surface below the high temperatures of reaction.

One good indicator for the mode of operation of a catalyst in carbon gasification is the ratio of CO and CO_2 formed, according to the reaction scheme in independent reactions, as simultaneous products. The possible post-oxidation of CO to CO_2 was excluded by a large number of observations, and is a basic ingredient in the formulation of the reaction scenario described above. In non-catalyzed gasification reactions, the product ratio is about unity, with a factor of five deviations for different carbons. The presence of Ca as a catalytic impurity on the carbon from a sooting flame reduced the product ratio to less than 0.01 [313]. A kinetic analysis also revealed that the addition of Ca enhanced the rate of CO formation, without affecting CO_2 production. This effect was attributed to a modification of the active sites of reaction step (6), which may be envisaged chemically as a Ca salt formation with a suitable surface complex. The suggested analytical value of the $CO:CO_2$ ratio should, however, be utilized with care, as other factors such as the basal : prism surface area ratio also affect the product ratio [37, 317, 329].

The oxidation of graphitic carbon is a complicated gas–solid interaction where topochemistry and its modification during burn-off [37, 79, 168, 330, 337, 343, 352, 354] controls the rate. Several independent factors, each with different dependencies on conversion, cooperate in the control of the overall reaction rate. The most difficult part to control is the strong influence of atomic intra-layer [318, 320] defects in the graphitic surface. It was made clear by using electron microscopy that these unpredictably occurring reaction initiators can dominate the overall reaction rate [336]. These defects lead to pitting corrosion and can enhance the available perimeter length where under-coordinated carbon atoms occur (site **C** in the reaction scheme) to a significant and, during burn-off, a varying [368] extent.

The reaction rate is shown schematically in Fig. 31, expressed as the normalized mass loss per time as a function of burn-off. In this figure, the main controlling factors and their dependencies on burn-off are indicated. The dominating influence is exerted by the changes in ASA and in fractional active prismatic faces surface area of the graphene stacks. In the initial phase of burn-off, the removal of surface defects and surface functional groups determines the rate. This was found experimentally

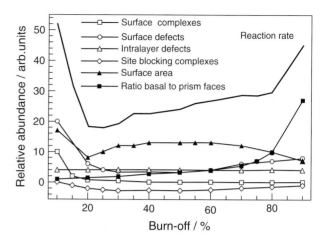

Fig. 31 Schematic representation of the convolution of key influences on the observable oxidation rate as function of carbon burn-off. The broad line for the rate indicates that fluctuations are superimposed on the slow trend. It is clear that a rate *constant* for carbon oxidation can only be a very crude approximation on this process.

by various techniques [333, 369] and is discussed with Fig. 29. The initially strongly changing rate led many research groups to exclude [351] the first 10–20% of the burn-off curve from a kinetic analysis, in order not to have to account for this initial effect. For catalytic purposes, however, this effect is highly relevant as the stability of the surface functional groups and surface defects will determine the catalytic reactivity of a carbon sample. In systems where active species are bonded to carbon, the surface structure and its stability will determine the dispersion and stability of the metal–support system. Some parameters shown in Fig. 31 are accounted for if the carbon reactivity is given as TOF based on RSA. The uncontrolled parameters remain the variation in intra-layer defect density, which is statistical and not measurable by any reaction-based technique (as the technique modifies the number of intra-layer defects), and the passivation by the formation of ultra-stable carbon–oxygen [370] groups. This information shows that the reactivity of a carbon material relevant in catalysis against oxygen (and other oxidizing gases such as CO_2 or NO and water) is not easily defined. As almost no experimental studies have ever been applied to the RSA–TOF concept, it is not surprising that severe inconsistencies regarding kinetic parameters exist within the literature. A recent review has highlighted the merits of the transient kinetic technique forming the basis for a unified description of carbon oxidation [371].

The issue of carbon oxidation reactivity is further complicated in cases where the oxidation of a carbonaceous deposit or a "coke" is involved. In these cases, the solid carbon oxidation kinetics is superimposed on entirely different oxidation pathways involving organic molecules and their fragments from the oxidative de-polymerization of hydrogen-poor hydrocarbons, and on the kinetics describing the carbonization reaction that occurs simultaneously with the gasification reaction. The carbonization reaction system described in Fig. 18 is a sequence of polymerization–dehydrogenation processes which leaves essentially pure carbon with widely differing degrees of structural ordering. This carbon, which is responsible for the frequently reported conversion from coke into graphite, is not oxidized at moderate temperatures. It is either a well-ordered graphite with high oxidation temperatures (see Fig. 30) due to the operation of a ordering mesophase [199, 200], or it is passivated by C–H atom terminations [83, 163] on disordered "glassy" carbon.

One particular case of carbon oxidation is the operation of NO abatement systems. It has been found that, during the oxidation of carbons with NO_x it is possible to induce catalytic reduction besides stoichiometric oxidation. As a result, an excess of nitrogen and oxygen is formed over the expected amount of CO_2 from direct reduction [187, 326, 372–374]. The presence of catalysts that are known to accelerate the dissociation of NO_x (e.g., Cu or iron), as well as the addition of alkali species found useful in the formation of nanostructured carbon surfaces [326, 375], were thought to create suitable oxygen functional groups. The complexity of the phenomenon [376], which involved several types of fractional surface that were either empty or covered with various types of oxygen group but still enhanced by the mode of operation of SCR using ammonia as coreagent [189, 377], provided limited, but conclusive, evidence of the origin of this interesting reactivity, and this is discussed further in the case studies below. It should be noted that the catalytic decomposition of NO_x is an adverse effect in applications where an upgrading of the sorption properties of activated carbon is sought by low-temperature oxidation, without damaging the high microporosity of the carbon structure. The high reactivity of NO_x with carbon, and in particular with their nanostructured forms, is the basis of the elimination of diesel soot particles [86, 153] by the deliberate generation of excess NO_x through combustion management. A detailed understanding of the structure–function relationship [378, 379] between diesel soot and oxidation kinetics [351] is required for a useful control of such a system to eliminate both detrimental emissions.

2.3.15.16 Surface Chemistry of Carbon

The surface chemistry of carbon is relevant to all aspects of catalytic carbon chemistry. The two most important heteroelements are hydrogen and oxygen, each of which can undergo a variety of chemically different coordination geometries, and these must be discriminated in analyses of the chemical function of the carbon–heteroatom bond. In the literature the term "surface complex" is often used [59, 148, 179, 200, 322, 343, 380] to describe a group of chemically different bonds which are not specified in detail.

The research field of functional surface groups on carbon is highly fragmented; for example, by using "carbon functional surface groups" as a keyword in the web of science database in 2006, a total of 1470 papers with approximately 3900 authors was identified. The research is distributed among the application areas of carbon in which catalysis plays only a minor role. Key issues besides catalysis [85, 135, 186, 188, 243, 285, 287, 372, 381–384] are polymer science [246, 385–387] and adsorption/purification applications [388–390]. An uncritical use of the methodology [391] and incomplete consideration of the fact that functional groups are dynamic entities on a carbon surface and hence change with chemical treatment and during evaluation [392], has led to a convoluted and sometimes conflicting situation in the literature. In this chapter it is impossible to analyze in an encyclopedic manner all ramifications of this fragmented situation. Neither can all aspects of chemical and biological [388] functionalization relevant to catalytic processes be analyzed. The strategy is rather, to use the information about structure and oxidation reactions collected in here, and to illustrate the consequences of exposing such materials to reactive environments. The dynamics of the research field is illustrated in the annual growth of peer-reviewed publications displayed in Fig. 32. The contribution from the emerging field of nanocarbons has resulted in only limited additions, as can be seen

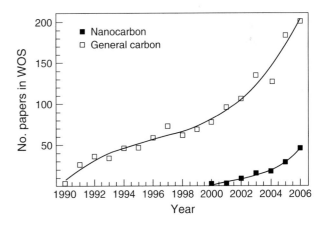

Fig. 32 Reflection of research in peer-reviewed journals about surface functionalization of carbon materials. The emerging character is clear. The studies on nanocarbons are additive to those of general carbon. No coherent view about the field can be deduced from these publications, despite a large agreement in methodology used.

References see page 419

from the ratio of general carbon-related papers to those of nanocarbon-related studies. The contribution of catalysis-related studies is too small to be estimated.

One basic consideration of the study of functional groups is the problem of their location. As far as oxygen groups are concerned, the fundamental process for their formation is either the binding of oxygen on a basal surface, followed by nucleophilic attack of water of the resulting epoxide into two OH groups. This reaction scenario was described first in the recipe of Boehm for generating basic [17, 392–394] functional groups, whereby it was assumed that, at a peripheral position of the initial oxidation, the incorporation of oxygen into the carbon skeleton would occur. This assumption and the resulting structures are discussed below. The difficult-to-reconcile statement that a basal plane of graphite should be non-reactive to activate oxygen [320] but active enough to dissociate water and adsorb protons [17] to generate a basic reaction of graphite in water, must be reconsidered in the light that the formation of an epoxide from a peroxide adsorption is a well-proven and plausible process. Theoretical studies [395] at a high level have shown that it is indeed not impossible on a flat surface, but that it is likely on a bent graphene surface (see e.g., Figs. 15–17) that oxygen reacts and forms a stable epoxide. A key result was that the interaction of a dissociated oxygen atom with a planar surface is exothermic by 176 kJ mol^{-1}, and on a single wall carbon nanotube outer surface by 310 kJ mol^{-1}. Care must be taken relating these numbers to reactivity, as the high activation barrier of dissociation of molecular oxygen must be overcome (as is also discussed in the paper). The traditional assumption of inertness of basal planes should be treated with caution if nanostructured surfaces are under consideration. The location of functional groups is thus not unambiguously clear, and in the presence of bent carbon surfaces and oxygen or even radical sources (such as naturally occurring peroxides in air or during illumination in air) a significant formation of spurious functional groups may occur, rendering carbons air-sensitive materials as they "age" and can oxidize organic molecules under ambient conditions [396]. Spontaneous explosions of carbon dusts and activated carbons immediately after activation may be explained, however, by this reactivity of non-planar surfaces with oxygen and water vapor.

The local connectivity of surface carbon atoms can be discriminated into sp^3, alkenic sp^2 and aromatic sp^2 for all types of carbon material of catalytic relevance. Each of the three connectivities can form with activated heteroatom several types of bond, allowing for a broad distribution of carbon heteroatom interactions coexisting at surfaces exhibiting structural defects (defined as deviation from the planar graphene surface, see Figs. 5 and 12). The surface chemistry of prismatic and basal planes of sp^2 carbons is fundamentally different (see Fig. 3), which renders the surface area ratio between the two orientations one of the dominating factors for the description of carbon reactivity. One way of obtaining an estimate for this ratio is to assume that the abundance of hydrophilic sites on a carbon is due to surface functional groups at the prism faces, whereas the basal planes should be hydrophobic. The existence of both properties on a carbon can be verified and quantitatively determined by selective chemisorption of long-chain aliphatic molecules (n-C$_{32}$H$_{66}$) or metal salts such as HAuCl$_4$ for the hydrophobic sites, and alcohol molecules such as n-butanol for the hydrophilic sites. A practical procedure using microcalorimetry has been developed and reported in the literature [397, 398].

The complexity of carbon materials, together with the surface chemical anisotropy and the modifying effects of defects, create a highly complex situation for surface chemistry, even when only one heteroelement is considered to be actively bonded. As the functional groups are covalently attached to the carbon, it is quite possible that multifunctionality occurs. The coexistence of acid and basic functional groups is frequently found, and will depend critically on the prehistory of a surface being an extrinsic and is not an intrinsic property of a material. It is a characteristic to describe such surfaces in terms of distributions of properties rather than in sharp numbers. Another issue to be addressed is the choice of the environment in which these surface properties are relevant. In most studies, it is explicitly assumed that an aqueous environment is present, as only then does a relation to the Brønsted acid–base scale make sense. When studies are carried out in ultra-high vacuum (UHV), for example in XPS analysis, or when samples are brought into an aprotic media (e.g., in polymer matrices), then strong modifications of the nature and number of oxygen functional groups may occur. The stated [17, 392] disagreement between instrumental and chemical analysis of functional groups may well be caused by the dynamic response of the carbon rather than by the inability of the analytical methodology to discriminate chemical structures.

The greatest attention in the literature has been paid to the carbon–oxygen–hydrogen interaction, because of their enormous chemical relevance and the existence of a suite of chemical and physical tools for their characterization. Carbon blacks have been studied extensively as they are of high elemental purity (no inorganic impurities) and exhibit sufficient specific surface area so as to allow a quantitative determination of the heteroelement content, which is the basis for any meaningful normalization of surface chemical properties.

In addition, the spectrum of industrial manufacturing processes allows the comparison of data from homogeneous samples with distinctly different [85, 152] microstructural properties.

Tab. 5 Heteroatoms in carbon black materials

Property	Philblack A furnace	Philblack E furnace	Spheron C channel	Spheron 9 channel	Mogul color
Surface area/m^2 g^{-1}	45.8	135.1	253.7	115.8	308
Total H/g 100 g^{-1} C	0.35	0.31	0.33	0.62	0.48
Total O/g 100 g^{-1} C	0.58	1.01	3.14	3.49	8.22
CO_2/g 100 g^{-1} C^a	0.187	0.401	0.575	0.536	2.205
CO/g 100 g^{-1} C^a	0.343	0.411	2.00	1.928	4.180
H_2O/g 100 g^{-1} C^a	0.00	0.435	0.600	0.710	1.440
H_2/g 100 g^{-1} C^a	0.209	0.152	0.152	11.5	0.132
H/µmol m^{-2}	38.2	11.5	6.5	7.7	7.7
O/µmol m^{-2}	7.9	4.7	7.7	18.0	16.7
CH_x	0.48	0.15	0.08	0.33	0.1
CO_x	0.1	0.06	0.10	0.23	0.21
C_xR	1.72	4.76	5.55	1.79	3.25
Ratio H : O	4.8	2.4	0.80	1.43	0.48
Ratio CO : CO_2	1.83	1.02	3.48	3.36	2.04

^{a}Integral from thermal desorption up to 1475 K.

2.3.15.17 Chemical Quantification of Oxygen Groups

A representative collection [210] of surface chemical data of several carbon blacks is compiled in Table 5. The spread of materials used in the classical studies of surface acidity [17, 399] is wider, but a comprehensive collection of characterization data is difficult to find. The specific surface area varies by an order of magnitude for the five samples. The total content of heteroatoms (as determined by classical microanalytical techniques) amounts to between about 1 and 5 wt.%, which is a sizeable quantity for a material which is nominally pure carbon.

One technique used often in chemical speciation of the surface functional groups is temperature-programmed desorption (TPD) of all functionalities. The strength of the surface chemical bonds requires maximum temperatures of usually 1273 K, with desorption profiles extending from 400 K to the upper limit of heating in inert atmosphere in a quartz tube furnace. The integral abundances of all possible decomposition products of C−O−H functionalities are listed in Table 5. These data show that both absolute amounts and product distributions (CO to CO_2 ratios) vary significantly, and exhibit no apparent correlation to surface area or synthesis method for the carbon. Reduction of the data per surface area reveals that carbon–hydrogen functions are abundant heterobonds on carbon black surfaces. Carbon–oxygen functions are always present, and in two samples (Spheron C and Mogul) are more frequent than hydrogen terminations. The importance of the heteroatoms for the description of the surface properties becomes apparent when the carbon to heteroatom ratio is considered. It seems that almost every other carbon atom exhibits in its connectivity one heteroatom.

The ratio of carbon to heteroatom may be used to estimate the size of the basic structural units. Assuming the microstructure of carbon to consist of small graphene platelets with few intralayer defects (see Figs. 11, 14, and 16), it is possible to estimate from the topology of *peri*-condensed aromatic structures the ratio of carbon to hypothetical oxygen terminations. These are all located at the prism faces of the graphene layers and saturate the dangling bonds in the carbon sp^2 configurations. A correlation between the carbon to heteroatom ratio and the size of hypothetical circular graphene layer is presented in Fig. 33. HRTEM images of carbon blacks exhibit 1.5–2.0 nm as characteristic diameters [192], compatible with $n \geq 5$ ring *peri*-condensed structures. The data in Table 5 agree reasonably well with this crude assumption,

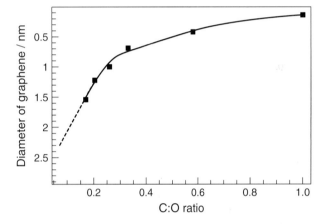

Fig. 33 Ratio between assumed carbon–oxygen terminations of hypothetical *peri*-condensed graphenes and their diameter. The dashed extrapolation falls well within the range of sizes seen by TEM and XRD, and agrees well with the data of Table 5.

References see page 419

as long as only oxygen functions are considered. The entry "C_xR" in Table 5 reveals, however, that significantly more heteroatomic bonds than perimeter sites may be present. In addition, the ratio between liberated CO and CO_2 indicates that a variety of oxygen functional groups is present rather than hypothetical carbonyl or hydroxy groups, which would be much too crowded from a synthetic organic standpoint on such a small graphene (aromatic) unit. It can be concluded that graphenes are richly decorated with oxygen functional groups at their perimeter. This location of heteroatom is, however, insufficient to explain the analytical results. There are several carboxylic and other more complex groups present as indicated by the CO/CO_2 ratio (the first two samples in Table 5), doubling the heteroatomic content per site. Even this is not sufficient to account for the analytical content. These considerations of Fig. 33 and the consistent structural data imply that other locations for functional groups must be considered. The rather irregular shape and the bending of many graphene units (as seen with TEM) and in the enlarged inter-planar distance point to the presence of a significant amount of in-graphene defects as location of functional groups. The other locations are hydrocarbon fragments between graphene units linking them covalently together, as discussed in Fig. 17.

It occurs, as discussed above, that the literature assumption about aromatic molecules being the carriers of functional groups does not account for the analytical results. It is thus a matter of simplification to deduce the chemical reactivity of functional groups only from aromatic substitution chemistry, as in the literature [192, 399]. It can be expected that a significant fraction of heteroatoms is bonded to sp^3 carbon centers, with a reactivity of oxo-alkanes adding to the expected spectrum of aromatic reactivity. Several sources of evidence from the literature support this notion. Using refined XRD techniques and by extracting the radial distribution function from molecular X-ray scattering, it has become possible to develop a model for a graphene layer [147, 228]. The model material was a coal sample before carbonization containing, as well as the main fraction of sp^2 centers, about 20% carbon atoms in aliphatic connectivity. This one-dimensional structure analysis represents a real example of the scheme displayed in Fig. 17C. The model predicts cluster sizes of between $n = 3$ and $n = 5$ for peri-condensed rings (coronene, hexaperibenzo-coronene), which is in full agreement with the electron microscopic (see Figs. 11 and 14) and NMR [400] data which led to the construction of Fig. 33. The data in Table 5 show clearly that a significant fraction of the total carbon surface which is chemically active (prism faces and defect sites) is covered by heteroatoms, and many of these sites are not passivated only by C—H bonds. This situation is also characteristic of carbonaceous deposits with C_xR values of 1.5 to 2.5 (corresponding to C : H ratios of 0.45 to 0.65), and pointing to truly smaller average diameters of the BSUs (see Fig. 19), as determined by chemical extraction and mass spectrometry [401, 402]. The situation is also typical for carbon materials resulting from dehydrogenation of organic polymers with a large abundance of prism faces and a significant presence of oxygen functional groups. For carbons with larger BSU diameters, the dominance of prism sites is gradually reduced and defect sites at crystallite boundaries and intra-plane defects become important locations for surface functional groups. The absolute abundance is, however, so much reduced that the accuracy of elemental analytical data as used above is too low to allow a meaningful chemical and structural interpretation. Figure 34 illustrates this situation, showing the deposit from oxidative dehydrogenation of butene over a carbon nanotube catalyst. The example was chosen to maintain the contrast between the deposit and the substrate such that structural details can be seen. The morphology may be compared to that of Fig. 14, bearing in mind that the C : H ratio of the deposit in Fig. 34 is close to 0.6.

As a critical evaluation of the size determination of BSUs is given here, it should be noted that the use of temperature-programmed analytical techniques is not free from error sources. In fact, in the present context it is of unproven but paramount importance

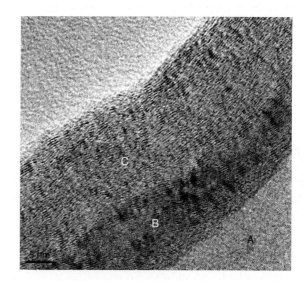

Fig. 34 High-resolution TEM image of a carbonaceous deposit from oxidative dehydrogenation of butene over a carbon nanotube catalyst. Region "A" is the hollow part of the nanotube, region "B" the wall of the nanotube containing at its outer side the active sites, and region "C" the deposit. The differing contrast between "B" and "C" is due to the reduced carbon density in the deposit explained by the hydrogen content. The functional groups cannot be seen at the present level of resolution.

that no readsorption of desorbed species may occur. At elevated temperatures, readsorbed gases will modify the chemical nature of the reacting surface. The presence of dangling bonds in the freshly created surface following the desorption of its top layer enhances such reactivity. Great care should be taken when data such as provided in Table 5 are analyzed in too-great detail, as most reported experiments have not been conducted in a vacuum but in a carrier gas stream with additional impurities of air and oxygen; moreover, no information is given about the pretreatment of the porous high-surface-area material. Despite its limitations, the chemical speciation of functional groups (as discussed in detail below) remains the most reliable – albeit the most complicated – way to learn about the chemical reactivity of a given carbon surface.

2.3.15.18 Non-Oxygen Heteroelements on Carbon Surfaces

The most important non-oxygen heteroelement is hydrogen bound to a variety of different reaction sites. Thermal desorption experiments showed that weakly held hydrogen desorbs at 1000 K, while more strongly bound species desorb at 1270 K and 1470 K. Only above 1900 K is all hydrogen released from a carbon material [403]. As this especially stable hydrogen can constitute as much as 30% of all hydrogen present, it was suggested that this species should be bonded to alkene and alkane linker structures within the graphene layers (see previous section), and represents hydrogen bonded to inner surfaces of carbon.

Hydrogen exerts a strong etching function to carbon by methanation that can well occur without the action of a catalytic metal, and which affects the stability of carbon under high temperature and under plasma conditions. Hydrogen atoms preferentially react with sp^2 carbon, leaving behind sp^3 carbon, a property that is used [63, 80, 162, 404, 405] to enhance the yield in diamond-like carbon and diamond CVD synthesis. It is also a great challenge to stabilize fusion reactor elements made from carbon against hydrogen etching, a process which has been studied in great detail from plasma physics [202, 405–407] and provides insight [408, 409] into the complex processes of CVD synthesis of hard carbons.

An important role was ascribed to the low-energy sorption of hydrogen [410] on carbon nanostructures once considered as a possible storage medium. The concept of liquefaction of hydrogen within the thin tubular opening of a nanotube proved not to be viable, as the chemisorption capacity at the outer surface was just below the required storage capacity per weight. For further details, the reader is referred to reviews in Refs. [410–412].

The chemical structure of the hydrogen groups is difficult to assess, as they can be present either directly bonded to carbon or as protons together with other functional groups. In addition, at low temperatures, all hydrophilic carbons carry a layer of molecular water that must first be desorbed to allow a quantitative analysis. Thermal desorption with the discrimination of water and carbon oxides can be used to control desorption of the unspecific water layer. It is observed, however, that the water desorption affects the surface chemistry of the carbon, as some labile groups are desorbed and the vacant surface sites can react with water to form C−O−H or C−H bonds [413, 414]. An observation in this direction was reported in the literature [17] where activated carbons were loaded with 200 µmol g^{-1} Pd and exposed to a humid atmosphere at 353 K. Under these very mild conditions, the abundance of acidic surface groups was more than doubled as compared to an exposure without Pd. The activated dissociation of water formed mainly phenolic groups, and likely also carbon hydrogen groups. Such observations are of relevance for the renaissance of carbon-supported catalysts in the processing of biofuels [190, 415, 416].

IR spectroscopy can be used to show that, even after mild desorption of water from an oxidized carbon black, a variety of hydrogen species is present at the surface. Figure 35 (upper panel) shows the assignment of absorption regions by functional groups using a typical carbon sample without any treatment. The corresponding transmission spectra of a technical carbon black with hydrophilic and hydrophobic reaction against liquid water after *in-situ* desorption of molecular water at 573 K are displayed in the lower panel of Fig. 35. The absorption from carbon–hydrogen and carbon–hydroxy groups overlap to form a complex peak structure. These data reveal, in agreement with TPD experiments, that hydrogen is present not only in strongly bound C−H groups, which coexist with other functional groups, but also in a variety of C−O−H groups which cannot be removed at moderate temperatures up to 675 K. For the hydrophilic sample, a broad structure of oligomeric OH group fundamental vibrations causes the strongly falling background at high wavenumbers. After more severe heat treatments the carbon surface is more uniform [417], but is modified in its defect properties and in its chemical reactivity.

The existence of several distinct C−H surface functional groups, and the modification of the carbon surface during adsorption of hydrogen, were studied in detail [418]. Four different kinetic regimes of coverage versus reaction time have been identified, indicating the structural heterogeneity of the carbon surface. The material studied (graphon) is a graphitized carbon black with no inorganic impurities. This allows the deduction of the heterogeneity of the adsorption sites being reproduced by a variation in

---

References see page 419

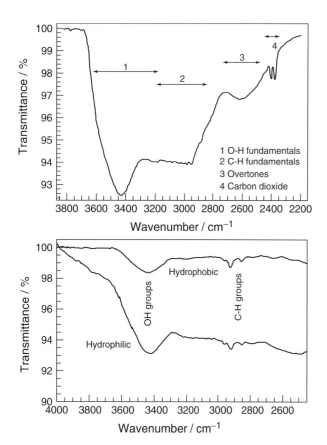

Fig. 35 Transmission FT-IR data for an untreated carbon black (upper panel) and two carbon blacks (lower panel) which were pretreated to desorb molecular water films. The hydrophilicity was induced by treatment with ozone at 300 K for 10 min. An in-situ reaction cell with specially prepurified KBr supports were used to accumulate the spectra in dry Ar. All sample handling was carried out in a glove box. A Bruker ISF 66 instrument was used to collect 1000 scans from each sample.

activation energy for adsorption as an intrinsic effect of the carbon material. The fact that temperatures above 700 K are required to saturate the carbon surface suggests that structural modifications (hydrogen etching) of the surface may significantly contribute to the surface reaction, being more complex than the reversible adsorption of hydrogen atoms on (pre-existing) chemisorption sites.

In order to exemplify the selective etching power of hydrogen, the SEM images are shown in Fig. 36 of a carbon–carbon composite before and after exposure to hydrogen plasma at nominally ambient temperature and without any sample bias. The strong preferential etching can clearly be seen of defective carbon, leaving behind graphite flakes partly oriented perpendicular to the initial surface. Chemically activated hydrogen will be much less aggressive at moderate temperatures, but will be as selective in modifying a heterogeneous carbon surface.

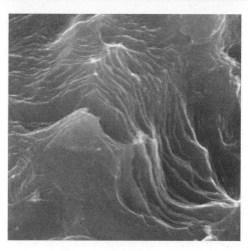

Fig. 36 SEM images (low-voltage) of a graphite-rich carbon–carbon composite before (A) and after (B) treatment with hydrogen plasma at 10^2 Pa pressure, nominally 300 K and without sample bias. The enlarged image (ca. 8× vs. B) at the bottom clearly reveals the preferential etching of the binder carbon leaving behind the graphite flakes.

Carbon–nitrogen surface groups are not abundant in carbons made from hydrocarbon precursors. Carbons from nitrogen-containing natural and synthetic polymers [419, 420] contain structural C–N–C groups in aliphatic and aromatic coordination. Surface complexes such as amines, amides, or nitro groups are not very stable, and are desorbed during heat treatment (manufacture, activation, precleaning). An extensive body of work has been assembled [413] concerning the intended incorporation of nitrogen from ammonia or cyanides using carbon surfaces prereacted with oxygen or halogens. Several structures of C−N and C−N−H have been identified with physical methods. However, the stability of the most abundant C−N−H groups is rather limited in terms of temperature.

Most significant is the observation that the catalytic behavior of a carbon (from various sources) remains affected by the amination treatment, even when the nitrogen is desorbed from the surface. The explanation was put forward that nitrogen binding (and removal) occurs on defective sites, and that the nitrogen activation is in fact a mild version of preferential removal of defective surface sites, leaving behind a carbon surface with a more graphitic electronic structure than was originally present. Some of the structural nitrogen (pyridine-like incorporation in the graphene layers) is thermally stable up to 1273 K [210, 413], without affecting the catalytic performance of the carbon material. This observation excludes the notion that the heterocyclic substitution injects additional electrons [421] into the conduction band of the semimetal graphite, and in doing so enhances its activity in electron-transfer reactions (e.g., activation of molecular oxygen in liquid phase oxidation catalysis [422]).

Carbon–sulfur groups are extremely stable surface compounds, which cannot be removed by thermal treatment up to 1470 K. Indeed, only by using a reductive treatment with hydrogen is it possible to clean carbon from sulfur adsorbates. One source of sulfur is the fuel used to generate the carbon from which about 90% are covalently bonded to the carbon and 10% are segregated as adsorbate, that can be removed by solvent extraction. The abundance of sulfur can amount to several wt.%. Removal of the structural sulfur is possible by hydrogen reduction to H_2S at about 1000 K. A collection of data on this subject is available [210]. In activated carbons, sulfur can also be present in an oxidized form as sulfate or as C−S=O compounds.

The generation of sulfur functional groups has been investigated in detail in order to prepare anchoring sites on surfaces used in rubber vulcanization (for tire manufacture). Only these sulfur groups which are not incorporated into the graphene layers can be chemically active for crosslinking polymerization in vulcanization. In a broad study [139, 423], carbon disulfide, hydrogen disulfide, thionylchloride and sulfur dioxide were used at elevated temperatures to generate sulfided carbon surfaces on a variety of sp^2 substrates. Structural data revealed in all cases the absence of graphitization phenomena, even though the reaction temperatures were up to 1173 K. Both, XPS and IR data indicated that two types of species were formed – one type should be C=S bonded to aromatic rings, and the other is a thiolactone.

Carbon–halogen groups are well known to exist for fluorine, chlorine and bromine. These groups can be attached by redox reactions involving the halogen elements, either in aqueous solution or in the gas phase. At elevated temperatures, bulk reactions occur under intercalation and eventually under formation of bulk covalent carbon–halogen materials ($CF_{1.1}$ solid, CCl_4 gaseous) [424]. At lower chemical potentials of the halogens, substitution reactions into C−H bonds [Eq. (8)] or addition reactions onto *in-situ*-formed unsaturated bonds [Eq. (9)] occur. These two reactions are among the most important analytical tools for describing the chemical reactivity of carbon materials. On the basis of extensive experiments with various carbons [425], the reaction of a carbon with aqueous bromine is suggested to serve as a reference experiment for the quantification of the unsaturated carbon atoms which are mostly masked by oxygen surface groups [Eq. (9)].

$$\text{Ar-H} + Br_2 \longrightarrow \text{Ar-Br} + HBr \quad (8)$$

The reaction with chlorine occurs, after high-temperature outgassing, directly as the addition of halogen on the unsaturated C=C bonds. The strength of the carbon–halogen bonds is determined by using halogen-saturated carbons for halogenation of nucleophiles [210]. It occurs that iodine groups are fully reversibly adsorbed, whereas bromine and chlorine are partly irreversibly adsorbed and can only be removed by high-temperature treatments under oxidizing or reducing conditions. This result led to the conclusion that aqueous iodine solutions should be used to determine the surface area of carbon as a selective adsorbent. Reaction with iodine vapor at elevated temperature (600 K) results in the same chemistry as shown in Eqs. (8) and (9), and is not suitable for carbon-selective surface area determination.

$$\text{C=C} + Br_2 \longrightarrow \text{CBr-CBr} \quad (9)$$

Care must be taken when interpreting the results of halogen uptake on carbon at low temperatures, in aqueous solutions, or at undefined surface states of the carbon (as received). A large number of reaction involving halogens and oxygen surface functional groups,

References see page 419

and of course between halogen hydrogen acids and basic functional groups, occur simultaneously with the reactions mentioned above. They may thus perturb the analytical data [in particular the amount of HX as quantitative measure for Eq. (8)] to such an extent that no meaningful result can be obtained. The only method is to define the surface chemistry of the carbon by preannealing and treatment in the gas phase prior to halogen exposure or by *in-situ* oxidation [426], thus forcing Eq. (9) to dominate.

Bromination is a powerful tool for purifying and functionalizing nanostructured [427–429] carbon. When bromine can react with carbon under catalytic conditions, it must be expected that strong segregation according to reactivity will occur, leaving well-graphitized structures unaffected but converting defective structures into species reacting with moisture or solvents leading to mobilization of individual graphenes. Traces of bromine (and to a lesser extent of chlorine) in systems operating with carbon-supported catalysts will be detrimental to the stability of the system, as the anchoring surface will be attacked and reactive sites will be preferentially [430] etched.

2.3.15.19 Chemical Analysis of Surface Oxygen Groups

The abundance of oxygen atoms on carbons varies over a wide range, with a maximum at about C_4O (see Table 5). The surface chemical – and hence many technical – properties of carbon materials (including diamond [404]) depend, however, crucially on the carbon–oxygen chemistry which controls the adsorption, adhesion, and oxidation activities of the surfaces. From the reaction with liquid water, acidic and basic surface complexes can be discriminated. Under almost all catalytic conditions both types of oxygen group are present simultaneously, and the pH value of an immersion of carbons in water is the result of protolysis of solid acids and solid base groups. The pH can be as low as weak mineral acids, indicating that a significant solid acidity can be present on carbon. The basic reaction is, due to a generally lower abundance of basic oxygen functions, rather more limited and rarely exceeds the value of pH 8. For this reason, an overall good quantitative correlation between pH in aqueous solution and analytical oxygen content for a carbon species has been established [431].

The possibility exists further, that the termination of a graphene perimeter occurs by changing its hybridization to localized, electron-rich configurations [37] termed "carbyne-like". Theoretical studies, together with some observations on the electronic properties of isolated graphene [37, 69, 432, 433], strongly imply that such terminations without functional groups should also exist. When searching the literature it is important to know that, in its discovery stage, the nanotubes were designated [434] "graphene tubules" – a structure of carbon that has, ideally, no termination and is thus not to be mixed with unconventional terminations. Under conditions relevant to and for catalysis, such self-termination should however, be a rare situation and the heteroatomic termination discussed here should be by far the most dominant situation, even for the samples discussed in the physics experiments.

A consequence of the Brønsted acid functionality is that many of the carbon–oxygen groups exhibit a cation-exchange capacity and a simultaneous anion-exchange capacity (from the basic groups). These functions are most relevant for the application of carbon as a catalyst support, and provide a method for anchoring metal ions also on carbons by the same method as used for oxide supports. It has been speculated that, for the special case of protons (exchange on basic groups), the graphene layer in its unperturbed aromatic state is sufficiently basic to allow direct adsorption [183]. This was supported by theoretical calculations on carbon cluster models [435]. Longer hydrocarbon molecules [436, 437] and molecular hydrogen [438] were verified to be chemisorbed on basal planes of graphene. It is, however, much more likely that protons are not adsorbed on graphene but rather on oxygen defects, forming hydroxyls or water clusters [439] with the resulting Brønsted-basic reaction being related to suitable defects and not to the bare graphene surface, the abundance of which would lead to a strongly basic reaction of graphene in water not being observed. The situation is much different when transition metals such as Pt are dispersed on carbon, and when hydrogen chemisorption is being used to determine the specific surface area. If a support with a substantial content of oxygen functional groups is used, it is quite likely that by spillover much too-high a dispersion will be reported. The application of CO as a probe molecule is not a suitable solution, as this molecule will also react with hydroxyl groups and thus lead to an overestimation of the dispersion. It is clear that the interference of defects and hydroxyl groups as acceptors for activated hydrogen play a role in hydrogenation catalysis [271, 440, 441] (carbon can be a source of active hydrogen) and in fuel cell applications [440] (carbon can store, and thus kinetically perturb, hydrogen evolution). The removal of all these interfering influences of oxygen functional groups is difficult, as the fixation of metal clusters will then become a problem [135, 383] on planar carbon.

The dominating importance of the oxygen functional groups has led to significant efforts aimed at preparation routes to obtain only part of the wide pK_a spectrum on the surface – that is, to prepare either only acidic or only basic functional groups. This was, however, only partly successful, with the result that in all circumstances a spectrum of neutralization equilibria [183, 323, 392] rather than a single pK value characterized a carbon surface. Basic functional groups can be obtained when carbons are annealed in vacuum or inert gas (Ar, not nitrogen) at 875 K for several hours, cooled down under inert conditions to 300 K, and then exposed to air, water, or oxygen [17, 394, 399]. The basic reactivity occurs only after contact with water molecules contributing to the formation of the basic groups.

Predominantly acidic groups are obtained by heating carbon in air or oxygen to 600–900 K with a decrease in acidity with increasing temperature (self-gasification). Very strongly acidic groups can be obtained with low-temperature liquid-phase oxidation [190] using hydrogen peroxide, ozone, chromic acid, or aqua regia. These groups also form unintentionally when carbons are used in liquid phase [190, 323, 439] for catalytic oxidations.

The characterization of the resulting surfaces has been problematic since the first attempts were made at applying chemical means [143, 394, 442]. Unfortunately, the advent of many surface physical methods [75, 85, 108, 152, 155, 321, 414, 439, 443] has not contributed significantly [392] more information than was obtained with chemical probe reactions (derivatization as in the early days of organic chemistry). Modern versions of chemical modifications can be used to detect and spatially resolve small amounts of oxygen functional groups by implementing [444] methods from molecular biology.

One reason for the relatively weak impact of surface analysis is the small abundance of a wide variety of chemically different species. Most surface spectroscopies are too insensitive for the variations in chemical environment, while others require single-crystal surfaces that are only functionalized under the loss of their well-defined structure, for example by sputtering or chemical etching [202, 445]. Another challenge is the need to perform surface analysis in UHV where, by definition, no water should exist and hence during sample preparation a major modification of chemically reactive groups will occur. This dynamic response of carbon leads to sample charging, line broadening, and many erroneous assignments of peak deconvolution, as discussed below. The argument [17, 392] is less valid that the electronegativity of oxygen would inherently preclude a meaningful chemical speciation. An example of spectroscopic multitechnique speciation on a complex substrate is found a report dealing with coal [446] functional groups.

Vibrational spectroscopy [59, 153, 167] is hampered by the strong bulk absorption of light in carbon. A comprehensive review [447] on nanoparticles of carbon and the characterization of reaction products under atmospheric conditions is given using FT-IR-coupled in-situ techniques. Raman spectroscopy [64, 92, 167, 448] is not particularly sensitive to highly polar carbon–heterobonds, and is not surface-sensitive at all. Nuclear magnetic resonance [43, 70, 173, 436, 449] has relaxation problems with non-hydrogen-carrying surface groups bound to aromatic graphene [450–452] structures.

For these reasons, the bulk of our knowledge on the structural details of carbon–oxygen functionalities stems from chemical experiments, as collected in several reviews [138, 175, 210, 388, 394, 453, 454]. A special series of publications is devoted to the structure and reactivity of basic oxygen groups [393, 399, 455, 456]. The most prominent analysis method – the "Boehm titration" – is described in great detail in the author's reviews. Additional hints on the application and usefulness of these chemical methods can be found in Ref. [454].

The analysis of the structural chemistry of oxygen groups using chemical methods is limited to stable configurations and misses, by definition (react until equilibrium is achieved), all metastable species that may be present under in-situ oxidation [39, 79] conditions. The chemical structures of the reaction products of the sequence provided by Eqs. (1–7) may be found in this description, but other less-stable configurations such as peroxides and hyperoxides [185, 413, 457] may also exit. Care should be taken to transfer the knowledge on oxygen groups essentially in contact with water reported here directly to the discussion about the atomic structure of oxidation reaction intermediates [78, 172, 327, 370, 371] outlined earlier in the chapter. It should be noted that hydrogen and water are strong inhibitors [344] of oxidation processes, and that the hydrogen spillover mentioned earlier is a known modifier [458] of oxygen groups.

The acidic functional groups can be characterized by their neutralization reactions with a selection of bases differing in their pK values. The usual reagents are sodium hydroxide, sodium carbonate, sodium bicarbonate and, for very weak acids, sodium ethoxide in alcohol. An additional dimension of information can be obtained by pretreating the carbon prior to neutralization analysis [59].

Figure 37 shows representations of the fundamental structures responsible for Brønsted acidity. The carbon substrate is represented, by intention, as partly aromatic and partly non-aromatic, without indicating the true local complexity of the "graphene" skeleton at and near its perimeter. This *caveat* is valid also for the following structural representations of functional groups.

Only carboxylic and anhydride groups are strictly acidic in water, with the anhydride hydrolyzing to two carboxylic acid functions. These two groups contribute mainly to the reaction with sodium hydroxide, but react with all weaker and stronger bases than hydroxide. These groups can occur frequently on carbon surfaces, in particular after partial gasification and cooling in oxygen-containing

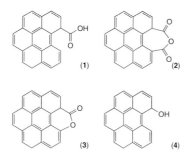

Fig. 37 Structures of acidic oxygen functional groups. (**1**) Carboxylic group; (**2**) acid anhydride group; (**3**) lactone; (**4**) phenol group. The acidity in water decreases in the order of presentation. Significant modifications occur from neighboring effects of multiple functional groups (see text).

atmospheres that include moisture. They will also result from liquid-phase oxidation with mineral acids, and in high abundance from peroxide treatments. The lactone groups must undergo ring opening in a basic medium, which can be achieved with 0.05 M sodium carbonate. Phenolic groups are thought to be frequent due to their structural simplicity [as the reaction product of Eq. (4)]; a rapid quench from gasification in weakly oxidizing atmospheres favors their formation. Their acidity is limited in water, and will depend strongly on neighboring functional groups. Sodium ethoxide is a selective reagent for their detection.

Some oxygen functional groups do not react with water molecules, and behave thus as neutral in aqueous media. They will react, however, in environments at higher temperatures, act as receptors for protons in gas-phase processes, and may also react in non-aqueous media. Their fundamental structures are shown in Fig. 38. Some carbonyl groups can still react with sodium ethoxide. Quinoidic structures result in an additional defect of an aromatic termination, whereby the initial radical site will either react with additional oxygen and give rise to gasification and likely further radical structures (a source of continuing reactivity), or it will react with hydrogen or an organic substrate and become quenched. It occurs that quinoidic groups are rare under oxidizing conditions as they will give rise to auto-oxidation, or that their generation is best accomplished with some sacrificial organic reagent such as a hydrocarbon sealing the primary defect with a non-reactive terminal group.

The ether or xanthene structure is fully inactive in aqueous media. Depending on its topology at internal or surface terminations, the relative stability in thermal or oxidation processes can be very high, indeed exceeding that of any other oxygen functional groups, and it may thus be an inhibitor of reactivity rather than an intermediate. Both neutral structures are believed to occur frequently due to their structural simplicity, and to the fact that in any oxidation reaction these structures must occur as intermediates in ring-destruction sequences. In quantitative studies where the oxygen content is compared to the sum of all neutralization reactions, a significant excess of chemically non-reactive oxygen is stated [323, 327, 370, 371] which is accounted for by these neutral functions.

Brønsted-basic groups are of a more complex structure, as can be seen from Fig. 39. They consist of two interacting oxygen centers [459, 460], one of which is active in proton adsorption and the other in anion exchange. The interrelationship has been demonstrated in a series of elegant chemical reactions [393, 399, 455, 456]. The structure shown in Fig. 39 represents a simplified configuration, and may have more complicated isomer structures. It is interesting to note that, during the formation of the basic groups, a non-stoichiometric amount of hydrogen peroxide is formed when the oxidized

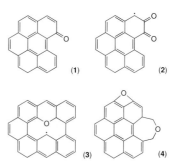

Fig. 38 Oxygen functional groups reacting "neutral" in liquid water. (**1**) Carbonyl; (**2**) quinoidic; (**3**) ether or xanthene in internal defect; (**4**) ether groups of varying stability on perimeter situations of a graphene. Significant modifications occur from neighboring effects of multiple functional groups (see text).

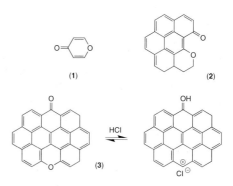

Fig. 39 Basic functional groups on carbons. (**1**) The γ-pyrone molecule giving the structures their name in literature. (**2**) represents the simple form of a pyrone structure where a resonance exists between the two oxygen atoms (alternating double bonds); numerous other configurations with this resonance exist and were drawn in the literature. (**3**) Pyrone structures react with acids (also with water) in an ionic equilibrium defining their base strength.

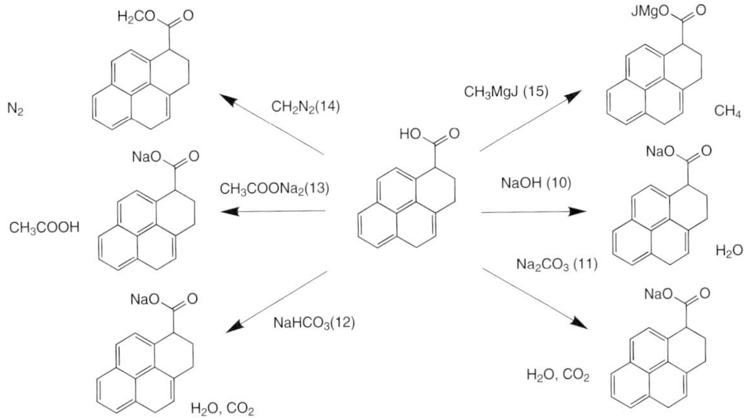

Fig. 40 Chemical probe reactions [Eqs. (10–15)] for carboxylic functional groups.

carbon is contacted with moisture or water. This reaction, which was also discussed in conjunction with an oxidation of ether structures to chromene groups [17, 394, 459], is reminiscent (see Fig. 39) of the anthraquinone catalysis of technical hydrogen peroxide synthesis. It should be taken into account that, in all situations in which dry oxidized carbons are contacted with water, a chemically detectable amount of hydrogen peroxide will be formed which may interfere with experimental observations.

Suitable reactions for the chemical identification of fundamental surface groups are collected in Eqs. (10–24); shown in the following figures, which summarize convenient reactions for chemical group identification. A large number of additional reactions with rather special applications can be found in the literature [17, 323, 388, 444, 454, 460]. When these reactions are used, it is advisable to optimize the reaction conditions by several different reactions characteristic of the same functional group. It occurs that the neutralization kinetics can be slow, in particular with hydrophobic and porous carbons. Although reaction times should not be under 24 h under ambient conditions, artifacts such as glass adsorption, reaction with traces of air and the intrinsic problem of conversion of the surface functional groups during chemical reaction, limit the reaction time to an optimum for complete but artifact-free determination. It is difficult to find reference values for well-characterized and accessible samples in the literature. From the studies of Boehm [17], data relating to carbon black CK3 can be used as characteristic values, although substantially different values for selected types of functional groups after highly specific treatments of incompletely characterized samples can be found in the original reports. The values are (in µmol g^{-1} carbon): basic: 40; carboxylate: 690; phenolic 320; and carbonyl: 210. These values illustrate the trends discussed in this section. Data from entirely other carbons [461–463] also support this trend, as much as the rigor of characterization of naturally occurring materials allows for a comparison of structure and reactivity.

Equations (12–15), which are shown collectively in Fig. 40, describe the standard neutralization reaction for an organic carboxylic acid. The discrepancy in neutralization capacity for the different reagents can be used to estimate the pK distribution [79, 464]. (For a discussion of the consequences of these differences in the assignment of functional groups, see below.) Amounts of between 50 and 500 µ Eq. acid per gram of carbon are characteristic values, indicating the high degree of experimental perfection required to determine such surface acidity data. Equations (14) and (15) form derivatives of the carboxylic acid, and can be used to volumetrically detect their presence by monitoring the gaseous reaction products.

Phenol groups are determined by the difference of the neutralization reactions between sodium hydroxide and sodium carbonate, which are common for carboxylic and phenolate groups. Oxidized carbons (typically in air at 700 K) contain equal amounts of these two functional groups. Activated carbons contain up to 800 µ Eq g^{-1} carbon of phenolate groups. The volumetric determinations are also sensitive to both carboxylic and phenolate groups, so that specific derivatization processes are needed to discriminate reliably between the two types of group. A selection of such derivatization reactions is provided in Fig. 41, with Eqs. (16–18). Since it is clear that Eq. (18) is not specific, the other two procedures allow the specific detection of isolated phenol groups by observing (via vibrational spectroscopy) the surface derivatization products, or by titrating the acids [Eq. (17)] formed as byproducts. Phenolic groups depend for their reactivity very strongly on the exact structure. Molecular substituents in phenol can change the pK_a

---

References see page 419

values by 10 orders of magnitude: a wide distribution of reactivities from almost Brønsted neutral for OH groups in non-aromatic structures to mineral acidic for OH groups concomitant with nitro-groups can be expected. Thus, care must be taken when reactivity–structure correlations [464, 465] are drawn.

With more complicated structures such as lactones, the identification problems become even more complex as their reaction in water results in the formation of a carboxylic acid group and a neighboring phenol group. This reaction is exemplified in Fig. 42. Derivatization with diazomethane allows quantification by volumetric nitrogen determination according to Eq. (19). As the coexistence of phenol and carboxylic acid will be very frequent, and much more frequent than the formation of a cyclic ether (lactone), it is necessary to perform the reaction also via Eq. (20) and, as a control for non-isolated lactones, via Eq. (21). The respective differences in nitrogen evolution allow very valuable distinctions about coexisting groups, and these data justify the substantial experimental effort necessary, including the strictly dry handling of high-surface-area material. Equation (21) creates new reactivity through the initially radical center that gives rise to the creation of new surface groups, depending on the nature and cleanliness of the reaction medium. Peroxides – and consequently basic OH groups – are the most likely candidates arising from traces of oxygen.

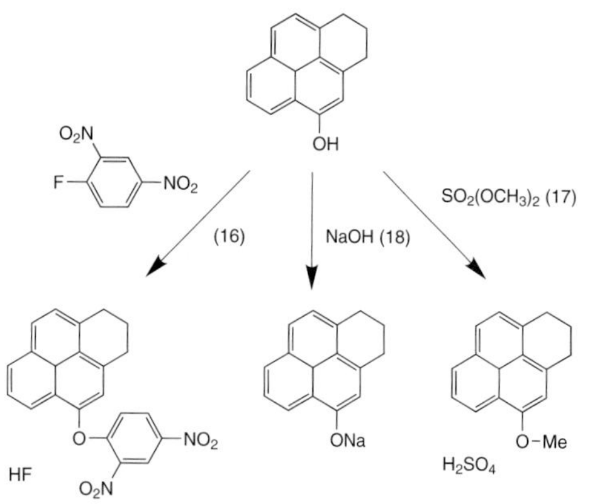

The complexity of the chemical situation increases further when combinations of different surface groups react with each other under the influence of a polar solvent such as water, when used for performing quantitative determinations. It should be noted that these reactions take time, and thus may also occur as a consequence of dynamic exchange reaction of hydrolysis products; these occur during the recommended time of 24 h time of reaction that is needed to exclude diffusion limitations in surface reactions. A prominent example, as shown in Fig. 43, is the lactol group originating from tautomeric equilibration [Eq. (22)] from the combination of a carboxyl group and a carbonyl group. This is a frequent situation due to the high abundance of both fundamental groups. The resulting structure, representing a complicated example of a "coadsorption effect", can undergo a whole spectrum of reactions and interfere with several derivation reactions, for example Eq. (23), which is aimed at determining the abundance of parent functional groups. This simple example illustrates the dynamics of a multifunctional surface at a very simple level.

Another type of complication arises from the geometry of a perimeter surface; the chemical inequivalence of (110) and (110) terminating surfaces – which are referred to as "zigzag" and "armchair" configurations of the graphene layers – leads to different reactivities of the same parent groups, depending on their location. An armchair configuration exhibiting two different functional groups at one six-membered ring is shown in Fig. 44. Upon reaction with base, the tautomerization [Eq. (24)] will be effective, with consequence that the keto form requires one equivalent for neutralization [Eq. (25)], whereas the enol form will consume two equivalents [Eq. (26)]. The equilibrium between the two forms depends upon the base concentration of the base used for detection. The extent of Eq. (24) will also depend on the time of contact of the surface with the base, and thus give rise to a slow kinetics of transformation and

Fig. 41 Chemical probe reactions [Eqs. (16–18)] for isolated phenol groups. Equation (16) is suitable for detection by vibrational spectroscopy; Eq. (17) for quantification by titration; Eq. (18) is non-specific and integrated the content of aromatic alcohols into the abundance of acidic groups.

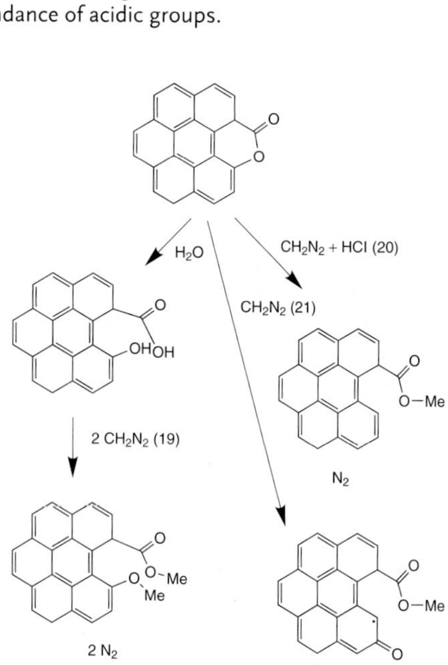

Fig. 42 Chemical probe reactions [Eqs. (19–21)] for lactones. Equation (21) requires an acidic proton, for example from a phenol group near the lactone group.

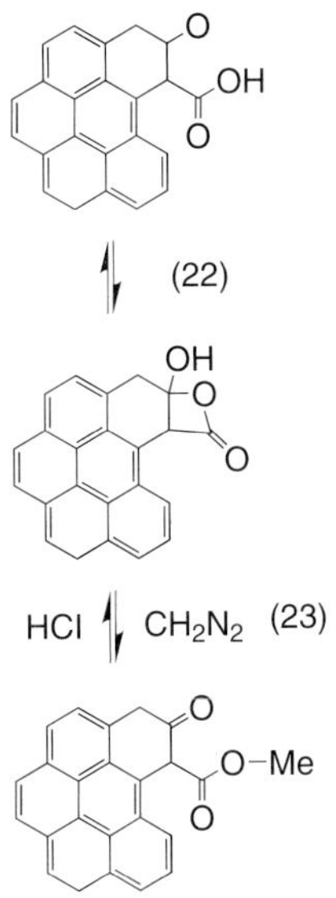

Fig. 43 Reaction of two groups independent from each other under influence of a solvent [Eq. (22)]. The resulting lactol group reacts with diazomethane [Eq. (23)] in the same way as the lactone group, although there would be no lactone detectable in vacuum-based analytical techniques.

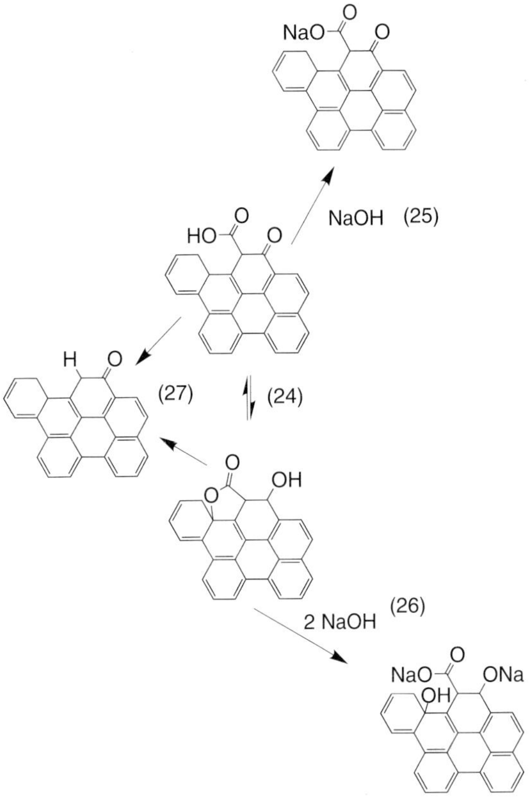

Fig. 44 The arbitrary situation of the lactol precursors (see Fig. 42) on an armchair configuration of the termination gives rise after tautomerism [Eq. (24)] to two ambiguous quantitative results using liquid-phase titration [Eqs. (25, 26)]. Vacuum-based thermal desorption [Eq. (27)] is insensitive to this phenomenon.

contributing to the experimental observation of "sloping" titration results with time of reaction.

Equation (24) is also a good example of the insensitivity of thermal desorption for such effects, because the same molecule results after decarboxylation [Eq. (27)]. Any quantitative correlation between neutralization and thermal desorption thus leads to inconsistencies, which become larger the more aggressive the chemicals being used for maximum complete determination of weakly reacting functional groups.

These few examples provide an impression of the chemical complexity of the carbon–oxygen surface reaction. They illustrate that surface acidity, as a single value for sample characterization, is a poor representation of the spectrum of reactivity already present in aqueous media. It becomes apparent why, in practical applications [59, 170, 181, 325, 461, 466–470], only poor correlations are observed between such simplified surface chemical parameters and the observed reactivity.

Conceptually, a simplifying attempt was made to overcome this complexity by analyzing carbon with TPD, integrating the total CO and CO_2 emission, and correlating the results with sample pretreatment and chemical reactivity [59, 210, 453]. The limited validity of such an approach was discussed above, and is also obvious from the examples presented in Figs. 41–44. In the applications of carbons as catalyst supports, it is immediately apparent that details of the carbon–metal interaction depend crucially on details of the surface chemistry. This explains the enormous number of carbon supports used commercially, numbering several thousands. To date, no systematic effort has been made to understand these relationships on the basis of modern analytical capabilities. Indeed, modification of the delicate surface chemistry at low and intermediate temperatures, and the sensitivity of functional groups to the presence of moisture or solvents, are not adequately represented in the destructive method of thermal analysis. Unfortunately, this *caveat* is also valid for most other spectroscopic methods.

2.3.15.20 Reactivity in Fluid Phase

Carbons in catalysis are frequently in contact with fluid (aqueous) phases. During synthesis, during modification, during loading with catalytic materials, and during service as either support, catalyst and/or sorbent, there is invariably present either a bulk fluid phase or a thin film of adsorbed water, forming the interface to the functionalized carbon surface. Only in the rare case

---

References see page 419

of full hydrophobicity [386, 471], which is achieved by complete hydrogenation of the surface, is there little contact. The previous sections provided an insight into the methodology of describing the functional group interface by a combination of structures derived from organic chemistry and analyzed using a static picture of their existence and reactivity. The tacit assumption, that functional groups can be characterized independently of each other (the basis of the categorization according to, for example the Boehm classification) was shown to be oversimplified, and strong coadsorption effects would prevail for samples held in reactive environments at ambient pressure and in the presence of water molecules. Both fundamental assumptions were shown to be approximations, however, and the carbon surface should be considered as a dynamic system with a flexible and multiple response of its reactivity to the reaction environment. Hence, it is not the categorization as "acid" or "basic" surfaces but rather the distribution of these properties (e.g., in the form of acid strength versus abundance plots) which provides a more adequate description of the surface. Acid–base reactivity should not be considered as a characteristic property of a material, but rather as a dynamic property, the extent of which will depend on the carbon structure that will respond flexibly (and sometimes irreversibly) to its environment.

In this situation, it is first adequate to monitor the modifications that are imposed on such a delicate system upon treatment with the test reagents of the chemical characterization toolbox. This can be achieved by performing a rigorous comparison (correlation) of analytical results obtained with different methods. As an example, for a group of typical carbons such a study [472, 473] on activated carbons is discussed in detail. Complementary methods of chemical analysis as described in the previous section are combined with calorimetric studies, elemental analysis, TPD experiments, and a method called "pH-drift". In this method, the buffering capacity of the solid acid–base is studied in an environment providing a constant electrolyte potential. The method is similar to electrokinetic experiments measuring the zeta potential of a sample. The exact definition and methodological discriminations have been discussed in detail [183] elsewhere.

The "bottom line" of this approach is that carboxylic functions can be well separated from other acidic functions. It seems also clear that oxygen functional groups are the only origin of acid reactivity, as shown by the clear structure–function correlation between surface oxygen content and total acid reactivity. Basic reactivity is conceptually more difficult to rationalize. First, there are conditions of the environment where acid and base reactivity are quantitatively the same. Whereas, in homogeneous media this would mean "neutralization", this is not the case in solid acid–base systems, where it means equal coexistence of groups. The possibility that the equality in response of acid and base groups to a neutralization agent arises from a reaction of the carbon with the neutralization agent cannot be excluded in particular, as pronounced kinetic effects are reported. The correlation of basicity with oxygen groups is much weaker, and can be traced back to the weak basicity of the species described in the previous section. Common for both acid and basic functions was that distributions of proton activities rather than well-defined material-dependent values were found. This raised, as in other studies, the question about the mechanism of formation [465, 474] of basic groups.

Indeed, in many studies it is inferred that reactions of water with "Lewis basic" sites must occur. An alternative idea, which is frequently found in the literature, assumes that the "pi-electron density" provides Lewis basicity. In practice, this would lead to a redox reaction between oxygen and water-forming OH groups:

$$O_2 + 4H^+ + 4e^- \rightleftharpoons 4OH^- + 4h^+ \quad (25)$$

This ties in with the observation by Boehm [17, 399] that the reaction of acids with carbon can only occur when oxygen is present. It should be noted here that care must be taken to verify these conditions in all attempts to titrate basic functional groups, as well as in porous systems, for example.

This reaction consumes graphitic carbon [formally for quenching the "hole states" in Eq. (25)], in analogy to its well-known property of being a reducing agent of oxides. The mild conditions lead to the reaction of "hole states" with water, forming the complicated oxygen functional groups shown in Figs. 41 to 44, including the heterocyclic incorporation of oxygen (pyrones) into the carbon lattice.

Such ideas can only be valid for the initial formation of basic groups from a pristine and elementally pure graphene perimeter, as would exist after high-temperature desorption under inert or vacuum conditions. The assumption that all basal planes of graphite would react continuously with oxygen and water to form hydroxyl groups cannot be correct, despite its frequent mention in the literature. Such reactivity does occur, however, under stoichiometric conditions at very harsh conditions of coking of coal or of coal gasification, but these processes are not described here. The role of defects in such reactivity [87, 153], for which theory [66, 475, 476] has produced evidence, has not yet been addressed in any systematic fashion. The process of the initial reaction of clean carbon (which will not be obtained after "conditioning" at mild temperatures [464, 465]) can be formally described via the formation of defects and free hydroxyls under the participation of Lewis-basic sites, or via the formation of peroxides [185, 413, 457, 467] (which seem to occur, as shown by the occasional spontaneous explosions of highly activated carbons). The symbolic equations are:

$$O_2 + 4e^- \longrightarrow 2O^{2-}$$
$$2LBS \longrightarrow 2LBS^{2+} + 4e^-$$
$$(LBS = \text{Lewis base site})$$
$$2LBS^{2+} + 2O^{2-} \longrightarrow 2LBS-O$$
$$2LBS-O + 2H_2O \longrightarrow 2LBS-OH + 2OH^- + 2h^+$$
$$O_2 + 2LBS + 2H_2O \longrightarrow 2LBS-OH + 2OH^- + 2h^+$$
(26)

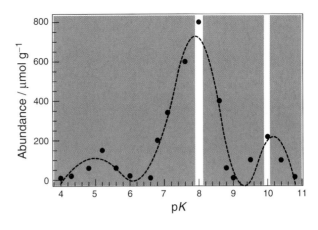

Fig. 45 Distribution of acid–base behavior of an activated carbon in a pK versus abundance plot. The data were replotted from the Ref. [477] to illustrate the rich information about reactivity that is available from the unusual surface analysis by continuous titration. The method of, for example, two point analyses with given reagents is indicated in the plot by the two "slits" in the hidden information represented by the dark areas of the plot. In the present example, the slits hit the two most relevant features but preclude the existence of the continuous distribution of reactivity over the pK scale.

The process described in Eq. (26) assumes that oxygen and water and the ill-defined Lewis-basic sites react to metastable hydroxo-complexes of the Lewis sites (their unification with the formal hole states would produce stoichiometric surface oxides as the final stable product of such a reaction) and hydroxyl species detectable by an initial pH jump when carbon is put into water in the presence of oxygen.

$$2C^* + O_2 \longrightarrow [C_2O_2] \quad (27)$$
$$[C_2O_2] + 2H_2O \longrightarrow 2C-OH + 2OH^- + 2h^+ \quad (28)$$

Equation (27) is part of the high-temperature oxidation cycle, where the peroxy-intermediate would occur as a short-lived surface transient. At low temperatures, however, the species can exist for long times and undergo chemical reactions. Equation (27) describes the recommended synthesis procedure for basic OH groups. The consecutive hydrolysis [Eq. (28)] produces the hydroxy-species and, as in the above formalism, also metastable species that will complete a redox reaction and end up in the pool of oxygen functional groups.

For the continuous reactivity of carbon resulting in basic properties these effects are, however, not relevant except as slow background processes using defect sites occurring during hydrolytic activation, as shown for example in Fig. 42. It must be concluded that the surface heterogeneity and local substituent effects at oxygen functional groups are the only factors that determine the acidity reactivity. For this reason, the beneficial effect of nitrogen substitution in redox catalysis [165, 189, 413, 443] with carbon is barely measurable by an increased basicity of the modified materials.

It must be concluded that all acid–base reactivity of carbon under steady-state conditions is associated with heteroatomic (oxygen) functional groups. The distribution of reactivity reflects, therefore, the complexity of multiple active structures [439] folded with the dynamics of neighboring effects, leading to unique fingerprints of carbons in acid–base reactivity despite a finite number of individual isolated functional groups. High sensitivity and chemical selectivity are characteristic of the titration methods used for such acid–base surfaces. The most prominent technique in a one- or two-point titration narrows down the information again to a small view on a large distribution. This is illustrated in Fig. 45, using published data [477]. The titration is usually carried out with two fixed concentrations of, for example, 0.05 M and 0.5 M bases as back-titration after an adsorption experiment. The value of the concentrated base is usually too high to perform neutralization reactions. Under these harsh conditions, hydrolysis and the creation of novel functional groups must be expected (see Fig. 46). More suitable values would be those indicated as windows in Fig. 45. These pK values can be reached by using not hydroxides but rather carbonates as bases. The results usually differ from each other, which reflects the distribution of acid sites over several orders of magnitude in pK values. Figure 45 provides an impression of the rich structure of this distribution, and the limited view resulting from the two values that are usually determined. Equations (12) and (18) are associated with these two values. For a variety of different carbons, the characteristic concentrations for the predominant analysis of carboxylic acids and phenols may differ due to variable distributions of local adsorption geometry. This affects, via neighboring group effects, the dissociation constants of the surface

References see page 419

groups, and hence modifies the pK distribution. In order to avoid possibly substantial errors, it is recommended that the complete pK distribution is determined using suitable techniques. If this is either impossible or impractical (e.g., due to wetting problems or to sensitivity issues of the samples), it may be very useful to employ a series of windows and to determine the adsorption isotherm of the neutralization reagent on the carbon; the resultant data may then be analyzed in terms of a multiple Langmuir chemisorption model.

The Langmuir model describes, for a uniform surface and a non-self-interacting adsorbate, the relationship between adsorbed amounts of a reagent and the exposure concentration. The parameters of the model are the maximum amount adsorbed as a full monolayer, and the equilibrium constant for the adsorption–desorption process. This quantity indirectly reflects the strength of the adsorbate–substrate interaction. For the present situation, the analysis is modified in the following ways:

- It is assumed that several independent site types exists which represent acids with different pK values.
- The neutralization process is seen as an equilibrium reaction which occurs in an additive way on the different site types.
- No interaction between different site types is assumed (no effects as described with Figs. 42 and 43).
- No complications with correlated equilibria for the generation of primary sites (e.g., hydrolysis of lactones or anhydrides; see Fig. 44) are taken into account.

Within these limitations, the method supplies information on the absolute abundance of the fundamental solid acid sites, and also provides a qualitative indication about the pK average of each site. The procedure works, in principle, for both acid and basic sites. In Fig. 46, an example is illustrated [79] for acid sites on a carbon black sample (FW-1 from Degussa). This sample was selected because the high purity allows an unambiguous assignment of the neutralization to carbon-bonded functional groups, while the high surface area provides sufficient chemical heterogeneity for a realistic model. In Fig. 46, the upper panel shows the experimental adsorption isotherms after 24 h equilibration in anaerobic conditions for the as-received sample and an extracted sample. Extraction was with xylene and ether to remove polycyclic aromatic molecules, which are chemisorbed in an abundance of less than one monolayer [192] on the carbon and passivate the genuine carbon surface. The strong effect of the extraction is clearly visible. The shape of the isotherm indicates, at least for the extracted material, some complications, as a stepped curvature should not occur from a single Langmuir isotherm. As the reaction should occur in an additive way, it is also not expected that the combination of several isotherms should give rise to such a feature, which is absent in the as-received sample. The discontinuity is ascribed to the generation of additional functional groups by chemical reactions between the NaOH with nominally non-reactive functional groups during the adsorption–equilibrium process.

The lower panel of Fig. 46 shows the analysis of the normalized isotherms. It occurs that two processes are superimposed on each isotherm. The break in the isotherm of the extracted carbon at 0.1 M base strength indicates that, in this regime, the extra reactivity consuming base (see e.g., Fig. 44) in addition to adsorption has occurred. These data may be quantified to provide the following information. The formal two adsorption sites differ in their acid strengths, as can be concluded from the equilibrium constants [$K = 14$ (17) for the majority site and $K = 64$ (107) for the minority

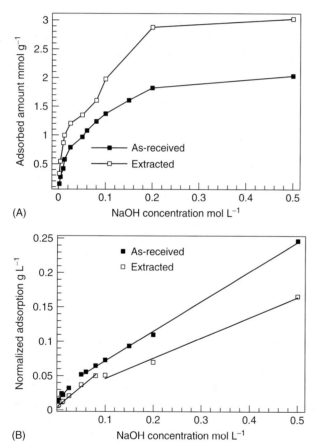

Fig. 46 Multi-point titration of the surface acidity of a carbon black. (A) Titration results in a Langmuir plot. The "extracted" sample was pretreated with xylene to remove PAH molecules covering part of the perimeter of the graphene BSUs. (B) The data in a linearized form to illustrate the partitioning of the action of the base reagent; at low concentrations, neutralization reactions dominate; at higher concentrations modification reactions creating new functional groups become relevant.

sites]. The limiting values of coverage are 2.32 mmol g^{-1} (3.41 mmol g^{-1}) and 1.24 mmol g^{-1} (1.74 mmol g^{-1}) for the two species. The data in parentheses are the values for the extracted surface, indicating the sensitivity of the method and the drastic effect of the surface purification process. The purification also has a detectable effect on the acid strength of the sites, which may be explained by the operation of neighbor group effects (adsorbate–adsorbate interaction).

The analysis of a multipoint titration clearly shows the danger of modifying the delicate surface of a carbon with too-strong probe reagents. With respect to catalytic applications, it must be noted that most metal salt solutions used to impregnate or deposit active species on carbon are at least as strong reagents as the present base. Great care must be taken therefore when carbons are prepared for deposition reactions, that desirable and eventually predetermined surface properties are not strongly modified [152] during the deposition process. A most critical effect is the *in-situ* modification of the abundance of acid groups, as shown in Fig. 46. This process will shift the zeta potential and the point of zero surface charge, and hence change the attractive or repulsive properties of the surface with respect to ions in solution to be deposited: the formation of agglomerates instead of molecularly dispersed species [11, 382, 391, 478–480] can easily be the consequence of neglecting the effect of metal ion concentration and hydrolysis reactions in the deposition [95, 179, 188]solution. Complicating effects from the admission of air with oxidizing and complexing components (oxygen, carbon dioxide) is not included here. It is no surprise, therefore, that correlations between the initial surface properties of carbon and final supported catalysts are difficult to evaluate, no matter what sophisticated initial characterization is applied. A series of successful preparations [382, 481–483] was carried out by several groups with nanocarbon supports, where a sufficient control over the dispersity of oxygen functional groups can be achieved, that is much more narrow [484–486] than on activated carbons or other highly defective carbon materials [85, 152] with nm-sized BSUs.

The quantitative aspects of surface acidic group determination by the Langmuir model (see Fig. 46) as much as other integral methods such as calorimetry [473] give no indication, however, about the chemical structure of the reactive sites. Only in combination with the chemical probe reactions is it possible to assign the two types of acid site to carboxylic acid and hydroxy groups, respectively. It is noted that such an approach can also be used to determine ion-exchange capacities for metal ion loading required for the generation of dispersed metal–carbon [372, 381, 487] catalyst systems.

The neglect of dynamic responses of highly reactive carbon surfaces leads frequently to disappointing results when surface analytical predictions are used to design synthesis procedures for dispersed catalysts.

For carbons with large amounts of surface groups per unit weight, the method of direct titration can be used. In this method a suitable arrangement for potentiometric titration with slow neutralization, anaerobic conditions and good temperature stability of the experiment is used to measure directly the titration curve of the solid acid. The first derivative shows the equivalence points of all stages of dissociation. The carbon is considered in this picture as an oligoprotic acid with a series of dissociation equilibria [138, 210, 488]. Figure 47 illustrates the information obtained from such a direct titration experiment with a carbon black sample (FW 1) that was oxidized at low temperature in liquid phase to obtain acidic functional groups. Large differences in the abundance (proportional to the peak positions on the abscissa) and in the dissociation constant (peak profile representing the shape of the discontinuity at the equivalence point) can be seen for the two oxidizing agents applied under identical conditions to the same carbon material. It is clear that ozone treatment resulted in the formation of a large amount of more strongly acidic surface functional groups, as seen from the sharp peak in the derivative plot. Three to four other types of functional group may be identified with widely varying abundance. Care must be taken when analyzing the derivative profiles, because for weaker acidic systems the effect of incomplete dissociation

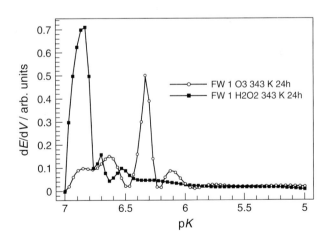

Fig. 47 Direct titration of acidic groups on a carbon black with many functional groups and excellent hydrophilicity. The curves represent the high-resolution first derivatives of the titration curves. Sodium hydroxide was used as reagent. Air was excluded and the temperature was kept strictly to 300 K. The ordinate scales are identical for both experiments.

References see page 419

and the formation of a buffering system which all broaden the derivative peaks must be taken into account. The data in Fig. 47 present a high-resolution view of a small section of the whole pK distribution, as can be seen by comparing the abscissa scales of Figs. 45 and 47. An impression about the strength of the surface groups is obtained by relating the data of Fig. 47 to the pK value of acetic acid being around 5, also at the far end of the scale in Fig. 47. The relatively weak carboxylic acid functions may be explained by the low temperature of the treatment and the high disorder of the carbon structure precluding the formation of functional groups with resonance stabilization using the "aromatic" parts of the surface.

The direct titration of the parent carbon black carrying significant amounts of functional groups was not possible due to wetting problems and the associated strong kinetic hindrance of the acid–base reaction. It is obvious that the two alternative chemisorption techniques of direct titration and of multipoint adsorption measurements (also with immersion calorimetry [180, 473]) both have their relevance in carbon chemistry.

A detailed numerical analysis procedure of the pK distribution functions based on direct titrations has been developed [464, 465, 467, 474, 477] that may be useful in all those cases where highly functionalized carbons are to be compared, and where dynamic responses of such reactive systems must be evaluated. The direct titration data need to be collected over a wide range of pK values, ideally between the lower and upper limit of the buffering curve of the solvent (water). These data need to be transformed into a continuous curve called the "proton binding isotherm PBI". The idea is that essentially the derivative of this curve contains the "pK spectrum" that is assumed to be composed of a series of Gaussian distributions of equal width (one pK unit) characterizing one type of binding site at the carbon surface. These deconvolutions become meaningful when reactions are also considered which involve acid–base couples where one element is outside the accessible pK curve determined by the solvent properties. Adequate corrections are made in the exact numerical analysis, and such treatment can thus be recommended within the limits mentioned above when quantitative comparisons of chemical modifications of carbon surfaces are required.

2.3.15.21 Spectroscopy of Functional Groups: XPS

One physical method that became popular in functional group analysis is photoelectron spectroscopy (XPS) [4, 85, 108, 152, 155, 182, 257, 281, 377, 414, 421, 439, 443, 446, 489] with X-ray excitation. The XPS or electron spectroscopy for chemical analysis (ESCA) method is used in this context as a fingerprinting tool using empirically derived tables for the chemical shift to analyze the data.

A methodologically consistent set of XPS data on carbon fibers is presented in a series of publications [492–494] discussing also the challenges [495] of analyzing chemically reactive modifications of carbon.

The issue of assigning carbon 1s data to chemical structures has been developed extensively in carbon polymer [490] science. The principle of the assignment resides on the following assumptions. All sp^2 and sp^3 C–C and C–H bonds give rise to one C 1s signal at 285 eV, with a tendency towards lower binding energies for pure graphite which was located for defect-poor samples at 284.6 eV [39, 67, 445, 496]. All carbon heterobonds shift the C 1s signal to a higher binding energy; this shift is the larger the more electronegative the heteroelement is relative to carbon. Further increments are brought about by carbon–heteroatom double bonds and by 1,1 substitution of the C–C or C–H bonds. A critical compilation of literature data on carbon–oxygen-induced shifts for several carbon materials is provided in Table 6. The data in the table confirm the general trend outlined above. The consistency about species assignment is high, as the sequences of chemical shifts are the same for all data sets. A substantial controversy exists about the shift range for oxygen spectra; some reports find only one binding energy highly reminiscent of water contamination of the polar sample surfaces. The modern studies with carbon fiber materials identify a variation in O 1s binding energy as would be expected [490] from theoretical calculations and semi-empirical model calculations using point charge models. The linewidth has also decreased with progress in instrumentation and vacuum conditions. Better sample preparations, less contamination and reduced beam damage through better X-ray sources (no heat load and electron emission) are reflected in these data.

Other tabulations in the literature [210] are physically inconsistent and contain erroneous data. These discrepancies arise from different sources of inconsistency:

- Different calibration of the binding energy scale for different experiments (see below).
- Influence of the defect structure on the absolute position of the main line which is often used as internal standard for calibration of the binding energy scale [445].
- Disregard of the asymmetric line profile of the graphitic carbon 1s signal. This asymmetry precludes a physically meaningful spectral deconvolution into a set of Gaussian peaks as is mostly done in the literature. The resulting inconsistencies in the positions of contributions and the variable line widths used in these deconvolution procedures arise from the varying contribution of the line asymmetry to the total spectral weight of the carbon 1s profile. The asymmetry is mainly caused by the coupling of the core hole

Tab. 6 Selected carbon 1s chemical shift data for carbon oxygen groups

Species	BE C1s/eV	BE O 1s/eV	FWHM C 1s/eV	FWHM O 1s/eV	Reference
Graphite	284.8	–	–	–	[490]
Alkanes	285.0	–	–	–	
C–OH	286.1 ± 0.1	533.2 ± 0.5	1.6	2.5	
C=O	287.9 ± 0.2	533.2 ± 0.5	2.9	2.0	
C=O(OR)	289.1 ± 0.2	533.2 ± 0.5	4.1	2.0	
Alkane	285.0	–	1.33	–	[491]
C–OH	286.5	533.4	1.33	1.75	
C=O	287.8	–	1.33	–	
C=O(OR)	289.2	–	1.33	–	
Graphite	284.6	–	–	–	[479]
C–OR	–	533.2	–	–	
C=O	–	532.0	–	–	
Metal oxides	–	530.0	–	–	
Alkanes	285.1 ± 0.25	–	–	–	[492]
C–OH	286.1 ± 0.25	532.5 ± 0.2	–	–	
C=O	287.6 ± 0.25	531.0 ± 0.1	–	–	
C=O(OR)	289.1 ± 0.25	533.8 ± 0.5	–	–	
Carbonate	290.6	–	–	–	
Graphite	284.6	–	1.25	–	Fig. 49
C–OH	–	532.5	–	1.75	
C=O	–	530.8	–	1.75	
C=O(OR)	–	533.8	–	1.75	

state created during photoemission to the semimetallic valence band states of graphite and by the convolution of the primary photoemission with the phonon and plasmon loss spectra [497, 498].

- Disregard of charging effects. Highly functionalized carbons are often not sufficiently metallic conducting to allow the photocurrent to flow through the sample without creating integral potential drops and affects the position of the binding energy scale. In difficult cases the charging up is locally variable (e.g., with a thick uneven specimen) and prevents correction by recalibration with an internal standard (the C–H + C–C line or the water line).

These effects shift all or part of the photoemission intensity for C–C and C–H species into the range of the carbon–oxygen chemical shifts, and can thus affect the interpretation in an undetected way. Oxygen functional groups can be analyzed by either the oxygen O 1s emission or the carbon C 1s emission. The carbon 1s range is used in almost all applications due to the larger shift sensitivity. As discussed with the data of Table 6, the presence of a film of molecular water, which is not removed in UHV, creates a large oxygen 1s signal at around 533 eV covering many structures in the functional group oxygen spectrum.

The same problems arise when carbon–nitrogen groups are analyzed. Nitrogen–oxygen groups (nitrate, nitrite, nitrosyl) are easily discernible (403–406 eV binding energy) from nitrogen–hydrogen bonds (ca. 401–400.5 eV for amino groups) and from carbon–nitrogen bonds (400.0–397 eV). The individual chemical configuration within the groups is, however, difficult to resolve [413]. In one report [443] on the reactivity of nitrogen species with oxidized activated carbons, a consistent series of nitrogen chemical shifts falling within the ranges specified above is established. This list of compounds is chemically self-consistent as the modifications used for their generation are fully in line with conventional nitrogen organic chemistry.

The C 1s line is less affected by the less-electronegative nitrogen heteroatom, and exhibits a much smaller shift range than for oxygen groups. All these problems are well documented in the literature aiming at the XPS analysis of the C_3N_4 material [499, 500]. The authors did not consider charging effects, nor did they account for oxidation of the sputtered surfaces during sample transfer. In addition, they used Gaussian line profiles and arrived at significantly high carbon 1s chemical shifts for two different C–N bonding geometries assigned as diamond-like (aliphatic) and polypyridine-like (heterocyclic aromatic).

In a detailed study of the chemical shifts of carbon 1s lines on the nature of the C–C bonding interaction [501] it was found that, using the well-determined C 1s value of 284.6 ± 0.3 eV [502], hydrogenated carbons exhibiting a binding energy of 285.3 ± 0.05 eV and carbidic carbon

References see page 419

in SiC were found to occur at 283.4 ± 0.05 eV. Diamond and graphite were found to be fully indistinguishable from their shift data, which is in full agreement with a different study in which compact [39] rather than film [502] samples were used. The significant uncertainty which is quoted for the value of graphite C 1s photoemission was clarified in a study on the dependence of the carbon 1s line position [445] on the crystalline quality (nature and abundance of surface defects). For hydrogenated amorphous carbon films, a complex influence of the substrate on the binding energy – and probably also on the detailed atomic structure – was detected in a study of a-C–H films on various GaAs surfaces. A carbon-rich species with a binding energy of 284.8 eV on GaAs and of 284.6 eV on Si was found in the near interface region of sputtered thin carbon overlayers [164]. These data indicate that a clear speciation of the chemical structure of a carbon species in a catalytic system can be quite ambiguous and that – contrary to chemical intuition – the connectivity of a carbon cannot be inferred from its carbon XPS spectrum. The reason for this failure of the ESCA effect lies in the complex convolution of ground state electronic properties of the carbons (which are all different [39, 87], as expected by intuition) with final state effects [497, 498] (relaxation of the core hole) which produce a diffuse but intense structure in each peak, accounting for their apparent similarity. This became apparent in studies aiming at a discrimination of the KVV Auger spectrum of carbon into s-derived and p-derived valence band states being suitable for the discrimination of sp^3 and sp^2 carbon connectivities by the Auger line shape [494, 503].

Care must be taken in interpreting the main component of the carbon 1s line as a fingerprint for the carbon connectivity. In the high binding energy side of the emission line, additional features arise from plasmon satellites occurring at 5.6 to 7.2 eV in sp^2 carbons and at 11.3 to 12.5 eV in sp^3 carbon above the main line [502]. These broad and weak structures interfere with chemical shift identification [257, 445, 489, 492, 494] of highly oxidized C–O functional groups.

The practical situation is illustrated with graphite in Figs. 48 and 49. A high-resolution laboratory-XP spectrum of natural purified graphite in its pristine state and after ball milling in hexane solvent introducing defects [325, 334] is compared in Fig. 48. The cleaned state (573 K in UHV) is characterized by the narrow asymmetric main line and the plasmon loss above 290 eV. Surface functional groups cannot be seen in this spectrum. The presence of numerous defects destroys the collective electron excitation loss and broadens the main line, without destroying the asymmetry indicating the local final state mechanism as the origin of the line profile. The corresponding O 1s data are presented in Fig. 49.

It should be noted that vacuum annealing does not remove the abundance of about 6 atom% oxygen from the surface, as already discussed for the formation mechanism of basic functional groups. The spectrum cannot be fully accounted for by chemical shifted components; the high-binding energy tail arises from differential charging, as discussed above. In contrast to many literature statements, but in agreement with the study on carbon fibers [492], three oxygen components are clearly seen, the assignment of which is given in Table 6. Defect formation removes many of the original peripheral functional groups and gives rise to a predominance of phenolic groups that also contribute to the line broadening of the carbon 1s peak displayed in Fig. 48.

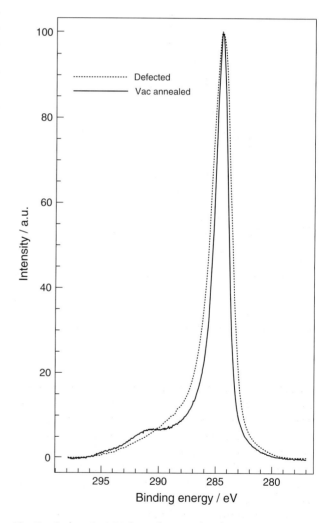

Fig. 48 Carbon 1s XPS line of a sample of well-crystalline and highly purified natural graphite. The pristine state was annealed in the vacuum of the spectrometer (10^{-8} Pa). The defected state was achieved by ball milling in hexane for 7 days, followed by vacuum drying at 373 K. The resolution of the spectrometer (SPECS EA 200 MCD) was set to the natural linewidth of the Mg k line.

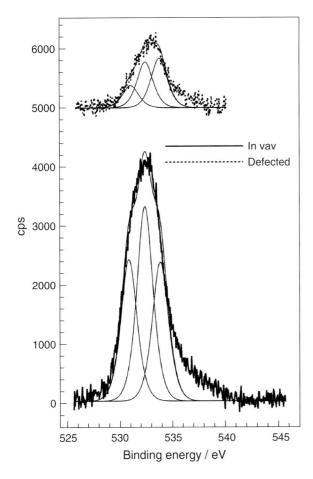

Fig. 49 Oxygen 1s data for the samples described in Fig. 48. The peak positions and profiles of the deconvolution were held fixed to predetermined values to explain the non-optimal fit to the line profile; this would be erroneous if freely optimized due to the presence of differential charging components. The sample surface contains patches that are non-electrically connected to the main material, as frequently observed [136] in STM artifacts. It should be noted that the ca. 5 atom% oxygen in the evacuated state do not give rise to discernible peaks in the C 1s spectrum shown in Fig. 48.

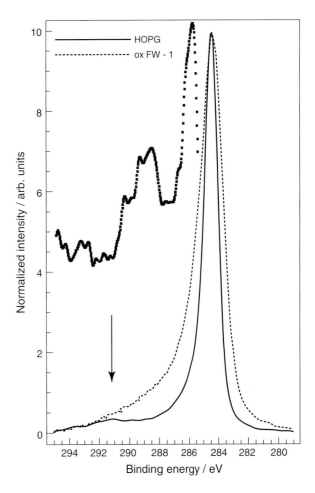

Fig. 50 XPS data in the carbon 1s region for a UHV cleaved graphite reference (HOPG) and a carbon black sample used in the titration experiments of Fig. 47. The arrow marks the position of the graphite surface plasmon. The ordinate-expanded spectrum in the high-binding energy side reveals several weak peaks for oxygen functional groups after removal of the asymmetric line profile from the main peak.

In Fig. 50, the carbon 1s data of HOPG graphite as reference are compared to those of a chemically etched graphitized carbon black sample. The defective graphitized carbon surface exhibits a significantly wider mainline (cf. Fig. 48) at almost the same position as the well-ordered graphite (see Table 6). It can be seen that neglect of the line asymmetry would invariably cause the erroneous detection of additional lines in the range 1 to 2 eV above the peak maximum. The difference spectrum is mainly caused by the broadened main line. Only after removing the asymmetric main contribution a small structure (expanded ordinate trace in Fig. 50) can be isolated revealing peaks at 286.0, 287.3, 288.7, and 290.0 eV that can be attributed to surface oxygen functional groups. Only 4.3% of the total intensity is due to the surface groups, in good agreement with 5.5 atom% oxygen content. The comparison of these numbers indicates [492, 494] that most functional groups can only carry one oxygen atom per carbon atom, and that carboxylic groups (with the largest shift) must be a small minority. These data agree fairly well with those from oxidized carbon fibers [495], and are in agreement with the values from Table 6. A maximum of three different oxygen functionalities can be resolved by XPS (these are the basic chemical structures mentioned in Fig. 37). Under UHV conditions, the resolving power of XPS for functional groups is limited to these groups, as more complex structures are known from theory to give rise to similar [490] peaks that cannot be recognized by

References see page 419

peak-fitting procedures. The widely varying line profiles, as given in Table 6, indicate very clearly that there is potential in better-resolved spectra [392] using synchrotron radiation studies under conditions limiting the beam damage. However, these studies have not yet been conducted.

A significant increase in the chemical specificity of XPS can be achieved when the surface sensitivity of the surface analytical detection is combined with the chemical specificity of the probe reactions discussed above. A highly specific analysis of the chemically reactive surface functional groups can be obtained using heteroatoms such as barium [504] from barium hydroxide, chlorine from HCl or $SOCl_2$ [505, 506], or nitrogen and sulfur tags from complex functional [507, 508] reagents. These data reveal that a large number of oxygen functions detected by integral elemental analysis are chemically inert and exhibit no acid–base activity. Such behavior is consistent with keto- functions or with oxygen heterocyclic functions (C−O−C in Table 6). From all chemical shift data in the literature, and for the spectra in Figs. 49 and 50, it appears that the heterocyclic function is dominating the functional group. This accounts for much of the analytical oxygen content without contributing to the low-temperature chemistry. From the sequence of events in carbon oxidation, this group is the primary reaction product and has thus also – from the mechanistic point of view – a high probability of survival in any oxidation treatment as the most abundant reaction intermediate.

2.3.15.22 Vibrational Spectroscopy of Functional Groups

At this point, a characterization technique with a higher chemical resolution is desirable because such functionalization plus surface analytical combination experiments are extremely difficult to perform in a clean and reproducible manner. Vibrational spectroscopy such as FT-IR has been developed into such a tool, following several methodical improvements with regards to sample preparation and detector sensitivity. In-situ oxidation experiments are still very difficult to conduct as heated black carbon is a perfect IR emission source and interferes with any conventional detection in the spectral range of carbon–oxygen fingerprint vibrations.

A compilation of fingerprint vibrations for characteristic carbon–oxygen functions is provided in Table 7. Data from a variety of organic molecules from spectra libraries are analyzed and grouped together. Only strong vibrations are tabulated, and extreme positions are removed. The compilation shows that several regions of interest exist:

- The OH^- valence region, which is due to combinations with molecular water is often difficult to resolve.
- The C=O region (1800–1650 wavenumbers).

Tab. 7 Characteristic group frequencies in IR absorption

Functional group	Absorption frequency/cm^{-1}	Assignment
Phenol groups		
O−H	3650–3590	Valence
O−H	3500 broad	Valence polymer
C−OH	1220–1180	Valence
C−OH	1350–1100	Valence aliphatic
O−H	1390–1330	Deformation
Keto groups		
C=O	1745–1715	Valence aliphatic
C=O	1700–1680	Valence aromatic
C=O	1730–1705	Valence diketo
Quinones		
C−O, para	1690–1655	Valence, two lines
C−O, ortho	1660	Valence, one line
Aldehyde groups		
C−H	2920–2880	Valence, combined
C=O	1720–1715	Valence, combined
arboxylic acids		
C=O (aliphatic)	1735–1700	Valence
C=O (aromatic)	1700–1680	Valence
C=O (anhydride)	1830, 1750	Valence, two lines
O−H	3000–2700	Valence
O−H	940–900	Deformation
Esters, lactones, ethers		
C=O	1730	Valence, benzoic
C=O	1740	Valence, lactones
C−O−C	1280–1100	Valence, ester, lactones
C−O−C	1320–1250	Valence, ethers

- The region at 1650 wavenumbers, of C=C vibrations which are isolated from the graphene network (the graphene vibrations are only Raman active; see Table 2).
- The region of C−O single bond vibrations at 1300–1100 wavenumbers, with a significant selectivity for different groups when combined with the simultaneously occurring C=O vibrations.

The practical observation of all these vibrations is difficult, as not all of the respective groups are present on a carbon with the same relative abundance.

As the quality of the spectra is generally poor, weak characteristic peaks of a minority group may not be recognized. Strong peaks at 2355 cm^{-1} and at about 2045 cm^{-1} arise from chemisorbed CO_2 and carbon monoxide during in-situ studies [447, 509, 510]. The diagnostic value of the data in Table 7 is enhanced by the search for combined absorptions for the more complex functions. Such combinations allow the resolution of the crowded spectral range between 1800 and 1700 cm^{-1}. The discussion of characteristic IR data on fullerenes and their oxidation products can be found [92, 95, 109, 112, 511] in the literature.

Two oxidized carbon samples are discussed as being representative for the many possible applications of Table 7. The main absorption frequencies for the two samples are collected in Table 8. One sample is a disc of glassy carbon formed under electrochemical oxidation; the other is a powder sample of a high-surface-area carbon black (FW-1, 235 m^2 g^{-1}). The *in-situ* electrochemical oxidation study of the glassy carbon [512] allows the tracing of several coexisting species by their respective fingerprints. It can be seen that several groups which remain undetected by chemical adsorption and XPS analysis can be found with the IR technique. No carbon–oxygen single bonds were found in this study. The oxidation potential may have been too high for the accumulation of a significant abundance of these still oxidizable groups. A technical reason may well be the poor spectral quality below 1600 cm^{-1}, which precluded the detection of any specific absorption besides the extremely strong background. The broad water hydroxy peak from the *in-situ* reaction environment also does not allow the detection of hydrogen groups. For these reasons, not all species are identified. This is required, however, to derive a plausible oxidation mechanism as suggested in this report.

The oxidation reaction by dry ozone at 335 K leads to the formation of a large number of strongly acidic surface groups [513]. A whole spectrum of chemical changes occurs, with the treatment leading finally to a breakdown of the carbon into soluble graphene fragments. The ozone reaction is also used to study the oxidation mechanism and its control by micromorphological effects of "hole burning" [514, 515], and also to elucidate structures in nanocarbon materials and composites [516, 517]. The high selectivity of ozonization on carbon structure should also be very useful in the diagnostics and application of regeneration procedures in catalysis. Finally, ozone oxidation is also of influence in atmospheric chemistry,

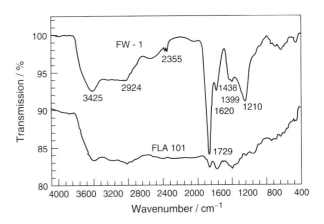

Fig. 51 FT-IR spectra of ozone-treated carbon blacks. The FLA sample is much more graphitic in the stacking order of the BSUs, and exhibits only 10% of the geometric surface area of the standard FW-1 sample. The treatments conditions were 343 K, aqueous suspension, and 10% ozone in oxygen bubbling through the suspension.

where it modifies carbon black particles to become efficient catalysts for the oxidation of NO and SO$_2$ under ambient conditions [518, 519].

Figure 51 illustrates FT-IR spectra from two carbon blacks after ozone treatment. The data from Table 8 and the spectra in Fig. 51 indicate that a wide variety of functional groups with single and double oxygen–carbon bonds are formed after mild ozone treatment at 335 K in aqueous medium.

The agreement between the assignments of the absorptions in both samples is quite remarkable, and agrees also with the reference data in Table 7. The influence of neighborhood effects on vibrational data

References see page 419

Tab. 8 Observed IR absorption bands for oxidized carbons. The data for the oxidized glassy carbon were taken [512] from the literature

Oxidized glassy carbon		Oxidized carbon black	
Absorption frequency/cm^{-1}	Assignment	Absorption frequency/cm^{-1}	Assignment
2045	Adsorbed CO	3425	C–OH
1780	Lactones	2924	C–H aliphatic
1722	Carboxylic acids	2355	CO gas phase
1696	Aromatic ketones	1760 sharp	Lactones
1678	Quinones	1729	Carboxylic acids
1611	Isolated C–C bonds	1620	Isolated C=C bonds
		1438	O–H deformation
		1399	C–O lactones
		1230 sharp	C–O aromatic ether
		1210	C–O phenol

(see for example complex structures of Figs. 42–44) and the uncertainty of locating the position of absorptions in complex bands explain the discrepancies in the numerical values of the peak positions. The regions of characteristic absorptions mentioned above can be identified easily in the top spectrum of Fig. 51. A distribution of absorption band positions for each chemical structure causes the broad envelopes, and indicates the structural heterogeneity of the oxidized carbon surface. The sharp double peak of gaseous CO_2 arises from the large micropore volume of the sample holding detectable amounts of CO_2 in relation to the natural abundance of this gas in the ambient air.

The nature of the functional groups seems not to depend on the structure of the carbon substrate, as even on oxidized fullerenes [117] the same bands can be found by IR spectroscopy. The relative abundance is, however, significantly different, as can be seen from the comparison of the two carbon black samples used for the data of Fig. 51. The FLA 101 sample is more graphitic, with a significantly reduced surface area. The abundance of carbon single bonds is greatly reduced, with the predominant species being carboxylates after the same treatment, giving rise to the broad variety of functional groups on FW-1. This is in full agreement with the published mechanistic picture [514, 515], which correlates velocity and depth of oxidation with microstructure and porosity of the carbon substrate.

When comparing FT-IR data of carbon materials, great care must be taken for the strong influence of the zero-band semiconducting properties of this material: the extent of IR absorption – and hence the background – depend critically on the ordering state of the carbon. The more graphitic the sample, the stronger will be the background and the less sensitive will be IR to the presence of functional groups. As IR spectroscopy is a non-surface-sensitive technique, it is the strong change in electronic structure between graphitic cores (absorbing all light) and "organic" perimeter structures (absorbing light selectively) that allows a sensitive detection of surface functional groups. Even if they are frequent, they may not be apparent due to the strongly absorbing bulk (as for example in the FLA 101 sample of Fig. 51).

Vibrational spectroscopy is a versatile and chemically well-resolving technique for the characterization of carbon functional groups. The immense absorption problems of earlier experiments seems to have been overcome in present times with modern FT-IR, diffuse reflectance infrared Fourier transform spectroscopy (DRIFTS), or photoacoustic [520–522] detection instruments. From an inspection of numerous published results, it is recommended for catalytic carbon materials that are no carbon fibers to rely on DRIFTS, as this method provides – with relatively facile sample preparation and dilution options – good spectra with a high content of surface information. The disadvantage of missing quantification can be overcome by using other quantification techniques for functional groups such as advanced thermal desorption.

2.3.15.23 Advanced Thermal Desorption

Desorption spectroscopy of oxygen functional groups is a frequently used physical method. The conceptually simple experiment, which uses a carrier gas and a non-dispersive IR detector for CO and CO_2, is not in line with the theoretical complexity of the experiment discussed partly in previous sections (see Figs. 30 and 31). One group of complications is of a technical nature, and refers to carrier gas purity, mass transport limitations, temperature gradients, non-linear heating ramps and temperature measurement problems. More severe is the problem that the method is destructive in two ways. First, it removes the functional groups and has to break all adsorbate–substrate bonds in a distribution of local geometries. Second, it is a high-temperature process, activating functional groups to change their structure during the experiment and to interconvert between them. This is particularly complicated in situations with carbon–oxygen single and double bonds located adjacently (e.g., see Figs. 40–44).

Despite these principal ambiguities, the thermal desorption method is a standard characterization technique in carbon surface chemistry. It has the advantage of being conceptually and, after correction of all technical challenges, a truly quantitative method that can be checked for internal consistency when executed as a TG-DTA-MS experiment. Various examples and data relating to desorption profiles for a selection of carbon treatments can be found in the literature [76, 79, 127, 331, 340, 386, 523]. A selection of references in which this methodology is used to study carbon deposition is also presented [356, 524–528], documenting that routine application and structural speciation are both well established. Less well documented is the use of the technique to study the reactivity of carbon as a function of its surface functional groups.

The method has its potential in qualitative analysis of carbons, as it is most valuable when samples are not amenable to quantitative analysis. The general features of a thermal desorption experiment is discussed with the data from Fig. 52. The sample was a mix of carbon nanotubes prepared from the plasma arc method [95] used for fullerene synthesis. The carbon was treated with ozone at 333 K in water and the dried product subjected to a temperature-programmed decomposition in flowing nitrogen using a special mass spectrometer (IMR-MS) as detector. This instrument [529] allows suppression of the nitrogen signal, and fully removes any fragmentation of CO_2 into CO. This problem is significant in conventional mass spectrometers, as it allows only estimates to be made of changing ratios of CO to CO_2.

The oxygen functional groups with acidic reactions shown in Fig. 37 (species 1, 2, 3) result in CO_2 as decomposition products, whereas all other surface groups produce mainly CO during their pyrolysis under non-oxidizing conditions. The more strongly oxidized α-carbon atoms in acidic groups desorb from the graphene surface at lower temperatures than the CO from the basic groups. This behavior can be recognized in the IMR-MS traces of Fig. 52. The CO_2 trace shows structures in three broad features (480 K, 620 K, 720 K), with the first peak exhibiting some internal structure which is ascribed to the decomposition of free carboxylic acid groups and structures with other functions next to the acidic group (see for example Fig. 44). The poor resolution of the main features indicates significant kinetic hindrance at a heating rate of 0.5 K s^{-1}, pointing to a distribution of binding energies rather than to a specific desorption event. The occurrence of post-desorption reactions of the CO_2 acting as oxidant for other functional groups must be taken into account at temperatures above 650 K [76]. This finds its expression in the crossing-over of the desorption traces (sensitivity-corrected) at around 650 K, coinciding with the onset of gasification, if the reaction is carried out with oxygen in the gas phase (data not shown). This is a confirmation of the general gasification mechanism discussed with Eqs. (1) to (7). The increased stability of the basic and neutral functional groups is seen in the high-temperature rise of the CO trace, which peaks only above 1000 K. At these temperatures, the structural rearrangement of the graphene layers (see Figs. 17 and 36) in amorphous carbon becomes an important process, as it prevents observation of the complete desorption of all oxygen functional groups without changing the bulk structure of the substrate. Structural changes in the substrate give rise to the noisy appearance of the CO trace, which indicates eruptions of gas due to the start of the hole-burning process [204] initiated by local concentration maxima of surface oxygen species which terminate quickly due to the lack of oxidant.

The highly reactive character of fullerene carbon present in the sample is indicated by the non-specific CO evolution following the dashed background line in Fig. 52. Entrapped oxygen [358] and polymerized fullerene molecules react in a secondary process besides the desorption reactions from the surface of more massive BSUs. Such situations are typical for low-temperature-generated carbon deposits on catalysts, but will be overlooked frequently due to the choice of too-rapid heating rates and the insensitivity of normal mass spectrometers to CO gas.

The relative contributions of different desorption processes changes with the average surface structure. Comparing the data in Fig. 52 with those of an analogous oxidation experiment with a different carbon substrate (carbon black FW-1; as shown in Fig. 53) reveals that the carbon black is much more robust against functionalization, as can be seen from the following observations. There is no continuous bulk decomposition fuelled by oxygen from the micropores (no low-temperature CO). The functionalized surface contains significantly more stable functional groups than the nanocarbon material, as can be seen from the steep increase of the CO trace above 800 K. The CO_2 curve, which is indicative of labile and acidic functions, is rather similar in the two experiments, indicating that the identical treatment caused the same induced surface modifications. The cross-over of the CO and CO_2 traces occurs at higher temperatures, which is in exact agreement with the enhanced stability of the carbon black. The presence of five-membered rings in the fullerene black, and the organization into stacks of planar graphene units in the carbon blacks, are consistent structural explanations. The high-temperature region of the desorption traces in Figs. 52 and 53 are rather similar, which points to the existence of similar stable oxygen functional groups, as should be expected if the description in Figs. 37 to 44 is of general validity. It should be pointed out that the use of detector allowing unambiguous

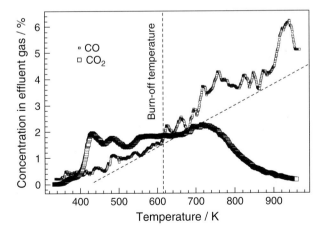

Fig. 52 Thermal desorption (heating in flowing nitrogen, heating rate 3 K s^{-1} of CO_2 and carbon monoxide) from heavily oxidized fullerene black (treated with 10% ozone in oxygen at 333 K in water). An IMR-MS detector was used for unperturbed gas analysis and simultaneous detection of other desorption products. The burn-off temperature in the figure is defined as the temperature where a weight loss of 3% had occurred in a gasification experiment using 5% oxygen in nitrogen.

References see page 419

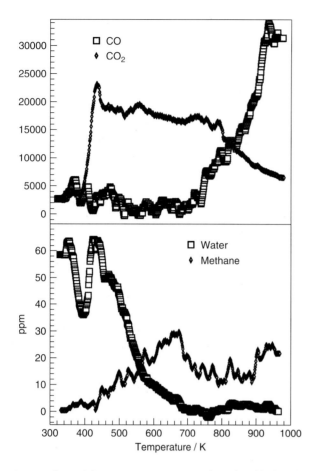

Fig. 53 Thermal desorption spectroscopy of a carbon black under the same conditions as used for the experiment in Fig. 52. Note the different scales of the ordinates.

recording of CO desorption without cross-sensitivity or fragmentation is vital to reach such conclusions.

The lower traces in Fig. 53 reveal that additional desorption processes occur with the decomposition of oxygen functional groups. The polar oxygen groups bind a layer of water molecules desorbing above 415 K, after the removal of water multilayers slightly above ambient temperature. Desorption of CO_2 from acidic groups leaves behind a C–H function. If this function is in an alkenic or aliphatic environment – that is, in a surface without a closed carbon hexagon termination – then desorption of methane can contribute to a stabilization of the surface by removal of the non-aromatic structures. The missing hydrogen atoms are taken from reactions of water with these defective carbon centers. The respectively small methane signal is clearly detectable in Fig. 53. No higher hydrocarbons are detectable in this experiment. The desorption of molecular hydrogen at temperatures above 900 K from the terminating hydrogen atoms with aromatic carbon structure [59, 210], leading to improvements of the crystalline order state (see Fig. 17), has been observed.

Thermal desorption data can be integrated and may be used to quantify the total number of oxygen functional groups on a carbon surface. When these data are compared with the sum of all chemically detected functional groups, a significant excess of desorption value over the chemical value is found. This excess serves as confirmation of the existence of chemically neutral oxygen functions that are bound strongly to the substrate. These groups act as side blockers in surface reactions, and are known in the carbon oxidation literature [189, 326, 376, 525] as "stable complexes" which inhibit [327] the oxidation.

The data in Figs. 52 and 53 reveal that it is not possible to fully detect the oxygen desorption without reaching temperatures at which structural changes occur in the substrates. This is a significant problem for hydrocarbon-rich carbons in catalytic deposits. The application of thermal characterization methods must be regarded with significant reservation, as severe carbonization and dehydrogenation will go along with the "analytical" desorption events. In subsequent adsorption or reaction experiments, the presence of a stable graphitic carbon as second phase is stated. The second phase is, in fact, an artifact [79, 331] of the preceding thermal treatment.

2.3.15.24 Synopsis: Carbon and Catalysis

The intensive and diversified research into carbon with respect to catalysis and surface chemistry reviewed here has brought about some general concepts. The extraction of these concepts attempted here should serve as guidelines to meet the many challenges posed by the action of carbon in catalysis.

Carbon exhibits an immense structural variability, given by its tendency to form homonuclear bonds in essentially two hybridizations. Depending on the availability of heteroatoms carbon has free principles of solving the termination issue: self-condensation until perfect graphene sheets or cubic diamond crystals exhibit the geometric minimum of free surfaces. These reactive perimeters stabilize by changing their hybridization (localized double bonds, in analogy to Si dimers), or a graphene layer terminating the diamond crystal. The few remaining vacancies from defects react with heteroatoms in the preference: H, O, N, halogen, S, B, P. The formation of defect into a once-crystallized stable modification is very difficult due to the high covalence of the carbon–carbon bonding. Thus, defects much more likely remain from the history of the material formation. Carbon formation always begins with molecules being subjected to homogeneous, radical or catalytic polycondensation processes. The primary formation of chains allows relaxation into the stable six-membered rings, or into metastable, non-six-membered rings. This cocondensation deforms planar carbon into all possible 3-D closed shapes from spheres to tubes, and always in accordance to minimize their perimeter length. An immense complexity of metastable intermediates (polymers, soot, carbon black, "deposits") exist along this pathway. These materials are potential catalysts and catalyst supports [430, 470, 530] as they can accommodate heteroatoms or active-phase particles and undergo a stable local restructuring. This synergy of two poorly reactive elements, namely carbon and gold, becoming very active at their inner interface when both constituents are nanostructured is exemplified in Fig. 54.

The generally preferred idea of anchoring active species by chemical bonding, and thus attempting to functionalize carbon surfaces, has brought about an immense factual knowledge about such groups. However, this knowledge is by and large cast in static descriptions of isolated structures usually derived from organic molecules. The fact that these structures are in electronic resonance to the semi-metal "graphene" has drastic consequences for their reactivity, their cross-coupling of reactivity, and their dynamic response to the chemical environment of a sample. Diamond carbon is either alkane-passivated or terminates in graphene layers, giving rise to the same surface chemistry as graphitic-type samples.

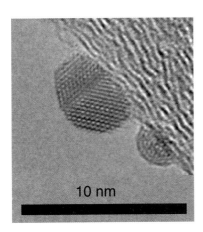

Fig. 54 Two Au nanoparticles on active carbon support. The high-resolution TEM was taken at 300 keV with a FEI Titan CS-corrected instrument. The sample was prepared from HAuCl$_4$ deposition on unmodified activated carbon followed by a mild reductive treatment in hydrogen. A large twinned particle with numerous high-energy terminations sits next to a metastable small particle with a greatly modified internal structure. The large particle interacts both on the bottom and on the side interfaces with graphene sheets, forming a tailored bed for the particle, plus it dissolves the graphene structure at the metal–carbon contacts where the carbon exhibits defects. The small particle "alloys" into the carbon surface, and it can be assumed that some carbon is also dissolved in the Au particle. No indication is found for an interlayer formation caused by surface functional groups that should form a reactive "carpet" on which the metal would be floating over the carbon.

In many cases, this dynamic response makes surface functional groups poor candidates for anchoring carbon with a matrix (polymers) or anchoring active species (metal particles) on them. The functional groups also determine the evolution of supramolecular structuring of carbon being relevant in pore-forming processes, in the formation of carbon–carbon composites, and in the use of nanocarbon in catalysts. Unintentional carbon also exerts these influences and thus adds considerably to changes of transport properties of many catalyst systems, including microporous systems.

The dynamics also give rise to a deep involvement of functional groups in activating small molecules on carbon surfaces. The most important case of this activation is the gasification process with oxygen-containing species requiring reductive activation. Besides a major proportion of present-day energy production being obtained from this reaction, there are numerous cases where this reaction should be inhibited, and this may be achieved by modifying the surface functionalization with stable ad-species or by changing the termination from heteronuclear to homonuclear by rehybridization. The dynamics also allows for the application of semi-metal carbon in heterogeneous catalysis [168, 190, 245, 251, 277, 324, 325, 373, 374, 384, 396, 457, 458, 466, 478, 485, 531–533] in its own right as an analogue to the molecular concept [534] of metal-free catalysis.

The questions raised in the Introduction, about the roles of carbon in catalysis, cannot be clearly answered with our present state of knowledge. Besides the obvious detrimental effects of deposition, and the problems of removing a highly energetic material with dynamic structural response from a delicate nanostructures catalyst system, carbon also has many beneficial roles when deposited during activation and being present in a partly carbonized form, resembling more a hydrogen-poor polymer rather than "carbon, soot or coke". The dynamic behavior of the surface carbon requires the adaptation of analytical tools that are only rarely applied. The static analysis hiding the dynamic response gives rise to the notion that (reactive) carbon is more a "black magic" than science, due to its apparent irritating complexity.

References

1. J. D. Bernal, *Proc. Royal Soc. A* **1924**, *106*, 749–773.
2. F. Atamny, J. Blöcker, B. Henschke, R. Schlögl, T. Schedel-Niedrig, M. Keil, A. M. Bradshaw, *J. Phys. Chem.* **1992**, *96*, 4522.
3. B. S. Itchkawitz, J. P. Long, T. Schedel-Niedrig, M. N. Kabler, A. M. Bradshaw, R. Schlögl, W. Hunter, *Chem. Phys. Lett.* **1995**, *243*, 211–216.
4. H. Werner, T. Schedel-Niedrig, M. Wohlers, D. Herein, B. Herzog, R. Schlögl, M. Keil, A. Bradshaw, J. Kirschner, *J. Chem. Soc. Faraday Trans.* **1994**, *90*, 403–409.

5. R. C. Cinti, B. S. Mathis, A. M. Bonnot, *Surf. Sci.* **1992**, *279*, 265–271.
6. G. Francz, P. Kania, P. Oelhafen, T. Schedel-Niedrig, R. Schlögl, U. Wild, *Surf. Sci.* **1996**, *365*, 825–830.
7. G. Post, V. Y. Dolmatov, V. A. Marchukov, V. G. Sushchev, M. V. Veretennikova, A. E. Sal'ko, *Russ. J. Appl. Chem.* **2002**, *75*, 755–760.
8. D. N. Belton, S. J. Schmieg, *J. Vac. Sci. Technol.* **1990**, *A8*, 2353–2362.
9. R. Berman, *Physical Properties of Diamond*, Academic Press, 1965.
10. A. V. Hamza, G. D. Kubiak, R. H. Stulen, *Surf. Sci.* **1990**, *237*, 35–52.
11. T. Braun, M. Wohlers, T. Belz, R. Schlögl, *Catal. Lett.* **1997**, *43*, 175–180.
12. R. Boschi, E. Clar, W. Schmidt, *J. Chem. Phys.* **1974**, *60*, 4404–4418.
13. E. Clar, J. M. Robertson, S. Schlögl, *J. Am. Chem. Soc.* **1981**, *103*, 1320–1328.
14. G. Monaco, R. G. Viglione, R. Zanasi, P. W. Fowler, *J. Phys. Chem. A.* **2006**, *110*(23), 7447–7452.
15. A. J. Groszek, *Carbon* **1987**, *25*, 717–722.
16. A. J. Groszek, S. Partyka, D. Cot, *Carbon* **1991**, *29*, 821–829.
17. H. P. Boehm, *Carbon* **1994**, *32*, 759–769.
18. H. P. Boehm, *Angew. Chem.* **1966**, *78*, 617–628.
19. D. D. Do, H. D. Do, *Fluid Phase Equilibria* **2005**, *236*, 169–177.
20. D. S. Rawat, L. Heroux, V. Krungleviciute, A. D. Migone, *Langmuir* **2006**, *22*, 234–238.
21. M. S. Dresselhaus, G. Dresselhaus, *Adv. Phys.* **1981**, *30*, 139–326.
22. R. Schlögl, R. Schöllhorn, *Graphite – A Unique Host Lattice*, Kluwer Academic Publishers Netherlands, 1994.
23. S. G. Chen, R. T. Yang, *J. Catal.* **1992**, *138*, 12–23.
24. M. Ishizuka, K. I. Aika, A. Ozaki, *J. Catal.* **1975**, *38*, 189–195.
25. F. Kapteijn, O. Peer, J. A. Moulijn, *Fuel* **1986**, *65*, 1371–1376.
26. M. Guraya, S. Sprenger, W. Rarog-Pilecka, D. Szmigiel, Z. Kowalczyk, M. Muhler, *Appl. Surf. Sci.* **2004**, *238*, 77–81.
27. H. W. Kroto, A. W. Allaf, S. P. Balm, *Chem. Rev.* **1991**, *91*, 1213–1235.
28. V. Pantin, J. Avila, M. A. Valbuena, P. Esquinazi, M. E. Davila, M. C. Asensio, *J. Phys. Chem. Solids* **2006**, *67*, 546–551.
29. W. T. Pong, C. Durkan, *J. Phys. D Appl. Phys.* **2005**, *38*, R329–R355.
30. D. H. Robertson, D. W. Brenner, C. T. White, *J. Phys. Chem.* **1992**, *96*, 6133–6135.
31. H. Schwarz, T. Weiske, D. K. Böhme, J. Hrusk, M. A. B. Ciufolini, *Exo- and Endohedral Fullerene Complexes in the Gas Phase*, VCH Publisher, New York, 1993.
32. H. S. Chen, A. R. Kortan, R. C. Haddon, D. A. Fleming, *J. Phys. Chem.* **1992**, *96*, 1016.
33. M. S. Dresselhaus, *Nature* **1992**, *358*, 195–196.
34. R. Saito, M. Fujita, G. Dresselhaus, M. S. Dresselhaus, *Mater. Sci. Eng. B* **1993**, *B19*, 185–191.
35. J. R. Fritch, K. P. C. Vollhardt, *Angew. Chem. Int. Ed. Engl.* **1980**, *19*, 559–561.
36. Vi. Kasatoch, V. V. Korshak, Yp. Kudryavt, A. M. Sladkov, Ie. Sterenbe, *Carbon* **1973**, *11*, 70–72.
37. L. R. Radovic, B. Bockrath, *J. Am. Chem. Soc.* **2005**, *127*, 5917–5927.
38. A. G. Whittacker, *Science* **1978**, *200*, 763–764.
39. F. Atamny, J. Blocker, B. Henschke, R. Schlögl, T. Schedelniedrig, M. Keil, A. M. Bradshaw, *J. Phys. Chem.* **1992**, *96*, 4522–4526.
40. P. Reinke, A. Knop-Gericke, M. Hävecker, T. Schedel-Niedrig, *Surf. Sci.* **2000**, *447*, 229–236.
41. T. Schedel-Niedrig, D. Herein, H. Werner, M. Wohlers, R. Schlögl, G. Francz, P. Kania, P. Oelhafen, C. Wild, *Europhys. Lett.* **1995**, *31*, 461–466.
42. T. Belz, A. Bauer, J. Find, M. Gunter, D. Herein, H. Mockel, N. Pfander, H. Sauer, G. Schulz, J. Schutze, O. Timpe, U. Wild, R. Schlögl, *Carbon* **1998**, *36*, 731–741.
43. C. Jager, T. Henning, R. Schlögl, O. Spillecke, *J. Non-Crystall. Solids* **1999**, *258*, 161–179.
44. H. Kanzow, A. Ding, J. Nissen, H. Sauer, T. Belz, R. Schlögl, *Phys. Chem. Chem. Phys.* **2000**, *2*, 2765–2771.
45. V. V. Roddatis, V. L. Kuznetsov, Y. V. Butenko, D. S. Su, R. Schlögl, *Phys. Chem. Chem. Phys.* **2002**, *4*, 1964–1967.
46. D. Roy, M. Portail, J. M. Layet, *Surf. Rev. Lett.* **2000**, *7*, 463–473.
47. R. A. Rosenberg, P. J. Love, V. Rehn, *Phys. Rev. B* **1986**, *33*, 4034–4037.
48. M. Kanowski, H. M. Vieth, K. Lüders, G. Buntkowsky, T. Belz, H. Werner, M. Wohlers, R. Schlögl, *Carbon* **1997**, *35*, 685–695.
49. H. Kroto, *Science* **1988**, *242*, 1139–1145.
50. J. Hrusük, H. Schwarz, *Chem. Phys. Lett.* **1993**, *205*, 187–190.
51. J. Kovac, G. Scarel, O. Sakho, M. Sancrotti, *J. Electron Spectrosc. Related Phenom.* **1995**, *72*, 71–75.
52. S. W. Biernacki, *Phys. Rev. B* **1997**, *56*, 11472–11476.
53. Y. K. Kim, J. H. Jung, J. Y. Lee, H. J. Ahn, *J. Mater. Sci.: Mater. Electron.* **1995**, *6*, 28–33.
54. V. L. Kuznetsov, M. N. Aleksandrov, I. Zagoruiko, A. L. Chuvilin, E. M. Moroz, V. N. Kolomiichuk, V. A. Likholobov, P. M. Brylykov, G. V. Sakovitch, *Carbon* **1991**, *29*, 665–668.
55. R. Sappok, *Chemie i. u. Zeit* **1970**, *4*, 145–151.
56. R. G. Cavell, S. P. Kowalcyk, L. Ley, R. Pollak, B. Mills, D. A. Shirley, W. Perry, *Phys. Rev. B* **1973**, *7*, 5313–5316.
57. S. Evans, J. M. Thomas, *Proc. R. Soc. Lond. A* **1977**, *353*, 103–120.
58. F. R. McFeely, S. P. Kowalczyk, L. Ley, R. G. Cavell, R. A. Pollak, D. A. Shirley, *Phys. Rev. B* **1974**, *9*, 5268–5278.
59. A. Dandekar, R. T. K. Baker, M. A. Vannice, *Carbon* **1998**, *36*, 1821–1831.
60. W. Engel, D. C. Ingram, J. C. Keay, M. Kordesch, *Diamond Relat. Mater.* **1994**, *3*, 1227–1229.
61. S. Matsumoto, H. Kanda, Y. Sato, M. Setaka, *Carbon* **1977**, *15*, 299–302.
62. F. Banhart, *Philos. Trans. R. Soc. Lond. Ser. A Math. Phys. Eng. Sci.* **2004**, *362*, 2205–2222.
63. F. Banhart, *J. Electron Microsc.* **2002**, *51*, S189–S194.
64. C. Castiglioni, F. Negri, M. Tommasini, E. Di Donato, G. Zerbi, *Carbon* **2006**, *100*, 381–402.
65. C. L. Lu, C. P. Chang, Y. C. Huang, J. M. Lu, C. C. Hwang, M. F. Lin, *J. Physics-Condens. Matter* **2006**, *18*, 5849–5859.
66. N. Ooi, A. Rairkar, J. B. Adams, *Carbon* **2006**, *44*, 231–242.
67. R. Schlögl, *Surf. Sci.* **1987**, *189*, 861–872.
68. S. Mrozowski, *Carbon* **1982**, *20*, 303–317.
69. N. M. R. Peres, F. Guinea, A. H. C. Neto, *Phys. Rev. B* **2006**, *73*.
70. M. Kanowski, G. Buntkowsky, H. Werner, M. Wohlers, R. Schlögl, H. M. Vieth, K. L. Åders, *Mol. Cryst. Liq. Cryst. A* **1994**, *244*, 271–275.
71. H. G. Karge, H. Darmstadt, A. Gutsze, H. M. Vieth, G. Buntkowsky, in *Zeolites and Related Microporous Materials: State of the Art 1994*, Vol. 84, 1994, pp. 1465–1474.
72. S. H. Yoon, C. W. Park, H. J. Yang, Y. Korai, I. Mochida, R. T. K. Baker, N. M. Rodriguez, *Carbon* **2004**, *42*, 21–32.

73. H. P. Boehm, R. W. Coughlin, *Carbon* **1964**, *2*, 1–6.
74. H. P. Boehm, U. Hofmann, *Z. Anorg. Allgem. Chem.* **1955**, *278*, 58–77.
75. E. L. Evans, Jd. Lopez-Gonzalez, A. Martin-Rodriguez, F. Rodriguez-Reinoso, *Carbon* **1975**, *13*, 461–464.
76. I. M. K. Ismail, P. L. Walker, Jr., *Carbon* **1989**, *27*, 549–559.
77. N. R. Laine, F. J. Vastola, P. L. Walker, Jr., *J. Phys. Chem.* **1963**, *67*, 2030–2034.
78. R. L. Radovic, P. L. Walker, R. G. Jenkins, *Fuel* **1983**, *62*, 849–856.
79. B. Henschke, H. Schubert, J. Blocker, F. Atamny, R. Schlögl, *Thermochim. Acta* **1994**, *234*, 53–83.
80. T. Schedel-Niedrig, D. Herein, H. Werner, M. Wohlers, R. Schlögl, G. Francz, P. Kania, P. Oelhafen, C. Wild, *Europhys. Lett.* **1995**, *31*, 461–466.
81. D. L. Wertz, M. Bissell, *Fuel* **1995**, *74*, 1431–1435.
82. P. Gallezot, C. Leclercq, M. Guisnet, P. Magnoux, *J. Catal.* **1988**, *114*, 100–111.
83. H. Kanzow, A. Ding, J. Nissen, H. Sauer, T. Belz, R. Schlögl, *Phys. Chem. Chem. Phys.* **2000**, *2*, 2765–2771.
84. R. B. Heimann, S. E. Evsyukov, Y. Koga, *Carbon* **1997**, *35*, 1654–1658.
85. P. Albers, K. Deller, B. M. Despeyroux, G. Prescher, A. Schafer, K. Seibold, *J. Catal.* **1994**, *150*, 368–375.
86. J. O. Muller, D. S. Su, R. E. Jentoft, U. Wild, R. Schlögl, *Environ. Sci. Technol.* **2006**, *40*, 1231–1236.
87. F. Atamny, A. Baiker, R. Schlögl, *Fresenius J. Anal. Chem.* **1997**, *358*, 344–348.
88. F. Atamny, T. F. Fässler, A. Baiker, R. Schlögl, *Appl. Phys. A: Mater. Sci. Process.* **2000**, *71*, 441–447.
89. A. W. Moore, *Chem. Phys. Carbon* **1981**, *17*, 233–286.
90. F. Rodriguezreinoso, *Carbon* **1989**, *27*, 305–312.
91. H. Knözinger, G. Mestl, *Top. Catal.* **1999**, *8*, 45–55.
92. A. C. Dillon, M. Yudasaka, M. S. Dresselhaus, *J. Nanosci. Nanotechnol.* **2004**, *4*, 691–703.
93. D. Porezag, T. Frauenheim, *Carbon* **1999**, *37*, 463–470.
94. A. M. S. V. I. Kasatochkin, Y. P. Kudryavtsev, M. M. Popov, J. E. Sterenberg, *Carbon* **1973**, *11*, 70–72.
95. T. Belz, R. Schlögl, *Synthetic Metals* **1996**, *77*, 223–226.
96. T. Belz, H. Werner, F. Zemlin, U. Klengler, M. Wesemann, E. Zeitler, A. Reller, R. Schlögl, B. Tesche, *Angew. Chem., Int. Ed. Engl.* **1994**, *33*, 1866–1869.
97. F. Cataldo, M. A. Pontier-Johnson, *Fullerenes Nanotubes Carbon Nanostructures* **2002**, *10*, 1–14.
98. J. D. Fitzgerald, G. H. Taylor, L. Brunckhorst, L. S. K. Pang, M. H. Terrones, A. L. Mackay, *Carbon* **1992**, *30*, 1251–1260.
99. M. S. Golden, M. Knupfer, J. Fink, J. F. Armbruster, T. R. Cummins, H. A. Romberg, M. Roth, M. Sing, M. Schmidt, E. Sohmen, *J. Phys.: Condens. Matter* **1995**, *7*, 8219–8247.
100. R. E. Haufler, J. Conceicao, L. P. F. Chibante, Y. Chai, N. E. Byrne, S. Flanagan, M. M. Haley, S. C. Obrien, C. Pan, Z. Xiao, W. E. Billups, M. A. Ciufolini, R. H. Hauge, J. L. Margrave, L. J. Wilson, R. F. Curl, R. E. Smalley, *J. Phys. Chem.* **1990**, *94*, 8634–8636.
101. D. Afanas'ev, I. Blinov, A. Bogdanov, V. Dyuzgev, V. Karataev, A. Kruglikov, *Tech. Phys.* **1994**, *39*, 1017–1024.
102. G. A. Dyuzhev, *Mol. Mater.* **1996**, *7*, 61–68.
103. D. Ugarte, *MRS Bull.* **1994**, *19*, 39–42.
104. H. W. Kroto, J. R. Heath, S. C. Obrien, R. F. Curl, R. E. Smalley, *Nature* **1985**, *318*, 162–163.
105. W. R. Creasy, J. T. Brenna, *Chem. Phys.* **1988**, *126*, 453–468.
106. T. Baum, S. Löffler, P. Weilmünster, K. H. Homann, *Ber. Bunsenges. Phys. Chem.* **1992**, *96*, 841–857.
107. K. H. Homann, *Angew. Chem. -Int. Ed.* **1998**, *37*, 2435–2451.
108. M. P. Johnson, R. W. Locke, J. B. Donnet, T. K. Wang, C. C. Wang, P. Bertrand, *Rubber Chem. Technol.* **2000**, *73*, 875–888.
109. H. Werner, D. Bublak, U. Göbel, B. Henschke, W. Bensch, R. Schlögl, *Angew. Chem. Int. Ed. Engl.* **1992**, *31*, 868–870.
110. Y. S. Grushko, T. Belz, T. Rühle, Y. Uchida, H. Werner, M. Wohlers, R. Schlögl, B. I. Smirnov, H. Kuzmany, J. Fink, M. Mehring, S. Roth, *Kinetics of Gas Phase Oxidation of Carbon Arc Cathode Deposits*, World Scientific, Kirchberg, 1996.
111. D. W. McKee, *Carbon* **1991**, *29*, 1057–1058.
112. M. Wohlers, A. Bauer, T. Rühle, F. Neitzel, H. Werner, R. Schlögl, *Fullerene Sci. Technol.* **1997**, *5*, 49–83.
113. M. Wohlers, H. Werner, T. Belz, T. Ruhle, R. Schlögl, *Mikrochim. Acta* **1997**, *125*, 401–406.
114. M. Wohlers, H. Werner, D. Herein, T. Schedelniedrig, A. Bauer, *Synthetic Metals* **1996**, *77*, 299–302.
115. C. Taliani, G. Ruani, R. Zamboni, R. Danieli, S. Rossini, V. N. Denisov, V. M. Burlakov, F. Negri, *J. Chem. Soc., Chem. Commun.* **1993**, 220.
116. H. J. Byrne, W. K. Maser, M. Kaiser, L. Akselrod, J. Anders, W. W. Ruhle, X. Q. Zhou, C. Thomsen, A. T. Werner, A. Mittelbach, S. Roth, *J. Mater. Process. Technol.* **1995**, *54*, 149–158.
117. M. Wohlers, A. Bauer, T. Rahle, F. Neitzel, H. Werner, R. Schlöglîgl, *Fullerenes Sci. Technol.* **1996**, *4*, 781–812.
118. F. Atamny, H. Kollmann, H. Bartl, R. Schlögl, *Ultramicroscopy* **1993**, *48*, 281–289.
119. J. B. Donnet, M. P. Johnson, D. T. Norman, T. C. Wang, *Carbon* **2000**, *38*, 1885–1902.
120. M. Kanowski, H. M. Vieth, K. Luders, G. Buntkowsky, T. Belz, H. Werner, M. Wohlers, R. Schlögl, *Carbon* **1997**, *35*, 685–695.
121. D. Ugarte, *Carbon* **1994**, *32*, 1245–1248.
122. H. Werner, M. Wohlers, D. Herein, D. Bublak, J. Blîcker, R. Schlöglîgl, *Fullerenes Sci. Technol.* **1993**, *1*, 199–219.
123. T. Belz, H. Werner, F. Zemlin, U. Klengler, M. Wesemann, B. Tesche, E. Zeitler, A. Reller, R. Schlögl, *Angew. Chem. Int. Ed. Engl.* **1994**, *33*, 1866–1869.
124. A. Keller, R. Kovacs, K. H. Homann, *Phys. Chem. Chem. Phys.* **2000**, *2*, 1667–1675.
125. H. Richter, E. deHoffmann, R. Doome, A. Fonseca, J. M. Gilles, J. B. Nagy, P. A. Thiry, J. Vandooren, P. J. VanTiggelen, *Carbon* **1996**, *34*, 797–803.
126. H. Richter, J. B. Howard, *Prog. Energy Combust. Sci.* **2000**, *26*, 565–608.
127. T. Belz, J. Find, D. Herein, N. Pfänder, T. Rühle, H. Werner, M. Wohlers, R. Schlögl, *Ber. Bunsen Ges. Phys. Chem.* **1997**, *101*, 712–725.
128. W. Ruland, *J. Appl. Physics* **1967**, *38*, 3585–3589.
129. W. Ruland, B. Smarsly, *J. Appl. Crystallogr.* **2002**, *35*, 624–633.
130. D. Joseph, A. Oberlin, *Carbon* **1983**, *21*, 559–564.
131. D. Joseph, A. Oberlin, *Carbon* **1983**, *21*, 565–571.
132. I. P. Batra, N. Garcia, H. Rohrer, H. Salemink, E. Stoll, S. Ciraci, *Surf. Sci.* **1987**, *181*, 126–138.
133. M. Kuwabara, D. R. Clarke, D. A. Smith, *Appl. Phys. Lett.* **1990**, *56*, 2396–2398.
134. D. Tomanek, S. G. Louie, H. J. Mamin, D. W. Abraham, R. E. Thomson, E. Ganz, J. Clarke, *Phys. Rev. B* **1987**, *35*, 7790–7793.
135. F. Atamny, A. Baiker, *Appl. Catal. A: General* **1998**, *173*, 201–230.
136. F. Atamny, O. Spillecke, R. Schlögl, *Phys. Chem. Chem. Phys.* **1999**, *1*, 4113–4118.
137. F. Stöckli, T. A. Centeno, J. B. Donnet, N. Pusset, E. Papirer, *Fuel* **1995**, *74*, 1592–1588.

138. A. Polania, E. Papirer, J. B. Donnet, G. Dagois, *Carbon* **1993**, *31*, 473–479.
139. J. B. Donnet, M. Brendle, T. L. Dhami, O. P. Bahl, *Carbon* **1986**, *24*, 757–770.
140. F. Atamny, T. Burgi, R. Schlögl, A. Baiker, *Surf. Sci.* **2001**, *475*, 140–148.
141. A. W. Moore, *Carbon* **1980**, *18*, 59–59.
142. H. P. Boehm, *Z. Anorg. Allgem. Chem.* **1958**, *297*, 315–320.
143. A. Clauss, H. P. Boehm, U. Hofmann, *Z. Anorg. Allgem. Chem.* **1957**, *290*, 35–51.
144. P. Pfeifer, F. Ehrburger-Dolle, T. P. Rieker, M. T. Gonzalez, W. P. Hoffman, M. Molina-Sabio, F. Rodriguez-Reinoso, P. W. Schmidt, D. J. Voss, *Phys. Rev. Lett.* **2002**, *8811*, 5502.
145. F. Atamny, R. Schlögl, A. Reller, *Carbon* **1992**, *30*, 1123–1126.
146. J. Lahaye, G. Prado, J. B. Donnet, *Carbon* **1974**, *12*, 27–35.
147. D. L. Wertz, M. Bissell, *Energy & Fuels* **1994**, *8*, 613–617.
148. L. R. Radovic, C. Moreno-Castilla, J. Rivera-Utrilla, *Chemistry Physics Carbon* **2001**, *27*, 227–405.
149. E. Raymundo-Pinero, D. Cazorla-Amoros, A. Linares-Solano, J. Find, U. Wild, R. Schlögl, *Carbon* **2002**, *40*, 597–608.
150. J. Lahaye, G. Prado, P. L. Walker Jr., P. A. Thrower, *Mechanisms of Carbon Black Formation*, Marcel Dekker, 1978.
151. A. I. Medalia, D. Rivin, *Carbon* **1982**, *20*, 481–492.
152. R. Burmeister, B. Despeyroux, K. Deller, K. Seibold, P. Albers, in *Heterogeneous Catalysis and Fine Chemicals Iii*, Vol. 78, 1993, pp. 361–368.
153. J. O. Müller, D. S. Su, R. E. Jentoft, J. Krohnert, F. C. Jentoft, R. Schlögl, *Catal. Today* **2005**, *102*, 259–265.
154. P. Ehrburger, F. Louys, J. Lahaye, *Carbon* **1989**, *27*, 389–393.
155. U. Wild, N. Pfänder, R. Schlögl, *Fresenius J. Anal. Chem.* **1997**, *357*, 420–428.
156. I. M. K. Ismail, S. L. Rodgers, *Carbon* **1992**, *30*, 229–239.
157. V. R. Choudhary, C. Sivadinarayana, P. Devadas, S. D. Sansare, P. Magnoux, M. Guisnet, *Micropor. Mesopor. Mater.* **1998**, *21*, 91–101.
158. M. Guisnet, P. Magnoux, *Catal. Today* **1997**, *36*, 477–483.
159. M. Guisnet, P. Magnoux, *Appl. Catal. A: General* **2001**, *212*, 83–96.
160. J. P. Lange, A. Gutsze, H. G. Karge, *J. Catal.* **1988**, *114*, 136–143.
161. J. Robertson, *Prog. Solid State Chem.* **1991**, *21*, 199–333.
162. J. Robertson, *Surface Coatings Technol.* **1992**, *50*, 185–203.
163. D. R. McKenzie, *Rep. Prog. Physics* **1996**, *59*, 1611–1664.
164. D. Ugolini, J. Eitle, P. Oelhafen, M. Wittmer, *Appl. Phys.* **1989**, *A48*, 549–558.
165. D. Mendoza, J. Aguilar-Hernandez, G. Contreras-Puente, *Solid State Commun.* **1992**, *84*, 1025–1027.
166. E. Meyer, H. Heinzelmann, P. Gratter, T. Jung, L. Scandella, H. R. Hidber, H. Rudin, H. J. Güntherodt, C. Schmidt, *Appl. Phys. Lett.* **1989**, *55*, 1624–1626.
167. A. C. Ferrari, S. E. Rodil, J. Robertson, *Physical Rev. B* **2003**, *67*.
168. J. Lahaye, *Fuel* **1998**, *77*, 543–547.
169. R. Arriagada, G. Bello, R. Garcia, F. Rodriguez-Reinoso, A. Sepulveda-Escribano, *Micropor. Mesopor. Mater.* **2005**, *81*, 161–167.
170. P. Ehrburger, O. P. Mahajan, P. L. Walker, Jr., *J. Catal.* **1976**, *43*, 61–67.
171. K. Lenghaus, G. G. H. Qiao, D. H. Solomon, C. Gomez, F. Rodriguez-Reinoso, A. Sepulveda-Escribano, *Carbon* **2002**, *40*, 743–749.
172. L. R. Radovic, *Surfaces Nanoparticles Porous Mater.* **1999**, *78*, 529–565.
173. M. S. Solum, R. J. Pugmire, M. Jagtoyen, F. Derbyshire, *Carbon* **1995**, *33*, 1247–1254.
174. A. R. Pande, *Fuel* **1992**, *71*, 1299–1302.
175. F. Rodriguezreinoso, M. Molinasabio, M. A. Munecas, *J. Phys. Chem.* **1992**, *96*, 2707–2713.
176. C. Almansa, M. Molina-Sabio, F. Rodriguez-Reinoso, *Micropor. Mesopor. Mater.* **2004**, *76*, 185–191.
177. E. Papirer, J. Dentzer, S. Li, J. B. Donnet, *Carbon* **1991**, *29*, 69–72.
178. J. B. Donnet, R. C. Bansal, M. J. Wang, *Carbon Black, Science and Technology*, Marcel Dekker Inc., 1993.
179. G. Bello, R. Garcia, R. Arriagada, A. Sepulveda-Escribano, F. Rodriguez-Reinoso, *Micropor. Mesopor. Mater.* **2002**, *56*, 139–145.
180. C. Gomez-de-Salazar, A. Sepulveda-Escribano, F. Rodriguez-Reinoso, *Carbon* **2000**, *38*, 1889–1892.
181. M. Molina-Sabio, M. J. Sanchez-Montero, J. M. Juarez-Galan, F. Salvador, F. Rodriguez-Reinoso, A. Salvador, *J. Phys. Chem. B* **2006**, *110*, 12360–12364.
182. F. Atamny, J. Blöcker, A. Dübotzky, H. Kurt, O. Timpe, G. Loose, W. Mahdi, R. Schlögl, *Mol. Physics* **1992**, *76*, 851–886.
183. J. A. Menendez, M. J. Illangomez, C. Leon, L. R. Radovic, *Carbon* **1995**, *33*, 1655–1657.
184. M. A. Rodriguez-Valero, M. Martinez-Escandell, M. Molina-Sabio, F. Rodriguez-Reinoso, *Carbon* **2001**, *39*, 320–323.
185. E. Ahumada, H. Lizama, F. Orellana, C. Suarez, A. Huidobro, A. Sepulveda-Escribano, F. Rodriguez-Reinoso, *Carbon* **2002**, *40*, 2827–2834.
186. M. Gurrath, T. Kuretzky, H. P. Boehm, L. B. Okhlopkova, A. S. Lisitsyn, V. A. Likholobov, *Carbon* **2000**, *38*, 1241–1255.
187. M. J. Illangomez, C. S. M. Delecea, A. Linaressolano, L. R. Radovic, *Energy Fuels* **1998**, *12*, 1256–1264.
188. L. B. Okhlopkova, A. S. Lisitsyn, V. A. Likholobov, M. Gurrath, H. P. Boehm, *Appl. Catal. A: General* **2000**, *204*, 229–240.
189. L. Singoredjo, F. Kapteijn, J. A. Moulijn, J. M. Martinmartinez, H. P. Boehm, *Carbon* **1993**, *31*, 213–222.
190. M. Besson, P. Gallezot, A. Perrard, C. Pinel, *Catal. Today* **2005**, *102*, 160–165.
191. K. S. W. S. S. J. Gregg, *Adsorption, Surface Area and Porosity*, Academic Press, London, 1982.
192. Fv. S. O. Vohler, E. Wege, H. von Kienle, M. Voll, P. Kleinschmidt, *Ullmann's Encyclopedia of Industrial Chemistry*, Wiley-VCH, 1986.
193. H. Bockhorn, F. Fetting, A. Heddrich, G. Wannemacher, *Phys. Chem. Chem. Phys.* **1987**, *91*, 819–825.
194. H. Bockhorn, H. Geitlinger, B. Jungfleisch, T. Lehre, A. Schon, T. Streibel, R. Suntz, *Phys. Chem. Chem. Phys.* **2002**, *4*, 3780–3793.
195. F. Mauss, T. Schafer, H. Bockhorn, *Combustion Flame* **1994**, *99*, 697–705.
196. Z. L. Wang, J. Wang, H. Richter, J. B. Howard, J. Carlson, Y. A. Levendis, *Energy Fuels* **2003**, *17*, 999–1013.
197. P. Gerhardt, S. Löffler, K. H. Homann, *Chem. Phys. Lett.* **1987**, *137*, 306–310.
198. K. H. Homann, H. G. Wagner, *Ber. Bunsen Ges. Phys. Chem.* **1965**, *69*, 20–&.
199. H. Marsh, P. L. Walker, Jr., P. A. Thrower, *The Formation of Graphitizable Carbons via Mesophase: Chemical and Kinetic Consideration*, Marcel Dekker, 1979.
200. P. Ehrburger, N. Pusset, P. Dziedzinl, *Carbon* **1992**, *30*, 1105–1109.

201. T. Belz, R. Schlögl, H. Kuzmany, J. Fink, M. Mehring, S. Roth, *Optical Emission Studies of Electric Arc Carbon Evaporation Plasmas*, World Scientific Publishing Co. Pte. Ltd., Singapore, 1995.
202. H. Grote, W. Bohmeyer, P. Kornejew, H. D. Reiner, G. Fussmann, R. Schlögl, G. Weinberg, C. H. Wu, *J. Nuclear Mater.* **1999**, *269*, 1059–1064.
203. D. Ugarte, *Carbon* **1995**, *33*, 989–993.
204. R. T. K. Baker, *J. Adhesion* **1995**, *52*, 13–40.
205. R. T. K. Baker, M. A. Barber, R. J. Waite, P. S. Harris, F. S. Feates, *J. Catal.* **1972**, *26*, 51–62.
206. O. C. Carneiro, N. M. Rodriguez, R. T. K. Baker, *Carbon* **2005**, *43*, 2389–2396.
207. N. M. Rodriguez, A. Chambers, R. T. K. Baker, *Langmuir* **1995**, *11*, 3862–3866.
208. H. Y. Wang, R. T. K. Baker, *J. Phys. Chem. B* **2004**, *108*, 20273–20277.
209. H. S. Bengaard, J. K. Norskov, J. Sehested, B. S. Clausen, L. P. Nielsen, A. M. Molenbroek, J. R. Rostrup-Nielsen, *J. Catal.* **2002**, *209*, 365–384.
210. R. C. B. J. B. Donnet, M. J. Wang, *Carbon Black, Science and Technology*, Marcel Dekker, New York, 1993.
211. I. Alstrup, M. T. Tavares, C. A. Bernardo, O. Sorensen, J. R. Rostrup-Nielsen, *Mater. Corrosion* **1998**, *49*, 367–372.
212. C. Bertrand, J. L. Delfau, *Combust. Sci. Technol.* **1985**, *44*, 29–45.
213. M. Endo, H. W. Kroto, *J. Phys. Chem.* **1992**, *96*, 6941–6944.
214. J. Lahaye, *Carbon* **1992**, *30*, 309–314.
215. H. F. Calcote, D. G. Keil, *Pure Appl. Chem.* **1990**, *62*, 815–824.
216. H. F. Calcote, D. G. Keil, *Combustion Flame* **1988**, *74*, 131–146.
217. J. Lahaye, *Polymer Degrad. Stability* **1990**, *30*, 111–121.
218. H. Richter, J. B. Howard, *Phys. Chem. Chem. Phys.* **2002**, *4*, 2038–2055.
219. J. W. M. Frenklach, *Combust. Sci. Technol.* **1987**, *51*, 265–283.
220. L. Baumgartner, H. Jander, H. G. Wagner, *Phys. Chem. Chem. Phys.* **1983**, *87*, 1077–1080.
221. J. H. Kent, H. G. Wagner, *Erdol Kohle Erdgas Petrochem.* **1985**, *38*, 543–549.
222. I. Glassman, J. G. Hansel, T. Eklund, *Combustion Flame* **1969**, *13*, 99–112.
223. M. Kasper, K. Sattler, K. Siegmann, U. Matter, H. C. Siegmann, *J. Aerosol Sci.* **1999**, *30*, 217–225.
224. V. V. Kislov, A. M. Mebel, S. H. Lin, *J. Phys. Chem. A* **2002**, *106*, 6171–6182.
225. J. P. Maier, *J. Phys. Chem. A* **1998**, *102*, 3462–3469.
226. L. B. Ebert, J. C. Scanlon, C. A. Clausen, *Energy Fuels* **1988**, *2*, 438–445.
227. M. Frenklach, L. B. Ebert, *J. Phys. Chem.* **1988**, *92*, 561–563.
228. D. L. Wertz, M. Bissell, *Fuel* **1995**, *74*, 1431–1435.
229. T. Cabioc'h, E. Thune, M. Jaouen, F. Banhart, *Philos. Mag. A – Phys. Condensed Matter Defects & Mech. Properties* **2002**, *82*, 1509–1520.
230. J. R. Rostrup-Nielsen, J. Sehested, in *Catalyst Deactivation 2001, Proceedings, Studies in surface science and Catalysis*, Vol. 139, Elsevier, 2001, pp. 1–12.
231. M. Loepfe, H. Burtscher, H. C. Siegmann, *Water Air Soil Pollution* **1993**, *68*, 177–184.
232. R. Engeln, G. von Helden, A. J. A. van Roij, G. Meijer, *J. Chem. Phys.* **1999**, *110*, 2732–2733.
233. G. von Helden, I. Holleman, G. Meijer, B. Sartakov, *Optics Express* **1999**, *4*, 46–52.
234. S. R. McDow, W. Giger, H. Burtscher, A. Schmidtott, H. C. Siegmann, *Atmosph. Environ. Part A – General Topics* **1990**, *24*, 2911–2916.
235. K. Siegmann, H. Hepp, K. Sattler, *Surf. Rev. Lett.* **1996**, *3*, 741–745.
236. T. Belz, R. Schlögl, H. Kuzmany, J. Fink, M. Mehring, S. Roth, *Diagnostics of Fullerene-Generating Plasmas by Optical Emission Spectroscopy*, World Scientific, 1996.
237. H. Bockhorn, *Soot formation in combustion: mechanisms and models*, Springer, Berlin, 1994.
238. X. X. Bi, M. Jagtoyen, M. Endo, K. Daschowdhury, R. Ochoa, F. J. Derbyshire, M. S. Dresselhaus, P. C. Eklund, *J. Mater. Res.* **1995**, *10*, 2875–2884.
239. J. Rodriguezmirasol, T. Cordero, L. R. Radovic, J. J. Rodriguez, *Chem. Mater.* **1998**, *10*, 550–558.
240. C. Thomazeau, P. Cartraud, P. Magnoux, S. Jullian, M. Guisnet, *Micropor. Mater.* **1996**, *5*, 337–345.
241. A. A. Bageev, A. P. Brosnik, V. V. Strelko, Y. Tarasenko, *Russ. J. Appl. Chem.* **2001**, *74*, 205–208.
242. R. Burmeister, B. Despeyroux, K. Deller, K. Seibold, P. Albers, *Stud. Surf. Sci. Catal.* **1993**, *78*, 361–368.
243. H. Jantgen, *Fuel* **1986**, *65*, 1436–1446.
244. J. Maruyama, K. Sumino, M. Kawaguchi, I. Abe, *Carbon* **2004**, *42*, 3115–3121.
245. E. Raymundo-Pinero, D. Carzorla-Amorós, A. Linares-Solano, J. Find, U. Wild, R. Schlögl, *Carbon* **2002**, *40*, 597–608.
246. R. Schlögl, *Abstr. Papers Am. Chem. Soc.* **2005**, *229*, U746–U747.
247. I. F. Silva, C. Palma, M. Klimkiewicz, S. Eser, *Carbon* **1998**, *36*, 861–868.
248. H. P. Boehm, *Carbon* **1973**, *11*, 583.
249. M. Audier, M. Coulon, L. Bonnetain, *Carbon* **1983**, *21*, 93–97.
250. M. Audier, M. Coulon, L. Bonnetain, *Carbon* **1983**, *21*, 105–110.
251. M. Audier, J. Guinot, M. Coulon, L. Bonnetain, *Carbon* **1981**, *19*, 99–105.
252. M. Audier, M. Coulon, *Carbon* **1985**, *23*, 317–323.
253. E. Boellaard, P. K. Debokx, A. Kock, J. W. Geus, *J. Catal.* **1985**, *96*, 481–490.
254. P. K. Debokx, A. Kock, E. Boellaard, W. Klop, J. W. Geus, *J. Catal.* **1985**, *96*, 454–467.
255. A. Kock, P. K. Debokx, E. Boellaard, W. Klop, J. W. Geus, *J. Catal.* **1985**, *96*, 468–480.
256. S. Helveg, C. Lopez-Cartes, J. Sehested, P. L. Hansen, B. S. Clausen, J. R. Rostrup-Nielsen, F. Abild-Pedersen, J. K. Norskov, *Nature* **2004**, *427*, 426–429.
257. N. M. Rodriguez, P. E. Anderson, A. Wootsch, U. Wild, R. Schlögl, Z. Paal, *J. Catal.* **2001**, *197*, 365–377.
258. I. Alstrup, *J. Catal.* **1988**, *109*, 241–251.
259. J. C. Charlier, S. Iijima, in *Carbon Nanotubes*, Vol. 80, 2001, pp. 55–80.
260. M. Morkel, G. Rupprechter, H. J. Freund, *Surf. Sci.* **2005**, *588*, L209–L219.
261. B. E. Bent, C. M. Mate, J. E. Crowell, B. E. Koel, G. A. Somorjai, *J. Phys. Chem.* **1987**, *91*, 1493–1502.
262. D. W. Blakely, G. A. Somorjai, *Surf. Sci.* **1977**, *65*, 419–442.
263. D. R. Kennedy, G. Webb, S. D. Jackson, D. Lennon, *Appl. Catal. A: General* **2004**, *259*, 109–120.
264. Z. Paal, A. Wootsch, R. Schlögl, U. Wild, *Appl. Catal. A: General* **2005**, *282*, 135–145.
265. L. A. West, G. A. Somorjai, *J. Chem. Phys.* **1971**, *54*, 2864–2873.

266. S. M. Davis, F. Zaera, B. E. Gordon, G. A. Somorjai, *J. Catal.* **1985**, *92*, 240–246.
267. G. C. Bond, G. Webb, P. B. Wells, *J. Catal.* **1968**, *12*, 157–165.
268. G. F. Taylor, S. J. Thomson, G. Webb, *J. Catal.* **1968**, *12*, 191–199.
269. S. J. Thomson, G. Webb, *J. Chem. Soc., Chem. Commun.* **1976**, 526–527.
270. D. Teschner, U. Wild, R. Schlögl, Z. Paal, *J. Phys. Chem. B* **2005**, *109*, 20516–20521.
271. Z. Paal, R. Schlögl, G. Ertl, *J. Chem. Soc. – Faraday Trans.* **1992**, *88*, 1179–1189.
272. J. M. Cogen, K. Ezaznikpay, R. H. Fleming, S. M. Baumann, W. F. Maier, *Angew. Chem. Int. Ed. Engl.* **1987**, *26*, 1182–1184.
273. M. Salmeron, G. A. Somorjai, *J. Phys. Chem.* **1982**, *86*, 341–350.
274. L. H. Dubois, D. G. Castner, G. A. Somorjai, *J. Chem. Physics* **1980**, *72*, 5234–5240.
275. A. Borodzinki, *Catal. Rev. -Sci. Eng.* **2006**, *48*, 91–144.
276. A. S. Canning, S. D. Jackson, A. Monaghan, T. Wright, *Catal. Today* **2006**, *116*, 22–29.
277. C. Glasson, C. Geantet, M. Lacroix, F. Labruyere, P. Dufresne, *J. Catal.* **2002**, *212*, 76–85.
278. M. Guisnet, P. Andy, N. S. Gnep, C. Travers, E. Benazzi, *J. Chem. Soc., Chem. Commun.* **1995**, 1685–1686.
279. J. Lojewska, W. Makowski, R. Dziembaj, *Chem. Eng. J.* **2002**, *90*, 203–208.
280. S. van Donk, J. H. Bitter, K. P. de Jong, *Appl. Catal. A: General* **2001**, *212*, 97–116.
281. Z. Paal, A. Wootsch, K. Matusek, U. Wild, R. Schlögl, *Catal. Today* **2001**, *65*, 13–18.
282. B. Hupfer, H. Schupp, J. D. Andrade, H. Ringsdorf, *J. Electron Spectrosc.* **1981**, *23*, 103–107.
283. J. Szaiman, R. C. C. Leckey, *J. Electron Spectrosc.* **1981**, *23*, 83–96.
284. H. Gabasch, E. Kleimenov, D. Teschner, S. Zafeiratos, M. Havecker, A. Knop-Gericke, R. Schlögl, D. Zemlyanov, B. Aszalos-Kiss, K. Hayek, B. Klotzer, *J. Catal.* **2006**, *242*, 340–348.
285. Z. Paal, D. Teschner, N. M. Rodriguez, R. T. K. Baker, L. Toth, U. Wild, R. Schlögl, *Catal. Today* **2005**, *102*, 254–258.
286. D. Teschner, E. Vass, M. Hävecker, S. Zafeiratos, P. Schnörch, H. Sauer, A. Knop-Gericke, R. Schlögl, M. Chamam, A. Wootsch, A. S. Canning, J. J. Gamman, S. D. Jackson, J. McGregor, L. F. Gladden, *J. Catal.* **2006**, *242*, 26–37.
287. A. Taguchi, F. Schuth, *Micropor. Mesopor. Mater.* **2005**, *77*, 1–45.
288. M. Guisnet, N. S. Gnep, F. Alario, *Appl. Catal. A: General* **1992**, *89*, 1–30.
289. H. G. Karge, J. P. Lange, A. Gutsze, M. Laniecki, *J. Catal.* **1988**, *114*, 144–152.
290. K. Moljord, P. Magnoux, M. Guisnet, *Appl. Catal. A: General* **1995**, *122*, 21–32.
291. D. Eisenbach, E. Gallei, *J. Catal.* **1979**, *56*, 377–389.
292. H. G. Karge, W. Niessen, H. Bludau, *Appl. Catal. A: General* **1996**, *146*, 339–349.
293. C. Li, P. C. Stair, *Catal. Today* **1997**, *33*, 353–360.
294. J. F. Haw, J. B. Nicholas, T. Xu, L. W. Beck, D. B. Ferguson, *Acc. Chem. Res.* **1996**, *29*, 259–267.
295. M. Hunger, T. Horvath, *J. Catal.* **1997**, *167*, 187–197.
296. J. P. Lange, A. Gutsze, J. Allgeier, H. G. Karge, *Appl. Catal.* **1988**, *45*, 345–356.
297. H. G. Karge, in *Introduction in Zeolite Science and Practice*, E. M. F. H. van Bekkum, J. C. Jansen (Eds.), Elsevier, Amsterdam, 1991, pp. 531–579.
298. S. M. Davis, Y. Zhou, M. A. Freeman, D. A. Fischer, G. M. Meitzner, J. L. Gland, *J. Catal.* **1993**, *139*, 322–325.
299. G. P. Handreck, T. D. Smith, *J. Catal.* **1990**, *123*, 513–522.
300. C. A. Henriques, A. M. Bentes, P. Magnoux, M. Guisnet, J. L. F. Monteiro, *Appl. Catal. A: General* **1998**, *166*, 301–309.
301. C. A. Henriques, J. L. F. Monteiro, P. Magnoux, M. Guisnet, *J. Catal.* **1997**, *172*, 436–445.
302. K. Moljord, P. Magnoux, M. Guisnet, *Appl. Catal. A: General* **1995**, *121*, 245–259.
303. Y. Shu, R. Ohnishi, M. Ichikawa, *Appl. Catal. A: General* **2003**, *252*, 315–329.
304. H. G. Karge, E.P., *Catal. Today* **1988**, *3*, 53–63.
305. H. Marsh, J. L. Figueiredo, J. A. Moulijn, *Structure in Carbons*, Martinus Nijhoff Publishers, 1986.
306. R. W. Henson, W. M. Reynolds, *Carbon* **1963**, *3*, 277–286.
307. J. Maire, J. Mering, *Chem. Phys. Carbon* **1970**, *6*, 125–190.
308. I. C. Lewis, *Carbon* **1982**, *20*, 519–529.
309. K. Hernadi, A. Gaspar, J. W. Seo, M. Hammida, A. Demortier, L. Forro, J. B. Nagy, I. Kiricsi, *Carbon* **2004**, *42*, 1599–1607.
310. M. Monthioux, J. G. Lavin, *Carbon* **1994**, *32*, 335–343.
311. G. Blyholder, H. Eyring, *J. Phys. Chem.* **1959**, *63*, 1004–1008.
312. J. M. Ranish, P. L. Walker, Jr., *Carbon* **1993**, *31*, 135–141.
313. Z. Du, A. F. Sarofim, J. P. Longwell, *Energy Fuels* **1991**, *5*, 214–221.
314. H. Marsh, T. E. O'Hair, R. Reed, *Trans. Faraday Soc.* **1965**, *61*, 285–293.
315. F. S. Feates, *Trans. Faraday Soc.* **1968**, *64*, 3093–3099.
316. D. W. McKee, *Chem. Phys. Carbon* **1991**, *23*, 173–232.
317. R. Schlögl, F. Atamny, W. J. Wirth, J. Stephan, *Ultramicroscopy* **1992**, *42–44*, 660–667.
318. J. M. Thomas, *Chem. Phys. Carbon* **1965**, *1*, 121.
319. C. Janiak, R. Hoffmann, P. Sjovall, B. Kasemo, *Langmuir* **1993**, *9*, 3427–3440.
320. E. L. Evans, R. J. M. Griffiths, J. M. Thomas, *Science* **1971**, *171*, 174–175.
321. H. Marsh, A. D. Foord, J. S. Mattson, J. Thomas, E. L. Evans, *J. Colloid Interface Sci.* **1974**, *49*, 368–382.
322. F. J. Vastola, P. J. Hart, P. L. Walker, *Carbon* **1964**, *2*, 65–71.
323. A. Pigamo, M. Besson, B. Blanc, P. Gallezot, A. Blackburn, O. Kozynchenko, S. Tennison, E. Crezee, F. Kapteijn, *Carbon* **2002**, *40*, 1267–1278.
324. H. P. Boehm, G. Mair, T. Stoehr, A. R. Derincon, B. Tereczki, *Fuel* **1984**, *63*, 1061–1063.
325. E. Sanchez, Y. Yang, J. Find, T. Braun, R. Schoonmaker, T. Belz, H. Sauer, O. Spillecke, Y. Uchida, R. Schlögl, in *Science and Technology in Catalysis 1998*, Vol. 121, 1999, pp. 317–326.
326. J. Yang, G. Mestl, D. Herein, R. Schlögl, J. Find, *Carbon* **2000**, *38*, 715–727.
327. A. A. Lizzio, H. Jiang, L. R. Radovic, *Carbon* **1990**, *28*, 7–19.
328. A. A. Lizzio, A. Piotrowski, L. R. Radovic, *Fuel* **1988**, *67*, 1691–1695.
329. A. A. Lizzio, L. R. Radovic, *Ind. Eng. Chem. Res.* **1991**, *30*, 1735–1744.
330. F. Atamny, R. Schlögl, W. J. Wirth, J. Stephan, *Ultramicroscopy* **1992**, *42*, 660–667.
331. R. Schlögl G. Loose, M. Wesemann, *Solid-State Ionics* **1990**, *43*, 183–192.
332. L. E. C. de Torre, J. L. Llanos, E. J. Bottani, *Carbon* **1991**, *29*, 1051–1061.
333. H. Marsh, T. E. O'Hair, *Carbon* **1969**, *7*, 702–703.

334. J. Yang, N. Pfänder, E. Sanchez-Cortezon, U. Wild, G. Mestl, J. Find, R. Schlögl, *Carbon* **2000**, *38*, 2029–2039.
335. H. Marsh, T. E. O'Hair, W. F. K. Wynne-Jones, *Trans. Faraday Soc.* **1965**, *61*, 274–283.
336. C. Wong, R. T. Yang, B. L. Halpern, *J. Chem. Physics* **1983**, *78*, 3325–3328.
337. G. A. Simons, *Carbon* **1982**, *20*, 117–118.
338. A. Cuesta, A. Martinez-Alonso, J. M. D. Tascon, *Energy Fuels* **1993**, *7*, 1141–1145.
339. D. Herein, Find, J., Herzog, B., Kollmann, H., Schmidt, R., Schlögl, R., *Abstracts, Papers Am. Chem. Soc.* **1996**, 148–156.
340. Y. J. Lee, L. R. Radovic, *Carbon* **2003**, *41*, 1987–1997.
341. D. W. McKee, C. L. Spiro, E. J. Lamby, *Carbon* **1984**, *22*, 507–511.
342. D. W. McKee, *J. Catal.* **1987**, *108*, 480–483.
343. O. W. Fritz, K. J. Hüttinger, *Carbon* **1993**, *31*, 923–930.
344. K. J. Hüttinger, W. F. Merdes, *Carbon* **1992**, *30*, 883–894.
345. D. W. McKee, *Fuel* **1983**, *62*, 170–175.
346. H. Marsh, N. Murdie, I. A. S. Edwards, H. Boehm, *Chem. Phys. Carbon* **1987**, *20*, 213.
347. A. A. Adjorlolo, Y. K. Rao, *Carbon* **1984**, *22*, 173–176.
348. E. Ferguson, R. Schlögl, W. Jones, *Fuel* **1984**, *63*, 1048–1058.
349. S. R. Kelemen, C. A. Mims, *Surf. Sci.* **1983**, *133*, 231–240.
350. S. R. Kelemen, H. Freund, *J. Catal.* **1986**, *102*, 80–91.
351. A. Messerer, R. Niessner, U. Pöschl, *Carbon* **2006**, *44*, 307–324.
352. F. Cangialosi, F. Di Canio, G. Intini, M. Notarnicola, L. Liberti, G. Belz, P. Caramuscio, *Fuel* **2006**, *85*, 2294–2300.
353. O. Senneca, P. Salatino, S. Masi, *Proc. Combustion Inst.* **2005**, *30*, 2223–2230.
354. P. S. Fennell, S. Kadchha, H. Y. Lee, J. S. Dennis, A. N. Hayhurst, *Chem. Eng. Sci.* **2007**, *62*, 608–618.
355. M. J. Nowakowski, J. M. Vohs, D. A. Bonnell, *Surf. Sci.* **1992**, *271*, L351–L356.
356. D. Gonzalez, O. Altin, S. Eser, A. B. Garcia, *Mater. Chem. Physics* **2007**, *101*, 137–141.
357. H. Werner, T. Schedelniedrig, M. Wohlers, D. Herein, B. Herzog, R. Schlögl, M. Keil, A. M. Bradshaw, J. Kirschner, *J. Chem. Soc., Faraday Trans.* **1994**, *90*, 403–409.
358. M. Wohlers, H. Werner, T. Belz, T. Råhle, R. Schlögl, *Microchim. Acta* **1997**, *125*, 401–406.
359. M. Wohlers, A. Bauer, R. Schlögl, *Microchim. Acta* **1997**, *14*, 267–270.
360. L. R. Radovic, J. Hong, A. A. Lizzio, *Energy Fuels* **1991**, *5*, 68–74.
361. J. A. Menendez, B. Xia, J. Phillips, L. R. Radovic, *Langmuir* **1997**, *13*, 3414–3421.
362. H. S. Chen, A. R. Kortan, R. C. Haddon, M. L. Kaplan, C. H. Chen, A. M. Mujsce, H. Chou, D. A. Fleming, *Appl. Phys. Lett.* **1991**, *59*, 2956.
363. R. T. Yang, C. Wong, *J. Catal.* **1984**, *85*, 154–168.
364. M. C. Bohm, J. Schulte, R. Schlögl, *Phys. Status Solid B: Basic Res.* **1996**, *196*, 131–144.
365. R. Schlögl, H. P. Bonzel, A. M. Bradshaw, G. Ertl, *Alkali Metals in Heterogeneous Catalysis*, Elsevier Science Publishers B. V., Amsterdam, 1989.
366. M. B. Cerfontain, R. Meijer, F. Kapteijn, J. A. Moulijn, *J. Catal.* **1987**, *107*, 173–180.
367. A. Simon, *J. Solid State Chem.* **1979**, *27*, 87–97.
368. A. Kavanagh, R. Schlögl, *Carbon* **1988**, *26*, 23–32.
369. M. Audier, A. Oberlin, M. Oberlin, M. Coulon, L. Bonnetain, *Carbon* **1981**, *19*, 217–224.
370. J. A. Moulijn, F. Kapteijn, *Carbon* **1995**, *33*, 1155–1165.
371. F. Kapteijn, R. Meier, S. C. van Eyck, J. A. Moulijn, in *Gasification of Carbon and Coal*, J. Lahaye, P. Ehrburger (Eds.), Kluwer, Dordrecht, 1991, pp. 221–237.
372. J. Ma, C. Park, N. M. Rodriguez, R. T. K. Baker, *J. Phys. Chem. B* **2001**, *105*, 11994–12002.
373. M. J. Illan-Gomez, S. Brandan, C. S. M. de Lecea, A. Linares-Solano, *Fuel* **2001**, *80*, 2001–2005.
374. A. Garcia-Garcia, M. J. Illan-Gomez, A. Linares-Solano, C. S. M. de Lecea, *Energy Fuels* **1999**, *13*, 499–505.
375. J. Yang, G. Mestl, D. Herein, R. Schlögl, J. Find, *Carbon* **2000**, *38*, 729–740.
376. Y. H. Li, L. R. Radovic, G. Q. Lu, V. Rudolph, *Chem. Eng. Sci.* **1999**, *54*, 4125–4136.
377. F. Kapteijn, J. A. Moulijn, S. Matzner, H. P. Boehm, *Carbon* **1999**, *37*, 1143–1150.
378. D. S. Su, J. O. Muller, R. E. Jentoft, D. Rothe, E. Jacob, R. Schlögl, *Top. Catal.* **2004**, *30–31*, 241–245.
379. D. S. Su, R. E. Jentoft, J. O. Muller, D. Rothe, E. Jacob, C. D. Simpson, Z. Tomovic, K. Mullen, A. Messerer, U. Pöschl, R. Niessner, R. Schlögl, *Catal. Today* **2004**, *90*, 127–132.
380. K. J. Huttinger, O. W. Fritz, *Carbon* **1991**, *29*, 1113–1118.
381. X. J. Xu, N. M. Rodriguez, R. T. K. Baker, *React. Kinet. Catal. Lett.* **2006**, *87*, 305–312.
382. T. Braun, M. Wohlers, T. Belz, G. Nowitzke, G. Wortmann, Y. Uchida, N. Pfänder, R. Schlögl, *Catal. Lett.* **1997**, *43*, 167–173.
383. F. Atamny, A. Baiker, *Surf. Interface Analysis* **1999**, *27*, 512–516.
384. B. S. Sherigara, W. Kutner, F. D'Souza, *Electroanalysis* **2003**, *15*, 753–772.
385. V. I. Povstugar, S. S. Mikhailova, A. A. Shakov, *J. Anal. Chem.* **2000**, *55*, 405–416.
386. N. Tsubokawa, *Bull. Chem. Soc. Jpn.* **2002**, *75*, 2115–2136.
387. N. Tsubokawa, *Polymer J.* **2005**, *37*, 637–655.
388. C. H. Cheng, J. Lehmann, J. E. Thies, S. D. Burton, M. H. Engelhard, *Org. Geochem.* **2006**, *37*, 1477–1488.
389. S. A. Snyder, P. Westerhoff, Y. Yoon, D. L. Sedlak, *Environ. Eng. Sci.* **2003**, *20*, 449–469.
390. C. Y. Yin, M. K. Aroua, W. Daud, *Sep. Purif. Technol.* **2007**, *52*, 403–415.
391. A. Braun, *J. Environ. Monitor.* **2005**, *7*, 1059–1065.
392. H. P. Boehm, *Carbon* **2002**, *40*, 145–149.
393. M. Voll, H. P. Boehm, *Carbon* **1971**, *9*, 481–488.
394. H. P. Boehm, W. Heck, R. Sappok, E. Diehl, *Angew. Chem. Int. Ed.* **1964**, *3*, 669–677.
395. D. C. Sorescu, K. D. Jordan, P. Avouris, *J. Phys. Chem. B* **2001**, *105*, 11227–11232.
396. M. Besson, A. Blackburn, P. Gallezot, O. Kozynchenko, A. Pigamo, S. Tennison, *Top. Catal.* **2000**, *13*, 253–257.
397. A. J. Groszek, S. Pratyka, D. Cot, *Carbon* **1991**, *29*, 821–829.
398. A. J. Groszek, *Carbon* **1987**, *25*, 712–722.
399. H. P. Boehm, M. Voll, *Carbon* **1970**, *8*, 227–240.
400. J. A. Franz, R. Garcia, J. C. Linehan, G. D. Love, C. E. Snape, *Energy Fuels* **1992**, *6*, 598–602.
401. M. Spiro, *Catal. Today* **1990**, *7*, 167–178.
402. R. C. Bansal, N. Bhatia, T. L. Dhami, *Carbon* **1978**, *16*, 65–68.
403. B. R. Puri, R. C. Bansal, *Carbon* **1964**, *1*, 451–455.
404. R. Sappok, H. P. Boehm, *Carbon* **1968**, *6*, 573–&.
405. G. Francz, P. Kania, P. Oelhafen, T. Schedel-Niedrig, R. Schlögl, R. Locher, *Surf. Sci.* **1996**, *365*, 825–830.
406. R. A. Causey, *J. Nuclear Mater.* **2002**, *300*, 91–117.
407. N. Yoshizawa, Y. Yamada, M. Shiraishi, *Carbon* **1993**, *31*, 1049–1055.

408. C. Casiraghi, J. Robertson, A. C. Ferrari, *Materials Today* **2006**, *10*, 44–53.
409. A. von Keudell, W. Jacob, *Prog. Surf. Sci.* **2004**, *76*, 21–54.
410. A. D. Lueking, R. T. Yang, N. M. Rodriguez, R. T. K. Baker, *Langmuir* **2004**, *20*, 714–721.
411. M. Hirscher, M. Becher, *J. Nanosci. Nanotechnol.* **2003**, *3*, 3–17.
412. J. Chen, F. Wu, *Appl. Physics A: Mater. Sci. Process.* **2004**, *78*, 989–994.
413. B. Stöhr, H. P. Boehm, R. Schlögl, *Carbon* **1991**, *29*, 707–720.
414. J. L. Hueso, J. P. Espinos, A. Caballero, J. Cotrino, A. R. Gonzalez-Elipe, *Carbon* **2007**, *45*, 89–96.
415. J. W. Shabaker, G. W. Huber, R. R. Davda, R. D. Cortright, J. A. Dumesic, *Catal. Lett.* **2003**, *88*, 1–8.
416. W. B. Kim, G. J. Rodriguez-Rivera, S. T. Evans, T. Voitl, J. J. Einspahr, P. M. Voyles, J. A. Dumesic, *J. Catal.* **2005**, *235*, 327–332.
417. R. C. Bansal, F. J. Vastola, P. L. Walker, Jr., *J. Colloid Interface Sci.* **1970**, *32*, 187–194.
418. R. C. Bansal, F. J. Vastola, P. L. Walker, *Carbon* **1971**, *9*, 185–192.
419. J. P. Boudou, P. Parent, F. Suarez-Garcia, S. Villar-Rodil, A. Martinez-Alonso, J. M. D. Tascon, *Carbon* **2006**, *44*, 2452–2462.
420. U. Zielke, K. J. Huttinger, W. P. Hoffman, *Carbon* **1996**, *34*, 999–1005.
421. S. Maldonado, S. Morin, K. J. Stevenson, *Carbon* **2006**, *44*, 1429–1437.
422. M. Zuckmantel, R. Kurth, H. Boehm, *Z. Naturforsch.* **1979**, *34*, 188–196.
423. C. H. Chang, *Carbon* **1981**, *19*, 175–186.
424. R. Schlögl, in *Progress in Intercalation Research, Physics and Chemistry of Materials with Low-dimensional Structures*, H. Müller-Warmuth (Ed.), Kluwer, Dordrecht, 1994, pp. 83–176.
425. B. R. Puri, R. C. Bansal, *Carbon* **1966**, *3*, 533–539.
426. W. O. Stacy, W. R. Imperial, P. L. Walker, Jr., *Carbon* **1966**, *4*, 343–352.
427. P. X. Hou, S. Bai, Q. H. Yang, C. Liu, H. M. Cheng, *Carbon* **2002**, *40*, 81–85.
428. Y. J. Chen, M. L. H. Green, J. L. Griffin, J. Hammer, R. M. Lago, S. C. Tsang, *Adv. Mater.* **1996**, *8*, 1012–1015.
429. A. I. Romanenko, O. B. Anikeeva, A. V. Okotrub, L. G. Bulusheva, N. F. Yudanov, C. Dong, Y. Ni, *Physics Solid State* **2002**, *44*, 659–662.
430. N. M. Rodriguez, M. S. Kim, R. T. K. Baker, *J. Phys. Chem.* **1994**, *98*, 13108–13111.
431. M. L. Studebaker, *Rubber Chem. Technol.* **1957**, *30*, 1400–1483.
432. Y. B. Zhang, Y. W. Tan, H. L. Stormer, P. Kim, *Nature* **2005**, *438*, 201–204.
433. K. S. Novoselov, A. K. Geim, S. V. Morozov, D. Jiang, M. I. Katsnelson, I. V. Grigorieva, S. V. Dubonos, A. A. Firsov, *Nature* **2005**, *438*, 197–200.
434. R. Saito, M. Fujita, G. Dresselhaus, M. S. Dresselhaus, *Appl. Physics Lett.* **1992**, *60*, 2204–2206.
435. A. Shimizu, H. Tachikawa, *J. Physics Chem. Solids* **2002**, *63*, 759–763.
436. M. D. Alba, M. A. Castro, S. M. Clarke, A. C. Perdigon, *Solid State Nucl. Magnet. Reson.* **2003**, *23*, 174–181.
437. D. Lennon, J. McNamara, J. R. Phillips, R. M. Ibberson, S. F. Parker, *Phys. Chem. Chem. Phys.* **2000**, *2*, 4447–4451.
438. H. M. Cheng, Q. H. Yang, C. Liu, *Carbon* **2001**, *39*, 1447–1454.
439. P. Albers, A. Karl, J. Mathias, D. K. Ross, S. F. Parker, *Carbon* **2001**, *39*, 1663–1676.
440. R. T. K. Baker, N. Rodriguez, A. Mastalir, U. Wild, R. Schlögl, A. Wootsch, Z. Paal, *J. Phys. Chem. B* **2004**, *108*, 14348–14355.
441. F. Atamny, D. Duff, A. Baiker, *Catal. Lett.* **1995**, *34*, 305–311.
442. U. Hofmann, G. Ohlerich, *Angew. Chemie* **1950**, *62*, 16–21.
443. R. J. J. Jansen, H. Bekkum, *Carbon* **1995**, *33*, 1021–1027.
444. X. Feng, N. Dementev, W. G. Feng, R. Vidic, E. Borguet, *Carbon* **2006**, *44*, 1203–1209.
445. R. Schlögl, H. P. Boehm, *Carbon* **1983**, *21*, 345–358.
446. G. Domazetis, M. Raoarun, B. D. James, J. Liesegang, P. J. Pigram, N. Brack, R. Glaisher, *Energy Fuels* **2006**, *20*, 1556–1564.
447. D. M. Smith, A. R. Chughtai, *Colloids Surfaces A: Physicochem. Eng. Aspects* **1995**, *105*, 47–77.
448. J. H. Lunsford, G. Mestl, *Raman Rev.* **1997**, 1–2.
449. K. V. Romanenko, J. B. D. de la Caillerie, J. Fraissard, T. V. Reshetenko, O. B. Lapina, *Micropor. Mesopor. Mater.* **2005**, *81*, 41–48.
450. J. Conard, H. Estrade, *Carbon* **1976**, *14*, 299–299.
451. P. Lauginie, A. Messaoudi, J. Conard, *Synthetic Metals* **1993**, *56*, 3002–3007.
452. C. Goze-Bac, S. Latil, P. Lauginie, V. Jourdain, J. Conard, L. Duclaux, A. Rubio, P. Bernier, *Carbon* **2002**, *40*, 1825–1842.
453. E. Papirer, S. Li, J. B. Donnet, *Carbon* **1987**, *25*, 243–247.
454. R. Schlögl, in F. Schüth, K. Sing, J. Weitkamp, *Handbook of Porous Solids*, Wiley-VCH, Weinheim, 2002, pp. 1863–1900.
455. M. Voll, H. P. Boehm, *Carbon* **1970**, *8*, 741–752.
456. M. Voll, H. P. Boehm, *Carbon* **1971**, *9*, 473–480.
457. S. Imamura, *Ind. Eng. Chem. Res.* **1999**, *38*, 1743–1753.
458. J. A. Menendez, L. R. Radovic, B. Xia, J. Phillips, *J. Phys. Chem.* **1996**, *100*, 17243–17248.
459. V. A. Garten, D. E. Weiss, *Austr. J. Chem.* **1957**, *10*, 309–328.
460. H. P. Boehm, H. Knoezinger, in *Catalysis Science and Technology*, Vol. 2, J. E. Anderson, M. Boudart (Eds.), Springer, Berlin, 1983.
461. Y. Chun, G. Y. Sheng, C. T. Chiou, B. S. Xing, *Environ. Sci. Technol.* **2004**, *38*, 4649–4655.
462. Z. J. Sui, T. J. Zhao, J. H. Zhou, P. Li, Y. C. Dai, *Chinese J. Catal.* **2005**, *26*, 521–526.
463. Y. N. Yang, Y. Chun, G. Y. Sheng, M. S. Huang, *Langmuir* **2004**, *20*, 6736–6741.
464. H. Benaddi, T. J. Bandosz, J. Jagiello, J. A. Schwarz, J. N. Rouzaud, D. Legras, F. Beguin, *Carbon* **2000**, *38*, 669–674.
465. A. Contescu, M. Vass, C. Contescu, K. Putyera, J. A. Schwarz, *Carbon* **1998**, *36*, 247–258.
466. Y. I. Matatov-Meytal, M. Sheintuch, *Ind. Eng. Chem. Res.* **1998**, *37*, 309–326.
467. T. J. Bandosz, J. Jagiello, J. A. Schwarz, A. Krzyzanowski, *Langmuir* **1996**, *12*, 6480–6486.
468. A. Bacaoui, A. Dahbi, A. Yaacoubi, C. Bennouna, F. J. M. Hodar, J. R. Utrilla, F. C. Marin, C. M. Castilla, *Environ. Sci. Technol.* **2002**, *36*, 3844–3849.
469. M. Baudu, G. Guibaud, D. Raveau, P. Lafrance, *Water Quality Res. J. Canada* **2001**, *36*, 631–657.
470. F. Rodriguez-Reinoso, *Carbon* **1998**, *36*, 159–175.
471. J. N. Barisci, G. G. Wallace, R. H. Baughman, *J. Electroanal. Chem.* **2000**, *488*, 92–98.
472. B. Buczek, A. Swiatkowski, S. Zietek, B. J. Trznadel, *Fuel* **2000**, *79*, 1247–1253.
473. M. V. Lopez-Ramon, F. Stoeckli, C. Moreno-Castilla, F. Carrasco-Marin, *Carbon* **1999**, *37*, 1215–1221.

474. T. J. Bandosz, J. Jagiello, C. Contescu, J. A. Schwarz, *Carbon* **1993**, *31*, 1193–1202.
475. E. A. Ustinov, D. D. Do, V. B. Fenelonov, *Carbon* **2006**, *44*, 653–663.
476. M. Grujicic, G. Cao, A. M. Rao, T. M. Tritt, S. Nayak, *Appl. Surf. Sci.* **2003**, *214*, 289–303.
477. A. Contescu, C. Contescu, K. Putyera, J. A. Schwarz, *Carbon* **1997**, *35*, 83–94.
478. A. S. Arico, A. K. Shukla, K. M. El-Khatib, P. Creti, V. Antonucci, *J. Appl. Electrochem.* **1999**, *29*, 671–676.
479. A. K. Shukla, A. S. Arico, K. M. El-Khatib, H. Kim, P. L. Antonucci, V. Antonucci, *Appl. Surf. Sci.* **1999**, *137*, 20–29.
480. A. S. Arico, A. K. Shukla, H. Kim, S. Park, M. Min, V. Antonucci, *Appl. Surf. Sci.* **2001**, *172*, 33–40.
481. M. K. van der Lee, A. J. van Dillen, J. H. Bitter, K. P. de Jong, *J. Am. Chem. Soc.* **2005**, *127*, 13573–13582.
482. M. L. Toebes, F. F. Prinsloo, J. H. Bitter, A. J. van Dillen, K. P. de Jong, *J. Catal.* **2003**, *214*, 78–87.
483. M. L. Toebes, M. K. van der Lee, L. M. Tang, M. H. H. in't Veld, J. H. Bitter, A. J. van Dillen, K. P. de Jong, *J. Phys. Chem. B* **2004**, *108*, 11611–11619.
484. C. Pham-Huu, N. Keller, V. V. Roddatis, G. Mestl, R. Schlögl, M. J. Ledoux, *Phys. Chem. Chem. Phys.* **2002**, *4*, 514–521.
485. D. S. Su, N. Maksimova, J. J. Delgado, N. Keller, G. Mestl, M. J. Ledoux, R. Schlögl, *Catal. Today* **2005**, *102*, 110–114.
486. J. J. Delgado, R. Vieira, G. Rebmann, D. S. Su, N. Keller, M. J. Ledoux, R. Schlögl, *Carbon* **2006**, *44*, 809–812.
487. C. A. Bessel, K. Laubernds, N. M. Rodriguez, R. T. K. Baker, *J. Phys. Chem. B* **2001**, *105*, 1115–1118.
488. A. S. Arico, P. L. Antonucci, M. Minutoli, N. Giordano, *Carbon* **1989**, *27*, 337–347.
489. J. A. Macia-Agullo, D. Cazorla-Amoros, A. Linares-Solano, U. Wild, D. S. Su, R. Schlögl, *Catal. Today* **2005**, *102*, 248–253.
490. D. T. Clark, H. R. Thomas, *J. Polymer Sci.* **1976**, *14*, 1671–1700.
491. S. Akhter, X. L. Zhou, J. M. White, *Appl. Surf. Sci.* **1989**, *37*, 201–216.
492. E. Desimoni, G. I. Casella, A. Morone, A. M. Salvi, *Surf. Interface Anal.* **1990**, *15*, 627–634.
493. E. Desimoni, G. I. Casella, A. M. Salvi, T. R. I. Cataldi, A. Morone, *Carbon* **1992**, *30*, 527–531.
494. E. Desimoni, G. I. Casella, A. M. Salvi, *Carbon* **1992**, *30*, 521–526.
495. E. Desimoni, G. I. Casella, T. R. I. Cataldi, A. M. Salvi, T. Rotunno, E. Dicroce, *Surf. Interface Anal.* **1992**, *18*, 623–630.
496. H. Werner, D. Herein, J. Blocker, B. Henschke, U. Tegtmeyer, T. Schedelniedrig, M. Keil, A. M. Bradshaw, R. Schlögl, *Chem. Physics Lett.* **1992**, *194*, 62–66.
497. Y. Baer, *J. Electron Spectrosc. Related Phenom.* **1981**, *24*, 95–100.
498. P. M. T. M. van Attekum, G. K. Wertheim, *Phys. Rev. Lett.* **1979**, *43*, 1896–1898.
499. M. C. dos Santos, F. Alvarez, *Phys. Rev. B* **1998**, *58*, 13918–13924.
500. A. Hoffman, I. Gouzman, R. Brener, *Appl. Physics Lett.* **1994**, *64*, 845–847.
501. D. N. Belton, S. J. Schmieg, *J. Vac. Sci. Technol.* **1990**, *A8*, 2353–2362.
502. F. R. McFeely, S. P. Kowalczyk, L. Ley, R. G. Cavell, R. A. Pollak, D. A. Shirley, *Phys. Rev. B* **1974**, *9*, 5268–5278.
503. A. Steiner, U. Falke, *Surf. Interface Anal.* **1995**, *23*, 789–793.
504. P. Denison, F. R. Jones, J. F. Watts, *Surf. Interface Anal.* **1986**, *9*, 431–435.
505. P. Denison, F. R. Jones, J. F. Watts, *Surf. Interface Anal.* **1988**, *12*, 455–460.
506. P. Denison, F. R. Jones, J. F. Watts, *J. Physics D: Appl. Physics* **1987**, *20*, 306–310.
507. L. J. Gerenser, J. F. Elman, M. G. Mason, J. M. Pochan, *Polymer* **1985**, *26*, 1162–1166.
508. J. M. Pochan, L. J. Gerenser, J. F. Elman, *Polymer* **1986**, *27*, 1058–1062.
509. M. Starsinic, Y. Otake, P. L. Walker, Jr., P. C. Painter, *Fuel* **1984**, *63*, 1002–1007.
510. L. S. K. Pang, A. M. Vassallo, M. A. Wilson, *Organic Geochem.* **1990**, *16*, 853–864.
511. S. Niyogi, M. A. Hamon, H. Hu, B. Zhao, P. Bhowmik, R. Sen, M. E. Itkis, R. C. Haddon, *Acc. Chem. Res.* **2002**, *35*, 1105–1113.
512. Y. Yang, Z. G. Lin, *J. Electroanal. Chem.* **1994**, *364*, 23–30.
513. J. B. Donnet, P. Ehrburger, A. Voet, *Carbon* **1972**, *10*, 737–746.
514. V. R. Deitz, J. L. Bitner, *Carbon* **1973**, *11*, 393–402.
515. V. R. Deitz, J. L. Bitner, *Carbon* **1972**, *10*, 145–157.
516. S. D. Razumovskii, V. N. Gorshenev, A. L. Kovarskii, A. M. Kuznetsov, A. N. Shchegolikhin, *Fullerenes Nanotubes Carbon Nanostructures* **2007**, *15*, 53–63.
517. F. Cataldo, O. Ursini, *Fullerenes Nanotubes Carbon Nanostructures* **2007**, *15*, 1–20.
518. A. Stohl, T. Berg, J. F. Burkhart, A. M. Fjaeraa, C. Forster, A. Herber, O. Hov, C. Lunder, W. W. McMillan, S. Oltmans, M. Shiobara, D. Simpson, S. Solberg, K. Stebel, J. Strom, K. Torseth, R. Treffeisen, K. Virkkunen, K. E. Yttri, *Atmos. Chem. Physics* **2007**, *7*, 511–534.
519. M. V. Martin, R. E. Honrath, R. C. Owen, G. Pfister, P. Fialho, F. Barata, *J. Geophys. Res. -Atmospheres* **2006**, *111*.
520. C. Q. Yang, J. R. Simms, *Fuel* **1995**, *74*, 543–548.
521. J. Ryczkowski, S. Pasieczna, *Journal De Physique Iv* **2003**, *109*, 79–88.
522. S. Pasieczna, J. Ryczkowski, T. Borowiecki, K. Stofecki, *Journal De Physique Iv* **2004**, *117*, 41–46.
523. J. L. Zimmerman, R. K. Bradley, C. B. Huffman, R. H. Hauge, J. L. Margrave, *Chem. Mater.* **2000**, *12*, 1361–1366.
524. L. Retailleau, R. Vonarb, V. Perrichon, E. Jean, D. Bianchi, *Energy Fuels* **2004**, *18*, 872–882.
525. D. Bianchi, *Energy Fuels* **2005**, *19*, 1453–1461.
526. M. Labaki, J. F. Lamonier, S. Siffert, F. Wymalski, A. Aboukais, *Thermochim. Acta* **2006**, *443*, 141–146.
527. B. R. Stanmore, J. F. Brilhac, P. Gilot, *Carbon* **2001**, *39*, 2247–2268.
528. T. Kim, G. Liu, M. Boaro, S. I. Lee, J. M. Vohs, R. J. Gorte, O. H. Al-Madhi, B. O. Dabbousi, *J. Power Sources* **2006**, *155*, 231–238.
529. D. Bassi, P. Tosi, R. Schlögl, *J. Vac. Sci. Technol. A: Vacuum Surfaces and Films* **1998**, *16*, 114–122.
530. L. R. Radovic, F. Rodriguez-Reinoso, in *Chemistry and Physics of Carbon*, Vol. 25, 1997, pp. 243–358.
531. N. F. Goldshleger, *Fullerene Sci. Technol.* **2001**, *9*, 255–280.
532. S. van Dommele, K. P. de Jong, J. H. Bitter, *Chem. Commun.* **2006**, 4859–4861.
533. J. D. D. Velasquez, L. A. C. Suarez, J. L. Figueiredo, *Appl. Catal. A: General* **2006**, *311*, 51–57.
534. J. W. Yang, M. T. H. Fonseca, B. List, *Angew. Chem. Int. Ed.* **2004**, *43*, 6660–6662.

2.4
Supported Catalysts

2.4.1
Preparation of Supported Catalysts by Deposition–Precipitation

John W. Geus* and A. Jos van Dillen

2.4.1.1 Introduction

Deposition–precipitation is a technique by which a catalytically active precursor may be applied to a support by precipitation. In order to deposit the precursor exclusively onto the surface of a support suspended in the solution in which the precipitation is executed, the precipitation is conducted such that nucleation in the bulk of the solution is prevented. This procedure enables one to apply an active precursor finely divided on the surface of a support, while the initially dissolved precursor is present in a solution the volume of which considerably exceeds the pore volume of the support. In this way, loadings of the active precursor significantly higher than would be possible by impregnation with a solution of an active precursor and drying can be achieved, while a small size of active particles (and hence a high active surface area per unit volume) is established. However, whilst impregnation and drying can be carried out with relatively large preshaped support bodies, a sufficiently rapid transport of the precipitating active precursor from the solution to the surface of the support usually calls for support bodies not larger than 2 mm. Although deposition–precipitation from a homogeneous solution is most suitable to effect precipitation exclusively on the surface of the support, the use of an inhomogeneous solution can be technically attractive and also provide excellent results. Whereas, originally, deposition–precipitation was developed to apply high loadings of a catalytically active precursor finely divided on the surface of a support, it has been employed more recently to deposit low loadings of extremely small gold or precious metal particles on the surface of supports such as titania and ceria.

Results obtained during the past few years have indicated the possibility of employing deposition–precipitation to control the reaction of the active precursor with the support. In this way, it is possible to control not only the distribution and size of the catalytically active particles, but also the porous structure of the catalyst. In addition, deposition–precipitation of a component that is not catalytically active can be used to produce new carriers for catalytically active components, the porous structure of which can be manipulated on a very small scale. These new supports are stable in alkaline solutions of a moderate strength – a property that at present is exhibited only by activated carbon and supports such as titania or zirconia, the porous and surface areas of which are difficult to control.

This chapter provides a general discussion of the properties and production of supported catalysts, as well as a brief survey of the theory of nucleation and growth of precipitating solids. Subsequently, the interaction of nuclei of precipitates with the surface of a solid suspended in solution, and the effect of such interaction on the precipitation process, will be discussed. An overview of deposition–precipitation procedures, together with some experimental results obtained with different procedures, will also be provided. The subject of the deposition of small gold particles onto supports such as ceria and titania is outlined, in addition to electrochemical deposition–precipitation procedures, and subsequently deposition–precipitation with pre-shaped support bodies. Finally, the new possibilities offered by controlling the reaction of either an active precursor or a more cheap inactive component during deposition–precipitation with silica will be discussed in detail.

2.4.1.2 Supported Catalysts

The activity of solid catalysts is usually proportional to the active surface area per unit volume of catalyst, provided that transport limitations are not present. A high activity per unit volume calls for small particles. As most active species sinter rapidly at the temperatures at which the thermal pretreatment and the catalytic reaction proceed [1], small particles of the active species alone generally do not provide thermostable, highly active catalysts. To arrive at solid catalysts of the desired shape, mechanical strength, porous structure, activity, and thermal stability, two different materials – viz., the support and the active material – provide the different functions that the catalyst must fulfill. The support, which is usually highly thermostable, furnishes the shape, mechanical strength and porous structure, while the catalytic activity and selectivity are due to the active component(s). As indicated in Fig. 1, the sintering of small particles alone leads to a low active surface area, whereas application of the active component on a support can stabilize the active surface area.

In view of the favorable bulk density, price and thermal stability, the most frequently used support is γ-alumina. Other supports often employed are silica and activated carbon. Silica has a lower bulk density and is more liable to sintering at temperatures above about 900 K than alumina. The volatility, as $Si(OH)_4$, with steam of elevated pressure, also limits the applicability of silica. Activated carbon supports usually offer very high surface areas, and in an environment which is inert to carbon

---
* Corresponding author.

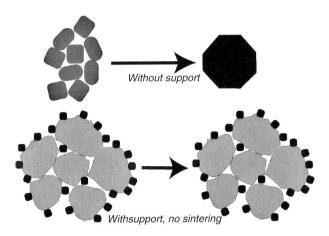

Fig. 1 Top: Rapid sintering of unsupported active particles. Bottom: Supported thermostable active material.

the thermal stability is extremely high. In an oxidizing environment, however, activated carbon will rapidly be oxidized; after application of most active precursors at lower temperatures than activated carbon alone. With some catalytically active components, reaction of the activated carbon with hydrogen to methane and with carbon dioxide to carbon monoxide can proceed.

Depending upon the nature of the active component, two different types of supported catalyst may be distinguished [2]. When the active component is very expensive (as with precious metals), the objective is a maximum active surface area per unit weight of the active component. With less costly active components, such as base metals or base metal oxides or sulfides, a maximum active surface area per unit volume is most attractive. The reason for this is that a high active surface area causes a relatively small catalytic reactor volume to be sufficient to achieve a technically satisfactory conversion. When the costs of the reactor dominate, a high loading of the support maximizing the active surface area per unit volume will therefore be most favorable. When, on the other hand, the costs of the active component are most important, the maximum active surface area per unit weight of active component is desired. Expensive active components, therefore, will be applied to supports at low loadings, leading to small or very small supported active particles and thus providing a high active surface area per unit weight of active component. As the support is diluting the active component, much higher loadings of the support are generally utilized with base metals or base metal compounds. However, the mobility of the atoms of the active species over the surface of the active component is usually high under the conditions of the catalytic reaction or the thermal pretreatment of the catalyst. As a result, contacting active particles will rapidly decrease their surface area by sintering, which can

lead at high loadings to very large bodies of the active component(s) and the exposure of a small active surface area. Contact between the supported active particles must consequently be limited as much as possible. An even, dense distribution of small particles of (the precursor) of the active component over the surface of the support is therefore desired with catalysts containing base metal (compounds) as the active component. Figure 2 shows, schematically, a supported precious metal catalyst and a catalyst in which a base metal (compound) has been applied at a high loading on a support.

When the surface of the support does not react with the active precursor and, hence, is not affected by the active precursor, the maximum active surface area that can be achieved is of the same order of magnitude as the surface area of the support. A surface of the support thus covered with the active component that the active particles are not quite touching will provide an active surface area approaching that of the support. Usually, such coverage calls for small particles of the active component, although the activity per unit surface area of the active component may vary with the size of the supported active moieties. When the activity per unit surface area drops with the size of the active species, the specific activity may be so low that the resulting activity is disappointing, in spite of a high catalytically active surface area per unit volume of catalyst.

Reactions for which the surface structure of the catalytically active component is affecting the activity and/or the selectivity are known as structure-sensitive or demanding [3]. With structure insensitive or facile reactions, the structure of the surface of the active component does not affect the catalytic properties. Small particles may not be favorable when the reaction is structure-sensitive, and larger particles may exhibit a higher activity and/or selectivity per unit surface area. Although the active surface area is lower, the higher

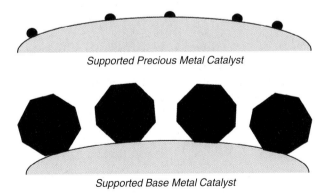

Fig. 2 Top: Characteristic supported precious metal catalyst. Bottom: Characteristic supported base metal catalyst.

References see page 465

activity per unit surface area of the active phase results in a higher overall activity. An example of this is supported platinum, when employed in oxidation reactions. Those platinum particles which are less than 3 nm in size disintegrate on reaction with oxygen, which means that the platinum–platinum distances disappear from the extended X-ray absorption fine structure (EXAFS) spectrum. The evidence indicates that small platinum particles react with oxygen to a platinum oxide in which the oxygen atoms are relatively strongly bonded, whereas larger platinum particles merely adsorb oxygen dissociatively on the surface. The more loosely bonded oxygen exhibits a much higher oxidation activity. With the selectivity of a catalytic reaction, the effect of the size of the active moieties may be even stronger, and may cause small supported active particles to be very unfavorable.

The most favorable size for supported active particles has only seldom been established, the reason being that it is difficult to vary systematically the size of the active particles deposited onto a support while maintaining a narrow particle size distribution. It is, consequently, desirable to apply the active component(s) uniformly and densely distributed over the surface of the support as particles the size of which can be controlled. This is one of the main goals when preparing supported catalysts by deposition–precipitation. A support surface which is very heavily loaded with nickel particles is shown in Fig. 3 (in fact, such loading of a support is difficult to achieve).

2.4.1.3 Production of Supported Catalysts

The procedures used to produce supported catalysts can be divided into two main groups: (i) the selective removal of one or more component out of usually non-porous bodies of a compound containing precursors of the support and the active component(s); and (ii) the application of (a precursor of) the active component(s) onto a separately produced support [2]. Both procedures are carried out extensively on a technical scale.

2.4.1.3.1 Selective Removal of One or More Component

With highly loaded supports, the first-mentioned procedure – selective removal – is often employed. One of the most successful examples is the technical ammonia synthesis catalyst, which is produced from magnetite (Fe_3O_4) containing some (up to about 4 wt.%) alumina [4]. Selective removal of the oxygen of the magnetite by reduction leads to a highly porous material, in which metallic iron particles of about 40 nm are separated by the remaining alumina. In the reduced catalyst, loading with the active component is no less than 96–98 wt.%, while the mechanical strength is extremely high. Copper-based catalysts are employed in methanol synthesis and the low-temperature CO shift conversion. The selective removal of water, carbon dioxide, and oxygen bonded to the copper from a coprecipitate of copper(II), aluminum(III), and zinc(II) is the procedure used to produce the catalysts. A final example is the removal of water and carbon dioxide from mixed oxalates, as with copper–magnesium or nickel–magnesium oxalates [5]. As shown in Fig. 4, such selective removal can result in a porous body (as with the ammonia synthesis catalyst) or in a powder of the support loaded with the active component (as with mixed oxalate precursors). Ultimately, the powder must be processed to bodies of the desired shape and dimensions.

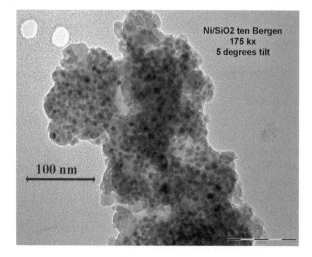 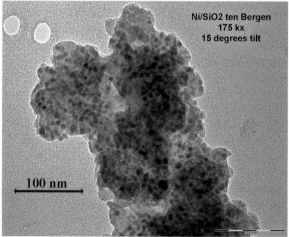

Fig. 3 A supported, highly loaded base metal catalyst containing small silica-supported nickel particles. As the silica does not display diffraction, the contrast is due to diffraction by the nickel particles. The orientation of the nickel particles with respect to the electron beam determines whether an individual particle is in a diffraction position. Upon tilting, the specimen nickel particles change, reversibly, their contrast from high (black) to low (gray). (Original magnification, ×175.)

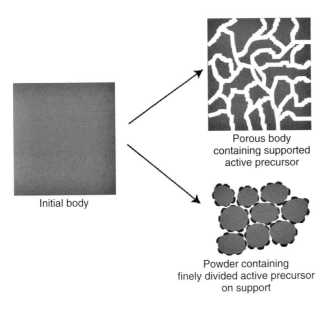

Fig. 4 Production of supported catalysts by selective removal of components from non-porous precursor leading to either porous bodies (top) or powder (bottom) of a supported active component.

Although when using the selective removal procedure the active component and support are usually intimately mixed, it is difficult to control the porous structure and/or the mechanical strength of the catalyst bodies that result after selective removal and the usually required shaping. Nonetheless, the procedure is difficult to beat for the production of highly loaded supports. The most well-known example of selective removal is that of Raney nickel (see Chapter 2.3.2), where aluminum is selectively removed from a nickel–aluminum alloy by dissolution of the aluminum in alkali. However, Raney metals do not involve a support, as pretreatment at elevated temperatures is not required to produce metallic nickel.

2.4.1.3.2 Application on Separately Produced Supports

Contacting with a solution of an active precursor and drying, precipitation of an active precursor in the presence of a support suspended in the solution, and decomposition of a gaseous active precursor on the surface of a support are procedures used to load separately produced supports. For the rapid development of a technical catalyst, starting from pre-shaped support bodies is the obvious route to take, as such bodies of the desired shape, size, surface area, porous structure, and mechanical properties are usually commercially available. As transport of the active precursor through the pores of pre-shaped support bodies can proceed too slowly to achieve an even distribution of the precursor over the surface of the support, preparation methods (such as precipitation) call for special measures. The most generally used preparations with separately produced supports are impregnation and drying of pre-shaped support bodies, and precipitation with suspended powdered support bodies, drying, and shaping.

A Impregnation and Drying of Supports The impregnation of a separately produced support with a solution of an active precursor can be executed with pre-shaped support bodies, or with powdered supports. Such impregnation can involve dissolved precursors that do not interact with the support, or that are removed from the solution by chemical interaction with the surface of the support. One technically attractive approach is that of incipient wetness or pore-volume impregnation and drying of pre-shaped support bodies. With this method waste water is not produced, and there is no risk of loss of active precursor. With precious metals, impregnation and drying of pre-shaped support bodies is therefore the procedure of choice.

With a precursor adsorbing onto the surface of the support, the distribution of active material within the support body can be controlled. The loadings that can be achieved with adsorbing active precursors are usually small, and hence procedures to apply adsorbing precious metal precursors onto the surface of supports are often employed. With loadings not sufficient to cover the surface of the support completely, impregnation with a solution of a precursor adsorbing on the surface of the support leads to deposition of the active component at the external edge of the support body (egg-shell). The addition of a more strongly adsorbing species results in deposition of the active precursor in more interior parts of the support bodies. An example is the application of platinum from a hexachloroplatinum acid solution into alumina support bodies. The addition of dibasic organic acids (e.g., oxalic acid) shifts the deposition of platinum from the edge into the alumina bodies (egg white). The impregnation of a powdered support with a solution of an adsorbing precursor of a volume larger than the pore volume of the support leads more readily to a homogeneous distribution of the active material throughout the support. Filtration or centrifugation separates the thus-loaded support from the remaining liquid.

With an active precursor adsorbing onto the surface of the support, the loading of the support is relatively low. Impregnation with a precursor that does not adsorb on the surface of the support can lead to higher loadings. Removal of the liquid by evaporation raises the concentration of the dissolved precursor and leads to crystallization of the precursor. Although a uniform filling of the pores of a support with the impregnated solution of the precursor can be easily achieved, it is difficult to maintain the uniform distribution when the

References see page 465

liquid is removed by drying. Evaporation of the liquid proceeds preferentially at the external edge of the support bodies, while capillary flow readily replenishes the liquid which is volatilized at the external edge. Catalyst supports generally do not contain cylindrical pores; rather, the pores in support bodies are much more irregularly shaped. The usual stacking of the elementary particles in support bodies brings about that, after emptying of the pore mouths, the liquid still flows as a film which connects the interstices of a small diameter in which the liquid is capillary-condensed to the edge of the support bodies. The small-diameter interstices are present between small interconnected particles of the support, and transport of the liquid to the external edge is much more rapid than that of water vapor to the edge. As a result, impregnation with a solution of a non-adsorbing precursor and drying leads generally to deposition of the active precursor exclusively on the external edge of the support bodies, as represented in Fig. 5. The deposition of an active precursor by drying impregnated support bodies occurs more uniformly within the support bodies with a bimodal pore size distribution, with both narrow and wider pores. After removal of the liquid film from the support surface by drying, the liquid remains present in the narrow pores. The water vapor is then transported through the pore system of the support to the gas flow, while the precursor dissolved in the remaining liquid crystallizes within the narrow pores.

To prevent migration of the liquid during drying of an impregnated solution of the active precursor, a solution has to be employed that interacts sufficiently strongly with the surface of the support and the viscosity of which rises during evaporation of the liquid. Incipient wetness impregnation of solutions of citric acid complexes of active precursors have proven to be very effective in achieving uniform distributions of active components within support bodies after drying. Also addition of hexaethyl cellulose and sugar appeared to be very effective in establishing uniform distributions. However, careful execution of the calcination of the loaded and dried supports is required to avoid formation of explosive mixtures particularly with nitrates.

Nevertheless, high loadings can often not be achieved by pore-volume impregnation, as the required amount of active precursor cannot be dissolved in a liquid volume equal to the pore-volume of the support. Contacting pre-shaped support bodies with a liquid volume which is larger than the pore-volume, and subsequent drying, leads to the preferential deposition of a significant fraction of the precursor on the external edge of the support bodies. Reasonable results can be achieved with a powdered, finely divided support that is continuously and intensively kneaded during evaporation of the liquid, but the distribution of the active component in the dried support is generally not homogeneous.

B Precipitation onto a Suspended Support Precipitation of an active precursor onto a support suspended in a solution of the precursor can provide high loadings, as the compound(s) dissolved in a liquid volume that is large compared to the support pore-volume, are concentrated onto the support. Besides coprecipitation of the precursors of the support and the active material and subsequent selective removal of some components, precipitation in the presence of a suspended support is also often carried out to achieve high loadings of the support. The apparatus and procedures do not differ significantly, but the precipitation usually proceeds where the precipitating liquid enters the suspension, and the precipitant necessarily does not enter the liquid present in the porous conglomerates of the support. When nucleation and growth of the precipitate of the precursor are rapid, large crystallites of the precursor result. With a rapid nucleation and slow growth (which is usually encountered with poorly soluble compounds), clusters of small particles of the active precursor outside the pore system of the support are often obtained. At high concentrations, at a point where the precipitant enters the suspension, the small particles rapidly and irreversibly flocculate, which leads to the clustering of small particles. Catalysts prepared in this way are liable to rapid deactivation during pretreatment or use at elevated temperatures, as the small elementary particles of the precursor are intimately connected and therefore sinter readily. A cluster which is often found with

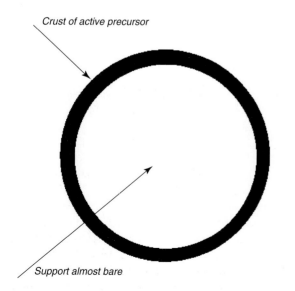

Fig. 5 Unfavorable distribution of active precursor that often results from impregnation and drying.

supported catalysts prepared by precipitation is shown, schematically, in Fig. 6.

The above discussion has indicated that precipitation onto a suspended support is often required to produce catalysts in which the support is highly loaded. Such supports are especially desirable with catalysts containing (compounds of) base metals, as mentioned above. In order to effect precipitation within the porous system of a suspended support, the precipitation process must first be modified so as to enable the still-soluble active precursor to migrate into the pores of the support. As the diffusion of colloidal particles is fairly slow, it is necessary to prevent the formation of colloidal particles of active precursor outside the pore system of the support. In order to determine the conditions needed to achieve transport into the pores of a suspended support, it is necessary to consider the nucleation of insoluble solids within (aqueous) solutions.

2.4.1.4 The Theory of Nucleation [13, 14]

Here, we will present only a simple review of the theory of nucleation and precipitation to assess the conditions needed to achieve precipitation exclusively on the support surface. First, we consider the formation of a spherical nucleus in the bulk of the solution. The change in the free enthalpy upon formation of the above nucleus is

$$\Delta G_{\text{tot}} = \Delta G_{\text{v}} + \Delta G_{\text{s}}$$

$$\Delta G_{\text{v}} = \frac{4}{3}\pi r^3 \frac{\rho}{M}(\mu_{\text{s}} - \mu_{\text{l}}) = \frac{4}{3}\pi r^3 \Delta v_{\text{sl}}$$

$$\Delta G_{\text{s}} = 4\pi r^2 \gamma$$

$$\Delta G_{\text{tot}} = \frac{4}{3}\pi r^3 \Delta v_{\text{sl}} + 4\pi r^2 \gamma$$

where ΔG_{v} is the volume change in free enthalpy, ΔG_{s} is the interfacial free enthalpy of the precipitating solid with the solution, $2r$ is the diameter of the nucleus of the precipitate, μ_{s} the molecular free enthalpy of the solid precipitate, and μ_{l} of the dissolved solid, while γ is the interfacial energy. Δv_{sl}, the difference in molecular free enthalpy per unit volume of solid between the solid and the dissolved compound, is negative with the bulk solid being stable.

A maximum in the free enthalpy is obtained when

$$\frac{d\Delta G_{\text{tot}}}{dr} = 0$$

$$\frac{d\Delta G_{\text{tot}}}{dr} = 4\pi r^2 \Delta v_{\text{ls}} + 8\pi r \gamma$$

$$r_C = \frac{2\gamma}{\Delta v_{\text{ls}}}$$

Note that in the latter expressions the negative quantity Δv_{sl} has been changed for Δv_{ls}, a positive value. When the size, $2r$, of the nucleus is greater than $2r_c$, the critical size for nucleation, the free enthalpy drops when the size of the nucleus rises. The critical size of the nucleus corresponds to the maximum in free enthalpy that must be crossed to arrive at a stable nucleus. Δv_{ls} strongly increases with the activity, and thus with the concentration of the dissolved compound. The size of the critical nucleus therefore falls rapidly at growing concentration of the dissolved compound. The change in free enthalpy upon formation of a critical nucleus is

$$\Delta G_{\text{tot}} = \frac{16}{3}\pi \frac{\gamma^3}{(\Delta v_{\text{ls}})^2}$$

The rate of nucleation is generally assumed to be proportional to the incident rate of elementary moieties of the crystallizing species onto the critical nuclei. Hence, the rate of nucleation is proportional to the concentration of critical nuclei, which is proportional to

$$\text{Exp}\left(-\frac{\Delta G_{\text{tot}}}{RT}\right)$$

and provided that the critical nucleus is stable and, hence,

$$r\Delta v_{\text{ls}} \geq 2\gamma$$

proportional to

$$\text{Exp}\left(-\frac{16}{3}\pi \frac{\gamma^3}{\Delta v_{\text{ls}}^2 RT}\right)$$

Insertion of the earlier-derived expression leads to the latter relationship, to which the density of critical nuclei and hence the rate of nucleation is proportional. It can be seen that a higher activity, and thus a higher concentration of the precipitating active precursor, which leads to a

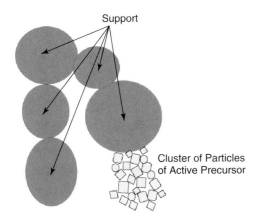

Fig. 6 Clusters of particles of active precursor within the support.

higher value of $\Delta\mu_{ls}$, causes the concentration of critical nuclei to increase exponentially. At a higher concentration the rate of nucleation thus rises sharply.

To account for the interaction with the support, we consider a hemispherical nucleus attached to the support. The change in free enthalpy upon crystallization to cap-shape nuclei on the support is

$$\Delta G_{tot} = \frac{2}{3}\pi r^3 \Delta v_{ls} + 2\pi r^2 \gamma + \pi r^2 \gamma_{ss}$$

where γ_{ss} is the interfacial energy with the support. The above equations indicate that interaction with the support causing a low value of γ_{ss} can decrease considerably the change in free enthalpy of the formation of a nucleus and hence the concentration of the dissolved active precursor at which a substantial rate of crystallization is exhibited.

We can survey the precipitation in the usual equilibrium diagram, in which the concentration and temperature of the saturated solutions are represented. In Fig. 7, the solubility of a solid as a function of the temperature is shown graphically, where it can be seen that, at increasing temperatures, the solubility increases. Since the rate of crystallization or precipitation increases rapidly over a narrow range of concentrations, as a function of the temperature, a concentration can be indicated where the precipitation abruptly becomes measurable. The plot of the above concentrations is referred to as the "supersolubility curve", and this is also represented in Fig. 7. It is important to note that, at concentrations between the solubility and supersolubility curves, precipitation within the bulk of the solution does not proceed, although large crystallites of the precursor are stable in the solution.

When the nuclei of the precursor to be precipitated interact significantly with the surface of the support, the rate of precipitation is measurable at much lower concentrations. Accordingly, it is possible to perform precipitation exclusively on the surface of the support by maintaining the concentration of the precursor between that of the solubility and supersolubility curves. Control of the concentration of catalytically active precursors within the above range is at the basis of the procedure of deposition–precipitation. With sparingly soluble solids, the concentration difference between the solubility and supersolubility curves is small, and consequently the concentration must be controlled fairly accurately.

The case where the active precursor reacts with the surface of the support to a compound of a lower solubility is shown in Fig. 8, where both the solubility and supersolubility curves have been shifted to lower concentrations. We will show that reaction to a compound with the support often takes place.

With the usual addition of a precipitant to a solution (as shown schematically in Fig. 9), the local concentration

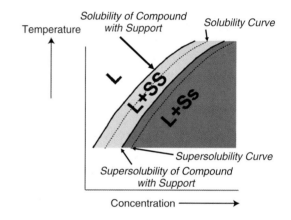

Fig. 8 Reaction of precipitating active precursor with support to a less-soluble compound. SS = large particles of precipitate interacting with the support stable; Ss = small particles of precipitate interacting with the support stable.

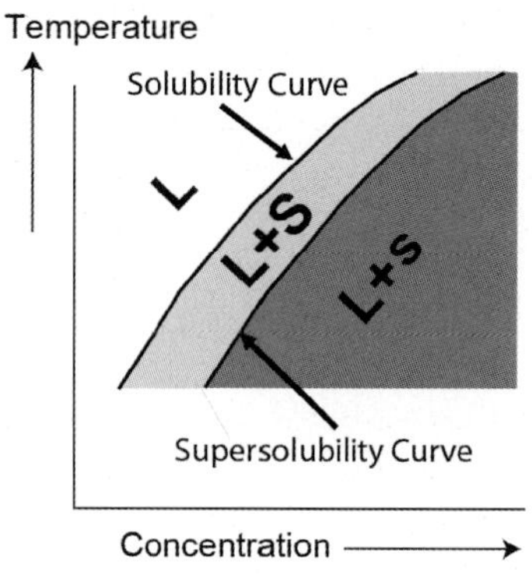

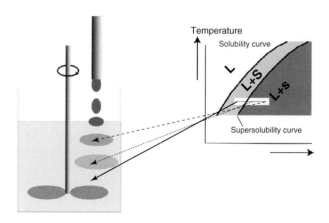

Fig. 7 Solubility plot indicating the solubility and supersolubility curves. S = large precipitated particles stable; s = small precipitated particles stable.

Fig. 9 Usual addition of precipitant to a suspension of a support in a solution of active precursor; rapid nucleation occurs outside the porous support bodies.

rises temporarily above that of the supersolubility curve, and this causes rapid nucleation of the precipitate. When, after homogenizing the liquid, the final concentration is lower than that of the supersolubility curve, nucleation has already proceeded and the crystallites precipitated are stable. Consequently, the concentration must be maintained continuously below that of the supersolubility curve. With deposition–precipitation, this is achieved by performing precipitation from a homogeneous solution [15–17]. A homogeneous solution is most easily obtained by separating the mixing and generation of the precipitant; this is usually brought about by carrying out the mixing at such a low temperature that the reaction generating the precipitant does not proceed markedly; thereafter, the temperature is raised to a level where the precipitant generation is rapid.

2.4.1.5 Deposition–Precipitation on Suspended Supports

2.4.1.5.1 Survey of Deposition–Precipitation Procedures

With deposition–precipitation, the interaction of nuclei of an insoluble active precursor with the surface of the suspended support is of decisive importance. First, the sign of the electrostatic charge of the support is important, as electrostatic repulsion usually prohibits the approach of ions or colloidal particles of a charge of the same sign as the surface. However, an electrostatic charge of an opposite sign of the species to be deposited is not sufficient to result in deposition–precipitation of the active precursor. Usually, a chemical interaction with the surface of the support is a prerequisite for deposition–precipitation.

Changing the pH level of a solution of an active precursor, changing the valency of a dissolved active precursor, or decreasing the concentration of a compound forming soluble complexes with the active precursor are procedures that can be used in deposition–precipitation. As many catalytically active precursors can be precipitated as hydroxides or basic salts, many deposition–precipitation procedures involve an increase in pH level of a suspension of the support in a solution of the active precursor. As a significant number of catalytically important metal compounds dissolve as oxy-anions, raising the pH level cannot be employed to bring about the precipitation of such metal ions, and consequently a procedure has been developed first to reduce the metal in the oxy-anions to a lower valency. The metal ions of a lower valency exhibit a basic behavior and are thus soluble at low pH levels and precipitate at high pH levels. With vanadium and molybdenum the procedure has provided very satisfactory results, with the reduction being performed most smoothly electrochemically. However, reduction by oxalate ions, for example, is also possible. Previous reduction of the oxy-anions allows one to apply the metal ions on negatively charged supports.

Deposition–precipitation by changing the valence of a dissolved active precursor involves either an increase in valency, such as from iron(II), which is more soluble, to iron(III), which is much less soluble, or a decrease in valency (e.g., chromate to chromium(III)), or reduction of a soluble species to the corresponding metal. Reduction to the metal is attractive with the deposition of noble or precious metals on suspended supports. Decreasing the concentration of a complexing agent is performed in a solution in which the (hydrated) metal ion alone is insoluble. Deposition–precipitation by decomplexing involves particularly the decomposition of soluble ammonia complexes. Ammonia is homogeneously removed by volatilization from a suspension of the support at elevated temperatures. The removal of an organic complexing agent by oxidation has also been effectively employed.

The best means of avoiding locally high concentrations of the precipitating agent is to separate the mixing of the different constituents of the solution and the generation of the precipitating agent. This separation of mixing and making available the precipitating agent is performed with precipitation from a homogeneous solution. Generally, the precipitating agent is generated by raising the temperature of a suspension of the support homogenized completely at a lower temperature. The procedures to execute deposition–precipitation on a technical scale were invented and developed during the 1960s. Indeed, the U.S. patent 4,113,658 of September 12, 1978 deals extensively with the different procedures and provides a large number of accurately described examples [18].

2.4.1.5.2 Deposition–Precipitation by Changing the pH

A Limited Subsequent Reaction of the Precipitated Precursor Raising the pH of the suspension of a support in a solution of the active precursor appears to be a fairly straightforward procedure with precursors such as manganese(II), nickel, and cobalt. The interaction of the precipitating species with the suspended support can be readily ascertained by recording the course of the pH during precipitation, with and without a suspended support. A sufficient interaction with the support is apparent from the precipitation with the suspended support to set on at significantly lower pH than without the support.

It has been found that the injection of an alkaline or acid solution below the level of the liquid of a suspension of the support can avoid concentrations so high that significant precipitation can proceed in the bulk of the solution. The reason for this is that high shear stresses can be established around the end of the injection tube,

References see page 465

which is not possible at the surface of the suspension, where the precipitant enters at a gas–liquid interface. The rate of injection appears to be very important, with a continuous slow flow being preferred as a rapid flow leads to elements of the injected solution ending up in regions in the suspension where no considerable shear stresses are present. The resulting locally high concentration leads to precipitation in the bulk of the solution. Although the scaling-up of procedures involving the mixing of liquids is normally difficult, scale-up of the injection procedure has been achieved successfully by performing the injection through a type of shower-head positioned beneath the surface of the suspension, close to a stirrer. In this way, a technically acceptable rate of injection can be established with a slow rate per injection point. Usually, a special stirrer is mounted in the precipitation vessel close to the injection point(s), while another stirrer is used to keep the suspension homogeneous. The precipitation has also been executed by injection of the alkaline or acid solution into a small vessel, the contents of which are vigorously agitated, while the suspension present in a large vessel is stirred less vigorously and recirculated through the small vessel.

The most interesting approach is to change the pH within a homogeneous solution; this readily avoids locally high concentrations and provides an embodiment, which can be scaled-up very smoothly. One of the best ways to raise the pH homogeneously is to use the hydrolysis of compounds, such as urea. The reason for this is that the mixing procedure is executed separately from the generation of hydroxyl ions. Urea can be mixed homogeneously with a suspension of the support in a solution of the active precursor at room temperature, and reacts only at temperatures above about 330 K according to:

$$CO(NH_2)_2 \rightleftharpoons NH_4^+ + CNO^-$$
$$CNO^- + 3H_2O \rightleftharpoons NH_4^+ + 2OH^- + CO_2$$

The rate of hydrolysis of urea, which is first order with respect to the concentration of urea, does not vary with the pH.

It is important to note that an early patent allocated to I.G. Farben with Stöwenert as inventor [19] mentions the use of urea for the production of supported catalysts. However, the I.G. Farben patent does not involve deposition–precipitation; rather, it refers to the impregnation of supports with a solution of urea or a range of analogously reacting compounds and an active precursor and raising the temperature in a single step to levels where the urea decomposes, the liquid volatilizes, and the support and the active precursor are calcined. The German patent furthermore does not identify the fact that precipitation from a homogeneous solution generally leads to large particles of a precipitate, and that interaction with the surface of the solid on which the precipitate should be deposited is a prerequisite to obtain the small particles required for an active catalyst. Instances where the hydrolysis of urea to increase the pH does not lead to deposition–precipitation include the suspension of silica in an iron(III) or aluminum solution. Stöwenert mentioned impregnation but did not recognize the possibility of employing a large liquid volume compared to the pore volume of the suspended support to arrive at highly loaded supports.

In order to account for the volatilization of ammonia, the amount of urea utilized is normally chosen as 1.5- to 2.5-fold the amount required stoichiometrically. As mentioned above, the urea first reacts to cyanate and subsequently to ammonium, carbon dioxide, and hydroxyl groups. It is the first of the above reactions that requires a temperature of at least 340 K, because ammonium cyanate reacts rapidly even at room temperature. When the precipitation is to be performed at lower temperatures, ammonium cyanate or a cyanate of an alkali metal may be used, but mixing must be conducted with the suspension temperature maintained at about 273 K to prevent the reaction from occurring during the mixing stage. Attempts have also been made to use an enzyme, urease, to hydrolyze the urea at low temperatures, but the enzyme was inhibited by the presence of dissolved, catalytically active precursors.

When high loadings of the support are to be applied with metal ions that can form soluble ammonia complexes, the formation of ammonium ions may cause problems. Examples of ions which form soluble ammonia complexes include nickel, cobalt, iron(II), and copper. At the higher pH levels needed to grow the metal hydroxides (which, due to surface loading of the support are no longer stabilized by interaction with the support), the ammonia concentration may become relatively high. As a result, the metal ions will remain dissolved as ammonia complexes. The ammonia can be volatilized by keeping the suspension for a prolonged period at a temperature close to 373 K.

To achieve high loadings of the support, therefore, it can be advantageous to use sodium nitrite, which does not lead to the formation of ammonium. Sodium nitrite reacts according to

$$3NO_2^- + H_2O \rightleftharpoons 2NO + NO_3^- + 2OH^-$$

Here, oxygen (air) must be excluded as oxidation of the NO released leads to the generation of hydrogen ions according to

$$2NO + O_2 \rightleftharpoons 2NO_2$$
$$(2/3)(3NO_2 + H_2O \rightleftharpoons NO + 2NO_3^- + 2H^+)$$

Consequently, the pH does change very much if the NO released is oxidized by oxygen.

One problem with using urea, cyanate and nitrite may be the production of waste water containing chemically bonded nitrogen, as this is difficult to dispose of. The NO released with the disproportionation of nitrite also calls for special measures to avoid emission. Another problem associated with the technical production of catalysts using the hydrolysis of urea is the relatively long time needed for such hydrolysis to occur. However, the reaction can be greatly accelerated by conducting it at (slightly) elevated pressures and temperatures above 373 K (as reported initially by Unilever). Since the suspension of the support can be thoroughly mixed at a temperature where the reaction does not proceed, highly concentrated solutions and correspondingly large amounts of highly viscous suspensions of support can be used with urea. Subsequently, the production of supported catalysts using urea hydrolysis has been scaled-up from volumes of approximately 1 L to more than 3 m^3. Notably, the loaded supports produced did not differ in terms of the scale of production.

Interaction with the support is prerequisite to achieving a dense, uniform coverage of the surface of the support with the precursor of the active component. Precipitation from a homogeneous solution without interaction with the support usually leads to relatively large precipitated crystallites. Interaction with the support can easily be assessed by measuring either the pH as a function of time or the amount of alkali injected, with and without a suspended support. If significant interaction with the support is found, the pH with a suspended support remains considerably below that measured without a suspended support. Without a support, the concentration of the supersolubility curve is often apparent by the pH curve passing through a maximum. When the concentration of the supersolubility curve has been attained, nucleation proceeds rapidly and subsequent growth of the nuclei proceeds faster than the generation or addition of hydroxyl ions.

Figure 10 shows the course of pH changes during the injection of alkali (NaOH) into a nickel nitrate solution kept at 293 K [20]. It can be seen that, without a suspended alumina support, the pH passes through the maximum, as mentioned above, but when alumina is suspended in the solution the pH remains below the level measured with nickel nitrate alone. At a low loading of the support (e.g., with a loading of 5% in Fig. 10), the pH remains considerably below the level found with only nickel nitrate. This low pH level indicates a reaction of the surface of the suspended alumina support with nickel(II) ions to a hydrotalcite structure. At higher loadings of the alumina support, the nickel ions can no longer be accommodated completely in the nickel–aluminum hydrotalcite formed

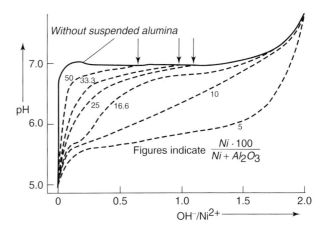

Fig. 10 Precipitation of nickel(II) from a homogeneous solution by injection of sodium hydroxide at 293 K. Measurements with and without suspended alumina are represented. The values on the curves indicate the final nickel loadings.

at the alumina surface, and nickel hydroxide grows, as indicated by the pH curve merging with the curve measured with nickel nitrate alone. In Fig. 10, vertical arrows indicate the points where the curves measured with suspended alumina (broken lines) reach the curve measured with nickel nitrate only (solid line). The reactivity of γ-alumina is known to be high. Accordingly, a patent application mentions the reaction of γ-alumina with a solution of magnesium acetate at 338 K to the corresponding hydrotalcite [21]. When hydroxyl ions are made available in the suspension, the nickel ions can smoothly react with the γ-alumina to the corresponding hydrotalcite. Using temperature-programmed reduction, De Bokx et al. [20] showed that the fraction of hydroxyl ions consumed at pH levels lower than that observed with nickel nitrate alone is indicative of the fraction of nickel hydrotalcite formed.

Figure 11 shows the course of pH changes during the precipitation of zinc(II), with and without suspended alumina [22]. The pH is recorded during the addition of alkali to the same volume of water, of a suspension of alumina in water, and of a solution of zinc nitrate. In Fig. 11 the curve obtained by the addition of hydroxyl ions consumed in the above three measurements is compared with the pH curve measured experimentally with suspended alumina. It is clear that the thus-calculated pH curve lies appreciably above the experimental curve. The interaction of the precipitating zinc ions with alumina is evident from the consumption of hydroxyl ions at a relatively low pH level.

Analogous experiments with silica as the support indicate the appreciable interaction of precipitating

---

References see page 465

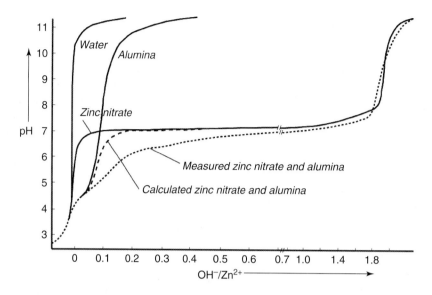

Fig. 11 Precipitation of zinc(II) from a homogeneous solution by injection of sodium hydroxide at 293 K. The pH was recorded during injection into water, an alumina suspension in water, a solution of zinc nitrate, and a suspension of alumina in a solution of zinc nitrate. From the curves measured with water, suspended alumina and the solution of zinc nitrate, the pH plot was calculated that would have been recorded had there been no interaction between the precipitating zinc(II) and the suspended alumina.

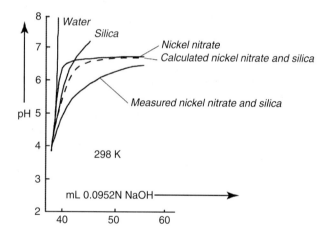

Fig. 12 Precipitation of nickel(II) from a homogeneous solution by injection of sodium hydroxide at 298 K. The pH was recorded during injection into water, a silica suspension in water, a solution of nickel nitrate, and a suspension of silica in a solution of nickel nitrate. From the curves measured with water, suspended silica and the solution of nickel nitrate, the pH plot was calculated that would have been recorded had there been was no interaction between the precipitating nickel(II) and the suspended silica.

nickel(II) ions with silica [23, 24]. These experiments were performed with silica produced by flame hydrolysis of silicon tetrachloride, Aerosil 200 and Aerosil 380 (Degussa, Germany; surface areas 200 and 380 m² g⁻¹, respectively). In the experiments represented in Fig. 12, the solutions or the suspension was kept at 298 K, and the curve measured with suspended silica was seen, again, to run considerably below that calculated from curves recorded with water, suspended silica, and nickel nitrate alone. The interaction with the silica support causes a higher consumption of hydroxyl ions at a low pH level. Burch and Flambard [25] performed the same experiments with both silica and titania supports, and observed an identical behavior; that is, a considerably lower pH level when the precipitation of nickel(II) was carried out in the presence of a silica or titania support.

Van der Lee et al. [26] investigated the deposition–precipitation process of nickel on carbon nanofibers. Previous oxidation with a (1:1) mixture of concentrated nitric and sulfuric acid produced oxygen-containing groups on the surface of the fibers, the surface area of which was found to be 150 m² g⁻¹ after treatment with the above mixture of acids. The amount of acid groups with a pK_a value less than 5 was 0.17 mmol g⁻¹, and the number with pK_a less than 8 was 0.32 mmol g⁻¹. The acid-treated carbon nanofibers are therefore highly interesting for the investigation of deposition–precipitation; the density of sites likely to interact strongly with precipitating nickel is accurately known, and the growth of nickel species deposited when the pH of a suspension of carbon nanofibers in nickel nitrate solution is raised may be initiated on the relatively strong acid sites.

For purposes of comparison, these authors measured the course of pH changes during the hydrolysis of urea with a nickel nitrate solution without a suspended support; the data showed a maximum at pH 6.4 and a subsequent plateau at pH ca. 6.2. At pH 5.5, the pH plot with dissolved nickel nitrate began to deviate from that measured with urea alone. The consumption of hydroxyl ions evident from the deviation from the urea-only plot indicates the onset of hydrolysis of nickel according to

$$Ni(H_2O)_6^{2-} + OH^- \rightleftharpoons Ni(H_2O)_5OH^+ + H_2O$$

Whereas the adsorption of nickel(II) ions on silica is insignificant at pH levels below 5.8, at pH 3.1 the oxidized carbon nanofibers took up 0.14 mmol Ni^{2+} g^{-1} carbon, or 0.9 mol Ni^{2+} mol^{-1} acid sites of $pK_a < 5$, which is close to one Ni^{2+} per strong acid site. The large uptake of Ni^{2+} is remarkable, as only 10% of the acid sites with $pK_a 4$ are dissociated at pH 3. It has been observed that $Al(H_2O)_6^{3+}$ reacts with urea to form complexes (viz., $Al(H_2O)_5(urea)$ and $Al(H_2O)_4(urea)_2$) in which urea is $CO(NH_2)_2$ [27]. It is possible that hydrated nickel ions also react to form a urea complex, which interacts more strongly with acid sites on the carbon nanofibers. An interesting question here is whether the nickel ions are adsorbed as an inner or outer sphere complex on the acid groups of the carbon surface.

Deposition–precipitation with the hydrolysis of urea at 363 K resulted in a pH course during hydrolysis that exhibited a faint maximum at pH 5.8 and an almost horizontal plateau at pH ca. 5.7. Consequently, nickel precipitates in the presence of oxidized carbon nanofibers at a significantly lower pH than within a solution containing only dissolved nickel nitrate and urea. The uptake of nickel by the carbon nanofibers during deposition–precipitation, which was experimentally determined, increased linearly with the deposition time. At the faint maximum in the pH plot the amount of nickel taken up corresponded to 3.8 Ni^{2+} per acid site of $pK_a < 5$. Investigation of the carbon nanofibers after deposition–precipitation of nickel using transmission electron microscopy (TEM) confirmed the presence of thin platelets attached to the fibers of lateral dimensions of 40–50 nm and a thickness of only 5–10 nm. The X-ray diffraction (XRD) pattern of the loaded carbon nanofibers after removal of the diffraction maxima due to the graphitic material of the fibers indicated the presence of α-$Ni(OH)_2$, a form of nickel hydroxide in which the stacking of the brucite layers is fairly disordered due to intercalation of anions, such as nitrate or carbonate ions. The strong broadening of the reflection corresponding to the d-spacing of 6.2 Å, was in agreement with a thickness of 5–10 nm observed using TEM.

Van der Lee et al. [26] assumed that the nickel ions adsorbed at pH ca. 3 are reacting to complexes containing two or three nickel ions from which nickel hydroxide nucleates at pH 5.8. These authors attributed this to the fact that the disordered α-$Ni(OH)_2$ grows in the presence of suspended oxidized carbon nanofibers at a much lower pH than without oxidized carbon nanofibers, to a kinetic phenomenon. The experimental data showed that the surface area of the edges of the nickel hydroxide platelets, from which the growth of nickel hydroxide can proceed more rapidly, is large due to the many nucleation sites present on the oxidized carbon nanofibers. The minimal thickness of the platelets of nickel hydroxide demonstrates preferential growth in the direction of the platelets, and a limited nucleation of new layers on the existing brucite-type platelets. As the surface area of the edges of the thin platelets grows with their lateral dimensions, the faint maximum in the pH curve may not in fact be due to nucleation calling for a certain pH level. When the surface area of the platelets increases, the equilibrium between generation and consumption of hydroxyl ions can be established at a slightly lower pH. That transport to growth sites is important with deposition–precipitation on oxidized carbon nanofibers is also evident from the observations of Van der Lee et al., that powdered clusters of carbon nanofibers are required to prevent the precipitation of bulk nickel hydroxide [26]. In precipitating nickel hydroxide by the hydrolysis of urea, without a suspended support, the number of nuclei generated will be small, as is usual for precipitation from a homogeneous solution. Accordingly, the surface area of the edges of the generated nuclei will also be small. It is possible that the maximum in the pH plot corresponds to the nucleation of new brucite-type layers on top of the existing platelets. As the flat bivalent carbonate ions can strongly bond a new layer to a nucleated platelet, it is possible that nickel hydroxide might only grow when the concentration of bivalent carbonate ions is sufficiently high. The significant consumption of hydroxyl ions before the maximum in the pH plot (as is evident from a comparison with the pH plot measured with only urea) may indicate the initial formation of plate-shaped nuclei on which subsequent brucite-type nickel hydroxide layers cannot be deposited at lower concentrations of carbonate ions.

Van der Lee et al. [26] also investigated deposition–precipitation on carbon nanofibers as grown, and on oxidized carbon nanofibers from which the acid groups had been removed by either thermal treatment at 873 K in nitrogen, or by treatment with lithium aluminum hydride ($LiAlH_4$). The pH plots measured with the as-grown fibers and with thermally treated oxidized fibers did not differ significantly from the plot recorded with only nickel nitrate and urea. With the oxidized fibers treated initially with $LiAlH_4$, some consumption of hydroxyl ions at a lower pH was observed, but the final pH coincided with that measured with nickel nitrate and urea alone. In agreement with the pH plots, the precipitation of nickel had proceeded, besides the carbon nanofibers, as relatively large platelets in which many brucite layers were stacked.

Earlier, Toebes et al. [28] had studied the deposition–precipitation of platinum from dissolved platinum

References see page 465

amminonitrate on carbon nanofibers. These authors established that the loading of very small platinum particles was higher when deposition–precipitation with urea hydrolysis was executed, than when adsorption of the ammine complex was performed. A platinum loading of about 4 wt.% could be achieved, whereas loading with the ion adsorption procedure achieved a level of less than 2 wt.%.

B Subsequent Reactions of the Initially Precipitated Precursor It is possible that the precipitate that initially results from a deposition–precipitation is not stable at higher pH levels. The decomposition of an initially precipitated basic salt, and the reaction of an initial precipitate with another anion, or with the support, can proceed. The reaction of basic salts of copper initially precipitated from a homogeneous solution by raising the pH level provides an interesting example of the reactions exhibited. As copper (oxide) catalysts are important in many catalytic reactions (e.g., methanol synthesis, CO shift conversion, selective hydrogenations and oxidations), it is highly desirable to modify the deposition–precipitation process so as to bring about a uniform and fine dispersion of the deposited copper species over the support. Thus, van der Meijden investigated in detail the precipitation of copper from a homogeneous solution [29].

Figure 13 illustrates the pH course during the precipitation of copper(II) from a copper(II) nitrate solution during the hydrolysis of urea at 283 K. The pH plots recorded with and without suspended silica are seen to be identical, and it is therefore necessary to shift the curve measured with suspended silica to exhibit the two plots.

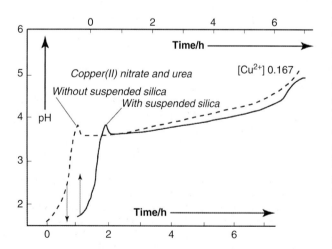

Fig. 13 Precipitation of copper(II) from a homogeneous solution of copper nitrate by hydrolysis of urea at 363 K, with and without suspended silica. The time scale of the two plots has been shifted to distinguish between the two plots.

Investigations with TEM and XRD each showed that large thin platelets of basic copper(II) nitrate, $Cu_2(OH)_3NO_3$, had precipitated besides the silica. As indicated already by the pH plots, the precipitating copper(II) species does not interact significantly with the suspended silica. Without any interaction between the precipitating species and the support, precipitation from a homogeneous solution leads to a low density of nuclei that grow to large crystallites. Accordingly, Labunakrom [30] prepared a large number of basic copper(II) salts by the hydrolysis of urea, and thus obtained crystallites sufficiently large to be investigated with optical microscopy. Thus, the data shown in Fig. 13 indicate that the use of urea does not necessarily lead to the desired uniform loading of the support with small particles of the active precursor.

The results of Martynenko [31] on the precipitation of basic copper salts are important, as this author established that nucleation of basic copper(II) salts calls for a higher supersaturation than that of copper hydroxide, $Cu(OH)_2$. Copper hydroxide initially precipitated by inhomogeneous addition of alkali can thus subsequently react to a basic salt. Previously Gauthier et al. [32] had observed that $Cu(OH)_2$ may react with anions such as sulfate, nitrate or chloride to form basic salts. These authors also found that, at higher pH levels, basic copper salts react to form copper hydroxide or copper(II) oxide. The reaction of the basic copper(II) salts is not confined to hydroxyl ions; Gauthier et al. additionally investigated the reaction of basic copper acetate, $Cu(CH_3CO_2)_2 \cdot 3Cu(OH)_2 \cdot 2H_2O$, with anions such as sulfate, chloride, and bromide. Gauthier's group assumed that the reaction can proceed without any intermediary dissolution of copper(II), and when the content of hydroxyl groups of the eventually precipitated basic copper salt was lower than that of the originally precipitated basic copper salt, a reaction to copper(II) oxide by dehydration of copper hydroxide was observed. However, in the presence of carbonate ions, the dissolution of copper(II) as $Cu(HCO_3)_2$ and $Cu(HCO_3)_4^{2+}$ was established. At high pH levels, the dissolution of copper(II) during the reaction of basic copper salts to copper hydroxide is not likely. Accordingly, Kohlschütter [33] found that the exposure of a basic copper salt to solutions containing a high concentration of hydroxyl ions results in a topotactic reaction of the basic copper salts. Without any change in the shape and size of the crystallites of the original basic copper salt, the reaction proceeds to small $Cu(OH)_2$ crystallites in a parallel orientation.

Van der Meijden [30] studied the precipitation of copper(II) without a suspended support from a homogeneous solution by the hydrolysis of urea and the injection of NaOH at 363 K, and at 305 K only by injection of NaOH. Carbonate-free NaOH, the concentration of which was varied from 0.0970 to 0.9575 mol L^{-1}, was injected continuously through a narrow poly(vinyl chloride) (PVC) tube attached to a Pyrex tube of internal diameter of 0.1 mm. PVC was employed to avoid any heterogeneous nucleation on the Pyrex tube. The rates of injection were 2.04 and 4.11 mL min^{-1}, with the tube ending below the surface of the liquid, just next to a stirrer blade. In this way, the precipitation from copper(II) chloride, copper(II) sulfate, copper(II) nitrate and copper(II) perchlorate was studied. Copper chloride precipitated initially as atacamite, $Cu_2(OH)_3Cl$, copper sulfate as brochantite, $Cu_4(OH)_6SO_4$, and copper nitrate as $Cu_2(OH)_3NO_3$. With the injection of NaOH, the proportion of the hydroxyl and copper(II) ions as a function of the pH is accurately known. The pH rose steeply when approximately 1.5 hydroxyls per copper ion were injected, these data agreeing well with the composition of the basic salts.

After passing through the maximum in the pH plot, the rate at which hydroxyl ions are made available and the rate of their consumption by the precipitation reaction are equal. The pH is seen to change only with the concentration of copper(II). During this stage, the pH should correspond to the solubility product of the basic copper salts. For basic copper chloride, published data refer to solubility products measured at 298 K and expressed as pS_k of 16.4 and 17.3 [34, 35]. Van der Meijden arrived at pS_k values of 15.7 and 16.1, measured at 305 K. If we take into account the value for pK_w falling from 13.8 to 12.4 when the temperature increased from 298 to 363 K, the experimentally recorded pH levels during steady state at 363 K led to pS_k values of 16.8 to 17.5, which was in good agreement with published values. For basic copper sulfate, published pS_k values of 15.4 and 17.1 are mentioned [34, 35], while the steady-state levels measured by van der Meijden at 303 K led to pS_k values of 15.0 and 15.1. The more numerous measurements at 363 K led, after correction for pK_w, to pS_k values ranging from 16.2 to 16.8, again well in the range of published values. The published pS_k values for basic copper nitrate are 15.6 and 16.4 [34, 35], while measurements at 305 K provided values of 15.3 and 15.4, and the many measurements at 363 K (after correction for temperature) to values ranging from 16.0 to 16.7. It is of interest that reducing the concentration of copper nitrate to about 0.020 mol L^{-1} led, after correction, to a much lower pS_k-value of 13.9. This low value indicates that at the low concentration, the growth of the nuclei cannot keep up with the generation of hydroxyl ions by the reaction of urea, which leads to the precipitation proceeding at a higher pH than corresponding to the solubility product.

As might be expected, the maxima of the pH curves were more pronounced with the hydrolysis of urea than with the injection of NaOH solutions. With the above three copper(II) salts, the maxima recorded with urea hydrolysis were 0.30 to 0.55 pH units above the pH exhibited at the subsequent steady state that corresponds to the solubility products. By using NaOH concentrations that allow the precipitation to be completed in a reasonable time, it is not possible to prevent the establishment of slightly higher concentrations at the injection point. The nucleation of basic copper salts proceeds where the local concentration of hydroxyl ions is higher. With the injection of alkaline solutions, the maxima in the pH curves were therefore only 0.05 to 0.20 above the pH of the following steady state. However, the pH during steady state still agreed with the above solubility products.

By injecting NaOH, the pH can be raised to >8, at which point the reaction to copper hydroxide or copper oxide will proceed. Figure 14 shows the pH course during the injection of NaOH into solutions of copper nitrate kept at 263 K. During one experiment in Fig. 14, a flow of carbon dioxide was passed through the solution. From the curve recorded without carbon dioxide, it can be seen that, upon injection of NaOH up to a proportion OH^-/Cu^{2+} of about 1.5, the pH steeply rises. However, after reaching pH 9, the pH abruptly and sharply fell due to the uptake of hydroxyl ions during the reaction of the basic nitrate to copper hydroxide or copper oxide. After injection of NaOH to a proportion of OH^-/Cu^{2+} of 2, as would be expected for the formation of CuO, the pH increased again. The reactions involved are

$$Cu_2(NO_3)(OH)_3 + OH^- \rightleftharpoons 2Cu(OH)_2 + NO_3^-$$

$$2Cu(OH)_2 \rightleftharpoons 2CuO + H_2O$$

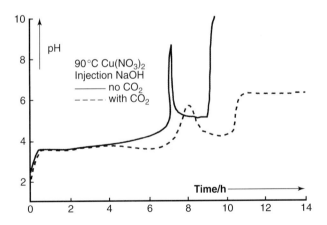

Fig. 14 Precipitation of copper(II) by injection of sodium hydroxide into a solution of copper(II) nitrate without suspended silica at 363 K. The pH course during two experiments is represented. The fully drawn curve indicates pH during injection in the copper nitrate solution. The first step rise in pH is shown after the injection of alkali to a proportion OH/Cu = 1.5, and the second step at OH/Cu = 2.0. Carbon dioxide was passed through the solution during pH monitoring (dashed line).

References see page 465

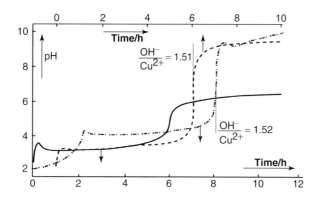

Fig. 15 Measurement of pH during injection of sodium hydroxide or hydrolysis of urea into a solution of copper(II) chloride without suspended silica. The solid curve represents pH during the hydrolysis of urea; the dashed curve shows pH during the injection of sodium hydroxide; the dash-point curve shows pH during the injection of sodium hydroxide into a solution of copper(II) chloride of a lower concentration.

With basic copper chloride and basic copper sulfate, the fall in pH is much less pronounced (or even absent), as shown in Fig. 15, which indicates that the reaction to copper hydroxide or copper oxide is slower. The size of the crystallites of the basic salts may account for the difference in the rate of reaction. Whereas the basic copper nitrate is present as very thin platelets, the basic copper chloride crystallizes as large bipyramidal crystallites, and the basic copper sulfate as needles.

Figure 16 shows that, after longer periods during the hydrolysis of urea, the pH remains lower than with the injection of NaOH, which is to be expected from the lower basic strength of ammonia. The pH course is different, however. Now, the pH plot exhibits no less than three maxima, the very first of which is related to the nucleation of basic copper nitrate. The structure of the different solid phases has been established with XRD. The results show that the second maximum is due to the nucleation of basic copper carbonate, malachite, $CuCO_3Cu(OH)_2$, and the third to nucleation of $Cu(OH)_2$ or CuO. At 363 K, $Cu(OH)_2$ is not stable and decomposes to CuO. It is not clear that the pH falls upon nucleation and growth of malachite in view of the reaction

$$Cu_2(NO_3)(OH)_3 + CO_3^{2-} \rightleftharpoons CuCO_3Cu(OH)_2 + NO_3^- + OH^-$$

It is felt that the fall in pH is due to an intermediary dissolution of copper as the hydrocarbonate complexes mentioned above [32]. Precipitation from the dissolved hydrocarbonate complexes results in consumption of hydroxyl ions according to

$$2Cu(HCO_3)_2 + 3OH^- \rightleftharpoons CuCO_3Cu(OH)_2 + 3HCO_3^- + H_2O$$

The above reaction rationalizes the consumption of hydroxyl ions during the reaction to copper oxide. It is finally important to note that, at the pH eventually obtained during urea hydrolysis at 363 K a significant fraction of the basic copper nitrate is still present. Figure 14 also compares the course of pH changes during the injection of NaOH, with and without passing a flow of carbon dioxide through the liquid. The effect of the carbon dioxide is evident from the difference in the pH plot. The presence of carbonate ions inferred that no formation of CuO was observed, but that the copper(II) precipitated completely as malachite, $CuCO_3Cu(OH)_2$.

Basic copper(II) perchlorate contains, analogously to basic copper(II) acetate, many hydroxyl groups according to the composition $Cu(ClO_4)_2 \cdot 6Cu(OH)_2$. A higher solubility of the basic salt might therefore be expected, as well as a higher reactivity with other anions and hydroxyl groups. During the injection of NaOH at 303 K in a solution of copper perchlorate, the pH formed a plateau at 4.8, after passing through a small maximum. The pH rose steeply at a OH^-/Cu^{2+} proportion of 1.66, which is close to the proportion of 1.71 predicted from the above composition of the basic perchlorate. A pS_k-value of 16.2 was calculated from the pH level of 4.8, which agrees satisfactorily with the pS_k-value of 16.6 measured at 278 K and lower concentrations by Nässänen and Tamminen [34]. The hydrolysis of urea represented in Fig. 17 leads to a pH plot which differs from the other plots; the maximum through which the pH passes is elevated and varies from experiment to experiment from 4.90 to 5.30. Next, a plateau is exhibited at pH 4.65 to 4.81, while a second fall in pH is subsequently shown. Whereas the precipitation of copper(II) perchlorate at 305 K leads to a grayish-green solid exhibiting a sharp, unknown XRD pattern, the hydrolysis of urea (at 363 K) leads only to basic copper carbonate. The maximum and first plateau of the pH plot are tentatively attributed to the formation of copper hydroxide, which reacts to form basic copper carbonate when the hydrolysis of urea has produced more carbonate ions.

The pH plot in Fig. 17, which was recorded during the hydrolysis of urea in the presence of suspended silica, does not exhibit a maximum. The copper species appeared to have been deposited finely divided on the silica. Neither did the loaded dried support show an XRD pattern. Thus, it may be concluded that precipitation from a homogeneous solution of copper(II) leads only to the desired deposition on silica when the copper precipitates at pH > 4. Although silica particles have a negative electrostatic charge above a pH of ca. 2.0 (the

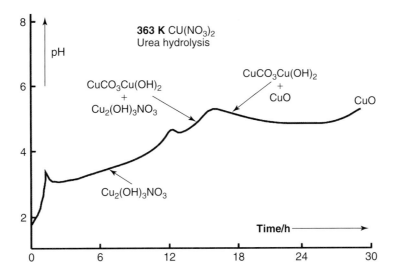

Fig. 16 The pH course during the hydrolysis of urea in a solution of copper(II) nitrate recorded for a relatively long period. X-ray diffraction was used to establish the structure of solids precipitated at different times. The results are indicated in the figure.

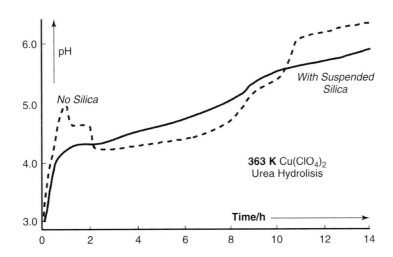

Fig. 17 pH course during the hydrolysis of urea in either a solution of copper(II) perchlorate or a suspension of silica in a copper(II) perchlorate solution.

isoelectric point), an appropriate electrostatic charge is not a sufficient condition for the interaction of copper(II) with the silica support. It is likely that the adsorption of copper(II) ions onto silica is a prerequisite for a sufficient interaction. From a pH of 5.0, the adsorption of copper on silica at room temperature is reported to rise with the pH. At lower pH levels, copper(II) does not adsorb onto silica. In view of the value of the water equilibrium of $pK_w = 13.8$ and 12.4 at 305 and 362 K, respectively, the adsorption of copper(II) could be expected to begin at about pH 4 at 362 K.

Figure 18 illustrates, graphically, two experiments in which copper nitrate was precipitated in the presence of suspended silica by the hydrolysis of urea. In contrast to experiments with higher copper concentrations, that in one of the experiments was only 0.04 mol L^{-1}. The pH plot recorded had no maximum, and the copper appeared to have deposited finely divided on the silica. This experiment showed also that the pH at which precipitation was initiated determined whether copper would be deposited onto the silica. Unfortunately, the use of a copper concentration of the order of 0.04 mol L^{-1} is technically not viable and, as the production of copper(II) perchlorate is not attractive, alternative procedures were sought.

As basic copper(II) sulfate is less stable than basic copper chloride, the reaction of the former at higher pH levels in the presence of suspended silica was investigated.

References see page 465

Basic copper sulfate was first precipitated by the injection of a NaOH solution into a suspension of silica in copper sulfate kept at 362 K. At an OH^-/Cu^{2+} proportion of 1.45, the pH rose rapidly, whereupon an automatic switch interrupted the NaOH injection at pH 7.2 and restarted it at pH 7.0. After reaching 7.2, the pH appeared to fall slowly when no alkali was injected. The copper from the basic copper sulfate was found to be deposited finely divided on the silica during this procedure, although about 90 h was required for complete deposition of copper using this procedure.

As basic copper nitrate is less stable than basic copper sulfate, it was investigated whether copper from the former might be deposited on silica at pH levels achievable by the hydrolysis of urea. Figure 18 also represents the pH curve recorded with a higher copper concentration of 0.19 mol L^{-1}. In addition to pH, the copper concentration was simultaneously monitored using an electrode specific for copper ions. On completion of the precipitation of copper as basic copper nitrate, the hydrolysis of urea was continued. It can be seen that, after 5.5, the pH begins to decrease abruptly. The data in Fig. 18 show that the amount of dissolved copper began to increase simultaneously with the fall in pH. At the final rise in pH, the copper concentration fell to very low levels. It is of interest that the maximum pH of ca. 5.0 that was displayed without suspended silica is now absent, whereas the maximum pH (5.5) is similar to that where the reaction to a mixture of basic copper carbonate, basic copper nitrate and copper oxide was indicated in the experiment without suspended silica. The copper appeared to have deposited in finely divided form onto the silica, and subsequent investigation of the precipitate showed that the copper(II) and silica had reacted to a clay mineral, *viz.*, chrysocolla [36–38]. The complete reaction is evident from the fact that reduction of the copper in thus-prepared catalyst precursors calls for temperatures of about 670 K – much higher than the value of 450 K which suffices to reduce usually deposited copper species. Although nitrates are not desirable in waste water, and copper nitrate is therefore not the compound of choice to produce copper catalysts, a technically acceptable procedure for the production of copper catalysts has been developed.

The relatively slow reduction of the copper present in chrysocolla is attractive. The reduction of copper oxide is highly exothermic, which can lead to high temperatures and severe sintering of the copper in technical reactors. The reduction of copper catalysts is therefore usually carried out with a gas flow containing only about 0.2% hydrogen in nitrogen. The heat of the reaction must be taken up by the gas flow, which calls for a very long period of reduction, often of two days or even a week. The rate of transport of hydrogen does not determine the reduction of copper within chrysocolla, but rather the slow decomposition of the mineral. Consequently, the reduction does not proceed within a narrow front within the reactor but at a low rate throughout the catalyst bed. The resulting metallic copper catalyst is, moreover, highly thermostable.

The highly relevant result of the above investigations is that the reaction of silica with copper cannot be confined to the surface of the silica, but can involve a considerable fraction of the silica. Depending upon the loading of the silica, even complete reaction of the silica can proceed.

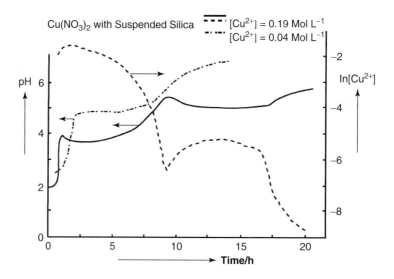

Fig. 18 pH course during the hydrolysis of urea in a suspension of silica in copper(II) nitrate solutions. Two different copper concentrations were employed (solid and dash-point curves). With a solution of 0.19 mol L^{-1}, the copper concentration in the solution was recorded simultaneously by means of an ion-sensitive electrode (dashed curve).

The decomposition of basic copper nitrate in the presence of suspended silica does not lead to a reaction of the silica. The injection of a KOH solution into a suspension of silica and basic copper nitrate kept at 363 K resulted in a black precipitate which exhibited a sharp XRD pattern of CuO [30]. It is likely that, at the point where the KOH solution enters the suspension, the basic copper nitrate decomposes without any reaction with silica. It is also possible that carbonate ions are required to intermediately dissolve the copper. The injection of a 0.51 mol L^{-1} copper nitrate solution into a suspension of silica in a urea solution kept at 363 K was also performed [30], with the rate of the injection controlled automatically so as to maintain the pH at 5.0; the time taken for the injection was also extended, to about 30 h. Although the copper was distributed fairly homogeneously throughout the silica support, the copper-containing particles which resulted after drying and reduction of the loaded support were not very small. It is likely that, at the injection point, copper(II) precipitates as malachite and not as basic copper nitrate. The solubility of copper present in malachite by reaction with bicarbonate ions may be less than that of the copper present in basic copper nitrate.

In Fig. 12, the interaction between precipitating nickel ions and suspended silica at room temperature was demonstrated, whereupon the pH curve measured in the presence of suspended silica could be seen to approach the curve recorded with nickel, but without suspended silica. When the precipitation of nickel is performed at higher temperatures (e.g., 363 K), the pH course is completely different. For example, Fig. 19 shows pH plots recorded during the precipitation of nickel(II) by the injection of a NaOH solution at 363 K. It can be seen that the pH plot measured with suspended silica remains far below the level recorded with nickel(II) alone. Following the injection of larger quantities of alkali, the pH also remains below the level exhibited with nickel nitrate alone.

Figure 20 represents measurements in which the hydroxyl ions were made homogeneously available by the hydrolysis of urea [24, 25]. The silica was non-porous, and had a high surface area of 380 m^2 g^{-1} (Aerosil 380; Degussa). Again, the curve measured with suspended silica in a nickel(II) solution at 363 K remained appreciably below that measured with nickel nitrate alone, and did not approach the curve measured without suspended silica. As the pH curve remains below the level at which nickel hydroxide precipitates, the formation of nickel hydroxide in the bulk of the solution cannot proceed. It can furthermore be seen that the pH curve measured with suspended silica exhibits a maximum, which indicates that a new solid nucleates, after which the consumption of hydroxyl ions by the growth of nucleated compound is temporarily more rapid than the generation of hydroxyl ions. The large difference in pH level between the curve measured with and without silica indicates that the nucleation does not involve nickel hydroxide stabilized by

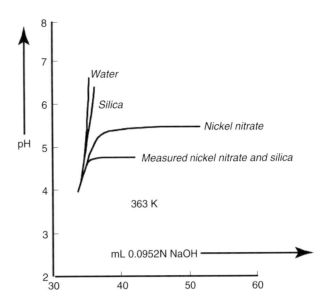

Fig. 19 Precipitation of nickel(II) from a homogeneous solution by injection of sodium hydroxide at 363 K. The pH was recorded during injection into water, a silica suspension in water, a solution of nickel nitrate, and a suspension of silica in a solution of nickel nitrate. From the curves measured with water, suspended silica and the solution of nickel nitrate, the pH plot was calculated that would have been recorded had there been no interaction between the precipitating nickel(II) and the suspended silica.

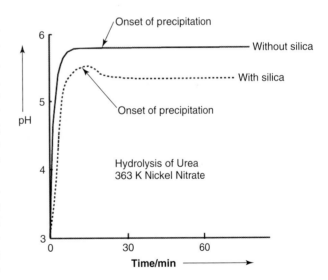

Fig. 20 pH course recorded during the hydrolysis of urea at 363 K in a solution of nickel nitrate (solid curve) and a suspension of silica in a solution of nickel nitrate (dashed curve).

References see page 465

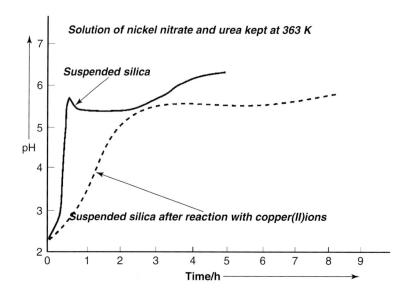

Fig. 21 pH course recorded during the hydrolysis of urea at 363 K. The solid curve represents pH during the hydrolysis of urea in a suspension of silica in a solution of nickel nitrate. The dashed curve represents an experiment in which silica previously reacted with copper(II) during the hydrolysis of urea was suspended in the nickel nitrate solution.

interaction by the unaffected silica surface. Richardson et al. [39] have reported precisely the same results with the deposition–precipitation of nickel(II) ions from a homogeneous solution by the hydrolysis of urea in the presence of a suspended silica (Cabosil) support, which was also produced by flame hydrolysis.

Nickel hydrosilicate grows without nucleation on silica that has reacted with copper(II) ions, as shown in Fig. 21 [40]. When nickel(II) ions are precipitated from homogeneous solution by urea hydrolysis at 363 K in the presence of suspended silica, the pH passes through a maximum (see Fig. 13). As the pH level remains much lower than with the precipitation of nickel(II) without suspended silica, the maximum in the pH curve indicates nucleation of a nickel phyllosilicate. In the presence of suspended silica that has reacted with copper, the pH remains substantially lower than with pure silica (see Fig. 13). Accordingly, the growth of a mixed phyllosilicate proceeds smoothly from copper(II) hydrosilicate. It is important to note that the platelets of the clay minerals do not fit on the surface of the essentially spherical silica particles of the suspended support. Rather, the fairly straight platelets are attached only at one end of the silica particles during growth. Consequently, a sufficient surface area of bare silica is available to provide silicon species to grow the clay minerals. When a sufficiently large amount of the metal ions is to be deposited, the growth of the clay minerals leads to complete consumption of the silica, as no trace of the initially present silica particles were observed in electron micrographs of the loaded carrier.

In analogy to copper(II) reacting to chrysocolla, the silica reacts with the precipitating nickel to form a new compound, in this case a phyllosilicate. The chemical reaction of the precipitating or precipitated nickel with the silica often leaves no silica particles behind, as mentioned above. The result is a new compound of a shape which is radically different from that of the initial silica particles. The silica may intermediately dissolve, or may migrate over the surface of the clay platelets.

Silica can react with the precipitating nickel to different compounds, namely phyllosilicates. As shown schematically in Fig. 22, the phyllosilicates have either a layer of $Si_2O_5(OH)$ units and a layer of hydroxyl ions around an octahedrally surrounded layer of bivalent metal ions [termed (1:1) phyllosilicates] or two layers of $Si_2O_5(OH)$ units around the bivalent metal ions [(2:1) phyllosilicates]. Different nickel-containing minerals with the above structures have been identified: antigorite, penouite, and lizardite [41] are (1:1) phyllosilicates, while willemseite [42] and stevensite [43] are (2:1) phyllosilicates. The most important evidence that the nickel deposited is not

$(OH)_3$	$Si_2O_5(OH)$	$Si_2O_5(OH)$
Ni_3	Ni_3	Ni_3
$(OH)_3$	$(OH)_3$	$Si_2O_5(OH)$
Nickel hydroxide	1:1 Phyllosilicate	2:1 Phyllosilicate

Fig. 22 Schematic representation of the structure of β-nickel hydroxide, a (1:1) nickel phyllosilicate, and a (2:1) nickel phyllosilicate.

nickel hydroxide but a phyllosilicate, is the fact that after the deposition–precipitation the nickel is not soluble in ammonia solution, whereas nickel hydroxide readily dissolves in ammonia. TEM images of the loaded silica show the presence of platelets that do not change upon calcination at temperatures above about 720 K. Figure 23 shows TEM images of a silica support loaded with nickel by deposition–precipitation employing urea hydrolysis at 363 K. The silica particles can be seen to have completely reacted with the platelets of the clay mineral. As nickel hydroxide alone precipitates also as platelets, it is important to distinguish the phyllosilicate platelets from platelets of pure nickel hydroxide. The widely different thermal stability can be employed to assess the presence of phyllosilicates. Teichner and coworkers have established that nickel hydroxide decomposes to cubic nickel oxide crystallites already at 470 K. Nickel hydroxide platelets therefore rearrange to arrays of cubic crystallites at temperature above 470 K; dark-field transmission electron microscopy readily exhibits the small cubic crystallites. The presence of nickel containing phyllosilicates is evident from the fact that the platelets are not broken up after thermal treatment at high temperature, e.g., at 723 K, as indicated in Fig. 23a. Another difference is that the nickel within the phyllosilicates can be reduced only at high temperatures, whereas with nickel hydroxide the dehydration and reduction proceed almost simultaneously from about 470 K. The metallic nickel particles resulting from reduction of nickel phyllosilicates are very small and highly thermostable as can be seen in Fig. 23b. The catalyst represented in this figure has been calcined at 723 K and subsequently reduced at the same temperature in a hydrogen-nitrogen flow. The fact that the nickel surface are as calculated from the extent of hydrogen adsorption is small as compared to the size of the nickel particles as evident from X-ray line broadening and electron microscopy, indicates that the nickel particles are partly buried with the silica remaining after the reduction. Since the silica thus completely prevents migration of the nickel particles, thus prepared nickel catalysts are highly thermostable.

Van Eijk van Voorthuijsen and Franzen [44, 45] were the first to investigate the interaction between nickel and silicate ions. These authors either combined boiling solutions of alkali silicate and nickel nitrate, or precipitated silica in the presence of freshly precipitated nickel hydroxide by the addition of acid to an alkali silicate solution, or precipitated nickel by addition of alkali to a suspension of silica in a nickel nitrate solution. The reaction products were characterized using XRD and differential thermal analysis. The nickel hydroxide was suspended in an alkali silicate solution kept at 373 K, and nitric acid slowly added. The nickel hydroxide did not react with the precipitating silica; this was in contrast to magnesia and magnesium hydroxide, both of which reacted to phyllosilicates. Combining alkali silicate and nickel nitrate solutions or precipitating nickel in a silica suspension resulted in nickel phyllosilicates. As

References see page 465

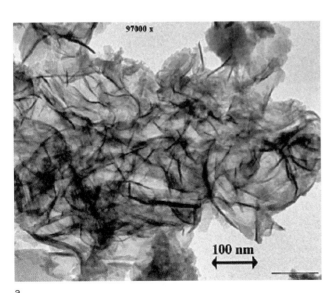 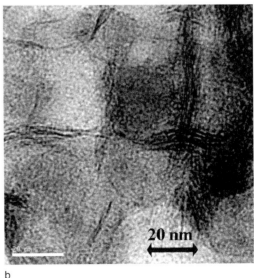

a b

Fig. 23 Transmission electron micrographs of silica loaded with nickel by deposition-precipitation (hydrolysis of urea at 363 K). Figure 23a catalyst calcined at 723 K in a flow of pure nitrogen. Figure 23b catalyst reduced at 723 K in a hydrogen-nitrogen flow imaged at a higher magnification. Where the (original) phyllosilicate layers are oriented perpendicularly to the electron beam, the layers either are represented greish (calcined catalyst, Fig. 23a) or exhibit the very small nickel particles (reduced catalyst, Fig. 23b).

the nickel:silica proportion of a (1:1) phyllosilicate is 3:2, and of a (2:1) phyllosilicate is 3:4, the nickel:silica ratio during the precipitation can determine which phyllosilicate will result. Unless the reaction products of nickel ions and silica were subjected to a hydrothermal treatment at 523 K, they were poorly crystallized. Whilst nickel hydroxide precipitated by the addition of alkali to a nickel nitrate solution exhibits a layer distance of 0.46 nm, the (1:1) phyllosilicate has a layer distance of about 0.71 nm and the (2:1) phyllosilicate ca. 0.95 nm, although willemseite shows a larger layer distance of about 1.2 nm. The kieselguhr used in the production of fat-hardening catalysts also reacts to phyllosilicates with nickel precipitated by the addition of alkali during the "hold", during which the loaded carrier is held in suspension for a certain period of time at about 365 K. The presence of phyllosilicates is evident not only from electron micrographs but also from the large increase in the surface area of the support, which is about 15 m^2 g^{-1} before being loading with nickel.

Che and coworkers have performed extensive investigations of the reaction products resulting from the interaction of nickel ions with silica [46–54]. Louis et al. [46] examined the interaction of solutions of nickel nitrate impregnated in silica during drying at 300 and 363 K; subsequent XRD analysis indicated hydrolysis of the nickel nitrate to $Ni(NO_3)_2 \cdot 2Ni(OH)_2$ and $NiNO_3 \cdot Ni(OH)_2 \cdot 2H_2O$. Drying for a very prolonged period led to a (1:1) nickel phyllosilicate.

Earlier, Clause et al. [50] investigated the interaction of nickel(II) ions with silica during deposition–precipitation by means of urea hydrolysis at 363 K. As XRD indicated only the presence of amorphous species, these authors employed EXAFS and temperature-programmed reduction (TPR) to characterize the reaction products. The times of the deposition–precipitations ranged from 0.3 to 72 h. Starting from a nickel nitrate solution that could provide a maximum loading of 10.66 wt.%, the nickel content after reaction for 72 h was 17.5 wt.%, which points to solution of some silica in the final stage of the precipitation process. TPR profiles of samples taken after 1 and 3 h showed a shoulder at a temperature of about 670 K and a broad peak centered at about 770 and 820 K, respectively. The shoulder was attributed to nickel hydroxide, and the main peak to a (1:1) nickel phyllosilicate. After urea hydrolysis for 72 h, the shoulder in the TPR profile had disappeared and the main peak had shifted to about 900 K, which points to a better-crystallized (1:1) nickel phyllosilicate or to reaction to a (2:1) nickel phyllosilicate, which is supposed to be reduced at higher temperatures.

The EXAFS data confirmed the TPR results. The spectra could be fitted by two structures, namely a (1:1) phyllosilicate structure with six neighboring nickel atoms and two silicon atoms surrounding a nickel atom, and a (2:1) phyllosilicate structure with six neighboring nickel atoms and four silicon atoms around a nickel atom. The authors did not consider the presence of a mixture of the two phyllosilicate structures, though it is well known that EXAFS cannot discriminate readily between a pure compound and a mixture of two compounds containing the same elements. Fitting the data with a (1:1) phyllosilicate structure led to 5.9, 5.8, and 6.0 neighboring nickel atoms, and 2.4, 2.4, and 3.1 silicon atoms around a nickel atom after 1, 3, and 72 h of deposition–precipitation, respectively. The lateral size of the platelets of the phyllosilicates calculated from the EXAFS data was larger than 5 nm. A (2:1) phyllosilicate structure resulted in 4.1, 4.0, and 4.3 neighboring nickel atoms and 4.0, 4.0, and 4.1 silicon atom around a nickel atom after 1, 3, and 72 h of deposition–precipitation, respectively. As was apparent from the smaller number of neighboring nickel atoms, the size calculated for the size of the platelets of the (2:1) phyllosilicate was smaller, viz., less than 3 nm. These authors used the Debye–Waller factor of the nickel atoms to discriminate between the two structures. The Debye–Waller factor of the nickel atoms of hydrothermally synthesized (1:1) and (2:1) phyllosilicates were both 0.09. As the ordering of the hydrothermally synthesized phyllosilicates was assumed to be higher than that of the species obtained by deposition–precipitation, the above Debye–Waller factor was considered to be the minimum value that is realistic for the precipitated species. Whereas the Debye–Waller factors of the (1:1) structure were 0.10, 0.10, and 0.09 after 1, 3 and 72 h of deposition–precipitation, the (2:1) structure led for all time periods to a Debye–Waller factor of 0.08. As the authors considered a factor of 0.08 not to be realistic, they favored the formation of a (1:1) phyllosilicate.

A subsequent report by Kermarec et al. [47] detailed the nickel species resulting from impregnation, deposition–precipitation, and contacting silica with an ammoniacal solution of nickel. In order also to identify badly crystallized species, the authors employed Fourier transform infrared spectroscopy (FTIR). Nickel catalysts were prepared from ammoniacal solutions by competitive cation exchange, which proceeded at pH values of 8.3 and 9.8. Deposition–precipitation was carried out in 50 mL of suspension containing 0.411 g nickel and 0.38 g silica and a urea:nickel ratio of 3. The supports were Spherosil (Rhône-Poulenc) with surface areas of 400 and 44 m^2 g^{-1}. The nickel loading of the 400 m^2 g^{-1} support was 9 and 17.5 wt.% in two experiments, and that of the 44 m^2 g^{-1} support only 2.8 wt.%. Generally, the surface area of the silica support is likely to have an appreciable effect on the rate of deposition. A later report by Che's group [52] confirmed the effect of the silica support's surface area on the initial rate of deposition: after 70 min, loading of the 400 m² g⁻¹ support was 5.2 wt.%, and of the 44 m² g⁻¹ only 2.8 wt.%. After 16 h of deposition–precipitation the loadings were more similar (44.6 and 40.8 wt.%, respectively). Kermarec et al. [47] prepared a reference sample of poorly crystallized α-Ni(OH)$_2$ by addition of ammonia to a nickel nitrate solution. The competitive cationic exchange produced either a (2 : 1) phyllosilicate or a mixture of (2 : 1) and (1 : 1) phyllosilicates. The higher solubility of silica at the relatively high pH levels of the exchange preparation promotes the formation of phyllosilicates; the high pH of 9.8 led to the (2 : 1) phyllosilicate of a higher silicon content, as might have been expected. The silica of lower surface area (50 m² g⁻¹; OX50, Degussa) displayed a mixture of a (1 : 1) phyllosilicate and nickel hydroxide; the characteristic peak of Ni(OH)$_2$ at 637 cm⁻¹ only appeared in a difference spectrum. The spectra of the samples prepared by deposition–precipitation on the high-surface-area silica indicated the presence of a (1 : 1) phyllosilicate. The highly loaded sample (17.5 wt.%) exhibited a broad band in the δ OH region, which could be deconvoluted in bands at 665 cm⁻¹, characteristic for a (1 : 1) phyllosilicate and a band at 644 cm⁻¹ attributed to nickel hydroxide. At high loadings, the growth of nickel hydroxide on top of a (1 : 1) phyllosilicate is not unreasonable. The spectrum of the sample of a loading of 2.8 wt.% showed a very faint indication of a band at 637 cm⁻¹.

In further reports from this group, Burattin et al. [51–54] employed XRD and EXAFS, in addition to infrared (IR) spectroscopy, to characterize the reaction products of the deposition–precipitation of nickel on silica. At this point, XRD produced important information. The first report, Burattin et al. [51], contained diffraction patterns of nickel hydroxide precipitated by urea hydrolysis without suspended silica, and patterns of phyllosilicates obtained by loading silica of surface areas 356 and 44 m² g⁻¹, employing different times of deposition–precipitation of nickel, followed by hydrothermal treatment. It is of interest to note that the nickel hydroxide precipitated without suspended silica recrystallized during urea hydrolysis after 48 h to an amorphous basic nickel carbonate. Nickel hydroxide exhibits two different structures, β-Ni(OH)$_2$ and α-Ni(OH)$_2$. β-Ni(OH)$_2$ has a brucite structure, characterized by layers of nickel ions sandwiched between two layers of hydroxyl ions. The hydroxyl ions of different layers are bonded by hydrogen bonds. The intercalation of foreign anions between the layers leads to the α-Ni(OH)$_2$ structure, where the stacking of the layers is less-ordered due to the foreign anions, which leads to the denotion "turbostratic" nickel hydroxide. The layer distance of 0.725 nm displayed by the initially precipitated nickel hydroxide is characteristic of turbostratic nickel hydroxide. β-Ni(OH)$_2$ exhibits a layer distance of only 0.46 nm. Apparently, the nucleation of a new layer of nickel hydroxide on existing layers does not proceed easily, as mentioned earlier in the discussion of the results of Van der Lee et al. [26]. The adsorption of foreign anions is required to nucleate a new layer, which brings about the turbostratic structure of the resulting nickel hydroxide. The reaction to amorphous basic nickel carbonate demonstrates that the initially precipitated nickel hydroxide is liable to further reactions. The flat bivalent carbonate ions are particularly strongly bonded to the layers. When the concentration of carbonate ions increases the reaction of nickel hydroxide to nickel carbonate can proceed. Nitrate and isocyanate ions taken up between the layers of nickel hydroxide are exchanged for carbonate ions. However, it has been observed that the removal of soluble salts from precipitates of platelets by filter-washing is very difficult, and three- to fourfold redispersion in water and filtration is required to produce pure solids. The presence of remaining nitrates and isocyanates in the samples of Burattin et al. [51] was evident from the fact that the bands in the IR spectra did not disappear after long-term deposition–precipitation, when the (1 : 1) nickel phyllosilicate had been formed. It is of interest that the planes of (2 : 1) phyllosilicates are usually negatively charged due to vacancies in the octahedral layer. Consequently, the layers adsorb cations and exhibit cation-exchange properties; this is in contrast to turbostratic nickel hydroxide, which displays the anion-exchange properties of hydrotalcites. When the platelets of (1 : 1) phyllosilicates in the samples prepared by the above authors were also negatively charged, the adsorption of ammonia ions from the solution in which the urea hydrolysis had proceeded was likely. That remaining ammonium salts were present in the samples prepared by Burattin et al. was apparent from the fact that, at high temperatures, part of the nickel was reduced to metallic nickel, without any consumption of hydrogen [53]. At high temperatures, ammonia can reduce nickel compounds to metallic nickel. The presence of ammonium may be due to the occlusion of ammonium salts between the platelets, or to adsorption on acid sites on the (1 : 1) phyllosilicates. Another important point is the separation of samples from the liquid during deposition–precipitation. When the suspension is cooled to room temperature, the concentration of hydrogen carbonate ions rises, which leads to an increase in the solubility of nickel as nickel hydrogen carbonate complexes [55]. Whereas the redissolution of nickel phyllosilicates during cooling to room temperature is not likely, the dissolution of β-nickel hydroxide almost certainly occurs. It is gratifying that Che and coworkers have published their experimental data accurately, and in detail.

The above authors also investigated the effect of urea hydrolysis on the silicas utilized. The surface area of the

References see page 465

Spherosil of 356 m^2 g^{-1} fell to about 230 m^2 g^{-1} when kept for 24 h in a urea solution hydrolyzing at 363 K but not containing dissolved nickel(II). The surface area of the 44 m^2 g^{-1} Spherosil did not change significantly during the same period. The effect on the surface area of the unloaded silica indicates the difference in reactivity of the silica. The silica of surface area 356 m^2 g^{-1} is liable to some Ostwald ripening by dissolution in the ammonia-containing solution, which is developed during urea hydrolysis. The surface of the 44 m^2 g^{-1} Spherosil is much less reactive, and consequently not affected. During the deposition of nickel the surface area of the 356 m^2 g^{-1} silica passes through a maximum of about 425 m^2 g^{-1} after 4 h, and then falls slowly to about 300 m^2 g^{-1}. The surface area of 44 m^2 g^{-1} Spherosil rises more rapidly during the initial 4 h of nickel deposition, and subsequently more slowly to about 150 m^2 g^{-1}. The effect on the silica of a low surface area indicates the reaction with nickel to platelets of phyllosilicates.

X-ray diffraction data abstracted from published sources and measured by Burattin et al. [51] on reference compounds of (2:1) phyllosilicates are listed in Table 1. The diffraction pattern of willemseite, a nickel-containing (2:1) phyllosilicate, has been included and two (2:1) phyllosilicates containing magnesium. The reference compounds of Burattin et al. involve nickel (2:1) phyllosilicates prepared hydrothermally at 298 and 423 K.

Table 2 contains details of reference compounds of (1:1) phyllosilicates. In addition to the diffraction maxima of the pattern of nickel hydroxide, the maxima of lizardite, a nickel-containing (1:1) phyllosilicate and two nickel

Tab. 1 Comparison of the diffraction maxima of (2:1) phyllosilicates

Talc (Mg)[a]	Stevensite (Mg)[b]	Willemseite[a]	Ni-prepared, 298 K[c]	Ni-prepared, 423 K[c]
	24 w		16.22 vs	
9.34 vs	11.2–12.4/ 15.5 vs	9.36 s		10.6 vs
4.56 s	4.54 vs	4.68 vw		4.51 m
	3.0–3.5 m	3.12 m	3.45 w	3.16 m
3.11 s				
2.63 wm	2.62 s			
2.60 m		2.503 m	2.54 m	2.53 ms
1.731 w	1.725 m			1.71 vw
1.527 ms	1.520 s	1.560 m	1.54 m	1.52 m
1.320 w	1.314 ms	1.170 m		
	0.994 mw			

[a] From Ref. [42].
[b] From Ref. [43].
[c] From Ref. [51].
vs: very strong; s: strong; m: medium; w: weak; vw: very weak.

Tab. 2 Comparison of diffraction maxima recorded for (1:1) phyllosilicates

α-Ni(OH)$_2$[a]	Lizardite[b]	Ni-prepared, 298 K[a]	Ni-prepared, 423 K[a]	With silica 356 m^2 g^{-1}
7.25 vs	7.216 s	11.2 s	7.68 s	8.4 vs
	4.628 m		4.45 w	4.5 m
	3.890 m			
3.36 w	3.624 s	3.61 w	3.63 m	3.7 m
	2.850 w			
2.67 m	2.668 w		2.63 s	2.66 s
2.50 m	2.506 vs	2.54 m	2.47 s	2.5 s
1.55 m	1.539 m	1.54 m	1.53 m	1.54 ms

[a] From Ref. [51].
[b] From Ref. [56].

(1:1) phyllosilicates prepared hydrothermally at 298 and 423 K [51] are included. The diffraction pattern of the material resulting from the hydrothermal treatment at 298 K may point to a (2:1) phyllosilicate. The right-hand column in Table 2 lists diffraction maxima measured on silica of 356 m^2 g^{-1} loaded with nickel during deposition–precipitation with urea for 100 h.

The analysis of the diffraction patterns of materials prepared by deposition–precipitation is complicated by the broad band of silica peaking at 0.4 nm, although during precipitation this gradually disappears. The agreement of the pattern exhibited by silica of 356 m^2 g^{-1} subjected to deposition–precipitation of nickel for 100 h with that of lizardite (as mentioned in the literature) is clear. As far as XRD is sufficiently sensitive, it can be concluded that silica with a surface area 356 m^2 g^{-1} reacts smoothly to a (1:1) phyllosilicate which exhibits the structure of lizardite. The diffraction patterns of samples obtained during reaction of the 44 m^2 g^{-1} silica are initially different from the finally recorded diffraction patterns. These two patterns are compared with those of lizardite and α-Ni(OH)$_2$ in Table 3.

As shown above for the initially precipitated basic copper salts, the nickel species initially precipitated during deposition–precipitation is also liable to further reactions, as is clearly evident from the extensive studies of Che and colleagues. As the pH levels at which the nickel(II) initially precipitates and subsequently reacts are not very different, establishment of the reactions calls for the identification of reaction products, and this has involved a wide range of identification techniques. Subsequent reaction with the silica, however, has been demonstrated beyond doubt. At pH values of between 5 and 6, and temperatures of approximately 363 K, small suspended silica particles have appeared to be fairly reactive. As mentioned above, the structure of

Tab. 3 Comparison of diffraction maxima of (1:1) phyllosilicates with those of silica of 40 m² g⁻¹ loaded with nickel by deposition–precipitation. (From Ref. [51])

α-Ni(OH)₂	Lizardite	Initial diffraction pattern	Final diffraction pattern
7.25 vs	7.216 s	7.25 vw	7.65 s
	4.628 m	4.0 (silica)	4.5 shoulder
	3.890 m		3.67 s
3.36 w	3.624 s		
	2.850 w		
2.67 m	2.668 w	2.6 w	
2.50 m	2.506 vs		
1.55 m	1.539 m	1.5 w	

the resulting phyllosilicates as apparent in electron micrographs suggests an intermediate dissolution of the silica or migration of a silica species over the surface of the phyllosilicate.

Louis [57], who continued the investigations of Che, concluded from the extensive results of the group in France that, initially, nickel hydroxide precipitates during deposition–precipitation with urea in the presence of suspended silica. The initially mechanism envisioned is that a monolayer of a (1:1) nickel phyllosilicate is formed on the silica surface, onto which subsequently brucite-type nickel hydroxide layers grow. The nickel hydroxide next reacts with the silica to form a (1:1) nickel phyllosilicate. The suggestion that transport into narrow pores can prohibit the deposition of nickel within support bodies is supported by evidence from another report by Louis and coworkers [58], in which the deposition–precipitation of nickel on zeolite-β bodies is described. The nickel is almost completely deposited on the external edge of the zeolite bodies.

The reactivity of the silica onto which the nickel(II) is deposited determines the rate at which the reaction to a phyllosilicate proceeds. Silica produced by flame hydrolysis (e.g., Aerosil and Cabosil silicas) which have a surface area in excess of 200 m² g⁻¹ were found to react rapidly to phyllosilicates. Silica obtained by the acidification of waterglass solutions also exhibits a very high reactivity. With silica of a high reactivity, (2:1) nickel phyllosilicates are often obtained. Nickel(II) can react to soluble bicarbonate complexes, as mentioned above for copper(II) [55], and consequently an intermediate solution of initially precipitated nickel hydroxide may occur. However, the fact that electron micrographs demonstrate unambiguously that the silica particles of the silica support produced by flame hydrolysis react completely to platelets of phyllosilicates, and that transport of silica must also proceed. The experimental results shown in Fig. 21, in which nickel(II) was deposition-precipitated on silica that had reacted with copper(II), are significant. The fact that the pH plot does not exhibit a maximum indicates that the nucleation of nickel phyllosilicate or nickel hydroxide on the surface of chrysocolla does not involve a nucleation barrier. The maximum displayed by the pH curves measured with deposition–precipitation by hydrolysis of urea on bare silica therefore involves restructuring of the silica surface.

The data in Fig. 24 indicate that iron(II) reacts analogously with silica upon precipitation from a homogeneous solution kept at 363 K [59]. Reaction to a new solid phase is also apparent from the maximum exhibited by the pH curve measured with suspended silica. The interaction of iron(II) with silica proceeds at pH >5.0, which is appreciably higher than the pH at which iron(III) precipitates. Dissolved iron(III) reacts to moieties of some nanometers of hydrated iron oxide, which is analogous to the reaction of aluminum to Al_{13} complexes, already at pH 2. When the pH is raised homogeneously by the hydrolysis of urea, the small moieties of hydrated iron(III) oxide do not deposit on suspended silica, but rather react to form larger clusters by olation and oxolation processes. Ultimately, the clusters of small moieties of iron(III) oxide are located between the suspended silica particles.

It is important to consider the interaction with the support more closely. Often, an electrostatic model has been used to account for the interaction of precipitating species with the support. It has been established that

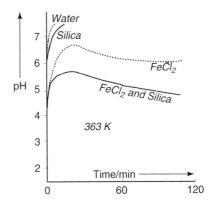

Fig. 24 Precipitation of iron(II) from a homogeneous solution by injection of sodium hydroxide at 363 K. The pH was recorded during injection into water, a silica suspension in water, a solution of iron(II) chloride, and a suspension of silica in a solution of iron(II) chloride. During the injection a flow of oxygen-free nitrogen was passed through the liquid to prevent oxidation of the iron(II). From the curves measured with water, suspended silica and the solution of iron (II) chloride the pH plot was calculated that would have been recorded had there been no interaction between the precipitating iron(II) and the suspended silica.

References see page 465

species with an electrostatic charge of the same sign as the support do not deposit onto the support. For example, the isoelectric point of silica is about 2, which implies that at lower pH levels the silica is positively charged due to an uptake of hydrogen ions, whereas at higher pH levels ionization of surface hydroxyl groups leads to a negative charge.

$$Si(OH_2^+) \rightleftharpoons SiOH = SiO^- + H^+$$

Hydrated iron oxide moieties are generated at pH 2.0. As there is consequently no attractive electrostatic interaction between the iron(III) moieties and the suspended silica, no small particles of hydrated iron(III) oxide will be deposited on the silica. If an electrostatic interaction alone were to determine the interaction of precipitating species with the surface of a suspended support, the interaction with precipitating hydrated iron(III) oxide would be small, whereas that with precipitating copper(II) hydroxide would be significant. However, the precipitation of copper(II), which proceeds at a pH of ca. 3 with most anions, is also not significantly affected by the presence of suspended silica. As shown by the experimental results presented above, copper(II) precipitates with many anions as a basic salt of the anion, being present during the precipitation without a significant interaction with a suspended silica support. Iron consequently must be applied onto silica from iron(II) solutions, as iron(II) precipitates at a much higher pH than iron(III) [60]. With iron(II), an extensive reaction with silica supports has been observed.

It is therefore felt that, besides an electrostatic interaction, a chemical interaction between the precipitating species and the support is also required. With colloidal silica solutions a reaction proceeds between the hydroxyl groups present on different silica particles. Condensation of water leads to the formation of oxygen bridges between the silica particles, which causes gelation of the silica sol; as a result, the viscosity of silica suspensions exhibits a maximum at a pH of about 5–6. Previously, it has been stated that hydroxyl ions catalyze the condensation of water from hydroxyl groups on contacting silica particles. An analogous reaction, which also is catalyzed by hydroxyl groups, will proceed between those hydroxyl groups present on the surface of silica particles and those of partially hydroxylated metal ions. The gelation of silica calls for a pH of at least 4, which agrees very well with the pH at which hydroxylated metal ions exhibit interaction with silica. Figure 25 illustrates, schematically, the reaction of the hydroxyl groups on contacting silica particles. It is relevant that, with recent discussions on the adsorption of ions onto (colloidal) particles, a chemical bond between the adsorbed ions and the surface of the particles must be assumed to account for the results of the adsorption of oxyanions on surfaces of metal oxides [61].

The extensive interaction of iron(II) with silica leads to precursors that are difficult to reduce to metallic iron. Silica loaded with iron(II) by deposition–precipitation therefore must be calcined previously at temperatures of approximately 1073 K in order to decompose the phyllosilicate generated during the deposition–precipitation. The loaded carrier assumes a red color during calcination, due to the formation of an iron(III) oxide. Reduction of the iron(III) oxide to metallic iron can proceed readily, provided that the water vapor is removed efficiently. When the silica loaded with iron(II) is kept in a reducing gas flow at temperatures as high as 1073 K, a green glassy solid results that does not contain any metallic iron.

A precursor in which the iron species can be reduced more easily can be produced by performing a precipitation of iron(III) at pH > 6. An acid iron(III) solution is injected into a suspension of silica kept at pH ≈ 6 by simultaneous urea hydrolysis or alkaline solution injection, and preferably at a temperature of ca. 363 K. Small hydrated iron(III) oxide particles are deposited onto the surface of suspended silica particles. The opposite sign of the electrostatic charge on the silica

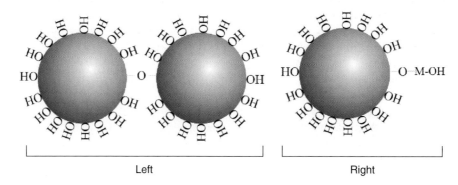

Left   Right

Fig. 25 Left: Two silica spheres connected by an oxygen bridge resulting from condensation of water from two SiOH groups. The condensation is catalyzed by hydroxyl groups. Right: A silica sphere that has bonded a (partially) hydrolyzed metal ion by an oxygen bridge resulting from the condensation of a water molecule.

and hydrated iron(III) oxide particles leads to mutual flocculation, while the higher reactivity of silica at pH > 6 also contributes to the bonding of the hydrated iron(III) oxide particles. As transport of the colloidal hydrated iron(III) oxide particles generated at the point where the acidified iron(III) solution enters the suspension into the pores of the silica support is slow, small suspended silica particles preferably prepared by flame hydrolysis (pyrogenic silica, "weisser Rusz") must be employed to achieve a uniform distribution of the iron(III) oxide over the support. Excellent results can also be obtained by oxidation of iron(II) in a homogeneous solution at a controlled pH level.

The precipitation of complex cyanides can be employed to produce interesting catalysts [62, 63]. The ability of the precursor metal ions to react with the support to a compound that is difficult to reduce is strongly suppressed by using complex cyanides. Furthermore, the cyanides react readily to the corresponding metals or alloys at not extremely high temperatures in an inert or reducing gas flow. A problem here is that the resulting metal particles are not anchored to oxidic supports; usually, the interaction of metal particles with the surface of supports proceeds via some metal ions of the precursor accommodated in the surface of the support. The size of the metal or alloy particles resulting from the thermal treatment of a support onto which complex cyanides have been deposited, can be controlled by previous oxidation of the applied cyanides to the corresponding oxides. Due to this oxidation, some metal ions of the complex cyanides are taken up into the surface of the support, and thus anchor the metal or alloy particles. However, when larger metal particles are desired, a direct reduction of the complex cyanides leads to the wanted result.

Another advantage of complex cyanides is in the preparation of alloy catalysts. As the mixing of two metals in complex cyanides is established on a molecular scale, alloy particles of a uniform chemical composition can readily be obtained. When the required selectivity of a catalytic reaction calls for active alloy particles of a uniform composition, complex cyanides are interesting precursors. Boellaard et al. have studied the deposition–precipitation of complex cyanides containing iron and iron and copper on alumina, titania, and silica supports [64]. Oxidation of the deposited cyanide precursors at a relatively low temperature (ca. 573 K) and subsequent reduction resulted in relatively small iron and copper–iron particles. The reduction proceeded very smoothly, considering the difficult reduction of iron-containing precursors applied in a conventional manner on supports. However, oxidation of the cyanide precursor on an industrial scale must be executed with a limited amount of oxygen in order to keep the temperature below the desired level.

Interesting iron–nickel alloys have been prepared by the deposition–precipitation of complex nickel–iron cyanide on titania [64]. The ratio of nickel and iron may be controlled by selecting the valency of the initial iron cyanide and employing a cyanide complex containing NO. However, a degree of caution is needed when utilizing complex cyanides. It was established that employing potassium ferrous cyanide resulted in catalysts containing much potassium. Subsequently, the uptake of potassium from potassium ferrous cyanide by titania-supported iron–nickel catalysts resulted in catalysts that were completely inactive in the Fischer–Tropsch synthesis. The use of ammonium ferrous cyanide, on the other hand, led to highly active catalysts.

C Reduction of Precursors Prior to Deposition–Precipitation A number of catalytically important metals cannot be applied smoothly onto silica, *viz.*, metals present in the most stable state as highly charged metal ions forming anions with oxygen. The most important examples of this group of metals are molybdenum, tungsten, and vanadium. The negatively charged MoO_4^{2-}, WO_4^{2-}, and VO_3^{-} ions cannot be deposited onto silica, which is also negatively charged. The electrostatic charges of the support, and of the species to be deposited being opposite, is a necessary – but not sufficient – condition to produce sufficient interaction with the support to enable deposition–precipitation to be performed. Vogt et al. [66–68] have established that decreasing the valence of the metal ions caused the metal ions to be present as hydrated cations in an aqueous solution. The electrochemical reduction of vanadium(V) and molybdenum(VI) to trivalent metal ions can be performed smoothly. Overbeek [69] has scaled-up the electrochemical reduction and achieved a very high current efficiency. With the lower-valent metal ions the usual deposition–precipitation can be performed by raising the pH of a suspension of silica in a solution of the lower-valent metal ion [66–68].

The behavior of vanadia and molybdena deposited in very finely divided form on silica is of great interest. Vogt et al. [66–68] showed that vanadia applied as vanadium(III) and subsequently oxidized to vanadium(V) by calcination at 623 K for 72 h was deposited on silica as a monolayer; the IR absorption due to vibrations of hydroxyl groups of the silica surface disappeared at a monolayer coverage of vanadia. Apparently, interaction of the precipitating vanadium(III) with the silica surface is strong. Later, de Boer [70] investigated the interaction of molybdena with silica. With molybdena deposited as molybdenum(III) and oxidized to molybdenum(VI),

References see page 465

the vibrations of hydroxyl groups of the silica surface did not disappear completely, which indicates that the molybdenum oxide species is not covering the silica surface completely. Low-energy ion-scattering, laser-Raman spectroscopy, and EXAFS have each revealed that the molybdena species are present as small hydrated clusters on the silica surface [71]. Upon dehydration, the molybdenum(VI) oxide moieties spread over the silica surface and cover the silica surface completely; rehydration leads again to clusters and to some bare silica. When the molybdenum(VI) oxide is finely divided, nucleation of MoO_3 cannot proceed and the molybdenum oxide interacts strongly with the silica. The uniform thin layer of molybdenum oxide can only be established by deposition from molybdenum(III) solutions.

Effecting deposition–precipitation by lowering the pH is of interest when metal ions are present in the stable state in aqueous solution as anions [72]. With silica, no interaction is observed, and this has led to the development of the electrochemical reduction procedure. In order to apply metal ions such as molybdenum or vanadium onto alumina, a homogeneous lowering of the pH is of interest. The pH may be decreased by the injection of either nitric acid or perchloric acid, or electrochemically. An electrochemical decrease in pH is attractive, as the hydronium ions are made available on the surface of an electrode, which is generally large as compared to the surface area of the injection tubes. In this way, a more homogeneous decrease in pH can be effected. The rate of crystallization of the hydrated oxides of vanadium(V) and molybdenum(VI) on the surface of suspended γ-alumina was, however, observed to be fairly low. Also, without suspended γ-alumina the rate of crystallization of hydrated oxides of molybdenic and vanadic acid is low, and no point in the pH curves could be distinguished at which crystallization proceeded at a significant rate. A higher pH level with a suspended support, indicating a positive interaction of the precipitating metal oxides with the support, could therefore not be measured. To prevent dissolution of the alumina supports, the pH could not be decreased to levels below about 3, at which point crystallization of the hydrated metal oxides may proceed more rapidly. Decreasing the pH of suspensions of a suitable support in a solution of an active precursor has therefore been studied much less often than raising the pH.

2.4.1.5.3 Changing the Valency of Precursor Ions

Precipitation from a homogeneous solution by changing the valence of the metal ions to be precipitated can be performed very smoothly in some cases. An increase in valency has been used with iron [18] and manganese. Iron(II) and manganese(II) are much more soluble than iron(III) and manganese(III) and (IV). Oxidation with a dissolved oxidizing agent therefore leads to precipitation from a homogeneous solution. With iron(II) the oxidation can be carried out with nitrate ions [18]. As hydrolysis of the resulting iron(III) ions releases hydrogen ions, the pH must be kept constant by simultaneous injection of an alkaline solution or by urea hydrolysis. Usually, the oxidation of iron(II) from a homogeneous solution leads to the precipitation of magnetite, Fe_3O_4. When other suitable bivalent metal ions are present in the solution, other ferrites, such as, nickel or cobalt ferrite, result.

$$3Fe^{2+} + NO_3^- + 4H^+ \rightleftharpoons 3Fe^{3+} + NO + 2H_2O$$

$$Fe^{2+} + 2Fe^{3+} + 4H_2O \rightleftharpoons Fe_3O_4 + 8H^+$$

$$Ni^{2+} + 2Fe^{3+} + 4H_2O \rightleftharpoons NiFe_2O_4 + 8H^+$$

Finely divided magnetite deposited on a support is liable to oxidation and hydrolysis, which leads to hydrated supported iron(III) oxide. To prevent oxidation and hydration, the loaded support must be dried in an inert gas. After drying, the supported magnetite is fairly stable on exposure to atmospheric air, though reaction to γ-Fe_2O_3 may proceed. Reduction to metallic iron is facilitated by preventing the formation of hydrated iron(III) oxide particles.

Precipitation by reduction of metal ions has many applications [73]. A procedure has been developed to precipitate molybdenum(VI) by reduction with hydrazine to a lower-valent insoluble molybdenum species. We will return to the reduction of molybdenum(VI) in Section 2.4.1.5.7. Precipitation from a homogeneous solution of chromium by reduction of chromium(VI), which is soluble at a pH of ≈ 7, to chromium(III), which is insoluble at this pH range, has also been carried out. The reduction of metal ions such as silver, palladium, or platinum to the metallic state can lead to very finely divided metal particles [74]. De Jong has studied the precipitation by reduction of silver ions in the presence of a suspended silica support [75, 76]. Usually, a bimodal particle size distribution results, with the silver complexes adsorbed on the support reacting with the reducing agent to form very small silver particles (≈ 1 nm), whereas in the bulk of the solution the more difficult nucleation of metallic silver leads to relatively large particles (≥ 10 nm). An analogous behavior has been observed with other precious metals. With suspended silica supports loaded at pH ≈ 7, it was shown that the electrostatic charge of the precursor complexes determines whether precipitation by reduction to the metal on the support can proceed. Only positively charged complexes, such as the silver ammine complex, resulted in small silver particles being deposited evenly over the silica support. The amount of silver complexes that can be adsorbed on the silica support, however, was also found to be limited. A higher concentration of silver simply resulted in large silver particles being formed in the bulk of the solution.

The controlled growth of previously deposited, very small silver particles can also be effected. The procedure involves first establishing the minimum temperature at which reduction of the metal proceeds in the bulk of the solution. When reducing silver ammine complexes with formaldehyde, the temperature to effect reduction in the bulk liquid is about 310 K. Metallic silver particles have been shown to catalyze the reduction of the silver ammine complexes. By suspending a silica support covered with small silver particles in a solution of $Ag(NH_3)_2^+$, cooling the suspension to 273 K, and subsequently injecting formaldehyde, a controlled growth of the silver particles can be effected. It is remarkable that the silver particles grow to an elongated shape; apparently, the electrostatic repulsion between the positively charged silver complexes brings about an approach from opposite sides. Thermal treatment at a moderate temperature in a reducing gas flow suffices to produce symmetric metal particles. An analogous procedure can be carried out with small particles of precious metals, such as platinum or palladium. It is also possible to grow different metals on small metallic nuclei, such as silver onto platinum particles.

2.4.1.5.4 Removal of a Complexing Agent

The best-known procedure in which a complexing agent is removed involves the volatilization of ammonia from ammine complexes. The first patent to mention the preparation of γ-alumina-supported copper catalysts by converting soluble or insoluble copper compounds into highly soluble ammine complexes was published in 1986 [77]. Copper compounds such as copper oxide or basic copper carbonate (malachite, $CuCO_3Cu(OH)_2$) are attractive starting compounds, the preference being for malachite. Usually, a mixture of ammonium carbonate and ammonia is used to obtain an intense blue-colored solution of the copper complex. Alumina support bodies are impregnated with the copper ammine solution, and the impregnated bodies are gradually heated to a temperature of 573 K. The procedure applies to the pore volume of the support bodies only, and does not involve deposition from a liquid volume larger than the pore volume on the support. The copper ammine complex thermally decomposes, which leads to precipitation of the copper on the alumina. The preparation of copper–zinc oxide–alumina catalysts is described in Ref. [77]. The patent process starts with metallic copper and zinc, which are dissolved in a solution of ammonium and ammonium carbonate by oxidation with air at ambient temperature. Subsequently, powdered alumina is suspended in the copper–zinc solution and the ammine complexes are decomposed at about 357 K. The resulting solid was filtered, calcined at 643 K and, after addition of 3 wt.% graphite, shaped into tablets and finally calcined at 633 K.

Later, Schaper et al. [78] employed the decomposition of a nickel ammonium complex to prepare alumina-supported nickel catalysts. With γ-alumina and ϑ-alumina, a nickel–aluminum hydrotalcite (a hydroxycarbonate) is deposited on the remaining alumina. As the reactivity of α-alumina is not sufficiently high, α-alumina is not suitable for this preparation procedure. The finely divided γ-alumina is so reactive that, with the precipitating nickel, a hydrotalcite develops. The catalyst (nickel content ca. 18 wt.%) resulting after calcination and reduction of the nickel contains nickel particles of 4–7 nm and is extremely stable at 1073 K in hydrogen–steam mixtures at 10 Pa (1 bar).

It is important to start from an insoluble compound and to dissolve the metal ions of the solid compound by complexing with ammonia. An example is nickel hydroxide, which is dissolved in a mixture of ammonium carbonate and ammonia; removal of the ammonia leads to precipitation of nickel carbonate. In this way it is difficult to achieve a sufficiently high concentration of dissolved nickel. Another problem is the relatively high pH required to bring the ammonium concentration at the required level to dissolve the nickel completely. At high pH due to the required ammonia concentration, supports such as alumina and silica dissolve. Upon desorption of the ammonia, hydrotalcites and phyllosilicates will precipitate.

A recent modification of the procedure involving volatilization of ammonia has been described [79]. A silica support is impregnated with a cobalt carbonate ammine complex and subsequently kept at a temperature where the ammonia desorbs from the silica bodies; this leads to the precipitation of basic cobalt carbonate. Suspending silica in a solution of cobalt ammine carbonate, and subsequent precipitation of basic cobalt carbonate, is also mentioned.

Procedures have also been developed to oxidize complexing agents such as EDTA (ethylene diamino tetraacetic acid) or ethylene diamine. The oxidation was performed with hydrogen peroxide (perhydrol) at a pH where the uncomplexed metal ions are insoluble [80]. The procedure was successful with a number of metal ions, but with copper a cautious operation was required. The oxidation of a copper–EDTA complex shows an induction period of about 30 min, after which copper oxide is deposited in very finely divided form onto the support, which provides a very efficient catalyst for the decomposition of hydrogen peroxide. As a consequence, the hydrogen peroxide decomposes rapidly to release a large volume of oxygen, which in turn causes a blow

---

References see page 465

out of the reactor contents. A film reactor was used to carry out the preparation involving the oxidation of EDTA complexes by hydrogen peroxide on a large scale, as this type of reactor allows the rapid evolution of gas.

2.4.1.5.5 Electrochemical Deposition–Precipitation

Base metals are generally cheaper than the corresponding metal salts. One reason for this is that the production and transport of the salt (which is heavier per unit weight of metal than the pure metal) can be avoided. When producing supported catalysts, it is therefore often advantageous to begin with the metal to be applied onto an appropriate support, and not from a salt of the metal. According to the usual procedure, the metal must be dissolved in a suitable acid in the catalyst production plant. At present, to minimize environmental problems, hydrochloric acid is most often used, although nitric acid can be more easily removed from the catalyst by a thermal treatment. Precipitation of the metal as a hydroxide, hydrated oxide, or as a basic salt, calls for the use of soda or other alkaline compounds and leads to the production of salt of the alkali metals used, which must be disposed of.

When a metal is applied without being completely dissolved intermediately, an attractive procedure can be utilized. Van Stiphout et al. have developed a precipitation method in which a metal such as copper, nickel, or cobalt can be deposited onto a suspended support by anodic dissolution of the metal [81–84]. For this, metal rods can be used, as can metal scrap which may be placed into a polymeric basket and connected to an electrode. When the metal is the anode and the voltage for dissolution is lower than that required to produce oxygen from water, dissolution of the metal proceeds readily. Deposition of the metal onto the cathode must be prevented, however. At the cathode, hydrogen and hydroxyl ions should be produced. Van Stiphout et al. investigated two set-ups to prevent deposition of the catalytically active metal onto a platinum cathode. With a polymer membrane capable of rapid transport of hydrogen ions and water, the migration of anodically dissolved metal atoms to the cathode, can be either prevented or impeded. As an alternative, a thin chromia layer may be deposited onto the platinum. Apparently, the nucleation of a metal on chromia proceeds with much greater difficulty than on platinum metal, because the deposition of metal on the thus-covered platinum electrode did not occur to any significant extent.

The above authors have established that the procedure with the polymeric membrane is unsatisfactory, because at high current densities metallic nickel nucleates on the polymeric membrane, and more nickel is subsequently deposited. Although a reasonable loading of a suspended silica carrier could be achieved, a marked fraction of the nickel was deposited on the polymeric membrane. The chromia layer was observed to yield much better results.

Silica-supported catalysts with nickel, copper, and copper–nickel as the active components have been prepared using the electrochemical precipitation procedure. The thus-prepared catalysts showed analogous characteristics as catalysts prepared by deposition–precipitation using urea. In particular, the copper–nickel catalysts are of great interest. Measurement of the saturation magnetization and temperature-programmed reduction experiments indicated the formation of copper–nickel alloys after reduction of supports loaded simultaneously or sequentially with copper and nickel. As the dissolution of nickel or copper is accompanied by the reaction of hydrogen ions to gaseous hydrogen, resulting in the formation of hydroxyl ions, the pH of the suspension does not vary as long as the hydroxides or phyllosilicates containing $Si_2O_5^{2-}$ species are generated. When no suspended silica is present, anodically dissolved copper initially reacts to basic copper salts, which leads to a rapid rise in pH, as the loss of hydrogen ions is not compensated by an equivalent consumption of hydroxyl ions. With suspended silica, however, the pH remains virtually constant, except when nitrate ions are present and reaction to nitrite or nitrogen oxide can proceed at the cathode. Both reactions consume hydrogen ions.

Electrochemical procedures thus can be used in the production of solid catalysts according to three different procedures:

- by homogeneously decreasing the pH of a suspension of a support in a solution of a precursor which precipitates at a lower pH by the consumption of hydronium ions at the cathode
- by reducing the higher-valent metal ions present (usually as oxyanions) to a lower-valent state, where they react less acidic and may be deposited by homogeneously raising the pH
- by anodic dissolution and subsequent precipitation of the metals, or combination of metals.

Although it has been established that the electrochemical production procedures may be very well controlled, they are rarely used in an industrial setting to produce heterogeneous catalysts, mainly because the scale-up of such processes is difficult.

2.4.1.5.6 Deposition–Precipitation of Gold

During the late 1980s, gold particles smaller than 5 nm appeared to be active catalysts for a number of oxidation reactions, for the oxidation of CO, the reduction of NO with CO, and for the CO shift conversion [85–88]. The oxidation activity is so high that, below 273 K the oxidation of CO is substantial also in the presence of water vapor. Thus, supported gold catalysts are employed in the removal of CO and odorous compounds at room temperature, in selective CO sensors, and in sealed CO_2 lasers. Haruta and coworkers were the first to report details of the high activity of small gold particles [85]. These authors noted that gold particles should be applied to reducible semiconducting metal oxides, to hydroxides of alkaline earth metals, or to amorphous zirconia. Initially, the supported gold catalysts were prepared exclusively by the coprecipitation of gold with (a precursor of) the support. Especially, catalysts prepared by the coprecipitation of hydrated iron oxide and gold have been investigated [87–89].

Since, with coprecipitation of precursors of the catalytically active component and the support, the porous structure and surface area of the final catalyst is difficult to control, the deposition of a gold precursor on the surface of a separately produced support is an attractive option. Thus, Haruta and coworkers developed a deposition–precipitation procedure for the deposition of small gold particles on commercial supports, namely magnesia, and two different titanias – an amorphous titania of a BET surface area of 110 m^2 g^{-1} and a crystalline titania (anatase) of 40 m^2 g^{-1} [86]. The supports were suspended in a solution of $HAuCl_4$. By adding Na_2CO_3, the pH of the titania suspensions was raised to 7, while the suspension of the more basic magnesia assumed a pH of 9.5. The results of XRD analyses showed that the magnesia reacted in solution to magnesium hydroxide. The aqueous suspensions were stirred for 2 h and subsequently filtered and washed. It appeared that the supports took up only a limited fraction of the gold present in the solution. The loading of the magnesia was 2 at.%, and that of the titania 1 at.%. The magnesia-supported catalysts were calcined at 523 K, and the titania-supported catalysts at 673 K for 5 h. The addition of magnesium citrate after dispersion of the solid in the gold solution was required to obtain the magnesia support with a desired gold particle size of $\approx$3 nm, whereas without the addition of Mg citrate the particle size was 10–14 nm. Consequently, the activity in the oxidation of CO of the magnesia-supported catalyst prepared with Mg citrate was much higher, as would be expected. It has been established that the drying of a support impregnated with a solution of a metal citrate and calcination leads to a thin layer containing very small particles of the corresponding metal oxide on the surface of the support. It is clear that the small magnesium oxide particles suppress the mobility of small gold particles over the surface of the magnesia. With the amorphous titania, an addition of Mg citrate or Mg nitrate was required to obtain small gold particles and to achieve an oxidation activity significantly higher than that of the magnesia-supported 3-nm gold particles. As crystallization of the titania leads to larger gold particles, magnesia (which suppresses the recrystallization of titania) is needed to obtain small gold particles.

With the crystalline titania, an addition of magnesium is not required; without magnesium addition a very high activity and small gold particles were also obtained. It is likely that there are oxygen vacancies on the titania surface; at oxygen vacancies one or more lower-valent titanium ions are present. The gold particles may be anchored by the oxygen vacancies, where the lower-valent titanium ions can chemically bond the metallic gold atoms.

In a subsequent report [90], the group of Haruta used sodium hydroxide to raise the pH of the solution of $HAuCl_4$. The support is then suspended in the solution, the pH readjusted, and the temperature of the suspension brought to 343 K; the suspension is then aged for 1 h. Although the amount of gold in solution may lead to a loading of 13 wt.%, the loading actually achieved is always lower. As a function of the pH of the suspension, the gold loading exhibited a sharp maximum of about 8 wt.% at pH 6. The loadings were lower at both lower and higher pH values. Others have employed a similar procedure to load the supports with gold. While the temperature of the suspension was not always noted, the gold solution is often held at 343 or 353 K during deposition of the gold. Although most publications deal with the application of gold on titania using the above deposition–precipitation procedure, the technique has also been employed with separately produced Fe_2O_3 and Co_3O_4 supports [87, 88]. It is important to realize that the supports used successfully with the deposition–precipitation procedure do not dissolve in alkaline solutions; titania is therefore an attractive support. Some authors have used gold catalysts prepared both by coprecipitation with the support and by deposition–precipitation [82–84] after drying at low temperatures [84], whereas others calcine the catalysts previously at 573 or 673 K in air [86, 88, 90] to convert the gold species deposited on the support into metallic gold. Evidence has been published that gold oxide is the active species in the catalysts [89], but it seems difficult to maintain an appreciable fraction of the gold in an oxidized state. Exposure to oxygen and CO also leads to a reduction of the gold species in the catalyst.

The titania supports employed are usually produced by flame hydrolysis of titanium tetrachloride of a surface area of about 40 to 50 m^2 g^{-1}. Normally, the isoelectric point of the titanias used is stated to be at pH 6, but Centeno et al. [92] showed the pH of the isoelectric point of the P25 titania produced by Degussa to be 7.2. The surface area was 56.9 m^2 g^{-1}, but after the deposition of gold and calcination for 2 h at 773 K this was lowered to 39.4 m^2 g^{-1}. The thus-produced titanias are usually a mixture of much anatase and less rutile. It is remarkable that only a limited fraction of the gold present in solution is deposited on the titania. Whereas, the amount of gold

---

References see page 465

in solution should lead to a loading of 13 wt.%, the actual loading of the titania support is often only about 3 wt.%; on occasion, however, the loading is even less at 1 wt.%. In addition, with a coprecipitation procedure the gold does not precipitate completely. For example, when coprecipitating gold and hydrated iron oxide by adding sodium carbonate to a solution of iron(III) nitrate and gold chloride, Khoudiakov et al. [89] noted that only 12–13% of the gold present in the solution was taken up within the precipitated solid. Haruta et al. [87], however, found a much higher fraction of the dissolved gold to be deposited within the solid during coprecipitation with iron oxide. A solution containing gold corresponding to a loading of 13.0 wt.% resulted in a gold content of 11.3 wt.% of the coprecipitated gold–iron(III) oxide solid.

In a solution of pH 7–8 at 353 K, gold is soluble as $AuCl_2(OH)_2^-$ and $AuCl(OH)_3^-$. Peck et al. [94] reported that gold is present as $Au(OH)Cl_3^-$ at pH levels close to 7. It has been established that the gold deposited on titania has no chloride ligands, and chemical analysis of the washed gold-loaded support does indeed not indicate the presence of chloride. Tsubota et al. [90], Zwijnenburg et al. [95], and Lin et al. [96] have all published evidence that the deposited gold does not contain chloride ligands. The deposition–precipitation is, therefore, likely to involve reaction of hydroxyl groups of the surface of the titania with the soluble gold complexes. In two publications, Louis and coworkers [97, 98] investigated in detail the nature of the deposited gold species. These authors described their preparation procedures accurately when investigating the adsorption of anionic $AuCl_4^-$ and cationic $Au(en)_2^{3+}$, and studied the incipient wetness impregnation with aqueous solutions of $HAuCl_4$, as well as the deposition–precipitation by sodium hydroxide of Haruta [90] and deposition–precipitation by the hydrolysis of urea. For Haruta's deposition–precipitation procedure, the period of suspension ageing was varied, and two different concentrations of gold chloride were employed at pH 7 and 8 of the suspension; the effect of magnesium citrate addition was then investigated. After 1 h, the loading was 1.8 wt.% gold, and after 2 h was 2.4 wt.%. Ageing for 16 h did not further increase the loading, although the concentration of gold was sufficient for a loading of 8 wt.%. The loaded titania support was dried *in vacuo* for 2 h at 373 K and subsequently calcined for 4 h at 573 K, which resultad in average size of the gold particles as determined by TEM of 1.5 to 1.8 nm. The particle size was smaller than that reported by others, although this may be due to the catalysts being stored in the dark, in desiccated form, at room temperature. Nevertheless, the size of the gold particles increased during storage, from 1.7 to 2.1 nm. Thus, the samples were stored as described after drying and calcined immediately before characterization.

The most important information regarding the gold species deposited during ageing of the suspension in solution at pH 7–10 was obtained with X-ray absorption spectroscopy. Louis and coworkers used samples that had been dried after loading, diluted with cellulose and pressed to create a pellet for X-ray absorption measurements. The X-ray absorption near edge spectroscopy (XANES) spectrum of the titania dried after loading with gold strongly resembled that of gold hydroxide, and differed from that of $HAuCl_4$. A comparison of the Fourier transform of the EXAFS signal of $Au(OH)_3$, Au foil, and $HAuCl_4$ with that of the dried titania support loaded with gold by deposition–precipitation with sodium hydroxide, showed that the transform of the loaded support was almost identical to that of gold hydroxide. The number of oxygen atoms around the gold atom was seen to be 4 at a distance of 0.198 nm, which was equal to the number and distance in gold hydroxide. Nevertheless, these authors hesitated to attribute the EXAFS spectrum of the loaded titania to gold hydroxide, as a feature at a distance of about 0.37 nm was lacking. This distance had been tentatively attributed to an Au–Au distance due to either gold hydroxide oligomers or to anhydrous Au_2O_3. Louis and coworkers therefore attributed the deposition to a reaction of hydroxyl groups of the titania surface with a gold chloride hydroxide complex according to

$$TiOH + [AuCl(OH)_3]^- \rightleftharpoons Ti-[O-Au(OH)_3]^- + H^+ + Cl^-$$

It is interesting that Haruta et al. [87], when studying the Fourier transform of the EXAFS profile of coprecipitated $Au-Fe_2O_3$, concluded that below 473 K the gold could exist as an oxidic species similar to hydrous gold oxide, Au_2O_3.

The number of hydroxyl groups at the surface of titania is limited, which in turn restricts the amount of gold that can be deposited. At pH 8, the density of undissociated hydroxyl groups is less than that at pH 6. Accordingly, Louis and coworkers identified a gold loading of 3.3 wt.% with a suspension of pH 7 and a loading of 1.8 wt.% with a suspension of pH 8 after ageing for 1 h. Haruta's group [85] also observed a gold loading that was lower with a suspension at pH 8 than with a suspension at pH 6. The distribution of the hydroxyl groups on the surface of titania is not evident from the results of Louis and coworkers. Akita et al. [100] employed a cleaved fragment of a single crystal of titania having the rutile structure for the deposition of gold according to Haruta's deposition–precipitation technique. Investigation at high resolution within a scanning electron microscope showed an inhomogeneous coverage of the titania with gold particles. The gold particles were deposited along (multi-atomic) steps on the surface. It is difficult to decide whether the distribution evident in the images had been established during the thermal treatment of the loaded support for 4 h at 573 K or during the deposition from the solution. Akita et al. [100], using a powerful SEM, published impressive results that may indicate the formation of a layer of titanium hydroxide on the titania surface during deposition in aqueous suspension. Dehydration in the vacuum of the electron microscope would lead to an amorphous layer which covered the titania crystallites, and in which the gold particles appear to be embedded.

Haruta's deposition–precipitation procedure has one of the characteristics of the general deposition–precipitation process, namely interaction of the support with the catalytically active precursor. One difference, however, is that with the general deposition–precipitation procedure, interaction with the support leads to nuclei that grow selectively and thus prevent nucleation and growth in the bulk of the solution. With Haruta's procedure, reactive surface hydroxyl groups on the support serve as the precipitating agent. The density of the reactive hydroxyl groups on the titania surface is such that only a fraction of the dissolved precursor can be deposited on the support.

It is of interest to investigate whether the usual deposition procedure using urea hydrolysis to raise the pH homogeneously can also raise the loading of gold on supports, such as titania, and also maintain the small size of the resultant gold particles. Dekkers et al. [101] performed the first deposition–precipitation of gold employing urea hydrolysis, carrying out the deposition at 353 K and obtaining a loading of 4.5 wt.%. However, the size of the resulting gold particles was disappointingly large (7.5 nm). Later, Louis and coworkers [97, 98] used the same procedure to deposit gold on titania, with much greater success. The latter authors employed aqueous solutions containing 1.1×10^{-3}, 1.6×10^{-3}, or 4.2×10^{-3} mol L^{-1} gold chloride and 0.42 or 0.84 mol L^{-1} urea (i.e., a very large excess of urea). The initial pH was about 2, and the suspension of titania in the gold chloride-urea solution was brought to 353 or 363 K. Samples were taken after 1, 2, 4, 16, and 90 h, keeping the suspension at 353 K.

It was surprising that, after only 1 h, with the pH 2.99, the gold had completely precipitated. The wet samples had taken on an orange color, in contrast to the white colors of the samples resulting from deposition–precipitation with sodium hydroxide. The average size of the gold particles obtained after calcination at 573 K was 5.6 nm (range 2.3 to 10.2 nm). After 4 h the pH was 7 and the average gold particle size after calcination had fallen to 2.7 nm (range 0.7 to 5.5 nm). Holding the suspension temperature at 353 K over 90 h increased the pH to 7.84 but did not significantly change the gold particle size. Holding the suspension temperature at 363 K led, after 4 h, to a pH of 7.8 and an average particle size of only 1.7 nm after calcination. The pH after 4 h at 353 K with a urea concentration of 0.84 mol L^{-1} was 7.2, and the average size of the gold particles 2.0 nm. Starting with gold concentrations of 1.1×10^{-3} and 1.6×10^{-3} mol L^{-1}, and holding the suspension at 353 K, resulted in particle sizes of 2.0 and 2.3 nm, respectively. Although the gold particle size obtained by deposition–precipitation with urea is slightly larger than that resulting from deposition–precipitation using sodium hydroxide, the particle size was much less than that obtained by Dekkers et al. [101]. It is important to note that deposition–precipitation with urea leads to a higher loading of gold, without a substantially larger size of gold particles and, hence, to more active catalysts. In separate studies, the authors investigated the influence of the conditions of thermal treatment and of storage of the gold-on-titania catalysts [99].

The structure of the gold species deposited on titania during deposition–precipitation with urea has not been elucidated. The Fourier transform of the EXAFS signal indicates four oxygen or nitrogen neighbors at a distance of 0.203 nm, which is slightly larger than the distance of the four oxygen neighbors around the gold atom with the deposition–precipitation with sodium hydroxide (i.e., 0.198 nm). One problem here is that oxygen and nitrogen cannot be distinguished in X-ray spectra.

Khoudiakov et al. [89] compared the preparation of gold-on-iron oxide catalysts by coprecipitation and by deposition–precipitation with urea. The latter approach was performed with hydrated iron oxide freshly precipitated with sodium carbonate at pH 9.5, thoroughly washed with warm water, separated from the liquid by centrifugation, and redispersed in a solution of urea and gold chloride. The authors executed the coprecipitation in analogy to the above preparation of the iron oxide support by adding the gold chloride to the iron nitrate solution. The gold was completely deposited with deposition–precipitation using urea, with gold recoveries of 99.2 and 99.8% being measured. With coprecipitation, the gold recovery was much less (11.8, 12.7, and 12.9%), as noted above. It is likely that, at the higher pH levels employed in coprecipitation, the gold dissolves as Au(OH)$_4$ species.

The XRD pattern of the dried catalysts displayed only two weak and very broad peaks indicating the presence of α-Fe$_2$O$_3$, hematite. After calcination at 573 K for 3 h, relatively sharp reflections of hematite could be seen, and two broadened peaks due to metallic gold. The broadening of the gold reflections points to a size of 5 nm of the gold particles, which was in agreement with that observed with TEM of the calcined catalysts (3–7 nm). It is interesting that X-ray photoelectron spectroscopy (XPS) measurements indicated that, in the dried catalysts prepared by deposition–precipitation with urea, 70% of the gold was present as metallic gold, compared to 85% in the coprecipitated catalysts. This contrasted with the findings of Louis and colleagues, who did not observe the presence of metallic gold in their dried samples. It is possible that, in the vacuum of the XPS apparatus, a large fraction of the gold species loses oxygen and becomes metallic. A relevant observation is that the dried catalyst prepared by coprecipitation exhibited initially a high activity in the oxidation of CO at room temperature, which gradually declined with time. The catalyst prepared by deposition–precipitation with urea, on the other hand, was initially not active, but subsequently developed an activity comparable with that of the coprecipitated catalyst after 35 min, which declined analogously with that of the coprecipitated catalyst. XPS measurements on the catalyst prepared by deposition–precipitation showed some nitrogen, but after sputtering of the sample the nitrogen signal had disappeared. The variation in catalytic activity of the differently prepared catalysts of Khoudiakov et al. [89] indicates that the gold species deposited by urea hydrolysis differs from that deposited at pH 7 to 10 adjusted by the addition of sodium carbonate or sodium hydroxide. The actual composition and structure of the gold complexes is, however, difficult to establish – the more so as the complexes may readily decompose to metallic gold.

2.4.1.5.7 Deposition–Precipitation within Pre-Shaped Support Bodies
The application of one or more precursor(s) uniformly over the internal surface of pre-shaped support bodies is an attractive option for the development of industrial catalysts within a short period of time. As impregnation and drying often leads to deposition more or less exclusively at the external edge of the support bodies, a better procedure would be highly desirable.

Knijff has extensively studied deposition–precipitation within the pore-systems of pre-shaped support bodies [6, 102]. Impregnation was performed with a solution of the active precursor and urea or ammonium or sodium nitrite. Preferably, the support bodies were evacuated before the impregnating fluid was admitted. In order to prevent evaporation of the liquid while the impregnated support bodies were maintained at a temperature of approximately 350 K, the open container in which the impregnated support bodies were present was inserted in a vessel in which an atmosphere of water saturated at 350 K was maintained. With the hydrolysis of urea and with the disproportionation of nitrite, gases are evolved. It is possible that gas bubbles generated within narrow pores may push the liquid out of the pores, thus preventing deposition of the active precursor onto the pore walls. However, some experiments have shown that removal of the liquid from (part of) the pore system by gas bubbles did not occur. Apparently, dissolution of the gases in the liquid present in the pore system effectively transported the evolved gases out of the pore system of the supports.

The results obtained were excellent, with extremely good distributions of the active precursor over the internal surface of the support bodies being observed. The difficulty, however, is the loading that can be achieved. The limited pore volume of pre-shaped support bodies and the solubility of the precursor to be deposited, and of the agent(s) to perform deposition–precipitation, cause multiple impregnations to be required to arrive at loadings characteristic of base metal (compounds). With urea and simple nitrates, however, highly concentrated solutions can be utilized, which allows high loadings of the active component to be applied. The difficulty here is to fill the pore system completely with the highly viscous concentrated solution. Thus, deposition–precipitation within the pores of pre-shaped support bodies is attractive for supports of a very high pore volume, and calls for additional research when high loadings are desired with supports of a low or intermediate pore volume.

De Jong [103] performed a very interesting study on the application of deposition–precipitation on pre-shaped support bodies by comparing the application of manganese(II) by homogeneously raising the pH level on silica powder and silica granules. The silica powder was Aerosil (Degussa; 281 $m^2\,g^{-1}$); the granules were from Grace Davison (0.6–1.4 mm, 1.2 mL g^{-1} pore volume and 310 $m^2\,g^{-1}$). The pH of suspensions of the supports in solution of manganese(II) nitrate was raised by the hydrolysis of urea executed at 363 K. The urea-to-manganese ratio was 5. The pH curves recorded during the precipitations indicated interaction of the precipitating manganese hydroxide with the silica surface; the pH with both silica powder and silica granules remained lower than that measured without suspended silica. In contrast to the pH curves recorded with suspended silica, the curve measured without suspended silica passed through a maximum. It is important that the pH curves measured with powdered silica and with silica granules did not show any significant difference. Apparently, transport into the pores of the granules did not affect the pH during the precipitation.

De Jong not only measured the pH curves, but also determined the manganese(II) concentration in solution during the precipitation. The kinetics of the urea hydrolysis were calculated from the increase in pH using a first-order rate with respect to the urea concentration. A first-order rate was also observed for the removal of manganese with suspended silica with respect to the manganese concentration. De Jong hesitated in assuming a first-order rate for the latter reaction, as the course of the manganese concentration with time suggested a zero-order dependence, and the amounts of manganese precipitated remained low in the absence of suspended silica. The rate of hydrolysis of urea appeared to be a factor of ≈5 higher than the removal of dissolved manganese from the solution, and without suspended silica, by a factor of 10. As the hydrolysis of urea is rate-determining in the precipitation process, a large fraction of the ammonia generated by urea hydrolysis is not effective in the precipitation of manganese(II), but is volatilized from the solution at 363 K.

De Jong furthermore assessed the rates of transport within the liquid to the surface of the granules, and within the pores of the granules. The rate of transport through the liquid phase was calculated from published data, and the rate of transport within the granules from the distribution of manganese within the granules, as determined by electron microprobe analysis. The ratio of the manganese concentration at the center of the granule and at the external edge enables the Thiele modulus to be calculated. The Thiele modulus, in combination with the diffusion coefficient and the geometry of the granules, provides a value for the first-order rate constant of the reaction of manganese with the silica surface. The deposition of manganese(II) into granules of silica may be considered as being controlled by three resistances connected in series, namely the reverse rates of urea hydrolysis, transport in the liquid to the granules surface, and transport within the pores of the granules. De Jong's quantitative results indicate a ratio of the three resistances of 90 : 1 : 10, which indicates that the rate of urea hydrolysis is also rate-determining with pre-shaped bodies. De Jong refers to values for the rate constants of the deposition reaction of rhodium and nickel on silica; these are a factor of about 3 and 12 lower than the deposition reaction of manganese(II) on silica. De Jong's results showed that it is difficult to achieve a homogeneous distribution of an active precursor within pre-shaped support bodies of 0.6 to 1.4 mm. However, when a preferential deposition of the active precursor at the external edge of support bodies is desired, deposition–precipitation can provide excellent results.

De Jong [103] also developed a new concept to arrive at a desired distribution of active precursors within pre-shaped support bodies – that is, employing the acid or basic properties of the support. De Jong employed reduction reactions depending on the pH of the solution. The first reaction was the reduction of molybdenum(VI) with hydrazine to molybdenum(III) [73], which occurs faster at a lower pH:

$$4MoO_4^{2-} + 3N_2H_4 + 4H_2O \rightleftharpoons 4Mo(OH)_3 + 3N_2 + 8OH^-$$

When silica support bodies are suspended into a solution of molybdate and hydrazine, and the solution penetrates into the pore system of the silica, the pH within the pores of the silica bodies is lower than that in the bulk of the solution. At the external edge the pH will rapidly rise, but in the center of the support bodies lower pH levels will be maintained for much longer periods of time. The lower pH at the center of the support bodies will bring about that the molybdenum(VI) is locally much more rapidly reduced to insoluble molybdenum(III) hydroxide. Three experimental procedures are used. In the first procedure, the support bodies are impregnated with a solution of ammonium molybdate, the pH of which had been installed at 8.7 by the addition of concentrated ammonia, after which the solution is cooled to 273 K. Subsequently, the silica support is suspended in the molybdate solution. Separately, a hydrazine solution is prepared, the pH of which is also installed at 8.7 by the addition of acetic acid. The hydrazine solution is also cooled to 273 K and added to the suspension of the silica spheres. The suspension is rotated in a round-bottomed vessel under nitrogen, and maintained at 273 K for 1 h; the temperature is then slowly raised over a 20-h period to 333 K. The second and the third procedures use a mixed molybdate–hydrazine solution at a temperature of 273–283 K. With the second procedure, the mixed solution is circulated through a small vessel containing the silica support bodies to be loaded. The silica bodies are held at a higher temperature than that of the circulating liquid (273–283 K). With the third procedure, the silica bodies are brought into a relatively small volume of a concentrated molybdate–hydrazine solution kept at 273 K. After 15 min, the reduction reaction is quenched by diluting the reaction mixture with water at 273 K. The circulation procedure can utilize a much larger liquid volume than the other two procedures, and is very versatile. In fact, it is possible to obtain a concave molybdenum distribution, while continuation of the reduction procedure with a solution of a sufficiently high concentration also allows a flat distribution or a distribution of a substantial level to be achieved, but with a heavier loading at the external edge of the support bodies.

De Jong [103] also envisioned two other reduction procedures, both of which exhibited a reaction producing hydrogen ions. The procedures involve the reduction of copper(II) to Cu_2O with hydrazine, and the reduction of

---

References see page 465

Ag(I) to metallic silver with formaldehyde. To cope with the hydrogen ions released by the reactions, γ-alumina support bodies were employed. An egg-yolk distribution of the copper and silver was indeed observed in the electron microprobe analysis. When the reaction of copper(II) with hydrazine was executed with silica bodies, the copper(I) oxide was deposited at the external edge, as would be expected with the more acidic silica. Based on his experimental findings, De Jong concluded that the pH gradients established within larger support bodies inhibited by solutions of pH differing from the isoelectric point, can exist for sufficiently long periods to be employed with deposition–precipitation processes.

Deposition–precipitation by the reduction of soluble silver compounds to metallic silver has also often been proposed for preparing silver catalysts for the production of ethylene oxide by the oxidation of ethene. In order to promote the rapid transport of ethylene oxide from the catalyst bodies, and to suppress subsequent undesired reactions of the ethene oxide, α-alumina of a very low surface area (<1 m^2 g^{-1}) is usually employed as a support. With powdered supports, good results can be obtained by a reduction of the silver ammine complex with, for example, formaldehyde. However, any reduction in the bulk of the solution, resulting in large silver particles, must be prevented. As silver nitrate solutions are liable to decomposition by (ultraviolet) light, they must be treated with nitric acid to dissolve the silver clusters generated by exposure to light. Furthermore, deionized water generally contains some organic residues that reduce silver(I) on the silver clusters; distilled water is therefore to be preferred. As the shaping of α-alumina is difficult and generally carried out by specialized companies, the application of finely divided silver on pre-shaped bodies of α-alumina is a technical requirement.

It is, therefore, attractive to perform an impregnation of pre-shaped bodies of α-alumina with both the silver compound and the species that reduces the silver. In order to prevent migration of the silver during the reduction, reducing agents are preferably utilized that reduce at higher temperatures. As the nitrate of silver nitrate consumes part of the reducing agent, the silver is often precipitated as oxalate or carbonate. After washing, the salts are dissolved in ethylene diamine solutions or in a solution of ethanol amine and ammonia. Carboxylic silver salts have also been mentioned in the patent literature [104]. At higher temperatures, ethylene diamine or ethanol amine reduces the silver. Other reducing agents from the patent literature include polyhydric alcohols such as ethylene glycol and glycerol. Carbonaceous residues from the reducing agent will be taken up by the silver and raise the activity of the silver to an undesired level. It has been proposed, therefore [105], that the catalyst is washed with methanol or ethanol after reduction with an organic amine. It has also been mentioned that, upon reduction at temperatures up to 573 K, the carbonaceous residues are less soluble in methanol or ethanol. Reduction with ethylene glycol or glycerol is therefore proposed at a temperature below 398 K. Use of the silver salt of lactic acid has also been mentioned in view of the high solubility.

Also of interest is a procedure [106] in which silver is first applied on the support and, after drying, a reducing agent which is immiscible with water is added; the reduction is then performed by raising the temperature. To prevent migration of the initially impregnated silver during drying, agar, gelatin, casein, pectin, starch, cellulose or polysaccharides are added to the impregnating silver solution. Reduction is subsequently executed with a high-boiling ester such as dioctyl sebacate or diethyl phthalate.

2.4.1.6 Production of Catalyst Supports and Synthetic Clay Minerals

The chemical interaction of metal ions precipitating from a homogeneous solution with silica at higher temperatures has been widely confirmed. With finely divided silica, reaction to synthetic clay minerals has been established. Consequently, the reaction is not confined to the surface of silica, but also involves the bulk of small silica particles. Reaction to another solid phase was already concluded from the maximum through which the pH curve often passes during the deposition–precipitation of nickel at a temperature of about 350 K, or higher. The complete reaction of the silica particles is also evident from electronmicrographs [2], which show a complete disappearance of the silica particles initially present. When the amount of bivalent metal ions is insufficient to convert the silica completely, the growth of platelets of clay minerals from the silica particles can easily be seen.

Based on the precipitation from homogeneous solution in the presence of suspended silica, a procedure for the production of synthetic clay minerals was achieved that is much more favorable than procedures used in the state of the art, which involves hydrothermal treatment at high temperatures and pressures for prolonged periods [27, 106–109]. The clay minerals from bivalent metal ions have the structure of the mineral stevensite, a (2:1) phyllosilicate (see above). With magnesium, very small platelets with a surface area of about 700 m^2 g^{-1} are obtained, while with zinc much larger platelets of surface area 200 m^2 g^{-1} are produced. The surface areas with nickel and cobalt are intermediate in value. These surface areas were considerably higher than those of the original silica supports. The reaction to clay minerals was also evident from the acid groups present within thus-loaded silica supports after partial reduction of reducible metal ions such as nickel or copper. The addition of tetrahedrally coordinated aluminum during the deposition–precipitation leads to reaction to saponite clay minerals, in which a fraction of the silicon ions present in the tetrahedral holes is substituted by aluminum ions. After exchange of the interlayer cations (usually ammonium), and, if required after thermal treatment, acid protons are present in the saponite clay minerals. The above-described preparation procedure leads reproducibly to clay minerals, the properties of which can be controlled accurately over a wide range, and offers interesting possibilities for the technical application of clay minerals in catalytic reactions.

The clay minerals obtained by deposition–precipitation of bivalent ions and silica can lead to highly attractive catalyst supports. It is important that the clay minerals are not affected by not-too-strong alkaline solutions, while the attack by acids is also limited at not too-low pH levels. Most interesting is that the porous structure and the surface area can be controlled to a level that is difficult to achieve with the most normally employed support alumina and silica. Figures 26 and 27 show electronmicrographs of

References see page 465

HAADF detector image

Secondary electron image

Magnification 79 kx

Fig. 26 Catalyst support synthesized from magnesium and zinc deposition-precipitated on silica from waterglass. Left: Image taken with a high-angle annular dark-field (HAADF) detector. The image indicated the areas in the specimen where the thickness of relatively heavy atoms is considerable. Right: Secondary electron image, indicating the profile of the specimen's surface. Both the platelets and stacking of the platelets are clearly visible.

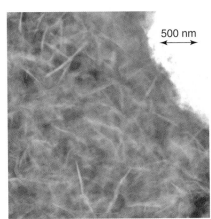
HAADF detector image

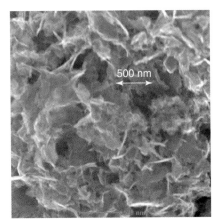
Secondary electron image

Magnification 56 kx

Fig. 27 Catalyst support synthesized from iron(II) deposition-precipitated on silica from waterglass. Left: Image taken with a high-angle annular dark-field (HAADF) detector: At the upper, right-hand edge, the specimen is fairly thick; this thickness does not affect the secondary electron image, but the number of electrons scattered within the thick area is large, such that the HAADF detector cannot exhibit suitable contrast. Right: Secondary electron image; the larger size of the platelets and their stacking is clearly visible.

thus-produced clay minerals. The left-hand micrographs were obtained with a high-angle annular dark-field (HAADF) detector; this detector only images the electrons scattered over large angles. Accordingly, areas of either a larger thickness or containing heavier elements are imaged with a higher contrast. The right-hand (secondary electron) images of Figs. 26 and 27 are of the same area of the specimen. As the depth of focus in TEM in scanning mode is limited, the secondary electron images are sharper than the HAADF images. The secondary electrons image the surface layer of the specimen, whereas the electrons scattered over large angles are from the bottom to the top of the specimen. It can be seen that the clay platelets form a "house-of-cards" structure of a high surface area and an elevated porosity, which leads to a very favorable mass transfer. The two sets of micrographs show that the size of the platelets can be varied by employing different metal ions. Whereas magnesium leads to very small platelets, zinc provides much larger platelets, and a mixture of magnesium and zinc results in platelets of an intermediate size. The stacking of the platelets can be controlled by the final ion strength of the solution. The surface of the clay platelets is negatively charged, and the edges positively. Washing to a very low content of electrolyte in the solution results in the edges of the clay platelets being attached to the surface of the plates due to positive and negative electrical charges. An almost complete removal of the electrolytes can only be achieved by multiple (three to five times) redispersion of the clay minerals, by washing and filtering. With a higher ionic strength, stacking of the clay platelets is preferred, and this leads to a lower surface area and a lower porosity. It may be concluded that the porous properties of the synthetic clay minerals can be installed within much larger ranges than those of usual silica and alumina.

The most obvious metal ions to produce phyllosilicate supports are zinc, magnesium, and iron(II). Whereas the application of phyllosilicate supports based on zinc and magnesium is obvious, the use of phyllosilicates of iron(II) may cause problems with hydrogenation catalysts. When the iron of the phyllosilicate is also reduced during reduction of the active hydrogenation catalyst (such as cobalt or nickel), an alloy with iron will result. As metallic iron is readily oxidized by water vapor, the segregation of iron oxide on the surface of the hydrogenation catalyst will proceed, and this leads to a catalyst of a relatively low activity (or even inactive). An excess of iron(II) with the preparation of the support has then to be prevented, and it must be assessed whether the iron of the phyllosilicate is reduced under the conditions of pretreatment of the hydrogenation catalyst.

2.4.1.7 Concluding Remarks

The above-described procedures for catalyst preparation have generally provided excellent results, with special importance applicable to surface-sensitive reactions. With supported catalysts in which the active components have a narrow particle size distribution, the optimum particle size for a demanding reaction can be established. Major improvements of supported catalysts – for example, with respect to carbon deposition and ammonia decomposition – can be achieved by preparing catalysts of a narrow particle size distribution. The size of the metal particles is especially important in the case of gold catalysts.

The preparation of catalysts in which the active components must have a uniform chemical composition is also extremely important. An example is the preparation of supported vanadium oxide phosphorous oxide (VPO) catalysts for the selective oxidation of n-butane to maleic anhydride, which has been carried out using vanadium(III) deposition onto silica. The preparation of catalysts in which the active component is also an alloy is an example where deposition–precipitation can be employed favorably. Controlling the chemical composition of alloy particles with dimensions of some nanometers is not a straightforward task. Particles without the desired chemical composition can adversely affect the selectivity of the catalytic reaction. In this chapter procedures have been presented to apply active precursor thus on supports that small alloy particles of a uniform chemical composition result.

The most technically desirable procedures are those that employ existing equipment for the commercial production of catalysts. Thus, the impregnation of pre-shaped support bodies, drying and subsequent calcination and/or reduction are therefore attractive techniques as they can be executed in this way. The scaling-up of processes that involve stirring is technically difficult, and it is important that deposition–precipitation can separate mixing of the ingredients for catalyst production and the generation of precipitating agents. Accordingly, the scaling-up of catalyst preparation procedures presents few problems. As the use of higher pressures considerably limits the time involved in catalyst preparation, the only remaining problem is the disposal of wastewater that results from the use of precipitating agents.

It is of interest that the inorganic chemistry and techniques used to produce advanced catalysts can also be used to produce magnetic materials, inorganic pigments, and finely divided metals. Much of the information outlined in this chapter has been published only in the patent literature, to which reference is made below. However, the patents contain accurate descriptions of the experiments involved, and these exemplify the results described in this chapter.

Acknowledgments

The authors are greatly indebted to Mr. J. J. M. G. Eurlings, who assisted at DSM in the initial experiments on which the deposition–precipitation techniques have been based. Thanks are also extended to the Ph.D. students of the Department of Inorganic Chemistry and Catalysis of Utrecht University, who contributed greatly to the development of the state of the art of deposition–precipitation.

References

1. J. W. Geus, in *Sintering and Catalysis, Material Science*, G. C. Kuczynski (Ed.), Vol. 10, Plenum Press, New York, 1975, p. 29.
2. J. W. Geus, in *Preparation of Catalysts III*, G. Poncelet, P. Grange, P. A. Jacobs (Eds.), *Studies in Surface Science and Catalysis*, Vol. 16, Elsevier, Amsterdam, 1983, p. 1.
3. C. N. Satterfield, *Heterogeneous Catalysis in Practice*, McGraw-Hill, New York, 1980, p. 131.
4. R. Schlögl, in *Catalytic Ammonia Synthesis Fundamentals and Practice*, J. R. Jennings (Ed.), Plenum Press, New York, 1991, p. 19.
5. W. Langenbeck, H. Dreyer, D. Nehring, *Naturwissenschaften* **1954**, *41*, 332; W. Langenbeck, H. Dreyer, D. Nehring, *Z. Anorg. Allg. Chem.* **1955**, *281*, 90.
6. L. M. Knijff, *The Application of Active Components into Catalyst Support Bodies*, Ph.D. Thesis, Utrecht University, 1992.
7. A. Q. M. Boon, *Catalytic Combustion of Methane in Fixed-Bed Reactors*, Ph.D. Thesis, Utrecht University, 1990.
8. P. J. van den Brink, *The Selective Oxidation of Hydrogen Sulfide to Elemental Sulfur on Supported Iron-Based Catalysts*, Ph.D Thesis, Utrecht University, 1992.
9. U. S. Patent 4,783,434 assigned to DOW Chemical Company, 1988.
10. G. R. Meima, B. G. Dekker, A. J. van Dillen, J. W. Geus, J. E. Bongaarts, F. R. van Buren, K. Delcour, J. M. Wigman, in *Preparation of Catalysts IV*, B. Delmon, P. Grange, P. A. Jacobs, G. Poncelet (Eds.), *Studies in Surface Science and Catalysis*, Vol. 31, Elsevier, Amsterdam, 1986, p. 83.
11. P. J. van den Brink, A. Scholten, A. van Wageningen, M. D. A. Lamers, A. J. van Dillen, J. W. Geus, in *Preparation of Catalysts V*, G. Poncelet, P. A. Jacobs, P. Grange, B. Delmon (Eds.), *Studies in Surface Science and Catalysis*, Vol. 63, Elsevier, Amsterdam, 1991, p. 2.
12. G. R. Meima, *Development of Supported Silver Catalysts*, Ph.D. Thesis, Utrecht University, 1987.
13. A. R. West, *Solid State Chemistry and its Applications*, Wiley & Sons, Chichester, 1984, p. 436.
14. D. Mealor, A. Townshend, *Talanta* **1966**, *13*, 1069.
15. H. H. Willard, N. G. Tang, *J. Am. Chem. Soc.* **1937**, *59*, 1190.
16. H. H. Willard, N. G. Tang, *Ind. Eng. Chem., Analyt. Ed.* **1937**, *9*, 357.
17. P. F. S. Cartwright, E. J. Newman, D. W. Wilson, *Analyst* **1967**, *9*, 7663.
18. U. S. Patent 4,113,658, assigned to Stamicarbon, 1978.
19. German Patent 740634, assigned to I. G. Farbenindustrie AG, 1943.
20. P. K. de Bokx, W. A. Wasserberg, J. W. Geus, *J. Catal* **1987**, *104*, 86.
21. PCT Patent WO 99/41196, assigned to AKZO-Nobel, 1999.
22. P. van Stiphout, *Prevention of Carbon Deposition on Silica-Supported Nickel-Based Catalysts*, Ph.D. Thesis, Utrecht University, 1987.
23. A. J. van Dillen, J. W. Geus, L. A. M. Hermans, J. van der Meijden, *Proceedings of the Sixth International Congress on Catalysis, London 1976*, G. C. Bond, P. B. Wells, F. C. Tompkins (Eds.), Chemical Society, Letchworth, UK, 1977, Vol. 2, p. 667.
24. L. A. M. Hermans, J. W. Geus, in *Preparation of Catalysts II*, B. Delmon, P. Grange, P. Jacobs, G. Poncelet (Eds.), *Studies in Surface Science and Catalysis*, Elsevier, Amsterdam, 1979, p. 113.
25. R. Burch, A. R. Flambard, in *Preparation of Catalysts III*, G. Poncelet, P. Grange, P. A. Jacobs (Eds.), *Studies in Surface Science and Catalysis*, Elsevier, Amsterdam, 1983, p. 311.
26. M. K. van der Lee, A. J. van Dillen, J. H. Bitter, K. P. de Jong, *J. Am. Chem. Soc.* **2005**, *127*, 13573.
27. R. J. M. J. Vogels, *Non-Hydrothermally Synthesized Trioctahedral Smectites*, Thesis, Utrecht University, 1996.
28. M. Toebes, M. K. van der Lee, Lai Mei Tang, M. H. Huis in 't Veld, J. H. Bitter, A. J. van Dillen, K. P. de Jong, *J. Phys. Chem. B* **2004**, *108*, 11611.
29. J. van der Meijden, *Preparation and Carbon Monoxide Oxidation Activity of Supported Copper (Oxide) Catalysts*, Ph.D. Thesis, Utrecht University, 1981.
30. T. Labunakrom, *Koll. Chem. Ber.* **1929**, *29*, 80.
31. L. J. Martynenko, *Zh. Neorg. Khim.* **1970**, 15.
32. M. Lemoinne, J. Gauthier, *Compt. Rend.* **1958**, *246*, 1994.
33. V. Kohlschütter, *Helv. Chim. Acta* **1929**, *12*, 512.
34. R. Nässänen, V. Tamminen, *J. Am. Chem. Soc.* **1949**, *71*, 1994.
35. W. Feitknecht, *Pure Appl. Chem.* **1963**, *6*, 130.
36. C. J. G. van der Grift, A. F. H. Wielers, A. Mulder, J. W. Geus, *Thermochim. Acta* **1990**, *171*, 95.
37. C. J. G. van der Grift, P. A. Elberse, A. Mulder, J. W. Geus, *Appl. Catal.* **1990**, *59*, 275.
38. C. J. G. van der Grift, A. Mulder, J. W. Geus, *Appl. Catal.* **1990**, *60*, 181.
39. J. T. Richardson, R. J. Dubus, J. G. Crump, P. Desai, U. Osterwalder, T. S. Cale, in *Preparation of Catalysts II*, B. Delmon, P. Grange, P. Jacobs, G. Poncelet (Eds.), *Studies in Surface Science and Catalysis*, Elsevier, Amsterdam, 1979, p. 131.
40. C. M. A. M. Mesters, *On the Preparation and Reactivity of Copper-Nickel Alloys*, Ph.D. Thesis, Utrecht University, 1984.
41. F. J. Wick, D. S. O'Hanley, in *Hydrous Phyllosilicates. Reviews in Mineralogy*, S. W. Bailey (Ed.), Mineralogical Society of America, 1988, Vol. 19, p. 126.
42. B. W. Evans, S. Guggenheim, in *Hydrous Phyllosilicates. Reviews in Mineralogy*, S. W. Bailey (Ed.), Mineralogical Society of America, 1988, Vol. 19, p. 239.
43. S. Shimoda, *Clay Miner.* **1971**, *9*, 185.
44. J. J. B. van Eijk van Voorthuijsen, P. Franzen, *Rec. Trav. Chim. Pays Bas* **1951**, *69*, 666.
45. J. J. B. van Eijk van Voorthuijsen, P. Franzen, *Rec. Trav. Chim. Pays Bas* **1951**, *70*, 793.
46. C. Louis, Z. X. Cheng, M. Che, *J. Phys. Chem.* **1993**, *97*, 5703.
47. M. Kermarec, J. Y. Carriat, P. Burattin, M. Che, A. Decarreau, *J. Phys. Chem.* **1994**, *98*, 12008.
48. P. Burattin, C. Louis, M. Che, *J. Chim. Phys. Physico-Chim. Biol.* **1995**, *92*, 1377.
49. C. Louis, in *Catalyst Deactivation, Studies in Surface Science and Catalysis*, Elsevier, Amsterdam, 1977, p. 617.
50. O. Clause, L. Bonneviot, M. Che, H. Dexpert, *J. Catal.* **1991**, *130*, 21.

51. P. Burattin, M. Che, C. Louis, *J. Phys. Chem. B* **1997**, *101*, 7060.
52. P. Burattin, M. Che, C. Louis, *J. Phys. Chem. B* **1998**, *102*, 2722.
53. P. Burattin, M. Che, C. Louis, *J. Phys. Chem. B* **1999**, *103*, 6171.
54. P. Burattin, M. Che, C. Louis, *J. Phys. Chem. B* **2000**, *104*, 10482.
55. *Gmelins Handbuch der Anorganische Chemie, Achte Auflage* Nr. 57, Verlag Chemie GmbH, Weinheim, 1966, p. 845.
56. F. J. Wicks, D. S. O'Hanley, in *Hydrous Phyllosilicates. Reviews in Mineralogy*, S. W. Bailey (Ed.), Vol. 19, Mineralogical Society of America, 1988, p. 126.
57. C. Louis, *Abstracts of Papers, 227th ACS National Meeting*, Anaheim CA, USA, March 28-April 1, 2004.
58. R. Nares, J. Ramirez, A. Gutierrez-Alejandre, R. Cuevas, C. Louis, R. Klimova, in *Studies in Surface Sciences and Catalysis*, E. Gaigneaux, D. E. de Vos, P. Grange, P. A. Jacobs, J. A. Martens, P. Ruiz, G. Poncelet (Eds.), *Scientific Bases for the Preparation of Heterogeneous Catalysts*, Vol. 143, Elsevier, Amsterdam, 2002, p. 537.
59. W. J. J. van der Wal, *Desulfurization of Process Gas by means of Iron Oxide-on-Silica Sorbents*, Ph.D. Thesis, Utrecht University, 1987.
60. J. W. Geus, *Appl. Catal.* **1986**, *25*, 313.
61. T. Hiemstra, W. G. van Riemsdijk, *J. Colloid Interface Sci.* **1996**, *178*, 488.
62. U. S. Patent 4,186,112, assigned to Hoechst Aktiengesellschaft, 1980.
63. EP Patent 447005, assigned to Shell International Research, 1991.
64. E. Boellaard, A. M. van der Kraan, J. W. Geus, in *Preparation of Catalysts VI*, G. Poncelet, J. Martens, B. Delmon, P. A. Jacobs, P. Grange (Eds.), *Studies in Surface Science and Catalysis*, Vol. 81, Elsevier, Amsterdam, 1995, p. 931.
65. J. van de Loosdrecht, A. J. van Dillen, J. van der Horst, A. M. van der Kraan, J. W. Geus, *Top. Catal.* **1995**, *2*, 29.
66. U. S. Patent 4,740,360, assigned to Harshaw Chemie, De Meern (NL), 1988.
67. E. T. C. Vogt, M. de Boer, A. J. van Dillen, J. W. Geus, *Appl. Catal.* **1988**, *40*, 255.
68. E. T. C. Vogt, *Preparation and Properties of Catalysts supported on Modified Silica*, Ph.D. Thesis, Utrecht University, 1988.
69. R. A. Overbeek, *New Aspects of the Selective Oxidation of n – Butane to Maleic Anhydride*, Ph.D. Thesis, Utrecht University, 1994.
70. M. de Boer, *On the Design and Properties of Silica-Supported Molybdenum Oxide Catalysts*, Ph.D. Thesis, Utrecht University, 1992.
71. M. de Boer, A. J. van Dillen, D. C. Koningsberger, J. W. Geus, M. A. Vuurman, I. E. Wachs, *Catal. Lett.* **1991**, *11*, 227.
72. Dutch Patent Appl. NL68/18691, December 25, 1968, Ger. Offen. 1,964,620, assigned to Stamicarbon, 1970.
73. Dutch Patent Appl. NL68/16771, assigned to Stamicarbon, 1968.
74. N.L. Patent 8201396, assigned to DOW Chemical Nederland, 1983.
75. K. P. de Jong, J. W. Geus, *Appl. Catal.* **1982**, *4*, 41.
76. K. P. de Jong, *Structural Aspects of Supported Monometallic and Bimetallic Catalysts*, Ph.D. Thesis, Utrecht University, 1982.
77. U. S. Patent 3,374,183, assigned to Ethyl Corp., 1986.
78. H. Schaper, D. J. Amesz, E. B. M. Doesburg, P. H. M. de Korte, J. M. C. Quartel, L. L. van Reijen, *Appl. Catal.* **1985**, *16*, 417.
79. U. S. Application 2003032684, assigned to Johnson & Matthey, 2003.
80. Dutch Patent Appl. 18337 December 20, 1968, Ger. Offen. 1,963,827 July 09, 1970, assigned to Stamicarbon.
81. U. S. Patent 4,869,792, assigned to Harshaw Chemie B.V., 1985.
82. P. C. M van Stiphout, H. Donker, C. R. Bayense, J. W. Geus, F. Versluis, in *Preparation of Catalysts IV*, B. Delmon, P. Grange, P. A. Jacobs, G. Poncelet (Eds.), *Studies in Surface Science and Catalysis*, Vol. 31, Elsevier, Amsterdam, 1987, p. 55.
83. P. C. M. van Stiphout, A. J. van Dillen, J. W. Geus, *Appl. Catal.* **1988**, *37*, 175.
84. P. C. M. van Stiphout, C. R Bayense, J. W. Geus, *Appl. Catal.* **1988**, *37*, 189.
85. M. Haruta, U. Hobayahsi, H. Sano, H. Yamada, *Chem. Lett.* **1987**, 405.
86. S. Tsubota, M. Haruta, T. Kobayshi, A. Ueda, Y. Nakahara, in *Preparation of Catalysts V*, G. Poncelet, P. A. Jacobs, P. Grange, B. Delmon (Eds.), *Studies in Surface Science and Catalysis*, Vol. 63, Elsevier, Amsterdam, 1991, p. 695.
87. M. Haruta, S. Tsubota, T. Kobayashi, H. Kageyama, M. J. Genet, B. Delmon, *J. Catal.* **1993**, *144*, 175.
88. D. Andreeva, T. Tabakova, V. Idakiev, P. Christov, R. Giovanoli, *Appl. Catal.* **1998**, *169*, 9.
89. M. Khoudiakov, M. C. Gupta, S. Deevi, *Appl. Catal. A General* **2005**, *291*, 151.
90. S. Tsubota, K. A. H. Cunningham, Y. Bando, M. Haruta, in *Preparation of Catalysts VI, Scientific Bases for the Preparation of Heterogeneous Catalysts*, G. Poncelet, J. Martens, B. Delmon, P. A. Jacobs, P. Grange (Eds.) *Studies in Surface Science and Catalysis*, Vol. 91, Elsevier, Amsterdam, 1995, p. 227.
91. M. Haruta, A. Ueda, S. Tsubota, R. M. Torrez Sanchez, *Catal. Today* **1996**, *29*, 443.
92. M. A. Centeno, I. Carrizosa, J. A. Odriozola, *Appl. Catal. A General* **2003**, *248*, 385.
93. D. A. H. Cunningham, W. Vogel, R. M. Torres Sanchez, K. Tanaka, M. Haruta, *J. Catal.* **1999**, *183*, 24.
94. J. A. Peck., G. E. Brown, *Geochim. Cosmochim. Acta* **1991**, *55*, 671.
95. A. Zwijnenberg, W. G. Goossens, M. W. J. Sloof, M. W. J. Crajé, A. M. van de Kraan, L. J. de Jong, M. Makkee, J. A. Moulijn, *J. Phys. Chem. B* **2002**, *106*, 9853.
96. C. H. Lin, S. H. Hsu, M. Y. Lee, S. D. Lin, *J. Catal.* **2002**, *209*, 62.
97. R. Zanella, L. Delannoy, C. Louis, *Appl. Catal. A General* **2005**, *291*, 62.
98. R. Zanella, S. Giorgio, C. R. Henry, C. Louis, *J. Phys. Chem.* **2002**, *106*, 7634.
99. R. Zanella, C. Louis, *Catal. Today* **2005**, *107–108*, 768.
100. T. Akita, M. Okumura, K. Tanaka, M. Haruta, *J. Catal.* **2002**, *212*, 119.
101. M. A. P. Dekkers, M. J. Lippits, B. E. Nieuwenhuys, *Catal. Today* **1999**, *54*, 381.
102. L. M. Knijff, P. H. Bolt, R. van Yperen, A. J. van Dillen, J. W. Geus, in *Preparation of Catalysts V*, G. Poncelet, P. A. Jacobs, P. Grange, B. Delmon (Eds.), *Studies in Surface Science and Catalysis*, Vol. 63, Elsevier, Amsterdam, 1991, p. 165.
103. K. P. de Jong, in *Preparation of Catalysts V. Scientific Bases for the Preparation of Heterogeneous Catalysts*, G. Poncelet, P. A. Jacobs, P. Grange, B. Delmon (Eds.), *Studies in Surface Science and Catalysis*, Vol. 63, Elsevier, Amsterdam, 1991, p. 17.

104. Fr. Patent 2,128,378, assigned to Shell International Research, 1972.
105. G. B. Patent 2045636, assigned to Halcon Res. & Dev., 1980.
106. U. S. Patent 4,350,616, assigned to DOW Chemical Company, 1980.
107. R. J. M. J. Vogels, M. J. H. V. Kerkhoffs, J. W. Geus, in *Preparation of Catalysts VI*, G. Poncelet, J. Martens, B. Delmon, P. A. Jacobs, P. Grange (Eds.), *Studies in Surface Science and Catalysis*, Vol. 91, Elsevier, Amsterdam, 1995, p. 1135.
108. Dutch Patent Application NL94.01433 Synthetic Clay Minerals, assigned to AKZO-Nobel, Shell, Engelhard.
109. Dutch Patent Application NL94.01431 Catalyst comprising at least a hydrogenation metal component and a synthetic clay, assigned to AKZO-Nobel.

2.4.2
Ion Exchange and Impregnation

Eric Marceau*, Xavier Carrier, Michel Che, Oliver Clause, and Christian Marcilly

Many practical catalysts consist of one or several catalytically active component(s) deposited on a high-surface-area support. The data in Fig. 1 indicate the number of reports published between 1967 and 2004 that were related to catalyst preparation involving various supports. The main purpose of using a support is to achieve an optimal dispersion of the catalytically active component and to stabilize it against sintering. In many reactions, the support is not catalytically inert and the overall process is actually the combination of several catalytic functions: those of the active component and that of the support.

There are many definitions of catalyst preparation, but the most revealing seem to have been provided by Richardson: "...catalyst preparation is the secret to achieving the desired activity, selectivity and life time" [1a] and by Baekeland: "... commit your blunders on a small scale and take your profits on a large scale" [1b]. Those definitions convey the idea that catalyst preparation is a strategic procedure which should not be disclosed or, if published, should be protected in the form of patents on the one hand, and which has some practical and economical consequences on the other hand.

Although the preparation of supported catalysts is one – if not the most – important step in the course of a catalytic process, it is striking to observe that for a long time little attention has been paid to this subject in most textbooks on catalysis. By contrast, the number of documents published on this subject is very large (Fig. 2a). This apparent contradiction lies in the fact that most of the papers and patents, if not all, describe preparation procedures rather than rationalizations and concepts, which makes any overview (that is usually required in textbooks) rather difficult. It is therefore no surprise that the term "catalyst design" was little used before the last decade.

It is interesting to observe that, while the number of papers steadily increases, the number of patents is, by contrast, almost constant (Fig. 2a). It is also worthwhile to note (Fig. 2b) that the literature on catalyst characterization follows the same trend as catalyst preparation but remains almost fourfold less abundant – which may explain why our understanding of the latter domain has needed time to be satisfactorily rationalized.

For all those reasons, a methodological approach of supported catalysts preparation requires one to:

- investigate the basic aspects of catalyst preparation from a phenomenological point of view
- identify the elementary steps and the related pertinent parameters so that new sequences of such steps can be proposed, making catalyst architecture possible
- establish procedures for choosing the most appropriate method to prepare industrial catalysts, including the problems associated with scaling-up.

In recent years, major advances have been made in techniques used for the physical and chemical characterization of supported catalysts [2] and on the quantitative and qualitative aspects of catalyst preparation [3–12], so that the *design* of supported catalysts has begun to become a feasible activity.

The term "design" is best appropriate when the precise nature of the active center for a given reaction is known and can be reproduced at the molecular level, including the oxidation state of the catalytically active element(s), the nature and symmetry of the(ir) environment(s). That is, the nature and number of the different ligands (in particular the type of support and number of bonds with the latter, the number and nature of ancillary ligands), and the number of coordination vacancies.

A properly designed catalyst should have the essential attributes of activity, stability, selectivity, and regenerability [13]. Such characteristics can be related to the physical and chemical properties of the catalyst, which in turn can be related to the many parameters inherent to the method used to prepare the catalyst. In the past, much of the literature on supported catalysts has not included this information.

2.4.2.1 Deposition of Active Component Precursors
There are two main steps in catalyst preparation. The first step consists of depositing the active component

---
* Corresponding author.

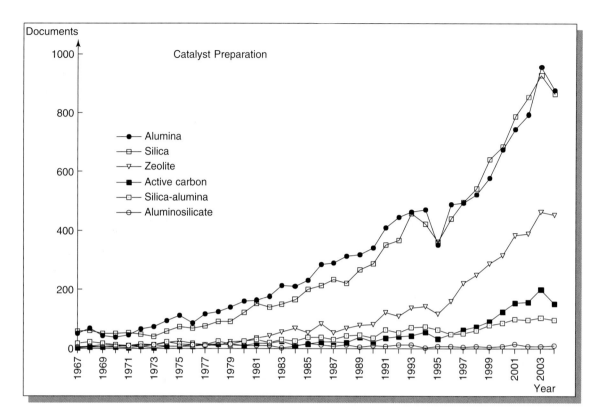

Fig. 1 Numbers of reports, published between 1967 and 2004, relating to catalyst preparation that involve the most frequently used supports. Database: SciFinder Scholar. Keywords: preparation of supported catalysts + support name.

precursor into a divided form on the support, and the second of transforming this precursor into the required active component(s) which, depending on the reaction to be catalyzed, can be found in the ionic, oxidic, sulfided, or metallic state. A large majority of deposition methods involve aqueous solutions and the liquid–solid interface, often referred to as "humid" interface. In some cases, deposition can be also performed from the gas phase and involves the gas–solid interface, often referred to as "dry" interface.

The methods most frequently used to achieve the deposition of the active component precursor are impregnation, ion exchange, anchoring, grafting, spreading and wetting, heterogenization of complexes, deposition–precipitation (homogeneous and redox), and adapted methods in the case of supported bimetallic catalysts. In some cases, the active component (not its precursor form) can be deposited directly onto the support. In the following sections only the case of a single active component will be discussed. When several active components are required, they can be deposited consecutively or simultaneously. Although the problem becomes more complicated because of possible interferences, it can be treated with the same basic concepts presented here.

When a support is placed in contact with a solution containing a precursor of the active component (in a dispersed state already, by the mere fact that it is dissolved), several phenomena can occur, among which electrostatic interactions, chemical bonding to the support surface, dissolution of the support, and the formation of surface compounds are the most important. If the experimental parameters are not properly adjusted, there is a fair probability that these phenomena may occur simultaneously. In the following, we will first discuss the physico-chemical principles on which two preparation procedures – ion exchange and impregnation – are based. The various possible interactions developing between support and precursor will then be analyzed from a chemical point of view.

The case of ion exchange is treated first, because it is simpler and involves only two types of chemical entities – the charged surface and the precursor complex ions of opposite charge. In fact, this preparation method also involves a subsequent step of washing, which eliminates the counter ions of the precursor complex ions. This is in sharp contrast to the impregnation method, which does not involve any washing step so that the precursor complex ions are left together with their counter ions on the oxide support.

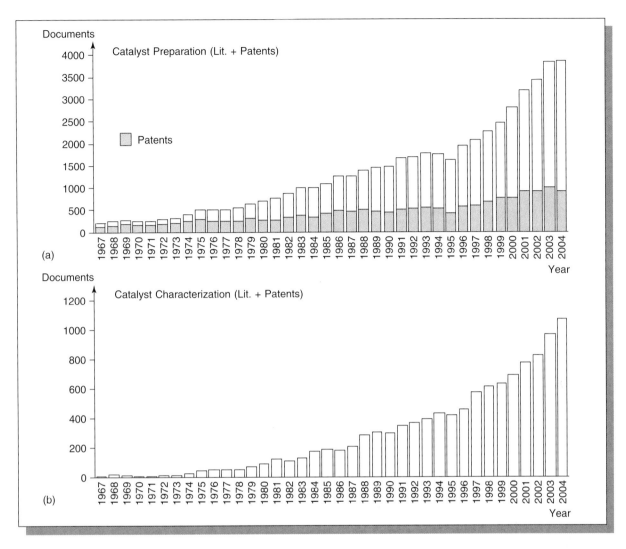

Fig. 2 Comparison of numbers of papers and patents identified in the literature between 1967 and 2004, related to catalyst preparation and characterization. Database: SciFinder Scholar. Keywords: preparation/characterization of supported catalysts.

2.4.2.2 Ion Exchange: Physicochemical Aspects

Ion exchange is an operation which consists of replacing an ion in an electrostatic interaction with the surface of a support by another ion species. The support containing ion A is plunged *into an excess volume* (much larger than the pore volume) of a solution containing ion B that is to be introduced. Ion B gradually penetrates into the pore space of the support and takes the place of ion A, which passes into the solution, until an equilibrium is established corresponding to a given distribution of the two ions between the solid and the solution. The solid is then washed, and finally separated by filtration or centrifugation.

A Surface Charges and their Consequences Almost all solid mineral supports are oxides; they behave like ion exchangers when their surface bears electric charges. Two categories may be distinguished: (i) natural ion exchangers; and (ii) amphoteric oxides.

Natural ion exchangers are composed of a framework bearing electric charges neutralized by ions with the opposite signs. In the case of zeolites, for example, these charges are negative and are due to the particular environment of aluminum. Aluminum, like silicon, is effectively situated in the center of a tetrahedron of four oxygen atoms, which provides it with four negative charges, whereas the aluminum itself has only three positive charges. The tetrahedron (AlO_4) is thus an overall bearer of a negative charge distributed over the oxygen atoms, and this charge is neutralized by the presence of various cations, Na^+, K^+, Ca^{2+}, etc.

References see page 482

These cations are not definitively linked to the framework but may be replaced by other cations during an ion-exchange operation. Whatever the exchange conditions, and in particular the pH, zeolites are cation exchangers and have a constant number of exchange sites, which is equal to the number of aluminum atoms in their framework.

There are natural ion exchangers other than zeolites. Clay silicates are cation exchangers while hydrotalcites are anion exchangers. As in the case of zeolites, the number of exchange sites is not pH-dependent.

Amphoteric oxides constitute the second category of ion exchangers. Oxide surfaces contacted with water are generally covered with hydroxyl groups which can be represented schematically by S—OH, where S represents Al, Si, Ti, Fe, etc. Some of these groups may behave as Brønsted acids or as Brønsted bases, giving rise to either one of the following two equations:

$$S-OH \rightleftharpoons S-O^- + H^+$$
$$S-OH + H^+ \rightleftharpoons S-OH_2^+ \quad (1)$$

The resulting surface charge which arises from the excess of one type of charged site over the other, is a function of the solution pH. A given value of pH exists for which the particle is overall not charged. This value, characteristic of the oxide, is called the point of zero charge (PZC) of the oxide, sometimes confused with the isoelectric point (IEP), determined by electrophoresis and influenced by the cloud of ions surrounding the oxide grains in solution [14]. The values of PZC of oxides frequently used as catalyst supports have been reviewed earlier [14, 15]. When oxide particles are suspended in aqueous solutions with pH > PZC, the oxide particles are negatively charged and adsorb cations. Conversely, at pH < PZC, anion adsorption is favored [16a]. Reported values of the PZC of γ-alumina range between 7 and 9, depending on thermal activation and impurities. This is illustrated in Fig. 3, which shows the experimental determination of PZC for silica and alumina [16b]. Thus, γ-alumina is amphoteric and may adsorb cations as well as anions. PZC values for silica range between 1.5 and 3. Silica may only adsorb cations. It should be mentioned that cation adsorption is actually significant only above pH 7 [16a, 17]. Such electrostatic adsorption (which is not ionic exchange *per se* [12c]) is valuable for the preparation of low-loaded catalysts containing expensive noble metals. However, ionic exchange is also possible on oxidic surfaces, for example between $PtCl_6^{2-}$ and an alumina previously dipped in an HCl solution and bearing $S-OH_2^+ \ Cl^-$ groups, where the chloride anion is exchangeable.

Many electrical double-layer and adsorption models have been proposed to account for experimental data dealing with the adsorption of ions on support oxides. Stern suggested to divide the solution region near the surface into two parts, the first part consisting of a layer of ions adsorbed at the surface (compact layer), and the second consisting of a diffuse Gouy layer [18]. More specifically, the compact region can be structured into an Inner Helmholtz Plane (IHP) located at the surface of the layer of Stern adsorbed ions, and an Outer Helmholtz Plane (OHP), located on the plane of centers of the next layer of ions [19, 20]. Specifically adsorbed ions are located in the IHP, whereas electrostatically adsorbed ions are located in the OHP (site binding models) [19, 21–26].

Assumptions underlying the adsorption models are not often discussed in the literature. In particular, lateral interactions between adsorbed ions, site heterogeneity as well as phenomena involving oxide dissolution or rehydration are not systematically taken into account. The latter phenomena will be discussed in Section 2.4.2.4.D. Lateral interactions between adsorbed ions (ion coadsorption) have been reported [27, 28] and make questionable the use of mass action equations at interfaces. The effect of surface structure, site heterogeneity and surface composition, in particular at the PZC value, were also pointed out [29, 30].

The adsorption of atomically dispersed ions generally allows for a much better dispersion of the active agent than does simple wetting without exchange [31–33]. Obviously, the high dispersions gained in the precursors must be kept during the subsequent drying, calcination or reduction operations, which is frequently observed. For example, the metal dispersion on silica resulting from impregnating $Pt(NH_3)_4^{2+}$ by ion exchange is excellent even when the platinum loading reaches several weight percent, whereas, incipient wetness impregnation with $PtCl_6^{2-}$ solutions at acidic pH leads to the formation of large metal crystallites after reduction [34].

B Elementary Background of Single Ion Exchange For the simple case of two monovalent cations (ion A^+ to be replaced on the solid Z, for instance a zeolite, by ion B^+ present in the solution S), the single ion exchange equilibrium can be written as:

$$A_Z^+ + B_S^+ \rightleftharpoons B_Z^+ + A_S^+ \quad (2)$$

in which the subscripts S and Z represent the solution and the solid, respectively.

An exchange is characterized by an isothermal curve representing the variation of the concentration C_{BZ} of ion B^+ in the solid as a function of the concentration C_{BS} of B^+ in the solution. The concentrations C_{BZ} and C_{BS} are expressed in ion-grams per unit of volume of the solid for the former and of the solution for the latter.

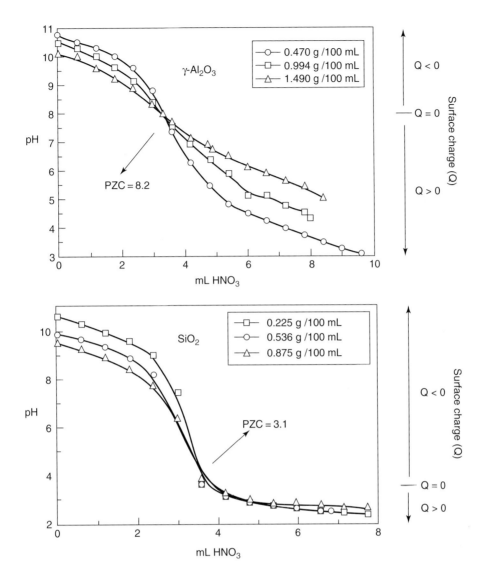

Fig. 3 Potentiometric mass titrations for the determination of point of zero charge (PZC). (Reproduced from Ref. [16b], by permission of The Royal Society of Chemistry.)

In the case of an ideal exchange (both, the exchanger and the solution are ideal), and of only a single type of site, the equilibrium constant K_a can be written as:

$$K_a = \frac{C_{BZ}C_{AS}}{C_{BS}C_{AZ}} \quad (3)$$

with $C_{AZ} + C_{BZ} = C_Z$ and $C_{AS} + C_{BS} = C_S$

It follows:

$$C_{BZ} = \frac{K_a C_Z C_{BS}}{C_S + (K_a - 1)C_{BS}} \quad (4)$$

If the exchange is no longer ideal, Eq. (4) must be expressed in terms of concentrations and activity coefficients. For non-monovalent ions and/or several types of site on the solid surface, experiments must be performed in order to obtain isothermal curves that can be used to predict the number and conditions of successive exchanges required to replace a given fraction of ions A by ions B.

A simple and standard example of an exchange is that used to replace Na^+ ions in zeolite NaY by NH_4^+ ions. Zeolite NaY, with the overall formula $Na_2O \cdot Al_2O_3 \cdot 5SiO_2$, contains 9.9 wt.% sodium. Available isothermal exchange curves show that, at room temperature, only about 73% of Na^+ ions can be exchanged and that, unless a very large volume of solution is used with an ammonium salt, a single exchange operation is not sufficient to eliminate all of the exchangeable sodium

References see page 482

ions. To reach a high exchange degree, one can adopt either a discontinuous technique, which involves several successive operations, or a continuous technique in which the exchange solution crosses the solid bed and is renewed by the progressive addition of fresh solution with the corresponding drain of the balanced solution.

C Multiple or Competitive Ion Exchange Involving a Single Metal Ion

This technique, which has long been known with regards to anions [35, 36], was indicated for the first time with regards to cations by Benesi [37]. It proves to be particularly useful when trying to introduce and homogeneously distribute small amounts of a noble metal on a support with a large surface area. This situation is examined in detail via a simple example, that of introducing $Pt(NH_3)_4^{2+}$ ions by cation exchange into a zeolite NH_4Y.

Statement of the Problem [38–40] In zeolite NH_4Y, about 73% of NH_4^+ ions are in large cavities and can be replaced by $Pt(NH_3)_4^{2+}$ ions. The saturation of all of these exchangeable sites corresponds to the fixing of more than 25 wt.% of platinum with respect to the anhydrous zeolite. Yet, for obvious reasons of cost, industrial catalysts must contain small amounts of noble metal (<1 wt.%), and the catalyst manufacturer will try to introduce in the zeolite all the metal contained in the solution. This amount of metal thus represents 3–4% at most of the exchangeable sites of the zeolite.

Let us consider a support composed of grains (beads, extrudates or others) several millimeters in diameter and containing a certain amount of zeolite Y (20 to 70 wt.%, for example). What is the situation that leads to impregnation with regards to:

- the degree of metal fixation by the solid and hence, the residual quantity of metal ions in solution?
- the macroscopic distribution of the metal inside the support?

The equilibrium between NH_4^+ and $Pt(NH_3)_4^{2+}$ ions can be written as follows:

$$Pt(NH_3)_4^{2+}{}_{(S)} + 2\,NH_4^+{}_{(Z)}$$
$$\rightleftharpoons Pt(NH_3)_4^{2+}{}_{(Z)} + 2NH_4^+{}_{(S)} \quad (5)$$

The isothermal exchange curve between NH_4^+ and $Pt(NH_3)_4^{2+}$ ions suggests that zeolite Y has a strong affinity for $Pt(NH_3)_4^{2+}$ ions [39, 40]. Since the exchange rate is limited by the diffusion inside the grain, $Pt(NH_3)_4^{2+}$ ions will be quantitatively and strongly linked to the first sites encountered – that is, on the periphery of the grains. The concentration of the solution, which is almost the sole driving force behind the diffusion (adsorbed-phase diffusion may be neglected), will thus decrease rapidly. A state of pseudoequilibrium is obtained which is characterized by a very heterogeneous metal deposit at the macroscopic scale (crust deposit) and which evolves only very slowly toward a state of true equilibrium corresponding to a homogeneous dispersion inside the volume of the grain [39, 40].

The only way to speed up this evolution towards true equilibrium is to maintain large amounts of residual metal in solution ($Pt(NH_3)_4^{2+}{}_{(S)} = Pt_S$), obviously while maintaining constant the total amount of platinum $Pt_T = Pt_S + Pt_Z$ in the system so as to accelerate diffusion. This result is obtained (as shown in Fig. 4) by adding an ionic agent, NH_4^+, whose role is to shift Eq. (5) towards the left, and hence to increase Pt_S. Such ionic agents are called "competitors", and the resulting effect the "competition effect" – hence the name "competitive ion exchange" given to the preparation method. It is also referred to as "adsorption equilibrium", though this term may be misleading [12c].

Thermodynamic Aspects: Fixation Degree of Metal Ions on the Zeolite [39, 40] The thermodynamic constant of Eq. (5) can be written in terms of concentrations of the various ionic species:

$$K_a = \frac{[Pt_Z]\cdot[(NH_4)_S]^2}{[Pt_S]\cdot[(NH_4)_Z]^2} \quad (6)$$

If concentrations are replaced by the amounts of the corresponding ions in the system, for example $[Pt_Z] = (Pt_Z/V_Z)$, Eq. (6) becomes:

$$K_a = \frac{Pt_Z \cdot (NH_4)_S^2}{Pt_S \cdot (NH_4)_Z^2} \cdot \frac{V_Z}{V_S} \quad (7)$$

where V_Z and V_S are the volumes of the zeolite crystals and of the solution, respectively. By replacing Pt_S by $(Pt_T - Pt_Z)$ and $(NH_4)_S$ by $((NH_4)_T - (NH_4)_Z)$, Eq. (7) becomes:

$$K_a = \frac{Pt_z}{(Pt_T - Pt_z)} \cdot \left(\frac{(NH_4)_T - (NH_4)_Z}{(NH_4)_Z}\right)^2 \cdot \frac{V_Z}{V_S} \quad (8)$$

with $(NH_4)_T$ equal to the total amount of NH_4^+ ions in the system.

If Pt_T is small compared to the total number N_Z of exchangeable sites of the zeolite, we can write $(NH_4)_Z \approx N_Z$. We can thus express the variation of the fraction of Pt attached to the zeolite as a function of the total amount $(NH_4)_T$ of competing ions:

$$\frac{Pt_T}{Pt_Z} \approx \frac{1}{K_a} \cdot \frac{V_z}{V_s}\left(\frac{(NH_4)_T}{(NH_4)_Z} - 1\right)^2 + 1 \quad (9)$$

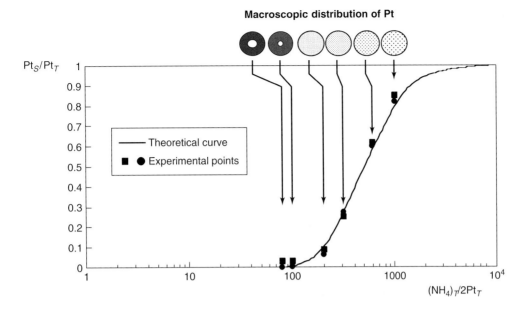

Fig. 4 Residual fraction of Pt in solution as a function of the $(NH_4)_T/2Pt_T$ ratio of competition between NH_4^+ and Pt^{2+} ions in the system.

It can thus be seen that the value of the fixation level (Pt_T/Pt_Z) can be adjusted to the desired value by tuning the total amount of the competing NH_4^+ ions in the system.

Practical Aspects The importance of the competition technique lies in the possibility of speeding up the impregnation kinetics and of reaching, within a reasonable period of time, a homogeneous macroscopic distribution of the metal on the support. This is particularly important when the catalyst grains are larger than several millimeters, such as a Y zeolite-based catalyst in the form of cylindrical pellets about 3 mm in diameter and 2 mm high. In the absence of NH_4^+ ions in the initial solution, the platinum is distributed in a peripheral crown about 0.5 mm thick. Castaing's microprobe shows that, inside this crown, the platinum concentration diminishes strongly from the outside to the inside. The choice of the proper amount of NH_4^+ ions per Pt^{2+} ion in the system must leave less than 5% of the platinum involved at the beginning of equilibrium in the final solution, but it must also bring about a quasi-homogeneous distribution of the metal within a few hours. Proper stirring of the medium is beneficial for the operation. In the absence of stirring, the kinetics is actually limited by diffusion in the bed of solid grains at the bottom of the recipient and of the boundary layer surrounding each grain.

The advantage of the competitive exchange technique is obviously much less evident when the zeolite is exchanged in the form of a fine powder. If no competition is used, various precautions must be taken:

- very effective stirring of the powder
- progressive addition of metal ions to the solution (for several dozen minutes)
- an exchange duration longer than or equal to about 5 h.

Slight competition also speeds up the operation appreciably.

2.4.2.3 Impregnation: Physical Aspects

Impregnation involves bringing the solution *into the pore space* of the support; by this technique, the volume of solution used is low and the whole precursor is expected to be retained on the support after drying (note that there is no intermediate washing step involved in this method), in contrast to the ion-exchange procedure. Two cases can be distinguished, depending on whether the pore space contains only ambient air at the start ("capillary" impregnation) or whether it is already filled by the solvent of the impregnation solution (usually water) or by another liquid ("diffusional" impregnation).

A Capillary Impregnation [41, 42] The operation consists of putting into contact with the previously dried support, with a pore volume V_{PT}, a volume V of solution containing the selected precursor. In almost all cases, $V = V_{PT}$, so that at the end of the operation no excess

References see page 482

solution remains outside the pore space. This is called "dry" or "incipient wetness" impregnation, and has the following characteristics.

Exothermicity Replacement of the solid–gas interface by a solid–liquid interface generally causes a considerable decrease in the free enthalpy of the system. Heat is released, which often has little effect on the quality of impregnation, except in the following specific cases:

- the precursor has retrograde solubility and its concentration is about the same as the saturation
- in a solution containing an unstable mixture of several metal precursors, the temperature rise may lead to precipitation of a mixed compound poorly distributed on the surface
- the impregnation solution is aggressive with regard to the support (high concentration of the precursor, very acid or basic pH), and a modification of the support surface may then occur, with accelerated alteration of its surface properties.

Solutions exist to avoid these drawbacks, such as:

- "deactivation" of the support by means of superheated steam to form a liquid surface film
- stabilization of the precursor(s), for example by complex formation.

Pressure Developed in Pores upon Impregnation [43–48]   As soon as the support is placed in contact with the solution, this latter is sucked up by the pores. Part of the air present in the pore space is imprisoned as bubbles and compressed under the effect of capillary forces. The pressure inside the bubbles depends on the radius r of the curve of the liquid–gas meniscus, and may reach several MPa (or tens of bars) when $r < 100$ nm, as the result of the Young–Laplace law:

$$\Delta P = P - P' = \frac{2\gamma}{r} \qquad (10)$$

where γ is the liquid–gas interfacial tension and ΔP refers to the difference of pressure inside the gas bubbles and the pressure outside the bubbles.

Considerable forces will thus be exerted on the portions of the pore walls in contact with these bubbles. The walls that are not strong enough will break down, causing degradation of the mechanical properties, and sometimes even bursting of the catalyst grains.

The development of high pressures is a transitory phenomenon. When highly compressed, air becomes dissolved and escapes progressively from the solid. The filling of the pore space is thus far from being instantaneous.

Solutions exist for preventing or limiting the bursting of the support upon impregnation, such as:

- operating in vacuum
- adding a surfactant to the solution.

Capillary Impregnation Rate [49–53]   Oversimplified calculations show that the time taken by the liquid to fill the pores is of the order of a few seconds. However, in practice determination of the impregnation time must take into consideration the imprisonment of air within the porosity and the elimination of this occluded air by dissolution and migration toward the outside of the grain [48]. The Young–Laplace law [Eq. (10)] shows that the last bubbles to disappear are those present in the largest pores, because they are subjected to the lowest pressures.

B Diffusional Impregnation   When the pore space of the support has first been filled up with pure solvent of the impregnation solution prior to being placed in contact with the latter, the characteristics defined above are valid only for the first step of saturation by the solvent.

The second step is generally an immersion that consists of plunging the solvent-saturated support into the impregnation solution. This step is no longer exothermic and does not cause the development of high pressure inside the pore space. The driving force of the progressive migration of the salt into the heart of the grains is the concentration gradient between the extragranular solution and the moving front of the soluble precursor in the intergranular solution. The migration time is obviously much longer than for capillary impregnation.

The case when the concentration of the solution outside the grains remains constant has been dealt with by various authors [54–59]. Weisz [56] defines a relaxation time constant τ for the system, which is representative of the actual time required to be very close to equilibrium:

$$\tau = \frac{R^2}{D} \qquad (11)$$

where R is a length parameter depending on the grain geometry (R is the radius of the grain in a sphere or an extrudate) and D is the diffusion coefficient of the precursor in the solvent.

When volume V is occupied by a porous solid, Eq. (11) is generally made to include a tortuosity coefficient β, the pore fraction ε (fraction of the support grain volume occupied by the pore space) and an interaction coefficient K between the precursor and the support ($K = 1$ if there is no interaction):

$$\tau = \frac{R^2}{D} \cdot \frac{\beta}{\varepsilon} \cdot K = \frac{R^2}{D_{\text{eff}}} \qquad (12)$$

in which $D_{\text{eff}} = (\varepsilon D/\beta K)$ is the effective or apparent diffusion coefficient of the solute in the porous solid.

The values of β are generally between 1.3 and 10, and more often between 2 and 6 (values determined various gases [60]). The values of ε are mostly between 0.3 and 0.7 [60]. The values of D in the liquid phase are generally about 10^{-5} cm^2 s^{-1} [55–60].

For example, let us consider a support made of beads with radius $R = 2$ mm. If we assume that $K = 1$, $D = 10^{-5}$ cm^2 s^{-1} and $\varepsilon = 0.5$, then time τ calculated from Eq. (11) is:

if $\quad \beta = 1.3 \quad \tau \sim 3$ h

if $\quad \beta = 5 \quad \tau \approx 12$ h

As the value of τ is proportional to the square of the radius, it can be appreciably reduced by diminishing the size of the particles. Thus, with beads of radius 1 to 0.5 mm, the above values are divided by 4 and 16, respectively.

Diffusional impregnation is almost never used for preparing catalysts when there is no appreciable interaction between the precursor and the support.

2.4.2.4 Support and Precursor during the Preparation Steps: Chemical Aspects

The precursors of the active phase involve transition metal ions (TMIs), complexed with ligands, which become catalytic sites at the end of preparation. From the characterization point of view, TMIs can be used to probe the various roles of the support, via their own interaction with the latter in the context of interfacial coordination chemistry [12, 61, 62]. Because TMIs have partly filled d orbitals, any change in their first coordination sphere immediately affects their optical and magnetic properties and can thus be detected by spectroscopy. The counterion associated with the TMI is also retained on the support when the catalyst is prepared by impregnation and, although less studied, its role should not be underestimated in the transformations taking place during preparation.

A The Oxide Surface as a Counter Ion The role of the support has been studied, essentially by diffuse reflectance UV-visible spectroscopy, in the case of the Ni/SiO$_2$ system prepared by the competitive cation exchange method (see Section 2.4.2.2.C) using nickel nitrate in water-ammonia solutions [12, 63]. In this method, the silica surface is first contacted with an ammonia solution at a pH above the ZPC of silica leading to a negatively charged surface:

$$\equiv\text{SiOH} + \text{NH}_3 \rightleftharpoons \{\equiv\text{SiO}^-, \text{NH}_4^+\} \quad (13)$$

Upon contact of silica with a nickel nitrate ammoniacal solution, it is observed that the purple blue color, which is characteristic of the nickel hexaammine Ni(NH$_3$)$_6^{2+}$ precursor complex, is transferred from the solution (Fig. 5, model I) to the silica support surface, while the diffuse reflectance spectrum of the resulting wet powder is the same as that of the initial solution. These phenomena suggest that the precursor Ni(II) complex has retained its integrity, and this can be explained in terms of the following competitive cation exchange:

$$2\{\equiv\text{SiO}^-, \text{NH}_4^+\} + [\text{Ni}(\text{NH}_3)_6]^{2+}$$
$$\rightleftharpoons \{2\equiv\text{SiO}^-, [\text{Ni}(\text{NH}_3)_6]^{2+}\} + 2\,\text{NH}_4^+ \quad (14)$$

Equation (14) shows that an ion-pair is formed with the oxide surface acting as a supermolecular counterion to which the TMI is bonded by ligand-screened electrostatic adsorption (Fig. 5, model II). In model II, as indicated by the arrows, the complex retains a certain degree of mobility [62a].

B The Oxide Surface as a Solid Solvent When complexes involving TMI are placed in microcontainers such as the pores of amorphous solids, or in the cavities of zeolites, their coordination is basically not changed, and is described by model I.

For pHs where surface functions are essentially OH groups, the oxide is able to enter the outer sphere of solvation of the complex in the same way as the solvent molecules (model I), and thus acts as a "solid" solvent (model III). When ligands have electronegative donor atoms (N, O, Cl, F, etc.), there may be some hydrogen bonds (model III, assuming L = NH$_3$, H$_2$O for instance) between the ligands and the OH or/and oxygen ions of the support oxide [64]. This aspect is different from that in solid solutions in oxides where TMIs come to rest after high-temperature diffusion to form doped or mixed oxides [65].

C From Electrostatic Adsorption to Grafting: the Oxide Surface as a Supermolecular Ligand Upon washing with ammonia solutions, in conditions of pH and NH$_3$ concentration corresponding to the stability range of the complex [Ni(NH$_3$)$_4$(H$_2$O)$_2$]$^{2+}$, it is shown that grafting occurs via substitution of 2 NH$_3$ by two surface $\equiv$SiO$^-$ ligands according to the successive reactions:

$$\{2\equiv\text{SiO}^-, [\text{Ni}(\text{NH}_3)_6]^{2+}\} + 2\text{H}_2\text{O}$$
$$\rightleftharpoons \{2\equiv\text{SiO}^-, [\text{Ni}(\text{NH}_3)_4(\text{H}_2\text{O})_2]^{2+}\} + 2\,\text{NH}_3 \quad (15)$$

$$\{2\equiv\text{SiO}^-, [\text{Ni}(\text{NH}_3)_4(\text{H}_2\text{O})_2]^{2+}\}$$
$$\rightleftharpoons [\text{Ni}(\text{NH}_3)_4(\equiv\text{SiO})_2] + 2\text{H}_2\text{O} \quad (16)$$

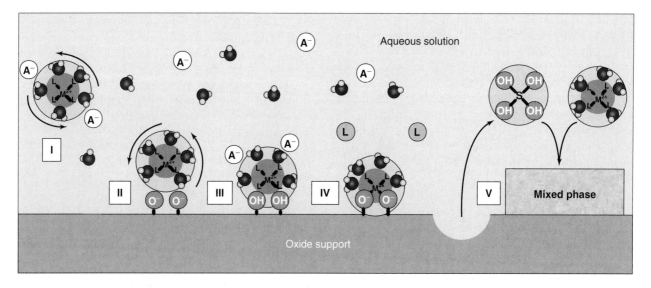

Fig. 5 Interaction models between a transition metal ion (TMI) of a precursor complex and an oxide support.

To date, the intermediate tetraammine complex has not been detected, and the actual process can in fact be written as the sum of the two preceding reactions as follows:

$$\{2\equiv SiO^-, [Ni(NH_3)_6]^{2+}\}$$
$$\rightleftharpoons [Ni(NH_3)_4(\equiv SiO)_2] + 2NH_3 \quad (17)$$

It appears that the ligand exchange NH_3/H_2O in the initial hexaammine complex weakens the coordination sphere, making grafting possible via two surface bonds (Fig. 5, model IV), with formation of a neutral *cis*-octahedral complex. The oxide, entering into the coordination sphere of Ni^{2+}, becomes a supermolecular bidentate ligand via vicinal $\equiv SiO^-$ groups. The latter have been found to be the most probable grafting sites, on the basis of geometric considerations [12, 63, 66] and extended X-ray absorption fine structure (EXAFS) studies [67].

The chelating effect of the surface appears to be the driving force for the formation of the surface *cis*-octahedral complex. Equation (17) is accompanied by an entropy increase with the release of water molecules and the disappearance of charged species, leading to a subsequent disordering of nearby solvent molecules [68, 69].

Once TMIs coming from complexes of the liquid phase (Fig. 5, model I) become bonded to the oxide surface (Fig. 5, model IV), they can remain attached to the support after the liquid phase is eliminated and the remaining gas–solid system heated at higher temperatures. It is then possible to produce vacant coordination sites. Their reactivity toward ligand addition or substitution is the driving force for adsorption and catalysis. This constitutes one of the major differences with coordination compounds in solution. Active ions can also form oligomeric/polymeric species, a monolayer or tridimensional oxide particles, with the surface acting as a ligand. In some other cases, intermediate compounds or solid solutions are formed when the support becomes a reactant, as shown in the next section.

D The Support as a Chemical Reactant The support/water system is thermodynamically metastable and evolves slowly towards the support precursor, namely the hydrous oxide or hydroxide. Thus, a realistic description of the oxide/water interfaces should take into account two phenomena: (i) the oxide hydration; and (ii) the support oxide dissolution.

On exposure to water, an anhydrous oxide can become hydrated by physical adsorption of water molecules without dissociation. It can also chemisorb water with dissociation leading to new hydroxyl groups, and finally to the formation of superficial oxyhydroxide or hydroxide, such as in the case of MgO [14a]. When silica surfaces are exposed to water for a long time, their hydroxylation produces polymeric chains of $-Si(OH)_2-O-Si(OH)_2-OH$ groups which can link up to form three-dimensional silica gel networks. Silica gel layers of about 2 nm thickness have been observed on silica surfaces prepared by evaporation on mica and exposed to humid air [70]. Thus, it may be postulated that surface groups are present not only in a two-dimensional oxide–liquid interface, but also in a phase of finite thickness extending from the surface into the interior of the solid [71].

Oxide dissolution has often been reported, in particular when impregnation is conducted at pH far from the PZC of the oxide, which is also the point of minimum solubility of the solid [72]. It is known that alumina dissolves in highly acidic impregnation solutions, and that a fraction of the dissolved aluminum may readsorb during impregnation or drying. Such an effect has been reported during the preparation of alumina-supported platinum [16a, 73–76], nickel [77], or palladium [78] catalysts. Silica solubility is very low from pH 2 to 8, but then increases very rapidly in basic medium [16a]. From pH 9 to 10.7, amorphous silica is in equilibrium with the neutral $Si(OH)_4$ monomers as well as silicate ions. Above pH 10.7, silica dissolves to form soluble silicates [17]. Silica dissolution in basic medium creates new hydroxyl groups and hence new sites with chemical affinity for cations, which may explain the observed sharp increase of cation adsorption on silica at pH higher than 10.5 [79, 80].

Interestingly, impurities such as aluminum, calcium, magnesium or zinc were reported to reduce both the rate of dissolution and the solubility of silica at equilibrium. In particular, nitric acid-cleaned silica immersed in solutions of Al was rendered insoluble at pH 9. This effect was attributed to the possible formation of a monolayer of insoluble silicate which lowered the silica solubility to that of the surface compound [17].

The formation of mixed extended phases by dissolution of the support oxide and reaction with TMI (Fig. 5, model V) has been reported in a variety of cases involving both cationic (e.g., $Ni(H_2O)_6^{2+}$) and anionic (e.g., MoO_4^{2-}) precursors.

In the former case, the formation of layer-type silicates was demonstrated during impregnation of silica with ammine Ni(II) ions in the pH range 8 to 9 [80]. No epitaxy was noted between the phyllosilicates and the carrier. This observation is consistent with experiments indicating the weathering at 100 °C of Pyrex vessels by aqueous solutions containing Ni(II) or Mg(II) salts, leading to the formation of antigorite-type silicates [81, 82]. Likewise, the formation of hydrotalcite-type coprecipitates involving Al ions extracted from the support was observed during impregnation of γ-alumina with Ni(II), Co(II) and Zn(II) complexes under mild conditions – that is, pH close to the PZC, ambient temperature and reasonable contact times [83, 84]. Anions whose presence was required for the electroneutrality of the hydrotalcite-like structures were adsorbed simultaneously with the metal cations. Thus, oxide supports should not be considered systematically as inert, even during impregnations conducted at pH close to the PZC.

In the case of anionic precursors, the formation of a mixed aluminometallic species in solution has been reported during the initial deposition step of Al_2O_3-supported Mo catalyst preparation by using selective adsorption procedures with ammonium heptamolybdate [85]. As a consequence, complexation of dissolved Al(III) by molybdenum resulted in extensive dissolution of the alumina support, even at pH-values close to neutrality. The *only* molybdic species present in solution at pH 4 after a few hours of contact was not the heptamolybdate initially introduced but rather the $Al(OH)_6Mo_6O_{18}^{3-}$ Anderson-type heteropolyanion: the transformation is essentially quantitative. Surface-selective ^{27}Al CP-MAS solid-state NMR showed that these species are not only formed in solution but also deposited *on the support*, and are thus the real precursors of the molybdic species present in the finished catalyst. It is essential to note that in this case, the mixed species proposed in Fig. 5 (model V) are not extended phases but deposited molecular compounds.

In all cases considered above, one cannot exclude the importance of gel-mediated pathways involving the diffusion of metal cations and anions through a hydrous gel layer, by analogy with phenomena occurring during the natural weathering of soils [86].

As shown above, the oxide support may play several different roles, and these may occur separately or simultaneously depending on the preparation method [12, 62].

E The Role of the Counter Ion Although attention has often been paid to the chemistry around the deposited TMI, it should be stressed that associated counter ions (such as NO_3^- for cationic complexes and NH_4^+ for anionic complexes) introduced in the impregnation solution also interact with the support, or with the TMI itself if they have coordinative properties [87]. They also determine the solubility of the precursor salt. During thermal treatment following impregnation, it is the chemistry of the overall system involving support, TMI, its ligands and the counter ions that must be considered: the oxidizing properties of NO_3^- toward ligands and/or TMI, or the reducing properties of NH_4^+ toward ligands and/or TMI may influence the final nature of the catalytic system.

2.4.2.5 Applications

The pertinent key parameters which may influence the interaction of a transition metal complex with an oxide support are first (and briefly) discussed, before an example of important application is given.

A Experimental Key Parameters When one deals with metal complexes in solution and a preformed support, the main problem is to know how to control the interaction between the complexes and the support. From the discussion above, six main parameters can be distinguished, which can tune the strength of this interaction. To these can be added the precursor

---

References see page 482

concentration, although in the case of impregnation this is a parameter determined by the quantity of metal to deposit on the support or the solubility of the precursor salt in the solvent.

Point of Zero Charge of the Oxide Support This is an essential parameter to initiate the electrostatic interaction between the metal complex and the oxide support. Two main methods of regulating the PZC of oxides have been developed. The first method involves the change of temperature of the impregnating solution, while the second involves selective doping of the support surface with various types of ion [88]. For instance, the PZCs of γ-Al_2O_3 and TiO_2 were found to increase upon raising the temperature of the impregnating solution, in contrast to that of SiO_2, which was found to decrease. Selective doping of γ-Al_2O_3 by very small amounts of Na^+ ions brought about an abrupt increase of PZC, while doping with F^- ions caused a shift to lower PZC values. It should be mentioned here that the selective doping method cannot be applied when the dopant leads to catalyst deactivation.

Solution pH and Precursor Speciation The pH appears to be an important parameter in the preparation process so as to achieve the desired ion–support interaction. It is worthwhile noting that this influences the liquid or/and solid part of the interface.

On the solid side, the pH allows one to select the sign of the surface charge (as shown in Fig. 3), and hence to determine the concentration of charged surface sites. Thus, the complex with electric charge of opposite sign must be chosen to obtain electrostatic interaction. Moreover, the pH controls the dissolution of the oxide support both thermodynamically and kinetically. The solubility (thermodynamic) and the dissolution rate (kinetic) of an oxide are all the more important as the pH of the solution departs from the PZC [16a, 72, 89].

On the liquid side, the pH determines the nature and nuclearity of the complex; that is, the speciation of the element involved (for a definition of speciation, see Ref. [90a]). For example, upon lowering the pH from 14, complexes gradually change from hexaammine to hexaaquametal(III) for many transition metals, or from mononuclear MoO_4^{2-} to polynuclear $Mo_8O_{26}^{4-}$ [90b]. The correct choice of pH allows one to select the type of the most abundant metal complex. Finally, the pH also allows control of the amount of monomers (such as $Al(H_2O)_6^{3+}$ in the case of alumina) produced in the solution upon oxide support dissolution [16a].

Having emphasized the importance of the pH, it is fair to recognize that its measurement and control are difficult tasks [62a]. When the surface is charged, the overall ion concentrations are not constant close to the oxide–fluid interface. Above the PZC (i.e., for negatively charged surfaces) there is an increase in cation and proton concentrations at the surface. Therefore, the pH is lower than in the bulk of the solution. The situation is reversed at a pH below the PZC. Moreover, when the volume of solution is low (impregnation), the pH in the pore space is often regulated by the acid-basic properties of the support ("buffer effect") [91].

In colloidal solutions, the measurement of pH is always delicate, as it depends on the average distance between the oxide particles and the pH electrode, and is then a function of the stirring rate. The best value, which is considered to be the pH of the bulk solution, is that obtained without stirring, the solid being far from the electrode and deposited at the bottom of the flask (centrifugation is sometimes necessary). This value, combined with the characteristics of the solution and the PZC of the oxide, allows the pH profile to be calculated from the theory developed by Davis et al. [19]. More recently, devices based on fiber-optic sensors [92] have been developed which might help to resolve the difficult problem of the determination of pH of oxides in suspension.

Type of Ligand The ligand type greatly affects both the nature and strength of the metal complex–surface interaction. Several structural and electronic aspects must be considered for each ligand:

- its charge (neutral molecules and/or anions), which determines the charge of the transition metal complex and hence the conditions such as support nature and pH range, for which ion exchange can occur
- its bulkiness, which tends to screen and decrease the electrostatic interaction between the transition metal complex and the charged surface
- ligands that are to be substituted in the grafting process by surface oxygens [see, e.g., Eq. (16)] should be sufficiently labile, while spectator polydentate ligands should be inert [64, 93a]. From a kinetic point of view, it is well known that the substitution of some aqua ligands for polydentate amine-type ligands in octahedral complexes increases the lability of the remaining aqua ligands [93a]. This lability has some practical consequences. It allows in particular to graft Ni(II) ions by one, two, or three bonds to the silica surface when the precursor complexes contain one, two, or three labile water ligands, e.g., $[Ni(en)(dien)(H_2O)]^{2+}$, $[Ni(en)_2(H_2O)_2]^{2+}$, and $[Ni(dien)(H_2O)_3]^{2+}$ respectively, (en = $NH_2-CH_2-CH_2-NH_2$ is the bidentate ligand ethylenediamine; dien = $NH_2-CH_2-CH_2-NH-CH_2-CH_2-NH_2$ is the tridentate ligand diethylenetriamine) [64, 93a].

- the presence of an adequate element (e.g., N, O, Cl, H) favors the formation of hydrogen bonds with the surface via its S−OH and/or S−O−S groups [64, 93b].
- the polydentate nature of the ligand, whether it is the organic ligand (e.g., ethylenediamine) or the surface (≡SiO), has a strong entropy effect on the stability of either the initial solution complex or the final grafted complex [61]
- polydentate ligands such as citrate produce complexes that deposit as amorphous compounds over the support surface, leading to a more dispersed phase than crystallized precursors would do [94]
- non-bridging ligands inhibit the formation of oligomers or mixed surface compounds such as silicates. Depending on their concentration, the pH chosen for the solution or imposed by the support, and the possibility of support dissolution, the precursors in solution can react with each other or with species released by the oxide (see Section 2.4.2.4.D). M−OH−M and M−O−M bonds can be formed via olation and oxolation reactions respectively which can be written as follows:

$$M-OH + M-H_2O \rightleftharpoons M-OH-M + H_2O \quad (18)$$

$$M-OH + M-OH \rightleftharpoons M-O-M + H_2O \quad (19)$$

where M is either the metal cation in the support monomer or in the complex. The result is that oligomers or intermediate surface compounds are formed [12, 95, 96] because both OH and H_2O are bridging ligands. By contrast, with ligands such as ammonia or ethylenediamine, Eqs. (18) and (19) cannot occur and isolated grafted complexes are favored.

The fact that the surface can be identified to a ligand, is an important aspect which has other practical consequences. For example, it is well known that the standard redox potential of a TMI can be largely modified depending upon the nature of its ligands [68, 69].

Competitive Ions   Examples of competitive ions were provided in Section 2.4.2.2.C. By use of competitive ions, it is possible not only to control [47, 97, 98] but also to model [9] the exchange process and the distribution profile of the active phase along the radius of the grain as a function of time. An important industrial application is described in Section 2.4.2.5.B.

Washing   It is known that washing must be avoided after the component precursor has been deposited by impregnation. When other deposition methods are involved, the washing step has been reported to have a variety of effects.

The fundamental aspects of washing have not been studied widely in the past, although washing may have important consequences. The first consequence is electrostatic in nature (see Section 2.4.2.4.A), in that washing can be accompanied by pH changes so that the sign of the surface can change, leading to repulsions with the adsorbed species [99]. The second consequence rests on the ligand nature of the surface. Changes of pH may lead to grafting of the component precursor (see Section 2.4.2.4.C) [63]. Yet another consequence of washing is connected with the role of reactant of the oxide support, as oxide dissolution will eventually lead to intermediate compounds formation at the surface of the oxide support (see Section 2.4.2.4.D) [80, 95].

Kinetics   The speciation of the deposited phase may be controlled to a large extent via the parameters that control the kinetics of interfacial phenomena – that is, the temperature and duration of preparation steps. Figure 6 illustrates this concept, classifying various steps of catalyst preparations in a $T = f(t)$ plot (T is the temperature and t the time of the corresponding preparation step). As an example, it is expected that the formation of mixed phases following support dissolution (see Fig. 5, model V) can be kinetically slowed down for low temperatures and their extent limited by short times and, conversely, accelerated for high temperatures and their extent increased by long times. As an application, it has been shown that incipient wetness impregnation immediately followed by freeze-drying effectively inhibits the formation of the Anderson-type $Al(OH)_6Mo_6O_{18}^{3-}$ heteropolyanion during alumina-supported Mo catalyst

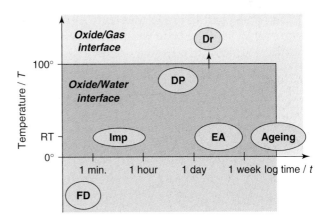

Fig. 6   Various steps of catalysts preparation in a $T = f(t)$ plot. FD = freeze-drying; Imp = impregnation; EA = equilibrium adsorption (ion exchange); DP = deposition–precipitation; Dr = drying; Ageing (maturation)

References see page 482

preparation, allowing to control the state of Mo after high-temperature calcination [85b].

B Case Study: Reforming Catalysts Pt/Al$_2$O$_3$–Cl

Reforming catalysts are usually constituted of well-dispersed platinum on alumina (η- and mainly γ-Al$_2$O$_3$) promoted by a halogen, usually chlorine [75]. The deposition of platinum is usually carried out by immersing the alumina in an aqueous solution of chloroplatinic acid and hydrochloric acid as competing agent [34a, 75].

Interactions between Acid Solutions and the Surface of Alumina
The hydroxyl groups present at the surface of alumina behave like bases with respect to the protons of the acids [14b, 16a]. According to various authors, the reaction is represented by two different mechanisms:

$$\text{—Al—OH} + \text{H}_3\text{O}^+, \text{X}^- \xrightleftharpoons[B]{A} \begin{array}{l} \text{—Al—OH}_2^+, \text{X}^- + \text{H}_2\text{O} \\ \text{—Al—X} + 2\text{H}_2\text{O} \end{array}$$

(20)

In mechanism A, the protonated OH group is in electrostatic interaction with the X$^-$ anion [16a, 19, 100–102], whereas in mechanism B [9, 13, 14b, 97, 103], the X$^-$ anion replaces the OH$^-$ group in the coordination sphere of the surface aluminum. Mechanism A is expected to predominate because the surface of transition aluminas is mainly covered by OH groups [104–107] in the presence of water.

Protonation of surface oxygen can also occur and lead to the scission of an Al–O bond according to the following reaction:

$$\text{—Al—O—Al—} + \text{H}_3\text{O}^+, \text{X}^- \rightleftharpoons \text{—Al—OH}_2^+, \text{X}^- + \text{—Al—OH}$$

(21)

According to Eqs. (20A) and (21), Al$_2$O$_3$ can behave as anion exchanger. The maximum quantity of anions that the surface can adsorb (i.e., the anion-exchange capacity) is related directly to the number of protonated OH groups [16a, 19, 100, 108]. Logically, it depends on the equilibrium, which shifts to the right as the pH decreases [16a, 19, 100, 108, 109]. The process of anion fixation on the surface is kinetically very fast [101], and rarely limits the overall kinetics. The rate-limiting step is expected to be the diffusion of ionic species in the pores of alumina particles.

Deposition of Platinum The maximum amount of platinum adsorbed on alumina from the acid H$_2$PtCl$_6$, without any significant attack of the support, differs widely according to the authors [16a, 34a, 35, 73, 98, 108, 110], varying between 2 and about 8 wt.%. Most of the values obtained at saturation of γ-aluminas with surface areas ranging between 150 and 250 m^2 g^{-1} are normally between 2 and about 4 wt.% [16a, 34a, 98, 110]. The platinum content of industrial catalysts is much lower, varying between 0.2 and 0.6 wt.%.

Many authors have observed that the interaction between the hexachloro complex of platinum H$_2$PtCl$_6$, and γ-Al$_2$O$_3$ is strong [16a, 34–36, 97, 98, 111]. At impregnation conditions approaching those used for the preparation of reforming catalysts (i.e., small amounts of platinum and low pH, hence high anion-exchange capacity), platinum tends to be adsorbed rapidly and strongly on the first sites encountered. The possibility of a heterogeneous macroscopic distribution of platinum on the surface is similar to that described above in the case of zeolite NH$_4$Y and Pt(NH$_3$)$_4^{2+}$.

In order to obtain a homogeneous distribution of platinum on alumina, it is necessary to add a competitor ion to the solution. The usual competitor is HCl [34a, 47, 112], whose role is to act on the equilibrium of the reaction:

$$[\text{PtCl}_6]^{2-}{}_{/\text{Al}} + 2\text{Cl}^-{}_{\text{aq}} \rightleftharpoons [\text{PtCl}_6]^{2-}{}_{\text{aq}} + 2\text{Cl}^-{}_{/\text{Al}}$$

(22)

where /Al represents the alumina surface.

The quantity of Cl$^-$ ions added to the system is usually adjusted to leave a sufficient quantity of PtCl$_6^{2-}$ in aqueous solution to ensure easy migration to the center of the grain so that, at the end of the operation, this residual non-adsorbed amount is only a small fraction of the total quantity of platinum involved.

However, chemical phenomena in this case are more complicated than in competitive cation exchange on NH$_4$Y zeolite. HCl acts not only as a competitor but also as a stabilizing agent for the coordination sphere of platinum. As a matter of fact, it has been shown [113, 114, 131], that the PtCl$_6^{2-}$ anion can lose two or four chloride ligands after a few hours in the absence of hydrochloric acid. Another level of complexity is caused by the chemical reactivity of the alumina surface.

Chemical Alteration of Alumina In an acidic medium, dissolution of the support occurs simultaneously with protonation of OH groups [Eq. (20A)] [16a, 31, 47, 101, 108, 111, 115]. Dissolution occurs below about pH 4, with the solubility of the support rising rapidly for decreasing pH [116b]. Actually, even at neutral pH, alumina is significantly altered upon hydration with formation of an hydroxide-like surface [116]. Equation (21) is probably the first step of formation of soluble compounds of the type $[Al(H_2O)_y(OH)_x]^{(3-x)+}$ with $x + y = 6$. This is expected to occur only after Eq. (20A) is complete. Several experimental observations [31] support these considerations. If alumina beads are contacted with an HCl solution, the following sequence of events occurs: rapid adsorption of Cl^- ions probably due to Eq. (20A); subsequent slowdown of this process; then a decrease in the quantity of Cl^- adsorbed, probably due to progressive dissolution of the external part of the alumina particle [Eq. (21)] where the pH is the lowest; and finally a slow adsorption of Cl^- ions. If the initial quantity of acid used is small (<4–5 wt.% Cl), the final event (slow adsorption of Cl^-) resumes after several minutes [31]. This happens because the situation at the particle scale is still far from equilibrium: the pH in the external solution is low, but higher in the internal solution near the center of the particle. The pH progressively evolves toward equilibrium, where the pH is slightly above 3. At such pH, chlorinated aluminum species which were solubilized at the lower pH during the beginning of the reaction will be redeposited on the support. This can be minimized by using low concentrations to prepare monometallic reforming catalysts, for which the solubility of γ-Al_2O_3 should be very low (a few hundred ppm).

Metal Concentration Profiles in Supports For the deposition of chloroplatinic acid on alumina, competitors other than HCl can be used [36, 97, 98, 111, 112, 116–123]. Acids such as HNO_3 and H_2SO_4 are usually more effective than the corresponding salts, such as ammonium, aluminum, and sodium nitrates [36]. Nitric acid is about fivefold more effective than acetic acid [120]. Hydrofluoric acid inhibits the adsorption of $PtCl_6^{2-}$ ions by blocking the sites [112]. Other competitors include CO_2 [119], phosphoric acid [117, 118], organic acids such as formic, oxalic, propionic, lactic, salicylic, tartaric, and citric acids [97, 116–118, 121–123], miscellaneous salts such as alkaline phosphate [98], alkaline halogenides [98, 118], alkaline benzoate [98] and amines, such as monoethanolamine [97].

Some of these "competitors" (e.g., monoethanolamine) do not act on adsorption via a competition effect, but rather via a pH effect [97]. In fact, a rise in pH tends to lower the exchange capacity of alumina, causing a more uniform occupation of the surface by the metallic complex.

Careful selection of metal precursors and competitors helps to obtain a large number of metal distribution profiles between the periphery and the center of the support grain [50, 95, 111, 124–127], depending on the strength of their interaction with the oxide support. Four main types of profile [50, 124–127] allow one to describe all the others: uniform, film or eggshell, internal ring or egg white, central or egg-yolk (Fig. 7).

HF [112] and oxalic, lactic, tartaric [121] and citric acids [121, 128], for example, help to obtain an egg-white or egg-yolk distribution of chloroplatinic acid. These competing agents, which all have a high affinity for alumina, become irreversibly adsorbed on the external sites of the support and force the platinum precursor to migrate farther toward the center of the grain in order to find free sites.

$PtCl_6^{2-}$ can undergo, in neutral or slightly acidic medium, various reactions of aquation and hydrolysis [16a, 49, 129–131] which modify the metal environment, the ion charge, and thus the affinity of the latter for the support, ending with grafting of the platinum ion to the oxide surface [131]. Indeed, the profile of the macroscopic distribution of the metal on the support also depends on the more or less strong ability of the metal ion to modify its sphere of coordination or to polymerize. Molybdic and tungstic ions, which are present as monomers MoO_4^{2-} or WO_4^{2-} in basic medium, polymerize in neutral or acidic medium. The polymerization of tungstic ions, for example, influences the distribution profile [132] via the diffusion rate of these ions. It can also lead to the blocking of part of the pores. Support dissolution followed by mixed-species formation may also influence the distribution profile of metallic precursors. Raman microscopy performed on cross-sections of alumina extrudates impregnated with molybdenum has shown that most of the Mo species are present as large clusters of alumino-molybdic mixed compounds precipitated on the outside of the pellets, while there is a low Mo concentration inside these pellets even after 24 h of ageing. The ageing period would thus favor an egg-shell distribution of Mo [133].

Precursor = metallic complex (MC), competitor (A)

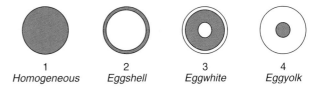

1
Homogeneous

2
Eggshell

3
Eggwhite

4
Eggyolk

1: roughly same affinities of MC and A
2: high affinity of MC; no competition
3: high affinity of A, low affinity of MC; low A/MC ratio
4: high affinity of A, low affinity of MC; high A/MC ratio

Fig. 7 The four main categories of macroscopic distribution of a metal within a support. A/MC (competitor/metallic complex) is the ratio representing competition (with a proportionality factor depending on the valence of the ions).

References see page 482

A number of theoretical models of migration and deposition of metallic ions in porous supports as a function of time, with or without the presence of competing agents, based on liquid-phase diffusion and adsorption mechanisms, have been proposed to account for the experimental results [56–58, 112, 134–138].

2.4.2.6 Conclusions

In this chapter we have described the principles and basic concepts underlying impregnation and ion exchange, the two most frequently employed methods used to prepare catalysts consisting of one or several catalytically active component(s) deposited on a high-surface-area oxide support. Oxide-supported catalysts, which form the most important class, represent about 75% of the ca. 10 billion worldwide demand of catalysts [139].

This chapter has emphasized the molecular and chemical aspects of catalyst preparation, particularly during the early steps involving the liquid–solid interface, the role of which in catalyst preparation has recently been reviewed [140]. Here, the case of ion exchange and impregnation has been presented in more detail, because of the possibilities of varying the "metal complex–support" interactions in a large domain, ranging from weak physical interactions to stronger chemical interactions.

Provided that certain experimental conditions are fulfilled, specific interactions can develop between the precursor metal complex and the oxide support, allowing the respective roles of the support (counter ion, solid solvent, supermolecular ligand, reactant), and of the counterion (ligand, oxidizing or reducing agent), to be visualized.

From the standpoint of applications, the main experimental key parameters which allow the tuning of "metal complex–support" interactions have been reviewed and their influence described for the particular case of Pt/Al$_2$O$_3$–Cl reforming catalysts, notably with regards to the metal concentration profiles in the support.

The main elementary processes [as indicated in Eqs. (1), (2), (5), and (13)–(22)] that occur during the steps of ion exchange and impregnation have been identified. An improved understanding of the processes that develop during successive unit operations (e.g., precursor deposition, ageing, filtration, washing, drying) provides help in improving catalyst design. By knowing the role of these key parameters, new sequences of preparation should be proposed to tune catalyst's ultimate properties, such that catalyst "architecture" will become common-place.

References

1. (a) J. T. Richardson, *Principles of Catalyst Development*, Plenum Press, New York, 1989, p. 134; (b) L. H. Baekeland, *J. Ind. Eng. Chem.* **1916**, *8*, 184.
2. (a) B. Imelik, J. C. Vedrine (Eds.), *Catalyst Characterization: Physical Techniques for Solid Materials*, Plenum Press, New York, 1994; (b) B. M. Weckhuysen (Ed.), *In Situ Spectroscopy of Catalysts*, ASP, Stevenson Ranch, 2004.
3. M. Campanati, G. Fornasari, A. Vaccari, *Catal. Today* **2003**, *77*, 299.
4. G. J. Hutchings, J. C. Védrine, *Springer Series in Chemical Physics* **2004**, *75*, 217.
5. J. M. Thomas, W. J. Thomas, *Principles and Practice of Heterogeneous Catalysis*, VCH, Weinheim, 1997.
6. K. Foger, *Catal. Sci. Technol.* **1984**, *6*, 227.
7. (a) Y. Iwasawa, in *Tailored Metal Catalysts*, Y. Iwasawa (Ed.), Reidel, Dordrecht, 1986; (b) Y. Iwasawa, *Adv. Catal.* **1987**, *35*, 187.
8. B. C. Gates, L. Guczi, H. Knözinger (Eds.), *Metal Clusters in Catalysis, Studies in Surface Science and Catalysis*, Vol. 29, Elsevier, Amsterdam, 1986, 648 pp.
9. L. L. Hegedus, R. Aris, A. T. Bell, M. Boudart, N. Y. Chen, B. C. Gates, W. O. Haag, G. A. Somorjai, J. Wei, *Catalyst Design*, Wiley, New York, 1987.
10. K. I. Tanaka, *Adv. Catal.* **1985**, *33*, 99.
11. M. Ichikawa, *Adv. Catal.* **1992**, *38*, 283.
12. (a) M. Che, *Stud. Surf. Sci. Catal.*, **1993**, *75A*, 31; (b) M. Che, *Stud. Surf. Sci. Catal.* **2000**, *130A*, 115; (c) J. F. Lambert, M. Che, in *Catalysis by Unique Metal Ion Structures in Solid Matrices*, G. Centi, B. Wichterlová, A. T. Bell (Eds.), NATO Science Series II: Mathematics, Physics and Chemistry, Vol. 13, Springer, Berlin, 2001, p. 1.
13. G. J. K. Acres, A. J. Bird, J. W. Jenkins, A. King, *Catalysis (London)* **1981**, *4*, 1.
14. (a) G. A. Parks, *Chem. Rev.* **1965**, *65*, 177; (b) W. Stumm, *Chemistry of the Solid-Water Interface: Processes at the Mineral-Water and Particle-Water Interface in Natural Systems*, J. Wiley & Sons, New York, 1992, 448 pp.
15. H. H. Kung, *Transition Metal Oxides: Surface Chemistry and Catalysis, Studies in Surface Science and Catalysis*, Vol. 45, Elsevier, Amsterdam, 1989, 343 pp.
16. (a) J. P. Brunelle, *Pure Appl. Chem.* **1978**, *50*, 1211; (b) J. Vakros, C. Kordulis, A. Lycourghiotis, *Chem. Commun.* **2002**, 1980.
17. R. K. Iler, *The Chemistry of Silica*, Wiley, New York, 1979, p. 56.
18. O. Stern, *Z. Elektrochem.* **1924**, *30*, 508.
19. J. A. Davis, R. O. James, J. O. Leckie, *J. Colloid Interface Sci.* **1978**, *63*, 480.
20. A. W. Adamson, *Physical Chemistry of Surfaces*, Wiley, New York, 1982, pp. 185–231.
21. D. E. Yates, S. Levine, T. W. Healy, *J. Chem. Soc., Faraday Trans. I* **1974**, *70*, 1807.
22. J. A. Davis, J. O. Leckie, *J. Colloid Interface Sci.* **1978**, *67*, 90.
23. T. W. Healy, L. R. White, *Adv. Colloid Interface Sci.* **1978**, *9*, 303.
24. J. Westall, H. Hohl, *Adv. Colloid Interface Sci.* **1980**, *12*, 265.
25. R. Sprycha, *J. Colloid Interface Sci.* **1984**, *102*, 173.
26. W. H. van Riemsdijk, J. C. M. de Wit, L. K. Koopal, G. H. Bolt, *J. Colloid Interface Sci.* **1987**, *116*, 511.
27. K. C. Williams, J. L. Daniel, W. J. Thomson, R. I. Kaplan, R. W. Maatman, *J. Phys. Chem.* **1965**, *69*, 250.
28. N. Spanos, A. Lycourghiotis, *Langmuir* **1993**, *9*, 2250.

29. W. H. van Riemsdijk, G. H. Bolt, L. K. Koopal, J. Blaakmeer, *J. Colloid Interface Sci.* **1986**, *109*, 219.
30. C. Contescu, J. Jagiello, J. A. Schwarz, *Langmuir* **1993**, *9*, 1754.
31. C. Marcilly, J. P. Franck, *Rev. IFP* **1984**, *3*, 337.
32. C. N. Satterfield, *Heterogeneous Catalysis in Industrial Practice*, McGraw-Hill, New York, *Chem. Eng. Series*, 1991, Chapter 7.
33. C. Perego, P. Villa, *The Catalytic Process from Laboratory to the Industrial Plant*, D. Sanfilippo, 1994, Chapter 2.
34. (a) J. F. Le Page, *Applied Heterogeneous Catalysis*, Technip, Paris, 1987, 12–122; (b) J. F. Lambert, E. Marceau, B. Shelimov, J. Lehman, V. Le Bel de Penguilly, X. Carrier, S. Boujday, H. Pernot, M. Che, *Stud. Surf. Sci. Catal.* **2000**, *130B*, 1043.
35. R. W. Maatman, C. D. Prater, *Ing. Eng. Chem.* **1957**, *49*, 253.
36. R. W. Maatman, *Ind. Eng. Chem.* **1959**, *51*, 913.
37. H. A. Benesi, US Patent 3.527.835, September 1970.
38. M. El Malki, J. P. Franck, C. Marcilly, R. Montarnal, *CR Acad. Sci. Paris* **1979**, *288C*, 173.
39. F. Ribeiro, C. Marcilly, *Rev. IFP* **1979**, *34*, 405.
40. F. Ribeiro, Ph.D. Thesis, *Préparation et propriétés catalytiques de Pt/zéolithe Hy et de Pt/zéolithe H. mordénite*, Poitiers, 1980.
41. L. I. Kheifets, A. V. Neimark, *Multiphase Processes in Porous Media*, Izd. Khimiya, Moscow, 1982, p. 320.
42. G. A. Aksel'rud, M. A. Al'tshuler, *Introduction to Capillary-Chemical Technology*, Izd. Khimiya, Moscow, 1983, p. 264.
43. R. Montarnal, *IFP* unpublished notes, 1973.
44. M. A. Al'tshuler, *Kolloidny. Zhurnal* **1961**, *23*, 646; M. A. Al'tshuler, *Kolloidny. Zhurnal* **1977**, *39*, 1142 (*Colloid J. USSR* **1977**, *39*, 1000).
45. K. M. Adam, *Water Resources Res.* **1969**, *5*, 840.
46. P. Leroux, Rhône-Poulenc, unpublished notes, 1978.
47. C. Marcilly, in *Catalyse par les Métau*, B. Imelik, G. A. Martin, A. J. Renouprez, (Eds.), CNRS, Paris 1984, p. 121.
48. V. V. Veselov, G. A. Chernaja, T. A. Levanjuk, *Kinet.Katal.* **1982**, *20*, 83.
49. J. R. Anderson, *Structure of Metallic Catalysts*, Academic Press, London, 1975, p. 172.
50. A. V. Neimark, L. I. Kheifets, V. B. Fenelonov, *Ind. Eng. Chem., Prod. Res. Dev.* **1981**, *20*, 439.
51. E. W. Washburn, *Phys. Rev* **1921**, *17*, 273.
52. C. H. Bosanquet, *Phil. Mag.* **1923**, *45*, 525.
53. M. A. Lauffer, *J. Chem. Educ.* **1981**, *58*, 250.
54. J. Crank, *Mathematics of Diffusion*, Clarendon Press, Oxford, 1964.
55. R. Paterson, *An Introduction to Ion Exchange*, Heyden and Son, London, 1970.
56. P. B. Weisz, *Trans. Faraday Soc.* **1967**, *63*, 1801.
57. P. B. Weisz, J. S. Hicks, *Trans. Faraday Soc.* **1967**, *63*, 1807.
58. P. B. Weisz, H. Zollinger, *Trans. Faraday Soc.* **1967**, *63*, 1815.
59. J. Crank, G. S. Park, *Diffusion in Polymers*, Academic Press, London, 1984, Chapter 1.
60. C. N. Satterfield, *Mass Transfer in Heterogeneous Catalysis*, MIT Press, London, 1970.
61. C. Lepetit, M. Che, *J. Mol. Catal.* **1995**, *100*, 147.
62. (a) J.-F. Lambert, M. Che, *J. Mol. Catal. A.* **2000**, *162*, 5; (b) J-F. Lambert, M. Che, *Stud. Surf. Sci. Catal.* **1997**, *109*, 91.
63. L. Bonneviot, O. Legendre, M. Kermarec, D. Olivier, M. Che, *J. Colloid Interface Sci.* **1990**, *134*, 534.
64. S. Boujday, J. F. Lambert, M. Che, *Chem. Phys. Chem.* **2004**, *5*, 1003.
65. A. Davidson, M. Che, *J. Phys. Chem.* **1992**, *96*, 9909.
66. G. W. Brindley, G. Brown, *Crystal Structures of Clay Minerals and their X-Ray Identification*, Mineralogy Society, London, 2nd edn., 1980, p. 2.
67. L. Bonneviot, O. Clause, M. Che, A. Manceau, H. Dexpert, *Catal. Today* **1989**, *6*, 39.
68. (a) K. F. Purcell, J. C. Kotz, *Inorganic Chemistry*, Saunders, Philadelphia, 1977; (b) F. A. Cotton, G. Wilkinson, C. A. Murillo, M. Bochmann, *Advanced Inorganic Chemistry*, 6th edn., Wiley, New York, 1999.
69. V. Vivier, F. Aguey, J. Fournier, J.-F. Lambert, F. Bedioui, M. Che, *J. Phys. Chem. B* **2006**, *110*, 900.
70. G. Vigil, Z. Xu, S. Steinberg, J. Israelachvili, *J. Colloid Interface Sci.* **1994**, *165*, 367.
71. J. Lyklema, *J. Electroanal. Chem.* **1968**, *18*, 341.
72. G. A. Parks, P. L. de Bruyn, *J. Phys. Chem.* **1962**, *66*, 967.
73. R. W. Maatman, P. Mahaffy, P. Hoekstra, C. Addink, *J. Catal.* **1971**, *23*, 105.
74. E. Santacesaria, S. Carrà, I. Adami, *Ind. Eng. Chem., Prod. Res. Dev.* **1977**, *16*, 41.
75. J. P. Boitiaux, J. M. Devès, B. Didillon, C. R. Marcilly, in *Catalytic Naphtha Reforming*, G. J. Antos, A. M. Aitani, J. M. Parera (Eds.), Marcel Dekker, New York, 1995, Chapter 4.
76. S. Parkash, S. K. Chakrabartty, T. Koanigawa, N. Berkowitz, *Fuel Process. Technol.* **1982**, *6*, 177.
77. S. L. Chen, H. L. Zhang, J. Hu, C. Contescu, J. A. Schwarz, *Appl. Catal.* **1991**, *73*, 289.
78. S. Subramanian, D. D. Obrigkeit, C. R. Peters, M. S. Chattha, *J. Catal.* **1992**, *138*, 400.
79. D. W. Fuerstenau, K. Osseo-Asare, *J. Colloid Interface Sci.* **1987**, *118*, 524.
80. O. Clause, M. Kermarec, L. Bonneviot, F. Villain, M. Che, *J. Am. Chem. Soc.* **1992**, *114*, 4709.
81. S. Henin, *C.R. Acad. Sci.* **1957**, *244*, 225.
82. S. Caillere, S. Henin, H. Besson, *C.R. Acad. Sci.* **1963**, *256*, 208.
83. J. L. Paulhiac, O. Clause, *J. Am. Chem. Soc.* **1993**, *115*, 11602.
84. J.-B. d'Espinose de la Caillerie, C. Bobin, B. Rebours, O. Clause, *Stud. Surf. Sci. Catal.* **1995**, *91*, 169.
85. (a) X. Carrier, J. F. Lambert, M. Che, *J. Am. Chem. Soc.* **1997**, *119*, 10137; (b) X. Carrier, J. F. Lambert, S. Kuba, H. Knözinger, M. Che, *J. Mol. Struct.* **2003**, *656*, 231.
86. (a) W. H. Casey, J. F. Banfield, H. R. Westrich, L. McLaughlin, *Chem. Geol.* **1993**, *105*, 1; (b) W. H. Casey, H. R. Westrich, J. F. Banfield, G. Ferruzzi, G. W. Arnold, *Nature* **1993**, *366*, 253.
87. F. Négrier, E. Marceau, M. Che, J. M. Giraudon, L. Gengembre, A. Löfberg, *J. Phys. Chem. B* **2005**, *109*, 2836.
88. A. Lycourghiotis, in *Acidity and Basicity of Solids: Theory, Assessment and Utility*, J. Fraissard, L. Petrakis (Eds.), NATO ASI Series, Series C, Math. Phys. Sci., 1994, 444, 415.
89. W. Stumm, *Colloids Surf. A* **1997**, *120*, 143.
90. (a) D. M. Templeton, F. Ariese, R. Cornelis, L. G. Danielsson, H. Muntau, H. P. van Leeuwen, R. Lobinski, *Pure Appl. Chem.* **2000**, *72*, 1453; (b) C. F. Baes, R. E. Mesmer, *The Hydrolysis of Cations*, Krieger, Malabar, 1986.
91. W. A. Spieker, J. R. Regalbuto, *Chem. Eng. Sci.* **2001**, *56*, 3491.
92. J. Lin, *Trends Anal. Chem.* **2000**, *19*, 541.
93. (a) J. F. Lambert, M. Hoogland, M. Che, *J. Phys. Chem. B* **1997**, *101*, 10347; (b) J. R. Bargar, S. N. Towle, G. E. Brown, Jr., G. A. Parks, *Geochim. Cosmochim. Acta* **1996**, *60*, 3541.
94. J. A. van Dillen, R. J. A. M. Terorde, D. J. Lensveld, J. W. Geus, K. P. de Jong, *J. Catal.* **2003**, *216*, 257.

95. M. Kermarec, J. Y. Carriat, P. Burattin, M. Che, A. Decarreau, *J. Phys. Chem.* **1994**, *98*, 12008.
96. K. Bourikas, C. Kordulis, J. Vakros, A. Lycourghiotis, *Adv. Colloid Interface Sci.* **2004**, *110*, 97.
97. A. K. Aboul-Gheit, *J. Chem. Tech. Biotechnol.* **1979**, *29*, 480.
98. Y.-S. Shir, W. Ernst, *J. Catal.* **1980**, *63*, 425.
99. C. Louis, M. Che, *J. Catal.* **1992**, *135*, 156.
100. J. A. Schwarz, *Catal. Today* **1992**, *15*, 395.
101. M. J. D'Anielo, Jr., *J. Catal.* **1981**, *69*, 9.
102. L. J. Jacimovic, J. Stevovic, S. Veljkovic, *J. Phys. Chem.* **1972**, *76*, 3625.
103. S. Sivasanker, A. V. Ramaswamy, P. Ratnasamy, *Stud. Surf. Sci. Catal.* **1979**, *3*, 185.
104. H. Knözinger, P. Ratnasamy, *Catal. Rev.* **1978**, *17*, 31.
105. J. B. Peri, *J. Phys. Chem.* **1965**, *69*, 211; J. B. Peri, *J. Phys. Chem.* **1965**, *69*, 220; J. B. Peri, *J. Phys. Chem.* **1965**, *69*, 231.
106. H. Knözinger, *Adv. Catal.* **1976**, *25*, 184.
107. Z. Vit, J. Vala, J. Malek, *Appl. Catal.* **1983**, *7*, 159.
108. E. Santacesaria, D. Gelosa, S. Carra, *Ind. Eng. Chem., Prod. Res. Dev.* **1977**, *16*, 45.
109. R. Poisson, J. P. Brunelle, P. Nortier, *Catalyst Supports and Supported Catalysts*, A. B. Stiles (Ed.), Butterworth, Boston, 1987, p. 11.
110. E. I. Gil'Debrand, *Intern. Chem. Eng.* **1966**, *6*, 449.
111. Th. Mang, B. Breitscheidel, P. Polanek, H. Knözinger, *Appl. Catal.* **1993**, *106*, 239.
112. L. L. Hegedus, T. S. Chou, J. C. Summers, N. M. Potter, *Stud. Surf. Sci. Catal.* **1979**, *3*, 17.
113. J. C. Summers, S. A. Ausen, *J. Catal.* **1978**, *52*, 445.
114. (a) H. Lieske, G. Lietz, H. Spindler, J. Völter, *J. Catal.* **1983**, *81*, 8; (b) G. Lietz, H. Lieske, H. Spindler, W. Hanke, J. Völter, *J. Catal.* **1983**, *81*, 17.
115. F. Umland, W. Fischer, *Naturwissenschaften* **1953**, *40*, 439.
116. (a) G. Lefèvre, M. Duc, P. Lepeut, R. Caplain, M. Fédoroff, *Langmuir* **2002**, *18*, 7530; (b) X. Carrier, E. Marceau, J. F. Lambert, M. Che, *J. Colloid Interf. Sci.* **2007**, *308*, 429.
117. V. Haensel, U. S. Patent 2, 840, 532, UOP, 1958.
118. M. S. Heise, J. A. Schwarz, *Stud. Surf. Sci. Catal.* **1987**, *31*, 1.
119. C. T. Kresge, A. W. Chester, S. M. Oleck, *Appl. Catal.* **1992**, *81*, 215.
120. G. N. Maslyanskii, B. B. Zharkov, A. Z. Rubinov, *Kinet. Katal.* **1971**, *12*, 699.
121. W. Jianguo, Z. Jiayu, P. Li, *Stud. Surf. Sci. Catal.* **1983**, *16*, 37.
122. T. A. Nuttal, CSIR Report CENG 182, CSIR, Pretoria, South Africa, 1977.
123. E. R. Becker, T. A. Nutall, *Stud. Surf. Sci. Catal.* **1979**, *3*, 159.
124. M. Komiyama, *Catal. Rev. Sci. Eng.* **1985**, *27*, 341.
125. A. Lekhal, B. Glasser, J. G. Khinast, *Chem. Eng. Sci.* **2001**, *56*, 4473.
126. A. Gavriilidis, A. Varma, *Catal. Rev. Sci. Eng.* **1993**, *35*, 399.
127. R. C. Dougherty, X. E. Verykios, *Catal. Rev. Sci. Eng.* **1987**, *29*, 101.
128. J. Papageorgiou, *J. Catal.* **1966**, *158*, 439.
129. (a) W. A. Spieker, J. Liu, J. T. Miller, A. J. Kropf, J. R. Regalbuto, *Appl. Catal. A* **2002**, *232*, 219; (b) W. A. Spieker, J. Liu, X. Hao, J. T. Miller, A. J. Kropf, J. R. Regalbuto, *Appl. Catal. A* **2003**, *243*, 53.
130. C. M. Davidson, R. F. Jameson, *Trans. Faraday Soc.* **1965**, *61*, 2462.
131. (a) B. Shelimov, J. F. Lambert, M. Che, B. Didillon, *J. Am. Chem. Soc.* **1999**, *121*, 545; (b) B. Shelimov, J. F. Lambert, M. Che, B. Didillon, *J. Catal.* **1999**, *185*, 462.
132. L. R. Pizzio, C. V. Caceres, M. N. Blanco, *Catal. Lett.* **1995**, *33*, 175.
133. J. A. Bergwerff, T. Visser, B. R. G. Leliveld, B. D. Rossenaar, K. P. de Jong, B. M. Weckhuysen, *J. Am. Chem. Soc*, **2004**, *126*, 14548.
134. P. Harriott, *J. Catal.* **1969**, *14*, 43.
135. R. C. Vincent, R. P. Merrill, *J. Catal.* **1974**, *35*, 206.
136. (a) M. Komiyama, R. P. Merrill, H. F. Harnsberger, *J. Catal.* **1980**, *63*, 35; (b) M. Komiyama, R. P. Merrill, *Bull. Chem. Soc. Jpn.* **1984**, *57*, 1169.
137. S. Y. Lee, R. Aris, *Stud. Surf. Sci. Catal.* **1983**, *16*, 35.
138. A. A. Castro, O. A. Scelza, E. R. Benvenuto, G. T. Baronetti, S. R. De Miguel, J. M. Parera, *Stud. Surf. Sci. Catal.* **1983**, *16*, 46.
139. G. Martino, *Stud. Surf. Sci. Catal.* **2000**, *130*, 83.
140. K. Bourikas, C. Kordulis, A. Lycourghiotis, *Catal. Rev.* **2006**, *48*, 363.

2.4.3
Solid-State Ion Exchange in Zeolites

*Hellmut G. Karge**

2.4.3.1 Introduction

The classical method for modification of zeolites with respect to their charge-compensating cations was, for a long time, ion exchange in an aqueous medium. Systematic studies of the potential of solid-state ion exchange commenced only in the mid-1980s, when two research groups began independent investigations into a broad variety of zeolites and compounds of ions to be incorporated into those microporous solids [1–7]. These studies were also independent of a few earlier observations reported during the 1970s by Rabo et al. [8] and Clearfield et al. [9], who used solid-state ion exchange to eliminate Brønsted acid-based catalytic properties from zeolites, and reacted mixtures of ammonium-containing zeolites with halides, respectively. Slinkin's group [1–4] focused on the solid-state introduction of transition metal cations into high-silica zeolites such as ZSM-5 and mordenite, using mainly electron spin resonance (ESR) spectroscopy for monitoring the reaction, whereas the present author's group began with quantitative investigations on solid-state exchange reactions between alkaline and alkaline earth halides and hydrogen or ammonium forms of zeolites such as NH_4-Y, H-ZSM-5 and H-MOR (for zeolite structures such as A, X, Y, MOR, L, BETA, SAPOs, MCM-41 and acronyms cf. Ref. [10]). In these latter studies, infrared (IR) spectroscopy and chemical analysis were mainly employed. During the subsequent decades, research into solid-state ion exchange was extended to as-synthesized sodium forms of zeolites as starting materials, and the incorporation of cations of possibly catalytic importance such as lanthanum, iron, copper, manganese, and noble metals. Similarly, the number of techniques suitable

---
* Corresponding author.

for monitoring solid-state ion exchange was considerably increased. In this chapter, the potential of the various methods will be demonstrated by appropriate examples. The systems studied with respect to modification via solid-state reaction are summarized in Table 4 (see below). An earlier extended and detailed review of the field has been published previously [11].

2.4.3.2 Comparison of Conventional and Solid-State Ion Exchange

In this section, the general chemistry of ion exchange in aqueous solution (conventional exchange, CE) will be contrasted with exchange in the solid state (solid-state ion exchange, SSIE). In some cases of CE, notably those with highly siliceous, acid-resistant zeolites, the cations of the as-synthesized form (usually Na^+ or K^+) may be exchanged for protons via a careful treatment with a mineral acid such as diluted HCl or HNO_3. However, CE is usually carried out by suspending the zeolite powder in aqueous solutions of salts which contain the desired in-going cation. The suspension is stirred for some time, frequently at temperatures higher than ambient. Basically, conventional ion exchange is described by Eq. (1) where, for the sake of simplicity, the chloride anion (Cl^-) is chosen as the counter ion of the in-going cation, M^{n+}, and the sodium form of the zeolite (Z) as the solid, is suspended in m moles of water:

$$a[M(H_2O)_w]^{n+} + naCl^- + bNa^+Z^- + mH_2O$$
$$\Longleftrightarrow xM^{n+}Z_n^- + (a-x)[M(H_2O)_w]^{n+} + naCl^-$$
$$+ nxNa^+ + (b-nx)Na^+Z^- + mH_2O \quad (1)$$

where M is the n-valent cation; Z^- is the monovalent negatively charged zeolite fragment; a, b, and x are stoichiometric numbers; m is the number of solvent (H_2O) moles; and w is the number of the solvating H_2O in the solvate shell of cation M.

Usually, the equilibrium does not lie heavily on the right hand side of Eq. (1). Thus, the (partially exchanged) solid and the solution must be separated. In order to achieve a high degree of exchange, the procedure must normally be repeated several times, and consequently ion exchange in aqueous solution is a time-consuming process. Moreover, the procedure produces large amounts of waste solutions which must either be regenerated or discarded in an environmentally friendly manner. Likewise, in several instances it is very difficult – or even impossible – to introduce all particular cations into a given zeolite by means of conventional ion exchange – that is, in aqueous solution. Examples are encountered where the cations are strongly solvated or available only in complexes, which are too bulky to enter the narrow pores of the respective zeolites (*vide infra*, e.g., Sections 2.4.3.5.7 and 2.4.3.6.2).

Because of the fundamental importance of ion exchange in zeolite chemistry, there exists a large body of literature on the thermodynamics of CE (see, e.g., Refs. [12–17] and references therein).

Solid-state ion exchange (SSIE) can be described by Eqs. (2a) or (2b), depending on whether the starting material is a cation-containing zeolite such as $Na^+ Z^-$ (as in the above case of CE) or a hydrogen (ammonium) form:

$$aM^{n+}Cl_n^- + bNa^+Z^- \Longleftrightarrow xM^{n+}Z_n^- + (a-x)M^{n+}Cl_n^-$$
$$+ (b-nx)Na^+Z^- + nxNa^+Cl^- \quad (2a)$$
$$aM^{n+}Cl_n^- + bH^+Z^- \longrightarrow xM^{n+}Z_n^- + (a-x)M^{n+}Cl_n^-$$
$$+ (b-nx)H^+Z^- + nxH^+Cl^- \uparrow \quad (2b)$$

With a stoichiometric mixture, as in the case of Eq. (2b), one may reach a complete exchange without any salt remaining in the product [cf. Eq. (2c)]:

$$aM^{n+}Cl_n^- + naH^+Z^- \longrightarrow aM^{n+}Z_n^- + naH^+Cl^- \uparrow$$
$$\quad (2c)$$

Thus, the solid-state reaction may be conducted in a mixture with the salt being present in a sub-stoichiometric ($b < na$), a stoichiometric ($b = na$), or an excess ($b > na$) amount. This provides us with the possibility of achieving a certain desired degree of exchange.

In the case of Eq. (2a) one arrives at an equilibrium, similar to the case of conventional exchange [cf. Eq. (1) and also Section 2.4.3.5.3]. In order to obtain a higher degree of exchange via solid-state ion exchange into a cation-containing form of a zeolite, one must remove the solid halide (e.g., NaCl) which has formed during the solid-state reaction (*vide infra*, e.g. Section 2.4.3.5.3). This can be achieved by a brief washing of the reaction product with a small amount of water. However, if the starting zeolite material is the hydrogen or ammonium form [cases of Eqs. (2b) and (2c)], the hydrogen halide (HCl) or a mixture of hydrogen halide and ammonia (HCl + NH_3) may be continuously removed in a stream of inert gas or into dynamic vacuum. Here, the equilibrium may be therefore completely shifted to the right-hand side of Eqs. (2b) and (2c), and a 100% exchange established (cf., e.g. Sections 2.4.3.5.1 and 2.4.3.5.3). In fact, in many other cases it has been reported that SSIE yielded a degree of exchange which was significantly higher than that reached by CE.

Whilst in the case of SSIE the main feature is that, usually at elevated temperatures, the *pure* solids are brought to reaction, in CE the solvent (water) is involved as a third component. Therefore, the in-going and out-going cations are usually solvated [cf. Eq. (1)]. Solvation of the cations in many cases impedes ion exchange, while

References see page 504

this cannot occur in SSIE. It should be mentioned here that, in several cases, SSIE was also observed when oxides instead of halides or similar compounds of the in-going cation were used [18], according to the following scheme:

$$M^{n+}O^{2-}_{n/2} + nH^+Z^- \longrightarrow M^{n+}Z^-_n + nH_2O \uparrow \quad (3)$$

In such a case, water is generated instead of a hydrogen halide.

To date, only some qualitative observations have been reported related to the thermodynamics of solid-state ion exchange [19, 20]. A few results have been obtained which seem to shed some light on the possible mechanism of solid-state ion exchange (cf. Section 2.4.3.5.4). Similarly, systematic investigations of the kinetics of SSIE are still rather scarce [21].

2.4.3.3 Experimental Procedure for Solid-State Ion Exchange in Zeolites

An efficient solid-state reaction between the starting zeolite and the salt, which contains the desired in-going cation, requires an intimate mixture of the solids. This can be achieved, for instance, by careful milling or grinding the two components together. In cases where an intense milling or grinding of the mixture may affect the integrity of the zeolite structure [22], it is preferable to prepare a suspension of the powdered salt and the zeolite in an inert solvent such as hexane. When the components have been thoroughly mixed by moving the suspension, the solvent may easily be removed [5]. The mixture obtained in either way is subsequently heated in a stream of inert gas or in high vacuum to remove volatile products such as hydrogen halides, ammonia, water, etc. A reaction temperature of 525–625 K and a reaction time of a few hours are usually sufficient to bring the process of SSIE to its maximum. In some cases, SSIE takes place at lower temperatures (e.g., at 400 K), but in other cases more severe conditions are required. This situation must be determined by appropriate monitoring of the reaction (see Section 2.4.3.4). In any case, it is advisable to check the crystallinity of the zeolite exchanged via SSIE after the reaction although, in the absence of water vapor, the evolution of the hydrogen halides (HCl, HF) does not affect the zeolite structure. The possible effect of water vapor can be reliably avoided if the mixture of the solids is carefully dried at about 400 K in a flow of inert gas, or a vacuum is employed prior to the solid-state reaction. It turns out that, in fact, chlorides of the in-going cations are most suitable for SSIE, while fluorides and bromides seem to be less reactive. In the case of iodides, elemental iodine is easily formed, and more complex cations such a CO_3^{2-} and SO_4^{2-} frequently decompose upon SSIE.

In some instances, the reaction between the solids (salts and zeolites) can be facilitated in the presence of an oxidizing agent (e.g., chlorine, cf. [23]) or reductive gases (see Section 2.4.3.7.4, e.g. Refs. [24, 25]). In the former case, the ion exchange may be mediated through the gas phase as a result of the formation of volatile reactants such as $PdCl_2$ [23].

2.4.3.4 Methods of Monitoring Solid-State Ion Exchange

In this section, the most important techniques will be described which enable the solid-state reaction between a zeolite and a salt to be monitored. The most prominent methods are:

- chemical analysis, including classical wet analysis (CA) and atomic absorption spectroscopy (AAS)
- temperature-programmed evolution (TPE) of gases
- re-exchange in solution
- catalytic test reactions
- infrared (IR) spectroscopy
- electron spin resonance (ESR) spectroscopy
- Mössbauer spectroscopy
- magic angle spinning nuclear magnetic resonance (MAS NMR)
- X-ray diffraction (XRD)
- extended X-ray absorption fine structure analysis (EXAFS)
- thermogravimetric analysis.

By no means does this overview pretend to be exhaustive, and several other methods might have the potential for analyzing SSIE, and may indeed have already been (or will be) employed in this field. For example, SSIE may be also monitored using a probe molecule such as pyridine (cf., Sections 2.4.3.5.2 and 2.4.3.6.3). However, the techniques discussed below are those most frequently applied, and their application in SSIE is illustrated in the following subsections.

2.4.3.4.1 Infrared Spectroscopy

Infrared spectroscopy was employed during the aforementioned early observations made by Rabo and coworkers [8, 26] as well as in the first systematic explorations of the new method of post-synthesis modification of zeolites [5, 6]. Essentially, two possible means exist of employing IR spectroscopy in order to disclose the occurrence and to determine the degree of SSIE, whereby wafers (platelets) of the salt/zeolite mixture in an appropriate IR cell are used (cf. Refs. [27, 28]). These two processes are in line with those described by Eqs. (2a) and (2b):

(i) A probe such as CO or pyridine may be utilized which, after adsorption, produces bands typical of the in-going cation (e.g., Be^{2+}, Ca^{2+}, La^{3+}, Cu^{2+}, etc.), or the replaced (i.e., out-going) cation of the starting zeolite (e.g., Na^+ of Na-Y).

(ii) If the salt containing the in-going cation reacts with OH groups of the hydrogen form of a zeolite, the reaction will result in the decrease in the corresponding OH band intensity, which can be measured via IR spectroscopy. Of course, the same procedure must be carried out with the same zeolite under the same conditions, but in the absence of the salt to ensure that a change in the OH band intensity is not merely a consequence of dehydroxylation [5, 29]. An additional check may be performed by using a suitable probe to verify the incorporation of the respective cation.

2.4.3.4.2 ESR Spectroscopy
This method was very successfully employed by Slinkin and his group [1–4, 30, 31] when studying the incorporation of transition metals into zeolites, where many of the cations give rise to typical ESR spectra. In most cases, an *ex-situ* heat-treatment of the respective mixtures of powdered salts or oxides and zeolites is carried out prior to running the spectra. However, the use of a particularly designed high-temperature ESR reactor (cf., e.g., Refs. [32, 33]) renders feasible the *in-situ* investigation of the kinetics of transition metal incorporation. Moreover, ESR spectroscopy enables us not only to detect and monitor solid-state reactions involving transition metals but also to obtain an insight into the particular type of coordination, which the incorporated transition metal cation has obtained.

2.4.3.4.3 Mössbauer Spectroscopy
Mössbauer spectroscopy is a valuable tool for investigating solid-state reactions of iron, which may be interesting in view of the metal's catalytic properties. Other promising candidates among the (unfortunately very limited) number of Mössbauer nuclei are Eu and Sn. Experiments conducted with mixtures of $FeCl_2$ and NH_4-Y zeolite have provided a wealth of information about the fraction of Fe^{2+} and Fe^{3+}, the degree of exchange, the chemical surroundings, and the coordination number of the incorporated species (cf. Section 2.4.3.5.5).

2.4.3.4.4 Solid-State NMR Spectroscopy
Several ways exist of employing solid-state NMR spectroscopy to investigate SSIE. In analogy to IR spectroscopy, ^1H MAS NMR could be used to determine the extent of reaction between salts and H- or NH_4-forms via the decrease in the respective signals of OH groups (cf. Chapter 3.2.4.4 of this Handbook, and Ref. [28]). The introduction of several cations such as Li^+, Na^+, Al^{3+}, which are directly "visible" in MAS NMR, can be studied in the case of microporous structures, even without assistance of probe molecules. The formation of tiny NaCl crystallites on the external surface of zeolite particles (*vide infra*) can be observed using ^{23}Na MAS NMR upon the replacement of Na^+ of the parent zeolite, according to Eq. (2a). Furthermore, in favorable cases MAS NMR may provide information concerning the location and chemical surroundings of the cations introduced through SSIE. Moreover, since the development of techniques which allow the heating of samples inside the spinning rotor [34–36], not only *ex-situ* but also *in-situ* MAS NMR experiments have become feasible.

2.4.3.4.5 X-Ray Diffraction
Monitoring the decrease of reflections of the compound of the in-going cation M^{n+} and, in favorable cases, the appearance of reflections originating from the corresponding compound of the out-going cation (e.g., NaCl), demonstrates that SSIE has in fact occurred. The resulting XRD pattern may be compared with that of the same, but conventionally exchanged, zeolite [37]. Using a heatable XRD chamber enables *in-situ* measurements of SSIE. *In-situ* XRD investigations reported to date have confirmed only on a qualitative basis that SSIE had occurred, and provided information of the required reaction temperature. However, future *in-situ* studies should also render possible relatively precise kinetic measurements. In addition, XRD should enable the determination of populations of various cation sites in the zeolite structure as a function of the reaction time.

2.4.3.4.6 Extended X-Ray Absorption Fine Structure (EXAFS) and X-Ray Absorption Near Edge Structure (XANES)
More recently, Hatje et al. [38, 39] and Förster and Hatje [40] succeeded in studying SSIE of Cu^+, Ni^{2+} Zn^{2+}, Pd^{2+} and Pt^{2+} into Y or ZSM-5 by EXAFS (cf. also Sections 2.4.3.5.4 and 2.4.3.5.7). This method not only provides evidence as to whether the solid-state reaction has taken place, or not: it also makes available data on the coordination of the introduced metal cations and, after reduction, on the size of produced metal particles.

2.4.3.4.7 Temperature-Programmed Evolution (TPE) of Volatile Gases
The compound of the cation, M^{n+}, may react with the starting zeolite under the formation of volatile gases, in particular when the hydrogen or ammonium form of the zeolite is used. The evolution of gaseous products such as HCl, NH_3, H_2O, etc. evolved into vacuum or a flow of a suitable carrier gas can be detected and/or quantitatively determined using mass spectrometry (MS), gas chromatography (GC), or (continuous) titration (in the case of H_2O via Karl Fischer's method). Different maximum temperatures in the TPE profiles of a homologous series of salt/zeolite systems

References see page 504

provide a measure of the ease of the respective solid-state reactions (cf., e.g., Section 2.4.3.5.1).

2.4.3.4.8 Thermogravimetric Analysis (TGA)
Frequently, the SSIE between salts and zeolites is accompanied by changes in weight of the reaction mixture. Therefore, the process may be studied using a thermal gravimetric analysis, perhaps utilizing a microbalance. Typically, the loss of weight, which occurs when a chloride reacts with the hydrogen form of a zeolite, provides a measure of the degree of exchange. Such measurement of changes in weight during temperature-programmed heating is combined advantageously with MS, GC or continuous titration to distinguish between the desorption of physisorbed volatiles (e.g., H_2O) and reaction products such as HCl or NH_3.

2.4.3.5 Systems Investigated for Solid-State Ion Exchange in Zeolites

Since the mid-1980s, many reports of studies in the field of SSIE have been made (cf. Refs. [11, 41]; see also Table 4). Thus, in the following section only a few examples of SSIE of some selected cations into selected zeolite structures will be discussed in greater detail, namely the incorporation of alkaline, alkaline earth, rare earth, some transition metals (Cu, Fe) and noble metals into ZSM-5, faujasite-type Y, and mordenite (MOR) zeolites.

2.4.3.5.1 Solid-State Ion Exchange with Salts of Alkaline Metals
In the systems M^+Cl/NH_4-ZSM-5, M^+Cl/H-ZSM-5, M^+Cl/NH_4-Y ($M^+ = Li^+, K^+, Rb^+, Cs^+$) and M^{2+}/H-MOR ($M^{2+} = Ca^{2+}, Mg^{2+}$) IR, TPE combined with MS or TGA were employed to confirm and quantitatively determine the effect of SSIE [5–7]. As an example obtained by application of IR, Fig. 1 shows – for the sake of comparison – the spectra of NH_4-Y in the absence of CsCl after treatment at 723 K (Fig. 1a), and the almost complete elimination of the acid OH groups resulting from the reaction of NH_4-Y with CsCl in high vacuum at 723 K. The latter is indicated by the almost complete elimination of absorbances of both the high-frequency (HF, 3640 cm^{-1}) and low-frequency (LF, 3550 cm^{-1}) band, which proves that the bulky Cs^+ cation enters both the large and small cavities (cf. Ref. [42]). As a result, SSIE yields an almost 100% exchange compared to a degree of only about 60% via CE in aqueous solution.

A high degree of incorporation of cesium is, *inter alia*, interesting in view of the generation of basicity in catalysts (see Refs. [43–46] and Chapter 3.2.4.1). The introduction of Rb^+ from Rb_2CO_3 via SSIE into H-Y (NH_4-Y) was studied by ^{87}Rb MAS NMR and ^{1}H MAS NMR [47]. Alkaline salt/NH_4-Y systems are also good examples to demonstrate the potential of TPE. TPE/MS profiles of CsCl/NH_4-Y are shown in Fig. 2.

One observes dehydration (molecular mass per electron, m/e, at 18) between 400 and 500 K. With the M^+Cl^-/NH_4- or H-zeolite series, one generally observes in TPE experiments a low- (LT) and a high- (HT) temperature peak of HCl ($m/e = 36$) around 500 and 860 K, respectively. In the LT region, there is in the case of NH_4-zeolites an overlap of NH_3 (evolved upon deammoniation) and HCl due to the reaction of M^+Cl^- with OH groups, which have intermittently formed via

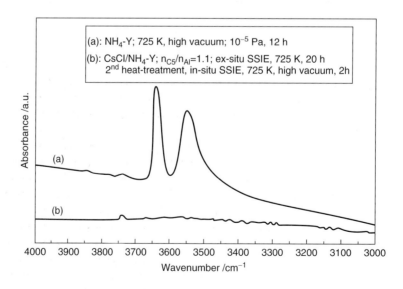

Fig. 1 Infrared spectra of the hydroxyl stretching frequency region after degassing of (a) NH_4-Y at 725 K for 12 h (final vacuum ca. 10^{-5} Pa) and of (b) CsCl/NH_4-Y ($n_{Cs}/n_{Al} = 1.1$) at 725 K for 20 h (*ex-situ* SSIE) and subsequently for 2 h in high vacuum (*in-situ* SSIE); the degree of exchange in NH_4-Y was 98%.

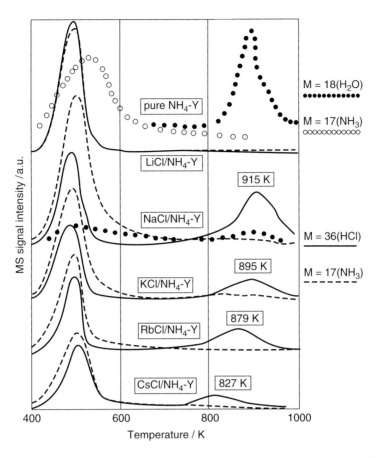

Fig. 2 Temperature-programmed evolution of gases ($m/e = 36$, HCl, ———; $m/e = 18$, H_2O, ···; $m/e = 17$, NH_3, oooo or -----) from pure NH_4-Y and MCl/NH_4-Y mixtures upon solid-state reaction. (After Ref. [48].)

NH_3 evolution. Less-accessible and/or less-reactive OHs react in the HT region. The LT region is particularly pronounced in the case of LiCl; its contribution decreases in the sequence Li > Na > K > Rb > Cs. Simultaneously, the temperature of the HT-HCl peak decreases in the same sequence. It is worth noting that with the exception of LiCl/NH_4-Y, where the HT peak is almost negligible, the sequence of the HT peak (i.e., Na > K > Rb > Cs) parallels the decrease in the lattice energies of the respective chlorides [5, 48]. This, however, seems not always to be the case with similar homologous series, and especially not with the salts of some transition metals investigated to date [18, 49]. The TPE results with NH_4-ZSM-5, H-ZSM-5 and H-BETA instead of NH_4-Y are very similar, and this also holds when, *vice versa*, the starting material NH_4-Cl was reacted with alkaline zeolite (BETA) [48, 50, 51]. Reaction in mixtures with zeolites exhibiting an M^+/Al_F ratio of 1 (Al_F = framework aluminum per unit cell, u.c., of the zeolite) led to an exchange degree of 100%, as derived from the elimination of the OH band intensities originating from acid OH groups. However, the seemingly non-acid so-called "silanol" groups (indicated by an IR band around $3740\ cm^{-1}$) could also react to some extent (especially if $M^+/Al_F > 1$), but they were easily restored via brief washing of the reaction product with water (hydrolysis), due to the low acid strength of these silanols.

The SSIE of NaCl/H-ZSM-5 eliminated the OH groups of H-ZSM-5, resulting in Na-ZSM-5. When this product was transformed via conventional exchange into NH_4-ZSM-5 and subsequently deammoniated, the original OH groups reappeared through this "re-exchange" showing that the integrity of the structure was not affected by the SSIE [5]; this was also confirmed by XRD. The re-exchange could be quantitatively determined by TPE of NH_3.

Solid-state ion exchange between NaCl and H-ZSM-5 was also monitored by a combination of TPE, TGA and continuous titration of the evolved HCl [5]. From the analysis of: (i) the starting zeolite material; (ii) the gases evolved; (iii) the aqueous extracts obtained from the salt/zeolite mixture after reaction; and (iv) the exchanged zeolite, the stoichiometry of the SSIE was determined.

References see page 504

Tab. 1 Starting materials, mixtures and results of solid-state ion exchange in the system NaCl/H-ZSM–5*

[1] zeolite	[2] $n_{Si}n_{Al}$	[3] Al	[4] NaCl (employed)	[5] HCl (evolved)	[6] Cl (extracted)	[7] Na (extracted)	[8] Na (irrev. held)	[9] δ (%)
H-ZSM-5 (I)	155	0.107	0.808	0.549	0.260	0.707	0.101	94
H-ZSM-5 (II)	23	0.691	1.306	0.838	0.478	0.670	0.636	92

*All data in mmol g^{-1} zeolite fired at 1273 K.
[1] Parent zeolite.
[2] n_{Si}/n_{Al} ratio of the parent zeolite.
[3] Al content of the parent zeolite.
[4] NaCl admixed to 1 g (dry) zeolite.
[5] HCl evolved on solid-state reaction.
[6] Cl$^-$ extracted after solid-state reaction.
[7] Na$^+$ extracted after solid-state reaction.
[8] Na$^+$ irreversibly held in zeolite after extraction.
[9] Degree of exchange; data of column [8] divided by data of column [3].

The results were very satisfying (cf., e.g., columns 3 and 8 of Table 1).

Finally, ^{23}Na MAS NMR [52] and ^{133}Cs MAS NMR [43] were used to determine the SSIE of Na$^+$ for Li$^+$, K$^+$, Be^{2+}, and Ca^{2+} in Na-Y and Cs$^+$ in NH$_4$-Y.

The easy solid-state reaction between alkaline chlorides (especially LiCl) and acid OH groups provides a convenient means of eliminating undesired Brønsted acidity, for example, acid OH groups which may form upon reduction of noble metal cations via reduction by H$_2$ as it occurs in the preparation of selective hydrogenation/dehydrogenation zeolite catalysts. Similar studies of SSIE between alkaline metal chlorides and NH$_4$-Y by XRD and IR, as described above, were conducted somewhat later also by Jiang et al. [53]. The same authors [54] also utilized SSIE (with KCl) to decrease the density of acid sites in potassium-containing zeolites loaded with Mo$_3$S$_4$ clusters.

2.4.3.5.2 Solid-State Ion Exchange with Salts of Alkaline Earth Metals

Some of the examples studied for SSIE in systems of alkaline earth/zeolite included CaCl$_2$/H-MOR, CaCl$_2$/NH$_4$-MOR, MgCl$_2$/H-MOR, and CaCl$_2$/H-ZSM-5. MgF$_2$ was also employed as a compound of an in-going cation, but it reacted only to a minor extent (*vide infra* and cf. Refs. [55, 56]). Again, IR spectroscopy (with and without pyridine as a probe, cf. Refs. [57, 58] and Section 2.4.3.4.1), chemical analysis and ^{23}Na MAS NMR (*vide supra*) were used to monitor the SSIE in such systems.

Figure 3 (spectrum 1a versus 2a) demonstrates the complete elimination of the band of acid OH groups (3610 cm^{-1}) and part of the silanol groups (band at 3750 cm^{-1}) upon SSIE in CaCl$_2$/H-MOR.

The adsorption of pyridine indicated that no pyridinium ions (band at 1540 cm^{-1}) formed because no acid OHs but only Ca^{2+} (band at 1446 cm^{-1}) and a few "true" Lewis acid sites (band at 1450 cm^{-1}) were left (Fig. 3, spectrum 2b). After short (2-min) contact with water vapor (1.3 kPa H$_2$O at 400 K) and subsequent degassing into high vacuum (10^{-5} Pa), the OH groups reappeared (spectrum 3a) due to rehydroxylation according to the Hirschler–Plank mechanism [Eq. (4)]:

$$Ca^{2+}Z_2^- + H_2O \longrightarrow H^+Z^- + Ca(OH)^+Z^- \quad (4)$$

Correspondingly, in spectrum 3b of Fig. 3 the band of pyridinium ions (1540 cm^{-1}) emerged. After SSIE, zeolite materials corresponding to spectrum 2 were almost inactive in acid-catalyzed reactions such as the disproportionation of ethylbenzene [58, 59], but became much more active catalysts after treatment with small amounts of water vapor (spectrum 3). This confirms that the bare Ca^{2+} ions were unable to act as carboniogenic centers in hydrocarbon reactions. Rather, the presence of acid OH groups (Brønsted acid sites) is an indispensable prerequisite for such reactions to occur (cf. the analogous observations in the case of SSIE with lanthanum salts; Section 2.4.3.5.3).

The mass balance for SSIE in the systems CaCl$_2$/H-MOR and CaCl$_2$/NH$_4$-MOR produced excellent results: 2.18 and 2.50 mmol Al g^{-1} water-free zeolite corresponded exactly to 1.09 and 1.26 mmol Ca^{2+} introduced via SSIE g^{-1} water-free zeolite MOR [6].

Especially in view of selective catalytic reduction (SCR), Ca-Y, Ca-X and Ba-Y (and, similarly, Na-Y, Na-X) were prepared via SSIE by Ebeling et al. [60] (cf. also Section 2.4.3.5.4) and investigated with respect to acidity, adsorption of NO$_2$, and stability as catalysts (coke formation).

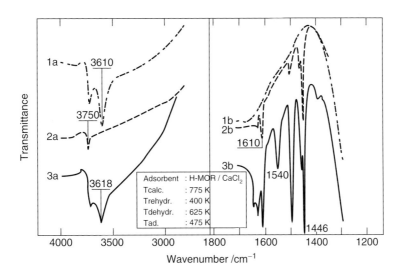

Fig. 3 Infrared spectra of H-MOR and the system $CaCl_2 \cdot 2H_2O$/H-MOR with $n_{Ca}/n_{Al} = 0.5$, after thermal treatment at 625 K and 10^{-5} Pa. Spectra 1a–3a in the OH stretching frequency region, 1b–3b in the region of pyridine ring deformation frequencies. 1a, 1b without and 2a, 2b with admixed $CaCl_2 \cdot 2H_2O$; 2b after pyridine adsorption and 3a after rehydroxylation, 3b after rehydroxylation and pyridine adsorption. (After Ref. [6].)

In the case of SSIE between $BeCl_2$ and Na-Y, it could be shown that indeed a temperature-dependent equilibrium was reached: at about 400 K the Na^+ cations were entirely replaced by Be^{2+} (indicated by a band of coordinatively bound pyridine at 1453 cm^{-1}). Correspondingly, the band of pyridine coordinated to Na^+ (at 1446 cm^{-1}) had completely disappeared. At higher temperatures the band at 1453 cm^{-1} decreased while the band at 1446 cm^{-1} re-appeared.

2.4.3.5.3 Solid-State Ion Exchange with Salts of Lanthanum

Rare earth cation-containing faujasite-type zeolites are important as components of acidic catalysts for a variety of hydrocarbon reactions. Therefore, $LaCl_3 \cdot 7H_2O$/NH_4-Y was also investigated with respect to SSIE [7, 61, 62]. An exchange degree of almost 100% was reached by only one exchange step, whereas in the case of CE this may be achieved only by repeatedly suspending the solid in an aqueous solution of an La salt, intermittent separations of the suspension and calcinations of the (partially) exchanged material. SSIE in La-salt/zeolite systems was studied by TPD (cf. Fig. 4, Ref. [7]), IR [7, 37], stoichiometric analysis [61, 63], TPE/continuous titration [7], and XRD [61].

For the sake of comparison, Fig. 4 displays the TPE/MS profile [7] obtained during temperature-programmed heating of a finely dispersed sample of the parent zeolite, NH_4-Y. SSIE was carried out with a mixture of $LaCl_3 \cdot 7H_2O$ and NH_4-Y (deammoniation: M = 16; dehydration or dehydroxylation: M = 18; evolution of HCl: M = 36). Similar to the TPE/MS profiles of alkaline salt/H-zeolites (vide supra), in Fig. 4 one can also distinguish between LT and HT peaks for the evolution of HCl. A comparison of the profiles obtained with NH_4-Y (Fig. 4, broken lines) and $LaCl_3 \cdot 7H_2O$/NH_4-Y (solid lines) shows that, in the latter case, no dehydroxylation (peak around 950 K) occurs, because the OH groups, which may have formed during heating in the LT region, were consumed by the solid-state reaction at higher temperatures.

The stoichiometry of SSIE in the $LaCl_3 \cdot 7H_2O$/NH_4-Y mixture based on chemical analysis is summarized in Table 2 for a ratio $n_{La}/n_{Al} = 0.33$, and in Table 3 for $n_{La}/n_{Al} = 0.67$. In both cases, the irreversibly held lanthanum (4.80 mequiv. g^{-1}) corresponded to the amount of framework aluminum (4.83 mequiv. g^{-1}) – that is, the maximum of bridging acid OH groups. Per gram, about 0.7 mequiv. Na^+ and 0.8 mequiv Cl^- – that is, approximately one NaCl molecule per β-cage – remained occluded in the SSIE product which, according to Rabo's studies [8, 26], may have enhanced the thermal stability of the structure of the exchanged zeolite.

It could be shown that a La-Y zeolite, prepared via SSIE, was catalytically active in disproportionation of ethylbenzene [7] and n-decane cracking [62] only after subsequent brief contact with water vapor (cf. Fig. 5, where the results of reactions in a flow-reactor IR cell are displayed [7]). Again, this demonstrates that bare polyvalent cations (such as La^{3+}) themselves are not catalytic centers but do provide catalytic activity for acid-catalyzed reactions only with the co-catalyst H_2O

References see page 504

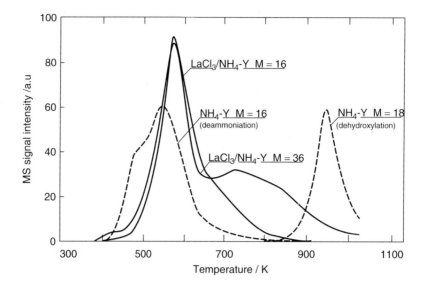

Fig. 4 Temperature-programmed evolution of H_2O ($m/e = 18$) from NH_4-Y and NH_4^+ ($m/e = 16$), HCl ($m/e = 36$) and H_2O ($m/e = 18$) from a $LaCl_3/NH_4$-Y mixture. (After Ref. [7].)

Tab. 2 Stoichiometry of solid-state ion exchange in the system $LaCl_3/NH_4$-Y*; $n_{La}/n_{Al} = 0.33$; heat-treatment at 850 K

	La^{3+}	Cl^-	Na^+	NH_4^+	Al
Parent zeolite	÷	÷	1.61	3.29	4.83
Admixed	1.61	4.83	–	–	–
Evolved as NH_4Cl/HCl	–	3.29	–	3.29	–
Extracted with water	0.06	0.72	0.94	–	–
Irreversibly held mmol g^{-1}	1.60	0.82	0.67	–	4.83
mequiv. g^{-1}	4.80	0.82	0.67	–	4.83

*All data given as mmol g^{-1}, except for last line (mequiv. g^{-1}).

Tab. 3 Stoichiometry of solid-state ion exchange in the system $LaCl_3/NH_4$-Y*; $n_{La}/n_{Al} = 0.67$; heat treatment at 850 K

	La^{3+}	Cl^-	Na^+	NH_4^+	Al
Parent zeolite	–	–	1.61	3.29	4.83
Admixed	3.22	9.65	–	–	–
Evolved as NH_4Cl/HCl	–	3.29	–	3.29	–
Extracted with water	1.57	5.51	0.94	–	–
Irreversibly held mmol g^{-1}	1.65	0.85	0.67	–	4.83
mequiv. g^{-1}	4.95	0.85	0.67	–	4.83

*All data given as mmol g^{-1}, except for last line (mequiv. g^{-1}).

according to the Hirschler–Plank scheme [see Eq. (4)]. Furthermore, when a conventionally exchanged La-Y catalyst with a high degree of exchange was compared with a catalyst of the same exchange degree (98%), but prepared via SSIE, then the latter was even superior with respect to activity and stability in the disproportionation of ethylbenzene [7].

In a comparative study, Jia et al. [63] prepared La-BETA zeolites through CE and SSIE, and characterized these using XRD, adsorption measurements (BET), Fourier transform IR (FTIR), transmission electron microscopy (TEM), energy dispersive spectroscopy (EDS), and chemical analysis. The La-content achieved by SSIE was as high as 100%, but was low even after several-fold repeated CE. Moreover, the samples obtained via SSIE were highly homogeneous and showed no loss of crystallinity, in contrast to CE materials.

Rare earth Y-type zeolites (RE = Ce, Nd, Sm, Eu, Yb) were obtained by Pires et al. [64] via SSIE from H-Y ($n_{Si}/n_{xsAl} = 2.6, 12.5, 28$) and the respective chlorides, $RECl_3$. The authors used the thus-obtained materials as catalysts for the liquid-phase oxidation of cyclohexane with tert-butyl hydroperoxide. The catalytic activity followed the sequences ($n_{Si}/n_{Al} = 12.5 > 2.6 > 28$) and Ce ≈ Yb > Sm > Eu > Nd. The main products were cyclohexene, n-hexanal, cyclohexanone and cyclohexanol.

Karge and Beyer [61] also succeeded in preparing catalytically active La-Y catalysts through solid-state reaction between $LaCl_3 \cdot 7H_2O$ and as-synthesized Na-Y (cf. also "contact-induced" solid-state ion exchange, cf. Section 2.4.3.7.1). The activity for acid-catalyzed reactions was checked by the disproportionation of ethylbenzene as a test reaction [58, 59]. Here, the solid-state cation exchange (La^{3+} for Na^+) was not only monitored via IR and ^{23}Na MAS NMR, but also with the help of XRD patterns: the reflections changed systematically when the SSIE proceeded and, furthermore, the reflections

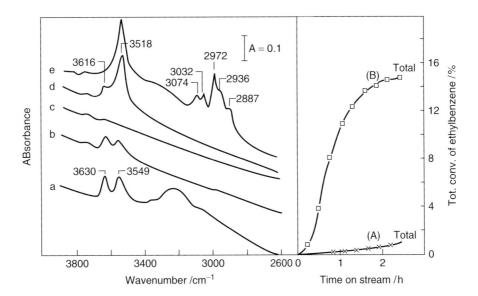

Fig. 5 In-situ infrared (IR) spectra of and ethylbenzene conversion over La,Na-Y obtained by heat treatment of a LaCl$_3$/NH$_4$-Y mixture in a flow reactor IR cell (A) prior to and (B) after brief contact with 10 Pa water vapor; (a), (b) and (c) pretreatment at 455, 575, and 675 K, respectively; (d) after short contact with 10 Pa water vapor; (e) after admission of the ethylbenzene/helium feed stream. (For details, see text; after Ref. [7].)

of the product NaCl appeared [60]. As reported by Difallah and Ginoux [65], in their study on the introduction of monovalent and polyvalent cations into zeolite Y and mordenite through SSIE (including La^{3+} and Cu^{2+}; see below), the incorporation of La^{3+} cations resulted in an increased adsorption capacity for CO.

2.4.3.5.4 Solid-State Ion Exchange with Salts of Copper

Copper-containing zeolites were, for some time, intensely studied because of their catalytic activity in DeNOx reactions ([66], cf. also Chapters 11.2 and 11.3 of this Handbook). The introduction of copper (e.g., from CuCl$_2$ or CuO) into NH$_4$-forms of various zeolites (A, X, Y) via SSIE was first reported and monitored via titration and ESR by Clearfield et al. [9] in 1973, and more than one decade later was investigated systematically through ESR experiments by Kucherow and Slinkin (besides cations of other transition metals [30, 67–69]) and Karge et al. [20]. As an example, Fig. 6 shows spectra obtained by SSIE in a mixture of CuO and H-ZSM-5 [69, 70].

This reaction required a relatively high temperature; typically, the degree of exchange was considerably enhanced when the temperature was increased from 793 to 1073 K. Spectrum b of Fig. 6 is in complete agreement with that obtained after the same treatment of a conventionally exchanged Cu-ZSM-5 (cf. spectrum a). The g-values and the hyperfine splitting (HFS) constants coincided. Two sets of g-values and HFS constants were observed: $g^1\| = 2.29$, $g^1_\perp = 2.05$, $A^1\| = 15.6$ mT, $A^1_\perp = 2.3$ mT, and $g^2\| = 2.31$, $g^2_\perp = 2.06$, $A^2\| = 15.3$ mT,

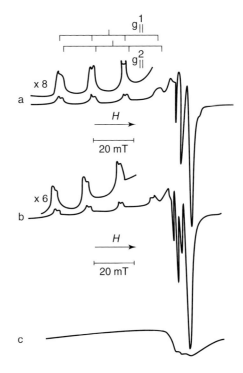

Fig. 6 Comparison of electron spin resonance (ESR) spectra of Cu^{2+}-containing ZSM-5 samples. Sample a prepared by solid-state ion exchange in a CuO/H-ZSM-5 mixture in vacuum at 1073 K; sample b obtained by conventional exchange, calcined in air at 1073 K and evacuated at 300 K; spectrum c obtained after contacting sample a with air. (After Refs. [1, 70].)

References see page 504

$A_\perp^2 = 2.25$ mT, these experimentally determined g-values being in good agreement with the computed values. Both Clearfield [9] and Kucherow and Slinkin [69, 70] concluded from their findings that the Cu^{2+} cations resided on two differently coordinated sites: set 1 was assumed to correspond to a square planar, and set 2 to a pyramidal and fivefold-coordinated state [31, 69]. SSIE in CuCl/H-ZSM-5 and $CuCl_2 \cdot 2H_2O$/H-ZSM-5 at 775 K was also studied by XRD [20].

In a study by Jiang and Fu [71], SSIE between CuCl and NH_4-Y was followed via IR and MS, and the resulting materials were compared with Cu(I)-Y obtained from Cu(II)-Y through auto-reduction or reduction by CO: in both cases, the characteristics of Cu^+ species were similar. In a subsequent publication, Borovkov et al. [72] reported on interesting diffuse reflectance FTIR spectroscopic (DR-FTIR) experiments on carbonyl formation on Cu(I)-zeolites produced through SSIE between CuCl and zeolites Y, ZSM-5, MOR, and L (for the acronyms, cf. Ref. [10]).

Hartmann and Boddenberg [73, 74] used SSIE of CuCl and NH_4,Na-Y with an exchange degree of 70% NH_4^+ to prepare a Cu,Na-Y(70) sample for a study of CO and Xe adsorption by ^{13}C and ^{129}Xe MAS NMR, respectively. SSIE provided by far the highest concentration of Cu^+ when compared with materials obtained by CE. The authors were able to determine quantitatively the distribution of the Cu^+ cations within the structure: about 70% resided in the supercages (27 of 29 per u.c.). Furthermore, it was shown that CO is a less-suitable probe than Xe, as the latter does not – in contrast to CO – change the distribution of the cations.

Cu(I)-Y zeolite catalysts prepared via SSIE were employed by King [75] in the oxidative carbonylation to dimethyl carbonate. In context with prospects for a "green chemistry", Anderson and Root [76] also studied the direct synthesis of dimethyl carbonate via carbonylation of methanol over a Cu(I)-X catalyst, which they had obtained by SSIE.

Selective catalytic reduction of NO in the presence of propene over Cu,Na-ZSM-5, which has formed upon SSIE of CuCl, $Cu(NO_3)_2$ or $Cu(CH_3COO)_2$ with H-ZSM-5/Na-ZSM-5, was investigated by Liese and Grünert [77]. These authors used X-ray photoelectron spectroscopy (XPS) and X-ray-induced Auger electron spectroscopy (AES) to monitor the transport of copper into the near-surface regions of the zeolite crystallites. Further contributions related to DeNOx catalysis, which employed zeolite catalysts prepared via SSIE, were provided by Varga et al. [78] (NO reaction over Cu-ZSM-5), Soria et al. [79] (adsorption of NO and/or O_2 on Cu-ZSM-5 studied by ESR), Schay et al. [80] (decomposition of NO, N_2O, and SCR of NO by C_3H_8 or CH_4 over Cu-[Al]TS-1 catalysts; for zeolite acronyms cf. Ref. [10]), Teraoka et al. [81] (NO decomposition over Cu-MFI catalysts), and Umamaheswari et al. [82] (ESR studies on Cu(I)-NO complexes in Cu-ZSM-5). As far as methods of catalyst preparation were compared, it was found in Ref. [81] that the performance of catalysts obtained by SSIE was superior to that of conventionally prepared materials, whereas the reverse was reported in Ref. [78].

2.4.3.5.5 Solid-State Ion Exchange with Salts of Iron Iron as an exchange cation is also interesting because of its potential in the formulation of catalysts. In fact, the introduction of Fe^{2+} and/or Fe^{3+} was systematically investigated by ESR [4, 9], XRD [4, 83] and TPE of NH_3 ($m/e = 17$), H_2O ($m/e = 18$) and HCl ($m/e = 36$). However, the incorporation of iron into NH_4-Y provides an excellent example for the application of Mössbauer spectroscopy. Thus, Fig. 7 displays a set of Mössbauer spectra obtained upon solid-state reaction of $FeCl_2 \cdot 4H_2O$ and NH_4-Y at various temperatures.

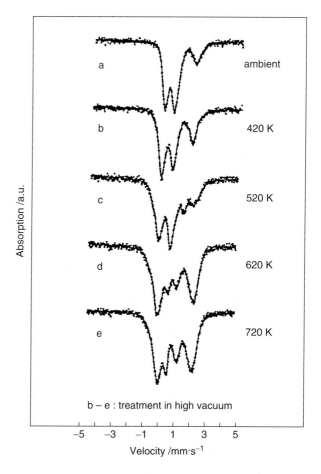

Fig. 7 Mössbauer spectra of a $FeCl_2 \cdot 4H_2O$/NH_4-Y mixture (a) ground in air at ambient temperature; material (a) after heat treatment in vacuum at (b) 420 K, (c) 520 K, (d) 620 K and (e) 720 K. (After Ref. [83].)

After decomposition of the spectra and determination of the Mössbauer parameters (isomer shift, IS; quadrupole splitting, QS; relative intensity, RI), the coordination and the fraction of Fe^{2+} and Fe^{3+} populating the respective sites were derived. It was found that after heat treatment at 720 K only about 4% Fe (III) species occurred and an almost pure Fe(II)-zeolite was produced, exhibiting Fe^{2+} in trigonal (30%) or octahedral (65%) coordination.

Čapek et al. [84] investigated Fe-zeolites of type MFI, FER, and BETA (cf. Ref. [10]) prepared *inter alia* through SSIE by UV-Vis, near infrared (NIR) and IR spectroscopy, as well as by voltammetry. These authors studied the formation of NO complexes and compared the results with those obtained using other preparation methods, for example exchange with $FeCl_3$ in acetyl acetonate. In the latter case, at low concentrations exclusively Fe cations at cation sites were detected, whereas at higher loadings dinuclear Fe–oxo complexes were preferably formed in MFI, but were absent in BETA structures. Both Fe^{2+} and Fe^{3+} salts (viz. $FeSO_4 \cdot 7H_2O$ and $FeCl_3$) were employed by Mohamed et al. [85] in SSIE with NH_4-MOR, H-MOR or Na-MOR; the Fe species were characterized by N_2 adsorption, XRD and Mössbauer spectroscopy. In the experiments with NH_4-MOR and H-MOR, 100% of Fe-loading was achieved, but simultaneously formation of a hematite phase was observed.

Similar to studies on SSIE with copper (*vide supra*), investigations of the introduction of iron (or other transition metal ions, *vide infra*) into zeolite structures were stimulated by the interest in decomposition of NO_x or SCR of nitrogen oxides. Thus, Szalay et al. [86] reported on the decomposition of NO_x and reduction of NO by CO over Fe-ZSM-5 catalysts obtained by SSIE. In most reactions, these materials were more active but less stable than conventionally prepared ones. Highly active Fe-ZSM-5 catalysts for N_2O decomposition were prepared by Rauscher et al. [87]; a hematite phase appeared when the ratio Fe^{2+}/Al was above 0.5. An attempt was made by Varga et al. [88] to evaluate the redox mechanism of NO decomposition over Fe- (or Cu)- containing ZSM-5. In contrast to Szalay et al. [86], these authors found catalysts produced through SSIE to be less active than those prepared by CE. However, this may be due to differences in the preparation and pretreatment procedures. Simultaneous reductive removal of NO and N_2O over Fe-MFI catalysts (cf. Ref. [10]) was described by Kögel et al. [89], who also found that the optimum of SSIE occurred at Fe^{2+}/Al = 0.5 (cf. Ref. [87]). Fe-ZSM-5 catalysts with systematically varied Fe/Al and Si/Al ratios were prepared via SSIE by Batista et al. [90, 91], and tested for the reduction of NO with *iso*-butane. The catalytic activity was found to depend strongly on the content and coordination of the cationic Fe-species compensating the negative charge of the framework.

SSIE experiments similar to those conducted with Fe-compounds were carried out with Ni^{2+} salts and BEA, L, MAZ, MFI, MOR, and Y zeolites (cf. e.g. [92, 93], see also Table 4; for the zeolite acronyms, cf. Ref. [10]). Incorporation of cobalt (and copper) into clinoptilolite via solid-state reaction was studied, *inter alia*, by Astrelin et al. [94] (cf. also Table 4).

2.4.3.5.6 Solid-State Ion Exchange with Salts of Manganese

The introduction of Mn^{2+} through SSIE was reported by Wichterlová et al. [95] and Beran et al. [18], who employed XPS, IR spectroscopy, TPD of ammonia, TPE, back-titration of evolved HCl and test reactions as methods of investigation. However, ESR was preferentially used in these studies. The following manganese-containing compounds were reacted with H-ZSM-5: $MnCl_2$, $MnSO_4$, Mn_3O_4 or $Mn(CH_3COO)_2$. A set of X-band ESR spectra of Mn^{2+} is shown in Fig. 8 [95].

The starting mixture of solid $MnSO_4$ and H-ZSM-5 did not exhibit any HFS of the signal at $g = 2.0$ originating from Mn^{2+} in crystalline $MnSO_4$. Only progressive heating up to 770 K and 870 K (Fig. 8, spectra b and c) and subsequent rehydration at ambient temperature made the

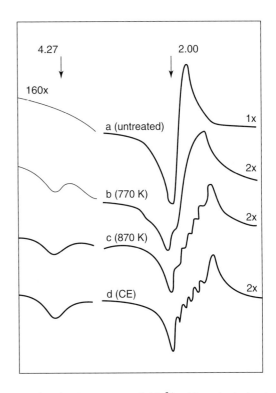

Fig. 8 X band ESR spectra of Mn^{2+}: (a) a physical mixture of $MnSO_4$ and H-ZSM-5; (b) sample (a) heat-treated at 770 K; (c) sample (a) heat-treated at 870 K; (d) Mn,H-ZSM-5 prepared by conventional ion exchange and calcined at 770 K. (After Ref. [95].)

Tab. 4 An overview over systems investigated for solid-state ion exchange

Compound of the introduced cation	Zeolite	References
LiCl; NaCl; KCl; RbCl; CsCl	H-ZSM-5; NH_4-ZSM-5; NH_4-Y, H-MOR; H-BETA; NH_4-BETA; 3A; 4A; Na-Y	[5, 8, 26, 44, 50, 52, 53, 60, 112, 120, 170, 171, 271–273]
	H-EMT; NH_4- EMT; Na-Y	[52, 172–174, 272]
	Na-X; Na-[Ge]X	[44, 45 60, 175, 176]
	NH_4-X; NH_4-Y; K-Y; K-L	[47, 53, 54]
RbCl; Rb_2CO_3; CsOH	NH_4-Y	[43, 47]
NaCl; KCl	H-MOR, MAPO-36; (H,Mo_3S_4)-Y; (H,Mo_3S_4)-L	[46, 54, 105, 177]
CsCl	H-Y; NH_4-BETA; Na-[Ge]X	[44, 46, 50]
$Cs_3[PW_{12}O_{40}]$	H-ZSM-5	[273]
NH_4Cl	H-BETA; H,Na-BETA; H,Cs-BETA	[51]
NR_4Cl (R: organoammonium)	Na-MM^a	[178, 225]
Na-MM^a, clay binders	H-ZSM-5; H-MOR; H-Y	[274, 280]
Li-A/Na-A	Na-A/Li-A	[118, 119]
Li-A/Na-Y; Li-A/Na-X; Li-A/Ca-X	Na-A/Li-Y; Na-A/Li-X; Ca-A/Li-X	[120]
$CaCl_2 \cdot 2H_2O$; $MgCl_2 \cdot 2H_2O$	H-MOR; NH_4-MOR	[6, 56]
$CaCl_2$; $BaCl_2$	NH_4-X; NH_4-Y	[60, 120, 210]
$BeCl_2$	Na-Y	[52]
$LaCl_3 \cdot 7H_2O$	NH_4-Y; H-ZSM-5; H-BETA; Na-Y	[7, 63, 64, 112]
$LaCl_3$ (water-free)	NH_4-Y; H-MOR; H-L; H-FER	[107]
$LaCl_3 \cdot 7H_2O$	Na-Y	[61, 179]
$CeCl_3$; $NdCl_3$; $SmCl_3$; $EuCl_3$; $YbCl_3$	NH_4-Y	[64]
$EuCl_3$	NH_4-Y	[181]
$CeCl_3$	H-ZSM-5	[180]
Cu^0 $(+O_2)$	H-MOR	[1, 142, 143, 276, 277]
$CuCl_2 \cdot 2H_2O$; CuF_2; $Cu_3[(OH)CO_3]_2$; $Cu_3(PO_4)_2$	NH_4-A; NH_4-X; NH_4-Y; H-MOR	[1, 9, 20, 40, 77, 183, 184]
$CuCl_2$	H-ZSM-5; Na-Y	[77, 78, 80, 88, 112, 165, 166, 185, 186]
CuCl	H-ZSM-5/NaZSM-5; NH_4-ZSM-5; 13X	[82, 112, 188–196]
CuCl	NH_4-Y, NH_4-X; H-BETA; H-L; H-[Al]TS-1	[20, 77, 80]
CuCl	H-BETA; H-CLIN	[40, 71–76, 197, 198]
CuCl	H-CLIN; Na-Y; Na-MOR; Na-Y	[94, 112, 199, 200]
CuO	H-[Ga]ZSM-5; Fe,H-ZSM-5	[21, 110]
	H-MOR; H-ZSM-5	[1, 3, 4, 20, 201, 202]
Cu_2O	NH_4-Y	[187, 203, 204]
Cu_2O; Cu_2S	H-ZSM-5	[20, 75, 205]
CuCl; $Cu(NO_3)_2$; $Cu(CH_3COO)_2$	Na-ZSM-5; Na-ZSM-5/H-ZSM-5	[1, 20, 81, 94]
$CuCl_2$	MCM-41	[1, 20, 77–79, 88]
$CuCrO_4$; $CuO + CrO_3$	H-ZSM-5	[3, 78, 79, 196, 206]
Ag^0 $(+O_2)$	H,Ag-MOR	[142, 143, 276]
AgCl	H-ZSM-5	[105]
$AuCl_3$	Na-Y	[207]
Zn^0	H-ZSM-5; NH_4-Y; NH_4-USY; H-MOR	[115, 134–137, 208]
$ZnCl_2$	NH_4-Y	[40, 134, 184, 209]
$ZnCl_2$	H-Y; Na-Y	[113, 210, 212]
$ZnCl_2$	NH_4-Y; Na-Y	[210–213]
$ZnCl_2$	H-ZSM-5	[130]
ZnO	NH_4-Y	[214]
ZnO	H-ZSM-5	[215–218]
Cd^0	NH_4-Y	[135–137, 208]

(continued overleaf)

Tab. 4 (continued)

Compound of the introduced cation	Zeolite	References
Cd(NO$_3$)$_2$; CdCl$_2$; CdO; CdS; CdSO$_4$	NH$_4$-X; NH$_4$-Y; NH$_4$-MOR	[210]
Hg$_2$Cl$_2$	H-ZSM-5	[105]
FeCl$_2 \cdot$ 4H$_2$O	NH$_4$-A; NH$_4$-X; NH$_4$-Y	[4, 9, 83, 84, 186, 220, 221]
FeCl$_2$; FeSO$_4$	H-BETA; NH$_4$-MOR; H-MOR; Na-MOR	[84, 85]
FeCl$_2 \cdot$ 4H$_2$O; FeSO$_4 \cdot$ 7H$_2$O	H-ZSM-5; H-MFI; H-FER	[84, 86–91, 193, 211, 222, 223]
FeCl$_3$	H-ZSM-5	[88, 90, 224, 283]
FeCl$_3$·	H-ZSM-5; Na-ZSM-5; H-[Ga]ZSM-5	[4, 202]
FeO; Fe$_3$O$_4$	H-ZSM-5	[4]
Fe$_2$O$_3$	H-ZSM-5; Na-ZSM-5	[95]
Fe(NO$_3$)$_3 \cdot$ 9H$_2$O	Na-MM^a	[225]
CoCl$_2$	NH$_4$-Y; H-ZSM-5	[185, 186, 192, 210, 226–234, 236, 283]
CoCl$_2$; Co(NO$_3$)$_2$; CoCl$_2 \cdot$ 6H$_2$O	NH$_4$-CLIN; H-CLIN; Na-Y	[94, 235, 282]
	H-BETA	[236]
	H-FER	[234]
Co$_2$(CO)$_8$	H-Y	[140]
NiCl$_2$	NH$_4$-Y	[40, 92, 237]
NiCl$_2$; NiSO$_4$; Ni(CH$_3$COO)$_2$; NiO	H-ZSM-5	[186, 192, 238]
NiCl$_2$	H-ZSM-5	[93, 165, 226, 227]
NiCl$_2$; NiCl$_2 \cdot$ 6H$_2$O; NiO	SAPO-42	[96]
NiCl$_2$	SAPO-n (n = 5, 8, 11, 34)	[219, 239–242]
NiCl$_2$	H-MOR; H-L; H-MAZ; H-BETA	[92, 93, 231]
NiCl$_2$	MCM-41	[244]
Raney nickel	Cu-ZSM-5	[278]
MnCl$_2$; MnSO$_4$; Mn(NO$_3$)$_2$; Mn$_3$O$_4$	H-ZSM-5	[18, 95]
Mn(CH$_3$COO)$_2$	H-ZSM-5	[18, 95]
MnCl$_2$; MnCl$_2 \cdot$ 4H$_2$O	NH$_4$-Y; Na-Y	[210, 282]
V$_2$O$_5$	H-MOR; H,Na-MOR; H-ZSM-5	[2, 19, 66]
	H,Na-ZSM-5	[246–248]
V$_2$O$_5$	H,Na-X; H-Y; H-ZSM-5; H-MOR	[179, 248]
V$_2$O$_5$	Na-Y	[249–251]
V$_2$O$_5$	H-[Ga]ZSM-5, [Al]BETA; [B]BETA	[201, 202, 281]
VO(NO$_3$)$_2$	AlPO$_4$-5	[133]
V$_2$O$_5$	NH$_4$-ZSM-5	[252]
VOCl$_3$; VCl$_3$	H-ZSM-5; Na-Y	[246, 282]
V$_2$O$_5$+CuO	H-ZSM-5	[3, 206]
Nb$_2$O$_5$	NH$_4$-Y; NH$_4$,Na-Y	[253]
Sb$_2$O$_3$	Na-Y; ZSM-5; La,Na-Y	[112, 254]
CrCl$_3$; CrCl$_3 \cdot$ 6H$_2$O	H-Y; Na-Y	[255, 279, 282]
CrO$_3$	H-MOR; H-ZSM-5	[2, 66, 255]
CuCrO$_4$	H-MOR; H-ZSM-5	[2, 3, 66, 255]
CrO$_3$	H-[Ga]ZSM-5; H-[Fe]ZSM-5	[4, 201, 202, 257]
Cr$_2$O$_3$	H-MOR; H-ZSM-5	[2, 66, 257, 258]
CrO$_2$Cl$_2$	MFI	[132]
CrO$_3$+CuO	H-ZSM-5	[206]
CrO$_3$	SAPO-11	[259]
MoCl$_5$; (MoOCl$_4$),	NH$_4$-Y; NH$_4$-DAY; Co,H-Y	[138, 139, 261]
MoCl$_3$	H-ZSM-5; H-MOR; H-ZSM-35; H-EU; H-FER; H-ZSM-48; H-L	[68, 96, 264]
MoO$_3$	H-ZSM-5; H-USY; H-FER; H-BETA	[2, 66, 138, 139, 261–264]
MoO$_3$	Na-Y; Na-ZSM-5	[251, 261]
WO$_3$	Na-Y	[251]
WO$_3$ (+CCl$_4$)	H-USY; H-ZSM-5; H-FER	[141]

(continued overleaf)

Tab. 4 (continued)

Compound of the introduced cation	Zeolite	References
Pt^0	K-L	[101]
Pd^0 (+Cl_2)	H-Y	[23]
$PdCl_2$; $Pd(NO_3)_2$; PdO; $PtCl_2$; $PtCl_4$; PtO_2	NH_4-Y; H-ZSM-5	[99, 100, 265]
$PdCl_2$	H-ALPHA; H-RHO; H-ZK-5; ZSM-48	[49, 96, 98]
PdO	H-SAPO-42	[45, 96, 231]
$PtCl_2$	H-RHO; H-ZK-5; H-ALPHA; H-ZSM-58; H-SAPO-42	[49, 96, 98]
$PtCl_2$; $PtBr_2$; PtO_2	NH_4-Y	[38, 39]
$RhCl_3$	ALPHA; H-SAPO-42; H-ZK-5	[49, 96, 98]
$RhCl_3$	H-$AlPO_4$; H-APO-11; H-APSO-11	[103]
$RhCl_3$	DAY	[102]
$PdCl_2+CaCl_2$; $PdCl_2+LaCl_3$	H-ZSM-5	[99, 100, 260, 265]
Ga^0	H-ZSM-5	[114, 116, 208]
Ga_2O_3	H-ZSM-5; MFI, BETA	[24, 25, 148–150, 156, 164–166]
$GaCl_3$	MFI; BETA	[156]
Ga_2O_3	NH_4,Na-Y; H-MOR; H-ZSM-5; H-BETA; H-Y	[155, 157, 266, 267]
Ga_2O_3	H-[Ga]ZSM-5	[25]
Ga_2O_3	H-SAPO-n ($n = 5, 34, 37$) (+template)	[167]
In^0	H-ZSM-5	[208]
In_2O_3	H-ZSM-5; NH_4-Y; H-MOR; H-OFF; H-BETA	[148, 157, 158, 165, 166, 187]
In_2O_3	H-ZSM-5; H,Na-Y; NH_4-MOR; NH_4-Y	[266, 267]
In_2O_3	H-BETA (TEA-BETA)	[160–163, 268, 269]
In_2O_3	H-SAPO-n ($n = 5, 34, 37$), (+template)	[151, 154, 163]
In_2O_3	MCM-41	[152, 167–169]
$InCl_3$	NH_4-ZSM-5	[159]

^{a} MM, montmorillonite.

hyperfine splitting appear. This pointed to a disintegration of the salt crystallites and migration of Mn^{2+} cations into the structure. The HFS constant of $A = 9.8$ mT was interpreted as being indicative of isolated Mn^{2+} cations in O_h coordination. In the parent H-ZSM-5, absorbance of the IR band at 3610 cm^{-1}, typical of the acid OH groups of the H-ZSM-5 sample used, corresponded to 0.91 mmol OHs g^{-1} dry zeolite. IR spectroscopic measurements evidenced a decrease in the absorbance of the band at 3610 cm^{-1} upon SSIE. Monitoring the solid-state reaction showed that in no case was a 100% degree of exchange reached. An increase in the amount of $MnCl_2$ applied from a substoichiometric ratio (0.33 mmol Mn^{2+} versus 0.91 mmol OHs g^{-1}) to a stoichiometric ratio (0.45 mmol Mn^{2+} versus 0.91 mmol OHs) did not lead to a significantly higher degree of exchange. An increase of the reaction temperature, from 570 to 770 K, enhanced the exchange degree from 21% to 57%. Measurements as a function of the reaction time revealed that most of the Mn^{2+} cations were introduced during the initial stage of the process, that is, within the first hour. With respect to the Mn-compounds employed in stoichiometric or slightly over-stoichiometric mixtures with H-ZSM-5, the degree of SSIE at 770 K increased in the order $MnSO_4$ (16%) < Mn_3O_4 (46%) < $MnCl_2$ (57%).

2.4.3.5.7 Solid-State Ion Exchange with Salts of Noble Metals Zeolites modified via ion exchange with salts of noble metals are, after reduction, interesting materials because of their potential as catalysts in hydrogenation/dehydrogenation reactions. It could be shown that SSIE is also a suitable route to noble metal-containing zeolites. In fact, SSIE in several cases offers the only route to post-synthesis modification of zeolites by incorporation of noble metal cations: with small-pore zeolites – and especially those with eight-membered oxygen rings (8-MR) as pore mouths – conventional ion exchange either fails or provides only a low degree of exchange. The reason is seen in geometric constraints: the solvated cations or complexes such as $Pt(NH_3)_4^{2+}$ are not able to penetrate the narrow-pore openings. Systematic studies of the introduction of noble metals (Pt, Pd, Rh) into 8-MR zeolites (ZSM-8, zeolite Rho, zeolite ZK-5, SAPO-42; cf. Ref. [10]) have been conducted by Bock [96] and

Weitkamp et al. [97, 98]. These authors demonstrated that the resulting materials were, after reduction, highly shape-selective catalysts in the competitive hydrogenation of slim n-hexene-(1) and bulky 2,4,4,-trimethylpentene-(1); the latter could react only on the external surfaces. Residual Brønsted acidity, which formed upon reduction of the noble metal cations introduced through SSIE, may be eliminated by subsequent SSIE with alkaline metal compounds (vide supra, Section 2.4.3.5.1).

However, for some types of reactions, such as hydroisomerization or hydrocracking of hydrocarbons, both hydrogenation/dehydrogenation and acidic functions of a catalyst are required. Such bifunctional catalysts were prepared via SSIE by Karge et al. [99, 100]. These authors used $PdCl_2$, PdO and $PtCl_2$ to incorporate the metal cations into H-ZSM-5. The resulting materials were, after reduction by H_2, successfully tested for hydrogenation of ethylbenzene. It was further shown that SSIE with Ca- or La-salts prior to (or simultaneously with) SSIE with noble metal compounds improved the performance of the final catalyst. The presence of Ca^{2+} or La^{3+} decreased the strength of Brønsted acidity. Moreover, the alkaline or rare earth cations play the role of anchors for the metal particles which form upon reduction and, thus, stabilize the particle size distribution.

Other interesting studies of SSIE with noble metal cations were carried out by Hatje et al. using temperature-programmed reduction (TPR), XRD, dispersive extended X-ray absorption fine structure (DEXAFS) and time-resolved X-ray absorption spectroscopy (XAS) [38, 39], by Mkombe et al. [101], who prepared Pt^0 clusters on K-L, by Schlegel et al. [102] (incorporation of Rh into DAY, studied by FTIR and CO adsorption under formation of $Rh(CO)_2^+$ complexes), and by Wasowicz and Kevan [103], who showed that a significantly higher loading of SAPO-11 with rhodium was achievable via SSIE than through CE and investigated the resulting Rh(II) and Rh(I) species by ESR.

2.4.3.5.8 Miscellaneous
A great number of further salt or oxide/zeolite systems have been studied with respect to SSIE, and it is beyond the limited size of this book to list and discuss all such systems described hitherto in the literature. Most such systems are listed in Table 4, together with pertinent references.

Only one type of solid-state reaction with zeolites should be briefly mentioned here, namely the SSIE with cations of high oxidation states, for example V^{5+} or V^{4+}, Cr^{6+} or Cr^{5+}, and Mo^{6+} or Mo^{5+} (cf. also Section 2.4.3.7.3). Important examples were described and discussed by Kucherov and Slinkin [68], and by Bock [96]. In this context, the interesting point is that not the bare cations of such high oxidation states but, frequently after partial reduction, complex cations such as $VO(OH)^+$, CrO_2^+, or $MoOCl_2^+$ were incorporated. This was elegantly confirmed by Bock [96] for the system $MoCl_3$/H-zeolites by carefully measuring the numbers of in-going and out-going Cl^- anions. (For further details concerning the introduction of cations with a high oxidation state via SSIE, see Refs. [11, 68, 96].)

2.4.3.6 Some Considerations on Thermodynamic and Kinetic Aspects of SSIE

2.4.3.6.1 Effect of Temperature and Salt Concentration in Salt/Zeolite Mixtures
Although reports on these effects are hitherto rather scarce, it has been observed on several occasions that an increase in reaction temperature (up to an optimum) shifted the SSIE towards a higher degree of exchange (cf. Section 2.4.3.5.4; e.g., SSIE in CuCl/H-ZSM-5, Ref. [104]; SSIE in $BeCl_2$/Na-Y, Ref. [52]; SSIE in $MnCl_2$/H-ZSM-5, Ref. [18]; and Section 2.4.3.6.3). In several cases the increase in the molar ratio of salt and zeolite in the mixtures also increased the degree of incorporation of the in-going cation. For instance, enhancing the Cu/OH molar ratio in mixtures of CuCl and H-ZSM-5 from 0.32 over 1.00, and 1.60 to 2.25, increased in corresponding steps the degree of exchange (OH consumption) by up to 86% [104]. However, this effect depends on the system under study: in the case of $MnCl_2$/H-ZSM-5, no such enhancement in the exchange degree was observed when, at the same reaction temperature, the amount of admixed $MnCl_2$ was increased (cf., e.g., Section 2.4.3.5.6; see Ref. [18]).

2.4.3.6.2 The Role of Water in Solid-State Ion Exchange
In general, the materials involved in reported SSIE experiments have been "dry", but this does not necessarily mean that H_2O was absolutely excluded. As the experiments were usually conducted under ambient conditions, strongly adsorbed water molecules or strongly held crystal water may have been present. Although the presence of water may not be detrimental to SSIE, arguments have been made that even traces of H_2O are not a prerequisite for this type of ion exchange to occur, as shown by the following three types of experiment:

(i) SSIE was successfully carried out with a great number of systems comprising compounds of the in-going cation which are insoluble in water: AgCl/H-ZSM-5 [105, 106], Hg_2Cl_2/H-ZSM-5 [105, 106], CuO/ H-ZSM-5 [104], Cu_2O/NH_4-Y [104], Mn_3O_4/H-ZSM-5 [18], and several others. This is

---
References see page 504

at least a strong hint that H$_2$O molecules are not required for SSIE.

(ii) A series of SSIE experiments was carefully designed by Sulikowski et al. [107] to secure that any traces of water were excluded. Each step of the procedure was carried out under exclusion of moisture, from the very initial steps and then throughout the whole subsequent process. The reactants were separately treated in high vacuum (10^{-6} Pa) at about 670 K in ampoules until water vapor was no longer detectable by MS and/or IR. The ampoules were then sealed and transferred into an efficiently working special glove box ($p_{H_2O} \leq 10^{-7}$ Pa). All subsequent steps of the sample preparation were carried out in this glove box: breaking the ampoules; mixing the powders of salts and zeolites; filling the powders into capillaries for XRD runs and sealing them; pressing wafers for the IR measurements, transferring these into sample holders and the sample holders then into an ultra-high vacuum-tight IR cell (cf. Ref. [108]). One experiment was conducted with NaCl/H-MOR [105]. After heating the IR cell, which was connected to an ultra-high vacuum system, IR spectroscopy revealed that the OH groups of the H-MOR had completely disappeared (cf. Section 2.4.3.5.1). This indicated a 100% exchange of the protons of the H-MOR for Na$^+$ from NaCl.

(iii) Another set of analogous experiments was carried out with water-free lanthanum chloride, which had been prepared from LaCl$_3 \cdot$ 7H$_2$O through a particular procedure in a flow of HCl. As zeolites H-L, H-MOR, ultra-stabilized (dealuminated) H-Y (i.e., H-USY), and H-FER (ferrierite) were employed. The IR measurements indicated that the acid OH groups of the zeolites had reacted, as the corresponding OH bands had disappeared (e.g., in the case of H-L) or were markedly decreased (H-MOR, H-USY). The fact that in the latter cases not all of the OH groups had reacted is easily understood in view of the fact that in these zeolites the distances between the acid OHs are frequently rather large and, therefore, it is more difficult to compensate the negative charge of three of those sites by one La^{3+} cation. In the case of H-FER no reaction at all occurred, most likely due to geometric constraints (*vide supra*).

2.4.3.6.3 Kinetics of Solid-State Ion Exchange

It should be possible to investigate the kinetics of SSIE, for example via time-resolved XRD patterns, through the intensity changes of the reflections of the salt (or oxide)/zeolite mixtures and/or of typical IR bands (framework vibrations, bands in the OH or NH region or bands due to interaction between the cations and

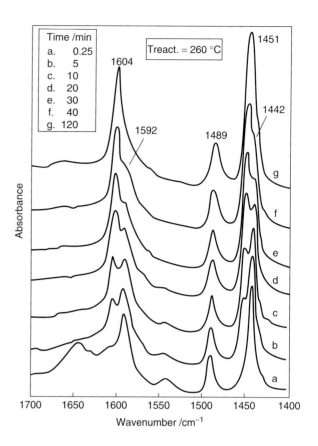

Fig. 9 Infrared spectra of pyridine adsorbed on a CuCl/Na-Y mixture at 533 K as a function of time. (For details, see text; after Ref. [21].)

probe molecules). In fact, in the latter case one tacitly assumes that the presence of the probe molecules does not affect the kinetics of SSIE. Here, results will be described which were obtained when the solid-state reaction between CuCl and Na-Y or Na-MOR was monitored by IR using pyridine (Py) as a probe [109, 110]. (For the experimental details, see Refs. [27, 108].) In order to illustrate such an experiment, Fig. 9 displays a set of IR spectra obtained during SSIE between CuCl and Na-Y at 533 K in the presence of pyridine (Py → Na$^+$). Initially, only those bands (at 1592 and 1442 cm^{-1}) were observed which indicate the interaction between pyridine and Na$^+$. However, after a few minutes additional bands at 1604 and 1451 cm^{-1} began to appear, originating from Py → Cu$^+$ complexes. While these bands increased with reaction time, concomitantly the bands of Py → Na$^+$ decreased.

From these changes of intensities with time, kinetic curves can be derived for various reaction temperatures (integrated absorbances versus reaction time; cf., e.g., Fig. 10). Clearly, the data must be corrected in order to account for the temperature-dependence of the adsorption equilibrium of pyridine. Figure 10 shows

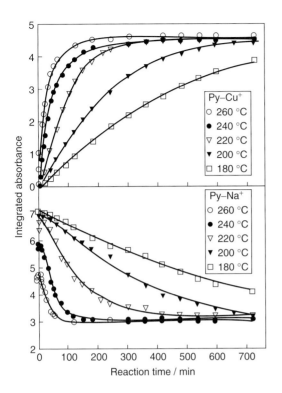

Fig. 10 Normalized and corrected integrated absorbances from pyridine adsorption on CuCl/Na-Y during solid-state ion exchange as a function of reaction time. (For details, see text; after Ref. [21].)

results for the kinetics of SSIE in the system CuCl/Na-Y; analogous graphs may be obtained with the bands of other Py → cation complexes. It is reasonable to assume that the absorbance, A, of the Py → Cu^+ complex at a given time, t, and at steady state (i.e., A_t and $A_{t\to\infty}$) are proportional to the amounts, M, of incorporated Cu^+ (i.e., to M_t and $M_{t\to\infty}$). With this assumption, plots of $M_t/M_{t\to\infty}$ versus $t^{1/2}$ (cf., e.g. Fig. 11) can be developed from figures such as Fig. 10. The data as shown in Fig. 11 can be described by solutions of Fick's second law for diffusion (see broken lines in Fig. 11), and suggest that SSIE in zeolites is possibly diffusion-controlled. Provided that this is correct, one would obtain diffusion coefficients of about 10^{-13} cm^2 s^{-1} at 500 K, and activation energies of 68–70 kJ mol^{-1} for the above systems CuCl/Na-Y and CuCl/Na-MOR.

2.4.3.6.4 A Possible Mechanism of Solid-State Ion Exchange In principle, the transport of the in-going cations may occur through the vapor phase or surface diffusion (cf. Refs. [111–113] and Chapter 2.4.7 of this Handbook). For example, in the case of several metals [114], such as metallic zinc and gallium, transport through the (intercrystalline) gas phase is most probably part of the exchange process. Schneider et al. [115] investigated the reaction between powders of elemental Zn° and H-ZSM-5 with various n_{Si}/n_{Al} ratios by TPE of hydrogen, which occurred above 620 K, XANES, and diffuse reflectance infrared Fourier transform spectroscopy (DRIFTS). Similarly, Altwasser et al. [116] reacted elemental Ga° with H-ZSM-5 and monitored the reaction by the TPE of H$_2$

References see page 504

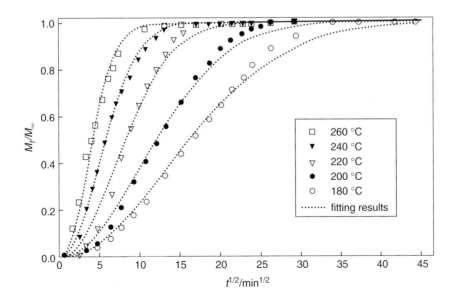

Fig. 11 Description of the kinetics of solid-state ion exchange in the system CuCl/Na-Y through a diffusion model; the symbols represent experimental data derived from the measured integrated absorbances of the probe (pyridine); the broken lines represent results of the fitting to the diffusion model. (For details, see text; after Ref. [21].)

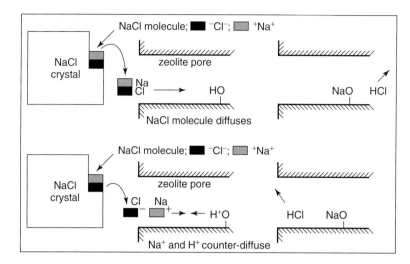

Fig. 12 Schematic representation of two possible models of the mechanism of solid-state ion exchange in microporous materials. Upper part, mechanism A: NaCl molecule diffuses. Lower part, mechanism B: Na^+ and H^+ counter-diffuse (see text).

peaks at about 400 K and 700 K, indicating that univalent Ga^+ was formed at the lower, trivalent Ga^{3+} at the higher temperature. This method enabled the introduction of gallium into zeolite structures with pore openings too small as to allow exchange in an aqueous solution of Ga-salts or SSIE with Ga_2O_3. However, many of the salts and oxides involved in SSIE have, even at the usual temperatures of solid-state reaction of about 530–730 K, a rather low vapor pressure. Thus, the second possibility of surface diffusion from the intimately contacted salt (or oxide) and to the zeolite particles seems to be more likely in SSIE. In any event, the question then remains as to whether ions or molecules of the respective compound diffuse into the zeolite pores after separation, for example from a kink ("Halbkristallage", after Stranski, i.e., the position of the "half-crystal", cf. Ref. [117]). Both possibilities are shown schematically in Fig. 12. If the anion and the cation migrated separately, then counter-diffusion of the in-going and out-going cations must proceed in such a way that no excessive electrical gradient occurs. Stranski [117] has computed that the separation of a NaCl molecule from a kink would require a lower energy than the separation of the anion and cation in a sequence. Moreover, in the case of the medium-pore zeolite ferrierite, H-FER, we have seen that no reaction occurred with $LaCl_3$. A simple explanation would be that the $LaCl_3$ molecule is too bulky as to be able to penetrate either the 10 MR pore openings (being smaller than in the case of ZSM-5) or the even narrower 8 MR pore openings of FER. This would mean that the mechanism of molecular diffusion (lower part of Fig. 12) is more likely to hold. This assumption was supported by experiments of SSIE of H-ZSM-5 with molecules containing bulky anions (e.g., salts of heteropoly acids such as $Cs_3[PW_{12}O_{40}]$ [48]). Compared to the exchange with CsCl (cf. Fig. 1), SSIE with $Cs_3[PW_{12}O_{40}]$ occurred only to a minor extent, due to a partial decomposition of the salt. Finally, the easiness of SSIE with highly covalent compounds (e.g., CuCl, cf. Section 2.4.3.5.4) suggested that the molecules did not dissociate but rather migrated as an intact species.

2.4.3.7 Modified Solid-State Ion Exchange and Related Processes

2.4.3.7.1 Contact-Induced Ion Exchange
In an interesting experiment, Kokotailo et al. [118] and Fyfe et al. [119] showed that, even at room temperature, cation exchange occurred when, for example, zeolite crystallites loaded with two different sorts of cations were brought into intimate contact. An example described was the simple mechanical mixing of equal amounts of Li-A and Na-A powders. The exchange was proven by ^{29}Si MAS NMR and XRD measurements. Initially, two well-separated ^{29}Si MAS NMR signals of Li-A and Na-A at -85.1 and -88.9 ppm (referenced to TMS) were observed, reflecting the different local surroundings of Si in both zeolites. Similarly, the XRD patterns showed split reflections. Thus, two different phases existed. After equilibrating at ambient conditions, only one sharp ^{29}Si MAS NMR line was left, and the splitting of the XRD reflections had disappeared. This indicated that the ion exchange had led to a homogeneous distribution of the Li^+ and Na^+ cations over all zeolite crystallites – that is, the existence of only one phase. Similar findings were reported by the same authors for the following pairs: Li-A/NH_4-A, Li-X/Na-X, Li-X/NH_4-X, and Li-A/Na-MOR. Huang et al. [120] described analogous results for Li-A/Na-Y, Li-A/NH_4-X, and Li-A/Ca-X. However, under the absolute exclusion of even of traces of H_2O (i.e., experimental conditions as

described in Section 2.4.3.6.2), no such contact-induced ion exchange occurred, as demonstrated by Koy and Karge [121]. Only when the capillaries containing the water-free equimolar mixtures of Li-A and Na-A powders were opened and exposed to moisture of the ambient air, did the splitting of the XRD reflections disappear, proving that the presence of adsorbed water vapor was required to make the Li$^+$ cations of Li-A crystallites migrate into the Na-A crystallites, and *vice versa* of Na$^+$ into Li-A, until an equilibrium was reached. Thus, contact-induced ion exchange is, in fact, mediated by water filling the zeolitic pores – in other words, the mechanism is essentially the same as that of conventional ion exchange. The possibility of contact-induced ion exchange was also investigated by Fraissard et al. [122] by applying ^{129}Xe NMR spectroscopy to the system Rb,Na-X/Na-Y. While no exchange was observed at 300 K, it did occur after heating the mixture to 673 K.

2.4.3.7.2 Incipient Wetness Technique of Impregnation

This technique comprises mixing a certain volume of dried (usually outgassed) zeolite powder with an equivalent volume of the aqueous solution containing a salt of the exchange cation (chloride, nitrate, carboxylates, etc.), thus producing a paste with a distribution of the cations as homogeneous as possible and subsequent heating [123–127]. For the first stage – which is conducted at lower temperatures when the zeolite powder is impregnated and the paste is produced – a fraction of the ions are exchanged in a similar manner as in CE, except that the intermittent washing steps are omitted, so that the out-going cations and the anions of the impregnated salt remain in the product of this procedure; this process resembles the contact-induced ion exchange. Subsequently, in a second stage the paste is heated to complete the exchange; that is, the mixture of the residual salt and zeolite powder is dried and, in fact, a real solid-state reaction occurs. However, if no volatile compounds are formed in the second stage an equilibrium with an exchange degree less than 100% will result. In essence, the incipient wetness method of impregnation lies between, or is a combination of, CE and SSIE. There are, however, some drawbacks associated with the procedure: in practice, the migration of the salt solution into the pores may be hindered, particularly when the pores are filled with gases, the wettability is low, and the pores are narrow. Furthermore, the salt molecules are usually not well-anchored on the internal surface, and therefore redistribution may take place during subsequent treatments (heating and, if desired, reduction; cf. Ref. [128]). Thus, the resulting products are often not homogeneously exchanged, and the concentration of the in-going cations is highest close to the external surface – that is, they exhibit a significant concentration gradient.

2.4.3.7.3 Gas-Phase-Mediated Processes Related to Solid-State Ion Exchange

The adsorption of elemental species or compounds of the in-going cation onto, and subsequent incorporation of the cation into, zeolite crystallites may occur in the case of volatile metals (*vide supra*), oxides or halides. Pertinent examples are the introduction of Zn^{2+}, Ga^{3+}, Fe^{3+} into hydrogen forms of zeolites via the interaction of them with vapors of the respective chlorides [129–131], the incorporation of V-, Ti-, and Cr-containing species from gaseous $VOCl_3$, $TiCl_4$, and $CrOCl_2$, respectively [132, 133], and the reaction of vapors of metallic zinc or cadmium with, for example, H-Y or H-ZSM-5 [134–137]. In some cases, the presence of additional molecules in the vapor phase may mediate SSIE at elevated temperatures. For instance, the presence of water vapor caused the ion exchange of MoO_3 with H,Na-Y to occur [138, 139], whereas attempts of SSIE with dry MoO_3 were unsuccessful [2]. It is possible that the interaction with H_2O prevented MoO_3 from polymerization and/or led to a disaggregation of initially formed Mo-oxide species which were too bulky to enter the porous framework of Y-zeolites. Similarly, the presence of CO in the gas phase enabled the incorporation of Co^{2+} and Rh^+ via the formation of carbonyl-complexes [102, 140]. Finally, Kucherov et al. [141] determined that the incorporation of Mo^{5+}-containing species such as $(MoOCl_2)^+$ or $MoCl_4^+$ into H-ZSM-5, H-BETA, H-FER, and H-USY was mediated by a flow of CCl_4-containing air stream.

2.4.3.7.4 Oxidative and Reductive Solid-State Ion Exchange

Previously, Beyer and colleagues [142, 143] observed during the 1970s that tiny silver and copper particles (Ag_n^0, Cu_n^0) formed on the external surfaces of zeolite crystallites upon reduction of Ag-Y and Cu-Y by H_2, respectively. Concomitantly, acid OH groups were restored. When the materials were then calcined in oxygen, however, Ag^0 and Cu^0 re-migrated into the zeolite structure, where they reacted with the protons of the OHs under formation of Ag^+ and Cu^+ on cation sites. Similarly, Kucherov et al. [144] claimed that, in the case of reduced Cu, H-ZSM-5 bulky Cu^0 aggregates on the external surface could also be re-oxidized ("oxidative SSIE"), and Feeley and Sachtler [23] reported that introduction of Pd^+ into H-Y proceeded through an analogous oxidative SSIE from Pd^0 particles produced on the outer zeolite surface in the presence of chlorine.

Reductive solid-state ion exchange (RSSIE) has been extensively studied, especially with the systems Ga_2O_3/H-ZSM-5, In_2O_3/H-ZSM-5 and CuO/H-ZSM-5, as well as in mixtures of oxides with H-MOR, H-BETA and H-Y (for details, see Ref. [145]). In fact, in systems such as

References see page 504

CuO/H-ZSM-5, so-called "auto-reduction" occurs in that a certain amount of the framework oxygen anions is oxidized to molecular oxygen under reduction of Cu^{2+} to Cu^+ [146, 147]. In contrast to RSSIE with Ga_2O_3 and In_2O_3, an additional reductive agent is, therefore, not necessary. Price and Kanazirev [24], Price et al. [25] and Kanazirev et al. [148–150] were the first to report on systematic studies of RSSIE. Temperature-programmed reduction monitored by a thermobalance (TPR/TG), XRD, transmission electron microscopy/energy dispersive analysis of X-rays (TEM/EDAX), XPS, and IR were each used to study RSSIE (cf. also Refs. [151–155]). The interest in such investigations was stimulated by the fact that Ga- and In-containing zeolites were assumed to be efficient catalysts for the aromatization of alkanes and redox reactions. Moreover, they appeared to be promising candidates for DeNOx reactions and SCR [156–159]. The stoichiometry of reductive ion exchange with Ga_2O_3 is, according to TPR experiments, most likely described by Eqs. (5) and (6):

$$Ga_2O_3 + 2H_2 \longrightarrow Ga_2O + 2H_2O \quad (5)$$

$$Ga_2O + 2H^+Z^- \longrightarrow 2Ga^+Z^- + H_2O \quad (6)$$

An analogous stoichiometry was proposed for the RSSIE with In_2O_3, even though no evidence for the intermediate formation of In_2O could be obtained; this suboxide is probably unstable and non-existent in the crystalline state. It was shown that reduction by CO or NH_3 is also possible. Studies on the incorporation of indium into zeolites (Na,H-Y, H-ZSM-5, H-MOR and H-BETA) were significantly amended and deepened by the investigations of Beyer and colleagues [154, 160–163], who employed IR (with and without probe molecules), TPR, temperature-programmed oxidation (TPO), and differential thermogravimetry (DTG) and a test reaction for acid sites (isomerization of m-xylene) to confirm and elucidate RSSIE of In_2O_3 in the above systems. For example, these authors clarified the formation and redox behavior of cationic InO^+ species and provided a consistent picture of the stoichiometric chemistry and mechanism of RSSIE in In_2O_3/zeolites. They also demonstrated that a closed (reversible) redox cycle is operative in the case of these systems, which makes indium-containing zeolite catalysts prepared by RSSIE promising candidates for redox reactions (*vide supra*; for details, see Ref. [145]). Besides the aromatization of alkanes (cf., e.g., Refs. [24, 150, 164, 165]), reactions of amines (cf., e.g., Ref. [166]) were investigated. Related studies concerned mixtures of Ga_2O_3 and In_2O_3 with SAPO-materials (SAPO-5, SAPO-34, SAPO-37 and the mesoporous material MCM-41 [152, 167–169] (for acronyms, cf. Ref. [10]), including the so-called template-induced introduction. In the latter case, decomposition products of templates used in the synthesis of the porous materials, act as reducing agents.

2.4.3.8 A Tabulated Survey of the Systems Studied in SSIE and Related Processes

As announced previously (see Sections 2.4.3.1, 2.4.3.5, 2.4.3.5.5, and 2.4.3.5.8), the systems studied with respect to modification via SSIE and related processes are summarized in Table 4. This should enable the interested reader to determine, conveniently, where research in this field has been already carried out and where knowledge is still lacking.

2.4.3.9 Concluding Remarks

In recent years, solid-state ion exchange has become a well-established and now routinely used method for the post-synthesis modification of zeolites, and is also applicable to other porous solids such as mesoporous materials, clays, and porous oxides. Hence, a large number of systems has been investigated and appropriate techniques of preparation and investigation developed that, without doubt, will continue to be extended and improved in the future. However, perhaps more important than studying additional systems might be an undertaking of efforts to obtain a deeper understanding of solid-state reactions of zeolites and related materials. As indicated in this chapter, a number of interesting problems remain to be solved systematically, notably the thermodynamics, kinetics, and mechanisms of solid-state ion exchange reactions.

References

1. A. V. Kucherov, A. A. Slinkin, *Zeolites* **1986**, *6*, 175.
2. A. V. Kucherov, A. A. Slinkin, *Zeolites* **1987**, *7*, 38.
3. A. V. Kucherov, A. A. Slinkin, *Zeolites* **1987**, *7*, 43.
4. A. V. Kucherov, A. A. Slinkin, *Zeolites* **1988**, *8*, 110.
5. H. K. Beyer, H. G. Karge, G. Borbély, *Zeolites* **1988**, *8*, 79.
6. H. G. Karge, H. K. Beyer, G. Borbély, *Catal. Today* **1988**, *3*, 41.
7. H. G. Karge, G. Borbély, H. K. Beyer, G. Onyestyák, in *Proceedings 9th International Congress on Catalysis*, M. J. Philips, M. Ternan (Eds.), Calgary, Alberta, Canada, June 26–July 1, 1988, Chemical Institute of Canada, Ottawa, 1988, p. 396.
8. J. A. Rabo, M. L. Poutsma, G. W. Skeels, in *Proceedings 5th International Congress on Catalysis*, J. W. Hightower (Ed.), Miami Beach, FL, USA, August 20–26, 1972, North-Holland Publishing Comp, New York, 1973, p. 1353.
9. A. Clearfield, C. H. Saldarriaga, R. C. Buckley, in *Proceedings 3rd International Conference on Molecular Sieves – Recent Research Reports*, J. B. Uytterhoven (Ed.), Zürich, Switzerland, September 3–7, 1973. University of Leuwen Press, Leuwen, Belgium, 1973, p. 241.
10. C. Baerlocher, W. M. Meier, D. H. Olson, *Atlas of Zeolite Framework Types*, 5th edn, Elsevier, Amsterdam, 2001, 302 pp.
11. H. G. Karge, H. K. Beyer, in *Molecular Sieves – Science and Technology*, H. G. Karge, J. Weitkamp (Eds.), Vol. 3, *Post-Synthesis Modification I*, Chapter 2, Springer, Berlin Heidelberg, 2002, p. 43.
12. D. W. Breck, *Zeolite Molecular Sieves*, Wiley, New York, 1974, Chapter 7, 529 pp.
13. L. V. C. Rees, *Chemistry and Industry* **1977**, 647.

14. R. P. Townsend, in *New Developments in Zeolite Science and Technology*, Y. Murakami, A. Iijima, J. W. Ward (Eds.), *Studies in Surface Science and Catalysis*, Vol. 28, Kodansha, Tokyo, Elsevier, Amsterdam, 1986, p. 273.
15. A. Dyer, *An Introduction to Zeolite Molecular Sieves*, Wiley, New York, 1988, Chapter 6, 63 pp.
16. G. H. Kühl, in *Catalysis and Zeolites, Fundamentals and Applications*, J. Weitkamp, L. Puppe (Eds.), Springer-Verlag, Heidelberg, 1999, Chapter 3, p. 81.
17. R. P. Townsend, E. N. Coker, in *Introduction to Zeolite Science and Practice*, H. van Bekkum, E. M. Flanigen, P. A. Jacobs, J. C. Jansen (Eds.), *Studies in Surface Science and Catalysis*, Vol. 137, Elsevier, Amsterdam, 2001, Chapter 11, p. 467.
18. B. Beran, B. Wichterlová, H. G. Karge, *J. Chem. Soc., Faraday Trans.* **1990**, *86*, 3033.
19. A. V. Kucherov, A. A. Slinkin, *Zeolites* **1987**, *7*, 583.
20. H. G. Karge, B. Wichterlová, H. K. Beyer, *J. Chem. Soc., Faraday Trans.* **1992**, *88*, 1345.
21. M. Jiang, J. Koy J, H. G. Karge, in *Proceedings 3rd International Symposium on the Synthesis of Zeolites, Expanded Layer Compounds and other Crystalline Microporous or Mesoporous Solids*, M. L. Occelli, H. Kessler (Eds.), ACS Meeting, Anaheim, California, USA, April 2–7, 1995. *Synthesis of Microporous Materials: Zeolites, Clays and Nanostructures*. Marcel Dekker Inc., New York, 1996, p. 335.
22. Q. Zhang, E. Kasai, H. Mimura, *Advanced Powder Technol.* **1994**, *5*, 289.
23. O. C. Feeley, W. H. M. Sachtler, *Appl. Catal.* **1991**, *75*, 93.
24. L. G. Price, V. Kanazirev, *J. Catal.* **1990**, *126*, 267.
25. G. L. Price, V. I. Kanazirev, K. M. Dooley, *Zeolites* **1995**, *15*, 725.
26. J. A. Rabo, P. H. Kasai, *Prog. Solid State Chem.* **1975**, *9*, 1.
27. H. G. Karge, W. Niessen, *Catal. Today* **1991**, *8*, 451.
28. H. G. Karge, M. Hunger, H. K. Beyer, in *Catalysis and Zeolites – Fundamentals and Applications*, J. Weitkamp, L. Puppe (Eds.), Springer, Berlin, Heidelberg, 1999, p. 209.
29. H. G. Karge, E. Geidel, in *Molecular Sieves – Science and Technology*, Vol. 4, *Characterization I*, H. G. Karge, J. Weitkamp (Eds.) Springer, Heidelberg, 2004, p. 73.
30. A. V. Kucherov, A. A. Slinkin, D. A. Kondrat'ev, T. N. Bondarenko, A. M. Rubinstein, K. M. Minachev, *Zeolites* **1985**, *5*, 320.
31. A. A. Slinkin, A. V. Kucherov, N. D. Chuvylkin, V. A. Korsunov, A. L. Kliachko, S. B. Nikishenko, *J. Chem. Soc., Faraday Trans.* **1989**, *85*, 3233.
32. J. P. Lange, Dissertation (PhD Thesis), Technische Universität, Berlin, 1986.
33. H. G. Karge, E. P. Boldingh, J.-P. Lange, A. Gutze, in *Proceedings, International Symposium on Zeolite Catalysis*, Siófok, Hungary, May 13–16, 1985, P. Fejes, D. Kalló, (Eds.), Petöfi Nyomda, Kecskemét, 1985, p. 639.
34. H. Ernst, D. Freude, T. Mildner, I. Wolf, in *Catalysis by Microporous Materials*, H. K. Beyer, H. G. Karge, I. Kiricsi, J. B. Nagy (Eds.), *Studies in Surface Science and Catalysis*, Vol. 94, Elsevier, Amsterdam, 1995, p. 413.
35. M. Hunger, T. Horvath, *J. Chem. Soc., Chem. Commun.* **1995**, 1423.
36. M. Hunger; T. Horvath, J. Weitkamp, in *Proceedings 11th International Zeolite Conference*, Scoul, Korea, August 12–17, 1996; *Progress in Zeolite and Microporous Materials*, H. Chon, S.-K. Ihm, Y. S. Uh (Eds.), *Studies in Surface Science and Catalysis*, Vol. 105, Part B, Elsevier, Amsterdam, 1997, p. 853.
37. H. G. Karge, G. Pál-Borbély, H. K. Beyer, *Zeolites* **1997**, *14*, 512.
38. U. Hatje, T. Ressler, S. Petersen, H. Förster, in *Proceedings European Symposium, Frontiers in Science and Technology with Synchrotron Radiation; J. de Physique IV, supplément au J. de Physique III*, 1994, p. 141.
39. U. Hatje, M. Hagelstein, T. Ressler, H. Förster, *Physica B* **1995**, *208/209*, 646.
40. H. Förster, U. Hatje, *Solid State Ionics* **1997**, *101–103*, 425.
41. H. G. Karge, in *Proceedings 11th International Zeolite Conference*, Scoul, Korea, August 12–17, 1996; *Progress in Zeolite and Microporous Materials*, H. Chon, S.-K. Ihm, Y. S. Uh (Eds.), *Studies in Surface Science and Catalysis*, Vol. 105, Part C, Elsevier, Amsterdam, 1997, p. 1901.
42. H. G. Karge, H. K. Beyer, in *Molecular Sieves – Science and Technology*. Vol. 3: *Post-Synthesis Modification I*, H. G. Karge, J. Weitkamp (Eds.), Chapter 2, Springer, Berlin, Heidelberg, 2002, p. 72.
43. J. Weitkamp, S. Ernst, M. Hunger, T. Röser, S. Huber, U. A. Schubert, P. Thomasson, H. Knözinger, in *Proceedings 11th International Congress on Catalysis*, J. W. Hightower, W. N. Delgass, E. Iglesia, A. T. Bell (Eds.), *Studies in Surface Science and Catalysis*, Vol. 101, Part B, Elsevier, Amsterdam, 1998, p. 731.
44. P. Conception-Heydorn, C. Jia, D. Herein, N. Pfänder, H. G. Karge, F. C. Jentoft, *J. Molec. Catal. A: Chemical* **2000**, *162*, 227.
45. Ch. Yang, Q.-H. Xu, *Huaxue Xuebao* **1997**, *55*, 562.
46. J. Xu, A. Yan, Q. Xu, *Wuli Huaxue Xuebao* **1998**, *14*, 422.
47. M. F. Ciarolo, J. C. Hanson, C. P. Grey, *Microporous Mesoporous Mater.* **2001**, *49*, 111.
48. H. G. Karge, in *Solid-State Reactions of Zeolites*, T. Hattori and T. Yashima (Eds.), *Studies in Surface Science and Catalysis*, Vol. 83, Kodansha, Tokyo, Elsevier, Amsterdam, 1994, p. 135.
49. J. Weitkamp, S. Ernst, T. Bock, A. Kiss, P. Kleinschmit, in *Catalysis by Microporous Materials*, H. K. Beyer, H. G. Karge, I. Kiricsi, J. B. Nagy (Eds.), *Studies in Surface Science and Catalysis*, Vol. 94, Elsevier, Amsterdam, 1995, p. 278.
50. V. P. Mavrodinova, *Microporous Mesoporous Mater.* **1998**, *24*, 1.
51. V. P. Mavrodinova, *Microporous Mesoporous Mater.* **1998**, *24*, 9.
52. G. Borbély, H. K. Beyer, L. Radics, P. Sandor, H. G. Karge, *Zeolites* **1989**, *9*, 428.
53. M. Jiang, S. Ye, S. Shan, Y. Fu, *Huaxue Xuebao* **1997**, *10*, 377.
54. M. Jiang, T. Tatsumi, *J. Phys. Chem. B* **1998**, *102*, 10879.
55. H. G. Karge, H. K. Beyer, in *Molecular Sieves – Science and Technology*, Vol. 3, *Post-Synthesis Modification I*, H. G. Karge, J. Weitkamp (Eds.), Chapter 2, Springer, Berlin Heidelberg, 2002, pp. 70, 83.
56. Y. Li, W. Xie, Y. Shen, *Shiyou Huagong* **1997**, *26*, 87.
57. H. G. Karge, E. Geidel, in *Molecular Sieves – Science and Technology*, Vol. 4, *Characterization I*, H. G. Karge, J. Weitkamp (Eds.), Springer, Berlin, Heidelberg, 2004, pp. 82, 131.
58. H. G. Karge, J. Ladebeck, Z. Sarbak, K. Hatada, *Zeolites* **1982**, *2*, 94.
59. H. G. Karge, K. Hatada, Y. Zhang, R. Fiedorow, *Zeolites* **1983**, *3*, 13.
60. A. Ebeling, A. G. Panov, D. E. McCready, M. L. Balmer, *Society of Automotive Engineers, SP* [Special Publication], 2000, SP-1639 (Non-Thermal Plasma Emission Control Systems), 111.
61. H. G. Karge, H. K. Beyer, in *C1-Chemie-Angewandte Heterogene Katalyse-C4-Chemie*, DGMK-Bericht 9101, ISBN No. 3-928164-07-4, DGMK, Hamburg, 1991, p. 191.
62. H. G. Karge, V. Mavrodinova, Z. Zheng, H. K. Beyer, *Appl. Catal.* **1991**, *75*, 343.

63. C. Jia, P. Beaunier, P. Massiani, *Microporous Mesoporous Mater.* **1998**, *24*, 69.
64. E. M. Pires, M. Wallau, U. Schuchardt, *J. Molec. Catal. A: Chemical* **1998**, *136*, 69.
65. M. Difallah, J. L. Ginoux, *J. Soc. Alger. Chim.* **1998**, *8*, 21.
66. M. Iwamoto, in *Zeolites and Related Microporous Materials: State of the Art 1994*, J. Weitkamp, H. G. Karge, H. Pfeifer, W. Hölderich (Eds.), Studies in Surface Science and Catalysis, Vol. 84, Part B, Elsevier, Amsterdam, 1994, p. 1395.
67. A. V. Kucherov, A. A. Slinkin, *Zeolites* **1987**, *7*, 583.
68. A. V. Kucherov, A. A. Slinkin, *J. Molec. Catal.* **1994**, *90*, 323.
69. A. V. Kucherov, A. A. Slinkin, *Zeolites* **1986**, *6*, 175.
70. A. V. Kucherov, A. A. Slinkin, D. A. Kondrat'ev, T. N. Bondarenko, A. M. Rubinstein, K. M. Minachev, *Zeolites* **1985**, *5*, 320.
71. M. Jiang, Y. Fu, *Wuli Huaxue Xuebao* **1997**, *13*, 822.
72. V. Y. Borovkov, M. Jiang, Y. Yilu, *J. Phys. Chem.* **1999**, *103*, 5010.
73. M. Hartmann, B. Boddenberg, in *Zeolites and Related Microporous Materials: State of the Art 1994*, J. Weitkamp, H. G. Karge, H. Pfeifer, W. Hölderich (Eds.), Studies in Surface Science and Catalysis, Vol. 84, Part A, Elsevier, Amsterdam, 1994, p. 509.
74. M. Hartmann, B. Boddenberg, *Microporous Mater.* **1994**, *2*, 127.
75. S. T. King, *J. Catal.* **1996**, *161*, 530.
76. S. A. Anderson, T. W. Root, *J. Catal.* **2003**, *217*, 396.
77. T. Liese, W. Grünert, *J. Catal.* **1997**, *172*, 34.
78. J. Varga, J. Halasz, I. Kiricsi, *Environ. Pollution* **1998**, *102*(Suppl. 1), 691.
79. J. Soria, A. Martínez-Arias, A. Martínez-Chaparro, J. C. Conesa, Z. Schay, *J. Catal.* **2000**, *190*, 352.
80. Z. Schay, L. Guczi, A. Beck, I. Nagy, V. Samuel, S. P. Mirajkar, A. V. Ramaswami, G. Pál-Borbély, *Catal. Today* **2002**, *75*, 393.
81. Y. Teraoka, H. Furukawa, I. Moriguchi, in *Science and Technology in Catalysis*, M. Anpo, M. Onaka, H. Yamashita, (Eds.), Studies in Surface Science and Catalysis, Vol. 145, Elsevier, Amsterdam, 2003, p. 231.
82. V. Umamaheswari, M. Hartmann, A. Poeppl, *J. Phys. Chem. B* **2005**, *109*, 19723.
83. K. Lázár, G. Pál-Borbély, H. K. Beyer, H. G. Karge, *J. Chem. Soc., Faraday Trans.* **1994**, *90*, 1329.
84. L. Čapek, V. Kreibich, J. Dědeček, T. Grygar, B. Wichterlová, Z. Sobalík, J. A. Martens, R. Brosius, V. Tokarová, *Microporous Mesoporous Mater.* **2005**, *80*, 279.
85. M. M. Mohamed, N. S. Gomaa, M. El-Moselhy, N. A. Eissa, *J. Colloid Interface Sci.* **2003**, *259*, 331.
86. D. Szalay, G. Horvath, J. Halasz, I. Kiricsi, Proceedings 5th International Symposium & Exhibition on Environmental Contamination in Central & Eastern Europe, Prague, Czech Republic, Sept. 12–14, 2000; Inst. Int. Cooperative Environmental Research, Florida State University, Tallahassee, Fla., 2000, p. 1989.
87. M. Rauscher, K. Kesore, R. Mönnig, W. Schwieger, A. Tissler, T. Turek, *Appl. Catal. A: General* **1999**, *184*, 249.
88. J. Varga, J. Halasz, D. Horvath, D. Mehn, J. B. Nagy, Gy. Schobel, I. Kiricsi, in *Preparation of Catalysts VII*, B. Delmon, P. A. Jacobs, R. Maggi, J. A. Martens, P. Grange, G. Poncelet (Eds.), Studies in Surface Science and Catalysis, Vol. 118, Elsevier, Amsterdam, 1998, p. 116.
89. M. Kögel, R. Mönnig, W. Schwieger, A. Tissler, T. Turek, *J. Catal.* **1999**, *182*, 470.
90. M. S. Batista, M. Wallau, E. A. Urquieta-Gonzalez, in *Recent Advances in the Science and Technology of Zeolites and Related Materials*, E. van Steen, L. H. Callanan, M. Claeys (Eds.), Studies in Surface Science and Catalysis, Vol. 154, Elsevier, Amsterdam, 2005, p. 2493.
91. M. S. Batista, E. A. Urquieta-Gonzalez, in *Impact of Zeolites and Other Porous Materials on the New Technologies at the Beginning of the New Millennium*, R. Aiello, F. Testa, G. Giordano (Eds.), Studies in Surface Science and Catalysis, Vol. 142, Elsevier, Amsterdam, 2002, p. 983.
92. M. A. Zanjanchi, A. Ebrahimian, *J. Molecular Structure* **2004**, *693*, 211.
93. G. Kinger, A. Lugstein, R. Swagera, M. Ebel, A. Jentys, H. Vinek, *Microporous Mesoporous Mater.* **2000**, *39*, 307.
94. I. M. Astrelin, T. Enhbold, M. Sychev, in Proceedings 12th International Zeolite Conference, Baltimore, Maryland, USA, July 5–10, 1998, Vol. 3, R. von Ballmoos, J. B. Higgins, M. M. J. Treacy (Eds.), Materials Research Soc., Warrendale, Pa., 1999, p. 2129.
95. B. Wichterlová, S. Beran, S. Bednárová, K. Nedomová, L. Dudíková, P. Jíru, in *Innovation in Zeolite Materials Science*, P. J. Grobet, W. J. Mortier, E. F. Vansant, G. Schulz-Ekloff (Eds.), Studies in Surface Science and Catalysis, Vol. 37, Elsevier, Amsterdam, 1988, p. 199.
96. T. Bock, Dissertation (PhD Thesis), University of Stuttgart, 1995.
97. J. Weitkamp, S. Ernst, T. Bock, A. Kiss, P. Kleinschmit, in *Catalysis by Microporous Materials*, H. K. Beyer, H. G. Karge, I. Kiricsi, J. B. Nagy (Eds.), Studies in Surface Science and Catalysis, Vol. 94, Elsevier, Amsterdam, 1995, p. 278.
98. J. Weitkamp, S. Ernst, T. Bock, T. Kromminga, A. Kiss, P. Kleinschmit, US Patent No 5,529,964, assigned to Degussa AG, 1996.
99. H. G. Karge, Y. Zhang, H. K. Beyer, *Catal. Lett.* **1992**, *12*, 147.
100. H. G. Karge, Y. Zhang, H. K. Beyer, in Proceedings 10th International Congress on Catalysis, L. Guczi, F. Solymosi, P. Tétényi, (Eds.), Akadémai Kladó, Budapest, Elsevier, Amsterdam, 1993, p. 257.
101. C. M. Mkombe, M. E. Dry, C. T. O'Connor, *Zeolites* **1997**, *19*, 175.
102. L. Schlegel, H. Miessner, D. Gutschick, *Catal. Lett.* **1994**, *23*, 215.
103. T. Wasowicz, L. Kevan, in *Modern Applications of EPR/ESR: From Biophysics to Materials Science; Proceedings 1st Asia Pacific EPR/ESR Symposium*, C. Z. Rudowict, P. K. N. Yu, H. Hiraoka (Eds.), Kowloon, Singapore, January 20–24, 1997, Springer-Verlag, Singapore, 1998, p. 279.
104. H. G. Karge, B. Wichterlová, H. K. Beyer, *J. Chem. Soc., Faraday Trans.* **1992**, *88*, 1345.
105. H. G. Karge, V. Mavrodinova, Z. Zheng, H. K. Beyer, in *Guidelines for Mastering the Properties of Molecular Sieves – Relationship between the Physico-chemical Properties of Zeolitic Systems and their Low Dimensionality*, D. Barthomeuf, E. G. Derouane, W. Hölderich (Eds.), Plenum Press, New York, p. 157; *NATO ASI Series B: Physics* **1990**, *221*, 157.
106. H. G. Karge, H. K. Beyer, in *Molecular Sieves – Science and Technology*. Vol. 3: *Post-Synthesis Modification I*, H. G. Karge, J. Weitkamp (Eds.), Chapter 2, Springer, Berlin, Heidelberg, 2002, p. 108.
107. B. Sulikowski, J. Find, H. G. Karge, D. Herein, *Zeolites* **1997**, *19*, 395.
108. H. G. Karge, M. Hunger, H. K. Beyer, in *Catalysis and Zeolites – Fundamentals and Applications*, J. Weitkamp, L. Puppe (Eds.), Springer, Berlin, Heidelberg, 1999, Chapter 4, p. 210.
109. M. Jiang, H. G. Karge, *J. Chem. Soc., Faraday Trans.* **1995**, *91*, 1845.

110. H. G. Karge, in *Proceedings 2nd Polish–German Zeolite Colloquium*, M. Rozwadowski (Ed.), Nicolaus Copernicus, University Press, Torún, 1995, p. 120.
111. H. Knözinger, E. Taglauer, in *Preparation of Solid Catalysts*, G. Ertl, H. Knözinger, J. Weitkamp (Eds.), Wiley-VCH, Weinheim, Germany, 1999, pp. 501–526.
112. Ch.-B. Wang, Y. Xie, Y. Tang, in *Proceedings 12th International Zeolite Conference*, Baltimore, Maryland, USA, July 5–10, 1998, R. von Ballmoos, J. B. Higgins, M. M. J. Treacy (Eds.), Materials Research Soc., Warrendale, Pa., 1999, Vol. 1, p. 175.
113. D. Yin, D. Yin, *Microporous Mesoporous Mater.* **1998**, *24*, 123.
114. H. Beyer, *Microporous Mesoporous Mater.* **1999**, *31*, 333.
115. E. Schneider, A. Hagen, J.-D. Grunwaldt, F. Roessner, in *Recent Advances in the Science and Technology of Zeolites and Related Materials*, E. van Steen, L. H. Callanan, M. Claeys (Eds.), *Studies in Surface Science and Catalysis*, Vol. 154, Elsevier, Amsterdam, 2005, p. 2192.
116. S. Altwasser, A. Raichle, Y. Traa, J. Weitkamp, *Chem. Ing. Tech.* **2004**, *76*, 140.
117. I. N. Stranski, *Z. Phys. Chem. (A)* **1928**, *136*, 259.
118. G. T. Kokotailo, S. L. Lawton, S. Sawruk, in *Molecular Sieves II, Proceedings 4th International Conference on Molecular Sieves*, J. Katzer (Ed.), *ACS Symposium Series*, Vol. 40, American Chemical Society, Washington, D.C., 1977, p. 439.
119. C. A. Fyfe, G. T. Kokotailo, J. D. Graham, C. Browning, G. C. Gobbi, M. Hyland, G. J. Kennedy, C. T. DeSchutter, *J. Am. Chem. Soc.* **1986**, *108*, 522.
120. Y. Huang, R. M. Paroli, A. H. Delgado, T. A. Richardson, *Spectrochim. Acta* **1998**, *54A*, 1347.
121. J. Koy, H. G. Karge, in *Molecular Sieves – Science and Technology*, Vol. 3, *Post-Synthesis Modification I*, H. G. Karge, J. Weitkamp (Eds.), Chapter 2, Springer, Berlin, Heidelberg, 2002, p. 164.
122. J. Fraissard, A. Gedeon, Q. Chen, T. Ito, in *Zeolite Chemistry and Catalysis*, P. A. Jacobs, N. I. Jaeger, L. Kubelková, B. Wichterlová (Eds.), *Studies in Surface Science and Catalysis*, Vol. 69, Elsevier, Amsterdam, 1991, p. 461.
123. D. J. Ostgard, L. Kustov, K. R. Poeppelmeier, W. M. H. Sachtler, *J. Catal.* **1992**, *133*, 342.
124. K. R. Poeppelmeier, T. D. Towbridge, J.-L. Kao, US Patent 4,568,656, 1986.
125. M. Che, O. Clause, Ch. Marcilly, in *Handbook of Heterogeneous Catalysis*, G. Ertl, H. Knözinger, J. Weitkamp (Eds.), VCH, Weinheim, Germany, 1997, p. 191.
126. A. V. Kucherov, A. A. Slinkin, S. S. Goryashenko, K. I. Slovetskaja, *J. Catal.* **1989**, *118*, 459.
127. A. V. Kucherov, T. N. Kucherova, A. A. Slinkin, *Catal. Lett.* **1991**, *10*, 289.
128. P. Gallezot, in *Molecular Sieves – Science and Technology*, Vol. 3, *Post-Synthesis Modification I*, H. G. Karge, J. Weitkamp (Eds.), Chapter 4, Springer, Berlin, Heidelberg, 2002, p. 262.
129. El-M. El-Malki, R. A. van Santen, W. M. H. Sachtler, *J. Phys. Chem B* **1999**, *103*, 4611.
130. M. Guisnet, N. S. Gnep, H. Vasques, F. R. Ribeiro, in *Zeolite Chemistry and Catalysis*, P. A. Jacobs, N. I. Jaeger, L. Kubelková, B. Wichterlová (Eds.), *Studies in Surface Science and Catalysis*, Vol. 69, Elsevier, Amsterdam, 1991, p. 321.
131. J. Liang, W. Tang, M.-L. Ying, S.-Q. Xhao, B.-Q. Zu, H.-Y. Li, in *Zeolite Chemistry and Catalysis*, P. A. Jacobs, N. I. Jaeger, L. Kubelková, B. Wichterlová (Eds.), *Studies in Surface Science and Catalysis*, Vol. 69, Elsevier, Amsterdam, 1991, p. 207.
132. B. I. Whittington, J. R. Anderson, *J. Phys. Chem.* **1991**, *95*, 3306.
133. B. I. Whittington, J. R. Anderson, *J. Phys. Chem.* **1993**, *97*, 1032.
134. A. Seidel, F. Rittner, B. Boddenberg, *J. Chem. Soc., Faraday Trans.* **1996**, *92*, 493.
135. A. Seidel, B. Boddenberg, *Chem. Phys. Lett.* **1996**, *249*, 117.
136. F. Rittner, A. Seidel, B. Boddenberg, *Microporous Mesoporous Mater.* **1997**, *24*, 127.
137. T. H. Sprang, A. Seidel, M. Wark, F. Rittner, B. Boddenberg, *J. Mater. Chem.* **1997**, *7*, 1429.
138. C. Yuan, J. Yao, K. Chen, M. Huang, *Shiyou Huagong (Petrochemical Technology)* **1992**, *21*, 724.
139. Q. Wang, D. Sayers, M. Huang, C. Yuan, S. Wei, *Am. Chem. Soc., Div. Fuel Chem.* **1992**, *37*, 283 (Preprint ISSN: 0569–3772).
140. G. A. Ozin, C. Gil, *Chem. Rev.* **1989**, *89*, 1749.
141. A. V. Kucherov, T. N. Kucherova, A. A. Slinkin, *Microporous Mesoporous Mater.* **1998**, *26*, 1.
142. H. K. Beyer, P. A. Jacobs, in *Molecular Sieves II, Proceedings 4th International Conference on Molecular Sieves*, J. Katzer (Ed.), *ACS Symposium Series*, Vol. 40, American Chemical Society, Washington, D.C., 1977, p. 493.
143. H. K. Beyer, P. A. Jacobs, J. B. Uytterhoeven, *J. Chem. Soc., Faraday Trans. I* **1976**, *72*, 674.
144. A. V. Kucherov, K. I. Slovetskaya, S. S. Goryaschenko, E. G. Aleshin, A. A. Slinkin, *Microporous Mater.* **1996**, *7*, 27.
145. H. G. Karge, H. K. Beyer, in *Molecular Sieves – Science and Technology*, Vol. 3, *Post-Synthesis Modification I*, H. G. Karge, J. Weitkamp (Eds.), Chapter 2, Springer, Berlin, Heidelberg, 2002, pp. 171–181.
146. P. A. Jacobs, W. De Wilde, R. A. Schoonheydt, J. B. Uytterhoeven, H. K. Beyer, *J. Chem. Soc., Faraday Trans.* **1976**, *I 72*, 1221.
147. P. A. Jacobs, J. B. Uytterhoeven, H. K. Beyer, *J. Chem. Soc., Faraday Trans. I* **1979**, *75*, 56.
148. V. Kanazirev, G. L. Price, in *Zeolites and Related Microporous Materials: State of the Art 1994*, J. Weitkamp, H. G. Karge, H. Pfeifer, W. Hölderich (Éds.), *Studies in Surface Science and Catalysis*, Vol. 84, Part C, Elsevier, Amsterdam, 1994, p. 1935.
149. V. Kanazirev, G. L. Price, K. M. Dooley, in *Zeolite Chemistry and Catalysis*, P. A. Jacobs, N. J. Jaeger, L. Kubelková, B. Wichterlová (Eds.), *Studies in Surface Science and Catalysis*, Vol. 69, Elsevier, Amsterdam, 1991, p. 277.
150. V. Kanazirev, Y. Neinska, T. Tsoncheva, L. Kosova, in *Proceedings 9th International Zeolite Conference*, R. von Ballmoos, J. B. Higgins, M. M. J. Treacy (Eds.), Butterworth-Heinemann, Boston, 1993, Vol. I, p. 461.
151. Y. Neinska, R. M. Mihályi, R. Magdolna, V. Mavrodinova, Ch. Minchev, H. K. Beyer, *Phys. Chem. Chem. Phys.* **1999**, *1*, 5761.
152. Y. Neinska, V. Mavrodinova, C. Minchev, R. M. Mihály, in *Porous Materials in Environmentally Friendly Processes*, I. Kiricsi, G. Pál-Borbély, J. B. Nagy, H. G. Karge (Eds.), *Studies in Surface Science and Catalysis*, Vol. 125, Amsterdam, 1999, p. 37.
153. G. L. Price, in *227th ACS National Meeting* (Abstracts of Papers), Anaheim, California, USA, March 28–April 1, 2004, American Chemical Society, Washington, D.C., 2004, COLL-172.
154. R. M. Mihályi, H. K. Beyer, V. Mavrodinova, Ch. Minchev, Y. Neinska, *Microporous Mesoporous Mater.* **1998**, *24*, 143.
155. R. M. Mihályi, H. K. Beyer, M. Keindl, in *Zeolites and Mesoporous Materials at the Dawn of the 21st Century*, A. Galarneau, F. Di Renzo, F. Fajula, J. Vedrine (Eds.), *Studies in Surface Science and Catalysis*, Vol. 135, Elsevier, Amsterdam, 2001, p. 1561.

156. J. Fuchsova, R. Bulanek, K. Novoveska, D. Cermak, V. Lochar, *Scientific Papers of the University of Pardubice*, Pardubice, Czech Rep., Series A: Faculty of Chemical Technology 2002, Vol. 2003, 8, 137.
157. M. Ogura, T. Ohsaki, E. Kikuchi, *Microporous Mesoporous Mater.* **1998**, *21*, 533.
158. C. Schmidt, T. Sowade, F.-W. Schuetze, M. Richter, H. Berndt, W. Gruenert, in *Zeolites and Mesoporous Materials at the Dawn of the 21st Century*, A. Galarneau, F. Di Renzo, F. Fajula, J. Vedrine (Eds.), *Studies in Surface Science and Catalysis*, Vol. 135, Elsevier, Amsterdam, 2001, p. 4973.
159. C. Schmidt, T. Sowade, E. Loeffler, A. Birkner, W. Gruenert, *J. Phys. Chem. B* **2002**, *106*, 4085.
160. H. K. Beyer, R. M. Mihályi, Ch. Minchev, Y. Neinska, V. Kanazirev, *Microporous Mesoporous Mater.* **1996**, *7*, 333.
161. R. M. Mihályi, G. Pál-Borbély, H. K. Beyer, Ch. Minchev, Y. Neinska, H. G. Karge, *React. Kinet. Catal. Lett.* **1997**, *60*, 195.
162. R. M. Mihályi, H. K. Beyer, in *Proceedings 3rd Polish-German Zeolite Colloquium*, M. Rozwadowski (Ed.), Torún, Poland, April 3–5, 1997, Nicholas Copernicus University Press, Torún, 1998, p. 309.
163. Y. Neinska, R. M. Mihályi, V. Mavrodinova, Ch. Minchev, H. K. Beyer, *Phys. Chem. Chem. Phys.* **1999**, *1*, 5761.
164. J. Halasz, Z. Konya, A. Fudala, I. Kiricsi, *Catal. Today* **1996**, *31*, 293.
165. G. L. Price, V. Kanazirev, K. M. Dooley, V. I. Hart, *J. Catal.* **1998**, *173*, 17.
166. V. I. Kanazirev, G. L. Price, K. M. Dooley, *J. Catal.* **1994**, *148*, 164.
167. Y. Neinska, Ch. Minchev, R. Dimitrova, N. Micheva, V. Minkov, V. Kanazirev, in *Zeolites and Related Microporous Materials: State of the Art 1994*, J. Weitkamp, H. G. Karge, H. Pfeifer, W. Hölderich (Eds.), *Studies in Surface Science and Catalysis*, Vol. 84, Part B, Elsevier, Amsterdam, 1994, p. 989.
168. Y. Neinska, Ch. Minchev, L. Kozova, V. Kanazirev, in *Catalysis by Microporous Materials, Proceedings ZEOCAT '95*, H. K. Beyer, H. G. Karge, I. Kiricsi, J. B. Nagy (Eds.), *Studies in Surface Science and Catalysis*, Vol. 94, Elsevier, Amsterdam, 1995, p. 262.
169. Y. G. Neinska, V. P. Mavrodinova, V. I. Kanazirev, C. I. Minchev, *Bulg. Chem. Commun.* **1998**, *30*, 357.
170. Ch. Jia, P. Massiani, P. Beaunier, D. Barthomeuf, *Appl. Catal. A: General* **1993**, *106*, L185.
171. P. R. Hari Prasad Rao, P. Massiani, D. Barthomeuf, in *Zeolites and Related Microporous Materials: State of the Art 1994*, J. Weitkamp, H. G. Karge, H. Pfeifer, W. Hölderich (Eds.), *Studies in Surface Science and Catalysis*, Vol. 84, Part B, Elsevier, Amsterdam, 1994, p. 1449.
172. E. Esemann, H. Förster, *J. Chem. Soc., Chem. Commun.* **1994**, 1319.
173. E. Esemann, H. Förster, *Z. Phys. Chem.* **1995**, *189*, 263.
174. E. Esemann, H. Förster, E. Geidel, K. Krause, *Microporous Mesoporous Mater.* **1996**, *6*, 321.
175. C. Yang, Q.-H. Xu, *Chem. J. Chinese Univ.* **1996**, *17*, 1336.
176. C. Yang, Q.-H. Xu, *Gaodeng Xuexiao Huaxue Xuebao* **1996**, *17*, 1336.
177. D. B. Akolekar, S. Bhargava, *J. Molec. Catal. A: Chemical* **1997**, *122*, 81.
178. M. Ogawa, T. Handa, K. Kuroda, C. Kato, *Chem. Letters* **1990**, 71.
179. S. Narayanan, A. Sultana, *Appl. Catal. A: General* **1998**, *167*, 103.
180. W. E. J. van Kooten, B. Liang, H. C. Krijsnen, O. L. Oudshoorn, H. P. A. Calis, C. M. van den Bleck, *Appl. Catal. B: Environmental* **1999**, *21*, 203.
181. E. J. Nassar, O. A. Serra, E. F. Souza-Aguiar, *Quim. Nova.* **1998**, *21*, 121.
182. R. G. Herman, J. H. Lunsford, H. Beyer, P. A. Jacobs, J. B. Uytterhoeven, *J. Phys. Chem.* **1975**, *79*, 2388.
183. R. M. Haniffa, K. Seff, *Microporous Mesoporous Mater.* **1998**, *25*, 137.
184. E. Esemann, H. Förster, *J. Molec. Struct.* **1999**, *483*, 7.
185. A. Auroux, A. Gervasini, C. Guimon, *J. Phys. Chem. B* **1999**, *103*, 7195.
186. J. Varga, A. Fudala, J. Halász, G. Schöbel, I. Kiricsi, in *Catalysis by Microporous Materials; Proceedings ZEOCAT '95*, H. K. Beyer, H. G. Karge, I. Kiricsi, J. B. Nagy (Eds.), *Studies in Surface Science and Catalysis*, Vol. 94, Elsevier, Amsterdam, 1995, p. 665.
187. V. I. Kanazirev, G. L. Price, *J. Catal.* **1994**, *148*, 164.
188. J. Varga, J. Halasz, I. Kiricsi, in *Proceedings 1st International Nitrogen Conference*, K. W. van der Hoek (Ed.), Elsevier, Oxford, UK, 1998, p. 691.
189. J. Varga, J. B. Nagy, J. Halasz, I. Kiricsi, *J. Molec. Struct.* **1997**, *410*, 149.
190. Z. Schay, H. Knözinger, L. Guczi, G. Pál-Borbély, *Appl. Catal. B: Environmental* **1998**, *18*, 263.
191. I. Halasz, G. Pál-Borbély, H. K. Beyer, *React. Kinet. Catal. Lett.* **1997**, *61*, 27.
192. J. Varga, A. Fudala, J. Halasz, I. Kiricsi, *J. Therm. Anal.* **1996**, *47*, 391.
193. I. Halasz, A. Brenner, *Catal. Lett.* **1998**, *51*, 23.
194. J. Halasz, J. Varga, G. Schobel, I. Kiricsi, K. Hernadi, I. Hannus, K. Varga, P. Fejes, in *Proceedings 3rd International Symposium on Catalysis Automotive Pollution Control*, Brussels, Belgium, April 20–22, 1994, Elsevier, Amsterdam, 1995, p. 675.
195. C. Setzer, D. Demuth, F. Schüth, *Chem. Ing. Tech.* **1997**, *69*, 79.
196. A. Poeppel, M. Newhouse, L. Kevan, *J. Phys. Chem.* **1995**, *99*, 10019.
197. S. T. King, *Catal. Today* **1997**, *33*, 173.
198. V. Y. Borovkov, M. Jiang, Y. Fu, *J. Phys. Chem. B* **1999**, *103*, 5010.
199. Q. Shi, C. Li, *Cuihua Xuebao* **2000**, *21*, 113.
200. C. Li, D. Li, Q. Shi, Q. Zhu, *Fenzi Cuihua* **1999**, *13*, 115.
201. A. V. Kucherov, A. A. Slinkin, H. K. Beyer, G. Borbély, *Kinet. Katal.* **1989**, *30*, 429. (Engl. Translat. *30*, 367.)
202. A. V. Kucherov, A. A. Slinkin, H. K. Beyer, G. Borbély, *J. Chem. Soc., Faraday Trans. I* **1989**, *85*, 2737.
203. V. I. Kanazirev, G. L. Price, *J. Molec. Catal. A: Chemical* **1995**, *96*, 145.
204. G. L. Price, V. I. Kanazirev, D. F. Church, *J. Phys. Chem.* **1995**, *99*, 864.
205. I. Jirka, B. Wichterlová, M. Maryska, in *Zeolite Chemistry and Catalysis*, P. A. Jacobs, N. I. Jaeger, L. Kubelková, B. Wichterlová (Eds.), *Studies in Surface Science and Catalysis*, Vol. 69, Elsevier, Amsterdam, 1991, p. 269.
206. A. V. Kucherov, A. A. Slinkin, *Kinet. Katal.* **1986**, *27*, 909. (Engl Translat. *27*, 789.)
207. T. M. Salama, T. Shido, R. Ohnishi, M. Ichikawa, *J. Phys. Chem.* **1996**, *100*, 3688.
208. H. K. Beyer, G. Pál-Borbély, M. Keindl, *Microporous Mesoporous Mater.* **1999**, *31*, 333.
209. B. Boddenberg, A. Seidel, *J. Chem. Soc., Faraday Trans.* **1994**, *90*, 1345.

210. G. Onyestyák, D. Kalló, J. Pap, Jr., in *Zeolite Chemistry and Catalysis*, P. A. Jacobs, N. I. Jaeger, L. Kubelková, B. Wichterlová (Eds.), *Studies in Surface Science and Catalysis*, Vol. 69, Elsevier, Amsterdam, 1991, p. 287.
211. D. Yin, D. Yin, Q. Li, Z. Fu, *Cuihua Xuebo* **2000**, *21*, 113.
212. A. Seidel, G. Kampf, A. Schmidt, B. Boddenberg, *Catal. Lett.* **1998**, *51*, 713.
213. D. Yin, D. Yin, Z. Fu, Q. Li, *Cuihua Xuebao* **1999**, *20*, 419.
214. E. Rojasova, A. Smieskova, P. Hudec, Z. Zidek, *Collect. Czech Chem. Commun.* **1999**, *64*, 168.
215. R. Salzer, in *Proceedings SPIE – Int. Opt. Eng. 1991; 8th International Fourier Transform Spectroscopy*, 1992, p. 50 (ISSN 0277-786X).
216. R. Salzer, U. Finster, F. Roessner, K.-H. Steinberg, P. Klaeboe, *Analyst* **1992**, *117*, 351.
217. F. Roessner, A. Hagen, U. Mroczek, H. G. Karge, K.-H. Steinberg, in *Proceedings International Congress on Catalysis*, L. Guczi, F. Solymosi, P. Tétényi (Eds.), Budapest, Hungary, July 19–24, 1992, Elsevier, Amsterdam, 1993, p. 1707.
218. A. Hagen, F. Roessner, in *Zeolites and Microporous Crystals*, T. Hattori, T. Yashima (Eds.), *Studies in Surface Science and Catalysis*, Vol. 83, Elsevier, Amsterdam, 1994, p. 313.
219. N. Azuma, M. Hartmann, L. Kevan, *J. Phys. Chem.* **1995**, *99*, 6670.
220. K. Lázár, G. Pál-Borbély, H. K. Beyer, H. G. Karge, in *Preparation of Catalysts VI, Scientific Bases for the Preparation of Heterogeneous Catalysts*, G. Poncelet, J. Martens, B. Delmon, P. A. Jacobs, P. Grange (Eds.), *Studies in Surface Science and Catalysis*, Vol. 91, Elsevier, Amsterdam, 1995, p. 51.
221. I. O. Liu, N. W. Nant, M. Kögel, T. Turek, *Catal. Lett.* **1999**, *63*, 241.
222. R. Giles, N. W. Cant, M. Kögel, T. Turek, D. L. Trimm, *Appl. Catal. B: Environmental* **2000**, *25*, L75.
223. M. Kögel, V. H. Sandoval, W. Schwieger, A. Tissler, T. Turek, *Catal. Lett.* **1998**, *51*, 23.
224. L. J. Lobree, I.-C. Hwang, J. A. Reimer, A. T. Bell, *J. Catal.* **1999**, *186*, 242.
225. M. Crocker, R. H. M. Herold, C. A. Emeis, M. Krijger, *Catal. Lett.* **1992**, *15*, 339.
226. A. Jentys, A. Lugstein, H. Vinek, *J. Chem. Soc., Faraday Trans.* **1997**, *93*, 4091.
227. A. Jentys, A. Lugstein, H. Vinek, *Zeolites* **1997**, *18*, 391.
228. El-M. El-Malki, D. Werst, P. E. Doan, W. M. H. Sachtler, *J. Phys. Chem. B* **2000**, *104*, 5924.
229. T. I. Korányi, N. H. Pham, A. Jentys, H. Vinek, in *Hydrotreatment and Hydrocracking of Oil Fractions, Proceedings 1st International Symposium/6th European Workshop*, G. F. Froment, B. Delmon, P. Grange (Eds.), *Studies in Surface Science and Catalysis*, Vol. 106, Elsevier, Amsterdam, 1997, p. 509.
230. A. Lugstein, A. Jentys, H. Vinek, *Appl. Catal. A: General* **1998**, *166*, 29.
231. A. Jentys, A. Lugstein, H. Vinek, *Acta Phys. Polon. A* **1997**, *91*, 969.
232. A. Jentys, A. Lugstein, O. E. Dusouqui, H. Vinek, M. Englisch, J. A. Lercher, in *Catalysts in Petroleum Refining and Petrochemical Industries 1995*, M. Absi-Halabi, J. Beshara, H. Qabazard, A. Stanislaus (Eds.), *Studies in Surface Science and Catalysis*, Vol. 100, Elsevier, Amsterdam, 1996, p. 525.
233. X. Wang, H.-Y. Chen, W. M. H. Sachtler, *Appl. Catal. B: Environmental* **2000**, *26*, L227.
234. Y.-K. Park, S. S. Goryashenko, D. S. Kim, S.-E. Park, in *Proceedings 12th International Zeolite Conference*, Baltimore, Maryland, USA, July 5–10, 1998, R. von Ballmoos, J. B. Higgins, M. M. J. Treacy (Eds.), Materials Research Soc., Warrendale, Pa., 1999, Vol. 2, p. 1157.
235. T. Enhbold, M. Sychev, I. M. Astrelin, M. Rozwadowski, R. Golembiewski, in *Proceedings 3rd Polish-German Zeolite Colloquium*, M. Rozwadowski (Ed.), Torún, Poland, April 3–5, 1997, Nicholas Copernicus University Press, Torún, 1998, p. 137.
236. Y. Li, J. N. Armor, *Appl. Catal. A: General* **1999**, *188*, 211.
237. R. M. Haniffa, K. Seff, *J. Phys. Chem.* **1998**, *102*, 2688.
238. B. Wichterlová, S. Beran, L. Kubelková, J. Nováková, A. Smiesková, R. Sebík, in *Zeolite as Catalysts, Sorbents and Detergent Builders – Applications and Innovations*, H. G. Karge, J. Weitkamp (Eds.), *Studies in Surface Science and Catalysis*, Vol. 46, Elsevier, Amsterdam, 1989, p. 347.
239. M. Hartmann, N. Azuma, L. Kevan, *J. Phys. Chem.* **1995**, *99*, 10988.
240. N. Azuma, Ch. W. Lee, M. Zamadics, L. Kevan, in *Zeolites and Related Microporous Materials: State of the Art 1994*, J. Weitkamp, H. G. Karge, H. Pfeifer, W. Hölderich (Eds.), *Studies in Surface Science and Catalysis*, Vol. 84, Part A, Elsevier, Amsterdam, 1994, p. 805.
241. M. Hartmann, N. Azuma, L. Kevan, in *Zeolites: A Refined Tool for Designing Catalytic Sites*, L. Bonneviot, S. Kaliaguine (Eds.), *Studies in Surface Science and Catalysis*, Vol. 97, Elsevier, Amsterdam, 1995, p. 335.
242. M. Hartman, L. Kevan, in *Progress in Zeolite and Microporous Materials, Proceedings 11th International Zeolite Conference*, Seoul, Korea, August 12–17, 1996, H. Chon, S. K. Ihm, Y. S. Uh (Eds.), *Studies in Surface Science and Catalysis*, Vol. 105, Elsevier, Amsterdam, 1997, p. 717.
243. M.-A. Djieugoue, A. M. Prakash, L. Kevan, *J. Phys. Chem. B* **1999**, *103*, 804.
244. M. Hartmann, A. Pöppl, L. Kevan, *J. Phys. Chem.* **1995**, *99*, 17494.
245. A. V. Kucherov, A. A. Slinkin, *Kinet. Katal.* **1986**, *27*, 678. (Engl. Translat. *27*, 585.)
246. M. Huang, S. Shan, C. Yuan, Y. Li, Q. Wang, *Zeolites* **1990**, *10*, 772.
247. S. Shan, S. Shen, K. Chen, M. Huang, *Acta Physico-Chimica Sinica; (Wuli Haxue Xuebao)* **1992**, *8*, 338.
248. S. Shen, S. Shan, K. Chen, M. Huang, *Cuihua Xuebao (Chin. J. Catal.)* **1993**, *14*, 455.
249. S. Narayanan, K. Deshpande, *Appl. Catal. A: General* **2000**, *199*, 1.
250. C. Marchal, J. Thoret, M. Gruia, C. Dorémieux-Morin, J. Fraissard, in *Symposium on Advances in FCC Technology II*, ACS, Catalysis Division (Ed.), Washington, USA, August 26–31, 1990, American Chemical Society, Washington D.C., 1990, p. 762.
251. C. Marchal, J. Thoret, M. G. Gruia, C. Dorémieux-Morin, J. Fraissard, in *Fluid Catalytic Cracking II; Concepts in Catalyst Design*, M. L. Ocelli (Ed.), *ACS Symposium Series*, Vol. 452, American Chemical Society, Washington D.C., 1991, p. 212.
252. J. Thoret, C. Marchal, C. Dorémieux-Morin, P. P. Man, M. Gruia, J. Fraissard, B. I. Whittington, *Zeolites* **1993**, *13*, 269.
253. M. Ignatovych, A. Gomenyuk, V. Ogenko, O. Chuiko, in *Porous Materials in Environmentally Friendly Processes; Proceedings 1st International FEZA Conference*, I. Kiricsi, G. Pál-Borbély, J. B. Nagy, H. G. Karge (Eds.), *Studies in Surface Science and Catalysis*, Vol. 125, Elsevier, Amsterdam, 1999, p. 261.
254. M. Ziłtek, I. Novak, H. G. Karge, in *Catalysis by Microporous Materials; Proceedings ZEOCAT '95*, H. K. Beyer, H. G.

Karge, I. Kiricsi, J. B. Nagy (Eds.), *Studies in Surface Science and Catalysis*, Vol. 94, Elsevier, Amsterdam, 1995, p. 270.
255. J. Thoret, P. P. Man, J. Fraissard, *Zeolites* **1997**, *18*, 152.
256. B. M. Weckhuysen, R. A. Schoonheydt, in *Zeolites and Related Microporous Materials: State of the Art 1994*, J. Weitkamp, H. G. Karge, H. Pfeifer, W. Hölderich (Eds.), *Studies in Surface Science and Catalysis*, Vol. 84, Part B, Elsevier, Amsterdam, 1994, p. 965.
257. A. V. Kucherov, A. A. Slinkin, *Kinet. Katal.* **1987**, *28*, 1199.
258. A. A. Slinkin, A. V. Kucherov, S. S. Goryashenko, E. G. Aleshin, K. I. Slovetskaya, *Kinet. Katal.* **1989**, *30*, 184. (*Engl. Translat. 30*, 158.)
259. A. A. Slinkin, A. V. Kucherov, S. S. Gorjachenko, E. G. Aleshin, K. I. Slovetskaja, *Zeolites* **1990**, *10*, 111.
260. Z. Zhu, T. Wasowicz, L. Kevan, *J. Phys. Chem. B* **1997**, *101*, 10763.
261. P.-S. E. Dai, J. H. Lunsford, *J. Catal.* **1980**, *64*, 173.
262. I. M. Harris, J. Dwyer, A. A. Garforth, C. H. McAteer, W. J. Ball, in *Zeolites as Catalysts, Sorbents and Detergent Builders–Applications and Innovations*, H. G. Karge, J. Weitkamp (Eds.), *Studies in Surface Science and Catalysis*, Vol. 46, Elsevier, Amsterdam, 1989, p. 271.
263. R. W. Borry, Y.-H. Kim, A. Huffsmith, J. A. Reimer, E. Iglesia, *J. Phys. Chem. B* **1999**, *103*, 5787.
264. W. Li, R. W. Borry, G. D. Meitzner, E. Iglesia, *J. Catal.* **2000**, *191*, 373.
265. Y. Zhang, Dissertation (PhD Thesis), Technical University of Berlin, 1992.
266. A. König, H. G. Karge, T. Richter, *Offenlegungsschrift DE*, 196 37 032 A1 6 : 25, assigned to Volkswagen AG, 1996.
267. E. Kikuchi, M. Ogura, I. Terasaki, Y. Goto, *J. Catal.* **1996**, *161*, 465.
268. M. Ogura, N. Aratani, E. Kikuchi, in *Progress in Zeolite and Microporous Materials, Proceedings 11th International Zeolite Conference*, Seoul, Korea, August 12–17, 1996, H. Chon, S.-K. Ihm, Y. S. Uh (Eds.), *Studies in Surface Science and Catalysis*, Vol. 105, Part B, Elsevier, Amsterdam, 1997, p. 1593.
269. R. M. Mihályi, H. K. Beyer, Y. Neinska, V. Mavrodinova, Ch. Minchev, *React. Kinet. Catalysis Letters* **1999**, *68*, 355.
270. R. M. Mihály, H. K. Beyer, *Chem. Commun.* **2001**, 2242.
271. H. G. Karge, H. K. Beyer, in *Zeolite Chemistry and Catalysis*, P. A. Jacobs, N. I. Jaeger, L. Kubelkova, B. Wichterlová (Eds.), *Studies in Surface Science and Catalysis*, Vol. 69, Elsevier, Amsterdam, 1991, p. 43.
272. E. Geidel, *Habilitation Thesis (Habilitationsschrift)*, University of Hamburg, 1997.
273. H. G. Karge, in *Zeolites and Microporous Crystals*, T. Hattori, T. Yashima (Eds.), *Studies in Surface Science and Catalysis*, Vol. 83, Elsevier, Amsterdam, 1994, p. 135.
274. P. Canizares, A. Duran, F. Dorado, M. Carmona, *Appl. Clay Sci.* **2000**, *16*, 273.
275. M. Hunger, G. Engelhardt, J. Weitkamp, in *Zeolites and Related Microporous Materials: State of the Art 1994*, J. Weitkamp, H. G. Karge, H. Pfeifer, W. Hölderich (Eds.), *Studies in Surface Science and Catalysis*, Vol. 84, Part A, Elsevier, Amsterdam, 1994, p. 725.
276. P. A. Jacobs, J. B. Uytterhoeven, H. K. Beyer, *J. Chem. Soc., Faraday Trans. I* **1997**, *73*, 1755.
277. J. Sárkány, W. M. H. Sachtler, *Zeolites* **1994**, *14*, 7.
278. A. Ma, M. Muhler, W. Gruenert, *Appl. Catal. B: Environmental* **2000**, *27*, 37.
279. B. M. Weckhuysen, H. J. Spooren, R. A. Schoonheydt, *Zeolites* **1994**, *14*, 450.
280. N. V. Choudary, S. G. T. Bhat, R. V. Jasra, *Adsorption Science Technology; Proceedings 2nd Pacific Basin Conference on Adsorption Science Technology*, D. D. Do (Ed.), Brisbane, Australia, May 14–18, 2000, World Scientific Publishing Co., Pte. Ltd. Singapore, 2000, p. 76.
281. R. P. Dimitrova, Y. G. Neinska, M. I. Spassova, N. G. Kostova, *Bulg. Chem. Commun.* **2002**, *34*, 261.
282. M. Zendehdel, M. Kooti, M. M. Amini, *J. Porous Mater.* **2005**, *12*, 143.
283. A. Feddag, A. Bengueddach, *J. Physique IV; Proc. Quatrieme Coll. Franco-Libanais sur la Science des Materiaux* **2005**, *124*, 243.

2.4.4
Metal Clusters in Zeolites

Wolfgang M. H. Sachtler* and Z. Conrad Zhang

2.4.4.1 Introduction

Although most industrial applications of zeolite catalysts make use of these materials in their acid mode, zeolites are also excellent supports for metals and their oxo-ions. Zeolites that contain both acid sites and clusters of a transition metal or multi-metallic clusters, act as bifunctional catalysts. Clusters inside zeolite cavities are always nanoparticles. Zeolite supported metals have been applied for shape-selective hydrogenations [1], hydroalkylation, isomerization, hydroisomerization, hydro-cracking, catalytic reforming [2], CO hydrogenation [3], and other reactions. For further details, the reader is referred elsewhere [4–6].

This chapter focuses on the chemistry of these catalysts. Zeolites are defined as alumosilicates having a channel or cage structure which is easily traversed by small molecules or ligated ions. Zeolites were discovered and described in 1756 by Cronsted [7], who coined the name because a powder of a ground zeolite releases its occluded water upon heating, giving the appearance of boiling soup. For details of the names of individual zeolites, the reader is referred to Ref. [8]. In this chapter, the word "metal" will be used in its restrictive sense, that is, for metal elements in their *zerovalent* state. Therefore, materials which use heavy metal ions in zeolites, such as the catalysts that convert light alkanes into aromatics [9–11], or the catalysts for NO_x abatement with oxo-ions of iron [12–14], copper, cobalt, palladium, or gallium inside zeolite cavities [15–19], and the group of Fe/MFI materials with Fe ions in the cavities of an MFI zeolite, that catalyze the one-step oxidation of benzene with N_2O to phenol [20–23], will not be discussed in this chapter.

---

* Corresponding author.

2.4.4.2 Nano Clusters of Metal versus Macroscopic Metals

When a macroscopic metal specimen is divided into clusters each of less than 50 atoms, some of the parameters that are characteristic of a macroscopic metal undergo dramatic changes, whereas others change only slightly. The first group of parameters are referred to as *dimension-sensitive*, and the second group is almost *dimension-insensitive*.

Clearly, typical collective electronic properties such as electric conductivity, thermal conductivity and ferromagnetic susceptibility are bound to change drastically when the size of the metal entity becomes smaller than the mean free path of the conduction electrons or the ferromagnetic domain. All properties related to the band structure will change for cluster sizes with low state density and discrete energy levels [24, 25]. The energy to remove an electron from a small cluster will be greater than the work function of the bulk metal, but less than the ionization potential of a metal atom. A striking proof of high dimension sensitivity is provided by results on Rh clusters. While bulk Rh is non-magnetic, Cox et al., using a modified Stern–Gerlach technique, found a very large magnetic moment for Rh_{10} clusters at 195 K [26]. The magnetic moment per Rh atom decreases with cluster size, but the moments of Rh_{15}, Rh_{16}, and Rh_{19} are anomalously large.

Molecular beam data show that, for alkali atom clusters, certain "magic numbers" characterize those clusters that are either very stable or are formed by favorable kinetics. Quantum mechanical calculations performed by Bonačić-Koutecký et al. predict which geometries lead to stable Na_n clusters for $n = 3$ to 20 [27]. These structures deviate from those which one would predict on the basis of maximum atom coordination; for example, the stable clusters Na_5 and Na_6 are planar.

Density functional theory (DFT) calculations of Pd_n clusters ($n = 1$ to 13) in their neutral, negatively, or positively charged states suggest that the stable structures of such clusters are dependent on their charge [28]. In particular, for clusters with $n \leq 6$, the stable structure of the anions is similar to that of the neutral clusters, which follows Stranski's basic concept of maximizing atom coordination. In contrast, positively charged Pd clusters favor a planar configuration. For Pd_n clusters with $n = 7$ or 13, the stable structural geometry converges for charged and neutral clusters. The isomeric cubo-octohedron is favored for $n = 13$.

Less is known at present with respect to the most stable shapes of other transition metal clusters. As their heats of sublimation are higher than those of alkali metals, it is probable that Stranski's basic concept of maximizing atom coordination remains valid for these metals. The preferred atom arrangements should then be given by the same "magic numbers" which have been found for rare gas clusters. For example, in the case of Pt or Rh, the smallest stable cluster would be a 13-atom icosahedron having one atom in the center and 12 atoms in equivalent positions surrounding it, or the isomeric cubo-octahedron. For a Pt_{13} cluster inside a cage, it is expected that the interaction between Pt atoms and cage walls are much weaker than the Pt–Pt bonds. Metal clusters in zeolites will, however, differ from those in molecular beams, if the interaction with the zeolite walls is significant.

Chemical characteristics and, by implication, the catalytic parameters of metal clusters appear less sensitive to sample size than some electronic effects. True quantum jumps in catalytic parameters are expected and observed only when dispersion is brought to the stage that the ensemble requirements of a catalytic reaction are no longer fulfilled, or when isolated metal atoms exist on a support.

In catalysis, nanoparticles of metals on a support have often been considered as very small pieces of metal, but exposing highly unsaturated atoms. The energy balance of a chemical process in which a metal reacts with another reactant contains the surface (free) energy of the metal. This term is negligible for a macroscopic specimen, but appears as a significant exothermic term in the energy of small metal particles. This has often been translated into the terminology of atomic processes by considering the extent of *coordinative unsaturation* of the surface atoms. Atoms involved in chemisorption are also coordinatively unsaturated in macroscopic samples, but the extent of coordinative unsaturation is greater for small clusters. The difference in coordinative unsaturation between clusters and macroscopic metals is comparable to that between crystal faces.

It has been known for several decades that chemisorption on metals is often accompanied by atom reconstruction at the surface. The group who discovered this phenomenon originally termed it "corrosive chemisorption" [29]. Later the term "surface reconstruction" was introduced. The condition is easily observed on atomically "rough" crystal faces [30, 31]. It thus stands to reason that reconstruction will be inherent to chemisorption and catalysis on nanosize metal clusters.

Restricting the dimension sensitivity of clusters in catalysis to the effects of coordinative unsaturation and ease of surface reconstruction might, however, be an oversimplification. Rosén et al. reported measurements on the sticking coefficient **S** of several diatomic gases on Rh, Co, Fe, Ni and Cu clusters in molecular beams [32, 33]. For clusters with fewer than 10 atoms, **S** is very low (except for CO on Rh), but it usually reaches a constant, high value for clusters with 25 atoms. For some systems the size-dependence of **S** is more complicated, reflecting

References see page 520

the change in ionization potential with cluster size and the occurrence of closed electron shells.

2.4.4.3 Preparation of Mono- or Bimetal Clusters in Zeolite Cavities

The methods used for depositing metal precursors into the cavities of a zeolite include: (i) ion exchange; (ii) impregnation of metal salts; (3) deposition of volatile or soluble complexes [34–38], and (iv) solid state ion exchange [39, 40]. It is possible to combine several of these techniques. For example, Pt or Rh can be brought into a Na-Y zeolite by ion exchange, followed by calcination and reduction in H_2. Subsequent chemical vapor deposition of the volatile carbonyl of a second metal creates bimetal combinations such as (Pt + Cu), (Pd + Fe) or (Pt + Re) with preferential deposition of the second metal onto the clusters of the first metal. Physical and catalytic characterization of Pt-Re/Na-Y [41] and Rh-Fe/Na-Y [42] confirms the predominant formation of bimetallic clusters. Stable monometal clusters of Ir and Rh have also been prepared through the reductive carbonylation of $Ir(CO)_2(acac)$ and $Rh(CO)_2(acac)$ precursors followed by decarbonylation [43, 44].

As catalyst preparation has been described extensively elsewhere, this chapter will merely summarize the chemistry of the most popular method, viz. ion exchange. Three major steps can be discerned, namely ion exchange, calcination, and reduction.

2.4.4.3.1 Ion Exchange
For the metals of high catalytic activity Pt, Pd, and Rh, the complex ions $[Pt(NH_3)_4]^{2+}$, $[Pd(NH_3)_4]^{2+}$, and $[Rh(NH_3)_5(H_2O)]^{3+}$ in aqueous solution are conveniently exchanged against alkali ions of the zeolite. Because of the rather bulky coordination shell of these ions relative to the pores of the zeolite, the exchange process often requires several days and temperatures above room temperature to achieve completion. If ion exchange is interrupted before equilibrium is reached, a metal ion concentration gradient will exist from the surface to the interior of the zeolite grains. If ion migration to the smaller cages at elevated temperature is not desired, the zeolite can be washed, after ion exchange, with a high-pH aqueous solution. This treatment converts transition metal ions to a hydroxygel which is later dehydrated to oxide particles. This procedure is recommended for Ni or Co in Na-Y, because reduction of oxide particles in supercages can be carried out under rather mild conditions, whereas the reduction of these ions located in the hexagonal prisms is very difficult [45].

2.4.4.3.2 Calcination
The calcination step is carried out to remove the water and the ligands from the exchanged ions. The metal ion-loaded zeolite is heated in a high flow of air or oxygen. Three phenomena often accompany the elimination of H_2O and NH_3:

1. Autoreduction, that is, the formation of metal clusters and zeolite protons due to the decomposing ammine ligands. Under calcination conditions, reduction to metal clusters is followed by their growth or formation of oxide particles. These particles, in turn, can react with protons to form metal ions and water.
2. Ion migration to smaller zeolite cages. A driving force for this process is the high negative charge density in small zeolite cages. Ion migration can be minimized by limiting the temperature of calcination. It is possible to initiate ion migration back to large zeolite cavities by offering them an attractive ligand (e.g., NH_3). This strategy is at the basis of catalyst rejuvenation [46], and has also been used to increase the Co content of zeolite supported $PdCo_x$ clusters [47].
3. Ion hydrolysis leads to dissipation of positive charge. Multivalent ions, such as Rh^{3+}, are converted to monopositive complexes (e.g., $(RhO)^+$) and protons. Each of these can find a location close to a negatively charged $[AlO_4]^-$ tetrahedron. In zeolites only the positive charge carriers are mobile, while the negative charge is localized in the Al-centered tetrahedra of the framework and is, therefore, immobile. For zeolites with high Si/Al ratios the negative charges will be separated far from each other; dissipation of the positive charge lowers the Coulomb energy of the system, because it enables the positive charge carriers to find positions close to the loci of the negative charge. This phenomenon is at the root of the gain in Gibbs Free Energy of all phenomena which dissipate multi-positive charges into mono-positive, mobile charge carriers. Hydrolysis [48], such as

$$Ga^{3+} + H_2O \longrightarrow [GaO]^+ + 2H^+ \quad (1)$$

or dissociative chemisorption of dinitrogen tetroxide on Ba/Y [49]:

$$Ba^{2+} + N_2O_4 \longrightarrow NO^+ + [Ba-NO_3]^+ \quad (2)$$

are well-documented examples.

Uncontrolled autoreduction is often undesired, because agglomeration of clusters to larger particles can take place. It can be minimized by carrying out the calcination step in pure oxygen, at a very high flow rate and a very slow temperature ramp (e.g., 0.5 K min^{-1}) [50]. However, no conditions are known to the present authors which will completely suppress the autoreduction of Pt or Rh.

2.4.4.3.3 Reduction
Reduction is typically carried out in a flow of H_2. The process:

$$M^{n+} + \frac{n}{2}H_2 = M^0 + nH^+ \quad (3)$$

creates protons of high Brønsted acidity in addition to metal atoms. If desired, the formation of protons can be minimized by limiting the temperature of calcination, thus retaining a sufficient concentration of NH_3 within the zeolite. Such ligands will neutralize the protons formed during the reduction step. Carbon monoxide, while unable to reduce unligated ions, can be used to reduce oxide particles or oxo-ions

$$(RhO)^+ + CO = Rh^+ + CO_2 \quad (4)$$

This chemistry is used to discriminate between bare ions and oxygen-containing entities [11, 50, 51]. Temperature-programmed reduction (TPR) also helps to identify ions in different locations; the consumed amounts of H_2 or CO provide the relative quantities in which various reducible species are present. Low H_2 consumption by calcined samples can indicate a high extent of autoreduction.

Bimetallic clusters are prepared in a similar manner. If ions of metals with widely different reducibility coexist in the same zeolite, the more easily reduced metal is reduced first; its clusters will then act as nuclei for H_2 dissociation and as anchors for the less easily reducible metal. As a consequence, the TPR peaks of the second element are often strongly shifted towards a lower temperature. The combinations (Pt + Cu) [52], (Pt + Co) [53], (Pd + Cu) [54], (Pd + Ni) [55], (Pd + Fe) [56], and (Rh + Fe) [57] have been studied in Y zeolite. In all cases, reducibility enhancement leads to the formation of bimetallic clusters. The combinations (Pt + Re) and (Rh + Fe) have also been prepared by the decomposition of a volatile carbonyl cluster of the second element onto the reduced zeolite-encaged metal clusters of the first element. With (Rh + Fe) NaY, Mössbauer data show that Fe^{3+} ions are reduced with H_2 to Fe^{2+} and Fe^0; no Fe^0 clusters are formed, as all Fe^0 is alloyed to Rh. Electron spin resonance (ESR) data indicate that $PdFe_x$ clusters in Na-Y give a ferromagnetic resonance signal, even if the cluster contains only one Fe atom [58].

2.4.4.4 Interaction of Metal Clusters and Zeolite Protons

2.4.4.4.1 Brønsted Acids in Zeolites
Protons are formed by the reduction of metal ions with H_2; a higher proton concentration can be achieved by first exchanging alkali ions against NH_4^+ ions which are converted to protons during calcination. The maximum achievable proton concentration is given by the molar Al/(Al + Si) ratio in the zeolite.

Although protons in zeolites act as Brønsted acid centers, the actual dipole moment of these OH groups is rather low. The oxygen in this dipole is bridging between a Si^{4+} and an Al^{3+} centered tetrahedron:

$$\underset{Al \quad\quad Si}{O{-}H}$$

The infra-red (IR) band at 3610 cm^{-1} in HZSM-5, the bands at 3560 and 3650 cm^{-1} in zeolites HX and HY, and the band at 3610 cm^{-1} in H-mordenite are assigned to this vibration [59]. Although olefins which are adsorbed on these sites are often called "carbenium ions", their Fourier transform IR (FTIR) signature identifies them as alkoxy-groups; that is, there exists a C–H bond with the "proton" and, in addition, a covalent bond between a C atom of the olefin and a basic oxygen of the zeolite wall. As the non-trivial problem of correctly determining the acid strength of these "protons" is beyond the scope of this chapter, we merely quote Kazansky here: "The only proper way to proceed is to follow the response of hydroxyl groups to their interaction with adsorbed bases" [60]. Solid-state NMR and IR spectroscopy are useful methods here; for example, the *change* in the MAS-^{1}H NMR chemical shift upon adsorbing a weak base, is an appropriate measure of the acid strength [61].

2.4.4.4.2 Metal Clusters as Lewis Bases
Protons are able to form chemical bonds with transition metal atoms. In the resulting "metal–proton adducts" the transition metal atoms act as Lewis bases. This type of chemical bond was first discovered with zeolite-supported metals [62–64]; the same type of chemical interaction has also been observed in metal–organic complexes. The group of van Koten showed, using ^{1}H NMR, that hydrogen-bridged Pt configurations with a chemical shift of ca. 20 ppm exist in metal ammine complexes [65]. Molecular orbital calculations of Pd complexes show that the Pd$\cdots$H-NH$_3$ configuration with, formally, Pd^{2+} is more stable than the tautomeric H-Pd$\cdots$NH$_3$ form with Pd^{4+}. The structure of zwitterionic complexes with a Pt$^{2+}\cdots$H-N$^+$ unit has been identified using X-ray diffraction (XRD) [66]. Hydrogen bonding to the metal centers of organometallic compounds was also identified by Kazarian et al. [67]; these authors reported Ir$\cdots$H-O hydrogen bonding between the 18-electron iridium center in CP* Ir(CO)$_2$ and a number of fluoroalcohols. In these complexes, the Ir atom clearly acts as a proton acceptor. The CO stretching frequency is shifted to higher wave

numbers by amounts increasing with the acid strength of the alcohol; the O–H frequency of the alcohols is broadened and shifted to lower wavenumbers. Complexes have also been detected in which *two* alcohol molecules are linked to the same Ir atom via C–O–H–Ir bridges. Complexes of Rh and Co have been found to form similar metal–proton bonds with acidic fluoroalcohols. It can thus be stated that transition metal atoms and their clusters act as Lewis bases towards protons of high acidity, forming chemical bonds M$\cdots$H$^+$.

The basicity of noble metal subnanometer clusters has been identified for Pt/γ-Al$_2$O$_3$. With Pt clusters <1.5 nm and predominantly <1 nm, it was found that exposure to an aqueous solution of saturated NH$_4$Cl resulted in immediate formation of NH$_3$, which was easily identified by its smell and by wet litmus paper. The size-specificity of the basicity was demonstrated by control experiments with a Pt/γ-Al$_2$O$_3$ catalyst having Pt clusters in the range of 3 to 8 nm on the same γ-Al$_2$O$_3$ support. No NH$_3$ was detectable in that case.

Evidently, clusters Pt$_n$ with small values of n exhibit a high basicity enabling the reaction:

$$\text{Pt}_n + m\text{NH}_4\text{Cl} \longrightarrow \text{H}_m\text{Pt}_n\text{Cl}_m + m\text{NH}_3 \quad (5)$$

The thermodynamics of this chemistry has been verified by density function calculations of Pt clusters of varying nuclearity for the simplest scenario of $m = 1$ [68] (see Fig. 1). The high exotherm of this process for Pt clusters with sizes in the nanometer and subnanometer range is evident from Fig. 1, which demonstrates the superior basicity of such Pt clusters relative to that of NH$_3$ molecules.

The high basicity of small Pt$_n$ clusters is also evident from their ability to modify the acid strength of the Lewis acid sites on the γ-Al$_2$O$_3$ support. Monitoring the adsorption of pyridine by using FTIR showed that the acid strength of the γ-Al$_2$O$_3$ was decreased due to the presence of the basic Pt$_n$ particles; this gave rise to a downward shift by 3 cm^{-1} of the band of adsorbed pyridine on the catalyst with the nano Pt clusters relative to the same adsorbate on Pt particles of 3 to 8 nm on the same alumina [69, 70].

The formation of metal proton adducts in zeolites leads to three important consequences for the use of metal/zeolites in heterogeneous catalysis:

- Protons act as "chemical anchors", impeding metal agglomeration, thus stabilizing high metal dispersion, even isolated metal atoms.
- As the partial positive charge of the "zeolite protons" is shared with the metal atoms of the anchored cluster, the cluster becomes "electron-deficient".
- The combination in one complex of the two ingredients of a bifunctional catalyst enables metal–proton adducts to act as *collapsed bifunctional sites*.

These consequences will be discussed in the following sections.

2.4.4.4.3 Chemical Anchoring Numerous examples are known showing that, in zeolite cavities, reduced metals such as Pt, Pd, or Rh are formed in higher dispersion on zeolites exposing Brønsted acid sites than on proton-free supports. This suggests that protons can act as "chemical anchors" for metal clusters. If protons are present, it has been shown that metal dispersion increases with proton concentration [71]. Likewise, transition metal ions and Lewis acids also function as chemical anchors. For Al$_2$O$_3$-supported metals, the size of Pt particles can be controlled by anchoring and de-anchoring them. Zhang and Beard reported pronounced changes for Pt/Al$_2$O$_3$ catalysts used for dehydrochlorination [72]. Pt clusters with sizes below 1.5 nm and exhibiting strong basicity are highly sensitive to acids such as HCl, which poisons these clusters by forming stable adducts H$_m$Pt$_n$Cl$_m$. In contrast, larger Pt particles with sizes in the 3- to 8-nm range are not susceptible to acid poisoning because their basicity is significantly lower, approaching that of bulk Pt metal.

Removal of the proton anchors mobilizes metal clusters, enabling them to agglomerate. Three possibilities exist for cutting the chemical anchor:

1. Neutralization of the protons: this is conveniently achieved by the adsorption of ammonia. The resulting increase in metal particle size is well documented by transmission electron microscopy (TEM) and extended X-ray absorption fine structure (EXAFS) data [73]. If hydrogen adsorption is used to monitor metal dispersion, then the adsorbed ammonia must first be removed.

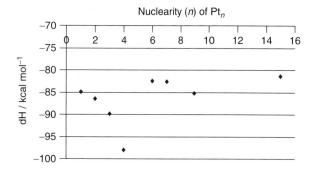

Fig. 1 Calculated enthalpy of the reaction Pt$_n$ + NH$_4$Cl $\rightarrow$ HPt$_n$Cl + NH$_3$ based on density function theory. (The *x*-axis shows the nuclearity n of Pt$_n$ clusters.)

2. Displacement of protons: this is often observed upon adding molecules, such as carbon monoxide, which are able to form stronger bonds with the metal. Direct evidence for proton displacement is the ensuing increase in the intensity of IR band for the zeolite O–H groups [74]. If protons are displaced by CO in the stage when metal clusters consist, for example of three or four atoms, then de-anchoring of these clusters induces their migration through zeolite channels, followed by coalescence to larger clusters [75, 76]. At low temperatures, where the zeolite lattice is essentially rigid, the size of the secondary clusters (formed upon CO admission) is controlled by the diameter of the apertures between zeolite cages [77]. For clusters that are too large to migrate through the cage windows at low temperature, the removal of some CO ligands and their replacement by zeolite protons is, however, reversible. With Pd in NaHY it has been demonstrated that mere pumping at room temperature suffices to transform the FTIR spectrum of the pure carbonyl cluster with a Pd_{13} core into that of the same core, but with mixed ligands (H^+ and CO). The process is reversible: upon admitting CO, the protons are displaced and the original spectrum is restored [63, 78].

3. Consumption of protons: olefins and aromatic molecules form carbenium ions with protons. This process deprives the metal cluster of its chemical anchor and enables it to migrate. As tertiary carbenium ions are more stable than secondary carbenium ions, metal agglomeration is more severe during catalysis with molecules containing tertiary C atoms. This has been demonstrated by TEM of initially identical Pd/ZSM-5 catalysts which were exposed to mixtures of H_2 with either cyclopentane or methyl-cyclopentane. With cyclopentane, little growth of the Pd clusters is detected, but methylcyclopentane – which has a tertiary C atom – induces a rather dramatic increase in Pd cluster size [72].

2.4.4.4.4 Heterogeneous Catalysis by Single Metal Atom Sites

As catalytic activity of metals is essentially confined to surface atoms, further investigations are required to achieve high metal dispersion. The *non plus ultra* in this respect is a metal divided into isolated atoms held by a support, but exposed to the gas phase. This goal has been achieved with platinum by using proton anchoring to the walls of an acidic zeolite with linear channels having "side-pockets". The support chosen by Lei et al. [79] and Zholobenko et al. [80] was H-mordenite At low metal loading and after reduction at low temperature, Pd and Pt were prepared in extremely high dispersions. For Pt/H-Mor the following experimental criteria support the conclusion that a large fraction of the platinum is present as Pt–proton complexes with only *one Pt atom*:

1. Zero Pt–Pt coordination, determined from EXAFS
2. A CO band in CO-FTIR at higher frequency than all CO bands known hitherto for CO adsorbed on Pt_n^0 clusters
3. An absence of any frequency shift in this band while the overall coverage with CO was varied
4. A total absence of this band in identical samples after neutralization of the protons
5. A unique product pattern in the H/D exchange of cyclopentane with D_2. It clearly shows that a *stepwise* exchange mechanism predominates over these Pt_1 catalysts, in contrast to the *multiple* exchange pattern that has always been observed for H/D exchange over conventional platinum catalysts with multi-atomic Pt_n sites.

The IR frequency shift of CO (2) reflects the high electron deficiency of Pt atoms held to the zeolite by anchoring protons. It is also evidence for the absence of adjacent adsorbed CO ligands, as dipole–dipole coupling of neighboring CO_{ads} would induce a frequency shift, as is always observed for CO_{ads} on multi-atomic Pt. Criterion (5) is unique: the H/D exchange of cyclopentane with D_2 has been studied by numerous groups using Pt catalysts with a variety of supports [81, 82]. Over such catalysts, the five H atoms at one side of the C_5 ring are exchanged during one residence of the adsorbed molecule, leading to a partial maximum of the $C_5H_5D_5$ product. A totally different product pattern is observed with Pt/H-MOR exposing isolated Pt atoms. In this case, a binomial distribution of the products is found which is the hallmark of a "stepwise" exchange [81, 82].

2.4.4.4.5 Electron Deficiency

Electron deficiency of platinum in zeolites was first reported by Dalla Betta and Boudart [83]. This condition is not limited to metal–proton adducts, but occurs whenever metal clusters interact with electron acceptors, including Lewis acids or metal ions. Electron deficiency is detected by several physical criteria (though some of them are not unambiguous). In X-ray absorption, the white line at the Pt LIII edge has been interpreted in terms of electron deficiency [84], but it was also attributed to the higher ionization potential of small clusters in comparison to the work function of larger metal particles [85]. Electron deficiency leads to a shift in binding energy in X-ray photoelectron spectroscopy (XPS); likewise, a shift of the IR vibration frequency of adsorbed CO has been quoted as evidence for a positive charge of the adsorbing metal cluster. Tri et al., using X-ray absorption spectroscopy,

References see page 520

observed a correlation between the electron deficiency of Pt in zeolites and Brønsted acidity [86]. Subsequently, Blackmond et al. confirmed that the electron-deficient character of the metal increases with the acidity of the support [87]. A quantum mechanical analysis of the interaction between zeolite protons and entrapped metal particles in zeolite cavities was reported by Zhidomirov et al. [88]. For a description of the electrical and electrochemical properties of metals in zeolite cavities, the reader is referred to Ref. [89]. The chemical reactivity of carbonyl groups on Ru and other clusters inside zeolite cavities and the mechanism of C–C bond formation have been discussed by Ichikawa et al. [90].

The results of an FTIR and XPS study of Pd/Y catalysts with varying proton concentrations show that the shift in both the CO frequency [82] and the Pd binding energy correlates with the proton concentration in the zeolite [72]. Remarkably, the XPS band width remains constant, while the band position shifts. This excludes an equilibrium distribution between coexisting pure metal clusters and metal–proton adducts in the same zeolite. The results suggest a high homogeneity of the metal–proton clusters for any given proton concentration. Where protons are scarce, metal clusters are linked each to one proton, but if the proton concentration is high, most metal clusters appear to be associated with several protons.

Electron deficiency affects the chemisorptive and catalytic characteristics of metal clusters. Dalla Betta et al. found that the catalytic activity of 1-nm Pt clusters in NaHY zeolite for neopentane isomerization and hydrogenolysis is 40-fold higher than that of Pt/SiO$_2$ and Pt/Al$_2$O$_3$. Other authors observed an increased activity for hydrogenation [91–94], changes in the competitive hydrogenation of benzene and toluene [95], changes in hydrogenolysis activity [96–98], and high resistance to sulfur poisoning [99–101]. Literature reports on changes in selectivity or catalyst life are more reliable than data on enhanced turnover frequencies, because the latter often depend on H$_2$ chemisorption data. It has been reported that small clusters of Ru [102], Pd [103], and Rh [51], stabilized by zeolite protons, adsorb little or no hydrogen. In the conversion of methylcyclopentane, electron-deficient palladium–proton [104] and platinum–proton [105] adducts exhibit not only strongly enhanced activity towards ring enlargement, but also markedly reduced activity towards ring opening. The former observation is ascribed to the bifunctional character of catalysts containing metal and acid sites, while the latter observation indicates that electron deficiency changes the catalytic properties of metal clusters. For instance, Rh/HY is more active for ring enlargement than Rh/NaY, but for ring opening Rh/NaY is more active, whereas this reaction is strongly suppressed over Rh/HY [106].

Two potential causes have been considered for the experimental fact that attaching zeolite protons to a metal cluster induces significant changes of its catalytic activity. Some authors assume that removing an electron from a metal atom will make this atom more similar to its left neighbor in the Periodic Table. Others reject this "alchymistic transmutation" of elements, but assume that the protons interact chemically with the adsorbate, transforming it, for instance, into a species resembling a carbenium ion which will interact with the metal [107].

Results obtained with neopentane have been quoted in favor of the former hypothesis, as neopentane is unable to form a carbenium ion in a single step. The enhanced catalytic activity of Pt [105], Pd [104], and Rh [108] in their electron-deficient state is in agreement with the fact that for each of these metals the left neighbor in the Periodic Table displays higher catalytic activity. Other observations, such as the formation of benzene in quantities exceeding the benzene/cyclohexane equilibrium ratio [109], have been ascribed to participation of the protons in the chemisorption complex.

2.4.4.4.6 Is Bifunctionality due to "Shuttling" between Sites or to "Collapsed Sites"?

Numerous hydrocarbon conversion reactions require a catalyst that contains both metal and acid sites. As Mills et al. showed [110], one can rationalize the product pattern obtained over such catalysts by assuming that an alkane molecule is first dehydrogenated on a platinum surface; the resulting alkene subsequently migrates to a Brønsted acid site forming an adsorbed carbenium ion which could, for instance, isomerize. Repetition of such steps in which intermediates shuttle between acid and metal sites could lead to alkane isomers, aromatic molecules, or cracking products and coke.

The metal–proton adduct model assumes that both catalytic functions are combined in one complex, acting as a "collapsed bifunctional site". The chemisorption of an alkane molecule on a metal site is dissociative; while the H atom can combine with the "proton" of the metal–proton adduct and leave as H$_2$, the positive charge of the remaining complex will, in one canonic form, be localized on the C atom of the alkyl group, thus opening the reaction channel for "chemisorbed carbenium ions". For instance, ring enlargement of methylcyclopentane to a six-membered ring becomes possible. In view of the facile dehydrogenation on the metal cluster, this adsorbed six-membered ring can leave the surface either as a benzene or as a cyclohexane molecule. The result will be similar to that predicted by the classical bifunctional mechanism, but all steps occur within one residence of the molecule at the metal–proton adduct. A comparison of Pd–proton adduct catalysts and a physical mixture of the corresponding monofunctional catalysts shows a strong gain in activity by combining protons and metal clusters in the same bifunctional site [104], thus eliminating the requirement for multiple shuttling of intermediates between sites.

2.4.4.4.7 Metal Oxidation by Protons, Selective "Leaching"

In the equilibrium of Eq. (3), at high proton concentration and low H_2 partial pressure, metal atoms will be oxidized by zeolite protons to metal ions. The entropy gain caused by the formation of H_2 favors the reverse of the reaction in Eq. (3) at high temperature, even for "noble" metals such as Pt or Rh [111, 112]. The H_2 release, which has been monitored by TPD, occurs at a higher temperature than desorption of chemisorbed hydrogen. It is possible to discriminate between the two processes by FTIR because only the oxidation process consumes protons, thus leading to a decrease in the intensity of the characteristic zeolite O–H bands. From Eq. (3) an equilibrium constant can be defined:

$$K = \frac{a_{Mo}(a_{H^+})^n}{(a_{M^{n+}})(a_{H_2})^{1/2n}} \quad (6)$$

where a represents the thermodynamic activity. Upon treating the clusters as a solid phase, of which the activity depends only on the temperature, one can include this activity in the equilibrium constant. The activities of the other constituents are replaced by their concentration or gas pressure. For fixed values of T and P_{H_2}, the proton concentration – and hence the extent of reoxidation – increases with metal loading.

For copper and rhodium, carbon monoxide forms stronger bonds with the positive ions than the zerovalent metal. For these metals, the equilibrium in Eq. (3) is, therefore, shifted towards the left side in the presence of CO. This was shown for Rh/NaY by FTIR detection of $Rh^+(CO)_2$, the concomitant emergence of the XPS peak for Rh^+, and by mass spectrometric identification of the H_2 evolution upon admission of CO [113]. With Cu/ZSM-5 the oxidation of Cu^0 by zeolite protons and the formation of $Cu^+(CO)$ has been monitored by FTIR as a function of time and temperature [114].

With bimetallic clusters in zeolites, the protons tend selectively to oxidize the less-noble alloy constituent. This "leaching" phenomenon was first observed by Tzou et al. [115], whose EXAFS data showed that in thoroughly reduced (Pt + Cu)/Na-Y samples some Cu was coordinated to zeolite oxygen; the concentration of this species increased upon heating in He. This suggests that Cu had been leached from $PtCu_x$ clusters, and that these Cu ions migrated into small zeolite cavities. This conclusion was confirmed by Zhang et al. [55] in their study of (Pd + Cu)/NaHY. These authors made use of the propensity of Pd to form hydrides that are easily identified with TPD by the temperature of their decomposition. As PdCu alloys do not form hydrides, the "leaching" of Cu out of the bimetallic clusters restores the ability of the cluster to form a hydride. Leaching of Fe from bimetallic $PdFe_x$ clusters was detected in the same manner by Xu et al. [57]. The H_2 release during leaching can be determined with considerable precision; this permits quantitative determination of the number of metal atoms which were initially present in the bimetallic cluster.

2.4.4.5 Effects of Zeolite Geometry on Catalysis

2.4.4.5.1 Transport and Transition State Restrictions

One reason for the high potential of zeolite-based catalysts in the chemical industry is the stereospecificity that is induced by the peculiar geometry of these systems. Zeolites are molecular sieves which only permit small kinetic diameter molecules to enter their pores. As a result, only one component in a mixture might react, while other components are unable to reach active sites in the pore walls or cavities. Likewise, when several isomers are formed in a catalytic process, one isomer may escape easily, whereas the other isomers are trapped inside the zeolite cages. These trapped species will leave only after undergoing secondary isomerization or cracking reactions. This phenomenon is exploited for octane upgrading. In catalytic processes where two reaction paths are possible – one path involving a monomolecular transition state, and the other requiring a bimolecular complex – the steric constraints of zeolite cavities will strongly favor the less-voluminous transition state. These effects are not restricted to reactions catalyzed by metals; they have also been demonstrated for acidic zeolites and are well described in the literature and in Chapter 5.5.3 of this Handbook. They will, therefore, not be discussed further in this chapter. However, two effects that are specific to zeolite-encaged metal clusters, namely oriented collision following collimation, and alteration of the diffusion type by metal particles inside channels, are discussed in Sections 2.4.4.5.2 and 2.4.4.5.4.

2.4.4.5.2 Collimation of Molecules in Pores

The diffusion of molecules through pores with a diameter comparable to the molecular dimensions is not of the Knudsen type. The term "configurational diffusion" was proposed by Weisz for this type of molecular transport [116]. Non-spherical molecules diffusing through linear pores will orient themselves with their long axis, essentially parallel to the pore axis. The first report that this *collimation* could induce stereospecific catalysis was

References see page 520

published by Tauster et al. [117]. Once the collimated molecule encounters a metal particle, a "head-on" collision results – that is, the molecule will be preferentially adsorbed at its front side. Methylcyclopentane is a non-spherical molecule, its length-to-width ratio being 1.7. When collimated in a narrow pore, it will collide with a metal particle either with its methyl group or with its flat, bottom end. Only the latter collision can result in opening of the C_5 ring, leading to the formation of 3-methylpentane (3MP). In contrast, for metals on an amorphous support, such as Al_2O_3 or SiO_2, the ring-opening products 3-methylpentane and 2-methylpentane (2MP) are formed in the statistical ratio of 2MP/3MP = 2, which is very close to the thermodynamic ratio. Numerous experimental data obtained by Gault et al. for metals on amorphous supports confirm this expectation [118]. Indications for a lower 2MP/3MP ratio in zeolites were obtained with Pt in Na-Y [119]. A much more pronounced collimation is expected for zeolites with linear, one-dimensional pores, and this was indeed confirmed for Pt inside the channels in zeolite L [120, 121] and in mordenite [122].

The same logic predicts that a collimated molecule moving through a one-dimensional pore can be attacked at its flanks, if small metal clusters are lurking in side pockets of the wall. This situation has been realized for Pt in mordenite, as small Pt clusters (even single atoms) can be accommodated in the side pockets of this zeolite. Due to the anchoring by protons, the H-form of the zeolite ensures a high metal dispersion, as required for this purpose. When prepared at low temperature and with low metal loading, the Pt clusters will be small enough not to protrude from their side pockets into the channels to provoke "head-on" collisions with collimated methyl-cyclopentane molecules. By varying the size of the Pt clusters, these authors succeeded in varying the 2MP/3MP ratio from 1.4 (for large Pt clusters) to 2.8 (for small clusters in side pockets) [123].

2.4.4.5.3 Types of Pore Diffusion in Zeolites; Effects on Apparent Activation Energy
Active sites are useful in heterogeneous catalysis only if they are accessible to reactants, and if the reaction products can readily leave the catalyst. As for all catalysts, an effectiveness factor η can be defined as the ratio of the actual reaction rate and that of a fictitious catalyst in which the same sites are accessible, without any diffusion limitation. Using this criterion, zeolites can be classified in two groups:

1. Materials with cages and intersecting channels, such as zeolite **Y**
2. Cage-free zeolites with one-dimensional, non-intersecting narrow channels (e.g., **L** and Mordenite).

For the first group, the ordinary Thiele modulus,

$$\Phi = \frac{L}{2}\left(\frac{k}{D}\right)^{1/2} \qquad (7)$$

where L is the pore length, k is the intrinsic rate constant and D is the effective diffusion coefficient can be used to calculate η. Even if channels are largely filled with physisorbed molecules, the reactants and products become mixed inside the cages and there are numerous paths connecting each site with the world outside the zeolite.

The situation is fundamentally different for the second group of zeolites, where one-dimensional pores permit molecules to diffuse in and out, but not to pass each other. This so-called "single file diffusion" is reminiscent of the restricted movement of pearls on a string [124]. When the pores are filled with reactant molecules, the reaction products that are formed inside such a pore are effectively trapped. Kärger et al. showed that a Thiele modulus, Φ_{sf}, can be defined for single file diffusion; the ratio of Φ_{sf} to the ordinary Thiele modulus Φ is given by:

$$\frac{\Phi_{sf}}{\Phi} = N\left\{\frac{(3/175)\theta^{3/2}}{(1-\theta)^{1/2}}\right\} \qquad (8)$$

where N is the number of adsorbing sites per pore and θ is the degree of their coverage. At high coverage and for typical zeolites with pores of 1 μm length, the Thiele modulus for single-file diffusion is two to three orders of magnitude higher than the normal Thiele modulus. The effectiveness factor for single-file diffusion is correspondingly lower. Only sites near the pore orifices contribute effectively to the catalytic reaction, as registered by the change in composition of the surrounding atmosphere. Another consequence of this phenomenon is that the depth of this effective surface zone increases with temperature, as θ decreases. In other words, the apparent activation energy for catalysts with single-file diffusion will be *higher* than the intrinsic activation energy in the absence of diffusion limitation. This is in contrast to the classical pore diffusion model, which leads to an apparent activation energy which is *lower* than the intrinsic value. It should be noted that the adsorption sites which must be considered for single-file diffusion will, in general, be different from the catalytically active sites; for instance, it has been found that the adsorption of hydrocarbons in zeolites is stronger on Na^+ ions than on protons [125].

Karpiński et al. studied, how the type of pore diffusion affects the apparent activation energy, by using the conversion of neopentane as a probe [126]. A variety of supported palladium catalysts was tested under identical conditions, including Pd/SiO_2, Pd/NaY, Pd/HY and a number of Pd/ML catalysts with M = Li, K, and Ca. The study confirmed that, for a given zeolite, the catalytic activity markedly increased with proton concentration, as would be expected on the basis of the metal–proton adduct model. More relevant to the present issue is that a very significant effect of the pore type was found. The activation energy on Pd/SiO$_2$ is 260 kJ mol^{-1}, which is presumably near the intrinsic value. All values for Y-type zeolites are either near this value or lower (184–278 kJ mol^{-1}), in agreement with the classical pore diffusion model, although the values for three Pd/ML catalysts, that were all reduced at 400 °C (to eliminate pore blocking by NH$_4^+$ ions) were higher; Pd/LiL: 370 kJ mol^{-1}; Pd/KL: 325 kJ mol^{-1}; Pd/CaL: 339 kJ mol^{-1}. All activation energies and pre-exponential factors measured in this study followed a classical "compensation" line. The data are presented in Fig. 2.

2.4.4.5.4 Beneficial Damage of Zeolite Matrix by Growing Metal Particles
The difference in catalytic effectiveness between one-dimensional and three-dimensional pore systems creates an incentive to introduce some three-dimensionality into zeolites with linear, non-intersecting pores. Several authors had reported that growing metal particles inside zeolite cavities can induce a local destruction of the matrix [127–129]. After reduction at high temperatures, a significant portion of the metal is present as particles larger than the zeolite cavities. These particles are located in voids created within the zeolite matrix. This phenomenon was utilized by Carvill et al. in a study of Pt/H-Mor catalysts that had been reduced at different temperatures [130]. When metal reduction was carried out at 500 °C instead of 350 °C, XRD data showed a significant loss in zeolite crystallinity, while TEM revealed Pt particles larger than the channel widths. However, upon probing these catalysts for n-heptane conversion, it appeared that the catalyst with the lower metal dispersion was twice as active as the sample that had been reduced at 350 °C. This non-trivial result clearly shows that inside the regime of single file diffusion control, catalyst activity can be increased by local destruction of the zeolite matrix, as depicted schematically in Figure 3.

2.4.4.6 Conclusions

Small clusters of transition metal atoms in zeolites are sites of high catalytic activity. For instance, an icosahedron of 13 palladium atoms has only one fully coordinated Pd atom in its center; the other 12 atoms are "coordinatively unsaturated" surface atoms and, therefore, sites of high catalytic activity. Chemisorption on such clusters is usually *corrosive* – that is, it involves a thorough *reconstruction* of the cluster.

Nanosized noble metal clusters display stronger basicity than ammonia, and therefore have strong affinity toward both the Brønsted and Lewis acid sites of a catalyst support. In a zeolite exposing Brønsted sites, this affinity results in strong chemical bonds between the nanosized metal clusters and the zeolitic protons. The protons effectively act as "chemical anchors" for transition metal clusters inside zeolite cavities. The resulting "metal–proton adducts" are "electron-deficient" – they combine the catalytic characteristics of an active transition metal cluster with those of a strong acid site, and they catalyze reactions which, on more conventional

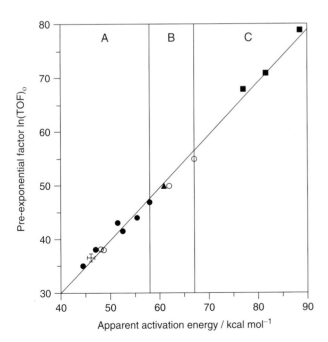

Fig. 2 Apparent activation energies and pre-exponential factors for neopentane conversion over supported Pd catalysts. ▲, Pd/SiO$_2$; ■, Pd/ML; ○, ●, Pd/Y.

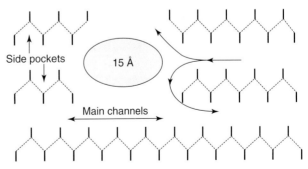

Fig. 3 Cartoon illustrating the transport paths opened by local destruction of mordenite channels caused by growing platinum.

References see page 520

catalysts, would require the "shuttling" of an adsorbed entity between physically separated metal sites and acid sites.

The anchoring effect for nanosized noble metal clusters also extends to transition metal ions in zeolites. In this case, the noble metal clusters are stabilized through the formation of an "electron-deficient" alloy with cations of the transition metal. The proximity of noble metal clusters and transition metal cations induces a significantly lowering of the reduction temperature for the transition metal ions in zeolites.

Complexes have been prepared in mordenite channels that consist of only *one* single Pt atom and one zeolite proton. These catalyze reactions that are typical for platinum, but the product distribution characteristically differs from that formed on multi-atomic Pt_n sites. Bimetallic clusters in zeolite cavities display peculiar catalytic effects, and their composition can be manipulated by selective *leaching* of one alloy component from the cluster.

The possibilities of manipulating the selectivity of catalytic reactions also include changing the geometry of the zeolite pores. Non-spherical molecules have been *collimated* in narrow zeolite pores, so that they hit a metal site with a particular part of the molecule, enforcing a chemical reaction at this spot. The diffusion of reactants and products through narrow zeolite pores is of the *single-file diffusion* type, in which case the measured activation energy of the catalytic process becomes larger than the intrinsic activation energy of the reaction in the absence of that diffusion effect.

Acknowledgments

The authors gratefully acknowledge the support provided by the EMSI program of the US National Science Foundation and the DOE Office of Science (CHE-9810378) at the Northwestern University Institute for Environmental Catalysis.

References

1. R. M. Dessau, *J. Catal.* **1982**, *77*, 304.
2. F. Ribeiro, C. Marcilly, M. Guisnet, *J. Catal.* **1982**, *78*, 275.
3. H. Treviño, W. M. H. Sachtler, *Catal. Lett.* **1994**, *27*, 251.
4. C. S. John, D. M. Clark, I. E. Maxwell, in *Perspectives in Catalysis*, J. A. Thomas, K. I. Zamaraev (Eds.), International Union of Pure and Applied Chemistry Publishers, 1992, 387.
5. W. M. H. Sachtler, Z. C. Zhang, in *Advances in Catalysis*, D. D. Eley, H. Pines, P. B. Weisz (Eds.), Academic Press, San Diego, Vol. 39, 1993, 29.
6. W. M. H. Sachtler, *Acc. Chem. Res.* **1993**, *26*, 383.
7. F. Cronsted, *Handl. Kon. Svensk. Acad. Vetensk.* **1756**, *17*, 120.
8. W. M. Meier, D. H. Olson, *Atlas of Zeolite Structure Types*, 2nd Revised Edn., Butterworths, London, 1987.
9. H. Kitagawa, Y. Sendoda, Y. Ono, *J. Catal.* **1986**, *101*, 12.
10. M. S. Scurrell, *Appl. Catal.* **1988**, *41*, 88.
11. B. S. Kwak, W. M. H. Sachtler, *J. Catal.* **1994**, *145*, 456.
12. H.-Y. Chen, W. M. H. Sachtler, *Catal. Lett.* **1998**, *50*, 125.
13. W. M. H. Sachtler, *Catal. Today* **1998**, *42*, 73.
14. H. Y. Chen, Q. Sun, B. Wen, Y.-H. Yeom, E. Weitz, W. M. H. Sachtler, *Catal. Today* **2004**, *96*, 1.
15. M. Iwamoto, In: *Zeolites and Related Microporous Materials: State of the Art 1994, Proceedings of the 10th International Zeolite Conference*, Garmisch-Partenkirchen, Germany, July 1994, J. Weitkamp et al. (Eds.), *Studies in Surface Science and Catalysis*, Vol. 84/B, Elsevier, Amsterdam, 1994, p. 1395.
16. Y. Li, J. N. Armor, *Appl. Catal. B, Environmental* **1993**, *2*, 239.
17. Y. Nishizaka, M. Misono, *Chem. Lett.* **1993**, *8*, 1295.
18. B. S. Kwak, W. M. H. Sachtler, *J. Catal.* **1993**, *141*, 729.
19. G. I. Panov, V. A. Sobolev, A. S. Kharitonov, *J. Mol. Catal.* **1990**, *61*, 85.
20. E. J. M. Hensen, Q. Zhu, M. M. R. M. Hendrix, A. R. Overweg, P. J. Kooyman, M. V. Sychev, R. A. van Santen, *J. Catal.* **2004**, *221*, 560.
21. Q. Zhu, R. M. van Teeffelen, R. A. van Santen, E. J. M. Hensen, *J. Catal.* **2004**, *221*, 575.
22. L. Kiwi-Minsker, D. A. Bulushev, A. Renken, *J. Catal.* **2003**, *219*, 273.
23. K. S. Pillai, J. Jia, W. M. H. Sachtler, *Appl. Catal. A.* **2004**, *264*, 133.
24. G. K. Wertheim, *Z. Phys. D. Atoms, Molecules and Clusters*, **1989**, *12*, 319.
25. S. N. Khanna, S. Linderoth, *Phys. Rev. Lett.* **1992**, *67*, 742
26. A. J. Cox, L. G. Louderback, S. E. Apsel, L. A. Bloomfield, *Phys. Rev. B* **1994**, *49*, 12295.
27. V. Bonačić-Koutecký, P. Fantucci, C. Fuchs, J. Koutecký, J. Pittner, *Z. Physik D* **1993**, *26*, 17.
28. Yu. Efremenko, M. Shteintuch, *J. Mol. Catal.* **2000**, *160*, 445.
29. A. A. Holscher, W. M. H. Sachtler, *Disc. Faraday Soc.* **1966**, *41*, 29.
30. R. Bouwman, W. M. H. Sachtler, *J. Catal.* **1970**, *19*, 127.
31. R. Bouwman, G. J. M. Lippits, W. M. H. Sachtler, *J. Catal.* **1972**, *25*, 350.
32. J. J. Persson, M. Andersson, A. Rosén, *Z. Phys. D., Atoms Molecules, Clusters* **1993**, *26*, 334.
33. M. Andersson, L. Holmgren, A. Rosén, *Surface Rev. Lett.* **1996**, *3*, 683.
34. B. S. Kwak, W. M. H. Sachtler, *J. Catal.* **1993**, *141*, 729.
35. Th. Bein, S. J. McLain, D. R. Corbin, R. D. Farlee, K. Moller, G. D. Stuckey. G. Woolery, D. Sayers, *J. Am. Chem. Soc.* **1988**, *110*, 1801.
36. C. Dossi, R. Psaro, R. Ugo, Z. C. Zhang, W. M. H. Sachtler, *J. Catal.* **1994**, *149*, 92.
37. O. C. Feeley, W. M. H. Sachtler, *Appl. Catal.* **1991**, *75*, 93.
38. G. Spoto, A. Zecchina, S. Bordiga, G. Ricchiardi, G. Martra, G. Leofanti, G. Petrini, *Appl. Catal. B* **1994**, *3*, 151.
39. H. G. Karge, H. K. Beyer, In: *Zeolite Chemistry and Catalysis, Proceedings of an International Symposium*, Prague, Czechoslovakia, September 1991, P. A. Jacobs, et al. (Eds.), *Studies in Surface Science and Catalysis*, Vol. 69, Elsevier, Amsterdam, 1991, p. 43.
40. B. Wichterlova, Z. Sobalik, M. Skokanek, *Appl. Catal.* **1993**, *103*, 269.
41. C. Dossi, J. Schaefer, W. M. H. Sachtler, *J. Molec. Catal.* **1989**, *52*, 193.
42. V. Schünemann, H. Treviño, W. M. H. Sachtler, K. Fogash, J. A. Dumesic, *J. Phys. Chem.* **1995**, *99*, 1317.
43. F. Li, B. C. Gates, *J. Phys. Chem. B* **2003**, *107*, 11589.

44. W. A. Weber, B. C. Gates, *J. Catal.* **1998**, *80*, 207.
45. M. Suzuki, K. Tsutsumi, H. Takahashi, Y. Saito, *Zeolites* **1988**, *8*, 284.
46. O. Feeley, W. M. H. Sachtler, *Appl. Catal.* **1990**, *67*, 141.
47. Y.-G. Yin, Z. Zhang, W. M. H. Sachtler, *J. Catal.* **1993**, *139*, 444.
48. B. S. Kwak, W. M. H. Sachtler, *J. Catal.* **1993**, *141*, 729.
49. M. Li, Y. Yeom, E. Weitz, W. M. H. Sachtler, *J. Catal.* **2005**, *235*, 201.
50. P. Gallezot, *Catal. Rev. - Sci. Eng.* **1979**, *20*, 121.
51. J. Sárkány, J. L. d'Itri, W. M. H. Sachtler, *Catal. Lett.* **1992**, *16*, 241.
52. D. C. Tomczak, V. L. Zholobenko, H. Treviño, G.-D. Lei, W. M. H. Sachtler, In: *Zeolites and Related Microporous Materials: State of the Art 1994. Proceedings of the 10th International Zeolite Conference*, Garmisch-Partenkirchen, Germany, July 1994, J. Weitkamp, et al. (Eds.), *Studies in Surface Science and Catalysis*, Vol 84/B, Elsevier, Amsterdam, 1994, p. 893.
53. G. Moretti, W. M. H. Sachtler, *J. Catal.* **1989**, *115*, 205.
54. Z. C. Zhang, W. M. H. Sachtler, *J. Chem. Soc., Faraday Trans.* **1990**, *86*, 2313.
55. Z. C. Zhang, L. Xu, W. M. H. Sachtler, *J. Catal.* **1991**, *131*, 502.
56. J. S. Feeley, W. M. H. Sachtler, *J. Catal.* **1991**, *131*, 573.
57. L. Xu, G. Lei, W. M. H. Sachtler, R. Cortright, J. A. Dumesic, *J. Phys. Chem.* **1993**, *97*, 11517.
58. V. Schünemann, H. Treviño, W. M. H. Sachtler, K. Fogash, J. A. Dumesic, *J. Phys. Chem.* **1995**, *99*, 317.
59. L. M. Kustov, V. Yu. Borovkov, V. B. Kazansky, *J. Catal.* **1981**, *72*, 149.
60. V. B. Kazansky, *Acc. Chem. Res.* **1992**, *24*, 379.
61. V. Adeeva, J. W. de Haan, J. Jänchen, G. D. Lei, V. Schünemann, L. J. M. van de Ven, W. M. H. Sachtler, R. A. van Santen, *J. Catal.* **1995**, *151*, 364.
62. L.-L. Sheu, H. Knözinger, W. M. H. Sachtler, *J. Am. Chem. Soc.* **1989**, *111*, 8125.
63. W. M. H. Sachtler, in *Chemistry and Physics of Solid Surfaces*, R. Vanselow, R. Howe (Eds.), Springer Series in Surface Sciences, No. 22, Springer-Verlag Berlin, 1990, p. 69.
64. W. M. H. Sachtler, A. Y. Stakheev, *Catal. Today* **1991**, *12*, 283.
65. I. C. M. Wehman-Ooyevaar, D. M. Grove, H. Kooijman, G. de Vaal, A. Dedieu, G. van Koten, *J. Am. Chem. Soc.* **1992**, *31*, 5484.
66. I. C. M. Wehman-Ooyevaar, D. M. Grove, H. Kooiijman, P. van der Sluis, A. L. Spek, G. van Koten, *J. Am. Chem. Soc.* **1992**, *114*, 9916.
67. S. G. Kazarian, P. A. Hamley, M. Poliakof, *J. Am. Chem. Soc.* **1993**, *115*, 9069.
68. Z. C. Zhang, J. Hare, B. Beard, In: *13th International Congress on Catalysis ICC*: Paris, France, July 2004, IACS, Paris, 2004, 02–017.
69. Z. C. Zhang, B. C. Beard, *Appl. Catal. A* **1999**, *188*, 229.
70. B. Beard, Z. C. Zhang, *Catal. Lett.* **2002**, *82*, 1.
71. L. Xu, Z. C. Zhang, W. M. H. Sachtler, *J. Chem. Soc., Faraday Trans.* **1992**, *88*, 2291.
72. Z. C. Zhang, B. C. Beard, *Appl. Catal. A* **1998**, *174*, 33.
73. Z. C. Zhang, B. Lerner, G.-D. Lei, W. M. H. Sachtler, *J. Catal.* **1993**, *140*, 481.
74. Z. Zhang, T. Wong, W. M. H. Sachtler, *J. Catal.* **1991**, *128*, 13.
75. Z. Zhang, H. Chen, W. M. H. Sachtler, *J. Chem. Soc., Faraday Trans.* **1991**, *87*, 1413.
76. Z. C. Zhang, H. Chen, L.-L. Sheu, W. M. H. Sachtler, *J. Catal.* **1991**, *127*, 213.
77. Z. C. Zhang, W. M. H. Sachtler, *J. Mol. Catal.* **1991**, *67*, 349.
78. L. L. Sheu, H. Knözinger, W. M. H. Sachtler, *J. Mol. Catal.* **1989**, *57*, 61.
79. G.-D. Lei, W. M. H. Sachtler, *J. Catal.* **1993**, *140*, 601.
80. V. L. Zholobenko, G.-D. Lei, B. T. Carvill, B. A. Lerner, W. M. H. Sachtler, *J. Chem. Soc., Faraday Trans.* **1994**, *90*, 233.
81. C. Kemball, in *Advances in Catalysis*, D. D. Eley, P. W. Selwood, P. B. Weisz (Eds.), Academic Press, San Diego, 1959, Vol. 11, p. 223.
82. N. Poole, D. A. Whan, in *Proceedings, 8th International Congress on Catalysis, Berlin 1984*, Vol. 4, Verlag Chemie, Frankfurt 1985, p. 345.
83. R. A. Dalla Betta, M. Boudart, in *Proceedings, 5th International Congress on Catalysis*, J. Hightower (Ed.), North-Holland, Amsterdam, 1973, p. 1329.
84. P. Gallezot, R. Weber, R. A. Dalla-Betta, M. Boudart, *Naturforschung* **1979**, *34A*, 40.
85. M. G. Samant, M. Boudart, *J. Phys. Chem.* **1991**, *95*, 4070.
86. T. M. Tri, J. P. Candy, P. Gallezot, J. Massardier, M. Primet, J. C. Védrine, B. Imelik, *J. Catal.* **1983**, *791*, 396.
87. D. G. Blackmond, J. G. Goodwin, Jr, *J. Chem. Soc., Chem. Commun.* **1981**, 125.
88. G. M. Zhidomirov, A. L. Yakovlev, M. A. Milov, N. A. Kachurovskaya, I. V. Yudanov, *Catal. Today* **1999**, *51*, 397.
89. N. Petranovi, D. Mini, *Ceramics Int.* **1996**, *22*, 449.
90. M. Ichikawa, P. W. Pan, Y. Imada, M. Yamaguchi, K. Isobe, T. Shido, *J. Mol. Catal.* **1996**, *107*, 23.
91. P. Gallezot, J. Datka, J. Massardier, M. Primet, B. Imelik, In: *Proceedings of the 6th International Congress on Catalysis*, Vol. 1: Imperial College, London, 1976, G. C. Bond et al. (Eds.), Chemical Society, London, 1997, p. 69.
92. J. Bandiera, *J. Chim. Phys.* **1980**, *77*, 303.
93. V. N. Romannikov, K. G. Ione, L. A. Pedersen, *J. Catal.* **1980**, *66*, 121.
94. F. Figueras, R. Gomes, M. Primet, *Adv. Chem. Ser.* **1973**, *121*, 480.
95. T. M. Tri, J. Massardier, P. Gallezot, B. Imelik, In: *Metal Support and Metal Additives Effects in Catalysis: Proceedings of an International Symposium* organized by the Institute de Recherches sur la Catalyse, CNRS, Villeurbanne, Ecully (Lyon) 1982, B. Imelik (Ed.), *Studies in Surface Science and Catalysis*, Vol. 11, Elsevier, Amsterdam, 1982, p. 141.
96. C. Naccache, N. Kaufherr, M. Dufaux, J. Bandiera, B. Imelik, in *Molecular Sieves - II*, J. R. Katzer (Ed.), ACS Symposium Series, American Chemical Society, Washington, D.C., 1977, p. 538.
97. S. T. Homeyer, Z. Karpinski, W. M. H. Sachtler, *J. Catal.* **1990**, *123*, 60.
98. S. T. Homeyer, Z. Karpinski, W. M. H. Sachtler, *Recl. Trav. Chim. Pays-Bas*, **1990**, *109*, 81.
99. T. M. Tri, J. Massardier, P. Gallezot, B. Imelik, In: *Catalysis by Zeolites: Proceedings of an International Symposium*, Ecully (Lyon) 1980, B. Imelik (Ed.), *Studies in Surface Science and Catalysis*, Vol 5, Elsevier, Amsterdam, 1980, p. 279.
100. J. A. Rabo, V. Schomaker, P. E. Pickert, In: *Proceedings of the 3rd International Congress on Catalysis*, Amsterdam, 1964, W. M. H. Sachtler et al. (Eds.), North Holland, Amsterdam, 1965, p. 612.

101. G. D. Chukin, M. V. Landay, V. Kruglikov, D. A. Agievskii, B. V. Smirnov, A. L. Belozerov, V. D. Asrieva, N. V. Goncharova, E. D. Radchenko, O. D. Konovalcherov, A. V. Agofonov, In: *Proceedings, 6th International Congress on Catalysis*, Vol. 1: Imperial College, London, 1976, G. D. Bond et al. (Eds.), Verlag Chemie, Weinheim, 1977, p. 621.
102. H. T. Wang, Y. W. Chen, J. G. Goodwin, Jr., *Zeolites* **1984**, *4*, 56.
103. L. Xu, Z. C. Zhang, W. M. H. Sachtler, *J. Chem. Soc., Faraday Trans.* **1992**, *88*, 2291.
104. X. L. Bai, W. M. H. Sachtler, *J. Catal.* **1991**, *129*, 121.
105. M. Chow, S. H. Park, W. M. H. Sachtler, *Appl. Catal.* **1985**, *19*, 349.
106. T. J. McCarthy, G.-D. Lei, W. M. H. Sachtler, *J. Catal.* **1996**, *159*, 90.
107. V. Ponec, G. C. Bond, in *Catalysis by Metals and Alloys*, V. Ponec, G. C. Bond (Eds.), *Studies in Surface Science and Catalysis*, Vol. 95, Elsevier, Amsterdam, 1995.
108. T. T. T. Wong, W. M. H. Sachtler, *J. Catal.* **1993**, *141*, 407.
109. B. A. Lerner, B. T. Carvill, W. M. H. Sachtler, *Catal. Lett.* **1993**, *18*, 227.
110. G. A. Mills, H. Heinemann, T. H. Millikan, A. G. Oblad, *Ind. Eng. Chem.* **1953**, 45134.
111. M. S. Tzou, B. K. Teo, W. M. H. Sachtler, *J. Catal.* **1988**, *113*, 220.
112. T. T. T. Wong, Z. Zhang, W. M. H. Sachtler, *Catal. Lett.* **1990**, *4*, 365.
113. T. T. T. Wong, A. Yu. Stakheev, W. M. H. Sachtler, *J. Phys Chem.* **1992**, *96*, 7733.
114. J. Sárkány, W. M. H. Sachtler, *Zeolites* **1994**, *14*, 7.
115. M. S. Tzou, M. Kusunoki, K. Asakura, H. Kuroda, M. Moretti, W. M. H. Sachtler, *J. Phys. Chem.* **1991**, *95*, 5210.
116. P. B. Weisz, *Chemtech* **1973**, *3*, 498.
117. S. J. Tauster, J. J. Steger, *Mater. Res. Soc. Symp. Proc.* **1988**, *111*, 418.
118. F. Gault, in *Advances in Catalysis*, D. D. Eley, H. Pines, P. B. Weisz (Eds.), Academic Press, San Diego, 1981, Vol. 30, p. 1.
119. G. Moretti, W. M. H. Sachtler, *J. Catal.* **1989**, *116*, 350.
120. W. E. Alvarez, D. E. Resasco, *Catal. Lett.* **1991**, *8*, 53.
121. D. J. Ostgard, L. Kustov, K. R. Poeppelmeier, W. M. Sachter, *J. Catal.* **1992**, *133*, 342.
122. B. A. Lerner, B. T. Carvill, W. M. H. Sachtler, *J. Mol. Catal.* **1992**, *77*, 99.
123. B. Lerner, B. Carvill, W. M. H. Sachtler, *Catal. Today* **1994**, *21*, 23.
124. J. Kärger, M. Petzold, H. Pfeifer, S. Ernst, J. Weitkamp, *J. Catal.* **1992**, *136*, 283.
125. P. E. Eberle Jr., *J. Phys. Chem.* **1963**, *67*, 2404.
126. Z. Karpiński, S. N. Gandhi, W. M. H. Sachtler, *J. Catal.* **1993**, 337.
127. N. I. Jaeger, J. Rathousy, G. Schulz-Ekloff, A. Svenson, A. Zukal, In: *Zeolites: Facts, Figures, Future: Proceedings of the 8th International Zeolite Conference*, Amsterdam, 1989, P. A. Jacobs et al. (Eds.), *Studies in Surface Science and Catalysis*, Vol. 49, Elsevier, Amsterdam, 1990, p. 1005.
128. N. I. Jaeger, A. L. Jaeger, G. Schulz-Ekloff, *J. Chem. Soc., Faraday Trans.* **1991**, *87*, 1251.
129. W. M. H. Sachtler, Z. Zhang, A. Y. Stakheev, J. S. Feeley, in *New Frontiers in Catalysis*, L. Guzci, F. Solymosi, P. Tetenyi (Eds.), Akademiai Kiado, Budapest, 1993, p. 271.
130. B. T. Carvill, B. A. Lerner, B. J. Adelman, D. C. Tomczak, W. M. H. Sachtler, *J. Catal.* **1993**, *144*, 1.

2.4.5
Grafting and Anchoring of Transition Metal Complexes to Inorganic Oxides

Frédéric Averseng, Maxence Vennat, and Michel Che

2.4.5.1 Introduction

In the past, many heterogeneous catalysts have been prepared by deposition on oxide supports of transition metal ions (TMIs) and transition metal complexes (TMCs), hereafter referred to as precursors of the catalytically active phase. The aim was to obtain a high dispersion of this phase and to strongly bind it to the oxide support. Originally, anchored compounds were engineered in order to replicate the properties of homogeneous catalysts with an easier recovery after the reaction.

"Grafting" and "anchoring" are terms which are widespread in the literature but which often have different meanings, most likely because they are used by communities belonging to different disciplines.

The present chapter aims at clarifying some points of terminology regarding the state of TMIs/TMCs interacting with oxide supports which has been coined a variety of terms, including supported, adsorbed, fixed, immobilized, dispersed, etc. [1]. The latter, which also include "grafted" and "anchored", are often ill-defined. A brief overview of the terms grafting and anchoring will show that there is still room for improvement, leading to definitions based on spectroscopic data and to a simple nomenclature. Following a brief description of the principles of grafting and anchoring, the characterization of the resulting complexes will be outlined. Finally, some typical examples will be provided of grafted and anchored complexes used as catalysts.

Although the basic concepts discussed here can apply to very different catalytic systems, only those catalysts which are prepared by the grafting and anchoring of TMCs to inorganic oxides will be considered.

2.4.5.2 Grafting versus Anchoring: Definitions, Nomenclature, and Basic Principles

2.4.5.2.1 A Brief Overview of Earlier Definitions Several attempts have been made in the past to distinguish between grafted and anchored species. Thus, for Murrell [2], an anchored catalyst is attached to a surface via its ligand(s), while a grafted catalyst results from "... the direct coupling reaction of an organometallic complex, which has reactive metal–carbon bonds, with the hydroxyl

---
* Corresponding author.

groups of an inorganic oxide surface". This definition does not include inorganic complexes which are, nevertheless, the precursors most often used in the industry to prepare supported catalysts.

Campbell [3] states that, "...an anchored catalyst is created by the binding of a species, without substantial change in its structure, to a solid surface. A grafted catalyst is produced when an initial structure bound to the surface is altered considerably by subsequent treatments". However, how substantial or considerable is the alteration is not defined. The nature of the precursor complex is not specified, although the example used by Campbell to illustrate grafting concerns $MoCl_5$ (Scheme 1): the formation of the grafted catalyst starts by reaction of the silanol groups of a silica previously dehydrated at 473 K with $MoCl_5$ such that an intermediate anchored species ($OMoCl_4$) is obtained. With subsequent heating to 773 K in oxygen, the remaining chlorines are removed to finally produce a grafted species with terminal molybdenyl bonds [4, 5]. Thus, $MoCl_5$ can be either anchored or grafted, making the definition dependent on preparation conditions.

The International Union of Pure and Applied Chemistry (IUPAC) recommendations [6] state that "...deposition involving the formation of a strong (e.g., covalent) bond between the support and the active element is usually described as grafting or anchoring. This is achieved through a chemical reaction between functional groups (e.g. hydroxyl groups) on the surface of the support and an appropriately selected inorganic or organometallic compound of the active element." This definition no longer distinguishes between grafting and anchoring.

Finally, for Cornils et al. [7], "...anchoring is the immobilization of homogeneous catalysts on suitable supports according to any bonding", specifying that considering several publications, "...anchored ought to mean complexes which are chemically bonded to the support surface". Grafting is defined as "...a process of equilibrium adsorption of an active catalytic species (metal or precursor) by covalent bonding to a solid support from a solution in which the support is suspended". These definitions are more involved than the preceding ones because they include catalytic species and not their precursors, and the method used to bind them to the support.

It appears from the above definitions, that grafted and anchored species cannot be unambiguously distinguished.

2.4.5.2.2 **Proposed New Definitions** [8] The cartoon shown in Fig. 1 depicts what common use defines as a "boat anchored normally to the seabed" – or here, for the sake of simplicity – to a pier, and also provides the chemical/molecular analogue. It should be noted that, there is no immobile common point between the complex and the support. The anchored complex still retains some degree of mobility within a space defined by the length of the linker, which anchors the metal center to the oxide surface.

By analogy with grafting defined in horticulture as "... a method of plant propagation where tissues of one plant are encouraged to fuse with those of another", it is easy to imagine a boat grafted to a pier via an immobile common point, the bollard. Figure 1 also provides the chemical/molecular analogue,, where the complex is grafted to the support surface via immobile and common donor atom(s). In this situation, the complex has lost most of its mobility.

According to these simple models, the following definitions, based on a molecular point of view, can be proposed, whatever the synthetic route used to bind the complex to the support:

- *Grafting* and *anchoring* both refer to chemical bonding of molecular species (often TMCs) to a surface, thus excluding physisorbed species (hydrogen-bonded, Van der Waals-sorbed or entrapped species, etc.) or purely organic species.
- *Grafting* refers to the situation where one/several group(s) of the support surface is/are chemically bonded to a metal center, these groups being part of its inner coordination sphere and thus acting as ligands. The grafting results in electron sharing between the surface and the metal center, a situation which modifies the features of the metal center (such as symmetry, coordination number, oxidation or even spin state).
- *Anchoring* refers to the situation where the metal center of a complex is connected to the surface by a linker via a series of covalent bonds. As a consequence, there cannot be any electron doublet shared between the metal center and the surface group. The ligand(s) involved in anchoring is/are most often referred to as "linker(s)", "tether(s)", or "spacer(s)". In anchored species, the coordination sphere of the metal center is not (or is only marginally) altered compared to

$$\underset{Silica}{\overset{OH}{|}} + MoCl_5 (g) \xrightarrow[-HCl]{473\ K} \underset{\underset{\text{anchored species}}{Silica}}{\overset{OMoCl_4}{|}} \xrightarrow[O_2]{773\ K} \underset{\underset{\text{grafted species}}{Silica}}{\overset{O}{\underset{|\ |\ |}{\overset{\|}{\underset{O\ O\ O}{M}}}}}$$

Scheme 1 Example for the preparation of a grafted catalyst. (Adapted from Ref. [3].)

References see page 537

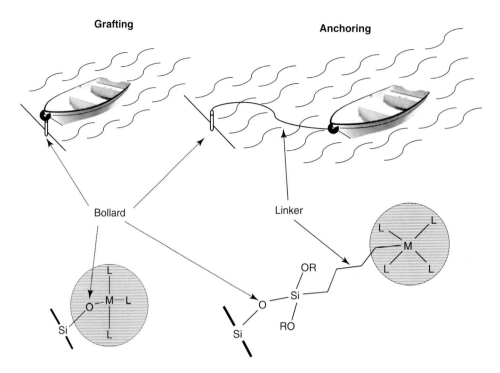

Fig. 1 Representation of "grafting" and "anchoring" of boats and complexes to a pier and an oxide support, respectively.

the precursor complex in solution or in the gas phase.

The above definitions allow a clear distinction to be made between grafting and anchoring in the case of mononuclear species, but this distinction becomes more difficult for polynuclear species. In such cases, as soon as at least one metal center is chemically bonded to the surface, the entire polynuclear species will be considered as grafted. For those particular situations, a nomenclature is proposed below to characterize the nature of grafting and/or anchoring.

2.4.5.2.3 Nomenclature, Basic Principles, and Oxide Support Characteristics Because of the importance of the metal center, initially in the definitions of grafting and anchoring and then in the catalytic process, the nomenclature of grafted/anchored species given below is essentially based on the metal center. Subsequently, the basic principles of grafting/anchoring to oxide support will be provided, the characteristics of which will be briefly discussed.

A Proposed Nomenclature for Grafted/Anchored Species
Proposed here a simple notation that may help rationalize the grafting or anchoring nature of a surface species. This notation completes that given in the IUPAC recommendations to describe TMCs [9]. It is a simple way to describe the points of attachment of species to the oxide surface, in terms of chemical nature and number.

A grafted species (Fig. 2) is noted $(aM/x, y, z \ldots)$, where a is the number of Metal centers of the species considered, $x, y, z \ldots$ are the number of chemical bonds between each Metal center and the surface (for the sake of simplicity, when there is only one metal center, $a = 1$ is omitted). Depending on the nature of the surface sites, geometrical restrictions (cf. Section 2.4.5.2.3B) and the number of labile ligands it possesses, the metal center of a TMC establishes up to three chemical bonds (thus, $x, y, z \leq 3$) with the surface, as illustrated in Fig. 2 for mono- and polynuclear species. This maximum number of three which refers to surfaces of conventional oxides (such as alumina, silica, titania, magnesia, etc.) can be exceeded for zeolites (e.g., up to six for internal sites such as S_I in faujasite-type zeolites), or for layered materials which can offer two parallel surfaces to sandwich a single TMC.

An anchored species is denoted $(aM/Lx, Ly, Lz \ldots)$, where a is the number of Metal centers, but $x, y, z \ldots$ now refers to the number of chemical bonds between each Linker and the surface. Ligands not involved as Linkers are not included. One example is given in Fig. 3. Due to the size and/or geometry of certain ligands, it should be noted that a single TMC may be anchored to the surface by more than three chemical bonds.

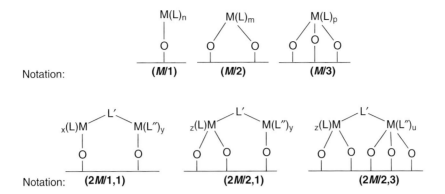

Fig. 2 Examples of mono- and polynuclear species grafted to an oxide support and proposed nomenclature.

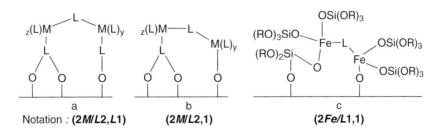

Fig. 3 Examples of polynuclear species anchored or grafted-anchored to a surface and proposed nomenclature. (Model (c) adapted from Ref. [10].)

This simple notation may be used to describe accurately some uncommon situations where grafting and anchoring are simultaneously observed for the same species, depending on the metal center being referred to. A general example is given in Fig. 3, and illustrated by a specific case taken from the literature [10].

As the preparation of either homogeneous supported or heterogeneous catalysts involves several steps which can strongly modify the interactions between the precursor and the support surface, this notation may also help describe each intermediate species encountered, as discussed in Section 2.4.5.3 and the following text.

B Basic Principles The previous notation can also be used to describe how specific TMC-support interactions can be produced via chemically engineered complexes (see Chapter 2.4.2 of this Handbook).

Thus, the octahedral Ni(II) complexes [Ni(en)(dien)(H$_2$O)]$^{2+}$, [Ni(en)$_2$(H$_2$O)$_2$]$^{2+}$, and [Ni(dien)(H$_2$O)$_3$]$^{2+}$ (where en = NH$_2$−CH$_2$−CH$_2$−NH$_2$ is the bidentate ligand ethylenediamine, and dien = NH$_2$−CH$_2$−CH$_2$−NH−CH$_2$−CH$_2$−NH$_2$ is the tridentate ligand diethylenetriamine) contain one, two, or three labile water ligands, respectively. From a kinetic point of view, it is known that the substitution of some aqua ligands for inert polydentate amine ligands in octahedral complexes increases the lability of the remaining aqua ligands [11]: the inertia of polydentate amine ligands is a consequence of their thermodynamic stability.

It has been shown that tuning the ratio of kinetically labile/inert ligands in the series of octahedral Ni(II) complexes shown in Fig. 4 allows one exclusively to obtain (M/1), (M/2) or (M/3) grafted species by replacing one, two, or three labile aqua ligands by one, two, or three Si−O surface groups, respectively [11, 12].

The previous results surprisingly observed with amorphous silica, which is formed from randomly distributed −[Si−O]$_n$−cycles, with $n \geq 3$ (Fig. 5), suggest that TMCs can act as probes towards specific adsorption sites of silica, hence leading to a similar phenomenon encountered in bio- and supramolecular chemistry, referred to as molecular recognition [12].

In the case of anchored species, the ligands, TMCs or oxide surface must at some stage be functionalized in order for anchoring to occur. Moreover, the functionalization of the ligand(s) must be conducted so that the geometry, steric hindrance and electronic properties of the metal center of the anchored TMC be as close as possible to its homogeneous counterpart.

References see page 537

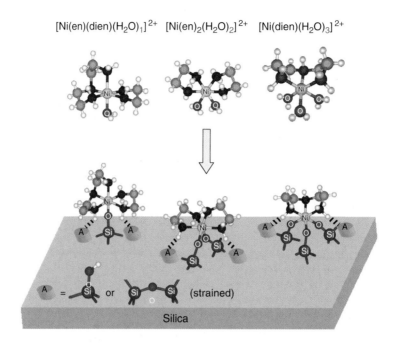

Fig. 4 Aqua/amine Ni(II) complexes presenting different aqua ligand/Ni stoichiometries for selective grafting. (Adapted from Ref. [12].)

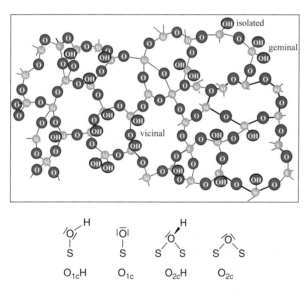

Fig. 5 Upper: Amorphous silica model with isolated, geminal or vicinal OH-groups (adapted from Ref. [24]). Lower: Terminal and bridging OH-groups and their deprotonated forms.

C Oxide Support Characteristics

The supports most frequently used in catalysis are inorganic oxides (alumina, silica, zeolites, magnesia, titania, zirconia, ceria, etc.) because they possess important physical features (such as high surface area, thermal stability, low cost, reproducibility, regenerability) [13, 14]. From the chemical point of view, they also exhibit intrinsic acid–base properties, which can be tuned in a wide range and used in catalytic reactions [15, 16]. Polymers and carbon-based materials have been less used as supports [17, 18].

The common feature of inorganic oxide supports is the presence on their surface of sites with donor atoms (most often oxygen) which act as weak σ and π donor ligands and are able to bind TMCs via ligand substitution reactions [19, 20]. These surface sites can either be naturally present on the support or the result of modification/addition of/to the surface by suitable thermal/chemical treatment. Qualitative and quantitative aspects must also be considered.

With regards to the qualitative aspect, the donor properties of surface oxygen atoms depend strongly on their surroundings. Consider, for example, the case of model systems such as MgO of the alkaline-earth series with a NaCl structure (Fig. 6). At the surface of dehydroxylated MgO, different types of oxide ion have been identified by spectroscopy [21], particularly by photoluminescence [22]. Such ions have been labeled O^{2-}_{LC}, where LC (low coordination) can take values of five, four, or three for oxide ions on planes, edges, or corners respectively, as shown in Fig. 6 [21]. Upon adsorption of water on MgO [23]:

$$Mg^{2+}_{LC} - O^{2-}_{L'C} + H_2O \rightleftharpoons Mg^{2+}_{LC} \overset{OH^-}{\underset{|}{}} - O^{2-}_{L'C} \overset{H^+}{\underset{|}{}} \quad (1)$$

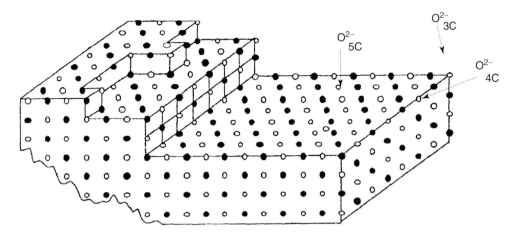

Fig. 6 Representation of a surface (100) plane of MgO showing surface imperfections such as steps, edges, kinks, corners, etc., which provide sites for ions of low coordination. (Adapted from Ref. [21].)

two main types of extraframework hydroxyl groups can be formed: $O_{L'C}H$ groups generated by protonation of surface oxide anions, and OH groups issued from the hydroxylation of surface Mg^{2+}_{LC} cations [Eq. (1)]. Both OH types can act as ligands towards TMCs: mono ($O_{1C}H$) and poly-coordinated ($O_{LC}H$ with L = 3, 4, or 5). The situation is more involved when neighboring oxygen atoms interact via hydrogen bonds.

For amorphous oxides, such as silica with a disordered structure (see Fig. 5) [24], the coordination of oxygen atoms can be defined in the same way. The latter, which can either be protonated or not, have a low coordination number, such as O_{1C} or O_{2C}, when they are in a terminal or bridging position, respectively. Depending on whether hydroxyl groups are linked to the same or two neighboring Si atom(s), they are referred to as either geminal or vicinal.

Finally, it is important to recall (see Chapter 2.4.2) that the pH of the aqueous solution in contact with the oxide support leads to either deprotonation or protonation following the equilibria:

$$S-OH \rightleftharpoons S-O^- + H^+ \quad (2)$$

$$S-OH + H^+ \rightleftharpoons S-OH_2^+ \quad (3)$$

With regards to the quantitative aspect, the concentration of TMCs grafted or anchored to the surface is dependent on the density of surface groups to which they bind, and thus on the preparation and further thermal treatment of the oxide support. In some cases, the surface of the oxide support may be chemically modified so as to obtain isolated surface groups. Thus, while treatment of MgO with water leads to the two types of hydroxyl groups described above, treatment with methanol will generate methoxy groups sitting on Mg^{2+}_{LC} cations and isolated hydroxyls of the type $O^{2-}_{L'C}H^+$.

2.4.5.3 Preparation: Context and Reactions

2.4.5.3.1 Grafting The definitions given above not only allow the distinction to be made between grafted and anchored species, but they are also in line with the different strategies used to prepare such species.

As described above, grafting can be achieved already at the deposition step at the liquid–solid interface when labile ligands are present in precursor TMCs, as shown in Section 2.4.5.2.3B for Ni(II) complexes. If this is not the case, or if precursor TMCs are not all grafted, then a thermal activation step (drying and/or calcination) may be required to initiate or complete the grafting and to make the grafted complexes catalytically active. This thermal activation alters the features of the initial species: the number and nature of ligands, the oxidation state and symmetry of the metal center environment with concomitant creation of vacant sites, for adsorption and catalysis to occur.

In contrast to anchoring, in the case of grafting the oxide support which is directly bonded to the metal center influences the catalytic properties of the latter. Grafted materials are prepared mostly via processes involving liquid phase (ion exchange, impregnation) and less via gas phase (chemical vapor deposition) processes.

A Grafting Reactions Grafting occurs by the insertion of one or several donor atoms of the surface into the coordination sphere of the TMI, resulting in chemical

References see page 537

bonding:

$$y[\text{supp}-\text{OH}] + \text{ML}_n$$
$$\longrightarrow [\text{supp}-\text{O}]_y - \text{ML}_{n-y} + y\text{HL} \quad (4)$$

This process does not normally change the metal oxidation state. When grafting involves a metal carbonyl complex, the metal center is oxidized as illustrated below:

$$[\text{supp}-\text{OH}] + \text{M}_3(\text{CO})_{12}$$
$$\longrightarrow [\text{supp}-\text{O}] - \text{M}_3\text{H}(\text{CO})_{10} + 2\text{CO} \quad (5)$$

Because both metal carbonyl and hydride complexes are air- and moisture-sensitive, a thermal pretreatment of the oxide support is required to remove air and water traces. The grafting reaction can take place either in inert solvents or in the gas phase via chemical vapor deposition (CVD) (see Chapter 2.4.6 of this Handbook).

After grafting, and in order to eliminate physisorbed and thus non-grafted TMCs, materials prepared in liquid phase are thoroughly washed with the solvent used for grafting, while those prepared by CVD are cleaned by purging under dry inert gas.

B Activation of the Grafted Material   In order to be catalytically active, the grafted material may require thermal and/or chemical activation, to remove the remaining ligands. This is best achieved by calcination, usually in the range 573 to 773 K in flowing air/oxygen in order to cancel the reducing properties of hydrogen-containing ligands such as ammonia by the following reaction:

$$2\,\text{NH}_3 + \tfrac{3}{2}\text{O}_2 \longrightarrow \text{N}_2 + 3\text{H}_2\text{O} \quad (6)$$

When activation is performed in a neutral atmosphere (e.g., nitrogen or argon), the thermal decomposition of the ligands (e.g., ethylenediamine $\text{H}_2\text{N}-\text{CH}_2-\text{CH}_2-\text{NH}_2$) can release hydrogen which in turn can reduce the metal ions into metal particles [25].

This can also be achieved by hydrolysis of the metal–ligand bonds with water, either in liquid or vapor phase in the case of moisture-sensitive species, leading to aqua or hydroxo species:

$$[\text{supp}-\text{O}^-]_y - \text{ML}_n + n\text{H}_2\text{O}$$
$$\longrightarrow [\text{supp}-\text{O}^-]_y - \text{M}(\text{OH}_2)_n + n\text{L} \quad (7)$$
$$[\text{supp}-\text{O}^-]_y - \text{ML}_n + n\text{H}_2\text{O}$$
$$\longrightarrow [\text{supp}-\text{O}^-]_y - \text{M}(\text{OH})_n + n\text{HL} \quad (8)$$

Mild thermal decomposition of the ligands under moist conditions may also undertaken leading to similar aqua or hydroxo species.

2.4.5.3.2 Anchoring   The idea lying behind the preparation of anchored complexes is to obtain materials which combine the properties of homogeneous catalysts (suited to fine and/or enantioselective chemistry) with an increased facility to recover the catalytic material from the reaction medium. The anchored complexes are expected to retain the same catalytic properties as their homogeneous counterparts. The support mainly acts as a point of anchoring and does not interfere in the catalytic process, except for instance if the anchored complexes are in high concentration and/or the oxide support exhibits narrow pores.

As for the support pretreatment and elimination of physisorbed TMCs, the same features given for grafted complexes hold true for anchored ones.

A Functionalization and Anchoring Reactions   Several strategies have been proposed for efficient and controlled anchoring of TMCs, but whatever the differences, the attachment of the TMC to the oxide surface necessarily involves the functionalization of one or more of the "anchoring actors" (namely, the surface, ligand or TMC). The process may be carried out in one step, by mixing all the reactants together, or performed in several steps. Two of the more common strategies are shown below (where L_f represents *functionalized* ligand):

Strategy 1 : $\quad x[\text{supp}-\text{OH}] + \text{M}(L_f)_y(\text{L})_z \longrightarrow$
$$[\text{supp}-\text{O}^-]_x(L_f)_y\text{M}(\text{L})_z + x\text{H}^+ \quad (9)$$

Strategy 2 : $\quad x[\text{supp}-\text{OH}] + yL_f \longrightarrow$
$$[\text{supp}-\text{O}^-]_x(L_f)_y + x\text{H}^+ \quad (10)$$
$$[\text{supp}-\text{O}^-]_x(L_f)_y + \text{MX}_z \longrightarrow$$
$$[\text{supp}-\text{O}^-]_x(L_f)_y\text{MX}_{z-x} + x\text{X} \quad (11)$$

It should be noted that the oxidation state of the metal does not change. These operations are most often carried out in non-aqueous, non-coordinating solvents or through gas-phase reactions by CVD. These reactions are thus limited by either the solvent boiling point or the decomposition temperature of the precursor TMCs.

B Activation of the Anchored Material   In contrast to grafted materials, once prepared, anchored materials are generally used as such for catalytic applications. It is important to note that they are most often very sensitive to any thermal or chemical treatment.

2.4.5.4 Characterization of Anchored/Grafted Complexes
Due to the low content of TMCs (often of the order of one percent) and the complexity of the catalytic

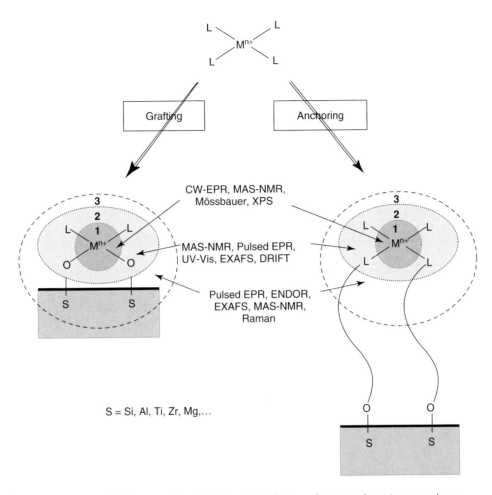

Fig. 7 Molecular spectroscopies and their range of analysis. 1, metal; 2, first coordination sphere; 3, outer sphere.

system, the combined use of surface techniques and spectroscopies is necessary to characterize the nature, quantity and dispersion of the species deposited [26–28]. Figure 7 shows a schematic representation of the "area of analysis" that a given technique covers, and how it helps to elucidate the nature of the bonds between TMCs and the oxide support.

Because surface groups are part of the coordination sphere of the metal center, grafted complexes exhibit different features from those of their precursors, in contrast to anchored complexes which retain most of their molecular specificities. Spectroscopies (e.g., Mössbauer [29], UV-visible [30a], continuous-wave electron paramagnetic resonance (EPR) [30a,b], metal center nuclear magnetic resonance NMR [27]) which provide information on the metal center and its close surroundings will thus help to distinguish between anchored and grafted species. The determination of other characteristics, such as metal dispersion and speciation on the oxide surface, requires techniques and spectroscopies [e.g., MS-coupled temperature programmed oxidation (TPO), extended X-ray absorption fine structure (EXAFS), multi-probe NMR] that cover a broader "area of analysis" or that focus (e.g., electron microscopy [31], X-ray photoelectron spectroscopy (XPS) [32]) on a larger than molecular scale.

2.4.5.5 Grafted and Anchored Materials as Heterogeneous Catalysts: Selected Overview

In this section, selected examples are illustrated of catalytic reactions promoted by TMCs obtained after grafting/anchoring. Various "inorganic support–transition metal complex" couples are presented which lead to grafted or anchored species labeled according to the nomenclature given in Section 2.4.5.2.3A.

2.4.5.5.1 Grafting

A Organometallics and TMCs

a One Metal Center – Two Oxygen Atoms of the Support: (M/2) Many examples of alkylidene ligands bound to

---

References see page 537

high-valent complexes of Mo, W, Ta, and Nb can be found in the literature. A large number have been characterized by X-ray diffraction (XRD) and spectroscopy [33, 34]. The need for well-defined complexes has been extensively demonstrated in the polymerization of alkynes [35] and the metathesis polymerization of alkenes [34, 36]. Alkylidene ligands can be formed on heterogeneous catalysts: when silica is employed as bidentate ligand in bis(alkyl) complexes of first-row transition metals, uncommon terminal alkylidene fragments are readily formed and stabilized [37]. These materials have the empirical formula $(\equiv SiO)_2M=CHEMe_3$ (E = C, Si).

In order to prepare silica-supported bis(alkyl) metal complexes, Beaudoin et al. [37] reported that homoleptic metal alkyls MR_4 (where M is Ti or Cr and R is CH_2CMe_3; M is V or Cr and R is CH_2SiMe_3) react at room temperature with surface hydroxyl groups of partially dehydroxylated Aerosil silica. Under controlled conditions, the reaction occurs with well-defined and reproducible stoichiometry. The amount of protonated ligand RH liberated into the gas phase represents two equivalents for each metal complex grafted onto the silica surface:

$$2(\equiv SiOH) + MR_4 \longrightarrow (\equiv SiO)_2MR_2 + 2RH \quad (12)$$

The 2 : 1 stoichiometry, leading to $M/2$ grafted species, was reported earlier by several groups by the reaction of MR_4 on pyrogenic silicas pretreated at 473 K [38–41], and differs from the 1 : 1 stoichiometry reported for silicas pretreated at higher temperatures [38, 42–46]. This led Ballard to speculate that, on silica pretreated at 473 K, most hydroxyl groups were not randomly distributed but rather occurred in pairs [39].

The formation of alkylidene ligands on heterogeneous catalysts, $(\equiv SiO)_2M=CHEMe_3$ (E = C, Si and M = Mo, W or Re) used for metathesis [47, 48] and polymerization (M = Cr) [49, 50] was suggested on doubly grafted complexes:

$$(\equiv SiO)_2M(CH_2EMe_3)_2$$
$$\longrightarrow (\equiv SiO)_2M=CHEMe_3 + EMe_4 \quad (13)$$

Zirconia, and its chemical modification by grafting heteroelements, have been studied to a lesser degree than silica and alumina, the advantages of zirconia's high thermal stability and acidic, basic, and redox properties being counterbalanced by its rather low surface area. However, since the development of SBA-15 and the success reported to synthesize mesoporous zirconia with surface areas exceeding 250 m^2 g^{-1}, interest in zirconia has grown somewhat [51].

Earlier reviews [52–54] reported that Fe/ZrO$_2$ systems were better catalysts than Fe/Al$_2$O$_3$ or Fe/SiO$_2$ in several reactions (NO–CO reaction, CO$_2$ hydrogenation, reduction of sulfates by H$_2$S, etc.). These few examples show a renewed interest in the use of the Fe/ZrO$_2$ catalytic system.

Van der Voort et al. [55] were the first to study the reaction of Fe(acac)$_3$ (where acac represents acetylacetonate) with zirconia (Fig. 8), which readily occurs both in liquid and gas phase. Fe(acac)$_3$ decomposes upon contact with zirconia: two of the ligands bind to zirconia, while the third remains bonded to Fe. The third ligand is, however, extremely unstable in the presence of water and hydrolyzes very quickly with the release of Hacac (Fig. 8). The thermally unstable Zr-acac sites transform into Zr-acetate species after thermal treatment in air at 383 K, or in vacuum above 473 K. After calcination, the organic ligands are removed and the zirconium hydroxyl groups restored. The grafting of iron onto zirconia is an interesting procedure to form either redox catalysts or solid-state fuel cells [56].

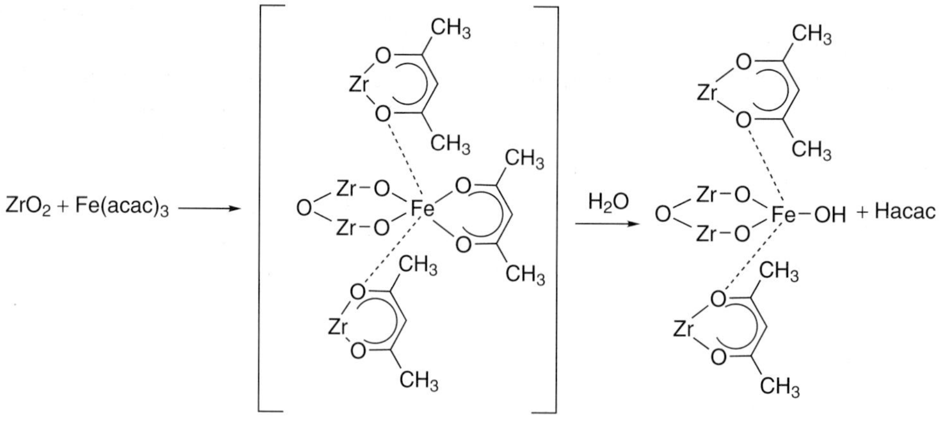

Fig. 8   Reaction of Fe(acac)$_3$ with coordinatively unsaturated Zr sites. (Adapted from Ref. [55].)

b One Metal Center – One, Two, or Three Oxygen Atoms of the Support: (M/1), (M/2) or (M/3) Basset and coworkers reported the preparation of vanadium(IV) surface complexes obtained via the reaction of tetrakis(dimethylamido)vanadium(IV), V[N(CH$_3$)$_2$]$_4$, with non-porous SiO$_2$, Al$_2$O$_3$, and TiO$_2$ as models for heterogeneous vanadium catalysts [57]. The reaction was performed either in pentane solution or in gas phase under inert conditions, and the density and number of hydroxyl groups were adjusted by thermal pretreatment at 333 K of the oxides in vacuum. Chemical analysis and EPR spectroscopy show that the coordination sphere of vanadium and the number of oxygen bridges to the surface depend heavily on the nature of the support. Indeed, it appears that all species obtained are of the type $M/1$, $M/2$, or $M/3$, on SiO$_2$, Al$_2$O$_3$, and TiO$_2$, according to elemental composition (Scheme 2).

B Metal Oxychlorides

a One Metal Center – One Oxygen Atom of the Support: (M/1) Heterometallic oxides find wide applications in materials science (e.g., in glasses and ceramics) and heterogeneous catalysis, because of unique properties conferred by the synergy between different metal sites. One particularly important heterometallic oxide is the ternary V–Ti–Si system. This is reported to be more active and selective than any of the possible bimetallic combinations (i.e., V–Si, Ti–Si, or V–Ti) for the oxidation of *o*-xylene to phthalic anhydride [58], and in the selective catalytic reduction (SCR) of NO$_x$ [59], for which it is used in a commercial process [60].

Anpo et al. [61a] and Rice and Scott [61b] used a non-hydrolytic, low-temperature route to ternary V–Ti–Si catalysts. The vanadium overlayer is deposited first, by complete reaction of VOX$_3$ (X = Cl, O*i*Pr) with surface hydroxyl groups of pyrogenic silica:

$$(\equiv SiOH) + VOCl_3 \longrightarrow (\equiv SiO)VOCl_2 + HCl \quad (14)$$

Ti(O*i*Pr)$_4$ migrates underneath the vanadium overlayer to produce a heterobinuclear complex containing one Ti and one V, bound to the silica surface by Si–O–Ti linkage [Eq. (15)]. The heterobinuclear complex is not formed when TiCl$_4$ is used [Eq. (16)], nor can it be generated by first depositing Ti(OR)$_4$ (note that the reaction is not written stoichiometrically but only lists the products formed):

$$(\equiv SiO)VOCl_2 + Ti(OiPr)_4 \longrightarrow$$
$$(\equiv SiO)Ti(O)_3V(OiPr)_2$$
$$+ \{iPrCl + iPrOH + CH_3CH=CH_2\} \quad (15)$$
$$(\equiv SiO)VOCl_2 + TiCl_4 \longrightarrow (\equiv SiO)TiCl_3 + VOCl_3 \quad (16)$$

The mixed V–Ti alkoxide complex on silica may serve as a structural model of the active site on dehydrated V–Ti–Si catalysts. The selectivity and activity of these materials have been linked to the high dispersion of V, the absence of exposed Ti, and the silica support, which contributes via its high surface area, as well as thermal and mechanical stability [62]. The sequential CVD technique holds promise for the future "design" of isolated multifunctional active sites in heterometallic oxide catalysts.

C Metal Alkoxides and Metal Oxo Complexes

a Two Metal Centers – Two Oxygen Atoms of the Support: (2M/1, 1) Epoxides are versatile reactive intermediates in organic synthesis because they readily transform into diols and polyols, polyethers, amino alcohols, etc., which form the basis for a wide variety of consumer products. As a consequence, the catalytic epoxidation of alkenes has been much investigated [63–65]. Homogeneous-based systems, including Ti(O*i*Pr)$_4$/*t*BuOOH, have been used for the asymmetric oxidation of allylic alcohols to epoxy alcohols. Shell developed a heterogeneous epoxidation catalyst for a wide range of alkene substrates [66], but unfortunately the catalyst rapidly deactivated in the presence of moisture. Indeed, moisture-tolerant zeolite-based titanium catalysts such as TS-1 have been used for epoxidation [67–71]. However, because zeolite cavities restrict access to the Ti sites, reactions are limited to unhindered alkenes. Recently, Mayoral and coworkers [72, 73] described the preparation of SiO$_2$-supported titanium catalysts derived from Ti(O*i*Pr)$_4$ which are active for the epoxidation of cyclohexene without necessary calcination. The active sites were suggested to be $(\equiv SiO)_2Ti(OiPr)_2$:

$$2(\equiv SiOH) + Ti(OiPr)_4$$
$$\longrightarrow (\equiv SiO)_2Ti(OiPr)_2 + 2iPrOH \quad (17)$$

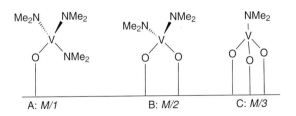

Scheme 2 Singly (A), doubly (B), and triply (C) grafted vanadium species to one, two, or three oxygen(s) of the support. (Adapted from Ref. [57].)

Bouh et al. [74] investigated the nature of active sites generated by gas-phase grafting of titanium isopropoxide onto Aerosil silica. On silica partially dehydroxylated at 473 K, the exclusive grafted product is [(≡SiO)Ti(OiPr)$_2$]$_2$(O) (note that the reaction is not written stoichiometrically, but only lists the products formed):

$$2(\equiv\text{SiOH}) + 2\text{Ti}(\text{O}i\text{Pr})_4$$
$$\longrightarrow [(\equiv\text{SiO})\text{Ti}(\text{O}i\text{Pr})_2]_2\text{O} + 3i\text{PrOH} + \text{C}_3\text{H}_6 \quad (18)$$

Sensarma et al. [75] reported that treatment of [(≡SiO)Ti(OiPr)$_2$]$_2$O with tBuOOH results in the quantitative displacement of the alkoxide ligand and formation of the titanium active sites (Fig. 9), with the grafting reaction being followed using in-situ infra-red (IR) spectroscopy.

Silica and alumina substrates have been much used to support transition-metal oxides for catalytic applications [76, 77], with micelle-templated inorganic oxides (in particular mesoporous silicas) have emerged as promising supports during the past decade. The grafting of organometallic or inorganic complexes onto such supports, followed by calcination, is known to produce isolated oxo-metal species. Ferreira et al. [78] recently reported a convenient one-step procedure to prepare Mo(VI) active sites on MCM-41 and MCM-48, involving the direct grafting of complexes [MoO$_2$X$_2$(thf)$_2$] (X = Cl, Br) onto purely siliceous supports using dichloromethane as solvent. Nunes et al. [79] used Mo EXAFS and IR to probe the local structure of supported Mo species in MCM-41 grafted from Mo(VI) complexes mentioned above. The high dispersion of active sites appears to promote good catalytic activity for the epoxidation of alkenes by $tert$-butyl hydroperoxide (TBHP). Using triethylamine to activate silanol groups, the material formed with the dichloro complex contains dimeric (Scheme 3) or polymeric species. Moreover, these oxo-bridged dimers were found to have the highest activity for the oxidation of alcohols.

b One Metal Center – Two Oxygen Atoms of the Support: (M/2) Under the same conditions as above, the material formed with the dibromo, instead of the dichloro complex, contains more mononuclear than dinuclear sites, revealing different reactivity and surface chemistry of the precursor complex (Scheme 4). Furthermore, the

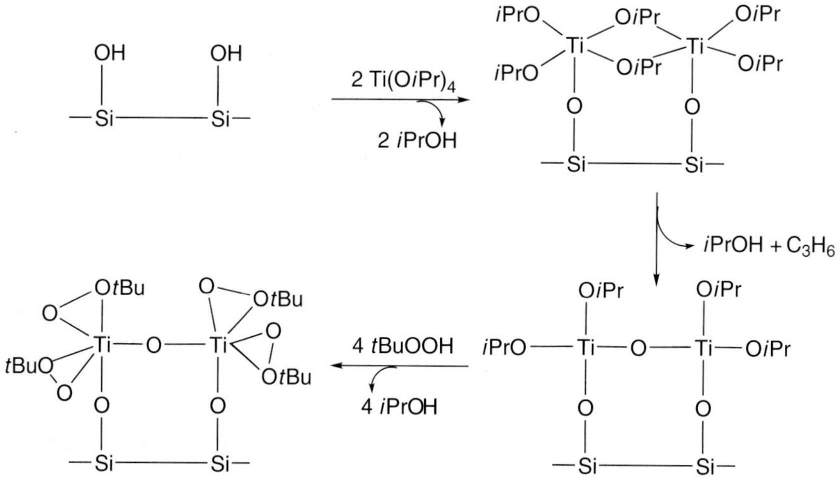

Fig. 9 Procedure for the preparation of silica-supported titanium catalyst. (Adapted from Ref. [75].)

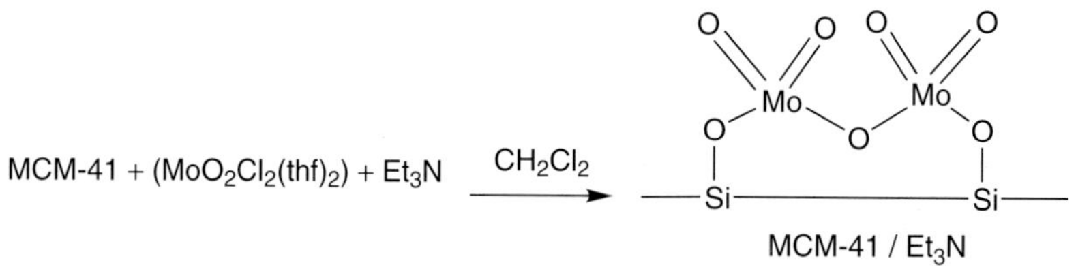

Scheme 3 Representation of the predominant species in MCM-41 in presence of Et$_3$N and [MoO$_2$Cl$_2$(THF)$_2$] as evidenced by Mo K-edge XAFS spectroscopy. (Adapted from Ref. [79].)

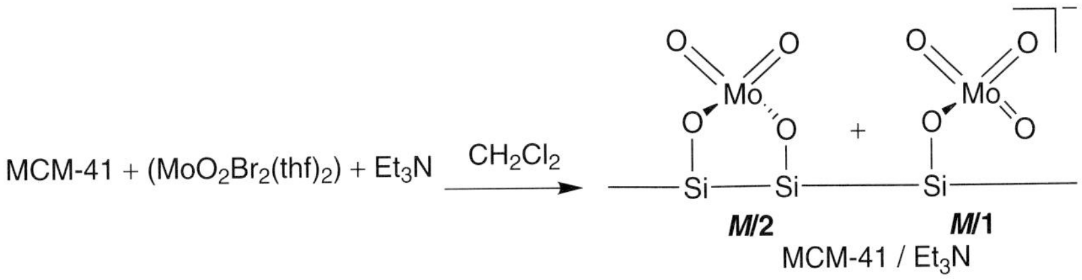

Scheme 4 Representation of the predominant species in MCM-41 in presence of Et$_3$N and [MoO$_2$Br$_2$(THF)$_2$] as evidenced by Mo K-edge XAFS spectroscopy. (Adapted from Ref. [79].)

preparation of isolated Mo(VI) active sites on ordered mesoporous silica is especially interesting, because the resulting materials are widely used for a variety of selective oxidative dehydrogenations, such as oxidation of methane to methanol and formaldehyde [80].

The conversion of methane is another reaction of industrial relevance; a widely studied case is the partial oxidation of methane (MPO) to formaldehyde and methanol. Recently, catalysts based on MoO$_3$ supported on zeolites, mainly HZSM-5, were found to be active in the dehydrogenation and aromatization of methane [81–83]. Both oxomolybdenum species, supported within the pore system, and acidic sites of the zeolite are implied in the reaction mechanism. Antiñolo et al. [84] reported a new MPO catalyst prepared by grafting MoO$_2$(acac)$_2$ onto HZSM-5 treated at 523 K under vacuum. By using FT-IR, the reaction was shown to take place between the acidic hydroxyl groups of the zeolite and the basic acac ligands. After grafting, activation of the solid leads to well-dispersed oxo Mo(VI) species:

$$\text{MoO}_2(\text{acac})_2 + 2(\equiv\text{SiOH})$$
$$\longrightarrow (\equiv\text{SiO})_2\text{MoO}_2 + 2\text{Hacac} \quad (19)$$

In comparison with the classical Mo/HZSM-5, the material prepared by Antiñolo exhibits enhanced activity.

c One Metal Center – One Oxygen Atom of the Support: (M/1) The FeZSM-5 zeolite has attracted considerable attention because of its high catalytic activity for the reduction of nitrogen oxides (NO$_x$) [85], and also for the selective oxidation of hydrocarbons with nitrous oxide (N$_2$O) as oxidant [86, 87]. Nozaki et al. [88] developed a new molecular precursor strategy to graft the Fe[OSi(OtBu)$_3$]$_3$(THF) complex onto SBA-15, leading to ≡SiO−Fe−[OSi(OtBu)$_3$]$_2$(THF) species (Scheme 5). By spectroscopy (particularly EPR, NMR, UV-vis, and *in-situ* IR) and elemental analysis, it has been shown that in the grafted species, the Fe(III) center is pseudotetrahedral.

2.4.5.5.2 Anchoring

A Metal Oxo complexes

a Two Metal Centers – One Linker – Two Oxygen Atoms of the Support: (2M/L2) Rhenium- and tungsten-based compounds are efficient catalysts for epoxidation using aqueous H$_2$O$_2$ [89]. Thus, Hoegaerts et al. [90] prepared a catalyst with tungsten anchored to mesoporous MCM-41. The method involves the reaction of tungsten peroxo anions ([PO$_4$(WO(O$_2$)$_2$)$_4$]$^{3-}$) with MCM-41 functionalized with covalently bonded phosphoramide groups (Fig. 10).

New signals appear in the solid-state ^{31}P-NMR spectrum, with the addition of peroxo-W complexes to phosphoramide-functionalized MCM-41, showing that the −PO$_3$H$_2$ groups react with peroxo-W compounds with the formation of P−O−W bonds.

b One Metal Center – One Linker – Two Oxygen Atoms of the Support: (M/L2) Jia and Thiel [91], in studying epoxidation catalysts, functionalized MCM-41 with a bidentate pyrazolylpyridine ligand to prepare an anchored complex of the type [MoO(O$_2$)$_2$(L-L)] [92]. The hybrid

---

References see page 537

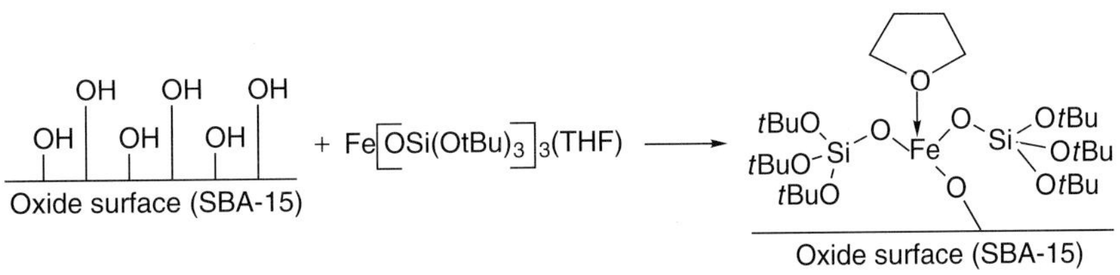

Scheme 5 Formation of grafted iron species, (≡SiO)−Fe−[OSi(OtBu)$_3$]$_2$(THF). (Adapted from Ref. [88].)

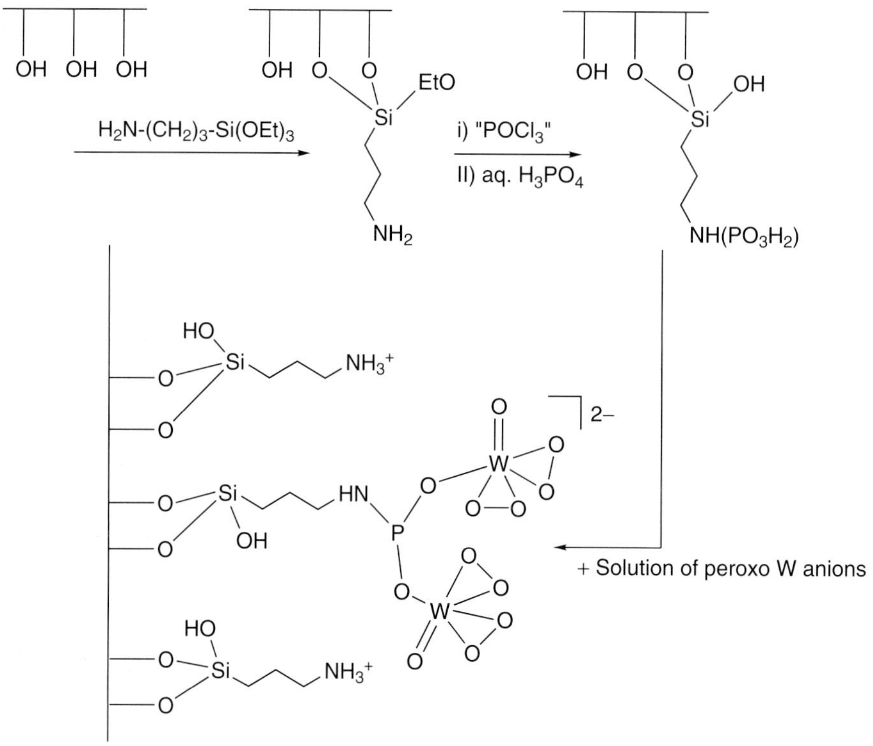

Fig. 10 Possible representation of the active anchored catalyst with phosphoramide linker. (Adapted from Ref. [90].)

system was found to be highly active for liquid-phase epoxidation of cyclooctene with TBHP as oxygen source.

The [MoO$_2$Cl$_2$(L^1)(L^2)] complexes of Mo(VI) have also been anchored to MCM-41 functionalized with monodentate nitrile and bidentate bipyridyl ligands [93, 94]. Unfortunately, they were found to be unstable toward molybdenum leaching, which accounts for their near-complete loss of activity in a second reaction cycle.

Nunes et al. [95] attempted to prepare more stable heterogeneous alkene epoxidation catalysts by anchoring of dioxo Mo(VI) complexes, [MoO$_2$Cl$_2$(L-L)], to MCM-41 and MCM-48 functionalized with a bidentate 1,4-diazabutadiene ligand, RN=C(Ph)−C(Ph)=NR [R≡(CH$_2$)$_3$Si(OEt)$_3$] (Fig. 11). By comparing the latter anchored materials with those synthesized by direct grafting of [MoO$_2$Cl$_2$(L-L)] onto MCM-41, it appears that they exhibit similar activities for the epoxidation of cyclooctene.

B Organometallics

a One Metal Center – One Linker – One Oxygen Atom of the Support: (M/L1) Petrucci and Kakkar [96] developed a new route that uses easily accessible starting materials and avoids some of the common problems encountered in the grafting of trialkoxysilane onto silica, including the presence of substantial amounts of surface-bound phosphorus oxide. These authors used a new acid–base hydrolytic approach [97] to synthesize two-dimensional thin film assemblies of functionalized long alkane chain donor ligands, HO−(CH$_2$)$_n$−PPh$_2$ ($n = 10$–12), on inorganic oxides. The long alkane chain hydrocarbons terminated with OH and PPh$_2$ groups were self-assembled on glass, quartz and single-crystal of Si via acid–base hydrolysis of surface-anchored [Si]−NEt$_2$ groups with terminal OH moieties (Fig. 12). The donor films were then used to introduce Rh(I) metal centers by bridge-splitting reactions to anchor metal complexes. The self-assembled organometallic thin films [Si]−O−(CH$_2$)$_n$−PPh$_2$[RhCl−(1,5-C$_8$H$_{12}$)] were found to be highly catalytically active in the hydrogenation of diphenylacetylene. The surface-bound Rh(I) catalysts were stable and could be recycled.

C Salen/Hydrosalen-Based Complexes

a One Metal Center – One Linker – Three Oxygen Atoms of the Support: (M/L3) The need for enantiomerically pure compounds in drug synthesis for the pharmaceutical industry has, during the past decade, triggered a tremendous interest in asymmetric catalysis and the search for suitable solid supports for heterogeneous stereoselective synthesis [98]. An approach to this end has been to encapsulate chromium salen complexes inside Y and EMT zeolites, or within the layers of K-10

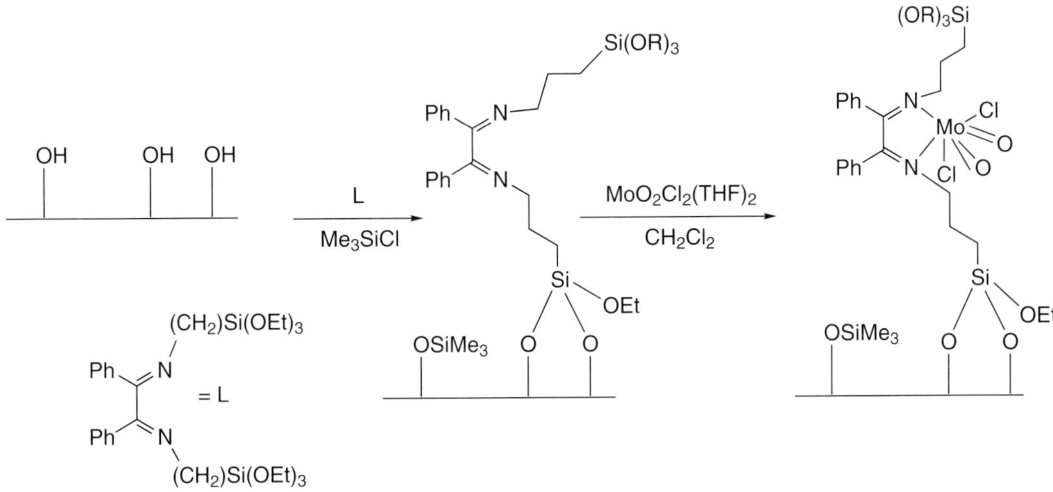

Fig. 11 Anchoring Mo(VI) complexes on mesoporous molecular sieves MCM-41 or MCM-48. (Adapted from Ref. [95].)

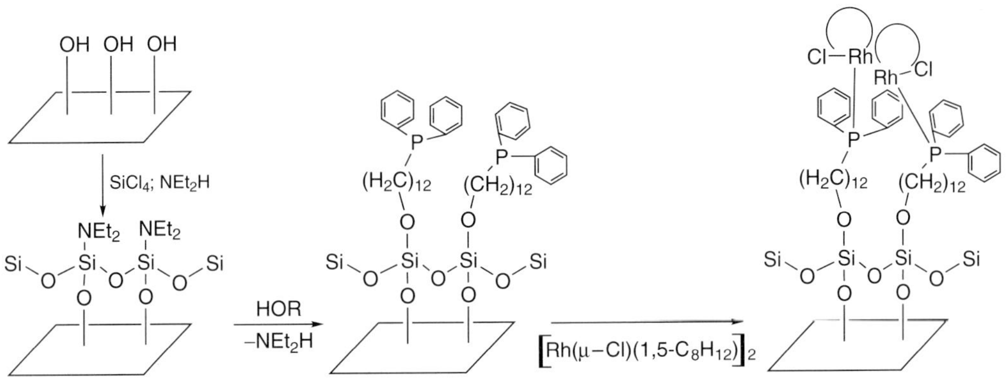

Fig. 12 Synthesis of self-assembled organometallics thin films, [Si]–O–$(CH_2)_n$–PPh_2[RhCl–(1,5-C_8H_{12})]. (Adapted from Ref. [96].)

montmorillonite. However, the major problem generally encountered is the lowering of the degree of asymmetric ability of the complexes supported on a solid.

Baleizão et al. [99] reported on the synthesis of two series of solid catalysts (Fig. 13) in which chiral chromium salen complexes are anchored through 3-aminopropyl linkers to functionalized MCM-41. For both series, the first step consists of reacting silanol groups of the support with 3-aminopropyltriethoxysiloxane leading to terminal amine group (Scheme 6). In the second step, chromium salen complexes are anchored to the aminopropyl-functionalized solid by two strategies. In the simplest route (Fig. 13; pathway A), the chromium salen complex is anchored to the solid by the metal through apical ligand coordination; this method involves the chromium Jacobsen catalysts prepared. In the second strategy (Fig. 13; pathway B), a chromium salen complex modified by chlorides is prepared in homogeneous phase in the absence of any supports. The coordination sphere of chromium ion is not involved in the consecutive step of anchoring of the complex to the support.

With regards to catalysis, heterogeneous materials were tested in the asymmetric ring opening of cyclohexenoxide. For catalysts in which the complex is anchored through coordination with the metal (Fig. 13; pathway A), a high enantiomeric excess (e.e.) was obtained (up to 70%), but the complex leached into the solution to a large extent. By contrast, for those solids in which the complex is covalently linked to the surface by the ligand (Fig. 13; pathway B), no leaching was observed, but the e.e.-value was modest (<20%) compared to those obtained in homogeneous catalysis under comparable conditions (>50%).

b One Metal Center – One Linker – Two Oxygen Atoms of the Support: (M/L2)   Sakthivel et al. [100] reported on the

---
References see page 537

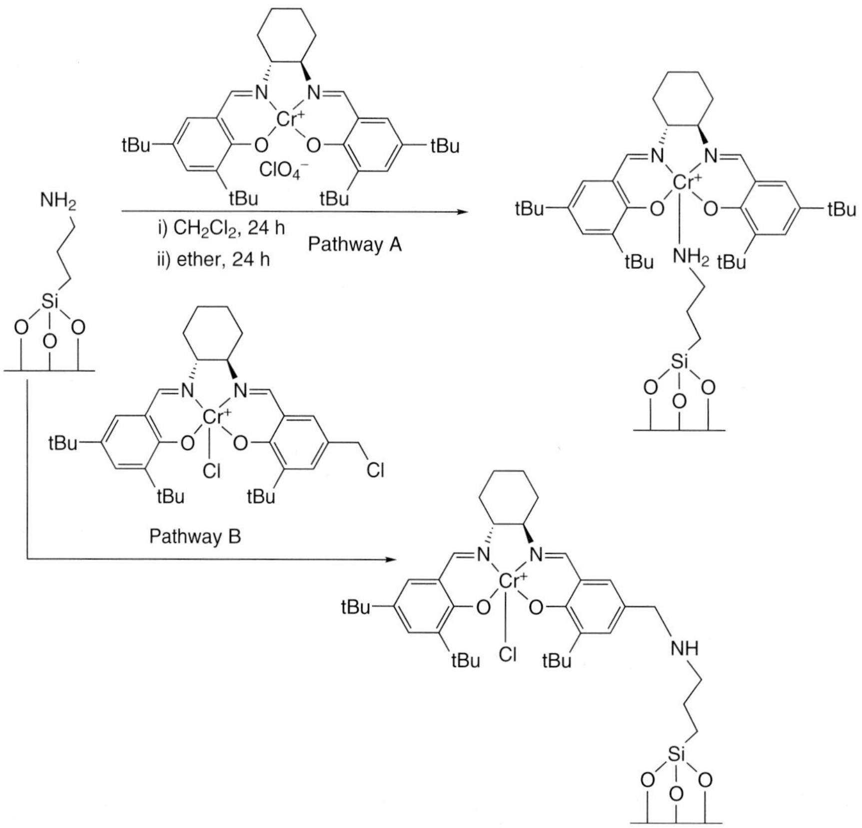

Fig. 13 Preparation of Cr(salen)-based catalysts. (A) Anchoring through surface functionalized amine linker. (B) Anchoring by functionalized salen ligand. (Adapted from Ref. [99].)

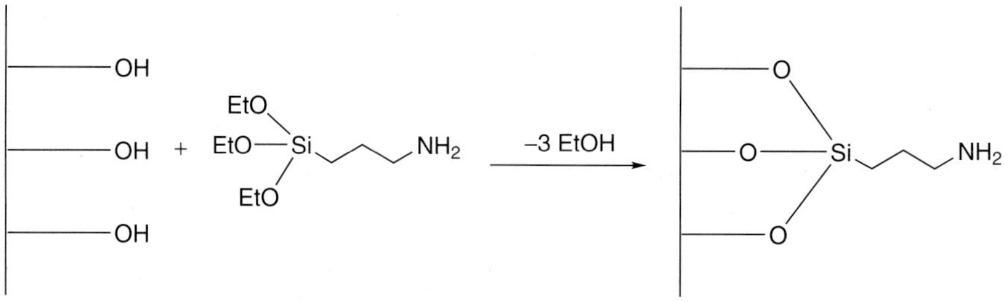

Scheme 6 Aminopropyl-functionalized solid supports. (Adapted from Ref. [99].)

anchoring of optically active Mo(VI) dioxo complexes bearing hydrosalen derivatives as ligands to MCM-41 and MCM-48, and examined their activities for asymmetric epoxidation. To prepare the catalyst, MCM-41 and MCM-48 are first modified with trimethoxy iodo propyl silane, after which the chiral catalysts are obtained by reaction of the iodo-groups on the silylated surface materials with the N-methyl groups present on the homogeneous chiral catalysts (Fig. 14). In asymmetric epoxidation, it appears that in the case of cis- and trans-β-methylstyrene, moderate e.e.-values of up to 31% can be reached when the reaction is carried out at room temperature.

2.4.5.6 Conclusions

In this chapter, an attempt has been made to review the state of TMCs bonded to inorganic oxide supports after either grafting or anchoring. Following an overview of earlier definitions which showed that the distinction between grafted and anchored species was blurred, new definitions were proposed for grafting and anchoring which led, with the help of typical examples, to a

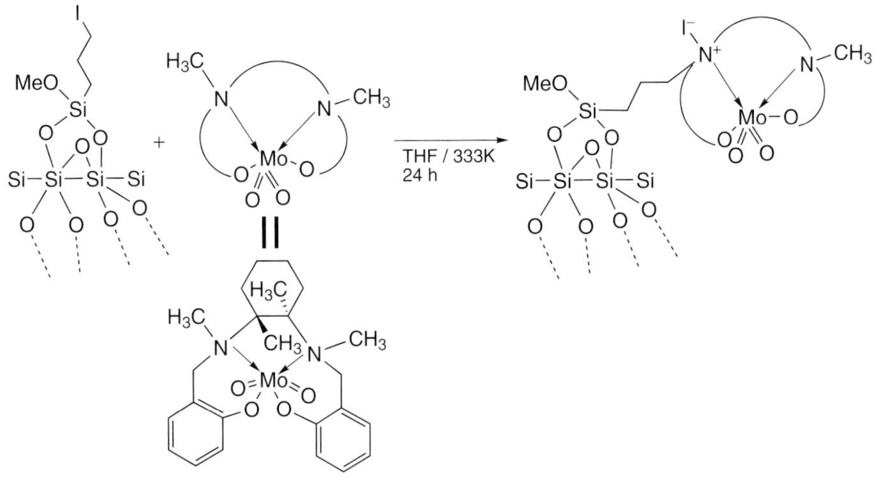

Fig. 14 Anchoring by reaction of surface-functionalized materials with the ligand of homogeneous chiral catalysts. (Adapted from Ref. [100].)

nomenclature for labeling TMCs as either "grafted" or "anchored".

The nature of the TMC depends largely on the type of precursor, and on whether a grafting or anchoring strategy has been used. Thus, in order to obtain a predetermined grafting (see Fig. 4) – that is, a metal center grafted by one, two or three chemical bonds to the oxide support – the coordination sphere of the precursor TMC must be specifically designed; in other words, it must contain one, two, or three labile ligands (usually water). In the case of anchoring, the oxide support and the TMC are connected via the linker. Hence, the attachment largely rests on the functionalization of the oxide, the precursor TMC, or its ligand(s).

The main characteristic of a grafted species is that the metal center is bonded directly to the surface of the oxide support and thus is essentially immobile. By contrast, the anchored species retain some degree of mobility within a space defined by the length of the linker, which anchors the metal center to the oxide support.

The concentration of both grafted and anchored species is limited by the amounts of surface hydroxyl groups (S–OH) or their derived forms available at the oxide surface. Unless special treatments of the oxide surface are used, it is therefore not possible to obtain a metal loading higher than that corresponding to the monolayer.

Typical examples have been given which describe how "inorganic support-transition metal complex" couples are prepared, leading to grafted or anchored species, labeled according to the notation detailed in Section 2.4.5.2, and further used in catalytic reactions.

Overall, because grafted species are often obtained after thermal activation at high temperature (several hundreds of degrees C), they usually involve high oxidation states and can withstand gas–solid reactions at such temperatures. By contrast, anchored materials that often are prepared in organic solvents involve lower oxidation states, are more fragile, and are used for liquid–solid reactions in which leaching may be a problem. On the other hand, they exhibit very interesting properties, for example in asymmetric catalysis.

References

1. J. F. Lambert, M. Che, in *Catalysis by Unique Metal Ion Structures in Solid. Matrices*, G. Centi, B. Wichterlová, A. T. Bell (Eds.), NATO Science Series, II: Mathematics, Physics and Chemistry, Vol. 13, Kluwer Academic Publishers, Dordrecht, 2001, p. 1.
2. L. L. Murrell, *Advanced Materials in Catalysis*, J. J. Burton, R. L. Garten (Eds.), Academic Press, New York, 1977, p. 235.
3. I. M. Campbell, *Catalysis at Surfaces*, Chapman & Hall, London, 1988, 250 pp.
4. C. Louis, M. Che, *J. Catal.* **1992**, *135*, 156.
5. C. Louis, M. Che, M. Anpo, *J. Catal.* **1993**, *141*, 453.
6. J. Haber, *Pure Appl. Chem.* **1991**, *63*, 1227.
7. B. Cornils, W. A. Herrmann, R. Schlögl, C. H. Wong, *Catalysis from A to Z*, 2nd Ed., Wiley-VCH, Weinheim, 2003, 840 pp.
8. F. Averseng, M. Vennat, M. Che, submitted for publication.
9. G. J. Leigh, *Nomenclature of Inorganic Chemistry, Recommendations 1990*, Blackwell Scientific Publications, Oxford, 1990, 289 pp.
10. A. W. Holland, G. Li, A. M. Shahin, G. J. Long, A. T. Bell, T. D. Tilley, *J. Catal.* **2005**, *235*, 150.
11. J. F. Lambert, M. Hoogland, M. Che, *J. Phys. Chem. B* **1997**, *101*, 10347.
12. S. Boujday, J. F. Lambert, M. Che, *Chem. Phys. Chem.* **2004**, *5*, 1003.
13. Y. Iwasawa, *Tailored Metal Catalysts*, D. Reidel Company, Dordrecht, 1986, 333 pp.

14. F. Schüth, K. S. W. Sing, J. Weitkamp, *Handbook of Porous Solids*, Wiley-VCH, Germany, 2002, Vol. 3, 3191 pp.
15. C. Marcilly, *Acido-Basic Catalysis. Application to Refining and Petrochemistry*, Editions Technip, Paris, 2005, Vol. 1, 432 pp, Vol. 2, 464 pp.
16. H. Pernot, *Catal. Rev.- Sci. Eng.* **2006**, *48*, 315.
17. R. Haag, S. Roller, *Top. Curr. Chem.* **2004**, *242*, 1.
18. K. P. De Jong, J. W. Geus, *Catal. Rev. Sci. Eng.* **2000**, *42*, 481.
19. L. Bonneviot, O. Legendre, M. Kermarec, D. Olivier, M. Che, *J. Coll. Interface Sci.* **1990**, *134*, 534.
20. C. Lepetit, M. Che, *J. Mol. Catal.* **1995**, *100*, 147.
21. M. Che, A. J. Tench, *Adv. Catal.* **1982**, *31*, 77 (AERE Report R-9971, November 1980).
22. M. Anpo, M. Che, *Adv. Catal.* **1999**, *44*, 119.
23. C. Chizallet, G. Costentin, M. Che, F. Delbecq, P. Sautet, *J. Phys. Chem. B* **2006**, *110*, 15878.
24. S. H. Garofalini, *J. Non-Crystall. Solids* **1990**, *120*, 1.
25. F. Négrier, E. Marceau, M. Che, J. M. Giraudon, L. Gengembre, A. Lofberg, *J. Phys. Chem. B* **2005**, *109*, 2836.
26. B. M. Weckhuysen, *In-Situ Spectroscopy of Catalysts*, Utrecht University, The Netherlands, 2004, 350 pp.
27. M. Hunger, *Adv. Catal.* **2006**, *50*, 149.
28. G. Leofanti, G. Tozzola, M. Padovan, G. Petrini, S. Bordiga, A. Zecchina, *Catal. Today* **1997**, *34*, 307.
29. J. W. Niemantsverdriet, W. N. Delgass, *Top. Catal.* **1999**, *8*, 133.
30. (a) Z. Sojka, F. Bozon-Verduraz, M. Che, see chapter of this Handbook; (b) K. Dyrek, M. Che, *Chem. Rev.* **1997**, *97*, 305.
31. J. M. Thomas, P. L. Gai, *Adv. Catal.* **2004**, *48*, 171.
32. A. M. Venezia, *Catal. Today* **2003**, *77*, 359.
33. R. R. Schrock, *Chem. Rev.* **2002**, *102*, 145.
34. J. Feldman, R. R. Schrock, *Prog. Inorg. Chem.* **1991**, *39*, 1.
35. R. R. Schrock, S. Luo, J. J. C. Lee, N. C. Zanetti, W. M. Davis, *J. Am. Chem. Soc.* **1996**, *118*, 3883.
36. R. R. Schrock, *Acc. Chem. Res.* **1990**, *23*, 158.
37. M. C. Beaudoin, O. Womiloju, A. Fu, J. A. N. Ajjou, G. L. Rice, S. L. Scott, *J. Mol. Catal. A* **2002**, *190*, 159.
38. V. A. Zakharov, Y. I. Yermakov, *Catal. Rev.-Sci. Eng.* **1979**, *19*, 67.
39. D. G. H. Ballard, *Adv. Catal.* **1973**, *23*, 263.
40. J. P. Candlin, H. Thomas, *Adv. Chem. Ser.* **1974**, *132*, 212.
41. J. Schwartz, M. D. Ward, *J. Mol. Catal.* **1980**, *8*, 465.
42. F. Quignard, C. Lecuyer, C. Bougault, F. Lefebvre, A. Choplin, D. Olivier, J. M. Basset, *Inorg. Chem.* **1992**, *31*, 928.
43. L. D'Ornelas, S. Reyes, F. Quignard, A. Choplin, J. M. Basset, *Chem. Lett.* **1993**, *11*, 1931.
44. J. A. N. Ajjou, S. L. Scott, *Organometallics* **1997**, *16*, 86.
45. S. A. Holmes, F. Quignard, A. Choplin, R. Teissier, J. Kervennal, *J. Catal.* **1998**, *176*, 173.
46. C. Rosier, G. P. Niccolai, J. M. Basset, *J. Am. Chem. Soc.* **1997**, *119*, 12408.
47. J. Engelhardt, J. Goldwasser, W. K. Hall, *J. Catal.* **1981**, *70*, 364.
48. V. B. Kazansky, B. N. Shelimov, *Res. Chem. Intermed.* **1991**, *15*, 1.
49. H. L. Krauss, K. Hagen, E. Hums, *J. Mol. Catal.* **1985**, *28*, 233.
50. M. Kantcheva, I. G. Dalla Lana, J. A. Szymura, *J. Catal.* **1995**, *154*, 329.
51. F. Chen, M. Liu, *J. Mater. Chem.* **2000**, *10*, 2603.
52. Y. Okamoto, T. Kubota, Y. Ohto, S. Nasu, *J. Catal.* **2000**, *192*, 412.
53. Z. H. Suo, Y. Kou, J. Z. Niu, W. Z. Zhang, H. L. Wang, *Appl. Catal. A* **1997**, *148*, 301.
54. E. Laperdrix, A. Sahibed-dine, G. Costentin, O. Saur, M. Bensitel, C. Nedez, A. B. Mohammed Saad, J. C. Lavalley, *Appl. Catal. B* **2000**, *26*, 71.
55. P. van der Voort, R. van Welzenis, M. de Ridder, H. H. Brongersma, M. Baltes, M. Mathieu, P. C. van de Ven, E. F. Vansant, *Langmuir* **2002**, *18*, 4420.
56. M. Guillodo, P. Vernoux, J. Fouletier, *J. Solid-State Ionics* **2000**, *127*, 99.
57. S. Grasser, C. Haessner, K. Köhler, F. Lefebvre, J. M. Basset, *Phys. Chem. Chem. Phys.* **2003**, *5*, 1906.
58. C. R. Dias, M. F. Portela, M. Galán-Fereres, M. A. Bañares, M. L. Granados, M. A. Peña, J. L. G. Fierro, *Catal. Lett.* **1997**, *43*, 117.
59. R. A. Rajadhyaksha, G. Hausinger, H. Zeilinger, A. Ramstetter, H. Schmelz, H. Knözinger, *Appl. Catal.* **1989**, *51*, 67.
60. M. J. Groeneveld, G. Boxhoorn, H. P. C. E. Kuipers, P. F. A. van Grinsven, R. Gierman, P. L. Zuideveld, *Proceedings 9th International Congress on Catalysis*, Calgary, 1988, M. J. Phillips, M. Ternan (Eds.), The Chemical Institute of Canada, Ottawa, 1988, p. 1743.
61. (a) M. Anpo, M. Sunamoto, M. Che, *J. Phys. Chem.* **1989**, *93*, 1187; (b) G. L. Rice, S. L. Scott, *Chem. Mater.* **1998**, *10*, 620.
62. M. G. Reichmann, A. T. Bell, *Appl. Catal.* **1987**, *32*, 315.
63. M. G. Finn, K. B. Sharpless, *J. Am. Chem. Soc.* **1991**, *113*, 106.
64. M. Palucki, N. S. Finney, P. J. Pospisil, M. L. Gueler, T. Ishida, E. N. Jacobsen, *J. Am. Chem. Soc.* **1998**, *120*, 948.
65. I. W. C. E. Arends, R. A. Sheldon, M. Wallau, U. Schuchardt, *Angew. Chem. Int. Ed. Engl.* **1997**, *36*, 1144.
66. H. P. Wulff, F. Wattimena, US Patent 4, 367,342, 1983.
67. M. G. Clerici, *Stud. Surf. Sci. Catal.* **1993**, *78*, 21.
68. U. Romano, A. Esposito, F. Maspero, C. Neri, M. G. Clerici, *Chimica e l'Industria* **1990**, *72*, 610.
69. M. A. Camblor, A. Corma, A. Martinez, J. J. Pérez-Pariente, *J. Chem. Soc., Chem. Commun.* **1992**, 589.
70. A. Corma, P. Esteve, A. Martinez, S. Valencia, *J. Catal.* **1995**, *152*, 18.
71. (a) J. M. Fraile, J. I. Garcia, J. A. Mayoral, E. Vispe, D. R. Brown, M. J. Naderi, *J. Chem. Soc., Chem. Commun.* **2001**, 1510; (b) P. Ratnasamy, D. Srinivas, H. Knözinger, *Adv. Catal.* **2004**, *48*, 1.
72. C. Cativiela, J. M. Fraile, J. I. Garcia, J. A. Mayoral, *J. Mol. Catal. A* **1996**, *112*, 259.
73. J. M. Fraile, J. I. Garcia, J. A. Mayoral, M. G. Proietti, M. C. Sánchez, *J. Phys. Chem.* **1996**, *100*, 19484.
74. A. O. Bouh, G. L. Rice, S. L. Scott, *J. Am. Chem. Soc.* **1999**, *121*, 7201.
75. S. Sensarma, A. O. Bouh, S. L. Scott, H. Alper, *J. Mol. Catal. A* **2003**, *203*, 145.
76. M. A. Banares, *Catal. Today* **1999**, *51*, 319.
77. N. C. Ramani, D. L. Sullivan, J. G. Ekerdt, J. M. Jehng, I. E. Wachs, *J. Catal.* **1998**, *176*, 143.
78. P. Ferreira, I. S. Gonçalves, F. E. Kühn, A. D. Lopes, M. A. Martins, M. Pillinger, A. Pina, J. Rocha, C. C. Romão, A. M. Santos, T. M. Santos, A. A. Valente, *Eur. J. Inorg. Chem.* **2000**, *10*, 2263.
79. C. D. Nunes, A. A. Valente, M. Pillinger, J. Rocha, I. S. Gonçalves, *Chem. Eur. J.* **2003**, *9*, 4380.
80. M. Seman, J. N. Kondo, K. Domen, R. Radhakrishnan, S. T. Oyama, *J. Phys. Chem. B* **2002**, *106*, 12965.
81. Y. Xu, S. Liu, I. Wang, M. Xie, X. Guo, *Catal. Lett.* **1995**, *30*, 135.

82. L. Chen, L. Lin, Z. Xu, X. Li, T. Zhang, *J. Catal.* **1995**, *157*, 190.
83. Y. Xu, X. Bao, L. Lin, *J. Catal.* **2003**, *216*, 386.
84. A. Antiñolo, P. Cañizares, F. Carrillo-Hermosilla, J. Fernández-Baeza, F. J. Fúnez, A. de Lucas, A. Otero, L. Rodríguez, J. L. Valverde, *Appl. Catal. A* **2000**, *193*, 139.
85. H. Y. Chen, W. M. H. Sachtler, *Catal. Today* **1998**, *42*, 73.
86. V. I. Sobolev, K. A. Dubkov, O. V. Panna, G. I. Panov, *Catal. Today* **1995**, *24*, 251.
87. G. I. Panov, A. K. Uriaete, M. A. Rodkim, V. I. Sobolov, *Catal. Today* **1998**, *41*, 365.
88. C. Nozaki, C. G. Lugmair, A. T. Bell, T. D. Tilley, *J. Am. Chem. Soc.* **2002**, *124*, 13194.
89. J. Rudolph, K. L. Reddy, J. P. Chiang, K. B. Sharpless, *J. Am. Chem. Soc.* **1997**, *119*, 6189.
90. D. Hoegaerts, B. F. Sels, D. E. de Vos, F. Verpoort, P. A. Jacobs, *Catal. Today* **2000**, *60*, 209.
91. M. Jia, W. R. Thiel, *J. Chem. Soc., Chem. Commun.* **2002**, 2392.
92. M. Jia, A. Seifert, W. R. Thiel, *Chem. Mater.* **2003**, *15*, 2174.
93. C. D. Nunes, A. A. Valente, M. Pillinger, A. C. Fernandes, C. C. Romão, J. Rocha, I. S. Gonçalves, *J. Mater. Chem.* **2002**, *12*, 1735.
94. P. Ferreira, I. S. Gonçalves, F. E. Kühn, A. D. Lopes, M. A. Martins, M. Pillinger, A. Pina, J. Rocha, C. C. Romão, A. M. Santos, T. M. Santos, A. A. Valente, *Eur. J. Inorg. Chem.* **2000**, *10*, 2263.
95. C. D. Nunes, M. Pillinger, A. A. Valente, J. Rocha, A. D. Lopes, I. S. Gonçalves, *Eur. J. Inorg. Chem.* **2003**, *21*, 3870.
96. M. G. L. Petrucci, A. K. Kakkar, *Chem. Mater.* **1999**, *11*, 269.
97. M. G. L. Petrucci, A. K. Kakkar, *Organometallics* **1998**, *17*, 1798.
98. J. Crosby, in *Chirality in Industry*, A. N. Collins, G. N. Sheldrake, J. Crosby (Eds.), Wiley, New York, 1992, p. 1.
99. C. Baleizão, B. Gigante, M. J. Sabater, H. Garcia, A. Corma, *Appl. Catal. A* **2002**, *228*, 279.
100. A. Sakthivel, J. Zhao, G. Raudaschl-Sieber, M. Hanzlik, A. S. T. Chiang, F. E. Kühn, *Appl. Catal. A* **2005**, *281*, 267.

2.4.6
Supported Catalysts from Chemical Vapor Deposition and Related Techniques

*Mizuki Tada and Yasuhiro Iwasawa**

2.4.6.1 Chemical Vapor Deposition (CVD) Technique

According to current IUPAC recommendations, deposition which occurs by adsorption or reaction from the gas phase is referred to as chemical vapor deposition (CVD) [1]. The term CVD is generally used without restriction with regards to the mode and mechanism of deposition of catalyst precursors in the field of catalysis.

The first practical use of the CVD technique was developed during the 1880s for the production of incandescent lamps, the goal being to improve the strength of filaments by carbon or metal deposition [2]. The carbonyl nickel process was developed by Mond and others during the same decade [3].

Today, applications of the CVD fabrication process represent key elements in many industrial products, such as semiconductors, optoelectronics, optics, cutting tools, refractory fibers, filters, corrosion applications, and many others [4]. CVD is also a very versatile process for the production of catalytic materials, and is a relatively simple and flexible technology which can accommodate many variations. For example, with CVD it is possible to deposit target species on materials of almost any shape and of almost any size; moreover, it is also possible to produce almost any metal and non-metallic element as well as compounds, such as oxides, sulfides, carbides, nitrides, and intermetallics [4]. The CVD processes link organometallic chemistry, inorganic synthetic chemistry, and surface chemistry. This has in turn led to an entirely new field of chemistry, where catalyst design can be considered as a promising, scientific means of preparing well-defined active surfaces with remarkable catalytic properties, as well as the elucidation of reaction mechanisms that include dynamic changes of active sites [5, 6]. Unfortunately, CVD also has several disadvantages, the most prominent being the requirement for chemical precursors with high vapor pressure that are often hazardous and, at times, also extremely toxic. In general, the byproducts of these precursors are also toxic and corrosive and must be neutralized, which may result in costly operations.

The chemical reactions used in CVD, besides adsorption, are numerous and include thermal decomposition (pyrolysis), hydrolysis, disproportionation, reduction, oxidation, carbonization, nitridation, and sulfurization. These can be used either singly or in combination [4]. A CVD reaction is governed both, by thermodynamics and by kinetics.

2.4.6.2 CVD Precursors and Equipment

In CVD, the correct choice of reactants (the precursors) is very important. These precursors fall into several general major groups, which include the halides, carbonyls, alkoxides, acetates, and organic complexes. The choice of a precursor is governed by certain general characteristics, including: (i) stability at room temperature; (ii) sufficient volatility at low temperature so that it can be easily transported to the reactor; (iii) a capability of being produced at very high degrees of purity; (iv) an ability

---

* Corresponding author.

to react cleanly on or with the support; and (v) an ability to react without producing side reactions or parasitic reactions [4]. Representative precursors are listed in Table 1. A branch of CVD known as metallo-organic CVD (MOCVD) is used extensively in semiconductor and opto-electronic applications [7].

Tab. 1 CVD precursors

Al	$AlCl_3$, $AlEt_3$
B	BCl_3, $B(OEt)_3$, BEt_2
Ba	$Ba(acac)_2$
Bi	$BiPh_3$
Ca	$Ca(DPM)_2$
Co	$Co_2(CO)_8$, $Co(CO)_3NO$, $Co(acac)_3$, $Co(allyl)_3$, $Co(AcO)_2$
Cr	$Cr(CO)_6$, $CrCl_2O_2$, $Cr(acac)_3$
Cu	$Cu(acac)_2$, $Cu(AcO)$, $Cu(DPM)_2$, $CuCl$, $Cu(OCHMeCH_2NMe_2)_2$
Fe	$Fe(CO)_5$, $Fe_3(CO)_{12}$, $Fe(acac)_3$, $FeCp_2$, $FeCl_3$
Ga	$Ga(acac)_3$, $GaMe_3$
In	$InMe_3$, $In(acac)_3$
Ir	$Ir_4(CO)_{12}$, $IrCl_2(CO)_2$, $Ir(acac)_2(CO)_2$, $Ir(allyl)_3$, $IrCp(1,5\text{-}COD)$, $IrCp^*(1,5\text{-}COD)$, $Ir_2(\mu\text{-}SCMe_3)_2(CO)_4$
Mn	$Mn_2(CO)_{10}$
Mo	$Mo(CO)_6$, $MoCl_5$, $Mo_2(AcO)_4$, $MoO_2(tmhd)_2$, MoO_3
Nb	$NbCl_5$, $Nb(OEt)_5$
Ni	$Ni(CO)_4$, $Ni(acac)_2$, $Ni(allyl)_2$, $NiCp_2$
Os	$Os_3(CO)_{12}$, $OsCl_2(CO)_3$
Pd	$Pd(acac)_2$, $Pd(AcO)_2$, $Pd(hfac)_2$, $Pd(allyl)(hfac)$, $Pd(allyl)_2$, $Pd(allyl)Cp$, $Pd(tmhd)_2$
Pt	$PtCl_2(CO)$, $PtCl_2(CO)_2$, $Pt(acac)_2$, $Pt(hfac)_2$, $Pt(allyl)_2$, $Pt(tmhd)_2$, $PtBu_2$, $PtCp^*Me_3$, $PtMe_2(1,5\text{-}COD)$
Re	$Re_2(CO)_{10}$, $MeReO_3$, Re_2O_7
Rh	$Rh_4(CO)_{12}$, $Rh_6(CO)_{16}$, $Rh_2Cl_2(CO)_4$, $Rh(acac)_3$, $Rh(acac)(CO)_2$, $Rh(AcO)_2$, $Rh(allyl)_3$, $Rh_2Cp_2(CO)_2$
Ru	$Ru(CO)_5$, $Ru_3(CO)_{12}$, $Ru(acac)_3$, $Ru(tmhd)_3$, $Ru(allyl)_2(1,5\text{-}COD)$, $Ru(1,5\text{-}COD)(toluene)$, $RuCp_2$
Sb	$Sb(OEt)_3$, $Sb(O_2CNMe_2)_3$, $SbEt_3$
Si	$Si(OMe)_4$, $Si(OEt)_4$, $Si(OMe)Pr_3$
Sm	$Sm(acac)_3$
Sn	$Sn(AcO)_2$, $SnMe_4$, $SnEt_4$
Ti	$TiCl_4$, $Ti(OEt)_4$, $Ti(OPr^i)_4$
V	$V(CO)_5$, VCl_4, $VOCl_3$, $VO(OPr^i)_3$
W	$W(CO)_6$, WCl_6
Zn	$ZnMe_2$, $ZnEt_2$, $Zn(OSiMe_3)_2$
Zr	$ZrCl_4$, $Zr(acac)_4$, $Zr(OEt)_4$, $Zr(OBu^t)_4$, $ZrCp_2Cl_2$, $ZrCp_2Me_2$

Acac, acetylacetonato; hfac, hexafluoroacetylacetonato; tmhd, 2,2,6,6-tetramethylheptane-3,5-dione; DPM, dipivaloylmethanate; Me, methyl; Et, ethyl; Pr, propyl; Pr^i, isopropyl; Bu, butyl; Bu^t, tert-butyl; Cp, cyclopentadienyl; Cp*, pentamethylcyclopentadienyl.

The equipment used to perform CVD is relatively simple, does not require ultra-high vacuum, and generally can be adapted to many process variations. Its flexibility is such that it allows many changes in composition during deposition, and the codeposition of compounds is readily achieved. A CVD reaction can occur either in a closed system or in a flow of inert gas [8].

The *closed reactor system* was the first type to be used for the purification of metals. As the name implies, the chemicals (solid at room temperature) are loaded into a reactor, in which the supports have been pretreated, and the reactor is then closed. CVD takes place upon heating to vaporize the chemicals with stirring. More conveniently, vapor precursors are used, which is easy to handle. The precursor vapor is interacted directly with pretreated supports in the reactor. In open-reactor CVD or "flowing gas CVD", the reactants of vapor are introduced continuously and flow through the reactor at given temperatures.

Many research groups have successfully expressed the flow dynamics and mass transport processes mathematically, and obtained a realistic model that could be used to predict CVD systems and in the design of reactors [8]. These models are designed to define the complex entrance effects and convection phenomena that occur in a reactor, and to solve the complex equations of heat, mass balance, and momentum. However, these problems may be of less importance in the case of catalyst preparation involving the adsorption and reaction of precursors and the subsequent chemical treatments.

2.4.6.3 CVD Catalysts

2.4.6.3.1 Gold (Au) $Au(CH_3)_2(acac)$ vapor was deposited on anatase and amorphous TiO_2 powders by a CVD technique at 306 K, followed by calcination in air to decompose the supported precursor to metallic gold particles. The CVD method was found to be advantageous in depositing gold particles less than 2 nm in diameter. The catalytic activity of Au/TiO_2 catalysts for CO oxidation was as high as that obtained for catalysts prepared by the liquid-phase deposition–precipitation method [9]. Au/Al_2O_3 was prepared by CVD of $Au(CH_3)_2(acac)$ on Al_2O_3, which provided finely dispersed Au/Al_2O_3 with $d_{Au} < 5$ nm [10].

2.4.6.3.2 Cobalt (Co) Cobalt species with different structures and oxidation states on inorganic oxides can be produced by using $Co_2(CO)_8$ as a precursor [11]. Scheme 1 illustrates the surface transformations of Co clusters which were obtained by interaction of $Co_2(CO)_8$ vapor with SiO_2 or Al_2O_3 [11]. These surface Co species have been characterized using extended X-ray absorption

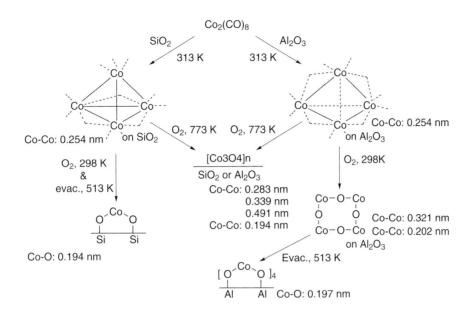

Scheme 1

fine structure (EXAFS), diffuse-reflectance UV/visible, Raman, composition analysis, etc. The $[Co_3O_4]_n$ species on SiO_2 and Al_2O_3, which are obtained as thin, small particles with an averaged diameter of 1.3 nm and less than 1.0 nm, respectively [12], were found to be remarkably active for the catalytic oxidation of CO with O_2 at 273 K, as shown in Fig. 1 [11, 13]. The traditional Co_3O_4/SiO_2 and Co_3O_4/Al_2O_3 catalysts prepared by a usual impregnation method were almost inactive under similar reaction conditions.

The $[Co(II)]_4$ species was also active for CO oxidation, as shown in Fig. 1. EXAFS Fourier transform of the species indicated that the Co assembly exists in a monomer form, without any Co–Co bonding [14]. However, the EXAFS Fourier transform measured under 0.1 MPa of O_2 exhibited a small new peak around 0.30 nm. The peak intensity increased under 1 MPa O_2. Curve-fitting analysis revealed that the peak is ascribed to a Co–Co arrangement at 0.330 nm. It is likely that the Co–Co peak appears by the linkage of an oxygen bridge (Co–O–Co) upon O_2 adsorption [14, 15]. The bridging oxygen linked to two Co atoms reacted almost instantaneously with CO to form CO_2 at room temperature, leading to the disappearance of the Co–Co peak.

SiO_2-supported Co catalysts were prepared by atomic layer epitaxy (ALE), using $Co(acac)_3$. The catalysts were used for gas-phase toluene hydrogenation in a microreactor system [16]. By using the $Co(acac)_3$ precursor, cubic spinel-type polycrystalline Co_3O_4 thin films were prepared on planar stainless steel. The Co_3O_4 catalyst was reported to be active for the combustion of propane and ethanol [17].

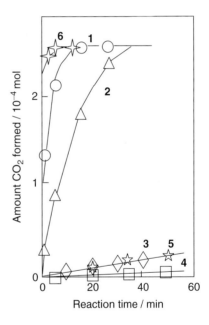

Fig. 1 Catalytic performances of CVD Co catalysts for CO oxidation. 1, $[Co_3O_4]n/Al_2O_3$; 2, $[Co^{(2+)}]_4/Al_2O_3$; 3, $[CoO]_4/Al_2O_3$; 4, Impregnated Co_3O_4/Al_2O_3; 5, $Co^{(2+)}/SiO_2$; 6, $[Co_3O_4]n/SiO_2$; Catalyst: 0.827 g (5 wt.% Co/support).

2.4.6.3.3 Chromium (Cr) Banks used $Cr(CO)_6$ on alumina or silica-alumina to polymerize ethene at 394–400 K [18, 19], while Brenner et al. noted that $Cr(CO)_6/Al_2O_3$ can be used to hydrogenate propene, with maximum activity being observed at 468 K [20].

References see page 553

NaX and LiX (57% ion-exchanged) zeolites were used as support for $Cr(CO)_6$. After evacuation at 673 K for 1 h, zeolite (12–40 mg) in a reactor was exposed to a $Cr(CO)_6$ vapor at room temperature for 1 h, providing a $Cr(CO)_6$ loading of 2.2/supercage of zeolite, as found by means of chemical analysis [21]. The tricarbonyl chromium(0) species encaged in LiX or NaX was found to be highly efficient and stereoselective for the hydrogenation of butadiene above 230 K. A reaction mechanism is proposed, in which cis-2-butene is formed via $(\eta^4\text{-}C_4H_6)Cr(CO)_3$, while 1-butene forms via $(\eta^2\text{-}C_4H_6)Cr(CO)_3$ and $(\eta^2\text{-}C_4H_6)Cr(CO)_3$ [21].

2.4.6.3.4 Copper (Cu)

Copper dipivaloylmethanate $Cu(DPM)_2$ was evaporated at 353 K onto SiO_2 [22]. Infrared (IR) bands of the adsorbed species resemble those of dipivaloylmethane, which indicates that the adsorption may originate from the interaction between the ligand and the surface OH groups, and not be related to the type of central metal atom. The Cu/ZnO system forms the basis of industrial methanol synthesis, and is also an important component of fuel-cell technology. Calcined MCM-41 was treated with the vapor of the blue-violet copper precursor $[Cu(OCHMeCH_2NMe_2)_2]$ at 340 K in a sealed Schlenk tube, which caused the original white siliceous support to turn light blue. The obtained sample was treated with diethyl zinc vapor $ZnEt_2$. Careful annealing of the $Cu/Zn(OCHMeCH_2NMe_2)_2/MCM$-41 at 623 K in vacuum provides a material which is free from CH_x contamination, but exhibits metallic Cu species and probably ZnO nanocrystallites, which was active for methanol synthesis [23]. A solid Cu catalyst was prepared by CVD of $Cu(acac)_2$ on porous activated carbon. The obtained monolayer Cu catalyst had a higher activity for catalytic wet oxidation of phenol solution than a conventional impregnated Cu/carbon catalyst [24].

Cu_2O–CuO and CuO thin films on a fiberglass substrate were prepared by CVD of $Cu(acac)_2$ in a flow of O_2 and catalytically evaluated by cyclohexane oxidation. The fiber with a CuO film exhibited a high activity, for example 48% at 623 K with a high yield of cyclohexanone (37% at 523 K) [25].

2.4.6.3.5 Iron (Fe)

Fe carbonyls interact with inorganic oxide surfaces through oxidative addition of acidic or amphoteric OH groups, and through coulombic interaction derived from acid–base reactions. The details of this chemistry have been reviewed elsewhere [11, 15, 26–28].

Highly dispersed iron particles in HY zeolite were produced by thermal and photochemical decomposition of $Fe(CO)_5$ [29]. Photochemical decomposition produced smaller particles because of the strong carbonyl–surface interaction due to electronic excitation of the carbonyl during decomposition. $Fe_3(CO)_{12}$ was deposited onto the SiR_x/SiO_2 surface using CVD at room temperature under vacuum, followed by heating at 313 K under vacuum [30]. The deposited Fe species has the composition $[Fe_3(CO)_6]$, and showed a Fe–Fe distance of 0.273 nm by EXAFS analysis; this is greater than that in $Fe_3(CO)_{12}$ (0.262 nm) and in Fe metal (0.248 nm). When the $[Fe_3(CO)_6]/SiR_x/SiO_2$ catalyst was heated at 368 K in vacuum, the Fe cluster showed a higher activity by a factor of 200 than the Fe species on SiO_2 for propene hydrogenation [30].

The structure of the iron species in mildly calcined over-exchanged Fe/ZSM-5 prepared by CVD of $FeCl_3$ was studied by in-situ Fe K-edge XANES and EXAFS [31]. The majority of iron was proposed (by XAFS study) to be present as Fe-binuclear complexes, where the closest Fe–O shell in the complexes was attributed to a $[HO-Fe-O-Fe-OH]^{2+}$ core. Heating to a moderate temperature in He as well as in a O_2/H_2 mixture resulted in the desorption of water from the Fe-coordination sphere. Performances of CVD Fe/ZSM-5 catalysts for benzene oxidation by N_2O to phenol were compared to those of iron-containing zeolites prepared by ion exchange and hydrothermal synthesis [32, 33], and the catalytic performances were found to be essentially similar. However, the performance of the CVD Fe/ZSM-5 samples was improved by steam treatment [34].

2.4.6.3.6 Gallium (Ga)

Gallium dispersed in acid materials can be applied to dehydrogenate alkanes. In combination with the shape-selective properties of a medium-pore zeolite such as ZSM-5, light paraffins can be converted to aromatic hydrocarbons including benzene, toluene, and xylenes. A commercial application of this is found in the Cyclar process. CVD of trimethylgallium in HZSM-5, followed by reduction or oxidation, led to well-defined reduced and oxidized species. The GaO^+ species exhibit a much higher H_2/D_2 exchange activity than reduced Ga^+ species [35].

2.4.6.3.7 Molybdenum (Mo)

A Mo Carbonyls Since the finding of Banks and Bailey in 1964 that $Mo(CO)_6/Al_2O_3$ is active for the metathesis reaction [36, 37], many metal carbonyls have been used to prepare improved catalytic materials [26, 27]. In 1969, Kemball et al. presented preliminary results from an IR spectroscopic investigation of the nature of $Mo(CO)_6$ surface bonded to alumina [38]. Howe et al. found that activation of supported $Mo(CO)_6$ catalysts involved carbonyl decomposition, the extent of which was a function of the basicity of support hydroxyl groups [39]. Brenner and Burwell proposed the species $Mo(CO)_5$,

Mo(CO)$_4$ and Mo(CO)$_3$ as resulting from Mo carbonyl decomposition with the component Mo(CO)$_3$ being metathesis-active [40]. This species was found to adsorb one molecule of oxygen at 298 K, ejecting a molecule of carbon monoxide to form adsorbed Mo(CO)$_2$O$_2$, a species which was 15-fold more active for metathesis than adsorbed Mo(CO)$_3$. The Mo(CO)$_{3ads}$ reacted further with surface hydroxyls (Z—OH) to evolve hydrogen with metal oxidation [41]. Surface-bonded Mo(CO)$_6$ exhibited greater activity for olefin hydrogenation than conventional MoO$_3$/Al$_2$O$_3$ catalysts [20].

It has been shown that Mo(CO)$_6$ encaged in Y-type zeolites shows very high activities for a stereoselective hydrogenation of butadiene or trans-1,3 pentadiene to the corresponding cis-2-olefin [42]. The structure and electronic state of thermally stable molybdenum subcarbonyl species entrapped in Y- and X-type zeolites were determined using EXAFS, X-ray photoelectron spectroscopy (XPS), and IR techniques. No Mo—Mo bonding was observed, suggesting a monomeric form. The ν (CO) band consists of three peaks at 1911(s), 1790(s) and 1760(sh) cm^{-1} for the NaY system. The IR peaks suggest a monomeric Mo(CO)$_3$(OZ)$_3$ (OZ : zeolite framework oxygen) with a fac configuration rather than a mer configuration [42]. In contrast, zeolite NaY adsorbs two Mo(CO)$_6$ molecules per supercage on exposure to the vapor at room temperature, and subsequent heating to 473 K causes complete decarbonylation, producing an average loading of two Mo per supercage [43]. EXAFS and ^{129}Xe NMR data are presented which show that adsorption of Mo(CO)$_6$ in zeolite NaY followed by decomposition at 473 K produces uniformly dispersed Mo$_2$ clusters in the zeolite supercages, as indicated by the Mo—Mo coordination number of 1.0 at 0.28 nm.

Time-resolved XAFS provides more precise information on the intermediate Mo species which are produced during temperature-programmed deposition (TPD) of Mo(CO)$_6$ in HY and NaY zeolites [44]. Mo(CO)$_6$ entrapped in HY pores decomposes to Mo monomers via a short-lived Mo(CO)$_3$(OZ)$_2$, while Mo(CO)$_6$ in NaY pores decomposes to Mo oxycarbide dimers [Mo$_2$(C)O$_x$] via a relatively more stable Mo intermediate.

B MoCl$_5$ Silica-supported molybdenum catalysts were prepared by the reaction of MoCl$_5$ vapor with surface OH groups on SiO$_2$ [45]. The samples were then dried in air at ca. 350 K with continuous stirring, and calcined in oxygen overnight at 773 K. A model of tetrahedral dioxo Mo species was found to account for the isolated Mo species of the CVD Mo/SiO$_2$ prepared by MoCl$_5$ after oxidizing treatment, on the basis of photoluminescence results [45, 46]. This is in agreement with the report by Iwasawa et al., who investigated Mo-allyl-based Mo/SiO$_2$ and Mo/Al$_2$O$_3$ catalysts and studied the systems with in-situ EXAFS and photoluminescence spectroscopy [11, 47–49]. The Raman investigations (Raman shift: 994–998 cm^{-1}) performed by Wachs et al. [50, 51] on Mo/SiO$_2$ catalysts prepared by different methods revealed the existence of polyoxomolybdates which were transformed into isolated octahedral monooxo Mo(VI) cations upon hydration [50, 51], whereas Cornac et al. observed the reverse phenomenon – that is, breaking of the Mo—O—X bridge (X = Mo or Si) [52]. Desikan et al. [53] have also characterized impregnated Mo/SiO$_2$ samples by in-situ Raman spectroscopy (Raman shift: 955 and 670 cm^{-1}). These authors found that the isolated surface Mo species present on their dehydrated catalysts is a distorted tetrahedron in C$_{2v}$ symmetry, with two short Mo=O bonds and two long Mo—O bonds.

C Allylmolybdenum Aigler et al. examined the oxidation states of allyl-based Mo/SiO$_2$ catalysts by XPS [8, 54]. The MOCVD Mo catalysts were prepared by a dry-mixing and sublimation method, which allowed the allyl complex to be loaded either onto 100- to 200-mesh particles of the SiO$_2$ support or onto pressed self-supporting wafers of the same material for XPS studies. An XPS Mo 3d spectrum was acquired for the Mo/SiO$_2$ catalysts after adsorption of Mo(η^3-C$_3$H$_5$)$_4$ on to SiO$_2$ (structure a in Scheme 2). The spectrum obtained showed a Mo 3d peak shape which suggests the presence of one oxidation state [8]. After reduction at 823 K for 12 h, the Mo 3d envelope shows a major contribution from a single oxidation state (Mo 3d$_{5/2}$ binding energy of 228.4 eV), and a minor contribution from Mo^{4+} (Mo 3d$_{5/2}$ binding energy of 230.2 eV). The Mo 3d$_{5/2}$ binding energy value at 228.4 eV is consistent with the presence of Mo^{2+}. This is in good agreement with structure b (see Scheme 2), as proposed by Iwasawa and co-workers [48, 49]. The Mo 3d$_{5/2}$ binding energies (223.0 and 236.2 eV) for the catalysts obtained following oxidation at 573 K are characteristic of Mo^{6+} [54]. In contrast, the conventional impregnation Mo^{6+}/SiO$_2$ catalyst results in mixed oxidation states, reflecting a heterogeneous feature.

Studies on single crystal oxide surfaces have extended this approach to include technologically interesting surface complexes that are not well-modeled by single-component metallic substrates. Hung et al. reported the use of organometallic surface deposition in ultra-high vacuum (UHV), followed by oxidation, to produce single-crystal models for mixed transition metal oxides under conditions which allow for their structural and compositional characterization [55]. Single-crystal Fe(100)

References see page 553

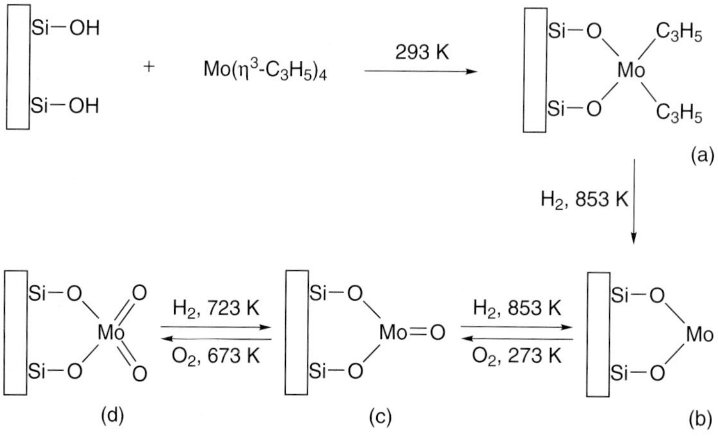

Scheme 2

was hydroxylated to give a p(2 × 1) overlayer (exposure to 1.2 L (1 Langmuir: 1.33× 10^{-4} Pa·s) of H_2O at 100 K followed by warming to 243 K). The Fe(100)–OH surface was exposed to Mo(η^3-allyl)$_4$ vapor for 2 min at a measured pressure below 1.3× 10^{-6} Pa. Fe(100)–Mo(allyl)$_2$ was treated at 573 K with H_2 for 15 min. Allyl ligand-related peaks at 3030, 1465, 1240, and 915 cm^{-1} were almost entirely lost by this treatment, and the HREELS loss peak assigned to the overlapping Fe–O–Mo units broadened and shifted to 425 cm^{-1} and a new loss peak at 980 cm^{-1} was assigned to v(Mo=O), in accord with similar assignments for MoO or MoO_2 complexes supported on bulk oxides and for soluble species. The O coverage increased to 1.2 ML (ML: monolayer), consistent with a Fe(100)–MoO_{1-2} unit stoichiometry [55].

D MoO$_3$ Vapor Metal oxide vapor synthesis (MOVS) has been developed for the preparation of unsupported and supported solid metal oxide systems [56]. Volatile metal oxides, such as MoO_3 (mp. 1068 K, b.p. 1530 K) or V_2O_5 (mp. 963 K, b.p. 2325 K), do not require high vacuum. The obtained MoO_3-vapor-based Mo/SiO_2 catalyst shows a unique nature and high efficiency in the catalytic conversion of methanol to formaldehyde [16]. The precursor vapor (MoO_3) was also directly contacted with pretreated HZSM-5 in a fused ampoule. The Mo/HZSM-5 catalyst obtained provided an excellent activity and stability for methane dehydrogenation to benzene [57].

2.4.6.3.8 Niobium (Nb) One-atomic layer niobium oxide catalysts were prepared on SiO_2 by a two-stage CVD procedure using Nb(OC$_2$H$_5$)$_5$ precursor. EXAFS analysis revealed the growth of a one-atomic-layer niobium oxide structure on SiO_2, showing the existence of a Nb–Si bond at 0.327 nm. The niobium oxide one-atom-layers on SiO_2 were found to be more active and selective (99%) for C_2H_4 formation (intramolecular dehydration) from ethanol at 373–573 K than a conventional impregnation catalyst [58].

2.4.6.3.9 Nickel (Ni) The Ni carbonyl CVD process was the first example of surface fabrication of advanced materials, to have an impact on industrial processes [2, 3]. Parkyns proposed that, at room temperature, Ni(CO)$_4$ initially physically adsorbs onto alumina, and then decomposes upon evacuation to small carbonylated metal clusters [59]. On partially dehydroxylated alumina, Ni(CO)$_4$ vapor reacts with the surface to form adsorbed Ni(CO)$_3$ at 298–323 K. Heating this material to 373 K results in further decarbonylation and agglomeration of nickel. Dispersion of the metallic nickel was maximized by isothermally decomposing the carbonyl in vacuum at 623 K, followed by passivation in nitrogen with traces of oxygen. Such catalysts were effective for benzene hydrogenation at 333–373 K [26].

A series of Ni/Al$_2$O$_3$ catalysts was prepared from vapor phase using the ALE technique [60]. Originally, ALE was used to grow single crystals and thin films [61], but today it is considered as a special mode of CVD, as strict demands are made upon the conditions under which the gas–solid reactions are carried out. In ALE, the surface is saturated with the volatile reactant, where metal halides, oxychloride, β-diketonates, metal alkoxides and acetates have been used [62]. The adsorbed Ni species were treated with air or H_2 at elevated temperatures. During the two to four sequences (5–10 wt.% Ni), nickel grows on the nucleation centers but also to some extent on unoccupied sites on the support, as characterized by low-energy ion scattering spectroscopy (ISS) and catalytic activity as a function of the number of preparation sequences. The catalytic hydrogenation of toluene to methylcyclohexane was examined on the obtained Ni/Al$_2$O$_3$ catalysts, and the highest activity in the hydrogenation of toluene was obtained at a Ni-loading of ca. 10 wt.% nickel [56].

2.4.6.3.10 Osmium (Os)
MOCVD has been used to immobilize Os species onto the internal porous structure of MCM-41. Os$_3$(CO)$_{12}$ was adsorbed on SiO$_2$ to form a trinuclear HOs$_3$(CO)$_{10}$(OSi) surface species. After heat treatment in air, these triangular sites break up to partially oxidized mononuclear surface species. In the presence of tert-butyl hydroperoxide, the monoculear species catalyzed the selective oxidation of trans-stilbene to the corresponding epoxide, showing selectivities of 70.5–92.6% at conversions of 1.1–27.4% at 328 K in CH$_3$CN-DMF [63].

2.4.6.3.11 Palladium (Pd) Films and Particles
Palladium, in the form of thin films and particles, is used in electronics, multilayer magneto-optical data storage, gas sensors, and catalysis. Few precursors have been studied for CVD. The bis(allyl) complexes [Pd(CH$_2$CRCH$_2$)$_2$](R=H, CH$_3$) gave palladium thin films with high purity (>99%) by thermal CVD. The complex [Pd(η^3-C$_3$H$_5$)Cp] is more stable but, as often occurs with cyclopentadienyl precursors, the deposited palladium films contained carbon impurities at the thermal CVD [64]. The precursor [Pd(η^3-C$_3$H$_5$)Cp] was deposited on a composite carbon nanotube–carbon microfiber system using a CVD technique. Only traces of Pd were detected after deposition on as-grown fibers, whereas a significantly higher concentration was found on plasma-treated fibers with oxygen-containing functional groups. The obtained Pd/C catalyst was active for the hydrogenation of cyclooctene [65]. A Pd/SiO$_2$ catalyst was prepared by CVD of Pd(allyl)Cp in a fluidized-bed reactor. The CVD of Pd(allyl)Cp was either carried out in one step by adding H$_2$ as reducing agent to the inert carrier gas, or in two steps using first N$_2$ as carrier gas (which resulted in the dissociative chemisorption of Pd(allyl)Cp on SiO$_2$), and then adding H$_2$ to obtain supported metallic Pd particles. The catalytic activity of the two-step CVD catalysts was very high in the hydrogenation of cyclooctene [66]. The CVD of palladium from [Pd(Cp)$_2$] and [Pd(acac)$_2$] has been patented [67], but films from [Pd(Cp)$_2$] are likely to suffer high carbon contamination and [Pd(acac)$_2$] has rather low volatility [68]. [Pd(η^3-CH$_2$CMeCH$_2$)(MeCOCHCOMe)], which is easily prepared by reaction of the sodium β-diketonate with the allylpalladium complex, is stable in air at room temperature for long periods of time, and this is advantageous when handling the material [68].

2.4.6.3.12 Pd/Zeolites
Zerovalent Pd particles in zeolites can be obtained through the reductive elimination of volatile ligands without formation of protons, unlike in the reduction of ion-exchanged Pd^{2+} in zeolites [69].

For chemical vapor deposition, both NaY and NaHY zeolites were first placed over a quartz frit in a U-tube reactor, and heated in flowing argon from 298 to 673 K over 2 h. After cooling to room temperature, a weighed amount of [Pd(η^3-C$_3$H$_5$)(η^5-C$_5$H$_5$)] was placed at the bottom of the U-tube under argon. Sublimation was carried out overnight in flowing argon (10 mL min^{-1}) at 298 K [70]. The supported material was heated in a H$_2$ (5%)/He flow from ambient temperature to 773 K at 3 K min^{-1}. The catalytic results in methylcyclopentane (MCP) conversion support the hypothesis of a total absence of protons. The Pd/NaY catalyst is 100% selective for the ring-opening reaction of MCP, in agreement with Gault's conclusions that this reaction is the sole one to be catalyzed by the metal function [71]. These catalysts compare well with other non-acidic Pd catalysts; in particular, a lower deactivation rate under catalytic condition is observed [72].

2.4.6.3.13 Platinum (Pt)
Platinum catalysts supported on alumina and silica were prepared by CVD of (trimethyl)methylcyclopentadienylplatinum (IV) as precursor. The catalysts were characterized by CO-chemisorption, temperature-programmed reduction (TPR) and XPS. The catalytic activity and selectivity were assessed by the hydrogenation of cinnamaldehyde [73]. Platinum hexafluoroacetylacetonate was selectively introduced inside KL zeolite channels via CVD in a flow of Ar at 343 K. Metal particles were then formed via thermal removal of the volatile organic ligands under H$_2$ at 623 K. In-situ diffuse reflectance infrared Fourier transform spectroscopy (DRIFTS) and EXAFS suggest the formation of very small Pt clusters with a nucleophilic character. These non-acidic Pt/KL catalysts show remarkably high activity and selectivity in the conversion of MCP to benzene at 773 K. A long catalyst life is achieved due to reduced coke formation and a very slow sintering rate [74]. A Pt/HL model catalyst was obtained by CVD of platinum hexafluoroacetylacetonate, Pt(hfac)$_2$, and characterized by EXAFS. In the conversion of MCP at 623 K, the selectivity is consistent with the typical bifunctional nature of the active sites. EXAFS results show that agglomeration of metal particles is greatly inhibited under catalytic reaction conditions [75]. The catalytic activity for MCP conversion, however, was not improved by the CVD preparation [76].

(Methylcyclopentadienyl)trimethyl platinum, MeCp-PtMe$_3$, was deposited on a TiO$_2$(110) in a self-limiting growth mode, which was imaged by scanning tunneling microscopy (STM). The precursor adsorbs

---
References see page 553

on five-fold coordinated Ti rows and decomposes to form two-dimensional-like low-profile Pt nanoparticles by annealing above 400 K. The inner structure of a specific low-profile particle was resolved by STM and site-dependent scanning tunneling spectroscopy (STS) was successfully measured. Based on these observations, a self-limiting growth mechanism was proposed, where two competing pathways of decomposition of the Pt precursor at the periphery of Pt particles and its blocking with TiO_x species, segregated from interstitial sites of the TiO_2 bulk, determine the particle size [77].

2.4.6.3.14 Rhenium (Re)
Rhenium oxide-encapsulated zeolites were prepared by CVD of methyl trioxo rhenium CH_3ReO_3 on H-ZSM-5 [78, 79]. The sample was pretreated under He at 673 K for 4 h *in situ* before use as catalyst. The catalyst exhibited high selectivity (88.2%) to acrolein(main) + acrylonitrile in propene-selective oxidation/ammoxidation at 648 K. The catalytic performance was much better than that of CVD catalysts obtained on HY and Na-ZSM-5, and also than that of an impregnated catalyst. Indeed, the coexistence of ammonia was found to be a prerequisite for the performance of the ReO_x/HZSM-5, where no selective oxidation proceeded in the absence of ammonia. The detailed characterization by XRD, ^{29}Si and ^{27}Al MAS NMR, X-ray absorption near edge structure (XANES), and EXAFS at the sequential stages of the catalyst preparation demonstrated the occurrence of structural changes of Re species during the preparation and under the catalytic reaction conditions. CH_3ReO_3 interacted with protons of HZ at 333 K, the subsequent treatment at 673 K formed tetrahedral [ReO_4] monomers, and ammonia promoted the formation of a new cluster, probably with a framework [Re_6O_{17}], at 673 K, having a Re—Re bond at 0.276 nm and terminal and bridge oxygen atoms (Re=O at 0.172 nm and Re—O at 0.203 nm, respectively). The active [Re_6O_{17}] clusters were converted to the inactive [ReO_4] monomers in the absence of ammonia [78, 79].

Phenol is one of the most important chemicals in industry, with annual worldwide production exceeding 7.2 megatons. Industrially, phenol has been produced from benzene by the three-step cumene process, which is not only energy-consuming but also relatively inefficient, producing many byproducts, such as acetone and α-methylstyrene. Direct phenol synthesis from benzene represents an alternative means of overcoming these problems, and O_2, H_2O_2, N_2O, $H_2 + O_2$, air/CO, and O_2/H_2O have been used as oxidants. Despite the good performances with N_2O and H_2O_2 as oxidants, no economically and environmentally favorable benzene–O_2 catalytic systems with high selectivity for phenol synthesis have been discovered to date, because molecular oxygen is difficult to activate for the selective achievement of benzene oxidation to phenol. A new Re cluster/HZSM-5 catalyst prepared by CVD of CH_3ReO_3 was found to be active for the selective oxidation of benzene to phenol with O_2, as shown in Scheme 3 [80]. The phenol selectivity at 553 K was tremendously high, reaching 88% in the steady-state reaction and 94% in the pulse reaction [80].

Zeolite-supported Re catalysts were prepared by CVD of CH_3ReO_3 with several zeolites such as H-ZSM-5, H-Beta, H-USY, and H-Mordenite. All catalysts were pretreated at 673 K in a He flow before use. The performances of the Re/zeolite catalysts for the selective oxidation of benzene with O_2 [80] are listed in Table 2. A Re-CVD/H-ZSM-5 catalyst ($SiO_2/Al_2O_3 = 19$) preferentially produced phenol with 87.7% selectivity in the presence of NH_3 (Table 2). It is to be noted that the coexistence of NH_3 is indispensable for the selective oxidation. The addition of H_2O and N_2O produced no positive effects on catalytic performance. Rather, the rate of phenol formation decreased in the order: H-ZSM-5 ($SiO_2/Al_2O_3 = 19$) > H-ZSM-5 ($SiO_2/Al_2O_3 = 23.8$) > H-ZSM-5 ($SiO_2/Al_2O_3 = 39.4$) ≫ H-Mordenite > H-beta > H-USY. The phenol selectivity was decreased slightly, from 87.7% to 82.4%, by increasing the Re loading (from 0.58 to 2.2 wt.%) and W/F (from 6.7 to 10.9 g_{cat} h mol^{-1}), while the conversion increased from 0.75% [turnover frequency (TOF): 65.6×10^{-5} s^{-1})] to 5.8% (TOF: 83.8×10^{-5} s^{-1}), as shown in Table 2 [80].

With CVD of methyl trioxo rhenium to HZSM-5, followed by He treatment at 673 K, two different Re—O bonds were observed at 0.173 ± 0.001 nm [coordination number (CN) = 2.5 ± 0.3] and 0.206 ± 0.001 nm (CN = 1.2 ± 0.5), which are attributed to Re=O and Re—O_{bridge} and/or Re—$O_{lattice}$, respectively, indicating the chemical attachment of Re species to the zeolite. The ammonia treatment of the Re species at the reaction temperature brought about the appearance of catalytic activity for the phenol synthesis. Indeed, the L_3-edge EXAFS data dramatically changed with the NH_3 treatment. The CN of Re—Re bonds increased from 1.2 ± 0.5 for the Re dimer to 5.2 ± 0.3 at 0.276 ± 0.002 nm, demonstrating the formation of Re clusters. Assuming a Re_6 octahedron as a cluster unit, the CN of Re—Re bonds indicates Re_{10} clusters edge-shared with two Re_6 octahedra. Furthermore, it was found that the NH_3-treated Re-cluster catalyst released N_2 molecules at 685 K in TPD spectra. The number of encapsulated nitrogen atoms was 0.12 N_2/Re, which corresponds to be one nitrogen atom per Re_6 octahedron. The Re—N(O) bonds in the NH_3-treated Re-CVD/HZSM-5 catalyst were observed at 0.204 ± 0.001 nm (CN = 2.8 ± 0.3) by EXAFS. Re=O double bonds at 0.172 ± 0.001 nm (CN = 0.3 ± 0.2) were also observed. The XPS Re 4f binding energy was 43.8 eV, which indicates a Re^{3+} state. The structure of the active Re clusters in the pore of

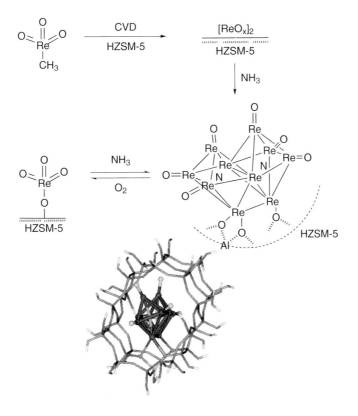

Scheme 3

Tab. 2 Catalytic performances of Re/zeolite catalysts for the direct phenol synthesis at 553 K[a]

Catalyst	SiO$_2$/Al$_2$O$_3$	Method	Re/wt.%	TOF/10^{-5} s^{-1}[b]	PhOH selec./%[c]
Re/HZSM–5[d]	19	CVD	0.58	Trace	0
Re/HZSM-5	19	CVD	0.58	65.6	87.7
Re/HZSM–5[e]	19	CVD	0.58	51.8	85.6
Re/HZSM–5[f]	19	CVD	2.2	83.8	82.4
Re/HZSM–5[g]	19	CVD	0.58	74.6	93.9
Re/HZM–5[h]	19	CVD	0.58	86.1	90.6
Re/HZSM-5	19	Imp.	0.6	11.8	27.7
Re/HZSM–5[d]	23.8	CVD	0.58	Trace	0
Re/HZSM-5	23.8	CVD	0.58	36.2	68.0
Re/HZSM-5	23.8	Imp.	1.2	18.5	15.3
Re/HZSM-5	39.4	CVD	0.59	31.0	48.0
Re/H-Beta	37.1	CVD	0.53	18.5	12.0
Re/H-USY	29	CVD	0.60	Trace	0
Re/H-Mordenite	220	CVD	0.55	26.3	23.4
Re/SiO$_2$-Al$_2$O$_3$	19	Imp.	1.2	Trace	0

[a] Catalyst = 0.20 g; W/F = 6.7 g$_{cat}$ h mol^{-1}; e/O$_2$/ NH$_3$/benzene = 46.4/12.0/35.0/6.6 (mol.%). The detailed carbon mass balance and material balance were examined in most of the experimental runs and the values were 97–99%.
[b] Consumed benzene/Re/s.
[c] Phenol selectivity in carbon%.
[d] In the absence of NH$_3$.
[e] W/F = 5.2 g$_{cat}$ h mol^{-1}.
[f] W/F = 10.9 g$_{cat}$ h mol^{-1}; He/O$_2$/NH$_3$/benzene = 46.4/12.0/35.0/6.6 (mol.%).
[g] Pulse reaction on the NH$_3$-pretreated catalyst (0.1 g); 1 pulse of benzene + O$_2$ [He/O$_2$/benzene = 81.4/12.0/6.6 (mol.%)].
[h] Pulse reaction on the NH$_3$-pretreated catalyst (1.0 g); 1 pulse of benzene + O$_2$ [He/O$_2$/benzene = 81.4/12.0/6.6 (mol.%)].
TOF, turnover frequency.

H-ZSM-5 was simulated by density functional theory (DFT) calculations, as illustrated in Scheme 3 [80]. The stable N-interstitial Re$_{10}$ cluster presented in Scheme 3 was modeled as a unique solution in the DFT calculations.

The Re$_{10}$ clusters with interstitial nitrogen atoms were highly selective for the catalytic phenol synthesis. It is to be noted that the selective benzene oxidation with molecular oxygen on the active Re$_{10}$ clusters in a pulsed reaction proceeds without ammonia. The Re$_{10}$ clusters were oxidized with O$_2$ to inactive ReO$_4$ monomers with Re^{7+}, as shown in Scheme 3. NH$_3$ reformed the active Re clusters in the zeolite pores under the working conditions.

2.4.6.3.15 **Rhodium (Rh)** The room-temperature interaction of Rh$_2$(CO)$_4$Cl$_2$ with TiO$_2$(110)–(1 × 1) and – (1 × 2) surfaces has been studied using STM. Dissociative adsorption of Rh$_2$(CO)$_4$Cl$_2$ at the (1 × 1) surface leads to the formation of an adsorbed Rh(CO)$_2$Cl species that coexists with particulate Rh^0. Adsorption of the precursor at the (1 × 2) surface results solely in the formation of metallic Rh [81].

2.4.6.3.16 **Ruthenium (Ru)** A supported cluster-derived Ru$_3$(CO)$_{12}$/2, 2′-bipyridine/SiO$_2$ catalyst has been found to be active both in water-gas shift reaction [82] and in 1-hexene hydroformylation [83]. The catalyst was prepared by vaporizing Ru$_3$(CO)$_{12}$ at 408 K in a separate sublimation oven and transferring the cluster in CO flow onto the silica bed in a fluidized-bed reactor. Sublimation can be carried out under low-pressure CO, but good mixing is achieved with a 200–300 mL min^{-1} flow of CO at normal pressure. 2,2′-Bipyridine was either added by a similar technique at 373 K or mixed with Ru$_3$(CO)$_{12}$/SiO$_2$ and activated separately at 373 K in a sealed glass cell under vacuum. Two types of Ru nanoparticle – round or flat, layered – were prepared by depositing Ru$_3$(CO)$_{12}$ vapor on a hole-modified highly oriented pyrolyzed graphite (HOPG) surface. N$_2$ adsorption/desorption behaviors on the round and flat Ru particles were seen to be different. The possibility of forming a high density of atomic N atoms on the surfaces may be a reason for the high activity of Ru/C catalysts in NH$_3$ synthesis [84]. MOCVD of [(1,5-cyclooctadiene))toluene)Ru^0] on copper and silicon substrates in a vertical cold-wall reactor led to the formation of thin metallic ruthenium films with low carbon content. Ru films could be deposited at temperatures as low as 423 K [85]. CVD of Ru(COD)(η^3-allyl)$_2$ on α-Al$_2$O$_3$ and oxidized Si(100) was conducted to provide Ru and RuO$_2$ nanostructured thin films [86].

2.4.6.3.17 **SiO$_2$** SiO$_2$ overlayers have been deposited onto the zeolite external surfaces to control the pore-opening size of zeolites. Pore-size control has usually been carried out by cation exchange, although an alternative method is silanation using SiH$_4$ [87]. These methods modify not only the acidity but also the pore structure, so an undesired change in catalytic activity may be introduced. In contrast, the pore-opening size of ZSM-5 was controlled by CVD of Si(OCH$_3$)$_4$. The kinetic diameter of the silicon alkoxide is supposed to be 0.89 nm, much larger than that of the H-ZSM-5 pore (0.54 × 0.55 nm). Alkylation of toluene with methanol and toluene disproportionation preferentially produced p-xylene over H-ZSM-5 modified by CVD of Si(OCH$_3$)$_4$ due to control of the pore-opening size [8, 88].

Katada, Niwa and colleagues have proposed a method for the CVD of tetramethoxysilane (Si(OCH$_3$)$_4$) using molecular templates on basic oxides such as Al$_2$O$_3$ and SnO$_2$. Aldehydes such as benzaldehyde adsorb strongly onto these basic surfaces. The obtained SiO$_2$/SnO$_2$ showed a chemisorption capacity and oxidation catalysis with specific selectivities probably based on the shapes of the cavities [89, 90].

A very elegant application of CVD to catalyst design has been performed in molecular imprinting catalysts. Recently, attempts to imprint metal complexes have been reported, and these constitute the current state of the art in this field [91, 92].

In most cases of metal-complex imprinting, ligands of the complexes are used as template molecules, with the aim of creating a cavity near the metal site. The molecular imprinting of metal complexes enables several features to be realized: (i) the attachment of the metal complex on robust supports; (ii) surrounding the metal complex by Si(OCH$_3$)$_4$ CVD, followed by a hydrolysis–polymerization to form the silica matrix; and (iii) the production of shape-selective cavities on the metal site by removal of a template ligand. Metal complexes imprinted in this way have been applied to molecular recognition, reactive complex stabilization, ligand-exchange reactions, and catalysis. Ligands of a metal complex not only influence its catalytic activity, but also provide an unsaturated, reactive metal site with a ligand-shape space following the removal of a ligand. Figure 2 shows a molecular imprinting Rh dimer catalyst for the hydrogenation of alkenes, where a ligand (P(OCH$_3$)$_3$) of the attached Rh complex is regarded as a template molecule with a similar shape to one of the half-hydrogenated species of 3-ethyl-2-pentene. By using the imprinting procedure, Rh 3d XPS intensity is reduced remarkably, indicating that the Rh species were embedded in the polymerized silica matrix overlayers that were characterized using ^{29}Si solid-state MAS NMR. The structures around the Rh atoms in the molecular imprinting catalyst at each step of the alkene hydrogenation were investigated by EXAFS [93, 94].

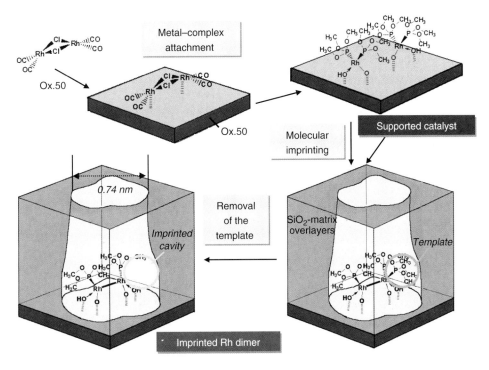

Fig. 2 Preparation steps for the molecular-imprinted Rh dimer catalyst.

The recognition of simple alkenes without any functional group is generally difficult, but the Rh imprinting catalyst demonstrated remarkable catalytic performances with both shape and size selectivity [95]. The method of combining metal-complex attachment and molecular imprinting on the surface demonstrates that the strategy can be used to both regulate and design chemical reactions [91–96].

2.4.6.3.18 Tin (Sn) CVD SnO_2/SiO_2 catalysts were prepared by repeating the following CVD cycle at 453 K: (i) the desiccation of silica gel (240 $m^2 g^{-1}$) in a flow of dry N_2; (ii) the deposition of $SnCl_4$ vapor on silica gel surface; (iii) purging of excess $SnCl_4$ by flowing dry N_2; and (iv) hydrolysis of the deposited Sn compound by water vapor [97]. The amount of deposited Sn was controlled by the number of CVD cycles repeated. The resulting materials were washed five times with 4 M NH_4OH to remove residual chloride ions, then with water to remove NH_4OH, and then dried and calcined at 773 K for 2 h in flowing air. The activity and selectivity in the oxidative dehydrogenation of ethylbenzene were measured in a conventional flow reactor. The high selectivity of the CVD catalyst was attributed to moderate acid strength due to the controlled dispersion of SnO_2 on SiO_2 [97].

2.4.6.3.19 Titanium (Ti) The surface or interface of two different metal oxide layers interacting chemically with each other through covalent bonds provides a potential means of producing new surface materials. A TiO_2 monolayer on SiO_2 was prepared by the CVD reaction between $Ti(O^iC_3H_7)_4$ vapor and OH groups of the SiO_2 surface (300 $m^2 g^{-1}$). XANES and EXAFS analyses revealed that the TiO_2 on SiO_2 exists as the anatase type. The shape of TiO_2 on SiO_2 was discussed on the basis of the coordination numbers of Ti–Ti nearest neighbors. The model which best fits the experimental results among the four model layer structures with different planes, (100), (001), (110), and (101), is the monolayer structure with (101) orientation [98]. The TiO_2 monolayer is not reduced with H_2 at 773 K, and surprisingly even in the presence of supported noble metals [98]. While the structure of the TiO_2 layers in $Pt/TiO_2/SiO_2$ were not changed by H_2 reduction at 773 K, a remarkable change was found in the structure of $Pd/TiO_2/SiO_2$, where the anatase structure was transformed to the rutile structure with the bond lengths of Ti–O, Ti–Ti, and Ti–Ti being 0.198, 0.294, and 0.357 nm, respectively; these values were almost the same as the corresponding bond lengths in rutile (0.197, 0.294, and 0.357 nm).

Highly dispersed TiO_2 supported on Vycor glass was also prepared by CVD using the reaction of $TiCl_4$ with the surface OH groups, followed by treatment with H_2O vapor to hydrolyze the supported compounds [99]. CH_4 and CH_3OH are produced from CO_2 and H_2O on the

References see page 553

CVD TiO$_2$/Vycor glass under UV irradiation at 275 K, with a good linear relationship between the yields of these products and the UV irradiation time.

The film structure, composition, and deposition kinetics of model Ziegler–Natta polymerization catalysts produced by CVD have been studied by TPD, Auger electron spectroscopy (AES), and XPS. Redox reactions of metallic Mg and TiCl$_4$ deposited from the vapor phase on a Au substrate produce the model catalysts composed of titanium chloride and magnesium chloride (TiCl$_x$/MgCl$_2$). After exposure to the AlEt$_3$ cocatalyst, the model catalyst films are active for propene polymerization [100].

Propene oxide is an important commodity chemical, with a present annual worldwide demand of about 5 million tons. The organic hydroperoxide process is currently a leading commercial method for PO production. This involves the reaction of propene with alkyl hydroperoxide, catalyzed by molybdenum compounds in solution or a solid titanium silica catalyst. The commercial titanium silica catalyst is prepared by impregnating silica with TiCl$_4$ or an organic titanium compound, followed by calcination and silylation [101]. The titanium silica samples with different titanium contents were prepared by CVD of TiCl$_4$ on silica gel in the temperature range of 673 to 1383 K. The resulting samples were used as catalysts for propene oxidation with *tert*-butyl hydroperoxide to propene oxide. The deposition temperature of TiCl$_4$ was found to have a strong effect on catalyst selectivity, but only a small effect on catalyst activity. With the increase in CVD temperature, the yield of propene oxide increased rapidly and passed through a maximum at 1173 K, corresponding to propene oxide yields of 92.7 to 94.4% in the Ti content range 0.89 to 1.07 wt.%. XPS data indicated that Si—OTiCl$_3$, formed from the reaction between TiCl$_4$ and a silanol site or a surface siloxane bridge site, was the active species for catalyzing the epoxidation [102].

Ti-containing mesoporous catalysts were prepared by CVD of TiCl$_4$ on silica MCM-41 at 973 to 1173 K. The samples were characterized by XRD, Fourier transform infra-red (FTIR), XPS, and transmission electron microscopy (TEM), and evaluated for the epoxidation of propene with *tert*-butyl hydroperoxide or ethylbenzene hydroperoxide. The best Ti/MCM-41 catalyst was prepared at a temperature of 1073 K, achieving a maximum propene oxide yield of 94.3% [103].

2.4.6.3.20 **Vanadium (V)** Bond et al. found that a V$_2$O$_5$/TiO$_2$ catalyst prepared by the reaction between VOCl$_3$ vapor and surface OH groups showed a high selectivity in the oxidation of *o*-xylene to phthalic anhydride [104, 105]. Hydroxyl groups on the surface of anatase react with the vapor of VOCl$_3$ at room temperature to give a partial monolayer of a vanadium species. The sample was heated at 670 K, then rehydroxylated by H$_2$O and outgassed at 410 K, and further treated with VOCl$_3$. After six such cycles, a monolayer catalyst is produced which contains 1.7 wt.% V$_2$O$_5$ and which is superior even to a doubly promoted catalyst prepared by conventional impregnation [105].

Oxide supports such as Al$_2$O$_3$, TiO$_2$, ZrO$_2$, and Nb$_2$O$_5$ have a high surface density of reactive surface hydroxyls, and tend to form a close-packed monolayer of the surface metal oxide phase, whereas, oxide supports such as SiO$_2$ have a lower density of reactive surface hydroxyls and do not form a close-packed overlayer of the surface vanadium oxide phase. The surface vanadium oxide species on the different oxide supports exhibit a Raman band at 1015–1040 cm^{-1} due to the terminal V=O bond of isolated VO$_4$ species [106, 107]. At low loadings (ca. 1 wt.% V$_2$O$_5$), a single sharp band is present at ca. 1030 cm^{-1} which is due to an isolated four-coordinated surface vanadium oxide species containing one terminal V=O bond and three bridging V—O—Ti bonds [108]. At intermediate loadings (2–6 wt.%), a second band is present at ca. 930 cm^{-1} which has been assigned to a polymerized, four-coordinated surface vanadium oxide species [108]. At high loadings (>6 wt.%), a third sharp band is present at 994 cm^{-1}, due to crystalline V$_2$O$_5$, which indicates that the close-packed surface vanadium oxide monolayer has been formed and essentially all the reactive surface hydroxyls are consumed.

The reactivity of the surface V$_2$O$_5$ overlayers on the different oxide supports was probed by the methanol oxidation reaction. The reactivity, in terms of TOF, of the surface vanadia species was found to depend dramatically on the specific oxide support [109]. The difference in reactivity may not be a structural factor as the structure is independent of the oxide support. One of the possible factors controlling the reactivity of various supported metal oxide catalysts as a function of oxide support is the difference in the terminal V=O bond or the bridging V—O-support bond. A plot of TOF versus Raman M=O position for supported vanadium oxide, molybdenum oxide, rhenium oxide, and chromium oxide catalysts suggests that no relationship exists between the catalyst reactivity and the terminal M=O bond strength. Wachs et al. proposed that the reactivity may be related to the bridging M—O-support bond [109].

Highly dispersed V$_2$O$_5$/SiO$_2$ materials were successfully synthesized using the atomic layer deposition by sequentially applying surface-saturating reactions of VO(O^{i}Pr)$_3$ and oxygen with a SiO$_2$ surface. The chemisorption-based growth of vanadia on SiO$_2$ occurred at 363–393 K, as suggested by elemental analysis and DRIFTS measurements. The maximum dispersion of vanadia (ca. 2.3 V nm^{-2}) was attained by two consecutive precursor binding–oxidation cycles on SiO$_2$ pretreated at 873 K. The catalysts with loadings 1.0–2.3 V nm^{-2} consisted of highly dispersed isolated VO$_4$ species, whereas in corresponding impregnated catalysts V$_2$O$_5$ crystallites were formed in addition to the monomeric vanadia species [110].

Highly dispersed supported vanadium oxide catalysts have also been prepared by using the reaction of the photochemically activated VOCl$_3$ with surface OH groups on Vycor glass at 273 K [111]. Photoactivation of adsorbed VOCl$_3$ was carried out by UV excitation of the charge-transfer absorption band of VOCl$_3$ at approximately 300–450 nm. The supported vanadium oxides exhibit a higher activity for the photocatalytic isomerization reactions of 2-butene, as compared with those obtained for the oxides prepared by a conventional impregnation method.

2.4.6.3.21 Tungsten (W) Tungsten nitride was synthesized on γ-alumina and quartz plate using a CVD method in a stream of WCl$_6$, NH$_3$, H$_2$, and Ar gases at 973 K under reduced pressure. The CVD tungsten nitride/alumina catalysts deposited for 30 and 60 min showed a constant activity for thiophene hydrodesulfurization (HDS) [112].

2.4.6.3.22 Zirconium (Zr) A Pd-supported sample was prepared by vapor deposition of Zr(OC$_2$H$_5$)$_4$. Zr(OC$_2$H$_5$)$_4$ powder was heated to ca. 423 K under vacuum and held for 13 h. The single crystal of Pd with the deposited Zr(OC$_2$H$_5$)$_4$ was oxidized at 673 K. The sample was subsequently reduced by treatment in H$_2$ at 773 K. When the sample was treated in O$_2$, the ZrO$_2$ appeared as a smooth, featureless overlayer of varying thickness wetting the Pd. After treatment in H$_2$, the ZrO$_2$ formed non-wetting particles on the Pd, with a sharp Pd/ZrO$_2$ interface [113]. Similar reversible structure transformation of a ZrO$_2$ overlayer has been found with the ZrO$_2$ overlayer on Pd black by EXAFS [114].

Single crystal Al(110) was hydroxylated in UHV to produce a disordered overlayer. Protolytically labile tetra-neopentylzirconium did not react with hydroxylated Al(110) in UHV, which was interpreted through a mechanistic proposal for proton-transfer-based surface chemical modification. According to this proposal, a zirconium alkoxide, although less basic than a zirconium alkyl, could be a kinetically viable substrate for proton-transfer-based organometallic CVD in UHV, and tetra-tert-butoxylzirconium did react with hydroxylated Al(110) to produce a surface-bound di-tert-butoxylzirconium species. The deposition reaction was readily studied by high-resolution electron energy loss spectroscopy (HREELS), XPS, and AES [115].

2.4.6.3.23 Bimetals Pt(acac)$_2$ and Ru(acac)$_3$ were deposited on carbon by CVD at 443–513 K to produce Pt–Ru nanoparticles (ca. 2 nm) on carbon. The obtained Pt–Ru/C catalyst was found to be active for electrocatalytic methanol oxidation [116].

Bimetallic Pt/Re particles entrapped inside supercages of NaY zeolite are easily obtained by CVD of Re$_2$(CO)$_8$ on prereduced Pt particles acting as nucleation sites for the decomposition of the metal carbonyl. The alloy particles were characterized by a high selectivity for deep hydrogenolysis of n-heptane to methane, which was much higher than that of the single Re and Pt metal catalyst [117].

Rh–Sn/SiO$_2$ catalysts were prepared by the selective CVD reaction between Sn(CH$_3$)$_4$ vapor and small Rh metallic particles supported on SiO$_2$ [118, 119]. Sn(CH$_3$)$_4$ deposition takes place much more rapidly on Rh metal surfaces than with OH groups of SiO$_2$. Thus, most of Sn(CH$_3$)$_4$ vapor reacts preferentially with Rh particles. The samples were finally reduced with H$_2$ at 573 K for 1 h, followed by evacuation *in situ* before use as catalysts, and also before each run. The CVD Rh–Sn/SiO$_2$ catalysts showed much higher catalytic activities for the NO–H$_2$ reaction and NO dissociation than Rh/SiO$_2$ and coimpregnated Rh/Sn/SiO$_2$ [118], showing a S-shape dependency on the amount of Sn [119]. The active bimetallic particles were characterized by CO adsorption, TEM, and EXAFS, which revealed a Rh–Sn ensemble surface structure with a relaxation of the first bimetallic layer [8].

The *in-situ* EXAFS technique was successfully applied to observe the molecular reaction intermediate for the NO–H$_2$ reaction on highly active Rh–Sn/SiO$_2$ catalysts. The EXAFS data, with the aid of FTIR, revealed the existence of the bent-type NO with a bond distance of 0.256 nm with Sn atoms (Rh–NO–Sn) on the bimetallic surface under steady-state reaction conditions. The tilted NO was suggested to dissociate to form an Sn–O bond at 0.205 nm [120].

Rh–Sn bimetallic particles entrapped in NaY cages were prepared by CVD of Sn(C$_2$H$_5$)$_4$ or Sn(C$_6$H$_5$)$_4$ onto reduced Rh/NaY samples, followed by H$_2$ reduction. Sn(C$_6$H$_5$)$_4$ leads to the formation of a higher Sn coverage than that obtained from Sn(C$_2$H$_5$)$_4$. In the selective hydrogenation of citral (3,7-dimethyl-2,6-octadienal), the promotional effect under working conditions is ascribed to the presence of non-ionic oxidized SnO$_x$ phases [121].

Particles of each of the single-phase Ni–Sn intermetallic compounds Ni$_3$Sn, Ni$_3$Sn$_2$, and Ni$_3$Sn$_3$ have been formed on silica by CVD of Sn(CH$_3$)$_4$ onto Ni/SiO$_2$ and subsequent hydrogen treatment, as characterized by XRD. The CVD method has resulted in the preparation of

References see page 553

silica-supported Ni−Sn intermetallic compound catalysts with higher activities than their unsupported counterparts, owing to the much higher specific surface areas, with retention of the high selectivity to benzene [122].

2.4.6.3.24 Mixed Metal Oxides and Sulfides

A Fe/Mo/DBH (partially deboronated borosilicate molecular sieve) catalyst was prepared by the CVD technique from FeCl$_3$ and chloride salts of molybdenum such as MoO$_2$Cl$_2$, MoOCl$_4$, and MoCl$_5$ with DBH [123]. The Fe/Mo/DBH catalyst, which contained 4.6 wt.%-Mo and 1.41 wt.%-Fe, Mo/Fe = 1.9, was found to be active for the gas-phase oxidation of *para*-xylene with oxygen to produce terephthaldehyde with high selectivity with sustained stability under relatively mild conditions. *Para*-tolualdehyde was also produced. The molar ratio of terephthaldehyde to *para*-tolualdehyde of up to 4:1 with rather limited burning (<10%) was realized under controlled conditions, despite the catalyst's large surface area of 250–325 m^2 g^{-1}. Combined techniques of TEM, analytical EM, Raman spectroscopy, XPS, XRD, and *in-situ* XANES were employed to characterize the catalyst. A reaction mechanism is postulated based on two different types of active sites, namely the pair sites (2 Mo^{6+}) for terephthaldehyde and the single sites (Mo^{6+}) for *para*-tolualdehyde [124]. The CVD Fe/Mo/DBH catalyst also exhibited better catalytic performance for the oxidation of benzene with N$_2$O to phenol than the impregnated counterpart [125]. The borosilicate molecular sieves themselves were also active. Two mechanistic pathways are postulated, based on reactive oxygen species such as O$^-$, which can be generated via interaction of N$_2$O with Fe sites, or OH$^+$, which can be generated by Brønsted sites on the borosilicate molecular sieve itself.

Considerable interest currently exists in developing highly active catalysts for deep HDS of dibenzothiophene (DBT) and 4,6-dimethyldibenzothiophene (4,6-DMDBT). Lee et al. prepared highly dispersed MoS$_2$ on fluorinated alumina (FAl) by a sonochemical method, then deposited Co on the MoS$_2$/FAl by a CVD method using Co(CO)$_3$(NO). The obtained catalyst exhibited 2.1- to 4.6-fold higher activity than a fluorine-free catalyst prepared by impregnation [126].

Okamoto et al. have reported the preparation of Co−Mo sulfide catalysts, in which the edge of MoS$_2$ particles is preferentially and fully covered by Co atoms forming the Co−Mo−S phase, by using Co(CO)$_3$NO as a precursor of Co. A selective formation of the Co−Mo−S phase in the CVD−Co/MoS$_2$/Al$_2$O$_3$ catalysts was evidenced by the Co 2p XPS, Co K-edge XANES, the magnetic susceptibility and effective magnetic moment of Co, a proportional correlation between Co/Mo and NO/Mo molar ratios, and a linear relationship between catalytic activity for thiophene HDS and the amount of Co accommodated [127, 128]. MoO$_3$/Al$_2$O$_3$ was prepared by an impregnation technique using (NH$_4$)$_6$Mo$_7$O$_{24}$·4H$_2$O, followed by calcination at 773 K for 5 h. The MoO$_3$/Al$_2$O$_3$ was sulfided in a 10% H$_2$S/H$_2$ flow at 673 K for 1.5 h. The MoS$_2$/Al$_2$O$_3$ was evacuated at 673 K for 1 h and then exposed to a vapor of Co(CO)$_3$NO, kept at 273 K, for 5 min at room temperature, followed by evacuation for 10 min at room temperature to remove physisorbed Co(CO)$_3$NO molecules and subsequent resulfidation at 673 K for 1.5 h in a 10% H$_2$S/H$_2$ flow. The Co atoms in CVD−Co/MoS$_2$/Al$_2$O$_3$ exhibited exclusively an antiferromagnetic property, indicating that even-numbered Co atoms are interacting with each other in the Co−Mo−S phase. On the basis of the NO adsorption behavior and magnetic property, it is proposed that the structure of the Co−Mo−S phase is represented as a Co sulfide dinuclear cluster located on the edge of MoS$_2$ particles (Fig. 3) [128].

2.4.6.4 CVD-Related Techniques

In plasma CVD (which is also known as plasma-enhanced or plasma-assisted CVD), the reaction is activated by a plasma. Thermal laser CVD occurs as a result of the thermal energy from the laser coming into contact with and heating an absorbing substrate. In photolytic CVD, the chemical reaction is induced by the action of light, specifically ultraviolet radiation, which has sufficient photon energy to break the chemical bonds in the reactant molecules. Fluidized-bed CVD is as special technique which is used primarily in the coating of powders, with the powder being provided with quasi-fluid properties in a flowing gas [129]. Microwave plasma technology has been used for the deposition of catalytically active chromia species on ZrO$_2$ and La-doped ZrO$_2$. The longer-term microwave plasma treatment causes a permanent hydrophobization of the surface. At low or medium microwave power and shorter treatment times, the surface chemical properties of supports are almost unchanged. The microwave plasma and radiation should have no negative effect on the plasma chemical preparation of supported catalysts [130].

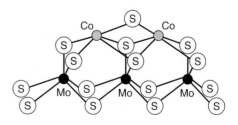

Fig. 3 A proposed structure of the Co−Mo−S phase in the CVD−Co/MoS$_2$/Al$_2$O$_3$.

The catalytic CVD method is a plasma-free deposition technique whereby deposition gases are decomposed by the catalytic or pyrolytic reactions with a heated catalyst (e.g., W wire) placed near the substrates. Thus, films are deposited at low temperatures without any aid from plasma or photochemical excitation [131].

2.4.6.5 Concluding Remarks

Today, CVD is no longer a laboratory curiosity but rather may be seen as a major technology which is on a par with other major disciplines such as conventional wet processing, sol–gel, and electrodeposition. Structurally well-defined zeolite-entrapped MOCVD- and CVD-type species such as $W(CO)_6$-Y, $Sn(CH_3)_2$-Y, $CdCH_3$-Y and Si_2H_5-Y have been used successfully for the self-assembly and organization of semiconductor nanocluster crystal lattices, including W_4O_{10}-Y, Sn_4S_6-Y, Cd_6Se_4-Y and Si_8-Y [132]. Materials of this type may have interesting catalytic, electronic, and luminescent properties. Currently, the method of chemical vapor deposition is opening a new area of catalyst preparation. These new types of materials, structures, and catalytic properties – which differ from those of the catalysts obtained by other, traditional methods of preparation – may make the application of CVD to the industrial-scale synthesis of catalytic materials an attractive proposition [80].

References

1. J. Haber, *Pure Appl. Chem.* **1991**, *63*, 1227–1246.
2. C. F. Powell, J. H. Oxley, J. M. Blocher Jr. (Eds.), *Vapor Deposition*, John Wiley & Sons, New York, 1966.
3. For example; L. Mond, US Patent 455230, 1891.
4. H. O. Pierson, *Handbook of Chemical Vapor Deposition – Principles, Technology and Applications*, Noyes, Park Ridge, New Jersey, 1992.
5. Y. Iwasawa, *Adv. Catal.* **1987**, *35*, 187.
6. R. Psaro, S. Recchina, *Catal. Today* **1998**, *41*, 139–147.
7. J. L. Zilko, *Handbook of Thin-Films Deposition Processes and Technologies*, K. K. Shuegraf (Ed.), Noyes Publications, New Jersey, 1988.
8. Y. Iwasawa, in G. Ertl, H. Knözinger, J. Weitkamp (Eds.), *Handbook of Heterogeneous Catalysis*, Vol. 2, Wiley-VCH, Weinheim, 1997, p. 853–873.
9. M. Okumura, K. Tanaka, A. Ueda, M. Haruta, *Solid State Ionics* **1997**, *95*, 143–149.
10. Y.-J. Chem, C.-T. Yeh, *J. Catal.* **2001**, *200*, 59–68.
11. Y. Iwasawa, *Adv. Catal.* **1987**, *35*, 187–264.
12. Y. Iwasawa, M. Yamada, Y. Sato, H. Kuroda, *J. Mol. Catal.* **1984**, *23*, 95–106.
13. Y. Iwasawa, M. Yamada, *Nikkashi* **1984**, 1042–1049.
14. K. Asakura, Y. Iwasawa, *J. Phys. Chem.* **1989**, *93*, 4213–4218.
15. Y. Iwasawa, *Catal. Today* **1993**, *18*, 21–72.
16. L. B. Backman, A. Rautiainen, M. Lindblad, O. Jylha, A. O. I. Krause, *Appl. Catal. A: General* **2001**, *208*, 223–234.
17. N. Bahlawance, E. F. Rivera, K. Kohse-Höinghaus, A. Brechling, U. Kleineberg, *Appl. Catal. B: Environmental* **2004**, *53*, 245–255.
18. R. L. Banks, US Patent 3 463 827, 1969.
19. R. L. Banks, Belgium Patent 633 418, 1963.
20. A. Brenner, D. A. Hucul, S. J. Hardwick, *Inorg. Chem.* **1979**, *18*, 1478–1484.
21. Y. Okamoto, H. Onimatsu, M. Hori, Y. Inui, T. Imanaka, *Catal. Lett.* **1992**, *12*, 239–244.
22. R. Sekine, M. Kawai, K. Asakura, Y. Iwasawa, *Proc. Mater. Res. Soc. Meeting*, Anaheim, **1991**, *222*, 333–338.
23. R. Becker, H. Parala, F. Hipler, O. P. Tkachenko, K. C. Klementiev, W. Grunert, H. Wilmer, O. Hinrichsen, M. Muhler, A. Birkner, C. Woll, S. Schafer, R. A. Fischer, *Angew. Chem. Int. Ed.* **2004**, *43*, 2839–2842.
24. H. P. Chu, L. Lei, X. Hu, P.-L. Yue, *Energy & Fuels* **1998**, *12*, 1108–1113.
25. J. Medina-Valtierra, J. Ramirez-Ortiz, V. M. Arroyo-Rojas, E. Ruiz, *Appl. Catal. A: General* **2003**, *238*, 1–9.
26. D. C. Bailey, S. H. Langer, *Chem. Rev.* **1981**, *81*, 109–148.
27. R. F. Howe, *The preparation of heterogeneous catalysts from mononuclear carbonyl complexes in inorganic supports*, in *Tailored Metal Catalysts*, Y. Iwasawa (Ed.), Reidel, Dordrecht, 1986, pp. 141–182.
28. M. Ichikawa, *Tailored Metal Catalysts*, Y. Iwasawa (Ed.), Reidel, 1986, pp. 183–263.
29. J. B. Nagy, M. van Eenoo, E. G. Derouane, *J. Catal.* **1979**, *58*, 230–237.
30. K. Asakura, K. Ooi, Y. Iwasawa, *J. Mol. Catal.* **1992**, *74*, 345–351.
31. A. A. Battiston, J. H. Bitter, W. M. Heijboer, F. J. F. de Groot, D. C. Koningsberger, *J. Catal.* **2003**, *215*, 279–293.
32. J. A. Z. Pieterse, S. Boonevelad, R. W. van den Brink, *Appl. Catal. B: Environmental* **2004**, *51*, 215–228.
33. G. D. Pirngruber, M. Luechinger, P. K. Roy, A. Cecchetto, P. Smirniotis, *J. Catal.* **2004**, *224*, 429–440.
34. Q. Zhu, R. M. van Teeffelen, R. A. van Santen, E. J. M. Hensen, *J. Catal.* **2004**, *221*, 575–583.
35. E. J. M. Hensen, M. Garcia-Sanchez, N. Rane, P. C. M. M. Magusin, P.-H. Liu, K.-J. Cao, R. A. van Santen, *J. Catal.* **2005**, *101*, 79–85.
36. R. L. Banks, G. C. Bailey, *Ind. Eng. Chem. Prod. Res. Dev.* **1964**, *3*, 170–173.
37. R. L. Banks, *CHEMTECH* **1986**, *16*, 112–117.
38. E. S. Davie, D. A. Whan, C. Kemball, *Chem. Commun.* **1969**, 1430–1431.
39. R. F. Howe, D. E. Davidson, D. A. Whan, *J. Chem. Soc., Faraday Trans. 1* **1972**, *68*, 2266–2280.
40. A. Brenner, R. L. Burwell, Jr., *J. Am. Chem. Soc.* **1975**, *97*, 2565–2566.
41. A. Brenner, R. L. Burwell, Jr., *J. Catal.* **1978**, *52*, 353–363; 364–374.
42. Y. Okamoto, T. Imanaka, K. Asakura, Y. Iwasawa, *J. Phys. Chem.* **1991**, *95*, 3700–3705.
43. J. M. Coddington, R. F. Howe, Y.-S. Yong, K. Asakura, Y. Iwasawa, *J. Chem. Soc., Faraday Trans. 1* **1990**, *86*, 1015–1016.
44. A. Yamaguchi, A. Suzuki, T. Shido, Y. Inada, K. Asakura, M. Nomura, Y. Iwasawa, *J. Phys. Chem. B* **2002**, *106*, 2415–2422.
45. M. Anpo, M. Kondo, S. Coluccia, C. Louis, M. Che, *J. Am. Chem. Soc.* **1989**, *111*, 8791–8799.
46. C. Louis, M. Che, M. Anpo, *J. Catal.* **1993**, *141*, 453–464.
47. Y. Iwasawa (Ed.), *Tailored Metal Catalysts*, Reidel, Dordrecht, 1986.
48. Y. Iwasawa, Y. Nakano, S. Ogasawara, *J. Chem. Soc., Faraday Trans. 1* **1978**, *74*, 2968–2981.

49. Y. Iwasawa, S. Ogasawara, *J. Chem. Soc., Faraday Trans. 1* **1979**, *75*, 1465–1476.
50. C. C. Williams, J. G. Ekerdt, J.-M. Jehng, F. D. Hardcastle, A. J. Turek, I. E. Wachs, *J. Phys. Chem.* **1991**, *95*, 8781–8791.
51. M. de Boer, A. J. van Dillen, D. C. Koningsberger, J. W. Geus, M. A. Wuurman, I. E. Wachs, *Catal. Lett.* **1991**, *11*, 227–239.
52. M. Cornac, A. Jeannin, J. C. Lavalley, *Polyhedron* **1986**, *5*, 183–186.
53. A. N. Desikan, L. Huang, S. T. Oyama, *J. Phys. Chem.* **1991**, *95*, 10050–10056.
54. J. M. Aigler, J. L. Brito, P. A. Leach, M. Houalla, A. Proctor, N. J. Cooper, W. K. Hall, D. M. Hercules, *J. Phys. Chem.* **1993**, *97*, 5699–5702.
55. W.-H. Hung, J. Schwartz, S. L. Bernasek, *Langmuir* **1994**, *10*, 2056–2059.
56. E. C. Alyea, K. F. Brown, K. J. Fisher, K. D. L. Smith, In: *New frontiers in catalysis, Proceedings of the 10th International Congress on Catalysis*, L. Guczi (Ed.), Budapest, 1992, *Studies in Surface Science and Catalysis*, Vol. 75, Elsevier, Amsterdam, 1993, pp. 503–514.
57. A. Malinowski, R. Ohnishi, M. Ichikawa, *Catal. Lett.* **2004**, *96*, 141–146.
58. K. Asakura, Y. Iwasawa, *J. Phys. Chem.* **1991**, *95*, 1711–1716.
59. N. D. Parkyns, In: *Proceedings of the 3rd International Congress on Catalysis*, W. M. H. Sachtler (E.d.), Amsterdam, 1964, Vol. 2, North-Holland, Amsterdam, 1965, pp. 914–927.
60. J.-P. Jacobs, L. P. Lindfors, J. G. H. Reintjes, O. Jylha, H. H. Brongersma, *Catal. Lett.* **1994**, *25*, 315–324.
61. T. Suntola, *Atomic layer epitaxy*, in *Handbook of Crystal Growth*, D. T. J. Hurle (Ed.), North-Holland, Amsterdam, 1993.
62. E.-L. Lakomaa, *Appl. Surf. Sci.* **1994**, *75*, 185–196.
63. V. Caps, S. C. Tsang, *Appl. Catal. A: General* **2003**, *248*, 19–31.
64. J. E. Gozum, D. M. Pollina, J. A. Jensen, G. S. Girolami, *J. Am. Chem. Soc.* **1988**, *110*, 2688–2689.
65. W. Xia, O. F.-K. Schlüter, C. Liang, M. W. E. van den Berg, M. Guruya, M. Mühler, *Catal. Today* **2005**, *102–103*, 34–39.
66. X. Mu, U. Bartmann, M. Guraya, G. W. Busser, U. Weckenmann, R. Fischer, M. Muhler, *Appl. Catal. A: General* **2003**, *248*, 85–95.
67. T. Kudo, A. Yamaguchi, Japanese Patent, JP 62 207 868, 1987.
68. Z. Yuan, R. J. Puddephatt, *Adv. Mater.* **1994**, *6*, 51–54.
69. C. Dossi, R. Psaro, R. Ugo, Z. C. Zhang, W. M. H. Sachtler, *J. Catal.* **1994**, *149*, 92–99.
70. C. Dossi, R. Psaro, A. Bartsch, A. Galasco, E. Brivio, P. Losi, *Catal. Today* **1993**, *17*, 527–538.
71. F. Gault, *Adv. Catal.* **1981**, *30*, 1–95.
72. R. Ugo, C. Dossi, R. Psaro, *J. Mol. Catal. A: Chemical* **1996**, *107*, 13–22.
73. M. Lashdaf, J. Lahtinen, M. Lindblad, T. Venäläinen, A. O. I. Krause, *Appl. Catal. A: General* **2004**, *276*, 129–137.
74. C. Dossi, R. Psaro, A. Bartsch, A. Fusi, L. Sordelli, R. Ugo, M. Bellatreccia, R. Zanoni, G. Vlaic, *J. Catal.* **1994**, *145*, 377–383.
75. C. Dossi, R. Psaro, L. Sordlli, M. Bellatreccia, R. Zanoni, *J. Catal.* **1996**, *159*, 435–440.
76. C. Dossi, A. Pozzi, S. Recchina, A. Fusi, R. Psaro, V. Dal Santo, *J. Mol. Catal. A: Chemical* **2003**, *204–205*, 465–472.
77. S. Takakusagi, K. Fukui, R. Tero, F. Nariyuki, Y. Iwasawa, *Phys. Rev. Lett.* **2003**, *91*, 066102-1-4.
78. N. Viswanadham, T. Shido, Y. Iwasawa, *Appl. Catal. A: General* **2001**, *219*, 223–233.
79. N. Viswanadham, T. Shido, T. Sasaki, Y. Iwasawa, *J. Phys. Chem. B*, **2002**, *106*, 10955–10963.
80. R. Bal, M. Tada, T. Sasaki, Y. Iwasawa, *Angew. Chem. Int. Ed.* **2006**, *45*, 448–452.
81. R. A. Bennett, M. A. Newton, R. D. Smith, M. Bowker, J. Evans, *Surf. Sci.* **2001**, *487*, 223–230.
82. U. Kiiski, T. Venalinen, T. A. Pakkanen, O. Krause, *J. Mol. Catal.* **1991**, *64*, 163.
83. L. Alvila, T. A. Pakkanen, O. Krause, *J. Mol. Catal.* **1993**, *84*, 145–156.
84. Z. Song, T. Cai, J. C. Hanson, J. A. Rodriguez, J. Hrbek, *J. Am. Chem. Soc.* **2004**, *126*, 8576–8584.
85. A. Schneider, N. Popvska, F. Holzmann, H. Gerhard, C. Topf, U. Zenneck, *Chem. Vap. Depos.* **2005**, *11*, 99–105.
86. D. Barreca, A. Buchberger, S. Daolio, L. E. Depero, M. Fabrizio, F. Morandini, G. A. Rizzi, L. Sangaletti, E. Tondello, *Langmuir* **1999**, *15*, 4537–4543.
87. G. Peeters, A. Thys, P. Devievre, E. F. Vansant, *Proceedings of the 6th International Zeolite Conference*, Reno, Butterworth, Guildford, 1983, s–49.
88. J.-H. Kim, A. Ishida, M. Okajima, M. Niwa, *J. Catal.* **1996**, *161*, 387–392.
89. T. Tanimura, N. Katada, M. Niwa, *Langmuir* **2000**, *16*, 3858.
90. N. Katada, S. Akazawa, N. Nishiaki, Y. Yano, S. Yamakita, K. Hayashi, M. Niwa, *Bull. Chem. Soc. Jpn.* **2005**, *78*, 1001–1007.
91. M. Tada, Y. Iwasawa, *J. Mol. Catal. A: Chemical* **2003**, *199*, 115–137.
92. M. Tada, Y. Iwasawa, *J. Mol. Catal. A: Chemical* **2003**, *204–205*, 27–53.
93. M. Tada, T. Sasaki, Y. Iwasawa, *Phys. Chem. Chem. Phys.* **2002**, *4*, 4561–4574.
94. M. Tada, T. Sasaki, T. Shido, Y. Iwasawa, *Phys. Chem. Chem. Phys.* **2002**, *4*, 5899–5909.
95. M. Tada, T. Sasaki, Y. Iwasawa, *J. Catal.* **2002**, *211*, 496–510.
96. M. Tada, T. Sasaki, Y. Iwasawa, *J. Phys. Chem. B* **2004**, *108*, 2918–2930.
97. T. Hattori, S. Itoh, T. Tagawa, Y. Murakami, In: *Preparation of catalysts: scientific bases for the preparation of heterogeneous catalysts IV*, B. Delmon, P. Grange, P. A. Jacobs, G. Poncelet (Eds.), Proceedings of the 4th International Symposium, Louvain-la-Neuve, September 1–4, 1986, *Studies in Surface Science and Catalysis*, Vol. 31, Elsevier, Amsterdam, 1987, pp. 113–123.
98. K. Asakura, Y. Iwasawa, *J. Phys. Chem.* **1992**, *96*, 829–834.
99. M. Anpo, K. Chiba, *J. Mol. Catal.* **1992**, *74*, 207–212.
100. S. H. Kim, G. A. Somorjai, *J. Phys. Chem. B* **2000**, *104*, 5519–5526.
101. Y. Z. Han, E. Marales, R. G. Gastinger, K. M. Carroll, U. S. Patent No. 6,114,552, 2000.
102. T.-T. Li, I.-C. Chen, *Ind. Eng. Chem. Res.* **2002**, *41*, 4028–4034.
103. K.-T. Li, C.-C. Lin, *Catal. Today* **2004**, *97*, 257–261.
104. G. C. Bond, K. Bruckman, *Faraday Discuss., Chem. Soc.* **1981**, *72*, 235–246.
105. G. C. Bond, P. Konig, *J. Catal.* **1982**, *77*, 309–322.
106. M. A. Vuurman, I. E. Wachs, *J. Phys. Chem.* **1992**, *96*, 5008–5016.
107. G. T. Went, S. T. Oyama, A. T. Bell, *J. Phys. Chem.* **1990**, *94*, 4240–4246.
108. I. E. Wachs, *J. Catal.* **1990**, *124*, 570–573.
109. I. E. Wachs, G. Deo, M. A. Vuurman, H. Hu, D. S. Kim, J.-M. Jehng, *J. Mol. Catal.* **1993**, *82*, 443–455.
110. J. Kerane, C. Guimon, E. Iiskola, A. Auroux, L. Niinisto, *J. Phys. Chem. B* **2003**, *107*, 10773–10784.

111. M. Anpo, M. Sunamoto, M. Che, *J. Phys. Chem.* **1989**, *93*, 1187–1189.
112. M. Nagai, T. Suda, K. Oshikawa, N. Hirano, S. Omi, *Catal. Today* **1999**, *50*, 29–37.
113. K. Asakura, Y. Iwasawa, S. K. Purnell, B. A. Watson, M. A. Barteau, B. C. Gates, *Catal. Lett.* **1992**, *15*, 317–327.
114. K. Asakura, Y. Iwasawa, *J. Phys. Chem.* **1992**, *96*, 7386–7389.
115. J. B. Miller, S. L. Bernasek, J. Schwartz, *J. Am. Chem. Soc.* **1995**, *117*, 4037–4041.
116. P. Sivakumar, R. Ishak, V. Tricoli, *Electrochim. Acta* **2005**, *50*, 3312–3319.
117. R. Ugo, C. Dossi, R. Psaro, *J. Mol. Catal. A: Chemical* **1996**, *107*, 13–22.
118. K. Tomishige, K. Asakura, Y. Iwasawa, *J. Chem. Soc., Chem. Commun.* **1993**, 184–185.
119. K. Tomishige, K. Asakura, Y. Iwasawa, *J. Catal.* **1994**, *149*, 70–80.
120. K. Tomishige, K. Asakura, Y. Iwasawa, *Chem. Lett.* **1994**, 235–238.
121. S. Recchina, C. Dossi, A. Fusi, L. Sordelli, R. Psaro, *Appl. Catal. A: General* **1999**, *182*, 41–51.
122. A. Onda, T. Komatsu, T. Yashima, *Chem. Commun.* **1998**, 1607–1508.
123. J. S. Yoo, J. A. Donohue, M. S. Kleefisch, P. S. Lin, S. D. Elfine, *Appl. Catal.* **1993**, *105*, 83–105.
124. J. S. Yoo, P. S. Lin, S. D. Elfline, *Appl. Catal. A* **1993**, *106*, 259–272.
125. J. S. Yoo, A. R. Sohail, S. S. Grimmer, C. Choi-Feng, *Catal. Lett.* **1994**, *29*, 299–310.
126. J. J. Lee, H. Kim, J. H. Koh, A. Jo, S. H. Moon, *Appl. Catal. B: Environmental* **2005**, *61*, 274–280.
127. Y. Okamoto, S. Ishihara, M. Kawano, M. Satoh, T. Kubota, *J. Catal.* **2003**, *217*, 12–19.
128. Y. Okamoto, M. Kawano, T. Kawabata, T. Kubota, I. Hiromitsu, *J. Phys. Chem. B* **2005**, *109*, 288–296.
129. J. L. Kaae, *Ceram. Eng. Sci. Proc.* **1988**, *9*, 1159–1168.
130. A. Dittmar, H. Kosslick, D. Herein, *Catal. Today* **2004**, *89*, 169–176.
131. J. L. Dupuie, E. Gulari, F. Terry, *J. Electrochem.* **1992**, *139*, 1151–1159.
132. G. A. Ozin, *Adv. Mater.* **1994**, *6*, 71–76.

2.4.7
Spreading and Wetting

Helmut Knözinger* and Edmund Taglauer

2.4.7.1 Introduction

Spreading and wetting are important phenomena in catalyst preparation. Many solid catalysts consist of several (at least two) solid phases which are brought into intimate contact. One of these phases is frequently the support, on the surface of which the catalytically active phases are dispersed. The most obvious examples of these types of heterogeneous catalyst are supported metals or oxides. Wetting and spreading frequently occur in such composite materials, and may critically influence or determine the structure and morphology of the active phases and, hence, their catalytic properties.

These processes may play an important role in several steps of catalyst syntheses, in catalyst ageing, and rejuvenation. One example is that of supported metal catalysts, which are typically prepared by impregnation from solutions containing suitable salt precursors. Dispersed metal particles are formed during subsequent reduction steps. The growth mechanisms of the metal particles, their sintering during use, and their redispersion during regeneration processes are strongly influenced by the wetting properties between the participating metal and oxide phases. Metal–support interactions (see also Chapter 3.2.5.1) may be discussed on the basis of wetting properties. The encapsulation of small metal particles by an overlayer of support oxide, which leads to the so-called strong metal-support interaction (SMSI) [1], is clearly due to the spreading of the support oxide material across the surface of the metal particles. Furthermore, promoter effects are frequently – if not always – due to the formation of an intimate contact between, for example, a metal particle and an oxide promoter. In bimetallic catalysts containing two metals that are immiscible in the bulk (e.g., Cu and Ru), one metal may still wet the other and thus form an overlayer (Cu on Ru) which ultimately may lead to complete encapsulation [2].

Another example is the synthesis of supported oxide catalysts by spreading of one oxide across the surface of another (support) oxide in physical mixtures. The phenomenon of solid-state ion exchange in zeolites may also be discussed within the framework of the wetting and spreading concept.

Despite the significance of wetting and spreading processes in heterogeneous catalysis, they have not been considered very frequently in the catalytic literature. In a recent review [3], Ko argues that this may be due to the fact that the term "wetting" is attributed to the liquid–solid interface by most researchers, and raises the question as to whether the term, in its strictest sense, can be used to describe solids that have low atomic mobility at ambient temperature. However, mobility may be induced in solid materials at catalytically relevant temperature conditions, and the general formalism of the thermodynamics of wetting can be applied to solid–solid interfaces [1]. This has been advocated by Haber and coworkers [4–7] in several articles, and the increasing awareness by catalytic chemists of the importance of wetting and spreading phenomena in heterogeneous catalysts is documented in three recent reviews [3, 8, 9].

---

* Corresponding author.

References see page 569

In this chapter, the formalism of the thermodynamics of wetting and spreading, and the dynamics of the spreading process, are briefly discussed in general terms. Experimental results related to supported metal catalysts, including sintering, redispersion, and encapsulation of a metal particle by oxide (SMSI) or by another metal, are then presented. Finally, the spreading of one oxide on the surface of another oxide is discussed in some detail, together with brief details of solid-state ion exchange in zeolites.

2.4.7.2 Theoretical Considerations

2.4.7.2.1 Thermodynamics of Wetting and Spreading

The thermodynamics of wetting of a solid by a liquid is well established, and has been discussed in detail in textbooks on colloid and interface chemistry. A schematic representation of wetting and spreading is shown in Fig. 1, where Θ is the contact angle between the two phases that are contacting each other. Θ is defined by Young's equation,

$$\gamma_{ag} \cos \Theta = \gamma_{sg} - \gamma_{as} \tag{1}$$

where γ_{ij} denotes the specific interface free energy between phases i and j, and the subscripts a, s, and g denote active phase, support and gas phases, respectively. If, under equilibrium conditions, the contact angle Θ is $>90°$, the supported active phase does not wet the support, whereas wetting occurs for $\Theta < 90°$. Spreading of the supported phase across the support surface may occur for the limiting case when Θ approaches zero.

When active-phase material is brought into contact with an uniform support, the overall change in interfacial free energy, ΔF, is given by Eq. (2)

$$\Delta F = \gamma_{ag} \Delta A_a - \gamma_{sg} \Delta A_s + \gamma_{as} \Delta A_{as} \tag{2}$$

The ΔA values are the changes in surface/interface areas. In order for spreading of the active phase across the support surface to occur, ΔF must be negative. In the opposite case, when $\Delta F > 0$, islands of the active phase will form on the support, which will tend to coalesce into larger particles in order to decrease the free energy of the system.

In order to fulfill the condition for spreading, namely $\Delta F < 0$, the inequality

$$(\gamma_{ag} \Delta A_a + \gamma_{as} \Delta A_{as}) < \gamma_{sg} \Delta A_s \tag{3}$$

or

$$\gamma_{ag} + \gamma_{as} < \gamma_{sg} \tag{4}$$

if $|\Delta A_a| = |\Delta A_s| = |\Delta A_{as}|$, must hold.

The specific interfacial free energy γ_{as} is given by [10–13]

$$\gamma_{as} = \gamma_{ag} + \gamma_{sg} - U_{as} \tag{5}$$

where

$$U_{as} = U_{int} - U_{strain} \tag{6}$$

Here, U_{int} is the interaction energy per unit interface area between the two phases and U_{strain} is the strain energy per unit area which is generated by the mismatch of the lattices of the two phases. Combining Eqs. (4) and (5) gives

$$U_{as} > 2\gamma_{ag} \tag{7}$$

that is, U_{as} must be more than twice the surface free energy γ_{ag} of the active phase for spreading to occur.

For predictions to be made as to whether or not solid–solid wetting can, in principle, take place in a given system, surface free energies must be known for the experimental temperature and environmental conditions applied. Surface free energies of metals and pure binary oxides have been compiled by Overbury et al. [14]. The available data are typically measured near the melting point of the material and the temperature coefficients of the γ values (which are of the order of magnitude of 10^{-8} J cm^{-2} K^{-1} [14]) are not known in most cases. Surface free energies will also vary with the nature and composition of the gas phase in an unknown form. Therefore, tabulated values can only be used for order-of-magnitude considerations. Surface free energies of several metals and oxides, which bear

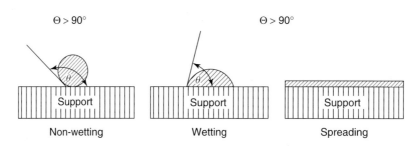

Fig. 1 The solid–solid–gas interphase.

Tab. 1 Surface free energies, bulk melting points (T_{melt}) and Tammann temperatures (T_{Tam}) of oxides [14]

Oxide	$\gamma/10^{-6}$ J cm^{-2}	T^a/K	T_{melt}/K	T_{Tam}/K
Al$_2$O$_3$	68–70	2323	2327	1163
	90	2123	–	–
MgO	110–115	298	3073	1536
	110	1870	–	–
SiO$_2$	60	298	1986	993
	39	2063	–	–
TiO$_2$	28–38	2125–2600	2173 (rutile)	1086
ZnO	90	–	2248	1124
ZrO$_2$	59–80	1423–2573	2988	1494
	113	<1423	–	–

^{a}Temperatures at which surface free energies were measured.

relevance in heterogeneous catalysis, are summarized in Tables 1 and 2, respectively, together with their bulk melting points T_{melt} and Tammann temperatures $T_{Tam} \approx 0.5\, T_{melt}$.

The ratio $U_{as}/2\,\gamma_{ag}$ is considered to be a measure for the tendency toward spreading of phase a across the surface of phases s [Eq. (7)]. In a microscopic picture, the same ratio must also be a measure of the relative strengths of interactions between the atoms of phase a and those of phase s in the interface, and among the atoms of phase a themselves. The interactions between the two phases are complex in nature and may involve dispersion, polar, and covalent interactions depending on the chemical nature of the two phases.

Van Delft et al. [16, 17] have discussed the tendency of monolayer formation versus crystallite growth of a metal phase a on a substrate s. Deliberately ignoring entropy effects, these authors calculated the difference ΔE in binding energy per atom for an infinite number of adsorbate atoms in either a monolayer or a cubic crystallite:

$$\Delta E = \left(\sum_{aa,as} E_{ij}\right)^{monolayer} - \left(\sum_{aa} E_{ij}\right)^{bulk} \quad (8)$$

If $\Delta E \leq 0$, monolayer formation (spreading) will occur, whereas $\Delta E \gg 0$ predicts the nucleation of crystallites (the Volmer–Weber growth mechanism). For a small positive value of ΔE, flat islands may be formed. For low coverages (finite number of adsorbate atoms), monolayer formation is favored over crystallite growth as the crystallite would have a finite size so that the unfavorable contributions of the binding energies of adsorbate atoms at the surface, and particularly at edges and corners of the crystallite, would become relevant [16]. The same authors also proposed that similar considerations may determine whether a layer-by-layer growth (Frank–van der Merwe mechanism) or growth of crystallites on top of a monolayer (Stranski–Krastanov mechanism) will occur. They also pointed out that this simple broken-bond model cannot be applied to systems of ionic solids (oxide–oxide) [17].

It must be noted that the thermodynamic treatment of wetting and spreading, as given above, is strictly correct only if the thickness of a film produced on a support surface by spreading is large enough to be considered as a bulk phase. As shown by Ruckenstein [18, 19], the free energy change becomes dependent on film thickness when the range of the interaction forces between an atom

References see page 569

Tab. 2 Surface free energies and Tammann temperatures of selected metals [15] and their oxides [14]

Metal	$\gamma/10^{-6}$ J cm^{-2}	T^a/K	T_{Tam}/K	Oxide	$\gamma/10^{-6}$ J cm^{-2}	T^a/K	T_{Tam}/K
Cu	135	1356	678				
Ir	225	2683	1342				
Fe	185	1808	904	FeO	59–73	1573–1693	846
					105	1683	
				Fe$_2$O$_3$	36–40	1811	906
Mo	225	2890	1445	MoO$_3$	5–7	1068	534
Ni	180	1726	863				
Pd	150	1827	914				
				Re$_2$O$_7$	3–4	600–800	285
Rh	200	2239	1120				
Ru	225	2583	1292				
				V$_2$O$_5$	8–9	963	482
				WO$_3$	10	1746	873
				ZnO	9		988

^{a}Temperature of measurement.

at the exposed surface of the film and the support is comparable to the thickness of the film. The continuum approach of these authors shows that there exists a critical film thickness for which the free energy change assumes a maximum negative value, which is typically of the order of interatomic distances and hence corresponds to the thickness of a monolayer.

Ruckenstein [20] also analyzed the stabilities of small crystallites relative to film formation. He showed that there is a minimum crystallite radius r_m above which the crystallite state is thermodynamically more stable than a film. This minimum radius depends on the contact angle and hence, on the degree of wetting of the support by the active phase:

$$r_m = \frac{h}{1 - \cos \Theta} \quad (9)$$

In Eq. (9), which is valid for thick films, h denotes the film thickness. Note that r_m becomes very large for $\cos \Theta = 1$ (wetting situation), suggesting that the film is always favored relative to crystallite formation. For situations relevant to supported catalysts, the loading of the support by an active phase would generally be small and, hence, the films would be thin. In this case, r_m becomes dependent on a parameter α, which is proportional to the difference of two quantities representing the strength of interactions between atoms of the active phase (a–a interactions) and the strength of interactions between a-atoms and s-atoms of the substrate (a–s interactions). For thin films, the minimum crystallite radius r_m is then given by

$$r_m = \frac{3h}{1 - \cos \Theta + \dfrac{\alpha}{h^2 \gamma_{ag}}} \quad (10)$$

If the a–a interactions are stronger than the a–s interactions, the parameter will most likely be negative, and if h is sufficiently small, then r_m can become negative. In this situation, the crystallite state is always thermodynamically favored over the film state. If, however, the denominator of Eq. (10) becomes positive, when a–s interactions are stronger than a–a interactions, then the radius r_m can become very large and positive. Under these conditions, the film state may be the thermodynamically preferred situation. For further information, the reader is referred to the detailed treatment by Ruckenstein [20].

Based on this thermodynamic treatment of wetting and spreading, Ruckenstein [21] concluded, that surface phase transformations may occur. Below a critical loading of the support, a submonolayer of active phase exists on the support. Above the critical loading, large crystallites will form in addition to the submonolayer (or monolayer). It is most likely that submonolayer films always exist in equilibrium with crystallites, although the fractional coverage can be very low. Obviously, these situations are closely related to the growth mechanisms of particles on the support surface.

2.4.7.2.2 **Qualitative Discussion of the Dynamics of Interfacial Processes** The foregoing discussions on the thermodynamics of wetting and spreading are limited to systems which are in their equilibrium state. However, this is not the case for many solid catalysts, and in particular not for supported catalysts in which the support material is supposed to stabilize small particles of the active phase. Because of the resulting high dispersions and high surface areas, such systems are typically in a state of high surface free energy. During heating or during catalytic reactions, these materials therefore have a high tendency to sinter, thus decreasing the exposed surface area and the surface free energy of the system. Especially, supported metals tend to age via crystallite growth. Under certain conditions, and for certain combinations of active-phase/support materials, the surface area of the active phase (especially oxides) being exposed to the chemical environment can be increased. Redispersion or spreading may occur. Sintering, redispersion and spreading are dynamic processes which drive a given system toward its equilibrium state. The extent of sintering or dispersion of a given catalytic material depends on several parameters, including time of treatment or use, temperature, gas-phase composition, presence of additional species on the support surface, and the properties of the support and of the active phase. Here, the surface free energies of the two components and the two-dimensional mobility (surface diffusion coefficients) of atoms, molecules or clusters of the active phase are of special importance.

The formation of highly dispersed particles or crystallites in the synthesis process of, for example, a supported metal catalyst, is governed by nucleation and growth mechanisms (*vide supra*) that have been described in the literature [16, 17, 22–24]. In order for sintering or redispersion (spreading and film formation) to occur, particles or atoms, molecules or clusters of the active phase must become mobile. As a rule of thumb, the Tammann temperature (in degrees K)

$$T_{Tam} \approx 0.5 T_{melt}^{bulk} \quad (11)$$

is considered to be sufficient to make atoms or ions of the bulk of a solid sufficiently mobile for bulk-to-surface migration, whereas the Hüttig temperature (which is approximately one-third of the bulk melting temperature) is enough to make the species already located on the surface sufficiently mobile to undergo agglomeration or sintering. Ruckenstein [13, 25] has shown that the enhanced mobility can be associated with the two-dimensional melting of the surface of a solid particle that leads to a liquid-like behavior of the surface layer. Kosterlitz and Thouless [26] have advanced a theory of two-dimensional melting which is based on the dislocation pairs model of melting. The two-dimensional melting temperature is given by

$$T_{\text{melt}}^{2-\text{dim}} = \left(\frac{mk}{8h^2}\right) a^2 \Theta_D^2 \quad (12)$$

where m is the atomic mass, k the Boltzmann constant, h is Planck's constant, a the lattice parameter, and Θ_D the Debye temperature. It follows that $T_{\text{melt}}^{2-\text{dim}}$ as given by Eq. (12) is proportional to the bulk melting temperature as obtained by Lindemann [27] with the proportionality constant being close to 0.5, the value used in the definition of the Tammann temperature. Values of bulk melting temperatures of catalytically relevant metals and oxides are listed in Tables 1 and 2, respectively.

Baker [28] observed mobilization of small particles of several metal oxides on graphite at a temperature (the so-called mobility temperature) that was identical to the Tammann temperature. Thus, at least in systems exhibiting relatively weak interactions between active phase and support surface, particle mobility may be induced at this temperature. The particle migration may perhaps be described as a floating of the active phase particle on the liquid-like surface layer.

For the sintering of small (metal) particles situated on a support surface, several simple mechanisms have been proposed:

- migration of crystallites and their coalescence [29, 30]
- emission of atoms from small crystallites and their capture by large ones by Ostwald ripening [31–35]
- a combination of the two above-mentioned mechanisms [33].

The theoretical treatment of these various mechanisms has been reviewed by Ruckenstein [36]. The essential features of sintering (of metal particles) can be qualitatively summarized on the basis of these simple processes. Consider a supported metal catalyst with a broad particle size distribution ranging from atoms to large particles. Single atoms may migrate across the surface of the support at sufficiently high temperatures and be captured by particles. Simultaneously, small particles can migrate across the surface and coalesce when they encounter each other, or large particles. This process can be diffusion-controlled if the interactions between two particles in contact are so strong that they form a single particle within a time that is short compared to the time of migration. If, however the coalescence of two particles into one is slow compared to the diffusion time, the process is said to be "coalescence-controlled" or "interface-controlled".

The migration rate of particles is size- (i.e., mass-) dependent. Small particles, being the most mobile, will therefore tend to be exhausted and further sintering can only occur if the remaining (slow or immobile) particles will emit atoms that migrate from smaller particles to larger ones by which they are captured. It can be shown that there exists a critical radius for the particles [33]. Those particles that have a radius r smaller than the critical radius, r_c, will lose atoms, while particles with $r > r_c$ will gain atoms and grow bigger. This is, of course, the mechanism of Ostwald ripening. As indicated in the previous section, monolayer or submonolayer films can coexist with particles. If there is a particle size distribution with small and large particles being present, atoms will be emitted to the film by the small particles and captured by the larger particles because the film is undersaturated relative to small particles, but supersaturated with respect to the large ones. This process can be considered as global ripening and should preferentially be operative at high metal loadings. Alternatively, direct ripening may occur at low metal loadings, whereby atoms are transferred directly from small particles to large ones without involving the two-dimensional film of single atoms.

It is clear that the sintering process can be globally dominated by either particle migration or single atom migration. The properties of the support and active phase of the considered catalyst system determine which of the two possible pathways will be dominantly chosen. Ruckenstein and Dadyburjor [33] considered a sintering mechanism in which single atoms and clusters of atoms are emitted from particles, particles of all sizes migrate at size-dependent rates, and two particles (including single atoms) that collide tend to coalesce. The theoretical treatment provided by Ruckenstein and Dadyburjor [33] takes into account the effects of metal loading, two-dimensional solubility, diffusional and interfacial processes for all sizes. None of these models can, however, fully describe the mechanisms of sintering because they still disregard the influence of wetting phenomena. Wetting determines the surface contact between support and active phase and therefore affects the migration of the active-phase particles across the support surface. Moreover, the two-dimensional film serving as a pool of single atoms in global ripening must be formed via the spreading of a submonolayer of atoms from the crystallites.

In all of the models discussed so far, the support surface has been assumed to be flat. In fact, very many case studies – particularly of supported metals – have used flat, low- surface-area substrates. However, Wynblatt and Ahn [37] have demonstrated that surface curvature does

---

References see page 569

affect the surface free energy, the growth of particles (sintering) via particle migration, and interparticle transport. Therefore, the sintering process of practical supported catalysts which frequently use high-surface-area, porous supports must be significantly more complex than described by the simple models.

Redispersion processes are very closely related to wetting and spreading phenomena. Inspection of the data accumulated in Table 2 shows that, typically, the surface free energies of metals are higher than those of the respective oxides. Hence, the thermodynamic condition for spreading [Eq. (7)] on the surface of an oxide support is more likely to be fulfilled for supported metal catalysts in an oxidizing atmosphere, particularly when strong chemical interactions (compound formation) are occurring, thus increasing the value of U_{as}. It appears that smaller particles spread on the support surface more easily than do larger particles. This phenomenon has been interpreted by Ruckenstein and Chu [38], by introducing a critical radius r_c for spreading (which clearly is very closely related to the critical radius for particle growth by emission and capture of atoms; vide supra). Smaller particles have a higher dissolution pressure than larger ones, and thus develop a greater tendency towards spreading. Particles with a radius smaller than r_c can spread completely across the support surface as a monolayer patch.

It is for the above reasons that rejuvenation of supported metal catalysts is typically performed in oxidizing atmospheres. However, redispersion mechanisms may be much more complex than indicated, as the fragmentation of particles may also occur during the thermal treatments in oxidizing atmospheres (see Section 2.4.7.3).

Several oxides which are used as active phases in supported oxide catalysts or catalyst precursors are also capable of spreading on typical support materials, such as alumina or titania [3, 8, 9]. Two driving forces may be principally involved in the spreading process. One force is the concentration gradient of the active phase, which might induce independent atom/molecule diffusion onto the support surface. Haber and co-workers [5, 39], however, argued that surface diffusion should be very slow in oxide systems in the temperature ranges frequently encountered in the solid-state synthesis of oxide systems, because of the typically high values of the lattice energies of oxides. Therefore, the variation of the surface free energy along the interface of a binary oxide system must be considered as the dominant driving force. The overall surface free energy in such a system can be minimized through the expansion of regions with a low surface free energy and contraction of those with high surface free energy. The gradient of the surface free energy along the surface creates a shear stress on the supported oxide phase, and the molecules of this phase may then migrate under the action of this stress from regions of lower surface free energy to those of higher surface free energy.

As discussed above, surface melting may occur at the Tammann temperature. The mechanism of spreading in powder mixtures of oxides has therefore been described [9] as a migration of species from a liquid-like surface layer of one solid across contact boundaries between grains onto the surface of the support, where they may be immobilized again if the interaction energy U_{int} is sufficiently high. A thin film (possibly a single atomic or molecular layer) may thus extend from the contact boundary between the particles onto the support surface. Further transport of the active oxide material can then be envisaged to occur via migration of the active species over the film surface toward the leading edge of the film, where they would ultimately be trapped again on the support surface. This process may be described as an "unrolling carpet mechanism" [40], and is shown schematically in Fig. 2. It might be expected that when a monolayer of active material has formed on the support, excess active phase is forming crystallites. Depending on the individual properties of the interacting oxide materials, the formation of thick films (several molecular layers) or islands (particles) with finite contact angle may also occur [13] (vide supra).

Yet another model for the spreading of an active oxide (V_2O_5) under oxidizing conditions onto the surface of a support (anatase) was proposed by Haber et al. [41]. In this model, the active-phase crystallite spreads spontaneously at the beginning of the thermal treatment (hot-plate effect), and this leads to amorphization of the active phase. Thereafter, the spreading is assumed to be diffusion-controlled. As migration of the active phase across the monolayers already formed was found to be highly improbable, it was proposed that spreading in this second stage occurs by diffusion of defects (vacancies) through the monolayer [40], these defects being refilled at the interface between the monolayer and the amorphous active phase (see Fig. 2). Most probably, this transport via vacancy diffusion can only occur if the interaction between active

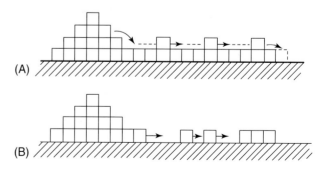

Fig. 2 Schematic representation of surface transport processes. (A) Unrolling carpet mechanism; (B) transport by defect diffusion.

phase and support is sufficiently weak for vacancies in the monolayer to be formed under the applied thermal conditions.

It may be of interest to note that Neiman [42] has recently proposed a mechanism for cooperative transport processes in complex oxide materials involving V, Nb, Mo, and W. In this model, it is assumed that the mobile species are MeO_x complexes (because of the high covalence of the Me—O bonds) which migrate by rearrangement of the Me—O bonds in regions of local amorphization and quasi-melting of the lattice. It may be speculated that similar processes also occur during spreading of two-dimensional films across the surface of supports.

2.4.7.3 Supported Metal Catalysts

The various stages of the preparation and thermal treatments of supported metal catalysts are shown schematically in Fig. 3. A similar presentation was provided earlier by van Delft et al. [17]. Typically, the support is impregnated with a metal salt (see Chapter 2.4.2), which serves as the metal precursor and should be well dispersed. Small metal particles may be formed by either direct reduction under mild conditions or by reduction after an intermediate oxidation step. Mild oxidation will lead to thin oxide films which spread out on the support, or to small oxide particles, where particles and film may also coexist. More severe treatments in oxidative atmospheres can lead to the formation of surface compounds (e.g., spinels) which again may coexist with oxide particles. Continued reduction or thermal treatment in a reducing atmosphere under more severe conditions can induce several processes, depending on the nature of the metal and of the support. If the metal is mobile and strong cohesive interactions are operative, then sintering will occur, whereas the metal particles may spread out and form flat islands (pill-box) if the metal–support interactions are strong. If, in contrast, the support is mobile and strong adhesive interactions come into play, the metal particles can become encapsulated and the so-called strong-metal-support-interaction (SMSI) may be induced. It is clear that wetting and spreading phenomena are significantly involved in all steps indicated in Fig. 3. Rather than report on the many studies on sintering and redispersion in practical supported metal catalysts, the following discussion briefly highlights some model studies which have been excellently reviewed by Ruckenstein et al. [1].

2.4.7.3.1 Sintering and Redispersion

Chen and Ruckenstein [43] have performed a mechanistic study of the sintering of Pd on alumina by transmission electron microscopy (TEM), using model catalysts. These consisted of thin self-supporting and electron-transparent γ-Al_2O_3 films onto which Pd metal was vacuum-evaporated. The resulting thin Pd film was transformed during treatment in flowing H_2 into crystallites in the diameter range 1 to 7 nm. Starting from this situation, the effects of additional heat treatments in H_2 at various temperatures, and for increasing periods of time, were studied using TEM. The time sequences of changes in size, shape and position of individual particles at 923 K indicated that the major processes occurring were large-scale crystallite migration and coalescence, because a large number of small crystallites were shown to disappear without change of size and with the average diameters of crystallites growing simultaneously. Interestingly, crystallites with diameters as large as 7.5 nm did migrate at 923 K, while some smaller ones appeared to be immobile. Thus, the mobilities of

---

References see page 569

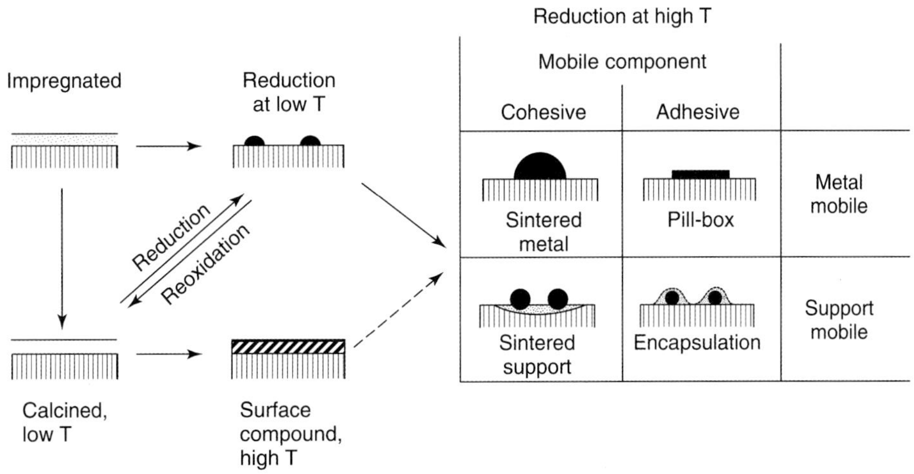

Fig. 3 Schematic representation of metal–support interactions. (Adopted from Ref. [17].)

crystallites are not only dependent on size. Small crystallites may be trapped in the valleys or pits of a rough surface, whereas larger ones, which contact only the tips of the ridges on the surface, may travel with greater ease. In addition, the local surface morphologies may affect the interaction forces between particles and the support surface. Although the size of small particles did not generally decrease before they disappeared, the possibility of Ostwald ripening could not be excluded. In fact, in localized regions where a few small crystallites were located in close vicinity to larger ones, the former decreased in size and then disappeared. This observation was interpreted as being due to direct ripening (*vide supra*).

Under more severe H_2 treatment, faceting of crystallites larger than 12.5 nm also occurred. Particle migration (particles up to 8 nm migrated over 25 nm at 773 K) was also observed for Pt on alumina [44]. The two major mechanisms of the sintering of supported Pt crystallites appeared to be: (i) short-distance, direction-selective migration of particles followed by either collision and coalescence or by direct transfer of atoms between the two approaching particles; or (ii) localized direct ripening between a few immobile, adjacent particles. Chen and Ruckenstein [45] also investigated the behavior of large Pd crystallites on γ-Al_2O_3 model supports in O_2 atmospheres in the temperature range 623 to 1193 K. The Pd crystallites are oxidized to form PdO at temperatures below the decomposition temperature of 1143 K. As the surface free energy of PdO is less than that of Pd metal (see Table 2), the crystallites extended onto the support surface to a smaller wetting angle, as expected. Spreading under oxidizing conditions occurs because the surface free energy of the metal oxide γ_{ag} and the interface free energy γ_{as} between the oxidized palladium and the alumina support are smaller. As can be seen from Eq. (5), γ_{as} is smaller because γ_{ag} is smaller, and the interaction energy U_{as} is larger for the metal oxide than for the metal.

Additional, much more complex, phenomena have been observed however. Depending on crystallite size and temperature conditions, the formation of porous structures, of torus-like particles, the extension of particles developing irregular leading edges with pits and cavities, and ultimately tearing and fragmentation was observed at 10^5 Pa O_2 over the temperature range mentioned. The pit formation was explained as a surface tension gradient-driven phenomenon (perhaps enhanced by a wetting-produced stress), and/or as a consequence of crack propagation caused by stress, which was induced by wetting and enhanced by the oxidation of the freshly exposed surface at the tip of the crack. The formation of the torus-like particles was attributed to incomplete oxidation of larger particles at lower temperatures.

It should be noted that, at temperatures higher than the decomposition temperature of bulk PdO at 10^5 Pa O_2, the palladium particles sintered and formed facetted crystallites.

Very similar phenomena were observed by Ruckenstein and Lee [46] when Ni supported on model alumina films was treated in an O_2 atmosphere. Extension of the particles (leading to a torus shape) during heating was reported, the torus being divided into interlinked subunits and containing small crystallites within the ring. The extension of the oxidized Ni particles is clearly due to the lower surface free energy of NiO and to the lower interfacial free energy between the NiO and the alumina support, as outlined above for the Pd–alumina system. However, the interfacial free energy γ_{as} can be extremely small (perhaps even negative under non-equilibrium conditions) when strong chemical interactions or surface compound formation increase the interaction energy U_{as} in Eq. (5). In contrast to PdO, NiO has a high tendency towards surface spinel formation in contact with alumina [47, 48], and hence U_{as} should be large. Therefore, under dynamic oxidation conditions, the large decrease in the dynamic value of γ_{as} and the simultaneous decrease of γ_{ag}, can very significantly increase the driving force for spreading. Spinel formation starts at the interface between the NiO particle and the alumina support, and propagates into the latter. The surface spinel layer grows until the rate of formation is slowed down by a large diffusional resistance. The interaction energy U_{as} is only large during the formation of the surface compound, and decreases again for the interaction of the NiO with the surface spinel. Therefore, when U_{as} is plotted against time, it passes through a maximum while, correspondingly, γ_{as} passes through a minimum (which can be negative), as shown schematically in Fig. 4. Although there may also be thermodynamic reasons

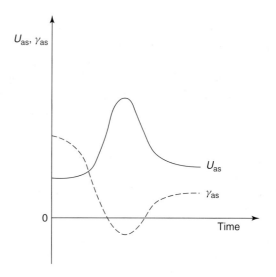

Fig. 4 Time-dependence of U_{as} and γ_{as}. (Adopted from Ref. [1].)

for the appearance of the torus-shaped NiO particles, Ruckenstein and Lee [46] also explained their formation as a kinetic phenomenon. These authors suggested that very rapid spreading takes place as soon as the surface spinel starts to form because of the resulting large driving force for wetting, and that this may lead to the formation of the torus shape by stress-stimulated fracture of the crystallites.

In a similar study, Sushumna and Ruckenstein [49] showed that the composition of the gas atmosphere may be of critical importance for the detailed behavior of Fe supported on alumina catalysts. Even trace impurities of O_2 and H_2O in H_2 in the ppm range had detectable effects on the sintering behavior of Fe particles.

Industrial regeneration techniques [50] of supported metal catalysts typically involve oxidation/reduction cycles. Redispersion is achieved during thermal treatments in oxidative atmospheres containing O_2, water, halogens, and others [51], often as mixtures as the U_{as}, interaction energies and, thus, the tendency towards wetting and spreading of the active phase onto the support surface, is strongly enhanced under those conditions. As a consequence, highly dispersed metal particles may be formed under subsequent mild reduction conditions.

An interesting case of a wetting–non-wetting transition under reaction conditions was recently described by Clausen et al. [52]. By measuring the changes of Cu–Cu coordination numbers by extended X-ray absorption fine structure (EXAFS) for a Cu/ZnO catalyst, these authors inferred that small metallic Cu particles dynamically changed morphology when the oxidation potential of the $H_2O/CO/CO_2/H_2$ gas phase was varied by changing the partial pressure of the components. The morphology changes were attributed to wetting–non-wetting phenomena which were due to gas-phase-induced changes of the interaction energy between Cu metal and the ZnO support. Essentially no changes in morphology were detected for Cu/SiO$_2$ catalysts. These findings could be explained by a dynamic microkinetic model that provides a good description of the kinetic measurements over a working methanol catalyst [53]. In-situ X-ray diffraction (XRD) and X-ray absorption fine structure (XAFS) measurements further support the model that reversible changes in wetting of ZnO by Cu occur upon changes in the reaction conditions [54].

Very recent developments in instrumentation provide the possibility of atom-resolved imaging of supported metal nanoclusters in the working state by high-resolution transmission electron microscopy (HRTEM) [55, 56]. This technique provides detailed insight into nanoparticle dynamics in general, and has already enabled a microscopic description to be provided of the dynamic shape changes in the supported Cu/ZnO catalysts mentioned above.

2.4.7.3.2 Strong Metal Support Interactions (SMSI) As shown schematically in Fig. 3, two extreme situations may occur during thermal treatment at high temperature (above ca. 770 K) in reducing atmospheres. If the support is immobile and strong interactions occur between the metal particle and support, the metal may spread out on the support surface and form flat islands (the pillbox model). If, however, the metal has a high melting point and the support is mobile, then encapsulation of the metal particle may occur by wetting of the metal by support oxide. Since Tauster and co-workers [57, 58] first reported that the chemisorption capacity for H_2 and CO decreased dramatically for supported metal catalysts in the SMSI state, there has been continuing discussion as to whether the phenomenon was caused by electronic or by geometric/morphological effects [1, 59, 60] (see Chapter 3.2.5.1). The encapsulation or decoration model was proposed because of the failure of the electron-transfer models to explain the behavior of large particles supported on titania [61–63]. However, electron-transfer effects may also be important contributions for the encapsulation to occur. The phenomenon of encapsulation can be explained in terms of wetting and spreading.

The SMSI effect has been observed for support materials such as TiO$_2$, V$_2$O$_5$, Nb$_2$O$_5$, and Ta$_2$O$_5$. Inspection of the surface free energies of these oxides (see Table 1) shows that they are smaller than those of other supports such as SiO$_2$, Al$_2$O$_3$, ZrO$_2$, MgO, which do not manifest SMSI. Considering the unrealistic situation of the formation of a thick oxide film on the metal particle, low γ_{as} values (the support must spread for encapsulation of the metal) would be favorable for spreading as the ratio $U_{as}/2\gamma_{sg}$ would increase. The driving force for spreading would increase even further if the interaction energy U_{as} between the metal and the support oxide were large. It should be noted that the oxides that manifest SMSI are reducible, and it is generally accepted that partial reduction (possibly catalyzed by the metal) of the support is a prerequisite for encapsulation to occur [1, 59, 60]. Electron transfer from the oxide support to the metal may in fact occur with greater ease if the oxide is in a reduced state. This would increase U_{as}, and significantly enhance the driving force for spreading. It should, however, also be noted that the surface free energies of reducible oxides increase if they become oxygen-deficient, and that their melting points simultaneously increase. Therefore, the driving force for spreading and the mobility of the oxide decrease in the reduced state. These effects may influence the tendency towards encapsulation, although the enhancement of the interaction term U_{as} seems to overcompensate for these effects.

References see page 569

In reality, the strong interactions are short-ranged, and thus the oxide film forming on the metal particle is expected to be a monolayer or submonolayer film. The driving forces involved are therefore the concentration gradient of the oxide and the gradient of the surface free energy along the metal surface, which leads to shear-stress-induced migration of oxide molecules across the metal surface, as discussed above.

Several studies have been reported in which model TiO_2 thin films were used as supports, and results have clearly demonstrated metal particle encapsulation [64, 65]. Linsmeier et al. [66] and Taglauer and Knözinger [67] have studied the behavior of Rh which had been evaporated onto an electrochemically produced TiO_2 (anatase) film using low energy ion scattering (LEIS) (see Chapter 3.2.2) as a surface-sensitive analytical technique. Figure 5A shows two models for supported metal catalysts, namely (a) a monolayer of atoms A on the support surface S; and (b) small islands of atoms A which are encapsulated by support S. Figure 5B represents the corresponding profiles that are expected for the two models when the scattering intensity ratio I_A/I_S is plotted versus the fluence (the total number of He^+ ions impinging on the surface). Experimental results for the Rh/TiO_2 model catalyst after mild H_2 treatment at 723 K (non-SMSI state) and after severe H_2 treatment at 773 K (SMSI state) are shown in Fig. 5C. The experimental profile obtained after low-temperature treatment resembles that of a highly dispersed metal film, whereas the profile after high-temperature treatment passes through a maximum and clearly indicates encapsulation of the Rh particles. A rough estimate showed that the fluence at the position of the maximum corresponds to the average value required to sputter one monolayer of oxide. This result relates well with the prediction made above for the expected film thickness.

2.4.7.3.3 Bimetallic Catalysts Surface segregation phenomena and the surface composition of bimetallic catalysts are controlled by the surface free energies of the constituents of the bimetallic particles. The deposition of metal on metal in relation to bimetallic catalysts has been discussed by Dodson [68]. Particularly interesting systems in the context of wetting are bimetallic supported catalysts containing two metals that are immiscible in the bulk, such as Ru and Cu. Sinfelt and co-workers [69, 70] first reported on a Cu-induced suppression of hydrogen chemisorption for SiO_2-supported RuCu catalysts; this observation was consistent with the structural model of the bimetallic particles consisting of a Ru core encapsulated by a Cu layer, which was developed by Sinfelt et al. [71], on the basis of EXAFS results. Several surface science studies have also demonstrated the formation of Cu films on the surface of Ru single crystal surfaces [72, 73] and infrared (IR) studies of CO chemisorption on $RuCu/SiO_2$ catalysts have supported the encapsulation model [74]. Thus, the available experimental evidence shows clearly that, although the two metals do not form binary alloys, Cu does wet the Ru. The surface free energy γ_{Cu} of Cu is relatively small (see Table 2) and lower than that of Ru. Moreover, surface science studies on single crystal model systems [75] and IR spectra of chemisorbed

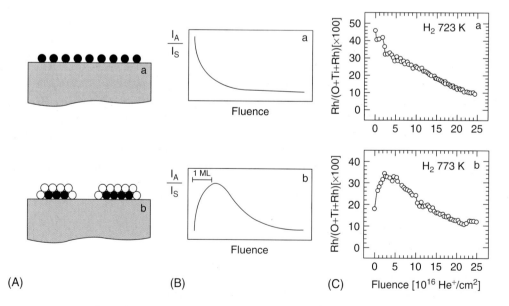

Fig. 5 Low-energy ion scattering from Rh/TiO_2 model catalysts. (A) Structure models; (B) expected intensity profiles; (C) experimental intensity profiles. (From Ref. [67].)

CO on RuCu/SiO$_2$ catalysts [74] showed that electron transfer occurs from Cu to Ru. This electron transfer contributes to the strength of interaction between the two metals at the interface, and presumably leads to a ratio $U_{RuCu}/2\gamma_{Cu} > 1$, so that the spreading condition for Cu across the surface of Ru is fulfilled.

2.4.7.4 Supported Oxide Catalysts

Supported oxides of transition metals, particularly of Groups Vb (V), VIb (Cr, Mo, W) and VIIb (Re), are widely used as catalysts or catalyst precursors for a variety of industrially important reactions. These so-called monolayer-type catalysts are formed when one metal oxide (mobile active phase) is dispersed on the surface of a second metal oxide (immobile support). Typical catalyst supports in industrial applications are transition aluminas, silica, and titania. Inspection of the data summarized in Table 1 indicates, that the surface free energies of these support oxides are higher than those of the above-mentioned transition metal oxides. Assuming the formation of thick films, the spreading condition of Eq. (7), namely $U_{as}/2\gamma_{as} > 1$, must be fulfilled. Unfortunately, the interaction energies U_{as} are practically always unknown, although it has been argued [76] that chemical contributions to U_{as} should be high if the two components have a tendency to form a ternary oxide via solid-state reactions. In this case the spreading condition should be fulfilled. Based on these arguments, the experimentally observed spreading of MoO$_3$ on alumina and titania could be explained [76], as both supports form bulk ternary oxide phases with MoO$_3$ (Al$_2$(MoO$_4$)$_3$ and Ti(MoO$_4$)$_2$, respectively), at higher temperatures, although these are not detected under the spreading conditions. In contrast, MoO$_3$ is not known to undergo a solid-state reaction with silica, and consequently spreading of MoO$_3$ on the surface of silica was not observed [76].

As mentioned above, the interaction forces between active phase and support are short range in nature and should be restricted to the interface. Therefore, the driving forces for spreading are expected to decrease significantly once a molecular monolayer is formed. As a consequence, oxide particles or crystallites of the active phase are expected to form when loading by the active oxide exceeds the theoretical monolayer capacity of the support. Experimental evidence for this situation has in fact been reported in the literature [9].

These principles can be applied for the preparation of supported oxide monolayer catalysts from mechanical mixtures, the advantage of this route being that the handling of solutions for impregnation can be avoided. However, in order for efficient spreading to occur, the two components of a powder mixture must be brought into intimate contact and optimal mixing is required so as to obtain a homogeneous product. Moreover, the active (and mobile) phase must be present as crystallites having diameters below a critical value [see Eqs. (9) and (10)], as otherwise the crystallite state may be thermodynamically more stable than the film. Grinding or milling is therefore usually applied to the powder mixture prior to thermal treatments. Often, these processes are not well controlled when powder mixtures are prepared for solid-state synthesis of bulk products or supported catalysts, although they must be expected to influence the reactivity of the powders very significantly.

Grinding will certainly influence the grain sizes and grain-size distributions, and thus the rates of spreading. The two-dimensional melting temperature should also be dependent on the grain size (curvature). Several additional phenomena occur in the very complex grinding process that must influence the spreading and reactivity behavior of powder mixtures [77–79]. During grinding, several particles are simultaneously and repeatedly subjected to stress application in the grinding zone, and with each stress application several fractures may occur in each particle. Cracks will be initiated and will propagate; flaw interaction in a particle, secondary breakage, and interaction of particles with each other will also occur. The physical and chemical interaction between particles and the grinding environment, as well as the transport of material through the grinding zone, will also affect the nature of the product obtained. Occasionally, material transport between chemically distinct particles may already lead to spreading during the grinding and milling procedure. This may occur especially when crystallites are involved that have layer structures (e.g., V$_2$O$_5$), so that stress application may lead to exfoliation processes. Even solid-state reactions in bulk phases can be induced by the mechanical activation of solid materials, and several tribochemical processes have found technological application [80]. For example, Angelov and Bonchev [81] have reported on the formation of a Cu-rich surface layer on Co$_3$O$_4$ by mechanically treating a powder mixture of CuO and Co$_3$O$_4$ in a friction grinder. These reactions are believed to occur due to strong local temperature increases which may lead to the melting of microscopic zones within contact regions. The effect of mechanical stress applied to MoO$_3$ by grinding in a planetary mill have recently been studied by Mestl and coworkers [82–84]. It was shown that the crystallite sizes decreased from an initial value of about 1 µm to 50 nm, with some ultrafine amorphous material also being produced. The BET surface area simultaneously increased from an initial value of 1.3 m^2 g^{-1} to 32 m^2 g^{-1}. A substoichiometric MoO$_{3-x}$ was formed during grinding. The presence of

---

References see page 569

shear defects was indicated by XRD, and Mo^{5+} was detected by electron spin resonance (ESR) and optical spectroscopy. When this milled MoO_3 was gently mixed with alumina, without applying additional mechanical stress, and then thermally treated in O_2 at 823 K (melting point of bulk MoO_3 is 1068 K), in-situ high-temperature Raman spectroscopy demonstrated the existence of a surface melt of molybdenum oxide which, on quenching to room temperature, transformed into a glassy MoO_3 surface phase [85]. The experiments described by Mestl et al. [85] clearly showed the higher reactivity of the milled MoO_3 and its more efficient spreading on the alumina surface as compared to the initial low-surface-area material.

As the reported literature has been summarized in several recent reviews [3, 8, 9], only a few experimental studies on spreading in oxide mixtures will be described briefly at this point.

2.4.7.4.1 Molybdenum-Based Catalysts

The most thoroughly studied supported oxide systems are those involving molybdenum oxide as the mobile phase. As mentioned above, spreading of MoO_3 in physical mixtures was observed on the surfaces of aluminas and titania (anatase) but not on silica; the reason for the different behavior of these support oxides was thought to be connected with the interaction energy U_{as} between MoO_3 and the support materials [76]. Highly dispersed molybdenum oxide surface phases were also detected on MgO [86, 87] and on SnO_2 [86]. It is interesting to note that MoO_3 and MgO undergo a solid-state reaction to form $MgMoO_4$.

The transport of MoO_3 across the surface of alumina was demonstrated in several model experiments. Hayden et al. [88] prepared molybdenum oxide on a graphite support. Because of the weak interaction between these two components, MoO_3 particles were formed; however, this material was then mixed with Al_2O_3 and studied with controlled atmosphere electron microscopy. MoO_3 particles were mobile on the graphite surface at 930 K and disappeared spontaneously when they contacted alumina grains. This observation was explained by rapid spreading of the MoO_3 over the Al_2O_3 surface, and related to the higher surface free energy of alumina as compared to graphite. It should, however, be emphasized that the large interaction energy U_{as} between MoO_3 and Al_2O_3 is crucial for the large driving force for spreading. Leyrer et al. [89] used model samples which consisted of an alumina wafer in contact with a MoO_3 wafer, forming a sharp dividing line. Concentration profiles across the dividing line between the two oxides were measured using Raman microscopy with a lateral resolution of approximately 20 μm. The results clearly demonstrated that MoO_3 was in fact transported onto the γ-Al_2O_3 surface at 800 K in streams of either dry or humid O_2 whereby the initially sharp concentration profile became diffuse and reached widths of several hundred (up to 1000 μm) micrometers after 100 h. It was also shown that the presence of water vapor increased the rate of spreading, and led to the formation of a surface polymolybdate (vide infra). These model experiments also enabled gas-phase transport and surface diffusion in a concentration gradient to be discounted as possible mechanisms for the migration of MoO_3, as no transport occurred onto a SiO_2 wafer in an analogous experiment. This result again emphasizes the crucial importance of the interaction energy in fulfilling the spreading conditions given by Eq. (7).

Spreading of polycrystalline MoO_3 in powder mixtures with γ-Al_2O_3 and TiO_2 (anatase) was indicated by the disappearance of the XRD pattern of MoO_3 after thermal treatment, provided that the MoO_3 content remained below a certain limit [8, 90, 91]. This limiting MoO_3 content was determined as 0.12 g MoO_3 per 100 m^2 of both alumina and titania, and interpreted as corresponding to the monolayer capacity. More direct evidence for spreading has been obtained by surface-sensitive techniques such as X-ray photoelectron spectroscopy (XPS) [8, 9, 90–92] and LEIS [8, 9, 67, 76, 93, 94]. As an example, Fig. 6 shows ion-scattering data

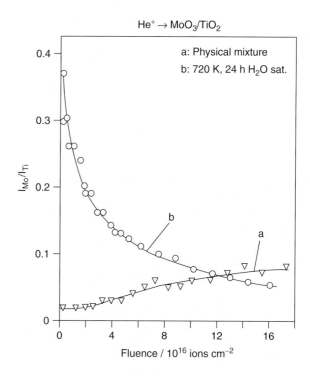

Fig. 6 Low-energy ion scattering (LEIS) intensity profiles for MoO_3/TiO_2 catalysts. (a) Physical mixture prior to thermal treatment. (b) After thermal treatment for 24 h at 720 K in an O_2 flow saturated with water vapor. (Adopted from Ref. [93].)

which clearly support the spreading of MoO_3 on TiO_2 during thermal treatment at 720 K.

In the figure, the intensity of He^+ ions backscattered from Mo atoms relative to that of He^+ backscattered from Ti atoms is plotted against the fluence (= total number of He^+ ions that have impinged on unit surface area). Profile (a) characterizes the physical mixture prior to thermal treatment; the second profile (b) was obtained after thermal treatment of the same mixture at 720 K for 5 h in a flow of oxygen saturated with 32×10^2 Pa (1 mbar = 10^2 Pa) H_2O vapor. The intensity ratios extrapolated to zero fluence are characteristic of the virgin surface being unaffected by sputtering processes. This value is very low for the starting physical mixture consisting of grains of Al_2O_3 and small crystallites of MoO_3. Because of the surface sensitivity of the LEIS technique, the Mo/Ti intensity ratio, when extrapolated to zero fluence, is expected to approach infinity if a monolayer of molybdenum oxide is formed on the TiO_2 surface. Profile (b) indeed rises steeply near zero fluence, and thus strongly supports the spreading of MoO_3 on TiO_2. Similar results were obtained with alumina as the support. LEIS experiments also verified that spreading of MoO_3 on both supports occurred in O_2 in the presence and absence of water vapor. Neither XPS nor LEIS provide structural information of the molybdenum oxide overlayers on the support surfaces. Therefore, laser Raman spectroscopy (LRS) [9, 76, 86, 93–95] and X-ray absorption spectroscopy (XRAS) [96, 97] have been applied to elucidate the structural characteristics of the two-dimensional oxide films. LRS emphasized the role of water vapor for the structure formation. As a representative example, Fig. 7 shows spectra of MoO_3–Al_2O_3 mixtures [94]. The bottom spectrum is characteristic for the oxide mixture prior to thermal treatment and represents the signature of polycrystalline MoO_3. When the mixture was thermally treated in water-free O_2, the same spectrum was obtained for the dispersed oxide film, suggesting that the X-ray-amorphous film still preserved the structural characteristics of MoO_3. Figure 7 also shows the spectra of the MoO_3–Al_2O_3 mixture after thermal treatment in humid O_2 obtained after 10 and 30 h. These spectra demonstrate the chemical transformation of MoO_3 into a surface heptamolybdate (spectrum after 30 h) with an intermediate monomeric MoO_4^{2-} anion being detectable by the characteristic band at 915 cm^{-1}. These conclusions were supported by XAFS [96] for the MoO_3–Al_2O_3 mixture and by the pre-edge and X-ray absorption near-edge (XANES) structure for the MoO_3–TiO_2 system [97].

The chemical transformation of MoO_3 into a surface polymolybdate on Al_2O_3 and TiO_2 was described as being due to the intermediate formation of $MoO_2(OH)_2$, a reaction which is known to occur under the experimental

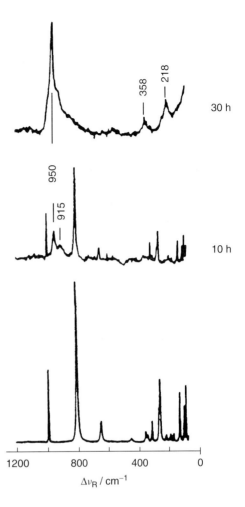

Fig. 7 Raman spectra of a physical mixture of 9 wt.% MoO_3–Al_2O_3 prior to thermal treatment (bottom spectrum) and after calcination at 723 K in a stream of O_2 saturated with water vapor for 10 and 30 h (top and middle spectra). (Adopted from Ref. [95].)

spreading conditions [98]:

$$MoO_3 + H_2O \longrightarrow MoO_2(OH)_2 \tag{13}$$

This oxyhydroxide was assumed to react with basic surface hydroxy groups $[OH^-]_s$ to yield the monomeric surface molybdate $[MoO_4^{2-}]_s$ that was detected by LRS as an intermediate (see Fig. 7):

$$MoO_2(OH)_2 + 2[OH^-]_s \longrightarrow [MoO_4^{2-}]_s + 2H_2O \tag{14}$$

These intermediates subsequently undergo condensation with formation of the surface polymolybdate if the local concentration of the monomer becomes sufficiently high. As gas-phase transport and diffusion in a concentration gradient could be excluded in this study (*vide supra* and *vide infra*), transport of the molybdenum oxide via the unrolling carpet mechanism was proposed [9] (see Fig. 2). However, migration via defect diffusion may still be an alternative. The results of recent *in-situ* LRS studies [85] indicated the presence of a liquid-like layer containing monomers and oligomers of MoO_3 which might be considered as the migrating species.

The development of the technique of photoelectron spectromicroscopy (SPEM) provides the possibility of directly studying spreading phenomena with high lateral resolution. Günther and co-workers [99] investigated the spreading of MoO_3 on titania and alumina with a spot size of 0.15 µm and an energy resolution of 0.5 eV in the imaging mode. Figure 8 [100] shows the spreading of molybdenum species on alumina due to heating in dry oxygen at an oxygen pressure of 10^5 Pa; spreading starting from the originally deposited crystallites is clearly visible. In this case, contributions from gas-phase transport – that is, evaporation into and redeposition from the gas phase – were discussed as a possible mechanism [100].

The influence of the crystallographic modification of the support on the dispersion of molybdena on titania (mixture of rutile and anatase) was recently studied by Zhu et al. [101] using FTIR. These authors found that molybdena preferentially dispersed on the rutile surface and, with high loadings, polymeric molybdena species were detected.

A systematic screening study of the spreading of 10 different metal oxides on four different supports was recently carried out by Bertinchamps and co-workers [102] using XPS and XRD. The group demonstrated a better monolayer spreading of the active phases on titania than on alumina and silica, with the latter inducing generally a poor dispersion. The results from the various active phases and supports are governed by differences in surface free energies, and are fully explained by the solid–solid wetting concept discussed in Section 2.4.7.2. The conversion of benzene by these catalysts progressively improved when the support changed from SiO_2 to Al_2O_3 to titania, thus demonstrating the importance of well-dispersed monolayers on the support surfaces.

2.4.7.4.2 Vanadium-Based Catalysts

The spreading of V_2O_5 on alumina and titania has been studied intensely because of the importance of these monolayer-type materials as catalysts for selective catalytic oxidation (see Chapter 14.11.1), and for the selective catalytic reduction of NO_x (see Chapter 11.3). V_2O_5 has a low melting point, and hence high mobility under mild temperature conditions. It also has a low surface free energy and probably undergoes strong interaction with supports such

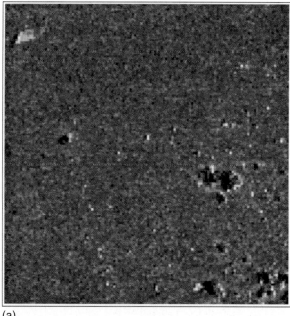

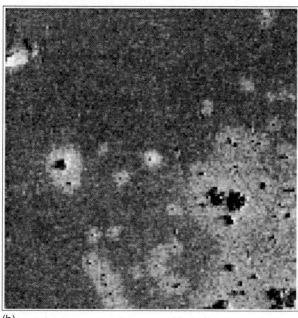

Fig. 8 Spreading of Mo species emanating from small MoO_3 crystallites on Al_2O_3 support (smooth area), imaged with the Mo 3d XPS intensity. (a) Before heat treatment; (b) after heating for 60 min at 660 K in dry oxygen. The gray scale represents the amount of spread of Mo species that can be clearly seen around the initially deposited crystallites. (Adopted from Ref. [100].)

as alumina as it forms the ternary oxide, $AlVO_4$. Thus, the spreading condition $U_{as}/2\gamma_{ag} > 1$ should be fulfilled. Moreover, in the presence of water vapor, highly volatile and reactive oxyhydroxides such as $V_2O_3(OH)_4$ [94] or

VO(OH)$_3$ [104] are formed which permit an interpretation of surface chemical transformations of V$_2$O$_5$. The many studies on vanadia-based materials have been reviewed [3, 9].

Haber and co-workers [4–6, 39, 41, 105, 106] have demonstrated, in several detailed investigations, that spreading occurred in V$_2$O$_5$/TiO$_2$ mixtures. These authors showed in particular [39] that the migration of V$_2$O$_5$ over the surface of anatase grains led to encapsulation of the latter by a thin overlayer, the properties of which were strongly modified by interaction with the anatase support. On top of this inner layer an outer layer was detected, its properties being similar to those of V$_2$O$_5$. Centi and co-workers [107, 108] also reported on the spreading of V$_2$O$_5$ over the surface of titania with formation of an oxide monolayer of V^{4+} oxide species [107] on top of which amorphous multilayer patches of V^{5+} oxide were believed to grow [108].

In contrast to Haber and co-workers [4–6, 39], Centi et al. [107, 108] observed the spreading of V$_2$O$_5$ not only on anatase but also on the rutile modification. Based on their results, Haber and co-workers [4–6, 39] proposed that spreading was dependent on the crystallographic modification of the support and on the type of exposed crystal plane. This interpretation is supported by the recent results on titania [101] mentioned above.

The migration mechanism of V$_2$O$_5$ on the surfaces of alumina and titania is not yet understood in detail. As already mentioned, however, results reported by Haber et al. [41] seem to indicate that a process of defect diffusion through the vanadia monolayer is involved.

2.4.7.4.3 Tungsten-Based Catalysts

The thermodynamic spreading condition may be fulfilled for WO$_3$ as its surface free energy is relatively low and it definitely forms a ternary oxide Al$_2$(WO$_4$)$_3$ with alumina. However, the Tammann temperature is 873 K, leading to low mobility. Nevertheless, spreading of WO$_3$ on γ-Al$_2$O$_3$, although slow, was observed at 820 K by LEIS [76] and LRS, while EXAFS [109] provided evidence for the formation of a surface polytungstate in the presence of water vapor via an intermediate oxyhydroxide WO$_2$(OH)$_2$. An analogous surface chemistry was also reported for physical mixtures of WO$_3$ with TiO$_2$ (anatase) [109].

2.4.7.4.4 Zeolites as Supports

A few studies involving zeolites and layer silicates as supports for oxides are reported in the literature. For example, Haase et al. [110] provided experimental evidence for the spreading of V$_2$O$_5$ on mordenite, while the migration of Ga$_2$O$_3$ in H-ZSM-5 zeolites was observed by Mériaudeau and Naccache [111]. Fierro et al. [112] proposed the incorporation of MoO$_3$ into the intracrystalline cavities of Y-type zeolites by means of vaporization of MoO$_3$ by reaction with water vapor at 623 K. It was claimed that the zeolite structure was largely retained under these conditions, although at higher temperatures a progressive loss of crystallinity was observed. Leyrer and Knözinger [113] have shown that complete degradation of the zeolite structure resulted, when incorporation of Mo into Y-zeolites was attempted via gas-phase transport of MoO$_2$(OH)$_2$ at 720 K.

2.4.7.5 Solid-State Ion Exchange of Zeolites

Solid-state ion exchange may be related to the phenomena of wetting and spreading in solid–solid systems. Rabo et al. [114, 115] first reported that proton-containing samples of zeolite Y reacted with sodium chloride under the evolution of HCl. Later, Karge and co-workers [116–121] developed solid-state ion exchange further as a synthetic route for the preparation of alkaline, alkaline earth, rare earth, and transition- and noble-metal-containing zeolites starting from parent materials that contain H$^+$, NH$_4^+$, or Na$^+$. In several cases a 100% cation incorporation could be achieved in a one-step, solid-state reaction. Even the large Cs$^+$ ions could be driven into S_I sites of zeolite Y via solid-state ion exchange [122]. Such degrees of exchange are difficult to obtain by conventional methods.

References

1. S. A. Stevenson, J. A. Dumesic, R. T. K. Baker, E. Ruckenstein, *Metal-Support Interactions in Catalysis, Sintering, and Redispersion*, Van Nostrand Reinhold, New York, 1987.
2. J. H. Sinfelt, *Bimetallic Catalysts: Discoveries, Concepts and Applications*, Wiley, New York, 1983.
3. E. I. Ko, in *Wettability*, J. C. Berg (Ed.), M. Dekker, New York, 1993, p. 431.
4. J. Haber, in *Surface Properties and Catalysis by Non-Metals*, J. P. Bonnelle, B. Delmon, E. Derouane (Eds.), Reidel, Dordrecht, 1983, p. 1.
5. J. Haber, *Pure Appl. Chem.* **1984**, 56, 1663.
6. J. Haber, *Proceedings of the 8th International Congress on Catalysis*, Berlin, Verlag Chemie, Weinheim, 1984, Vol. 1, p. 85.
7. J. Haber, J. Ziolkowski, in *Proceedings of the 7th International Symposium Reaction Solids*, Bristol, 1972, Chapman & Hall, London, 1972, p. 782.
8. Y. C. Xie, Y. Q. Tang, *Adv. Catal.* **1990**, 37, 1.
9. H. Knözinger, E. Taglauer, in *Catalysis*, J. J. Spivey, S. K. Agarwal (Eds.), Royal Society of Chemistry, Cambridge, 1993, Vol. 10, p. 1.
10. E. Ruckenstein, S. H. Lee, *J. Catal.* **1987**, 104, 259.
11. I. Sushumna, E. Ruckenstein, *J. Catal.* **1985**, 94, 239.
12. S. A. Stevenson, J. A. Dumesic, R. T. K. Baker, E. Ruckenstein, *Metal-Support Interactions in Catalysis, Sintering, and Redispersion*, Van Nostrand Reinhold, New York, 1987, p. 141.
13. E. Ruckenstein, in *Sintering and Heterogeneous Catalysis*, G. C. Kuczinski, A. E. Miller, G. A. Sargent (Eds.), Plenum Press, New York, London, 1984, p. 199.

14. S. H. Overbury, P. A. Bertrand, G. A. Somorjai, *Chem. Rev.* **1975**, *75*, 547.
15. R. Andersen, *Structure of Metallic Catalysts*, Academic Press, New York, 1975.
16. F. C. M. J. M. van Delft, A. D. van Langefeld, B. E. Nieuwenhuys, *Thin Solid Films* **1985**, *123*, 333.
17. F. C. M. J. M. van Delft, A. D. van Langefeld, B. E. Nieuwenhuys, *Solid State Ionics* **1985**, *16*, 233.
18. E. Ruckenstein, *Metal-Support Interactions in Catalysis, Sintering, and Redispersion*, S. A. Stevenson, J. A. Dumesic, R. T. K. Baker, E. Ruckenstein, Van Nostrand Reinhold, New York, 1987, p. 241.
19. E. Ruckenstein, *J. Cryst. Growth* **1979**, *47*, 666.
20. E. Ruckenstein, in *Metal-Support Interactions in Catalysis, Sintering, and Redispersion*, S. A. Stevenson, J. A. Dumesic, R. T. K. Baker, E. Ruckenstein, Van Nostrand Reinhold, New York, 1987, p. 236.
21. E. Ruckenstein, in *Metal-Support Interactions in Catalysis, Sintering, and Redispersion*, S. A. Stevenson, J. A. Dumesic, R. T. K. Baker, E. Ruckenstein, Van Nostrand Reinhold, New York, 1987, p. 247.
22. G. E. Rhead, M. G. Barthes, C. Argile, *Thin Solid Films* **1981**, *82*, 201.
23. E. Bauer, H. Poppa, *Thin Solid Films* **1972**, *12*, 167.
24. E. Bauer, H. Poppa, G. Todd, P. R. Davis, *J. Appl. Phys.* **1977**, *48*, 3773.
25. E. Ruckenstein, in *Metal-Support Interactions in Catalysis, Sintering, and Redispersion*, S. A. Stevenson, J. A. Dumesic, R. T. K. Baker, E. Ruckenstein, Van Nostrand Reinhold, New York, 1987, p. 153.
26. (a) J. M. Kosterlitz, D. J. Thouless, *J. Phys. C* **1972**, *5*, L 124; (b) J. M. Kosterlitz, D. J. Thouless, *J. Phys. C* **1973**, *6*, 1181.
27. F. A. Lindemann, *Z. Phys.* **1910**, *11*, 609.
28. R. T. K. Baker, *J. Catal.* **1982**, *78*, 473.
29. E. Ruckenstein, B. Pulvermacher, *J. Catal.* **1973**, *29*, 224.
30. E. Ruckenstein, B. Pulvermacher, *AIChE J.* **1973**, *19*, 356.
31. B. K. Chakraverty, *J. Phys. Chem. Solids* **1967**, *28*, 2401.
32. P. C. Flynn, S. E. Wanke, *J. Catal.* **1974**, *33*, 233.
33. E. Ruckenstein, D. B. Dadyburjor, *J. Catal.* **1977**, *48*, 73.
34. E. Ruckenstein, D. B. Dadyburjor, *Thin Solid Films* **1978**, *55*, 89.
35. H. H. Lee, *J. Catal.* **1980**, *62*, 129.
36. E. Ruckenstein, in *Metal-Support Interactions in Catalysis, Sintering, and Redispersion*, S. A. Stevenson, J. A. Dumesic, R. T. K. Baker, E. Ruckenstein (Eds.), Van Nostrand Reinhold, New York, 1987, pp. 156, 187.
37. P. Wynblatt, T. M. Ahn, in *Sintering and Catalysis*, G. C. Kuczynski (Ed.), Plenum Press, New York, 1975, p. 83.
38. E. Ruckenstein, Y. F. Chu, *J. Catal.* **1979**, *59*, 109.
39. J. Haber, T. Machej, T. Czeppe, *Surf. Sci.* **1985**, *151*, 301.
40. P. von Blanckenhagen, in *Structure and Dynamics of Surfaces II. Phenomena, Models, and Methods*, W. Schommers, P. von Blanckenhagen (Eds.), Springer-Verlag, Berlin, 1987, p. 73.
41. J. Haber, T. Machej, E. M. Serwicka, I. E. Wachs, *Catal. Lett.* **1995**, *32*, 101.
42. A. Ya. Neimann, *Solid State Ionics* **1996**, *83*, 263.
43. J. J. Chen, E. Ruckenstein, *J. Catal.* **1981**, *69*, 254.
44. I. Sushumna, E. Ruckenstein, *J. Catal.* **1988**, *709*, 433.
45. J. J. Chen, E. Ruckenstein, *J. Phys. Chem.* **1981**, *85*, 1606.
46. E. Ruckenstein, S. H. Lee, *J. Catal.* **1984**, *86*, 457.
47. M. Lo Jacono, M. Schiavello, A. Cimino, *J. Phys. Chem.* **1971**, *75*, 1044.
48. Y. Chen, L. Zhang, *Catal. Lett.* **1992**, *72*, 51.
49. I. Sushumna, E. Ruckenstein, *J. Catal.* **1985**, *94*, 239.
50. E. Ruckenstein, D. B. Dadyburjor, *Rev. Chem. Eng.* **1983**, *7*, 251.
51. M. J. D'Aniello, Jr., D. R. Monroe, C. J. Carr, M. H. Knieger, *J. Catal.* **1988**, *709*, 407.
52. B. S. Clausen, J. Schiotz, L. Gräbaek, C. V. Oveson, K. W. Jacobsen, J. K Norskov, H. Topsoe, *Top. Catal.* **1994**, *7*, 367.
53. C. V. Ovesen, B. S. Clausen, J. Schiøtz, P. Stolze, H. Topsøe, J. K. Nørskov, *J. Catal.* **1997**, *168*, 133.
54. J.-D. Grunwaldt, A. M. Molenbroek, N.-Y. Topsøe, H. Topsøe, B. S. Clausen, *J. Catal.* **2000**, *194*, 452.
55. P. L. Hansen, J. B. Wagner, S. Helveg, J. R. Rostrup-Nielsen, B. S. Clausen, H. Topsøe, *Science* **2002**, *295*, 2053.
56. P. L. Hansen, S. Helveg, A. K. Datye, *Adv. Catal.* **2006**, *50*, 77.
57. S. J. Tauster, S. C. Fung, R. L. Garten, *J. Am. Chem. Soc.* **1978**, *100*, 170.
58. S. J. Tauster, S. C. Fung, *J. Catal.* **1978**, *55*, 29.
59. G. L. Haller, D. E. Resasco, *Adv. Catal.* **1989**, *36*, 173.
60. R. T. K. Baker, S. J. Tauster, J. A. Dumesic (Eds.), *Strong Metal Support Interactions*, American Chemical Society, Washington, 1986.
61. S. Engels, B. Freitag, W. Mörke, W. Röschke, M. Wilde, *Z. Anorg. Allg. Chem.* **1981**, *474*, 209.
62. P. Mériaudeau, J. F. Dutel, M. Dufaux, C. Naccache, *Stud. Surf. Sci. Catal.* **1982**, *77*, 95.
63. J. Santos, J. A. Dumesic, *Stud. Surf. Sci. Catal.* **1982**, *77*, 43.
64. B.-H. Chen, J. M. White, *J. Phys. Chem.* **1983**, *87*, 1327.
65. A. D. Logan, E. J. Braunschweig, A. K. Datye, D. J. Smith, *Langmuir* **1988**, *4*, 827.
66. Ch. Linsmeier, H. Knözinger, E. Taglauer, *Nucl. Instr. Methods B* **1996**, *118*, 533.
67. E. Taglauer, H. Knözinger, *Phys. Stat. Sol. (B)* **1995**, *792*, 465.
68. B. W. Dodson, *Surf. Sci.* **1987**, *184*, 1.
69. J. H. Sinfelt, Y. L. La, J. A. Cusumano, A. E. Barnett, *J. Catal.* **1976**, *42*, 227.
70. E. B. Prestridge, G. H. Via, J. H. Sinfelt, *J. Catal.* **1977**, *50*, 115.
71. J. H. Sinfelt, G. H. Via, F. W. Lytle, *J. Phys. Chem.* **1980**, *72*, 4832.
72. (a) K. Christmann, G. Ertl, H. Shimizu, *J. Catal.* **1980**, *61*, 397; (b) K. Christmann, G. Ertl, H. Shimizu, *Thin Solid Films* **1979**, *57*, 247.
73. J. T. Yates, C. H. F. Peden, D. W. Goodman, *J. Catal.* **1985**, *94*, 321.
74. R. Liu, B. Tesche, H. Knözinger, *J. Catal.* **1991**, *729*, 402.
75. J. E. Houston, C. H. F. Peden, P. J. Feibelman, *Surf. Sci.* **1987**, *192*, 457.
76. J. Leyrer, R. Margraf, E. Taglauer, H. Knözinger, *Surf. Sci.* **1988**, *201*, 603.
77. P. Somasundaram, in *Ceramic Processing before Firing*, G. Y. Onoda, Jr., L. L. Hench (Eds.), Wiley, New York, 1978, p. 78.
78. K. Shinohara, in *Powder Technology Handbook*, K. Jinoya, K. Gotoh, K. Higashitani (Eds.), M. Dekker, New York, 1991, p. 481.
79. K. Miyanami, in *Powder Technology Handbook*, K. Jinoya, K. Gotoh, K. Higashitani (Eds.), M. Dekker, New York, 1991, p. 595.
80. G. Heinicke, *Tribochemistry*, Akademie-Verlag, Berlin, 1984.
81. S. A. Angelov, R. P. Bonchev, *Appl. Catal.* **1986**, *24*, 219.
82. G. Mestl, B. Herzog, R. Schlögl, H. Knözinger, *Langmuir* **1995**, *11*, 3027.
83. G. Mestl, N. F. D. Verbruggen, H. Knözinger, *Langmuir* **1995**, *11*, 3055.

84. G. Mestl, T. K. K. Srinivasan, H. Knözinger, *Langmuir* **1995**, *11*, 3795.
85. G. Mestl, N. F. D. Verbruggen, F. C. Lange, B. Tesche, H. Knözinger, *Langmuir* **1996**, *12*, 1817.
86. S. R. Stampfl, Y. Chen, J. A. Dumesic, C. Niu, C. G. Hill, Jr., *J. Catal.* **1987**, *105*, 445.
87. J. M. M. Llorente, V. Rives, *Solid State Ionics* **1990**, *38*, 119.
88. T. F. Hayden, J. A. Dumesic, R. D. Sherwood, R. T. K. Baker, *J. Catal.* **1987**, *105*, 299.
89. J. Leyrer. D. Mey, H. Knözinger, *J. Catal.* **1990**, *124*, 349.
90. Y. Xie, L. Gui, Y. Liu, B. Zhao, N. Yang, Y. Zhang, Q. Guo, L. Duan, H. Huang, X. Cai, Y. Tang, in *Proceedings of the 8th International Congress on Catalysis*, Berlin 1984, Dechema, Frankfurt, 1984, Vol. V, p. 147.
91. Y. Xie, L. Gui, Y. Liu, Y. Zhang, B. Zhao, N. Yang, Q. Guo, L. Duan, H. Huang, X. Cai, Y. Tang, in *Adsorption and Catalysis on Oxide Surfaces*, M. Che, G. C. Bond (Eds.), Elsevier, Amsterdam, 1985, p. 139.
92. B. M. Reddy, K. Narsimha, P. Kanta Rao, *Langmuir* **1991**, *7*, 1551.
93. R. Margraf, J. Leyrer, E. Taglauer, H. Knözinger, *Surf. Sci.* **1987**, *189/190*, 842.
94. R. Margraf, J. Leyrer, E. Taglauer, H. Knözinger, *React. Kinet. Catal. Lett.* **1987**, *35*, 261.
95. J. Leyrer, M. L. Zaki, H. Knözinger, *J. Phys. Chem.* **1986**, *90*, 4775.
96. G. Kisfaludi, J. Leyrer, H. Knözinger, R. Prins, *J. Catal.* **1990**, *124*, 349.
97. L. M. J. von Hippel, F. Hilbrig, H. Schmelz, B. Lengeler, H. Knözinger, *Collect. Czech. Chem. Commun.* **1992**, *57*, 2456.
98. O. Glemser, H. G. Wendland, *Angew. Chem.* **1973**, *75*, 949.
99. S. Günther, M. Marsi, A. Kolmakov, M. Kiskinova, M. Noeske, E. Taglauer, G. Mestl, U. A. Schubert, H. Knözinger, *J. Phys. Chem.* **1997**, *101*, 10004.
100. S. Günther, F. Esch, L. Gregoratti, A. Barinov, M. Kiskinova, E. Taglauer, H. Knözinger, *J. Phys. Chem.* **2004**, *108*, 14223.
101. H. Zhu, M. Shen, Y. Wu, X. Li, J. Hong, B. Liu, X. Wu, L. Dong, Y. Chen, *J. Phys. Chem. B* **2005**, *109*, 11720.
102. F. Bertinchamps, C. Grégoire, E. M. Gaigneaux, *Appl. Catal. B* **2006**, *66*, 1.
103. O. Glemser, A. Müller, *Z. Anorg. Allg. Chem.* **1963**, *325*, 220.
104. N. L. Yannopoulos, *J. Phys. Chem.* **1968**, *72*, 3293.
105. M. Gasior, J. Haber, T. Machej, *Appl. Catal.* **1987**, *33*, 1.
106. T. Machej, J. Haber, A. M. Turek, I. E. Wachs, *Appl. Catal.* **1991**, *70*, 115.
107. G. Centi, E. Giamello, D. Pinelli, F. Trifiro, *J. Catal.* **1991**, *130*, 220.
108. G. Centi, D. Pinelli, F. Trifiro, D. Ghoussoub, M. Guelton, L. Gengembre, *J. Catal.* **1991**, *130*, 238.
109. F. Hilbrig, H. E. Göbel, H. Knözinger, H. Schmelz, B. Lengeler, *J. Phys. Chem.* **1991**, *95*, 6973.
110. R. Haase, J.-G. Jerschkewitz, G. Öhlmann, J. Richter-Mendau, J. Scheve, in *Book of Abstracts, 2nd International Symposium on Scientific Bases for the Preparation of Heterogeneous Catalysts*, Louvain-la-Neuve, Belgium, 1978, paper F5.
111. P. Mériaudeau, C. Naccache, *Appl. Catal.* **1991**, *73*, L13.
112. J. L. G. Fierro, J. C. Conesa, A. Lopez Agudo, *J. Catal.* **1987**, *108*, 334.
113. J. Leyrer, H. Knözinger, unpublished results.
114. J. A. Rabo, P. H. Kasai, *Prog. Solid State Chem.* **1975**, *9*, 1.
115. J. A. Rabo, in *Zeolite Chemistry and Catalysis*, J. A. Rabo (Ed.), American Chemical Society, Washington, 1976, p. 332.
116. H. K. Beyer, H. G. Karge, G. Borbely, *Zeolites* **1988**, *8*, 79.
117. H. G. Karge, H. K. Beyer, G. Borbely, *Catal. Today* **1988**, *3*, 41.
118. H. G. Karge, V. Mavrodinova, Z. Zheng, H. K. Beyer, in *Guidelines for Mastering the Properties of Molecular Sieves*, D. Barthomeuf, E. G. Derouane, W. Hölderich (Eds.), Plenum Press, New York, 1990, p. 157.
119. H. G. Karge, H. K. Beyer, in *Zeolite Chemistry and Catalysis*, P. A. Jacobs, N. I. Jaeger, L. Kubelkova, B. Wichterlova (Eds.), Elsevier, Amsterdam, 1991, p. 43.
120. H. G. Karge, B. Wichterlova, H. K. Beyer, *J. Chem. Soc., Faraday Trans.* **1992**, *88*, 1345.
121. H. G. Karge, H. K. Beyer, in *Molecular Sieves*, H. G. Karge, J. Weitkamp (Eds.), Vol. 3, Springer-Verlag, Berlin, 2002, p. 43.
122. J. Weitkamp, S. Ernst, M. Hunger, T. Röser, S. Huber, U. A. Schubert, P. Thomasson, H. Knözinger, in *Proceedings of the 11th International Congress on Catalysis*, J. W. Hightower, W. N. Delgass, E. Iglesia, A. T. Bell (Eds.), Baltimore 1996, Elsevier, Amsterdam, 1996, p. 731.

2.4.8
Mechanochemical Methods

Bernd Kubias*, Martin J. G. Fait, and Robert Schlögl

2.4.8.1 Introduction

Since the beginning of the last century it has been known that the impact of mechanical energy on a solid leads, besides its comminution and the formation of new free surfaces, to its mechanical activation (MA) and, with increasing activation, to mechanochemical reactions [1, 2].

The term "activation" includes both enhanced activity with respect to reactions of the solid itself and changed physico-chemical properties concerning adsorption and catalytic activity and selectivity. Mechanical activation and mechanochemical reactions are the subject of mechanochemistry, a special branch of chemistry which was first defined by Ostwald at the end of the 19th century [3]. An extensive literature has been devoted to this subject during the past four decades [1, 4–8], whereby it was shown that the occurrence of these reactions cannot be explained only by thermal influence on the system by the exerted mechanical treatments. In contrast, it was proven that this phenomenon is of an independent nature (see Section 2.4.8.2).

During the past years, special domains of mechanochemistry and mechanical activation have been developed

---

References see page 581
* Corresponding author.

rapidly [9], such as mechanical alloying and amorphization of metals as well as mechanofusion and the manufacture of new, nanosized materials (see, e.g. Refs. [9–12]). Successful attempts have been made to apply mechanochemistry to pharmacy, synthesis, and to the manufacture of pigments. New approaches to the manufacture of novel or improved catalysts, and for the activation and "selectivation" of known catalysts, have been reported [13–15]. Thus, in addition to "classic" materials processing in engineering, the use of mechanochemistry in catalysis has gained increasing scientific and industrial interest. Many systematic basic and applied investigations have been conducted in this field during the past two decades, especially in the former Soviet Union and Russia, respectively [13–17], based on former fundamental and applied investigations by German and Japanese groups.

The mechanochemically influenced catalytic ("tribocatalytic") reactions known up to 1984 are listed by Heinicke in the monograph *Tribochemistry*, which deals with all aspects of mechanochemistry (in the older German literature the term tribochemistry/Tribochemie was used synonymously for the mechanochemistry of solids). Results from investigations conducted during the past 50 years concerning the mechanochemical preparation of catalysts have been collected in a recently published review by Molchanov and Buyanov [13].

In this chapter, an overview of the methods of MA in the activation and selectivation of catalytic materials, and of the contribution of mechanochemistry to the preparation of catalysts, is presented. Due to its minor importance, the influence of mechanical impact on a running reaction is mentioned only briefly. The first part of the chapter includes some outline information concerning the state of the art of MA. Similarities exist in the use of ultrasonic irradiation, which is also a viable method for the preparation and treatment of solid catalysts (cf. Chapter 8.5).

Mechanical activation and mechanochemical reactions are carried out predominantly in mills, where the milling processes play an important role in chemical industry and are responsible for a large proportion of the process costs [8]. The action of mills and the comminution of solids have formed the basis of many investigations in engineering science for decades, and the findings have led to a deeper understanding of the comminution processes and the development of highly efficient mills [9, 18].

2.4.8.2 The Effect of Mechanical Activation on the Reactivity of Solids

When mechanical energy is applied to a solid by means of pressure and shear, the resulting physicochemical changes are the consequence of relaxation of the field of stresses in the solid during and after the mechanical treatment [13]. The strain field manifests itself by shifts of atoms from the stable equilibrium positions, by changes of bond length and angles and, in some cases, by excitation of the electron subsystem [19]. These changes result in an activated state of the solid. The accumulated energy can lead to destruction of the crystal and finally to amorphization or transition to a metastable polymorphous state. If relaxation of the strain field results in the rupture of chemical bonds, a mechanochemical reaction occurs [2]. Thus, the thermal excitation of chemical reactions, which is also induced by mechanical exertions, is not the only reason for the observed MA results [20].

In all cases the thermodynamic potentials of the treated compounds are changed, and free energy and enthalpy are enhanced. These depend on the size of particles and on the presence of defects in their crystal structure. Often, a direct relationship between the mechanical energy added, the concentration of defects, and the reactivity of the solids has been observed. These relationships have been widely studied for brittle substances, but are also described in some investigations with ductile materials (e.g., Refs. [21–24]). The imperfections are categorized as point defects and extended defects (one- or two-dimensional defects) [25, 26], and are commonly assumed to act as the catalytically active sites.

The non-equilibrium structural and phase states of catalysts arising in the process of their mechanochemical preparation, or by their treatment by MA, tend to anneal during the use of catalysts. As a consequence, the catalytic properties of the treated solids may change to a less-active steady state during their application at higher temperature. Thus, to maintain the enhanced activation for catalytic purposes a permanent addition of mechanical energy may be necessary. However, there are numerous examples of rather high stability of the metastable activated states of solid catalysts at elevated temperatures under real reaction conditions [13, 27]. This is explained by the fact that the thermal annealing of defects is also an activated process. Therefore, some structures with higher reactivity may exist for a rather long time, and the increase in catalytic activity may be used for practical purposes.

On occasion, mechanical treatment is even necessary for the creation of sufficient catalytic activity, as in the case of the activation of Ziegler–Natta catalysts by milling [28]. These catalysts and other mechanically activated catalysts, which are used at lower temperatures, remain stable during their use.

A close correlation between the nature and the concentration of lattice defects and the specific catalytic activity has been found [29–33], and consequently defects such as dislocations [21, 23, 30, 34–37], planes of crystallographic shear [34, 38], and steps in the facets [39] are assumed to be catalytically active sites. However, according to Molchanov and Buyanov [13, 40], only a few studies [23, 41] have provided direct experimental evidence for the role of well-defined defects in the increase in the catalytic activity and allowed a description of the structure of active sites at the atomic level.

Systematic studies of the effects of different types of stress during MA showed [42] that the activation effects depend on the types of stress valid for the different comminution machines. Whereas impact stress changes the microstructure in the whole bulk of the particle, pressure and shear stresses cause a significant enhancement of defects at the surface of the treated particles. Among the factors controlling defect development, the energy input of the mills and the efficiency of the energy transfer in the mills are the most important [43]. In the case of complex reactions, MA-effected changes in the catalytic properties can also include altered properties of selectivity (see Section 2.4.8.5.2).

The MA-effected changes of the structure of solids may be characterized by the usual methods of structural and morphological characterization of solids such as X-ray diffraction (XRD) (real structure analysis: crystallite size, amorphous fraction, microstructure), laser diffraction (particle size distribution), BET and porosity measurements (specific surface area, pore structure), density measurements, calorimetry and thermal analysis, and transmission and scanning electron microscopy (TEM, SEM). Vibrational spectroscopy [Raman, diffuse reflectance infrared Fourier transform spectroscopy (DRIFTS), DR-UV-visible] and electron paramagnetic resonance (EPR) spectroscopy were also used to characterize the defect structures of mechanically activated materials (e.g., Refs. [44–48]). A detailed description of the methods mentioned with respect to MA can be found in Refs. [1, 49], and in the respective chapters of this Handbook.

2.4.8.3 Mills

In general, mills are used for the comminution and mechanochemical treatment of solids. It should be pointed out that although normally used mills are developed for optimal comminution, and not for optimum activation, they have nonetheless proved to be useful tools for mechanical activation.

In the case of a given uniform chemical compound, comminution by milling results in the formation of a middle particle size of the material with increasing milling duration, independent of the particle size of the starting material; in other words, the so-called Hüttig milling equilibrium is reached [50]. Here, typical particle sizes are ca. 10 μm in case of fine grinding with ball mills, and ca. 1 μm (super-fine grinding with attritors). Under special conditions, comminution may be used for the synthesis of a wide range of nanopowders with mean particle sizes as small as 4 nm [51]. As shown by Rehbinder et al., the addition of surfactants to the treated materials leads to a diminished strength of the milled metals [19, 52, 53] (the Rehbinder effect; see Section 2.4.8.3.3) and, thus, enables an efficient comminution.

Due to the importance of comminution for industrial materials processing, the types of crushers and mills available are many and diverse. Despite the progress, which has been made during the past decade in understanding the milling process, the choice of the best suited mill for any given comminution or activation task and, of the "right" conditions, has been until now a question of experiment and experience. This is due mainly to the fact that today the description of a comminution process is possible, but only to a limited extent using appropriate models [54].

The comminution process may be traced by means of standard methods for solids characterization, as listed in Section 2.4.8.2.

As mentioned above, mechanical activation comprises the complex structural changes of the solid in the micro and macro scale. For MA in the laboratory, a number of excellent mills have particularly proved their worth. The different types of stress mechanism exerted in these different mills lead to different grinding results and different structural changes, depending on the material's behavior. A short description of the different stress types is presented below, and an overview is also provided of the main characteristics of selected types of laboratory mill from the point of view of an experimentalist. Some examples will be used to illustrate the mode of action and use of mills in the treatment of solids.

2.4.8.3.1 **Stress Types** The micro processes which occur during the comminution can be classified after Rumpf, according to the types of exerted stresses [55, 56].

A Stress Between Two Surfaces (cf. Fig. 1) In this case, the material is stressed individually or collectively between two surfaces. If the influence is vertical, then a compressive stress takes place resulting in pressure and percussion stress. With an additional tangential movement of the milling areas, shearing and friction stress takes place. In practice both stresses occur together. Whereas pressure and percussion dominate at crushing and rolling, shearing dominates at trituration and cutting. At distinct parameters of the mill the transition to percussion stress is adjustable, which is valid especially for planetary mills if their construction allows the independent adjustment of the relevant components.

References see page 581

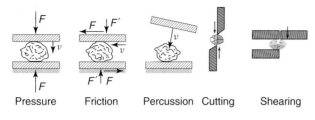

Pressure Friction Percussion Cutting Shearing

Fig. 1 Comminution principles according to Rumpf [57]: stress between two surfaces (from Ref. [56]). F = force; F' = force directed perpendicular to F; v = stress velocity.

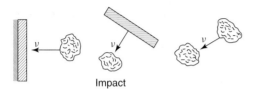

Impact

Fig. 2 Comminution principles according to Rumpf [57]: stress at a single surface; v = stress velocity (From Ref. [56]).

B Stress at a Single Surface (cf. Fig. 2) This type of stress is exerted by a single surface and appears as impact stress. The material is stressed by its striking against the face of the grinding medium, by the striking of the grinding medium against the material, or by mutual impact of the material. The stress intensity is a function of the velocity. This type of stress occurs mainly in a high peripheral-speed pin mill (disintegrator), a jet mill, and an attritor (see Section 2.4.8.3.2).

C Stress Without Milling Faces Energy can also be transmitted to solids by surrounding liquid and gaseous media, for example, by a jet of gases. However, this type of energy input is of minor interest for the comminution and mechanical activation and, therefore, is not mentioned in the following.

2.4.8.3.2 Types and Mode of Action of Mills In the following section, the mode of action of some selected laboratory mills are described in order to provide an impression of their application, as well as the problems connected with their practical use. The family of crushers will not be described at this point (see Ref. [56]). A detailed description of a variety of mills from the viewpoint of process engineering can be found in Refs. [56, 58, 59], while the monograph of Höffl [60] concerns the constructive characteristics of different industrial grinding machines.

Well-known members of the family of grinders with two grinding surfaces are the vibratory mill, the mortar grinder, and the planetary ball mills, which are selected here as examples to demonstrate the possible function of mills.

The *pinned disc mill*, as a representative of a comminution machine with one grinding surface, utilizes a high-speed rotor with one or more rows of rods that propel particles into stationary pins that mesh with the rotor. In *jet mills*, the material is accelerated by means of propellant air or gas jets; comminution then occurs by particle impact with a surface (target), or by interparticle collision. The *attritor* is a mill containing balls in liquid media, which is internally agitated by a stirrer; however, the liquid must be removed after grinding and this may affect the properties of the material treated.

A Vibratory Mills Vibratory mills achieve size reduction through the combination of impact and friction. An electromagnetically powered mortar generates vibrations (see Fig. 3), which are transferred to the grinding ball via the grinding material. The impact energy of the grinding ball can be controlled by adjusting the amplitude. The vibrating mill has proved its worth as a good mechanical activation machine, and is often used for the investigations of mechanical activation effects.

B Mortar Grinder The principle of action of a mortar grinder (Fig. 4) is similar to that of a hand mortar, in that the material to be ground is reduced in size between the pestle and the mortar bowl through pressure, friction, shearing, and cutting. Pressure is applied by the pestle on the mortar wall in horizontal as well as vertical directions. Pressure arises by the pestle's own weight and by preloaded adjustable springs, or by an additional

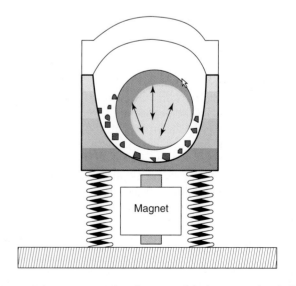

Fig. 3 Schematic view of a vibratory mill with one grinding ball. (Illustration reproduced with permission from Fritsch GmbH.)

and by selecting suitable rotation times, accelerations of 600 to 1000 m s^{-2} (ca. 60–100 g) can be achieved, ensuring very high mechanical activation effects.

An adjustment of different types of stress is possible in a specially designed planetary mill (type FIA; UVR-FIA GmbH, Freiberg, Germany) by appropriate choice of the characteristic machine parameters; this allows the transfer from friction stress via a transition region to impact stress, as demonstrated in the case of mechanical alloying of binary mixtures of metals [9]. In a similar manner, the creation of new surfaces and structural distortions, respectively, can be influenced specifically in the same mill. This effect was studied, for example with quartz [61], using a vibration mill and by varying the revolutions and the amplitude.

Nowadays, both a gas pressure and temperature measuring system (GTM system, Fritsch GmbH) and a means of providing an inert atmosphere are available for ball mills. Such systems enable a planetary mill to be converted, for example, into an *in-situ* measuring system. Without any modification of the mill itself, the grinding bowl is substituted by a bowl with a transmitter integrated into the lid. The data are transferred by wireless mode to a computer, where they can be visualized and analyzed.

Direct observation of the self-combustion of elemental powders during the course of ball-milling treatment by using quartz, transparent reaction vessels and a high-speed image-acquisition system has been described [62].

2.4.8.3.3 The Practical Use of Mills: Some Comments

The different comminution machines vary by three orders of magnitude in their maximum power input related to the milling body mass [63]. The respective maximum values are in the range of 40 W kg^{-1} (vibration mill), ca. 100 W kg^{-1} (planetary ball mill; type "Pulverisette 5", Fritsch GmbH, Germany), and up to 4 kW kg^{-1} (planetary ball mill; type FIA, UVR-FIA GmbH, Germany). An especially powerful high-energy mill is the Spex mill 8000 (SPEX CertiPrep, USA), which enables a power input of up to 30 kW kg^{-1}. The energy input is correlated directly to the milling time; mills transferring a very high energy permit short milling times in the range of minutes, whereas mills with lower power input require longer milling durations that range from several hours up to days.

The key points of carrying out effective grinding, such as optimum filling, cleaning procedures of grinding sets, and safety requirements, are summarized with respect to analytical aspects in Ref. [64]. In this section, only the

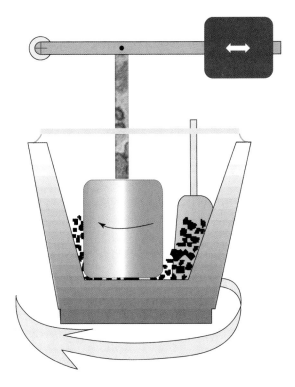

Fig. 4 Schematic view of a mortar grinder. (Illustration reproduced with permission from Fritsch GmbH.)

weight acting on the pestle axis. A fixed scraper turns the material and guides the whole sample amount back to the pestle. The mortar grinder seems to be suited especially for materials with a layered structure, and especially if the layers must be spread in order to expose the catalytically active and selective surfaces.

C Planetary Ball Mill In the planetary ball mill (Fig. 5), the sample material is primarily comminuted by the high-energy impact of grinding balls, together with friction between the balls and the wall of the grinding bowl. The grinding bowls, when filled with the grinding material and balls, rotate around their own axis on a counter-rotating supporting disc (the sun disc). The centrifugal forces caused by the rotation of the grinding bowls and supporting disc work on the grinding material. The force resulting from rotation of the grinding bowl when the mill is started causes the rotating balls to rub against the inside wall of the bowl (sliding regime; Fig. 6a). With increasing centrifugal force, comminution by percussion is reached (throwing regime; Fig. 6b). Depending on the revolution of the grinding bowls and supporting disc, turbulence and circle regimes may occur (Fig. 6c,d). The various milling regimes result in different main stress mechanisms of percussion, shearing, or pressure. In these mills, by using an appropriate geometric arrangement of the trajectories,

References see page 581

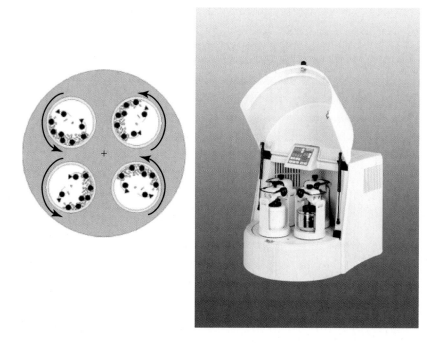

Fig. 5 Left: a schematic view of a planetary ball mill. Right: A "Pulverisette 5", with four grinding bowls. (Illustration reproduced with permission from Fritsch GmbH.)

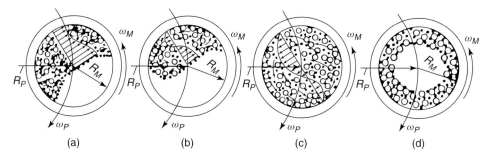

Fig. 6 Different milling regimes depending on the operating parameters in a planetary ball mill (from Ref. [60]). R_P = radius of the supporting disc; R_M = radius of the grinding bowl; ω_P = angular velocity of the supporting disc; ω_M = angular velocity of the grinding bowl. (a) Sliding regime; (b) throwing regime; (c) turbulence regime; (d) circle regime

problems of contamination and agglomeration will be detailed.

In general terms, contamination of the milled materials must be considered, mainly as a result of abrasion of the mill's grinding parts. For example, it could be expected that a milled material would contain iron if the grinding part were to be made from stainless steel [65]. In order to overcome this problem, the vessels should be lined with appropriate materials (e.g., semiprecious stone). Moreover, for catalytically relevant studies, grinding media with high abrasion resistance should be used, such as agate, sintered corundum, and zirconia.

One undesired effect during grinding is that of agglomeration or "caking"; this may be caused by electrostatic charge accumulation and/or the fusion of particles under pressure or humidity, and in turn leads to an inhibition of further particle size reduction. The multichoice of grinding aids available (slurry grinding) makes this technique an "art". The theoretical background of efficient grinding in suspension is termed the "Rehbinder effect", which consists of a decrease in the surface tension of a solid under the action of an ambient medium such as water, alcohol, or other liquids. These media are added to the sample before grinding begins, and are removed afterwards. The effect results in an abated strength of the solid, and thus favors the evolution of defects and comminution at a lower level of mechanical stress.

In order to achieve reproducible results, it is necessary to know the ratio of the mass of balls to the mass of material to be milled, as well as the filling degree of the grinding vessels. These data, together with the velocity, determines the amount of energy which is transferred into the grinding material [9].

Before choosing and using a distinct laboratory mill for MA, the following points must be clarified [66]:

- the type of grinding material (i.e., its chemical and physical properties, especially hardness according to the classification hard/tough, hard/brittle, medium hard/tough, medium hard/brittle, soft/tough, soft/brittle)
- the initial sample characteristics (i.e., initial particle size, quantity, and number of samples)
- the final sample characteristics (i.e., fine particle size, particle size range, contamination, abrasion)
- the universal applicability (i.e., milling in a slurry medium or in an inert atmosphere and under vacuum, sample cooling)
- the processing time (i.e., milling time, time for cleaning of grinding sets).

2.4.8.4 The Use of Mechanochemistry for the Synthesis of Catalysts

In this section, selected examples of the application of mechanochemistry in the synthesis of catalysts, and for the control of their performance, are presented. Today, catalyst synthesis represents a specialized area of mechanochemistry in the production of new or improved materials.

According to Molchanov and Buyanov [13], mechanochemical syntheses can be successfully carried out if: (i) the crystalline structures of the solids to be converted are similar; (ii) in the case of an interaction of acid–base type, one of the reactants is a strong acid or base; or (iii) in the formation of pseudomelts, an intensive mechanical treatment increases the mobilities of molecules of at least one reagent to the level of their mobilities in a melt.

One of the most impressive examples of improving catalytic properties by mechanical treatment is the preparation of polymerization catalysts of the Ziegler–Natta type. The ball milling of $TiCl_3$ was found to cause a large increase in catalyst activity in the stereoselective production of isotactic propylene [28]. The reason for this is that, under the conditions of milling, formation of the δ-$TiCl_3$ phase is favored, which shows a higher activity than the α- and γ-phases and proved to be necessary for high stereospecificity [67]. The processes occurring during the manufacture of Al–Ti-chloride catalysts by the mechanical treatment of $Al/TiCl_4$ or $Ti/AlCl_3$ or chloride/alkylaluminum mixtures were investigated by Tornquist et al. [68–70], who showed that activation was due to the formation of a solid solution of the δ-$TiCl_3$ structure [68].

An additional example of the successful use of MA has been the synthesis of hydrides of intermetallic Mg compounds as selective hydrogenation catalysts [71, 72]. These catalysts are used for the purification of olefin monomers from undesired alkine and diene contaminations. As yet, the known noble metal hydrogenation catalysts have too-low selectivities in these reactions but, as was shown during the 1980s, the discovered hydrides of intermetallic compounds proved to be active and selective in these hydrogenation reactions due to their highly mobile hydrogen [73–75]. In fact, the mechanically prepared alloys were found to be readily hydrogenated without any long-lasting hydrogenation/dehydrogenation treatments. As shown by Molchanov et al., under the conditions of high hydrogen pressure, new Mg intermetallic compounds such as Mg_2FeH_6, Mg_3CoH_5, and Mg_2CoH_5 can easily be prepared via the stage of mechanical alloying [40]. These mechanochemically obtained catalysts show 100% selectivities in the hydrogenation of butadiene to butenes and acetylene to ethylene, and also proved to be active in the hydrogenation of nitrogen to ammonia.

Novel waste-free methods for the manufacture of heteropoly acids (HPA) by the mechanical treatment of the constituents, allow for the preparation of these compounds [76], which are efficient catalysts for acid–base and oxidation reactions [77]. A direct synthesis from the oxides of tungsten, molybdenum, and vanadium was possible due to an increase in their reactivity by suitable milling procedures. More recently, new methods developed by Molchanov et al. for the synthesis of P–Mo–V– and P–Mo–HPA's comprise more the MA of the respective oxides and their mixtures in interaction with phosphoric acid [76].

According to Maksimov et al. [78], mechanochemically prepared HPA catalysts proved successful (instead of oleum) for the synthesis of diacetone L-sorbose, an intermediate product in the synthesis of vitamin C. This new technology is also said to include a method for the recovery of spent catalysts, thereby allowing the easy recycling of catalytic materials and avoiding the loss of expensive constituents.

Mechanical activation may also be successfully applied in the preparation of hydrocarbon decomposition catalysts for the manufacture of filamentous carbons. These catalysts are normally obtained by the catalytic decomposition of hydrocarbons on iron subgroup metals or their alloys with other metals such as Cu. The traditional method of catalyst preparation is to coprecipitate the hydroxides of metals and supports to obtain metal or alloy particles with a size of less than 50 nm [79]. The MA of Ni oxide

References see page 581

or its mixture with other oxides, together with suitable lamellar-structured supports in planetary mills, yielded catalysts with better activities compared to the conventionally prepared materials [80, 81]. Moreover, their good long-term stability makes them attractive for use in the production of hydrogen and carbonaceous materials from suitable hydrocarbons.

Mechanical activation proved also to be helpful in the manufacture of skeletal catalysts for the hydrogenation of organic compounds, as shown by Boldyrev et al. [82] and Fasman et al. [83]. Thus, the mechanical alloying of mixtures of powders of Ni and Al metals by means of planetary-centrifugal and vibratory mills leads to Raney catalysts with improved activity, which can be easily leached [82, 83]. In addition, energy-consuming pyrometallurgical steps for the preparation of alloys are avoided. Mechanical activation may also be successfully applied in the regeneration of spent skeletal catalysts, leading to catalysts with higher activity compared to the original catalysts. This result is attributed to the formation of a metastable compound with a Ni–Al structure, whereas conventional Raney catalysts contain Ni as a face-centered cubic lattice [84].

An impressive example of the advantage of mechanical treatment in the synthesis of supported metal catalysts is the milling of a mixture of metallic Ni with quartz in a vibratory mill. As shown by Schrader et al. [85], this procedure led to a coating of quartz with Ni and, consequently, to an improvement of the hydrogenation activity with respect to benzene. The increased activity is assumed to be due to distortions of the crystal lattice of Ni. If Cr, V, and Mo on silica gel were treated in a vibratory mill in different media, the catalysts obtained showed increased activities for ethylene polymerization [86], propylene metathesis [87], and the selective oxidation of methane to formaldehyde [88].

A further example of the successful application of mechanochemical methods for the manufacture of catalysts, namely the preparation of supported oxide monolayer catalysts from mechanical mixtures of the respective oxides, is discussed in Chapter 2.4.7. The preparation of catalysts for total oxidation on the basis of Co/Cu oxides, the synthesis of phosphate catalysts with improved strength, the preparation of alumina-based catalysts and of supports including hydroxyaluminates, alumosilicates, and zeolites have been described by Molchanov and Buyanov [13, and references cited therein].

It follows from the above-mentioned examples that mechanochemistry might help to identify new solutions with reduced expenditures for the purification of liquid and gaseous emissions, and that this will in turn make a major contribution to reducing the costs associated with catalyst production. As some of these mechanochemical syntheses show great potential with regards to not only energy saving but also improved safety, this technology should prove to be highly attractive when substituting conventional catalyst preparations.

2.4.8.5 The Influence of Mechanical Activation on Catalytic Properties

Evidence suggests that MA of catalytic materials prior to their use as catalysts may have a major influence on both the activity and selectivity. Whereas the influence of mechanical treatment on catalytic activity was realized during the initial investigations on catalytic effects caused by MA, the improvements in selectivity in complex reactions has become increasingly important only during the past 20 years.

2.4.8.5.1 The Influence of Mechanical Activation on Catalytic Activity
The effect of MA on the catalytic activity of metals and metal oxides has been demonstrated in numerous experiments, and is well documented in the literature. As a rule, mechanical treatments result in a considerable increase in the catalytic activity of solids – in some cases by several orders of magnitude [89].

Such an effect on the catalytic activity of metals was first realized during the 1930s [90], when Eckell observed that the rolling of a Ni foil led to an increase in its activity for ethylene hydrogenation. However, this enhanced activity was seen to disappear following annealing of the foil at temperatures in excess of 200 °C.

From later studies with different noble and other metal catalysts, it was concluded that dislocations and point defects were responsible for the observed increase in catalytic activity [91, 92]. Further, it was recognized that thermal annealing resulted in a decrease in activity to a level identical with the activity of the untreated metal.

Systematic studies of the influence of MA on the catalytic properties of metals were conducted by Uhara et al. and Kishimoto and Nishioka during the 1960s. These groups demonstrated increases in the catalytic activities of Cu [21, 30], Ni [35, 36], Ag [37], Au [93], and Pt [94] in a series of reactions such as the decomposition of diazonium chloride, the dehydrogenation and oxidation of ethanol, the hydrogenation of cinnamic acid, and the decomposition of formic acid. Dislocations were concluded to act as active sites, although in some cases the increase in activity was also attributed to the occurrance of vacancies [35, 36, 94].

At the same time, Schrader et al. [41, 95–99] showed that the activity of Co and Ni powder in the hydrogenation of benzene and ocenol, as a model reactant for fats, depended heavily on the duration of treatment in the applied ball and vibratory mills, and correlated well with the presence of defects and their density, as well as distortions of the crystal lattice. In the case of cobalt, Schottky defects were identified as active sites [99]. Again, the deactivation of catalysts by heating was attributed to the annihilation of vacancies [99].

The effects of MA on the catalytic activity of oxides were first observed by Clark and Rowan in 1941, while studying the catalytic activity of lead oxide in the decomposition of hydrogen peroxide [100]. Later studies by Paudert [101, 102] and by Schrader et al. [95, 96, 103, 104] concerning the activation behavior of different oxides (e.g., α-Fe$_2$O$_3$ in the oxidation of SO$_2$ and CO) revealed that the increase in activity caused by grinding the oxide in a vibratory mill was due rather to distortions of the crystal lattice than to the observed enlargement of the surface area. Analogously, the activation of α-Fe$_2$O$_3$ in a ball mill led to an enhanced activity in the selective catalytic reduction of NO with CO, which was shown to be caused by defects in the oxygen lattice [27]. Other types of mechanical impacts, such as compacting pressure and shock waves, were also found to cause higher activities of the treated mixed oxides or solid acids (e.g., tungstic acid and potassium hydrogensulfate) in different types of reactions such as polymerization, hydration/dehydration, hydrogenolysis, and oxidation [105–108].

The responsibility of defects for this increased activity was also clearly evidenced by Avvakumov et al. in a study of the MA of TiO$_2$ in the oxidation of CO. These authors' results revealed a quantitative dependence of the specific rate of CO oxidation on the concentration of the defects such as the concentration of the planes of crystallographic shear, which emerged on the surface of TiO$_2$ [38].

The above-mentioned examples show clearly that several effects may be responsible for the mechanochemically effected changes in the activity of a catalytic material. In most cases, the high concentration of defects in the solids as the result of treatments led to a decrease in activation energy and, therefore, to an increased activity. If, at the same time, an inhibitory effect caused by adsorbed molecules on the catalyst surface was to be lifted by the formation of fresh surfaces, a particularly strong increase in catalytic activity would be expected to appear [109].

There are also examples, however, when mechanical treatment leads to a *reduction* in activity; an example is the case of aluminum oxide which, after milling, no longer possess any activity for coke formation [110]. According to Buyanov et al. [15], the MA of rutile in air resulted in a decrease in specific activity for the oxidation of CO by one order of magnitude, whereas in the presence of Ar, MA led to almost complete deactivation. Another example of the somewhat surprising effects of MA was that such treatment of α-Fe$_2$O$_3$ in a ball mill, with up to 90% amorphization, caused a negative effect on the solid-state reactivity for reduction by hydrogen; this inhibitory effect was thought to be caused by diffusion of the Fe ions [27].

2.4.8.5.2 The Influence of Mechanical Activation on Catalytic Selectivity During the past 20 years, an increasing number of studies have been devoted to the effect of MA on the selectivity of catalysts. One such investigation involved the influence of MA on the selectivity of vanadyl pyrophosphate [(VO)$_2$P$_2$O$_7$] catalysts in the oxidation of butane to maleic anhydride; this effect is of special interest due to the industrial importance of this oxidation process. A study of the behavior of defects in the treated catalyst precursor, vanadyl hydrogen phosphate hemihydrate, VOHPO$_4 \cdot$ 0.5H$_2$O, showed that structural changes in the precursor induced by MA were preserved due to the topotactic nature of the conversion of the hemihydrate into the pyrophosphate (the "heredity effect") [43]. According to Horowitz et al., the treatment of a suspension of VOHPO$_4 \cdot$ 0.5H$_2$O in isopropanol in a (not-specified) mill for 24 h improved the maleic anhydride selectivity by about 6% at a butane conversion of 80% [111]. This improvement was attributed to an increased exposure of the catalytically selective face of the (VO)$_2$P$_2$O$_7$ catalyst relative to other crystallographic surfaces, this being due to shear forces provided by the wet milling. A rise in the catalytic selectivity and activity of the vanadyl pyrophosphate catalyst was also described by Zazhigalov et al. [112], independent of the fact, whether the samples were obtained after pretreatments of the raw material V$_2$O$_5$, the precursor, or of the catalyst itself. These authors suggested that such an effect might be due to an enhanced exposure of the (100) planes of the vanadyl pyrophosphate crystals formed by topotactic transformation of the vanadyl planes (001) of the precursor. In contrast, other groups [113–115] attributed the increase in selectivity of maleic anhydride formation to an increased surface area and the generation of fresh and reactive surfaces. Fait et al. [116] studied the influence of grinding the precursor and catalyst in different mills, thus allowing different mechanisms of mechanical exertions on the real structure parameters. Subsequently, a 5% increase was observed in the selectivities of maleic anhydride, though the area-specific rates did not change significantly. According to these authors, the improved selectivity cannot be explained by catalytic anisotropy but rather by induced lattice imperfections connected with a particle size effect. After conditioning, the changes in selectivity and activity were found to remain constant for more than 70 h while on stream.

A further example of a selectivity-enhancing effect of MA was found in the grinding of MoS$_2$ in a vibratory mill, with selectivity in thiophene hydrodesulfurization being markedly changed. As shown by Stevens and Edmonds [117], the composition of the medium in the mill had a major influence on the direction of this change. For example, if the grinding was performed

References see page 581

in heptane, a ratio of the surfaces of basal and edge planes of molybdenum disulfide of ca. 3 : 1 was found, but when ground in air the ratio amounted to 1 : 19. The edge plane was found to be a strong site for thiophene adsorption and for activated hydrogen adsorption, and also appeared to be a strong hydrogenating center. The basal plane of MoS_2 proved to be active for thiophene conversion and was also more selective, producing low levels of thiophene and hydrogen adsorption and low levels of product hydrogenation. The sulfided commercial CoMo–Al_2O_3 catalyst resembled heptane-ground MoS_2, which suggested that the main active site of the CoMo catalyst is the MoS_2 basal plane.

As shown by Trovarelli et al. [118], nanophase carbides obtained by the ball milling of elemental Fe and graphite powders at room temperature were found to be more active and selective in the hydrogenation of CO_2 than Fe/C mixtures and coarse-grained, conventionally synthesized carbides. Depending on the milling duration and the formation of iron carbide, the selectivity towards C_{2+} hydrocarbons varied markedly. The alkene/alkane ratio in C_2–C_4 products decreased, indicating that secondary hydrogenation of alkenes was promoted on these samples, with the product distributions following Anderson–Schulz statistics.

As observed by Molchanov [17], grinding may also lead to a decrease in selectivity compared to the behavior of the untreated material. Thus, Molchanov reported a decrease in the selectivity of butadiene formation by dehydrogenation of n-butenes on a mechanically activated Fe/K catalyst.

In most cases, the observed changes in selectivity were related to changes in the exposure of special crystallographic planes caused by grinding. However, as mentioned above, studies performed under the conditions of varying exertions of pressure and shear showed clearly that induced imperfections of the catalyst crystals may also have a selectivity-directing effect.

2.4.8.6 Pseudocatalytic and Catalytic Reactions During Mechanical Activation

According to Molchanov and Buyanov [13, 15], heterogeneous reactions, which occur exclusively under the permanent influence of mechanical impacts to a solid, can be discriminated either as "pseudocatalytic" or "catalytic".

Pseudocatalytic reactions are characterized by the repeated transformation of chemical reactants with the continuously activated solid. In this case, the catalyst acts as the initiator of the chemical reaction and is consumed in the process. The driving force of these reactions is the permanent generation of active centers by mechanical impact.

In contrast, in *catalytic reactions* the active sites are regenerated as a result of cyclic chemical transformations. With regards to activation of the reactants, it was postulated [13] that the necessary activation energy is supplied only partially by the transformation of mechanical energy into thermal energy, thus enabling the activation of the molecules in the usual manner. Subsequently, a direct transfer of excess energy concentrated in short-living centers formed by mechanical impact to the reactants was proposed as a more probable means of activation, termed "mechanochemical catalysis".

According to the above definition, examples of pseudocatalytic reactions include the oxidation of C, H_2, and CO during the mechanical activation of quartz [119–122]. Polymerization reactions of vinyl monomers occurring under the influence of vibromilling of Al, Fe, and SiO_2 powders in the medium of the respective monomers are also assumed to belong to this type [123]. Radicals or charged centers of freshly formed surfaces are thought to start these polymerization reactions [124].

Catalytic reactions were first observed by Heinicke et al. [119, 125], who studied the formation of NH_3 during the treatment of an industrial catalyst with carborundum particles in either a jet mill [119, 125] or in a ball mill [126] at room temperature. The catalytic reaction [126] was found to take place only during the treatment process. As shown by Lischke and Heinicke, the hydrogenation of benzene on Ni metal under mechanical exertion took place even in the presence of the strong catalyst poison thiophene if corundum particles were added to Ni [127, 128]. Studies at elevated temperature comprised the oxidation of SO_2 on V_2O_5 in a vibratory mill at 350–500 °C [101]. In addition to these reactions, Molchanov et al. [129] highlighted the fact that the pyrolysis of butane on MgO, the hydrogenolysis of butadiene on intermetallides of Mg, and the synthesis of boric esters on zeolites also take place under MA conditions.

2.4.8.7 Conclusions

As shown in this chapter, the mechanochemistry of catalysts and catalysis has developed rapidly during the past few decades. Comprehensive reviews characterizing the state of the art and providing detailed information and access to the original papers are available. Today, mechanochemistry provides a valuable tool, with multiple choice, for the syntheses of catalytic materials and for the activation and selectivation of catalysts. A variety of novel, environmentally friendly approaches to the manufacture of catalysts has been developed, with the MA of known catalysts proving to be most successful in the preparation of more active catalysts with sufficient long-term stability and enhanced selectivity. Notably, the investigation of the latter possibility should open new routes to the development of improved catalysts.

In addition, MA has contributed to a better understanding of catalytic processes by providing deeper insights into the nature of active sites. With further development of the methods of characterization, it is expected that open questions concerning mechanisms at the atomic level will be answered within the next few years. In particular, knowledge of the types, concentration, and distribution of induced imperfections in solids should be expanded in order to gain a more detailed understanding of the relationships between structure and activity/selectivity for mechanically activated solids. This includes the development and application of methods for the investigation of single defects and the separation of amorphous, partly crystalline, and crystalline parts of the treated solids [2].

For the reader's information, current manufacturers of mills for laboratory applications include:

- Fritsch GmbH, Industriestraße 8, D-55743 Idar-Oberstein, Germany, www.fritsch.de
- Retsch GmbH, Rheinische Straße 36, D-42781 Haan, Germany, www.retsch.de
- SPEX CertiPrep, 203 Norcross Avenue, Metuchen, NJ 08840, USA, www.spexcsp.com
- UVR-FIA GmbH, Chemnitzer Straße 40, D-09599 Freiberg, Germany
- Pascall Engineering, Capco Test Equipment Ltd., Wickham Market, Suffolk, IP13 OTA UK, United Kingdom, www.capco.co.uk
- Tema B.V. Chemical & Mineral Process Equipment, Steenplaetsstraat 22–26, 2288 AA Rijswijk, The Netherlands, www.tema.nl
- Herzog Maschinenfabrik GmbH & Co., Auf dem Gehren 1, D-49086 Osnabrück, Germany, www.herzog-maschinenfabrik.de
- Bühler AG, CH-9240 Uzwil, Switzerland, www.buhlergroup.com
- Siebtechnik GmbH, Platanenallee 46, D-45478 Mülheim an der Ruhr, Germany, www.siebtechnik-gmbh.de
- Bico Braum International, 3116 Valhalla Dr, Burbank, CA 91505, USA, www.bicoinc.com

References

1. G. Heinicke, *Tribochemistry*, Akademie-Verlag, Berlin, 1984, 495 pp.
2. V. Boldyrev, K. Tkacova, *J. Mater. Synth. Proc.* **2000**, *8*, 121.
3. W. Ostwald, *Lehrbuch der Allgemeinen Chemie*, Engelmann, Leipzig, 1891, 1163 pp.
4. E. G. Avvakumov, *Mechanicheskie Metody Activazii Chimicheskich Prozessov*, Izd. Nauka, Novosibirsk, 1986, 305 pp.
5. K. Tkacova, *Mechanical Activation of Minerals*, Elsevier, Amsterdam, 1989, 155 pp.
6. P. A. Thießen, K. Meyer, G. Heinicke, *Grundlagen der Tribochemie*, Akademie-Verlag, Berlin, 1967, 194 pp.
7. P. Baláž, *Extractive Metallurgy of Activated Minerals*, Elsevier Science B.V., Amsterdam, 2000, 278 pp.
8. R. Schrader, *Technik* **1969**, *24*, 88.
9. H. Heegn, *Chem. Ing. Tech.* **2001**, *73*, 1529.
10. R. Bormann, in *Mechanical alloying: Fundamental Mechanisms and Applications, Material by Powder Technology*, DGM Informationsgesellschaft, Oberursel, 1993.
11. T. Yokoyama, K. Urayama, M. Naito, M. Kato, T. Yokoyama, *KONA* **1987**, *5*, 59.
12. E. Arzt, L. Schultz, in *New Materials by Mechanical Alloying Techniques*, DGM Informationsgesellschaft, Oberursel, 1989.
13. V. V. Molchanov, R. A. Buyanov, *Russian Chem. Rev.* **2000**, *69*, 435.
14. V. V. Molchanov, R. A. Buyanov, *Kinetics Catal.* **2001**, *42*, 366.
15. R. A. Buyanov, B. P. Solotovskii, V. V. Molchanov, *Izvest. Sibirsk. Otdel. Russ. Akad. Nauk, (Sib. J. Chem.)* **1992**, 5.
16. R. A. Buyanov, V. V. Molchanov, *Khim. Promst.* **1996**, *3*, 7.
17. V. V. Molchanov, *Khim. Promst.* **1992**, 386.
18. V. Boldyrev, V. A. Pavlov, V. A. Poluboyarov, A. V. Duzhkin, *Inorg. Mater.* **1995**, *31*, 1128.
19. P. Yu. Butyagin, *Russ. Chem. Rev.* **1994**, *63*, 1031.
20. P. Yu. Butyagin, *Colloid J.* **1999**, *61*, 581.
21. I. Uhara, S. Yanagimota, K. Tani, G. Adachi, *Nature* **1961**, *39*, 867.
22. H. J. Fecht, *Materials Transactions, JIM* **1995**, *36*, 777.
23. R. Schrader, W. Staedter, H. Oettel, *Z. Phys. Chem.* **1972**, *249*, 87.
24. U. Steinike, K.-P. Hennig, *KONA* **1992**, *10*, 15.
25. H. Schmalzried, *Chemical Kinetics of Solids*, VCH Verlagsgesellschaft, Weinheim, 1995, 433 pp.
26. F. Agullo-Lopez, C. R. A. Catlow, P. D. Townsend, *Point Defects in Materials*, Academic Press, London, 1988.
27. T. Rühle, O. Timpe, N. Pfänder, R. Schlögl, *Angew. Chem.* **2000**, *112*, 4551.
28. G. Natta, I. Pasquon, *Adv. Catal.* **1959**, *11*, 2.
29. U. Steinike, K. Tkacova, *J. Mater. Synth. Proc.* **2000**, *8*, 197.
30. I. Uhara, S. Yanagimota, K. Tani, G. Adachi, S. Teratani, *J. Phys. Chem.* **1962**, *66*, 2691.
31. R. Schrader, W. Staedter, H. Oettel, *Chem. Techn.* **1971**, *23*, 363.
32. S. Kishimoto, *J. Phys. Chem.* **1962**, *66*, 2694.
33. R. Schrader, J. Deren, B. Fritsche, J. Ziolkowski, *Z. Anorg. Allg. Chem.* **1970**, *379*, 25.
34. L. E. Cratty, A. V. Granato, *J. Chem. Phys.* **1957**, *26*, 96.
35. I. Uhara, T. Hikino, Y. Numata, H. Hamada, Y. Kageyama, *J. Phys. Chem.* **1962**, *66*, 1374.
36. I. Uhara, S. Kishimoto, T. Hikino, Y. Kageyama, H. Hamada, Y. Numata, *J. Phys. Chem.* **1963**, *67*, 996.
37. I. Uhara, S. Kishimoto, Y. H. T. Yoshida, *J. Phys. Chem.* **1965**, *69*, 880.
38. E. G. Avvakumov, V. V. Molchanov, R. A. Buyanov, V. V. Boldyrev, *Dokl. Akad. Nauk SSSR* **1989**, *306*, 367.
39. R. Schrader, W. Staedter, *Acta Chim. Acad. Sci. Hung.* **1968**, *55*, 39.
40. V. V. Molchanov, A. A. Stepanov, R. A. Buyanov, I. G. Konstanchuk, V. Boldyrev, V. V. Goidin, SU Patent 1638865, assigned to Institut Kataliza SO AN SSSR and Institut Chimii Tvordovo Tela i Mineralnovo Syrya SO AN SSSR, 1988.
41. R. Schrader, H. Grund, G. Tetzner, *Z. Chem.* **1963**, *3*, 365.
42. C. Bernhardt, H. Heegn, *Silikattechnik* **1978**, *29*, 373.
43. U. Steinike, B. Müller, A. Martin, *Mater. Sci. Forum* **2000**, *321–324*, 1078.

44. G. Mestl, B. Herzog, R. Schlögl, H. Knözinger, *Langmuir* **1995**, *11*, 3027.
45. G. Mestl, N. F. D. Verbruggen, H. Knözinger, *Langmuir* **1995**, *11*, 3035.
46. G. Mestl, T. K. K. Srinivasan, H. Knözinger, *Langmuir* **1995**, *11*, 3795.
47. G. Mestl, N. F. D. Verbruggen, F. C. Lange, B. Tesche, H. Knözinger, *Langmuir* **1996**, 1829.
48. G. Mestl, N. F. D. Verbruggen, E. Bosch, H. Knözinger, *Langmuir* **1996**, *2961*.
49. V. Boldyrev, in *Experimental methods in the mechanochemistry of inorganic solids*, H. Herman (Ed.), *Treatise on materials science and technology*, Vol. 19B, Academic Press Inc., New York, 1983, p. 185.
50. G. F. Hüttig, *Dechema Monographien* **1952**, *21*, 96.
51. P. G. McCormick, T. Zsuzuki, J. S. Robinson, J. S. Ding, *Adv. Mater.* **2001**, *13*, 1008.
52. W. I. Lichtman, P. A. Rehbinder, G. W. Karpenko, *Der Einfluss grenzflächenaktiver Stoffe auf die Deformation von Metallen*, Akademie Verlag, Berlin, 1964, 237 pp.
53. P. Butyagin, *Colloids Surf., A: Physicochemical and Engineering Aspects* **1999**, *160*, 107.
54. K. Husemann, *Chem. Ing. Tech.* **2005**, *77*, 205.
55. H. Rumpf, *Chem. Ing. Tech.* **1959**, *31*, 323.
56. H. Schubert, *Aufbereitung fester mineralischer Rohstoffe*, VEB Deutscher Verlag für Grundstoffindustrie, Leipzig, 1989, p. 77.
57. H. Rumpf, *Chem. Ing. Tech.* **1965**, *37*, 187.
58. E. C. Blanc, *Technologie des appareils de fragmentation et de classement dimensionnel*, Editions Eyrolles, Paris, 1974, Vol. I, 219 pp, Vol. II, 190 pp, Vol. III, 202 pp.
59. C. L. Prasher, *Crushing and Grinding Process Handbook*, John Wiley, Chichester, 1987, 474 pp.
60. K. Höffl, *Zerkleinerungs- und Klassiermaschinen*, VEB Deutscher Verlag f. Grundstoffindustrie, Leipzig, 1985, p. 253.
61. C. Bernhardt, H. Heegn, *Dechema Monographien.* **1976**, *79*, 213.
62. M. Monagheddu, S. Doppiu, C. Deida, G. Cocco, *J. Phys. D: Appl. Phys.* **2003**, *36*, 1917.
63. H. Heegn, *Chem. Ing. Tech.* **1990**, *62*, 458.
64. R. H. Obenauf, D. Nash, J. Martin, R. Bostwick, M. DeStefano, J. Akers, K. Tucker, *Handbook of sample preparation and handling*, SPEX SamplePrep, L.L.C., Metuchen, 2005, 224 pp.
65. K. Tkacova, N. Stevulova, J. Lipka, V. Sepelak, *Powder Techn.* **1995**, *83*, 163.
66. W. Mutter (Fritsch GmbH), Personal Communication, 2005.
67. G. Natta, P. Corradini, G. Allegra, *J. Polymer Sci.* **1961**, *51*, 399.
68. E. G. M. Tornquist, J. T. Richardson, Z. W. Wilchinsky, R. W. Looney, *J. Catal.* **1967**, *8*, 189.
69. W. R. Carradine, H. F. Rase, *J. Appl. Polym. Sci.* **1971**, *15*, 889.
70. Z. W. Wilchinsky, R. W. Looney, E. G. M. Tornquist, *J. Catal.* **2005**, *28*, 351.
71. E. Yu. Ivanov, I. G. Konstanchuk, A. A. Stepanov, V. V. Boldyrev, *Dokl. Akad. Nauk SSSR.* **1986**, *286*, 385.
72. V. V. Molchanov, A. A. Stepanov, I. G. Konstanchuk, R. A. Buyanov, V. Boldyrev, V. V. Goidin, *Dokl. Akad. Nauk SSSR.* **1989**, *305*, 1406.
73. J. W. Ward, *J. Less-Common Met.* **1980**, *73*, 183.
74. V. V. Lunin, A. Z. Khan, *J. Phys. Chem.* **1984**, *25*, 317.
75. I. R. Konenko, E. V. Starodubtseva, E. A. Fedorovskaya, E. I. Klabunovskii, I. S. Sazonova, V. P. Mordovin, *Izv. Akad. Nauk SSSR, Ser. Khim.* **1984**, *25*, 754.
76. V. V. Molchanov, V. V. Goidin, S. M. Kulikov, O. M. Kulikova, I. V. Koshevnikov, R. A. Buyanov, R. I. Maksimovskaya, L. M. Plyasova, O. B. Lapina, RU Patent 2076071, assigned to Institut Kataliza SO RAN, 1994.
77. Y. Ono, *Perspective in Catalysis: A Chemistry for the 21st Century*, Blackwell Science, Oxford, 1992, p. 431.
78. G. M. Maksimov, V. V. Goidin, M. N. Timofeeva, R. I. Maksimovskaya, RU Patent 2080923, assigned to Institut Kataliza SO RAN, 1995.
79. F. Benissard, P. Godella, M. Coulon, L. Bonnetain, *Carbon* **1988**, *26*, 61, 425.
80. V. V. Chesnokov, N. A. Prokudina, R. A. Buyanov, V. V. Molchanov, Russkaya Federaziya Patent 2042425, assigned to Institut Kataliza SO RAN, 1992.
81. V. V. Chesnokov, R. A. Buyanov, V. V. Molchanov, N. A. Prokudina, Russkaya Federaziya Patent 2071932, assigned to Institut Kataliza SO RAN, 1993.
82. V. Boldyrev, G. V. Golubkova, T. F. Grigorjeva, E. Yu. Ivanov, O. T. Kalinina, S. D. Mikhailenko, A. B. Fasman, *Dokl. Akad. Nauk SSSR.* **1987**, *297*, 1181.
83. A. B. Fasman, S. D. Mikhailenko, O. T. Kalinina, E. Yu. Ivanov, T. F. Grigorjeva, V. Boldyrev, G. V. Golubkova, *Izvest. Sibirsk. Otdel. Russ. Akad. Nauk, Ser. Khim. Nauk* **1988**, 83.
84. G. V. Golubkova, E. Yu. Ivanov, T. F. Grigorjeva, *Izvest. Sibirsk. Otdel. Russ. Akad. Nauk* **1990**, 60.
85. R. Schrader, P. Nobst, G. Tetzner, D. Petzold, *Z. Anorg. Allg. Chem.* **1969**, *365*, 255.
86. A. A. Bobyshev, V. B. Kazanskii, I. R. Kibardina, V. A. Radtsig, B. N. Shelimov, *Kinetics Catal.* **1989**, *30*, 1427.
87. A. A. Bobyshev, V. B. Kazanskii, I. R. Kibardina, B. N. Shelimov, *Kinetics Catal.* **1992**, *33*, 363.
88. O. V. Krylov, A. A. Firsova, A. A. Bobyshev, V. A. Radtsig, D. P. Shashkin, L. Ya. Margolis, *Catal. Today* **1992**, *13*, 381.
89. V. Boldyrev, A. S. Kolosov, M. V. Chaikina, E. G. Avvakumov, *Dokl. AN SSSR.* **1977**, *223*, 892.
90. J. Eckell, *Z. Elektrochem. Angew. Phys. Chem.* **1933**, *39*, 433.
91. G. Rienäcker, *Z. Elektrochem. Angew. Phys. Chem.* **1940**, *46*, 369.
92. G. Rienäcker, J. Völter, *Z. Anorg. Allg. Chem.* **1958**, *296*, 210.
93. S. Kishimoto, M. Nishioka, *J. Phys. Chem.* **1972**, *76*, 1907.
94. S. Kishimoto, *J. Phys. Chem.* **1963**, *67*, 1161.
95. R. Schrader, G. Tetzner, H. Grund, *Z. Anorg. Allg. Chem.* **1966**, *342*, 204.
96. R. Schrader, G. Tetzner, H. Grund, *Z. Anorg. Allg. Chem.* **1966**, *342*, 212.
97. R. Schrader, G. Tetzner, H. Grund, *Z. Anorg. Allg. Chem.* **1966**, *343*, 308.
98. G. Tetzner, R. Schrader, *Z. Anorg. Allg. Chem.* **1974**, *407*, 227.
99. G. Tetzner, R. Schrader, *Z. Anorg. Allg. Chem.* **1974**, *409*, 77.
100. G. L. Clark, R. Rowan, *J. Am. Chem. Soc.* **1941**, *63*, 1302.
101. R. Paudert, *Chem. Techn.* **1965**, *17*, 449.
102. R. Paudert, *Monatsber. Deutschen Akad. Wissenschaft.* **1967**, *9*, 719.
103. R. Schrader, G. Jacob, *Chem. Techn.* **1966**, *18*, 414.
104. R. Schrader, W. Vogelsberger, *Z. Anorg. Allg. Chem.* **1969**, *368*, 187.
105. Y. Ogino, T. Kawakami, K. Tsurumi, *Bull. Chem. Soc. Jpn.* **1966**, *39*, 639.
106. Y. Ogino, T. Kawakami, T. Matsuoka, *Bull. Chem. Soc. Jpn.* **1966**, *39*, 359.
107. Y. Ogino, S. Nakajima, *J. Catal.* **1967**, *9*, 251.

108. Y. Saito, M. Ichimura, Y. Ogino, *J. Chem. Soc. Jpn.* **1970**, *73*, 266.
109. P. A. Thießen, G. Heinicke, K. Meyer, *III. Internationaler Kongreß für grenzflächenaktive Stoffe*, 1961, p. 514.
110. V. V. Chesnokov, V. V. Molchanov, E. A. Paukshtis, T. A. Konovalova, *Kinetics Catal.* **1995**, *36*, 759.
111. H. S. Horowitz, C. M. Blackstone, A. W. Sleight, G. Teufer, *Appl. Catal.* **1988**, *38*, 193.
112. V. Zazhigalov, J. Haber, J. Stoch, A. I. Kharlamov, I. V. Bacherikova, L. V. Bogutskaya, in *Alternative methods to prepare and modify vanadium-phosphorus catalysts for selective oxidation of hydrocarbons*, R. K. Grasselli, S. T. Oyama, A. M. Gaffney, J. E. Lyons (Eds.), *Studies in Surface Science and Catalysis*, Vol. 110, Elsevier, Amsterdam, 1997, p. 337.
113. C. B. Hanson, C. R. Harrison, EP Patent 0098065A2, assigned to ICI PLC, 1984.
114. K. Shima, M. Ito, M. Murayama, M. Hatano, *Sci. Technol. Catal.* **1995**, *92*, 355.
115. G. J. Hutchings, R. Higgins, *Appl. Catal. A* **1997**, *154*, 103.
116. M. Fait, B. Kubias, H.-J. Eberle, M. Estenfelder, U. Steinike, M. Schneider, *Catal. Lett.* **2000**, *68*, 13.
117. G. C. Stevens, T. Edmonds, *J. Less-Common Met.* **1977**, *54*, 321.
118. A. Trovarelli, P. Matteazzi, G. Dolcetti, A. Lutman, F. Miani, *Appl. Catal., A* **1993**, *95*, L9.
119. R. Schrader, G. Glock, K. Köhnke, *Z. Chem.* **1969**, *9*, 156.
120. A. V. Bystrikov, A. I. Streletzkii, P. Yu. Butyagin, *Kinet. Catal.* **1980**, *21*, 1148.
121. I. V. Kolbanev, I. V. Berestetskaya, P. Yu. Butyagin, *Kinet. Catal.* **1980**, *21*, 1154.
122. J. I. Jarim-Agajev, P. Yu. Butyagin, *Dokl. Akad. Nauk SSSR.* **1972**, *207*, 892.
123. Y. Tamai, S. Mori, *Z. Anorg. Allg. Chem.* **1981**, *476*, 221.
124. S. Schönner, R. Schrader, K.-H. Steinert, *Z. Anorg. Allg. Chem.* **1977**, *432*, 215.
125. G. Heinicke, K. Meyer, U. Senzky, *Z. Anorg. Allg. Chem.* **1961**, *312*, 180.
126. P. A. Thießen, G. Heinicke, N. Bock, *Z. Chem.* **1974**, *14*, 76.
127. G. Heinicke, I. Lischke, *Z. Chem.* **1963**, *3*, 355.
128. G. Heinicke, I. Lischke, *Z. Chem.* **1971**, *11*, 332.
129. V. V. Molchanov, R. A. Buyanov, V. V. Goidin, *Abstracts of Reports of the 2nd International Conference on Mechanochemistry and Mechanical Activation (INCOME-2)*, Novosibirsk, 1997, p. 125.

2.4.9
Immobilization of Molecular Catalysts

*Reiner Anwander**

2.4.9.1 Introduction and Scope

The beneficial effects of molecular catalyst immobilization (IMC; immobilized molecular catalyst) have been substantiated by numerous scientific reports in the fields of homogeneous, heterogeneous, and bio(enzyme)-catalysis [1, 2]. Such IMCs are also referred to as heterogenized homogeneous catalysts, anchored catalysts, supported catalysts, or hybrid catalysts [3–9]. Conceptually, molecular catalysts are immobilized in order to evoke synergistic effects between homogeneous catalysis, heterogeneous catalysis, and enzyme catalysis (Fig. 1).

IMCs provide access to environmentally benign chemical syntheses due to the ease of product separation and concomitant waste decrease; these prominent "classical" issues also comprise catalyst recovery and reusability, as well as the application of such IMCs in both batch and fluidized-bed processes [10–12]. Moreover, catalyst immobilization facilitates combinatorial library generation via simplification and automation of synthesis and characterization procedures ("high-throughput screening") [12]. Finally, modern catalyst immobilization techniques not only aim at the superior performance and facile reactivity control of molecularly well-defined catalytic sites [13]. The quest for tailor-made catalysts featuring *improved* catalytic performance (enhanced reaction rate, conversion, and selectivity) and/or new reactivity pathways are the ultimate challenge. This can be achieved by emulating and adapting the exceptional catalytic performance of metalloproteins which is imparted by site isolation, coordination confinement [keywords: entatic state (Greek "entasis" = under tension), local environment] and the intriguing protein environment of the active center (keywords: lock-and-key principle, micro- and mesoenvironment) [13, 14]. Importantly, site isolation and limited mobility of the catalytically active moieties counteract bimolecular decomposition pathways/collisional deactivation [13].

Accordingly, the "active" promoter role of the support material is no longer determined exclusively by its chemical composition. The implications of the topology ("architecture") and prefunctionalization of the (hybrid) organic/inorganic host for the catalytic performance are additional delicate issues (keywords: "molecular factories", "nanoreactors", "zeozymes", "mesozymes") [15]. The formation of precise reaction chambers can be accomplished via multicomponent self-assembly or via the use of nanoscopic host materials (e.g., periodically nanoporous silica or dendrimers) [16–18]. Such advanced concepts of catalyst immobilization involve nanostructured catalysts which are supposed to narrow the gap between nature and innovation [19, 20]. Catalyst encapsulation can counteract catalyst deactivation, and simultaneously enhance product selectivity. Furthermore, the molecular precursor route to tailor-made nanostructured host materials according to a soft chemistry approach ("chimie douce") facilitates bioinspired catalysis [21] and advanced methods

* Corresponding author.

References see page 608

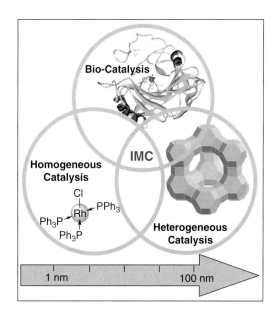

Fig. 1 Immobilized molecular catalysts (IMCs): aiming at synergisms between bio-catalysis (shown: zinc-dependent carbonic anhydrase), homogeneous catalysis (shown: Wilkinson's catalyst), and heterogeneous catalysis (shown: supercage of zeolite Y). The abscissa indicates the dimensions (order of magnitude) in nm units.

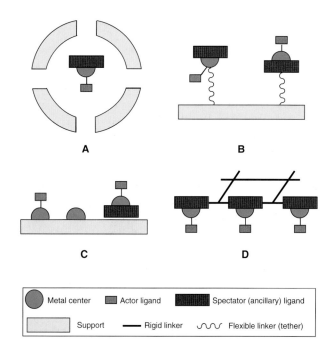

Fig. 2 Methods of metal complex immobilization: **A**, encapsulation; **B**, covalent tethering; **C**, adsorption and grafting; **D**, metal–organic assembly.

of surface functionalization for the fabrication of modular catalysts [22].

By nature, this chapter will provide a rather subjective survey on *metal*-containing (= dependent) IMCs; IMCs with functional organic groups as catalytic sites are not included [23]. The selected examples will be discussed according to the various approaches toward catalyst immobilization, as outlined schematically in Fig. 2. Intrinsic features resulting from the heterogenization are briefly addressed – that is, comparison is drawn to the "corresponding" molecular pendant.

More detailed information about related topics may be found in other chapters of this Handbook: anchoring and grafting of coordination metal complexes onto oxide surfaces (Chapter 2.4.5), the immobilization of metal clusters (Chapters 2.4.4 and 4.2) [24], relevant methods of synthesis including impregnation, ion exchange (Chapter 2.4.2), chemical vapor deposition techniques (Chapter 2.4.6), and routine methods of characterization (Chapter 3). Information about the immobilization of biocatalysts and reactions on immobilized biocatalysts may be found in Chapters 2.4.12 and 16 of this Handbook.

2.4.9.2 Metal Complex–Support Interaction

Functionalized organopolymers (e.g., resins, polystyrene, dendrimers) [19, 25], high-surface-area carbons (e.g., activated carbon, charcoal, carbon nanotubes; see Chapter 2.3.15) [26], and inorganics (e.g., silica, metal oxides such as alumina, magnesia, and titania, as well as metals) [27–29] are the most prominent materials used for the immobilization of molecular catalysts. Primary particle size, morphology, durability, agglomeration, and solubility are important criteria for support classification which qualify such IMCs for industrial applications. However, surface properties such as polarity, hydrophobicity/hydrophilicity, surface area, type and population of functional groups (e.g., polymer: phosphino groups; silica: silanol groups) are "molecular-level" criteria which directly affect the catalytic transformation/catalyst performance in a multifaceted manner. Depending on the surface functionalities/reactivity, the molecular catalyst can interact with the support according to the entire repertoire of chemical bonding [30]; that is: (i) adsorption via weak dipolar and van der Waals interactions; (ii) hydrogen bonding; (iii) ionic (electrostatic) interactions; (iv) covalent attachment (σ-bonding); (v) π-bonding interactions; and (vi) supported-phase immobilization. The "high-surface-area" criterion favors nanoporous solids including zeolites [31, 32], zeotypes [33], ordered mesoporous materials [17, 34] and layered materials (clays) [35] as superior inorganic supports (see also Chapters 2.3.5, 2.3.6, 2.3.7, and 2.3.9). Compared to organopolymers inorganic supports exhibit favorable shape stability and a non-swelling behavior in solvents, and hence are applicable at elevated and low temperatures as well as under pressure [36].

One special class of supports includes high-surface-area periodic nanoporous inorganic materials.

Porous oxidic materials featuring rigidity, thermal stability as well as regular, adjustable, nanosized cage and pore structures are commonly classified as promising catalyst supports. The instrinsic structural and morphological properties of semicrystalline periodic mesoporous silica (PMS) materials have been summarized in several reviews [15–17, 34]. PMSs are unique as they combine the favorable properties of crystalline microporous materials such as structural variety [31], with those of amorphous silica such as a high number of surface silanol groups [37]. Due to their intrinsic zeolite-like pore architecture [38] and thermal stability, channel-like MCM-41 [39], MCM-48 [40], FSM-16 [41], and SBA-15 [42] silica as well as cage-like SBA-1/2/16 [43] provide a unique platform for efficient intrapore chemistry comprising complex immobilization, surface-mediated ligand exchange, and catalytic applications. Surface areas as high as 1500 m^2 g^{-1} and uniformly arranged mesopores with pore volumes as high as 3 cm^3 g^{-1} ensure both a higher guest loading and a more detailed characterization by means of nitrogen adsorption/desorption, powder X-ray diffraction (PXRD), and high-resolution transmission electron microscopy (HRTEM) compared to conventional silica support materials. Pore-confinement of the catalytic center can have a favorable effect with some respects, including: (i) protection of the reactive site from deactivation processes [44]; (ii) stabilization of labile reaction intermediates (encapsulation phenomenon) [45]; and (iii) control of product selectivity (regioselectivity, enantioselectivty) [13] and morphology [15]. More detailed information about PMSs may be found in Chapter 2.3.6 of this Handbook. Molecular catalyst immobilization within the intracrystalline space of traditional microporous materials such as zeolites, zeotypes, and pillared clays is limited to small-sized metal complexes [32–34], thus, excluding the heterogenization of most of the known efficient homogeneous catalysts. Note that even small-sized organometallics such as AlMe$_3$ (ca. 6.15 × 3.60 Å) [46] and Fe(η^5-C$_5$Me$_5$)$_2$ (ca. 5.73 × 7.19 Å) [47] are prone to pore blockage [48].

Figure 3 outlines the immobilization strategies which are not dealt with in the following chapters, demonstrating the diversity of molecular catalyst–support interactions [49]. Surface attachment via hydrogen bonding as shown for Rh and Ru hydrogenation catalysts **1–3** can successfully counteract IMC leaching compared to extremely weak van der Waals interactions [50–53].

Hydrogen-bonding-promoted immobilization can occur via sulfonic/phosphonic acid-modified phosphino ligands [50–52], and was also exploited for the heterogenization of cationic Rh complexes (IMC **3**, trifluoromethanesulfonate (= triflate) counteranion-support interaction) [53] and for Rh(phosphine)–PTA–support hybrid catalysts (PTA = heteropolyacid, e.g., phosphotungstic acid; support = montmorillonite K, carbon, alumina, lanthana; not shown) [54]. IMC **2** is a rare example of a magnetically recoverable chiral catalyst on a nanosized support [55, 56], with the catalyst separation being performed by magnetic decantation [52, 57]; the superparamagnetic magnetite nanoparticles were synthesized by thermal decomposition or by the coprecipitation method.

Examples **4**, **5**, and **6** represent IMCs featuring electrostatic (ionic) interactions of the molecular catalyst precursor with the support (Fig. 3) [58, 59]. Chiral manganese SALEN complexes can be attached to the anionic surface of mesoporous aluminosilicate via various methods of ion exchange. Both, reaction of complex [Mn^{III}(SALEN)][PF$_6$] with Na$^+$-type MCM-41 and reaction of H$_2$SALEN with Mn^{2+}-exchanged MCM-41 followed by oxidation with Cp$_2$FePF$_6$ gave IMC **4**, which exhibited higher enantioselectivities for the epoxidation of styrene as compared with the homogeneous analogues (e.g., $ee_{max}[-80\,°C] = 84$ versus 78%) [60]. The anionic mercaptoethanoate dioxo Mo complex [Mo^{VI}(O$_2$CC(S)Ph$_2$)$_2$]$^{2-}$ was ion-exchanged into a Zn(II)–Al(III) layered double hydroxide (LDH) host affording an extension of the interlayer spacing from 4.2 Å to close to 13 Å (IMC **6**) [61]. Application of IMC **6** in oxygen-transfer reactions proved that intercalation prevents catalyst deactivation via formation of dimeric Mo(V) species. Superacidic sulfonated zirconia (SZR) was used for the cationization of an organometallic zirconocene complex, (C$_5$H$_5$)$_2$Zr^{IV}(CH$_3$)$_2$, for alkene polymerization (IMC **5**) [62].

The concept of supported aqueous-phase organometallic catalysis [63] was successfully applied to supported liquid phase (SLP) catalysis for oxidation and asymmetric hydrogenation [64, 65]. In the latter case (IMC **7**, Fig. 3), the organometallic Ru^{II} catalyst was dissolved in a hydrophilic liquid (ethylene glycol) which forms a thin film on a porous hydrophilic support, thus: (i) providing a large interfacial area between the catalytic phase and the solvent; and (ii) facilitating the separation of the product from the hybrid catalyst [64]. A similar approach is utilized in supported ionic liquid catalysis (SILC) [66–70]. The molecular catalyst is dissolved in an ionic liquid which is confined to the surface either by covalent interactions or hydrogen bonding (IMC **8**, Fig. 3). Catalytic applications comprise Lewis acid (Friedel–Crafts) reactions [66], hydroformylation [68], and enantioselective hydrogenation [70].

References see page 608

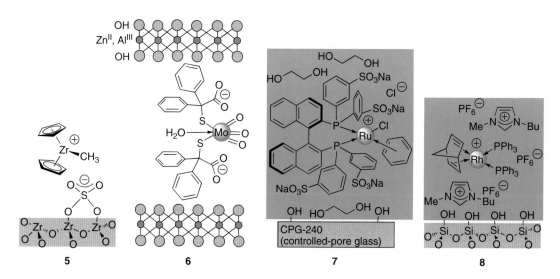

Fig. 3 Non-covalent interactions of homogeneous catalysts with inorganic supports.

More detailed information about metal support interactions and supported liquid catalysts may be found in Chapters 3.2.5.1 and 2.4.11 of this Handbook, respectively.

2.4.9.3 Physically Immobilized ("Entrapped") Metal Complexes

2.4.9.3.1 Intrazeolite Assembly: Ship-in-a-Bottle Catalysts

Molecular catalysts entrapped in well-defined (super)cages of crystalline microporous materials such as zeolites Y (faujasite, $d_{\text{supercage}} = 1.3$ nm), EMT, EMC-2, MCM-22, and VPI-5 are often termed "true IMCs", assuming the absence of any interaction with the support [71–74]. Such intrazeolite assembly is anticipated to retain much of the solution properties of the physically entrapped metal complex, while at the same time aiming at synergistic effects emerging from the intrinsic shape selectivity, charge density, and acid–base properties of the zeolitic host ("nanoreactor"-confined reactivity). DuPont chemist Norman Herron first coined the term "ship-in-a-bottle" to describe Ni(CO)$_{n-4}$L$_n$ complexes encapsulated in zeolite X [75]. Later on, such inorganic-matrix(zeolite)-encapsulated metalorganic complexes were shown to act as viable enzyme mimics (*zeozyme* = *ze*olite and en*zyme*) [76, 77]. Two main synthetic methods are applied to generate metal-containing ship-in-a-bottle species:

- The *flexible ligand method* exploits the conformational flexibility of the free ligand precursor (often a SALEN ligand) which can pass the restricting windows of the larger cages, but after complexation to the *d*-transition metal of the ion-exchanged zeolite it becomes too large to exit.

- The *template synthesis* approach ("ship-in-a-bottle method") involves introduction of the metal via ion-exchange in aqueous solution or preadsorption of labile metal carbonyl or metallocene complexes and subsequent treatment with molecular building blocks such as 1,2-dicyanobenzene or pyrrole/acetaldehyde to produce entrapped metal phthalocyanine (Pc, $d =$ ca. 1.4 nm) and porphyrin (Por) complexes, respectively.

Extensive Soxhlet extractions are performed to purify the zeolite from the uncomplexed ligand and side-products. A third method uses chemically and thermally robust transition metal complexes as a template synthesizing the zeolite around the metal complex ("build-bottle-around-ship"). Depending on the charge of the chelating ligand (e.g., [*bi*pyridine]^0, [bis(oxazoline)]^0, [SALEN]$^{2-}$, [phthalocyanine]$^{2-}$, discrete cationic or neutral metal complexes are formed which are unable to escape from the cavities of the zeolite host matrix [73]. Representative examples include [Co(SALEN)@Y] **9** (oxygen carrier/*hemoglobin* mimic) [76], [Fe^{II}(Pc)@Y@PDMS] **10** (alkane oxidation/*cytochrome P-450* mimic; PDMS = polydimethylsiloxane = mimic of *phospholipid membrane*) [78], [*cis*-Mn^{II}(BIPY)$_2^{2+}$@Y] **11** (selective alkene oxidation) [79], [(TiO$_2$)$_{nano}$@Fe^{II}(BIPY)$_3^{2+}$@Y] **12** [80], [Mn^{II}(SALEN)@EMT] **13** (enantioselective epoxidation of aromatic alkenes, $ee_{max} = 88\%$; Fig. 4) [81], [Pd^{II}(SALEN)@Y] **14** (selective hydrogenation) [82], and [Cu{bis(oxazoline)}$^{n+}$@Y] **15** (asymmetric aziridation of styrene, $ee_{max} = 95.2\%$; Fig. 4) [83]. A striking feature of the latter IMC was much higher *ee*-values as compared with the homogeneous catalyst Cu(OTf)$_2$/{bis(oxazoline)}, even when sterically less demanding bis(oxazolines) were used [83].

The steric constraints imparted by the zeolite cage [Y versus EMT versus MCM-22 (super cage = 0.71 × 1.82 nm) versus highly dealuminated Y] were also crucial for achieving high *ee*-values in the epoxidation of alkenes with zeolite-immobilized Mn^{III}–SALEN complexes (Jacobsen–Katsuki catalyst) [81, 84–86]. Here, a precise "side-on" approach of the alkene to the manganese center is a prerequisite for an efficient asymmetric induction. [Pd^{II}(SALEN)@Y] **14** not only was more active than the molecular counterpart (2 mol.% versus 0.3 mol.%, after 24 h, 400 torr H$_2$, 25 °C) but also displayed shape selectivity in hex-1-ene/cyclohexene hydrogenation, preferentially converting the linear alkene [82]. TiO$_2$ nanoclusters act as an electron relay for zeolite Y-encapsulated Fe^{II}(BIPY)$_3^{2+}$ **12**, as evidenced by the enhanced photocatalytic activity of the intrazeolite iron guest in the test system 2,4-xylidine/H$_2$O$_2$ [80]. The increased catalytic activity and selectivity of [Ru^{II}(F$_{18}$Pc)@NaX] **16**, featuring a chemically very robust fluorinated phthalocyanine ligand, in the oxidation of cyclohexane to cyclohexanone was ascribed to site isolation of the intrazeolite guest [turnover number (TON): 3000 versus 250, after 1 day, 25 °C; deactivation: >20 000 (ca. 70% conversion); cyclohexane/cyclododecane: 2933 versus 295 TON day^{-1}] [87]. IMC **16** was synthesized according to the "build-bottle-around-ship" method [88].

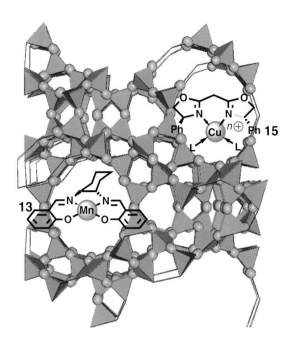

Fig. 4 Enantioselective copper and manganese ship-in-the-bottle catalysts.

More detailed information about ship-in-a-bottle catalysts may be found in Chapter 2.4.10 of this Handbook.

2.4.9.3.2 Entrapment in Sol–Gel and Polysiloxane Matrices
Similar to the zeolite-based "build-bottle-around-ship" concept (see Section 2.4.9.3.1), various homogeneous catalysts have been encapsulated in less-ordered inorganic networks according to the sol–gel process [89, 90]. In addition to phosphino complexes such as Ru^{II}Cl$_2$(PPh$_3$)$_3$, Rh^{I}Cl(PPh$_3$)$_3$, and Ir^{I}Cl(CO)(PPh$_3$)$_3$, quaternary ammonium ion pair variants have been occluded (Scheme 1: IMCs **17** and **18**) and used as extremely stable, efficient and recyclable catalysts in alkene isomerization, hydrogenation, and hydroformylation [90–92].

The catalyst performance was markedly affected by the nature of the matrix, which varied according to different condensation conditions (pH value, silica

References see page 608

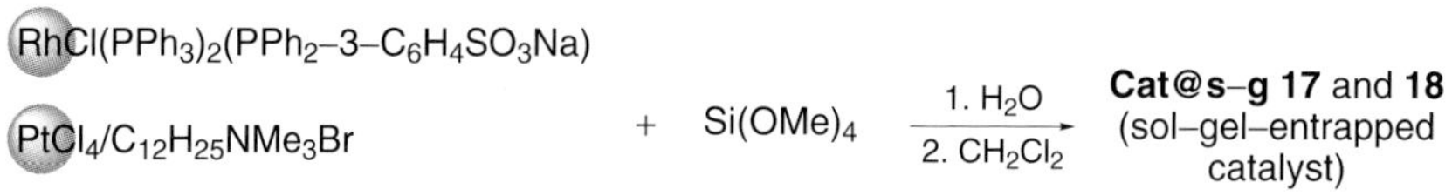

Scheme 1 Sol–gel occlusion of sulfonated and quaternary ammonium-functionalized transition metal ion pairs.

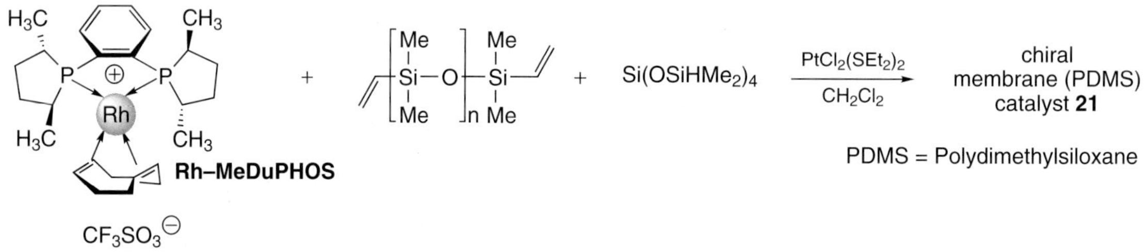

Scheme 2 Heterogenization of a chiral rhodium complex into a PMDS membrane (approximate thickness: 1000 μm).

source ($Si(OMe)_4$ versus $Si(OEt)_4$)). This sol–gel entrapment methodology can also be successfully applied for one-pot reaction sequences with "opposing reagents" mutually destroying each other [93–95]. While the enzyme/catalyst pair lipase@s–g/Rh^{I}Cl(PPh$_3$)$_3$@s–g **19** was used for one-pot esterification and C–C double bond hydrogenation reactions (substrates: 10-undecenoic acid, 1-pentanol) [94], the base/acid pair H$_2$N(CH$_2$)$_2$ NH(CH$_2$)$_3$@s–g/H$_4$SiMo$_{12}$O$_{40}$@s–g **20** simultaneously catalyzed the dehydrohalogenation of 2-iodoethylbenzene to styrene (base catalysis) and the alkylation of anisole with styrene (acid catalysis) [95].

Solvent evaporation from a mixture of PDMS solution and chiral catalyst component produced chiral catalytic membranes (e.g., IMC **21**, Scheme 2) [96–98]. The PDMS solution was prepared from vinyldimethyl-terminated silicone polymers and silane crosslinkers, such as $Si(OSiHMe_2)_4$. The examined chiral molecular catalysts include Jacobsen–Katsukis catalysts, N,N'-bis(3,5-di-*tert*-butylsalicylidene)-1,2-cyclohexanediaminemanganese chloride (for asymmetric alkene epoxidation) [96] and Noyoris Ru^{II}–BINAP complex [2,2′-bis(diphenylphosphanyl)-1,1′-binaphthyl](*p*-cymol)ruthenium chloride [97] and Rh^I–MeDuPHOS (for asymmetric hydrogenation of methyl acetylacetate) [98].

2.4.9.3.3 Microencapsulation and Polymer-Incarcerated Methods

Microencapsulated catalysts (MCs) are a special type of organopolymer-supported homogeneous catalysts which enable environmentally benign and powerful high-throughput organic syntheses [99]. These IMCs are synthesized by adding the metal precursor as a solid core to a polymer solution (solvent: e.g., cyclohexane, THF). Upon cooling this mixture slowly to 0 °C, the dispersed metal complex is physically enveloped by the polymers (coaservates) and phase separation occurs. At this stage, the MC–IMC are physically soft and the capsule walls can be hardened by the addition of hexane. Depending on the presence of copolymer functional groups (e.g., epoxide), the microcapsule can be crosslinked according to the polymer-incarcerated (PI) method (Scheme 3) [99].

The formation of MC–IMC is particularly feasible for polystyrene-based polymer backbones (M_w = 300 000; polystyrene ≫ polydienes ≫ polyethene), and it was proposed that the microcapsules are stabilized by interactions of the metal center with the phenyl groups of the polystyrene. According to scanning electron microscopy (SEM) and metal energy dispersive X-ray (EDX) maps, the metal complexes are located all over the surface of the microcapsules. The MC method was initially developed for scandium triflate and the resulting MC–Sc^{III}(OTf)$_3$ **22** successfully used in several fundamental Lewis acid-catalyzed carbon–carbon bond-forming reactions (e.g., aldol, aza Diels–Alder, Mannich-type and Strecker reactions) [100]. Other examples comprise MC–Os^{VIII}O$_4$ **23** (asymmetric dihydroxylation of alkenes) [101], MC–Pd^0(PPh$_3$) **24** (Suzuki coupling) [102], MC–Ru^{II}Cl$_2$(PR$_3$) **25** (R = Ph, cyclohexyl; ring-closing alkene metathesis) [103], MC(PI)–Pd^0 **26** (hydrogenation, allylation) [104], and MC(PI)–Ru^0 **27** (oxidations of alcohols to aldehydes and ketones) [105]. The latter ruthenium catalyst was more active than the original non-immobilized RuCl$_2$(PPh$_3$)$_3$ ((*R*)-3-[(*tert*-butyldiphenylsilyl)oxy]-2-methylpropanal as a substrate: 84% (97% *ee*) versus 77% (81% *ee*), after 2 h, 30 °C) and was applied in a flow system revealing excellent conversions and yields as well as no leaching of Ru [105].

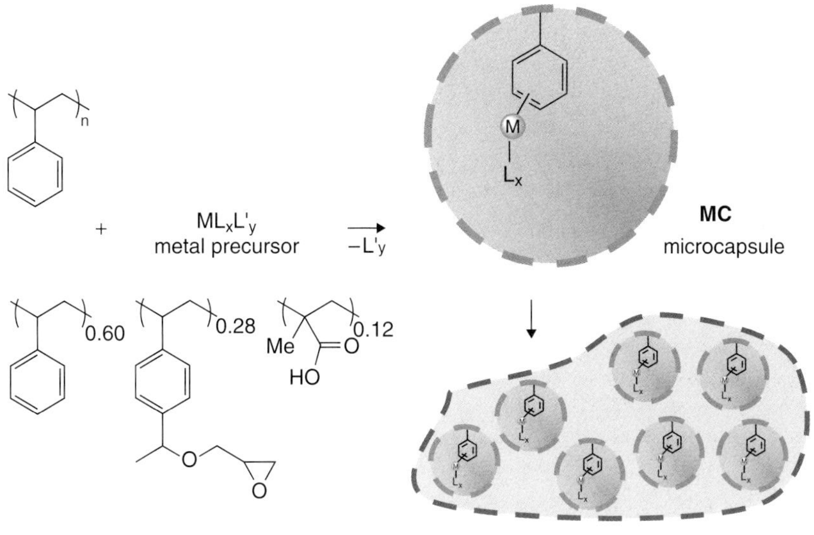

Scheme 3 The "polymer-incarcerated method" (PI method; according to reference [99]).

2.4.9.3.4 Supramolecular Complexation: Mechanical Immobilization
A supramolecular coordination chemistry approach [106, 107] gave access to molecular square encapsulated catalysts (Fig. 5). A pyridine-functionalized manganese porphyrin complex was placed into the cavity of a large preorganized porphyrin-derived Re_4Zn_4 octametallic ring structure, affording IMC **28** [108].

Such directed assembly originates from the interaction of the peripheral Lewis bases with the Lewis acidic Zn^{II} receptor sites. By analogy to protein superstructures, this catalyst encapsulation concept imparted both stability and reaction selectivity in the epoxidation of alkenes (*artificial enzyme*). For example, the functional stability of the Mn complex precursor (against bimolecular degradation pathways such as formation of oxo-bridged Mn−O−Mn dimers) could be enhanced 10- to 100-fold through supramolecular complexation. Moreover, substrate selectivity could be improved by half-cavity-tuning (empty half-cavity: ca. 9 × 18 Å), as shown for the addition of 3,5-dinicotinic acid dineomenthyl ester (Fig. 5) [108]. Similarly, the catalytic activity (3-fold) and catalyst stability (>20-fold) of a pyridine-functionalized Jacobsen–Katsuki Mn^{III}(SALEN)a catalyst system could be increased in the presence of the bulky Lewis acid zinc tetraphenylporphyrin (IMC **29**; epoxidation of 2,2-dimethylchromene: TON 181 versus 63, after 3 h, 25 °C) [109].

2.4.9.4 Tethered Metal Complexes
The covalent attachment of molecular catalysts to organic or inorganic support materials via a linker ("tether") is certainly the most popular and extensively studied concept of catalyst immobilization [10]. Particularly in fine chemical synthesis (asymmetric catalysis) and polymerization catalysis, numerous reports witness on this prolific field of catalysis research [110–113]. While the early studies on molecular catalyst immobilization of phosphino-functionalized polymer resins were aimed at preserving the efficiency of the molecular catalyst (e.g., $RhCl(PPh_3)_3$) [114–117], current topics of covalent IMCs include synergistic effects gained via the support topology (nanoporous materials) and bi(multi)functional surfaces [13, 118]. This section will provide selected examples of the latter topics, simultaneously paying tribute to different synthesis strategies (Fig. 6: *sequential* versus *convergent* approach) [118] and privileged chiral catalysts [119].

Further information concerning anchored chiral complexes (see Chapter 14.16.2) and oxidations on immobilized molecular catalysts (see Chapter 14.17.3) may be found elsewhere in this Handbook.

2.4.9.4.1 Multifunctional Surfaces
The TCSM catalyst **30** (TCSM = *t*ethered *c*omplex on a *s*upported *m*etal) was obtained by immobilizing an alkoxy-functionalized rhodium complex on a silica-supported palladium catalyst Pd−SiO₂, as illustrated in Scheme 4 (convergent approach) [120].

The heterobimetallic catalysts showed an intrinsic catalytic activity in the hydrogenation of aromatic substrates which was substantially higher than that of the tethered complex [toluene-to-methylcyclohexane: turnover frequency (TOF): 5.5 versus 0.7 mol H_2 mol_{Rh}^{-1} min^{-1}; TON: 2420 versus 143, after 8.5 h, 40 °C] or the supported metal separately. This was ascribed to a "metal–metal cooperativity" effect.

References see page 608

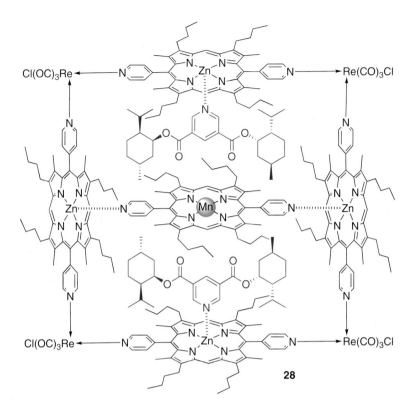

Fig. 5 Supramolecular embedding of a manganese-based epoxidation catalyst and cavity fine-tuning via addition of 3,5-dinicotinic acid dineomenthyl ester (purple).

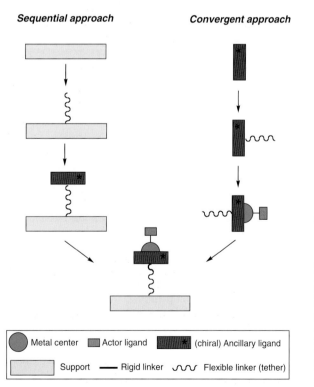

Fig. 6 Anchoring strategies for chiral metal complexes: *sequential* versus *convergent*.

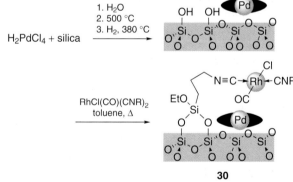

Scheme 4 Synthesis of a TCSM catalyst.

2.4.9.4.2 Spatial Restrictions Imparted by Periodic Mesoporous Silica A sequential approach was used to immobilize a chiral Rh^I catalyst derived from (R, R)-diphenylethylenediamine (DPEN) on the inner walls of a pre-functionalized MCM-41 silica (Fig. 7, upper left) [121]. IMC **31** exerted a significant increase in enantioselectivity for the asymmetric hydrogenation of phenylcinnamic acid relative to the homogeneous counterpart and to a closely related non-porous carbosil-supported catalyst (93% versus 81% versus 79% *ee*). This distinct catalytic behavior

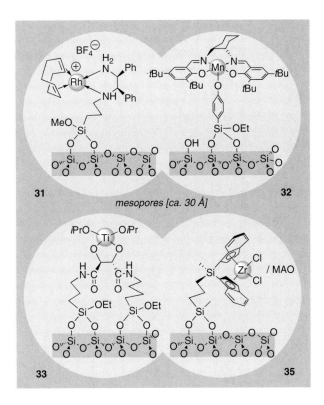

Fig. 7 Mesoporous silica-confined tethered IMCs: implications for stereoselectivity.

was ascribed to a chiral confinement of the catalytic center on the concave silica surface featuring a deliberately restricted spatial freedom in its vicinity [122].

Enhanced enantiomeric excess relative to the "free" molecular catalyst has also been observed in the asymmetric epoxidation of α-methylstyrene catalyzed by anchored Mn^{III}SALEN complex IMC **32** (Fig. 7, upper right; 72 versus 56% ee) [123]. Covalent attachment of the catalytic metal center to the spacer via a Mn–O linkage (sequential approach) ensured stability and reusability. A heterogenized variant of the Sharpless–Katsuki chiral catalyst system (Ti(OiPr)$_4^+$ chiral dialkyl tartrate) for the asymmetric epoxidation of allylic alcohols with *tert*-butyl hydroperoxide was obtained according to the sequential approach [124]. When a commercially available silica ($a_s = 375$ m^2 g^{-1}) was used as a support, the corresponding organic–inorganic chiral hybrid catalyst **33** (Fig. 7, lower left) displayed similar TONs (14 versus 15, after 48 h, 0 °C) and *ee*-values (86 versus 83%) as the non-immobilized catalyst. The slightly lower activity of a MCM-41-based hybrid catalysts (TON 11, after 48 h; 84% ee) was ascribed to hindered diffusion of the substrate and product molecules within the mesopores. A silica-tethered Ti^{IV}–TADDOLate complex IMC **34** obtained by a sequential approach was employed as another privileged chiral complex in the Lewis acid-promoted enantioselective addition of ZnEt$_2$ to benzaldehyde ($ee_{max} = 98$%; no influence of the pore size (20 versus 50 nm) of the CPG (= controlled pore glass) support; TADDOL = $\alpha,\alpha,\alpha',\alpha'$-tetraaryl-1,3-dioxalane-4,5-dimethanol) [125].

Mesoporous silicas MCM-41, SBA-15, and MCM-48 were used as supports for tethering Brintzinger-type bis(indenyl)zirconocene complexes according to a convergent approach [126–128]. When activated with methylalumoxane (MAO), IMC **35** (Fig. 7, lower right) produced polyethylenes and polypropylenes with very high molecular weights, low polydispersities and, in the case of polypropylene, higher levels of isotacticity were obtainable than with the molecular counterpart. The lower activities were attributed to deactivation of metal sites upon reaction with surface silanol groups, and alternatively to pore-blocking (ethene polymerization: 234.00 (MCM-48) versus 603.72 kg polyethylene (PE) mol$_{Zr}^{-1}$ h^{-1} bar^{-1}; propene polymerization: 4.93 (MCM-41) versus 567.30 kg PP mol$_{Zr}^{-1}$ h^{-1} bar^{-1}).

2.4.9.4.3 Site Isolation via Spatial Patterning of the Silica Surface

Surface silylation not only is an efficient tool for the passivation ("deactivation") of the outer surface area ("exterior walls") of mesoporous silica and endcapping of surface OH groups [129]. An eight-step sequential approach involving surface prefunctionalization with a bulky alkoxysilane patterning molecule and hexamethyldisilazane-based endcapping was applied for the synthesis of a single-site ethene polymerization catalyst on mesoporous silica SBA-15 (d_p = ca. 50 Å; Fig. 8) [130]. When activated with tris(pentafluorophenyl)borane (BARF) and tri*iso*butylaluminum (TIBA), the patterned constrained geometry Ti^{IV} complex **36** was shown to be more active than covalently tethered catalysts prepared via traditional techniques, even when compared to the

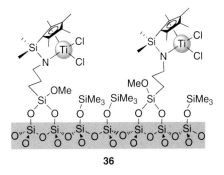

Fig. 8 Constrained geometry complex, immobilized on a mesoporous silica SBA-15 via a patterning methodology.

References see page 608

non-anchored catalyst (ethene polymerization: 28.7 (patterned) versus 5.1 (traditional) versus 19.8 (homogeneous) kg PE mol$_{Ti}^{-1}$ h^{-1}(60 psi)$^{-1}$, 25 °C). More efficient assembly of the surface species, uniform and "more" isolated surface sites, ease of activation and monomer incorporation, and disfavored surface–metal interactions were proposed as advantages of such a patterning protocol.

2.4.9.5 Grafted Metal Complexes: Surface Organometallic Chemistry (SOMC)

Surface organometallic chemistry (SOMC) is the immobilization ("grafting", "chemisorption") of molecularly well-defined organometallic compounds (e.g., molecular catalyst precursors) on chemically and thermally robust condensed solid materials [131–133]. SOMC – this term was coined by Jean-Marié Basset [134] – is performed predominantly on silica and alumina materials, and represents a unique approach to the generation of novel hybrid materials that are relevant to heterogeneous catalysis. SOMC changes the composition of the organometallic complex by single or multiple ligand displacement, with the surface acting as a new conformationally rigid non-innocent (multifunctional) ligand. SOMC produces catalytic surface sites which are easily accessible and the distribution of which is easily controllable over a wide range.

Such vapor- or solution-phase-grafting reactions are routinely performed under anaerobic conditions on highly dehydrated support materials in order to avoid deactivation and hydrolysis and hence molecular complex agglomeration (e.g., by the formation of metal hydroxide). The spacing and dispersion of the metal centers can be controlled by: (i) the ligand size (steric bulk) of the molecular complex [135, 136]; (ii) thermal pretreatment of the oxidic support [132]; (iii) prefunctionalization (alkylation, silylation) of the oxidic support [130]; (iv) the morphology (concave/convex surface) of the oxidic support [15]; and (v) the synthesis protocol of the support (e.g., porous material: template/surfactant removal via calcination or extraction) [15]. For silica, the various types of silanol groups (isolated, vicinal, geminal, and hydrogen-bridged moieties) can be thermally transformed into predominantly two "reactive sites": isolated silanol groups and strained siloxane bridges [37, 132]. The ratio of the latter depends on the dehydration conditions (temperature/vacuum; silica gel: ca. 2.6 SiOH nm^{-2} (200 °C), 1.2 (500), 0.7 ± 0.2 (700)) [132j].

SOMC can produce IMCs with unforeseen reactivity patterns. Enhanced reactivity can originate from sterically unsaturated metal centers featuring a highly distorted coordination environment, and from a strongly electron-withdrawing effect of the support (type of M−O−X(support) linkage/connectivity) [15]. Moreover,

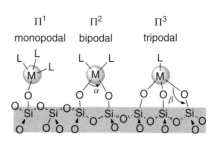

Fig. 9 Classification of surface bonding by podality Π^n. Angles α and β indicate the extent of surface confinement.

surface confinement facilitates the stabilization of unusual oxidation states and a "small-ligand" chemistry of the metal centers. Numerically, the surface confinement of a metal center can be explained by its podality Π^n; that is, the number of covalent M−O(support) bonds, and the angles O−M−O (α_n) and M−O−X(support) (β_n) (Fig. 9) [20]. While Π^n and the M−O(support) distances can in theory be determined by quantification of the released protonated ligand and X-ray absorption spectroscopy (EXAFS, extended X-ray absorption fine structure; XANES, X-ray absorption near-edge structure), respectively, the two sets of angles remain unknown. Π^n depends on the silanol population which can be controlled by thermal pretreatment (i.e., dehydration/dehydroxylation) of the support material and its surface curvature. The intrinsic monodisperse pore structure of PMS materials allows for the variation of the inner surface curvature – that is, the concavity of the pore walls, via the pore diameter. The local environment and, hence, the reactivity of the metal center can be further fine-tuned by the stereoelectronic properties and asymmetry of the remaining (chiral) ligand L [15].

The extent of surface reaction/ligand displacement is also determined by the reactivity of the molecular complex – that is, that of the M−R bond of a homoleptic MR$_x$ or heteroleptic MR$_x$L$_y$ complex. Assuming a pK_a value of ca. 5–7 of the SiOH groups of a dehydrated silica surface [137], it is not only real organometallics such as alkyl, allyl, cyclopentadienyl, and carbonyl derivatives that can be employed for SOMC. Also, *pseudo*-organometallics such as amide and alkoxide complexes that contain no direct M−C linkage but otherwise readily protolyzable M−X bonds can undergo surface reactions. Accordingly, the basicity of the metal-bonded ligand (i.e., the pK_a value of the protonated ligand) gives a reasonable measure of the surface reactivity (Fig. 10). Furthermore, the non-displaced ligands can either be exploited as "probe ligands" for spectroscopic investigations or subjected to secondary ligand exchange reactions [15].

Detailed information relating to metal (carbonyl) clusters in zeolites may be found in Chapters 2.4.4 and 4.2

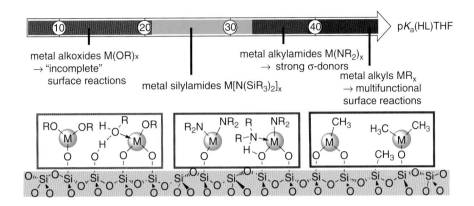

Fig. 10 The pKa criterion as a measure for surface reactivity.

of this Handbook. Further information about elementary steps and mechanisms of grafting and chemisorption may be found in Chapters 2.4.5 and 5.1, respectively.

2.4.9.5.1 Alkyl-Based SOMC Alkyl complexes MR_x of oxophilic and electrophilic main group and early transitions metals often violently (over)react with all of the surface silanol sites and even strained siloxane bridges according to a multifunctional surface reaction (Fig. 10, lower right; e.g., $AlMe_3$@silica) [138]. Equally, surface-bonded alkyl species display highly reactive surface sites which are exploited for many catalytic reactions. Prominent examples include organozirconium and organochromium-modified silica/alumina materials (see Fig. 11). The Union Carbide catalyst derived from Cp_2Cr and silica (IMC 37) [139, 140] and silica-supported Ziegler–Natta systems such as $[(C_9H_7)_2C_2H_4]Zr^{IV}Cl_2$/MAO (IMC 38, MAO = methylalumoxane) [141] are industrially applied catalysts for the synthesis of high-density polyethylene (HDPE) and stereospecific polypropylene, respectively.

Monopodally silica-anchored rhodium allyl species $(\equiv Si-O)Rh^{III}(C_3H_5)_2$ 39 [142, 143] undergo hydrogenolysis ("secondary ligand exchange") to form $(\equiv Si-O)Rh^{III}(C_3H_5)H$ which catalyzes the interaction of ethane with D_2 via sequential H/D exchange and formation of deuterated ethanes rich in ethane-d_1 [142c]. Activation of methane was observed with $(\equiv Si-O)Rh^{III}ClH$ 40 (chlorination of methane) [132a].

Hydrogenolysis of monopodally grafted $(\equiv Si-O)Zr^{IV}(CH_2tBu)_3$ alkyl species and concomitant surface-mediated "ligand redistribution" afforded a tripodal zirconium hydride surface species and surface silanes (Fig. 11, IMC 41) [144–146]. The sterically highly unsaturated $(\equiv Si-O)_3Zr^{IV}H$ surface species catalyze alkene hydrogenation, isomerization, and polymerization. Such catalytic cleavage of C–H and C–C bonds of alkanes was also exploited for novel, unprecedented reaction pathways such as the depolymerization (retro-polycondensation) of polyalkenes polyethylene (PE) and polypropylene (PP) at 150 °C by the action of H_2 [144].

Low-valent tantalum (IMC 42) and tungsten (IMC 43) hydride surface species are tailor-made alkane methathesis catalysts, obtainable from mixed alkylidene/alkyl precursors such as IMC 44 [147–151]. The catalytic conversion of propane depends not only on the metal center ($Ta_{@silica} > W_{@silica}$: TON = 1.2 mol versus 6.1 propane per mol metal after 120 h, 150 °C) but also on the support ($W_{@alumina} > Ta_{@silica}$: TON = 18 versus 8.2 propane per mol metal after 120 h) [149]. Linear alkanes are favored over branched alkanes, and the selectivities for higher homologues were $C_4 > C_5 \gg C_6$; the selectivity for methane was higher for the Ta-based catalyst (ca. 10% versus 3%). These studies corroborate that alkane metathesis proceeds via carbene and metallacyclobutane intermediates [149]. Moreover, stoichiometric alkane cross-metathesis was observed with the hydrocarbyl ligands of the surface complexes $(\equiv Si-O)_xTa^V(=CHtBu)(CH_2tBu)_{(3-x)}$ 45 [152]. The monopodal molybdenum alkylidene surface complex $(\equiv Si-O)Mo^{VI}(=NAr)(=CHtBu)(CH_2tBu)$ 47 displayed a four-fold higher activity in propene (alkene) metathesis than Re^{VII} surface species IMC 46 (TOF: 1.0 versus 0.25 mol $mol_{Mo/Re}^{-1}$ s^{-1}, 25 °C) [153, 154]. Furthermore, the IMC had a longer lifetime under catalytic conditions than the molecular model complex $([(c-C_5H_9)_7Si_7O_{12}SiO]Mo^{IV}(=NAr)(=CHtBu)(CH_2tBu))$, which was attributed to active-site isolation preventing deactivation for example, via the dimerization of reactive intermediates.

Given the enormous commercial interest, comprehensive studies exist for the generation of supported organometallic α-alkene polymerization catalysts on various oxidic supports (see Chapter 15 of this Handbook) [155]. In general, three strategies provide active

References see page 608

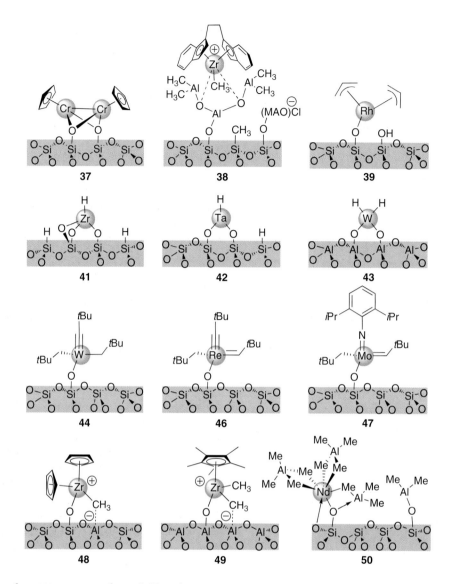

Fig. 11 Reactive surface sites originating from alkyl-based SOMC.

species. Method I involves the interaction of metal alkyl complexes (metal = Zr, Ac) with Lewis-acidic supports such as alumosilicates, alumina, and sulfonated zirconia which implies metal cation formation ("cationization") and formation of the active catalyst without any addition of cocatalysts (IMCs **6**, **48**, and **49**) [156–158]. According to this methodology, grafted halfzirconocene complex, IMC **49**, displayed much higher activity than the zirconocene congeners **48** in ethylene polymerization, with alumina supports giving the best results (ethene polymerization: 20 versus 3 kg PE mol_{Zr}^{-1} h^{-1} atm^{-1}, 50 °C) [157]. Alternatively, pretreatment/derivatization of the support with organoaluminum cocatalysts (e.g., MAO (IMC **38**), AlMe$_3$) or borate species (e.g., formation of an anilinium borato surface complex [($\equiv$Si—O)B(C$_6$F$_5$)$_3$)$^-$ (HNMe$_2$Ph)$^+$] [159]) and subsequent immobilization of the transition metal complex also leads to cationization (= activation; method II). Finally, the reverse procedure – consecutive immobilization of the "neutral" zirconium complex (e.g., on a non-functionalized silica surface) and of cocatalyst (e.g., MAO or B(C$_6$F$_5$)$_3$) – represents method III [155].

Surface organometallic chemistry on periodic mesoporous silica, SOMC@PMS, was exploited for the nanofabrication of polymeric materials [160]. A functionalized mesoporous silica film (MSF) material prepared by coating a mica plate with a PMS layer ($P6_3$/mmc structure with the c-axis perpendicular to the film), was sequentially grafted with titanocene Cp$_2$Ti^{IV}Cl$_2$ and MAO (method III), and shown to act as a nanoextruder for crystalline nanofibers of linear PE with an ultrahigh-molecular weight (6 200 000) and a diameter of 30 to 50 nm (IMC **51**,

Fig. 12) [161]. It was proven by small-angle X-ray scattering (SAXS) and differential scanning calorimetry (DSC) that the mesoscopic pores in MSF serve as a template to produce extended-chain crystals of PE. Formation of the kinetically favored chain folding process, typical of ordinary HDPE, is suppressed as the pore diameter (ca. 30 Å) is almost one-order of magnitude smaller than the lamellar thickness (270 Å) of the folded chain crystals of ordinary PE.

Such extrusion polymerization mechanistically mimics the biosynthesis of cellulose.

Heterogeneous single-site alkene polymerization catalyst [rac-C_2H_4(Ind)$_2$]Zr^{IV}Cl$_2$@MAO@MCM-41 **52** (method II) (Scheme 5) was also reported to affect the polymer morphology [162–164]. Propene extrusion polymerization produced spherulite PP particles of approximate diameter 10 μm, featuring a distinct core and shell arrangement [163]. Although displaying one-quarter the activity (ca. 1400 kg PP mol$_{Zr}^{-1}$ atm^{-1} h^{-1}), these catalysts produced a four-fold increase in the number average molecular mass of PP (similar polydispersity 1.9) and a higher isotacticity PP (83%) relative to the homogeneous system (78%). Better stereocontrol was associated with a lower monomer concentration at the active sites disfavoring 1,3-insertion, and could be even enhanced by utilization of the extra large pore zeotype VPI-5 as a support material [162].

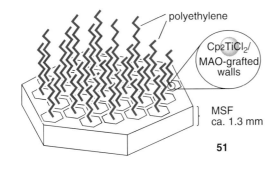

Fig. 12 Nanoextrusion of crystalline nanofibers of linear polyethylene promoted by SOMC@PMS.

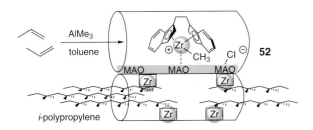

Scheme 5 Nanofabrication of spherulite polypropylene particles (d_P = 10 mm) promoted by SOMC@PMS.

In a preliminary study, cubic MCM-48 featuring a three-dimensional mesopore system was applied to generate a heterogeneous rare-earth metal-based diene polymerization initiator [165]. Two synthesis protocols were described, Nd^{III}(AlMe$_4$)$_3$@Et$_2$AlCl@MCM-48 (method II) and Et$_2$AlCl@Nd^{III}(AlMe$_4$)$_3$@MCM-48 (method III, IMC **50**, Fig. 11; before activation with Et$_2$AlCl), which seemed to have minor implications for the catalyst performance. The proposed surface species were modeled with hexane-soluble lanthanide alkyl and siloxide complexes [165]. All of the neodymium-*grafted* materials performed as efficient single-component catalysts in the slurry polymerization of isoprene. Polymer analysis revealed high-*cis*-1,4-stereospecificities (>99% *cis*), with higher molecular weights ($M_w \approx 1 \times 10^6$ g mol^{-1}) and significantly smaller polydispersities (PDI) relative to the homogeneous binary Nd^{III}(AlMe$_4$)$_3$/Et$_2$AlCl catalysts (PDI = 1.33–1.88 versus 2.78–3.45). The narrow molecular weight distributions were attributed to the absence of any organoaluminum cocatalyst dissociation/reassociation processes at the heterogenized active neodymium centers.

2.4.9.5.2 Alkoxide-Based SOMC Metal alkoxide complexes [166], particularly homoleptic derivatives M(OR)$_x$ (Al^{III}(OiPr)$_3$, Nd^{III}(OCtBu$_3$)$_3$, Ti^{IV}(OiPr)$_4$, Zr^{IV}(OnPr)$_4$, Mo$_2^V$(OEt)$_{10}$, W^V(OEt)$_5$), have been widely used to tailor silica materials for catalytic applications, particularly oxidation catalysis [167, 168], often via thermal degradation and oxo-layer formation. Depending on their substituents R, the pK_a value of alcohols HOR can range from 5 to 20 (see Fig. 10). Although being the most inexpensive and easy-to-handle precursors for SOMC, metal alkoxide complexes reveal several disadvantages, including incomplete surface silanol consumption [167b], release of strongly surface-coordinating alcohols [169], agglomeration of alkoxide surface species [169], and limited secondary ligand exchange [15]. As a consequence, the endcapping of non-reacted OH groups is often required for catalytic applications [170].

As an example, aluminum alkoxide-grafted MCM-41, Al^{III}(OiPr)$_3$@MCM-41 **53** (Scheme 6), revealed a remarkably enhanced catalytic activity in the Meerwein–Ponndorf–Verley (MPV) reduction of cyclic ketones compared to the homogeneous system [Al(OiPr)$_3$]$_4$ (conversion: 86% versus < 1%, after 5 h, 25 °C) [171]. A detailed ^{27}Al MAS NMR study revealed that the enhanced catalytic activity can be ascribed to the formation of low coordinated (4-, 5-), geometrically distorted aluminum species. It was proposed that surface confinement prevents the aluminum alkoxide moieties from agglomeration, while

References see page 608

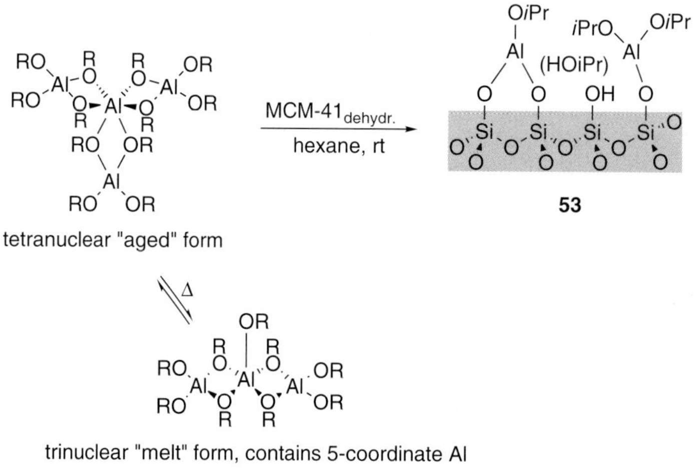

Scheme 6 Generation of reactive aluminum via grafting of [Al(OiPr)$_3$]$_4$.

the silicate material simultaneously acts as an electron-withdrawing matrix.

Interestingly, a previous kinetic study showed that the "melt" form of Al(OiPr)$_3$ (Scheme 6), which consists of the predominantly trimeric form with 4- and 5-coordinated aluminum centers, was found to be 10^3-fold more reactive in MPV reductions than the tetrameric form containing 4- and 6-coordinated aluminum [172]. These findings could be corroborated by grafting 4-coordinate Al^{III}[N(SiHMe$_2$)$_2$]$_3$(THF), which produces lower-coordinated aluminum surface species exclusively [171]. Subsequent silylamide/HOiPr ligand exchange produced a material which displayed similar catalytic activity as IMC **53**.

Other O-ligand-based SOMC involves siloxide (e.g., Ti^{IV}[OSi(OtBu)$_3$]$_4$, Fe^{III}[OSi(OtBu)$_3$]$_3$(THF), {Cu^I[OSi(OtBu)$_3$]}$_4$, {Rh^I[μ-OSi(OtBu)$_3$](diene)}$_2$, Ta^V(OiPr)$_2$[OSi(OtBu)$_3$]$_3$, {Fe^{II}[OSi(OtBu)$_3$]$_2$}$_2$, M^{VI}O[OSi(OtBu)$_3$]$_4$ (M = Mo, W), V^{V}O[OSi(OtBu)$_3$]$_3$ [173], Mo^{IV}(NtBu$_2$)$_2$(OSiMe$_3$)$_2$, Ti^{IV}(OSiMe$_3$)$_4$ [174]) and β-diketonato precursors (e.g., Cr^{III}(acac)$_3$ [175]). The grafting of such oxygen-rich metal–[OSi(OtBu)$_3$] siloxide complexes was successfully applied on PMS SBA-15 for the preparation of site-isolated metal species displaying remarkable catalytic activity and/or selectivity in hydrocarbon oxidation [173].

2.4.9.5.3 Amide-Based SOMC Similar to metal alkoxides, SOMC with metal alkyl amide derivatives M(NR$_2$)$_x$ [176] (e.g., V^{III}(NiPr$_2$)$_3$ [136], Nd^{III}(NiPr$_2$)$_3$(THF)$_x$ [177], Ti^{IV}(NMe$_2$)$_4$ [178], Ti^{IV}(NEt$_2$)$_4$ [169]) generates protonated ligands (amines) which can strongly interact with electrophilic surface sites (see Fig. 10). In contrast, silylamide complexes M[N(SiR$_3$)$_2$]$_x$ display a favorable surface reaction featuring: (i) mild reaction conditions; (ii) formation of thermodynamically stable metal siloxide bonds; (iii) surface hydrophobization (endcapping) via concomitant surface silylation; (iv) favorable atom economy; (v) suppression of extensive complex agglomeration due to the steric bulk of the silylamide ligands; (vi) release of weakly coordinated and hence easily separable silylamines; (vii) the absence of any insoluble byproducts; and (viii) ease of silylamide ligand functionalization (e.g., with spectroscopic probe moieties as shown for the Si–H group of the dimethylsilylamide ligand N(SiHMe$_2$)$_2$). Additionally, silylamide ligands provide a stabilizing environment (low coordination numbers, low metal oxidation states) for most of the main group and transition metals except for the noble metals [176]. Various conceptional approaches of this heterogeneously performed silylamine elimination have been initially reported for the lanthanide elements (Scheme 7) [45, 177, 179–181] and later applied for magnesium [182], aluminum [177], bismuth [183], titanium [183], and zinc [184], as outlined in Scheme 7.

In Scheme 7, approach **a** involves the sequential grafting of the silylamide complexes (IMC **54**) and a surface-mediated secondary ligand exchange reaction (IMC **55**) [179]. The latter can be an alkylation of the rare-earth metal center with organoaluminum reagents for applications in polymerization catalysis [185–187] or ligand protonolysis reactions (e.g., introduction of chiral ligands) [188]. As an example, Hfod@Ln^{III}[N(SiHMe$_2$)$_2$]$_3$(THF)$_x$@MCM-41 **57** [Fig. 13; Ln = Sc, Y, La; (Hfod = 1, 1, 1, 2, 2, 3, 3-heptafluoro-7,7-dimethyl-4,6-octanedione)] are IMCs obtained via sequential silylamide grafting and protonolysis [181]. IMCs **57** showed a highly selective reaction behavior in the Danishefsky transformation, a special hetero-Diels–Alder reaction starting from Danishefsky's diene (*trans*-1-methoxy-3-trimethylsilyloxy-1,3-butadiene) and benzaldehyde, to form the cycloaddition product exclusively

2.4.9 Immobilization of Molecular Catalysts 597

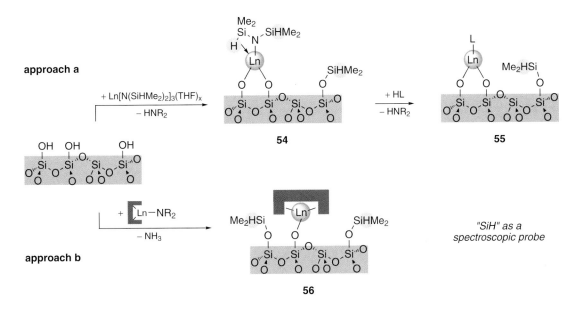

Scheme 7 Two variants of the heterogeneously performed silylamide route.

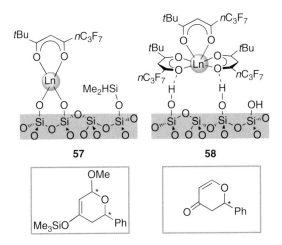

Fig. 13 The effect of a favorable microenvironment (= surface silylation) on product selectivity in a hetero-Diels–Alder reaction [181].

(TOF = 70 : Ln = Y). For comparison, hybrid catalyst Y^{III}(fod)$_3$@MCM-41 **58**, obtained by contacting a dehydrated MCM-41 sample directly with Y(fod)$_3$, gave an elimination product from the beginning. However, IMC **58** displayed enhanced initial catalytic activity comparable to that of molecular Y(fod)$_3$.

The selective behavior of IMC **57** was ascribed to a favorable microenvironment, which resulted from the heterogeneously performed silylamide route; that is, an *in situ* silylation due to the released silylamine ensuring the complete end-capping of all of the Brønsted acidic surface silanol groups [189]. IMC **57**(Y) also exhibited reusability and long-time stability, while the molecular Y(fod)$_3$ underwent deactivation after 1 h. The decreased initial catalytic activity of IMC **57** was attributed to diffusion limitation of the substrates and the product (TOF: 30 versus 180 mol mol$_Y^{-1}$ h^{-1}, 25 °C).

In Scheme 7, approach **b** utilizes tailor-made heteroleptic molecular precursors exhibiting reactive docking positions (a silylamide moiety) and strongly chelating ancillary ligands such as SALEN and disulfonamides which disfavor protonolysis and counteract oligomerization reactions [180]. IMCs **56** were also successfully employed in the above Danishefsky transformation [180].

SOMC appears to be a prolific method for the development of biomimetic catalysts based on mesoporous host materials ("mesozymes"). The application of SOMC@PMS to emulate the incredible cooperativity between metal site, protein tertiary structure, and substrate molecule in natural enzymes is certainly challenging.

2.4.9.5.4 Molecular Model Oxo–Surfaces Modern methods of surface characterization contribute powerfully to a better understanding of the formation, appearance, and catalytic action of SOMC-derived active surface sites (see Chapter 3 of this Handbook). Due to their structural order and mesoporosity, PMS materials provide access to a more diverse physico-chemical characterization which makes them especially well-suited as a *model* support for more ordinary forms of silica [137]. Multivalent molecular model oxo–surfaces are also investigated for a better understanding of the surface reactivity and attachment

References see page 608

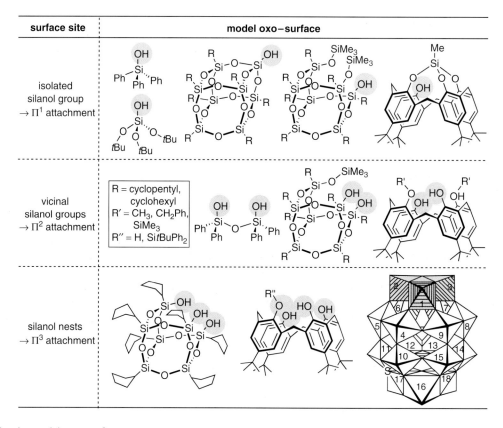

Fig. 14 Molecular model oxo–surfaces.

of the organometallic compounds/fragments and silylating reagents [20, 132j, 190–193]. Accordingly, ordinary silanols such as triphenylsilane, HOSiPh$_3$ [180, 194], and tri-*tert*-butoxysilanol, HOSi(O*t*Bu)$_3$ [165], are employed as models for a monopodal (π^1) surface attachment of metal ligand fragments to isolated surface hydroxy groups. Polyhedral oligosilsesquioxane (POSS) [191] and calix[4]arene derivatives [192, 195] seem to qualify as superior mimics, and represent oxo–surfaces with various sets of OH groups (Fig. 14). Control of POSS condensation and post-synthesis functionalization (alkylation, silylation) enables the modeling/manipulation of monopodal (Π^1), bipodal (Π^2), and tripodal (Π^3) surface attachments.

Incompletely condensed silsesquioxanes of type (*c*-C$_5$H$_9$)$_7$Si$_7$O$_9$(OH)$_3$ (cyclopentyl-T$_7$(OH)$_3$; according to the conventional silicone nomenclature, T denotes a trifunctional unit of siloxane structure) [196] implicate a "realistic" electronic and steric situation as revealed by the strongly electron-withdrawing character of the trisilanol "SiO-framework". Of particular significance are the short-range structural similarities between cyclopentyl-T$_7$(OH)$_3$ and geometrically comparable trisilanol sites available on idealized surfaces of SiO$_2$ polymorphs such as β-cristobalite and β-tridymite [191]. Furthermore, the pK_a value of the hydroxyl groups in calix[4]arene [197] is similar to the pK_a value found for SiOH groups on silica surfaces, and the distance of 2.8–3.2 Å [195a] between two neighboring phenolic OH moieties is comparable to the density of silanols in MCM-41 dehydrated at ca. 300 °C (1.69 OH nm^{-2}). Polyoxyanion-supported catalysts such as [Cp*Rh^{I}P$_2$W$_{15}$Nb$_3$O$_{62}$]$^{7-}$ and [(Cp$_2$U^{IV})$_2$(μ-κ^2O-TiW$_5$O$_{19}$)$_2$]$^{4-}$ can also provide models of atomically dispersed metal complexes (Fig. 14, lower right) [193].

Combined examination of the soluble model complexes via high-resolution solution spectroscopy and diffraction methods (e.g., X-ray structure analysis) help in the development of a detailed qualitative understanding of immobilization processes and the appearance of surface species. However, such investigations are usually hampered by agglomeration processes in solution (e.g., monomer–dimer equilibria), which produce higher-coordinated metal centers and, hence, imply changed (decreased) reactivity.

2.4.9.6 Metal–Organic Assemblies: Self-Immobilized Homogeneous Catalysts

In this section, a representative selection of IMCs will be provided where the catalytically relevant metal–ligand moiety simultaneously constitutes the support, or is an integral part of the support as a result of tailor-made ligand topology and functionality, respectively [198, 199].

2.4.9.6.1 Metal–Organic Frameworks
More detailed information concerning the preparation of metal–organic frameworks [200, 201] may be found in Chapter 2.3.8 of this Handbook.

A Achiral Metal–Organic Assemblies Heterogeneous catalysis promoted by an IMC coordination network was initially described for a Cd^{II}–(4,4′-bipyridine) square-grid cationic network (Scheme 8) [202]. The porous six-coordinate "grid complex" **59** catalyzed the cyanosilylation of imines more efficiently (N-phenylbenzimin as a substrate: 98% conversion after 1 h, 0 °C) than moderately soluble $Cd(pyridine)_4(NO_3)_2$ (no infinite framework; 85% conversion after 1 h) and eight-coordinate $Cd(NO_3)_2 \times 4H_2O$. The enhanced catalytic activity was attributed to hydrophobic grid cavities (efficient substrate binding) and a more cationic/Lewis-acidic metal center in IMC **59**, respectively.

It is noteworthy, that the one-dimensional nanochannels of porous coordination polymers of type $[Cu_2^{II}(pzdc)_2(L)]_n$ **60** (pzdc = pyrazine-2,3-dicarboxylate; L = pillar ligand such as pyrazine or 4,4′-bipyridine) [203] can be exploited for the fabrication of single-chain polymers via radical (polystyrene) [204] and anionic polymerization (polymethylpropiolate) [205].

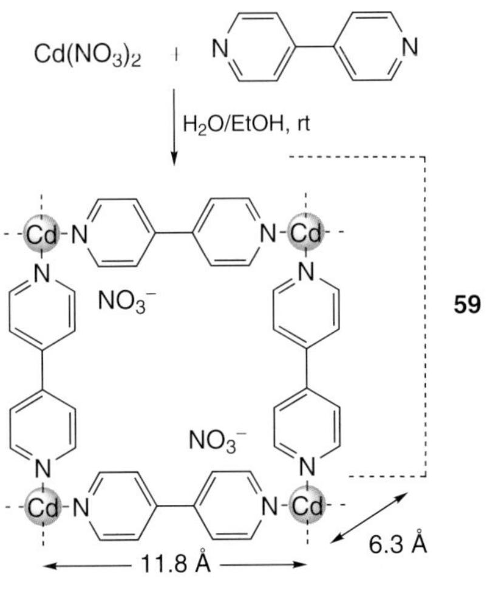

Alkoxide complexes of Group III, IV, and XIII metals easily undergo transalcoholysis reactions with an anthracene bisresorcinol H_4L_1 ("apohost") to afford highly insoluble microporous polyphenoxides **61–64** with a neutral framework structure (Scheme 9) [206]. In marked contrast to sol–gel processes, such precipitation reactions proceed very rapidly. Compared to the hydrogen-bonded apohost H_4L_1 (7 m^2 g^{-1}), such functional organic zeolite analogues display specific surface areas a_S of ca. 200–250 m^2 g^{-1} and a pore size of ca. 0.7 nm, albeit with fairly small micropore volumes (<100 mm^3 g^{-1}). The dialkoxyzirconium polyphenoxide $[L_1^{4-} \times 2[Zr^{IV}(OtBu)_2]^{2+}]$ **62** was used as a Lewis acid catalyst for the Diels–Alder reaction of acrolein with 1,3-cyclohexadiene and revealed better performance than soluble $Zr(OtBu)_4$ and apohost H_4L_1 (TOF: 40 versus 0.1 versus 0.3 h^{-1}) [206]. IMC **62**, which could be employed repeatedly without deactivation, was also used in a flow system.

The titanium $[L_1^{4-} \times 2[Ti^{IV}(OiPr)Cl]^{2+}]$ **61** and aluminum derivatives $[L_1^{4-} \times 2[Al^{III}(CH_3)]^{2+}]$ **64** catalyzed the Diels–Alder reaction in a similar highly stereoselective manner (endo/exo product < 99/1) [207]. The La^{III}-immobilized organic solid $[L_1^{4-} \times 2[La^{III}(OiPr)]^{2+}]$ **63** was investigated as a microporous enolase mimic after aqueous work-up [208]. The guest/host complex obtained from water adduct $[L_1^{4-} \times 2[La^{III}(OH)(H_2O)_6]^{2+}]$ and cyclohexanone gave facile deuterium incorporation at the α-positions of the carbonyl group, in D$_2$O. H/D exchange went to completion with a half-life of $\tau = 1.8$ h, while neither soluble $LaCl_3$ nor insoluble $La(OH)_3$ as a reference were active. A Michaelis–Menten-type kinetic behavior proposed a reversible substrate prebinding and activation in an enzyme-mimetic manner. Such lanthanum polyphenoxides performed also as solid Brønsted-base catalysts in the aldol reaction of cyclohexanone with aldehydes [208].

B Chiral Metal–Organic Assemblies Oxo-bridged trinuclear metal carboxylates $[M_3^{II}(\mu_3-O)(O_2CR)_6(H_2O)_3]^{2-}$

Scheme 8 Synthesis of a Cd^{II}–(4,4′-bipyridine) square-grid cationic network with nitrate ions located inside the cavities.

References see page 608

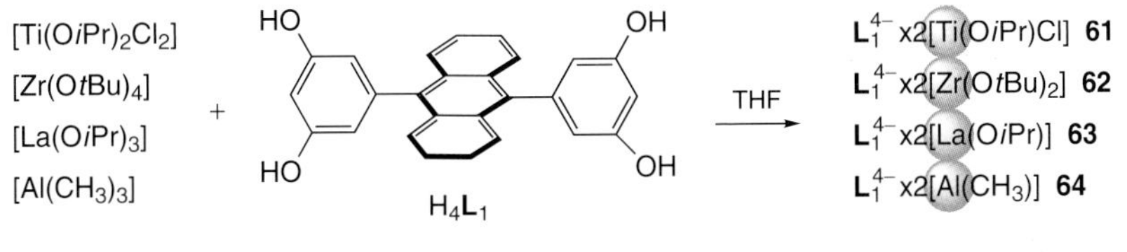

Scheme 9 Formation of microporous phenoxide-based metal–organic frameworks.

("basic carboxylates") are known to form rigid, thermally robust neutral homochiral metal–organic frameworks (MOF) [201]. This strategy was exploited for achieving homochiral framework structures with large chiral voids, as outlined in Scheme 10 [209, 210].

Reaction of D-tartaric acid with 4-aminopyridine gave the enantiopure building block HL$_2$, which reacted with Zn^{II} to produce D-POST-1 **65**, the void volume of which (ca. 47%) is filled with water molecules [209]. While the pyridyl group-catalyzed transesterification reactions of 2,4-dinitrobenzoic acid methyl ester with alcohols of varying bulk proposed size-selectivity (i.e., the transformation mainly occurred in the channels), the use of racemic 1-phenyl-2-propanol gave the corresponding esters only with poor enantiomeric excesses (ca. 8% ee) [209]. The hydrothermal synthesis of [Zn^{II}$_2$(bdc)(L-lac)(dmf)](DMF) **66** utilizing 1,4-benzenedicarboxylic acid and L-lactic acid as a rigid spacer and chiral auxiliary, respectively, also gave a homochiral MOF material with open framework structure (Langmuir surface area, ca. 190 m^2 g^{-1}) [210]. The catalytic oxidation of thioethers with urea hydroperoxide revealed size- and chemoselectivity (formation of sulfoxide rather than sulfons), although no asymmetric induction was observed for homochiral material IMC **66**.

Similar to the anthracene bisresorcinol-derived polyphenoxides (see Scheme 9) [206–208], rigid BINOL-based ligands (BINOL = 1, 1′-bi-2-naphthol) spontaneously coordinate with metal ions to form homochiral metal–ligand assemblies (Figs. 15 and 16) [211–213].

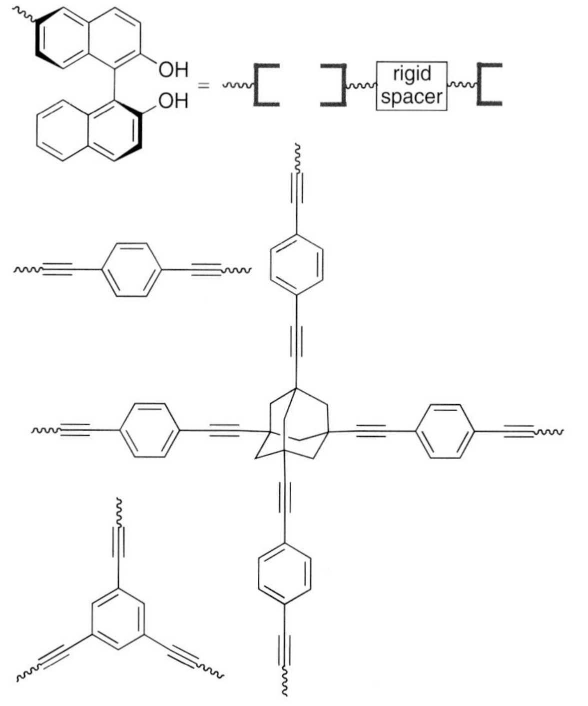

Fig. 15 Design of chiral rigid multitopic (= multifunctional) BINOL ligands.

This self-supporting strategy was used for the heterogenization of Shibasaki's La/BINOL catalyst by simply adding the corresponding multitopic ligand precursor to

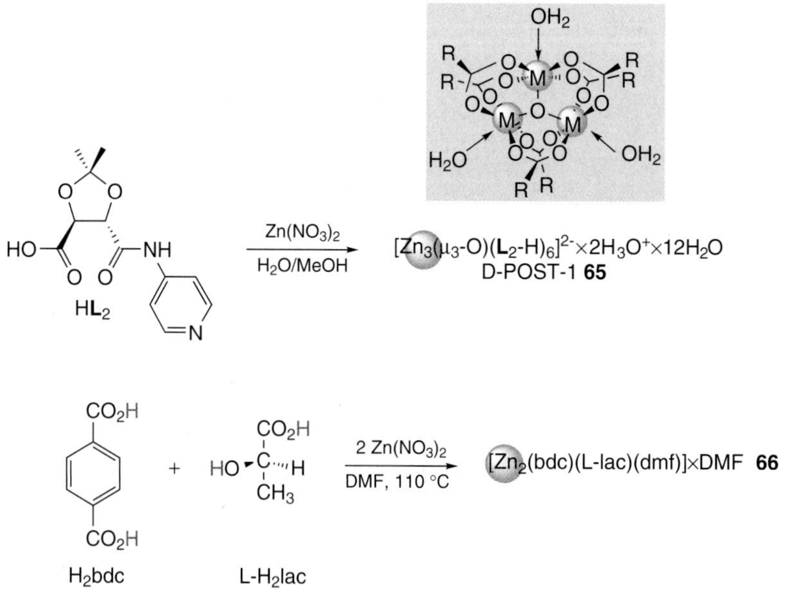

Scheme 10 Exploiting Yaghi's concept of carboxylate-based neutral metalorganic frameworks (MOFs) for the construction of homochiral porous IMCs.

a solution of [La(OiPr)₃] in THF in the presence of triphenylphosphine oxide [214]. Albeit not single-crystalline, such self-supported IMCs **67** display multicomponent asymmetric catalysts with high enantiocontrol. In the epoxidation of α, β-unsaturated ketones with cumene hydroperoxide (CMHP), reusable IMCs **67** gave ee-values ranging from 84.9 to 97.6%, depending on the substitution of the substrate and the bridging spacer of the multitopic ligands [214].

The generality of this concept has earlier been demonstrated for the synthesis of [LiAl^{III}–BINOLate] (**68**) and [Ti^{IV}–BINOLate] (**69**) metal-bridged polymers from LiAlH₄ and Ti(OiPr)₄, respectively [215]. The use of phenylene-bridged BINOLate ligands revealed that a location of the phenolic hydroxy groups at the opposite sides in the multidentate ligand affords high enantioselectivity in the Michael reaction of cyclohexenone with dibenzyl malonate (ee_{max} = 96%).

The Ti-based IMC **69** (Fig. 16) was also used in an asymmetric carbonyl-ene reaction of aldehyde with α-methylstyrene [215]. The initial ee-value of 88% was maintained after six catalytic cycles.

A similar concept was used for the programmed assembly of Ferringa's catalyst and Noyori-type catalysts as non-crystalline "cationic frameworks" (Fig. 17). The self-supported chiral rhodium(I) catalyst **70**, obtained from [Rh^I(cod)₂]BF₄ and bridged monophosphoramidite ligands, gave high enantioselectivities in the hydrogenation of β-aryl- or alkyl-substituted dehydro-α-amino acids and enamide derivatives (94.3–97.3% ee) [216]. A homocombination of bridged BINAP and bridged DPEN ligands was exploited for the complexation of [(C₆H₆)Ru^{II}Cl₂]₂. This "mixed ligand" IMC **71** exhibited a remarkable performance in the asymmetric hydrogenation of aromatic ketones (ee_{max} = 97.4%) [217]. Both hydrogenation catalysts **70** and **71** failed to show any significant metal leaching, and an improved enantioselectivity was demonstrated for special substrates as compared with the homogeneous analogues.

Two successful approaches comprising (i) the assembly of single-crystalline pyridine-based neutral MOFs, and (ii) the enantiocontrol of BINOL-based catalysts, were prolifically combined in the homochiral porous IMC **72** (Scheme 11) [218]. Removal of all solvent molecules from the single-crystalline Cd-based MOF revealed permanent porosity with a specific surface area of 600 m² g⁻¹ and a pore volume of 0.26 cm³ g⁻¹. Reaction of the framework dihydroxy groups with Ti(OiPr)₄ gave heterobimetallic MOF **72**, which was used in the catalytic addition of ZnEt₂ to aromatic aldehydes. For example, the addition of ZnEt₂ to 1-naphthaldehyde afforded (R)-1-(1-naphthyl)propanol with 93% ee, rivaling that of the homogeneous counterpart under similar conditions (94% ee). Moreover, IMC **72** displayed size-selectivity in the presence of different-sized aldehydes, demonstrating that the catalytic sites are accessed via the open chiral one-dimensional nanochannels (ca. 1.6 × 1.8 nm) [218].

2.4.9.6.2 Matrix-Embedded Catalysts

A Sol–Gel-Processed Interphase Catalysts Nondestructive sol–gel immobilization exploits the incorporation of homogeneous catalysts into oxidic networks via a condensable ligand functionality [36, 219–224]. The inorganic matrix precursor (e.g., Si(OEt)₄, or simple metal alkoxides) are hydrolyzed in the presence of a functionalized organometallic complex. The resulting hybrid organic–inorganic catalysts display textural stability, porosity, high surface areas, and a uniform distribution of catalytic sites [36]. Moreover, these IMCs often exhibit increased catalytic efficiency

References see page 608

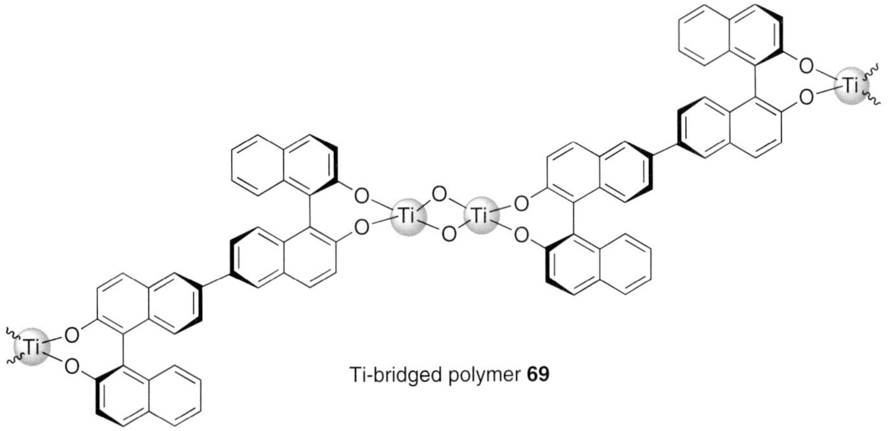

Ti-bridged polymer **69**

Fig. 16 Spontaneous assembly of a homochiral self-supported [Ti^{IV}–BINOLate] polymer.

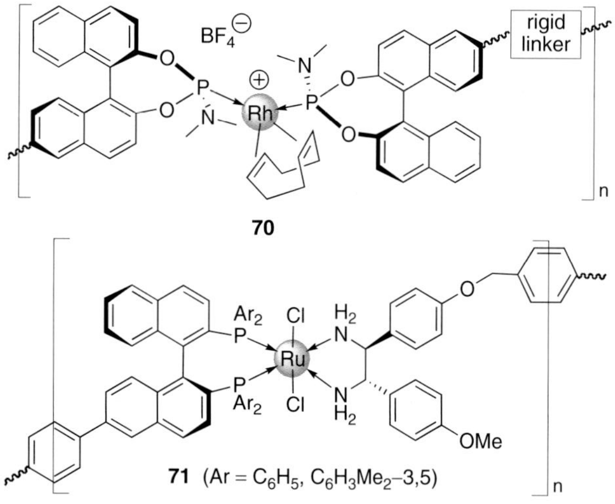

Fig. 17 Self-supported chiral rhodium and ruthenium hydrogenation catalysts.

(improved stability, better performance, better handling, higher TON) as compared with their homogeneous counterparts [221]. Figure 18 shows four molecular catalyst precursors **P73–P76** which have been efficiently incorporated into silica matrices. The Ru^{II}–IMC **73** was employed in the catalytic synthesis of N,N-diethylformamide from CO_2, H_2, and diethylamine [225]. A mesoporous aerogel ($V_p = 1.74$ cm^3 g^{-1}, $a_s = 1000$ m^2 g^{-1}) afforded a TOF of 18400 h^{-1}, almost 10-fold that of a corresponding microporous xerogel ($V_p = 0.44$ cm^3 g^{-1}, $a_s = 670$ m^2 g^{-1}), which was attributed to the absence of any significant intraparticle diffusion limitations in the aerogel. The TOF for the synthesis of dimethylformamide from carbon dioxide reached remarkable 1860 h^{-1} [226]. Iron and manganese metalloporphyrinosilicas were discussed as selective biomimetic oxidation catalysts (*cyctochrome P450* mimic) [227]. Iron-based IMC **74** ($a_s = 690$ m^2 g^{-1}) catalyzed the hydroxylation of cyclohexane and heptane with PhIO in ca. 50% yields. Furthermore, a size-selective hydroxylation was observed for adamantane/cyclohexane mixtures when the "homopolymer" of **P74** ($a_s = 60$ m^2 g^{-1}) was used as an IMC [227].

Cocondensation of alkoxy-functionalized precursors **P75** with tetraethoxysilane (TEOS) in the presence of cetyltrimethylammonium bromide (CTAB) under acidic conditions and a subsequent solvent extraction/silylation (Me$_3$SiCl or Me$_2$SiCl$_2$) sequence led to a mesoporous inorganic–organometallic hybrid material with the organometallic complex integrated into the pore walls [228]. Rh^I-containing SBA-3 silica **75** (1.7 wt% Rh) displays pore diameters in the range of 2.0 to 2.2 nm and a BET specific surface area of 930 m^2 g^{-1}. IMC **75** was evaluated in the hydrogenation of alkenes such as styrene, cyclohexene, acrolein, and crotonaldehyde, and exhibited similar activities as a hybrid catalyst obtained from postsynthesis grafting of mesoporous silica SBA-15. For styrene hydrogenation, the activity was in the same range as observed for the homogeneous catalyst [RhCl(PPh$_3$)$_3$] (TOF: 2650 versus 3300 h^{-1}) [228]. Similarly, the first chiral mesostructured organosilica (ChiMO) with MCM-41-like topology was obtained from TEOS and **P76** (molar ratio 85:15–95:5) and CTAB as a structure-directing reagent [229]. After solvent extraction, IMC **76** revealed a pore diameter of 4.5 nm and a BET specific surface area of 900 m^2 g^{-1}. Chiral V^V-IMC **76** gave a lower enantiomeric excess in the cyanosilylation of benzaldehyde with TMSCN compared to a MCM-41 hybrid material obtained via a postsynthesis grafting sequence (30 versus 63% *ee*; TON: 320 versus 287).

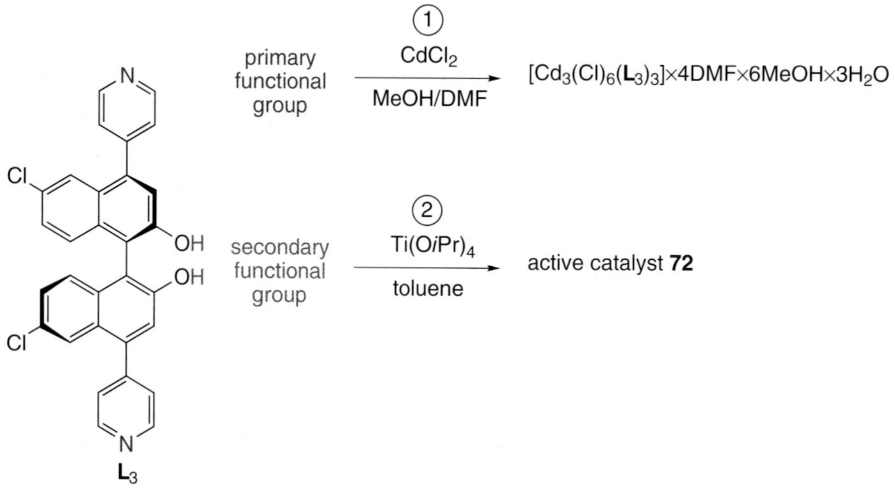

Scheme 11 Design of a multifunctional MOF for enantioselective and size-selective catalysis.

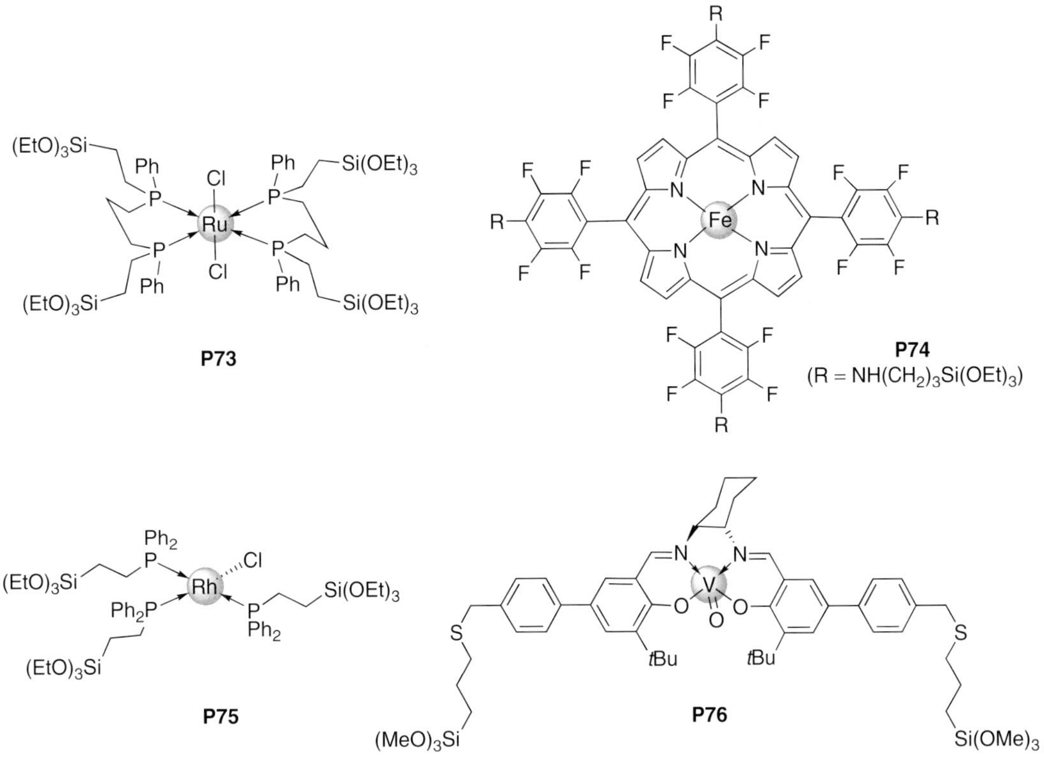

Fig. 18 Metal–organic precursors for sol–gel immobilization.

Chiral porous zirconium phosphonates **77** containing Ru^{II}–BINAP–DPEN moieties were generated by a condensation reaction of ruthenium complexes carrying phosphonic acid-substituted BINAP ligands with Zr(OtBu)$_4$ [230, 231]. Both the porosity and enantiocontrol depended on the bis(phosphonic acid) substitution of the BINAP ligand. 4,4′-substitution, as shown in Scheme 12, afforded highly porous high-surface-area hybrid materials ($a_{s,max}$(BET) = 475 m^2 g^{-1}, $V_{p,max}$ = 1.02 cm^3 g^{-1}) which catalyzed the asymmetric hydrogenation of aromatic ketones with high activity (TOF ca. 700 h^{-1} at 70% conversion) and enantioselectivity (ee_{max} = 99.2%) [230]. For example, acetophenone was hydrogenated to 1-phenylethanol with 96.3% ee, which was significantly higher than that observed for the parent Ru^{II}–BINAP–DPEN homogeneous catalyst under similar conditions (ca. 80% ee). Such IMCs could be readily recycled/reused and also successfully employed for the asymmetric hydrogenation of β-keto esters (ee_{max} = 95.0%) [231].

B Organic–Polymer-Embedded Catalysts Copolymerization of suitably functionalized privileged ligands with common monomers gave access to a series of uniform polymer-embedded IMCs [232–234]. Scheme 13 shows part of the reaction sequence which produced the first chiral polymer(polystyrene)-bonded metathesis catalyst **78** [232].

Application of Mo^{VI}-based IMC **78** for the asymmetric ring-closing metathesis (ARCM) and ring-opening/cross metathesis (AROM/CM; 7-oxynorbornene substrates and styrene) revealed excellent levels of enantioselectivity (ee_{max} = 95 and 98%, respectively). However, the transformations were less efficient than those of the parent homogeneous catalyst, this being attributed to limited diffusion of the substrate molecules into the polymers.

A soluble polymer-bonded catalyst for asymmetric hydrogenation was prepared *in situ* from polyester-supported BINAP ligands – obtained by polycondensation of 5,5′-diamino-BINAP, terephthaloylchloride, and (2S,4S)-pentanediol – and [Ru^{II}(cymene)Cl$_2$]$_2$ [233]. Under otherwise identical conditions, chiral IMC **79** showed higher activity and enantioselectivity in the hydrogenation of 2-(6′-methoxy-2′-naphthyl)acrylic acid (rt, 69.0 kg cm^{-2} H$_2$ pressure, methanol : toluene, 9 : 1, v/v) as compared to the corresponding homogeneous catalyst (conversion: 95.5%/4 h versus 37.5%/60 h; 87.7 versus 80.5% ee). It was proposed that this superior performance might originate from a polyester backbone–BINAP ligand steric "cooperativity effect".

References see page 608

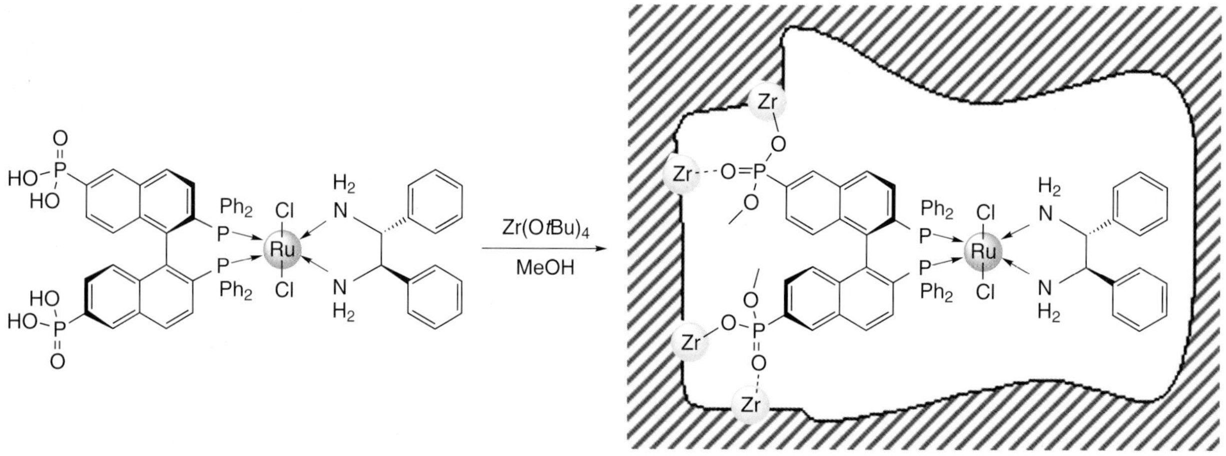

Scheme 12 Synthesis of a chiral porous zirconium phosphonate (according to Ref. [230]).

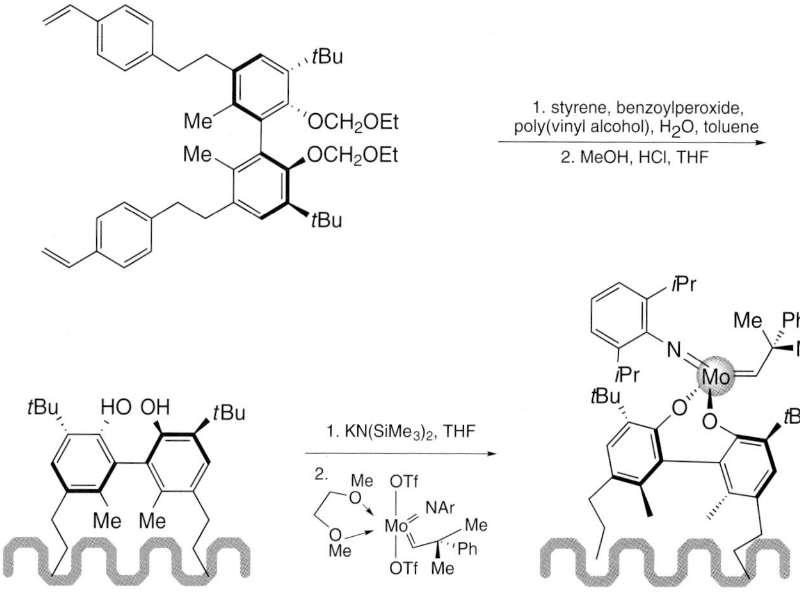

Scheme 13 Immobilization of a chiral molybdenum complex for enantioselective olefin metathesis.

Acrylester-functionalized, Si-tethered binaphthol ligands were used as monomers for the copolymerization with methyl methacrylate according to a "catalyst analogue" route (AIBN and ethylene glycol dimethylacrylate (EGDMA) were used as starter and crosslinker, respectively) [234]. In a sequential reaction involving (i) hydrolysis of the hybrid polymer and (ii) AlMe$_3$/*t*BuLi addition, the Si atom was displaced by Al−Li to yield a polymer-supported multicomponent asymmetric catalyst (MAC) **80**. IMC **80** promoted the Michael reaction of 2-cyclohexen-1-one with dibenzylmalonate, with 91%

ee. For comparison, a reference material obtained by skipping the Si-tethering ("place-marker") exhibited poor enantiocontrol (21% *ee*).

C Self-Immobilizing Polymerization Catalysts Various metalorganic polymerization catalyst precursors carrying a functional alkenyl group were successfully copolymerized with α-alkenes to generate multinuclear networks of active sites (Scheme 14) [235, 236]. This approach is based on the idea that the catalyst precursor acts simultaneously as a monomer and that, upon activation with a suitable

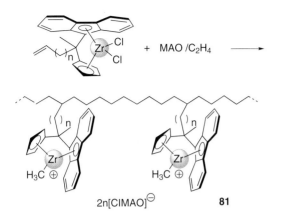

Scheme 14 Self-immobilization of a homogeneous ansa-metallocene catalyst.

co-catalyst (e.g., MAO), the active center copolymerizes monomers (e.g., ethene, propene, styrene) and other catalyst molecules [237]. Such organometallic–organic polymeric networks can be generated in situ and act as heterogeneous catalysts for α-alkene polymerization. The polymerizable metalorganic complexes comprise metallocenes and constrained geometry catalysts of titanium, zirconium and hafnium as well as postmetallocenes of the late transition metals (e.g., iron, nickel) [235, 236, 238]. In a similar way, β-H-elimination in metallacyclic metallocenes can be exploited to copolymerize active sites [239]. Furthermore, alkenyl group transfer from a zirconocene dialkenyl complex to the MAO cocatalyst was used to generate a "self-immobilizing cocatalyst" [240]. IMCs such as **81** give rise to polyethenes (resins) with new material properties that cannot be achieved with other catalysts, such as evenly distributed ethyl branches, excellent clarity, and tensile strength [235b].

D Molecularly Imprinted Polymers The molecular imprinting technique was originally applied in catalytic antibody technology for producing ligand-selective recognition sites in synthetic polymers [241, 242]. Accordingly, molecularly imprinted metal complex catalysts can be synthesized with polymerizable metal complexes that contain a combination of polymerizable (L_p) and non-polymerizable ligands [243, 244]. By copolymerization with organic or inorganic monomers, often in the presence of a porogen (e.g., $CHCl_3$), the three-dimensional structure of the metallomonomer is transferred into the surrounding rigid matrix. Subsequent removal of the non-polymerizable ligand provides a functional shape-selective reaction space for molecules exhibiting similar shape/functionality as compared to the removable ligand (the imprint or template or pseudo-substrate). Such IMCs operate in an enzyme-inspired active site by utilizing reaction intermediate or transition-state analogues as imprints. Examples include $Co^{II}(Py_p)_2$@poly-4-vinylpyridine–styrene–divinylbenzene (**82**: aldol condensation) [245], $Rh^I(diamine_p)_2$@polyurea (**83**, enantioselective hydride transfer reduction of prochiral ketones: $ee_{max} = 70\%$) [246], $Ti^{IV}(OAr_p)_2Cl_2$@polystyrene (**84**: Diels–Alder reaction) [247], $Ru^{II}(p$-cumene)(sulfonamide–amine$_p$)Cl@poly-EGDMA (**85**: transfer hydrogenation of ketones, EGDMA = ethylene glycol dimethylacrylate) [248], $Pd^{II}(phosphine_p)_2Cl_2$@polystyrene–divinyl benzene (**86**: Suzuki and Stille reactions) [249, 250], $Rh^I(OSi\equiv_p)[P(OMe)_3]$@silica (**87**: hydrogenation of alkenes; surface imprinting) [251], and $Ru^I(Por_p)(CO)$@poly-EGDMA (**88**: oxidation of alcohols and alkanes) [252]. The molecularly imprinted IMCs displayed substrate (regio)selectivity (**82–87**), shape-selectivity (**87**), and rate enhancement as compared to the corresponding molecular complexes (**82, 86, 88**) and to complexes polymerized in the absence of the imprint (**82, 85–88**). This increase in reactivity was ascribed to the imprint cavity and rigid attachment of the metal complex to the polymer backbone. Other beneficial effects of molecular imprinting include higher yields as compared to the molecular catalyst (**86, 88**) and reusability (**82, 86**).

Scheme 15 shows the synthesis of molecularly imprinted polymer (MIP) **89** via copolymerization of the metallomonomer (vinyl–dppe$_p$)Pt^{II}[(R)-tBu$_2$BINOL] with EDGMA and the subsequent displacement (ca. 70%) of the imprinting ligand by acidic phenol α, α, α-trifluoro-m-cresol to yield **89a** [253]. Consideration was given to the rebinding of rac-BINOL, an imprinting analogue, and it was found that the imprinted enantiomer preferentially rebound via ligand exchange, up to 69% ee and 58% Pt sites. Measurement of the kinetic selectivity revealed that the least reactive (= least accessible) Pt sites showed the highest recognition of the imprinted monomer, with selectivities up to 94% ee. Therefore, utilization of such IMPs for asymmetric catalysis might require poisoning of the "hyperactive" sites to reveal the more selective ones.

2.4.9.6.3 Liquid Crystal Assemblies Surfactant molecules were shown to act as useful additives for the efficient immobilization of molecular catalysts via sol–gel entrapment (see Scheme 1: IMC **18**) [90]. In addition, neutral or anionic amphiphilic molecules can be employed as "structure-directing" heterogenizing ligands by implying the formation of Langmuir–Blodgett thin films or micelles via molecular self-assembly [22, 254]. IMCs **90** revealed the impact of molecular order in Langmuir–Blodgett films (glass surface) on the catalytic activity (Fig. 19) [255]. Catalytically active films of

References see page 608

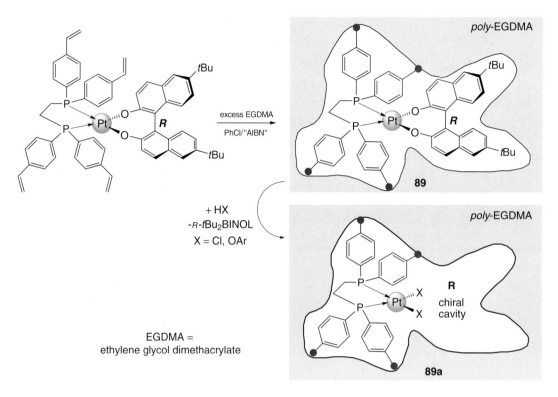

Scheme 15 Molecular imprinting of a platinum complex–chiral cavity entity (according to Ref. [253]).

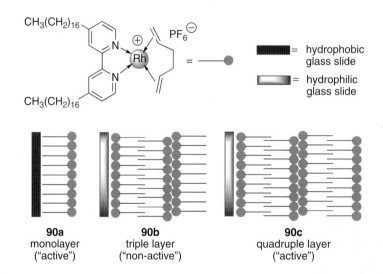

Fig. 19 Schematic presentation of different types of self-assembled metal-containing Langmuir–Blodgett films.

a water-stable rhodium complex [(4,4'-diheptadecyl-2,2'-bipy)Rh^I(hexadiene)][PF$_6$] were obtained with the polar metal-center-containing headgroups oriented in a way to terminate the film (e.g., **90a, 90c**). The active layers exhibited dramatically increased catalytic activity as compared to the molecular counterpart in the hydrogenation of aliphatic ketones and aldehydes (acetone as a substrate: TON$_{max}$: 60 000 (**90c**) versus 500; after 48 h, 25 °C, 73 psi H$_2$). Furthermore, high substrate selectivity as evidenced by the selective conversion of an acetone–butanone mixture (product ratio = isopropanol : sec-butanol > 100 : 1) and an unusual temperature effect corroborated the importance of molecularly ordered structures [255].

Other "nano-organometallic" thin films were obtained via sequential functionalization of a support (glass, quartz, silicon) involving: (i) covalent attachment of a monolayer with terminal phosphine donor groups; and (ii) complexation of a metal–organic complex [255–257]. Examples include {Ni^0(CO)$_2$(PPh$_3$)[(PPh$_2$)(CH$_2$)$_3$O]}@[Si] (**91**, oligomerization of phenylacetylene) [256] and {Rh^I(Cl)(1,5-COD)[(PPh$_2$)(CH$_2$)$_{12}$O]}@[Si] (**92**, hydrogenation of diphenylacetylene) [257]. IMC **91** displayed enhanced catalytic activity (TOF: 1656 versus 0.001 h^{-1}, 80 °C) and selectivity compared to the molecular analogue, suggesting a beneficial effect of the orientation of the surface-bonded organometallic species. IMC **92** revealed enhanced stability and could be recycled while the molecular analogue decomposed under similar catalytic reaction conditions.

Sulfate-based amphiphiles were employed for the self-assembly (heterogenization) of homogeneous catalysts of both late and early transition metals (Fig. 20). Sodium dodecylsulfate (SDS) was found to accelerate hydrogen transfer from cis-1,2-cyclohexanedimethanol to (E)-4-phenyl-3-buten-2-one, catalyzed by cationic [Ru^{II}Cl(S)-(binap)(benzene)]Cl **P93** [258]. Rhodium-containing amphiphiles **94** [259] and **95** [260] were obtained in water from the corresponding anionic surfactants SDS and diphosphine 2,7-bis(SO$_3$Na)xantphos with complexes [Rh^I{(R,R)-HO-diop}(cod)]BF$_4$ and Rh^{I}H(CO)(PPh$_3$)$_3$, respectively.

References see page 608

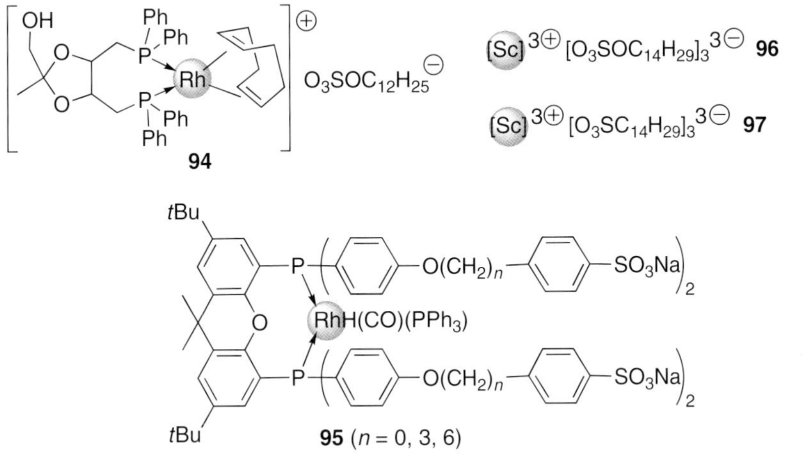

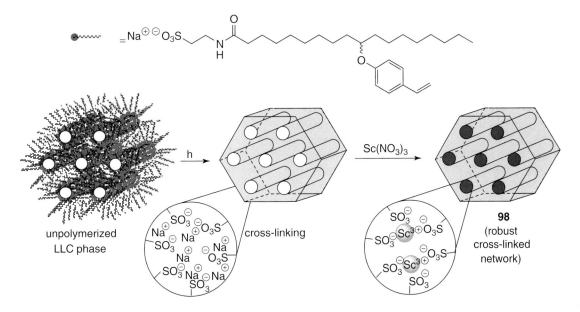

Fig. 20 Formation of metal-containing micelles and nanovesicles in the presence of anionic amphiphiles.

Scheme 16 Crosslinking of LLC phases for the synthesis of nanostructured heterogenized Lewis acid catalysts (according to Ref. [268]).

For IMC **95**, spontaneous formation of thermostable vesicles (average size ca. 140 nm, stable up to 290 °C) was reported if the hydrophobic part (tail) of the ligand is large enough ($n = 3, 6$) [260]. IMC **94** exhibited an impressive enhancement of the enantioselectivity in the asymmetric hydrogenation of alkene substrates ($ee_{max} = 76.6\%$, relative enantioselectivity $Q_{a/b(max)} = 7.3$) [259, 261]. Vesicular IMCs **95** were applied in the accelerated biphasic hydroformylation of 1-octene revealing TOFs which were 14-fold higher compared to ligands that do not form aggregates (e.g., TOF: 21 versus 5 mol aldehyde $mol_{Rh}^{-1} h^{-1}$, at 120 °C). Moreover, catalyst recycling showed no decrease of the highest reported regioselectivities (linear : branched = 99 : 1) [260].

The synthesis of defined Lewis acid–surfactant-combined catalysts (LASC) of type **96** and **97** [262] was prompted by an improved performance of scandium triflate [Sc(OTf)$_3$] in Lewis acid-catalyzed transformations in the presence of SDS [263]. IMCs **96** and **97** are easily prepared from ScCl$_3$ and SDS, and rapidly form stable colloidal suspensions (spherical particles of ca. 1 μm size) in the presence of organic substrates in water [262]. LASC-catalyzed reactions include aldol, allylation, and Mannich-type reactions. The surfactant-type catalysts can be removed from the reaction mixture via centrifugation, without using organic solvents. The catalyst performance depended on the type of the anionic surfactant (alkyl chain length, alkane sulfate versus alkene sulfonate), and was best in water.

The reactive microemulsion approach [263–267] was exploited for the synthesis of "organic zeolites" [Scheme 16; cf., the polymer incarcerated (PI) method; Scheme 3] [268].

Accordingly, nanostructures self-assembled from lyotropic liquid crystals (LLC) were stabilized by polymerization/crosslinking of the functionalized surfactant tails [268, 269]. Such organic analogues to zeolites and PMS materials display catalytically active head groups which are localized at the inner surfaces of microscopic "water pools" [263]. The catalytic viability of these reactive microemulsions was initially shown for the Cu(II)-catalyzed phosphate ester hydrolysis [263] and Pd(II)-promoted vinylation and hydrogenation reactions [264, 265]. The crosslinked inverted hexagonal MCM-41-type LLC phase, as shown in Scheme 16, features closely packed, monodisperse water channels ca. 15 Å in diameter [268]. The sodium cations could be quantitatively exchanged by Sc^{III} with retention of the hexagonal structure to afford IMC **98** for the Mukaiyama aldol reaction of aldehydes with silyl enolates and Mannich-type reactions [270].

References

1. B. Cornils, W. A. Herrmann, W. Schlögl, C.-H. Wong (Eds.), *Catalysis from A to Z*, Wiley-VCH, Weinheim, 2003.
2. B. Cornils, W. A. Herrmann (Eds.), *Applied Homogeneous Catalysis with Organometallic Compounds*, Wiley-VCH, Weinheim, 2002.
3. D. G. H. Ballard, *Adv. Catal.* **1973**, *23*, 263.
4. (a) F. R. Hartley, P. N. Vezey, *Adv. Organomet. Chem.* **1977**, *15*, 189; (b) F. R. Hartley, *Supported Metal Complexes*, D. Reidel, Dordrecht, 1985.
5. D. C. Bailey, S. H. Langer, *Chem. Rev.* **1981**, *81*, 109.
6. Yu. Yermakov, B. N. Kuznetsov, V. A. Zakharov, *Catalysis by Supported Complexes*, Elsevier, Amsterdam, 1981.
7. U. Schubert, *Phys. Uns. Zeit* **1987**, *18*, 137.
8. G. Alberti, T. Bein (Eds.), *Comprehensive Supramolecular Chemistry*, Vol. 7, Elsevier, Oxford, 1996.
9. D. E. De Vos, I. F. J. Vankelecom, P. A. Jacobs (Eds.), *Chiral Catalyst Immobilization and Recycling*, Wiley-VCH, Weinheim, 2000.
10. For a special issue on recoverable catalysts and reagents, see: *Chem. Rev.* **2002**, *102*, 3215.
11. A. Kirschning, H. Monenschein, R. Wittenberg, *Angew. Chem. Int. Ed.* **2001**, *40*, 650.
12. G. Jas, A. Kirschning, *Chem. Eur. J.* **2003**, *9*, 5708.
13. J. M. Thomas, R. Raja, D. W. Lewis, *Angew. Chem. Int. Ed.* **2005**, *44*, 6456.
14. P. McMorn, G. J. Hutchings, *Chem. Soc. Rev.* **2004**, *33*, 108.
15. R. Anwander, *Chem. Mater.* **2001**, *13*, 4419.
16. For review articles on periodically mesoporous silicas as catalyst support, see: (a) J. H. Clark, D. J. Macquarrie, *Chem. Commun.* **1998**, 853; (b) K. Moller, T. Bein, *Chem. Mater.* **1998**, *10*, 2950; (c) J. Y. Ying, C. P. Mehnert, M. S. Wong, *Angew. Chem. Int. Ed. Engl.* **1999**, *38*, 56; (d) J. M. Thomas, R. Raja, *J. Organomet. Chem.* **2004**, *689*, 4110.
17. (a) J. M. Thomas, *Angew. Chem. Int. Ed.* **1999**, *38*, 3588; (b) T. J. Barton, L. M. Bull, W. G. Klemperer, D. A. Loy, B. McEnaney, M. Misono, P. A. Monson, G. Pez, G. W. Scherer, J. C. Vartuli, O. M. Yaghi, *Chem. Mater.* **1999**, *11*, 2633.
18. For review articles on dendrimers as catalyst supports, see: (a) J. W. Kriesel, T. D. Tilley, *Adv. Mater.* **2001**, *13*, 1645; (b) R. van Heerbeek, P. C. J. Kamer, P. W. N. M. van Leeuwen, J. N. H. Reek, *Chem. Rev.* **2002**, *102*, 3717; (c) J. N. H. Reek, S. Arevalo, R. van Heerbeek, P. C. J. Kamer, P. W. N. M. van Leeuwen, *Adv. Catal.* **2006**, *49*, 71.
19. A. T. Bell, *Science* **2003**, *299*, 1688.
20. S. L. Scott, C. M. Crudden, C. W. Jones (Eds.), *Nanostructured Catalysts*, Kluwer Academic/Plenum Publishers, New York, 2003.
21. R. J. P. Corriu, A. Mehdi, C. Reyé, *J. Mater. Chem.* **2005**, *15*, 4285.
22. For examples, see: (a) M. G. L. Petruccei, A. K. Kakkar, *J. Chem. Soc., Chem. Commun.* **1995**, *8*, 1577; (b) M. G. L. Petruccei, A. K. Kakkar, *Adv. Mater.* **1996**, *8*, 251.
23. A. P. Wright, M. E. Davis, *Chem. Rev.* **2002**, *102*, 3589.
24. P. J. Dyson, *Coord. Chem. Rev.* **2004**, *248*, 2443.
25. For organopolymer supported catalysts, see: (a) D. C. Sherrington, *Chem. Commun.* **1998**, 2275; (b) D. C. Sherrington, *Catal. Today* **2000**, *57*, 87; (c) C. A. McNamara, M. J. Dixon, M. Bradley, *Chem. Rev.* **2002**, *102*, 3275; (d) D. E. Bergbreiter, *Chem. Rev.* **2002**, *102*, 3345.

26. For the functionalization of carbon materials and carbon supported catalysts, see: (a) J. M. Planeix, N. Coustel, B. Coq, V. Brotons, P. S. Kumbhar, R. Dutartre, P. Geneste, P. Bernier, P. M. Ajayan, *J. Am. Chem. Soc.* **1994**, *116*, 7935; (b) C. Pham-Huu, N. Keller, L. J. Charbonniere, R. Ziessel, M. J. Ledoux, *Chem. Commun.* **2000**, 1871; (c) A. Hirsch, *Angew. Chem. Int. Ed.* **2002**, *41*, 1853; (d) D. Tasis, N. Tagmatarchis, A. Bianco, M. Prato, *Chem. Rev.* **2006**, *106*, 1105.
27. For oxidic supports other than silica and alumina, see: (a) R. Andrés, M. Galakhov, A. Martin, M. Mena, C. Santamaria, *J. Chem. Soc., Chem. Commun.* **1995**, 551; (b) K. I. Hadjiivanov, D. G. Klissurski, *Chem. Soc. Rev.* **1996**, *25*, 61; (c) P. Serp, P. Kalck, R. Feurer, *Chem. Rev.* **2002**, *102*, 3085.
28. For metallic supports, see: (a) T. R. Lee, G. M. Whitesides, *J. Am. Chem. Soc.* **1991**, *113*, 2576; (b) C. E. Lee, S. H. Bergens, *J. Phys. Chem. B* **1998**, *102*, 193; (c) J.-P. Candy, B. Didillon, E. L. Smith, T. B. Shay, J.-M. Basset, *J. Mol. Catal.* **1994**, *86*, 179.
29. For phosphonate-based supported catalysts, see: C. Maillet, P. Janvier, M. Pipelier, T. Praveen, Y. Andres, B. Bujoli, *Chem. Mater.* **2001**, *13*, 2879.
30. L. Pauling, *The Nature of the Chemical Bond*, 3rd edn, Ithaca, 1960.
31. (a) J. V. Smith, W. J. Dytrych, *Nature* **1984**, *309*, 607; (b) J. V. Smith, *Chem. Rev.* **1988**, *88*, 149; (c) M. E. Davis, *Acc. Chem. Res.* **1993**, *26*, 111; (d) C. S. Cundy, P. A. Cox, *Chem. Rev.* **2003**, *103*, 663.
32. (a) G. A. Ozin, C. Gil, *Chem. Rev.* **1989**, *89*, 1749; (b) G. A. Ozin, S. Özkar, R. A. Prokopowicz, *Acc. Chem. Res.* **1992**, *25*, 553; (c) M. Ichikawa, *Adv. Catal.* **1992**, *38*, 283; (d) W. M. H. Sachtler, Z. Zhang, *Adv. Catal.* **1993**, *39*, 129; (e) W. M. H. Sachtler, *Acc. Chem. Res.* **1993**, *26*, 383; (f) G. A. Ozin, *Adv. Mater.* **1994**, *6*, 71; (g) T. Bein, in G. Alberti, T. Bein (Eds.), *Comprehensive Supramolecular Chemistry*, Vol. 7, Elsevier, Oxford, 1996, p. 579; (f) A. M. Schneider, P. Behrens, *Chem. Mater.* **1998**, *10*, 679.
33. F. Lefebvre, A. de Mallmann, J.-M. Basset, *Eur. J. Inorg. Chem.* **1999**, 361.
34. For review articles, see: (a) F. Schüth, *Ber. Bunsenew. Phys. Chem.* **1995**, *99*, 1306; (b) A. Sayari, *Chem. Mater.* **1996**, *8*, 1840; (c) X. S. Zhao, G. Q. Lu, G. J. Millar, *Ind. Eng. Chem. Res.* **1996**, *35*, 2075; (d) A. Corma, *Chem. Rev.* **1997**, *97*, 2373; (e) S. Biz, M. L. Occelli, *Catal. Rev. -Sci. Eng.* **1998**, *40*, 329; (f) U. Ciesla, F. Schüth, *Microporous Mesoporous Mater.* **1999**, *27*, 131; (g) G. J. de A. A. Soler-Illia, C. Sanchez, B. Lebeau, J. Patarin, *Chem. Rev.* **2002**, *102*, 4093.
35. (a) D. O'Hare, in *Inorganic Materials*, D. W. Bruce, D. O'Hare (Eds.), Wiley, Chichester, 1992, p. 166; (b) D. O'Hare, *New J. Chem.* **1994**, *18*, 989; (c) Y. Morikawa, *Adv. Catal.* **1993**, *39*, 303; (d) J. T. Kloprogge, *J. Porous Mater.* **1998**, *5*, 5; (e) A. Gil, L. M. Gandía, M. A. Vicente, *Catal. Rev.-Sci. Eng.* **2000**, *42*, 145.
36. E. Lindner, T. Schneller, F. Auer, H. A. Mayer, *Angew. Chem. Int. Ed.* **1999**, *38*, 2155.
37. E. F. Vansant, P. Van Der Voort, K. C. Vrancken (Eds.), *Characterization and Chemical Modification of the Silica Surface, Studies in Surface Science and Catalysis*, Vol. 93, Elsevier, Amsterdam, 1995.
38. M. W. Anderson, C. C. Egger, G. J. T. Tiddy, J. L. Casci, K. A. Brakke, *Angew. Chem. Int. Ed.* **2005**, *44*, 3243.
39. (a) C. T. Kresge, M. E. Leonowicz, W. J. Roth, J. C. Vartuli, J. S. Beck, *Nature* **1992**, *359*, 710; (b) J. S. Beck, J. C. Vartuli, W. J. Roth, M. E. Leonowicz, C. T. Kresge, K. D. Schmitt, C. T.-W. Chu, D. H. Olson, E. W. Sheppard, S. B. McCullen, J. B. Higgins, J. L. Schlenker, *J. Am. Chem. Soc.* **1992**, *114*, 10834.
40. M. S. Morey, A. Davidson, G. D. Stucky, *J. Porous Mater.* **1998**, *5*, 195.
41. (a) T. Yanagisawa, T. Shimizu, K. Kuroda, C. Kato, *Bull. Chem. Soc. Jpn.* **1990**, *63*, 988; (b) S. Inagaki, Y. Fukushima, K. Kuroda, *J. Chem. Soc., Chem. Commun.* **1993**, 680; (c) S. Inagaki, A. Koiwai, N. Suzuki, Y. Fukushima, K. Kuroda, *Bull. Chem. Soc. Jpn.* **1996**, *69*, 1149.
42. (a) D. Zhao, J. Feng, Q. Huo, N. Melosh, G. H. Fredrickson, B. F. Chmelka, G. D. Stucky, *Science* **1998**, *279*, 548; (b) D. Zhao, Q. Huo, J. Feng, B. F. Chmelka, G. D. Stucky, *J. Am. Chem. Soc.* **1998**, *120*, 6024.
43. (a) C. Huo, R. Leon, P. M. Petroff, G. D. Stucky, *Science* **1995**, *268*, 1324; (b) C. Huo, D. I. Margolese, G. D. Stucky, *Chem. Mater.* **1996**, *8*, 1147; (c) Y. Sakamoto, M. Kaneda, O. Terasaki, D. Y. Zhao, J. M. Kim, G. Stucky, H. J. Shin, R. Ryoo, *Nature* **2000**, *408*, 449; (d) T.-W. Kim, R. Ryoo, M. Kruk, K. P. Gierszal, M. Jaroniec, S. Kamiya, O. Terasaki, *J. Phys. Chem. B* **2004**, *108*, 11480.
44. For an example, see: K. Mukhopadhyay, B. R. Sarkar, R. V. Chaudhari, *J. Am. Chem. Soc.* **2002**, *124*, 9692.
45. For an example, see: I. Nagl, M. Widenmeyer, S. Grasser, K. Köhler, R. Anwander, *J. Am. Chem. Soc.* **2000**, *122*, 1544.
46. (a) R. de Ruiter, J. C. Jansen, H. van Bekkum, in *Molecular Sieves*, M. L. Occelli, H. E. Robson (Eds.), Van Nostrand Reinhold, New York, 1992, p. 167; (b) J. Murray, M. J. Sharp, J. A. Hockey, *J. Catal.* **1970**, *18*, 52.
47. (a) Z. Li, T. E. Mallouk, *J. Phys. Chem.* **1987**, *91*, 643; (b) G. A. Ozin, J. Godber, *J. Phys. Chem.* **1989**, *93*, 878.
48. A. Théolier, E. Custodero, A. Choplin, J.-M. Basset, F. Raatz, *Angew. Chem., Int. Ed. Engl.* **1990**, *29*, 805.
49. D. J. Cole-Hamilton, *Science* **2003**, *299*, 1702.
50. C. Bianchini, D. G. Burnaby, J. Evans, P. Frediani, A. Meli, W. Oberhauser, R. Psaro, L. Sordelli, F. Vizza, *J. Am. Chem. Soc.* **1999**, *121*, 5961.
51. C. Bianchini, P. Barbaro, V. Dal Santo, R. Gobetto, A. Meli, W. Oberhauser, R. Psaro, F. Vizza, *Adv. Synth. Catal.* **2001**, *343*, 41.
52. A. Hu, G. T. Yee, W. Lin, *J. Am. Chem. Soc.* **2005**, *127*, 12486.
53. F. M. de Rege, D. K. Morita, K. C. Ott, W. Tumas, R. D. Broene, *Chem. Commun.* **2000**, 1797.
54. R. Augustine, S. Tanielyan, S. Anderson, H. Yang, *Chem. Commun.* **1999**, 1257.
55. M. Tamura, H. Fujihara, *J. Am. Chem. Soc.* **2003**, *125*, 15742.
56. B. M. Choudary, M. L. Kantam, K. V. S. Ranganath, K. Mahendar, B. Sreedhar, *J. Am. Chem. Soc.* **2004**, *126*, 3396.
57. P. D. Stevens, G. Li, J. Fan, M. Yen, Y. Gao, *Chem. Commun.* **2005**, 4435.
58. R. Raja, J. M. Thomas, M. D. Jones, B. F. G. Johnson, D. E. W. Vaughan, *J. Am. Chem. Soc.* **2003**, *125*, 14982.
59. C. Simons, U. Hanafeld, I. W. C. E. Arends, R. A. Sheldon, T. Maschmeyer, *Chem. Eur. J.* **2004**, *10*, 5829.
60. G.-J. Kim, S.-H. Kim, *Catal. Lett.* **1999**, *57*, 139.
61. (a) A. Cervilla, A. Corma, V. Formes, E. Llopis, P. Palanca, F. Rey, A. Ribeira, *J. Am. Chem. Soc.* **1994**, *116*, 1595; (b) A. Corma, F. Rey, J. M. Thomas, G. Sankar, G. N. Greaves, A. Cervilla, E. Llopis, A. Ribeira, *Chem. Commun.* **1996**, 1613.

62. C. P. Nicholas, H. Ahn, T. J. Marks, *Organometallics* **2002**, *21*, 1788.
63. (a) J. P. Arhancet, M. E. Davis, J. S. Merola, B. E. Hanson, *Nature* **1989**, *339*, 454; (b) M. E. Davis, *CHEMTECH* **1992**, 498.
64. (a) K. T. Wan, M. E. Davis, *Nature* **1994**, *370*, 449; (b) K. T. Wan, M. E. Davis, *J. Catal.* **1995**, *152*, 25.
65. R. Neumann, M. Cohen, *Angew. Chem. Int. Ed.* **1997**, *36*, 1738.
66. M. H. Valkenberg, C. deCastro, W. F. Hölderich, *Green Chemistry* **2002**, *4*, 88.
67. C. P. Mehnert, *Chem. Eur. J.* **2005**, *11*, 50.
68. C. P. Mehnert, R. A. Cook, N. C. Dispenziere, M. Afeworki, *J. Am. Chem. Soc.* **2002**, *124*, 12932.
69. C. P. Mehnert, E. J. Mozeleski, R. A. Cook, *Chem. Commun.* **2002**, 3010.
70. H. H. Wagner, H. Hausmann, W. F. Hölderich, *J. Catal.* **2001**, *203*, 150.
71. D. E. De Vos, F. Thibault-Starzyk, P. P. Knops-Gerrits, R. F. Parton, P. A. Jacobs, *Macromol. Symp.* **1994**, *80*, 157.
72. G. H. Hutchings, *Chem. Commun.* **1999**, 301.
73. A. Corma, H. Garcia, *Eur. J. Inorg. Chem.* **2004**, 1143.
74. For the first example of metal complexes in zeolite hosts, see: W. DeWilde, G. Peeters, J. H. Lunsford, *J. Phys. Chem.* **1980**, *84*, 2306.
75. N. Herron, G. D. Stucky, C. A. Tolman, *Inorg. Chim. Acta* **1985**, *100*, 135.
76. N. Herron, *Inorg. Chem.* **1986**, *25*, 4714.
77. N. Herron, *CHEMTECH* **1989**, 542.
78. R. F. Parton, I. F. J. Vankelecom, M. J. A. Casselman, C. P. Bezoukhanova, J. B. Uytterhoeven, P. A. Jacobs, *Nature* **1994**, *370*, 541.
79. P.-P. Knops-Gerrits, D. De Vos, F. Thibault-Starzyk, P. A. Jacobs, *Nature* **1994**, *369*, 543.
80. S. H. Bossmann, N. Shahin, H. L. Thanh, A. Bonfill, M. Wörner, A. M. Braun, *Chem. Phys. Chem.* **2002**, *3*, 401.
81. (a) S. B. Ogunwumi, T. Bein, *Chem. Commun.* **1997**, 901; (b) M. J. Sabater, S. García, M. Alvaro, H. García, J. C. Scaiano, *J. Am. Chem. Soc.* **1998**, *120*, 8521.
82. S. Kowalak, R. C. Weiss, K. J. Balkus, Jr., *J. Chem. Soc., Chem. Commun.* **1991**, 57.
83. S. Taylor, J. Gullick, P. McMorn, D. Bethell, P. C. Bolman Page, F. E. Hancock, F. King, G. J. Hutchings, *J. Chem. Soc., Perkin Trans. 2* **2001**, 1714.
84. M. J. Sabater, A. Corma, A. Domenech, V. Fornés, H. García, *Chem. Commun.* **1997**, 1285.
85. G. Gbery, A. Zsigmond, K. J. Balkus, Jr., *Catal. Lett.* **2001**, *74*, 77.
86. C. Schuster, W. F. Hölderich, *Catal. Today* **2000**, *60*, 193.
87. K. J. Balkus, Jr., M. Eissa, R. Levado, *J. Am. Chem. Soc.* **1995**, *117*, 10753.
88. B.-Z. Zhan, X.-Y. Li, *Chem. Commun.* **1998**, 349.
89. L. L. Hench, J. K. West, *Chem. Rev.* **1990**, *90*, 33.
90. D. Avnir, *Acc. Chem. Res.* **1995**, *28*, 328.
91. A. Rosenfeld, D. Avnir, J. Blum, *Chem. Commun.* **1993**, 583.
92. J. Blum, A. Rosenfeld, N. Polak, O. Israelson, H. Schumann, D. Avnir, *J. Mol. Catal. A* **1996**, *107*, 217.
93. (a) F. Gelman, J. Blum, D. Avnir, *New J. Chem.* **2003**, *27*, 205; (b) F. Gelman, J. Blum, D. Avnir, *J. Am. Chem. Soc.* **2000**, *122*, 11999.
94. F. Gelman, J. Blum, D. Avnir, *J. Am. Chem. Soc.* **2002**, *124*, 14460.
95. F. Gelman, J. Blum, D. Avnir, *Angew. Chem. Int Ed.* **2001**, *40*, 3647.
96. I. Vankelecom, D. Tas, R. F. Parton, V. Van de Vyver, P. A. Jacobs, *Angew. Chem. Int Ed. Engl.* **1996**, *35*, 1346.
97. A. Wolfson, S. Janssens, I. Vankelecom, S. Geresh, M. Landau, M. Gottlieb, M. Herskowitz, *Chem. Commun.* **1999**, 2407.
98. I. Vankelecom, A. Wolfson, S. Geresh, M. Gottlieb, M. Herskowitz, *Chem. Commun.* **2002**, 388.
99. (a) S. Kobayashi, *Top. Organomet. Chem.* **1999**, *2*, 285; (b) S. Kobayashi, R. Akiyama, *Chem. Commun.* **2003**, 449.
100. S. Kobayashi, S. Nagayama, *J. Am. Chem. Soc.* **1998**, *120*, 2985.
101. S. Kobayashi, M. Endo, S. Nagayama, *J. Am. Chem. Soc.* **1999**, *121*, 11229.
102. R. Akiyama, S. Kobayashi, *Angew. Chem. Int. Ed.* **2001**, *40*, 3469.
103. R. Akiyama, S. Kobayashi, *Angew. Chem. Int. Ed.* **2002**, *41*, 2602.
104. R. Akiyama, S. Kobayashi, *J. Am. Chem. Soc.* **2003**, *125*, 3412.
105. S. Kobayashi, H. Miyamura, R. Akiyama, T. Ishida, *J. Am. Chem. Soc.* **2005**, *127*, 9251.
106. S. R. Seidel, P. J. Stang, *Acc. Chem. Res.* **2002**, *35*, 972.
107. P. H. Dinolfo, J. T. Hupp, *Chem. Mater.* **2001**, *13*, 3113.
108. M. L. Merlau, M. del Pilar Mejia, S. T. Nguyen, J. T. Hupp, *Angew. Chem. Int. Ed.* **2001**, *40*, 4239.
109. G. A. Morris, S. T. Nguyen, J. T. Hupp, *J. Mol. Catal. A* **2001**, *174*, 15.
110. H.-U. Blaser, C. Malan, B. Pugin, F. Spindler, H. Steiner, M. Studer, *Adv. Synth. Catal.* **2003**, *345*, 103.
111. C. E. Song, S.-G. Lee, *Chem. Rev.* **2002**, *102*, 3495.
112. M. Heitbaum, F. Glorius, I. Escher, *Angew. Chem. Int. Ed.* **2006**, *45*, 4732.
113. J. R. Severn, J. C. Chadwick, R. Duchateau, N. Friederichs, *Chem. Rev.* **2005**, *105*, 4073.
114. R. H. Grubbs, L. C. Kroll, *J. Am. Chem. Soc.* **1971**, *93*, 3062.
115. M. Capka, P. Svoboda, M. Cerny, J. Hetflejs, *Tetrahedron Lett.* **1979**, *50*, 4787.
116. J. P. Collman, L. S. Hegedus, M. P. Cooke, J. R. Norton, G. Dolcetti, D. N. Marquardt, *J. Am. Chem. Soc.* **1972**, *94*, 1789.
117. W. Dumont, J.-C. Poulin, T.-P. Dang, H. B. Kagan, *J. Am. Chem. Soc.* **1973**, *95*, 8295.
118. J. M. Thomas, B. F. G. Johnson, R. Raja, G. Sankar, P. A. Midgley, *Acc. Chem. Res.* **2003**, *36*, 20.
119. T. P. Yoon, E. N. Jacobsen, *Science* **2003**, *299*, 1691.
120. (a) H. Gao, R. J. Angelici, *J. Am. Chem. Soc.* **1997**, *119*, 6937; (b) H. Gao, R. J. Angelici, *Organometallics* **1999**, *18*, 989.
121. M. D. Jones, R. Raja, J. M. Thomas, B. F. G. Johnson, D. W. Lewis, J. Rouzaud, K. D. M. Harris, *Angew. Chem. Int. Ed.* **2003**, *42*, 4326.
122. For more examples, see: (a) B. F. G. Johnson, S. A. Raynor, D. S. Shephard, T. Maschmeyer, J. M. Thomas, G. Sankar, S. Bromley, R. Oldroyd, L. Gladden, M. D. Mantle, *Chem. Commun.* **1999**, 1167; (b) A. Bleloch, B. F. G. Johnson, S. V. Ley, A. J. Price, D. S. Shephard, J. M. Thomas, *Chem. Commun.* **1999**, 1907; (c) S. A. Raynor, J. M. Thomas, R. Raja, B. F. G. Johnson, R. G. Bell, M. D. Mantle, *Chem. Commun.* **2000**, 1925; (d) D. S. Shephard, W. Zhou, T. Maschmeyer, J. M. Matters, C. L. Roper, S. Parson, B. F. G. Johnson, M. J. Duer, *Angew. Chem. Int. Ed.* **1998**, *37*, 2719.
123. S. Xiang, Y. Zhang, Q. Xin, C. Li, *Chem. Commun.* **2002**, 2696.
124. S. Xiang, Y. Zhang, Q. Xin, C. Li, *Angew. Chem. Int. Ed.* **2002**, *41*, 821.
125. A. Heckel, D. Seebach, *Angew. Chem. Int. Ed.* **2000**, *39*, 163.

126. C. J. Miller, D. O'Hare, *Chem. Commun.* **2004**, 1710.
127. H. C. L. Abbenhuis, *Angew. Chem. Int. Ed.* **1999**, *38*, 1058.
128. For an example of an immobilized hybrid catalyst for atom transfer radical polymerization (ATRP) of vinyl monomers, see: S. C. Hong, K. Matyjaszewski, *Macromolecules* **2002**, *35*, 7592.
129. C. Zapilko, R. Anwander, *Chem. Mater.* **2006**, *18*, 1479.
130. (a) M. W. McKittrick, C. W. Jones, *J. Am. Chem. Soc.* **2004**, *126*, 3052; (b) M. W. McKittrick, C. W. Jones, *Chem. Mater.* **2005**, *17*, 4758.
131. For early work on SOMC, see: (a) Y. I. Yermakov, A. M. Lazutkin, E. A. Demin, Y. P. Grabovskii, V. A. Zakharov, *Kinet. Katal.* **1972**, *13*, 1422; (b) V. A. Zakharov, V. K. Dudchenko, E. A. Paukshtis, L. G. Karakchiev, Y. I. Yermakov, *J. Mol. Catal.* **1977**, *2*, 421; (c) V. A. Zakharov, Y. I. Yermakov, *Catal. Rev. Sci. Eng.* **1979**, *19*, 67.
132. For review articles, see: (a) J. Schwartz, *Acc. Chem. Res.* **1985**, *18*, 302; (b) Y. Iwasawa, *Adv. Catal.* **1987**, *35*, 187; (c) H. H. Lamb, B. C. Gates, H. Knözinger, *Angew. Chem., Int. Ed. Engl.* **1988**, *27*, 1127; (d) B. C. Gates, H. H. Lamb, *J. Mol. Catal.* **1989**, *52*, 1; (e) Y. Iwasawa, B. C. Gates, *CHEMTECH* **1989**, 173; (g) T. J. Marks, *Acc. Chem. Res.* **1992**, *25*, 57; (f) A. Zecchina, C. O. Areán, *Catal. Rev. -Sci. Eng.* **1993**, *35*, 261; (h) J. Evans, *Chem. Soc. Rev.* **1997**, 11; (i) E. Cariati, D. Roberto, R. Ugo, E. Lucenti, *Chem. Rev.* **2003**, *103*, 3707; (j) C. Copéret, M. Chabanas, R. P. Saint-Arroman, J.-M. Basset, *Angew. Chem. Int. Ed.* **2003**, *42*, 156.
133. For a special issue on surface and interfacial organometallic chemistry, see: *Top. Organomet. Chem.* **2005**, 16.
134. J.-M. Basset, B. C. Gates, J. P. Candy, A. Choplin, H. Leconte, F. Quignard, C. Santini (Eds.), *Surface Organometallic Chemistry, Molecular Approaches to Surface Chemistry*, Kluwer, Dordrecht, 1988.
135. (a) T. Maschmeyer, F. Rey, G. Sankar, J. M. Thomas, *Nature* **1995**, *378*, 159; (b) R. D. Oldroyd, J. M. Thomas, T. Maschmeyer, P. A. MacFaul, D. W. Snelgrove, K. U. Ingold, D. D. M. Wayner, *Angew. Chem., Int. Ed. Engl.* **1996**, *35*, 2787.
136. K. Köhler, J. Engweiler, H. Viebrock, A. Baiker, *Langmuir* **1995**, *11*, 3423.
137. (a) G. A. Parks, *Chem. Rev.* **1965**, *65*, 177; (b) R. K. Iler, *The Chemsitry of Silica*, Wiley-Interscience, New York, 1979; (c) K. K. Unger, *Porous Silica*, Elsevier, Amsterdam, 1979.
138. R. Anwander, C. Palm, O. Groeger, G. Engelhardt, *Organometallics* **1998**, *17*, 2027.
139. M. P. McDaniel, *Adv. Catal.* **1985**, *33*, 47.
140. For examples, see: (a) F. J. Karol, G. L. Karapinka, C. Wu, A. W. Dow, R. N. Johnson, W. L. Carrick, *J. Polym. Sci. A* **1972**, *10*, 2621; (b) F. J. Karol, C. Wu, W. T. Reichle, N. J. Maraschin, *J. Catal.* **1979**, *60*, 68; (c) A. Zecchina, G. Spoto, S. Bordiga, *Faraday Diss. Chem. Soc.* **1989**, *87*, 149; (d) M. Schnellbach, F. H. Köhler, J. Blümel, *J. Organomet. Chem.* **1996**, *520*, 227; (e) J. A. N. Ajjou, A. L. Scott, *J. Am. Chem. Soc.* **2000**, *122*, 8968.
141. For reviews, see: (a) J. B. P. Soares, A. E. Hamielec, *Polym. React. Eng.* **1995**, *3*, 131; (b) G. Fink, R. Mülhaupt, H. H. Brintzinger (Eds.), *Ziegler Catalysts*, Springer Verlag, Berlin, 1995; (c) H.-H. Brintzinger, D. Fischer, R. Mülhaupt, B. Rieger, R. M. Waymouth, *Angew. Chem., Int. Ed. Engl.* **1995**, *34*, 1143; (d) M. Bochmann, *J. Chem. Soc., Dalton Trans.* **1996**, 255; (e) W. Kaminsky, *J. Chem. Soc., Dalton Trans.* **1998**, 1413.
142. (a) M. D. Ward, J. Schwartz, *J. Mol. Catal.* **1981**, *11*, 397; (b) M. D. Ward, J. Schwartz, *J. Am. Chem. Soc.* **1981**, *103*, 5253; (c) M. D. Ward, J. Schwartz, *Organometallics* **1982**, *1*, 1030; (d) S. A. King, J. Schwartz, *Inorg. Chem.* **1991**, *30*, 3771.
143. H. C. Foley, S. J. DeCanio, K. D. Tau, K. J. Chau, J. H. Onuferko, C. Dybowski, B. C. Gates, *J. Am. Chem. Soc.* **1983**, *105*, 3074.
144. (a) J. Corker, F. Lefebvre, C. Lécuyer, V. Dufaud, F. Quignard, A. Choplin, J. Evans, J.-M. Basset, *Science* **1996**, *271*, 966; (b) V. Dufaud, J.-M. Basset, *Angew. Chem. Int. Ed.* **1998**, *37*, 806.
145. F. Rataboul, A. Baudouin, C. Thieuleux, L. Veyre, C. Copéret, J. Thivolle-Cazat, J.-M. Basset, A. Lesage, L. Emsley, *J. Am. Chem. Soc.* **2004**, *126*, 12541.
146. W. Kaminsky, F. Hartmann, *Angew. Chem. Int. Ed.* **2000**, *39*, 331.
147. (a) V. Dufaud, G. P. Niccolai, J. Thivolle-Cazat, J.-M. Basset, *J. Am. Chem. Soc.* **1995**, *117*, 4288; (b) V. Vidal, A. Theolier, J. Thivolle-Cazat, J.-M. Basset, J. Corker, *J. Am. Chem. Soc.* **1996**, *118*, 4595.
148. (a) V. Vidal, A. Theolier, J. Thivolle-Cazat, J.-M. Basset, *Science* **1997**, *276*, 99; (b) O. Maury, L. Lefort, V. Vidal, J. Thivolle-Cazat, J.-M. Basset, *Angew. Chem. Int. Ed.* **1999**, *38*, 1952.
149. E. Le Roux, M. Taoufik, C. Copéret, A. De Mallmann, J. Thivolle-Cazat, J.-M. Basset, B. M. Maunders, G. J. Sunley, *Angew. Chem. Int. Ed.* **2005**, *44*, 6755.
150. (a) E. Le Roux, M. Chabanas, A. Baudouin, A. De Mallmann, C. Copéret, E. A. Quadrelli, J. Thivolle-Cazat, J.-M. Basset, W. Lukens, A. Lesage, L. Emsley, G. J. Sunley, *J. Am. Chem. Soc.* **2004**, *126*, 13391; (b) J.-M. Basset, C. Copéret, L. Lefort, B. M. Maunders, O. Maury, E. Le Roux, G. Saggio, S. Soignier, D. Soulivong, G. J. Sunley, M. Taoufik, J. Thivolle-Cazat, *J. Am. Chem. Soc.* **2005**, *127*, 8604.
151. H. Ahn, T. J. Marks, *J. Am. Chem. Soc.* **2002**, *124*, 7103.
152. C. Copéret, O. Maury, J. Thivolle-Cazat, J.-M. Basset, *Angew. Chem. Int. Ed.* **2001**, *40*, 2331.
153. F. Blanc, C. Copéret, J. Thivolle-Cazat, J.-M. Basset, A. Lesage, L. Emsley, A. Sinha, R. R. Schrock, *Angew. Chem. Int. Ed.* **2006**, *45*, 1216.
154. M. Chabanas, A. Baudouin, C. Copéret, J.-M. Basset, W. Lukens, A. Lesage, S. Hediger, L. Emsley, *J. Am. Chem. Soc.* **2003**, *125*, 492.
155. (a) G. G. Hlatky, *Chem. Rev.* **2000**, *100*, 1347; (b) G. Fink, B. Steinmetz, J. Zechlin, C. Przybyla, B. Tesche, *Chem. Rev.* **2000**, *100*, 1377; (c) E. Y.-X. Chen, T. J. Marks, *Chem. Rev.* **2000**, *100*, 1391.
156. C. P. Nicholas, H. Ahn, T. J. Marks, *J. Am. Chem. Soc.* **2003**, *125*, 4325.
157. M. Jezequel, V. Dufaud, M. J. Ruiz-Garcia, F. Carrillo-Hermosilla, U. Neugebauer, G. P. Niccolai, F. Lefebvre, F. Bayard, J. Corker, S. Fiddy, J. Evans, J.-P. Broyer, J. Malinge, J.-M. Basset, *J. Am. Chem. Soc.* **2001**, *123*, 3520.
158. For examples of SOAcC, see: (a) R. G. Bowman, R. Nakamura, P. J. Fagan, R. L. Burwell, Jr., T. J. Marks, *J. Chem. Soc., Chem. Commun.* **1981**, 257; (b) M.-Y. He, G. Xiong, P. J. Toscano, R. L. Burwell, Jr., T. J. Marks, *J. Am. Chem. Soc.* **1985**, *107*, 641; (c) P. J. Toscano, T. J. Marks, *J. Am. Chem. Soc.* **1985**, *107*, 653; (d) D. Hedden, T. J. Marks, *J. Am. Chem. Soc.* **1988**, *110*, 1647; (e) W. C. Finch, R. D. Gillespie, D. Hedden, T. J. Marks, *J. Am. Chem. Soc.* **1990**, *112*, 6221; (f) X. Yang, C. L. Stern, T. J. Marks, *Organometallics* **1991**, *10*, 840; (g) M. S. Eisen, T. J. Marks, *J. Mol. Catal.* **1994**, *86*, 23.
159. N. Millot, A. Cox, C. C. Santini, Y. Molard, J.-M. Basset, *Chem. Eur. J.* **2002**, *8*, 1438.

160. K. Tajima, T. Aida, *Chem. Commun.* **2000**, 2399.
161. K. Kageyama, J. Tamazawa, T. Aida, *Science* **1999**, *285*, 2113.
162. Y. S. Ko, T. K. Han, J. W. Park, S. I. Woo, *Macromol. Rapid. Commun.* **1996**, *17*, 749.
163. (a) J. Tudor, D. O'Hare, *Chem. Commun.* **1997**, 603; (b) S. O'Brien, J. Tudor, T. Maschmeyer, D. O'Hare, *Chem. Commun.* **1997**, 1905.
164. L. K. Van Looveren, D. F. M. C. Geysen, K. A. L. Vercruysse, B. H. J. Wouters, P. J. Grobet, P. A. Jacobs, *Angew. Chem., Int. Ed. Engl.* **1998**, *37*, 517.
165. A. Fischbach, M. G. Klimpel, M. Widenmeyer, E. Herdtweck, W. Scherer, R. Anwander, *Angew. Chem. Int. Ed.* **2004**, *43*, 2234.
166. D. C. Bradley, R. C. Mehrotra, I. P. Rothwell, A. Singh, *Alkoxo and Aryloxo Derivatives of Metals*, Academic Press, London, 2001.
167. For examples, see: (a) R. Mokaya, W. Jones, *Chem. Commun.* **1997**, 2185; (b) R. Anwander, C. Palm, *Stud. Surf. Sci. Catal.* **1998**, *117*, 413; (c) Z. Luan, E. M. Maes, P. A. W. van der Heide, D. Zhao, R. S. Czernuszewicz, L. Kevan, *Chem. Mater.* **1999**, *11*, 3680; (d) M. S. Morey, S. O'Brien, S. Schwarz, G. D. Stucky, *Chem. Mater.* **2000**, *12*, 898; (e) M. S. Morey, G. D. Stucky, S. Schwarz, M. Fröba, *J. Phys. Chem. B* **1999**, *103*, 2037; (f) M. S. Morey, J. D. Bryan, S. Schwarz, G. D. Stucky, *Chem. Mater.* **2000**, *12*, 3435.
168. M. Morey, A. Davidson, H. Eckert, G. D. Stucky, *Chem. Mater.* **1996**, *8*, 486.
169. (a) A. O. Bough, G. L. Rice, S. L. Scott, *J. Am. Chem. Soc.* **1999**, *121*, 7201; (b) G. L. Rice, S. L. Scott, *J. Mol. Catal. A* **1997**, *125*, 73.
170. For examples, see: (a) T. Tatsumi, K. A. Koyano, N. Igarashi, *Chem. Commun.* **1998**, 325; (b) M. B. D'Amore, S. Schwarz, *Chem. Commun.* **1999**, 121; (c) S. Klein, W. F. Maier, *Angew. Chem., Int. Ed. Engl.* **1996**, *35*, 2230.
171. R. Anwander, C. Palm, G. Gerstberger, O. Groeger, G. Engelhardt, *Chem. Commun.* **1998**, 1811.
172. (a) L. Horner, U. B. Kaps, *Liebigs Ann. Chem.* **1980**, 192; (b) V. J. Shiner, Jr., D. Whittaker, *J. Am. Chem. Soc.* **1969**, *91*, 394.
173. (a) T. D. Tilley, *J. Mol. Catal. A: Chem.* **2002**, *182/183*, 17; (b) J. Jarupatrakorn, T. D. Tilley, *J. Am. Chem. Soc.* **2002**, *124*, 8380; (c) C. Nozaki, C. G. Lugmair, A. T. Bell, T. D. Tilley, *J. Am. Chem. Soc.* **2002**, *124*, 13194; (d) K. L. Fujdala, T. D. Tilley, *J. Catal.* **2003**, *216*, 265; (e) I. J. Drake, K. L. Fujdala, S. Baxamusa, A. T. Bell, T. D. Tilley, *J. Phys. Chem. B* **2004**, *108*, 18421; (f) J. Jarupatrakorn, T. D. Tilley, *Dalton Trans.* **2004**, 2808; (g) R. L. Brutchey, C. G. Lugmair, L. O. Schebaum, T. D. Tilley, *J. Catal.* **2005**, *229*, 72; (h) A. W. Holland, G. Li, A. M. Shahin, G. L. Long, A. T. Bell, T. D. Tilley, *J. Catal.* **2005**, *235*, 150; (i) J. Jarupatrakorn, M. P. Coles, T. D. Tilley, *Chem. Mater.* **2005**, *17*, 1818; (j) D. A. Ruddy, N. L. Ohler, A. T. Bell, T. D. Tilley, *J. Catal.* **2006**, *238*, 277.
174. C. Roveda, T. L. Church, H. Alper, S. L. Scott, *Chem. Mater.* **2000**, *12*, 857.
175. For examples, see: (a) B. M. Weckhuysen, R. R. Rao, J. Pelgrims, R. A. Schoonheydt, P. Bodart, G. Debras, O. Collart, P. Van Der Voort, E. F. Vansant, *Chem. Eur. J.* **2000**, *6*, 2960; (b) P. Van der Voort, M. Baltes, E. F. Vansant, *J. Phys. Chem. B* **1999**, *103*, 10102.
176. M. F. Lappert, P. P. Power, A. R. Sanger, R. C. Srivastava, *Metal And Metalloid Amides*, Ellis Horwood, Chichester, 1980.
177. R. Anwander, R. Roesky, *J. Chem. Soc., Dalton Trans.* **1997**, 137.
178. M. Widenmeyer, S. Grasser, K. Köhler, R. Anwander, *Microporous Mesoporous Mater.* **2001**, *44/45*, 327.
179. R. Anwander, O. Runte, J. Eppinger, G. Gerstberger, E. Herdtweck, M. Spiegler, *J. Chem. Soc., Dalton Trans.* **1998**, 847.
180. R. Anwander, H. Görlitzer, C. Palm, O. Runte, M. Spiegler, *J. Chem. Soc., Dalton Trans.* **1999**, 3611.
181. G. Gerstberger, C. Palm, R. Anwander, *Chem. Eur. J.* **1999**, *5*, 997.
182. C. Zapilko, R. Anwander, *Stud. Surf. Sci. Catal.* **2005**, *158*, 461.
183. M. Widenmeyer, PhD thesis, Technische Universität München, 2000.
184. Y. Liang, R. Anwander, *Dalton Trans.* **2006**, 1909.
185. (a) I. Nagl, M. Widenmeyer, E. Herdtweck, G. Raudaschl-Sieber, R. Anwander, *Microporous Mesoporous Mater.* **2001**, *44–45*, 311; (b) R. Anwander, I. Nagl, C. Zapilko, M. Widenmeyer, *Tetrahedron* **2003**, *59*, 10567.
186. R. M. Gauvin, A. Mortreux, *Chem. Commun.* **2005**, 1146.
187. T. J. Woodman Y. Sarazin, G. Fink, K. Hauschild, M. Bochmann, *Macromolecules* **2005**, *38*, 3060.
188. G. Gerstberger, R. Anwander, *Microporous Mesoporous Mater.* **2001**, *44/45*, 303.
189. (a) W. Hertl, M. L. Hair, *J. Phys. Chem.* **1971**, *75*, 2181; (b) R. Anwander, I. Nagl, M. Widenmeyer, G. Engelhardt, O. Groeger, C. Palm, T. Röser, *J. Phys. Chem. B* **2000**, *104*, 3532.
190. (a) Y. T. Struchkov, S. V. Lindeman, *J. Organomet. Chem.* **1995**, *488*, 9; (b) R. Murugavel, A. Voigt, M. G. Walawalkar, H. W. Roesky, *Chem. Rev.* **1996**, *96*, 2205; (c) V. Lorenz, A. Fischer, S. Gießmann, J. W. Gilje, Y. Gun'ko, K. Jacob, F. T. Edelmann, *Coord. Chem. Rev.* **2000**, *206–207*, 321.
191. For review articles on silsesquisiloxane model oxo surfaces, see: (a) F. J. Feher, T. A. Budzichowski, *Polyhedron* **1995**, *14*, 3239; (b) H. C. L. Abbenhuis, *Chem. Eur. J.* **2000**, *6*, 25; (c) R. Duchateau, *Chem. Rev.* **2002**, *102*, 3525; (d) R. W. J. M. Hanssen, R. A. van Santen, H. C. L. Abbenhuis, *Eur. J. Inorg. Chem.* **2004**, 675.
192. For review articles on calix[4]arene model oxo surfaces, see: (a) C. Floriani, *Chem. Eur. J.* **1999**, *5*, 19; (b) C. Floriani, R. Floriani-Moro, *Adv. Organomet. Chem.* **2001**, *47*, 167.
193. (a) V. W. Day, C. W. Earley, W. G. Klemperer, D. J. Maltbie, *J. Am. Chem. Soc.* **1985**, *107*, 8261; (b) M. Pohl, D. K. Lyon, N. Mizuno, K. Nomiya, R. G. Finke, *Inorg. Chem.* **1995**, *34*, 1413.
194. For examples, see: (a) W. A. Herrmann, A. W. Stumpf, T. Priermeier, S. Bogdanovic, V. Dufaud, J.-M. Basset, *Angew. Chem., Int. Ed. Engl.* **1996**, *35*, 2803; (b) J.-Y. Piquemal, S. Halut, J.-M. Brégeault, *Angew. Chem., Int. Ed. Engl.* **1998**, *37*, 1146.
195. For examples, see: (a) S. Shang, d. V. Khasnis, J. M. Burton, C. J. Santini, M. Fan, A. C. Small, M. Lattman, *Organometallics* **1994**, *13*, 5157; (b) R. Anwander, J. Eppinger, I. Nagl, W. Scherer, M. Tafipolsky, P. Sirsch, *Inorg. Chem.* **2000**, *39*, 4713; (c) E. Hoppe, C. Limberg, B. Ziemer, C. Mügge, *J. Mol. Catal. A: Chem.* **2006**, *251*, 34.
196. J. F. Brown, Jr., L. H. Vogt, Jr., *J. Am. Chem. Soc.* **1965**, *87*, 4313.
197. (a) K. Araki, K. Iwamoto, S. Shinkai, T. Matsuda, *Bull. Chem. Soc. Jpn.* **1990**, *63*, 3480; (b) S. Shinkai, *Tetrahedron* **1993**, *40*, 8933.
198. L.-D. Dai, *Angew. Chem. Int. Ed.* **2004**, *43*, 5726.
199. C. L. Bowes, G. A. Ozin, *Adv. Mater.* **1996**, *8*, 13.

200. For recent reviews on porous metal–organic frameworks, see: (a) P. J. Hagrman, D. Hagrman, J. Zubieta, *Angew. Chem. Int. Ed.* **1999**, *38*, 2638; (b) S. L. James, *Chem. Soc. Rev.* **2003**, *32*, 276; (c) O. M. Yaghi, M. O'Keeffe, N. W. Ockwig, H. K. Chae, M. Eddaoudi, J. Kim, *Nature* **2003**, *423*, 705; (d) C. Janiak, *Dalton Trans.* **2003**, 2781; (e) S. Kitagawa, R. Kitaura, S. Noro, *Angew. Chem. Int. Ed.* **2004**, *43*, 2334; (f) C. N. R. Rao, S. Natarajan, R. Vaidhyanathan, *Angew. Chem. Int. Ed.* **2004**, *43*, 1466; (g) I. Goldberg, *Chem. Commun.* **2005**, 1243; (h) J. L. C. Rowsell, O. M. Yaghi, *Angew. Chem. Int. Ed.* **2005**, *44*, 4670; (i) N. W. Ockwig, O. Delgado-Friedrichs, M. O'Keefe, O. M. Yaghi, *Acc. Chem. Res.* **2005**, *38*, 176; (j) G. Férey, C. Mellot-Draznieks, C. Serre, F. Millange, *Acc. Chem. Res.* **2005**, *38*, 217; (k) D. Bradshaw, J. B. Claridge, E. J. Cussen, T. J. Prior, M. J. Rosseinsky, *Acc. Chem. Res.* **2005**, *38*, 273; (l) R. J. Hill, D.-L. Long, N. R. Champness, P. Hubberstey, M. Schröder, *Acc. Chem. Res.* **2005**, *38*, 335.
201. For examples, see: (a) H. Li, M. Eddaoudi, M. O'Keeffe, O. M. Yaghi, *Nature* **1999**, *402*, 276; (b) S. S.-Y. Chui, S. M.-F. Lo, J. P. H. Charmant, A. G. Orpen, I. D. Williams, *Science* **1999**, *283*, 1148; (c) M. Eddaoudi, J. Kim, N. Rosi, D. Vodak, J. Wachter, M. O'Keeffe, O. M. Yaghi, *Science* **2002**, *295*, 469; (d) H. K. Chae, D. Y. Siberio-Pérez, J. Kim, Y. Go, M. Eddaoudi, A. J. Matzger, M. O'Keeffe, O. M. Yaghi, *Nature* **2004**, *427*, 523.
202. (a) M. Fujita, Y. J. Kwon, S. Washizu, K. Ogura, *J. Am. Chem. Soc.* **1994**, *116*, 1151; (b) O. Ohmori, M. Fujita, *Chem. Commun.* **2004**, 1586.
203. R. Matsuda, R. Kitaura, S. Kitagawa, Y. Kutoba, R. V. Belosludov, T. C. Kobayashi, H. Sakamoto, T. Chiba, M. Takata, Y. Kawazoe, Y. Mita, *Nature* **2005**, *436*, 238.
204. T. Uemura, K. Kitagawa, S. Horike, T. Kawamura, S. Kitagawa, M. Mizuno, K. Endo, *Chem. Commun.* **2005**, 5968.
205. T. Uemura, R. Kitaura, Y. Ohta, M. Nagaoka, S. Kitagawa, *Angew. Chem. Int. Ed.* **2006**, *45*, 4112.
206. T. Sawaki, T. Dewa, Y. Aoyama, *J. Am. Chem. Soc.* **1998**, *120*, 8539.
207. T. Sawaki, Y. Aoyama, *J. Am. Chem. Soc.* **1999**, *121*, 4793.
208. (a) T. Saiki, Y. Aoyama, *Chem. Lett.* **1999**, 797; (b) T. Dewa, T. Saiki, Y. Aoyama, *J. Am. Chem. Soc.* **2001**, *123*, 502.
209. J. S. Seo, D. Whang, H. Lee, S. I. Jun, Y. J. Jeon, K. Kim, *Nature* **2000**, *404*, 982.
210. Y. Cui, O. R. Evans, H. L. Ngo, P. S. White, W. Lin, *Angew. Chem. Int. Ed.* **2002**, *41*, 1159.
211. D. N. Dybtsev, A. L. Nuzhdin, H. Chun, K. P. Bryliakov, E. P. Talsi, V. P. Fedin, K. Kim, *Angew. Chem. Int. Ed.* **2006**, *45*, 916.
212. (a) L. Pu, *Chem. Rev.* **1998**, *98*, 2405; (b) L. Pu, *Chem. Eur. J.* **1999**, 2227.
213. (a) S. Matsunaga, T. Oshima, M. Shibasaki, *Adv. Synth. Catal.* **2002**, *344*, 3; (b) M. Shibasaki, S. Matsunaga, *Chem. Soc. Rev.* **2006**, *35*, 269.
214. X. Wang, L. Shi, M. Li, K. Ding, *Angew. Chem. Int. Ed.* **2005**, *44*, 6362.
215. (a) S. Takizawa, H. Somei, D. Jayaprakash, H. Sasai, *Angew. Chem. Int. Ed.* **2003**, *42*, 5711; (b) X. Wang, H. Guo, Z. Wang, K. Ding, *Chem. Eur. J.* **2005**, *11*, 4078.
216. X. Wang, K. Ding, *J. Am. Chem. Soc.* **2004**, *126*, 10524.
217. Y. Liang, Q. Jing, X. Li, L. Shi, K. Ding, *J. Am. Chem. Soc.* **2005**, *127*, 7694.
218. C.-D. Wu, A. Hu, L. Zhang, W. Lin, *J. Am. Chem. Soc.* **2005**, *127*, 8940.
219. (a) U. Schubert, *New J. Chem.* **1994**, *18*, 149; (b) B. Breitscheidel, J. Zieder, U. Schubert, *Chem. Mater.* **1991**, *3*, 559.
220. J. J. E. Moreau, M. W. C. Man, *Coord. Chem. Rev.* **1998**, *178/180*, 1073.
221. Z.-L. Lu, E. Lindner, H. A. Mayer, *Chem. Rev.* **2002**, *102*, 3543.
222. K. J. Shea, D. A. Loy, *Chem. Mater.* **2001**, *13*, 3306.
223. G. Cerveau, R. J. P. Corriu, E. Framery, *Chem. Mater.* **2001**, *13*, 3373.
224. C. Sanchez, G. J. De. A. A. Soler-Illia, F. Ribot, D. Grosso, *C. R. Chimie* **2003**, *6*, 1131.
225. (a) L. Schmid, M. Rohr, A. Baiker, *Chem. Commun.* **1999**, 2303; (b) E. F. Murphy, L. Schmid, T. Bürgi, M. Maciejewski, A. Baiker, D. Günther, M. Schneider, *Chem. Mater.* **2001**, *13*, 1296.
226. (a) O. Kröcher, R. A. Köppel, A. Baiker, *Chem. Commun.* **1996**, 1497; (b) O. Kröcher, R. A. Köppel, M. Fröba, A. Baiker, *J. Catal.* **1998**, *178*, 284.
227. P. Battioni, E. Cardin, M. Louloudi, B. Schöllhorn, G. A. Spyroulias, D. Mansuy, T. G. Traylor, *Chem. Commun.* **1996**, 2037.
228. V. Dufaud, F. Beauchesne, L. Bonneviot, *Angew. Chem. Int. Ed.* **2005**, *44*, 3475.
229. C. Baleizao, B. Gigante, D. Das, M. Alvaro, H. Garcia, A. Corma, *Chem. Commun.* **2003**, 1860.
230. A. Hu, H. L. Ngo, W. Lin, *Angew. Chem. Int. Ed.* **2003**, *42*, 6000.
231. A. Hu, H. L. Ngo, W. Lin, *J. Am. Chem. Soc.* **2003**, *125*, 11490.
232. (a) K. C. Hultzsch, J. A. Jernelius, A. H. Hoveyda, R. R. Schrock, *Angew. Chem. Int. Ed.* **2002**, *41*, 589; (b) S. J. Dolman, K. C. Hultzsch, F. Pezet, X. Teng, A. H. Hoveyda, R. R. Schrock, *J. Am. Chem. Soc.* **2004**, *126*, 10945.
233. (a) T. Arai, T. Sekiguti, K. Otsuki, S. Takizawa, H. Sasai, *Angew. Chem. Int. Ed.* **2003**, *42*, 2144; (b) T. Sekiguti, Y. Iizuka, S. Takizawa, D. Jayaprakash, T. Arai, H. Sasai, *Org. Lett.* **2003**, *5*, 2647.
234. Q.-H. Fan, C.-Y. Ren, C.-H. Yeung, W.-H. Hu, A. S. C. Chan, *J. Am. Chem. Soc.* **1999**, *121*, 7401.
235. (a) H. G. Alt, *J. Chem. Soc., Dalton Trans.* **1999**, 1703; (b) H. G. Alt, *Dalton Trans.* **2005**, 3271.
236. J. Zhang, X. Wang, G.-X. Jin, *Coord. Chem. Rev.* **2006**, *250*, 95.
237. B. Peifer, W. Milius, H. G. Alt, *J. Organomet. Chem.* **1998**, *553*, 205.
238. D. Zhang, G.-X. Jin, N.-H. Hu, *Eur. J. Inorg. Chem.* **2003**, 1570.
239. A. I. Licht, H. G. Alt, *J. Organomet. Chem.* **2003**, *687*, 142.
240. C. E. Denner, H. G. Alt, *J. Appl. Polym. Sci.* **2003**, *89*, 3379.
241. (a) G. Wulff, *Trends Biotechnol.* **1993**, *11*, 85; (b) K. J. Shea, *Trends Polym. Sci.* **1994**, *2*, 166; (c) K. Mosbach, *Trends Biochem. Sci.* **1994**, *19*, 9; (d) M. Yan, O. Ramströhm (Eds.), *Molecularly Imprinted Materials – Science and Technology*, Marcel Dekker, New York, 2005.
242. (a) R. A. Lerner, S. V. Benkovic, P. G. Schultz, *Science* **1991**, *252*, 659; (b) G. Vlatakis, L. I. Andersson, R. Müller, K. Mosbach, *Nature* **1993**, *361*, 645.
243. For reviews, see: (a) M. E. Davis, A. Katz, W. R. Ahmad, *Chem. Mater.* **1996**, *8*, 1820; (b) K. Severin, *Curr. Opin. Chem. Biol.* **2000**, *4*, 710; (c) G. Wulff, *Chem. Rev.* **2002**, *102*, 1; (d) C. Alexander, L. Davidson, W. Hayes, *Tetrahedron* **2003**, *59*, 2025; (e) M. Tada, Y. Iwasawa, *J. Mol. Catal. A: Chem.* **2003**, *199*, 115; (f) M. Tada, Y. Iwasawa, *J. Mol. Catal. A: Chem.* **2003**, *204/205*, 27; (g) J. J. Becker, M. R. Gagné, *Acc. Chem. Res.* **2004**, *37*, 798.

244. (a) K. M. Padden, J. F. Krebs, C. E. MacBeth, R. C. Scarrow, A. S. Borovik, *J. Am. Chem. Soc.* **2001**, *123*, 1072; (b) S. C. Zimmerman, M. S. Wendland, N. A. Rakow, I. Zharov, K. S. Suslick, *Nature* **2002**, *418*, 399; (c) Z. Cheng, L. Zhang, Y. Li, *Chem. Eur. J.* **2004**, *10*, 3555.
245. J. Matsui, I. A. Nicholls, I. Karube, K. Mosbach, *J. Org. Chem.* **1996**, *61*, 5414.
246. F. Locatelli, P. Gamez, M. Lemaire, *J. Mol. Catal. A: Chem.* **1998**, *135*, 89.
247. B. P. Santora, A. O. Larsen, M. R. Cagne, *Organometallics* **1998**, *17*, 3138.
248. (a) K. Polborn, K. Severin, *Chem. Commun.* **1999**, 2481; (b) K. Polborn, K. Severin, *Chem. Eur. J.* **2000**, *6*, 4604.
249. A. N. Cammidge, N. J. Baines, R. K. Bellingham, *Chem. Commun.* **2001**, 2588.
250. F. Viton, P. S. White, M. R. Cagné, *Chem. Commun.* **2003**, 3040.
251. (a) M. Tida, T. Sasaki, Y. Iwasawa, *J. Catal.* **2002**, *211*, 496; (b) M. Tida, T. Sasaki, Y. Iwasawa, *Phys. Chem. Chem. Phys.* **2002**, *4*, 4561; (c) M. Tida, T. Sasaki, T. Shido, Y. Iwasawa, *Phys. Chem. Chem. Phys.* **2002**, *4*, 5899.
252. E. Burri, M. Öhm, C. Dauenet, K. Severin, *Chem. Eur. J.* **2005**, *11*, 5055.
253. N. M. Brunkan, M. R. Cagné, *J. Am. Chem. Soc.* **2000**, *122*, 6217.
254. A. K. Kakkar, *Chem. Rev.* **2002**, *102*, 3579.
255. K. Töllner, R. Popovitz-Biro, M. Lahav, D. Milstein, *Science* **1997**, *278*, 2100.
256. M. G. L. Petrucci, A. K. Kakkar, *Organometallics* **1998**, *17*, 1798.
257. M. G. L. Petrucci, A. K. Kakkar, *Chem. Mater.* **1999**, *11*, 269.
258. K. Nozaki, M. Yoshida, H. Takaya, *J. Organomet. Chem.* **1994**, *473*, 253.
259. (a) R. Selke, J. Holz, A. Riepe, A. Börner, *Chem. Eur. J.* **1998**, *4*, 769; (b) T. Dwars, U. Schmidt, C. Fischer, I. Grassert, R. Kempe, R. Fröhlich, K. Drauz, G. Oehme, *Angew. Chem. Int. Ed.* **1998**, *37*, 2851.
260. M. Schreuder Goedheijt, B. E. Hanson, J. N. H. Reek, P. C. J. Kamer, P. W. N. M. van Leeuwen, *J. Am. Chem. Soc.* **2000**, *122*, 1650.
261. For the enantioselective hydrogenation of enamides and itaconic acid in the presence of a water-soluble Rh(I) catalyst and SDS, see: K. Yonehara, K. Ohe, S. Uemura, *J. Org. Chem.* **1999**, *64*, 9381.
262. K. Manabe, Y. Mori, T. Wakabayashi, S. Nagayama, S. Kobayashi, *J. Am. Chem. Soc.* **2000**, *122*, 7202.
263. (a) S. Kobayashi, T. Wakabayashi, S. Nagayama, H. Oyamada, *Tetrahedron Lett.* **1997**, *38*, 4559; (b) S. Kobayashi, T. Wakabayashi, H. Oyamada, *Chem. Lett.* **1997**, 831; (c) S. Kobayashi, T. Busujima, S. Nagayama, *Synlett.* **1999**, 545.
264. F. M. Menger, T. Tsuno, *J. Am. Chem. Soc.* **1990**, *112*, 6723.
265. E. Ruckenstein, L. Hong, *Chem. Mater.* **1992**, *4*, 122.
266. M. J. Sundell, E. O. Pajunen, O. E. O. Hormi, J. H. Näsman, *Chem. Mater.* **1993**, *5*, 372.
267. H. Huang, E. E. Remsen, K. L. Wooley, *Chem. Commun.* **1998**, 1415.
268. (a) D. L. Gin, W. Gu, B. A. Pindzola, W.-J. Zhou, *Acc. Chem. Res.* **2001**, *34*, 973; (b) D. L. Gin, W. Gu, *Adv. Mater.* **2001**, *13*, 1407.
269. (a) R. C. Smith, W. M. Fischer, D. L. Gin, *J. Am. Chem. Soc.* **1997**, *119*, 4092; (b) H. Deng, D. L. Gin, R. C. Smith, *J. Am. Chem. Soc.* **1998**, *120*, 3522; (c) S. A. Miller, E. Kim, D. H. Gray, D. L. Gin, *Angew. Chem. Int. Ed.* **1999**, *38*, 3022; (d) M. A. Reppy, D. H. Gray, B. A. Pindzola, J. L. Smithers, D. L. Gin, *J. Am. Chem. Soc.* **2001**, *123*, 363.
270. W. Gu, W.-J. Zhou, D. L. Gin, *Chem. Mater.* **2001**, *13*, 1949.

2.4.10
Zeolite-Entrapped Metal Complexes

*Stefan Ernst**

2.4.10.1 Introduction

Since the publication of the first edition of this Handbook [1], knowledge concerning the preparation of zeolite-entrapped metal complexes, the methods for their characterization, and their application in catalytic transformations have been gradually extended. In particular, whilst the examples of catalytic applications of these host/guest compounds have considerably broadened, the principally available methods for their preparation and characterization have been extended only in an "evolutionary" manner. Thus, the first two sections of this chapter, which describe the synthesis and characterization of zeolite-encaged metal complexes, have been only slightly modified and/or extended, and more recent references have been added. Overall, the subject matter of this chapter is restricted to physically entrapped complexes, while immobilization by anchoring or grafting is detailed in Chapters 2.4.5 and 2.4.9 of this Handbook. The third section of the chapter, which describes catalysis by zeolite-entrapped transition metal complexes, has been extended and updated in order to cover more recent developments in the field.

2.4.10.2 Synthesis of Zeolite-Entrapped Metal Complexes

A variety of monodentate as well as bidentate and polydentate ligands has been used for the preparation of transition metal complexes in zeolites. The early studies on monodentate-based complexes were originally extensively reviewed by Lunsford [2, 3] and later updated by Mortier and Schoonheydt [4] and by Ozin and Gil [5]. The present chapter will, therefore, focus on the preparation of zeolite-entrapped transition metal complexes with bidentate and polydentate ligands.

Until about 1996, only relatively few bidentate and polydentate ligands had been investigated for the synthesis of transition metal complexes in zeolites. These included ethylenediamine, tetraethylenepentamine, dimethylglyoxime, phenanthroline, bi- and terpyridine, various Schiff bases, phthalocyanines, and porphyrins. From a catalytic point of view,

---

* Corresponding author.

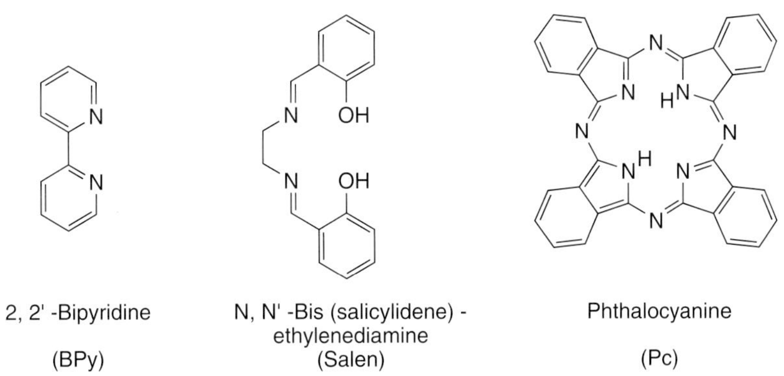

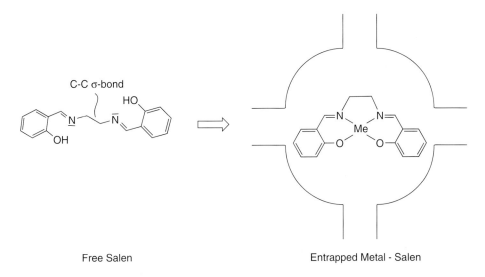

Fig. 1 Selected examples for bidentate and tetradentate ligands used to synthesize zeolite-entrapped transition metal complexes.

Fig. 2 Synthesis of zeolite-entrapped metal–salen complexes via the flexible-ligand method. (Adapted from Ref. [19].)

the three ligand types depicted in Fig. 1 – viz. the bidentate bipyridine (Bpy), and the tetradentate ligands N,N′-bis(salicylidene)ethylenediamine (salen) and its derivatives, as well as phthalocyanines (Pc) and substituted phthalocyanines – have attracted by far the most attention [6–15]. More recently, the catalytic potential of poly-aza macrocycle complexes encapsulated in zeolites was also explored [14–17].

Essentially, three methods have been developed for the preparation of zeolite-encaged metal complexes:

- reaction of the preformed ligand with the transition metal previously introduced into the zeolite cages (flexible-ligand method)
- assembling the ligand from smaller species inside the zeolite cavities (ship-in-the-bottle technique)
- synthesis of the zeolite structure around the preformed transition metal complex (zeolite synthesis method).

These three basic approaches are discussed in more detail in the following sections.

2.4.10.2.1 Flexible-Ligand Method This approach is based on the principle that the free ligand guests can easily enter the cavities of the zeolite host material because they are flexible enough to pass through the restricting windows, giving access to the larger cages. As depicted in Fig. 2 for the Schiff base salen, the free unfolded ligand is able to enter the pores of the zeolite (in the present case faujasite) due the possible free rotation around the carbon–carbon σ-bond connecting the two salicylidene moieties of the salen molecule. However, once the ligand has entered the zeolite cage and chelated with the previously exchanged transition metal ions, the complex adopts a square–planar configuration which is unable to escape from the zeolite host matrix via the 12-membered ring windows. Hence, the metal–salen complex is physically entrapped in the supercages of faujasite. This principle was first exploited by Herron [20] to prepare

References see page 628

cobalt–salen complexes in the supercages of faujasite. Later, the transition metal ion as well as the structure of the Schiff base ligand were varied extensively [21–37]. The resulting host–guest compounds have been shown to act as oxygen carriers mimicking hemoglobin [21–25], as selective oxidation catalysts [21, 26, 29–31, 33, 34, 36, 37], or as catalysts for selective hydrogenations [23, 24, 26, 32, 35]. In addition, they possess interesting electrochemical properties, as observed by investigations using cyclic voltammetry [28, 38].

The principal experimental approach for preparing zeolite-entrapped transition metal Schiff base complexes is as follows. The Na^+ form of the selected zeolite is partially exchanged with the envisaged transition metal ion. In order to avoid precipitation of hydrolyzed transition metal species during the exchange or the subsequent high-temperature dehydration, all precautions generally taken during ion exchange of zeolites with transition metal ions should be followed. In particular, dilute exchange solutions of the transition metal salts (e.g., $5 - 20 \times 10^{-3}$ mol L^{-1}) and low heating rates during dehydration should be applied. An alternative to ion exchange in aqueous suspension is solid-state ion exchange [39], especially for those transition metals which are easily hydrolyzed in water or which do not form water-soluble salts. The ion-exchanged and dried zeolite is mixed in an inert atmosphere (i.e., in a glove box) with the predried Schiff base ($n_{ligand} : n_{metal}$ usually from 2 to 4), filled in a glass tube, evacuated and sealed, and finally heated for several hours at a temperature slightly above the melting point of the ligand (e.g., 140 °C for salen) until a bright color is observed. After the reaction, a Soxhlet extraction with acetone, acetonitrile, or dichloromethane is usually applied to purify the zeolite from the uncomplexed ligand. During the complexation step, the Schiff base loses its two ionizable protons to the zeolite framework upon coordination with a divalent transition metal ion. This has two consequences: (i) the resulting transition metal complex is electroneutral; and (ii) Brønsted-acid sites are formed inside the zeolite cavities. Therefore, care must be taken that these acid sites do not interfere in subsequent catalytic reactions where they could potentially catalyze undesirable side reactions. Hence, measures must be taken to remove these acid sites, for example via an ion exchange with monovalent metal ions.

The principal synthetic approach used for the preparation of zeolite-encapsulated Schiff base complexes also holds for the complexation of transition metal ions with bipyridine, and has been reviewed in detail [10]. By carefully adjusting the ligand-to-metal ion ratio (e.g., 2:1 or 3:1) and the temperature of the synthesis (e.g., 90 °C or 200 °C), the complexation can be easily directed with high selectivity to either the bis- or the tris-coordinated complex [10, 40, 41]. Recently, *cis*-manganese(II)-bis-2,2′-bipyridine complexes encaged in faujasite-type zeolites have found particular interest in industrially relevant catalytic applications, that is, for selective alkene oxidations (see below) [42, 43].

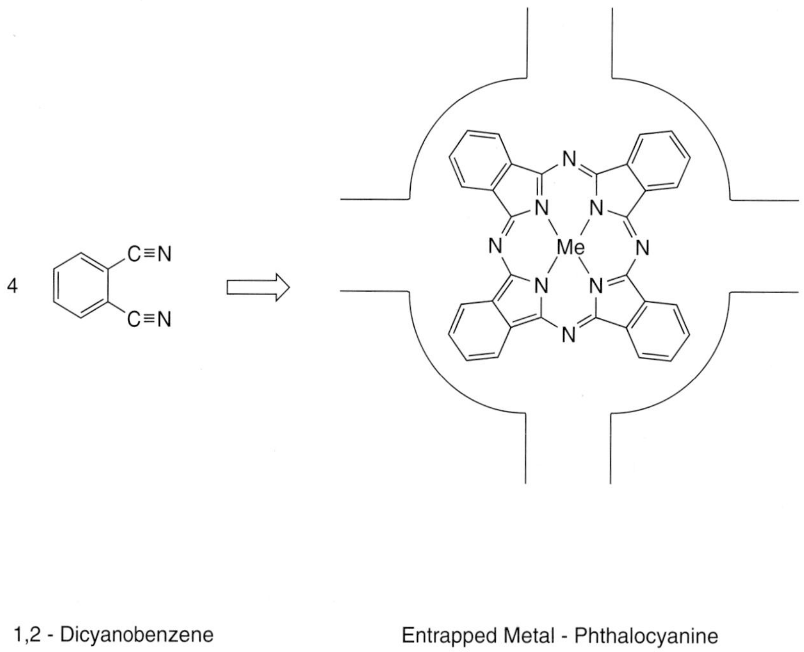

1,2 - Dicyanobenzene   Entrapped Metal - Phthalocyanine

Fig. 3 Ship-in-the-bottle synthesis of zeolite-encaged metal–phthalocyanines. (Adapted from Ref. [19].)

2.4.10.2.2 **Ship-in-the-Bottle Method** This approach is generally used to synthesize transition metal-phthalocyanines encapsulated in zeolites. The method was first described by Romanowski and coworkers [44–46], and its principle is depicted in Fig 3. It involves the introduction of the metal via ion exchange or preadsorption of a labile metal complex, such as a carbonyl or a metallocene, followed by reaction with 1,2-dicyanobenzene at temperatures from 250 °C to 350 °C. This synthesis principle was later adopted by others to prepare phthalocyanine complexes of, for example, cobalt [47–50], copper [48, 51], iron [52–55], manganese [56], nickel [47, 57], osmium [57], rhodium [58], and ruthenium [57] in the large cavities of faujasite, EMC-2 (EMT) [59] and in the aluminophosphate VPI-5 [56, 60].

The introduction of the transition metal ions into the zeolite by ion exchange in aqueous solution requires the same precautions as already described for Schiff base complexes. After loading of the ion-exchanged zeolite with 1,2-dicyanobenzene (DCBz), tetramerization is induced by heating in vacuum at temperatures between ca. 250 °C and 350 °C [47]. Differential scanning calorimetry has been shown to be a useful tool for determining the optimum complexation temperature. The latter obviously varies with the type of transition metal ion and the structure of the zeolite. Complexation occurs according to the stoichiometry given in Eq. (1) [10, 47].

$$Me^{2+}Y + 4DCBz + H_2O \longrightarrow MePcY + \tfrac{1}{2}O_2 + 2H^+ \quad (1)$$

The phthalocyanine precursor must be reduced for the formation of the transition metal complex. Since it has been found that complexation does not occur if the ion-exchanged zeolite was strongly dehydrated before the complexation step, water is thought to be the electron source in this process [47]. After complexation, the zeolite is usually purified by three successive Soxhlet extractions using acetone, pyridine, or dimethylformamide, and again acetone to dissolve unreacted dicyanobenzene and the side product phthalimide, and to remove phthalocyanine or metal–phthalocyanine which was formed on the outer surface of the zeolite crystallites. Typically, the individual extraction steps must be performed for a period from ca. 2 to 5 days in order to achieve complete extraction. Detailed chemical analyses and mass balances revealed that only a certain fraction of the introduced transition metal ions is complexed, the remainder occupying cation positions in the zeolite [47, 59]. It is to be noted that both Brønsted acid sites produced in the complexation step and uncomplexed metal ions can interfere in catalytic applications by catalyzing undesirable side reactions.

Another possible method of ship-in-the-bottle synthesis involves introduction of the transition metal by adsorption of the corresponding carbonyl complex in the zeolite, followed by synthesis of the phthalocyanine ligand around the transition metal. In this case, two alternatives exist: (i) the carbonyl is first decomposed thermally or photochemically whereby small metal clusters are formed; or (ii) a direct ligand exchange of CO by 1,2-dicyanobenzene at lower temperature (e.g., from 100 °C to 200 °C). In both cases, large fractions of the transition metal remain uncomplexed. In the former case the transition metal is present as a metal cluster in the supercages of faujasites, and only one phthalocyanine complex can be formed per supercage, whereas in the latter case the decomposition of the carbonyl complexes is faster than the rate of complex formation with 1,2-dicyanobenzene. The reaction equation for the "carbonyl route" [Eq. (2)] reveals that no protons – and hence no acid sites – are produced during complex formation since the valence of the introduced transition metal is zero.

$$Me(CO)_n + Y + 4DCBz \longrightarrow MePcY + nCO \quad (2)$$

When using metallocenes as the transition metal-supplying agents, again small amounts of water are required during phthalocyanine synthesis [Eq. (3)]. In this case, however, the produced protons are not bound as Brønsted acid sites to the zeolite framework, but rather are required to convert the cyclopentadienyl fragments (Cp) of the metallocene to cyclopentadiene (Cpen) [10].

$$Me(Cp)_2 + Y + 4DCBz + H_2O \longrightarrow MePcY + 2Cpen + \tfrac{1}{2}O_2 \quad (3)$$

The host–guest compounds produced by the metal-locene route have the advantage that they are virtually free from uncomplexed metal species. However, unlike with the two previously described procedures, large amounts of the free phthalocyanine base are also formed inside the zeolite [55]. Potentially, these entrapped free ligands may hinder the diffusion of reactants and/or products in catalytic reactions.

The synthesis of *substituted* metal-phthalocyanines in the supercage of faujasite-type zeolites via the ship-in-the-bottle technique has also been claimed. The incentive to encapsulate substituted complexes is at least twofold:

- to enhance the stability of the complex via bulky substituents which protect the meso-atoms of the phthalocyanine ring
- to increase the catalytic activity of the metal by introducing electron-withdrawing groups.

Examples published so far are tetra-*tert*-butyl-substituted iron-phthalocyanine [61], perfluorophthalocyanines of iron [62], cobalt [63, 64], copper [63, 64] and manganese [63], and iron-tetranitrophthalocyanine [65]. The starting materials for the *in-situ* ligand synthesis were 4-*tert*-butyl-1,2-dicyanobenzene, tetrafluoro-1,2-dicyanobenzene and 4-nitro-1,2-dicyanobenzene, respectively. In the case of the nitro-substituted phthalocyanine it was shown unambiguously that the complex is only formed at the outer surface of the zeolite crystallites [65]. By contrast, there seems to be sufficient experimental evidence for true encapsulation of the perfluorated phthalocyanines [63, 64]. Insufficient data are available on the tetra-*tert*-butyl-substituted system to allow for a final conclusion as to whether or not this very bulky complex is really encapsulated.

The principle of ship-in-the-bottle synthesis was also applied for the *in-situ* preparation of porphyrin-type ligands in the supercages of Y-type zeolites. Nakamura et al. [66] claimed the encapsulation of iron- and manganese-tetramethylporphyrins in the supercages of zeolite NaY by refluxing the transition metal-exchanged zeolite with pyrrole and acetaldehyde in methanolic solution. Unfortunately, however, apart from one UV/visible spectrum no further experimental evidence for true intrazeolitic complex formation was provided, and the catalytic results obtained with the synthesized materials were poor. The synthesis of encapsulated tetraphenylporphyrins from pyrrole and benzaldehyde has been claimed by Chan and Wilson [67]. However, no experimental evidence has been presented for true complex encapsulation. Furthermore, there are certainly severe size limitations for these large complexes, at least in the supercages of faujasite-type zeolites. As a whole, the topic of encapsulated porphyrin-type complexes seems still to be underdeveloped. Nevertheless, this interesting field deserves further systematic research, especially in view of the steadily increasing number of large- and super-large-pore zeolites and molecular sieves which may serve as host matrices.

2.4.10.2.3 **Zeolite Synthesis Method** Incorporation of transition metal complexes by synthesizing the molecular sieve structure around the preformed complexes has also been claimed [68–71]. Prerequisites for this method are the stability of the complex under the conditions of zeolite synthesis (pH, temperature, hydrothermal conditions) and a sufficient solubility in the synthesis medium to enable random distribution of the complex in the synthesis mixture as well as in the final zeolite. The synthesis of zeolite mordenite from gels containing bipyridine, phenanthroline, or phthalocyanine complexes has been reported by Rankel and Valyocsik [68]. However, encapsulation of the complexes in the final product was neither claimed nor proved. Balkus and coworkers [69–71] claimed that homogeneous encapsulation of transition metal complexes in zeolites is possible by careful design of the zeolite synthesis procedure to avoid aggregation of the complexes in the aqueous synthesis medium. According to these authors, phthalocyanine complexes can be incorporated during crystallization into zeolite X [69] and into the aluminophosphates $AlPO_4$-5 and $AlPO_4$-11 [70]. At least in the two latter cases, the formation of mesopores must be assumed in order to allow for the accommodation of the relatively spacious complexes in the narrow one-dimensional pore systems of $AlPO_4$-5 or even $AlPO_4$-11. In addition, these two molecular sieves require the presence of further organic templates for their synthesis which must be removed by calcination before an application in adsorption or catalysis is possible. It is anticipated that, during this calcination step, the transition metal complex is also destroyed and/or burnt. Hence, the zeolite synthesis method seems to be restricted to those cases, where the synthesis of the zeolite matrix is possible without the use of further organic templates.

More recently, the encapsulation of ruthenium-hexadecafluorophthalocyanine (RuF10Pc) in zeolite NaX [72], of bis(di(2-aminoethyl)amine) nickel(II) in $AlPO_4$-5 [73] and of chromium (III) porphyrins [74] and nickel (II)-tetrakis-(*N*-methyl-4-pyridyl) porphyrin complexes [75] in zeolite NaY has also been reported.

Gbery et al. used a modified zeolite synthesis method to encapsulate the so-called Jacobsen catalyst [(*R*, *R*)-*N*, *N*-bis(3,5-di-*tert*-butylsalicylidene)-1,2-cyclohexane-diamine-manganese(II) chloride] in the large intracrystalline cages of zeolite MCM-22, which possess dimensions of ca. 0.71×1.82 nm [76]. In essence, these authors made use of the fact that the synthesis of MCM-22 proceeds through a layered precursor [77]. At this stage of the synthesis, the Jacobsen catalyst could be intercalated between the layers. After drying and heating to 280 °C, the layers condense and form the crystalline MCM-22 structure. The metal complex becomes encapsulated during this process, with no possibility to escape through the relatively small 10-membered ring windows in the cages. The authors were able to show that the immobilized complex is active for enantioselective epoxidation reactions [76].

2.4.10.3 **Characterization**

The characterization of zeolite-encapsulated transition metal complexes is carried out with the objective to answer questions about:

- the distribution of the complex over the host crystal
- the degree of complexation
- the type of coordination compared to that in a liquid solvent
- the influence of a host–guest interaction on the structure

- the stability compared to that in the solid or in dissolved states
- the ability to exhibit free coordination sites for chemisorption and catalysis.

In the following sections, methods for characterization are described which have provided largely unambiguous information about these most important questions.

2.4.10.3.1 Electron Paramagnetic Resonance Spectroscopy

Electron paramagnetic resonance (EPR) spectroscopy is a powerful tool for the study of square–planar complexes having Co^{II} as central metal ion due the low-spin configured doublet ($S = 1/2$) ground states in which the unpaired electron resides in the Co $3d_{z^2}$ orbital, where z is perpendicular to the chelate plane. The spin Hamiltonian parameters of such ground states are, thus, extremely sensitive to changes in axial ligation. This property can be used to follow the processes of complexation, hydration, and oxygen bonding by means of a theoretical analysis of the spin Hamiltonian parameters and a comparison of the individual ligand-field effects with the experimental spectra.

A relatively clear way to follow the complexation process is the evaluation of intensity changes in EPR spectra, as has been demonstrated for the formation of the tetrakis(pyridine)copper(II) complex [77]. The expected stoichiometry $[Cu^{II}(Py)_4]^{2+}$ for the square–planar complex, exhibiting a superhyperfine spectrum, is adjusted in the supercage of the zeolite Y (Fig. 4).

A low-spin state appears in faujasite Y using bipyridine as a ligand for the formation of tris(bipyridine) complexes of Co^{II}, Fe^{III}, or Ru^{III} [40, 41, 78]. The low-spin state is complete at 77 K, but a temperature-dependent equilibrium conversion into the octahedral high-spin state appears at higher temperatures according to

$$(^6t_{2g}{}^1e_g) \quad \rightleftharpoons \quad (^5t_{2g}{}^2e_g)$$
low-spin complex $\qquad$ high-spin complex

The multiplicity of the hyperfine structure indicates whether monomeric $[Co^{III}L_xO_2^-]^{2+}$ or dimeric $[Co^{III}L_xO_2^{2-}Co^{III}L_x]^{4+}$ complexes are formed from methylamine ligands, since the high-spin state of the latter μ-peroxo complex can be converted into the low-spin μ-superoxo species $[Co^{III}L_xO_2^-Co^{III}L_x]^{5+}$ by oxidation, for example with chlorine or H_2O_2 [79]. The preferential location of the unpaired electron on the oxygen molecule is concluded from the small ^{59}Co hyperfine splitting constants.

From the EPR spectra of a variety of zeolite-entrapped Co^{2+} Schiff base complexes, it could be concluded that planar complexes of Co^{2+} with tetradentate salen (bis(salicylidene)ethylenediamine) and acacen (bis(acetylacetonyl)ethylenediamine) are only formed in low concentrations in faujasites, but that the pentadentate analogues facilitate the elimination of the zeolite from the primary coordination sphere of the metal ion. In some cases, the spectral characteristics of the entrapped complexes are identical to those of the corresponding solution complexes; in other cases, small axial coordination strength is found, presumably due to the influence of the zeolite host [26].

Charge-compensating protons, which are formed during the chelating process, have been shown to interact with the pyrrole nitrogen atoms of CoPc, resulting in a non-equivalency of these N atoms [48]. As a consequence, two nitrogen atoms with their nuclear spins $I = 1$ couple to the low-spin Co^{2+} ($S = 1/2$; $m_s = \pm 1/2$; $m_I = 0, \pm 1, \pm 2$; $M_s = 1$; $M_I = 0$). This results in five-component spectra interfering with the octet of the ^{59}Co hyperfine structure (Fig. 5).

2.4.10.3.2 Electron Excitation by Ultraviolet/Visible Light

The spectroscopy of ultraviolet (UV) and visible light absorption provides information about the d-orbital splitting through the d–d transitions and the ligand–metal interaction through the ligand-to-metal charge-transfer transitions. Information about the coordinative bond is obtained either qualitatively by comparison of the spectra with spectra of known complexes, or quantitatively by ligand field or angular overlap model calculations on the d–d spectrum.

Generally, diffuse reflectance spectroscopy is applied to follow the chelating process in zeolite cages. For example, the formation of tris(ethylenediamine) in cobalt

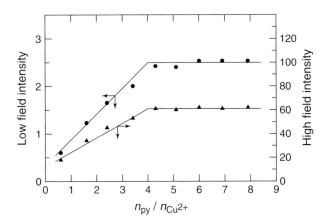

Fig. 4 Intensity of the $[Cu^{II}(Py)_4]^{2+}$ EPR spectrum versus the molecules of Py adsorbed per Cu^{2+} ion. Left ordinate: intensity based on the $m_I = -3/2$ component of the spectrum. Right ordinate: intensity based on the high-field minimum. (From Ref. [77].)

References see page 628

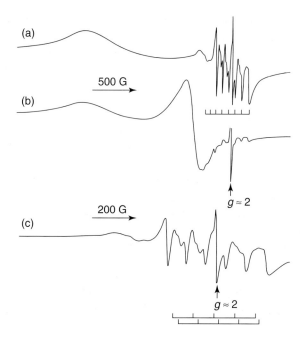

Fig. 5 X-band EPR spectra of CoPc–NaX (b) at 130 K and 1 mPa and (a,c) at 10^5 Pa (=1 bar) in air. The identified features are inserted. The arrows indicate the increase of the magnetic field H and the abscissa extensions. (From Ref. [48].)

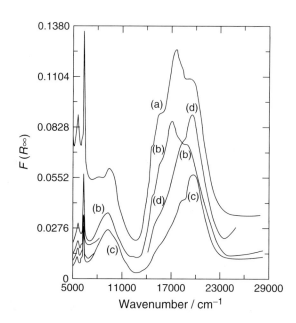

Fig. 6 Dynamics of Co chelation by ethylenediamine on hydrated (a) CoNaY. The chelation proceeds by successive removal of water at (b) 329 K, (c) 428 K, and (d) 477 K. (From Ref. [80].)

ion-exchanged faujasite Y is indicated by an increase of the relative intensity of the band at 20 650 cm^{-1} (Fig. 6), which is characteristic for a d–d transition of an octahedral complex [80]. Since only relative intensities of the bands, assigned to six-fold coordination, change during the chelate formation the process must occur according to

$$[Co(H_2O)_6]^{2+} + 3en \longrightarrow [Co(en)_3]^{2+} + 6H_2O$$

That is, it represents the replacement of the weaker ligand H$_2$O by the stronger one. This complex is stable up to 473 K. At higher temperature, a stepwise decomposition

$$[Co(en)_3]^{2+} \longrightarrow [Co(en)_2]^{2+} \longrightarrow [Co(en)]^{2+}$$

proceeds. The intermediate [Co(en)$_2$]$^{2+}$ is identified by an EPR spectrum of the monomeric superoxocomplex exhibiting the presence of a low-spin species, and the monoligand complex is identified by the appearance of charge-transfer bands in the UV region, i.e. above 20 000 cm^{-1} [80].

The dynamics of the formation of the dimethylglyoximato–Co(II) complex by uptake of dimethylglyoxime in a cobalt ion-exchanged zeolite NaX single crystal has been followed by *in-situ* UV/VIS microscope spectrometry in transmission [81]. The temporal changes of the spectra (Fig. 7) indicate the tetrahedral coordination of the starting complex [Co^{II}(Zeo)$_3$OH]$^+$ concluded from the band quadrupole between 450 and 700 nm, which originates from the splitting of the transition $^4F(\Gamma_2) \rightarrow {}^4P(\Gamma_4)$ due to (L, S) coupling effects. The shift of the red band from 655 to 625 nm is due to an increasing distortion of the quasi-tetrahedral complex, presumably due to the formation of the intermediate [Co^{II}(dmgH)$_2$(H$_2$O)$_2$]. This model is supported by EPR studies [82].

Using UV/VIS spectroscopy in transmission with large single crystals (50 µm size), NaX-entrapped cobalt phthalocyanine complexes have been analyzed [83]. The spectra exhibit a splitting of the Q-band, which is expected for a lowering of symmetry from D$_{4h}$ to D$_{2h}$ [84]. Partial protonation of metal-phthalocyanine (MePc) can result in such a lowering to a two-fold symmetry [85]. The charge-compensation protons which are generated during the chelate formation are the source of the partial protonation.

The analysis of luminescence spectra in the UV/VIS region can provide information about the distribution of zeolite-encapsulated complexes, or it can be used as a probe for processes causing acceleration or quenching of emission intensity. For tris(2,2'-bipyridine)ruthenium(II) complexes in zeolite Y, concentration quenching starts at three [Ru^{II}(bpy)$_3$]$^{2+}$ complexes per unit cell; that is, at distances of 2–3 nm [40, 41]. Small amounts of water can increase the emission intensity, whereas oxygen acts always as a quencher due to singlet oxygen formation.

For zinc porphyrins incorporated in AlPO$_4$-5 molecular sieves by crystallization inclusion, an average distance

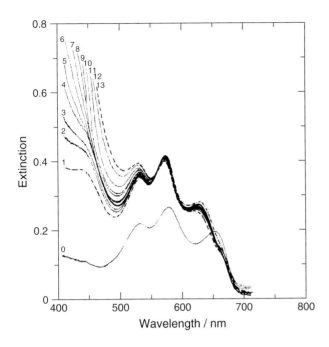

Fig. 7 Dynamics of Co chelation by dimethylglyoxime via vapor-phase deposition in dehydrated (673 K, 5 h) CoNaX at 373 K. Spectrum 0 is pure CoNaX. (From Ref. [81].)

for Förster quenching by inductive resonance is found to be 6 nm – that is, less than spacings valid for dissolved complexes [86].

2.4.10.3.3 X-Ray Photoelectron Spectroscopy

X-ray photoelectron spectroscopy (XPS), which can be applied to the analysis of zeolites due to increased electron escape depths, is used to answer questions on:

- the homogeneity of the distribution of the encaged complexes, i.e., either uniform throughout the zeolite crystal or enriched in a shell close to the crystal surface
- the degree of complexing
- the structure of the complex
- the interaction with the framework [87–90].

The detailed analysis of Me : Si, Me : Na or N : Si ratios, where Me denotes the complexing metal ion, provides evidence about the preferential location of the complex; that is, strong deviations from the expected bulk compositions to higher values for these ratios indicate an enrichment at or close to the external surface of the zeolite crystals. High degrees of complexing result in a shift of the binding energy E_b for the complexing metal to lower values by ca. 1 eV (e.g., for Ni 2p3/2 from 855.8 for the nickel ion-exchanged faujasite Y to 854.9 for the zeolite-encaged nickel–phthalocyanine) [87]. Another proof for the formation of the phthalocyanine complex is the decrease in the intensity of the satellite shake-up due to the reduction of the magnetic moment; that is, the decrease in the Ni unpaired spin density during the course of the formation of diamagnetic NiPc. The degree of binding of the metal ions to the Pc ligand can also be followed via the N : Me ratio; when the ratio approaches the maximum value of 8, this indicates an increase in the degree of complex formation.

The structure of zeolite-encaged metal–phthalocyanine can strongly deviate from that of the free complex. This is indicated by a change of the spectra of N 1s (Fig. 8). Instead of a narrow N 1s singlet with $E_b = 399$ eV and full width of half maximum (FWHM) = 1.9 eV, either strongly broadened peaks (FWHM = 3.5 eV) or poorly resolved doublets are observed. Deconvolution of the N 1s peaks yields two components with binding energies of about 399 and 400.5 eV. Obviously, the eight nitrogen atoms of the Pc molecules are no longer equivalent in the zeolite-encaged state. Presumably, the charge-compensating protons, which are formed during the

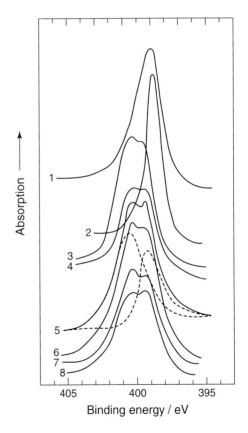

Fig. 8 XPS spectra of N 1s in MePc–NaY samples and reference spectra. (1) CoPc; (2) NiPc; (3) NiPc–NaY(60); (4) NiPc–NaY(40); (5) NiPc–NaY(30) with spectrum synthesis; (6) CoPc–NaY(60); (7) CoPc–NaY(30); (8) CuPc–NaY(60). (From Ref. [89].)

References see page 628

complex formation from the metal ions and the 1,2-dicyanobenzene, cause a protonation of the pyrrole N atoms and a corresponding shift of the signal.

2.4.10.3.4 Infrared Spectroscopy
Infrared spectroscopy can provide information on whether ligand molecules have coordinated to transition metal cations, if different patterns appear in the free or in the chelated state, or if characteristic bands exhibit defined shifts upon chelation.

A striking example, where both conditions are fulfilled, is given for faujasite-hosted transition metal complexes of 1,4,8,11-tetraazacyclotetradecane (cyclam) [26]. The high-frequency domain of the infra-red (IR) spectra shows a shift of the most intense N–H stretching vibration in the order

$$(L)\text{-NaY} > [Cr^{III}(L)]^{3+}\text{-NaY} > [Mn^{II}(L)]^{2+}\text{-NaY}$$
$$> [Co^{II}(L)]^{2+}\text{-NaY} > [Ni^{II}(L)]^{2+}\text{-NaY}$$

and the shoulders at 3273 and 3321 cm^{-1} are lost (Fig. 9).

2.4.10.3.5 Nuclear Magnetic Resonance Spectroscopy
High-resolution magic angle spinning nuclear magnetic resonance spectroscopy (HR MAS NMR) has been used to study the interaction of phthalocyanines with the molecular sieve host [91]. For FePc in the aluminophosphate molecular sieve VPI-5, ^{27}Al double rotation (DOR) spectroscopy and ^{1}H ^{27}Al cross-polarization (CP) DOR NMR spectroscopy of hydrated and intercalated VPI-5 have been applied (Fig. 10). From the ^{27}Al line broadening and the CP enhancement of the line intensity in the Pc-intercalated VPI-5 a strong interaction between Al atoms in the pore walls and the protons of encapsulated Pc were deduced. The assumption of a local deformation of the framework by host–guest interaction is supported by the loss of resolution in the ^{31}P spectrum (Fig. 10).

2.4.10.3.6 Raman Spectroscopy
Raman spectroscopy was used by Zhan and Li [92] to study bis- and tris-(1,10-phenanthroline)manganese(II) complexes encapsulated in zeolite Y. These authors suggest that the benefit of Raman spectroscopy in this context is at least threefold:

- The Raman spectra are relatively simple and quite well defined; vibrations coming from the zeolite host matrix can be easily identified and subtracted from the measured spectrum.
- Almost the whole range of mid-IR and far-IR can be used for characterizing the occluded transition metal complexes.
- A well-defined Raman peak at 500 cm^{-1}, attributed to the T–O–T bending of the zeolite framework, can be exploited as an internal standard for quantitative studies, i.e. to determine the concentration of occluded species [92].

Knops-Gerrits et al. [93] used Raman spectroscopy to investigate in more detail iron phthalocyanines in zeolites.

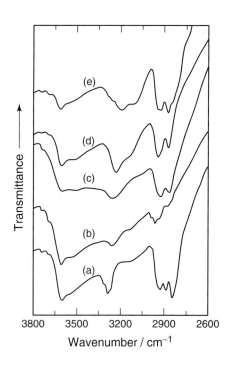

Fig. 9 High-frequency domain of the IR spectra of transition metal complexes of 1,4,8,11-tetraazacyclotetradecane (cyclam) in NaY. (a) cyclam; (b) [Cr^{III}(cyclam)]$^{3+}$; (c) [Mn^{II}(cyclam)]$^{2+}$; (d) [Co^{II}(cyclam)]$^{2+}$; (e) [Ni^{II}(cyclam)]$^{2+}$. (From Ref. [26].)

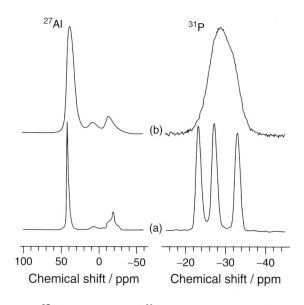

Fig. 10 ^{27}Al DOR NMR and ^{31}P MAS NMR spectra of VPI-5. (a) As-synthesized and (b) after incorporation of FePc. (From Ref. [41].)

From their results, they proposed a novel model for the conformation of encapsulated metal–phthalocyanines in the supercages of zeolite Y, where two of the aromatic rings of the phthalocyanine are pushing against the supercage walls [93].

2.4.10.3.7 Stability Analysis

The chemical stability of faujasite-encaged metal phthalocyanines has been studied by thermal analysis under air [94], oxygen, hydrogen, and helium [88]. Under helium, the complexes remained stable up to 873 K. Maximum rates of decomposition appear in hydrogen around 700 K (presumably due to catalytically assisted hydrogenolysis), in air around 700 K (related to catalytic self-oxidation), and in oxygen even below 500 K. Consequently, poor stabilities are found in catalytic gas-phase reactions on zeolite-entrapped complexes [94]. In liquid-phase oxidations at relatively low temperatures (e.g., the oxidation of ethylbenzene by O_2 at 409 K), a relatively rapid destruction of the CoPc complex has also been reported [48]. Clearly, oxyfunctionalization reactions of alkanes or alkenes using *tert*-butylhydroperoxide or iodosobenzene at room temperature do not significantly affect the stability of zeolite-anchored metal complexes [52, 56].

The strong increase in the photostability of metal–porphyrin compounds upon incorporation into molecular sieves has been used [95] to judge on the location of the complex – that is, at either the external surface or in the voids of the porous mineral crystals (Fig. 11).

2.4.10.3.8 Oxygen Adsorption

Oxygen adsorption is applied to titrate the fraction of Co^{II} ions in the zeolite framework which has been chelated successfully to such complexes exhibiting the capability to bind dioxygen. The oxygen pulse technique has been applied to titrate this fraction for a $[Co^{II}(bipyridine)(terpyridine)]^{2+}$-LiY zeolite [96]. Attempts to prepare cobalt complexes of bis(salicylidene)ethylenediamine in faujasite-Y for reversible oxygen uptake failed [20] because the formation of a quadratic–planar chelate is impeded [25]. Reversible uptake of oxygen has been successfully realized with cobalt complexes of bis(salicylaldehyde)methyldipropyltriamine (smdpt) in zeolites NaY and NaEMT [25]. The molar ratio bonded O_2: total Co^{2+} has been determined by quantitative evaluation of EPR signal intensities. Such ratios amounted to 0.25 for $[Co^{II}(smdpt)]$-NaEMT and to 0.05 for cobalt complexes of bis(2-pyridinecarboxylaldehyde)ethylenediimide in NaY at room temperature [25, 26]. Presumably, the sorption kinetics is controlled by diffusion processes.

2.4.10.3.9 Cyclovoltammetry

Cyclovoltammetry was proposed as a method to recognize the preferential location of a zeolite-anchored transition metal complex – that is, at the external surface of the molecular sieve crystal or encapsulated in the void structure. Moreover, discrimination between different types of accommodation (i.e., physically entrapped or ion-exchanged) has been claimed [38]. However, the nature of electron-transfer processes and the role of electrolyte cations in the charge compensation are not fully understood, and are debated controversially in the literature. Further success in the application of zeolite-encaged complexes for selective electroassisted catalytic reactions requires new strategies to enable ready electron transfer between external electrodes and internal chelates [38, 97].

2.4.10.4 Catalysis by Zeolite-Entrapped Transition Metal Complexes

Zeolite-encapsulated catalytically active transition metal complexes offer a number of advantages over their counterparts in homogeneous solution. The general benefits of the heterogenization of homogeneous catalysis are an easy separation from the reaction medium and the possibility of using a large variety of different solvents and reaction conditions [1, 6–15, 98]. Examples of the advantages introduced by using a microporous zeolite host matrix are the possibility of shape-selective catalysis due to the constrained intrazeolitic environment, a higher catalytic activity, and a higher stability of the encapsulated complexes. The latter arises from the spatial isolation which prevents dimerization of the monomeric complexes and also strongly suppresses an oxidative self-destruction.

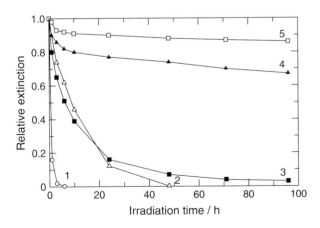

Fig. 11 Stability of zinc(II)-tetrakis(*N*-methylpyridyloxy)-phthalocyanine under irradiation with 20 mW cm^{-2} of visible light in air in: (1) DMF; (2) water; (3) adsorbed on the surface of AlPO$_4$-5; (4) incorporated in AlPO$_4$-5; and (5) latter sample in vacuum. (From Ref. [95].)

Zeolite-encapsulated complexes have also been suggested as model compounds for enzyme mimicking [7, 45]. In this respect, the term "zeozymes" (zeolite-based enzyme mimics) has been coined to describe a catalytic system in which the zeolite replaces the protein mantle of the enzyme and the entrapped metal complex (e.g., a metal–phthalocyanine) mimics the active site of the enzyme (e.g., an iron–porphyrin) [7].

The catalytic properties of zeolite-encapsulated transition metal complexes have been explored in a variety of selective oxidation and hydrogenation reactions. Such host–guest compounds were even demonstrated to be able to catalyze oxidations or hydrogenations in an enantioselective manner, if the immobilized complexes possessed chiral properties.

2.4.10.4.1 Selective Oxidations on Zeolite-Encaged Transition Metal Complexes

In the case of oxidation reactions over zeolite-encaged complexes, the choice of the correct oxidant was in general identified as a critical step with respect to catalytic activities and selectivities. In homogeneously catalyzed oxidations with, for example porphyrins, many different oxygen-supplying reagents have been used (e.g., iodosobenzene, NaClO, KHSO$_5$, peracids, perchlorates, organic peroxides, hydrogen peroxide, molecular oxygen) [99]. However, only few of these are useful and practical, especially in combination with zeolite-encapsulated complexes, for the following reasons [8]:

- they are too large to diffuse through the zeolite pores (e.g., certain aromatic peracids)
- they require a phase-transfer catalyst which, however, is too bulky to enter the pores (as in the case of NaClO)
- they are too expensive
- they are nowadays no longer acceptable under environmental aspects.

For oxidation reactions on zeolite-immobilized complexes, iodosobenzene, organic peroxides (in particular *tert*-butylhydroperoxide, TBHP), H$_2$O$_2$ and O$_2$ have been used. While iodosobenzene is known as a good oxygen atom donor because it does not form radicals [99], it has certain disadvantages if used in heterogeneously catalyzed oxidations. The two main drawbacks are reduced reaction rates due to a relatively low mobility in the pores [21, 52, 53] and the intrazeolitic formation of iodoxybenzene by disproportionation of iodosobenzene [52, 53]. The former leads to relatively fast deactivation of the catalyst due to pore clogging [52, 53]. Therefore, peroxides are by now generally recommended as the preferred oxidants. In this case, however, care must be taken in order to minimize the amount of residual (i.e., uncomplexed) metal cations in the zeolite which could induce decomposition of the peroxide [100].

Zeolite-encapsulated iron–phthalocyanine (FePc) complexes are able to catalyze the selective oxidation of alkanes to alcohols and ketones under ambient conditions with iodosobenzene [52, 53] or *tert*-butylhydroperoxide as oxygen atom donor [55]. Some illustrative examples are depicted in Fig. 12 for the oxidation of the homologous series of C$_5$ through C$_{10}$ *n*-alkanes. It can be seen that free FePc (dissolved in dichloromethane) is much less active than the same complex immobilized in the supercages of zeolite Y. Moreover, whereas the encaged complex remains intact during the reaction, the homogeneous FePc catalyst is completely destroyed, which is indicated by a loss in activity and a color change from the typical blue-green to light yellow [55]. Turnover numbers between 20 and 30 were observed for the free complex (depending on the feed alkane), while they were one order of magnitude higher for the immobilized complex. The occurrence of a maximum in catalytic activity at C$_6$/C$_7$ has been interpreted in the present case as the increasing reactivity of the alkane being superimposed by a decreasing diffusivity in the zeolite pores [55].

Not only activity but also selectivity changes have been observed upon immobilizing iron–phthalocyanines in a microporous environment: Herron [53] reported *shape selectivity effects* when oxidizing a mixture of cyclohexane and cyclododecane over FePc in zeolite Y. By contrast to the free complex, where the conversion of cyclododecane is much faster, cyclohexane is preferentially oxidized over the encaged complex. This has been interpreted as an example of reactant shape selectivity. After exchanging the zeolite with alkali cations of increasing size (i.e., K$^+$, Rb$^+$), the effective pore width is further reduced and the discrimination between reactants of different size is further enhanced [53].

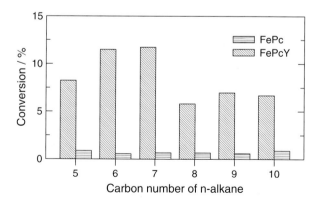

Fig. 12 Oxidation of the homologous C$_5$–C$_{10}$ *n*-alkanes on FePc and FePcY under ambient conditions with *tert*-butylhydroperoxide as oxygen donor; conversions are given after 2 h of reaction. (Adapted from Ref. [56].)

Regioselectivity has been observed during the oxidation of n-octane on FePc encaged in zeolite Y or in the molecular sieve VPI-5. Over both catalysts, oxidation towards the end of the alkane chain is favored [53, 55, 101], while over the free FePc oxidation occurs with equal rates [101].

Early examples of *stereoselective oxidations* have been reported by Herron [53] for the oxidative conversion of a mixture of methylcyclohexane and norbornane. Later, Bowers and Dutta [21] and Ichikawa et al. [61] reported that *trans*-stilbene is preferentially oxidized over Mn-SalenY [21], FePcY, and Fe(*tert*-butyl)PcY [61], while *cis*-stilbene is more easily oxidized in homogeneous catalysis.

Zeolitic host–guest systems for the direct synthesis of adipic acid from cyclohexane or cyclohexene have been reported by Jacobs and coworkers. Suitable catalysts are *cis*-manganese(II)bis-2,2'-bipyridyl (cis-[Mn(Bpy)$_2$)]$^{2+}$ [44] or iron–phthalocyanine [102] in zeolite Y. A reaction sequence for the oxidation of cyclohexene with aqueous hydrogen peroxide obtained from detailed catalytic studies is depicted in Fig. 13. The primary oxidation product of cyclohexene conversion is cyclohexene oxide, which is then converted by hydration to 1,2-cyclohexanediol and further oxidized to 1,2-cyclohexanedione and adipic acid. In the hydration step, as well as in the cleavage of the cyclohexane ring, residual acidity created during the preparation of the encapsulated complex is supposed to assist. Side reactions include the formation of cyclohexenone via allylic oxidation of cyclohexene on uncomplexed Mn^{2+}, and of cyclohexanol from the acid-catalyzed hydration of cyclohexene. After turnovers of ca. 800 [15, 102], the pores of the zeolitic host are reported to be filled with adipic acid. It seems that – if a practical application of this type of catalyst is envisaged – a medium (solvent) is required in which the strongly polar product(s) (e.g., adipic acid) are continuously removed from the zeolite phase.

The oxidation of cyclohexane or cyclohexene was also studied by other groups using a broad variety of zeolite-entrapped transition metal complexes. Balkus et al. studied cyclohexane oxidation using *tert*-butylhydroperoxide as oxidant over hexadecafluoro-phthalocyanine complexes of ruthenium(II) which were introduced into the supercages of zeolite NaX via the synthesis method ("build bottle around ship-method") [72]. These authors report complete conversion of their feed to cyclohexanol and cyclohexanone (with ca. 3000 turnovers per day), without the apparent formation of adipic acid and/or concomitant catalyst deactivation. Farzaneh et al. [103] investigated the influence of the transition metal ion [e.g., Cr(III), Mn(II), Co(II)] and the nature of the ligand (e.g., 2,2'-bipyridine, tetramethylethylenediamine, tetramethyl-1,8-naphthylenediamine) in the oxidation of cyclohexane with *tert*-butylhydroperoxide (TBHP). The best results were obtained with a relatively large surplus of cyclohexane with respect to TBHP (ca. 6.25) using a [Mn(bpy)$_2$]$^{2+}$-containing Y-type zeolite [103]. Quite similar results were obtained by Knops-Gerrits et al. with vanadyl bis-bipyridine complexes encaged in zeolite NaY [104]. More recently, Salavati-Niasari reported the encapsulation of nickel(II)-containing hexaazacyclohexadecane complexes in the supercages of zeolite Y [105]. This author claims "superior" catalytic activity for the oxidation of cyclohexene with molecular oxygen to 2-cyclohexene-1-ol and 2-cyclohexene-1-one [105].

As a whole, it seems that a broad variety of zeolite-immobilized transition metal complexes has been explored in the oxidation of cyclohexane or cyclohexene with the idea in mind, to identify active/selective and environmentally friendly catalysts for adipic acid production. Unfortunately, however, much of these catalytic data

References see page 628

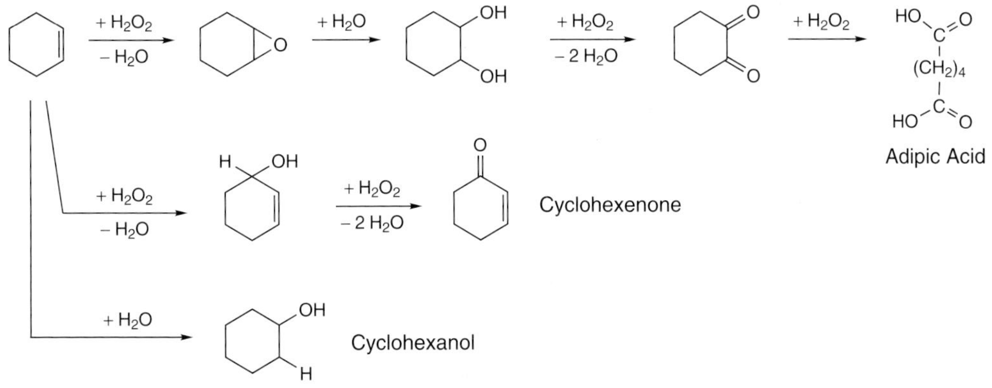

Fig. 13 Consecutive and competing reactions in the oxidation of cyclohexene with hydrogen peroxide over faujasite-encapsulated *cis*-[Mn(Bpy)$_2$]$^{2+}$ complexes. (After Ref. [43].)

have been obtained under different experimental conditions, using different oxidants and catalysts which were characterized to different degrees. Hence, a standardized catalytic testing and standardized methods for a meaningful minimum characterization would be highly desirable in the future to allow for a comparison of different catalysts on a sound experimental basis.

In an attempt to prepare real zeozymes, viz. a mimic of cytochrome P-450, Parton et al. [106] used a very interesting concept, whereby iron–phthalocyanine was first encapsulated in the supercages of zeolite Y, after which this inclusion compound was embedded in a polydimethyl siloxane membrane. The resulting system is reported to oxidize alkanes at room temperature at rates comparable to those of the true enzyme. To date, unfortunately, this principle could obviously not be generalized or transferred to other catalytic systems.

The oxidation of styrene with tert-butylhydroperoxide on copper–phthalocyanine-containing faujasites at temperatures around 60 °C yields the epoxide, as well as phenylacetaldehyde and benzaldehyde as the main products [107]. It has been reported that the turnover frequency (TOF) (per Cu or per CuPc) for the oxidation reaction is superior (ca. $50\,h^{-1}$) for the encapsulated CuPc complex as compared to the free CuPc complex (TOF ca. $8\,h^{-1}$) or copper-exchanged zeolite Y (TOF ca. $4-5\,h^{-1}$). Benzaldehyde is probably derived through a secondary oxidation of a radical type, while phenylacetaldehyde seems to be an isomerization product of benzaldehyde formed in the primary step. The presence of residual protons (most likely generated during copper exchange, followed by dehydration of the zeolite) seems to favor consecutive isomerization. In this context, a report by Raja and Ratnasamy [108] is exciting as it claims that $CuCl_{17}Pc$ encapsulated in zeolite Y is able to oxidize primary carbon atoms using molecular oxygen in a very selective manner, viz. without the formation of typical products appearing with a free radical pathway. Unfortunately, however, no further report on this surprising result has been published to date.

The hydroxylation of phenol with hydrogen peroxide over zeolite encapsulated metal–phthalocyanines has also attracted considerable attention [109, 110]. One attractive catalytic system comprises copper- or cobalt-containing perchlorinated phthalocyanines ($Cl_{14}Pc$) or tetranitro-substituted phthalocyanines (($NO_2)_4Pc$). It was reported that the selectivities for the products formed depend on the size of the ligand: over the bulkier (immobilized) complex, the slimmer hydroquinone is preferentially formed, while with the less bulky complex ($Cl_{14}PcCu(II)$) catechol and benzoquinone are to a considerable extent also observed as products [109].

More recently, Maurya et al. have shown that hydroxylation of phenol with hydrogen peroxide is also possible using chromium(III), iron(III) or bismuth(III) complexes of the amidate ligand 1,2-bis-(2-hydroxybenzamido)ethane encapsulated in the supercages of zeolite Y [110]. These authors report the formation of catechol as major and of hydroquinone as minor product. As the conditions of the catalytic reaction in this latter case were not too different from those in the former example, this is also an interesting example for the possibility to fine-tune catalytic selectivity in the field of zeolite-encaged transition metal complexes.

2.4.10.4.2 Selective Hydrogenations on Zeolite-Encaged Transition Metal Complexes Beside the many examples of oxidation reactions over zeolite-encapsulated transition metal complexes, only relatively few examples of hydrogenation reactions have been reported [23, 24, 54, 111, 112]. Shape selectivity effects were observed in the hydrogenation of an equimolar mixture of 1-hexene and cyclohexene over Pd–salen encapsulated in zeolites X and Y [23]. It was reported that both reactants are hydrogenated with comparable rates over the free Pd–salen in homogeneous reaction. Over the encapsulated complex, the linear complex is still hydrogenated, whereas the cycloalkane is not hydrogenated at all. Double-bond isomerization of 1-hexene to 2-hexene occurs over the zeolite catalyst as a side reaction. The effect is more pronounced over the aluminum-rich X-type zeolite [23]. Similar selectivities were also reported for Ni–salen in zeolite Y [112].

An interesting approach has been used by Kimura et al. [54], who prepared an electron donor–acceptor complex $(Na^+)_4(FePc)^{4-}$-NaY from iron–phthalocyanine (FePc) encapsulated in zeolite NaY by reaction with sodium-naphthalene, $Na^+(C_{10}H_8)^-$, and this complex gave unusually high ratios of trans: cis-butene-2 in the selective hydrogenation of butadiene.

More recently, the immobilization of Pd–salen complexes in the intracrystalline voids of zeolites with the FAU and the EMT framework was investigated by Ernst et al. [113–116]. It has been shown that the resulting host–guest compounds possess catalytic activity for the hydrogenation of unsaturated hydrocarbons (i.e., species containing a carbon–carbon double bond or/and a carbonyl group), both in the liquid phase [113–115] and in the gas phase [116]. Kahlen et al. prepared Pt(II)-salen complexes encapsulated in ultrastable zeolite Y and explored their catalytic properties in the hydrogenation of compounds bearing carbonyl groups [117]. The encaged Pt(II)–salen was able to catalyze the hydrogenation of methylpyruvate with almost 100% selectivity at conversions above 70% and of methyl acetoacetate with selectivities around 92% at ca. 50% conversion. Bulkier substrates such as isopropylpyruvate and tert-butylpyruvate gave less or no conversion [117].

Zhan et al. reported on the hydrogenation of cyclooctadiene (COD) on nickel(II)-tetrakis-(N-methyl-4-pyridyl)-porphyrin complexes immobilized in zeolite Y via the zeolite synthesis method [75]. The catalyst exhibited high activity for the hydrogenation of COD to cyclooctene with only little cyclooctane formed. A shape selectivity effect was observed with larger substrate molecules: the rate of hydrogenation for COD was ca. 33 times higher than that for the hydrogenation of the larger cyclododecene [75].

2.4.10.4.3 Catalysis by Zeolite-Encapsulated Chiral Complexes

The catalytic chemistry of chiral complexes entrapped in zeolites has been the subject of recent reviews [118–121]. By far the most attention was paid to the immobilization of complexes bearing the so-called Jacobsen ligand (cf. Fig. 14), or derivatives of it. Zeolite-encapsulated manganese-containing complexes of this type were first prepared by Ogunwumi and Bein [29] and by Sabater et al. [30]. In principle, these catalysts showed some enantioselectivity with monooxygen atom donors such as NaClO or iodosobenzene as oxidant. In no case, however, could the ee-values for the free complexes be observed with the immobilized ones. This observation was recently rationalized by Jacobs [15] and De Vos and Jacobs [119] in the following way: The initial (homogeneous) system consisting of the chiral Jacobsen catalyst and the oxidant (sodium hypochlorite, "bleach") is typically a two-phase system, in which the catalyst resides in the organic phase together with the prochiral substrate and the enantiomerically enriched products. In this way, the products are protected from an overoxidation through aqueous sodium hypochlorite. If, however, the complex is immobilized in an (aluminum-rich) zeolite, the resulting catalyst becomes hydrophilic and, hence, resides in the wrong phase. As a result, the (conventionally) immobilized chiral complexes could, in principal, never reach the enantioselectivity of the free complexes.

In this context, the studies of Hölderich and coworkers deserve attention. In an earlier investigation, these authors prepared, via consecutive treatment with $SiCl_4$

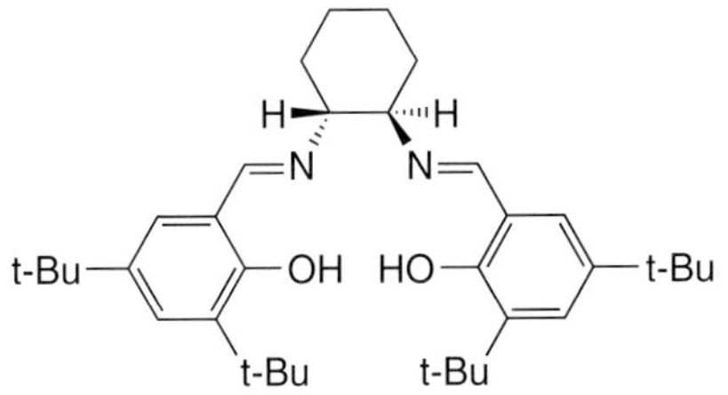

Fig. 14 Structure of the chiral salen-type ligand (R, R)-N',N'-bis-(3,5-di-tert-butylsalicylidene)-1,2-cyclohexanediamine used for the preparation of enantioselective transition metal complexes in zeolite host materials.

vapors and steaming, a high-silica faujasite host material with mesopores which are completely surrounded by micropores [122, 123]. Cobalt Schiff base complexes in these mesopores were able to epoxidize (−)-α-pinene using oxygen and a sacrificial aldehyde with 100% conversion, 96% selectivity, and 91% diastereoselectivity. With a similar catalyst, the same group reported the transhydrogenation of acetophenone to 1-phenylethanol with an enantiomeric excess as high as obtained over the dissolved complex. Moreover, the high ee-values for the immobilized complex were obtained at higher temperature (25 °C) than for the same species in solution (−10 °C). This was tentatively explained by assuming that the zeolite host favors a certain structure of the occluded cobalt–salen complex and preserves this conformation at elevated temperatures [123].

Another example of chiral catalysis on immobilized transition metal complexes in zeolites has been reported by Zsigmond et al. [124, 125]. These authors prepared [Rh(COD)L]$^+$ complexes with L = L-prolinamide or N-tert-butyl-L-prolinamide in zeolite NaY and explored the catalytic properties of the resulting host–guest compounds in the enantioselective hydrogenation of (Z)-methyl-(2-acetamido-3-phenylacrylate). The prepared solid catalysts not only possessed much higher specific activities for the hydrogenation of alkenes (e.g., 1-hexene or cyclohexene) as compared to those of the corresponding complexes in solution, but the enantiomeric excess in the hydrogenation of (Z)-methyl-(2-acetamido-3-phenylacrylate) was also much higher with the immobilized complexes.

2.4.10.4.4 Miscellaneous Reactions

Beside selective oxidations and hydrogenations on zeolite-encaged transition metal complexes, examples for only a few other reaction types are known from the literature also to be catalyzed by zeolite inclusion compounds.

One example for the oligomerization of ethylene over certain nickel chelates has been described by Keim [126]. Whereas in homogeneous solution and supported on amorphous carriers much higher carbon numbers are observed, the length of the oligomers could be restricted to C_{20} when the complexes were incorporated into a faujasite host matrix.

Cyclopropanation reactions were conducted by Alcon et al. [127]. Thus, the reaction of ethyl or tert-butyl diazoacetate with styrene and 1,2-dihydropyrane was catalyzed by rhodium- and copper-complexes encaged in an ultrastabilized zeolite Y. It was observed that high activity and selectivity for cyclopropanation as compared to

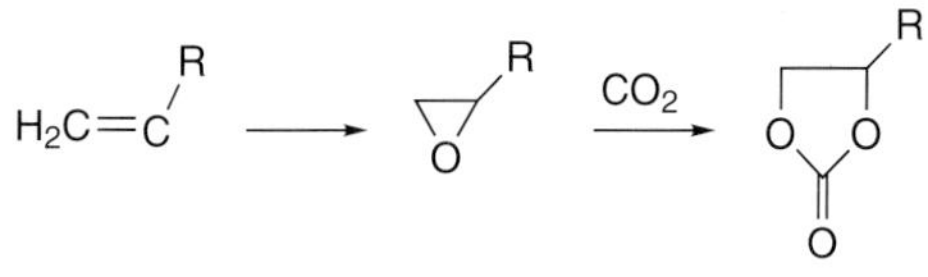

Fig. 15 Carboxylation of epoxides to organic carbonates on zeolite-encapsulated CuPc complexes.

the competing dimerization of styrene could be achieved in this way.

The hydrolysis of nitriles to amides was reported for chromium(III)porphyrin complexes incorporated via the zeolite synthesis method into zeolite Y [74]. It was shown that such host–guest compounds possess catalytic activity for the highly efficient hydrolysis of acetonitrile to acetamide.

Finally, Srivastava et al. [128] found that MePc complexes with Me = Cu^{2+}, Co^{2+}, Ni^{2+}, Al^{3+} entrapped in zeolite Y are active for the cycloaddition of CO_2 to propylene oxide, resulting in the formation of the corresponding cyclic carbonates (Fig. 15). The catalyst (in particular CuPcY) is reported to be stable and much more active than the same complex in solution or supported on silica.

2.4.10.5 Conclusions

Zeolites have been used successfully for many years as host materials for the immobilization of catalytically active transition metal complexes. Particularly suitable are zeolites possessing large intracrystalline cavities which are accessible only via smaller windows. So far, almost exclusively zeolites Y, EMT and, to a minor extent, MCM-22 have been used to entrap a large variety of transition metal complexes. The encapsulated catalyst can be obtained by assembling the transition metal complex within the pores or by building the zeolite framework (host) around the complex (guest). A multitude of physico-chemical methods is required to prove true encapsulation and to explore the chemical nature of the encapsulated complexes. The catalytic properties of zeolite-encapsulated transition metal complexes have been demonstrated in many selective oxidation and hydrogenation reactions. If the immobilized complexes possess chiral properties, they are even able to catalyze oxidations and hydrogenations in an enantioselective manner. This offers good prospects for catalytic applications of such host–guest compounds in the future. The major challenges which must be dealt with in this context are the stability of the inclusion compound under reaction conditions [temperature, solvent, oxidant (if used)] and the minimization of diffusion limitations for reactants/products in the intracrystalline pores of the host material.

Acknowledgments

The author gratefully acknowledges financial support of his research on catalysis and adsorption on zeolites and zeolite-related porous solids by Deutsche Forschungsgemeinschaft, Fonds der Chemischen Industrie, Max-Buchner-Forschungsstiftung, Volkswagenstiftung and Stiftung Rheinland-Pfalz für Innovation.

References

1. G. Schulz-Ekloff, S. Ernst, in *Handbook of Heterogeneous Catalysis*, G. Ertl, H. Knözinger, J. Weitkamp (Eds.), Part 1, VCH, Weinheim, 1997, p. 374.
2. J. H. Lunsford, *Catal. Rev.-Sci. Eng.* **1975**, *12*, 137.
3. J. H. Lunsford, in *Molecular Sieves-II*, J. R. Katzer (Ed.), American Chemical Society Symposium Series, Vol. 40, American Chemical Society, Washington, DC, 1977, p. 473.
4. W. J. Mortier, R. A. Schoonheydt, *Prog. Solid State Chem.* **1985**, *16*, 1.
5. G. A. Ozin, C. Gil, *Chem. Rev.* **1989**, *89*, 1749.
6. J. H. Lunsford, *Rev. Inorg. Chem.* **1987**, *9*, 1.
7. R. Parton, D. De Vos, P. A. Jacobs, in *Zeolite Microporous Solids: Synthesis, Structure and Reactivity*, E. G. Derouane, F. Lemos, C. Naccache, F. Ramoa Ribeiro (Eds.), Kluwer Academic Publishers, Dordrecht, The Netherlands, 1992, p. 555.
8. D. E. De Vos, F. Thibault-Starzyk, P. P. Knops-Gerrits, R. F. Parton, P. A. Jacobs, *Macromol. Symp.* **1994**, *80*, 157.
9. B. V. Romanovsky, *Macromol. Symp.* **1994**, *80*, 185.
10. D. E. De Vos, P. P. Knops-Gerrits, R. F. Parton, B. M. Weckhuysen, P. A. Jacobs, R. A. Schoonheydt, *J. Incl. Phenom.* **1995**, *21*, 185.
11. K. J. Balkus, Jr., A. G. Gabrielov, *J. Incl. Phenom.* **1995**, *21*, 159.
12. K. J. Balkus, Jr., in *Phthalocyanines – Properties and Applications*, Vol. 4, C. C. Leznoff, A. B. P. Lever (Eds.), Academic Press, 1996, p. 289.
13. D. E. De Vos, P. A. Jacobs, in *Introduction to Zeolite Science and Practice*, 2nd edn., H. van Bekkum, E. M. Flanigen, P. A. Jacobs, J. C. Jansen (Eds.), *Studies in Surface Science and Catalysis*, Vol. 137, Elsevier, Amsterdam, 2001, p. 957.
14. A. Corma, H. Garcia, *Eur. J. Inorg. Chem.* **2004**, 1143.
15. P. A. Jacobs, in *Zeolites and Ordered Mesoporous Materials*, J. Cejka, H. van Bekkum (Eds.), *Studies in Surface Science and Catalysis*, Vol. 157, Elsevier, Amsterdam, 2005, p. 289.
16. D. E. De Vos, P. A. Jacobs, *J. Organomettalic Chem.* **1996**, *520*, 195.
17. D. E. De Vos, T. Bein, *J. Am. Chem. Soc.* **1997**, *119*, 9460.
18. D. E. De Vos, J. Meinershagen, T. Bein, in *Progress in Zeolite and Microporous Materials*, H. Chon, S.-K. Ihm, U. S. Uh (Eds.), *Studies in Surface Science and Catalysis*, Vol. 105, Part B, Elsevier, Amsterdam, 1997, p. 1069.
19. J. Weitkamp, in *Proceedings from the Ninth International Zeolite Conference*, Vol. 1, R. von Ballmoos, J. B. Higgins, M. M. J. Treacy (Eds.), Butterworth-Heinemann, Boston, 1993, p. 13.
20. N. Herron, *Inorg. Chem.* **1986**, *25*, 4714.
21. C. Bowers, P. K. Dutta, *J. Catal.* **1990**, *122*, 271.
22. K. J. Balkus, Jr., A. A. Welch, B. E. Gnade, *Zeolites* **1990**, *19*, 722.

23. S. Kowalak, R. C. Weiss, K. J. Balkus, Jr., *J. Chem. Soc., Chem. Commun.* **1991**, 57.
24. D. E. De Vos, P. A. Jacobs, in *Proceedings from the Ninth International Zeolite Conference*, Vol. 2, R. von Ballmoos, J. B. Higgins, M. M. J. Treacy (Eds.), Butterworth-Heinemann, Boston, 1993, p. 615.
25. D. E. De Vos, F. Thibault-Starzyk, P. A. Jacobs, *Angew. Chem., Int. Ed. Engl.* **1994**, *33*, 432.
26. D. E. De Vos, PhD Thesis, Katholieke Universiteit Leuven, Leuven, Belgium, 1994.
27. D. E. De Vos, E. J. P. Feijen, R. A. Schoonheydt, P. A. Jacobs, *J. Am. Chem. Soc.* **1994**, *116*, 4746.
28. F. Bedioui, L. Ruoe, J. Devynck, K. J. Balkus, Jr., in *Zeolites and Related Microporous Materials; State of the Art 1994*, J. Weitkamp, H. G. Karge, H. Pfeifer, W. Hölderich (Eds.), *Studies in Surface Science and Catalysis*, Vol. 84, Part B, Elsevier, Amsterdam, 1994, p. 917.
29. S. B. Ogunwumi, T. Bein, *Chem. Commun.* **1997**, 901.
30. M. J. Sabater, A. Corma, A. Domenech, V. Fornes, H. Garcia, *Chem. Commun.* **1997**, 1285.
31. A. Zsigmond, F. Notheisz, Z. Frater, J. E. Bäckvall, in *Heterogeneous Catalysis and Fine Chemicals IV*, H. U. Blaser, A. Baiker, R. Prins (Eds.), *Studies in Surface Science and Catalysis*, Vol. 108, Elsevier, Amsterdam, 1997, p. 453.
32. S. Ernst, H. Disteldorf, X. Yang, *Microporous Mesoporous Materials* **1998**, *22*, 457.
33. C. R. Jacob, S. P. Varkey, P. Ratnasamy, *Appl. Catal. A: General* **1998**, *168*, 353.
34. D. Chatterjee, A. Mitra, *J. Molec. Catal. A: Chemical* **1999**, *144*, 363.
35. S. Ernst, E. Fuchs, X. Yang, *Microporous Mesoporous Materials* **2000**, *35–36*, 137.
36. K. O. Xavier, J. Chacko, K. K. Mohammed Yusuff, *Appl. Catal. A: General* **2004**, *258*, 251.
37. C. Jin, W. Fan, Y. Jia, B. Fan, J. Ma, R. Li, *J. Molec. Catal. A: Chemical*, **2006**, *249*, 23.
38. F. Bedioui, L. Roue, E. Briot, J. Devynck, S. L. Bell, K. J. Balkus, Jr., *J. Electroanal. Chem.* **1994**, *373*, 19.
39. H. G. Karge, H. K. Beyer in *Zeolite Chemistry and Catalysis*, P. A. Jacobs, N. I. Jaeger, L. Kubelková, B. Wichterlová (Eds.), *Studies in Surface Science and Catalysis*, Vol. 69, Elsevier, Amsterdam, 1991, p. 43.
40. W. H. Quale, G. Peeters, G. L. De Roy, E. F. Vansant, J. H. Lunsford, *Inorg. Chem.* **1982**, *21*, 2226.
41. W. De Wilde, G. Peeters, J. H. Lunsford, *J. Phys. Chem.* **1980**, *84*, 2306.
42. P. P. Knops-Gerrits, D. De Vos, F. Thibault-Starzyk, P. A. Jacobs, *Nature* **1994**, *369*, 543.
43. P. P. Knops-Gerrits, F. Thibault-Starzyk, P. A. Jacobs, in *Zeolites and Related Microporous Materials; State of the Art 1994*, J. Weitkamp, H. G. Karge, H. Pfeifer, W. Hölderich (Eds.), *Studies in Surface Science and Catalysis*, Vol. 84, Part B, Elsevier, Amsterdam, 1994, p. 1411.
44. V. Y. Zakharov, B. V. Romanovsky, *Vest. Mosk. Univ., Khim. Ser. 2* **1977**, *18*, 143.
45. B. V. Romanovsky, in *Proceedings 8th International Congress on Catalysis*, Vol. 4, Verlag Chemie, Weinheim, 1984, p. 657.
46. B. V. Romanovsky, A. G. Gabrielov, in *New Developments in Selective Oxidation by Heterogeneous Catalysts*, P. Ruiz, B. Delmon (Eds.), *Studies in Surface Science and Catalysis*, Vol. 72, Elsevier, Amsterdam, 1992, p. 443.
47. G. Meyer, D. Wöhrle, M. Mohl, G. Schulz-Ekloff, *Zeolites* **1984**, *4*, 30.
48. G. Schulz-Ekloff, D. Wöhrle, V. Iliev, E. Ignatzek, A. Andreev, in *Zeolites as Catalysts, Sorbents and Detergent Builders – Applications and Innovations*, H. G. Karge, J. Weitkamp (Eds.), *Studies in Surface Science and Catalysis*, Vol. 46, Elsevier, Amsterdam, 1989, p. 315.
49. K. J. Balkus, Jr., J. P. Ferraris, *J. Phys. Chem.* **1990**, *94*, 8019.
50. E. Páez-Mozo, N. Gabriunas, F. Lucaccioni, D. D. Acosta, P. Patrono, A. La Ginstra, P. Ruiz, B. Delmon, *J. Phys. Chem.* **1993**, *97*, 12819.
51. J. P. Ferraris, K. J. Balkus, Jr., A. Schade, *J. Incl. Phenom.* **1992**, *14*, 163.
52. N. Herron, G. D. Stucky, C. A. Tolman, *J. Chem. Soc., Chem. Commun.* **1986**, 1521.
53. N. Herron, *J. Coord. Chem.* **1988**, *19*, 25.
54. T. Kimura, A. Fukuoka, M. Ichikawa, *Catal. Lett.* **1990**, *4*, 279.
55. R. F. Parton, L. Uytterhoeven, P. A. Jacobs, in *Heterogeneous Catalysis and Fine Chemicals II*, M. Guisnet, J. Barrault, C. Bouchoule, D. Duprez, G. Pérot, R. Maurel, C. Montassier (Eds.), *Studies in Surface Science and Catalysis*, Vol. 59, Elsevier, Amsterdam, 1991, p. 395.
56. Z. Jiang, Z. Xi, *Fenzi Cuihua* **1992**, *6*, 467.
57. B. V. Romanovsky, A. G. Gabrielov, *J. Molec. Catal.* **1992**, *74*, 293.
58. K. J. Balkus, Jr., A. A. Welch, B. E. Gnade, *J. Incl. Phenom.* **1991**, *10*, 141.
59. R. F. Parton, C. P. Bezoukhanova, F. Thibault-Starzyk, R. A. Reynders, P. J. Grobet, P. A. Jacobs, in *Zeolites and Related Microporous Materials; State of the Art 1994*, J. Weitkamp, H. G. Karge, H. Pfeifer, W. Hölderich (Eds.), *Studies in Surface Science and Catalysis*, Vol. 84, Part B, Elsevier, Amsterdam, 1994, p. 813.
60. S. Ernst, Y. Traa, U. Deeg, in *Zeolites and Related Microporous Materials; State of the Art 1994*, J. Weitkamp, H. G. Karge, H. Pfeifer, W. Hölderich (Eds.), *Studies in Surface Science and Catalysis*, Vol. 84, Part B, Elsevier, Amsterdam, 1994, p. 925.
61. M. Ichikawa, T. Kimura, A. Fukuoka, in *Chemistry of Microporous Crystals*, T. Inui, S. Namba, T. Tatsumi (Eds.), *Studies in Surface Science and Catalysis*, Vol. 60, Elsevier, Amsterdam, 1991, p. 335.
62. A. G. Gabrielov, K. J. Balkus, Jr., S. L. Bell, F. Bedioui, J. Devynck, *Microporous Materials* **1994**, *2*, 119.
63. F. Bedioui, L. Roué, L. Gaillon, J. Devynck, S. L. Bell, K. J. Balkus, Jr., *Preprints, Div. of Petroleum Chemistry, American Chemical Society* **1993**, *38*(No. 3), 529.
64. K. J. Balkus, Jr., A. G. Gabrielov, S. L. Bell, F. Bedioui, L. Roué, J. Devynck, *Inorg. Chem.* **1994**, *33*, 67.
65. R. F. Parton, C. P. Bezoukhanova, J. Grobet, P. J. Grobet, P. A. Jacobs, in *Zeolites and Microporous Crystals*, T. Hattori, T. Yashima (Eds.), *Studies in Surface Science and Catalysis*, Vol. 83, Elsevier, Amsterdam, 1994, p. 371.
66. M. Nakamura, T. Tatsumi, H. Tominaga, *Bull. Chem. Soc. Jpn.* **1990**, *63*, 3334.
67. Y.-W. Chan, R. B. Wilson, *Preprints, Div. of Petroleum Chemistry, American Chemical Society* **1988**, *33*(No. 3), 453.
68. L. A. Rankel, E. W. Valyocsik, US Patent 4 500 503, assigned to Mobil Oil Corp., 1985.
69. K. J. Balkus, Jr., S. Kowalak, K. T. Ly, D. C. Hargis, in *Zeolite Chemistry and Catalysis*, P. A. Jacobs, N. I. Jaeger, L. Kubelková, B. Wichterlová (Eds.), *Studies in Surface Science and Catalysis*, Vol. 69, Elsevier, Amsterdam, 1991, p. 93.
70. S. Kowalak, K. J. Balkus, Jr., *Collect. Czech. Chem. Commun.* **1992**, *57*, 774.

71. K. J. Balkus, Jr., S. Kowalak, PCT Intern. Appl. WO 92/09527, assigned to the Board of Regents, The University of Texas Systems, 1992.
72. K. J. Balkus, Jr., M. Eissa, R. Lavado, in *Catalysis by Microporous Materials*, H. K. Beyer, H. G. Karge, I. Kiricsi, J. B. Nagy (Eds.), *Studies in Surface Science and Catalysis*, Vol. 94, 1995, p. 713.
73. N. Rajic, D. Stojakovic, A. Meden, V. Kaucic, in *Proceedings of the 12th International Zeolite Conference*, M. M. J. Treacy, B. K. Marcus, M. E. Bisher, J. B. Higgins (Eds.), Part II, Materials Research Society, Warrendale, Pennsylvania, 1999, p. 1765.
74. B.-Z. Zhan, P. A. Jacobs, X.-Y. Li, in *Proceedings of the 12th International Zeolite Conference*, M. M. J. Treacy, B. K. Marcus, M. E. Bisher, J. B. Higgins (Eds.), Part IV, Materials Research Society, Warrendale, Pennsylvania, 1999, p. 2905.
75. B.-Z. Zhan, P. A. Jacobs, X.-Y. Li, in *Proceedings of the 12th International Zeolite Conference*, M. M. J. Treacy, B. K. Marcus, M. E. Bisher, J. B. Higgins (Eds.), Part IV, Materials Research Society, Warrendale, Pennsylvania, 1999, p. 2897.
76. G. Gbery, A. Zsigmond, K. J. Balkus, Jr., *Catal. Lett.* **2001**, *74*, 77.
77. P. S. E. Dai, J. H. Lunsford, *Inorg. Chem.* **1980**, *19*, 262.
78. K. Mizuno, J. H. Lunsford, *Inorg. Chem.* **1983**, *22*, 3483.
79. R. F. Howe, J. H. Lunsford, *J. Am. Chem. Soc.* **1975**, *97*, 5156.
80. R. A. Schoonheydt, J. Pelgrims, *J. Chem. Soc. Dalton* **1981**, 914.
81. H. Diegruber, P. J. Plath, in *Metal Microstructures in Zeolites*, P. A. Jacobs, N. I. Jaeger, P. Jiru, G. Schulz-Ekloff (Eds.), *Studies in Surface Science and Catalysis*, Vol. 12, Elsevier, Amsterdam, 1982, p. 23.
82. W. Lubitz, C. J. Winscom, H. Diegruber, R. Möseler, *Z. Naturforsch.* **1987**, *42A*, 970.
83. H. Diegruber, P. J. Plath, *Z. Phys. Chem. (Leipzig)* **1985**, *266*, 641.
84. L. Edwards, M. Gouterman, *J. Molec. Spectr.* **1970**, *33*, 292.
85. U. Ahrens, H. Kuhn, *Z. Phys. Chem. (Frankfurt)* **1963**, *37*, 1.
86. M. Ehrl, F. W. Deeg, C. Bräuchle, O. Franke, A. Sobbi, G. Schulz-Ekloff, D. Wöhrle, *J. Phys. Chem.* **1994**, *98*, 47.
87. S. V. Gudkov, B. V. Romanovsky, E. S. Shpiro, G. V. Antoshin, Kh. M. Minachev, *Izv. AN SSSR, Ser. Khim.* **1980**, 2448.
88. E. S. Shpiro, G. V. Antoshin, O. P. Thachenko, S. V. Gudkov, B. V. Romanovsky, Kh. M. Minachev, in *Structure and Reactivity of Modified Zeolites*, P. A. Jacobs, N. I. Jaeger, P. Jiru, V. B. Kazansky, G. Schulz-Ekloff (Eds.), *Studies in Surface Science and Catalysis*, Vol. 18, Elsevier, Amsterdam, 1984, p. 31.
89. Kh. M. Minachev, E. S. Shpiro, *Catalyst Surface: Physical Methods of Studying*, CRC Press, Boca Raton, 1990, p. 201.
90. J. Strutz, H. Diegruber, N. I. Jaeger, R. Möseler, *Zeolites* **1983**, *3*, 102.
91. R. F. Parton, C. P. Bezoukhanova, F. Thibault-Strazyk, R. A. Reynders, P. J. Grobet, W. Sun, Y. Wu, P. A. Jacobs, *J. Molec. Catal.* **1984**, *24*, 115.
92. B.-Z. Zhan, X.-Y. Li, in *Progress in Zeolite and Microporous Materials*, H. Chon, S.-K. Ihm, Y. S. Uh (Eds.), *Studies in Surface Science and Catalysis*, Vol. 105, Part A, Elsevier, Amsterdam, 1997, p. 615.
93. P.-P. H. J. M. Knops-Gerrits, F. Thibault-Starzyk, R. Parton, in *Impact of Zeolites and Other Porous Materials on the New Technologies at the Beginning of the New Millennium*, R. Aiello, G. Giordano, F. Testa (Eds.), *Studies in Surface Science and Catalysis*, Vol. 142, Part C, Elsevier, Amsterdam, 2002, p. 1809.
94. H. Diegruber, P. J. Plath, G. Schulz-Ekloff, M. Mohl, *J. Molec. Catal.* **1984**, *24*, 115.
95. D. Wöhrle, G. Schulz-Ekloff, *Adv. Mater.* **1994**, *4*, 875.
96. S. Imamura, J. H. Lunsford, *Langmuir* **1985**, *1*, 326.
97. C. Senaratne, J. Zhang, M. D. Baker, C. A. Bessel, D. R. Rolison, *J. Phys. Chem.* **1996**, *100*, 5849.
98. B. V. Romanovsky, *Acta Phys. Chem.* **1985**, *31*, 215.
99. B. Meunier, *Chem. Rev.* **1992**, *92*, 1411.
100. K.-H. Bergk, F. Wolf, B. Walter, *J. Prakt. Chem.* **1979**, *321*, 529.
101. R. F. Parton, D. R. C. Huybrechts, Ph. Buskens, P. A. Jacobs, in *Catalysis and Adsorption by Zeolites*, G. Öhlmann, H. Pfeifer, R. Fricke (Eds.), *Studies in Surface Science and Catalysis*, Vol. 65, Elsevier, Amsterdam, 1991, p. 47.
102. F. Thibault-Starzyk, R. F. Parton, P. A. Jacobs, in *Zeolites and Related Microporous Materials; State of the Art 1994*, J. Weitkamp, H. G. Karge, H. Pfeifer, W. Hölderich (Eds.), *Studies in Surface Science and Catalysis*, Vol. 84, Part B, Elsevier, Amsterdam, 1994, p. 1419.
103. Z. Farzaneh, M. Majidian, M. Ghandi, *J. Molec. Catal. A: Chemical* **1999**, *148*, 227.
104. P. P. Knops-Gerrits, C. A. Trujillo, B. Z. Zhan, X. Y. Li, P. Rouxhet, P. A. Jacobs, *Topics Catal.* **1996**, *3*, 437.
105. M. Salavati-Niasari, *Microporous Mesoporous Materials* **2006**, *92*, 173.
106. R. F. Parton, I. F. J. Vankelecom, M. J. A. Casselman, C. P. Bezoukhanova, J. B. Uytterhoeven, P. A. Jacobs, *Nature* **1994**, *370*, 541.
107. S. Seelan, A. K. Sinha, D. Srinivas, S. Sivasanker, *J. Molec. Catal. A: Chemical* **2000**, *157*, 163.
108. R. Raja, P. Ratnasamy, in *11th International Congress on Catalysis – 40th Anniversary*, J. W. Hightower, W. N. Delgass, E. Iglesia, A. T. Bell (Eds.), *Studies in Surface Science and Catalysis*, Vol. 101, Part A, Elsevier, Amsterdam, 1996, p. 181.
109. S. Seelan, A. K. Siha, D. Srinivas, S. Sivasanker, *Bull. Catal. Soc. India* **2002**, *1*, 29.
110. M. R. Maurya, S. J. J. Titinchi, S. Chand, *J. Molec. Catal. A. Chemical* **2004**, *214*, 257.
111. A. N. Zakharov, *Mendeleev Commun.* **1991**, 80.
112. D. Chatterjee, H. C. Bajaj, A. Das, K. Bhatt, *J. Molec. Catal.* **1994**, *92*, L235.
113. S. Ernst, S. Ost, X. Yang, in C_4 *Chemistry – Manufacture and Use of* C_4 *Hydrocarbons*, W. Keim, B. Lücke, J. Weitkamp (Eds.), DGMK, Hamburg, 1997, p. 279.
114. S. Ernst, H. Disteldorf, X. Yang, *Microporous Mesoporous Materials* **1998**, *22*, 457.
115. S. Ernst, S. Sauerbeck, X. Yang, in *Proceedings of the 12th International Zeolite Conference*, M. M. J. Treacy, B. K. Marcus, M. E. Bisher, J. B. Higgins (Eds.), Part III, Material Research Society, Warrendale, Pennsylvania, 1999, p. 2155.
116. S. Ernst, O. Batréau, in *Catalysis by Microporous Materials*, H. K. Beyer, H. G. Karge, I. Kiricsi, J. B. Nagy (Eds.), *Studies in Surface Science and Catalysis*, Vol. 94, Elsevier, Amsterdam, 1995, p. 479.
117. W. Kahlen, A. Janssen, W. F. Hölderich, in *Heterogeneous Catalysis and Fine Chemicals IV*, H. U. Blaser, A. Baiker, R. Prins (Eds.), *Studies in Surface Science and Catalysis*, Vol. 108, Elsevier, Amsterdam, 1997, p. 469.
118. C. Li, *Catal. Rev. – Science Eng.* **2004**, *46*, 419.
119. D. De Vos, P. A. Jacobs, in *Recent Advances in the Science and Technology of Zeolites and Related Materials*, E. van Steen, L. H. Callanan, M. Claeys (Eds.), *Studies in Surface Science and Catalysis*, Vol. 154, Part A, Elsevier, Amsterdam, 2004, p. 66.

120. A. Cornejo, J. M. Fraile, J. L. Garcia, M. J. Gil, C. L. Herrerias, G. Legaretta, V. M. Martinez-Merino, J. A. Mayotal, *J. Molec. Catal. A: Chemical* **2003**, *196*, 101.
121. I. F. Vankelecom, P. A. Jacobs, in *Chiral Catalyst Immobilization and Recycling*, D. E. De Vos, I. F. Vankelecom, P. A. Jacobs (Eds.), Wiley-VCH, Weinheim, 2000, p. 19.
122. W. Kahlen, H. H. Wagner, W. F. Hölderich, *Catal. Letters* **1998**, *54*, 85.
123. E. Möllmann, P. Tomlinson, W. F. Hölderich, *J. Molec. Catal. A: Chemical* **2003**, *206*, 253.
124. A. Zsigmond, K. Bogar, F. Notheisz, *Catal. Letters* **2002**, *83*, 55.
125. A. Zsigmond, K. Bogar, F. Notheisz, *J. Catal.* **2003**, *213*, 103.
126. W. Keim, in *Homogeneous and Heterogeneous Catalysis*, Y. Yermakov, V. Likholobov (Eds.), VNU Press, Utrecht, The Netherlands, 1986, p. 499.
127. M. Alcon, A. Corma, M. Iglesias, F. Sanchez, *J. Molec. Catal. A: Chemical* **1999**, *144*, 337.
128. R. Srivastava, D. Srinivas, P. Ratnasamy, *Catal. Letters* **2003**, *89*, 81.

2.4.11
Supported Liquid Catalysts

Anders Riisager, Rasmus Fehrmann, and Peter Wasserscheid*

2.4.11.1 Introduction

The development and applications of supported liquid catalyst systems can generally be divided into three groups: (i) supported molten salt catalysts (or supported ionic liquid-phase (SILP) catalysts if the melting point of the applied salt is below 100 °C); (ii) supported organic liquid-phase systems; and (iii) supported aqueous-phase (SAP) systems.

Historically, the development of supported molten salt catalysts started first, and investigations in this field can be grouped into periods which are closely related to the liquid temperature range of the used ionic salts (Fig. 1). Since low-melting ionic liquids have been used for catalytic applications only quite recently, the early applications using supported ionic salt catalysts have been restricted to high-temperature reactions (most often >200 °C), and to catalyst systems which are stable under these reaction conditions.

2.4.11.2 Historical Development

The record of supported molten salt catalysis began in 1914, when an early claim of a supported liquid salt catalyst system was filed by BASF [1]. This catalyst system was the silica-supported V_2O_5-alkali pyrosulfate sulfur dioxide oxidation catalyst used to prepare sulfuric

---
* Corresponding author.

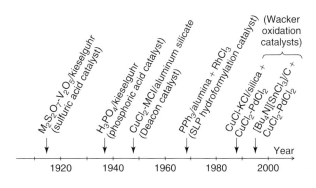

Fig. 1 Important steps in the historical development of supported molten salt catalysts.

acid and which, even today (in slightly modified form) is still the preferred catalyst for sulfuric acid production [2]. It was some years later however, during the 1940s, before it was realized [3, 4], that the catalyst system actually was a supported molten catalyst system which is best described as a mixture of vanadium alkali sulfate/hydrogensulfate/pyrosulfate complexes at reaction conditions in the temperature range of 400 to 600 °C, with the vanadium complexes playing a key role in the catalytic reaction [2] (for details, see Section 2.4.11.3).

Between the late 1930s and the late 1970s, several other molten salt catalyst systems containing alkali metal sulfate-pyrosulfate, copper(I) chloride, zinc(II) chloride or aluminum(III) chloride or eutectic salt mixtures of the salts have, among others, been prepared and applied for various organic reactions. These include commercial reactions, such as condensations, cracking reactions, isomerizations, halogenations and oxidations (see, e.g. Ref. [5] and references cited therein for a comprehensive review). In addition, in many cases these ionic salts provided, besides being high-melting (typically 200–500 °C), highly reactive and corrosive catalyst phases. Some of the most widely examined and important supported catalyst systems developed during this period include the industrial Deacon catalyst [6] and the related oxychlorination catalyst systems [5] used for the oxidation of hydrogen chloride and oxidative chlorination of unsaturated hydrocarbons (e.g., ethylene), respectively. The solid silica- or kieselguhr-supported phosphoric acid catalyst used for petrochemical reactions such as, for example, olefin oligomerization and alkylations, was also developed during these years following its initial invention in the 1930s (for more details, see Section 2.4.11.4).

An analogous catalyst reported in 1970 [7] describes a system where supported copper oxide and mixed CuO/M-oxide (M = Cr, Co, Mn, or V) hydrocarbon oxidation catalysts were generated by thermal treatment of various

References see page 642

precatalyst liquid salt systems prepared by supporting an eutectic mixture of alkali metal salts containing copper salts (e.g., $CuSO_4$-$NaHSO_4$-$KHSO_4$, melting point <160 °C), on alumina or silica. Noticeably, only the calcined catalysts were applied for oxidation reactions and not the supported liquid salt precatalysts themselves. A few years later, however, the use of analogous supported eutectic salt catalyst systems containing alkali metal halides and transition metal halides, such as KCl–$LiCl$–$MnCl_2$ supported on alumina, were applied by Monsanto [8] for the formation of C_2–C_4 alkenes by gas-phase dehydrogenation of the corresponding alkanes at temperatures above the temperature of fusion of the eutectics of the systems.

An early example of a supported catalyst system consisting of a relatively low-melting liquid salt comprising transition metal catalysts was reported for the aerial partial oxidation of ethylene to acetaldehyde (Wacker oxidation) [9]. The catalyst contained $CuCl_2$–$PdCl_2$ dissolved in a eutectic $CuCl/KCl$ melt (65 : 35 mol%) on a porous silica gel support, and proved stable during 150 h continuous reaction (195 °C, 1.6 MPa), providing a product selectivity of acetaldehyde of 95%. In a later study, both the $CuCl/KCl$ molten salt Wacker oxidation system and an analogous ionic liquid tetra-n-butylammonium trichlorostannate $[Bu_4N][SnCl_3]$ system (melting point 60 °C) was applied for the electrocatalytic generation of acetaldehyde from ethanol by the cogeneration of electricity in a fuel cell [10]. In addition to the Wacker oxidation catalysts, supported eutectic molten salt $CuCl/KCl$-based catalyst systems have also been examined for other processes including, for example, the production of synthesis gas from methanol for use as on-board combustion fuels in vehicles (unpromoted/$ZnCl_2$-promoted on silica/alumina/zinc silicate, 350–550 °C, 0.2–1.5 MPa) [11] providing high activity (in some cases the methanol conversion was >99%) and excellent selectivity (up to 95%). Another example of these catalyst systems proved to be suitable for the quantitative combustion of chlorinated hydrocarbons to CO_x and HCl/Cl_2 at ambient pressure (200–500 °C) with silica-based systems [12, 13].

Silica-supported catalyst systems comprised of low-melting tetra-n-butylammonium ionic liquids and $PdCl_2$ with $CoCl_2/CuCl_2$ promoters (melting points around 60 °C), have also been used for the hydrodechlorination of chloroform with hydrogen at 90–150 °C, yielding mixtures of short-chain paraffins and olefins [14].

Since the mid-1990s, imidazolium-based, room temperature ionic liquids have been used to prepare supported ionic liquid systems for catalytic applications. Depending on the role of the ionic liquid, the terms "supported ionic liquid catalyst" (SILC) and "supported ionic liquid-phase" (SILP) catalysts were coined. While the first term indicates a catalytic activity of the supported ionic liquid itself, SILP is used for systems in which the ionic liquid is acting as a non-volatile, inert, immobilizing, liquid film for the catalytic component (mostly a transition metal complex). SILC systems using chloroaluminate ionic liquids immobilized on supports have been disclosed in the patent literature by the IFP research team of Olivier-Bourbigou and Chauvin in 1993. These authors applied such systems in the alkylation of olefins to generate hydrocarbon fuels [15–17]. Later, other groups employed similar concepts for Friedel–Crafts alkylations in liquid- and gas-phase reactions [18, 19].

In contrast, SILP catalysts have only been investigated during the past three years. Mehnert's group at Exxon Mobil pioneered this field with studies of Rh-catalyzed hydrogenation and hydroformylation reactions [20]. Other applications of this technique in Rh-catalyzed hydroformylation followed later, among them the first continuous applications using SILP catalysts [21], in addition to extended studies proving the homogeneous nature of the catalyst dissolved in the ionic liquid films both by spectroscopic [22] and kinetic investigations [23]. Furthermore, SILP catalyst systems have been successfully applied for hydrogenations [24] (Rh-catalyzed), for Heck reactions [25] (Pd-catalyzed), and for hydroaminations [26] (Rh-, Pd-, and Zn-catalyzed).

However, the concept of supported liquid catalysis is not restricted to liquid salts. Since the pioneering studies of Acres et al. [27] and Rony [28, 29], supported liquid-phase (SLP) catalysis has also been investigated using high-boiling organic solvents for continuous gas-phase reactions. By transferring traditional homogeneously catalyzed reactions into these supported systems, the authors intended to take advantage of the larger gas/liquid transfer area per unit of reactor volume and reduced transport paths of reactants. Hjortkjaer et al. [30, 31], Strohmeier et al. [32], and especially Scholten's group [33–42], published during the late 1970s and 1980s an impressive number of hydroformylation studies using this concept. Typical systems under investigation contained a hydridocarbonyltris(triphenylphosphine)-rhodium(I) catalyst dissolved in excess triphenylphosphine which was capillary-condensed in the pores of α-Al_2O_3. In some of these studies, catalyst stabilities of more than 800 h time on stream were attained [33], and even a pre-design of a large-scale SLP plant for propylene hydroformylation has been proposed [43, 44]. However, later studies raised some doubts concerning the long-term stability of these SILP catalysts, as it was found that the evaporation of the loaded liquid cannot be avoided completely during operation in open reactors [45].

During the late 1980s, an alternative supported liquid catalyst concept was developed, namely a technique that involves absorption of an aqueous solution of a catalytically active organometallic complex onto a high-surface-area hydrophilic support material. The technique, often referred to as "Supported Aqueous-Phase Catalysis" (SAPC), offers a high interfacial area between the catalytic phase and the reactants, and is especially suitable for slurry-phase reactions with hydrophobic reaction mixtures. The SAPC concept has shown to be effective for many transition metal-catalyzed reactions [46–48], and has been especially well developed for many hydroformylation reactions using Rh- [49–54], Pt- [55], and Co-complexes [56].

In the following sections, the aim is to demonstrate in more detail important aspects of selected examples of supported liquid-phase catalysts. The selection has been made such as to include both well-established, technically most relevant examples and very recent developments with great technical potential. Attention will be focused on the preparation, properties and application of the sulfuric acid catalyst (Section 2.4.11.3), on the solid phosphoric acid (SPA) catalyst (Section 2.4.11.4), and on the most recently developed SILP catalyst systems (Section 2.4.11.5).

2.4.11.3 The Sulfuric Acid Catalyst

Sulfuric acid is the most important chemical by weight produced in a chemical reaction, with total production amounting to around 170×10^6 tons annually on a worldwide basis. This corresponds to about 25 kg sulfuric acid produced per year per human being on Earth.

The catalyst used for sulfuric acid production and catalytic flue gas cleaning, catalyzing the oxidation of SO_2 to SO_3 [Eq. (1)], is a supported liquid-phase catalyst, usually made by calcination of diatomaceous earth, vanadium pentoxide (or other V salts) and alkali salt promoters (usually in the form of sulfates) with an alkali-to-vanadium molar ratio ranging from 2 to 5.

$$SO_2(g) + \tfrac{1}{2}O_2(g) \longrightarrow SO_3(g) \qquad (1)$$

During the activation process, large quantities of sulfur oxides are taken up by the catalyst, forming molten alkali pyrosulfates [3, 4, 57] which dissolve the vanadium salts. Thus, the molten salt gas system $M_2S_2O_7-MHSO_4-V_2O_5/SO_2-O_2-SO_3-H_2O-N_2$ (M = Na, K, Cs) is considered to be a realistic model of the working industrial catalyst. A major problem in the SO_2 oxidation process is the sudden drop in activity which is experienced for all commercial catalysts at an operating temperature below 420 °C. At this temperature, the equilibrium of the reaction in Eq. (1) is not sufficiently shifted to the right, leading to a content of SO_2, typically around 2000 ppm – too high to be emitted directly to the atmosphere. Interstage (and costly) absorption of SO_3 before the last catalyst bed in the four-bed converter has thus become unavoidable in order to shift Eq. (1) to the right and to obtain a sufficiently low SO_2 content in the stack gas, as shown in Fig. 2.

The exothermic reaction in Eq. (1) ($\Delta H^\circ = -96$ kJ mol^{-1}) needs to be carried out at around 350 °C in order to make the interstage absorption of SO_3 unnecessary, but the activity of the catalyst decreases steeply in this temperature region and becomes far too low at 350 °C.

Previously [5, 6], very little has been known about the complex and compound formation in the catalyst. However, this fundamental knowledge is essential for the understanding of the reaction mechanism and the severe deactivation of the catalyst below ca. 420 °C; it also provides probably the best basis for the systematic improvement of the low-temperature activity of the catalyst. Unfortunately, a direct study of the species formed in the liquid phase, which is dispersed in the small pores of the industrial catalyst, is very difficult, and only methods such as electron spin resonance (ESR) and nuclear magnetic resonance (NMR) can be applied. Applications of magnetic resonance techniques to study vanadium catalysts have been initiated by Mastikhin and coworkers [58, 59]. Indeed, these authors were the first to show directly that, under reaction conditions (i.e., at 400–500 °C), the active component exists as a melt forming a thin liquid layer on the surface of the

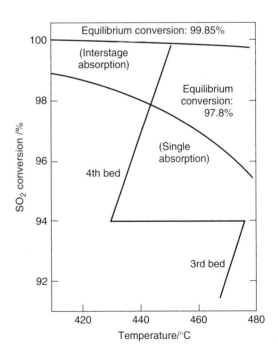

Fig. 2 Conversion of SO_2 in typical single- and double-absorption sulfuric acid plants with a volumetric feed gas composition of 10% SO_2, 11% O_2, 79% N_2.

References see page 642

support [60]. ESR spectra measured at temperatures up to 500 °C revealed the precipitation of V(IV) compounds, causing deactivation of the catalysts at around 420 °C, where a breakpoint was observed on the activity versus temperature plot [60, 61].

This *in-situ* ESR study is probably the first investigation using operando spectroscopy on supported catalysts. Since the early 1980s, joint international efforts have focused on obtaining a deep understanding of the chemistry of the catalyst. This still ongoing study deals with: (i) the complex and redox chemistry of vanadium; (ii) the formation of V(III), V(IV) and V(V) compounds; and (iii) the physico-chemical properties of the catalyst model-system [62–65 and references therein]. The project strategy is to study both the working industrial catalysts and model systems in order to check if their chemistries can be linked together. In addition to ESR and NMR spectroscopy, studies of the catalysts and model systems have included methods such as UV/visible, Fourier transform IR (FTIR), and Raman spectroscopy, electrical conductivity, potentiometry, extended X-ray absorption fine structure (EXAFS), X-ray diffraction (XRD), neutron diffraction, thermal analysis, differential enthalpic analysis, and differential scanning calorimetry.

The original BASF catalyst from 1914 [1] contained only vanadium and potassium salts supported on porous kieselguhr. By uptake of SO_3, the catalyst is well described by the binary molten salt system $K_2S_2O_7$–V_2O_5 distributed in the pores of the support. The phase diagram of this system has attracted much controversy, but by combining electrical conductivity and high-temperature NMR spectroscopy the diagram shown in Fig. 3 could be obtained [66]. Also shown in Fig. 3 is the ORTEP drawing of the structure of the dimeric V(V) compound formed at $x_{V_2O_5} = 0.33$ (unpublished but analogous to $Cs_4(VO)_2O(SO_4)_4$ [64]). This most likely plays a central role in the reaction mechanism (see discussion later in the chapter).

The molar K/V ratio of the BASF catalyst was in the range 1 to 1.5 (mole fraction $x_{V_2O_5} = 0.4$–0.5). The BASF catalyst was operated at 440 °C, well above the liquidus temperature of around 375–425 °C. During the last century, catalysts with higher activity at lower temperature have been developed, usually with a mixture of Na- and K-salts as promoters and a considerably lower content of V_2O_5 (i.e., $x_{V_2O_5}$ ca. 0.20). As judged from Fig. 3, this is close to the lowest melting eutectic of the phase diagram found at $x_{V_2O_5} = 0.17$ and $T_{fus} = 314$ °C (temperature of fusion). Thus, the catalyst should be molten and active at 350 °C, the desired low temperature of operation. Furthermore, the addition of Na, as in the industrial catalyst VK-38 (Haldor Topsøe A/S) (i.e., 20% Na + 80% K and $x_{V_2O_5} = 0.21$) further reduces the melting point of the catalyst according to the phase diagram [67]. Although expected to be molten and active at 350 °C, all industrial catalysts deactivate rapidly below the breakpoint temperature, usually in the range 400 to

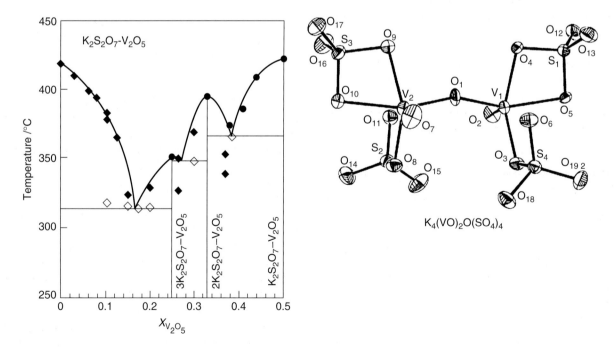

Fig. 3 Phase diagram of the $K_2S_2O_7$–V_2O_5 catalyst model system (left) [66] and the structure of the dimeric V(V) compound formed at $x_{V_2O_5} = 0.33$ (right) [64].

440 °C, depending on the catalyst and gas composition. *In-situ* ESR measurements on VK-38 converting a traditional sulfuric acid synthesis gas (i.e., 9% SO_2, 11% O_2, 80% N_2) revealed [68] that the V(IV) compound $K_4(VO)_3(SO_4)_5$ begins to precipitate at the breakpoint temperature, thus depleting the melt gradually of the active component by further cooling. This compound and several other weakly soluble V(IV) and V(III) compounds have been isolated from catalyst model melts (i.e., $M_2S_2O_7-V_2O_5$; M = Na, K, Cs) in partly converted $SO_2-O_2-N_2$ synthetic gas, and their crystal structure and ESR-spectra were determined. These compounds are listed in Table 1.

Thorough investigations by, for example, ESR [69, 70] of industrial VK-38, VK-58 and VK-WSA (Haldor Topsøe A/S) and model catalysts in sulfuric acid synthesis gas and simulated power plant flue gas, revealed that similar compounds are formed in the supported catalysts under operation and deactivation. Based on these investigations, it could be concluded that the primary effect of mixing of other alkali ions in addition to K in the catalyst leads to a lower temperature of precipitation (and deactivation) as the solubility product for the salts is decreased by dilution of the alkali ion in question by other ions. Cs is preferred over Li as additive to Na-K-promoted catalysts as Cs^+ has much less polarizing power than Li^+, thus stabilizing the $S_2O_7^{2-}$ and $S_3O_{10}^{2-}$ solvent species which further stabilizes the catalytically active V(V) instead of inactive V(IV)- and V(III)-complexes. In addition, it has been shown [71] that decreasing the pore size of the support also lowers the temperature of precipitation as smaller crystals have a higher lattice energy (i.e., they are less stable) than larger crystals.

In summary, the low-temperature activity of sulfuric acid catalysts is enhanced by: (i) the increased atomic number of the alkali promoter; (ii) mixing of the alkali promoters; (iii) dilution of the melt with respect to vanadium (which, however, decreases the overall activity); and (iv) decreasing the pore size of the support. Thus,

Tab. 1 Possible compounds responsible for SO_2 oxidation catalyst deactivation [2]

V(IV) compounds	V(III) compounds
$Na_2VO(SO_4)_2$	$NaV(SO_4)_2$
$Na_3(VO)_2(SO_4)_4$[a]	$Na_3V(SO_4)_3$
$K_4(VO)_3(SO_4)_5$	$KV(SO_4)_2$
$K_3(VO)_2(SO_4)_4$[a]	
$Cs_2(VO)_2(SO_4)_3$	$CsV(SO_4)_2$
β-$VOSO_4$	
$VOSO_4(SO_2,SO_3)_x$[b]	

[a] Mixed valence V(IV)–V(V) compound.
[b] $VOSO_4$-like lattice with incorporated SO_2 and/or SO_3 molecules.

in-situ ESR spectroscopy is a key method for the understanding of the chemistry of the sulfuric acid catalysts. In combination with all the other spectroscopic, thermal, electrochemical, and X-ray crystallographic methods mentioned above, the chemistry of the sulfuric acid catalysts and model systems has been largely elucidated, and a possible catalytic cycle as shown in Fig. 4 has been proposed [2, 72].

The dimeric V(V) complex $(VO)_2O(SO_4)_4^{4-}$ seems to play a central role, initially activating the rather inert oxygen molecule for reaction with SO_2. It should be noted that the vanadium complexes of the catalytic cycle remain in the oxidation state +V. The presence of the dimeric complex $(VO)_2O(SO_4)_4^{4-}$ in a large composition range of the binary catalyst model system $M_2S_2O_7-V_2O_5$ and in supported industrial catalysts has been confirmed by potentiometric [73] and Raman spectroscopic measurements [74], respectively. The side reactions lead to deactivation by either SO_3, forming the monomeric V(V) complex $VO(SO_4)_2^-$ or by SO_2, reducing V(V) to less-soluble V(IV) compounds, here exemplified by $VOSO_4$.

Although some incremental steps towards the creation of an industrial catalyst working at the desired low temperature at around 350 °C have been taken, the goal has not yet been achieved, and double absorption plants have still to be applied for traditional sulfuric acid production. Thus, room exists for further research and development in this area.

2.4.11.4 Solid Phosphoric Acid (SPA) Catalyst

One supported liquid acid catalyst which has been used widely in industry for more than 60 years is the so-called "solid phosphoric acid" (SPA) catalyst [75]. This consists most often of a silicate support material (generally silica gel or a diatomaceous earth such as kieselguhr) which is impregnated with large amounts of phosphoric acid. However, related materials such as montmorillonite and bentonite [76] have also been applied to improve the mechanical properties of SPA catalysts. In 1993, the annual use of industrial SPA catalysts was estimated at approximately 15×10^3 tons [77].

The catalytic activity of the SPA catalyst is linked to carbocation chemistry [78], which is associated with the Brønsted acidity of the deposited phosphoric acid on the catalyst. This acidity depends largely on the water content, as different acid forms (i.e., ortho-, pyro-, tri- and polyphosphoric acids), with increasing degrees of condensation and decreasing acid strengths and amounts may be present, depending on the P_2O_5 to water ratio [79]. In continuous processing, the necessary concentration of

References see page 642

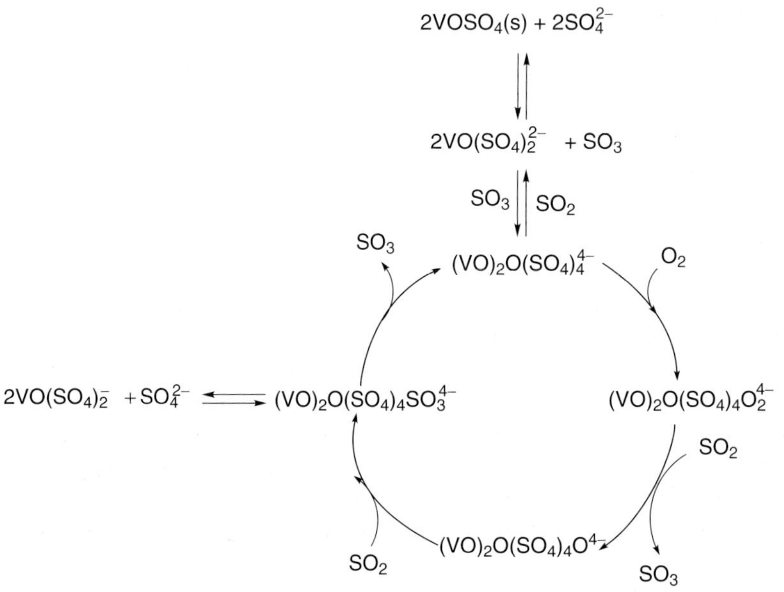

$$2VOSO_4(s) + 2SO_4^{2-}$$

$$2VO(SO_4)_2^{2-} + SO_3$$

$$(VO)_2O(SO_4)_4^{4-}$$

$$2VO(SO_4)_2^- + SO_4^{2-} \rightleftarrows (VO)_2O(SO_4)_4SO_3^{4-}$$

$$(VO)_2O(SO_4)_4O_2^{4-}$$

$$(VO)_2O(SO_4)_4O^{4-}$$

Fig. 4 Proposed mechanism of SO_2 oxidation on the vanadium-based sulfuric acid catalyst [72].

Brønsted acidic sites is maintained by continuously co-feeding water. In the case of propene oligomerization and cumene synthesis by propylation of benzene, as examples (see below) [77], the water content in the reactor inlet typically ranges from about 50 to 1000 ppm.

The concentration and type of acidic sites in the SPA catalyst is frequently found to depend further on the nature of the support. Lorenzelli and coworkers [80, 81], for example, have examined catalysts consisting of a monolayer of phosphoric acid on different carrier materials or mixed carrier materials of silica, titania, and alumina by thermogravimetric-differential thermal analysis (TG-DTA), XRD and FTIR with probe molecules (e.g., CO and amines). It could be shown that phosphoric acid on silica is pure Brønsted acidic in nature, resulting in only weak interaction between supported phosphoric acid and silica, producing mainly liquid-like species, together with covalently bonded phosphate and hydrogen phosphate species. In contrast, a dual Brønsted and Lewis acid character was displayed on titania and alumina, which resulted in a significantly different state (described as "superacidic" [82]) of the supported phosphoric acid having much stronger support interactions, producing supported hydrogen phosphate and phosphate species.

During the early stages of their development, SPA catalyst systems were occasionally considered as a purely solid catalyst, even though immobilization of phosphoric acid on silica (amorphous or quartz crystals) rarely led to the complete formation of silicon phosphates with compositions such as the pyrophosphate SiP_2O_7 [83]. Indeed, Mitsutani and Hamamoto [84] showed that, after treatment at high temperature, the support in SPA catalysts forms silicon phosphates only when extremely small amounts of water are present, while the remaining phosphoric acid appears to consist of a mixture of pyro-, meta- and poly-phosphates with a phosphorus(V) oxide content which is considerably higher than that present in H_3PO_4. Recently, magic-angle spinning nuclear magnetic resonance (MAS-NMR) and X-ray powder diffraction studies performed by Krawietz et al. [85] established that the acid function of silica SPA is due to a liquid or glassy solution of phosphoric acid oligomers supported on mixed crystalline silicon phosphate phases, such as, $Si_5O(PO_4)_6$, *hexagonal*-SiP_2O_7, $Si(HPO_4)_2 \cdot H_2O$, and $SiHP_3O_{10}$. Therefore, SPA should be considered as a multicomponent solid system consisting of silica, silicon phosphates, and a liquid phosphoric acid film (i.e., an SLP catalyst).

The development of SPA catalysts was pioneered by Egloff and Ipatieff et al. during the 1930s [86–89], primarily for the production of liquid gasoline fractions with high octane numbers by oligomerization and hydrogenation of C_3–C_5 olefin feedstocks obtained from thermal cracking processes. Later, SPA catalysts were applied for a wide variety of related acid-catalyzed reactions, most of which were described in the patent literature. These include, for example, cracking, alkylation/acylation, polymerization, rearrangements, (de)hydration and condensation [6, 75, 90, 91]. Today, the industrial application of SPA is essentially restricted to oligomerization of propene [92], ethylene hydration for ethanol production [93], and alkylation of benzene with propene to produce cumene [94], of which essentially all is further converted to phenol or α-methylstyrene [95].

The main advantage of SPA [77] is its very low operating cost, which originates from both its low price and its very high selectivity to the desired products (e.g., cumene; see below). Important disadvantages, on the other hand, are mainly related to disposal of the spent catalyst, which is not regenerable, and the need constantly to replace entrained or evaporated phosphoric acid (and water) during processing in order to maintain the catalytic performance. In addition, constant exposure of the plant construction materials to the highly corrosive vaporized phosphoric acid requires that these materials are lined with copper, copper alloys, or carbon bricks [96] for protection. Furthermore, in certain applications (e.g., ethylene hydration [97]) the silica support deteriorates during prolonged operation due to phase transformation of the carrier material.

For these reasons, SPA has today been largely replaced by different types of solid acids, mainly zeolites [98, 99], thus significantly reducing the use of SPA catalysts. In the case of cumene production, for example, the introduction of zeolite-based catalysts in 1996 led, within only three years, to a decrease in SPA-based processes from 90% to 55% [95]. Zeolitic catalysts also replaced SPA in many processes involving the oligomerization of propene, as they generally provide higher activity. Nevertheless, SPA remains important in industrial catalysis, as indicated in the data listed in Table 2.

In summary, SPA is a supported liquid-phase catalyst which, traditionally, is based on silica supports, and comprises a highly concentrated phosphoric acid solution on supports of various crystalline silicon phosphate phases formed during preparation of the catalyst. It follows that an SPA catalyst is in fact a complex material, and not simply phosphoric acid supported on silica.

2.4.11.5 Supported Ionic Liquid-Phase (SILP) Catalysts

When a substantial amount of an ionic liquid is immobilized on a support, a film of free ionic liquid on the carrier may act as an inert reaction phase, dissolving various molecular catalysts [102]. This further implies that, although the resulting material appears as a solid, the active species dissolved in the ionic liquid phase on the support comprise the attractive features of ionic liquid molecular catalysts, for example high selectivity and dispersion.

In the first reported example using this approach, supported hydroformylation catalysts were prepared by impregnation of a surface-modified silica gel containing covalently anchored ionic liquid fragments with ionic liquid solutions of the precursor [Rh(acetylacetonate)(CO)$_2$] and the ligand tri(m-sulfonyl)triphenyl phosphine as either trisodium salt (TPPTS) or as the more "ionic liquidphilic" tris(1-butyl-3-methylimidazolium) salt (TPPTI)

References see page 642

Tab. 2 Industrial applications of solid phosphoric acid (SPA) catalysts [76, 93, 95, 100, 101]

Reaction type	Process	Process description	Process efficiency[a] /% (source)	Production capacity[b] / 10^6 t·a^{-1} (year)
Oligomerization	Propene to isononenes and isododecenes	Vapor phase, 170–220 °C, 40–60 bar (UOP)	90	1.1 (1997)
Alkylation of aromatics	Benzene and ethene to ethylbenzene	Vapor phase, 300 °C, 70 bar (UOP)	97 (benzene), 90 (propene)	25 (1999)
	Benzene and propene to cumene	Liquid phase, 200–250 °C, 15–35 bar (UOP, BF$_3$-promoted)	96 (benzene) 91 (propene)	10.8 (2000)
	Phenol and C$_{10}$–C$_{18}$ olefins to alkylphenols	a) Liquid phase, 20–100 °C, 1–20 bar		
		b) Vapor phase, 300–400 °C, 1–30 bar		0.3 (1992)
Hydration	Ethene and water to ethanol	Vapor phase, 250–300 °C, 60–80 bar (BP, Shell, UCC, USI, Hüls)	4 (ethene)	2.6 (1998)
	Propene and water to 2-propanol	Vapor phase, 180 °C, 25–45 bar (Hüls)	5–6 (propene)	2.1 (2000)

[a] Process efficiency corresponds to the lower or average amounts of product obtained from the reactants.
[b] Total production capacity including all processes.

(molar ligand to Rh ratio of 10) [20]. The ligand TPPTI was found to dissolve in both 1-butyl-3-methylimidazolium tetrafluoroborate ([BMIM][BF$_4$]) and hexafluorophosphate ([BMIM][PF$_6$]), while the corresponding sodium salt only dissolved in [BMIM][BF$_4$].

The initial preparation of the catalyst involved the modification of a pre-dried silica gel support by treatment in an immobilizing step with functionalized ionic liquids containing the cation 1-butyl-3-(3-triethoxysilylpropyl)imidazolium and the desired anions [BF$_4$]$^-$ or [PF$_6$]$^-$. Analysis of the surface coverage revealed an average of 0.4 ionic liquid fragments per nm^2, corresponding to the involvement of approximately 35% of all the hydroxyl groups of the pretreated silica gel. Treatment of the obtained monolayer of ionic liquid with additional ionic liquid resulted in a multiple layer of free ionic liquid on the support corresponding to an ionic liquid phase loading of 25 wt.%. When the solvent was removed under reduced pressure, a yellowish powder was obtained.

The prepared catalysts were investigated in the hydroformylation of 1-hexene to produce heptanal in a batch-wise, liquid-phase reaction, and all results obtained were compared to the identical reaction in the liquid–liquid biphasic process design. A comparison between the supported [BMIM][BF$_4$]-based catalyst and the biphasic ionic liquid reaction showed that the supported system exhibited a slightly enhanced activity (turnover frequency (*TOF*) = 65 min^{-1} versus 23 min^{-1}) due to the higher concentration of rhodium complexes at the interface, and the generally larger interface of the supported system. However, at high aldehyde concentrations the ionic liquid [BMIM][BF$_4$] was found to dissolve partly in the organic phase, and this led to a considerable rhodium loss of up to 2.1 mol.%. Moreover, pronounced catalyst deactivation was found, even at lower conversion, during recycling of the catalyst, independently of presilylation of the support [103].

Later, the hydroformylation of 1-hexene has also been performed using an MCM-41-supported Rh-TPPTS catalyst based on the ionic liquid 1,1,3,3-tetramethylguanidinium lactate (TMGL) [104]. Here, the catalyst exhibited practically unchanged performance in 12 consecutive runs, providing about 50% conversion and low *n/iso* product ratios of about 2.5, as was also found with the [BMIM][BF$_4$] ionic liquid (see above).

The concept was later extended to continuous-flow processes. The reactions studied were Rh-phosphine catalyzed hydroformylation of propene and 1-octene using fixed-bed reactions [21, 22, 105] (a more detailed description of the reaction set-ups may be found in Refs. [106, 107] for the gas-phase reactions, and in Ref. [108] for the liquid-phase reactions). The supported ionic liquid-phase catalyst systems were prepared by immobilizing [Rh(acetylacetonate)(CO)$_2$] and the applied phosphine ligands in a film of either [BMIM][PF$_6$] or [BMIM][*n*-octylsulfate] ionic liquid on silica gel. Three different phosphine ligands were used for the preparation of the supported catalysts, all being modified with charged groups to increase the ionic liquid solubility (Fig. 5).

In 1-octene hydroformylation using silica-supported [BMIM][PF$_6$] with a dissolved Rh-NORBOS catalyst, a steady catalyst performance (TOF 44 h^{-1} and *n/iso* ratio 2.6) was achieved after 3 to 4 h of reaction. Furthermore, no leaching of rhodium metal could be detected by inductively coupled plasma-atomic emission spectrometry (ICP-AES) analysis of outlet samples, at least not after this relatively short reaction time.

Interestingly, in the continuous, fixed-bed, gas-phase hydroformylation reaction the silica-supported catalysts based on the different ligands were found essentially not to be influenced by the type of ionic liquid, as no significant differences were observed when the reaction was carried out with films of [BMIM][PF$_6$] or [BMIM][*n*-octylsulfate] (except for the catalyst preformation which appeared to depend on the solubility of the ligand and precursor). The same insensitivity of the SILP-catalyzed hydroformylation was observed when alternative oxide supports such as titania, alumina and zirconia, were tested [109].

The use of catalyst systems containing the bidentate phosphine ligand sulfoxantphos proved particular interesting, as excellent *n/iso* ratios of 23 (i.e., linear product selectivities up to 96%) were attained. Moreover, it suggested that the original principles of homogeneous catalysis hold true in supported ionic liquid-phase hydroformylation, with chelating ligands being considerably more selective towards the linear products than monodentate ones. Also, a tendency of decreased activity with higher ligand/rhodium molar ratio supported this conclusion. At low ligand/rhodium molar ratios around 2.5,

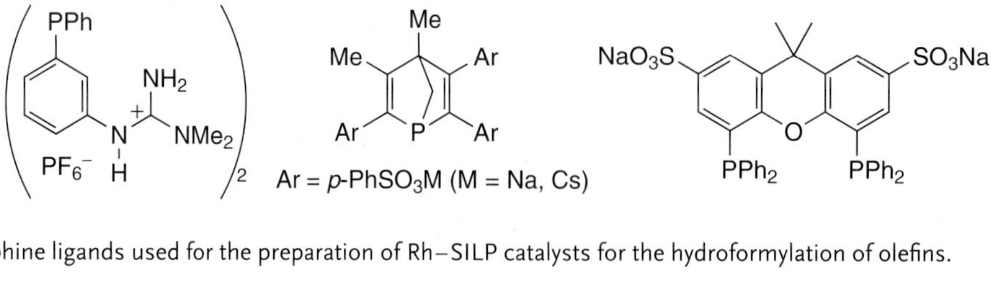

Fig. 5 Ionic phosphine ligands used for the preparation of Rh–SILP catalysts for the hydroformylation of olefins.

however, the selectivity of the Rh-sulfoxantphos system was significantly lower and comparable to the best selectivity obtained with the monodentate phosphines. In comparison to analogous experiments performed with the catalyst system in the ionic liquid–liquid biphasic mode where high selectivity was obtained [110], this result suggested the occurrence of some reaction between the ligand and the solid surface, possibly causing ligand loss under these reactions.

Importantly, it was also realized in this initial study that the catalysts deactivated in prolonged use with a simultaneous decrease in catalytic activity and selectivity independent of the type of ionic liquid, its loading, and the ligand/rhodium molar ratio of the catalyst. Furthermore, the beneficial effect of the ionic liquid solvent in the Rh-sulfoxantphos catalyst systems was not entirely clear in this initial study, as the catalyst without ionic liquid was active and only slightly less selective than the ionic liquid-containing catalysts.

Later, a suitable pretreatment of the silica support involving dehydroxylation gained by thermal treatment was devised and used to prepare Rh-sulfoxantphos catalysts with [BMIM][n-octylsulfate] or without ionic liquid; these were then tested in 60 h continuous propene hydroformylation reactions using similar conditions as previously reported (Fig. 6) [22].

In the reaction using the Rh-sulfoxantphos catalyst without ionic liquid, the activity and selectivity decreased sharply after 5 to 10 h on stream (reaching $TOF = 14$ h^{-1} and $n/iso = 7$ after 50 h), while the supported ionic liquid catalyst system ($\alpha = 0.1$) reached its maximum activity only after 30 h ($TOF = 44$ h^{-1}) and maintained this level stable up to 60 h (i.e., $TON > 2400$) along with a high selectivity corresponding to an n/iso ratio of 21–23. Also, it can be deduced that the apparently negative ionic liquid effect on the catalyst activity measured in these short-term reactions may partly be explained by a delayed formation of the catalytically active species.

Besides the importance of the ionic liquid solvent, it was also shown that a relatively large excess of sulfoxantphos ligand is required to obtain a stable catalyst system, even in reactions using a thermally pretreated support. This aspect was further recognized by solid-state ^{29}Si- and ^{31}P-MAS-NMR measurements to be directly related to an irreversible reaction of the ligand with the acidic silanol surface groups during catalysis. Moreover, FTIR measurements on the supported catalysts under synthesis gas at catalytically relevant conditions, verified the complex formation in the supported ionic liquid-phase to be similar to that observed for Rh-sulfoxantphos and analogous xanthene ligand-based rhodium complexes in organic and ionic liquid solvents [111] (Table 3). This proved definitively that the studied reactions were indeed homogeneously catalyzed.

Finally, the FTIR examinations established the catalyst instability as being associated with degradation of the co-existing isomeric [HRh(CO)$_2$(sulfoxantphos)] complexes, formed by the amount of ligand being available for complex formation. Thus, the prerequisites for obtaining an active, highly selective, and durable supported ionic liquid hydroformylation catalyst, were shown to involve both the presence of ionic liquid solvent and a relatively large excess of bisphosphine ligand to compensate for some detrimental surface reactions.

In another early example, the first silica-based SILP hydrogenation catalyst was prepared by impregnating a [BMIM][PF$_6$] ionic liquid phase containing [Rh(norbornadiene)(PPh$_3$)$_2$][PF$_6$] complexes onto the carrier [24a] (25 wt.% ionic liquid corresponding to an average ionic liquid catalyst layer of 6 Å). The catalyst was used for liquid-phase hydrogenation of 1-hexene, cyclohexene and 2,3-dimethyl-2-butene under various reaction conditions in a batch reactor, and compared to analogous reactions performed in ionic liquid and organic media with the same complex.

Evaluation of the catalysts revealed a superior activity for SILP catalysts in comparison to both the homogeneous and biphasic ionic liquid reaction systems giving up to 10^2 times higher average TOFs under similar reaction conditions. The enhanced activity of the rhodium complex in ionic liquid reactions was explained by the absence of any coordinating solvent, while the higher activity obtained for reactions using supported catalysts compared to the biphasic reactions was attributed to the higher interfacial complex concentration gained by using the

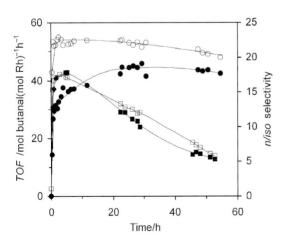

Fig. 6 Continuous hydroformylation of propene with silica Rh–sulfoxantphos/[BMIM][n-C$_8$H$_{17}$OSO$_3$] catalysts (molar ligand/rhodium ratio = 10) having ionic liquid loadings of $\alpha = 0$ (□, ■) and $\alpha = 0.1$ (○, ●). Activity (closed symbols) and selectivity (n/iso ratio, open symbols) [22].

References see page 642

Tab. 3 Infrared $v(CO)$ bands in systems with $[HRh(CO)_2(L)]$ complexes

Complex	Solvent	ea-isomer $v(CO)/cm^{-1}$	ee-isomer $v(CO)/cm^{-1}$	Reference
SILP Rh-sulfoxantphos	[BMIM][n-octylsulfate]	1994, 1948	2035(w), 1964	[22]
$[HRh(CO)_2(sulfoxantphos)]$	$[BMIM][PF_6]$	1985, 1935	2032, 1967	[111]
$[HRh(CO)_2(xantphos)]$	benzene	1991, 1941	2036, 1969	[111]
$[HRh(CO)_2(thixantphos)]$	cyclohexane	1999, 1953	2040, 1977	[111]

supported catalysts having a highly dispersed catalyst phase. Moreover, the SILP catalyst also showed excellent durability and was reused for 18 consecutive runs, without any significant loss of activity and indication of rhodium metal leaching (the level remained below the detection limit of 33 ppb); this may be attributed to the preservation of all active Rh-complexes as cationic species during the catalytic reaction. Furthermore, the isolated organic phases after each run did not exhibit any further reactivity, suggesting full retention of the active species.

In a different approach, supported ionic liquid poly(vinylidene fluoride) filter membranes containing an active phase of $[Rh(norbornadiene)(PPh_3)_2][PF_6]$ complex dissolved in imidazolium-based perfluoroanion ionic liquids (e.g., [RMIM][X] with R = ethyl, butyl and X = BF_4, PF_6, $(CF_3SO_2)_2N$ and CF_3SO_3), were employed to examine hydrogenation of propene and ethene [112]. In order to accomplish the hydrogenation reactions, the olefin was maintained at atmospheric pressure on the feed side of the membrane, while a flow of hydrogen gas at the same pressure was swept over the permeate side to remove hydrogenated products when these passed through the catalytic ionic liquid bulk phase of the membrane. The maximum hydrogenation rates for the catalytic membrane materials based on the different ionic liquids were found to follow the order $[EMIM][(CF_3SO_2)_2N)] > [EMIM][CF_3SO_3] > [EMIM][BF_4] > [BMIM][PF_6]$ (EMIM = 1-ethyl-3-methylimidazolium). This order of reactivity was in line with results previously found for some of the ionic liquids in the analogous biphasic liquid–liquid hydrogenation of 1-pentene [113], which showed the importance of the possible coordinating ability of the ionic liquid perfluoroanions on reaction rates and of the purity of the solvent.

SILP hydrogenation catalysts based on a polymeric poly(allyldimethylammonium chloride) support were also prepared by attachment of $[BMIM][PF_6]$ solutions of $[RhCl(PPh_3)_3]$ (Wilkinson catalyst) and [RuCl(S-BINAP)]Cl (chloro-[(S)-(−)-2,2′-bis(diphenylphosphino)-1,1′-biphenyl](isopropylbenzene)-ruthenium(II)-chloride), respectively [24b]. The supported catalysts were applied for liquid-phase hydrogenations of 2-cyclohexen-1-one and 1,3-cyclooctadiene and asymmetric hydrogenation of methyl acetoacetate (Fig. 7), and compared to similar reactions performed under biphasic and homogeneous reaction conditions.

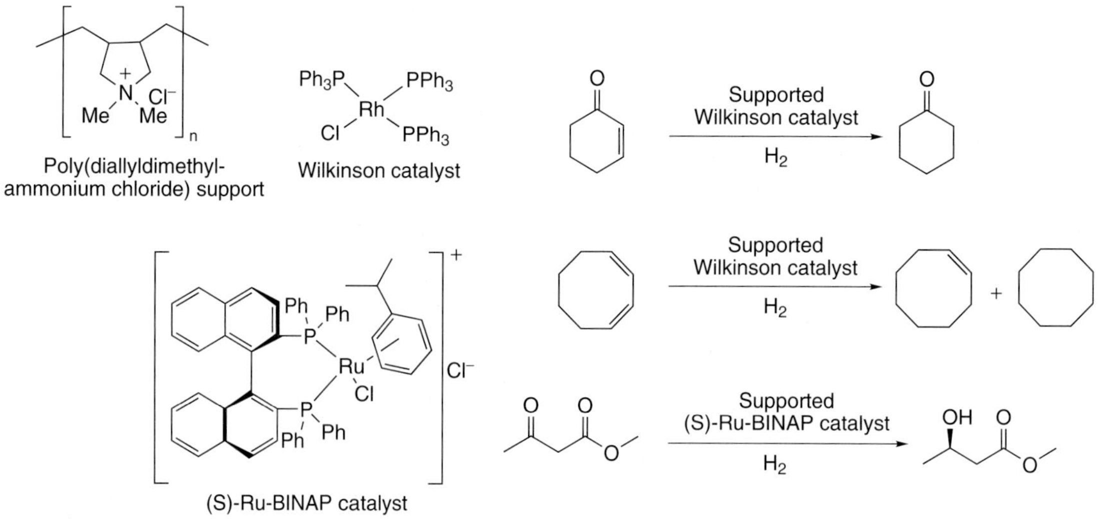

Fig. 7 Hydrogenations using poly(diallyldimethylammonium chloride) supported ionic liquid-phase catalysts. (Illustration reproduced from Ref. [24b].)

For all reactions studied, the activity of the supported catalysts was higher than for the similar biphasic ionic liquid system, and ascribed to improved mass transfer between the substrates and the ionic liquid phase. In addition, the observed product selectivities of 64–87% and an enantioselectivity of 97% for the supported Ru-(S)-BINAP-catalyzed reaction equated with those of the homogeneous reference reactions. Whereas the recyclable system based on Ru-(S)-BINAP was slower than the homogeneous reference system, the supported Wilkinson catalyst system showed activities that were comparable to those of the homogeneous reaction. Also, in all reactions where the supported catalysts were used in combination with ionic liquid-immiscible solvents, no indication of rhodium metal leaching was observed by atomic absorption spectrometry (AAS) analysis of the reaction filtrate.

The synthesis of imines using catalytic hydroamination is another type of homogeneously catalyzed reaction, which has recently been tested using the concept of SILP catalysis containing late transition metal complexes [26]. In this study, hydroamination of phenylacetylene was performed by the direct addition of 4-isopropylamine (Fig. 8), using different supported catalysts obtained by shock-freezing (followed by freeze-drying) a mixture of [EMIM][CF$_3$SO$_3$]-dichloromethane solutions of cationic rhodium(I), palladium(II), copper(I) and zinc(II), respectively, containing a macroporous, diatomaceous earth support (Chromosorb P, pre-modified by dichlorodimethylsilane treatment, i.e., silylated) in liquid nitrogen.

For the supported catalyst containing the complexes [Rh(DPPF)(2,5-norbornadiene)]ClO$_4$, [Pd(DPPF)] [(CF$_3$SO$_3$)$_2$] or Zn(CF$_3$SO$_3$)$_2$ (DPPF: 1,1′-bis-(diphenylphosphino)-ferrocene), the rate of reaction was found to be two- to six-fold higher than in the corresponding homogeneously catalyzed reactions. In contrast, [Cu$_2$(C$_6$H$_5$CH$_3$)][CF$_3$SO$_3$]$_2$ showed a significantly lower catalytic activity in the supported catalysts than in the non-supported catalyst, possibly due to competing coordination of the copper(I) center by the ionic liquid and the substrate. It is noteworthy that also the product selectivity (85–100%) was significantly increased for all four supported catalysts compared to the homogeneous reactions.

Palladium complexes have also been impregnated on an amorphous silica support with the aid of a solution containing [BMIM][PF$_6$] dissolved in tetrahydrofuran, and applied as highly efficient catalysts for promoting liquid Heck coupling reactions between various aryl halides and cyclohexyl acrylate in alkanes, without the presence of additional ligands [25] (Fig. 9).

With optimized reaction conditions using a supported catalyst containing palladium(II) acetate, a high TOF of 8000 h^{-1} was reached in the reaction of iodobenzene with cyclohexyl acrylate in n-dodecane using tri-n-butylamine as base. Moreover, the catalyst could readily be reused up to six times giving yields of 89–98% and without taking any precautions (e.g., avoidance of air), if surface-deposited tri-n-butylammonium iodide salt was removed by washing with aqueous sodium hydroxide at an intermediate stage during the consecutive reactions.

In another interesting approach, an oxime carbapalladacycle ionic liquid catalyst was attached to an aluminosilicate support (Al/MCM-41, Si : Al = 13) via electrostatic interaction by impregnation, and tested for base-promoted Suzuki cross-coupling of halobenzenes and phenylboronic acid in toluene and dimethylformamide (DMF), respectively [114] (Fig. 10).

References see page 642

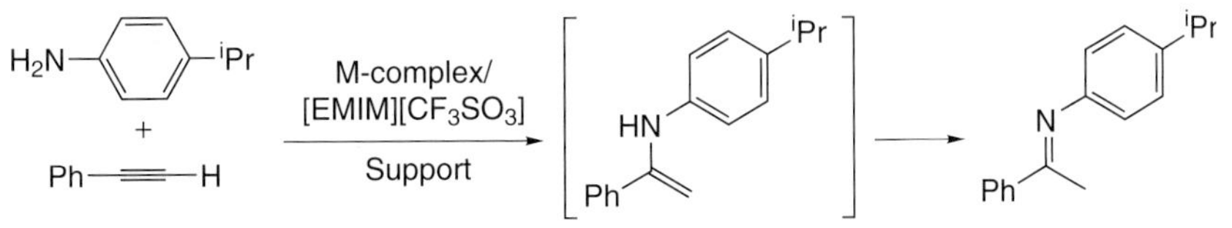

Fig. 8 Hydroamination of phenylacetylene with 4-isopropylamine using supported ionic liquid-phase metal catalysts [26].

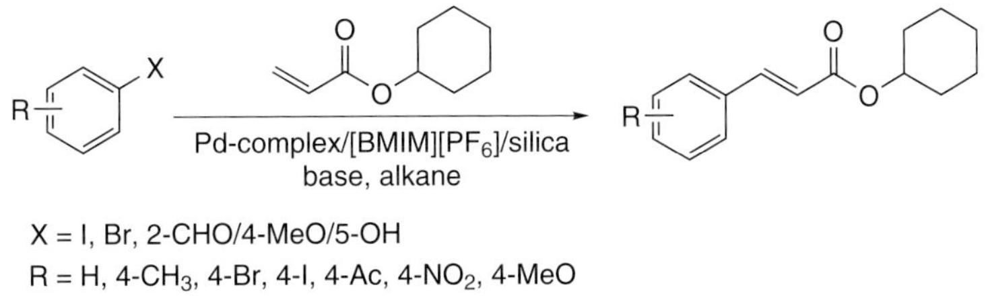

X = I, Br, 2-CHO/4-MeO/5-OH
R = H, 4-CH$_3$, 4-Br, 4-I, 4-Ac, 4-NO$_2$, 4-MeO

Fig. 9 Heck coupling reactions between aryl halides and cyclohexyl acrylate in alkanes using silica/[BMIM][PF$_6$] supported ionic liquid-phase catalysts [25].

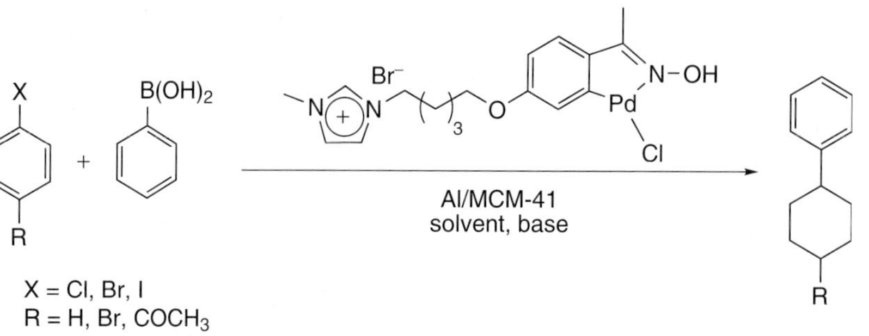

Fig. 10 Suzuki cross-coupling of halobenzenes and phenylboronic acid using Al/MCM-41-supported oxime carbapalladacycle ionic liquid catalyst [114].

The catalyst performance of the supported palladium ionic liquid catalyst was found to depend heavily on the organic solvent, revealing higher activity (comparable to analogous homogeneously catalyzed reactions) and higher product selectivity towards Suzuki-coupled products compared to homocoupled products in a polar solvent such as DMF (no homocoupled products observed) than in toluene (1:2 to 1:5 distribution of products). Unfortunately, it was also shown that the catalyst could only be reused for a few consecutive runs with similar activity (resulting in 86–90% conversion) and selectivity, due to a partial instability of the oxime carbapalladacycle complex under the reaction conditions, resulting in the formation of palladium metal particles which were detached from the support into the liquid phase (up to 30% of the conversion originating from dissolved palladium metal).

In a different approach based on membrane technology, oligomerization of ethene has been examined using a poly(ethersulfone)-supported ionic liquid membrane containing [EMIM]-chloroaluminates ionic liquids in the presence or absence of a $[NiCl_2P(cyclohexyl)_3]_2$ dimerization catalyst and dichloroethylaluminate as acid scavenging cocatalyst [115].

In summary, SILP catalysis combines well-defined catalyst species, non-volatile ionic liquids, and porous solid supports in a manner which offers advantages over regular biphasic ionic liquid–organic liquid systems, namely substantially reduced amounts of catalyst and ionic liquid, higher turnovers, no loss of organic solvent, and no catalyst leaching. In particular, the advantage of the non-volatility of ionic liquids compared to traditional supported catalysts comprising organic solvents with low vapor pressure (SLP, supported liquid-phase) or water (SAP, supported aqueous-phase) on supports, which are clearly limited by solvent evaporation, makes it possible to perform gas-phase continuous processes using fixed-bed reaction technology.

2.4.11.6 Conclusions

It is likely that supported liquid-phase catalysts are second only to zeolites in terms of the numbers of materials, processes, patents, and journal publications relating to them. Nonetheless, they can be very straightforward materials, such as liquid acids (e.g., phosphoric acid) or simple metal oxides (e.g., V_2O_5) dispersed on common porous support materials (e.g., kieselguhr). However, the chemistry of the working supported catalyst may be very complicated, as exemplified by the SO_2 oxidation catalyst.

The recent use of ionic liquids as solvents for supported liquid-phase catalysts has demonstrated a remarkable potential for the SILP concept. This will most likely initiate the development of more continuously operated applications and new reactions, and help to accelerate their introduction into commercial processes. In addition, combinations of SILP methodology with research disciplines such as nanotechnology, separation techniques and biocatalysis hold great promise for the future.

References

1. E. Blum, Swiss Patent CH71326, assigned to BASF, 1914.
2. O. B. Lapina, B. S. Balzhinimaev, S. Boghosian, K. M. Eriksen, R. Fehrmann, *Catal. Today* **1999**, *51*, 469–479.
3. J. H. Frazer, W. J. Kirkpatrick, *J. Am. Chem. Soc.* **1940**, *62*, 1659–1660.
4. H. F. A. Topsøe, A. Nielsen, *Trans. Danish Acad. Tech. Sci.* **1948**, *1*, 3–17.
5. C. N. Kenney, *Catal. Rev. -Sci. Eng.* **1975**, *11*, 197–224.
6. J. Villadsen, H. Livbjerg, *Catal. Rev. -Sci. Eng.* **1978**, *17*, 203–272.
7. British Patent 1351802, assigned to Societé Cooperative Metachimie (Belgium), 1970.
8. D. B. Fox, E. H. Lee, *CHEMTECH* **1973**, *3*, 186–189.
9. V. Rao, R. Datta, *J. Catal.* **1988**, *114*, 377–387.
10. S. Malhotra, R. Datta, *Proc. Electrochem. Soc.* **1994**, *94*, 773–780.
11. A. D. Schmitz, D. P. Eyman, *Energy & Fuels* **1994**, *8*, 729–740.
12. R. M. Lago, M. L. H. Green, S. C. Tsang, M. Odlyha, *Appl. Catal. B: Environmental* **1996**, *8*, 107–121.

13. M. L. H. Green, R. M. Lago, S. C. Tsang, *J. Chem. Soc., Chem. Commun.* **1995**, 365.
14. X. Wu, Y. A. Letuchy, D. P. Eyman, *J. Catal.* **1996**, *161*, 164–177.
15. E. Benazzi, A. Hirschauer, J. F. Joly, H. Olivier, J. Y. Bernhard, European Patent 553009, assigned to IFP, 1993.
16. E. Benazzi, H. Olivier, Y. Chauvin, J. F. Joly, A. Hirschauer, *Abstr. Pap. Am. Chem. Soc.* **1996**, *212*, 45.
17. E. Benazzi, Y. Chauvin, A. Hirschauer, N. Ferrer, H. Olivier, J. Y. Bernhard, US Patent 5,69,3585, assigned to IFP, 1997.
18. F. G. Sherif, L.-J. Shyu, WO9903163, assigned to Akzo Nobel Inc., 1999.
19. C. deCastro, E. Sauvage, M. H. Valkenberg, W. F. Hölderich, *J. Catal.* **2000**, *196*, 86–94.
20. C. P. Mehnert, R. A. Cook, N. C. Dispenziere, M. Afeworki, *J. Am. Chem. Soc.* **2002**, *124*, 12932–12933.
21. A. Riisager, P. Wasserscheid, R. van Hal, R. Fehrmann, *J. Catal.* **2003**, *219*, 252–255.
22. A. Riisager, R. Fehrmann, S. Flicker, R. van Hal, M. Haumann, P. Wasserscheid, *Angew. Chem., Int. Ed.* **2005**, *44*, 815–819.
23. A. Riisager, R. Fehrmann, M. Haumann, B. S. K. Gorle, P. Wasserscheid, *Ind. Eng. Chem. Res.* **2005**, *44*, 9853–9859.
24. (a) C. P. Mehnert, E. J. Mozeleski, R. A. Cook, *Chem. Commun.* **2002**, 3010–3011; (b) A. Wolfson, I. F. J. Vankelecom, P. A. Jacobs, *Tetrahedron Lett.* **2003**, *44*, 1195–1198.
25. H. Hagiwara, Y. Sugawara, K. Isobe, T. Hoshi, T. Suzuki, *Org. Lett.* **2004**, *6*, 2325–2328.
26. S. Breitenlechner; M. Fleck; T. E. Müller; A. Suppan, *J. Mol. Catal. A: Chemical* **2004**, *214*, 175–179.
27. G. J. K. Acres, G. C. Bond, B. J. Cooper, J. A. Dawson, *J. Catal.* **1966**, *6*, 139–141.
28. P. R. Rony, *J. Catal.* **1969**, *14*, 142–147.
29. P. R. Rony, *Chem. Eng. Sci.* **1968**, 1021–1034.
30. J. Hjortkjaer, M. S. Scurell, P. Simonsen, *J. Mol. Catal.* **1979**, *6*, 405–420.
31. J. Hjortkjaer, M. S. Scurell, P. Simonsen, *J. Mol. Catal.* **1981**, *12*, 179–195.
32. W. Strohmeier, R. Marcec, B. Graser, *J. Organomet. Chem.* **1981**, *221*, 361–366.
33. L. A. Gerritsen, A. van Meerkerk, M. H. Vreugdenhill, J. J. F. Scholten, *J. Mol. Catal.* **1980**, *9*, 139–155.
34. L. A. Gerritsen, J. M. Herman, W. Klut, J. J. F. Scholten, *J. Mol. Catal.* **1980**, *9*, 157–168.
35. L. A. Gerritsen, J. M. Herman, J. J. F. Scholten, *J. Mol. Catal.* **1980**, *9*, 241–256.
36. L. A. Gerritsen, W. Klut, M. H. Vreugdenhill, J. J. F. Scholten, *J. Mol. Catal.* **1980**, *9*, 257–264.
37. L. A. Gerritsen, W. Klut, M. H. Vreugdenhill, J. J. F. Scholten, *J. Mol. Catal.* **1980**, *9*, 265–274.
38. N. A. de Munck, M. W. Verbruggen, J. J. F. Scholten, *J. Mol. Catal.* **1981**, *10*, 313–330.
39. N. A. de Munck, M. W. Verbruggen, J. E. de Leur, J. J. F. Scholten, *J. Mol. Catal.* **1981**, *11*, 331–342.
40. H. L. Pleit, G. van der Lee, J. J. F. Scholten, *J. Mol. Catal.* **1985**, *29*, 319–334.
41. H. L. Pleit, J. J. J. J. Brockhus, R. P. J. Verburg, J. J. F. Scholten, *J. Mol. Catal.* **1985**, *31*, 107–118.
42. H. L. Pelt, P. J. Gijsman, R. P. J. Verburg, J. J. F. Scholten, *J. Mol. Catal.* **1985**, *33*, 119–128.
43. J. J. F. Scholten, R. van Hardeveld, *Chem. Eng. Commun.* **1987**, *52*, 75–92.
44. J. M. Herman, A. P. A. F. Rocourt, P. J. van den Berg, P. J. van Krugten, J. J. F. Scholten, *Chem. Eng. J.* **1987**, *35*, 83–103.
45. R. Brüsewitz, D. Hesse, *Chem. Eng. Technol.* **1992**, *15*, 385–389.
46. M. E. Davis, in *Aqueous Phase Organometallic Chemistry*, B. Cornils, W. A. Herrmann (Eds.), Chapter 4.7, Wiley-VCH, Weinheim, 1998, pp. 241–251.
47. M. E. Davis, *Chemtech* **1992**, *22*, 498–502.
48. M. S. Anson, M. P. Leese, L. Tonks, J. M. J. Williams, *J. Chem Soc., Dalton Trans.* **1998**, 3529–3538.
49. J. P. Arhancet, M. E. Davis, J. S. Merola, B. E. Hanson, *Nature* **1989**, *339*, 454–455.
50. J. P. Arhancet, M. E. Davis, J. S. Merola, B. E. Hanson, *J. Catal.* **1990**, *121*, 327–339.
51. J. P. Arhancet, M. E. Davis, B. E. Hanson, *J. Catal.* **1991**, *129*, 94–99.
52. J. P. Arhancet, M. E. Davis, B. E. Hanson, *J. Catal.* **1991**, *129*, 100–105.
53. H. Zhu, Y. Ding, H. Yin, L. Yan, J. Xiong, Y. Lu, H. Luo, L. Lin, *Appl. Catal. A: General* **2003**, *245*, 11–117.
54. A. Riisager, K. M. Eriksen, J. Hjortkjaer, R. Fehrmann, *J. Mol. Catal. A: Chemical* **2003**, *193*, 259–272.
55. I. Tóth, L. Guo, B. E. Hanson, *J. Mol. Catal.* **1997**, *116*, 217–229.
56. T. Bartik, B. Bartik, I. Guo, B. E. Hanson, *J. Organomet. Chem.* **1994**, *480*, 15–21.
57. G. K. Boreskov, *Catalysis in the Sulphuric Acid Production*, Goskhinizdat, Moskva, 1954 (in Russian).
58. O. B. Lapina, V. M. Mastikhin, A. A. Shubin, V. N. Krasilnikov, K. I. Zamaraev, *Prog. NMR Spectroscopy* **1992**, *24*, 457–525.
59. V. M. Mastikhin, O. B. Lapina, Vanadium Catalysts: Solid State NMR, in *Encyclopedia of Nuclear Magnetic Resonance*, Vol. 8, John Wiley & Sons, 1996, pp. 4892–4904.
60. G. K. Boreskov, L. P. Davydova, V. M. Mastikhin, G. N. Polyakova, *Dokl. Akad. Nauk SSSR, Ser. Khim.* **1966**, *171*, 648–651 (in Russian).
61. V. M. Mastikhin, G. N. Polyakova, J. Ziolkowsky, G. K. Boreskov, *Kinet. Katal.* **1970**, *1*, 1463–1468 (in Russian).
62. S. V. Kozyrev, B. S. Bal'zhininiaev, G. K. Boreskov, A. A. Ivanov, V. M. Mastikhin, *React. Kinet. Catal. Lett.* **1982**, *20*, 53–57.
63. S. Boghosian, R. Fehrmann, N. J. Bjerrum, G. N. Papatheodorou, *J. Catal.* **1989**, *199*, 121–134.
64. K. Nielsen, R. Fehrmann, K. M. Eriksen, *Inorg. Chem.* **1993**, *32*, 4825–4828
65. D. A. Karydis, K. M. Eriksen, R. Fehrmann, S. Boghosian, *J. Chem. Soc. Dalton Trans.* **1994**, 2151–2157.
66. G. E. Folkmann, G. Hatem, R. Fehrmann, M. Gaune-Escard, N. J. Bjerrum, *Inorg. Chem.* **1991**, *30*, 4057–4061.
67. D. A. Karydis, S. Boghosian, R. Fehrmann, *J. Catal.* **1994**, *145*, 312–317.
68. K. M. Eriksen, R. Fehrmann, N J Bjerrum, *J. Catal.* **1991**, *132*, 263–265.
69. K. M. Eriksen, D. A. Karydis, S. Boghosian, R. Fehrmann, *J. Catal.* **1995**, *155*, 32–42.
70. S. G. Masters, A. Chrissanthopoulos, K. M. Eriksen, S. Boghosian, R. Fehrmann, *J. Catal.* **1997**, *166*, 16–24.
71. C. Oehlers, R. Fehrmann, S. G. Masters, K. M. Eriksen, D. E. Sheinin, B. S. Balzhinimaev, V. I. Elokhin, *Appl. Catal.* **1996**, *147*, 127–144.
72. G. Hatem, K. M. Eriksen, M. Gaune-Escard, R. Fehrmann, *Topics Catal.* **2002**, *19*, 323–331.

73. S. B. Rasmussen, K. M. Eriksen, R. Fehrmann, *J. Phys. Chem.* **1999**, *103*, 11282–11289.
74. A. Christodoulakis, S. Boghosian, *J. Catal.* **2003**, *215*, 139–150.
75. E. Weisang, P. A. Engelhard, *Bull. Soc. Chim. Fr.* **1968**, 1811–1820.
76. A. Farkas, in *Ethanol, Ullmann's Encyclopedia of Industrial Chemistry*, 6th edn, Wiley-VCH, Weinheim, 2002.
77. F. Cavani, G. Girotti, G. Terzoni, *Appl. Catal. A: General* **1993**, *97*, 177–196.
78. L. Schmerling, V. N. Ipatieff, *Adv. Catal.* **1950**, *2*, 21–80.
79. A. V. Slack, *Phosphoric Acid*, Marcel Dekker, New York, 1968, 992 pp.
80. G. Ramis, P. F. Rossi, G. Busca, V. Lorenzelli, *Langmuir* **1989**, *5*, 917–923.
81. G. Busca, G. Ramis, V. Lorenzelli, P. F. Rossi, *Langmuir* **1989**, *5*, 911–916.
82. A. Krzywicky, M. Marczewski, *J. Chem. Soc. Faraday Trans 1* **1980**, *76*, 1311–1322.
83. P. A. Obraztsov, V. S. Malinsky, *J. Mol. Catal.* **1989**, *55*, 285–292.
84. A. Mitsutani, Y. Hamamoto, *Kogyo Kagaku Zasshi (Japanese)* **1964**, *67*, 1231.
85. T. R. Krawietz, P. Lin, K. E. Lotterhos, P. D. Torres, D. H. Barich, A. Clearfield, J. F. Haw, *J. Am. Chem. Soc.* **1998**, *120*, 8502–8511.
86. G. Egloff, *Ind. Eng. Chem.* **1936**, *28*, 1461–1467.
87. V. N. Ipatieff, B. B. Corson, *Ind. Eng. Chem.* **1938**, *28*, 1316–1317.
88. V. N. Ipatieff, R. E. Schaad, *Ind. Eng. Chem.* **1938**, *30*, 596–599.
89. V. N. Ipatieff, R. E. Schaad, *Ind. Eng. Chem.* **1948**, *40*, 78–80.
90. W. H. Corkern, A. Fry, *J. Am. Chem. Soc.* **1967**, *89*, 5888–5894.
91. R. Vetrivel, B. Viswanathan, *Surface Tech.* **1984**, *22*, 1–8.
92. E. K. Jones, in *Advances in Catalysis and Related Subjects*, D. D. Eley, W. G. Frankenberg, V. I. Komarewsky (Eds.), Vol. 8, Academic Press, New York, 1956, pp. 219–238.
93. K. Weissermel, H.-J. Arpe, in *Alcohols, Industrial Organic Chemistry, Fourth, Completely Revised Edition*, Wiley-VCH, Weinheim, 2003, pp. 93–215.
94. M. F. Bentham, G. J. Gajda, R. H. Jensen, H. A. Zinnen, *Erdöl Erdgas Kohle* **1997**, *113*, 84–88.
95. K. Weissermel, H.-J. Arpe, in *Benzene Derivatives, Industrial Organic Chemistry, Fourth, Completely Revised Edition*, Wiley-VCH, Weinheim, 2003, pp. 337–385.
96. S. R. Bethea, J. H. Karchmer, *Ind. Eng. Chem.* **1956**, *48*, 370–377.
97. C. M. Fougret, M. P. Atkins, W. F. Hölderich, *Appl. Catal. A: General* **1999**, *181*, 145–156.
98. A. Corma, *Chem. Rev.* **1995**, *95*, 559–614.
99. W. F. Hölderich, D. Heinz, *Res. Chem. Intermed.* **1998**, *24*, 337–348.
100. M. Röper, E. Gehrer, T. Narbeshuber, W. Siegel, in *Acylation and Alkylation, Ullmann's Encyclopedia of Industrial Chemistry*, 6th edn., Wiley-VCH, Weinheim, 2002, pp. 1–41.
101. K. Weissermel, H.-J. Arpe, in *Olefins, Industrial Organic Chemistry, Fourth, Completely Revised Edition*, Wiley-VCH, Weinheim, 2002, pp. 59–89.
102. C. P. Mehnert, R. A. Cook, US Patent 6,673,737, assigned to Exxon Mobil, 2004.
103. C. P. Mehnert, R. A. Cook, E. J. Mozeleski, N. C. Dispenziere, M. Afeworki, *Abstr. Pap. Am. Chem. Soc.* **2003**, *226*, 277.
104. Y. Yang, H. Lin, C. Deng, J. She, Y. Yuan, *Chem. Lett.* **2005**, *34*, 220–221.
105. A. Riisager, K. M. Eriksen, P. Wasserscheid, R. Fehrmann, *Catal. Lett.* **2003**, *90*, 149–153.
106. B. Heinrich, Y. Chen, J. Hjortkjaer, *J. Mol. Chem.* **1993**, *80*, 365–375.
107. A. Riisager, K. M. Eriksen, J. Hjortkjaer, R. Fehrmann, *J. Mol. Catal. A: Chemical* **2003**, *193*, 259–272.
108. J. Hjortkjaer, Y. Chen, B. Heinrich, *Appl. Catal.* **1991**, *67*, 269–278.
109. A. Riisager, R. Fehrmann, P. Wasserscheid, R. van Hal, in *Ionic Liquids IIIB: Fundamentals, Progress, Challenges, and Opportunities – Transformations and Processes*, R. D. Rogers, K. R. Seddon (Eds.), Vol. 902, Chapter 23, ACS, 2005, pp. 334–349.
110. H. Bohnen, J. Herwig, D. Hoff, R. van Hal, P. Wasserscheid, European Patent 1400504-A1, assigned to Celanese Chemical Europe GmbH, 2003.
111. S. M. Silva, R. P. J. Bronger, Z. Freixa, J. Dupont, P. W. N. M. van Leeuwen, *New J. Chem.* **2003**, *27*, 1294–1296.
112. T. H. Cho, J. Fuller, R. T. Carlin, *High Temp. Mat. Process.* **1998**, *2*, 543–558.
113. Y. Chauvin, L. Mussmann, H. Olivier, *Angew. Chem. Int. Ed. Engl.* **1995**, *34*, 2698–2700.
114. A. Corma, H. García, A. Leyva, *Tetrahedron* **2004**, *60*, 8553–8560.
115. R. T. Carlin, T. H. Cho, J. Fuller, *Proc. Electrochem. Soc.* **1998**, *98*, 180–185.

2.4.12
Immobilization of Biological Catalysts

Marion B. Ansorge-Schumacher[*]

2.4.12.1 Introduction

With the increasing importance of biological catalysts, particularly isolated enzymes and microbial cells, in the synthesis of fine and commodity chemicals [1], the development of methods for their immobilization has grown into an important and challenging field of research. Basically, this is due to the possible advantageous effects of the "physical confinement or localization of biocatalysts in a certain defined region of space with retention of catalytic activities" [2] on the economy of biocatalyzed reactions. These effects mainly result from the robustness and recyclability, high catalyst density, facilitated downstream processing, and increased stability that can be achieved by use of immobilized biocatalysts [3].

By definition, the immobilization of biological catalysts encompasses a wide variety of possible methods which, in accordance with the suggestions of the international "Enzyme Engineering Conference" at Henniker, New

---
[*] Corresponding author.

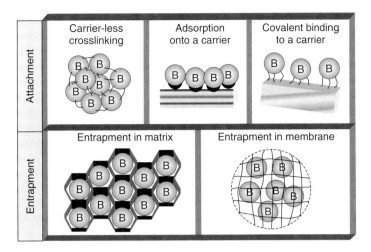

Fig. 1 Methods for the immobilization of biological catalysts.

Hampshire (USA) in 1973, are usually divided into the main categories of "attachment" and "entrapment" [4] and their subdivisions, as illustrated in Fig. 1. The suitability of any of these methods for the application of biological catalysts in synthetic reactions depends on a variety of factors including the nature of the catalyst, the participating reactants and process conditions, and all the micro-environmental effects and limitations in mass transfer resulting from the transformation of a dissolved to a solid or solid-like catalyst. Direct influences on stability and catalytic properties, such as pH and temperature optima, kinetics, and substrate specificity can be exerted due to reactions with or at the active site of the biocatalyst or conformational changes induced by hydrophobicity, porosity, charge, or other properties of an immobilization support [5]. Indirectly, the activity and stability of an immobilized biocatalyst can be influenced by the density of active sites [6], the occlusion of active sites in the complex [7], the prohibition of conformational changes of biocatalysts [8], restrictions of the spatial rotation of substrates [9], partitioning effects, and mass transfer limitations [6, 10]. Finally, the performance during application also depends on several non-catalytic parameters such as mechanical and chemical stress, biodegradation, loading capacity, retention of biocatalysts, growth control in live-cell systems, additional costs, and waste disposal.

2.4.12.2 Immobilization by Attachment

The immobilization of biocatalysts by attachment involves the formation of macroscopic insoluble aggregates either by directly linking the biocatalysts to each other, or by binding them to the surface of an organic or inorganic support. The pros and cons of these two principles have been discussed in detail by various authors [11, 12], particularly emphasizing the possible severe loss of activity [13] at high enzyme loadings [14], as well as the extra costs and waste resulting from the use of insoluble carriers. However, use of a carrier can also provide easy control over the physical properties of an immobilized biocatalyst or positively influence its catalytic features [15]. Moreover, the internal mobility and flexibility of the biocatalysts are considerably reduced by either form of attachment.

2.4.12.2.1 Carrier-Less Crosslinking Many biological catalysts possess a natural tendency for aggregation, particularly when suspended or dissolved at high density [16]. This results from physico-chemical interactions of their surfaces, sometimes mediated by macromolecules [17]. For technical purposes, the aggregation of microbial cells can be artificially induced by polyelectrolytes, such as chitosan or derivatives of polymethacrylate [18], while isolated enzymes precipitate in the presence of many different organic solvents, electrolytes, non-ionic polymers, or acids [19]. The procedure often exerts only minor effects on the catalytic activity, which makes aggregation an obvious and promising method for immobilization. Because of the weak binding forces, however, only few such aggregates, such as the flocculated cells of *Zymomonas mobilis* and *Saccharomyces cerevisiae* [20], or coaggregates of lipases with dialkyl ether-type phospholipid analogues [21], are sufficiently stable for technical application.

The stability of aggregated biocatalysts towards chemical and mechanical stress is usually increased when intermolecular covalent crosslinks are introduced into the complex. The functional groups on the surface of the biocatalysts involved in these crosslinks are mostly

References see page 652

nucleophilic amino acid residues, such as the ε-amino function of lysine, the γ- or δ-carboxy groups of aspartic and glutamic acid, and the sulfhydryl function of cysteine [22, 23]. The residues are interlinked by the formation of plain peptide bonds [24] or the use of bifunctional reagents, the backbone of which may reveal favorable characteristics, such as a certain chain length or bulkiness [23]. Although the most nucleophilic residue – and thus best target for covalent linking – is the sulfhydryl function of cysteine, most bifunctional crosslinkers currently in use react with the ε-amino group of lysine. This is due to the frequent average presence of this residue in protein molecules [22]. Important types of crosslinking reagents are summarized in Fig. 2.

Among the many possible crosslinkers, glutaraldehyde is the most prominent for the crosslinking of both whole cells and isolated enzymes. It reacts rapidly and almost irreversibly at neutral to slightly basic pH [25], and is more efficient than other dialdehydes in generating thermally and chemically stable links [26]. The reactivity of glutaraldehyde is targeted at ε-amino functions, α-amino functions, guanidinyl functions, and secondary amino groups in decreasing order, and side reactions with hydroxyl functions and histidine residues can occur occasionally [27]. The mechanism of the crosslinking reaction is not fully understood to date, which might partly be due to the diverse mono- and polymeric forms of glutaraldehyde in commercially available preparations. Apart from this diversity, it seems that not only one mechanism governs the reaction with amino acid residues (for a review, see Ref. [28]). Nonetheless, the reaction conditions optimal for the crosslinking of biocatalysts with bifunctional reagents must be individually adapted to every application, as it has been demonstrated that

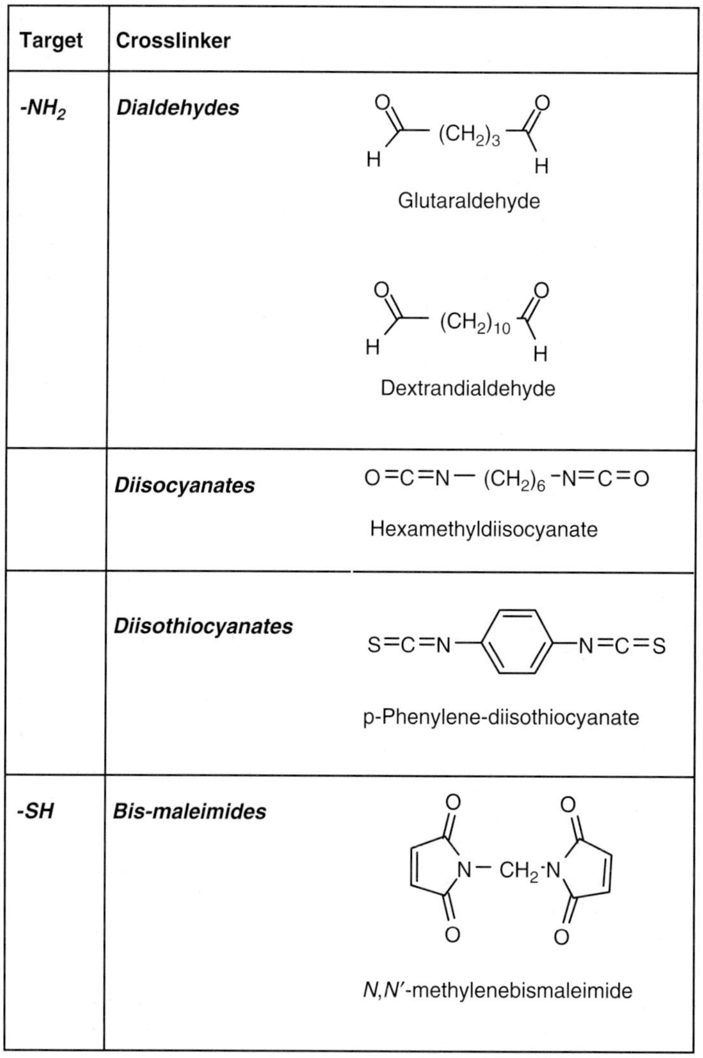

Target	Crosslinker	
-NH$_2$	Dialdehydes	Glutaraldehyde
		Dextrandialdehyde
	Diisocyanates	Hexamethyldiisocyanate
	Diisothiocyanates	p-Phenylene-diisothiocyanate
-SH	Bis-maleimides	N,N'-methylenebismaleimide

Fig. 2 Bifunctional reagents for covalent crosslinking of biocatalysts.

the success depends on a delicate balance of type [29–31] and concentration [32] of the biocatalyst, preparation of the reagent [30], pH [33] and ionic strength [34] of the solution, temperature [35], and time of reaction [36].

While covalent crosslinking of whole-cell biocatalysts usually starts from cell suspensions mixed with spacer molecules such as polyethyleneimine [37], gelatin, or albumin [38–40], solutions [30, 41, 42] as well as spray-dried [43], crystalline [44, 45], or pre-aggregated preparations [46] can be used for the crosslinking of isolated enzymes. The procedure often results in an increase of biocatalyst stability but a decrease in activity; in the case of enzymes this effect has been ascribed to the fixing of the bioactive conformation by intermolecular and intramolecular patches [28].

Aggregates resulting from the crosslinking of dissolved enzymes gain only a gelatinous consistency, while the residual activity of spray-dried enzymes is rather low. Neither of these preparations has therefore become important for technical application. In contrast, crosslinked enzyme crystals, which were originally invented to facilitate X-ray analysis [44], usually reveal both reasonable residual activity [47] and stability towards mechanical forces [48], extreme pHs and temperatures, and organic solvents [49]. As such, crosslinked enzyme crystals of some technically important enzymes, such as penicillin acylase, have been commercialized under the trademark CLEC® (Altus Biologics, Cambridge MA, USA). Crosslinked enzyme crystals generally consist of a solid microporous structure (1–100 μm) with uniform solvent-filled channels that traverse the body of the crystal [50]. Catalytic and technical properties can be influenced by the properties of the crosslinking reagent, but mainly depend on the conditions of crystallization preceding the actual crosslinking (for a review, see Ref. [51]). Thus, the preparation of optimal enzyme crystals is the most crucial step of this immobilization technique, and it has to be considered that the development of a reproducible crystallization protocol can be extremely time-consuming [52]. Crosslinked enzyme crystals also have disadvantages in the catalysis of fast diffusion-controlled reactions and in the conversion of macromolecular compounds [47].

The covalent crosslinking of physico-chemical aggregates instead of crystalline enzymes has been introduced as a promising alternative technique for immobilization only recently [46, 53]. The catalytic performance of the resulting crosslinked enzyme aggregates (CLEAs) is influenced by the employed crosslinking reagent [54], but mainly depends on the preparation and properties of the basic enzyme precipitates [55]. However, a major drawback of CLEAs can be found in their low mechanical stability [56], which is almost comparable to aggregates from dissolved enzymes. To date, this could only be overcome by stabilizing the aggregates by entrapment in polyvinyl alcohol [56], overlaying with polyelectrolytes [57], or attachment to microporous polymeric membranes [58].

2.4.12.2.2 Adsorption onto a Carrier

The adsorption of biocatalysts onto solid materials is a spontaneous process which results from hydrophobic, polar, electrostatic, and chelating interactions of the surfaces [59, 60]. Consequently, immobilization can be achieved by simply contacting a biocatalyst with a suitable support. The strength of the attachment depends on the type and amount of the physical interactions, which in turn are strongly influenced by the physical characteristics of biocatalyst and support, surface area, porosity, incubation time, temperature, pH, and ionic strength [61].

Many different types of carrier have already been applied to adsorb biocatalysts (Table 1), ranging from complex biological material such as wood shavings [62] to highly functionalized organic polymers of natural or synthetic origin, such as polysaccharides [63] or polyacrylate derivatives [64], and inorganic supports such as zeolites [65] or transition metals [59]. For synthetic application, the various carriers have been modified in many ways in order to achieve a better performance under particular reaction conditions. In this way, the chemical stability of the organic polymers dextran and agarose was considerably improved by internal crosslinking with epichlorhydrin or hydrophilic polymer chains [66], obtaining the now well-known and frequently used carriers "Sephadex" and "Sepharose" (Pharmacia, Uppsala, Sweden). The flow-through properties of many synthetic carriers such as styrene-divinylbenzene copolymers, acrylates, methacrylates, vinylpyridines, vinylpyrrolidone, vinyl acetate, and polystyrene (Cavilink™) have been improved by an increase in, and control of, their pore sizes [66, 67], while as inorganic carriers, zeolites, ordered mesoporous materials such as MCM-41 [68], or hierarchical bimodal porous silicas such as UVM-7 and organosilicas such as HPNO [69] are now frequently employed. For the adhesion of microbial cells, porous materials with voids that allow organisms to penetrate and grow into large colonies (e.g., cotton or nylon mesh or cloth, metallic mesh and various types of foam or sponge) are most advantageous [70].

Retention of the biocatalyst on the carrier is optimized by maximizing the number and strength of physico-chemical interactions. For lipases, which usually have large hydrophobic regions on their surfaces [71], this can be achieved by increasing the hydrophobicity of a carrier, as in octyl-agarose 4BCL [15]. Stable adsorption of most other native biocatalysts requires the introduction

References see page 652

Tab. 1 Carriers used for adsorption of isolated enzymes (E) and microbial cells (C) in biocatalysis

Organic carriers		Inorganic carriers	
Complex	Wood shavings (C)	**Metallics**	Alumina (E, C) Steel (C)

Natural polymers	Polysaccharides:
	  Agarose (E)
	  Cellulose, Dextran (E, C)
	  Chitosan (E, C)
	  Hemicellulose (C)
	  Starch (E)
	Proteins:
	  Collagen (E)
	  Gelatin (C)
	  Gluten (E)
		Minerals	Brick (C)
			Caolinite (E)
			Celite (C)
			Charcoal (E, C)
			Clay (E, C)
			Bentonite (E)
			Kieselguhr (C)
			Stone (C)

Synthetic polymers	Polyesters (C) Poly(meth)acrylates (E) Polyphenylacetylene (E) Polypropylene (E) Polystyrenes (E) Polyurethanes (E) Poly(vinyl alcohol) (E)	Ceramics	Glass (E, C) Silicate (E) Zeolite (E)

of multi-point electrostatic interactions, either by covering biocatalysts [72] and/or non-activated supports with derivatives of phosphoric acid [73] or charged polymers such as poly-L-lysine [74], polyethyleneimine [75], and alternating layers of polyethyleneimine and poly(styrene sulfonate) [76], or by covalently linking charged functional groups onto the surface of the carriers and operating the biological catalysts at pHs well above or below their isoelectric points. Many such carriers are commercially available for protein purification; examples include the anion exchangers DEAE (diethylaminoethyl)-cellulose and -Sephadex, Octylamino-Sepharose and Sephadex, Amberlite IRA, Dowex, or Ionex, or the cation exchangers CM-(carboxymethyl-)cellulose, cellulose phosphate, Amberlite IRC and IR, and dextran sulfate. A good binding capacity for enzymes and whole cells of *Saccharomyces cerevisiae* has also been reported for aminated and sulfonated poly(vinyl alcohol) (SFF) [77].

An equally stable alternative to adsorption by ionic interactions is the formation of metal complexes between functional groups of biocatalysts, primarily histidyl residues and adjacent aromatic side chains of tryptophan, phenylalanine, and tyrosine [78], and ions of transition metals or aluminum [72, 79] embedded in or bound to organic [80, 81] or inorganic [81, 82] carriers. This has become particularly interesting since modern molecular technologies enable the introduction of histidine clusters ("His-tags") to the N- or C-terminus of recombinant proteins [83], thus enhancing the adsorption of any protein to Ni^{2+}-containing nitrilotriacetate (NTA) matrices [84] (Fig. 3) or Cu^{2+}-activated silicate [85]. However, a major drawback of the method comprises the leakage of metal ions during use [86], as this can contaminate and destabilize the final product [87].

Finally, molecular techniques are now frequently applied to achieve "bioaffinity-adsorption" of biocatalysts onto a carrier, including protein–ligand or protein–protein interactions such as enzyme–inhibitor, avidin–biotin, protein A–IgG, or cell–protein interactions [88]. Glycosylated enzymes and whole-cell biocatalysts can be immobilized on supports exposing a covalently bound recombinant lectin [89], such as Concavalin A, or hydrazides [90]. Suitable carriers for the specific adsorption of recombinant proteins exposing a Strep-tag [91], FLAG-tag [92] or SBP-tag [93] are becoming increasing available on a commercial basis (for a review, see Ref. [94]), while

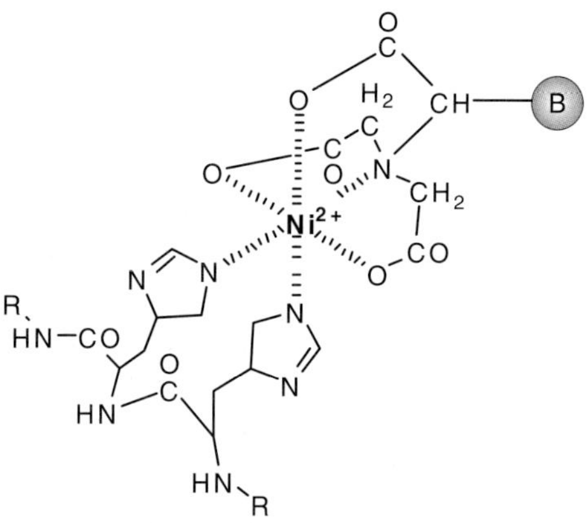

Fig. 3 Attachment of His-tagged proteins to Ni^{2+}-agarose.

at the same time lectins anchored to the surface of recombinant whole cells [95] determine a strong interaction with glycosphingolipids [96] as a definite class of standard supports.

In spite of all specific physico-chemical interactions between biocatalysts and carriers, high sensitivity of the attachment towards changes in the surrounding remains a general feature of immobilization by adsorption. While this offers the unique possibility of recovery and repeated use of a carrier after deactivation of the biocatalysts [97], it complicates the technical application, as working conditions such as pH, ionic strength, or substrate and product concentrations may change during the course of a reaction and thus lead to the gradual detachment of the immobilized biocatalysts.

2.4.12.2.3 **Covalent Binding to a Carrier** Biocatalysts can be attached covalently to most of the carriers suitable for adsorption (see Table 1), although the formation of stable multi-point links is facilitated by use of carriers with a defined surface composition. This is due to the specific activation that is usually required to create sufficient reactivity for the coupling with the biocatalyst or appropriate spacer molecules. The activation procedure depends on the functional groups available on the carrier surface, mostly hydroxyl, amino, amide, and carboxyl functions, and many useful derivatives have been described to date. For the same reasons as described in Section 2.4.12.2.1, they are mainly targeted at the ε-amino functions of lysine residues on the surfaces of biocatalysts. A detailed survey on supports and activation protocols has been provided by Cabral and Kennedy [22].

For technical purposes, commercially available carriers based on polysaccharides and synthetic polymers have become the most important and will therefore be briefly described at this point.

Mechanically stable, but non-derivatized polysaccharide supports such as Sephadex or Sepharose are best activated by reacting their hydroxyl functions with cyanogen bromide to cyanate esters or cyclic imidocarbonate intermediates [98, 99] (Fig. 4). However, as the stability of these reactive functions in aqueous solution is low, and isocyanates released during the coupling of the biocatalyst may affect the residual activity of the obtained immobilizates [100], polysaccharides functionalized for protein purification (such as CM-cellulose or DEAE-Sephadex) are preferable for covalent attachment. Carboxy and amino functions present in these carriers can be activated with carbodiimide [98] and glutaraldehyde [101], respectively.

Synthetic polymers are available in many compositions, and therefore contain a variety of reactive functions suitable for activation and subsequent covalent binding. Amino functions in Cavilink™ beads (Polygenetics, Los Gatos, CA, USA), a commercial support based on porous crosslinked polystyrene [67], can again be activated with glutaraldehyde. In contrast, the carriers Eupergit® C (Röhm, Darmstadt, Germany), which is a macroporous copolymer of methacrylamide, glycidyl methacrylate and allyl glycidyl ether, crosslinked with N,N'-methylene-bis(methacrylamide) [102], Amberzyme™ (Röhm & Haas, Darmstadt, Germany), which consists of porous spheres

References see page 652

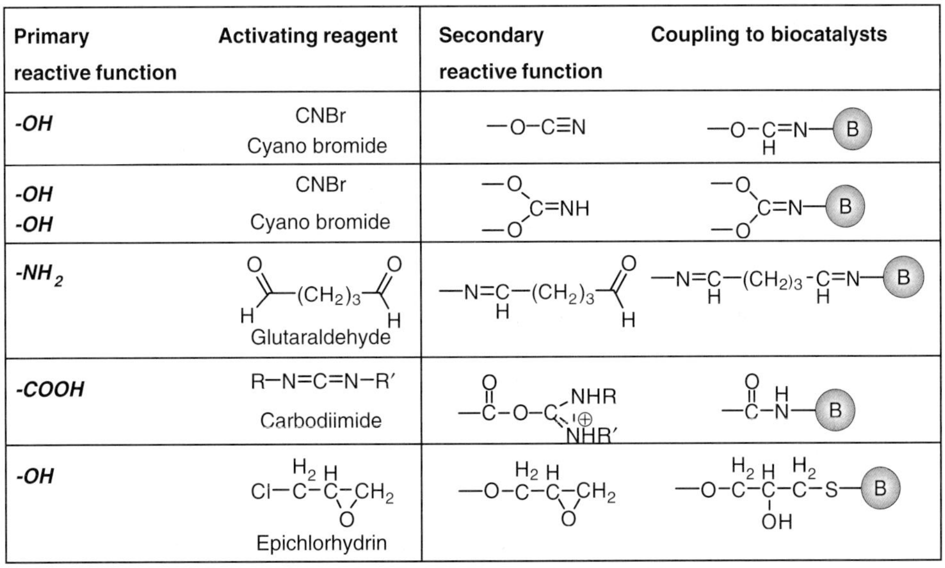

Primary reactive function	Activating reagent	Secondary reactive function	Coupling to biocatalysts
-OH	CNBr Cyano bromide	—O—C≡N	—O—C(H)=N—B
-OH -OH	CNBr Cyano bromide	—O\\C=NH / —O	—O\\C=N—B / —O
-NH$_2$	OHC-(CH$_2$)$_3$-CHO Glutaraldehyde	—N=C(H)—(CH$_2$)$_3$—CHO	—N=C(H)—(CH$_2$)$_3$—C(H)=N—B
-COOH	R—N=C=N—R' Carbodiimide	—C(=O)—O—C(NHR)(⊕NHR')	—C(=O)—N(H)—B
-OH	Cl—CH$_2$—CH—CH$_2$(O) Epichlorhydrin	—O—CH$_2$—CH—CH$_2$(O)	—O—CH$_2$—CH(OH)—CH$_2$—S—B

Fig. 4 Activation of carrier functional groups and coupling to biocatalysts.

of a polymer composition similar to Eupergit® C [http://www.advancedbiosciences.com], and Sepabeads FP-EP (Resindion, Milan, Italy), as well a polymethacrylate-based resin [103], all expose a dense layer of oxirane functions which directly couple to sulfhydryl and amino functions of the biocatalysts. In this way, stable multi-point attachment is allowed [103, 104], while undesired reactions of the support with protein can be prevented by blocking ("end-capping") the remaining oxirane groups with reagents such as mercaptoethanol, ethanolamine, or glycine. The stability of the attachment can be further improved by underlaying the oxirane functions with ethylenediamine [105] or metal affinity adsorption of biocatalysts onto the support prior to covalent linking [106]. Additionally, multi-point attachment can be enhanced by partially functionalizing the epoxy groups with sodium sulfide, yielding thiol-epoxy supports that allow thiol-disulfide interchanges [107]. A noteworthy new development in synthetic carriers for covalent immobilization comprises so-called "smart" polymers (e.g., poly-N-isopropylacrylamide), which have the ability to reversibly alter their swelling degree upon changes in the reactions conditions [108]. In this way they offer the unique possibility of conducting a reaction on a soluble carrier, while separation of biocatalysts can easily be achieved by deliberately insolubilizing the carrier after reaction [109].

In contrast to organic polymers, inorganic supports are relatively inert towards derivatization. A certain activation of free hydroxyl functions can be achieved by use of cyanobromide, but most often the carriers are silanized with trialkoxy silane derivatives containing organic functional groups [110]. These functional groups can directly be reacted with the biocatalysts when exposing an oxirane function – that is, after the use of glycidoxypropyl-trimethoxysilane and N,N'-diisopropyltrimethoxysilane for silanization [74]. Alternatively, the primary functional groups of the silane derivative are further converted to activated intermediates which, in turn, are reacted with the biocatalysts (for a review, see Ref. [111]). The most popular and versatile silane compound for the latter protocol is aminopropyltriethoxy silane, which forms an alkylamine derivative and is then activated with glutaraldehyde [74, 112]. The residual activity of biocatalysts immobilized in this way can sometimes be improved by the presence of albumin during silanization [113]. An attractive new inorganic carrier for the covalent attachment of biocatalysts are Maghemite nanocrystals (Fe_2O_3) silanized with aminopropyltriethoxysilane which, due to their magnetic properties, allows easy recovery of immobilizates despite their very small particle size [114].

The major drawback of immobilization by covalent attachment is the often low residual activity of biocatalysts resulting from modifications of essential amino acid residues in the active sites. However, it has been demonstrated that these can successfully be diminished by: (i) covalent attachment of the enzyme in the presence of a competitive inhibitor or substrate; (ii) a reversible covalently linked enzyme–inhibitor complex; (iii) a chemically modified soluble enzyme which is linked to the matrix by newly incorporated residues; or (iv) a Zymogen precursor [32].

2.4.12.3 Immobilization by Entrapment

The immobilization of biocatalysts by entrapment involves embedding in a matrix and encapsulation in semi-permeable membranes. With both methods, the biocatalysts have considerably less contact with the carrier than after immobilization by attachment, and usually retain a very high mobility and flexibility. Thus, the catalytic activity is hardly affected, while technical problems often arise from the leakage of biocatalysts during application. This can be reduced by controlling the pore sizes in matrices and membranes, but only at the cost of a restricted mass transfer of substrates and products.

2.4.12.3.1 Entrapment in a Matrix
The embedding of biocatalysts in the cavities of a polymer network is a very simple process, involving the polymerization or crosslinking of precursor molecules in the presence of a solution or suspension of the catalysts. As precursors, a variety of natural and synthetic polymers [77, 115–118], synthetic monomers [119, 120] or inorganic compounds [119–121] are suitable. A concise overview including the specification of the respective types of matrix formation is provided in Table 2.

Matrices achieved by thermal gelation (i.e., by heating and subsequent cooling of dissolved polymers) are usually characterized by a low thermal and mechanical stability [122], even if the salt concentration in the polymer solution can sometimes influence the melting temperature [117]. An exception can only be found in gels from the synthetic polymer poly(vinyl alcohol), which combines a high elasticity with a long-term resistance against temperatures up to 65 °C [123]. For application in organic synthesis, lens-shaped gels made from poly(vinyl alcohol) have been commercialized as Lenticats® (geniaLab, Braunschweig, Germany), while other thermoreversible gels are most often shaped into beads by an emulsification/dispersion process involving gelation in the presence of an oil as a discontinuous phase [124].

Alginate beads, the most popular material for entrapment of biocatalysts [125], are obtained when an aqueous solution of this natural polymer is dropped into a solution of divalent cations, mostly Ca^{2+}, which crosslink the alginate macromolecules by ionotropic interactions [116]. The physical properties of the resulting gels depend heavily on

Tab. 2 Processes and materials for entrapment of biocatalysts in a matrix

Process of matrix formation	Material
Thermal gelation	Agar, agarose, κ-carrageenan, gelatin, gellan gum Poly(vinyl alcohol)
Ionotropic gelation	Alginate, chitosan, pectate
Photo-induced crosslinking	Photo-sensitive resin prepolymers
Precipitation	Cellulose, cellulose triacetate
Polymerization	Polyacrylamide, polymethacrylate, polyacrylamide-hydrazide Photo-crosslinkable resin prepolymers
Polycondensation	Polyurethane, epoxy resin

the composition, sequential structure, molecular size and concentration of the polymer, as well as on the concentration and type of the crosslinking ion [126–128]. The major disadvantage of calcium-alginate is its sensitivity towards chelating compounds such as phosphate, citrate, EDTA, and lactate, or anti-gelling cations such as Na^+ or Mg^{2+}.

Networks from synthetic polymers are generally superior to those from natural polymers in terms of mechanical strength and longevity. Additionally, the porosity of the carrier, as well as the ionic and hydrophobic or hydrophilic properties, can be influenced by deliberately adjusting the molecular composition [119]. Gel formation usually starts from a mixture of suitable synthetic monomers, including small amounts of a crosslinker which determines the density of the network. Polymerization can be initiated by radiation, ultraviolet light, or chemical catalysts, depending on the participating monomers and solvents. Each initiation yields free radicals which, in turn, initiate the polymerization of monomers. The synthesis of polyacrylamide and polymethacrylamide usually involves N,N,N',N'-tetramethylethylenediamine (TEMED) as chemical initiator, and ammonium persulfate as oxidizing reagent [129]. For efficient polymerization, the use of an oxygen-free medium is most important. One drawback of the method is that the toxicity of monomers [130] and initiator [119], in addition to the heat evolved during polymerization, often results in a severe loss of biocatalytic activity. Initiating the polymerization by radiation can lower the reaction temperature [131], but does not decrease inactivation by the initiator itself [119].

Inorganic porous wet gels or xerogels can be obtained by the hydrolysis and condensation–polymerization of metal and semi-metal alkoxides, mainly based on SiO_2 [132]. This has been successfully applied by Reetz et al. [133] for the entrapment of lipases, yielding biocatalysts with a ca. 100-fold increased activity and stability in synthesis reactions. The resulting activity was influenced by the selected alkylsilane precursor, and the use of additives such as isopropyl alcohol, 18-crown-6, Tween 80®, methyl-β-cyclodextrin, and KCl [134]. Unfortunately, the entrapment of other enzymes in sol–gels usually leads to a severe loss in catalytic activity [135], unless a hybrid nanocomposite consisting of THEOS [tetrakis(2-hydroxyethyl)orthosilicate] and polysaccharides is employed [136]. This composite prevents the use of organic solvents during the entrapment process due to the complete solubility of THEOS in water, as well as the use of an additional catalyst because of the catalytic effect of polysaccharides on the sol–gel process. Thus, the entrapment can be performed at any *pH* suitable for structural integrity and functionality of the biocatalysts, and the gel can be prepared at reduced concentrations of THEOS, thereby preventing notable heat release during the course of gel formation. Alternatively, silicones – a diverse group of copolymers containing alkylated silicates and siloxanes as basic components – combine promising activities of the biological catalysts with an advantageous elasticity of the matrix [120]. For an emulsion of dissolved lipases entrapped in such silicone elastomers, an activity increase comparable to that resulting from entrapment in sol–gels has recently been reported [137].

In general, immobilization by entrapment in a matrix has mainly been applied to whole-cell biocatalysts (for a review, see Ref. [138]) because of the aforementioned continuous leakage of the considerably smaller isolated enzymes during use. Without severe restriction of the transfer of substrates and products through the matrix, this leakage can only be prevented by crosslinking the enzymes [56] or by attaching them to a support [139] prior to entrapment. However, the method can be advantageous for application of isolated enzymes in organic solvents, because leakage is of no importance in these media, while the entrapment protects the biocatalysts from detrimental effects of the surrounding solvent [122, 140–142].

2.4.12.3.2 Entrapment in Membranes The entrapment of biocatalysts in semi-permeable membranes involves either the formation of membrane capsules in the presence of the biocatalysts, or insertion of the biocatalysts into preformed membrane modules. The first method yields microscopically small catalytic units with very

References see page 652

good properties with regard to mass transfer during reaction, but can have negative effects on the activity of the encapsulated biocatalysts. The second method provides comparably large catalytic units with high mechanical strength during technical application, but severely restricted mass transfer and longevity.

Three principal methods are usually employed to produce membrane microcapsules: coacervation (phase separation); interfacial polymerization; and liquid drying [143].

- For membrane formation by *coacervation*, an aqueous solution of biocatalysts is dispersed in a water-immiscible organic solvent containing the membrane-forming polymer. On addition of a second water-immiscible solvent to this mixture, phase separation of the polymer occurs, resulting in its association around small aqueous droplets and subsequent fusion to a continuous membrane. For the entrapment of biocatalysts, it is particularly important that the polymer separates as a concentrated phase without precipitation [144]. Polymers employed for this type of membrane formation include cellulose nitrate, ethyl cellulose, nitrocellulose, polystyrene, polyethylene, polyvinyl acetate, polymethalmethacrylate, and polysiobutylene.
- *Interfacial polymerization* involves an aqueous mixture of biocatalysts and hydrophilic monomers dispersed in a water-immiscible organic solvent. A hydrophobic monomer solution dissolved in the same water-immiscible organic solvent is then added and a polymeric membrane is formed at the interface of the aqueous and organic phase. The size of the resulting microcapsules can be varied by adjustment of the stirring rates during polymerization. The method has typically been used for the formation of nylon membranes, but other materials such as polyurethane, polyester and polyurea can also be employed [144]. A related method, which avoids use of organic solvents, is the formation of polyelectrolytic complexes as hollow spheres by ionic interactions between cellulose sulfate and polydimethyldiallylammoniumchloride (PDMDAAC) or similar compounds [145]. These are formed at the boundary surface when a suspension of biocatalysts in a solution of cellulose sulfate is dropped into a bath containing PDMDAAC [125].
- For membrane formation by *liquid drying*, a polymer such as polystyrene or ethylcellulose is dissolved in a water-immiscible solvent which has a boiling point lower than water (e.g., benzene or chloroform). An aqueous solution of the biocatalyst is then emulsified in the organic phase, and the emulsion is dispersed in an aqueous phase containing protective colloidal substances, such as gelatin, to form a second emulsion. Membranes form upon removing the organic solvent from the emulsion by heating under vacuum. The size of the resulting microcapsules can be adjusted by polymer concentration and stirrer speed [146].

Among the many available preformed membrane modules, hollow fibers and ultrafiltration devices are most often used for the encapsulation of biocatalysts [147]. The membranes typically consist of polysulfones, cellulose acetate, and acrylic copolymers [148], with a range of molecular weight cut-off values between 200 and 100 000 Daltons. For reaction, the biocatalysts are introduced into the lumen of the fibers or on one side of an ultrafiltration chamber, while the reactants are supplied in the reaction container or on the other side of the chamber, respectively. The transfer of substrates and products over the membrane improves with increasing pore size, while the retention of the biocatalysts decreases. For a suitable synthetic performance and operational stability, the use of carrier-enlarged biocatalysts can therefore be advantageous [149]. In any case, this strategy is mandatory for retention of small cofactor molecules such as NADH in a membrane reactor [150].

2.4.12.4 Conclusion

The vast number of different methods available for the immobilization of biological catalysts, including all possible combinations of the basic types, makes the selection of a suitable method for a specific application rather difficult. With increasing knowledge relating to structural properties and structure–activity relationships, and the emerging methods in molecular biology and polymer sciences, attempts at producing a rational design for immobilization methods have increased in recent years [151]. The promising perspectives of this approach are particularly evident from the progress that has been made in the immobilization of lipases for synthetic applications. These use the characteristics of the surfaces of the enzymes not only for stable adsorption onto a support [15, 21], but also to enhance enzyme activity in sol–gels [133] and silicone elastomers [137]. It must be admitted, however, that for most other classes of biocatalysts, the relationships between the nature of a carrier and the performance of an immobilized biocatalyst in any given application are not yet understood [152]. Thus, the design of immobilized biocatalysts still relies largely on empirical and laborious screening procedures [153].

References

1. A. Liese, K. Seelbach, C. Wandrey, *Industrial Biotransformations: A Collection of Processes*, Wiley-VCH, Weinheim, 2000, 423 pp.; A. Schmid, J. S. Dordick, B. Hauer, A. Kiener, M. Wubbolts, B. Witholt, *Nature* **2001**, *409*, 258; K. M. Koeller, C.-H. Wong, *Nature* **2001**, *409*, 232; A. Zaks, *Curr. Opin. Chem. Biol.* **2001**, *5*, 130; A. J. J. Straathof,

S. Panke, A. Schmid, *Curr. Opin. Biotechnol.* **2002**, *13*, 548; A. Schmid, F. Hollmann, J. B. Park, B. Bühler, *Curr. Opin. Biotechnol.* **2002**, *13*, 359; S. Panke, M. Held, M. Wubbolts, *Curr. Opin. Biotechnol.* **2004**, *15*, 272; E. Garcia-Junceda, J. F. Garcia-Garcia, A. Bastida, A. Fernanez-Mayoralas, *Bioorg. Med. Chem.* **2004**, *12*, 1817; S. Panke, M. Wubbolts, *Curr. Opin. Chem. Biol.* **2005**, *9*, 188; T. Ishige, K. Honda, S. Shimizu, *Curr. Opin. Chem. Biol.* **2005**, *9*, 174.
2. E. Katchalski-Katzir, *Trends Biotechnol.* **1993**, *11*, 471.
3. A. M. Klibanov, *Anal. Biochem.* **1979**, *93*, 1; V. Bihari, S. K. Basu, *J. Sci. Ind. Res.* **1984**, *43*, 679.
4. P. V. Sundaram, E. K. Pye, in *Enzyme Engineering*, E. K. Pye, L. B. Wingard Jr., (Eds.), Vol. 2, Plenum Press, New York and London, 1974, p. 449.
5. J. M. Engasser, *Appl. Biochem. Bioeng.* **1976**, *1*, 127; L. Goldstein, *Methods Enzymol.* **1976**, *44*, 397; M. D. Trevan, *Immobilized Enzymes. An Introduction and Applications in Biotechnology*. John Wiley & Sons, Chichester, New York, Brisbane, Toronto, 1980, 152 pp.; B. E. Dale, D. H. White, *Enzyme Microb. Technol.* **1983**, *5*, 227.
6. W. Halwachs, *Process Biochem.* **1979**, *14*, 25; D. D. Do, D. S. Clark, J. E. Bailey, *Biotechnol. Bioeng.* **1982**, *24*, 1527.
7. T. Cha, A. Guo, X.-J. Zhu, *Proteomics* **2005**, *5*, 416.
8. V. V. Mozhaev, K. Martinek, *Enzyme Microb. Technol.* **1982**, *4*, 299.
9. I. A. Webb, *Biotechnol. Lett.* **1983**, *5*, 207.
10. F. H. Verhoff, *Biotechnol. Bioeng.* **1982**, *24*, 703; M. R. Riley, F. J. Muzzio, C. Sebastian, *Appl. Biochem. Biotechnol.* **1999**, *80*, 151.
11. W. Tischer, V. Kasche, *Trends Biotechnol.* **1999**, *17*, 326.
12. L. Cao, L. van Langen, R. A. Sheldon, *Curr. Opin. Biotechnol.* **2003**, *14*, 387.
13. J. Bryarz, B. N. Kolarz, *Biochemistry* **1998**, *33*, 409.
14. M. H. A. Jansen, L. M. van Langen, S. R. M. Perreira, F. van Rantwijk, R. A. Sheldon, *Biotechnol. Bioeng.* **2002**, *78*, 425.
15. A. Bastida, P. Sabuquillo, P. Armisen, R. Fernandez-Lafuente, J. Huguet, J. M. Guisan, *Biotechnol. Bioeng.* **1998**, *58*, 486.
16. A. W. Bunch, *Biotechnol. Genet. Eng. Rev.* **1994**, *12*, 535; K. J. Verstrepen, G. Derdelinckx, H. Verachtert, F. R. Delvaux, *Appl. Microbiol. Biotechnol.* **2003**, *61*, 197.
17. R. H. Harris, R. Mitchell, *Annu. Rev. Microbiol.* **1973**, *27*, 27.
18. S. Barany, A. Szepesszentgyoergyi, *Adv. Colloid Interface Sci.* **2004**, *111*, 117; T. G. van de Ven, *Adv. Colloid Interface Sci.* **2005**, *114–115*, 147.
19. C. W. Nystrom, US Patent 3935068, assigned to Reynolds R. J., Tobacco Co., 1976; F. Rothstein, in *Protein Purification Process Engineering*, R. G. Harrison (Ed.), Marcel Dekker, New York, 1994, p. 115.
20. C. D. Scott, *Ann. N. Y. Acad. Sci.* **1983**, *413*, 448; J. D. Bu'lock, D. M. Comberback, C. Ghommidh, *Biochem. Eng. J.* **1984**, *29*, B9; C. B. Netto, A. Destruhaut, G. Goma, *Biotechnol. Lett.* **1985**, *7*, 355; H. Kuriyama, Y. Seiko, T. Murakami, H. Kobayashi, Y. Sonada, *J. Ferment. Technol.* **1985**, *63*, 159.
21. A. Hiroyuki, I. Umezawa, H. Matsukura, T. Oishi, *Chem. Pharm. Bull.* **1992**, *40*, 318.
22. J. M. Cabral, J. E. Kennedy, *Bioprocess Technol.* **1991**, *14*, 73.
23. S. S. Wong, L.-C. Wong, *Enzyme Microb. Technol.* **1992**, *14*, 866.
24. G. Talsky, G. Gianitsopoulos, *3rd European Congress of Biotechnology* **1984**, *1*, 299.
25. K. Okuda, I. Urabe, Y. Yamada, H. Okada, *J. Ferment. Eng.* **1991**, *71*, 100.
26. M. E. Nimni, D. Cheung, B. Strates, M. Kodama, K. Sheikh, *J. Biomed. Mater. Res.* **1987**, *21*, 741.
27. G. Alexa, D. Chisalita, G. Chirita, *Rev. Tech. Ind. Cuir.* **1971**, *63*, 5.
28. I. Migneault, C. Dartiguenave, M. J. Bertrand, K. C. Waldron, *BioTechniques* **2004**, *37*, 790.
29. S. Avrameas, T. Ternynck, *Immunchem.* **1969**, *6*, 53.
30. G. B. Broun, *Methods Enzymol.* **1976**, *44*, 263.
31. J. F. Kennedy, J. M. S. Cabral, in *Solid Phase Biochemistry: Analytical and Synthetic Aspects*, W. H. Scouten (Ed.), Wiley & Sons, New York, 1983, p. 253.
32. O. R. Zaborsky, *Immobilized Enzymes*, CRC Press, Cleveland, OH, 1973, 175 pp.
33. E. F. Jansen, Y. Tomimatsu, A. C. Olson, *Arch. Biochem. Biophys.* **1971**, *144*, 394.
34. Y. Tomimatsu, E. F. Jansen, W. Gaffield, A. C. Olson, *J. Colloid Interface Sci.* **1971**, *36*, 51.
35. M. Ottesen, B. Svensson, *Comptes-rendus des Travaux du Laboratoire Carlsberg* **1971**, *38*, 171.
36. E. F. Jansen, A. C. Olson, *Arch. Biochem. Biophys.* **1969**, *129*, 221.
37. S. Becka, F. Shrob, K. Plkackova, P. Kujan, P. Holler, *Biotechnol. Lett.* **2003**, *25*, 277.
38. M. E. Reichlin, *Methods Enzymol.* **1980**, *70*, 159.
39. F. B. Kolot, *Process Biochem.* **1981**, *5*, 2.
40. S. F. D'Souza, *Curr. Sci.* **1999**, *77*, 69.
41. Y. Rajput, M. N. Gupta, *Biotechnol. Bioeng.* **1988**, *31*, 220.
42. S. E. D'Souza, W. Altekar, S. F. D'Souza, *World J. Microbiol. Biotechnol.* **1997**, *13*, 561.
43. S. Arnotz, US Patent 4,665,028, assigned to Novo Industri A/S, 1987.
44. F. A. Quiocho, F. M. Richards, *Proc. Natl. Acad. Sci. USA* **1964**, *52*, 833.
45. D. Häring, P. Schreier, *J. Am. Chem. Soc.* **1992**, *114*, 7314.
46. L. Cao, L. M. van Langen, M. H. A. Janssen, R. A. Sheldon, EP1088887A1, assigned to Technische Universiteit Delft, 1999.
47. C. P. Govardhan, *Curr. Opin. Biotechnol.* **1999**, *10*, 331.
48. M. Baust-Timpson, P. Seufer-Wasserthal, *Specialty Chemicals* **1998**, *18*, 248.
49. G. DeSantis, J. B. Jones, *Curr. Opin. Biotechnol.* **1999**, *10*, 324.
50. L. Z. Vilenchik, J. P. Griffith, N. St Clair, M. N. Navia, A. L. Margolin, *J. Am. Chem. Soc.* **1998**, *120*, 4290.
51. J. J. Roy, T. E. Abraham, *Chem. Rev.* **2004**, *104*, 3705.
52. J. D. Vaghjiani, T. S. Lee, G. J. Lye, M. K. Turner, *Biocat. Biotrans.* **2000**, *18*, 151.
53. P. Lopez-Serrano, L. Cao, F. van Rantwijk, R. A. Sheldon, PCT International Application WO 2002061067, assigned to Technische Universiteit Delft, 2002; L. Cao, J. Elzinga, European Patent Application EP1333087, assigned to Avantium International B.V., 2003.
54. R. Schoevaart, M. W. Wolbers, M. Golubovic, M. Ottens, A. P. G. Kieboom, F. van Rantwijk, L. A. M. van der Wielen, R. A. Sheldon, *Biotechnol. Bioeng.* **2004**, *87*, 754; C. Mateo, J. M. Palomo, L. M. van Langen, F. van Rantwijk, R. Sheldon, *Biotechnol. Bioeng.* **2004**, *86*, 273.
55. L. Cao, F. van Rantwijk, R. A. Sheldon, *Org. Lett.* **2000**, *2*, 1361; P. Lopez-Serrano, L. Cao, F. van Rantwijk, R. A. Sheldon, *Biotechnol. Lett.* **2002**, *24*, 1379; L. Wilson, A. Illanes, O. Albian, B. C. Pessela, R. Fernandez-Lafuente, J. M. Guisan, *Biomacromolecules* **2004**, *5*, 852; F. Lopez-Gallego, L. Betancor, A. Hidalgo, N. Alonso, R. Fernandez-Lafuente, J. M. Guisan, *Biomacromolecules* **2005**, *6*, 1839.

56. L. Wilson, A. Illanes, C. Benevides, C. Pessela, O. Abian, R. Fernández-Lafuente, J. M. Guisán, *Biotechnol. Bioeng.* **2004**, *86*, 558.
57. N. G. Balabushevitch, G. B. Sukhorukov, N. A. Moroz, D. V. Volodkin, N. I. Larionova, E. Donath, H. Mohwald, *Biotechnol. Bioeng.* **2001**, *76*, 207.
58. N. Hilal, R. Nigmatullin, A. Alpatova, *J. Membrane Sci.* **2004**, *238*, 131.
59. R. A. Messing, *Methods Enzymol.* **1976**, *44*, 148.
60. D. R. Oliveira, *Appl. Sci.* **1993**, *223*, 45.
61. J. Klein, H. Ziehr, *J. Biotechnol.* **1990**, *16*, 1; P. S. Rouxhet, *Ann. N. Y. Acad. Sci.* **1990**, *613*, 265.
62. G. Durand, J. M. Navarro, *Proc. Biochem.* **1978**, *14*, 14.
63. S. Dumitriu, E. Chornet, in *Polysaccharides*, S. Dumitriu (Ed.), Dekker, New York, 1998, p. 629.
64. M. Sadar, R. Agarwal, A. Kumar, M. N. Gupta, *Enzyme Microb. Technol.* **1997**, *20*, 361.
65. A.-X. Yan, X.-W. Li, Y.-H. Ye, *Appl. Biochem. Biotechnol.* **2002**, *101*, 113.
66. R. Arshady, *J. Chromatogr. A* **1991**, *586*, 181.
67. N. H. Li, J. R. Benson, US Patent 5,583,162, assigned to Polygenetics Inc., 1998.
68. H. Takahashi, B. Li, T. Sasaki, C. Miyazaki, T. Kajino, S. Inagaki, *Microporous Mesoporous Mater.* **2001**, *44–45*, 755; D. Moelans, P. Cool, J. Baeyens, E. F. Vansant, *Catal. Commun.* **2005**, *6*, 307.
69. M. Tortjada, D. Ramón, D. Beltrán, P. Amorós, *J. Mater. Chem.* **2005**, *15*, 3859.
70. C. D. Scott, *Enzyme Microb. Technol.* **1987**, *9*, 66.
71. U. Derewenda, A. M. Brozowski, D. M. Lawson, Z. S. Derewenda, *Biochem.* **1992**, *31*, 1532.
72. V. P. Lewis, S.-T. Yang, *Biotechnol. Bioeng.* **1992**, *40*, 465.
73. M.-A. Coletti-Previero, A. Previdero, *Anal. Biochem.* **1989**, *180*, 1.
74. A. Subramanian, S. J. Kennel, P. I. Oden, J. B. Jacobson, J. Woodward, M. J. Doktycz, *Enzyme Microb. Technol.* **1999**, *24*, 26.
75. S. F. D'Souza, *Biotechnol. Bioeng.* **1983**, *25*, 1661; S. F. D'Souza, J. S. Melo, A. Deshpande, G. B. Nadkarni, *Biotechnol. Lett.* **1986**, *8*, 643; B. Champluvier, B. Kamp, P. G. Rouxhet, *Appl. Microbiol. Biotechnol.* **1988**, *27*, 464; S. F. D'Souza, N. Karnath, *Appl. Microbiol. Biotechnol.* **1988**, *29*, 136; N. Karnath, J. S. Melo, S. F. D'Souza, *Appl. Biochem. Biotechnol.* **1988**, *19*, 251; S. S. Godbole, B. S. Kubal, S. F. D'Souza, *Enzyme Microb. Technol.* **1990**, *12*, 214; R. Bahulekar, N. R. Ayyangar, S. Ponrathanam, *Enzyme Microb. Technol.* **1991**, *13*, 858; J. C. Melo, B. S. Kubal, S. F. D'Souza, *Food Biotechnol.* **1992**, *6*, 175; N. Albayrak, S.-T. Yang, *Biotechnol. Prog.* **2002**, *18*, 240.
76. J. P. Santos, E. R. Welsh, B. P. Gaber, A. Singh, *Langmuir* **2001**, *17*, 5361.
77. H. Ichijo, N. Nagasawa, A. Yamauchi, *J. Biotechnol.* **1990**, *14*, 169.
78. F. H. Arnold, *Bio/Technology* **1991**, *9*, 150.
79. J. F. Kennedy, J. M. S. Cabral, *Art. Cells Blood Subst. Immob. Biotechnol.* **1995**, *23*, 231.
80. S. A. Barker, A. N. Ermery, J. M. Novais, *Process Biochem.* **1971**, *6*, 11; J. F. Kennedy, B. Kalogerakis, J. M. S. Cabral, *Enzyme Microb. Technol.* **1984**, *6*, 68; S. Afaq, J. Iqbal, *Electronic J. Biotechnol.* **2001**, *4*, 120.
81. A. N. Emery, J. S. Hough, J. M. Novais, P. T. Lyons, *Chem. Eng. (London)* **1972**, *259*, 71.
82. M. Charles, R. W. Coughlin, E. K. Panuchuri, B. R. Allen, F. X. Hasselberger, *Biotechnol. Bioeng.* **1975**, *17*, 203.
83. E. Hochuli, H. Doebeli, A. Schachter, *J. Chromatogr.* **1987**, *411*, 177; E. Hochuli, W. Bannwarth, H. Doebeli, R. Gentz, D. Stueber, *Bio/Technology* **1988**, 1321.
84. B. M. Brena, L. G. Ryden, J. Porath, *Biotechnol. Appl. Biochem.* **1994**, *17*, 217; E. Ordaz, A. Garrido-Pertierra, M. Gallego, A. Puyet, *Biotechnol. Prog.* **2000**, *16*, 287; Y. Wang, F. Caruso, *Chem. Commun.* **2004**, 1528.
85. L.-F. Ho, S.-Y. Li, S.-C. Lin, W.-H. Hsu, *Process Biochem.* **2004**, *39*, 1573.
86. T. Oswald, G. Hornbostel, U. Rinas, F. B. Anspach, *Biotechnol. Appl. Biochem.* **1997**, *25*, 109.
87. V. Gabarec-Porekar, V. Menart, *J. Biochem. Biophys. Methods* **2001**, *49*, 335.
88. W. H. Scouten, *Curr. Opin. Biotechnol.* **1991**, *2*, 37; M. Glad, P. O. Larsson, *Curr. Opin. Biotechnol.* **1991**, *2*, 413; M. Saleemuddin, *Adv. Biochem. Eng. Biotechnol.* **1999**, *64*, 203.
89. B. Mattiasson, P. A. Johansson, *Immunol. Methods* **1982**, *52*, 233; B. Mattiasson, *Appl. Biochem. Biotechnol.* **1982**, *7*, 121; F. B. Viera, B. B. Barragan, B. L. Busto, *Biotechnol. Bioeng.* **1988**, *31*, 711; M. Saleemuddin, Q. Hussain, *Enzyme Microb. Technol.* **1991**, *13*, 290; M. A. Montero, A. Remeu, *Biochem. Mol. Biol. Int.* **1993**, *30*, 685; P. Vrabel, M. Plakovic, S. Godo, V. Bales, P. Docolomansky, P. Gemeiner, *Enzyme Microb. Technol.* **1997**, *21*, 196.
90. Z. Bilkova, M. Slovakova, D. Horak, J. Lenfeld, J. Churacek, *J. Chromatogr. B* **2002**, *770*, 177.
91. T. G. M. Schmidt, A. Skerra, *Protein Eng.* **1993**, *6*, 109; F. Jafri, S. Husain, M. Saleemuddin, *Biotechnol. Techniques* **1995**, *9*, 117.
92. T. P. Hopp, K. S. Pricket, V. L. Price, R. T. Libby, C. J. March, D. P. Ceretti, D. L. Urdal, P. J. Conlon, *Bio/Technology* **1988**, *6*, 1204.
93. D. S. Wilson, A. D. Keefe, J. W. Szostak, *Proc. Natl. Acad. Sci. USA* **2001**, *98*, 3750.
94. K. Terpe, *Appl. Microbiol. Biotechnol.* **2003**, *60*, 523.
95. L. G. J. Frenken, P. de Geus, F. M. Klis, H. Y. Toschka, C. T. Verrips, PCT Int. Appl. WO 9418330, assigned to Unilever PLC, UK; Unilever N. V., 1994.
96. B. E. Collins, L. J.-S. Yang, R. L. Schnaar, *Methods Enzymol.* **2000**, *312*, 438.
97. A. Deshpande, K. Sankaran, S. F. D'Souza, *Enzyme Microb. Technol.* **1989**, *11*, 617; T. Fossati, M. Colombo, C. Castiglioni, G. Abbiati, *J. Chromatogr. B* **1994**, *656*, 59; G. Massolini, E. Talleri, E. de Lorenzi, M. Pregnolato, M. Terreni, G. Felix, C. Gandini, *J. Chromatogr. A* **2001**, *921*, 147; E. Calleri, C. Temporini, S. Furlanetto, F. Loiodice, G. Facchiolla, G. Massolini, *J. Pharm. Biomed. Anal.* **2003**, *32*, 715.
98. R. Axen, J. Porath, S. Ernback, *Nature* **1967**, *214*, 1302.
99. J. Kohn, M. Wilchek, *Enzyme Microb. Technol.* **1982**, *4*, 161.
100. R. F. Taylor, *Bioprocess Technol.* **1991**, *14*, 139–160.
101. Y. Kosugi, Q.-L. Chang, K. Kanazawa, H. Nakanishi, *J. Am. Oil Chem. Soc.* **1997**, *74*, 1395.
102. E. Katchalsky-Katzir, D. M. Kraemer, *J. Mol. Catal. B.: Enzymatic* **2000**, *10*, 157.
103. C. Mateo, O. Abian, R. Fernandez-Lafuente, J. M. Guisan, *Enzyme Microb. Technol.* **2000**, *26*, 509; A. Tam, D. Re, P. Caimi, M. Daminati, Eur. Pat. Appl. EP 2003–7881 20030407, assigned to Resindion S.r.L., 2003.
104. C. Mateo, O. Abian, G. Fernandez-Lorente, J. Pedroche, R. Fernandez-Lafuente, J. M. Guisan, A. Tam, M. Daminati, *Biotechnol. Prog.* **2002**, *18*, 629.
105. C. Mateo, R. Torres, G. Fernandez-Lorente, C. Ortiz, M. Fuentes, A. Hidalgo, F. Lopez-Gallego, O. Abian,

J. M. Palomo, L. Betancor, B. C. C. Pessela, J. M. Guisan, R. Fernandez-Lafuente, *Biomacromol.* **2003**, *4*, 772.
106. B. C. C. Pessela, C. Mateo, A. V. Carrascosa, A. Vian, J. L. Garcia, G. Rivas, C. Alfonso, J. M. Guisan, R. Fernandez-Lafuente, *Biomacromol.* **2003**, *4*, 107.
107. V. Grazu, O. Abian, C. Mateo, F. Batista-Viera, R. Fernandez-Lafuente, J. M. Guisan, *Biotechnol. Bioeng.* **2005**, *90*, 597.
108. E. Kokufuta, *Progr. Polymer Sci.* **1992**, *17*, 647.
109. I. Y. Galaev, B. Mattiasson, *Trends Biotechnol.* **1999**, *17*, 335.
110. H. H. Weetall, A. M. Filbert, *Methods Enzymol.* **1974**, *34*, 59.
111. H. H. Weetall, *Appl. Biochem. Biotechnol.* **1993**, *41*, 157.
112. L. C. Shriver-Lake, W. B. Gammeter, S. S. Bang, M. Pazirandeh, *Analyt. Chem. Acta* **2002**, *470*, 71.
113. C. M. F. Soares, M. H. A. Santana, G. M. Zanin, H. F. de Castro, *Biotechnol. Prog.* **2003**, *19*, 803.
114. H. M. R. Gardimalla, D. Mandal, P. D. Stevens, M. Yen, Y. Gao, *Chem. Commun.* **2005**, 4432.
115. T. Heinze, D. Klemm, F. Loth, B. Philipp, *Acta Polym.* **1990**, *41*, 259; R. M. Banik, B. Kanari, S. N. Upadhyay, *World J. Microbiol. Biotechnol.* **2000**, *16*, 407.
116. F. van de Velde, N. D. Lourenco, H. M. Pinheiro, M. Bakker, *Adv. Synth. Catal.* **2002**, *344*, 815.
117. K. Nilsson, P. Brodelius, K. Mosbach, *Methods Enzymol.* **1987**, *135*, 222.
118. C. Bucke, *Methods Enzymol.* **1987**, *135*, 175.
119. W. R. Gombotz, A. S. Hoffman, *Hydrogels Med. Pharm.* **1986**, *1*, 95.
120. I. Gill, A. Ballesteros, *Trends Biotechnol.* **2000**, *18*, 469.
121. I. Gill, A. Ballesteros, *Trends Biotechnol.* **2000**, *18*, 282.
122. M. B. Ansorge-Schumacher, G. Pleß, D. Metrangolo, W. Hartmeier, *Landbauforschung Völkenrode* **2002**, *SH 241*, 99.
123. V. Lozinsky, F. Plieva, *Enzyme Microb. Technol.* **1998**, *23*, 227.
124. R. J. Neufeld, D. Poncelet, *Focus Biotechnol.* **2004**, *8A*, 311.
125. N. Gerbsch, R. Buchholz, *FEMS Microbiol. Rev.* **1995**, *16*, 259.
126. A. Martinsen, G. Skjåk-Bræk, O. Smidsrød, *Biotechnol. Bioeng.* **1989**, *33*, 79.
127. M. P. J. Kierstan, C. Bucke, *Biotechnol. Bioeng.* **1977**, *19*, 387.
128. F. Paul, P. M. Vignais, *Enzyme Microb. Technol.* **1980**, *2*, 281; W. E. Rochefort, T. Regh, P. C. Chau, *Biotechnol. Lett.* **1986**, *8*, 115.
129. H. Nilsson, R. Mosbach, K. Mosbach, *Biochim. Biophys. Acta* **1972**, *268*, 253; A. C. Johansson, K. Mosbach, *Biochim. Biophys. Acta* **1974**, *370*, 339.
130. K. Mosbach, R. Mosbach, *Acta Chem. Scand.* **1966**, *20*, 2807; Y. Degani, T. Miron, *Biochim. Biophys. Acta* **1970**, *212*, 362.
131. H. Maeda, A. Yamauchi, H. Suzuki, *Biochim. Biophys. Acta* **1973**, *315*, 18.
132. D. Avnir, S. Braun, O. Lev, M. Ottolenghi, *Chem. Mater.* **1994**, *6*, 1605; D. Avnir, L. C. Klein, D. Levy, U. Schubert, in *The chemistry of organic silicon compounds*, Vol. 2, John Wiley & Sons Ltd, New York, 1998, p. 2317.
133. M. Reetz, A. Zonta, J. Simpelkamp, *Angew. Chem. Int. Ed. Engl.* **1995**, *34*, 301; M. T. Reetz, *Adv. Mater.* **1997**, *9*, 943.
134. M. T. Reetz, P. Tielmann, W. Wiesenhöfer, W. Könen, A. Zonta, *Adv. Synth. Catal.* **2003**, *345*, 717.
135. A. Basso, L. de Martin, C. Ebert, L. Gardossi, A. Tomat, M. Casarci, O. Li Rosi, *Tetrahedron Lett.* **2000**, *41*, 8630.
136. Y. A. Shipunov, T. Y. Karpenko, I. Y. Bakunina, Y. V. Burtseva, T. N. Zvyagintseva, *J. Biochem. Biophys. Methods* **2004**, *58*, 25.
137. A. Buthe, A. Kapitain, W. Hartmeier, M. B. Ansorge-Schumacher, *J. Mol. Catal. B: Enzymatic* **2005**, *35*, 93.
138. V. Bihari, S. K. Basu, *J. Sci. Ind. Res.* **1984**, *43*, 679; K. Mosbach, *Ann. N. Y. Acad. Sci.* **1984**, *434*, 239; R. G. Willaert, G. V. Baron, *Rev. Chem. Eng.* **1996**, *12*, 1.
139. M. P. J. Kierstan, A. McHale, M. P. Coughlan, *Biotechnol. Bioeng.* **1982**, *24*, 1461; B. Hahn-Hägerdal, *Biotechnol. Bioeng.* **1984**, *26*, 771.
140. S. Fukui, A. Tanaka, T. Iida, in *Biocatalysis in Organic Media*, C. Laane, J. Tramper, M. D. Lilly (Eds.), Elsevier Science Publishers B. V., Amsterdam, 1987, p. 21.
141. D. Metrangolo-Ruìz de Temiño, W. Hartmeier, M. B. Ansorge-Schumacher, *Enzyme Microb. Technol.* **2005**, *36*, 3.
142. T. Hischer, D. Gocke, M. Fernández, P. Hoyos, A. Alcántara, J. V. Sinisterra, W. Hartmeier, M. B. Ansorge-Schumacher, *Tetrahedron* **2005**, *61*, 7378.
143. J. K. Park, H. N. Chang, *Biotechnol. Adv.* **2000**, *18*, 303.
144. I. Chibata, *Immobilized enzymes*, John Wiley & Sons, New York, London, Sidney, Toronto, 1978, 384 pp.
145. J. Mansfeld, H. Dauzenberg, *Methods Biotechnol.* **1997**, *1*, 309.
146. M. P. Kierstan, M. P. Coughlan, *Bioprocess Technol.* **1991**, *14*, 13.
147. D. S. Inloes, D. P. Taylor, S. N. Cohen, A. S. Michaelis, C. R. Robertson, *Appl. Environ. Microbiol.* **1983**, *46*, 264; V. I. Baranov, I. Y. Morosov, S. A. Ortlepp, A. S. Spirin, *Gene* **1989**, *84*, 463.
148. A. Artmains, J. Neufeld, T. M. S. Chang, *Enzyme Microb. Technol.* **1984**, *6*, 135.
149. R. P. Chambers, W. Cohen, W. H. Baricos, *Methods Enzymol.* **1976**, *44*, 291; S. P. O'Neill, J. R. Wykes, P. Dunnill, M. D Lilly, *Biotechnol. Bioeng.* **1971**, *13*, 319.
150. A. F. Bückmann, M. Morr, M.-R. Kula, *Biotechnol. Appl. Biochem.* **1987**, *9*, 258.
151. L. Cao, *Curr. Opin. Chem. Biol.* **2005**, *9*, 217.
152. T. Boller, C. Meier, S. Menzler, *Org. Process Res. Develop.* **2002**, *6*, 509.
153. J. van Roon, R. Beeftink, K. Schroen, H. Tramper, *Curr. Opin. Biotechnol.* **2002**, *13*, 398.

2.5
Formation of the Final Catalyst

2.5.1
Reactions During Catalyst Activation

*Bernard Delmon**

2.5.1.1 Introduction

Catalysts are rarely prepared in the reactor where they will be used or tested, the exceptions being some grafted and heterogenized complexes, some heteropolyacids and their salts, and suspensions of nanoparticles prepared *in situ*. In the majority of cases, catalysts are not used as delivered by the manufacturer, nor directly as they

---

References see page 673
* Corresponding author.

were prepared in a separate section of the laboratory. First, they must acquire the desired activity by their being transformed to a different material. According to a generally accepted terminology, the corresponding process or procedure is termed "activation" or sometimes "formation of the final catalyst" (making a more active catalyst using dopes is generally referred to as "promotion"). Typical examples are the transformation of hydroxides to oxides, the reduction of metal oxides to dispersed metals particles, for example for hydrogenation reactions, for reduction-sulfidation in hydrodesulfurization (HDS) or, more generally, hydrotreating.

Activation is a crucial step. Indeed, the whole succession of steps during the preparation of a precursor has an impact on activity, selectivity, and resistance to deactivation. The IUPAC defines activation as the transformation of a solid precursor to the material immediately active for the desired reaction [1]. However, this definition is not – nor cannot be – precise. Strictly speaking, the material called the "precursor" should be specified among the succession of intermediates obtained before introduction into the reactor (e.g., dried or wet, contaminated by foreign substances or not, etc.). On the other hand, surface science shows that the contact with any one of the reactants already alters the catalyst (e.g., surface structure, gradient of elemental composition in the near-surface regions, etc.). The reaction, or at least the initial rate and selectivity, sometimes depend on the order in which the catalyst is contacted with the different reactants. In some cases, a catalyst has to catalyze the desired reaction for an appreciable length of time – a sort of "running-in" period – before reaching stable activity and selectivity. This is the case of the so-called vanadium phosphate catalysts for the oxidation of benzene or butane to maleic anhydride. Almost all solid catalysts are materials that undergo transformations during their life, and for this reason a single chapter in a Handbook such as this cannot examine all cases of "activation" or "formation of the catalytic activity". Thus, the objective is to provide a very brief presentation of the fundamental mechanisms that operate during activation and in the development of catalytic activity.

Many catalysts are supported on a more or less inert material ("support" or "carrier"), and in this case the chemical reactivity of a solid is fundamentally altered. Compared to that of a non-supported compound, the kinetics of the reaction can even lead not only to different textures but also to different compositions – that is, to a different catalytic material. Some of these changes are intuitively expected, and this is certainly the case when a chemical reaction has taken place between the active component and the support, when the reactivity reflects that of the new compound. Some other modifications are often not properly recognized, however, and one of the most common errors in this respect can be pinpointed.

The problem results from the ignorance that, beyond a certain threshold, the mere fact that a substance is more highly dispersed may lead to a decrease, rather than an increase of reactivity.

Only a small proportion of practical (industrial) catalysts is used unsupported. Although the activation of such catalysts is also important, a much larger amount of book space must be dedicated to the reactivity of supported precursors, mainly because of the particular behavior of supported substances. Principally, two factors must be taken into account:

- The state of dispersion of the precursor; this becomes increasingly important with the current trend to nanotechnology and the hope of applying the corresponding techniques for catalyst preparation.
- The interaction of this precursor with the support.

These two aspects will be examined in Section 2.5.1.2, while the case of catalysts where activation is generally recognized as a specific step in the preparation of the "final catalyst" – that is, the material introduced into the reactor – will be examined subsequently. The other steps include calcination (or decomposition at a more or less high temperature), reduction, and reduction–sulfidation.

Today, progress in technology and fundamental research is giving increasing importance to the processes that occur in the reactor when catalysts are initially contacted with the feed (this is the "running-in" period mentioned above). Recently, the use of elaborate physicochemical techniques provided some insight into these processes. However, the example of hydrodesulfurization (HDS) catalysis taught us that the need for a better control of activity, selectivity and long life required some knowledge of the "running-in" mechanisms. The reason for this is that, formerly, HDS catalysts were introduced as supported oxides into the reactor and directly contacted with the feed, often enriched in sulfur compounds. In the corresponding "running-in" process, the sulfur contained in the feed, together with hydrogen, progressively transformed the oxides to sulfides, these constituting the catalytically active material. Later, a series of observations and studies led to sulfurization of the precursor oxide form in special plants outside the reactor. Observations showed that these sulfided materials possessed better characteristics in many respects, and demonstrated the need to allude, very shortly, to the "running-in" processes with regards to activation. Today, oxidation reactions and related processes are becoming increasingly important in the modern petroleum chemistry, and recent investigations have begun to unveil the mechanisms that occur in catalysts when freshly introduced into the catalytic reactor. It is therefore logical to speculate that catalysts and activation procedures will be modified in such a way as to take into account these recent findings, thereby justifying some of the comments made at the end of this chapter.

2.5.1.2 Particular Features of Chemical Reactions of Supported Solids

2.5.1.2.1 Characteristics of Reactions of Solids: Possible Consequences in the Kinetics of Activation

It is necessary to recall briefly the fundamental processes involved in the reactions of solids, which comprise two main families:

- *Solid-state diffusion-controlled* (ssDC) reactions, which take place as the result of diffusion of a chemical species through the bulk of one of the solids involved (reactant or product). Typical examples are those of transition metal oxides, such as NiO or Co_3O_4, with an alumina support, forming a solid solution of the transition metal in γ-Al_2O_3, or even leading to the formation of a spinel, $NiAl_2O_4$ or $CoAl_2O_4$.
- *Interface-controlled* (IC) reactions, reactions which take place at the interface between the solid and the other reactant and kinetically limited by the chemical process occurring at this interface. This corresponds to the case of many dehydroxylations (transformations of hydroxides obtained by precipitation to oxides), reductions of oxides to metals (NiO to Ni) or reduction–sulfidation processes leading to sulfides (e.g., the "cobalt–molybdate" bi-layer transforming to $Co_9S_8 + MoS_2$). For this reason, the specific features of IC reactions are briefly described here.

Interface-Controlled (IC) Reactions IC reactions may, in turn, be of two types:

- In *nucleation-controlled interface reactions* (NCIR), the nucleation of the new phase (e.g., Ni produced by the reduction of NiO) may be rate-determining, in which case separate nuclei of the new phase can be detected.
- In *pure IC reactions*, all the surface of the initial solid reacts, and a continuous interface entirely covers the solid reactant. With respect to kinetics, the interfacial process entirely controls the reaction in this last case.

For IC processes, the rate of reaction at the interface, r_i is proportional to the area of the interface S_i, with a proportionality factor k_i, determined by the reaction conditions, which is the rate or velocity of the interface movement:

$$r_i = k_i S_i$$

For IC reactions in constant conditions, r_i continuously decreases during the whole process, because the area of the interface between the product and the reactant steadily decreases. This is not the case for NCIR, where r_i first increases and later decreases. The overall phenomenon thus depends both on the rate of nucleation and number of nuclei forming per unit surface area per unit time and k_i [2, 3].

It is possible to calculate the variation of the overall rate due to the convolution of the kinetics of nucleation and interface movement [4–7]. The corresponding variation of the degree of transformation, α, can be calculated from k_i and S_i; this can be performed for different particle shapes [7]. Starting from large particles, a diminution of their size initially leads to higher rates of the overall reaction. However, contrary to intuitive expectation, a maximum is reached and the rate diminishes when the particles are very small. This is shown in Fig. 1, which is a plot of α against time (common scale for all curves) for samples composed of spherical particles of diameter d (by convention, $d = 1$ for the sample reacting with the highest rate at very low values of α). The rate of reaction decreases dramatically, by a factor of approximately 20, when the particles are just one order of magnitude smaller ($d = 0.1$). On the other hand, particles with very different d-values, for example 0.46 and 21.5, surprisingly have comparable reactivity. For small particles, nucleation becomes rate-limiting. At this point it should be noted that the various experimental parameters control nucleation on the one hand, and interface reaction on the other hand, but in different ways (for further details, see Ref. [7]).

It should also be noted that this phenomenon must also be taken into account in temperature-programmed reactions [e.g., temperature-programmed reduction (TPR),

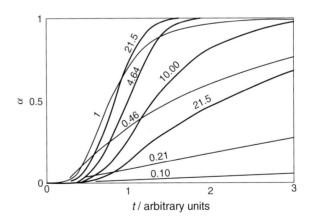

Fig. 1 Comparison of the course of reaction (degree of reaction, α, versus time) of samples composed of identical spherical particles of various diameter. Conventionally, the dimension unit for diameter has been chosen so that it corresponds to 1 for the curve at the left. (Adapted from Ref. [7].)

References see page 673

temperature-programmed sulfidation, thermal analysis, etc.] for correct interpretations. A shift of a reaction peak to higher temperature should not automatically be attributed to a change in the nature or composition of the reactant, or the fact that it is less dispersed [8]. Rather, the contrary may be true. Similarly, changes in activation processes of supported substances compared to pure unsupported ones are frequently attributed to physical or chemical interactions between the supported phase and the support. This error may have serious consequences with regards to an understanding of the activation mechanism and the efficient search for experimental parameters permitting the control of activation.

2.5.1.2.2 Interaction Between Supported Phases and Supports

The cases where active phase–support interactions exist must be analyzed in detail. It is clear that a change in reactivity will be observed if the supported phase reacts chemically with the carrier. For example, nickel can combine with silica to produce a hydroxysilicate compound when the deposition–precipitation method is used for the preparation. The reactivity of this hydroxysilicate with hydrogen during activation by reduction is very different (usually lower) compared to that of nickel oxide. In general, quite different situations possibly exist, according to the mutual arrangement and the type of physico-chemical or chemical interaction between the supported material and the support. However, a given precursor–support system may correspond to different cases, depending on the whole succession of steps in the preparation. The classification presented below corresponds to increasingly stronger chemical interactions (schematic illustrations were presented in the First Edition of this Handbook: see Section 2.1.2, pp. 265–271).

Case 1: Very Weak Forces This is the case of van der Waals forces or hydrogen bonds. Such weak forces are probably insufficient for retaining an active phase attached to the support during industrial processes, but may be sufficient for handling a precursor. Organic precursors deposited from organic solutions are actually retained on the support by this type of interaction in most cases. For hydrated salts, hydrogen bonds may also contribute. In both cases, the role of the support is only to spread or disperse the precursor. Graphite [9] and sometimes silica [10] may be examples of carriers that exert only a weak action on the supported substance. However, care must be taken that activated carbon generally possesses a functionalized surface (see Chapter 2.3.15), and that the hydroxyls of silica can react with precursors to form Si−O−R bridges (where R may represent many different organic or inorganic moieties: acyl or alkyl species, coordination compounds, etc.).

Case 2: Electronic Interaction In principle, semiconducting or conducting substances deposited on a semiconducting support may form with the latter a *junction* in the electronic sense, even if no chemical bond (as defined conventionally) is formed. Electron transfer across the boundary can, in principle, change the electron density of the deposited aggregates and their reactivity. The effect is expected to be more pronounced for small aggregates. The present developments in the synthesis of nanoparticles mean that this type of interaction (among others) can be used for deposition on supports.

Cases 3 to 7: Attachment Through a Transition Layer In the adhesion or gluing of two solids by more or less strong forces, a whole spectrum of mechanisms may explain the mutual interaction across the interface. However, except for weak forces (these are predominantly active in advanced glues) and electronic interaction, the existence of a transition layer is assumed. This transition layer may be a mixed compound of molecular thickness joining the solids, or it may involve a thicker layer across which the composition progressively changes from that of one of the solids to that of the other [11]. This leads to different situations. If the surface energy of the deposited precursors is higher than the interface energy of the interface with the support, the formation of crystallites rather than that of covering layers is more favorable. This corresponds to *Case 3*. A good example of deposited precursor is cobalt oxide (obtained from the nitrate) on silica. There is a relatively strong interaction, as shown by some tendency to form a monolayer, but the stable situation corresponds to small crystallites deposited on silica. The interface probably has the structure of cobalt silicate but is certainly only a few Angstroms or nanometers thick. With regards to reactivity in the reaction leading to the activated catalyst, the main difference with unsupported precursors results from the higher dispersion.

The formation of patches of monolayers ("rafts" or islands) or complete monolayers corresponds to *Case 4*. This category of interaction corresponds to the ideal situation where all the atoms or molecules of the precursor are exposed on the surface. The upper limit is encountered when a continuous monolayer is formed. A typical example is the monolayer of MoO_3 on the surface of γ-Al_2O_3. Formally at least, this corresponds to the reaction of one surface hydroxyl of alumina with one hydroxyl of hydrated molybdenum oxide, forming H_2O and a $[\gamma\text{-}Al_2O_3]-O-(MoO_xH_y)$ bond, with x and y depending on the composition of the gaseous phase during calcination. This is a real surface reaction, leading to the formation of a layer of molecular thickness, actually a real two-dimensional (2-D) aluminum molybdate with specific characteristics. This is a "bi-layer", for which much evidence has been presented in the literature [12–17].

The most conspicuous cases concern electronic effects [18–24]. Special structural features were detected using a variety of spectroscopic techniques (see, e.g., Ref. [25]; for a comprehensive review, see Ref. [26]). An excellent example is constituted by a shift of about 0.5 eV towards higher binding energies of the $3d_{3/2}$ and $3d_{5/2}$ levels of molybdenum, as compared to MoO_3. As long as the amount of material corresponds to less than a monolayer, no "bulk-like" three-dimensional (3-D) aluminum molybdate forms. The reactivity of monolayers is very different from that of the pure supported precursor (here MoO_3) or those of the mixed compound of composition similar to the monolayer (here $Al_2(MoO_4)_3$ (see Ref. [26])).

The formation of a "bi-layer" constitutes *Case 5*. The most conspicuous example concerns the 2-D cobalt–molybdenum and nickel–molybdenum oxide bi-layers on γ-Al_2O_3 [26–28]. While much of the structure of these bi-layers is known [26, 28–30], they should be considered as 2-D cobalt or nickel molybdates. Ion scattering spectroscopy (ISS) studies have shown that the Group VIII element may, according to circumstances, sit either on the outside (exposed to the exterior) or the inside (between γ-Al_2O_3 and the layer) [29, 30]. As generally expected with oxides containing two or several metallic elements [44], the reactivity of bi-layers is increased by the presence of the element most easily reacted (e.g., Co or Ni in Co–Mo or Ni–Mo oxide bi-layers with respect to reduction).

A more extensive reaction between a precursor and the support leads to the formation of solid solutions of supported elements (*Case 6*), compounds of ill-defined stoichiometry, or new phases (*Case 7*). This can occur during one of the preparation stages, especially during impregnation or when precipitation–deposition or similar techniques are used (see Ref. [31]), and during calcination. The result may range from an ionic exchange confined to the near-surface regions of the support, to the formation of a multiple-layer thick solid solution or to a new compound with a different crystallographic structure.

The reactivity of solid solutions and compounds containing several elements has been discussed in many cases that are not related to supported catalysts but, nevertheless, are quite illustrative. The example of the NiO–CuO system [3] illustrates the case of bulk solid solutions, as discussed in Ref. [32]. Instances of the formation of solid solutions involving substances used as supports (*Case 6*) are abundant in the literature. For example, Cr_2O_3 can react with an alumina support [33], and a similar effect is observed when CoO is contacted with magnesia [34]. A very important case for industry was (and still is) the partial dissolution of cobalt (or nickel) in γ-Al_2O_3 in HDS catalyst precursors. For cobalt, this is explained by the fact that Co_3O_4 and γ-Al_2O_3 both have the spinel structure.

The support is necessarily inert with respect to activation conditions. The element trapped in the solid solution is therefore much less reactive. The partial loss of cobalt or nickel by reaction with γ-Al_2O_3 is of concern in hydrotreating reactions. Regeneration by calcination, which increases the proportion of the active element trapped in the support as solid solution, further decreases the amount of cobalt or nickel able to play a catalytic role. This occurs readily upon calcination when the oxide precursor contains transition metal elements such as copper or manganese (and also cobalt or nickel) [35–38]. Because of their very high lattice energy and correlatively low reactivity, compounds such as spinels and some perovskites can trap part of the active elements in a near-irreversible manner. Nickel hydroxysilicate is formed in the preparation of the precursor of Ni/SiO_2 catalysts. At high loadings of MoO_3 (e.g., 20% MoO_3) on γ-Al_2O_3, in particular in catalysts used for very difficult hydrotreating reactions, $Al_2(MoO_4)_3$ crystallites are formed [39, 40].

Case 8: Grafted Precursors Currently, organometallic complexes are very frequently used. These are anchored on the inorganic surface (the terms "grafted" and perhaps "tethered" are also used, the latter almost exclusively for polymerization catalysts). Although these anchored systems may be used as catalysts without any further treatment (see Chapter 2.4.5), an increasing number of attempts have been made to use the atomically dispersed species as precursors for all sorts of catalysts (oxides, bare metal crystallites, sulfides). This implies that these organometallic complexes undergo an activation procedure. In parallel with these developments, other molecules able to attach to surfaces through a chemical bond are also used. This is the case for chlorides which, by reacting with surface hydroxyls while releasing HCl, form a bond between one remaining oxygen atom and the metal. This has been used for dispersing vanadium pentoxide on anatase, and also for preparing other highly dispersed catalysts [41–43]. One direction that nanotechnology is taking is to develop increasingly more sophisticated grafting procedures. Highly dispersed supported precursors have been successfully prepared using grafted alkoxides as precursors, and a comparison made between this grafting and that of chlorides [41–43].

Such precursors do not behave as solid reactants, but rather as molecular species. However, the resulting molecular product (bare metal atoms in the zero-valent state, oxide or sulfide "monomers") are generally not

References see page 673

stable, in particular because of their extremely high surface energy. They aggregate very rapidly in most cases, leading to the formation of tiny crystallites. Although the chemical mechanisms concerning grafting are currently under investigation, it seems that the *transformation* for activation of the grafted species has not yet been explored in detail. This is unfortunate because the remainders of organometallic complexes potentially play a role in determining the properties of the supported species. These remainders may be carbon or other residues for preventing sintering, or may act as dopes.

Intermediate Cases There are two major difficulties for describing supported oxides or, more generally, precursors (oxychlorides, salts of various oxides or other compounds) that undergo activation. One problem is the fact that the same precursor may be involved in different types of reaction according to the nature of the support. The second problem is that several situations described above may coexist in a given catalyst. MoO_3 offers an excellent illustration of these difficulties [44] when considering the interaction between MoO_3 and various supports or oxides [44–64]. The whole range of situations can be identified. At the one end of the scale is the extremely weak interaction with α-Sb_2O_4 [45–47]. The MoO_3–α-Sb_2O_4 interface energy is much lower than the surface energy of MoO_3, so that MoO_3 detaches spontaneously from α-Sb_2O_4. $BiPO_4$ behaves in a very similar way [48, 49]. The interaction becomes progressively stronger with Co_3O_4 [50], SiO_2 [51–57], TiO_2 [50, 53, 54, 58–62], ZnO_2 [54], MnO_2 [50], SnO_2 [53], MgO [53, 54], Al_2O_3 [53], and innumerable other oxides. The MoO_3–γ-Al_2O_3, system, with a strong tendency of MoO_3 to form a monolayer, stands almost at the other end of the scale. Other oxides, such as bismuth molybdates, behave as γ-Al_2O_3, or even more strongly, in the sense that they may form compounds with other stoichiometries [63, 64]. Additional details on the different cases mentioned above are provided in Ref. [44].

The coexistence of several types of interaction between given partners (*Cases 3 to 7* above) is illustrated by the case of MoO_3 dispersed on silica. Silica is generally considered to be an inert carrier, and is expected to exert only weak interactions with the supported phase and to act merely as a dispersing agent. However, this is not the case [65, 66], as shown by a series of MoO_3/SiO_2 catalysts prepared by pore volume impregnation, with drying at 383 K and calcination at 773 K. A progressive shift towards higher values of the binding energy (BE) of the Si_{2p} line was observed with increasing active phase loading from 0 to 20 wt.% of MoO_3. The shift reached the considerable value of 1 eV. The value at low loading corresponded to a strong interaction, which was substantiated by the formation of a monolayer [52, 56, 57]. The reason for this is that the monolayer structure has a limited stability, and crystallites form by spontaneous detachment of the MoO_3 monolayer. As a consequence, the following species are found in MoO_3/SiO_2 precursors [66]: MoO_3 (*Case 3* above); silicomolybdic acid (SMA) (*Case 7*); dimolybdate (DMA), polymolybdates (PMAs) and molybdenum bound to the surface of the support as a monolayer [14]. The proportion of the various species changes according to composition [66]. DMA and some PMAs are probably included in the monolayer (*Case 4*).

The additional complication due to the presence of several supported elements was illustrated very early in the precursors (oxide form) of HDS catalyst supported on alumina [35, 39, 67], and confirmed later by innumerable authors. The system contains, or may contain, according to cases: cobalt pseudo-aluminate or $CoAl_2O_4$ with the (approximately) normal spinel structure; free MoO_3; a Co_x–Mo_yO_z bilayer; and possibly $Al_2(MoO_4)_3$ at high loading. The case of alumina-supported hydrodenitrogenation NiMo catalysts is very similar to that of CoMo [68], and this type of situation is also encountered in most hydrotreating catalysts containing one or several of the elements currently used (Mo, W, Co, Ni, Fe, etc.).

Although studied in less detail, the case of perovskites deposited on a CeO_2–ZrO_2 solid solution also illustrates the complexity of the interactions between supported materials and the carrier [69]. Most of the types of interaction mentioned above can also be observed in systems involving, as precursors, many different salts (to be later decomposed to a catalytically active oxide) or a metal (for example to be sulfided), and complex supports.

2.5.1.3 Activation of Supported Catalysts by Calcination

The calcination step is crucial in the case of oxides used as catalysts and supports (Al_2O_3, SiO_2, TiO_2, ZrO_2, silica-alumina and many other oxides or mixed oxides). The term "calcination" corresponds to the action of burning or the action of oxygen or air on a material, and also to a simple decomposition to an oxide. Elaborate procedures are often used (this concerns bulk, non-supported systems, and is detailed in Chapters 2.3.3, 2.3.4, 2.3.7, and 2.3.12 of this Handbook). The case of SiO_2 *mixed* with active phases (e.g., in catalytic oxidation) has little relevance to the subject of this section, because it seems that silica does not play the role of a real support, but rather that of a dispersing agent or spacer [70]. Few cases have been reported of truly supported catalysts in which the final active form is obtained by calcination. Important examples are V_2O_5 deposited on the anatase form of TiO_2, a catalyst used for the oxidation of *o*-xylene to phthalic anhydride, and V_2O_5 deposited on various modified supports (in particular TiO_2) used for the selective catalytic reduction of NO_x. In these cases, the publications have paid much attention to the details of impregnation or grafting (notably to hydroxyls of the support), and to pretreatment of the support and support modifiers, but very little information has related to changes during calcination. Heating in the presence of air or oxygen most frequently destroys monolayers, and still more easily, the fragile attachment of grafted species. This leads to the production of supported crystallites, that often are highly dispersed (e.g., V_2O_5 deposited on the anatase form of TiO_2) [71–73].

The supported initial precursor and the supported final oxide each correspond to one of the cases mentioned in Section 2.5.1.2.2. The general tendency is to reach a higher thermal stability, with account taken of the surface energy and energy of the interface. If there is only a very weak interaction between the oxide and the support, the oxide tends to detach from the support. This will occur irrespective of the nature of the precursor, except if the remainders of some of these precursors create specific links. In spite of their advantage of being easily spread on many supports, nitrates may be inferior to precursors that leave some carbon, because that carbon acts as an "anchoring" site and prevents the growth of crystallites. In the case of α-Sb_2O_4 supported on MoO_3, or vice versa, such an anchoring seems difficult and detachment of the supported oxide seems unavoidable. In principle, the existence of monolayers or a high dispersion of the precursor constitute favorable conditions for transformation to a dispersed oxide in the calcined catalyst. The oxide/support interface energy is sometimes so high that spontaneous spreading will take place, even when the precursor is not perfectly dispersed, although complete breakdown may also occur. The extensive formation of solid solutions or new compounds contributes to anchor the active oxide crystallites to the support. This is apparently the case of V_2O_5 deposited on the anatase form of TiO_2. Very highly dispersed V_2O_5 species are observed (perhaps as patches of monolayers), together with small "rafts" which are several lattice units thick, and some dissolution of vanadium in anatase occurs.

A correct identification of supported phase–support interaction suggests many ways to control the calcination step. These include: favoring the spreading of precursors thanks to an adequate choice of these precursors; the use of additives; and inhibiting (or, conversely, promoting) the formation of solid solutions or new compounds with the support by adequate doping of the latter. A ramping rate increase or decrease in temperature or a modification of the calcination atmosphere also represent ways to modify the interactions between the active phase and the support. One reason for this is that the oxidation state of the different elements in oxides depends heavily on temperature and the surrounding atmosphere. This oxidation state modifies lattice defects in solids, and consequently interdiffusion in oxides. This can be critical for the formation (or not) of oxide structures such as spinels, perovskites, and garnets, or for modifying the active phase–support interactions and all structures in general.

The thermal decomposition of unsupported compounds, notably in the presence of air or oxygen, has been the object of extensive studies conducted since the early 1940s. A wealth of information can be found in classic text books [2, 3, 74], in the proceedings of International Symposia on the Reactivity of Solids, and in journals such as *Solid State Ionics* or *Thermochimica Acta*. With respect to the symposia, the references to the 7th, 8th, 9th and 10th are listed in Refs. [75–78]; the details of two subsequent symposia (1988 and 1992) were published in *Solid State Ionics*, while other topics acquired prominence at later meetings.

Without forgetting the factors mentioned in Section 2.5.1.1, analogies with reactions of bulk compounds might suggest modifications of the calcination procedures, such as previous handling in vacuum or using a reducing atmosphere (which often limits the sintering of oxides), or the use of a succession of different reaction conditions in order to influence nucleation. The role of water vapor in the calcination furnace is very important. According to circumstances, it can be either detrimental (in particular to dispersion) or favorable (e.g., if the introduction of vacancies in the lattice of the activated supported oxide is desirable). Another source of suggestions is that of studies concerning MoO_3 and V_2O_5, or the preparation of supported oxide *precursors* (oxides to be further activated to produce active metals or sulfides). Infrared and Raman spectroscopies, electron microscopy and microanalysis with high spatial resolution, ISS, X-ray photoelectron spectroscopy (XRPS), EXAFS and similar techniques have all provided data concerning the various steps in the transformation of precursor compounds to oxides, and are increasingly used in current research investigations.

2.5.1.4 Activation of Supported Catalysts by Reduction

In the majority of cases, the final step in the preparation of catalytically active metals is the reduction of an oxide. However, an oxychloride is the real precursor of active platinum and some noble metals, if complexes such as salts of chloroplatinic acid or similar molecules are used. Yet other precursors are used, however, with several experiments having shown that the direct reduction

References see page 673

without an intermediate oxide can lead to a higher dispersion of metals. On the other hand, nearly all catalytic metals are used as supported catalysts. A notable exception is iron used for ammonia synthesis and, together with other metals, for Fischer–Tropsch synthesis. It is sufficient to cite one detailed study concerning these catalysts [79]. The other important unsupported metals are Raney nickel and similar "skeletal" catalysts (see Chapter 2.3.2), platinum sponge or platinum black, and similar catalysts described elsewhere in this Handbook (see Chapter 2.3.2). The preparation of borohydrides is a special case.

The importance of supported metals in catalysis highlights the need to understand the mechanisms involved in activation by reduction. Control of the kinetics of reduction is essential to determine the optimal compromise between an acceptable duration for activation and high activity. Reduction is a typical reaction of solids, not only in the metals industry but also in catalysis, and many reductions have been investigated in detail with unsupported materials. Details of the necessary background for the present section, namely that concerning *unsupported* oxides, can be found elsewhere [3, 7, 80, 81], in the latter case for more chemical engineering-oriented problems. The guiding line for discussion will be provided by results on the activation of supported nickel catalysts. The interpretation can very much benefit from results concerning *unsupported* nickel oxide, this reaction having been the object of extensive fundamental investigations. (Although the reduction of iron oxides has been examined in numerous studies, the potential contribution to this section is minimal.)

2.5.1.4.1 General: Effect of Precursor Dispersion
A more dispersed precursor allows a higher rate of reduction and the production of a more disperse activated metal, provided that the reaction is not nucleation-limited and interactions do not cause side effects. This actually is the trend, although care should be taken that diffusion limitations in pores may hinder activation in the interior of catalyst particles, beads, pellets, or tablets. The situation is more complicated in NCIR (Section 2.5.1.2.1), where extreme dispersion brings about a diminution of reactivity. This effect is confirmed in Fig. 2, where the continuous curves correspond to the theoretical situation [7]. Electron microscopy provides an estimated average diameter of 550 nm for bulk NiO particles (nearly spherical or slightly cubic), and 80–100 nm for NiO in NiO/SiO$_2$ [82]. The data in Fig. 2 [83] indicate that the diminution of reactivity of NiO/SiO$_2$ is due to a diminution in size of the NiO particles, by a factor of about 6 (cube root of the ratio of parameter $A = 0.5$

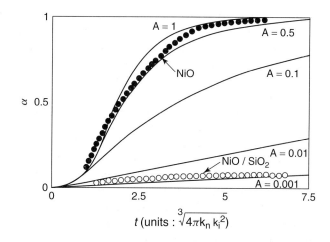

Fig. 2 Quantitative comparison of the reduction (α = fraction reduced) of bulk and silica-supported NiO at 538 K in a flow of pure H$_2$ [78]. Parameter A represents the relative rate of nucleation (constant k_n) and interface progress (constant k_i). Assuming identical spherical particles, A corresponds to the average number of nuclei being formed on each NiO particle in the time necessary for the interface to travel a distance equal to particle radius a: $A = 4\pi k_n \cdot k_i^{-1} \cdot a^3$ [7].

for bulk to $A = 0.002$ for supported NiO). Calculations from electron microscopy data gives approximately the same value (a factor of 5.0–6.3). The conclusion that nucleation limits the rate of activation by reduction of precursors was simultaneously proposed following the above deductions [84], and also by Coenen, who used an impressive series of converging arguments [85].

2.5.1.4.2 Overall Effects: Influence of Calcination Temperature in Precursor Preparation
Figure 3 illustrates,

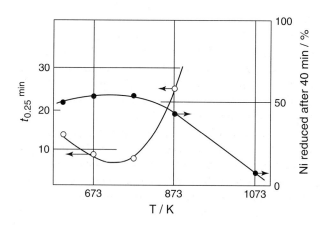

Fig. 3 Influence of calcination temperature on the reduction of NiO/SiO$_2$ at 598 K by pure H$_2$ [86]. SiO$_2$ surface area 334 m^2 g^{-1}; 18.9 wt.% Ni, calcined at 773 K, 5 h; $t_{0.25}$ is the time necessary to reduce 25% of NiO; the degree of reduction obtained after 40 min is indicated on the right-hand axis.

graphically, the important influence of calcination temperature on the reactivity of NiO supported on silica [86]. The curve giving the fraction of NiO reduced after 40 min suggests the general trends. Between 570 and 770 K, the dispersion of NiO diminishes, and the reaction rate increases, as indicated in the Section 2.5.1.4.1. Above 770 K, NiO reacts with silica, so that the rate of reduction (more precisely the reciprocal of time necessary for 25% reduction, $1/t_{0.25}$) and the amount reduced both decrease. Although such effects are frequently found [87], they are even more dramatic if the support has not been calcined before impregnation [68]. Calcination may influence several characteristics of the supported precursor, including dispersion and the extent of formation of a compound by reaction with the support [88, 89].

2.5.1.4.3 **Role of Precursors** The overall reactivity, as well as the intensity of interaction with the support, depends on the precursor. The changes concern purely chemical reactivity (e.g., as measured by the rate of interface progress), and the intervention (or not) of rate-limiting nucleation processes. Difficult nucleation processes actually play an important role in the transformations to metals. As a consequence, the characteristics of the activated catalysts are also different. Nickel catalysts prepared by incipient wetness impregnation of silica with an aqueous solution of nickel nitrate constitute good examples. One sample was simply dried at 393 K, and another was calcined at 723 K. Nickel silicate was not detected in either case, but the calcined sample exhibited very sharp X-ray diffraction (XRD) peaks of NiO [90]. Figure 4 shows the TPR profiles of these samples, together with that of a "bulk" (non-supported) NiO, similar to that of curve 1 of Fig. 1. The lower reactivity of supported "calcined" NiO compared to bulk NiO is due to high dispersion. The TPR peak consists of two components which very likely correspond to two different collections of crystallites of different size (bimodal distribution). As shown independently, nickel combined with silica would reduce at temperatures 160–180 K higher [90, 91]. The dried sample reacts 70–90 K lower and, with a much higher dispersion than the calcined sample, reacts 25 K lower than bulk NiO (peak at the highest temperature), because the interaction with silica is weaker. The dispersion of reduced nickel was measured by hydrogen chemisorption, on the assumption of two atoms per surface Ni atom, and a surface area of 6.33 Å^2 occupied by one Ni atom. After reduction at 773 K, the dispersion of supported nickel was 20% for the dried sample, but only 2% for the calcined one. The result underlines the difference of reactivity of supported precursors: Ni(OH)$_2$ or partially decomposed Ni(NO$_3$)$_2$, on the one hand, and NiO, on the other hand. If a deeper reduction and a greater dispersion are required,

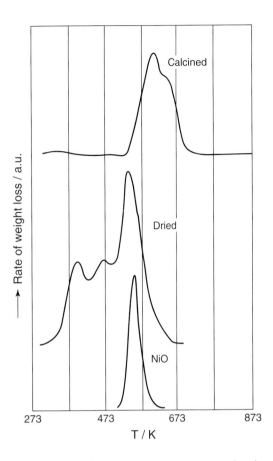

Fig. 4 Comparison of the temperature-programmed reduction (TPR) profiles of samples of NiO/SiO$_2$: SiO$_2$ surface area 275 m^2 g^{-1}; pore volume 0.80 cm^3 g^{-1}; pore distribution maximum 8 nm; incipient wetness impregnation by Ni(NO$_3$)$_2$ solution; drying at 393 K, 16 h (*Dried*); calcination at 773 K (*Calcined*); NiO weight content, on the basis of SiO$_2$ + NiO, 10.6%. TPR, measured as weight loss, was performed in deoxygenated, pre-dried H$_2$, with a heating rate of 10 K min^{-1}. The peaks at 403 and 468 K for the dried samples probably correspond, respectively, to dehydroxylation of Ni(OH)$_2$ and decomposition-reduction of ill-defined compounds containing nitrate ions [90].

it is often advantageous to start from non-calcined samples. In a less-detailed study on cobalt catalysts supported on silica, interesting differences were found when either nitrates, or acetates, or chlorides were used [87].

2.5.1.4.4 **Influence of the Amount of Deposited Precursor Loaded on the Support** When precursors mainly form crystallites, their size tends to increase with the amount deposited (Fig. 5) [68, 84]. The effect on the rate of reduction by hydrogen may easily be inferred from the data in Fig. 1.

When the supported phase can form both partial monolayer patches and crystallites, as MoO$_3$/SiO$_2$, the

References see page 673

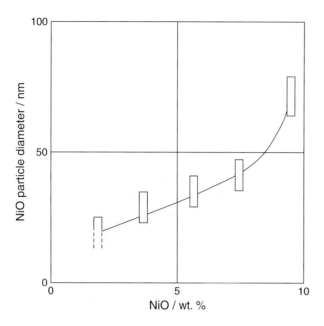

Fig. 5 Increase in size of crystallites supported on SiO_2 with increasing amounts of NiO [83]. SiO_2 had a surface area of 415 m^2 g^{-1}, and a pore volume of 1.12 cm^3 g^{-1}. The samples were obtained by pore volume impregnation with an aqueous solution of nickel nitrate. Calcination was at 773 K for 4 h. Crystallite size was determined by electron microscopy.

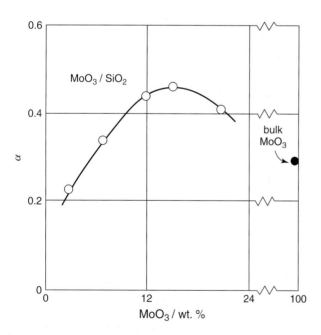

Fig. 6 Degree of reduction (α) of MoO_3/SiO_2 to MoO_2/SiO_2 at 673 K, after 10 h, H_2 pressure 100 kPa [83].

situation is more complicated, as shown in Fig. 6. The decrease in reactivity for low loading is due to the fact that: (i) the reduction of the MoO_3 crystallites to MoO_2 is nucleation-limited (see Section 2.5.1.4.1); and (ii) the monolayer patches that coexist with crystallites may have a lower reactivity than the latter. The decrease of reactivity for loadings exceeding 14% is due to the increase in size of the MoO_3 crystallites over a range of dimensions where the role of interface progress is more rate-limiting than that of nucleation. These effects have been analyzed in some detail [3, 92].

2.5.1.4.5 Effect of the Formation of Compounds between the Precursor and the Support

Impregnation of SiO_2 by nickel nitrate brings about only very little interaction when calcination is made at low temperatures (<770 K). However, the need to increase dispersion leads to the use of other methods [85, 93, 94], for example the so-called deposition–precipitation technique (see Chapter 2.4.1) that creates large amounts of nickel hydroxysilicate in the form of filaments [85, 90, 94]. This corresponds to a considerable decrease in reducibility, as shown in Fig. 7 (e.g., see also Refs. [90, 91]). Samples impregnated with nickel hexaammine $[Ni(NH_3)_6]^{2+}$ also show a dramatic decrease in the reduction rate and ultimate degree of reduction [84]. The explanation for this is the formation of some nickel silicate (or hydroxysilicate) and a much higher dispersion. For a NiO loading of 9.43%, the ratio of the Ni_{2p} to Si_{2p} XPS bands is then multiplied by a factor of about 14 [84].

Other effects, which sometimes are beneficial, may occur when a strong interaction exists. One such example is the case of Ni/Al_2O_3 catalysts for synthesis gas production from carbon dioxide and methane [88]. The formation of nickel aluminate in the precursor was favored by heating at 1073 K in a CH_4/O_2 mixture, and this yielded a very stable catalyst after activation. The reduction certainly gives a finely interdispersed mixture of nickel metal and alumina particles. A detailed study of the reduction by hydrogen of 10% NiO/α-Al_2O_3 doped by CaO or MgO showed that NiO was transformed directly to Ni [89]. Although the nickel crystallites were smaller than the starting NiO grains, some migration of nickel (atoms?) took place, presumably leading to an increase of the size of the nickel nuclei. The additives inhibited nucleation, but oxidation of the activated catalysts meant that the following reduction almost suppressed this benefit. It may be speculated that, if CO were to be used as a reducing agent, the possible deposition of some carbon could assist alumina in preventing nickel sintering.

2.5.1.4.6 Nature of the Support

Comparison between SiO_2, TiO_2 and Nb_2O_5 in the case of nickel [93] shows the influence of the support. This is also shown in the reduction of NiO supported on a range of

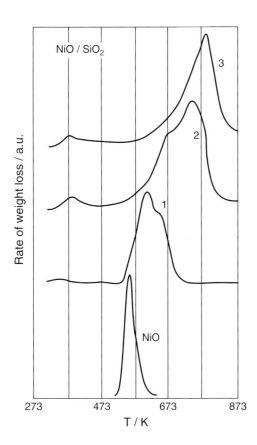

Fig. 7 Comparison of temperature-programmed reduction (TPR) profiles of NiO/SiO$_2$ samples prepared by nitrate impregnation (see legend of Fig. 4) or by deposition–precipitation on the same support [83]. The NiO loadings were almost identical (10.6 and 10.9 wt.%, on the basis of SiO$_2$ + NiO). Curve 1: nitrate-impregnated, calcined at 723 K (same as Fig. 4). Curve 2: deposition–precipitation, dried at 393 K for 16 h. Curve 3: same precursor, calcined under the same conditions as the nitrate-impregnated sample. The curve for bulk NiO is added for reference.

silica-aluminas [95]. Reducibility reflects changes in the chemical interactions between NiO and the support; this is indicated by: (i) the UV reflection spectra, specifically the presence of Ni^{2+} in an octahedral environment on alumina-rich samples; and (ii) XPS, with a strong change in the binding energy of nickel, silicon, and oxygen. In addition, the dispersion of nickel oxide on the surface increases as expected when the alumina content increases, as indicated also by XPS [95, 96].

2.5.1.4.7 Effect of Modifiers (Promotors)

Common additives, such as alkali metals, are used for depressing acidity, while others oppositely promote acidity (e.g., fluorine, chlorine). To a certain extent, their role can be inferred from results obtained with a series of studies using alumina as a support, and cobalt or molybdenum oxides as supported phase [97–116]. The interpretation of this is complex. Some dissolution of surface species, often including some elements of the support itself, occur during the impregnation step, and this leads to the migration of species inside pores, the formation of compounds involving the supported phase and the modifiers, and also changes of dispersion. This considerably alters the reducibility, as well as the dispersion and distribution inside the pellets of the reduced metal. One such example is the role of Ba and La for Ni/α-Al$_2$O$_3$ [117].

One case deserves special consideration at this point, namely the role of metal additives (e.g., platinum, copper) and other donors of spillover hydrogen used in small amounts in comparison with the supported active element. The various effects of spillover are described in Chapter 5.3.2. The maximum degree of reduction, α_{max}, for NiO/SiO$_2$ increases very much when 0.4 wt.% copper is added (Fig. 8) [86]. The data in Fig. 9 compare the amount of NiO reducible at 598 K without and with copper, for samples of different NiO loading containing 5.3 wt.% Cu on the basis of the Ni content. Copper brings about a substantial increase of the (near-ultimate) reducibility. The explanation for this is that Cu (or Pt, or Pd) produces spillover hydrogen, which considerably accelerates the formation of nickel metal nuclei in the

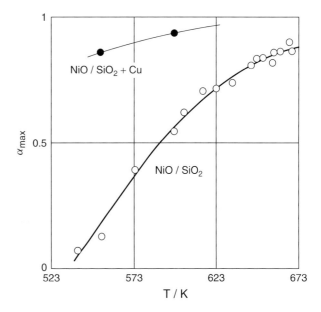

Fig. 8 Influence of copper on the reduction of NiO/SiO$_2$. The NiO/SiO$_2$ sample is the same as for the experiments of Fig. 3. Part of the sample was subjected to a second impregnation with copper nitrate (total Cu added 0.4%, calculated on the total NiO + SiO$_2$ weight) and calcined at 773 K in air for 5 h. The apparent (or conventional) maximum degree of reduction, α_{max}, corresponds to that obtained after 40 min. Reduction in pure H$_2$ (100 kPa) [86].

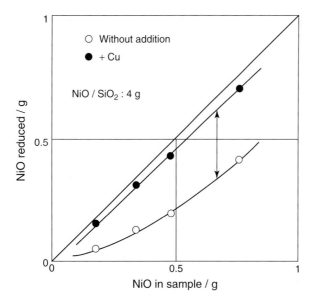

Fig. 9 Additional reduction of NiO due to copper in NiO/SiO$_2$ catalysts. Samples of different NiO loadings (weight in grams in a 4 g NiO/SiO$_2$ sample) prepared on the same silica and with the same procedure as the samples used for Figs. 3 and 8. Copper, in the proportion Cu:Ni = 5.3 wt.%, was introduced as explained in the legend of Fig. 8. Reduction was in pure H$_2$ (100 kPa) at 598 K. The weight of NiO reduced corresponds to α_{max} after 40 min. The double-headed arrow indicates the magnitude of the additional reduction due to copper [86].

reduction conditions. At low loading, the high dispersion of nickel means that the nucleation is rate-limiting in the case of pure NiO. Copper initiates nucleation and consequently accelerates reduction for high loading. The effect is proportionally smaller, and the part of NiO that becomes reducible is diminished [86]. The NiO crystallites are larger, and Ni can nucleate spontaneously at an appreciable rate, even in the absence of copper (see Section 2.5.1.2 and Ref. [3]), but the amount reduced in a pure IC reaction in 40 min diminishes. This effect of spillover has been recognized with unsupported oxides, and a review of the results obtained can be found in Refs. [3, 86, 118]; further data are accessible in Refs. [112–115]. Some publications mention the use of platinum to promote the activation of supported catalysts by reduction [111, 112].

Experiments concerning the influence of other promotors and additives on the reduction of bulk oxides [3, 118–120] can sometimes provide useful information to help understand the behavior of supported oxides.

2.5.1.4.8 Influence of the Activation Conditions

Many studies have been conducted to examine the influence of the reduction conditions in which activation was carried out. Although an overview of the trends is difficult, two studies concerning NiO/SiO$_2$ can be used to illustrate some important aspects [85, 123]. A higher temperature always brings about a higher degree of reduction [124], and this is accompanied by a loss in dispersion of the metal, unless this increase permits the reduction of a *compound* of the active metal. For example, nickel from hydroxysilicates is usually more dispersed and less prone to sintering than nickel from NiO. Extremely high temperatures may lead to the so-called strong metal-support interaction (SMSI) effect (see Chapter 3.2.5.1), or alloying of the active metal with some reduced metallic element of the support, for example nickel with silicon [125, 126]. While the majority of activation procedures use hydrogen as the reducing gas, other gases (e.g., CO) can also be used, with the advantage or disadvantage of producing residues (such as carbon deposits) that are supposed to prevent sintering in certain cases. At present, it does not appear that any systematic studies of the influence of hydrogen pressure have been conducted. The influence of water seems always to be detrimental [85, 127], for two main reasons. The first reason is that water has an inhibitory effect, and working in its presence necessitates higher temperatures, with the consequence of an increasing danger of the loss of metal surface area by purely thermal sintering. Water, independently of temperature, also promotes sintering for reasons that are not yet clear.

2.5.1.4.9 Final Remarks

One short article [128], and another cited earlier [86], suggest that the fundamental approach presented in here is useful for understanding the overall phenomenon and for controlling the activation process. No single method can provide all of the necessary information needed to optimize activation by reduction. TPR used in previous examples, and also in many other studies [91, 129], is valuable for providing preliminary data, but it must be complemented by techniques adapted to the system under study (e.g., Mössbauer spectroscopy if iron is present in the active phase [130]). Although nickel catalysts have been used as examples here, many articles dealing with other metals have shown that the same concepts do apply. Indeed, this is the case of cobalt deposited on silica or alumina [128], Cu/γ-Al$_2$O$_3$ [131] or Ag/η-Al$_2$O$_3$ and Ag/TiO$_2$ [132], for example. The preparation of bimetallic catalysts is more complicated, because of a possible preferential reduction of one metal before the other, a phenomenon which is well-recognized with bulk oxides [3]. An example of complex processes that may occur is provided in a study of the reduction of a silica-supported Pt–Ag catalyst [133]. A few other studies have suggested that approaches similar to those presented here may also be used [129]. It should be mentioned at this point that controlled partial reduction can increase the thio-resistance of Ni/Al$_2$O$_3$ [134].

2.5.1.5 Reduction–Sulfidation

Hydrotreating reactions (hydrodesulfurization, hydrogenation, hydrodenitrogenation, hydrodeoxygenation, hydrodemetallation, etc., and presently a large part of hydrocracking reactions) use sulfided catalysts (see Chapter 13.2). Until recently, these catalysts contained, as the main active species, Group VI metals – that is, molybdenum or sometimes tungsten, on the one hand, and nickel and/or cobalt (or sometimes iron) on the other hand. Environmental constraints relative to the sulfur content of fuels led to the introduction of noble metals into the formulation for deep hydrodesulfurization and de-aromatization. (These latter systems will not be examined here because too few data concerning their activation are accessible.) The precursors of the sulfided form of traditional catalysts contain one molecule-thick layers or "bi-layers" of the active metals in the oxide state. The formation of active sulfided catalysts necessitates both a reduction and a deep sulfidation of this oxide precursor. This operation, which always is referred to as *activation* in the field of hydrotreating, is absolutely crucial in the sense that changes in the activation procedure can bring about variations of activity by factors in excess of 2 or 3, they may modify selectivity, and also either extend or diminish catalyst life [135]. Two remarkable articles must be cited in this context [136, 137].

During the early days of the industrial development of hydrotreating, the precursors were simply contacted with hydrogen and a sulfur-containing feed, which usually was an ordinary petroleum fraction to which sulfur-containing compounds had been added (a so-called "spiked" feed). Later, the catalysts were charged in the reactor and treated with a special mixture containing, in addition to hydrogen, either H$_2$S or a compound that easily released reactive sulfur (e.g., dimethyl sulfide or mercaptans). Unfortunately, the operation in a large reactor was difficult to control, especially with regards to temperature, and for this reason a pre-sulfidation outside the rector is becoming common practice. The reactions are identical or similar in all cases, however, and correspond to the reaction of hydrogen and sulfur-containing compounds with the oxide precursor (see Section 2.5.1.5.6). Innumerable recipes have been proposed in industry and described in literature. In addition to Refs. [136] and [137], one article (among many) provides a valuable (though still non-exhaustive) overview of possible procedures [138]. An understanding of the reduction–sulfidation process is essential for the reasons indicated at the beginning of section, and also because these catalysts must be regenerated (either *in situ* or in special plants). This regeneration is carried out almost exclusively by carefully controlled oxidation, whereby the process removes residual carbonaceous material and oxidizes the metal sulfides. A reactivation must therefore take place. This will be done through an approach different from the first activation and possible previous reactivations. This implies that the new activation is equally crucial to the first, and potentially may produce a catalyst of higher activity – an often-explored matter in catalysis. While *regeneration* (examined in Chapter 7.1) is not the topic of the present section, the second, third, etc., activation episodes are. Consequently, we should briefly mention some references illustrating the modifications that catalysts may undergo after a regeneration–activation step [139–141].

2.5.1.5.1 Fundamental Data

Reduction–sulfidation corresponds to a drastic change of the supported material. The *starting* bi-layer is a 2-D oxide adhering to the support (usually alumina) and forming a continuous interface. The *final* catalyst is constituted of tiny crystallites attached to this support by small patches of interfaces. These patches have a composition, and probably also a structure, achieving a transition between both partners (Fig. 10). The whole reduction–sulfidation reaction involves many simultaneous or successive processes, including: reduction of MoO^{6+} to MoO^{4+}; sulfidation of MoO^{4+} to MoS$_2$; sulfidation of the oxides of Co, Ni or Fe to sulfides; destruction of the bi-layer; and the formation of crystallites. The Group VI and VIII metals that initially are associated together in the bi-layer segregate to a large extent in the sulfide state, though the actual extent of segregation remains the subject of debate [143–150]. The question here concerns the existence of compound sulfide species, dubbed "CoMoS", or similar species for Ni or Fe (see Chapter 13.2). In principle, the activation procedure might favor the formation of this association or, conversely, lead to the segregation of highly dispersed MoS$_2$ and cobalt sulfide. A slow increase

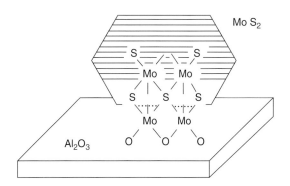

Fig. 10 Schematic representation of a tiny crystallite of MoS$_2$ (diameter, typically 5–10 nm) attached to the alumina surface in a sulfided hydrotreating catalyst.

References see page 673

in temperature in a poorly reducing atmosphere would be expected to hinder segregation. Conversely, the contact with a more reducing atmosphere at a slightly higher temperature would favor segregation by producing Co_9S_8 or a slightly more reduced cobalt sulfide.

Leaving aside the uncertainty concerning the real form of the freshly activated catalyst, there is clear evidence that a high dispersion of the various sulfide species and some proximity between the crystallites of different partners are crucial for high activity, and may control selectivity and perhaps also stability. Reduction–sulfidation involves strong coupling effects, as this generally occurs in reactions of solids [3, 151]. This has a strong influence on the structure and texture of the final sulfided catalysts. The most striking coupling phenomena concern the reduction and sulfidation reactions. MoO_3 is reduced at much lower temperatures (sometimes 200 K lower) in a H_2/H_2S mixture than in pure H_2 [152, 153]. Starting from already reduced molybdenum (namely Mo^{4+} in MoO_2), sulfidation takes place at a lower rate than when reduction and sulfidation occur simultaneously (that is, when starting from MoO_3). The presence of cobalt (or nickel) probably has a slightly promoting effect on the reduction [154–156], because the corresponding oxides are more easily reduced and sulfided than molybdenum and tungsten.

The explanation of the coupling effect in reduction–sulfidation is most likely as follows. The reduction of MoO_3 to MoO_2 is nucleation-limited [157], with spillover hydrogen accelerating the reaction [118, 158, 159]. Because it adsorbs dissociatively [157, 160], H_2S probably also produces a reactive H species, promoting nucleation to lower molybdenum oxides at low temperature and, as a consequence, further reduction. Because of its high state of dispersion, the thus-formed suboxide reacts more easily than MoO_2 produced by separate reduction. These results are based on the fundamental studies of Steinbrunn et al. [157, 161]. Upon reduction–sulfidation with pure H_2S, the surface of MoO_3 is first reduced to MoO_2, and then sulfided at a much slower rate to MoS_2 [161]. However, during the reaction with H_2/H_2S mixtures, a more complicated mechanism sets in, that leads to the formation of highly dispersed MoS_2 clusters [151, 162]. MoS_2 crystallites are formed only at a later stage, this being due to the fact that the reduction step with H_2/H_2S is slower than with H_2S. Some refinements were obtained using thin films [163], although studies with composite films of MoO_3 evaporated on a freshly cleaned CoO surface indicated that these observations were still valid in the presence of cobalt [164]. The success of the reduction–sulfidation is due to a better balance between reduction and sulfidation rates. On the basis of this improved balance, the formation of much smaller MoS_2 domains (clusters, instead of crystallites) is possible. The fundamental analysis of the process thus suggests several ways by which HDS catalyst preparation and activation could be improved. Progress might come from independent control of the different processes, using parameters other than just temperature, pressure and the composition of the mixtures used for activation. This may include the use of promoters of reduction, such as donors of spillover, or conversely inhibitors, or inhibitors of sintering. Incidentally, it is possible that part of the beneficial role of noble metals in hydrotreating catalysts might also be to promote reduction during activation, because of their capacity to emit spillover hydrogen. As the height/diameter ratio of MoS_2 crystallites has been claimed to control HDS/hydrogenation selectivity, additives favoring the lateral growth of clusters at the expense of stacking of MoS_2 layers (or vice versa) could perhaps be found by following this basic idea.

A comparison with studies on supported catalysts containing only either cobalt or molybdenum [19, 20, 165–167] shows that the simultaneous presence of both elements profoundly alters the process. This is linked to the existence of the CoMo oxide bi-layer, and is partly due to the interaction of both elements in the reduction–sulfidation and the segregation of cobalt from molybdenum in this process. The reduction–sulfidation of cobalt molybdate ($CoMoO_4$) seems of little relevance for an understanding of the reduction–sulfidation mechanism [162, 168]; the reason for this is, perhaps, the fact that $CoMoO_4$ becomes sulfided *before* sufficient reduction, as shown by thermogravimetric analysis [169].

2.5.1.5.2 Influence of the Sequence of Steps in Reduction-Sulfidation

The importance of coupling between reduction and sulfidation can be shown by comparing experiments in which the precursor is either reduced and sulfided simultaneously, or is reduced first and sulfided afterwards (Fig. 11). Sulfidation of the pre-reduced catalyst becomes less complete as the reaction temperature increases. Presumably, this is due to the growth of MoO_2 crystallites that makes them less reactive. When carried out at low temperature (below ca. 600 K), a preliminary reduction step is much less harmful, although a limited pre-reduction has been recommended. The most likely reason for this is that the Group VIII metal is reduced, segregates and, in the form of sulfide, enhances the production of spillover hydrogen in the subsequent reduction–sulfidation. The crucial experiment of Fig. 11 can be analyzed in more detail [156]. Mo^{5+} species can be detected by XPS in series (R + RS), but not in series (RS), indicating incomplete reduction. Oxygen is removed in (RS), thus compensating by a weight loss the weight gain due to the introduction of sulfur. Nevertheless, the gain in weight due to sulfidation takes place more rapidly in (RS) than in the (RS) step of (R + RS) experiments.

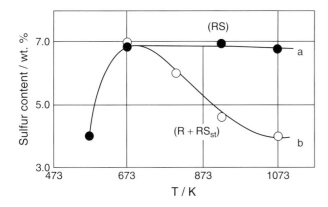

Fig. 11 Comparison of the sulfur content of a commercial CoMo catalyst (Procatalyse HR 306) activated by simultaneous reduction–sulfidation or by reduction followed by sulfidation (R + RS$_{st}$). (a) Simultaneous reduction–sulfidation by a H$_2$S/H$_2$ mixture (15% H$_2$S/H$_2$ by volume). Except for the experiment at 573 K, the samples were first reacted for 4 h at 673 K, then progressively heated to the reaction temperature indicated in the figure and maintained at that temperature for 4 h. (b) Reduction at the indicated temperature for 4 h, followed by sulfidation with the H$_2$S/H$_2$ mixture at 673 K for 4 h (the subscript in the RS$_{st}$ symbol indicates that this step was conducted at a standard temperature [156]).

This shows that sulfidation is more rapid when starting from Mo^{6+} than from reduced forms of molybdenum. NO adsorbed on Mo ions exhibits a special infrared band at 1680–1700 cm^{-1}, the position of which is characteristic of the reduction state of Mo. Following the (R + RS) treatment, some deeply reduced Mo ions are present on the surface, whereas in (RS) catalysts, only Mo^{4+} ions are observed. This confirms that sulfidation is difficult when starting from Mo^{4+} ions [152, 156].

2.5.1.5.3 Influence of the Activation Temperature

The influence of temperature is always considerable. All *in-situ* activation procedures stipulate that the temperature be raised stepwise, following a procedure that depends on the catalyst, the nature of the sulfur-containing compound used for sulfidation, and all the operating conditions [156]. The objective of the present discussion is to outline the fundamental mechanisms that could justify these procedures. Figures 12 and 13 correspond to activity in hydrodesulfurization (HDS) and hydrogenation, respectively, in experiments in which both reactions (HDS of thiophene and hydrogenation of cyclohexene) took place simultaneously [156]. In addition to the procedures used for the results of Fig. 11, samples were also successively reduced and then sulfided at the same temperature (S + RS). In addition to rates, the HDS/hydrogenation selectivity also changed. For the (RS) series, the selectivity varied by a factor higher than 2. Similar results are observed for both 573 K and 1073 K [170].

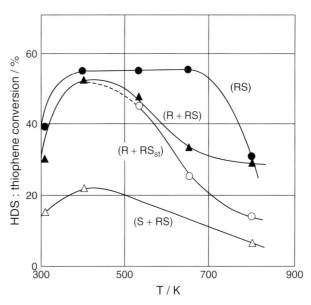

Fig. 12 Hydrodesulfurization activity of catalysts (Procatalyse HR 306) activated according to different procedures at different temperatures. (RS): Simultaneous reduction–sulfidation (15% H$_2$S in H$_2$); except for the experiment at 573 K, the samples were first reacted for 4 h at 673 K, then progressively heated to the temperature indicated in the figure and maintained at this temperature for 4 h. (R + RS): Successive reduction in H$_2$ (R) and reduction–sulfidation, as above (4 h) both at the temperature indicated. (R + RS$_{st}$): Reduction at the temperature indicated (4 h) followed by reduction–sulfidation with the H$_2$S/H$_2$ mixture at the "standard" temperature 673 K (4 h). (S + RS): Sulfidation with pure H$_2$S (4 h) followed by reduction–sulfidation as above, with both steps conducted at the temperature indicated. The reaction with hydrogen and a feed containing 0.5% thiophene and 30% cyclohexene in cyclohexane was conducted in a continuous-flow equipment under a total pressure of 3 MPa [148, 156].

The MoS$_2$ crystallites grow larger when the activation temperature increases, and their distribution in size is modified (Fig. 14), although the ratio of the total edge-planes area to that of the basal planes changes little (by ca. ±20%). However, the area of edge planes relative to the volume of crystallites decreases considerably, actually by a factor of 5 [170]. Infrared studies of adsorbed NO [136, 137, 139, 141], ISS measurements [145, 156, 171], and ^{57}Co Mössbauer emission studies [136, 137, 149] have also been made (see Chapter 13.2). When the activation temperature increases, the signal of cobalt pseudoaluminate (or aluminate-like cobalt species) diminishes, indicating that part of the cobalt strongly bound to alumina reacts. The signal attributed to the so-called CoMoS species strongly diminishes at 923 K, but the catalytic activity remains high (Fig. 15) [148–150]. Among many others, three additional systematic studies of the influence of temperature must be cited [172–174].

References see page 673

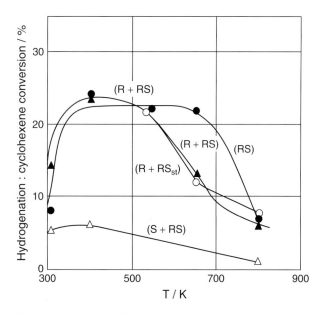

Fig. 13 Hydrogenation activity of catalysts activated according to different procedures. The experimental conditions are as described for Fig. 12 [148, 156].

2.5.1.5.4 Influence of the Sulfiding Molecule Figures 16 and 17, which relate to the HDS of thiophene and the hydrogenation of cyclohexene [175], confirm that the nature of the feed, the "spiking" of the feed, and the nature of sulfiding molecules generally speaking, are of considerable importance [138, 176, 177]. Changing the activating molecule modifies activity and selectivity more strongly than the procedures illustrated in Figs. 12 and 13. There is no safe explanation yet concerning the relation between changes in selectivity and changes in the texture of the crystallites and their arrangement. The origin may be found mainly in modifications in the very subtle coupling effects between elementary reactions and/or the possible formation of highly dispersed carbon left behind by the destruction of some organic molecules.

2.5.1.5.5 Other Results Other notable data on the reduction–sulfidation of $CoMo/Al_2O_3$ catalysts can be found in articles not yet mentioned [177, 178]. In a study using supports calcined at various temperatures, with incomplete reduction or reduction–sulfidation according to several procedures, and characterization of used samples, an attempt was made to obtain a more in-depth understanding of activation. The merit of the study was to show that accepted conclusions still hold when the activation is studied in a wider investigation range than previously [164]. The activation of HDS or hydrodenitrogenation catalysts other than CoMo has also been the object of a variety of studies: NiMo [165–168], NiW [147, 178, 184], and FeMo [185].

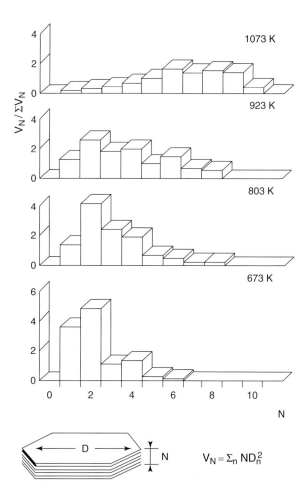

Fig. 14 Variation of the size of MoS_2 crystallites in supported CoMo catalysts (Procatalyse HR 306) obtained after activation at different temperatures (procedure (RS); see legends in preceding figures) as determined by high-resolution electron microscopy. The figure takes into account the collection of crystallites (number n), with the same number N of MoS_2 layers ("thickness" N). The total volume of each collection can be calculated as indicated in the figure. The fraction in volume (or weight) that each collection represents in the overall sample $V_N/\Sigma V_N$ gives the height of the corresponding bar in the figure [170].

2.5.1.5.6 Outlook The tendency at present is to presulfide catalysts before loading the reactor in order to control the activation process in a better way. Currently, however, a new technology is emerging that essentially consists of coating the precursor catalyst (namely the oxide) with a mixture of organic sulfides and polysulfides. The technique has its origin in the *in-situ* soaking technique developed during the late 1970s [138, 186] and "liquid phase sulfiding" (e.g., see Ref. [186]). A limited amount of data are available on a few of these new techniques [172, 187], and the effect of this type of presulfiding with alkylpolysulfides has been discussed [172, 188]. The decomposition of polysulfides releases active species that permit sulfidation at low

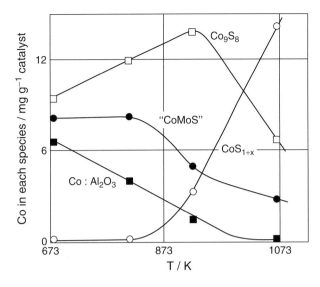

Fig. 15 Distribution of cobalt among the various species of an activated HDS catalyst, as determined by Mössbauer emission spectroscopy. The catalyst, prepared by impregnation of ammonium heptamolybdate and ^{57}Co nitrate, was as similar as possible to the commercial one used for obtaining the results presented in the previous figures (3 wt.% as CoO, 13 wt.% as Mo on γ-alumina). All precautions were taken to avoid possible errors due to non-instantaneous charge compensation in the sequence of nuclear events in the ^{57}Co decay (this should be taken into account in the non-conducting MoS_2 matrix, but this was not always the case in literature) [149, 150].

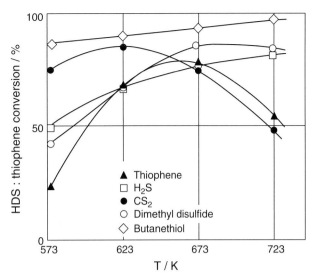

Fig. 16 Influence of the nature of the sulfiding agent on HDS activity. The catalyst was Procatalyse HR 306. The amount of sulfiding agent was calculated in order to obtain a sulfur content of 3 wt.% in the charge (hexane). The precursor samples (2 g) were heated in argon (36 L h^{-1}, 3 MPa) up to the desired activation temperature. Argon was then replaced by hydrogen (36 L h^{-1}, same pressure) and the feed introduced (liquid hourly velocity: 42 h^{-1}). The activation time was 4 h. Activity measurements were made as described for Fig. 12 [175].

temperature. A comparison between activation with H_2S and with these polysulfides suggests that results are very similar [189–191]. In addition, there is a suspicion (and also some proof) that carbon initially contained in the active sulfides may be beneficial for activity. This carbon presumably forms highly dispersed particles separating the crystallites of the different sulfides. As carbon is known to permit easy circulation of hydrogen (in particular spillover hydrogen), these particles may facilitate the circulation of hydrogen [192].

2.5.1.6 Activation of Other Catalysts

On completion of the manufacture of other catalysts, some steps correspond to activation, as defined previously in this chapter. This is the case for Raney nickel, where a Ni–Al alloy is leached by an alkali, leaving catalytically active nickel, though this is a very special case. A situation which is more similar to the discussion above is the spontaneous increase in performance which occurs when precursors are contacted with the reacting mixture or feed under the normal conditions of the catalytic reactions. This was the situation during the early development of hydrotreatments. Another conspicuous example is the activation of vanadium phosphate catalysts used in the oxidation of butane to maleic anhydride. Here, the

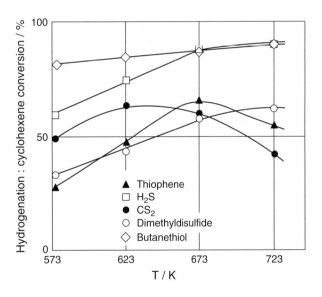

Fig. 17 Influence of the nature of the sulfiding agent on hydrogenation activity. Conditions were as indicated for Fig. 16 [175].

fixed-bed catalyst must be contacted for several days with an air–butane mixture before reaching normal activity. Until now, the vanadium phosphate catalysts

References see page 673

used seem to have been unsupported, and therefore do not strictly coincide with the situation of the other catalysts described in this chapter. This remark does, however, deserves mention here because such transient periods (i.e., of "running-in") cannot be formally distinguished from activation. Although future technologies may well use supported catalysts for MAA manufacture, reports published to date have not indicated whether a new cyclic two-step scheme – the "rising reactor process", whereby the catalyst is "stuffed" with oxygen and subsequently contacted with butane in the absence of air or oxygen – also requires a "running-in" period.

2.5.1.7 Conventional Activation and the Real State of Catalysts during Catalysis

Examples highlighting the previous sections demonstrate the difficulties encountered in making a clear distinction between activation, as it is usually understood, and the progressive transformation of catalysts (or precursors) during the first minutes, hours or days of operation under catalytic conditions. For example, during the first days of progressive transformation of a vanadium phosphate catalyst (VPO) contacting butane and air, no change in composition could be detected using either "normal" procedures or advanced techniques, despite both yield and selectivity changing. Today, the demand for smooth running of industrial plants means that such changes represent an inconvenience for the practical control of production processes, and lead to substantial costs. An unveiling of these still partly unknown changes, in order to control them, would be logical by considering a "jump" in present practice, from a largely empirical "running-in" step to controlled "activation". It is clear that advanced activation procedures will be developed in the near future, many of which will be inspired by comparative studies of catalysts when introduced into the reactor and when functioning under reaction conditions. Some examples demonstrating the development of such development are cited in the following.

A reconstruction of the VPO catalyst occurs in near-surface regions during the synthesis of maleic anhydride. In addition, the surface composition is changed, and there is some reduction of the catalytic oxides and an increase in oxygen mobility. A number of publications have indicated the extension of these phenomena [193–195]. In the oxidative dehydrogenation of propane on VMgO catalysts, the structure of near-surface layers changes, with VO_4^{3-} moieties that interact with MgO reorganizing in different ways according to the reaction conditions [196]. In the selective oxidation of isobutene to methacrolein on MoO_3 in the presence of Sb_2O_4, in addition to a conspicuous reconstruction of the apparent surface at the scale of several nanometers, the development of special oxide structures (actually molybdenum suboxides) is observed [197]. Still more elaborate pictures of the composition of near-surface layers of VPO catalysts during toluene ammoxidation have been proposed [198–201], while a recent review has provided similar evidence for other reactions [201]. The present intensive studies of complex catalysts, when explored for the ammoxidation of propane, may prove the existence of the same type of feature [202–204]. Incidentally, these latter results may be related to those obtained with heteropolyacids (see, e.g., Refs. [205, 206]).

In these cases, the creation of an optimal surface structure based on a previous treatment may not prove helpful because this occurs spontaneously, inside the reactor. Nevertheless, a new form of "activation", brought about by approaches that differ greatly from established procedures, suggests that innovative procedures will be identified in the near future to inspire investigation in a variety of directions.

2.5.1.8 Conclusions

Although crucial for activity, selectivity and sustained activity (catalyst life), the activation of catalysts has rarely been investigated in systematic manner. Here, we suggest that a methodical approach is used, leading to the adoption of improved procedures, but despite clear scientific explanation often being impossible. Today, the equipment available in most catalysis laboratories, whether industry, university, or research institutes, is sufficient for the gathering of necessary information. However, while classical mechanisms can be used to explain the phenomena observed, the science of solid reactivity is often ignored, which not only hampers interpretation but also limits future development. In recent years, contacts between industry and academia have greatly improved, and collaborative programs now provide access to elaborate physico-chemical techniques. Unfortunately, today's huge variety of catalysts creates difficulties in providing a comprehensive overview, while the fascination for new techniques causes fewer investigators to show interest in the mechanisms of catalyst activation. Despite increasing numbers of publications reflecting greater numbers of research projects, the data produced are often not really explained. As a consequence, very few fundamental investigations are initiated, and reports correspond to specific problems rather than to major scientific advances. The situation, therefore, is tending towards empiricism, with great interest in the applications of catalysis, but with a rapid attainment of objectives. Although the overall situation might appear problematic, we should in fact rejoice at the extreme interest shown by industry for catalysis. In this chapter, we have tried to show that a better multidisciplinary approach, perhaps even inspired by bottlenecks in industry, may provide a much clearer picture of the activation of supported catalysts.

References

1. J. Haber, *Pure Appl. Chem.* **1991**, *63*, 1227.
2. A. K. Galwey, *Chemistry of Solids*, Chapman & Hall, London, 1967, Chapter 5.
3. V. V. Boldyrev, M. Bulens, B. Delmon, *The Control of the Reactivity of Solids, Studies in Surface Science and Catalysis*, Vol. 2, Elsevier, Amsterdam, 1979, Chapter 2.
4. W. A. Johnson, R. Mehl, *Trans. Am. Inst. Min. Metal. Engrs.* **1939**, *135*, 416.
5. K. L. Mampel, *Z. Phys. Chem.* **1940**, *A187*, 43.
6. K. L. Mampel, *Z. Phys. Chem.* **1940**, *A187*, 235.
7. B. Delmon, *Introduction à la Cinétique Hétérogène*, Ed. Technip, Paris, 1969, Chapter 11.
8. J. L. Lemaître, in *Characterization of Heterogeneous Catalysts*, F. Delannay (Ed.), Marcel Dekker, New York, 1984, p. 29.
9. G. C. Stevens, in *Preparation of Catalysts II*, B. Delmon, P. Grange, P. A. Jacobs, G. Poncelet (Eds.), *Studies in Surface Science and Catalysis*, Vol. 3, Elsevier, Amsterdam, 1978.
10. A. Cimino, B. A. Angelis, *J. Catal.* **1975**, *36*, 11.
11. B. Delmon, in *Interfaces in New Materials*, P. Grange, B. Delmon (Eds.), Elsevier, London, 1991, p. 1.
12. J. Sonnemans, P. Mars, *J. Catal.* **1973**, *31*, 209.
13. J. M. J. G. Lipsch, G. C. A. Schuit, *J. Catal.* **1969**, *15*, 174.
14. M. Dufaux, M. Che, C. Naccache, *J. Chim. Phys.* **1970**, *67*, 527.
15. V. H. J. de Beer, H. M. van Sint Fiet, J. F. Engelen, A. C. van Haandel, M. V. J. Wolfs, C. H. Hamberg, G. C. A. Schuit, *J. Catal.* **1972**, *27*, 357.
16. G. C. A Schuit, B. C. Gates, *AIChe. J.* **1973**, *19*, 417.
17. W. K. Hall, M. LoJacono, in *Proceedings 6th International Congress on Catalysis*, G. C. Bond, P. B. Wells, E. C. Tompkins (Eds.), The Chemical Society, London, 1977, p. 246.
18. P. Gajardo, R. I. Declerck-Grimée, G. Delvaux, P. Olodo, J. M. Zabala, P. Canesson, P. Grange, B. Delmon, *J. Less Common Met.* **1977**, *54*, 311.
19. R. I. Declerck-Grimée, P. Canesson, R. M. Friedman, *J. Phys. Chem.* **1978**, *82*, 885.
20. R. I. Declerck-Grimée, P. Canesson, R. M. Friedman, *J. Phys. Chem.* **1978**, *82*, 889.
21. P. Canesson, C. Defossé, R. I. Declerck-Grimée, B. Delmon, *5th Ibero-American Symposium on Catalysis*, Lisbon, 1976, paper A3-8.
22. G. C. Stevens, T. Edmonds, in *Chemistry and Uses of Molybdenum*, P. C. H. Mitchell, A. Seamen (Eds.), Climax, Ann Arbor (Mich.), 1976, p. 155.
23. A. Miller, W. Atkinson, M. Barber, P. Swift, *J. Catal.* **1971**, *22*, 140.
24. A. W. Armour, P. C. H. Mitchell, B. Folkesson, R. Larsson, *J. Less Common Met.* **1974**, *36*, 361.
25. A. Lycourghiotis, C. Defossé, F. Delannay, B. Delmon, *J. Chem. Soc., Faraday Trans. I* **1980**, *76*, 2052.
26. H. Knözinger, in *Proceedings 9th International Congress on Catalysis*, M. J. Phillips, M. Ternan (Eds.), Vol. 1, The Chemical Institute of Canada, Ottawa, 1988, p. 201.
27. B. Delmon, P. Grange, M. A. Apecetche, P. Gajardo, F. Delannay, *C. R. Acad. Sci., Ser. C* **1978**, *287*, 401.
28. P. Gajardo, P. Grange, B. Delmon, *J. Catal.* **1980**, *63*, 201.
29. F. Delannay, E. N. Haeussler, B. Delmon, *J. Catal.* **1980**, *66*, 469.
30. H. Knözinger, H. Jeziorowski, E. Taglauer, in *Proceedings 7th International Congress on Catalysis, Tokyo*, T. Seiyama, K. Tanabe (Eds.), Vol. 1, Kodansha and Elsevier, Tokyo and Amsterdam, 1981, p. 604.
31. J. B. Espinosa de la Callère, C. Robin, B. Rebours, O. Clause, in *Preparation of Catalysts VI*, G. Poncelet, J. Martens, B. Delmon, P. A. Jacobs, P. Grange (Eds.), *Studies in Surface Science and Catalysis*, Vol. 91, Elsevier, Amsterdam, 1995, p. 169.
32. P. Grange, H. Charcosset, Y. Trambouze, *J. Thermal Anal.* **1969**, *1*, 311.
33. C. Marcilly, B. Delmon, *J. Catal.* **1972**, *24*, 336.
34. A. P. Hagan, C. O. Arean, F. S. Stone, in *Reactivity of Solids*, J. Wood, O. Lindqvist, C. Helgessen, N. G. Vannerberg (Eds.), Plenum Press, New York, 1977, p. 69.
35. M. LoJacono, A. Cimino, G. C. A. Schuit, *Gazz. Chim. Ital.* **1973**, *103*, 1281.
36. M. LoJacono, M. Schiavello, A. Cimino, *J. Phys. Chem.* **1971**, *75*, 1044.
37. M. LoJacono, M. Schiavello, D. Cordischi, *Gazz. Chim. Ital.* **1975**, *105*, 1165.
38. M. LoJacono, M. Schiavello, in *Preparation of Catalysts I, Scientific Bases for the Preparation of Heterogeneous Catalysts*, B. Delmon, P. A. Jacobs, G. Poncelet (Eds.), *Studies in Surface Science and Catalysis*, Vol. 1, Elsevier, Amsterdam, 1976, p. 473.
39. N. Giordano, J. C. J. Bart, A. Vaghi, A. Castellan, G. Matinotti, *J. Catal.* **1975**, *36*, 81.
40. G. N. Asmolov, O. V. Krylov, *Kinet. Katal.* **1970**, *11*, 1028.
41. X. L. Xiong, L. T. Weng, B. Zhou, B. Yasse, E. Sham, L. Daza, F. Gil-Llambías, P. Ruiz, B. Delmon, in *Preparation of Catalysts VI*, G. Poncelet, J. Martens, B. Delmon, P. A. Jacobs, P. Grange (Eds.), *Studies in Surface Science and Catalysis*, Vol. 91, Elsevier, Amsterdam, 1995, p. 537.
42. R. Castillo, B. Koch, P. Ruiz, B. Delmon, *J. Mater. Chem.* **1994**, *4*, 903.
43. R. Castillo, B. Koch, P. Ruiz, B. Delmon, in *Preparation of Catalysts VI*, G. Poncelet, J. Martens, B. Delmon, P. A. Jacobs, P. Grange (Eds.), *Studies in Surface Science and Catalysis*, Vol. 91, Elsevier, Amsterdam, 1995, p. 291.
44. B. Delmon, *J. Mol. Catal.* **1990**, *59*, 179.
45. B. Zhou, B. Doumain, B. Yasse, P. Ruiz, B. Delmon, in *Proceedings 9th International Congress on Catalysis*, M. J. Phillips, M. Ternan (Eds.), Vol. 4, The Chemical Institute of Canada, Ottawa, 1988, p. 1850.
46. B. Zhou, E. Sham, P. Bertrand, T. Machej, P. Ruiz, B. Delmon, *J. Catal.* **1991**, *132*, 157.
47. L. T. Weng, B. Zhou, B. Yasse B. Doumain, P. Ruiz, B. Delmon, in *Proceedings 9th International Congress on Catalysis*, M. J. Phillips, M. Ternan (Eds.), Vol. 4, The Chemical Institute of Canada, Ottawa, 1988, p. 1609.
48. M. V. E. Rodriguez, B. Delmon, J. P. Damon, in *Proceedings 7th International Congress on Catalysis, Tokyo*, T. Seiyama, K. Tanabe (Eds.), Vol. 1, Kodansha and Elsevier, Tokyo and Amsterdam, 1981, p. 1141.
49. J. M. D. Tascón, P. Bertrand, M. Genet, B. Delmon, *J. Catal.* **1986**, *56*, 300.
50. J. Haber, *Pure Appl. Chem.* **1984**, *56*, 1663.
51. F. Delannay, P. Gajardo, P. Grange, *J. Microsc. Electron.* **1978**, *3*, 411.
52. P. Gajardo, D. Pirotte, C. Defossé, P. Grange, B. Delmon, *J. Electron Spectrosc. Relat. Phenom.* **1979**, *17*, 121.

53. S. R. Stampfl, Y. Chen, J. A. Dumesic, C.-M. Niu, *J. Catal.* **1987**, *105*, 445.
54. D.-S. Kim, *J. Catal.* **1994**, *146*, 268.
55. M. A. Befiares, *J. Catal.* **1994**, *150*, 407.
56. P. Gajardo, P. Grange, B. Delmon, *J. Phys. Chem.* **1979**, *83*, 1771.
57. P. Gajardo, D. Pirotte, P. Grange, B. Delmon, *J. Phys. Chem.* **1979**, *83*, 1780.
58. K. Y. S. Ng, E. Gulari, *J. Catal.* **1985**, *92*, 108.
59. Y. C. Liu, G. L. Griffin, S. S. Chan, I. E. Wachs, *J. Catal.* **1985**, *94*, 108.
60. G. C. Bond, S. Flamerz, J. Nowotny, *Catal. Today* **1987**, *1*, 229.
61. G. J. Eon, E. Bordes, A. Vejux, P. Courtine, in *Proceedings 9th International Symposium on the Reactivity of Solids*, K. Dyrek, J. Haber, J. Nowotny (Eds.), Vol. 2, PWN, Warsaw, 1982, p. 603.
62. T. Machej, B. Doumain, B. Yasse, B. Delmon, *J. Chem. Soc., Faraday Trans. I* **1988**, *84*, 3905.
63. M. El Jamal, Ph.D. Thesis, Université Claude Bernard, Lyon, 1984.
64. M. El Jamal, M. Forissier, G. Coudurier, J. C. Védrine, in *Proceedings 9th International Congress on Catalysis*, M. J. Phillips, M. Ternan (Eds.), Vol. 4, The Chemical Institute of Canada, Ottawa, 1988, p. 1617.
65. P. Biloen, G. T. Pott, *J. Catal.* **1973**, *30*, 169.
66. A. Castellan, J. C. J. Bart, A. Vaghi, N. Giordano, *J. Catal.* **1976**, *42*, 162.
67. V. H. J. de Beer, M. J. M. van der Aalst, C. J. Machiels, G. C. A. Schuit, *J. Catal.* **1976**, *43*, 78.
68. S. Narayanan, K. Uma, *J. Chem. Soc., Faraday Trans. I* **1985**, *81*, 2733.
69. A. M. Alifanti, N. Blangenois, M. Florea, B. Delmon, *Appl. Catal. A* **2005**, *280*, 255.
70. F. Delannay, *Characterization of Heterogeneous Catalysts*, Marcel Dekker, New York, 1984, pp. 113–115.
71. T. Machej, M. Remy, P. Ruiz, B. Delmon, *J. Chem. Soc., Faraday Trans. I* **1990**, *86*, 715.
72. T. Machej, M. Remy, P. Ruiz, B. Delmon, *J. Chem. Soc., Faraday Trans. I* **1990**, *86*, 723.
73. T. Machej, M. Remy, P. Ruiz, B. Delmon, *J. Chem. Soc., Faraday Trans. I* **1990**, *86*, 731.
74. W. E. Garner, *Chemistry of Solids*, Butterworth, London, 1955.
75. J. S. Anderson, M. W. Roberts, F. S. Stone, *Reactivity of Solids*, Chapman & Hall, London, 1973.
76. J. Wood, O. Lindqvist, C. Helgesson, N. G. Vannerberg, *Reactivity of Solids*, Plenum Press, New York and London, 1977.
77. K. Dyrek, J. Haber, J. Nowotny, *Proceedings 9th International Symposium on the Reactivity of Solids*, Elsevier, Amsterdam, 1982.
78. P. Barret, L.-C. Dufour, *Reactivity of Solids*, Elsevier, Amsterdam, 1985.
79. K. Sudsakorn, J. G. Goodwin, Jr., A. A. Adeyga, *J. Catal.* **2003**, *213*, 204.
80. P. Barret, L.-C. Dufour, *Cinétique Hétérogène*, Gauthier-Villars, Paris, 1973.
81. J. Szekeley, J. W. Evans, H.-Y. Sohn, *Gas-Solid Reactions*, Academic Press, New York, 1976.
82. M. Montes, Ph.D. Thesis, Université Catholique de Louvain, 1984.
83. B. Delmon, M. Houalla, in *Preparation of Catalysts II*, B. Delmon, P. Grange, P. A. Jacobs, G. Poncelet (Eds.), Studies in Surface Science and Catalysis, Vol. 3, Elsevier, Amsterdam, 1978, p. 439.
84. M. Houalla, F. Delannay, I. Matsuura, B. Delmon, *J. Chem. Soc. I* **1980**, *7*, 2128.
85. J. W. E. Coenen, in *Preparation of Catalysts II*, B. Delmon, P. Grange, P. A. Jacobs, G. Poncelet (Eds.), *Studies in Surface Science and Catalysis*, Vol. 3, Elsevier, Amsterdam, 1978, p. 81.
86. A. Roman, B. Delmon, *J. Catal.* **1973**, *30*, 333.
87. M. P. Rosynek, C. A. Polansky, *Appl. Catal.* **1991**, *73*, 97.
88. M. V. M. de Souza, L. Clavé, V Dubois, C. A. C. Perez, M. Schmal, *Appl. Catal. A* **2004**, *272*, 133.
89. J. T. Richardson, R. M. Scates, M. V. Twigg, *Appl. Catal. A* **2004**, *267*, 35.
90. M. Montes, C. Penneman de Bosscheyde, B. K. Hodnett, F. Delannay, P. Grange, B. Delmon, *Appl. Catal.* **1984**, *12*, 309.
91. J. A. Moulijn, P. W. N. M. van Leeuwen, R. A. van Santen, *Catalysis*, Elsevier, Amsterdam, p. 355.
92. J. Masson, J. Nechtschein, *Bull. Soc. Chim. Fr.* **1968**, 3933.
93. E. I. Ko, J. M. Hupp, F. H. Rogan, N. J. Wagner, *J. Catal.* **1983**, *84*, 85.
94. L. A. M. Hermans, J. W. Geus, in *Preparation of Catalysts II*, B. Delmon, P. Grange, P. A. Jacobs, G. Poncelet (Eds.), *Studies in Surface Science and Catalysis*, Vol. 3, Elsevier, Amsterdam, 1978, p. 113.
95. M. Houalla, B. Delmon, in *Reactivity of Solids*, K. Dyrek, J. Haber, J. Nowotny, (Eds.), Elsevier, Amsterdam, 1982, p. 923.
96. M. Houalla, B. Delmon, *J. Phys. Chem.* **1980**, *84*, 2194.
97. M. Houalla, B. Delmon, *C.R. Acad. Sci. Paris, Sér. C* **1980**, *290*, 301.
98. M. Houalla, J. Lemaître, B. Delmon, *J. Chem. Soc., Faraday Trans. I* **1982**, *78*, 1389.
99. M. Houalla, B. Delmon, *C.R. Acad. Sci. Paris, Sér. C* **1979**, *289*, 77.
100. A. Lycourghiotis, C. Defossé, B. Delmon, *Rev. Chim. Minérale* **1979**, *16*, 473.
101. A. Lycourghiotis, C. Defossé, F. Delannay, J. Lemaître, B. Delmon, *J. Chem. Soc., Faraday Trans. I* **1980**, *76*, 1677.
102. A. Lycourghiotis, C. Defossé, F. Delannay, B. Delmon, *J. Chem. Soc., Faraday Trans. I* **1980**, *76*, 2052.
103. A. Lycourghiotis, C. Defossé, B. Delmon, *Bull. Soc. Chim. Belg.* **1980**, *89*, 929.
104. M. Houalla, F. Delannay, B. Delmon, in *Preprints 7th Canadian Symposium on Catalysis*, S. E. Wanke, S. E. Chakrabartty (Eds.), The Chemical Institute of Canada, Ottawa, 1980, p. 158.
105. M. Houalla, F. Delannay, B. Delmon, *J. Phys. Chem.* **1981**, *85*, 1704.
106. J. M. Porto López, S. Ceckiewicz, C. Defossé, P. Grange, B. Delmon, *Appl. Catal.* **1984**, *12*, 331.
107. F. M. Mulcahi, M. Houalla, D. M. Hercules, *J. Catal.* **1987**, *106*, 210.
108. C. Papadopoulou, A. Lycourghiotis, P. Grange, B. Delmon, *Appl. Catal.* **1988**, *38*, 255.
109. H. Matralis, A. Lycourghiotis, P. Grange, B. Delmon, *Appl. Catal.* **1988**, *38*, 273.
110. M. Houalla, B. Delmon, *Appl. Catal.* **1981**, *1*, 285.
111. M. Houalla, F. Delannay, B. Delmon, *J. Electron. Spectrosc. Relat. Phenom.* **1982**, *25*, 59.
112. H. Lafiteau, E. Néel, C. Clément, in *Preparation of Catalysts I*, B. Delmon, P. A. Jacobs, G. Poncelet (Eds.), *Studies in Surface Science and Catalysis*, Vol. 1, Elsevier, Amsterdam, 1976, p. 393.

113. J. P. R. Vissers, S. M. A. M. Bouwens, V. H. J. de Beer, R. Prins, in *Proceedings, Symposium on Fundamental Chemistry of Promoters and Poisons in Heterogeneous Catalysis*, American Chemical Society, New York, April 13–18, 1986, p. 227.
114. A. Stanislaus, M. Habsi-Halabi, K. Al-Dolama, *Appl. Catal.* **1988**, *39*, 239.
115. P. Atanasova, R. Halachev, *Appl. Catal.* **1988**, *38*, 235.
116. M. M. Ramírez de Agudelo, A. Morales, in *Proceedings 9th International Congress on Catalysis*, M. J. Phillips, M. Ternan (Eds.), Vol. 1, The Chemical Institute of Canada, Ottawa, 1988, p. 42.
117. S. Narayanan, K. Uma, *J. Chem. Soc., Faraday Trans. I* **1988**, *84*, 521.
118. H. Charcosset, B. Delmon, *Ind. Chim. Belg.* **1973**, *38*, 481.
119. W. Verhoeven, B. Delmon, *Bull. Soc. Chim. Fr.* **1966**, 3065.
120. W. Verhoeven, B. Delmon, *Bull. Soc. Chim. Fr.* **1966**, 3073.
121. E. J. Novak, R. M. Koros, *J. Catal.* **1967**, *7*, 50.
122. E. J. Novak, *J. Phys. Chem.* **1969**, *73*, 3790.
123. J. T. Richardson, R. J. Dubus, J. G. Crump, P. Desai, U. Osterwalder, T. S. Cale, in *Preparation of Catalysts II*, B. Delmon, P. Grange, P. A. Jacobs, G. Poncelet (Eds.), Studies in Surface Science and Catalysis, Vol. 3, Elsevier, Amsterdam, 1978, p. 131.
124. E. B. Doesburg, S. Orr, J. R. H. Ross, L. L. van Rijen, *J. Chem. Soc., Chem. Commun.* **1977**, *20*, 734.
125. H. Praliaud, G. A. Martin, *J. Catal.* **1981**, *72*, 394.
126. R. Frety, L. Tournayan, M. Primet, G. Bergeret, M. Guénin, J. B. Baumgartner, A. Borgna, *J. Chem. Soc., Faraday Trans. I* **1993**, *89*, 3313.
127. G. A. Martin, C. Mirodatos, H. Praliaud, *Appl. Catal.* **1981**, *1*, 367.
128. J. Cosyns, M.-T. Chénebaux, J. F. Le Page, R. Montarnal, in *Preparation of Catalysts I*, B. Delmon, P. A. Jacobs, G. Poncelet (Eds.), Studies in Surface Science and Catalysis, Vol. 1, Elsevier, Amsterdam, 1976, p. 459.
129. S.-H. Chen, H.-W. Huang, K.-L. Lu, *Bull. Inst. Chem. Acad. Sinica* **1985**, *32*, 9.
130. B. S. Clausen, H. Topsøe, S. Morup, *Appl. Catal.* **1989**, *48*, 327.
131. J. M. Dumas, C. Géron, K Kribii, J. Barbier, *Appl. Catal.* **1989**, *47*, L9.
132. L. M. Strubinger, G. L. Geoffroy, M. A. Vannice, *J. Catal.* **1985**, *96*, 72.
133. H. Zea, K. Lester, A. K. Datye, E. Rightor, R. Culotty, W. Waterman, M. Smith, *Appl. Catal. A* **2005**, *282*, 237.
134. D. Duprez, M. Mendez, in *Catalyst Deactivation*, B. Delmon, G. F. Froment (Eds.), Elsevier, Amsterdam, 1987, p. 525.
135. D. S. Thakur, M. G. Thomas, *Appl. Catal.* **1985**, *15*, 197.
136. S. Eijsbouts, *Appl. Catal. A* **1997**, *158*, 53.
137. S. Eijsbouts, *Appl. Catal. A* **1997**, *158*, 69.
138. H. Hallie, *Oil Gas J.* **1982**, December 20, 6.
139. A. Arteaga, J. L. G. Fierro, F. Delannay, B. Delmon, *Appl. Catal.* **1986**, *26*, 227.
140. A. Arteaga, J. L. G. Fierro, P. Grange, B. Delmon, in *Catalyst Deactivation*, B. Delmon, G. F. Froment (Eds.), Studies in Surface Science and Catalysis, Vol. 34, Elsevier, Amsterdam, 1987, p. 245.
141. A. Arteaga, J. L. G. Fierro, P. Grange, B. Delmon, *Appl. Catal.* **1987**, *34*, 89.
142. B. Delmon, in *Catalysts in Petroleum Refining*, D. L. Trimm, S. Akashah, M. Absi-Halabi, A. Bishara (Eds.), Elsevier, Amsterdam, 1990, p. 1.
143. B. Delmon, *Latin American Appl. Res.* **1995–1996**, *26*, 87.
144. B. Delmon, *Bull. Soc. Chim. Belg.* **1995**, *104*, 173.
145. K. Inamura, R. Prins, in *Science and Technology in Catalysis 1994*, Y. Izumi, H. Arai, M. Iwamoto (Eds.), Studies in Surface Science and Catalysis, Vol. 92, Elsevier, Amsterdam, 1994, p. 111.
146. D.-H. Zuo, M. Vrinat, H. Nie, F. Maugé, Y.-H. Shi, M. Lacroix, D.-D. Li, *Catal. Today* **2004**, *93–95*, 751.
147. M. W. J. Crajé, V. H. J. de Beer, J. A. R. van Veen, A. M. van der Kraan, *J. Catal.* **1993**, *143*, 601.
148. R. Prada Silvy, P. Grange, B. Delmon, in *Catalysts in Petroleum Refining*, D. L. Trimm, S. Akashah, M. Absi-Halabi A. Bishara (Eds.), Elsevier, Amsterdam, 1990, p. 1233.
149. J. Ladrière, R. Prada Silvy, *Hyperfine Interactions* **1988**, *41*, 653.
150. R. Prada Silvy, Ph.D. Thesis, Université catholique de Louvain, 1987.
151. B. Delmon, in *Reactivity of Solids*, K. Dyrek, J. Haber, J. Nowotny (Eds.), Elsevier Amsterdam, 1982, p. 327.
152. J. M. Zabala, P. Grange, B. Delmon, *C. R. Acad. Sci. Paris, Sér. C* **1974**, *279*, 725.
153. H. Zahradnikova, V. Kamik, L. Beránek, *Coll. Czech. Chem. Commun.* **1985**, *50*, 1573.
154. J. M. Zabala, P. Grange, B. Delmon, *C. R. Acad. Sci. Paris, Sér. C* **1974**, *279*, 563.
155. J. M. Zabala, P. Grange, B. Delmon, *C. R. Acad. Sci. Paris, Sér. C* **1975**, *280*, 1129.
156. R. Prada Silvy, J. L. G. Fierro, P. Grange, B. Delmon, in *Preparation of Catalysts IV, Scientific Bases for the Preparation of Heterogeneous Catalysts*, B. Delmon, P. Grange, P. A. Jacobs, G. Poncelet (Eds.), Studies in Surface Science and Catalysis, Vol. 31, Elsevier, Amsterdam, 1987, p. 605.
157. A. Steinbrunn, C. Lattaud, *Surf. Sci.* **1985**, *155*, 279.
158. D. K. Lambiev, T. T. Tomova, G. Samsonov, *Powder Metall. Intern.* **1972**, *4*, 17.
159. J. Masson, B. Delmon, J. Nechtschein, *C. R. Acad. Sci. Paris, Sér. C* **1968**, *266*, 428.
160. A. Steinbrunn, J. C. Colson, C. Lattaud, C. G. Gachet, L. de Mourgues, M. Vrinat, J. P. Bonnelle, in *Reactivity of Solids*, Elsevier, Amsterdam, 1985, p. 1079.
161. A. Steinbrunn, C. Lattaud, H. Reteno, J. C. Colson, in *Physical Chemistry of the Solid State: Applications in Metals and their Compounds*, P. Lacombe (Ed.), Elsevier, Amsterdam, p. 551.
162. G. L. Schrader, C. P. Cheng, *J. Catal.* **1984**, *85*, 488.
163. P. A. Spevak, N. S. McIntyre, *J. Phys. Chem.* **1993**, *97*, 11031.
164. A. Steinbrunn, M. Bordignon, *Bull. Soc. Chim. Belg.* **1987**, *96*, 941.
165. R. Thomas, E. M. van Oers, V. H. J. de Boer, J. A. Moulijn, *J. Catal.* **1983**, *84*, 275.
166. B. Scheffer, P. Arnoldy, J. A. Moulijn, *J. Catal.* **1988**, *112*, 516.
167. B. Scheffer, N. J. J. Dekker, P. J. Mangnus, J. A. Moulijn, *J. Catal.* **1990**, *121*, 31.
168. T. I. Korányi, I. Manninger, Z. Paál, O. Marks, J. R. Günter, *J. Catal.* **1989**, *116*, 422.
169. D. Pirotte, P. Grange, B. Delmon, in *Reactivity of Solids*, K. Dyrek, J. Haber, J. Nowotny (Eds.), Elsevier, Amsterdam, 1982, p. 973.
170. R. Prada Silvy, F. Delannay, B. Delmon, *Indian J. Technol.* **1987**, *25*, 627.
171. R. Prada Silvy, J. M. Beuken, J. L. J. Fierro, P. Bertrand, B. Delmon, *Surf. Interf. Anal.* **1986**, *8*, 167; R. Prada Silvy, J. M. Beuken, J. L. J. Fierro, P. Bertrand, B. Delmon, *Surf. Interf. Anal.* **1986**, *9*, 247.
172. V. Stuchly, K. Klusacek, *Appl. Catal.* **1987**, *34*, 263.
173. S.-I. Kim, S.-I. Woo, *Appl. Catal.* **1991**, *74*, 109.

174. P. J. Mangnus, E. K. Poels, J. A. Moulijn, *Ind. Eng. Chem. Res.* **1993**, *32*, 1818.
175. R. Prada Silvy, P. Grange, F. Delannay, B. Delmon, *Appl. Catal.* **1989**, *46*, 113.
176. T. I. Korányi, I. Manninger, Z. Paál, *Solid State Ionics* **1989**, *32–33*, 1012.
177. M. W. J. Crajé, V. H. J. de Beer, J. A. R. van Veen, A. M. van der Kraan, *J. Catal.* **1993**, *143*, 601.
178. P. J. Mangnus, A. Bos, J. A. Moulijn, *J. Catal.* **1994**, *146*, 437.
179. T. I. Korányi, M. Schikorra, Z. Paál, R. Schlögl, J. Schütze, M. Wesemann, *Appl. Surf. Sci.* **1993**, *68*, 307.
180. R. Badilla-Ohlbaum, D. Chadwick, in *Proceedings 7th International Congress on Catalysis, Tokyo*, T. Seiyama, K. Tanabe (Eds.), Vol. 2, Kodansha and Elsevier, Tokyo and Amsterdam, 1981, p. 1126.
181. J. Abart, E. Delgado, G. Ertl, H. Jeziorowski, H. Knözinger, N. Thiele, X.-Zh. Wang, E. Taglauer, *Appl. Catal.* **1982**, *2*, 155.
182. F. Maugé, J. C. Duchet, J. C. Lavalley, S. Houssenbay, E. Payen, J. Grimblot, S. Kasztelan, *Catal. Today* **1991**, *10*, 561.
183. V. L. S. Teixeira da Silva, R. Frety, M. Schmal, *Ind. Eng. Chem. Res.* **1994**, *33*, 1692.
184. M. Breysse, M. Cattenot, T. Decamp, R. Frety, C. Gachet, M. Lacroix, C. Leclercq, L. de Mourgues, J.-L. Portefaix, M. Vrinat, M. Houari, J. Grimblot, S. Kasztelan, J.-P. Bonnelle, S. Housni, J. Bachelier, J. C. Duchet, *Catal. Today* **1988**, *4*, 3.
185. W. L. T. M. Ramselaar, M. W. J. Crajé, E. Gerkema, V. H. J. de Beer, A. M. van der Kraan, *Bull. Soc. Chim. Belg.* **1987**, *96*, 931.
186. G. Marroquín, J. Ancheyta, J. A. I. Díaz, *Catal. Today* **2004**, *98*, 75.
187. S. R. Murff, E. A. Karlisle, P. Dufresne, H. Rabehasaina, *Symposium on Regeneration, Reactivation and Reworking of Spent Catalysts*, Div. Petrol. Chem. ACS, Denver, Preprints, ACS 1993, Vol. 34, p. 81.
188. J. van Gestel, J. Léglise, J.-C. Duchet, *J. Catal.* **1994**, *145*, 429.
189. R. Lebreton, S. Brunet, G. Pérot, V. Harlé, S. Kasztelan in *Catalyst Deactivation 1999*, B. Delmon, G. F. Froment (Eds.), Elsevier, Amsterdam, 1999, p. 195.
190. P. Dufresne, N. Brahma, F. Labruyère, M. Lacroix, M. Breysse, *Catal. Today* **1996**, *29*, 251.
191. S. Texier, G. Berhault, G. Pérot, V. Harlé, F. Diehl, *J. Catal.* **2004**, *223*, 404.
192. C. Glasson, C. Geantet, M. Lacroix, F. Labruyère. P. Dufresne, *J. Catal.* **2002**, *212*, 76.
193. G. J. Hutchings, A. Desmartin-Chomel, R. Olier, J. C. Volta, *Nature (Lett.)* **1994**, *368*, 41.
194. K. Aït-Lachgar-Ben Abdelouahad, M. Roullet, M. Brun, A. Burrows, C. J. Kiely, J. C. Volta, M. Abon, *Appl. Catal. A* **2001**, *210*, 235.
195. K. C. Waugh, Y.-H. Taufiq-Yap, *Catal. Today* **2003**, *81*, 215.
196. A. Pantazidis, A. Burrows, C. J. Kiely, C. Mirodatos, *J. Catal.* **1998**, *177*, 235.
197. E. M. Gaigneaux, M. J. Genet, P. Ruiz, B. Delmon, *J. Phys. Chem. B* **2000**, *104*, 5724.
198. A. Brückner, A. Martin. N. Steinfeldt, G.-U. Wolf, B. Lücke, *J. Chem. Soc., Faraday Trans.* **1996**, *92*, 4257.
199. A. Brückner, A. Martin, B. Kubias, B. Lücke, *J. Chem. Soc., Faraday Trans.* **1998**, *94*, 2221.
200. A. Brückner, *Appl. Catal.* **2000**, *200*, 287.
201. A. Brückner, *Catal. Rev.* **2003**, *45*, 97.
202. E. K. Novakova, J. C. Védrine, E. G. Derouane, *J. Catal.* **2002**, *211*, 235.
203. W. Ueda, D. Vitry, T. Katou, *Catal. Today* **2004**, *96*, 235.
204. M. O. Guerrero-Pérez, M. A. Bañares, *Catal. Today* **2004**, *96*, 265.
205. L. Marosi, G. Cox, A. Tenten, H. Hibst, *J. Catal.* **2000**, *194*, 140.
206. L. Marosi, C. Otero Areán, *J. Catal.* **2003**, *213*, 235.

2.5.2
Catalyst Forming

Ferdi Schüth* and Michael Hesse

2.5.2.1 Introduction

Catalyst forming processes are, in most cases, not important in academic catalysis research. In academic studies, it is mostly pressed and subsequently split catalyst with the correct particle size fraction that is used to ensure proper fluid dynamics in the reactors, but otherwise shaping of the catalyst is not considered. However, when a process is scaled up to pilot or commercial scale, the form in which a catalyst is used is decisive for the ultimate performance of the unit. In general, the size and shape of the catalyst pieces is a compromise between the wish to minimize pore diffusion effects in the catalyst particles, for which small particles are necessary, and the pressure drop across the reactor, which requires large particle sizes [1]. In addition to balancing these demands, particle shaping has to serve other needs: The particles must have sufficient strength and abrasion resistance to survive transport, filling of the reactor and use in operation, while reactor charging and discharging must be easy, dust build up must be minimized and, if it occurs, it must proceed in a uniform manner. Moreover, shaping needs to provide reasonably cheap routes to the final catalyst [2–4]. The typical synthetic procedures for the catalytic materials themselves do not lead to particle shapes which meet the requirements of the process, and thus shaping operations are necessary to produce the catalyst in a form in which it can be charged to the reactor. As the shaping methods are often proprietary – and also more often an art than a science – information on exact procedures is rarely disclosed, and patents typically provide a range of processing conditions while rarely disclosing details of the most successful formulation. A general treatment of this topic, which also includes actual recipes for the synthesis of catalysts, can be found in the excellent book by Stiles and Koch [5].

Several different forming methods are known, the most important being spray drying, extrusion, pelleting (also called pilling or tableting), and granulation (also called

---
* Corresponding author.

spherudizing). These will be discussed in more detail in this chapter. A survey of the different shapes, sizes and areas of application for these catalysts is provided in Table 1. It should be borne in mind however that, when utilizing this table, the sizes at the fringes of the ranges may pose special problems. For example, with 50-mm extrudates drying and calcination without producing cracks can be a major challenge. In the following sections, some physico-chemical aspects will be described before the procedures for shaping of powders are discussed in more detail. This discussion follows to some extent the treatment given in the *Handbook of Porous Solids*, which covers the basics of creating porous materials by compaction [6]. Useful information on the physico-chemical aspects of particle agglomeration can also be found in Ref. [7].

2.5.2.2 The Physico-Chemical Background of Forming Processes

During a forming process in the manufacture of catalysts, the properties of the active phase should be maintained as well as possible, together with the primary porosity which exists on the level of the primary powder. In addition to this primary porosity, compaction of the powder creates additional porosity which can be very important for mass transfer. The range of pore sizes which can be achieved by compaction of the primary powder is determined by the size of the primary particles and of the aggregates or agglomerates to be compacted and their particle size distribution. The porogens, which are included in the formulation and burnt out in the final calcination stage, may also have an adverse effect. In order to obtain a rough estimate of the porosity which can be expected, the void volumes obtained by using different ways to pack the particles are briefly discussed in the following. Experimental investigations of this topic, together with modeling to predict the packing of particles are active fields of research (for a review, see Ref. [8]). Although such studies are often aimed at the production of materials with as little porosity as possible (as are needed in powder metallurgy or ceramics), a number of studies has been carried out in which the creation of porous solids has been investigated. Most models were developed for spherical particles of identical or differing sizes (e.g., Refs. [9–13]), but some models also considered irregularly shaped particles (e.g., Ref. [14]). For equally sized particles with a close-to-spherical shape and dense packing, the space filling is 74%, tetrahedral voids can accommodate spheres with a radius which is 0.225 times that of the constituting spheres, and octahedral holes those with a radius which is 0.414 times that of the constituting spheres [15].

Random close packings of identically sized spheres have substantially lower densities (63.7% [16, 29]). The packing of equally sized particles will lead to a narrow pore size distribution, but if the particle size distribution is broad then a broadened pore size distribution results, although this depends crucially on the particle size distribution. It has been shown that, for Gaussian-type particle size distributions, Gaussian-type pore size distributions will result. However, the void sizes are typically not proportional to the particle sizes and more complex dependencies exist [12]. For the distributions investigated in Ref. [12] the ratio between mean void size and mean particle size was around 0.25.

In the forming of catalysts, tests of the compaction of powders indicate that it is desirable for good compaction behavior to have powders with mixed particle size [17]; that is, in a coarse powder a substantial fraction of fines is normally beneficial. The fines fill the voids and tend to agglomerate and strengthen the larger particles in the final compression operations. However, there are cases (e.g., with copper chromite, nickel chromite, and some molybdates) where excessively fine powder will cause severe pilling problems. Even if a distribution of powder particle sizes may be desirable, it could be difficult to mix particles of all sizes in a random fashion, as the smaller particles will always tend to move to the bottom of a container while the larger particles tend to accumulate at the top. Attempts have been made to model the packing of particles under the influence of various forces to create different types of packing (e.g., Refs. [9, 18, 19]), and the degree of "randomness" of these packings has been investigated on a theoretical basis [20]. The effect of "shaking" was studied previously in a model [9], although a satisfactory and general solution appears still to be lacking. The presence of a liquid, which binds particles together via capillarity, partly solves the problem of size fractionation, as the capillary forces will counterbalance gravity.

Tab. 1 Different types of catalyst shape and the reactor types in which they are used. (After Ref. [4])

Shape	Size[a]	Type of reactor
Extrudate	$d = 1$–50 mm $l = 3$–30 mm	Fixed bed
Pellet	$d = 3$–15 mm $l = 3$–15 mm	Fixed bed
Granule, bead	$d = 1$–20 mm	Fixed bed
Sphere	$d = 1$–10 mm	Fixed bed Moving bed
Microspheroidal	$d = 20$–100 μm	Fluidized bed Slurry

[a] d = diameter; l = length.

References see page 698

Crucial to the success of a forming method are the forces between the primary particles that are compacted. These forces are different in the forming stage itself and in the final shaped body. Whilst in the forming stage and green body the liquid bridges and capillarity are the most important mechanisms (as they bind the powder particles together), in the final shaped body – which typically is obtained after a calcination step – it is the solid bridges and chemical bonding between primary particles that are the predominant factors.

The mechanisms of cohesion between loose particles are summarized in Table 2. Typically, it is not only one force which controls the adhesion between particles, but rather several forces which act simultaneously to provide the strength of the formed particle. In the simplest case, the particles can be compacted simply by exerting pressure, for example in a pellet press or in a tableting machine. However, in some cases covalent bonds might then form between particles during such processes. Silica particles, for instance, have many hydroxyl groups on their external surface (four to six per nm^2), and these can condense to form a siloxane bridge at the contact point by eliminating water (Fig. 1a). If several such bonds can be formed, this may impart substantial strength to the formed body. Likewise, if particles flow plastically under pressure, then the individual particles may also weld at the contact area during the deformation. Another important contribution can be derived from liquid bridges between particles; these may even be formed in dry

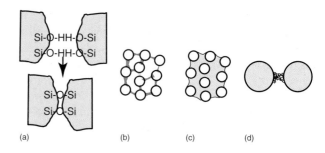

Fig. 1 Mechanisms of agglomeration of solid particles. (a) Chemical bonds for the example of a siloxane bond; (b) liquid bridges; (c) complete filling of granule volume resulting in strong binding due to capillary forces; (d) solid bridges formed by the crystallization of salt from evaporating pore liquid. (From Ref. [6].)

powders, as capillary condensation can lead to water condensation from the atmosphere, if particles approach each other closely. The forces caused by liquid bridges may be substantial, and although analytical solutions are available for special cases under certain assumptions [21], it seems better to solve numerically the Laplace–Young equation which describes the problem [22]. The forces are even stronger if not only bridges are formed but also the void space between particles is almost completely filled with liquid due to capillary forces (Fig. 1b,c). At a liquid filling of about 90% of the total pore volume, the tensile strength is at a maximum – a situation which can be demonstrated experimentally and also predicted in theoretical terms [23].

Liquid bridges or capillary filling may serve as the starting point for another adhesion mechanism, the formation of solid bridges (Fig. 1d). If salts are dissolved in the liquid between the particles, these will crystallize upon evaporation of the solvent. As the solvent retracts to the bridges between particles during evaporation, and thus the solution there becomes increasingly concentrated, there are localized spots where the salts precipitate and this will lead to the formation of a solid bridge between two adjacent particles. The strength of these connections is difficult to predict, as normally the size of the individual salt particles and their intergrowth (which will determine the overall strength of the bridge) is not known. In the formation of catalysts by extrusion, additives such as nitric or hydrochloric acid are often used to serve for peptization, but this may also lead to a minor dissolution of the powder. During the subsequent drying and calcination steps the dissolved salt may be reprecipitated and serve as binder, thereby creating solid bridges.

Compared with the other contributions to the strength of a compacted particle, the two weakest forces are dispersion forces (van der Waals forces) and electrostatic forces. However, even these can become very important in certain particle size ranges. For example, in the case of very

Tab. 2 Mechanisms for particle agglomeration and relative strength of the forces. Liquid bridges can even be possible in "dry" powders due to capillary condensation in the necks between small particles

Mechanism	Strength	Comment
Van der Waals forces	Medium at short distances, rapid decay with distance	Magnitude depends on the interaction potentials
Electrostatic forces	Weak, but dominant for distances approaching μm	Can be repulsive, if charging of particles with same sign occurs; different for conductors and insulators
Liquid bridges	Strong	
Capillary forces	Very strong	Full saturation of granule with liquid, situation (c) in Fig. 1c
Solid bridges	Variable	Depends very much on conditions of solvent evaporation and crystallizing solid in bridge
Covalent bonds	Very strong	

small particles the van der Waals forces can contribute substantially to the strength of the agglomerate, because as the particles approach each other closely there will be many points where the inter-particle distances are small enough for the dispersion forces to manifest themselves. Coulombic interactions, in contrast, are insignificant for small particles, because they typically do not carry high charges, although they may become significant for larger particles exceeding micrometer dimensions. In some cases, however, electrostatic forces rather have an adverse effect, as particles of the same type tend to charge identically; consequently, repulsive forces occur that may make the handling of the particles difficult.

The relative contributions of the different forces between particles are combined in Fig. 2 for two simple model systems, one of which also considers the effect of roughness, which is important for the liquid bridges and the van der Waals interaction. As can be seen, for short distances, liquid bridges are typically the most significant forces between particles.

The strong forces exerted by the liquids between small particles are the main reason why mostly wet powders are used in compaction processes. In addition to creating a high initial attraction between particles, the liquid can also facilitate the formation of solid bridges, as discussed above.

If the particles do not form stable bodies in a shaping process, then binders are added. These may be oxide powders (such as alumina, silica, steatite, cements, or liquids) which form chemical bonds between the particles to be shaped. It is important that the pellets are not heated to elevated temperatures after shaping, as any liquid binders (which mostly are polymers, such as polyethylene oxide) would be removed during heating. The binders (and often also lubricants) are added before the compaction process is started in a kneader in order to provide intimate mixing between the ingredients. The types of additive used are discussed in more detail with the specific methods used to form the particles.

2.5.2.3 Spray Drying

Spray drying is a forming technology which can serve both for the production of the final shapes of the catalyst and for the creation of suitable powders which are then further

References see page 698

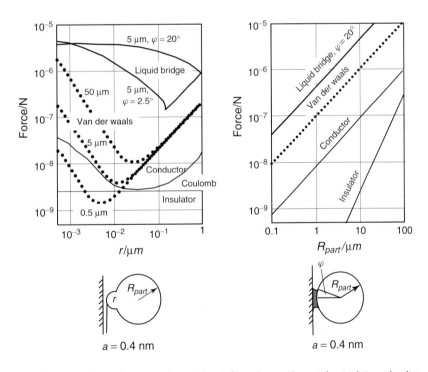

Fig. 2 Dependence of forces between a flat wall and rough particles (left) and smooth particles (right) on the distance from the wall. The different contributions to the total force are plotted separately. Bold solid line: liquid bridges; dotted lines: van der Waals attraction; fine solid lines: Coulomb interactions for insulator or conductor. The left-hand graph is plotted as dependence on the radius of the surface roughness, R_{part} (radius of the whole particle) is given as a parameter. For liquid bridges, two bridge angles are given as parameters. For $\varphi = 20°$ the roughness is submerged in the liquid. For small angles, the bridge to the large sphere is broken from a certain roughness (discontinuity at ca. 100 nm roughness). For van der Waals interaction, the Lifshitz–van der Waals constant is fixed at 5 eV. For Coulomb interactions between conductors the contact potential is assumed to be 0.5 V, for insulators the surface charge density is assumed to be $10^{-2} e\ \mu m^{-2}$, with e being the elementary charge. The right-hand graph side is plotted for an ideally smooth sphere. (From Ref. [6].)

processed by another shaping method. Spray drying, as a method for shaping the final product, is important in the synthesis of catalysts used for fluid catalytic cracking (FCC) units; although these catalysts can be prepared via several different routes, they all use a spray-drying step in their shaping. A correctly operated spray-drying process results in the formation of particles that range in size from 20 to 100 μm, and these are ideal for use in fluidized-bed operation. Some basic considerations for spray dryers are compiled in Ref. [24].

2.5.2.3.1 **The Spray-Drying Process** Spray drying has been labeled as more of an art than a science [25], and this is certainly true to some extent. The spray drying of industrial catalysts is difficult to emulate on the laboratory scale, as bench-top spray dryers are limited with respect to the particle sizes which can be produced; typically, the use of a bench-top spray dryer will result in spherulites of less than 10 μm diameter. Over the years a substantial degree of empirical and modeling knowledge has been accumulated, and today the processes may be reasonably well designed. Unfortunately, although much knowledge has been acquired – especially for the spray drying of zeolites – it is not widely published in the open, nor even in the patent literature.

Spray drying typically starts with slurries containing small particles of the catalyst to be formed, possibly together with matrix components which bind the catalyst particles and impart the attrition resistance necessary for their use in fluidized beds. A slurry, with a solids content of 5 to 50 wt.% (with the preferred range being 10–30 wt.%) is sprayed through a nozzle into a chamber into which hot gas is flowing, mostly cocurrently, with the spray (Fig. 3). As the spray falls downwards in the dryer, the solvent evaporates and microspheres are formed. Depending on the temperature of the gas phase, only drying will occur in the spray dryer, although at high gas temperatures the solid can be simultaneously calcined. Since the dimensions of the spray dryer are among the crucial factors that determine the size of the final particles, it is clear that a bench-scale spray dryer with a diameter of some 10 cm cannot produce the same particle sizes as can an industrial-scale system with a diameter of several meters and a height of some ten of meters. Fines are always produced to some extent, but these can be separated in cyclones and recycled for the slurry preparation.

The particle sizes for the slurries are typically in the range of some hundred nanometers, and these particles are obtained, for example, by milling the filter cake of the previously produced catalyst. Some reports have been published where sols with primary particle sizes of only some tens of nanometers are shaped by spray drying [26], thus combining spray drying with a sol–gel process. Generally, materials which are either gel-forming or film-forming can easily be spray dried to form a satisfactory product. If the material to be shaped does not easily form stable microspheres in a spray-drying process, then the catalytic material can be included in a matrix composed of a gel- or film-forming precursor which is added to the slurry to be sprayed. Such matrices (e.g., silica or

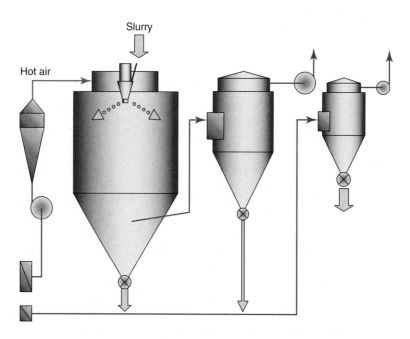

Fig. 3 Schematic drawing of a spray dryer with rotary atomizer and cocurrent air flow. (Illustration courtesy of Niro A/S.)

alumina) are obtained from the corresponding sols. In order to create highly attrition-resistant particles, it may be advantageous to include a small abrasion-resistant species (subcolloidal particles preferably of <5 nm diameter) into the slurry. These move to the outer shell of the droplets with the evaporating liquid in the initial drying stage, where they form a hard coating (shell) that will protect the active phase against attrition. An example of this was when 10% polysilicic acid with particle sizes of 3 to 4 nm was added to the slurry of a vanadium phosphorous oxide (VPO) catalyst for butane oxidation. When sprayed, the VPO microspheres were coated with abrasion-resistant silica (Fig. 4) [27].

The drying process in spray dryers can qualitatively be described as follows [28]. As soon as the droplet with many finely dispersed solid particles leaves the atomizer, it contacts the hot air in the drying tower, whereupon evaporation from the droplet surface is rapid. The solid material is left behind and forms a skin or crust on its surface. Due to the heat of evaporation, the droplet stays relatively cool during this stage. In the next, slower stage the liquid from the interior of the particle must diffuse outwards through the surface skin/crust, which simultaneously grows towards the interior. As the droplets do not rotate, one side of the forming particle is always exposed to the hot gas stream, and on this side the evaporation is faster and hence the skin/crust grows thicker. On the rearwards side the particle stays softer and, due to the evaporation of liquid on the hot side, this section is pulled inwards, often creating a dimpled morphology of spray-dried particles. One undesirable effect which may occur is that of "ballooning". Here, if the initially formed skin is relatively strong and dense, the liquid may evaporate in the center of the forming particle and the pressure may cause the particle to expand. In the extreme case, the particle may even explode, leading to the formation of small fragments.

The density of the particle produced can be adjusted by a number of different measures. Normally, the bulk density increases with increasing solids content of the slurry, fewer solubles, and slower drying through lower gas temperatures [28].

2.5.2.3.2 Atomizers One of the most crucial points in a spray dryer is the nozzle (also called the atomizer), which generates the spray. This part of the spray dryer creates a fine mist of the slurry when it enters the main chamber. Three major types of atomizer are used in spray dryers for the production of catalysts: rotary atomizers; one-fluid (pressure or hydraulic) nozzles; and two-fluid (pneumatic) atomizers [28–31]. Design equations for the different atomizers can be found in Ref. [30]. All three types have their specific advantages and disadvantages. The various types of atomizer are shown, schematically, in Fig. 5.

A Rotary Atomizers A rotary atomizer employs a wheel which rotates at high speed (up to 20 000 revolutions min^{-1}) to spray the slurry mechanically. The slurry is fed through the center into the hollow wheel, which is perforated around its circumference. The centrifugal force projects the liquid through the perforations, and as the liquid leaves the orifices it is sprayed into droplets; the drying process then starts. As the rotary atomizer is rotating horizontally in the drying tower, the droplets are sprayed towards the walls of the drying tower. It is crucial

References see page 698

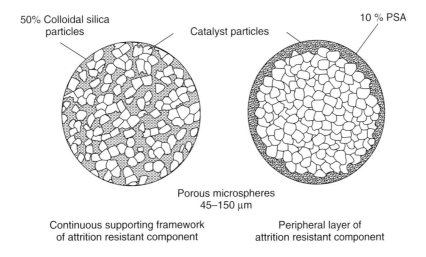

Fig. 4 Schematic drawing of two different concepts to confer attrition resistance to porous microspheres produced by spray drying. (Reproduced with permission from Ref. [27].)

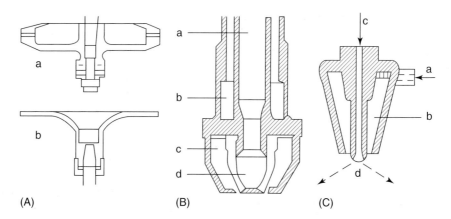

Fig. 5 Devices for spraying. (A) Disk atomizer: (a) suspended installation; (b) standing installation. (B) Pressure nozzle: (a) fluid delivery; (b) heating or cooling jacket; (c) nozzle support; (d) nozzle orifice. (C) Two-fluid nozzle: (a) tangential high-pressure air supply; (b) conical expansion and swirl chamber; (c) liquid supply line; (d) spray zone. (From Ref. [30].)

that the drops are at least superficially dried before they hit the wall. Thus, rotary atomizers cannot be used in bench-scale spray dryers but only in production plants. The droplet diameter in dependence of the chamber diameter for typical slurries used to produce FCC catalysts is shown in Fig. 6 [29]. Rotary atomizers produce rather narrow particle size distributions, do not require high-pressure pumps (unless very thick pastes are sprayed), do not plug easily, can handle relatively abrasive slurries, and can achieve very high capacities. The disadvantages are the mechanically more demanding design compared to the other nozzle types, and the size limitations of the produced spheres, which relates to the diameter of the drying chamber.

B Hydraulic Nozzles Hydraulic nozzles convert hydraulic pressure, at which the fluid enters the nozzle, into velocity. Different designs of pressure nozzle are available, which are distinguished by the inlet and outlet geometries of the nozzles [31]. *Sheet nozzles*, which are most often employed in spray dryers, first produce a conical liquid film which then breaks up into individual droplets. The thinner the conical film, the finer the mist created. Pressure nozzles are mechanically relatively simple and do not have moving parts. However, they require a high-pressure pump for feeding the slurry. They also are more easily plugged and cannot handle highly abrasive media. Although they have a limited throughput, many nozzles can be simultaneously operated in one drying chamber. As the mist is predominantly vertically directed, larger particles can be produced, since the drying can proceed over the full height of a drying tower of up to several tens of meters in height. However, operational flexibility is limited: if the throughput is to be increased, the pressure must be increased according to

$$\Delta p \propto \dot{V}^2 \qquad (1)$$

where Δp is the pressure difference and $\dot{V}$ the volume flow. However, this leads simultaneously to a change in droplet size and thus in particle size according to

$$d \propto \dot{V}^{-2/3} \qquad (2)$$

where d is the Sauter mean diameter [31]. Thus, if throughput is to be changed in a spray dryer with a pressure nozzle, typically the nozzles must be changed, although with well-designed nozzles some degree of control over product quality is possible even with pressure nozzles [32].

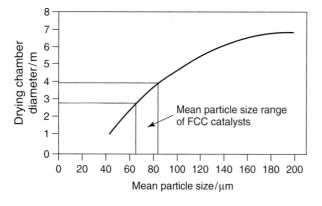

Fig. 6 Dependence of particle size on drying chamber diameter for rotary atomizers. (Reproduced with permission from Ref. [29].)

C Two-fluid Atomizers In two-fluid nozzles, the spraying of the fluid is achieved by means of a spraying gas which provides the atomizing energy. Such atomizers are relatively simple in their operation, and with the additional parameter of air flow they have more operation parameters, allowing a wider variation in drop size with one nozzle. In addition, as the spray leaves the nozzle at relatively low velocity, the drying tower can be built rather compactly. With regards to disadvantages, the consumption of compressed air is relatively high (up to the same consumption on a weight basis as the slurry), the spraying air must be heated by the drying air (thus needing additional energy), and the particle size distribution produced in two-fluid nozzles is fairly wide. Two-fluid nozzles are also typically limited to small plants.

2.5.2.3.3 Dryer Configuration The drying process in the drying chamber is also influenced by the dryer configuration, namely cocurrent, countercurrent, or mixed flow (fountain nozzle) operation. The hot air in the drying chamber must be distributed in such a way that the flow is homogeneous over the cross-section of the chamber. Eddy currents, recirculations and short-circuiting must be avoided. This is achieved by means of an air distributor, the design of which is normally considered proprietary information by spray-dryer manufacturers. The temperature profiles to which the particles are exposed are also very important for product quality. Excessive temperatures for long periods of time should be avoided in order to prevent deterioration of product quality. In this respect, it should be borne in mind that a spray-drying chamber is a moist environment, and materials which are hydrothermally unstable may suffer from partial degradation, if temperatures are too high. The most gentle temperature profiles are generally achieved with cocurrent air flow. This is due to the fact that, in this configuration, the hottest air is contacted with the droplets at a stage when the majority of the water is still present. Evaporative cooling prevents excessive temperatures in the droplets. However, as the droplets and forming particles are entrained by the gas, the chambers tend to be larger with cocurrent flow. If it is important that the chamber is compact, the so-called "mixed flow" or "fountain nozzle" arrangement is chosen. Here, the hot gas enters the chamber from the top and the nozzle is placed near the bottom, with the spray directed upwards. Thus, the flow is countercurrent initially, when the spray leaves the nozzle, but turns to cocurrent in later stages of the drying process. In order to produce a stable spray, a minimum cone angle at the nozzle is needed, which requires a minimum diameter of about 2.5 m for such flow configurations.

The residence time of gas in the chamber is another important parameter which determines the quality of the product. This parameter normally must be determined in test runs, with initial values chosen from experience; the time is typically on the order of seconds.

2.5.2.4 Extrusion

In extrusion processes a paste of the particles, typically with several additives, is pressed through a die. This may result in the formation of a variety of different shapes, in the simplest case cylinders, but rings, trilobates, stars, star-rings, or monolithic honeycombs can also be produced by using a suitable die. After leaving the die, the extrudate is either left to break, resulting in inhomogeneously sized extrudates, or it is cut off (e.g., by a rotating blade), which leads to extrudates which are all approximately identical. The extrusion method has several advantages, including possible high throughput, relatively low cost, a variety of possible extrudate shapes, and the possibility to work porogens into the paste which later can be removed to create additional porosity. However, extrudates are normally less uniform and less resistant to abrasion compared to tableted material [33]. The most formidable extruded material is probably the cordierite honeycomb used as the catalyst support structure in automotive catalysts (see Chapter 11.2 of this Handbook). This is extruded on a scale of millions of honeycombs per year with channel diameters of only 1 mm and a wall thickness of 150 μm over lengths of several tens of centimeters for the standard monolith having about 62 cells per cm^2, as used by most car manufacturers. The most advanced commercial systems have 140 cells per cm^2 and a wall thickness of only about 75 μm [34].

After extrusion, the obtained product must be dried to remove water from the green body, and typically also calcined to impart the final mechanical properties. While this step is less critical for small structure sizes (which will have short diffusion pathways for the water), it can prove to be a highly sophisticated process for larger structures present in the extruded titania-based monoliths used in selective catalytic reduction (SCR) processes in power plants.

An excellent survey of the basic principles of extrusion is provided by Benbow and Bridgwater [35].

2.5.2.4.1 Basic Considerations Many parameters govern an extrusion process. Among the most important is the flow behavior of the paste: if the paste is too viscous it can block the extruder, whereas if it is too thin the extrudates will be unstable. In order to better understand the basics of the extrusion process, some information regarding the rheology of pastes (i.e., liquids with a high solids content) may be helpful. The simplest behavior which a fluid can have is that of a Newtonian fluid, where the shear stress τ is directly proportional to the strain tensor e, with the viscosity as the proportionality constant. For a simple flow pattern, such as the shearing motion

References see page 698

between two parallel plates, which is often used for characterizing rheological behavior, the strain tensor has only one non-zero component, corresponding to the velocity gradient between the two plates, and is often called the shear rate $\dot{\gamma}$:

$$\tau = 2\eta \cdot \dot{\gamma} \qquad (3)$$

where τ is the shear stress, η the viscosity, and $\dot{\gamma}$ the shear rate. Figure 7 illustrates graphically the types of flow behavior which a fluid may have in the form of flow curves [36] – that is, for one-dimensional flow. Unfortunately, Newtonian behavior is almost never encountered when dealing with pastes. Typically, the viscosity is not independent of the shear rate, but rather becomes a function of the shear rate. If the shear rate is varied over several decades, viscosity may change by several orders of magnitude:

$$\tau = 2\eta(\dot{\gamma}) \cdot \dot{\gamma} \qquad (4)$$

Most pastes used for extrusion are either pseudoplastic (shear thinning) or viscoelastic (see below, the viscous flow contribution also shows shear thinning in this case) – that is, their viscosity decreases when the shear rate is increased. Such behavior can be attributed to the reduction of interactions on the colloidal level; that is, weak bonds between particles present in the paste or the entanglement of polymer additives are broken up with increasing shear rate and flow becomes easier. Dilatant (shear thickening) behavior is less often encountered, but it may occur at high solids content and very homogeneous particle sizes. This has been reported for titania suspensions with titania concentrations exceeding 40% [37]. In some cases, the time over which the shear strain is exerted also influences viscosity. Such systems are termed "thixotropic" or "rheopectic" (Fig. 7, lower graph). Thixotropic behavior is occasionally observed for pastes used in extrusion processes.

Finally, fluids may also be viscoelastic; that is, in addition to the viscous flow component, in which the energy required for the deformation is dissipated as heat, an elastic component is also present, by which part of the energy is stored in the fluid and recovered when the force is released. Thus, depending on the flow, a viscoelastic fluid may exhibit, in varying degrees, the elastic properties of solids and the viscous properties of fluids [36]. In an extrusion process, viscoelasticity could lead to so-called "extrudate swelling", where the fluid stream leaving a capillary may swell to several times the capillary diameter.

The preferred flow behavior for pastes in extrusion processes is pseudoplastic. In this situation, the energy required for the deformation is reasonably low, as the viscosity is low at high shear rates, as occurs in an extruder. However, when the paste leaves the die, at either a low or zero shear rate, the viscosity increases so that a form-stable extrudate is obtained. Ideally, extrudable materials are those which actually have plateaus in the flow curves – that is, over a certain range of shear rates the shear stress does not change. An example of this can be seen in the flow behavior of plastic clay [38], which can be extruded without any internal deformation (ideal plug flow with slip along the wall). However, most catalyst pastes do not behave in this ideal manner and demonstrate some shear deformation which can be tolerated if not too severe [39]. Whilst thixotropic behavior is acceptable, the process is more difficult to control, as the viscosity of the

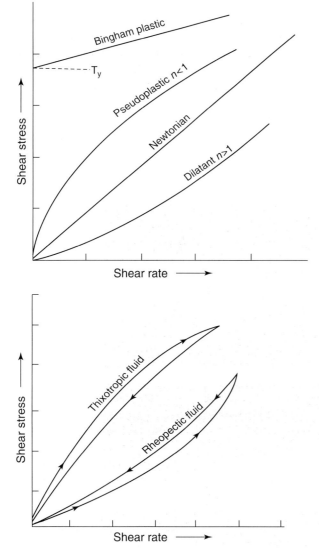

Fig. 7 Schematic flow diagrams for fluids with different rheological behavior. Upper graph: fluids with time-independent flow characteristics. Lower graph: fluids with time-dependent flow characteristics. (From Ref. [36].)

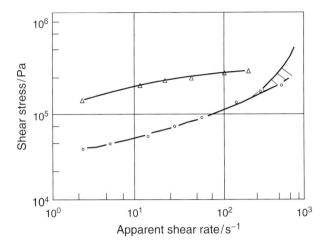

Fig. 8 Flow curves for two different alumina pastes. Triangles: γ-alumina + Pural SB + ammoniumalginate; this mass shows good extrusion behavior. Circles: Pural SB + Luviskol (polyvinylpyrrolidone); this mass shows poor extrusion behavior. (From Ref. [38].)

paste then becomes dependent on its thermal and its flow history.

The importance of the flow characteristics of the paste on extrudability can be illustrated with an example of the extrusion of alumina monoliths. Figure 8 shows the flow curves for two different alumina pastes, based on Pural SB, but with different plasticizers. In Fig. 9, the photograph of the extrudate of the paste with slightly dilatant behavior and wall sticking at high shear rates shows that the production of a shaped body with this mass is impossible, whereas Pural SB with γ-alumina and ammoniumalginate as plasticizer is pseudoplastic and can be extruded very wel.

2.5.2.4.2 Equipment Before being extruded through a die to produce the shaped body, the paste must be homogenized. Insufficient homogenization of the paste can lead to severe extrusion problems and damaging of the shaped bodies. The homogenization process can be carried out in a separate device such as a kneader. In the kneader, the base materials, additives and water are combined and kneaded to produce a homogeneous paste. Alternatively, the dry catalyst material, together with additives and water, can be fed directly into the extruder, which has a kneading section and also serves as a homogenization device. In the latter case, screw extruders must be used which possess different zones

References see page 698

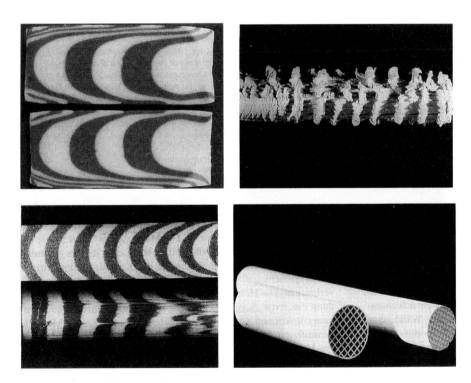

Fig. 9 Extrudates formed from the two masses for which flow curves are given in Fig. 8. Upper images: Extrudates from paste with poor extrusion characteristics. Lower images: Extrudates from paste with good extrusion characteristics. The images on the left-hand side show cross-sections of an extrudate in which the paste was alternately colored to visualize the flow. In the lower image, deformation in the inner part of the paste is much smaller than in the upper figure. (From Ref. [38].)

along the screw, for blending and for the final extrusion. Kneading should be carried out under well-controlled conditions of temperature, pH, and agitation.

Today, many different types of extruder are available. In the laboratory, *piston extruders* are normally used, whereby a piston forces the paste through the die. As a piston extruder does not require a large amount of paste for reproducible operation, it is often preferred for laboratory operation. However, on an industrial scale, press extruders with a rotating pressing cylinder, roller presses or gear presses, or screw extruders, are used (Fig. 10), which allow the continuous production of material. Both require rather large amounts of paste before they can operate under steady-state conditions; hence, they are not often used in the laboratory to prepare shaped catalyst bodies.

In a *press extruder*, the catalyst paste is exposed to the pressing force over only a relatively short distance, for example in the gap between the pressing and screening cylinder, or in the die of the press extruder (see Fig. 10), where the degree of shear deformation is less than in screw extruders. Thus, the flow behavior of the paste is less critical than in screw extruders, where it must be moved along the length of the barrel by the screw; consequently, press extruders allow the relatively easy extrusion of viscous pastes. The force in press extruders is exerted by counter-rotating wheels (see Fig. 10), by rollers which press the paste through dies in a flat disk on which they roll, by cog-wheels in gear presses (where the cogs have the dies between the teeth of the cog wheel), or by various other methods [40]. Finally, the extruded paste is detached from the dies mostly under the action of gravity.

In a *screw extruder*, the screw is fed with the paste from a hopper at the rearside, after which the paste is conveyed towards the die whilst being compressed. In general, single screw extruders are used for inorganic pastes, while twin screws are often preferred for the melt extrusion of polymers. Another difference is seen in the ratio of screw length to barrel diameter, which is typically much shorter when extruding inorganic materials such as catalysts. When the extrudate has left the die, it is cut off directly by a rotating blade and either falls from the die by the action of gravity in irregular segments, or it can be transferred onto a conveyor belt for drying and solidification before being cut or broken into segments, depending on the solidification properties of the paste. Screw extruders allow more complex shapes to be formed compared to press extruders, in which the production of hollow or monolithic profiles would be very difficult. In order to produce complex shapes, it is often mandatory that the paste is free from gas bubbles, and in such cases a vacuum section is used for degassing before the final extrusion stage. Some examples of extruders during catalyst extrusion are shown in Fig. 11; a selection of different catalyst shapes which can be prepared by extrusion, together with the industrial production of sulfuric acid catalysts by extrusion, are shown in Fig. 12.

For both, press and screw extruders, the choice of the die material can be important. As wall slipping is a favorable flow mode in the die, a low-friction die is desirable and consequently polished stainless steel, nylon or Teflon may be used as the die material.

2.5.2.4.3 Paste Composition and Additives The paste formulation is the single most important factor that determines the success of an extrusion process. Some of the formulation parameters have been discussed by Forzatti et al., for the example of the extrusion of titania-based monoliths [41]. If complex shapes are to be extruded, such as the monolith for automotive exhaust gas treatment with its delicate structures, it is mandatory that no large solid particles are present in the paste, and that the solids must be carefully screened before the paste is formulated. Although the different additives are normally classified according to function, this is not a strict categorization. Many additives have two or more functions. For example, clay minerals can function as both binder and plasticizer, while organic polymers may serve as binders for the green body, plasticizers, and also lubricants. The additives will be discussed, on the basis of their primary function in the paste, in the following sections.

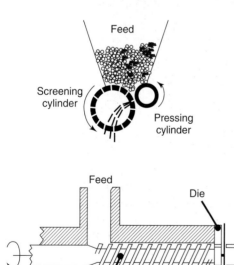

Fig. 10 Schematic drawing of two types of extruder. Press extruder (top) and screw extruder (bottom). (From Ref. [6].)

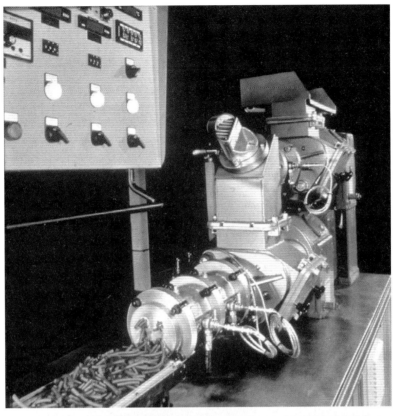

Fig. 11 Top: A laboratory screw extruder used for catalyst production. Bottom left: Extruder head for simultaneous extrusion of many thin extrudates. Bottom right: The same head in operation. (Illustration courtesy of BASF AG.)

A Water Content The water content of the paste must be adjusted precisely. If it is substantially lower than the pore volume of the solid, the paste is too dry and may stick in the extruder; however, if it is much higher the paste will not settle after leaving the extruder [4]. For the extrusion of Cu/ZnO catalysts in an alumina matrix, differences as low as ±1 wt.% in water content were shown to be decisive for a successful extrusion [42]; the importance of water content on the stability of resultant pillared clay monoliths was also demonstrated [43]. The required water content to produce a paste with good rheological properties always depends on the properties of the solid to be extruded, although water contents of between 20 and 40 wt.% in the final paste are typical for a variety of pastes.

B Peptization Agents The rheological behavior of concentrated slurries is determined on a molecular level by the colloid–chemical interactions between particles, and can thus be strongly pH-dependent. This is due to the colloid chemistry of the dispersions. At most pH values, particles in aqueous solution are charged. However, if the charges on the surface of a solid are balanced (i.e., no net

References see page 698

Fig. 12 Top: Samples of extrudates. (Illustration courtesy of Hosokawa-Bepex GmbH, Leingarten, Germany.) Bottom: Industrial-scale extrusion of sulfuric acid catalyst in the form of star rings. (Illustration courtesy of BASF AG.)

charge is present), one speaks of the point of zero charge (PZC), which is mainly dependent on the material. At lower pH, the particles are positively charged, because the surface oxide or hydroxide groups are protonated. At pH values above the PZC, surface OH-groups are deprotonated and the particles are negatively charged. Thus, if the pH deviates widely from the PZC, the particles are charged and repulsion forces keep them separated. This normally reduces the viscosity, since at low shear rates the viscosity of a paste is mainly determined by the solid–solid and solid–liquid interaction, whereas at high shear rates the viscosity of the paste is primarily governed by solvent viscosity. Around the PZC of the solid, the viscosity is highest, and at this pH possibly an otherwise liquid paste may settle into a solid one [44]. A valuable compilation of the PZC values for many different oxides has been published by Parks [45]. Normally, for extrusion well-dispersed particles are desired in the paste, and thus the pH value is often adjusted somewhat different to the PZC in order to break up agglomerates. Agents used to adjust the colloid chemical properties of the paste are called "peptizing" or "peptization" agents.

Typically, the pH is selected to be slightly acidic to deviate from the PZC, and nitric acid, formic acid or acetic acid solutions are often used as the peptizing agent. In addition to protonating the surface which stabilizes the sol, acid peptizing agents can also hydrolyze oxo-bridges between particles which are aggregated by bond formation and thus further assist in deagglomeration. The above-mentioned acids have the advantage that they are decomposed tracelessly in the later calcination step. Since the presence of dissolved salts (i.e., the ionic strength) also influences the screening of surface charges and any ions present compete with the particles for water molecules, ionic strength is an additional factor which should be taken into account when preparing the paste.

C Binders Most catalytic materials do not themselves possess properties which would give them sufficient strength after extrusion, and most behave rather like sand [3]. Therefore, it is important to work binders into the paste, as these give the extrudates sufficient strength after drying and calcination. As most extrudates are calcined at relatively high temperatures after their production, organic binders are not suitable for the stabilization of extrudates. They are helpful in maintaining the integrity of the paste and the green body, but the final strength must be derived from inorganic binders, the most-often used of which are aluminas, silica sols, or clays. *Alumina* is typically added in the form of boehmite or pseudoboehmite which, upon calcination, convert to transition aluminas. The acicular crystal morphology of boehmite helps in its binding ability, providing a felt-like needle network. During calcination, chemical bonds (oxo-bridges) are formed between the primary particles, and this provides the final stability. Silica sols used as binders gelate during processing and calcination, also resulting in the formation of a strongly chemically bonded network. Clay minerals (bentonite is often used) have layered structures which can swell and delaminate to some extent in the paste. Upon drying and calcination, they form agglomerates of these delaminated fragments which give the stability to the extrudate. On some oxides they also may form fibers which could entangle and thus contribute to mechanical strength [46]. In addition to their function as binder, clays may also act as plasticizers, improving the workability of the paste. Finally, glass fibers have also been added to the paste for the production of titania monoliths [41]; these align with the shear flow during extrusion and thus provide additional axial strength.

D Plasticizers   Plasticizers are additives which improve the rheological behavior of the paste, and may be either inorganic (e.g., clays) or organic. Clay has ideal plastic behavior in itself, and its addition to the paste provides some of the characteristics of pure clay pastes. In addition, clays act as a binder in the final extrudate, and thus have a double functionality.

In the case of organic plasticizers, there is a wide variety available, including starch, sugars, polyethylene oxide, polyethylene glycol, polyvinyl alcohol, polyacrylamide, polyvinylpyrrolidone, alginates, or different types of cellulose, such as methyl cellulose, methylhydroxyethyl cellulose, carboxymethylcellulose, and many others. Several plasticizer systems have been compared (see Ref. [47]). Water-soluble plasticizers, such as cellulose ethers or polyethylene oxides, can precipitate at only moderately high ionic strength. Consequently, they may lose their plasticizing properties in the presence of alkali salts [48]. For extrudability, it appears that polymer plasticizers with too-low glass temperatures (below working temperature) are unfavorable, as they settle quite rapidly causing the paste to become brittle [48]. The choice of plasticizer is often empirically driven and governed by the compatibility with the paste. Often, rather than a single plasticizer a proprietary mix of different plasticizers is used. When using organic plasticizers, it must be borne in mind that these will be burned in the final calcination step to create porosity. However, whilst this may be beneficial for mass transfer in the catalyst, the additional pores may also lead to deterioration in mechanical stability.

E Lubricants   Lubricants help the extrusion process, and their action can often not clearly be discriminated from that of the plasticizers. In many extrusion processes it is favorable if there is no internal shear in the paste and the movement occurs predominantly by slipping along the wall, resulting in a piston flow through the die. Such flow is aided by lubricants, the additives used including glycerin, ethylene glycol, propylene glycol, polyethylene oxide, or mineral oil. It should also be borne in mind that the organic material is removed during calcination to leave voids in the extrudate.

F Porogens   The remarks about the voids created during removal of organic plasticizers or lubricant point to the use of another possible additive, namely porogens. For this, organic material can be added with the purpose of creating porosity during the calcination step. In principle any organic material can be used, such as carbon black, starch, and sawdust.

2.5.2.4.4 Drying and Calcination   The primary product of the extrusion process is the green body which, if produced from a correctly formulated paste, has some mechanical stability in its own right. However, the green body still contains water and the organic additives used for the paste formulation, which must be removed during the drying and calcination stages. Calcination is also necessary to induce the solid-state reactions necessary for the final stability of the material [44]. Since in these stages substantial shrinkage of the green body occurs which easily leads to crack formation or breaking, the drying and calcination must be carried out under well-controlled conditions, especially with defined temperature ramps and controlled humidity during the drying. The larger an extrudate is, the more carefully it must be dried and calcined, because in larger bodies the stresses are higher than in smaller extrudates, in which less steep gradients develop. Thus, the 50 mm upper size limit of extrudates listed in Table 1 can be achieved only in fortunate cases. Whilst, for the delicate structures of automotive exhaust catalysts the extrusion is most difficult step, the drying and calcination steps are crucial for the fabrication of titania-based monoliths used in SCR units, with their larger feature sizes. For some systems, larger monoliths are individually wrapped during drying in order to prevent

References see page 698

too-rapid drying of the outer parts of the monoliths, whereby the resultant drying stress might lead to crack formation.

2.5.2.5 Tableting

Tableting (pelleting, pilling) produces the most regularly shaped bodies (called pellets, tablets or pills) which typically also have high mechanical strength. Although the technology is most widespread in the pharmaceutical industry, the physico-chemical basis of tableting is the same if catalysts are tableted, even if the material properties are quite different. The basics of tableting are covered in substantial detail in reviews which focus on pharmaceutically relevant substances [49, 50]. As the quality of the tablets created depends to a large extent on the mechanical properties of the powders to be tableted, some of the basic concepts are discussed at this point.

2.5.2.5.1 Basic Considerations
The deformation behavior of a solid has a major influence in a tableting process, as the response of the material to applied stress is the main factor controlling the properties of the resulting tablets. The elastic deformation of a solid, which is fully reversible if the stress is released, is given by Hooke's law

$$\sigma = E\varepsilon \quad (5)$$

where σ is the stress (force divided by surface area), E is Young's modulus of elasticity, and ε the resulting strain (i.e., the magnitude of dimensional change). Typically, a material is elastic over only a certain range of deformation; if the stress exceeds a critical limit, the so-called "yield stress", then the material is either deformed permanently by flow (plastic deformation), or it breaks. Plastic deformation is important in the tableting process as it facilitates the formation of permanent particle–particle contacts over larger areas, which in turn will lead to increased adhesive forces between particles and possibly even to a welding of particles. However, plastic deformation depends not only on the stress applied but also on factors such as the overall time of compression, rate of compression, and the time for which the material is under maximum stress. The plastic response to stress is crucial for compaction, and must be evaluated for the powders to be tableted. The elastic component of the response to the applied pressure is stored in the material, which relaxes again when the pressure has been released. If this elastic component is too high, then the tablet tends to be unstable and may "cap" after pressure release; that is, a thin section at the end of the tablet will become detached. Calorimetrically, it has been estimated for pharmaceutical substances that 75–80% of the energy is converted to heat by plastic flow, while the remainder of the energy is retained in the tablet, mostly in form of mechanical tension [51]. The properties of tablets, the associated parameters or material constants, the technique for analysis, and the type of information derived for the tableting process are summarized in Table 3 [49].

Most catalytic materials do not flow plastically (or perhaps only to a limited extent), but rather break as a result of brittle fracturing. If brittle fracturing is the only deformation mode of a powder, the production of stable tablets is not possible. There exists a crude empirical relationship between ductility, melting point, elastic modulus and Mohs hardness, which allows the classification of solids into "tabletable" and "non-tabletable" (Fig. 13) [52]. Many catalytic materials fall into the class of refractory oxides, and thus are expected to be very difficult to pelletize in neat form. Scanning electron microscopy (SEM) images of different powders before and after tableting reveal the relative importance of brittle fracture and plastic deformation, as can be seen in Fig. 14 [53]. It can be seen that the corn starch is plastically deformed, and it also becomes clear that such a soft, plastic material is unsuitable for the production of catalyst pellets, due to the almost complete absence of porosity. $CaCO_3$ has an intermediate situation between brittle fracture and plastic flow, and good tablets can be made, while barium sulfate – which would lie on the upper left side of the pelletable range in Fig. 13 – almost exclusively fractures, resulting in small contact areas between the particles and therefore weak tablets. In such cases, additives must be used which plastically deform and provide bonding of the tablet.

For catalysts, the porosity developed in the pellet is, in addition to the mechanical strength, the most important property. The development of the apparent density of the compact in the die can be approximated by the Heckel equation [54]:

$$\ln\left(\frac{1}{1-D}\right) = KP + A \quad (6)$$

where D is the relative density of the compact in the die at pressure P, and K and A are regression constants. $1/K$ can be identified with the mean yield pressure P_y.

Alternatively, the Kawakita equation [55] is used to describe the compaction of powders under pressure:

$$\frac{P_a}{C} = \frac{1}{ab} + \frac{P_a}{a} \quad (7)$$

where P_a is the applied axial pressure, C the relative volume decrease $C = (V_0 - V)/V_0$, and a and b are constants. Both compaction equations have been shown to be equivalent under certain conditions [56], and describe the compaction curve often quite well over a wide range of pressures. According to the Heckel equation,

Tab. 3 Characterization of mechanical properties of powdered materials by mechanical constants/parameters and techniques. (From Ref. [49])

Property	Mechanical parameters or constants	Technique or method for determination	Information derived
Plasticity or ductility	Yield strength	Density pressure profiles (Heckel equation)	Minimum pressure to form coherent compact
	Yield pressure	Indentation hardness compression cycles	Local plasticity of materials
			Capping potential, plastic/elastic deformation
	Network of plastic deformation	Force displacement curves (compaction simulator)	Precompression possibilities
			Work of die wall friction
			Simulated final scale compaction
Brittleness	Brittle fracture index	Hiestand tableting indices (Instron universal tester)	Laminating tendencies
	Critical stress	Notched beam fracture, double torsion, radial edge cracked tablet or disc	Fracture toughness
	Intensity factor	Vicker's indentation	
Elasticity	Young's modulus	Beam bending, indentation testing, Compression testing	
	Brinell hardness number		
	Elastic recovery (%)	Force displacement curves	Work of elastic deformation
		Work done on the lower punch in a second compression	
	Strain index	Hiestand tableting index	
Visco elasticity	Stress relaxation		Plastic flow, die sticking potential, lamination tendency
	Viscoelastic slope		
	Strain rate sensitivity	Heckel analysis	Effect of scaling-up to high-speed tablet presses
		Compaction simulator	
	Creep compliance		
	Elastic and viscous moduli		
Compactibility	Tablet strength	Compression force versus crushing strength profile	Laminating tendencies
	Deformation hardness	Leuenberger equation, indentation hardness	Maximum compaction pressures
Compressibility	Bonding index	Hiestand tableting indices	Capping and sticking potential
	Compressibility	Leuenberger equation	
		Heckel equation	

the development of porosity of the tablet follows an exponential decay curve with pressure; that is, the major loss of porosity occurs at relatively low pressure, which normally does not lead to stable tablets. This substantial loss of porosity must be accepted if catalysts are to be tableted. If high porosity is desired, then porogens will need to be added.

2.5.2.5.2 Equipment Today, tableting machines are highly developed and widely used, with productivities reaching several hundreds of thousands of tablets per hour with modern machines. Different designs of tableting machines are available, which may function on a stroke principle or by sinking a piston into a die. However, for high output a rotary tableting machine is preferred.

A rotary tableting machine is illustrated schematically in Fig. 15. The powder is fed to the die table from a hopper and filled into the die using a distributor. At this point the lower punch is in the lower part of the die, but the upper punch is then brought down into the die, compacting the powder. After the desired time, the upper punch is pulled back, and the lower punch moved upwards to eject the tablet, and then pulled back again so that the cycle can start anew.

References see page 698

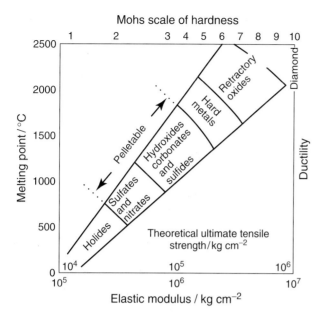

Fig. 13 Schematic correlation between melting point, elastic modulus, hardness, and ductility for different solids. The production of pills is only easy for materials in the middle part of the figures. (Reproduced with permission from Ref. [52].)

The typical tablet form is a simple cylinder with roughly the same height as diameter, but with special dies and punches, rings may also be produced. It is clear that, in order to produce good tablets, the machines must be manufactured and maintained at very high precision. If the dies are not exactly cylindrical but barrel-shaped as a result of wear, then the tablets are squeezed during ejection, and the stress built up during this process can lead to weakening of the pill or, in the extreme case, to complete fragmentation [57]. This is a more severe problem for catalyst production than in the pharmaceutical industry, as pilling pressures are mostly higher and the ingredients much more abrasive than the rather soft organic solids used in pharmacology. Especially hard coatings are therefore necessary, both on the surface of the die and on the punches.

An industrial tableting machine for catalyst production in a partly disassembled state is shown in Fig. 16.

2.5.2.5.3 Formulation of the Feed In tableting processes, dry feeds are used. For successful operation, the feed must meet specifications with respect to density, moisture content, and particle sizes and size distribution. Compaction tests indicate that it is advantageous for tableting that the feed consists of mixed particle sizes, which give the tablet high mechanical stability – just as a concrete is best if it contains sand and gravel [58]. Some catalyst powders are relatively easy to pill, such as magnesium oxide, copper oxide, zinc oxide and aluminum oxide, and thus rather forgiving with respect to particle sizes and

Fig. 14 Scanning electron micrographs of: (a) $CaCO_3$ powder and (b) the surface of a tableted material made from that powder which had been exposed to the punch. (c) Corn starch powder and (d) the surface of a tableted material made from that powder which had been exposed to the punch. (e) $BaSO_4$ powder and (f) the surface of a tableted material made from that powder which had been exposed to the punch. (From Ref. [53].)

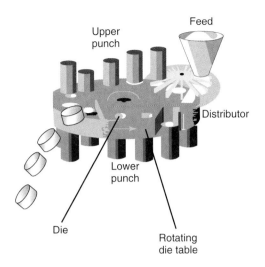

Fig. 15 Schematic drawing of a rotary tablet press. (From Ref. [6].)

other parameters. Others, such as nickel chromite, copper chromite, and some molybdates, are very difficult to process, and the feed needs to be exactly according to the specifications in order to successfully produce stable tablets [58].

The typical synthetic procedures for catalytic solids, such as precipitation and calcination, do not lead directly to the formation of sufficiently dense powders for direct feeding into a tableting machine, and thus the solids must be precompacted. For densification, the as-made catalyst powder can be worked to a thick paste in a kneader or an extruder under the addition of water, with the optimum final water content typically being around 25 to 35 wt.%. After sufficient kneading, the paste is dried, crushed again and screened, for example, through an 8-mesh sieve to produce a pillable solid [59]. During this precompaction, additives as pilling aids can also be worked into the mass. Alternatively, pilling aids can also be mixed with the catalyst powder after precompaction in mixing devices.

Instead of wet compaction as described above, dry compaction routes are also possible, in which the powder may be briquetted in compactors in which the press rolls are equipped with pockets, thus producing briquettes under pressure. These are subsequently crushed again to the desired particle size fraction. Compaction using flat or profiled roll tires is also possible. The main advantage of using a dry route is the elimination of the drying step, which of course would be required after wet compaction processes.

Although shaped bodies can be produced by briquetting, this method is not generally used for the production of the final catalyst shapes, but only for precompaction. This is due to the fact that the homogeneity of pellets made by briquetting is not very high, and the pellets are normally denser on the outside and less dense in the interior.

References see page 698

Fig. 16 An industrial tableting machine used for the production of catalysts. The machine is shown in partly disassembled state for better visibility of the punches. In operation, the machine is encased in a housing. (Illustration courtesy of BASF AG.)

The pilling aids used include graphite, polyvinyl alcohol, polyethylene, talc, mineral oil or stearin as lubricants; other pilling aids include silicates, aluminates, magnesium hydroxide, magnesium oxide, and some chromates. These solids are easier to pill than many other oxides, and thus act as binders. Some oxides, which are difficult to pill, can be shaped into pellets, if some of the carbonates or hydroxides, which serve as precursors of the oxides, are left in the feed for the tableting machine [52].

Graphite is one of the most frequently used pilling aids, and normally is added at concentrations between 0.5 and 1%. However, on occasion larger proportions of up to 3% are necessary. Graphite has been labeled a "highly objectionable" pilling aid, but a necessary evil to achieve high stability of the tablets [60]. The graphite would only be needed on the surfaces of the die and on the punches to facilitate ejection of the tablet. However, its application exclusively by mechanical means is almost impossible to achieve, and consequently the graphite must be added to the feed, preferably to the precompacted powder, as graphite within the powder particles would not help the ejection process. If the graphite is detrimental to catalytic performance, it may be necessary to remove part of it by gentle oxidation. However, care must be taken as the graphite may be oxidized by oxygen from the catalyst, which in turn may be difficult to reoxidize to the same level of intrinsic activity. If other organic lubricants are to be used, they typically must be added in larger amounts than graphite (about twofold), but these can be removed by oxidation at lower temperatures. Other organic lubricants can also lead to a reduction of the catalyst during combustion. Compared to graphite, organic lubricants may burn much more rapidly, leading to higher local heating; they may also partly decompose, which may become problematic due to volatile, badly smelling decomposition products. As problematic as the removal of the organic pilling aids may be in terms of stability of the tablets, they have the positive effect of creating voids, which are beneficial for mass transfer during the later catalytic reaction, even if the porosity is not very pronounced due to the low levels at which they are used. Boron nitride circumvents the problems associated with graphite or organic lubricants, because it can typically be left in the tablets without any negative effect on catalyst performance. However, the level needed is about twice that of graphite, and it is more expensive.

Altogether, the feed for a tableting machine should ideally be prepared to a state that it has the flow properties of sand [61].

2.5.2.6 Granulation

Granulation (also known as "spherudizing") is a method of producing shaped particles which can, in simple terms, be described like the formation of a snowball. Small, moist particles are agitated and stick to each other, and thus grow to larger aggregates. The underlying principles have been treated in an over 30-year-old, but still excellent, classic article by Rumpf [23]. A more recent issue of the journal *Powder Technology* provides a survey on agglomeration processes [62], containing an excellent review which also addresses fundamental questions related to the compaction of particles [21]. A treatment of the granulation process in connection with application to the shaping of catalysts has recently been provided by Holt [63].

Granulation results in less dense and strong particles compared to tableting or extrusion. Also, the shape and size of the particles is more irregular. Granulation is probably the cheapest forming process, provided that a catalyst can be produced continuously in a granulation pan and is needed in constant quality. Low catalyst volumes, or catalysts demanding varying quality, are less favorably produced by granulation, because the start-up and shut-down of the process leads to substantial losses, and start-up may require appreciable time until satisfactory stationary quality is produced. Granulation is ideal for the large-scale production of support spheres, as they can be classified and sold in specific size fractions for further processing. Specialty products are less amenable to production by granulation, since for small-scale applications typically only selected size ranges can be used, and the remainder must be either recycled or discarded.

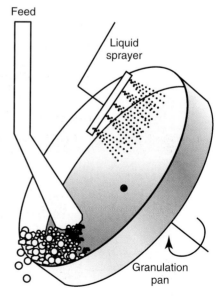

Fig. 17 Schematic drawing of a granulation pan for continuous production of granules. (From Ref. [6].)

One of the most often used apparatus for granulation is the granulation pan (Fig. 17). In order to produce granules in such a device the powder is placed on a rotating pan, which is tilted to a certain degree, and the powder is optionally wetted, for instance using spray nozzles. During rotation of the pan, the particles agglomerate and become bigger. As the bigger particles tend to accumulate on the surface of the granulating solid bed in the pan, and additional powder or slurry is constantly fed into the pan, they will fall over the rim as soon as they have achieved a certain size. Pan granulation is not the only method used to produce agglomerates by granulation procedures. Instead of a pan, a tumbling drum can be used, and the use of agglomeration in a fluidized bed is also possible to achieve granulation. The major effects governing granulation processes are extensively reviewed in Ref. [21], and only some key points will be repeated here.

2.5.2.6.1 Basic Considerations
During a granulation process there are many elementary events which can be treated either separately or in a more integral fashion (Fig. 18). In the integral description, several possible elementary events are grouped together to provide a description with only three rate processes: (i) wetting and nucleation (nucleation here does not mean the initial formation of a new phase, but the initiation of the formation of a new sphere); (ii) consolidation and growth (by collision between two granules, granules and feed powder, or granules and the equipment); and (iii) attrition and breakage, which destroy or diminish the size of already formed granules. If a stationary state is reached at some point between the accretion of matter on granules and attrition and breaking, the size of the granules is limited by this balance.

As the liquid bridges between particles and capillarity provide the major binding force in granulation, it is essential that the particles are easily wettable by the liquid used. This can be especially difficult, if different types of particle are to be granulated together. For a narrow granule size distribution a homogeneous liquid distribution is essential. The liquid can be delivered by pouring, spraying, or melting, if a solid binder is used with a relatively low melting point. Although the features of the granulation process depend heavily on the particular type of equipment used, spraying – preferably with a controlled droplet size distribution – seems to provide the best control over the product (see Ref. [21] for further references). How quickly the nuclei of granules will form, and whether one binder drop leads to the formation of only one nucleus or more than one, depends on several factors. These include the droplet size compared to the powder particle size, and the penetration kinetics of the binder liquid into the powder.

Growth of particles occurs whenever material collides and sticks together, be it by the collection of small

References see page 698

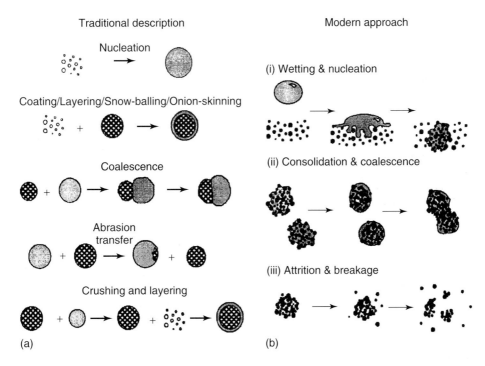

Fig. 18 Processes occurring during granulation compared for two different ways of describing the overall process. (From Ref. [21].)

particles (layering) or by sticking of two already formed granules (coalescence); the distinction here is, to some degree, quite arbitrary. If the granules become bigger, coalescence must be avoided as this may lead to the cementing of several granules together, which would be highly undesirable for later applications. The build-up of the granules may occur in a layer-by-layer fashion, and the process must be operated in such a way as to avoid the formation of non-coherent layers, as in an onion where successive layers can be peeled off. Whether sticking takes place depends on a number of factors, such as the mechanical properties of the granules, the presence of binder liquid, and the angle and relative velocities during collision. Collisions among particles and with the apparatus, however, do not only lead to particle growth but also to the consolidation (densification) of the particles, by which porosity is changed. Typically, a low degree of consolidation means a low stability of the particle but a high pore volume, and vice versa. Depending on the desired application, a compromise has to be made between the strength of particles and the porosity. In many cases, porosity during granulation decays rapidly initially and then levels off, which can be described by an empirical relationship [64]

$$\frac{\varepsilon^g - \varepsilon^g{}_{\min}}{\varepsilon^g{}_0 - \varepsilon^g{}_{\min}} = \exp(-k_{con} N_{drum}) \tag{8}$$

where ε^g is the average granule porosity after N_{drum} drum revolutions, $\varepsilon^g{}_0$ is the initial average porosity of the feed, $\varepsilon^g{}_{\min}$ is the final average porosity reached at infinite number of revolutions, and k_{con} is a consolidation rate constant.

Whilst collisions among particles, and of particles with the apparatus, can lead to agglomeration, it might also have the reverse effect and lead to attrition. Although these processes are difficult to predict, they are less critical if they occur in the granulator with the wet particles. While this may lead to an altered particle size distribution, it may even be beneficial in limiting the upper size of particles. A more serious effect is the attrition and breaking of the final, dried particles. The mechanisms by which particles break are reasonably well understood, and the materials and process parameters that control granule breakage have been identified. Currently, although models are under development to predict granule breakage behavior, a substantial amount of investigation is still necessary in this field [21].

2.5.2.6.2 Equipment
A variety of different processes, using different equipment, can be used for granulation. The most frequently used systems are the granulation pan and the granulation drum. The design rules for the aggregates have been described by Sommer and Herrmann [65].

A The Granulation Pan A granulation pan is flat, has a rim, and rotates on an axis, in practice with a tilt angle of 45° to 55° against the floor. The diameter of the pan may be up to several meters, and the rim may be higher than 1 m for high-diameter pans. The rotation speed of the pan is typically adjusted between 0.6 and 0.75 of the critical rotation speed – this is the speed at which the centrifugal force just compensates the weight (i.e., at higher speeds the particles would be pressed against the rim and would not roll down, even from the highest point). The critical rotation speed can be calculated according to [65]

$$\frac{n_{krit}}{\min^{-1}} = \frac{42.3}{\sqrt{D/m}}\sqrt{\cos \alpha} \tag{9}$$

where n_{crit} is the critical rotation speed, D is the diameter of the pan, and α the tilt angle. The critical speed for a pan of 3 m diameter and a tilt angle of 50° is calculated at about 20 rpm; that is, such a pan would be operated at between 12 and 15 rpm. The productivity of a granulation pan scales with the square of the pan's diameter, and the proportionality constant (also called the granulation factor) depends on the materials' properties. For a 3-m pan, the factor can range from several thousands to several tens of thousands of kilograms per hour.

Since liquid provides the binding forces between particles, it must be sprayed on the granulation pan onto which the dry powder is fed from a hopper. Alternatively, small particles can be placed in the pan at the start of the granulation process, and additional solid is fed together with the liquid in form of a slurry. As a sprayable slurry is typically too wet for the granulation process, the liquid must evaporate in this case to some extent in order to allow accretion of the solid on the growing spheres. A binder, such as cement or polyvinyl pyrrolidone, may be added to either the solid or the liquid to improve the mechanical properties of the granules.

In a granulation pan, the nuclei and the initially formed small particles move at the bottom of the pan. Due to friction the initially irregularly shaped particles move upwards, with the larger, more rounded particles rolling more easily on the smaller particles and thus staying on top of the particle bed. Ultimately, the large particles will roll over the rim of the pan with a relatively defined size, so that classification is often unnecessary. If the primary product is classified, then the smaller and crushed larger granules are fed back to the granulation pan. This not only maximizes the use of the raw materials but also reduces the batch times by changing the way in which nucleation proceeds [63].

B The Granulation Drum The granulation drum represents an alternative to the granulation pan. This is a cylindrical drum into which the granulation mass is fed on one side. The mass is transported through the drum, which is slightly tilted, via rotation of the drum. In a granulation drum the tilt is not used to control the granulation process, but simply to transport the mass. Granule size distribution is broader than for a granulation pan, so that a classification step is often used after granulation. The drum diameters range typically from 0.5 to 1 m, and the lengths up to 3 m.

A less-often used technique – at least for the spherudation of catalysts – is granulation in fluidized-bed granulators; these may be designed as conventional fluidized beds [66], as rotary fluidized beds [67], or as so-called spouted beds [67]. The density of granules prepared in fluidized beds tends to be lower than those of similar materials granulated in either pans or drums. It appears, however, that fluidized bed granulation does not seem to have found wide acceptance for the production of catalysts.

An industrial-scale spherudizer for the batchwise production of granules is shown in Fig. 19. The feed is loaded into the drum and then tumbled until granules with the target properties have formed, and these are then discharged. The lower part of Fig. 19 shows catalyst granules prepared using a granulation process.

2.5.2.7 Miscellaneous Techniques

2.5.2.7.1 Oil Drop Coagulation
Gelation itself may be considered as a shaping process. If gel blocks are prepared, these can be cut into pieces in the wet state and, after drying, crushed and screened to specific particle sizes. This process is in fact used for the production of some qualities of silica gel.

One special forming method for materials which easily form a gel is that of oil drop coagulation. Here, drops of a sol (either silica or alumina) are dripped into an immiscible liquid, such as mineral oil or silicon oil, which is heated to temperatures around 100 °C. The surface tension leads to the formation of spherical droplets, which solidify via a sol–gel process. Two process options are available [68]:

- Small spheres are superficially solidified by passing through the hot oil layer and then enter an aqueous ammonia solution. This induces further gelation, which is completed in a final aging step in order to produce rigid spheres [69].
- Bigger spheres cannot be externally gelled, but need a homogeneous change of pH to induce gelation throughout the droplet. In order to achieve this,

Fig. 19 An industrial spherudizer used for the batchwise production of catalyst granules (illustration courtesy of BASF AG) and of catalyst granules made by spherudizing (illustration courtesy of Hosokawa-Bepex GmbH, Leingarten, Germany).

the sol itself contains compounds such as urea or hexamethylenetetramine, which decompose in the hot oil to produce hydroxide ions, thus causing gelation [70].

References see page 698

As drop sizes can be relatively precisely adjusted, the sizes of the resulting spheres may also have a rather narrow distribution. Oil drop coagulation is very suitable for producing regular spheres with diameters in the millimeter range, or below. The main disadvantage of the process is the possible presence of organic impurities in the shaped bodies, as these may have to be removed by calcination. This problem may be solved by allowing the gelating drops to fall through air instead of through oil before they enter the aqueous phase. However, this requires a very rapidly gelling sol, and such a procedure does not seem to be generally applicable.

2.5.2.7.2 **Pastillation** Pastilles are produced by suspending the catalyst in a melt, followed by solidification of the melt. This process is used in order to facilitate handling of air-sensitive catalysts, such as nickel powder used in the hydrogenation of fats or Fischer–Tropsch catalysts [71, 72]. Such catalyst pastilles are typically manufactured by suspending the catalyst precursor in a fat (in the case of the nickel catalyst) or a molten paraffin or wax (in the case of the Fischer–Tropsch catalyst), reducing the catalyst to the active metal state, and then dropping the catalyst suspension onto a belt cooler, thus producing pastilles of the catalyst embedded in fat or wax. The catalyst can now be handled without deterioration in air for limited periods of time, but is released under process conditions when the matrix melts. As the matrix is either the raw material or a product of the catalytic process, it does not interfere with the reaction.

2.5.2.7.3 **Coating of Preshaped Bodies** Finally, a number of different technical catalysts are produced by coating a preshaped carrier with the catalytically active mass. Such preshaped bodies are often used in the automotive industry, for example, cordierite monoliths, which form the basis of many three-way catalysts, or rolled-up corrugated steel sheets. Various different coating methods are possible, such as washcoating of the cordierite monolith with the alumina/zirconia/ceria support (see Chapter 11.2), chemical vapor deposition (CVD) from gaseous catalyst precursors, or plasma spraying.

Important in this context is also the coating of non-porous carriers or the coating of the inside of metal tubes with catalysts for partial oxidation reactions, such as those used in the production of phthalic anhydride [73–75] or acrylic acid [76]. Occasionally, the use of coated monolithic structures or ceramic foams has also been reported for this purpose [77]. In these partial oxidation reactions, it is mandatory that the initial product leaves the pore system rapidly in order to prevent further oxidation, and the use of a thin coating (on the order of 100 μm) of the active catalyst on a non-porous support, such as steatite or α-alumina, are suitable measures to ensure this. The methods used for these coating processes resemble those applied in granulation, and coating drums are often used into which slurries of the catalyst and various binders are sprayed. Fluidized-bed coating procedures have also been reported in the patent literature [78]. Crucial points here are the sticking of the catalyst on the carrier body, and the homogeneity of the coating. The methods and conditions are rather specific for each of the catalyst systems, and a full discussion of these miscellaneous techniques is beyond the scope of this chapter.

2.5.2.8 **Perspectives**

Forming is an important step in catalyst manufacture, and the correct shape and forming method can support the action of the active mass. However, in spite of the progress made in the understanding of the underlying physico-chemical principles, forming operations remain strongly governed by practical knowledge. In order for a forming procedure to be successful, a wide range of experience with a given system – and often also a number of forming experiments with different additives and different equipment – is necessary. Hence, this approach will most likely continue to dominate the shaping of catalysts for a number of years to come.

References

1. M. S. Spencer, in *Catalyst Handbook*, M. V. Twigg (Ed.), 2nd Ed., Wolfe Publishing, London, 1989, p. 17.
2. M. Campanatti, G. Fornasari, A. Vaccari, *Catal. Today* **2003**, *77*, 299.
3. C. Perego, P. L. Villa, *Catal. Today* **1997**, *34*, 281.
4. E. B. M. Doesberg, J. H. C. van Hooff, in *Catalysis: An integrated approach*, R. A. van Santen, P. W. M. N. van Leuwwen, J. A. Moulijn, B. A. Averill (Eds.), Studies in Surface Science and Catalysis, Vol. 123, Elsevier, Amsterdam, 1999, p. 433.
5. A. B. Stiles, T. A. Koch, *Catalyst Manufacture*, 2nd Ed., Marcel Dekker, New York, 1995, 291 pp.
6. F. Schüth, in *Handbook of Porous Solids*, F. Schüth, K. S. W. Sing, J. Weitkamp (Eds.), Vol. 1, Wiley-VCH, Weinheim, 2002, p. 533.
7. K. Sommer, *Size Enlargement*, in *Ullmann's Encyclopedia of Industrial Chemistry*, Wiley-VCH, Weinheim, 2005, available on CD or as Web-Edition: http://jws-edck.interscience.wiley.com:8087/index.html.
8. D. J. Cumberland, R. J. Crawford, in *Handbook of Powder Technology*, J. C. Williams, T. Allen (Eds.), Vol. 6, Elsevier, Amsterdam 1987, 150 pp.
9. W. M. Visscher, M. Bolsterli, *Nature* **1972**, *239*, 504.
10. D. He, N. N. Ekere, L. Cai, *Mater. Sci. Eng.* **2001**, *A298*, 209.
11. J. Zheng, W. B. Carlson, J. S. Reed, *J. Eur. Ceram. Soc.* **1995**, *15*, 479.
12. Y. Rouault, S. Assouline, *Powder Technol.* **1998**, *96*, 33.
13. G. Mason, D. W. Mellor, *J. Colloid Interf. Sci.* **1995**, *176*, 214.
14. L. N. Smith, P. S. Midha, *Comp. Mater. Sci.* **1997**, *7*, 377.
15. A. F. Wells, *Structural Inorganic Chemistry*, 5th Ed., Clarendon Press, Oxford, 1984, p. 151.

16. J. L. Finney, *Proc. Roy. Soc. London A* **1970**, *319*, 479.
17. A. B. Stiles, T. A. Koch, *Catalyst Manufacture*, 2nd Ed., Marcel Dekker, New York, 1995, p. 76.
18. E. L. Hinrichsen, J. Feder, T. Jossang, *J. Stat. Phys.* **1986**, *44*, 793.
19. G. Mason, *J. Colloid Interf. Sci.* **1976**, *56*, 483.
20. Z. P. Zhang, A. B. Yu, R. B. S. Oakeshott, *J. Phys. A: Math. Gen.* **1996**, *29*, 2671.
21. S. M. Iveson, J. D. Litster, K. Hapgood, B. J. Ennis, *Powder Technol.* **2001**, *117*, 3.
22. C. D. Willet, M. J. Adams, S. A. Johnson, J. P. K. Seville, *Langmuir* **2001**, *16*, 9396.
23. H. Rumpf, *Chem. Ing. Tech.* **1974**, *46*, 1–11.
24. R. H. Snow, J. F. Hasbrouck, R. E. Barnes, M. Castro, E. M. Cook, M. Roisum, P. Schmidtchen, *Spray Dryers. A Guide to Performance Evaluation*, 2nd Ed., AIChE, New York, 2003, 68 pp.
25. A. B. Stiles, T. A. Koch, *Catalyst Manufacture*, 2nd Ed., Marcel Dekker, New York, 1995, p. 99.
26. J. Kim, K. C. Song, O. Wilhelm, S. E. Pratsinis, *Chem. Ing. Tech.* **2001**, *73*, 461.
27. H. E. Bergna, in *Characterization and Catalyst Development*, S. A. Bradley, M. J. Gattuso, R. J. Bertolacini (Eds.), ACS Symposium Series, Vol. 411, ACS, Washington, 1989, p. 55.
28. F. V. Shaw, in *Symposium on Catalyst Supports: Chemistry, Forming and Characterization*, ACS Division of Petroleum Chemistry, New York City Meeting, August 25–30, 1991, p. 524
29. T. Roberie, D. Hildebrandt, J. Creighton, J. P. Gilson, in *Zeolites for Cleaner Technologies*, M. Guisnet, J. P. Gilson (Eds.), Imperial College Press, London, 2002, p. 57.
30. E. Tsotsas, V. Gnielinski, E. U. Schlünder, *Drying of Solid Materials*, in *Ullmann's Encyclopedia of Industrial Chemistry*, Wiley-VCH, Weinheim, 2005, available on CD or as Web-Edition: http://jws-edck.interscience.wiley.com:8087/index.html.
31. P. Walzel, *Spraying and Atomizing of Liquid*, in *Ullmann's Encyclopedia of Industrial Chemistry*, Wiley-VCH, Weinheim, 2005, available on CD or as Web-Edition: http://jws-edck.interscience.wiley.com:8087/index.html.
32. J. C Hein, R. Rafflenbeul, M. Beckmann, *Chem. Ing. Tech.* **1982**, *54*, 787.
33. A. B. Stiles, T. A. Koch, *Catalyst Manufacture*, 2nd Ed., Marcel Dekker, New York, 1995, p. 95.
34. http://www.corning.com/docs/environmentaltechnologies/celcor-ultra-thin-wall.pdf.
35. J. Benbow, J. Bridgwater, *Paste Flow and Extrusion*, Clarendon Press, Oxford, 1993, 153 pp.
36. D. V. Boger, Y. Leong Yeow, *Fluid Mechanics*, in *Ullmann's Encyclopedia of Industrial Chemistry*, Wiley-VCH, Weinheim, 2005, available on CD or as Web-Edition: http://jws-edck.interscience.wiley.com:8087/index.html.
37. A. B. Metzner, M. Whitlock, *Trans. Soc. Rheol.* **1958**, *2*, 239.
38. W. Gleißle, J. Graczyk, *Chem. Ing. Tech.* **1993**, *65*, 1206.
39. J. Graczyk, W. Gleißle, *Erdöl und Kohle-Erdgas-Petrochemie* **1990**, *43*, 27.
40. J. F. LePage, *Applied Heterogeneous Catalysis – Design, Manufacture, Use of Solid Catalysts*, Editions Technip, Paris, 1987, 516 pp.
41. P. Forzatti, D. Ballardini, L. Sighicelli, *Catal. Today* **1998**, *41*, 87.
42. S. P. Müller, M. Kucher, C. Ohlinger, B. Kraushaar-Czarnetzki, *J. Catal.* **2003**, *218*, 419.
43. F. Mohino, A. B. Martin, P. Salerno, A. Bahamonde, S. Mendioroz, *Appl. Clay Sci.* **2005**, *29*, 125.
44. P. Avila, M. Montes, E. E. Miró, *Chem. Eng. J.* **2005**, *109*, 11.
45. G. A. Parks, *Chem. Rev.* **1965**, *65*, 177.
46. S. K. Kawatra, S. J. Ripke, *Miner. Eng.* **2001**, *14*, 647.
47. V. Lyakhova, G. Barannyk, Z. Ismagilov, in *Preparation of Catalysts VI*, G. Poncelet, J. Martens, B. Delmon, P. A. Jacobs, P. Grange (Eds.), Studies in Surface Science and Catalysis, Vol. 91, Elsevier, Amsterdam, 1999, p. 775.
48. W. P. Addiego, W. Liu, T. Boger, *Catal. Today* **2001**, *69*, 25.
49. S. Jain, *Pharm. Sci. Technol. Today* **1999**, *2*, 20.
50. S. Patel, A. M. Kaushal, A. K. Bansal, *Crit. Rev. Ther. Drug Carrier Systems* **2006**, *23*, 1.
51. C. Führer, in *Formulation and Preparation of Dosage Forms*, J. Polderman (Ed.), Elsevier/North Holland, Amsterdam, 1977, p. 289.
52. S. P. S. Andrew, *Chem. Eng. Sci.* **1981**, *36*, 1431.
53. K. H. Sartor, H. Schubert, *Chem. Ing. Tech.* **1978**, *50*, 708.
54. R. W. Heckel, *Trans. Metal Soc. AIME* **1961**, *221*, 671, 1001.
55. K. Kawakita, K. H. Ludde, *Powder Technol.* **1970–1971**, *4*, 61.
56. P. J. Denny, *Powder Technol.* **2002**, *127*, 162.
57. A. B. Stiles, T. A. Koch, *Catalyst Manufacture*, 2nd Ed., Marcel Dekker, New York, 1995, p. 94.
58. A. B. Stiles, T. A. Koch, *Catalyst Manufacture*, 2nd Ed., Marcel Dekker, New York, 1995, p. 76.
59. A. B. Stiles, T. A. Koch, *Catalyst Manufacture*, 2nd Ed., Marcel Dekker, New York, 1995, p. 77.
60. A. B. Stiles, T. A. Koch, *Catalyst Manufacture*, 2nd Ed., Marcel Dekker, New York, 1995, p. 88.
61. N. Pernicone, *Catal. Today* **1997**, *34*, 535.
62. *Powder Technology* **2001**, *117*, 1–171.
63. E. M. Holt, *Powder Technol.* **2004**, *140*, 194.
64. D. Ganderton, B. M. Hunter, *J. Pharm. Pharmacol.* **1971**, *23*, 1.
65. K. Sommer, W. Herrmann, *Chem. Ing. Tech.* **1978**, *50*, 518.
66. A. M. Metha, M. J. Valazza, S. E. Able, *Pharm. Technol.* **1986**, *10*, 46.
67. P. Gauthier, J. M. Aiache, *Pharmaceut. Technol. Europe* 2001, September issue, 22.
68. A. J. Kamphuis, J. R. Walls, in *Preparation of Catalysts VII*, B. Delmon, P. A. Jacobs, R. Maggi, J. A. Martens, P. Grange, G. Poncelet (Eds.), Studies in Surface Science and Catalysis, Vol. 118, Elsevier, Amsterdam, 1998, p. 775.
69. M. E. A. Hermans, *Sci. Ceram.* **1970**, *5*, 523
70. R. M. Cahen, J. M. Andre, H. R. Debus, in *Preparation of Catalysts IV*, B. Delmon, P. Grange, P. A. Jacobs, G. Poncelet (Eds.), Studies in Surface Science and Catalysis, Vol. 3, Elsevier, Amsterdam, 1979, p. 585.
71. S. R. Mohedas, R. L. Espinoza, R. Ajoy, K. L. Coy, US Patent 182145, assigned to Conocophillips, 2005.
72. D. C. Wolfe, P. D. Schneider, R. O'Brien, X. D. Hu, J. Bradne, P. McLaughlin, J. Stack, US Patent 157731, assigned to Süd-Chemie, 2004.
73. K. Ishida, M. Umatachi, R. Yamamoto, JP 49041035, assigned to Mitsui Toatsu Chemicals Inc., 1974.
74. E. Ricker, O. Goehre, DE 2232448, assigned to BASF AG, 1974.
75. V. Nikolov, D. Klissurski, A. Anastasov, *Catal. Rev. -Sci. Eng.* **1991**, *33*, 319.
76. S. Unverricht, H. Arnold, A. Tenten, O. Machhammer, P. Zehner, WO 9941011, assigned to BASF AG, 1999.
77. B. Schimmoeller, H. Schulz, S. E. Pratsinis, A. Bareiss, A. Reitzmann, B. Kraushaar-Czarnetzki, *J. Catal.* **2006**, *243*, 82.
78. S. Neto, W. Rummer, S. Storck, J. Zuehlke, F. Rosowski, WO 2005030388, assigned to BASF AG, 2005.

2.6
Standard Catalysts

2.6.1
Non-Zeolitic Standard Catalysts

Geoffrey C. Bond*

2.6.1.1 Introduction

It is one of the fundamental requirements of research in the physical sciences that any observation reported in the literature should be capable of repetition, anywhere, and at any time. Authors of scientific papers are therefore under an obligation to describe their materials and procedures in sufficient detail to make this possible. Nowhere is this requirement more necessary than in the field of heterogeneous catalysis where, by reason of the complexity of the material used and the subtlety of the procedures applied in their pretreatment, the results obtained often depend critically on the variables involved. Full, detailed and accurate descriptions of what has been done are therefore needed, as it is sometimes the case that the really critical variable is not recognized by the operator, and is therefore not controlled. Such adequate descriptions are by no means always to be found in published works.

It therefore follows that, ideally, a literature description should permit anyone to copy the preparation and pretreatment of a catalyst, and to reproduce the measurement of its activity and other reaction characteristics within acceptable limits. It is uncertain whether this has ever been attempted as a conscious exercise, but from the fragmentary comparisons that are possible from the literature the outcome would be uncertain to say the least. Thus, while one may feel confident of the self consistency of results within a given paper, it does not automatically follow that another operator in another laboratory would reach the same conclusions. It is, however, of overriding importance for the advancement of the subject that one should be able to regard *all* observations, wherever made, as being equally valid.

The control that is needed to achieve good reproducibility of results from one laboratory to another extends from the composition and structure of the catalyst and the conditions of its pretreatment to the apparatus in which this, and the ensuing catalytic reaction, are performed. Experience teaches that factors such as reactor dimensions and the material of its construction are critical for reproducibility, especially in the case of exothermic reactions (as shown in Section 2.6.1.3.1). It is of course supposed that, when comparison between different pieces of equipment is attempted, all other controllable factors such as temperature, reactant pressures, and purification of reactants, are held constant to within whatever limits are practicable.

In order to ensure comparability, it is possible to advance in one of two directions: (i) either to use perfectly standard apparatus of a type possessed by all; or (ii) to calibrate one's home-made apparatus by use of a standard or reference catalyst available to all. There is much to be said in favor of the former approach, but it implies the use of commercially available equipment. However, experience teaches that there is little difficulty in reproducing measurements concerning the physical properties of catalytic materials [X-ray photoelectron spectroscopy (XPS), X-ray diffraction (XRD), transmission electron microscopy (TEM), etc.] when, as is usually the case, standard instrumental methods are used. The alternative is to accept that for some time to come many laboratories will continue to use home-made facilities for chemisorption and catalysis, and that there will therefore be a need for a number of standard catalysts that may be tested to ensure the proper functioning of the equipment (and of the operator!).

There is an additional reason for wishing to see a number of standard catalysts, prepared on a quite large scale, available for widespread use. In many research laboratories, experimental catalysts are often prepared on a scale of only a gram or two, most of which will be used for the tests required by the project. It is therefore impossible to repeat the work at a later date, or to extend it, or to provide colleagues elsewhere with portions of the material for their own experimentation.

There are of course understandable reasons for making small amounts – the method may be truly novel and uncertain of success, or one may wish to economize in the use of costly reagents – but it is worth noting that, in the field of molecular biology, anyone publishing data on specific genes or proteins is under an obligation to provide samples to others, so that the work can be confirmed and developed.

While not denying that work on small-scale preparations may have value, its worth will be limited by the small chance of accurate repetition and by the constraints imposed on the amount of work that can be carried out on such a material. It is notoriously difficult to reproduce exactly the conditions under which preparations on a small scale are conducted: in much research it is adequate or even preferable to use a standard material where possible. Work performed on these materials in different laboratories will accumulate, and eventually there will result a catalyst having a broad profile of factual information. Here at least will be one rock in the shifting sands of uncertainty that are literature of heterogeneous catalysis.

---

* Corresponding author.

It was with these and similar considerations in mind that a group of European scientists, having constituted a Research Group on Catalysis under the leadership of Professor Derouane, decided to prepare two standard supported metal catalysts, one a Pt/SiO_2 and the other a Ni/SiO_2. At a meeting of the group held at Imperial College, London, during the Sixth International Congress on Catalysis, it was agreed to ask the Johnson Matthey Company to prepare 6 kg of a 5% Pt/SiO_2 (EUROPT-1), while Professors Geus and Coenen would collaborate in preparing about 4 kg of about 25% Ni/SiO_2 (EURONI-1) at the University of Nijmegen. Later, the research group became inaugurated under Belgian Law as the European Association of Catalysis, known by the acronym EUROCAT.

Extensive work carried out over a period of some years, mainly by research workers in addition to their normal duties, led finally to broad agreement on their major characteristics, and to a deeper understanding of the processes of chemisorption and catalysis. What was achieved is reviewed in Section 2.6.1.2.

As investigations with these first catalysts progressed, it became clear that there was a demand for standardized platinum catalysts more akin to those used in industry. EUROCAT therefore selected two typical petroleum-reforming catalysts manufactured by AKZO, namely, 0.3% Pt/Al_2O_3 and 0.3% $Pt-0.3\%$ Re/Al_2O_3. These were called EUROPT-3 and -4, respectively (see Section 2.6.1.2.3). More recently still, EUROCAT decided to expand its range of standard catalysts to cover oxides (Section 2.6.1.3.1), sulfides (Section 2.6.1.3.3), and zeolites (Section 2.6.1.3.2).

A major development since the first edition of this Handbook was published has been in the field of catalysis by gold. It is now established that gold has outstanding catalytic properties in a number of oxidations, but only when the particle size is very small, and to obtain this it has been necessary to invent new methods of preparation. To assist in this work, four supported gold "reference" catalysts have been made and distributed by the World Gold Council (Section 2.6.1.4).

A number of other organizations and groups initiated programs of work on standard catalysts and supports, but they were mainly parochial in their scope and not intended to produce internationally recognized standards. Their intentions and accomplishments are briefly summarized in Section 2.6.1.5.

The performance of a catalyst in industrial usage is likely to be determined by its pore structure; that is to say, by its total pore volume and its pore size distribution. In cases where the active phase is mounted on a porous support, its pore characteristics may affect the accessibility of the active phase to the reactants, as well as other features of the catalyst's performance. For these and other reasons it is important to have agreed on reliable procedures for the measurement of these and related quantities: progress in this direction is surveyed in Section 2.6.1.4.7.

2.6.1.2 EUROCAT Metal Catalysts

2.6.1.2.1 EUROPT-1

EUROPT-1 is a silica-supported platinum catalyst containing 6.3% Pt. [1–7]. This unusually high metal loading was chosen in order to facilitate the use of as many techniques of characterization as possible; in particular, it is difficult to apply electron microscopy if the metal content is less than 1%. Samples were distributed to some 20 members of EUROCAT, and detailed results on its physical characterization were reported in five papers [1–5]. Results concerning its catalytic behavior were published later [6], and most recently a comprehensive review of work performed by other users of the catalyst has appeared [7].

The catalyst was prepared by Johnson Matthey Chemicals plc by an ion-exchange procedure, using SORBOSIL grade AQ U30 silica as support and $Pt(NH_3)_4Cl_2$ as the source of metal: it was reduced in H_2 at 473 K for 0.5 h [2]. It has a somewhat broad distribution of grain sizes, with about 60% being in the range 250 to 500 μm. Weight losses on heating up to 473 K were minimal (≈0.25%); BET surface area measurement derived from full N_2 physisorption isotherms (Type II) gave pleasingly uniform results (185 ± 5 m^2 g^{-1}); mercury porosimetry gave a similar area, and both methods afforded pore volumes of 0.72–0.77 cm^3 g^{-1}, corresponding to a mean pore diameter of about 14 nm. Single-point BET measurements showed that variation of the temperature used for outgassing (to constant weight) did not affect the result [2].

More disturbing were the measurements on the support itself. Outgassing at 573 K led to a surface area of 364 m^2 g^{-1}, and it was concluded that the process of catalyst preparation had decreased both the area and the pore volume by about one-half, the loss of porosity being mainly in the mesopore region. It was concluded that hydrothermal sintering of the silica had occurred; as the literature usually gives only the physical characteristics of the support, and not the finished catalyst, it is sensible to recommend that *reported physical properties should always refer to the catalyst as used.* The possibility that further changes may occur during use needs also to be kept in mind.

EUROPT-1 was subjected to chemical analysis by a number of methods [2]; all gave closely concordant results; namely 6.3 ± 0.1 wt.% Pt. Grains larger than 500 μm, however, contained only 5.7% metal, so care needed to be exercised in taking representative samples. There

References see page 714

is considerable evidence from X-ray absorption fine-structure spectroscopy (XAFS), XPS [2], and temperature-programmed reduction (TPR) [8] measurements that the metal in the catalyst as supplied is substantially oxidized, with at least 90% being in the form of a disordered oxide phase. This is, however, easily reduced by H_2 [9, 10], and providing this is not done above 623 K the dispersion remains high.

Particle size distributions were measured by TEM in a number of laboratories. Some confusion arose because different workers used different size intervals (Fig. 1), but all agreed that the virgin catalyst contained particles between 0.9 and 3.5 nm in size, of which 75% were less than 2 nm; these are distributed quite homogeneously throughout the support grains. A number average size of 1.8 nm has been proposed [3]. This paper gives a lengthy discussion on how the TEM results might be converted into a degree of dispersion, so that turnover frequencies (TOFs) might be calculated, but the procedure adopted assumes (probably incorrectly) that the Pt remains as an oxide: it is common experience that platinum group metal oxides are reduced *in vacuo* under the influence of ionizing radiation. Nevertheless, it seems safe to take a dispersion value of 60% for the reduced catalyst [3]. More recent TEM investigations [7] have suggested a somewhat larger number average size (2.1 nm), and there are indications that the result obtained depends on both the method of sample preparation and the accelerating voltage used [7]. The XAFS spectrum obtained after reduction at 520 K has been interpreted [3] to give a Pt–Pt distance of 0.277 nm and a Pt–Pt metallic coordination number of 9.6 ± 1.0. This agrees with the TEM dispersion, and a structure consistent with the data is a 7 × 7 array of Pt atoms on which a further eight atoms reside [7]. Structural information can also be derived by Debye function analysis of the XRD pattern [9]; superficial oxidation of the as-received material was confirmed, and the reduced catalyst appeared to consist of regular face-centered cubooctahedra, 90% of which were in the form of 55-atom crystallites. The Debye temperature (147 K) was lower than that of bulk Pt^0 (234 K), and there was an increase in static disorder following evacuation. The statistical improbability of creating particles having exactly 55 atoms has been noted, as has the contradiction between the XAFS- and XRD-derived structures [7]. NMR studies on the ^{195}Pt nucleus confirm the dispersion to be about 60% [7]. It must, however, be concluded that these techniques of physical characterization are not yet sufficiently refined for highly quantitative analysis of well-dispersed supported metal catalysts.

Heating in H_2 at temperatures up to 1273 K led to particle growth and broadening of the distribution [3], but the corresponding loss in H_2 adsorption capacity [10] may be due in part to the formation of a Pt silicide.

Selective chemisorption of H_2 is probably the most widely practiced means of assessing dispersion, particle size and metal area in supported metal catalysts. The process has been investigated with EUROPT-1 by several groups [4, 7, 11, 12], with results that have general importance for the validity of the method. It has been widely assumed that surface stoichiometry H/Pt_s may be taken as unity irrespective of particle size, despite evidence to the contrary with other metals. Volumetric H_2 chemisorption isotherms of good quality are obtained (Fig. 2), with low slopes at pressures above about 10^3 Pa (1 bar = 10^5 Pa) pressure; these appear to fit the Langmuir equation for dissociative chemisorption [12] (Fig. 3), which when applied in the linearized form

$$\frac{P_e^{0.5}}{n} = (K^{0.5} n_m)^{-1} + \left(\frac{P_e^{0.5}}{n_m}\right)$$

(where P_e is the equilibrium pressure, n and n_m are respectively the moles of H_2 adsorbed at pressure P_e and at saturation, and K is the adsorption equilibrium constant) enables n_m to be obtained both from the slope and (less accurately) the intercept (Table 1). The difficulty with the often-used "extrapolation to zero pressure" method for estimating monolayer volume is that the intercept increases as the pressure range is extended [11] and thus has no real significance. The data also seem to obey the Temkin equation over very wide pressure ranges [4, 11], although the logarithmic form of the equation does not admit of a saturation limit. It is, however, concluded that in practice adsorption is complete at about 3.3×10^4 Pa pressure, the uptake being 200 µmol g^{-1} (H:Pt$_s$ = 1.24),

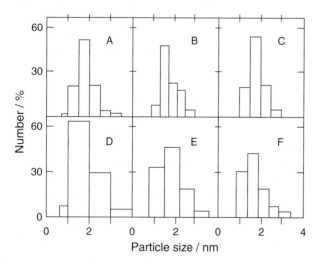

Fig. 1 Particle size distributions of EUROPT-1 by transmission electron microscopy [3]: (a–e) material as received examined in five different laboratories; (f) after re-reduction at 623 K. Numbers of particles counted: A, B, D, F = 500; C, E = 1000.

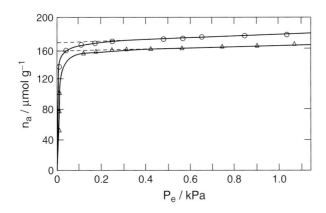

Fig. 2 H$_2$ chemisorption isotherms on EUROPT-1 [12] (△) after reduction at 758 K; (○) oxidized before reduction at 758 K.

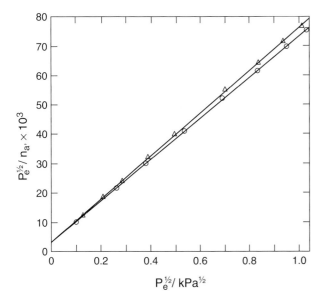

Fig. 3 H$_2$ chemisorption isotherms on EUROPT-1 plotted according to the Langmuir equation [12] (see also Table 1). Symbols as in Fig. 2.

Tab. 1 Monolayer capacities (μmol g^{-1}) for H$_2$ on EUROPT-1 derived in various ways

Measurement[b]				Reference
n^0_m	n^{40}	n^L_m	H:Pt^L	
166	174	–	–	[4]
166 ± 4	168	≈192	1.19	[7]
159.5	162	171	1.06	[12]
173.5[a]	178[a]	188[a]	1.17[a]	[12]
–	200	–	–	[11]

[a] Catalyst oxidised at 623 K for 1 h before reduction.
[b] n^0_m = monolayer capacity by extrapolation to zero pressure; n^{40} = H$_2$ uptake at 53×10^2 Pa (53 mbar) equilibrium pressure; n^L_m = monolayer capacity according to Langmuir equation; H:Pt^L = corresponding H:Pt ratio.

with little further adsorption occurring up to 6.5×10^5 Pa pressure. A value derived by the Langmuir equation, provided the data cover a sufficient pressure range, is therefore acceptable (Table 1).

However, even if one takes a value of ~180 μmol g^{-1}, corresponding to the uptake at 5×10^3 Pa pressure (Table 1), the number of H atoms adsorbed exceeds the *total* number of Pt atoms by some 10%, so there is a flat contradiction between the greater than 100% dispersion that this implies, and the 60% dispersion seen by TEM. The dilemma is partially resolved by the results of temperature-programmed desorption studies [4]. At least three peaks are recognized in the spectrum; the amount adsorbed in the most populated state, together with the assumption that H:Pt$_s$ equals unity, gives a dispersion of 65%, in fair agreement with the TEM value. The most strongly held state is assigned to H spillover, and some of the H$_2$ "chemisorbed" is thought to be used to break Pt–O–Si bridges formed between the metal particles and the support. Three groups have reported heats of adsorption for H$_2$ [7, 11], but the results are not in exact agreement. ^{1}H NMR has also been used to identify states of H$_2$ adsorption [4, 7]. The study of this system is a salutary reminder of exactly how complex an apparently simple process can be.

Carbon monoxide chemisorption on EUROPT-1 has been the subject of several studies [5, 7]. Isotherms of low slope in higher pressure ranges have been obtained, and monolayer volumes at zero pressure of 165–198 μmol g^{-1} have been reported. Infra-red (IR) spectroscopy shows the linear form to predominate, with band frequencies increasing from 2061 to 2075 cm^{-1} with increasing coverage; weak features at 1878–1849 and 1720 cm^{-1} were also seen [5] (Fig. 4). Diffuse reflectance infrared Fourier transform (DRIFT) measurements have found bands for the linear species at somewhat higher frequencies [7]. The heat of adsorption is given as 160 kJ mol^{-1}, independent of coverage up to $\theta = 0.7$. To a first approximation, the surface stoichiometry is the same as for H atoms, namely CO:Pt$_s$ is unity.

Oxygen chemisorption has also been followed using a number of techniques [5, 7]. Because of the great strength of its adsorption, prolonged and rigorous outgassing is not required, and zero pressure uptakes at room temperature were in the region 79–100 μmol g^{-1}. This implies a surface stoichiometry O:Pt$_s$ also of unity. However, O$_2$ uptakes increase with increasing temperature and with

References see page 714

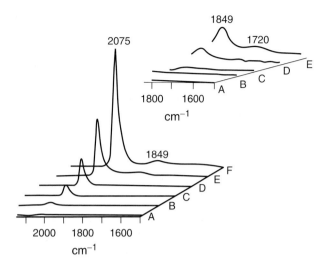

Fig. 4 Infrared spectra of carbon monoxide on H$_2$-depleted EUROPT-1 surface [5]. Spectra A–F represent 3-min exposure at progressively increasing pressures of from 1.3×10^{-5} to 15 mbar (1.3×10^{-3} to 15×10^2 Pa).

hydrogenation, and five different types of site have been recognized and quantified.

A clearer and more penetrating identification of the catalytic sites on the surface of the Pt particles is, however, likely to be provided by reactions that are classified as being "structure-sensitive". In contrast to those of the "structure-insensitive" type – to which most of the reactions cited in the last paragraph belong – these seem to require an ensemble of several atoms, with perhaps some added requirement of specified coordination numbers. The rates and products formed in "structure-sensitive" reactions may therefore vary with: (i) insertion of an inactive (e.g., Group 11) element as in an alloy; (ii) the arrangement of atoms in single crystal surfaces; and (iii) particle size in supported metal catalysts. Since the theoretical infrastructure of these observations is not yet explicated, it is well to apply the terms ensemble-size sensitivity, plane-geometry sensitivity, and particle-size sensitivity, respectively, to these phenomena. The final caveat must be added that these effects do not have a black-and-white character, and that *degrees* of sensitivity to the relevant variable should be considered.

Extensive studies have been described of the reactions of linear alkanes with H$_2$ over EUROPT-1 [6, 7, 10, 13, 14]. Table 2 summarizes the kinetic parameters reported for ethane, propane, and *n*-butane. Skeletal isomerization is of course possible with this last molecule and selectivity to isobutane is between 10 and 30% at 573 K, using a 10-fold excess of H$_2$. Its activation energy is higher than for hydrogenolysis, and its rate and kinetic parameters are somewhat less reproducible. A perpetual problem in the study of these reactions, and of alkane transformations in general, is the formation of strongly held carbonaceous deposits, which are (at

time, and great care is needed to obtain reliable and meaningful results for O$_2$ chemisorption. It is possible that measurements at sub-ambient temperature would be the most useful.

Turning now to the catalytic properties of EUROPT-1, very extensive measurements of all types of hydrocarbon transformation have been undertaken, and an impressive body of information is now available [6, 7, 12–14]. Only the briefest of surveys is possible here; the available reviews [6, 7] and the references cited therein provide additional information.

A number of supposedly structure-insensitive reactions have been examined: these include methane–D$_2$ exchange, hydrogenation of *n*-hexene isomers, of isoprene (2-methyl-1,3-butadiene), of benzene and toluene, the exchange of benzene and of ethene with D$_2$, and epimerization of *cis*-1,2-dimethylcyclohexane. The results were in the main as expected for a platinum catalyst; while several laboratories examined the reactions of benzene, precise comparison of TOF values between laboratories and with the literature is difficult, because experimental conditions were rarely exactly the same. It was observed that the TOF passed through a maximum as the temperature was raised, and that the temperature at which this occurred increased with the H$_2$ pressure. It is therefore reasonable to relate this effect to the changing concentration of H atoms on the surface. With the ethene–D$_2$ exchange, minor differences in product distributions have been correlated with changes in particle size produced by sintering, but the possibility of other effects obtruding, as noted above, must be remembered. The so-called "single turnover" (STO) technique has been applied to 1-butene

Tab. 2 Kinetic parameters* for the hydrogenolysis of linear alkanes on EUROPT-1: TOF and selectivities at 573 K [6, 15]

Alkane	E/kJ mol^{-1}	TOF/h^{-1}	S_2[a]	S_3[a]	Reference
C$_2$H$_6$	210	0.22	–	–	[6]
C$_2$H$_6$	189.5	0.15	–	–	[13]
C$_3$H$_8$	181–189	1.5	–	–	[6]
C$_3$H$_8$	[b]	2.2[b]	0.995[b]	–	[13]
C$_3$H$_8$	146–164[c]	1.45[c]	0.995[c]	–	[13]
n-C$_4$H$_{10}$	114–142	3.5.7.3	0.72	0.58	[6]
n-C$_4$H$_{10}$	[b]	≈4.5[b]	0.67[b]	0.64[b]	[13]
n-C$_4$H$_{10}$	118–127[c]	≈1.8[c]	0.77[c]	0.60[c]	[13]
n-C$_4$H$_{10}$	109	10.2	–	–	[6]

[a] For C$_3$H$_8$, $S_1 + 2S_2 = 3$; for *n*-C$_4$H$_{10}$, $S_1 + 2S_2 + 3S_3 = 4$, where the subscript is the number of C atoms in the product.
[b] Initial state of catalyst: for C$_3$H$_8$, the Arrhenius plot is non-linear.
[c] Steady state of catalyst.
* Kinetics measured with H$_2$:alkane ratios of ≈10.

least initially) extensively dehydrogenated forms of the reactant. Attempts have been made to minimize their formation, for example by using short reaction periods, and to identify their effects on the residual reaction. In the case of n-butane [13], they increase the activation energy and direct the reaction towards greater central C—C bond fission; isomerization is selectively suppressed.

The following conclusions may be drawn from Table 2:

- Activation energies decrease and TOFs increase with the length of the carbon chain.
- As is commonly observed, ethane and methane are formed in almost exactly equimolar amounts from propane.
- The two types of bond in n-butane are broken with almost equal probabilities.

With linear alkanes having five or more carbon atoms, cyclization becomes possible as well as isomerization and hydrogenolysis. With n-pentane, cyclization is minimal and with n-hexane it does not exceed 25% in the range 470–570 K [6]; with the latter molecule, isomerization predominates above 520 K.

Many investigations have also been carried out on the reactions in the presence of H_2 of branched alkanes, and on the mechanism of their skeletal isomerization [6, 7]. The use of ^{13}C-labeled molecules permits alternative reaction pathways to be distinguished. Thus, for example, most of the 3-methylpentane formed from 2-methylpentane has followed the bond-shift route, but most of the n-hexane has resulted from the cyclic mechanism. Labeled molecules also allow mechanisms of aromatization of C_7 and C_8 alkanes to be followed [6].

In the context of these mechanistic considerations, the way in which the C_5 ring is split by H_2 is also of interest; numerous studies of substituted cyclopentanes have been performed [7]. The mode of breaking in methylcyclopentane is particle-size sensitive: EUROPT-1 affords 44% n-hexane, 38% 2-methylpentane, and 18% 3-methylpentane at 520 K.

Hungarian workers have made extensive studies with EUROPT-1 and other platinum catalysts of the dependence of rates or product selectivities on H_2 pressure [7]. Yields of alkenes understandably decrease, and yields of hydrogenolysis products increase, with increasing H_2 pressure; those of other products pass through maxima which move to higher H_2 pressures with increasing temperature. There is a lack of any comprehensive and quantitative theoretical framework to interpret all these observations, but it is clear that the concentration of adsorbed H atoms is crucial in determining to what extent each available pathway is followed. A principal factor in the effect of increasing temperature on product yields is the *decrease* in H atom concentration.

It appears in the main that EUROPT-1 behaves as expected for a Pt catalyst having particles of about 2 nm in size. Compared to other catalysts, however, its hydrogenolysis selectivity is low, as is its tendency to form carbonaceous residues [7]; these characteristics may be connected. An attempt has been made [15] to show how the various reactive intermediates might be accommodated on the faces of a 55-atom cubooctahedron.

EUROPT-1 has also proved to be of exceptional value as a model catalyst to which systematic modifications may be made. The effects on its catalytic properties of the addition of silver, titania and alumina [14], chlorine, oxygen, sulfur, ammonia, and carbon monoxide have all been examined [7], but of particular interest is its extensive use in studies of enantioselective hydrogenation [16–25], where its long-term stability has been an asset over some 20 years of experimentation. Its metallic function has always maintained the same dispersion and other particle characteristics, while being modified in a number of different ways to impart enantioselectivity. Although the Pt particles have no nett chiral quality of themselves, chiral environments can be created at the Pt surface by the adsorption of cinchona, morphine, and strychnos alkaloids [20, 22, 25]; that is, sites having a chiral quality are created adjacent to these adsorbed alkaloid molecules. Selective enantioface adsorption of pro-chiral molecules may occur at these sites providing, on hydrogenation, preferential formation of one enantiomer of the product. Pro-chiral molecules studied include α-ketoesters such as methyl and ethyl pyruvates, methylbenzoyl formate ($R_1COCOOR_2$) and α,β-diketones (R_1COCOR_2). The best performance recorded for EUROPT-1 has been observed in methyl pyruvate hydrogenation in ethanolic solution at 293 K and 3×10^6 Pa pressure over catalyst modified by 10,11-dihydrocinchonidine where the product contained 90% R-methyl lactate and 10% S-methyl lactate at 100% conversion [17]. The opposite chiral outcome is obtainable by replacing cinchonidine by its near-stereoisomer cinchonine. The coadsorption of another strongly adsorbing molecule such as either oxygen or an alkyne favored the desired reaction by governing the number of desired sites and removing some that gave non-selective behavior [23]. A remarkable feature of this work has been the rate enhancement that accompanies enantioselective reaction; factors of up to 100 were recorded, attributable to the role of the modifier in assisting the chemisorption of the reactant in the best possible way [24]. Recent studies, involving doping by Bi, have shown that the enantioselectivity achieved at sites at

References see page 714

or near steps in the Pt surface of EUROPT-1 is higher than that achieved at terrace sites [25b].

Very small amounts of EUROPT-1 may be available from Professor Peter Wells if required (peter@wells19.fsnet.co.uk).

2.6.1.2.2 EUROPT-2
It was intended to prepare a supported platinum catalyst by the inverse micelle method, hoping that this would yield metal particles of uniform size; the designation EUROPT-2 was provisionally allocated to this material, but unfortunately it proved impossible to prepare on sufficiently large scale to have utility as a standard catalyst. However by the time this decision had been taken, the next two standards had been acquired, and it was thought inadvisable to change their designations (EUROPT-3 and -4). No information of any value is therefore available on EUROPT-2.

2.6.1.2.3 EUROPT-3 and EUROPT-4
As noted above, these were typical petroleum-reforming catalysts. EUROPT-3 was 0.3% Pt:Al$_2$O$_3$, and EUROPT-4 contained in addition 0.3% Re. Their surface areas were about 185 m^2 g^{-1}, their N$_2$ pore volumes about 0.5 cm^3 g^{-1}, and both contained about 1% Cl [26].

It was appreciated at the outset that the characterization of these materials would be an altogether more difficult problem than that of EUROPT-1, by reason of the lower metal content, and the bimetallic nature of EUROPT-4. It was hoped that study of the latter would help resolve some of the questions concerning the intimacy of mixing of the two metals in the functioning catalyst, but these queries have been only partially answered. TPR studies have confirmed the sensitivity of the reduction profile to experimental conditions and sample pretreatment but, notwithstanding reports in the literature concerning the selective chemisorption of H$_2$ and of other simple molecules, members of EUROCAT have been unable to measure adsorption characteristics as precisely as they had wished. Contributing factors to the difficulties may be the presence of the Cl, and H spillover. It is believed that most of the platinum is in the form of particles of about 1 nm in size, although by TEM it is only possible to see a few very large particles [28]. For the bimetallic catalyst, various models have been proposed. For example, Re atoms (or ions) may "decorate" the surface of the Pt crystallites, eliminating the large ensembles which can cause hydrogenolysis. It must be recalled that in the working state the catalyst is sulfided, and a model involving decoration by ReS$_x$ species is widely accepted. In the unsulfided state, however, surface energy considerations would determine that rhenium should go to the core, but the particles are so small as to have almost no inside

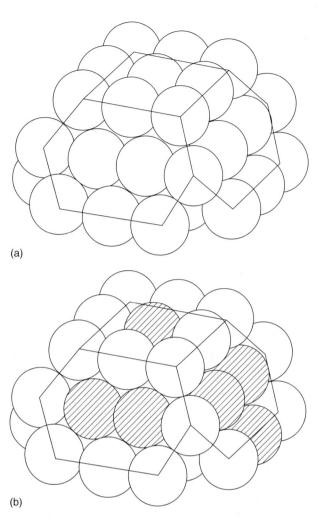

Fig. 5 Models of metal particles containing (a) 46 Pt atoms (size ~1.2 nm, 72% dispersion) or (b) 23 atoms each of Pt and Re (shaded). Of the 13 interior atoms, all are Re, and of the remaining 10 only six are visible [29].

(Fig. 5). It is highly likely that some of the rhenium remains on the Al$_2$O$_3$ surface as Re^{4+} ions occupying octahedral holes; it is certain that there are no pure Re^0 particles or large clusters of Re atoms, because these are extremely active for deep hydrogenolysis, which is not observed. Great care was exercised to use closely defined pretreatment conditions before making catalytic measurements, to maximize the chance of obtaining reproducible effects [26].

Few studies on "structure-insensitive" reactions have been reported [26]. Both catalysts were very active for ethene hydrogenation, and rapid deactivation occurred even at 176 K. Ethyne and 1,3-butadiene react in a more controlled manner: a study of ethyne hydrogenation using both ^{14}C-labeled ethyne and ethene showed that ethane formation took place directly from adsorbed ethyne, without the intervention of gas-phase ethene.

Tab. 3 Kinetic parameters for the hydrogenolysis of linear alkanes on EUROPT-3 and EUROPT-4 rates and selectivities at 603 K [26, 29, 32]

Alkane	Catalyst	E/kJ mol^{-1}	r^a/mmol g$_{Pt}^{-1}$ h^{-1}	S_2^a	S_3^a
C_2H_6	EP-3	230	0.70	–	–
C_3H_8	EP-3	187	1.42	0.992	–
n-C_4H_{10}	EP-3	148	3.62	0.581	0.703
C_2H_6	EP-4	195	7.35	–	–
C_3H_8	EP-4	185	0.76	0.969	–
n-C_4H_{10}	EP-4	142	8.70	0.787	0.523

^{a} See footnote to Table 2.

Hydrogenolysis of alkanes and related reactions on these catalysts have been studied by a number of groups [26, 27, 30–32]. Kinetic parameters for ethane, propane, and n-butane on both catalysts are summarized in Table 3. With propane, selectivities to ethane on EUROPT-3 are greater than 99%, as with EUROPT-1, but with EUROPT-4 they are lower ($S_2 = 0.84$–0.97 at 603 K) and rates are not well reproducible; faster rates correlate with lower values of S_2. Activation energies for both catalysts lie in the range 170 to 200 kJ mol^{-1}. With the Pt/Al$_2$O$_3$, as with the EUROPT-1, Pt/SiO$_2$, rates increase and activation energies decrease with increasing carbon chain length, but a major difference is seen in isomerization selectivities, which with EUROPT-3 are only about 2%. The effect of rhenium on the n-butane reaction is to raise S_2 and depress S_3, these changes being interpreted to mean that adsorbed intermediate C$_3$ species has a much lower chance of desorbing as propane when rhenium is present.

The kinetics of alkane hydrogenolysis – that is, the dependence of rate on reactant concentration – have been the subject of numerous studies, and much effort has been devoted to devising rate expressions based on the Langmuir–Hinshelwood formalism to interpret them. Reactions of ethane, propane, and n-butane with H$_2$ on EUROPT-3 and -4 have been carefully studied, with the surfaces in either as clean a state as possible, or deliberately carbided [29, 31].

Detailed consideration has been given to interpreting the changes in apparent activation energy for and rates (or TOFs) of hydrogenolysis of the C$_2$ to C$_4$ alkanes as a function of their chain length (Tables 2 and 3). Determination of rate as a function of H$_2$ pressure at various temperatures for each alkane revealed that: (i) the apparent activation energy was inevitably a function of the pressure [32, 34] (Fig. 6); and (ii) at a given temperature the pressure at which the rate was maximal increased with chain length [32]. The relative activities of the three alkanes therefore depended on the H$_2$ pressure at which

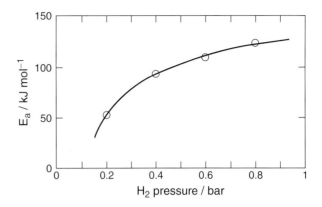

Fig. 6 Apparent activation energy for n-butane hydrogenolysis on EUROPT-3 as a function of hydrogen pressure [32, 34].

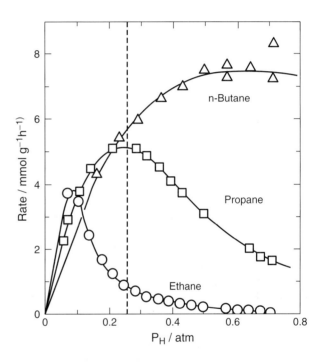

Fig. 7 Rates of hydrogenolysis of C$_2$H$_6$, C$_3$H$_8$ and n-C$_4$H$_{10}$ over EUROPT-3 at 608 K [32, 35].

they were measured [32, 35] (Fig. 7). Thus, the variation in kinetic parameters was a direct consequence of the different hydrogen coverages obtaining at a fixed pressure in consequence of the different tendencies of the alkanes to undergo dissociative chemisorption.

The situation may be summarized diagrammatically [34–36] (Fig. 8) by a plot of rate r, apparent activation energy and "order" in H$_2$ versus the hydrogen coverage, the hatched zones A, B and C representing typical ranges of coverage experienced when the alkanes n-butane,

References see page 714

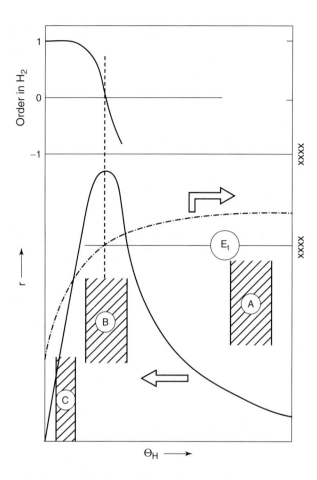

Fig. 8 Schematic diagram showing the dependence of rate, "order" in H_2 and apparent activation energy as a function of hydrogen coverage θ_H. The hatched zones are located around the dashed line in Fig. 7.

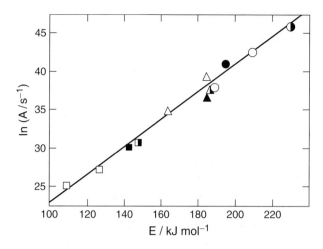

Fig. 9 Compensation plot of the results shown in Tables 2 and 3. Symbols: circles, C_2H_6; triangles, C_3H_8; squares, n-C_4H_{10}. The open points are for EUROPT-1, the half-filled for EUROPT-3 and the filled points for EUROPT-4.

propane and ethane respectively are hydrogenolyzed using normal H_2 pressure ranges. The activation energy found in the region of maximum rate approximates to the true value, but this value differs for each alkane, as it reflects the energetics of the transition state, and hence the manner of bonding of the reactive hydrocarbon intermediate. Fitting the kinetic curves to an appropriate rate expression derived from a reaction model allowed estimates to be made of true activation energies and enthalpies of adsorption, although the quality of fit was surprisingly insensitive to the form of the rate expression [21, 38]. We may note also that this analysis provides an explanation of one of the longstanding puzzles of heterogeneous catalysis, namely, the linear correlation of activation energy with pre-exponential factor, commonly referred to as "compensation" [34, 35]. This is now seen to arise from the use of *apparent* kinetic parameters, and following the Temkin equation the variability of apparent activation energy is attributable [32] to variable enthalpies of adsorption of the alkane. The results in Tables 2 and 3 are presented as compensation plots in Fig. 9; the use of more extensive results for these catalysts than are contained in those tables only serve to reinforce the remarkable linearity of such plots as activation energy changes over a very wide range [34, 36, 37]. For a more detailed presentation of these arguments, see Refs. [34] and [35].

Results obtained for the reactions of n-hexane; methylcyclopentane, methylpentanes, and n-octane with H_2 on these catalysts have been summarized [18]. On the whole, the manner in which product selectivities change with temperature is not well reproducible, due no doubt to the use of different pretreatment and the operation of other factors already mentioned. Fairly consistent results were however obtained with methylcyclopentane, product selectivities between 560 and 610 K being 2-methylpentane $\approx 40\%$, n-hexane $\approx 34\%$, and 3-methylpentane $\approx 26\%$. Increasing the surface Cl^- concentration increases the isomerization selectivity with 2,2-dimethylbutane, while its removal by water decreases it [19].

The amount of work performed with these catalysts is less than with EUROPT-1, and it has been more difficult to achieve consistent and reproducible results with them. What has been done does, however, serve to emphasize the extreme care that is needed to eliminate spurious effects; of these, the formation of carbonaceous deposit is undoubtedly the most serious. It is least troublesome with ethane, and becomes worse as chain length increases.

2.6.1.2.4 **EURONI-1** This catalyst [39–41] was formulated as a 25% Ni/SiO_2, typical of those used for fat-hardening and other large-scale hydrogenations. One of the major problems with catalysts of this type arises from the extensive interaction, amounting to compound formation, that occurs between the precursor components during the preparation, and the consequential difficulty of obtaining complete reduction of the nickel. This was a major focus of effort in the study of this catalyst.

EURONI-1 was prepared by the homogeneous precipitation rate route developed by Geus. An aqueous solution of $Ni(NO_3)_2 \cdot 6H_2O$ in which Degussa Aerosil 180 was suspended was hydrolyzed by hydroxyl ions liberated during the slow hydrolysis of urea at 363 K over a period of 20 h; the products of two separate preparations were mixed, and spray-dried [26]. The nickel content was estimated by several analytical methods, which in the main gave consistent results in the range 24.0–25.5 wt.% (mean 24.6%); only X-ray fluorescence and atomic absorption spectroscopy gave values outside this limit. The unreduced material was hygroscopic; its BET area was about 270 m^2 g^{-1}.

Temperature-programmed reduction (TPR) was followed both by thermogravimetry and conventionally; that is, by measuring the change in composition of a 5% H_2 in Ar mixture. Slight reduction occurred between 570 and 670 K, due perhaps to the presence of small amounts of $Ni(OH)_2$ or basic carbonate; reduction was rapid above 670 K but H_2 consumption continued until at least 1000 K. In isothermal studies, high degrees of reduction were achieved at 900–920 K over periods of 4 to 26 h, and 90% reduction was obtained in 24 h at 700 K. Much depended on the dryness of the gas stream as water (which was liberated in the reduction) might reoxidize the metal. Reproducible reduction rates could only be obtained if bed geometry was tightly controlled. Degrees of reduction were measured by measuring the volume of H_2 liberated on treating the reduced catalyst with H_2SO_4 [40].

Careful studies of the unreduced precursor gave no indication of the presence of $Ni(OH)_2$, the principal component being a poorly crystallized nickel antigorite, a compound having a layer structure. Thus, much – if not all – of the SiO_2 had undergone a reaction involving the Ni^{2+} ions and the base, forming a new structure in which the layers are not in register. This was stable in air up to 770 K.

Samples of EURONI-1 reduced under various conditions were examined by XRD, XPS, magnetic methods, TEM and H_2 chemisorption in order to ascertain the mean size and size distribution of the nickel particles [41]. According to TEM, in a sample reduced at 770 K most of the Ni particles were less than 5 nm in size, but some sintering occurred on raising the temperature to 920 K, after which particles in the range 5–8 nm were detected (Fig. 10). The reduced catalyst was found to be superparamagnetic, so the Ni particles consisted of single domains. The decrease in saturation magnetization following H_2 chemisorption was 1.43 μ_B per H_2 molecule. Considerable difficulty was experienced in the measurement of H_2 chemisorption isotherms, with factors such as variable degree of reduction, outgassing efficiency, reoxidation of Ni particles, and uncertain equilibration times, being held responsible for differences in monolayer volume of up to two-fold between different laboratories. All monolayer volumes were assessed by the extrapolation method, the hazards of which have been described above; no attempt has been made to fit the experimental data to any of the established adsorption equations. Despite these difficulties, the most reliable monolayer volume was stated to be 53 cm^3 g$_{Ni}^{-1}$ for degrees of reduction (obtained at 700–770 K) of 85–90%.

Finally an attempt was made to reconcile the various techniques for particle size estimation (Table 4). To do this it was necessary to assume a particle shape, and the results were evaluated both for spherical and hemispherical models, the former showing the greatest consistency. The particle size was defined as the cube root of the volume. The results obtained by the various methods are in quite close agreement (Table 4), and Coenen was justified in concluding "... that the agreement between the methods is probably better than reported anywhere in the literature, which is the more remarkable when we

---

References see page 714

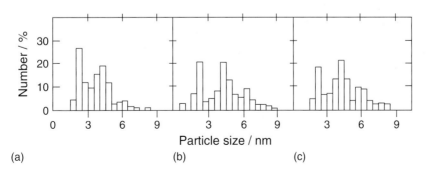

Fig. 10 Particle size distributions for EURONI-1 reduced under various conditions [41]; (a) 16 h at 633 K + 6.5 h at 773 K; (b) 3 h at 923 K; (c) 19 h at 923 K.

Tab. 4 Mean nickel particle sizes (nm) in EURONI-1 determined by various methods [41]

Model[a]	T_{red}/K	XRD	H_2-chemisorption	XPS	Magnetic	TEM[b]
HS	723–773	2.2	2.1	2.4	–	2.2
HS	903–923	–	3.7	–	5.0	3.0
S	723–773	2.2	2.6	2.4	–	3.5
S	903–923	–	4.6	–	5.0	4.6

[a] HS = hemisphere, S = sphere.
[b] For the HS model, TEM means sizes are divided by 1.56 because the projected area is greater than that corresponding to the cube root of the volume.

note that measurements were carried out in ten different laboratories." [41].

2.6.1.3 Other EUROCAT Catalysts

2.6.1.3.1 Vanadia-Titania EUROCAT Oxides

When the decision was made to extend the EUROCAT program of standard catalysts to other types, there was no dispute as to which oxide catalyst should be selected: the V_2O_5/TiO_2 system was the unanimous choice. The chief reasons were the following:

- it was already the subject of attention in a number of European laboratories
- it had important industrial uses for selective oxidations, especially of *o*-xylene to phthalic anhydride
- it was active for the selective catalytic reduction (SCR) of nitrogen oxides (NO_x) by NH_3 and prototypical of those to be used in practice
- determination of its structure and modes of operation presented a major intellectual challenge
- a great variety of analytical techniques could be directed to the problems and could be expected to contribute to their solution.

There was already a considerable body of knowledge on catalysts of this type [42]. For those used for selective oxidations, there was much evidence to show that the active phase was a monolayer of oxovanadium species chemically bonded to the TiO_2 surface; such a material would have about 1 wt.% V_2O_5 for a TiO_2 area of $10 \, m^2 \, g^{-1}$, but technical catalysts usually contained substantially larger amounts. The excess appeared to be in the form of V_2O_5 microcrystals which neither helped nor hindered in selective oxidation; rather, it seemed to serve as a reserve supply to replenish the monolayer, should the latter become depleted. There was also evidence that an uncovered TiO_2 surface was harmful, in that it could cause deep oxidation to carbon oxides. In these applications, the anatase form of TiO_2 was generally used and, unless the contrary is stated, the formula TiO_2 will imply this form.

Based on the collaboration of Rhône-Poulenc, CRA 4-kg quantities of two catalysts were prepared containing about 1% and 8% V_2O_5; the corresponding precursors were also made available (i.e., the uncalcined materials, and also the support). Portions of the catalysts were distributed to some 25 laboratories, and the findings were reported in a special issue of *Catalysis Today* [43].

The catalysts were coded V1 and V8, and the support (an uncoated pigmentary anatase) EL10. They were prepared by mixing vanadyl oxalate solution with the support to form a paste that was kneaded before drying and calcinations (4 h at 723 K) [43b]. Mean values for surface areas were EL10 10.5 $m^2 \, g^{-1}$, V1 10.1 $m^2 \, g^{-1}$, and V8 11.0 $m^2 \, g^{-1}$; these are not significantly different. The mean grain size of the anatase was about 0.5 µm, and the pore volume 0.85 $cm^3 \, g^{-1}$; the grains were non-microporous [43]. XRD measurements showed the presence of crystalline V_2O_5 in V8, but probably not in V1; no other vanadium oxides were present, and no vanadium had dissolved in the anatase [43].

Some points of interest arose in the chemical analysis. Total vanadium contents, estimated by complete dissolution of the catalysts in H_2SO_4 or aqua regia, were close to the expected values of 0.975% for V1 and 7.82% for V8. Attempts were made to distinguish between "soluble" and "insoluble" vanadium, the former being crystalline or weakly held species, and the latter chemically linked to the TiO_2 surface; it was also hoped to estimate the relative amounts of V^{IV} and V^V present. Treatments involving extraction with boiling isobutanol, with NH_4OH and with H_2SO_4 were used; the kinetics of the reaction with NH_4OH was studied in detail. An approximately constant 0.12–0.16 wt.% vanadium was designated as insoluble, which was much less than the amount forming a monolayer. Paradoxically, V^V was reduced to V^{IV} during the H_2SO_4 extraction [43].

Thermal analysis of the precursors showed decomposition in two stages: first, an endothermic loss of water, and then an exothermic loss of CO_2. With V8, the melting of V_2O_5 was detected at higher temperature [43]. TPR is a well-established technique for distinguishing between different reducible species. In the present system, the monolayer species were more readily reducible than the crystalline form, and with V8 two peaks of comparable size were observed (Fig. 11). This required some explaining; perhaps there was an intermediate disordered V_2O_5 state, the reduction of which was not distinguishable from that of the monolayer species, and that together they accounted for about half the total. Reproducibility of TPR profiles required exact control over a number of variables, and a recommended schedule has been given [43]. The

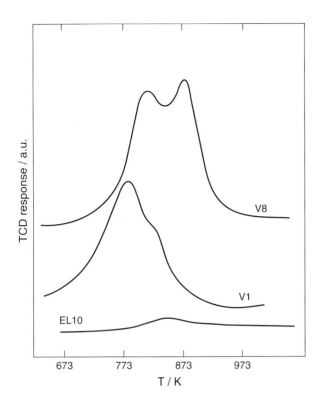

Fig. 11 Temperature-programmed reduction profiles of V_2O_5/TiO_2 catalysts V8 (8% V_2O_5), V1 (1% V_2O_5) and of the support EL 10 [43g].

easier reduction of V1 was also demonstrated in work designed to assess literature claims that the dispersion of the V_2O_5 could be measured by chemisorption of O_2 after reduction [43].

A major effort was devoted to the characterization of these catalysts by vibrational and electronic spectroscopies [43]. The use of IR and Raman spectroscopies confirmed that the monolayer consisted of vanadyl complexes, exhibiting both Lewis and Brønsted sites, and suppressing nucleophilic sites on the support. According to UV-VIS spectra, most of the vanadium in the calcined catalysts was in the V^V state, but part was easily reducible. Evidence for electronic interaction between vanadium and titanium ions was also presented. The monolayer was effective in suppressing Raman bands due to the anatase support, and an explanation for this effect has been offered [44].

Solid-state ^{51}V NMR spectroscopy in both the static (wide line) and magic angle spinning modes has been employed [43]. Electron paramagnetic resonance (EPR) measurements [43] failed to detect significant quantities of V^{IV} species, but confirmed their formation in attack by H_2SO_4; their appearance in solid solution with the TiO_2 following the phase transformation to rutile above about 970 K was also confirmed. TEM studies [43] detected enrichment in vanadium at the edge of the TiO_2 particles, and some unattached V_2O_5 platelets in V8; electron-beam damage on prolonged examination was also noted. XPS measurements by different laboratories yielded widely different results, especially in respect of V : Ti ratios, and recommended conditions for obtaining consistent results have been given [43]. Secondary-ion mass spectrometry (SIMS) revealed the presence of numerous trace impurities, especially niobium, and provided some evidence for removal from the surface of species containing both V and Ti atoms [43].

Electrical conductivity measurements posed the greatest difficulty, as the conclusions they led to were difficult to reconcile with the models generated by the other techniques [43]. These seemed to show that the calcined catalysts contained V^V ions in the TiO_2 lattice, as they increased conductivity: their amount increased from V1 to V8. It was not entirely clear whether they were distinct from the monolayer species, or whether the latter were so firmly bonded to the TiO_2 that they in effect became part of it, and hence increased its conductivity. The presence of anionic vacancies in the TiO_2 was also detected.

The catalytic activity of the EUROCAT oxides toward the selective oxidation of o-xylene was examined in several laboratories, with some discrepancies being apparent [43]. In particular, the temperatures required to achieve a given conversion varied somewhat and, since the reaction is extremely exothermic, the differences could be attributed to poor temperature control, or to the use of various bed dimensions and reactor materials. These uncertainties underline the hazards of measuring catalytic activities in home-made apparatus, and emphasize the need to have standard catalysts to serve as a bench-mark.

The decomposition of isopropanol proceeded either via (A) dehydration to propene or (B) dehydrogenation to acetone. Both continuous- and pulse-flow measurements have shown that the reactions had similar rates on V1 and V8, but (A) was faster than (B) on V8. Reaction (A) required acid centers; these were entirely removed when the soluble vanadium species were removed by isobutanol or NH_4OH [43].

An attempt has been made to synthesize the observations into a model that portrays their structures [43]. The main outline is clear: with V1 the anatase surface was largely covered by a single layer of oxovanadium (vanadyl) groups, firmly attached, and perhaps interacting with the support so as to modify its electrical properties. This monolayer remained intact in V8, but the additional V_2O_5 formed separate crystallites, perhaps as "towers" that obscured only a small part of the surface, and which were easily removed by chemical means. No doubt considerable refinement of this model is possible, especially in terms of "soluble" and "insoluble" forms. The information in

References see page 714

this volume [43] provides much food for thought. Stocks of these materials are held by Mme Pascale Mascunan at the IRC, Lyon, France.

2.6.1.3.2 EUROTS-1 Zeolite
In a further extension of its interest to develop a range of standard catalysts, EUROCAT authorized the preparation and distribution of the titanium silicalite-1 (TS-1); in this zeolite, some of the silicon atoms in the silicalite structure are replaced by titanium. TS-1 is noted for its ability to catalyze the selective partial oxidation of organic substances by H_2O_2; it thus represents new catalytic chemistry which forms the basis of important innovations in partial oxidation technology. More than 100 scientific publications and patents on its use appeared in the decade following its discovery in 1983.

EUROTS-1 was synthesized (150 g) by hydrothermal treatment of a gel derived from ethyl orthosilicate and ethyl orthotitanate, in the presence of tetrapropylammonium hydroxide as template [45]. Each participating laboratory performed its own calcination (3 h or 16 h at a maximum temperature of 823 K, other conditions being variable). The Ti content was 2.8%, and the calcined product consisted of a single phase of orthorhombic symmetry, with a BET surface area of 435 $m^2\,g^{-1}$. The IR and diffuse reflectance spectra (DRS) of EUROTS-1 have been recorded; the material had the form of small, rounded cubes of about 0.15 µm size.

The chosen catalytic test reaction was the oxidation of phenol, which yields a mixture of catechol, hydroquinone, and 1,4-benzoquinone (Scheme 1). The reaction was conducted at atmospheric pressure by continuously adding aqueous H_2O_2 to a mixture of catalyst, phenol, water, and a solvent (either methanol or acetone) at the reaction temperature (usually 373 K); reaction times were 1 to 4 h. Conversions and product selectivities depended on the composition of this mixture; under the best conditions, H_2O_2 conversion was 100%, phenol conversion 27%, and phenol hydroxylation selectivity 91%. The ratio of $o:p$-substituted products (Scheme 1) was usually about unity. It was concluded that catalytic performance depended critically on calcination conditions – that is, on the completeness of removal of the template.

2.6.1.4 Gold Reference Catalysts
Perhaps the most surprising and interesting new development in heterogeneous catalysis has been the realization that gold catalysts are able to show exceptional activity and specificity in certain reactions, especially those in which oxygen is one of the reactants [46, 47]. Following initial hints during the 1970s that activity might depend importantly on particle size, M. Haruta and his colleagues in Japan reported in 1987 that very small gold particles (<5 nm) supported on certain transition metal oxides (e.g., Fe_2O_3, TiO_2) were active for the oxidation of CO well below ambient temperature. This discovery has resulted in much research on catalysts for this reaction [46, 47]; it transpires that only coprecipitation and deposition–precipitation of an hydroxyl-precursor from $HAuCl_4$ solution at raised pH (~9) are able to create the necessary catalyst structure. Au^{III} species are readily reduced to Au^0, but catalyst precursors are very sensitive to conditions of storage and to light, and activity depends critically on chloride concentration and thermal treatment. It has subsequently also been found that gold on carbon catalysts show excellent behavior for the selective oxidation of reducing sugars and related reactions.

Such has been the degree of interest shown in gold catalysts, including potential applications too numerous to mention here, that there have already been several international conferences in which catalysis by gold has featured largely, and a substantial monograph has been published [47]. However, because of the unusual sensitivity of gold catalysts to their method of preparation and conditions of storage, a decision was taken following a conference held in Cape Town in 2003 to authorize the preparation of several supported gold catalysts, and to offer these to interested persons for a nominal charge through the agency of the World Gold Council (www.gold.org). Three such catalysts were manufactured by Süd-Chemie Catalysts Japan Inc., following procedures devised by M. Haruta and S. Tsubota [48]; a fourth catalyst (Au/activated carbon) was made by M. Rossi (Università di Milano). Their compositions, physical characteristics and some catalytic activity measurements are listed in Table 5.

Types A and C were prepared some 12 times in 400-g batches, with closely reproducible properties; the

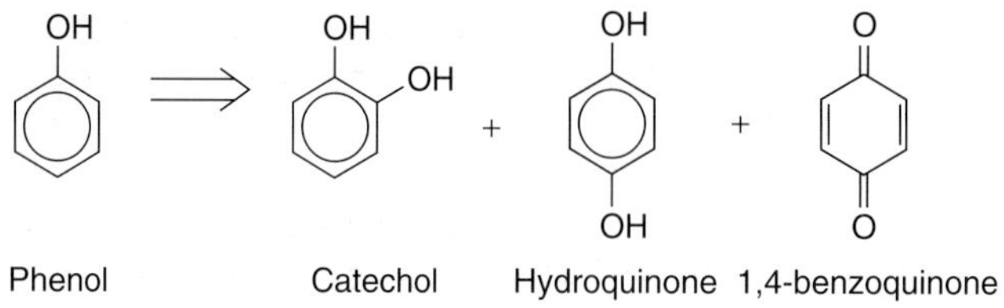

Scheme 1

Tab. 5 Compositions and properties of supported gold "reference" catalysts

Type	Method[1]	Nominal composition	[Au]/%	[Na]/%	d_{av}/nm	$T_{50}(CO)/K$[2]	$T_{50}(H_2)/K$[2]	Notes
A	DP	3% Au/TiO$_2$	1.51	0.042	3.8	228	313	a
B	DP[3]	0.3% Au/Fe$_2$O$_3$/Al$_2$O$_3$	0.28	0.75	–	311	–	b
C	COPPT	5% Au/Fe$_2$O$_3$	4.48	0.019	3.7	232	317	c, d
D	Sol	1% Au/C	0.8	–	10.5	–	–	–

[1] DP = deposition–precipitation; COPPT = coprecipitation.
[2] Temperature (K) for 50% conversion of 1% CO or H$_2$ in air using 100 mg catalyst and a flow-rate of 33 cm^3 min^{-1}.
[3] The Au/Fe$_2$O$_3$ is mounted on Al$_2$O$_3$ beads, [Fe] = 11.6%.
[a] Ref. 51 gives [Au] = 1.38%, [Na] = 0.053%, [Cl] = 47 ppm.
[b] Oxidation of CO under "wet" conditions complete at 273 K; T_{50} = 311 K is for "dry" conditions.
[c] Ref. 49 gives d_{av} = 2.8 nm, BET area = 39 m^2 g^{-1}.
[d] Ref. 50 gives d = 2–5 nm, BET area = 38.7 m^2 g^{-1}, α-Fe$_2$O$_3$ by Mössbauer spectroscopy.

compositions listed in Table 5 are mean values. Type B was made in one batch of 5 kg, while Type D was prepared by deposition of colloidal gold onto activated carbon, and is supplied as a wet paste containing 40% water; results of activity tests for the selective oxidation of ethane-1,2-diol and of D-glucose are available (r.holliday@gold.org).

There is evidence in the literature that unreduced precursors to gold catalysts can suffer changes at room temperature, especially when exposed to light, that lead on reduction to increases in particle size. Information on the long-term stability of the catalysts shown in Table 5 is lacking, and they are therefore described as "reference" rather than "standard" catalysts. They have however been quite widely used, and there are for example reports on their use for the selective hydrogenation of benzalacetone [49], for the water-gas shift [50], for the selective oxidation of CO in H$_2$ [51], as well as for the oxidation of CO alone [52]. In a particularly interesting application, suspensions of Types A and D in aqueous H$_3$PMo$_{12}$O$_{40}$ were active at 298 K for the oxidation of CO in a pressurized batch reactor (15 × 10^5 Pa); as part of an electrical cell, this reaction was used to produce electricity in the manner of a fuel cell [53]. A detailed study of the Type C Au/Fe$_2$O$_3$ has been carried out in connection with its use for the water-gas shift [50]. In TPR, there is a sharp feature (T_{max} ∼ 523 K) as the support is reduced to magnetite, followed by a much broader feature (T_{max} ∼ 873 K) as it is reduced to metallic iron. Chemisorption of CO leads to a sharp band at 2114 cm^{-1} characteristic of the linear form on Au^0, as well as a plethora of bands between 1200 and 1650 cm^{-1}, due to carbonate and bicarbonate-like species.

2.6.1.5 Other Programs

In addition to the investigations summarized above on catalysts initiated by the EUROCAT organization, several other countries have launched similar programs having somewhat related aims. However, none of these appears still to be active, and so at this point a brief mention of their intentions and accomplishments will suffice.

The Catalysis Society of Japan began its program in 1978; its organization and method of working has been described [54]. The materials selected included a number of oxides that were either solid acid catalysts or were usable as supports for metals or other active components. Their surface areas and porosities were measured in several laboratories, and the importance of prior outgassing (minimum 15 min at 700 K) was stressed. Acidities were also measured in a variety of ways, including those of six zeolites [55]. Supported metal catalysts containing either platinum, palladium, or rhodium were prepared by impregnation or ion exchange, and a variety of methods was tried for estimating metal area and particle size. The results from various laboratories appeared to be somewhat unsatisfactory, although a pulse method using CO with H$_2$ as carrier gas met with some success. There is no evidence that these materials were widely employed for reference purposes, and they were not used outside Japan.

Three Russian laboratories have collaborated in the preparation and characterization of Pt/SiO$_2$ and Pt/Al$_2$O$_3$ catalysts. Emphasis was placed on distinguishing between metallic and ionic Pt (i.e., that which could not be reduced at 773 K). After such reduction, much of the Pt in the Pt/Al$_2$O$_3$ catalysts remained in ionic form (∼ 60%), and this complicated attempts to apply traditional methods to estimating metal dispersion. Metal concentrations on various sizes of support grains were measured, and the amount of metal in pores of various diameters was estimated by O$_2$ chemisorption after progressively filling the pores with water. This type of information is not commonly available, but a variation in metal content with support grain size was also observed with EUROPT-1. Catalysts were tested for benzene hydrogenation and

References see page 714

cyclohexane dehydrogenation, there being a marked inverse correlation between metal loading and TOF for the latter reaction, perhaps caused by selective deactivation of larger particles by coke formation. No published account of this work was traced, but some results were given in Chapter 11.1.4.2 of the First Edition of this Handbook.

The Chemistry Department of Northwestern University, Evanston, Illinois, USA, under the leadership of the late Professor Robert L. Burwell, devised a number of supported metal catalysts pretreated in various ways so as to obtain a wide range of dispersions. These ranged between 10 and 150%, the latter figure being given by a Rh/SiO$_2$ catalyst containing particles so small that the H:Rh$_s$ was much greater than unity. Investigations with these catalysts were described in a series of reports [56–58], and they were used by certain other laboratories, but were not made systematically available to the wider catalytic fraternity.

Other organizations have contributed to the orderly and accurate evaluation of the physical characterization of solid catalysts. The American Society for Testing Materials (ASTM) has developed specifications for analytical methods relevant to catalysts, and the British Standards Institute (BSI), which has a longstanding reputation for standardizing measurements of all kinds, has drafted a specification for determining metal surface areas by gas adsorption [59]. The International Union of Pure and Applied Chemistry (IUPAC) has also been active in promoting standard procedures for measuring surface area and porosity of solids by physical adsorption of gases. These studies have been described [60], and presented in Chapter 11.2 of the First Edition of this Handbook.

2.6.1.6 Summary and Conclusions

The value of having well-characterized materials as standards in studies of catalysts is appreciated by all who have used them. In this field – perhaps more than in any other part of chemistry, where performance depends critically on so many variables – it is important to have available standard materials against which the behavior of equipment and operators can be judged. Results obtained with commercial equipment are most easily reproduced, except where, as in the outgassing of porous materials, an element of personal judgment remains. Experience gained in attempts to obtain reliable and consistent rate measurements with the catalysts described above strongly suggest that very great care is needed in controlling all relevant features, and indeed the problem often is to be aware of what these features are. It is worth stressing also that in many published works there has been inordinate time and money devoted to characterizing materials that turn out to have little catalytic value, and, even when they have, the effort devoted to the catalysis is often scant. The properties of a catalyst are not defined by a single rate measurement under a single set of conditions, and comparison between members of a series of catalysts under such limited conditions causes one to wonder whether use of other conditions would have resulted in the same hierarchy.

References

1. G. C. Bond, P. B. Wells, *Appl. Catal.* **1985**, *18*, 221–224.
2. G. C. Bond, P. B. Wells, *Appl. Catal.* **1985**, *18*, 225–230.
3. J. W. Geus, P. B. Wells, *Appl. Catal.* **1985**, *18*, 231–242.
4. A. Frennet, P. B. Wells, *Appl. Catal.* **1985**, *18*, 243–257.
5. P. B. Wells, *Appl. Catal.* **1985**, *18*, 259–272.
6. G. C. Bond, F. Garin, G. Maire, *Appl. Catal.* **1988**, *41*, 313–335.
7. G. C. Bond, Z. Paál, *Appl. Catal. A.* **1992**, *86*, 1–35.
8. G. C. Bond, M. R. Gelsthorpe, *Appl. Catal.* **1987**, *35*, 169–176.
9. V. Gutzmann, W. Vogel, *J. Phys. Chem.* **1990**, *94*, 4991.
10. G. C. Bond, Xu Yide, *J. Chem. Soc. Faraday Trans. 1* **1984**, *80*, 969–980.
11. C. Hubert, A. Frennet, *Catal. Today* **1993**, *17*, 469–482.
12. G. C. Bond, Lou Hui, *J. Catal.* **1994**, *147*, 346–348.
13. G. C. Bond, Lou Hui, *J. Catal.* **1992**, *137*, 462–472.
14. G. C. Bond, Lou Hui, *J. Catal.* **1993**, *142*, 512–530.
15. Z. Paál, *Catal. Today* **1992**, *12*, 297.
16. I. M. Sutherland, A. Ibbotson, R. B. Moyes, P. B. Wells, *J. Catal.* **1990**, *125*, 77–88.
17. P. A. Meheux, A. Ibbotson, P. B. Wells, *J. Catal.* **1991**, *128*, 387–396.
18. J. A. Slipszenko, S. P. Griffiths, P. Johnston, K. E. Simons, W. A. H. Vermeer, P. B. Wells, *J. Catal.* **1998**, *179*, 267–276.
19. G. Bond, K. E. Simons, A. Ibbotson, P. B. Wells, D. A. Whan, *Catal. Today* **1992**, *12*, 421–425.
20. K. E. Simons, P. A. Meheux, S. P. Griffiths, I. M. Sutherland, P. Johnston, P. B. Wells, A. F. Carley, M. K. Rajumon, M. W. Roberts, A. Ibbotson, *Recueil Trav. Chim. Pays-Bas* **1994**, *113*, 465–474.
21. S. P. Griffiths, P. B. Wells, K. Griffin, P. Johnston, in *Catalysis of Organic Reactions*, F. E. Herkes (Ed.), Marcel Dekker, New York, 1998, pp. 89–100.
22. P. B. Wells, K. E. Simons, J. A. Slipszenko, S. P. Griffiths, D. F. Ewing, *J. Molec. Catal. A: Chemical* **1999**, 159–166.
23. S. P. Griffiths, P. Johnston, P. B. Wells, *Appl. Catal. A: General* **2000**, *191*, 193–204.
24. W. A. H. Vermeer, A. Fulford, P. Johnston, P. B. Wells, *J. Chem. Soc. Chem. Commun.* **1993**, 1053–1054.
25. (a) S. P. Griffiths, P. Johnston, W. A. H. Vermeer, P. B. Wells, *J. Chem. Soc. Chem. Commun.* **1994**, 2431–2432; (b) D. J. Jenkins, A. M. S. Alabdulrahman, G. A. Attard, K. G. Griffin, P. Johnston, P. B. Wells, *J. Catal.* **2005**, *234*, 230–239.
26. G. C. Bond, *J. Molec. Catal.* **1993**, *81*, 99–118.
27. R. Burch, Z. Paál, *Appl. Catal. A* **1994**, *114*, 9–33.
28. Z. Huang, J. R. Fryer, C. Park, D. Stirling, G. Webb, *J. Catal.* **1994**, *148*, 478–492.
29. G. C. Bond, R. H. Cunningham, *J. Catal.* **1996**, *163*, 328–337.
30. G. C. Bond, M. R. Gelsthorpe, *J. Chem. Soc., Faraday Trans. 1* **1989**, *85*, 3767–3783.
31. G. C. Bond, R. H. Cunningham, E. L. Short, in *Proceedings 10th International Congress on Catalysis*, L. Guzci, F. Solymosi, P. Tétényi (Eds.), Akadémiai Kiadó, Budapest, 1993, Vol. A, p. 849.

32. G. C. Bond, R. H. Cunningham, *J. Catal.* **1997**, *166*, 172–185.
33. G. C. Bond, R. H. Cunningham, J. C. Slaa, *Top. Catal.* **1994**, *1*, 19–24.
34. G. C. Bond, *Metal-Catalysed Reactions of Hydrocarbons*, Springer, New York, 2005.
35. G. C. Bond, M. A. Keane, H. Kral, J. A. Lercher, *Catal. Rev.-Sci. Eng.* **2000**, *42*, 323–384.
36. G. C. Bond, *Appl. Catal. A: General* **2000**, *191*, 23–34.
37. G. C. Bond, *Catal. Today* **1999**, *49*, 41–48.
38. G. C. Bond, *Ind. Eng. Chem. Res.* **1997**, *36*, 3173–3179.
39. J. W. E. Coenen, *Appl. Catal.* **1989**, *54*, 59–64.
40. J. W. E. Coenen, *Appl. Catal.* **1989**, *54*, 65–78.
41. J. W. E. Coenen, *Appl. Catal.* **1991**, *75*, 193–223.
42. G. C. Bond, S. F. Tahir, *Appl. Catal.* **1991**, *71*, 1–31.
43. EUROCAT Oxide: *Catal. Today* **1994**, *20*; (a) G. C. Bond, J. C. Védrine, *Catal. Today* **1994**, *20*, 1–6; (b) E. Garcin, *Catal. Today* **1994**, *20*, 7–10; (c) J. Haber, *Catal. Today* **1994**, *20*, 11–16; (d) P. Ruiz, B. Delmon, *Catal. Today* **1994**, *20*, 17–22; (e) J. Ph. Nogier, *Catal. Today* **1994**, *20*, 23–34; (f) V. Rives, *Catal. Today* **1994**, *20*, 37–44; (g) R. A. Koeppel, J. Nickl, A. Baiker, *Catal. Today* **1994**, *20*, 45–52; (h) F. Majunki, M. Baerns, *Catal. Today* **1994**, *20*, 53–60; (i) G. Busca, A. Zecchina, *Catal. Today* **1994**, *20*, 61–76; (j) C. Fernandez, M. Guelton, *Catal. Today* **1994**, *20*, 77–86; (k) A. Aboukais, *Catal. Today* **1994**, *20*, 87–96; (l) M. de Boer, *Catal. Today* **1994**, *20*, 97–108; (m) J. Ph. Nogier, M. Delamar, *Catal. Today* **1994**, *20*, 109–124; (n) G. C. Bond, *Catal. Today* **1994**, *20*, 125–134; (o) J.-M. Hermann, *Catal. Today* **1994**, *20*, 135–152; (p) G. Golinelli, F. Trifirò, *Catal. Today* **1994**, *20*, 153–164; (q) B. Grzybowska-Świerkosz, *Catal. Today* **1994**, *20*, 165–170; (r) G. C. Bond, J. C. Védrine, *Catal. Today* **1994**, *20*, 171–178.
44. D. N. Waters, *Spectrochim. Acta* **1994**, *50A*, 1833–1840.
45. J. A. Martens, Ph. Buskens, P. A. Jacobs, A. Van der Pol, J. H. C. van Hooff, C. Ferrini, H. W. Kouwenhoven, P. J. Kooyman, H. van Bekkum, *Appl. Catal. A* **1993**, *99*, 71–84.
46. G. C. Bond, D. T. Thompson, *Catal. Rev.-Sci. Eng.* **1999**, *41*, 319–388.
47. G. C. Bond, C. Louis, D. T. Thompson, *Catalysis by Gold*, Imperial College Press, London, 2006.
48. S. Tsubota, A. Yamaguchi, M. Daté, M. Haruta, *Gold 2003*, Vancouver, 2003.
49. C. Milone, R. Ingoglia, A. Pitstone, G. Neri, F. Frusteri, S. Galvagno, *J. Catal.* **2004**, *222*, 348–356.
50. J. A. Moulijn, M. Makkee, B. Silberova, unpublished results.
51. C. Rossignol, S. Arrii, F. Morfin, L. Piccolo, V. Caps, J.-L. Rousset, *J. Catal.* **2005**, *230*, 476–483.
52. S. Carrettin, P. Concepcion, A. Corma, J. M. Lopez Nieto, V. F. Puntes, *Angew. Chem. Int. Ed.* **2004**, *43*, 2538–2540.
53. W. B. Kim, T. Voith, G. J. Rodriguez-Rivera, J. A. Dumesic, *Science* **2004**, *305*, 1280–1283.
54. Y. Murakami, in *Preparation of Catalysts III*, G. Poncelet, P. Grange, P. A. Jacobs (Eds.), Elsevier, Amsterdam, 1993, pp. 775–784.
55. M. Niwa, M. Iwamoto, K. Segawa, *Bull. Chem. Soc. Japan* **1986**, *59*, 3735–3739.
56. Z. Karpinski, T.-K. Chuang, H. Katsuzawa, J. B. Butt, R. L. Burwell, Jr., J. B. Cohen, *J. Catal.* **1986**, *99*, 184–197.
57. J. B. Butt, R. L. Burwell Jr., *Catal. Today* **1992**, *12*, 177–188.
58. D. K. Takehara, J. B. Butt, R. L. Burwell, Jr., *J. Catal.* **1992**, *133*, 279–293, 294–308.
59. British Standards Institute BS 4359, Part 4, 1995.
60. D. H. Everett, G. D. Parfitt, K. S. W. Sing, R. Wilson, *J. Appl. Chem. Biotechnol.* **1974**, *24*, 199–219.

2.6.2
Zeolite Standard Catalysts and Related Activities of the International Zeolite Association

*Michael Stöcker and Jens Weitkamp**

2.6.2.1 Introduction

During the early 1960s, zeolite catalysts were introduced on a commercial scale into fluid catalytic cracking of heavy petroleum fractions (see, e.g., Ref. [1] and Chapter 13.5 of this Handbook). The improvement of the process and its economics brought about by the new class of solid catalysts was truly spectacular, and subsequently zeolites conquered an impressive number of additional large-scale processes, mainly in petroleum refining (see Chapter 13.1 of this Handbook) and basic petrochemistry (see Chapters 14.3 and 14.4 of this Handbook). Today, while these traditional fields continue to be domains of zeolite catalysts, they have also found important applications in the chemical and environmental industries. In general terms, zeolites are looked upon as one of the main classes of catalytic materials the future of which looks very bright [2], the more so since the family of zeolitic materials is steadily growing. As a matter of fact, numerous research groups dealing with heterogeneous catalysis, either in industry or at the academia, are building up expertise in zeolite materials science and the use of zeolites in heterogeneous catalysis.

The International Zeolite Association (IZA) [3], together with some of its commissions, supports and assists the catalysis community. The IZA does so, *inter alia*, by providing and disseminating a straightforward zeolite nomenclature approved by the International Union for Pure and Applied Chemistry (IUPAC) [4], up-to-date collections of framework types of so-far discovered zeolitic materials [5, 6], proven recipes for the hydrothermal syntheses of zeolites [7], as well as established procedures for converting as-synthesized zeolites into active catalysts [8] and testing their performance in selected catalytic reactions [8]. The aim of this chapter is to draw the reader's attention to the services of the IZA in as far as they are relevant to heterogeneous catalysis.

2.6.2.2 Atlas of Zeolite Framework Types

"The Atlas" [5, 6] is a compilation of zeolite framework types which were approved by the IZA Structure Commission. Upon approval, each framework type is assigned a code consisting of three capital letters (to be printed in bold face type). Zeolite framework types

---
References see page 719
* Corresponding author.

should not be confused or equated with real zeolitic materials, as they only describe the network of the corner-sharing tetrahedrally coordinated framework atoms [5]. An example for a three-letter code is **FAU** (derived from the mineral faujasite, but also describing the framework of the synthetic zeolites X and Y). The three-letter code can be used in the IUPAC crystal chemical formula [4] of a real zeolitic material; for example, $|Na_{56}|$ $[Al_{56}Si_{136}O_{384}]$-**FAU** represents zeolite Y with an n_{Si}/n_{Al}-ratio of 2.4 (note that the guest species and the framework host are given in vertical brackets and square brackets, respectively).

At the time of writing this chapter, a total of 176 framework types had been approved by the IZA Structure Commission [6], of which 133 are compiled in the last printed version of the Atlas [5]. There, each framework type is described on two pages of which one page is listing the information that characterizes the framework type (crystallographic data, coordination sequences, vertex symbols and loop configurations), and the second page presents data for real materials of this framework type. Figure 1 shows that the crystallographic pore diameters of existing materials with the 176 approved framework types cover a wide range, from less than 0.2 nm to approximately 1.3 nm. It is moreover seen that there is a particularly broad choice of framework types with eight-membered-ring (or, synonymously, narrow-pore) materials, ten-membered-ring (medium-pore) materials and twelve-membered-ring (large-pore) materials, from which the zeolite catalyst for a given application can be selected.

Listed in Table 1 are some selected framework type codes and real materials that are especially relevant to catalysis. More extended lists of this form may be found in Refs. [5, 9].

2.6.2.3 Verified Syntheses of Zeolitic Materials

Once a zeolite with an appropriate framework type has been identified for a given catalytic application, a real material must be prepared by hydrothermal synthesis. Countless recipes for zeolites have been published in the patent and scientific literature, but for various reasons the experimentalist may encounter reproducibility problems in some instances. With its "Verified Syntheses" [7], the IZA Synthesis Commission provided a most valuable compilation of proven recipes for the synthesis of more than 70 zeolitic materials. Many of these are of a framework type listed in Table 1. All recipes published in

Tab. 1 Selected three-letter framework type codes and some real zeolitic materials with relevance to catalysis. (Source: IZA Structure Commission [5, 6])

Three-letter code	Material(s)
AEL	AlPO-11, SAPO-11
AFI	AlPO-5, SAPO-5, CoAPO-5, SSZ-24
*BEA	Beta
CHA	Chabazite, AlPO-34, SAPO-34
DON	UTD-1
EMT	EMC-2
ERI	Erionite
EUO	EU-1, ZSM-50
FAU	Faujasite, Linde X, Linde Y, LSX, SAPO-37
FER	Ferrierite, ZSM-35
ISV	ITQ-7
KFI	ZK-5
LTA	Linde Type A, Alpha, SAPO-42, ZK-4
LTL	Linde Type L
MAZ	Mazzite, Omega, ZSM-4
MEL	ZSM-11, silicalite-2, TS-2
MFI	ZSM-5, silicalite-1, TS-1
MOR	Mordenite
MTT	ZSM-23
MTW	ZSM-12, NU-13, Theta-3
MWW	MCM-22, ITQ-1, PSH-3, SSZ-25
NES	NU-87
OFF	Offretite
OSI	UiO-6
RHO	Rho
SFE	SSZ-48
STF	SSZ-35, ITQ-9
STT	SSZ-23
TON	Theta-1, ISI-1, NU-10, ZSM-22
VFI	VPI-5, AlPO-54, MCM-9

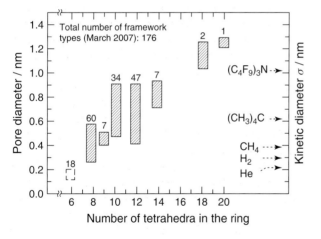

Fig. 1 The range of crystallographic pore diameters of materials with framework types approved by the IZA Structure Commission, as of March 2007 [6]. By pore diameter, the largest diameter of the largest pore of a given n-membered-ring material is meant ($n = 8, 9, 10, 12, 14, 18, 20$). The kinetic diameters, σ, of some selected molecules are shown for comparison. The numbers on top of the bars give the number of framework types into which the known n-ring materials can be classified. The dotted bar for six-membered-ring materials is to indicate that no pore diameters are given in the Atlas for these materials. Moreover, since they are dense rather than porous, they are of limited value from the viewpoint of catalysis.

the Verified Syntheses were cross-checked by experts who are experienced in hydrothermal synthesis. Given in each recipe are the starting materials, the gel composition and preparation, details of the crystallization process as well as product recovery and characterization. In particular, a powder X-ray diffractogram is included for each sample.

In addition to the specific synthesis recipes, the book offers brief introductory articles describing basic skills that are relevant to zeolite synthesis and characterization. All in all, the Verified Syntheses represent a most valuable source of information for those who plan to prepare, in their own laboratories, zeolite catalysts for whatever reaction.

2.6.2.4 Preparation of the Standard Large-Pore, Acidic Zeolite Catalyst La,Na-Y

In almost all instances, the zeolite in its as-synthesized form is not yet the active catalyst aimed at, but a catalyst precursor which must be treated by one or several modification procedures, so as to arrive at an active and selective catalyst for the envisaged reaction. Frequently involved techniques at this stage are ion exchange, either in aqueous suspension [10] or in the solid state [11], dealumination of the framework [12], introduction of metal guests into the pores [13], and activation of the zeolite by treatment at elevated temperatures, in the absence or presence of a reducing or oxidizing atmosphere. Especially for newcomers in the field, it is desirable to compare their own results obtained on a home-made catalyst and/or in a home-built apparatus with those achieved by experienced experts in the field.

With this in mind, the IZA Catalysis Commission identified a suitable test reaction catalyzed by large-pore, acidic zeolites, *viz.* the disproportionation of ethylbenzene into benzene and the three isomeric diethylbenzenes and a standard large-pore, acidic zeolite catalyst, *viz.* La,Na-Y [8]. In terms of the approved IUPAC and IZA nomenclature [4], the designation of this zeolite is $|La_{13.4}Na_{15.8}|[Al_{56}Si_{136}O_{384}]$-**FAU**. This is zeolite Y with an n_{Si}/n_{Al}-ratio of 2.4 and a degree of lanthanum exchange of ca. 72%. The standard catalyst is prepared from $|Na_{56}|[Al_{56}Si_{136}O_{384}]$-**FAU** (i.e., zeolite Na-Y) in the following manner [8]: Na-Y zeolite is stored in a desiccator over a saturated aqueous solution of $Ca(NO_3)_2$, until the water content referenced to the moist zeolite is 24.2 wt.%. For the ion exchange with $La(NO_3)_3$, 166.5 g of the moist zeolite (corresponding to 126.2 g of the water-free zeolite) are suspended in 400 mL of water. To this solution, 41.0 g $La(NO_3)_3 \cdot 6H_2O$ dissolved in 50 mL of water are added. The *pH* should not be below 5.5, and if necessary it is adjusted with a dilute aqueous solution of NaOH. Upon adjusting the volume with water to 500 mL, the suspension is heated at 353 K and stirred for 18 h. The suspension is then filtered while hot, and the filter cake washed with 1.5 L of water. After resuspension of the zeolite cake in 400 mL of water, a second La^{3+} ion exchange is performed under identical conditions, followed by a third and a fourth ion exchange. After the fourth ion exchange, the zeolite is washed until nitrate-free, as checked with Lunge's reagent (zinc dust, sulfanilic acid, α-naphthylamine), air-dried and dried for 24 h in an oven at 373 K. At this point, the degree of lanthanum exchange amounts to about 72% of the total ion-exchange capacity. The zeolite is then granulated by pressing the powder without a binder at a maximum pressure of 400 MPa, crushed and sieved to obtain particles with a diameter between 0.2 and 0.3 mm. The water content of these particles is determined by thermogravimetric analysis (TGA), and a mass of catalyst corresponding to 0.29 g of dry zeolite is diluted with 3 mL washed sand or chips from fused silica with a particle size of 0.1 to 0.3 mm. The dilution with the catalytically inert solid results in an appropriate bed height in the fixed-bed reactor, even if a small amount of zeolite catalyst is used.

2.6.2.5 Apparatus and Procedure for Catalyst Testing

Scheme 1 shows the disproportionation of ethylbenzene. Ideally, just four product hydrocarbons are formed, *viz.* benzene and *ortho*-, *meta*-, and *para*-diethylbenzene. This renders the product analysis simple. Note, however, that the diethylbenzenes formed tend to undergo acid-catalyzed consecutive reactions, for example transalkylation with ethylbenzene into benzene and triethylbenzenes, especially at elevated ethylbenzene conversions. Such product mixtures can be readily analyzed by capillary gas chromatography. Suitable stationary phases and temperature programs are provided in Ref. [8].

Prior to its use in a catalytic experiment, ethylbenzene must be purified from potential catalyst poisons, such as ethylbenzene hydroperoxide, for example by passing the liquid hydrocarbon over a chromatographic column filled with activated alumina [8].

The IZA Catalysis Commission recommends testing of the La,Na-Y zeolite in a continuously operated apparatus with a fixed-bed reactor. Nitrogen with a flow rate of 40 mL min^{-1} is first passed through a saturator containing the liquid feed hydrocarbon thermostated at 294 K. Appropriate saturator constructions are discussed

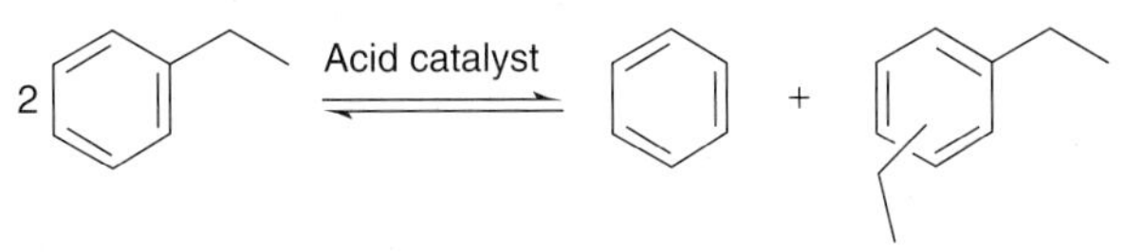

Scheme 1 The disproportionation of ethylbenzene.

References see page 719

in Chapter 9.2 of this Handbook. Given the vapor pressure of ethylbenzene (E-Bz) at 294 K, the molar feed rate into the reactor amounts to 1.0 mmol h^{-1}. The reactor consists of a tube from glass or fused silica with an internal diameter of 10 mm. Inside this tube, the zeolite catalyst diluted with an inert solid (*vide supra*) is held in place by a glass frit or a plug from silica glass wool. To measure the reaction temperature, a thermocouple is placed at an axial position in the catalyst bed. The N$_2$/E-Bz gas mixture flows through the catalyst bed from top to bottom. Prior to the experiment, the catalyst is activated inside the reactor in a flow of nitrogen (50 mL min^{-1}) by heating at a rate of 2 K min^{-1} up to 523 K, followed by a constant temperature of 523 K for 12 h. For heating, the reactor is housed in an electrical oven [8].

As shown by Karge et al. [14, 15], and discussed in more detail by Weiß et al. [16], ethylbenzene disproportionation in large-pore zeolites may show an induction period – that is, initially the catalyst activity and hence the ethylbenzene conversion may increase with time-on-stream (*TOS*). After a sufficiently long *TOS*, the ethylbenzene conversion either remains constant or declines slightly (due to the formation of carbonaceous deposits). In order to detect such instationarities, the product gas leaving the reactor is sent to a six-way sampling valve which is connected with the gas chromatograph. By means of the sampling valve, product samples are injected into the gas chromatograph for analysis at regular intervals. From the peak areas of the various products and unconverted ethylbenzene, the conversion (X_{E-Bz}) and product yields (Y_j) are calculated, as demonstrated in Ref. [8].

2.6.2.6 Comparison of the Results Achieved in Five Different Laboratories

Ethylbenzene was converted on the standard La,Na-Y zeolite catalyst by the members of the IZA Catalysis Commission in five different laboratories [8]. The reaction conditions applied by all groups were identical:

- Mass of dry catalyst: $W = 290$ mg
- Molar flux of feed: $F_{E-Bz} = 1.0$ mmol h^{-1}
- Saturator temperature: $T_{Sat.} = 294$ K
- Reaction temperature: $T_{React.} = 453, 523,$ or 593 K.

Figure 2 shows, as an example, the ethylbenzene conversions (X_{E-Bz}) measured as a function of *TOS* at a reaction temperature of 453 K in the different laboratories. Clearly, an induction period was observed by all groups, but its duration varied considerably (from ca. 4 to 24 h). It was found that the duration of the induction period in ethylbenzene disproportionation could be markedly influenced by minor deviations in the reaction conditions (which were nominally identical in all five experiments). By contrast, the agreement in X_{E-Bz} at the end of the

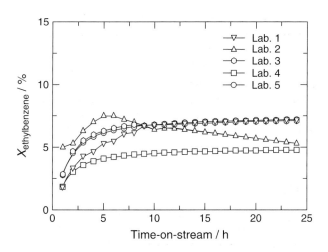

Fig. 2 Conversion of ethylbenzene (E-Bz) on the standard La,Na-Y zeolite catalyst at $T_{React.} = 453$ K and $W/F_{E-Bz} = 290$ g h mol^{-1} in nitrogen as carrier gas measured in five different laboratories (Lab. 1 to Lab. 5). (Adapted from Ref. [8].)

induction period is much better, and amounts to ca. 6% (except for the results of one laboratory).

In Ref. [8], the reader will find a much more comprehensive discussion of the results achieved in the laboratories of the five groups. This discussion includes the ethylbenzene conversion at temperatures higher than 453 K, the yields of benzene and ethylbenzenes, the isomer distribution of *ortho*-, *meta*-, and *para*-diethylbenzene, and a statistical evaluation of the results obtained in the five participating laboratories. Moreover, a list of eight potential pitfalls and sources for aberrant results is discussed which will considerably help newcomers to improve and optimize their catalytic equipment.

2.6.2.7 Conclusions

In hindsight, the disproportionation of ethylbenzene on large-pore acidic zeolite catalysts is not as simple a reaction as the stoichiometric equation (see Scheme 1) might suggest. The occurrence of an initial induction period, the deviation of the yield ratio of benzene and diethylbenzenes from unity (as described in Ref. [8]), and the occurrence or not of catalyst deactivation after termination of the induction period (Fig. 2) all point to an important role of phenomena related to sorption of the reactant, the diethylbenzene products and byproducts such as triethyl- or even more highly alkylated benzenes. The potential role of such phenomena in ethylbenzene disproportionation has been discussed in more detail in the literature (e.g., Refs. [16, 17]). If one takes into account the complexity of the catalytic cycle of this seemingly simple hydrocarbon reaction [16], the agreement achieved in the five different laboratories is reasonable. In any event, the meticulously described procedures for preparing the La,Na-Y standard catalyst and for conducting the catalytic experiment will be of considerable value for newcomers who wish to control their experimental skills.

The IZA Catalysis Commission is currently investigating on simpler gas-phase reactions, for example on bifunctional zeolite catalysts in which an acidic and a hydrogenation/dehydrogenation function are combined, and on liquid-phase test reactions [8], whereby the services to the catalysis community will be significantly extended.

References

1. P. B. Venuto, E. T. Habib, Jr., *Fluid Catalytic Cracking with Zeolite Catalysts*, Marcel Dekker, New York, 1979, 156 pp.
2. A. Corma, *J. Catal.* **2003**, *216*, 298–312.
3. http://www.iza-online.org/.
4. L. B. McCusker, F. Liebau, G. Engelhardt, *Micropor. Mesopor. Mater.* **2003**, *58*, 3–13.
5. Ch. Baerlocher, W. M. Meier, D. H. Olson, *Atlas of Zeolite Framework Types*, 5th Ed., Elsevier, Amsterdam, 2001, 302 pp.
6. http://www.iza-structure.org/databases/.
7. H. Robson (Ed.), (K. P. Lillerud, XRD patterns), *Verified Syntheses of Zeolitic Materials*, 2nd Ed. Elsevier, Amsterdam, 2001, 266 pp.
8. D. E. De Vos, S. Ernst, C. Perego, C. T. O'Connor, M. Stöcker, *Micropor. Mesopor. Mater.* **2002**, *56*, 185–192.
9. F. Schüth, K. S. W. Sing, J. Weitkamp (Eds.), *Handbook of Porous Materials*, Vol. 5, Wiley-VCH, Weinheim, 2002, p. 3140–3141.
10. R. P. Townsed, R. Harjula, in *Molecular Sieves – Science and Technology*, Vol. 3, *Post-Synthesis Modification I*, H. G. Karge, J. Weitkamp (Eds.), Springer, Berlin, Heidelberg, New York, 2002, p. 1–42.
11. H. G. Karge, H. K. Beyer, in *Molecular Sieves – Science and Technology*, Vol. 3, *Post-Synthesis Modification I*, H. G. Karge, J. Weitkamp (Eds.), Springer, Berlin, Heidelberg, New York, 2002, p. 43–201.
12. H. K. Beyer, in *Molecular Sieves – Science and Technology*, Vol. 3, *Post-Synthesis Modification I*, H. G. Karge, J. Weitkamp (Eds.), Springer, Berlin, Heidelberg, New York, 2002, p. 203–255.
13. P. Gallezot, in *Molecular Sieves – Science and Technology*, Vol. 3, *Post-Synthesis Modification I*, H. G. Karge, J. Weitkamp (Eds.), Springer, Berlin, Heidelberg, New York, 2002, p. 257–304.
14. H. G. Karge, Y. Wada, J. Weitkamp, S. Ernst, U. Girrbach, H. K. Beyer, in *Catalysis on the Energy Scene*, S. Kaliaguine, A. Mahay (Eds.), *Studies in Surface Science and Catalysis*, Vol. 19, Elsevier, Amsterdam, 1984, p. 101–111.
15. H. G. Karge, S. Ernst, M. Weihe, U. Weiß, J. Weitkamp, in *Zeolites and Related Microporous Materials: State of the Art 1994*, J. Weitkamp, H. G. Karge, H. Pfeifer, W. Hölderich (Eds.), *Studies in Surface Science and Catalysis*, Vol. 84, Part C, Elsevier, Amsterdam, 1994, p. 1805–1812.
16. U. Weiß, M. Weihe, M. Hunger, H. G. Karge, J. Weitkamp, in *Progress in Zeolite and Microporous Materials*, H. Chon, S.-K. Ihm, Y. S. Uh (Eds.), *Studies in Surface Science and Catalysis*, Vol. 105, Part B, Elsevier, Amsterdam, 1997, p. 973–980.
17. N. Arsenova-Härtel, H. Bludau, W. O. Haag, H. G. Karge, *Micropor. Mesopor. Mater.* **2000**, *35–36*, 113–119.